T0074050

**Springer Handbook
of Bio-/Neuro-Informatics**

Springer Handbook provides a concise compilation of approved key information on methods of research, general principles, and functional relationships in physical and applied sciences. The world's leading experts in the fields of physics and engineering will be assigned by one or several renowned editors to write the chapters comprising each volume. The content is selected by these experts from Springer sources (books, journals, online content) and other systematic and approved recent publications of scientific and technical information.

The volumes are designed to be useful as readable desk reference book to give a fast and comprehensive overview and easy retrieval of essential reliable key information, including tables, graphs, and bibliographies. References to extensive sources are provided.

Springer Handbook

of Bio-/Neuro-Informatics

Nikola Kasabov (Ed.)

With 490 Figures and 92 Tables

Editor
Nikola Kasabov
Auckland University of Technology
KEDRI – Knowledge Engineering and Discovery Research Institute
120 Mayoral Drive
Auckland, New Zealand

ISBN: 978-3-642-30573-3   e-ISBN: 978-3-642-30574-0
DOI 10.1007/978-3-642-30574-0
Springer Dordrecht Heidelberg London New York

Library of Congress Control Number: 2013951342

Production and typesetting: le-tex publishing services GmbH, Leipzig
Senior Manager Springer Handbook: Dr. W. Skolaut, Heidelberg
Typography and layout: schreiberVIS, Seeheim
Illustrations: Hippmann GbR, Schwarzenbruck
Cover design: eStudio Calamar Steinen, Barcelona
Cover production: WMXDesign GmbH, Heidelberg
Printing and binding: Stürtz GmbH, Würzburg

Printed on acid free paper

Springer is part of Springer Science+Business Media (www.springer.com)

89/3180/YL   5 4 3 2 1 0

For my mother Kapka Nikolova Mankova-Kassabova (1920–2012)
who wrote in Bulgarian, among many others, the little poems:

Не ще ми стигнат дните	I will run out of time
Да разгадая тайните на битието	To unlock all the secrets,
Да излекувам човешките рани	To heal all the wounds,
И да изкажа любовта си	To express all my love.

From the editor of this Springer Handbook, Nikola Kasabov

Foreword

Life is a striking state of material that emerged 3.6 billion years ago on our earth. It has the ability of self-reproduction, by keeping information about itself in itself. The marriage of information and matter makes it possible to implement the evolutional mechanism. Thus, life has developed to achieve a variety of amazing structures and functions. In particular, we humans have appeared, with the characteristics of intelligence and consciousness, constructing societies and culture.

Shun-ichi Amari

Now the time is ripe to understand the mechanisms of living systems, where information is kept and processed efficiently and robustly. These mechanisms are highly complex and exquisite; their structures range from the molecular level to the higher level of the brain function. Advanced information science/technology is required to decipher the information mechanisms of life. This research field is composed of bioinformatics and neuroinformatics, which are closely related. They include a wide range of hierarchical levels of molecules, cells, systems, and individuals. We need a good perspective of these extremely wide fields in a systematic and understandable way.

The present Handbook is intended to provide not only an overview of the state of the art in bioinformatics and neuroinformatics and the related research problems, but also detailed informatics methods and techniques to address them. In this Handbook one can easily find important theoretical informatics tools necessary for bioinformatics and neuroinformatics research, including the necessary details.

I am highly pleased to see that such a handbook, covering all aspects of bioinformatics and neuroinformatics together with their tools and methods, is being published.

Brain Science Institute
RIKEN, Saitama

Shun-ichi Amari

Professor Shun-ichi Amari is IEEE Life Fellow, foreign member of the Polish Academy of Science and Professor-Emeritus at the University of Tokyo. He is past-Director of he RIKEN Brain Science Institute and is now a senior advisor. He has published more than 300 papers, including initiation of information geometry and pioneering works on neural networks. He is the past President of International Neural Networks Society (INNS), the Institute of Electronic, Information and Communication Engineers, Japan (IEICE) and founding President of Asian-Pacific Neural Networks Assembly (APNNA). He is a founding co-editor-in-chief of *Neural Networks*. He received the Emanuel A. Piore Award and Neural Networks Pioneer Award from the IEEE, the Japan Academy Award, the Japanese Order of Cultural Merit, and the INNS Gabor Award, among many others.

Foreword

From Data to Synthesis: The Human Brain Project

Brain research is entering its maturing stage. Enormous amounts of data, information, and knowledge have been accumulated and are being accumulated every day. The challenge now is to put all these into integrated models and systems and to facilitate a better understanding of the brain. This *Springer Handbook of Bio-/Neuroinformatics* is very timely in this respect.

The European Commission has recently approved the *Human Brain Project* (HBP) as a flagship in the *Future Emerging Technologies* (FET) program (www.humanbrainproject.eu). The HBP commenced a 30 months ramp-up phase in 2013 and is expected to continue for a total of 10 years. The overall goal of the project is to build a new European research infrastructure for brain related research with a strong emphasis on state-of-the-art information and communication technologies (ICT). Through the development of methods and tools, the integration of data, and the construction of technology platforms the HBP will follow the successful model of large-scale research infrastructures in other scientific fields to tackle one of the most fundamental challenges of basic research, linking the results directly to medical and computing applications.

The HBP addresses three major research areas: neuroscience, neuromedicine, and future computing. In neuroscience the project aims at an accelerated understanding of the human brain by building and simulating unifying brain models. Progress in this challenging area of basic research is expected by increasing the value of past and future experiments through integration into publicly available brain atlases; filling open knowledge gaps using predictive software tools and strategically selected experiments, which will be prioritized and optimized for the use in simulations. In neuromedicine the goal is to accelerate the understanding of brain diseases by increasing the value of existing clinical records and data with modern data storage and machine learning technologies. By systematically understanding the similarities and differences of brain diseases the HBP expects to provide tools for pharmaceutical and nutrition companies to prevent, diagnose, and treat brain diseases. The future computing research area aims at an accelerated development of brain inspired computing technologies. Those technologies will be based on the knowledge generated from the new insights of brain structure and function obtained in the HBP. Specifically, the HBP will develop a high performance, visually interactive exa-scale computing infrastructure, novel low power, robust, self-adapting computing, and communication devices and with this lay the groundwork for a new fundamentally different paradigm of computing.

In order to achieve these goals, the HBP has started to build six technology platforms, which will serve the scientific community in Europe and beyond in addressing the goals described above. The six platforms will be hosted by partner groups of the HBP and dedicate their work to the following tasks:

Foto: F. Hentschel

Karlheinz Meier
Universität Heidelberg
Kirchhoff-Institut
für Physik
Co-Director of the
Human Brain Project

1. Neuroinformatics platform: The task is to aggregate neuroscience data and to deliver brain atlases.
2. Medical informatics platform: the task is to aggregate clinical records and to classify brain diseases.
3. Brain simulation platform: the task is to develop software tools and to run closed loop brain simulations.
4. High performance computing platform: the task is to develop and operate HPC systems optimized for brain simulations.
5. Neuromorphic computing platform: the task is to develop and operate novel brain derived computing hardware.
6. Neurorobotics platform: the task is to develop virtual robotic systems for closed loop cognitive experiments.

The platforms will be complemented by a new *European Institute for Theoretical Neuroscience* to integrate basic research on information processing in the brain and to use the HBP platforms to inspire new theoretical approaches to the study of the brain. Also, the project will have a strong ethics group to monitor on-going research and to the enable a permanent discourse with the public.

Heidelberg, July 2013   Karlheinz Meier

Curriculum Vitae

- *since 2011*: Co-Director of the EU Human Brain Project, ET Flagship Project
- *since 2011*: Invited professor at EPFL (Lausanne, Switzerland)
- *since 2010*: Coordinator of the EU BrainScaleS Consortium
- *since 2009*: Coordinator of the Marie-Curie Network FACETS-ITN
- *2005–2010*: Coordinator of the EU FACETS Consortium
- 1999: Founding Director of the Kirchhoff Institut für Physik
- 1994–2012: Project leader of the LHC-ATLAS Trigger PreProcessor
- *since 1992*: Chair Experimental Physics, Ruprecht-Karls-Universität Heidelberg
- *1990–1992*: Scientific Staff (tenure) at DESY (Hamburg)
- *1988–1990*: Scientific Staff at CERN (Geneva, Switzerland)
- *1984–1988*: Research Fellow at CERN (Geneva, Switzerland)
- 1984: PhD in Physics from Hamburg University
- 1981: Diploma in Physics from Hamburg University

Fields of Interest

Experimental particle physics, microelectronics instrumentation, physics foundations of information processing, brain-inspired computing

Preface

One of the greatest scientific achievements of the twentieth century was the deciphering of the human genome and the genomes of many other species. The Human Genome Project revealed the "mystery" of the human genome and provided us with an enormous quantity of data and information that still needs to be properly processed in order to obtain new knowledge about the human race and its living environment on this planet, their biological co-evolution, their current state of development, and possibly, their future. These scientific and technological advancements led to the firm establishment of the science discipline *bioinformatics*. New biological data, information, and knowledge have been produced in many institutes from all over the world, for example: the European Bioinformatics Institute, the National Centre for Biotechnology Information, and the RIKEN Bioinformatics Institute.

It is agreed that the twenty-first century will be devoted to the deciphering of the human brain. Several projects and institutes have already addressed this challenge. The Blue Brain Project aimed at the computer simulation of brain functions at a low level of brain information processing. The Allen Brain Science Institute has already mapped tens of thousands of human genes to the brain structures and functions of mice and humans, providing an enormous amount of data (see Chap. 62). The goals of the RIKEN Brain Science Institute for the twenty-first century are to understand, to create, and to protect the brain (see the Foreword by Prof. Amari). There is an international coordination of neuroinformatics activities in the International Neuroinformatics Coordinating Facilities, in which Japan has taken a leading role (see Part J). All these activities have led to the firm establishment of another scientific discipline – *neuroinformatics*.

Recently, several new large-scale world projects have started to generate by the hour new, large-scale data and information about the human brain, including data related to biology, physiology, psychology, cognitive studies, and brain diseases. The Human Brain Project was founded in 2013 by the European Union. It includes more than 80 participants from Europe. It aims at collecting data at each level of brain functioning – molecular, genetic, proteomic, synaptic, neuronal ensembles, and cognitive, and to simulate brain func-

tions in a computer system. This will enable a better understanding of many cognitive processes and brain diseases, such as Alzheimer's disease, stroke, epilepsy, Parkinson's disease, mental retardation, and aging. The BRAIN project, founded in the USA also in 2013, aims at the development of a broad set of activities for the study of the human brain including: key investments to jumpstart the effort, strong academic leadership, public-private partnerships, and maintaining highest ethical standards.

The active research in bio- and neuroinformatics and the vast amount of data and information, much of it overlapping and strongly related, requires a unified approach to the analysis of this data for the sake of new knowledge discovery. This is to be provided by information science as demonstrated in this Springer Handbook of Bio-/Neuroinformatics.

The Springer Handbook was a challenging project, as for the first time it puts together both the foundations and the state-of-the art of three major science disciplines in their interaction and mutual relationship, namely: information science (IS), bioinformatics (BI), and neuroinformatics (NI).

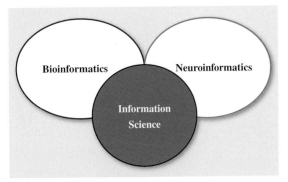

The text contains 62 peer reviewed chapters organized in 12 parts, 6 of them covering topics from IS and BI, and 6 topics from IS and NI. Each chapter consists of three main sections: introduction to the subject area, presentation of methods, and advanced and future developments.

Some of the IS methods are presented separately, some are used to address problems in BI or/and NI, but there are also specific methods designed for BI or NI only. The BI and NI areas also share IS methods, as more BI methods are now being used in NI and vice versa. As the overlap between the two areas of BI and NI is increasing, with the increased molecular information accumulated in NI, more IS and BI methods are being used in NI.

Despite the large number of IS methods developed and used so far, both generic and specific, some of them presented in the Handbook, new information processing methods will be needed in the future in order to better solve the challenging problems in BI and NI. As all the three areas of IS, BI, and NI are developing rapidly, the scope is open for future development and for new editions.

The Springer Handbook of Bio-/Neuroinformatics can be used as both a textbook in the mentioned three disciplines, and as a research or reference book for postgraduate study and advanced research in these areas. The target audience includes students, scientists, and practitioners from the areas of information, biological, and neurosciences.

It was indeed a long journey to complete this book, and an exciting one! I was offered this grand project by Springer. During the preparation of the Handbook I received tremendous support of a high professional quality from the Springer team. I would like to especially acknowledge Werner Skolaut and also Anne Strohbach, Veronika Hamm, Thomas Ditzinger, Leontina Di Cecco, Ronan Nugent, and the whole editorial technical team.

I would like to acknowledge the Part Editors of the Handbook Heike Sichtig, Chris Brown, Irwin King, Kaizo Huang, Francesco Masulli, Danilo Mandic, Lubica Benuskova, Shiro Usui, and Raphael Ritz. Some of them participated in the organization of two or more Parts and communicated with many of the authors from the large team of more than 100. Professors Shin-ichi Amari and Karlheinz Meier were kind to write a memorable Foreword. Many thanks to all authors who did the work in extra time that they theoretically did not have!

In my communications with the authors, I was helped by Joyce D'Mello. My wife Diana Kassabova was actively involved from the very beginning of this project. She is the co-author of the Glossary and she proofread the early versions of several chapters. Some of the editing work on this Handbook I did as part of my EU FP7 Marie Curie Fellowship at the Institute for Neuroinformatics at the ETH/University of Zurich.

My sincere thanks go to all who were involved and helped with *our* Springer Handbook! Many thanks!

Auckland, NZ, October 2013   Nikola Kasabov

About the Editor

Professor Nikola Kasabov is a world leading scientist in the area of computer and information sciences, more specifically in computational intelligence, knowledge engineering, neuroinformatics and bioinformatics. He is Fellow of IEEE (2010) and Fellow of the Academy of the Royal Society of New Zealand, RSNZ (2001) and Distinguished Visiting Fellow of the Royal Academy of Engineering, UK (2013). He obtained his Masters degree in computing and electrical engineering (1971) and PhD in mathematical sciences (1975) from the Technical University of Sofia and his secondary school education in Pavlikeni, Bulgaria. He is the Director and the Founder of the Knowledge Engineering and Discovery Research Institute (KEDRI, www.kedri.info) and Professor of Knowledge Engineering at the School of Computing and Mathematical Sciences at the Auckland University of Technology. Before that he worked at the Technical University of Sofia, University of Essex in the UK, University of Otago. He has published more than 500 papers, that include 15 books, 150 journal papers, 28 patents and many conference publications in the areas of computational intelligence, neural networks, bioinformatics, neuroinformatics. His books *Foundations of neural networks, fuzzy systems and knowledge engineering* (MIT, 1996) and *Evolving connectionist systems* (Springer, 2003, 2007) are well cited and used in many universities as text and research books. Prof. Kasabov was the President of the International Neural Network Society (INNS) (2009, 2010) and currently is a Governor of the INNS. He is also a Past President of the Asia Pacific Neural Network Assembly (APNNA) (2008 and 1997) and currently a member of the Governing Board. He is a Distinguished IEEE CIS Lecturer (2011-2013). Kasabov is an EU Marie Curie Fellow and a Visiting Professor at the Institute for Neuroinformatics of the University of Zurich and ETH and a Guest Professor at the Shanghai Jiao Tong University. Among his awards are the APNNA *Outstanding achievements Award* (2012), INNS Gabor Award (2012), the AUT VC Individual Research Excellence Award (2010), the Bayer Innovation Award (2007), DAAD Visiting Professorship in Germany – University of Kaiserslautern (2006); APPNA Excellence Award (2005), the RSNZ Science and Technology Medal (2002), the Dutch Science Visiting Professorship – University of Maastricht (1999), Full Professorship at the University of Trento (1996), Leverhulme Trust Scholarship UK (1990), numerous IEEE best paper awards. He has given more than 60 keynote and plenary talks at international conferences and served as a chair and a committee member of numerous IEEE, ICONIP, ANNES and other international conferences. He is the General Chair of biannual international conferences in New Zealand *Neuro-Computing and Evolving Intelligence* (NCEI). More than 35 PhD students have graduated under his supervision. More information about Prof. Kasabov and his KEDRI Institute can be obtained from http://www.kedri.aut.ac.nz.

About the Part Editors

Lubica Benuskova

University of Otago
Department of Computer Science
Dunedin, New Zealand
lubica@cs.otago.ac.nz

Parts I, K, L

Lubica Benuskova is a Senior Lecturer at the Department of Computer Science at the University of Otago, Dunedin, New Zealand. She has become a Professor at Comenius University, Bratislava, Slovakia, where she also obtained her PhD degree in Biophysics in 1994. She has a Master degree in Psychology from Vanderbilt University, Nashville, TN, USA and is a Member of the IEEE Neural Networks Technical Committee. Her research activities are in computational neuroscience, cognitive science, neuroinformatics and bioinformatics. Dr. Benuskova is a Principal Investigator of the Brain Health Research Center and a Member of the AI Research Group. She is associated with the University of Otago Research Theme on Memory: Mechanisms Processes and Applications and Bioinformatics and Computational Molecular Biology.

Chris M. Brown

University of Otago
Department of Biochemistry
Dunedin, New Zealand
chris.brown@otago.ac.nz

Parts B, D

Dr. Chris Brown received his PhD from the University of Otago, New Zealand in Biochemistry. He published his first solely computational research papers as part of those studies of gene expression. He was awarded a Human Frontier Science Long Term Fellowship to investigate gene expression in viral genomes at Iowa State University, then returned to NZ to research and teach Biochemistry and Genetics at the Otago School of Medical Sciences. His current studies aim to understand the regulation of gene expression in complex organisms and viruses, along with collaborators worldwide. He continues to take two synergistic approaches to this-blending Bioinformatics and Cell Biology.

Kaizhu Huang

Xi'an Jiaotong
Department of Electrical and Electronic
Engineering
Jiangsu, China
Kaizhu.Huang@xjtlu.edu.cn

Parts C, D

Kaizhu Huang received the PhD degree from Chinese University of Hong Kong (CUHK) in 2004. He is currently a Senior Lecturer at Xi'an Jiaotong-Liverpool University. Before that, he was an Associate Professor at the National Laboratory of Pattern Recognition, CAS. He worked as a researcher in Fujitsu R&D Center from 2004 to 2007. During 2008 and 2009, he was a research fellow at CUHK and a research assistant at the University of Bristol, UK. His research interests include machine learning, pattern recognition, and data mining. He is the recipient of APNNA Young Researcher Award 2011. He has published a book and over 70 research papers including 22 journal articles and many top conference papers.

Irwin King

The Chinese University of Hong Kong
Shenzhen Research Institute (SZRI)
Rich Media and Big Data Key Laboratory
Department of Computer Science and
Engineering
Shatin, NT, Hong Kong
king@cse.cuhk.edu.hk

Parts C, D

Dr. King is Professor at the Department of Computer Science and Engineering, The Chinese University of Hong Kong. He received his BSc degree in Engineering and Applied Science from California Institute of Technology, Pasadena and his MSc and PhD degree in Computer Science from the University of Southern California, Los Angeles. Recently, he was a Visiting Scholar at University of California, Berkeley and taught two courses on Social Computing and Data Analytics. Dr. King's research interests include machine learning, social computing, Big Data, web intelligence, data mining, and multimedia information processing. In these research areas, he has over 210 technical publications in journals. In addition, he has contributed over 30 book chapters and edited volumes. Moreover, Dr. King has over 30 research and applied grants resulted in five patents.

Danilo P. Mandic

Imperial College London
Department of Electrical and Electronic
Engineering, Communication and Signal
Processing Research Group
London, UK
d.mandic@imperial.ac.uk

Parts F, H, L

Danilo Mandic (PhD) is a Professor of Signal Processing at Imperial College London, UK. He has been working on multivariate and nonlinear signal processing, data-driven signal analysis, and adaptive multidimensional learning systems. He has authored two research monographs, has been a Guest Professor at KU Leuven, Belgium, and a frontier researcher at RIKEN Japan. Prof Mandic has also served as a member of the IEEE Technical Committee on Signal Processing Theory and Methods. He has received several best paper awards for his work on brain-computer interface, and has pioneered Ear-EEG, the ultimate noninvasive EEG recording system. He is a Fellow of the IEEE.

Francesco Masulli

Università di Genova
DIBRIS – Dipartimento di Informatica,
Bioingegneria, Robotica e Ingegneria dei
Sistemi
Genova, Italy
francesco.masulli@unige.it

Parts D, E, F, K

Francesco Masulli is an Associate Professor of Computer Science with the Dipartimento di Informatica, Bioingegneria, Robotica e Ingegneria dei Sistemi (DIBRIS) of the University of Genova, Italy, and Adjunct Associate Professor at the Sbarro Institute for Cancer Research and Molecular Medicine, Center, Center for Biotechnology of Temple University – Philadelphia (PA, USA).

He is a Lecturer of courses on Computer Science Foundations, Computer Science for Medical Doctors, and Machine Learning at the University of Genoa. Professor Masulli is the recipient of the 2008 Pattern Recognition Society Award for the paper: M. Filippone, F. Camastra, F. Masulli, S. Rovetta, *A survey of kernel and spectral methods for clustering* (2008). He is author of more than 200 scientific papers on clustering, machine learning, neural networks, fuzzy systems and bioinformatics.

Raphael Ritz

MPI for Plasma Physics
Garching Computing Centre of the Max
Planck Society
Garching, Germany
raphael.ritz@gmail.com

Part J

Raphael Ritz received his PhD in Theoretical Physics from the Technical University of Munich in 1995 for a study on spatio-temporal patterns in models of associative neuronal networks. At the Salk Institute for Biological Studies in San Diego he turned to biophysical neuronal modeling. Later he extended his activities into neuronal data analysis and neuroinformatics at the Institute of Advanced Studies (Wissenschaftskolleg) and the Humboldt-University in Berlin. From 2006 to 2013 he was the Scientific Officer of the International Neuroinformatics Coordinating Facility (INCF) – an international endeavor to foster neuroinformatics at a global scale. Since October 2013 he is heading the Data Service Group within the Garching Computing Centre of the Max Planck Society at the MPI for Plasma Physics, Germany.

Heike Sichtig

U.S. Food and Drug Administration
CDRH/OIR/Microbiology Devices
Silver Spring, MD, USA
Heike.Sichtig@fda.hhs.gov

Parts A, C, D, G, I

Dr. Sichtig is a subject matter expert and scientific reviewer within the Division of Microbiology Devices in the Office of In-vitro Diagnostics and Radiological Health(OIR), Centre for Devices and Radiological Health at the U.S. Food and Drug Administration (FDA). She is the Lead Technical Scientist for microbial diagnostic devices related to high throughput sequencing and bioinformatics. She is developing approaches to use alternative analytical models for assessing safety and effectiveness for these novel sequence-based diagnostic devices. Prior to joining the FDA, Dr. Sichtig developed and assessed a novel adaptive computational modeling platform for transcription factor binding site detection and discovery using machine learning techniques (artificial spiking neural networks and genetic algorithms) at the Department of Molecular Genetics and Microbiology at University of Florida. She is an IEEE senior member and the current Computational Intelligence Society (CIS) chapter subcommittee chair.

Shiro Usui

Toyohashi University of Technology
EIIRIS
Toyohashi, Japan
usui@eiiris.tut.ac.jp; usuishiro@riken.jp

Part J

Professor Shiro Usui holds a PhD from University of California, Berkeley (1974) and a is Team-Leader of Neuroinformatics at the RIKEN Brain Science Institute, Japan. His contribution is not limited to the academic activities for understanding the visual system, but also a leadership in establishing a world organization of Neuroinformatics: INCF (2006). He guided the board to promote Neuroinformatics worldwide. He also worked actively for the establishment of the INCF Japan-node where he is the first Director. In addition, he served as the organizing chair of Neuroinformatics Congress of INCF in 2010 at Kobe, Japan.

Since 2012 until today Professor Usui is Project Professor in Computational Retinal Physiology at the Electronics-Inspired Interdisciplinary Research Institute (EIIRIS), Toyohashi University of Technology. From 2007 – 2012 he was Director of the Neuroinformatics Japan Center, RIKEN BSI after having served as President of the Japanese Neural Network Society from 2005 to 2007.

List of Authors

Mones S. Abu-Asab
National Institutes of Health
National Eye Institute
10 Center Drive
Bethesda, MD 20892, USA
e-mail: *mones@mail.nih.gov*

Kamal Abuhassan
University of Ulster
Intelligent Systems Research Centre
Northland Road
Derry, Northern Ireland BT48 7JL, UK
e-mail: *Abuhassan-K@email.ulster.ac.uk*

Kazuyuki Aihara
The University of Tokyo
Institute of Industrial Science
4-6-1 Komaba, Meguro-ku
153-8505 Tokyo, Japan
e-mail: *aihara@sat.t.u-tokyo.ac.jp*

Federico Ambrogi
University of Milan
Department of Clinical Sciences and Community
Health, Section of Medical Statistics, Biometry and
Bioinformatics "Giulio A. Maccacaro"
via A.Vanzetti 5
20133 Milan, Italy
e-mail: *federico.ambrogi@unimi.it*

Periklis Andritsos
University of Toronto
Faculty of Information (iSchool)
140 St. George Street
Toronto, ONT M5S 3G6, Canada
e-mail: *periklis.andritsos@utoronto.ca*

Alessandro Astolfi
Imperial College London
Electrical and Electronic Engineering
London, SW7 2AZ, UK
e-mail: *a.astolfi@ic.ac.uk*

Annalisa Barla
University of Genova
DIBRIS
via Dodecaneso 35
16146 Genova, Italy
e-mail: *annalisa.barla@unige.it*

Niccolò Bassani
University of Milan
Department of Clinical Sciences and Community
Health
via Vanzetti, 5
20133 Milano, Italy
e-mail: *niccolo.bassani@unimi.it*

José A. Becerra Permuy
University of A Coruña
Department of Computer Science
Mendizábal s/n
15403 Ferrol, Spain
e-mail: *jose.antonio.becerra.permuy@udc.es*

Francisco Bellas
Universidade da Coruña
Escola Politecnica Superior, Department of
Computer Science
15405 Ferrol, Spain
e-mail: *francisco.bellas@udc.es*

Lubica Benuskova
University of Otago
Department of Computer Science
133 Union Street East
Dunedin, 9016, New Zealand
e-mail: *lubica@cs.otago.ac.nz*

Basabdatta S. Bhattacharya
University of Lincoln
School of Engineering
Brayford Pool
Lincoln, LN6 7TS, UK
e-mail: *bbhattacharya@lincoln.ac.uk*

Elia M. Biganzoli
University of Milan
Department of Clinical Sciences and Community
Health, Section of Medical Statistics, Biometry and
Bioinformatics "Giulio A. Maccacaro"
via A.Vanzetti 5
20133 Milan, Italy
e-mail: *elia.biganzoli@unimi.it*

Veselka Boeva
Technical University of Sofia, Branch Plovdiv
Department of Computer Systems and
Technologies
Str. Tsanko Dyustabanov 25
Plovdiv, 4000, Bulgaria
e-mail: *vboeva@tu-plovdiv.bg*

Patrizia Boracchi
University of Milan
Department of Clinical Sciences and Community
Health, Section of Medical Statistics, Biometry and
Bioinformatics "Giulio A. Maccacaro"
via A.Vanzetti 5
20133 Milano, Italy
  and
Fondazione IRCCS Istituto Nazionale Tumori
e-mail: *patrizia.boracchi@unimi.it*

Carol Brayne
University of Cambridge
Department of Public Health and Primary Care
Forvie Site, Robinson Way
Cambridge, CB2 0SR, UK
e-mail: *cb105@medschl.cam.ac.uk*

Chris M. Brown
University of Otago
Department of Biochemistry
710 Cumberland St.
Dunedin, 9016, New Zealand
e-mail: *chris.brown@otago.ac.nz*

Mónica F. Bugallo
Stony Brook University
Electrical and Computer Engineering
Stony Brook, NY 11794-2350, USA
e-mail: *monica@ece.sunysb.edu*

Matteo Cacciola
Uuniversity Mediterranea of Reggio Calabria
DICEAM
via Graziella Feo di Vito
89060 Reggio Calabria, Italy
e-mail: *matteo.cacciola@unirc.it*

Colin Campbell
University of Bristol
Department of Engineering Mathematics
Bristol, BS8 1UB, UK
e-mail: *c.campbell@bristol.ac.uk*

Mario Cannataro
University Magna Graecia of Catanzaro
Department of Medical and Surgical Sciences
viale Europa (Località Germaneto)
88100 Catanzaro, Italy
e-mail: *cannataro@unicz.it*

Mariella Caputo
Università di Salerno
Dipartimento di Farmacia
via Ponte Don Melillo
84084 Fisciano, Italy
e-mail: *macaputo@unisa.it*

Hyeygjeon Chang
Kookmin University
School of Electrical Engineering
77 Jeongneung-ro, Seongbuk-gu
Seoul 136-702, Korea
e-mail: *hchang@kookmin.ac.kr*

Badong Chen
Xi'an Jiaotong University
Institute of Artificial Intelligence and Robotics
Xi'an, 710049, P. R. China
e-mail: *chenbd@mail.xjtu.edu.cn*

Sixue Chen
University of Florida
Department of Biology
2033 Mowry Roa
Gainesville, FL 32610, USA
e-mail: *schen@ufl.edu*

Girija Chetty
University of Canberra
Faculty of Information Sciences and Engineering
Bruce Canberra, ACT 2601, Australia
e-mail: *girija.chetty@canberra.edu.au*

Yoonjoo Choi
Dartmouth College
Department of Computer Science
6211 Sudikoff Lab
Hanover, NH 03755, USA
e-mail: *yoonjoo@cs.dartmouth.edu*

Feng Chu
Siemens Ptd Ltd
The Siemens Center
60 MacPherson Road
348615 Singapore
e-mail: *chufeng67@gmail.com*

Lore Cloots
KU Leuven
Centre of Microbial and Plant Genetics
Kasteelpark 20
3001 Heverlee, Belgium
e-mail: *lore.cloots@biw.kuleuven.be*

Danila Coradini
University of Milan
Department of Clinical Sciences and Community
Health, Fondazione IRCCS Istituto Nazionale Tumori
via A.Vanzetti 5
20133 Milano, Italy
e-mail: *danila.coradini@libero.it*

Damien Coyle
University of Ulster
Intelligent System Research Centre
Northland Rd.
Derry, Northern Ireland BT48 7JL, UK
e-mail: *dh.coyle@ulster.ac.uk*

Alexandra E. D'Agostino
Stony Brook University
Department of Neurobiology
Stony Brook, NY 11794, USA
e-mail: *alexandra.dagostino@stonybrook.edu*

Joyce D'Mello
Auckland University of Technology
Knowledge Engineering and
Discovery Research Institute
120 Mayoral Drive
Auckland, 1010, New Zealand
e-mail: *joyce.dmello@aut.ac.nz*

Shaojun Dai
Northeast Forestry University
Alkali Soil Natural Environmental Science Center
Hexing Road No.26
Harbin, 150040, China
e-mail: *daishaojun@hotmail.com*

Chinh Dang
Allen Institute for Brain Science
Department of Technology
34th St.
Seattle, WA 98103, USA
e-mail: *chinhda@alleninstitute.org*

Dries De Maeyer
KU Leuven
Department of Microbial and Molecular Systems
Kasteelpark 20
3001 Heverlee, Belgium
e-mail: *dries.demaeyer@biw.kuleuven.be*

Charlotte M. Deane
University of Oxford
Department of Statistics
1 South Parks Road
Oxford, OX1 3TG, UK
e-mail: *deane@stats.ox.ac.uk*

Jim DeLeo
National Institutes of Health Clinical Center
Laboratory for Informatics Development
10 Center Drive
Bethesda, MD 20892, USA
e-mail: *jdeleo@nih.gov*

Charles DeLisi
Boston University
Program in Bioinformatics
24 Cummington St.
Boston, MA 02215, USA
e-mail: *delisi@bu.edu*

Patricia M. Di Lorenzo
Binghamton University
Department of Psychology
Binghamton, NY 13902-6000, USA
e-mail: *diloren@binghamton.edu*

Petar M. Djurić
Stony Brook University
Department of Electrical and
Computer Engineering
Stony Brook, NY 11794-2350, USA
e-mail: *djuric@ece.sunysb.edu*

Michele Donato
Wayne State University
Department of Computer Science
408 State Hall
Detroit, MI 48202, USA
e-mail: *dw2237@wayne.edu*

Stijn van Dongen
EMBL – European Bioinformatics Institute
Wellcome Trust Genome Campus
Hinxton, Cambridge CB10 1SD, UK
e-mail: *stijn@ebi.ac.uk*

Sorin Drăghici
Wayne State University
Department of Computer Science
408 State Hall
Detroit, MI 48202, USA
e-mail: *sorin@wayne.edu*

Richard J. Duro
Universidade da Coruña
Department of Computer Science
EPS, Mendizabal s/n
15403 Ferrol, Spain
e-mail: *richard@udc.es*

Anton J. Enright
EMBL – European Bioinformatics Institute
Wellcome Trust Genome Campus
Hinxton, Cambridge CB10 1SD, UK
e-mail: *aje@ebi.ac.uk*

Yue Fan
Boston University
Department of Mathematics and Statistics
111 Cummington Street
Boston, MA 02215, USA
e-mail: *yue@bu.edu*

Valery Feigin
AUT University
National Institute for Stroke and
Applied Neurosciences
90 Akoranga Drive
Northcote, Auckland 0627, New Zealand
e-mail: *valery.feigin@aut.ac.nz*

David Feng
Allen Institute for Brain Science
Department of Technology
34th St.
Seattle, WA 98103, USA
e-mail: *davidf@alleninstitute.org*

Michele Filosi
Fondazione Bruno Kessler
Predictive Models for Biomedicine and
Environment
via Sommarive, 18
38123 Povo, Italy
e-mail: *filosi@fbk.eu*

Alexandru G. Floares
SAIA, OncoPredict
Cancer Institute Cluj-Napoca
Str. Republicii, Nr. 34–36
Cluj-Napoca, 400015, Romania
e-mail: *alexandru.floares@oncopredict.com;*
alexandru.floares@saia-institute.org

Yan Fu
Chinese Academy of Sciences
Academy of Mathematics and Systems Science,
Haidian District
Beijing, 100190, China
e-mail: *yfu@amss.ac.cn*

Kunihiko Fukushima
634-3, Miwa Machida
195-0054 Tokyo, Japan
e-mail: *fukushima@m.ieice.org*

Cesare Furlanello
Fondazione Bruno Kessler
via Sommarive, 18
38123 Povo, Italy
e-mail: *furlan@fbk.eu*

Matthias E. Futschik
University of Algarve
IBB/Centre for Biomedical and
Structural Biomedicine
Campus de Gambelas
8000 Faro, Portugal
e-mail: *mfutschik@ualg.pt*

Carlo Gambacorti-Passerini
University of Milano Bicocca
Department of Health Science
via Cadore 48
20900 Monza, Italy
e-mail: *carlo.gambacorti@unimib.it*

Petia Georgieva
University of Aveiro
Department of Electronics Telecommunications
and Informatics (DETI)
Campus Universitario
3800–193 Aveiro, Portugal
e-mail: *petia@ua.pt*

Muriel Gevrey
University of Maryland
Chesapeake Biological Laboratory
1 Williams St.
Solomons, MD 20688-0038, USA
e-mail: *mgevrey@gmail.com*

Antonio Giordano
Temple University
Biology – Center for Biotechnology
Biolife Sciences Bldg., 1900 N. 12th Street
Philadelphia, PA 19122, USA
e-mail: *antonio.giordano@temple.edu*

Janice I. Glasgow
Queen's University
School of Computing
557 Goodwin Hall
Kingston, ONT. K7L 3N6, Canada
e-mail: *janice@cs.queensu.ca*

Pietro H. Guzzi
University Magna Graecia of Catanzaro
Surgical and Medical Sciences
viale Europa
88100 Catanzaro, Italy
e-mail: *hguzzi@unicz.it*

Michael Hawrylycz
Allen Institute for Brain Science
Department of Modeling, Analysis, and Theory
551 N. 34th St.
Seattle, WA 98103, USA
e-mail: *mikeh@alleninstitute.org*

Miguel A. Hernandez-Prieto
University of Algarve
IBB/Centre for Biomedical and
Structural Biomedicine
Campus de Gambelas
8000 Faro, Portugal
e-mail: *mprieto@ualg.pt*

Yoshito Hirata
The University of Tokyo
Institute of Industrial Science
4-6-1 Komaba, Meguro-ku
153-8505 Tokyo, Japan
e-mail: *yoshito@sat.t.u-tokyo.ac.jp*

Mika Hirvensalo
University of Turku
Department of Mathematics
Turku 20014, Finland
e-mail: *mikhirve@utu.fi*

Zeng-Guang Hou
Chinese Academy of Sciences
Institute of Automation, State Key Laboratory of
Management and Control for Complex Systems
Beijing, 100190, China
e-mail: *zengguang.hou@ia.ac.cn*

Yingjie Hu
Auckland University of Technology
Knowledge Engineering and
Discovery Research Institute
Auckland, 1142, New Zealand
e-mail: *rhu@aut.ac.nz*

Xiao Huang
Room 1214, City Center Plaza, 555110th St. NE
Bellevue, WA 98155, USA

Sally Hunter
University of Cambridge
Public Health and Primary Care
Forvie Site, Robinson Way
Cambridge, CB2 0SR, UK
e-mail: *seh66@medschl.cam.ac.uk*

Hitoshi Iba
The University of Tokyo
Graduate School of Information, Science and
Technology
Hongo 3, Bunkyo-ku
Tokyo, Japan
e-mail: *iba@iba.t.u-tokyo.ac.jp*

Takayoshi Ikeda
University of Otago
Dean's Department
Wellington, Wellington South 6021, New Zealand
e-mail: *tak.ikeda@otago.ac.nz*

Hiroko Imasato
Fuzzy Logic Systems Institute (FLSI)
1-5-204 Hibikino, Wakamatsu, Kitakyushu
808-0135 Fukuoka, Japan
e-mail: *imasato@flsi.or.jp*

Grant H. Jacobs
BioinfoTools
PO Box 6129
Dunedin, New Zealand
e-mail: *gjacobs@bioinfotools.com*

Igor Jurisica
University of Toronto
Department of Computer Science
6 King's College Road
Toronto, Ontario M5S 3H5, Canada
e-mail: *juris@ai.utoronto.ca*

Giuseppe Jurman
Fondazione Bruno Kessler
Predictive Models for Biomedicine and
Environment
via Sommarive, 18
38123 Povo, Italy
e-mail: *jurman@fbk.eu*

Ravi K.R. Kalathur
University of Algarve
DIBB/Centre for Biomedical and Structural
Biomedicine
Campus de Gambelas
8000 Faro, Portugal
e-mail: *rkkalathur@ualg.pt*

Nikola Kasabov
Auckland University of Technology
KEDRI – Knowledge Engineering and
Discovery Research Institute
120 Mayoral Drive
Auckland, New Zealand
e-mail: *nkasabov@aut.ac.nz*

Diana A. Kassabova
Knowledge Engineering Consulting Ltd.
47C Nihill Crescent
Auckland, 1071, New Zealand
e-mail: *diana.kassabova@gmail.com*

Sebastian Kelm
University of Oxford
Department of Statistics
1 South Parks Road
Oxford, OX1 3TG, UK
e-mail: *kelm@stats.ox.ac.uk*

Alvin T. Kho
Boston Children's Hospital
320 Longwood Avenue
Boston, MA 02115, USA
e-mail: *alvin_kho@hms.harvard.edu*

Alistair Knott
University of Otago
Department of Computer Science
Dunedin, 9016, New Zealand
e-mail: *alik@cs.otago.ac.nz*

Hiroshi Kojima
Tamagawa University
Intelligent Information Systems
6-1-1 Tamagawa Gakuen, Machida
194-8610 Tokyo, Japan
e-mail: *hkojima@lab.tamagawa.ac.jp*

Mark Kon
Boston University
Department of Mathematics and Statistics
111 Cummington St.
Boston, MA 02215, USA
e-mail: *mkon@bu.edu*

Elena Kostadinova
Technical University of Sofia, Plovdiv Branch
Department of Computer Systems and
Technologies
Str. Tsanko Dyustabanov 25
Plovdiv, 4000, Bulgaria
e-mail: *elli@tu-plovdiv.bg*

Rita Krishnamurthi
AUT University
National Institute for Stroke and Applied
Neurosciences
90 Akoranga Drive
Auckland, 0627, New Zealand
e-mail: *rita.krishnamurthi@aut.ac.nz*

Eric J. Lang
New York University
Department of Physiology & Neuroscience
School of Medicine
550 First Avenue
New York, NY 10016, USA
e-mail: *eric.lang@nyumc.org*

Gwenaël Leday
Vrije Universiteit Amsterdam
Department of Mathematics
De Boelelaan 1081a
1081 HV, Amsterdam, The Netherlands
e-mail: *g.g.r.leday@vu.nl*

Lin Li
Philips Research North America
Briarcliff Manor, NY 10510, USA
e-mail: *lin-li@philips.com*

Qingling Li
China University of Mining & Technology, Beijing
Department of Mechanical & Electrical Engineering
No. Ding 11, Xueyuan Road, Haidian District
Beijing, China
e-mail: *doudouhit@163.com*

Xingfeng Li
University of Ulster
Department of Computing and Engineering
Northland Road
Derry, Northern Ireland BT48 7JL, UK
e-mail: *x.li@ulster.ac.uk*

Wen Liang
Auckland University of Technology
School of Computing and Mathematical Science
120 Mayoral Drive
Auckland, 1142, New Zealand
e-mail: *lliang@aut.ac.nz*

Hongye Liu
Harvard Medical School/Boston Children's Hospital
Informatics Program
300 Longwood Avenue
Boston, MA 02115, USA
e-mail: *hongye.liu@gmail.com;*
hongye.liu@childrens.harvard.edu

Weifeng Liu
Jump Trading
600 W Chicago Ave.
Chicago, IL 60654, USA
e-mail: *weifeng@ieee.org*

David Looney
Imperial College London
Communication and Signal Processing Research
Group, Department of Electrical and
Electronic Engineering
Exhibition Road
London, SW7 2BT, UK
e-mail: *david.looney06@imperial.ac.uk*

Irina Luludachi
SAIA Institute
Bulevardul Nicolae Titulescu 4
Cluj-Napoca, 400420, Romania
e-mail: *irina.luludachi@saia.ro*

Marcella Macaluso
Temple University
Biology – Center for Biotechnology
1900 N. 12th Street
Philadelphia, PA 19122, USA
e-mail: *macaluso@temple.edu*

Liam Maguire
University of Ulster
Intelligent Systems Research Centre
Northland Road
Derry, Northern Ireland BT48 7JL, UK
e-mail: *lp.maguire@ulster.ac.uk*

Danilo P. Mandic
Imperial College London
Department of Electrical and Electronic
Engineering, Communication and Signal
Processing Research Group
Exhibition Road
London, SW7 2BT, UK
e-mail: *d.mandic@imperial.ac.uk*

Kathleen Marchal
Ghent University
Department of Plant Biotechnology and
Bioinformatics
Technologiepark 927
9052 Gent, Belgium
e-mail: *kamar@psb.ugent.be*

Manuela Marega
Justus-Liebig-Universität
Institut für Neuropathologie
Aulweg 123
35392 Gießen, Germany
e-mail:
manuela.marega@patho.med.uni-giessen.de

T. Martin McGinnity
University of Ulster
School of Computing and Intelligent Systems,
Computer Science Research Institute
Derry, Northern Ireland BT48 7JL, UK
e-mail: *tm.mcginnity@ulster.ac.uk*

Mariofanna Milanova
University of Arkansas at Little Rock
Department of Computer Science
University Ave.
Little Rock, AR 72204, USA
e-mail: *mgmilanova@ualr.edu*

Angela Mogavero
Università di Milano Bicocca
Department of Health Science
via Cadore 48
20900 Monza, Italy
e-mail: *angela.mogavero@gmail.com*

Micaela Montanari
Temple University
Biology-Center for Biotechnology
1900 N. 12th Street
Philadelphia, PA 19122, USA
e-mail: *montanar@temple.edu*

Francesco C. Morabito
University Mediterranea
DICEAM
via Graziella – Feo di Vito
89127 Reggio Calabria, Italy
e-mail: *morabito@unirc.it*

Giuseppe Morabito
University of Pavia
via Ferrata
Pavia, Italy
e-mail:
giuseppe.morabito01@universitadipavia.it

Tamás Nepusz
Eötvös Loránd University
Department of Biological Physics
Pázmány Péter sétány 1/a
1117 Budapast, Hungary
e-mail: *nepusz@hal.elte.hu*

Lydia Ng
Allen Institute for Brain Science
Department of Technology
551 N. 34th Street
Seattle, WA 98103, USA
e-mail: *lydian@alleninstitute.org*

Nasimul Noman
University of Tokyo
Graduate School of Engineering
IBA Laboratory
Department of Electrical Engineering and
Information Systems
Tokyo, Japan
e-mail: *noman@iba.t.u-tokyo.ac.jp,*
noman@univdhaka.edu

Gianluigi Occhiuto
University Mediterranea
DICEAM
via Graziella – Feo di Vito
89127 Reggio Calabria, Italy
e-mail: *gianluigi.occhiuto@unirc.it*

Alberto Paccanaro
University of London
Department of Computer Science
Royal Holloway
Egham, Surrey TW20 0EX, UK
e-mail: *alberto@cs.rhul.ac.uk*

Leon Palafox
The University of Tokyo
School of Electrical Engineering
Hongo 3, Bunkyo-ku
Tokyo, Japan
e-mail: *leon@iba.t.u-tokyo.ac.jp*

Cheolsoo Park
University California, San Diego
Department of BioEngineering
9500 Gilman Drive
La Jolla, CA 92093, USA
e-mail: *charles586@gmail.com*

David Parry
Auckland University of Technology
School of Computing and Mathematical Sciences
Auckland, 1142, New Zealand
e-mail: *dave.parry@aut.ac.nz*

Michael G. Paulin
University of Otago
Department of Zoology
Dunedin, 9022, New Zealand
e-mail: *mike.paulin@otago.ac.nz*

Russel Pears
Auckland University of Technology
Department of Computing and
Mathematical Sciences
2–14 Wakefield Street
Auckland, 1010, New Zealand
e-mail: *rpears@aut.ac.nz*

Rocco Piazza
University of Milano-Bicocca
Department of Health Science
via Cadore 48
20900 Monza, Italy
e-mail: *rocco.piazza@unimib.it*

Leonie Z. Pipe
Auckland University of Technology
Knowledge Engineering and Discovery
Research Institute (KEDRI)
Auckland, 1142, New Zealand
e-mail: *leonie.pipe@gmail.com*

Alessandra Pirola
University of Milano-Bicocca
Department of Health Science
via Cadore 48
20900 Monza, Italy
e-mail: *alessandra.pirola@unimib.it*

Joel Pitt
Lincoln University
c/0 Bio-Protection Research Centre
P.O. Box 84
Lincoln, 7674, New Zealand
e-mail: *joel@joelpitt.com*

José C. Príncipe
University of Florida
Department of Electrical and
Computer Engineering
Gainesville, FL 32611, USA
e-mail: *principe@cnel.ufl.edu*

Hima Raman
University of Milano Bicocca
Department of Health Science
via Cadore, 48
20900 Monza, Italy
e-mail: *ram.hima@gmail.com*

Naveed ur Rehman
COMSATS Institute of Information Technology
Park Road, Chak Shahzad
Islamabad, Pakistan
e-mail: *naveed.rehman@comsats.edu.pk*

Samantha Riccadonna
Fondazione Edmund Mach
Computational Biology Department
via E. Mach, 1
38010 S. Michele all'Adige, Italy
e-mail: *samantha.riccadonna@fmach.it*

Raphael Ritz
MPI for Plasma Physics
Garching Computing Centre of the
Max Planck Society
Boltzmannstrasse 2
85748 Garching, Germany
e-mail: *raphael.ritz@gmail.com*

Alberto Riva
University of Florida
Molecular Genetics and Microbiology
Gainesville, FL 32610, USA
e-mail: *ariva@ufl.edu*

Roberta Rostagno
University Milano Bicocca
Department of Health Science
via Cadore 48
20900 Monza, Italy
e-mail: *roberta.rostagno@unimib.it*

Reinhard Schliebs
University of Leipzig
Medical Faculty, Paul Flechsig Institute for
Brain Research
Jahnallee 59
04109 Leipzig, Germany
e-mail: *schre@medizin.uni-leipzig.de*

Stefan Schliebs
Auckland University of Technology
School of Computing and Mathematical Sciences
Auckland, 1142, New Zealand
e-mail: *stefan.schliebs@aut.ac.nz*

Filipe Silva
University of Aveiro, Campus Universitário de
Santiago
Department of Electronics, Telecommunications
and Informatics
3810-193 Aveiro, Portugal
e-mail: *fmsilva@ua.pt*

Lavneet Singh
University of Canberra
Bruce Canberra, ACT 2601, Australia
e-mail: *lavneet.singh@canberra.edu.au*

Snjezana Soltic
Manukau Institute of Technology
Engineering Centre of Excellence
South Auckland mail Centre
Manukau, 2241, New Zealand
e-mail: *ssoltic@manukau.ac.nz*

Yu Song
Beijing Jiaotong University
Department of Automation
No.3 Shang Yuan Cun, Hai Dian District
Beijing, 100044, China
e-mail: *songyu@bjtu.edu.cn*

Roberta Spinelli
University of Milano-Bicocca
Department of Health Science
via Cadore, 48
20900 Monza, Italy
e-mail: *roberta.spinelli@unimib.it*

Andrea Splendiani
Rothamsted Research
Harpenden, Hertfordshire AL5 2BG, UK
e-mail: *andrea.splendiani@intellileaf.com,
andrea.splendiani@rothamsted.ac.uk*

Margherita Squillario
University of Genova
DIBRIS
via Dodecaneso 35
16146 Genova, Italy
e-mail: *margherita.squillario@unige.it*

Stewart G. Stevens
University of Otago
Department of Biochemistry
Cumberland Street
Dunedin, 1019, New Zealand
e-mail: *stewart.stevens@otago.ac.nz*

Susan Sunkin
Allen Institute for Brain Science
Scientific Program Management
551 N. 34th Street
Seattle, WA 98103, USA
e-mail: *susans@alleninstitute.org*

Aaron Szafer
Allen Institute for Brain Science
Department of Technology
551 N. 34th Street
Seattle, WA 98103, USA
e-mail: *aarons@alleninstitute.org*

Mario F. Tecce
Università di Salerno
Dipartimento di Farmacia
via Ponte Don Melillo
84084 Fisciano, Italy
e-mail: *tecce@unisa.it*

Shoba Tegginmath
Auckland University of Technology
Computing and Mathematical Sciences
2-14 Wakefield Street
Auckland, 1142, New Zealand
e-mail: *stegginm@aut.ac.nz*

Giuseppe Tradigo
University Magna Graecia of Catanzaro
Department of Medical and Surgical Sciences
viale Europa (Località Germaneto)
88100 Catanzaro, Italy
e-mail: *gtradigo@unicz.it*

Elena Tsiporkova
Sirris, The Collective Center for the Belgian
Technological Industry
Department of ICT & Software Engineering
A. Reyerslaan 80
1030 Brussels, Belgium
e-mail: *elena.tsiporkova@sirris.be*

Shiro Usui
Toyohashi University of Technology
EIIRIS
Hibarigaoka, Tempaku
441-8580 Toyohashi, Japan
  and
INCF Japan-node (NIJC)
RIKEN BSI
Hirosawa, Wako
Saitama, 351-0198, Japan
e-mail: *usui@eiiris.tut.ac.jp; usuishiro@riken.jp*

Simona Valletta
University of Milano Bicocca
Department of Health Science
via Cadore, 48
20900 Monza, Italy
e-mail: *s.valletta@campus.unimib.it*

Pierangelo Veltri
University Magna Graecia of Catanzaro
Department of Medical and Surgical Sciences
viale Europa (Località Germaneto)
88100 Catanzaro, Italy
e-mail: *veltri@unicz.it*

Roberto Visintainer
Fondazione Bruno Kessler
Predictive Models for Biomedicine and
Environment
via Sommarive, 18
38123 Povo, Italy
e-mail: *visintainer@fbk.eu*

Lipo Wang
Nanyang Technological University
School of Electrical and Electronic Engineering
50 Nanyang Avenue
639798 Singapore
e-mail: *elpwang@ntu.edu.sg*

Juyang Weng
Michigan State University
Department of Computer Science and Engineering
East Lansing, MI 48824, USA
e-mail: *weng@cse.msu.edu*

KongFatt Wong-Lin
University of Ulster
Intelligent Systems Research Centre
Northland Road
Derry, Northern Ireland BT48 7JL, UK
e-mail: *k.wong-lin@ulster.ac.uk*

Susan P. Worner
Lincoln University
Bio-Protection Research Centre
Lincoln, 7674, New Zealand
e-mail: *worner@lincoln.ac.nz*

Wei Xie
Institute for Infocomm Research
1 Fusionopolis Way, #21-01 Connexis
138632 Singapore
e-mail: *wxie@i2r.a-star.edu.sg*

Takeshi Yamakawa
Fuzzy Logic Systems Institute (FLSI)
808-0135 Kitakyushu, Japan
e-mail: *yamakawa@flsi.or.jp*

Zhengyou Zhang
Microsoft
Microsoft Research
One Microsoft Way
Redmond, WA 98052, USA
e-mail: *zhang@microsoft.com*

Hylde Zirpoli
Università di Salerno
Dipartimento di Farmacia
via Ponte Don Melillo
84084 Fisciano, Italy
e-mail: *hzirpoli@unisa.it*

Xin Zou
The Universities of Greenwich and Kent at Medway
Medway School of Pharmacy
Central Avenue, Chatham Maritime
Chatham, Kent CT2 7NZ, UK
e-mail: *x.zou@kent.ac.uk*

Contents

Part A Understanding Information Processes
in Biological Systems

Part B Molecular Biology, Genome and Proteome Informatics

Part C Machine Learning Methods for the Analysis, Modeling and Knowledge Discovery from Bioinformatics Data

Part D Modeling Regulatory Networks: The Systems Biology Approach

Part E Bioinformatics Databases and Ontologies

Part F Bioinformatics in Medicine, Health and Ecology

Part G Understanding Information Processes in the Brain and the Nervous System

36 Information Processing in Synapses

37 Computational Modeling with Spiking Neural Networks

38 Statistical Methods for fMRI Activation and Effective Connectivity Studies

39 Neural Circuit Models and Neuropathological Oscillations

Part I Information Modeling of Perception, Sensation and Cognition

44 Modeling Vision with the Neocognitron

45 Information Processing in the Gustatory System

46 EEG Signal Processing for Brain–Computer Interfaces

47 Brain-like Information Processing for Spatio-Temporal Pattern Recognition

Part J Neuroinformatics Databases and Ontologies

Part K Information Modeling for Understanding and Curing Brain Diseases

Part L Nature Inspired Integrated Information Technologies

List of Abbreviations

β-CTF	β-carboxy-terminal fragment
1-D	one-dimensional
1-DE	one-dimensional electrophoresis
2-D	two-dimensional
2-D-DIGE	two-dimensional fluorescence difference gel electrophoresis
2-DE	two-dimensional electrophoresis
3C	chromosome conformation capture
3-D	three-dimensional

A

AAR	adaptive auto regressive
AC	adenylate cyclase
AC	alternating current
aCGH	array-comparative genomic hybridization
Ach	achromatic
ACh	muscarinic acetylcholine
AChE	acetylcholinesterase
aCML	atypical chronic myeloid leukemia
ACO	ant colony optimization
ACSF	artificial cerebrospinal fluid
ACTB	actin cytoplasmic
AD	activating domain
AD	Alzheimer's disease
ADAM	a disintegrin and metalloproteinase
ADCA	autosomal dominant cerebellar ataxias
ADDL	amyloid-derived diffusible ligand
ADHD	attention-deficit hyperactivity disorder
AE	absolute error
AER	address event representation
AFP	across-fiber pattern
AGEA	Anatomic Gene Expression Atlas
AHP	after-hyperpolarization
AI	artificial intelligence
AIC	Akaike information criterion
AICD	intracellular domain of APP
AIDS	acquired immunodeficiency syndrome
ALD	approximate linear dependency
ALS	amyotrophic lateral sclerosis
AM	aesthetic measure
Aβ	β-amyloid peptide
AMPA	α-amino-3-hydroxy-5-methyl-4-isoxazolepropionic acid
AMPAR	(amino-methylisoxazole-propionic acid) receptor
ANN	artificial neural network
ANNOVAR	annotation of genetic variants
ANOVA	analysis of variance
AOC	acoustooptical crystal
AOCD	acoustooptical crystal deflector

APACHE	acute physiology, age, chronic health evaluation
APH-1	anterior pharynx-defective-1
API	application programming interface
AP-MS	co-affinity purification followed by mass spectrometry
APP	amyloid precursor protein
APR	average overall ranking precision
AR	auto regressive
ARACNE	algorithm for the reconstruction of accurate cellular network
ARE	AU-rich element
ARMA	autoregression and moving average
ARm	alpha rhythm model
ART	adaptive resonance theory
ASI	adaptive subspace iteration
ASRS	automatic speech recognition system
ATP	adenosine triphosphate
AUC	area under curve
AUROC	area under the ROC curves

B

BACE	β-site APP cleaving enzyme
BALL	lymphotropic leukemia cell line
BAM	binary alignment format
BBD	brain-based device
BBN	Bayesian belief network
BC	blast crisis
BCGA	bacterial colony growth algorithm
BCI	brain-computer interface
BD	binding domain
BDGP	Berkeley *Drosophila* genome project
BDNF	brain-derived neurotrophic factor
BEMD	bivariate EMD
BER	bit error rate
BFA	bacterial foraging algorithm
BF	beamforming
BFO	basic formal ontology
BGO	brain-gene ontology
BI	bioinformatics
BIND	biomolecular interaction network database
BiNGO	biological networks gene ontology
BioGRID	biological general repository for interaction dataset
BioPAX	biological pathways exchange
BL	Burkitt lymphoma
BLAST	basic local alignment search tool
BLI	broad long-inhibition

BLOSUM	blocks of amino acids substitution matrix		ChIP	chromatin immunoprecipitation
BLR	brain-like robotics		CHS	cardiovascular health study
BLyS	B-lymphocyte stimulator		CI	computational intelligence
BNST	bed nucleus of the stria terminalis		CI	cross intensity
BOLD	blood oxygen level-dependent		CID	collision-induced dissociation
BP	biological process		CIRCOS	circular visualization of tabular data
BPRG	Biomedical Proteomics Research Group		CiTO	citation typing ontology
BRCA	breast cancer-associated gene		CLC	chloride channel
BSED	binding site enrichment detection		CLL	chronic lymphocytic leukaemia
BSI	broad short-inhibition		CLR	context likelihood of relatedness
BSS	blind source separation		CM	covariance model
BWA	Burrows–Wheeler alignment		CML	chronic myeloid leukemia
BWT	Burrows–Wheeler transform		CNGM	computational neurogenetic model
BY	blue–yellow		CNS	central nervous system
			CNV	copy number variation
			CO	carbon monoxide
			CoA	coenzyme A

C

CAA	cerebral amyloid angiopathy
CAE	childhood absence epilepsy
CAGE	cap analysis gene expression
CAM	cell adhesion molecule
CaMKII	calcium/calmodulin-dependent protein kinase II
cAMP	cyclic adenosine monophosphate
CARMEN	Code Analysis Repository and Modeling for eNeuroscience
CART	classification and regression trees
CASP	critical assessment of techniques for protein structure prediction
CATH	class, architecture, topology, and homologous
Cb	cerebellum
CBF	cerebral blood flow
CbN	cerebellar nuclei
CBR	case-based reasoning
CC	cellular component
CCC	chromosome conformation capture
CDK1	cyclin-dependent kinase 1
CDK5	cyclin-dependent kinase 5
cDNA	complementary DNA
CDO	chronic disease ontology
CDR	cognitive developmental robotics
CDS	coding sequence
CDSS	clinical decision support system
CE	capillary electrophoresis
CeA	central nucleus of the amygdala
cEAP	coevolutionary based algorithm for personalized modeling
CEC	Congress on Evolutionary Computation
CEMD	complex EMD
CG	compatibility grade
ChAT	choline acetyltransferase
ChIA-PET	chromatin interaction analysis using paired-end tag sequencing
ChIP-3C	ChIP-loop

COG	centre-of-gravity defuzzification method
Co-IP	coimmunoprecipitation
COPASI	complexpathway simulator
C-OWL	context ontology web language
CPT	continuous performance test
CRIPT	cysteine-rich interactor of PDZ3
CRM	cis-regulatory module
cRNA	complementary RNA
CSC	cancer stem cell
CSIM	circuit simulator
CSP	common spatial pattern
CSV	comma-separated values
CT	chorda tympani
CT	computer tomography
CTF	carboxy-terminal fragment
CTP	cytidine triphosphate
ctree	conditional tree
CUI	concept unique identifier
CV	cross validation
CVS	clustering-wise sensitivity

D

DAG	diacylglycerol
DAG	directed acyclic graph
DAUB4	Daubechies-4
DBM	low nutrient/soft agar conditions
DBN	dynamic Bayesian network
DC	direct current
DCM	dynamic causal modeling
DDBJ	DNA Data Bank of Japan
DDE	delay differential equation
ddNTP	dideoxynucleotide
DE	differential evolution
DEAD	asp-glu-ala-asp
DEG	differentially expressed gene
DENFIS	dynamic neuro-fuzzy inference system
DEP	dielectrophoresis
DFT	discrete Fourier transform

DG	dentate gyrus
DHA	docosahexaenoic acid
DICOM	digital imaging and communications in medicine
DIGE	difference in gel electrophoresis
DIP	database of interaction proteins
DLA	low nutrient/hard agar conditions
DLBCL	diffuse large B-cell lymphoma
DM	data mining
dmdo	diminuendo
DMH	differential methylation hybridization
DNA	deoxyribonucleic acid
DNMT	DNA methyltransferase
DOLCE	descriptive ontology for linguistics and cognitive engineering
DR6	death receptor 6
DRIM	discovery of rank-imbalanced motifs
dsDNA	double-strand DNA
dsRNA	double-strand RNA
DSSP	define secondary structure of proteins
DTI	diffusion tensor imaging
DTW	dynamic time warping
DWT	discrete wavelets transform
DYRK1A	dual specificity tyrosine-phosphorylation-regulated kinase 1A

E

EA	evolutionary algorithm
EBI	European Bioinformatics Institute
EC	entorhinal cortex
EC	evolutionary computation
ECD	electron-capture dissociation
ECF	evolving classification function
ECG	electrocardiogram
ECIST	expressed CpG island sequence tag
ECM	evolving clustering method
ECM	extracellular matrix
ECoG	electrocorticographic
ECOS	evolving connectionist system
ECR	energy conservation relation
ED	edit distance
EEG	electroencephalography
EEMD	extended empirical mode decomposition
EF	elongation factor
EF-G	elongation factor G
EFuNN	evolving fuzzy neural network
EGF	epidermal growth factor
EGFR	epidermal growth factor receptor
EGG	electrogastrogram
EHR	electronic health record
eIN	excitatory interneuron
EIS	evolving intelligent systems
EIV	Experiment Image Viewer
EM	expectation-maximization
EMBL	European Molecular Biology Laboratory

EMD	empirical mode decomposition
EMG	electromyogram
EMSE	excess mean square error
EMT	epithelial-to-mesenchymal transition
ENU	N-ethyl-N-nitrosourea
EOG	electrooculography
EPA	eicosapentaenoic acid
EPP	end-plate potential
EPR	Einstein–Podolsky–Rosen
EPSC	excitatory postsynaptic current
EPSP	excitatory postsynaptic potential
eQED	eQTL electrical diagram
eQTL	expression quantitative trait loci
ER	endoplasmic reticulum
ER−	estrogen receptor-negative
ER+	estrogen receptor-positive
ERA	epigenetic robotics architecture
ERD	event related desynchronization
ERK2	extracellular signal-related kinase 2
ERK	extracellular signal-regulated kinase
ERS	event-related synchronization
ES	evolutionary strategy
ESI	electrospray ionization
eSNN	evolving spiking neural network
ESST	environment-specific substitution table
EST	expressed sequence tag
ETD	electron-transfer dissociation
EWS	Ewing family of tumors
EX-RLS	extended RLS

F

FAD	familial AD
FASIC	fuzzy adaptive subspace iteration-based two-way clustering
FBN3	fibrilin 3
FCA	formal concept analysis
FDA	Federal Drug Administration
FDR	false discovery rate
FFT	fast Fourier transformation
FGF	fibroblast growth factor
FGN	fractional Gaussian noise
fIN	fast inhibitory interneurons
FIRE	finding informative regulatory element
FISH	fluorescence in situ hybridization
FL	follicular lymphoma
fMRI	functional magnetic resonance imaging
fNIR	functional near-infrared system
FNN	fuzzy neural network
FOA	focus of attention
FP	false positive
FP	fixation point
FPE	final prediction error
FPGA	field-programmable gate array
FPR	type one error rate α

FPS	frames per second		GTPBP	GTP binding protein
FRSAC	fuzzy-rough supervised attribute clustering		GWAS	genome-wide association scan
FST	follistatin			

H

FT	Fourier transform	HAT	histone acetyltransferase
FT-ICR	Fourier-transform ion cyclotron resonance	HCA	hierarchical cluster analysis
		HCN	hyperpolarization-activated, cyclic nucleotide-gated current
FT-IR	Fourier transform infrared	HCQ	hydroxychloroquine
ftp	file transfer protocol	HDAC	histone deacetylase
FWE	familywise error rate	HDACi	histone deacetylase inhibitor
FWHM	full-width at half-maximum	HeLa	Henrietta Lacks
		HiAcc	high accuracy

G

		HiCov	high coverage
GA	genetic algorithm	HILIC	hydrophilic interaction chromatography
GABA	gamma-aminobutyric acid	Hip	hippocampus
GABA$_A$R	GABA$_A$ receptor	HIPPA	Health Insurance Portability and Privacy Act
GABA$_B$R	GABA$_B$ receptor		
GABRA	GABA$_A$ receptor	HiTS-CLIP	highthroughput sequencing with crosslinking and immunoprecipitation
GABRB	GABA$_B$ receptor		
GBM	glioblastoma multiforme	HIV	human immunodeficiency virus
GC	Granger causality	H-KNN	K-local hyperplane
GCM	Granger causality model	HL-60	human promyelocytic leukemia cell
GC–MS	gas chromatography–mass spectrometry	HL7	Health Level 7
gDNA	genomic DNA	HMG	β-hydroxy-β-methyl-glutaryl
GDP	guanosine diphosphate	HMM	hidden Markov model
GE-EPI	gradient-echo, planar images	HMMR	hyaluronan-mediated motility
GEO	gene expression omnibus	HMT	histone methyltransferase
GePS	genomatix pathway system	HOSF	high-oleic acid sunflower oil
GFP	green fluorescent protein	HPLC	high-performance liquid chromatography
GFS	global field synchronization	HPRD	human protein reference database
GIS	geographical information system	HRF	hemodynamic response function
GKAP	guanylate kinase-associated protein	HRG	hierarchical random graph,
GLAM	gapped local alignment of motifs	HuEMAP	homolog of echinoderm microtubule associated protein EMAP
GLM	general linear model		
GLMM	general linear mixed model	HUPO	Human Proteome Organization
GMD	galvanometer deflector	HUPO-PSI	HUPO–Proteomics Standards Initiative
GNS	global navigation system	HWHM	half-width at half-maximum
GO	gene ontology		
GOA	GO annotation		

I

GO-id	GO category identification		
GP	Gaussian process	IAC	intelligent adaptive Curiosity
GP	genetic programming	IAO	information artifact ontology
GP	globus pallidus	IBM	individual-based model
GPCR	G protein coupled receptor	IC	information content
GPM	Global Proteome Machine	ICA	independent component analysis
GPRN	gene–protein regulatory network	ICAT	isotope-coded affinity tag
GRG	geometric random graph	ICD	international classification of disease
GRIP	glutamate receptor-interacting protein	ICR	imprinting control region
GRK	G protein-coupled receptor kinase	ICR	ion cyclotron resonance
GRN	gene regulatory network	ICR-FT	ion-cyclotron resonance Fourier transform
GSEA	gene set enrichment analysis		
GSK	glycogen synthase kinase	IDA	intelligent discovery assistant
GSO	Gram–Schmidt orthogonalization	IDFT	inverse discrete Fourier transform
GSP	greater superficial petrosal	IEX	ion exchange
GTP	guanosine triphosphate		

IEETA	Institute of Electrical Engineering and Telematics of Aveiro
IEF	isoelectric focusing
IEMG	integral of EMG
IF	initiation factor
IFM	incremental fuzzy mining
IFM	integrate-and-fire model
IGV	integrative genomics viewer
IHDR	incremental hierarchical discriminant regression
IIR	infinite impulse response
IL-1β	interleukin-1β
IMA	intelligent machine architecture
IM	images line
IMAC	ion-affinity chromatography
IMF	intrinsic mode function
IMPM	integrated method for personalized modeling
IN	interneuron
INCF	International Neuroinformatics Coordinating Facility
INDEL	insertion or a deletion
IOR	inhibition of return
IPA	ingenuity pathway analysis
IPG	immobilized pH gradient
IPSC	inhibitory postsynaptic current
IPSP	inhibitory postsynaptic potential
IR	infrared
IRE	iron responsive element
IS	information science
IS	intelligent system
ISH	in situ hybridization
ISPM	integrated optimization system for personalized modeling
IT	infero temporal cortex
iTRAQ	isobaric tags for relative and absolute quantitation
IUPAC	International Union of Pure and Applied Chemistry
IV	intravenous

J

JAK/STAT	Januskinase/signal transducers and activators of transcription
JSON	JavaScript object notation

K

KAF	kernel adaptive filtering
KAPA	kernel affine projection algorithm
KA-R	kainate receptor
KBNN	knowledge-based neural network
KBP	knotted branching pattern
KCN	kalium (potassium) voltage-gated channel

KDD	knowledge discovery in databases
KDDONTO	knowledge discovery in databases ontology
KEGG	Kyoto encyclopedia for genes and genomes
KF	Kalman filtering
KFDA	kernel Fisher discriminant analysis
KH	K homology
KINARM	kinesiological instrument for normal and altered reaching movement
K–L	Karhunen–Loeve
KL	Kullback–Leibler
KLMS	kernel least mean squares
KNN	K nearest neighbor
KPCA	kernel principal component analysis
KPI	Kunitz protease inhibitor
KRLS	kernel recursive least squares
KS	Kolmogorov–Smirnov
KSDP	kernel spectral dot product

L

LARS	least angle regression
LC	liquid chromatography
LCA	lowest common ancestor
LCD	low-calorie diet
LC-MS	liquid chromatography-mass spectrometry
LC-MS/MS	liquid chromatography tandem mass spectrometry
LCMV	linearly constrained minimum variance
lda	linear discriminant analysis
LDA	linear discriminant analysis
LFP	local field potential
LGN	lateral geniculate nucleus
LH	lateral hypothalamus
LHMM	layered hidden Markov model
LIBSVM	library for support vector machines
LIF	leaky integrate-and-fire neuron
LIFM	leaky IFM
LL	labeled line
LM	likelihood multiplier
LMA	ligation-mediated amplification
LMA	ligation-mediated annealed
LM MDR	log-linear model-based multifactor dimensionality reduction
LMS	least mean square
LOD	linked open data
log	logistic regression
LOH	loss of heterozygosity
LOI	loss of imprinting
LOO	leave-one-out
LOOCV	leave-one-out cross validation
LPC	linear predictive coefficient
LPD	latent process decomposition
LR	likelihood ratio

LRP	lipoprotein receptor-related protein
LRP1	lipoprotein receptor-related protein 1
LRT	linear regression technique
LSID	life science identifier
LSM	liquid state machine
LT	lingual-tonsillar branch of the glossopharyngeal nerve
LTD	long-term depression
LTM	long-term memory
LTP	long-term potentiation
LTQ	linear ion trap
LVQ	learning vector quantization

M

mAChR	cholinergic muscarinic receptor
MAF	minor allele frequency
MAF	multiple alignment format
MAGE-ML	microarray and gene expression markup language
MAGE-OM	microarray gene expression object model
MALDI	matrix assisted laser desorption/ionization
MANOVA	multivariate analysis of variance
MAP	maximum a posteriori
MAPK	mitogen-activated protein kinase
MAR	multivariate autoregression
MAST	motif selection and validation algorithm
MC	memory complainer
mCI	memoryless CI kernel
MCI	mild cognitive impairment
MCL	Markov clustering
MCMC	Markov chain Monte Carlo
MCODE	molecular complex detection
MDA	melanoma differentiation-associated
MDB	multi-level Darwinist brain
MDIG	modular dispersal in GIS
MDP	Markov decision process
MEG	magnetoencephalography
MEMD	multivariate EMD
MEMD-CSP	multivariate EMD-common spatial pattern
MEME	multiple expectation maximization for motif elicitation
MEMERIS	multiple EM for motif elucidation in RNAs including secondary structures
mEPP	miniature end-plate potential
MeSH	medical subject heading
MET	met proto-oncogene
MF	median frequency
MF	membership function
MF	molecular function
MFCC	mel-frequency cepstral coefficient
MFE	minimum free energy
MI	molecular interaction

MI	mutual information
MIAME	minimum information about a microarray experiment
MIAPE	minimum information about a proteomics experiment
MIMO	multiple-output
MINC	medical imaging NetCDF
MiNET	mutual information networks package
MINT	Molecular INTeraction
MIPS	Munich Information Center for Protein Sequences
miRNA	microRNA
mirSVR	micro support vector regression
MKL	multiple kernel learning
ML	machine learning
ML	maximum likelihood
MLE	maximum-likelihood estimator
MLP	multilayer perceptron
MLR	multiple linear regression
mmCIF	macromolecular crystallographic information file
MMP	matrix metalloproteinase
MMR	maximum matching ratio
MNI	4-methoxy-7-nitroindolinyl
MOC	metal oxide chromatography
MOE4CBR	mixture of experts for CBR
MP	maximum parsimony
MPF	mean power frequency
MPPI	Mammalian Protein–Protein Interaction Database
MP-RAGE	magnetization prepared rapid gradient echo
MQAP	model quality assessment program
MR	magnetic resonance
MRI	magnetic resonance imaging
mRNA	messenger RNA
MS	mass spectrometry
MSE	mean square error
MSG	monosodium glutamate
MS/MS	tandem mass spectrometry
MT	master line
MT	middle temporal cortex
MTG	middle temporal gyrus
MU	motor unit
MUAP	motor unit action potential
MudPIT	multidimensional protein identification technology
MUSIC	multi-simulation coordinator
mzData	mass spectrometry data

N

nAChR	nicotinic acetylcholine receptor
NADPH	nicotinamide adenine dinucleotide phosphate
NA-MEMD	noise-assisted MEMD

NanoESI	nanoelectrospray ionization
NAPP	N-terminal fragment of amyloid precursor protein
NARX	autoregressive with exogenous inputs
NARMAX	autoregressive moving-average model with exogenous input
nb	naive Bayes
NB	neuroblastoma
NBC	naive Bayesian classifier
NC	novelty criterion
NCBI	National Center for Biotechnology Information
NCBO	National Center for Biomedical Ontologies
nCI	nonlinear cross intensity kernel
NC-KLMS	NC kernel least mean square
ncRNA	noncoding RNA
NDEI	nondimensional error index
NFT	neurofibrillary tangle
NGF	nerve growth factor
NGS	next-generation sequencing
NHL	non-Hodgkin's lymphoma
NHS	N-hydroxysuccinimidyl
NI	neuroinformatics
NIEpi	normal epithelial
NIH	National Institutes of Health
NIJC	Neuroinformatics Japan Center
NINDS	National Institute of Neurological Disorders and Stroke
NLM	US National Library of Medicine
NLP	natural-language processing
NMDA	N-methyl-D-aspartate
NMDAR	(N-methyl-D-aspartate acid) NMDA receptor
NMDR	(N-methyl-D-aspartate acid) NMDA receptor
NMJ	neuromuscular junction
NMR	nuclear magnetic resonance
NN	neural network
NNET	neural network
NO	nitric oxide
NP	neuritic plaque
NP-hard	nondeterministic polynomial-time hard
NPRC	Neuroscience Peer Review Consortium
NSIM	nonlinear system identification method
NTRK	neurotrophic tyrosine receptor kinase
NTS	nucleus of the solitary tract

O

OBI	ontology of biomedical investigation
OBO	open biology ontology
OBO	open biomedical ontology
ODE	ordinary differential equation

ODMDP	observation-driven Markov decision process
OLM	oriens lacunosum moleculare
ORC	origin recognition complex
ORF	open reading frame
OVTK	Ondex Visual Tool Kit
OWL	ontology web language
OWL-DL	OWL description logic

P

p38-MAP	P38 mitogen-activated protein
p75NTR	p75 neurotrophin receptor
PAGE	polyacrylamide gel electrophoresis
par-CLIP	photoactivatable ribunocleoside enhanced-CLIP
PBMC	peripheral blood mononuclear cell
PbN	parabrachial nucleus of the pons
PBS	phosphate-buffered saline
PC	posterior cingulate cortex
PC12	pheochromocytoma 12
PCA	principle component analysis
PCM	possibilistic c-mean
pCNGM	probabilistic computational neurogenetic model
PCR	polymerase chain reaction
PCV	phase coherence value
PD	Parkinson's disease
PDAPP	PD amyloid precursor protein
PDB	protein data bank
PDE	partial differential equation
PDGFR	platelet-derived growth factor receptor
PDSP	Psychoactive Drug Screening Program
PE	persistence of excitation
PEN-2	presenilin-enhancer-2
PET	positron emission tomography
PF	particle filter
PF	particle filtering
PFC	prefrontal cortex
PFK	phosphofructokinase
PGD2	prostaglandin D2
PGH2	prostaglandin H2
Ph	Philadelphia chromosome
PHAT	predicted hydrophobic and transmembrane
PHF	paired helical filament
PhI	phosphorylation interaction
PI3K	phosphatidylinositol 3-kinase
pI	isoelectric point
PINV	Moore–Penrose pseudoinverse
PIR	protein information resource project
PKA	protein kinase A
PKC	protein kinase C
PKD1L3	polycystic kidney disease 1-like 3
PKD2L1	polycystic kidney disease 2-like 1
PKU	phenylketonuria

PLDE	piecewise-linear differential equation
PM	prophylactic mastectomy
PMF	personalized modeling framework
PML-RAR	promyelocytic leukemia-retinoic acid receptor
PMS	personalized modeling system
PNGase F	N-glycosidase F
POMPD	partially observable MDP
PP	Poisson process
PP	posterior parietal cortex
PPARα	peroxisome proliferator-activated receptor-α
PPI	protein–protein interaction
PPV	positive predictive value
PRESS	predicted sum of squares
PRIDE	Proteomics Identifications Database
PRN	protein regulatory network
PROM	probabilistic regulation of metabolism
proNGF	NGF precursor
PRP	pre-mRNA processing
PS	presenilin
PSD	positive semidefinite
PSD	postsynaptic density
PSD	power spectral density
PSI	proteomics standard initiative
pSILAC	protein stable labeling by amino acids in cell culture
pSNN	probabilistic spiking neural network
PSO	particle swarm optimization
PSP	post-synaptic potential
PSSM	position-specific scoring matrix
PSWM	position-specific weight matrix
PTEN	phosphatase and tensin homolog
PTM	posttranslational modification
PTZ	pentylenetetrazol
Pu	putamen
PUFA	polyunsaturated fatty acid
PV	parvalbumin
PWM	position weight matrix
PY	pyramidal cell

Q

QA-R	quisqualate receptor
QD	quantum dot
qda	quadratic discriminant analysis
QDA	quadratic discriminant analysis
QDE	qualitative differential equation
QI	quantum inspired
QIEC	quantum inspired methods of evolutionary computation
QiSNN	quantum-inspired spiking neural network
QKLMS	quantized KLMS
QKLMS-GU	QKLMS with global update
QMEAN	qualitative model energy aNalysis
QoS	quality of service

QP	quadratic programming
QReg	quantized regressor
QSAR	quantitative structure-activity relationship
QTOF	quadrupole TOF
QTRAP	quadrupole ion-trap

R

R	R programming language
RAAM	recursive auto-associative memory
RAGE	receptor for advanced glycation end products
RAHC	rheumatoid arthritis, healthy control
RAOA	rheumatoid
RAS	rat sarcoma
Rb	retinoblastoma
RBC	erythrocytes
RBF	radial basis function
RBF	radical basis function
RBPDb	RNA-Binding Protein DataBase
RCT	randomized controlled trial
RDF	resource description framework
RDFS	resource description framework schema
RE	restriction endonuclease
RE	restriction enzyme
REDUCE	regulatory element detection using correlation with expression
REML	restricted maximum-likelihood
REST	representational state transfer
ReSuMe	remote supervised method
RF	releasing factor
RG	red–green
RHAMM	hyaluronan-mediated motility receptor
RI-EMD	rotation-invariant EMD
RIM	Rab3-interacting module
RIP-chip	RNA immunoprecipitation chip
RKHS	reproducing kernel Hilbert space
RKL	rank of the last relevant item
RLS	recursive least-squares
RM	reduction mammoplasty
RMP	resting membrane potential
RMS	root-mean-square
RMSD	root-mean-square deviation
RMSE	root mean squared error
RNA	ribonucleic acid
RN	red nucleus
RN	regularization network
RNAi	RNA interference
RNA-seq	next-gen sequencing
RNAseq	next-gen sequencing
RNN	recurrent neural network
RNSC	restricted neighborhood search clustering

rNTS	rostral nucleus of the solitary tract
RO	relations ontology
ROC	receiver operating characteristic
RODES	reversing ordinary differential equation system
ROI	region of interest
ROL	relative operating level
ROS	reactive oxygen species
RP	reversed-phase
R&R	gage repeatability reproducibility
RRM	RNA recognition motif
rRNA	ribosomal RNA
RSA	Rivest, Shamir, Adleman
RSAT	regulatory sequence analysis tools
RSS	residual sum of squares
RST	random set theory
RSVM	recursive SVM
RTA	real time analyzer

S

SA	simulated annealing
SAD	sporadic AD
SADI	semantic automated discovery and integration
SAGE	serial analysis of gene expression
SAM	sequence alignment map
SAM	surface adhesion molecule
SAMtools	tools for sequence alignment maps
SAP-102	synapse-associated protein-102
SASE	self-aware and self-effecting architecture
SBEAMS	Systems Biology Experiment Analysis Management System
SBGN	Systems Biology Graphical Notation
SBML	systems biology markup language
SBW	Systems Biology Workbench
SC	Schwartz's criterion
SCX	strong cation exchange
SCFG	stochastic context-free grammar
SCG	subcallosal gyrus
SCI	structure conservation index
SC-KLMS	Schwartz's criterion kernel least mean squares
SCN	sodium voltage-gated channel
SCOP	structural classification of proteins
SCS	sequencing control software
SD	standard deviation
SDM	species distribution model
SDP	semidefinite programming
SDP	spectral dot product
SDS	sodium dodecyl sulfate
SDSP	spike driven synaptic plasticity
SDS-PAGE	sodium dodecyl sulfate-polyacrylamide gel electrophoresis
SECIS	selenocysteine insertion sequence

SELDI	surface-enhanced laser desorption ionization
SELEX	systematic evolution of ligands by exponential enrichment
SEM	scanning electron microscopy
sEMG	surface electromyography
SF	straight filament
SFG	superior frontal gyrus
SFS	sequential sequential feature selection
SFS	sequential forward selection
SGS	squid giant synapse
SI	silhouette index
SI	substantia innominata
SI	swarm intelligence
SIB	Swiss Institute of Bioinformatics
SIFT	sorts intolerant from tolerant
SILAC	stable-isotope labeling with amino acids in cell culture
sIN	slow inhibitory interneuron
siRNA	small interfering RNA
SKOS	simple knowledge organization system
SLE	systemic lupus erythematosus
SLIM	scorematrx leading intramembrane
SLN	superior laryngeal branch of the vagus nerve
SM	sensory memory
SMC	sequential Monte Carlo
SMR	sensorimotor rhythm
SNAP	soluble NSF attachment protein
SNARE	SNAP receptor
SNN	spiking neural network
SNOMED	Systematized Nomenclature of Medicine-Clinical Terms
SNOP	Systematized Nomenclature of Pathology
snoRNA	small nucleolar RNA
SNP	single-nucleotide polymorphism
SNR	signal-to-noise ratio
snRNA	small nuclear RNA
SNV	single nucleotide variant
SOM	self-organizing map
SORL	sortilin-related sorting receptor
SOS	sum of squares
SPARQL	a query system for the Semantic Web
SPECT	single photon emission computerized tomography
SPF	standard PF
SPINE	safe programmable and integrated network environment
SPM	Statistical Parametric Mapping
SQL	structured query language
SQUID	super-conducting-quantum-interference-device
SRBCT	small round blue cell tumor
SRDA	spectral regression discriminant analysis

SRM	spike response model		TRC	taste receptor cell
SRN	simple recurrent network		TRI	integrated transcriptional
sRNA	small RNA		trkA	high-affinity receptor
SRS	sequence retrieval system		tRNA	transfer RNA
SRT	saccadic reaction time		TRN	thalamic reticular nucleus
SSAP	sequential structure alignment program for protein structure comparison		TS	T-score
			TSA	test set accuracy
ssDNA	single-stranded DNA		t-SNARE	target soluble NSF attachment receptor
SSE	squares due to error			
SSI	selective short-inhibition		TSS	true skill statistic
SSTD	spatio and spectro-temporal data		TWNFI	transductive weighted neuro-fuzzy inference engine
SSVEP	steady state visual evoked potential			
STD	spatio-temporal data		TWRBF	transductive inference based radial basis function
STDP	spike-timing dependent plasticity			
STFT	short-time Fourier transform			
STM	short-term memory			
STPR	spatio-temporal pattern recognition			
STRING	search tool for the retrieval of interacting genes			

U

STS	superior temporal sulcus		UCSC	University of California Santa Cruz
SUMO	suggested upper merged ontology		UML	unified medical language
SV	support vector		UMLS	unified medical language system
SVD	singular value decomposition		UniHI	Unified Human Interactome database
svm	support vector machine		UPLC	ultra-performance liquid chromatography
SVM	support vector machine			
SWB	systems biology workbench		URI	uniform resource identifier
SWD	slow-wave discharge		URL	unified resource locator
SynGAP	synaptic GTPase-activating protein		UTP	uridine triphosphate
			UTR	untranslated regions
			UV	ultraviolet

T

T1R1	taste receptor type 1 member 1		VA	visual attention
T1R2	taste receptor type 1 member 2		VAChT	vesicular acetylcholine transporter
T1R3	taste receptor type 1 member 3		VAMP	vesicle associated membrane protein
TAP	tandem affinity purification		VC	Vapnik–Chervonenkis
TC	thalamo-cortical		VCX	primary visual cortex
TCR	thalamocortical relay cell		VCL	vinculin
TE	Tsallis entropy		VEGF	vascular endothelial growth factor
TEMD	trivariate EMD		VEP	visual evoked potential
TF	transcription factor		VEP	visually evoked potential
TFBS	transcription factor binding site		VF	visual field
tf-idf	term-frequency, inverse document-frequency		VIP	vasoactive intestinal peptide
			VOR	vestibulo-ocular reflex
Tg	transgenic		VPL	ventral posterolateral nucleus
TG	triacylglyceride		VPMpc	parvocellular subdivision of the ventral posteromedial nucleus of the thalamus
TGF	transforming growth factor			
Th1	T-helper cell		VQ	vector quantization
TLE	temporal lobe epilepsy		v-SNARE	vesicle SNARE
TM	transmembrane		VTA	ventral tegmental area
TMS	transcranial magnetic stimulation			
TNF	tumor necrosis factor			

W

TOF	time-of-flight		WAMP	Willison amplitude
TOF-TOF	tandem time-of-flight		WBC	leukocytes
TPA	third party annotation		WEKA	Waikato environment for knowledge analysis
TR	repetition time			
TRANSFAC	transcription factor database			

WEP	weight error power
WGCNA	weighted gene co-expression network analysis
WGCN	weighted gene co-expression network
WGN	white Gaussian noise
WHO	World Health Organization
WHS	Waxholm space
WKNN	weighted nearest neighbor
WPT	wavelet packet transform
WTA	winner-take-all
WT	wavelet transform
WVD	Wigner–Wille distribution
WWKNN	weighted distance and weighted variables K nearest neighbor

X

XIAPE	XML information about a proteomics experiment
XML	extensible markup language

Y

Y2H	yeast two hybrid
YMF	yeast motif finder

Z

ZC	zero crossing

1. Understanding Nature Through the Symbiosis of Information Science, Bioinformatics, and Neuroinformatics

Nikola Kasabov

This chapter presents some background information, methods, and techniques of information science, bio- and neuroinformatics in their symbiosis. It explains the rationale, motivation, and structure of the Handbook that reflects on this symbiosis. For this chapter, some text and figures from [1.1] have been used. As the introductory chapter, it gives a brief overview of the topics covered in this *Springer Handbook of Bio-/Neuroinformatics* with emphasis on the symbiosis of the three areas of science concerned: information science (informatics) (IS), bioinformatics (BI), and neuroinformatics (NI). The topics presented and included in this Handbook provide a far from exhaustive coverage of these three areas, but they clearly show that we can better understand nature only if we utilize the methods of IS, BI, and NI, considering their integration and interaction.

1.1 Nature as the Ultimate Inspiration and Target for Science

Science aims at understanding nature. Scientific methods are inspired by principles from nature, too. The beauty of our world is that, along with the fascinating geographical formations, it has life as a variety of biological species. Most importantly, it has the highest level of life: the conscious brain. Nature and life have been the ultimate inspirations and targets for three important areas of science: information science (IS, also called informatics), bioinformatics (BI), and neuroinformatics (NI).

IS deals with *generic methods* for information processing. BI applies these methods to biological data and also develops its own *specific* information processing methods. NI applies the methods of IS to brain and

nervous system data and also develops its own *specific* methods.

Nature evolves in time. The most obvious example of an evolving process is life. Life is defined in the *Concise Oxford English Dictionary* as follows:

> *a state of functional activity and continual change peculiar to organized matter, and especially to the portion of it constituting an animal or plant before death; animate existence; being alive.*

Continual change, along with certain stability, is what characterizes life. Modeling living systems requires that the continuous changes are represented in the model,

i. e., that the model adapts in a lifelong mode and at the same time preserves some features and principles that are characteristic to the process. This stability–plasticity dilemma is a well-known principle of life. In a living system, evolving processes are observed at different levels (Fig. 1.1).

At the quantum level, particles are in a complex evolving state all the time, being at several locations at the same time, which is defined by probabilities. General evolving rules are defined by several principles, such as entanglement, superposition, etc. [1.2–6]. The discovery of atomic structure by physics and chemistry revolutionized understanding of these quantum principles. Among the scientists who contributed to this is Ernest Rutherford (1871–1937) (Fig. 1.2).

At a molecular level, ribonucleic acid (RNA) and protein molecules, for example, evolve and interact in a continuous way based on the deoxyribonucleic acid (DNA) information and on the environment. The central dogma of molecular biology constitutes a general evolving rule, but what are the specific rules for different species and individuals? The area of science that deals with the information processing and data manipulation at this level is BI. At the cellular level (e.g., neuronal cells) all the metabolic processes, cell growth, cell division, etc., are evolving processes [1.7].

At the level of cell ensembles, or at a neural network level, an ensemble of cells (neurons) operates in concert, defining the function of the ensemble or the network through learning, for instance, perception of sound, perception of an image, or learning languages [1.8].

In the human brain, complex dynamic interactions between groups of neurons can be observed when certain cognitive functions are performed, e.g., speech and language learning, visual pattern recognition, reasoning, and decision making [1.9–11].

At the level of a population of individuals, species evolve through evolution (Charles Darwin). A bio-

logical system evolves its structure and functionality through both lifelong learning of an individual and the evolution of populations of many such individuals. In other words, an individual is a result of the evolution of many generations of populations, as well as a result of its own developmental lifelong learning processes [1.11, 12].

Processes at different levels from Fig. 1.1 show general characteristics, such as those listed below:

1. *Frequency*: Frequency, denoted as F, is defined as the number of signal/event repetitions over a period of time T (seconds, minutes, centuries). Some processes have stable frequencies, but others change their frequencies over time. Different processes from Fig. 1.1 are characterized by different frequencies defined by their physical parameters. Usually, a process is characterized by a spectrum of frequencies. Different frequency spectrums characterize brain oscillations (e.g., delta waves), speech signals, image signals, and quantum processes.

2. *Energy*: Energy is a major characteristic of any object or organism. Albert Einstein's most celebrated formula defines energy E as depending on the mass of the object m and the speed of light c as $E = mc^2$. Defining the energy in a living system is more complicated. The energy of a protein, for example, depends not only on the DNA sequence that is translated into this protein, but on the three-dimensional (3-D) shape of the protein and on external factors.

3. *Information*: Information is a characteristic that can be defined in different ways, as discussed in Sect. 1.2.

4. *Interaction*: There are many interactions within each of the six levels from Fig. 1.1 and across these levels. Interactions are what make a living organism complex. Understanding them is also a challenge for BI and NI; For example, there are complex interactions between genes in a genome, and between

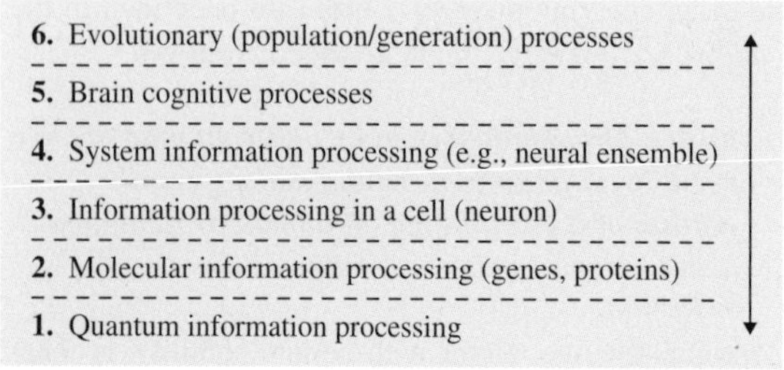

6. Evolutionary (population/generation) processes

5. Brain cognitive processes

4. System information processing (e.g., neural ensemble)

3. Information processing in a cell (neuron)

2. Molecular information processing (genes, proteins)

1. Quantum information processing

Fig. 1.1 Six levels of evolving processes in a higher-level living organism (after [1.1])

Fig. 1.2 Ernest Rutherford (1871–1937)

proteins and DNA. There are complex interactions between the genes and the functioning of each neuron, a neural network, and the whole brain [1.13].

Abnormalities in some of these interactions are known to cause brain diseases, and many of them remain unknown at present [1.14].

1.2 Information Science (IS)

Information science is the area of science that develops generic methods and systems for information and knowledge processing, regardless of the domain specificity of this information.

1.2.1 The Scope of IS

IS incorporates the following subject areas:

- Data collection and data communication (sensors and networking)
- Information storage and retrieval (database systems)
- Methods for information processing (information theory)
- Creating computer programs and information systems (software engineering and system development)
- Acquisition, representing, and processing knowledge (knowledge engineering)
- Creating intelligent systems and machines (artificial intelligence).

Generally speaking, *data* are raw entities: numbers, symbols etc., e.g., 36.

Information is labeled, understood, interpreted data, e.g., the temperature of the human body is 36 °C.

Knowledge is the understanding of a human, the way we do things, the interpretable in different situations, general information; e.g., IF the human temperature is between 36 °C and 37 °C degrees, THEN the human body is in a healthy state.

Some basic ways to represent data, information, and knowledge are presented in Sect. 1.2.2.

1.2.2 Probability, Entropy, and Information

The formal theory of probability relies on the following three axioms, where $p(E)$ is the probability of an event E to happen and $p(\neg E)$ is the probability of an event not to happen. $E_1, E_2, \ldots, E_k$ is a set of mutually exclusive events that form a universe U:

Axiom 1.1
$0 \le p(E) \le 1$.

Axiom 1.2
$\sum p(E_i) = 1$, $E_1 \cup E_2 \cup \ldots \cup E_k = U$, U-problem space.

Corollary 1.1
$p(E) + p(\neg E) = 1$.

Axiom 1.3
$p(E_1 \vee E_2) = p(E_1) + p(E_2)$, where E_1 and E_2 are mutually exclusive events.

Probabilities are defined as:

- Theoretical: some rules are used to evaluate the probability of an event.
- Experimental: probabilities are learned from data and experiments – throw a die 1000 times and measure how many times the event "getting a 6" has happened.
- Subjective: probabilities are based on common-sense human knowledge, such as defining that the probability of getting a 6 after throwing a die is (1/6)th, without really throwing it many times.

A random variable x is characterized at any moment of time by its uncertainty in terms of what value this variable will take in the next moment – its *entropy*. A measure of uncertainty $h(x_i)$ can be associated with each random value x_i of a random variable x, and the total uncertainty $H(x)$, called the *entropy*, measures our lack of knowledge, the seeming disorder in the space of the variable x

$$H(X) = \sum_{i=1,\ldots,n} p_i h(x_i), \qquad (1.1)$$

where p_i is the probability of the variable x taking the value x_i.

The following axioms for the entropy $H(x)$ apply:

- Monotonicity: if $n > n'$ is the number of events (values) that a variable x can take, then $Hn(x) > Hn'(x)$; so, the more values x can take, the greater the entropy.

Fig. 1.3 Claude Shannon (1916–2011)

- Additivity: if x and y are independent random variables, then the joint entropy $H(x, y)$, meaning $H(x$ AND $y)$, is equal to the sum of $H(x)$ and $H(y)$.

The following log function satisfies these two axioms:

$$h(x_i) = \log(1/p_i) . \qquad (1.2)$$

If the log has base 2, the uncertainty is measured in [bit], and if it is the natural logarithm ln, then the uncertainty is measured in [nat],

$$H(x) = \sum_{i=1,\dots,n} [p_i h(x_i)] = -c \sum_{i=1,\dots,n} (p_i \log p_i) , \qquad (1.3)$$

where c is a constant.

Based on *Shannon's* (Fig. 1.3) measure of uncertainty – *entropy* – we can calculate an overall probability for a successful prediction for all states of a random variable x, or the predictability of the variable as a whole

$$P(x) = 2^{-H(x)} . \qquad (1.4)$$

The maximum entropy is calculated when all the n values of the random variable x are equiprobable, i. e., have the same probability $1/n$ – a uniform probability distribution

$$H(x) = -\sum_{i=1,\dots,n} p_i \log p_i \le \log n . \qquad (1.5)$$

The *joint entropy* between two random variables x and y (for example, an input and an output variable in a system) is defined by the formulas

$$H(x, y) = -\sum_{i=1,\dots,n} p(x_i \text{ AND } y_j)$$
$$\times \log p(x_i \text{ AND } y_j) , \qquad (1.6)$$

$$H(x, y) \le H(x) + H(y) . \qquad (1.7)$$

The *conditional entropy*, i. e., the uncertainty of a variable y (output variable) after observing the value of a variable x (input variable), is defined as

$$H(y|x) = -\sum_{i=1,\dots,n} p(x_i, y_j) \log p(y_j|x_i) , \qquad (1.8)$$

$$0 \le H(y|x) \le H(y) . \qquad (1.9)$$

Entropy can be used as a measure of the information associated with a random variable x, its uncertainty, and its predictability.

The *mutual information* between two random variables, also simply called the *information*, can be measured as

$$I(y; x) = H(y) - H(y|x) . \qquad (1.10)$$

The process of online information entropy evaluation is important, as in a time series of events; after each event has happened, the entropy changes and its value needs to be reevaluated.

Information models based on probability include:

- Bayesian classifiers
- Hidden Markov models (HMM).

A *Bayesian classifier* uses a conditional probability estimated to predict a class for a new datum (1.11), which is represented as the conditional probability between two events C and A, known as the Bayes formula (Tamas Bayes, 18th century)

$$p(A|C) = p(C|A)p(A)/p(C) . \qquad (1.11)$$

It follows from (1.11) that

$$p(A \wedge C) = p(C \wedge A) = p(A|C)p(C)$$
$$= p(C|A)p(A) . \qquad (1.12)$$

Problems with the Bayesian learning models relate to unknown prior probabilities and the requirement of a large amount of data for more accurate probability calculation. This is especially true for a chain of events $A, B, C, \dots$, where the probabilities $p(C|A, B), \dots$, etc. need to be evaluated. The latter problem is addressed in techniques called hidden Markov models (HMM).

HMM [1.15] is a technique for modeling the temporal structure of a time series or sequence of events. It is a probabilistic pattern-matching approach which models a sequence of patterns as the output of a random process. The HMM consists of an underlying Markov chain

$$P[q(t+1)|q(t), q(t-1), q(t-2), \dots, q(t-n)]$$
$$\approx P[q(t+1)|q(t)] , \qquad (1.13)$$

where $q(t)$ is state q sampled at time t.

1.2.3 Statistical Information Methods

Correlation coefficients represent possible relationships between variables. For every variable x_i ($i = 1, 2, \ldots, d_1$), its correlation coefficients $\mathrm{Corr}(x_i, y_j)$ with all other variables y_j ($j = 1, 2, \ldots, d_2$) are calculated. Equation (1.14) is used to calculate the Pearson correlation between two variables x and y based on n values for each of them

$$\mathrm{Corr} = \mathrm{SUM}_i \frac{(x_i - \mathrm{M}x)(y_i - \mathrm{M}y)}{(n-1)\mathrm{Std}x\mathrm{Std}y} , \tag{1.14}$$

where $\mathrm{M}x$ and $\mathrm{M}y$ are the mean values of the two variables x and y, and $\mathrm{Std}x$ and $\mathrm{Std}y$ are their respective standard deviations.

The *t*-test and the signal-to-noise ratio (SNR) evaluate how important a variable is to discriminate samples belonging to different classes. For the case of a two-class problem, the SNR ranking coefficient for a variable x is calculated as an absolute difference between the mean value $\mathrm{M}1x$ of the variable for class 1 and the mean $\mathrm{M}2x$ of this variable for class 2, divided by the sum of the respective standard deviations

$$\mathrm{SNR\_}x = \mathrm{abs} \frac{\mathrm{M}1x - \mathrm{M}2x}{\mathrm{Std}1x + \mathrm{Std}2x} . \tag{1.15}$$

A similar formula is used for the *t*-test:

$$t\text{-test}\_x = \mathrm{abs} \frac{\mathrm{M}1x - \mathrm{M}2x}{\mathrm{Std}1x^2/N_1 + \mathrm{Std}2x^2/N_2} , \tag{1.16}$$

where N_1 and N_2 are the numbers of samples in class 1 and class 2, respectively.

Principal component analysis (PCA) aims at finding a representation of a problem space X defined by its variables $X = \{x_1, x_2, \ldots, x_n\}$ into another orthogonal space having a smaller number of dimensions defined by another set of variables $Z = \{z_1, z_2, \ldots, z_m\}$, such that every data vector x from the original space is projected onto a vector z of the new space, so that the distance between different vectors in the original space X is maximally preserved after their projection into the new space Z.

Linear discriminant analysis (LDA) is a transformation of classification data from the original space into a new space of LDA coefficients that has an objective function to preserve the distance between the samples, using also the class label to make them more distinguishable between the classes.

Multiple Linear Regression Methods (MLR)

The purpose of MLR is to establish a quantitative relationship between a group of p independent variables

(X) and a response y

$$y = XA + b , \tag{1.17}$$

where p is the number of independent variables, y is an $n \times 1$ vector of observations, X is an $n \times p$ matrix of regressors, A is a $p \times 1$ vector of parameters, and b is an $n \times 1$ vector of random disturbances. The solution to the problem is a vector, A', which estimates the unknown vector of parameters.

The *least-squares* solution is used so that the linear regression formula or another model approximates the data with the least root-mean-square error (RMSE) as

$$\mathrm{RMSE} = \mathrm{SQRT} \left\{ \underset{i=1,2,\ldots,n}{\mathrm{SUM}} \frac{(y_i - y_i')^2}{n} \right\} , \tag{1.18}$$

where y_i is the desired value from the dataset corresponding to an input vector x_i, y_i' is the value obtained through the regression formula for the same input vector x_i, and n is the number of samples (vectors) in the dataset.

Another error measure is also used to evaluate the performance of the regression and other models – the nondimensional error index (NDEI) – the RMSE divided by the standard deviation of the dataset:

$$\mathrm{NDEI} = \mathrm{RMSE}/\mathrm{Std} . \tag{1.19}$$

1.2.4 Machine-Learning Methods

Machine learning is an area of IS concerned with the creation of information models from data, the representation of knowledge, and the elucidation of information and knowledge from processes and objects. Machine learning includes methods for feature selection, model creation, model validation, and knowledge extraction.

One of the widely used machine-learning method is artificial neural networks (ANNs) [1.9, 11, 16–22].

ANNs are computational models that mimic the nervous system in its main function of adaptive learning. An ANN consists of small processing units – artificial neurons – connected with each other. It has two major functions: learning, which is the process of presenting examples to the ANN and changing the connection weights, and recall, which is the process of presenting new examples to the trained ANN and examining its reaction (output). The connection between the neurons are analogized to the synaptic weights in the nervous system. Most of the known ANNs learning algorithms are influenced by a concept introduced by *Hebb* [1.23]. He proposed a model for unsupervised learning in which

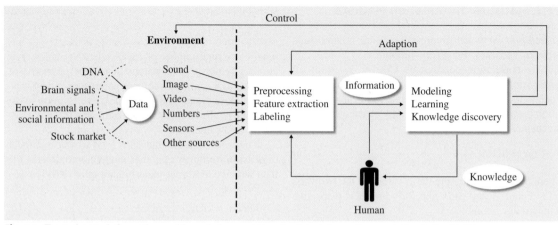

Fig. 1.4 From data to information and knowledge, and then back to information modeling (after [1.1])

the synaptic strength (weight) is increased if both the source and the destination neurons become simultaneously activated. It is expressed as

$$w_{ij}(t+1) = w_{ij}(t) + c o_i o_j \,, \qquad (1.20)$$

where $w_{ij}(t)$ is the weight of the connection between the ith and jth neurons at moment t, and o_i and o_j are the output signals of neurons i and j at the same moment t. The weight $w_{ij}(t+1)$ is the adjusted weight at the next time moment $(t+1)$.

In general terms, a learning system $\{S, W, P, F, L, J\}$ is defined by its structure S, its parameter set P, its variable (e.g., connections) weights W, its function F, its goal function J, and a learning procedure L. The system learns if the system optimizes its structure and its function F when observing events $z_1, z_2, z_3, \ldots$ from a problem space Z. Through a learning process, the system improves its reaction to the observed events and captures useful information that may be later represented as knowledge.

Another class of machine-learning methods are inspired by evolution of biology in nature, being called *evolutionary computation* (EC) [1.24, 25]. Here, learning is concerned with the performance not only of an individual system but of a population of systems that improve their performance through generations. The best individual system is expected to emerge and evolve from such populations. EC methods, such as genetic algorithms (GA), utilize ideas from Darwinism.

Another popular machine-learning method is called *support vector machines* (SVM). It was first proposed by *Vapnik* and his group at AT&T Bell Laboratories [1.26]. For a typical learning task defined as probability estimation of output values y depending on

input vectors x

$$P(\boldsymbol{x}, y) = P(y|\boldsymbol{x})P(\boldsymbol{x}) \,, \qquad (1.21)$$

a SVM classifier is used to build a decision function

$$f_\mathrm{L} : \boldsymbol{x} \rightarrow \{-1, +1\} \qquad (1.22)$$

based on a training set

$$f_\mathrm{L} = L(S_\mathrm{train}) \,, \qquad (1.23)$$

where $S_\mathrm{train} = (\boldsymbol{x}_1, y_1), (\boldsymbol{x}_2, y_2), \ldots, (\boldsymbol{x}_n, y_n)$.

In SVM, the primary concern is to determine an optimal separating hyperplane that gives a low generalization error. Usually, the classification decision function in the linearly separable problem is represented by

$$f_{\boldsymbol{w},b} = \mathrm{sign}(\boldsymbol{w} \cdot \boldsymbol{x} + b) \,. \qquad (1.24)$$

In SVM, this optimal separating hyperplane is determined by giving the largest margin of separation between vectors that belong to different classes. It bisects the shortest line between the convex hulls of the two classes.

1.2.5 Knowledge Representation

The ultimate goal of information processing is the creation of knowledge. The process of knowledge acquisition from nature is a continuous process that will never end. This knowledge is then used to understand nature, to preserve it, to model it, and to predict events. From data to information and knowledge and then back: this is what information science is concerned with (Fig. 1.4).

Different types of knowledge can be used in machine-learning systems, some of them being [1.1, 21]:

- Propositional logic rules (Aristotle)
- First-order logic rules
- Fuzzy logic rules [1.27]
- Semantic maps
- Schemata
- Metarules
- Finite automata
- Higher-order logic.

Fuzzy logic is an extension of propositional logic. It was first introduced by *Zadeh* in 1965 [1.27]. It deals with fuzzy propositions that can have any truth value between true (1) and false (0). Fuzzy propositions are used to represent fuzzy rules that better represent human knowledge [1.21, 28], e.g., IF the gene expression of gene G is High AND the age of the person is Old THEN the risk of cancer is High, where fuzzy membership functions (rather than Yes or No values) are used to represent the three propositions.

Modeling and knowledge can be [1.26, 29]:

- Global: valid for the whole population of data
- Local: valid for clusters of data [1.1, 30]
- Personalized: valid only for an individual [1.1, 26, 31].

1.3 Bioinformatics

1.3.1 Biology Background

Bioinformatics brings together several disciplines – molecular biology, genetics, microbiology, mathematics, chemistry and biochemistry, physics, and of course informatics, with the aim of understanding life. The *theory of evolution through natural selection* (Charles Darwin, 1809–1882, Fig. 1.5) was a significant step towards understanding species and life.

With the completion of the first draft of the human genome and the genomes of some other species, the task is now to be able to process this vast amount of ever-growing dynamic information and to discover new knowledge.

Deoxyribonucleic acid (DNA) is a chemical chain, present in the nucleus of each cell of an organism; it consists of ordered double-helix pairs of small chemical molecules (bases), adenine (A), cytosine (C), guanine (G), and thymine (T), linked together by a sugar phosphate nucleic acid backbone (Fig. 1.6).

The *central dogma of molecular biology* states that DNA is transcribed into RNA, which is translated into *proteins*.

DNA contains millions of base pairs, but only 5% or so is used for production of proteins, and these are the segments of the DNA that contain genes. Each gene is a sequence of base pairs that is used in the cell to produce RNA and/or proteins. Genes have a length of hundreds to thousands of bases.

Ribonucleic acid (RNA) has a similar structure to DNA, but here thymine (T) is substituted by uracil (U). In pre-RNA, only segments that contain genes are extracted from the DNA. Each gene consists of two types of segments: exons, which are segments translated into proteins, and introns, which are segments that are considered redundant and do not take part in protein production. Removing the introns and ordering only the exon parts of the genes in a sequence is called splicing, and this process results in the production of a messenger RNA (mRNA) sequences.

Fig. 1.5 Charles Darwin (1809–1882)

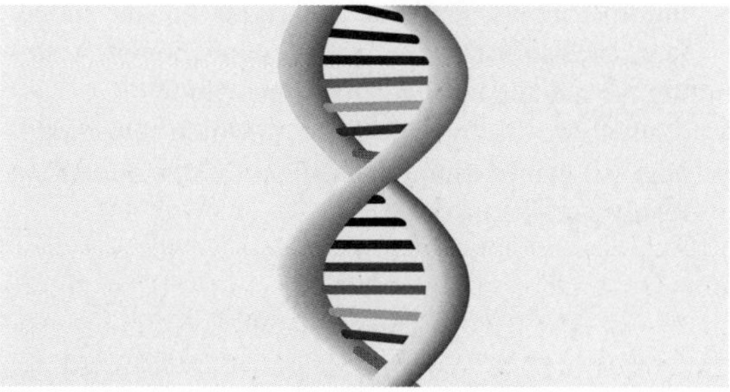

Fig. 1.6 DNA is organized as a double helix

mRNAs are directly translated into proteins. Each protein consists of a sequence of amino acids, each of them defined by a base triplet, called a codon. From one DNA sequence, many copies of mRNA produced, the presence of a certain gene in all of them defining the level of gene expression in the cell and indicating what and how much of the corresponding protein will be produced in the cell.

Genes are complex chemical structures that cause dynamic transformation of one substance into another during the whole life of an individual, as well as the life of the human population over many generations. When genes are *in action*, the dynamics of the processes in which a single gene is involved are very complex, as this gene interacts with many other genes and proteins, and the process is influenced by many environmental and developmental factors.

Modeling these interactions, learning about them, and extracting knowledge are major goals for bioinformatics.

Bioinformatics is concerned with the application of the methods of information science for the analysis, modeling, and knowledge discovery of biological processes in living organisms.

1.3.2 Data Analysis and Modeling in Bioinformatics

There are five main phases of information processing and problem solving in most bioinformatic systems:

1. Data collection, e.g., collecting biological samples and processing them
2. Feature analysis and feature extraction – defining which features are more relevant and therefore should be used when creating a model for a particular problem (e.g., classification, prediction, decision making)
3. Modeling the problem, which consists of defining the inputs, outputs, and type of the model (e.g., probabilistic, rule-based, connectionist), training the model, and statistical verification
4. Knowledge discovery in silico, in which new knowledge is gained through analysis of the modeling results and the model itself

5. Verifying the discovered knowledge in vitro and in vivo – biological experiments in both laboratory and real life to confirm the discovered knowledge.

When creating models of complex processes in molecular biology, the following issues must be considered:

- How to model complex interactions between genes and proteins, between the genome and the environment.
- Both stability and repetitiveness are features that need to be modeled, because genes are relatively stable carriers of information.
- Dealing with uncertainty; For example, when modeling gene expressions, there are many sources of uncertainty, e.g., alternative splicing (a splicing process of the same RNAs resulting in different mRNAs); mutation in genes caused by ionizing radiation (e.g., x-rays); chemical contamination, replication errors, viruses that insert genes into host cells, etc. Mutated genes express differently and cause the production of different proteins.

There are many problems in bioinformatics that require solution through data analysis and modeling. Typical problems are:

- Discovering patterns from DNA and RNA sequences (e.g., promoters, binding sites, splice junctions)
- Analysis of gene expression data and gene profile creation
- Protein discovery and protein function analysis
- Modeling interaction networks between genes, proteins, and other molecules
- System biology approach to modeling the whole dynamic process of biological functioning
- Creating biological and medical prognostic and decision-support systems

All the above tasks require different information methods, both generic (taken from IS) and specific, being created for the particular analysis and modeling of a specific problem and type of biological data.

1.4 Neuroinformatics

1.4.1 Human Brain and Nervous System

The human brain can be viewed as a dynamic, evolving information-processing system, probably the most complex one. Processing and analysis of information recorded from brain and nervous system activity, and modeling of perception, brain functions, and cognitive processes, aim at understanding the brain and creating brain-like intelligent systems. This is a subject of *neuroinformatics*.

The brain evolves initially from stem cells. It evolves its structure and functionality from an embryo to a sophisticated biological information processing system (Fig. 1.7).

In an embryo, the brain grows and develops based on genetic information and nutritional environment. The brain evolves its functional modules for vision, speech and language, music and logic, and many other cognitive tasks.

There are *predefined* areas of the brain that are *allocated* for language and visual information processing, for example, but these areas may change during the neuronal evolving processes. The paths of the signals traveling in, and the information processes of, the brain are complex and different for different types of information. Even at the age of 3 months, some functional areas are already formed, but identical embryos with the same genetic information can develop in different ways to reach the state of an adult brain. This is because of the environment in which the brain evolves. Both the genetic information (nature) and the environment (nurture) are crucial factors. They determine the *evolving rules* for the brain. The challenge is how to reveal these rules and eventually use them in brain models. Are they the same for every individual?

A significant step in understanding the brain and the nervous system was the discovery of the structure of the neural system by Ramón y Cajal (1852–1934, Figs. 1.8, 1.9).

A neuron, which receives signals (spikes) through its *dendrites* and emits output signals through its *axon*, is connected to thousands other neurons through *synapses*. The synaptic connections are subject to *adaptation and learning* and represent the *long-term memory* of the brain.

Neurons can be of different types according to their main functionality [1.11]. There are, for example: sensory neurons, motor neurons, local interneurons, projection interneurons, and neuroendocrine cells.

It is through the organization of neurons into ensembles that functional compartments emerge. Neurosciences provide a very detailed picture of the organization of the neural units in the functional compartments (functional systems). Each functional system is formed by various brain regions that are responsible for processing of different types of information. It is shown that the paths which link different components of a functional system are hierarchically organized.

It is mainly in the *cerebral cortex* where the cognition functions take place. Anatomically, the cerebral cortex is a thin, outer layer of the cerebral hemisphere with thickness of around 2–4 mm. The cerebral cortex is divided into four lobes: frontal, parietal, temporal, and occipital.

Several principles of the evolving structure, functions, and cognition of the brain are listed below:

- Redundancy; i. e., there are many redundant neurons allocated to a single stimulus or a task; e.g., when a word is heard, there are hundreds of thousands of neurons that are immediately activated.
- Memory-based learning; i. e., the brain stores exemplars of facts that can be recalled at a later stage. Some studies suggest that all human actions, including learning and physical actions, are based on memorized patterns.
- Learning is achieved through interaction of an individual with the environment and with other individuals.
- Inner processes take place, e.g., information consolidation through sleep learning.

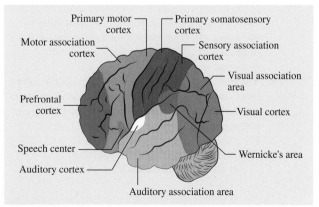

Fig. 1.7 The brain evolves its structure and functionality through genetic information and developmental learning (after [1.32])

Fig. 1.8 Santiago Ramón y Cajal (1852–1934) (after [1.33])

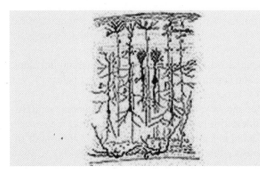

Fig. 1.9 A drawing by Ramón y Cajal of a neuronal circuitry (after [1.33])

- The learning process is continuous, evolving, and lifelong.
- Learning and memory are of three main types: short term (within hundreds of milliseconds), which is manifested in the synapses, the neuronal membrane potentials, and the spiking activity of neurons; long term (within minutes or days), which is manifested in the synaptic weights; and genetic (within months, years, or millions of years), which is manifested in the genes and their expressions). These three types of memory and learning interact in a dynamic way in the brain.
- Through the process of evolving brain structures (neurons, connections), higher-level concepts emerge; these are embodied in the structure and represent a level of abstraction.

1.4.2 Data Analysis and Modeling in Neuroinformatics

The brain is the most complex information processing machine. It processes data, information, and knowledge at different levels. Modeling the brain as an informa-

tion processing machine has different results depending on the goals of the models and the detail with which the models represent the genetic, biological, chemical, physical, physiological, and psychological rules and the laws that govern the functioning and behavior of the brain.

Several levels of brain data analysis and modeling can be distinguished.

Molecular/Genetic Level
At the genetic level, the genome constitutes the input information, while the phenotype constitutes the output result, which causes:

1. Changes in the neuronal synapses (learning)
2. Changes in DNA and its gene expression [1.12].

Neurons from different parts of the brain, associated with different functions, such as memory, learning, control, hearing, and vision, function in a similar way, and their functioning is genetically defined. This principle can be used as a unified approach to building different neuronal models to perform different functions, such as speech recognition, vision, learning, and evolving. The genes relevant to particular functions can be represented as a set of parameters of a neuron. These parameters define the way the neuron functions and can be modified through feedback from the output of the neuron.

Single Neuronal Level
There are many information models of neurons that have been explored in neural network theory (for a review, see [1.11]). Among them are:

1. Analytical models. An example is the Hodgkin–Huxley model (1952) [1.34].
2. McCulloch and Pitts-type models (1943) [1.35].
3. Spiking neuronal models (Maas, Gerstner, Kistler, Izhikevich, Thorpe, Wysoski et al.) [1.29, 36–45].
4. Neurogenetic models, where a gene regulatory network is part of the neuronal model ([1.8, 32, 38, 40]).

Neural Network (Ensemble) Level
Information is processed in ensembles of neurons that form a functionally defined area, such as sensory modalities [1.46]. The human brain deals mainly with five sensory modalities: vision, hearing, touch, taste, and smell. Each modality has different sensory receptors. After the receptors perform the stimulus transduction, the information is encoded through the excitation of neural action potentials. The information is encoded us-

ing the average of pulses or the time interval between pulses. This process seems to follow a common pattern for all sensory modalities, however there are still many unanswered questions regarding the way the information is encoded in the brain.

Cognitive Level

Information is processed in the whole brain through many interacting modules. Many neuronal network modules are connected together to model a complex brain structure and learning algorithms [1.47, 48]. To date, the most effective means available for brain activ-

ity measurements are electroencephalography (EEG), magnetoencephalography (MEG), and functional magnetic resonance imaging (fMRI). Once the data from these measurement protocols has been transformed into an appropriate state-space representation, an attempt to model different dynamic brain functions can be made.

Modeling the entire brain is far from having been achieved, and it will take many years to achieve this goal, but each step in this direction is useful towards understanding the brain and towards the creation of intelligent machines that will help people [1.49].

1.5 About the Handbook

This Springer Handbook includes 12 parts, 6 of them covering topics from BI and 6 from NI. Each part includes chapters, and each chapter introduces topics that integrate BI and IS, or NI and IS, or BI, NI, and IS together.

1.5.1 Bioinformatics

Part A is about *understanding information processes in biological systems*. It includes chapters that reveal the information processing at cellular level, genomics level, proteomics level, and evolutionary molecular biology point of view.

Part B covers the *methods of molecular biology*, including: analysis of DNA sequences, analysis and discovery of microRNA signatures, discovery of regulatory elements in RNA, protein data modeling, and protein structure discovery.

Part C presents different *machine-learning methods for analysis, modeling, and knowledge discovery from bioinformatics data*. It includes chapters that review the applications of different methods, such as Bayesian classifiers and support vector machines (SVM); case-based reasoning; hybrid clustering; fuzzy logic; and phylogenetic cladograms.

Part D presents more sophisticated methods for *integrated, system biology analysis* and modeling in BI, including chapters on: inferring interaction network from Omics data; inferring gene transcription networks; analysis of transcriptional regulations, inferring genetic networks using differential evolution; pattern discovery in protein–protein networks; visual representation of molecular networks; and a pipeline model for identi-

fying somatic mutations with examples from leukemia and colon cancer.

Part E presents *databases and ontologies* that contain structured bioinformatics data to enable worldwide research and study in bioinformatics. It includes chapters on bioinformatics databases and bioinformatics ontology systems.

Part F is about *applications of bioinformatics in medicine, heath, and ecology*. It includes chapters on modeling cancer stem formation, epigenetics, immune system control, nutrigenomics, nanomedicine, personalized medicine, health informatics, and ecological informatics.

1.5.2 Neuroinformatics

Part G is about understanding *information processes in the brain and the nervous system*. It includes chapters on information processes at a lower, synaptic level, spiking neural networks that represent and model information processes at a neuronal ensemble level; brain connectivity study based on fMRI data; and information processes at the level of the whole brain.

Part H introduces *advanced signal processing methods for brain signal analysis and modeling*. The methods are applicable to study spatiotemporal spiking activities of single neurons and neuronal ensembles along with spiking activity of the whole cortex. This part includes chapters on adaptive filtering in kernel spaces for spike train analysis, analysis and visualization of multiple spike trains, and the multivariate empirical mode decomposition method for time–frequency analysis of EEG signals.

Part I is concerned with modeling *perception, sensation, and cognition*. It includes chapters on modeling vision, modeling the gustatory system, perception and motor control modeling based on EEG with application for brain–computer interfaces, spiking neural network and neurogenetic systems for spatio- and spectrotemporal brain data analysis and modeling, and models of natural language.

Part J presents *neuroinformatics databases and systems* to help brain data analysis and modeling. It includes chapters on brain-gene ontology systems, neuroinformatics databases, and worldwide organizations.

Applications of neuroinformatics methods for understanding and curing of brain diseases is presented in Part K. It contains chapters on Alzheimer disease

genetic regulatory networks, integrating data and prior knowledge for understanding Alzheimer disease, a system biology approach to modeling and understanding Parkinson and Alzheimer disease, modeling gene dynamics in epilepsy, predicting outcome of stroke, and surface electromyography methods for nerve–muscle system rehabilitation using the case study of stroke.

Nature-inspired integrated information technologies, presented in the last Part L, combine different principles from the biology, brain, and quantum levels of information processing (Fig. 1.1). It includes chapters on brain-like robotics, interactive, developmental multimodal robotic systems, quantum and biocomputing integration, and integrated brain-, gene-, and quantum-inspired computational intelligence.

1.6 Conclusion

This chapter presents a brief overview of the topics covered in this *Springer Handbook of Bio-/Neuroinformatics*, with emphasis on the symbiosis of the three areas of science concerned: information science (informatics), bioinformatics, and neuroinformatics. The

topics presented and included in the Handbook provide a far from exhaustive coverage of these three areas, but they show clearly that we can better understand nature only if we utilize the methods of IS, BI, and NI, considering their integration and interaction.

References

1.1 N. Kasabov: *Evolving Connectionist Systems: The Knowledge Engineering Approach* (Springer, London 2007)

1.2 R.P. Feynman, R.B. Leighton, M. Sands: *The Feynman Lectures on Physics* (Addison–Wesley, Redding 1965)

1.3 R. Penrose: *The Emperor's New Mind* (Oxford Univ. Press, Oxford 1989)

1.4 R. Penrose: *Shadows of the Mind. A Search for the Missing Science of Consciousness* (Oxford Univ. Press, Oxford 1994)

1.5 C.P. Williams, S.H. Clearwater: *Explorations in Quantum Computing* (Springer, Berlin 1998)

1.6 M. Brooks: *Quantum Computing and Communications* (Springer, Berlin, Heidelberg 1999)

1.7 D.S. Dimitrov, I.A. Sidorov, N. Kasabov: Computational biology. In: *Handbook of Theoretical and Computational Nanotechnology*, Vol. 1, ed. by M. Rieth, W. Sommers (American Scientific Publisher, New York 2004), Chap. 21

1.8 N. Kasabov, L. Benuskova: Computational neurogenetics, Int. J. Theor. Comput. Nanosci. **1**(1), 47–61 (2004)

1.9 F. Rosenblatt: *Principles of Neurodynamics* (Spartan Books, New York 1962)

1.10 W. Freeman: *Neurodynamics* (Springer, London 2000)

1.11 M. Arbib (Ed.): *The Handbook of Brain Theory and Neural Networks* (MIT, Cambridge 2003)

1.12 H. Chin, S. Moldin (Eds.): *Methods in Genomic Neuroscience* (CRC, Boca Raton 2001)

1.13 J.J. Hopfield: Neural networks and physical systems with emergent collective computational abilities, Proc. Natl. Acad. Sci. USA **79**, 2554–2558 (1982)

1.14 National Center for Biotechnology Information (US): *Genes and Disease [Internet]* (NCBI, Bethesda 1998), available online at http://www.ncbi.nlm.nih.gov/books/NBK22183/

1.15 L.R. Rabiner: A tutorial on hidden Markov models and selected applications in speech recognition, Proc. IEEE **77**(2), 257–285 (1989)

1.16 S. Grossberg: On learning and energy – Entropy dependence in recurrent and nonrecurrent signed networks, J. Stat. Phys. **1**, 319–350 (1969)

1.17 D.E. Rumelhart, G.E. Hinton, R.J. Williams (Eds.): Learning internal representations by error propagation. In: *Parallel Distributed Processing: Explorations in the Microstructure of Cognition* (MIT/Bradford, Cambridge 1986)

1.18   T. Kohonen: *Self-Organizing Maps* (Springer, Berlin, Heidelberg 1997)

1.19   S. Haykin: *Neural Networks – A Comprehensive Foundation* (Prentice Hall, Engelwood Cliffs, 1994)

1.20   C. Bishop: *Neural Networks for Pattern Recognition* (Oxford Univ. Press, Oxford 1995)

1.21   N. Kasabov: *Foundations of Neural Networks, Fuzzy Systems and Knowledge Engineering* (MIT, Cambridge 1996)

1.22   S. Amari, N. Kasabov: *Brain-like Computing and Intelligent Information Systems* (Springer, New York 1998)

1.23   D. Hebb: *The Organization of Behavior* (Wiley, New York 1949)

1.24   X. Yao: Evolutionary artificial neural networks, Int. J. Neural Syst. **4**(3), 203–222 (1993)

1.25   D.B. Fogel: *Evolutionary Computation – Toward a New Philosophy of Machine Intelligence* (IEEE, New York 1995)

1.26   V. Vapnik: *Statistical Learning Theory* (Wiley, New York 1998)

1.27   Z.A. Zadeh: Fuzzy sets, Inf. Control **8**, 338–353 (1965)

1.28   T. Yamakawa, H. Kusanagi, E. Uchino, T. Miki: A new effective algorithm for neo fuzzy neuron model, Proc. Fifth IFSA World Congress (IFSA, 1993) pp. 1017–1020

1.29   N. Kasabov: Global, local and personalized modeling and profile discovery in Bioinformatics: An integrated approach, Pattern Recognit. Lett. **28**(6), 673–685 (2007)

1.30   M. Watts: A decade of Kasabov's evolving connectionist systems: A review, IEEE Trans. Syst. Man Cybern. C **39**(3), 253–269 (2009)

1.31   Q. Song, N. Kasabov: TWNFI – Transductive neural-fuzzy inference system with weighted data normalization and its application in medicine, IEEE Trans. Fuzzy Syst. **19**(10), 1591–1596 (2006)

1.32   L. Benuskova, N. Kasabov: *Computational Neuro-Genetic Modeling* (Springer, New York 2007)

1.33   http://www.wikipedia.org (last accessed April 4 2012)

1.34   A.L. Hodgkin, A.F. Huxley: A quantitative description of membrane current and its application to conduction and excitation in nerve, J. Physiol. **117**, 500–544 (1952)

1.35   W. McCullock, W. Pitts: A logical calculus of the ideas immanent in nervous activity, Bull. Math. Biophys. **5**, 115–133 (1943)

1.36   W. Gerstner: Time structure of the activity of neural network models, Phys. Rev. **51**, 738–758 (1995)

1.37   E. Izhikevich: Simple model of spiking neurons, IEEE Trans. Neural Netw. **14**(6), 1569–1572 (2003)

1.38   N. Kasabov, L. Benuskova, S. Wysoski: A Computational Neurogenetic Model of a Spiking Neuron, IJCNN 2005 Conf. Proc., Vol. 1 (IEEE, New York 2005) pp. 446–451

1.39   N. Kasabov, R. Schliebs, H. Kojima: Probabilistic computational neurogenetic framework: From modeling cognitive systems to Alzheimer's disease, IEEE Trans. Auton. Ment. Dev. **3**(4), 1–12 (2011)

1.40   N. Kasabov: To spike or not to spike: A probabilistic spiking neuron model, Neural Netw. **23**(1), 16–19 (2010)

1.41   G. Kistler, W. Gerstner: *Spiking Neuron Models – Single Neurons, Populations, Plasticity* (Cambridge Univ. Press, Cambridge 2002)

1.42   W. Maass, C.M. Bishop (Eds.): *Pulsed Neural Networks* (MIT, Cambridge 1999)

1.43   S. Thorpe, A. Delorme, R. Van Rullen: Spike-based strategies for rapid processing, Neural Netw. **14**(6/7), 715–725 (2001)

1.44   S. Wysoski, L. Benuskova, N. Kasabov: Evolving spiking neural networks for audiovisual information processing, Neural Netw. **23**(7), 819–835 (2010)

1.45   S. Guen, S. Rotter (Eds.): *Analysis of Parallel Spike Trains* (Springer, New York 2010)

1.46   E. Rolls, A. Treves: *Neural Networks and Brain Function* (Oxford Univ. Press, Oxford 1998)

1.47   J.G. Taylor: *The Race for Consciousness* (MIT, Cambridge 1999)

1.48   R. Koetter (Ed.): *Neuroscience Databases: A Practical Guide* (Springer, Berlin, Heidelberg 2003)

1.49   D. Tan, A. Nijholt (Eds.): *Brain-Computer Interfaces* (Springer, London 2010)

Introduction

Part A
Understa

Part A Understanding Information Processes in Biological Systems

Ed. by Heike Sichtig

2. Information Processing at the Cellular Level: Beyond the Dogma

Alberto Riva

The classical view of information flow within a cell, encoded by the famous *central dogma of molecular biology*, states that the instructions for producing amino acid chains are read from specific segments of DNA, just as computer instructions are read from a tape, transcribed to informationally equivalent RNA molecules, and finally *executed* by the cellular machinery responsible for synthesizing proteins. While this has always been an oversimplified model that did not account for a multitude of other processes occurring inside the cell, its limitations are today more dramatically apparent than ever. Ironically, in the same years in which researchers accomplished the unprecedented feat of decoding the complete genomes of higher-level organisms, it has become clear that the information stored in DNA is only a small portion of the total, and that the overall picture is much more complex than the one outlined by the dogma.

The cell is, at its core, an information processing machine based on molecular technology, but the variety of types of information it handles, the ways in which they are represented, and the mechanisms that operate on them go far beyond the simple model provided by the dogma. In this

chapter we provide an overview of the most important aspects of information processing that can be found in a cell, describing their specific characteristics, their role, and their interconnections. Our goal is to outline, in an intuitive and nontechnical way, several different views of the cell using the language of information theory.

The central dogma of molecular biology was originally proposed 50 years ago by Francis Crick, who called it a dogma mostly as a joke. It was indeed a conceptual milestone since it explicitly described for the first time the flow of information within living cells. The central point of the theory, updated with the discoveries of the 1970s and 1980s, is that information can flow from DNA to DNA (replication), DNA to RNA (transcription), RNA to DNA (reverse transcription), and from RNA to protein (translation), but once the information is converted into the form of protein, it does not go back to RNA or DNA (Fig. 2.1). Today, 50 years later, it has

become clear that this model only applies to the portion of the genome (less than 5%) that encodes for proteins, and therefore, while still being valid and extremely important, it is somewhat reductive, since the information content of the noncoding portion of the genome, although at present difficult to quantify, is certainly very significant. In fact, the flow of information within the cell is much more complex than the simple conversion of a series of triplets into a chain of amino acids.

Information can be defined as a force that transforms chaos into organized structures and processes, and indeed the technical definition of information is

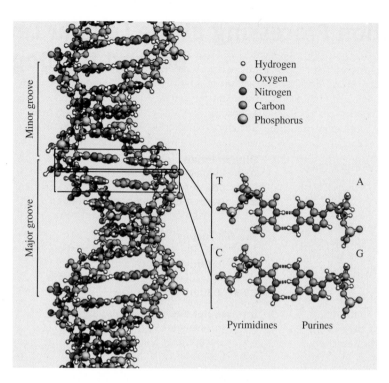

Fig. 2.1 The double-helix structure of DNA

○ Hydrogen
◉ Oxygen
● Nitrogen
◉ Carbon
◯ Phosphorus

T   A

C   G

Pyrimidines   Purines

based on measuring the extent to which a system or phenomenon under observation can be distinguished from chaos (i. e., total randomness). The cell is a highly structured and organized object, and its information content is therefore extremely high. However, the elucidation of the genetic code and the consequent formulation of the dogma represented the first, and so far only, discovery of a form of symbolic encoding of information in a natural system. The realization that living matter had evolved a mechanism to permanently store information about its own structure and function in digital form (using an alphabet consisting of only four different symbols) surely represents one of the most revolutionary discoveries in the history of the natural sciences.

On the other hand, its undisputed importance should not lead to the belief that the information flow described by the central dogma is the only one taking place in the cell, or even the only one necessary to explain protein synthesis. As we will see, the cell is a complex, highly integrated system, whose components are constantly exchanging information with each other employing a multitude of different *languages*, which are embodied into physical entities in widely different ways. No part of the cell could exist and function separately from the surrounding environment produced by its other parts, and the system can only develop and evolve as a whole: you cannot build a cell by assembling separately produced parts. The scope of any discussion about information flows at the cellular level must therefore be widened to include all aspects of cellular activities. This is what we do in the following sections, describing various examples of information processing taking place in the cell and showing how they are structured, regulated, and interconnected.

2.1 Fundamentals

In this section we provide an overview of the basic concepts of molecular biology, and specifically of those entities and processes that are involved in the central dogma, from the point of view of information theory.

2.1.1 Representing Information

In information theory, the basic unit of information is the bit. A bit is a variable that can assume only one of two states, usually represented as 0 and 1. Using multiple bits one can represent variables that assume an arbitrary number of states; for example, there are eight possible combinations of 3 bit (000, 001, 010, ..., 111); a set of 3 bit can therefore be used to represent all integer numbers between 0 and 7 (or any other set of eight distinct entities). In general, n bit encode 2^n different values. Given a sufficient number of bits and an appropriate encoding, anything that can be represented in numerical form can be expressed as a sequence of bits.

Given an alphabet of S distinct, mutually exclusive symbols, each symbol on average carries $\log 2(S)$ bit of information. If the probability of occurrence of the different symbols is not uniform (e.g., in the English language the letter "e" occurs far more often than "q"), then the information conveyed by each symbol is $-\log 2(pi)$, where pi is the probability of occurrence of symbol i. Intuitively, the information content of a symbol in an alphabet represents the increase in the amount of knowledge that we gain by observing that symbol. More generally, given a series of mutually exclusive events (such as picking a letter at random from an English text), events that occur with higher probability will convey a small amount of information, and conversely, rare events will convey a higher amount of information: the letter "e" in English, occurring at frequency of 12%, conveys 3.06 bit of information, while the letter "q," with frequency of 0.095%, conveys over 10 bit of information.

In other words, information can be viewed as a measure of unexpectedness: observing an event that occurs with absolute certainty does not give us any information, since we already knew it was going to happen, while observing a rare event provides us with a large amount of information.

Finally, it is important to note that information is not simply a theoretical concept, but has very practical applications: information can in fact be directly linked to the physical concept of energy. In a complex system such as an ideal gas, the number of microstates (e.g., the position and velocity of each particle) that produce the same macrostate (e.g., the gas temperature and pressure) increases with the amount of energy in the system. Therefore, a gas at a high temperature will convey a larger amount of information than one at a low temperature, because describing its exact state at any instant implies selecting from a much larger number of possibilities.

2.1.2 DNA and RNA

DNA is a double-stranded molecule that encodes genetic information as a sequence of nucleotides. The four nucleotides found in a DNA sequence are adenine, cytosine, guanine, and thymine, commonly represented by the letters A, C, G, and T. Since there are only four possible nucleotides, each position in the DNA sequence conveys 2 bit of information. The human genome contains approximately 3.2 billion nucleotides, divided into 23 chromosomes of variable lengths, located in the cellular nucleus. The information content of the human genome is therefore normally indicated as being of the order of 6.4×10^9 bit. This, however, does not take into account the fact that there are two copies of each chromosome (except for the Y chromosome), and they are not exactly identical to each other.

The Structure of DNA

A DNA molecule is composed of two long polymers coiled around a common axis in the well-known double-helix structure (Fig. 2.1). Each polymer consists of a long sequence of repeating elements, the nucleotides, attached to a common backbone. The two DNA strands are held together by the pairing between complementary nucleotides: adenine forms hydrogen bonds with thymine, and cytosine with guanine. The sequence of nucleotides on one strand is therefore complementary to the sequence on the other strand, and each strand can be reconstructed using the other strand as a template (as takes place, for example, during cell replication). The A–T and C–G pairing is known as Watson–Crick base pairing, or canonical pairing.

To a first approximation, we can assume that the majority of differences between the two copies of a chromosome (or between chromosomes of different individuals) consist of single-nucleotide polymorphisms (SNPs). It is estimated that there are about 3 million differences between any two individuals in the entire genome (many other types of genetic polymorphisms exist, but they occur at much lower frequencies). If we consider that at each position of the DNA sequence we can have either one of the four nucleotides or one of the six possible pairs of nucleotides, the number of bits necessary to represent each position increases to approximately 3.3, and the total information content rises to over 10^{10} bit.

RNA is a single-stranded molecule composed of four nucleotides, like DNA (except that thymine is replaced by uracil). RNA is normally produced by transcription, a process that copies a region of DNA to synthesize an equivalent RNA molecule. The information content of an RNA is therefore, at first glance, the same as that of the DNA fragment it was transcribed from, and indeed until recently it was thought that its only function was to convey this information outside of the nucleus. In fact, RNA is an active molecule capable of catalyzing reactions, and its three-dimensional structure, partially determined by its sequence, is critically important towards determining its properties. Measuring the information content of an RNA molecule should therefore take its structure into account, or at least the presence of structural features that confer specific properties.

2.1.3 Transcription and Gene Expression

The main purpose of transcription is to copy the DNA sequence of genes into a messenger RNA (mRNA) molecule. The mRNA is then exported from the nucleus to the cytoplasm, where it is translated into a sequence of amino acids by the ribosome. Transcriptional activity (in other words, which genes are transcribed and how much of each mRNA is produced) is very carefully regulated by a highly complex network of molecular interactions. The main actors in this process are a class of proteins called transcription factors (TFs). TFs bind to the DNA molecule at specific sites (binding sites), generally located upstream of genes in a region called the promoter, and control the expression of the corresponding genes in a combinatorial fashion. In other words, the expression of a gene is dynamically determined by the pattern of TFs that bind to its promoter. The *function* that determines gene expression on the basis of the bound TFs can be a simple one (binding of an activator TF starts transcription, binding of a repressor TF suppresses it), or a complex combinatorial one, in which the exact expression level is determined by the interplay of multiple positive and negative signals. At a basic level, binding of a TF to the DNA molecule is a binary event (either the TF is bound, or it is not), and the locations of binding sites in the promoter of a gene are fixed, since they are mainly determined by the nucleotide sequence. Therefore, describing the possible arrangements of TFs in a gene promoter would require a number of bits that is equal to the number of binding sites it contains. In other words, a promoter containing b binding sites can be in any of 2^b possible states, each one of which may lead to a different expression level for the downstream gene. In reality, the picture is more complex because of a number of additional factors: for example, different TFs may recognize the same binding site, and they will therefore bind in an antagonistic fashion (if one of them binds to the DNA, it prevents the other one from doing so). TFs are, themselves, proteins that are produced by the expression of specific genes, and the gene they reg-

ulate may in turn act as a TF for other genes. Moreover, TF binding is a dynamic process, since TFs continually bind DNA and dissociate from it; time therefore represents a critically important aspect of the regulation process. The pattern of TFs binding a gene promoter is not static but changes over time, and the true *input* that determines the expression of the gene is the sequence of patterns with their extremely precise timings. The overall regulatory network can therefore be viewed as a very large set of interconnected transfer functions, dynamically mapping the expression levels of all input TFs into the expression level of the RNA produced as output.

In most higher organisms, there is a further step between transcription and protein synthesis. After transcription, the newly synthesized RNA molecule undergoes splicing, a process that removes segments called introns and produces the mature mRNA by joining together the remaining segments (the exons). This process is guided by a dedicated regulatory mechanism, that (similarly to transcriptional regulation) relies on the presence of appropriate signals at the sequence level, and on the action of specific classes of proteins (RNA binding proteins, splicing factors).

2.1.4 Translation and Protein Synthesis

The process of translating an mRNA into an amino acid sequence takes place inside the ribosome, a highly complex organelle composed of RNA and proteins. During translation, the ribosome interprets the RNA sequence in blocks of three nucleotides (called triplets). Out of the 64 possible triplets, one is interpreted as a start signal (ATG), three are interpreted as stop signals (TAA, TAG, TGA), and the remaining 60 encode the 20 different amino acids. The genetic code is redundant, meaning that some amino acids are encoded by more than one triplet. The translation process therefore results in an apparent loss of information, since representing 20 amino acids requires only 4.3 bit of information, on average, instead of the 6 encoded by three nucleotides ($\log 2(20) = 4.32$). In fact, the opposite is true. Proteins are highly complex molecules, characterized by a secondary structure (describing the presence of specific structural features such as α-helices or β-sheets), a tertiary structure (its exact three-dimensional structure, in terms of the spatial coordinates of each atom), and in some cases by a quaternary structure (the arrangement of multiple proteins in a complex). These structures, in turn, largely determine the protein's properties and function. Although the number of possible three-dimensional structures for a typical amino acid chain is astronomically high, the protein spontaneously folds into the correct one, in a way that is ultimately determined by the amino acid sequence together with environmental conditions (temperature, concentration, etc.) and, in some cases, the presence of *assisting* proteins (chaperone proteins). Protein folding can therefore be viewed as a process that amplifies the information content of the amino acid sequence by many orders of magnitude.

By selecting the one correct structure out of the innumerable possible ones, protein folding gives rise to a complex object whose information content is much larger than the one encoded in the original amino acid sequence. There is an apparent paradox here, since information, just like energy, cannot be created out of nowhere. How can a relatively short sequence composed of at most 20 different symbols describe in perfect detail an extremely complex three-dimensional molecular structure? The solution to this paradox lies in

the role played by natural selection: the amino acid sequence that produces a properly folded, fully functional protein is the result of a trial-and-error process that has taken place over millions of years and billions of replication events. DNA provided evolution with both an alphabet for representing evolved protein sequences in a symbolic, simplified way, and with the *long-term storage* device for permanently recording them. The ability to retain the information produced in the course of this extremely long, massively parallel biochemistry experiment (that has effectively being going on since the origin of life on Earth) is what ultimately explains the information amplification that takes place during protein synthesis: the process is not generating new information, but is *replaying* a single successful event selected out of an almost infinite number of possible ones (and as we saw in Sect. 2.1.1, rare events convey a very large amount of information). This idea can be extended from the protein level to the entire cell, as we will see in Sect. 2.2.

2.2 The Cell Factory

The cell, and in particular the animal cell, is probably the most complex object on Earth. It can be imagined as a factory in which millions of parts move, interact, and transform in a coordinated way to perform metabolic functions, to respond to external stimuli, and, ultimately, to ensure its own survival. Of course, a fundamental property of the cell is its ability to produce two identical copies of itself, a process during which the entire information content of the cell must be duplicated almost exactly. Keeping in mind the equivalence between information and energy, it is easy to understand why cell replication mobilizes a large part of the cell's resources.

Like any factory, its functioning requires energy and a blueprint of instructions. The energy supply that allows the cell to live derives from high-energy compounds. Discussing how this fuel is produced by the cell itself is beyond the scope of this chapter; suffice it to say that it derives from products of the life processes (life feeds on life) and ultimately from the energy of solar radiation.

Since we are living in the *digital era*, it is common to imagine DNA as a sort of biological hard disk, storing information in digital form, with the rest of the cell functioning like an *operating system* that enables the execution of the program encoded in it. Once again,

we have a model in which DNA-encoded information plays the lead role, with the cellular environment as the backdrop.

In fact, this vision is profoundly misleading. DNA and its harboring cell cannot be viewed as separated entities: they are fully integrated, having evolved together during millions of years of evolution. In other words, one cannot say that the DNA sequence contains the blueprint to create a cell, because a new cell can only be generated from an existing one: DNA in isolation, outside of the cellular environment, is useless, just as a cell cannot survive without DNA. DNA should instead be viewed as a *historical record* of the evolutionary process that produced the cell, a record that can be used to repeat some parts of that process whenever necessary.

The informative content of the cell is distributed in many layers; the information stored in the DNA polymer in the form of a sequence of nucleotides is just one of them, although it is the one that can be most easily decoded and interpreted in a symbolic, deterministic way. In fact, all the components of the cell carry an information content that is, in general, difficult to quantify, but nevertheless critically important. Starting from the smaller components (amino acids, nucleotides, etc.) to the more complex macromolecules (enzymes, RNAs, lipids, polysaccharides, etc.), what makes them function is the stereochemistry of the weak chemical bonds they can engage in. This confers extreme specificity and, at the same time, flexibility on the intracellular interactions. In other words, every weak bond, of any nature (van der Waals, ionic bonds, hydrogen bonds, etc.) carries information, since these bonds are responsible for the properties and behavior of the cellular complexes and components they create.

Moving from here to a higher level, the other cellular nucleic acid, RNA, deserves special attention because of its key role in the dynamic behavior of the cell.

2.2.1 RNA: The Key Mediator of Cellular Information Flow

RNA differs from DNA in one critical aspect: the ribose sugar has a chemically reactive 2′-OH (instead of H) which changes the conformation of the sugar ring and alters the chemistry of the neighboring phosphate. The reactive 2′-OH confers unique properties on RNA. While complementary RNAs form a double helix (although one different from DNA), single-stranded RNA polymers can fold into biologically active secondary and tertiary structures through Watson–Crick and non-Watson–Crick base pairing.

The RNA World

It is commonly accepted that life on Earth started around 4 billion years ago with the appearance of small RNA molecules (formed from simpler molecules) capable of catalyzing simple reactions and, most importantly, of self-replicating, evolving, and undergoing natural selection. The hypothesis of an RNA world that preceded the DNA world is now widely accepted and is supported by the existence of molecules such as ribozymes and of self-splicing introns in some modern organisms.

In modern cells, RNA plays a key role in the processing of cellular information. RNA can be viewed as the *messenger* of DNA since its main task is to transport the information content, which would be otherwise immobilized on DNA, to other cellular compartments. It is now known that the entire genome (both its coding and noncoding portions) is transcribed into a variety of RNA species (Table 2.1). While RNA potentially carries an amount of information equivalent to the DNA sequence, the overall amount of RNA in cells is much greater. The effective amount of information carried by RNA is therefore amplified by the relative concentrations

Table 2.1 Some of the most common RNA types

Name	Description	Function
mRNA	Messenger RNAs	Carry protein sequence information to ribosomes
tRNA	Transfer RNAs	Transfer a specific amino acid to a polypeptide during protein synthesis
rRNA	Ribosomal RNAs	The catalytic components of ribosomes
miRNA	MicroRNAs	Downregulators of gene expression
siRNA	Small interfering RNAs	Downregulators of gene expression
snRNA	Small nuclear RNAs	Involved in splicing, transcription, telomere maintenance
snoRNA	Small nucleolar RNAs	Perform modifications of RNA nucleotides

of expressed transcripts, and is further expanded by the structural features of RNA (secondary and tertiary structures) and its elaborate metabolism (processing, splicing, editing, degradation, etc.).

Moreover, as a legacy from its ancient role in the RNA world, RNA operates with a large amount of autonomy in interpreting and amplifying the information contained in the DNA. The most striking example of such amplification is the ribosome, i. e., the organelle where translation of mRNA into polypeptides takes place. Ribosomes are complexes of ribosomal RNAs and proteins. However, the catalytic activity of the ribosome resides primarily in its RNA component, with ribosomal proteins having only an ancillary role; in other words, the ribosome is an RNA machine. How is this achieved? On paper, ribosomal RNA can fold in an almost infinite number of ways (over 10^{80}), yet only one catalytically active structure (Fig. 2.2) is adopted.

Where does the information for such *quality control* come from? In part, it comes from the nucleotide sequence (hence, from DNA), but most of it comes from the intrinsic capacity of ribonucleotides to engage in canonical and noncanonical pairings.

Similar considerations apply to the other cellular RNAs (tRNAs, snRNAs, microRNAs, etc.), and the same concepts apply to the folding of polypeptides into a biologically active structure.

Also in this case, the informational content of a linear array of amino acids is amplified by the intrachain bonds between residues to generate secondary and tertiary structures (α-helices, β-sheets, etc.) that con-

Fig. 2.2 The three-dimensional structure of the large ribosomal subunit (50S) of *Haloarcula marismortui*, facing the 30S subunit. The ribosomal proteins are shown in *blue*, the rRNA in *ochre*, and the active site (A 2486) in *red*. Data were taken from Protein Data Bank (PDB) 3CC2, rendered with PyMOL

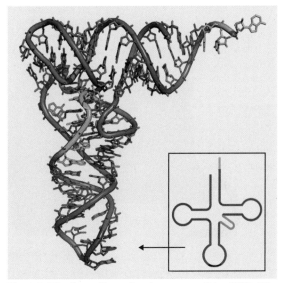

Fig. 2.3 The three-dimensional structure of the tRNA. The three nucleotides in *grey at the bottom* of the figure are the anticodon, and are complementary to the triplet recognized by this tRNA. The *yellow portion at the top right* is the acceptor stem that binds to the appropriate amino acid. It is important to note how Watson–Crick base pairings and other bonds result in a three-dimensional structure that determines the tRNA's behavior, beyond what could be inferred from the *logical* structure in the *lower right box*. In particular, this structure exposes the anticodon and the acceptor stem, making them accessible for interaction with the translation machinery

fer their unique biologically active conformations on proteins.

2.2.2 Molecular Recognition Information

The information driving specific recognition between biologically active molecules resides in the stereochemistry of weak bonds. In the case of macromolecules (DNA, RNA, proteins, etc.) the key factors are the shape of their surfaces, the nature of exposed residues, and the configuration of active sites. A striking example of how such specificity is attained can be found in the tRNA charging reaction (aminoacylation), that is catalyzed by a family of enzymes called aminoacyl-tRNA synthetase. tRNAs are small adaptor molecules that directly implement the genetic code, during the translation of mRNA into proteins. Each tRNA carries a triplet of nucleotides at one end, and receives the corresponding amino acid at the other end thanks to highly specific interactions with aminoacyl-tRNA synthetases. This molecular recognition, performed by elements located in different parts of the enzyme, is so finely tuned that it is able to distinguish between two closely related amino acids and to charge the tRNA with the correct one according to the triplet it carries (Fig. 2.3). The information that guides this reaction is carried by

the structural, chemical, and molecular features of the three partners (enzyme – tRNA – amino acids), and the strong evolutionary conservation of this reaction suggests a primordial origin of the tRNA recognition code. This fact, and the critical importance of this process to all living cells, have led some authors to describe it as a *second genetic code*.

2.2.3 Spatial Information

Just as in real factories the various types of tools and equipment are distributed in the available space in an organized way, the cell volume is organized into discrete compartments in which cellular molecules are recruited and specific reactions are performed. The nucleus and the cytoplasm are the main subdivisions, inside which many other specialized *bodies* are found (Fig. 2.4). The function of such bodies is to separate and concentrate reagents in order for them to reach critical concen-

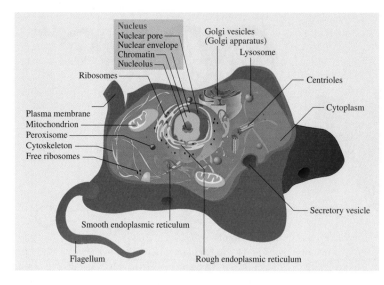

Fig. 2.4 Diagram of an animal cell, showing the most important compartments and organelles

trations for reactions. Such structural organization is highly dynamic, with extensive trafficking of molecules between nucleus and cytoplasm and within their compartments.

A major nuclear compartment is the chromatin, i.e., the assembly of DNA and proteins that compacts the DNA polymer to various degrees, up to the extreme compaction of the mitotic chromosomes. Chromatin is an important example to illustrate how multiple dynamic information layers overlap the information encoded by the DNA sequence, itself static and linear. The uneven distribution of histones and other proteins along the DNA chain and their chemical modifications (the epigenetic structure) lead to differences in accessibility to transcription factors. Therefore, a key step in gene expression regulation is guided by an additional informational code (the so-called histone code or epigenetic code) that precedes and affects sequence recognition: The chromatin state of a specific region of a chromosome determines whether expression of genes located in that region is allowed or suppressed. The chromatin state is determined, to a first approximation, by the positions of histones and other chromatin proteins on the chromosome. Another striking example of positional information can be found in the fertilized egg. The first step of embryonic development (such as anterior–posterior patterning) is dictated only by the uneven concentration gradients of maternal RNAs in the egg space, without any new transcription or DNA intervention.

In a more abstract fashion, the cell can be seen as a very large set of compartments, each of which is characterized by the concentrations of the reagents it contains. Concentrations change over time in response to the concentrations of the *input* reagents, and determine the rate of change of the concentrations of other *output* reagents, in the same or in different compartments, according to laws that can be modeled as differential equations. This view of the cell gives rise to an entirely new approach for analyzing its information content: using methods derived from the field of system dynamics, the entire cell could be modeled through a very large system of differential equations whose parameters describe the behavior of each specialized compartment and whose solutions represent the instantaneous concentration of each reagent in each compartment.

If we denote the number of compartments by C, the average number of reagents in each compartment by R, and the average number of *inputs* to each compartment by U, the number of parameters required to represent the whole cellular system is $P = CRU$, and the amount of information required to describe it would be Pb, where b is the number of bits used to encode the value of each parameter; For example, if we assume $C = 100$, $R = 100\,000$, $U = 5$, and $b = 20$, the amount of information required would be of the order of 100 million bit, to which should be added the CRb bit necessary to specify the initial values of the concentrations of the reagents in all cellular compartments.

Given this description, the whole system could in theory be simulated to predict its behavior and evolution. If the concentrations of all reagents in all compartments at time t_0 are known, solving the system of differential equations would produce the values of all concentrations at time t_1. In practice, this is not yet feasible for a variety of reasons:

1. In general, measuring the initial concentrations with sufficiently high accuracy is extremely difficult, or even impossible, since most compartments are not directly accessible.
2. Solving a system of a very large number of differential equations is computationally very hard.

3. Given the nonlinearity of the system, the solution critically depends on the initial values, meaning that very small errors in their estimation will result in dramatically different final states.
4. Finally, our knowledge of the compartmental structure of the cell (e.g., which compartments it contains, which inputs affect each compartment, etc.) is still very far from being complete enough to perform a meaningful simulation.

2.3 Intracellular and Extracellular Signaling

The cell has sophisticated mechanisms to receive and process signals coming not only from within itself, but also from other cells or from the external environment. While signals inside the cell are transported exclusively by chemical signals, signals from outside the cell are received through specialized proteins or, in the case of cells that belong to the nervous system, electrochemical signals. In most cases, the effect of the incoming signal is not due to the properties of the signaling molecule directly, but is the result of a long chain of chemical reactions that involve many other active elements. Again, we see here how evolution has evolved an abstract *language* through which cells can communicate with each other or react to the outside environment.

A prominent role in cellular signaling networks is played by a class of proteins known as protein kinases. Protein kinases are enzymes that transfer phosphate groups from high-energy donor molecules to other proteins (a process called phosphorylation), thereby changing their chemical properties. In particular, a large number of enzymes and receptors are *activated* by phosphorylation and *deactivated* by the opposite process, dephosphorylation.

Phosphorylation can therefore be viewed as a process that transmits a single bit of information across the cell. On the other hand, proteins may contain multiple phosphorylation sites, and their activation state can therefore be affected by many different incoming signals.

2.3.1 Intracellular Signaling

An interesting example of intracellular signaling that plays a key role in cell survival concerns genome stability and hence the maintenance of genetic information. DNA is continuously exposed to damage both from external sources (such as radiation) and from mutagens and toxic products of cellular metabolism (such as free radicals). Damage to DNA can occur in the form of chain breaks and/or base modifications; it is estimated that human cells may be affected by up to 1 million molecular lesions per cell per day. Since they essentially happen at random, these modifications reduce the amount of information carried by DNA, and despite the high redundancy present in the cell's genetic complement, their cumulative effect can become deleterious. To prevent this from happening, the cell employs elaborate DNA repair mechanisms that restore DNA integrity. How does the cell know that its DNA has been damaged and needs to be repaired before proceeding with replication? The signal is the damage itself, which is recognized by proteins belonging to the so-called *check point* control system. These proteins (that belong to the protein kinase group) cause a temporary arrest in the progression of the cell cycle, thus allowing the intervention of the DNA repair system that restores the correct sequence. Whenever possible, the damaged DNA strand is repaired using information from the undamaged strand.

2.3.2 Extracellular Signaling

Cells communicate with each other using specialized molecules, such as growth factors, hormones, or cytokines. Their action results in the activation or repression of selected genes, usually accompanied by epigenetic modifications of chromatin. This effect is achieved thanks to proteins embedded in the cell membrane, called *receptors*. Thanks to their characteristic three-dimensional structure, receptors cross the lipidic double layer of the membrane, putting the external world in communication with the internal milieu. They therefore act not just as receptors, but also as transductors that relay the information content of an external signal to the cell. When the receptor is activated, it activates a cascade of kinases that, through phosphorylation of critical proteins, modifies chromatin and in turn activates other enzymes and transcription factors. This may also cause a change in their nucleocytoplasmic distribution, thus carrying the signal's information all the way to the cell's inner compartments. Finally, this cascade of events usually results in the activation or inactivation of genes, leading to changes in the concentrations of the proteins they encode.

From the information point of view, the signaling process can therefore be viewed as a mechanism that amplifies the incoming information, translating an event that can be represented as a single bit (the binding of an external agent to a receptor) into a set of high-level events affecting a potentially large number of genes and proteins in very complex ways. The energy that powers the amplification process comes from the high-energy compounds involved in phosphorylation, and, as in all amplifiers, the presence of negative feedback loops maintains the stability and equilibrium of the whole system.

2.3.3 Computer Cells: Neurons

Any discussion of information processing at the cellular level would of course not be complete without a mention of neurons. Neurons are specialized information processing cells, that are able to communicate with each other directly through electrochemical signals. They are composed of a central body (the soma, containing the nucleus), out of which depart the dendrites and the axon. Dendrites are thin, branching structures that connect the neuron to a large number of other neurons, while the axon is a thicker structure that can extend over very long distances (up to 1 m in humans). Although many exceptions exist, in general neurons communicate with each other through synapses, which connect a neuron's axon with the dendrites of other neurons. When a neuron becomes excited, it generates an electrical signal (action potential) that travels through its axon; the signal is received by all other neurons whose dendrites are synaptically connected to the axon, and can in turn switch them to an excited state.

A single neuron can be connected to tens of thousands of other neurons, resulting in a neuronal network that in higher organisms can reach astonishing complexity. The human brain contains approximately 10^{11} neurons, each one having on average 7000 synaptic connections to other neurons. This results in a total number of synapses in the range 10^{14}–10^{15}, depending on age (the number of synapses reaches its maximum around 3 years of age, and then declines until adult age is reached). By comparison, the most advanced microprocessor designed so far contains approximately 2×10^9 logic elements, each of which is connected to a very small number of other components. Finally, several specialized types of neurons exist; For example, sensory neurons convey information from tissues and other organs to the central nervous system, while motor neurons transmit information to the muscles.

A detailed description of neurons and their properties is beyond the scope of this chapter and can be found in Chaps. 36 and 37. It is nevertheless interesting to analyze the information processing behavior of neurons at a high level. From this perspective, a neuron is essentially a time integrator followed by an all-or-nothing logical gate. Incoming signals from other cells (received through its dendrites) are summed together, and if the sum of the amplitudes of the signals received over a set time interval is higher than the activation threshold, the neuron will fire (i. e., generate a signal of its own). If the sum of the incoming signals does not reach the activation threshold, the neuron will remain in its basic state and the incoming signals will be lost. When a neuron is excited, the amplitude of its response does not depend on the amplitude of the incoming signals; what may change as a function of the magnitude of input signals is the frequency at which the neuron fires. It should also be noted that the interaction between neurons can be excitatory or inhibitory: the effect of an incoming excitatory signal can therefore be suppressed by a simultaneous, stronger inhibitory signal.

The neuronal network is therefore a massively parallel, highly interconnected computing system, composed of an enormous number of simple processing elements. The basic unit of information transmitted in the neuronal network is the spike, which is a time-discrete signal, in other words, an event that, to a first approximation, occurs at a definite instant of time, while its amplitude can vary over a continuous range. This is completely different in nature from all the other forms of information processing we have seen so far. We are still far from being able to understand how the architecture and complexity of neuronal networks can produce the high-level cognitive functions typical of higher organisms; but without any doubt, the development of neuronal networks represents one of the most outstanding accomplishments of life on Earth.

2.4 Conclusions

Cells are not simply biochemical machines, but are able to store, process, and communicate information through a wide variety of mechanisms and approaches. The DNA double helix stores genetic information permanently, in symbolic form. mRNA molecules represent a transient copy of a portion of the information contained in DNA, which is then converted into an amino acid sequence through a deterministic translation process. The binding of a transcription factor to DNA can be seen as a binary event (either the factor binds, or it does not) with an associated temporal span (the factor can dissociate from DNA after binding, in a stochastic fashion). The concentrations of reagents in different cellular compartments are continuous variables, whose changes can be modeled by differential equations. Cell-to-cell signaling allows cells to communicate with each other using signals expressed in symbols from a chemical alphabet.

In all these cases information flows along complex and highly interconnected networks of chemical signals. Genetic information and chemical signaling can therefore be viewed as components of highly sophisticated cellular software, whose language and algorithms are derived from the laws of molecular recognition. Similarly to computer software, it includes integrated debugging and error-correcting functionality that ensure its correct and consistent behavior.

Just as the operation of a computer is best understood at the software level rather than at the level of individual transistors, understanding the ways in which information is encoded and manipulated inside the cell can lead to a more profound understanding of its complexity, structure, and architecture, opening the way to more accurate predictions about its behavior.

Further Reading
- B. Alberts, A. Johnson, J. Lewis, M. Raff, K. Roberts, P. Walter: *Molecular Biology of the Cell*, 5th edn. (Garland Science, New York 2008)
- H. Lodish, A. Berk, P. Matsudaira, C.A. Kaiser, M. Krieger, M.P. Scott, S.L. Zipursky, J. Darnell: *Molecular Cell Biology*, 5th edn. (Freeman, New York 2004)
- J.D. Watson, T.A. Baker, A.B. Bell, A. Gann, M. Levine, R. Losick: *Molecular Biology of the Gene*, 6th edn. (Cummings, San Francisco 2008)

3. Dielectrophoresis: Integrated Approaches for Understanding the Cell

Takeshi Yamakawa, Hiroko Imasato

The complex permittivity of a biological cell reflects its substance and structure and thus seems to reflect its function, activity, abnormality, life/death, age, and life expectancy. Although it may be very difficult to measure the complex permittivity of each cell, the movement or behavior of the cell as affected by its complex permittivity can be observed under the microscope. The *dielectrophoretic force* (DEP force) generated on a particle in a nonuniform electric field causes movement of the particle in accordance with its complex permittivity or polarizing characteristics. Thus, differences in the substance or structure of biological cells lead to differences in their movement or behavior in a nonuniform electric field. The principle of dielectrophoresis (DEP) and the estimation of the DEP force are described in this chapter. The distinctive features of DEP are applied in the separation of biological cells, e.g., leukocytes from erythrocytes, leukemia cells from normal leukocytes. This cell separation ability is affected by the frequency and amplitude of the applied voltage. To estimate the DEP force generated on a single cell, the terminal velocity of the cell in the medium should be measured without taking it out of the DEP device. The procedure to measure the terminal velocity is also described.

Biological tissue is a large population of cells interconnected with extracellular matrix, performing a similar function within an organism. Tissue is studied by histology or histopathology. The classical tools for studying tissue are the wax block for fixation and cutting, tissue stain for highlighting particular structures of interest, and the optical microscope. Functional analysis and discussion of tissue can be achieved by employing these classical tools.

Blood cells, e.g., leukocytes (white blood cells), erythrocytes (red blood cells), and platelets, exhibit and serve individual functions in peripheral blood. Thus, pathologists, pharmacologists, hematologists, internal medicine specialists, etc. are eager to know the function, activity, abnormality, dead/alive status, age, and life expectancy of a single blood cell rather than the cell population. These factors are significantly important from the viewpoint of bioinformatics and are difficult to determine using classical tools.

This chapter describes a promising tool to extract biological and morphological information from a single blood cell by employing *dielectrophoresis* [3.1], which is based on the polarization characteristics of the cell.

3.1 Complex Permittivity as Bioinformation

Polarization characteristics of solid, liquid or gaseous materials give us important information about the substance and structure of the target. When the target is uniform or quasi-uniform material (whether crystalline, polycrystalline, or amorphous), the polarization characteristics can be measured by shaping the target material into a cuboid and attaching parallel-plate electrodes as shown in Fig. 3.1a, and described in the form of an impedance, e.g., a parallel connection of a resistor and a capacitor as shown in Fig. 3.1b. The frequency

response of the impedance implies a polarizing mechanism and polarizing domain.

When we want to measure the dielectric property of a small particle or a biological cell floating in water as shown in Fig. 3.1c, it is impossible to adopt the impedance description of the particle or the cell, because the impedance between terminals a and b does not reflect the dielectric property of only the target particle or cell, but mostly that of the water.

The dielectric property of a single particle (or a biological cell) in water can be measured as a force generated on the target particle, or from the behavior of the particle, without taking the particle out of the water. One possibility is dielectrophoresis, which reflects the dielectric property of the particle and the medium that generates a mechanical force on the particle and makes it move in some direction.

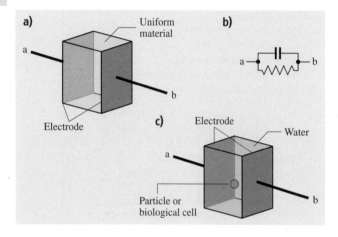

Fig. 3.1a–c How can the dielectric property of a material be measured? (**a**) In case of a uniform material, it can be measured by shaping the material into a cuboid and attaching two parallel-plate electrodes facing each other, being described as (**b**) the impedance between the two electrodes. (**c**) In case of a single particle or a biological cell in water, how the dielectric property of the particle or cell can be measured.

3.2 Dielectrophoresis (DEP)

Charged particles in a medium where a static electric field is generated by a direct current (DC) voltage are forced to move along the field. This phenomenon is known as *electrophoresis* and is utilized for estimation of molecular mass and polarity of ions. Electrophoresis cannot be employed for neutral particles such as biological cells in body fluid or blood. On the other hand, under a nonuniform alternating electric field, charged or uncharged particles are forced to move in accordance with the relative difference of the dielectric properties between the particles and the medium in which they are floating. This phenomenon is called *dielectrophoresis*.

3.2.1 Principle of Dielectrophoresis

Dielectrophoresis (DEP) has attracted much interest because it is an effective label-free method (enabling noninvasive approaches) to manipulate and separate

a particle such as a living cell, virus, DNA, or biomolecule [3.2–4]. The basic theory and applications were comprehensively presented by *Pohl* [3.1].

When a material is placed in an electric field, all the molecules constructing the material are polarized. Since the polarized charges of molecules in the material cancel those of adjacent molecules, only charges at boundaries between different materials should be considered, for instance, the boundary between the electrodes and the medium, and that between the particle and the medium. When a particle in a given medium is placed in an electric field E, both the particle and the medium are simultaneously polarized. In case of higher permittivity of the particle than that of the medium ($\varepsilon_p^* > \varepsilon_m^*$), the electric field induces charges at the particle side of the boundary as shown in Fig. 3.2a,c. On the other hand, in case of lower permittivity of the particle than that of the medium ($\varepsilon_p^* < \varepsilon_m^*$), the electric field

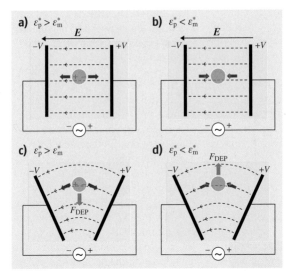

Fig. 3.2a–d Line distribution of electric force generated by an external voltage and the polarization induced by an electric field in case of parallel-plate electrodes ((**a**) $\varepsilon_p^* < \varepsilon_m^*$, (**b**) $\varepsilon_p^* > \varepsilon_m^*$). Non-parallel-plate electrodes ((**c**) $\varepsilon_p^* < \varepsilon_m^*$, (**d**) $\varepsilon_p^* > \varepsilon_m^*$)

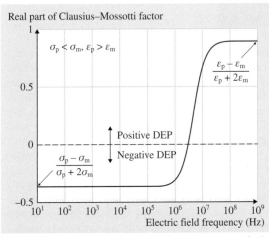

Fig. 3.3 Frequency response of the real part of the Clausius–Mossotti factor characterizing the direction and magnitude of the dielectrophoretic force

induces charges at the medium side of the boundary as shown in Fig. 3.2b,d.

In a uniform electric field, as shown in Fig. 3.2a,b, the particle experiences equal opposing forces (blue arrows) and thus the net force is zero. On the other hand, in a nonuniform electric field with $\varepsilon_p^* > \varepsilon_m^*$ as shown in Fig. 3.2c, the particle is forced to move toward a re-

gion of higher electric field by the resultant force F_{DEP} (red arrow) of two component forces (blue arrows), and in a nonuniform electric field with $\varepsilon_p^* < \varepsilon_m^*$ as shown in Fig. 3.2d, the particle is forced to move toward a region of lower electric field by the resultant force F_{DEP} (red arrow) [3.1, 5, 6]. Each resultant force is the so-called dielectrophoretic force (DEP force).

The DEP force F_{DEP} depends on the dielectric properties of the particle and of the medium, the radius of the particle, and the electric field as described by (3.1) [3.1, 5, 6]

$$F_{DEP} = 2\pi r^3 \cdot \varepsilon_m \cdot \mathrm{Re}[F_{CM}(\omega)] \cdot \nabla E_{rms}^2 , \qquad (3.1)$$

where r is the radius of the particle,

$$F_{CM}(\omega) = \frac{\varepsilon_p^* - \varepsilon_m^*}{\varepsilon_p^* + 2\varepsilon_m^*} \text{ is the Clausius–Mossotti factor ,}$$

$$(3.2)$$

$$\varepsilon^* = \varepsilon - i\frac{\sigma}{\omega} \text{ is the complex permittivity ,}$$

$$(3.3)$$

ε and σ are the permittivity and conductivity, respectively, $\mathrm{Re}[F_{CM}(\omega)]$ is the real part of the complex function of angular frequency ω, and E_{rms} is the root-mean-square value of the electric field (subscripts "p" and "m" indicate particle and medium, respectively).

The frequency response of the real part of the Clausius–Mossotti factor can be calculated from (3.2) and (3.3) and is shown in Fig. 3.3 for $\sigma_p < \sigma_m$ and $\varepsilon_p > \sigma_m$. When $\mathrm{Re}[F_{CM}(\omega)] > 0$, the particle moves towards the region of higher electric field (positive DEP). When $\mathrm{Re}[F_{CM}(\omega)] < 0$, the particle moves towards the region of lower electric field (negative DEP). Consequently, the actual direction of particle movement is determined by both $\mathrm{Re}[F_{CM}(\omega)]$ and the shape of the electrodes.

3.2.2 Creek-Gap Electrode and DEP Device

Dielectrophoresis can be achieved using various kinds of electrodes, such as a pin–pin electrode [3.7,8], a pin–plate electrode [3.8], a castellated interdigitated electrode [3.9], an extruded-quadrupole electrode [3.10], etc. These are useful for separation of particles by the sign of the DEP force, but not for measuring the DEP force. The creek-gap electrode [3.11–13] is adopted here for measuring the DEP force of the particle, as shown in Fig. 3.4. Figure 3.4a shows a side view of the DEP device, and Fig. 3.4b a top view of the DEP device, where only a partial view of two electrodes is magnified and shown as well as electric terminals. Figure 3.4c

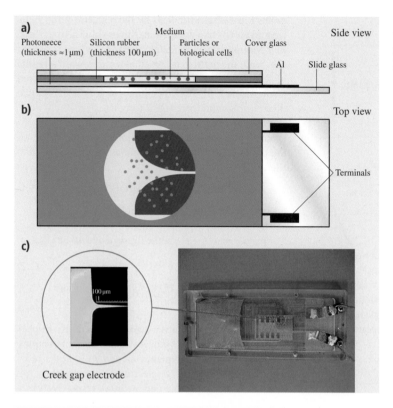

Fig. 3.4a–c The DEP device: (**a**) side view, (**b**) top view, and (**c**) photograph of the creek-gap electrode and the DEP device

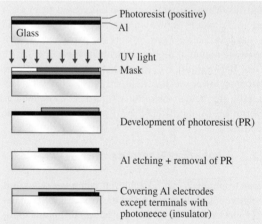

Fig. 3.5 Photolithography of the creek-gap electrode employed in the DEP device

rubber spacer of 100 μm in thickness is placed between a cover glass and a slide glass to construct a thin chamber of 100 μm in height and about 6 mm in diameter, which is filled with the particle or biological cell suspension where the DEP force is to be generated. The DEP device possesses two electric terminals for alternating current (AC) voltage application.

The DEP device is fabricated by photolithography as shown in Fig. 3.5. An aluminum film is so firmly formed on the surface of a slide glass by sputtering that it may not to be removed by frictioning with a finger. The electrode is covered by photosensitive polyimide (Photoneece, about 1 μm in thickness; Tray Industries, Inc.) as an insulating film. The Photoneece film is exposed with the mask to form two terminals. A side view of the accomplished DEP device is shown in Fig. 3.4a.

3.2.3 Measurement of DEP Force by Inclined Gravity

To measure the DEP force generated on a particle or biological cell in a medium, the so-called null method is adopted here. In the inclined floor environment where the DEP force F_{DEP} is upward along the floor inclined

shows a photograph of the whole shape of the electrodes and the DEP device. A sputtered aluminum film (about 1 μm in thickness) is shaped to two electrodes which construct an electrode gap like a creek. Thus, the pair of electrodes is called a creek-gap electrode. A silicon

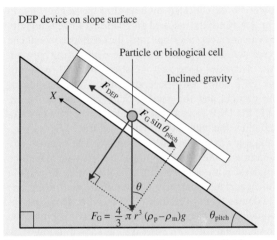

Fig. 3.6 The DEP force canceled by the controllable gravitational force as a reference force

radius r moving at velocity $\dot{x}$ is affected by a viscous drag force of $6\pi r \eta \dot{X}$ due to the fluid viscosity η, where X is the sloping coordinate. The particle is affected by a force from the external environment, leading to a total force of

$$\frac{4}{3}\pi \rho_p r^3 \ddot{X} = F_{DEP} - F_G \sin\theta_{pitch} - \mu_s F_G \cos\theta_{pitch}$$
$$- \mu_r F_G \cos\theta_{pitch} - 6\pi r \eta \dot{X}, \quad (3.4)$$

where ρ_p is the mass density of the particle, μ_s is the coefficient of static friction, and μ_r is the coefficient of rolling friction. Let us consider (3.4) under the condition of negligibly small friction terms $\mu_s F_G \cos\theta_{pitch}$ and $\mu_r F_G \cos\theta_{pitch}$. If the particle is stationary ($\dot{X}=0$), the acceleration is also zero ($\ddot{X}=0$), and thus $4/3\pi r \rho_p r^3 \ddot{X}=0$ and $6\pi r \eta \dot{X}=0$. Therefore, (3.4) is reduced to the following simple equation:

$$F_{DEP} = F_G \sin\theta_{pitch} = \frac{4}{3}\pi(\rho_p - \rho_m)r^3 g \sin\theta_{pitch}, \quad (3.5)$$

where ρ_m is the mass density of the medium, g is the gravitational acceleration, and $4/3\pi \cdot (-\rho_m)r^3 g$ is the effective gravitational force applied to the particle considering the buoyancy force. Equation (3.5) implies that F_{DEP} can be obtained by measuring the angle θ_{pitch}, if ρ_p and ρ_m are known.

at an angle of θ_{pitch}, the component of the gravitational force parallel to the floor $F_G \sin\theta_{pitch}$ acts oppositely to the DEP force as shown in Fig. 3.6. The normal component of the gravitational force $F_G \cos\theta_{pitch}$ causes static friction or dynamic friction depending upon whether the particle is stationary or moving. A spherical particle of

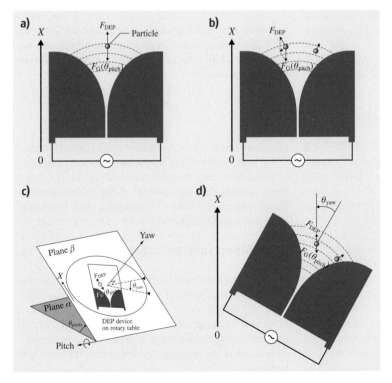

Fig. 3.7a–d Principle of the null method. (a) The DEP force generated on the particle in the center line between two electrodes can be canceled by the adjustable gravitational force. (b) The DEP force on a particle off the center line which cannot be canceled by the adjustable gravitational force. The particle moves to the right. (c) Device rotation around the yaw axis by the angle θ_{yaw} enables the balance between F_{DEP} and the reference force $F_G(\theta_{pitch}) = F_G \sin\theta_{pitch}$ without changing the amount of the reference force. (d) The DEP force on a particle off the center line canceled by the adjustable gravitational force by changing the direction

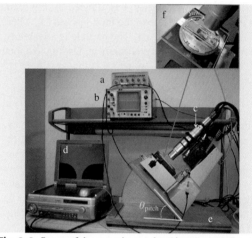

Fig. 3.8 Setup of the experimental equipment; (*a*) function generator, (*b*) oscilloscope, (*c*) digital microscope, (*d*) display of digital microscope, (*e*) microscope holder, (*f*) DEP device on the rotary table of the digital microscope

The electric field in the upper part of the gap between two electrodes is very small. The differential electric field in the lower part of the gap is also very small, because two electrodes are designed so that the gap may be constant at the lower part. For these two reasons the profile of the DEP force is almost zero at the upper and lower parts of the gap and maximum halfway.

The two forces which are canceled in the null method should be exactly equal to each other ($F_{\mathrm{DEP}} = F_{\mathrm{G}} \sin \theta_{\mathrm{pitch}}$), as shown in Fig. 3.7a. The upward DEP

force can be canceled by adjusting the angle θ_{pitch} in Fig. 3.6. Since the inclined gravitational force is directly downward along the slope face, the DEP force should be generated directly upward on the central line between two electrodes. However, the strength and direction of the DEP force vary depending upon the position of the particle and the electric field distribution. Therefore, the DEP force cannot be always directly upward, and thus the inclined gravitational force $F_{\mathrm{G}} \sin \theta_{\mathrm{pitch}}$ cannot always cancel the DEP force. The strength and direction of the canceling force $F_{\mathrm{G}} \cdot \sin \theta_{\mathrm{pitch}}$ should be adjusted in accordance with the DEP force. To achieve this cancelation properly in the null method, we rotate the device by θ_{yaw} as shown in Fig. 3.7c. Thus, the upward left DEP force is possibly canceled by this inclined gravitational force $F_{\mathrm{G}} \cdot \sin \theta_{\mathrm{pitch}}$, as shown in Fig. 3.7d. In a similar manner, the upward right DEP force can be canceled by the inclined gravitational force by adjusting the opposite angle θ_{yaw}. Accordingly, accurate measurement of the DEP force can be realized by inclining the plane of the DEP device and rotating the device around its yaw axis.

Figure 3.8 shows a setup of the experimental equipment to measure the DEP force at any place between the electrodes (clear field). The amount of the inclined gravitational force can be easily adjusted by changing the angle θ_{pitch}. If the particle is not on the center line between the electrodes, the balance between the DEP force and the inclined gravitational force is accomplished by rotating the rotary table (*f* in Fig. 3.8) of the digital microscope.

3.3 Separation of Blood Cells

Blood cells form in bone marrow, which is the soft material in the center of most bones. Normally, bone marrow begins by producing immature blood cells, called stem cells and blasts. As these blood cells mature, they form into one of three types of mature blood cells:

1. Erythrocytes (red blood cells), which carry oxygen and other materials to all tissues of the body
2. Leukocytes (white blood cells), which fight infection and disease
3. Platelets, which help prevent bleeding by causing blood clots to form.

Mature blood cells are created from one of two types of stem cells, i.e., lymphoid stem cells or myeloid stem cells. As lymphoid stem cells mature, they become

lymphoid blasts (so-called lymphoblasts), then lymphocytes, and finally white blood cells. As myeloid stem cells mature, they become myeloid blasts (so-called myeloblasts), then myelocytes (also called granulocytes), and finally either white blood cells, red blood cells, or platelets depending on the type of granulocyte.

3.3.1 Blood Cells and Leukemia

In leukemia [3.14], bone marrow produces abnormal cells such as abnormal lymphoid cells or myeloid cells. When abnormal lymphoid cells are created, it is known as lymphocytic leukemia (or lymphoblastic leukemia). When abnormal myeloid cells are created, it is known as myeloid leukemia (or myelogenous leukemia).

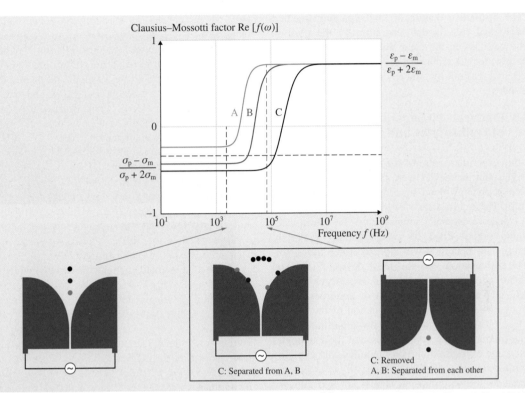

Fig. 3.9a–c Separation of biological cells by dielectrophoresis (DEP). (**a**) Separation of cells A, B, and C exhibiting negative DEP. (**b**) Separation of a cell C with negative DEP from cells B and C with positive DEP. (**c**) Separation of cells A and B, both exhibiting positive DEP

These blood cells are different from each other, whether they are normal or abnormal, in substance and structure, and thus in dielectric property or polarizing characteristics. This may cause different behavior of these blood cells in nonuniform electric fields.

3.3.2 Principle of Cell Separation by DEP

The Clausius–Mossotti factor (CM factor) factor $F_{CM}(\omega)$ characterizes the dependency of the polarity and the magnitude of the DEP force on frequency. It is assumed that there are three kinds of cells (cell A, cell B, cell C) which have characteristic features as shown in Fig. 3.9a. Let us consider the following three cases for separating blood cells using the equipment shown in Fig. 3.8 [3.15]. Let us consider the following two situations at 1 kHz and 100 kHz:

1. The CM factors of the three cell types (A, B, C) exhibit the same negative sign at 1 kHz, and thus

separation can be achieved by the magnitude of DEP (Fig. 3.9b).
2. The CM factors of two cell types (A, B) are positive and that of the other last one (C) is negative at 100 kHz. The first separation can be achieved by the sign of DEP (Fig. 3.9c). Only cell C with negative DEP is separated from the other two cells A and B with positive DEP. Cells A and B are attracted to the edge of the electrodes because of the convergence of the electric field, which is much more significant than the attractive force to the narrow gap area. Sequentially the DEP device is turned around by 180° at the same frequency. Then, cell C falls down and disappears from view due to the inclined gravitational force and the additional negative DEP force. At the same time, cells A and B are separated by the amount of the positive DEP force (Fig. 3.9d). The applied voltage or the pitch angle may be carefully adjusted to release cells A and B from the electrodes.

The distinctive feature of this separation method of biological cells is the separation by the position of the particles in the clear field (space in the electrode gap) for the same sign of DEP force as shown in Fig. 3.9b,d, but not by adherence to or repulsion from electrodes.

3.3.3 Practical Cell Separation of Erythrocytes and Leukocytes by DEP

For this experiment, almost the same amounts of erythrocytes and leukocytes were taken from heparinized human peripheral blood by centrifugation (2000 rpm, 30 min). These cells were washed in a mixture of 8.75% (w/v) sucrose isotonic solution and 1% phosphate-buffered saline (PBS) by centrifugation (1500 rpm, 10 min). This operation was repeated three times, followed by injection into the DEP device [3.15] and application of an AC voltage to the Al electrodes. Figure 3.10 shows the separation of erythrocytes and leukocytes 23 min after the application of the voltage (10 V_{pp}, 60 MHz). Figure 3.10a shows the initial state of the mixture of the erythrocytes and the leukocytes of the same concentration. Forty minutes after applying the voltage (10 V_{pp}, 60 MHz), almost all of the cells are separated into two groups, as shown in Fig. 3.10b. We can observe the erythrocytes moving toward the weak electric field (negative DEP) and the leukocytes adhering to the edge of the electrodes (positive DEP) at which the strong electric field converges. These two kinds of cells can also be identified by watching their shapes in the microscope.

3.3.4 Practical Separation of Leukemia Cells (BALL-1, HL-60)

Two kinds of leukemia cells, i. e., the human B-cell acute lymphotropic leukemia cell line (BALL-1) and human promyelocytic leukemia (HL-60) cell line, were examined. These test cells were provided by the RIKEN BRC through the National Bio-Resource Project of MEXT, Japan. BALL-1 cells and HL-60 cells were cultured in RPMI 1640 medium (Wako Pure Chemical Industries, Ltd.) with 10% fetal bovine serum (Thermo Scientific) under 5% CO_2/95% air at 37 °C. For the experiment, they were washed three times in 8.75% (w/v) sucrose isotonic solution with 1% PBS (18.9 mS/m).

Normal leukocytes (mononuclear cells and granulocytes) for the experiment were separated from heparinized human peripheral blood sample by density-

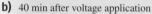
a) Initial state

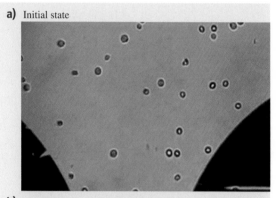

b) 40 min after voltage application

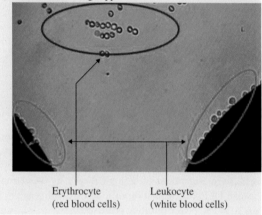

Erythrocyte   Leukocyte
(red blood cells)   (white blood cells)

Fig. 3.10 Separation of erythrocytes (RBC) from leukocytes (WBC). Applied voltage: 10 V_{pp}, frequency: 60 MHz, time required for separation: 40 min

gradient centrifugation (700 g, 30 min) as shown in Fig. 3.11. Then, the normal leukocytes were washed three times in 8.75% (w/v) sucrose isotonic solution with 1% PBS.

BALL-1 cells and normal leukocytes (mononuclear cells) were mixed together so that the concentration of each cell type might be almost the same, followed by injection into the DEP device. Next, HL-6 cells and normal leukocytes (granulocytes) were also mixed in the same way and injected into the DEP device.

An AC voltage was applied to the electrodes. Figure 3.12 shows the separation of the leukemia cells from the normal leukocytes. We can observe the normal leukocytes moving toward the weak electric field (negative DEP), and the leukemia cells (BALL-1 cells) adhering to the edge of the electrodes (positive DEP) where a strong electric field is generated.

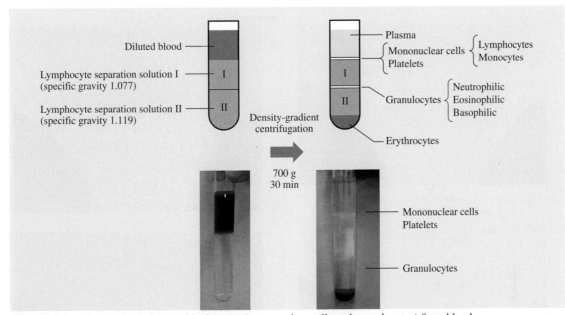

Fig. 3.11 Separation method of normal leukocytes (mononuclear cells and granulocytes) from blood

This separation was observed 30 min after AC voltage application (14 V_{pp}, 37 kHz) as shown in Fig. 3.12. On the other hand, separation of HL-60 cells from normal leukocytes was observed about 40 min after voltage application (14 V_{pp}, 45 kHz), as shown in Fig. 3.13.

While separation of inorganic materials is relatively easy, separation of living cells is sophisticatedly difficult because of the complexity of the substance and its structure. Control of the osmotic pressure and electric conductivity of the solution is necessary for effective separation of living biological cells.

3.4 Estimation of the DEP Force

Separation of cells may be accomplished by the DEP device with a creek-gap electrode on an inclined plane without knowing the actual DEP force generated on each cell, because the relative DEP force or its relative magnitude between cells is sufficient to separate them. However, estimation of the actual DEP force generated on a single biological cell may be significant to obtain biological information of the cell through its polarizing characteristics [3.16].

In (3.5), g is a constant, θ_{pitch} can be measured, and r can also be measured under the microscope. Therefore, the DEP force F_{DEP} can be estimated by (3.5), if the specific gravity difference between the target cell and the medium ($\rho_p - \rho_m$) is obtained *without taking the cell out of the device*. One way to obtain the value ($\rho_p - \rho_m$) of a single biological cell is through the *terminal velocity*.

3.4.1 Terminal Velocity of a Single Biological Cell

The *terminal velocity* is the velocity which the particle reaches in a viscous medium under falling condition as shown in Fig. 3.14. When a small particle of volume V is falling in a fluid of viscosity η, the net force F_{net} acts on the particle downward, consisting of two upward forces of buoyancy F_b and viscous drag F_D and one downward force of gravity F_G. Thus,

$$F_{net} = F_G - F_b - F_D = V\rho_p g - V\rho_m g - F_D . \quad (3.6)$$

If the Reynolds number Re of the flow satisfies Re $\ll 1$, this is called Stokes flow (creeping flow). A spherical particle in Stokes flow experiences a viscous drag force F_D in (3.6) of $6\pi\eta(d/2)v$, where v is the velocity of the

a) Initial state

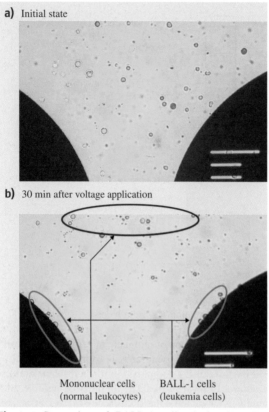

b) 30 min after voltage application

Mononuclear cells   BALL-1 cells
(normal leukocytes)   (leukemia cells)

Fig. 3.12 Separation of BALL-1 cells (leukemia cells) from mononuclear cells (normal leukocytes). Applied voltage: 14 V$_{pp}$, frequency: 37 kHz, time required for separation: 30 min

a) Initial state

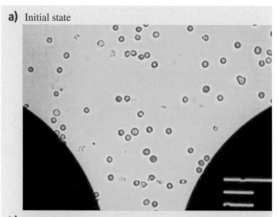

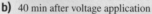

b) 40 min after voltage application

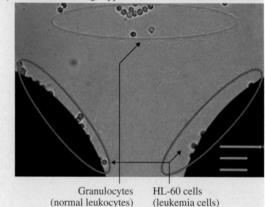

Granulocytes   HL-60 cells
(normal leukocytes)   (leukemia cells)

Fig. 3.13 Separation of HL-60 leukemia cells from normal granulocytes

particle. Thus, (3.6) can be rewritten as

$$F_{net} = \tfrac{4}{3}\pi r^3 \rho_p g - \tfrac{4}{3}\pi r^3 \rho_m g - 6\pi\eta r v \,. \quad (3.7)$$

When $F_{net} = 0$, the particle falls at a constant speed in a steady state. This velocity is called the terminal velocity v_t and satisfies

$$0 = \tfrac{4}{3}\pi r^3 \rho_p g - \tfrac{4}{3}\pi r^3 \rho_m g - 6\pi\eta r v_t \,,$$
$$(3.8)$$

$$\therefore \quad \rho_p - \rho_m = \frac{9\eta \cdot v_t}{2g \cdot r^2} \,. \quad (3.9)$$

Equations (3.5) and (3.9) result in the following simple equation:

$$F_{DEP} = 6\pi r\eta v_t \cdot \sin\theta_{pitch} \,. \quad (3.10)$$

Thus, the dielectrophoretic force can be estimated by (3.10), if the viscosity of the medium has been measured by viscometer in advance.

3.4.2 Measuring the Terminal Velocity

The terminal velocity v_t of a single particle in a medium in the thin space (e.g., $100\,\mu m$) of the DEP device has to be measured. Figure 3.15a,b shows the scheme for raising the DEP device to the vertical (Fig. 3.15b) from the recumbent position (Fig. 3.15a). This may seem a promising approach, at first sight, to measure the terminal velocity. However, the falling behavior of a polystyrene particle in water exhibits fluctuation and no steady state, as shown in Fig. 3.15c.

Although the flow around the particle in a wide space is symmetric (Fig. 3.16a), the flow around the particle falling near a wall is asymmetric (Fig. 3.16b). Therefore, the latter exhibits strange behavior, e.g., repeated contact with and peel-off from the wall. Consequently, the velocity fluctuates and does not exhibit falling in free space.

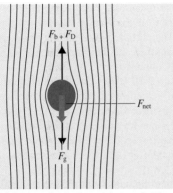

Fig. 3.14 Free fall of a spherical particle in a fluid under the condition of Re ≪ 1, i. e., so-called Stokes flow (creeping flow)

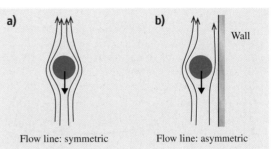

Flow line: symmetric   Flow line: asymmetric

Fig. 3.16 Flow around the particle with and without a close wall

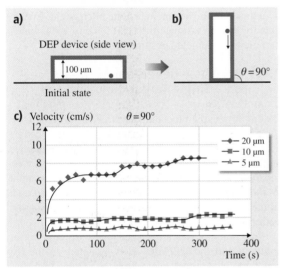

Fig. 3.15a–c Falling behavior of a polystyrene particle in water. (**a**) Initial state of the DEP device, where the particle is on the floor. (**b**) The DEP device is rapidly raised up to vertical. (**c**) Time response of the falling velocity of the particle for different sizes

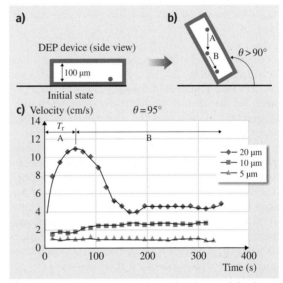

Fig. 3.17a–c The procedure to make the particle depart from the wall. (**a**) Initial state of the DEP device, where the particle is on the floor. (**b**) The DEP device is rapidly raised up in excess of $\theta = 90°$. (**c**) Time response of the falling velocity of the particle for different sizes

To avoid this problem, it is necessary to make the particle depart from the wall and preferably fall through the central space between the walls of the DEP device (50 μm from the wall). When the DEP device is raised by more than 90°, as shown in Fig. 3.17a,b, the particle departs from one wall (where it was stationary) to fall down until reaching the other sloping face, i. e., the other wall (stage A in Fig. 3.17b). After that, the particle rolls down the sloping face (stage B in Fig. 3.17b). Fig-

ure 3.17c shows the experimental results for $\theta = 95°$. The velocity gradually increases to reach a maximum in the first stage A, and then decreases in the next stage B. The maximum point is the time when the particle reaches the opposite wall. Thus, the time to reach the maximum is the *reach time* T_r. Half of this is the optimum time to maintain the angle of device inclination so that the particle falls down through the intermediate space between the walls. This time is the *release time* $T_{1/2}$. Although $\theta = 95°$ is adequate for a large particle, it is not enough for other smaller particles which cannot achieve the reach time.

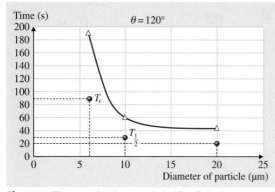

Fig. 3.18 The reach time and half of it in case of $\theta = 120°$

Figure 3.17 also shows the release times as 90 s for a 6 μm particle, 30 s for a 10 μm one, and 20 s for a 20 μm one. Figure 3.18 shows the relationship between the diameter of the particle and the reach time T_r (triangles) together with the release time $T_{1/2}$ (filled circles) in case of $\theta = 120°$.

Considering the issue described above, the terminal velocities of various sizes of particle can be measured by the procedure shown in Fig. 3.19. First, the DEP device is raised to 120° to examine the reach time of the target particle. Next, the DEP device is turned back by 60° (i. e., to attitude of 60° from the initial state), and one waits for half the reach time, i. e., the release time. Then, the target particle is located near the midpoint between the two walls. Finally, the DEP device is raised by 30° (i. e., to the vertical) to measure the terminal velocity. The particle is now falling down near the center of the chamber, and thus the stable terminal velocity can be easily obtained.

Figure 3.20 shows the experimental results for the terminal velocities of polystyrene particles in distilled water for three diameters, i. e., 21.5 μm/s, 5.0 μm/s, and 2.0 μm/s for particles of 24 μm, 10 μm, and 6 μm in diameter, respectively. Since these values do not exhibit fluctuation and reach the steady state, they are reliable.

Earlier, the radius r of the target particle was measured under a microscope, the viscosity η of the medium was also measured by viscometer, and the pitch angle θ_{pitch} of the target particle at the target position in the clear field of the creek-gap electrode was determined by adjusting the yaw angle and the pitch angle. Consequently we can estimate the DEP force of the targeted particle at the targeted position by (3.10).

3.4.3 DEP Force Generated on a Single Target Biological Cell

By employing the procedure described above, the DEP force generated on a single target biological cell can be measured. The target cells in this case are a granulocyte and a lymphocyte, which are both normal leukocytes, separated by the density-gradient centrifugation method as shown in Fig. 3.11. The medium is sucrose isotonic solution with a small amount of PBS, so-called isotonic buffer solution.

The granulocytes are suspended in isotonic buffer solution, and the DEP device is mounted on the rotary table of a digital microscope. A voltage (6.5 V$_{pp}$, 1 MHz) is applied to the device, and the yaw angle

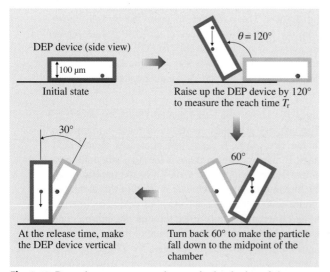

Fig. 3.19 Procedure to measure the terminal velocity of the target particle without fluctuation and with steady-state velocity

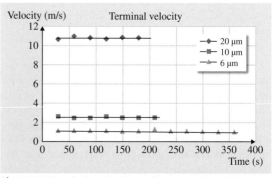

Fig. 3.20 Terminal velocities of target polystyrene particles of three sizes in distilled water

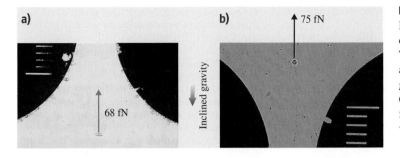

Fig. 3.21a,b DEP forces estimated by the DEP device with creek-gap electrode and inclined gravity, using the terminal velocity. The medium is sucrose isotonic solution with a small amount of PBS. (**a**) Positive DEP force generated on a granulocyte. Applied voltage: 6.5 V_{pp}, frequency: 1 MHz. (**b**) Negative DEP force generated on a lymphocyte. Applied voltage: 10 V_{pp}, frequency: 20 kHz

θ_{yaw} and pitch angle θ_{pitch} are adjusted so that a granulocyte may stay at the target position. Figure 3.21a shows two granulocytes staying at the target position in the clear field, forming a *pearl chain*. These exhibit the same DEP force of 68 fN. Other granulocytes are all attracted to the edge of the electrode, because the creek-gap electrode is a film electrode and thus significant convergence of the electric field occurs at the edge of the electrode.

Now, lymphocytes are suspended in isotonic buffer solution, and the DEP device is mounted on the rotary table of a digital microscope. A voltage (10 V_{pp}, 20 kHz) is applied to the device, and the yaw angle θ_{yaw} and pitch angle θ_{pitch} are adjusted so that a lymphocyte may stay at the target position. Figure 3.21b shows a single lymphocyte staying at the target position in the clear field. The magnitude of the negative electrophoretic force is 75 fN.

3.5 Conclusions

Dielectrophoresis (DEP) is discussed from the viewpoint of bioinformation retrieval from biological cells as well as inorganic particles. Several kinds of biological cells can be separated by the positive or negative DEP force in accordance with the polarizing characteristics based on their substance and structure. Measurement of not only the sign of the DEP force but also its magnitude is discussed, employing the terminal velocity of the particle in a fluid. This method facilitates estimation of the DEP force generated on a single

biological cell rather than a cell population, in a nonuniform electric field without taking it out of the DEP device. Since the behavior of the cell based on the DEP force reflects its polarization characteristics, its frequency response seems to reflect the substance and structure of the target cell. Consequently, the frequency response of the DEP force is expected to be useful for monitoring the immature/mature status, apoptosis, necrosis, and other activities of individual biological cells.

References

3.1   H.A. Pohl: *Dielectrophoresis: The Behavior of Neutral Matter in Nonuniform Electric Fields* (Cambridge Univ. Press, Cambridge 1978)

3.2   F.F. Becker, X.B. Wang, Y. Huang, R. Pethig, J. Vykoukal, P.R.C. Gascoyne: Separation of human breast cancer cells from blood by differential dielectric affinity, Proc. Natl. Acad. Sci. USA **92**, 860–864 (1995)

3.3   S. Nedelcu, J.H.P. Watson: Size separation of DNA molecules by pulsed electric field dielectrophoresis, J. Phys. D **37**, 2197–2204 (2004)

3.4   C.P. Luo, A. Heeren, W. Henshel, D.P. Kern: Nanoelectrode arrays for on-chip manipulation of

biomolecules in aqueous solutions, Microelectron. Eng. **83**, 1634–1637 (2006)

3.5   T.B. Jones: *Electromechanics of Particles* (Cambridge Univ. Press, Cambridge 1995) pp. 5–33

3.6   H. Morgan, N.G. Green: *AC Electrokinetics: Colloids and Nanoparticles* (Research Studies, London 2003)

3.7   H.A. Pohl, J.S. Crane: Dielectrophoresis of cells, Biophysical J. **11**, 711–727 (1971)

3.8   Y. Huang, R. Pethig: Electrode design for negative dielectrophoresis, Meas. Sci. Technol. **2**, 1142–1146 (1991)

3.9   R. Pethig, X.B. Wang, Y. Huang, J.P.H. Burt: Positive and negative dielectrophoretic collection of colloidal particles using interdigitated castellated microelectrodes, J. Phys. D **24**, 881–888 (1992)

3.10   J. Voldman, M. Toner, M.L. Gray, M.A. Schmidt: Design and analysis of extruded quadrupolar dielectrophoretic traps, J. Electrostatics **57**, 69–90 (2003)

3.11   T. Yamakawa, H. Imasato: Dielectrophoresis device and method, Japanese Patent Application, TOKUGAN2007-262058, 5 October (2007)

3.12   T. Yamakawa, H. Imasarto: Dielectrophoresis device and method, International Patent Application, PCT/JP2008/068114 (United States, Germany, United Kingdom, France, Italy, China, Korea, Japan) 26 September (2008)

3.13   H. Imasato, T. Yamakawa: Measurement of dielectrophoretic force by employing controllable gravitational force, J. Electrophoresis **52**(1), 1–8 (2008)

3.14   Clinaero, Inc.: Leukemia, eMedTV (Clinaero, San Antonio 2008) available online from http://leukemia.emedtv.com/leukemia-cells/leukemia-cells.html

3.15   H. Imasato, T. Yamakawa, M. Eguchi: Separation of leukemia cells from blood by employing dielectrophoresis, Intell. Autom. Soft Comput. **18**(2), 121–137 (2012)

3.16   H. Imasato, T. Yamakawa: Measuring the mass density of a target blood cell to calculate its dielectrophoretic force, World Autom. Congr. 2012 (WAC2012) (2012)

4. Information Processing at the Genomics Level

Alvin T. Kho, Hongye Liu

A central objective in biology is to identify and characterize the mechanistic underpinnings (e.g., gene, protein interactions) of a biological phenomenon (e.g., a phenotype). Today, it is technologically feasible and commonplace to measure a great number of biomolecular features in a biological system at once, and to systematically investigate relationships between the former and the latter phenotype or phenomenological feature of interest across multiple spatial and temporal scales. The canonical starting point for such an investigation is typically a real number valued data matrix of N genomic features $\times M$ sample features, where N and M are integers, and N is often orders of magnitude greater than M. In this chapter we describe and rationalize the broad concepts and general principles underlying the analytic steps that start from this data matrix and lead to the identification of coherent mathematical patterns in the data that represent potential and testable mechanistic associations. A key challenge in this analysis is how one deals with false positives that largely arise from the high dimensionality of the data. False positives are mathematical patterns that are not coherent (from a technical or statistical standpoint) or coherent patterns that do

not correspond to a true mechanistic association (from a biological standpoint).

A biological system is composed of multiscalar features organized in space within a specific environment. The state of the system is all its features at a specific time point. It may be conceptually informative to separate these features into two groups based largely on their spatial scales: microscopic and macroscopic. Examples of microscopic scale features include genotypes and protein interactions in a cell cytoplasm that exist or occur at the molecular scale. Examples of macroscopic feature scales include phenotypes and morphological transformations during the course of organ development that might be observed by the naked eye or optical imaging devices.

4.1 Characterizing a Biological System State, the Central Dogma of Molecular Biology, and the Advent of Bioinformatics

A central objective in biology is to identify and characterize associations between features of a system within and across multiple spatial and temporal scales. Consider the following example of a feature association: In a human study population, all subjects with a phenotype P1 are found to have a specific sequence mutation or genotype G1 of a particular gene relative to the non-mutated reference genotype G0, but not every subject with the genotype G1 has the phenotype P1. The association between any two features F1 and F2 may be weak (e.g., correlative where F1 and F2 are, say, similarly regulated by another feature F3) or strong (e.g., causal where F1 regulates F2, or vice versa). Associations between features may be conceptualized as logical truth tables, conditional probabilities, or state transition matrices in cases where features are represented as binary or discrete valued objects, or systems of difference or differential equations in cases where features are represented as continuous valued objects.

Suppose that one is given a specific phenotype to investigate in a study population. The set of microscopic scale candidate features that one could interrogate for an association with the given phenotype could be orders of magnitude larger than the size of the study population; For example, the set of biomolecules alone includes diverse classes of molecules such as genes and proteins, and distinct classes of epigenetic molecular modifications such as deoxyribonucleic acid methylation and histone deacetylation. In certain cases, the so-called central dogma of molecular biology may be assumed in order to simplify and focus the interrogation on the subset of microscopic scale features that contain nucleic acids (the primary building blocks of genes) and proteins [4.1]:

> The central dogma of molecular biology deals with the detailed residue-by-residue transfer of sequential information. It states that such information cannot be transferred from protein to either protein or nucleic acid.

In its most simplistic interpretation, the central dogma states that biological information is transcribed from deoxyribonucleic acid (DNA) to ribonucleic acid (RNA), and RNA is translated into proteins that interact at the microscopic or molecular scale to engender a phenotype, the ultimate expression of the biological information at the macroscopic scale. It is critical to note that the complete transformation from geno-type to phenotype is not instantaneous. Clearly it is a time-dependent and cumulative process that can span anywhere from a fraction of a second to days or longer, and may be affected by environmental factors. The central dogma assumption is often used to reduce the biological state of a system to its transcriptome (i.e., the set of all expressed genes) at a specific time point. Obviously this assumption is limited for identifying feature associations in cases where a significant time interval exists between earlier causal genotypic events and later phenotypic manifestations during which non-causal events and environmental factors might cooccur or interject.

The original definition of a gene is the abstract concept of a molecular unit of heredity [4.2]. Genes became physically embodied and localized to segments of the DNA following a series of landmark experiments in the 1900s [4.3–7]. The identification of genes to the DNA led naturally to questions about their sequential constituents, i.e., nucleotide base pairs. This in turn led to the ethos and simplifying assumption that the sequence of the complete DNA or genome of an organism was necessary, though as it turned out not always sufficient, for understanding its biology. Subsequently, big sequencing enterprises such as the Human Genome Project (1990–2003) arose that needed computational and mathematical modeling techniques to synthesize the high-throughput fragmented sequence data of diverse qualities from different sequencing technologies at multiple sequencing centers [4.8]. Not surprisingly, some of these techniques were originally developed in nonbiological quantitative domains that shared similar abstracted problems. Frequently, novel techniques had to be developed to tackle the unique problems that arose from sequencing such as assembly of many small DNA fragments [4.9]. This entry of nontrivial applications of computer science and mathematics into the biological domain marks the advent of bioinformatics.

From a technical standpoint, the identification of abstract gene to physical DNA paved the way for high-throughput parallel gene expression measurement technologies such as microarrays and serial analysis of gene expression (SAGE). The resulting genomic and transcriptomic data in conjunction with the reductionistic assumption of the central dogma made it convenient and feasible to simplify many basic biological inquiries into questions about gene sequences and their expression. This simplification enabled one to abstract and reformulate a biological inquiry into an equivalent computational or mathematical problem. In this chapter we review the bioinformatics assumptions, methods, and common practices related to transcriptomic data analyses.

It is important to note that, through experimentation, logic, and serendipity, researchers have since discovered features in the so-called epigenome such as noncoding RNA/DNA and chromatin modifications that are heritable and do not adhere to the transcription–translation machinery of the central dogma to be biologically functional, i. e., to be associated with or engender a phenotype. We should note that, in genome-wide association (GWAS) studies, it is commonplace to represent gene expression as a phenotype and to identify correlations between expression and single-nucleotide polymorphisms (SNPs) or haplotypes, i. e., an expression quantitative trait loci (eQTL) [4.10].

4.2 High-Throughput Gene Expression Measurement Principles and Technologies

The most common high-throughput gene expression technologies are based upon massively parallel implementations of two basic principles along with minor modifying assumptions: nucleotide base pair complementarity and nucleotide sequencing. A common modifying assumption is to claim the existence of a short nucleotide sequence that uniquely identifies each RNA or gene species. The optimal sequence lengths to achieve this objective vary according to the technical context. To complicate matters, there exist genes that are transcribed into distinct RNA isoforms and therefore translated into distinct protein isoforms in a process known as alternative splicing [4.11]. A library of expressed sequence tags (ESTs) is used to map each short identifier sequence to a known RNA or gene, and by extension to its biological ontology. Recall that genes are sequences of DNA bases – adenine (A), cytosine (C), guanine (G), and thymine (T). The DNA molecule is physically stabilized as a double helix of complementary DNA base pairs A-T and C-G. DNA is transcribed into RNA, which is comprised of bases adenine (A), cytosine (C), guanine (G), and uracil (U), with base pair complementarity A-U and C-G.

Total RNA is first extracted from a biological system of interest and then processed (e.g., fragmented, transcribed into single-strand complementary RNA/DNA, or chemically labeled) to produce so-called target RNA/DNA. Targets are quantified using the following representative technologies based on the two aforementioned basic principles.

4.2.1 Microarrays

Short, uniquely identifying single-strand DNA/RNA probe sequences are placed at predefined locations on a physical substrate in order to hybridize (by nucleotide base pair complementarity) with fluorescent labeled single-strand target DNA/RNA fragments from a system of interest. The fluorescence intensity at specific locations is assumed to be proportional to the abundance of a respective target. The readout is a real valued continuous variable corresponding to this fluorescence intensity, and therefore to the abundance of specific RNA species, which is in turn a proxy for gene expression. Notable disadvantages of this approach include cross-hybridization, and the effect of percentage GC content on the rate of hybridization, which becomes prominent as probe or target sequence lengths increase [4.12].

4.2.2 Serial Analysis of Gene Expression (SAGE)

Short, uniform length, uniquely identifying DNA/RNA target sequences (SAGE tags) are cut at predefined positions in each RNA molecule from a system of interest using a restriction enzyme. These tags are concatenated and sequenced. The readout is an integer count of the number of SAGE tags observed which correspond to the expression of specific genes. SAGE tags are identified with known genes using a SAGE tag gene library. Notable disadvantages include the availability of a comprehensive SAGE library to identify resulting tags, and the temperature-dependent sensitivity/specificity of the restriction enzyme [4.13].

4.2.3 Sequencing Techniques

The numerous next-generation sequencing techniques that are becoming commonplace today are largely based upon the foregoing two basic principles. We will not describe them except to refer the interested reader to key references [4.14–19].

Regardless of the technologies and their principles, the summary readout from a generic high-throughput gene expression assay is a data matrix of size N rows $\times M$ columns, where N is the number of genes or biomolecules measured, and M is the number of samples or conditions under investigation. Conceptually the rows and columns represent microscopic and macroscopic scale features, respectively. N is commonly orders of magnitude greater than M. The i, j-th data matrix entry is typically a real number corresponding to the expression level (or RNA abundance) of the i-th gene in the j-th sample. This data matrix is the canonical starting point for transcriptomic data analysis, whose broad objective is to identify coherent mathematical patterns in the matrix, and to determine whether these patterns correspond to nontrivial biological associations between the N genes and M samples. In datasets where N or M is large, there exists a nonzero likelihood of identifying a mathematical pattern that superficially seems coherent (i.e., a nonrandom or statistically rare event) due to multiple statistical comparisons [4.20]. We call these events technical or statistical false positives. Characterizing the false-positive rate – both technical and biological – is a critical universal issue in high-throughput data analyses in particular, and in biological investigations in general. We will survey the more common guiding principles for characterizing and quantifying technical false positives later. Biological false positives cannot generally be characterized using computational/mathematical methods alone. There exist numerous examples of genes and proteins whose expression are superficially correlated with, but are mechanistically independent of, a phenotype of interest. Biological false positives typically require additional hypothesis-testing experiments for clarity.

In summary, empirical data from almost all biological investigations can be summarized in the form of a data matrix. For transcriptomic investigations, the data matrix provides dual and related broad perspectives or representations of a system under investigation: genes (rows) in sample (columns) space, and samples in gene space. Mathematically, these perspectives correspond to genes as vectors with sample components, and samples as vectors with gene expression components, respectively. The computational and mathematical methods that are applicable in one perspective will typically be applicable in the other, with minor modifications that account for the differences in dimensionality.

4.3 Typical Workflow in Generic Transcriptomic Data Analysis

Starting with a real-valued data matrix of N gene $\times M$ sample features, the typical steps in transcriptomic data analysis are:

1. Data representation and visualization
2. Modeling noise and normalization
3. Identifying coherent (nonrandom) mathematical patterns
4. Assessing the association between coherent mathematical patterns and the biology of the system
5. Generating new testable hypotheses concerning the existence and characteristics of feature associations.

We graphically summarize these broad steps in Fig. 4.1. Our objective in the sections below is to describe and rationalize the broad concepts and general principles underlying these analytic steps. We do not intend for this to be an encyclopedia of the numerous, and daily growing, bioinformatics methods.

4.4 Data Representation and Visualization

Data have to be transformed from the original state into a representation that is suitable for manipulation to enable visualization, analysis, and reporting. These representations must be derivable from the original data, and retain as much of its information content and related context as possible. The underlying structure of a representation may in fact be more complex and dimensionally larger than the original data itself, and must be transformed into lower-dimensional representations for visualization and human intelligibility. A representation frequently highlights specific properties rather than complete details of the original data so that they are useful for data abstraction, visualization, and ultimately analysis [4.21]; For example, the dual perspectives of

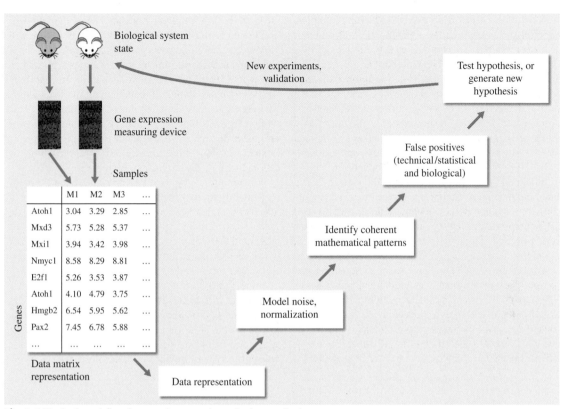

Fig. 4.1 Typical workflow in generic transcriptomic data analysis

the N gene $\times M$ sample data matrix described earlier constitute two broad representation classes: genes in sample space, and samples in gene space.

Data representation is essentially the mathematical formulation of a physical problem or system into formal structures suitable for mathematical operations and transformations. At the most basic level, it maps data features into a suitable mathematical space where one can formally define and quantify dis/similarity between features; For example, working in mathematical spaces that are equipped with either Pearson correlation or Euclidean distance as measures of dis/similarity will lead to different outcomes and conclusions about the data. In these formal spaces, the data representation of a physical problem can be analyzed (conceptualized, modeled, resolved) using suitable standard mathematical methods and theorems (e.g., statistical tests, clustering algorithms) to derive a formal result or outcome. After analysis, the investigator will test this formal result for physical (biological) validity and meaning.

Consider the following Example 1 of data representation and visualization (Fig. 4.2). Suppose that a researcher is investigating the expression level of two genes G1 and G2 in a human study population consisting of 50 subjects with disease X and 50 control subjects O who are sex and age matched. The collected data can be represented as a real-valued data matrix of two genes (rows) $\times 100$ subjects (columns). Let G1 and G2 denoted the expression values of genes G1 and G2, respectively. Inspecting the expression level of each gene separately, i. e., applying a two-group statistical test comparing the G1 (or G2) values of groups X versus O, the researcher concludes that the statistics of G1 (or G2) in groups X and O are not significantly different. Note that Euclidean distance is the implicit measure of dissimilarity in this analysis. When the data are plotted (Fig. 4.2a) to represent each subject as a point in G1 versus G2 space, it is observed that there is a coherent separation by disease status along the $G1 + G2 = 0$ line when visually zooming in on the points. $G1 + G2$ appears to be an indicator of disease state. When $G1 + G2 < 0.04$ for a subject, the subject is more likely to be in group X. When $G1 + G2 > 0.04$, the subject is more likely to be in group O. This insight would have been less ob-

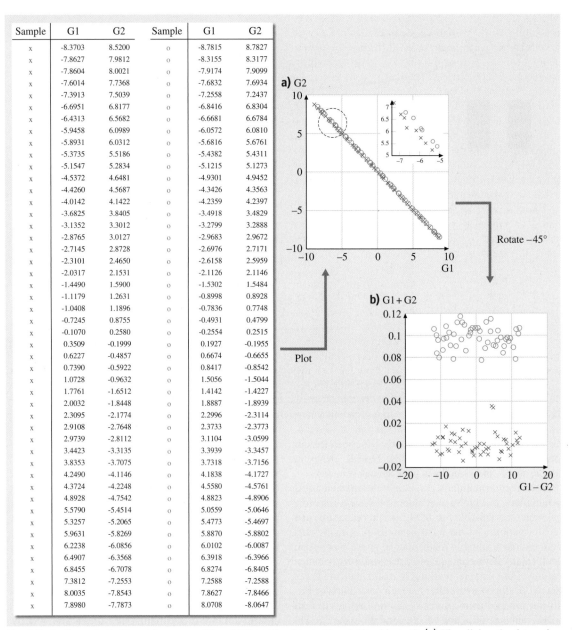

Sample	G1	G2	Sample	G1	G2
x	-8.3703	8.5200	o	-8.7815	8.7827
x	-7.8627	7.9812	o	-8.3155	8.3177
x	-7.8604	8.0021	o	-7.9174	7.9099
x	-7.6014	7.7368	o	-7.6832	7.6934
x	-7.3913	7.5039	o	-7.2558	7.2437
x	-6.6951	6.8177	o	-6.8416	6.8304
x	-6.4313	6.5682	o	-6.6681	6.6784
x	-5.9458	6.0989	o	-6.0572	6.0810
x	-5.8931	6.0312	o	-5.6816	5.6761
x	-5.3735	5.5186	o	-5.4382	5.4311
x	-5.1547	5.2834	o	-5.1215	5.1273
x	-4.5372	4.6481	o	-4.9301	4.9452
x	-4.4260	4.5687	o	-4.3426	4.3563
x	-4.0142	4.1422	o	-4.2359	4.2397
x	-3.6825	3.8405	o	-3.4918	3.4829
x	-3.1352	3.3012	o	-3.2799	3.2888
x	-2.8765	3.0127	o	-2.9683	2.9672
x	-2.7145	2.8728	o	-2.6976	2.7171
x	-2.3101	2.4650	o	-2.6158	2.5959
x	-2.0317	2.1531	o	-2.1126	2.1146
x	-1.4490	1.5900	o	-1.5302	1.5484
x	-1.1179	1.2631	o	-0.8998	0.8928
x	-1.0408	1.1896	o	-0.7836	0.7748
x	-0.7245	0.8755	o	-0.4931	0.4799
x	-0.1070	0.2580	o	-0.2554	0.2515
x	0.3509	-0.1999	o	0.1927	-0.1955
x	0.6227	-0.4857	o	0.6674	-0.6655
x	0.7390	-0.5922	o	0.8417	-0.8542
x	1.0728	-0.9632	o	1.5056	-1.5044
x	1.7761	-1.6512	o	1.4142	-1.4227
x	2.0032	-1.8448	o	1.8887	-1.8939
x	2.3095	-2.1774	o	2.2996	-2.3114
x	2.9108	-2.7648	o	2.3733	-2.3773
x	2.9739	-2.8112	o	3.1104	-3.0599
x	3.4423	-3.3135	o	3.3939	-3.3457
x	3.8353	-3.7075	o	3.7318	-3.7156
x	4.2490	-4.1146	o	4.1838	-4.1727
x	4.3724	-4.2248	o	4.5580	-4.5761
x	4.8928	-4.7542	o	4.8823	-4.8906
x	5.5790	-5.4514	o	5.0559	-5.0646
x	5.3257	-5.2065	o	5.4773	-5.4697
x	5.9631	-5.8269	o	5.8870	-5.8802
x	6.2238	-6.0856	o	6.0102	-6.0087
x	6.4907	-6.3568	o	6.3918	-6.3966
x	6.8455	-6.7078	o	6.8274	-6.8405
x	7.3812	-7.2553	o	7.2588	-7.2588
x	8.0035	-7.8543	o	7.8627	-7.8466
x	7.8980	-7.7873	o	8.0708	-8.0647

Fig. 4.2a,b Example 1: Measuring genes G1 and G2 in 50 subjects X and 50 subjects O. (**a**) Visualizing the data using given basis vectors {G1,G2}. The subplot is a visual magnification of the data inside the broken circle. (**b**) A different but equivalent representation of the data relative to new basis vectors identified by principal component analysis $\{G1, G2\} \rightarrow \{G1 - G2, G1 + G2\}$

vious had the researcher not visualized or plotted the data.

Each subject was originally represented as a point in the (G1, G2) feature space where the measure of dissim-

ilarity is Euclidean distance. In this original representation, the expression level of each gene for a subject corresponds to its projections onto the G1 (horizontal) or G2 (vertical) axes, where we see that X and O subjects are uniformly intermixed – this is a visual representation of the two-group statistical test result that the researcher performed earlier. Principal component analysis (or more generally, singular value decomposition) of the 2×2 covariance matrix (or more generally, dis/similarity matrix) of the column vectors of the data matrix was used to identify a new representation (formally an algebraic basis) in the $(G1 - G2, G1 + G2)$ feature space [4.22]. The covariance matrix represents the subject variance in the original $(G1, G2)$ space. When the subjects are represented as points in the new feature space, we see a coherent separation between groups X and O along the $G1 + G2$ (vertical) axis (Fig. 4.2b).

The starting point for transcriptomic data analysis is almost always a real-valued data matrix of N gene $\times M$ sample features. Unless otherwise stated, this representation carries an implicit measure of dis/similarity between features, frequently Euclidean distance. It is not uncommon for the researcher to overlook the question of whether the implicit measure of dis/similarity is suited to the biological problem under investigation. The initial data representation is a critical step in the analysis, and the choice of a measure of dis/similarity should depend on the problem. Consider another Example 2 (Fig. 4.3). Suppose that we have features that are time-series expression profiles of three genes G1, G2, and G3. Let us consider the question of similarity between these features in mathematical spaces equipped with either a Euclidean distance (a dissimilarity measure) or Pearson correlation (a similarity measure). In Euclidean space, G1 and G2 are more similar than are G1 and G3 – note the difference between their areas under the curves. On the other hand, in Pearson correlation space, G1 and G2 are less similar than G1 and G3. This example illustrates how oppo-

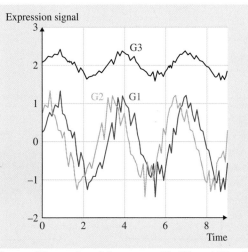

Fig. 4.3 Time series of three genes G1, G2, and G3, where G2 = G1 phase-shifted by 0.5 radians and G3 = 0.3G1 + 2 (Example 2)

site conclusions can be obtained from one dataset by simply changing the measure of dis/similarity in its representation.

Let us consider a more complex Example 3 of a transcriptomic dataset of the time series of the mouse developing whole cerebellum (Fig. 4.4) [4.23]: Starting from the data matrix of ≈ 7000 time points (postnatal days 1–60, P1–P60), we consider data representations and visualizations (using principal component analysis) from both genes in sample (time points) space, and samples in gene space perspectives. Note the distinct visualizations and their subsequent conclusions that can arise from one data matrix by simply changing the underlying measure of dis/similarity in its representations.

4.5 Noise and Normalization

Empirical data from physical experiments and observations inevitably contain *noise*; i. e., real-world data are *noisy*. "Noise" is a term that every researcher thinks he or she intuitively understands, yet both its formal and informal definitions are nontrivial and vary greatly from one domain to another. It is generally understood to be a lack of deterministic reproducibility of measurements of a system state that one assumes to be static relative to some reference framework that can be physical or theoretical; For example, if we measure the tail length of a single mouse multiple times in a 10 min time window using the same measure-

ment device and protocol, then we expect these readings to be identical or closely similar to one other. This is because it is not extraordinary to assume that the tail length is constant within that time window, and presumably static environmental variables that might affect tail length. These variations in tail length measurements are noise, and could depend on the dynamic range of the measuring device. A device that can measure in the (minimum) centimeter range is more likely to yield more uniform measurement values than one capable of measuring in the (minimum) nanometer range.

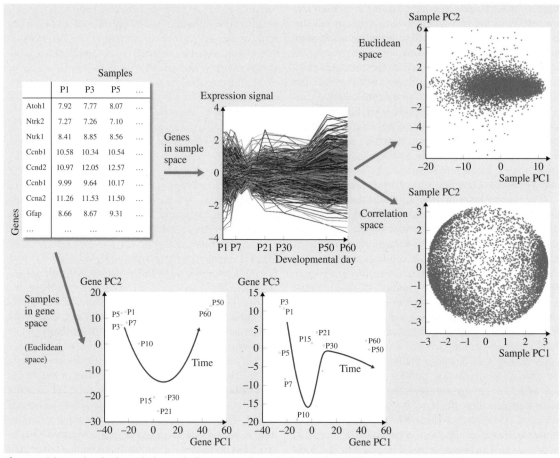

Fig. 4.4 Mouse developing whole cerebellum transcriptomic time series. Different data representations visualized using principal component analysis. In the genes in sample space perspective, each *grey dot* represents a gene. (Example 3)

For this discussion, it suffices to say that noisy data are those that lack deterministic reproducibility but possess statistical regularity or coherence [4.24]. For this reason, a common noise reduction strategy is to average over numerous repeated measurements of the system state or measurements of numerous replicate system states. In the mouse tail length example, we expect that averaging over a larger number of tail measurements in that time window leads us closer to the true length. The critical assumption underpinning this averaging strategy is that the law of large numbers and/or the central limit theorem, or generalizations or variations of them, hold for a large enough number of repeats or replicates [4.25]. We say that two system states A and B are replicates of one another, with respect to a set of predefined properties P, if A and B are identical in terms of P. Clearly no two system states are identical when P is the universal set of all properties, as the philosopher Heraclitus of Ephesus elegantly stated: *One cannot step into the same river twice* [4.26].

Noise can arise from technical (e.g., assaying instrument or protocol) and biological (e.g., different test subjects) sources. Therefore, technical and biological replicates are needed to characterize and model noise comprehensively. Technical replicates could be more than one independent measurement of a quantity of interest from one biological test subject. Biological replicates could be measurements of a quantity of interest from more than one independent biological test subject with an identical set of predefined properties. The term "independence" here is informally used to refer to the notion that one measurement (or test subject or event) has no relation to, or influence on, another [4.25]. Note that noise sources are not necessarily linearly additive, and gage repeatability and reproducibility (R&R) analysis is a standard way to quantify the different sources of noise, specifically measurement variations [4.27].

Normalization is technically a mathematical transformation of the data that puts features from different scales into a common scale to enable physically meaningful comparisons and mathematical operations [4.28–30]. The choice of transformations is a function of local and universal assumptions, as well as properties of the dataset and the system from which the data were derived; For example, microarrays may contain housekeeping probes that are assumed to give a predefined readout under universal experimental conditions, and the data from these are used as pivots for background normalization transformations. Assumptions about and models of noise in the data can influence normalization transformations. A common assumption with regards to the data matrix representation described earlier is that each feature vector (the sample profile of a gene, or the transcriptome profile of a sample) arose from a common probabilistic distribution (e.g., all samples or transcriptome profiles have a constant mean and/or variance across all genes or samples), and specific mathematical transformations are applied to the column or rows of the data matrix to normalize the data to this effect. One common normalization transformation involves standardizing a feature vector whereby the mean (over all feature vector components) is subtracted from each feature vector component, which is then divided by the standard deviation (over all feature vector components). In this case, what is happening geometrically is that feature vectors in N dimensions are transformed into vectors on a hypersphere of radius $\sqrt{(N-1)}$ centered at the origin in $N-1$ dimensions. Note that, for points on a hypersphere, the Euclidean distance and Pearson correlation are equivalent measures; For example, recalling the earlier Example 2, suppose that G1, G2, and G3 live on a hypersphere. If G1 and G2 are more similar than are G1 and G3 in the Euclidean sense, then G1 and G2 are also more similar than are G1 and G3 in the Pearson correlation sense, and vice versa. Normalization methods are as numerous as there are mathematical transformations, and the analyst would be prudent to not simply apply them following blind custom or unverified assumptions about the data at hand.

4.6 Identifying Coherent Mathematical Patterns: Technical/Statistical False Positives

Data representation and normalization render data into a form such that one can apply mathematical operations and transformations on the data in a physically meaningful and rational way. Recall that the data matrix representation of N gene $\times M$ sample features is the canonical starting point for generic transcriptomic data analyses, whose broad objective is to identify coherent mathematical patterns in the data, and to determine whether these patterns correspond to nontrivial biological associations between the N genes and M samples. The term "coherent mathematical pattern" is used here to informally refer to a nonrandom or statistically rare event. The pattern is used to validate an existing hypothesis, or to generate new hypotheses about the system from which the data were derived for further experiments. From a machine learning perspective, methods for identifying these patterns fall into two broad categories: supervised and unsupervised. A supervised method uses the feature label as an input parameter, whereas an unsupervised method does not. Examples of a feature label are the sample (column in data matrix) phenotype such as a particular disease state, or the gene (row in data matrix) ontological class such as its role in a specific biological process. Supervised methods are generally used to identify features that are related to a reference set of other features or their labels, for example, to identify genes (row features) significantly differently expressed between samples with a disease state X and samples with a disease state O (column feature labels). The methods include Student's t test, Mann–Whitney–Wilcoxon test, analysis of variance (ANOVA), and many standard statistical and inference tests [4.31]. Unsupervised methods are typically used to discover relationships between features that are independent of feature labels, for example, clusters of genes that have similar (e.g., correlated) sample expression profiles. The methods include principal component analysis (PCA), self-organizing maps (SOM), and numerous feature clustering algorithms [4.30, 32]. A cluster refers to a set of features that are deemed similar to one another based upon a predefined threshold value of a specific measure of dis/similarity between features.

A central problem in transcriptomic data analysis is determining whether a derived or identified mathematical pattern is coherent (i. e., nonrandom or statistically rare event), also known as a true positive. The false-positive rate or type 1 statistical error in transcriptomic data analysis is prominent due to the high dimensionality of the feature space (notably the high number of genes queried) in a typical dataset, and therefore the repeat applications of a statistical test and its parameters for each feature, i. e., multiple comparisons or multiple hypothesis testing; For example, suppose that we statistically test whether gene expression is significantly different between sample populations of two groups X versus O in a dataset of 100 genes whose average expression in the two groups we know a priori to be similar, i. e., the null hypothesis (average expression in group X equals average expression in group O) being true for each gene. After checking that initial conditions for correct application of Student's t test hold, we use the customary threshold probability value < 0.05 for significance to identify significantly differentially expressed genes. The likelihood of one gene being not significant (i. e., one null hypothesis being true) with this t test is 0.95. Assuming that all t tests are mutually independent, the likelihood of N genes being not significant (i. e., N null hypotheses being true) is 0.95^N. The logical complement or the likelihood of rejecting any one of the 100 null hypotheses is $1 - 0.95^{100} = 0.9941$; i. e., by chance alone, 1 of 100 of these genes will be found to be significant (i. e., a false positive) at the $p < 0.05$ threshold.

There are intense current debates and discussions in the bioinformatics community about how one deals with multiple comparisons. Some argue that traditional multiple-comparison adjustment strategies control false positives while potentially inflating false negatives, which defeats the goals of exploratory empirical research [4.33, 34]. On the other hand, the easy availability of large datasets in transcriptomic studies often leads to testing a large number of hypotheses

with no a priori expectation of the number of true hypotheses, as shown in the 100-gene example above. Multiple comparison is handled differently depending upon the nature of the transcriptomic investigation: hypothesis driven versus exploratory [4.20, 35]. When a study is testing a predefined set of hypotheses, one typically controls the probability of a false discovery or the family-wise error rate, where family refers to the set of hypotheses being tested. One assigns an acceptable type 1 error rate, customarily 0.05, to each family and controls them jointly using techniques such as Bonferroni correction. When a study is exploratory and its results (coherent patterns) can be retested in an independent study, one typically controls the false discovery rate (FDR), which is the proportion of false positives among all rejected hypotheses; For example, if 100 genes were experimentally predicted to be differentially expressed with an FDR of 0.20, then 20 $(0.20 \cdot 100)$ of these are false positives. The FDR is often determined using permutation tests where features of the original dataset are permuted and reanalyzed for coherent patterns to model the likelihood of obtaining such patterns in a *typical* (permuted) dataset generated by the stochastic process that generated the original (unpermuted) dataset.

We should note here the increasingly commonplace application of Bayesian modeling in transcriptomic data analysis [4.36]. In contrast to the more orthodox frequentist perspective where data are regarded to be repeatable random variables whose underlying parameters (e.g., average, variance) remain fixed, the Bayesian perspective considers data to be fixed observations generated from processes whose parameters are unknown and described probabilistically. The frequentist is concerned with the overall rate of false positives (e.g., confidence interval), whereas the Bayesian is concerned with the rate of error among positives (e.g., credibility interval) [4.37].

4.7 General Strategies to Minimize Biological False Positives

As mentioned earlier, false positives may be technical/statistical (as described in Sect. 4.6) or biological in character. We use the term "biological false positive" to informally refer to a coherent mathematical pattern that does not correspond to an actual physical–biological phenomenon; For example, suppose that we identify a gene G whose genotype or expression is correlated with tail length in a mouse transcriptome study. How-

ever, if in a mouse model where G is modified (e.g., knocked out or mutated), we do not observe any measurable relationship between genetic variation and tail length, then G is a biological false positive.

Any researcher who has ever attempted to distinguish correlation from causation in a study will undoubtedly be familiar with biological features that are superficially correlated with but mechanistically independent of (or noncausal with respect to) a phenotype of interest. On the other hand, there exist many examples of distinct and independent biological processes and molecular pathways that converge to a common phenotypic endpoint such as inflammation. It should be evident that biological false positives cannot generally be characterized or unraveled using mathematical methods alone. Additional experiments that test more refined follow-up hypotheses are necessary for a definitive resolution.

The resulting coherent mathematical pattern from a generic transcriptomic data analysis is often a set of candidate genes that pass some predefined mathematical criterion for coherence. A perennial and somewhat open-ended question that follows is what one can do next with this set of genes to further the study. This next step obviously depends upon the main objectives and context of the study. In particular, one often asks whether and how one might rank or prioritize the genes rationally for further investigation, since the scientific method typically only allows for construction and testing of one hypothesis (i. e., one gene) at a time.

One common strategy is to identify additional and potentially biologically relevant patterns within the set of candidate genes, such as determining whether there is a gene subset that is enriched (i. e., statistically over-represented) for a specific ontological property, e.g., elements of a specific biochemical pathway, localization to a particular cellular component, or share a common transcription factor binding sequence motif. Practical implementations of this strategy include gene ontology (GO) enrichment analyses [4.38], and gene–protein–disease interaction networks [4.39].

Another common strategy – which is sometimes used as in silico validation prior to further experimental validation – is informally known as *integrative genomics*. Here one performs a meta-analysis involving biological systems and datasets that are cognate to one's main system and dataset of interest to look for common coherent patterns. Deciding which systems and datasets are cognate is often a nontrivial exercise requiring extensive technical and biological background knowledge, and can seem ad hoc in actual practice. As we have stated earlier, no two systems (or study designs) are ever exactly identical. The central assumption underlying this strategy is that the identification of a common coherent pattern across independent cognate studies (which includes different measurement modalities such as genes, proteins, epigenetic modifications) endows the pattern with greater technical and biological merit (i. e., less likely to be a false positive by either chance or individual subject variation) and higher prioritization for further experimental investigation. Notable applications of integrative genomics include a meta-analysis of breast cancer signatures across independent studies [4.40], and the correlation of gene expression and epigenetic modifications in cancer [4.41].

4.8 Generalizations and Conclusion

Starting from an N gene $\times M$ sample real-valued data matrix, there is a great diversity of methods for analysis – representing, transforming, visualizing, in silico validating – with the ultimate goal of identifying coherent mathematical patterns in the data. The resulting coherent pattern is typically a set of candidate genes that often has additional internal structure such as a subset with a specific ontological enrichment. The choice of methods should be guided firstly by one's question about the system being investigated with a well-formulated and well-defined objective, and secondly by assumptions (which ought to be checked) about the underlying data structure for the valid application of the selected method. A standard tool for assessing the sensitivity and specificity of different methods or modeling parameters with respect to the underlying data structure is receiver operating characteristic (ROC) curve analysis [4.42].

With minor modifications, the analytic methods are often generalizable to data matrices whose rows and columns represent features other than genes, e.g., proteins and other genotypic or phenotypic traits that can be represented as discrete or continuous valued variables. The distinction between microscopic and macroscopic scaled features can be fuzzy. As we have noted earlier, the expression of a gene is commonly treated as a quantitative trait known as an expression quantitative trait (eQTL) in studies that investigate the effect of a genomic locus (e.g., a SNP) on gene expression [4.43,44].

References

4.1   F. Crick: Central dogma of molecular biology, Nature **227**(5258), 561–563 (1970)

4.2   B. Lewin: *Genes VII* (Oxford Univ. Press, Oxford 2000)

4.3   T.H. Morgan: Sex limited inheritance in *Drosophila*, Science **32**(812), 120–122 (1910)

4.4   H.J. Muller: Artificial transmutation of the gene, Science **66**(1699), 84–87 (1927)

4.5   H.B. Creighton, B. McClintock: A correlation of cytological and genetical crossing-over in *Zea mays*, Proc. Natl. Acad. Sci. USA **17**(8), 492–497 (1931)

4.6   B. McClintock: The order of the genes C, Sh and Wx in *zea mays* with reference to a cytologically known point in the chromosome, Proc. Natl. Acad. Sci. USA **17**(8), 485–491 (1931)

4.7   A.H. Sturtevant: The linear arrangment of six sex-linked factors in *Drosophila*, as shown by their mode of association, J. Exp. Zool. **14**, 39–45 (1927)

4.8   R.J. Robbins: Challenges in the Human Genome Project, IEEE Eng. Biol. Med. **11**(1), 25–34 (1992)

4.9   M. Pop, S.L. Salzberg, M. Shumay: Genome sequence assembly: Algorithms and issues, IEEE Computer **35**(7), 47–54 (2002)

4.10   E.E. Schadt, J. Lamb, X. Yang, J. Zhu, S. Edwards, D. Guhathakurta, S.K. Sieberts, S. Monks, M. Reitman, C. Zhang, P.Y. Lum, A. Leonardson, R. Thieringer, J.M. Metzger, L. Yang, J. Castle, H. Zhu, S.F. Kash, T.A. Drake, A. Sachs, A.J. Lusis: An integrative genomics approach to infer causal associations between gene expression and disease, Nat. Genet. **37**(7), 710–717 (2005)

4.11   P.A. Sharp: Splicing of messenger RNA precursors, Science **235**(4790), 766–771 (1987)

4.12   D.J. Duggan, M. Bittner, Y. Chen, P. Meltzer, J.M. Trent: Expression profiling using cDNA microarrays, Nat. Genet. **21**(1), 10–14 (1999)

4.13   E.H. Margulies, S.L. Kardia, J.W. Innis: Identification and prevention of a GC content bias in SAGE libraries, Nucleic Acids Res. **29**(12), E60–60 (2001)

4.14   M.L. Metzker: Sequencing technologies – The next generation, Nat. Rev. Genet. **11**(1), 31–46 (2010)

4.15   B. Wold, R.M. Myers: Sequence census methods for functional genomics, Nat. Methods **5**(1), 19–21 (2008)

4.16   D. Branton, D.W. Deamer, A. Marziali, H. Bayley, S.A. Benner, T. Butler, M. Di Ventra, S. Garaj, A. Hibbs, X. Huang, S.B. Jovanovich, P.S. Krstic, S. Lindsay, X.S. Ling, C.H. Mastrangelo, A. Meller, J.S. Oliver, Y.V. Pershin, J.M. Ramsey, R. Riehn, G.V. Soni, V. Tabard-Cossa, M. Wanunu, M. Wiggin, J.A. Schloss: The potential and challenges of nanopore sequencing, Nat. Biotechnol. **26**(10), 1146–1153 (2008)

4.17   T.D. Harris, P.R. Buzby, H. Babcock, E. Beer, J. Bowers, I. Braslavsky, M. Causey, J. Colonell, J. Dimeo, J.W. Efcavitch, E. Giladi, J. Gill, J. Healy, M. Jarosz, D. Lapen, K. Moulton, S.R. Quake, K. Steinmann,

E. Thayer, A. Tyurina, R. Ward, H. Weiss, Z. Xie: Single-molecule DNA sequencing of a viral genome, Science **320**(5872), 106–109 (2008)

4.18   J. Eid, A. Fehr, J. Gray, K. Luong, J. Lyle, G. Otto, P. Peluso, D. Rank, P. Baybayan, B. Bettman, A. Bibillo, K. Bjornson, B. Chaudhuri, F. Christians, R. Cicero, S. Clark, R. Dalal, A. Dewinter, J. Dixon, M. Foquet, A. Gaertner, P. Hardenbol, C. Heiner, K. Hester, D. Holden, G. Kearns, X. Kong, R. Kuse, Y. Lacroix, S. Lin, P. Lundquist, C. Ma, P. Marks, M. Maxham, D. Murphy, I. Park, T. Pham, M. Phillips, J. Roy, R. Sebra, G. Shen, J. Sorenson, A. Tomaney, K. Travers, M. Trulson, J. Vieceli, J. Wegener, D. Wu, A. Yang, D. Zaccarin, P. Zhao, F. Zhong, J. Korlach, S. Turner: Real-time DNA sequencing from single polymerase molecules, Science **323**(5910), 133–138 (2009)

4.19   D.R. Bentley, S. Balasubramanian, H.P. Swerdlow, G.P. Smith, J. Milton, C.G. Brown, K.P. Hall, D.J. Evers, C.L. Barnes, H.R. Bignell, J.M. Boutell, J. Bryant, R.J. Carter, R. Keira Cheetham, A.J. Cox, D.J. Ellis, M.R. Flatbush, N.A. Gormley, S.J. Humphray, L.J. Irving, M.S. Karbelashvili, S.M. Kirk, H. Li, X. Liu, K.S. Maisinger, L.J. Murray, B. Obradovic, T. Ost, M.L. Parkinson, M.R. Pratt, I.M. Rasolonjatovo, M.T. Reed, R. Rigatti, C. Rodighiero, M.T. Ross, A. Sabot, S.V. Sankar, A. Scally, G.P. Schroth, M.E. Smith, V.P. Smith, A. Spiridou, P.E. Torrance, S.S. Tzonev, E.H. Vermaas, K. Walter, X. Wu, L. Zhang, M.D. Alam, C. Anastasi, I.C. Aniebo, D.M. Bailey, I.R. Bancarz, S. Banerjee, S.G. Barbour, P.A. Baybayan, V.A. Benoit, K.F. Benson, C. Bevis, P.J. Black, A. Boodhun, J.S. Brennan, J.A. Bridgham, R.C. Brown, A.A. Brown, D.H. Buermann, A.A. Bundu, J.C. Burrows, N.P. Carter, N. Castillo, E. Chiara, M. Catenazzi, S. Chang, R. Neil Cooley, N.R. Crake, O.O. Dada, K.D. Diakoumakos, B. Dominguez-Fernandez, D.J. Earnshaw, U.C. Egbujor, D.W. Elmore, S.S. Etchin, M.R. Ewan, M. Fedurco, L.J. Fraser, K.V. Fuentes Fajardo, W. Scott Furey, D. George, K.J. Gietzen, C.P. Goddard, G.S. Golda, P.A. Granieri, D.E. Green, D.L. Gustafson, N.F. Hansen, K. Harnish, C.D. Haudenschild, N.I. Heyer, M.M. Hims, J.T. Ho, A.M. Horgan, K. Hoschler, S. Hurwitz, D.V. Ivanov, M.Q. Johnson, T. James, T.A. Huw Jones, G.D. Kang, T.H. Kerelska, A.D. Kersey, I. Khrebtukova, A.P. Kindwall, Z. Kingsbury, P.I. Kokko-Gonzales, A. Kumar, M.A. Laurent, C.T. Lawley, S.E. Lee, X. Lee, A.K. Liao, J.A. Loch, M. Lok, S. Luo, R.M. Mammen, J.W. Martin, P.G. McCauley, P. McNitt, P. Mehta, K.W. Moon, J.W. Mullens, T. Newington, Z. Ning, N.B. Ling, S.M. Novo, M.J. O'Neill, M.A. Osborne, A. Osnowski, O. Ostadan, L.L. Paraschos, L. Pickering, A.C. Pike, A.C. Pike, D. Chris Pinkard, D.P. Pliskin, J. Podhasky, V.J. Quijano, C. Raczy, V.H. Rae, S.R. Rawlings,

A. Chiva Rodriguez, P.M. Roe, J. Rogers, M.C. Rogert Bacigalupo, N. Romanov, A. Romieu, R.K. Roth, N.J. Rourke, S.T. Ruediger, E. Rusman, R.M. Sanches-Kuiper, M.R. Schenker, J.M. Seoane, R.J. Shaw, M.K. Shiver, S.W. Short, N.L. Sizto, J.P. Sluis, M.A. Smith, J. Ernest Sohna Sohna, E.J. Spence, K. Stevens, N. Sutton, L. Szajkowski, C.L. Tregidgo, G. Turcatti, S. Vandevondele, Y. Verhovsky, S.M. Virk, S. Wakelin, G.C. Walcott, J. Wang, G.J. Worsley, J. Yan, L. Yau, M. Zuerlein, J. Rogers, J.C. Mullikin, M.E. Hurles, N.J. McCooke, J.S. West, F.L. Oaks, P.L. Lundberg, D. Klenerman, R. Durbin, A.J. Smith: Accurate whole human genome sequencing using reversible terminator chemistry, Nature **456**(7218), 53–59 (2008)

4.20   Y. Benjamini, Y. Hochberg: Controlling the false discovery rate: A practical and powerful approach to multiple testing, J. R. Stat. Soc. B **57**(1), 289–300 (1995)

4.21   J.J. Thomas, K.A. Cook (Eds.): *Illuminating The Path: The Research and Development Agenda for Visual Analytics, National Gov. Pub* (IEEE Computer Society, Los Alamitos 2005)

4.22   R.O. Duda, P.E. Hart, D.G. Stork: *Pattern Classification* (Wiley, New York 2001)

4.23   A.T. Kho, Q. Zhao, Z. Cai, A.J. Butte, J.Y. Kim, S.L. Pomeroy, D.H. Rowitch, I.S. Kohane: Conserved mechanisms across development and tumorigenesis revealed by a mouse development perspective of human cancers, Genes Dev. **18**(6), 629–640 (2004)

4.24   E. Parzen: *Modern Probability Theory and its Applications* (Wiley, New York 1992)

4.25   M.H. DeGroot, M.J. Schervish: *Probability and Statistics*, 4th edn. (Addison-Wesley, Boston 2012)

4.26   D.W. Graham: Heraclitus. In: *The Stanford Encyclopedia of Philosophy*, ed. by E.N. Zalta (SEP, Stanford 2011)

4.27   NIST/SEMATECH: *e-Handbook of Statistical Methods* (NIST, 2012) available online at http://www.itl.nist.gov/div898/handbook/

4.28   J. Quackenbush: Microarray data normalization and transformation, Nat. Genet. **32**, 496–501 (2002)

4.29   P. Stafford (Ed.): *Methods in Microarray Normalization* (CRC, Boca Raton 2008)

4.30   I.S. Kohane, A.T. Kho, A.J. Butte: *Microarrays for an Integrative Genomics* (MIT, Cambridge 2003)

4.31   R.A. Johnson, D.W. Wichern: *Applied Multivariate Statistical Analysis*, 5th edn. (Prentice Hall, Upper Saddle River 2002)

4.32   B. Everitt: *Cluster Analysis*, 5th edn. (Wiley, Chichester 2011)

4.33   T.D. Wu: Analysing gene expression data from DNA microarrays to identify candidate genes, J. Pathol. **195**(1), 53–65 (2001)

4.34   J. Taylor, R. Tibshirani, B. Efron: The *miss rate* for the analysis of gene expression data, Biostatistics **6**(1), 111–117 (2005)

4.35   S. Dudoit, J.P. Shaffer, J.C. Boldrick: Multiple hypothesis testing in microarray experiments, Stat. Sci. **18**(1), 71–103 (2003)

4.36   D.K. Dey, S. Ghosh, B.K. Mallick (Eds.): *Bayesian Modeling in Bioinformatics* (Chapman Hall/CRC, New York 2010)

4.37   K. Winstein: *Styles of Inference: Bayesianness and Frequentism* (CSAIL MIT, Cambridge 2011), available online at http://groups.csail.mit.edu/mac/users/gjs/6.945/readings/winstein-bayes-frequentist-2011.pdf

4.38   D.W. Huang, B.T. Sherman, R.A. Lempicki: Systematic and integrative analysis of large gene lists using DAVID bioinformatics resources, Nat. Protoc. **4**(1), 44–57 (2009)

4.39   O. Vanunu, O. Magger, E. Ruppin, T. Shlomi, R. Sharan: Associating genes and protein complexes with disease via network propagation, PLoS Comput. Biol. **6**(1), e1000641 (2010)

4.40   L. Ein-Dor, I. Kela, G. Getz, D. Givol, E. Domany: Outcome signature genes in breast cancer: Is there a unique set?, Bioinformatics, **21**(2), 171–178 (2005)

4.41   M. Esteller: Cancer epigenomics: DNA methylomes and histone-modification maps, Nat. Rev. Genet. **8**(4), 286–298 (2007)

4.42   T. Fawcett: An introduction to ROCanalysis, Pattern Recognit. Lett. **27**(8), 861–874 (2006)

4.43   Y. Gilad, S.A. Rifkin, J.K. Pritchard: Revealing the architecture of gene regulation: The promise of eQTL studies, Trends Genet. **24**(8), 408–415 (2008)

4.44   W. Cookson, L. Liang, G. Abecasis, M. Moffatt, M. Lathrop: Mapping complex disease traits with global gene expression, Nat. Rev. Genet. **10**(3), 184–194 (2009)

5. Understanding Information Processes at the Proteomics Level

Shaojun Dai, Sixue Chen

All living organisms are composed of proteins. Proteins are large, complex molecules made of long chains of amino acids. Twenty different amino acids are usually found in proteins. Proteins are produced on protein-synthesizing machinery directed by codons made of three deoxyribonucleic acid (DNA) bases. DNA is an information storage macromolecule. With the fast advancement of DNA sequencing technology, more and more genomes have been sequenced. Sequence analysis of this exploding genomic information has revealed a lot of novel genes for which molecular and/or biological functions are to be determined. The huge genomic information stored in DNA and genes is stationary and heritable. At cellular level, genomic information flows selectively from DNA to messenger RNA (mRNA) through transcription and from mRNA to proteins through translation for biological functions, such as response to changes in the environment. Different large-scale, high-throughput studies have been performed to investigate the information flow, e.g., transcriptomic profiling using microarray or RNAseq technologies. As a complementary approach to genomics and transcriptomics, proteomics has been fast developing to investigate gene expression at protein levels including quantitative changes, posttranslational modifications, and interactions with other molecules. These protein-level events represent a global view of information processing at the proteomics level. In this chapter, we focus on the description of technological and biological aspects of the information flow from the static genome to the dynamic proteome through gene transcription, protein translation, posttranslational modification, and protein interactions.

5.1 Genetic Information Transfer from DNA to Protein by Transcription and Translation

Genetic information is transferred from DNA to protein by transcription and translation. The most important molecules involved are messenger RNA (mRNA), transfer RNA (tRNA), and ribosome.

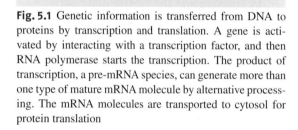

Fig. 5.1 Genetic information is transferred from DNA to proteins by transcription and translation. A gene is activated by interacting with a transcription factor, and then RNA polymerase starts the transcription. The product of transcription, a pre-mRNA species, can generate more than one type of mature mRNA molecule by alternative processing. The mRNA molecules are transported to cytosol for protein translation

5.1.1 mRNA and Transcription

The intermediate molecule in charge of transferring genetic information to protein is mRNA, which is synthesized from a DNA template. The process in which a double-strand DNA (dsDNA) provides the information for the synthesis of a single-strand RNA (pre-mRNA) is called transcription (Fig. 5.1). The nucleotide sequence of the pre-mRNA is complementary to that of the piece of DNA (i. e., the gene) from which it is transcribed. Therefore, pre-mRNA retains the same information as the gene itself. After processing, mature mRNA transfers the stored genetic information from

nucleus to cytoplasm. The mRNA molecules in the cytoplasm serve as a template to direct the incorporation of amino acids.

Transcription is catalyzed by an RNA polymerase, which requires a DNA template and the four precursor ribonucleotides: adenosine triphosphate (ATP), guanosine triphosphate (GTP), cytidine triphosphate (CTP), and uridine triphosphate (UTP). It includes initiation, elongation, and termination stages. In the initiation stage, RNA polymerase binds to a specific DNA sequence called the promoter. The promoter sequence is located at the 5′-end (upstream) of the coding region. There are some short conserved sequences in promoters, which serve as the binding sites for the polymerase and other DNA-binding proteins (such as transcription factors). The dsDNA helix can be locally unwound at the promoter site to which RNA polymerase binds. One of the two strands of DNA is the template for RNA synthesis. This template strand is known as the antisense strand, whereas the other stand is the sense strand. The polymerase then initiates the synthesis of the RNA strand at a specific nucleotide location, called the start site. This site is defined as position +1 of the gene sequence (Fig. 5.2). The polymerase covalently adds ribonucleotides to the 3′-end of the RNA chain, extending the growing RNA chain in the 5′ → 3′ direction to sustain the RNA elongation. The enzyme itself moves in the 3′ → 5′ direction along the antisense DNA strand, and locally unwinds the DNA helix to expose the template strand for the RNA strand elongation. The helix is reformed behind the polymerase (Fig. 5.2). Finally, termination of transcription occurs at a specific DNA sequence known as a terminator, which often contains self-complementary regions forming a stem-loop or hairpin secondary structure in the primary RNA product.

A pre-mRNA species can generate more than one type of mature mRNA molecule by alternative mRNA processing. Alternative mRNA processing is a very important information processing step following the

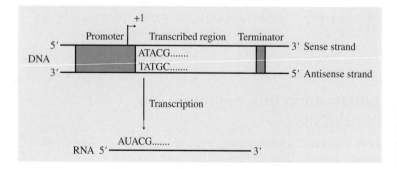

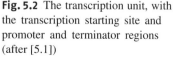

Fig. 5.2 The transcription unit, with the transcription starting site and promoter and terminator regions (after [5.1])

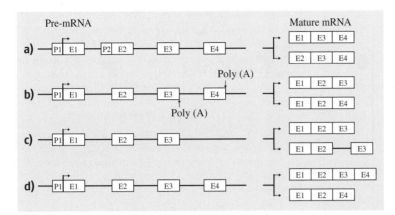

Fig. 5.3a–d Modes of alternative mRNA splicing. (**a**) Alternative selection of promoters P1 and P2; (**b**) alternative selection of cleavage/polyadenylation sites; (**c**) retention of an intron; (**d**) exon skipping (after [5.1])

formation of pre-mRNA. It can occur by varying the use of 5′- to 3′-splice sites in different ways, e.g., by using different promoters or different poly(A) sites, or by retaining certain introns, and retaining or removing certain exons (Fig. 5.3). In addition, RNA editing is another form of mRNA processing in which the nucleotide sequence of pre-mRNA is altered by changing, inserting, or deleting nucleotide residues at specific sequence sites.

5.1.2 tRNA, Ribosome, and Protein Synthesis

Transfer RNAs (tRNAs) are the adaptor molecules that decode the information in mRNA by delivering different amino acids to ribosomes. The primary structure of tRNA is a 60–95 nt (commonly 76 nt) long sequence. Conventionally, position 1 is numbered at the 5′-end and 76 is at the 3′-end. The secondary structure of tRNAs is a cloverleaf structure consisting of stems and loops by base pairing of different regions (Fig. 5.4). At the 3′-end, there are invariant residues 74–76 (5′-CCA-3′) that are not included in this base pairing region. This stem functions as an amino acid acceptor. tRNAs are joined to amino acids to become aminoacyl-tRNAs by aminoacyl-tRNA synthetases.

Another important structure consists of a 5 bp stem and a seven-residue loop including three adjacent nucleotides (anticodons). The anticodon is complementary to the codon sequence (a triplet in the mRNA) that the tRNA recognizes. The tRNA molecule folds into an L-shape three-dimensional (3-D) structure by base pairing between several invariant bases in the D- and T-arms. In the 3-D structure, the anticodon is at one end and the amino acid acceptor is at the other end. This structure enables tRNA to transfer amino acids to ribosomes ac-

cording to the specific mRNA sequence during protein synthesis.

Ribosome is the protein synthesis factory that can translate the information from mRNA molecules into specific polypeptide sequences. It can translate about 10–20 codons per second. Ribosomes in bacteria, eukaryotes, and the Archaea share this fundamental functionality, although they differ in structural details. Each ribosome contains a small and a large subunit. The small subunit functions to decode mRNA by interacting codons in mRNA with anticodons in aminoacyl-tRNAs. The large subunit has peptidyl transferase activity that is

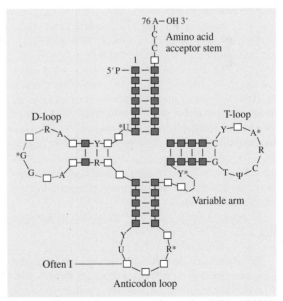

Fig. 5.4 Cloverleaf structure of transfer RNA (tRNA), consisting of amino acid acceptor stem, variable arm, D-loop, T-loop, and anticodon loop

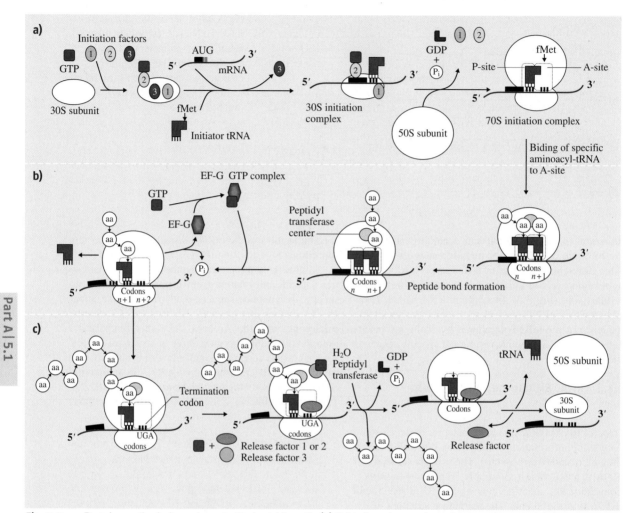

Fig. 5.5a–c Protein synthesis in prokaryote *Escherichia coli*. (**a**) Initiation. A complete ribosome is assembled to an mRNA molecule at the initiation codon site and an initiator tRNA is bound. (**b**) Elongation. The peptide chain is growing with the help of elongation factor EF-Tu, GTP/GDP, and translocator (EF-G). (**c**) Termination. A new protein molecule is released after the releasing factors (RF1 or RF2) recognizing the stop codons, and with the help of RF3 and peptidyl transferases

needed for peptide bond formation, and hydrolase activity required to release the completed polypeptide from the ribosome complex.

Protein synthesis has three stages in prokaryotes (Fig. 5.5).

1. Initiation stage. In this stage, a complete ribosome is assembled to an mRNA molecule at the initiation codon site and an initiator tRNA is bound. The components involved include large and small ribosome subunits, mRNA, the initiator tRNA, three initiation factors (IF1, IF2, and IF3), and GTP. First, IF1 and

IF3 bind to a free 30S subunit, which can prevent the large ribosome subunit from binding, and then IF2 complexed with GTP binds to the small subunit to assist initiator tRNA binding. Then, mRNA uses its ribosome-binding site to bind to the ribosome subunit. The initiator tRNA binds to the complex by base pairing of its anticodon with the AUG codon on the mRNA, and IF3 is released. This forms the 30S initiation complex. Finally, the 50S subunit is bound to displace IF1 and IF2, and the GTP is hydrolyzed. This is called the 70S initiation complex. There are two tRNA-binding sites (A- and P-sites) in the small subunit of the complex, for the binding of aminoacyl-tRNA molecules and the growing polypeptide chain, respectively.

2. Elongation stage. A charged aminoacyl-tRNA is delivered to the A-site as a complex with elongation factor EF-Tu and GTP. Then, GTP is hydrolyzed and EF-Tu·GDP is released, which can be reused with the help of EF-Ts and GTP. A peptide bond between the two adjacent amino acids is formed, depending on the 50S subunit peptidyl transferase activity. Finally, translocator (EF-G) with energy from GTP moves the ribosome one codon along the mRNA, ejecting the uncharged tRNA and transferring the growing peptide chain to the P-site.

3. Termination stage. Releasing factors (RF1 or RF2) recognize the stop codons and, with the help of RF3 and peptidyl transferases, join the polypeptide

chain to a water molecule rather than to the usual aminoacyl-tRNA. Thus, the new protein is released. Ribosome releasing factors help to dissociate the ribosome subunits from the mRNA.

To sum up, genetic information is transferred from DNA to protein by transcription and translation, which include the following processes.

● A pre-mRNA is transcribed from a DNA template.
● The pre-mRNA species can generate more than one type of mature mRNA molecule through alternative mRNA splicing.
● Amino acids are delivered to ribosomes by tRNAs, which are the adaptor molecules that decode the genetic information in mRNA.
● The genetic information is translated from mRNA to protein on ribosomes.

5.2 Protein Information Processing by Translational and Posttranslational Control

The delivery of genetic information to the expression of functional protein information is controlled by translational and posttranslational processes.

5.2.1 Translational Control

Protein translation is well controlled using different mechanisms in prokaryotes and eukaryotes. In eukaryotes, there are multiple copies of 5'-AUUUA-3' in the 3'-noncoding region of a gene that signal the mRNA for rapid degradation and thus limited translation. In addition, some proteins can bind to the mRNA directly to inhibit translation. When the conditions are appropriate, these proteins dissociate from the mRNA and the mRNA can be translated again. Prokaryotes control translation by the inhibition of exonucleases, formation of specific mRNA tertiary structure, and formation of duplexes between antisense RNA and the mRNA at its ribosome-binding site. For details, refer to *Turner* et al. [5.1].

5.2.2 Protein Targeting/Sorting

One gene can encode many different protein species, partly because of alternative mRNA splicing and translational control. Numerous protein species need to be localized to different subcellular compartments to carry out their functions. The ultimate subcellular location of

proteins is usually determined by specific amino acid sequences within the proteins themselves. Newly synthesized proteins are imported into different organelles through three main pathways.

1. Soluble nuclear proteins enter the nucleus through nuclear pores. Particular amino acids in the proteins serve as the nuclear localization signal to bind to nuclear import receptor protein. The proteins are transported into the nucleus by interacting with the nuclear pore structures in the nuclear envelope.

2. Proteins can enter an organelle by passing through the organelle membrane. Most proteins of mitochondria, chloroplast, and peroxisomes have signal sequences in their N-termini. The signaling sequences can be recognized by receptor proteins in the membrane of the organelles. Thus, the unfolded proteins can move across the organelle membrane through protein translocators in the membrane.

3. Proteins enter an organelle by vesicle transport and fusion. Proteins for endoplasmic reticulum (ER), Golgi body, vacuoles, lysosomes, plasma membrane, and extracellular space can be directed across the ER membrane as they are being synthesized. The proteins from ER can then move to other endomembrane systems through transporting vesicles derived from ER and Golgi apparatus.

5.2.3 Protein Folding and Modification

Newly synthesized polypeptides are usually not functional. They need to be folded into various secondary structures, such as right-handed α-helix and parallel/antiparallel β-pleated sheets stabilized by hydrogen bonds. Different sections of secondary structure and connecting regions are further folded into a well-defined tertiary structure that is stabilized by noncovalent interactions and sometimes disulfide bonds. Besides, there are a number of other alterations required for activity, including cleavage, various covalent modifications, and formation of quaternary structures; For example, the signal sequences at the N-terminal are cleaved off in secreted proteins. Many proteins are composed of two or more polypeptide chains, enabling large protein molecules to have sophisticated functionality. In addition, there are more than 300 chemically known modifications that affect the protein functions, such as acetylation, hydroxylation, phosphorylation, methylation, glycosylation, and even addition of lipids and nucleotides. Among the modifications, phosphorylation and glycosylation have drawn much attention; For example, the activities of many enzymes, such as glycogen phosphorylase, transcription factors, and G proteins, are regulated by phosphorylation. The stability and signal recognition of many proteins are dependent on glycosylation.

To sum up, the information expressed in protein is processed by translational and posttranslational control:

- The information translated from mRNA to protein is adjusted by mRNA noncoding region and mRNA binding proteins.
- The information expressed and protein functionality are dependent on protein targeting and subcellular localization.
- Protein folding and posttranslational modifications play important roles in the regulation of protein functions.

5.3 High-Throughput Platforms for Protein Separation and Characterization

5.3.1 Two-Dimensional Gel Electrophoresis (2-DE)

2-DE is a powerful technology for separation of complex protein samples extracted from cells, tissues, and organisms. With the recent development in sample solubilization and the development of immobilized pH gradient (IPG) gel strips, 2-DE offers high resolution and reproducibility in protein separation. The technology is based on two independent physicochemical parameters of proteins: isoelectric point (pI) and molecular weight. The first dimension of 2-DE, termed as isoelectric focusing (IEF), separates proteins according to their different pIs. In the electrical field, proteins rehydrated into the first-dimension IPG gel matrix move along the pH gradient and accumulate at their respective pIs, where the net charges will be zero. The differences in protein pIs depend on their amino acid composition and modifications, especially the number of acidic and basic amino acids within the proteins. Therefore, different proteins with different pIs are separated at different positions along the pH gradient gel. There are two methods to create the pH gradient. At the beginning of 2-DE, prefocusing of carrier ampholytes was required to generate the pH gradient. During IEF, the pH gradient can collapse or drift, creating variation and irreproducibility. In the early 1990s, IPG gels were invented as a big technological breakthrough. The IPG gel is more stable than carrier ampholytes as the pH gradient is immobilized, leading to a significant improvement in reproducibility. The second step of 2-DE is to separate proteins based on their molecular weights using sodium dodecyl sulfate-polyacrylamide gel electrophoresis (SDS-PAGE). Proteins are treated with SDS along with other reagents so that they are unfolded and bind large amounts of SDS proportional to their masses. The SDS coating of the proteins allows all of the proteins to be negatively charged and have similar mass-to-charge ratios. The polyacrylamide gel acts like a molecular sieve when the current is applied, separating the proteins on the basis of their molecular weights, with larger proteins being retained in the upper portions of the gel and smaller proteins passing through the sieves and reaching the lower regions of the gel. After staining with Coomassie Blue, silver or Sypro Ruby, protein spots can be visualized and digitized. Proteins exhibiting differences in abundance between samples are revealed by comparing the patterns and densitometric intensities of the spots in different gels [5.2].

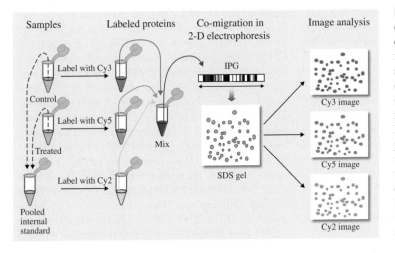

Samples · Labeled proteins · Co-migration in 2-D electrophoresis · Image analysis

Label with Cy3

Control

Label with Cy5

Treated · Mix

IPG

Cy3 image

SDS gel

Cy5 image

Label with Cy2

Pooled internal standard

Cy2 image

Fig. 5.6 Two-dimensional fluorescence difference in gel electrophoresis (2-D-DIGE) workflow. Protein samples are labeled with spectrally resolvable fluorescent dyes (synthetic *N*-hydroxysuccinimidyl (NHS) ester derivatives of the cyanine dyes Cy3 and Cy5, respectively). Equal aliquots of the samples are pooled to form an internal control, which is Cy2. Then the samples and the internal control are mixed and co-separated on one 2-D gel, followed by image scanning at different wavelengths and image analysis of differentially expressed spots

Despite considerable improvement, regular 2-DE still has the limitations of reproducibility between gels, dynamic range, and sensitivity. Two-dimensional fluorescence difference gel electrophoresis (2-D-DIGE) was invented to circumvent some of these limitations (Fig. 5.6) [5.2]. For DIGE, various protein samples are prelabeled with spectrally resolvable fluorescent dyes [synthetic *N*-hydroxysuccinimidyl (NHS) ester derivatives of the cyanine dyes Cy2, Cy3, and Cy5] before pooling them at equal amounts. The cyanine dye reagents react with the ε-amine of lysine, which is prevalent in proteins. Because different samples are labeled with different cyanine dyes, samples can be pooled and separated in the same 2-D gel. Protein spots from different samples on the 2-D gel are detected by different fluorescence signals and quantified using spot volumes. In minimal DIGE experiments, the fluorescent dyes are used at low stoichiometry relative to protein quantities. Usually the samples are labeled with Cy3 or Cy5. The Cy2 dye is used to label a pooled internal standard, which consists of an equal aliquot of each test sample. The internal standard not only serves as a loading control, but also allows for normalization of Cy3 : Cy2 and Cy5 : Cy2 ratios for each protein across all the gels in a large experiment. Because proteins from different samples are separated in one gel and the international standard allows normalization across many DIGE gels, DIGE has greatly enhanced 2-D gel reproducibility and throughput for large-scale proteomic studies. In addition, DIGE provides superior sensitivity and dynamic range afforded by fluorescence detection. In spite of these advantages, DIGE does not resolve some of the inherent problems with 2-DE, such as resolving hydrophobic proteins, very large and very small proteins, and proteins with extreme pIs.

5.3.2 High–Performance Liquid Chromatography (HPLC)

HPLC is a widely used chromatographic technique that can separate a mixture of molecules. HPLC consists of various types of stationary phases, mobile phases, pumps to move the mobile phases and analytes through the column, and detectors that detect different molecules on a retention-time scale. Analytes are separated depending on their strength of interaction with the stationary phase (column), the composition of solvents used, and the flow rate of the mobile phases [5.3]. Usually, proteins and peptides are separated by reversed-phase (RP) HPLC. RP-HPLC is carried out with hydrophobic stationary phases (e.g., alkylated silica gel or hydrophobic organic polymers) in combination with gradients of increasing concentration of nonpolar organic solvents in aqueous solutions. Separation of proteins and peptides in RP-HPLC is based on the differences of their hydrophobic properties. Compared with RP-HPLC, ion-exchange (IEX)-HPLC has the advantage of maintaining protein three-dimensional structure during the separation process. IEX-HPLC is based on electrostatic interaction of analyte molecules with positively or negatively charged groups on the stationary phase. These groups are immobilized on the stationary phase and are in equilibrium with exchangeable counterions in the mobile phase. In the adsorption process, the mobile counterions are exchanged by charged analyte molecules. Both cation- and anion-exchange

HPLC can be used to separate proteins and peptides, depending on their isoelectric points and the pH condition at which separation is carried out. For shotgun proteomics, strong cation exchange (SCX) is often used in combination with RP-HPLC for separation of complex peptide mixtures, followed by tandem mass spectrometry (MS/MS) for peptide identification and characterization. Recent studies have shown that hydrophilic interaction chromatography (HILIC) offers a unique separation mechanism that is based on retention by hydrophilicity and by charge. Since HILIC is a type of normal-phase liquid chromatography, volatile solvent can be used and desalting steps for downstream applications are not necessary. HILIC may provide an excellent alternative to SCX in shotgun proteomics [5.4].

5.3.3 Mass Spectrometry

The Edman degradation technique is a classical protein sequencing method. However, it is a low-throughput and insensitive technique (requiring microgram levels of protein) that only works for about 50% of proteins because the N-termini of the rest of proteins are often blocked. Thus, it cannot meet the needs of proteomics. MS is a reliable, sensitive, and high-throughput method for protein identification and characterization. Mass spectrometers measure the mass-to-charge ratio (m/z) of ions. The m/z data of the peptides (MS data) and/or their fragments (MS/MS data) can be used to search against protein or nucleotide databases for protein identification. Mass spectrometers consist of three major parts: an ionization source that converts molecules into ions, a mass analyzer to separate the ions based on m/z, and a detector to detect the ions. Two types of ionization source: matrix-assisted laser desorption and ionization (MALDI) and electrospray ionization (ESI), are commonly used in proteomics (Fig. 5.7). In MALDI, peptide molecules to be ionized are cocrystallized with a crystalline matrix and a pulsed laser beam causes excitation of the matrix and ionization of the peptides. MALDI-MS is normally used to analyze relatively simple peptide mixtures. ESI creates ions by high-voltage induction of charged droplet formation and evaporation. Nanoelectrospray ionization (NanoESI) involves the use of a miniaturized electrospray source consisting of a metal-coated glass capillary with an opening of $1-10\,\mu$m. The tip of the capillary is held at a potential difference of $1-2\,$kV with respect to the orifice of the mass analyzer, which results in creation of charged droplets that are about 100

times smaller in volume than those produced by conventional electrospray sources. NanoESI has the advantage of low solvent suppression and thus high sensitivity. ESI is often coupled to liquid-based separation tools. Liquid chromatography ESI MS systems (LC-MS) are the preferred technology for analysis of complex proteomic samples; For example, in shotgun proteomics, SCX-, RP- and HILIC-HPLC can be used in combination with MS. The mass analyzer functions to separate ions according to their m/z ratios. Mass range, sensitivity, resolution, and mass accuracy are key parameters of a mass analyzer. There are five basic types of mass analyzer: time-of-flight (TOF), quadrupole, 3-D and linear ion trap, Orbitrap, and Fourier-transform ion cyclotron resonance (FT-ICR) (Fig. 5.7) [5.4]. Each mass analyzer can be used by itself or configured in tandem with another mass analyzer to produce a hybrid instrument. The TOF analyzer measures ions according to the time of travel from source to detector, usually being coupled with a MALDI ion source to measure the mass of peptides and proteins. The quadrupole mass analyzer consists of four parallel rods, and ions pass through the middle axis in between the rods. The combination of the constant voltage and radiofrequency voltage determines that ions with a certain mass-to-charge ratio (m/z) can fly through the center towards the detector. In ion-trap analyzers, ions are captured or *trapped* for a certain time interval and are then subjected to MS or MS/MS analysis. A linear ion trap can trap ions longer and has a high capacity for holding ions compared with 3-D traps. FT-ICR is a high-end instrument that captures the ions under high vacuum in a high magnetic field. All charged ions in the magnetic field experience circular movement (cyclotron motion). The radius of the circle depends on the field strength and the ion energy. The frequency of the cyclotron motion depends on the m/z value. FT-ICR MS offers ultrahigh resolution and mass accuracy. Orbitrap is a new mass analyzer that works similarly to FT-ICR, but without the strong magnetic field. Ions are trapped in motion with circular elements by means of an electrostatic field along a central electrode. Orbitrap offers very good resolution and mass accuracy close to those of FT-ICR, and it is robust and needs low maintenance. Trap, triple quadrupole, Orbitrap, and FT-ICR analyzers are often coupled to an ESI ion source and used to generate fragment ion spectra (MS/MS spectra) of selected precursor ions. In the last decade, hybrid instruments have been well developed for proteomics. These include TOF/TOF MS, quadrupole TOF (QTOF) MS, quadrupole ion-trap (QTRAP) MS, and linear ion trap Orbitrap. In such configurations, ions of a particu-

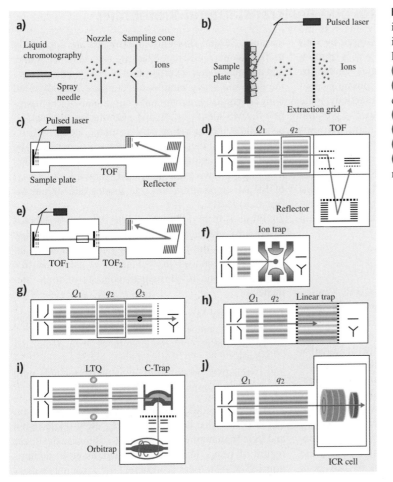

Fig. 5.7a–j Mass spectrometers used in proteome research: (**a**) electrospray ionization (ESI); (**b**) matrix-assisted laser desorption/ionization (MALDI); (**c**) reflector time-of-flight (TOF); (**d**) quadrupole time-of-flight; (**e**) tandem time-of-flight (TOF-TOF); (**f**) ion trap; (**g**) triple quadrupole; (**h**) quadrupole linear ion trap; (**i**) linear ion trap (LTQ) Orbitrap; (**j**) Fourier-transform ion cyclotron resonance (FT-ICR) (after [5.5])

lar m/z ratio are selected in a first mass analyzer (e.g., quadrupole or Orbitrap) and fragmented in a collision cell, and the fragmented ions are analyzed by the second analyzer (e.g., TOF or trap). Different mass spectrometers are equipped with different detectors to detect the ion signals. TOF MS utilizes a multiple-channel plate detector in combination with a scintillator plate and photomultiplier. Quadrupole and ion trap MS often use an electron multiplier as the mass detector. Interestingly, Orbitrap and FT-ICR MS detect ions differently without destruction of ions. Ions in motion generate a minuscule current, which can be measured over time. Change in this current in the time domain is subjected to FT for transformation into m/z values and the intensities of the ions.

In summary, proteins can be analyzed using high-throughput proteomics platforms as follows:

- 2-DE can simultaneously separate thousands of protein species in complex mixtures.
- 2-D-DIGE improves 2-D gel reproducibility and throughput for large-scale proteomic studies.
- HPLC, e.g., RP-HPLC, IEX-HPLC, and HILIC, can separate proteins and peptides from a mixture of compounds depending on their interactions with the stationary phase and the mobile-phase solvents.
- MS is a sensitive and high-throughput method for protein identification and characterization. In shotgun proteomics, RP- and HILIC-HPLC can be used in combination with different MS instruments.

5.4 Quantitative Proteomics and Posttranslational Proteomics

Most biological questions cannot be answered by simply determining whether a given protein is present or not. Current proteomics technologies enable us to measure the amounts of individual proteins, peptides, protein modifications, and interactions. The exciting research areas include comparative expression proteomics, protein modification, and interaction proteomics. Expression proteomics aims to identify and quantify proteins in various samples. Protein modification proteomics focuses on analyzing posttranslational modifications (e.g., phosphorylation and glycosylation) of proteins from different samples. Interaction proteomics aims to characterize the network of protein interactions [5.6].

5.4.1 Quantitative Proteomics

The strategies of expression proteomics, often called quantitative proteomics, include 2-DE-based, MS-based stable isotope labeling, and label-free quantification (Fig. 5.8) [5.7]. Technologies of stable isotope labeling of proteins have been well developed for quantitative proteomics. In the case of metabolic labeling, which exploits in vivo incorporation of stable isotopes during protein synthesis in cells, two populations of cells are grown in isotopically distinct media, being labeled with, for example, ^{15}N and ^{14}N, respectively (Fig. 5.9). The two samples are then combined before being separated by 2-DE. The differential expression of specific peptides is determined by comparison of MS peak intensities between the two samples. However, each nitrogen atom in each amino acid will be replaced by ^{15}N, resulting in varied mass shifts that depend on the length and amino acid composition of the peptide. This makes result interpretation difficult. An alternative approach that overcomes this problem is known as stable-isotope labeling with amino acids in cell culture (SILAC). In SILAC, natural variants of essential amino acids are left out of the media and replaced by ^{13}C, ^{15}N, or ^{2}H

variants. Several essential amino acids are commonly used to achieve efficient labeling, e.g., leucine, lysine, and arginine. After incorporation of heavy amino acids, the light and heavy samples are combined and digested with trypsin to generate an isotopic pair of each peptide for MS identification and quantitation. Because of near 100% incorporation of the isotope labels, SILAC reduces the number of manipulations and increases the accuracy of quantitation. Recently, SILAC mouse and *Arabidopsis* have been generated, showing the power of SILAC application in large tissues and organisms (Fig. 5.9) [5.8].

Proteins from different samples can also be labeled in vitro by chemical reactions. Isotope-coded affinity tag (ICAT) is the first technology designed to quantify protein changes in two samples using gel-free approaches. The ICAT reagent consists of three elements:

1. An affinity tag (biotin), which is used to isolate ICAT-labeled peptides.
2. A tag linker containing eight or zero deuterium atoms.
3. A reactive group with specificity toward thiol groups (cysteines) (Fig. 5.10).

Due to the different linker, the ICAT reagent exists in two forms: heavy (containing eight deuteriums) and light (containing no deuteriums). Because the linker region of heavy ICAT reagent contains eight deuteriums (d8 ICAT), there is sufficient m/z spacing in mass spectra. Thus, it is better to detect different labeled peptide from two samples. However, the d0 (light) and d8 (heavy) ICAT-tagged peptides exhibit different retention times on reversed-phase HPLC columns, and the retention of the biotin group complicates interpretation of MS/MS spectra. To solve these problems, an advanced version of ICAT (cleavable ICAT, cICAT) has been introduced. Carbon-13 instead of deuterium is used for the tag linker, and an acid-cleavable moiety is

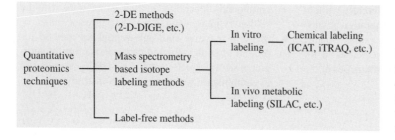

Fig. 5.8 Quantitative proteomics techniques, including 2-D electrophoresis-based methods, mass spectrometry based on stable isotope labeling methods, and label-free methods

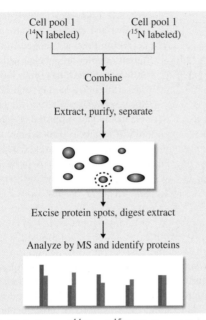

Fig. 5.9 Technique of ^{14}N and ^{15}N labeling. After labeled with ^{14}N and ^{15}N, respectively, the cells are combined. Proteins are extracted and separated on 2-D gels. Then, the protein spots are excised and digested for mass spectrometry, which will identify different proteins based on MS/MS and quantify them based on peak areas of precursor peptides labeled with different isotopes

introduced ahead of the biotin group. Nevertheless, the ICAT method is still limited to two samples and proteins containing cysteine residues.

To overcome these problems, a new technology called isobaric tags for relative and absolute quantitation (iTRAQ) has been developed. It is based on labeling of all primary amines of peptides derived from a protein digest with isobaric tags containing reporter groups of varying masses. iTRAQ labeling combined with HPLC-MS/MS enables multiplexed quantitation of all the peptides via reporter signals in the low mass region of MS/MS spectra and thus quantitation of protein changes across four (4-plex) or eight different samples (8-plex). The reagents are designed as isobaric tags consisting of a charged reporter that is unique to each of the reagents, a neutral balance group to maintain an overall mass of 145 for 4-plex and 305 for 8-plex, and a peptide primary amine reactive group. Upon MS/MS fragmentation of peptide precursors, the neutral balance groups undergo neutral loss and peptides from different samples give rise to four ($m/z = 114$–117) or eight unique reporter ions ($m/z = 113$–119 and 121) that are used to quantify the protein abundance in different samples (Fig. 5.11). The peptide reactive group is designed to react with all primary amines, including the N-terminus and the ε-amino group of the lysine side-chain, to label all the peptides in up to eight different biological samples, thus enhancing peptide coverage for any given protein while allowing for retention of other important structural information such as posttranslational modifications (PTMs).

All the aforementioned isotope labeling-based technologies have limitations, such as expensive isotope tags, variation in labeling efficiency, and the limited number of experiments that can be compared. Label-free approaches can overcome these limitations. One of the label-free approaches is based on the observation and comparison of MS peak areas, because they are proportional to the corresponding peptide concentrations. Thus, highly reproducible LC-MS and careful chromatographic peak alignment using automated methods and software are critical when comparing across various samples. However, this method is limited by unequally detectable peptides in different samples because of dynamic-range limitation, instrument sensitivity, and matrix ion suppression. Spectral counting is another label-free method to circumvent these problems. It is based on the correlation between protein abundance and the number of MS/MS spectra of the peptides de-

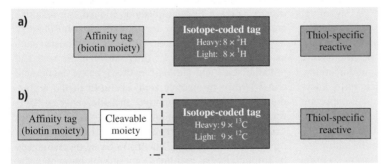

Fig. 5.10a,b The ICAT reagent consists of three elements: biotin, linker, and thiol-specific reactive group. (a) Heavy reagent d8-ICAT, or light reagent d0-ICAT; (b) heavy reagent C13-ICAT, or light reagent C12-ICAT

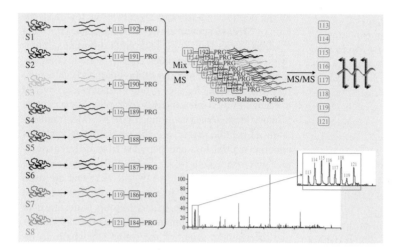

Fig. 5.11 Eight-plex isobaric tags for relative and absolute quantitation (iTRAQ). Peptides derived from different samples are labeled with eight different iTRAQ tags (m/z 113, 114, 115, 116, 117, 118, 119, and 121), respectively. After labeling, the samples are mixed and subjected to MS and MS/MS analysis. Proteins are identified based on MS/MS, and quantified via reporter tag signals in the low mass region of the MS/MS spectra ($m/z = 113–119$ and 121)

rived from the protein. Therefore, this method exhibits good technical reproducibility and dynamic range, and can be used as a simple method for relative protein quantification. In addition, Synapt MS is a novel label-free quantification system. This system is composed of ultraperformance LC and quadrupole time-of-flight MS.

Protein identification and quantification data were collected in LC–ion mobility–MS^E acquisition by alternating the energy applied to a gas cell between a low-energy state and an elevated-energy state. The low-energy scan mode is used to obtain accurate precursor ion mass and intensity data for quantification, while the elevated collision energy mode generates multiplex peptide fragments of all peptide precursors for identification. The Synapt system enables quantitative analysis of protein changes in complex mixtures.

5.4.2 Posttranslational Modification Proteomics

Protein posttranslational modification is an indispensable part of protein information processing and plays important roles in regulating a suite of signaling, developmental, and metabolic processes. There are more than 360 known chemical modifications of proteins [5.9], and many more remain to be discovered. Here, we focus on two important known modifications: phosphorylation and glycosylation. Protein phosphorylation is one of the most widespread and important posttranslational modifications. Dynamic phosphorylation/dephosphorylation at specific amino acid residues serves as a switch of signaling and metabolic pathways. This reversible process is mediated by various protein

kinases and phosphatases. Large-scale identification and quantification of phosphoproteins and their phosphorylation sites under various conditions are essential to fully understand the functions of protein phosphorylation and dephosphorylation. Gel-free LC-MS/MS and 2-DE are two major proteomic technologies for phosphoprotein characterization. The latter has been used to quantify the changes of phosphorylated proteins by monitoring the quantitative shifts in pIs of protein spots, usually being combined with Cy dye labeling, ^{32}P incorporation, and immunoblotting of proteins with phosphospecific antibodies. Because most phosphoproteins have low abundance, low stoichiometry, and high dynamics, effective prefractionation and enrichment of phosphoproteins or phosphopeptides are essential. One widely used enrichment method is immobilized metal ion-affinity chromatography (IMAC). IMAC can enrich phosphopeptides and phosphoproteins from lysates of cells and tissues. The IMAC method is based on the high affinity of phosphates to certain trivalent metal ions, such as Fe^{3+}, Ga^{3+}, Al^{3+}, and Zr^{3+}. Another widely used technique is metal oxide chromatography (MOC) based on titanium dioxide (TiO_2) or zirconium dioxide (ZrO_2). TiO_2 has been developed as a selective and robust technology for large-scale phosphoproteomics. In this method, competitive binders such as aromatic modifiers (2,5-dihydroxybenzoic acid and phthalic acid) and aliphatic hydroxy acids (lactic acid and 3-hydroxypropanoic acid) are used in the buffers during enrichment to improve the selectivity of TiO_2 for phosphopeptides. In addition, the phosphopeptides can be fractionated using SCX and HILIC chromatography (Sect. 5.2). HILIC chemistry based on charge and hydrophilicity has exhibited superior retention and separation of phosphopeptides. The combination of SCX or HILIC with IMAC or MOC has been successfully applied to several large-scale phosphoproteomic studies. Furthermore, immunoprecipitation with the antiphosphotyrosine antibody followed by SDS-PAGE or LC-MS/MS analysis has been used to detect phosphoproteins, although this method has not shown utility in mapping protein phosphorylation sites. Mapping protein phosphorylation sites is important and challenging. In addition to general data-dependent MS/MS acquisition, selective mass spectrometry scanning functions such as precursor ion scan and neutral loss scan are very useful; For example, phosphopeptides can be analyzed in a linear ion trap Orbitrap using the neutral loss-dependent MS3 mode. This mode has been developed into a multistage activation method, in which peptide ions resulting from neutral losses by collision-induced dissociation (CID) are subjected to MS/MS analysis. Because phosphopeptide precursor and neutral lost species are all fragmented to generate the MS/MS spectra, the intensities of fragment ions for sequencing tend to be high, which is helpful for phosphorylation site mapping. In addition to CID, electron-transfer dissociation (ETD) and electron-capture dissociation (ECD) are two new methods of fragmentation that preserve sidechains and modifications; for example, phosphorylation on certain amino acid residues is left intact. ETD and ECD have been shown to be very effective in protein phosphorylation analysis [5.10].

Glycosylation is another important posttranslational modification. Glycoproteins are key molecules for cell signaling, structure, and metabolism. Thus, glycoproteomics has been well developed in recent years. Because glycoproteins are usually masked by other high-abundance nonglycosylated proteins (e.g., albumin and fibrinogen in human plasma), lectin or boronate affinity chromatography is often used to enrich or fractionate glycoproteins prior to one-dimensional gel electrophoresis (1-DE) or 2-DE. After gel separation, glycoproteins are processed to release N- and/or O-linked oligosaccharides, followed by separation using graphitized carbon liquid chromatography and MS analysis (Fig. 5.12). Lectin affinity chromatography is designed based on the capability of lectin to bind to sugars. Lectins are immobilized on a chromatographic support (e.g., agarose); then, they can specifically recognize and reversibly bind to glycoproteins without altering their covalent structures. The specificity of the chromatography is dependent on both the specificity of the lectin and the experimental conditions used to bind and elute glycoproteins. After washing unbound

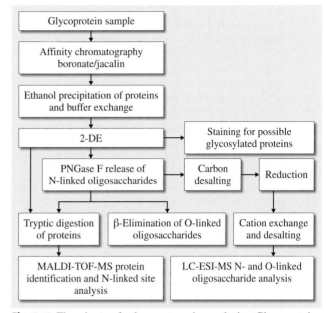

Fig. 5.12 Flowchart of glycoproteomic analysis. Glycoproteins can be enriched and then separated by 2-D gel electrophoresis. *N*-glycans and *O*-glycans can be released by PNGase and beta-elimination. After trypsin digestion, the peptides and oligosaccharides can be analyzed by mass spectrometry

materials, the bound glycoproteins can be displaced by addition of a competing oligosaccharide. Boronate affinity chromatography appears to be superior to lectin affinity chromatography, with increased stability and affinity for the majority of glycoproteins. A coplanar conformation of carbohydrate hydroxyl groups with adjacent carbon atoms is needed for boronate interaction to occur with oligosaccharides; For example, sugars such as fucose with *cis*-diol groups are prime targets for boronate affinity chromatography, which enables fucose-containing glycoproteins to be separated from the complex mixture. In alkaline solution (pH 8), the boronate ligand is ionized and covalently interacts with the appropriate hydroxyl geometry by releasing two H_2O molecules. Under acidic conditions or using a soluble diol-containing agent such as sorbitol, glycoproteins can be displaced from the column. So, glycoproteins are enriched by changing the pH or content of the elution solution. Glycoproteins and the attached oligosaccharide structures can be analyzed by nanoflow LC-MS. N-linked oligosaccharides attached to the asparagine residue on the glycoprotein backbone can be enzymatically released by using *N*-glycosidase F (PNGase F). The released N-linked oligosaccharides can then be analyzed directly or reduced prior to MS/MS analysis. However, there are no universal enzymes available to cleave the linkage of O-linked oligosaccharides attached via the hydroxyl group of serine/threonine on the protein backbone. Thus, chemical β-elimination in a reducing environment is widely used to liberate the O-linked oligosaccharide structures. MALDI-TOF, Orbitrap, and other mass spectrometers can be used for glycoprotein identification and N-linked site analysis. Both N- and O-linked reduced oligosaccharides can be analyzed by LC-ESI-MS in negative ion mode. Various reduced oligosaccharides can be separated at their different chromatographic retention-time points. The MS/MS data provide m/z and intensity data of oligosaccharide parent ions and their fragments. The detailed sequence and linkage structural information can be interpreted using software and databases, such as GlycosidIQ (Proteome Systems Ltd.) and Glycosuit-eDB (Proteome Systems Ltd.).

5.4.3 Protein Complex and Protein Interactions

Proteins rarely act alone. They always interact with other molecules to exert functions. Proteomics research has started to focus on detection of protein interactions, including protein–protein, protein–DNA, protein–lipid, and protein–metabolite interactions. There are many molecular, biochemical, and proteomics methods to investigate these interactions. The yeast two-hybrid screen is a molecular biology technique for discovery of protein–protein interactions. It is based on the principle that the activating domain (AD) and binding domains (BD) in most eukaryotic transcription factors can function in close proximity to each other without direct binding. Even though the transcription factor is split into two parts, it can still activate transcription when the two fragments are indirectly connected. The yeast two-hybrid assay system utilizes genetically engineered nutrient mutants of yeast. Bait and prey plasmids are simultaneously introduced into the mutant yeast to produce BD-fused protein (bait protein) and AD-fused protein (prey protein), respectively. The protein fused to the AD can be either a single known

protein or a library of known or unknown proteins. If the bait and prey proteins interact, the transcription factor is indirectly connected and transcription of the reporter gene can occur. Thus, protein–protein interaction is reported by a change in cell growth and metabolic phenotype. It is evident that the yeast two-hybrid screen detects binary interactions and works best for proteins in the nucleus. Another technique for large-scale analysis of protein interactions is tandem affinity purification (TAP). This method involves creating a fusion protein of the protein of interest with a designed TAP tag. The protein with TAP tag first binds to beads coated with IgG, the first TAP tag is then hydrolyzed by an enzyme, and the second TAP tag binds reversibly to beads of a different type. After the protein of interest has been washed through two affinity columns, it can be examined for its binding partners. Because the TAP tag method requires two successive steps of protein purification, it cannot readily detect transient protein–protein interactions. In addition to yeast two-hybrid and TAP, co-immunoprecipitation is a gold-standard assay for verifying interactions between suspected interaction partners. The protein of interest is isolated with a specific antibody, and its direct or indirect interaction partners are subsequently identified by Western blotting and/or mass spectrometry. Pull-down assay is a variant of immunoprecipitation, where bait protein is tagged and captured on an immobilized affinity ligand specific for the tag. Affinity electrophoresis is used for estimation of binding constants. For example, lectin affinity electrophoresis can be used to characterize molecules with specific features such as glycan content. Another technology worth mentioning is surface plasmon resonance [5.11], which has clear benefits of sensitive affinity capture and characterization of binding and dissociation kinetics of interacting molecules, including protein–protein, protein–DNA, and protein–metabolite interactions. Recently, another instrument called Octet has been developed to bring the detection surface directly into a 96-well sample plate in a simple dip-and-read format [5.12]. This is a high-throughput system for label-free quantitation of antibodies, proteins, peptides, DNA, and other biomolecules and provides kinetic characterization of biomolecular interactions.

5.5 Summary and Future Perspectives

Information processes at the proteomic level determine cellular physiological reactions and phenotypes. There-

fore, it is essential to investigate and achieve a better understanding of these information processes. Modern proteomics has provided unprecedented tools for studying protein identity and structure, protein expression changes, modifications, protein–protein interactions, and protein interactions with other molecules. Here are some important technologies covered in this chapter.

2-D-DIGE, stable isotope labeling (such as SILAC, ICAT, and iTRAQ), and label-free quantification technologies have been well developed for analysis of differential expression of specific proteins/peptides in different samples.

Protein posttranslational modifications, represented by phosphorylation and glycosylation, are very important for protein functions in various signaling and metabolic processes. The two main technologies for phosphorylated protein analysis are LC-MS/MS and 2-DE combined with phosphoprotein labeling/staining, and immunoblotting.

Phosphoprotein/peptide enrichment and fractionation methods (e.g., IMAC, MOC, and SCX/HILLIC chromatography) are essential for phosphoprotein analysis by MS. In addition, CID, ETD, and ECD are effective methods for phosphorylation site mapping.

Glycoproteins can be enriched or fractionated by lectin or boronate affinity chromatography, separated by 1-DE, 2-DE, and LC, followed by MS characterization.

Proteins always interact with other molecules. Among such interactions, protein–protein interaction is important and can be analyzed by yeast two-hybrid screen, TAP, co-immunoprecipitation, and pull-down assay. In addition, protein interactions with other molecules can be analyzed using surface plasmon resonance technology.

Looking back on the development of proteomics, studies in the 1990s and early 2000s were mostly focused on protein identification and cataloging. Since the early 2000s, large-scale quantitative proteomic analyses of different samples, e.g., wild-type versus mutant, control versus treatment or disease, and different genotypes and ecotypes, have dominated the field. In the meantime, research has increasingly focused on organelle proteomes, protein posttranslational modifications, protein–protein interactions, and protein interactions with other molecules. This trend of development has been greatly facilitated by the advancement in genome sequencing, high-sensitivity and high-throughput mass spectrometers, and proteomics software tools. In the coming decades, systematic characterization of protein modifications (such as phosphorylation, glycosylation, acetylation, and redox modification), protein interactions and complexes, organelle and single-cell proteomics, and whole-proteome coverage, together with spatial and temporal dynamics, and correlation of protein information with physiological output and phenotypes can be predicted to be the hot areas of research. As these technologies continue to advance quickly, more and more information processes at the proteomic level will be discovered, reconstructed, and integrated with information processes at other levels for holistic understanding of cellular molecular networks and systems.

References

5.1 P.C. Turner, A.G. Mclennan, A.D. Bates, M.R.H. White: *Instant Notes in Molecular Biology* (Bios Scientific, Milton Park 1997)

5.2 G.B. Smejkal, A. Lazareu: *Separation Methods in Proteomics* (CRC, Boca Raton 2006)

5.3 D.S. Sem: *Spectral Techniques in Proteomics* (CRC, CRC, Boca Raton 2007)

5.4 G. Marko-Varga: *Proteomics and Peptidomics New Technology Platforms Elucidating Biology*, Comprehensive Analytical Chemistry, Vol. 46 (Elsevier, Amsterdam 2005)

5.5 R. Aebersold, M. Mann: Mass spectrometry-based proteomics, Nature **422**, 198–207 (2003)

5.6 S. Chen, A.C. Harmon: Advances in plant proteomics, Proteomics **6**, 5504–5516 (2006)

5.7 Y. Wang, H.Y. Li, S. Chen: Advances in quantitative proteomics, Front. Biol. **5**, 195–203 (2010)

5.8 M. Krüger, M. Moser, S. Ussar, I. Thievessen, C.A. Luber, F. Forner, S. Schmidt, S. Zanivan, R. Fässler,

M. Mann: SILAC mouse for quantitative proteomics uncovers kindlin-3 as an essential factor for red blood cell function, Cell **134**, 353–364 (2008)

5.9 ABRF Delta Mass: A database of protein post translational modifications, available online at http://www.abrf.org/index.cfm/dm.home

5.10 H. Kosako, K. Nagano: Quantitative phosphoproteomics strategies for understanding protein kinase-mediated signal transduction pathways, Expert Rev. Proteomics **8**, 81–94 (2011)

5.11 A. Kausaite, A. Ramanaviciene, V. Mostovojus, A. Ramanavicius: Surface plasma resonance and its application to biomedical research, Medicina **43**, 355–365 (2007)

5.12 J. Li, A. Schantz, M. Schwegier, G. Shankar: Detection of low-affinity anti-drug antibodies and improved drug tolerance in immunogenicity testing by octet biolayer interferometry, J. Pharm. Biomed. Anal. **54**, 286–294 (2011)

6. Pattern Formation and Animal Morphogenesis

Michael G. Paulin

The millions of species of animals on Earth can be divided into only about 35 phyla based on underlying morphology. Animal bodies are constructed using a small set of structural motifs that, as 19th century embryologists recognized, can be generated spontaneously by nonliving physico-chemical processes. The discovery of genes early in the 20th century, and of their molecular identity a few decades later, led to the view that morphology is a consequence of patterned gene expression during development. Advances in mathematical theory and numerical methods in the second half of the 20th century have made it possible to analyze, classify, and simulate patterns that emerge spontaneously in nonlinear dynamical systems.

The body of this chapter is in three sections. The first section (Sect. 6.1) introduces mathematical models and methods of dynamical systems theory. Section 6.2 explains principles and mechanisms of dynamical pattern formation using this theory, while Sect. 6.3 discusses the possible role of these mechanisms in the evolution and development of animal morphology. The mathematical notation is loose and the presentation avoids technicalities, in order to make the chapter more accessible to its intended audience: biologists who have not yet mastered nonlinear dynamical systems theory, and mathematical engineers and physicists seeking opportunities to apply their skills in biology.

The theory shows that macromolecular reaction networks are capable in principle of generating a larger class of patterns than actually occurs. This raises an interesting puzzle: Why do developmental genes only build structures that could build themselves? The question lies at the heart of *evo-devo*, an emerging scientific program that aims to synthesize evolutionary molecular biology and developmental mechanics. Dynamical models suggest that metazoan developmental genes may have evolved not as generators of morphology, but to stabilize and coordinate self-organizing mechanical and physicochemical processes. Simple

Part A | 6

simulations show how molecular patterns that now presage anatomical patterns in development may have been a consequence rather than a cause of those patterns in early animal evolution.

6.1 Historical Overview

In the Aristotelian world view, every effect has a cause. This principle can be formally expressed by

$$x' = f(x), \tag{6.1}$$

according to which the rule f describes how the cause x generates the effect x'. In the 16th century, Newton introduced a new way to quantify causality in terms of continuously changing states. The Newtonian world view can be encapsulated in the same formal equation, except that x' represents the rate of change of state x at an instant, rather than a new state at a subsequent time.

Newtonian mechanics allowed us to supersede the primitive, common-sense view that patterns in structure and behavior must reflect patterns in an underlying cause. Newton showed that patterns can be generated autonomously in a physical system when the rate of change of state is a function of state.

The iconic example is Newton's model of the orbits of the planets. Seen from the Earth, planets move in complex patterns among the stars. Copernicus explained that these paths would look simple if we could view them from the sun. Newton, building on Kepler's mathematical model of the paths that planets take around the sun, explained how they result from a simple rule of the form (6.1). The cause in Newton's model is a radial force acting towards the sun, but the effect is qualitatively different, a periodic elliptical orbit around the sun. Newton's beautifully simple, accurate predictive model displaced the beautifully simple ancient explanation that planets perform a complex dance in the heavens because an intelligent designer employed angels to make it so (Fig. 6.1a).

Medieval thinkers had a very simple explanation of animal form: preformation. They proposed that a miniature human, a homunculus, is folded into each human egg. Development merely unfolds a structure that already exists. In the late 19th century the Newtonian scientific revolution began to have an influence on developmental biology. In 1874, His demonstrated how the development of anatomical structures can be mimicked by nonliving materials. Shortly thereafter, Roux coined the term *Entwicklungsmechanik* (or *developmental mechanics*) to describe this approach to explaining animal form. The approach was eloquently championed by Darcy Wentworth Thompson in the early 20th century. His epic tome, *On Growth and Form* [6.1], was described by Sir Peter Medawar as the greatest work of scientific literature ever written in the English language [6.2]. Its frontispiece showing a drop of milk

splashing onto a surface (Fig. 6.1) has become an iconic image in biology. Thompson pointed out that the beautiful, regular, and reproducible pattern is not present in the falling milk drop. We must seek its creator in the dynamics of the splash, not in angels that intervene just at that moment.

However, the early insights of developmental mechanics were overtaken by developments elsewhere. While Thompson was writing his book in Scotland, Thomas Hunt Morgan was studying the inheritance of fruit fly body parts in New York. Building on Mendel's earlier work, *Morgan* made the Nobel Prize-winning discovery that body parts are inherited as if instructions for building them are laid out in lines *like beads on a string* [6.3]. He called these instructions genes. They remained theoretical objects until the middle of the 20th century, when they were identified with DNA nucleotide sequences [6.4]. This discovery ushered in a period of spectacular productivity in molecular biology, as the beautiful ideas of developmental mechanics

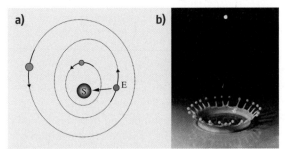

Fig. 6.1 (a) Viewed from Earth (E) other planets follow idiosyncratic complex paths. Newton showed that the different paths can be predicted by the same simple rule. In Newton's model planets move nearly at right angles to the forces that move them (*arrow*). Common sense is not merely useless, but misleading in trying to understand pattern formation in this simplest of dynamical systems. **(b)** Howard Edgerton's famous photograph of a milk splash, which formed the frontispiece of Darcy Wentworth Thompson's *On Growth and Form*. In violation of common sense, a complex, lifelike form is generated from a simple, egg-shaped precursor, by dropping it from a height onto a formless constraint. Does animal morphogenesis employ analogous, autonomous pattern-forming mechanisms? Thompson thought so, but died before mathematical methods capable of explaining crown splash formation, and computer simulation methods capable of replicating it, were developed

were swept away by the ugly facts: Morphogenesis is controlled by patterns of gene expression [6.5]. Genomes evolve by blind evolutionary tinkering to construct organisms that perpetuate the genes [6.6, 7]. Any resemblance between the living and the dead is merely coincidental.

As predicted by the theory that genes determine morphology, disrupting genes or gene expression can disrupt morphological development. However, genetic determination of form has turned out not to be as simple as first suggested by Mendel's peas and Morgan's flies. Many of the claims that have been made about the effects of genes can be made, using comparably good evidence, about the effects of star signs. The zodiacal sign of one's birth predicts height, susceptibility to mental and physical illness, career choices, sporting ability, and various other personal attributes and life outcomes [6.8–11]. These observations are real, but it requires some critical thought and a little understanding of statistical theory to understand why these correlations occur. Ought not the same standard be applied in developmental genetics? Morphological patterns *are* predictable from gene expression patterns during development. The question is, why?

As *Newman* and *Forgacs* [6.12] point out, the idea of a *genetic program* for building an organism took hold among molecular biologists despite the fact that no convincing model of a causal link between gene expression and the three-dimensional form of an animal had ever been presented. The Human Genome Project, whose goal was to print out the *instruction book for human biology*, marked the high point for the paradigm that every problem in organismal biology can be reduced to the problem of finding a gene for it [6.13]. However, the failure of genomics to live up to its early promise has not dampened enthusiasm for more of the same. Instead, molecular biologists are extending the paradigm to incorporate multiple genes for a trait, multiple effects of a gene, interactions between genes, and feedback loops between genes and gene products [6.13].

In the late 20th century, molecular genetics became eerily reminiscent of the last gasps of Ptolemaic astronomy. A proliferation of epicycles leads to an increasingly accurate description of rapidly accumulating data, and ever more accurate predictions [6.14]. We can only thank our lucky stars – irony intended – that Kepler's contemporaries did not have computers. With even a modest 21st century desktop computer, 17th century astrologers could have developed *astroinformatics* and bequeathed to us an ability to describe everything in the world with arbitrary precision, while leaving us

utterly ignorant of our place in it. It is worth noting that the Ptolemaic model is not inherently false, just a different way of looking at the data. It remains to this day capable of describing observations with greater precision than modern astronomy, because Ptolemaic epicycles are constrained only by a need to fit the data while modern astronomers are obliged to ensure that their models obey certain constraints now called *the laws of physics.*

It is true that large numbers of genes and gene products interact in complex networks during development, and there is nothing inherently false in the theory that there are systematic relationships between patterns of gene expression and patterns of morphology, physiology, and behavior. The problem is that developmental systems are clearly dynamical systems. As illustrated above, even the simplest dynamical systems are beyond human comprehension if we attempt to model them as Aristotelian causal chains or networks; For example, while on the one hand we cannot help but be impressed by the prodigious effort, skill, and technology that has gone into mapping out the molecular genetic network underlying sea urchin morphogenesis [6.5], we also cannot help noticing that the result tells us nothing about how sea urchin morphology arises in this simplest of animal developmental systems. From a computational modeler's point of view a molecular network map makes it possible to simulate the molecular network that operates while the animal is developing, without giving us the slightest hint about how to simulate the development of the animal.

This chapter is a call for developmental genetics to become reacquainted with developmental mechanics. Modern mathematical methods and computational tools make it possible to analyze the dynamics of complex networks, and morphogenesis in expanding soft matter (i.e., the idea formerly known as *growth and form*). This chapter will lay out some ideas that may be fundamental to understanding the relationship between genes and proteins at the microscopic level, and morphology, physiology, and behavior at the macroscopic level. These ideas are currently not a standard part of molecular geneticists' training. One of the key ideas that will be explained is that, in general, there are infinitely many sets of components and interactions that will generate a particular pattern. As a consequence of this, attempts to understand how the pattern arises by cataloguing the underlying components and their interactions are likely to be hindered by large variability at the microscopic level; For example, what appears to be the same macroscopic process in two individuals could be accompanied by overexpression of a particular gene in one individual and reduced expression in the other. The good news is that the global characteristics of a network that specify its behavior are generally much fewer than the number of components, and very much fewer than the number of interactions. This means that the cost of learning abstract mathematical concepts is repaid by a simpler reality.

A second important concept is that, while the dynamics of genetically regulated macromolecular re-action networks does have the potential to account for the diversity of pattern and form in biology, it struggles to explain the lack of it. Why, among all of the patterns that could be generated by such networks, do they appear to restrict themselves to instructing tissues to develop in ways that they could develop without instructions? This chapter will outline a possible solution, drawing on recent progress in *evo-devo*, a research program that is producing a new synthesis of 19th century developmental mechanics and modern molecular genetics.

6.2 Models

6.2.1 Ordinary Differential Equations

Ordinary differential equations quantify how the rate of change of some variable(s) depends on some other variable(s), which may include the state variables themselves.

This section introduces three simple dynamical systems that will subsequently be used to illustrate principles of dynamical pattern formation. These are systems whose behavior can be analyzed using two-dimensional plots. The methods of analysis are general, but are difficult to visualize for more complex systems.

It is not necessary to follow this section in detail in order to get the key points, which can be summarized as follows.

1. There is a general mathematical model, called the state space model, for any dynamical system with a finite number of components.
2. We can draw a map showing the kinds of behavior that a system can generate – the *topology of its trajectories* – by examining mathematical properties of its state space model.
3. There is, in general, an infinite class of state space models whose trajectories have a given topology.
4. Correspondingly, on the one hand we can design infinitely many networks of interacting components that exhibit any specified behavior, while on the other hand it is impossible to determine how a network will behave by examining its components and their interactions, unless you know all of them.
5. The ideas extend to systems with an infinite number of components, in particular to the mechanics of continuous materials.

6.2.2 Pendulum

Pendulums oscillate spontaneously at a frequency depending on their effective length. The oscillations die away because energy is dissipated by friction and drag. The equation of motion for a simple pendulum including velocity-dependent drag is

$$\ddot{\theta} + \mu\dot{\theta} + \frac{g}{r}\sin\theta = 0 \,, \tag{6.2}$$

where θ is the angular deviation from the vertical equilibrium position, g is the acceleration due to gravity, r is the length of the pendulum, and μ is a drag parameter. The dot notation represents differentiation with respect to time; i. e., $\dot{\theta}$ is angular velocity and $\ddot{\theta}$ is angular acceleration. This equation is nonlinear because the restoring force, the component of gravity driving the pendulum back towards the vertical equilibrium position $\theta = 0$, depends on the sine of the angle.

6.2.3 van der Pol Oscillator

Formally, the van der Pol oscillator resembles the pendulum. It has a linear restoring force, but has nonlinear *drag* that switches to *antidrag* when $\theta < 1$. It generates periodic behavior that does not die away. Energy dissipated by drag in some parts of the cycle is replaced by work done by antidrag in other parts of the cycle.

$$\ddot{\theta} + \mu(1 - \theta^2)\dot{\theta} + \theta = 0 \,. \tag{6.3}$$

6.2.4 Lotka–Volterra

The Lotka–Volterra equations are often used to model predator–prey systems. Using r to represent the number of rabbits and f to represent the number of foxes,

$$\dot{r} = \alpha r - \beta r f , \qquad (6.4a)$$
$$\dot{f} = -\gamma f + \delta r f . \qquad (6.4b)$$

The first equation states that rabbits are born at a constant birth rate, and die at a rate proportional to the number of foxes. The second says that foxes die at a constant rate and are born at a rate proportional to the number of rabbits.

This model has a simple realistic interpretation in terms of the probability that foxes and rabbits will encounter each other. If each species is randomly distributed over a region, then the probability that one will encounter the other is proportional to the product of the population densities. These densities are proportional to population numbers in a fixed region. The coefficients of the product terms are different in each equation because the effects of predator–prey encounters are not symmetric. The predator gains a little while the prey loses a lot when they meet (the *life–lunch principle*).

The Lotka–Volterra equations generate periodic fluctuations in rabbit and fox numbers. These patterns are qualitatively similar to patterns generated by the pendulum and van der Pol oscillator equations, and yet (6.4) seem qualitatively different from (6.2) and (6.3). Subsequently, we shall see that these equations are not as different as they seem to be at first sight.

6.2.5 Linearization

Because $\sin\theta \approx \theta$ for small θ, for small swing angles we can approximate the nonlinear pendulum model with the linear model

$$\ddot{\theta} + \mu\dot{\theta} + \frac{g}{r}\theta = 0 . \qquad (6.5)$$

The same equation (6.5) is obtained by linearizing the van der Pol oscillator at $\theta = 0$.

A linear approximation to the Lotka–Volterra equation at $f = 0, r = 0$ is

$$\dot{r} = \alpha r , \qquad (6.6a)$$
$$\dot{f} = -\gamma f . \qquad (6.6b)$$

The linearized pendulum and the linearized van der Pol system are damped oscillators, but the linearized Lotka–Volterra system (6.6) does not oscillate. We will examine this in more detail now.

6.2.6 State Space Models

The pendulum can be rewritten as a pair of first-order equations like the Lotka–Volterra system, by introducing state variables $x_1 = \theta$ and $x_2 = \dot{\theta}$. The pendulum equation becomes

$$\dot{x}_1 = x_2 , \qquad (6.7a)$$
$$\dot{x}_2 = -\mu x_2 - \frac{g}{r}\sin x_1 \qquad (6.7b)$$

and the van der Pol equation becomes

$$\dot{x}_1 = x_2 , \qquad (6.8a)$$
$$\dot{x}_2 = \mu\left(1 - x_1^2\right)x_2 - x_1 . \qquad (6.8b)$$

In each case, state variable x_1 specifies the configuration while state variable x_2 specifies the rate of change of configuration.

This trick illustrated for single second-order equations can be used to convert any set of differential equations, of any order, into a system of first-order differential equations in a set of state variables. The resulting state space form is a general model for a finite-dimensional dynamical system,

$$\dot{x}_k = f_k(\boldsymbol{x}) , \qquad (6.9)$$

where x_k is the kth state variable and $\boldsymbol{x}$ is the state vector, containing all of the state variables.

The generality of the state space model for finite-dimensional nonlinear dynamical systems means that we can analyze arbitrary systems in terms of this model. Because all of the relevant variables and their rates of change are treated on the same footing, conceptually we can consider any finite-dimensional nonlinear dynamical system to be an ecosystem of interacting species: foxes, rabbits, etc. In different dynamical systems the state variables may be chemical reagent concentrations, mechanical configuration variables, or any properties of interacting components in a network.

Nonlinear dynamical systems can be difficult to understand because everything is connected to everything and everything is always changing. You generally cannot see what a dynamical system will do next by looking at its current configuration, even if you have its equation of motion. However, state space models make it possible to represent and visualize dynamical systems geometrically. Some *species* in a dynamical model may correspond to directly observable or measurable quantities – like the *species* in the Lotka–Volterra model – while others may correspond to abstract or unobservable properties that are much more troublesome to

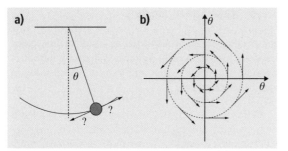

Fig. 6.2a,b In a snapshot of a pendulum (**a**) it is impossible to tell which way the pendulum is moving. As illustrated in (**a**), it can be impossible to see how a simple mass will move even if you know the forces acting on it. (**b**) In contrast, the pendulum's future behavior is easy to predict and visualize from a snapshot in state space. At any point the state space model specifies a vector showing how the state changes at that point. A trajectory starting at any point can be determined by following the flow of the vector field

human intuition. These different kinds of variables are treated on the same footing in state space models, which quantify how the state changes as a function of the current state. It is possible to visualize this across the state space (Fig. 6.2). This is a major conceptual advantage of the state space model; it lets us freeze arbitrary nonlinear dynamics in a *snapshot* that characterizes what happens next. We will explore how to use state space maps to analyze dynamical systems in more detail below.

Another advantage of the state space model is that it is relatively straightforward for a numerical algorithm to map out trajectories from arbitrary starting positions in state space, given such a model. The most commonly used numerical integration routines for solving ordinary differential equations require the equations to be specified in state space form [6.15].

6.2.7 Linear State Space Models

We derived linear approximations to three nonlinear differential equations in Sect. 6.2.5. It is generally easy to linearize a state space model at any specified point in state space by taking partial derivatives of the functions f_k in (6.9) at that point. The state space model linearized at x_0 is

$$\dot{x} = \mathbf{F}x ,$$ (6.10)

where $\mathbf{F}$ is the matrix of partial derivatives $\mathbf{F}_{kj} = \partial f_k / \partial x_j |_{x=x_0}$; For example, the state space model (6.6)

for the pendulum becomes

$$\dot{x}_1 = x_2$$ (6.11a)

$$\dot{x}_2 = -\mu x_2 - \frac{g}{r}x_1 ,$$ (6.11b)

which can be written in the form (6.10) with

$$\mathbf{F} = \begin{pmatrix} 0 & 1 \\ -g/r & -\mu \end{pmatrix} .$$ (6.12)

The linearized state space model describes the local behavior of a smooth dynamical system near x_0. The behavior can be characterized in terms of the properties of the matrix $\mathbf{F}$; that is, analyzing the local behavior of a dynamical system comes down to matrix algebra. If the system is linear then its global behavior can be determined by matrix algebra (Sect. 6.3).

6.2.8 Critical Points

A critical point is a point in state space at which the state derivatives are zero,

$$f_k(x) = 0 , \quad \text{for each } k .$$ (6.13)

At a critical point, the linear approximation to a smooth nonlinear system is

$$\dot{x}_k = 0 ,$$ (6.14)

which implies that the system will freeze up if it reaches a critical point. This can happen, but a critical point may be unstable, meaning that arbitrarily small perturbations cause the system to move away.

Critical points of nonlinear systems are important for understanding the kinds of behavior that they can generate, because near unstable critical points small perturbations of the state can cause large changes in a system's behavior. Near stable critical points, small perturbations have little or no effect.

In one dimension there are three kinds of critical point (Fig. 6.3). The critical point is either stable, in which case the system will converge to the critical point from nearby states, or unstable, in which case the system will diverge away from the critical point.

Nonlinear systems are *qualitatively linear* away from critical points, in the sense that small perturbations have proportionately small effects on trajectories.

6.2.9 Autonomy

The nonlinear state space model (6.9) describes how the rate of change of state of a system depends on its

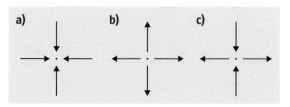

Fig. 6.3a–c Critical points in two-dimensional state space. (a) Stable critical point, all state derivative vectors point inwards. (b) Unstable critical point, states diverge away. (c) Saddle point, some trajectories lead in to the critical point, while others lead away

current state. It conspicuously fails to include external inputs that may also influence the evolution of the state. Equation (6.9) should be modified to include external influences,

$$\dot{x}_k = f_k(x, u) \,, \tag{6.15}$$

where $u(t)$ is an external signal acting on the system.

However, by introducing time as a state variable,

$$x_{n+1} = t \tag{6.16}$$

and an additional equation of motion for this state variable,

$$\dot{x}_{n+1} = 1 \,, \tag{6.17}$$

(6.15) can be transformed into (6.9).

The equivalence of models (6.9) and (6.15) means that technically it makes no difference at all whether we regard system and environment as separate entities that interact, or as parts of one larger system. From a mathematical point of view, then, the *nature–nurture* debate is epistemological (about how we describe things) rather than ontological (about things). The mathematical solution is unambiguous: Choose the model that is simpler to analyze (rather than getting caught up in the nature–nurture debate – the map is not the territory).

6.2.10 Partial Differential Equations

We want to consider dynamics of continua, such as spatially inhomogeneous chemical reactions and the mechanics of continuous materials. In these systems the states are functions of location and time $x(r, t)$, not just functions of time $x(t)$. Models of continuum dynamics require partial differential equations (PDEs), rather than the ordinary differential equations (ODEs) that we have been considering thus far.

We can model a continuum approximately by considering states at grid points. In two dimensions we can choose an array of locations r_{kj} and replace the spatially distributed state with a finite set of state variables,

$$x_{kj}(t) \triangleq x(r_{kj}, t) \,. \tag{6.18}$$

In this way we can model a continuum using a large set of ordinary differential equations instead of one partial differential equation. As we have seen in Sect. 6.2.5 this set of ODEs can be rewritten in state space form, so (6.9) is a general model for continuum systems.

This observation that continuum systems can be modeled as very large dynamical networks of discrete interacting components may seem a little simplistic. However, the truth is that, under the hood, many numerical methods for solving partial differential equations explicitly solve systems of equations like (6.18). Conversely, materials that have classically been modeled as continua are in fact very large collections of very small interacting components. As computing technology advances, we are increasingly able to simulate macroscopic phenomena explicitly in terms of microscopic mechanisms, and as we will see below there are modern computing environments such as NetLogo that make it remarkably easy to do this. I am not trying to suggest that PDE models have nothing to contribute to developmental biology, only pointing out that for present purposes we can avoid the complexities of PDEs and treat everything as a network.

6.2.11 Networks

In integral form, (6.9) becomes

$$x_k = \int f_k(x) \, dt \,. \tag{6.19}$$

It follows that, via a state space model, any dynamical system can be modeled using an array of integrators whose outputs loop back to the inputs via a transformation. It is worth the effort to understand how the networks in Fig. 6.4 can be drawn by inspection of the corresponding ODEs in state space form.

In the light of preceding theory, Fig. 6.4 shows that networks of integrators and static transformations can mimic arbitrary dynamical systems. This result is applied, for example, in analog circuit design. Given a state space model, an engineer can design a circuit whose behavior mimics any dynamical system. The task is made easy in electronics by the availability of components designed to implement standard mathematical

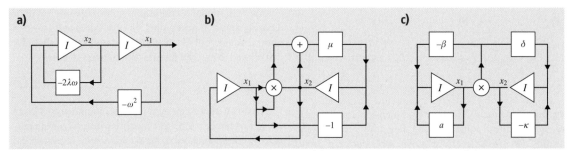

Fig. 6.4a–c Network implementations of differential equations. (**a**) Linear pendulum with drag, (**b**) van der Pol oscillator, (**c**) Lotka–Volterra system. These networks can be implemented as analog electronic circuits using off-the-shelf components. A state space model of a dynamical system can be directly interpreted as a circuit diagram for a network of interacting operators that mimics the system. The reader is encouraged to check the correspondence between signals and operations in the illustrated networks, and the operands and operators in the differential equation models

operations. Given the ability to select from a sufficiently diverse set, circuits could be constructed using other kinds of components; For example, the state variables could be implemented using reagent concentrations in a macromolecular reaction network, or spiking probabilities in a neural network [6.16].

6.3 Where Patterns Come From

6.3.1 Oscillations in Linear Systems

It is easy to verify by substitution that the function

$$\theta(t) = A e^{-t/\tau} \cos \omega t \tag{6.20}$$

satisfies the pendulum equation (6.2), when $\tau = 2/\mu$ and $\omega = \sqrt{g/r - 1/\tau^2}$. This function describes an oscillation that may decay or grow (Fig. 6.5c). For a real pendulum, μ is positive, corresponding to a velocity-dependent drag force, and in this case the oscillation decays exponentially with time constant τ. The decay reflects energy dissipated by the drag force.

If there is no drag ($\mu = 0$), as in a vacuum, then the solution is simple harmonic oscillation (Fig. 6.5b),

$$\theta(t) = A \cos \sqrt{\frac{g}{r}} t . \tag{6.21}$$

Linear oscillations occur in a mechanical system when there is a force driving its configuration towards a point, proportional to how far away it is from that point; For example, such restoring forces can be approximately generated by gravity acting through a rotational constraint, as in a pendulum, or by springs. Oscillations can be generated by second-order dynamical loops in other kinds of physical systems. In a second-order loop, the rates of change of two state variables are coupled in a closed chain.

6.3.2 Feedback and Dynamic Stability

Suppose we use an actuator to apply a force on the pendulum proportional to its velocity. This is called feedback, because the applied force is a function of state. Adding this term to the linearized model (6.5), we obtain the closed-loop equation of motion

$$\ddot{\theta} + \mu \dot{\theta} + \frac{g}{r}\theta = \lambda \dot{\theta} . \tag{6.22}$$

By choosing $\lambda = \mu$, we create a pendulum that oscillates periodically.

A pattern is *dynamically stable* if it persists when the state is perturbed. The feedback-controlled pendulum oscillator is not stable because small perturbations of θ and/or $\dot{\theta}$ alter its amplitude. However, it is not unstable either, because while a perturbation will move the pendulum off one trajectory, it will move it onto a similar, neighboring trajectory. This is an example of neutral stability.

When $\lambda < \mu$ in the feedback system, so that $\mu - \lambda > 0$, the pendulum's state moves onto the trajectory $(\theta, \dot{\theta}) = (0, 0)$; i.e., it comes to rest at the origin. This trajectory is dynamically stable, but not very interesting from a biological pattern-formation point of view.

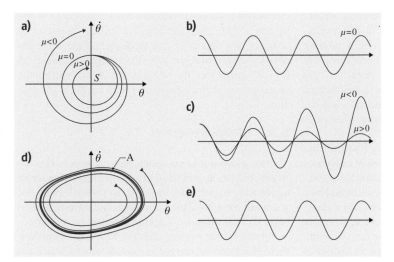

Fig. 6.5a–e Oscillations due to second-order coupling between state variables and their rates of change. (**a**) State trajectories of a pendulum with different values of the drag parameter μ. A closed cycle appears when $\mu = 0$. There is a stable point attractor S at the origin when $\mu > 0$. (**b,c**) Configuration of the pendulum over time with different values of μ. Persistent sinusoidal oscillation occurs when $\mu = 0$. (**d**) State trajectories of a van der Pol oscillator, showing trajectories that start near the periodic attractor A converging onto it. (**e**) Configuration θ of the van der Pol oscillator over time. When μ is small, oscillations generated by the van der Pol oscillator closely resemble an undamped pendulum (simple harmonic motion)

6.3.3 External Pattern Generators

Persistent oscillations in the feedback-controlled pendulum are not free. The actuator must use energy to do the work necessary to compensate for dissipation due to drag. This means that the controlled system must have an external energy source, powering an actuator that delivers a periodic force to the pendulum. Without the illumination provided by dynamical systems theory, intelligent observers might agree that this periodic external force causes the pattern of movement. However, from a dynamical systems point of view the *external* feedback loop is a component added to an autonomous pattern-generating system in order to select and stabilize a desirable pattern. The feedback element simply defends a naturally occurring pattern from the ravages of the second law of thermodynamics.

6.3.4 Structural Stability

Dynamical stability, considered in Sect. 6.2.8, is about whether a pattern persists when the state is perturbed. Structural stability is about whether a system's behavior persists when its structure is perturbed.

The controlled pendulum (6.22) is structurally unstable. Arbitrarily small errors in the feedback parameter λ destroy its periodic oscillation. If $\lambda < \mu$, the oscillation decays and the pendulum comes to a halt at the origin in state space. This behavior is structurally stable. When there is net drag, small changes in its magnitude do not qualitatively alter this behavior. They only affect how long it takes the pendulum to stop swinging.

The growing oscillation that occurs when $\lambda > \mu$ is also structurally stable. If the feedback is too strong then small changes in its strength only affect how rapidly the oscillations grow. Note that *structural stability* is a formal mathematical property of the model (6.20). Excessive feedback in a real oscillating system will eventually result in some kind of physical breakdown that invalidates the model. Investigating that breakdown would require a more sophisticated model.

6.3.5 Attractors

An attractor is a locus in state space onto which a system's trajectories converge from nearby trajectories. Closed loops in the state space of the feedback-controlled pendulum when net drag is zero (Fig. 6.5a) are not attractors, because the pendulum will shift onto a neighboring closed-loop trajectory if the state is perturbed. The origin is an attractor when net drag is positive, however, because a damped pendulum will slow to a halt at the origin and stay there in the face of perturbations.

However, as noted before, attractors in pendulum dynamics are not very interesting from a biological pattern-generating point of view. The periodic trajectories of a linear pendulum are not attractors (they are not dynamically stable), and not structurally stable. The only stable attractor of (6.5) is a point at the origin, where it goes to die.

The van der Pol oscillator, on the other hand, has a structurally stable periodic attractor (Fig. 6.5d). When its parameter μ is small, the behavior of a van der Pol oscillator closely resembles the behavior of an undamped pendulum (Fig. 6.5e), but this pattern resists perturbations in the state and structure of the oscillator. This stability incurs a design cost and a running cost. The oscillator must incorporate a mechanism that provides appropriate nonlinear state feedback, and this mechanism must draw power from an external source because the nonlinear term dissipates energy when the coefficient of $\dot{\theta}$ is positive and does work when it is negative.

We have seen on the one hand that pattern formation is easy to analyze in linear systems using analytical solutions of ODE models, but this is not directly relevant to biological pattern formation because the patterns that linear systems generate are unstable and/or uninteresting. On the other hand, the van der Pol example illustrates how nonlinear ODE models can generate structurally and dynamically stable patterns, but it is usually impossible to solve nonlinear ODEs analytically.

Fortunately, it is straightforward in principle to characterize and map the attractors of a nonlinear system from a state space model. We need not consider the details of how to do this, because our present concern is not to know *how* it is done so much as to know *that* it can be done. The technical procedure is very clearly explained by *Strogatz* [6.17]. The result is a map of the state space showing critical points and attractors, and how the system's trajectories flow around them, i. e., diagrams such as Fig. 6.5a for the simple pendulum and Fig. 6.5d for the van der Pol oscillator.

Dynamical systems are said to be topologically equivalent if their critical points and attractors can be matched up by continuously warping the state space. Topologically equivalent systems have essentially the same sets of behaviors.

6.3.6 Bifurcations

By adding a parameter to the van der Pol model we obtain a model that can be smoothly modified into a damped linear pendulum model,

$$\ddot{\theta} + \mu(1 - \lambda^2\theta^2)\dot{\theta} + \theta = 0 \,. \tag{6.23}$$

This model becomes (6.3) when $\lambda = 1$ and (6.5) when $\lambda = 0$. Assuming that μ is small and positive, this system can have either a stable point attractor or a stable periodic attractor. This change happens suddenly as the parameter λ changes gradually.

This sudden qualitative change in global dynamics, from actively holding still to oscillating periodically, is a Hopf bifurcation. In general, a bifurcation oc-

curs when a parameter change causes a change in the system's dynamical topology. A critical point or an attractor may appear or disappear, and although the underlying cause may be a small continuous change in a model parameter, the effect is the sudden emergence or extinction of some pattern(s) of behavior in the system.

6.3.7 Global Dynamics

There is a trivial sense in which each network in Fig. 6.4 is just one of an infinite set of networks that generate a particular pattern; For example, in network Fig. 6.4a, we could note that $2\lambda\mu = \lambda\mu + \lambda\mu$ and have two pathways each feeding $\lambda\mu$ back around the integrator for x_2 instead of one pathway feeding back $2\lambda\mu$. However, there is a more subtle and important way in which infinite families of networks are functionally equivalent. Suppose that we construct new state variables y_1 and y_2 by transforming the original x_1 and x_2,

$$y = Ax \,, \tag{6.24}$$

so that

$$x = A^{-1}y \,. \tag{6.25}$$

Expressed in terms of the new state variables, (6.10) becomes

$$\dot{y} = AFA^{-1}y \,. \tag{6.26}$$

Equations 6.25 and 6.26 define an infinite family of dynamical networks whose outputs x – what we actually observe – are indistinguishable.

In general, the dynamics of a linear system with an N-dimensional state vector are characterized not by the N^2 coefficients of its dynamical matrix F but by the N eigenvalues of this matrix. Correspondingly, the behavior of a linear integrator network depends not on the components and their interactions but on a relatively small number of global characteristics of the network.

This result means that engineers have considerable flexibility in analog circuit design. They can use the similarity transform, $F(A) = AFA^{-1}$, which leaves eigenvalues unchanged, to change circuit components and layout without altering the function of the circuit [6.16]. Systems related by a similarity transform are said to be similar. This is not about having free parameters that are unconstrained by the function of the circuit, but about the ability to transform signals and operations to achieve the same function in different ways.

In biology, evolution selects dynamical networks according to their function. Different molecular components of these networks may have very similar properties, and superficially very different networks may generate similar patterns. This implies that what must be conserved across organisms to achieve common goals is not particular molecules or pathways but global characteristics of the molecular reaction networks. As in engineering, there are likely to be tasks for which particular components and circuit topologies tend to be used commonly or even universally, for reasons other than that they produce an advantageous behavior; For example, it may be particularly easy and cheap to produce certain components, they may draw less power to perform the task, or it may simply be that one design became standard many years ago and it would be too disruptive to introduce a new design now, even if it would be better in the long run.

It is possible that the role played by a particular gene product in one species could be carried out by an unrelated gene product in a related species. For example, *bicoid* expression seems to be essential for establishing the anterior–posterior axis in *Drosophila* embryos, but other insects appear to use different gene products for the task [6.18]. From a dynamical systems point of view, insect development would appear to require *something* to be expressed near one pole of the embryo so that its concentration gradient can guide anterior–posterior differentiation during development. The particular molecule that is selected for this task may be just one of many different gene products capable of performing it.

In complex networks where many similar components are available, even if different individuals employ the same molecular components, they do not need to exhibit the same molecular concentration patterns (internal state variables) to achieve the same outcomes. In particular, a *pathological* perturbation of any pathway can be compensated by adjustments in other pathways to maintain global function. This is straightforward from an engineering perspective; For example, given a functioning network F plus a constraint such as a maximum allowable value for some coefficient (maximum possible reaction rate on the corresponding pathway), it is a simple exercise in algebra to find a similar network that satisfies the constraint – if there is one. If there is not, then the constraint is fatal to the operation of the network. If similarity is merely difficult or expensive to achieve under the constraint, then we might label the constraint *pathological*.

Dynamical systems theory suggests that we should not necessarily be surprised or concerned about substitution of unrelated gene products in homologous pathways, or large interindividual variation in molecular profiles even within a species. By contrast, the natural prediction of the *genetic program* theory of biological organization is that there is some optimal level for each gene product, that selection acts to tune gene expression to these optimal levels, and that pathology can be identified by abnormally large deviations from population norms. This strategy does work in some cases, but, as noted above, it has not lived up to early expectations. Extending the same idea to look at multiple gene products may simply be a more difficult and expensive path to the same disappointment.

The *genetic program* paradigm encourages scientists to try to correlate the expression of particular genes to outcomes in morphology, physiology or behavior, rather than asking how a gene product operates as a network element and how the network is constructed and regulated to generate the outcomes; For example, the function of a resistor is to limit current flow in an electronic pathway. However, if we try to determine function by observing the consequences of removing or modifying resistors in functioning circuits, we would discover that they are *pleiotropic* devices with multiple roles. They can prevent a system from overheating or generating blue smoke, alter the loudness of sounds or the brightness of lights. Their roles can be contradictory: Removing a resistor from an amplifier can make it a siren, while removing a resistor from a siren can silence it. In addition, it might be observed that manufacturers may substitute resistors with different sizes and shapes, made from different materials, without altering the function of a device. This would be an excellent indoor game to play if there was a reward for discovering and classifying patterns in the relationship between circuit structure and function, because it occupies the mind without stressing it and there is always something to do next. An exponent of this game might learn how to make copies of simple electronic devices and to repair more complex ones, but this would resemble primitive folklore and witchcraft more than modern physical science and engineering.

6.4 Evolution and Development of Morphology

6.4.1 The Origin of Order

In cellular metabolism, chemical reactions among thousands of nucleic acid and protein species are coordinated to produce simple, stable patterns in relation to changing conditions at the cell boundary. Metabolism is not a collection of many simple, independent reactions but a molecular ecosystem in which all species interact in a single giant network. Extending the methods illustrated before by simple two-species networks to analyze and design networks with thousands of species seems beyond the capacity of finite intelligence. However, *Kauffman* [6.20] has shown that, under mundane conditions, large networks of interacting macromolecules are almost certain to form spontaneously.

The basic principle of Kauffman's model is simple. A diverse population of macromolecules presents a diverse set of potential reactions and catalytic interactions. Given some probability that macromolecules will react, and some probability that a macromolecule will catalyze the reaction, the probability that some catalyzed reactions will occur in a collection of macromolecular species grows as the number of species grows. Kauffman demonstrated that, given some small initial probability of catalyzed reactions, increasing the number of molecular species eventually leads to a *tipping point* at which all of the species join a single large reaction network with probability approaching 1. At this critical point there is a phase transition where a large set of macromolecules containing small subsets of interacting species suddenly coalesces into a single network of interactions.

Kauffman's model shows that complex interacting networks inevitably crystalize out of macromolecular soup containing a sufficiently large number of ingredients. He has discussed it in the context of a *metabolism first* model of the evolution of living cells. This is beyond the brief of the current chapter, but it may be noted in passing that network phase transitions, first identified mathematically some decades earlier [6.21], provide a simple potential explanation for the *irreducible complexity* of the very complex and apparently very finely tuned molecular machinery in living cells. Natural selection does not have to construct such machines by gradual modification of simpler ones. It only has to select and modify complex whole networks that necessarily occur for thermodynamic reasons.

While Kauffman's work has primarily been theoretical, using mathematical models and computer simulations, examples of self-catalyzing molecular reaction networks have been generated in laboratory experiments.

6.4.2 Turing Patterns

In 1952 the British mathematician *Alan Turing* developed a theory of how chemical reactions can create spatial patterns [6.22]. The mechanism is a chemical reaction in which one reagent catalyzes the formation of the other, while the second inhibits the formation of the first. This is a molecular analog of a predator–prey system. We have already seen how temporal oscillations can arise when two quantities are dynamically coupled in this way.

In Turing's reaction–diffusion model, the reaction takes place in a thin layer of solute, in which the reagents diffuse at different rates. Initial small spatial variations in relative reagent concentrations are amplified into spatial patterns whose wavelengths are determined by the reaction–diffusion kinematics.

The ideas put forward by Turing have been picked up and extended by others [6.23–28]. Various patterns can be generated by Turing's mechanism, most commonly stripes and spots resembling patterns on animal coats (Fig. 6.6). The patterns are affected by the size and shape of the surface in which the reaction occurs. The Turing–Murray theorem famously asserts that a spotty animal may have a stripy tail but a stripy animal cannot have a spotty tail, a prediction that appears to be confirmed in nature [6.29].

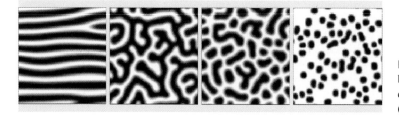

Fig. 6.6 Turing patterns, generated by simulating Turing's reaction–diffusion equations using NetLogo (after [6.19])

Despite many examples of two-dimensional patterns on surfaces of organisms that bear a compelling resemblance to Turing patterns, it took the best part of half a century to demonstrate that there really is a Turing-like chemical reaction–diffusion process underlying any of these patterns [6.30, 31]. Other patterns that superficially resemble Turing patterns have been shown not to be generated by this mechanism.

6.4.3 Segments: Growth Transforms Time into Space

In 1894 *Bateson* noted that periodic patterns in animal form, such as vertebrate somites and earthworm segments, could be generated by oscillating processes coupled to growth [6.32]. In 1976, *Cooke* and *Zeeman* formalized this idea in the mathematical *clock-wavefront model* [6.33] (Fig. 6.7). An oscillating reaction generates periodic fluctuations in reagent concentrations, while the tissue grows steadily. The reaction is activated when the concentration of another chemical produced at one end of the tissue is above a threshold level. As the tissue grows steadily, the reagent concentrations are frozen when the level of the activating substance falls below the required threshold, thus converting predator–prey-like temporal oscillations into a periodic spatial pattern.

About 10 years ago, *Pourquie* and colleagues observed that certain genes are expressed with a temporal periodicity that matches the time taken for one somite to form in chick embryos [6.34, 35]. They and others have subsequently identified the gene products involved and confirmed the clock-wavefront model for vertebrate somite generation, more than a century after the basic idea was put forward and three decades after the dynamics of the pattern-forming mechanism were formulated in a mathematical model [6.12].

6.4.4 Sticking Together: The Invisible Hand of Adhesion

The drop in electrostatic potential energy that occurs when oppositely charged entities approach each other means that it is energetically favorable for them to be adjacent to each other and work must be done to pull them apart. This microscopic effect is responsible for the macroscopic phenomenon of adhesion, or stickiness.

In a fluid of polarized and nonpolarized particles, the polarized particles will tend to clump together. Surface tension arises because there is free energy associated with unmatched charge on the surface. In the absence of other forces and constraints the clump will contract into a sphere; For example, water molecules are polarized and air molecules are not. It costs 7.3×10^{-8} J to increase the surface area of a drop of water in air by $1\,\mathrm{mm}^2$. At small scales this overwhelms other forces, and this is why small water droplets are spherical.

In a mix containing particles with different adhesivity, less adhesive particles will form layers around more adhesive particles. If localized charges (the sticky bits) are restricted to parts of larger particles then the minimum energy configurations can be sheets, tubes or shells rather than spheres [6.12].

Cells have polarized molecules, appropriately called cell adhesion molecules (CAMs) and surface adhesion molecules (SAMs), embedded in their membranes [6.36]. CAMs stick cells to each other, while SAMs stick cells to the extracellular matrix. CAMs and SAMs are gene products, and so genes can exploit adhesion to create tissues that spontaneously self-organize into layered structures by modulating the expression of these proteins [6.12, 36]. In contrast, according to the *genetic program* model, cell adhesion molecules are simply labels that tell cells where they should be in the body and how they should choose their neighbors, analogous to color coding on joints and fasteners of kitset furniture.

Surface tension–adhesion effects dominate other forces at small scales, and therefore probably play a major role in organizing the overall form of organisms in early development. As the organism grows, adhesion takes on the more mundane role of just keeping it together. The freshwater predator *Hydra* can be dissociated into single cells that can spontaneously reassemble into an intact animal. This effect appears

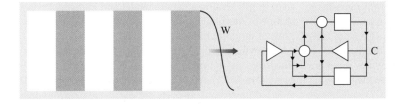

Fig. 6.7 Clock-wavefront model for segmentation. A periodic reaction network of genes and gene products forms a *clock* (C), periodically changing cell states. Meanwhile, gene products diffuse along the tissue. Behind the advancing wavefront (W), the periodic reaction stops, leaving alternating stripes of cell states

to be due to differential cell adhesion [6.37]. *Hydra* is apparently small and simple enough that the state space of position and movement of isolated floating cells has a single point attractor, a minimum-energy configuration corresponding to the adult morphology. More complex organisms appear to require specific developmental pathways to reach their stable adult morphologies. They tend to drop into nonviable configurations corresponding to local energy minima if these pathways are disrupted.

6.4.5 Making a Splash

Thompson and *White* [6.38] famously noted the morphological similarity of splashes and certain animals (Fig. 6.1). There is still technical debate about the physically correct mathematical model for crown formation in splashing fluid droplets [6.39–41], but relevant fluid-dynamical principles can be explained in simple terms.

The kinetic, gravitational, and adhesion energies of molecules in a body of water are dynamically coupled. Consequently, mechanical work that locally accelerates water molecules or alters their height causes ripples to propagate radially across the surface. The dynamical coupling is such that any kinetic energy in bulk flow (*near field*) is quickly converted to surface energy in the

form of ripples that increase the surface area. Energy is dissipated by friction between water molecules as their kinetic, gravitational, and adhesion energies exchange periodically, and the surface eventually returns to the flat, minimum-energy configuration. Water molecules bounce up and down like masses suspended by springs. Waves radiate outwards but water molecules do not. As in a pendulum, the wavelengths of these ripples are characteristics of the fluid, not of the perturbing force.

If the perturbation is sufficiently energetic, a second set of ripples can emerge around the crest of a radiating wave. Random fluctuations along the ridge are amplified as kinetic energy is transferred into surface energy by rippling. The number of peaks in the crown is determined by the wavelength of these ripples. As the radiating wave continues to expand outwards, if there is still enough energy, the peaks will pinch off and form droplets, transferring additional kinetic energy into surface energy.

This informal description of crown splash formation illustrates that the crown morphology depends on the dynamical properties of the fluid: viscosity, density, and surface energy. In particular, other things being equal, because the number of points in the crown depends on the wavenumber of the secondary ripples around a circular wave crest, the number of points depends on the dynamical parameters of the fluid. This suggests an interesting thought experiment: It ought to be possible to selectively breed or genetically engineer cows to alter the number of points in milk droplet splash crowns, not by selecting a molecular *program* that runs during splash formation, but by selecting for gene products that affect the viscosity, density, and surface energy of the milk.

6.4.6 Buckling the Trend

Metazoan development does not involve being dropped from a height, and in any case our tissues are too viscous for the crown splash mechanism to be a realistic model of embryogenesis. Metazoan tissues are *soft matter*, viscoelastic materials that can be shaped by applied forces. Kinetic energy is negligible in soft matter dynamics, but elastic strain energy may play an important role. In this section a simple model and numerical simulation of morphogenesis by buckling when two tissues grow at different rates is presented.

Growth-driven morphogenesis in soft matter is illustrated by a simple two-dimensional model implemented in MATLAB (Fig. 6.8). The model consists of a two-dimensional viscoelastic *mesoderm* surrounded

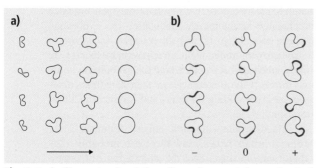

Fig. 6.8a,b Computer simulation of morphogenesis by mechanical symmetry breaking in two dimensions. (**a**) When tissue growth rates match so that there is no stress on either tissue, the *organisms* develop as circles (*right column*). If the mesoderm grows at a slower rate, they collapse into *frozen splashes*. The specific morphology depends on the relative growth rate, which increases from *left to right*. Columns are replicates with the same relative growth rate. (**b**) A single mutant cell is introduced into an initially circular embryo. Its descendents are labelled by *heavy line segments*. If mutant cells are less stiff than normals (−) then they are more likely to end up in the *mouths* of the adults, while if they are stiffer they are more likely to end up near the tips of the arms (+). *Mutants* with normal stiffness (0) end up at random locations on the adult. Details in text

by a line of viscoelastic *ectoderm*. Mesoderm is modeled as a continuum that stores strain energy if it is compressed or dilated. Ectoderm is modeled as a closed chain of linear springs, representing cells, connected at cell junctions by angular springs. These springs store energy when the ectoderm is stretched, compressed or bent.

The structure adopts the shape that minimizes total energy or, as Newton would say, balances the net forces of the tissues against each other. The *visco* in viscoelastic means that inertia is negligible, so this adjustment takes place as a gradual smooth movement. During slow tissue growth the *embryo* will track the minimum-energy configuration.

The embryo is initialized so that the unstressed area of mesoderm equals the area enclosed by a regular polygon formed by ectodermal cells at their rest length. Ectodermal cell lengths and junction angles are then randomly perturbed, and changes that result in lower total energy are selected until the embryo settles into the minimum-energy configuration. Initially, because the system was initialized so that the unstressed area of the mesoderm equals the area enclosed by a regular polygon of unstressed ectoderm, it morphs into that regular polygon.

Now the two tissues begin to grow. On each cycle, a cell is added to the ectoderm and the unstressed area of the mesoderm is increased. Ectodermal cell positions are adjusted by random perturbations to reduce the total strain energy of the organism.

If the area of mesoderm grows so that it remains equal to the area of the regular polygon enclosed by the current number of ectodermal cells at their rest lengths, then the embryo develops as a regular polygon, ageing gracefully into a circle. However, if mesoderm grows at a slower rate, the ectoderm buckles.

The underlying cause of ectodermal buckling in this model is that the energetic cost of bending the ectoderm is smaller than the energetic benefit of reducing mesodermal stretching and ectodermal compression. Buckling tends to form uniform ripples around the organism, rather than sharp folds, because it is energetically favorable to distribute strain energy uniformly in the ectoderm rather than concentrate it at a point.

The morphology of these embryo models can be systematically modified by adjusting relative growth rates and elastic parameters of the tissues. In Fig. 6.8a, the columns are outcomes of repeated runs using the same relative growth rates for the two tissues. If the ectoderm grows much faster than the mesoderm, the embryos quickly collapse into two-armed critters. As the relative growth rate of the mesoderm increases, the collapse is delayed and the number of arms tends to increase. Finally, when the mesodermal growth rate is such that its area always equals the unstressed area enclosed by the unstressed ectoderm, the critters grow up to be circles.

Other tissue parameters were fixed for these simulations, which were able to generate two-, three-, and four-armed critters. The number of arms in the adult morphology is consistent given a particular relative growth rate of the tissues. Although this parameter is continuous, four qualitatively distinct morphologies are generated as the parameter is varied, with sudden switching from one morph to the next as the parameter gradually increases.

6.4.7 Genes for Regional Specification

In the model of Sect. 6.4.6 (Fig. 6.8a), buckling is initiated by amplification of small random perturbations in the ectoderm. As a consequence, the adult morphs are randomly oriented, i. e., arms are produced at random locations on the body.

One can imagine that arms might confer certain advantages on a creature that evolved the capacity to develop them. However, to exploit those advantages it might be quite handy to be able to coordinate developmental processes so that tissues and organs are arranged in repeatable configurations rather than sprouting in random locations with respect to each other.

In Fig. 6.8b, mutant ectodermal cells are introduced at random locations in the initial embryos. These mutants have either stiffer or softer angular springs than other ectodermal cells. In real organisms, such differences could be due to altered amounts or types of CAM gene expression.

If the mutant cells are *soft* their descendants tend to end up in the *mouths* of the adult creatures, but if they are *stiff* their descendants tend to end up near the ends of the arms. The descendants of control mutants, with normal stiffness, tend to end up in random parts of the adult body (Fig. 6.8b).

This simulation illustrates a simple principle. The mutant cell breaks the mechanical symmetry of the early embryo, so that morphogenesis tends to be aligned with the locations of that cell's descendants. Tissues that develop from the mutant tend to be in particular locations in the body in the adult.

In general there is no need for the *organizer* of the morphology-generating process to be the progenitor of tissues destined for particular location in the adult. It is only necessary for there to be some kind of local marker that the two different processes, tissue specialization and mechanical buckling, can align with; For example, if the embryos in Fig. 6.8 were generated by budding off from an adult, there might be some cytological differences in the cells at the budding point that could on the one hand affect mechanical properties and influence buckling, and on the other hand affect gene expression in those cells.

In this model a local alteration of gene expression precedes and predicts the appearance of an arm (or a mouth) at that location, but it is not a *gene for an arm* (or a mouth). As Fig. 6.8a shows, arms are perfectly capable of building themselves without the help of such genes. The expression of the mutant gene in a particular location does, however, predict that an arm will appear at that location. The symmetry-breaking signal occurs early in development, long before the morphology emerges, making it possible in principle for other genes to be activated in spatial patterns that align with the as-yet invisible adult morphology. This makes it possible in principle for evolution to capitalize on morphogenesis by coordinating other developmental processes around it.

In this model the signal breaks the mechanical symmetry directly; i.e., the mutant cell has different mechanical properties. For evolution to be able to take advantage of such a signal by systematically altering the expression of other genes, it would be simpler if the signalling molecule was a transcription factor rather than a structural protein such as a CAM or a SAM. A transcription factor is simply a gene product that influences gene expression by influencing DNA transcription. In particular, developmental genes, such as the *HOX* genes that appear to sketch out the morphology of the adult before it appears, produce transcription factors. Modulated CAM expression leading to mechanical symmetry breaking could then be just one of a number of developmental processes that could be localized to specific regions of the developing embryo marked by prior expression of transcription factors.

6.5 Genes and Development

6.5.1 Morphology First

The facts of molecular biology show that morphology is presaged by spatial patterns of gene products, and the relationship is evidently causal because disrupting expression of these genes or manipulating the concentrations of their products disrupts morphogenesis. However, there is something rather odd about this picture: Genes apparently instruct embryos to develop in ways that they would develop without instructions.

As outlined in Sect. 6.3, macromolecular reaction networks could in principle generate arbitrary spatial and temporal patterns. However, pattern formation in animal morphogenesis is actually restricted to a small repertoire of basic motifs that, as *Thompson* and *Whyte* [6.38] observed, occur spontaneously in growing materials. As hinted at by the simple model in Sect. 6.4.6, one kind of possible explanation for this observation is that the patterns generated by *patterning genes* are a consequence of, not a cause of, morphogenesis due to physical properties of expanding soft matter.

Newman et al. [6.42] proposed that morphogenesis in the first metazoans may have been determined by mechanical properties of growing tissues, which were subsequently stabilized and elaborated by genetic mechanisms.

Fossil evidence and comparative morphology both indicate that animals evolved from sponge larvae that developed into pelagic suspension feeders. Sponges, or poriferans, evidently evolved by aggregation of choanoflagellates, unicellular organisms that can express cell adhesion molecules and spontaneously aggregate into clumps that cooperate as suspension-feeding colonies [6.43].

Poriferans are benthic suspension feeders with simple, variable morphologies including hollow blobs, barrels, and cylinders. Fragments of a sponge, even when completely dissociated into cells, can spontaneously reorganize into their species-specific form. Sponge morphology and development seem to be largely, if not entirely, due to the self-organizing properties of differential adhesion among particles in a viscous fluid [6.44, 45].

In addition to the ability to reproduce and disperse by simply fragmenting and floating away, sponges can reproduce sexually and produce larvae. The simplest of these are spherical cell aggregates that disperse passively in ocean currents. Larvae of some species have streamlined shapes, and the ability to actively respond to environmental cues and thereby increase the probability of settling in a favorable habitat [6.46–48].

Ctenophores or comb jellies are the closest living relatives to poriferans, and both fossil and comparative evidence suggest that ctenophores evolved directly from sponges. According to Nielsen's trochaea theory, the first eumetazoans – animals with distinct tissues and organs – were derived poriferan larvae that matured and started feeding in the water column as pelagic suspension feeders, instead of settling onto the benthos [6.43].

The trochaea, the hypothetical ancestor of ctenophores and all other eumetazoans, is morphologically a collapsed spherical shell of tissue (Fig. 6.9). On one hand, this morphology results from mechanical symmetry breaking in an expanding shell of viscoelastic material [6.49, 50]. On the other hand, this is the basic morphology of ctenophores, the simplest eumetazoans.

Nielsen examines trochaea evolution from a Darwinian perspective, detailing the adaptive advantages of its morphological features and associated tissue specializations. In brief these are that it is somewhat larger than its larval ancestors, giving it a lower Reynolds number [6.51] that enables it to move, and therefore feed, more efficiently in the water column; it is radially symmetric and streamlined, favoring motion in one direction; anterior–posterior tissue differentiation takes advantage of this hydrodynamic asymmetry; sensory cells at the anterior pole detect environmental cues correlated with higher nutrient density; ciliated motor cells along the sides propel and steer the organism, under the influence of neural signals from the anterior sensors; the caudal invagination collects food particles because of hydrodynamic eddy currents as the trochaea

moves forwards; and tissues in this *mouth/gut* region are specialized for capturing and digesting the particles.

There is nothing new in any of the specialized tissue functions of the trochaea, relative to its pre-Cambrian predecessors. Colonial choanoflagellates are now, and presumably were prior to the Cambrian explosion, capable of sensing and moving, and of capturing and digesting food. What is new among Cambrian eumetazoans is not only that these capabilities have been delegated to specialized subgroups of cells, but that these subgroups are systematically arranged within the organism in relation to its overall morphology.

Figure 6.9a is a diagrammatic representation of a symmetrical pelagic suspension feeder whose epithelial cells are all sensors, motors, and feeders. Figure 6.9b shows a suspension feeder with a collapsed morphology. Its morphological asymmetry means that it now has a front and a back and it moves more efficiently forwards than backwards. Information about what is ahead is more valuable than information about what is behind, and nutrients naturally accumulate in the caudal invagination [6.43]. Thus, there would be a selective advantage for these organisms if they could evolve some mechanism(s) so that anterior cells (*i*) specialize for sensing, lateral cells (*ii, iii*) specialize for propulsion, and cells in the invagination or *mouth* specialize for feeding. These tissue specializations, coordinated with morphogenesis, appear to be the crucial steps that marked the transition from brainless ancestors with variable morphology to eumetazoans, animals with regular, reproducible morphology, and tissues and organs including a nervous system [6.43].

6.5.2 Post Hox ergo Propter Hox?

Nielsen discusses the adaptive significance of morphological features and tissue specializations of the trochaea, and gives a detailed explanation of how these could have arisen in a sequence of small steps by random modification of prior structures. However, at the end of that fateful day, 543 million years ago at the onset of the Cambrian explosion, we have an organism whose morphology is a predictable consequence of the morphology and material of its ancestors (blobs of viscoelastic soft matter consisting of replicating sticky particles) under selection pressure to get larger, because being bigger makes suspension feeding more efficient [6.51, 52].

It is evidently possible to believe that natural selection could gradually sculpt random perturbations of morphology into arbitrary forms, given that genera-

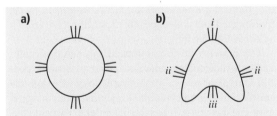

Fig. 6.9 (a) Eumetazoans evidently evolved from sponge larvae, the simplest of which are spheroidal masses of ciliated cells, represented diagrammatically here. **(b)** An expanding shell can spontaneously collapse, creating an asymmetrical mass with an invagination. This morphology has the potential to provide certain advantages to a pelagic suspension feeder, but only if tissue differentiation can be coordinated to align with the mechanical symmetry break. See text for details

tions of biologists seem to have believed that story. However, as outlined above, viscous fluids of sticky particles are self-organizing. With given boundary conditions an aggregate of differentially adhesive cells will have a preferred, energetically favored morphology. Because it will be energetically expensive to maintain any slightly different morphology, natural selection cannot produce new morphologies by accumulating small random morphological changes. Developmental processes may steer morphogenesis towards particular stable morphologies while avoiding nonviable forms, but they cannot arbitrarily create new ones.

In Darwinian terms the evolution of the basic morphology of the hypothetical trochaea and ac-
tual ctenophores can be explained in terms of soft matter dynamics with initial selection for increased size, rather than selection for morphology. This implies that *Newman* and colleagues are right [6.12, 42]. The spatial patterns of molecular concentration that presage morphogenesis in development must have *followed* morphogenesis in evolution. The simple model outlined in Sects. 6.4.6 and 6.4.7 explains how developmental genes could evolve to coordinate tissue differentiation with morphogenesis, leading to the situation that we see today in which transcription factors are expressed in spatial patterns that predict the adult morphology before it starts to appear.

6.6 Discussion and Future

During the second half of the 20th century explanations of biological pattern formation and animal morphogenesis were dominated by the theory that spatial patterns and structures are a result of patterned gene expression during development. According to the *new synthesis* of evolutionary theory and molecular genetics, patterns arise at random and the forms that we see are simply those that survived Darwinian natural selection. Compelling evidence in favor of this theory accumulated over the century. After the development of technology that permitted spatial patterns of gene expression preceding the appearance of corresponding morphology to be clearly visualized in developing embryos, the case seemed closed.

In spite of this, a small group of researchers continued to build on the ideas of 19th century developmental mechanics. Thompson's beautiful exposition of those ideas was recognized as great literature, but his arguments were based on analogy and esthetics rather than rigorous mathematical models. In fact, Thompson was an excellent mathematician in his day, but the mathematics of the day was not up to the task. There has been considerable progress in dynamical systems theory especially in the last quarter of the 20th century [6.17, 49, 50]. Whether this theoretical framework is now adequate to complete Thompson's program is unclear, but at the start of the 21st century we have a mathematical language and the computational capacity to develop and test self-organizing dynamical systems models of pattern formation and morphogenesis.

It is clear from 20th century advances in molecular biology that 19th century developmental mechanics cannot be an alternative to evolutionary molecular genetics; the truth must be a synthesis of the two. This *newer synthesis*, called *evo-devo*, is now gathering steam. This chapter outlined the mathematical principles of self-organizing dynamical systems and proposed how such systems, containing reaction networks of genes and gene products as well as soft matter components, may generate patterns and forms in biology. More comprehensive treatments of ideas that will form the framework for evo-devo in the coming century may be found in books by *Strogatz* [6.17], *Stewart* [6.49, 50], *Raff*, *Raff*, and *Kauffmann* [6.53, 54], *Kauffman* [6.20, 55], *Newman* and colleagues [6.12, 42, 56], *Hall* [6.57], and *Carroll* [6.58, 59].

The promise of developmental mechanics synthesized with molecular genetics is that it may become possible to explain not only the morphology of extant organisms, but to predict morphologies that might have, or might one day, exist. It has the potential to explain phylogenesis in terms of a mathematical taxonomy of form, demoting random mutation from the creator to a mere explorer of animal morphology. We might then be less surprised by the anatomy of the first alien beings that we encounter than the first Europeans to arrive in Australia were by kangaroos, platypus, and black swans.

References

6.1   D.A.W. Thompson: *On Growth and Form* (Cambridge Univ. Press, Cambridge 1917) p. 793

6.2   R.D.A. Thompson: *D'Arcy Wentworth Thompson, The Scholar-Naturalist, 1860-1948* (Oxford Univ. Pres, London 1958) p. 244

6.3   T.H. Morgan: *The Mechanism of Mendelian Heredity* (Holt, New York 1915) p. 262

6.4   J.D. Watson, F.H.C. Crick: Molecular structure of nucleic acids – A structure for deoxyribose nucleic acid, Nature **171**(4356), 737–738 (1953)

6.5   I.S. Peter, E.H. Davidson: Genomic control of patterning, Int. J. Dev. Biol. **53**(5/6), 707–716 (2009)

6.6   R. Dawkins: *The Selfish Gene*, 2nd edn. (Oxford Univ. Press, Oxford 1976)

6.7   R. Dawkins: *The Blind Watchmaker* (Norton, New York 1986)

6.8   G. Doblhammer, J.W. Vaupel: Lifespan depends on month of birth, Proc. Natl. Acad. Sci. USA **98**(5), 2934–2939 (2001)

6.9   J. Verhulst: World cup soccer players tend to be born with sun and moon in adjacent zodiacal signs, Br. J. Sports Med. **34**(6), 465–466 (2000)

6.10   E. Salib: Astrological birth signs in suicide: Hypothesis or speculation?, Med. Sci. Law **43**(2), 111–114 (2003)

6.11   P. Castrogiovanni, S. Iapichino, C. Pacchierotti, F. Pieraccini: Season of birth in psychiatry – A review, Neuropsychobiology **37**(4), 175–181 (1998)

6.12   S.A. Newman, G. Forgacs: Complexity and self-organization in biological development and evolution. In: *Complexity in Chemistry, Biology, and Ecology*, ed. by D. Bonchev, D.H. Rouvray (Springer, New York 2005) pp. 49–95

6.13   F.S. Collins, V.A. McKusick: Implications of the human genome project for medical science, Jama-J. Am. Med. Assoc. **285**(5), 540–544 (2001)

6.14   P. Feyerabend: *Against Method* (Verso, New York 1988) p. 296

6.15   C.B. Moler: *Numerical Computing with MATLAB2004* (SIAM, Philadelphia 2008)

6.16   R.E. Kalman, P.E. Falb, M.A. Arbib: *Topics in Mathematical Systems Theory* (McGraw-Hill, New York 1969)

6.17   S.H. Strogatz: *Nonlinear Dynamics and Chaos* (Perseus, New York 1994)

6.18   P. Dearden, M. Akam: Developmental evolution: Axial patterning in insects, Curr. Biol. **9**(16), R591–R594 (1999)

6.19   U. Wilensky: NetLogo (Center for Connected Learning and Computer-Based Modeling, Northwestern University, Evanston 1999) http://ccl.northwestern.edu/netlogo/

6.20   S.A. Kauffman: *The Origins of Order: Self-Organization and Selection in Evolution* (Oxford Univ. Press, Oxford 1993)

6.21   P. Erdös, A. Rényi: On the evolution of random graphs, Bull. Int. Stat. Inst. **38**(4), 343–347 (1960)

6.22   A.M. Turing: The chemical basis of morphogenesis, Bull. Math. Biol. **52**(1-2), 153–197 (1990)

6.23   A. Gierer, H. Meinhardt: A theory of biological pattern formation, Kybernetik **12**(1), 30–39 (1972)

6.24   A.J. Koch, H. Meinhardt: Biological pattern-formation – From basic mechanisms to complex structures, Rev. Mod. Phys. **66**(4), 1481–1507 (1994)

6.25   H. Meinhardt: Models of biological pattern-formation in the development of higher organisms, Biophys. J. **47**(2), A355–A355 (1985)

6.26   H. Meinhardt: Mechanisms of biological pattern-formation, Biol. Inspir. Phys. **263**, 279–293 (1991)

6.27   H. Meinhardt: Pattern-formation in biology – A comparison of models and experiments, Rep. Prog. Phys. **55**(6), 797–849 (1992)

6.28   H. Meinhardt: Models of biological pattern formation: From elementary steps to the organization of embryonic axes, Multiscale Model. Dev. Syst. **81**, 1–63 (2008)

6.29   J.D. Murray: *Mathematical Biology*, 3rd edn. (Springer, New York 2002)

6.30   A.D. Economou, A. Ohazama, T. Porntaveetus, P.T. Sharpe, S. Kondo, M.A. Basson, A. Gritli-Linde, M.T. Cobourne, J.B.A. Green: Periodic stripe formation by a Turing mechanism operating at growth zones in the mammalian palate, Nat. Genet. **44**(3), 348–351 (2012)

6.31   S. Kondo, T. Miura: Reaction-diffusion model as a framework for understanding biological pattern formation, Science **329**(5999), 1616–1620 (2010)

6.32   W. Bateson: *Materials for the Study of Variation* (MacMillan, London 1894)

6.33   J. Cooke, E.C. Zeeman: Clock and wavefront model for control of number of repeated structures during animal morphogenesis, J. Theor. Biol. **58**(2), 455–476 (1976)

6.34   O. Pourquie: The segmentation clock: Converting embryonic time into spatial pattern, Science **301**(5631), 328–330 (2003)

6.35   O. Pourquie: Segmental patterning of the vertebrate axis, Mech. Dev. **122**, S3–S3 (2005)

6.36   G.M. Edelman: *Topobiology: An Introduction to Molecular Embryology* (Basic Books, New York 1988) p. 240

6.37   U. Technau, C. Cramer von Laue, F. Rentzsch, S. Luft, B. Hobmayer, H.R. Bode, T.W. Holstein: Parameters of self-organization in Hydra aggregates, Proc. Nat. Acad. Sci. **97**(22), 12127–12131 (2000)

6.38   D.A.W. Thompson, L.L. Whyte: *On Growth and Form* (Cambridge Univ. Press, Cambridge 1942) p. 1116

6.39   R.F. Allen: The mechanics of splashing, J. Colloid Interface Sci. **124**(1), 309–316 (1988)

6.40   L.J. Leng: Splash formation by spherical drops, J. Fluid Mech. **427**, 73–105 (2001)

6.41   L.V. Zhang, P. Brunet, J. Eggers, R.D. Deegan: Wavelength selection in the crown splash, Phys. Fluids **22**(12), 122105–122114 (2010)

6.42   S.A. Newman, G. Forgacs, G.B. Müller: Before programs: The physical origination of multicellular forms, Int. J. Dev. Biol. **50**(2/3), 289–299 (2006)

6.43   C. Nielsen: Six major steps in animal evolution: Are we derived sponge larvae?, Evol. Dev. **10**(2), 241–257 (2008)

6.44   P.S. Galtsoff: Regeneration after dissociation (An experimental study on sponges) II Histogenesis of *Microciona prolifera*, J. Exp. Zool **42**(1), 223–255 (1925)

6.45   G. Korotkov: Comparative morphological study of development of sponges from dissociated cells, Cah. Biol. Mar. **11**(3), 325 (1970)

6.46   R. Collin, A.S. Mobley, L.B. Lopez, S.P. Leys, M.C. Diaz, R.W. Thacker: Phototactic responses of larvae from the marine sponges *Neopetrosia proxima* and *Xestospongia bocatorensis* (*Haplosclerida*: *Petrosiidae*), Invertebr. Biol. **129**(2), 121–128 (2010)

6.47   S. Mariani, M.J. Uriz, X. Turon, T. Alcoverro: Dispersal strategies in sponge larvae: Integrating the life history of larvae and the hydrologic component, Oecologia **149**(eq1), 174–184 (2006)

6.48   G. Jekely, J. Colombelli, H. Hausen, K. Guy, E. Stelzer, F. Nédélec, D. Arendt: Mechanism of phototaxis in marine zooplankton, Nature **456**(7220), 395–399 (2008)

6.49   I. Stewart: *Life's Other Secret: The New Mathematics of the Living World* (Wiley, New York 1998) p. 285

6.50   I. Stewart: *Mathematics of Life* (Basic Books, New York 2011) p. 358

6.51   E.M. Purcell: Life at low Reynolds number, Am. J. Phys. **45**(1), 3–11 (1977)

6.52   S. Vogel: *The Physical World of Animalsand Plants*, Life's Devices (Princeton Univ. Press, Princeton 1988) p. 367

6.53   R.A. Raff, T.C. Kaufman: *Embryos, Genes, and Evolution* (Indiana Univ. Press, Bloomington 1991) p. 395

6.54   R.A. Raff: *The Shape of Life: Genes, Development, and the Evolution of Animal Form* (Univ. Chicago Press, Chicago 1996) p. 520

6.55   S.A. Kauffman: *Investigations* (Oxford Univ. Press, Oxford 2002)

6.56   G. Forgaìcs, S. Newman: *Biological Physics of the Developing Embryo* (Cambridge Univ. Press, Cambridge 2005) p. 337

6.57   B.K. Hall: *Evolutionary Developmental Biology*, 2nd edn. (Chapman Hall, London 1998)

6.58   S.B. Carroll: *Endless Forms Most Beautiful: The New Science of Evo Devo and the Making of the Animal Kingdom* (Norton, New York 2005) p. 350

6.59   S.B. Carroll, J.K. Grenier, S.D. Weatherbee: *From DNA to Diversity: Molecular Genetics and the Evolution of Animal Design*, 2nd edn. (Blackwell, Malden 2005) p. 258

7. Understanding Evolving Bacterial Colonies

Leonie Z. Pipe

Microbial colonies are collections of cells of the same organism (in contrast to biofilms, which comprise multiple species). Within an evolving colony, cells communicate, pass information to their daughters, and assume roles that depend on their spatiotemporal distribution. Thus, they possess a collective intelligence which renders them model systems for studying biocomplexity. Since the early 1990s, a plethora of models have been proposed to investigate and understand bacterial colonies. The majority of these are based on continuum equations incorporating physical and biological phenomena, such as chemotaxis, bacterial diffusion, nutrient diffusion and consumption, and cellular reproduction. Continuum approaches have greatly advanced our knowledge of the likely drivers of colony evolution, but are limited by the fact that diverse methods yield the same or similar solutions. Some researchers have turned instead to agent-based, heuristic approaches, which provide a natural description of complex systems. Yet others have recognized that chemotaxis constitutes an optimization problem, as bacteria weigh nutrient requirement against competition and energy expenditure. This chapter begins with a brief introduction to bacterial colonies and why they have attracted research interest. The experiments

on which many of the published models have been based, and the modeling approaches used, are discussed (Sect. 7.1). In Sect. 7.2 a wide cross-section of published models for comparison and contrast is presented. Limitations of existing models are discussed in Sects. 7.3–7.7, and the chapter concludes with current and future trends in this important research area (Sect. 7.8).

Part A | 7

A pure bacterial colony comprises billions of clonal cells derived from a single cell. The initial stage of colony formation, in which the cells are monolayered and can still be individually recognized, is known as a microcolony [7.1]. The arrangement of cells in microcolonies (networks, flat plates, rosettes or square clumps) is affected by properties such as cell shape, cell division processes, and polymer secretion [7.1]. Gliding, sheet-forming, or filamentous organisms form flat microcolonies, while raised colonies result from passive cell division and/or movement in three dimensions. Rough colonies arise from cells lacking surface polymers; cells that secrete polymers produce smooth, shiny colonies [7.1]. Many microorganisms growing in colonies undergo periodic oscillations in metabolic processes such as pigment production, which manifest themselves as rings of intense color alternating with rings of virtually no color [7.2].

Cells within colonies interact in complex and often highly structured ways. Embedded within colonies,

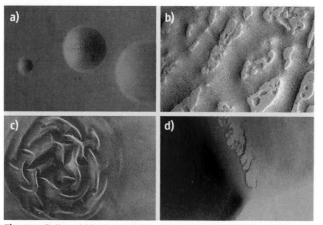

Fig. 7.1 Cells within bacterial colonies are anything but homogeneous. This figure shows the diversity of shapes, colors, and textures exhibited by microbial colonies in nature (colonies of unspecified organisms photographed at the laboratory at the School of Biological Sciences, University of Auckland, New Zealand, under 40 × magnification)

bacteria are shielded from drying, predation, and sudden changes in their environment [7.1]. *Saier* [7.3] asserted that, by understanding colonies, we can better understand how complex organisms formed at the very dawn of life. This is because colonies differentiate as they grow, as do embryonic cells. Saier attributed this to differential gene expression within the cells, which forces them into different roles depending on their position inside the colony. Indeed, bacterial colonies are increasingly being recognized as model systems by which to understand how biocomplexity emerges and evolves.

As a member of a colony, an undifferentiated bacterium becomes part of a collective intelligence, assuming a role that depends on its spatiotemporal location. Epigenetic studies suggest that bacteria possess a rudimentary *memory*, as daughter cells may inherit the genetic modifications that determine whether particular genes will be expressed or not [7.4]. The ability of colonial cells to sense, remember, and respond to their environment has secured their reputation as in-

telligent agents which will inspire new computational developments in fields ranging from data mining to robotics [7.5].

Microbiologists have recognized that growth on solid surfaces is very different from broth culture growth, in which cells are assumed homogeneous (Fig. 7.1). By studying surface microbial growth, researchers have developed new insights into food safety and microbial ecology; For example, a study by *Abellana* et al. [7.6] led to predictions of how a yeast fungus will affect bakery products. *Thomas* et al. [7.7] and her colleagues investigated how a harmless bacterium could be added to food, to prevent growth of a toxigenic one, by modeling spatial colony interactions. *Newman* and *Bowen* [7.8] found specific patterns in the distribution of microcolonies on the root surfaces of grassland plants, hinting at possible higher-level interactions between the bacteria and their host plant.

The earliest bacterial colony growth models recognized that colonies are three-dimensional structures, particularly when growing under favorable conditions [7.9–11]. Pioneering researchers, however, were daunted by the intricacies of simulating an evolving profile, and they restricted their colony geometries to simple shapes such as hemispheres (as in [7.11]). Succeeding researchers were drawn to the similarity between colonial surface-spreading behaviors and the physical behaviors of nonliving systems, whose mechanisms are generally understood. Their models view colonies as essentially two-dimensional entities that exhibit different morphologies under different environmental conditions. As a result of this focal shift, modeling of three-dimensional colony growth remains underdeveloped to this day. My colleague and I are currently developing a model in which the colony geometry is not fixed, but evolves in response to the changing physical dynamics of the system, such as nutrient diffusion from the underlying substrate and cell growth/reproduction (which gives rise to a macroscopic velocity field). In this chapter, however, the focus is on the morphological aspects of spreading colonies, which are more pertinent to the problems of cooperative group behavior and artificial intelligence.

7.1 Lessons Learned from Experiment

When surface colonies are grown on a range of nutrient and agar concentrations, a series of morphology transitions arise, which have been studied in great de-

tail [7.12, and references therein]. *Ohgiwari* et al. [7.13] identified five broad morphology classes for bacteria grown on agar surface (Fig. 7.2). As the agar softens, or

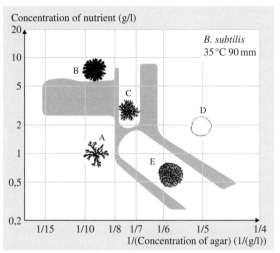

Fig. 7.2 The five morphology classes for the growth of bacterial colonies on agar surfaces identified by *Ohgiwari* et al. (After [7.13] with permission)

as nutrient level increases, the trend is toward smoother, more compact growth. The five regions identified by Ohgiwari et al. are:

- Region A (diffusion-limited aggregation or DLA) – low nutrient/hard agar conditions
- Region B (Eden-like) – high nutrient/hard agar conditions
- Region C (intermediate) – medium nutrient/medium agar conditions
- Region D (compact) – medium–high nutrient/soft agar conditions
- Region E (dense branching morphology or DBM) – low nutrient/soft agar conditions.

As mentioned above, colony morphologies share many characteristics of nonliving ordered forms, such as snowflakes and crystalline aggregates. The fractal structures of DLA result from small-scale self-organization in the absence of surface tension. In DBM, branched structures are confined within a clearly defined envelope, as the smoothing effects of surface tension compete with macroscopic diffusion. Eden-like structures are compact with rough edges, reflecting interplay between surface tension, Gaussian noise, and growth normal to the interface. Compact forms are smooth and regularly shaped, and may be superficially uniform. Interpretation of the C region varies. *Ohgiwari* et al. [7.13] described colony growth in the C region as DBM with a rough, DLA or Eden-like envelope (the patterns becoming denser with rising nutrient level).

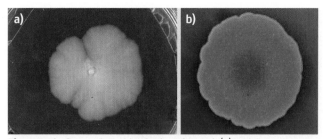

Fig. 7.3a,b Seven-day-old colonies of *E. coli* (**a**) and *S. marcescens* (**b**) grown on high-nutrient plate count agar. *E. coli* colonies tend to form irregular lobes, which appear to be clonal variants of higher transparency and frontal speed than the rest of the colony. *S. marcescens* colonies remain smooth, undifferentiated, and relatively circular throughout extremely long incubation periods

Wakita et al. [7.14] and later workers have identified concentric ring formation in this region, which they use as the C region prototype. These latter workers identified the concentric rings for higher values of nutrient than those used in Ohgiwari's study.

The phase boundaries in Ohgiwari et al.'s morphology diagram are flexible, and differ between strains. Furthermore, the morphology diagram was derived from observations of a particular species, *Bacillus subtilis*, a motile rod-shaped bacterium with a wide range of morphology types. It should not be assumed that the diagram fits all bacterial species [7.13].

On poor nutrient and very soft agar ($\approx 1.5\,\mathrm{g}/100\,\mathrm{ml}$ or less), the developing tip-splitting morphology in some species dissolves into bursts of curved branches, which rapidly populate the agar surface [7.15, 16]. These branches are chiral forms. Chirality, the distinction between right- and left-handed forms of an object, is an intrinsic property of nature. It is found in such diverse entities as DNA, climbing plants, seashells, and human beings [7.17].

Ben-Jacob et al. [7.16] proposed that friction of the cells with the soft agar surface enables the bacteria to distinguish between up and down, an important property of chiral systems. They ruled out gravitational attraction (the most obvious means) by observing that growth in an upside-down Petri dish does not reverse the direction of twist.

On hard agar, bacterial movement requires considerable effort and the bacteria may form vortex modes [7.15]. Here, trails or streams of bacteria are formed, with leading droplets of collectively moving cells. Each droplet contains a swarm of bacteria orbiting a center at velocities highly dependent on the growth conditions. Vortices may rotate clockwise or anticlock-

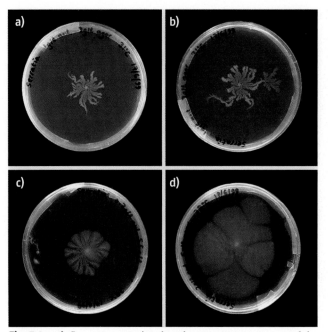

Fig. 7.4a–d *S. marcescens* incubated at temperatures around its growth optimal (28 °C), but subjected to nutrient stress. Fractal (**a**,**b**) and dense (**c**) branching morphologies emerge on hard and soft substrate, respectively. As the agar softens further, a morphology intermediate between diffuse spreading and DBM is observed (**d**)

wise, and may be single- or multilayered. In general, the larger leading droplets enter new regions of agar, while the smaller droplets colonize the empty areas left behind. This is a nice example of active cooperation [7.18].

Under the microscope, colonies growing on hard agar show no sign of active movement; colony growth occurs by cell division and population increase alone [7.13]. Sporulated cells in the colony center remain in that state and are static thereafter. Repeating their experiments with a flagella-less mutant, Ohgiwari et al. established that cell movement is necessary for both DBM and rapid homogeneous spreading over soft agar surfaces. Immotile bacteria do not produce these forms at any agar concentration. Increasing nutrient supply merely induces a crossover from DLA to Eden-like patterns.

In sections of compact colonies, visible morphology changes sometimes arise, which are usually attributed to mutations or phase variants. Clonal variants may manifest as wedges of different transparency or spreading speed. However, different phenotypes within a bacterial colony are not always visibly distinct (Fig. 7.3). Using

a well-known staining technique, *Shapiro* [7.19] found sharp regions of heterogeneity within young compact *Escherichia coli* colonies, and confirmed that new cells were added to the periphery of the colony rather than to the center. (Note that he was studying the surface of the colony, not the region of active growth, which is thought to lie below the surface.) Shapiro suggested that physical and chemical gradients alone are not sufficient to explain the diversity of colony development, and that coordination of cell activity plays a major role.

Developing DLA-like colonies are characterized by random branching, the outer branches growing while the inner ones terminate, suggesting a screening effect [7.20]. DLA patterning in *B. subtilis* (which is hydrophobic compared with species such as *E. coli*) may be related to colony surface tension (ibid). Supporting this idea, *Pipe*[7.21] found no instances of *E. coli* developing DLA-like forms, whereas the other species investigated (*Serratia marcescens*, which is strongly hydrophobic [7.22]) formed classic examples (Figs. 7.4 and 7.5).

Mendelson et al. [7.23] demonstrated that motile cells in *B. subtilis* colonies move in organized whirls and jets. In wet colonies, cell organization occurs at three levels: individual cell movement, whirl and jet motion involving groups of cells, and the organization of whirls and jets into an organized macrostructure. Whirl motion is periodic, being interrupted by chaotic motion and reversal of whirl direction, suggesting that the cells retain a memory of their previous orientation. These organized macrostructures, with their chaotic interludes, are difficult to explain in terms of traditional swimming and chemotaxis [7.23]. Mendelson and colleagues suggested that colony morphology depends on the way in which whirls and jets strike the colony boundary, and the subsequent behavior of the cells deposited there. They assert that understanding these complex short-term phenomena is crucial to correctly interpreting colony development.

E. coli cells have been grown on intermediates of the tricarboxylic acid cycle (such as succinate, fumarate, and malate), an important metabolic pathway in aerobic organisms which, besides yielding energy, aids in providing carbon for amino acid synthesis. In semisolid agar, a migrating band appears, leaving regularly spaced stationary arrays of spots or stripes in its wake [7.24]. These spots or stripes are aggregates of cells, which intensify and later fade as many of their members escape to rejoin the swarm ring [7.25]. *Budrene* and *Berg* [7.24] proposed a plausible mechanism for pattern formation. Buildup of toxic respiratory wastes such as peroxide

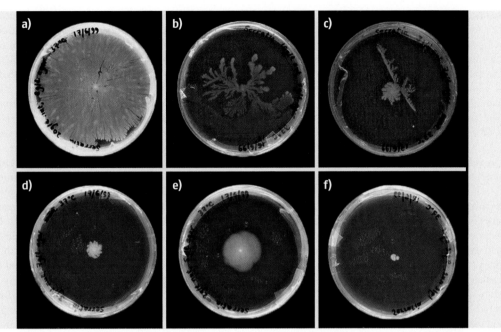

Fig. 7.5a–f *S. marcescens* incubated at 37 °C causes a loss of pigment production but does not significantly affect colony growth. On hard agar surface, extremely high nutrient induces rapid spreading (**a**), while very low levels of nutrient result in tiny colonies after several days (**f**). Varying nutrient level and agar hardness between these extremes induces fractal (**b**), pinwheel (**c,d**) and diffuse spreading (**e**) forms. In (**b**) and (**c**), the medium appears to have been incompletely mixed, and bacteria have responded by altering their morphologies. In (**c**), the bacteria possibly encounter and follow a strong nutrient gradient

may trigger cellular secretion of aspartate or one of its analogs. Aspartate, a chemoattractant, aggregates the cells, rendering them less sensitive to the local effects of toxic byproducts [7.24].

Zoomed-in, a traveling swarm ring resembles a diffuse band of densely packed cells [7.26]. Prior to aggregation, this band curls up into a cylindrical structure. The transition causes inhomogeneities in cell density, and hence in chemoattractant, as chemoattractant concentration is highest in regions of greatest cell density. Bacteria flow along the toroidal axis in response to these gradients, thus forming aggregates. Collapse into the aggregate precursor state occurs when the linear density of cells in the migrating ring exceeds a critical value [7.26].

As succinate concentration increases from low to high, the arrangement of the spots alters from spoke-like to hexagonal [7.25]. These shifting array patterns may be attributed to variation in the secretion rate of aspartate, which depends on cell density and nutrient availability. At low succinate concentrations, the aspartate chemical gradient is insufficient to overcome the

radial effect of the advancing swarm ring, and successive sets of aggregates are radially aligned [7.25]. At higher concentrations, regions of highest swarm-ring density appear roughly half a cycle out of phase between *odd* and *even* cycles, because significant numbers of cells have been left behind in previous aggregates; that is, the chemoattractant effect competes with the forward progression of the swarm ring.

Blat and *Eisenbach* [7.27], repeating Budrene and Berg's experiment with a strain of *Salmonella typhimurium*, found that pattern formation depends on glucose, rather than fumarate, in this species. This implies that a different metabolic pathway, using a different receptor, is employed by these organisms. That the pathway is still sensitive to oxidative stress led Blat and Eisenbach to conclude that pattern aggregation is a general response to oxidative stress.

Woodward et al. [7.28], however, reported that *S. typhimurium* colonies do form regular arrays on succinate-infused soft agar. At first the colony is featureless, but after about 40 h of growth, concentric rings emerge, rippling outwards from the colony center. The mechanism of ring formation is identical to that of aggregate formation in *E. coli*; circumferential spreading of points at specific radii occurs at a rate dependent on succinate concentration. At low succinate concentration, the rings become discrete spots or arcs [7.28]. Cells within newly formed rings are actively motile; those in older rings are nonmotile.

The patterns described above appear closely related to three-dimensional *rosette* structures described in [7.1]. These structures are the result of polar adhesion by cells to a central spot and are apparently, at least in some cases, triggered by release of a chemoattractant [7.1].

Swarming and gliding bacteria have also attracted considerable interest, on account of their highly coordinated group behavior. The model swarming organism is *Proteus mirabilis*. This intriguing member of the *Enterobacteriaceae* intersperses periodic swarming with immotile growth, leading to regularly spaced ridges in colonies of the organism. Swarming is probably not triggered by exhausted nutrient supply, but by a population threshold in the outer ridge [7.29].

The myxobacteria are a fascinating example of predatory bacteria, occurring naturally in aerobic soil environments. Their lifecycle is similar to that of cellular slime molds [7.30]: as nutrient is consumed, the cells aggregate into fruiting bodies containing myxospores. Myxobacteria possess no flagella, but glide in highly organized swarms towards a food source. The structure of the matrix in which they are embedded, as well as the intercellular interactions, may largely govern the social behavior of the cells [7.31].

Myxobacterial colonies exhibit a unique rippling behavior, so called because it imitates the movement of ripples on a pond. Rippling occurs when myxobacteria are introduced to a prey species. The presence of rippling cells is thought to be associated with broken-down cell-wall components of the prey species [7.31]. It appears to be largely independent of fruiting body formation. The mechanisms of directed movement seem to involve more than simple chemotaxis, and likely involve the orientation or composition of molecules in the extracellular matrix (for a full review of this unique group of bacteria see [7.31]).

7.2 Which Modeling Approach Should I Use?

Modeling biological systems requires a delicate balance between realism and tractability. A model that attempts to incorporate every phenomenon rapidly sinks in its own complexity, while an oversimplified model is unlikely to adequately represent the system. *Golding et al.* [7.12] favor the generic approach, which incorporates known biological information but ignores those features not easily incorporated into a simplified mathematical model.

When deciding upon a modeling approach, the nature of the problem being addressed should be considered; For instance, in deterministic models, the state of a variable at a particular time is governed by the previous state of the variable. The global behavior of the system emerges, while random effects are ignored. If random effects play a strong role in the system, deterministic approaches should be used with caution, or not at all [7.32, 33]. Discrete, random temporal events such as birth/death processes in populations, or fluctuations in environmental factors such as light intensity and temperature, require a stochastic approach [7.32]. The archetypal technique for modeling such processes is the Monte Carlo method, in which a random number generated by a computer is used to decide the fate of the next event.

Generic models can be grouped into two categories: discrete and continuous [7.12]. Discrete models represent bacteria or groups of bacteria as separate entities. These individual entities can reproduce, diffuse, consume nutrients, and respond to chemical gradients. Continuous models (also called reaction–diffusion models) describe the time evolution of the system in terms of bacterial density coupled to chemical fields, such as nutrient density.

In the majority of published studies, morphological colony diversity is simulated by varying parameters in standard continuum equations. However, because different models yield the same or similar emergent patterns, it is difficult to know which continuum approaches, if any, have accurately captured the biological phenomena. Recently, some researchers have taken a heuristic approach to bacterial colony growth modeling, using discrete approaches to treat cells as autonomous entities.

Common discrete modeling approaches, such as agent-based models and coupled-map lattices, may be considered bottom-up approaches; that is, the macroscopic evolution of the system emerges from the interaction of individual elements. They thus provide a natural description of a system. In agent-based models, each individual entity (agent) exists in isolation, senses its local environment, and makes independent decisions based on a set of specific rules. This approach is commonly used to model social, political, and economic systems whose outcomes cannot be reliably predicted [7.34, 35]. The ability of agent-based models to produce useful outputs depends on the quality of the input data, which are often difficult to quantify [7.34]. Another limitation of agent-based modeling is that the agents are truly individual, and cannot be treated as aggregates. Describing the interactions and behavior of each entity can be extremely computer intensive if the system is large [7.34].

Agent-based toolkits have become freely available as a result of the interest, and subsequent investment in, this approach to modeling. Among these is the Swarm toolkit, employed by *Kreft* et al. [7.36] in their individual-based model of bacterial colony growth [7.35]. Otherwise, one generates an agent-based model by writing an object-oriented program in a suitable language such as Java or C++ [7.35].

In coupled-map lattice models, a range of spatiotemporal behaviors can be generated by the tuning of a few parameters. Typically, these consist of a local parameter that controls the chaoticity of individual agents, and a coupling parameter that connects the agents [7.37]. A simple example is the difference form of the logistic equation [7.38]

$$x(n+1) = cx(n)[1-x(n)], \quad n = 0, 1, 2, \ldots,$$

which, when iterated on a lattice of regularly spaced sites x, exhibits exponential growth or decay, oscillatory behavior, multiple steady states, or chaotic behavior, de-

pending on the choice of the chaoticity parameter c. Introduction of a coupling parameter into this system causes the value at a single site to spread its influence across the lattice with successive iterations [7.38]. Coupled-map lattice models successfully mimic dendritic and needle-like patterns in crystal growth by varying diffusion, surface tension, and anisotropy parameters [7.39].

Coupled-map lattice models are an extension of cellular automata, in which individual cells interact only with their nearest neighbors. Coupled-map lattices are less strict about spatial discretization than are cellular automata; this is the essential difference between them. The state of a given cell at a particular time, selected from a range of specified values defined by simple rules, depends upon the state of its neighbors at that time. Highly complex behaviors, mimicking many natural phenomena, emerge as the system is iterated in time. The rules governing the behaviors range from deterministic to fuzzy, may be standard (e.g., diffusion) or tailored to a particular application, and may alternate between successive time steps [7.40]. The number of states that a cell may assume at any given time, and the number of neighbors surrounding each cell, can be varied if desired [7.40]. An example is described in [7.41], in which the initial state of the system is a set, or *window*, of sharp Boolean states (0 or 1) embedded in a background of noninteger states selected from the interval $[0, 1]$.

The cellular automaton approach is not suitable for high-density populations [7.42, 43]. Moreover, it can be difficult to select, from among the rich set of rules available, those that are appropriate to the physical behaviors of interest [7.43].

7.3 Modeling Morphology Transitions in Bacterial Colonies

Matsushita et al. [7.44] expressed the need for models that successfully reproduced all five pattern types on the morphology diagram. In particular, the movement of the interface differs markedly between colony forms. In Eden-like colonies, the interfacial cells form chains which advance slowly in bundles, whereas the interface of disk-like colonies moves via the active motion of cells. As a result, the D interface travels much faster than the B interface [7.44].

Wakita et al. [7.14] found that bacterial density of *B. subtilis* cells on bacto-peptone agar in the D region is low compared with that in the DLA and Eden regions, but that cells are active out to the interface.

Wakita et al. concluded that diffusion in the D region is a result of active two-dimensional movement. DLA patterns, they propose, are governed by nutrient diffusion, while Eden-like dynamics are determined by local bacterial growth.

Matsushita et al. [7.44] introduced a variable for each cell that they called the *internal state*. Essentially, a cell can be active or inactive. Their model generates the superficial aspects of all five pattern types by varying just two parameters: initial nutrient concentration and agar concentration. However, the simplicity of their model left the colony fine structure (such as the spacing and periodicity of rings in the

C-type colonies) unexplored. Later, these researchers assumed that bacterial diffusion is constant on soft agar, but a function of bacterial density on hard agar. The improved model could produce concentric rings, by treating the growth/consolidation phases typical of such formations as a density-dependent switch between moving and nonmoving states. Compact, rough-edged colonies are produced by the model only when bacterial diffusion is nonlinear [7.45]. *Mimura* et al. [7.45] concluded that the essential ingredient in their model is the rate of switching between active and inactive states.

The communicating walkers models of *Ben-Jacob* et al. [7.46] have been developed over a number of years, successively incorporating nutrient diffusion, chemotaxis, bacterial motility, reproduction, and dormant states into a comprehensive simulator of complex bacterial systems. Chemotactic responses are of three types: food chemotaxis, which dominates at certain nutrient levels, long-range repulsive chemotaxis, and short-range attractive chemotaxis. Food chemotaxis is expected to hasten the colony growth velocity, as bacteria actively respond to the nutrient gradient. Repulsive chemotaxis is a warning signal from starved bacteria deep within the colony. Being a long-range signal, it stabilizes colony growth, confining it to a smooth circular envelope as observed in the experiments. Short-range attraction is a response to signals from bacteria at the colony edge, operating during growth in rich media when toxic waste secretion is high [7.46].

Eventually, a *lubricating* fluid was incorporated into the communicating walkers model [7.12]. This fluid is secreted by the cells, is dragged along by the cells as they move, and hence flows by diffusion and convection. It is closely coupled to the bacterial movement. Since it is metabolically expensive, its production depends on nutrient level [7.12]. Under certain assumptions, lubrication can be built into the bacterial diffusion term. Additionally, because bacterial density cannot become vanishingly small (at least one bacterium is required for the system to grow), a cutoff in the bacterial density is imposed [7.12].

The communicating walkers model leads to true branching patterns, and allows the cutoff to be arbitrarily chosen. Thus, it has the advantage over the Mimura model, in which the threshold is fixed at ≈ 0.1 bacteria/μm^2 [7.12]. However, a maximum density cutoff might also be expected, due to physical constraints. Once the bacteria are as tightly packed as possible, they will pile up in layers (in the simplest case), or push each other out of the way (as described by *Kreft* et al. [7.36]). The Golding model gives no upper boundary for the

density and no apparent restriction in colony height, though a maximum height is imposed on the lubrication fluid.

Using a lubricant-producing strain of *E. coli*, *Zorzano* et al. [7.47] observed that colonies of motility-limited cells expanded irregularly and produced thick, dense branches, but that more motile cells produced round, compact colonies with copious fluid. Zorzano et al. developed a set of reaction–diffusion equations to model colony growth in this specialized class of organism. The model's treatment of the lubricating field is quite different from that of Ben-Jacob et al.; in particular, fluid is not dragged by the cells as they move. An important distinction of Zorzano et al.'s model is that differentiation into static, reproducing cells is triggered by the amount of fluid exceeding a threshold, and that swarming is a response to critically low fluid densities. This does not conflict with models that use bacterial density as the triggering factor, since lubricating fluid density is expected to positively correlate with bacterial density. Bacterial density, however, is the more generic triggering factor, since not all strains will produce lubricant.

Yeasts, unlike filamentous fungi, always form compact colonies [7.48]. Like many bacteria, yeast cells live in a self-produced liquid, which they carry around as they grow [7.48]. Sams et al. conducted studies of the yeast *Pichia membranaefaciens* at 30 °C and 100% relative humidity. They proposed that metabolites secreted by the yeast cells are analogous to impurities in an undercooled metal melt subjected to a temperature gradient. The crystallizing front is preceded by a field of diffusing impurities which control the evolving front morphology. Tip-splitting morphologies arise naturally from metabolic buildup in dense regions, but the introduction of a noise term, representing statistical fluctuations in front velocity, is necessary to reproduce experimentally observed irregularities. When the metabolic and yeast fields are completely decoupled, rough, Eden-like structures emerge. The envelope then grows by random deposition [7.48]. The model is interesting because it uses buildup of waste, rather than chemotaxis, as a basis for inducing morphology transition.

The relevance of motility and nutrient diffusion has been challenged by some workers. *Li* et al. [7.49] ignored the long-range diffusion of nutrient and bacteria, and asserted that the reproduction ability of the bacteria and the *occupation rate* adequately describe the growth process. The occupation rate defines the rate of cell growth, given that the number of cells in a local region is limited by the specified nutrient level in that region. The reproduction rate depends on the environmental conditions such as agar concentration and moisture. By varying the reproduction rate, Li et al. obtained transitions between compact, DBM, and fractal growth. They argued that motility and nutrient diffusion are less important than the multiplication process and nutrient concentration. Given that a plausible diffusivity of nutrient through bacterial mass is of order $100 \, \mu m^2 \, s^{-1}$ (i.e., occurring very rapidly from a bacterium's viewpoint), this is a reasonable assumption at the microscopic scales assumed by the model. At the whole-colony scale, however, it is unlikely that either bacterial motility or nutrient diffusion can be disregarded. To account for such scaling differences, nutrient diffusion and uptake can be allowed to occur many times during each reproductive time step (as was implemented by *Kreft* et al. [7.36]).

The Li model bears similarities to *Azbel*'s model [7.50], based on the resemblance between bacterial colony growth and growth phenomena in nonliving systems such as flame and chemical reactions. The microscopic dynamics of the cell – movement into an unoccupied adjacent site, consumption of nutrient at a site, and death of starved bacteria – lead to ensemble averaging and spatial homogeneity on a macroscopic scale. Such behavior is typical of a broad class of natural systems, Azbel argues, though the parameters are specific to bacterial growth.

Azbel considers only the dynamics of live bacteria. Rapid multiplication leads to compact, very efficient

colonies in which regions of untouched nutrients are sparse. Unfavorable conditions (rapid death rate) ensure that cells die before they have a chance to reach more nutrients. The result is DLA patterning, with large areas of untouched nutrient between dying cells. Food diffusion is disregarded, on the basis that its inclusion would complicate spore formation. In the event of rapid spreading, where cells at the colony periphery consume nutrient rapidly enough to prevent its entry to the interior, this approach might be justified. However, sporulation could be readily accommodated in a model that included food diffusion, if one imposed a cutoff in nutrient density below which cells could not maintain metabolic activity.

Lega and *Passot* [7.51], recognizing that substrate moisture plays a role in colony morphology, developed the first fully hydrodynamic model, one which includes water concentration as well as nutrient and bacterial density. The model is based on experimental conditions set up by *Mendelson* [7.23]: wet agar containing nutrient, and an upper layer of water containing bacteria. The water/bacterial mixture is modeled as a dense, viscous Newtonian fluid, whose dynamics can be approximated by the Navier–Stokes equations. Bacteria are driven away from overpopulated regions; coupled with nutrient diffusion, this effect is analogous to chemotaxis [7.51]. Although the equations proposed are different from those incorporating nonlinear diffusion and lubricating fluid, they produce equivalent structures [7.51].

7.4 Dynamics of Bacterial Diffusion

The movement of bacteria within colonies is generally regarded as diffusive. *Ohgiwari* et al. [7.13] reported that actively swarming cells on soft agar become sluggish as they approach the colony periphery. Deep within the colony, however, cells are nonmotile as on hard agar. Guided by these results, *Kawasaki* et al. [7.52] proposed that bacteria switch their status from active to inactive as either bacterial or nutrient density declines. They provided a nonlinear, nutrient-density-dependent description of bacterial diffusion, which naturally leads to DBM. Cells in nutrient-depleted inner zones become immobile and hence cannot fill the valleys developing in an incipient DBM formation. An analogous nonlinear bacterial diffusion term, designated the *death term*, was introduced by *Ben-Jacob* and colleagues to generate the same effect in their communicating walkers model [7.12].

Nonlinear bacterial diffusion was proposed also by *Kitsunezaki* [7.53], who introduced it to account for the immobility of nonswarming bacteria on hard agar. They set the bacterial diffusion coefficient $D(b)$ proportional to b^k, where b is the bacterial density and $k = 0$ corresponds to linear bacterial diffusion. When $k = 0$, the growing colony interface is stable and propagates at constant speed. This speed depends on D and on μ, the rate of switching between the active and inactive states [7.53]. Interface instabilities, leading to finger-like patterns and ultimately to DLA, arise from increasing μ and k, and by decreasing D. Unlike normal viscous fingering, surface tension effects play only a second-order role in the dynamics [7.53]. Kitsunezaki's model can generate simple disks, Eden-like patterns, and DLA branching patterns in accordance with experiment but, like Matsushita et al.'s early

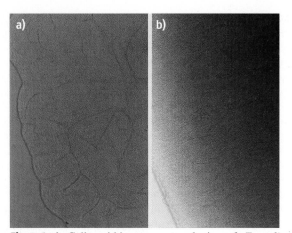

Fig. 7.6a,b Cells within compact colonies of *E. coli* (**a**) and *S. marcescens* (**b**) appear to grow by cell division alone. The images show the edges of the colonies (marked by a thin film in the case of *S. marcescens*). Viewed at 40 × magnification, there is no sign of active cellular movement. The mesostructure of the colony, however, is convoluted in *E. coli*, and radially aligned in *S. marcescens*. This mesostructure forges macroscopic irregularity in *E. coli* colonies, and smooth uniformity in those of *S. marcescens*. Does bacterial diffusion offer the best strategy for modeling these emerging structures?

model, does not produce the concentric rings seen at medium agar concentrations.

The idea that nonlinear diffusion is essential in physically realistic DLA is now entrenched in bacterial colony growth modeling. The original DLA synthesis model, proposed by *Witten* and *Sander* [7.54], begins with a single particle at the origin of a lattice, and a free particle which wanders randomly (diffuses) until it reaches the seed. Clusters grow by adding diffusing particles to neighboring clusters at some average rate. As large clusters grow from a germ cluster, the germ cluster is buried within the ensemble and disappears, also at some average rate. In DLA, diffusion of free particles is modified by adding strong absorption at cluster perimeter sites. The local rate of absorption is then equal to the rate of local cluster growth [7.54].

At this point, it is pertinent to ask whether DLA structures emerging from cell growth are the same as those formed by particle deposition. In the latter, aggregates grow from the very particles that are diffusing near and around them, whereas growth of bacterial clusters occurs from within the cluster. It transpires that the two processes are entirely equivalent. DLA emerges naturally from a model of dielectric breakdown (of which

lightning is a well-known example), in which charge is transferred from a central point electrode to an outer circular one via probabilistic spatiotemporal linking of lattice points [7.55].

Brener et al. [7.56] emphasized the main flaw in the Witten and Sander model, namely that it allows amplification of any small fluctuation ahead of the growing front. In true DLA, there should be a finite threshold in aggregate density ρ below which cluster growth cannot occur, or at least the reduction in cluster growth rate as $\rho \to 0$ should be faster than linear. *Brener* et al. [7.56] replaced the original formulation of *Witten* and *Sander* [7.54]

$$\frac{\partial \rho}{\partial t} = u(\rho + a^2 \nabla^2 \rho)$$

with

$$\frac{\partial \rho}{\partial t} = u(\rho^\gamma + a^2 \nabla^2 \rho),$$

where $\gamma > 1$, u is the walker density, and a is the lattice spacing.

Numerical investigations by *Uwaha* and *Saito* [7.57] show that the Witten and Sander model undergoes phase transitions as the particle density increases. When the particle density is small, diffusing particles are rare around the aggregate periphery, and the aggregate develops as a fractal structure. At high particle densities, movement is heavily restricted, and the growth of the cluster is driven by interface kinetics. In the high-density limit, no diffusion is possible and growth occurs by the Eden process [7.57]. Importantly, the modified model also undergoes phase transitions, though dendritic forms arise if and only if $\gamma = 1$ [7.58].

In the original Witten and Sander formulation, the interface velocity can accelerate towards infinity [7.56]. Imposing a cutoff on cluster density prevents this. However, by Brener et al.'s own admission, the correction factor γ is one of several ways of introducing the cutoff, and as such, is somewhat arbitrary. A similar approach is described in [7.59], where the term $D_s = D_1 s^{-\gamma}$ is used to describe diffusion of already formed clusters (here, D_1 is the diffusion coefficient of a single particle, s is the cluster size, and γ characterizes the relationship between D_s and cluster size).

Although nonlinear bacterial diffusion has been universally embraced by the bacterial colony modeling community, the biological mechanisms underlying the process remain elusive. Furthermore, diffusion, regardless of linearity, may insufficiently describe colonies that grow by cell division alone (Fig. 7.6). Merely setting the diffusion coefficient to zero does not account for the fact that cells within such colonies can move, whether by dividing or pushing by other cells, and are perhaps best described by a velocity field. It is not even clear that the correct form of the nonlinear diffusion has been identified. It has been argued that the most

popular form of bacterial diffusivity (that employed by *Kitsunezaki* [7.53]) allows arbitrarily high diffusion at large cell densities [7.60]. Czirók and colleagues favor a tanh(ρ) function, which imposes an asymptotic limit of 1 on swarmer diffusivity.

7.5 Modeling Ring Formations in Colonies of *Proteus mirabilis* and *B. subtilis*

The formation of rings and more complex morphologies in bacterial colonies has received considerable attention. Most models specific to ring formations are based on the rapidly spreading *Proteus mirabilis*. Cells of this species growing on agar surfaces can exist in two states: swimmers, which undergo typical growth/division cycles, and swarmers, elongated, highly flagellated forms which grow but do not divide, and which aggregate in parallel arrays as they migrate outwards. In most situations, contact with other cells is required for mass migration; that is, swarming behavior depends on bacterial density. Swarming is triggered by amassing of enough cells in the swarmer state, and is terminated by an abundance of swimmers [7.61]. Cells in an expanding *P. mirabilis* colony alternate between the two phases.

The *P. mirabilis* swarming/consolidation cycle is insensitive to changes in velocity and duration of the swarming phase [7.29]. A model based on bacterial differentiation and interactions can duplicate this robustness, provided that all cells septate at the same age [7.62]. Such a model simulates the clock-like nature of the *P. mirabilis* colony, with the fixed lifetime of the cells setting a lower bound for the cycle period. The swarming/consolidation period is equally insensitive to nutrient availability and chemotactic substances [7.62, and references therein, [7.61]]. *Esipov and Shapiro*'s [7.62] model introduces a nonlinear diffusivity term (which dampens small deviations in the timing of individual swarmer groups), to represent organization of swarmer cells into mesostructures, and sharing of extracellular polysaccharides.

In a more tractable variant of the Esipov–Shapiro formulation, both swimmer and swarmer cell populations evolve by growth, conversion to the other cell type, and gain from the other cell type [7.63, and references therein]. A lower swarmer density threshold controls the swarming/consolidation cycle [7.63], and is crucial to ring formation in the model. The threshold represents decoherence of bacterial cell interactions,

and consequent disruption of the motility mechanism, if the distance between cells becomes too great. By additionally incorporating nutrient dynamics, the model can simulate the branched ring structures observed in *B. subtilis*.

Czirók et al. [7.60] developed their model in the hindsight of Itoh et al.'s work. They replaced the age distribution of the swarmer population in Esipov and Shapiro's model with a density measure and a constant swarmer decay rate. Motility of swarmers is controlled by cell density, and no growth occurs in these cells. No interaction is assumed to exist between swarmers and swimmers, in contrast to the Esipov–Shapiro model. Experimentally, the reproduction time of a swarmer cell is typically 20 times that of a swimming cell, and may therefore exceed the consolidation period. In Czirók et al.'s model, the first cohort of swarmers does not emerge until the swarmer reproduction time has elapsed.

The model of *Lacasta* et al. [7.64] was not designed specifically for periodic behavior, but is another approach to modeling morphology transitions in *B. subtilis*. The model accounts for the proportion of long (i.e., swarmer) cells in a population, and assigns a different diffusivity to cells in this state. Unlike the Esipov–Shapiro formulation, nutrient concentration, as well as bacterial density, plays a role in initiating conversion into long cells. Diffusivity of cells in either state can be switched off, or diffusion can be made to alternate between both states, mimicking the cyclic behavior of migration/consolidation phases. DLA-like patterns emerge from the model when both bacterial diffusivity and nutrient availability are low. As nutrient level increases, the number of randomly diffusing cells increases and the colony becomes homogeneous. The model is economical, condensing the three to five equations required by other models into just two equations. However, it is not certain that two different forms of bacterial movement actually occur in *B. subtilis* [7.65].

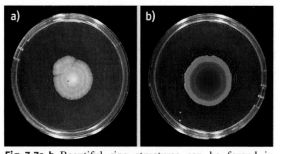

Fig. 7.7a,b Beautiful ring structures can be forged in colonies of *E. coli* (**a**) and *S. marcescens* (**b**) subjected to alternating 24 h cycles of room temperature ($\approx 23\,^{\circ}\mathrm{C}$) and 37 °C. In *S. marcescens*, these are caused by pigment loss at the higher temperature, but *E. coli* is known to not produce pigment. Perhaps the rings represent density changes or some unknown coordinated reaction in cells of this species

Cellular automata approaches may be superior to reaction–diffusion equations when the system is spatially inhomogeneous. *Badoual* et al. [7.43] developed a cellular automata model that successfully reproduces all five regions of the morphology diagram, without resorting to esoteric features not borne out by experiment. Each cell contains a bacterium, or is empty. A full cell can migrate, proliferate, or do nothing. Switching between the three states is controlled by two characteristic times: a proliferation time and a migration time; in an improved version of the model these were replaced by bacterial density thresholds [7.66]. Other parameters in the model are the agar and nutrient concentrations. The model nicely demonstrates how complex behavior emerges from simple rules; however, many of the biological principles widely considered of prime impor-

tance, such as chemotaxis and the presence of lubricant, are not considered in the model.

Østergård and *Sørensen* [7.67] defined a set of rules that reproduced swarming/consolidation behavior of an unspecified periodic ring-forming species. Besides the usual swarmer and vegetative cell densities and nutrient supply, these authors incorporated casamino acid concentration (a nutrient that controls speed of growth). The important processes (diffusion, nutrient consumption by both cell types, and differentiation of swarming into vegetative state and vice versa) are modeled via density difference between a cell and its neighbor, Michaelis–Menten kinetics, and Heaviside functions, respectively. These rules are all sensible and biologically plausible, yet differ in both selection and means of application from those of Badoual and colleagues. It should be noted that *Badoual* et al. [7.43] used periodic ring formation as a basis for the entire morphology diagram, whereas *Østergård* and *Sørensen*'s [7.67] approach was directed at periodicity only.

In conclusion, modeling of the C region section of the morphology diagram is dogged by controversy. Some researchers assert that the same periodic growth mechanism governs both *P. mirabilis* and *B. subtilis* colonies ([7.65, 66]). In *P. mirabilis* colonies, migration/consolidation cycles are known to be independent of nutrient concentration. On the other hand, colonies of *B. subtilis* develop ring morphologies only at specific nutrient and agar concentrations [7.60, 63], and *Mimura* et al.'s [7.45] model regards periodic behavior as interplay between cell movement and nutrient dynamics [7.65]. Ring formations may also be induced by exposing colonies to cyclic environmental changes (Fig. 7.7).

7.6 Modeling Arrays and Aggregates

Tsimring et al. [7.68] considered that both waste products and chemotaxis contribute to pattern organization of the type described by *Budrene* and *Berg* [7.25]. Recall that, in these patterns, an advancing ring of motile cells deposits nonmotile spotted or striped aggregates in its wake. Differentiation into the nonmotile form is assumed to result from starvation. Aggregation is triggered by waste accumulation, which itself induces chemotactic signaling. Tsimring et al. found that, in order to mimic the complex structures reported by Budrene and Berg, a cutoff in chemotactic emission must be imposed. Mathematically, the cutoff intro-

duces a strong nonlinearity into the system, sensitizing the wake of the outer ring to tiny changes in waste and chemoattractant densities. Where chemoattractant emission is favored, spots or stripes form depending on whether the chemotactic response is high or low, respectively.

Rather than impose limits on chemoattractant levels, *Ben-Jacob* et al. [7.15] suggested that a hexagonal array of spots can be converted to a radial structure by adding a chemorepellent emitted by starved bacteria. They claim that chemorepulsive signaling is a biologically sound alternative.

Chemotactic signaling is also thought responsible for the swarming rings that develop when colonies of *B. subtilis* are exposed to ultraviolet radiation. *Nelson* and *Schnerb* [7.69] began work in this area by considering the effect of local ultraviolet (UV) irradiation and the migration of the colony as the UV source is moved; this model, as pointed out by *Zhang* et al. [7.70], does not apply to irradiation of an entire colony by UV. Exposure of a normally growing colony to uniform UV triggers a morphology change [7.71], in which the bacteria, as though in a futile rush to escape the threat, migrate to the colony edge. Arresting the growth of the bacteria (i. e., exposing the colony to the damaging radiation) causes production of chemoattractant in regions where toxic waste concentration is low enough to accommodate viable cells. Cells diffuse without filling the vacant areas with new cells (because growth has been halted), and thus become concentrated at the colony margin [7.71].

Recognizing that vastly different models produce the same or similar outputs, *Brenner* et al. [7.26] sought a minimal theory, that is, the simplest model based on known biological data that predicts the main features of experiment. They found that bacterial density, chemoattractant, and a triggering agent for the chemoattractant were sufficient. In addition, to maintain realistic aggregate density, some means of halting positive feedback between bacterial density and chemoattractant production is required; this is most likely provided by oxygen availability [7.26].

Brenner et al. considered that depletion of succinate, which is used to make chemoattractant, triggers migration of the swarm ring. This mechanism differs from that of *Tsimring* et al. [7.68], in which movement is triggered by competition between cell division and nutrient [7.26]. Brenner et al. argue that their model, which predicts a decrease in front velocity as substrate concentration is increased, and an increase in front velocity with increasing cell numbers, is far more in accord with experiment than *Tsimring* et al.'s [7.68] and similar models, which predict a front velocity independent of both factors.

An important consequence of Brenner et al.'s formulation is that aggregate formation is a result of nonlinear singularities in the equations for bacterial density and chemoattractant flux. Changes in diffusion constants and consumption rates should not affect the form of the solution. This is in contrast to the linear instabilities present in other models, which render the models sensitive to both of these factors, affecting both the stable thresholds and the pattern formation [7.26].

7.6.1 Aggregation in *Bacillus circulans* Colonies

Bacterial density triggers aggregate formation by colonies of *Bacillus circulans*. The trajectories of these aggregates as the cells inside them move, grow, and reproduce form so-called knotted branching patterns (KBPs), similar to those seen in *Paenibacillus vortex* [7.72]. Aggregate formation is associated with a rapid increase in colony expansion rate, as aggregates race across the agar surface. An aggregate is typically a few millimeters in radius.

Wakano et al. [7.73] considered this phenomenon as an analog of nucleation theory in physical chemistry. Support for the analogy comes from experimental knowledge of the chemical forces existing on bacterial cell surfaces, such as van der Waals attractive forces, repulsion between negatively charged cells, and the close-packing of aggregates [7.73]. In this scenario, small aggregates forming on hard agar surface cannot overcome the viscous drag, which traps them within the enveloping lubricant fluid. Thus, they remain locked within the parent envelope. As an aggregate grows, the flagellated cells move in unison, generating a propulsion force which overcomes the drag and allows the aggregate to move towards "greener pastures." Coordinated motion is presumably triggered by chemotactic signaling. The aggregate continues growing as it propels, encouraged by the higher nutrient availability, until it becomes large enough to be halted by friction. It continues to grow following motion cessation.

Wakano et al. report one major inconsistency between their model's output and experimental observation: that, experimentally, aggregate size is a function of initial nutrient concentration. They link this discrepancy to two-dimensional modeling of a nutrient field which is really three-dimensional (that is, nutrient is being replenished from the agar), then state that the three-dimensional process is left for future work.

On very soft agar, colonies of *B. circulans* penetrate the agar as well as traverse it [7.74]. In such a three-dimensional growth mode the bacteria are largely separated, and colony form is independent of nutrient and agar concentration. Growth undergoes a sharp transition to two-dimensional spreading across the agar as the agar concentration is increased, after which changes in substrate conditions do affect colony patterns [7.74]. It is noteworthy that, in two-dimensional mode, bacteria are tightly packed [7.74]. Bacterial density is therefore seemingly responsible not only for the formation of aggregates in this species, but also for the switch to two-dimensional spreading [7.74]. This supports the notion that chemotaxis is a primary driver of complex pattern formation in microbial colonies, since sparsely spaced cells are unlikely to communicate effectively.

7.7 Limitations of Bacterial Colony Growth Models

Numerous efforts to understand bacterial colony dynamics have yielded divergent results; For example, the nonlinear chemoattractive field proposed by *Woodward* et al. [7.28], and the triggering of chemoattractant by waste production [7.15, 46], produce similar output [7.26]. Researchers generally agree that chemotaxis is required for complex pattern formation by bacteria, but disagree with respect to factors such as nutrient diffusion and cell motility. A model recently developed by myself and colleagues suggests that, in the development of compact colonies, nutrient diffusion plays a primary role, while the effects of chemotaxis are minimal at best. Further discrepancies exist between the models of *Lacasta* et al. [7.64], *Czirók* et al. [7.60], *Mimura* et al. [7.45], and *Wakita* et al. [7.65] regarding the relative importance of nutrient density, the universality of the mechanisms governing periodicity in ringed colonies, and swimmer–swarmer interaction. We note that *Badoual* et al. [7.43] and *Arouh* [7.63]

have successfully generated *B. subtilis* patterns by coupling the essentials of *P. mirabilis* growth (periodicity triggered by time and/or swimmer/swarmer density thresholds) with nutrient density fields.

All of the models described here, without exception, assume that the evolution of bacterial density is correctly modeled by diffusion. Modern bacterial colony growth models further insist that the diffusivity term describing bacterial movement be nonlinear. As discussed earlier, setting $D(b)$ proportional to b^k to generate physically acceptable solutions in DLA is quite arbitrary, and may not accurately describe the motion of dense cell populations.

Another common trend in bacterial colony growth models is to assume that compact colonies with smooth interfaces evolve by the Fisher equation. The one-dimensional Fisher equation, originally formulated to describe the spread of advantageous genes through populations [7.76], takes the form

$$\frac{\partial u}{\partial t} = u(1-u) + \frac{\partial^2 u}{\partial x^2} ,$$

where the first term on the right represents the increase of a favorable mutant gene u at the expense of its parents, and the second term describes the diffusion of the gene (x and t are the spatial and temporal coordinates, respectively). The Fisher equation is applicable to any system that involves reaction–diffusion. However, when applied to the movement and growth of bacterial cells on agar, it predicts linear increase in colony radius with time, as first described by *Pirt* [7.9], and not challenged until *Pipe* [7.77] and independently *Panikov* et al. [7.78] found that the linear growth phase is a component of a more inclusive, single nonlinear growth phase that holds over a much longer period.

I would argue that the D region of the morphology diagram (occupied by compact colonies) has been unduly neglected in bacterial colony growth modeling. Apart from the dubious application of diffusion to a mass of cells which moves by growth alone, and the assumption that radial growth is linear in these colonies, it is doubtful that other factors, such as chemotaxis, play a primary role in compact growth (as is often as-

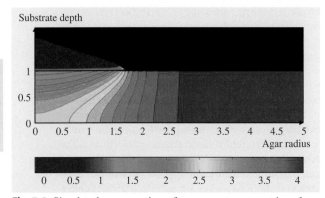

Fig. 7.8 Simulated cross-section of a compact, symmetric colony growing on substrate after 12 500 time steps. The *color bar* delineates nutrient density, the horizontal axis is radial distance, and the vertical axis is substrate depth (0 = bottom of dish; 1 = top). The colony is about 0.57 units high (simulations conducted in MATLAB 6.5). Growth is restricted to a narrow band at the colony base, as observed by *Reyrolle* and *Letellier* [7.75]. Although the vertical nutrient gradients are smaller than the horizontal gradients, they constitute a nutrient source term at each point on the substrate surface which is not considered in strictly two-dimensional models

sumed). On the other hand, nutrient diffusion in the z-direction, which is difficult to incorporate into two-dimensional growth models and which has hence been completely disregarded, is a likely driver of compact radial growth (Fig. 7.8). The height dynamics of evolving compact colonies have been similarly ignored. We have investigated vertical nutrient dynamics by assuming that compact colonies are symmetric around the z-axis. Thus, a planar cross-section of the colony plus agar dish is representative of the whole system. By exploring the parameter space of the model, it may be possible to predict vertical nutrient responses to surface nutrient levels. In the two-dimensional (2-D) colony plane, this response would constitute a nutrient source term at each grid point. In this way, one could account not only for nutrient consumption by the cells, but also for nutrient replenishment from the substrate.

Some of the above concerns have been echoed by *Müller* and *van Saarloos* [7.79]. They point out the lack of quantitative arguments supporting many mod-els and the questionable relevance of the arguments to nonmotile cells, and even that their relevance to motile cells is debatable. Although solutions to the simple reaction–diffusion equation with nonlinear diffusivity are unstable (i. e., can generate complex morphologies), the extent to which such models capture the true dynamics of bacterial colony growth remains unclear [7.79].

It is likely that no one model can capture the enormous diversity of natural bacterial colonies – fractals, dense-branching morphologies, spots, stripes, vortices, compact, chiral – all of which depend on complex and still poorly understood interactions and physical processes. On the other hand, all of the above studies have enhanced our understanding of the group behavior of cells in colonies, by providing biological insights or by encouraging debate and providing innovative ways of viewing colony dynamics. From this understanding has emerged a new, previously unsuspected role for bacterial colonies in the development of computational biology and robotics.

7.8 Recent Trends and Future Directions

Recent trends in bacterial colony growth modeling are towards chemotaxis as an optimization problem. Researchers in the 1990s became excited by the analogies between bacterial organization and that of nonliving systems; today's researchers perceive opportunities for developing biologically inspired machines and algorithms. Chemotaxis constitutes an optimization problem because it represents a compromise between resource exploitation (limited by local nutrient availability) and cost (energy expended in seeking new food sources [7.80]).

The bacterial foraging algorithm (BFA) likens bacteria to organisms that search cooperatively for resources. Its essence is that flagellated bacteria move by a series of runs and tumbles. During a run, flagella movement is tightly coordinated, while during tumbling it is desynchronized. Chemotaxis is distinguished from normal swim/tumble motility by an increase (repulsive chemotaxis) or decrease (attractive chemotaxis) in tumbling frequency [7.81]. In the BFA, bacteria seek to maximize their individual fitness via chemotaxis, reproduction, and dispersal [7.82]. Bacteria with large runs between tumbles are better equipped to explore the wider environment; those with small runs utilize their immediate surrounds, and thus attain the advantage in high-quality local environments. The original

BFA may preclude bacteria from finding their global optima, because of trapping by initial conditions or local optima [7.80]. An improved BFA allows the bacteria to forage with progressively shorter run times, such that they initially perform a coarse-scale assessment of their environment, before homing in on local optima [7.80]. Alternatively, the search space n may be split into $n/2$ subspaces, with the fittest bacterium from each subspace becoming the updated solution for that subspace [7.80].

Optimal foraging may be defined as maximum food gain for the least effort. Thus, the BFA can be applied to real-world optimization problems, such as minimizing identification errors during training of a direct current (DC) servo motor to a plant model [7.83]. Another bacterial-inspired algorithm, the bacterial colony growth algorithm (BCGA), aims to solve the localization problem in robotics [7.84]. How a robot senses and reacts to its environment is considered analogous to a bacterium making choices based on nutrient availability and population density. In essence, a robot assesses the suitability of its environment based on how closely its measurements match the group consensus. Both BFA and BCGA, however, focus on separate aspects of microbial cooperative behavior, and neither appreciates the rich set of capabilities exhibited by bacterial ag-

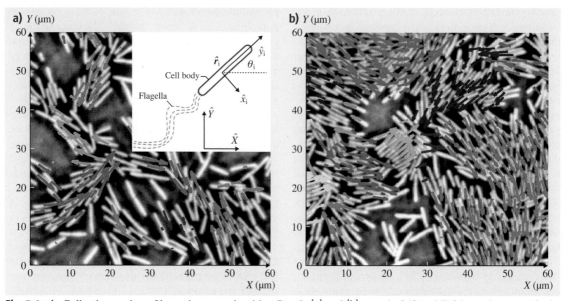

Fig. 7.9a,b Collective motion of bacteria at two densities. Panels (**a**) and (**b**) contain 343 and 718 bacteria, respectively. Velocity of a bacterium is indicated by the *length of an arrow*, and cohorts of moving cells are delineated by *color*. The *inset* in (**a**) shows the ith bacterium's local coordinates $(\hat{x}_i, \hat{y}_i)$ with respect to the laboratory coordinates $(\hat{X}, \hat{Y})$. The coordinate $\boldsymbol{r}_i$ represents the center of mass of the bacterium's reference frame. (After [7.86] with permission)

gregations. Similar sentiments have been expressed by *Xavier* et al. [7.5] with regard to BFA. Recently, an extended BFA which allows for varying populations, and which incorporates additional bacterial behaviors such as metabolism and quorum sensing, has been proposed [7.85]. The performance of the model is superior to that of conventional BFA, because quorum sensing naturally introduces population diversity and prevents local optima trappings in mixed-species populations [7.85].

Kerr et al. [7.87] used a different approach – that of game theory – to investigate bacterial ecology dynamics. They seeded a flask solution (homogeneous environment) and an agar plate (environment allowing spatial separation) with three different strains of *E. coli*. In the homogeneous environment, one strain outcompeted the others, but in the spatially discrete environment, all three strains survived. The authors suggested that interaction, dispersal, and competition generate spatiotemporal diversity within populations, provided they act over small spatial scales. The authors also addressed the issue of cost versus fitness (i. e., is it worth producing a toxin, which is expensive to make, if it reduces local competition), an issue which seems to have been neglected in conventional models of bacterial colony growth.

Grimm and *Railsback* [7.88] recognized that, to gain realistic insight into swarming and flocking behavior, one must investigate structural organization at different spatial scales and by varying multiple parameters; For instance, flocking in fish schools depends not only on nearest-neighbor density but also on how the fish are angled relative to each other (ibid). In both homogeneous and heterogeneous bacterial populations, *Ben-Jacob* and colleagues found that flocking can naturally lead to vortex formation [7.89]. Bacterial propulsion is moderated by friction with the substrate, attractive forces (representing physical interaction), and short-range repulsion (which prevents overcrowding). Given appropriate interaction parameters, a random mixture of two cell types settles into two homogeneous vortices [7.89].

Lack of experimental data has been a major impediment to properly understanding swarm behavior [7.86]. By comprehensive tracking of individual cell movement within *B. subtilis* colonies, *Zhang* et al. [7.86] demonstrated the existence of densely packed clusters whose collective movements are tightly aligned and distinct from those of other clusters (Fig. 7.9). Bacterial populations constrained to a surface, therefore, indeed exhibit the same swarming behaviors as do birds and fish, with the advantage that bacterial colonies are readily grown and controlled in the laboratory [7.86]. Bacterial colonies thus provide an excellent opportunity to understand collective movement in a wider context [7.86].

We are now in a position to investigate bacterial colony growth at a more intimate level. Some preliminary efforts are currently being made in this direction [7.5, 90]. *Xavier* et al. [7.5] envisaged a two-layer hierarchy of bacterial information processing: micro (the genetically controlled signal transduction intracellular network) and macro (cellular communication and environmental responses). *Schultz* et al. [7.90] went one step further. They modeled the complex signal transduction network that controls transitions between sporulation and competency in *B. subtilis*, with a view to linking these into the wider framework of colony organization. Actually, three hierarchical layers exist in bacterial colonies: microscale (gene switching, epigenetic memory inheritance, gene products), mesoscale (small-range chemical signaling, local organization), and macroscale (long-range signaling, pattern formation, maximizing the fitness of the colony as a whole). *Shklarsh* et al. [7.91] regarded the mesoscale in branching colonies as the synchronized movement of myriads of bacteria inside a single branch.

To model the system in its entirety, different modeling approaches might be required. A heuristic (i.e., agent-based) approach might be better suited to the modeling of individual behavior at the micro- and mesoscales, yet the rules should be chosen to be consistent with the macroscopic evolution of the colony, which may be governed by physical processes such as diffusion, surface tension, and internal forces. At the same time, one must not lose track of biological plausibility; For instance, fractal patterns are likely associated with nearest-neighbor communication and local sensing of the environment. In other environments, the bacteria fare better by collective communication.

To develop a complete understanding of colonial dynamics, we need to fully identify the physical and biological drivers of colony evolution. Researchers generally agree on the importance of chemotaxis, but what about factors such as nutrient diffusion, drag forces, lubricant production, and the range over which signaling occurs? Which of these exert primary influence? How do genetics affect the macroscopic development of the colony, and under what conditions do genetic effects become significant? For example, within the dynamic branches of *P. vortex* colonies, bacterial motion is rapid and highly coordinated [7.72], and occurs on a different time scale from that of cellular reproduction. To bring genetics into bacterial colony growth models, perhaps we should first consider compact colonies. In such colonies the genetic component is likely to dominate, as the immotile cells must adapt to their changing physical environment (frequently by sporulation). The experiments of *Shapiro* [7.19, 92] could provide an ideal testing ground for models linking genetics to the macroscopic via mesoscale differentiation and organization. Furthermore, dense three-dimensional colonies might inspire a different class of algorithms for multitasking machines and robots, in which tasks such as touching and locating could be assigned to different parts.

Finally, is there a role for epigenetics in biological complexity? It has been suggested that bacteria can collectively remember their past encounters with toxins such as antibiotics [7.91]. Links between epigenetic marks and memory are becoming increasingly recognized; For instance, in the malaria parasite, memory is retained long enough for the parasite to establish within the host, but gene expression switches before the host immune system can adjust [7.93]. In this way the parasite can assume long-term residence in the host and cause extensive damage. In colonies, epigenetic-controlled gene switching and memory might ensure that the greatest number of individuals survive for the longest time. This suggests that the convoluted mechanisms governing colony evolution are underpinned by a simple and fundamental principle, namely that of global optimization.

To conclude, the history of bacterial colony growth modeling spans four decades, and much has been learned since *Pirt* [7.9] first realized that cells growing in colonies must experience changing environments, which will lead to their spatiotemporal differentiation. When I first embarked upon the study of bacterial colonies, a common reaction was *Very interesting, but why should we care?* In this chapter, an attempt was made to show why we should care. Far from being a scientific curiosity, microbial colonies have taught us extensively about complex organization, optimization, and group behavior. By investigating these structures we can exploit and emulate them, but can we truly match what Nature has been doing for billions of years?

References

7.1 P. Hirsch: Microcolony formation and consortia. In: *Microbial Adhesion and Aggregation*, ed. by K.C. Marshall (Springer, Berlin, Heidelberg 1984) pp. 373–393

7.2 C. Friend-Norton: *Microbiology* (Addison-Wesley, Boston 1981)

7.3 M.H. Saier Jr: Bacterial diversity and the evolution of differentiation, Austr. Soc. Microbiol. News **66**(6), 337–343 (2000)

7.4 J. Casadesús, D. Low: Epigenetic gene regulation in the bacterial world, Microbiol. Mol. Biol. Rev. **70**(3), 830–856 (2006)

7.5 R.F. Xavier, N. Omar, L. Nunes de Castro: Bacterial colony: Information processing and computational behavior, 3rd World Congr. Nat. Biol. Inspired Comput. (2011) pp. 439–443

7.6 M. Abellana, J. Benedí, V. Sanchis, A.J. Ramos: Water activity and temperature effects on germination and growth of *Eurotium amstelodami, E. chevalieri* and *E. herbariorum* isolates from bakery products, J. Appl. Microbiol. **87**, 371–380 (1999)

7.7 L.V. Thomas, J.W.T. Wimpenny, G.C. Barker: Spatial interactions between subsurface bacterial colonies in a model system: A territory model describing the inhibition of *Listeria monocytogenes* by a nisin-producing lactic acid bacterium, Microbiology **143**, 2575–2582 (1997)

7.8 E.I. Newman, H.J. Bowen: Patterns of distribution of bacteria on root surfaces, Soil Biol. Biochem. **6**, 205–209 (1974)

7.9 S.J. Pirt: A kinetic study of the mode of growth of surface colonies of bacteria and fungi, J. Gen. Microbiol. **47**, 181–197 (1967)

7.10 J.W.T. Wimpenny: The growth and form of bacterial colonies, J. Gen. Microbiol. **114**, 483–486 (1979)

7.11 R.S. Kamath, H.R. Bungay: Growth of yeast colonies on solid media, J. Gen. Microbiol. **134**, 3061–3069 (1988)

7.12 I. Golding, Y. Kozlovsky, I. Cohen, E. Ben-Jacob: Studies of bacterial branching growth using reaction-diffusion models for colonial development, Physica A **260**, 510–554 (1998)

7.13 M. Ohgiwari, M. Matsushita, T. Matsuyama: Morphological changes in growth phenomena of bacterial colony patterns, J. Phys. Soc. Japan **61**(3), 816–822 (1992)

7.14 J. Wakita, K. Komatsu, A. Nakahara, T. Matsuyama, M. Matsushita: Experimental investigation on the validity of population dynamics approach to bacterial colony formation, J. Phys. Soc. Japan **63**(3), 1205–1211 (1994)

7.15 E. Ben-Jacob, I. Cohen, O. Shochet, I. Aranson, H. Levine, L. Tsimring: Complex bacterial patterns, Nature **373**, 566–567 (1995)

7.16 E. Ben-Jacob, I. Cohen, I. Golding, Y. Kozlovsky: Modeling branching and chiral patterning of lubricating bacteria, Proc. IMA Workshop Pattern Form. Morphogenet. (1998) (1999) pp. 1–50

7.17 R.A. Hegstrom, D.K. Kondepudi: The handedness of the universe, Sci. Amer. **Jan**, 108–115 (1990)

7.18 E. Ben-Jacob, O. Shochet, A. Tenenbaum, I. Cohen: Communication, regulation and control during complex patterning of bacterial colonies, Fractals **2**(1), 15–44 (1994)

7.19 J.A. Shapiro: The use of Mudlac transposons as tools for vital staining to visualise clonal and non-clonal patterns of organisation in bacterial growth on agar surfaces, J. Gen. Microbiol. **130**, 1169–1181 (1984)

7.20 H. Fujikawa, M. Matsushita: Fractal growth of *Bacillus subtilis* on agar, J. Phys. Soc. Japan **58**(11), 3875–3878 (1989)

7.21 L.Z. Pipe: unpublished observations

7.22 R. Bar-Ness, N. Avrahamy, T. Matsuyama, M. Rosenberg: Increased cell surface hydrophobicity of a *Serratia marcescens* NS 38 mutant lacking wetting activity, J. Bacteriol. **179**(9), 4361–4364 (1988)

7.23 N.H. Mendelson, A. Bourque, K. Wilkening, K.R. Anderson, J.C. Watkins: Organized cell swimming motions in Bacillus subtilis colonies: Patterns of short-lived whirls and jets, J. Bacteriol. **181**(2), 600–609 (1999)

7.24 E.O. Budrene, H.C. Berg: Complex patterns formed by motile cells of *Escherichia coli*, Nature **349**, 630–633 (1991)

7.25 E.O. Budrene, H.C. Berg: Dynamics of formation of symmetrical patterns by chemotactic bacteria, Nature **376**, 49–53 (1995)

7.26 M.P. Brenner, L.S. Levitov, E.O. Budrene: Physical mechanisms for chemotactic pattern formation by bacteria, Biophys. J. **74**(4), 1677–1693 (1998)

7.27 Y. Blat, M. Eisenbach: Tar-dependent and-independent pattern formation by *Salmonella typhimurium*, J. Bacteriol. **177**, 1683–1691 (1995)

7.28 D.E. Woodward, R. Tyson, M.R. Myerscough, J.D. Murray, E.O. Budrene, H.C. Berg: Spatio-temporal patterns generated by *Salmonella typhimurium*, Biophys. J. **66**, 2181–2189 (1995)

7.29 O. Rauprich, M. Matsushita, C.J. Weijer, F. Siegert, S.E. Esipov, J.A. Shapiro: Periodic phenomena in *Proteus mirabilis* swarm colony development, J. Bacteriol. **178**(22), 6525–6538 (1996)

7.30 L.M. Prescott, J.P. Harley, D.A. Klein: *Microbiology*, 3rd edn. (McGraw-Hill, Columbus 1996)

7.31 L.J. Shimkets: Social and developmental biology of the myxobacteria, Microbiol. Rev. **54**(4), 473–501 (1990)

7.32 R.E. Keen, J.D. Spain: *Computer Simulation in Biology – A BASIC Introduction* (Wiley-Liss, New York 1992)

7.33 J. Baranyi: Comparison of stochastic and deterministic concepts of bacterial lag, J. Theor. Biol. **192**, 403–408 (1998)

7.34 E. Bonabeau: Agent-based modeling: Methods and techniques for simulating human systems, Proc. Natl. Acad. Sci. USA **99**(Suppl. 3), 7280–7287 (2002)

7.35 C.M. Macal, M.J. North: Tutorial on agent-based modeling and simulation Part 2: How to model with agents, Proc. Winter Simul. Conf. 2006 (2006) pp. 73–83

7.36 J. Kreft, G. Booth, J.W.T. Wimpenny: BacSim, a simulator for individual-based modeling of bacterial colony growth, Microbiology **144**, 3275–3287 (1998)

7.37 E. Katzav, L.F. Cugliandolo: From coupled map lattices to the stochastic Kardar-Parisi-Zhang equation, Physica A **371**(1), 96–99 (2006)

7.38 C. Ormerod, N.S. Bordes, B.A. Pailthorpe: Characterising coupled map lattices (2006), available online at www.ms.unimelb.edu.au/~cormerod/pailthorpe.pdf

7.39 H. Sakaguchi, M. Ohtaki: A coupled map lattice for dendritic patterns, Physica A **272**, 300–313 (1999)

7.40 N. Ganguly, B.K. Sikdar, A. Deutsch, G. Canright, C. Pal Chaudhuri: *A survey on cellular automata*, Technical Report (Centre for High Performance Computing, Dresden University of Technology, Dresden 2003), available online at www.cs.unibo.it/bison/publications/CAsurvey.pdf

7.41 G. Cattaneo, P. Flocchini, G. Mauri, C. Quaranta Vogliotti, N. Santoro: Cellular automata in fuzzy backgrounds, Physica D **105**, 105–120 (1997)

7.42 E. Bettelheim, B. Lehmann: Microscopic simulation of reaction-diffusion processes and applications to population biology and product marketing, Annu. Rev. Comput. Phys. **7**, 311–339 (1999)

7.43 M. Badoual, P. Derbez, M. Aubert, B. Grammaticos: Simulating the migration and growth patterns of *Bacillus subtilis*, Physica A **338**, 549–559 (2009)

7.44 M. Matsushita, J. Wakita, H. Itoh, I. Ráfols, T. Matsuyama, H. Sakaguchi, M. Mimura: Interface growth and pattern formation in bacterial colonies, Physica A **249**, 517–524 (1998)

7.45 M. Mimura, H. Sakaguchi, M. Matsushita: Reaction-diffusion modeling of bacterial colony patterns, Physica A **282**, 283–303 (2000)

7.46 E. Ben-Jacob, I. Cohen, D.L. Gutnick: Cooperative organisation of bacterial colonies: From genotype to morphotype, Annu. Rev. Microbiol. **52**, 779–806 (1998)

7.47 M.-P. Zorzano, D. Hochberg, M.-T. Cuevas, J.-M. Gómez-Gómez: Reaction-diffusion model for pattern formation in *E. coli* swarming colonies with slime, Phys. Rev. E **71**, 031908 (2005)

7.48 T. Sams, K. Sneppen, M.H. Jensen, C. Ellegaard, B.E. Christensen, U. Thrane: Morphological instabilities in a growing yeast colony: Experiment and theory, Phys. Rev. Lett. **79**(2), 313–316 (1997)

7.49 B. Li, J. Wang, B. Wang, W. Liu, Z. Wu: Computer simulations of bacterial-colony formation, Europhys. Lett. **30**(4), 239–243 (1995)

7.50 M.Y.A. Azbel: Survival-extinction transition in bacteria growth, Europhys. Lett. **22**(4), 311–316 (1993)

7.51 J. Lega, T. Passot: Hydrodynamics of bacterial colonies: A model, Phys. Rev. E **67**, 031906 (2003)

7.52 K. Kawasaki, A. Mochizuku, M. Matsushita, T. Umeda, N. Shigesada: Modeling spatio-temporal patterns generated by *Bacillus subtilis*, J. Theor. Biol. **188**, 177–185 (1997)

7.53 S. Kitsunezaki: Interface dynamics for bacterial colony formation, J. Physical Soc. Japan **66**(5), 1544–1550 (1997)

7.54 T.A. Witten, L.M. Sander: Diffusion-limited aggregation, Phys. Rev. B **27**(9), 5686–5697 (1983)

7.55 L. Niemeyer, L. Pietronero, H.J. Wiesmann: Fractal dimension of dielectric breakdown, Phys. Rev. Lett. **52**, 1033–1036 (1984)

7.56 E. Brener, H. Levene, Y. Tu: Mean-field theory for diffusion-limited aggregation in low dimensions, Phys. Rev. Lett. **66**(15), 1978–1981 (1991)

7.57 M. Uwaha, Y. Saito: Aggregation growth in a gas of finite density: Velocity selection via fractal dimension of diffusion-limited aggregation, Phys. Rev. A **40**(8), 4716–4723 (1989)

7.58 Y. Tu, H. Levine, D. Ridgway: Morphology transitions in a mean-field model of diffusion-limited growth, Phys. Rev. Lett. **71**(23), 3838–3841 (1993)

7.59 P. Jensen, A.-L. Barabási, H. Larralde, S. Havlin, H.E. Stanley: Model incorporating deposition, diffusion, and aggregation in submonolayer nanostructures, Phys. Rev. E **50**(1), 618–621 (1994)

7.60 A. Czirók, M. Matsushita, T. Vicsek: Theory of periodic swarming of bacteria: Application to *Proteus mirabilis*, Phys. Rev. E **63**, 031915-1–031915-11 (2001)

7.61 H. Itoh, J. Wakita, K. Watanabe, T. Matsuyama, M. Matsushita: Periodic colony formation of bacteria due to their cell reproduction and movement, Prog. Theor. Phys. **139**, 139–151 (2000)

7.62 S.E. Esipov, J.A. Shapiro: Kinetic model of *Proteus mirabilis* swarm colony development, J. Math. Biol. **36**, 249–268 (1998)

7.63 S. Arouh: Analytic model for ring pattern formation by bacterial swarmers, Phys. Rev. E **63**, 031908 (2001)

7.64 A.M. Lacasta, I.R. Cantalapiedra, C.E. Auguet, A. Peñaranda, L. Ramírez-Piscina: Modeling of spatio-temporal patterns in bacterial colonies, Phys. Rev. E **59**(6), 7036–7041 (1999)

7.65 J. Wakita, H. Shimada, H. Itoh, T. Matsuyama, M. Matsushita: Periodic colony formation by bacterial species *Bacillus subtilis*, J. Phys. Soc. Japan **70**(3), 911–919 (2001)

7.66   A. Nishiyama, T. Tokihiro, M. Badoual, B. Grammaticos: Modeling the morphology of migrating bacterial colonies, Physica D **239**, 1573–1580 (2010)

7.67   J. Østergård, L. Sørensen: Modeling bacterial colony growth with cellular automata (2006), available online at ftp://ftp.diku.dk/diku/image/publications/ostergaard.sorensen.060925.pdf

7.68   L. Tsimring, H. Levine, I. Aranson, E. Ben-Jacob, I. Cohen, O. Shochet, W.N. Reynolds: Aggregation patterns in stressed bacteria, Phys. Rev. Lett. **75**(9), 1859–1862 (1995)

7.69   D.R. Nelson, N.M. Schnerb: Non-hermitian localisation and population biology, Phys. Rev. E **58**(2), 1383–1403 (1998)

7.70   S. Zhang, L. Zhang, R. Liang, E. Zhang, Y. Liu, S. Zhao: Lubricating bacteria model for the growth of bacterial colonies exposed to ultraviolet radiation, Phys. Rev. E **72**, 051913 (2005)

7.71   A.M. Delprato, A. Samadani, A. Kudrolli, L.S. Tsimring: Swarming ring patterns in bacterial colonies exposed to ultraviolet radiation, Phys. Rev. Lett. **87**(15), 158102-1–158102-4 (2001)

7.72   E. Ben-Jacob: Learning from bacteria about natural information processing, Nat. Genet. Eng. Nat. Genome Ed. Ann. Acad. Sci. N. Y. **1178**, 78–90 (2009)

7.73   J.Y. Wakano, S. Maenosono, A. Komoto, N. Eiha, Y. Yamaguchi: Self-organised pattern formation of a bacteria colony modeled by a reaction-diffusion system and nucleation theory, Phys. Rev. E **90**, 258102 (2003)

7.74   N. Eiha, A. Komoto, S. Maenosono, J.Y. Wakano, K. Yamamoto, Y. Yamaguchi: The mode transition of the bacterial colony, Physica A **313**, 609–624 (2002)

7.75   J. Reyrolle, F. Letellier: Autoradiographic study of the localization and evolution of growth zones in bacterial colonies, J. Gen. Microbiol. **111**, 399–406 (1979)

7.76   R.A. Fisher: The wave of advance of advantageous genes, Ann. Eugen. **7**, 353–369 (1937)

7.77   L.Z. Pipe: The influence of temperature on the form and development of bacterial colonies, N. Z. Microbiol. **6**(2), 28–32 (2000)

7.78   N.S. Panikov, S.E. Belova, A.G. Dorofeev: Nonlinearity in the growth of bacterial colonies: Conditions and causes, Microbiology **71**(1), 50–56 (2002)

7.79   J. Müller, W. van Sarloos: Morphological instability and dynamics of fronts in bacterial growth models with nonlinear diffusion, Phys. Rev. E **65**, 061111 (2002)

7.80   H. Chen, Y. Zhu, K. Hu: Cooperative bacterial foraging optimization, Discrete Dynam. Nat. Soc. **2009**, 815247 (2009)

7.81   Y. Shi, T. Duke: Cooperative model of bacterial sensing, Phys. Rev. E **58**(5), 6399–6406 (1998)

7.82   Y. Liu, M. Passino: Biomimicry of social foraging bacteria for distributed optimization: Models, principles, and emergent behaviors, J. Optim. Theory Appl. **115**(3), 603–628 (2002)

7.83   B. Bhushan, M. Singh: Adaptive control of DC motor using bacterial foraging algorithm, Appl. Soft Comput. **11**, 4913–4920 (2011)

7.84   A. Gasparri, M. Prosperi: A bacterial colony growth algorithm for mobile robot localization, Auton. Robots **24**, 349–364 (2008)

7.85   M.S. Li, T.Y. Ji, W.J. Tang, Q.H. Wu, J.R. Saunders: Bacterial foraging algorithm with varying population, Biosystems **100**, 185–197 (2010)

7.86   H.P. Zhang, A. Be'er, E.-L. Florin, H.L. Swinney: Collective motion and density fluctuations in bacterial colonies, Proc. Natl. Acad. Sci. USA **107**(31), 13626–13630 (2010)

7.87   B. Kerr, M.A. Riley, M.W. Feldman, B.J. Bohannan: Local dispersal promotes biodiversity in a real-life game of rock-paper-scissors, Nature **418**, 171–174 (2002)

7.88   V. Grimm, S.F. Railsback: Agent-based models in ecology: Patterns and alternative theories of adaptive behavior,. In: *Agent-Based Computational Modeling*, ed. by F.C. Billari, T. Fent, A. Prskawetz, J. Scheffran (Physica, Heidelberg 2006) pp. 139–152, (2006)

7.89   H. Levine, E. Ben-Jacob, I. Cohen, W.-J. Rappel: Swarming patterns in microorganisms: Some new modeling results, Proc. 45th IEEE Conf. Decis. Control (2006) pp. 5073–5077

7.90   D. Schultz, P.G. Wolynes, E. Ben-Jacob, J.N. Onuchic: Deciding fate in adverse times: Sporulation and competence in *Bacillus subtilis*, Proc. Natl. Acad. Sci. USA **106**(50), 21027–21034 (2009)

7.91   A. Shklarsh, O. Kalishman, C.J. Ingham, E. Ben-Jacob: Bacteria self-organization and swarming intelligence, Lecture Presented at the 8th Agents Multi-agent Syst., AAMAS (2008)

7.92   J.A. Shapiro: Organisation of developing Escherichia coli colonies viewed by scanning electron microscopy, J. Bacteriol. **169**(1), 142–156 (1987)

7.93   J.J. Lopez-Rubio, A.M. Gontijo, M.C. Nunes, N. Issar, R.H. Rivas, A. Scherf: 5' flanking region of var genes nucleate histone modification patterns linked to phenotypic inheritance of virulence traits in malaria parasites, Mol. Microbiol. **66**(6), 1296–1305 (2007)

Part B
Molecula

Part B Molecular Biology, Genome and Proteome Informatics

Ed. by Chris Brown

8. Exploring the Interactions and Structural Organization of Genomes

Grant H. Jacobs

Bioinformatics typically treats genomes as linear DNA sequences, with features annotated upon them. In the nucleus, genomes are arranged in space to form three-dimensional structures at several levels. The three-dimensional organization of a genome contributes to its activity, affecting the accessibility and regulation of its genes, reflecting the particular cell type and the epigenetic state of the cell.

The majority of the cell cycle occurs during interphase. During metaphase and meiosis, chromosomes are highly condensed. By contrast, interphase chromosomes are difficult to visualize by direct microscopy. Several attempts have been made to understand the nature of metaphase chromosomes and genome structures. Approaches to indirectly derive the spatial proximity of portions of a genome have been devised and applied (Fig. 8.1, Table 8.1).

This chapter reviews these approaches briefly and examines early methods used to investigate the structure of a genome from these data. This research involves taking experimental data, processing them with variants of existing bioinformatic DNA sequence analyses, then analyzing the proximity data derived using biophysical approaches. This chapter emphasizes the background to the biological science and the latter, biophysics-oriented, analyses. The processing of the genomic data is outlined only briefly, as these approaches draw on established bioinformatic methods covered elsewhere in this Handbook. The main focus is on the methods used to derive three-dimen-

sional (3-D) structural information from the interaction data.

8.1 The Structure of Chromosomes, Genomes, and Nuclei

For readers new to the biology of the cell nucleus, this section gives a very brief overview of current knowledge of genome structure. This description best relates to animal genomes; each of the major classes of

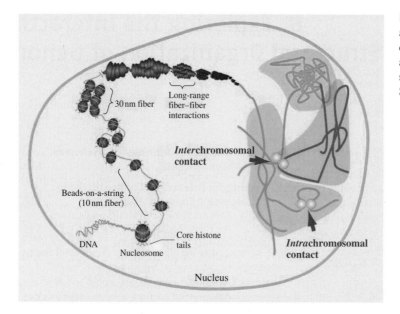

Fig. 8.1 Chromosomes are organized at a variety levels, several of which can bring portions of the genome that are sequentially separated into close spatial proximity, as introduced in Sect. 8.1 (after *Fraser* et al. [8.1])

life have some differences that are not examined here. Those working on computational methods in this area are strongly encouraged to develop a full and detailed understanding of the structure of DNA within the eukaryotic nucleus. This is a substantial topic with several textbooks devoted to it.

The description below roughly works in order from the smallest structural elements to the largest, as outlined in Fig. 8.1, aiming to note key terms and structural organizations that can serve as starter keywords for researchers venturing into this arena to explore further.

DNA bases can be decorated with a range of covalently bonded additions. A considerable number of these extra or alternative DNA bases are known, with their distribution differing in different classes of life. While many covalent modifications of DNA bases have been identified, most are rare. The best known base modification, and the dominant modification found in eukaryotes, is methylation of cytosine (5-hydroxy-methylcytosine), typically in CpG steps.

These can denote epigenetic states. DNA methylation (or other modifications) can impede the binding of DNA-binding proteins, or serve as recruiting points for protein–DNA complexes and through this define the state of the gene, accessible to be used or not. (Note how this differs from the classical control of the rate of transcription of a gene.)

Little of the DNA in a cell is naked. The most consistently exposed DNA in a genome is near the start of a gene and the immediate promoter; the majority of

the remainder of the DNA in the nucleus is bound into nucleosome–DNA complexes. Coding regions (both introns and exons) have nucleosomes that are displaced and reassembled as the DNA passes the transcriptional machinery.

Nucleosomes are made from an octamer of histone proteins (composed of two of each of the H2A, H2B, H3, and H4 proteins) around which approximately 147 base pairs of DNA is wrapped, taking approximately 1 and 7/8ths turns around the histone octamer. A so-called linker, typically 80 base pairs and usually bound by histone H1 or its isoforms, spans each bound octamer. The typical linker length varies between different species.

Histones have a compact core with extended, disordered N- and C-terminal tails that protrude from the histone octamer core of nucleosomes. These exposed tails are covalently modified in what has been termed the histone code.

Different histone modifications have been associated with different transcriptional states, in particular if chromatin containing the gene is open or closed. In active genes, the chromatin is open (euchromatin) with the nucleosome-bound DNA forming a relatively extended structure. In closed chromatin (heterochromatin), typically located near the periphery of the nucleus, nucleosomes are packed against one another to condense unused portions of the genome. A number of models of higher-order structures – so-called nucleosomal array models – have been proposed. The detailed models proposed for these higher-order structures are

Exploring the Interactions and Structural Organization of Genomes | 8.1 The Structure of Chromosomes, Genomes, and Nuclei   117

Part B | 8.1

Table 8.1 Outline of the experimental protocols (Fig. 8.1). An outline of the steps taken to prepare sequence data representing regions of spatial proximity. This table should be read with Fig. 8.1 and Sect. 8.2. Some of the methods can also be assessed using (so-called) gene-array profiling; for simplicity only the sequencing options are presented here

	ChIP-PET	3C	4C	5C	Hi-C
Cross-link the complexes	Cross-link chromatin using formaldehyde				
	Sonicate to shear DNA	Cleave DNA with restriction enzyme (e.g., *Hin*dIII)			
	Immunoprecipitate complexes				Fill in sticky ends with biotin-labeled DNA
Ligate	Ligate, using conditions that favor ligation (joining) of DNA ends from the same cross-linked complex, not between different cross-linked complexes.				
Prepare for sequencing, and purify DNA	Linearize DNA			Circularize DNA, LMA	Sonciate to shear DNA; purify using streptavidin beads
Sequencing	Direct sequencing	PCR	Inverse PCR	Sequence *copies*	End-pair sequencing

somewhat controversial, as is the extent to which they are well defined in vivo. These arrays of nucleosomes are considered to form the extended fiber, 30 nm fiber, and higher-order packing of DNA, such as that seen in densely packed chromatin.

In interphase, the active or growth phase of the cell cycle, the packing of chromatin varies along the length of the chromosomes, reflecting the accessibility of each portion of the chromosome to gene expression.

The genome is anchored to a number of substrates in the periphery of the nucleus (e.g., DNA–lamina interactions), the nuclear matrix, and complexes associated with the genome. Insulator or boundary elements (two separable but related activities) are among the attachment points.

Loops within chromosomes form, determined by DNA methylation, histone modification and the binding proteins that organize the genome into higher-order structural units. These loops affect how regulatory factors act on the genes, for example, depending on if the regulatory element is in the same DNA loop as the gene it might regulate, or not.

Related to this is the well-established observation that many regulatory sites are tens, or even hundreds, of thousands of bases away from the genes they regulate. While they might be distant in linear sequence, they may be in close proximity in space. Thus, we can think of linear (along the genome) and spatial (through space) distances with respect to the genome. The methods described in this chapter identify pairs (or more) of loci that have large linear but short spatial distances.

At a higher level, chromosome territories define large portions of a chromosome that tend to occupy one region of the volume of the nucleus. (That is, regions of chromosomes tend to occupy their own space in the nucleus rather than mixing or tangling with other chromosomes or other parts of the same chromosome.)

In addition to these layers, or hierarchy, of structure are specialized regions of the genome, such as telomeric regions, ribosomal DNA (e.g., nucleolus), and a number of nuclear bodies. Each of these regions have specialized genome structures.

These observations have been informed, in part, by the techniques reviewed in this chapter. A key question is how the three-dimensional organization of the genome relates to control of gene expression. Related questions are how genome structures define what genes are made accessible to be used in a particular cell type and how or if alterations in these structures impact on disease.

Careful interpretation of the interaction data is important. One point to remember as you read about these methods is that the structure of a genome is dynamic, changing over time, varying with environmental conditions and cell type. Another is that, while the regions of chromatin between complexes bound to a matrix or lamina or other substrate may have structural elements and structural properties and be (relatively) well defined, the loop regions may not have any particularly well-defined structure.

The remainder of this chapter proceeds by outlining the main experimental techniques used, briefly examining the data analysis requirements of these methods (Table 8.1), then exploring the methods used to derive spatial information from the proximity data. It closes by briefly mentioning complementary methods and thoughts for the future.

An important complementary field to this endeavor is (theoretical) polymer modeling, which aims to build models of chromatin structure from an understanding of polymer physics. While this strongly impacts on what is presented here, this is not covered. Readers approaching this field would benefit from gaining an understanding of the polymer science and biophysical aspects, as this field is a translational field, moving from linear sequence-based genomics to a three-dimensional, biophysical genome (or four-dimensional if time is considered).

8.2 Testing for Interacting Chromatin Using Chromosome Conformation Capture

When applied to a modest number of loci, chromosome conformation capture (3C or CCC) can identify interacting loci. Initially developed in 2002 [8.2], a range of variants have since been developed aiming to allow high-throughput processing (e.g., 5C [8.3] or Hi-C [8.4]) or relating chromatin-associated proteins to chromosome conformation (e.g., 4C [8.5] or ChIP-PET (chromatin immunoprecipitation using paired-end tags)) (Fig. 8.2). Some sources point to the work of *Cullen* et al. [8.6] as an early precursor of these methods, with its proximity ligation concept.

This section outlines the conceptual basis of these methods. Experimental details are left for the interested reader to pursue via the references supplied.

These methods do not determine a structure, as such, but identify interactions between different portions of the genome through DNA-bound proteins. With these in hand, one can infer what modeled structures might be consistent with the interactions observed.

A common feature of these techniques of note for a computational biologist or bioinformatician is that, the final 3-D analysis aside, the data are familiar DNA sequencing products. This means that much of the initial data processing has similarities to other (high-throughput) sequence analysis projects.

For all of the methods described, the first step is to cross-link the DNA-bound proteins. Usually formaldehyde-based cross-linking is used. This yields a portion of DNA, its bound proteins from one portion of the genome cross-linked to a portion of DNA, and its bound proteins from another portion of the genome. The different portions may lie on the same chromosome or different chromosomes. The cross-linked complex may capture more than two protein–DNA complexes.

Next, excess DNA from around the complexed protein–DNA structure is removed. For the 3C method, restriction endonucleases (REs) are used. REs of different cutting frequency can be chosen. REs with smaller DNA recognition sites will typically cut the genome into more pieces of smaller size. Alternatively, sonication can be used, as in the ChIP-PET and Hi-C protocols. (Sonication is ultrasound vibra-

tion of the mixture aimed at physically fracturing the DNA.)

The free ends of the DNA strands from the complexes are then ligated. Conditions used are chosen to favor ligation within each protein–DNA complex, rather than joining different protein–DNA complexes together (e.g., by using highly dilute solutions of the ligation agent). This favoring of ligating complexes held together by protein–DNA interactions, over those that are not, is key to the approach.

The cross-links are reversed (broken), protein removed, DNA extracted and purified, and the DNA ligation products quantified (e.g., by polymerase chain reaction (PCR)) to measure the frequency at which the different interactions occur.

Which sequencing and quantitation methods can be applied depends on how the DNA is cut and ligated. A number of different approaches have been developed with a view to automating the sequencing of the joins (if possible) to allow the material to be sequenced en masse or profiled against a gene array. The cutting method also affects the number of complexes found and the extent to which complexes of weaker interactions are retained in the sample, as discussed below.

Adding chromatin immunoprecipitation (ChIP) biases the sample to contain a particular protein by using an antibody for the particular protein to select complexes with the protein.

Figure 8.1 shows the key steps discussed above, using as an illustration a complex of only a pair of DNA fragments. In practice, complexes can be of several protein–DNA complexes cross-linked, with more complex DNA products to be considered. To the left are the key steps; arrayed along the bottom are the different methods.

Estimates of the typical distance between loci is derived from the argument that the frequency of ligation is approximately inversely proportional to the typical distance of separation of the ligated elements.

The different techniques have different merits.

Because 3C DNA sequences from specific cut ends, for any one sequencing reaction it can only

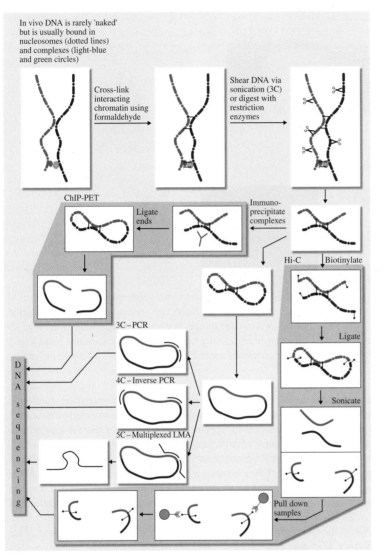

In vivo DNA is rarely 'naked' but is usually bound in nucleosomes (dotted lines) and complexes (light-blue and green circles)

Cross-link interacting chromatin using formaldehyde

Shear DNA via sonication (3C) or digest with restriction enzymes

ChIP-PET

Ligate ends

Immuno-precipitate complexes

Hi-C | Biotinylate

3C – PCR

4C – Inverse PCR

5C – Multiplexed LMA

Ligate

Sonicate

Pull down samples

DNA sequencing

Fig. 8.2 A schematic outline of the *conceptual* differences in the different chromosome conformation capture (CCC) methods published at the time of writing. (Further differences occur throughout each method.) All CCC methods rely on cross-linking spatially adjacent chromatin, breaking the linearly adjacent DNA (either through sonication or use of restriction enzymes), ligating the ends formed, then sequencing across these ends or screening against gene arrays if desired and appropriate. ChIP-PET uses sonication to fragment the cross-linked DNA, then immunoprecipitation to extract the cross-linked material, before ligating the ends, isolating and purifying the DNA from chromatin, then sequencing it. The immunoprecipitation selects from the sample particular proteins of interest, thus the method is able to investigate the interactions formed by particular proteins. Hi-C uses restriction enzymes (*Hin*dIII) to cleave the cross-linked DNA. The sticky ends are filled in with biotinylated bases, then ligated. After fragmenting the DNA using sonication, the ligated ends are isolated using using streptavidin-coated beads and the DNA is sequenced. A particular value of this approach is that it scales well. The 3C, 4C, and 5C methods mainly differ in setting up different later stages, enabling high-throughput analysis. Note in particular that, for 5C, the DNA sequenced is ligation-mediated annealed (LMA) copy. As 5C uses standardized primers, this scales better than 3C or 4C. (These standardized primer sites are depicted in the figure as tails on the LMA copy)

investigate one particular ligation. 3C is also limited to intramolecular interactions, i.e., within one chromosome, whereas the variants can also examine interactions between chromosomes (needed for whole-genome studies).

4C ensures the ligated ends form circular products. Any one end is circularized with all the other ends in the complex, resulting in a mixture of circular products. As a result, one sequencing reaction can investigate all the ligation products of a given end.

5C, with its carbon-copy step that adds universal sequencing primers, can capture all combinations of the ligated ends into the sequencing mixture, with the result

that all (many) pairs of ligated ends can be sequenced from one sequencing mixture.

In addition, to add selection for a particular protein in the complexes examined, the ChIP-based variants can, in principle, be more sensitive, too.

Below, each of the methods are individually examined.

8.2.1 Chromosome Conformation Capture for Case Studies: 3C

Conceptually, the idea behind 3C is straightforward. Chromosomal DNA has many proteins associated with

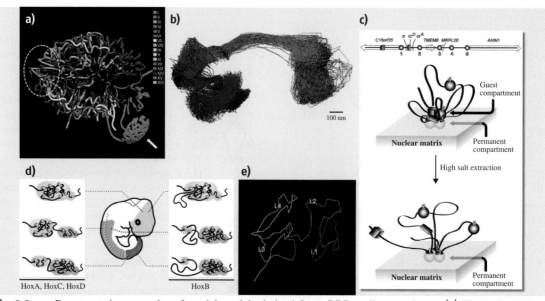

Fig. 8.3a–e Representative examples of spatial models derived from CCC studies are shown. (**a**) Three-dimensional (3-D) model of the yeast genome (after Fig. 5a [8.7]). (**b**) 3-D model of the human α-globin locus (after Fig. 2b [8.8]). (**c**) Structure around the TMEM8 gene of chicken erythroid cells under two different salt conditions (after Fig. 5 [8.9]). (**d**) Schematic of the structures of the mouse Hox gene clusters (after Fig. 3 [8.10]). (**e**) 3-D model of human chromosome 14 (after Fig. 10 [8.11])

it, in particular nucleosomes. Formaldehyde (methanal, CH_2O) is a small, reactive and diffusible chemical that can cross-link nearby protein (and DNA) amino groups through formation of a $-CH_2-$ linkage. Once cross-linked, the DNA in the protein–DNA complexes is cleaved using restriction enzymes with known specificity. The cleaved ends of the DNA in the protein–DNA complexes are then ligated (joined) under dilute conditions that favor ligation of DNA ends from the same protein–DNA complex. The ligated DNA is isolated, amplified (e.g., via PCR), and sequenced. The sequences obtained are mapped onto a reference genome sequence to identify the two portions of the genome that the formaldehyde-based reaction cross-linked. The frequency with which a particular pair of sequences is found to be cross-linked is considered to reflect their proximity in space in the nucleus. (In the opinion of the author, this founding assumption would benefit from closer examination.) From this, spatial models of (a region of) the genome are constructed that satisfy the proximity measures derived (Fig. 8.3).

A key limitation of the original 3C protocol is that each annealed sequence pair must be individually amplified, through PCR, using specific primer pairs from the target sequence and potential interacting sequences.

(Because this targets specific sequences, the method is applied to previously known or suspected sites, rather than as a screen to locate interacting sites.) Furthermore, each amplification needs to be individually controlled. As a consequence this approach is not practicable for studies involving a large number of sites. Typically it is used for regions up to a several hundred kilobases in size to explore the finer structure of particular model genes or genetic loci.

Another issue is that the high level of local noise means that results are of low resolution. The number of nonspecific interactions is related to the inverse square of the linear distance separating two loci. As a consequence, putative interactions between sites of less than some distance (e.g., 100 kb) are typically discounted.

The 3C approach has been applied to a number of individual genes, revealing the looping structure of the gene under different methylation states or cell types.

8.2.2 Finding All Interactions with a Single Locus: 4C

Four laboratories independently developed variants of what are now grouped under the label 4C, circular chromosome conformation capture (circular 3C: *Zhao* et al. [8.5]; 3C-on-chip: *Simonis* et al. [8.12]; open-ended 3C: *Wurtele* and *Chartrand* [8.13]; olfactory receptor 3C: *Lomvardas* et al. [8.14]).

This approach determines the interactions formed by a single site or locus. The 3C products are circularized (e.g., using a frequent-cutting restriction enzyme followed by ligation), selectively amplified using inverse PCR, then passed on to microarray analysis or (deep) sequencing. Inverse PCR uses only a single primer, rather than two, allowing amplification then sequencing of all of the sequences present on the other side of the ligation from a known sequence (unique to the chosen site being studied).

8.2.3 Towards High-Throughput Chromosome Conformation Capture: 5C

A high-throughput variant of 3C, this method enables larger-scale screening of many potential interactions in parallel (see [8.15] for an introduction). The key innovation is that, once the 3C ligation step is complete, probes with standardized PCR primers (T7 and its complement T3) are hybridized to the ligated region. This enables quantitative capture of all the individual ligations within DNA fragments with standardized primers at either end.

This process, known as ligation-mediated amplification (LMA), creates a copy of the original 3C library that is based on standard primers, allowing many genome interactions to be surveyed at once in parallel. Because the copied library is amplified, the products can be examined using (so-called) deep sequencing or microarray analysis. Furthermore, all pairs of ligation products can be examined to generate a matrix of all interacting loci and their frequencies without many rounds of analysis.

Deep sequencing may be preferred over gene arrays, as it is more capable of handling very large datasets. (The complexity of the datasets increases exponentially with the number of primers.)

The automation and large scale of this approach make it useful to look at much larger portions of chromosomes (e.g., Mbps in length) or explore 3-D structure or complex interactions (e.g., promoter–enhancer interactions [8.16]).

8.2.4 Adding Paired-End Tag Sequencing: 6C

6C and some other variants, e.g., ChIP-loop (also called ChIP-3C) and ChIA-PET (chromatin interaction analysis using paired-end tag sequencing, e4C), combine 3C with chromatin immunoprecipitation (ChIP) through first enriching the chromatin for the protein of interest before ligation. ChIP is a popular way of identifying the location of protein–DNA complexes, genome-wide, annotating the linear sequence of the genome with the positions of the protein–DNA interactions identified. Through applying chromosome conformation capture to ChIP-seq libraries, the long-range interactions with which the protein–DNA complexes are associated may be investigated.

Fullwood and *Ruan* [8.17] provide good coverage of the issues associated with these approaches. In their discussion, they recommend sonication of the cross-linked product, rather than use of restriction enzymes (REs), arguing that 3C and 4C methods are noisy (capturing many nonspecific chromatin interactions) and that sonication shakes apart weak or nonspecific interactions, reducing the noise in the data. (It would be interesting to see if computational methods that combine both levels of interactions can be put to good effect.) Fraser et al. argue that addition of ChIP reduces the noise further and makes the results specific for the protein of interest.

8.2.5 High-Throughput Chromosome Conformation Capture: Hi-C

Used to examine whole genome structure, this method labels the ligated regions of 3C constructs with biotin before shearing the DNA, then capturing the biotin-labeled fragments using streptavidin beads. The captured DNA is then sequenced. The biotin is incorporated by choosing restriction enzymes that leave a $5'$ overhang that is filled in using nucleotides that include a biotinylated nucleotide. The blunt-ended fragments are ligated.

This approach collects only the ligated products (i. e., those labeled with biotin) for sequencing. Furthermore, shearing the DNA does not restrict the interactions to DNA regions containing the sites of the restriction enzymes used in other variants.

The particular value of this method is that it can be applied to large datasets, i. e., whole genomes.

8.3 Processing Chromosome Conformation Capture Sequence Data

As the focus of this chapter is the methods to construct models of three-dimensional structures of genomes, this section is limited to lightly covering conceptual issues of the main steps typically involved in a CCC-based study, raising a few of the key issues and pointing to some of the software developed to address them. Excellent coverage of the public tools for working with 3C data can be found in *Fraser* et al. [8.15] (Table 8.2).

It is useful to first appreciate the limitations of the experimental elements that impact on the computational work. As one example, *Fullwood* and *Ruan* [8.17] point out that the use of restriction enzymes (versus sonication) is not truly genome-wide, as it is biased to where the RE sites are in the genome. Also, there is the issue of eliminating oversampling of the same sequence by removal of nonunique sequences, as the methods in use rely on multiple overlapping unique signals to correct oversampling. (Likewise for clustering methods.)

In the case of RE-based digestion of the isolated cross-linked DNA, a first step is to choose which REs to use. RE digest patterns are readily generated using a wide variety of software. The choice of REs balances the frequency of cutting, fragment size, evenness of cutting across the studied (portion of the) genome, where they cut (especially with respect to low-complexity regions), etc. Advice (e.g., Fraser et al.) suggests that is that it is useful at this point to create files of all annotation data, e.g., locations of genes and other features, for reference through the project.

Next is primer design. In the case of the 5C method, with its carbon-copy step, the 5C forward and reverse primers need to be designed. The Dostie lab offer software via a WWW interface for this purpose (Table 8.2). One typical concern is avoiding primers that are homologous to low-complexity regions within the genome (region) under investigation. Another is to determine the uniqueness of the primers (e.g., via BLAST searches or a hash lookup). For other CCC methods, except those

Table 8.2 Software for analysis of chromosome conformation capture data. Software for tasks specifically related to chromosome conformation capture data are listed. Not listed is software for tasks related to the wider scenario of high-throughput sequence analysis, which is considered elsewhere in this Handbook. Currently, some of the software available is only described in the methods sections of experimental chromosome conformation capture research literature rather than in separate publications. Listed are those that can be obtained outside of this context, i.e., are readily available through conventional means on the Internet

Package/Website	URL	Contents	References
Genome3D	http://genomebioinfo.musc.edu/Genome3D/Index.html	Visualization of data using three-dimensional models. Able to integrate data from different levels of resolution	[8.18]
My5C server	http://3dg.umassmed.edu/welcome/welcome.php http://3dg.umassmed.edu/my5Cprimers/5C.php http://3dg.umassmed.edu/my5Cuploads/upload.php http://3dg.umassmed.edu/my5Cheatmap/heatmap.php	A collection of packages from the Dekker laboratory presented online, including: my5C.primers, my5C.uploads, my5C.heatmap	[8.16]
3C and 5C projects from the Dostie laboratory	http://dostielab.biochem.mcgill.ca/	Includes standalone versions of 3Cprimer, 5Cprimer, the 5C Program collection [8.1] (including 5C arrayBuilder, 5C3D and Microcosm), and MCMC5C	[8.1, 11]
The integrative modeling platform	http://www.integrativemodeling.org/	Although intended for molecular interactions, some groups are using this software for genome structures as it can accept the interaction data generated by chromosome conformation capture experiments	[8.8, 19]
Circos	http://circos.ca/	Not genome structure software, but a general-purpose visualization tool that can be used to represent genomes as a circle with relationships between loci (e.g., interactions) illustrated as lines between the loci. Similar or related tools include *Circoletto* and *Gremlin*	[8.20–22]
3PD	http://www.pristionchus.org/3CPrimerDesign/	Web form-based server offering primer design for 3C experiments	[8.23]

Exploring the Interactions and Structural Organization of Genomes | 8.4 Calculating Genomic Loci Proximity or Interaction Data   123

Part B | 8.4

using sonication to break the DNA, similar needs for primer design have to be addressed.

If a gene array is to be used to (semi)quantitate the frequency of the interaction events, this will need to be designed. One approach, used in the Dostie lab's 5CArrayBuilder program, is to determine a series of probes of increasing length, where the shorter length is used to assess the background signal when calculating interaction frequencies.

Once the raw sequencing or array data are available, interaction frequencies for each set of interacting loci are derived. This considers background noise and controls; For example, using different length probes for each loci, all probe interaction frequencies that are too close to the background probe interaction frequencies can be discarded and the useful values averaged. Suitable quality control steps might (and should) be undertaken, e.g., serial dilution and PCR amplification of a sample of the probes.

As one can see, with the exception of the determination of the interaction frequencies, these steps are variants on existing molecular-genetics problems and largely do not introduce any new concepts. Further details on these types of tasks can be found elsewhere in this Handbook.

During the course of writing this chapter, *Yaffe* and *Tanay* presented an examination of factors biasing the initial data in Hi-C experiments, offering an (in their words) integrated probabilistic model for analyzing this data [8.24]. They draw attention in particular to

ligation products that are likely to have arisen from non-specific cleavage sites rather than restriction fragment ends, the length of restriction fragments (with respect to ligation efficiency), the nucleotide composition of the genome (chromosome) being investigated, and issues with uniquely mapping the interactions back onto the reference genome (chromosome) sequence.

For 5C data, the My5C server [8.16] offers primer design facilities, in several different ways. (See Table 8.2 for related software and access to them.) Given the large genomic regions covered, primer design needs to be automated.

My5C also provides online generation of heatmaps of interaction data, which can be manually examined for particular interactions. Comparisons of two heatmaps is provided, using difference, ratio, or log ratio comparisons. Smoothing functions can be applied to examine larger interaction patterns, e.g., averaging over a sliding window. Similarly, sliding window plots of the data can be examined.

My5C interacts with the UCSC genome browser, in the sense that data can be moved between the two, with genome annotation data placed on the My5C data and custom tracks prepared to be presented in the UCSC browser alongside genome data.

Although My5C offers to generate pairwise interaction data that Cytoscape (used to visualize complex networks) can present, this treats these data as static, whereas in practice they are averages over different cells and potentially different chromatin conformations.

8.4 Calculating Genomic Loci Proximity or Interaction Data

8.4.1 General Observations

The previous section dealt with processing steps that have much in common with other high-throughout genomic studies, with familiar issues of primer design, PCR amplification, sequencing error, etc.

These data yield collections of loci which have been identified as (putative) sites of interactions. It is assumed that the frequency with which a particular loci pair is observed reflects the relative proximity or opportunity to interact of that loci pair compared with other loci pairs in the dataset. (Note that proximity and opportunity to interact are not synonyms but alternative ways of viewing interaction frequencies. Consider that two loci can be spatially close, but not interact because they are constrained; by contrast, two loci that

are typically well separated spatially might occasionally interact through large-scale movements.)

As the data are from a population of cells, and accepting that chromosomes are (perhaps highly) flexible structures, one may not be able to, or perhaps cannot, derive a meaningful single structure from these interaction data.

Related to this is that the methods deriving structural models from interaction data are probabilistic and yield an ensemble of possible models, rather than a single model. This is familiar ground to those working in molecular modeling and other probabilistic areas (e.g., phylogenetics), but may be new to those working on DNA sequence data. In some senses, the resulting models are perhaps best considered to represent structural properties, rather than structures per se.

Interactions with very low interaction frequencies have very few constraints in modeling. These points should not be overinterpreted in the final models. It may be possible to infer that they are likely to be distant from the other points, provided false-negative rates are low.

Models derived may reflect cell growth conditions (or cell type). In additional to computational and experimental protocol repeatability, there is biological repeatability to consider (*Fraser* et al. report that they are investigating this issue [8.15]). This requires that all environmental factors, reagents, and cell conditions be standardized (e.g., the point on the growth curve, synchronizing cell growth, etc.).

8.4.2 My5C, 5C3D, and Microcosm (*Dostie* et al.)

Dostie and colleagues have developed software to identify chromatin conformation signatures in 5C data [8.1], which they have applied to their work on the HoxA gene cluster. Their overall approach is to generate a series of randomized starting conformations, which they minimize against the root-mean-square (RMS) deviation of their interloci data, being the inverse of the interaction frequencies. The computational tools are presented online on the My5C website (Table 8.2).

Fraser et al. [8.15] describe the Dostie group's program 5C3D, which performs a gradient descent minimization approach. Euclidian distances are set to be the inverse of the interaction frequencies. Starting models are a chain set on N points randomly distributed within a cube. Gradient descent is applied in a conventional way to minimize the overall difference in the distance matrix and the Euclidian distances until convergence is achieved.

While this approach is perhaps applicable to smaller models with few interaction points, and will be very fast to execute, it is likely to be too naïve to tackle larger-scale models or genome-wide modeling. In particular, gradient descent used alone is well known to become trapped away from the global minima if local minima are present. Having said this, this general approach may be of use to range-find complex models using a reduced set of interactions as the method will be fast.

Their approach concludes with a separate inference of the best fitting model(s) using their *Microcosm* program. (Their description is not especially clear, but this appears to be a best-fit procedure by inspecting if spheres of appropriate size placed around each interac-

tion locus capture most of the data points of collections of data selected at random from the original interaction frequency data, under the distributions associated with the interaction frequency of each locus. The size of the spheres chosen appears to be arbitrary in that is it set by the user by manual experiment.)

The local density of interactions is represented and plotted. This approach can also be used to present a comparison of two models as a graph with assessment by deriving p-values for the differences (being the probability of incorrectly predicting a difference, assuming normally distributed differences).

8.4.3 Modeling the Yeast Genome

A model of the yeast genome [8.7] is based on a simple polymer model where each 130 bp of DNA is set to occupy 1 nm with chromosomes treated as a string with 10 kb of these 1 nm units assigned to each RE fragment from the experiment. The model was then constructed by minimizing these 10 kb fragments against the distances derived from the experiments.

This mixes something that might otherwise be close to a pure experimental approach with one that draws on past work on chromatin structure models and polymer theory. Given the noisiness of the data over shorter linear distances, it seems worth exploring incorporation of some model for structure over the shorter distances; further studies on appropriate models for the shorter distances will be useful.

8.4.4 Using the Integrative Modeling Platform

Baù and *Marti-Renom* [8.25] suggest adopting a modeling platform (IMP) intended for protein assemblies (Table 8.2), illustrating their example by examining the 3-D structure of the α-globin locus, which corresponds to the Enm008 ENCODE region.

Before describing this work, it should be noted that the concepts used are based on those used for molecular simulations, e.g., of proteins. These will be new and different concept for those whose work resolves solely around DNA sequences.

The IMP software allows users to develop a representation of spatial data and a scoring scheme, and then generates models fitting the criteria of the model, as well as providing some analysis facilities.

Initial data are normalized to generate an interaction matrix. Nominally each RE site is where, or close to where, the interactions occur. *Baù* and *Marti-Renom*

[8.25] represented the α-globin locus as a polymer of 70 particles, one for each *Hind*III RE fragment, using a sphere with an exclusion volume proportional to the size of the particle. This was modeled in part on a canonical 30 nm fiber, with length to base-pair ratio of 0.01 nm/base. (There is some controversy over the extent to which 30 nm fiber is present in vivo.)

This step is a common theme in these modeling efforts: how to simplify the initial model (i. e., every base pair) in a way that reduces the complexity of the modeling without incorrectly representing the data so as to render the model void. Compared with molecular modeling, one may think of these fragments as being the residue, or repeated unit, used for the modeling.

Calculating the energy (cost, favorability) of a model draws on polymer simulation norms, with neighbor and nonneighbor interaction properties and restraints being defined. (One can consider these restraints to be springs attached to a point, restraining the residue if it is pulled away from the point.)

Neighboring (adjacent) residues were constrained to have an approximate distance between them, to lie within what was considered a reasonable range of the distribution of interaction distances derived from the interaction frequencies (i. e., upper- and lower-bound harmonic restraints were applied to the dis-

tances of neighboring residues based on the interaction data).

Neighbors with no interaction data were set to remain bound to their neighboring residue through an upper-bound harmonic constraint.

Nonneighboring residues were defined to be at least a minimum distance apart, using a lower-bound harmonic (essentially defining a sphere of exclusion that defines the volume occupied by each residue).

Modeling seeks to minimize the sum of the violations of the individual restraints using simulated annealing, with a subset of positions in each Monte Carlo step moved in probabilistic proportion to the objective function score of the model before and after the move, given the temperature of the system. (Higher temperatures allow greater dynamic movement.) Optimization used 500 Monte Carlo steps with 5 steps of local optimization (minimization) from 50 000 random starting models, yielding 50 000 potential models of the α-globin locus.

While more detailed, and perhaps involving more labor, this approach builds on many years of developments in molecular modeling (but see the discussion in the following section for possible caveats). Discussion of the analysis and interpretation of the ensembles generated can be found in the next section.

8.5 Interpreting the Data

An issue common to these methods is that they identify interactions observed in a (large) population of cells. It is an average picture that is obtained, not data specific to any one cell or one cell type within that population.

By contrast, microscopy studies including methods such as FISH (fluorescence in situ hybridization) that use single cells, may only represent a subset of the interactions of that one cell through time but are able to provide absolute measures of distances – as opposed to inferred probabilities of proximity from population studies – and some idea of the true frequency with which (subsets of) the interactions occur. Subsets are important in that some subgroups of interactions may work as a unit in the presence or absence of specific transcription factors.

Another way of looking at this is to compare with what is used for protein nuclear magnetic resonance (NMR) data, which similarly uses distance restraints to derive structural models. A key difference is that secondary structural elements in proteins are relatively

stable. While the detailed interactions with a particular side-chain might differ in different individual proteins in the sample, the overall relationship of the secondary structural elements, or fold, of the protein will be relatively constant (breathing motions excepted). Furthermore, these stable structures show many interactions that in effect triangulate the positions of the structural features, leading to relatively robust structural models.

By contrast, the chromatin structure equivalent of protein secondary structural elements – the different nucleosome array conformations – may be too flexible to sustain stable structures in the way that protein secondary structures do. If so, there may be no real way to compute an overall structure for the genome, as it would be a constantly moving target within any one cell, and/or varying from cell to cell. However, one could still examine *properties* of the structure.

Likewise, you might identify the tethers, or anchoring points, of loop structures that are defined by well-bound protein complexes and hence stable as a complex forming the base of a loop, even if the body of the loop itself does not form a stable structure (and hence interactions that the portions of the loop make with the rest of the genome vary constantly over time).

Thus, although it is tempting to compare this problem with protein modeling (e.g., using NMR restraints), as, for example, *Marti-Renom* and *Mirny* [8.26] do, caution is needed not to take such comparison too far.

As these methods generate ensembles of models, the collection of models generated need to be examined as a dataset itself. A simple strategy, such as that taken in *Baù* et al. [8.8], is to take the better models generated (the better 10 000 of the 50 000 in total in their case) and apply cluster analysis to identify common themes in the models generated. In these authors' case, a Markov cluster algorithm was used based on rigid-body superimposition, but other approaches might be adopted.

Their 10 000 models yielded 393 clusters, the largest two clusters having 483 and 314 models, respectively, revealing a dumbbell-shaped structure with two relatively globular domains spanned by an extended domain (Fig. 8.3b).

This author has some concern over the very large number of clusters reported. This would suggest either that the cluster criteria are too fine, splitting models into clusters that might otherwise resemble a larger group, or that the modeling process was unable to yield consistent models. One possible avenue that future groups might explore is to use structure comparisons that allow for movement, i. e., non-rigid-body comparison. Another approach would be to identify structural motifs or domains that can be found within the simulation models. (For example, without examining Baù and Marti-Renom's data first-hand, it is difficult to know if the actual results are that there are two relatively globular domains spanned by an extended segment, but that the conformation within each globular domain taken over all the models is not well defined. Using a mixture of examining domains and non-rigid-body comparisons might resolve this.)

Considerable effort has been expended on methods such as this for protein structure comparison over many years; it would be worth those working in the genome structure field examining the possible ways of comparing the ensembles of structures more closely.

Different subclasses of solutions are not in themselves errors or the result of a lack of constraints – they may indicate different structure in different transcription states, or different subtypes of cells in the population studied.

These concerns indicate a need to examine carefully if the data are adequately explored in the simulations (e.g., the degree to which the constraints have been tested). Independent verification by experimental data may assist, e.g., FISH experiments.

Sanyal et al. [8.27] point out that current methods, despite being able to collect data over whole genomes, are perhaps limited to domains of several Mb, as the larger structures are expected to be dynamic in nature and vary between cells. Related to this concern is that, to model whole genomes computationally, very large numbers of long-range interactions would be required to yield stable (meaningful) structural models.

8.6 Future Prospects

While an exciting area, there is clearly a lot to do.

Annotation of the interaction data and three-dimensional aspects of genome structure will be required. *Reed* et al. [8.28] briefly touch on the then upcoming need for this. This may require tools beyond the current approach of annotating genomes using linear tracks.

Visualization tools are needed to present the interactions and the resulting three-dimensional models. Heat maps and circular plots (Table 8.2) offer static presentations of interaction data. While giving overviews of the datasets, these static plots have limited capacity to represent the scale and complexity of the data, which will likely involve careful examination of patterns and subgroups within the datasets. Thus, we can expect further development of tools to aid examination of these datasets.

An initial tool to visualize genome models in three dimensions is Genome3D [8.18]; we can expect further development in this direction.

The data analysis itself will further evolve.

With larger datasets, higher resolution might be possible provided that background noise does not prove problematic. Optimistically, it may be possible to detect if 30 nm fiber-like structures are present in vivo, although if there are too many short-range random collisions, this nonspecific background noise may make detection of smaller-scale regularities such as a 30 nm structure difficult or impossible [8.29]. Developing methods that cope with noisy data might be one useful avenue for exploration. Related to this may be further examination of what controls might be applied to these data.

Another avenue might be approaches to compare different datasets, say from different cell types or stages of the cell cycle.

Better support for, and understanding of, the main underlying premise – that the frequency of interactions between two loci is inversely proportional to their proximity in space – would be useful. As far as the author is aware, at the time of writing, few imaging studies are available to offer support for this premise (e.g., *Lieberman-Aiden* et al. [8.4] and *Miele* et al. [8.30]).

The lack of (high-resolution) true-positive *truth* sets against which these computational methods might be compared is a concern. While one can test for self-consistency, there needs to be some way to assess the accuracy of the models generated and what can be inferred from the interaction data.

To conclude, an important overriding conclusion is that biophysics matters.

While you can argue about the specifics [8.31], it should be clear to anyone perusing this field that this work involves a shift to a physical genome: physical both in the sense of dealing with a physical substrate (rather than abstract information) and in the sense of physics.

References

8.1   J. Fraser, M. Rousseau, S. Shenker, M.A. Ferraiuolo, Y. Hayashizaki, M. Blanchette, J. Dostie: Chromatin conformation signatures of cellular differentiation, Genome Biol. **10**, R37 (2009)

8.2   J. Dekker, K. Rippe, M. Dekker, N. Kleckner: Capturing chromosome conformation, Science **295**, 1306–1311 (2002)

8.3   J. Dostie, T.A. Richmond, R.A. Arnaout, R.R. Selzer, W.L. Lee, T.A. Honan, E.D. Rubio, A. Krumm, J. Lamb, C. Nusbaum, R.D. Green, J. Dekker: Chromosome conformation capture carbon copy (5C): A massively parallel solution for mapping interactions between genomic elements, Genome Res. **16**, 1299–1309 (2006)

8.4   E. Lieberman-Aiden, N.L. van Berkum, L. Williams, M. Imakaev, T. Ragoczy, A. Telling, I. Amit, B.R. Lajoie, P.J. Sabo, M.O. Dorschner, R. Sandstrom, B. Bernstein, M.A. Bender, M. Groudine, A. Gnirke, J. Stamatoyannopoulos, L.A. Mirny, E.S. Lander, J. Dekker: Comprehensive mapping of long-range interactions reveals folding principles of the human genome, Science **326**, 289–293 (2009)

8.5   Z. Zhao, G. Tavoosidana, M. Sjölinder, A. Göndör, P. Mariano, S. Wang, C. Kanduri, M. Lezcano, K.S. Sandhu, U. Singh, V. Pant, V. Tiwari, S. Kurukuti, R. Ohlsson: Circular chromosome conformation capture (4C) uncovers extensive networks of epigenetically regulated intra- and interchromosomal interactions, Nat. Genet. **38**, 1341–1347 (2006)

8.6   K.E. Cullen, M.P. Kladde, M.A. Seyfred: Interaction between transcription regulatory regions of prolactin chromatin, Science **261**, 203–206 (1993)

8.7   Z. Duan, M. Andronescu, K. Schutz, S. McIlwain, Y. Kim, C. Lee, J. Shendure, S. Fields, C.A. Blau, W.S. Noble: A three-dimensional model of the yeast genome, Nature **465**, 363–367 (2010)

8.8   D. Baù, A. Sanyal, B.R. Lajoie, E. Capriotti, M. Byron, J.B. Lawrence, J. Dekker, M.A. Marti-Renom: The three-dimensional folding of the α-globin gene domain reveals formation of chromatin globules, Nat. Struct. Mol. Biol. **18**, 107–114 (2011)

8.9   A.A. Gavrilov, I.S. Zukher, E.S. Philonenko, S.V. Razin, O.V. Iarovaia: Mapping of the nuclear matrix-bound chromatin hubs by a new M3C experimental procedure, Nucleic Acids Res. **38**, 8051–8060 (2010)

8.10   D. Noordermeer, M. Leleu, E. Splinter, J. Rougemont, W. De Laat, D. Duboule: The dynamic architecture of hox gene clusters, Science **334**, 222–225 (2011)

8.11   M. Rousseau, J. Fraser, M.A. Ferraiuolo, J. Dostie, M. Blanchette: Three-dimensional modeling of chromatin structure from interaction frequency data using Markov chain Monte Carlo sampling, BMC Bioinformatics **12**, 414 (2011)

8.12   M. Simonis, P. Klous, E. Splinter, Y. Moshkin, R. Willemsen, E. de Wit, B. van Steensel, W. de Laat: Nuclear organization of active and inactive chromatin domains uncovered by chromosome conformation capture-on-chip (4C), Nat. Genet. **38**, 1348–1354 (2006)

8.13   H. Wurtele, P. Chartrand: Genome-wide scanning of HoxB1-associated loci in mouse ES cells using and open-ended chromosome conformation capture methodology, Chromosome Res. **14**, 477–495 (2006)

8.14   S. Lomvardas, G. Barnea, D.J. Pisapia, M. Mendelsohn, J. Kirkland, R. Axel: Interchromosomal interactions and olfactory receptor choice, Cell **126**, 248–250 (2006)

8.15   J. Fraser, M. Rousseau, M. Blanchette, J. Dostie: Computing Chromosome Conformation. In: *Computational Biology of Transcription Factor Binding*, Methods in Molecular Biology, Vol. 674, ed. by I. Ladunga (Humana, Totowa 2010) pp. 251–268

8.16   B.R. Lajoie, L.N. van Berkum, A. Sanyal, J. Dekker: My5C: Web tools for chromosome conformation capture studies, Nat. Methods **6**, 690–691 (2009)

8.17   M.J. Fullwood, Y.J. Ruan: ChIP-Based methods for the identification of long-range chromatin interactions, Cell. Biochem. **107**, 30–39 (2009)

8.18   T.M. Asbury, M. Mitman, J. Tang, W.J. Zheng: A viewer-model framework for integrating and visualizing multi-scale epigenomic information within a three-dimensional genome, BMC Bioinformatics **11**, 444 (2010)

8.19   D. Russel, K. Lasker, B. Webb, J. Velázquez-Muriel, E. Tjioe, D. Schneidman-Duhovny, B. Peterson, A. Sali: Putting the pieces together: Integrative modeling platform software for structure determination of macromolecular assemblies, PLoS Biology **10**(1), e1001244 (2012)

8.20   N. Darzentas: Circoletto: Visualizing sequence similarity with Circos, Bioinformatics **26**, 2620–2621 (2010)

8.21   M. Krzywinski, J. Schein, I. Birol, J. Connors, R. Gascoyne, D. Horsman, S.J. Jones, M.A. Marra: Circos: An information aesthetic for comparative genomics, Genome Res. **19**, 1639–1645 (2009)

8.22   T.M. O'Brien, A.M. Ritz, B.J. Raphael, D.H. Laidlaw: Gremlin: An interactive visualization model for analyzing genomic rearrangements, IEEE Trans. Vis. Comput. Graph. **16**, 918–926 (2010)

8.23   S. Fröhler, C. Dieterich: 3PD: Rapid design of optimal primers for chromosome conformation capture assays, BMC Bioinformatics **10**, 635 (2009)

8.24   E. Yaffe, A. Tanay: Probabilistic modeling of Hi-C contact maps eliminates systematic biases to characterize global chromosomal architecture, Nat. Genet. **43**, 1059–1063 (2011)

8.25   D. Baù, M.A. Marti-Renom: Structural determination of genomic domains by satisfaction of spatial restraints, Chromosome Res. **19**, 25–35 (2011)

8.26   M.A. Marti-Renom, L.A. Mirny: Bridging the resolution gap in structural modeling of 3D genome organization, PLoS Comput. Biol. **7**, 2125 (2011)

8.27   A. Sanyal, D. Baù, M.A. Martí-Renom, J. Dekker: Chromatin globules: A common motif of higher order chromosome structure?, Curr. Opin. Cell Biol. **23**, 325–331 (2011)

8.28   J.L. Reed, I. Famili, I. Thiele, B.O. Palsson: Towards multidimensional genome annotation, Nat. Rev. Genet. **7**, 130–141 (2006)

8.29   P.J. Shaw: Mapping chromatin conformation, F1000, Biology Rep. **2**, 18 (2010)

8.30   A. Miele, J. Dekker: Mapping cis- and transchromatin interaction networks using chromosome conformation capture (3C). In: *The Nucleus. Volume 2: Chromatin, Transcription, Envelope, 105 Proteins, Dynamics, and Imaging*, Methods in Molecular Biology, Vol. 468, ed. by R. Hancock (Humana, Totowa 2008)

8.31   J. Langowski: Chromosome conformation by crosslinking: Polymer physics matters, Nucleus **1**, 37–39 (2010)

9. Detecting MicroRNA Signatures Using Gene Expression Analysis

Stijn van Dongen, Anton J. Enright

Small RNAs such as microRNAs (miRNAs) have been shown to play important roles in genetic regulation of plants and animals. In particular, the miRNAs of animals are capable of downregulating large numbers of genes by binding to and repressing target genes. Although large numbers of miRNAs have been cloned and sequenced, methods for analyzing their targets are far from perfect. Methods exist that can predict the likely binding sites of miRNAs in target transcripts using sequence alignment, thermodynamics or machine learning approaches. It has been widely illustrated that such de novo computational approaches suffer from high false-positive and false-negative error rates. In particular these approaches do not take into account expression information regarding the miRNA or its target transcript. In this chapter we describe the use of miRNA seed enrichment analysis approaches to this problem. In cases where gene or protein expression data are available, it is possible to detect the signature of miRNA binding events by looking for enrichment of microRNA seed binding motifs in sorted gene lists. In this chapter we introduce the concept of miRNA target analysis, the background to motif enrichment analysis, and a number of programs designed for this purpose. We focus on the Sylamer algorithm for miRNA seed enrichment analysis and its applications for miRNA target discovery with examples from real biological datasets.

Complete genome sequencing has revolutionized biology and our understanding of the structure and function of the genome. The scale and scope of noncoding RNA transcription was surprising to many, and a number of classes of functional noncoding RNAs have since been characterized. There are likely many more classes, as yet, undiscovered. One class of small noncoding RNA is the microRNA (miRNA). These short (19–22 nt) molecules were first identified in *Caenorhabditis elegans* and subsequently found to be highly conserved among other animals. Plants too possess miRNAs, although they function differently from animal miRNAs. These small molecules bind to target messenger RNAs (mRNAs) through complementarity between their sequence and the sequence of their target. This binding causes a number of significant effects, including translational inhibition and transcript destabilization through de-adenylation and subsequent

decapping. A single miRNA can regulate the expression of hundreds or thousands of target transcripts, and they have been shown to have significant roles in tissue identity, development, response to stress, pluripotency, and a number of other areas of biology. Indeed, already a large number of miRNAs have been associated with diseases. A major goal of miRNA research is to accurately predict the likely target sites of known miRNAs. A number of purely computational methods were initially developed, but these methods have high false-positive and false-negative detection rates. More recently, approaches have been developed to detect miRNA binding from gene expression or proteomics datasets. This chapter discusses detection of miRNA binding sites by combining computational prediction and statistical analysis with gene expression data to accurately predict miRNA targets using seed enrichment analysis.

9.1 Background

Currently there are 1424 confirmed miRNAs in human [9.1]. Initially miRNAs were discovered through cloning and capillary sequencing. The advent of next-generation sequencing accelerated the discovery of small RNAs. One expects miRNAs to have multiple targets. However, few have been experimentally confirmed (1094 so far according to the TarBase webserver [9.2]). The ideal solution would likely be an accurate high-throughput experimental approach that can directly confirm binding between miRNA and target transcript. A number of approaches look promising (e.g., highthroughput sequencing with crosslinking and immunoprecipitation (HiTS-CLIP) and photoactivatable ribunocleoside enhanced-CLIP (par-CLIP) [9.3, 4]), however such approaches are still extremely challenging and require significant effort and expense. Hence, it is vital that computational techniques be developed to unravel the regulatory effects of miRNAs and to explore their implications for health and disease.

Prediction of miRNA targets has been ongoing ever since the 3′ untranslated regions (UTRs) of transcripts were determined to contain their complementary binding sites [9.5]. The efficacy of computational approaches to locate and rank potential genomic binding sites was initially supported by the apparent high degree of miRNA complementarity to experimentally determined binding sites. Despite the later identification of hundreds of miRNAs in a variety of species through large-scale sequencing projects [9.1], only a handful of targets were

experimentally identified for an even smaller number of miRNAs [9.2]. Given the laborious nature of experimental target validation, it became imperative that computational approaches be developed. A number of different approaches for de novo computational miRNA target prediction were developed over the years, including miRanda, TargetScan, PicTar [9.6–10], and many others [9.11]. However, there are a number of challenges for purely computational prediction to overcome.

9.1.1 Target Prediction Basics

Initially, researchers determined the target transcript for an miRNA through experiment (usually a reporter 3′ UTR construct attached to green fluorescent protein (GFP) or luciferase), then identified potential sites by manually searching the target transcript for matching locations, and finally confirmed the site through site-directed mutagenesis or other techniques. When the first few target sites had been identified for miRNAs, such as let-7 and lin-4 [9.12, 13], it was obvious that miRNAs had relatively clearly defined patterns of complementarity to the 3′ UTRs of their target transcripts [9.12]. These first few targets detected for *C. elegans* had enough similar features that it became apparent that computational techniques might be able to allow their discovery in silico. However, computational analysis is hampered by a number of significant issues described below.

9.1.2 miRNA Size

The apparent complementarity between miRNA and target could have been seen as an advantage for computational analysis. However, other features of miRNA–UTR association make matters more complicated. The usual sequence alignment algorithms assume longer sequences than the 19–22 nt of miRNAs. This short length makes ranking and scoring of targets very difficult, as statistical techniques for sequence matching (such as Karlin–Altschul statistics [9.14]) require longer sequences. Binding sites actually consist of regions of complementarity, bulges, and mismatches [9.7, 15, 16]. As standard sequence analysis tools were designed for sequences with longer stretches of matches and less gaps, they are much less useful for miRNA target prediction. Nucleotides 2–9 of miRNAs, the so-called *seed region*, have been described as a key specificity determinant of binding and require perfect complementarity [9.16] (Fig. 9.1). If one ignores GC content and performs a simple order-of-magnitude calculation, then a perfect match for a six-nucleotide seed region of an miRNA should occur roughly once every 4 kb in a genome, in other words on average approximately once in every three human 3′ UTRs. However, it would not seem realistic for a single miRNA to regulate more than a few hundred targets. Effective regulation of transcript translation requires that miRNAs and their targets be located in the same cellular compartment. Hence, most of these theoretical targets correspond to false positives, as the spatial and temporal context of miRNA and mRNA target are not considered.

9.1.3 Identification of 3′ UTRs

In order to identify miRNA targets in a given species, acquiring an accurate set of 3′ UTRs for this species is vital. Despite the accumulation of genome sequences for many species, the location, extent, or splice variation of 3′ UTRs is still poorly characterized for many organisms. Some species-specific projects, such as the Berkeley *Drosophila* Genome Project (BDGP), produce high-quality transcript information that enables accurate determination of a 3′ UTR from a stop codon to a polyadenylation site. The Ensembl database [9.17] uses alignment of complementary DNAs (cDNAs) and expressed sequence tags (ESTs) to genomic sequences and also human curation to identify 3′ UTR regions where evidence is available for human, mouse, and zebrafish genomes. However, for many sequenced genomes (especially those of low coverage), definitive 3′ UTR boundaries are not described. One can estimate these regions by taking downstream flanking sequence to the stop codon corresponding to the length of an average 3′ UTR from that species. Experimental techniques such as tiling arrays [9.18], ditag/cap analysis gene expression (CAGE) tagging, and RNA sequencing are promising approaches toward the generation of high-quality 3′ UTR datasets and have already dramatically improved the quality of 3′ UTR sequences for a number of model organisms. The availability of reliably annotated and verified 3′ UTR datasets will potentially benefit target prediction more than small incremental improvements to existing computational prediction methods.

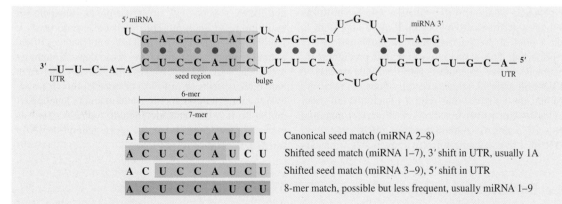

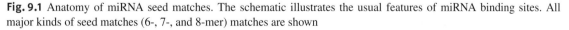

Fig. 9.1 Anatomy of miRNA seed matches. The schematic illustrates the usual features of miRNA binding sites. All major kinds of seed matches (6-, 7-, and 8-mer) matches are shown

9.1.4 Conservation Analysis

Solutions to reduce the number of false positives in target predictions include filtering out those binding sites that do not appear to be conserved across species. Predicted binding sites conserved across orthologous 3′ UTRs in multiple species are considered more likely [9.6]. However, recently evolved miRNAs, such as miR-430 in zebrafish (recently and significantly expanded), have large numbers of nonconserved targets [9.19] in the scope of the currently available set of fish genomes. One caveat of conservation analysis is the set of species that are compared. Conservation of targets between human and chimp is hardly significant given that at least 99% of the entire transcript is conserved [9.20] on average. Other species might seem more relevant to compare with human transcripts (e.g., mouse, rat, dog, etc.), but the fact is that sequenced genomes are not evenly sampled from the phylogenetic tree. As a result, the number of *false positives* (targets predicted which turn out to be incorrect) can effectively be greatly reduced, but this will be done at the expense of increased *false negatives* (true valid targets which are not successfully detected by a prediction method).

9.1.5 False Positives and False Negatives

Typically when performing large-scale prediction of targets across a whole genome, one wants to ensure a high degree of specificity (few false positives) over sensitivity (few false negatives) to ensure fewer predictions of better quality. Many of the published algorithms and released databases choose such an approach. However, for an individual researcher interested in a single gene or pathway, wanting to investigate a role for miRNAs in their system, sensitivity becomes more important so that a pool of predicted targets for individual testing is preferred. The major issue affecting prediction inaccuracies is one of context. Genome-wide prediction of possible miRNA targets typically does not take into account the temporal and spatial expression of either the miRNA or its predicted target. A perfect match to the seed region of a brain-specific miRNA in the 3′ UTR of a liver-specific mRNA transcript is irrelevant if the miRNA and target are never expected to be in the same place at the same time. Transcripts within the liver will evolve or avoid target sites for miRNAs they have the potential to encounter, but are free to contain possible matches to miRNAs they will never encounter [9.21]. Hence, taking into account the expression context of both miRNA and target mRNA is vital to accurate miRNA target prediction.

9.1.6 The Influence of miRNAs on Gene Expression

One key breakthrough in this field was the discovery that miRNAs can significantly affect gene expression levels of target transcripts [9.19, 22]. Previously, it had been thought that miRNA binding was predominantly a translational control mechanism and did not directly affect transcript levels or stability. An experiment involving the transfection of miRNAs into HeLa cells followed by gene expression analysis clearly showed that miRNAs exert a significant and direct effect on gene expression levels mediated by mRNA binding [9.22]. Subsequent work proved that miRNA binding stimulated both de-adenylation and decapping of the target mRNA, triggers for mRNA destabilization [9.19, 23]. These experiments clearly illustrated that miRNAs do not just affect translation but have a profound impact on their target mRNA molecule that can be measured by both protein and mRNA expression levels. These discoveries paved the way for gene expression to be used as a readout of miRNA binding. Hence, it is possible to detect the activity of a miRNA by its global effect on gene expression or protein levels. Gene expression technologies (including microarrays and RNA sequencing) are widespread, inexpensive, and simple to analyze. They provide a genome-wide readout, and there have been significant advances in statistical approaches to analyzing these data. Proteomics approaches are extremely powerful, as they can measure direct effects of miRNAs on translation. However, even new techniques such as protein stable labeling by amino acids in cell culture (pSILAC) remain relatively expensive and cumbersome to apply genome-wide for miRNA target discovery [9.24].

9.2 Enrichment Analysis

The use of gene expression or proteomics, as a proxy for measuring miRNA effects, addresses a number of the problems associated with pure computational prediction. Such approaches implicitly take into account the context of both miRNA and mRNA/protein and do not rely on genome conservation. However, detected shifts in gene or protein expression levels in response to a miRNA comprise both primary and secondary effects. Primary effects are directly bound targets of a miRNA being downregulated, while secondary effects are other genes, which are not directly bound, whose expression shifts as a response to the primary effect. Previous experiments had shown that direct effects are usually associated with strong enrichment of miRNA seed matching sites among those genes/proteins whose expression is significantly altered. Enrichment analysis is commonly used in biology for detection of overrepresented annotations in experimental gene lists [9.25]. The goal here is to use enrichment to detect a significant overrepresentation of likely miRNA seed matching sites in genes which have responded in an expression experiment. This section discusses enrichment analysis and how it is applied to the problem of miRNA target site analysis.

9.2.1 Basics of Enrichment Analysis

Enrichment analysis is used to answer questions of the type:

1. I have collected a set S of 50 genes, 25 of which have property P. I know that in the full universe of 20 000 genes there are 2000 genes with this property. Property P seems thus enriched in my set. What is the probability associated with this enrichment?
2. I have collected a set of 50 genes, and a set of gene annotations. I would like to find out whether my set is enriched for any of those annotations.
3. I have a ranked list of genes. Is there evidence of miRNA regulation, by considering at a cutoff in the list whether the part to the left is enriched or depleted for putative target sites for a given miRNA?

The first two examples are basic yet commonly occurring examples and serve as an introduction to enrichment analysis, in particular to introduce the *hypergeometric distribution*. The third example is of particular interest in this chapter, and requires more careful use of this distribution (Sects. 9.2.4, 9.2.5). It can be noted that in scenario (1) the ratio of genes with property P is 0.5 in the sample set S, compared with a ratio of 0.1 in the full universe of genes, clearly indicating that S is enriched for property P. The statistical framework ideally suited to such situations is the hypergeometric distribution, which associates an exact P value with the event under a natural model commonly described as drawing balls from an urn without replace-

ment. The model posits an urn with a total of U balls, of which W are white and $U - W$ are black. A sample of k balls is drawn from the urn *without replacement*. It is observed that w balls in the sample are white (and thus $k - w$ are black). The chance of this happening, $P(X = w)$ is described by a binomial formula $f(w; k, W, U)$ on the parameters of the model

$$P(X = w) = f(w; k, W, U) = \frac{\binom{W}{w}\binom{U-W}{k-w}}{\binom{U}{k}}.$$

(9.1)

One is generally not interested in the exact probability, but rather in the one-tailed tests $P(x \geq w)$ if testing for enrichment and $P(x \leq w)$ if testing for depletion. These are simply obtained by summing the appropriate P values, e.g., the enrichment P value

$$P(X \geq w) = \sum_{i=w}^{W} P(X = i).$$

(9.2)

Fast software exists to compute hypergeometric probabilities to high precision, making this a fast exact test. This allows scenario (1) above to be dealt with satisfactorily, noting that the correct statement for a one-tailed test is *what is the probability associated with this or better levels of enrichment*. In the example one would compute the cumulative probability of *finding at least 25 genes with property P when drawing 50 genes randomly from the universe*, and obtain the value 9.12×10^{-13}. The null hypothesis is thus that the set of 50 genes was obtained by random drawing, and the P value describes the probability of observing at least 25 genes with property P under this model.

9.2.2 Hypergeometric Enrichment in a Set Is Equivalent to Depletion in the Complement Set

The hypergeometric distribution has several symmetry properties; For example, it is possible to count the number of black marbles drawn rather than the number of white marbles drawn, leading to the identity $f(w; k, W, U) = f(k - w; k, U - W, U)$. More importantly, by interchanging the roles of which balls are considered drawn, it follows that the probability of drawing *at least w white balls in a sample of size k* is equal to the probability of *drawing fewer than W − w white balls in a sample of size U − k*. This has attractive implications when applying the hypergeometric distribution to ranked lists, as it implies that depletion events are automatically detected in this model. It has been established [9.21] that tissue-specific genes try to avoid regulation by miRNAs active in the same tissue, so that upregulated genes in a tissue–tissue comparison are depleted in the corresponding word (i. e. sequence complementary to the *seed*). The hypergeometric distribution will bring such events to light.

9.2.3 The Hypergeometric Distribution Applied to GO Enrichment Analysis

Scenario (2) commonly occurs, for example, when a wet lab experiment returns a candidate set of genes and the researcher would like to see whether the set is enriched for some type of functional annotation. An often used approach is that of gene ontology (GO) term enrichment analysis [9.26]. Various packages exist implementing this functionality. In this case property P from scenario (1) translates to *the gene is annotated by GO term g*, and the natural approach then is to test for many GO terms whether they are enriched in set S. This immediately establishes the fact of multiple testing. If one assumes a model where GO terms are distributed randomly across genes, then repeating a hypergeometric test for different GO terms will increase the likelihood of a significant result. The normal approaches to multiple testing correction apply, such as Bonferroni or Benjamini–Hochberg correction. Additionally, a researcher may want to restrict the number of GO terms tested a priori, for example, by omitting one of the GO hierarchies (molecular function, cellular component, and biological process), by omitting terms that describe only few genes in the universe, and by taking into account the hierarchical nature of GO terms and by collapsing and restricting terms to a certain level in the GO tree.

9.2.4 The Hypergeometric Distribution Applied to Motif Discovery

Scenario (3) describes the situation in motif discovery. The urn model still applies, if requiring a somewhat more detailed exposition, and this is the model essentially used by discovery of rank-imbalanced motifs (DRIM) [9.27], oligo-analysis [9.28–30], and Sylamer [9.31] (Sect. 9.4). Later we describe the approach used by Sylamer and highlight the differences from DRIM and oligo-analysis. Each particular word that is examined leads to new values for W and w in the hypergeometric formula $f(w; k, W, U)$. For convenience

we assume that the RNA hexamer TACCTC, complementary to the seed region (described in Fig. 9.1) of the let-7 family, is the word under consideration. A list of genes is given, ranked according to some experiment, and a UTR is available for each gene. The UTRs may differ in length. The universe is constructed by considering the UTRs themselves as a ranked set, and by sliding a window of six nucleotides across all the UTRs in the entire ranked set. The number of different windows thus found defines the universe size U. A given cutoff in the ranked list of genes now divides the UTRs into two sets accordingly. At this point we count the total number of six-nucleotide windows in the first set using the sliding window technique. This defines the sample size k. We also count the number of occurrences of TACCTC in both the first set and in the two sets combined. These are associated with the number of white balls drawn (w) and the total number of white balls (W), respectively. At this point we are able to compute the hypergeometric P value of seeing at least w instances of TACCTC in a sample of k nucleotide windows drawn from a universe of size U, knowing that there are W instances of TACCTC in the universe. As previously mentioned, this probability is identical to *the probability of seeing fewer than $W - w$ instances of TACCTC in a sample of $U - k$ nucleotide windows*. In other words, an enrichment event in the one set corresponds to a depletion event in the other set with identical P value. There is a minor technicality in that the sliding window technique leads to a universe count U that cannot even in theory be attained by the count W of instances of TACTTC, due to the fact that sliding windows may overlap. In practice, using a more realistic universe count changes results only very slightly.

The model just described, due to the way words are counted, is not affected by UTR length. Briefly, Sylamer then applies this model at regular cutoffs in the gene list and plots the result in a landscape plot (Sect. 9.4.2). Oligo-analysis uses the same approach in counting words, but proceeds with the binomial distribution and requires an a priori chosen set S rather than proceeding down a ranked list. DRIM uses a different approach to counting. In its urn model, a ball represents a single sequence rather than a sliding window of nucleotides, and the presence and absence of a motif under consideration (such as TACCTC) in the sequence constitute the two different ball (sequence) colors. In this case, the length of the sequences under consideration may have an effect on the efficacy of the model. UTRs can vary greatly in length, and UTR length biases do occur in gene expression data. Hence, Sylamer uses by default the sliding window approach, and offers the absence/presence model as an option. On the other hand, DRIM was applied to analyze motif enrichment of promoter regions, where one typically uses upstream sequence segments of identical lengths. As in the previous example, it is necessary to correct for multiple testing in scenario (3), if multiple motifs are tested. There are, for example, approximately 1100 different hexamers corresponding to the complement of the 2–7 nt miRNA seed region of all human miRNAs. A more stringent selection of human miRNAs requiring them to be conserved in mammals still leaves approximately 400 different hexamers.

9.2.5 Biases in Sequence Composition Require Parameter Estimation

A common problem in sequence analysis is the occurrence of sequence composition biases. When Sylamer was created, the hypergeometric distribution had already been employed to discover motifs in ranked lists [9.27]. Similarly, Markov models had been used to address the issue of nucleotide composition bias [9.21, 28, 30]. The latter is important if one imagines a ranked list of genes derived from expression data where GC-rich sequences have a slight propensity to be upregulated compared with AT-rich sequences. This happens surprisingly often in practice and will hamper methods that do not address it. Large numbers of GC-rich motifs will be very significantly enriched in the upregulated part of the gene list (considering that the remainder of the list is relatively GC-poor), yet this is explained by a background trend in the data and will obfuscate any miRNA effect. Conveniently, a well-established approach exists to apply *background correction* and has already been applied in this field [9.21, 29]. The basic tenet of this approach is that the global count W is replaced by an *expected global count*, derived from the nucleotide composition in the part of the gene list under consideration. This is done using a Markov model on occurrences of smaller word size. This expected global count is computed for each word separately, based on a model that predicts the expected count from the occurrences of smaller words in the part of the gene list under consideration. The intuition is that, if this part is, for example, GC-rich, the expected global counts will be inflated for GC-rich words and diminished for AT-rich words, in such a manner that the overall apparent enrichment of GC-rich words and depletion of AT-rich words is decreased. The smallest word size taken

into account is called the order of the Markov model. This affects the stringency of background correction, higher orders being more stringent; For example, using a Markov model of order 3 we compute the expected frequency of the hexamer TACTTC as

$$
\begin{aligned}
P(TACTTC) &= P(TACT)P(ACTT|ACT) \\
&\quad \times P(CTTC|CTT) \\
&= P(TACT)\frac{P(ACTT)}{P(ACT)}\frac{P(CTTC)}{(PCTT)} \\
&= \frac{P(TACT)P(ACTT)P(CTTC)}{P(ACT)P(CTT)} .
\end{aligned}
$$

(9.3)

The frequencies for words of smaller length are obtained by computing the observed frequencies in the considered part of the gene list. Thus,

$$
\begin{aligned}
P(TACTTC) &= \text{Expected}(TACTTC) = \\
&\frac{\text{Observed}(TACT)\text{Observed}(ACTT)\text{Observed}(CTTC)}{\text{Observed}(ACT)\text{Observed}(CTT)} .
\end{aligned}
$$

(9.4)

It is important to note that the observed frequencies are derived from the subset under consideration (i.e., the set of sequences to the left of the cutoff chosen). The result is the *expected* frequency of occurrence of the hexamer TACTTC based on observing this *subset* of sequences. The expected global count of occurrences of TACTTC is now computed as a proportion of the universe size U using this computed expected frequency. If the latter is close to the *observed* frequency in the subset under consideration, then background correction will cancel out any previously observed enrichment. In other words, enrichment can only be detected if a word occurs more often in the subset than would be expected if its constituent words combined randomly. A higher-order Markov model implies a more stringent background correction; For example, using a Markov model of order 5 when analyzing hexamer enrichment requires the hexamer to be enriched significantly above the level where its two constituent pentamers combine randomly, which is a much stronger requirement compared with the analogous statements for constituent words of, say, length 3.

Replacing universe counts by expected global counts introduces a further complication. As the query sets grow in size, the expected global counts may become unrealistically small or large. This issue can be addressed by moderating the expected global counts with a normalization factor based on the expected global counts based on the observed sequences in the *entire* universe [9.21]. This approach, also taken by Sylamer (Sect. 9.5), dampens the discrepancies between expected global counts and the actual global counts as the query sets approach the entire universe.

9.3 Existing Enrichment Methodologies

9.3.1 Degenerate and Stringent Motifs

A large and varied body of research exists related to the discovery of motifs in biological sequences. A significant amount of this work has been devoted to discovering the binding site specificity of transcription factors and other nucleotide-binding proteins, and this work preceded and influenced the development of methods for miRNA binding site enrichment detection (BSED). Transcription factor binding sites reflect binding properties between peptide chains and nucleotides. Importantly, these sites are most often characterized by so-called *degenerate* sequence motifs; that is, the peptide chain can bind to different nucleotide sequences that share certain positional characteristics. Such a motif is commonly represented by a position-specific weight matrix (PSWM), that is, a matrix containing the observed frequencies of specific nucleotides at each

position in the motif (Fig. 9.2). These PSWMs are commonly represented graphically as sequence logos [9.32] (Fig. 9.2), showing the preferred affinity of each nucleotide at each position in the motif. In the case of miRNA binding, the interaction of interest is between two RNA molecules and it has been shown that a region of 6−8 consecutive nucleotides of perfect complementarity at the 5′ end of the miRNA (seed region) is critical. Such a region is thus a fixed hexamer, heptamer, or octamer, and is referred to simply as a *word*, a *stringent motif* or a *k-mer*. For both degenerate and stringent motifs, the basic model is that a *query* set of sequences is compared with a *universe* or *background* set of sequences. A motif is counted in both sets, and both counts as well as the sizes of the query set and the universe are taken into account to derive a *P* value, for example, the *hypergeometric P* value (Sect. 9.2.4). Degenerate motifs are often sought by finding small

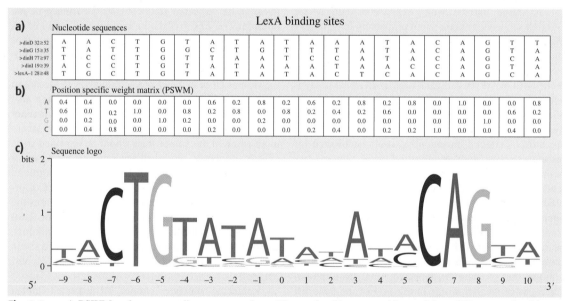

Fig. 9.2a−c A PSWM and corresponding sequence logo. The nucleotide compositions of four different nucleotide sequences (binding sites for lexA) are shown in the topmost matrix (**a**). The middle matrix (**b**) represents the same data as a position-specific weight matrix, which represents the frequencies of each of four nucleotides at each position of the four sequences above. The plot (**c**) is a sequence logo obtained using the weblogo tool (http://weblogo.berkeley.edu), which is a visual representation of the PSWM (**a,b**), where the height of nucleotides reflects their individual frequencies together with the information content in bits of that position

stringent motifs and using these as seeds to detect extended and degenerate motifs.

9.3.2 Enrichment Analysis of Stringent Motifs Corresponding to miRNA–mRNA Binding

The specific characteristics of miRNA–mRNA binding have allowed the miRNA BSED methods to forgo some of the steps required when the aim is to detect enrichment of degenerate motifs, as well as to integrate new approaches or improve existing methodology. Examples of the latter are

1. The detection of depletion signals
2. (More) thorough approaches towards coping with sequence composition bias
3. Length normalization
4. A potentially more accurate delineation of the sequences under consideration
5. The use of exact statistics obviating the need for computationally demanding sampling approaches
6. The ability to handle ranked data
7. Correct handling of large sets and large universes. Importantly, stringent (i. e., nondegenerate) motifs allow much faster methods.

9.3.3 Enrichment Analysis of Sets and Ranked Lists

An important categorization of BSED methods is whether they query a set of sequences of fixed size, or a ranked list instead. It is often the case that sequences that are queried for enriched motifs are derived from experimental data and come equipped with a score or P value that imparts a ranking on them. Examples are microarray expression data, Chip-Seq data, and DNA methylation data. Older methods generally required a fixed set of sequences to be analyzed, for example, the topmost up- or the topmost downregulated genes in a microarray experiment, necessitating a choice of cutoff. It is usually difficult to establish guidelines to find the most appropriate cutoff, hampering analysis. Later methods operated directly on a ranked list, finding for each individual motif a cutoff dividing the list such that one part is optimally enriched for the motif. Such methods are both more powerful and easier to use. A well-established example of such a method, albeit not in the context of sequence analysis, is gene set enrichment analysis (GSEA) [9.25]. This method accepts a ranked list. It then queries a large reposi-

tory of gene annotations, asking at different positions in the ranked lists whether the genes to the left of that position are enriched in a given annotation category when compared with the genes to the right of that position.

9.3.4 Different Approaches Used by Motif Search Methods

Search methods have been developed that target the challenges specific to the detection of degenerate motifs. A common approach is to represent such a motif by a weight matrix, such as a PSWM (Sect. 9.3.1), reflecting the frequency of occurrence of each nucleotide at each position of known or predicted binding sites (Fig. 9.2). In this approach probabilistic learning techniques such as expectation maximization are employed to find and repeatedly update PSWMs. Examples are multiple expectation maximization for motif elicitation (MEME) [9.33] and MdScan [9.34]. These algorithms require the input of a single set of sequences to analyze for motif enrichment.

A method that is conceptually close to miRNA BSED algorithms is oligo-analysis [9.29]. It uses binomial statistics to assign a binomial P value to the frequency of occurrence of a word in a given set of sequences, compared with a background expectation. Although it operates on stringent rather than degenerate motifs, it is very flexible in that it allows different background models, and was proven effective for extracting motifs from yeast (*Saccharomyces cerevisiae*) regulatory families. Background models can be computed from the sequence data themselves using Markov models, computing expected occurrences of k-mers from word occurrences for smaller (than k) word sizes (Sect. 9.2.5). It does not operate on ranked lists, however, and the binomial distribution assumes an infinite universe. This assumption becomes limiting in the setting of a ranked list of UTRs where miRNA binding effects may extend into a large fraction of the list. A more natural choice in this scenario, employed by Sylamer [9.31], is to use the hypergeometric distribution, as it requires a finite universe of sequences and accounts for the depletion effect in the complement of a set when analyzing enrichment. DRIM [9.27] is a motif discovery method that employs the hypergeometric distribution (Sect. 9.2.4). Like Sylamer, oligo-analysis is available either on the web or as a standalone suite of programs [9.30].

Linear models have been applied in both degenerate and stringent motif discovery, in particular to associate inhibiting and activating terms with sequence motifs in a linear model to explain changes in genome-wide expression data [9.35]. The method regulatory element detection using correlation with expression (REDUCE) [9.36] iteratively selects motifs to optimize their independence and constructs PSSMs from groups of significant dimers. This method was later applied to find miRNA signals in expression data [9.37]. The method repeatedly fits a linear model to the motifs, at each stage removing the motif that reduces the difference between the model and the expression data the most. It can be remarked that the methods previously described differ from this method in that the former are descriptive and do not seek to explain the data from the motifs. REDUCE goes beyond simply ranking the data and factors the logarithm of the expression ratio into the linear model. However, motifs are not tracked across the ranked data, and hence no information is available that correlates rank cutoffs with motif enrichment.

The Kolmogorov–Smirnov (KS) test has also been used in motif discovery [9.38] to detect sequence patterns correlated with sequences ranked in accordance with whole-cell expression measurements. Other groups [9.21] used a modified KS test for detecting enriched and depleted words corresponding to miRNA binding sites (i. e., complementary to the miRNA seed region). They used both control cohorts and fitting of the asymptotic KS test statistic tail to assign P values. To account for sequence composition biases, they subtracted the number of expected targets in each bin from the number of observed targets and used the largest cumulative negative difference along a ranked list of sequences as the KS test statistic. The number of expected targets was computed using a Markov model (Sect. 9.2.5) similar to oligo-analysis and Sylamer. The Sylamer method for miRNA seed enrichment analysis is described more fully in the following sections.

9.4 Sylamer: A Tool for miRNA Seed Enrichment Analysis

Sylamer is a fast program to detect miRNA and siRNA off-target effects from ranked lists [9.31]. Customarily, expression analysis is used to produce a list of genes ranked according to fold change or t-statistic. This is not imposed by the program, and it simply accepts ranked lists obtained by any means. The only other input required is a FASTA file with the sequences to be analyzed for enrichment. Sylamer is written in C and optimized for speed and memory usage. It utilizes highly optimized scientific numeric code from the GNU scientific library for the computation of hypergeometric P values, and is itself also freely available under the GNU general public license.

9.4.1 Sylamer Program Input and Output

Sylamer takes as input a file containing one gene identifier per line, and a file in FASTA format containing sequences for these genes. The identifiers in the rank file and the FASTA file should be exact matches. Sylamer will order the sequences in the FASTA file according to the ordering of the genes in the rank file. In the usual mode of operation, sequences that are present in the FASTA file for which the identifier is absent in the rank file are ignored, as are identifiers in the rank file for which no sequence is found in the FASTA file. Sylamer accepts a size argument specifying the size of the words to be analyzed. By default all possible

DNA/RNA words [composed of ACG(T/U)] of that size are analyzed, but it is possible to specify a smaller set of words by supplying them in a word file. When the number of words is large, the issue of multiple testing should be considered (see below). Sylamer tests the ranked gene list using multiple cutoffs, at each cutoff asking whether a particular word is more or less abundant in the top of the list then expected when compared with the rest of the list. By default this is done until the cutoff includes the entire set of sequences to be analyzed. Cutoffs are constructed using a *stepsize*, determining the granularity of measurements along the ranked gene list. The n-th cutoff taken thus defines the initial n^* *stepsize* sequences. Such a set of sequences is called a leading bin (of the ordered sequences). The Sylamer significance values (see later) are passed to a drawing program. This can be an R script (an example is included with Sylamer) or a wrapper program. A significance curve is now constructed for each word separately, plotting significance scores on the y-axis for each cutoff on the x-axis. The x-axis thus tracks the number of genes in the top of the list. Significance is calculated using hypergeometric (default setting) or binomial statistics. At each cutoff, for each word, this yields two P values, one for depletion of the word and one for enrichment. The smallest is taken and then negative log-transformed if the word is enriched in the top of the list, or log-transformed if the word is depleted. The

Detecting MicroRNA Signatures Using Gene Expression Analysis | 9.4 Sylamer: A Tool for miRNA Seed Enrichment Analysis 139

Part B | 9.4

transformed value is plotted on the y-axis at the relevant cutoff position on the x-axis. All the plotted points for a single word are connected to form a curve.

For larger word sizes (i. e., 8–15) restricted word lists speed up Sylamer significantly. Running Sylamer on 40 bins takes less than 1 min using the full word lists, but takes approximately 5 s using word lists consisting of miRNA seed matches only. Tracking and plotting all possible words has the advantage that sequence composition biases are revealed. These biases can be corrected, and low-complexity sequences are removed by filtering (see later).

9.4.2 Landscape Plot

An intuitive way to visualize Sylamer results is to generate a landscape plot showing for each word its associated log-transformed P values (Sect. 9.6). Over- and underrepresentation are plotted on the positive and negative y-axis, respectively. The curves generated by Sylamer in the landscape plot describe, for each word, the enrichment or depletion for that word across the ranked gene list. It can be observed that all curves start and end at a y-value of zero, corresponding with a P value of one. This corresponds on the right side of the plot with a sample that is equal to the entire universe. One necessarily draws *all* balls in the urn model, and the probability of drawing all white balls is accordingly one. On the left side of the plot one may consider the consequences of drawing an empty sample, but it is easiest to use the symmetry properties of the hypergeometric distribution and consider the *complement* of the sample, which is again the entire universe (Sect. 9.2.2).

Consider a word W and its associated curve in the Sylamer landscape plot. For a *peak* occurring on the *positive* y-axis, W is overrepresented in the 3′ UTRs for the genes to the left of that peak, and W is underrepresented in the genes to the right. The P values associated with these two events are identical, as discussed in Sect. 9.2.2. If we suppose that the maximum deviation is found as a *trough* on the *negative* y-axis, then W is underrepresented in the genes to the left of that trough and W is overrepresented in the genes to the right of that trough. If W is the complement of a miRNA seed (seed match) and a significant peak is found on the positive y-axis, it means that the set of genes to the left of the peak is overrepresented for that word, likely as a result of an underlying biological trait of the system being examined. This is consistent with an experiment in which the miRNA was knocked out or knocked down, where the genes that it would normally suppress are up-regulated.

If W is the seed match of a miRNA and a trough is found on the negative y-axis, it means that the set of genes to the right of the trough is overrepresented for that word. This is consistent with an experiment where a miRNA is overexpressed, reinjected in a null model, or possibly occurring as a nonnative miRNA (e.g., with a site mutation in the seed). In this case the miRNA suppresses genes which are found in the downregulated part of the gene list. In this case, if we reverse the gene ranking, the signal will now show as a peak on the positive y-axis, with the genes on the left of the peak downregulated and enriched for the seed match. Another condition may arise where a ranking is constructed by comparing mRNA expression for a particular tissue against, for example, average expression in a range of other tissues, and genes are again ranked from upregulated to downregulated. In this case genes that are upregulated in the tissue of interest are likely to avoid regulation by a miRNA that is highly expressed in that tissue. This effect has been demonstrated to exist [9.21] and is revealed as a depletion signal for the seed match in the upregulated genes. In the Sylamer landscape plot this will show as a trough on the negative y-axis on the left side of the gene ranking.

It can be remarked that Sylamer does not attempt to explain the expression data in terms of these biases, and does not assume any particular quantitative relationship between the two. It simply describes the biases found in a robust manner, associating a P value with the event of finding a given number of sites in a set of sequences, each of a certain length. For random rankings of the gene list, one typically finds maximum enrichment peaks consistent with the number of words tested.

9.4.3 Correction of Composition Biases

Composition biases in a section of the ranked sequences (e.g., % GC, di- or trinucleotide content) may cause word occurrence biases. Sylamer uses background correction as discussed in Sect. 9.2.5 to combat this issue (Fig. 9.3). It is possible to choose the order of the Markov model to be employed by using the −m parameter. For historical reasons, the parameter supplied is the order of the Markov model plus one, that is, the larger of the word sizes considered in the model.

9.4.4 Low-Complexity and Redundant Sequences

Low-complexity sequences such as monomer or dimer repeats can cause drastic biases in counts, particu-

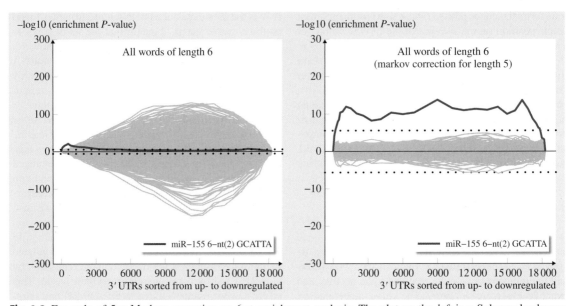

Fig. 9.3 Example of 5 nt Markov correction on 6 nt enrichment analysis. The plot on the *left* is a Sylamer landscape enrichment analysis of a miR-155 knockout in Th2 cells (similar to the Th1 experiment, Sect. 9.6.1) without Markov correction for sequence composition biases. The results are clearly and significantly biased, making it almost impossible to distinguish the effect of the miRNA (shown in *brown*). The plot on the *right* shows the same analysis performed using Markov correction of length 5 nt

larly for words that are part of the repeat. Sylamer will by default collapse any monomer or dimer repeat into a single occurrence. For more complex patterns, we suggest using DUST (*Tatusov* and *Lipman* [9.39]) prior to Sylamer analysis. In higher organisms, many alternative transcripts contain identical or overlapping 3′ UTR sequences. Paralogous genes can also encode for transcripts with very similar 3′ UTRs. If these sequences end up at similar positions in the gene list, it can lead to inflated estimations of significance of any word contained (or depleted) in them. To avoid this, we suggest masking repetitive and redundant sequences by using, for example, the regulatory sequence analysis tools (RSAT) [9.30] purge-sequence interface (http://rsat.ulb.ac.be/rsat/) to Vmatch (http://www.vmatch.de/).

9.4.5 Multiple Testing Correction

Sylamer generates a P value for word occurrences for all DNA/RNA words of a given length K, for each bin in the ranked sequence universe. Assuming b bins and word length K, this produces $b4K$ P-values in total. Additionally, a user may test multiple word lengths. Firstly, it can be noticed that findings for different word

lengths should be expected to corroborate one another. Different word lengths should thus be used to create a bigger picture of an overall finding, and do not contribute to multiple testing. A good example in this respect is miRNA seed match analysis where one uses words of length six, seven, and eight. In this particular case, we have additionally developed a method to integrate results for words of different lengths (see below). Secondly, a typical result takes the form of a significant incline or decline towards one of the ends of the ranked-sequence universe. The number of bins does not play a large role in this respect, merely determining the granularity of measurement across the ranked universe. Consequently it is recommended to apply Bonferroni correction solely by the number of words tested.

9.4.6 Integrating Sylamer Results and Increasing Sensitivity

A recent development increases sensitivity and simplifies analysis by integrating Sylamer results for words of different length when analyzing miRNA signals. For a miRNA that is differentially expressed in a particular comparison and has an effect on mRNA profiles, we typically find that plots of words complementary to the extended seed region (1–8 nt) follow a coordinated pattern. These are the two heptamers, the core hexamer, and the encompassing octamer. However, no single element among these exhibits always the strongest significance, potentially requiring inspection and a decision procedure. A certain degree of correlation can be expected for arbitrary octamers, as different elements are not independent. This correlation is not strictly implied however, and a strongly significant result for all elements simply increases confidence that one of the associated miRNAs is differentially expressed.

A Sylamer result for a single word represents a normalized result, in that it is no longer associated with word frequency (which in itself varies greatly with word length). Instead it signifies the enrichment or depletion of that word according to the hypergeometric model, and these significance values are comparable even between different word sizes. These observations led us to devise a simple model for combining Sylamer results for different elements complementary to the extended seed [9.40]. This entails summing the significance curves for these elements, boosting coordinate peaks. Each seed is then associated with the maximum absolute value of its associated summed curve. The resulting scores follow an extreme value distribution by the nature of the scoring criterion. By fitting this distribution, a single P value can be associated to a seed region independent of the number of cutoffs tested, taking into account results of different word sizes. This method has been employed to recognize dysregulation of the miR-371–373 and miR-302 clusters in malignant germ cell tumors [9.40].

9.4.7 Additional Facilities Including Exploratory Options

Sylamer has several modes that enable different methods and different outputs. It is possible to generate tables and listings just with word counts, optionally filtering sequences based on the absence or presence of specified words (*k*-mers). Words (to be tested) can be filtered based on nucleotide bag content, entropy, and with a pattern language that supports wild cards and the International Union of Pure and Applied Chemistry (IUPAC) extended DNA/RNA alphabet. It is possible to:

1. Use the binomial distribution
2. Test for enrichment using Mann–Whitney–Wilcoxon statistics
3. Treat sequences as units with absence/presence coloring rather than nucleotide windows
4. Analyze both strands from the sequences
5. Perform shuffling and random trials
6. Analyze enrichment of dyads (words of identical length separated by a variable spacer).

9.4.8 Other Signals Found in UTRs

Transcript UTRs are a hotbed of regulation. AU-rich elements (AREs) in UTRs are recognized by RNA-binding proteins. Although thought to be potentially dyadic and degenerate, these motifs may cause enrichment of purely stringent motifs. Another class of signals associated with UTRs are the polyadenylation signals, the most common of which is AATAAA. This signal is often seen in gene lists derived from expression data contrasts. This is almost always seen accompanied by a skew towards shorter UTRs at one end of the gene list. Shorter UTRs will generally carry the polyadenylation signal, thus contributing towards enrichment (noting that the expected occurrence of a hexamer is roughly once every 4 kb). It is not clear whether such a skew towards shorter UTRs could have a biological underpinning or is more likely to arise from experimental factors. A signal that arises from either class of motif and is stronger than a concomitant miRNA signal is by itself not reason to discard the miRNA signal. It is advisable to judge the miRNA signal on its own merits, and in case of ARE enrichment it is additionally recommended to use Markov background correction.

9.5 Using Sylamer and Related Tools

Sylamer has been implemented as a core algorithm written in C. This serves as the key component of a number of implementations of tool developed for a variety of users. These range from pure command-line UNIX usage of the system to fully web-based use. This section discusses the three main ways of using Sylamer.

9.5.1 Sylamer Command Line

The command-line tool *sylamer* is the program implementing the core Sylamer method, as well as other modes as described in Sect. 9.4.8 and Fig. 9.4. It is a versatile tool that reads and writes well-established file formats. It is easy to embed in bioinformatic scripts and

```
prompt> sylamer -fasta mouse.fa -universe wt-mut.rank -k 7 -m 3 -grow 200 -o wt-
mut.k7m3

read 11809 query ids
read 11809 sequences
searched/found/skipped 11809/11809/6121 ids
bases 13006517, masked 814429, counted 12039352, rejected 89203, spaced 1730966,
r2-skipped 4008, universe 12039352
.....................................
the window count exceeded the expected background times 8924 times

Please cite:
        Stijn van Dongen, Cei Abreu-Goodger & Anton J. Enright,
        Detectiong microRNA binding and siRNA off-target effects
        from expression data, Nature Methods 2008 (PMID 18978784).

prompt>
```

Fig. 9.4 The command-line version of Sylamer. Example command-line invocation of the Sylamer algorithm, analyzing words of 7 nt with Markov correction set to 3 nt

```
1> sylamer -fasta mouse.fa -universe wt-mut.rank -k 7 -m 0 -grow 200 -o wt-mut.k7m0
2> sylamer -fasta mouse.fa -universe wt-mut.rank -k 7 -m 3 -grow 200 -o wt-mut.k7m3
3> sylamer -fasta mouse.fa -universe wt-mut.rank -words mouse.7 -grow 200 -o wt-
mut.k7m3.mirna-words
4> sylamer -fasta mouse.fa -universe wt-mut.rank -w .......A -m 3 -grow 200 -o wt-mut.k8m3
5> sylamer -fasta mouse.fa -universe wt-mut.rank  --word-count -o universe.word-counts
6> R --vanilla --args --table=wt-mut.k7m3 --pdf=wt-mut.k7m3.pdf < plot-sylamer.txt
```

Fig. 9.5 A screen shot of results from the experimental program from Fig. 9.4

pipelines. It is also very fast and thus suitable for interactive use; a typical *sylamer* run should take less than half a minute. A number of command-line invocations showing typical usage are given in Fig. 9.5. All cases share some of the same parameters, identifying the data files to be read.

A number of remarks can be made. First, *sylamer* does not place any requirements on file names; in particular there is no requirement that they have a specific suffix. In the examples the suffix "rank" was used for the file containing ranked gene identifiers and the suffix "fa" was used for the FASTA file, but this is not enforced by the program, and the user is free to develop their own conventions. The argument to the −fasta option should be a FASTA file, and the argument to the −universe option should be a line-based format. The first white-space-delimited field on each line is taken and used as a sequence identifier. The FASTA file is queried for sequences for all identifiers. It is allowed for identifiers to be missing from the FASTA file, in which case they will be ignored, and any additional identifier–

sequence pairs in the FASTA file are ignored as well. In invocations (1) and (2) (Fig. 9.5) the word length is set using the −k option, causing each possible word of the specified length to be analyzed. In both invocations Markov correction (Fig. 9.3) is used with −m 3 (corresponding to a Markov model of order one less, i.e., 2). It is often but not always necessary to do this. In (3) a subset of words to be analyzed is specified using the −words option. The argument to this is the name of a file for which the first white-space-delimited field on each line should be a valid DNA word, with all words of the same length. A possible use is to limit the analysis to words of a certain class, for example, heptamers that are found to be complementary to seed regions of conserved mouse miRNAs. In (4) a set of words of length eight is specified with the −w pattern option, the only requirement being that the last nucleotide is an adenosine. A similar option −xw exists that specifies removal of words. The options −words, −w, and −xw may all be combined freely. In (5) *sylamer* is used to simply store all word counts for all identifiers in the list. It is

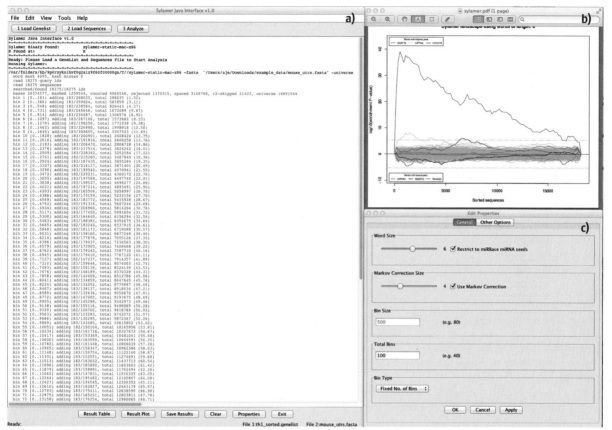

Fig. 9.6a–c The interface of the *jSylamer* application. A *jSylamer* application window (**a**) is shown with an example output plot (**b**) and the *jSylamer* preferences panel (**c**)

again possible to filter the set of words using the options previously discussed (not shown). Finally, (6) utilizes an R script shipped with *sylamer* to convert the table output into a landscape enrichment plot. Figure 9.4 shows an example command-line invocation of *sylamer*.

9.5.2 *jSylamer* Graphical User Interface

The Sylamer command-line tool is useful for bioinformaticians and other users familiar with UNIX-style interfaces. In order to produce a more straightforward graphical interface to the tool we developed a JAVA interface for Windows, Macintosh, and Linux (Fig. 9.6). This tool contains the latest release of the *sylamer* command-line tool together with an interface [9.31]. Using *jSylamer*, one can select a sorted gene list from a high-throughput proteomics or gene expression experiment together with a FASTA file containing 3′ UTR sequences. The tool assumes that gene identifiers in

the gene list correspond directly with sequence identifiers in the FASTA file. A preferences panel allows the user to choose the word size of interest, control the type of Markov chain correction, and specify individual miRNA seeds of interest to highlight in the output plot (Fig. 9.6). Like the command-line version, *jSylamer* requires the user to have the R system installed to produce enrichment landscape plots.

9.5.3 *SylArray*

Although *jSylamer* sought to make application of large-scale enrichment analysis more straightforward, it did not address one significant challenge most users face, i.e., how to download and link high-quality 3′ UTR sequences to the probes or gene identifiers from their experimental data. With the *SylArray* project [9.41] we developed a web-based interface to Sylamer that requires only a single gene list from a gene expres-

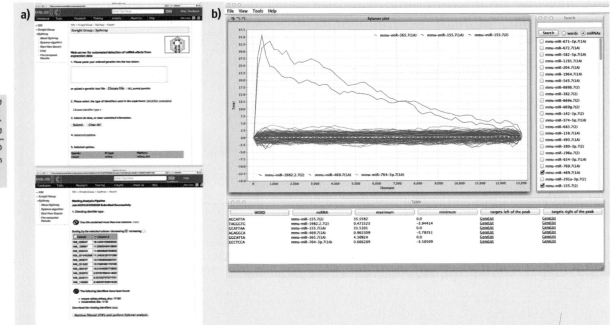

Fig. 9.7a,b Example of the web-based *SylArray* system and its graphical interface. The basic web-interface pages for starting a Sylamer analysis are shown (**a**) together with the graphical JAVA-based analysis toolkit window that is produced for visualizing results (**b**)

sion experiment. The user chooses which platform their gene list originated from and then automatically builds a set of high-quality 3′ UTR sequences specifically for that platform by interfacing directly with the Ensembl database [9.17]. The system goes further by reducing redundancy and biases within the gene list by automatically filtering cases where multiple array probes match the same 3′ UTR or where one array probe may match a large number of possible 3′ UTR sequences.

The *SylArray* website (Fig. 9.7, left), then automatically runs Sylamer on this set of 3′ UTR sequences, or on coding sequences or 5′ UTRs if desired. We

have built a graphical data-analysis environment around the results from Sylamer that allows the user to accurately assess the enrichment landscape plots produced (Fig. 9.7). A *peak-selection* mode allows the user to identify significant peaks, determine the sequence behind the enrichment peak, and then identify the likely target molecules from their original input gene list, which can be downloaded as a target result file. The graphical interface allows word enrichment analysis for words of length 6, 7, and 8 nt to be displayed simultaneously and provides a rich system for selecting or deselecting words corresponding to miRNAs of interest.

9.6 Use of Sylamer on Real Biological Examples

In this section we describe the successful application of miRNA seed enrichment analysis with a number of biological examples from published literature. In many cases, direct perturbation of miRNAs has been used to elicit a measurable response at the gene expression level. However, it is also possible to detect miRNA signatures in gene expression experiments that do not involve exogenous miRNA perturbation but instead rely

on natural variation of miRNA levels in the sample being profiled.

9.6.1 The Murine Immune System and miR-155

One of the first applications of miRNA seed enrichment analysis was on the role of miR-155 and the

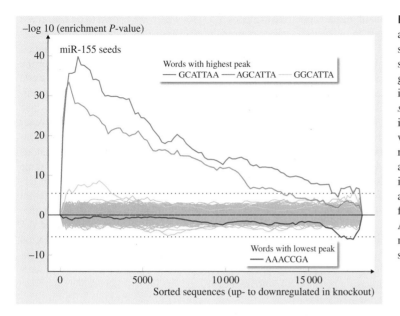

Fig. 9.8 Result for miR-155 seed enrichment analysis in blood cells. Sylamer landscape plot showing significance for all possible miRNA seed matches of length 7 nt across the sorted gene list (*x*-axis). In this case the gene list is sorted from most upregulated (*left-hand side*) to most downregulated (*right-hand side*) in ΔmiR-155 mutant samples as compared with wild-type samples of Th1 cells. Each line represents a 7-mer word, and colored lines are deemed significant when multiple testing is taken into account (*dotted lines*). The *y*-axis represents the $-\log(P)$ of significance for each word at that position in the gene list. All enriched seeds in upregulated genes in the mutant correspond to the canonical or shifted seeds of miR-155

murine immune system [9.42]. This miRNA was the first for which a mouse knockout model was produced. This miRNA (miR-155) was chosen as it was highly expressed in spleen and blood. Initially, it appeared that these mice possessed a mild phenotype. Later analysis demonstrated that these mice had severe inflammation of the gastrointestinal tract and lung and were significantly immunocompromised. The growth and function of T-cells, B-cells, and dendritic cells were significantly impaired. The functional roles for miR-155 and its targets were completely unknown, although a small number of predicted miRNA targets did appear to have roles in immunity. It was decided to focus on T-cells in the initial analysis, specifically T-helper cells (Th1).

An Affymetrix gene chip analysis was performed on T-cells isolated from wild-type mice compared with Th1 cells isolated from miR-155$^{-/-}$ mice. Two biological replicates were taken for each. The resulting expression data were normalized (RMA normalization) and filtered for nonexpressed probes. A linear model and empirical Bayes analysis (R/Bioconductor) was used to determine the log fold-changes of each transcript between wild-type versus miR-155$^{-/-}$. This gene list was sorted according to log fold-change and analyzed using Sylamer with 3′ UTR sequences derived directly from Ensembl (Fig. 9.8). In this experiment we compare wild-type cells with those where an miRNA has been knocked out. Hence, one would expect overexpression of likely targets of the miRNA to present

themselves as an enrichment in the upregulated set of genes from the gene list. Indeed, the results clearly showed massive enrichment for the seed match to miR-155 in the most upregulated genes in the mutant Th1 cells. The significance of this enrichment was 1×10^{-40}, one of the highest target enrichments observed so far with this approach (Fig. 9.8). Interestingly, the plot also indicates a potential match in the downregulated genes for AAACCGA. This does not match any known mouse miRNA seed region, but is however identical to the consensus enhancer motif for Elf-1, an important transcription factor involved in T-cell regulation, perhaps indicating that loss of miR-155 targeting causes important secondary effects on integral transcription factor regulatory networks. A total of 55 targets were identified using this approach, and a number have subsequently been validated in both T- and B-cells [9.42, 43].

9.6.2 Deafness and miR-96

A large number of mouse models exist for deafness, especially forms of the disease which resemble early-onset deafness in humans. A number of these models came from large-scale mutagenesis screens using *N*-ethyl-*N*-nitrosourea (ENU). Such screens aimed to identify mice which fail to display the correct preyer reflex (involuntary response to sound) when subjected to sonic pressure of a specified level. One such mutant, *diminuendo* (dmdo), was shown to have early-onset deafness in mice, and its genetic basis was unclear [9.44]. Once genotyped, these mice were shown to have an A to T transversion, but not within a coding sequence. It was subsequently noticed that this A–T was directly within the seed region of the miR-96 miRNA [9.44]. At that stage the function of miR-96 was poorly understood, although it was known to be highly expressed in sensory tissues including the eye and ear. Electron microscopy clearly showed that the inner-ear hair cell bundles were malformed compared with wild-type mice that have very regular patterning of stereocilia within the inner ear. It was decided to attempt gene expression profiling on these hair bundles and to compare expression directly between wild-type and dmdo mice.

In this case, Illumina bead arrays were used for profiling, as they required a low quantity of input RNA material. This is important as RNA was difficult to harvest from the inner ear without pooling mice.

The goal here again is to determine the likely targets of miR-96 in the inner ear and the loss-of-function effects in mutant dmdo mice. An important issue in this example is also that of gain of function, as mutant mice have lost a wild-type copy of miR-96 but gained a mutant miR-96 with a different seed region. Hence, one would expect the observed phenotype to be a mixture between genes upregulated in the mutant which have lost their cognate wild-type miR-96 and genes downregulated in the mutant which are being targeted by the acquired mutant miR-96. Again the gene expression data obtained were background-corrected and normalized, and a linear model analysis using empirical Bayes was used to determine differential expression as fold-changes and significances. Sylamer enrichment analysis of the derived gene list (sorted according to log fold-change and filtered for expressed probes) shows an extremely interesting result (Fig. 9.9). A very clearly defined peak (Fig. 9.9) is observed in the upregulated genes for the seed match to wild-type miR-96. This is consistent with the expected loss of targeting of miR-96 in mutant animals where its seed no longer recognizes wild-type targets. Additionally, we observe another peak of significant enrichment in the downregulated genes for matches to the mutated miR-96 dmdo seed region (Fig. 9.9). Both peaks are significant, and this result clearly indicates that, at least at the level of mRNA expression, the observed effect is a roughly equal mixture of large-scale downregulation and upregulation of genes in the mutant animals. Again one must question whether the observed phenotype is due to loss of function or gain of function. Interestingly, if

one excludes 3′ UTRs which are known to be conserved across mammals, one finds that wild-type miR-96 targets are indeed conserved, whereas the acquired mutant targets do not appear to be conserved. Evidence for the phenotype to be mostly a loss-of-function effect came from a human family identified in Spain with dominant hereditary deafness [9.45]. These individuals also had mutations in the seed region of miR-96, however in different nucleotides of the seed region. These individuals have a similar phenotype, where endogenous targeting of miR-96 is lost, affecting largely the same set of genes observed in mice. However, acquired targets of their mutant miR-96 would be entirely different as they contain differing mutations in the seed region. This experiment was one of the first and clearest descriptions of the causal nature of a miRNA mutation relevant to human disease. The experiment also highlights the usefulness and importance of mouse models of human disease.

9.6.3 Red Blood Cell Development and the miR–451/144 Cluster

The previous examples have involved single miRNAs with strong and measurable effects on gene expression levels. In this case we focus on the knockout of a miRNA cluster containing a pair of miRNAs whose expression is usually linked. The miR-451/144 cluster is expressed during hematopoesis. A knockout mouse for this cluster displayed anaemia and defects in maturation of red blood cells and a significant inability to cope with chemotoxic stress [9.46]. Again gene expression analysis was used to compare erythroblasts from wild-type and miR-451/144$^{-/-}$ mice using Illumina bead arrays, similarly to the miR-96 experiment above. In this case a pair of miRNAs have been removed in the mutant mice, and one observes four significant peaks among the most upregulated genes in the mutant animals (Fig. 9.10). A strong peak for both the canonical seed and 1 nt-shifted seed of miR-451 is observed above the multiple-testing threshold. A second set of peaks (canonical and 1 nt shifted) are observed just at or above threshold for the other miRNA miR-144. This result clearly shows a strong influence of miRNA activity on the levels of expression of genes upregulated in mutant animals and also suggests that one miRNA has a stronger effect on gene expression levels than the other [9.46]. This result was later confirmed experimentally when individual copies of each cluster member were knocked out. The miR-451$^{-/-}$ phenotype was identical to the miR-451/144$^{-/-}$ phenotype, while miR-144$^{-/-}$ mice had no obvious phenotype. Analysis of cooperativity between these miRNAs suggests that, while they share few direct targets, a small number of mRNAs are targeted by both miRNAs and those cooperative targets are very significantly down-regulated in wild-type animals. In total, 228 transcripts had binding sites for miR-451 and 98 had miR-144 sites, with 57 transcripts having both seed matching sites and showing significant expression effects from Sylamer analysis.

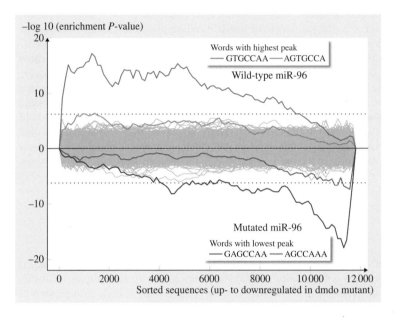

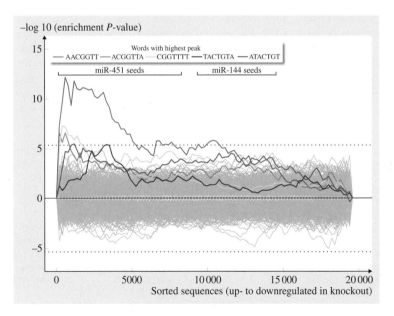

Fig. 9.9 Result for the diminuendo (dmdo) miR-96 deafness causative mutation. Sylamer landscape plot showing significance for all possible miRNA seed matches of length 7 nt across the sorted gene list (x-axis). In this case the gene list is sorted from most upregulated (*left-hand side*) to most downregulated (*right-hand side*) in dmdo mutant samples as compared with wild-type samples from mouse inner ear. Each line represents a 7-mer word, and colored lines are deemed significant when multiple testing is taken into account (dotted lines). The y-axis represents the significance for each word at that position in the gene list. All enriched seeds in upregulated genes in the mutant correspond to the canonical or shifted seeds of wild-type miR-96, consistent with loss of function. All peaks on the right-hand side of the plot correspond to enrichment of mutant miR-96 seed matches in those genes downregulated in mutant samples, consistent with gain of function

Fig. 9.10 Result for the knockout of the miR-451/144 cluster in mouse. Sylamer landscape plot showing significance for all possible miRNA seed matches of length 7 nt across the sorted gene list (x-axis). In this case the gene list is sorted from most upregulated (*left-hand side*) to most downregulated (*right-hand side*) in ΔmiR-451/144 mutant samples as compared with wild-type samples of erythrocytes. Each line represents a 7-mer word, and colored lines are deemed significant when multiple testing is taken into account (*dotted lines*). The y-axis represents the $-\log(P)$ of significance for each word at that position in the gene list. The enriched seeds in upregulated genes in the mutant correspond to the canonical and shifted seeds of miR-451 and miR-144. The most significant peaks are for miR-451 with the seed matches to miR-144 being just at the level of significance, indicating that, while both miRNAs are involved in this response, miR-451 is the key player

9.6.4 Detection of RNAi Off-Target Effects

In the examples described above we have been searching for miRNA signatures in datasets where we have perturbed a miRNA or expect its expression to significantly change and induce mRNA target level effects. Small interfering RNAs (siRNAs) are frequently used as tools to knock down mRNA levels for a single, specific target gene. They are designed using specific rules for specificity. It has been known for some time that a poorly designed siRNA may act like a miRNA and additionally target many hundreds or thousands of *off-targets* unintentionally [9.47]. Hence, the miRNA seed enrichment approaches described above can be used in this context as a quality control step to assess the likelihood of off-target effects in RNA interference (RNAi) experiments that have used gene expression or proteomics to profile genome-wide effects [9.48].

A previous off-target study used microarrays to measure the effects of transfecting 12 different siRNAs into HeLa cells [9.49]. These data were reanalyzed using Sylamer [9.31]. For each transfection experiment, a gene list was ranked according to fold-change, starting with the most downregulated genes (likely to be direct off-targets). In most cases, there was significant enrichment of words matching the 5′ end of the siRNA. The effect on the expression profile was due to a miRNA-like effect, as the only significant words were those that matched to the beginning of the siRNA. In agreement with previous results [9.50], a positive correlation was observed between the maximum enrichment value caused by each siRNA and the total number of its seed matches in human 3′ UTRs. These results suggest that use of seed enrichment analysis can be a useful control to perform on genome-wide RNAi experiments where expression data are available.

9.7 Conclusions

Use of gene expression analysis for miRNA target prediction has been significant. This approach has been used to elucidate the role of miRNAs in hundreds of different experiments across a multitude of systems and organisms to date. Typically, miRNA targets predicted from this kind of seed enrichment analysis have validation rates above 85%, a significant improvement over de novo computational methods (approximately 30–50%). There are a number of approaches and statistical frameworks for performing this kind of analysis. For this chapter we have focused on the Sylamer system

that we have developed as it is specifically designed for miRNA seed enrichment analysis, freely available, and simple to use. The advent of high-throughput proteomics and next-generation sequencing is encouraging, and it is envisaged that such datasets will also become widely available for miRNA analysis. We believe that seed and motif enrichment analysis will continue its success when applied to these datasets in the future, helping to improve our understanding of miRNAs, their regulatory targets, and their potential importance for health and disease.

References

9.1   A. Kozomara, S. Griffiths-Jones: miRBase: Integrating microRNA annotation and deep-sequencing data, Nucleic Acids Res. **39**, D152–D157 (2011)

9.2   P. Sethupathy, B. Corda, A.G. Hatzigeorgiou: Tar-Base: A comprehensive database of experimentally supported animal microRNA targets, RNA **12**, 192–197 (2006)

9.3   S.W. Chi, J.B. Zang, A. Mele, R.B. Darnell: Argonaute HITS-CLIP decodes microRNA–mRNA interaction maps, Nature **460**, 479–486 (2009)

9.4   M. Hafner, M. Landthaler, L. Burger, M. Khorshid, J. Hausser, P. Berninger, A. Rothballer, M. Ascano, A.C. Jungkamp, M. Munschauer, A. Ulrich, G.S. Wardle, S. Dewell, M. Zavolan, T. Tuschl: PAR-CliP –

A method to identify transcriptome-wide the binding sites of RNA binding proteins, J. Vis Exp. **41**, e2034 (2010), video article

9.5   R.C. Lee, V. Ambros: An extensive class of small RNAs in Caenorhabditis elegans, Science **294**, 862–864 (2001)

9.6   A.J. Enright, B. John, U. Gaul, T. Tuschl, C. Sander, D.S. Marks: MicroRNA targets in *Drosophila*, Genome Biol. **5**, R1 (2003)

9.7   B.P. Lewis, I.-H. Shih, M.W. Jones-Rhoades, D.P. Bartel, C.B. Burge: Prediction of mammalian microRNA targets, Cell **115**, 787–798 (2003)

9.8   A. Krek, D. Grün, M.N. Poy, R. Wolf, L. Rosenberg, E.J. Epstein, P. MacMenamin, I. da Piedade,

K.C. Gunsalus, M. Stoffel, N. Rajewsky: Combinatorial microRNA target predictions, Nat. Genet. **37**, 495–500 (2005)

9.9   J. Brennecke, D.R. Hipfner, A. Stark, R.B. Russell, S.M. Cohen: Bantam encodes a developmentally regulated microRNA that controls cell proliferation and regulates the proapoptotic gene hid in *Drosophila*, Cell **113**, 25–36 (2003)

9.10   B. John, A.J. Enright, A. Aravin, T. Tuschl, C. Sander, D.S. Marks: Human MicroRNA targets, PLoS Biol. **2**, e363 (2004)

9.11   P. Mazière, A.J. Enright: Prediction of microRNA targets, Drug Discov. Today **12**, 452–458 (2007)

9.12   B.J. Reinhart, F.J. Slack, M. Basson, A.E. Pasquinelli, J.C. Bettinger, A.E. Rougvie, H.R. Horvitz, G. Ruvkun: The 21-nucleotide let-7 RNA regulates developmental timing in *Caenorhabditis elegans*, Nature **403**, 901–906 (2000)

9.13   P.H. Olsen, V. Ambros: The lin-4 regulatory RNA controls developmental timing in Caenorhabditis elegans by blocking LIN-14 protein synthesis after the initiation of translation, Dev. Biol. **216**, 671–680 (1999)

9.14   S. Karlin, S.F. Altschul: Methods for assessing the statistical significance of molecular sequence features by using general scoring schemes, Proc. Natl. Acad. Sci. USA **87**, 2264–2268 (1990)

9.15   J. Brennecke, A. Stark, R.B. Russell, S.M. Cohen: Principles of microRNA-target recognition, PLoS Biol. **3**, e85 (2005)

9.16   B.P. Lewis, C.B. Burge, D.P. Bartel: Conserved seed pairing, often flanked by adenosines, indicates that thousands of human genes are microRNA targets, Cell **120**, 15–20 (2005)

9.17   P. Flicek, M.R. Amode, D. Barrell, K. Beal, S. Brent, Y. Chen, P. Clapham, G. Coates, S. Fairley, S. Fitzgerald, L. Gordon, M. Hendrix, T. Hourlier, N. Johnson, A. Kähäri, D. Keefe, S. Keenan, R. Kinsella, F. Kokocinski, E. Kulesha, P. Larsson, I. Longden, W. McLaren, B. Overduin, B. Pritchard, H.S. Riat, D. Rios, G.R. Ritchie, M. Ruffier, M. Schuster, D. Sobral, G. Spudich, Y.A. Tang, S. Trevanion, J. Vandrovcova, A.J. Vilella, S. White, S.P. Wilder, A. Zadissa, J. Zamora, B.L. Aken, E. Birney, F. Cunningham, I. Dunham, R. Durbin, X.M. Fernández-Suarez, J. Herrero, T.J. Hubbard, A. Parker, G. Proctor, J. Vogel, S.M. Searle: Ensembl 2011, Nucleic Acids Res. **39**, D800–D806 (2011)

9.18   L. David, W. Huber, M. Granovskaia, J. Toedling, C.J. Palm, T. Bofkin, T. Jones, R.W. Davis, L.M. Steinmetz: A high-resolution map of transcription in the yeast genome, Proc. Natl. Acad. Sci. USA **103**, 5320–5325 (2006)

9.19   A.J. Giraldez, Y. Mishima, J. Rihel, R.J. Grocock, S. Van Dongen, K. Inoue, A.J. Enright, A.F. Schier: Zebrafish MiR-430 promotes deadenylation and clearance of maternal mRNAs, Science **312**, 75–79 (2006)

9.20   S. Griffiths-Jones, H.K. Saini, S. van Dongen, A.J. Enright: miRBase: Tools for microRNA genomics, Nucleic Acids Res. **36**, D154–D158 (2008)

9.21   K.K.-H. Farh, A. Grimson, C. Jan, B.P. Lewis, W.K. Johnston, L.P. Lim, C.B. Burge, D.P. Bartel: The widespread impact of mammalian MicroRNAs on mRNA repression and evolution, Science **310**, 1817–1821 (2005)

9.22   L.P. Lim, N.C. Lau, P. Garrett-Engele, A. Grimson, J.M. Schelter, J. Castle, D.P. Bartel, P.S. Linsley, J.M. Johnson: Microarray analysis shows that some microRNAs downregulate large numbers of target mRNAs, Nature **433**, 769–773 (2005)

9.23   A.J. Giraldez, R.M. Cinalli, M.E. Glasner, A.J. Enright, J.M. Thomson, S. Baskerville, S.M. Hammond, D.P. Bartel, A.F. Schier: MicroRNAs regulate brain morphogenesis in zebrafish, Science **308**, 833–838 (2005)

9.24   M. Selbach, B. Schwanhäusser, N. Thierfelder, Z. Fang, R. Khanin, N. Rajewsky: Widespread changes in protein synthesis induced by microRNAs, Nature **455**, 58–63 (2008)

9.25   A. Subramanian, P. Tamayo, V.K. Mootha, S. Mukherjee, B.L. Ebert, M.A. Gillette, A. Paulovich, S.L. Pomeroy, T.R. Golub, E.S. Lander, J.P. Mesirov: Gene set enrichment analysis: A knowledge-based approach for interpreting genome-wide expression profiles, Proc. Natl. Acad. Sci. USA **102**, 15545–15550 (2005)

9.26   M. Ashburner, S. Lewis: On ontologies for biologists: The Gene Ontology – Untangling the web, Novartis Found. Symp. **247**, 66–80 (2002), discussion 80–3, 84–90, 244–52

9.27   E. Eden, D. Lipson, S. Yogev, Z. Yakhini: Discovering motifs in ranked lists of DNA sequences, PLoS Comput. Biol. **3**, e39 (2007)

9.28   J. van Helden: Regulatory sequence analysis tools, Nucleic Acids Res. **31**, 3593–3596 (2003)

9.29   M. Defrance, R. Janky, O. Sand, J. van Helden: Using RSAT oligo-analysis and dyad-analysis tools to discover regulatory signals in nucleic sequences, Nat. Protoc. **3**, 1589–1603 (2008)

9.30   M. Thomas-Chollier, M. Defrance, A. Medina-Rivera, O. Sand, C. Herrmann, D. Thieffry, J. van Helden: RSAT 2011: Regulatory sequence analysis tools, Nucleic Acids Res. **39**, W86–91 (2011)

9.31   S. van Dongen, C. Abreu-Goodger, A.J. Enright: Detecting microRNA binding and siRNA off-target effects from expression data, Nat. Methods **5**, 1023–1025 (2008)

9.32   G.E. Crooks, G. Hon, J.-M. Chandonia, S.E. Brenner: WebLogo: A sequence logo generator, Genome Res. **14**, 1188–1190 (2004)

9.33   T.L. Bailey, M. Boden, F.A. Buske, M. Frith, C.E. Grant, L. Clementi, J. Ren, W.W. Li, W.S. Noble: MEME SUITE: Tools for motif discovery and searching, Nucleic Acids Res. **37**, W202–W208 (2009)

9.34   X.S. Liu, D.L. Brutlag, J.S. Liu: An algorithm for finding protein-DNA binding sites with applications to chromatin-immunoprecipitation microarray experiments, Nat. Biotechnol. **20**, 835–839 (2002)

9.35   H.J. Bussemaker, H. Li, E.D. Siggia: Regulatory element detection using correlation with expression, Nat. Genet. **27**, 167–171 (2001)

9.36   B.C. Foat, A.V. Morozov, H.J. Bussemaker: Statistical mechanical modeling of genome-wide transcription factor occupancy data by MatrixREDUCE, Bioinformatics **22**, e141–e149 (2006)

9.37   P. Sood, A. Krek, M. Zavolan, G. Macino, N. Rajewsky: Cell-type-specific signatures of microRNAs on target mRNA expression, Proc. Natl. Acad. Sci. USA **103**, 2746–2751 (2006)

9.38   L.J. Jensen, S. Knudsen: Automatic discovery of regulatory patterns in promoter regions based on whole cell expression data and functional annotation, Bioinformatics **16**, 326–333 (2000)

9.39   Tatusov, Lipman: unpublished

9.40   R.D. Palmer, M.J. Murray, H.K. Saini, S. van Dongen, C. Abreu-Goodger, B. Muralidhar, M.R. Pett, C.M. Thornton, J.C. Nicholson, A.J. Enright, N. Coleman: Children's Cancer and Leukaemia Group: Malignant germ cell tumors display common microRNA profiles resulting in global changes in expression of messenger RNA targets, Cancer Res. **70**, 2911–2923 (2010)

9.41   N. Bartonicek, A.J. Enright: SylArray: A web server for automated detection of miRNA effects from expression data, Bioinformatics **26**, 2900–2901 (2010)

9.42   A. Rodriguez, E. Vigorito, S. Clare, M.V. Warren, P. Couttet, D.R. Soond, S. van Dongen, R.J. Grocock, P.P. Das, E.A. Miska, D. Vetrie, K. Okkenhaug, A.J. Enright, G. Dougan, M. Turner, A. Bradley: Requirement of bic/microRNA-155 for normal immune function, Science **316**, 608–611 (2007)

9.43   E. Vigorito, K.L. Perks, C. Abreu-Goodger, S. Bunting, Z. Xiang, S. Kohlhaas, P.P. Das, E.A. Miska, A. Rodriguez, A. Bradley, K.G. Smith, C. Rada, A.J. Enright, K.M. Toellner, I.C. Maclennan, M. Turner: microRNA-155 regulates the generation of immunoglobulin class-switched plasma cells, Immunity **27**, 847–859 (2007)

9.44   M.A. Lewis, E. Quint, A.M. Glazier, H. Fuchs, M.H. De Angelis, C. Langford, S. van Dongen, C. Abreu-Goodger, M. Piipari, N. Redshaw, T. Dalmay, M.A. Moreno-Pelayo, A.J. Enright, K.P. Steel: An ENU-induced mutation of miR-96 associated with progressive hearing loss in mice, Nat. Genet. **41**, 614–618 (2009)

9.45   A. Mencía, S. Modamio-Høybjør, N. Redshaw, M. Morín, F. Mayo-Merino, L. Olavarrieta, L.A. Aguirre, I. del Castillo, K.P. Steel, T. Dalmay, F. Moreno, M.A. Moreno-Pelayo: Mutations in the seed region of human miR-96 are responsible for nonsyndromic progressive hearing loss, Nat. Genet. **41**, 609–613 (2009)

9.46   K.D. Rasmussen, S. Simmini, C. Abreu-Goodger, N. Bartonicek, M. Di Giacomo, D. Bilbao-Cortes, R. Horos, M. Von Lindern, A.J. Enright, D. O'Carroll: The miR-144/451 locus is required for erythroid homeostasis, J. Exp. Med. **207**, 1351–1358 (2010)

9.47   A.L. Jackson, S.R. Bartz, J. Schelter, S.V. Kobayashi, J. Burchard, M. Mao, B. Li, G. Cavet, P.S. Linsley: Expression profiling reveals off-target gene regulation by RNAi, Nat. Biotechnol. **21**, 635–637 (2003)

9.48   I. Sudbery, A.J. Enright, A.G. Fraser, I. Dunham: Systematic analysis of off-target effects in an RNAi screen reveals microRNAs affecting sensitivity to TRAIL-induced apoptosis, BMC Genomics **11**, 175 (2010)

9.49   A. Birmingham, E.M. Anderson, A. Reynolds, D. Ilsley-Tyree, D. Leake, Y. Fedorov, S. Baskerville, E. Maksimova, K. Robinson, J. Karpilow, W.S. Marshall, A. Khvorova: 3′ UTR seed matches, but not overall identity, are associated with RNAi off-targets, Nat. Methods **3**, 199–204 (2006)

9.50   E.M. Anderson, A. Birmingham, S. Baskerville, A. Reynolds, E. Maksimova, D. Leake, Y. Fedorov, J. Karpilow, A. Khvorova: Experimental validation of the importance of seed complement frequency to siRNA specificity, RNA **14**, 853–861 (2008)

10. Bioinformatic Methods to Discover *Cis*-regulatory Elements in mRNAs

Stewart G. Stevens, Chris M. Brown

Cis-regulatory elements play a number of important roles in determining the fate of messenger RNAs (mRNAs). Due to these elements, mRNAs may be translated with remarkable efficiency, or destroyed with little translation. Untranslated regions cover over a third of a typical human mRNA and often contain a range of regulatory elements. Some elements along with their RNA or protein binding partners are well characterized, though many are not. These require different types of bioinformatic methods for identification and discovery. The most successful techniques combine a range of information and search strategies. Useful information may include conservation across species, prior biological knowledge, known false positives, or noisy high-throughput experimental data. This chapter focuses on current successful methods designed to discover elements with high sensitivity but low false-positive rates.

10.1 The Importance of *Cis*-regulatory Elements in mRNAs

Cells respond to the environment by changing gene expression. In human cells, gene expression is often regulated at both the transcriptional and translational levels. Steady-state levels of mRNAs and their proteins will alter due to the combined effects of this regulation. Regulation at the posttranscriptional level is critical for rapid responses to environmental factors.

While there is correlation with the levels of mRNA, translational control is key to determining cellular levels of specific protein [10.1]. *Cis*-regulatory elements commonly affect mRNA stability or translational efficiency of the mRNA [10.2]. Figure 10.1 shows a simplified schematic of the regulatory elements found in mRNAs.

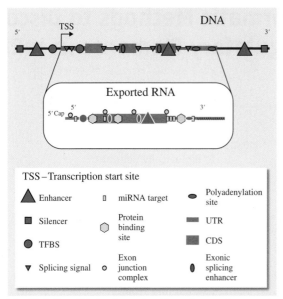

Fig. 10.1 Simplified schematic of regulatory elements in DNA and an mRNA transcribed from the corresponding genomic region. *White outlined* elements (e.g., TFBS) are present but nonfunctional in the mature mRNA (CDS: coding sequence)

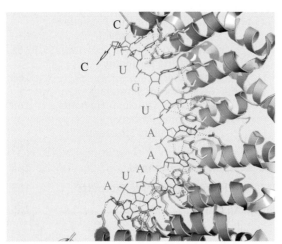

Fig. 10.2 PUF3p in complex with the Cox17 mRNA. The PUF3 binding site is not base paired. (Figure rendered using pyMol, PDB, 3K49)

Many different proteins potentially bind to RNAs. In yeast there are estimated to be over 600 RNA-binding proteins (Fig. 10.2) [10.3], and in humans, based on the occurrence of RNA binding motifs in proteins, it is likely that this number is substantially higher [10.4].

The most common domains are K homology (KH, Fig. 10.3c), RNA recognition motif (RRM, Fig. 10.3d), and double-stranded RNA (dsRNA, Fig. 10.3a) binding domains [10.5]. Many of these proteins are part of other structures, for example, ribosomes or splicing complexes, but some of these have additional roles in mRNA binding. In addition, there are many proteins that do not have obvious canonical RNA binding domains but bind specifically to groups of RNAs [10.6].

Some of these *trans*-acting binding proteins bind to specific *cis*-regulatory elements in mRNAs, providing a regulatory mechanism determining transcript fate. The translation, stability, and localization of an mRNA may vary in response to these interactions.

Interactions with microRNAs (miRNAs) are of particular importance in determining mRNA stability. Predicted target sites for miRNAs have been reported for over 30% of human genes. However, only a small proportion of these predictions are experimentally confirmed.

The *cis*-regulatory elements encoded in the transcript are most often found in the untranslated regions (UTRs). This bias may partly be because study of these regions is more tractable to experimental and bioinformatic analysis. There is however evidence that *trans*-acting factors such as miRNAs are easily displaced by the translational machinery and so operate more readily in the 3′ UTR [10.7]. Translation can be entirely repressed immediately after nuclear export, and so this displacement is not an issue for elements such as those found in the coding region of the mRNA for *ASH1* [10.8]. Some *cis*-regulatory elements such as the binding site for iron regulatory proteins certainly occur in the 5′ UTR [10.9].

Regulatory elements that act in the DNA may be also present in the mRNA. RNAs exported from the nucleus contain some of these elements that are non-functional in the mature transcript. This is particularly confounding for de novo element discovery in mRNAs. Knowledge of the genomic elements acting at the transcriptional level can help to resolve this.

Many types of structured and unstructured *cis*-regulatory mRNA elements act at the posttranslational level. Examples of such elements are the selenocysteine insertion sequence (SECIS), Histone3, PUF3, and iron responsive element (IRE). Computational methods have been developed to find each of these [10.10–12]. It is also useful to consider such well-characterized elements in order to estimate the variation that may exist in novel elements.

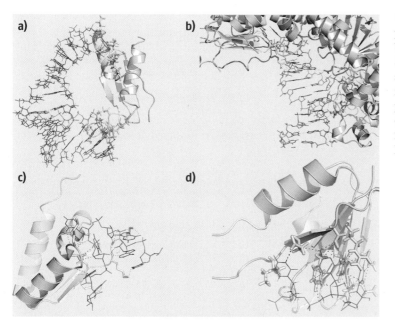

Fig. 10.3a–d Examples of RNA–protein interactions. (**a**) dsRNA binding domain in complex with Staufen (PDB 1EKZ). (**b**) Iron responsive element of ferritin mRNA in complex with an iron regulatory protein (PDB 2IPY). (**c**) KH domain (PDB 2ANR). (**d**) RRM domain (PDB 2L41). (All figures rendered using PyMol)

Methods developed for genes encoding structured noncoding RNAs (ncRNAs) may be applied to *cis*-regulatory elements [10.13, 14].

Most importantly, these well-characterized elements may be used for benchmarking the performance of prediction algorithms that may then be applied in the search for novel elements. For this purpose a subset of the Rfam database [10.15] (e.g., CisReg), or small parts of databases of known RNA secondary structures, e.g., RNA STRAND [10.16] or CompaRNA [10.17], can be used.

In the past, the limitations of methodology for determining *cis*-regulatory elements have meant that there was a large role for bioinformatic prediction. In the last few years new high-throughput RNA-Seq techniques combined with bioinformatics are now being developed [10.18–21].

10.2 Searching for *Cis*-regulatory Elements Using Bioinformatics

Elements in mRNAs are necessary for regulation of stability, translational control, and localization. They may have structural motifs critical to their function or be characterized by primary sequence alone. These *cis*-regulatory elements are the targets for miRNAs and protein binding sites. Bioinformatic analysis of mRNA can be useful in proposing new models and hypotheses for experimental testing and interpreting existing data. The aim of this analysis is to discover those regulatory sequences that regulate the fate of the mRNAs containing them.

There are many challenges in identifying *cis*-regulatory elements within RNAs. Their primary sequence patterns are often sparse. The binding sites for proteins may depend on just a few nucleotides with critical secondary structure. Determination of RNA structure is experimentally difficult, and prediction tools are often inaccurate. Regulatory elements are only sometimes conserved, and there are many distracting signals, such as elements operating at the transcriptional level. Some regulatory elements may be unique to a particular mRNA, but others such as the iron responsive element certainly operate within many mRNAs with divergent regulatory outcomes [10.12]. Even when effective bioinformatic models exist for a *cis*-regulatory element, the application of this model for the purposes of discovery will inevitably yield some false positives.

Some methods have been developed that attempt to discover new regulatory elements in mRNA sequences using only limited biological knowledge. However, the most successful methods utilize as much biological knowledge as is available in order to refine and inform the predictions. Despite the development and publication of several hundred methods (particularly for

Table 10.1 Commonly used and current tools and research directions

Name	Purpose	Type	Reference	URL
General tools				
UCSC Genome Browser	Visualize and download many different data for various genomes	Web	[10.22]	http://genome.ucsc.edu/
ncRNA Genome Browser	Version of the UCSC browser that includes many data tracks particularly aimed at RNA analysis	Web	[10.23]	http://www.ncrna.org/
Galaxy Suite	Web-based service that eases data acquisition, processing, and visualization by integrating many different tools	Web	[10.24]	http://galaxy.psu.edu/
Narrowing the search space				
RepeatMasker	Identify repeat elements	Command line via UCSC	[10.22, 25]	http://www.repeatmasker.org/
Transfac	Identify TFBS	Web	[10.26]	http://www.gene-regulation.com/
Jaspar		Web	[10.27]	http://jaspar.genereg.net/
STAMP	Binding motif/matrix comparison tool	Web Command line	[10.28]	http://www.benoslab.pitt.edu/services.html
Refseq	Gene annotation database that allows focus on UTRs	Via UCSC	[10.29]	http://www.ncbi.nlm.nih.gov/RefSeq/
Known regulatory elements				
Rfam	Contains covariance models of many known regulatory elements	Web	[10.15]	http://rfam.sanger.ac.uk/
Transterm	Contains pattern-based models of known regulatory elements	Web	[10.30]	http://mrna.otago.ac.nz/Transterm.html
UTRsite		Web	[10.31]	http://utrsite.ba.itb.cnr.it/
RBPDb	Database of protein binding sites	Web	[10.32]	http://rbpdb.ccbr.utoronto.ca/
TargetScan	Tools and database of predicted miRNA binding sites	Web Via UCSC	[10.33]	http://www.targetscan.org/
PicTar	Database of predicted miRNA binding sites	Web	[10.34]	http://pictar.mdc-berlin.de/
miRbase	Database of miRNAs	Web	[10.35]	http://www.mirbase.org/
ElMMo	Database of predicted miRNA binding sites – also allows searching based on mRNA expression profiles	Web	[10.36]	http://www.mirz.unibas.ch/ElMMo2/
Primary sequence analysis				
MEME	Tools for finding overrepresented patterns in primary sequences	Web Command line	[10.37]	http://meme.nbcr.net/
Weeder		Web Command line	[10.38]	http://www.pesolelab.it/
TEIRESIAS		Web Command line	[10.39]	http://cbcsrv.watson.ibm.com/Tspd.html

miRNA sites) only a few are commonly used (or cited). This may sometimes be because the expected or actual utility of the software is outweighed by the difficulty of installing and using it. In other cases the benefits of a particular tool may be outweighed by the familiarity of commonly used tools that do a similar job.

10.2.1 Summary of Tools and Data Sources

The most promising commonly used and current tools and research directions are discussed here (listed in Table 10.1). Lists with different foci can be found in the literature [10.40–42].

Table 10.1 (continued)

Name	Purpose	Type	Reference	URL
Secondary structure prediction [10.42]				
mfold/UNAfold	Predict secondary structure from primary sequences	Command line	[10.43]	http://mfold.rna.albany.edu/
RNAfold and RNAplFold		Web Command line	[10.44]	http://www.tbi.univie.ac.at/RNA/
RNAalifold	Predict secondary structure from alignments of primary sequences	Web Command line	[10.45]	http://www.tbi.univie.ac.at/RNA/
Dynalign/Multilign	Simultaneous alignment and folding of multiple similar RNAs to predict structure	Command line GUI	Mathews, 2010 #28239}	http://rna.urmc.rochester.edu/ RNAstructure.html
Turbofold		Command line GUI	Mathews, 2010 #28239}	http://rna.urmc.rochester.edu/ RNAstructure.html
Comparing secondary and tertiary structures				
RNAdistance	2-D structure comparison	Command line	[10.44]	http://www.tbi.univie.ac.at/~ivo/RNA/
RNAforester		Command line	[10.46]	http://bibiserv.techfak.uni-bielefeld.de/ rnaforester/
iPARTS	3-D structure comparison	Web	[10.47]	http://bioalgorithm.life.nctu.edu.tw/ iPARTS/
Searching for secondary structures				
RNAMotif	Search method to identify motifs that may be described structurally and/or by sequence	Command line	[10.48]	http://casegroup.rutgers.edu/
CMFinder	A tool that finds conserved motifs based on predicted structures using covariance models	Web Command line	[10.49]	http://wingless.cs.washington.edu/htbin-post/unrestricted/CMfinderWeb/ CMfinderInput.pl
Combining primary and secondary structural search methods				
Infernal/cmsearch	Search method using covariance models built from sequence alignments to a consensus structure	Command line	[10.50]	http://infernal.janelia.org/
Scan_for_matches	Pattern-based search method	Command line	[10.51]	http://blog.theseed.org/servers/2010/07/ scan-for-matches.html
Evidence for common regulation and tissue-specific expression				
GEO	Database of gene expression experiments	Web	[10.52]	http://www.ncbi.nlm.nih.gov/geo/
Publicly available combinatorial methods				
MEMERIS	An extension of MEME guided by predicted secondary structure	Command line	[10.53]	http://cs.stanford.edu/people/hillerm/Data/ MEMERIS/
RNAz	A tool that finds conserved structural motifs in aligned sequences	Web Command line	[10.54]	http://www.tbi.univie.ac.at/~wash/RNAz/
FIRE	A tool that combines detection of overrepresented primary sequence patterns with other biological data	Web Command line	[10.55]	https://iget.princeton.edu/

Part B | 10.2

10.3 Obtaining High-Quality mRNA Sequence Data to Analyze

Individual mRNA sequences can be obtained from the University of California, Santa Cruz (UCSC), Ensembl, or the National Center for Biotechnology Information (NCBI). Refseq annotations of UTRs are useful to focus on, as these avoid many elements acting at the transcriptional level and also patterns in coding sequences. The Refseq annotations are more conservatively curated than the Ensembl ones, which contain many predicted transcripts. The UTRs may be readily obtained from Ensembl using the web-based Biomart interface. Refseq transcripts along with their annotations can be obtained from NCBI or UCSC – the annotations must be processed to produce UTR sequences. Downstream statistical processing may be necessary to remove duplicate sequences and to remove redundant, very similar sequences (e.g., using CD-HIT-EST).

Multiple sequence alignments for several species can be obtained from the UCSC genome browser. The output is supplied in multiple alignment format (MAF). The alignments can be filtered to include only those sequences for which there is reasonable conservation. This process is simplified by using the online tool Galaxy [10.24].

10.3.1 Evidence for Common Regulation and Tissue-Specific Expression

mRNA Expression

In order to discover regulatory elements in a gene of interest it is useful to look for common sequences and structures in similarly regulated genes. Also, in the scenario where a regulatory element is known for a particular gene, it is sensible to search for similar elements that may be identified in coregulated genes. The gene expression omnibus [10.52] (GEO) contains mRNA expression data for multiple species and tissues under many experimental conditions. GEO also includes data from RNA immunoprecipitation chip (RIP-chip) experiments, e.g., with Staufen1 (GSE8438) or HuR (GSE29778) and more recently RIP-RNA-Seq data (e.g., Tdrd1 GSE29418).

The web interface provided by GEO allows the researcher to identify coregulated mRNAs via a link to *profile neighbors*. Expression data may also be downloaded and other statistical tools used to identify and quantify these neighbors, which is worth doing if a study of special interest is available. The profile neighbors provided by GEO are simply the 200 most closely expressed genes.

There are also some relevant sets of publicly available data from published work that are not included in GEO, e.g., co-localized genes for fruit-fly embryos [10.56] or the data from yeast RIP-Chip [10.3], which are provided as websites or supplemental data.

RNA Binding Protein Expression. The expression of RNA binding proteins in specific tissues can also be obtained from GEO. On the GEO website the protein expression data may be readily accessed using the advanced search option and specifying "Protein" for the "Sample Type" field.

miRNA Expression. Coexpression of miRNA and target may be an indication that a functional interaction occurs. Many miRNAs, like mRNAs, are expressed in specific tissues. Studies and methods that consider coexpression in the same tissues have been useful in identifying biologically relevant pairs. Large studies have determined the expression of most small RNAs in many tissues, for example, over 250 small RNA libraries from 26 different organ systems [10.57]. Several databases provide access to these expression data [10.58]. Some databases such as ElMMo combine both miRNA and mRNA data.

10.3.2 Narrowing the Search Space to Biologically Relevant Regions

In the search for novel *cis*-regulatory elements that operate at the posttranscriptional level, sequence elements that are likely to be false positives must be avoided. These include repetitive elements and elements acting purely at the transcriptional level.

When de novo element discovery is being pursued these distracting elements may overwhelm pattern prediction algorithms and so must be masked in input sequence. When searching for known elements these distractions should be considered at a later stage when assessing the likelihood of a putative hit.

Coding sequence contains confounding patterns arising from protein constraints and common combinations of amino acids in the translated product. They may be masked from analysis in a straightforward fashion where sufficient gene annotation is available. If this is not the case but sufficient protein information is available, it is possible to use tblastn (protein–nucleotide 6-frame translation) to map and mask out likely coding regions. It must be noted that this approach will inevitably fail when the interesting *cis*-regulatory regions are actually in the coding region.

10.3.3 Distinguishing and Avoiding Repetitive Elements

Repetitive sequences arising from events such as viral retrotransposition are abundant in many genomes; For example, the Alu-element forms 10% of the human genome; furthermore, it is concentrated in gene regions and overlaps mRNAs [10.59]. Therefore, many human 3' UTRs contain detectable Alu remnants of the 7S RNA.

This and other repeat elements are catalogued in Repbase [10.25]. These sequences may be masked out at an early stage of analysis using the RepeatMasker [10.60] program. Alternatively, the UCSC genome browser [10.22] may be used when assessing putative elements – this has a RepeatMasker track and allows intersection of uploaded candidate *cis*-regulatory elements to report overlaps.

It is possible that a novel common repeat element may be discovered as a putative regulatory element in a set of coregulated mRNAs. The choice of an appropriate negative control set of real RNAs (rather than simulated sequences) from the same species is important to avoid this.

10.3.4 Distinguishing and Avoiding Elements Acting at the Transcriptional Level and Other RNA Features

Elements acting in the DNA such as transcription factor binding sites (TFBSs) and enhancers may be misidentified as elements acting posttranscriptionally. If the genome of interest is well annotated or sufficient transcript sequence data are available, a genomic search can be avoided. This will go some way to avoiding transcriptionally acting elements.

The methods discussed in this chapter may be usefully restricted to untranslated regions (UTRs). This will avoid many false positives from genomic sequences and patterns associated with coding sequence.

TFBSs in Transcription Promoters that May Overlap 5' Regions of the mRNA. TFBS databases such as Transfac [10.26] and Jaspar [10.27] contain both experimentally defined and computational predictions. As these databases contain many false positives it is inadvisable to mask or remove predicted TFBSs from an early stage of analysis.

The difficulties with predicted TFBSs are reviewed here [10.61]. One approach to improving TFBS prediction is to use conservation information, although this is controversial as binding sites may not be conserved in multispecies alignments [10.62].

The web-based tool STAMP [10.28] may be employed to determine whether a set of aligned predicted *cis*-regulatory elements coincides with any of those TFBSs in the public databases.

Enhancers that May Overlap 3' Regions

ChIP-seq studies can be used to identify binding sites for known enhancers [10.63]. The cited review points out that there is conflicting evidence regarding the conservation of enhancer regions and TFBS; although many sites are conserved, there are a large number of nonconserved sites. Bearing these limitations in mind, the available data should still be considered; For example, a recent study first identified highly conserved noncoding sequences and then tested associated genes for tissue-specific expression during mouse embryogenesis [10.64]. Although the identified enhancer sites are clearly a subset of all such elements and very few occur within gene transcripts, putative *cis*-regulatory elements acting in mRNAs should be cross-checked against known enhancer regions.

10.4 Known Regulatory Elements

When a *cis*-regulatory element is characterized by a primary sequence, pairwise alignment methods or multiple ones, including hidden Markov models (HMMs), may be used to search for it in other RNAs. Pairwise alignment methods are best established and have many fast, readily available implementations such as the basic local alignment search tool (BLAST) [10.65]. On the other hand, HMMs which probabilistically model state transitions and thereby account for gaps in a nonarbitrary manner can have increased sensitivity, albeit at a computational cost [10.66].

An example where bioinformatic primary sequence analysis has been successful is in a study of the mRNA for the *Vg1* localized mRNA. This mRNA was shown to be localized to the vegetal cortex of *Xenopus laevis* oocytes by parts of a 340-base region in the 3′ UTR. Subsequently, this was shown to be bound by proteins that contain multiple KH domains. *Schnapp*'s group first identified using bioinformatics four repeated sequence elements, but experimental deletion showed that only one of these elements, *UUCAC* (E2), was critical for function [10.67].

Mowry's group also studied the localization of the mRNA for *Vg1* and took a different approach [10.68]. They systematically deleted sequences over the entire 340-nt region. Interestingly, the result they found was quite different – a different sequence, UUUCUA (VM1), was identified as critical, and this was supported using site-specific mutagenesis [10.68]. Although this sequence corresponded to one of the three (E1) elements identified in the Schnapp laboratory, their results showed those deletion constructs had reduced but not abolished localization.

The UUCAC, E2 element was also found to be required for localization of the mRNA for *vegT* [10.69]. Interestingly, both the *vegT* and the *Vg1* mRNAs had multiple (five) copies of this element. Subsequently a shorter more generalized motif, *CAC*, was postulated, repeats of which are present in the majority of RNAs localized to the vegetal cortex of *Xenopus laevis* oocytes [10.70].

This demonstrates the utility of bioinformatic analysis, although the requirement to find repeated clusters of short sequence required the development of a utility specialized to this task. For such a short motif (3–6 bases) multiple copies may be found by chance in any mRNA, and statistical tools have been developed to analyze this [10.71]. Functionally, multiple small dispersed E2 and VM1 elements provide, in combination, a specific binding site for the RNA-binding proteins in this case.

Notably many of the experimentally determined elements collected in the RBPDb are short (4–8 bases) and alone would not provide specificity. Computational tools that consider combinations of weakly informative sites have been used in other systems, e.g., for TFBS and miRNA targets. However, the functional relationships between multiple instances of the same or different sites are not usually known. For some sites, notably the important classes of AU-rich and CU-rich elements, programs to detect these sites operate by weighting multiple nearby repeats.

10.4.1 RNA Binding Protein Target Sites

Known protein binding sites are available from several public databases. The Rfam [10.15] database contains a growing group of covariance models for many *cis*-regulatory elements including RNA binding protein sites. Transterm [10.30] also contains patterns and descriptions of many known *cis*-regulatory elements – particularly protein binding sites. Nonredundant sets of sequences from NCBI or other uploaded sequences of interest may be searched for matches against these patterns. UTRsite [10.31] is a similar database of patterns. The RNA-Binding Protein DataBase [10.32] (RBPDb) catalogs proteins and their binding sites curated from the literature. Unstructured sequence motifs of binding sites may be downloaded and filtered by experiment type, species, and/or binding protein affinity.

These databases (Rfam, Transterm, UTRsite, and RBPDb) are useful in identifying known elements and may be used to find interesting candidates for testing. The reasoning behind this approach is to direct the identification of putative elements by their identity or similarity to known elements that have been experimentally demonstrated in other mRNAs. Another important usage of these databases is as a source of benchmarking datasets for any algorithms or pipelines designed to predict *cis*-regulatory elements de novo.

A note of caution is warranted. Database entries depend on manual curation based on literature review to remain current; For example, we have recently updated the Rfam model for the IRE [10.12] – the previous model was out of date and could not be used to identify many IREs that had been more recently experimentally demonstrated. It must also be considered that the models in Rfam, Transterm, and UTRsite inevitably vary in their sensitivity and specificity when identifying elements in target sequences.

Most of the entries in RBPDb are characterized by single sequences alone, and so if these data are to be used, the user must currently confine their search to be based on primary sequence only or build their own models to include predicted/demonstrated secondary structure. RBPDb does, however, contain some position weight matrices (PWMs) from systematic evolution of ligands by exponential enrichment (SELEX) experiments – the web interface may be employed to search mRNA sequences using these.

For some types of RNA–protein interactions it may be possible to predict either the binding from the protein structure, or the reverse. For a few classes of proteins this has been possible (e.g., PUF domain-containing proteins), but research in this area is beyond the scope of this chapter [10.72, 73].

10.4.2 miRNA Target Sites

miRNAs have short ungapped seed sequences complementary to their target sites and act to downregulate expression. These targets are characterized by primary sequence with secondary structure normally impairing binding. Higher eukaryotic genomes may contain many hundreds of miRNAs (1,424 Human miRNAs in MiR-Base release 17, 4/2011), and each one of those tested affects the expression of several hundred mRNAs. Some of the changes in expression are undoubtedly the result of indirect regulation; For example, transcription factors are regulated by miR-34a, so changes in miR-34a expression certainly have effects beyond the immediate targets [10.74].

There are numerous predicted miRNA binding sites, mainly in 3′ UTRs. It has been estimated that over 30% of human genes contain target sites. However, relatively few target sites have been validated experimentally. miRTarBase [10.75] listed the greatest number (2,819) of verified targets for the 269 human miRNAs that had been tested (4/2011). Databases of less reliable high-throughput data map over 150 000 targets to genomes [10.75].

Prediction of target sites for miRNAs is more straightforward than for proteins. However, there are many methods available which will detect different subclasses of sites with different accuracies; these have been recently reviewed [10.76], and ensemble methods that combine several different tools are available [10.77, 78]. Several examples of predictive tools are outlined below.

TargetScan [10.33] predicts miRNA target sites. Base pairing between the seed sequence at the 5′ end of the miRNA and the target mRNA and evolutionary conservation of the sites are the primary consideration. The software has been developed to additionally account for conservation at the seed region, minimum free energy of

the hybridization including additional 3′ binding, flanking AU-rich sequence, proximity to additional miRNA target sites, and the position of the target site within the UTR [10.79, 80]. The UCSC genome browser is a straightforward way of accessing the TargetScan predictions. These have been calculated using conservation information based on multiple species alignment. The inclusion of this information reduces false positives and is probably a good idea for genome-wide analysis. If the target for analysis is restricted to several genes it may be worthwhile to consider nonconserved miRNA target sites by running the TargetScan program on unaligned sequence.

An alternative method is PicTar [10.34]. It allows the identification of target sites that have reduced conservation but are within mRNAs coexpressed with the targeting miRNA. The authors report that 30–50% of such sites are functional [10.81].

Another method, miRanda/mirSVR [10.82], has the distinction of allowing noncanonical G–U base pairs in the seed sequence; miRanda is also available on the web and has recently been extended to include a support vector regression algorithm – mirSVR [10.83]. This incorporates the relevant biological data, including expression data, into a scoring system and avoids a strict filter based on conservation.

An alternative method offering greater sensitivity to nonconserved target sites is ElMMo [10.36]. The algorithm uses Bayesian methods to assign priors calculated from the phylogenetic distribution of target sites for *each* miRNA. This allows miRNA-dependent adjustment of posterior probabilities for target sites with similar conservation patterns. The benefit of this is that conservation information is automatically tuned to each specific miRNA. The disadvantage is that it will be hard to match a nonconserved target site of an miRNA that has many widely conserved target sites. ElMMo is available via a web server and includes convenient filtering of mRNA targets to search using expression levels from numerous datasets, also providing Gene Ontology enrichment information on the identified targets.

10.5 De Novo Element Discovery

The discovery of new regulatory elements is a key goal for improved understanding of gene regulation. Recurring patterns in sequence and predicted structure may be detected and assessed for statistical significance. Some approaches for dealing with false positives caused by distracting sequence patterns

have already been discussed. In de novo detection these steps are of particular importance – there is no model of even low specificity with which to begin.

There are many tools available for the detection of patterns in primary sequence, usually because they have been developed for finding DNA regulatory regions and this is then applied to RNA.

This is despite the demonstrated importance of structure in many regulatory elements. The reason for this lies in the difficulty of secondary structure prediction and that primary sequence patterns are still characteristic of structured elements.

10.5.1 Primary Sequence Analysis for Elements Lacking Significant Secondary Structure

A number of enumerative methods are available for de novo detection of primary sequence patterns. The patterns being sought are generally far shorter than the genomic and transcript sequences in which they

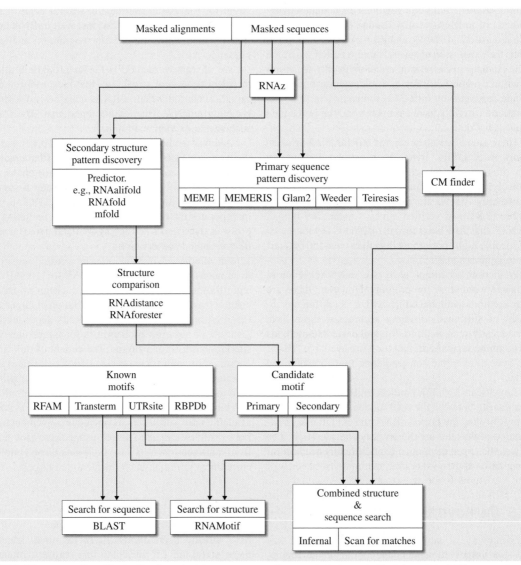

Fig. 10.4 Discovery of motifs starting from input sequences or alignments that are similarly regulated. The general goal of each component is shown, with some specific examples of currently available software named

are contained. Real primary sequence elements may be degenerate, gapped, redundant, and repetitive. Several differently patterned elements may be responsible for similar regulatory outcomes in different transcripts. All these factors contribute to the difficulty of detection.

Overrepresented patterns in a set of unaligned input sequences may be identified by multiple expectation maximization for motif elicitation (MEME) [10.37]. A background model may be provided for this analysis. The output motifs consist of position weight matrices (PWMs) showing the probability of a particular nucleotide at each position within the motif. MEME will not consider gaps in a motif. GLAM2 [10.84] allows gaps in the matched primary sequence, but it does not include these gaps in the output motifs.

An alternative approach offered by *Weeder* [10.38] involves building suffix trees from a set of input sequences. These are used to find all patterns of a set length, occurring in at least a certain number of sequences, with an upper limit on the number of mismatches (mutations). The program may be run in an automated way multiple times to detect patterns of different lengths.

The TEIRESIAS [10.39] algorithm is not restricted to searching for patterns of specific length and can detect gapped patterns. This is computationally more intensive, in both memory and processing requirements. The large number of results requires further processing for statistical significance.

The application of these methods is shown in overview in Fig. 10.4. Further methods are reviewed and benchmarked elsewhere [10.85].

10.5.2 Secondary Structure Prediction for Structured Element Discovery

Relatively few confirmed secondary or tertiary structures for *cis*-regulatory elements are available. Therefore, predictions of RNA structures are made computationally. High-throughput methods that may allow more structures to be determined experimentally are becoming available [10.86]. However, these methods are limited, and a combination of bioinformatic and high-throughput experiment has been successful [10.18].

In addition, it may be possible to predict the three-dimensional structures of RNAs using bioinformatics, which will become increasingly feasible as the number of known structures increases [10.87]. Packages are available to assist in tertiary structural predic-

tion [10.88, 89]. These may be used with sequences of interest alone or in combination with available experimental data on similar structures.

Single Sequences
Predicting folding on individual sequences is a common technique. This may be done globally (for the entire mRNA) or locally on windows within a biologically relevant section (e.g., 80-base windows in 3′ UTRs). A global RNA fold prediction algorithm from the *Zuker* laboratory is implemented by the mfold [10.43] program. These methods are commonly used, and the paper associated with this program has over 700 citations in the literature. The Vienna RNA package provides a similar program, RNAfold [10.44]. Like mfold, this calculates predicted secondary structure for RNAs based on minimum free energies (MFE) using conformations derived from published values for stacking and destabilizing energies.

The UNAFold [10.90] software is a development of mfold and further predicts hybridization and melting profiles. Like RNAfold, it produces dotplots showing pairing probabilities over the sequence (Fig. 10.5). The dotplots from both programs include the pairing probabilities corresponding to suboptimal (predicted) folding of the input sequence. Suboptimal structures are discussed later.

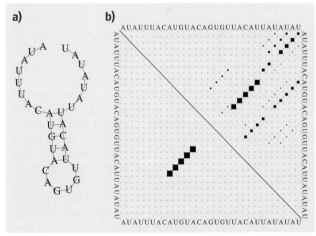

Fig. 10.5a,b Predicted secondary structures for the IRE in human *CDC14A*. (**a**) Optimal MFE structure from RNAfold (-2.9 kcal/mol). The dot plot (**b**) represents the ensemble of structures. Some suboptimal structures contain predicted pairs in a lower stem, their presence being more consistent with other IRE structures. The C–G base pair in the loop observed in the experimental structure (**b**) is not predicted by MFE methods

Local folding methods are likely to be more appropriate for small *cis*-regulatory elements. Several algorithms which are particularly suitable have been developed. These perform better on known *cis*-regulatory elements [10.91]. This is an area of active development and testing; current methods include RNAplfold and Rfold [10.92, 93]. In addition, methods that combine structure prediction with comparison with known secondary structures of homologous sequences are being developed [10.94].

RNA structure is dynamic. Binding interactions with a structured *cis*-regulatory element are likely to further influence the structure of that element. A limited set of thermodynamic measurements provides the basis of the RNA structure predicting algorithms. Furthermore, these methods do not account for pseudoknots, G-quartets, and other structures. The predicted structures cannot be expected to entirely model the complex molecular interactions found in the in vivo environment. Approaches using homology to known secondary and tertiary structures that are not necessarily the MFE structures should assist in this [10.87, 94, 95].

Some elements may contain pseudoknots, but the prediction of pseudoknots remains computationally slow and is only tractable for shorter RNA sequences. The pknots [10.96] program and HotKnots [10.97] are methods that may be employed on targeted regions if such structures are expected [10.98]. There are *cis*-regulatory elements containing pseudoknots, for example, frameshift elements. Further examples may be found in general and specific databases [10.15, 99].

Predicted structures do not always match natively observed structures. An example is the iron responsive element (IRE) found in the human mRNA for *CDC14A*. Both the UNAfold and RNAFold program find the same MFE. Additional base pairs are seen in the dotplot, some of which would make a lower stem observed in all IREs tested. Some pairs are not predicted at all by MFE approaches, e.g., the C–G in the apical loop (Fig. 10.5).

When there is existing information about the structure of a particular *cis*-regulatory element, these constraints may be provided to the folding programs. This allows the estimation of the MFE in a candidate element even when this is not the lowest resulting from prediction. Other sources, such as gene expression and phylogenetic information, may then be used in combination to arrive at strong candidates for experimental testing.

Benchmarking on ncRNA genes has shown both sensitivity and specificity of RNA structure prediction using MFE methods to be limited (22–63% and 20–60%, respectively) [10.100]. Newer algorithms improve on this [10.101]. Although secondary and tertiary structure is a factor in RNA interaction, the difficulty of experimentally determining these structures and of accurately predicting them must be always borne in mind.

These programs do not allow noncanonical bases, e.g., U–U or A–G, which have been observed in many experimentally determined RNA structures. Several algorithms do allow predictions that include these noncanonical pairs [10.89, 101]. Though these are considerably slower, they may be used with short structured *cis*-regulatory elements.

Multiple Sequences

Calculating the consensus structure for aligned sequences can overcome some of the shortcomings in the accuracy of MFE calculations for single sequences. This approach depends on the ability to obtain a reasonable alignment of the primary sequences. Methods available include RNAalifold [10.45] – part of the Vienna RNA package.

Covariance (e.g., an A–U base pair being exchanged by some other pairing) or compatible mutation (e.g., a G–C base pair being exchanged by a G–U pair) help tools such as RNAalifold to provide an optimal structure consistent with the alignment. However, too many variations in the primary sequence will make alignment at this level impossible.

A further class of algorithm simultaneously folds and aligns input sequences. The approach is computationally more intensive, though may be useful where the primary sequence alignments have limited similarity [10.41]. Dynalign, Multilign, and Turbofold are all part of the RNAstructure package [10.14]. The original Dynalign works with two sequences – Multilign operates on multiple sequences. Turbofold does not produce an alignment but presents separate structures for each of the sequences in the input, rather than one consensus structure. Structures are predicted based on pairing probabilities in each sequence severally, combined with the pairing probabilities in a consensus model [10.102].

Foldalign and FoldalignM [10.103] will produce local and global alignments along with structure predictions. An interesting feature of Foldalign is that it will attempt local alignments over the input sequences (based on structure and primary sequence) and then report the best alignments found. It is therefore also an element detection tool and not limited to structural prediction. Additional discussion and comparison of RNA fold prediction algorithms can be found in recent reviews [10.14, 87, 104].

10.5.3 Comparing Secondary Structures

It is useful to compare two structures to assess their similarity. In some cases the best information about a regulatory element may be its secondary structure, and so this is key in finding similar elements. RNAdistance, RNAforester, and Cofolga2mo [10.105] allow these comparisons for simple secondary structures. A simplification of the problem is to compare overall shapes, e.g., stem loop or cloverleaf [10.106].

The Vienna RNA package provides RNAdistance [10.44], which allows not only comparison of pairs of structures but also simultaneous comparison of multiple structures, providing a comparison matrix for all input structures. The output quantifies the differences between the structures. A pipeline on the Vienna RNA servers, *Structure Conservation Analysis*, includes this method. Input is an alignment and RNAalifold is used to predict MFE structures that are compared with predicted structures of individual sequences.

The RNAforester [10.46] algorithm builds tree-like data structures that represent RNA secondary structure. These can then be used to build multiple alignments of different RNA structures. Thresholds may be applied determining whether a particular structure is sufficiently similar to form part of an aligned group. This allows the degree of similarity between structures to be assessed as well as the grouping of RNAs into structurally determined families. Another useful output of this tool is an alignment of input sequences that is wholly determined by the given structures. This can be useful to build a seed for a covariance model.

10.5.4 Searching for Secondary or Tertiary Structural Elements

When there is good evidence for structure but the specificity of the primary sequence in the regulatory element is largely or completely unknown, a search based on structure alone is required. RNAMotif [10.48] allows the creation of a pattern which has no or little information about primary sequence. The resulting matches may be used to find additional sequences for testing; alignments of these will hopefully allow the incorporation of primary sequence information into the model for the regulatory element in question. Three-dimensional (3-D) motifs (for example, *G-bulges* from a lysine riboswitch can be searched for with RMdetect [10.95].

10.5.5 Combining Primary and Secondary Structural Search Methods

A covariance model (CM) is a stochastic context-free grammar (SCFG) that can be used to model the consensus sequence and structure of RNAs. Given an alignment to a consensus secondary structure, not only nucleotide residues at single-stranded positions but also base pairings, insertions, and deletions are scored. The Infernal [10.50] software package provides the tools to build covariance models and to search for matches to the model over a target sequence. The resulting *bit score* is the log-odds ratio of the probability of the target matching the model to the probability of target matching random sequence. This methodology is key to the Rfam database which catalogs RNA families using these models, showing their paralogs and homologs [10.15].

When sufficient information exists about a *cis*-regulatory element such that known examples may be meaningfully aligned to a consensus structure, a covariance model may be constructed. This model may be used to search for matches within other mRNA sequences – as has been done in the case of the IRE [10.12] and other *cis*-regulatory elements in the Rfam database. This can result in new candidates for experiment testing.

An alternative to using covariance models is to build a pattern corresponding to a motif. This is the approach taken by Transterm and UTRsite. A useful tool for interpreting and searching using such patterns is scan_for_matches [10.51]. The pattern descriptions used by this tool can incorporate structural information. Such a pattern does not depend on being able to construct an alignment of known sequence elements. However, complex patterns can be difficult to construct, and the ability to make a good pattern depends on prior knowledge of a motif, including permissible variations at different points. The output from scan_for_matches does not include the statistical information provided by the Infernal software's method.

10.6 Combinatorial Methods

Figures 10.6 and 10.4 give an overview of how the tools discussed here may be used together in the search for *cis*-regulatory elements. Some combinations of tools are sufficiently novel to be considered new methods in their own right. These methods are discussed here.

Unpaired bases are more likely to be involved directly with RNA interaction, and certain protein interactions. Consequently, primary sequence patterns found in unpaired regions are of particular interest. This can be seen for example in the IRE, where the unpaired nucleotides have been shown to interact with the iron regulatory proteins. MEMERIS [10.53] is an extension of the MEME algorithm. A script that comes with the package first uses the Vienna RNA tools to predict secondary structure across the input sequences (which must be in FASTA format with all sequence on one line). Altering the prior probabilities for putative motif start sites directs the MEME algorithm towards predicted single-stranded positions. The weakness of not being able to detect gapped patterns is inherited from MEME.

Given the importance of structure in *cis*-regulatory elements, another useful approach is to identify sequence regions likely to have conserved structure. RNAz [10.54] was originally developed to detect structured ncRNAs in genomic sequence, but this method is also of potential interest in *cis*-regulatory element

discovery and can successfully detect several known elements. The software takes a sequence alignment as input and uses RNAalifold to calculate the MFE of consensus structures arising over a sliding window. MFE values for these same sequences are also severally calculated using RNAfold. The consensus MFE in ratio to the average single sequence MFE gives a structure conservation index (SCI). Additionally, a z-score is calculated, which represents the deviation of the MFE score from random sequence of similar composition and length. The z-score and the SCI are used as inputs to a support vector machine together with the number of aligned sequences and the pairwise identity of the sequences – this results in a probability value for the occurrence of a conserved structural motif. RNAz may be run from the command line or on the Vienna RNA servers. Also the ncRNA [10.23] site has a version of the UCSC genome browser that includes a track for sites predicted by RNAz.

For multiple sequences with common function, automated production of covariance models for elements with similar sequence and structure is provided by the CMFinder [10.49] program, which takes short unaligned sequences ($< 500\,\text{bp}$) as input. Based on MFE, the algorithm first selects a number of candidates from within these sequences. The candidates are

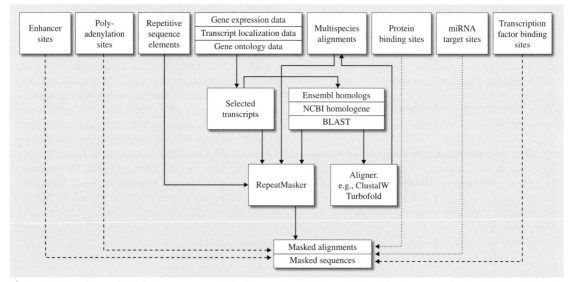

Fig. 10.6 Preparing and obtaining sequences and alignments. *Solid lines* indicate basic approach. *Dashed lines* indicate additional steps appropriate to genome-scale analysis or de novo element discovery. *Dotted lines* indicate additional processes dependent on the specific problem

aligned according to predicted secondary structure and an expectation-maximization algorithm used to refine a covariance model that identifies elements within the candidate sequences distinct from a background distribution. The input sequences are then rescanned using this covariance model, and the top hits are included as candidates. The authors of CMFinder note that identifying larger motifs is problematic and go some way to addressing this by attempting to merge smaller motifs as a final step.

The importance of incorporating biological data into the element discovery process has already been discussed. Selection of sequence for pattern discovery together with appropriate background models is an important first step in many analyses. The finding informative regulatory elements (FIRE) [10.55] pipeline automates this process by combining the detection of primary sequence patterns with other biological data – either discrete or continuous data (e.g., gene expression data). FIRE starts by looking for 7-mer *seed* motifs, but these can be extended one base in either direction. The initial *seeds* are systematically modified using degenerate International Union of Pure and Applied Chemistry (IUPAC) codes to arrive at a motif most significantly associated with the other biological data. This pipeline also offers the convenience of displaying gene ontology terms with enriched association to the identified motifs.

10.7 Conclusions and Future Prospects

The importance of posttranscriptional regulation is becoming increasingly apparent as large-scale proteomic data become available. Transcripts are translated with a wide range of efficiencies, giving differing numbers of functional proteins per message.

It has now become almost routine to measure transcript levels at a genome scale using microarrays or next-generation sequencing. The expectation was that transcript abundance would provide good estimates of protein abundance in the cell. However, the early studies done over a decade ago that suggested that mRNA levels might predict less than 45% of protein levels have been reiterated by recent studies. This indication of widespread posttranscriptional control is seen in many organisms.

High-throughput *wet-lab* studies and analysis of regulatory elements will facilitate discovery of elements with widely conserved functions. However, it should also be noted that some key elements might only be found in a small number of messages or species, e.g., human-specific miRNA targets, or the targets of other noncoding RNAs. These exceptional elements are a challenge for both bioinformatic and wet-lab studies and may also be of critical importance for cell growth and development, and have applications in biotechnology.

At least some of the variation in the amount of protein translated from individual mRNAs will be mediated by *cis*-regulatory elements in the mRNAs. This chapter has outlined current bioinformatic methods available for their discovery. Development of new methods and the use of high-throughput data on a genome-wide scale, particularly comparative genomic, proteomic, and high-throughput transcriptomic data, will facilitate this.

References

10.1   C. Vogel: Translation's coming of age, Mol. Syst. Biol. **7**, 498 (2011)
10.2   S.A. Tenenbaum, J. Christiansen, H. Nielsen: The post-transcriptional operon, Methods Mol. Biol. **703**, 237–245 (2011)
10.3   D.J. Hogan, D.P. Riordan, A.P. Gerber, D. Herschlag, P.O. Brown: Diverse RNA-binding proteins interact with functionally related sets of RNAs, suggesting an extensive regulatory system, PLoS Biology **6**, e255 (2008)
10.4   M.J. Moore: From birth to death: The complex lives of eukaryotic mRNAs, Science **309**, 1514–1518 (2005)
10.5   P.A. Galante, D. Sandhu, R. de Sousa Abreu, M. Gradassi, N. Slager, C. Vogel, S.J. de Souza, L.O. Penalva: A comprehensive in silico expression analysis of RNA binding proteins in normal and tumor tissue: Identification of potential players in tumor formation, RNA Biology **6**, 426–433 (2009)
10.6   N.G. Tsvetanova, D.M. Klass, J. Salzman, P.O. Brown: Proteome-wide search reveals unexpected RNA-binding proteins in *Saccharomyces cerevisiae*, PLoS One **5**, e12671 (2010)
10.7   D.P. Bartel: MicroRNAs: Target recognition and regulatory functions, Cell **136**, 215–233 (2009)

Part B | 10

10.8   P. Chartrand, X.H. Meng, S. Huttelmaier, D. Donato, R.H. Singer: Asymmetric sorting of ash1p in yeast results from inhibition of translation by localization elements in the mRNA, Mol. Cell. **10**, 1319–1330 (2002)

10.9   M.W. Hentze, S.W. Caughman, T.A. Rouault, J.G. Barriocanal, A. Dancis, J.B. Harford, R.D. Klausner: Identification of the iron-responsive element for the translational regulation of human ferritin mRNA, Science **238**, 1570–1573 (1987)

10.10   S. Castellano, V.N. Gladyshev, R. Guigo, M.J. Berry: SelenoDB 1.0: A database of selenoprotein genes, proteins and SECIS elements, Nucleic Acids Res. **36**, D332–338 (2008)

10.11   M. Davila Lopez, T. Samuelsson: Early evolution of histone mRNA 3' end processing, RNA **14**, 1–10 (2008)

10.12   S.G. Stevens, P.P. Gardner, C. Brown: Two covariance models for iron-responsive elements, RNA Biology **8**, 792–801 (2011)

10.13   R. Backofen, S.H. Bernhart, C. Flamm, C. Fried, G. Fritzsch, J. Hackermüller, J. Hertel, I.L. Hofacker, K. Missal, A. Mosig, S.J. Prohaska, D. Rose, P.F. Stadler, A. Tanzer, S. Washietl, S. Will: RNAs everywhere: Genome-wide annotation of structured RNAs, J. Exp. Zool. B **308**, 1–25 (2007)

10.14   D.H. Mathews, W.N. Moss, D.H. Turner: Folding and finding RNA secondary structure, Cold Spring Harb. Perspect. Biol. **2**, a003665 (2010)

10.15   P.P. Gardner, J. Daub, J. Tate, B.L. Moore, I.H. Osuch, S. Griffiths-Jones, R.D. Finn, E.P. Nawrocki, D.L. Kolbe, S.R. Eddy, A. Bateman: Rfam: Wikipedia, clans and the "decimal" release, Nucleic Acids Res. **39**, D141–145 (2011)

10.16   M. Andronescu, V. Bereg, H.H. Hoos, A. Condon: RNA STRAND: The RNA secondary structure and statistical analysis database, BMC Bioinformatics **9**, 340 (2008)

10.17   K. Rother, M. Rother, M. Boniecki, T. Puton, J.M. Bujnicki: RNA and protein 3-D structure modeling: Similarities and differences, J. Mol. Model. **17**, 2325–2336 (2011)

10.18   J.G. Underwood, A.V. Uzilov, S. Katzman, C.S. Onodera, J.E. Mainzer, D.H. Mathews, T.M. Lowe, S.R. Salama, D. Haussler: FragSeq: Transcriptome-wide RNA structure probing using high-throughput sequencing, Nat. Methods **7**, 995–1001 (2010)

10.19   D.P. Riordan, D. Herschlag, P.O. Brown: Identification of RNA recognition elements in the *Saccharomyces cerevisiae* transcriptome, Nucleic Acids Res. **39**, 1501–1509 (2011)

10.20   Y. Wan, M. Kertesz, R.C. Spitale, E. Segal, H.Y. Chang: Understanding the transcriptome through RNA structure, Nat. Rev. Genet. **12**, 641–655 (2011)

10.21   S. Kishore, S. Luber, M. Zavolan: Deciphering the role of RNA-binding proteins in the post-transcriptional control of gene expression, Brief Funct. Genomics **9**, 391–404 (2010)

10.22   W.J. Kent, C.W. Sugnet, T.S. Furey, K.M. Roskin, T.H. Pringle, A.M. Zahler, D. Haussler: The human genome browser at UCSC, Genome Res. **12**, 996–1006 (2002)

10.23   T. Mituyama, K. Yamada, E. Hattori, H. Okida, Y. Ono, G. Terai, A. Yoshizawa, T. Komori, K. Asai: The Functional RNA Database 3.0: Databases to support mining and annotation of functional RNAs, Nucleic Acids Res. **37**, D89–92 (2009)

10.24   J. Goecks, A. Nekrutenko, J. Taylor: Galaxy: A comprehensive approach for supporting accessible, reproducible, and transparent computational research in the life sciences, Genome Biol. **11**, R86 (2010)

10.25   J. Jurka: Repbase update: A database and an electronic journal of repetitive elements, Trends Genet. **16**, 418–420 (2000)

10.26   V. Matys, O.V. Kel-Margoulis, E. Fricke, I. Liebich, S. Land, A. Barre-Dirrie, I. Reuter, D. Chekmenev, M. Krull, K. Hornischer, N. Voss, P. Stegmaier, B. Lewicki-Potapov, H. Saxel, A.E. Kel, E. Wingender: TRANSFAC and its module TRANSCompel: Transcriptional gene regulation in eukaryotes, Nucleic Acids Res. **34**, D108–110 (2006)

10.27   J.C. Bryne, E. Valen, M.H. Tang, T. Marstrand, O. Winther, I. da Piedade, A. Krogh, B. Lenhard, A. Sandelin: JASPAR, the open access database of transcription factor-binding profiles: New content and tools in the 2008 update, Nucleic Acids Res. **36**, D102–106 (2008)

10.28   S. Mahony, P.V. Benos: STAMP: A web tool for exploring DNA-binding motif similarities, Nucleic Acids Res. **35**, W253–258 (2007)

10.29   K.D. Pruitt, T. Tatusova, W. Klimke, D.R. Maglott: NCBI Reference Sequences: Current status, policy and new initiatives, Nucleic Acids Res. **37**, D32–36 (2009)

10.30   G.H. Jacobs, A. Chen, S.G. Stevens, P.A. Stockwell, M.A. Black, W.P. Tate, C.M. Brown: Transterm: A database to aid the analysis of regulatory sequences in mRNAs, Nucleic Acids Res. **37**, D72–76 (2009)

10.31   G. Grillo, A. Turi, F. Licciulli, F. Mignone, S. Liuni, S. Banfi, V.A. Gennarino, D.S. Horner, G. Pavesi, E. Picardi, G. Pesole: UTRdb and UTRsite (RELEASE 2010): A collection of sequences and regulatory motifs of the untranslated regions of eukaryotic mRNAs, Nucleic Acids Res. **38**, D75–80 (2009)

10.32   K.B. Cook, H. Kazan, K. Zuberi, Q. Morris, T.R. Hughes: RBPDB: A database of RNA-binding specificities, Nucleic Acids Res. **39**, D301–308 (2010)

10.33   B.P. Lewis, C.B. Burge, D.P. Bartel: Conserved seed pairing, often flanked by adenosines, indicates that thousands of human genes are microRNA targets, Cell **120**, 15–20 (2005)

10.34 A. Krek, D. Grun, M.N. Poy, R. Wolf, L. Rosenberg, E.J. Epstein, P. MacMenamin, I. da Piedade, K.C. Gunsalus, M. Stoffel, N. Rajewsky: Combinatorial microRNA target predictions, Nat. Genet. **37**, 495–500 (2005)

10.35 A. Kozomara, S. Griffiths-Jones: miRBase: Integrating microRNA annotation and deep-sequencing data, Nucleic Acids Res. **39**, D152–157 (2010)

10.36 D. Gaidatzis, E. van Nimwegen, J. Hausser, M. Zavolan: Inference of miRNA targets using evolutionary conservation and pathway analysis, BMC Bioinformatics **8**, 69 (2007)

10.37 T.L. Bailey, N. Williams, C. Misleh, W.W. Li: MEME: Discovering and analyzing DNA and protein sequence motifs, Nucleic Acids Res. **34**, W369–373 (2006)

10.38 G. Pavesi, G. Mauri, G. Pesole: An algorithm for finding signals of unknown length in DNA sequences, Bioinformatics **17**(Suppl. 1), S207–214 (2001)

10.39 I. Rigoutsos, A. Floratos: Combinatorial pattern discovery in biological sequences: The TEIRESIAS algorithm, Bioinformatics **14**, 55–67 (1998)

10.40 A.D. George, S.A. Tenenbaum: Web-based tools for studying RNA structure and function, Methods Mol. Biol. **703**, 67–86 (2011)

10.41 R.S. Hamilton, I. Davis: Identifying and searching for conserved RNA localisation signals, Methods Mol. Biol. **714**, 447–466 (2011)

10.42 Wikipedia: List of RNA structure prediction software (2012), available at http://en.wikipedia.org/wiki/List_of_RNA_structure_prediction_software

10.43 M. Zuker, P. Stiegler: Optimal computer folding of large RNA sequences using thermodynamics and auxiliary information, Nucleic Acids Res. **9**, 133–148 (1981)

10.44 I.L. Hofacker, W. Fontana, P.F. Stadler, L.S. Bonhoeffer, M. Tacker, P. Schuster: Fast folding and comparison of RNA secondary structures, Monatsh. Chem./Chem. Mon. **125**, 167–188 (1994)

10.45 S.H. Bernhart, I.L. Hofacker, S. Will, A.R. Gruber, P.F. Stadler: RNAalifold: Improved consensus structure prediction for RNA alignments, BMC Bioinformatics **9**, 474 (2008)

10.46 M. Hochsmann, B. Voss, R. Giegerich: Pure multiple RNA secondary structure alignments: A progressive profile approach, IEEE/ACM Trans. Comput. Biol. Bioinform. **1**, 53–62 (2004)

10.47 C.W. Wang, K.T. Chen, C.L. Lu: iPARTS: An improved tool of pairwise alignment of RNA tertiary structures, Nucleic Acids Res. **38**, W340–347 (2010)

10.48 T.J. Macke, D.J. Ecker, R.R. Gutell, D. Gautheret, D.A. Case, R. Sampath: RNAMotif, an RNA secondary structure definition and search algorithm, Nucleic Acids Res. **29**, 4724–4735 (2001)

10.49 Z. Yao, Z. Weinberg, W.L. Ruzzo: CMfinder – A covariance model based RNA motif finding algorithm, Bioinformatics **22**, 445–452 (2006)

10.50 E.P. Nawrocki, D.L. Kolbe, S.R. Eddy: Infernal 1.0: Inference of RNA alignments, Bioinformatics **25**, 1335–1337 (2009)

10.51 M. Dsouza, N. Larsen, R. Overbeek: Searching for patterns in genomic data, Trends Genet. **13**, 497–498 (1997)

10.52 T. Barrett, D.B. Troup, S.E. Wilhite, P. Ledoux, C. Evangelista, I.F. Kim, M. Tomashevsky, K.A. Marshall, K.H. Phillippy, P.M. Sherman, R.N. Muertter, M. Holko, O. Ayanbule, A. Yefanov, A. Soboleva: NCBI GEO: Archive for functional genomics data sets – 10 years on, Nucleic Acids Res. **39**, D1005–1010 (2010)

10.53 M. Hiller, R. Pudimat, A. Busch, R. Backofen: Using RNA secondary structures to guide sequence motif finding towards single-stranded regions, Nucleic Acids Res. **34**, e117 (2006)

10.54 S. Washietl, I.L. Hofacker, P.F. Stadler: Fast and reliable prediction of noncoding RNAs, Proc. Natl. Acad. Sci. USA **102**, 2454–2459 (2005)

10.55 O. Elemento, N. Slonim, S. Tavazoie: A universal framework for regulatory element discovery across all genomes and data types, Mol. Cell. **28**, 337–350 (2007)

10.56 E. Lecuyer, H. Yoshida, N. Parthasarathy, C. Alm, T. Babak, T. Cerovina, T.R. Hughes, P. Tomancak, H.M. Krause: Global analysis of mRNA localization reveals a prominent role in organizing cellular architecture and function, Cell **131**, 174–187 (2007)

10.57 P. Landgraf, M. Rusu, R. Sheridan, A. Sewer, N. Iovino, A. Aravin, S. Pfeffer, A. Rice, A.O. Kamphorst, M. Landthaler, C. Lin, N.D. Socci, L. Hermida, V. Fulci, S. Chiaretti, R. Foà, J. Schliwka, U. Fuchs, A. Novosel, R.U. Müller, B. Schermer, U. Bissels, J. Inman, Q. Phan, M. Chien, D.B. Weir, R. Choksi, G. De Vita, D. Frezzetti, H.I. Trompeter, V. Hornung, G. Teng, G. Hartmann, M. Palkovits, R. Di Lauro, P. Wernet, G. Macino, C.E. Rogler, J.W. Nagle, J. Ju, F.N. Papavasiliou, T. Benzing, P. Lichter, W. Tam, M.J. Brownstein, A. Bosio, A. Borkhardt, J.J. Russo, C. Sander, M. Zavolan, T. Tuschl: A mammalian microRNA expression atlas based on small RNA library sequencing, Cell **129**, 1401–1414 (2007)

10.58 D. Betel, M. Wilson, A. Gabow, D.S. Marks, C. Sander: The microRNA.org resource: Targets and expression, Nucleic Acids Res. **36**, D149–153 (2008)

10.59 M.A. Batzer, P.L. Deininger: Alu repeats and human genomic diversity, Nat. Rev. Genet. **3**, 370–379 (2002)

10.60 A. Smit, R. Hubley, P. Green: *RepeatMasker Open-3.0.* (1996–2010), available at http://www.repeatmasker.org

10.61 S. Hannenhalli: Eukaryotic transcription factor binding sites–modeling and integrative search methods, Bioinformatics **24**, 1325–1331 (2008)

10.62 D. Schmidt, M.D. Wilson, B. Ballester, P.C. Schwalie, G.D. Brown, A. Marshall, C. Kutter, S. Watt,

C.P. Martinez-Jimenez, S. Mackay, I. Talianidis, P. Flicek, D.T. Odom: Five-vertebrate ChIP-seq reveals the evolutionary dynamics of transcription factor binding, Science **328**, 1036–1040 (2010)

10.63   A. Visel, E.M. Rubin, L.A. Pennacchio: Genomic views of distant-acting enhancers, Nature **461**, 199–205 (2009)

10.64   L.A. Pennacchio, N. Ahituv, A.M. Moses, S. Prabhakar, M.A. Nobrega, M. Shoukry, S. Minovitsky, I. Dubchak, A. Holt, K.D. Lewis, I. Plajzer-Frick, J. Akiyama, S. De Val, V. Afzal, B.L. Black, O. Couronne, M.B. Eisen, A. Visel, E.M. Rubin: In vivo enhancer analysis of human conserved noncoding sequences, Nature **444**, 499–502 (2006)

10.65   S.F. Altschul, W. Gish, W. Miller, E.W. Myers, D.J. Lipman: Basic local alignment search tool, J. Mol. Biol. **215**, 403–410 (1990)

10.66   S.R. Eddy: Profile hidden Markov models, Bioinformatics **14**, 755–763 (1998)

10.67   J.O. Deshler, M.I. Highett, B.J. Schnapp: Localization of Xenopus Vg1 mRNA by Vera protein and the endoplasmic reticulum, Science **276**, 1128–1131 (1997)

10.68   D. Gautreau, C.A. Cote, K.L. Mowry: Two copies of a subelement from the Vg1 RNA localization sequence are sufficient to direct vegetal localization in Xenopus oocytes, Development **124**, 5013–5020 (1997)

10.69   S. Kwon, T. Abramson, T.P. Munro, C.M. John, M. Kohrmann, B.J. Schnapp: UUCAC- and vera-dependent localization of VegT RNA in Xenopus oocytes, Curr. Biol. **12**, 558–564 (2002)

10.70   S. Choo, B. Heinrich, J.N. Betley, Z. Chen, J.O. Deshler: Evidence for common machinery utilized by the early and late RNA localization pathways in Xenopus oocytes, Dev. Biol. **278**, 103–117 (2005)

10.71   B.B. Andken, I. Lim, G. Benson, J.J. Vincent, M.T. Ferenc, B. Heinrich, L.A. Jarzylo, H.Y. Man, J.O. Deshler: 3'-UTR SIRF: A database for identifying clusters of short interspersed repeats in 3' untranslated regions, BMC Bioinformatics **8**, 274 (2007)

10.72   P.P. Tam, I.H. Barrette-Ng, D.M. Simon, M.W. Tam, A.L. Ang, D.G. Muench: The Puf family of RNA-binding proteins in plants: Phylogeny, structural modeling, activity and subcellular localization, BMC Plant Biol. **10**, 44 (2010)

10.73   I. Tuszynska, J.M. Bujnicki: DARS-RNP and QUASI-RNP: New statistical potentials for protein-RNA docking, BMC Bioinformatics **12**, 348 (2011)

10.74   M. Kaller, S.T. Liffers, S. Oeljeklaus, K. Kuhlmann, S. Roh, R. Hoffmann, B. Warscheid, H. Hermeking: Genome-wide characterization of miR-34a induced changes in protein and mRNA expression by a combined pulsed SILAC and microarray analysis, Mol. Cell. Proteomics **10**(M111), 010462 (2011)

10.75   S.D. Hsu, F.M. Lin, W.Y. Wu, C. Liang, W.C. Huang, W.L. Chan, W.T. Tsai, G.Z. Chen, C.J. Lee, C.M. Chiu, C.H. Chien, M.C. Wu, C.Y. Huang, A.P. Tsou, H.D. Huang: miRTarBase: A database curates experimentally validated microRNA-target interactions, Nucleic Acids Res. **39**, D163–169 (2011)

10.76   M. Thomas, J. Lieberman, A. Lal: Desperately seeking microRNA targets, Nat. Struct. Mol. Biol. **17**, 1169–1174 (2010)

10.77   F. Xiao, Z. Zuo, G. Cai, S. Kang, X. Gao, T. Li: miRecords: An integrated resource for microRNA-target interactions, Nucleic Acids Res. **37**, D105–110 (2009)

10.78   H. Dweep, C. Sticht, P. Pandey, N. Gretz: miRWalk-database: Prediction of possible miRNA binding sites by "walking" the genes of three genomes, J. Biomed. Inform. **44**, 839–847 (2011)

10.79   R.C. Friedman, K.K. Farh, C.B. Burge, D.P. Bartel: Most mammalian mRNAs are conserved targets of microRNAs, Genome Res. **19**, 92–105 (2009)

10.80   A. Grimson, K.K. Farh, W.K. Johnston, P. Garrett-Engele, L.P. Lim, D.P. Bartel: MicroRNA targeting specificity in mammals: Determinants beyond seed pairing, Mol. Cell. **27**, 91–105 (2007)

10.81   K. Chen, N. Rajewsky: Natural selection on human microRNA binding sites inferred from SNP data, Nat. Genet. **38**, 1452–1456 (2006)

10.82   B. John, A.J. Enright, A. Aravin, T. Tuschl, C. Sander, D.S. Marks: Human MicroRNA targets, PLoS Biol. **2**, e363 (2004)

10.83   D. Betel, A. Koppal, P. Agius, C. Sander, C. Leslie: Comprehensive modeling of microRNA targets predicts functional non-conserved and non-canonical sites, Genome Biol. **11**, R90 (2010)

10.84   M.C. Frith, N.F. Saunders, B. Kobe, T.L. Bailey: Discovering sequence motifs with arbitrary insertions and deletions, PLoS Comput. Biol. **4**, e1000071 (2008)

10.85   M. Tompa, N. Li, T.L. Bailey, G.M. Church, B. De Moor, E. Eskin, A.V. Favorov, M.C. Frith, Y. Fu, W.J. Kent, V.J. Makeev, A.A. Mironov, W.S. Noble, G. Pavesi, G. Pesole, M. Regnier, N. Simonis, S. Sinha, G. Thijs, J. van Helden, M. Vandenbogaert, Z. Weng, C. Workman, C. Ye, Z. Zhu: Assessing computational tools for the discovery of transcription factor binding sites, Nat. Biotechnol. **23**, 137–144 (2005)

10.86   E. Westhof, P. Romby: The RNA structurome: High-throughput probing, Nat. Methods **7**, 965–967 (2010)

10.87   E. Westhof, B. Masquida, F. Jossinet: Predicting and modeling RNA architecture, Cold Spring Harb. Perspect. Biol. **3**, a003632 (2011)

10.88   F. Jossinet, T.E. Ludwig, E. Westhof: Assemble: An interactive graphical tool to analyze and build RNA architectures at the 2-D and 3-D levels, Bioinformatics **26**, 2057–2059 (2010)

10.89 M. Parisien, F. Major: The MC-Fold and MC-Sym pipeline infers RNA structure from sequence data, Nature **452**, 51–55 (2008)

10.90 N.R. Markham, M. Zuker: UNAFold: Software for nucleic acid folding and hybridization, Methods Mol. Biol. **453**, 3–31 (2008)

10.91 S.J. Lange, D. Maticzka, M. Mohl, J.N. Gagnon, C.M. Brown, R. Backofen: Global or local? Predicting secondary structure and accessibility in mRNAs, Nucleic Acids Res. **40**, 5215–5216 (2012)

10.92 S.H. Bernhart, U. Muckstein, I.L. Hofacker: RNA accessibility in cubic time, Algorithms Mol. Biol. **6**, 3 (2011)

10.93 H. Kiryu, G. Terai, O. Imamura, H. Yoneyama, K. Suzuki, K. Asai: A detailed investigation of accessibilities around target sites of siRNAs and miRNAs, Bioinformatics **27**, 1788–1797 (2011)

10.94 M. Hamada, K. Yamada, K. Sato, M.C. Frith, K. Asai: CentroidHomfold-LAST: Accurate prediction of RNA secondary structure using automatically collected homologous sequences, Nucleic Acids Res. **39**, W100–W106 (2011)

10.95 J.A. Cruz, E. Westhof: Sequence-based identification of 3-D structural modules in RNA with RMDetect, Nat. Methods **8**, 513–519 (2011)

10.96 E. Rivas, S.R. Eddy: A dynamic programming algorithm for RNA structure prediction including pseudoknots, J. Mol. Biol. **285**, 2053–2068 (1999)

10.97 J. Ren, B. Rastegari, A. Condon, H.H. Hoos: HotKnots: Heuristic prediction of RNA secondary structures including pseudoknots, RNA **11**, 1494–1504 (2005)

10.98 S. Bellaousov, D.H. Mathews: ProbKnot: Fast prediction of RNA secondary structure including pseudoknots, RNA **16**, 1870–1880 (2010)

10.99 M. Bekaert, A.E. Firth, Y. Zhang, V.N. Gladyshev, J.F. Atkins, P.V. Baranov: Recode-2: New design, new search tools, and many more genes, Nucleic Acids Res. **38**, D69–D74 (2010)

10.100 P.P. Gardner, R. Giegerich: A comprehensive comparison of comparative RNA structure prediction approaches, BMC Bioinformatics **5**, 140 (2004)

10.101 C.H. zu Siederdissen, S.H. Bernhart, P.F. Stadler, I.L. Hofacker: A folding algorithm for extended RNA secondary structures, Bioinformatics **27**, i129–136 (2011)

10.102 A.O. Harmanci, G. Sharma, D.H. Mathews: TurboFold: Iterative probabilistic estimation of secondary structures for multiple RNA sequences, BMC Bioinformatics **12**, 108 (2011)

10.103 E. Torarinsson, J.H. Havgaard, J. Gorodkin: Multiple structural alignment and clustering of RNA sequences, Bioinformatics **23**, 926–932 (2007)

10.104 C.M. Reidys, F.W. Huang, J.E. Andersen, R.C. Penner, P.F. Stadler, M.E. Nebel: Topology and prediction of RNA pseudoknots, Bioinformatics **27**, 1076–1085 (2011)

10.105 A. Taneda: An efficient genetic algorithm for structural RNA pairwise alignment and its application to non-coding RNA discovery in yeast, BMC Bioinformatics **9**, 521 (2008)

10.106 S. Janssen, R. Giegerich: Faster computation of exact RNA shape probabilities, Bioinformatics **26**, 632–639 (2010)

11. Protein Modeling and Structural Prediction

Sebastian Kelm, Yoonjoo Choi, Charlotte M. Deane

Proteins perform crucial functions in every living cell. The genetic information in every organism's DNA encodes the protein's amino acid sequence, which determines its three-dimensional structure, which, in turn, determines its function. In this postgenomic era, protein sequence information can be obtained relatively easily through experimental means. Sequence databases already contain millions of protein sequences and continue to grow. Structural information, however, is harder to obtain through experimental means – we currently know the structure of about 75 000 proteins. Knowledge of a protein's structure is extremely useful in understanding its molecular function and in developing drugs that bind to it. Thus, computational techniques have been developed to bridge the ever-increasing gap between the number of known protein sequences and structures.

In addition to proteins in general, this chapter discusses the specific importance of membrane proteins, which make up about one-third of all known proteins. Membrane proteins control communication and transport into and out of every living cell and are involved in many medically important processes. Over half of current drug targets are membrane proteins.

Part B | 11

A brief introduction to protein sequence and structure is followed by an overview of common techniques used in the process of computational protein structure prediction. Emphasis is put on two particularly hard problems, namely protein loop modeling and the structural prediction of membrane proteins.

11.1 Proteins

Proteins play important roles in living organisms. They perform crucial functions in all biological processes. They serve as structural building blocks as well as performing most of the cell's molecular functions [11.1]. In general, a protein's function is specified by its three-dimensional structure.

11.2 Protein Structure

It has been widely accepted since the early 1970s that in most cases a protein's sequence directly determines its three-dimensional (3-D) structure [11.3], which in turn determines its function [11.1] (Fig. 11.1). A protein's function can also be investigated without knowledge of its structure (e.g., using binding assays and enzyme kinetics assays). However, knowledge of a protein's

structure allows a fuller exploration of how a specific function is obtained.

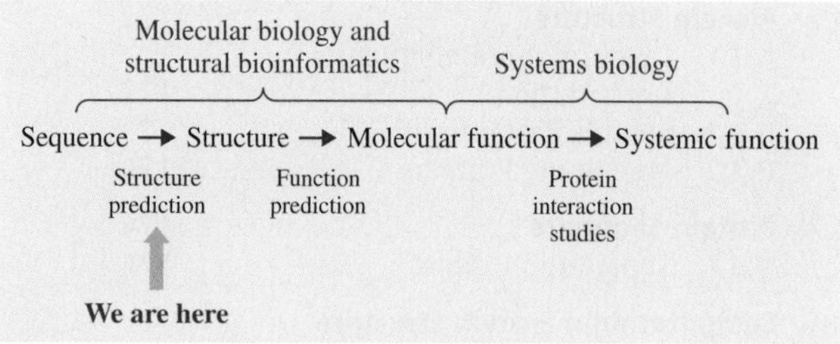

The building blocks of proteins are amino acids. The central atom of an amino acid is its α carbon, which has four distinctive groups linked to it: an amine group containing nitrogen and hydrogen atoms, a carboxyl group consisting of a carbon atom and an oxygen atom, an R group (side-chain) that specifies the amino acid, and a hydrogen atom. These form a tetrahedral structure.

Due to its tetrahedral nature and the four different substituents, two isomers are possible, L and D, which are mirror images of one another. Natural proteins consist of only the L isomer.

Amino acids are linked by peptide bonds, which are planar in character. All the bond lengths and angles in the backbone of a protein are tightly constrained.

Since all bond lengths and angles in the main chain of a protein are almost fixed, the main degrees of freedom can be described by dihedral rotations around the bonds. There are two possible rotations per amino acid of the backbone. These dihedral angles are denoted ϕ ($C_{i-1} - N_i - C\alpha_i - C_i$) and ψ ($N_i - C\alpha_i - C_i - N_{i+1}$). If $\phi = \psi = 180°$, then two adjacent residues are coplanar.

Ramakrishnan and *Ramachandran* described the Ramachandran plot (Fig. 11.2), which displays combinations of the ϕ and ψ dihedral angles in a two-dimensional space [11.4]. Due to steric clashes among adjacent side-chains, some Ramachandran areas are disallowed (the white blank areas in Fig. 11.2); For example, the large disallowed region from roughly 70° to 170° in ϕ dihedral angle is due to the steric clash between side-chain atoms and the backbone oxygen atom. Further distinctions can be calculated by examining the dihedral angles observed in known protein structures, which allow identification of favorable conformations. The Ramachandran plot also shows local patterns of a protein, i. e., secondary structures.

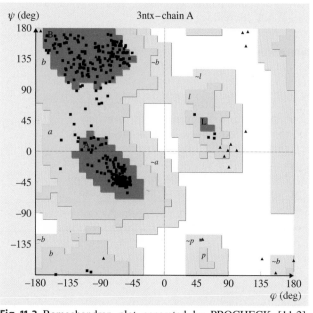

Fig. 11.1 The context of protein structure prediction in understanding biology. Structure prediction is an essential step in moving from the sequence of a gene to the understanding of the gene product's (i. e., the corresponding protein's) role in the organism

Fig. 11.2 Ramachandran plot generated by PROCHECK [11.2]. The *white areas* are disallowed dihedral angle regions. Favorable angles are colored in *red*. *a* indicates the α-helical region, *b* indicates β-sheets, and the *l* region is for left-handed α-helices. *Black dots* represent dihedral angles of the protein structure (PDB code 3NTX)

11.2.1 Levels of Protein Structure Organization

Primary Structure

The primary structure of a protein is the unfolded peptide chain of that protein. It is described by the sequence of residues along the protein. By convention, the primary structure reads from the amine group (N terminal) to the carboxyl group (C terminal) (Fig. 11.3a).

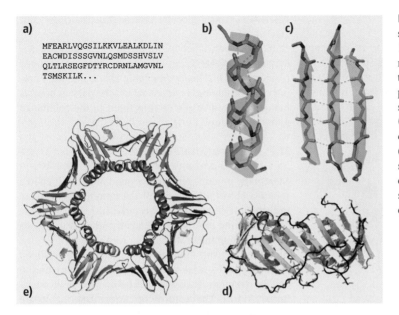

Fig. 11.3a–e The levels of protein structure organization (PDB code 1AXC). (**a**) A sequence is the primary structure of the protein. The three-dimensional structure of the protein consists of several regular secondary structures such as helices (**b**) and β-sheets (**c**). (**d**) The secondary structures are linked by loops (*colored black*) and form a tertiary structure. (**e**) Multiple folded proteins can be arranged together and construct a multisubunit complex, the quaternary structure

The primary structure also refers to the network of covalent bonds between amino acids that holds a protein together. The primary structure is critically important in protein folding. Anfinsen performed a series of experiments with bovine pancreatic RNase and demonstrated that a protein structure is in thermodynamic equilibrium with its environment and the interactions between its residues determine its three-dimensional structure [11.3]. This hypothesis is the basic assumption of most computational protein modeling.

Secondary Structure

Secondary structure in proteins is a repeating local pattern in three-dimensional space. There are two main patterns: α-helix and β-strand (Fig. 11.3b,c). Secondary structures are formally defined by hydrogen bonds and can be annotated by computational methods such as define secondary structure of proteins (DSSP) [11.5].

Helices are rod-like helical structures where the side-chains point outwards. They involve periodic hydrogen bonds between residue i and residue $i + 3$ (3_{10} helices), residue $i + 4$ (α-helices), or residue $i + 5$ (π-helices). Helices are mostly right-handed due to steric preferences; β-strands have an extended conformation and do not form local hydrogen bonds. Instead, they interact with other strands further along in the sequence to form larger structures known as β-sheets. According to the directions of adjacent β-strands, a β-sheet can be parallel or antiparallel.

All amino acid subsequences not fitting into these two categories are commonly referred to as random coils or loops.

Tertiary Structure

The tertiary structure of a protein is the overall shape of the polypeptide chain in three-dimensional space. A tertiary structure usually consists of several elements of secondary structure. Secondary structure elements (helices and strands) interact with each other to form compact structures, called *domains* (Fig. 11.3d).

Quaternary Structure

The quaternary structure is formed when several protein chains, called subunits, assemble into a larger multimeric complex (Fig. 11.3e). Multiple chains can form homomers (identical subunits), or can associate with other proteins to form heteromers. Assembly into a complex allows more complicated functions to be carried out, e.g., transcription of DNA to RNA by the RNA polymerase complex, which involves a multitude of different proteins and cofactors.

11.2.2 Soluble Proteins

Soluble proteins often adopt a globular (ball-like) conformation (Fig. 11.4). In their natural environment they are completely surrounded by water. Their structure is usually optimized to have hydrophobic residues in the protein core, while the water-exposed surface

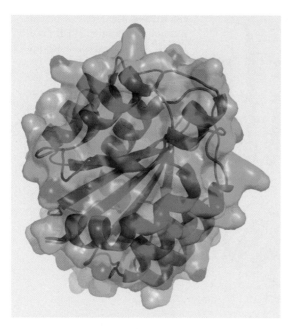

Fig. 11.4 A typical globular protein structure (PDB code 1QCY) ◄

contains a higher concentration of polar or charged residues [11.6]. This *hydrophobic effect*, i. e., the burying of hydrophobic residues inside the protein core, is thought to be a major driving force in the folding of globular proteins [11.6].

11.2.3 Membrane Proteins

Membrane proteins represent around 30% of all known proteins [11.7]. They control the communication and exchange of chemicals between the inside and outside of every living cell. Many medically relevant processes involve membrane proteins, e.g., recognition of foreign substances by the immune system, adhesion of blood cells to blood vessel walls, electron transport during cellular respiration, and recycling of used neurotransmitters [11.8]. The majority of current drug targets are membrane proteins [11.7].

11.3 Protein Sequence

Two proteins are said to be *homologous* if they have a common evolutionary ancestor. Homologous proteins are found to have the same or very similar molecular functions and structures.

The most common way to quantify homology is by sequence alignment. Each protein sequence is represented by a string of letters, each of which encodes a particular type of amino acid (e.g., "A" for alanine). If each sequence is one row in a matrix, the goal of sequence alignment is to shift the sequences left or right such that the maximum number of matrix columns contain two similar amino acids. In addition, gaps can be inserted, e.g., if the two sequences are of different length, or if they are only locally similar. Two components are required for sequence alignment:

1. A measure of how similar each type of amino acid is to the others
2. A scoring system to control the location and size of gaps.

Once the sequences have been aligned, the similarity between the two sequences can be computed (Fig. 11.5). Sequence alignment can be generalized to multiple sequence alignment or can involve structural information, e.g., in sequence-to-structure alignment.

11.3.1 Substitution Tables

Substitution tables are typically 20-by-20 matrices, each row and column representing one of the 20 amino acid types. Each value in such a table represents the likelihood (usually as a log-odds score) of observing the substitution from one particular type of amino acid to another in a pair of homologous proteins.

The most commonly used type of log-odds matrix is known as the blocks of amino acids substitution matrix (BLOSUM) [11.9]. Matrices of this type are built from the frequencies of amino acid substitutions observed in conserved blocks of sequences in local multiple alignments of closely related proteins. The most popular version is BLOSUM62 (Fig. 11.5b). The number in the name signifies the redundancy level of the sequences used to build the table. For BLOSUM62, proteins with no more than 62% sequence identity were used. This is done to avoid biasing the score matrix towards common sequences.

BLOSUM-like matrices have been constructed specifically for membrane proteins: JJT [11.10], PHAT [11.11], and SLIM [11.12]. These matrices mainly differ from BLOSUM matrices in that they were

Fig. 11.5a,b Sequence alignment. (**a**) An example sequence alignment of globular proteins (the last 39 residues are shown). One conserved leucine (L) residue is highlighted in *red*. One of the sequences shows a valine (which is similar in size and hydrophobicity) at the same position. (**b**) The BLOSUM62 substitution matrix. The substitutions from leucine to leucine (*red*) and from leucine to valine (*yellow*) are highlighted. Scores are log-odds scores (computed from substitution frequencies), where higher numbers indicate more favorable substitutions. Substitutions to self (e.g., L to L) indicate how conserved a residue tends to be. Note that BLOSUM matrices are symmetric ▶

a)

Name	Sequence
2sli_A	QVGLLYEKYDSWSRNELH-LKDILKFEKYSISELTGQA-
gi|109070532	QAPQLYVLYEKGRNHYTE-SISMAKISVYGTL-------
gi|109101506	LFGCLYEANDYEEIVFLM-FTLKQAFPAEYLPQ------
gi|170712512	TF--VLTSYGYWEKDYNKPYIKSLRVTLKEIDEIVREMV
gi|110674221	NIGLLYEGTPSEEMSYIE---MNLKYLESGANK------

b)

	G	A	V	L	M	I	F	Y	W	S	T	C	P	N	Q	K	R	H	D	E
G	6	0	-3	-4	-3	-4	-3	-3	-2	0	-2	-3	-2	0	-2	-2	-2	-2	-1	-2
A	0	4	0	-1	-1	-1	-2	-2	-3	1	0	0	-1	-2	-1	-1	-1	-2	-2	-1
V	-3	0	4	1	1	3	-1	-1	-3	-2	0	-1	-2	-3	-2	-2	-3	-3	-3	-2
L	-4	-1	1	4	2	2	0	-1	-2	-2	-1	-1	-3	-3	-2	-2	-2	-3	-4	-3
M	-3	-1	1	2	5	1	0	-1	-1	-1	-1	-1	-2	-2	0	-1	-1	-2	-3	-2
I	-4	-1	3	2	1	4	0	-1	-3	-2	-1	-1	-3	-3	-3	-3	-3	-3	-3	-3
F	-3	-2	-1	0	0	0	6	3	1	-2	-2	-2	-4	-3	-3	-3	-3	-1	-3	-3
Y	-3	-2	-1	-1	-1	-1	3	7	2	-2	-2	-2	-3	-2	-1	-2	-2	2	-3	-2
W	-2	-3	-3	-2	-1	-3	1	2	11	-3	-2	-2	-4	-4	-2	-3	-3	-2	-4	-3
S	0	1	-2	-2	-1	-2	-2	-2	-3	4	1	-1	-1	1	0	0	-1	-1	0	0
T	-2	0	0	-1	-1	-1	-2	-2	-2	1	5	-1	-1	0	-1	-1	-1	-2	-1	-1
C	-3	0	-1	-1	-1	-1	-2	-2	-2	-1	-1	9	-3	-3	-3	-3	-3	-3	-3	-4
P	-2	-1	-2	-3	-2	-3	-4	-3	-4	-1	-1	-3	7	-2	-1	-1	-2	-2	-1	-1
N	0	-2	-3	-3	-2	-3	-3	-2	-4	1	0	-3	-2	6	0	0	0	1	1	0
Q	-2	-1	-2	-2	0	-3	-3	-1	-2	0	-1	-3	-1	0	5	1	1	0	0	2
K	-2	-1	-2	-2	-1	-3	-3	-2	-3	0	-1	-3	-1	0	1	5	2	-1	-1	1
R	-2	-1	-3	-2	-1	-3	-3	-2	-3	-1	-1	-3	-2	0	1	2	5	0	-2	0
H	-2	-2	-3	-3	-2	-3	-1	2	-2	-1	-2	-3	-2	1	0	-1	0	8	-1	0
D	-1	-2	-3	-4	-3	-3	-3	-3	-4	0	-1	-3	-1	1	0	-1	-2	-1	6	2
E	-2	-1	-2	-3	-2	-3	-3	-2	-3	0	-1	-4	-1	0	2	1	0	0	2	5

constructed from datasets containing only the transmembrane domains of membrane proteins.

A further refinement of substitution matrices are environment-specific substitution tables (ESST) [11.13]. If the structure of a protein is known, then the protein can be partitioned into distinct structural environments, based on environmental factors; For example, if one considers only the environmental factors of accessibility and secondary structure, then one particular structural environment might be defined as residues on the protein surface and in an α-helix. Given enough proteins of known structure, one can generate substitution tables for each distinct structural environment. These can be used in sequence alignment and allow for more accurate alignments. This principle has been applied to membrane proteins by using substitution tables specific to membrane environments to improve alignment quality [11.14].

11.4 Computational Protein Structure Prediction

The Protein Data Bank (PDB) contains experimentally determined three-dimensional structures of proteins.

Although experimental determination is the most accurate way to obtain protein structures, it is costly and time-consuming. There were 71 158 protein structures deposited by December 2010. There were 20 333 nonredundant structures (sequence identity < 30%) and 3427 human protein structures in that set. Compared with the number of protein sequences known (about 12 million nonredundant sequences), only a small fraction of proteins have solved structures. Computational protein structure prediction is an alternative method to close this gap (Fig. 11.6).

Current protein structure modeling techniques are examined at the critical assessment of techniques for protein structure prediction (CASP) [11.15–17], a biannual community-wide experiment on protein structure prediction. The main goal of CASP is to examine the ability and limitation of current computational modeling methods. Protein modelers perform blind predictions of given protein sequences while experimentalists determine the structures. Broadly, computational protein structure prediction can be divided into two main categories: template-based modeling and template-free modeling [11.17].

11.4.1 Template-Free Modeling

Modeling without the use of templates is called ab initio or de novo prediction. It aims to simulate the protein folding process in silico. This is generally implemented by encoding the rules of chemistry and physics, combined with empirical knowledge about typical protein structures, in an energy function and then exploring a protein chain's conformational space while minimizing its energy. Such methods have, so far, produced useful results only for small proteins [11.17], below about 150 residues in length. The larger the protein, the more degrees of freedom need to be explored. A typical membrane protein is over 300 amino acids long (up to about 1000), with approximately 10 atoms per amino acid.

More recently, ab initio methods have begun to incorporate the principle of cotranslational folding [11.18, 19], which tends to decrease the simulation time until biologically plausible structures are observed [11.20]. While these approaches show interesting results in terms of their general folding behavior, their practical application in structure prediction of large proteins is still in its infancy.

Currently, the most successful template-free method is ROSETTA [11.21]. Its algorithm is based on covering the entire length of the protein sequence with small overlapping fragments, representing all possible local 3-D conformations. These fragments are combined in many different ways to form candidate tertiary structures, which are then scored using a statistical energy function based on empirical knowledge of protein structure. Candidate models are clustered in terms of overall similarity, and the program returns one representative per cluster. ROSETTA has consistently performed well [11.22–24]. Nevertheless, it cannot beat the accuracy of template-based methods, especially for molecules larger than about 150 residues, and requires large amounts of processing time.

11.4.2 Template-Based Modeling

Template-based modeling uses the assumption that similar sequences share similar structures. For a given target sequence to model, the first step is to identify similar sequences in databases; then, if one or more of these have a known structure, they will act as modeling templates.

While this use of previous knowledge is the major advantage of template-based methods, it is also the major drawback. The complete reliance on a database of known structures means that it is possible to accurately model structures similar to those already known, but it is impossible to discover entirely new protein folds in this fashion.

When most currently used modeling methods were developed, the number of available 3-D structures of membrane proteins was vanishingly small and their experimental elucidation was thought to be near impossible. The lack of available modeling templates meant that template-based methods were not applicable. Therefore, membrane proteins were all but ignored during the design of the popular modeling approaches available today.

The two subtypes of template-based modeling are comparative (or homology) modeling and fold recognition. In practice, the two approaches differ mainly in the group of template proteins that the methods aim to identify.

Template Identification and Alignment

Comparative modeling relies on the presence of templates that are highly sequence-similar to the target protein, which are assumed to have a direct evolutionary relationship with the target and a highly similar 3-D structure. Such homologous proteins can be identified using fast sequence alignment methods such as the basic local alignment search tool (BLAST) [11.25]. The target and template sequences are aligned, coordinates from corresponding residues are copied from the template, and unaligned regions filled using loop modeling techniques (see later).

In the absence of such highly sequence-similar template proteins, fold recognition is used instead. This approach aims to identify template proteins that are evolutionarily distant. They share a similar structure (or fold) with the target but their sequence is so different that these proteins are hard to detect with classical sequence alignment methods. Identifying such distantly related proteins usually involves the use of

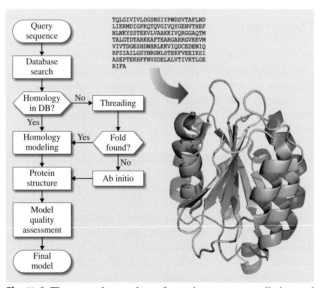

Fig. 11.6 The general procedure of protein structure prediction and the sequence and three-dimensional structure (PDB code 1QCY)

sequence-to-structure alignment (threading), based on a scoring function that determines how likely a particular sequence is to take up the candidate template's structure. Some popular implementations of threading use ESSTs [11.13] or profile–profile alignment via hidden Markov models [11.26].

No matter how the templates were identified, they need to be correctly aligned to the target protein. This may happen in the same step as template identification, or in a separate step thereafter. Alignment quality is one of the major contributors to the accuracy of the final model [11.28].

The target–template alignment is then converted into a 3-D model of the target, in a step termed *coordinate generation*.

Coordinate Generation

One of the most popular coordinate generation programs available today is MODELLER [11.29], which combines information gained from the 3-D structure of the template(s) with previous knowledge gained from a database of soluble proteins, encoded in an energy function.

Two main approaches exist for template-based coordinate generation: assembly of rigid bodies [11.30–35] and satisfaction of spatial restraints [11.29]. All these approaches result in a 3-D model, which may contain gaps (missing residues) and often only the backbone atoms (N, Cα, C, O) of the protein chain. Some methods model all atoms, but may sometimes generate low-quality side-chain coordinates, while other programs specialize in remodeling only the side-chains on a given model [11.36].

The classical method for coordinate generation is the assembly of rigid bodies (COMPOSER [11.30], 3-D-JIGSAW [11.31], SWISS-MODEL [11.32]). Here, a *framework* is calculated from the templates by finding conserved regions within the multiple alignment and averaging the Cα coordinates across the templates (Fig. 11.7). Main-chain atom coordinates of each core segment of the target model are then obtained from one of the templates and superimposed onto the framework. Segments which still have no structure are considered to be loops. These are filled in by searching a loop database for a loop with matching anchor regions (structured regions immediately adjacent to the loop) and similar sequence (Sect. 11.4.3). Side-chains are finally added to the model, and their conformation optimized using energy minimization or molecular dynamics.

Another type of approach is modeling by satisfaction of spatial restraints (MODELLER [11.29]). First,

many constraints on the structure of the target sequence are generated from the initial alignment of the target sequence to the template structures. The assumption made is generally that distances (in 3-D space) between corresponding atoms in the alignment are similar in the target and template proteins. Homology-derived restraints are usually combined with stereochemical restraints on bond lengths and angles, as well as noncovalent interactions obtained using a classical molecular mechanics force field. The final model is obtained by

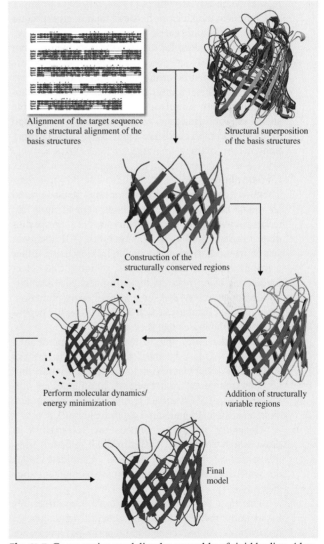

Alignment of the target sequence to the structural alignment of the basis structures

Structural superposition of the basis structures

Construction of the structurally conserved regions

Addition of structurally variable regions

Perform molecular dynamics/ energy minimization

Final model

Fig. 11.7 Comparative modeling by assembly of rigid bodies. Algorithm used in homology modeling by assembly of rigid bodies (see text for details) (after [11.27])

Part B | 11.4

an optimization process that minimizes the violations of these restraints.

11.4.3 Protein Loop Structure Prediction

Protein loops are a class of structure which connects regular secondary structures (helices or strands).

A typical globular protein has on average a third of its residues in loops [11.37]. They tend to be located on the surface of the protein and show far more variation between homologous protein structures than regular secondary structures [11.38, 39]. Due to larger differences in loop regions between homologous protein structures, difficulties arise in their prediction. Loop regions tend to account for the largest errors in any given model.

Protein loop structure prediction generally consists of three stages: sampling, filtering, and ranking. Candidate loops are sampled while avoiding implausible structural conformations and redundancy. A method ranks the sampled candidate loops using scoring functions.

Sampling

There are two methods of sampling: database and ab initio. Database search methods depend upon the assumption that similarities between local properties may suggest similar local structures. All database search methods work in a similar fashion using either a complete set or a classified set of loops and select predictions using features including sequence similarity, anchor geometry, and some form of energy function. The quality of database search methods is highly dependent on the databases that they use.

Many database search methods use predefined loops [11.40–45]. A database can contain entire protein structures without classifying loops in order to take into account ambiguous anchor regions in a modeling situation [11.34, 46]. Candidate loops can also be sampled from an artificially generated database [11.47].

Generally, ab initio methods are referred to as those which do not use solved protein structure fragments per se for loop prediction. Candidate loops are generated and optimized against scoring functions (Sect. 11.4.4).

The most common way to sample loop structures in ab initio methods is to use dihedral angle (ϕ, ψ) propensities. As backbone bond lengths and angles are nearly fixed, the main degrees of freedom are dihedral angle rotations. Each amino acid has distinct dihedral angle propensities, and a loop is sampled based on

the propensities of the sequence. Instead of using the standard dihedral angle set (ϕ, ψ), it has been proposed [11.48, 49] to use (ψ_i, ϕ_{i+1}) dihedral angles to reflect neighborhood residues.

The sampled fragment may well not connect the anchors. Hence, a step to close the gap is required. This process is called loop closure. Loop closure is a mathematical technique to find dihedral angle rotations that steer a polypeptide chain to a desired position. Each dihedral angle is optimized [11.50], or all dihedral angles along the loop are optimized at the same time [11.51–53].

Filtering

While loops are sampled, fragments which violate steric clashes are removed in both database search and ab initio methods. Due to the dihedral angle propensities, for a given loop sequence, ab initio methods may produce many highly similar structure conformations. For faster optimization, this redundancy is removed using a simple structure similarity cutoff value [11.54] or a clustering algorithm [11.55]. Database search methods use local properties, such as local sequence similarity and anchor orientation, as filters during sampling.

Ranking

Even if near-native structures are obtained during the sampling and filtering steps, ranking is needed to select the best structure among the sampled models. The models are ranked using scoring functions. The basic assumption is that the native structure is in the global minimum of a scoring function.

The scoring functions are in general categorized into three types: physics-based energy functions, statistical potential functions, and local similarity measures. A method can have one or more scoring functions. However, as ab initio methods do not use fragments from databases, the local similarity measures are used only by database search methods.

11.4.4 Model Quality Assessment and Model Selection

Usually, coordinate generation (whether in template-based or ab initio approaches) results in the creation of more than one model. The quality of the models can be assessed using objective functions, which may test for problems such as steric clashes and unfavorable bond angles and distances, as well as checking the agreement with typical (empirical) local structures. These functions generally return a score, which should ideally correlate with the similarity of a model to the *true* protein structure. Various functions of this type have been implemented in model quality assessment programs (MQAPs), e.g., QMEAN [11.56, 57]. Using MQAP scores, the models may be ranked and/or filtered before returning the result to the user.

11.4.5 Homology Modeling of Transmembrane Proteins

The current computational structure prediction methods may not be ideal for transmembrane proteins, designed as they are for water-soluble proteins [11.59]. The physical differences between water-soluble and membrane proteins may mean that many of the steps in structure prediction should be approached differently.

Membrane Protein–Specific Information
Soluble proteins often adopt a globular conformation, with their hydrophobic residues mainly in the protein core and their polar and charged residues predominantly on the water-exposed surface. Membrane proteins, on the other hand, sit in a lipid bilayer and thus contain stretches of residues that are exposed to the hydrophobic environment of the membrane. Such transmembrane (TM) segments usually have one of two structure types: α-helices or β-strands. These structural differences between membrane and soluble proteins have been used in various computational methods [11.60], for example, to identify membrane proteins from sequence alone.

In a typical template-based modeling pipeline, the database of known protein structures is searched for modeling templates. Membrane protein structures are gathered and annotated in specialized databases [11.61–63], which provide information about each protein's orientation within the lipid bilayer (*membrane insertion*). This information is not directly available from experiments and is therefore predicted computationally, either from the surface hydrophobicity of a static 3-D structure, or using coarse-grained molecular dynamics simulations. The latter approach gives the most detail, including knowledge about each residue's contact with either the hydrophobic lipid tails or the polar head groups of the membrane lipids. It is, however, also the most intensive to compute. Its use has been extended through the projection of the generated annotation onto homologous protein structures [11.64].

Informing the Modeling Process
Such membrane protein-specific information can be used to inform the model building process. It has

been shown that more accurate target–template alignments can be achieved when using membrane protein-specific substitution tables. This improvement in alignment quality directly results in better 3-D models [11.14]. A recent coordinate generation approach, MEDELLER [11.58], uses membrane protein-specific

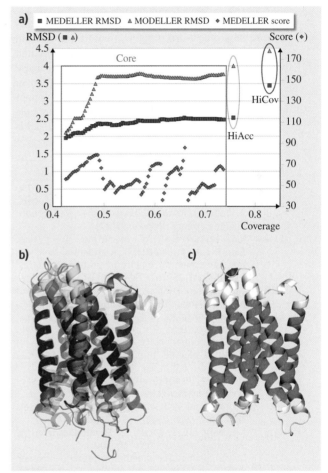

Fig. 11.8a–c Example model of the membrane protein human adenosine A2A receptor (PDB code 3EML, chain A). (**a**) Progression of model accuracy through the modeling process. The modeling phases corresponding to the different MEDELLER models [core, high accuracy (HiAcc) and high coverage (HiCov)] are labeled. (**b**) MEDELLER's high-coverage model, colored by modeling confidence in a *blue-to-red* spectrum, aligned to the native x-ray structure (*transparent orange*). (**c**) MEDELLER's high-coverage model, colored by membrane insertion using iMembrane [*red*, middle hydrophobic (tail group) layer; *white*, peripheral polar (head group) layers; *blue*, aqueous (nonmembrane) layers] (after [11.58])

Part B | 11.4

substitution tables to identify a reliable *core* structure, shared by the template and the target proteins, from the input alignment. This is achieved by first identifying the part of the template structure located in the middle layer of the membrane (marked red in Fig. 11.8c). This is typically the most conserved part of a transmembrane protein's structure [11.65]. This *core* model is then grown outwards, away from the membrane, until the substitution score for each transmembrane segment drops below a cutoff value. From this *core* model, typically only the loops are missing, which are then modeled using the FREAD database search algorithm [11.46]. MEDELLER has been shown to achieve improved accuracy, compared with the popular program MODELLER, which uses no membrane protein-specific information. An example 3-D model is shown in Fig. 11.8b, superimposed with the same protein's native x-ray structure. A plot of the MEDELLER model's root-mean-square difference (RMSD) and substitution score is shown in Fig. 11.8a, along with the RMSD of the corresponding coordinates in a model produced by the popular coordinate generation program MODELLER. Figure 11.8c illustrates the model's membrane insertion, as predicted by iMembrane.

Model Refinement

The largest errors in membrane protein models remain the loop regions, as in soluble proteins. Additional membrane-specific problems include helix kinks, reentrant loops, and helices that span only a part of the membrane. Solutions for the refinement of such problematic regions in the model are under active research but have, as of yet, not been successfully implemented in any automated modeling tools.

References

11.1 B. Alberts, A. Johnson, J. Lewis, M. Raff, K. Robets, P. Walter: *Molecular Biology of the Cell*, 4th edn. (Garland Science, New York 2002)

11.2 R.A. Laskowski, M.W. MacArthur, D.S. Moss, J.M. Thornton: PROCHECK: A program to check the stereochemical quality of protein structures, J. Appl. Cryst. **26**, 283–291 (1993)

11.3 C.B. Anfinsen: Principles that govern the folding of protein chains, Science **181**(96), 223–230 (1973)

11.4 C. Ramakrishnan, G.N. Ramachandran: Stereochemical criteria for polypeptide and protein chain conformations: II. Allowed conformations for a pair of peptide units, Biophys. J. **5**, 909–933 (1965)

11.5 W. Kabsch, C. Sander: Dictionary of protein secondary structure: Pattern recognition of hydrogen-bonded and geometrical features, Biopolymers **22**(12), 2577–2637 (1983)

11.6 J.M. Berg, J.L. Tymoczko, L. Stryer: *Biochemistry*, 5th edn. (Freeman, New York 2002)

11.7 G. von Heijne: The membrane protein universe: What's out there and why bother?, J. Intern. Med. **261**(6), 543–557 (2007)

11.8 D.J. Müller, N. Wu, K. Palczewski: Vertebrate membrane proteins: Structure, function, and insights from biophysical approaches, Pharmacol. Rev. **60**(1), 43–78 (2008)

11.9 S. Henikoff, J.G. Henikoff: Amino acid substitution matrices from protein blocks, Proc. Natl. Acad. Sci. USA **89**(22), 10915–10919 (1992)

11.10 D.T. Jones, W.R. Taylor, J.M. Thornton: A model recognition approach to the prediction of all-helical membrane protein structure and topology, Biochemistry **33**(10), 3038–3049 (1994)

11.11 P. Ng, J. Henikoff, S. Henikoff: PHAT: A transmembrane-specific substitution matrix, Bioinformatics **16**(9), 760–766 (2000)

11.12 T. Müller, S. Rahmann, M. Rehmsmeier: Non-symmetric score matrices and the detection of homologous transmembrane proteins, Bioinformatics **17**(1), S182–S189 (2001)

11.13 J. Shi, T.L. Blundell, K. Mizuguchi: FUGUE: Sequence-structure homology recognition using environment-specific substitution tables and structure-dependent gap penalties, J. Mol. Biol. **310**(1), 243–257 (2001)

11.14 J.R. Hill, S. Kelm, J. Shi, C.M. Deane: Environment specific substitution tables improve membrane protein alignment, Bioinformatics **27**(13), i15–i23 (2011)

11.15 A. Kryshtafovych, C. Venclovas, K. Fidelis, J. Moult: Progress over the first decade of CASP experiments, Proteins **61**(7), 225–236 (2005)

11.16 J. Moult, K. Fidelis, A. Kryshtafovych, B. Rost, T. Hubbard, A. Tramontano: Critical assessment of methods of protein structure prediction – Round VII, Proteins **69**, 3–9 (2007)

11.17 J. Moult, K. Fidelis, A. Kryshtafovych, B. Rost, A. Tramontano: Critical assessment of methods of protein structure prediction – Round VIII, Proteins **9**(77), 1–4 (2009)

11.18 C.M. Deane, M. Dong, F.P. Huard, B.K. Lance, G.R. Wood: Cotranslational protein folding – fact or fiction?, Bioinformatics **23**(13), i142–i148 (2007)

11.19 J.J. Ellis, F.P.P. Huard, C.M. Deane, S. Srivastava, G.R. Wood: Directionality in protein fold prediction, BMC Bioinformatics **11**(1), 172 (2010)

11.20 B.R. Jefferys, L.A. Kelley, M.J.E. Sternberg: Protein folding requires crowd control in a simulated cell, J. Mol. Biol. **397**(5), 1329–1338 (2010)

11.21 R. Das, D. Baker: Macromolecular modeling with rosetta, Annu. Rev. Biochem. **77**(1), 363–382 (2008)

11.22 K.T. Simons, R. Bonneau, I. Ruczinski, D. Baker: Ab initio protein structure prediction of CASP III targets using ROSETTA, Proteins **37**(3), 171–176 (1999)

11.23 P. Bradley, L. Malmström, B. Qian, J. Schonbrun, D. Chivian, D.E. Kim, J. Meiler, K.M.S. Misura, D. Baker: Free modeling with Rosetta in CASP6, Proteins **61**(7), 128–134 (2005)

11.24 R. Das, B. Qian, S. Raman, R. Vernon, J. Thompson, P. Bradley, S. Khare, M.D. Tyka, D. Bhat, D. Chivian, D.E. Kim, W.H. Sheffler, L. Malmström, A.M. Wollacott, C. Wang, I. Andre, D. Baker: Structure prediction for CASP7 targets using extensive all-atom refinement with Rosetta@home, Proteins **69**(S8), 118–128 (2007)

11.25 S.F. Altschul, W. Gish, W. Miller, E.W. Myers, D.J. Lipman: Basic local alignment search tool, J. Mol. Biol. **215**(3), 403–410 (1990)

11.26 J. Söding, A. Biegert, A. Lupas: The HHpred interactive server for protein homology detection and structure prediction, Nucleic Acids Res. **33**(2), W244–W248 (2005)

11.27 C.M. Deane: *Protein Structure Prediction: Amino Acid Propensities and Comparative Modelling* (Univ. of Cambridge, Cambridge 2000)

11.28 R. Sánchez, A. Sali: Advances in comparative protein-structure modelling, Curr. Opin. Struct. Biol. **7**(2), 206–214 (1997)

11.29 A. Sali, T.L. Blundell: Comparative protein modelling by satisfaction of spatial restraints, J. Mol. Biol. **234**(3), 779–815 (1993)

11.30 M.J. Sutcliffe, I. Haneef, D. Carney, T.L. Blundell: Knowledge based modelling of homologous proteins, Part I: Three-dimensional frameworks derived from the simultaneous superposition of multiple structures, Protein Eng. **1**(5), 377–384 (1987)

11.31 P.A. Bates, L.A. Kelley, R.M. MacCallum, M.J. Sternberg: Enhancement of protein modeling by human intervention in applying the automatic programs 3D-JIGSAW and 3D-PSSM, Proteins **45**(5), 39–46 (2001)

11.32 T. Schwede, J. Kopp, N. Guex, M.C. Peitsch: SWISS-MODEL: An automated protein homology-modeling server, Nucleic Acids Res. **31**(13), 3381–3385 (2003)

11.33 D. Petrey, Z. Xiang, C.L. Tang, L. Xie, M. Gimpelev, T. Mitros, C.S. Soto, S. Goldsmith-Fischman, A. Kernytsky, A. Schlessinger, I.Y. Koh, E. Alexov, B. Honig: Using multiple structure alignments, fast model building, and energetic analysis in fold recognition and homology modeling, Proteins **53**(6), 430–435 (2003)

11.34 C.M. Deane, T.L. Blundell: CODA: A combined algorithm for predicting the structurally variable regions of protein models, Protein Sci. **10**(3), 599–612 (2001)

11.35 P. Koehl, M. Delarue: A self consistent mean field approach to simultaneous gap closure and side-chain positioning in homology modelling, Nat. Struct. Biol. **2**(2), 163–170 (1995)

11.36 G.G. Krivov, V.M. Shapovalov, L.R. Dunbrack: Improved prediction of protein side-chain conformations with SCWRL4, Proteins **77**(4), 778–795 (2009)

11.37 L.E. Donate, S.D. Rufino, L.H.J. Canard, T.L. Blundell: Conformational analysis and clustering of short and medium size loops connecting regular secondary structures: A database for modeling and prediction, Protein Sci. **5**, 2600–2616 (1996)

11.38 G.D. Rose: Prediction of chain turns in globular proteins on a hydrophobic basis, Nature **272**, 586–590 (1978)

11.39 I.P. Crawford, T. Niermann, K. Kirschner: Prediction of secondary structure by evolutionary comparison: Application to the α subunit of tryptophan synthase, Proteins **2**(2), 118–129 (1987)

11.40 J. Wojcik, J.-P. Mornon, J. Chomilier: New efficient statistical sequence-dependent structure prediction of short to medium-sized protein loops based on an exhaustive loop classification, J. Mol. Biol. **289**, 1469–1490 (1999)

11.41 N. Fernandez-Fuentes, B. Olivia, A. Fiser: A supersecondary structure library and search algorithm for modeling loops in protein structures, Nucleic Acids Res. **34**, 2085–2097 (2006)

11.42 N. Fernandez-Fuentes, J. Zhai, A. Fiser: ArchPRED: A template based loop structure prediction server, Nucleic Acids Res. **34**, W173–W176 (2006)

11.43 E. Michalsky, A. Goede, R. Preissner: Loops in proteins (LIP) – a comprehensive loop database for homology modeling, Protein Eng. **16**, 979–985 (2003)

11.44 A. Hildebrand, M. Remmert, A. Biegert, J. Söding: Fast and accurate automatic structure prediction with HHpred, Proteins **77**(S9), 128–132 (2009)

11.45 H. Peng, A. Yang: Modeling protein loops with knowledge-based prediction of sequence-structure alignment, Bioinformatics **23**, 2836–2842 (2007)

11.46 Y. Choi, C.M. Deane: FREAD revisited: Accurate loop structure prediction using a database search algorithm, Proteins **78**(6), 1431–1440 (2010)

11.47 C.M. Deane, T.L. Blundell: A novel exhaustive search algorithm for predicting the conformation of polypeptide segments in proteins, Proteins **40**, 135–144 (2000)

11.48 S. Sucha, R.F. Dubose, C.J. March, S. Subashini: Modeling protein loops using a ϕ_{i+1}, ψ_i dimer database, Protein Sci. **4**, 1412–1420 (1995)

11.49 V.Z. Spassov, P.K. Flook, L. Yan: LOOPER: A molecular mechanics-based algorithm for protein loop prediction, Protein Eng. **21**, 91–100 (2008)

11.50 A.A. Canutescu, R.L. Dunbrack Jr.: Cyclic coordinate descent: A robotics algorithm for protein loop closure, Protein Sci. **12**, 963–972 (2003)

11.51 P.S. Shenkin, D.L. Yarmush, R.M. Fine, H. Wang, C. Levinthal: Predicting antibody hypervariable loop conformation. I. Ensembles of random conformations for ringlike structures, Biopolymers **26**, 2053–2085 (1987)

11.52 J. Lee, D. Lee, H. Park, E.A. Coutsias, C. Seok: Protein loop modeling by using fragment assembly and analytical loop closure, Proteins **78**(16), 3428–3436 (2010)

11.53 T. Hurst: Flexible 3D searching: The directed tweak technique, J. Chem. Inf. Comput. Sci. **34**, 190–196 (1994)

11.54 M.A. DePristo, P.I.W. de Bakker, S.C. Lovell, T.L. Blundell: Ab initio construction of polypeptide fragments: Efficient generation of accurate, representative ensembles, Proteins **51**, 41–55 (2003)

11.55 M.P. Jacobson, D.L. Pincus, C.S. Rapp, T.J.F. Day, B. Honig, D.E. Shaw, R.A. Friesner: A hierarchical approach to all-atom protein loop prediction, Proteins **55**, 351–367 (2004)

11.56 P. Benkert, S.C.E.C. Tosatto, D. Schomburg: QMEAN: A comprehensive scoring function for model quality assessment, Proteins **71**(1), 261–277 (2007)

11.57 P. Benkert, M. Kunzli, T. Schwede: QMEAN server for protein model quality estimation, Nucleic Acids Res. **37**(2), W510–514 (2009)

11.58 S. Kelm, J. Shi, C.M. Deane: MEDELLER: Homology-based coordinate generation for membrane proteins, Bioinformatics **26**(22), 2833–2840 (2010)

11.59 A. Elofsson, G. von Heijne: Membrane protein structure: Prediction versus reality, Annu. Rev. Biochem. **76**(1), 125–140 (2007)

11.60 M. Punta, L.R. Forrest, H. Bigelow, A. Kernytsky, J. Liu, B. Rost: Membrane protein prediction methods, Methods **41**(4), 460–474 (2007)

11.61 M.A. Lomize, A.L. Lomize, I.D. Pogozheva, H.I. Mosberg: OPM: Orientations of proteins in membranes database, Bioinformatics **22**(5), 623–625 (2006)

11.62 G.E. Tusnády, Z. Dosztányi, I. Simon: PDB: Selection and membrane localization of transmembrane proteins in the protein data bank, Nucleic Acids Res. **33**(1), D275–D278 (2005)

11.63 K.A. Scott, P.J. Bond, A. Ivetac, A.P. Chetwynd, S. Khalid, M.S. Sansom: Coarse-grained MD simulations of membrane protein-bilayer self-assembly, Structure **16**(4), 621–630 (2008)

11.64 S. Kelm, J. Shi, C.M. Deane: iMembrane: Homology-based membrane-insertion of proteins, Bioinformatics **25**(8), 1086–1088 (2009)

11.65 L. Forrest, C. Tang, B. Honig: On the accuracy of homology modeling and sequence alignment methods applied to membrane proteins, Biophys. J. **91**(2), 508–517 (2006)

Part C Machine

Part C Machine Learning Methods for the Analysis, Modeling and Knowledge Discovery from Bioinformatics Data

Ed. by Irwin King and Kaizhu Huang

12. Machine Learning Methodology in Bioinformatics

Colin Campbell

Machine learning plays a central role in the interpretation of many datasets generated within the biomedical sciences. In this chapter we focus on two core topics within machine learning, *supervised* and *unsupervised* learning, and illustrate their application to interpreting these datasets. For supervised learning, we focus on *support vector machines* (SVMs), which is a subtopic of *kernel-based learning*. Kernels can be used to encode many different types of data, from continuous and discrete data through to graph and sequence data. Given the different types of data encountered within bioinformatics, they are therefore a method of choice within this context. With *unsupervised learning* we are interested in the discovery of structure within data. We start by considering hierarchical cluster analysis (HCA), given its common usage in this context. We then point out the advantages of *Bayesian* approaches to unsupervised learning, such as a principled approach to model selection (how many clusters are present in the data) through to confidence measures for assignment of datapoints to clusters. We outline five case studies illustrating these methods. For supervised learning we consider prediction of disease progression in cancer and protein fold prediction. For unsupervised learning we apply HCA to a small colon cancer dataset and then illustrate the use of Bayesian unsupervised learning applied to breast and lung cancer datasets. Finally we consider

network inference, which can be approached as an unsupervised or supervised learning task depending on the data available.

In this chapter we consider the application of modern methods from machine learning to the analysis of biomedical datasets. There are a substantial number of machine learning methods which could be used in the context of bioinformatics, and so we are necessarily selective. We focus on the two commonest themes within machine learning, namely *supervised* and *unsupervised* learning. Many methods have been proposed for supervised learning, and so, in Sect. 12.1, we choose to concentrate on *kernel-based methods*, specifically *support vector machines* (SVMs). SVMs are a popular approach to classification and have several advantages in handling datasets from bioinformatics. In particular, biomedical data can appear in many forms from continuous-valued and discrete data to network structures and sequence data. These different types of data can be encoded into *kernels* which quantify the similarity of data objects. We start by introducing classification and the support vector machine for binary classification. In Sect. 12.1.1 we then extend this approach to multiclass classification, learning in the presence of noise, the association of confidence measures to class labels, and regression (i. e., using continuously valued labels). In Sect. 12.1.3 we consider simple kernels, complex kernels for graphs, strings, and sequences, and *multiple kernel learning*, where we build a decision function for prediction using multiple types of input data.

The second area of machine learning we consider is *unsupervised learning*, where we are interested in

the discovery of structure in data. In Sect. 12.2.1 we start with *hierarchical cluster analysis* (HCA), given the common usage of this unsupervised learning approach in the biomedical community. We then point out that Bayesian approaches can have certain advantages over HCA. We consider *variational* approaches to Bayesian unsupervised learning and illustrate the use of this approach in finding novel disease subtypes in cancer research. We then consider *Markov chain Monte Carlo* (MCMC) approaches, which can be more accurate than variational methods, and illustrate their use in cancer research for finding the most probable pathway structure from a set of candidate pathway topologies.

12.1 Supervised Learning

Many bioinformatics problems involve prediction over two classes; For example, we may want to predict whether a tumor is benign or malignant, based on genetic data. An abstract *learning machine* will learn from *training data* and attempt to *generalize* and thus make predictions on novel input data. For the training data we have a set of input vectors, denoted x_i, with each input vector having a number of component *features*. These input vectors are paired with corresponding *labels*, which we denote y_i, and there are m such pairs ($i = 1, \ldots, m$). Thus, for our cancer example, $y_i = +1$ may denote malignant and $y_i = -1$ benign. The matching x_i are input vectors encoding the genetic data derived from each patient i. Typically, we would be interested in quantifying the prediction performance before any practical usage, and so we would evaluate a *test error* based on a *test set* of data.

The training data can be viewed as labeled datapoints in an input space, which we depict in Fig. 12.1. For two classes of well-separated data, the learning task amounts to finding a *directed hyperplane*, i.e., an oriented hyperplane such that datapoints on one side will be labeled $y_i = +1$ and those on the other side as $y_i = -1$. The directed hyperplane found by a *support vector machine* is intuitive: it is that hyperplane which is maximally distant from the two classes of labeled datapoints. The closest such points on both sides have most influence on the position of this separating hyperplane and are therefore called *support vectors*. The separating hyperplane is given as $w \cdot x + b = 0$ (where "·" denotes the inner or scalar product). b is the *bias* or offset of the hyperplane from the origin in input space, and x are points located within the hyperplane. The normal to the hyperplane, the *weight vector* w, determines its orientation.

Of course this picture is too simple for many applications. The two clusters could be highly intermeshed with many overlapping datapoints: the dataset is then *not linearly separable*. This situation is one motivation for introducing the concept of *kernels* later in this chapter. We can also see that stray datapoints could have a significant impact on the orientation of the hyperplane, and so we need a mechanism for handling anomalous datapoints and noise.

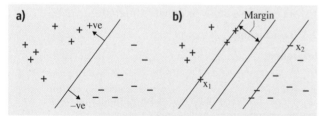

Fig. 12.1 (a) The argument inside the decision function of our SVM classifier is $w \cdot x + b$. The separating hyperplane corresponding to $w \cdot x + b = 0$ is shown as a line on this plot. This hyperplane separates the two classes of data, with points on one side labeled $y_i = +1$ ($w \cdot x + b \geq 0$) and points on the other side labeled $y_i = -1$ ($w \cdot x + b < 0$). **(b)** The perpendicular distance between the separating hyperplane and a hyperplane through the closest points (the support vectors) is called the *margin*, γ. x_1 and x_2 are examples of *support vectors* of opposite sign. The hyperplanes passing through the support vectors are *canonical hyperplanes*, and the region between the canonical hyperplanes is the *margin band*. The projection of the vector ($x_1 - x_2$) onto the normal to the separating hyperplane ($w / ||w||_2$) is 2γ

Statistical learning theory is the theoretical study of learning and generalization. From the perspective of statistical learning theory, the motivation for considering binary classifier SVMs comes from a theoretical upper bound on the *generalization error*, that is, the theoretical prediction error when applying the classifier to novel, unseen instances. This generalization error bound has two important features.

1. The bound is minimized by maximizing the *margin*, γ, i.e., the minimal distance between the hyperplane separating the two classes and the closest datapoints to the hyperplane, and
2. The bound does not depend on the dimensionality of the space.

Suppose we consider a binary classification task with datapoints x_i ($i = 1, \ldots, m$) having corresponding labels $y_i = \pm 1$ and a *decision function*

$$f(x) = \text{sign}\, (w \cdot x + b) \,, \tag{12.1}$$

where $\cdot$ is the inner product. From the decision function we see that the data is correctly learnt if $y_i(w \cdot x_i + b) > 0 \forall i$, since $(w \cdot x_i + b)$ should be positive when $y_i = +1$ and it should be negative when $y_i = -1$. The decision function is invariant under a positive rescaling of the argument inside the sign-function, leading to an ambiguity in defining a distance measure and therefore the margin. Thus, we implicitly define a scale for the (w, b) by setting $w \cdot x + b = 1$ for the closest points on one side and $w \cdot x + b = -1$ for the closest on the other side. The hyperplanes passing through $w \cdot x + b = 1$ and $w \cdot x + b = -1$ are called *canonical hyperplanes*, and the region between these canonical hyperplanes is called the *margin band*. Let x_1 and x_2 be two points inside the canonical hyperplanes on both sides of the separating hyperplane (Fig. 12.1b). If $w \cdot x_1 + b = 1$ and $w \cdot x_2 + b = -1$, we deduce that $w \cdot (x_1 - x_2) = 2$. For the separating hyperplane $w \cdot x + b = 0$, the normal vector is $w/\|w\|_2$ (where $\|w\|_2$ is the square root of $w^T w$). Thus, the distance between the two canonical hyperplanes is equal to the projection of $x_1 - x_2$ onto the normal vector $w/\|w\|_2$, which gives $(x_1 - x_2) \cdot w/\|w\|_2 = 2/\|w\|_2$. As half the distance between the two canonical hyperplanes, the margin is therefore $\gamma = 1/\|w\|_2$. Maximizing the margin is therefore equivalent to minimizing

$$\frac{1}{2}\|w\|_2^2 \,, \tag{12.2}$$

subject to the constraints

$$y_i\,(w \cdot x_i + b) \geq 1 \quad \forall i \,. \tag{12.3}$$

This is a constrained optimization problem in which we minimize an *objective function* (12.2) subject to the *constraints* (12.3).

As a constrained optimization problem, the above formulation can be reduced to minimization of a *Lagrange function*, consisting of the sum of the objective function and the m constraints multiplied by their respective *Lagrange multipliers*, denoted α_i. We will call this the *primal* formulation

$$L(w, b) = \frac{1}{2}(w \cdot w) - \sum_{i=1}^{m} \alpha_i \left[y_i(w \cdot x_i + b) - 1 \right] \,, \tag{12.4}$$

where α_i are Lagrange multipliers, and thus $\alpha_i \geq 0$ (a necessary condition for the Lagrange multiplier). At the optimum, we can take the derivatives with respect to b and w and set these to zero

$$\frac{\partial L}{\partial b} = -\sum_{i=1}^{m} \alpha_i y_i = 0 \,, \tag{12.5}$$

$$\frac{\partial L}{\partial w} = w - \sum_{i=1}^{m} \alpha_i y_i x_i = 0 \,. \tag{12.6}$$

Substituting w from (12.6) back into $L(w, b)$ we then get a *dual formulation* (also known as a *Wolfe dual*)

$$W(\alpha) = \sum_{i=1}^{m} \alpha_i - \frac{1}{2} \sum_{i,j=1}^{m} \alpha_i \alpha_j y_i y_j \,(x_i \cdot x_j) \,, \tag{12.7}$$

which must be *maximized* with respect to the α_i subject to the constraints

$$\alpha_i \geq 0, \quad \sum_{i=1}^{m} \alpha_i y_i = 0 \,. \tag{12.8}$$

The objective function in (12.7) is *quadratic* in the parameters α_i, and thus it is a constrained *quadratic programming* (QP) problem. QP is a standard problem in optimization theory, so there are a number of resources available, such as *QUADPROG* in *MATLAB*, *MINOS*, and *LOQO*. In addition there are a number of packages specifically written for SVMs such as *SVM-light*, *LIBSVM*, and *SimpleSVM* [12.1].

So far we have not considered point (2), i.e., that the generalization bound does not depend on the dimensionality of the space. For the objective (12.7) we notice that the data x_i only appear inside an inner product. To get an alternative representation of the data, we could therefore map datapoints into a space with different dimensionality, called *feature space*, through

$$x_i \cdot x_j \to \Phi\,(x_i) \cdot \Phi(x_j) \,, \tag{12.9}$$

where $\Phi(\cdot)$ is the mapping function. Data which are not separable in input space can always be separated in a space of high enough dimensionality. A consequence of the generalization bound, given in (2), is that there is no loss of generalization performance if we map to a feature space where the data are separable and a margin can be defined.

Surprisingly, the functional form of the mapping $\Phi(x_i)$ *does not need to be known in general*, since it is implicitly defined by the choice of the *kernel*: $K(x_i, x_j) = \Phi(x_i) \cdot \Phi(x_j)$ or inner product in feature space. For nonseparable continuous-valued data a common choice is the *Gaussian kernel* (Sect. 12.1.3)

$$K(x_i, x_j) = e^{-(x_i - x_j)^2 / 2\sigma^2} . \tag{12.10}$$

The introduction of a kernel with its implied mapping to feature space is known as *kernel substitution*. The class of mathematical functions which can be used as kernels is very general. Apart from continuous-valued data we can also consider many data objects which appear in bioinformatics such as graphs (representing *networks* and *pathways*), strings, and sequences (such as genetic or protein sequence data).

For binary classification with a given choice of kernel the learning task therefore involves maximization of

$$W(\boldsymbol{\alpha}) = \sum_{i=1}^{m} \alpha_i - \frac{1}{2} \sum_{i,j=1}^{m} \alpha_i \alpha_j y_i y_j K(x_i, x_j) , \tag{12.11}$$

subject to the constraints (12.8). The bias, b, is found separately. For a datapoint with $y_i = +1$,

$$\min_{\{i|y_i = +1\}} [\boldsymbol{w} \cdot \Phi(x_i) + b] =$$

$$\min_{\{i|y_i = +1\}} \left[\sum_{j}^{m} \alpha_j y_j K(x_i, x_j) \right] + b = 1 , \tag{12.12}$$

using (12.6), with a similar expression for datapoints labeled $y_i = -1$. We thus deduce that

$$b = -\frac{1}{2} \left\{ \max_{\{i|y_i = -1\}} \left[\sum_{j}^{m} \alpha_j y_j K(x_i, x_j) \right] \right.$$

$$\left. + \min_{\{i|y_i = +1\}} \left[\sum_{j}^{m} \alpha_j y_j K(x_i, x_j) \right] \right\} . \tag{12.13}$$

Thus, to construct an SVM binary classifier, we place the data (x_i, y_i) into (12.11) and maximize $W(\alpha)$ subject to the constraints (12.8). From the optimal values of α_i, which we denote $\alpha_i^\star$, we calculate the bias b from (12.13). For a novel input vector z, the predicted class is then based on the sign of

$$\phi(z) = \sum_{i=1}^{m} \alpha_i^\star y_i K(x_i, z) + b^\star , \tag{12.14}$$

where $b^\star$ denotes the value of the bias at optimality.

12.1.1 Multiclass Classification and Other Extensions

Multiclass Classification

Many bioinformatics problems involve multiclass classification, and a number of schemes have been outlined. If the number of classes is small then we can use a *directed acyclic graph* (DAG) [12.2] with the learning task reduced to binary classification at each node. The idea is illustrated in Fig. 12.2. Suppose we consider a three-class classification problem. The first node is a classifier making the binary decision label 1 versus label 3, say. Depending on the outcome of this decision, the next steps are the decisions 1 versus 2, or 2 versus 3. We could also use a series of one-against-all classifiers [12.3]. Thus, we construct C separate SVMs, with the c-th SVM trained using data from class c as the positively labeled samples and the remaining classes as the negatively labeled samples. Associated with the c-th SVM we have $f_c(z) = \sum_i y_i^c \alpha_i^c K(z, x_i) + b^c$, and the novel input z is assigned to class c such that $f_c(z)$ is largest. Other schemes have been suggested [12.4–7].

Learning with Noise: Soft Margins

Many biomedical datasets are intrinsically noisy, and a learning machine could fit to this noise, leading to poor generalization. As remarked earlier, outliers can have an undue influence on the position of the separating hyperplane used by an SVM (Fig. 12.1). Potential

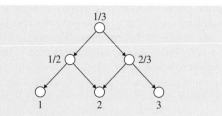

Fig. 12.2 A multiclass classification problem reduced to a series of binary classification tasks

noise in a dataset can be handled by the introduction of a *soft margin* [12.8]. Two schemes are commonly used. With an L_1 *error norm*, the learning task is the same as in (12.11, 12.8) except for the introduction of the *box constraint*

$$0 \le \alpha_i \le C. \tag{12.15}$$

On the other hand, for an L_2 *error norm*, the learning task is (12.11, 12.8) except for the addition of a small positive constant to the leading diagonal of the kernel matrix

$$K(\boldsymbol{x}_i, \boldsymbol{x}_i) \leftarrow K(\boldsymbol{x}_i, \boldsymbol{x}_i) + \lambda. \tag{12.16}$$

The appropriate values of these parameters can be found by means of a *validation study*. With sufficient data we would split the dataset into a *training set*, a *validation set*, and a *test set*. With regularly spaced values of C or λ, we train the SVM on the training data and find the best choice for this parameter based on the validation error. With more limited data, we may use *cross-validation*, or rotation estimation, in which the data are randomly partitioned into subsets and rotated successively as training and validation data.

With many biomedical datasets there is an imbalance between the amount of data in different classes, or the significance of the data in the two classes can be quite different; For example, for the detection of tumors on magnetic resonance imaging (MRI) scans, it may be best to allow a higher number of false positives if this improved the true-positive detection rate. The balance between the detection rate for different classes can be easily shifted by introducing *asymmetric soft margin parameters* [12.9]. Thus, for binary classification with an L_1 error norm we use $0 \le \alpha_i \le C_+$ ($y_i = +1$) and $0 \le \alpha_i \le C_-$ ($y_i = -1$), but $K(\boldsymbol{x}_i, \boldsymbol{x}_i) \leftarrow K(\boldsymbol{x}_i, \boldsymbol{x}_i) + \lambda_+$ (if $y_i = +1$) and $K(\boldsymbol{x}_i, \boldsymbol{x}_i) \leftarrow K(\boldsymbol{x}_i, \boldsymbol{x}_i) + \lambda_-$ (if $y_i = -1$) for the L_2 error norm.

Introducing a Confidence Measure

Suppose we are using a support vector machine for diagnostic categorization or prediction of disease progression; then, it would be plainly useful to have a confidence measure associated with the class assignment. A clinician could be expected to plan differently with a high confidence prediction over a low confidence one. A SVM has an in-built quantity that could provide a confidence measure for the class assignment, i. e., the distance of a new point from the separating hyperplane (Fig. 12.1). A new datapoint with a large distance from the separating hyperplane should be assigned a higher degree of confidence than a point which lies close to the

hyperplane. Before thresholding, the output of a SVM is given by

$$\phi(z) = \sum_i y_i \alpha_i K(\boldsymbol{x}_i, z) + b. \tag{12.17}$$

One approach is to fit a *probability* measure $p(y|\phi)$ directly [12.10]. A good choice for the mapping function is the *sigmoid*

$$p(y = +1|\phi) = \frac{1}{1 + \exp(A\phi + B)}, \tag{12.18}$$

with the parameters A and B found from the training set $(y_i, \phi(\boldsymbol{x}_i))$. Let us define t_i as the target probabilities

$$t_i = \frac{y_i + 1}{2}, \tag{12.19}$$

so for $y_i \in \{-1, 1\}$ we have $t_i \in \{0, 1\}$. We find A and B by performing the following minimization over the entire training set:

$$\min_{A, B} \left[-\sum_i t_i \log(p_i) + (1 - t_i) \log(1 - p_i) \right], \tag{12.20}$$

where p_i is simply (12.18) evaluated at $\phi(\boldsymbol{x}_i)$. This is a straightforward two-dimensional minimization problem which can be solved using a variety of optimization methods. Once the sigmoid has been found using the training set, we can use (12.18) to calculate the probability that a new test point belongs to either class. Figure 12.3 shows the margins of training data points

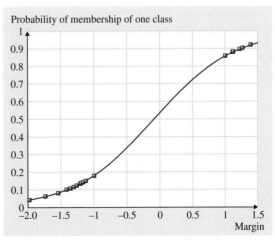

Fig. 12.3 Probability of membership of one class (y-axis) versus margin (x-axis). The plot shows the training points and fitted sigmoid for an ovarian cancer dataset

(*x*-axis) and fitted sigmoid for a microarray dataset for ovarian cancer. The distinction is ovarian cancer versus normal. There are no datapoints present in a band between $+1$ and -1 since this corresponds to the margin band of the SVM.

Regression

Some bioinformatics applications involve *regression* and the construction of models with real-valued labels y_i. To model the dependency between the input vectors x_i and the y_i we could use a linear function of the form

$$g(x_i) = w^T x_i \qquad (12.21)$$

to approximate y_i. Thus, we could minimize the following function in w:

$$L(w) = \frac{1}{2} \sum_{i=1}^{m} \left[y_i - g(x_i) \right]^2 \qquad (12.22)$$

to get a solution $y_i \approx g(x_i)$. If we make a mapping to feature space $x_i \to \Phi(x_i)$ and introduce a variable $\xi_i = y_i - w^T \Phi(x_i)$, then (12.22) could be reformulated as

$$\min_{w, \xi} \left\{ L = \sum_{i=1}^{m} \xi_i^2 \right\}, \qquad (12.23)$$

subject to the constraints

$$y_i - w^T \Phi(x_i) = \xi_i \forall i \qquad (12.24)$$

$$w^T w \leq B^2. \qquad (12.25)$$

The latter constraint (12.25) is a *regularization* constraint on the w; that is, it is used to avoid an overcomplex solution which fits to noise in the data, leading to poor generalization. As a constrained optimization problem with objective function (12.23) and two constraint conditions (12.24) and (12.25), we derive a Lagrange function

$$L = \sum_{i=1}^{m} \xi_i^2 + \sum_{i=1}^{m} \beta_i \left[y_i - w^T \Phi(x_i) - \xi_i \right]$$
$$+ \lambda \left(w^T w - B^2 \right) \qquad (12.26)$$

with Lagrange multipliers β_i and λ for these two constraint conditions. If we take derivatives of L with respect to ξ_i and w, we get

$$\xi_i = \frac{1}{2} \beta_i, \qquad (12.27)$$

$$w = \frac{1}{2\lambda} \sum_{i=1}^{m} \beta_i \Phi(x_i). \qquad (12.28)$$

Substituting these back into L gives the *dual* formulation

$$W = \sum_{i=1}^{m} \left(-\frac{1}{4} \beta_i^2 + \beta_i y_i \right)$$
$$- \frac{1}{4\lambda} \sum_{i,j=1}^{m} \left(\beta_i \beta_j K(x_i, x_j) \right) - \lambda B^2, \qquad (12.29)$$

which with a redefined variable $\alpha_i = \beta_i / 2\lambda$ (a positive rescaling, since $\lambda \geq 0$) gives the following restatement:

$$\max_{\alpha_i, \lambda} \left\{ W = -\lambda^2 \sum_{i=1}^{m} \alpha_i^2 + 2\lambda \sum_{i=1}^{m} \alpha_i y_i \right.$$
$$\left. - \lambda \sum_{i,j=1}^{m} \alpha_i \alpha_j K(x_i, x_j) - \lambda B^2 \right\}. \qquad (12.30)$$

In contrast to a SVM for binary classification, where we must solve a constrained QP problem, (12.30) gives a direct solution

$$\alpha = (K + \lambda I)^{-1} y. \qquad (12.31)$$

This gives a means for finding w in (12.21) via (12.28). The above is one of the simplest kernel-based approaches to regression [12.11, 12]. However, for prediction on a novel datapoint we implicitly use *all* the datapoints via the kernel matrix K. Sample sparsity is desirable since it reduces the complexity of the model. For this reason, it not the best approach to finding a regression function, and thus a number of approaches which involve constrained quadratic programming have been developed [12.13–17]. These minimize the number of support vectors favoring sparse hypotheses and giving smooth functional approximations to the data.

12.1.2 Case Study 1: Predicting Disease Progression

As an example of the use of SVMs within the context of bioinformatics, we consider an application to predicting disease progression. In this example, the objective is to predict relapse versus nonrelapse for Wilm's tumor, a cancer which affects children and young adults [12.18]. This tumor originates in the kidney, but it is a curable disease in the large majority of affected children. However, there is a recognized aggressive subtype with a high probability of relapse within a few years. It is therefore clini-

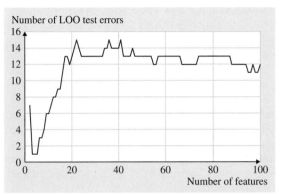

Fig. 12.4 The number of LOO test errors (y-axis) versus number of top-ranked features (x-axis) remaining (with features ranked by a t-test statistic) for predicting relapse or nonrelapse for Wilm's tumor

cally important to predict risk of relapse when the disease is first diagnosed, with an alternative treatment regime if risk of relapse is high. In this study we used microarray data as input to the support vector machine. The microarray had 30 720 probes, each measuring *gene expression*, roughly quantifying the amount of protein produced per gene. A number of these readings were poor quality, so quality filtering was used, reducing the number of *features* to 17 790. The dataset consisted of 27 samples, of which 13 samples were from patients who relapsed with the disease within 2 years and the remainder labeled as nonrelapse due to long-term survival without disease recurrence.

The task is binary classification, and the large number of features means the relatively small number of datapoints are embedded in a high-dimensional space. As a consequence, the dataset is separable and we use a *linear kernel* $K(x_i, x_j) = x_i \cdot x_j$. Since we do not have large training and test sets we use *leave-one-out* (LOO) testing; i. e., we train on 26 samples and evaluate classifier performance on the single left-out datapoint, successively rotating the test datapoint through the data. The vast majority of features are likely to be irrelevant, so we use *feature selection* to remove uninformative features and thus improve performance.

Using a t-test statistic to rank features [12.19] we can obtain a minimal LOO test error of 1 from 27 (the t-test must be run separately per LOO partitioning to avoid corrupting the test statistic). However, this tentative prediction performance is only achieved with a small set of the most informative features, and, if all features are included, the vast number of uninforma-

tive and noisy features is sufficient to overwhelm this predictive genetic signature.

12.1.3 Different Types of Kernels

Biomedical data can appear in many different formats, and in this section we consider how to construct kernels representing these different types of data.

Permissable Kernels
If a proposed kernel matrix is *positive semidefinite* (PSD) then it is an allowable kernel. For any arbitrary set of real-valued variables $a_1, \ldots, a_m$, a PSD kernel satisfies

$$\sum_{i=1}^{m} \sum_{j=1}^{m} a_i a_j K(x_i, x_j) \geq 0. \tag{12.32}$$

This type of kernel is symmetric, $K(x_i, x_j) = K(x_j, x_i)$, with positive components on the diagonal, $K(x, x) \geq 0$. An example is the *linear kernel* introduced earlier

$$K(x_i, x_j) = x_i \cdot x_j. \tag{12.33}$$

It is symmetric, $K(x, x) \geq 0$, and it satisfies (12.32) since

$$\sum_{i=1}^{m} \sum_{j=1}^{m} a_i a_j (x_i \cdot x_j) = \left\| \sum_{i=1}^{m} a_i x_i \right\|^2 \geq 0. \tag{12.34}$$

We can determine if a proposed kernel matrix is PSD by determining its spectrum of eigenvalues: if the matrix has at least one negative eigenvalue λ with corresponding eigenvector v, say, then $v^\mathrm{T} K v = \lambda v^\mathrm{T} v < 0$, so it is not PSD. There are, indeed, strategies for handling non-PSD kernel matrices [12.20–24].

From permissable kernels we can construct other permissable kernels; For example,

1. If $K_1(x_i, x_j)$ is a kernel then so is

$$K(x_i, x_j) = c K_1(x_i, x_j), \tag{12.35}$$

   where c is a positive constant.
2. If $K_1(x_i, x_j)$ and $K_2(x_i, x_j)$ are two kernels then the sum $K(x_i, x_j) = K_1(x_i, x_j) + K_2(x_i, x_j)$ and the product $K(x_i, x_j) = K_1(x_i, x_j) K_2(x_i, x_j)$ are both permissable kernels.
3. If $K_1(x_i, x_j)$ is a kernel and $f(x)$ is any function of x, then the following is a permissable kernel:

$$K(x_i, x_j) = f(x_i) K_1(x_i, x_j) f(x_j). \tag{12.36}$$

4. If $K_1(x_i, x_j)$ is a kernel then so is

$$K(x_i, x_j) = p\left[K_1(x_i, x_j)\right], \tag{12.37}$$

where $p(\cdot)$ is a polynomial with nonnegative coefficients. As an example, the following is a kernel:

$$K(x_i, x_j) = \exp[K_1(x_i, x_j)], \qquad (12.38)$$

since $\exp(\cdot)$ can be expanded in a Taylor series with positive coefficients.

A further manipulation we can apply to a kernel matrix is *normalization*. This is achieved using a modified mapping function to feature space: $x \to \Phi(x)/\|\Phi(x)\|_2$. The normalized kernel is then

$$
\begin{aligned}
\widehat{K}(x_i, x_j) &= \frac{\Phi(x_i) \cdot \Phi(x_j)}{\|\Phi(x_i)\|_2 \, \|\Phi(x_j)\|_2} \\
&= \frac{\Phi(x_i) \cdot \Phi(x_j)}{\sqrt{\Phi(x_i) \cdot \Phi(x_i)} \sqrt{\Phi(x_j) \cdot \Phi(x_j)}} \\
&= \frac{K(x_i, x_j)}{\sqrt{K(x_i, x_i) K(x_j, x_j)}} \, . \qquad (12.39)
\end{aligned}
$$

Thus, consider the Gaussian kernel introduced in (12.10). Since

$$
\begin{aligned}
K(x_i, x_j) &= \exp\left(-\frac{(x_i - x_j)^2}{2\sigma^2}\right) \\
&= \exp\left(\frac{x_i \cdot x_j}{\sigma^2} - \frac{x_i \cdot x_i}{2\sigma^2} - \frac{x_j \cdot x_j}{2\sigma^2}\right) \\
&= \frac{\exp\left(\frac{x_i \cdot x_j}{\sigma^2}\right)}{\sqrt{\exp\left(\frac{x_i \cdot x_i}{\sigma^2}\right) \exp\left(\frac{x_j \cdot x_j}{\sigma^2}\right)}} \, , \qquad (12.40)
\end{aligned}
$$

it is a normalized kernel. Since we know that the linear kernel $K(x_i, x_j) = x_i \cdot x_j$ is a permissable kernel, validity of a Gaussian kernel follows from properties (1), (2), and (4) above.

For the Gaussian kernel, and various other kernels, there is a *kernel parameter* (the σ for the Gaussian). The value for this parameter needs to be found, and there are several ways to do this. If we have enough data we can split it into a *training set*, a *validation set*, and a *test set*. We then pursue a *validation* study in which the learning machine is trained at regularly spaced choices of the kernel parameter, and we use that value which minimizes the validation error (the error on the validation data). If insufficient data are available and we are considering classification, then we can estimate the kernel parameter using generalization bounds with no recourse to using validation data [12.25–29].

Kernels for Strings and Sequences

Strings appear in many bioinformatics contexts; For example, we could be considering DNA genetic sequences composed of the four DNA bases A, C, G, and T, or we could be considering proteinogenic amino acid sequences composed of the 21 amino acids found in eukaryotes. Strings can be defined as ordered sets of symbols drawn from an alphabet. We can evidently see a degree of similarity between strings. Thus, suppose we consider the genetic sequences ACTGA, CCACTG, and CTGACT. They have the string CTG in common, and a matching algorithm should pick up this similarity irrespective of the differing prefixes and suffixes. Strings could differ by *deletions* or *insertions*, thus ACGA differs from ACTGA by a gap consisting of a single deletion.

We can consider two distinct categories when matching ordered sets of symbols [12.30]. The first we will call *string matching*: in this case contiguity of the symbols is important. For the second category, *sequence matching*, only the order is important. Thus, for our example with ACGA and ACTGA, there are only two short contiguous strings in common: AC and GA. On the other hand, A, C, G, and A are ordered the same way in both words. When matching the same genes between two individuals, there will be extensive commonality, interrupted by occasional mutations and rare deletions and insertions. In this section we consider the p-spectrum kernel (contiguity is necessary), the subsequence kernel (contiguity is not necessary and the order of symbols is important), and a gap-weighted kernel which is influenced by both contiguity and order.

The p-Spectrum Kernel. Two strings can be compared by counting the number of contiguous substrings of length p which are in common. A p-spectrum kernel [12.31–33] is based on the set of frequencies of all *contiguous substrings of length p*; For example, suppose we wish to compute the 2-spectrum of the string $s = \text{CTG}$. There are two contiguous substrings of length $p = 2$, namely $u_1 = \text{CT}$ and $u_2 = \text{TG}$, both with frequency of 1. As a further example, let us consider the set of strings $s_1 = \text{CTG}$, $s_2 = \text{ACT}$, $s_3 = \text{CTA}$, and $s_4 = \text{ATA}$. The 2-spectrum mapping function Φ is given in Table 12.1, where each entry is the number

Table 12.1 A mapping function Φ for the p-spectrum kernel

Φ	CT	AT	TG	TA
CTG	1	0	1	0
ATG	0	1	1	0
CTA	1	0	0	1
ATA	0	1	0	1

Table 12.2 The p-spectrum kernel matrix from the mapping function in Table 12.1

K	CTG	ATG	CTA	ATA
CTG	2	1	1	0
ATG	1	2	0	1
CTA	1	0	2	1
ATA	0	1	1	2

of occurrences of the substring u (say $u = CT$) in the given string (say $s = CTG$). The corresponding kernel is shown in Table 12.2.

Thus, to compute the (CTG, ATG) entry in the kernel matrix, we sum the products of the corresponding row entries under each column in the mapping function (Table 12.1). Only the pair of entries in the TG substring column both have nonzero entries of 1, giving $K(CTG, ATG) = 1$. For an entry on the diagonal of the kernel matrix, we take the sum of the squares of the entries in the corresponding row in Table 12.1. Thus, for $K(CTG, CTG)$, there are nonzero entries of 1 under CT and TG, and so $K(CTG, CTG) = 2$.

The All-Subsequence Kernel. With this kernel the implicit mapping function is taken *over all contiguous and noncontiguous ordered subsequences of a string*, which includes the empty set. As an example let us consider two sequences $s_1 = CTG$ and $s_2 = ATG$. Let Ω represent the empty set, then the mapping function is given in Table 12.3.

The off-diagonal terms in the kernel matrix are then evaluated as the sum across all columns of the products of the two entries in each column. The diagonal terms are the sum of squares of all entries in a row (Table 12.4).

Finally, we can consider a *gap-weighted subsequence kernel*. With this kernel a penalization is used so that the length of the intervening gap or insertion decreases the score for the match. As an example, CTG is a subsequence of CATG and CAAAAATG. However, CTG differs from CATG by one deletion, but CTG differs from CAAAAATG by a gap of five symbol deletions. By appropriately weighting the penalty associated with the gap or insertion, we can interpolate between a p-spectrum kernel and the all-subsequence kernel.

Table 12.4 The all-subsequences kernel matrix for the mapping function in Table 12.3

K	CTG	ATG
CTG	8	4
ATG	4	8

Kernels for Graphs

Graphs appear in many settings in bioinformatics; For example, we can use a graph to represent a transcriptional regulatory network with the genes as the nodes and the edges representing functional connections between genes. We can consider two types of similarity. For a given graph we may be interested in the similarity of two nodes within the same graph. On the other hand we may be interested in constructing a measure of similarity between two different graphs. We can construct kernels for both cases, and we refer to *kernels on graphs* [12.34, 35] when constructing a within-graph kernel between nodes and *kernels between graphs* [12.36, 37] for the latter comparison.

12.1.4 Multiple Kernel Learning

Many bioinformatics prediction tasks involve different types of data. In the preceding sections we saw that kernels are available for encoding a variety of different data types. Thus, we now consider prediction based on multiple types of input data. Plainly, if we build a predictor which can use all available relevant information, then it is likely to be more accurate than a predictor based on one type of data only.

As an example, for our case study 2, considered shortly, we predict protein fold class based on the use of 12 different types of data. This potentially includes sequence data derived from RNA sequences but also continuous-valued data from various physical measurements. Later, in case study 5, we consider *network completion*. Based on a training set of known links and nonlinks we attempt to predict possible links to new nodes in an incomplete network. For the case of network completion with protein–protein interaction data there are a variety of informative data types such as gene expression correlation, protein cellular localization, and phylogenetic profile data. Individually, these

Table 12.3 A mapping function Φ for the all-subsequences kernel

Φ	Ω	A	C	G	T	CT	AT	TG	CG	AG	CTG	ATG
CTG	1	0	1	1	1	1	0	1	1	0	1	0
ATG	1	1	0	1	1	0	1	1	0	1	0	1

types of data may be weakly informative for the purposes of prediction. However, taken together, they may yield a stronger predictive signal.

This type of problem is called *multiple kernel learning* (MKL). The most common approach to MKL is to use a linear combination of candidate kernels, with these kernels representing different types of input data. Let $\mathcal{K}$ be such a set of candidate kernels, then the objective of MKL is to simultaneously find the optimal weighted combination of these kernels *and* the best classification function. With such a linear combination of p prescribed kernels $\{K_\ell : \ell = 1, \ldots, p\}$, we have a composite kernel

$$K = \sum_{\ell=1}^{p} \lambda_\ell K_\ell , \qquad (12.41)$$

where $\sum_{\ell=1}^{p} \lambda_\ell = 1$, $\lambda_\ell \geq 0$, and the λ_ℓ are called the *kernel coefficients*.

There are a number of criteria for learning the kernel. For SVM binary classification an appropriate criterion would be to maximize the margin with respect to all the kernel spaces capable of discriminating different classes of labeled data. In particular, for a given kernel $K \in \mathcal{K}$ with $K(x_i, x_j) = \Phi(x_i) \cdot \Phi(x_j)$, we have seen that maximizing the margin involves minimizing $\|w\|^2$ in feature space subject to

$$y_i (w \cdot \Phi(x_i) + b) \geq 1, \quad i = 1, \ldots, m . \qquad (12.42)$$

As we have seen, maximizing the margin subject to these constraints gives the following dual problem:

$$\omega(K) = \max_{\alpha} \left\{ \sum_{i=1}^{m} \alpha_i - \frac{1}{2} \sum_{i,j=1}^{m} \alpha_i \alpha_j y_i y_j K(x_i, x_j) : \right.$$
$$\left. \sum_{i=1}^{m} \alpha_i y_i = 0, \alpha_i \geq 0 \right\} . \qquad (12.43)$$

If $\mathcal{K}$ is now a linear combination of kernel matrices, this maximum margin approach to kernel combination learning reduces to

$$\min_{\lambda} \max_{\alpha} \left\{ \mathcal{L}(\alpha, \lambda) = \sum_{i=1}^{m} \alpha_i \right.$$
$$\left. - \frac{1}{2} \sum_{i,j=1}^{m} \alpha_i \alpha_j y_i y_j \left[\sum_{\ell=1}^{p} \lambda_\ell K_\ell(x_i, x_j) \right] \right\}$$
$$\qquad (12.44)$$

subject to

$$\sum_{i=1}^{m} \alpha_i y_i = 0 \quad 0 \leq \alpha_i \leq C , \qquad (12.45)$$

$$\sum_{\ell=1}^{p} \lambda_\ell = 1 \quad \lambda_\ell \geq 0 . \qquad (12.46)$$

$\mathcal{L}(\alpha, \lambda)$ is *concave* with respect to α and *convex* with respect to λ. A number of approaches have been proposed for dealing with this type of optimization problem. One of the earliest approaches [12.38] proposed a *semidefinite programming* (SDP) approach. SDP is computationally intensive, so more efficient methods were developed subsequently [12.38–43].

12.1.5 Case Study 2: Protein Fold Prediction Using Multiple Kernel Learning

During *folding* a protein forms its final three-dimensional structure. Understanding a protein's structure gives important insights into function. Its structure can lead to an understanding of protein–protein interaction or likely biological function. A knowledge of protein structure is important in the design of small molecular inhibitors for disabling the function of target proteins. We can use machine learning methods to predict protein structure based on sequence and other types of data: indeed this is an obvious application domain for MKL techniques. In this case study we only consider a subproblem of structure prediction in which the predicted label is over a set of *fold classes*. The fold classes are a set of structural components, common across proteins, which give rise to the overall three-dimensional structure. In this study we will use 27 fold classes with 313 proteins used for training and 385 for testing.

There are a number of relevant types of data which can be used to predict fold class, and in this study we use 12 different types of data, each encoded into a kernel. These data types included sequence and physical measurements such as hydrophobicity, polarity, and van der Waals volume. In Fig. 12.5 we illustrate the performance of a MKL algorithm on this dataset. In Fig. 12.5a the vertical bars indicate the test set accuracy based on using one type of data only: for example, H is hydrophobicity, P is polarity, and V is van der Waals volume. The horizontal line indicates the performance of the MKL algorithm with all data types included. Thus, we get an improvement in per-

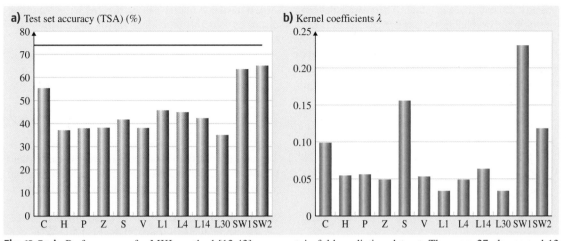

Fig. 12.5a,b Performance of a MKL method [12.43] on a protein fold prediction dataset. There are 27 classes and 12 types of data. (**a**) Test set accuracy (TSA, in %) based on individual data types (*vertical bars*) and using MKL (*horizontal line*). (**b**) Kernel coefficients λ_ℓ, which indicate the relative significance of individual types of data

formance if we use all available relevant sources of data over just using the single most informative data source.

Figure 12.5b gives the values of the kernel coefficients λ_ℓ based on using a linear combination (12.41). The relative height of the peaks indicates the relative significance of different types of input data. This algorithm indicates that all 12 types of data are relevant, though some types of data are more informative than others (the most informative, SW1 and SW2, are based on sequence alignments). MKL methods have been successfully demonstrated on other bioinformatics problems requiring integration of heterogeneous datasets [12.38, 43–46].

12.2 Unsupervised Learning

Having considered supervised learning, we now discuss unsupervised learning and the discovery of structure in biomedical datasets. Hierarchical cluster analysis is a commonly used approach to unsupervised learning in the biomedical literature, and so we briefly review this topic in Sect. 12.2.1. One of our main objectives in this section will be to show that there are some advantages to using more contemporary methods, for example, from Bayesian statistics. In Sects. 12.2.3–12.2.5 we introduce *variational* methods. These are not as accurate as the *Markov chain Monte Carlo* (MCMC) methods discussed in Sect. 12.2.7. However, they are fast in practice and thus suited to some of the large datasets which appear in many bioinformatics applications. After introducing variational methods we consider an application to the interpretation of gene expression array datasets in cancer research. After introducing MCMC, we illustrate its use with network inference in Sect. 12.2.8 with an application to find-ing the most probable network topology given a set of candidate network topologies.

12.2.1 Hierarchical Cluster Analysis

In the biomedical literature, hierarchical cluster analysis (HCA) is a commonly used technique for unsupervised learning. This approach can be divided into *agglomerative methods*, which proceed through a series of successive fusions of m samples into larger and larger clusters, and *divisive methods*, which systematically separate clusters into smaller and smaller groupings. HCA methods are usually represented by a *dendrogram* which illustrates the successive fusions or divisions produced by this approach. In this section we only consider agglomerative methods. In this case we start with m single sample clusters, representing each data point. At each stage in the clustering procedure we fuse those samples or groups of samples which are currently closest to each other. There are a number of criteria for evaluating the similarity or closeness of data points. Thus, if x_{id} corresponds to the i-th sample ($i = 1, \ldots, m$) with d the corresponding *feature* index ($d = 1, \ldots, p$), then a commonly used similarity measure is the *squared distance*

$$D(x_i, x_j) = \sum_{d=1}^{p} (x_{id} - x_{jd})^2 .$$

Other criteria are used such as the *correlation coefficient*,

$$C(x_i, x_j) = \frac{\sum_d (x_{id} - \bar{x}_i)(x_{jd} - \bar{x}_j)}{\sqrt{\sum_d (x_{id} - \bar{x}_i)^2 \sum_d (x_{jd} - \bar{x}_j)^2}} ,$$

where $\bar{x}_i = (\sum_d x_{id})/p$. If each sample vector is standardized to zero mean and unit standard deviation, then clustering based on the correlation coefficient becomes equivalent to use of a squared distance. Using the chosen distance measure, we then derive a *distance matrix* encoding the distances between all data points.

Just as there are a number of ways of quantifying similarity, there are a number of criteria for deciding which clusters to fuse at each stage. Six methods are commonly used in practice: single linkage, complete linkage, average linkage, the centroid and median methods, and Ward's method. We will illustrate the approach with *single linkage clustering* as one of the simplest of these methods. In this case the distance between groupings is defined as the shortest distance between any pair of samples. This method is best illustrated with an example. Thus, suppose we have a set of five samples with a corresponding initial distance matrix given by

$$D_1 = \begin{array}{c} 1 \\ 2 \\ 3 \\ 4 \\ 5 \end{array} \begin{array}{c} 1\ 2\ 3\ 4\ 5 \end{array} \left(\begin{array}{ccccc} 0 & 4 & 8 & 9 & 7 \\ 4 & 0 & 6 & 5 & 6 \\ 8 & 6 & 0 & 3 & 8 \\ 9 & 5 & 3 & 0 & 2 \\ 7 & 6 & 8 & 2 & 0 \end{array} \right) .$$

In this matrix it is evident that the smallest distance is that between samples 4 and 5. Thus, we place these two samples into a new cluster. The new distances between this cluster, which we label (45) and the other three samples is obtained from the above distance matrix as

$$d_{1(45)} = \min (d_{14}, d_{15}) = d_{15} = 7 ,$$
$$d_{2(45)} = \min (d_{24}, d_{25}) = d_{24} = 5 ,$$
$$d_{3(45)} = \min (d_{34}, d_{35}) = d_{34} = 3 .$$

A new distance matrix can be derived based on these distances and the remaining set of pre-existing distances, thus

$$D_2 = \begin{array}{c} 1 \\ 2 \\ 3 \\ (45) \end{array} \begin{array}{c} 1\ \ 2\ \ 3\ (45) \end{array} \left(\begin{array}{cccc} 0 & 4 & 8 & 7 \\ 4 & 0 & 6 & 5 \\ 8 & 6 & 0 & 3 \\ 7 & 5 & 3 & 0 \end{array} \right) .$$

The smallest entry in this new distance matrix is then between sample 3 and the cluster (45), and so we form a new three-member cluster and a new set of distances

$$d_{1(345)} = \min (d_{13}, d_{14}, d_{15}) = d_{15} = 7 ,$$
$$d_{2(345)} = \min (d_{23}, d_{24}, d_{25}) = d_{24} = 5 ,$$

leading to a new 3×3 distance matrix. This process is iterated until all data points belong to one cluster. The sequence of clusterings is therefore

Step	Clusters
1	(1), (2), (3), (4), (5)
2	(1), (2), (3), (45)
3	(1), (2), (345)
4	(12), (345)
5	(12345)

Of course, using the closest points within each cluster is not necessarily a good fusion criterion. With *complete linkage clustering* the distance between two groupings is defined as the most distant pair of data points, with each pair consisting of one sample from each of the two groups. Again, this type of method can be influenced by outliers within each cluster, and so a better method is to use the average of a cluster instead. With *average clustering* the distance between two groupings is defined as the average of the distances between all pairs of samples, with one sample in a pair from each group. Thus, with our example above, after merging data points 4 and 5 into the first cluster, we would compute the new set of distances from this cluster to the other three data points through

$$d_{1(45)} = \tfrac{1}{2} (d_{14} + d_{15}) = 8.0 ,$$
$$d_{2(45)} = \tfrac{1}{2} (d_{24} + d_{25}) = 5.5 ,$$
$$d_{3(45)} = \tfrac{1}{2} (d_{34} + d_{35}) = 5.5 .$$

With *centroid clustering* each grouping is represented by a mean vector, that is, a vector composed of the mean values of each feature taken over all samples within the grouping. The distance between two clusters is then the distance between the corresponding mean vectors. However, centroid clustering has the following disadvantage: if the number of members in one cluster is very different from the other, then, when they are fused, the centroid of the new cluster will be most heavily influenced by the larger grouping and may remain within that cluster. Clustering can be made independent of group size by assuming that the two groups are of equal size. *Median clustering* is a variant on centroid clustering in which the cluster arising from the fusion lies at such a midpoint between the two clusters. Finally, with *Ward's method*, the two clusters chosen for fusion are the two which result in the least increase in the sum of the distances of each sample to the centroid of its originating cluster.

Of course, the above methods provide an approach to finding a cluster structure but they do not indicate the most likely number of clusters in the data. However, various criteria have been proposed to indicate the most probable number of clusters [12.48–50].

12.2.2 Case Study 3: An HCA Application to Colon Cancer

As an illustration, we now apply HCA to a gene expression microarray dataset from a cancer study. A DNA microarray has a series of microscopic probes, each constructed from strings of nucleotides. A microarray has tens of thousands of such probes, each of which can hybridize to a particular target strand of complementary DNA (cDNA). Probe–target hybridization can be measured by using fluorophore labels, for example, with altered levels of gene expression quantified

by the level of fluorescence. In Fig. 12.6 we depict the corresponding dendrogram (using *average linkage* and a *correlation coefficient* distance measure) for a small microarray dataset consisting of 17 colon tumor and 18 normal samples. The dendrogram has been largely successful in separating normal from cancer samples.

12.2.3 Bayesian Unsupervised Learning

HCA has been the method of choice for cluster analysis for many biomedical publications. However, HCA does have some drawbacks. Firstly, there is an implicit assumption that each sample is associated to a particular cluster. This may not be realistic in many situations where a sample should be better represented as overlapping several clusters. Thus, tumors can be genetically heterogeneous; i.e., tissue regions in different regions of the tumor may have different genetic signatures. The models we describe below are *mixed membership models* with each sample represented as a combinatorial mixture over clusters. With the clinical assignment of patient samples to clusters or disease subtypes it would also be useful to associate a confidence measure with the cluster assignment. This can be achieved using the Bayesian methods outlined here. Further advantages of a Bayesian approach are the use of sound methods to objectively assess the number of sample clusters and the means to penalize overfitting (at the top levels of the dendrogram in Fig. 12.6 we are finding structure, but, toward the leaves at the base, we are starting to fit to noise in the data).

Let us suppose that M is a model and D is the data, then $p(M|D)$ represents the probability of a model given the data. Intuitively we should seek to maximize

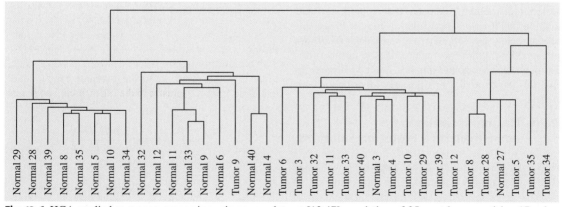

Fig. 12.6 HCA applied to a gene expression microarray dataset [12.47] consisting of 35 samples comprising 17 colon cancer and 18 normal colon samples

the probability of the model given the data. Given an assumed model structure, a fit to data is achieved through the adjustment of *model parameters*, which we denote Θ as a set and represent as the components of a vector $\boldsymbol{\theta}$. *Bayes's theorem* then implies that we should maximize

$$p(\Theta|D) = \frac{p(D|\Theta)p(\Theta)}{p(D)} \ , \tag{12.47}$$

where $p(\Theta|D)$ is the *posterior*, $p(D|\Theta)$ is the *likelihood*, and $p(\Theta)$ is the *prior*. Thus, $p(\Theta|D) \sim p(D|\Theta)p(\Theta)$ states that multiplying the likelihood by our prior beliefs about the parameters, $p(\Theta)$, will give us posterior beliefs about the parameters, having observed the data, $p(\Theta|D)$. The normalization term $p(D)$ is called the *evidence* and can be found through an integration over the $\boldsymbol{\theta}$ (we will refer to this as *marginalizing* out the $\boldsymbol{\theta}$)

$$p(D) = \int p(D|\boldsymbol{\theta})p(\boldsymbol{\theta})\mathrm{d}\boldsymbol{\theta} \ . \tag{12.48}$$

Maximizing the probability of a model, given the data, would require finding the optimal set of values for the model parameters, which we denote $\widehat{\Theta}$. We will call this a set of *point estimates* for these parameters.

However, we cannot make inferences which are not justified by the data. As a consequence, given some set of data, the most we can really say is that there is a spectrum of models which fit the data and some of these models are more probable than others. We call this a *posterior distribution over models*. This posterior distribution carries information beyond a model based on point estimates since a relatively flat posterior distribution means that many models fit the data well and the most probable model is not particularly unique. On the other hand, a sharply peaked posterior distribution indicates that a point estimate solution might be a sound solution to use. In the discussion below we will start with *maximum likelihood* and *maximum a posteriori* approaches to unsupervised learning which use point estimates. Later we discuss approaches which give a full posterior distribution over models.

12.2.4 Maximum Likelihood and Maximum a Posteriori Solutions

With a *maximum likelihood* (ML) approach we derive a set of parameters that maximize the likelihood of the data given the model parameters, $p(D|\Theta)$. With the *maximum a posteriori* (MAP) approach we find the set of parameters that maximize the posterior,

$p(\Theta|D)$, given the data. Thus, in terms of parameter-dependent probabilities, the MAP and ML solutions are related through Bayes rule $p(\Theta|D) \sim p(D|\Theta)p(\Theta)$. The MAP solution therefore enables us to include any prior knowledge we may have about the distribution of the parameter values.

The ease with which we may calculate $p(\Theta|D)$ depends on the functional forms of the likelihood and the prior. Also, if Θ is high dimensional, the evidence $p(D)$ may be difficult to evaluate. With a MAP solution only point estimates are used for the model parameters. These are denoted Θ_{MAP}, and they are based on the mode of the posterior distribution. Since the mode is unaffected when the distribution is multiplied by a constant, we can ignore the evidence in the denominator and only use the unnormalized distribution, which we denote $\widehat{p}(\Theta|D) = p(D|\Theta)p(\Theta)$. In general, though, we may need to evaluate the evidence, and this motivates our discussion of Monte Carlo methods below.

12.2.5 Variational Bayes

With a *variational Bayes* approach we can determine a posterior distribution over models. In this approach we approximate a posterior distribution $p(R|D)$ by a parameterized distribution $q(R)$. In this case D is the data, as before, and R is a parameter vector consisting of the model parameters Θ and any hidden variables within the model (for example, hidden labels assigning data components to clusters). Thus, we wish to make $q(R)$ as close as possible to the posterior $p(R|D)$. Since $p(R, D) = p(R|D)p(D)$ and $\int_{-\infty}^{\infty} q(R)\mathrm{d}R = 1$, we can write

$$\log\left[p(D)\right] = \log\left(\frac{p(R, D)}{p(R|D)}\right) \tag{12.49}$$

so

$$\log\left[p(D)\right] = \int_{-\infty}^{\infty} q(R)\log\left(\frac{q(R)p(R, D)}{q(R)p(R|D)}\right)\mathrm{d}R$$

$$= \int_{-\infty}^{\infty} q(R)\log\left(\frac{p(R, D)}{q(R)}\right)\mathrm{d}R$$

$$+ \int_{-\infty}^{\infty} q(R)\log\left(\frac{q(R)}{p(R|D)}\right)\mathrm{d}R$$

$$= F[R] + KL[q||p] \ .$$

The second term

$$KL\left[q||p\right] = \int q(R) \log\left(\frac{q(R)}{p(R|D)}\right) dR \quad (12.50)$$

is a *Kullback–Leibler* (KL) divergence which quantifies the similarity of the two distributions $q(R)$ and $p(R|D)$. The key observation is that $\log\left[p(D)\right]$ does not depend on R. Since we want to minimize the KL divergence between $q(R)$ and $p(R|D)$, we can achieve this by optimizing $F[R]$, which is called the *variational free energy*. After making distributional and other assumptions we can derive an *expectation-maximization* (EM) algorithm which optimizes $F[R]$ and gives an effective approximation to the posterior distribution $p(R|D)$.

Though we do not give more detail here, the ML, MAP, and variational Bayes methods outlined above can be extended in many ways; For example, we could consider a *marginalized variational Bayes approach* in which we marginalize, or integrate out, further model parameters [12.51, 52]. With fewer parameters to estimate, this approach gives higher cluster accuracy but at the cost of model interpretability. We could consider *semisupervised* clustering [12.53] if *side information*, in the form of some sample labels, is present. Then again, whereas HCA may not be amenable to data integration, Bayesian methods could provide a route to unsupervised learning using multiple types of data; For example, with a *correspondence model* [12.54, 55], we can consider two types of data with one type of data believed dependent on the other. In a bioinformatics context an example would be gene expression array data which we believe may be partially influenced by microRNA expression [12.55].

12.2.6 Case Study 4: An Application to the Interpretation of Expression Array Data in Cancer Research

To illustrate an application of these methods we consider the use of ML, MAP, and variational Bayes methods to interpret gene expression array data derived in cancer research. We will start with a maximum likelihood or MAP solution giving point estimates for the model parameters. We start by specifying a model which includes making certain distributional assumptions for the data and model parameters. This leads us through to a likelihood bound which is optimized using an algorithmic procedure (an EM or expectation-maximization method).

As an example we consider latent process decomposition (LPD) [12.56]. We first make assumptions about the type of data we are considering. For expression array data, a reasonable assumption is that the gene expression measurements follow an approximate Gaussian distribution which will have a mean μ and standard distribution σ. Each sample in the data has a set of features labeled by an index $g = 1, \ldots, G$. Then, for feature g we draw a cluster index k ($k = 1, \ldots, K$) with a probability denoted β_k which selects a Gaussian with parameters μ_{gk} and σ_{gk}. Next we make assumptions about the probability distributions involved. For the β we assume a Dirichlet distribution, which is a standard assumption in this context. This Dirichlet distribution is parameterized by a K-dimensional vector α. The expression array data D consist of a set of samples, indexed $a = 1, \ldots, A$, each with a set of features labeled by g, and we thus denote the experimental measurements by e_{ga}. It is normal to work with the log of the likelihood so that we deal with summations rather than products. This is sound since the log-function is monotonic, so maximizing the log-likelihood is equivalent to maximizing the likelihood. The log-likelihood is then $\log p(D|\mu, \sigma, \alpha)$, which can be factorized over the individual samples as

$$\log p(D|\mu, \sigma, \alpha) = \sum_{a=1}^{A} \log p(a|\mu, \sigma, \alpha) . \quad (12.51)$$

We can rewrite this as a *marginalized* integral over the β, that is,

$$\log p(D|\mu, \sigma, \alpha)$$
$$= \sum_{a=1}^{A} \log \int p(a|\mu, \sigma, \beta) p(\beta|\alpha) d\beta , \quad (12.52)$$

with $p(\beta|\alpha)$ being the Dirichlet distribution. We introduce the Gaussian distributional assumption for the data

$$p(a|\mu, \sigma, \beta) = \prod_{g=1}^{G} \sum_{k=1}^{K} N\left(e_{ga}|k, \mu_{gk}, \sigma_{gk}\right) \beta_k .$$
$$(12.53)$$

Exact inference with this model is intractable, so to increase the likelihood and therefore, implicitly, the probability of the model given the data, we follow the indirect route of deriving a lower bound on this log-likelihood via *Jensen's inequality*. We then maximize this lower bound using an iterative procedure based on an expectation-maximization algorithm. Thus we get

$$\sum_{a=1}^{A} \log \left[p(a|\boldsymbol{\mu}, \boldsymbol{\sigma}, \boldsymbol{\alpha}) \right] =$$

$$\sum_{a=1}^{A} \log \int_{\boldsymbol{\beta}} \left[\prod_{g=1}^{G} \sum_{k=1}^{K} N(e_{ga}|k, \mu_{gk}, \sigma_{gk})\beta_k \right] p(\boldsymbol{\beta}|\boldsymbol{\alpha})\, d\boldsymbol{\beta} \,.$$

$$(12.54)$$

If we define $E_p(z) = \int z p(z) dz$, then Jensen's inequality for a concave function $f(x)$ states that

$$f(E_{p(z)}[z]) \geq E_{p(z)}[f(z)] \,. \qquad (12.55)$$

We have assumed a Dirichlet distribution for the $\boldsymbol{\beta}$, so we could introduce a sample-specific (a-dependent) distribution for the $p(\boldsymbol{\beta}|\boldsymbol{\gamma}_a)$ as

$$E_{p(\boldsymbol{\beta}|\boldsymbol{\gamma}_a)}[f(\boldsymbol{\beta})] = \int_{\boldsymbol{\beta}} f(\boldsymbol{\beta})p(\boldsymbol{\beta}|\boldsymbol{\gamma}_a)\, d\boldsymbol{\beta} \,, \qquad (12.56)$$

giving

$$f\left[\int_{\boldsymbol{\beta}} p(\boldsymbol{\beta}|\boldsymbol{\gamma}_a)\boldsymbol{\beta}\, d\boldsymbol{\beta} \right]$$
$$= f(E_{p(\boldsymbol{\beta}|\boldsymbol{\gamma}_a)}[\beta]) \geq E_{p(\boldsymbol{\beta}|\boldsymbol{\gamma}_a)}[f(\boldsymbol{\beta})]$$
$$= \int_{\boldsymbol{\beta}} f(\boldsymbol{\beta})p(\boldsymbol{\beta}|\boldsymbol{\gamma}_a)\, d\boldsymbol{\beta} \qquad (12.57)$$

and so

$$\sum_a \log \left[p(a|\boldsymbol{\mu}, \boldsymbol{\sigma}, \boldsymbol{\alpha}) \right]$$

$$= \sum_a \log \left[\int_{\boldsymbol{\beta}} p(a|\boldsymbol{\mu}, \boldsymbol{\sigma}, \boldsymbol{\beta})p(\boldsymbol{\beta}|\boldsymbol{\alpha})d\boldsymbol{\beta} \right]$$

$$= \sum_a \log \left[\int_{\boldsymbol{\beta}} \left\{ p(a|\boldsymbol{\mu}, \boldsymbol{\sigma}, \boldsymbol{\beta}) \frac{p(\boldsymbol{\beta}|\boldsymbol{\alpha})}{p(\boldsymbol{\beta}|\boldsymbol{\gamma}_a)} \right\} p(\boldsymbol{\beta}|\boldsymbol{\gamma}_a)d\boldsymbol{\beta} \right]$$

$$= \sum_a \log \left[E_{p(\boldsymbol{\beta}|\boldsymbol{\gamma}_a)} \left\{ p(a|\boldsymbol{\mu}, \boldsymbol{\sigma}, \boldsymbol{\beta}) \frac{p(\boldsymbol{\beta}|\boldsymbol{\alpha})}{p(\boldsymbol{\beta}|\boldsymbol{\gamma}_a)} \right\} \right]$$

$$\geq \sum_a E_{p(\boldsymbol{\beta}|\boldsymbol{\gamma}_a)} \left[\log \left\{ p(a|\boldsymbol{\mu}, \boldsymbol{\sigma}, \boldsymbol{\beta}) \frac{p(\boldsymbol{\beta}|\boldsymbol{\alpha})}{p(\boldsymbol{\beta}|\boldsymbol{\gamma}_a)} \right\} \right]$$

$$= \sum_a E_{p(\boldsymbol{\beta}|\boldsymbol{\gamma}_a)} \left[\log \left\{ p(a|\boldsymbol{\mu}, \boldsymbol{\sigma}, \boldsymbol{\beta}) \right\} \right]$$
$$+ \sum_a E_{p(\boldsymbol{\beta}|\boldsymbol{\gamma}_a)} \left[\log \left\{ p(\boldsymbol{\beta}|\boldsymbol{\alpha}) \right\} \right]$$
$$- \sum_a E_{p(\boldsymbol{\beta}|\boldsymbol{\gamma}_a)} \left[\log \left\{ p(\boldsymbol{\beta}|\boldsymbol{\gamma}_a) \right\} \right] \,.$$

It is this lower bound which we optimize using a two-step expectation-maximization algorithm [12.56]. A variational Bayes approach leads to a similar iterative procedure to maximize the free energy term in (12.50).

We applied ML, MAP, and variational Bayes methods [12.55] to the interpretation of expression array data

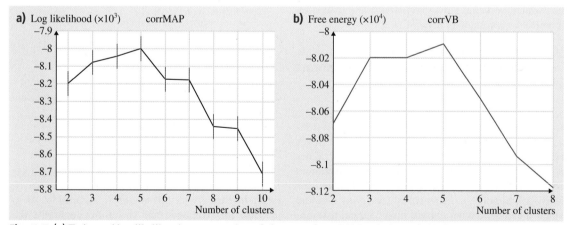

Fig. 12.7 (a) Estimated log-likelihood versus number of clusters using a MAP solution [12.55] for a breast cancer expression array dataset [12.57]. (b) Variational Bayes solution for the same dataset. For both methods the peak at five clusters indicates five principal subtypes of breast cancer. If more data were used, higher resolution may be achieved and we may observe more subtypes

derived from 78 primary breast cancer samples [12.57]. Using a MAP approach we obtained a solution for the model parameters using an EM algorithm. If we split the data into training and validation data, parameters in the log-likelihood can be estimated from the training set. Using this estimated log-likelihood and the validation data, we can then estimate model complexity, i. e., how many clusters are apparently present in the data – in this case how many subtypes of breast cancer are indicated. In Fig. 12.7a we show the estimated log-likelihood on left-out validation data for this breast cancer dataset. The peak indicates a minimum of five subtypes.

We also used a variational Bayes method with the same dataset [12.55]. In this case the maximum of the free energy versus number of clusters indicates the appropriate model complexity. In Fig. 12.7b the peak is also at 5, indicating that this is the most appropriate number of subtypes to consider. Interestingly, with variational Bayes, we do not need to use validation data.

Having found the most appropriate number of subtypes, we can use these methods to find those genes which are abnormally functioning within subtypes. The parameters μ_{gk} and σ_{gk} can be used to model the data distribution for gene g in cluster k. This is illustrated in Fig. 12.8 for two genes, *FOXA1* and *FOXC1*, which appear to be operating abnormally within one of these subtypes (cluster 5, denoted Cl5). The Gaussian distributions are determined by the (μ_{gk}, σ_{gk}), and the actual distribution of data values is shown below these Gaussian distributions. Whereas *FOXA1* and *FOXC1* appear to function normally in the other subtypes, *FOXA1* appears to be underexpressing and *FOXC1* overexpressing within this subtype.

As pointed out at the beginning, one further aspect in which these Bayesian unsupervised learning methods differ from HCA is that they allow for mixed membership. As a second example (Fig. 12.9), we apply a variational Bayes method to a lung cancer gene expression array dataset derived from 73 patient samples [12.59]. The peaks indicate the confidence that a particular patient belongs to a particular cluster. Many peaks are 1, but a number overlap several clusters, possibly indicating an unclear assignment of patient to subtype. We have used lung cancer as an illustration because there are a number of clinically established subtypes for this disease, such as small cell lung cancer and adenocarcinoma of the lung. The clinical assignments are indicated by the boundary markers in this plot, and there is reasonable agreement between clinical assign-

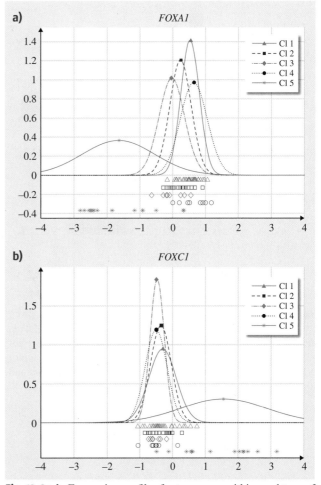

Fig. 12.8a,b Expression profiles for two genes within a subtype of breast cancer: *FOXA1* (**a**) and *FOXC1* (**b**). This subtype can be identified with the basal-like or *basaloid* subtype of breast cancer. These two genes show a strong reciprocal anticorrelated expression profile within this subtype (after [12.58])

ment to subtype and those assignments made by the variational Bayes method.

12.2.7 Monte Carlo Methods

Monte Carlo methods are potentially more exact than variational methods, and they are commonly used in probabilistic inference. If θ is high dimensional, evaluating the evidence $\int p(D|\theta)p(\theta)\,d\theta$ and finding the posterior $p(\theta|D)$ is a difficult task. For the evidence, we could perform a *Monte Carlo integration*. Taken over an arbitrary distribution $h(\theta)$, we can perform Monte Carlo

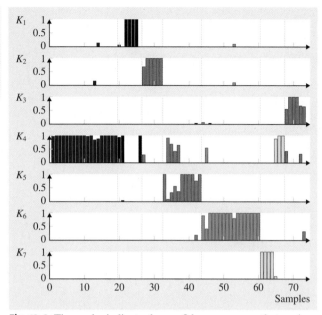

Fig. 12.9 The peaks indicate the confidence measure that a given sample is assigned to a particular cluster (see [12.58] for details about the derivation of this measure). The *fourth band down* has mainly adenocarcinoma of the lung (samples 1–20), and the plot indicates some confusion of assignment between this category and other subtypes of lung cancer

integration by writing $h(\theta) = f(\theta)g(\theta)$ as

$$\int h(\theta)\,d\theta = \int f(\theta)g(\theta)\,d\theta = E_{g(\theta)}\left[f(\theta)\right]$$

$$\approx \frac{1}{M}\sum_{m=1}^{M} f\left(\theta^{(m)}\right), \tag{12.58}$$

so that the integration becomes an expectation of $f(\theta)$ over $g(\theta)$. By deriving a number of observations $\theta^{(m)}$ ($m = 1, \ldots, M$), sampled from the distribution $g(\theta)$, and using these samples in $f(\theta)$, the original integral can be approximately evaluated. We could use this approach to evaluate the evidence and the posterior distribution by sampling from the unnormalized distribution $\widehat{p}(\theta|D) = p(D|\theta)p(\theta)$ to find the normalization $\int p(\theta|D)\,d\theta$. To use this method of integration, we must be able to draw samples reliably from a distribution of choice, such as $\widehat{p}(\theta|D)$. This is not easy, and here we briefly describe *Markov chain Monte Carlo* (MCMC) methods [12.60–62], based on the *Metropolis* algorithm, for performing this task.

A *Markov chain* is a sequence of samples $\{\theta_1, \theta_2, \ldots\}$ such that each sample is only dependent

on the previous sample, i. e., $p(\theta_t|\theta_{t-1}, \theta_{t-2}, \ldots, \theta_1) = p(\theta_t|\theta_{t-1})$, where t is the iteration index. In the MCMC approach, a *proposal* or *jump* distribution $Q(\theta_t|\theta_{t-1})$ is used to generate θ_t from θ_{t-1}. With the *Metropolis algorithm* we assume that this distribution is symmetric, i. e., that $Q(\theta_{t+1}|\theta_t) = Q(\theta_t|\theta_{t+1})$. Our objective is to reliably draw samples from a distribution $g(\theta)$, such as $p(\theta|D)$. Thus, we start the procedure with some θ_0 such that $\widehat{g}(\theta_0) > 0$. After a number of iterations this procedure will tend toward the stationary distribution $g(\theta)$ so that θ_t represents a random draw from $g(\theta)$. In the standard approach to MCMC, to arrive at this stationary distribution, at each step a sample is selected from Q and either accepted or rejected based on a comparison with $g(\theta)$. Specifically, a candidate sample $\theta^\star$ is sampled from $Q(\theta_t|\theta_{t-1})$ and accepted with probability α_t given by

$$\alpha_t = \min\left(\frac{\widehat{g}(\theta^\star)Q(\theta_{t-1}|\theta^\star)}{\widehat{g}(\theta_{t-1})Q(\theta^\star|\theta_{t-1})}, 1\right) \tag{12.59}$$

or rejected. If $\theta^\star$ is accepted, then $\theta^\star$ is assigned to θ_t. Otherwise, if $\theta^\star$ is rejected, then θ_{t-1} is assigned to θ_t instead. The Metropolis Hastings algorithm [12.63–65] extends the Metropolis algorithm to jump distributions which are not symmetric, and both of these methods are members of a broad class of approaches to sampling from high-dimensional distributions.

12.2.8 Case Study 5: Network Inference

The understanding of pathways and networks is crucial to our understanding of the functional organization of genes and proteins. At an abstract level a network can be viewed as a set of *nodes*, together with a set of directed or undirected *edges* between these nodes. Biological networks of interest include, for example, *transcriptional regulatory networks*. In this case a gene may express a protein that functions as a transcriptional inhibitor or, alternatively, as an activator of one or more target genes. In this case the genes can be viewed as the nodes of a network with the edges representing direct regulatory connections to other genes. With *signal transduction networks* the proteins are viewed as the nodes and the edges are corresponding protein–protein interactions. Further networks of interest are *metabolic networks*, where the metabolites are the nodes.

We could use unsupervised learning to determine the network structure, a task we could call *global inference of network topology*. However, suppose we consider a fully connected network with n nodes, then we are attempting to find $n(n-1)/2$ possible connec-

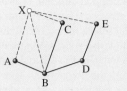

Fig. 12.10 Network completion. Using a training set of known links (*bold lines*) and nonlinks for nodes A–E we use supervised learning to predict links or nonlinks to a new node X (*dashed lines*)

tions given limited amounts of data. A cell may have tens of thousands of genes, whereas, in most experiments, we are only considering a few hundred samples with measurements corrupted by noise. Thus, the inference problem is typically highly underdetermined.

To make network inference more amenable as a machine learning task, a more tractable approach is *network completion* [12.66]. In this case we have an established pathway of interest and consequently we know certain links and nonlinks between pairs of nodes. The problem is therefore to determine whether a given node, perhaps representing a gene, has a link to this pathway or not. Since we have a training set of known links and nonlinks we can train a classifier via supervised learning and then predict a link to the pathway or otherwise, for the node of interest (Fig. 12.10). Various types of data are informative as to whether a functional link exists, and hence we can cast this supervised learning task as an application of the multiple kernel learning techniques of Sect. 12.1.4. As a supervised learning problem and with a smaller set of linkages to infer, this problem can give more reliable results than global unsupervised inference of network topology.

We can improve accuracy by reducing the search space further. Thus, as a third approach, we can consider the use of Bayesian methods to decide *the most probable network structure given a small set of candidate network topologies*. In this case it is assumed that a biologist has partially determined a network but remains undecided over a set of candidate topologies. Thus, the task is to determine which of these proposed network topologies is most probable, given the

data. As an example, *Calderhead* et al. [12.67, 68] used Bayesian unsupervised learning to decide over alternative topologies proposed for the extracellular signal-regulated kinase (ERK) pathway. This signaling pathway regulates growth and is defective in many cancers. Binding of epidermal growth factor (EGF) to the epidermal growth factor receptor (EGFR) at the cell surface membrane activates ERK through a chain of proteins. Four candidate topologies were proposed. To compare individual pairs, representing different topologies, we use the *Bayes factor*, that is, the likelihood ratio

$$\frac{p(D|M_i)}{p(D|M_j)} \qquad (12.60)$$

stated in terms of the *marginal likelihood* for model M to generate data D

$$p(D|M_i) = \int p(D|M_i, \boldsymbol{\theta}_i) p(\boldsymbol{\theta}_i) \, d\boldsymbol{\theta}_i , \qquad (12.61)$$

where the $\boldsymbol{\theta}_i$ are the set of model parameters marginalized or integrated out. Each protein in the chain is modeled using an ordinary differential equation (ODE), and we use optimization methods with a least-squares parameter fitting approach to find the optimal parameter values in these ODEs. We do not describe the ODEs here but refer to the original paper [12.67, 68]. Thus, a model can be written as $M = \{S, \boldsymbol{\theta}\}$, where S is the system of differential equations and $\boldsymbol{\theta}$ is the set of parameters. The marginal likelihood is therefore a nonlinear function based on the solution of these ODEs. It cannot be computed analytically, and so we must use MCMC-based methods. Because the posterior distribution is generated by complex dynamical systems models, it is necessary to use more sophisticated sampling methods than those mentioned in Sect. 12.2.7. Indeed, rather than static sampling distributions we use populations of annealed (temperature-dependent) distributions in an approach commonly referred to as population MCMC. Applied to ERK pathway models it was possible to use population MCMC to compute marginal likelihoods and thus Bayes factors, lending support to one of the proposed topology models for the ERK pathway [12.68].

12.3 Conclusions

Progress in the biomedical sciences can be furthered by improved data interpretation, in addition to the acquisition of more data. In this chapter we have seen that

contemporary methods from machine learning can offer many advantages over more long-established data analysis methods. There are an increasing number of studies where multiple types of data are acquired from the same sample. An example is the Cancer Genome Atlas [12.69] project, where multiple types of data are derived from the same tumor sample. In Sect. 12.1.3 we saw that many different types of data can be encoded into corresponding kernels and prediction can be achieved using multiple kernel learning. In Sect. 12.2 we saw that contemporary Bayesian unsupervised learning methods compare favorably with hierarchical cluster

analysis, providing a confidence measure for assigning datapoints to clusters (Fig. 12.9) and an effective approach to model selection (Fig. 12.7). In the last section we saw that Bayesian methodology is the most effective approach to other tasks such as determining the most probable network topology given a set of candidate network topologies. Further innovation in machine learning will improve and expand the interpretability of many datasets from the biomedical sciences.

References

12.1   L. Bottou, O. Chapelle, D. DeCoste, J. Weston: *Large-Scale Kernel Machines*, Neural Information Processing Series (MIT Press, Cambridge 2007)

12.2   J. Platt, N. Cristianini, J. Shawe-Taylor: Large margin DAGS for multiclass classification, Adv. Neural Inform. Proces. Syst. **12**, 547–553 (2000)

12.3   Y. Lee, Y. Lin, G. Wahba: *Multicategory support vector machines*, *Technical Report 1043* (Univ. Madison, Wisconsin 2001)

12.4   T. Hastie, R. Tibshirani: Classification by pairwise coupling, Ann. Stat. **26**, 451–471 (1998)

12.5   T.G. Dietterich, G. Bakiri: Solving multiclass learning problems via error-correcting output codes, J. Artif. Intell. **2**, 263–286 (1995)

12.6   E.L. Allwein, R.E. Schapire, Y. Singer: Reducing multiclass to binary: A unifying approach for margin classifiers, J. Mach. Learn. Res. **1**, 133–141 (2000)

12.7   K.-B. Duan, S.S. Keerthi: Which is the best multiclass SVM Method? An empirical study, Proc. 6th Int. Workshop Multiple Classifier Syst. (2005), Vol. 3541 (Springer, Berlin, Heidelberg 2006) pp. 278–285

12.8   C. Cortes, V. Vapnik: Support vector networks, Mach. Learn. **20**, 273–297 (1995)

12.9   K. Veropoulos, C. Campbell, N. Cristianini: Controlling the sensitivity of support vector machines, Proc. Int. Joint Conf. Artif. Intell. (IJCAI) (1999)

12.10 J. Platt: Probabilistic outputs for support vector machines and comparison to regularised likelihood methods, Adv. Large Margin Classifiers (MIT Press, Cambridge 1999) pp. 61–74

12.11 A.E. Hoerl, R. Kennard: Ridge regression: Biased estimation for nonorthogonal problems, Technometrics **12**, 55–67 (1970)

12.12 C. Saunders, A. Gammermann, V. Vovk: Ridge regression learning algorithm in dual variables, Proc. Fifteenth Int. Conf. Mach. Learn. (ICML), ed. by J. Shavlik (Morgan Kaufmann, 1998)

12.13 V. Vapnik: *The Nature of Statistical Learning Theory* (Springer, New York 1995)

12.14 V. Vapnik: *Statistical Learning Theory* (Wiley, New York 1998)

12.15 B. Schölkopf, A.J. Smola: *Learning with Kernels* (MIT Press, Cambridge 2002)

12.16 J. Weston, A. Gammerman, M. Stitson, V. Vapnik, V. Vovk, C. Watkins: *Support vector density estimation*, Advances in Kernel Methods: Support Vector Machines (MIT Press, Cambridge 1998) pp. 293–306

12.17 A.J. Smola, B. Schölkopf: A tutorial on support vector regression, Stat. Comput. **14**, 199–222 (2004)

12.18 R.D. Williams, S.N. Hing, B.T. Greer, C.C. Whiteford, J.S. Wei, R. Natrajan, A. Kelsey, S. Rogers, C. Campbell, K. Pritchard-Jones, J. Khan: Prognostic classification of relapsing favourable histology Wilms tumour using cDNA microarray expression profiling and support vector machines, Genes Chromosom. Cancer **41**, 65–79 (2004)

12.19 I. Guyon, A. Elisseeff: An Introduction to Variable and Feature Selection, J. Mach. Learn. Res. **3**, 1157–1182 (2003)

12.20 T. Graepel, R. Herbrich, P. Bollmann-Sdorra, K. Obermayer: Classification on pairwise proximity data, Adv. Neural Inform. Proces. Syst. **11**, 438–444 (1998)

12.21 E. Pekalska, P. Paclik, R.P.W. Duin: A generalized kernel approach to dissimilarity based classification, J. Mach. Learn. Res. **2**, 175–211 (2002)

12.22 V. Roth, J. Laub, M. Kawanabe, J.M. Buhmann: Optimal cluster preserving embedding of nonmetric proximity data, IEEE Trans. Pattern Analys. Mach. Intell. **25**, 1540–1551 (2003)

12.23 R. Luss, A. d'Aspremont: Support vector machine classification with indefinite kernels, Adv. Neural Inform. Proces. Syst. **20**, 953–960 (2008)

12.24 Y. Ying, C. Campbell, M. Girolami: Analysis of SVM with Indefinite Kernels, Adv. Neural Informat. Proces. Syst. **22**, 2205–2213 (2009)

12.25 N. Cristianini, C. Campbell, J. Shawe-Taylor: Dynamically adapting kernels in support vector machines, Adv. Neural Inform. Proces. Syst. **11**, 204–210 (1999)

12.26 T. Joachims: Estimating the generalization performance of an SVM efficiently, Proc. 17th Int. Conf. Mach. Learn. (Morgan Kaufmann, 2000) pp. 431–438

12.27 O. Chapelle, V. Vapnik: Model selection for support vector machines, Adv. Neural Inform. Proces. Syst. **12**, 673–680 (2000)

12.28   V. Vapnik, O. Chapelle: Bounds on error expectation for support vector machines, Neural Comput. **12**, 2013–2036 (2000)

12.29   P. Sollich: Bayesian methods for support vector machines: Evidence and predictive class probabilities, Mach. Learn. **46**, 21–52 (2002)

12.30   J. Shawe-Taylor, N. Cristianini: *Kernel Methods for Pattern Analysis* (Cambridge Univ. Press, Cambridge 2004)

12.31   H. Lodhi, C. Saunders, J. Shawe-Taylor, N. Cristianini, C. Watkins: Text classification using string kernels, J. Mach. Learn. Res. **2**, 419–444 (2002)

12.32   C. Leslie, R. Kuang: Fast kernels for inexact string matching, 16th Ann. Conf. Learning Theory 7th Kernel Workshop, Vol. 2777 (Springer, Berlin, Heidelberg 2003) pp. 114–128

12.33   S. Vishwanathan, A. Smola: Fast Kernels for String and Tree Matching, Adv. Neural Inform. Proces. Syst. **15**, 569–576 (2003)

12.34   I.R. Kondor, J.D. Lafferty: Diffusion kernels on graphs and other discrete structures, Proc. Int. Conf. Mach. Learn. (Morgan Kaufmann, San Francisco, 2002) pp. 315–322

12.35   A.J. Smola, I.R. Kondor: Kernels and regularization on graphs, Conf. Learning Theory (COLT), Vol. 2777 (Springer, Berlin, Heidelberg 2003) pp. 144–158

12.36   T. Gartner, P. Flach, S. Wrobel: On graph kernels: Hardness results and efficient alternatives, Proc. Annu. Conf. Computational Learning Theory (COLT) (Springer, Berlin, Heidelberg 2003) pp. 129–143

12.37   S.V.N. Vishwanathan, K.M. Borgwardt, I.R. Kondor, N.N. Schraudolph: Graph Kernels, J. Mach. Learn. Res. **9**, 1–41 (2008)

12.38   G.R.G. Lanckriet, N. Cristianini, P. Bartlett, L. El Ghaoui, M.I. Jordan: Learning the kernel matrix with semidefinite programming, J. Mach. Learn. Res. **5**, 27–72 (2004)

12.39   F. Bach, G.R.G. Lanckriet, M.I. Jordan: Multiple kernel learning, conic duality and the SMO algorithm, Proc. 21st Int. Conf. Machine Learning (ICML) (Morgan Kaufmann, New York 1998)

12.40   S. Sonnenburg, G. Rätsch, C. Schäfer, B. Schölkopf: Large scale multiple kernel learning, J. Mach. Learn. Res. **7**, 1531–1565 (2006)

12.41   A. Rakotomamonjy, F. Bach, S. Canu, Y. Grandvalet: SimpleMKL, J. Mach. Learn. Res. **9**, 2491–2521 (2008)

12.42   Z. Xu, R. Jin, I. King, M.R. Lyu: An extended level method for multiple kernel learning, Adv. Neural Inform. Proces. Syst. **22**, 1825–1832 (2008)

12.43   Y. Ying, K. Huang, C. Campbell: Enhanced protein fold recognition through a novel data integration approach, BMC Bioinf. **10**, 267–285 (2009)

12.44   T. Damoulas, M. Girolami: Probabilistic multi-class multi-kernel learning: On protein fold recognition and remote homology detection, Bioinformatics **24**, 1264–1270 (2008)

12.45   G.R.G. Lanckriet, T. De Bie, N. Cristianini, M.I. Jordan, W.S. Noble: A statistical framework for genomic data fusion, Bioinformatics **20**, 2626–2635 (2004)

12.46   M. Kloft, U. Brefeld, S. Sonnenburg, P. Laskov, K.-R. Müller, A. Zien: Efficient and accurate lp-norm multiple kernel learning, Adv. Neural Inform. Proces. Syst. **22**, 997–1005 (2009)

12.47   U. Alon, N. Barkai, D.A. Notterman, K. Gish, S. Ybarra, D. Mack, A.J. Levine: Broad patterns of gene expression revealed by clustering analysis of tumor and normal colon tissues probed by oligonucleotide arrays, Proc. Natl. Acad. Sci. USA **96**(12), 6745–6750 (1999)

12.48   B. Everitt: *Cluster Analysis* (Arnold, New York 1993)

12.49   L. Kaufman, P.J. Rousseeuw: *Finding Groups in Data* (Wiley, New York 2005)

12.50   R.O. Duda, P.E. Hart, D.G. Stork: *Pattern classification* (Wiley, New York 2001)

12.51   Y.W. Teh, D. Newman, M. Welling: A collapsed variational Bayesian inference algorithm for latent dirichlet allocation, Adv. Neural Inform. Proces. Syst. **19**, 1353–1360 (2006)

12.52   Y. Ying, P. Li, C. Campbell: A marginalized variational Bayesian approach to the analysis of array data, BMC Proc. **2**(4), S7 (2008)

12.53   P. Li, Y. Ying, C. Campbell: A variational approach to semi-supervised clustering, Proc. ESANN2009 (2009) pp. 11–16

12.54   D.M. Blei, M.I. Jordan: Modeling annotated data, Proc. 26th Annu. Int. ACM SIGIR Conf. Res. Dev. Inf. Retr. (ACM Press, New York 2003) pp. 127–134

12.55   P. Agius, Y. Ying, C. Campbell: Bayesian Unsupervised Learning with Multiple Data Types, Stat. Appl. Genet. Molec. Biol. **8**, 27 (2009)

12.56   S. Rogers, M. Girolami, C. Campbell, R. Breitling: The latent process decomposition of cdna microarray datasets, IEEE/ACM Trans. Comput. Biol. Bioinforma. **2**, 143–156 (2005)

12.57   C. Blenkiron, L.D. Goldstein, N.P. Thorne, I. Spiteri, S.F. Chin, M.J. Dunning, N.L. Barbosa-Morais, A.E. Teschendorff, A.R. Green, I.O. Ellis, S. Tavaré, C. Caldas, E.A. Miska: MicroRNA expression profiling of human breast cancer identifies new markers of tumour subtype, Genome Biol. **8**(10), R214-1–R214-16 (2007)

12.58   L. Carrivick, S. Rogers, J. Clark, C. Campbell, M. Girolami, C. Cooper: Identification of prognostic signatures in breast cancer microarray data using Bayesian techniques, J. R. Soc. Interf. **3**, 367–381 (2006)

12.59   E. Garber, O.G. Troyanskaya, K. Schluens, S. Petersen, Z. Thaesler, M. Pacyna-Gengelbach, M. van de Rijn, G.D. Rosen, C.M. Perou, R.I. Whyte, R.B. Altman, P.O. Brown, D. Botstein, I. Petersen: Diversity of gene expression in adenocarcinoma of the lung, Proc. Natl. Acad. Sci. USA **98**, 13784–13789 (2001)

12.60   C. Andrieu, N. De Freitas, A. Doucet, M.I. Jordan: An introduction to MCMC for machine learning, Mach. Learn. **50**, 5–43 (2003)

12.61   W.R. Gilks, S. Richardson, D.J. Spiegelhalter: *Markov Chain Monte Carlo in Practice* (Chapman Hall/CRC, New York 1996)

12.62   C.P. Robert, G. Casella: *Monte Carlo Statistical Methods* (Springer, Berlin, Heidelberg 2004)

12.63   S. Chib, E. Greenberg: Understanding the Metropolis Hastings Algorithm, Am. Stat. **49**(4), 327–335 (1995)

12.64   B.A. Berg: *Markov Chain Monte Carlo Simulations and Their Statistical Analysis* (World Scientific, Singapore 2004)

12.65   W.M. Bolstad: *Understanding Computational Bayesian Statistics* (Wiley, New York 2010)

12.66   K. Bleakley, G. Biau, J.-P. Vert: Supervised reconstruction of biological networks with local models, Bioinformatics **23**, i57–i65 (2007)

12.67   B. Calderhead, M. Girolami: Estimating Bayes factors via thermodynamic integration and population MCMC, Comput. Stat. Data Anal. **53**, 4028–4045 (2009)

12.68   T.R. Xu, V. Vyshemirsky, A. Gormand, A. von Kriegsheim, M. Girolami, G.S. Baillie, D. Ketley, A.J. Dunlop, G. Milligan, M.D. Houslay, W. Kolch: Inferring signaling pathway topologies from multiple perturbation measurements of specific biochemical species, Sci. Signal. **3**(113), ra20:1–ra20:10 (2010)

12.69   Cancer Genome Atlas: Available at http://cancergenome.nih.gov

13. Case-Based Reasoning for Biomedical Informatics and Medicine

Periklis Andritsos, Igor Jurisica, Janice I. Glasgow

Case-based reasoning (CBR) is an integral part of artificial intelligence. It is defined as the process of solving new problems through their comparison with similar ones with existing solutions. The CBR methodology fits well with the approach that healthcare workers take when presented with a new case, making its incorporation into a clinical setting natural. Overall, CBR is appealing in medical domains because a case base already exists, storing symptoms, diagnoses, treatments, and outcomes for each patient. Therefore, there are several CBR systems for medical diagnosis and decision support. This chapter gives an overview of CBR systems, their lifecycle, and different settings in which they appear. It also discusses major applications of CBR in the biomedical field, the methodologies used, and the systems that have been adopted. Section 13.1 provides the necessary background of CBR, while Sect. 13.2 gives an overview of techniques. Section 13.3 presents different systems in which CBR has been successfully applied, and Sect. 13.4 presents biomedical appl-

ications. A concluding discussion closes the chapter in Sect. 13.5.

Many areas of bioinformatics have benefited from artificial intelligence problem-solving techniques. Computational biology approaches have been applied and offered enormous advances to a wide breadth of medical applications including diagnosis, prognosis, etc. [13.1, 2]. These fields are often characterized by complex data, many unknowns, incomplete theories, and rapid evolution. In decision-making, reasoning is often based on experience, rather than on general knowledge. In this chapter we consider one such machine-learning approach, *case-based reasoning* (CBR), and discuss how it has been applied to problems in bioinformatics.

CBR using analogy-based reasoning is a multidisciplinary area of research that deals with the reuse of experiences, called *cases* [13.3, 4]. CBR is defined as a plausible, high-level model for cognitive processing [13.5] as well as a computational paradigm for problem-solving [13.3]. This paradigm uses a bottom-up approach by exploiting knowledge gathered after solving specific problem situations. In contrast, top-down approaches start with the problem domain theory, which is usually expressed using a specific language, e.g., rule-based language, frames, first-order logic, and semantics networks. In brief, CBR is well suited for capturing both objective details as well as contextual ones [13.6].

Our goal is to present the specific challenges involved in analyzing the underlying datasets. Hence, we will explore CBR techniques as they are applied to image data, as well as the diagnosis of diseases using numerical and textual data.

The rest of this chapter is organized as follows. Section 13.1 provides the necessary background of CBR, while Sect. 13.2 gives an overview of techniques used to perform feature selection on data to be handled by CBR systems. Section 13.3 presents different systems in which CBR has been successfully applied, while Sect. 13.4 focuses on applications in the biomedical domain. Finally, Sect. 13.5 presents concluding discussions.

13.1 Case–Based Reasoning

When building a CBR system, one of the fundamental components is how cases are represented, i. e., the exemplar solutions that we have stored from previous experience. These are solutions that will be used together with the problem description and possibly evaluation of the solution. At design time, one must decide on the following two issues:

1. The model (data structure) used to store the case-base content,
2. The organization of the case memory.

13.1.1 CBR Content Modeling

Case-base content is usually application dependent. Typically, there are three main ways to represent the cases:

1. As a set of features in a vector, usually called the vector-space model,
2. As text, structured (e.g., inside a database management system) or semistructured [e.g., in extensible markup language (XML) documents],
3. As complex objects, such as graphs.

There are also hybrid approaches that mix, for instance, text-based and vector-based models.

Vector-based modeling is of particular interest since this is a representation inherent in most machine-learning techniques, such as feature selection and extraction as well as clustering and classification methods [13.7]. Therefore, most CBR systems have adopted this representation so that these techniques can be applied on available datasets. Objects in these datasets are modeled as vectors, and therefore they often need to be converted into a space where relevant similarity measures are defined and can be incorporated as part of the system; For example, one of the typical transformations of document corpuses into vectors of decimal numbers (or scores) is done by means of the *term-frequency, inverse document-frequency* (tf-idf) weighting method. This method computes the frequency of a term t in a document $d \in D$ as

$$\mathrm{tf}(x) = \|t \in t \in d\| \,,$$

as well as the inverse document frequency of x in the corpus D as

$$\mathrm{idf}(x) = \log \frac{\|D\|}{\|d : t \in d\|} \,.$$

Then, the tf-idf of term x is defined as $\mathrm{tf\text{-}idf}(x) = \mathrm{tf}(x) \times \mathrm{idf}(x)$, giving higher scores to terms that appear many times in a document and have a low document frequency, i. e., that are considered highly informative with respect to others. After computing these scores for each term inside the documents, we can represent each document by means of the tf-idf scores of their terms.

Finally, many CBR approaches view problem-solving as automatic classification or function approximation tasks [13.8, 9].

Regardless of the case representation, all cases include the problem description, the solution, and the outcome. The first refers to the set of features that are matched when we have a new problem in the system. This must include all the information needed to first discover that a case can be successfully reused for solving a similar problem. The solution models the information for which we are searching, e.g., the diagnosis of a disease or the plan to perform protein crystallization. Finally, the outcome provides an evaluation of the applicability or quality of the solution for the given problem.

13.1.2 CBR System Lifecycle

The problem-solving cycle of a typical CBR system is shown in Fig. 13.1.

Retrieve (1)
Given the description of a new problem, the CBR system retrieves a set of cases stored in the case base. The retrieval uses a similarity metric to compare the problem component of the new case that we are about to build with the problem descriptions of the cases in the base.

In a problem where documents are represented as vectors of tf-idf scores, as described above, one common similarity metric that is used is called *cosine similarity* and calculates the angle between corresponding vectors of documents. Given the tf-idf vectors v_i and v_j of two documents $d_i \in D$ and $d_j \in D$, respectively, the *cosine similarity*, $\approx (v_i, v_j)$ is given by the expression

$$\approx (v_i, v_j) = \frac{v_i \cdot v_j}{\|v_i\| \cdot \|v_j\|} \; ,$$

where the nominator is the inner product between the two document vectors and the denominator the arithmetic product of their norms. For very large case bases, other techniques such as indexing and clustering of cases may need to be incorporated to make this step more efficient [13.10]. This is useful in particular when the context does not change. By context we mean the constraints that are involved in the problem-solving without intervening in it [13.11].

Reuse (2)

This step includes the testing of the solved case in the real dataset. Cases can be reused through evaluation against a domain expert, a simulated model, or known solutions (*test set*). In other words, this step is based on the ability to associate concepts to facts by analogy, called *analogy-based reasoning* [13.12]. The similar cases that are retrieved are reused to build the best solution. This solution could simply be the solution of the most similar case in the base, or an integration of solutions that are extracted from the retrieved cases to build a new candidate solution.

Revise (3)

The candidate solution is then adapted to fit any specific constraints of the current situation; For instance, an extracted therapy should be adapted for a new patient suffering from a particular disease.

Review (4)

The solution that the system builds should now be evaluated by applying it (or simulating the application) to the current problem. If we detect failure, we have to go back and revise the solution or keep the solution as a negative example for future use [13.13]. The *reuse*, *revise*, and *review* stages are also called *case adaptation* [13.5].

Retain (5)

The new case may or may not be added to the case base, depending on its similarity to existing cases and poten-

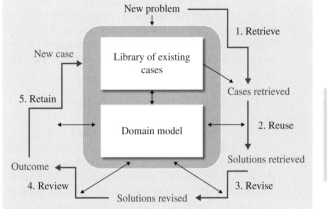

Fig. 13.1 Problem-solving cycle of a CBR system

tial value during problem-solving, since there would be little or no value to add identical or highly similar cases to the case base.

13.1.3 Case Maintenance

The last component of the CBR cycle has attracted particular attention, as unnecessary growth in the size of the case base may negatively affect its performance. Case-based maintenance methods have been proposed in response to this problem through deletions, additions of selected cases, and merging of similar cases. *Yang* and *Wu* [13.14] introduced a density-based clustering and information theory-based approach that results in case bases of smaller sizes where retrieval is guided by comparing the information contents of new problems and the clusters built. However, this approach is moving away from the CBR strategy (lazy learning) toward traditional machine-learning approaches, and thus would be applicable in more stable domains. *Lawanna* and *Daengdej* [13.15] give a concise set of case deletion and addition heuristics. Their heuristics evaluate the utility of candidate cases by measuring their *coverage* and *reachability*. Coverage is the set of problems that the case can solve, while reachability is the set of cases that can be used to provide a solution for a target problem. Given a value for both measures, cases are retained or deleted. Finally, *Arshadi* and *Jurisica* [13.16] approach the problem of case maintenance from a machine-learning point of view. The goal of their method, termed mixture of experts for CBR (MOE4CBR), is to increase the prediction accuracy of CBR classifiers in high-dimensional domains by using a mixture of experts where an ensemble of CBR systems is integrated with clustering and feature selection to improve performance. They employ spectral clustering to group samples, and each group forms a separate CBR system. Logistic regression is then applied to select those features that more accurately predict the class labels.

13.1.4 Adaptation of CBR in Medicine

The CBR cycle described above fits well with the approach that a healthcare worker takes when presented with a new case, making its incorporation into a clinical setting natural. Overall, CBR is appealing in medical domains because a case base already exists, storing symptoms, diagnoses, treatments, and outcomes for each patient. Therefore, there are several CBR systems for medical diagnosis and decision support [13.17–21]. These systems increasingly use knowledge engineering techniques [13.22]. As more complex domains are tackled by CBR systems, where representing cases

and adapting the solutions of retrieved cases become difficult, systematic approaches to CBR development are needed [13.23, 24]. This is important in these domains because it elucidates knowledge that aids in the construction of a meaningful case representation, meaningful in the sense that it allows for retrieved cases to be matched as closely as possible to the target case in order to reuse their solutions with little adaptation. CBR still has clear benefits in these domains as long as the knowledge engineering efforts required to construct such a case representation are less than what would be required to construct an entire general model [13.24]. Overall, the effectiveness of CBR depends on the domain stability, coverage, as well as the quality and quantity of cases in a case base. With an increased number of unique cases, the problem-solving capabilities of CBR systems may improve at the expense of a decrease in efficiency. In addition, the scalability of a system depends on the model used and the similarity-based retrieval algorithm.

13.2 Feature Selection for CBR

13.2.1 Feature Subset Selection

As one of the key components of CBR, a similarity measure is used to assess closeness of a given *problem case* to cases in the case base, considering any specific conditions (context). This similarity is computed over sets of features that are considered important for the problem at hand. The choice of case features that best distinguish classes of instances has a large impact on the similarity measure and has become an important preprocessing step of case-based reasoning for different domains, termed *feature selection* [13.25]. Techniques that involve the selection of features according to how well they predict given class labels are called *supervised* [13.26]. They usually perform an exhaustive search of all possible subsets of features, and therefore proper heuristics may be used to reduce the complexity. On the other hand, *unsupervised* techniques do not incorporate given class labels but instead employ importance or proximity measures in order to select appropriate sets of features [13.27]

There are two types of feature selectors:

1. *Wrapper* selectors, which use the learning algorithm as a black box with the goal of minimizing the fitting error for a particular problem [13.28], and,

2. *Filter* selectors, which choose features by evaluating some preset criteria independent of the learning algorithm.

In practice, filter selectors have much lower complexity than wrappers, and the features selected often yield comparable prediction errors [13.29].

13.2.2 Feature Ranking

Several feature selection search algorithms have been proposed specifically for case-based reasoning, such as exhaustive search, hill-climbing, and forward sequential selection [13.30, 31]. Such techniques employ objective functions that can be used for *feature ranking*, such as the tf-idf criterion discussed above, Fisher's criterion, *t*-test, and logistic regression [13.7, 9, 32]. Highly ranked features are deemed more *valuable*, as opposed to features that have lower ranks. The latter ones are of importance in certain domains, such as in the search for prognostic gene signatures for clinical outcomes [13.33].

Depending on the application domain, a different feature selection approach may need to be applied so that appropriate similarity measures can be employed. Feature selection is highly dependent upon the type of data with which we are dealing. Numerical data stored as cases of CBR systems have well-defined geometrical characteristics where similarity measures, such as Euclidean distance, can be applied. On the other hand, in a text-based CBR system a case includes the information extracted from the text. This information comprises a set of important keywords that can also be ranked using, for example, the tf-idf score and stored in CBR cases. This approach usually integrates CBR with information retrieval [13.34–36] and leverages statistical information inherent in the documents.

One domain in which we are interested in building case bases is that of protein–protein interactions (PPI) from text contained in publication abstracts [13.37]. In this application, features are keywords that are related to the way that PPIs are expressed in collections of papers that deal with specific biological experiments for their detection. As an input to the system, we use sets of terms related to PPI experiments, such as *mass spectrometry*. We use clustering to separate the expressions that are very similar to each other and dissimilar to other PPI expressions. We finally perform keyword scoring in order to measure the utility of each keyword using a scoring function called *category utility* [13.38, 39]. Intuitively, this function measures the *utility* of the individual keywords in the clusters of similar PPI experiments. If we have a corpus D of sentences relevant to PPIs and a cluster C that has been created from this corpus, the utility of a keyword $k \in C$ is calculated as [13.40]

$$\mathrm{CU}(k, C) = P(k|C)^2 - P(k)^2 \, ,$$

which calculates the difference between the expectation of the keyword k when the clustering C is given with its expectation in the full corpus. Using this measure, the keywords that are more responsible for the creation of the cluster will have a higher probability of appearance in it and, hence, a higher CU score.

More complex objects such as images often require more advanced preprocessing before storing them in a CBR. Implicit image features are extracted using different techniques such as morphological characterization that applies image segmentation and feature extraction to determine and quantify image texture, distinctive objects contained within images, intensity, presence or absence of straight lines, light distribution, dark points, etc. [13.41]. Image and nonimage features can be combined, and similarity between stored and raw cases can be measured [13.42]. If the purpose of using a CBR system is to classify the images, many of the features can become redundant if they have similar predictive and expressive power. Hence, the challenge is to select a minimal number of features from each image.

The following is a list of techniques proposed for feature extraction in protein crystallization:

- Laplacian pyramid filter [13.43]: The Laplacian filter is used to decompose the image into three different levels. The Laplacian filter is used to extract the boundary information and image features. The multiscale representation is capable of extracting the following useful features of the image (invariant to orientation): mean, standard deviation, skewness, kurtosis, energy, entropy, autocorrelation, and power.
- Extraction of contours [13.44]: Using edge-detection techniques, proper contours are extracted, and the identification of the type of line segments helps in the classification of images. Useful features for which these techniques produce scores are the maximum length and number of line segments as well as the ratio of linear regions.
- Texture features [13.45].
- Gabor wavelet decompositions for edge detection, noise filtering, image compression, texture analysis, and synthesis [13.46].
- Fourier and wavelet analysis [13.47].

Feature selection can be an inherent component of a CBR system, or it can be performed as a preprocessing step. Systems such as eXiT*CBR [13.48] incorporate basic preprocessing and feature selection methods to facilitate experimentation.

13.3 Case–Based Reasoning Systems

Early systems were mainly applied in datasets with a low number of symbolic attributes that contained discrete values [13.49]. With richer heterogeneous real-world data sources available, appropriate treatment and interpretation of large and complex types of data constitute key issues for developing CBR systems.

CBR systems are being developed in both industrial and scientific applications. Moreover, a variety of systems have been developed to solve problems in design [13.50], cost estimation [13.51], business process planning and customer support call-centers [13.52], finance [13.53], and legal reasoning [13.54]. Most of the systems follow the standard problem-solving lifecycle described above. Depending on the data that they need to handle, they differ in the way they preprocess and store the cases, as well as the algorithmic approach used to assess the similarity of existing cases with new problems. *Göker* et al. [13.49] suggest a software engineering methodology for CBR. Similar to the development of software applications, they propose a workflow model, and their methodology deems the user, the organization, and the domain at hand as its main characteristics, while it provides a classification of each one.

CBR systems have also been incorporated as modules within other decision-making processes; For instance, *Ahmed* et al. [13.55] describe a system that uses intelligent agents that deploy CBR systems whose role is to assist these agents to gain experience by collecting past solved cases and adapt them to the current context. This enables flexible and modular maintenance systems where different suppliers can deliver agents that eventually develop to become experts in specific tasks. Their framework is geared towards knowledge transfer in complex technical fields, cost reduction, and faster response times.

External knowledge is often included in CBR systems as part of their main knowledge source and can be used from retrieval to reasoning. *D'Aquin* et al. [13.56] have integrated the C-OWL context ontology with a CBR system. Within this system, semantic relations between contexts and the associated reasoning mechanisms in a particular problem are reused and shared in other problems. In a similar fashion, *Sauer* et al. [13.57] discuss the advantages of integrating linked open data (LOD) with a CBR system. In particular they use the DBpedia ontology to retrieve information and similarities of diseases to be used in a system that is concerned with the prevention, management, and research of health problems associated with travel, and covers all medical aspects a traveler has to take care of before, during, and after a journey. DBpedia elements are queried and used in the construction of cases as well as when existing cases are compared against new problems. Their goal is to achieve better semantic similarity of the results and simplify the task of manually filling the case base with existing knowledge.

Case-based reasoning has also been applied to facilitate job runtime estimation [13.13, 58]. In this work, past performance acts as a good indicator for job

scheduling optimization in a grid environment and a CBR system is used to predict the runtime of long-term applications in heterogeneous systems. Using the *TA*3 CBR system, *Xia* et al. [13.13] have investigated job characteristics and ranked them according to their runtime statistics; i.e., those characteristics with low runtime standard deviation are ranked higher. In a similar fashion they rank machine characteristics and build cases that are stored in the CBR system. Novel similarity measures for job and machine characteristics are defined to be used when new runtimes need to be predicted.

13.3.1 The *TA*3 CBR System

The last application mentioned above uses the *TA*3 case-based reasoning system [13.59]. *TA*3 represents cases as attribute–value pairs whose domains are defined in what is called a case description. More formally, a case C is represented as

$$C = \langle a_0 : V_0 \rangle, \langle a_1 : V_1 \rangle, \ldots, \langle a_n : V_n \rangle \,,$$

where a_i represent attributes and V_i their corresponding values, $0 \leq i \leq n$. Given this representation, a case for which we already know the solution is represented by C_{source} while a cases that constitutes a new problem is denoted by C_{input}. Finally, a set of source cases $\{C_{\text{source}}^1, C_{\text{source}}^2, \ldots, C_{\text{source}}^k\}$ constitutes a case base, i.e., the search space of cases in the CBR system.

There are three classes of data defined in a case description:

1. Description: the nonpredictive data,
2. Problem: the predictive data,
3. Solution: the classification, diagnosis, or outcome.

Focusing on the problem class, attributes are grouped into categories. The advantage of grouping attributes is that it allows the assignment of different constraints and priorities depending on the relevance of an attribute or collection of attributes (i.e., their value in matching similar cases). During retrieval an explicit context is used for similarity assessment, and the process is guided by incremental transformations of the context. A context is simply a subset of the problem part of the case description with constraints applied to the attribute–value pairs. More formally, a context T is defined by a finite set of attributes and related constraints on their values

$$T = \langle a_0 : CV_0 \rangle, \langle a_1 : CV_1 \rangle, \ldots, \\ \langle a_m : CV_m \rangle \,,$$

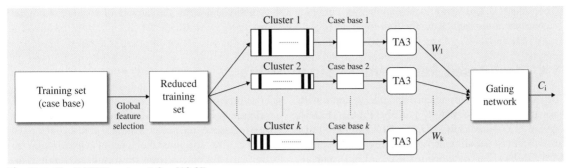

Fig. 13.2 The E4CBR system (after [13.8])

where a_i denotes the attributes and CV_i is the set of values that the attribute can have. This minimizes the effect that irrelevant or less relevant attributes may have when trying to match similar cases. Category membership can be assigned either by an expert with domain knowledge of the relevance of different attributes or by a machine-learning approach.

The retrieval process uses modified nearest-neighbor matching: predictive attributes are grouped to allow different priorities and constraints, an explicit context is used during similarity assessment, and the retrieval algorithm is guided by incremental transformations of the context. A context is simply a subset of the problem part of the case description with constraints applied to the attribute–value pairs.

The flexible nature of *TA3* lies in the fact that retrieval strategies can be dynamically defined for a particular domain and the specific application. The system also includes a genetic algorithm for knowledge discovery purposes; given two or more test sets representing different classes of cases, this functionality maximizes the distances between different classes and minimizes the distances within the same class. The distance between two cases is defined as the amount of relaxations needed to make the two cases similar. The information gained by this process may not only determine previously unknown relations in the data, but may provide a new context with which to guide the retrieval process with greater prediction accuracy.

The *TA3* system uses a structured query language (SQL) database to store the case base and thus handles very large inputs; For instance, it is being used to store and analyze protein crystallization experiments. There are 12 000 experiments, each of which has 9216 attributes, and the data are derived from 110, 592, 000 images. The repository grows at a rate of 200 experiments per month [13.41]. Before *TA3* can suggest crystallization strategies for a new protein, we need to compute 12 375 features from the images and classify them into 10 categories [13.60].

Finally, the *TA3* system can also be used as a classifier. The attribute–value pairs can be accompanied by class labels, and nearest-neighbor techniques can be modified to classify new cases. To extend the capabilities of CBR, *Arshadi* and *Jurisica* [13.8] implemented E4CBR, where an ensemble of CBR classifiers is combined with clustering and feature selection. A set of case features is selected first, and then clustering of the cases into disjoint groups is employed, where each group of cases forms the case base of one of the member classifiers (Fig. 13.2). In each case base a subset of features is *locally* selected individually. To predict the label of an unseen case, each classifier in the ensemble provides a prediction, and the aggregation component of E4CBR combines the predictions by weighting each classifier using a CBR approach; a classifier with more cases similar to the test case receives a higher weight.

Similarly, *Spasic* et al. [13.61] have proposed MaSTerClass, a system that classifies terms by using natural-language processing (NLP) to perform feature selection, comparison of terms, and classification in order to update existing semantic networks with new terms. More precisely, they used a string-based similarity measure called *edit distance* (ED) to find similarities between individual terms and keyphrases. Intuitively, ED counts the insertions, deletions, and updates that should be performed to convert one string (or keyphrase) to another [13.7]. In addition to ED, they incorporate a tree similarity measure to take into account the similarity of corresponding terms in a semantic hierarchy or ontology. This measure is defined as

$$\mathrm{ts}(C_1, C_2) = \frac{2 \cdot \mathrm{common}(C_1, C_2)}{\mathrm{depth}(C_1) + \mathrm{depth}(C_2)},$$

where C_1 and C_2 are the two given concepts, while

- common(C_1, C_2) denotes the number of common concepts in the paths from the root to C_1 and C_2
- depth(C_i) denotes the number of concepts from the root to C_i, $i = 12$.

13.4 Case–Based Reasoning in Biomedicine

CBR systems in health science include systems with tasks in diagnosis (SHRINK [13.62], Protos [13.63], CASEY [13.64], MEDIC [13.65], BOLERO [13.66]), assessment test planning and clinical research planning (MNAOMIA [13.67]), tutoring (CADI [13.68]), and image analysis (MacRad [13.69], ImageCreek [13.70]).

Nilsson and *Sollenborn* provide a classification of developments in systems applied in the medical domain, according to their purpose [13.71]. They observe that the majority of these systems belong to the areas of *diagnosis*, *classification*, and *planning*. At the same time, systems such as *TA3*, which we described in Sect. 13.3.1, are *domain independent* and may be applied in the medical domain.

13.4.1 The eXiT*CBR System

Several domain-independent CBR systems have been proposed, including eXiT*CBR [13.48]. eXiT*CBR is a framework that supports the development and experimentation with CBR systems in general. The system supports a classification component tailored to diagnostic tasks. According to the authors, the advantage of using eXiT*CBR lies in the fact that it is modular and helps the user to preprocess the data and visualize the results of different parameter settings. The system includes the following components:

- *Experiment interpreter:* This is the core of the framework that interprets a configuration file given by the user and applies one of the two methods that the system currently supports: *batch processing* and *cross-validation*. In the first method the interpreter reads a set of training data, generates a case base, and uses a different test dataset to obtain the results. Results are interpreted according to a performance measure given as input by the user. The cross-validation method performs multiple runs of a CBR configuration with different datasets (split into training and test data). The performance of the system is averaged at the end of the processing.
- *Preprocessor:* The system accepts a configuration file and the type of preprocessing to be performed. Currently, it supports discretization, normalization, and feature selection.

- *Postprocessor:* This module starts when the CBR engine has finished and involves the application of performance measures that the user is interested in. It also specifies the type of visualization to be used for interpreting the results. Default visualization involves receiver operating characteristic (ROC) curves for the diagnosis system, while it can also calculate an area under curve (AUC) value. The latter measures the accuracy under a ROC curve value and assesses the accuracy of the results. The closer the results are to a value of 1.0, the better the accuracy is.

The system can also assist in the generation of datasets, which can be created according to different user preferences.

13.4.2 Diagnosis with CBR Systems

Other CBR frameworks that have been used in the medical domain are jCOLIBRI [13.72] and MyCBR [13.73]. The former is similar to eXiT*CBR in that it is modular, but instead of configuration files, it incorporates an ontology to be used in the different phases of CBR. The ontology has been developed as a plug-in to the PROTEGE (http://protege.stanford.edu/) editor for ontologies. In this system, ontologies are used to define properties of the features to be used.

The eXiT*CBR tool has been used in a breast cancer diagnosis scenario. Data from patients and healthy population are stored in a database. The 1199 attributes represent habits, such as smoker or not, sport activities and eating, disease characteristics, such as type and size of tumor, as well as the gynecological history of the women involved. The authors describe their experiments with the eXiT*CBR tool when it is given different input parameters, such as the number of cases to be retrieved or the number of attributes to be used [13.48]. They also demonstrate how small changes in the initial configuration and preprocessing help in rapid visualization and assessment of the results.

Breast cancer decision support has also been addressed in the KASIMIR system [13.74]. The focus of this work is a methodology that adopts knowledge from experts. The authors describe adaptation patterns such as cases of inapplicable decisions and consequences of particular decisions. They also take into account missing data during the retrieval phase. Clusters also help in the discovery of patterns, and they have been used in recent work on the diagnosis of melanoma by *Armengol* [13.75]. In this work explanations produced by the system are used to describe clusters and these explanations become part of the system's domain theory, which can be valuable to domain experts.

The *TA*3 CBR system has been used in two medical diagnostic case studies, one in attention-deficit hyperactivity disorder (ADHD) and one in stroke diagnosis. ADHD is a problem with symptoms of inattentiveness, overactivity, impulsivity, or a combination thereof. It is a neuropsychiatric disorder that appears in both children and adults of different ages. One of the problems with ADHD is the lack of objective tests for its proper diagnosis. *Brien* et al. [13.20] have tackled this problem by proposing a methodology that incrementally improves a CBR system, namely the *TA*3 system that we described above.

Standard diagnoses of ADHD usually include interviews with parents and teachers as well as rating scales of hyperactivity and impulsivity, clinical history, cognitive assessments, and neurological examinations. When other disorders are present, diagnosis of ADHD becomes harder and the aforementioned tests less accurate. Therefore, their validity and reliability are questionable.

Brien et al. explore a very specific symptom that can lead to more accurate diagnosis of ADHD, that of *saccadic eye movement*. Saccades are rapid movements of the eye that bring new visual targets onto the fovea of the retina. They take place either consciously or automatically as a response to stimuli that appear in front of the person. ADHD can be diagnosed through saccadic eye movements since the regions in the brain responsible for controlling them are well understood and related to the regions that cause ADHD. The data collected for each subject in the saccade tasks include the diagnosis group (ADHD or control), any drugs that were used, the age, sex, handedness, hyperactivity, impulsivity, and saccadic reaction time (SRT) [13.20]. Two particular tasks, i.e., the prosaccade and antisaccade tasks, have been used to investigate whether the eye movement was performed voluntarily or as a response to stimuli. In both tasks the subjects are looking at fixed objects called fixation points (FP). A new visual target (T) appears to the left or right side of the fixation point, and the subject is asked to look toward the target (prosaccade) or away from it (antisaccade). The sac-

Description	
Subject code: aba	Hyperactivity: 87
Handedness: 10	Impulsivity: 76

Problem

Priority 0

Age: 8   Sex: Male

Task variables

Task	Mean SRT (ms)	CV	Dir. error (%)	Exp. (%)
Anti/Gap/Left	363.00	39.35	93.33	0.00
Anti/Gap/Right	458.00	50.01	75.76	12.50
Anti/Over/Left	351.67	42.07	77.78	16.67
Anti/Over/Right	483.33	44.50	62.50	0.00
Pro/Gap/Left	338.00	59.82	12.50	7.14
Pro/Gap/Right	343.06	58.70	5.88	9.38
Pro/Over/Left	431.50	46.88	3.45	7.14
Pro/Over/Right	471.24	44.73	9.38	0.00

Solution

Diagnosis: ADHD

Fig. 13.3 Example case description from model 1. SRT = Saccadic reaction time. CV = coefficient of variation in SRT. Dir. error = percentage of direction error. Exp. = percentage of express saccades. Anti = antisaccade task. Pro = prosaccade task. Over = overlap condition. Gap = gap condition (after [13.20])

cadic reaction time (SRT) is measured during these tests and compared against controls to make a more objective decision about ADHD.

An example of the results from the *TA*3 system is given in Fig. 13.3.

The data used involve saccadic eye movement from adults and children. In both categories, measurements of people with ADHD are compared against controls, and the accuracy of the results is 72% for children and 76% for adults. The results of the overall study using *TA*3 are comparable to those of the continuous performance test (CPT) [13.76], the most objective clinical laboratory test for assessing attention and vigilance.

In the diagnosis and prognosis of stroke, conventional assessment techniques are carried out by clinicians, who attempt to measure the degree of impairment with subjective measures. Normally, an early diagnosis is made by assessing the symptoms, reviewing medical history, conducting tests to confirm the occurrence of a brain attack, and measuring the degree of impairment. Conventional stroke assessment scales convert motor status to a score in an ordinal (nonnumeric) scale. Typically, each patient performs a certain task where the main emphasis is laid on task completion rather than specific details. Therefore it is a nonqualitative scoring. In qualitative scoring, other factors are considered as well, such as measurement of the amount of assistance required, alteration in the normal (gross) position, and time utilized to complete a test.

CBR can be used to create a repository of data from stroke patients who have an explicit diagnosis and prognosis and are receiving subsequent rehabilitation. For a new stroke patient, whose diagnosis is yet to be confirmed and who has an indefinite prognosis, CBR retrieves similar cases from the case base, which may provide useful information to the clinicians, hence facilitating them in reaching a potential solution for stroke diagnosis [13.21]. The work by *Baig* exploits data that are collected via the kinesiological instrument for normal and altered reaching movement (KINARM) [13.77], a robotic device that monitors and manipulates upper body movements. It records quantifiable kinetic/kinematic measures, such as reaction time, velocity, joint torque, and hand trajectories of both stroke and control subjects for specific motor and sensory tasks. The stroke diagnosis framework involving KINARM and the *TA*3 CBR systems is shown in Fig. 13.4. After collecting stroke patient data using KINARM, the data are preprocessed to find errors and perform feature selection using data mining through the WEKA tool [13.78]. WEKA also classifies the data into three contexts, A, B, and C, which are defined on different sets of attributes and will be used for retrieval. The data are then stored in a persistent database management system from which cases of known problems, symptoms, and outcomes are extracted and passed to the *TA*3 CBR system. The final stage is the decision regarding the stroke diagnosis and prognosis of patients. A detailed list of all the attributes used in the case base can be found in [13.21].

Given a new problem (data from a new patient) the case is first classified into one of the three contexts and then compared against the known cases. The main objectives when experimenting with stroke data and the CBR system are:

- To differentiate a *stroke* subject from a *control*,
- To classify the type of stroke as *hemorrhagic* or *ischemic*,
- To classify the stroke subject as *right-brain affected* or *left-brain affected*,
- To determine the prognosis of a stroke patient in terms of *affected vascular territory* and identify the *lesion location*.

The evaluation of the system showed good performance in terms of sensitivity (51%), specificity (98%), and accuracy (82%) [13.21]. A sensitivity of 100% would mean that the test recognizes all sick people as such, whereas a specificity of 100% would mean that the test recognizes all healthy people as healthy.

13.4.3 Medical Imaging and CBR Systems

All the aforementioned systems handle textual and numerical data in medical diagnosis applications. CBR systems have also been found useful at handling bioinformatics data that have been produced as the result of image analysis. Tasks such as interpretation, classification of images, and planning of experiments have been guided by their use; For example, the ImageCreek system [13.70] dealt with the problem of image interpretation. More precisely, this system was used to interpret computer tomography (CT) images, which can then be used in disease diagnosis. The ImageCreek system includes two case-base reasoners, one for segment identification, called *Segment ImageCreek*, and one for image interpretation, called *Wholistic ImageCreeek*. This image segment-based CBR system has been used in a study of abdominal CT images that come with a set of hypotheses. The system uses these hypotheses to interpret new images or, according to domain expertise, change the existing ones.

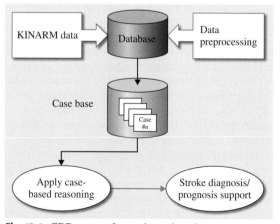

Fig. 13.4 CBR system for stroke patient data

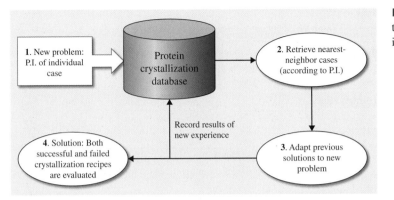

Fig. 13.5 Case-based planning of protein crystallization (P.I. – precipitation index)

13.4.4 Protein Crystallization Planning with CBR

Images are also useful for understanding how proteins acquire their three-dimensional structure, and protein crystallization techniques assist in this task. The fragile nature of crystals as well as the many environmental factors make the process of protein crystallization inherently difficult. Therefore, attempting to crystallize a protein without a proven protocol can be challenging and time consuming. CBR systems have been proposed to aid in the planning of protein crystallization experiments, since one of the difficulties in planning crystal growth experiments is that the history of experiments is not well known [13.80].

Figure 13.5 illustrates the process of using a CBR system in the planning of protein crystallization experiments, where solubility experiments give a quantitative score of similarity for the proteins. Hence, planning strategies that are used in one protein could also be ap-

plied to another. New crystallization problems can then be approached by execution and analysis of a set of precipitation reactions, followed by automated identification of similar proteins and analysis of the recipes used to crystallize them (that is, crystal growth method, temperature and pH ranges, concentration of a protein, crystallization agent, etc.) [13.79].

Case retrieval involves a modified k-nearest-neighbor similarity matching that compares the precipitation indexes of an existing case with a new problem that is given as input [13.79, 81]. Adaptation follows as a next step, and previous solutions are modified to address the new problem. Combined with domain knowledge, the system acts both as an adviser to the crystallographer by suggesting parameter settings for further experiments, as well as an evaluator of potential experiments that the user might propose. The adaptation module constitutes a dynamic process that evolves over time as new knowledge becomes available. Once a plan (in the form of a set of experiments) has been derived

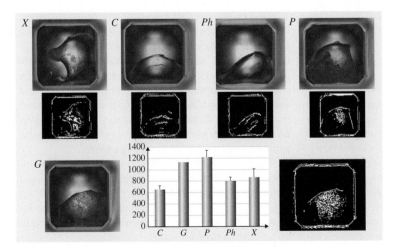

Fig. 13.6 Image analysis and classification (for explanation see text, after [13.79])

Table 13.1 Domain-independent CBR systems

System	Case representation	Preprocessing	New case processing
TA3	Attribute–value pairs	Feature selection	k-NN[a] search
eXiT*CBR	CSV[b] files of attribute descriptions and weights	Discretization, normalization, feature selection	Case classification

[a] k-NN: k-nearest neighbor; [b] CSV: comma-separated values

Table 13.2 Different biomedical tasks in CBR systems

Condition/experiment	Type	System	Case comparison
Breast cancer	Diagnosis Prognosis	eXiT*CBR, KASIMIR	Preclassified cases are compared with new ones
ADHD	Diagnosis	TA3	Similarity of saccade eye movement measurements
Stroke	Diagnosis	TA3	Similarity of kinetic/kinematic measures from KINARM
Image analysis	Segmentation	ImageCreek	Existing CT scan hypotheses compared against new
Protein crystallization	Planning	TA3	k-NN compares precipitation indexes of existing/new cases

and executed for a novel protein, the results are recorded as a new case that reflects this experience. Cases with both positive and negative outcomes are equally valuable for future decision-making processes and are also required for the application of data-mining techniques to the case base.

Figure 13.6 shows example results of image analysis and classification. After performing image segmentation, multiple classes of crystallization results can be detected. The figure demonstrates example images that have been classified as crystal (X), phase separation (Ph), precipitate (P), clear (C), and gel (G). The corresponding contour images have been utilized to compute the Euler number, which, in turn, has been utilized to cluster similar images. The bar graph also shows that crystal and phase separation overlap, but can be separated from clear drops and precipitates. The results have been shown to have 89% accuracy [13.79, 82].

Finally, for a summary of the two main domain-independent CBR systems discussed, as well as the different biomedical cases where they are being used, see Tables 13.1 and 13.2.

13.5 Conclusions

This chapter presents an overview of CBR systems and their use. We discuss their architecture and the process they follow to solve new problems as well as the challenges involved. Especially in healthcare applications, CBR systems can help solving problems that would otherwise be too difficult to manage using other methods and techniques.

The systems and applications we review illustrate the different roles that CBR systems can play, from diagnosis and prognosis of diseases to image classification and experiment planning. CBR systems can play a significant role in supporting medical decisions nowadays. As the available real-life datasets appear in many different sources and formats, such as data coming from sensors, image, and video as well as data related to semantic properties of entities, e.g., ontologies, CBR systems have been adapted to fit these diverse needs. New data characteristics call for new data preprocessing techniques, new proximity measures, as well as aggregation and evaluation of results.

Due to the size and heterogeneity of data sources, there is a trend for integration of CBR systems with existing knowledge discovery tools in order to improve the data quality of the cases and their efficiency. Recently CBR has been used in text mining, information retrieval, and natural-language processing. There are also problems in which CBR integrates existing semantic information as part of the domain, which can help address entity resolution problems as well as semantic similarity among new and past experiences.

Finally, new CBR systems facilitate experimentation with different datasets and parameter settings, which allow more effective quality assessment of the results. Combined with advanced visualization techniques, CBR systems can improve healthcare delivery by optimizing decision-support processes.

Part C | 13

References

13.1   M.N. Nguyen, J.M. Zurada, J.C. Rajapakse: Toward better understanding of protein secondary structure: Extracting prediction rules, IEEE/ACM Trans. Comput. Biol. Bioinforma. **8**, 858–864 (2011)

13.2   Z. Zhu, S. Sun, M. Bern: Classification of protein crystallization imagery, IEEE EMBS Annu. Conf. (2004) pp. 1628–1631

13.3   A. Aamodt, E. Plaza: Case-based reasoning: Foundational issues, methodological variations, and system approaches, AI Commun. **7**(1), 39–59 (1994)

13.4   I. Jurisica, D. Wigle: *Knowledge Discovery in Proteomics* (Chapman Hall, London 2004)

13.5   J. Kolodner: *Case-based Reasoning* (Morgan Kaufmann, New York 1993)

13.6   S.V. Pantazi, J.F. Arocha, J.R. Moehr: Case-based medical informatics, BMC Med. Inform. Decis. Mak. **4**, 1–23 (2004)

13.7   R. Baeza-Yates, B. Ribeiro-Neto: *Modern Information Retrieval* (Addison-Wesley-Longman, Amsterdam 1999)

13.8   N. Arshadi, I. Jurisica: An ensemble of case-based classifiers for high-dimensional biological domains, Int. Conf. Case-Based Reason. (2005) pp. 21–34

13.9   N. Arshadi, I. Jurisica: Data mining for case-based reasoning in high-dimensional biological domains, IEEE Trans. Knowl. Data Eng. **17**(8), 1127–1137 (2005)

13.10   Z.-J. Huang, B.-Q. Wang: A novel swarm clustering algorithm and its application for CBR retrieval, 2nd Int. Conf. Inf. Eng. Comput. Sci. (ICIECS) (2010) pp. 1–5

13.11   S. Montani: How to use contextual knowledge in medical case-based reasoning systems: A survey on very recent trends, Artif. Intell. Med. **51**(2), 125–131 (2011)

13.12   S.J. Russell: *Use of Knowledge in Analogy and Induction* (Morgan Kaufmann, New York 1989)

13.13   E. Xia, I. Jurisica, J. Waterhouse, V. Sloan: RuNtime estimation using the case-based reasoning approach for scheduling in a grid environment, Int. Conf. Case-Based Reason. (2010) pp. 525–539

13.14   Q. Yang, J. Wu: Keep it simple: A case-base maintenance policy based on clustering and information theory, Can. Conf. AI (2000) pp. 102–114

13.15   A. Lawanna, J. Daengdej: Methods for case maintenance in case-based reasoning, Int. J. Comput. Inform. Eng. **4**, 10–18 (2010)

13.16   N. Arshadi, I. Jurisica: Maintaining case-based reasoning systems: A machine learning approach, Eur. Conf. Case-Based Reasoning (2004) pp. 17–31

13.17   K.-D. Althoff, R. Bergmann, S. Wess, M. Manago, E. Auriol, O.I. Larichev, A. Bolotov, Y.I. Zhuravlev, S.I. Gurov: Case-based reasoning for medical decision support tasks: The INRECA approach, Artif. Intell. Med. **12**(1), 25–41 (1998)

13.18   M. Frize, R. Walker: Clinical decision-support systems for intensive care units using case-based reasoning, Med. Eng. Phys. **22**, 671–677 (2000)

13.19   I. Jurisica, J. Mylopoulos, J.I. Glasgow, H. Shapiro, R.F. Casper: Case-based reasoning in IVF: Prediction and knowledge mining, Artif. Intell. Med. **12**(1), 1–24 (1998)

13.20   D. Brien, J.I. Glasgow, D. Munoz: The application of a case-based reasoning system to attention-deficit hyperactivity disorder, Int. Conf. Case-Based Reason. (2005) pp. 122–136

13.21   M. Baig: *Case-Based Reasoning: An Effective Paradigm for Providing Diagnostic Support for Stroke Patients*, Master Thesis (Queen's University, Kingston 2008)

13.22   R.G. Ross, D. Hommer, D. Breiger, C. Varley, A. Radant: Eye movement task related to frontal lobe functioning in children with attention deficit disorder, J. Am. Acad. Child Adolesc. Psychiatry **33**, 869–874 (1994)

13.23   A. Aamodt: Modeling the knowledge contents of CBR systems, Int. Conf. Case-Based Reason. (2001) pp. 32–37, Naval Res. Note AIC-01-003

13.24   P. Cunningham, A. Bonzano: Knowledge engineering issues in developing a case-based reasoning application, Knowl. Based Syst. **12**(7), 371–379 (1999)

13.25   N. Xiong, P. Funk: Combined feature selection and similarity modelling in case-based reasoning using hierarchical memetic algorithm, IEEE Congr. Evol. Comput. (2010) pp. 1–6

13.26   L. Gazendam, C. Wartena, R. Brussee: Thesaurus based term ranking for keyword extraction, DEXA Workshops (2010) pp. 49–53

13.27   Y. Matsuo, M. Ishizuka: Keyword extraction from a single document using word co-occurrence statistical information, Int. J. Artif. Intell. Tools **13**(1), 157–169 (2004)

13.28   R. Kohavi, G.H. John: Wrappers for feature subset selection, Artif. Intell. **97**(1-2), 273–324 (1997)

13.29   H. Peng, F. Long, C.H.Q. Ding: Feature selection based on mutual information: Criteria of max-dependency, max-relevance, and min-redundancy, IEEE Trans. Pattern Anal. Mach. Intell. **27**(8), 1226–1238 (2005)

13.30   K. Gopal: Efficient case-based reasoning through feature weighting, and its application in protein crystallography. Ph.D. Thesis (Texas A & M University, College Station 2007)

13.31   C. Kirsopp, M.J. Shepperd, J. Hart: Search heuristics, case-based reasoning and software project effort prediction, Genet. Evol. Comput. Conf. (2002) pp. 1367–1374

13.32　I. Guyon, A. Elisseeff: An introduction to variable and feature selection, J. Mach. Learn. Res. **3**, 1157–1182 (2003)

13.33　P.C. Boutros, S.K. Lau, M. Pintilie, N. Liu, F.A. Shepherd, S.D. Der, M.-S. Tsao, L.Z. Penn, I. Jurisica: Prognostic gene signatures for non-small-cell lung cancer, Proc. Natl. Acad. Sci. USA **106**(8), 2824–2828 (2009)

13.34　K. Börner, E. Pippig, E.-C. Tammer, C.-H. Coulon: Structural similarity and adaptation, Eur. Winter Conf. Brain Research (1996) pp. 58–75

13.35　E.L. Rissland, J.J. Daniels: The synergistic application of CBR to IR, Artif. Intell. Rev. **10**(5-6), 441–475 (1996)

13.36　M. Lenz, H.-D. Burkhard, P. Pirk, E. Auriol, M. Manago: CBR for diagnosis and decision support, AI Commun. **9**(3), 138–146 (1996)

13.37　Y. Niu, D. Otasek, I. Jurisica: Evaluation of linguistic features useful in extraction of interactions from PubMed; Application to annotating known, high-throughput and predicted interactions in I^{2}D, Bioinformatics **26**(1), 111–119 (2010)

13.38　P. Andritsos, P. Tsaparas, R.J. Miller, K.C. Sevcik: LIMBO: Scalable clustering of categorical data, Eur. Conf. Case-Based Reasoning (2004) pp. 123–146

13.39　P. Andritsos, R.J. Miller, P. Tsaparas: Information-theoretic tools for structure discovery in large data sets, SIGMOD (2004) pp. 731–742

13.40　M. Gluck, J. Corter: Information, uncertainty, and the utility of categories, Proc. 7th Annu. Conf. Cogn. Sci. Soc. (COGSCI) (1985) pp. 283–287

13.41　C.A. Cumbaa, I. Jurisica: Protein crystallization analysis on the world community grid, J. Struct. Funct. Genomics **11**, 61–69 (2010)

13.42　P. Perner: Similarity-based image segmentation for determination of brain/liquor ratio in CT image by Alzheimer dementia, Bildverarb. für die Med. (1998)

13.43　G. Xu, C. Chiu, E.D. Angelini, A.F. Laine: An incremental and optimized learning method for the automatic classification of protein crystal images, IEEE EMBS Annu. Int. Conf. (2006) pp. 6526–6529

13.44　K. Kawabata, M. Takahashi, K. Saitoh, H. Asama, T. Mishima, M. Sugahara, M. Miyano: Evaluation of crystalline objects in crystallizing protein droplets based on line-segment information in greyscale images, Acta Crystallogr. D **62**, 239–245 (2006)

13.45　K. Saitoh, K. Kawabata, H. Asama, T. Mishima, M. Sugahara, M. Miyano: Evaluation of protein crystallization states based on texture information derived from greyscale images, Acta Crystallogr. D **61**, 873–880 (2005)

13.46　S. Pan, G. Shavit, M. Penas-Centeno, D.-H. Xu, L. Shapiro, R. Ladner, E. Riskin, W. Hol, D. Meldrum: Automated classification of protein crystallization images using support vector machines with scale-invariant texture and gabor features, Acta Crystallogr. D **62**, 271–279 (2006)

13.47　J. Wilson: Automated Classification of Images from Crystallization Experiments, Ind. Conf. Data Min. (2006) pp. 459–473

13.48　B. López, C. Pous, P. Gay, A. Pla, J. Sanz, J. Brunet: eXiT*CBR: A framework for case-based medical diagnosis development and experimentation, Artif. Intell. Med. **51**(2), 81–91 (2011)

13.49　M. Göker, C. Baudin, M. Manago: Development of industrial knowledge management applications with case-based reasoning. In: *Successful Case-Based Reasoning Applications*, ed. by S. Montani, L. Jain (Springer, Berlin, Heidelberg 2010) pp. 53–82

13.50　C.L. Cobb, A.M. Agogino: Case-based reasoning for evolutionary mems design, J. Comput. Inform. Sci. Eng. **10**(3), 39–48 (2010)

13.51　K.J. Kim, K. Kim: Preliminary cost estimation model using case-based reasoning and genetic algorithms, J. Comput. Civ. Eng. **24**(6), 499–505 (2010)

13.52　C.-H. Leon Lee, A. Liu, H.-H. Huang: Using planning and case-based reasoning for service composition, J. Adv. Comput. Intell. Intell. Inform. **14**, 540–548 (2010)

13.53　M. Tingyu, M. Biao: Case based reasoning applied in personal financing: Representing cases based on XML, J. Adv. Comput. Intell. Intell. Inform. **14**, 540–548 (2010)

13.54　A. Wyner, T. Bench-Capon: Argument schemes for legal case-based reasoning, Proc. 2007 Conf. Leg. Knowl. Inform. Syst. (2007) pp. 139–149

13.55　M.U. Ahmed, S. Begum: Case-based reasoning for medical and industrial decision support. In: *Successful Case-Based Reasoning Applications*, ed. by S. Montani, L. Jain (Springer, Berlin, Heidelberg, 2010) pp. 7–52

13.56　M. d'Aquin, J. Lieber, A. Napoli: Decentralized case-based reasoning for the semantic web, Int. Seman. Web Conf. (2005) pp. 142–155

13.57　C.S. Sauer, K. Bach, K.-D. Althoff: Integration of linked-open data in case-based reasoning systems, LWA Workshop (2010) pp. 269–274

13.58　L.N. Nassif, J. Marcos, S. Nogueira, A. Karmouch, M. Ahmed, F. de Andrade: Job completion prediction using case-based reasoning for grid computing environments, Concurr. Comput. Pract. Exp. **19**(9), 1253–1269 (2007)

13.59　I. Jurisica: *TA*3: Theory, implementation and applications of similarity-based retrieval for case-based reasoning. Ph.D. Thesis (University of Toronto, Toronto 1998)

13.60　V.K. Chaudhri, I. Jurisica, M. Koubarakis, D. Plexousakis, T. Topaloglou: The KBMS project and beyond, Concept. Model. (2009) pp. 466–482

13.61　I. Spasic, S. Ananiadou, J. Tsujii: MaSTerClass: A case-based reasoning system for the classification of biomedical terms, Bioinformatics **21**(11), 2748–2758 (2005)

13.62   J. Kolodner, R.M. Kolodner: Using experience in clinical problem solving: Introduction and framework, IEEE Trans. Syst. Man Cybern. **17**, 420–431 (1987)

13.63   R. Bareiss, B.W. Porter, C.C. Wier: Protos: An exemplar-based learning apprentice, Int. J. Man-Mach. Stud. **29**(5), 549–561 (1988)

13.64   P. Koton: Reasoning about evidence in causal explanations, AAAI (1988) pp. 256–263

13.65   R.M. Turner: Using schemas for diagnosis, Comput. Methods Programs Biomed. **30**, 199–207 (1989)

13.66   B. López, E. Plaza: Case-based planning for medical diagnosis, Int. Symp. Methodol. Intell. Syst. (ISMIS) (1993) pp. 96–105

13.67   I. Bichindaritz: A case-based reasoner adaptive to different cognitive tasks, Int. Conf. Case-Based Reason. (1995) pp. 390–400

13.68   K.D. Fenstermacher: CADI – An intelligent, multimedia tutor for cardiac auscultation, AAAI/IAAI, Vol. 2 (1996) p. 1387

13.69   R.T. Macura, K.J. Macura: MacRad: Radiology image resource with a case-based retrieval system, Int. Conf. Case-Based Reason. (1995) pp. 43–54

13.70   M. Grimnes, A. Aamodt: A two layer case-based reasoning architecture for medical image understanding, Eur. Winter Conf. Brain Res. (1996) pp. 164–178

13.71   M. Nilsson, M. Sollenborn: Advancements and trends in medical case-based reasoning: An overview of systems and system development, Proc. 17th Int. Florida Artif. Intell. Res. Soc. Conf. (AAAI, Menlo Park, 2004) pp. 178–183

13.72   J.A. Recio-García, B. Díaz-Agudo, M.-A. Gómez-Martín, N. Wiratunga: Extending jCOLIBRI for textual CBR, Int. Conf. Case-Based Reason. (2005) pp. 421–435

13.73   A. Stahl, T. Roth-Berghofer: Rapid prototyping of CBR applications with the open source tool myCBR, Eur. Conf. Case-Based Reasoning (2008) pp. 615–629

13.74   M. d'Aquin, J. Lieber, A. Napoli: Adaptation knowledge acquisition: A case study for case-based decision support in oncology, Comput. Intell. **22**(3-4), 161–176 (2006)

13.75   E. Armengol: Classification of melanomas in situ using knowledge discovery with explained case-based reasoning, Artif. Intell. Med. **51**(2), 93–105 (2011)

13.76   H.E. Rosvold, A.F. Mirsky, I. Sarason, E.D. Bransome Jr., L.H. Beck: A continuous performance test of brain damage, J. Consult. Clin. Psychol. **20**(5), 343–350 (1956)

13.77   S.H. Scott: Apparatus for measuring and perturbing shoulder and elbow-joint positions and torques during reaching, J. Neurosci. Methods **89**, 119–127 (1999)

13.78   I.H. Witten, E. Frank, L. Trigg, M. Hall, G. Holmes, S.J. Cunningham: Weka: Practical Machine Learning Tools and Techniques with Java Implementations (1999), available online at http://www.cs.waikato.ac.nz/

13.79   I. Jurisica, J.I. Glasgow: Applications of case-based reasoning in molecular biology, AI Mag. **25**(1), 85–96 (2004)

13.80   A. Ducruix, R. Giege: *Crystallization of Nucleic Acids and Proteins. A Practical Approach* (Oxford Univ. Press, Oxford 1992)

13.81   I. Jurisica, P. Rogers, J.I. Glasgow, R.J. Collins, J.R. Wolfley, J.R. Luft, G.T. DeTitta: Improving objectivity and scalability in protein crystallization: Integrating image analysis with knowledge discovery, IEEE Intell. Syst. **16**(6), 26–34 (2001)

13.82   I. Jurisica, C.A. Cumbaa, A. Lauricella, N. Fehrman, C. Veatch, R. Collins, J. Luft, G. DeTitta: Automatic classification of protein crystallization screens on 1536-well plates, Proc. Annu. Conf. Am. Crystallogr. Assoc. (ACA03) (2003)

14. Analysis of Multiple DNA Microarray Datasets

Veselka Boeva, Elena Tsiporkova, Elena Kostadinova

In contrast to conventional clustering algorithms, where a single dataset is used to produce a clustering solution, we introduce herein a MapReduce approach for clustering of datasets generated in multiple-experiment settings. It is inspired by the *map-reduce* functions commonly used in functional programming and consists of two distinctive phases. Initially, the selected clustering algorithm is applied (mapped) to each experiment separately. This produces a list of different clustering solutions, one per experiment. These are further transformed (reduced) by portioning the cluster centers into a single clustering solution. The obtained partition is not disjoint in terms of the different participating genes, and it is further analyzed and refined by applying formal concept analysis.

Part C | 14

Gene clustering is one of the most important top-down microarray analysis techniques when it comes to extracting meaningful information from gene expression profiles. In general, clustering is the process of grouping data objects into sets of disjoint classes called clusters, so that objects in the same cluster are more similar to each other than objects in the other clusters, given a reasonable measure of similarity. In the context of microarray analysis, clustering algorithms have been used to divide genes into groups according to the degree of their expression similarity. Such a grouping may suggest that the respective genes are correlated and/or coregulated, and subsequently indicates that the genes could possibly share a common biological role.

14.1 Rationale

Presently, with the increasing number and complexity of available gene expression datasets, the combination of data from multiple microarray studies addressing a similar biological question is gaining high importance. There are several approaches in the literature [14.1–4] devoted to the combination of the information contained in different gene representations within a single clustering process. These different representations may refer to gene expression data produced within a single mi-croarray experiment or to gene ontology representation containing knowledge about, e.g., gene functions. Other studies [14.1, 4, 5] propose several different ways to combine gene information representations at the level of similarity matrices.

Methods for the combination of the partitions derived for each representation separately have also been considered [14.6–8]. For instance, the algorithm proposed in *Johnson* and *Kargupta* [14.6] first generates local cluster models and then combines them to generate the global cluster model of the data. The study in *Topchy* et al. [14.8] focuses on clustering ensembles, that is, seeking a combination of multiple partitions that provides improved overall clustering of the given data. The combined partition is found as a solution to the corresponding maximum-likelihood problem using the *expectation-maximization* (EM) algorithm [14.9]. *Strehl* and *Ghosh* [14.7] consider the problem of combining multiple partitions of a set of objects into a single consolidated clustering without accessing the features or algorithms that determined these partitions. The cluster ensemble problem is then formalized as a combinatorial optimization problem in terms of shared mutual information.

Kostadinova et al. [14.10] study two microarray data integration techniques that can be applied to the problem of deriving clustering results from a set of microarray experiments. They consider initially a cluster integration approach, which combines the information contained in multiple microarray experiments at the level of expression or distance matrices and then applies a partitioning algorithm on the combined matrix. Furthermore, a technique for the integration of partitioning results derived from multiple microarray datasets is introduced. It is shown that the application of a partition algorithm on the integrated data yields better performance compared with the approach of aggregating multiple partitions.

In this chapter, we describe a clustering method that merges partitioning results derived separately from a set of microarray experiments studying the same biological phenomenon. It is inspired by the *map-reduce* functions commonly used in functional programming and consists of two distinctive steps. The selected clustering algorithm is initially applied (i. e., performing the *map* step) to each experiment separately. This produces a list of different clustering solutions, one per experiment, which is further transformed by portioning the cluster centers into a single clustering solution (i. e., performing the *reduce* step). This approach has certain advantages:

1. It uses all data by allowing potentially each experiment to have a different set of genes; i. e., the total set of studied genes is not restricted to those contained in all datasets.
2. It is better tuned to each experimental condition by identifying the initial number of clusters for each experiment separately depending on the number, composition, and quality of the gene profiles.
3. It avoids the problem with ties, i. e., a case when a gene is randomly assigned to a cluster because it belongs to more than one cluster, by merging similar clusters into partitions that are not necessarily disjoint.

The latter interesting artifact of the proposed mapreduce clustering (Sect. 14.2.4) implies that it may occur that some genes are assigned to more than one cluster.

The overlapping partition produced by the reduce step is further analyzed by employing formal concept analysis (FCA) [14.11], which allows one to extract valuable insights from the data and further refine the partition into a disjoint one. FCA produces a concept lattice where each concept represents a subset of genes that belong to a number of clusters. The concepts compose the final disjoint clustering partition.

FCA or the *concept lattice approach* has been applied for extracting local patterns from microarray data [14.12, 13] or for performing microarray data comparison [14.14, 15]. For example, the FCA method proposed in [14.15] builds a concept lattice from the experimental data together with additional biological information. Each vertex of the lattice corresponds to a subset of genes that are grouped together according to their expression values and some biological information related to gene function. It is assumed that the lattice structure of the gene sets might reflect biological relationships in the dataset. In [14.16], a FCA-based method is proposed for extracting groups or classes of coexpressed genes. A concept lattice is constructed where each concept represents a set of coexpressed genes in a number of situations. A serious drawback of the method is the fact that the expression matrix is transformed into a binary table (the input for the FCA step), which may lead to possible introduction of biases or information loss.

14.2 Data and Methods

We first describe the multiexperiment microarray dataset used to demonstrate the proposed clustering method and provide a short overview of the common clustering approaches and the FCA framework. Then, we proceed with the detailed description of our combined MapReduce and FCA clustering approach. In addition, we also consider the cluster validation measures that are used to evaluate the performance of the proposed algorithm.

14.2.1 Microarray Datasets

The clustering results of the proposed hybrid algorithm are evaluated on *benchmark* (where true clustering is known) and *real* (where true clustering is unknown) gene expression time-series data obtained from a study examining the global cell-cycle control of gene expression in fission yeast *Schizosaccharomyces pombe* [14.17]. The study includes eight independent time-course experiments synchronized respectively by

1. elutriation (three independent biological repeats)
2. cdc25 block-release (two independent biological repeats, of which one in two dye-swapped technical replicates and one experiment in a sep1 mutant background)
3. a combination of both methods (elutriation and cdc25 block-release as well as elutriation and cdc10 block-release).

Thus, nine different expression test sets are available. In the preprocessing phase the rows with more than 25% missing entries have been filtered out from each expression matrix and any other missing expression entries have been imputed by the dynamic time warping (DTW) impute algorithm [14.18]. In this way nine complete matrices which are our real test datasets have been obtained.

Rustici et al. [14.17] identified 407 genes as cell-cycle regulated. These have been subjected to clustering, which resulted in the formation of four separate clusters. Subsequently, the time expression profiles of these genes have been extracted from the complete data matrices, and thus nine new matrices, which form our benchmark datasets with a known clustering solution, have been constructed. Note that some of these 407 genes were removed from the original matrices during the preprocessing phase; i.e., each benchmark dataset may have a different set of genes. Thus, a set of *386 different genes* are present in the nine benchmark datasets in total.

The test datasets have been additionally normalized by applying a data transformation method aiming at multipurpose data standardization and inspired by gene-centric clustering approaches, proposed in [14.19].

14.2.2 Clustering Algorithms

Clustering techniques can be broadly divided into density-based, hierarchical, and partitioning methods [14.20]. *Density-based* algorithms implement the so-called local principle to group neighboring objects into clusters based on density conditions, and thus they are capable of discovering clusters of arbitrary shape [14.21]. *Hierarchical* clustering methods generate a set of nested clusters by either merging smaller clusters into larger ones, or by splitting larger clusters in a hierarchical manner [14.22]. In contrast to these approaches, *partitioning* algorithms decompose the dataset into a set of k disjoint clusters such that the within-cluster sum of distances between each object in a given cluster and the corresponding cluster center is minimized. Density-based and hierarchical approaches have certain shortcomings that may be obstructive for the usage of these methods in the microarray analysis field. Density-based clustering exhibits some lack in interpretability, i.e., the results may appear difficult to understand and comprehend by human domain experts, while the hierarchical methods do not foresee improvement of already constructed clusters, which may significantly affect the quality of the resulting clustering solution. In contrast, partitioning algorithms gradually attempt at each step to improve the constructed clusters, which is their general advantage and the reason they are among the most competitive for gene expression data [14.23]. Partitioning methods have also attracted great interest in the literature because of their straightforward implementation and small number of iterations.

Three partitioning algorithms are well known and commonly used for microarray analysis to divide data objects into k disjoint clusters [14.24]: k-means clustering, k-medians clustering, and k-medoids clustering. All three methods start by initializing a set of k cluster centers, where k is preliminarily determined. Then, each object of the dataset is assigned to the cluster whose center is the nearest, and the cluster centers are recomputed. This process is repeated until the objects inside every cluster become as close to the center as possible and no further object item reassignments take place. The EM algorithm is commonly used for that purpose, i.e., to find the optimal partitioning into k disjoint groups. The three partitioning methods in question differ in how the cluster center is defined. In k-means clustering, the cluster center is defined as the mean data vector averaged over all objects in the cluster, while the median is calculated for the k-medians clustering, instead. For k-medoids clustering [14.25], which is a robust version of the k-means method, the cluster center is defined as the object which has the smallest sum of distances to the other objects in the cluster, i. e., the most centrally located point in a given cluster.

The quality of the clustering solutions obtained from the above partitioning algorithms is sensitive to the choice of the initial cluster centers. Two simple approaches to cluster center initialization are either to select the initial values randomly, or to choose the first k objects. As an alternative, different sets of initial values are considered, and the set that is closest to optimal is chosen. However, testing different initial sets is considered not very practical, especially for large numbers of clusters. Therefore, different methods for cluster initialization have been proposed in literature [14.26–28]; For instance, *Moth'd Belal* and *Al-Daoud* [14.28] propose a cluster initializing algorithm which is based on finding a set of medians extracted from the dimension with maximum variance.

14.2.3 Formal Concept Analysis

Formal concept analysis (FCA) [14.11] is a mathematical formalism allowing the derivation of a concept lattice from a formal context constituted of a set of objects O, a set of attributes A, and a binary relation defined on the Cartesian product $O \times A$. The context is described as a table [14.29, Fig. 12.8]: the rows correspond to objects and the columns to attributes or properties, and a cross in a table cell means that *an object possesses a property*. FCA can be used for a number of purposes, including knowledge formalization and acquisition, ontology design, and data mining.

The *concept lattice* is composed of formal concepts, or simply concepts, organized into a hierarchy by a partial ordering (a subsumption relation allowing comparison of concepts). Intuitively, a concept is a pair (X, Y), where $X \subseteq O$, $Y \subseteq A$, and X is the maximal set of objects sharing the whole set of attributes in Y and vice versa. The set X is called the *extent* and the set Y the *intent* of the concept (X, Y). The subsumption (or subconcept–superconcept) relation between concepts is defined as

$$(X_1, Y_1) \prec (X_2, Y_2) \Leftrightarrow X_1 \subseteq X_2 \quad (\text{or } Y_2 \subseteq Y_1) .$$

Relying on this subsumption relation $\prec$, the set of all concepts extracted from a context is organized within a complete lattice, which means that for any set of concepts there is a smallest superconcept and a largest subconcept, called the *concept lattice*.

14.2.4 Hybrid Clustering Approach for Analysis of Multiple Microarray Datasets

We propose here a MapReduce algorithm for deriving a clustering result from multiple microarray datasets. The main idea is inspired by the *map-reduce* functions commonly used in functional programming. Google also exploited the map-reduce principle when developing their MapReduce software framework to support distributed computing on large datasets on clusters of computers, although the purpose of the map-reduce functions in the Google MapReduce framework is not the same as their original functional programming forms.

Assume that a particular biological phenomenon is monitored in several high-throughput experiments under n different conditions. Each experiment i ($i = 1, 2, \ldots, n$) is supposed to measure the gene expression levels of m_i genes in n_i different experimental conditions or time points. Thus, a list of n different data matrices $M_1, M_2, \ldots, M_n$ will be produced, one per experiment. Suppose that N different genes are in total monitored by all the different experimental datasets.

Initialization Step

Initially, the number of cluster centers is identified for each experiment separately. As discussed in [14.30, 31], this can be performed by running the selected clustering algorithm on each dataset for a range of different numbers of clusters. Subsequently, the quality of the obtained clustering solutions needs to be assessed in some way in order to identify the clustering scheme which best fits the dataset in question; For example, some of the internal validation measures that are presented in Sect. 14.2.5 can be used as validity indices to identify the best clustering scheme. Suppose that k_i cluster centers are determined for each experiment i ($i = 1, 2, \ldots, n$).

Map Step

The selected clustering algorithm is applied (*Map* function) to each expression matrix (experiment) M_i ($i = 1, 2, \ldots, n$) separately. The latter will generate a list of n different clustering solutions, one per experiment. Clearly, the Map function takes a pair of number of clusters and expression matrix, and returns a list of clusters.

It is applied to each experiment in the input data, i. e., $\mathrm{Map}(k_i, M_i) \rightarrow [C_{i1}, C_{i2}, \ldots, C_{ik_i}]$ for $i = 1, 2, \ldots, n$.

Reduce Step

To integrate the different clustering results into a final clustering solution, we propose the following merge (*Reduce* function) schema. Consider the cluster centers (genes) of the obtained clusters represented by their expression profiles extracted from the original data matrices. Subsequently, these profiles (genes) can be divided into k groups (clusters) according to the degree of their expression similarity. The number of groups is identified by integration analysis of the quality of the obtained clustering solutions generated on the involved datasets for a range of different numbers of clusters.

In this way a partition matrix $P = \left\{ p_{ij}^r \right\}$ is generated, where p_{ij}^r can be considered as the membership of cluster C_{ij} ($i = 1, 2, \ldots, n$; $j = 1, 2, \ldots, k_i$) to the r-th ($r = 1, 2, \ldots, k$) group. The expression matrices may eventually need to be normalized or standardized (e.g., by applying a suitable transformation technique) prior to performing the clustering.

Subsequently, the clusters whose centers belong to the same partition are merged in order to obtain the final partitioning, i. e.,

$$C_r = \bigcup_{i=1}^{n} \bigcup_{j=1}^{k_i} p_{ij}^r C_{ij}, \quad r = 1, 2, \ldots, k.$$

Obviously, the *Reduce* function is applied to a list of clustering solutions and produces the final list of clusters, i. e.,

$$\mathrm{Reduce} \left(\mathrm{list}_{i=1}^{n}[C_{i1}, C_{i2}, \ldots, C_{ik_i}] \right)$$
$$\rightarrow [C_1, C_2, \ldots, C_k].$$

Thus, the MapReduce framework transforms a list of clustering results into a single clustering solution. This behavior is similar to the functional programming map-reduce combination, which accepts a list of arbitrary values and returns one single value that combines all the values returned by Map.

The above procedure will generate a clustering partition which is disjoint in terms of the gene expression profiles produced in the different experiments. However, the partition is not guaranteed to be disjoint in terms of the different participating genes; i. e., there will be genes which will belong to more than one cluster. This is because it is likely to occur that some genes will exhibit quite divergent expression profiles among the different experiments due to noise, poor data quality, and other experimental artifacts.

The resulting overlapping gene partition can easily be turned into a *fuzzy* one by assigning to each (gene, cluster) pair a confidence score expressing the membership degree of the gene to the cluster, defined as the ratio of the number of expression profiles of the gene belonging to this cluster divided by the total number of expression profiles available for this gene across the different experiments. Thus, a gene will have a membership degree of 1 to a certain cluster only in case all the expression profiles of this gene are assigned to the cluster in question. In that way, each cluster r ($r = 1, 2, \ldots, k$) can be considered as composed by two types of genes: *core* genes (membership degree 1), which belong with all their multiexperiment expression profiles to the same cluster, and *periphery* genes (membership degree < 1), which have their multiexperiment expression profiles split, being distributed between several different clusters.

The degree of cluster overlap can also be used to derive some insights about the consistency of the gene expression profiles over the different experiments and the data quality as a whole.

Formal Concept Analysis (FCA) Step

As discussed before, the N studied genes are grouped during the *Reduce* step into k clusters that are not guaranteed to be disjoint; i. e., some genes may be assigned to more than one cluster. This overlapping partition is further analyzed and refined into a disjoint one by applying FCA. As mentioned in Sect. 14.2.3, FCA is a principled way of automatically deriving a hierarchical conceptual structure from a collection of objects and their properties. The approach takes as input a matrix (referred to a formal context) specifying a set of objects and the properties thereof, called attributes. The genes are the objects, and the clusters are the attributes.

In our case, a (formal) *context* consists of the set G of the N studied genes, the set of clusters $C = \{C_1, C_2, \ldots, C_k\}$ produced by the reduce step, and an indication of which genes belong to which clusters. Thus, the context is described as a matrix, with the genes corresponding to the rows and the clusters corresponding to the columns of the matrix, and a value 1 in cell (i, j) whenever gene i belongs to cluster C_j. Subsequently, a (formal) *concept* for this context is defined to be a pair (X, Y) such that:

- $X \subseteq G$ and $Y \subseteq C$ and every gene in X belongs to every cluster in Y

Part C | 14.2

- For every gene in G that is not in X, there is a cluster in Y that does not contain that gene,
- For every cluster in C that is not in Y, there is a gene in X that does not belong to that cluster.

The family of these concepts obeys the mathematical axioms defining a *concept lattice*. The built lattice consists of concepts where each one represents a subset of genes belonging to a number of clusters. The set of all concepts partitions the genes into a set of disjoint clusters.

14.2.5 Cluster Validation Measures

One of the most important issues in cluster analysis is the validation of clustering results. Essentially, cluster validation techniques are designed to find the partitioning that best fits the underlying data, and should therefore be regarded as a key tool in the interpretation of clustering results. The data mining literature provides a range of different cluster validation measures, which are broadly divided into two major categories: external and internal [14.32]. External validation measures have the benefit of providing an independent assessment of clustering quality, since they validate a clustering result by comparing it with a given external (gold) standard.

However, an external gold standard is rarely available, and for this particular reason, external validation methods are not considered here. Internal validation techniques, on the other hand, avoid the need for using such additional knowledge, but have the alternative problem that the validation is based on the same information used to derive the clusters themselves.

Additionally, some authors consider a third approach of clustering validity, which is based on relative criteria [14.30,31]. The basic idea is to evaluate the clustering result by comparing clustering solutions generated by the same algorithm but with different input parameter values. A number of validity indices have been defined and proposed for each of the above approaches.

Furthermore, internal measures can be split with respect to the specific clustering property they reflect and assess to find an optimal clustering scheme: compactness, separation, connectedness, and stability of the cluster partitions. *Compactness* evaluates the cluster homogeneity, which is related to the closeness within a given cluster. *Separation* demonstrates the opposite trend by assessing the degree of separation between individual groups. The third type of internal validation measure (*connectedness*) quantifies the extent to which nearest-neighboring data items are placed into the same cluster.

The *stability* measures evaluate the consistency of a given clustering partition by clustering from all but one experimental condition. The remaining condition is subsequently used to assess the predictive power of the resulting clusters by measuring the within-cluster similarity in the removed experiment. A detailed summary of different types of validation measures can be found in [14.33].

Since none of the clustering algorithms performs uniformly best under all scenarios, it is not reliable to use a single validation measure, but instead to use a few that reflect various aspects of a partitioning. In this sense, we have implemented two different validation measures for estimating the quality of clusters: *silhouette index* (*SI*) for assessing compactness and separation properties of a partitioning, and *connectivity* for assessing connectedness.

Connectivity

Connectivity captures the degree to which genes are connected within a cluster by keeping track of whether neighboring genes are put into the same cluster [14.33]. Define $m_{i(j)}$ as the j-th nearest neighbor of gene i, and let $x_{im_{i(j)}}$ be zero if i and j are in the same cluster and $1/j$ otherwise. Then for a particular clustering solution $C_1, C_2, \ldots, C_k$ of matrix M, which contains the expression values of m genes (rows) in n different experimental conditions or time points (columns), the *connectivity* is defined as

$$\text{Conn}(k) = \sum_{i=1}^{m} \sum_{j=1}^{n} x_{im_{i(j)}} . \tag{14.1}$$

The connectivity has a value between zero and infinity and should be minimized.

Silhouette Index (SI)

The silhouette index [14.34] is a cluster validity index that is used to judge the quality of any clustering solution (partition) $C_1, C_2, \ldots, C_k$. Suppose a_i represents the average distance of gene i from the other genes of the cluster to which the gene is assigned, and b_i represents the minimum of the average distances of gene i from genes of the other clusters. Then the SI of matrix M, which contains the expression profiles of m genes, is defined as

$$s(k) = \frac{1}{m} \sum_{i=1}^{m} \frac{b_i - a_i}{\max\{a_i, b_i\}} . \tag{14.2}$$

The values of SI vary from -1 to 1, and higher value indicates better clustering results. It reflects the compactness and separation of clusters.

14.3 Results and Discussion

In this section, we discuss the performance of the MapReduce clustering method on the benchmark datasets described in Sect. 14.2.1. The results on the real datasets are presented in the supplementary material [14.29]. The map step employs the k-medoids clustering algorithm in order to generate clustering partitions on each of the considered nine microarray matrices. The quality of these partitions is evaluated using two cluster validation measures: connectivity (14.1) and SI (14.2). A modified version of the k-medoids clustering algorithm based on dynamic time warping distance has been implemented by using publicly available open-source clustering software [14.35]. The reduce step of the MapReduce algorithm and the cluster validation measures have been implemented in C++.

14.3.1 Map Step

The clustering algorithm is first tested on the benchmark datasets. Initially, the number of cluster centers is identified for each dataset separately. This is performed by running the k-medoids clustering algorithm on each experiment for values of k between 2 and 10. Subsequently, the quality of the obtained clustering solutions is assessed by using *connectivity* and *SI* as validity indices (Figs. 14.1, 14.2 [14.29]). The clustering scheme which best fits each separate dataset is determined by identifying the values of k for which a significant local change in the index value occurs [14.30]. The selected optimal number of clusters for the different experiments is presented in Table 14.1, ranging between 4 and 8 and once more confirming the strong experiment dependency of the data. A total set of 48 clusters is generated across the whole data compendium.

Table 14.1 The selected optimal number of clusters per experimental dataset

Experiment	Number of clusters
elu1	4
elu2	5
elu3	4
cdc25-1	4
cdc25-2.1	6
cdc25-2.2	8
cdc25-sep1	6
elu-cdc10	6
elu-cdc25	5

Figure 14.1 depicts the *connectivity* (Fig. 14.1a) and *SI* (Fig. 14.1b) values generated by applying the k-medoids clustering algorithm on the benchmark matrices using the selected optimal number of clusters. The clustering performance varies significantly between the different experiments; For instance, according to *connectivity*, the best clustering performance is recorded for the elu3 experiment and the worst for cdc25-2.2, while *SI* has the best value for cdc25-1 and the worst for the cdc25-2.1 experiment, respectively.

14.3.2 Reduce Step

The reduce step is executed next, and as a result the cluster centers (genes) of the obtained 48 clusters represented by their expression profiles extracted from the normalized test data matrices are divided into four final groups (clusters). Figure 14.2 depicts the expression profiles of these 48 cluster medoids as partitioned in the final four groups. Group 1 is composed of profiles expressing mostly fluctuating noise, which is probably a result of some stress response phenomenon. The expression profiles in groups 2 and 3 are clearly periodic

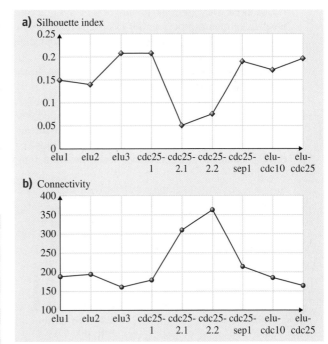

Fig. 14.1a,b Connectivity (**b**) and *SI* values (**a**) generated on the benchmark datasets using the optimal k

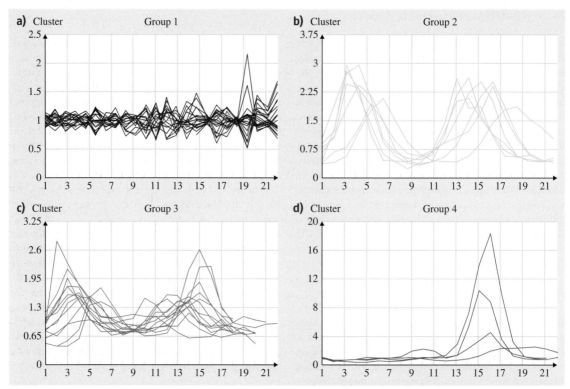

Fig. 14.2a–d The expression profiles of the cluster medoids as partitioned in the final four groups. Group 1: Profiles expressing fluctuating behaviour (**a**); group 2: periodic profiles with two distinctive expression peaks (**b**); group 3: periodic profiles with two distinctive expression peaks (**c**); and group 4: periodic profiles with a single expression peak (**d**)

with two distinctive peaks, while those in group 4 exhibit one single peak later in time.

Each of the four cluster groups generated by the reduce step contains a subset of the 48 cluster medoids generated by the map step. Subsequently, it is assumed that each group is expanded to contain the clusters corresponding to these medoids. Thus, the final four

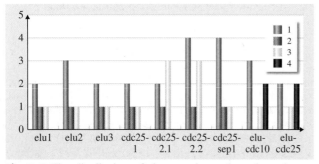

Fig. 14.3 The distribution of the experiment-dependent clusters among the final four groups

cluster groups form a disjoint partition of all the gene expression profiles contained in the nine experimental datasets. Figure 14.3 presents how the expression profiles of the nine different experiments are distributed between the four different reduce clusters. It is interesting to observe that the expression profiles of each experiment are partitioned in exactly three groups. Groups 1 and 3 are obtained by merging clusters from all the experiments, group 2 is not presented in experiments elu-cdc10 and elu-cdc25, while group 4 is formed only by clusters of these two experiments. Consequently, groups 2 and 4 separate the experiments into two complementary groups: one containing experiments synchronized by either elutriation or cdc methods, and the other consisting of those produced by the combination of both methods.

Figure 14.4 compares the connectivity (Fig. 14.4a) and SI (Fig. 14.4b) values generated on the clustering solutions obtained by the k-medoids clustering as performed during the map step and those produced by the merge algorithm during the reduce step. The map results are generated by applying the k-medoids clustering algorithm on the benchmark matrices using the selected optimal number of clusters (Table 14.1), while the reduce ones are based on the final partitions of the individual datasets obtained by the reduce step (Fig. 14.3). The traditional k-medoids approach (as applied in the map step) exhibits worse performance than the proposed MapReduce algorithm under both validation measures. This has also been supported by the experiments conducted on the real test datasets [14.29, Fig. 12.6].

Note that the partition produced by the reduce step is only disjoint in terms of expression profiles, but is not disjoint in terms of the different participating genes; e.g., about 70% of genes belong to more than one group. The overlapping gene partition can further be turned into a fuzzy one by assigning to each (gene, cluster) pair a confidence score, calculated as the ratio of the number of expression profiles of the gene falling in the cluster divided by the total number of expression profiles available for this gene across the different experiments. Figure 14.5 depicts the cumulative distribution of the confidence scores generated for each cluster group. It is interesting to observe that, while only 30% of the genes in group 1 have confidence score lower than 0.5 (i.e., at most half of the multiexperiment expression profiles per gene belong to group 1), this is the case for more than 80% of the genes in group 3 and about 90% in group 2. Even more, there are no genes in group 4 with confidence score higher than 0.3. It is clear that group 1 is the most strongly represented in terms of confidence score levels. The latter is also supported by the fact that 115 genes from the all 117 core genes (membership degree 1) are in group 1.

14.3.3 FCA Step

The (overlapping) partition produced by the reduce step is further analyzed and refined to a disjoint one by applying FCA using a publicly available tool (http://www.iro.umontreal.ca/~galicia/features.html). We have created a context that consists of the set of 386 studied genes and the set of four clusters (groups) produced by the reduce step. It is described as a binary matrix, with the genes corresponding to the rows and the clusters corresponding to the columns. Subsequently, a lattice of 15 concepts for this context is generated (Fig. 14.6). The list of genes belonging to each concept can be found in the supplementary material [14.29, Tab. 12.2]. The FCA step partitions the benchmark gene set into 11 disjoint clusters (concepts) in total, since four concepts appear to be empty.

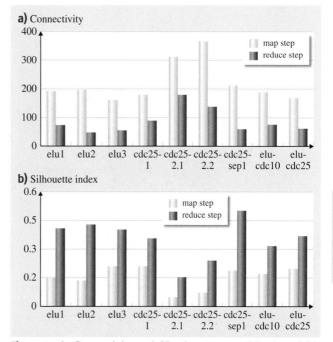

Fig. 14.4a,b Connectivity and SI values generated by k-medoids (map step) and the modified clustering (reduce step)

As explained in Sect. 14.2.1, the benchmark datasets have been built by using the time expression profiles of genes identified as cell-cycle regulated from *Rustici* et al. [14.17]. The identified periodic genes have been grouped into four main clusters, defining successive waves of transcription. Cluster 1 contains 87 genes whose expression peaks during mitosis and whose gene ontology (GO) associations are enriched for the terms M phase, cytokinesis, and chromosome condensation. Cluster 2 contains 78 genes whose expression peaks

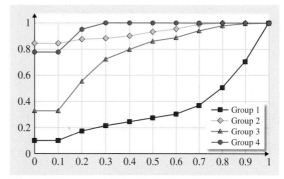

Fig. 14.5 The cumulative distribution of confidence scores calculated for the four fuzzy clusters

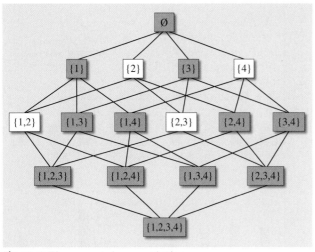

Fig. 14.6 The generated concept lattice

a little later, around anaphase, cytokinesis, and entry into G1 phase. This cluster contains many highly regulated genes and includes the highest number of known periodic genes, most of which function in cell-cycle control or regulation of DNA replication, as reflected in the associated GO terms. Cluster 3 contains only 46 genes whose expression peaks during DNA replication (S phase). The ten most highly regulated genes in this cluster all encode histones, which form a tight sub-cluster. Accordingly, the GO term *Chromatin* was overrepresented among GO associations of cluster 3 genes. Cluster 4 contains 147 genes, but most are only weakly regulated (low-amplitude genes). This is the most heterogeneous cluster, with genes peaking at different times during G2 phase, although a majority peak in early G2.

Each concept consisting of a set of genes generated by the FCA algorithm was subjected to analysis with the BiNGO tool [14.36] to determine which gene ontology categories are statistically overrepresented in each concept. The results are generated for a cutoff p-value of 0.05 and Benjamini and Hochberg (false discovery rate) multiple testing correction. For each gene concept a table is generated consisting of five columns:

1. The GO category identification (GO-id)
2. The multiple testing corrected p-value (p-value)
3. The total number of genes annotated to that GO term divided by the total number of genes in the test set (cluster frequency)
4. The number of selected genes versus the total GO number (total frequency)
5. A detailed description of the selected GO categories (description).

Only six FCA concepts were associated GO categories by the BiNGO tool: C4, C7, C8, C12, C13, and C15. Concretely:

- Concept C4 contains 2 genes annotated to 12 GO categories (9 have total frequency $> 0.0\%$),
- Concept C7 contains 173 genes connected with 27 GO categories,
- Concept C8 contains 5 genes associated with 29 GO categories (22 have total frequency $> 0.0\%$),
- Concept C12 contains 10 genes that are annotated to 15 GO categories,
- Concept C13 contains 5 genes related to 3 GO categories (all 3 have total frequency $= 0.0\%$),
- Concept C15 has 31 genes that are associated with more than 50 GO categories.

Table 14.2 The pairwise coverage values between the clusters of the known clustering solution and the concepts of the gene partition generated by the FCA step

Original clusters FCA concepts	Cluster 1	Cluster 2	Cluster 3	Cluster 4
C2 = {Group 1}	0.1478261	0.090498	0.263736	**0.516129**
C4 = {Group 3}	0	0.018519	0.028986	0
C7 = {Group 1, Group 3}	**0.4930556**	**0.344086**	0.2	**0.367953**
C8 = {Group 1, Group 4}	0	0.018018	0.055556	0.035503
C10 = {Group 2, Group 4}	0	0.037037	0	0
C11 = {Group 3, Group 4}	0.0169492	0.036697	0.028571	0
C12 = {Group 1, Group 2, Group 3}	0.096	0.034483	0	0.022989
C13 = {Group 1, Group 2, Group 4}	0.0166667	0.072072	0	0
C14 = {Group 1, Group 3, Group 4}	0.0559441	0.149254	0.126316	0.135417
C15 = {Group 2, Group 3, Group 4}	0.0958904	0.233577	0.163265	0
C16 = all groups	0.0629921	0.118644	0.025316	0

It is interesting to observe that some of the genes distributed to the above concepts are unknown, e.g., gene *prl36* in C4. The latter implies that the proposed FCA clustering can be useful for making hypotheses about the function of these genes based on the known GO annotations of genes that are located in the same concept.

Further, we compare the final disjoint gene partition generated by the FCA step with the known clustering solution found by *Rustici* et al. [14.17]. Table 14.2 presents the pairwise coverage values calculated between the clusters of the two gene partitions. The best coverage (in bold) is recorded for pairs: (C2, cluster 4),

(C7, cluster 1), (C7, cluster 4), and (C7, cluster 2). Evidently, the genes of concept C7 are mainly distributed between cluster 1, cluster 2, and cluster 4, while those of concept C2 are allocated to cluster 4. This is not surprising since concepts C2 and C7 are the largest ones and their intents are {Group 1} and {Group 1, Group 3}, respectively. The latter two cluster groups are shown to have high overlap with cluster 1, cluster 2, and cluster 4. In addition, it is interesting to observe the high coverage between concept C15, containing 31 genes associated with more than 50 GO categories, and cluster 2, found by *Rustici* et al. [14.17], containing the highest number of known periodic genes.

14.4 Conclusions

We have discussed a MapReduce technique for deriving clustering results from information presented in multiple gene expression matrices. It has been also demonstrated how the obtained overlapping partition can be further analyzed and refined into a disjoint one by using formal concept analysis. Future work includes further empirical evaluation and validation

of the methodology proposed. The aim is to investigate the applicability of MapReduce and FCA approaches to other types of microarray data and to study the obtained clustering results and corresponding concept lattices in order to gain a biological insight into possible functions of previously unknown genes.

References

14.1   T.C. Havens, J.M Keller, M Popescu, J.C Bezdek, E. MacNeal Rehrig, H.M Appel, J.C Schultz: Fuzzy cluster analysis of bioinformatics data composed of microarray expression data and gene ontology annotations, Proc. North Am. Fuzzy Inf. Process. Soc. (2008) pp. 1–6

14.2   D. Huang, W. Pan: Incorporating biological knowledge into distance-based clustering analysis of microarray gene expression data, Bioinformatics **22**(10), 1259–1268 (2006)

14.3   J. Kasturi, R. Acharya: Clustering of diverse genomic data using information fusion, Bioinformatics **21**(4), 423–429 (2005)

14.4   G. Li, Z. Wang: Incorporating heterogeneous biological data sources in clustering gene expression data, Health **1**, 17–23 (2009)

14.5   R. Kustra, A. Zagdanski: Incorporating gene ontology in clustering gene expression data, Proc. 19th IEEE Symp. Comput.-Based Med. Syst. (2006) pp. 555–563

14.6   E. Johnson, H. Kargupta: Collective hierarchical clustering from distributed, heterogeneous data, LNCS **1759**, 221–244 (1999)

14.7   A. Strehl, J. Ghosh: Cluster ensembles – A knowledge reuse framework for combining multiple partitions, J. Mach. Learn. Res. **3**, 583–617 (2002)

14.8   A. Topchy, K. Jain, W. Punch: Clustering ensembles: Models of consensus and weak partitions, IEEE Trans. Pattern Anal. Mach. Intell. **27**, 1866–1881 (2005)

14.9   A.P. Dempster, N.M. Laird, D.B. Rubin: Maximum likelihood from incomplete data via the EM algorithm, J. Roy. Stat. Soc. B **39**(1), 1–38 (1977)

14.10   E. Kostadinova, V. Boeva, N. Lavesson: Clustering of multiple microarray experiments using information integration, LNCS **6865**, 123–137 (2011)

14.11   B. Ganter, G. Stumme, R. Wille (Eds.): *Formal Concept Analysis: Foundations and Applications*, Lect. Notes Artif. Intell., Vol. 3626 (Springer, Berlin, Heidelberg 2005)

14.12   J. Besson, C. Robardet, J.-F. Boulicaut: Constraint-based mining of formal concepts in transactional data, LNCS **3056**, 615–624 (2004)

14.13   J. Besson, C. Robardet, J.-F. Boulicaut, S. Rome: Constraint-based concept mining and its application to microarray data analysis, Intell. Data Anal. **9**(1), 59–82 (2005)

14.14   D.P. Potter: A combinatorial approach to scientific exploration of gene expression data: An integrative method using formal concept analysis for the comparative analysis of microarray data. Ph.D. Thesis (Department of Mathematics, Virginia Tech 2005)

14.15 V. Choi, Y. Huang, V. Lam, D. Potter, R. Laubenbacher, K. Duca: Using formal concept analysis for microarray data comparison, J. Bioinf. Comput. Biol. **6**(1), 65–75 (2008)

14.16 M. Kaytoue-Uberall, S. Duplessis, A. Napoli: *Using Formal Concept Analysis for the Extraction of Groups of Coexpressed Genes CCIS 14* (Springer, Berlin, Heidelberg 2008) pp. 445–455

14.17 G. Rustici, J. Mata, K. Kivinen, P. Lió, C.J. Penkett, G. Burns, J. Hayles, A. Brazma, P. Nurse, J. Bähler: Periodic gene expression program of the fission yeast cell cycle, Nat. Genet. **36**, 809–817 (2004)

14.18 E. Tsiporkova, V. Boeva: Two-pass imputation algorithm for missing value estimation in gene expression time series, J. Bioinf. Comput. Biol. **5**(5), 1005–1022 (2007)

14.19 V. Boeva, E. Tsiporkova: A multipurpose time series data standardization method. Intelligent systems: From theory to practice, Stud. Comput. Intell. **299**, 445–460 (2010)

14.20 A.K. Jain, M.N. Murty, P.J. Flynn: Data clustering: A review, ACM Comput. Surv. **31**(3), 264–323 (1999)

14.21 M. Ester, H.P. Kriegel, J. Sander, X. Xu: A density-based algorithm for discovering clusters in large spatial databases with noise, Proc. 2nd ACM SIGKDD, Portland (1996) pp. 226–231

14.22 M. Eisen, P.T Spollman, P.O Brown, D. Botstein: Cluster analysis and display of genome-wide expression patterns, Proc. Natl. Acad. Sci. USA **95**, 14863–14868 (1998)

14.23 S. Datta, S. Datta: Comparisons and validation of statistical clustering techniques for microarray gene expression data, Bioinformatics **19**, 459–466 (2003)

14.24 J.B. MacQueen: Some methods for classification and analysis of multivariate observations, Proc. 5th Berkeley Symp. Math. Stat. Prob. **1**, 281–297 (1967)

14.25 L. Kaufman, P.J. Rousseeuw: *Fitting Groups in Data: An Introduction to Cluster Analysis* (Wiley, New York 1990)

14.26 G. Babu, M. Murty: A near optimal initial seed value selection in k-means algorithm using a genetic algorithm, Pattern Recognit. Lett. **14**, 763–769 (1993)

14.27 S.S. Khan, A. Ahmad: Cluster center initialization algorithm for k-means clustering, Pattern Recognit. Lett. **25**, 1293–1302 (2004)

14.28 M. Al-Daoud: A new algorithm for cluster initialization, World Acad. Sci. Eng. Technol. **4**, 74–76 (2005)

14.29 V. Boeva, E. Tsiporkova, E. Kostadinova: Analysis of multiple DNA microarrays (2012), available online at http://cst.tu-plovdiv.bg/bi/ SupplementaryMaterial_MapReduce-FCA.pdf

14.30 M. Halkidi, Y. Batistakis, M. Vazirgiannis: On clustering validation techniques, J. Intell. Inf. Syst. **17**(2/3), 107–145 (2001)

14.31 S. Theodoridis, K. Koutroubas: *Pattern Recognition* (Academic, New York 1999)

14.32 A.K. Jain, R.C. Dubes: *Algorithms for Clustering Data* (Prentice Hall, Englewood Cliffs 2006)

14.33 J. Handl, J. Knowles, D.B. Bell: Computational cluster validation in post-genomic data analysis, Bioinformatics **21**, 3201–3212 (2005)

14.34 P. Rousseeuw: Silhouettes: A graphical aid to the interpretation and validation of cluster analysis, J. Comput. Appl. Math. **20**, 53–65 (1987)

14.35 M. de Hoon: Open clustering software (Laboratory of DNA Information Analysis, Human Genome Center, Institute of Medical Science, University of Tokyo, 2012) available at http://bonsai.hgc.jp/~mdehoon/software/cluster/software.htm

14.36 S. Maere, K. Heymans, M. Kuiper: BiNGO: A Cytoscape plugin to assess overrepresentation of gene ontology categories in biological networks, Bioinformatics **21**, 3448–3449 (2005)

15. Fuzzy Logic and Rule-Based Methods in Bioinformatics

Lipo Wang, Feng Chu, Wei Xie

This chapter reviews some fuzzy logic and rule-based approaches in bioinformatics. Among the fuzzy approaches, we emphasize fuzzy neural networks (FNN), which have advantages from both fuzzy logic (e.g., linguistic rules and reduced computation) and neural networks (e.g., ability to learn from data and universal approximation). After the overview in Sect. 15.1, the structure and algorithm of the FNN are reviewed in Sect. 15.2. In Sect. 15.3, we describe a t-test-based gene importance ranking method followed by a description of how we use the FNN to classify three important microarray datasets, for lymphoma, small round blue cell tumor (SRBCT), and ovarian cancer (Sect. 15.4). Section 15.5 reviews various fuzzy and rule-based approaches to microarray data classification proposed by other authors, while Sect. 15.6 reviews fuzzy and rule-based approaches to clustering and prediction in microarray data. We discuss and draw some conclusions in Sect. 15.7.

Part C | 15

Within the field of bioinformatics, here we are especially interested in microarray data processing, including cancer diagnosis from microarray gene expression data, gene expression data clustering, and gene function prediction. In particular, we describe a FNN that we proposed earlier for cancer classification. This FNN contains three valuable aspects, i.e., automatic generation of fuzzy membership functions, parameter optimization, and rule-base simplification. One major obstacle in microarray dataset classification is that the number of features (genes) is much larger than the number of objects (samples). We therefore use a fea-

ture selection method based on the t-test to select more significant genes before applying the FNN. To demonstrate our approach, we use three well-known microarray databases, i.e., the lymphoma dataset, the small round blue cell tumor (SRBCT) dataset, and the ovarian cancer dataset. In all cases we obtain 100% accuracy with fewer genes in comparison with previously published results. Our result shows that the FNN classifier not only improves the accuracy of cancer classification but also helps biologists to find a better relationship between important genes and development of cancers.

15.1 Overview

Since *Zadeh*'s proposal of fuzzy set theory in 1965 [15.1], fuzzy logic has enjoyed tremendous success in terms of theoretical development and practical applications [15.2, 3]. Fuzzy logic enables linguistic interpretations of logic operations and reduces computational load, similar to the way in which we humans carry out computations. Neural networks are able to learn from datasets [15.2, 3], and trained neural networks can perform a wide variety of tasks, such as classification and control. Fuzzy neural networks combine the best of both fuzzy logic and neural networks, consisting of fuzzy rules generated and modified during training [15.2, 3]. Fuzzy logic has been used extensively in solving various bioinformatics problems [15.4]. This chapter is mainly concerned with microarray data processing.

Accurate diagnosis of different types of cancers is an important and challenging job to be accomplished before selecting a proper treatment. However,

similar morphological appearances of some types of cancers are a main challenge for traditional diagnostic methods to accurately classify cancers. Gene expression data obtained from microarrays have brought new promise for addressing the fundamental problems relating to cancer diagnosis. The DNA microarray technique has made simultaneous monitoring of thousands of gene expressions possible. This abundance of gene expression data has attracted great attention due to the possibility of differentiating cancers at molecular level [15.5–7]. Quite a number of machine learning or statistical methods have been proposed for cancer diagnosis. Some recent approaches include neural networks [15.8], support vector machines [15.9, 10], nearest shrunken centroids [15.11], and so on. In this chapter, we use a fuzzy neural network (FNN) proposed earlier by *Frayman* and *Wang* [15.12, 13] to classify cancers based on their gene expression profiles [15.14, 15].

15.2 Fuzzy Neural Network: Structure and Training

In this section, we review the structure and the training algorithm of the FNN proposed by *Frayman* and *Wang* [15.12, 13, 16, 17], which has been successfully applied in classification, rule extraction, and control. The structure of the FNN is shown in Fig. 15.1. The network consists of four layers, i.e., the input layer, the input membership function (MF) layer, the rule layer, and the output layer.

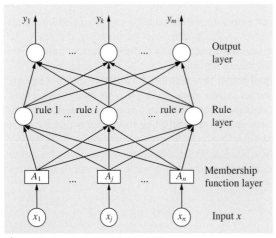

Fig. 15.1 The structure of the FNN

Input data are either numerical or categorical. The input membership function layer generates input membership functions for numerical inputs; i.e., numerical values are converted to categorical values.

Each rule node is connected to each input membership function node and each output node. Each rule node performs a product of its inputs. The input membership functions act as fuzzy weights between the input layer and the rule layer. Links between the rule layer, the output layer, and the input membership functions are tuned during the learning process. In the output layer, each node receives inputs from all the rule nodes connected to this output node and produces the actual output of the network. The structure generation and learning algorithm of the FNN are shown in Fig. 15.2.

15.2.1 FNN Initialization

Firstly, we create n nodes for the input layer and m nodes for the output layer, where n and m are the numbers of input variables (attributes) and output variables (classes), respectively. The rule layer is empty; i.e., there are initially no rules in the rule base.

Two equally spaced triangular membership functions are added along the operating range of each input

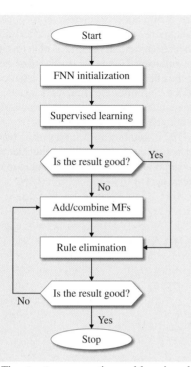

Fig. 15.2 The structure generation and learning algorithm of the FNN

variable. The piecewise-linear triangular membership function is chosen for computational efficiency [15.18].

Then we create the initial rule-base layer using the following form for rule i:

rule i :   if x_1 is $A^i_{x1}, \cdots$, and x_n is A^i_{xn},

then $y_1 = \omega^i_1, \cdots , y_m = \omega^i_m$,

where x_j $(j = 1, 2, \ldots, n)$ and y_l $(l = 1, 2, \ldots, m)$ are the inputs and the outputs, respectively. ω^j_l is a real number. A^i_q $(q = x_1, x_2, \cdots , x_n)$ is the membership function of the antecedent part of rule i for node q in the input layer.

The membership value μ_i of the premise of the i-th rule is calculated with a fuzzy AND using the product operator

$$\mu_i = A^i_{x1}(x_1) \times A^i_{x2}(x_2) \times \cdots \times A^i_{xn}(x_n) . \quad (15.1)$$

The output y_l of the fuzzy inference is obtained using the weighted average [15.19]

$$y_l = \frac{\sum_i \mu_i \times \omega^i_l}{\sum_i \mu_i} . \quad (15.2)$$

15.2.2 FNN Training

The network is trained using the general learning rule [15.3]

$$y^i_l(k+1) = y^i_l(k) - \eta \frac{\partial \varepsilon_l}{\partial y^i_l} . \quad (15.3)$$

The learning rules for ω^i_l and A^i_q are

$$\omega^i_l(k+1) = \omega^i_l(k) - \eta \frac{\partial \varepsilon_l}{\partial \omega^i_l} , \quad (15.4)$$

$$A^i_q(k+1) = A^i_q(k) - \eta \frac{\partial \varepsilon_l}{\partial A^i_q} , \quad (15.5)$$

where η is the learning rate. The objective is to minimize the error function

$$\varepsilon_l = \tfrac{1}{2} \times (y_l - y_{dl})^2 , \quad (15.6)$$

where y_l is the current output, and y_{dl} is the target output.

We tune the learning rate η to improve the speed of convergence, as well as the learning performance (accuracy). We update η according to the following heuristic rules:

1. If the error undergoes five consecutive reductions, increase η by 5%.
2. If the error undergoes three consecutive combinations of one increase and one reduction, decrease η by 5%.
3. Otherwise, keep η unchanged.

Furthermore, due to this dynamical updating strategy, the initial value of η is usually not critical as long as it is not too large. The learning error ε_l is reduced towards zero or a prespecified small value $\varepsilon_{\text{def}} > 0$ as the iteration number k increases.

15.2.3 Rule-Base Modification

An additional membership function is added to each input at the point of the maximum output error, following *Higgins* and *Goodman* [15.18]. One vertex of the additional membership function is placed at the point of the maximum output error, and it must have the membership value unity; the other two vertices lie at the centers of the two neighboring regions, respectively, and they have membership values zero. As the output of the network is not a binary 0 or 1, but a number ranging from 0 to 1, we can speed up the convergence of the network substantially by eliminating the error whose deviation from the target value is the greatest.

We then evaluate the accuracy and the simplicity of the rules generated in Sect. 15.2.2. We use a weighting parameter between accuracy and simplicity, namely the compatibility grade (CG) of each fuzzy rule. The CG of rule j is calculated by the product operator as

$$\mu_j(x) = \mu_{j1}(x_1) \times \mu_{j2}(x_2) \times \cdots \times \mu_{jn}(x_n), \quad (15.7)$$

when the system provides the correct classification result.

Each rule whose CG falls below a predefined threshold is deleted. Elimination of rule nodes is rule by rule;

i.e., when a rule node is deleted, its associated input membership nodes and links are deleted as well. By adjusting the CG threshold, the user is able to specify the degree of rule-base compactness. The size of the rule base can thus be kept minimal. If the classification accuracy of the FNN after the elimination of rule nodes is below the requirement, we add another rule as described above; otherwise we stop the process.

This FNN combines the powerful features of initial fuzzy model self-generation, parameter optimization [15.3], and rule-base simplification to achieve good performance.

15.3 Ranking Importance of Genes for Feature Selection

Machine learning has found numerous applications in bioinformatics, especially in microarray data processing. One major obstacle in microarray data classification is that the number of features (genes) is very large and can range from a few thousands to tens of thousands. Classifiers, such as fuzzy neural networks [15.3] and support vector machines [15.20], usually are not able to handle so many input features. Furthermore, most genes are irrelevant to cancer distinction and may therefore act like noise if included in the input. We use feature selection algorithms to obtain the most important genes related to diagnosis. Here, we use t-test feature selection. First, we rank genes via their T-score (TS). Then, we select the most important genes with the highest rankings. After that, we are ready to apply the FNN classifier that we shall describe in Sect. 15.4. In this section, we describe how the genes can be ranked.

The TS of gene i is defined as [15.21, 22]

$$TS_i = \max\left\{ \left| \frac{\bar{x}_{ik} - \bar{x}_i}{m_k s_i} \right|, k = 1, 2, \ldots, K \right\}, \quad (15.8)$$

$$\bar{x}_{ik} = \sum_{j \in C_k} \bar{x}_{ij} / n_k, \quad (15.9)$$

$$\bar{x}_i = \sum_{j=1}^{n} x_{ij} / n, \quad (15.10)$$

$$s_i^2 = \frac{1}{n - K} \sum_{k} \sum_{j \in C_k} (x_{ij} - \bar{x}_{ik})^2, \quad (15.11)$$

$$m_k = \sqrt{1/n_k + 1/n}, \quad (15.12)$$

where K is the number of classes, C_k refers to class k that includes n_k samples, x_{ij} is the expression value of gene i in sample j, $\bar{x}_{ik}$ is the mean expression value in class k for gene i, n is the total number of samples, $\bar{x}_i$ is the general mean expression value for gene i, and s_i is the pooled within-class standard deviation for gene i. In fact, the TS used here is a t-statistic between a specific class and the overall centroid of all the classes [15.21, 23].

15.4 Demonstration with Benchmarking Data

15.4.1 Lymphoma Data

Several major microarray datasets are available in the public domain. Here, we describe applications of the FNN to three well-known microarray datasets. The lymphoma data set [15.24, 25] contains 42 samples derived from diffuse large B-cell lymphoma (DLBCL), 9 samples from follicular lymphoma (FL), and 11 samples

from chronic lymphocytic leukaemia (CLL). The entire dataset includes the expression data of 4026 genes. In this dataset, a small part of the data is missing. A k-nearest neighbor algorithm was applied to fill those missing values [15.26].

First, we randomly divided the 62 samples into two parts: 31 samples for training, and 31 samples for testing. We ranked the entire 4026 genes according to their

Table 15.1 Lymphoma gene importance ranking: 174 genes with the highest TSs, in decreasing order (Gene ID is defined in [15.24])

Rank	Gene ID	Gene description
1	GENE2307X	(CD23A = low affinity II receptor for Fc fragment of IgE; Clone = 1352822)
2	GENE3320X	(Similar to HuEMAP = homolog of echinoderm microtubule associated protein EMAP; Clone = 1354294)
3	GENE708X	*Ki67 (long type); Clone = 100
4	GENE2393X	*MDA-7 = melanoma differentiation-associated 7 = anti-proliferative; Clone = 267158
5	GENE1622X	*CD63 antigen (melanoma 1 antigen); Clone = 769861
6	GENE1641X	*Fibronectin 1; Clone = 139009
7	GENE2391X	(Unknown; Clone = 1340277)
8	GENE1636X	*Fibronectin 1; Clone = 139009
9	GENE1644X	(cathepsin L; Clone = 345538)
10	GENE1610X	*Mig = Humig = chemokine targeting T cells; Clone = 8
11	GENE707X	(Topoisomerase II alpha (170 kD); Clone = 195630)
12	GENE689X	*lamin B1; Clone = 1357243
13	GENE695X	*mitotic feedback control protein Madp2 homolog; Clone = 814701
14	GENE1647X	*cathepsin B; Clone = 261517
15	GENE537X	(B-actin,1099-1372; Clone = 143)
...	...	...
165	GENE1539X	*lysophospholipase homolog (HU-K5); Clone = 347403
166	GENE2385X	*Unknown UG Hs.124382 ESTs; Clone = 1356466
167	GENE719X	(Myt1 kinase; Clone = 739511)
168	GENE2415X	(Unknown; Clone = 1289937)
169	GENE527X	*glutathione-S-transferase homolog; Clone = 1355339
170	GENE1598X	*Similar to ferritin H chain; Clone = 1306027
171	GENE1192X	*Interferon-induced guanylate-binding protein 2; Clone = 545038
172	GENE731X	*Chromatin assembly factor-I p150; Clone = 1334875
173	GENE769X	*14-3-3 epsilon; Clone = 266106
174	GENE724X	(Hyaluronan-mediated motility receptor (RHAMM); Clone = 756037)

TS with the 31 training samples. Then we picked out the 174 genes with the highest TS values (Table 15.1). We subsequently input the selected 174 genes one by one into the FNN according to their TS ranks, starting with the gene ranked 1 in Table 15.1. That is, we first used only a single gene that is ranked 1 as the input to the FNN. We trained the FNN with the training data, and subsequently tested the FNN with the test data. We repeated this process with the first two genes in Table 15.1, then the first three genes, and so on. We found that the FNN performed very well: its training error and testing error both decreased to 0 when only the first nine genes in Table 15.1 were input into the FNN.

15.4.2 Small Round Blue Cell Tumors (SRBCT) Data

The SRBCT data [15.8, 27] contains the expression data of 2308 genes. There are totally 63 training samples and 25 testing samples provided; 5 of the testing samples are not SRBCTs. The 63 training samples contain 23 Ewing family of tumors (EWS), 20 rhabdomyosarcoma (RMS), 12 neuroblastoma (NB), and 8 Burkitt lymphomas (BL), and the 20 SRBCT testing samples contain 6 EWS, 5 RMS, 6 NB, and 3 BL.

We followed the same procedure as for the lymphoma dataset. We first ranked the entire 2308 genes

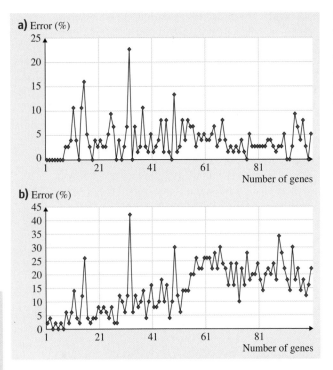

Fig. 15.3a,b The training (**a**) and testing (**b**) results for the ovarian cancer dataset ◄

TSs. We input the selected 96 genes one by one into the FNN according to their TSs, in decreasing order. Both the training error and the testing error decreased to 0 when the top eight genes were input into the FNN.

15.4.3 Ovarian Cancer Data

The ovarian cancer data [15.28] contains 125 samples, including 68 samples derived from breast cancer and 57 samples derived from ovarian cancer. The entire dataset includes the expression data of 3363 genes.

Similarly, we first randomly divided the data into two parts: 75 samples for training, and 50 samples for testing. We ranked the entire 3363 genes according to their TSs with the 75 training samples. Then we picked out the 100 genes with the highest TSs. We subsequently input the selected 100 genes one by one into the network according to their TSs, in decreasing order. Figure 15.3 shows the training and the testing result. Form these results, we found that our FNN also performed very well: both the training error and the testing error decreased to 0 when only the top three genes were input into the FNN.

according to their TSs with the 63 training samples. Then, we picked out the 96 genes with the highest

15.5 Additional Fuzzy and Rule-Based Approaches

Schaefer and *Nakashima* [15.29] presented a fuzzy rule-based classifier consisting of a set of fuzzy if–then rules that enable accurate nonlinear classification of input patterns. Rule i is

$$\text{rule } i: \quad \text{if } x_1 \text{ is } A^i_{x1}, \cdots, \text{ and } x_n \text{ is } A^i_{xn},$$
$$\text{then class } C_i \text{ with } CF_i, i = 1, 2, ..., N_r,$$

where C_i is the consequent class (i.e., one of the C given classes), and CF_i is the grade of certainty of the fuzzy if–then rule i. N_r is the total number of rules. Rules are first generated from training patterns. A new pattern is classified by the rule that has the maximum output. The number of rules created in such a way is high. *Schaefer* and *Nakashima* [15.29] then presented a hybrid fuzzy classification scheme in which a small number of fuzzy if–then rules are selected through a genetic algorithm (GA), leading to a compact classifier for gene expression analysis. The fitness of a rule is the number of training patterns correctly classified by this

rule. Extensive experimental results on various well-known gene expression datasets confirm the efficacy of these approaches. The datasets are the colon cancer dataset [15.30], leukemia dataset [15.6], and lymphoma dataset [15.24]. In particular, for the lymphoma dataset, 100% classification accuracy was achieved using 50 and 100 genes.

Maji and *Pal* [15.31] used the concept of the fuzzy equivalence partition matrix based on the theory of fuzzy-rough sets to approximate the true marginal and joint distributions of continuous gene expression values, where each row of the matrix can automatically be derived from the given expression values. The performance of the discussed approach was compared with that of existing approaches using the class separability index and the support vector machine. It was shown to be effective for selecting relevant and nonredundant continuous-valued genes from microarray data. The datasets used in this paper are for breast cancer [15.32], colon cancer [15.30], leukemia [15.6], RAOA (rheumatoid arthritis, osteoarthritis) [15.33], and RAHC (rheumatoid arthritis, healthy control) [15.34]. Later, *Maji* [15.35] proposed fuzzy-rough supervised attribute clustering (FRSAC) to find relevant genes. A new quantitative measure was introduced that incorporates the information of sample categories to measure the similarity among genes, whereby redundancy among the genes was removed. The clusters were refined incrementally based on sample categories. The naive Bayes classifier, the *k*-nearest neighbor rule, and the support vector machine were used as classifiers.

Tang et al. [15.36] proposed a fuzzy granular support vector machine recursive feature elimination algo-rithm (FGSVM-RFE) to select multiple highly informa-tive gene subsets for cancer classification and diagnosis. As a hybrid algorithm of statistical learning, fuzzy clus-tering, and granular computing, the FGSVM-RFE elim-inates irrelevant, redundant, or noisy genes in different granules at different stages and selects highly informa-tive genes. The colon cancer dataset [15.30], leukemia dataset [15.6], and prostate cancer dataset [15.37] were used to demonstrate the algorithm. In particular, the best result for the testing accuracy for the prostate cancer dataset was 100% with only 8 genes, which is signifi-cantly better than the previous best result of 86% with 16 genes. The identified genes were annotated by Onto-Express to be biologically relevant.

15.6 Fuzzy and Rule-Based Approaches to Clustering and Prediction in Microarray Data

Du et al. [15.38] used fuzzy logic to effectively model gene regulation and interaction, which accurately reflect the underlying biology. Specifically, they applied a new multiscale fuzzy clustering method that allows genes to interact between regulatory pathways and across differ-ent conditions at different levels of detail. Fuzzy cluster centers showed causal relationships between groups of coregulated genes. Du et al. demonstrated their ap-proach using gene expression data from an experiment on carbohydrate metabolism in the model plant *Ara-bidopsis thaliana* [15.39, 40]. Various gene regulatory relationships were evaluated using the gene ontology database.

Maraziotis et al. [15.41] studied reverse-engineering problems concerning the reconstruction and identi-fication of gene regulatory networks through gene expression data and proposed an approach for inferring the complex causal relationships among genes from mi-croarray data based on a neural fuzzy recurrent network. The method derived information on the gene interac-tions in fuzzy rules and took into account the dynamical aspects of gene regulation. Microarray data from two experiments on *Saccharomyces cerevisiae* [15.42] and *Escherichia coli* [15.43–45] were used to test gene expression time course prediction. Many biologically valid relationships among genes were discovered in the study. For a recent survey on generating genetic networks using fuzzy and other related approaches, see [15.46].

Hu et al. [15.47] proposed a method to model a biomolecular network. A biomolecular network was first clustered into various subnetworks. Computa-tional models were then generated for the subnetworks and simulated to predict their behavior in the cellu-lar context. Modeling approaches included state-space modeling, probabilistic Boolean network modeling, and fuzzy logic modeling. The modeling and simulation results were tested against biological datasets (mi-croarrays and/or genetic screens) under normal and perturbation conditions.

Shaik and *Yeasin* [15.48] introduced fuzzy member-ship to transform the hard adaptive subspace iteration (ASI) algorithm [15.49] into a fuzzy-ASI algorithm to perform two-way clustering and then applied this fuzzy adaptive subspace iteration-based two-way clus-tering (FASIC) approach to microarray data to find differentially expressed genes (DEGs). The approach assigned a relevance value to genes associated with each cluster, and each gene cluster was ranked based on its potential to provide a correct classification of the sample classes. These ranks were converted into *P* values using the *R*-test, and the significance of each gene was determined. A fivefold validation was performed on the genes selected. The gastric cancer dataset [15.50], leukemia dataset [15.6], and colon cancer dataset [15.30] were used to demonstrate the approach and visualize clusters.

Sjahputera et al. [15.51] studied methylation-expression relational data generated in the simultaneous analysis of promoter methylation using differential methylation hybridization (DMH) and its associated gene expression using expressed CpG island sequence

Part C | 15.6

tag (ECIST) microarrays. They proposed various algorithms based on fuzzy set theory, especially possibilistic c-means (PCM) and cluster fuzzy density. For each gene, these algorithms calculate measures of confidence of various methylation-expression relationships in each non-Hodgkin's lymphoma (NHL) subclass. Thus, these tools can be used as a means of high-volume data exploration to better guide biological confirmation using independent molecular biology methods.

Ma and *Chan* [15.52] proposed an incremental fuzzy mining (IFM) technique to effectively capture heterogeneity in expression data for pattern discov-

ery. The method uses a fuzzy measure to determine if interesting association patterns exist between the linguistic gene expression levels. Each gene can be allowed to belong to more than one functional class with different degrees of membership. IFM can be used either as a classification technique or together with existing clustering algorithms to improve the cluster groupings discovered for greater prediction accuracies. The approach was tested with real expression datasets for both classification and clustering tasks. Simulation results showed that IFM can effectively discover hidden patterns for accurate gene function predictions.

15.7 Summary and Discussion

Cancer diagnosis from huge microarray gene expression data is both important and challenging. In this chapter, we showed how to use a fuzzy neural network (FNN) that we proposed earlier for cancer classification, an important and challenging bioinformatics problem. This FNN contains three valuable aspects, i.e., automatic generation of fuzzy membership functions, parameter optimization, and rule-base simplification. One major obstacle in microarray dataset classification is that the number of features (genes) is much larger than the number of objects (samples). We used a feature selection method based on the t-test to select more significant genes before applying the FNN. To demonstrate our approach, we used three well-known microarray databases, i.e., the lymphoma dataset, the small round blue cell tumor (SRBCT) dataset, and the ovarian cancer dataset. In all cases we obtained 100% accuracy with fewer genes in comparison with previously published results.

Our results showed that the microarray data classification problem can be solved with a much smaller number of genes. Compared with the method of nearest shrunken centroids using 48 genes, our FNN leads to 100% accuracy using only 9 genes for the lymphoma dataset. For the SRBCT data, the best known result is from the evolutionary algorithm reported by *Deutsch* [15.54]. He used 12 genes to obtain 100% accuracy. However, our FNN requires only eight genes to obtain the same accuracy. A comparison of the number of genes required for the methods, all with 100% accuracy, is given in Table 15.2. For the ovarian cancer data, *Schaner* et al. [15.28] used at least 61 genes to classify the ovarian cancer and the breast cancer with

the nearest shrunken centroids methods. In comparison with *Schaner* et al.'s work [15.28], our FNN also greatly reduced the number of genes required to obtain an accurate result (to four). In view of the smaller number of genes required by our FNN and its high accuracy, we conclude that our FNN classifier not only helps biological researchers differentiate cancers that are difficult to classify using traditional clinical methods, but can also help researchers focus on a small number of important genes to find the relationship between those important genes and the development of cancers.

Other researchers have also demonstrated significant gene reduction in other gene expression datasets using other approaches. Despite the great progress already made, the following issues need to be addressed in future studies. The semi-exhaustive search, divide-and-conquer [15.10], and semi-unsupervised gene selection via spectral biclustering [15.55] approaches that we proposed earlier should be tested with other datasets, e.g., for breast cancer [15.32], colon cancer [15.30],

Table 15.2 Comparisons of results, all with 100% accuracy, for the SRBCT data

Method	Number of genes required
MLP (multi-layer perceptron) neural network [15.8]	96
Nearest shrunken centroids [15.11]	43
Support vector machine (SVM) [15.53]	20
Evolutionary algorithm [15.54]	12
L. Wang, F. Chu, W. Xie FNN	8

leukemia [15.6], RAOA (arthritis) [15.33], prostate cancer [15.37], gastric cancer [15.50], and RAHC [15.34]. Other feature selection algorithms have been proposed and should be tested in gene selection, for example, class-dependent and GA-based feature selection [15.56–58].

References

15.1 L.A. Zadeh: Fuzzy sets, Inf. Control **8**(3), 338–353 (1965)

15.2 V. Kecman: *Learning and Soft Computing, Support Vector machines, Neural Networks and Fuzzy Logic Models* (MIT, Cambridge 2001)

15.3 L. Wang, X. Fu: *Data Mining with Computational Intelligence* (Springer, Berlin, Heidelberg 2005)

15.4 S. Mitra, Y. Hayashi: Bioinformatics with soft computing, IEEE Trans. Syst. Man Cybern. C **36**, 616–635 (2006)

15.5 M. Schena, D. Shalon, R.W. Davis, P.O. Brown: Quantitative monitoring of gene expression patterns with a complementary DNA microarray, Science **2**, 467–470 (1995)

15.6 T.R. Golub, D.K. Slonim, P. Tamayo, C. Huard, M. Gaasenbeek, J.P. Mesirov, H. Coller, M.L. Loh, J.R. Downing, M.A. Caligiuri, C.D. Bloomfield, E.S. Lander: Molecular classification of cancer: Class discovery and class prediction by gene expression monitoring, Science **286**, 531–537 (1999)

15.7 S. Dudoit, J. Fridlyand, T.P. Speed: Comparison of discrimination methods for the classification of tumors using gene expression data, J. Am. Stat. Assoc. **97**, 77–87 (2002)

15.8 J.M. Khan, J.S. Wei, M. Ringner, L.H. Saal, M. Ladanyi, F. Westermann, F. Berthold, M. Schwab, C.R. Antonescu, C. Peterson, P.S. Meltzer: Classification and diagnostic prediction of cancers using gene expression profiling and artificial neural networks, Nat. Med. **7**, 673–679 (2001)

15.9 M.P. Brown, W.N. Grundy, D. Lin, N. Cristianini, C.W. Sugnet, T.S. Furey, M. Ares Jr., D. Haussler: Knowledge-based analysis of microarray gene expression data by using support vector machines, Proc. Natl. Acad. Sci. USA **97**, 262–267 (2000)

15.10 L. Wang, F. Chu, W. Xie: Accurate cancer classification using expressions of very few genes, IEEE-ACM Trans. Comput. Biol. Bioinform. **4**(1), 40–53 (2007)

15.11 R. Tibshirani, T. Hastie, B. Narashiman, G. Chu: Diagnosis of multiple cancer types by shrunken centroids of gene expression, Proc. Natl. Acad. Sci. USA **99**, 6567–6572 (2002)

15.12 Y. Frayman, L. Wang: Data mining using dynamically constructed recurrent fuzzy neural networks, Research and Devevelopment in Knowledge Discovery and Data Mining, Vol. 1394 (Springer, Berlin, Heidelberg 1998) pp. 122–131

15.13 Y. Frayman, L. Wang: A Dynamically-constructed fuzzy neural controller for direct model reference adaptive control of multi-input-multi-output nonlinear processes, Soft Comput. **6**, 244–253 (2002)

15.14 F. Chu, W. Xie, L. Wang: Gene selection and cancer classification using a fuzzy neural network, Proc. North-Am. Fuzzy Inf. Process. Conf. (NAFIPS 2004), Vol. 2 (2004) pp. 555–559

15.15 W. Xie, F. Chu, L. Wang, E.T. Lim: A fuzzy neural network for intelligent data processing, SPIE Proc. **5812**, 283–290 (2005)

15.16 Y. Frayman, L.P. Wang, C. Wan: Cold rolling mill thickness control using the cascade-correlation neural network, Control Cybern. **31**(2), 327–342 (2002)

15.17 L. Wang, Y. Frayman: A Dynamically generated fuzzy neural network and its application to torsional vibration control of tandem cold rolling mill spindles, Eng. Appl. Artif. Intell. **15**(6), 541–550 (2002)

15.18 C.M. Higgins, R.M. Goodman: Fuzzy rule-based networks for control, IEEE Trans. Fuzzy Syst. **2**, 82–88 (1994)

15.19 M. Sugeno, G.T. Kang: Structure identification of fuzzy model, Fuzzy Sets Syst. **28**, 15–33 (1988)

15.20 L. Wang (Ed.): *Support Vector Machines: Theory and Applications* (Springer, Berlin, Heidelberg 2005)

15.21 J. Devore, R. Peck: *Statistics: The Exploration and Analysis of Data*, 3rd edn. (Duxbury, Pacific Grove 1997)

15.22 V.G. Tusher, R. Tibshirani, G. Chu: Significance analysis of microarrays applied to the ionizing radiation response, Proc. Natl. Acad. Sci. USA **98**, 5116–5121 (2001)

15.23 R. Tibshirani, T. Hastie, B. Narasimhan, G. Chu: Class prediction by nearest shrunken centroids with applications to DNA microarrays, Stat. Sci. **18**, 104–117 (2003)

15.24 A.A. Alizadeh, M.B. Eisen, R.E. Davis, C. Ma, I.S. Lossos, A. Rosenwald, J.C. Boldrick, H. Sabet, T. Tran, X. Yu, J.I. Powell, L. Yang, G.E. Marti, T. Moore, J. Hudson Jr., L. Lu, D.B. Lewis, R.T. Tibshirani, G. Sherlock, W.C. Chan, T.C. Greiner, D.D. Weisenburger, J.O. Armitage, R. Warnke, R. Levy, W. Wilson, M.R. Grever, J.C. Byrd, D. Botstein, P.O. Brown, L.M. Staudt: Distinct types of diffuse large B-cell lymphoma identified by gene expression profiling, Nature **403**, 503–511 (2000)

15.25 Web supplement to reference 15.24: http://llmpp.nih.gov/lymphoma

15.26 O. Troyanskaya, M. Cantor, G. Sherlock, P. Brown, T. Hastie, R. Tibshirani, D. Botstein, R.B. Altman:

Missing value estimation methods for DNA microarrays, Bioinformatics **17**, 520–525 (2001)

15.27   Web supplement to reference 15.8: http://research. nhgri.nih.gov/microarray/Supplement/

15.28   M.E. Schaner, D.T. Ross, G. Ciaravino, T. Sorlie, O. Troyanskaya, M. Diehn, Y.C. Wang, G.E. Duran, T.L. Sikic, S. Caldeira: Gene expression patterns in ovarian carcinomas, Mol. Biol. Cell **14**, 4376–4386 (2003), supplementary material available at   http://genome-www.stanford.edu/ovarian_cancer/

15.29   G. Schaefer, T. Nakashima: Data mining of gene expression data by fuzzy and hybrid fuzzy methods, IEEE Trans. Inf. Biomed. **14**, 23–29 (2010)

15.30   U. Alon, N. Barkai, D. Notterman, K. Gish, S. Ybarra, D. Mack, A. Levine: Broad patterns of gene expression revealed by clustering analysis of tumor and normal colon tissues probed by oligonucleotide arrays, Proc. Natl. Acad. Sci. USA **96**, 6745–6750 (1999)

15.31   P. Maji, S.K. Pal: Fuzzy-rough sets for information measures and selection of relevant genes from microarray data, IEEE Trans. Syst. Cybern. B **40**, 741–752 (2010)

15.32   M. West, C. Blanchette, H. Dressman, E. Huang, S. Ishida, R. Spang, H. Zuzan, J.A. Olson, J.R. Marks, J.R. Nevins: Predicting the clinical status of human breast cancer by using gene expression profiles, Proc. Natl. Acad. Sci. USA **2**(20), 11462–11467 (2001)

15.33   T.C.T.M. van der Pouw Kraan, F.A. van Gaalen, P.V. Kasperkovitz, N.L. Verbeet, T.J.M. Smeets, M.C. Kraan, M. Fero, P.-P. Tak, T.W.J. Huizinga, E. Pieterman, F.C. Breedveld, A.A. Alizadeh, C.L. Verweij: Rheumatoid arthritis is a heterogeneous disease: Evidence for differences in the activation of the STAT-1 pathway between rheumatoid tissues, Arthritis Rheum. **48**(8), 2132–2145 (2003)

15.34   T.C.T.M. van der Pouw Kraan, C.A. Wijbrandts, L.G.M. van Baarsen, A.E. Voskuyl, F. Rustenburg, J.M. Baggen, S.M. Ibrahim, M. Fero, B.A.C. Dijkmans, P.P. Tak, C.L. Verweij: Rheumatoid arthritis subtypes identified by genomic profiling of peripheral blood cells: Assignment of a type I interferon signature in a subpopulation of patients, Ann. Rheum. Dis. **66**(8), 1008–1014 (2007)

15.35   P. Maji: Fuzzy-rough supervised attribute clustering algorithm and classification of microarray data, IEEE Trans. Syst. Cybern. B **41**, 222–233 (2011)

15.36   Y. Tang, Y.-Q. Zhang, Z. Huang, X. Hu, Y. Zhao: Recursive fuzzy granulation for gene subsets extraction and cancer classification, IEEE Trans. Inf. Technol. Biomed. **12**, 723–730 (2008)

15.37   D. Singh, P.G. Febbo, K. Ross, D.G. Jackson, J. Manola, C. Ladd, P. Tamayo, A.A. Renshaw, A.V. D'Amico, J.P. Richie, E.S. Lander, M. Loda, P.W. Kantoff, T.R. Golub, W.R. Sellers: Gene expression correlates of clinical prostate cancer behavior, Cancer Cell **1**(2), 203–209 (2002)

15.38   P. Du, J. Gong, E.S. Wurtele, J.A. Dickerson: Modeling gene expression networks using fuzzy logic, IEEE Trans. Syst. Cybern. B **35**, 1351–1359 (2005)

15.39   L. Mendoza, E.R. Alvarez-Buylla: Dynamics of the genetic regulatory network for Arabidopsis thaliana flower morphogenesis, J. Theor. Biol. **193**, 307–319 (1998)

15.40   L. Mendoza, D. Thieffry, E.R. Alvarez-Buylla: Genetic control of flower morphogenesis in Arabidopsis thaliana: A logical analysis, Bioinformatics **15**, 593–606 (1999)

15.41   I.A. Maraziotis, A. Dragomir, A. Bezerianos: Gene networks reconstruction and time-series prediction from microarray data using recurrent neural fuzzy networks, IET Syst. Biol. **1**, 41–50 (2007)

15.42   P.T. Spellman, G. Sherlock, M.Q. Zhang, V.R. Iver, K. Anders, M.B. Eisen, P.O. Brown, D. Botstein, B. Futcher: Comprehensive identification of cell cycle-regulated genes of the yeast Saccharomyces cerevisiae by microarray hybridization, Mol. Biol. Cell **9**, 3273–3297 (1998)

15.43   M. Ronen, R. Rosenberg, B.I. Shraiman, U. Allon: Assigning number to the arrows: parameterizing a gene regulation network by using accurate expression kinetics, Proc. Natl. Acad. Sci. **99**, 10555–10560 (2002)

15.44   S.S. Shen-Orr, R. Milo, S. Mangan, U. Alon: Network motif in the transcriptional regulation network of Escherichia coli, Nat. Genet. **31**, 64–68 (2002)

15.45   F.C. Neidhardt, M.A. Savageau: Regulation beyond the operon. In: *Escherichia coli and Salmonella: Cellular and Molecular Biology*, 2nd edn., ed. by F.C. Neidhardt (Am. Soc. of Microbiology, Washington 1996) pp. 1310–1324

15.46   S. Mitra, R. Das, Y. Hayashi: Genetic Networks and Soft Computing, IEEE-ACM Trans. Comput. Biol. Bioinform. **8**, 616–635 (2011)

15.47   X. Hu, M. Ng, F.-X. Wu, B.A. Sokhansanj: Mining, modeling, and evaluation of subnetworks: From large biomolecular networks and its comparison study, IEEE Trans. Inf. Technol. Biomed. **13**, 184–194 (2009)

15.48   J. Shaik, M. Yeasin: Fuzzy-adaptive-subspace-iteration-based two-way clustering of microarray data, IEEE-ACM Trans. Comput. Biol. Bioinform. **6**, 244–259 (2009)

15.49   T. Li, S. Ma, M. Ogihara: Document clustering via adaptive subspace iteration, Proc. ACM SIGIR'04 (2004) pp. 218–225

15.50   X. Chen, S.Y. Leung, S.T. Yeuen, K.M. Chu, J. Ji, R. Li, A.S.Y. Chan, S. Law, O.G. Troyanskaya, J. Wong, S. So, D. Botstein, P.O. Brown: Variation in gene expression patterns in human gastric cancers, Mol. Biol. Cell **14**, 3208–3215 (2003)

15.51   O. Sjahputera, J.M. Keller, J.W. Davis, K.H. Taylor, F. Rahmatpanah, H. Shi, D.T. Anderson,

S.N. Blisard, R.H. Luke, M. Popescu, G.C. Arthur, C.W. Caldwell: Relational analysis of CpG islands methylation and gene expression in human lymphomas using possibilistic c-means clustering and modified cluster fuzzy density, IEEE-ACM Trans. Comput. Biol. Bioinform. **2**, 176–189 (2007)

15.52 P.C.H. Ma, K.C.C. Chan: Incremental fuzzy mining of gene expression data for gene function prediction, IEEE Trans. Biomed. Eng. **58**, 1246–1252 (2011)

15.53 Y. Lee, C.K. Lee: Classification of multiple cancer types by mulitcategory support vector machines using gene expression data, Bioinformatics **19**, 1132–1139 (2003)

15.54 J.M. Deutsch: Evolutionary algorithms for finding optimal gene sets in microarray prediction, Bioinformatics **19**, 45–52 (2003)

15.55 B. Liu, C. Wan, L. Wang: An efficient semi-unsupervised gene selection method via spectral biclustering, IEEE Trans. Nano-Biosci. **5**(2), 110–114 (2006)

15.56 X.J. Fu, L.P. Wang: A GA-based novel RBF classifier with class-dependent features, Proc. 2002 IEEE Congr. Evol. Comput. (CEC 2002), Vol. 2 (2002) pp. 1890–1894

15.57 X.J. Fu, L.P. Wang: Rule extraction from an RBF classifier based on class-dependent features, Proc. 2002 IEEE Congr. Evol. Comput. (CEC 2002), Vol. 2 (2002) pp. 1916–1921

15.58 L. Wang, N. Zhou, F. Chu: A general wrapper approach to selection of class-dependent features, IEEE Trans. Neural Netw. **19**(7), 1267–1278 (2008)

16. Phylogenetic Cladograms: Tools for Analyzing Biomedical Data

Mones S. Abu-Asab, Jim DeLeo

This chapter provides an introduction to phylogenetic cladograms – a systems biology evolutionary-based computational methodology that emphasizes the importance of considering multilevel heterogeneity in living systems when mining data related to these systems. We start by defining intelligence as the ability to predict, because prediction is a very important objective in mining data, especially biomedical data (Sect. 16.1). We then give a brief review of artificial intelligence (AI) and computational intelligence (CI) (Sects. 16.2, 16.3), provide a conciliatory overview of CI, and suggest that phylogenetic cladograms which provide hypotheses about speciation and inheritance relationships should be considered to be a CI methodology. We then discuss heterogeneity in biomedical data and talk about data types, how statistical methods blur heterogeneity, and the different results obtained between more traditional CI methodologies (phenetic) and phylogenetic techniques. Finally, we give an example of constructing and interpreting a phylogenetic cladogram tree.

Part C | 16

Those of us who have worked for many years in biomedical research environments cannot escape the recurring issues that bedevil the field. The problems that are not divulged in seminars and symposia come out during personal conversations. For the last decade, with the adoption of the new technology of microarrays and the new generation of mass spectrometers, biomedical scientists have found themselves sitting on mounds of data worth tens of millions of dollars but with inadequate tools with which to make sense of these data.

What they thought was going to help them decipher the mysteries of disease has turned into a bioinformatic nightmare.

Recent rapid advances in machine technology have led to the massive accumulation of datasets in gen*omics*, metabol*omics*, and prote*omics*, areas collectively referred to as *omics*. However, these significant advances in the technical aspects of science have not been matched by sufficient advances in applied analytical and bioinformatic methodologies. As a result, data interpretation, data mining, and knowledge discovery from these data have not kept pace with the data pileup [16.1, 2]. Consequently, we now face the challenge of discovering ways to properly analyze the massive and extremely heterogeneous datasets associated with biological specimens, and to assemble these rapidly growing datasets via integrative modeling of systems biology – an integral synthesis of all the available data [16.1, 3].

Moreover, computer scientists and bioinformaticians are struggling to develop the most biologically compatible methods for analyzing, mining, and modeling an astounding amount of variability in the new omics data types from gene-expression microarrays, mass spectrometry of proteomics and metabolomics, as well as genome-wide association studies (GWAS), copy number variation (CNV), and single-nucleotide polymorphism (SNP) surveys. Because these techniques allow the quantitation of total amounts of RNA, microRNA (miRNA), proteins, and metabolites within cells and tissues, as well as genome associations between specimens and populations to be analyzed at the same time, we are now confronted with an avalanche of heterogeneous data that awaits meaningful analysis [16.4, 5]. Attempting to figure out the biological significance of thousands of gene-expression values with incongruent distribution among a number of specimens is a formidable task for biologists, bioinformaticians, and computer scientists.

This problem is amplified by two other factors that are contributing to the generation of *toxic assets* in biomedical research. These are the researchers' underestimation, or sometimes lack of awareness, of population heterogeneities, and the application of unsuitable analytical paradigms when dealing with such heterogeneous data [16.6]. These two factors are hampering progress on some of the major issues in current biomedical research such as disease definition and its molecular boundaries, class discovery (i. e., subtyping

of disease), early detection, omics biomarker discovery, profiling (from genetic, metabolic, and proteomic data), sorting out clonal from nonexpanded mutations, genetic versus epigenetic driving events, posttreatment assessment, and figuring out the primary origin of some cancers, e.g., Ewing, synovial, and chondromyxoid sarcomas [16.6, 7]. Progress on these issues will greatly benefit from effective data mining that accounts for the heterogeneous nature of biological data.

Reliance on averaging, statistical, and phenetic approaches (such as clustering) rarely produces results that are fully biologically relevant [16.4, 8]. Heterogeneity is a product of the individual's interaction with the environment and is sustained by selective evolutionary pressures. Heterogeneity exists at various levels, namely (1) within the individual, (2) between individuals within a population, and (3) between populations.

We envision a way out of this logjam by finding or designing tools that are more suitable for dealing with heterogeneity. As we demonstrate in this chapter, the phylogenetic paradigm of data analysis and modeling is in our opinion an important step for making sense out of heterogeneity. This is especially true for high-throughput data with tens of thousands of data variables. Furthermore, as has been demonstrated, it offers a multidimensional analysis that provides predictability that is superior to other methods presently being used [16.9, 10]. As we detail the theoretical basis and practical steps of our premise, it is important for the reader to keep in mind that the area of biomedical phylogenetics is young and there is ample room for improvement and expanding on currently existing ideas. Phylogenetics is a way of bringing intelligent insight into analysis of high-throughput data related to phenomena in which it is clear that microevolutionary forces are at work. We think that bringing phylogenetics under the umbrella of computational intelligence (CI) may in fact enhance both phylogenetics and CI. Before we discuss that, let us first consider what we mean by *intelligence*.

16.1 On Intelligence

In computer science, CI emerged as a follow-up to classical artificial intelligence (AI). AI and CI are types of machine intelligence. AI attempts to emulate human knowledge by having knowledge engineers interview human subject-matter experts to capture the implicit or tacit experiential knowledge of these experts and emulate it in a computerized knowledge base in the form of

rules, semantic networks, and frames. AI software engineers would also develop computer programs called inference engines that would apply the knowledge in the knowledge base to specific individual cases, such as in diagnosing an individual patient. At some point in the history of computer science it was conceived that, since humans derive knowledge by processing lots of data, why not bypass the humans and let the computer extract the knowledge directly from the data. This thinking gave rise to what is generally known as machine intelligence; later, the term *computational intelligence* arose.

Today's inductees in the IEEE Intelligent Systems Hall of Fame include John McCarthy, Marvin Minsky, Noam Chomsky, Nils Nilsson, Raj Reddy, Doug Engelbart, Ed Feigenbaum, Tim Berners-Lee, Judea Pearl, and Lotfi Zadeh. Before we explore AI and CI further, it is of interest to note that, until the recent induction of Professor Zadeh (*the father of fuzzy logic*), there was almost total commitment in AI and CI to classical Aristotelian bivalent logic with no room for other logical systems. Fuzzy logic can be understood to be a superset of conventional Boolean logic. It gives us a precise tool for coping with imprecision, particularly the imprecision that arises in natural language.

Even before considering CI and AI machine intelligence further, it is of interest to note that it is ironic and humorous that the very word *intelligence* has been defined in many different and vague ways, including the ability for abstract thought, understanding, communication, reasoning, learning, planning, emotional knowing, and problem solving. *Intelligence* (whatever it is) has been studied in humans, animals, and plants; so let us then refer to this kind of intelligence as *biological-based intelligence*. It is also ironic and humorous that the computer has been anthropomorphized by attributing intelligence to it, given that there is no consensus about what intelligence means. The pioneers of AI (e.g., Marvin Minsky) were well aware of this, and they obviously had a sense of humor when they defined AI in terms of machines exhibiting any behavior which, if performed by humans, would be considered to be intelligent [16.11]. To get a little more precise about the notion of machine intelligence, *Alan Turing* introduced the Turing test in his 1950 paper [16.12]. His test is a test of a machine's ability to exhibit intelligent behavior. In Turing's original version of the test, a human judge engages in a natural-language conversation via a computer keyboard and screen with a human and a machine designed to generate performance indistinguishable from that of a human being in a particular circumscribed domain of knowledge. In our modern-day rendering of the Turing test (Fig. 16.1), the young lady in the center is the *human judge*, and she is online with a human man on her left and a young robot couple on her right. All three workstations are physically and sensorily separated from one another. If the human judge (the young lady) cannot reliably tell the machine

(the robots) from the human (the dude), the machine is said to have passed the Turing test, and therefore is considered to be *intelligent*. The test is not concerned with the correctness of the answers but rather how closely the answers resemble typical human answers. John von Neuman, another great contributor to computer science, claimed computers could think. When challenged on this, von Neuman told the challenger to write down specifically what he thought a computer could not do and he (von Neuman) would build him a computer that would do just that. More recently, *Searle* has offered us the Chinese room thought experiment to tease us some more about what we mean by this fuzzy and vague term *intelligence* [16.13]. If you write an English sentence on a piece of paper and slip it under the door of a room and a few minutes later your sentence comes back under the door perfectly translated into a Chinese language and you did this repeatedly, would you say that the room is exhibiting intelligence? Most likely you would. Well, suppose you later learned that there was a human in the room who knew nothing about Chinese languages and he simply took the paper and looked into translation books to get the translations, would you now consider the room to be intelligent? Perhaps there is no right or wrong answer. Perhaps the point of the Chinese room experiment is to again point out that we do not know what we mean by *intelligence*, or, more forgivingly, that it is a matter of semantics.

The AI approach to have machines emulate biological-based intelligence is a top-down approach in which knowledge base and inference engine paradigms are built, whereas the CI approach has been to build problem-specific numerically based computational methodologies – a bottom-up approach. This will be discussed further in Sect. 16.2.

So what is intelligence after all? For our present purposes we are content with *Hawkins* and *Blakeslee*'s prac-

Fig. 16.1 A modern-day illustration of the Turing test (artist Ann Aiken)

tical definition of intelligence, which is the capability to make predictions about the future [16.14]. We want to explore how the convergence of CI and biologically based phylogenetics can be brought to the problem of reliably predicting things in medicine: things such as diagno-sis, patient treatment outcome, efficacy of treatment, and suitable candidates for specific clinical research proto-cols and medical regimens. We believe that, once we get some deep experience in our work, our learning can be leveraged to other subject matter domains.

16.2 Computational Intelligence and Cladograms

CI is a branch of computer science that addresses problems for which no effective solutions have been found, either because it is difficult to formulate them or because they are NP-hard (nondeterminis-tic polynomial-time hard), i. e., they require enormous computational resources. CI is a new aspect of com-puter science. It is a diverse field that encompasses several methodologies including neural, fuzzy, evolu-tionary, probabilistic, artificial life, multiagent systems, and swarm intelligence computing. CI is also a style of computing in which computer solutions are designed to meet the specific needs called for by a specific problem. This bottom-up style is in contrast to devel-oping broad top-down robust schemas and solutions such as is done in classical artificial intelligence (AI). In AI the focus is on developing and using robust canonical schemas for building knowledge bases and inference engines that mimic human- and animal-style intelligence. CI also focuses on problems that require human- and animal-like intelligence; however, CI sys-tems deal directly with raw numerical and categorical data and do not use biology-based knowledge the way AI does.

CI is becoming well established. It is concerned with building intelligent computational agents that rec-ognize patterns. New theories with sound biological and cognitive metaphors have been evolving. *Engelbrecht* defines CI as *the study of adaptive mechanisms to en-able or facilitate intelligent behavior in complex and changing environments* [16.15]. Current work in CI is directed towards developing improved knowledge rep-resentations, retrieval models, and pattern recognition algorithms appropriate for specific purposes. CI is what AI guru *Minsky* calls a suitcase (umbrella) term [16.11]. We believe that phylogenetic thinking and cladogram algorithms belong in the CI suitcase. Modern phylo-genetics has evolved in biology and not in computer science and informatics. It has been recently addressing primarily problems associated with omics data, e.g., ge-nomics, proteomics, and metabolomics data, which are mostly derived from microarray, mass spectrometry, and sequencing technologies. As phylogenetics methods are brought into the CI suitcase, both CI and phylogenet-ics will mutually benefit – CI can enhance phylogenetics methods, and phylogenetics methods can be extended to broader applications beyond omics data applications.

16.3 Computational Intelligence: A Conciliatory View

To begin to understand how phylogenetics and clado-grams are associated with CI, it may be helpful to establish a bird's eye, conciliatory view of computa-tional intelligence. Figure 16.2 provides such a view. As this figure suggests, the main topics of concern in CI are feature extraction, data preparation, regres-sion, clustering, and classification. Feature extraction is concerned with methods for drawing out appropriate features (variables) to be used in regression, classifica-tion, and clustering problems [16.16]. Data preparation is usually a huge effort in CI applications. It includes such tasks as gathering and normalizing data; dealing with missing, messy, and noisy data; and determining the most appropriate methods for scaling data. Regres-sion has to do with using a set of feature values to predict some value (e.g., a missing cholesterol value for an individual patient). In both clustering and clas-sification, the idea is to group entities (e.g., patients, specimens, diseases, analyte values) in a way that is meaningful. Clustering methods assume that data are unlabeled, meaning there are no assigned gold-standard confirmed classes associated with the data, whereas classification methods work with labeled data, mean-ing reliable class values are present as features in the data [16.17, 18].

In both clustering and classification applications, the work starts with extracting an appropriate set of feature-variable values suitable for solving the prob-lem at hand. These feature variables can be any mix of continuous, discrete, and nominal variables. In a medical example, continuous feature variables might be age, weight, blood and serum values, and intelligence test scores. Discrete variables are those with a finite set of values (e.g., good–better–best, low–medium–high). Discrete variables also include bivalent variables that have only two values, e.g., true–false, yes–no, male–female, gene under- or overexpressed. Nominal variables could be ethnicity, eye color, and disease subtype. A conciliatory view of variables is that they are all subsumed under nominal variables. This is obvious for discrete variables. The reason that (measured) continuous variables can be seen as nominal variables is that all measured continuous data have particular granularity, limited in accuracy by the machine or process producing them, and are therefore discrete.

Feature extraction is then followed by the arduous task of data preparation, which was described above.

Clustering problems can be understood to subsume classification problems if we consider that the class feature in a classification problem is just one of the feature values. Conversely, any feature in a clustering problem can be viewed as the classifying feature to be predicted. Classification, as depicted in Fig. 16.2, is seen as being subsumed by regression. Again, regression methods build models to predict missing or unknown values for new cases. The value to be predicted can be a class or a value, which when thresholded provides class assignment. The back error propagation artificial neural network (ANN) is a CI method that can be used for both regression and classification. For classification, receiver operating characteristic (ROC) methodology may be employed to determine appropriate threshold values [16.19]. The criticism that ANNs do not provide

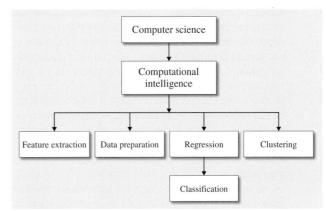

Fig. 16.2 A bird's eye perspective of computational intelligence

explanations for their classifications has been addressed in *Dayhoff* and *DeLeo* [16.20].

Phylogenetics fits within the CI suitcase because it provides multidimensional data visualization, descriptive and clustering tools for suggesting hypotheses, as well as classification capability. Conversely, CI can augment phylogenetics by providing new optional and improved methods for data preparation and parameter selection as well as in other ways yet to be discovered. As we will see, phylogenetics offers trees called cladograms that represent data relationships. CI offers trees in the form of classification and regression trees (CART) and random forests, for example. These CI tree methods use subsets of the features, while phylogenetic parsimony techniques use all the features. Furthermore, phylogenetic trees reveal imputed inheritance patterns, which are generally not a feature of traditional CI tree methods.

16.4 Heterogeneity Within Biological Data

16.4.1 Sources and Levels of Heterogeneity

In short, heterogeneity means the lack of synchronicity and concordance within and between variables. In biology, heterogeneity refers to variation within a variety of structural, anatomical, functional, behavioral, and molecular traits among organisms. Since all biological systems and their related data are characterized by heterogeneity, it seems reasonable to begin data analysis by examining the sources and levels of heterogeneity occurring within specimens to determine an optimal analytical approach. The type of heterogeneity that manifests in a biological system depends on the

source from which the specimen's data are gathered; For example, there are inter- and intrapopulation variations, within-individual variations (tumors from the same individual), or cellular components (DNA from nuclear, mitochondrial, or chloroplast sources); and sometimes according to the disease pathology (e.g., cancer, degenerative illnesses, etc.).

Analysis and interpretation of an interpopulational study where individuals are sampled in order to compare populations differs from a study of cancer where tumors from the same individual are sampled, and the two differ from a study where the mitochondria are sampled. Here is where the level of heterogeneity makes a difference; For example, when comparing the microarray gene expression of a cancer type between individuals, the data encompass three levels of variation: intratumoral (within the tumor), intertumoral within the individual (tumors within the same individual), and intertumoral from different individuals [16.21, 22]. Furthermore, the importance of each level is determined by its rate of evolution (i. e., the speed of change, such as the introduction and establishment of new mutations) and whether the question at hand deals with finding the shared similarities or the unique differences between the specimens.

Evolutionary forces acting on individuals are the source of heterogeneity at the level of individuals and below, such as in genes, chromosomes, genomes, epigenomes, and tissues. While under normal conditions the individual is the ultimate expression of the genetic variation that exists at all levels within their cells, the case is different in cancers, as discussed later. Variations constituting heterogeneity are maintained within the population by natural selection, and it is assumed that such variation confers selective advantage and facilitates the survival of individuals with the right combinations during catastrophic events. However, cross-breeding of individuals from different populations produces interpopulational heterogeneity [16.23]. The case is slightly different in certain pathologies such as cancer tumors, where one tumor has usually more than one clonal lineage [16.24], and therefore the tumor in this case resembles a population with several subpopulations. This notion of intratumoral variation, which was hypothesized in the 1950s, has been detected recently by the new technology of deep sequencing [16.25–28]. The same is true for the hundreds of mitochondria within one cell, where each mitochondrion differs from the rest of the mitochondrial population, a phenomenon termed heteroplasmy [16.28]. The situation is also similar when considering the ontogenies of a number of diseases where the disease phenotype encompasses several genotypes or developmental pathways, i. e., multiple etiologies leading to the same phenotype–disease heterogeneity.

At the population level, heterogeneity can be detected within the population as an intrapopulational variation among individuals, and between populations as interpopulational differences. However, the level of heterogeneity increases and populational boundaries become blurred as a result of random interpopulational mating in areas of high immigrant influx; For example, in the USA, the implication of interracial marriages

and their successive generations will produce individuals that do not fit into the current racial definitions. By 2050, it is anticipated that 50% of the population will share a Hispanic, African American, or Asian ancestry. This would render as invalid the current racial categories used in government population stratification and statistics as well as in biomedical trials [16.29]. The obsolete racial definitions will then have to be replaced by a new profiling method that will take into consideration the new heterogeneous make-up of the population.

Furthermore, our current understanding of biological heterogeneity and its implications on pathogenesis remain rudimentary and are far from complete [16.26, 27, 30, 31]. Unfortunately, the effect of our having only a rudimentary understanding of biological heterogeneity on biomedical research has hampered meaningful progress on many aspects of research and has not provided a good return on our investments. Researchers working on a good number of diseases, especially in cases using high-throughput equipment that produces what is collectively called omics, are struggling to make sense of their data. Moreover, biological heterogeneity is observed in strains of microorganisms – bacteria, fungi, and viruses – and is also blamed for the variable response of patients to drug treatments and their adverse effects.

16.4.2 Data Types

Knowing the level of data variation is not sufficient for selecting the analytical tool to use. Settling on the most suitable method or strategy for the data also requires a decision regarding whether to treat the data as qualitative or quantitative. Conversion of quantitative to qualitative, and vice versa, is easily accomplished; For example, mapping quantitative microarray gene expression and mass spectrometry data into qualitative data by discretization is straightforward, as is converting qualitative immunohistochemistry data into a quantitative representation [16.1]. Table 16.1 presents the various types of the most commonly encountered omics data listed according to their natural (i. e., generated) qualitative or quantitative form. In phylogenetic computational intelligence techniques, conversion to qualitative form is required. It may be worth mentioning here that to date the application of phylogenetic methods has been in work with omics data. However, phylogenetic methods can also be applied to more traditional biomedical quantitative and qualitative data. This is one of the main objectives of our exposing phylogenetic techniques here. Another objective of our writing is to suggest that

Table 16.1 Common qualitative and quantitative types of biological data. A listing of the various biological data types that are currently generated by researchers

Qualitative data
Aneuploidy
Chromosomal translocation, inversion, duplication
Chromosome fragmentation
Copy number variation (CNV) of genes
DNA and RNA sequences
Epigenetic changes
Immunohistochemical labeling
Loss of heterozygosity (LOH)
Morphological discrete characters
Polyploidy
Protein sequences
Restriction sites
Single-nucleotide polymorphisms (SNP)
Quantitative data
Gene frequencies
Mass spectrometry metabolomic data
Mass spectrometry proteomic data
Microarray data
Morphological measurements

phylogenetics methods may be enhanced by incorporating appropriate CI influences, such as conversions from qualitative to quantitative and using ensembling (multiple models) to assess variance.

As has become clearly obvious during the last 10 years to bioinformaticians analyzing omics quantitative data of mass spectrometry and microarrays, as well as the qualitative omics data of single-nucleotide polymorphism (SNP), copy number variation (CNV), and other genome-wide association studies, it is a formidable task to analyze such heterogeneous data in order to identify biological patterns or biomarkers. Some researchers have crashed their careers by publishing results based on custom algorithms that later turned out to be erroneous and irreproducible by others. The field of biomarker discovery for diagnosis is notorious for its many disappointments, and it is littered with discarded biomarkers [16.1, 32–35]. These problems emanate from poor experimental design, eagerness to beat the competition, the lack of widely acceptable methods of analysis suitable for these types of datasets, and above all, severe lack of understanding and appreciation of heterogeneity [16.31, 36].

Similar results should be reproducible by different objective methods for the same dataset; For example, two methods claiming early detection of a cancer type should identify identical biomarkers or profiles, and the same individuals that are susceptible within a study group. Concordance of results between two or more methods is important to producing valid results. If both methods give disparate results, then one of them or both are not valid. Reliable methods of analysis for high-throughput omics data are needed in the biomedical sciences in order to resolve the current issues discussed in the introduction and to bring new and more appropriate computational methodologies to clinical medicine.

One of the characteristics of quantitative mass spectrometry data and gene-expression microarray data is bivalence (two-state data) [16.8, 31, 36]. This pattern ex-

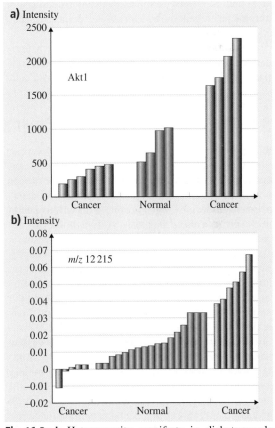

Fig. 16.3a,b Heterogeneity manifests in dichotomously expressed genes and proteins as in the prostate cancer specimens of this figure. (**a**) Quantitation of gene-expression microarrays of Akt1 in 10 cancerous and 4 normal specimens. (**b**) Intensity of mass spectrometry at $m/z = 12\,215$ of 11 cancerous and 17 normal specimens (after [16.9])

ists in a group of cancerous specimens, thus far, where for certain gene expressions or proteins their data are bimodal and fall outside the normal range, i. e., above and below the maximum and minimum normal limits (Fig. 16.3). The dichotomous data form represents a challenge for statistical and computational methods, since these methods do not include such data in their analysis. Statistical data analyses use comparison of sample means to reveal population-level differentially expressed genes of one type, thus discarding other population-level markers such as those dichotomously expressed.

16.5 Statistical Methods and Heterogeneity

In order to estimate the efficacy of clinical tests, drugs, and procedures, statisticians often analyze data for researchers by first eliminating noise, which sometimes translates into reducing heterogeneity. Researchers usually accept this because they are either ambivalent about it and like to tune it out [16.8, 37], or unaware of its extent and its impact on the analysis. Statisticians circumvent heterogeneity by calculating point estimates for groups and not for individuals.

More recently, recognizing the ubiquity of heterogeneity in complex systems and the negative effects of ignoring it, statisticians and researchers have been call-

ing for a two-stage study design. In the first stage the study group is stratified into well-defined yet broad populations using traditional experimental methods. In the second stage subgroups are constructed [16.8, 9, 25, 38]. Although this two-stage approach is easy to achieve in a small simple case with a few variables, it becomes difficult to use when the variables are in the tens, hundreds, or thousands. Thus, the two-stage approach will not work when the data are high-throughput omics data such as those obtained from microarray and mass spectrometry metabolomic and proteomic studies.

16.6 The Phenetic Versus the Phylogenetic Approaches to Analysis

Objective methods of biological data analysis can be classified into the phenetic and phylogenetic approaches. Objectivity here means that similar methods of analysis should produce closely similar or identical results, especially when dealing with serious health issues such as cancer diagnosis. Although the two approaches aim to construct the most accurate hypothesis of relationships among specimens, they differ conceptually and algorithmically in their use of data to perform classification, and thus rarely produce similar results. Except for those that are strictly phylogenetic (see the list of phylogenetic methods below), all other methods are phenetic.

The current phenetic methods applied to omics data include clustering [16.37], pattern recognition (finding known patterns in new data) [16.39], data mining (discovery of previously unknown patterns) [16.40], machine learning algorithms (learning from experience/new data) [16.41], and neural networks (modeling data to find patterns) [16.42].

The general methods of phylogenetics are Bayesian, compatibility, distance, least-squares method, likelihood, neighbor joining, and parsimony. These methods differ by virtue of the type of data they handle (numeric or categorical); however, data can be adapted to run

on different programs. For a summery of phylogenetic methods and algorithms, see *Felsenstein*'s book [16.43].

The phenetic algorithms mentioned above also differ from the phylogenetic ones with regard to their original purpose. While the phylogenetics algorithms are designed to reconstruct evolutionary history of organisms at various taxonomic levels to infer the relationships among the organisms on the basis of known evolutionary mechanisms and principles, the phenetic algorithms were not developed with the evolutionary aspect in mind. This difference between the two approaches usually makes the phylogenetic methods more appropriate for biological data. This is because normal variations and disease processes in most pathologies develop and progress in an evolution-based manner. Therefore, mining biological data for pattern discovery, specimen classification, clinical applications, or any other purpose means that the analytical tool when evolution based will be biologically compatible and give more meaningful and more useful results [16.6].

One main goal for mining biological data is to infer the processes affecting the biological system [16.9]. Thus, in order for an analytical tool to be useful in assessing the dynamics of a biological system, it is appropriate that it be evolution based and biologically compatible. Such a tool should be designed with evolutionary theory as a conceptual framework and built into a software package to process any mixture of biological omics (genomic, proteomic, metabolomic, etc.) data for the purpose of circumscribing biological phenotypes (taxonomic classes and subclasses, as well as disease classes and subclasses) [16.6].

16.7 Phylogenetics: Biologically Compatible Computational Intelligence Tools

16.7.1 What Is Phylogenetics?

During the 1950s, German entomologist *Hennig* codified evolutionary systematics into what has become known as phylogenetic systematics or cladistics [16.44]. The field has since evolved into an analytical paradigm applied in major fields of biology such as botany, microbiology, and zoology; and just recently it is being applied in biomedical research. Phylogenetics groups specimens on the basis of shared derived similarity rather than overall similarity; i.e., it sorts out similarity in derived and ancestral traits before constructing a classification. Thus, it produces a hypothesis of relationships among the study specimens that summarizes the data matrix – the distribution of the derived and ancestral states among the specimens. The resulting hypothesis of relationships is presented in a graphical format called a cladogram, or simply a tree (Figs. 16.4, 16.5b).

Only a few phylogenetic terms need to be explained here. A shared derived state among two or more specimens is called a synapomorphy, and the group of specimens delimited by a synapomorphy or more is termed a clade. Plesiomorphy is an ancestral state, and a symplesiomorphy is a shared ancestral state. The cladogram, which is a dichotomously branching tree, is the map in the form of a tree that shows the specimens and the location of the synapomorphies and symplesiomorphies.

There are a few types of phylogenetic analysis algorithms that differ from each other in their assumptions and in the type of data they process (numerical or categorical variables), including distance methods (usually applied to DNA and protein sequences, for example, neighbor joining method), probability methods (such as likelihood and Bayesian methods, applied to quantitative datasets), and parsimony methods (such as Wagner [16.45] and Camin–Sokal methods [16.46], applied to qualitative discrete characters [16.6]).

Parsimony is the oldest of the phylogenetic methods [16.43]. It functions better than the other two methods when handling data related to diseases for which the genomic, epigenetic, and metabolic rates of change are fast and heterogeneous, such as in cancer [16.47–49]. These methods have been compared, and parsimony has turned out to be the most suitable for the purposes of dealing with heterogeneous high-throughput data of various diseases [16.6].

16.7.2 What Is Parsimony?

In practice, parsimony refers to choosing the simplest explanation, or choosing the hypothesis with the least number of assumptions among all plausible ones. In other fields of study, it is called Occam's razor or the *principle of simplicity*. However, in a precise phylogenetic sense, the preferred phylogenetic hypothesis (or cladogram, tree) is the one that requires the least number of steps to construct (i. e., the shortest tree). The parsimonious paradigm can also be interpreted as a dynamic multidimensional analytical and modeling tool, which is data based, not specimen based, and integrates heterogeneous data, the nature of the biological data type (such as high-throughput microarrays), and the principles of evolution [16.6, 9].

In biomedicine as in other fields of research, it is important that the generated cladogram has high

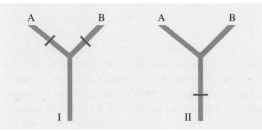

Fig. 16.4 Using parsimony to choose between two competing hypotheses. For two specimens, A and B, sharing a derived gene-expression value, there are two possible explanations: (I) they have two independent pathways/origins/parallel evolution, or (II) they have one common pathway/origin. The most parsimonious cladogram is the one with fewest assumptions (II)

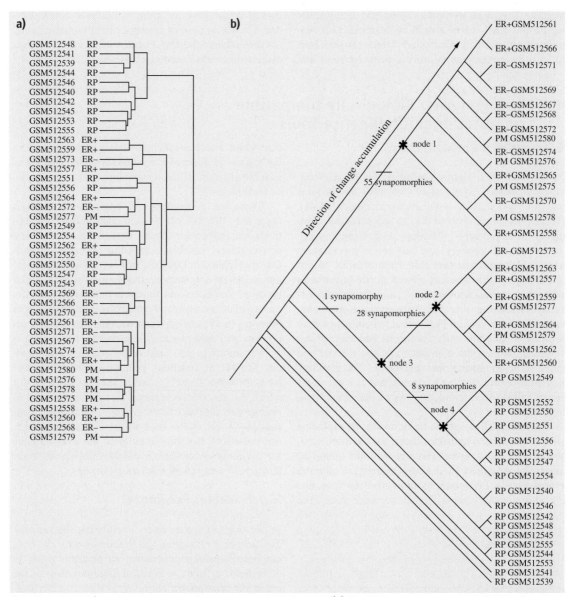

Fig. 16.5a,b A comparison between a dendrogram and cladogram. (**a**) Dendrogram as constructed [16.50] based on the authors' 98 differentially expressed gene probes. (**b**) Cladogram constructed with MIX of the PAUP package using the 22 283 gene probes from microarray chip [HG-U133A] Affymetrix Human Genome U133A Array [16.50, 51]. The cladogram shows the number of synapomorphies for several clades as well as the direction of change accumulation, two features that are specific to cladograms that the dendrogram does not show

predictivity, which in a practical sense means high objective accuracy and low margin of error regarding its hypothesis. Predictivity has practical implications beyond stratification of populations [16.31]; For example, a highly predictive parsimonious analysis can be used in a clinical setting in distinguishing between healthy and diseased individuals, in identifying the metabolic changes associated with disease, and in locating transitional cases between the healthy and diseased, i.e., individuals susceptible to developing disease.

It is important to explain how parsimony sorts out the data and selects among competing hypotheses, because the question of why the most parsimonious cladogram is preferred is often asked. We will use a simple example to illustrate how the principle of parsimony is used in phylogenetics. In Fig. 16.3, specimens (a) and (b) both have a similar state (such as a gene expression or a metabolite value). The state is represented here by crossing bars on the tree branches. There are two competing hypotheses that explain the distribution of the state among the two specimens. The first hypothesis (I) shows two steps (or locations) on the cladogram, one associated with each of the specimens (i.e., it shows two origins for the state), while the second hypothesis (II), shows one step (or one location). Both cladograms explain the presence of the same state among the two specimens, and both cladograms convey the same information, but one has one step fewer than the other. The cladogram with the single step is considered the most parsimonious, and is selected as a more valid explanation. Now, you can extrapolate from this example to larger more complex cladograms with tens of specimens and hundreds of characters.

16.8 Analyzing, Modeling, and Mining Biological Data

Working at large research institutions where various omics technologies are used to study the significant issues in biomedical research has influenced our reasoning concerning the type of analysis that is most appropriate for the understanding of the biological significance of these data. Over the past few years, we have witnessed colleagues struggle to analyze omics data, and we have been aware of the efforts of bioinformaticians in attempting to arrive at creative solutions for demanding biomedical scientists. Aside from our advocacy for parsimony phylogenetics, to date we are not aware of any new promising solutions whose proponents claim an ability to address data heterogeneity. The recent clinical trials of targeted treatments to counteract cancer are examples of reductionist-style thinking which has produced disappointing results [16.52]. However, the irony about the targeted treatment approach is that it brought the issue of heterogeneity to the forefront and increased awareness of the need to deal with this issue [16.26, 53].

When dealing with large datasets that contain thousands of variables, especially datasets obtained from high-throughput microarrays or mass spectrometry, there are two steps in carrying out parsimony analysis; First, polarity assessment of data points through outgroup comparison into either derived (abnormal in case of disease phenotypes) or ancestral (normal) must be carried out. The second step is the processing of the polarized values through a maximum-parsimony algorithm to classify the specimens into a cladogram [16.43, 54]. Polarity assessment transforms the continuous quantitative data points into discrete entities of zeros (0s) and ones (1s), where zero indicates that the value is ancestral (normal) and one indicates that the value is different and therefore assumed to be derived in an evolutionary sense. So, the new data matrix of polarized bivalent values has only zeros and ones, and it is this matrix that will be processed in a parsimony algorithm such as MIX of the PAUP package [16.54].

16.8.1 Dendrogram Versus Cladogram

To illustrate the parsimony analysis of a heterogeneous dataset we selected a publicly available microarray dataset GDS3716 from the National Center for Biotechnology Information (NCBI [16.55]). Then we followed the steps outlined in the above paragraph. The GDS3716 dataset contains gene expression of four sets of histologically normal epithelial (NIEpi) breast specimens from

1. 18 reduction mammoplasty (RM) surgeries
2. 6 prophylactic mastectomy (PM) surgeries
3. 9 biopsies from estrogen receptor-positive (ER+) breast cancer patients, and
4. 9 biopsies from estrogen receptor-negative (ER−) breast cancer patients [16.1, 50].

Graham et al. [16.50] hypothesized and demonstrated that altered gene expressions within NIEpi of breast cancer patients occur in both ER+ and ER− cancer patients and are similar to gene expressions in breast tissues of high breast cancer risk women (the PM group) [16.50]. Our analysis below has also shown the same conclusion, but with a few additional implications.

Part C | 16.8

Table 16.2 Synapomorphies of four main clades as denoted by nodes 1–4 of the cladogram in Fig. 16.5b. List produced by the MIX program of the PHYLIP package through the application of maximum parsimony as assigned to the nodes

Synapomorphies of node 1

ARPC3, ATP5E, ATP5SL, BLCAP, C19orf10, C2CD2, CALD1, CAPN7, CCDC53, CTSB, FAM65A, FLOT1, FNDC3B, GANAB, GPX1, H2AFX, HNRNPA1, HSPD1, ILF2, KIAA0195, LOC440926, LONP2, LPAL2, MED28, MKNK1, MKRN1, MSH6, MY-CBP2, NEDD4L, NGLY1, ORC2L, PLSCR3, PPCDC, R02172, RAB7A, RARS, RBM16, RBX1, RPIA, SERTAD2, SF3B5, SLC25A14, SLC35D2, SNRPB, ST7, TARDBP, TEAD4, TIMP1, TMEM106C, TRADD, UBXN7, UQCRC2, WDR92, YIPF3, ZNF451

Synapomorphies of node 2

AF117899, ANK1, ARMC9, ATP6V0E1, C2orf42, C2orf43, C6orf106, CCDC90B, CSE1L, ERCC6, FOXG1, GTF2H1, KPNB1, LSM6, MVK, POLR3B, PSMF1, RNF13, RPS21, RRN3P1, SRI, TSHB, UBE2K, VAMP4, VGLL4, WIZ, WNT6, ZMYND11

Synapomorphy of node 3

SIK1

Synapomorphies of node 4

AF198444, C9orf114, CITED2, FOS, HIST1H2BC, HSPC157, JUN, LRRC37A2

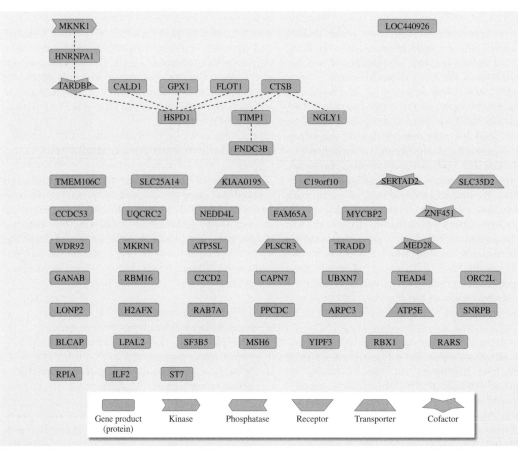

Fig. 16.6 Network pathway analysis of the 55 synapomorphies of node 1 of the cladogram (Fig. 16.5b) with the genomatix pathway system (GePS) [16.56]. Eleven of the 55 synapomorphies are connected by several pathways, while the rest of the genes are not, thus indicating a punctuated pattern of gene-expression irregularity that contributes to the initiation of carcinogenesis

In their study, *Graham* et al. [16.50] identified 98 differentially expressed probes in the expression data by comparing the gene expressions of the last three groups (PM, ER+, ER−) with those of RM, and constructed a clustering dendrogram (Fig. 16.5a). However, when we examined some of their differentially expressed genes (*Graham* et al.'s Table 2 [16.50]), we noticed that none of these genes were universally up- or downregulated for any of the study groups but rather up- and downregulated within each group, i.e., heterogeneously. Similar to *Graham* et al. [16.50], we used their RM group as the outgroup to carry out the polarity assessment of the three other NIEpi groups (there was no a priori selection of genes), and processed the polarized matrix with the parsimony program MIX [16.54]. MIX produced one most parsimonious cladogram (Fig. 16.5b).

The dendrogram of *Graham* et al. [16.50] differs from our cladogram in several aspects. The most obvious difference is the branching of each that determines the relationships of the main groupings. Membership of their groups also differs from that of our clades. On the cladogram, we have listed the number of synapomorphies for each clade. These synapomorphies define a clade of specimens (except where there is a homoplasy). Therefore, parsimony makes it easier to identify events that lead to carcinogenesis. Also, the clade at the upper side of the cladogram is the one with highest number of aberrations (Fig. 16.5b, Table 16.2). Thus, in addition to mapping the shared derived changes (the synapomorphies), the cladogram possesses a directionality for the change accumulation; both of these properties are lacking in the dendrogram.

16.8.2 Interpreting the Results of a Parsimony Analysis

In order to interpret the parsimony analysis result we need to dissect the cladogram. There are three aspects to look for in a cladogram: the groupings, the synapomorphies, and the topology of the cladogram. In our example (Fig. 16.5b), the first aspect we will consider is the distribution of the four sets that form the study collection, i.e., histologically NIEpi breast specimens of RM, PM, as well as ER+ and ER− biopsies from breast cancer patients. The first group formed several small clades at the lower part of the cladogram and a well-defined clade at node 4, while the other three sets were mixed between two clades at nodes 1 and 2. The RM clade at node 4 differed from the other RM clades in having eight synapomorphies that are exclusive to the clade and one synapomorphy that it shared with the clade of node 2 comprising PM, ER+, and ER− specimens (Fig. 16.5b, Table 16.2). The clade of node 1 has 55 synapomorphies, the largest number of synapomorphies for any clade in this study group, signifying a larger number of aberrant gene expressions (Fig. 16.5b, Table 16.2).

By providing synapomorphies for each clade, parsimony allows us to connect the perturbations in gene expression to their cellular pathways, a feature that is lacking in other types of analysis. When the 55 synapomorphies of node 1 were processed for pathway analysis with the genomatix pathway system (GePS), 11 of the 55 synapomorphies appeared to be connected by several related pathways, while the rest of the genes were not, thus indicating a punctuate pattern of gene-expression irregularity (Fig. 16.6).

16.9 Applications of Parsimony to Various Biomedical Issues

There are a few recent examples of advocating and applying phylogenetics to biomedical issues. *Ewald* [16.57] argued that advances in molecular techniques should provide the tools for evaluating viral causation of cancer, and advocated the use of molecular phylogenies to test the viral association with carcinogenesis [16.57]. The cladograms in this case can be used to map the association of viral mutations with the phases of oncogenesis.

Abu-Asab et al. [16.1] applied maximum parsimony to a microarray dataset and showed that metastatic prostate cancer arose from different cell lineages than the primary prostate cancer, and that the two phenotypes had different shared derived changes (clonal expressions, synapomorphies) [16.1].

Furthermore, *Greig* et al. called for the application of phylogenetics to proteomics, genomics, and metabolomics of marine mammal health evaluations due to the limitations of using conventional hematology and serum chemistry panels alone to assess health status [16.58].

16.10 Implications of Applying Parsimony Phylogenetics

There are three sets of implications for applying parsimony to biological data: bioinformatic, biological, and clinical (Table 16.3). From a bioinformatic point of view, parsimony is seamless and capable of incorporating very large datasets with tens of thousands of variables; the only bottleneck that we are aware of thus far is the computational power. In addition to the tens of thousands (or more) of variables, it also can use hundreds of specimens, and the pooling of multiple experiments from various datasets (mass spectrometry, microarrays, morphological data, etc.). It produces higher concordance, for example, between gene lists from different datasets from similar specimens (e.g., a type of cancer) than statistical methods (F- and t-statistics and fold-change) [16.31]. Furthermore, it is possible to carry out intra- and intercomparability analysis using parsimony; this can be done by polarizing

each dataset to its own baseline, then combining the sets in one analysis.

Since it defines only shared derived states (the synapomorphies) as the basis of similarity between specimens and groups of specimens (the clades), it uses them to delineate clades (a process of class discovery); these synapomorphies are also the potential biomarkers. Parsimony has the capacity to incorporate gene and protein expressions that violate normal distribution in a set of specimens, e.g., dichotomously expressed genes and peptides (Fig. 16.3a,b) [16.9].

Another advantage of parsimony that many bioinformaticians are not aware of is the reduction in the number of homoplasies in the produced cladogram. Homoplasy encompasses independent multiple origins of the same state in different specimens. This biological phenomenon is due to the evolutionary processes

Table 16.3 Advantages and implications of applying maximum parsimony (MP) to heterogeneous biological data

Bioinformatic implications
• MP defines only shared derived states (synapomorphies) as the basis of similarity between specimens, and uses them to delineate clades (class discovery); synapomorphies are also the potential biomarkers for diseases
• Reduces experimental noise by transforming quantitative datasets into polarized qualitative points
• Allows intra- and intercomparability of data
• Produces higher concordance between gene and protein lists than statistical methods (F- and t-statistics and fold-change)
• Seamless, permits the use of tens of thousands (or more) of variables and the pooling of multiple experiments from various datasets
• MP analysis seeks to minimize as much as possible multiple origins of similar states – the homoplasies (due to convergence, parallelism, or reversal)

Biological implications
• MP accurately reveals the biological pattern(s) and subtypes of the disease
• Offers a data-based, not specimen-based, listing of derived genes, metabolites, and proteins
• Produces lists of simultaneously deregulated genes, metabolites, and proteins, therefore permits establishing of linkage criteria
• Offers a qualitative assessment of gene, metabolite, and protein expressions by showing their directionality
• Efficiently models the heterogeneous expression profiles of the diseased specimens. Those with fast mutation rate such as cancer and degenerative diseases
• Incorporates gene, metabolite, and protein expressions that violate normal distribution in a set of specimens, e.g., dichotomously expressed genes and peptides
• Dynamic, can be used for modeling disease, comparing between two states for the same individual such as before and after treatment. A stable cladogram incorporates novel specimens without major alterations of initial hypothesis of relationships/ classification
• Predictive, ability to list (predict) a specimen's characteristics when its class becomes known
• Elucidates the direction of accumulation of synapomorphies among a number of specimens that leads to their molecular and cellular diversity, e.g., the presence of one or more developmental pathway

Clinical implications
• MP can be translated into a clinical setting for early detection, diagnosis, prognosis, and posttreatment evaluation
• Can be used to stratify population(s) ahead of clinical trial and trace response within homogeneous clades and between clades

of convergence, parallelism, or reversal. Thus, the best cladogram is the one with least amount of homoplasies, i.e., that minimizes the effect of the homoplasies on the classification hypothesis.

The process of polarity assessment that is used on some quantitative datasets such as mass spectrometry and gene expression reduces the noise in the data, since polarity assessment cancels out the noise when it transforms the continuous quantitative data to discrete bivalent data.

The approach is also dynamic; it incorporates novel specimens without major alterations of the initial hypothesis of relationships or classification. The parsimony classification is highly predictive; it lists (predicts) a specimen's characteristics when its position within the clade becomes known (i.e., its location on the cladogram), and correctly classifies biological classes, as we have seen in the above example and other published ones [16.1].

Additionally, polarization of the data (i.e., converting it into bivalent 0–1 data) allows pooling of multiple experiments, and therefore facilitates intra- and inter-comparability of data. In this regard, the analysis is a systems biology approach; it can be seamless, permitting also pooling of data from related diseases to determine common changes among them, e.g., several cancer types [16.31].

The biological advantages of parsimony over other methods are numerous. Parsimony accurately reveals biological patterns among a group of specimens, for example, distinguishing between primary and metastatic cancer types. Because it is data based, and not specimen based, it allows tracing of variables on the cladogram and interpretation of their biological significance, such as whether they are synapomorphies or homoplasies. It also allows quantification and linkage of synapomorphies as mapped on the cladogram.

Linkage is inferred when the synapomorphies codefine a clade.

Parsimony is best suited for data that include a high rate of mutations, since it efficiently models heterogeneous expression, and profiles specimens such as those from cancerous specimens and degenerative diseases with hundreds of mutations. Furthermore, parsimony can incorporate gene, metabolite, and protein expressions that violate normal distribution in a set of specimens, e.g., dichotomously expressed genes and peptides.

Biological systems are dynamic and not static. Therefore, they need to be studied with a tool that can account for and model this characteristic. Phylogenetics in general and parsimony in particular are suitable tools for modeling dynamic systems. The cladogram presents that change as a continuous spectrum from one side to the other (Fig. 16.5b). Therefore, the cladogram can be used for modeling a disease to show the changes from normal, through the transitional states, to the disease; and for comparing between two states for the same individual, such as before and after treatment. Additionally, a stable cladogram incorporates novel specimens without major alterations of the initial hypothesis of relationships/classification.

The cladogram produced by parsimony has directionality; thus it elucidates the direction of accumulation of synapomorphies among a number of specimens that leads to their molecular and cellular diversity, e.g., the presence of one or more developmental pathways.

The power of parsimony can be applied in a clinical setting. The cladogram can be utilized for early detection, diagnosis, prognosis, and posttreatment evaluation. Additionally, a parsimonious cladogram can be used to stratify population(s) ahead of a clinical trial and trace response within homogeneous clades and between clades.

16.11 Closing Thoughts

In general, heterogeneity means the lack of synchronicity and concordance among variables. In biology, heterogeneity refers to variation within structural, anatomical, functional, behavioral, and molecular traits among organisms. Since all biological systems and their related data are characterized by heterogeneity, it seems reasonable to examine the sources and levels of heterogeneity when designing an analytical approach to data mining. Heterogeneity is a plausible reason for the ineffectiveness of targeted treatment of cancer, as

well as for the unfruitful search for biomarkers for early detection, diagnosis, and posttreatment assessment of cancer and other diseases. It may also explain limited results in disease definition and profiling using genetic, metabolic, and proteomic data (collectively known as omics data). Other areas of research perhaps found wanting because of their negligence of heterogeneity include sorting out clonal from nonexpanded mutations, differentiating between genetic versus epigenetic driving events, determining the primary origins

Part C | 16.11

of cancers and other diseases, subtyping of diseases (class discovery), and side-effect susceptibility. Phylogenetics provides tools that consider heterogeneity and address important issues in biomedical research such as those just mentioned. These tools show promise because they provide a multidimensional data-based modeling paradigm that features improved considera-

tion of heterogeneity to address and resolve difficulties encountered with current computational methods. We believe that the convergence of CI and phylogenetics will provide even better tools for addressing problems such as the ones just mentioned, as well as other data mining problems related to other kinds of data.

References

16.1   M.S. Abu-Asab, M. Chaouchi, S. Alesci, S. Galli, M. Laassri, A.K. Cheema, F. Atouf, J. VanMeter, H. Amri: Biomarkers in the age of *omics*: Time for a systems biology approach, OMICS **15**(3), 105–112 (2011)

16.2   H.H. Heng, G. Liu, J.B. Stevens, S.W. Bremer, K.J. Ye, B.Y. Abdallah, S.D. Horne, C.J. Ye: Decoding the genome beyond sequencing: The new phase of genomic research, Genomics **98**(4), 242–252 (2011)

16.3   L. Hood, J.R. Heath, M.E. Phelps, B. Lin: Systems biology and new technologies enable predictive and preventative medicine, Science **306**(5696), 640–643 (2004)

16.4   M. Eklund, O. Spjuth, J.E. Wikberg: An eScience-Bayes strategy for analyzing *omics* data, BMC Bioinformatics **11**, 282 (2010)

16.5   A. Galvan, J.P. Ioannidis, T.A. Dragani: Beyond genome-wide association studies: Genetic heterogeneity and individual predisposition to cancer, Trends Genetics **26**(3), 132–141 (2010)

16.6   M.S. Abu-Asab, H. Amri: Analyzing heterogeneous complexity in CAM research: A systems biology solution through parsimony phylogenetics, Forsch. Komplementärmed./Res. Complement. Med. **19**(1), 42–48 (2012)

16.7   C.L. Sawyers: The cancer biomarker problem, Nature **452**(7187), 548–552 (2008)

16.8   F. Davidoff: Heterogeneity is not always noise: Lessons from improvement, JAMA **302**(23), 2580–2586 (2009)

16.9   M. Abu-Asab, M. Chaouchi, H. Amri: Evolutionary medicine: A meaningful connection between *omics*, disease, treatment, Proteomics Clin. Appl. **2**(2), 122–134 (2008)

16.10   M. Abu-Asab: Microarrays need phylogenetics. Science STKE e-Lett (2009) available online from http://stke.sciencemag.org/cgi/eletters/sigtrans;1/51/eg11

16.11   M.L. Minsky: *The Emotion Machine* (Simon Schuster, New York 2006)

16.12   A. Turing: Computing machinery and intelligence, Mind **50**, 433–460 (1950)

16.13   J. Searle: Minds, Brains and Programs, Behav. Brain Sci. **3**(3), 417–457 (1980)

16.14   J. Hawkins, S. Blakeslee: *On Intelligence* (Times Books, New York 2004)

16.15   A.P. Engelbrecht: *Computational Intelligence: An Introduction* (Wiley, Hoboken 2002)

16.16   I. Guyon: *Feature Extraction: Foundations and Applications*, Studies in Fuzziness and Soft Computing (Springer, Berlin, Heidelberg 2006)

16.17   V.S. Cherkassky, F. Mulier: *Learning from Data: Concepts, Theory, Methods*, 2nd edn. (Wiley-Interscience, Hoboken 2007)

16.18   R. Xu, D.C. Wunsch: *Clustering*, IEEE Press Series on Computation Intelligence (Wiley, Hoboken 2009)

16.19   J.M. DeLeo: Receiver operating characteristic laboratory (ROCLAB): Software for developing decision strategies that account for uncertainty, Proc. 2nd Int. Symp. Uncertain. Modeling Anal. (1993)

16.20   J.E. Dayhoff, J.M. DeLeo: Artificial neural networks: Opening the black box, Cancer **91**(8), 1615–1635 (2001)

16.21   G.H. Heppner, B.E. Miller: Tumor heterogeneity: Biological implications and therapeutic consequences, Cancer Metastasis Rev. **2**(1), 5–23 (1983)

16.22   F. Michor, K. Polyak: The origins and implications of intratumor heterogeneity, Cancer Prev. Res. **3**(11), 1361–1364 (2010)

16.23   K. Puniyani, S. Kim, E.P. Xing: Multi-population GWA mapping via multi-task regularized regression, Bioinformatics **26**(12), i208–i216 (2010)

16.24   J.J. Berman: *Precancer: The beginning and the End of Cancer* (Jones Bartlett, Sudbury 2010)

16.25   W. Liu, W. Zhao, M.L. Shaffer, N. Icitovic, G.A. Chase: Modelling clinical trials in heterogeneous samples, Stat. Med. **24**(18), 2765–2775 (2005)

16.26   H.H. Heng, S.W. Bremer, J.B. Stevens, K.J. Ye, G. Liu, C.J. Ye: Genetic and epigenetic heterogeneity in cancer: A genome-centric perspective, J. Cell Physiol. **220**(3), 538–547 (2009)

16.27   J. McClellan, M.C. King: Genetic heterogeneity in human disease, Cell **141**(2), 210–217 (2010)

16.28   M.M. Holland, M.R. McQuillan, K.A. O'Hanlon: Second generation sequencing allows for mtDNA mixture deconvolution and high resolution detection of heteroplasmy, Croat. Med. J. **52**(3), 299–313 (2011)

16.29   United States Census Bureau: http://www.census.gov/population/www/projections/usinterimproj/

16.30 W. Liu, N. Icitovic, M.L. Shaffer, G.A. Chase: The impact of population heterogeneity on risk estimation in genetic counseling, BMC Med. Genetics **5**, 18 (2004)

16.31 M.S. Abu-Asab, M. Chaouchi, H. Amri: Phylogenetic modeling of heterogeneous gene-expression microarray data from cancerous specimens, OMICS **12**(3), 183–199 (2008)

16.32 K. Hotakainen, U.H. Stenman: Will emerging prostate cancer markers redeem themselves?, Clin. Chem. **56**(8), 1212–1213 (2010)

16.33 R. Jones: Biomarkers: Casting the net wide, Nature **466**(7310), S11–S12 (2010)

16.34 M. May: Biomarkers still off the mark for detecting breast cancer, Nat. Med. **16**(1), 3 (2010)

16.35 G. Poste: Bring on the biomarkers, Nature **469**(7329), 156–157 (2011)

16.36 J. Lyons-Weiler, S. Patel, M.J. Becich, T.E. Godfrey: Tests for finding complex patterns of differential expression in cancers: Towards individualized medicine, BMC Bioinformatics **5**, 110 (2004)

16.37 R. Xu, S. Damelin, B. Nadler, D.C. Wunsch 2nd: Clustering of high-dimensional gene expression data with feature filtering methods and diffusion maps, Artif. Intell. Med. **48**(2-3), 91–98 (2010)

16.38 M. Abu-Asab, M. Chaouchi, H. Amri: Phyloproteomics: What phylogenetic analysis reveals about serum proteomics, J. Proteome Res. **5**(9), 2236–2240 (2006)

16.39 R.S. Varghese, H.W. Ressom: LC-MS data analysis for differential protein expression detection, Methods Mol. Biol. **694**, 139–150 (2011)

16.40 H. Tsugawa, Y. Tsujimoto, M. Arita, T. Bamba, E. Fukusaki: GC/MS based metabol*omics*: Development of a data mining system for metabolite identification by using soft independent modeling of class analogy (SIMCA), BMC Bioinformatics **12**, 131 (2011)

16.41 B.A. Goldstein, A.E. Hubbard, A. Cutler, L.F. Barcellos: An application of Random Forests to a genome-wide association dataset: Methodological considerations & new findings, BMC Genetics **11**, 49 (2010)

16.42 A. Thakur, V. Mishra, S.K. Jain: Feed forward artificial neural network: Tool for early detection of ovarian cancer, Sci. Pharm. **79**(3), 493–505 (2011)

16.43 J. Felsenstein: *Inferring Phylogenies* (Sinauer, Sunderland 2004)

16.44 W. Hennig: *Phylogenetic systematics* (Univ. of Illinois Press, Urbana 1966)

16.45 R. Eck, M. Dayhoff: *Atlas of Protein Sequence and Structure* (National Biomedical Research Foundation, Silver Spring 1966)

16.46 J. Camin, R. Sokal: A method for deducing branching sequences in phylogeny, Evolution **19**, 311–326 (1965)

16.47 S. Sridhar, F. Lam, G.E. Blelloch, R. Ravi, R. Schwartz: Direct maximum parsimony phylogeny reconstruction from genotype data, BMC Bioinformatics **8**, 472 (2007)

16.48 D. Stefankovic, E. Vigoda: Pitfalls of heterogeneous processes for phylogenetic reconstruction, Syst. Biol. **56**(1), 113–124 (2007)

16.49 D. Stefankovic, E. Vigoda: Phylogeny of mixture models: Robustness of maximum likelihood and non-identifiable distributions, J. Comput. Biol. **14**(2), 156–189 (2007)

16.50 K. Graham, A. de las Morenas, A. Tripathi, C. King, M. Kavanah, J. Mendez, M. Stone, J. Slama, M. Miller, G. Antoine, H. Willers, P. Sebastiani, C.L. Rosenberg: Gene expression in histologically normal epithelium from breast cancer patients and from cancer-free prophylactic mastectomy patients shares a similar profile, Br. J. Cancer **102**(8), 1284–1293 (2010)

16.51 Affymetrix: http://www.affymetrix.com/ (2012)

16.52 D.H. Roukos: Mea Culpa with cancer-targeted therapy: New thinking and new agents design for novel, causal networks-based, personalized biomedicine, Expert Rev. Mol. Diagn. **9**(3), 217–221 (2009)

16.53 H.H. Heng, J.B. Stevens, S.W. Bremer, K.J. Ye, G. Liu, C.J. Ye: The evolutionary mechanism of cancer, J. Cell Biochem. **109**(6), 1072–1084 (2010)

16.54 J. Felsenstein: PHYLIP: Phylogeny inference package (version 3.2), Cladistics **5**, 164–166 (1989)

16.55 National Center for Biotechnology Information: GEO DataSets, available online at http://www.ncbi.nlm.nih.gov/gds/ (2012)

16.56 Genomatix: www.genomatix.de/en/index.html (2012)

16.57 P.W. Ewald: An evolutionary perspective on parasitism as a cause of cancer, Adv. Parasitol. **68**, 21–43 (2009)

16.58 D.J. Greig, F.M. Gulland, C.A. Rios, A.J. Hall: Hematology and serum chemistry in stranded and wild-caught harbor seals in central California: Reference intervals, predictors of survival, and parameters affecting blood variables, J. Wildl. Dis. **46**(4), 1172–1184 (2010)

17. Protein Folding Recognition

Lavneet Singh, Girija Chetty

Protein folding recognition is a complex problem in bioinformatics where different structures of proteins are extracted from a large amount of harvested data including functional and genetic features of proteins. The data generated consist of thousands of feature vectors with fewer protein sequences. In such a case, we need computational tools to analyze and extract useful information from the vast amount of raw data to predict the major biological functions of genes and proteins with respect to their structural behavior. In this chapter, we discuss the predictability of protein folds using a new hybrid approach for selecting features and classifying protein data using support vector machine (SVM) classifiers with quadratic discriminant analysis (QDA) and principal component analysis (PCA) as generative classifiers to enhance the performance and accuracy. In one of the applied methods, we reduced the data dimensionality by using data reduction algorithms such as PCA. We compare our results with previous results cited in the literature and show that use of an appropriate feature selection technique is promising and can result in a higher recognition ratio compared with other competing methods proposed in previous studies. However, new approaches are still needed, as the problem is complex and the results are far from satisfactory. After this introductory section, the chapter is or-

ganized as follows: In Sect. 17.1 we discuss the problem of protein fold prediction, protein database, and its extracted feature vectors. Section 17.2 describes feature selection and classification using SVM and fused hybrid classifiers, while Sect. 17.4 presents the experimental results. Section 17.5 discusses experimental results, including conclusions and future work.

Part C | 17

Protein folding recognition is one of the challenging and most important problems in the area of bioinformatics. The structure of a protein plays an important role in its biological and genetic function [17.1]. Thus, to know about protein and genetic sequences, which are basically sequences of various amino acids in protein molecules, we first need to know its structure, and then how it is folded. With the increase in available computational power, several research works have been reported in the genome sequencing area with the objective of amino acid sequencing, although there are still challenges in determining the three-dimensional (3-D) structure of proteins. Several machine-learning methods have been introduced in previous studies to predict protein folding structures and amino acid sequences. *Ding* and *Dubchak* [17.2] used support vector machine

(SVM) and artificial neural network (ANN) classifiers to extract features and various properties, based on which certain predefined protein folds were predicted. *Shen* and *Chou* [17.3] proposed a model based on a nearest-neighbor algorithm and its modification called the K-local hyperplane (H-KNN) method [17.4]. *Nanni* [17.5] proposed a model using Fisher's linear classifier and H-KNN classifier. *Eddy* [17.6] used hidden Markov models (HMM) for protein folding recognition. Their model predicted the most accurate rate of protein folding. However, the disadvantage of using HMM is that it needs high computational power working on large datasets of protein folding for training and testing, although the reduced state-space HMM method with smaller architecture can also be used [17.7, 8].

Basically, in protein folding problems, classification is done by using two types of classifiers. The probabilistic approach uses training data to map the probability estimates of each class and find the posterior probability of test data. The second type of classifier uses the likely neighborhood or weights between different classes based on instances of training data. In our study, we use the support vector machine (SVM) with specific parameters, with generative quadratic discriminant analysis (QDA) for classification of protein folds.

Many researchers have used the fusion of different classifiers to increase the performance and accuracy of the recognition rate in bioinformatics [17.9]. *Shen* and *Chou* [17.10] proposed a fused classifier for large-scale human protein subcellular location prediction, and *Nanni* et al. [17.11] used SVM classifiers fused with max rule algorithms. The challenge in terms of the classification and accuracy in the protein folding problem is that, after converting data into $m \times n$ matrices, the number of features is large (sometimes thousands) with lesser training data, which makes it harder to implement learning models. To address this problem, a better approach is to reduce the high dimensionality of the features by using some feature selection techniques so that, with n observations and instances, we have n features for the learning or training phase. Reducing the features can be achieved by using statistical or probabilistic approaches. There are two methods for feature reduction: feature selection and feature transformation, to convert high-dimensional data into a new space with reduced dimensionality.

The major challenge in the above-mentioned methods and approaches lies in the complexity of the data, due to the large number of folding classes with only a small number of training samples and the multiple, heterogeneous feature groups, making it harder to discover patterns. Due to this problem, the classification accuracy achieved in protein folding prediction problems in most previously reported work is no more than 60%, which is in general lower than typical pattern recognition problems in other application domains.

In this study, we examined several learning classifiers, such as SVM, linear regression classifiers, ANN with multilayer perceptron (MLP), and random forest, and then finally we compared the performance of these classifiers with that of our proposed hybrid classifier using a fusion of QDA and principal component analysis (PCA) with SVM. Experimental validation on publicly available protein databases showed that the recognition accuracy achieved with our proposed classifiers is significantly higher than achieved with the methods proposed in previous work. SVM is a binary classifier, whereas protein folding recognition is a multiclass problem. Thus, in this study, we use the strategy of assigning weights using a discriminant function based on QDA.

17.1 Protein Folding Recognition Problem

Proteins are macromolecules composed of 20 different amino acids linked by peptide bonds in a linear order [17.12]. The linear polypeptide chain is called the primary structure of the protein. The primary structure can be represented as a sequence of 20 different letters, where each letter denotes an amino acid. In the native state, the amino acids (or residues) of a protein fold into local secondary structures including alpha helix, beta sheet, and nonregular coil. The secondary structure elements are further packed to form tertiary structure due to hydrophobic forces and side-chain interactions between amino acids. The tertiary structures of several related proteins can bind together to form a protein complex called the quaternary structure. In a cell, proteins with native tertiary structures interact to carry out all kinds of biological functions, including enzymatic catalysis, transport and storage, coordinated motion, mechanical support, immune protection, generation and transmission of nerve impulses, and control of growth and differentiation. Extensive biochemical experiments in the past have shown that a protein's function is determined by its structure.

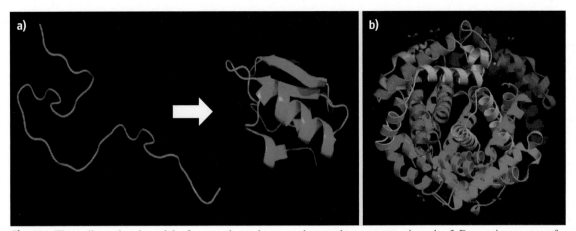

Fig. 17.1 Three-dimensional model of converting primary and secondary structure into the 3-D protein structure for various folds

Thus, elucidating a protein's structure is the key to understand its function, which has fundamental significance in biological and medical sciences. Currently, there are about 30 000 proteins with determined structures deposited in the protein data bank (PDB) [17.13]. These diverse and abundant structures provide invaluable data to understand how a protein folds into its unique 3-D structure and to predict the structure from its sequence.

17.1.1 Protein Sequence, Structure, and Function

Since the pioneering experiments [17.12] showed that a protein's structure is dictated by its sequence, predicting the structure of a protein from its sequence has become one of the most fundamental problems in structural biology. In the postgenomic era, with the application of high-throughput DNA and protein sequencing technologies, the number of protein sequences has increased exponentially, whereas experimental determination of protein structure remains very expensive, time consuming, labor intensive, and sometime impossible. Currently, only about 1.5% of protein sequences (about 30 000 out of 2 million) have solved structures, and the gap between proteins with known and unknown structures is still increasing.

Thus, predicting the structure of a protein from its sequence is increasingly imperative and useful. Protein structure prediction is becoming a vital tool for understanding phenomena in modern molecular and cell biology and has important applications in medical sciences, e.g., in drug design [17.14].

17.1.2 1-D, 2-D, and 3-D Protein Structure Prediction

Protein structure prediction is often classified into three levels: one-dimensional (1-D), two-dimensional (2-D), and 3-D [17.15]. One-dimensional prediction is used to predict structural features such as the secondary structure and the solvent accessibility of each residue along the one-dimensional protein sequence. Two-dimensional prediction is used to predict the relationship between residues (e.g., contact map prediction and disulfide bond prediction). Three-dimensional prediction is used to predict the 3-D coordinates of all residues or all atoms in a protein. Although the ultimate goal is to predict the 3-D structure, 1-D and 2-D prediction are of great interest to biologists and represent important steps toward 3-D structure prediction. The existing, rather practical, method for 3-D and 2-D structure prediction as shown in Fig. 17.1 is the template based approach including comparative modelling (or homology modelling) and fold recognition (or threading). This approach is based on the observation that nature tends to reuse existing structures/folds to accommodate new protein sequences and functions during evolution.

More than half a century ago, evidence began to accumulate that a major part of most proteins' folded structure consists of two regular, highly periodic arrangements, designated α and β. The key to both structures is the hydrogen bond. A hydrogen atom is nothing more than a proton with a surrounding electron cloud. When one of these atoms is chemically bonded to an electron-withdrawing atom such as nitrogen or oxygen, much of the electron cloud moves toward the

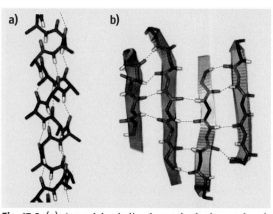

Fig. 17.2 (a) A model α-helix shows the hydrogen bonds (*dotted lines*) between oxygen and hydrogen atoms of the fourth amino acid up the chain. (**b**) β-Sheets are also held together by hydrogen bonds. *Transparent arrows* show the direction of individual β-strands. Chains running in the same direction (*left pair*) are called parallel β-sheet; strands running in opposite directions (*right pair*) are said to be antiparallel β-sheet. Atom coloring is as follows: carbon = *green*, oxygen = *red*, nitrogen = *blue*, and hydrogen = *white*. (Courtesy: Stanley Krystek, Bristol-Myers Squibb, Pharmaceutical Research Institute)

nitrogen or oxygen. The proton is thus left almost bare, with its positive charge largely unshielded. If it comes close to a another atom with a small extra negative charge, typically an oxygen or nitrogen atom, the partial positive and negative charges will attract each other. It is this attraction that produces the hydrogen bond and stabilizes the α and β structures. The structure now called an α-helix is a right-handed spiral stabilized by hydrogen bonds between each amino acid's nitrogen atom and the oxygen atom of the fourth one up the chain. This means that there are 3.6 amino acids for each turn of the helix. The main part of the amino acid (the side-chain, designated R in Fig. 17.2) sticks out from this spiral backbone like the bristles on a bottle brush. The β structure is now called a β-sheet. It is essentially flat, with the side-chains sticking out on alternate sides. β-Sheet

is also stabilized by hydrogen bonds between nitrogen and oxygen atoms. In this case, however, the hydrogen-bonded atoms belong to different amino acid chains running alongside each other. The sheets are *parallel* if all the chains run in the same direction and *antiparallel* if alternate chains run in opposite directions. Antiparallel sheets are often, but not always, formed by a single chain looping back upon itself. When a single chain loops back on itself to form an antiparallel β-sheet, the one to three amino acids linking the two strands are known as a β-turn. Today, scientists recognize the β-turn as one of the fundamental elements of protein structure [17.11]. All other local arrangements of amino acids are described as *random coil*, although they are random only in the sense of not being periodic.

Thus, although the protein sequence space (the number of protein sequences) is very large, the protein structure space (the number of unique protein folds) is relatively small and expected to be limited. Currently, millions of protein sequences have been collected, but the number of unique structures (folds) in protein classification databases such as SCOP (structural classification of proteins) and CATH (protein structure classification database developed by University College, London), which are the publically available databases for protein structures classified domains based on their structures and amino acid sequences, among them only about 1000 (out of 30 000 protein structures). Moreover, among the protein structures newly determined in structural genomics projects, the novel folds account for only a small portion (about 10%), and the overall fraction of new folds has continued to decrease over the past 15 years. Thus, most protein sequences, particularly similar protein sequences within the same family and superfamily evolving from common ancestors, have structures similar to other proteins. So, given a query protein without a known structure, template-based prediction is used first to identify a template protein (if one exists) with a solved, similar structure to the query protein. The structure of the template protein is then used to model the structure of the query protein based on the alignment between the query sequence and the template structure.

17.2 Protein Database and Its Features

17.2.1 Protein Database

In this study, we used datasets derived from the structural classification of proteins (SCOP) database [17.11]. Details of these protein sets have being clearly de-

scribed in [17.16] and are presented in Table 17.1. All feature vectors are standardized and normalized to the range of $[-1; +1]$ before applying any classifiers. The proteins in both the training and test sets belong to 27 different protein folds, corresponding to four major structural classes: α, β, α/β, and α + β. In this study, we compare classification results of protein folds using the overall accuracy Q, defined as the percentage of correctly recognized proteins among all proteins in the test dataset, which can be expressed as $Q = c/n$, where c is the number of query proteins whose folds have been correctly recognized and n is the total number of proteins in the test dataset. Table 17.1 presents the parameters extracted from the protein sequence with a 125-dimensional feature vector for each protein in this dataset.

17.2.2 Feature Vectors

In our experiments, we used the features described by *Ding* and *Dubchak* [17.2]. These feature vectors are based on six parameters: amino acid composition (C), predicted secondary structure (S), hydrophobicity (H), normalized van der Waals volume (V), polarity (P), and polarizability (Z). Each parameter consists of 21 features, except for the amino acid composition

Table 17.1 Six features extracted from the protein sequence with their dimensions

Protein feature	Dimension
Amino acid composition	20
Predicted secondary structure	21
Hydrophobicity	21
van der Waals volume	21
Polarity	21
Polarizability	21

(C), which consists of 20 features. We then concatenate all features into one feature vector (C, S, H, V, P, Z) containing 125 features. All feature vectors are standardized and normalized to the range of $[-1; +1]$ before applying any classifiers. Standardizing and normalizing the features results in features with smaller numeric ranges, making it much easier for classifiers to do the classification task.

Section 17.3 describes the feature selection and classification approaches examined for this problem.

17.3 Feature Selection and Classification

17.3.1 Support Vector Machine Classifiers

The support vector machine (SVM) is a well-known large-margin classifier proposed by *Vapnik* [17.17]. The basic concept of the SVM classifier is to find an optimal separating hyperplane that separates two classes. The decision function of the binary SVM is

$$f(x) = \left[\sum_{i=1}^{N} \alpha_i y_i K(x_i, x) + b \right], \qquad (17.1)$$

where b is a constant, $y_i \in \{-1, 1\}$, $0 \le \alpha_i \le C$, $i = 1, 2, \ldots, N$ are nonnegative Lagrange multipliers, C is a cost parameter that controls the trade-off between allowing training errors and forcing rigid margins, x_i are the support vectors, and $K(x_i, x)$ is the kernel function.

Here, we use SVM for the multiclass problem using the one-against-one method. It was first introduced in [17.18], and the first use of this strategy for SVM was in [17.19, 20]. This method constructs $k(k-1)/2$ classifiers, where each one trains data from two classes. For training data from the i-th and j-th classes, we solve the following binary classification problem. In this study, we use the max win voting strategy suggested in [17.21]. If $\text{sign}[(w_{ij})T\phi(x) + (b_{ij})]$ says x is in the i-th class, then the vote for the i-th class is increased

by 1. Otherwise, the j-th vote is increased by 1. Then, the largest vote will be given to specific protein structure class domains on variable x.

We used the LIBSVM software library to perform the experiments. LIBSVM is a general library for support vector classification and regression, being available at http://www.csie.ntu.edu.tw/~cjlin/libsvm. As mentioned before, there are various functions to map data to higher-dimensional spaces, and for this we need to select the kernel function $K(x_i; x_j) = \phi(x_i)T\phi(x_j)$. There are several types of kernels that can be used for different problems. Based on the specific problem, each kernel has different parameters; For example, some well-known problems with large numbers of features, such as text classification [17.22], DNA problems [17.23], brain abnormality analysis [17.24, 25], and speech and image analysis [17.26–28], are classified more correctly with a linear kernel. In our study, we use the radial basis function (RBF) kernel, which is a real value function whose value depends on the distance from origin. A learner with the RBF kernel usually performs no worse than others do, in terms of generalization ability. In this study, we carried out some simple comparisons and observed that, when using the RBF kernel, the performance was better than for the linear kernel $K(x_i; x_j) = \phi(x_i)T\phi(x_j)$ for all the problems

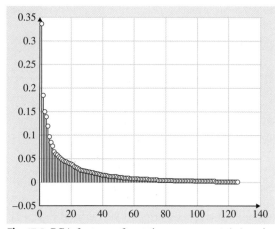

Fig. 17.3 PCA features of protein sequences. *x-Axis*: principle components (eigenvectors); *y-axis*: the weights (importance) of the principle components to discriminate different data points (protein sequences) from the original feature space

we studied. Therefore, for the three datasets, instead of staying in the original space, a nonlinear mapping to a higher-dimensional space seems useful. Another important issue is the selection of parameters. For SVM training, a few parameters such as the penalty parameter C and the kernel parameter of the RBF function must be determined in advance. Choosing the optimal parameters for use in support vector machines is an important step in SVM design.

17.3.2 The PCA/LDA–Based Support Vector Machine Algorithm

Step 1: Use PCA on training and testing dataset

$$\hat{S} = \frac{1}{N} \sum_{j=1}^{K} \sum_{i=1}^{N_j} (x_{ji} - \hat{\mu})(x_{ji} - \hat{\mu})^{\mathrm{T}} .$$

Step 2: Use Linear Discriminant Analysis (LDA) on training and testing dataset

$$\hat{S}_{\mathrm{W}} = \frac{1}{N} \sum_{j=1}^{K} \sum_{i=1}^{N_j} (x_{ji} - \hat{\mu})(x_{ji} - \hat{\mu})^{\mathrm{T}} ,$$

$$\hat{S}_{\mathrm{B}} = \sum_{j=1}^{K} N_j (\mu_j - \mu)(\mu_j - \mu)^{\mathrm{T}} ,$$

$$J(T) = \frac{|\hat{S}_{\mathrm{B}}|}{|\hat{S}_{\mathrm{W}}|} ,$$

where $J(T) =$ linear discriminant function.

Step 3: Calculate the decision function of SVM as

$$f(x) = \left[\sum_{i=1}^{N} \alpha_i y_i K(x_i, x) + b \right] .$$

We used RBF, sigmoid (Sig), and linear (Lin) activation functions in the SVM. In Sect. 17.5, a comparative study on multikernel SVM is presented with respect to their learning classification accuracy rates.

17.3.3 Quadratic Discriminant Analysis

Quadratic discriminant analysis (QDA) [17.29] describes the likelihood of a class as a Gaussian distribution and then uses the posterior distribution estimates to estimate the class for a given test vector. This approach leads to the function

$$d_k(x) = (x - \mu_k)^{\mathrm{T}} \sum_{k}^{-1} (x - \mu_k)$$
$$+ \log \sum k - 2 \log p(k) , \qquad (17.2)$$

where $\sum k$ is the covariance matrix, x is the test vector, μ_k is the mean vector, and $p(k)$ is the prior probability of class k. The Gaussian parameters for each class can be estimated from the training dataset, so the values of $\sum k$ and μ_k are replaced in (17.2) by their estimates $\sum \hat{k}$ and $\hat{\mu}_k$. However, when the number of training samples is small compared with the number of dimensions of the training vector, the covariance estimation may be ill-posed. The approach to resolve this ill-posed estimation is to regularize the covariance matrix $\sum k$.

To apply QDA, our first goal is to reduce the dimension of the data by finding a small set of important features that can result in good classification performance. Feature selection algorithms can be roughly grouped into two categories: filter methods and wrapper methods. Filter methods rely on general characteristics of the data to evaluate and select feature subsets without involving the chosen learning algorithm (QDA in this case). Filters are usually used as a preprocessing step, since they are simple and fast. A widely used filter method for bioinformatics data is to apply a univariate criterion separately on each feature, assuming that there is no interaction between features. We applied the t-test on each feature and compared the p-value (or the absolute values of the t-statistic) for each feature as a measure of how effective it is at separating groups.

17.4 Illustrative Experiments

This experiment illustrates the complexity of the problem and the difficulty in obtaining good solutions. Our experiments include implementation of SVM and other classifiers on the prescribed protein dataset and a comparative study to find the higher accuracy rate. To work on these experiments, firstly we chose certain parameters. We used the cross-validation technique to avoid overfitting. We used k-fold cross-validation with $k = 7$, because there must be at least 7 samples of each class in the training dataset. Then, the experiments are tested and classified by using different kernels for SVM clas-

sifier. The RBF kernel is

$$K(x_i, x) = -\gamma(x - x_i)^2, \quad \gamma > 0. \quad (17.3)$$

The model uses various kernels such as linear, polynomial, and Gaussian. The RBF kernel gave the best results in our experiments. The parameters C from Sect. 17.3.1 and γ from (17.3) have certain parametric values. Both values were experimentally chosen by using a cross-validation procedure on the training dataset. The best recognition ratio was achieved using parameter values of $\gamma = 0.7$ and $C = 300$.

17.5 Exemplar Results

In this study we used SVM and the proposed hybrid classifier to increase performance and accuracy. Accuracy was measured in terms of the percentage recognition ratio. Suppose that there are $N = n_1 + n_2 + n_3 + \cdots + n_p$ test proteins, where n_i is the number of proteins belonging to class i. Suppose that c_i of the n_i proteins are correctly recognized (as belonging to class i). Then, a total number of $C = c_1 + c_2 + c_3 + \cdots + c_p$ proteins are correctly recognized. Therefore, the total accuracy is $Q = C/N$.

The SVM classifier works with the feature vector for each class representing the minimum value of the discriminant function. So, for every instance, we calculate the discriminant function value of each class. In this way we create $m \times n$ matrices of two different feature vectors. The SVM classifiers were used with 126-dimensional feature vectors. All binary classifiers were trained, and in the testing phase, each instance was assigned to each class. The recognition ratio using the SVM classifiers was 63.89%.

The final step was to examine the performance of three feature selection algorithms. The results are presented in Table 17.2. There are three different voting tables and weights for the three different matrices with different selected feature vectors. The recognition ratio was about 66.43%, which is slightly higher than when using the SVM classifiers alone. Table 17.3 describes the comparative performance of using various classifiers, including SVM and our proposed hybrid SVM classifiers. Table 17.4 defines the comparative study of multikernels with SVM in terms of their learning classification accuracy rates. The results are still not sat-

isfactory, suggesting that new methods for this problem are needed.

Table 17.2 Recognition ratio obtained using various feature selection algorithms

Selection method (number of features)	Recognition ratio SVM (%)
Generalized linear model (28)	55.36
SFS[a] (69)	60.19
QDA (28)	65.17
PCA (60)	66.43

[a] SFS: sequential feature selection

Table 17.3 Comparison among different methods

Method	Recognition ratio (%)
SVM	63.75
H-KNN	57.4
Bayesian naive	52.30
Random forest	53.72
MLP	54.72
SVM-QDA (proposed method)	65.17
SVM-PCA (proposed method)	66.43

Table 17.4 Comparison among different methods using multiple kernels

Method	Sigmoid kernels	RBF kernel	Linear kernel
QDA-SVM	57.23	65.17	52.11
PCA-SVM	63.78	66.43	51.89

Part C | 17.5

17.6 Conclusions and Future Work

In this chapter, we discuss the problem of protein folding recognition. We have applied different feature selection and classification methods to illustrate the complexity of the problem. As seen from the experimental results, the feature selection algorithms along with an appropriate classifier approach can result in better outcomes and may be promising for future investigations in high-dimensional protein databases. All our experiments were done on the feature vector previously developed by *Ding* and *Dubchak* [17.2], allowing us to assess the improvements achieved by the new approach. Further work will be done on developing new feature selection algorithms for other high-dimensional protein sequences, resulting in reduced computational power, and better accuracy and recognition rate. Also, a novel classifier approach based on SVM combined with binary decision trees and other algorithms to make hybrid classifiers will be investigated. In conclusion, new computational methods will be needed for future research on this complex problem.

References

17.1   H.S. Chan, K. Dill: The protein folding problem, Phys. Today **46**(2), 24–32 (1993)

17.2   C.H. Ding, I. Dubchak: Multi-class protein folds recognition using support vector machines and neural networks, Bioinformatics **17**, 349–358 (2001)

17.3   H.B. Shen, K.C. Chou: Ensemble classifiers for protein fold pattern recognition, Bioinformatics **22**, 1717–1722 (2006)

17.4   O. Okun: Protein fold recognition with K-local hyperplane distance nearest neighbor algorithm, Proc. 2nd Eur. Workshop Data Min. Text Min. Bioinform. (Pisa 2004) pp. 51–57

17.5   L. Nanni: A novel ensemble of classifiers for protein folds recognition, Neurocomputing **69**, 2434–2437 (2006)

17.6   S.R. Eddy: Hidden Markov models, Curr. Opin. Struct. Biol. **6**, 361–365 (1995)

17.7   M. Madera, J. Gough: A comparison of profile hidden Markov model procedures for remote homology detection, Nucl. Acids Res. **30**(19), 4321–4328 (2002)

17.8   C. Lampros, C. Papaloukas, T.P. Exarchos, Y. Golectsis, D.I. Fotiadis: Sequence-based protein structure prediction using a reduced state-space hidden Markov model, Comput. Biol. Med. **37**, 1211–1224 (2007)

17.9   C. Lampros, C. Papaloukas, K. Exarchos, D.I. Fotiadis: Improving the protein fold recognition accuracy of a reduced state-space hidden Markov model, Comput. Biol. Med. **39**, 907–914 (2009)

17.10   H.B. Shen, K.C. Chou: Hum-mPLoc: An ensemble classifier for large-scale human protein subcellular location prediction by incorporating samples with multiple sites, Biochem. Biophys. Res. Commun. **355**, 1006–1011 (2007)

17.11   L. Nanni, A. Lumini: MppS: An ensemble of support vector machine based on multiple physicochemi-cal properties of amino acids, Neurocomputing **69**, 1688–1690 (2006)

17.12   F. Sanger, E.O. Thompson: The amino–acid sequence in the glycyl chain of insulin. I. The identification of lower peptides from partial hydrolysates, J. Biochem. **53**(3), 353–366 (1953)

17.13   H.M. Berman: The protein data bank, Nucl. Acids Res. **28**, 235–242 (2007)

17.14   M. Jacobson, A. Sali: Comparative protein structure modeling and its applications to drug discovery. In: *Annual Reports in Medicinal Chemistry*, Vol. 39, ed. by J. Overington (Academic, London 2004) pp. 259–276

17.15   J. Cheng, A. Randall, P. Baldi: Prediction of protein stability changes for singlesite mutations using support vector machines, Protein Struct. Funct. Bioinform. **62**(4), 1125–1132 (2006)

17.16   C.X. Zhang, J.S. Zhang: RotBoost: A technique for combining rotation forest and adaboost, Pattern Recognit. Lett. **29**, 1524–1536 (2008)

17.17   V. Vapnik: *The Nature of Statistical Learning Theory* (Springer, New York 1995)

17.18   S. Knerr, L. Personnaz, G. Dreyfus: Single-layer learning revisited: A step-wise procedure for building and training a neural network. In: *Neurocomputing: Algorithms, Architectures and Applications*, ed. by J. Fogelman (Springer, Berlin, Heidelberg 1990)

17.19   J. Friedman: Another approach to polychotomous classification, Technical Report, Department of Statistics (Stanford University, Stanford 1999), available online from http://www-stat.stanford.edu/~jhf/ftp/poly.pdf

17.20   U. Krebel: Pair-wise classification and support vector machines. In: *Advances in Kernel Methods – Support Vector Learning*, ed. by B. Scholkopf, C.J.C. Burges, A.J. Smolapages (MIT Press, Cambridge 1999) pp. 255–268

17.21 C.-J. Lin: Formulations of support vector machines: A note from an optimization point of view, Neural Comput. **13**(2), 307–317 (2001)

17.22 T. Joachims: *The maximum-margin approach to learning text classifiers: Methods, theory, and algorithms*, Dissertation (Universität Dortmund, Dortmund 2001)

17.23 L. Singh, G. Chetty: Hybrid approach in protein folding recognition using support vector machines, Proc. Int. Conf. Mach. Learn. Data Min. (MLDM 2012) (Berlin, LNCS, Springer 2012)

17.24 C.-H. Yeang, S. Ramaswamy, P. Tamayo, S. Mukherjee, R.M. Rifkin, M. Angelo, M. Reich, E. Lander, J. Mesirov, T. Golub: Molecular classification of multiple tumor types, Bioinform. Discov. Note **1**(1), 1–7 (2001)

17.25 L. Singh, G. Chetty: Review of classification of brain abnormalities in magnetic resonance images using pattern recognition and machine learning, Proc.

Internat. Conf. Neuro Comput. and Evolving Intelligence, NCEI 2012 (Auckland, LNCS Bioinformatics, Springer 2012)

17.26 A. Mishra, L. Singh, G. Chetty: A novel image water marking scheme using extreme learning machine, Proc. IEEE World Congr. Computat. Intell. (WCCI 2012) (Brisbane 2012)

17.27 L. Singh, G. Chetty, S. Singh: A novel algorithm using MFCC and ERB gammatone filters in speech recognition, J. Inf. Syst. Commun. **3**(1), 365–371 (2012)

17.28 L. Singh, G. Chetty: A comparative study of recognition of speech using improved MFCC algorithms and Rasta filters, Int. Conf. Inf. Intell. Syst. Technol. Management ICISTM 2012 (Springer, 2012) pp. 304–314, , Communications in Computer and Information Science Ser., Vol. 285

17.29 K. Fukunaga: *Introduction to Statistical Pattern Recognition*, 2nd edn. (Academic, New York 1990)

18. Kernel Methods and Applications in Bioinformatics

Yan Fu

The kernel technique is a powerful tool for constructing new pattern analysis methods. Kernel engineering provides a general approach to incorporating domain knowledge and dealing with discrete data structures. Kernel methods, especially the support vector machine (SVM), have been extensively applied in the bioinformatics field, achieving great successes. Meanwhile, the development of kernel methods has also been strongly driven by various challenging bioinformatic problems. This chapter aims to give a concise and intuitive introduction to the basic principles of the kernel technique, and demonstrate how it can be applied to solve problems with uncommon data types in bioinformatics. Section 18.1 begins with the product features to give an intuitive idea of kernel functions, then presents the definition and some properties of kernel functions, and then devotes a subsection to a brief review of kernel engineering and its applications to bioinformatics. Section 18.2 describes the standard SVM algorithm. Finally, Sect. 18.3 illustrates how kernel methods can be used to address the peptide identification

and the protein homology prediction problems in bioinformatics, while Sect. 18.4 concludes.

Kernel methods are a large class of pattern analysis methods that use the kernel technique to implicitly map input patterns to a feature space [18.1]. As the dot product in the feature space, a kernel function can be incorporated into any computational processes that exclusively involve dot product operations on input patterns. Kernel functions have been most successful in statistical learning algorithms, exemplified by the support vector machine (SVM) [18.2]. They can also be viewed as similarity measures of input patterns and used in all distance-based algorithms.

When the input patterns or the variables taken by a kernel function are vectors, the mapping induced by the kernel function is usually nonlinear, and all dot-product-based linear algorithms can be directly extended to their nonlinear versions by simply replacing the dot product with the kernel function. When the input patterns are not vectors, such as strings, trees, and graphs, kernel functions provide a general approach to making vector-based algorithms applicable to these nonvector or discrete data structures.

Kernel methods, especially the SVM, have been extensively applied in bioinformatics, achieving great successes (see [18.3–5] for reviews). Meanwhile, the development of kernel methods has also been strongly driven by various challenging bioinformatic problems;

Part C | 18

For example, bioinformatics is a field full of discrete data structures, such as nucleic acid and protein sequences, phylogenetic trees, and molecular interaction networks. Engineering proper kernel functions for these biological data has considerably expanded the application scope of kernel methods.

This chapter aims to give readers a concise and intuitive introduction to the basics of the kernel technique, and demonstrate how it can help construct a nonlinear algorithm and how it can be used to solve challenging problems in bioinformatics. Although the well-known SVM algorithm will be briefly described as an example of kernel methods, the emphasis of this chapter is on more extensive usages of kernel functions, such as kernel engineering and kernel functions for similarity search.

18.1 Kernel Functions

18.1.1 Product Features

To have an intuitive idea of kernel functions, let us start with the product features [18.6]. In pattern classification, the input patterns are usually from the real vector space, i.e., $x \in \mathbb{R}^N$, and linear classification algorithms use a hyperplane in the vector space to classify the patterns. Different ways to search for the hyperplane result in different algorithms. However, it is often the case in practice that the patterns cannot be classified by a hyperplane in the original space; For example, the essential features of patterns may be all the d-order products (called product features) of dimensions of the input space

$$x_{j_1} \cdot x_{j_2} \cdot \cdots \cdot x_{j_d},$$
$$j_1, j_2, \ldots, j_d \in \{1, \ldots, N\}. \tag{18.1}$$

All product features constitute a new space F (called the feature space), in which patterns are linearly separable; i.e., they can be separated by a hyperplane in F. For example, when $N = 2$ and $d = 2$, the dimensionality N_F of the feature space F is 3

$$(x_1, x_2) \rightarrow \left(x_1^2, x_2^2, x_1 x_2\right). \tag{18.2}$$

In general, we have

$$N_F = \frac{(N+d-1)!}{d!(N-1)!}. \tag{18.3}$$

N_F will increase rapidly with N and d; For example, for images with 16×16 pixels, we have $N = 256$, and when $d = 2$, the order of magnitude of N_F is as large as 10^{10}. Such high dimensionality will lead to serious computational difficulties.

However, if an algorithm only involves dot product operations on input vectors, then we need not actually work in the feature space. We can construct a version of the algorithm for the feature space from the input space, as long as we can compute the dot product in the feature space from the original space. Consider the feature space with dimensions being the ordered product features. When $N = 2$ and $d = 2$, the input patterns are transformed from two dimensions to four dimensions

$$C_2 : (x_1, x_2) \rightarrow \left(x_1^2, x_2^2, x_1 x_2, x_2 x_1\right). \tag{18.4}$$

The dot product of two vectors in this feature space is then

$$C_2(\boldsymbol{x}) \cdot C_2(\boldsymbol{y}) = x_1^2 y_1^2 + x_2^2 y_2^2 + 2x_1 x_2 y_1 y_2 = (\boldsymbol{x} \cdot \boldsymbol{y})^2. \tag{18.5}$$

In general, if C_d maps $\boldsymbol{x} \in \mathbb{R}^N$ to vector $C_d(\boldsymbol{x})$ with elements consisting of all ordered d-order products, then we have

$$k(\boldsymbol{x}, \boldsymbol{y}) = C_d(\boldsymbol{x}) \cdot C_d(\boldsymbol{y}) = (\boldsymbol{x} \cdot \boldsymbol{y})^d. \tag{18.6}$$

Further, if the feature space consists of all product features of up to d-orders, we have

$$k(\boldsymbol{x}, \boldsymbol{y}) = (\boldsymbol{x} \cdot \boldsymbol{y} + 1)^d, \tag{18.7}$$

which is the commonly used polynomial kernel function in kernel methods.

18.1.2 Definition and Properties of Kernel Functions

A general definition of kernel functions is given as: A *kernel function*, or a *kernel* for short, is a binary function k such that, for all $\boldsymbol{x}, \boldsymbol{y} \in A$, we have

$$k(\boldsymbol{x}, \boldsymbol{y}) = \phi(\boldsymbol{x}) \cdot \phi(\boldsymbol{y}), \tag{18.8}$$

where ϕ is some mapping from the input space A to a feature space B. Usually, ϕ is a nonlinear mapping, and the feature space B has very high or even infinite dimensionality (Hilbert space). According to this definition of kernel functions, any computations based on dot products in the feature space can be accomplished by the kernel function from the input space, thus avoiding the explicit mapping from the input space to the feature space. Typical examples of kernel functions are

$$\text{Linear kernel: } \boldsymbol{x} \cdot \boldsymbol{y}$$
$$\text{Polynomial kernel: } [a(\boldsymbol{x} \cdot \boldsymbol{y}) + \gamma]^d$$
$$\text{RBF kernel: } \exp(-\lambda ||\boldsymbol{x} - \boldsymbol{y}||^2)$$
$$\text{Sigmoid kernel: } \tanh[a(\boldsymbol{x} \cdot \boldsymbol{y}) + \gamma] . \quad (18.9)$$

RBF stands for radical basis function. Note that the sigmoid kernel satisfies the definition of kernel functions only for certain parameter values [18.2]. Some operations on kernel functions lead to still valid kernel functions; For example, if $k_1(\boldsymbol{x}, \boldsymbol{y})$ and $k_2(\boldsymbol{x}, \boldsymbol{y})$ are valid kernels, then the following functions are also valid kernels:

$$k(\boldsymbol{x}, \boldsymbol{y}) = a_1 k_1(\boldsymbol{x}, \boldsymbol{y}) + a_2 k_2(\boldsymbol{x}, \boldsymbol{y}) , \quad \text{for } a_1, a_2 > 0 ,$$
$$k(\boldsymbol{x}, \boldsymbol{y}) = k_1(\boldsymbol{x}, \boldsymbol{y}) k_2(\boldsymbol{x}, \boldsymbol{y}) ,$$
$$k(\boldsymbol{x}, \boldsymbol{y}) = \text{pol}^+[k_1(\boldsymbol{x}, \boldsymbol{y})] ,$$
$$k(\boldsymbol{x}, \boldsymbol{y}) = \exp[k_1(\boldsymbol{x}, \boldsymbol{y})] , \quad (18.10)$$

where pol^+ indicates polynomials with positive coefficients.

According to the definition of kernel functions, all operations that exclusively involve dot products in the feature space can be implicitly done in the input space by an appropriate kernel. A nonlinear version of an algorithm can be readily obtained by simply replacing the dot products with kernels. This is called the *kernel technique*. The underlying mathematical conclusions of it were derived almost one century ago [18.7], but it was very recently that the kernel technique was widely used. The most successful application of kernels is to extend various linear learning algorithms to their nonlinear versions. The first such attempt was made in 1964 [18.8]. However, the great success of kernel methods is due to the SVM algorithm, introduced in the early 1990s [18.2, 9]. Other famous kernel-based nonlinear learning algorithms include kernel principal analysis [18.10], kernel canonical correlation analysis [18.11], kernel discriminant analysis [18.12], kernel independence analysis [18.13], etc.

However, kernels are not limited to learning algorithms. As the dot product in the feature space, a kernel is a measurement of similarity, and defines the metrics of the feature space [18.6]; For example, we have the following kernel-based cosine and Euclidian distances in the feature space:

$$\cos[\phi(\boldsymbol{x}), \phi(\boldsymbol{y})] = \frac{k(\boldsymbol{x}, \boldsymbol{y})}{\sqrt{k(\boldsymbol{x}, \boldsymbol{x}) \cdot k(\boldsymbol{y}, \boldsymbol{y})}} , \quad (18.11)$$
$$d[\phi(\boldsymbol{x}), \phi(\boldsymbol{y})] = \sqrt{k(\boldsymbol{x}, \boldsymbol{x}) + k(\boldsymbol{y}, \boldsymbol{y}) - 2k(\boldsymbol{x}, \boldsymbol{y})} . \quad (18.12)$$

Therefore, in a more general sense, all distance-based algorithms can be kernelized [18.14]. Kernels can even be directly used for similarity search, e.g., nearest-neighbor search [18.15, 16] and information retrieval [18.17–19].

18.1.3 Kernel Engineering and Applications in Bioinformatics

In many real-world problems, the input patterns are not given in the form of vectors but are nonvector, discrete structures, such as strings, trees, graphs, or objects in any form. Usually, these discrete structures are very difficult, if not impossible, to represent as vectors. Thus, vector-based algorithms cannot be directly applied to these structures. Even if these structures can be vectorized in some way, the essential information may be irretrievably lost, and therefore the discriminative power of algorithms may be greatly reduced.

The common kernels given in (18.9) require the input patterns to be vectors. However, in fact, in the definition of kernel functions the input space can be the set of any types of patterns or objectives, as long as the kernel can be represented as the dot product in some space. If we can construct an appropriate kernel for the discrete structure of interest, all vector-based algorithms can be directly applied to the discrete structure by simply replacing the dot product with the kernel. In such cases, a kernel can be considered as a similarity measurement between input patterns [18.6].

A string kernel has been proposed for text classification, which implicitly maps strings into the feature space of all possible substrings [18.20]. *Watkins* showed that the conditionally symmetrically independent joint probability distributions can be written as dot products and therefore are valid kernels [18.21]. *Haussler* proposed a more general scheme for kernel engineering, named convolution kernels [18.22]. The Fisher kernel is a method for constructing kernels for classifiers using generative models [18.23]. *Kondor* and *Lafierty* proposed the diffusion kernels on graphs [18.24].

Kernel engineering has been successfully applied to bioinformatics. The mismatch kernels, a class of string kernels, showed good performance in the protein classification problem [18.25, 26]. A kernel on fixed-length strings was proposed and used for prediction of signal peptide cleavage sites [18.27]. It is based on the fact that strings are similar to each other if they possess many common rare substrings. *Zien* et al. [18.28] incorporated the local correlation in DNA sequences into a kernel function, and obtained better results than previous methods for translation initiation site prediction. The Fisher kernel based on the hidden Markov model (HMM) was used to detect protein homologies and performed better than other methods including HMM [18.29]. As pointed out by *Watkins* [18.21], the pair HMM used for scoring sequence alignment is in fact a valid kernel. Diffusion kernels were used for microarray data analysis [18.30]. A tree kernel on phylogenetic profiles was engineered and used together with SVM and principle component analysis (PCA) for biological function pre-

diction [18.31]. Kernels on graphs were engineered and used to predict protein functions [18.32].

In summary, as the dot products in the feature space, kernels can be used in all dot-product-based algorithms. We can directly use a kernel without knowing the specific form of the feature space induced by the kernel. This advantage makes kernels a powerful tool to deal with problems on discrete structures. There are plenty of discrete structures in the biological domain, such as DNA or protein sequences, protein three-dimensional structures, phylogenetic trees, interaction networks, etc. These data are hard to convert into vectors for analysis by vector-based methods. The kernel technique is a promising solution to these problems. Research in this direction is still in its infancy. Engineering more powerful kernels on more discrete structures is an open problem.

18.2 Support Vector Machines

This section illustrates how the kernel function can be incorporated into SVM [18.2, 9, 33], the first and most successful kernel method.

18.2.1 Maximum–Margin Classifier

When two classes of patterns (represented as vectors) are linearly separable, multiple hyperplanes that can classify the patterns exist in the vector space. Different linear classifiers use different criteria to find the hyperplane. The so-called maximum-margin hyperplane is the one that not only separates the patterns but also has the largest margin to the patterns. According to the statistical learning theory [18.2], the maximum-margin hyperplane has low Vapnik–Chervonenkis (VC) dimension and better generalizability (ability to classify unseen patterns). The algorithm searching for the maximum-margin hyperplane is called a support vector machine (SVM). As we will see later, SVM only involves dot product operations on input vectors and therefore can be kernelized. In addition to the statistical foundation, the kernel technique is another factor contributing to the great success of SVM. For this reason, SVM is sometimes called a kernel machine. A third advantage of SVM is the global optimality of its solution. Multiple ways to formulate SVM have been presented, and there are SVMs for different purposes, e.g., classification and regression. Below, we introduce the standard SVM for classification.

Let H be a hyperplane that can separate two classes of samples, and H_1 and H_2 be two hyperplanes parallel to H and passing through the samples closest to H. The distance between H_1 and H_2 is called the classification margin. Suppose that the equation for hyperplane H is $x \cdot w + b = 0$, where $x \in \mathbb{R}^N$, w is the weight coefficient vector, and b is the displacement. We can normalize this equation so that a linearly separable two-class training sample set, (x_i, y_i), $i = 1, 2, \ldots, n$, $x \in \mathbb{R}^N$, $y \in \{+1, -1\}$, satisfies

$$y_i(x \cdot w + b) - 1 \geq 0, \quad i = 1, 2, \ldots, n. \quad (18.13)$$

Under this condition, we have that the classification margin is $2/\|w\|$. Maximizing the margin is equivalent to minimizing $\|w\|$. The hyperplane that satisfies the condition in (18.13) and minimizes $\|w\|$ is the maximum-margin hyperplane. Figure 18.1 shows an example in two-dimensional space. To find the

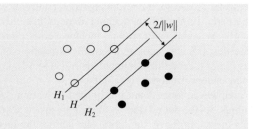

Fig. 18.1 Maximum margin

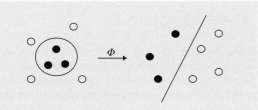

Fig. 18.2 Nonlinear mapping

maximum-margin hyperplane, we need to solve the following optimization problem

$$\min_{\boldsymbol{w},b} \ ||\boldsymbol{w}||^2$$

$$\text{with} \quad y_i(\boldsymbol{x} \cdot \boldsymbol{w} + b) - 1 \geq 0, \quad i = 1, 2, \ldots, n.$$

$$(18.14)$$

The Lagrangian dual problem of this problem is

$$\max_{\alpha} \sum_{i=1}^{n} \alpha_i - \frac{1}{2} \sum_{i=1}^{n} \sum_{j=1}^{n} y_i y_j \alpha_i \alpha_j (\boldsymbol{x}_i \cdot \boldsymbol{x}_j)$$

$$\text{with} \ \sum_{i=1}^{n} y_i \alpha_i = 0,$$

$$\forall i : \alpha_i \geq 0, \quad i = 1, 2, \ldots, n. \qquad (18.15)$$

Solving this optimization problem leads to a classification function

$$f(x) = \text{sgn} \left[\sum_{i=1}^{n} \alpha_i^* y_i (\boldsymbol{x} \cdot \boldsymbol{x}_i) + b^* \right]. \qquad (18.16)$$

When the training samples are not linearly separable, we cannot find a solution to the optimization problem in (18.15). However, if the samples are nonlinearly separable, we can map the input samples to a higher-dimensional space with a properly chosen nonlinear mapping Φ so that the samples become linearly separable and the problem in (18.15) is solvable, as shown in Fig. 18.2. However, this may dramatically increase the dimensionality of the feature space, resulting in the so-called dimension disaster. The answer lies in the kernel technique introduced above. Notice that the problem in (18.15) and the optimal classification function in (18.16) exclusively involve the dot product operation on input vectors. We can avoid an explicit nonlinear mapping by using some appropriate kernel function $k(\boldsymbol{x}_i, \boldsymbol{x}_j)$ to replace the dot product $\boldsymbol{x}_i \cdot \boldsymbol{x}_j$, where $k(\boldsymbol{x}_i, \boldsymbol{x}_j) = \Phi(\boldsymbol{x}_i) \cdot \Phi(\boldsymbol{x}_j)$. In this way, we have the nonlinear maximum-margin classifier.

18.2.2 Soft Margin

Above, we have shown how kernel functions can be used to implicitly map the samples into a feature space in which the samples become linearly separable. However, it may be problematic if we always rely on the mapping induced by a kernel function to achieve perfect separation of samples. This is because the samples in practice are usually imperfect and are subject to interferences of various noises, and the linear inseparability of samples in the input space may have resulted from the noises and not be an inherent characteristic of the patterns. Further, even if the patterns are not linearly separable in the input space, there are still noises in the samples. A strong nonlinear mapping seeking perfect separation may lead to potential overfitting to training samples and reduced generalizability of classifiers. Therefore, a good algorithm should be able to tolerate some training errors. In SVM, this is achieved by introducing the concept of soft margin. The idea of this is to augment the margin by allowing some misclassified samples in the training process. The optimization problem then becomes

$$\min_{\boldsymbol{w},b} \ \frac{1}{2} ||\boldsymbol{w}||^2 + C \sum_{i=1}^{n} \xi_i$$

$$\text{with} \quad y_i(\boldsymbol{x} \cdot \boldsymbol{w} + b) - 1 + \xi_i \geq 0, \quad i = 1, 2, \ldots, n,$$

$$\xi_i \geq 0, \quad i = 1, 2, \ldots, n, \qquad (18.17)$$

where ξ_i are slack variables, C is the parameter controlling the tradeoff between the classification margin and training errors, and the coefficient $1/2$ is introduced for representation convenience. The Lagrangian dual problem of this problem is

$$\max_{\alpha} \ \sum_{i=1}^{n} \alpha_i - \frac{1}{2} \sum_{i=1}^{n} \sum_{j=1}^{n} y_i y_j \alpha_i \alpha_j (\boldsymbol{x}_i \cdot \boldsymbol{x}_j)$$

$$\text{with} \ \sum_{i=1}^{n} y_i \alpha_i = 0,$$

$$\forall i : 0 \leq \alpha_i \leq C, \quad i = 1, 2, \ldots, n. \quad (18.18)$$

Again, the problem exclusively involves the dot product operation on input vectors. Therefore, the dot product can be replaced by a kernel function; That is,

$$\max_{\alpha} \ \sum_{i=1}^{n} \alpha_i - \frac{1}{2} \sum_{i=1}^{n} \sum_{j=1}^{n} y_i y_j \alpha_i \alpha_j k(\boldsymbol{x}_i, \boldsymbol{x}_j)$$

$$\text{with} \ \sum_{i=1}^{n} y_i \alpha_i = 0,$$

$$\forall i : 0 \leq \alpha_i \leq C, \quad i = 1, 2, \ldots, n. \quad (18.19)$$

Solving this optimization problem leads to the nonlinear soft margin classifier

$$f(x) = \text{sgn}\left[\sum_{i=1}^{n} \alpha_i^* y_i k(x, x_i) + b^*\right], \qquad (18.20)$$

in which α_i^* is the solution to the problem in (18.19), and b^* can be calculated with

$$b^* = \frac{1}{y_i} - \sum_{j=1}^{n} \alpha_i^* y_j k(x_i, x_j), \qquad (18.21)$$

where i is an arbitrary index satisfying $0 < \alpha_i < C$.

Only the samples with $\alpha_i^* > 0$ contribute to the final maximum-margin hyperplane, being called support vectors (SVs). SVs are those samples that are on or between the hyperplanes H_1 and H_2.

18.3 Applications of Kernel Methods to Bioinformatics

Kernel methods, especially the SVM, have been successfully used to address various biological problems, e.g., gene prediction [18.34], gene expression data analysis [18.35], RNA splicing site prediction [18.36], protein classification [18.26], protein structure prediction [18.37], protein function prediction [18.38], protein mass-spectrometric data analysis [18.39, 40], etc. This section does not list all applications of kernel methods in bioinformatics (some reviews already exist, e.g., [18.3–5]), but gives some detailed research results on two specific problems, namely peptide identification from tandem mass spectra and protein homology prediction.

18.3.1 Kernel Spectral Dot Product for Peptide Identification

Peptide identification from tandem mass spectra is the foundation of bottom-up proteomics [18.41]. In tandem mass spectrometry, peptides are ionized, isolated, and fragmented. Roughly speaking, the tandem mass spectrum of a peptide is a mass (or more accurately, mass-to-charge ratio m/z) histogram of fragment ions of this peptide [18.42]. To identify the peptide responsible for a tandem mass spectrum, the database search approach is commonly used, in which theoretical mass spectra are predicted from the peptide sequences in a database for comparison with the input experimental spectrum [18.43]. However, the mass spectrum of a peptide cannot in general be accurately predicted. Therefore, the scoring function that measures the similarity between the theoretical and the experimental spectra is the core of a peptide identification algorithm.

The spectral dot product (SDP) is the most widely used similarity measure for mass spectra [18.43, 44]. In SDP, mass spectra are represented as vectors and the SDP is simply the dot product of spectral vectors. A disadvantage of SDP is that it, as a linear function, totally

ignores possible correlations among fragment ions. According to expert knowledge, when positively correlated fragment ions are matched together, the matches are more reliable as a whole than as individuals. Therefore, such matches should be scored higher than separate matches.

Kernel Spectral Dot Product

A straightforward idea is to extend the SDP with a kernel to incorporate the correlations between fragment ions. Since not all fragment ions are correlated, common kernel functions, e.g., the polynomial kernel, do not directly apply here. The central problem becomes how to find a kernel that only emphasizes co-occurring matches of truly correlated fragment ions while ignoring others. Inspired by the locally improved polynomial kernel [18.28, 45], we solved this problem by exerting kernels separately on all possible groups of correlated fragment ions, and then summing them up [18.46]. This is achieved by arranging the predicted fragment ions into a correlative matrix as shown in Fig. 18.3, and grouping the correlated fragment ions into local correlative windows, e.g., the dotted rectangles in Fig. 18.3.

All fragment ions in a theoretical spectrum are assumed to possess unique m/z values. Under this assumption, all nonzero dimensions in the theoretical spectral vector t can be arranged into a matrix $T = (t_{pq})_{m \times n}$ according to their fragmentation positions and fragment ion types, where m is the number of fragment ion types for theoretical fragmentation and $n + 1$ is the peptide sequence length; For example, $t_{2,3}$ corresponds to the fragment ion b_3^* in Fig. 18.3. For the experimental spectral vector c, the dimensions at the m/z value corresponding to t_{pq} are also extracted and arranged into a matrix $C = (c_{pq})_{m \times n}$. It follows that

$$\text{SDP} = c \cdot t = \sum_{p=1}^{m} \sum_{q=1}^{n} c_{pq} t_{pq}. \qquad (18.22)$$

Given the definition of correlative windows, the general form of the kernel spectral dot product (KSDP) is defined as

$$\text{KSDP} = \sum_{j} k(\boldsymbol{c}_j, \boldsymbol{t}_j), \qquad (18.23)$$

where $k(\boldsymbol{c}_j, \boldsymbol{t}_j)$ is a kernel function, $\boldsymbol{c}_j$ is the vector with elements c_{pq}, and $\boldsymbol{t}_j$ is the vector with elements t_{pq}, in which the subscript (p, q) traverses all the elements in the j-th correlative window in the matrices $\boldsymbol{C}$ and $\boldsymbol{T}$.

The KSDP given in (18.23) is also a kernel function. If $k(\boldsymbol{c}_j, \boldsymbol{t}_j)$ is the dot product kernel, the KSDP reduces to the SDP. For a properly chosen kernel function $k(\boldsymbol{c}_j, \boldsymbol{t}_j)$, the KSDP implicitly maps the spectral space to a high-dimensional space where the new dimensions correspond to the combinations of correlated fragment ions.

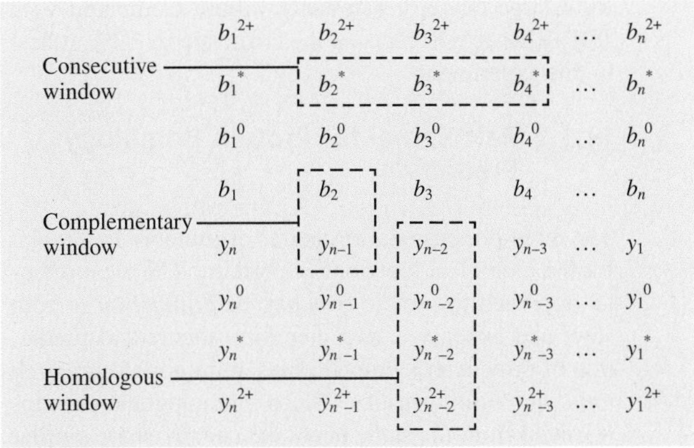

KSDP for Consecutive Fragment Ions

In our earlier work [18.46], we developed the computation of the KSDP for consecutive fragment ions using the locally improved polynomial kernel, given by

$$\sum_{i=1}^{m} \sum_{j=1}^{n} \left[\sum_{k=j-l_1}^{j+l_2} (c_{ik} t_{ik})^{\frac{1}{d}} \right]^d, \qquad (18.24)$$

where the positive integers l_1 and l_2 are equal to $\lfloor (l-1)/2 \rfloor$ and $\lceil (l-1)/2 \rceil$, respectively, the integer l is the size of the correlative window, and c_{ik} and t_{ik} are set to zero for $k \le 0$ and $k > n$.

Here, we build on our previous work by developing a radial basis function (RBF) version of the KSDP for consecutive fragment ions as

$$\sum_{i=1}^{m} \sum_{j=1}^{n} \exp \left[-\gamma \sum_{k=j-l_1}^{j+l_2} (c_{ik} - t_{ik})^2 \right], \qquad (18.25)$$

where γ is the parameter in the RBF kernel. The RBF-KSDP can be computed in $O(mn)$ time, similarly to the polynomial KSDP.

Performance of KSDP

To show the effectiveness of the KSDP and to explore its parameters, the polynomial KSDP given in (18.24) and the RBF-KSDP given in (18.25) were directly used as the scoring function of the pFind search engine [18.46, 47]. Spectra from a standard protein dataset were searched [18.46, 48]. The error rates of the two KSDP implementations are given in Fig. 18.4. Both implementations significantly outperformed SDP

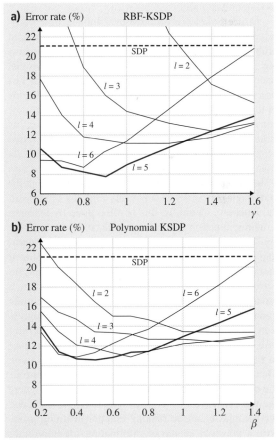

Fig. 18.3 Correlative matrix and examples of correlative windows. *Horizontal direction* is the fragmentation position in peptides, and *vertical direction* is the ion type

Fig. 18.4a,b Error rates of KSDP for consecutive fragment ions (**a**). In the polynomial KSDP, the parameter d is equal to $1 + \beta(l-1)$ (**b**)

for a large range of parameter values. Compared with SDP, RBF-KSDP reduced the error rate by 13% at best in this experiment.

18.3.2 Pair Kernel for Protein Homology Prediction

The three-dimensional structures of proteins are crucial for the biological functions of proteins. The experimental approach to protein structure determination is both slow and expensive, and therefore, theoretical prediction of protein structure becomes an important research topic in bioinformatics. Since homologous proteins (evolved from the same ancestor) usually share similar structures, predicting protein structures based on protein homologies has been one of the most important bioinformatic problems [18.49]. Protein homology prediction is a key step in template-based protein structure prediction methods.

In terms of information retrieval and machine learning, protein homology prediction is a typical ranking problem [18.50]. In this problem, the database objects are protein sequences with known three-dimensional structures, and the query is a protein sequence with unknown structure. The objective is to find those proteins in the database that are homologous to the query protein so that the homologous proteins can be used as structural templates. The homology or match of two proteins can be captured by multiple features, which, as in other retrieval problems, need to be integrated into a single score (called a ranking function) in an intelligent manner in order to accurately rank the proteins in the database. Currently, this is often done by learning a ranking function from a training dataset.

A major characteristic of the ranking-function learning problem is that each feature vector is computed based on a query. Therefore, all feature vectors are partitioned into groups by queries. Here, we call each group of data associated with a query a block. Unlike traditional learning tasks, e.g., classification and regression, in which data are assumed to be independently and identically distributed, the ranking data belonging to the same block are correlated via the same query. This block structure of the data is a unique feature of the ranking-function learning problem. Although ignoring the existence of the block structure would reduce the complexity of the ranking-function learning problem, a potential source of information for improving the ranking performance would also be ignored. In the past, the block structure was not fully explored by most ranking-function learning algorithms.

Pair Kernel

Here, we explore a kernel engineering approach to making use of the block structure information and give some preliminary results. The approach is quite general. We propose the following kernel, named a *pair kernel*, for learning to rank

$$k_{\mathrm{p}}(\langle d_i, q_u\rangle, \langle d_j, q_v\rangle) = k_{\mathrm{s}}(s_{iu}, s_{jv}) + k_{\mathrm{q}}(q_u, q_v),$$

(18.26)

where d_i and d_j are two database items, q_u and q_v are two queries, $\langle d_i, q_u\rangle$ and $\langle d_j, q_v\rangle$ are item–query pairs whose relevances are of interest, s_{iu} and s_{jv} are feature vectors that measure the similarities between database item i and query u and between item j and query v, respectively, k_{s} is a kernel on the similarity feature vectors, and k_{q} is a kernel on queries.

The pair kernel defined above says that, when we are predicting the relevance or similarity of an item–query pair according to another (training) pair, we should not only consider the pairs as a whole but also consider the similarity between queries alone. When $k_{\mathrm{q}} \equiv 0$, the pair kernel reduces to the common kernel used for ranking problems in information retrieval. Different implementations of k_{q} lead to different query kernels. Here, we give an implementation of k_{q} as

$$k_{\mathrm{q}}(q_u, q_v) = k_{\mathrm{b}}(B_u, B_v) = k[\Phi(B_u), \Phi(B_v)],$$

(18.27)

where B_u and B_v are the data blocks associated with query q_u and query q_v, respectively, $\Phi(B_u)$ and $\Phi(B_v)$ are feature vectors describing block B_u and block B_v, respectively, k_{b} is a kernel defined on blocks, and $k[\Phi(B_u), \Phi(B_v)]$ is a kernel defined on vectors. Further, we define

$$\Phi(B_u) = \langle \mu_{u1}, \sigma_{u1}, \mu_{u2}, \sigma_{u2}, \ldots, \mu_{ud}, \sigma_{ud}\rangle,$$

(18.28)

where μ_{uk} and σ_{uk} are the mean and the standard deviation, respectively, of the k-th feature in block B_u.

Performance of Pair Kernel

The dataset used to validate the pair kernel is from the ACM KDD Cup 2004 competition [18.51]. ACM KDD Cup is the annual data mining and knowledge discovery competition organized by the ACM special interest group on data mining and knowledge discovery, the leading professional organization of data miners. One of the tasks in KDD Cup 2004 was to predict protein

Table 18.1 Performances of SVM with the common kernel and the pair kernel for protein homology prediction

	TOP1 (maximize)	RKL (minimize)	APR (maximize)	RMS (minimize)
Common kernel	0.8497	54.78	0.8033	0.0373
Pair kernel	0.8497	47.26	0.8253	0.0363

homologies. The best results on this task were mostly obtained with SVMs [18.52–56]. The success of SVMs in the competition demonstrated that kernel methods are among the best methods for solving complicated bioinformatic problems.

In this dataset, homology between a database protein and the query protein is described by 74 features (constituting the input vector s of the k_s kernel). These features were generated by a protein fold recognition program named LOOPP [18.57] and include various scores of sequence alignments, scores of threading features, measures of secondary structure fitness, etc. [18.58]. There are a total of 153 training queries (test queries are without published labels and are not used here). For each query, a data block consisting of about 1000 samples is given. A sample is a 74-dimensional vector measuring the homology between the query and a protein in the database.

Four metrics were used to evaluate the performance of ranking algorithms, including the frequency of a relevant item being ranked highest (TOP1), the average rank of the last relevant item (RKL), the average overall ranking precision (APR), and the average root-mean-square error (RMS). For model selection, the leave-one-block-out strategy was used [18.52]. SVM was used as the learning algorithm. Table 18.1 presents the results obtained with the pair kernel and the common kernel. In both cases, the dot product kernel was used. It is demonstrated that the proposed pair kernel can significantly improve the ranking performance, in comparison with the commonly used kernel. This improvement is owing to the query kernel k_q added to the common kernel. Finally, it should be noted that the pair kernel proposed here is a general scheme, and the constituent query kernel implemented here is only a simple demonstration of this methodology.

18.4 Conclusions

The kernel technique is a powerful tool for constructing new algorithms, especially nonlinear ones. Kernel engineering provides a promising methodology for addressing nonvector data structures. Over the past two decades, kernel methods have become very popular for pattern analysis, especially in the bioinformatics field. It is neither possible nor necessary to enumerate all applications of kernel methods in bioinformatics in this short

chapter, because they have actually been applied to almost all bioinformatic problems that one can imagine. This chapter, instead of giving a comprehensive review or focusing on a specific research, presents basic principles of kernel methods and focuses on kernel engineering, in the hope that readers, after reading this chapter, can explore their own kernel methods to address various problems with uncommon data types in bioinformatics.

References

18.1   J. Shawe-Taylor, N. Cristianini: *Kernel Methods for Pattern Analysis* (Cambridge University Press, Cambridge 2004)

18.2   V.N. Vapnik: *The Nature of Statistical Learning Theory* (Springer, Berlin Heidelberg 1995)

18.3   B. Schölkopf, K. Tsuda, J.-P. Vert (Eds.): *Kernel Methods in Computational Biology* (MIT Press, Cambridge 2004)

18.4   A. Ben-Hur, C.S. Ong, S. Sonnenburg, B. Schölkopf, G. Rätsch: Support vector machines and kernels for computational biology, PLoS Comput. Biol. **4**(10), e1000173 (2008)

18.5   K.M. Borgwardt: Kernel methods in bioinformatics,. In: *Handbook of Statistical Bioinformatics*, ed. by H.H. Lu, B. Schölkopf, H. Zhao (Springer, Berlin Heidelberg 2011) pp. 317–334

18.6   B. Schölkopf: Support Vector Learning. Dr. Thesis (Technische Universität Berlin, Berlin 1997)

18.7   J. Mercer: Functions of positive and negative type and their connection with the theory of integral equations, Philos. Trans. R. Soc. **A209**, 415–446 (1909)

18.8   M.A. Aizerman, E.M. Braverman, L.I. Rozonofier: Theoretical foundations of the potential function method in pattern recognition learning, Autom. Remote Control **25**, 821–837 (1964)

18.9   B.E. Boser, I.M. Guyon, V.N. Vapnik: A training algorithm for optimal margin classifiers, Proc. 5th Annu. ACM Workshop Comput. Learn. Theory (1992) pp. 144–152

18.10   B. Schölkopf, A.J. Smola, K.R. Müller: Nonlinear component analysis as a kernel eigenvalue problem, Neural Comput. **10**, 1299–1319 (1998)

18.11   P.L. Lai, C. Fyfe: Kernel and nonlinear canonical correlation analysis, Int. J. Neural Syst. **10**, 365–377 (2000)

18.12   S. Mika, G. Rätsch, J. Weston, B. Schölkopf, K.-R. Müller: Fisher discriminant analysis with kernels, Neural Networks for Signal Processing, Vol. 9, ed. by Y.-H. Hu, J. Larsen, E. Wilson, S. Douglas (IEEE, 1999) pp. 41–48

18.13   F.R. Bach, M.I. Jordan: Kernel independent component analysis, J. Mach. Learn. Res. **3**, 1–48 (2003)

18.14   B. Schölkopf: The kernel trick for distances. In: *Advances in Neural Information Processing Systems*, Vol. 13, ed. by T.K. Leen, T.G. Dietterich, V. Tresp (MIT Press, Cambridge, MA 2001) pp. 301–307

18.15   K. Yu, L. Ji, X. Zhang: Kernel nearest-neighbor algorithm, Neural Process. Lett. **15**, 147–156 (2002)

18.16   J. Peng, D.R. Heisterkamp, H.K. Dai: Adaptive quasiconformal kernel nearest neighbor classification, IEEE Trans. Pattern Anal. Mach. Intell. **26**, 656–661 (2004)

18.17   Y. Fu: Machine Learning Based Bioinformation Retrieval, Dissertation (Chinese Academy of Sciences 2007)

18.18   J. Xu, H. Li, C. Zhong: Relevance ranking using kernels, Proc. 6th Asian Inf. Retr. Soc. Symp. (2010) pp. 1–12

18.19   W. Wu, J. Xu, H. Li, S. Oyama: Learning a robust relevance model for search using kernel methods, J. Mach. Learn. Res. **12**, 1429–1458 (2011)

18.20   H. Lodhi, C. Saunders, J. Shawe-Taylor, N. Cristianini, C. Watkins: Text classification using string kernels, J. Mach. Learn. Res. **2**, 419–444 (2002)

18.21   C. Watkins: Dynamic alignment kernels,. In: *Advances in Large Margin Classifiers*, ed. by A.J. Smola, P.L. Bartlett, B. Schölkopf, D. Schuurmans (MIT, Cambridge 1999) pp. 39–50

18.22   D. Haussler: Convolution kernels on discrete structures, Technical Report UCSC-CRL-99-10 (1999)

18.23   T.S. Jaakkola, D. Haussler: Exploiting generative models in discriminative classifiers,. In: *Advances in Neural Information Processing Systems*, Vol. 11, ed. by M.S. Kearns, S.A. Solla, D.A. Cohn (MIT Press, Cambridge, MA 1999) pp. 487–493

18.24   R.I. Kondor, J. Lafierty: Diffusion kernels on graphs and other discrete input spaces, Proc. 9th Int. Conf. Mach. Learn. (2002) pp. 315–322

18.25   C. Leslie, E. Eskin, W.S. Noble: The spectrum kernel: A string kernel for SVM protein classification,

Proc. 7th Pac. Sympos. Biocomput. (2002) pp. 564–575

18.26   C. Leslie, E. Eskin, J. Weston, W.S. Noble: Mismatch string kernels for discriminative protein classification, Bioinformatics **20**(4), 467–476 (2004)

18.27   J.P. Vert: Support vector machine prediction of signal peptide cleavage site using a new class of kernels for strings, (2002) pp. 649–660

18.28   A. Zien, G. Rätsch, S. Mika, B. Schölkopf, T. Lengauer, K.-R. Muller: Engineering support vector machine kernels that recognize translation initiation sites, Bioinformatics **16**, 799–807 (2000)

18.29   T. Jaakkola, M. Diekhans, D. Haussler: A discriminative framework for detecting remote protein homologies, J. Comput. Biol. **7**, 95–114 (2000)

18.30   J.P. Vert, M. Kanehisa: Graph-driven features extraction from microarray data using diffusion kernels and kernel CCA. In: *Advances in Neural Information Processing Systems*, Vol. 15, ed. by S. Becker, S. Thrun, K. Obermayer (MIT Press, Cambridge, MA 2003) pp. 1425–1432

18.31   J.P. Vert: A tree kernel to analyze phylogenetic profiles, Bioinformatics **18**, S276–S284 (2002)

18.32   K.M. Borgwardt, C.S. Ong, S. Schönauer, S.V.N. Vishwanathan, A.J. Smola, H. Kriegel: Protein function prediction via graph kernels, Bioinformatics **1**(Suppl.), 47–56 (2005)

18.33   N. Cristianini, J. Shawe-Taylor: *An Introduction to Support Vector Machines and Other Kernel-based Learning Methods* (University Press, Cambridge 2000)

18.34   G. Schweikert, A. Zien, G. Zeller, J. Behr, C. Dieterich, C.S. Ong, P. Philips, F. De Bona, L. Hartmann, A. Bohlen, N. Krüger, S. Sonnenburg, G. Rätsch: mGene: Accurate SVM-based gene finding with an application to nematode genomes, Genome Res. **19**(11), 2133–2143 (2009)

18.35   I. Guyon, J. Weston, S. Barnhill, V. Vapnik: Gene selection for cancer classification using support vector machines, Mach. Learn. **46**, 389–422 (2002)

18.36   Y. Sun, X. Fan, Y. Li: Identifying splicing sites in eukaryotic RNA: Support vector machine approach, Comput. Biol. Med. **33**(1), 17–29 (2003)

18.37   J. Gubbi, A. Shilton, M. Palaniswami: Kernel methods in protein structure prediction. In: *Machine Learning in Bioinformatics*, ed. by Y.-Q. Zhang, J.C. Rajapakse (Wiley, Hoboken 2008)

18.38   G.R.G. Lanckriet, M. Deng, N. Cristianini, M.I. Jordan, W.S. Noble: Kernel-based data fusion and its application to protein function prediction in yeast, Proc. 9th Pac. Symp. Biocomput. (2004) pp. 300–311

18.39   H. Wang, Y. Fu, R. Sun, S. He, R. Zeng, W. Gao: An SVM scorer for more sensitive and reliable peptide identification via tandem mass spectrometry, Proc. 11th Pac. Symp. Biocomput. (2006) pp. 303–314

18.40   Y. Li, P. Hao, S. Zhang, Y. Li: Mol cell proteomics, feature-matching pattern-based support vector machines for robust peptide mass fingerprinting, Mol. Cell. Proteomic. **10**(12), M110.0057852011 (2011)

18.41 R. Aebersold, M. Mann: Mass spectrometry-based proteomics, Nature **422**, 198–207 (2003)

18.42 H. Steen, M. Mann: The ABC's (and XYZ's) of peptide sequencing, Nat. Rev. Mol. Cell **5**, 699–711 (2004)

18.43 J.K. Eng, A.L. McCormack, J.R. Yates: An approach to correlate tandem mass spectral data of peptides with amino acid sequences in a protein database, J. Am. Soc. Mass Spectrom. **5**, 976–989 (1994)

18.44 K.X. Wan, I. Vidavsky, M.L. Gross: Comparing similar spectra: From similarity index to spectral contrast angle, J. Am. Soc. Mass Spectrom. **13**, 85–88 (2002)

18.45 B. Schölkopf, P. Simard, A. Smola, V. Vapnik: Prior knowledge in support vector kernels, Adv. Neur. Inf. Proces. Syst., Vol. 10, ed. by M. Jordan, M. Kearns, S. Solla (MIT, Cambridge 1998) pp. 640–646

18.46 Y. Fu, Q. Yang, R. Sun, D. Li, R. Zeng, C.X. Ling, W. Gao: Exploiting the kernel trick to correlate fragment ions for peptide identification via tandem mass spectrometry, Bioinformatics **20**, 1948–1954 (2004)

18.47 D. Li, Y. Fu, R. Sun, C. Ling, Y. Wei, H. Zhou, R. Zeng, Q. Yang, S. He, W. Gao: pFind: A novel database-searching software system for automated peptide and protein identification via tandem mass spectrometry, Bioinformatics **21**, 3049–3050 (2005)

18.48 A. Keller, S. Purvine, A.I. Nesvizhskii, S. Stolyar, D.R. Goodlett, E. Kolker: Experimental protein mixture for validating tandem mass spectral analysis, Omics **6**, 207–212 (2002)

18.49 Y. Zhang: Progress and challenges in protein structure prediction, Curr. Opin. Struct. Biol. **18**, 342–348 (2008)

18.50 T. Liu: *Learning to Rank for Information Retrieval* (Springer, New York 2011)

18.51 R. Caruana, T. Joachims, L. Backstrom: KDD Cup 2004: Results and analysis, SIGKDD Explorations **6**, 95–108 (2004)

18.52 Y. Fu, R. Sun, Q. Yang, S. He, C. Wang, H. Wang, S. Shan, J. Liu, W. Gao: A block-based support vector machine approach to the protein homology prediction task in KDD Cup 2004, SIGKDD Explorations **6**, 120–124 (2004)

18.53 C. Foussette, D. Hakenjos, M. Scholz: KDD-Cup 2004 – Protein homology task, SIGKDD Explorations **6**, 128–131 (2004)

18.54 B. Pfahringer: The Weka solution to the 2004 KDD cup, SIGKDD Explorations **6**, 117–119 (2004)

18.55 Y. Tang, B. Jin, Y. Zhang: Granular support vector machines with association rules mining for protein homology prediction, Artif. Intell. Med. **35**, 121–134 (2005)

18.56 Y. Fu, R. Pan, Q. Yang, W. Gao: Query-adaptive ranking with support vector machines for protein homology prediction. In: *ISBRA 2011*, Lecture Notes in Bioinformatics, Vol. 6674, ed. by J. Chen, J. Wang, A. Zelikovsky (Springer, Berlin Heidelberg 2011) pp. 320–331

18.57 D. Tobi, R. Elber: Distance dependent, pair potential for protein folding: Results from linear optimization, Proteins Struct. Funct. Genet. **41**, 40–46 (2000)

18.58 O. Teodorescu, T. Galor, J. Pillardy, R. Elber: Enriching the sequence substitution matrix by structural information, Proteins Struct. Funct. Bioinform. **54**, 41–48 (2004)

Part D
Modeling

Part D Modeling Regulatory Networks: The Systems Biology Approach

Ed. by Chris Brown

19. Path Finding in Biological Networks

Lore Cloots, Dries De Maeyer, Kathleen Marchal

Understanding the cellular behavior from a systems perspective requires the identification of functional and physical interactions among diverse molecular entities in a cell (i. e., DNA/RNA, proteins, and metabolites). The most straightforward way to represent such datasets is by means of molecular networks of which nodes correspond to molecular entities and edges to the interactions amongst those entities. Nowadays with large amounts of interaction data being generated, genome-wide networks can be created for an increasing number of organisms. These networks can be exploited to study a molecular entity like a protein in a wider context than just in isolation and provide a way of representing our knowledge of the system as a whole. On the other hand, viewing a single entity or an experimental dataset in the light of an interaction network can reveal previous unknown insights in biological processes.

In this chapter we focus on different approaches that have been developed to reveal the functional state of a network, or to find an explanation for the observations in functional data through paths in the network. In addition we give an overview of the different omics datasets and data-integration techniques that can be used to build integrated biological networks.

19.1 Background

With the advent of new molecular profiling techniques, genome-wide datasets that describe interactions between molecular entities (i. e., mRNA, proteins, metabolites, etc.) are being generated at an ever increasing pace. These datasets each measure a specific type of interaction that is active under certain conditions within the cell or that occurs as a response to specific environmental signals. The distinct nature of these datasets often brings about complementary views on cellular behavior. A network-based representation of various biological systems captures many of the essential characteristics of these data and integrating complementary molecular interaction layers into a single network thus provides a way of representing our knowledge of the system as a whole. Application of well-established tools and concepts developed in fields such as graph theory on such networks can provide valuable insights into the system's mode of action and functionalities [19.1]. The identification of motifs, which are statistically significant reoccurring characteristic patterns in a network [19.2], has for instance shown that specific types of motifs carry out specific information-processing functions within cells [19.3].

An integrated interaction network can also reveal previous unknown insights in biological processes or functional behavior by explicitly interrogating it with independent functional data sets. Methodologies that identify and explore paths in networks between given input and output nodes have gained much interest. Such a path in a network can be seen as a mechanistic representation of the way information propagates through the network. Identifying biologically meaningful paths in the network between nodes of interest,

nodes which can be defined from functional data sets that are independent from the network itself, can unveil previously uncovered signal flow mechanisms that are responsible for the observed functional behavior or define a measure for relatedness of two nodes in the network.

In this chapter, we highlight diverse omics datasets and data-integration techniques that can be used to build integrated biological networks and discuss several categories of network-based path finding methodologies that aim at obtaining a more functional understanding of cellular behavior.

19.2 Inferring Interaction Networks from Omics Data

Understanding the cellular behavior from a systems perspective requires the identification of functional and physical interactions among diverse molecular entities in a cell (i.e., DNA/RNA, proteins and metabolites). The most straightforward manner of capturing interactions between molecular entities is by representing them as an interaction network. Here, molecular entities are represented by nodes and the interactions between them by edges. In this section, we elaborate on different types of networks that can be constructed from omics data and present supervised learning strategies to assign reliabilities to interactions.

19.2.1 Network Representations

Classically, a distinction is made between a functional network in which nodes usually correspond to proteins or genes, and edges represent functional relations between the nodes and a physical network where edges represent direct physical interactions (Fig. 19.1). Proteins connected in a functional network can be interpreted as being active in the same pathway or being needed together to mediate a specific function, but they do not necessarily physically interact. Examples of specific functional networks are, for instance, genetic interaction networks and coexpression networks. Within a physical network, different molecular layers can be distinguished: intracellular signal transduction, which transmits information from the surface to the nucleus, for instance by means of protein phosphorylation, and protein interactions that propagate this signal. In addition, (post-) transcriptional regulation processes comprise transcription factor (TF) proteins or sRNAs regulating the expression of genes and finally metabolic

reactions catalyzed by enzymes, where metabolites are converted into energy and building blocks. Each of these different layers in a physical network can be deduced from their own specific datasets and will have their own characteristics.

Overview of Different Datasets

Small-scale laboratory experiments alone are impractical for creating a genome-scale network of different types of interactions, mainly for reasons of cost and time. Recently, advances in experimental methods made it possible to generate interaction datasets in a high throughput manner. Such datasets, like for instance, protein–protein interactions (PPI), have been generated for model organisms such as *Saccharomyces cerevisiae* [19.4–8], *Caenorhabditis elegans* [19.9], and *Drosophila melanogaster* [19.10, 11], as well as *Homo sapiens* [19.12, 13] by genome-wide yeast two hybrid (Y2H) screens and large-scale affinity purification/mass spectrometry. Technologies such as ChIP-chip and ChIP-seq, make it possible to measure the TF-DNA interactions at a genomic scale [19.14–16], and mass spectrometry (MS)-based proteomics have enabled the large-scale mapping of in vivo phosphorylation sites [19.17].

Due to the large flood of experimental interaction data becoming available, several efforts have been made to store and centralize these datasets through the construction of databases. Some databases capture data about a specific organism or research topic, like, for instance, transcriptional regulation, while others integrate data from specific organisms and/or different interaction types in a standardized manner. Gradually more and more specific databases are merged into these integrated

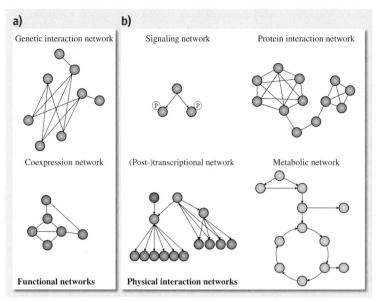

Fig. 19.1a,b Overview of molecular networks that can be inferred from omics data. (**a**) Functional networks: molecular entities represented by nodes in the network share a functional relation that does not require their physical contact. Genetic interaction network: edges reflect the phenotype that is observed when both nodes (genes) are inactivated simultaneously (double mutant). *Coexpression network*: nodes correspond to genes and edges represent the mutual similarity in expression profiles between connected nodes. (**b**) Physical interaction networks. Physical contact occurs amongst members (nodes) of the network. *Signaling network*. Nodes are proteins and edges represent signaling events (e.g., phosphorylation). *Protein interaction network*. Edges represent physical interactions between proteins represented by the nodes. *(Post-)transcriptional network*. Nodes represent either regulators or target genes and the directed edges reflect the physical regulator (transcription factor, sRNA)-target interactions. *Metabolic network*. Edges correspond to metabolic reactions catalyzed by enzymes represented by the nodes

databases. Table 19.1 gives an overview of some frequently used databases that provide physical interaction data for several eukaryotic model organisms, categorized by the type(s) of data they provide.

The fragmentation of data over a rising amount of databases [19.33] makes an integrated and comprehensive use very difficult. To reduce this problem, *Bader*, *Cary*, and *Sander* [19.34] provide an extensive overview of several databases (at the time of this writing this number was equal to 328) spanning different interaction layers across a multitude of organisms in a meta-database named Pathguide.

Assessing the Confidence in Interactions Derived from High-Throughput Experimental Data

Experimental datasets generated by high-throughput methods like Y2H or ChIP-chip are not only prone to

Table 19.1 Overview of databases containing physical interaction data for eukaryotic model organisms

Type of interactions	Database
Metabolic pathways	KEGG [19.18], MetaCyc [19.19], BiGG [19.20]
Protein–protein interactions	BioGRID [19.21], BIND [19.22], DIP [19.23], MINT [19.24], MIPS [19.25], STRING [19.26], HPRD [19.27] (*H. sapiens*)
Protein complexes	MIPS [19.25], CORUM [19.28] (mammalian organisms)
Transcription factor–DNA interactions	TRANSFAC [19.29], modENCODE [19.16] (*D. melanogaster, C. elegans*), YPD [19.30] (*S. cerevisiae*)
Signaling interactions	PhosphoPOINT [19.31], PhosphoSite [19.32]

high rates of false positive interactions [19.15, 35], low overlap between them also indicates that the current interaction maps are far from complete [19.36]. Assessing the quality of the data obtained is useful both for deciphering the correct molecular mechanisms underlying given biological functions, and for intelligent future experiment design [19.37].

Von Mering et al. [19.38] addressed the problem of extracting highly confident interactions between proteins from high throughput data sources by using the intersection of direct high throughput experimental results. Although they were able to achieve low false positive rates, the coverage in the number of retrieved interactions was also very low. Increasing the coverage of a network is especially useful for humans since the protein interaction map, for instance, was estimated to be only 10% complete [19.36]. Predicting interactions augments the current knowledge of the relationship between distinct cellular processes and underlying mechanisms of diseases.

Several machine-learning methodologies were suggested to assign reliabilities to interactions identified by experimental data, as well as to predict de novo interactions. In Sect. 19.2.2, we, therefore, introduce the concept of supervised learning to assess interactions.

19.2.2 Integration Frameworks Based on Supervised Techniques to Predict Interactions

Supervised learning methods [19.39] infer a function from a training set. Such a training set consists of pairs of input vectors and their corresponding known output. When the output is discrete, the supervised method is called a classifier. The learned function between input and output can then be used to predict the output of any valid input of which the output is not yet known. Applied here, a classifier would exploit known interactions to infer novel interactions from omics data. They can learn the set of data characteristics (features) that allow distinguishing true from false interactions from a set of known interactions (training set). A novel interaction is then predicted to be true or false, depending on the extent to which it shares similar features with the interactions in the training set.

Since classifiers are commonly used to assess the confidence in interactions from omics data, some guidelines for the choice of features and training sets are discussed in the next section, as well as two types of classifiers that are frequently used to stratify many candidate interactions by confidence or predict novel interactions, namely Bayesian approaches and logistic regression. A short case study in predicting PPI and functional interactions is presented in the last part of this section, as an enormous amount of high throughput experimental data for PPI is nowadays freely available in several databases. However, the integration process is general and can be used for assessing interactions at other network levels as well, using the standard frameworks described below, together with a set of features and training set that is specific for the dataset being assessed (e.g., [19.35, 40] for assessing TF-DNA interactions obtained from ChIP-chip).

Features

The set of features provided to the classifier are measurable entities or *evidences* that characterize an interaction or a noninteraction. The classifier then learns which of the provided features are predictive for the interactions at hand. Such measurable entities can be direct information (e.g., the interaction was seen in an experiment) or indirect information (e.g., the expression correlation of two proteins could indicate that they are members of the same complex). Examples for predicting PPI are, for instance, network topology-based features [19.41] and GO biological process similarity [19.42,43], amongst others. Nucleosome occupancy [19.40], DNA binding motifs [19.35, 40], and shared phylogenetic profiles (i.e., occurrence of the interaction in multiple species) [19.35] have been shown to be predictive for TF-DNA interactions. Coexpression between genes [19.35,40,42–45] has been used both for predicting TF-DNA and PPIs.

Training Sets

The prediction quality of a classification scheme stands or falls with the choice of a golden standard training set. This training dataset usually consists of positive and negative examples and is used to discover a predictive relationship between several features and the positive and negative examples.

An ideal golden standard should be independent of the data sources serving as features, sufficiently large for reliable statistics, and free of systematic bias [19.42]. Moreover, the choice of training set also depends on the prediction task at hand: positive and negative examples should reflect the same entities as the ones one would like to predict. This means, for instance, that a golden standard for predicting protein complexes should consist of proteins belonging or not belonging to the same complex, while positive and negative examples for predicting a functional network, on the other hand, should reflect functional and nonfunctional relationships, respectively.

A set of positive examples is usually based on a curated, literature-derived dataset, containing only high-confidence interactions. For predicting physical protein interactions, a high quality subset of the Database of Interaction Proteins (i.e., DIP [19.23]) discovered by small-scale experiments or data from individually performed experiments listed in The Munich Information Center for Protein Sequences (i.e., MIPS [19.25]) can, for instance, be applied. TF–DNA interactions from the Incyte YPD database [19.30] could serve as a positive set for transcriptional interactions in yeast [19.35, 40]. Positive examples of functional relations between proteins can be extracted from the gene ontology (i.e., GO [19.46]) database annotations. Usually, proteins are considered functionally related if they share a specific biological process GO term (e.g., contain less than 200 annotations [19.47]).

A good set of negative examples is harder to define, since noninteracting pairs cannot be observed. Negative training sets can, for instance, consist of randomly combined pairs [19.35], 44], randomly observed interactions in a high throughput dataset [19.45, 48], or, in the case of protein interactions, proteins occurring in different subcellular components [19.42–44], and proteins not sharing any specific GO term [19.47, 49–53].

Bayesian Approaches

Different sources of evidence can be probabilistically combined to predict interactions using Bayesian formalism. This learning framework allows for combining highly dissimilar types of data in a model that is easy to interpret and that can readily accommodate missing data.

The posterior odds of interaction between two molecular entities (O_{post}) represent the probability that an interaction occurs given the presence of several genomic features, divided by the probability that such an interaction will not occur given the presence of these features. This can be formalized using Bayes' theorem,

$$O_{\text{post}} = \frac{P(I|f_1, \ldots, f_N)}{P(\sim I|f_1, \ldots, f_N)}, \tag{19.1}$$

$$= \frac{\frac{P(f_1, \ldots, f_N|I) \cdot P(I)}{P(f_1, \ldots, f_N)}}{\frac{P(f_1, \ldots, f_N|\sim I) \cdot P(\sim I)}{P(f_1, \ldots, f_N)}}, \tag{19.2}$$

$$= \frac{P(I)}{P(\sim I)} \cdot \frac{P(f_1, \ldots, f_N|I)}{P(f_1, \ldots, f_N| \sim I)}, \tag{19.3}$$

$$= O_{\text{prior}} \cdot \text{LR}. \tag{19.4}$$

The posterior odds of an interaction can thus be calculated as the product of the prior odds (O_{prior}) of interaction and the likelihood ratio (LR) of an interaction (19.4).

The prior odds of interaction are defined as the probability of encountering an interaction among all pairs, divided by the probability of observing no interaction between a pair. The likelihood ratio represents the probability of observing the values in the predictive datasets, given that a pair of molecular entities interacts, divided by the probability of observing these values given that the pair does not interact.

A naive Bayesian classifier makes the assumption that the genomic features (denoted by $f_1, \ldots, f_N$) are independent. In this case, the LR can be calculated as the product of the individual likelihood ratios from the respective genomic features (19.5)

$$\text{LR} = \prod_{i=1}^{N} \left[\frac{P(f_i|I)}{P(f_i| \sim I)} \right]. \tag{19.5}$$

The likelihood ratio for every genomic feature can be estimated by counting the frequency of occurrence of interacting and noninteracting pairs in the golden standard that possess a particular value of the feature.

In the case of features with correlated evidence, the likelihood ratio cannot be factorized in this way and all possible combinations of all states of the features must be considered, which can be computationally intensive. The prior odds are more difficult to assess, since not all true interactions are known. For PPI this parameter, for instance, be estimated by examining the average number of interactions per protein for which all known interactions have been identified in the literature [19.43, 44].

After deriving likelihood ratios for independent features from the golden standard, the likelihood ratio for every protein pair can be determined by combining the likelihood ratios for every independent evidence source [19.43, 44]. An interacting pair is then predicted as positive if its likelihood ratio exceeds a certain cut off [19.42, 49, 51].

Logistic Regression

A logistic regression is a generalized linear model that is used to calculate the probability of the outcome of an event, e.g., the probability of observing an interaction between two proteins. The relationship between the response variable (e.g., observing an interaction ($= 1$) or not ($= 0$)) and the predictor variables (i.e., genomic features and/or experimental observations) is given by a logistic function

$$P(I) = \frac{1}{1 + e^{-(\beta_0 + \beta_1 f_1 + \dots + \beta_N f_N)}} \ . \qquad (19.6)$$

The logistic function can take as an input (i.e., evidence features $f_1, \dots, f_N$) any value from negative infinity to positive infinity, whereas the output (i.e., probability of an interaction $P(I)$) is confined to values between 0 and 1. Parameters $\beta_1, \dots, \beta_N$ can be estimated by using a set of positive and negative interaction pairs as output (yielding output values of 1 and 0, respectively) and their corresponding features as input in a maximum likelihood approach [19.39]. The estimated parameters can then be used together with evidence features corresponding to an interaction between two molecular entities to predict the probability that these entities truly interact.

Case Study: Inferring Protein Interaction Networks from Omics Data

Examples of PPI and Functional Networks. Using the Bayesian framework or slight variations of it and specific sets of genomic features or experimental datasets, both functional networks for *S. cerevisiae* [19.49, 50], *C. elegans* [19.54], human [19.53], *Arabidopsis thaliana* [19.52], mouse [19.47] and protein-complex [19.42], and PPI networks [19.43, 44] for human were developed.

Methods based on logistic regression have been used to assess the confidence of interactions observed in experimental data [19.41, 45, 48] by integrating experimental information, topological measures and/or expression correlation, and have been used in several path finding approaches to assign a confidence score to the PPI in a yeast network [19.55–58] or a human network [19.59], and to assess the reliability of experimentally determined protein interactions in *D. melanogaster* [19.11].

Performance. Many supervised classification methods have been developed to integrate direct and indirect information on protein interactions. They each differ in the collection of integrated data sources, approach, and implementation. *Qi* et al. [19.60] independently investigated the performance of different classifiers and the importance of different biological datasets, together with various golden standards. They concluded that a classifier based on Random Forests performed best among the classifiers, followed by a logistic regression. However, *Suthram* et al. [19.61] assessed the performance of six approaches, each with their own combination of features and classification method, and showed that a rather complex approach based on Random Forests [19.62] had lower overall performance compared to other methods tested. Both authors could conclude that including many input variables does not necessarily result in a better prediction performance, and in some cases even the opposite can be true. However, utilizing any probability scheme turned out to be better than considering all interactions observed to be true or equally probable.

19.3 Using Interaction Networks to Interpret Functional Data

Nowadays with large amounts of interaction data being generated, genome-wide networks can be created for an increasing number of organisms. These networks can be exploited to study a molecular entity like a protein in a wider context than just in isolation. However, the inferred physical networks are static and do not reveal which parts of the networks are active under certain conditions and how perturbations are propagated through the network. Integrating physical interaction networks with functional data, like gene expression data makes it possible to reveal relevant active paths or substructures in the network.

High-throughput techniques now allow genome-wide views of the molecular changes that occur in cells as they respond to stimuli. However, data derived from these high-throughput techniques unveil that our understanding of cellular systems is still fragmentary, even of well-characterized model systems. In humans, for instance, only about 30–40% of all differentially expressed genes for transcription factors NF-kB and STAT1 appear to be direct targets [19.63]. *Yeger–Lotem* et al. [19.64] observed that the results of genetic screenings (i.e., identifying genetic hits, or genes whose individual manipulation alters the phenotype of stimulated cells) and mRNA profiling (i.e., identifying differentially expressed genes following stimuli) often hardly overlap and provide a limited and biased view of cellular responses.

Exploiting the network structure can help in gaining a comprehensive picture of the functioning of a cell,

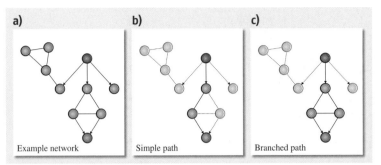

Fig. 19.2a–c Definition of a path (**a**) example of an interaction network, with an input node (*red node*), output node (*blue node*), and other interacting entities (*green nodes*) mapped on the network. *Arrows* between nodes represent directed interactions, *lines* between nodes represent undirected interactions (**b**) a simple path from input to output node is highlighted in the network, *dashed lines*, *arrows*, and transparent nodes do not belong to the selected path (**c**) a branched path from input to output node, containing multiple simple paths is highlighted in the network. *Dashed lines*, *arrows*, and *transparent nodes* do not belong to the path

by providing a mechanistic explanation that links the observed effects to the perturbation or cause exerted. Such an underlying mechanism is unlikely to be discoverable when looking at all datasets separately. Several approaches exist for mining the information embedded in integrated networks, dependent on the specifics of the problem statement. Network clustering strategies for instance, which search for highly connected subnetworks, have been successfully used to distinguish cancer-causing mutations from neutral mutations [19.65] or to assess the structure of the yeast genetic interaction network, revealing insights in gene function and modular organization [19.66].

In this chapter we focus on different approaches that have recently been developed to reveal the functional state of a network or to find an explanation for the observations in functional data through paths in the network. A simple path in a network is illustrated in Fig. 19.2b and is defined as a collection of edges that connect a source node (i.e., gene causing an effect or input gene of interest) and a target node (i.e., affected gene or output gene of interest) in an interaction network, such that each selected edge is connected to one other selected edge and the information spread by the source node can reach the target node without interruption. A path can be a collection of simple paths, containing several branches connecting a source with a target node (Fig. 19.2c). There is no further constraint that the nodes within a path should be densely connected to each other, which would refer to a cluster in a graph and would comprise a different problem statement.

In this second part of the chapter, an extensive overview of path finding methodologies, illustrated with

several applications, is given. Different approaches are categorized according to the underlying goal they try to accomplish. These goals are represented in an abstract way in Figs. 19.3–19.6, and are further clarified at the beginning of each category.

19.3.1 Connecting One or Several Causes to Their Effect(s) by Unveiling the Underlying Active Paths

The common objective of methods described in this paragraph is to reveal the underlying pathways transmitting a signal from one or several causes to their corresponding observed effect(s) by adopting a network-based approach (Fig. 19.3). The cause could, for instance, be a membrane protein and the observed effect a DNA binding protein that receives the signal, but the intermediate molecular interactions through which the signal was transduced from cause to effect is unknown.

Several of the methods developed for this purpose use, in addition to the given cause and effect pairs, other functional data like gene expression to extract biologically relevant paths from the network. This either by using the extent to which a network node is differentially expressed as an indication of its contribution to a plausible signaling path [19.64, 67] or by using a measure of expression correlation between edge nodes [19.55, 68], between edge nodes and the source and target nodes [19.69], or between edge nodes and target node [19.70] to indicate the confidence we have in an edge contributing to a causal path.

The reconstruction of signaling pathways by overlaying PPI data with cause–effect pairs has received

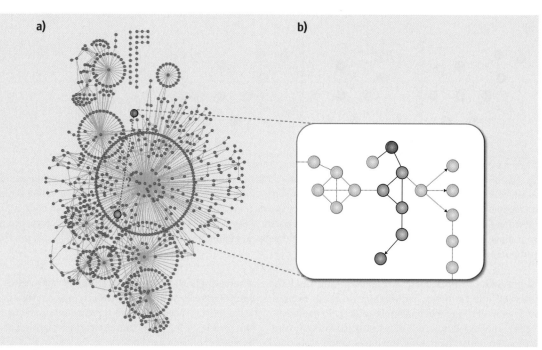

Fig. 19.3a,b Connecting one or several causes to their effect(s) by unveiling the underlying active paths. (**a**) Example of an interaction network on which a known causal gene (*red node* or input) and its affected gene (*blue node* or output) were mapped. The underlying path responsible for transferring the information from input to output is unknown (*blue dashed line*). (**b**) The underlying mechanism that explains the observed effect is highlighted in the network, *dashed lines*, *arrows*, and *transparent nodes* do not belong to the selected path. *Arrows* between nodes represent directed interactions, *lines* between nodes represent undirected interactions

a great deal of attention. *Steffen* et al. [19.71] were one of the first to model simple paths of a specified length through a physical protein interaction network, starting at a membrane protein and ending on a DNA binding protein in a procedure called NetSearch. Paths were ranked based on a statistical scoring metric, reflecting how many path members clustered together according to their expression profiles. Simple paths that had common starting points and endpoints and the highest ranks among each other were then combined into the final model of branched networks.

In reality, simple paths cannot capture the full complexity of signaling pathways since there may be multiple interaction paths within a pathway. *Scott* et al. [19.55], therefore, adapted the color coding technique and allowed the identification of more complicated substructures such as trees and series-parallel graphs. A number of candidate paths are firstly found with a score assigned to each candidate and the top scoring paths are then assembled into a signaling network. *Lu* et al. [19.69] extracted nonlinear path structures

from the network and potential interactions between related paths were taken into account.

However, most of these methods generally cannot directly find a signaling network as a whole, i.e., they first identify separate paths and then heuristically assemble them into a signaling network. Other approaches like those of *Zhao* et al. [19.68], *Yosef* et al. [19.58], *Yeger-Lotem* et al. [19.64], and *Ren* et al. [19.70] infer active paths immediately as a subnetwork from the whole network. The methods have in common that they try to explain cause–effect pairs in a particular set of experiments by solving an optimization problem which typically balances the reliability of the edges used by the length and complexity of the possible paths. A third category of methods uses the frequency of occurrence of a path with a predefined form (i.e., a motif) in the network to explain cause–effect pairs on a more statistical basis. An example of this category is the method of *Joshi* et al. [19.72], which is discussed in more detail in the case study at the end of this section.

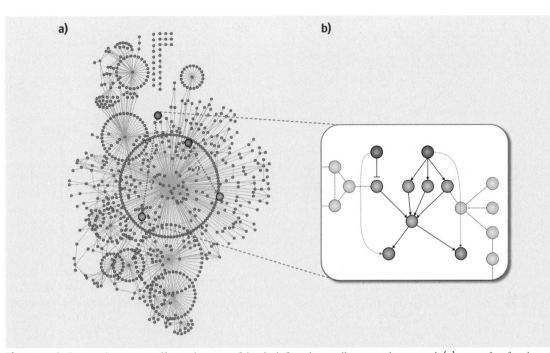

Fig. 19.4a,b Integrating cause–effect pairs to confidently infer edge attributes on the network (**a**) example of an interaction network on which different known cause (*red nodes*)–effect (*blue nodes*) pairs were mapped. The type of observed effect is also taken into account: a *regular arrow* represents an activating effect from input to output; a *cut arrow* represents an inhibiting effect from input to output. The underlying path responsible for transferring the information from input to output is unknown (*blue dashed line*). Also the type of effect (i. e., activating or inhibiting) for each edge on the path must be inferred consistently by making use of the cause–effect pairs (**b**) the underlying path that explains the observed effect is highlighted in the network, whereby also a type of effect to each edge in the path is assigned. *Dashed lines*, *arrows*, and *transparent nodes* do not belong to the selected path, the *blue dashed lines* show the observed cumulative effect from input to output as in (**a**). *Arrows* between nodes represent directed interactions, *lines* between nodes represent undirected interactions

The majority of these methods use one or more MAP kinase signaling pathways involved in pheromone response, filamentous growth, maintenance of cell wall integrity, and high osmolarity as their benchmark, since these are among the best studied signaling networks. These pathways are activated by G protein-coupled receptors and characterized by a core cascade of MAP kinases that activate each other through sequential binding and phosphorylation reactions. A method comparison performed by *Zhao* et al. [19.68] demonstrated that most methods can to a large extent uncover the known signaling paths, which confirm the effectiveness and prediction power of the approaches. On the other hand, the results also show that there is no single method that can perform the best in all cases, and different models are complementary to each other.

While previous methods concentrate on the reconstruction of signaling cascades between a membrane receptor protein and a target protein, *Yeger–Lotem* et al. [19.64] focus on identifying molecular interaction paths connecting several related genetic hits (sources) and differentially expressed genes (targets), revealing the underlying response pathways. They hypothesize that some of the genetic hits, which are enriched for regulators of cellular response, will be connected via regulatory paths to the differentially expressed genes, which are the outputs of such paths, via components of the response that were not detected by either the genetic or the mRNA profiling assays themselves. To identify these undetected path components, the authors developed a flow algorithm called ResponseNet. *Huang* and *Fraenkel* [19.67] reconsidered this problem by taking into account that both the input data from experimen-

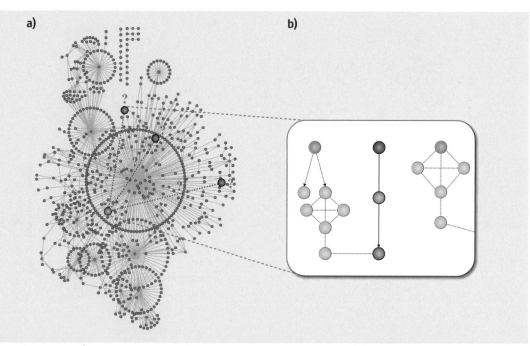

Fig. 19.5a,b Identifying (an) unknown causal input(s) for (an) observed effect(s). (**a**) Example of an interaction network on which several candidate causal genes (*red nodes* or possible inputs) and an affected gene (*blue node* or output) were mapped. The underlying path responsible for transferring the information from input to output is unknown (*blue dashed line*). Also the most likely input for the observed output should be identified (*red question marks*). (**b**) the most likely causal gene (*red node*) together with the underlying mechanism that explains the observed effect (*blue node*) is highlighted in the network, *dashed lines* and *arrows* and *transparent nodes* do not belong to the selected path. *Arrows* between nodes represent directed interactions, *lines* between nodes represent undirected interactions

tal observations and the interactome can contain noise. They treat the goal of connecting data as a constraint that is attempted to be satisfied through an optimization procedure, resulting in a subnetwork that contains mainly reliable edges while excluding possibly false positive source or target nodes.

Other application examples of this problem formulation can be found in the reconstruction of metabolic pathways [19.73], connecting a source metabolite to a target metabolite, and the reconstruction of transcriptional regulation [19.74], connecting regulators to their target module consisting of coexpressed genes.

Case Study
Previously mentioned techniques do not search for general mechanisms or path structures that are common between different cause and effect pairs, nor include a significance analysis that assesses the statistical significance of the inferred paths. In this way, it is difficult to assess if a given network model truly reflects under-

lying regulation mechanisms or appears just by chance due to the inevitable noise inherent in the perturbation data as well as in the physical interaction networks.

Joshi et al. [19.72], therefore, propose an alternative strategy by searching for *regulatory path motifs* in an integrated transcriptional (TRI), protein–protein (PPI) and phosphorylation (PhI) interaction network. Regulatory path motifs are defined as paths of length up to three, which connect a causative gene (for example a transcription factor) to a set of effect genes which are differentially expressed after perturbation of the causative gene, and occur significantly more often than expected by chance in an integrated physical network. The method was tested by searching regulatory motifs between 157 deleted [19.75] and 55 overexpressed [19.76] TFs in *S. cerevisiae*, together with their corresponding differentially expressed genes.

The significance of the regulatory path motifs is determined by a randomization strategy: the cause and

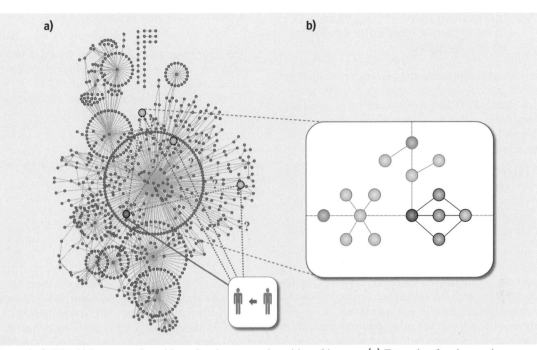

Fig. 19.6a,b Identifying network entities related to network entities of interest. (**a**) Example of an interaction network on which a known disease-related gene (*red node* and *full blue line* to the disease) and candidate disease-related genes (*orange nodes* and *dashed blue lines* to the disease) were mapped. To infer the most likely candidate disease-gene(s), their relatedness to the known disease-related gene (*dashed orange lines*) is examined through interactions on the network. (**b**) The most likely candidate disease-related gene and its relation to the known disease-related gene is highlighted on the network. *Dashed lines* and *arrows* and *transparent nodes* do not belong to the selected paths

effect pairs were permuted for 10 000 times, keeping the number of perturbed genes for each transcription factor constant. Next, the frequency of occurrence of each path motif in the randomized data sets was calculated. If the number of paths in the real perturbation data lies at the right tail of this random distribution (using a z-test statistic), the path was considered significant.

Out of all possible paths of length up to three, the algorithm identified eight regulatory path motifs, of which five where enriched in both deletion and overexpression data. These eight motifs explain 13% of all genes differentially expressed in deletion data and 24% in overexpression data, a more than five to tenfold increase compared to using directional transcriptional links only, confirming that perturbational microarray experiments contain mostly indirect regulatory links.

Like static network motifs [19.2, 77, 78], regulatory path motifs were found to aggregate into modular structures where the differentially expressed targets of a transcription factor reached by the same path through the same intermediate nodes, form a module. Many path

modules showed a high coexpression and were overrepresented in a particular functional category, validating the biological relevance of the regulatory path motifs.

This approach is not only more likely to reflect reflect general regulatory strategies used in biological networks, but also the specificity of a TF to a particular regulatory path can hint towards its mode of action. It was found for instance, that 75% of the genes being perturbed after MET4 overexpression, can be explained by TRI, PPI-TRI, and PPI-TRI-TRI motifs, indicating that MET4 acts together with different combinations of auxiliary factors. In addition they observed that many network motifs that were significantly enriched in response to DNA damage in yeast were shorter than those enriched during cell cycle, exemplifying that environmental responses prefer fast signal propagation while developmental processes progress through multiple stages of interconnecting TFs [19.72, 79]. Thus regulatory path motifs can be used to characterize the condition-dependency of the response mechanisms across multiple integrated networks.

19.3.2 Integrating Cause–Effect Pairs to Confidently Infer Edge Attributes on the Network

While previous approaches limit themselves to inferring common underlying paths connecting related cause–effect pairs, others integrate several (even unrelated) known cause–effect pairs in order to assign specific attributes to the interaction graph (Fig. 19.4). Such attributes are, for instance:

1. The presence or absence of an edge in the path connecting cause and effect
2. The regulatory effect of a node, i. e., activating or repressing
3. The direction of information flow through an edge.

By integrating several cause–effect pairs more assignments of attributes can be made than when considering each pair in isolation. This because the assignment of attributes explaining a particular cause–effect pair should also be able to explain causes and observed effects that occur downstream of this pair in the network. Or in other words, the objective is trying to explain as many cause–effect pairs as possible such that the biological constraints on the network are consistent.

The inferred models of *Yeang* et al. [19.80], called physical network models, are annotated molecular interaction graphs. In this framework the presence of a certain edge in the physical network, the directionality of signal transduction in PPIs, and the regulatory effect of the interaction are determined if their combination is able to explain observed differential expression upon single gene knockouts. These strategies where further explored to investigate the mechanisms of the coupling between regulatory and metabolic networks [19.81]. The MTO algorithm of *Medvedovsky* et al. [19.82] and the method of *Gitter* et al. [19.83] limit themselves to determining a single direction for each edge, so that a maximum number of pairs have a directed path from the cause to the effect.

Case Study
SPINE [19.57] improved the physical network models [19.80] by assigning an activation/repression attribute with each protein so as to explain (in expectation) a maximum number of knockout effects observed. They do not explicitly model the direction of the edges, but most PPIs appeared in one direction only in the inferred consistent pathways. The goal of the algorithm is to infer regulatory pathways in the network that provide a consistent explanation for the input set of knockout pairs.

A path is a consistent explanatory path (Fig. 19.4b) if:

1. The aggregate sign of the path is equal to the observed expression direction (upregulated or activated versus downregulated or inhibited)
2. If every subpath connecting another knockout pair is also consistent.

The optimization problem is defined as that of finding an assignment that will maximize the expected number of pairs that have at least one consistent path, given by

$$E\left(\sum_{(s,t)\in X} K_{s,t}\right) = \sum_{(s,t)\in X} E(K_{s,t})$$
$$= \sum_{(s,t)\in X} p(K_{s,t}=1), \qquad (19.7)$$

where $K_{s,t}$ is a variable that indicates if there exists at least one regulatory path consistent with a knockout pair (s, t) out of a collection of knockout pairs X; $p(K_{s,t}=1)$ corresponds to the probability that at least one consistent path exists for knockout pair (s, t). The optimization problem is reformulated and solved as an integer linear program.

The authors evaluated their method by applying it on a genome-wide integrated yeast network consisting of PPI and TF-DNA data, in order to explain the effects observed in gene expression under different single-gene knockouts [19.84]. Here, a significant overlap between the model's prediction and the known signs was seen. Moreover, increasing the path length from one edge (i. e., only a direct TF-DNA link) up to three edges (i. e., one TF-DNA link and two other PPI/TF-DNA links) in different runs clearly showed the importance of looking at paths rather than considering direct edges only, since the amount of explained knockout pairs increased accordingly.

19.3.3 Identifying (an) Unknown Causal Input(s) for (an) Observed Effect(s)

In this section, we discuss applications where several effects are observed, but the true cause of these effects is unknown (Fig. 19.5). The common objective here is thus to infer this unknown cause or causes by use of a network. Examples of such problems can be found in the domain of expression quantitative trait loci (eQTL) mapping. With the availability of complete genomes of single strains, identifying which alterations in the DNA sequence (i. e., causes) are responsible for observed changes in gene expression (i.e., the quantitative trait, or effects) becomes increasingly important. Usually, in a first step the association between a gene's expression level (i.e., eQTL mapping) and each genomic region (i.e., the expression quantitative locus or eQTL) is examined by a statistical method [19.85]. In the case of multifactorial traits, multiple loci can be associated with the gene's expression behavior, which complicates eQTL analysis. In addition, due to linkage disequilibrium each of the associated loci can contain several genes, which limits the localization of the true causal gene. Even when the causal gene can be identified, the molecular mechanism through which the association is exerted often remains elusive [19.86].

Path finding methods can here be used to identify the causal gene within an associated genomic locus and the underlying pathways that transmit signals from the locus to the affected target. The genes that altered their expression levels are considered as effects, while all the genes in the associated loci are defined as possible causes.

Tu et al. [19.87] proposed a random walk approach to infer the causal gene in a locus and the underlying pathways from a physical interaction network consisting of protein phosphorylation, PPI, and TF-DNA interactions. They assumed that the pathway starts with one causal gene in an associated locus and ends at the transcription factors regulating the target gene such that the expression of the genes on the pathway are correlated with the target gene. For each affected gene and each of its associated eQTLs their stochastic algorithm is performed separately to identify the causal gene. During the walks initiated on the network, different genes will be visited with different frequencies depending on their expression profile. The genes with higher frequencies are then assumed to be more likely to be the causal gene, and the most frequently traveled paths are regarded as the underlying regulatory pathways. For 239 out of 585 eQTLs identified in a study of 112 yeast segregants of *Brem* et al. [19.88] a causal gene could be significantly predicted. The authors highlighted GPA1 as causal regulator for target gene PRP39, a result that was experimentally verified by *Yvert* et al. [19.89].

Suthram et al. [19.86] further adapted the method of Tu by considering the analogy between random walks and electric circuits in a new method, named eQTL electrical diagrams (eQED). eQED models the flow of information from a locus to a target gene as electric currents through the protein network. The authors consider all loci influencing the target simultaneously, allowing multiple loci to reinforce each other when they fall along a common regulatory pathway. The causal gene in each locus is then predicted as the one with the highest current running through it. By validating the eQED model on the eQTL data set of Brem and Kruglyak against a golden standard of knockout expression profiles, the multilocus model indeed showed a highly improved accuracy compared to the single-locus model and the method of Tu et al. (80% versus 72% and 50%, respectively).

Case Study

Inspired by the eQED electric circuit model, *Kim* et al. [19.90] developed a method for the identification of candidate causal genes and dysregulated pathways that are potentially responsible for the altered expression of target genes associated to glioblastoma multiforme (GBM), the most common and most aggressive malignant primary brain tumor in humans. Applied on gene expression and genomic alteration (in this case copy number variations or CNVs) profiles of 158 GBM patients, their methodology comprises four steps:

1. In a first step a set of genes is selected that show differential expression in the patients while taking into account disease heterogeneities among different patients, thus extracting sets of differentially expressed genes that are specific to a subgroup of patients. These differentially expressed genes are hereafter called target genes.
2. Subsequently, an eQTL mapping is performed by a linear regression analysis to determine the association between the expression of each target gene and copy number alterations of tag loci. A liberal p-value was chosen to retain most of potentially interesting relationships.
3. Then, to filter out false positives and to determine the most likely causal genes within each region of associated CNVs for each target gene, a physical network-based approach based on the electric circuit diagrams of *Suthram* et al. [19.86] was applied. Each node in the circuit represents a gene and holds a certain voltage to be determined. Each edge represents an interaction between node entities and has a conductance (i.e., how easily electricity flows along a path) defined by the mean expression correlation of its nodes with the target gene. As such, the authors ensured that a single noncorrelated node reduced but not completely interrupted the current flow, while a cluster of noncorrelated nodes put a considerable resistance to the current flow. Using Ohm's law and Kirchhoff's current law, the amount

of current through a node was calculated. Candidate causal genes for each target gene where then selected based on a permutation test to estimate the statistical significance of the current flow through the nodes.

4. Finally, this resulting set of causal genes was further reduced by imposing another filter: a minimum set of causal genes was selected that could explain all disease cases except for a few outliers.

Assessing the significance of the identified set of causal genes by determining overlap with sets of known GBM/glioma specific genes showed that their approach could uncover more cancer relevant genes than a simple association approach and demonstrated the increased predictive power of the model.

The authors also assessed the importance of genes in the paths from putative causal genes to their target genes and observed the emergence of hubs, genes that appeared in a disproportionally large number of paths. Such a set of hubs contained important transcription factors such as MYC and E2F1, and oncogenes such as JUN and RELA, and was enriched in genes that appeared in cancer pathways, the cell cycle, and several important signaling pathways. While such hub genes were clearly related to cancer, they would hardly have been identified by analyzing differentially expressed genes alone, demonstrating the advantages of a pathway-based approach.

Moreover, a GO biological process enrichment analysis of the uncovered subnetworks revealed frequently re-occurring classical cancer related pathways like insulin receptor signaling pathways, RAS signaling, as well as a glioma-associated regulation of transforming growth factor-b2 production and SMAD pathway. Such pathways can then be considered as *GO biological process hubs* or *highways*, connecting many different causal genes with their targets. Such an observation supports the hypothesis of a pathway-centric view of complex disease, namely that many different genomic alterations potentially dysregulate the same pathways in complex diseases.

Among the discovered set of putative causal genes and pathways, an influence of PTEN and CDC2 was observed on the expression of WEE1 through transcription factors TP53 and E2F4. This tyrosine kinase in turn phosphorylates the protein product of CDC2 (i. e., CDK1), a signaling event that is crucial for the cyclin-dependent passage of various cell cycle checkpoints and suggested as an important feedback mechanism for cancer by the authors.

19.3.4 Identifying Network Entities Related to Network Entities of Interest

In this fourth and last category, the objective is to identify entities related to a set of entities of interest (hereafter named seeds), by exploring the paths (e.g., long versus short paths, paths through highly connected nodes versus through very specific nodes, one simple path versus multiple paths connecting cause and effect, . . .) that connect them in the network, rather than inferring the underlying paths that transfer the signal from a cause to affected genes like in the previously described approaches (Fig. 19.6). This approach is useful when one is looking for the cause of an observed effect, but they cannot both be mapped on a physical or functional molecular network. A domain where this strategy, for instance, has been proven useful is when dealing with diseases as observed effects.

For most diseases only a limited number of causal genes is currently known [19.91]. Genome-wide association studies, whereby genomic variation are associated to a certain phenotype, typically result in one or more linked chromosomal regions, which in turn can contain several genes. Since the elucidation of disease mechanisms can improve diagnose or medical care, several approaches have been developed to identify novel disease genes.

Motivated by the observation that genes causing a specific or similar disease phenotype tend to lie close to each other in a protein–protein network [19.92], several network-based approaches have been developed. These methods have a common approach in the sense that they try to score candidate disease genes based on the assumption that good candidates reside in the neighborhood of certain a priori determined genes. These a priori determined genes are, for instance, genes known to be involved in the disease [19.93–95] or in related phenotypes [19.59, 96, 97], or differentially expressed genes upon the phenotype [19.98]. These a priori determined disease related genes are called *seed genes* in what follows.

Several types of measures can be applied to score candidate genes. A first, intuitive way of identifying disease related genes is based on direct neighborhood: candidate genes that are directly connected to one or more seed genes are then predicted to be potentially causative [19.93]. However, it is possible that two disease related genes do not interact with each other directly, but are, for instance, part of the same pathway, and disruption in either one of them leads to the same disease. These cases will be missed by direct neighborhood counting. To account for indirect interactions, one can use the shortest path length between a seed node and a candidate gene as a measure of its relatedness to the seed gene: if a certain candidate node lies at most k edges away from the seed node, it is considered as a disease gene. *George* et al. [19.94] observed that as the shortest path length between a seed node and a candidate node increases, the sensitivity of identified disease genes improves, but the number of false positives increases exponentially and reduces the specificity.

Protein–protein networks, however, possess the *small world* property, meaning that the average path length between any two nodes in the network is rather short [19.99]. One consequence of this for methods relying on direct interactions or shortest paths is that it is not very unlikely to observe genes interacting with the disease seed genes but that are unrelated to the disease as such [19.95]. Moreover, methods based on shortest path length ignore the fact that there might be multiple shortest paths or also other paths with longer lengths, which could point to a higher relatedness to the seed gene than when only one path is present between seed and candidate node.

To overcome these limitations, several methods have been developed that consider the topology of the entire network (i. e., global distance measure versus local) to define a distance measure between two nodes. These methods generally propose a strategy based on random walks on graphs (i. e., the interaction network). A random walk of a certain length k on a graph represents a stochastic process starting at a seed node and each subsequently visited node is chosen uniformly at random from the neighbors of each previous node. The steady state probability $p_{x,y,k}(G)$ is then the probability that a random walk of length k, starting at node x would end in node y.

Kohler et al. [19.95] proposed a variant of the random walk, namely the random walk with restart, to identify disease genes in a human PPI network. Here, the walk is allowed to restart in every time step at a known disease gene (i. e., seed gene) with a certain probability. All candidate disease genes are then ranked according to the probability of the random walker reaching the candidate gene from a (set of) seed node(s), which reflects their global relatedness to the known disease seed genes. Applied on 110 disease gene families, together with their known associated genes in a leave-one-out cross-validation setting, they clearly outperformed local measures as direct neighborhood or shortest path, or approaches not based on any network.

Vanunu et al. [19.59] extended this approach by using causal genes both from the disease of interest or similar ones as seed genes in the random walk with restart. This approach can be very useful when no causal genes of the disease are known. Diseases, however, can be very heterogeneous, meaning that they can result in different phenotypes and encompass various subtypes. Exploiting all known disease genes that have been related to a heterogeneous disease might not have a sufficiently high resolution to predict novel genes for a specific subtype of the disease. To overcome this, two other methods proposed to identify gene-phenotype relationships rather than finding the gene-disease relationship directly. This strategy decomposes a disease in phenotypes and tries to identify novel phenotype-related genes by using genes related to one or more phenotypes of interest and related ones as seed genes. *Li* and *Patra* [19.96] build upon the random walk with restart strategy, but perform it on a heterogeneous network connecting the gene network (i. e., consisting of PPI) with the phenotype network (i. e., a k nearest neighbor graph presentation of phenotypes similarity) through gene-phenotype relations. *Yang* et al. [19.97], on the other hand, make use of the associations between protein complexes and the phenotypes of interest to perform a random walk with restart on a protein complex network.

Case Study
The aforementioned methods rank candidate genes based on their connections to known disease genes or to known causal genes of related phenotypes in a protein–protein network. However, these methods are usually ineffective when little is known about the molecular basis of the phenotype (e.g., no confirmed causal genes), or when the observed phenotypes are very specific. To this end, *Nitsch* et al. [19.98] developed a computational method to prioritize candidate disease genes for which limited or no prior knowledge is available, by using experimental data on differential gene expression between healthy and affected individuals for a phenotype of interest.

Genes for which significant differential expression was measured in an affected tissue compared to wild type are usually considered as promising candidates being involved in the disease. However, not necessarily the expression levels of the disease gene are affected, but rather expression of genes downstream of this causal gene. Therefore, by mapping differential expression levels on a gene network, one expects to observe a disrupted expression module around the disease gene. Other candidate genes that are not causally related to the phenotype should not be part of such a module. For this reason, the relevance of a candidate gene is scored by considering the level of differential expression in its neighborhood in a protein network instead of only taking its own expression level into account, under the assumption that strong candidates tend to be surrounded by differentially expressed neighbors.

In this work, a functional network was created using human protein associations obtained from the STRING database [19.26], since protein interaction networks are still far from complete and according to the authors might give suboptimal results due to many missing components and pathways. For each gene, differential expression values are determined from microarray experiments that measure wild type versus diseased cell lines, for a phenotype of interest. The prioritization can be performed on a list of candidate genes on a chromosomal region of interest (e.g., determined from a linkage study) or genome-wide when no list of candidate genes are available (although this will probably result in many more false positives).

The neighborhood of a candidate gene is then determined by using a graph kernel, namely the Laplacian exponential diffusion kernel. This gives a weight to each gene in the network, which decreases as a function of the distance from the causal gene, taking into account that there might be multiple paths between the causal gene and each gene in the network. It can be seen as a random walk, starting from a node and transitioning to a neighboring node with a certain probability.

Finally, each candidate gene is scored by summing up the levels of differential expression (measures by absolute fold changes) of each gene, weighted by its network distance from the candidate. Higher differential expression of neighboring genes will, therefore, result in higher scores. The significance of a candidate gene is determined by randomly distributing the differential expression data on the network and computing an empirical p-value from the random distribution of scores.

Besides benchmarking, the methodology on several monogenic diseases for which the causal gene is known, the authors also applied their method on the polygenic disorder *Stein–Leventhal* [19.100] for which currently no disease gene is known. They highly ranked two genes on two different chromosomal regions that were previously assigned a possible role in this disorder, namely fibrilin 3 (FBN3) and follistatin (FST). Another gene, DEAD box4, was found to be the best scoring gene and was suggested as a new candidate gene potentially involved in this disease. Although little is known about the molecular function of DEAD box 4 in mammals, the authors found several indications in literature that indicate a plausible role in the Stein–Leventhal syndrome, for instance because of its association with stem cell recruitment to the ovaries, interaction with the mRNA processing machinery, and impact on apoptosis.

Although the expression levels of genes can be determined by multiple genes together in the case of a polygenic order, making it difficult to determine the true causes of the effects, the approach has nevertheless shown to provide plausible candidates when little knowledge is available for the disease at hand.

19.4 Conclusion

Despite the emerging amount of data and the integration strategies presented at the beginning of this chapter, eukaryotic molecular networks are still low in coverage due to their size and complexity. We expect that the increasing number of available data sets will continue to expand current networks and further refine our knowledge on the included interactions, ultimately resulting in a better understanding of the cell's behavior. In this chapter we have presented several path finding methodologies to interrogate networks with functional data and showed how they can be used to predict disease genes, unveil hidden signaling paths, ... The use of interaction networks for unveiling mode of actions will only gain in importance with functional omics datasets profiling human diseases growing steadily in number and size.

References

19.1   U. Alon: Biological networks: The tinkerer as an engineer, Science **301**(5641), 1866–1867 (2003)

19.2   R. Milo, S. Shen-Orr, S. Itzkovitz, N. Kashtan, D. Chklovskii, U. Alon: Network motifs: Simple building blocks of complex networks, Science **298**(5594), 824–827 (2002)

19.3   U. Alon: Network motifs: Theory and experimental approaches, Nat. Rev. Genet. **8**(6), 450–461 (2007)

19.4   Y. Ho, A. Gruhler, A. Heilbut, G.D. Bader, L. Moore, S.L. Adams, A. Millar, P. Taylor, K. Bennett, K. Boutilier, L. Yang, C. Wolting, I. Donaldson, S. Schandorff, J. Shewnarane, M. Vo, J. Taggart, M. Goudreault, B. Muskat, C. Alfarano, D. Dewar, Z. Lin, K. Michalickova, A.R. Willems, H. Sassi, P.A. Nielsen, K.J. Rasmussen, J.R. Andersen, L.E. Johansen, L.H. Hansen, H. Jespersen, A. Podtelejnikov, E. Nielsen, J. Crawford, V. Poulsen, B.D. Sørensen, J. Matthiesen, R.C. Hendrickson, F. Gleeson, T. Pawson, M.F. Moran, D. Durocher, M. Mann, C.W. Hogue, D. Figeys, M. Tyers: Systematic identification of protein complexes in Saccharomyces cerevisiae by mass spectrometry, Nature **415**(6868), 180–183 (2002)

19.5   T. Ito, T. Chiba, R. Ozawa, M. Yoshida, M. Hattori, Y. Sakaki: A comprehensive two-hybrid analysis to explore the yeast protein interactome, Proc. Natl. Acad. Sci. USA **98**(8), 4569–4574 (2001)

19.6   N.J. Krogan, G. Cagney, H. Yu, G. Zhong, X. Guo, A. Ignatchenko, J. Li, S. Pu, N. Datta, A.P. Tikuisis, T. Punna, J.M. Peregrín-Alvarez, M. Shales, X. Zhang, M. Davey, M.D. Robinson, A. Paccanaro, J.E. Bray, A. Sheung, B. Beattie, D.P. Richards, V. Canadien, A. Lalev, F. Mena, P. Wong, A. Starostine, M.M. Canete, J. Vlasblom, S. Wu, C. Orsi, S.R. Collins, S. Chandran, R. Haw, J.J. Rilstone, K. Gandi, N.J. Thompson, G. Musso, P. St Onge, S. Ghanny, M.H. Lam, G. Butland, A.M. Altaf-Ul, S. Kanaya, A. Shilatifard, E. O'Shea, J.S. Weissman, C.J. Ingles, T.R. Hughes, J. Parkinson, M. Gerstein, S.J. Wodak, A. Emili, J.F. Greenblatt: Global landscape of protein complexes in the yeast Saccharomyces cerevisiae, Nature **440**(7084), 637–643 (2006)

19.7   P. Uetz, L. Giot, G. Cagney, T.A. Mansfield, R.S. Judson, J.R. Knight, D. Lockshon, V. Narayan, M. Srinivasan, P. Pochart, A. Qureshi-Emili, Y. Li, B. Godwin, D. Conover, T. Kalbfleisch, G. Vijayadamodar, M. Yang, M. Johnston, S. Fields, J.M. Rothberg: A comprehensive analysis of protein-protein interactions in Saccharomyces cerevisiae, Nature **403**(6770), 623–627 (2000)

19.8   A.C. Gavin, M. Bösche, R. Krause, P. Grandi, M. Marzioch, A. Bauer, J. Schultz, J.M. Rick, A.M. Michon, C.M. Cruciat, M. Remor, C. Höfert, M. Schelder, M. Brajenovic, H. Ruffner, A. Merino, K. Klein, M. Hudak, D. Dickson, T. Rudi, V. Gnau, A. Bauch, S. Bastuck, B. Huhse, C. Leutwein, M.A. Heurtier, R.R. Copley, A. Edelmann, E. Querfurth, V. Rybin, G. Drewes, M. Raida, T. Bouwmeester, P. Bork, B. Seraphin, B. Kuster, G. Neubauer, G. Superti-Furga: Functional organization of the yeast proteome by systematic analysis of protein complexes, Nature **415**(6868), 141–147 (2002)

19.9   S. Li, C.M. Armstrong, N. Bertin, H. Ge, S. Milstein, M. Boxem, P.O. Vidalain, J.D. Han, A. Chesneau, T. Hao, D.S. Goldberg, N. Li, M. Martinez, J.F. Rual, P. Lamesch, L. Xu, M. Tewari, S.L. Wong, L.V. Zhang, G.F. Berriz, L. Jacotot, P. Vaglio, J. Reboul, T. Hirozane-Kishikawa, Q. Li, H.W. Gabel, A. Elewa, B. Baumgartner, D.J. Rose, H. Yu, S. Bosak, R. Sequerra, A. Fraser, S.E. Mango, W.M. Saxton, S. Strome, S. van den Heuvel, F. Piano, J. Vandenhaute, C. Sardet, M. Gerstein, L. Doucette-Stamm, K.C. Gunsalus, J.W. Harper, M.E. Cusick, F.P. Roth, D.E. Hill, M. Vidal: A map of the interactome network of the metazoan *C. elegans*, Science **303**(5657), 540–543 (2004)

19.10   E. Formstecher, S. Aresta, V. Collura, A. Hamburger, A. Meil, A. Trehin, C. Reverdy, V. Betin, S. Maire, C. Brun, B. Jacq, M. Arpin, Y. Bellaiche, S. Bellusci, P. Benaroch, M. Bornens, R. Chanet, P. Chavrier, O. Delattre, V. Doye, R. Fehon, G. Faye, T. Galli, J.A. Girault, B. Goud, J. de Gunzburg, L. Johannes, M.P. Junier, V. Mirouse, A. Mukherjee, D. Papadopoulo, F. Perez, A. Plessis, C. Rossé, S. Saule, D. Stoppa-Lyonnet, A. Vincent, M. White, P. Legrain, J. Wojcik, J. Camonis, L. Daviet: Protein interaction mapping: A *Drosophila* case study, Genome Res. **15**(3), 376–384 (2005)

19.11   L. Giot, J.S. Bader, C. Brouwer, A. Chaudhuri, B. Kuang, Y. Li, Y.L. Hao, C.E. Ooi, B. Godwin, E. Vitols, G. Vijayadamodar, P. Pochart, H. Machineni, M. Welsh, Y. Kong, B. Zerhusen, R. Malcolm, Z. Varrone, A. Collis, M. Minto, S. Burgess, L. McDaniel, E. Stimpson, F. Spriggs, J. Williams, K. Neurath, N. Ioime, M. Agee, E. Voss, K. Furtak, R. Renzulli, N. Aanensen, S. Carrolla, E. Bickelhaupt, Y. Lazovatsky, A. DaSilva, J. Zhong, C.A. Stanyon, R.L. Finley Jr., K.P. White, M. Braverman, T. Jarvie, S. Gold, M. Leach, J. Knight, R.A. Shimkets, M.P. McKenna, J. Chant, J.M. Rothberg: A protein interaction map of *Drosophila melanogaster*, Science **302**(5651), 1727–1736 (2003)

19.12   J.F. Rual, K. Venkatesan, T. Hao, T. Hirozane-Kishikawa, A. Dricot, N. Li, G.F. Berriz, F.D. Gibbons, M. Dreze, N. Ayivi-Guedehoussou, N. Klitgord, C. Simon, M. Boxem, S. Milstein, J. Rosenberg, D.S. Goldberg, L.V. Zhang, S.L. Wong, G. Franklin, S. Li, J.S. Albala, J. Lim, C. Fraughton, E. Llamosas, S. Cevik, C. Bex, P. Lamesch, R.S. Sikorski, J. Vandenhaute, H.Y. Zoghbi, A. Smolyar, S. Bosak, R. Sequerra, L. Doucette-Stamm, M.E. Cusick, D.E. Hill, F.P. Roth, M. Vidal: Towards a proteome-scale map of the human protein-protein interaction network, Nature **437**(7062), 1173–1178 (2005)

19.13   U. Stelzl, U. Worm, M. Lalowski, C. Haenig, F.H. Brembeck, H. Goehler, M. Stroedicke, M. Zenkner, A. Schoenherr, S. Koeppen, J. Timm, S. Mintzlaff, C. Abraham, N. Bock, S. Kietzmann, A. Goedde, E. Toksöz, A. Droege, S. Krobitsch, B. Korn, W. Birchmeier, H. Lehrach, E.E. Wanker: A human protein-protein interaction network:

A resource for annotating the proteome, Cell **122**(6), 957–968 (2005)

19.14 T.I. Lee, N.J. Rinaldi, F. Robert, D.T. Odom, Z. Bar-Joseph, G.K. Gerber, N.M. Hannett, C.T. Harbison, C.M. Thompson, I. Simon, J. Zeitlinger, E.G. Jennings, H.L. Murray, D.B. Gordon, B. Ren, J.J. Wyrick, J.B. Tagne, T.L. Volkert, E. Fraenkel, D.K. Gifford, R.A. Young: Transcriptional regulatory networks in *Saccharomyces cerevisiae*, Science **298**(5594), 799–804 (2002)

19.15 C.T. Harbison, D.B. Gordon, T.I. Lee, N.J. Rinaldi, K.D. Macisaac, T.W. Danford, N.M. Hannett, J.B. Tagne, D.B. Reynolds, J. Yoo, E.G. Jennings, J. Zeitlinger, D.K. Pokholok, M. Kellis, P.A. Rolfe, K.T. Takusagawa, E.S. Lander, D.K. Gifford, E. Fraenkel, R.A. Young: Transcriptional regulatory code of a eukaryotic genome, Nature **431**(7004), 99–104 (2004)

19.16 S.E. Celniker, L.A. Dillon, M.B. Gerstein, K.C. Gunsalus, S. Henikoff, G.H. Karpen, M. Kellis, E.C. Lai, J.D. Lieb, D.M. MacAlpine, G. Micklem, F. Piano, M. Snyder, L. Stein, K.P. White, R.H. Waterston, modENCODE Consortium: Unlocking the secrets of the genome, Nature **459**(7249), 927–930 (2009)

19.17 R. Aebersold, M. Mann: Mass spectrometry-based proteomics, Nature **422**(6928), 198–207 (2003)

19.18 M. Kanehisa, S. Goto, M. Furumichi, M. Tanabe, M. Hirakawa: KEGG for representation and analysis of molecular networks involving diseases and drugs, Nucleic Acids Res. **38**(Database issue), D355–360 (2010)

19.19 R. Caspi, T. Altman, J.M. Dale, K. Dreher, C.A. Fulcher, F. Gilham, P. Kaipa, A.S. Karthikeyan, A. Kothari, M. Krummenacker, M. Latendresse, L.A. Mueller, S. Paley, L. Popescu, A. Pujar, A.G. Shearer, P. Zhang, P.D. Karp: The MetaCyc database of metabolic pathways and enzymes and the BioCyc collection of pathway/genome databases, Nucleic Acids Res. **38**(Database issue), D473–479 (2010)

19.20 J. Schellenberger, J.O. Park, T.M. Conrad, B.O. Palsson: BiGG: A biochemical genetic and genomic knowledgebase of large scale metabolic reconstructions, BMC Bioinformatics **11**, 213 (2010)

19.21 C. Stark, B.J. Breitkreutz, A. Chatr-Aryamontri, L. Boucher, R. Oughtred, M.S. Livstone, J. Nixon, K. Van Auken, X. Wang, X. Shi, T. Reguly, J.M. Rust, A. Winter, K. Dolinski, M. Tyers: The BioGRID interaction database: 2011 update, Nucleic Acids Res. **39**(Database issue), D698–704 (2011)

19.22 G.D. Bader, D. Betel, C.W. Hogue: BIND: The biomolecular interaction network database, Nucleic Acids Res. **31**(1), 248–250 (2003)

19.23 I. Xenarios, L. Salwinski, X.J. Duan, P. Higney, S.M. Kim, D. Eisenberg: DIP, the Database of interacting proteins: A research tool for studying cellular networks of protein interactions, Nucleic Acids Res. **30**(1), 303–305 (2002)

19.24 A. Ceol, A. Chatr Aryamontri, L. Licata, D. Peluso, L. Briganti, L. Perfetto, L. Castagnoli, G. Cesareni: MINT, the molecular interaction database: 2009 update, Nucleic Acids Res. **38**(Database issue), D532–539 (2010)

19.25 H.W. Mewes, A. Ruepp, F. Theis, T. Rattei, M. Walter, D. Frishman, K. Suhre, M. Spannagl, K.F. Mayer, V. Stumpflen, A. Antonov: MIPS: Curated databases and comprehensive secondary data resources in 2010, Nucleic Acids Res. **39**(Database issue), D220–224 (2011)

19.26 C. von Mering, M. Huynen, D. Jaeggi, S. Schmidt, P. Bork, B. Snel: STRING: A database of predicted functional associations between proteins, Nucleic Acids Res. **31**(1), 258–261 (2003)

19.27 R. Goel, B. Muthusamy, A. Pandey, T.S. Prasad: Human protein reference database and human proteinpedia as discovery resources for molecular biotechnology, Mol. Biotechnol. **48**(1), 87–95 (2011)

19.28 A. Ruepp, B. Waegele, M. Lechner, B. Brauner, I. Dunger-Kaltenbach, G. Fobo, G. Frishman, C. Montrone, H.W. Mewes: CORUM: The comprehensive resource of mammalian protein complexes – 2009, Nucleic Acids Res. **38**(Database issue), D497–501 (2010)

19.29 E. Wingender: The TRANSFAC project as an example of framework technology that supports the analysis of genomic regulation, Brief Bioinf. **9**(4), 326–332 (2008)

19.30 P.E. Hodges, A.H. McKee, B.P. Davis, W.E. Payne, J.I. Garrels: The yeast proteome database (YPD): A model for the organization and presentation of genome-wide functional data, Nucleic Acids Res. **27**(1), 69–73 (1999)

19.31 C.Y. Yang, C.H. Chang, Y.L. Yu, T.C. Lin, S.A. Lee, C.C. Yen, J.M. Yang, J.M. Lai, Y.R. Hong, T.L. Tseng, K.M. Chao, C.Y. Huang: PhosphoPOINT: A comprehensive human kinase interactome and phosphoprotein database, Bioinformatics **24**(16), i14–20 (2008)

19.32 P.V. Hornbeck, I. Chabra, J.M. Kornhauser, E. Skrzypek, B. Zhang: PhosphoSite: A bioinformatics resource dedicated to physiological protein phosphorylation, Proteomics **4**(6), 1551–1561 (2004)

19.33 G.R. Cochrane, M.Y. Galperin: The 2010 nucleic acids research database issue and online database collection: A community of data resources, Nucleic Acids Res. **38**(Database issue), D1–4 (2010)

19.34 G.D. Bader, M.P. Cary, C. Sander: Pathguide: A pathway resource list, Nucleic Acids Res. **34**(Database issue), D504–506 (2006)

19.35 A. Beyer, C. Workman, J. Hollunder, D. Radke, U. Moller, T. Wilhelm, T. Ideker: Integrated assessment and prediction of transcription factor binding, PLoS Comput. Biol. **2**(6), e70 (2006)

19.36 G.T. Hart, A.K. Ramani, E.M. Marcotte: How complete are current yeast and human protein-protein interaction networks?, Genome Biol. **7**(11), 120 (2006)

19.37 F. Ramirez, A. Schlicker, Y. Assenov, T. Lengauer, M. Albrecht: Computational analysis of human protein interaction networks, Proteomics **7**(15), 2541–2552 (2007)

19.38 C. von Mering, R. Krause, B. Snel, M. Cornell, S.G. Oliver, S. Fields, P. Bork: Comparative assessment of large-scale data sets of protein-protein interactions, Nature **417**(6887), 399–403 (2002)

19.39 C.M. Bishop: *Pattern Recognition and Machine Learning* (Springer, New York 2006)

19.40 D. Ucar, A. Beyer, S. Parthasarathy, C.T. Workman: Predicting functionality of protein-DNA interactions by integrating diverse evidence, Bioinformatics **25**(12), i137–i144 (2009)

19.41 J.S. Bader, A. Chaudhuri, J.M. Rothberg, J. Chant: Gaining confidence in high-throughput protein interaction networks, Nat. Biotechnol. **22**(1), 78–85 (2004)

19.42 R. Jansen, H. Yu, D. Greenbaum, Y. Kluger, N.J. Krogan, S. Chung, A. Emili, M. Snyder, J.F. Greenblatt, M. Gerstein: A Bayesian networks approach for predicting protein-protein interactions from genomic data, Science **302**(5644), 449–453 (2003)

19.43 D.R. Rhodes, S.A. Tomlins, S. Varambally, V. Mahavisno, T. Barrette, S. Kalyana-Sundaram, D. Ghosh, A. Pandey, A.M. Chinnaiyan: Probabilistic model of the human protein-protein interaction network, Nat. Biotechnol. **23**(8), 951–959 (2005)

19.44 M.S. Scott, G.J. Barton: Probabilistic prediction and ranking of human protein-protein interactions, BMC Bioinformatics **8**, 239 (2007)

19.45 R. Sharan, S. Suthram, R.M. Kelley, T. Kuhn, S. McCuine, P. Uetz, T. Sittler, R.M. Karp, T. Ideker: Conserved patterns of protein interaction in multiple species, Proc. Natl. Acad. Sci. USA **102**(6), 1974–1979 (2005)

19.46 M.A. Harris, J. Clark, A. Ireland, J. Lomax, M. Ashburner, R. Foulger, K. Eilbeck, S. Lewis, B. Marshall, C. Mungall, J. Richter, G.M. Rubin, J.A. Blake, C. Bult, M. Dolan, H. Drabkin, J.T. Eppig, D.P. Hill, L. Ni, M. Ringwald, R. Balakrishnan, J.M. Cherry, K.R. Christie, M.C. Costanzo, S.S. Dwight, S. Engel, D.G. Fisk, J.E. Hirschman, E.L. Hong, R.S. Nash, A. Sethuraman, C.L. Theesfeld, D. Botstein, K. Dolinski, B. Feierbach, T. Berardini, S. Mundodi, S.Y. Rhee, R. Apweiler, D. Barrell, E. Camon, E. Dimmer, V. Lee, R. Chisholm, P. Gaudet, W. Kibbe, R. Kishore, E.M. Schwarz, P. Sternberg, M. Gwinn, L. Hannick, J. Wortman, M. Berriman, V. Wood, N. de la Cruz, P. Tonellato, P. Jaiswal, T. Seigfried, R. White, Gene Ontology Consortium: The gene ontology (GO) database and informatics resource, Nucleic Acids Res. **32**(Database issue), D258–261 (2004)

19.47 Y. Guan, C.L. Myers, R. Lu, I.R. Lemischka, C.J. Bult, O.G. Troyanskaya: A genomewide functional net-

work for the laboratory mouse, PLoS Comput. Biol. **4**(9), e1000165 (2008)

19.48 T. Shlomi, D. Segal, E. Ruppin, R. Sharan: QPath: A method for querying pathways in a protein-protein interaction network, BMC Bioinformatics **7**, 199 (2006)

19.49 I. Lee, S.V. Date, A.T. Adai, E.M. Marcotte: A probabilistic functional network of yeast genes, Science **306**(5701), 1555–1558 (2004)

19.50 I. Lee, Z. Li, E.M. Marcotte: An improved, bias-reduced probabilistic functional gene network of baker's yeast, *Saccharomyces cerevisiae*, PLoS One **2**(10), e988 (2007)

19.51 I. Lee, B. Lehner, C. Crombie, W. Wong, A.G. Fraser, E.M. Marcotte: A single gene network accurately predicts phenotypic effects of gene perturbation in *Caenorhabditis elegans*, Nat. Genet. **40**(2), 181–188 (2008)

19.52 I. Lee, B. Ambaru, P. Thakkar, E.M. Marcotte, S.Y. Rhee: Rational association of genes with traits using a genome-scale gene network for *Arabidopsis thaliana*, Nat. Biotechnol. **28**(2), 149–156 (2010)

19.53 B. Linghu, E.S. Snitkin, Z. Hu, Y. Xia, C. Delisi: Genome-wide prioritization of disease genes and identification of disease-disease associations from an integrated human functional linkage network, Genome Biol. **10**(9), R91 (2009)

19.54 I. Lee, U.M. Blom, P.I. Wang, J.E. Shim, E.M. Marcotte: Prioritizing candidate disease genes by network-based boosting of genome-wide association data, Genome Res. **21**(7), 1109–1121 (2011)

19.55 J. Scott, T. Ideker, R.M. Karp, R. Sharan: Efficient algorithms for detecting signaling pathways in protein interaction networks, J. Comput. Biol. **13**(2), 133–144 (2006)

19.56 G. Bebek, J. Yang: PathFinder: Mining signal transduction pathway segments from protein-protein interaction networks, BMC Bioinformatics **8**, 335 (2007)

19.57 O. Ourfali, T. Shlomi, T. Ideker, E. Ruppin, R. Sharan: SPINE: A framework for signaling-regulatory pathway inference from cause-effect experiments, Bioinformatics **23**(13), i359–366 (2007)

19.58 N. Yosef, L. Ungar, E. Zalckvar, A. Kimchi, M. Kupiec, E. Ruppin, R. Sharan: Toward accurate reconstruction of functional protein networks, Mol. Syst. Biol. **5**, 248 (2009)

19.59 O. Vanunu, O. Magger, E. Ruppin, T. Shlomi, R. Sharan: Associating genes and protein complexes with disease via network propagation, PLoS Comput. Biol. **6**(1), e1000641 (2010)

19.60 Y. Qi, Z. Bar-Joseph, J. Klein-Seetharaman: Evaluation of different biological data and computational classification methods for use in protein interaction prediction, Proteins **63**(3), 490–500 (2006)

19.61 S. Suthram, T. Shlomi, E. Ruppin, R. Sharan, T. Ideker: A direct comparison of protein interaction confidence assignment schemes, BMC Bioinformatics **7**, 360 (2006)

19.62   Y. Qi, J. Klein-Seetharaman, Z. Bar-Joseph: Random forest similarity for protein-protein interaction prediction from multiple sources, Pac. Symp. Biocomput. (2005) pp. 531–542

19.63   X. Zhu, M. Gerstein, M. Snyder: Getting connected: Analysis and principles of biological networks, Genes Dev. **21**(9), 1010–1024 (2007)

19.64   E. Yeger-Lotem, L. Riva, L.J. Su, A.D. Gitler, A.G. Cashikar, O.D. King, P.K. Auluck, M.L. Geddie, J.S. Valastyan, D.R. Karger, S. Lindquist, E. Fraenkel: Bridging high-throughput genetic and transcriptional data reveals cellular responses to alpha-synuclein toxicity, Nat. Genet. **41**(3), 316–323 (2009)

19.65   E. Cerami, E. Demir, N. Schultz, B.S. Taylor, C. Sander: Automated network analysis identifies core pathways in glioblastoma, PLoS One **5**(2), e8918 (2010)

19.66   J. Bellay, G. Atluri, T.L. Sing, K. Toufighi, M. Costanzo, P.S. Ribeiro, G. Pandey, J. Baller, B. VanderSluis, M. Michaut, S. Han, P. Kim, G.W. Brown, B.J. Andrews, C. Boone, V. Kumar, C.L. Myers: Putting genetic interactions in context through a global modular decomposition, Genome Res. **21**(8), 1375–1387 (2011)

19.67   C.Y. Huang, E. Fraenkel: Integration of proteomic, transcriptional, and interactome data reveals hidden signaling components, Sci. Signal. **2**, ra40 (2009)

19.68   X.M. Zhao, R.S. Wang, L. Chen, K. Aihara: Uncovering signal transduction networks from high-throughput data by integer linear programming, Nucleic Acids Res. **36**(9), e48 (2008)

19.69   S. Lu, F. Zhang, J. Chen, S. Sze: Finding pathway structures in protein interaction networks, Algorithmica **48**, 363–374 (2007)

19.70   X. Ren, X. Zhou, L.Y. Wu, X.S. Zhang: An information-flow-based model with dissipation, saturation and direction for active pathway inference, BMC Syst. Biol. **4**, 72 (2010)

19.71   M. Steffen, A. Petti, J. Aach, P. D'Haeseleer, G. Church: Automated modelling of signal transduction networks, BMC Bioinformatics **3**, 34 (2002)

19.72   A. Joshi, T. Van Parys, Y.V. Peer, T. Michoel: Characterizing regulatory path motifs in integrated networks using perturbational data, Genome Biol. **11**(3), R32 (2010)

19.73   J. Pey, J. Prada, J.E. Beasley, F.J. Planes: Path finding methods accounting for stoichiometry in metabolic networks, Genome Biol. **12**(5), R49 (2011)

19.74   N. Novershtern, A. Regev, N. Friedman: Physical Module Networks: An integrative approach for reconstructing transcription regulation, Bioinformatics **27**(13), i177–i185 (2011)

19.75   Z. Hu, P.J. Killion, V.R. Iyer: Genetic reconstruction of a functional transcriptional regulatory network, Nat. Genet. **39**(5), 683–687 (2007)

19.76   G. Chua, Q.D. Morris, R. Sopko, M.D. Robinson, O. Ryan, E.T. Chan, B.J. Frey, B.J. Andrews, C. Boone, T.R. Hughes: Identifying transcription factor functions and targets by phenotypic activation, Proc. Natl. Acad. Sci. USA **103**(32), 12045–12050 (2006)

19.77   S.S. Shen-Orr, R. Milo, S. Mangan, U. Alon: Network motifs in the transcriptional regulation network of *Escherichia coli*, Nat. Genet. **31**(1), 64–68 (2002)

19.78   L.V. Zhang, O.D. King, S.L. Wong, D.S. Goldberg, A.H. Tong, G. Lesage, B. Andrews, H. Bussey, C. Boone, F.P. Roth: Motifs, themes and thematic maps of an integrated *Saccharomyces cerevisiae* interaction network, J. Biol. **4**(2), 6 (2005)

19.79   N.M. Luscombe, M.M. Babu, H. Yu, M. Snyder, S.A. Teichmann, M. Gerstein: Genomic analysis of regulatory network dynamics reveals large topological changes, Nature **431**(7006), 308–312 (2004)

19.80   C.H. Yeang, T. Ideker, T. Jaakkola: Physical network models, J. Comput. Biol. **11**(2/3), 243–262 (2004)

19.81   C.H. Yeang, M. Vingron: A joint model of regulatory and metabolic networks, BMC Bioinformatics **7**, 332 (2006)

19.82   A. Medvedovsky, V. Bafna, U. Zwick, R. Sharan: An algorithm for orienting graphs based on cause-effect pairs and its applications to orienting protein networks, LNCS **5251**, 222–232 (2008)

19.83   A. Gitter, J. Klein-Seetharaman, A. Gupta, Z. Bar-Joseph: Discovering pathways by orienting edges in protein interaction networks, Nucleic Acids Res. **39**(4), e22 (2011)

19.84   T.R. Hughes, M.J. Marton, A.R. Jones, C.J. Roberts, R. Stoughton, C.D. Armour, H.A. Bennett, E. Coffey, H. Dai, Y.D. He, M.J. Kidd, A.M. King, M.R. Meyer, D. Slade, P.Y. Lum, S.B. Stepaniants, D.D. Shoemaker, D. Gachotte, K. Chakraburtty, J. Simon, M. Bard, S.H. Friend: Functional discovery via a compendium of expression profiles, Cell **102**(1), 109–126 (2000)

19.85   Y. Gilad, S.A. Rifkin, J.K. Pritchard: Revealing the architecture of gene regulation: The promise of eQTL studies, Trends Genet. **24**(8), 408–415 (2008)

19.86   S. Suthram, A. Beyer, R.M. Karp, Y. Eldar, T. Ideker: eQED: An efficient method for interpreting eQTL associations using protein networks, Mol. Syst. Biol. **4**, 162 (2008)

19.87   Z. Tu, L. Wang, M.N. Arbeitman, T. Chen, F. Sun: An integrative approach for causal gene identification and gene regulatory pathway inference, Bioinformatics **22**(14), e489–496 (2006)

19.88   R.B. Brem, G. Yvert, R. Clinton, L. Kruglyak: Genetic dissection of transcriptional regulation in budding yeast, Science **296**(5568), 752–755 (2002)

19.89   G. Yvert, R.B. Brem, J. Whittle, J.M. Akey, E. Foss, E.N. Smith, R. Mackelprang, L. Kruglyak:

Trans-acting regulatory variation in *Saccharomyces cerevisiae* and the role of transcription factors, Nat. Genet. **35**(1), 57–64 (2003)

19.90 Y.A. Kim, S. Wuchty, T.M. Przytycka: Identifying causal genes and dysregulated pathways in complex diseases, PLoS Comput. Biol. **7**(3), e1001095 (2011)

19.91 J. Amberger, C.A. Bocchini, A.F. Scott, A. Hamosh: McKusick's online Mendelian inheritance in man (OMIM), Nucleic Acids Res. **37**(Database issue), D793–796 (2009)

19.92 M. Oti, H.G. Brunner: The modular nature of genetic diseases, Clin. Genet. **71**(1), 1–11 (2007)

19.93 M. Oti, B. Snel, M.A. Huynen, H.G. Brunner: Predicting disease genes using protein-protein interactions, J. Med. Genet. **43**(8), 691–698 (2006)

19.94 R.A. George, J.Y. Liu, L.L. Feng, R.J. Bryson-Richardson, D. Fatkin, M.A. Wouters: Analysis of protein sequence and interaction data for candidate disease gene prediction, Nucleic Acids Res. **34**(19), e130 (2006)

19.95 S. Kohler, S. Bauer, D. Horn, P.N. Robinson: Walking the interactome for prioritization of candidate disease genes, Am. J. Hum. Genet. **82**(4), 949–958 (2008)

19.96 Y. Li, J.C. Patra: Genome-wide inferring gene-phenotype relationship by walking on the heterogeneous network, Bioinformatics **26**(9), 1219–1224 (2010)

19.97 P. Yang, X. Li, M. Wu, C.K. Kwoh, S.K. Ng: Inferring gene-phenotype associations via global protein complex network propagation, PLoS One **6**(7), e21502 (2011)

19.98 D. Nitsch, L.C. Tranchevent, B. Thienpont, L. Thorrez, H. Van Esch, K. Devriendt, Y. Moreau: Network analysis of differential expression for the identification of disease-causing genes, PLoS One **4**(5), e5526 (2009)

19.99 C. Boone, H. Bussey, B.J. Andrews: Exploring genetic interactions and networks with yeast, Nat. Rev. Genet. **8**(6), 437–449 (2007)

19.100 M. Cortón, J.I. Botella-Carretero, A. Benguria, G. Villuendas, A. Zaballos, J.L. San Millan, H.F. Escobar-Morreale, B. Peral: Differential gene expression profile in omental adipose tissue in women with polycystic ovary syndrome, J. Clin. Endocrinol. Metab. **92**(1), 328–337 (2007)

Part D | 19

20. Inferring Transcription Networks from Data

Alexandru Floares, Irina Luludachi

Reverse engineering of transcription networks is a challenging bioinformatics problem. Ordinary differential equation (ODEs) network models have their roots in the physicochemical base of these networks, but are difficult to build conventionally. Modeling automation is needed and knowledge discovery in data using computational intelligence methods is a solution. The authors have developed a methodology for automatically inferring ODE systems models from omics data, based on genetic programming (GP), and illustrate it on a real transcription network. The methodology allows the network to be decomposed from the complex of interacting cellular networks and to further decompose each of its nodes, without destroying their interactions. The structure of the network is not imposed but discovered from data, and further assumptions can be made about the parameters' values and the mechanisms involved. The algorithms can deal with unmeasured regulatory variables, like transcription factors (TFs) and microRNA (miRNA or miR). This is possible by introducing the *regulome probabilities* concept and the techniques to compute them. They are based on the statistical thermodynamics of regulatory molecular interactions. Thus, the resultant models are mechanistic and theoretically founded, not merely data fittings. To our knowledge, this is the first reverse engineering approach capable of dealing

with missing variables, and the accuracy of all the models developed is greater than 99%.

20.1 Introduction and Background

The final goal of any reverse engineering of biochemical networks approach is to decipher the reactions and interactions involved and their mechanisms from available data and domain knowledge. Whenever is possible, a mathematical model, describing not only the structure but also the dynamics of the network and its regulation, is preferred. This allows the quantitative understanding of the network and opens the possibility of controlling it. Thus, maximizing a desired product in a biotechnological context or designing in silico drug dosages regimens, capable of restoring the physiological dynamics of the network, perturbed in various diseases, become possible.

Unfortunately, building dynamic mathematical models of biochemical networks is a difficult task. Time series data of all variables involved are needed, but time course experiments are rare at present. Besides, in most available time series data only some variables are measured, but all variables are needed to infer transcriptome dynamics. Most often, only mRNA is measured, but TFs and microRNAs are also needed. Conventional mathematical modeling is a time-consuming, hypothesis-driven approach, usually needing unavailable knowledge (for example, all parameter values) and advanced mathematical skills.

As a consequence, some challenging goals are:

1. To automate the most difficult steps of network model development
2. To discover the networks' structure from data and knowledge, instead of imposing it to the data
3. To estimate automatically the unknown parameters of the model from data
4. To discover the mechanisms of the networks' reactions and interactions from data, instead of just fitting the data
5. To infer automatically networks' dynamic models even if key variables are unmeasured or even unknown.

These goals are ambitious but some of the concepts and methods proposed in this chapter are promising.

The rate at which the concentration of a particular molecular species inside a cell changes depends on the rate of reactions producing it and consuming it (i.e., degradation and dilution). For example, the rate of change of a protein depends mainly on the rate at which its mRNA is produced and degraded, the rates at which the mRNA molecules are translated, and the rate at which the protein itself degrades. Because all these are biochemical reactions the corresponding rates are naturally described by kinetic equations. A deterministic and a stochastic framework have been developed to model and simulate biochemical reactions.

Deterministic modeling is based on constructing a set of rate equations to describe the biochemical reactions. These rate equations are nonlinear ODEs with concentrations of chemical species as variables. Deterministic simulation produces concentrations by integrating the ODEs.

The stochastic modeling involves forming a set of chemical master equations with probabilities as variables [20.1]. Stochastic simulation produces counts of molecules of the chemical species, as realizations of random variables, drawn from the probability distribution, described by the chemical master equations.

Besides, in addition to purely deterministic or stochastic time course simulations for ODE models, software packages like COPASI [20.2] can also use hybrid methods. These methods split the ODE model into two segments according to the number of particles participating in a reaction. The reactions with many particles are simulated deterministically. The reactions with only a few particles are simulated stochastically. The hybrid simulation can lead to significant simulation speedup, compared to purely stochastic simulation, while still being more accurate than purely deterministic simulation for small particle numbers.

This chapter focuses on large networks and only the deterministic approach scales well to high-dimensional data; thus, the concepts and methods are based on this approach. They are by no means restricted to transcription networks. The same framework can be used reverse engineering of any biochemical network.

Conventional mathematical modeling starts with a set of hypotheses and simplifying assumptions. Computational intelligence based reverse engineering starts from data, discovers a model from data, and only then explores what are the hypotheses and assumptions compatible with the model discovered. However, we will combine the two approaches, which complement each other. Thus, we will present first the mathematical models that we expect to discover from data and their theoretical foundations.

20.2 The Dynomics and Regulomics of Transcription Networks

The vector of mRNAs concentrations gives the state of a transcription network. TFs are proteins regulating the production rates of some mRNAs like transcription activators or repressors. MicroRNA (miRNA or miR) molecules regulate some target mRNAs degradation rates, usually increasing these degradation rates [20.3]. Thus, inside the cells of an organism, many dynamic omics networks act and interact together in a concerted way, coupled by intercellular communication. We proposed the self-explanatory term *dynomics* for the field dealing with the dynamics of omics networks. The term *regulomics* has already been introduced in the literature for omics network regulation.

Using the proposed techniques, we can decompose the omics networks to focus only to the transcription network and its regulation. Further, we can decompose the transcription networks to focus only on one mRNA variable or node. It is crucial to preserve all the interactions in the decomposing process.

20.2.1 The Generic Equation of Transcription Networks

We will describe the most complex generic ODE model of mRNA transcription. This equation describes the whole transcription network, but we omit indices for simplicity. The reverse engineering task is to discover a similar ODE or a simpler version, if not all possible regulatory interactions are present.

The rate of change in a specific mRNA concentration is

$$\frac{\text{dmRNA}}{\text{d}t} = \beta_0 + \beta P - \gamma \text{mRNA} - \lambda_f \cdot \text{mRNA} \cdot \text{miR}$$
$$+ \lambda_r \cdot \text{mRNA:mi} , \qquad (20.1)$$

where β_0 is a basal mRNA transcription rate, which could be zero, β represents the maximal transcription rate, γ is the mRNA degradation constant, λ_f is the forward rate constant of the mRNA-microRNA complex (mRNA : miR) formation, λ_r is the reverse rate constant of the mRNA-microRNA complex degradation, and P is a function describing mRNA transcription regulation by TFs, which can be interpreted as a probability (Sect. 20.2.2).

This nonlinear ordinary differential equation belongs to a coupled system of ODEs describing the interacting omics networks. The mRNA transcription network interacts with the microRNA transcription network, regulating mRNA degradation, and with the TF proteins regulating mRNA synthesis. We have two different types of regulation: 1) regulatory interactions, where the regulators are not consumed in the regulatory process, performed by transcription factors, and 2) regulatory reactions, where the regulators are consumed in the regulatory process, performed by microRNAs.

To be able to focus only on the transcription of a single mRNA, some techniques (see below) must be developed to decompose first the cellular network and then the mRNA transcription network. This is a challenging problem even if all variables are known, but the coupling terms in (20.1), $\beta \cdot P$, $\lambda_f \cdot \text{mRNA} \cdot \text{miR}$, and $\lambda_r \cdot \text{mRNA:miR}$, contain an unknown number of missing variables: the TFs activating and repressing mRNA synthesis in P, the miR usually increasing target mRNA degradation, and the mRNA-microRNA complex.

20.2.2 Regulome Probabilities and Regulome Functions

Genetic programming, the central computational intelligence tool of our GP RODES (reversing ordinary differential equation systems) algorithm (this is described in detail in Sect. 20.3), cannot deal directly with missing variables. To discover an accurate model from data it needs all the variables involved. In our case, missing variables are unmeasured concentrations' profiles of some molecular species. Most often, in microarray time course experiments, only mRNA is measured. Thus, regulators like TFs and microRNAs, elements of the mRNA transcription network *regulome*, are not measured. Besides, for most mRNAs not all regulatory factors are known.

We introduced the term *regulome*, which we define as the set of regulatory factors (like TFs and microRNAs as mentioned above) intervening in mRNA transcription.

We further introduce the terms of regulome probabilities and regulome functions. To enable GP RODES to cope with missing variables, we developed a set of regulome probabilities, based on a well-established statistical thermodynamics foundation, which was first proposed in [20.4] (see also [20.5] for a recent introduction). The *regulome probabilities* are components of the *regulome functions* and they can be estimated without knowledge of the regulators' concentration profiles. The regulome functions consist in a combination of regulome probabilities and some usually unknown constants.

Typically, these are monotonic S-shaped functions that increase for activators and decrease for repressors [20.6]. One such function is the *Hill function*, which accurately describes empirical data of regulatory interactions. However, it can also be derived from the statistical thermodynamic analysis of equilibrium binding of TFs to its site on the promoter [20.4, 5, 7]. This does not mean considering the network in chemical equilibrium, not even in the steady state. It just takes into account that the binding and dissociation of the molecules composing the transcription apparatus is orders of magnitude faster than transcription and translation.

For an activator, the Hill function is a curve rising from zero and approaching a maximal saturation level

$$f_a(\text{TF}) = \beta P_a = \beta \frac{\text{TF}^n}{K^n + (\text{TF})^n} , \qquad (20.2)$$

where K is termed the *activation coefficient* for the TF activating the gene and has units of concentration, β is the *maximal expression level* (also (20.1)), and n is the *Hill coefficient*, typically in the range $n = 1-4$. The value of K is related to the chemical affinity between the TF and its promoter on the gene and some other factors.

It is important to note that this activation regulome function has two terms – a constant β and an activation probability, $P_a = \frac{\text{TF}^n}{K^n + \text{TF}^n}$. To be able to compute this probability, without knowing the TF concentration profile, TF was scaled to K, forming the variable X, known to have typical values in the continuous range $[0, 2]$. Thus, four activation probabilities of the form

$$P_a = \frac{X^n}{1 + X^n}, \quad \text{with} \quad X = \frac{\text{TF}}{K} \qquad (20.3)$$

were computed for $n = 1-4$.

As we will show, this enables GP RODES to automatically discover such activation functions (20.2) from data.

For a repressor TF, the Hill function is a decreasing S-shaped function, again composed by a constant and a repression probability, where the scaled variable X is used,

$$f_r(X) = \beta P_r = \beta \frac{1}{1 + X^n}, \quad \text{with} \quad X = \frac{\text{TF}}{K} . \qquad (20.4)$$

Since gene transcription is high when the repressor is not bound to the promoter, this Hill function can be derived by considering the probability that the promoter is unbound by TF. Similarly, the probability is computed taking TF/K in the biologically plausible ranges of $[0, 2]$ and $n = 1-4$, and GP RODES will discover the appropriate constant β. Moreover, giving the activation and repression regulome probabilities as inputs to GP RODES, we expect the algorithm to discover all the regulatory interactions. Ideally, for any activation/repression interaction it will discover a term containing the activation/repression regulome probability and the corresponding β constant.

A similar statistical thermodynamics approach can be applied to the mRNA microRNA interaction [20.8, 9]. The details of the derivation of these probabilities and their integration in (20.1) will be presented else-

where (manuscript in preparation). Here, it is sufficient to say that we can use exactly the same regulome probabilities as above for GP RODES inputs.

As it will be explained in detail later, giving the estimated regulome probabilities as inputs to GP RODES, together with the measured variables (mRNA), we expect it to find automatically the proper constants, thus reconstructing the regulome functions. The regulome functions discovered in this way could represent any number of potential regulatory interactions, either activatory or inhibitory, and any potential regulatory mechanism.

The same approach can be used for any number of regulatory inputs. *Setty* et al. [20.10] found that the following function describes data very well experimentally,

$$f(\text{TF}_1, \ldots, \text{TF}_n) = \frac{\sum_i \beta_i \cdot (\text{TF}_i / K_i)^{n_i}}{1 + \sum_i (\text{TF}_i / K_i)^{m_i}} , \qquad (20.5)$$

where K_i is the activator or repressor coefficient for the TF TF_i, β_i is the maximal contribution to expression, and the Hill coefficients are $n = m$ for activators and $n = 0$, $m > 0$ for repressors. Again, the probabilities can be computed and given as inputs to GP RODES, together with the measured variables.

20.2.3 Challenging Problems of Reverse Engineering of Transcription Networks

Modeling the dynomics and regulomics of transcription networks with ordinary differential equations has a strong theoretical foundation. This approach does not exclude the stochastic processes taking place in gene regulatory networks. The stochastic and deterministic approaches can be integrated both conceptually and technically. There are algorithms, implemented in software tools like COPASI [20.2] and Systems Biology Workbench (SBW) [20.11], that can combine them or even can transform a differential equations system in its stochastic counterpart.

Any algorithm inferring an ODE system from omics data, either by conventional mathematical modeling techniques or by computational intelligence techniques, faces the following problems:

1. The measured time series omics data are corrupted by noise.
2. The number of measured time points is too small, usually less than 10, sometimes even 2.

3. The sampling time points are rarely chosen such as to properly uncover the real dynamic of molecular concentrations.
4. Most of the reaction and interaction parameters are unknown.
5. Important regulatory molecules, like TFs or microRNAs, are not measured, but they should be variables of the complete ODE system model.
6. The ODE models are usually large systems of coupled, stiff nonlinear differential equations.
7. Most optimization criteria of the proposed algorithm are computationally too intensive, requiring large ODE system integration at each iteration.
8. A scalable algorithm should decompose the ODE system (network) without destroying the coupling between equations (network interactions).
9. Most conventional and computational intelligence algorithms impose a predefined structure to the data.
10. Algorithms capable of discovering the structure from data are needed to mechanistically understand gene regulatory networks.

It is hoped that the noise can be removed by smoothing techniques (Sect. 20.3.1).

Data with two time points can only be fitted using linear interpolation, thus, the methodology described in this chapter is not suited for this kind of data. Most data have less than 10 points, but the real problem is the choice of the sampling times. A proper choice usually requires some knowledge about the actual dynamic of the concentrations of the molecular species measured. For example, a typical dynamic is that of a perturbed mRNA by a drug. The perturbation is followed by a decrease/increase of the mRNA concentration, which is then relaxed to a steady state, which could be different from the initial concentration. It is difficult to capture the real dynamics if the sampling points are clustered merely in the decreasing/increasing part of the dynamic. Few existing data have more than 10 points and are sampled such as to uncover the real molecular profiles.

Fitting and resampling can generate any desired number of concentration (or number of molecules) values at the desired time points. Equidistant time points should be preferred as they are required by many algorithms dealing with time series. Sometimes, if proper domain knowledge is available, manual curve rectification could be useful. These preprocessing steps will be described in the next section.

Dealing with missing variables or nodes is a challenging problem, which is more difficult than the well-known problem of missing data – missing values of a measured variable. Several techniques developed for dealing with missing variables in the ODE system, or equivalently nodes and interactions in the network, were introduced in Sect. 20.2.2 and will be illustrated in Sect. 20.4.

All existing algorithms for inferring gene network models, including ODE system models, use optimization criteria to fit the models to data. Most of them rely on the error between the measured molecular profiles and the predicted ones. This implies the integration of the complex ODE system at each iteration of the optimization process, which is computationally very intensive.

Genetic programming is, in principle, capable of discover the model (network) structure from data. Using GP, which is computationally intensive, with an optimization criterion (fitness function in GP) requiring the integration of complex ODE systems, is not scalable to large gene networks. Optimization criteria which make ODE system integration unnecessary will be introduced in Sect. 20.3.3. This effectively decomposes the ODE system or the network, without destroying its regulome, and reduces the computation time [20.12] by order of magnitude, making it scalable.

20.3 The GP RODES Algorithm

Reverse engineering transcription networks with GP RODES algorithm consists in the following steps:

1. Smooth and fit the experimental data; resample the fitted data to increase the number of the available time points.
2. Decompose the transcription network or ODE system model without destroying the network's regulome.
3. Automatically discover the structure of the networks or their ODE systems model.
4. Automatically estimate the parameters of the ODE system's model.
5. Identify the biochemical and pharmacological mechanisms involved.

The GP RODES algorithm starts from time-series data. The name of the algorithm is related to its results, not to the biological or any of the physical systems investigated. This is because we successfully applied it not only to biochemical networks, but also to complex biological neural networks, for example, the subthalamopallidal neural network of the basal ganglia [20.12]. The result is an ODE system, $dX/dt = f(X)$ with n equations, one for each variables or network nodes X_i, $i = 1, \cdots, n$. (note here that X refers to vector of input variables for GP RODES, and not to the scaled variable introduced in Sect. 20.2.2) In the following, we give details for each of the aforementioned five steps.

20.3.1 Data Fitting and Time Derivative Computing

Data fitting implies determining a function that approximates (is close to) the given discrete data points. Interpolation refers to an exact fit, when the function passes through the given data points. Polynomial interpolation is preferred when a function is to be approximated locally, and it is sufficiently smooth. One drawback of this approach appears when the function needs to be approximated on a rather large interval. In such a case, the degree of the interpolating polynomial may become unacceptably large.

An alternative solution is spline interpolation [20.13]. The interval is split in a sufficient number of smaller intervals, and a good approximation of the function on each smaller interval can be obtained with a low degree polynomial.

We identified a problem when using cubic spline interpolation on noisy time-series data. By performing an exact fit of the measured data, the resulting function exhibits many oscillations. These are unlikely to appear in reality in the gene's temporal profile and are most likely due to the noise in the data. This problem can be overcome by applying a smoothing spline technique ([20.13]; Sect. 20.4.2, Fig. 20.8a). A smoothing spline performs the fitting of the data based on a smoothing parameter α situated in the interval $[0, 1]$. This parameter influences the fitted function making it smoother or closer to the data. For example, for $\alpha = 1$, one obtains a cubic spline interpolation, while for $\alpha = 0$, the straight line fitting by least squares is determined.

Another problem was identified in the distribution of the time points. On a large time interval where no measured data is available the fitted function took values unlikely to appear in reality (Sect. 20.4.3, Fig. 20.8b).

Fitting time series data results in a function that is used to resample the data. In this way, values can be obtained for a desirable, larger, number of time points. The time derivative of the resampled time series data is obtained by computing the first-order derivative of the fitted function and evaluating it on the desired time points.

20.3.2 Decomposing ODE System Models of Transcription Networks

Each differential equation of transcription network is a relationship between the time series of the time derivative of a variable and the time series of some of the variables of the system (network). This relationship should be valid for any given discrete time point t_j, where $j = 0, 1, 2, \cdots, T$ time points. Thus, the equations of the ODE system can be reconstructed one by one via a simple data mining approach, as an algebraic relation $f_i(X)$ between some of the inputs from X and the output dX_i/dt (Fig. 20.1).

As will be shown, for decomposing the network or ODEs system model, it is crucial to develop optimization criteria (fitness function) based on the error between the model predicted time derivative and the data time derivative. Using the error between the predicted and actual concentration profile requires the integration of a large and complex system of ODEs. To be able to integrate the ODEs system all equations must be known. Using, for example, GP, the ODEs structure can be discovered from data, but this implies integrating a population of hundreds of candidate model systems each generation, for thousands of generations. This is a regression problem, because we want to determine an algebraic relation between continuous inputs and outputs (discrete numbers in computer representation). A well-known problem, which plagues mining omics data with computational intelligence, the small number of samples, is solved by resampling the fitted data (Sect. 20.3.1).

20.3.3 Structure and Parameter Discovery from Omics Time Series Data

With GP, both the structure and the parameters of the mathematical model can be determined, unlike traditional regression where the structure of the model must be given. For this reason, the regression performed with GP is also known as symbolic regression.

The GP RODES algorithm was already implemented in [20.14, 15] using linear genetic programming [20.16]. Here, tree-based GP [20.17] will be used for the first time in GP RODES. Using LGP or tree GP is a matter of investigators' preference, both implementations giving good results.

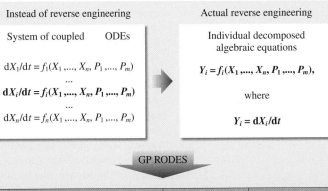

Fig. 20.1 Main features of GP RODES: GP RODES with no missing variables uses as values of input variables on discrete time points $X_i(t)$, $i = 1, \cdots, n$. If variables are missing, GP RODES will also use as inputs the regulome probabilities $P_j(t)$, $j = 1, \cdots, m$. The output to be predicted is a variable's first-order derivative; thus GP RODES performs reverse engineering on individual algebraic decomposed ODEs instead of reverse engineering of a whole system of ODEs

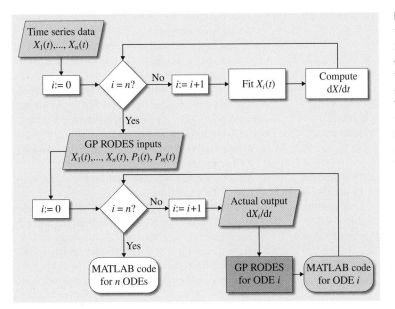

Fig. 20.2 GP RODES workflow: time series values for each variable X_i, $i = 1, \cdots, n$ are fitted resulting in a function whose first-order derivative is then computed. The fitted function and the first-order derivative are evaluated on the desired number of discrete time points. These steps are repeated n times, i.e., for each variable X_i, $i = 1, \cdots, n$. The input variables for GP RODES will then be: the variables X_i, $i = 1, \cdots, n$ (containing values of the evaluated corresponding fitted function) if there are no missing variables; the variables X_i, $i = 1, \cdots, n$ and the regulome probabilities P_j, $j = 1, \cdots, m$ if there are missing variables. The actual output for GP RODES will be the first-order derivative dX_i/dt. The result of GP RODES is a mathematical model in MATLAB that actually represents the i-th ODE. To obtain all n decomposed ODEs, the process of assigning an actual output for GP RODES and performing the GP RODES algorithm needs to be repeated n times, i.e., for each dX_i/dt, $i = 1, \cdots, n$

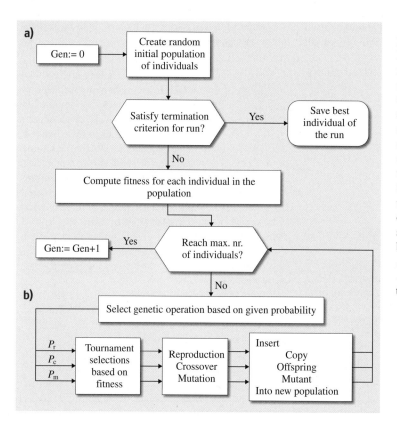

Fig. 20.3a,b GP workflow inside GP RODES. A run consists of the following steps: I. Create generation 0: random population of a maximum number of individuals. II. Perform the following substeps (i.e., create a generation of a maximum number of individuals) until the termination criterion for the run is met (i.e., a maximum number of generations or the fitness desired has been reached). **(a)** Assess fitness of each individual. **(b)** Perform genetic operations on individuals chosen by tournament selection: reproduction (with probability P_r); crossover (with probability P_c) and mutation (with probability P_m); III. Save the best individual of the run

The tree-based GP RODES method consists of the following steps (Figs. 20.2, 20.3):

1. Assume that for some variables X_i their values are measured at given discrete time points $t_j, j = 0, \cdots, T$.
   a) Apply a smoothing method to reduce the noise.
   b) Fit the data using a suitable fitting method.
   c) Differentiate each fitted variable X_i with respect to time to obtain $dX_i/dt(t_j)$, for $j = 0, \cdots, T$ and $i = 1, \cdots, n$.
   d) Resample the data to obtain values for X_i and $dX_i/dt(t_j)$ at desired time points in the interval $[t_0, t_T]$; T is the number of desired time points.
2. Build input–output pairs $(X, dX_i/dt)$ for each time point t_j.
   a) Use all variables of X supposed to belong to the right-hand side of the reconstructed ODE as inputs.
   b) Include the computed regulome probabilities P_i values for each time point t_j as inputs, for the unmeasured regulatory variables.

c) Use the time derivative dX_i/dt of one of the variables as output.
3. Build training, validation (optional, to avoid overfitting), and testing sets from the input–output pairs to perform symbolic regression.
4. Initialize a population of randomly generated trees, coding the mathematical relationship between the inputs $X_k, k \in \{1, \cdots, n\}$ and the output dX_i/dt.
5. Perform tournament selections.
   a) Randomly select a number of trees to take part in tournament selection.
   b) Choose the tree with the best fitness as the winner of the tournament.
   c) Randomly mutate the winner by changing subtrees to create a new tree or directly reproduce the winner (both with a given probability).
   d) Repeat steps 5(a) and (b) to obtain another winner for crossover.
   e) Perform crossover to the copies of the two winners:
      i. One-point crossover: select a single gene in each winner and swap the subtrees between them.

ii. Two-point crossover: select two genes in each winner and swap the subtrees between those genes.

f) If the population size has not reached the maximum specified:
i. Insert the trees resulted from two-point crossover in the population.
ii. Replace the winners with the trees obtained from their one-point crossover.

6. Repeat tournament selections until the desired fitness of a tree is achieved or the maximum number of generations is reached.

7. Extract the ODE for the variable X_i from the tree with the best fitness.

8. Repeat steps 2–7 for each variable to infer the whole ODE system model.

It is often the case that the initial experimental data time points are insufficient for symbolic regression, and the problem becomes more difficult if we wish to split the data into two (or three) datasets. However, as mentioned before, steps 1(b) and (c) allows us to extract more data from the experimental data.

Steps 1–3 reduce the problem of reversing a system of coupled ODEs, $dX/dt = f(X)$, to that of reversing individual, decomposed, algebraic equations, $dX_i/dt = f_i(X)$. Even though the output is in reality a time derivative, the algorithm is simply searching for an algebraic equation relating the inputs to the output, at each discrete time point t. The corresponding relation is the predicted function for the right-hand side of each differential equation of the system.

This approach drastically reduces the CPU time of the algorithm, by orders of magnitude, because in step 5(b) the fitness evaluation does not require the integration of the ODE system. Moreover, the ordinary differential equation system is not even known. The fitness function is simply based on the difference between the actual derivative dX_i/dt computed at step 1(c) and the predicted output of the GP RODES model, $d\hat{X}_i/dt$ (these notations are shown in Fig. 20.1 as well).

For the examples presented in this chapter, we use the root mean squared error fitness function. Thus, the fitness of a tree was evaluated as

$$\text{RMSE} = \sqrt{\frac{1}{T} \sum_{j=0}^{T} \left[\frac{d\hat{X}_i}{dt}(t_j) - \frac{dX_i}{dt}(t_j) \right]^2}. \quad (20.6)$$

To perform symbolic regression, we used an implementation of tree based genetic programming from a freely available MATLAB toolbox called GP-TIPS [20.7]. The tool allows the use of multigene symbolic regression [20.18], i. e., a mathematical model is not given by a single tree, but by a linear combination of trees (called an individual). In this case, the term of *gene* refers to a tree in the linear combination and the term of *tree* to an individual. The coefficients in the linear combination are determined from the training data using a least squares method.

To perform GP RODES, several settings of the evolutionary algorithm were made (Sect. 20.4). The data were split into a training set used to evolve the (multigene) regression model, a testing set used to determine the performance of the model on new data, and a supplementary validation set, used to test the fitness of the best tree in each generation. Neither the testing, nor the validation data set is used in developing the model. Choosing an additional validation data set helps to avoid overfitting.

The accuracy of a model obtained with GP RODES for a variable X_i was given by the percent of variation in the predicted variable, explained by the model (in short, *variation explained*), i. e., multiplying by 100 the following R^2 coefficient,

$$R^2 = 1 - \frac{\sum_{j=0}^{T} \left[\frac{d\hat{X}_i}{dt}(t_j) - \frac{dX_i}{dt}(t_j) \right]^2}{\sum_{j=0}^{T} \left[\frac{dX_i}{dt}(t_j) - \overline{\frac{dX_i}{dt}} \right]^2}, \quad (20.7)$$

where, as before, dX_i/dt is the actual derivative and $d\hat{X}_i/dt$ is the predicted output of the GP RODES model; by $\overline{dX_i/dt}$ we denoted the mean of dX_i/dt.

20.4 Results

To illustrate the methodology behind the GP RODES, we used published protein and mRNA time-series data [20.3]. Protein data are available free of charge at http://pubs.acs.org/, while the mRNA data are accessible at the NCBI GEO database [20.19] with

the accession GSE17708. The study explores how TGF-β treatment induces epithelial-to-mesenchymal transition (EMT) [20.20], related to cancer progression and metastases, in A549 lung adenocarcinoma cells.

20.4.1 Genomics and Proteomics Data

Protein concentration was assessed at 0, 2, 4, 8, 16, 24, 48, and 72 h for a first biological replicate. The time point 0 h corresponds to untreated (control) samples. Since major differences in expression were not observed at the 2 and 4 h time points, a second biological replicate experiment was performed by replacing the 2 and 4 h time points with two 72 h biological replicates. The mRNA data was measured at 0, 0.5, 1, 2, 4, 8, 16, 24, and 72 h using three biological replicates at each time point. Again, at time point 0 h, the samples were untreated. Both the protein and the mRNA data were $\log_{10}$ transformed and values for each sample were made relative to a sample chosen at the 0 h time point (see [20.3] for details).

The authors identified 66 proteins that showed differential expression at least at one time point assessed. A similar expression profile was observed between these proteins and their corresponding mRNA in a proportion of 79%.

We based our examples on two protein coding mRNAs: The actin cytoplasmic 1 (ACTB) mRNA, identified as upregulated and considered one of the most important biomarkers for the epithelial-to-mesenchymal transition and the AGR2 mRNA, identified as downregulated.

The expression values of the mRNAs were taken from [20.3]. Since the data were made relative to a time point 0 h sample, the mRNAs initial values at time point 0 h are 1. The time series values of the mRNAs used for modeling with GP RODES are presented in the following sections (Tables 20.1 and 20.2).

20.4.2 Reverse Engineering the Dynomics and Regulomics of Upregulated mRNAs

The time series values for the ACTB mRNA are presented in Table 20.1.

The first step in applying GP RODES is to fit the given discrete time points. We used the smoothing spline technique (Sect. 20.3.1), implemented in the MATLAB curve fitting toolbox function *csaps*.

For the mRNA discrete data, the smoothing parameter was set at $\alpha = 0.573$. The upper plot in Fig. 20.4 shows the fit that we obtained through smoothing spline

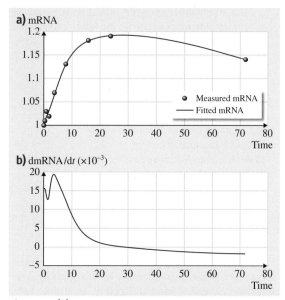

Fig. 20.4 (**a**) ACTB mRNA fit using the smoothing spline technique with $\alpha = 0.573$. Goodness of fit: $R^2 = 0.9938$, SSE = 0.0002839, RMSE = 0.0105. (**b**) First-order derivative of mRNA with respect to time (RMS validation set error: 0.00047849). Variation explained: 99.2657%

(the continuous curve) and the discrete data (represented by * signs). The goodness of fit was evaluated by computing the R-square coefficient, which measures how successful the fit is in explaining the variation of the data, the sum of squares due to error (SSE), and the root mean squared error (RMSE). A value for R-square closer to 1 and values for SSE and RMSE closer to 0 indicate a good fit and in our case the values were: R-square $= 0.9938$; SSE $= 0.0002839$ and RMSE $= 0.0105$.

After obtaining this fit, we differentiated the function with respect to time to obtain dmRNA/dt; its plot can be observed in the bottom graphic of Fig. 20.4. We then evaluated the fitted function and its first-order derivative on time points chosen at an equal distance of 0.1 in the interval [0 h, 72 h]. Thus, we obtained 721 discrete values, corresponding to 721 time points in the interval [0 h, 72 h] for ACTB mRNA and dmRNA/dt.

We applied GP RODES to obtain an ODE for dmRNA/dt. The experiments were performed with

Table 20.1 ACTB mRNA time series values

Time (h)	0	0.5	1	2	4	8	16	24	72
mRNA	1	1.01	1.03	1.02	1.07	1.13	1.18	1.19	1.14

GPTIPS, and the multigene symbolic regression method was preferred. After testing various settings, the following run parameters were used for all the experiments presented in this contribution:

1. Population size: 500
2. Number of generations: 5000
3. Number of trees in tournament: 7
4. Parsimony pressure: *Luke* and *Panait* [20.21]
5. Maximum tree depth: 5
6. Maximum number of genes: 4
7. Probability of tree mutation: 0.1
8. Probability of tree crossover: 0.85
9. Probability of tree direct copy: 0.05
10. Function set: $\{+, -, *\}$
11. Fitness (RMSE) to be achieved: 0.0005.

The data were split into training, testing, and validation. The data contained 721 values for the input and output variables, and it is important to note that the mRNA (one of the inputs) and the dmRNA/dt (output) variables are time-dependent. Thus splitting the data into three sets was done so that in each data set, the values corresponded to time points on the whole interval $[0, 72]$ h.

First, we considered only one input and one output, the mRNA time series and its time derivative, respectively. Mechanistically, only the known linear degradation of the mRNA is taken into account by this experiment. Thus, the treatment effects were deliberately ignored, just to see if GP RODES is capable of detecting that variables are missing from its input set. The initial experiment provided, as expected, a low accuracy: on the validation dataset, the RMSE was 0.0025 and the variation explained (based on R^2) by the model was 79.33%. This shows that GP RODES is able to detect missing variables.

The next step was to consider three input variables: an activation regulome probability P_a, a repression regulome probability P_r (Sect. 20.2.2), and the ACTB mRNA; the output was dmRNA/dt. We computed eight regulome probabilities, four activation probabilities $X^n/(1+X^n)$, and four repression probabilities $1/(1+X^n)$, taking $n = 1, 2, 3, 4$. A suitable range for X was determined by making assumptions based on the data and common choices in the literature [20.6]:

1. The regulome activation probability $X^n/(1+X^n)$ is supposed to be related to the regulation of mRNA production, as ACTB mRNA is upregulated. We further suppose that the activation is performed by one

or more TFs, possibly also upregulated. Nevertheless, inhibitory interactions are also possible, but the net effect is an activation and we intended to discover the main effects of the TGF β treatment. Thus, $X = \text{TF}/K_a$, with TF denoting a generic TF, because TFs are not known/measured, and k_a is a generic activation constant, because we do not know its value. To determine a plausible range for a possible upregulated unknown TF, from the available upregulated protein data in [20.3], we estimated a maximum value for TF, $\text{TF}_{\max} = 2.35$. The minimum value for TF is $\text{TF}_{\min} = 1$, because the values are relative to the untreated initial value.

2. The repression regulome probability $1/(1+X^n)$ is supposed to be related to the experimentally demonstrated effect of TGF β treatment to increase the ACTB mRNA stability, decreasing its degradation [20.3]. Thus, supposing further that microRNAs are involved in this effect, $X = \text{miR}/K_r$, where miR represents a generic unmeasured microRNA, and K_r a generic unknown repression constant. To determine a plausible range for the unmeasured microRNAs, the maximum value for miR was estimated from the available mRNA data, corresponding to over-expressed proteins, $\text{miR}_{\max} = 4.24$, while the minimum is $\text{miR}_{\min} = 1$.

To be able to compute the regulome probabilities, which are dimensionless, dimensional analysis should be performed. Note that mRNA, TF, and miR do not have concentration units in the data [20.3]. They are dimensionless, being relative to the time point 0 h sample – $\text{mRNA}/\text{mRNA}_0$, TF/TF_0, and miR/miR_0, where mRNA_0, TF_0 and miR_0 are the mRNA, TF, and miR concentrations at time point 0 h. Thus, we also need dimensionless constants and the natural choice is: K_a/TF_0 and K_r/miR_0.

We decided to set a maximum value for X of $X_{\max} = 2$ [20.6] and thus compute the constants $K_a = \text{TF}_{\max}/X_{\max}$ and $K_r = \text{miR}_{\max}/X_{\max}$, which yielded: $K_a = 2.35/2$, so $K_a = 1.175$ and $K_r = 4.24/2$, i.e., $K_r = 2.12$. Using this information, a minimum value for X: $X_{\min}$ was computed and used to determine the regulome probabilities as follows:

1. For P_a, $X_{\min} = \text{TF}_{\min}/K_a$, i.e., $X_{\min} = 1/1.175$, which resulted in $X_{\min} = 0.8511$. The four activation regulome probabilities were evaluated on 721 evenly spaced points in the interval $[X_{\min}, X_{\max}] = [0.8511, 2]$ for $n = 1, 2, 3, 4$.

2. For P_r we had $X_{min} = miR_{min}/K_r$, so $X_{min} = 1/2.12$ from which $X_{min} = 0.4717$. Again, the four repression regulome probabilities were evaluated on 721 evenly spaced points in the interval $[X_{min}, X_{max}] = [0.4717, 2]$ for $n = 1, 2, 3, 4$.

All the combinations of one activation regulome probability P_a and one repression regulome probability P_r were used as input data. More precisely, all the combinations of the following input variables were used in experiments:

1. An activation probability P_a computed with $n = i$, where $i = 1, 2, 3, 4$.
2. A repression probability P_r computed with $n = j$, where $j = 1, 2, 3, 4$.
3. The mRNA time series values, which was the same for all the combinations.

The resulting models (individuals in the population) were analyzed considering their accuracy, their complexity, and the duration of the run. The fitness criterion was reached very quickly, below 50 generations being computed before the run was completed. The accuracy of the model was estimated on the validation dataset, i.e., the model was chosen as the individual in the population which provided the best fitness on the validation dataset, and the variation explained was between 99.20 and 99.39%. This suggests that the regulome probabilities successfully replace the missing variables in model development. More precisely, GP RODES discovered the constants that, together with the regulome probabilities, composed the regulome functions (Sect. 20.2.2).

An example of a good model, from both an accuracy and a complexity viewpoint, was obtained using a P_a activation regulome probability obtained for $n = 3$ and a P_r repression regulome probability computed with $n = 1$. The plots of the predicted versus actual output, as well as the variation explained by the model (99.26%) on the validation dataset and the fitness (RMSE = 0.00047), can be seen in Fig. 20.5.

To prove that the regulome probabilities P_a and P_r are important variables for GP RODES and that the model we obtained above was not just a simple fit of data, we assigned GP RODES random inputs instead of P_a computed with $n = 3$ and P_r computed with $n = 1$. For this, we first determined the intervals that the values of P_a (with $n = 3$) and P_r (with $n = 1$) belonged too, thus obtaining:

1. An interval $I_a = [\min(P_a), \max(P_a)]$, i.e. $I_a = [0.3814, 0.8889]$; we then generated 721 uniformly distributed random values in I_a to obtain a random input variable for GP RODES as a substitute for P_a.
2. An interval $I_r = [\min(P_r), \max(P_r)]$, i.e. $I_r = [0.3333, 0.6795]$; again, we generated 721 uniformly distributed random values in I_b to obtain a random input variable for GP RODES as a substitute for P_r.

Then, the inputs for GP RODES were: two random variables as substitutes for P_a and P_r and the run was executed with the same parameters as for the above experiments. The run was complete after the maximum number of 5000 generations was reached, and the duration of the run was approximately 2 h. The accuracy (variation explained) was 77.19%. The best fitness achieved was 0.0026 for the individual evaluated as the best on the validation dataset (Fig. 20.6).

The models obtained represent the ODE for mRNA, i.e., dmRNA/dt. By integrating the ODE obtained with respect to time on the interval [0 h, 72 h] we obtain the predicted mRNA values on the given interval. In Fig. 20.7, we have the plots of the measured mRNA values (solid curve), the predicted mRNA values (mRNA1, dashed curve) computed by integrating the model obtained with the activation regulome probability for $n = 3$ and the repression regulome probability with $n = 1$, and the predicted mRNA values (mRNA2, dash-dotted curve) computed by integrating the model obtained with a random variable with values between the minimum and the maximum value of P_a for $n = 3$

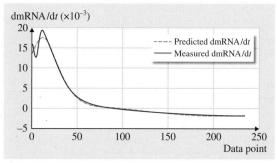

Fig. 20.5 Measured dmRNA/dt and predicted dmRNA/dt values on the validation dataset used in GP RODES. The model was given by the individual with the best fitness (RMSE) on the validation dataset. The inputs for GP RODES were P_a computed with $n = 3$, P_r computed with $n = 1$, mRNA (RMS validation set error: 0.0026665 Variation explained: 77.1939%)

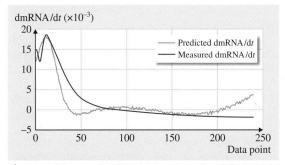

Fig. 20.6 Measured dmRNA/dt and predicted dmRNA/dt values on the validation dataset used in GP RODES. The model was given by the individual with the best fitness (RMSE) on the validation dataset. The inputs for GP RODES were: random variable with values between the minimum and the maximum value of P_a computed for $n = 3$ and random variable with values between the minimum and the maximum value of P_r computed for $n = 1$; mRNA

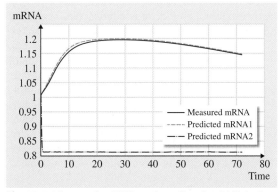

Fig. 20.7 Measured mRNA and predicted mRNA values (*solid curve*). Predicted values were obtained by integrating the models inferred with GP RODES using as inputs. For mRNA1 (*dashed curve*): P_a computed with $n = 3$; P_r computed with $n = 1$; mRNA. For mRNA2 (*dash-dotted curve*): random variable with values between the minimum and the maximum value of P_a computed with $n = 3$; random variable with values between the minimum and the maximum value of P_r computed with $n = 1$; mRNA

and a random variable with values between the minimum and the maximum value of P_r for $n = 1$.

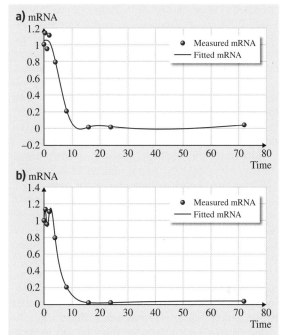

Fig. 20.8 (a) AGR2 mRNA fit using the smoothing spline technique with $\alpha = 0.573$. Goodness of fit: $R^2 = 0.9876$, SSE $= 0.02497$, RMSE $= 0.0984$. **(b)** AGR2 mRNA fit using the smoothing spline technique with $\alpha = 1$ (cubic spline interpolation). Goodness of fit: $R^2 = 1$ and SSE $= 4.815 \times 10^{-35}$. These two fits are examples of problems that may occur during interpolation (negative values for mRNA as in plot **(a)** or unlikely oscillations as in plot **(b)**) in the fitted function. Data points from these two fits were further used to create an acceptable fit for AGR2 mRNA

20.4.3 Reverse Engineering of the Dynomics and Regulomics of Downregulated mRNAs

We also applied GP RODES on the mRNA corresponding to the AGR2 protein identified by the authors as downregulated [20.3]. The values of the AGR2 mRNA are presented in Table 20.2.

To fit the given measured data points, we used the smoothing spline technique. A first fit was obtained by setting the smoothing parameter $\alpha = 0.573$.

Table 20.2 mRNA AGR2 time series values

Time (h)	0	0.5	1	2	4	8	16	24	72
mRNA	1	1.14	0.96	1.11	0.80	0.21	0.02	0.02	0.04

The resulted fitted function (Fig. 20.8a), having SSE = 0.02497, $R^2 = 0.9876$ and RMSE = 0.09846, was acceptable on the interval [0 h, 4 h], but afterwards, the function's values decreased below 0, which was no longer a valid option for mRNA data. Therefore, a second fit was performed with $\alpha = 1$, which returned a perfect fit of the data (Fig. 20.8b), with SSE = 4.815×10^{-35} and $R^2 = 1$. AGR2 mRNA values have some oscillations between $t = 0$ h and $t = 4$ h suspected to be caused by noise.

To remove this segment, it was decided to create new data to be fitted. We thus combined the data points of the first fitted function on the interval [0 h, 4 h], which was smoothly decreasing, and the data points of the second fit on the interval [4 h, 72 h], which were greater than 0. These new data points were further fitted using the smoothing spline technique with $\alpha = 0.529$ and the goodness of fit results were: SSE = 0.004861, $R^2 = 0.998$; RMSE = 0.0185. Figure 20.9 shows this final fitting of the data (Fig. 20.9a, solid curve) and the time series mRNA values (Fig. 20.9a, asterisks). The first-order derivative dmRNA/dt is represented in Fig. 20.9b.

We then evaluated the fitted function and its first-order derivative on time points chosen at an equal distance of 0.1 in the interval [0 h, 72 h]. Thus, we obtained 721 discrete values, corresponding to 721 time points in the interval [0 h, 72 h] for AGR2 mRNA and dmRNA/dt.

Because TGF-β induced a downregulation of mRNA, the following assumptions were made to compute the regulome probabilities:

1. The production of mRNA is stimulated through downregulated TFs. Therefore, an activation regulome probability was considered with $X = \mathrm{TF}/k_a$, and the observation that its values will have to decrease in time. Because downregulated TFs intervene, the maximum value was set at $\mathrm{TF}_{max} = 1$; the minimum value was taken from the downregulated protein data in [20.3]: $\mathrm{TF}_{min} = 0.13$. The constant $k_a = \mathrm{TF}_{max}/X_{max}$ i.e., $k_a = 1/2$ resulting in $k_a = 0.5$. Then, the minimum value for X was $X_{min} = \mathrm{TF}_{min}/k_a$, so $X_{min} = 0.13/0.5$, giving $X_{min} = 0.26$. The activation regulome probabilities were computed choosing X in the interval [0.26, 2] for different sigmoidicity coefficients.

2. The production of mRNA is inhibited, probably by upregulated TFs with an inhibitory role. Therefore, a repression regulome probability was considered with: $X = \mathrm{TF}/k_r$; the maximum value for X was considered as before $X_{max} = 2$. The maximum value for TF was taken from the upregulated protein data in [20.3]: $\mathrm{TF}_{max} = 2.35$ and $\mathrm{TF}_{min} = 1$. We computed k_r to be $k_r = \mathrm{TF}_{max}/X_{max}$, i.e., $k_r = 2.35/2$ resulting in $k_r = 1.175$. The minimum value for X was computed as $X_{min} = \mathrm{TF}_{min}/k_r$, so $X_{min} = 1/1.175$ which gave $X_{min} = 0.8511$. The repression regulome probabilities were computed choosing X in the interval [0.8511; 2], for different sigmoidicity coefficients.

3. The degradation of mRNA is accelerated by upregulated miRNAs. Thus, an activation regulome probability intervenes with $X = \mathrm{miR}/k_a$. The value for X_{max} remains the same: $X_{max} = 2$. The maximum value for miR, as before, is taken from the data of mRNA in [20.3] corresponding to upregulated proteins: $\mathrm{miR}_{max} = 4.24$ and $\mathrm{miR}_{min} = 1$. We computed k_a to be $k_a = \mathrm{miR}_{max}/X_{max}$, i.e., $k_a = 4.24/2$ resulting in $k_a = 2.12$. The minimum value for X was computed as $X_{min} = \mathrm{miR}_{min}/k_a$, so $X_{min} = 1/2.12$, which yielded $X_{min} = 0.4717$. The repression regulome probabilities were computed choosing X in the interval [0.47171, 2] for different sigmoidicity coefficients.

The regulome probabilities were evaluated on 721 points evenly spaced in the intervals: [0.8511, 2] for re-

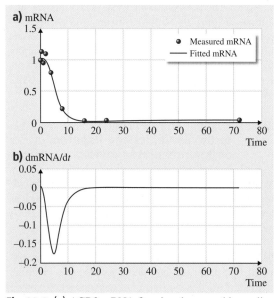

Fig. 20.9 (a) AGR2 mRNA fit using the smoothing spline technique with $\alpha = 0.529$. Goodness of fit: $R^2 = 0.9980$, SSE = 0.004861, RMSE = 0.0185. **(b)** First-order derivative of mRNA with respect to time

pression of production; [0.26, 2] for the activation of production, and [0.47171, 2] for activation of degradation. The input data for GPTIPS consisted in:

1. A repression regulome probability with its values decreasing in time
2. An activation regulome probability with its values decreasing in time
3. An activation regulome probability with its values increasing in time
4. The mRNA AGR2 values obtained by evaluating the fitted function for mRNA.

The desired output represented the dmRNA/dt values on the 721 time points.

We experimented with combinations of a regulome probability with values decreasing in time (stimulation of production by downregulated TFs) and an activation regulome probability with increasing values in time (stimulation of mRNA degradation by upregulated miRs) for different sigmoidicity coefficients. The same settings for GP RODES were used as for the mRNA ACTB experiments. The fitness to be reached was set as before at 0.0005. Many experiments computed the maximum number of generation (5000 generations) without actually reaching the desired fitness, but having the best fitness value of the order of 10^{-4}. The accuracy (variation explained) was about 99.9% for all experiments.

Figure 20.10 shows the measured and the predicted mRNA values on the validation dataset for the best

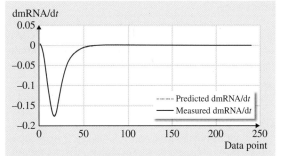

Fig. 20.10 Measured dmRNA/dt and predicted dmRNA/dt values on the validation dataset used in GP RODES. The model was given by the individual with the best fitness (RMSE) on the validation dataset. The inputs for GP RODES were P_a computed with $n = 3$ decreasing in time, P_a computed with $n = 1$ increasing in time mRNA (RMS validation set error: 0.00056012; variation explained: 99.9784%)

model obtained with a decreasing activation regulome probability computed with $n = 3$ and an increasing activation regulome probability computed with $n = 1$. The accuracy (variation explained) was of 99.97% and the best fitness achieved after 5000 generations was RMSE = 0.00056.

By introducing conceptually and technically the activation and repression regulome probabilities, the accuracy is no longer a problem. However, model selection and interpretation should be developed, as many different models have similar accuracy.

20.5 Conclusion

We proposed a methodology for reverse engineering of biochemical networks, illustrated on a transcription network using real data. The automatically inferred models are systems of ordinary differential equations. The algorithms developed, based on genetic programming, can decompose the interacting networks and each node of a network, without destroying the network interactions. Decomposition makes the algorithms fast and scalable to large omics networks. The structure of the network is discovered from data, not imposed, and further hypothesis about the parameters and the mechanisms involved may be formed. The main challenging problem solved by this methodology is that of unmeasured variables in omics time course experiments. Unknown or unmeasured variables, like TFs or microRNAs, are replaced by *regulome probabilities*, a concept introduced here for the first time. The regulome probabilities have a theoretical foundation based on the statistical thermodynamics of regulatory molecular interactions. Thus, the resulting models are not just simple data fittings. To our knowledge, this is the first approach capable of dealing with missing variables, and the accuracy of the models is greater than 99%.

References

20.1   H. Guo, N.T. Ingolia, J.S. Weissman, D.P. Bartel: Mammalian microRNAs predominantly act to decrease target mRNA levels, Nature **466**(7308), 835–840 (2010)

20.2   J.R. Koza: *Genetic Programming: On the Programming of Computers by Means of Natural Selection* (MIT Press, Cambridge 1992)

20.3   V.G. Keshamouni, P. Jagtap, G. Michailidis, J.R. Strahler, R. Kuick, A.K. Reka, P. Papoulias, R. Krishnapuram, A. Srirangam, T.J. Standiford, P.C. Andrews, G.S. Omenn: Temporal quantitative proteomics by iTRAQ 2D-LC-MS/MS and corresponding mRNA expression analysis identify post-transcriptional modulation of actin-cytoskeleton regulators during TGF-β-induced epithelial-mesenchymal transition, J. Proteome Res. **8**(1), 35–47 (2009)

20.4   G.K. Ackers, A.D. Johnson, M.A. Shea: Quantitative model for gene regulation by lambda phage repressor, Proc. Natl. Acad. Sci. USA **79**(4), 1129–1133 (1982)

20.5   L. Bintu, N.E. Buchler, H.G. Garcia, U. Gerland, T. Hwa, J. Kondev, R. Phillips: Transcriptional regulation by the numbers: Models, Curr. Opin. Genet. Dev. **15**(2), 116–124 (2005)

20.6   U. Alon: *An Introduction to Systems Biology: Design Principles of Biological Circuits* (Chapman Hall/CRC, New York 2006)

20.7   M.A. Shea, G.K. Ackers: The OR control system of bacteriophage lambda: A physical-chemical model for gene regulation, J. Mol. Biol. **181**(2), 211–230 (1985)

20.8   D. Searson: GPTIPS: Genetic programming and symbolic regression for MATLAB (2009) available from http://gptips.sourceforge.net/

20.9   D.P. Searson, D.E. Leahy, M.J. Willis: GPTIPS: An open source genetic programming toolbox for multigene symbolic regression, Proc. Int. Multiconf. Eng. Comput. Sci. (IMECS 2010) (2010) pp. 77–80

20.10 Y. Setty, A.E. Mayo, M.G. Surette, U. Alon: Detailed map of a cis-regulatory input function, Proc. Natl. Acad. Sci. USA **100**, 7702–7707 (2003)

20.11 H.M. Sauro, M. Hucka, A. Finney, C. Wellock, H. Bolouri, J. Doyle, H. Kitano: Next generation simulation tools: The systems biology workbench and BioSPICE integration, OMICS **7**(4), 353–370 (2003), SBW latest version available free from http://sourceforge.net/projects/jdesigner/

20.12 G. Greenburg, E.D. Hay: Epithelia suspended in collagen gels can lose polarity and express characteristics of migrating mesenchymal cells, J. Cell Biol. **95**(1), 333–339 (1982)

20.13 C. de Boor: *A Practical Guide to Splines* (Springer, Berlin, Heidelberg 1978)

20.14 A.G. Floares: Automatic reverse engineering algorithm for drug gene regulating networks, Proc. 11th IASTED Int. Conf. Artif. Intell. Soft Comput. (2007)

20.15 A.G. Floares: A reverse engineering algorithm for neural networks, applied to the subthalamopallidal network of basal ganglia, Neural Netw. Spec. Issue **21**, 379–386 (2008)

20.16 M. Brameier, W. Banzhaf: *Linear Genetic Programming* (Springer, Berlin, Heidelberg 2007)

20.17 U. Mückstein, H. Tafer, J. Hackermüller, S.H. Bernhart, P.F. Stadler, I.L. Hofacker: Thermodynamics of RNA–RNA binding, Bioinformatics, **22**(10), 1177–1182 (2006)

20.18 N.G. van Kampen: *Stochastic Processes in Physics and Chemistry* (North-Holland, Amsterdam 1992)

20.19 T. Barrett, D.B. Troup, S.E. Wilhite, P. Ledoux, C. Evangelista, I.F. Kim, M. Tomashevsky, K.A. Marshall, K.H. Phillippy, P.M. Sherman, R.N. Muertter, M. Holko, O. Ayanbule, A. Yefanov, A. Soboleva: NCBI GEO: Archive for functional genomics data sets – 10 years on, Nucleic Acids Res., **39**(1), D1005–D1010 (2011)

20.20 S. Hoops, S. Sahle, C. Lee, J. Pahle, N. Simus, M. Singhal, L. Xu, P. Mendes, U. Kummer: COPASI – A COmplex PAthway SImulator, Bioinformatics **22**(24), 3067–3074 (2006)

20.21 S. Luke, L. Panait: Lexicographic parsimony pressure, Proc. GECCO-2002 (Morgan Kaufmann, San Fancisco 2002) pp. 829–836

21. Computational Methods for Analysis of Transcriptional Regulation

Yue Fan, Mark Kon, Charles DeLisi

Understanding the mechanisms of transcriptional regulation is a key step in understanding many biological processes. Many computational algorithms have been developed to tackle this problem by identifying (1) the binding motifs, (2) binding sites, and (3) regulatory targets of given transcription factors. In this chapter, we survey the scope of currently used methods and algorithms for solving each of the above subproblems. We also focus on the newer subarea of machine learning (ML) methods, which have introduced a framework for a new set of approaches to solving these problems. The connections between these machine learning algorithms and conventional position weight matrix (PWM)-based algorithms are also highlighted, with the suggestion that ML algorithms can often generalize and expand the capabilities of existing methods.

Part D | 21

21.1 Background

Individual genes can be thought of as the fundamental units of information encoded in a biological genome. Roughly speaking, a gene can be described as a specific region of double-stranded DNA which is transcribed into a single-stranded RNA sequence, which is in turn translated into a protein [21.1]. This process is known as *gene expression*. The expression of genes in tissue and in biological pathways is primarily controlled by *transcription factors* (TFs) [21.2, 3], which activate or inhibit transcription of a gene by binding, with different levels of affinity, to specific binding sites, which are collections of short (usually 6–15 base pairs)

and degenerate (similar but not identical) oligonucleotides [21.4]. These sites are located in upstream promoter regions (near the transcription start site of the target gene) or enhancers/silencers (regions further from the target gene), which are DNA regions hundreds to thousands of base pairs long [21.3, 5].

In this chapter we survey existing methodologies for solving transcriptional networks at three levels (described below in detail). These are, for a given TF, the discovery of its binding motifs, the location of its binding sites, and the identification of its gene targets. We further describe some recently developed methods for solving these problems using a family of machine learning approaches which have achieved close to optimal power in all three of the above problems.

Understanding the mechanisms of transcription regulation is still a major challenge in biology. For a fixed transcription factor t, the study of its DNA binding is largely focused on the following three problems:

1. Identification of binding motifs (characteristic DNA sequence patterns of length approximately 10 to which it binds)
2. Identification of functional binding sites (binding positions identified at the nucleotide level)
3. Identification of gene targets of t.

Without describing each of them in detail at this point, we should point out that the above three problems are interrelated. Approaches to solving them also interact with one another in the process of achieving the ultimate goal of identifying gene targets of TFs, and thus constructing the complete transcription regulatory network of an organism. Thus, an important intermediate step toward the goal of solving the gene regulatory network (among others) is the problem considered here of analyzing the details of TF binding [21.6]. The transcriptional regulatory network is in essence a directed bipartite gene network that describes the causal regulatory relationship between TF genes (those coding transcription factors) on one side and target genes on the other [21.7]. An edge from a TF gene to a target gene indicates that the first transcriptionally regulates the second (via its TF).

To be specific, the above three problems have the following relationship with the goal of identifying transcriptional interactions and transcriptional regulation. From partial knowledge of a transcription network in the form of a group of known targets of the given TF t, problem (1) for the TF t can be solved, yielding a motif (a characteristic pattern for DNA regions to which t binds). This is done by identifying and incorporating similar sequence patterns from the promoter sequences of known targets of t [21.5, 8]. With identified binding motifs for t, it is then possible to explore potential binding sites for t genome-wide in silico by finding the sequences that match the predicted motif pattern from (1). Such potential candidate binding sites provide the starting points for solving problem (2). Although computationally identifying functional (actual) binding sites with high confidence is difficult, experimenters are able to validate predicted binding sites and also the corresponding TF–gene associations [21.9] and so complete more of the transcriptional regulatory network; That

is, the result of solving problems (1) and (2) provides one solution to problem (3), which is the prediction of the existence or absence of any edge in the transcriptional network. The process can interactively grow the network, since the improved network in turn can be used to refine the predicted binding motifs and recalibrate the predicted binding sites. In addition to this canonical loop for completion of transcription regulatory networks, there are many alternatives for solving problems (2) and (3) independently [21.7,9–11], a number of which will be mentioned later on.

Identifying binding motifs (problem 1) plays the most important role in the above-mentioned loop; this may be why, in the last two decades, this problem has attracted the most attention of the above three [21.5,8,12]. Computationally, the motif finding problem can be formulated as follows. Assume that, for the given TF t, there is a set of known large DNA regions (e.g., upstream regions of known target genes, often more than 1000 base pairs long, the positive examples) likely to be bound in vivo by t. These (search) regions can be regions identified by chromatin immunoprecipitation chip (ChIP-chip) [21.13, 14] or ChIP-seq [21.15] experiments, or be the promoter regions of potentially coregulated genes (for example, genes with high coexpression, or correlated expression) [21.16–19]. The function of a motif finding algorithm is to find a common pattern of candidate DNA motif subsequences of length approximately 10 (up to a factor of 2) which appear in most of these regions.

This problem is challenging in two respects. First, the signal-to-noise ratio is small. Not all sequences in the positive example set (in vitro binding regions) may actually be bound by the TF in vivo (low enrichment of the positive set). Also, the binding sites are narrow compared with the search region, which is usually hundreds to thousands of base pairs long. Second, the actual binding site patterns are usually similar but not identical sequences (degeneracy). Because of these two challenges, enrichment and degeneracy are often used as a goodness measure of the predicted binding motifs.

Problem (2) of identifying functional binding sites at the nucleotide level is often difficult [21.9]. One commonly used approach is to scan the predicted motif pattern obtained from solving (1) through potential binding regions and identify approximately matching DNA strings as potential binding sites [21.20,21]. This method is computationally simple but generates many false positives (i.e., accidental matches to the pattern that are not biologically functional binding sites) [21.22,23]. To improve prediction accuracy, workers have incorporated other sources of information for identification of binding sites which can complement the above sequence information; For example, phylogenetic information can be used to restrict the search for binding sites to highly conserved regions among related species, based on the knowledge that the functional regulatory elements are more conserved than other regions [21.24, 25]. Some motif search algorithms have used this information to improve the accuracy of their predictions [21.26–32]. Nucleosome positioning information can also be leveraged, since it has been observed that a TF is more accessible to DNA regions that are not occupied by nucleosomes [21.33–35]. Some other physicochemical information (e.g., hydroxyl radical cleavage patterns [21.36, 37]) has also been found to be promising for predicting binding sites of some TF families [21.10, 38].

Identifying target genes (problem 3) and thus expanding the transcriptional regulatory network is the final stage of this process. Beside the standard approaches forming the canonical loop mentioned above, many algorithms have been developed to solve this problem directly. Because transcriptional regulation directly influences the expression of target genes, many computational methods for identifying TF targets start from gene expression profile data, which measure the level of messenger RNA (mRNA) transcripts of genes across different experimental conditions. The success of these algorithms is based on the observation that correlation (similarity) between the gene expression profiles of a TF and of its target genes is often high [21.11, 39]. Unsupervised methods (in which examples of known targets of t are not utilized as above) use gene–gene expression correlation or mutual information (another measure of correlated expression) to associate TF genes and their targets.

Another approach takes advantage of partial knowledge of a regulatory network structure as a set of examples, after which supervised classifiers [21.7] are trained to predict regulatory relationships between new gene pairs just from the training set of known regulatory relationships. These so-called learning-based methods are suitable for expression profiles which are believed to be high dimensional and noisy. Beside these static methodologies, other reverse-engineering methodologies can be applied to time series of gene expression profiles to capture the dynamics of transcriptomic responses (based on relative times of activations of TF genes and regulated genes). Useful mathematical formalisms have included Boolean networks [21.40], Bayesian networks [21.41, 42], and ordinary or partial differential equation approaches [21.43–47]. These methods for constructing the regulatory network through estimating process dynamics are often computationally expensive and data intensive.

21.2 Binding Motif Discovery Algorithms

In this section we introduce some commonly used motif discovery algorithms used for the solution of the binding motif problem (problem 1) mentioned in Sect. 21.1. Even though some of these were developed 20 years ago, they are still the standard tools used nowadays by bioinformaticians and biologists. We describe two major types of motif discovery algorithms, alignment-based and word-based algorithms, in Sects. 21.2.2 and 21.2.3, respectively. Section 21.2.4 introduces learning-based algorithms, which search motifs that distinguish binding targets from a set of nontargets based on an effective training set consisting of known targets and possibly known (or inferred) nontargets. Finally, Sect. 21.2.5 outlines machine-learning-based ensemble algorithms.

21.2.1 Position Weight Matrix and Consensus

To identify binding motifs, we first need a model to describe them. A common motif model is the *position weight matrix* (PWM, also known as *position-specific scoring matrix* or PSSM) model [21.20, 21, 48]. A PWM is a $4 \times l$ matrix $\mathbf{M} = (\theta_{ij})$ whose j-th column $\boldsymbol{\theta}_j$ defines the probability distribution of $\{A, C, G, T\}$, respectively, appearing at position j of the aligned common pattern of a binding site of a given TF t. It is usually generated empirically from a large number of likely binding sites of t by counting frequencies of DNA bases at aligned positions.

A DNA sequence $\alpha = (a_1 a_2 \ldots a_k)$ of size k, known as a k-mer, is considered a significant (highly probable) instance of the motif $\mathbf{M}$ if the probability of generating α under the probability model defined by $\mathbf{M}$ (assuming bases are independent) is significantly higher than the probability of generating α from a background DNA model $\mathbf{B}$; That is, the log-ratio $\log_2\{[P(\alpha|\mathbf{M})]/[P(\alpha|\mathbf{B})]\}$ is significantly larger than 1, where $P(\mathbf{A})$ denotes the probability of an event $\mathbf{A}$.

Computationally, the probability matrix $\mathbf{M}$ is first converted into a log-ratio scoring matrix $\mathbf{N}$ with entries

$$\mathbf{N}_{ij} = \log_2 \left(\frac{\theta_{ij}}{b_i} \right) ,$$

measuring the probabilities in $\mathbf{M}$ against a certain background distribution assigning an a priori independent probability of b_i to base i. The log-ratio score, the *PWM scanning score*, of α with respect to $\mathbf{M}$ (or $\mathbf{N}$) can be calculated as

$$s^{(r)}(\alpha; \mathbf{M}) = \log_2 \left(\frac{P(\alpha|\mathbf{M})}{P(\alpha|\mathbf{B})} \right) = \sum_{j=1}^{l} \mathbf{N}_{a_j, j} ,$$

where the superscript (r) in $s^{(r)}$ denotes use of the PWM scanning score. We will use $s^{(w)}$ and $s^{(e)}$ to represent w-scanning score and ensemble scanning score in later sections.

A higher score suggests that α is more likely to be a binding site. In practice, a threshold τ is often used to identify significant binding k-mers. A promoter position with $s^{(r)}(\alpha; \mathbf{M}) > \tau$ is reported as a candidate binding site and may suggest a new target; more details are presented in Sect. 21.3.

Two commonly used goodness measures of PWMs are the *information content* (IC, or *relative entropy*) and *maximum a posteriori score* (MAP score, or the *log-likelihood ratio*). These are defined by

$$\text{IC}(\mathbf{M}) = \sum_{i=1}^{4} \sum_{j=1}^{k} \theta_{ij} \log_2 \left(\frac{\theta_{ij}}{b_i} \right) = \sum_{i=1}^{4} \sum_{j=1}^{k} \mathbf{M}_{ij} \mathbf{N}_{ij} .$$

The MAP score is defined similarly as

$$\text{MAP}(\mathbf{M}) = n \, \text{IC}(\mathbf{M}) = \sum_{i=1}^{4} \sum_{j=1}^{k} n \mathbf{M}_{ij} \mathbf{N}_{ij} ,$$

where n is the number of k-mer instances used to construct the PWM $\mathbf{M}$.

The IC measures the *degeneracy* of the PWM. The minimum IC value is obtained when the PWM $\mathbf{M}$ implies the same distribution as the background, thus not providing any additional information. On the other hand, a high IC indicates the PWM/motif is rarely observable (significant) by chance.

The MAP score is often computed as a result of solving the problem of identifying common motif patterns from a set $\mathscr{S}$ of binding targets of TF t. It does not only score the PWM/motif by its degeneracy, but also scores it by how often it is observed in binding targets (*enrichment*). The more often a motif is observed in $\mathscr{S}$, the more likely it is the true binding motif of t.

The *consensus* is another commonly used profiling model to characterize a motif. Instead of showing the probabilistic distribution of nucleotide bases at each position of a motif, a consensus displays only the most frequently occurring base or bases. Standard codes created by the International Union of Pure and Applied Chemistry (IUPAC) [21.49] are often used to allow some variations in certain positions; For example, TGASTCA can be used to indicate either TGAGTCA or TGACTCA. Even with the IUPAC codes, the consensus model is not as flexible or accurate as the PWM model. However, it can summarize information in a way that is easy to visualize.

21.2.2 Alignment–Based Algorithms for Finding PWM

The uses of PWMs mentioned above include assistance in both TF binding site and target identification. Algorithms for the identification of PWM motifs for TFs are known as *motif discovery algorithms*. Such algorithms take a number of forms, one of the most common of which is the so-called alignment-based algorithm. As its name implies, the idea is to find the best-aligned blocks among the set of positive sequences (i. e., promoters of known targets of the TF). The PWM is then used to model the aligned blocks. The enrichment (matching frequency) and degeneracy (matching goodness) are usually used to measure the quality of the identified PWM.

In theory, it has been proved that exhaustively evaluating all possible alignments of the set of positive sequences is an NP-hard problem [21.8, 50], meaning that the exact solution is not computationally feasible. Therefore, some local approximations have been developed to prune the search space in order to avoid this computational obstacle, resulting in algorithms such as greedy optimization [21.51], expectation maximization [21.52, 53], and Gibbs sampling [21.54–57], which have been used quite successfully in practice.

Greedy Optimization

Given a set of potential target sequences, $\mathscr{S} = \{S_1, S_2, \ldots, S_N\}$, the basic idea of greedy optimization is to dynamically identify a PWM as well as its optimal width in order to describe a common sequence pattern from the increasing series of sequences $\{S_1\}$, $\{S_1, S_2\}, \{S_1, S_2, S_3\}, \ldots$. The PWM width is generally determined before the program is run. The algorithm is summarized as follows:

1. The program forms the PWMs for the most common subsequences of size k (known as *k-mers*, *strings*, or *biopolymers*) in S_1.
2. *Greedy search*: For each candidate PWM from sequence S_1, the algorithm combines best-matching k-mers of S_2 into the profile and rebuilds the candidate PWM. At this second iteration every candidate PWM is scored based on its information content. Only the best PWMs are saved to continue to the next step.
3. The procedure continues in the next iteration with all k-mers of S_i considered for combination with each saved PWM from the previous step. The updated PWMs are scored and selected based on the newly added information. This procedure is repeated until all sequences S_i in the original set are incorporated.

These steps are the essential core of the CONSENSUS motif discovery tool [21.51]. For this procedure we note that a potential problem is that some factors may affect the final result systematically. First note that the result depends on the order of presentation of the sequences S_i in $\mathcal{S}$, and in particular on the choice of the first two sequences. This potential oversensitivity is solved in a later version of CONSENSUS known as WCONSENSUS [21.58], by considering all pairs of sequences at the beginning of the algorithm. Second, the number of PWMs that are allowed to continue to the next iteration is also critical to the performance of the algorithm and is a parameter input into the program. In practice, different values can be tried in order to obtain the best results. In WCONSENSUS, a *p*-value is calculated for the information content of each reported PWM, providing a measure to help the user select significant PWMs. The previous requirement of providing PWM size as an input is also dropped, because the program can automatically try a range of sizes and select optimal PWMs based on their *p*-values.

Expectation Maximization

A commonly used algorithm which implements so-called expectation maximization is multiple expectation maximization for motif elicitation (MEME) [21.52,53]. Given a dataset of potential binding target regions $\mathcal{S} = \{S_1, S_2, \ldots, S_N\}$, MEME *guesses* for each k-mer in every sequence whether it is generated from the motif model or from the background model. The expectation-maximization (EM) algorithm is employed to fit the most likely two-component finite mixture model that would have generated the sample $\mathcal{S}$.

Denote the k-mer starting from position j of sequence S_i as α_{ij}. The finite mixture model assumes that some of these k-mers are generated from the motif of interest, modeled by a PWM $\mathbf{M} = (\theta_{ij})$ (foreground model), while the rest are generated by a background model $\mathbf{B} = (b_i)$, where b_i denotes an (independent) background probability for nucleotide base i. A set of auxiliary variables $\mathbf{Z} = (\mathbf{Z}_{ij})$ is also introduced to indicate the model-based annotations of each α_{ij}; For example, in the two-component case, $\mathbf{Z}_{ij} = 1$ indicates that k-mer α_{ij} is an instance of motif $\mathbf{M}$ and $\mathbf{Z}_{ij} = 0$ indicates not. The entire dataset $\mathcal{S}$ is viewed as arising from a mixture of these two models weighted by global constants λ_1 and λ_0, i.e.,

$$P(\mathcal{S}) = \lambda_1 \cdot P(\mathcal{S}|\mathbf{M}) + \lambda_0 \cdot P(\mathcal{S}|\mathbf{B}) .$$

Therefore, it can be shown that the model parameters θ_{ij} and λ_k can be obtained from maximizing the logarithm of the likelihood function L (the log-likelihood function), defined as

$$\ln(L(\mathbf{M}, \lambda | \mathcal{S}, \mathbf{Z})) = \sum_{ij} \mathbf{Z}_{ij} \cdot \ln(\lambda_1 P(\alpha_{ij}|\mathbf{M})) + (1 - \mathbf{Z}_{ij}) \cdot \ln(\lambda_0 P(\alpha_{ij}|\mathbf{B})) .$$

Because the annotations $\mathbf{Z}_{ij}$ are not observable, the EM component of the algorithm is used to obtain a maximum-likelihood estimate of the model parameters θ_{ij} and λ_k by iteratively calculating the expected value of $\mathbf{Z}$ (i.e., of the components $\mathbf{Z}_{ij}$ – this is the E-step or expectation step) and reestimating the parameters θ_{ij} and λ_k (this is the M-step or maximization step). Intuitively, the expected value of $\mathbf{Z}_{ij}$ is calculated by comparing which model (foreground or background) is more likely generating the α_{ij} (E-step), based on the current estimation of θ_{ij} and λ_k. Then the PWM $\mathbf{M} = (\theta_{ij})$ is constructed by aligning the α_{ij} with larger $\mathbf{Z}_{ij}$ values, since they are more likely generated by the foreground model. Some pseudocounts are also added when constructing $\mathbf{M}$, which is in turn equivalent to a Dirichlet prior distribution over the θ_{ij} obtained from a Bayesian point of view. The iteration is stopped when the maximum a posteriori (MAP) score of the predicted PWM has converged.

Similarly to the CONSENSUS algorithm, the size of the motif is a required parameter in the model. Thus, in practice, a researcher needs to run MEME several times with different motif size parameters to obtain a set of candidate PWMs. To overcome this limitation, a motif selection and validation algorithm, MAST [21.59], has also been developed by the same group as a complement to compare models with different width parameters. This extended motif selection algorithm can help search multiple databases to identify meaningful motifs as well as identify the transcription factors corresponding to them [21.52]. In addition, the algorithm can be generalized to mixture models with more than two components. However, the difficulty is that the maximum-likelihood estimator (MLE) obtained from the algorithm is then more likely to converge to local optima when the dimensionality is large. Currently, the MEME algorithm removes k-mers of a given motif completely from the analysis of the promoter regions in $\mathcal{S}$ and reruns the algorithm with the remaining data to obtain the next PWM [21.52].

Gibbs Sampling

One of the most successful approaches to motif discovery has been the Gibbs sampling strategy. It was first introduced for motif discovery in protein sequences by *Lawrence* et al. [21.54]. There are several successful algorithms implementing Gibbs sampling for DNA sequences, including Motif Sampler [21.60, 61], AlignACE [21.57], and BioProspector [21.55]. The idea of the basic Gibbs sampling algorithm is to replace the EM deterministic optimization with a stochastic search, so as to avoid the problem of premature convergence to local optima of either the IC or MAP functions.

Given a set of input sequences $\mathcal{S}$, the goal is again to identify the most significant aligned block of size k in the set $\mathcal{S}$. The simplest version of the Gibbs sampler assumes only one binding site in each sequence. Similarly to the procedure in MEME, the motif (aligned pattern) is modeled with a PWM $\mathbf{M}$ and the background is represented as a model $\mathbf{B}$. Instead of the auxiliary indicators $\mathbf{Z}$ used in MEME, a set of position indices $\{a_i\}$ (indicating the starting position index of the current most highly matched motif in sequence S_i) is used to help in the estimation of $\mathbf{M}$. Similarly to the procedure in the EM algorithm, the Gibbs sampling algorithm consists of a predictive update step (analogous to the M-step) and a sampling step (analogous to the E-step). These can be summarized as follows:

1. A set of initial binding site positions a_i is randomly chosen in each sequence S_i (one for each sequence).
2. *Predictive update step*: At iteration m, one sequence, say S_z, is chosen at random or from a specified ordering. A PWM $\mathbf{M}^{(m)}$ summarizes the k-mers at the positions a_i, including the starting positions a_i in all sequences S_i excluding the currently chosen S_z.

3. *Sampling*: Every k-mer α_{zj} starting at position j in the selected sequence S_z is considered as a candidate to replace the currently selected k-mer α_{za_z} starting at the currently selected position a_z in sequence S_z. For each α_{zj}, the probability Q_{zj} of generating α_{zj} from the currently updated PWM $\mathbf{M}^{(m)}$ versus the probability P_{zj} based on the background model $\mathbf{B}$ are computed. The weight $\mathbf{A}_{zj} = Q_{zj}/P_{zj}$ is assigned to each k-mer α_{zj}, and one of the k-mers is selected at random according to the normalized weights (probabilities) $\mathbf{A}_{zj}/\sum_l \mathbf{A}_{zl}$. The starting position j of the selected k-mer α_{zj} becomes the new value of the starting position a_z in S_z.
4. The above predictive update step is repeated with the newly updated position variables a_z. A new PWM $\mathbf{M}^{(m+1)}$ is formed and is used to draw a new position $a_{z'}$ from a newly selected sequence $S_{z'}$.

These steps are iterated a minimum number of times or until convergence of the MAP score is reached.

The main difference between the Gibbs sampling approach and MEME is how the k-mers for updating the estimated PWM $\mathbf{M}^{(k)}$ are selected: while MEME always picks a k-mer deterministically according to how well it fits the current motif, the Gibbs sampler chooses the k-mer randomly according to a weighting distribution defined by the current motif. Due to this stochasticity, the reported PWM could be different in multiple runs, even with the same set of parameters. This property is a double-edged sword. On one hand, with this stochasticity the Gibbs sampler is less likely to become stuck in local optima. On the other hand, however, it is more difficult for practitioners to select the most informative predictions. In practice, workers can run the program multiple times with the same or different size parameters, and then for quality control compare a subset of the results with other algorithms, other information, or with database searches.

21.2.3 Word-Based Algorithms

So-called word-based algorithms were essentially the first approach introduced to solve the motif problem starting from the mid-1980s [21.62]. Conceptually, these algorithms first enumerate all possible k-mers that could be a part of the binding motif of interest and put them in a set $\mathbf{A} = \{\alpha_1, \alpha_2, \ldots, \alpha_n\}$. Without any prior knowledge, the candidate set $\mathbf{A}$ must generally contain all 4^k possible k-mers. The key step is to measure how much a candidate k-mer is overrepresented in the set of potential target sequences (positives), relative to the entire genome. The measure of this is usually developed upon counting the absolute frequency of occurrence of each given k-mer (with no mismatches), or the frequency of occurrence with at most e mismatches, both in potential target sequences and associated background sequences. A ranked list is then generated based on this measure, and the highest ranking k-mers are reported as potential binding k-mers. Various subsequent steps are then also developed to identify either a consensus sequence or to form a PWM from this set of high-ranking (potential binding) k-mers, in order to use this identified pattern to find new gene targets for the given TF.

The computational difficulty of this method initially limited its development because of limited computational resources. In particular, exhaustively computing frequency measures for all k-mers, in general with 4^k candidates, was considered too computationally intensive, as the dimension of the problem increases exponentially with the size k. More recently, some new techniques have been developed to work around this complexity problem. The exhaustive search can be significantly speeded up by organizing the input sequences in a suitable structure, such as the suffix tree used by the Weeder algorithm [21.63], which yields an execution time only exponential in the number e of substitutions allowed. On the other hand, the initial set of candidates can be downsized in many different ways; For example, the MDscan algorithm [21.64] restricts the initial candidate set to those k-mers appearing in sequences that are most likely bound by the transcription factor (i. e., with highly significant p-values in chromatin immunoprecipitation, or ChIP-chip, experiments). Thus, improved quality of biological information can be used to improve the efficiency of this computational method.

A simple and illustrative statistical measure of overrepresentation of a k-mer α is computed from the z-scores [21.65] of frequency of k-mer α, defined as

$$\mathbf{Z}_\alpha = \frac{\mathbf{N}_\alpha - EN_\alpha}{\sigma_\alpha} \, .$$

$\mathbf{N}_\alpha$ is the observed frequency of α in a sequence, and EN_α is the expected frequency of α under the background model. Their difference is divided by the standard deviation σ_α of the random variable $\mathbf{N}_\alpha$. The calculation of EN_α and σ_α are detailed in [21.65, 66]. This measure is used by the yeast motif finder (YMF) [21.67, 68], Weeder [21.63], and similar algorithms with some extensions; For example, Weeder computes this score with a suffix tree for measuring the overrepresentation of k-mers with several mutations (e-mismatches). Variations of the algorithm are also im-

plemented using different combinations of (k, e), and their resulting lists of top-ranked k-mers.

21.2.4 Learning–Based Methods

All of the aforementioned algorithms use only the sequences of a set of potential targets as input data, and the significance of the motif is measured against a generic background model that characterizes the nucleotide distribution of the genome. On the other hand, some learning-based algorithms have also been quite successful in the past few years. In addition to taking only putative binding targets into account, a learning-based method generally also considers negatives (genes that are unlikely bound by the transcription factor) in training its discrimination mechanism. Such presumed negatives play the role of characterizing nonbinding k-mers relative to the given TF, rather than using the entire genome to form a background, as is done in some of the previously mentioned methods. When known negatives are not available, using random sequences generated from the background model can also be used. These random sequences thus provide information similar to that from the background model in non-learning-based methods. Therefore, the unavailability of explicit negative examples does not need to limit the use of a learning-based method. As previously, learning-based methods can again be categorized into alignment-based versus word-based algorithms.

Instead of using enrichment of positives with respect to a putative motif as one of its goodness measures, a learning-based method uses discriminative power, i. e., the ability to distinguish positives (binding target genes) from negatives (others). Hence, the objective functions (in the above-mentioned alignment-based algorithms) and the k-mer scoring functions (of the word-based algorithms) are effectively modified to include the information obtained from binary labeling of positive and negative examples. Two examples of learning-based algorithms in our sense are MoAn [21.69] and SVMotif [21.70], which will be mentioned in some more detail in this section.

Learning Alignment–Based Methods

The distinguishing characteristic of learning-based algorithms is the use of the labels (classes) of examples in the objective function. For instance, for the nonlearning expectation-maximization-based algorithms (e.g., MEME), the objective function is the log-likelihood

$$L(\mathbf{M}, \Phi; \mathcal{S}) = P(\mathcal{S}|\mathbf{M}, \Phi) \, ,$$

where $\mathcal{S} = \{S_1, S_2, \ldots, S_N\}$ is the set of sequences of interest, $\mathbf{M} = (\theta_{ij})$ is the PWM, and Φ are other model nuisance parameters. In the class of learning-based algorithms, the log-likelihood to be optimized is the joint probability of observing both sequences $\mathcal{S}$ and their labels $Y = (y_1, y_2, \ldots, y_N)$, where $y_i = \pm 1$ depending on whether sequence S_i is a positive or negative example of targets of the TF of interest. In general, the objective function has the form

$$L(\mathbf{M}, \Phi; \mathcal{S}, Y) = P(\mathcal{S}, Y | \mathbf{M}, \Phi)$$
$$\propto P(\mathcal{S} | Y, \mathbf{M}, \Phi) P(Y).$$

We consider MoAn [21.69] as an example of a learning-based algorithm. The objective function has the form

$$\log_2 L(\mathbf{M}, \tau; \mathcal{S}, Y) = \sum_i \log_2 P(y_i | S_i, \mathbf{M}, \tau)$$
$$\propto \sum_i \log_2 P(S_i | y_i, \mathbf{M}, \tau) P(y_i),$$

where τ is a threshold to be used in determining the binding site.

The probability of a sequence $S \in \mathcal{S}$ conditioned on its label y and motif model (PWM) $\mathbf{M}$ is represented by

$$P(S | y, \mathbf{M}, \tau) = q(S) \left(v_{0,\text{eff}}^y + v_1^y r(S, \mathbf{M}, \tau) \right),$$

where

$$q(S) = P(S | \mathbf{B})$$

is the probability of S under a background model and

$$r(S, \mathbf{M}, \tau) = n^{-1} \sum_{z: s^{(r)}(\alpha_z; \mathbf{M}) > \tau} 2^{s^{(r)}(\alpha_z; \mathbf{M})};$$

$s^{(r)}(\alpha_z; \mathbf{M})$ is the scanning score (Sect. 21.3) of PWM $\mathbf{M}$ against the k-mer α_z starting at position z in sequence S. Note that $q(S) 2^{s^{(r)}(\alpha_z; \mathbf{M})}$ is equivalent to $P(\alpha_z | \mathbf{M})$ and condition $s^{(r)}(\alpha_z; \mathbf{M}) > \tau$ indicates that α_z matches the pattern defined by $\mathbf{M}$ (using threshold τ). Thus, $q(S) r(S, \mathbf{M}, \tau)$ is the probability of observing sequence S conditioned on the k-mers matching $\mathbf{M}$ (having a scanning score above threshold τ) being generated from the motif model $\mathbf{M}$ and the rest of the sequences being generated from the background model $\mathbf{B}$. The v^y are the prior probabilities of the matching k-mers α_z (scanning score $s^{(r)}(\alpha_z; \mathbf{M}) > \tau$) being generated from either the background (v_0^y) or motif model (v_1^y) given y, the label of the sequence S. Even if its scanning score is above the threshold, there is still a chance that a given k-mer is from the background. The

prior probabilities are different for positives and negatives; For example, $v_1^1 = 0.99$ and $v_1^0 = 0.8$ mean that 99% of the matches (with scanning score over threshold) found in the positives are considered true, but that only 80% of matches found in the negatives are considered true. This objective function is then optimized using simulated annealing (SA) with respect to the parameters defined in the PWM $\mathbf{M} = (\theta_{ij})$. The width of the PWM and the scanning threshold τ are also optimized simultaneously.

Word-Based Learning Methods

Learning methods have also been developed in a word-based framework. We will also refer to these as machine learning methods, given the usually very high-dimensional feature spaces they involve. SVMotif [21.70], as an example, delivers good results in revealing *Saccharomyces cerevisiae* motifs. Instead of measuring the overrepresentation of each k-mer feature (word) as in nonmachine word-based algorithms (e.g., using normalized z-score of word counts), SV-Motif measures the discriminatory power of each k-mer by fitting the problem into a support vector machine (SVM) feature selection framework. A gene's promoter sequence S_i is represented by a feature vector $x_i = (x_1, x_2, \ldots, x_n) \in F$, whose elements count frequencies of a list $\mathbf{A} = \{\alpha_1, \alpha_2, \ldots, \alpha_n\}$ of k-mers. F is then called the k-mer feature space. A support vector machine (SVM) then uses this class of feature vectors x_i as input to predict the labels y_i of a training set of sequences ($y_i = 1$ for positives/putative binding targets, $y_i = -1$ for negatives/nontargets). More specifically, a classification function

$$f(x) = \sum_j w_j x_j + b$$

that minimizes the Lagrangian

$$\mathcal{L} = C \cdot \sum_{i=1}^N \text{Loss}(y_i, f(x_i)) + \frac{1}{2} \|w\|^2$$

is learned from the training sample $\mathcal{S}$, where the above used function Loss is a measure of the error between the prediction $f(x_i)$ and the true class y_i.

Using the recursive SVM (RSVM) algorithm [21.71], the feature importance score I_j of k-mer α_j is defined by

$$I_j = w_j (\bar{x}_j^+ - \bar{x}_j^-),$$

where $\bar{x}_j^+$ and $\bar{x}_j^-$ are average frequencies of α_j in the positive and negative training examples, respectively.

Table 21.1 Top 10 k-mers in the output for sample yeast transcription factors GCN4, UME6, MIG1, and STE12. True (standard) binding motifs, which are retrieved from YeastGenome [21.73], are listed in *upper case* at the top. k-Mers matching the corresponding motifs are *highlighted in bold*. The feature importance scores I_j for k-mer features are listed next to the corresponding k-mers

Rank	GCN4 (TGACTCA)		UME6 (TAGCCGCCSA)		MIG1 (WWWWSYGGGG)		STE12 (TGAAACA)	
1	**gagtca**	8.099	**gccgcc**	5.262	cccgc	0.858	**gtttca**	2.764
2	**agtcat**	4.434	**agccgc**	5.103	ggggaa	0.658	cgagaa	1.084
3	**gactca**	4.094	**cggcta**	4.253	accca	0.622	**gaaaca**	1.014
4	**agtca**	1.679	**gccgc**	3.04	ccccgc	0.616	cattcc	0.934
5	**gactc**	1.66	**ccgcc**	2.718	ccgga	0.604	tcctaa	0.79
6	**agtcac**	1.127	**ccgccg**	2.05	accc	0.603	agtatg	0.708
7	cattag	0.933	**cgccga**	1.857	ccgg	0.597	acattc	0.649
8	cttatc	0.886	**agccg**	1.634	ccccac	0.568	**aaacag**	0.609
9	**actca**	0.832	**cgccg**	1.124	ccgta	0.556	**atgaaa**	0.562
10	catgac	0.734	**gcgcc**	0.845	gcaaca	0.524	taggaa	0.556

This importance score I_j describes the contribution of the j-th k-mer to the final classifier's differentiation of positives from negatives. The k-mer features are then ranked by their feature importance scores I_j from the largest down, and a dimensionally reduced set consisting of the top half of the ranked k-mer features is kept to train the SVM (again on the training data) for another iteration. This feature ranking and elimination process is iterated until only the top m k-mers remain. The remaining features (the k-mers with the best discriminatory power) are most closely related to the true binding motif (Table 21.1). In principle, any feature selection/ranking method for any machine learning algorithm can also be used in this implementation; For example, using the *random forest* algorithm [21.72] to rank k-mer features (using the permutation test on so-called out-of-bag samples) yields comparable performance to that of SV-Motif.

Table 21.1, which was generated by running SV-Motif on yeast data, also shows that k-mers which are not related to known binding motifs can also be ranked close to the top. There are at least two reasons for this. First, such k-mers may be related to secondary binding motifs of the TF t of interest which are active under different conditions from those which activate the standard motif. Alternatively, they may be related to binding motifs of cofactors of t. Cofactors are additional TFs functioning together with the TF of interest which form so-called *cis*-regulatory modules, tightly packaged molecular TF complexes that act together to initiate transcription at some gene locations [21.6]. Thirdly, low signal-to-noise ratios can cause the feature selection methods to select false-positive features. The signal-to-noise ratio can be affected by many factors,

including the size of the sought motif versus the size of the searching region, the specificity of the motif versus the background, or the quality of the negatives in the training dataset.

The mentioned problems can be addressed using a *k-mer agglomeration step*. After ranking the top m k-mers based on their feature importance scores, SV-Motif then greedily agglomerates top k-mers into k-mer clusters based on how well and with what alignment these k-mers overlap with each other. By clustering top k-mers, we can filter out the k-mers which are incidental to the motif and reveal the overlapping k-mer clusters which are consistent with the true motif PWM, and are typically ranked consistently at the top. In addition, because of the limitations involved in handling the high dimensionality of spaces generated by long k-mers (e.g., greater than 8 in length), the PWM is still preferred as a low-dimensional profiling tool for characterizing binding motifs. Thus, SVMotif operates by discovering longer motif PWMs based on the w-vector in the 6-mer space $F^{(6)}$ as listed here:

1. Rank the k-mers (e.g., 6-mers) α_j based on their SVM training weights w_j obtained from the training dataset. The top ranked k-mers are more likely to be part of the full binding motif.
2. Construct an initial size k PWM $\mathbf{M}_1$ deterministically to reflect the top k-mer α_1 only.
3. Add the next ranking k-mer α_2 into $\mathbf{M}_1$ if the scanning score $s^{(r)}(\alpha_2, \mathbf{M}_1)$ is above a certain threshold. Otherwise, construct a new PWM $\mathbf{M}_2$ for α_2 only.
4. For each successive k-mer α_i, check whether this k-mer should be incorporated into any previously formed PWM matrix or be used to form a new

PWM, and so either modify the previous PWM or create a new one.

5. Repeat until the top m k-mers are agglomerated into PWMs (effectively k-mer clusters).

6. The best PWM (k-mer clusters) are then selected based on a goodness measure (e.g., IC, MAP), yielding a final ranked list of PWM.

21.2.5 Ensemble Algorithms

With dozens of motif discovery tools developed, a study of many commonly used algorithms was performed under standardized dataset and performance metrics [21.22]. These datasets contained real as well as simulated sequences from several species, among which yeast had the simplest genome. The performance of each method was measured at two different levels, the site level and the nucleotide level. Metrics for both levels assessed the performance of predicted binding sites (problem 2) rather than predicted binding motifs (problem 1). This is a more stringent standard for assessing the motif discovery tools since the ultimate goal of such motifs is to identify regulatory targets via identifying binding sites.

The result shows that none of these algorithms performs consistently well across different datasets (Table 21.2). This suggests that ensemble methods taking advantage of several different algorithms can be useful. Some ensemble methods, which combine the predictions from different algorithms, have been developed recently to refine predicted motifs. One type of ensemble approach is built to score each motif in a candidate pool of predicted motifs by measuring its *goodness*. WebMOTIFS (the web interface of TAMO) [21.75, 76] assesses each candidate motif with several statistics (e.g., hypergeometric enrichment score [21.14]), and a ranked list is reported for the user. Another type of ensemble approach starts with predicted binding sites on the relevant promoter sequences from single or multiple algorithms. The locations agreed to by the most algorithms are reported as predicted binding sites [21.77–80]. The motifs are then formed by compiling the subsequences at those locations, and a scoring scheme is then used to rank predicted motifs. Essentially, the scoring function consists of two main components, enrichment and degeneracy (Sect. 21.2.1). Some algorithms then optimize the motif score by locally adjusting the motif locations to improve accuracy; these include BEST [21.81], GAME [21.82], and BioOptimizer [21.83].

As an extension of the SVMotif machine learning algorithm, an ensemble method has also been developed within a feature selection framework [21.84]. The idea of this approach is that, instead of using exact k-mers,

Table 21.2 Performance (F_1 score) of 16 algorithms on individual Tompa datasets [21.22]. *Columns* 2 and 6 indicate the number of datasets on which the algorithm produces nonzero nucleotide- and site-level F_1 score. *Columns* 3–5 and 7–9 indicate the number of datasets on which the algorithm performs the best among the 16 (in terms of F_1 score [21.74])

Algorithms	Nucleotide level				Site level			
	Nonzero	Top	Top 2	Top 3	Nonzero	Top	Top 2	Top 3
AlignACE	10	2	2	7	8	2	2	6
ANN-Spec	25	5	8	12	20	2	6	7
Consensus	7	1	1	2	6	1	2	4
GLAM	15	0	2	4	11	0	3	5
Improbizer	22	1	4	4	21	1	3	5
MEME	23	3	7	12	20	5	9	13
MEME3	19	3	5	7	15	3	5	8
MITRA	16	1	3	4	10	1	4	5
MotifSampler	21	4	8	10	18	4	7	11
oligodyad-analysis	15	2	3	5	12	2	3	4
QuickScore	14	1	5	6	7	0	1	1
SesiMCMC	19	4	6	11	19	4	8	12
Weeder	18	4	6	7	18	7	9	11
YMF	20	3	8	10	18	3	5	7
SVMotif	20	5	11	17	19	2	6	8
Ensemble	24	11	15	16	22	9	12	13

k-mer profiles are used as features, with a PWM used to characterize the profile. This approach can also be considered a generalization from the matching of k-mers with at most e-mismatches.

The PWM-based feature map on promoters is defined similarly to the previously mentioned map using k-mer features. In particular, the promoter sequence S of a gene is mapped into a metafeature space F which contains feature vectors $x = \psi(S)$ based on two types of feature maps $\psi^{(r)}$ and $\psi^{(w)}$ as follows:

- The *PWM scanning feature* map $x^{(r)} = \psi^{(r)}(S)$, defined by components

$$x_i^{(r)} = s^{(r)}(S; \mathbf{M}_i) = \sum_{\alpha \subset S} s^{(r)}(\alpha; \mathbf{M}_i) \,,$$

where

$$s^{(r)}(\alpha; \mathbf{M}_i) = \sum_{j=1}^{k} \log_2 \left(\frac{\theta_{a_j,j}}{b_{a_j}} \right)$$

is a PWM scanning score for k-mer $\alpha = (a_1 a_2 \ldots a_k)$ against PWM $\mathbf{M} = (\theta_{ij})$, defined in detail below in Sect. 21.3. Intuitively, this feature is similar to counting the matching frequencies of PWM $\mathbf{M}_i$ in sequence S.
- The *PWM subspace feature* map $x^{(w)} = \psi^{(w)}(S)$, defined by

$$x_i^{(w)} = \mathbf{w}_i \cdot x_i \,,$$

where $\mathbf{w}_i$ is the w-vector of the SVM trained in a restricted PWM subspace F_i (Sect. 21.5.2 below) and x_i is the feature vector to which S is mapped in F_i. Intuitively, the feature $x_i^{(w)}$ scores a gene using the classifier formed by restricting to a subspace F_i of the full k-mer feature space, with a basis consisting

of a random selection of k-mers consistent with the motif matrix $\mathbf{M}_i$ (see Sect. 21.5 for a more detailed discussion of the w-vector, and the PWM and PWM subspaces).

We call a feature space F of either type above a *synopsis feature space*, whose features x_i indicate a likelihood that promoter sequence S is a target of the TF t based on the motif matrix $\mathbf{M}_i$. The SVMs built on feature vectors $x^{(r)} \in F^{(r)}$, on $x^{(w)} \in F^{(w)}$, or on their combination $x^{(e)} = (x^{(r)}, x^{(w)}) \in F^{(e)}$ are called the scanning ensemble, the subspace ensemble, and the comprehensive ensemble SVMs, respectively. Feature selection methods can also be applied to one of the synopsis feature spaces ($F^{(r)}$, $F^{(w)}$, or $F^{(e)}$) in order to select the most informative PWM features. Then, top-ranked PWMs will be agglomerated into PWM clusters [21.84] similarly to the way this is done in SVMotif. This approach is considered an ensemble method because the individual PWM-based features x_i are calculated from predicted PWMs obtained from various independent motif tools such as MEME [21.52], Weeder [21.63], etc. This ensemble algorithm shows a very large predictive improvement over its individual components [21.84]. Eighty-eight yeast transcription factors whose binding motifs had been previously identified [21.13, 14, 85, 86] were used as a benchmark dataset. Symmetrized Kullback–Leibler divergences were used as a similarity measure to compare the predicted motifs with benchmarks (previously identified motifs). An ideal ensemble should generally be able to reproduce the best (most similar to the standard) motif predicted by component algorithms. The ensemble-recommended top predictions reproduced equal or better similarity compared with all of its components for 46 out of 88 TFs [21.84].

21.3 Binding Site Identification

The second problem in the original list in Sect. 21.2.5 (problem 2) is binding site identification. After a motif pattern becomes available, a standard subsequent task is to find matched locations of the pattern along any genetic regions of interest. A traditional algorithm, the PWM scanning algorithm, is presented in Sect. 21.3.1 below. This algorithm, which locates k-mers in DNA that match the probability distribution implied by a given PWM, works statistically, and identifying functional binding sites with it is difficult. Therefore, some confirmatory algorithms (Sect. 21.3.2) have been developed to refine the identifications of

given PWMs and their scanning-based binding site predictions. At the end of this section (Sect. 21.3.3), we once again introduce an ensemble machine method which integrates multiple motif discovery tools as mutually confirmatory sources.

21.3.1 PWM Scanning

Once a motif for a given TF t has been identified (e.g., in the form of a PWM), it can be used as a primary tool for finding potential binding sites of t in candidate promoter sequences in $\mathcal{S}$. Recall (as mentioned in Sect. 21.2.1) that a $4 \times k$ PWM matrix ([21.20, 21, 48]) gives a position-based probability distribution of nucleotides at n successive DNA locations. Thus, $\mathbf{M} = (\theta_{ij})$ has entries θ_{ij} equaling the probability of observing DNA base i (A, C, G, or T) at position j relative to the start of a putative binding site. PWM scanning is a procedure which scores each position z in S with a motif PWM $\mathbf{M}$ by calculating the probability of the length-k string starting at z under the probability model $\mathbf{M}$.

For the k-mer $\alpha_z = (a_1 a_2 \ldots a_k)$ starting at position z in sequence S, its PWM scanning score is defined as

$$
\begin{aligned}
s^{(r)}(\alpha_z; \mathbf{M}) &= \log_2 P(\alpha_z|\mathbf{M}) - \log_2 P(\alpha_z|\mathbf{B}) \\
&= \sum_{j=1}^{k} \log_2 \left(\theta_{a_j, j} \right) - \sum_{j=1}^{k} \log_2 \left(b_{a_j} \right) \\
&= \sum_{j=1}^{k} \log_2 \left(\frac{\theta_{a_j, j}}{b_{a_j}} \right) .
\end{aligned}
$$

Statistically the scanning score $s^{(r)}(\alpha_z; \mathbf{M})$ is the logarithm of the likelihood ratio between the motif model and background model for generating α_z. A larger score indicates that k-mer α_z is more likely to be a match for the motif. In practice, a threshold τ is used to define significant binding site predictions; that is, the PWM scanning binding site identification algorithm predicts α_z as a binding site if $s^{(r)}(\alpha_z; \mathbf{M}) > \tau$.

The performance of such PWM scanning algorithms depends not only on the quality of the motif PWM $\mathbf{M}$ but also on the chosen background model $\mathbf{B}$ in computing the scanning score. The purpose of a background model is to characterize what random (specifically, nonbinding) sequences look like in the region being scanned (e.g., the promoter). A simple choice of a background distribution is one where the probability of each background nucleotide is calculated empirically based on its overall frequency, and nucleotides are treated as independent (as in the above definition of $s^{(r)}(\alpha_z; \mathbf{M})$). With this background model, however, the scanning method often produces large numbers of false positives [21.9]. Instead of using an independent nucleotide base model, some higher-order Markov models together with context information (upstream or downstream) have been investigated for this purpose [21.87, 88]; For example, the calculation of $\log_2 P(\alpha_z|\mathbf{B})$ depends on the sequences (α_{z-k}) located upstream and sequences (α_{z+k}) located downstream, as well as the higher-order Markov score of α_z itself. It has been shown that such improvements in the background model can improve the specificity of PWM scanning methods in binding site identification [21.88].

Beside the above-mentioned approaches to improving the specificity of binding site discovery methods, some confirmatory methods which use supporting evidence to eliminate nonfunctional putative binding sites have also been developed by various groups. Some examples are methods using additional information such as phylogenetic footprinting, *cis*-regulatory module-based analysis, and nucleosome occupancy-based analysis. We introduce these in more detail in the following sections.

21.3.2 Confirmatory Methods Using External Information

Phylogenetic Footprinting

Identification of sequences conserved between related species is termed *phylogenetic footprinting* [21.89]. Evolutionary conservation of functional regulatory elements in orthologous promoter regions of closely related genomes is considered informative in confirmation of putative TF binding sites [21.32, 90]; That is, we expect PWM scanning hits in highly conserved regions to be more likely to represent functional binding sites.

Some motif discovery algorithms [21.26, 27, 31] find binding motifs and binding sites interactively in order to incorporate phylogenetic information better. The usefulness of phylogenetic footprinting [21.91] has been shown in a number of studies, for example, in the reconstruction of the yeast regulatory map by *MacIsaac* and collaborators [21.32]. This work involved conservation analysis using PhyloCon [21.27] and Converge [21.14], which compared the yeast genome with that of closely related species in order to improve specificity of binding site identification. Compared with previous binding site maps [21.13, 14] using only information from *Saccharomyces cerevisiae*, the new map discovers a number of binding sites for 36 new TFs [21.32].

Cis-regulatory Module–Based Analysis

Transcription factors often work cooperatively as *cis*-regulatory modules (CRM), and in these cases their binding sites tend to be in close proximity (i.e., within a few tens of bases) along DNA sequences [21.92, 93]. Based on this knowledge, given a fixed TF t, the observed cooccurrence of binding sites for TFs belonging to a *cis*-regulatory module containing t within a predefined window about a putative binding site increases the confidence of binding site predictions for t [21.9]. Recently, computational approaches have been developed to identify putative *cis*-regulatory modules, based on the repetitive presence in various promoters of cooccurring motifs [21.94–96], on site clustering information [21.97–100], and on other related observations [21.9]. Some algorithms have also been developed to identify binding motifs and binding sites simultaneously from given sets of coregulated targets [21.93, 101]; For example, the method presented by *Gupta* and *Liu* in [21.101] employs an explicitly specified hidden Markov model (HMM) to describe the structure of *cis*-regulatory modules (in terms of numbers of binding motifs, their binding site locations, and binding order preference) for a set of coregulated genes. Additional parameters such as the number of component TFs, their PWMs, distance between neighboring TF binding sites, transition probabilities between different TFs' binding sites, and the like, can be estimated by optimizing the probability of observing a set of coregulated genes under this full probability model.

Competitive Occupancy by Other Types of DNA Binding Factors

Beside transcription factors, DNA sequences can also be bound by a number of other types of DNA binding factors, including nucleosomes as well as other proteins and protein complexes, including, for example, the origin recognition complex (ORC) [21.35]. These alternative DNA binding factors typically compete with transcription factors for binding along the genome. Therefore, high DNA occupancy levels by another binding factor reduce the likely functionality of a transcription factor binding site at the same location. Many experimental and computational methods are used to identify DNA occupancy by other factors [21.33, 102–105]. Nucleosomes, for example, are coiled structures binding stretches of about 147 DNA base pairs, which loop around a histone octomer [21.106] (a linked group of eight histone proteins). These can occupy as much as 75–90% of the genome [21.107]. An interesting algorithm, COMPETE [21.35], used HMM [21.108] to model the competition between nucleosomes and 88 TFs in yeast. The results showed that explicitly modeling the competition among DNA binding factors improves both TF binding and nucleosome binding predictions genome-wide [21.35].

21.3.3 Machine-Learning Methods for Binding Site Identification

In this section, we highlight one type of machine learning method used in the discovery of binding sites, known as w-scanning [21.84]. In Sect. 21.5, we discuss the connection between this use of the SVM w-vector and PWM-based scanning for binding sites in detail. Based on this discussion, we will see w-scanning as an improved alternative to PWM scanning methods. The confirming evidence below suggests a capability for improving specificity.

w-Scanning

In an earlier section, we introduced an SVM-based motif discovery algorithm, SVMotif [21.70]. This utilizes the ranking of k-mer features in the w-vector $\mathbf{w} = (w_1, w_2, \ldots, w_n)$ used in the classification function $f(\boldsymbol{x}) = \mathbf{w} \cdot \mathbf{x} + b$ to rank k-mers. In SVMotif, a feature importance score $I_j = w_j(\bar{x}_j^+ - \bar{x}_j^-)$, which is a variant of the w-vector weight w_j suggested by RSVM [21.71], is used to rank k-mers. Some other feature-ranking scores, such as $I_j = w_j^2$, are also suggested in other literature [21.109]. For this discussion we will use $I_j = w_j$ in describing the w-scanning algorithm. The w-scanning procedure identifies potential binding sites for the TF t of interest by scoring each k-mer α_z starting at position z along the sequence S and by assigning it the score $s^{(w)}(\alpha_z; \mathbf{w}) = w_{\alpha_z}$. Here, w_{α_z} is the entry corresponding to k-mer α_z in the w-vector. Similarly to the PWM scanning method, a threshold τ is also used to identify significant binding site predictions.

The w-scanning method can be considered an improvement over PWM rescanning because it does not involve an assumption regarding the statistical independence of successive bases, which is implicitly made in a PWM-based motif model. To see this, note that, given a PWM $\mathbf{M}$, using the PWM scanning method, a k-mer $\alpha = (a_1 a_2 \ldots a_k)$ has a score $s^{(r)}(\alpha; \mathbf{M})$ equal to the logarithm of the ratio of its probability $\prod_j \mathbf{M}_{a_j, j}$ under the model $\mathbf{M}$ and its probability $\prod_j \mathbf{B}_{a_j, j}$ under the background model $\mathbf{B}$. On the other hand, scoring k-mer α using the w-rescanning method uses the above score w_α, by which we mean the w-component corresponding to k-mer α. Now note that, in the special case where (using the above notation) $w_\alpha = s^{(r)}(\alpha; \mathbf{M})$, we can effectively reduce the PWM scanning score to a w-scanning score, as can easily be shown through a simple calculation. More general choices of w, on the other hand, are equivalent to scoring α according to a fully general probability model (as opposed to a PWM-based model implicitly assuming independent bases). Such a fully general model might also be considered a $(k-1)^{th}$-order Markov model, since when restricted to words of size k, such a $(k-1)^{th}$-order model is equivalent to a general probability distribution on all words of length k.

Ensemble–Based PWM Scanning and w–Scanning

Scanning by a single PWM or single w-vector can typically produce large numbers of nonfunctional binding sites (false positives). In addition to the confirmatory validation methods introduced in Sect. 21.3.2, in this section we consider an ensemble method which integrates scanning scores from multiple PWMs and w-vectors as mutually confirmatory information. For each k-mer α_z starting at position z along the sequence S, the ensemble scanning score $s^{(e)}$ is defined as a linear combination of multiple PWM-scanning scores $s_i^{(r)} = s^{(r)}(\alpha_z; \mathbf{M}_i)$ and w-scanning scores $s_i^{(w)} = s^{(w)}(\alpha_z; \mathbf{w}_i)$ [21.84], i.e.,

$$s^{(e)}(\alpha_z) = \sum_i \beta_i s_i^{(r)} + \sum_i \gamma_i s_i^{(w)} \ .$$

Once again, a threshold τ is necessary, and here is learned from the score distribution on negative/background sequences. If $s^{(e)}(\alpha_z) > \tau$, then α_z is predicted as a potential binding site.

The coefficients β_i and γ_i are determined by the SVM built on the synopsis feature space $F^{(e)}$ (Sect. 21.2.5). PWMs and w-vectors with strong discriminatory power (in separating binding targets and nontargets) will be assigned large coefficients by the machine. Therefore, if k-mer α_z matches multiple good PWMs and w-vectors, its ensemble score $s^{(e)}$ will be large. One the other hand, accidental matching k-mers are less likely to be confirmed by multiple PWMs or w-vectors, and their ensemble scores $s^{(e)}$ will be relatively small. Under the general conditions required below, this algorithm will produce reliable ensemble scanning scores:

1. The candidate PWM pool should be formed to contain multiple PWMs which are similar to the true motif.
2. PWMs should be learned from varying tools so that the false PWMs are dissimilar to each other and their scanning scores will not add up to high values accidentally.

The ensemble was tested on *Tompa*'s benchmark dataset [21.22] (see Sect. 21.5 for more details).

21.4 Target Gene Identification and Construction of the Gene Regulatory Network (GRN)

In this section, we introduce several methodologies for identifying transcriptional regulatory target genes for a given TF t, with a view toward using this approach in the construction of high-confidence transcriptional regulatory backbones for more general gene regulatory networks. This involves problem (3). mentioned in the introduction, identifying transcriptional regulatory targets of TFs. In this section we consider different methods based on the information sources they use. Primary information sources are expression (Sect. 21.4.1) and sequence (Sect. 21.4.2) information. Other data (Sect. 21.4.2) are also explored as complements to these. Machine learning methods are used to integrate information from different sources.

21.4.1 Using Coexpression Information

Unsupervised Methods

Some unsupervised algorithms for identifying TF gene targets which are introduced below were originally developed for identifying gene coexpression clusters or gene–gene networks in general. Based on the assump-

tion that TF genes are more likely to be coexpressed with their target genes, these algorithms have also been applied to identify transcriptional regulatory relationships.

The core of these algorithms is based on calculating the similarity between the expression values of any gene pair (in our case a TF gene and potential target genes). Many similarity measures have been tried with relatively successful results over the past few years. Let the expression profiles of two genes (measured at different times or under different conditions) be $X = (x_1, x_2, \ldots, x_n)$ and $Y = (y_1, y_2, \ldots, y_n)$. Here x_i and y_i are expression levels of gene X and gene Y at the i-th time or under the i-th condition.

● The *correlation* (with adjustment [21.110]) is $\hat{r}^2 = \text{sgn}(r) \cdot r^2$. Here, r is the standard correlation coefficient between the gene expression profiles X and Y, i.e.,

$$r = \frac{\sum_i \left(x_i - \overline{X}\right)\left(y_i - \overline{Y}\right)}{\sqrt{\sum_i \left(x_i - \overline{X}\right)^2 \left(y_i - \overline{Y}\right)^2}} \ ,$$

where $\overline{X}$ and $\overline{Y}$ represent the mean values of the vectors X and Y.

- The *mutual information* (MI [21.39]) is $I(X, Y) = \sum p(x, y) \ln p(x, y)/p(x)p(y)$. Here, x and y range over possible values of x_i and y_i, respectively, and probabilities are observed empirical probabilities of individual and paired values over the dataset. Because the expression values are continuous, in practice, mutual information is often calculated after discretization into K bins [21.39]. Some nonparametric algorithms have also been developed to accommodate continuous values [21.111].

After obtaining the pairwise similarity information (e.g., correlation or mutual information), the final step in compiling the transcriptional regulatory network relies on the choice of a correlation/mutual information threshold τ (for r or I) for determining above-threshold regulatory relationships between genes. Thus, a TF gene and target gene are predicted to interact if the similarity between their expression profiles is above τ. The choice of τ involves a trade-off between potential true-positive and false-positive rates in identifying significant regulatory interactions. A high threshold results in a sparse network with fewer links, in which genes typically form disjoint small clusters. The predicted interactions are more accurate, but the potential to identify novel interactions is limited. Conversely, a low threshold will often yield a dense network with many predicted links which will be more likely to capture novel interactions. However, there will also be many false-positive interactions in such a network, corresponding to background correlations and misannotations of indirect dependences as direct interactions. Thus, choosing the correct threshold often requires additional analysis, which can frustrate practitioners applying these algorithms to specific datasets.

Some postanalysis algorithms have been developed to eliminate insignificant edges (false positives) from predictions. As mentioned above, a correlation does not always suggest a direct interaction between two genes. A pair of genes separated in the regulatory network by one or more intermediaries (indirect interactions) may be highly coregulated without implying an irreducible interaction between the two; this can produce large numbers of false positives. The ARACNe algorithm [21.112] uses a well-known information-theoretic property referred to as the *data processing inequality* to eliminate some indirect interactions. The algorithm iteratively examines the triangular structures in the network and removes the edge with the smallest MI score among each of the three edges in a triangular interaction. On synthetic datasets, ARACNe achieves low error rates and outperforms the relevance network [21.39] and Bayesian network [21.42] algorithms [21.112].

The CLR algorithm [21.11] is an extension of the relevance network approach using mutual information [21.39] as the pairwise similarity measure. After computing the MI between a TF gene and its potential targets, CLR normalizes the MI by calculating the statistical likelihood of each MI value within its network context. The network context or *background* of a pair (t, g) consisting of the TF gene t and its potential target g includes all possible such pairs that include either the TF t or the putative target g. This step is designed to remove so-called promiscuous cases in which the TF gene weakly covaries with a large number of genes, or one putative target gene weakly covaries with many TF genes. Such promiscuity arises when the assayed conditions are inadequately or unevenly sampled, thus failing to distinguish direct interactions from indirect influences, or when microarray normalization fails to remove false background correlations due to interlaboratory variations in methodology [21.11].

Supervised Methods

Supervised methods are developed to leverage partial knowledge of regulatory networks in the form of a list of known TF–gene interactions as well as expected noninteracting pairs as training labels (positives and negatives) to build classifiers. Expression profiles have been successfully used as features of genes [21.7]. The idea is that, if a gene A has an expression profile similar to a gene B known to be regulated by a given TF, then gene A is likely to be also regulated by this TF. Thus, for a given transcription factor t, a classifier f (e.g., SVM) is built upon the data $D = \{x_i, y_i\}_i$. The x_i are the expression profile feature vectors of known targets of t (positives) and some (randomly selected) presumed nontargets (negatives). The y_i label the targets as 1 and nontargets as -1. The learned classifier is applied to the expression profiles of other genes in the same genome with feature vectors z, defining z to be a new regulatory target if the classification function $f(z) = 1$. Because of the effective use of the negatives/background genes, this method does not need any postprocessing step in order to select a threshold. Compared with the CLR algorithm, which is one of the most successful unsupervised algorithms (i. e., one using no training negatives), this supervised method predicts six times more known regulations at a 60%

precision level. The recall is 44.5% at the same precision level, meaning that this algorithm is able to retrieve almost half of the existing transcriptional regulatory relationships and that 60% of the newly predicted ones are correct.

21.4.2 Using Sequence and Other Information Sources

Even though the above coexpression-based algorithms have obtained significant results, high levels of noise in expression microarrays limit these algorithms' accuracy. In addition, because the biochemical interactions of the transcription factors with their target genes are determined by the DNA binding domain of the TF and the nucleotides of its binding sites [21.113], identification of functional TF binding sites naturally provides useful and reliable information on transcriptional regulation relationships between TFs and genes. Because computationally identifying specific functional binding sites is difficult, machine learning methods have been developed to use k-mer count signatures of the promoter of a gene, along with those of its other potential binding locations. These can then be used to identify target genes directly from such k-mer count features, rather than from analysis of specific putative binding sites. The innovation involved in these algorithms is that no motif needs to be formed in intermediate steps, preventing information loss related to summarizing binding site information via PWM or consensus sequences.

PWM Scanning

In Sect. 21.2 we discussed motif discovery algorithms – an accurate motif model consists of significant information usually obtained from sequences and expected functional binding sites. As mentioned earlier, the standard algorithm for the extraction of gene targets from PWM information has been the PWM scanning algorithm [21.20, 21, 48]. A motif (in the form of a PWM) for a transcription factor t is used to score each k-mer in the promoter region $S(g)$ of a candidate gene g (note that PWM scanning scores for k-mers are defined in Sects. 21.2.1 and 21.3.1). We define the PWM scanning score of a long sequence $S(g)$ as a function of all individual k-mer scanning scores (Sect. 21.2.1), i. e.,

$$s^{(r)}(S(g); \mathbf{M}) = \phi \left\{ \left[s^{(r)}(\alpha_z; \mathbf{M}) \right]_{\alpha_z \subset S(g)} \right\} .$$

Several forms for ϕ are available to summarize k-mer scores at the gene level; For example,

- *Linear scanning score*

$$\left[s^{(r)}(S(g); \mathbf{M}) \right] = \sum_{\alpha_z \subset S(g)} s^{(r)}(\alpha_z; \mathbf{M}) ,$$

- *Maximum scanning score*

$$\left[s^{(r)}(S(g); \mathbf{M}) \right] = \max_{\alpha_z \subset S(g)} s^{(r)}(\alpha_z; \mathbf{M}) ,$$

- *Top-m scanning score*

$$\left[s^{(r)}(S(g); \mathbf{M}) \right] = \sum_{\text{top } m \; \alpha_z \subset S(g)} s^{(r)}(\alpha_z; \mathbf{M}) .$$

The latter score represents a sum of the top m k-mer scanning scores for a fixed m.

If the PWM scanning score of the promoter $S(g)$ of g is greater than a predefined threshold τ, i. e., $s^{(r)}(S(g); \mathbf{M}) > \tau$, then g is predicted as a regulatory target of t. In machine learning language, the above-mentioned decision rule can be translated into the classifier

$$f(g) = \text{sgn}\left(s^{(r)}(S(g); \mathbf{M}) - \tau \right) .$$

This method is simple and fast, but in practical terms does not always give good results. First, the accuracy of the classifier not only depends on the quality of the motif PWM $\mathbf{M}$, but also on the background model chosen in computing the rescanning score. The function of a background model is characterizing what random (specifically, nonbinding) sequences look like in the region being scanned (e.g., the gene promoter). A simple choice of the background is one in which the probability of each background nucleotide is assumed independent and is calculated empirically based on its overall frequency. With this background model, the scanning method often produces large numbers of false positives [21.9]. Instead of using an independent nucleotide base model, some higher-order Markov models also including context information (e.g. upstream or downstream) have been investigated for this purpose [21.87, 88]. It has been shown that such improvements in the background model can also improve binding site identification.

Ensemble Methods for Target Identification

One of the ensemble methods described above in the discussion of motif identification (Sect. 21.2.5) is also useful in the identification of binding targets of a TF. As a single feature in determining whether gene g is a target of TF t, one can use the scanning score for a particular PWM $\mathbf{M}$ over the gene's promoter $S(g)$ agglomerated

Computational Methods for Analysis of Transcriptional Regulation | Target Gene Identification and Construction of GRN 343

Part D | 21.4

using one of the above three scoring methods over all k-mers (successive base sequences of length k) in $S(g)$. The corresponding classification rule is a simple weak classifier, $f(g) = s^{(r)}(S(g); \mathbf{M}) - \tau$. However, even if the choice of background model can improve its quality, by itself the scanning score $s^{(r)}$ cannot perform well, again partly because the motif $\mathbf{M}$ may not be optimized.

Rather than using only single motif scores as features, we consider an ensemble method which uses the multiple scores calculated from scanning by several predicted motifs from multiple motif discovery algorithms, for example, AlignACE [21.57], MEME [21.52], etc. The larger feature vector

$$\boldsymbol{x}^{(r)} = \Big\{ s^{(r)} \big[S(g); \mathbf{M}_1 \big], s^{(r)} \big[S(g); \mathbf{M}_2 \big], \ldots,$$
$$s^{(r)} \big[S(g); \mathbf{M}_m \big] \Big\}$$

can provide more complete and reliable information regarding the transcription binding pattern. In our analysis, the ensemble SVM-based classifier which uses as features the motif scanning scores of motifs from five algorithms, including AlignACE [21.57], Bio-Prospector [21.55], MEME [21.52], Weeder [21.63], and SVMotif [21.70], outperforms the best components for 75 of 88 yeast transcription factors [21.84]. The F_1 score [21.74] ($F_\beta = (1 + \beta^2) \times \text{Precision} \times \text{Recall}/(\beta^2 \text{Precision} + \text{Recall})$) is improved from approximately 0.55 (MEME) to approximately 0.7 averaged over all 88 TFs under investigation in this study (Table 21.3). At a fixed 80% precision level, the above F_1 scores translate to approximately 40% recall versus approximately 60% recall, respectively. Thus, the ensemble method is able to retrieve 20% more known targets from the genome with the same precision.

Integrating Other Information

Not all of the factors determining TF binding to DNA sequences are fully documented or understood. Using machine learning kernel methods, *Holloway*, et al. [21.10] integrated 26 different information sources including DNA sequence, gene coexpression features, and physicochemical information (e.g., local DNA melting temperatures) to predict binding targets. Machine learning methods, especially kernel-based ones, play an important role in integrating information of different dimensions, different quality, and different consistencies together, and from this making state-of-the-art predictions.

The machine learning paradigm generally starts with a *training set* consisting of a data collection $D = \{\boldsymbol{x}_i, y_i\}_{i=1}^N$, with each $\boldsymbol{x}_i$ a feature vector representing a well-defined collection of numbers describing the i-th training sample, while y_i represents the class of the same sample (e.g., $+1$ for target gene and -1 for non-target gene). The learning machine is presented with the full dataset D, and based on this is automatically trained to make further predictions y for feature vectors x of novel examples based on the training with D (see Sect. 21.4.1 for a special case using expression profile features).

At the center of many machine learning approaches is the associated kernel matrix $\mathbf{K}$, an $\mathbf{N} \times \mathbf{N}$ matrix whose entry $\mathbf{K}_{ij} = \boldsymbol{x}_i \cdot \boldsymbol{x}_j$ is the inner product between the feature vectors $\boldsymbol{x}_i$ and $\boldsymbol{x}_j$ (essentially their unnormalized pairwise similarity). Different classes of feature vectors (based on different information sources) are then combined through addition of the corresponding kernel matrices. Specifically, let $\mathbf{K}_{(i)}$ be the kernel matrix based on the i-th information source; then the weighted sum $\mathbf{K} = \sum_i \beta_i \mathbf{K}_{(i)}$ is known as a combined kernel and can now be similarly used to train the classifier, now using combined information sources.

Table 21.3 F_1 scores by the number of positives in the training set. The table shows the performance of five component algorithms, three ensembles, and the benchmark (PWM scanning classifier using PWMs retrieved from the UCSC [21.85, 86] genome browser). The three ensemble algorithms had similar performance to each other, but had consistently better performance than other methods

No. pos	BioProspector	Align-ACE	MEME	SVMotif	Weeder	$F^{(r)}$	$F^{(w)}$	$F^{(e)}$	Benchmark
20–40	0.60	0.49	0.59	0.56	0.58	0.72	0.68	0.72	0.64
40–60	0.55	0.47	0.55	0.54	0.53	0.68	0.66	0.68	0.61
60–100	0.54	0.47	0.53	0.51	0.58	0.66	0.67	0.66	0.53
100–200	0.54	0.49	0.51	0.51	0.55	0.67	0.66	0.68	0.62
200+	0.61	0.53	0.59	0.59	0.56	0.75	0.73	0.75	0.63
All	0.57	0.49	0.55	0.54	0.56	0.70	0.68	0.70	0.61

Among the different information sources for determining whether a promoter $S(g)$ corresponding to a gene g is a target of the fixed TF t, sequence information in the form of k-mer frequency counts has been shown to be the most effective [21.10].

21.5 Machine Learning Approaches

We have introduced several machine learning algorithms in Sects. 21.2 and 21.4. In this section we present a more comprehensive picture of the uses of machine learning and feature space-based methods for solving the previously mentioned three problems (those of motif, binding site, and target discovery). A number of papers have shown that machine-learning-type algorithms can have a number of advantages over traditional PWM-based algorithms.

Using SVM methods, our group has developed a series of algorithms, TFSVM [21.10], SVMotif [21.70], and SVMotifM [21.114], to solve these three problems, individually and also in the larger context of solving and better understanding the gene regulatory network. A key concept in these machine methods is the previously discussed k-mer feature map, which we will denote as the *k-mer spectrum map ϕ*. This maps each sequence S (e.g., the promoter/upstream sequence of a potential target gene g) into a k-mer (spectrum) feature vector x [21.115] whose component x_i, indexed by the k-mer label α_i (α_i is a fixed consecutive DNA substring of length k), simply counts the number of distinct occurrences of the k-mer string α_i in S. Each feature vector x_i in the dataset $D = \{x_i, y_i\}_{i=1}^{N}$ has label $y_i = \pm 1$ indicating whether training gene g_i is a target ($y_i = +1$) or nontarget ($y_i = -1$).

A machine classifier f is trained on $D = \{x_i, y_i\}_{i=1}^{N}$ via optimization of a certain Lagrangian functional $\mathcal{L}(f)$ which grades f with respect to both its accurate classification of the training data D and its generalizability to classification of novel data. The optimizing f is then used to classify an incoming promoter sequence $S(g)$ from gene g based on whether the value of the classifier $x = \phi(S(g))$ is positive or negative [21.10]. For a linear SVM the classification function has the form $f(\mathbf{x}) = \mathbf{w} \cdot \mathbf{x} + b$ for a fixed vector $\mathbf{w}$ and scalar b. It can be learned by minimizing a Lagrangian $\mathcal{L} = C \cdot \sum_i \text{Loss}(y_i, f(\mathbf{x_i})) + \frac{1}{2}\|\mathbf{w}\|^2$. The so-called w-vector $\mathbf{w} = (w_j)$, which is a vector normal to the classification hyperplane $f(\mathbf{x}) = 0$, separates predicted positives (targets) from negatives (nontargets). Each component w_j of the w-vector gives a weight to the j-th k-mer feature, and so measures the discriminative power of that k-mer.

21.5.1 Connection Between PWM and SVM

We note that there is a connection between the above-mentioned SVM feature space methods and PWM scanning methods for identification of target genes, in the following sense. First, with linear scanning methods (Sect. 21.4.2), there is an equivalent SVM (a classification function with the form $f(\mathbf{x}) = \mathbf{w} \cdot \mathbf{x} + b$) which selects the same target genes, assuming that the w-vector is selected properly and the feature space consists of components measuring counts of k-mers of the same size as the PWM being compared. Specifically, for a given PWM $\mathbf{M}$, if w_j is chosen to be exactly the PWM scanning score of the j-th k-mer α_j, i.e., $s^{(r)}(\alpha_j; \mathbf{M})$, then an SVM with this w-vector will in the testing stage give exactly the same testing score for a given gene promoter as the linear PWM scanning score [21.84],

$$s^{(r)}(S) = \sum_{\alpha \subset S} s^{(r)}(\alpha) = \sum_j n_{\alpha_j} s^{(r)}(\alpha_j) = \mathbf{w} \cdot \mathbf{x} ,$$

where α in the sum ranges over all k-mers in sequence S, and n_{α_j} is the number of appearances of k-mer α_j. To the extent above, the SVM method is a strict generalization of such linear scanning methods, given that an SVM w-vector exists which gives the same scores to promoters as does linear PWM-based scoring. It should be noted that, if training is done by constraining the potential w-vectors to fall into this class of vectors derived from PWMs (which we will call PWM-compatible w-vectors), then a w which optimizes the Lagrangian $\mathcal{L}(f)$ (see before) may correspond to a PWM which is not the same as that which would be obtained through standard PWM methods. Nevertheless, given that the optimization of $\mathcal{L}$ corresponds to an optimal choice of $\mathbf{w}$ on the training set, such a machine-selected PWM-compatible w-vector can in some cases correspond to a PWM which does a better job than the standard one for discriminating between positives and negatives; That is, the Lagrangian-optimizing PWM may be a better one based on the training set for the machine learning method.

Though the above method may allow the claim that SVM generalizes linear scanning methods, it should be noted that most PWM scanning methods use nonlinear systems of scoring. Specifically, a scanning score on a given promoter usually considers only the maximum scanning matches of k-mers against the given PWM along the promoter, rather than a linear combination of all matches as above.

We note, however, that it is also possible to construct an SVM-based classifier whose scoring emulates this type of maximum score obtained from PWM rescanning. Specifically, if we form a w-vector as above corresponding to a given PWM, we can form a linear classifier based on this vector with the replacement of each component w_i by w_i^p with p an integer (here superscript p denotes a selected fixed high power of w_i). If we simultaneously apply a feature map $\phi(\mathbf{x})_p = (x_i)^p$ to the feature vectors, again with p fixed but sufficiently large, the linear classifier $f(x) = \phi(\mathbf{w}) \cdot \phi(\mathbf{x}) + b$ will emulate the scoring system obtained from the PWM scoring mentioned above. This follows from the fact that the p-norm $\|\mathbf{v}\|_p \equiv (\sum_i v_i^p)^{1/p}$ of a vector $\mathbf{v}$ approaches its ∞-norm $\|\mathbf{v}\|_\infty \equiv \sup_i v_i$ as $p \to \infty$.

More specifically, if we consider the class of PWM-based classifiers which use maximum matching scores in the PWM scanning, we will show that there exists an SVM-based classifier which effectively emulates such scanning-based classifiers as well. For this we consider a restricted class of feature vectors $\mathbf{x} = (x_i)$ for which the components x_i are defined by the values

$$x_i = \begin{cases} 1, & \text{if the } i\text{-th } k\text{-mer } \alpha_i \text{ appears in } S, \\ 0, & \text{if } \alpha_i \text{ does not appear in } S. \end{cases}$$

We note that this restricted class of feature vectors has proved to be often more effective in practical SVM classifiers for TF binding as compared with feature vectors using standard k-mer counts ([21.70]). Then, if we again form the above-defined w-vector determined by a particular PWM $\mathbf{M}$, the classifier

$$f(\mathbf{x}) = \phi(\mathbf{w}) \cdot \phi(\mathbf{x}) + b = \mathbf{w}^{\cdot p} \cdot \mathbf{x}^{\cdot p} + b$$

(where $\mathbf{w}^{\cdot p}$ by definition has components w_i^p) for large p will have values which are largest when the maximum PWM scanning score for the sequence S is the largest. Note above that $\mathbf{x}^{\cdot p} = \mathbf{x}$ (since $x_i = 0$ or 1), and the p-norm $\mathbf{w}^{\cdot p} \cdot \mathbf{x}^{\cdot p}$ is effectively an ∞-norm (taken to the power p) when p is large. Thus, $\mathbf{w}^{\cdot p} \cdot \mathbf{x}$ computes the maximum component of the vector $\mathbf{w}^{\cdot p} \cdot \mathbf{x} = (w_i^p x_i) = w_i^p$ (since $x_i = 1$). Since this high power w_i^p is monotonic in w_i, it follows that the highest values for $f(\mathbf{x})$ will correspond, as $p \to \infty$, to (the p-th power of) the highest maximum PWM rescanning scores $w_i x_i = w_i$ for feature vectors $\mathbf{x}$, yielding

for a given threshold choice b of $f(\mathbf{x})$ an equivalent TF target classifier using this SVM as is obtained using the (nonlinear) maximum PWM scanning score (with threshold $\tau = -b^{1/p}$) as a criterion.

Thus, we conclude that, for any maximum scanning score PWM-based classifier, there exists a weight vector $\mathbf{w}' = \mathbf{w}^{\cdot p}$ (for sufficiently large p) which for the above (0–1) feature vectors $\mathbf{x}$ emulates the PWM maximum scanning score-based classifier to arbitrary accuracy. However, the above modified SVM algorithm with restricted binary (0–1) feature vectors $\mathbf{x}$ searches the full space of w-vectors in order to optimize the classification of the dataset D of positive and negative examples. Therefore (using the above Lagrangian to optimize the SVM-based classifier on the dataset D over all $\mathbf{w}$ and the binary feature vectors $\mathbf{x}$), this SVM searches a set of classifiers which forms a superset of all of the above PWM maximum scanning score-based classifiers in its quadratic programming optimization algorithm. Though this particular optimization criterion may choose a vector $\mathbf{w}$ which is not equal to $\mathbf{w}'^{\cdot p}$ (the choice which for sufficiently large p is equivalent to use of a PWM maximum scanning score classification criterion), the vector $\mathbf{w}'^{\cdot p}$ is in the search space of the Lagrangian optimization, and so it may be argued that the SVM finds a vector which in some cases may be better than the canonical vector based on PWM methods. We summarize these arguments as follows (using feature vectors $\mathbf{x} = (x_i)$ defined as above with only 0 or 1 components):

Theorem 21.1

Assume we are given a prior choice of a PWM as a classifier of targets of a given TF t, and a training dataset $D = \{x_i, y_i\}_{i=1}^{N}$ of known positive and negative examples of targets of t using only binary features in $\mathbf{x}$. Assume that the linear SVM algorithm based on the dataset D determines a classifier $f(\mathbf{x}) = \mathbf{w} \cdot \mathbf{x} + b$ which is optimized as a target classifier for t according to standard SVM quadratic programming criteria from the dataset D. Then, this optimization is done by searching a space of $\mathbf{w}$ which includes a particular $\mathbf{w}$ for which the score $f(\mathbf{x})$ is equivalent to (i. e., is a monotonic function of) the PWM maximum scanning score for each gene with feature vector $\mathbf{x}$.

More specifically, as shown above, the SVM score based on this choice of $\mathbf{w}$ vector is asymptotically for large p a monotonic function (the p-th power) of the above-mentioned PWM maximum scanning score, and so a w vector matched in this way with the PWM gives

Table 21.4 Area under the ROC curves (AUROC) for different classifiers. The column headed "PWM" presents the AUROC of the PWM maximum scanning score-based classifier. Columns 2–4 present the AUROC of the above-mentioned SVM-based classifier using the p-th powers of PWM-compatible w-vectors ($p = 1, 4, 10$). The two columns on the right present the AUROC for two linear SVMs trained on full 7-mer feature space (F) and $\mathbf{M}_0$-derived 7-mer subspace ($F_{\mathbf{M}_0}$)

PWM	SVM-based (1)	SVM-based (4)	SVM-based (10)	SVM (F)	SVM ($F_{\mathbf{M}_0}$)
0.8428	0.6691	0.8282	0.8441	0.6833	0.8195

Table 21.5 Number of matched predictions. The row "Top" indicates the numbers of motifs matched with the standards [21.32, 85, 86] out of the predictions ranked first by each algorithm. The row "Top 3" indicates the total numbers out of the top three predictions from each algorithm

Algorithm	AlignACE	BioProspector	Weeder	SVMotif	SVMotif(B)
Top	19	27	33	29	33
Top 3	34	33	45	43	47

the same binding site selection criteria as the PWM maximum scanning score criterion.

We present one numerical example assembled by the authors to illustrate this. The PWM $\mathbf{M}_0$ for a well-studied yeast TF, GCN4, was retrieved from the University of California, Santa Cruz (UCSC) database [21.85, 86], and positive and negative samples of targets and nontarget promoters were obtained (based on ChIP-chip experiment data [21.13]). The methods tested included PWM scanning classifiers (using maximum scanning scores), the above-mentioned SVM-based classifiers ($f(S) = \mathbf{w}^{.p} \cdot \mathbf{x} - \tau$ with $p = 1, 4, 10$), linear SVM classifiers (trained on full k-mer feature space), and SVM subspace-based classifiers (trained on $\mathbf{M}_0$-derived subspaces; see Sect. 21.5.2 and [21.84]).

Note that the above SVM-based classifier with a PWM-compatible w-vector does not involve training, since the w-vector is obtained strictly using PWM-based k-mer scanning scores, i.e., $w_j = s^{(r)}(\alpha_j; \mathbf{M}_0)$ (with α_j the k-mer corresponding to component w_j of $\mathbf{w}$). The two SVM classifiers were assessed under a five-fold cross-validation protocol. Table 21.4 presents the area under the receiver operating characteristic (ROC) curve for each algorithm, showing that for sufficiently large p the SVM-based classifier has approximately the same performance as the PWM-scanning classifier. Another interesting finding is that the dimension-reduced SVM classifier built on the subspace $F_{\mathbf{M}_0}$ (the subspace of k-mers generated by $\mathbf{M}_0$, here with only 35 dimensions) outperforms the SVM classifier built on the full space (with 4^7 dimensions), which suggests the effectiveness of dimension reduction through PWM subspaces (Sect. 21.5.2).

The agglomeration algorithm used in SVMotif [21.70] (mentioned earlier) can be used to help convert the optimized w-vector back into the form of a PWM. The benefit of doing this has been illustrated earlier. We tested this application of SVMotif using Boolean feature vectors (SVMotif(B); feature $x_i = 1$ if k-mer α_i appears once or more in the gene promoter) and SVMotif [21.70] using standard features on a benchmark yeast dataset consisting of target information for 85 yeast TFs [21.14, 32] (we eliminated 10 TFs with gapped motifs for this dataset). We compared the PWMs abstracted using SVMotif from the w-vectors with the PWMs predicted by BioProspector, AlignACE, and Weeder. The result shows that the SVMotif(B) algorithm is quite promising and can perform better than the other (standard) ones (Table 21.5).

21.5.2 Dimension Reduction and Ensemble Methods

From the discussion in Sect. 21.5.1, given a dataset D, the optimal linear classifier based on SVM is obtained through a search which includes PWM scanning methods involving both linear and maximum scanning score-based classifiers. Because the size of the k-mer features is the same as the width of the PWM, the dimension of the above-mentioned PWM-based w-vector will increase exponentially with the size of the PWM. For long binding motifs, computationally identifying the optimal w-vector therefore becomes difficult. One strategy, used successfully by the MoAn algorithm [21.69] and several other discriminative motif discovery algorithms, is to restrict the search to the class of PWM-compatible w-vectors; recall that such vectors assign the weight w_i to the exact PWM score (for a fixed choice of PWM) of the corresponding k-mer α_i. In other words, the entries of the PWM are effectively the only free parameters used to optimize the Lagrangian. The search space thus has $(4-1) \times k$ dimensions, rather than 4^{k-1} dimensions for the case of arbitrary w-vectors in the search space. Another strategy for dimension reduction is the use of certain standard dimension reduction techniques in machine learning as discussed below.

Beside traditional dimension reduction techniques, we outline an ensemble machine method [21.84] using a collection of algorithms (motif discovery algorithms such as BioProspector, MEME, and the like) used initially to identify informative subspaces F_i of a large feature space F as a dimensional reduction tool. In machine learning language, individual algorithms in such an ensemble are denoted as weak learners. Because the component algorithm i learns a dimensionally reduced subspace F_i, we further denote the component algorithms as subspace-valued weak learners. Before discussing the details of the algorithm, we first introduce the concept of *PWM subspaces*.

Recall from above that an equivalent SVM representation of the maximum PWM scanning score classifier is $f(x) = \sum_j w_j^p x_j$ for large p (assuming that we restrict x_j to be 0 or 1). This can be approximated by the score $\hat{f}(x) = \sum_j \left(I(w_j > c) w_j \right)^p x_j$, where $I(w_j > c) = 1$ if $w_j > c$ and 0 otherwise; That is, the PWM scanning classifier can be restricted onto the subspace spanned by k-mer features with $w_j > c$. Since w_j is the scanning score of the PWM against the

j-th k-mer, we define the subspace spanned by those k-mer features that are best matched with the PWM (i. e., have the highest PWM scanning scores) as the *PWM subspace*.

Use of such a PWM score-based subspace usually reduces the dimension substantially. By restricting SVM training to within this PWM subspace, the new restricted k-mer subspace w-vector can be optimized much more efficiently than the original w-vector (the full PWM-based vector) using the same training samples to split positives and negatives.

Because the above PWM-based subspace is determined by the initial PWM, the quality of the corresponding w-vector again depends on the quality of the initial PWM. Without knowing the true PWM, the resulting w-vector may be trapped into local maxima around this perhaps nonoptimal initial PWM. To minimize the chances of this, in [21.84] five different algorithms are integrated, providing a total of 29 candidate PWMs $\{M_i\}_{i=1}^{29}$ for the fixed TF t. Each PWM from this ensemble will define a candidate PWM-based subspace F_i and a trained w-vector $\mathbf{w}_i \in F_i$. A linear combination of this ensemble of w-vectors, $\mathbf{w} = \sum \beta_i \mathbf{w}_i$ (where the coefficients β_i are selected by training in *synopsis space* [21.84]), is then chosen as the final optimized w-vector and used to classify positives and negatives. The synopsis space essentially defines the final ensemble-based feature vector of a potential

Table 21.6 Performance of 15 algorithms on 56 Tompa datasets [21.22]. The definitions of all performance metrics are listed in [21.22]. Some brief definitions of acronyms: $xSn = $ sensitivity (recall), $xPPV = $ precision, $xF_1 = F_1 = 2 \cdot$ Recall $\cdot$ Precision/(Recall + Precision), where $x = n$ or s indicates that the metric is at nucleotide or site level

Algorithm	Nucleotide level						Site level			
	nSn	nPPV	nSp	nPC	nCC	nF1	sSn	sPPV	sASP	sF1
AlignACE	0.055	0.112	0.991	0.038	0.066	0.074	0.088	0.123	0.105	0.103
ANN-Spec	0.087	0.088	0.982	0.046	.	0.088	0.155	0.085	0.120	0.110
Consensus	0.021	0.113	0.997	0.018	0.040	0.035	0.040	0.133	0.087	0.062
GLAM	0.026	0.038	0.987	0.016	.	0.031	0.046	0.048	0.047	0.047
Improbizer	0.069	0.070	0.982	0.036	.	0.069	0.123	0.084	0.103	0.099
MEME	0.067	0.107	0.989	0.043	0.071	0.083	0.111	0.139	0.125	0.124
MEME3	0.078	0.091	0.985	0.044	.	0.084	0.125	0.135	0.130	0.129
MITRA	0.031	0.062	0.991	0.021	0.031	0.042	0.050	0.063	0.057	0.056
MotifSampler	0.060	0.107	0.990	0.040	0.067	0.077	0.098	0.101	0.100	0.100
oligodyad-analysis	0.040	0.154	0.996	0.033	0.069	0.063	0.073	0.121	0.097	0.091
QuickScore	0.017	0.030	0.989	0.011	0.008	0.022	0.033	0.019	0.026	0.024
SeSiMCMC	0.061	0.037	0.969	0.024	0.049	0.046	0.080	0.075	0.078	0.078
Weeder	0.086	0.300	0.996	0.072	0.152	0.134	0.161	0.289	0.225	0.207
YMF	0.064	0.137	0.992	0.046	0.082	0.087	0.121	0.120	0.120	0.120
Ensemble	0.146	0.119	0.979	0.070	0.113	0.131	0.194	0.154	0.174	0.171

Part D | 21.5

target gene g as a 29-dimensional vector v whose i-th component v_i ($i = 1, \ldots, 29$) represents the SVM score $f_i(g)$ in the feature space F_i determined by the i-th PWM of the 29 [21.84].

We compare this ensemble with the performance of its five component PWM generators, namely AlignACE [21.57], BioProspector [21.55], MEME [21.52], SVMotif [21.70], and Weeder [21.63], in identifying the target genes of 88 yeast TFs [21.84]. For 75 out of 88 TFs, the ensemble w-vector outperforms the best PWM-based classifier using PWMs predicted by component algorithms. The average F_1 score over all TFs is about 10 percentage points higher than the best component algorithm (Table 21.3).

We tested the ensemble w-vector also to identify individual binding sites on the well-known Tompa benchmark datasets for binding site identification [21.22]. Because the dataset was not designed for learning-based algorithms, we generated putative negative sequences by randomly selecting sequences from the same genome of the test dataset (the test dataset consists of 56 individual gene sets from four species, including human, mouse, *Drosophila melanogaster*, and *Saccharomyces cerevisiae*). A w-scanning method was applied to score

each k-mer along the promoter sequence $S(g)$ of each candidate gene g by the corresponding k-mer score in the ensemble w-vector. The regions with a total score above a certain threshold τ were then predicted to be binding sites.

Averaged over all 56 datasets from four species, the ensemble w-vector method yields the best sensitivity (recall) at both the nucleotide and site levels (14.6% and 19.4%; Table 21.6) using a threshold of $\tau_0 = 3$ (τ_0 is selected to achieve a < 0.01 significance level based on the empirical distribution of the normalized w-scanning scores on background sequences). This indicates that the ensemble w-vector is able to retrieve the most known binding sites. The precision (positive predictive value – PPV) (17.4% at site level) is lower only than the best contender, Weeder (28.8% at site level), which utilizes an eighth-order Markov model to capture the background distribution for each species. The overall F_1 scores at the nucleotide and site levels are 13.1% and 17.1%, respectively, which ranks in second place among the 15 (14 tested by [21.22]).

The above performance level occurs partially because the training sample sizes for Tompa datasets are in general too small for machine learning methods.

Table 21.7 F_1 scores by minimal number of sample target sequences, m, in each of the *Tompa* [21.22] datasets. The total number of Tompa datasets that have more than m samples are listed in the second column and denoted as n. The F_1 scores at nucleotide and site level are listed for four algorithms: AlignACE, MEME, Weeder, and the ensemble machine method. The correlation coefficient between the F_1 scores and m are computed and listed in the bottom row. This table illustrates that the correlation coefficients between performance metrics and minimal sample size m are much higher for the ensemble machine method than the other methods

| m | n | F_1 at nucleotide level | | | | F_1 at site level | | | |
		Align-ACE	MEME	Weeder	Ensemble	Align-ACE	MEME	Weeder	Ensemble
4	42	0.084	0.088	0.143	0.138	0.117	0.135	0.223	0.181
5	35	0.090	0.088	0.148	0.148	0.123	0.134	0.231	0.194
6	27	0.091	0.097	0.168	0.161	0.126	0.152	0.271	0.211
7	23	0.098	0.096	0.161	0.165	0.136	0.152	0.265	0.218
8	19	0.101	0.086	0.161	0.187	0.136	0.141	0.27	0.241
9	16	0.104	0.950	0.155	0.198	0.143	0.161	0.266	0.249
10	11	0.133	0.113	0.189	0.238	0.176	0.194	0.321	0.287
Corr. coef.		0.8889	0.6544	0.7427	0.9607	0.8993	0.8193	0.8777	0.9792

Table 21.8 Performance of the ensemble and some of its components on human and mouse datasets

| Algorithm | Nucleotide level | | | | | | Site level | | | |
	nSn	nPPV	nSp	nPC	nCC	nF1	sSn	sPPV	sASP	sF1
AlignACE	0.037	0.085	0.992	0.026	0.044	0.051	0.063	0.096	0.079	0.076
MEME	0.047	0.076	0.989	0.030	0.045	0.058	0.081	0.106	0.093	0.092
Weeder	0.056	0.239	0.996	0.048	0.107	0.091	0.111	0.230	0.171	0.150
Ensemble	0.133	0.114	0.979	0.066	0.104	0.123	0.179	0.146	0.163	0.161

In order to show the sample size effects, we measure the performance only on those datasets with at least m sample sequences. For $m = 4$–10, Table 21.7 shows that the F_1 scores (see definition) of the ensemble at both nucleotide and site levels steadily increase as m increases. The correlation coefficients between the F_1 scores and m are 0.96 and 0.98. These are much higher than those for AlignACE, MEME, and Weeder (selected because they are used as components of the ensemble).

In particular, on the mammalian subset of the Tompa datasets (human and mouse), which has larger sample sizes on average, the ensemble ranks at the top in terms of F_1 scores (Table 21.8) [21.84].

21.6 Conclusions

In this chapter, we have presented an overall picture of algorithms for the three problems of finding transcription factor motifs, binding sites, and gene targets, the latter being useful in the construction of the transcriptional regulatory network. Particular attention has been given to new machine learning methodologies for these purposes, a number of which have been developed by the authors' group [21.10, 70, 84, 114]. We have also illustrated some connections between machine-learning-based algorithms and conventional PWM-based algorithms for the three problems. From several perspectives, we can claim that machine-learning-based algorithms form an alternative which is often superior to that of PWM-based algorithms. Computational difficulties in dealing with high dimensionality of feature spaces related to analysis using long k-mer words can be partially solved by dimensional reduction methods and ensemble methods, the latter leveraging the capabilities of individual weak learners, each learner yielding a dimensionally reduced subspace. Tested on some standard benchmark datasets for the above three problems, the machine learning methods achieve state-of-the-art performance. More importantly, these machine learning methods show great potential in integrating information from different sources. With the growing amount of information available from biological datasets, such methods have greatest promise for consistent improvement based on integration of new information sources and data types.

References

21.1 M.B. Gerstein, C. Bruce, J.S. Rozowsky, D. Zheng, J. Du, J.O. Korbel, O. Emanuelsson, Z.D. Zhang, S. Weissman, M. Snyder: What is a gene, post-ENCODE? History and updated definition, Genome Res. **17**(6), 669–681 (2007)

21.2 E.H. Davidson, D.R. McClay, L. Hood: Regulatory gene networks and the properties of the developmental process, Proc. Natl. Acad. Sci. USA **100**(4), 1475–1480 (2003)

21.3 M. Levine, R. Tjian: Transcription regulation and animal diversity, Nature **424**(6945), 147–151 (2003)

21.4 J.B. Warner, A.A. Philippakis, S.A. Jaeger, F.S. He, J. Lin, M.L. Bulyk: Systematic identification of mammalian regulatory motifs' target genes and functions, Nat. Methods **5**(4), 347–353 (2008)

21.5 M. Das, H.K. Dai: A survey of DNA motif finding algorithms, BMC Bioinformatics **8**(7), S21 (2007)

21.6 W.W. Wasserman, A. Sandelin: Applied bioinformatics for the identification of regulatory elements, Nat. Genet. **5**(4), 276–287 (2004)

21.7 F. Mordelet, J.-P. Vert: SIRENE: Supervised inference of regulatory networks, Bioinformatics **24**(16), i76–i82 (2008)

21.8 G. Pavesi, G. Mauri, G. Pesole: In silico representation and discovery of transcription factor binding sites, Brief. Bioinform. **5**, 217–236 (2004)

21.9 L. Elnitski, V.X. Jin, P.J. Farnham, S.J.M. Jones: Locating mammalian transcription factor binding sites: A survey of computational and experimental techniques, Genome Res. **16**(12), 1455–1464 (2006)

21.10 D.T. Holloway, M.A. Kon, C. DeLisi: Machine learning methods for transcription data integration, IBM J. Res. Dev. **50**(6), 631–644 (2006)

21.11 J.J. Faith, B. Hayete, J.T. Thaden, I. Mogno, J. Wierzbowski, G. Cottarel, S. Kasif, J.J. Collins, T.S. Gardner: Large-Scale mapping and validation of *Escherichia coli* transcriptional regulation from a compendium of expression profiles, PLoS Biology **5**(1), 54–66 (2007)

21.12 E. van Nimwegen: Finding regulatory elements and regulatory motifs: A general probabilistic framework, BMC Bioinformatics **8**(6), S4 (2007)

21.13 T.I. Lee, N.J. Rinaldi, F. Robert, D.T. Odom, Z. Bar-Joseph, G.K. Gerber, N.M. Hannett, C.T. Harbison, C.M. Thompson, I. Simon, J. Zeitlinger, E.G. Jennings, H.L. Murray, D.B. Gordon, B. Ren, J.J. Wyrick,

Part D | 21

J.-B. Tagne, T.L. Volkert, E. Fraenkel, D.K. Gifford, R.A. Young: Transcriptional regulatory networks in *Saccharomyces cerevisiae*, Science **298**(5594), 799–804 (2002)

21.14   C.T. Harbison, D.B. Gordon, T.I. Lee, N.J. Rinaldi, K.D. MacIsaac, T.W. Danford, N.M. Hannett, J.-B. Tagne, D.B. Reynolds, J. Yoo, E.G. Jennings, J. Zeitlinger, D.K. Pokholok, M. Kellis, P.A. Rolfe, K.T. Takusagawa, E.S. Lander, D.K. Gifford, E. Fraenkel, R.A. Young: Transcriptional regulatory code of a eukaryotic genome, Nature **431**(7004), 99–104 (2004)

21.15   G. Robertson, M. Hirst, M. Bainbridge, M. Bilenky, Y. Zhao, T. Zeng, G. Euskirchen, B. Bernier, R. Varhol, A. Delaney, N. Thiessen, O.L. Griffith, A. He, M. Marra, M. Snyder, S. Jones: Genome-wide profiles of STAT1 DNA association using chromatin immunoprecipitation and massively parallel sequencing, Nat. Methods **4**(8), 651–657 (2007)

21.16   M. Schena, D. Shalon, R.W. Davis, P.O. Brown: Quantitative monitoring of gene expression patterns with a complementary DNA microarray, Science **270**(5235), 467–470 (1995)

21.17   M. Chee, R. Yang, E. Hubbell, A. Berno, X.C. Huang, D. Stern, J. Winkler, D.J. Lockhart, M.S. Morris, S.P. Fodor: Accessing genetic information with high-density DNA arrays, Science **274**(5287), 610–614 (1996)

21.18   J.L. DeRisi, V.R. Iyer, P.O. Brown: Exploring the metabolic and genetic control of gene expression on a genomic scale, Science **278**(5338), 680–686 (1997)

21.19   H. Yu, N.M. Luscombe, J. Qian, M. Gerstein: Genomic analysis of gene expression relationships in transcriptional regulatory networks, Trends Genet. **19**(8), 422–427 (2003)

21.20   G.Z. Hertz, G. Hartzell, G.D. Stormo: Identification of consensus patterns in unaligned DNA sequences known to be functionally related, Comput. Appl. Biosci. **6**(2), 81–92 (1990)

21.21   G.D. Stormo: DNA binding sites: Representation and discovery, Bioinformatics **16**(1), 16–23 (2000)

21.22   M. Tompa, N. Li, T.L. Bailey, G.M. Church, B.D. Moor, E. Eskin, A.V. Favorov, M.C. Frith, Y. Fu, W.J. Kent, V.J. Makeev, A.A. Mironov, W.S. Noble, G. Pavesi, G. Pesole, M. Régnier, N. Simonis, S. Sinha, G. Thijs, J. van Helden, M. Vandenbogaert, Z. Weng, C. Workman, C. Ye, Z. Zhu: Assessing computational tools for the discovery of transcription factor binding sites, Nat. Biotechnol. **23**, 137–144 (2005)

21.23   J. Hu, B. Li, D. Kihara: Limitations and potentials of current motif discovery algorithms, Nucleic Acids Res. **33**, 4899–4913 (2005)

21.24   P. Kheradpour, A. Stark, S. Roy, M. Kellis: Reliable prediction of regulator targets using 12 *Drosophila* genomes, Genome Res. **17**(12), 1919–1931 (2007)

21.25   P. Cliften, P. Sudarsanam, A. Desikan, L. Fulton, B. Fulton, J. Majors, R. Waterston, B.A. Cohen, M. Johnston: Finding functional features in saccharomyces genomes by phylogenetic footprinting, Science **301**(5629), 71–76 (2003)

21.26   S. Sinha, E. van Nimwegen, E.D. Siggia: A probabilistic method to detect regulatory modules, Bioinformatics **19**(1), i292–i301 (2003)

21.27   T. Wang, G. Stormo: Combining phylogenetic data with co-regulated genes to identify regulatory motifs, Bioinformatics **19**(18), 2369–2380 (2003)

21.28   L.A. McCue, W. Thompson, C.S. Carmack, M.P. Ryan, J.S. Liu, V. Derbyshire, C.E. Lawrence: Phylogenetic footprinting of transcription factor binding sites in proteobacterial genomes, Nucleic Acids Res. **29**(3), 774–782 (2001)

21.29   W. Thompson, S. Conlan, L.A. McCue, C.E. Lawrence: Using the Gibbs Motif Sampler for phylogenetic footprinting, Methods Mol. Biol. **395**, 403–424 (2007)

21.30   L.A. Newberg, W.A. Thompson, S. Conlan, T.M. Smith, L.A. McCue, C.E. Lawrence: A phylogenetic Gibbs sampler that yields centroid solutions for cis-regulatory site prediction, Bioinformatics **23**(14), 1718–1727 (2007)

21.31   R. Siddharthan, E.D. Siggia, E. van Nimwegen: Phylogibbs: A Gibbs sampling motif finder that incorporates phylogeny, PLoS Comput. Biol. **1**(7), e67 (2005)

21.32   K.D. MacIsaac, T. Wang, D.B. Gordon, D.K. Gifford, G.D. Stormo, E. Fraenkel: An improved map of conserved regulatory sites for Saccharomyces cerevisiae, BMC Bioinformatics **7**(1), 113 (2006)

21.33   E. Segal, Y. Fondufe-Mittendorf, L. Chen: A genomic code for nucleosome positioning, Nature **442**(7104), 772–778 (2006)

21.34   L. Narlikar, R. Gordân, A.J. Hartemink: A nucleosome-guided map of transcription factor binding sites in yeast, PLoS Comput. Biol. **3**(11), e215 (2007)

21.35   T. Wasson, A.J. Hartemink: An ensemble model of competitive multi-factor binding of the genome, Genome Res. **19**(11), 2101–2112 (2009)

21.36   J.A. Greenbaum, B. Pang, T.D. Tullius: Construction of a genome-scale structural map at single-nucleotide resolution, Genome Res. **17**(6), 947–953 (2007)

21.37   The ENCODE Project Consortium: Identification and analysis of functional elements in 1% of the human genome by the ENCODE pilot project, Nature **447**(7146), 799–816 (2007)

21.38   J.A. Greenbaum, S.C. Parker, T.D. Tullius: Detection of DNA structural motifs in functional genomic elements, Genome Res. **17**(6), 940–946 (2007)

21.39   A.J. Butte, I.S. Kohane: Mutual information relevance networks: Functional genomic clustering using pairwise entropy measurements, Pac. Symp. Biocomput. (2000) pp. 418–429

21.40 T. Akutsu, S. Miyano, S. Kuhara: Algorithms for identifying Boolean networks and related biological networks based on matrix multiplication and fingerprint function, J. Comput. Biol. **7**(3-4), 331–343 (2000)

21.41 N. Friedman, M. Linial, I. Nachman, D. Pe'er: Using Bayesian networks to analyze expression data, J. Comput. Biol. **7**(3-4), 601–620 (2000)

21.42 D. Heckerman, D. Geiger, D.M. Chickering: Learning Bayesian networks: The combination of knowledge and statistical data, Mach. Learn. **20**, 197–243 (1995)

21.43 M. Bansal, G. Della Gatta, D. di Bernardo: Inference of gene regulatory networks and compound mode of action from time course gene expression profiles, Bioinformatics **22**(7), 815–822 (2006)

21.44 K.-C. Chen, T.-Y. Wang, H.-H. Tseng, C.-Y.F. Huang, C.-Y. Kao: A stochastic differential equation model for quantifying transcriptional regulatory network in *Saccharomyces cerevisiae*, Bioinformatics **21**(12), 2883–2890 (2005)

21.45 T.S. Gardner, D. di Bernardo, D. Lorenz, J.J. Collins: Inferring genetic networks and identifying compound mode of action via expression profiling, Science **301**(5629), 102–105 (2003)

21.46 D. di Bernardo, M.J. Thompson, T.S. Gardner, S.E. Chobot, E.L. Eastwood, A.P. Wojtovich, S.J. Elliott, S.E. Schaus, J.J. Collins: Chemogenomic profiling on a genome-wide scale using reverse-engineered gene networks, Nat. Biotechnol. **23**(3), 377–383 (2005)

21.47 J. Tegner, S.K. Yeung, J. Hasty, J.J. Collins: Reverse engineering gene networks: Integrating genetic perturbations with dynamical modeling, Proc. Natl. Acad. Sci. USA **100**(10), 5944–5949 (2003)

21.48 G.Z. Hertz, G.D. Stormo: Identifying DNA and protein patterns with statistically significant alignments of multiple sequences, Bioinformatics **15**(7), 563–577 (1999)

21.49 A.D. Johnson: An extended iupac nomenclature code for polymorphic nucleic acids, Bioinformatics **26**(10), 1386–1389 (2010)

21.50 T. Akutsu, H. Arimura, S. Shimozono: On approximation algorithms for local multiple alignment, Proc 4th Annu. Int. Conf. Comput. Mol. Biol. RECOMB '00 (ACM, New York 2000) pp.1–7

21.51 G.Z. Hertz, G.W. Hartzell, G.D. Stormo: Identification of consensus patterns in unaligned DNA sequences known to be functionally related, Comput. Appl. Biosci. **6**, 81–92 (1990)

21.52 T.L. Bailey, N. Williams, C. Misleh, W.W. Li: MEME: Discovering and analyzing DNA and protein sequence motifs, Nucleic Acids Res. **34**(2), W369–W373 (2006)

21.53 T.L. Bailey, C. Elkan: Unsupervised learning of multiple motifs in biopolymers using expectation maximization, Mach. Learn. **21**, 51–80 (1995)

21.54 C.E. Lawrence, S.F. Altschul, M.S. Boguski, J.S. Liu, A.F. Neuwald, J.C. Wootton: Detecting subtle sequence signals: A Gibbs sampling strategy for multiple alignment, Science **262**, 208–214 (1993)

21.55 X. Liu, D.L. Brutlag, J.S. Liu: Bioprospector: Discovering conserved DNA motifs in upstream regulatory regions of co-expressed genes, Proc. 6th Pac. Symp. Biocomput. (2001) pp.127–138

21.56 J.D. Hughes, P.W. Estep, S. Tavazoie, G.M. Church: Computational identification of cis-regulatory elements associated with groups of functionally related genes in Saccharomyces cerevisiae, J. Mol. Biol. **296**(5), 1205–1214 (2000)

21.57 F.P. Roth, J.D. Hughes, P.W. Estep, G.M. Church: Finding DNA regulatory motifs within unaligned noncoding sequences clustered by whole-genome mrna quantitation, Nat. Biotechnol. **16**, 939–945 (1998)

21.58 G.D. Stormo: Wconsensus website (2002) available online from http://ural.wustl.edu/consensus/html/Html/main.html

21.59 T.L. Bailey, M. Gribskov: Combining evidence using p-values: Application to sequence homology searches, Bioinformatics **14**, 48–54 (1998)

21.60 W.A. Thompson, L.A. Newberg, S. Conlan, L.A. McCue, C.E. Lawrence: The Gibbs centroid sampler, Nucleic Acids Res. **35**(2), W232–W237 (2007)

21.61 W. Thompson, E.C. Rouchka, C.E. Lawrence: Gibbs Recursive Sampler: Finding transcription factor binding sites, Nucleic Acids Res. **31**(13), 3580–3585 (2003)

21.62 J.R. Sadler, M.S. Waterman, T.F. Smith: Regulatory pattern identification in nucleic acid sequences, Nucleic Acids Res. **38**(1), D613–619 (2010)

21.63 G. Pavesi, G. Mauri, G. Pesole: An algorithm for finding signals of unknown length in DNA sequences, Bioinformatics **17**(1), S207–214 (2001)

21.64 X.S. Liu, D.L. Brutlag, J.S. Liu: An algorithm for finding protein-DNA binding sites with applications to chromatin-immunoprecipitation microarray experiments, Nat. Biotech. **20**, 835–839 (2002)

21.65 M. Tompa: An exact method for finding short motifs in sequences, with application to the ribosome binding site problem, Proc. Int. Conf. Intell. Syst. Mol. Biol. (ISMB) (1999) pp.262–271

21.66 S. Sinha, M. Tompa: Discovery of novel transcription factor binding sites by statistical overrepresentation, Nucleic Acids Res. **30**(24), 5549–5560 (2002)

21.67 S. Sinha, M. Tompa: A statistical method for finding transcription factor binding sites, Proc. the 8th Int. Conf. Intell. Syst. Mol. Biol. (AAAI, Menlo Park 2000) pp.344–354

21.68 S. Sinha, M. Tompa: YMF: A program for discovery of novel transcription factor binding sites by statistical overrepresentation, Nucleic Acids Res. **31**(13), 3586–3588 (2003)

Part D | 21

21.69   E. Valen, A. Sandelin, O. Winther, A. Krogh: Discovery of regulatory elements is improved by a discriminatory approach, PLoS Comput. Biol. **5**(11), e1000562 (2009)

21.70   M.A. Kon, Y. Fan, D. Holloway, C. DeLisi: SVMotif: A machine learning motif algorithm, ICMLA '07: Proc. 7th Int. Conf. Mach. Learn. Appl. (IEEE Computer Soc., Washington 2007) pp. 573–580

21.71   X.G. Zhang, X. Lu, Q. Shi, X.Q. Xu, H.C.E. Leung, L.N. Harris, J.D. Iglehart, A. Miron, J.S. Liu, W.H. Wong: Recursive SVM feature selection and sample classification for mass-spectrometry and microarray data, BMC Bioinformatics **7**(1), 197 (2006)

21.72   L. Breiman: Random forests, Mach. Learn. **45**(1), 5–32 (2001)

21.73   J.M. Cherry, E.L. Hong, C. Amundsen, R. Balakrishnan, G. Binkley, E.T. Chan, K.R. Christie, M.C. Costanzo, S.S. Dwight, S.R. Engel, D.G. Fisk, J.E. Hirschman, B.C. Hitz, K. Karra, C.J. Krieger, S.R. Miyasato, R.S. Nash, J. Park, M.S. Skrzypek, M. Simison, S. Weng, E.D. Wong: Saccharomyces Genome Database: The genomics resource of budding yeast, Nucleic Acids Res. **40**, D700–D705 (2012)

21.74   C.J. van Rijsbergen: *Information Retrieval* (Butterworth, London 1979)

21.75   K.A. Romer, G.-R. Kayombya, E. Fraenkel: Webmotifs: Automated discovery, filtering and scoring of DNA sequence motifs using multiple programs and Bayesian approaches, Nucleic Acids Res. **35**(2), W217–W220 (2007)

21.76   D.B. Gordon, L. Nekludova, S. McCallum, E. Fraenkel: TAMO: A flexible, object-oriented framework for analyzing transcriptional regulation using DNA-sequence motifs, Bioinformatics **21**(14), 3164–3165 (2005)

21.77   J. Hu, Y. Yang, D. Kihara: EMD: An ensemble algorithm for discovering regulatory motifs in DNA sequences, BMC Bioinformatics **7**(1), 342 (2006)

21.78   T. Reddy, B. Shakhnovich, D. Roberts, S. Russek, C. Delisi: Positional clustering improves computational binding site detection and identifies novel cis-regulatory sites in mammalian GABAA receptor subunit genes, Nucleic Acids Res. **35**(3), e20 (2007)

21.79   T. Reddy, C. Delisi, B. Shakhnovich: Binding site graphs: A new graph theoretical framework for prediction of transcription factor binding sites, PLoS Comput. Biol. **3**(5), e90 (2007)

21.80   C. Yanover, M. Singh, E. Zaslavsky: M are better than one: An ensemble-based motif finder and its application to regulatory element prediction, Bioinformatics **25**(7), 868–874 (2009)

21.81   D. Che, S.T. Jensen, L. Cai, J.S. Liu: Best: Binding-site estimation suite of tools, Bioinformatics **21**(12), 2909–2911 (2005)

21.82   Z. Wei, S.T. Jensen: GAME: Detecting cis-regulatory elements using a genetic algorithm, Bioinformatics **22**(13), 1577–1584 (2006)

21.83   S.T. Jensen, J.S. Liu: BioOptimizer: A Bayesian scoring function approach to motif discovery, Bioinformatics **20**(10), 1557–1564 (2004)

21.84   Y. Fan, M. A. Kon, C. DeLisi: Ensemble machine methods for analysis of transcription factor and DNA interactions (2013), preprint

21.85   D. Karolchik, R. Baertsch, M. Diekhans, T.S. Furey, A. Hinrichs, Y.T. Lu, K.M. Roskin, M. Schwartz, C.W. Sugnet, D.J. Thomas, R.J. Weber, D. Haussler, W.J. Kent: The UCSC Genome Browser Database, Nucleic Acids Res. **31**(1), 51–54 (2003)

21.86   B. Rhead, D. Karolchik, R.M. Kuhn, A.S. Hinrichs, A.S. Zweig, P.A. Fujita, M. Diekhans, K.E. Smith, K.R. Rosenbloom, B.J. Raney, A. Pohl, M. Pheasant, L.R. Meyer, K. Learned, F. Hsu, J. Hillman-Jackson, R.A. Harte, B. Giardine, T.R. Dreszer, H. Clawson, G.P. Barber, D. Haussler, W.J. Kent: The UCSC Genome Browser database: Update 2010, Nucleic Acids Res. **38**(1), D613–619 (2010)

21.87   G. Thijs, M. Lescot, K. Marchal, S. Rombauts, B.D. Moor, P. Rouzé, Y. Moreau: A higher-order background model improves the detection of promoter regulatory elements by Gibbs sampling, Bioinformatics **17**(12), 1113–1122 (2001)

21.88   N.-K. Kim, K. Tharakaraman, J.L. Spouge: Adding sequence context to a Markov background model improves the identification of regulatory elements, Bioinformatics **22**(23), 2870–2875 (2006)

21.89   W.W. Wasserman, J.W. Fickett: Identification of regulatory regions which confer muscle-specific gene expression1, J. Mol. Biol. **278**(1), 167–181 (1998)

21.90   M. Kellis, N. Patterson, M. Endrizzi, B. Birren, E.S. Lander: Sequencing and comparison of yeast species to identify genes and regulatory elements, Nature **423**, 241–254 (2003)

21.91   M. Blanchette, B. Schwikowski, M. Tompa: Algorithms for phylogenetic footprinting, J. Comput. Biol. **9**(2), 211–223 (2002)

21.92   E.H. Davidson: *The Regulatory Genome: Gene Regulatory Networks In Development And Evolution* (Academic, Burlington 2006)

21.93   W. Thompson, M.J. Palumbo, W.W. Wasserman, J.S. Liu, C.E. Lawrence: Decoding human regulatory circuits, Genome Res. **14**(10A), 1967–1974 (2004)

21.94   G. Kreiman: Identification of sparsely distributed clusters of cis-regulatory elements in sets of co-expressed genes, Nucleic Acids Res. **32**(9), 2889–2900 (2004)

21.95   Q. Zhou, W.H. Wong: CisModule: De novo discovery of cis-regulatory modules by hierarchical mixture modeling, Proc. Natl. Acad. Sci. USA **101**(33), 12114–12119 (2004)

21.96   Z. Zhu, J. Shendure, G.M. Church: Discovering functional transcription-factor combinations in the human cell cycle, Genome Res. **15**(6), 848–855 (2005)

21.97   E. Segal, R. Sharan: A discriminative model for identifying spatial cis-regulatory modules, J. Comput. Biol. **12**(6), 822–834 (2005)

21.98   L. Marsan, M.-F. Sagot: Extracting structured motifs using a suffix tree – algorithms and application to promoter consensus identification, RECOMB (2000) pp. 210–219

21.99   W.W. Wasserman, M. Palumbo, W. Thompson, J.W. Fickett, C.E. Lawrence: Human–mouse genome comparisons to locate regulatory sites, Nat. Genet. **26**(2), 225–228 (2000)

21.100 M.C. Frith, J.L. Spouge, U. Hansen, Z. Weng: Statistical significance of clusters of motifs represented by position specific scoring matrices in nucleotide sequences, Nucleic Acids Res. **30**(14), 3214–3224 (2002)

21.101 M. Gupta, J.S. Liu: De novo cis-regulatory module elicitation for eukaryotic genomes, Proc. Natl. Acad. Sci. USA **102**(20), 7079–7084 (2005)

21.102 W. Lee, D. Tillo, N. Bray, R.H. Morse, R.W. Davis, T.R. Hughes, C. Nislow: A high-resolution atlas of nucleosome occupancy in yeast, Nat. Genet. **39**(10), 1235–1244 (2007)

21.103 I. Whitehouse, O.J. Rando, J. Delrow, T. Tsukiyama: Chromatin remodelling at promoters suppresses antisense transcription, Nature **450**(7172), 1031–1035 (2007)

21.104 S. Shivaswamy, A. Bhinge, Y. Zhao, S. Jones, M. Hirst, V.R. Iyer: Dynamic remodeling of individual nucleosomes across a eukaryotic genome in response to transcriptional perturbation, PLoS Biology **6**(3), e65 (2008)

21.105 N. Kaplan, I.K. Moore, Y. Fondufe-Mittendorf, N. Kaplan, I.K. Moore, Y. Fondufe-Mittendorf, A.J. Gossett, D. Tillo, Y. Field, E.M. LeProust, T.R. Hughes, J.D. Lieb, J. Widom, E. Segal: The DNA-encoded nucleosome organization of a eukaryotic genome, Nature **458**(7236), 362–366 (2009)

21.106 K.E. van Holde, J.R. Allen, K. Tatchell, W.O. Weischet, D. Lohr: DNA-histone interactions in nucleosomes, Biophys. J. **32**(1), 271–282 (1980)

21.107 K.E. van Holde: *Chromatin* (Springer, New York 1989)

21.108 L.R. Rabiner: A tutorial on hidden Markov models and selected applications in speech recognition, Proc. IEEE **77**(2), 257–286 (1989)

21.109 I. Guyon, J. Weston, S. Barnhill, V. Vapnik: Gene selection for cancer classification using support vector machines, Mach. Learn. **46**(1), 389–422 (2002)

21.110 A.J. Butte, P. Tamayo, D. Slonim, T.R. Golub, I.S. Kohane: Discovering functional relationships between RNA expression and chemotherapeutic susceptibility using relevance networks, Proc. Natl. Acad. Sci. USA **97**(22), 12182–12186 (2000)

21.111 J. Beirlant, E.J. Dudewicz, L. Györfi, E.C. Meulen: Nonparametric entropy estimation: An overview, Int. J. Math. Stat. Sci. **6**, 17–39 (1997)

21.112 A. Margolin, I. Nemenman, K. Basso, C. Wiggins, G. Stolovitzky, R. Favera, A. Califano: ARACNE: An algorithm for the reconstruction of gene regulatory networks in a mammalian cellular context, BMC Bioinformatics **7**(1), S7 (2006)

21.113 P.J. Wittkopp: Variable transcription factor binding: A mechanism of evolutionary change, PLoS Biology **8**(3), e1000342 (2010)

21.114 M.A. Kon, Y. Fan, C. DeLisi: Ensemble machine methods for DNA binding, ICMLA '08: Proc. 6th Int. Conf. Mach. Learn. Appl. (IEEE Computer Soc., San Diego 2008) pp. 709–716

21.115 R. Kuang, E. Ie, K. Wang, K. Wang, M. Siddiqi, Y. Freund, C. Leslie: Profile-based string kernels for remote homology detection and motif extraction, CSB '04: Proc. 2004 IEEE Comput. Syst. Bioinform. Conf. (IEEE Computer Soc., Washington 2004) pp. 152–160

22. Inferring Genetic Networks with a Recurrent Neural Network Model Using Differential Evolution

Nasimul Noman, Leon Palafox, Hitoshi Iba

In this chapter, we present an evolutionary approach for reverse-engineering gene regulatory networks (GRNs) from the temporal gene expression profile. The regulatory interaction among genes is modeled by the recurrent neural network (RNN) formalism. We used the differential evolution (DE) algorithm with a random restart strategy for inferring the underlying network structure as well as the regulatory parameters. The random restart mechanism is particularly useful for avoiding premature convergence and is hence expected to be valuable in recovering important regulations from noisy gene expression data. The algorithm has been applied for inferring regulation by analyzing gene expression data generated from both in silico and in vivo networks. We also investigate the effectiveness of the method in obtaining an acceptable network from a limited amount of noisy data.

Part D | 22

22.1 Background and History

Over the last decade, systems biology has been at the center of attention throughout many branches of biology. At the turn of the twenty-first century, *Kitano* created timely awareness and enhanced interest for the systems approach to biology through his seminal papers [22.1, 2]. Although it is argued that the systems approach to biology has a long history [22.3], the formal strategy, borrowing ideas from the theory of system identification, to understand the collective cellular organization or the dynamic behavior in changeable environment developed not long ago.

Biological organisms are systems of a large number of functional units. These components cooperatively render the complex structure and coherent behaviors through selective and nonlinear interactions [22.2]. Because of the inherent hierarchy in biology, cells, tissues,

organs, organisms and the ecological community could all be the target of the systems approach. Depending on what we call a system the constituent parts may range from individual molecules to whole organisms.

Gene regulation is believed to be one of the elementary processes that pulses the functionalities of the molecular machines and their interplay in living organisms [22.4]. A comprehensive understanding of this process can be very useful in making the dreams of synthetic biology come true. Therefore, decoding the regulatory interactions between genes and their products has been regarded as one of the most important objectives of systems biology and considerable research effort has been expended over the last decade [22.5–10].

The research to unravel the molecular mechanism behind gene regulation became possible primarily for two key reasons. First, with technological progress, we gained access to the molecular level of cells and we are able to monitor, investigate, and experiment how different metabolites interact with each other. With this enhanced molecular knowledge, we were able to model the activity of a single gene [22.11]. The second reason is the massive amount of gene expression data that became available with the advent of the microarray technology. This paradigm shifting technology, providing us with the molecular snapshots of the cell, made the task of identifying the gene network feasible [22.12].

The DNA microarray technology functions based on the basic principle that a given mRNA (messenger RNA) molecule will pair with, or hybridize to, the complementary DNA strand from which it came. With cutting edge technology it is possible to spot thousands of DNA fragments or oligonucleotides, which represent specific coding regions, in a tiny slide or membrane. Thus scientists are able to measure precisely the amount of mRNA bound to the spots on the microarray in a single experiment [22.13]. In this way, microarrays enable us to measure the expression levels of hundreds of genes under various experimental conditions and to generate the gene expression profile of the cell. With rapid advancement of DNA microarray technology, the global cellular patterns of gene expression can be monitored as either time series or steady-state data. These gene expression profiles can be considered as snapshots of the dynamics of the underlying system over a time or under certain conditions. Therefore, theoretically it is possible to uncover the underlying regulatory system of gene regulation when we have enough sets of snapshots [22.14].

Analysis of gene expression data to decipher the underlying gene interactions is done in two major approaches: clustering and reverse-engineering. Cluster analysis is a useful exploratory technique for gene expression data that identifies groups of similar genes or experiments or both. Such an analysis helps us to determine potential relationships among the iden-

tified groups. Clustering techniques have proven to be helpful to understand gene function, gene regulation, cellular processes, and subtypes of cells [22.15]. There are a variety of algorithms available for clustering microarray data, including k-means, hierarchical clustering [22.16], self-organizing maps (SOMs) [22.17], and biclustering [22.18]. Cluster analysis can extract valuable information out of large-scale gene expression data regarding the co-regulated genes. However, it often fails to fulfill the ultimate target of gene network inference – to identify the system-wide causal relationship among genes. Reverse engineering the network using a computational model is a better approach to reveal the characteristics of the regulatory architecture [22.19].

The reverse engineering approach to gene networks involves two major components other than the expression data. The first component is a mathematical model that adequately represents the gene regulation process. The other component is a computational method to learn the model parameters from the gene expression profile. In fact this is the system identification process that tries to estimate a set of system parameters that reproduces the observed response [22.20]. Both of these components will be discussed in more detail in subsequent sections of this chapter.

In this chapter, we discuss the model based reconstruction of gene regulatory network (GRN) using an evolutionary algorithm. Considering the large number of modeling approaches to gene networks and various reverse engineering algorithms available to each model we keep our discussion limited to one specific model – the recurrent neural network (RNN) [22.21]. We present a reverse engineering algorithm for inferring transcriptional regulation from expression profile using the RNN model. We try to analyze the strength and weakness of the algorithm in identifying correct regulation from various amount of expression data with different levels of noise. The reconstruction method was also tested in real microarray data analysis.

22.2 Modeling Approaches to Gene Regulatory Networks

Like any other complex dynamic system, in order to fully understand the gene regulation process, we need to mathematically model it as promulgated in systems biology. Therefore, a large number of modeling approaches have been applied to gene networks [22.22–33]. The modeling band extends from the abstract Boolean representation to detailed differential equation

based model, where every formalism has its own benefits and limitations.

The Boolean network model, based on Boolean logic, assumes that each gene is in either of the two states: *fully expressed* or *not expressed* [22.14, 24]. The state change of a gene is given by a Boolean function with the states of influencing genes as inputs. This coarse representation offers certain advantages in terms of search space, computational effort and data requirement. However, the two-state discretization of gene expression is a crude abstraction of what we observe in nature and inevitably results in information loss. Additionally, the synchronous transition of genes is not biologically plausible. Nevertheless, these simple models and some of their advanced variants have been used successfully to obtain a first representation of large-scale gene networks with a reasonable computational effort [22.34].

Murphy and *Mian* [22.35] and *Friedman* et al. [22.23] have suggested using Bayesian network models of gene expression networks. In Bayesian network formalism the structure of a genetic network is modeled by a directed acyclic graph in which vertices represent genes and/or other components and the edges capture the conditional dependence relation and represent the interactions between genes. The Bayesian network approach to GRN has several advantages, such as its solid statistical basis, inherent capability to handle the stochastic aspects of gene expression and noisy measurements, the ability to work with incomplete knowledge, etc. However, static Bayesian networks cannot handle temporal information and are unable to consider the dynamical aspects of gene regulation. Therefore, dynamic Bayesian networks (DBNs) [22.36] and state space models [22.37] have become a more reasonable choice to overcome problems like hidden variables, prior knowledge, etc.

In the differential equation based approach, the continuous changes in the concentration levels of different reactants are modeled by differential equations [22.38]. If a network consists of N genes, an ordinary differential equation model will represent the change in the expression level of the i-th gene using an equation of the form

$$\frac{de_i}{dt} = f_i(e_1, e_2, \ldots, e_N), \qquad (22.1)$$

where f_i describes how the transcription rate of the i-th gene is influenced by other genes. The concentration level of the product of the i-th gene will increase, decrease, or remain unchanged [22.39]. Naturally, f_i may be positive or negative depending on the combinations of $e_1, e_2, \ldots, e_N$. The biological interpretation of this phenomenon is that in some states the genes of the network act to switch on the i-th gene and in other states they act to switch it off.

Now, the change in transcription rate of the i-th gene can be modeled by a linear or nonlinear function. A nonlinear model is more desirable because in vivo gene regulatory networks contain significant nonlinearities [22.40]. Therefore, any model of the system derived using purely linear approaches will not accurately represent the true nonlinear behavior of the real gene network [22.41]. On the contrary, GRN inference using a nonlinear f_i involves estimation of a large number of model parameters and hence will need a greater volume of data [22.19]. Among the nonlinear models for genetic networks, the S-system [22.42] is a very flexible and popular one and offers an excellent compromise between accuracy and mathematical flexibility. However, the regression task for this model becomes difficult and time consuming as there is a large parameter space to be optimized with this nonlinear formalism [22.43, 44]. Consequently, only small scale network was successfully reconstructed using S-system model. In this study we used a particular differential equation based formalism called recurrent neural network (RNN) model [22.28]. The details of this model will be described in next section. Some other models for GRN exist, which use special cases of the rate equation of (22.1) such as the *piecewise-linear differential equation* (PLDE), the *qualitative differential equation* (QDE), and the *partial differential equation* (PDE). More about these models can be learned from [22.10].

Many other gene network models have been proposed, such as the Petri net based model [22.45], finite state linear models [22.46], and many other hybrid models [22.11]. There exist some excellent review articles [22.8, 10, 19] from which more can be read about these and other modeling approaches to GRN.

22.3 Recurrent Neural Network Model for GRN

The recurrent neural network (RNN) model for a gene network is proposed based on the assumption that the dynamic behavior of a gene network can be represented as a neural network [22.47, 48]. Each node of the network represents a particular gene and the connection between nodes represents the regulatory interaction of one gene on the other.

Like in any other continuous dynamic system, the RNN formalism represents the gene regulatory network using a set of coupled differential equations. In the RNN model, the description of the generalized rate equation of (22.1) becomes

$$\tau_i \frac{de_i}{dt} = g\left(\sum_{j=1}^{N} w_{ij}e_j + \sum_{k=1}^{K} v_{ik}u_k + \beta_i\right) - \lambda_i e_i , \tag{22.2}$$

where e_i represents the gene expression level for the i-th gene ($i \le N$, N is the total number of genes in the network). w_{ij} represents the nature and strength of the causal interaction from the j-th gene towards the i-th gene. A positive value of w_{ij} represents the activation control (i. e., the j-th gene activates the i-th gene) and a negative value indicates inhibitory relationship (i. e., the j-th gene represses the i-th gene). When w_{ij} is zero, there is no regulatory control on the i-th gene from the j-th gene. Similarly, u_k ($1 \le k \le K$) represents external entities such as externally added chemicals, metabolites, signals or some other exogenous agents. Moreover, the variable v_{ik} represents the influence of the k-th ($1 \le k \le K$) external variable on the i-th gene. β_i indicates the bias term that can be interpreted as the basal expression level of the i-th gene. High negative and/or positive values of β result in a low influence of regulatory factors [22.28]. λ_i represents the decay rate parameter and τ_i^{-1} represents the rate constant; $g(\cdot)$ represents the nonlinear activation function that can be selected in many different ways. One of the most common choices for g, used in this work, is

$$g(z) = \frac{1}{(1 + e^{-\kappa z})} . \tag{22.3}$$

We used $\kappa = 1$ in this work. The choice of κ is not very significant, since for a different setting of κ we can rescale the model parameters and obtain the same dynamics [22.47].

In this work, we have focused on the regulatory interaction among the genes. Therefore, it is justified to ignore the influence of the external entities $\sum_{k=1}^{K} v_{ik}u_k$. Moreover, it is not easy to measure these external variables and, therefore, they are usually ignored in similar work [22.33, 47, 49, 50]. With this simplification the RNN model becomes

$$\tau_i \frac{de_i}{dt} = g\left(\sum_{j=1}^{N} w_{ij}e_j + \beta_i\right) - \lambda_i e_i . \tag{22.4}$$

For computational convenience, the RNN model is generally expressed in discrete form as follows:

$$\frac{e_i(t + \Delta t) - e_i(t)}{\Delta t}$$
$$= \frac{1}{\tau_i}\left\{ g\left[\sum_{j=1}^{N} w_{ij}e_j(t) + \beta_i\right] - \lambda_i e_i(t)\right\} \tag{22.5}$$

or

$$e_i(t + \Delta t) = \frac{\Delta t}{\tau_i} \cdot g\left[\sum_{j=1}^{N} w_{ij}e_j(t) + \beta_i\right]$$
$$+ \left(1 - \frac{\lambda_i \Delta t}{\tau_i}\right) e_i(t) . \tag{22.6}$$

In this form the RNN model for GRN can be described by the following set of parameters $\Omega = \{w_{ij}, \beta_i, \lambda_i, \tau_i\}$, where $1 \le i, j \le N$. Figure 22.1 illustrates the architecture of a fully connected RNN network and the details of the neuron that realizes the simplified form of (22.6). The delay element was inserted in Fig. 22.1b to emphasize that the state of the neuron is updated in a stepwise manner; the state of the gene expression at time $(t + \Delta t)$ is determined by the state at time t.

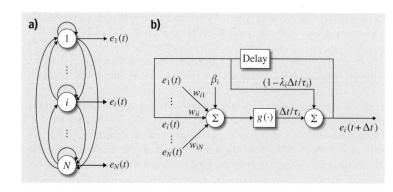

Fig. 22.1 (a) Structure of a fully connected RNN. **(b)** Details of a single neuron in RNN model

22.4 Relevant Research on GRN Reconstruction Using RNN

Once we have selected a model for gene regulation, we can focus on the other component of reverse engineering a GRN – the reconstruction algorithm. A large spectrum of reconstruction algorithms such as analytical methods [22.14], Bayesian networks [22.51], clustering and classification [22.52, 53], statistical methods [22.27, 54], and various kinds of machine-learning algorithms [22.17, 55, 56] have been applied to reconstruct GRNs. However, the choice of algorithm remains largely dependent on the model selected. Evolutionary algorithms (EAs) are preferred as the reconstruction method for different types of GRN models, especially for differential equation based models [22.50, 57–60]. A complete survey of the overwhelming number of reconstruction algorithms really goes beyond of the scope of this study, hence, we review the reconstruction algorithms that use the RNN model.

In his work, *Vohradský* suggested that any of the existing methods of *training* the neural networks such as gradient descent methods (e.g., back propagation) can be used for reconstructing the weight matrix from known initial and terminal conditions [22.28, 48]. He also suggested the use of stochastic optimization algorithms to search the parameter space. Moreover, others also recommended EA as a preferred method for both identifying the structure and estimating the connection weights of neural networks [22.61].

Wahde and *Hertz* used a genetic algorithm (GA) to find the parameters of the RNN that modeled the gene network [22.21, 47]. They investigated the effectiveness of their GA in reconstructing the target network both from single and multiple time series from artificial networks. They also applied the algorithm to a coarse-grained representation of a real data set consisting of measurements from a rat's central nervous system.

Keedwell and *Narayanan* used a neural genetic hybrid algorithm for reverse engineering GRN using RNN formalism [22.62]. In their approach they used GA for evolving the structure of the network and estimated the weights of the structure using the gradient descent method. Regulatory interactions present in the maximum number of reconstruction trials or repeated runs, were taken as a significant connection. Their approach was validated using an artificial data set, a rat spinal cord data set, and a *Saccharomyces cerevisiae* data set.

Ressom et al. used an evolutionary computation approach combining two swarm intelligence (SI) methods to infer regulatory interactions with the RNN model [22.63]. In their algorithmic framework they used ant colony optimization (ACO) to identify the network architecture and particle swarm optimization (PSO) was applied to optimize the weights for the ACO evolved structures. The proposed SI-RNN method was validated using an artificial gene network and a subnetwork involved in cell-cycle regulation of *S. cerevisiae*.

Xu et al. used a bi-level architecture of PSO for inferring RNN models of gene networks [22.50]. Their proposed PSO/RNN method used a binary version of PSO in the outer layer to choose a subset of network connection. In the inner layer they used a conventional PSO algorithm for weight estimation. They applied their method both on a synthetic data set and the real gene expression data collected from the SOS DNA repair network in *Escherichia coli* (*E. coli*).

Mondal et al. applied the canonical differential evolution (DE) algorithm for estimating RNN parameters of GRNs [22.64]. They supported their proposal by reconstructing the gene network both from in silico gene expression data and in vivo gene expression data from SOS DNA repair network of *E. coli*.

22.5 A Modified DE Algorithm for Inferring GRN with RNN

Because of the success of EAs in reverse engineering GRNs, as found in our survey of recent literature, here we present another evolutionary approach for discovering regulatory relationships from expression profiles. We used a modified version of differential evolution (DE) with the random restart strategy to estimate the model parameters of RNN. In this section we describe our reconstruction algorithm after a brief introduction of the canonical DE algorithm.

22.5.1 Canonical Differential Evolution Algorithm

Storn and *Price* proposed DE as a family of algorithms for real-parameter optimization more than a decade ago [22.65, 66]. The variants of DE are distinguished from each other by varying the mutation and/or the recombination operations of the algorithm within a common framework. However, in this chapter we use a modified version of the classic DE algorithm (DE/rand/1/bin variant). Therefore, we describe the canonical DE algorithm with this variant.

DE searches for a global optimum point in D dimensional search space of real parameters. It works with a population of individuals x_G^i, $i = 1, 2, \ldots, P$, each representing a solution to the problem. DE individuals, also called chromosomes, are encoded as real vectors of size D, the dimension of the problem. The number of individuals in a population is called population size and is denoted by P, and the generation number is denoted by G where $G = 1, 2, \ldots, G_{\max}$. The initial population, $\mathcal{P}_1$ is created by randomly creating the vectors in appropriate search ranges. Then the fitness score of each individual is calculated through evaluation.

DE practices random parent selection regardless of the fitness values. In every generation, each individual x_G^i has a chance to become the principal parent and to breed its own offspring mating with other randomly chosen auxiliary parents. Formally, for every principal parent x_G^i, $i = 1, 2, \ldots, P$, three other auxiliary parents x_G^{r1}, x_G^{r2}, x_G^{r3} are selected randomly such that $r1, r2, r3 \in \{1, 2, \ldots, P\}$ and $i \neq r1 \neq r2 \neq r3$. Then, these three auxiliary parents participate in a *differential mutation* operation to create a mutated individual x_G^{mut} as follows:

$$x_G^{\mathrm{mut}} = x_G^{r1} + F\left(x_G^{r2} - x_G^{r3}\right), \tag{22.7}$$

where F is the *amplification factor*, a real-valued control parameter chosen from $[0.1, 1.0]$ [22.67]. Subsequently, the mutated vector, x_G^{mut}, participates in a *binomial crossover* operation with the principal parent x_G^i to generate the trial individual or offspring x_G^{child}.

The genes of x_G^{child} are inherited from x_G^i and x_G^{mut}, determined by a parameter called *crossover probability* (CR $\in [0, 1]$), as

$$x_{G,j}^{\mathrm{child}} = \begin{cases} x_{G,j}^{\mathrm{mut}} & \text{if } r(j) \leq \text{CR or } j = \mathrm{rn}(i), \\ x_{G,j}^i & \text{if } r(j) > \text{CR and } j \neq \mathrm{rn}(i), \end{cases} \tag{22.8}$$

where $j \ (= 1, \ldots, D)$ denotes the j-th element of individual vectors; $r(j) \in [0, 1]$ is the j-th evaluation of a uniform random number generator, and $\mathrm{rn}(i) \in \{1, \ldots, D\}$ is a randomly chosen index that ensures that x_G^{child} obtains at least one element from x_G^{mut}.

The selection scheme used in DE is also known as *knock-out competition*. As the name suggests, DE plays a one-to-one competition between the principal parent, x_G^i, and its offspring, x_G^{child}, to select the survivor for the next generation. The DE selection scheme can be described as

$$x_{G+1}^i = \begin{cases} x_G^{\mathrm{child}} & \text{if } f\left(x_G^{\mathrm{child}}\right) \text{ is better than } f\left(x_G^i\right) \\ x_G^i & \text{otherwise} \end{cases}. \tag{22.9}$$

Repeating the aforementioned mutation and crossover operations on each individual of the current generation, DE creates a new generation of population, which replaces the current generation. This generation alternation process is iterated until the termination criteria are satisfied. The control parameters of DE (F, CR, and P) are chosen beforehand and are kept constant throughout the search in this canonical version of the algorithm.

DE is one of the most elegant new-generation EAs for solving real-parameter optimization problems [22.67]. The algorithm has many attractive characteristics compared to other EAs, such as its simple and easy-to-understand structure, few controlling parameters, superior convergence characteristics, and robust performance. DE has proven to be very effective in solving nonlinear, nondifferentiable, nonconvex, and multimodal optimization problems. The variants of DE have secured high ranks in various competitions arranged at the IEEE Congress on Evolutionary Computation (CEC) conferences over the last few years. Due to its robust performance, DE has been successfully applied in solving many real world problems. Because of its impressive performance, we have chosen DE to learn the RNN parameters in GRN reconstruction problems.

22.5.2 Modified DE for Reconstructing GRN

The model based reconstruction algorithm for GRN is a strongly nonlinear problem with numerous local optima. The model flexibility poses a huge challenge for the optimization algorithm that easily gets trapped in suboptimum solutions. Consequently, though the inferred model parameters can reproduce the observed dynamics of the gene regulation system very closely, the actual model parameters remain undetected [22.68]. In order to overcome the problem of premature convergence a long existing method in EAs is random restart. Random restart is a heuristic, plugged in a meta search algorithm, which is applied when the search becomes stagnant in some local minima. It has been shown that if random restart is combined with a deterministic descent algorithm, then the hybrid algorithm converges asymptotically to the optimal solution [22.69]. Hybridization of random restart with genetic algorithms (GA) has also been found to be valuable [22.70]. Therefore, to avoid premature convergence to some local optimum, especially in the case of noise corrupted gene expression data, we embedded the random restart mechanism in a differential evolution (DE) algorithm. The details of the modified DE for reconstructing RNN based gene networks are given below.

Individual Structure

As mentioned before, each individual of DE represents a candidate solution of the problem – here the RNN model for the gene network. Therefore, the individuals of the presented DE algorithm contain $N(N+3)$ parameters of the RNN model, where N denotes the number of genes in the network. In other words, an individual consists of the array of decision variable vectors as

$$x^i_G = \begin{pmatrix} w_{1,1} & w_{1,2} & \dots & w_{1,N} & \beta_1 & \lambda_1 & \tau_1 \\ w_{2,1} & w_{2,2} & \dots & w_{2,N} & \beta_2 & \lambda_2 & \tau_2 \\ \vdots & \vdots & \ddots & \vdots & \vdots & \vdots & \vdots \\ w_{N,1} & w_{N,2} & \dots & w_{N,N} & \beta_N & \lambda_N & \tau_N \end{pmatrix}.$$

However, for convenience of implementation we serialized these parameters into a single dimensional vector

$$x^i_G = (w_{1,1} \dots w_{1,N}\ w_{2,1} \dots w_{2,N}\ w_{N,1}$$
$$\dots w_{N,N}\ \beta_1 \dots \beta_N\ \lambda_1 \dots \lambda_N\ \tau_1 \dots \tau_N).$$

Therefore, in this reverse engineering problem we will optimize parameters in a $N(N+3)$ dimensional search space using the random restart DE algorithm.

Initialization

In DE all individuals of the initial population are generated randomly. Generally, for each parameter of the RNN model, there is a feasible range within which the value of the parameter should be searched. This search range indicates the appropriate bound for the parameter value, e.g., the regulatory interaction, $w_{i,j}$, could be positive or negative to represent activation or repression, respectively; on the other hand, the rate constant τ_i should always be positive. Naturally, the search for the rate constants should be restricted to the range of positive real numbers. In order to find the global optimum values for these parameters, the initial population should be created covering this range as much as possible. This requirement is satisfied by choosing the value of each parameter x by uniformly sampling it within the maximum bound (x_{max}) and minimum bound (x_{min}) of

its search range as follows

$$x = x_{min} + (x_{max} - x_{min}) \times U_j(0, 1), \qquad (22.10)$$

where $U_j(0, 1)$ is a uniform random number between 0 and 1. The index j indicates that the distribution of the uniform random number is sampled anew for each parameter of each individual. After initialization, each individual is evaluated as the candidate model for the target GRN using evaluation criteria. The process of model evaluation, described later, assigns a fitness score to the evaluated individual.

Parent Selection

In DE each individual of the current population becomes the principal parent for breeding its own offspring. The auxiliary parents, which are mutually exclusive from each other and from the principle parent, are then selected from the rest of the population. Note here that, unlike the fitness proportionate selection, DE does not impose selection pressure during parents selection – all auxiliary parents are selected randomly. Therefore, each individual has equal probable participation in producing the new generation of individuals.

Offspring Generation

After the parents are selected, they participate in DE mutation and crossover operations to generate the offspring. In other words, the genetic operations of (22.7) and (22.8) are used to breed the offspring. However, the parameters of the offspring generated in this way may fall beyond the respective search ranges. To keep the GRN model parameters within the appropriate search space, the model parameters that violate the boundary constraints are adjusted using the reflection mechanism. In this process, adjustment of the violating parameter is done by reflecting back from the bound by the amount of violation

$$x_G = \begin{cases} 2 \cdot x_{min} - x_G, & \text{if } x_G < x_{min} \\ 2 \cdot x_{max} - x_G, & \text{if } x_G > x_{max} \end{cases}.$$

Selection

After a valid offspring is generated, it is evaluated using the fitness evaluation criteria to determine its fitness value. Then its fitness is compared with that of its principal parent. The better one is promoted to the next generation. When all the individuals of the current generation finish breeding, the new generation replaces the current generation.

Algorithm 1 Random Restart DE

1: Select P, F and CR and Set $G = 1$
2: **for** $i = 1$ to P **do**
3:     Initialize the i-th individual, $\varkappa_G^i$, in population, P_G, randomly
4:     Evaluate $\varkappa_G^i$
5: **end for**
6: **while** termination criteria not satisfied **do**
7:     **for** each individual $\varkappa_G^i$ in P_G **do**
8:         Select auxiliary parents $\varkappa_G^{r1}$, $\varkappa_G^{r2}$ and $\varkappa_G^{r3}$
9:         Create offspring $\varkappa_G^{child}$ using mutation and crossover
10:        Evaluate $\varkappa_G^{child}$
11:        $P_{G+1} = P_{G+1} \cup \text{Best}(\varkappa_G^{child}, \varkappa_G^i)$
12:     **end for**
13:     Set $G = G+1$
14:     **if** $(G \bmod G_{theta} = 0)$ AND $((f_{worst} - f_{best}) < (\delta \cdot f_{best}))$ **then**
15:         Reinitialize each $\varkappa_G^i$ in P_G except $\varkappa_G^{best}$
16:     **end if**
17: **end while**

Fig. 22.2 Pseudo-code description of random restart differential evolution for RNN model simulation

Random Restart

As discussed before, the random restart mechanism is used to get out of a local optimum. It also helps to improve the population diversity and thereby increases the probability of locating the global optimum. Generally, there are three decisions involved with the random restart mechanism: i) when ii) which, and iii) how. The first criterion *when* indicates how to decide when the random restart mechanism should be invoked. The most commonly used mechanism is to track the fitness of the best individual (or the average fitness of the population). If the tracked fitness does not change over a certain generation, then it is assumed that the population has converged and random restart is triggered. In this work, we have used a slightly different mechanism to detect the convergence. We took the difference between the best fitness (f_{best}) and the worst fitness (f_{worst}) of the current population as the decisive criterion. Based on this difference the random restart decision is taken as

$$\text{Random Restart if } (f_{worst} - f_{best}) < \delta \cdot f_{best},$$
$$(22.11)$$

where a smaller fitness value indicates a better individual. This condition is checked after every G_θ generation and, if satisfied, the random restart is triggered. The second criteria, *which*, is about the choice of the individuals who will undergo reinitialization. Here, we have reini-

tialized every individual of the population except the best individual. Finally, the *how* criteria concerns how to reinitialize the selected individuals. One possibility is to reinitialize in an absolute random manner as initialized in the first generation. The other possibilities are to add some random variation to every or some selected parameters of every individuals or to reinitialize individuals using random variation to the elite individual. In this work, we used the first choice – total random initialization. After the random initialization, the algorithm proceeds in a regular fashion.

Termination Criteria

The process of reproduction and selection continues generation after generation until a maximum number of generations, G_{max} have elapsed. At that point the optimization is terminated and the best individual in the final generation is selected as the GRN model for the target network.

The pseudo-code description of random restart DE for RNN model estimation is presented in Algorithm 1 (Fig. 22.2).

22.5.3 Model Evaluation Criteria

As we search for the optimum model for the target gene regulatory network using the presented algorithm, different candidate models are generated by the reconstruction algorithm. As the algorithm works with a fixed number of candidate models (population size P is fixed), we need to evaluate the alternate models and filter them based on some criterion. The most widely used criterion for GRN model evaluation is difference between the dynamic response generated by the candidate model and the observed network dynamics. Therefore, the fitness score for a candidate RNN model (Ω) is given by

$$f(\Omega) = \frac{1}{TN} \sum_{t=1}^{T} \sum_{i=1}^{N} \left[e_i^{cal}(t) - e_i^{exp}(t) \right]^2, \quad (22.12)$$

where $e_i^{exp}(t)$ and $e_i^{cal}(t)$ represent the expression levels of the i-th gene at time t in experimentally collected and model simulated data, respectively.

In general, for a dynamic system, if we can provide multiple trajectories of its responses then the reverse engineering task can be done with higher accuracy. This has also been found true for reconstructing GRN from gene expression data [22.21, 43, 47]. The fitness evaluation function for RNN models using multiple sets of time dynamics becomes

$$f(\Omega) = \frac{1}{TNM} \sum_{k=1}^{M} \sum_{t=1}^{T} \sum_{i=1}^{N} \left[e_{i,k}^{\text{cal}}(t) - e_{i,k}^{\text{exp}}(t)\right]^2 ,$$

(22.13)

where $e_{i,k}^{\text{exp}}(t)$ and $e_{i,k}^{\text{cal}}(t)$ represent the expression levels of the i-th gene in k-th set of time courses at time t in experimental and simulated data, respectively. This fitness function of (22.13) was used in this work for evaluating the candidate network models.

22.6 Reconstruction Experiments

In this work, the effectiveness of the proposed approach was mainly validated using synthetic networks since the actual complete structure is unknown for most real networks. The performance of any reconstruction algorithm is challenged by the amount of expression profile used for learning the parameters and also the level of noise present in the expression data. Therefore, we tried our algorithm with various quantities and/or qualities of the gene expression data. Finally, we verified the proposed method using real gene expression data collected from the SOS DNA repair network of *Escherichia coli*.

22.6.1 In Silico Network Inference

First, we investigated the effectiveness of the proposed method in capturing the dynamics of a gene network as well as estimating the correct kinetic parameter. For this purpose we employed the reverse engineering algorithm in reconstructing a simplified synthetic genetic network with four genes. The RNN model parameters for this network are displayed in Table 22.1. Note that in this model $\lambda_i = 1$ was used for all genes and is therefore not displayed in Table 22.1. The same network model was studied in [22.47, 50].

In order to generate synthetic gene expression data we simulated this RNN network model with the random initial state for each gene. We generated 50 time samples for each gene. In order to experiment with various amounts of gene expression data we generated ten sets of gene expression data for the target network. We also varied the quality of the gene expression data sets by adding various amounts of measurement noise with the expression data. Besides the noise free ideal gene expression data set, we generated expression profiles

with 5 and 10% Gaussian noise, respectively, to simulate measurement error. In our experimental study we fed the algorithm with gene expression data of different quality and quantity.

We performed all the experiments for this synthetic network reconstruction under the same experimental conditions. The algorithmic setup was as follows: $F = 0.5$, $CR = 0.9$, $P = 100$ $G_{\text{max}} = 10\,000$, $G_{\theta} = 10$, $\delta = 1 \times 10^{-3}$. The search regions for the RNN parameters were: $w_{ij} \in [-30.0, 30.0]$, $\beta_i \in [-10.0, 10.0]$, $\tau_i \in [0.0, 20.0]$. Since $\lambda_i = 1.0$ was fixed in the target model, we kept those fixed and did not include them in our search. Our algorithm was implemented in Java language and the experiments were run on an Intel Core $i7$ CPU 2.67 GHz computer with 6 GB RAM.

We applied a structure skeletalization scheme to decrease the computational effort, as done in other methods [22.43, 44, 59]. In this process if the magnitude of a RNN parameter becomes less than 1×10^{-3} then we ignore it by resetting it to zero. We performed ten repetitions of each reconstruction algorithm to confirm the reliability of our stochastic algorithm.

In our initial group of experiments, we estimated the target RNN parameters in an ideal noise-free environment. In the first experiment of this group we reverse engineered the target GRN from a single noise free gene expression data set. The results of this experiment are summarized in Table 22.2a. Each cell of the table represents the average and standard deviation of the estimated values for the corresponding parameter. In summary, the algorithm was able to estimate the correct network structure with a reasonable estimate of the target parameter values. However, the magnitudes of the false positive regulations were too high to ignore. In the subsequent experiments in this group, we gradually increased the number of gene expression data sets for reconstructing the GRN. We reverse engineered the network with three, five, and ten sets of gene expression profiles; the results are shown in Table 22.2b–d, respectively. It is obvious that as we give more system response to the reverse engineering process, the accuracy will increase. Moreover, we observed

Table 22.1 RNN model of the target synthetic network

w_{ij}	1	2	3	4	β_i	τ_i
1	20.0	−20.0	0.0	0.0	0.0	10.0
2	15.0	−10.0	0.0	0.0	−5.0	5.0
3	0.0	−8.0	12.0	0.0	0.0	5.0
4	0.0	0.0	8.0	−12.0	0.0	5.0

Table 22.2a–d Results from RNN model parameter estimation from noise-free gene expression data (**a**) one data set, (**b**) three data sets, (**c**) five data sets, (**d**) ten data sets

(a)

w_{ij}	1	2	3	4
1	18.557±8.020	−17.300±6.367	1.250±10.015	−1.190±7.527
2	13.174±2.513	−6.151±3.636	2.073±8.772	−8.644±6.872
3	15.896±11.178	−12.721±17.336	9.621±12.630	−5.886±16.650
4	−0.744±2.148	2.513±4.379	6.357±7.233	−15.888±6.865
β_i	−1.163±8.947	−2.570±7.713	−0.943±7.909	3.073±7.295
λ_i	9.318±1.847	5.652±0.419	5.184±0.046	5.116±0.310

(b)

w_{ij}	1	2	3	4
1	22.682±3.401	−22.590±3.728	−4.800±6.005	1.905±3.047
2	14.997±0.095	−10.000±0.083	−0.005±0.040	0.035±0.046
3	−5.024±8.144	−8.263±8.721	13.172±2.802	0.005±12.859
4	−0.044±0.058	0.047±0.055	8.107±0.145	−12.210±0.286
β_i	3.092±3.838	−5.014±0.042	4.122±5.613	0.030±0.051
λ_i	10.077±0.114	4.992±0.015	5.003±0.019	5.030±0.047

(c)

w_{ij}	1	2	3	4
λ_i	10.077±0.114	4.992±0.015	5.003±0.019	5.030±0.047
1	20.087±0.192	−20.177±0.370	−0.129±0.260	0.211±0.422
2	14.973±0.058	−9.973±0.058	0.016±0.039	−0.035±0.081
3	−3.431±5.292	−9.612±7.020	12.416±1.752	2.436±10.079
4	0.000±0.002	−0.008±0.012	7.991±0.023	−11.992±0.028
β_i	0.033±0.067	−4.990±0.020	3.105±4.733	0.007±0.013
λ_i	10.002±0.014	5.000±0.002	5.018±0.032	5.001±0.003

(d)

w_{ij}	1	2	3	4
1	20.000±0.000	−20.000±0.000	0.000±0.000	0.000±0.000
2	15.000±0.000	−10.000±0.000	0.000±0.000	0.000±0.000
3	0.000±0.001	−8.000±0.000	12.000±0.001	0.000±0.001
4	0.000±0.000	0.000±0.000	8.000±0.000	−12.000±0.000
β_i	0.000±0.000	−5.000±0.000	0.000±0.000	0.000±0.000
λ_i	10.000±0.000	5.000±0.000	5.000±0.000	5.000±0.000

the same trend in our experiments. With the increase of the number of data sets the estimation of the parameter values becomes more accurate and the values of the false positive diminish towards zero. In particular, when we reconstructed the network from ten data sets we obtained the exact model with zero error (i. e., fitness score = 0) in a couple of runs. This justifies that if sufficient time dynamics is given, then the presented algorithm can estimate very accurate RNN model parameters for the target GRN.

Table 22.3 shows the average and standard deviation of the fitness scores of the best GRN models found in different trial runs. This table also shows the required time for reconstructing the GRN. According to this table, the reconstruction of the network was possible within few seconds with a single gene expression data set. However, with an increase in the amount of gene expression data the reconstruction time naturally increases. Yet, it was possible to reconstruct the network within few minutes from ten sets of gene expression data.

After verifying the success of the algorithm in an ideal environment, we tried the algorithm with the noise corrupted data to understand the effect of the measurement noise on the reverse engineering process. The

Table 22.3 Statistics of the fitness scores and required times in different experiments

Data set	Noise level (%)	Score	Time (s)
1	0	8.70E-09±1.20159E-08	20.30±0.04
3	0	7.53E-08±6.54317E-08	57.78±0.08
5	0	4.93E-08±1.15222E-07	96.40±0.15
10	0	2.02E-12±3.24571E-12	191.16±0.21
1	5	6.97E-04±2.00188E-06	20.96±0.06
3	5	9.76E-04±8.70371E-06	60.30±0.07
5	5	9.85E-04±1.85125E-06	98.70±0.11
10	5	1.05E-03±9.46458E-06	192.80±0.23
1	10	4.28E-03±6.92098E-05	21.00±0.28
3	10	3.51E-03±6.95777E-07	58.89±0.06
5	10	3.45E-03±6.21808E-06	96.46±0.24
10	10	3.98E-03±8.54596E-07	191.30±0.25

reconstruction results from 5% noise corrupted data are displayed in Table 22.4. When noise was present in the data, the algorithm failed to predict some of the regulation types correctly (e.g., $w_{4,4}$) from a single time series data or from three sets of time series. Moreover, the estimation of the parameters was less accurate and the false positives were too large to neglect. However, with the increase of data sets, it was possible to reduce the effect of noise. With five or ten sets of data, the algorithm predicted all the regulation types correctly and the estimation of the regulation strength was reasonably accurate. With more gene expression data, it was also easier to identify most of the possible false positives as those values became very small. However, as Table 22.3 shows, with the noise present in the data the fitness scores for the predicted models became larger,

which indicates higher discrepancy between the target and the simulated gene expression profile. Nevertheless, it is good in the sense that the method was able to avoid overfitting to the noisy dynamics.

Finally, we evaluated the algorithm for the same network but with increased noise levels. The results of reconstructing the network from 10% noise corrupted expression data are summarized in Table 22.5. With a increase in the noise level the overall performance of the reconstruction algorithm degraded another level: poorer estimation of the regulatory strengths and higher values for the false positives. However, once again with the increase of the number of different dynamics, the effect of noise could be reduced, as we can observe from Table 22.5a–d. Nevertheless, in general, the accuracy of the estimated GRN model degraded with an increased level of noise.

In order to present a joint perspective on the effect of noise and number of data sets we compared those in terms of error in model estimation. The error $\epsilon_{m,n}$ in model estimation from $n\%$ noise corrupted m data sets is defined as

$$\epsilon_{m,n} = \sum_{i=1}^{N} \sum_{j=1}^{N} (w_{i,j}^{T} - w_{i,j}^{EA}) + \sum_{i=1}^{N} (\beta_i^{T} - \beta_i^{EA})$$

$$+ \sum_{i=1}^{N} (\tau_i^{T} - \tau_{i,j}^{EA}), \qquad (22.14)$$

where $w_{i,j}^{T}$, β_i^{T}, and τ_i^{T} represent the target model parameters for the GRN and $w_{i,j}^{EA}$, β_i^{EA}, and τ_i^{EA} represent the averaged estimated parameters from the gene expression data, respectively. We plot this error in the graph shown in Fig. 22.3. The graph clearly shows that

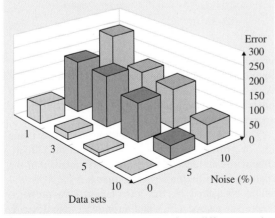

Fig. 22.3 Errors in model prediction from different number of data sets and error levels

Table 22.4a–d Results from RNN model parameter estimation from 5% noisy gene expression data, (**a**) one data set, (**b**) three data sets, (**c**) five data sets, (**d**) ten data sets

(a)

w_{ij}	1	2	3	4
1	29.624±0.302	−29.844±0.161	−6.171±18.712	−3.758±5.426
2	25.513±0.955	−29.787±0.238	28.841±0.488	29.687±0.319
3	2.124±14.821	−10.005±9.926	4.292±10.445	2.562±19.169
4	−2.433±8.159	6.539±7.062	13.170±13.772	3.537±15.089
β_i	−3.241±0.684	−9.118±0.828	−1.895±7.216	−5.375±3.928
λ_i	0.050±0.000	4.942±0.016	5.326±1.213	5.022±0.228

(b)

w_{ij}	1	2	3	4
1	−11.581±5.834	8.050±2.985	1.804±0.254	29.165±0.593
2	15.189±3.173	−21.260±7.328	6.184±7.033	−3.440±2.243
3	−3.290±2.846	−7.191±2.553	17.060±8.458	−2.395±1.520
4	−3.348±0.284	−0.577±0.188	4.031±0.169	−2.137±0.414
β_i	2.773±6.113	1.922±3.191	3.513±4.029	2.082±0.292
λ_i	10.493±0.106	4.846±0.158	2.985±0.624	3.598±0.124

(c)

w_{ij}	1	2	3	4
1	10.095±3.764	−14.164±7.227	−2.696±5.082	29.697±0.215
2	11.216±0.556	−5.654±0.515	0.804±0.048	1.348±0.054
3	0.398±2.540	−9.617±1.720	13.555±0.513	−0.434±1.536
4	−0.725±0.083	0.684±0.096	5.672±0.059	−6.052±0.168
β_i	−3.049±4.161	−5.518±0.136	0.514±3.810	−0.726±0.034
λ_i	10.762±0.065	4.613±0.076	4.075±0.064	4.397±0.009

(d)

w_{ij}	1	2	3	4
1	28.038±1.939	−23.157±5.432	−7.251±6.142	2.135±4.761
2	21.432±2.283	−18.219±3.071	0.264±0.129	−2.749±0.922
3	0.415±0.042	−7.121±0.122	11.031±0.115	−0.114±0.080
4	0.017±0.022	0.140±0.053	6.768±0.073	−9.769±0.147
β_i	2.513±3.760	−4.254±0.447	−0.537±0.110	−0.318±0.063
λ_i	9.927±0.744	5.270±0.021	4.250±0.017	4.495±0.007

for the same amount of data sets, the error increases with noise level, and for the same level of noise, the error decreases with the increase in data set. So it is possible to have a reasonable estimate of the target GRN parameters if we can increase the number of data sets or keep the level of noise within a reasonable limit.

22.6.2 In Vivo Network Inference

Finally, we verified our method by analyzing real gene expression data. We employed the algorithm presented to reconstruct the SOS DNA repair network in *E. coli* bacterium. The DNA repair system in *E. coli* is trig-

gered when the bacterium is exposed to agents or conditions that damage or interfere with its DNA replication process [22.71]. The complete repair system consists of about 40 genes that produce many defense proteins for the repair task [22.72].

The function of the SOS system in *E. coli* operates, in general, as proposed by *Little* et al. [22.73]. Although many genes and proteins are involved in the repair mechanism, the key functioning of the SOS system is controlled by the interplay of two proteins: a repressor named LexA and an inducer called RecA. During normal growth, i.e., in an un-induced cell, the LexA repressor binds to a specific sequence – the

Table 22.5a–d Results from RNN model parameter estimation from 10% noisy gene expression data (**a**) one data set, (**b**) three data sets, (**c**) five data sets, (**d**) ten data sets

(a)

w_{ij}	1	2	3	4
1	17.023 ± 13.551	7.571 ± 13.195	7.891 ± 10.349	7.838 ± 15.758
2	27.564 ± 2.369	-11.394 ± 3.290	-27.190 ± 1.963	28.667 ± 1.598
3	23.068 ± 2.869	-9.180 ± 1.806	3.803 ± 0.238	-10.628 ± 2.590
4	-20.030 ± 9.793	19.214 ± 5.480	25.365 ± 3.544	-10.287 ± 12.385
β_i	3.101 ± 4.989	-6.978 ± 2.544	-9.578 ± 0.429	-9.111 ± 1.089
λ_i	7.803 ± 0.101	4.662 ± 0.055	0.016 ± 0.002	8.635 ± 0.396

(b)

w_{ij}	1	2	3	4
1	11.308 ± 0.783	-29.508 ± 0.434	11.059 ± 7.094	-3.482 ± 4.436
2	29.416 ± 0.311	-25.073 ± 0.335	-6.821 ± 0.443	5.494 ± 1.065
3	2.168 ± 0.486	-3.992 ± 0.810	3.029 ± 0.521	4.039 ± 1.021
4	-3.978 ± 1.580	4.183 ± 1.661	24.890 ± 2.217	-25.629 ± 2.867
β_i	9.727 ± 0.207	-5.999 ± 0.193	-2.595 ± 0.095	-5.765 ± 1.356
λ_i	8.904 ± 0.042	4.780 ± 0.023	1.942 ± 0.136	7.264 ± 0.030

(c)

w_{ij}	1	2	3	4
1	13.972 ± 3.723	-20.557 ± 2.941	17.471 ± 5.142	-25.024 ± 4.430
2	28.013 ± 1.209	-24.085 ± 1.112	-7.210 ± 0.471	8.121 ± 0.523
3	6.735 ± 0.480	-9.020 ± 0.631	-0.022 ± 0.413	9.781 ± 0.894
4	0.359 ± 0.739	-0.363 ± 0.938	4.746 ± 2.035	-5.992 ± 3.743
β_i	7.768 ± 1.727	-6.218 ± 0.287	-3.818 ± 0.182	-0.715 ± 0.421
λ_i	9.070 ± 0.041	5.010 ± 0.023	2.516 ± 0.078	3.510 ± 0.541

(d)

w_{ij}	1	2	3	4
1	14.448 ± 2.019	-22.340 ± 2.517	5.683 ± 0.533	-14.690 ± 1.505
2	13.923 ± 0.420	-9.127 ± 0.361	0.085 ± 0.119	1.819 ± 0.198
3	2.545 ± 0.097	-2.844 ± 0.097	4.517 ± 0.035	3.534 ± 0.112
4	-0.645 ± 0.116	0.219 ± 0.098	5.953 ± 0.122	-8.424 ± 0.305
β_i	8.325 ± 0.851	-5.548 ± 0.130	-3.782 ± 0.052	0.073 ± 0.113
λ_i	10.391 ± 0.071	4.900 ± 0.030	1.890 ± 0.045	4.292 ± 0.029

Part D | 22.6

SOS box, present in the operator sites of the SOS genes and repress their transcription. Depending on the characteristics of their SOS boxes, the SOS genes are repressed to different levels under normal growth. When the cell experiences an increased level of DNA damage, an inducing signal that leads to the expression of the SOS regulon is generated. One of the SOS proteins, RecA, serves as the sensor of the DNA damage by binding to single-stranded DNA (ssDNA). The binding of RecA to ssDNA in the presence of a nucleoside triphosphate forms a nucleoproteins filament and converts RecA to an activated form (often referred to as RecA*) [22.72]. The interaction of the activated RecA protein with the LexA protein induces the self-cleavage reaction of the LexA repressor. With the decrease of the LexA pool, various SOS genes, repressed by LexA, are de-repressed. Genes with SOS boxes that bind LexA weakly are the first to be expressed. If the induced treatment is very high, more molecules of RecA become activated, resulting in self-cleavage of more LexA repressors. Eventually, genes with a strong binding of LexA at their SOS box become expressed [22.74].

Once the DNA damage is repaired, the cell begins to return to its normal state and the RecA molecules return to their inactive state. In the absence of the RecA protease, the amount of LexA molecule continues to rise as a result of continuous synthesis. With the increase of the LexA pool, the repression of SOS genes is re-instigated and the cell returns to its normal un-induced state.

Gene Expression Data Set

Ronen et al. [22.75] developed a system for real-time monitoring of the transcriptional activity of operons by using low-copy reporter plasmids in which a promoter controls green fluorescent protein (GFP). This process complements, at higher accuracy, the genomic-scale perspective given by DNA microarrays. It was shown that this approach can be used to determine the order of genes in an assembly pathway [22.75].

Using this method, the kinetics of eight genes (uvrD, lexA, umuD, recA, uvrA, uvrY, ruvA and polB) of the SOS DNA repair network were collected. Measurements were done after irradiation of the DNA at the initial time with UV light. Four experiments were done for various light intensities (experiments 1, 2: $5\,J/m^2$, experiments 3, 4: 20 $5\,J/m^2$). Each experiment consisted of 50 measurements, evenly sampled every 6 min. The data sets are publicly available from the home page of *Uri Alon Lab* [22.76].

Since the expression level at the first time point was zero for all genes in all data sets, we ignored it and normalized the data in the interval [0, 1]. We experimented with all data sets together and with each data set independently. Based on the experimental results from the previous section, we expected better results from multiple data sets. Unfortunately, the best results were obtained from the first data set and are presented here. We assume that the possible presence of the high level measurement errors in the expression data prevented the algorithm from predicting uniform regulatory interactions from multiple time series.

Predicted Network

The algorithmic settings for the SOS repair network was the same as used in the previous section. However, we changed the search ranges for the parameters of the RNN model. The search ranges were: $w_{ij} \in [-10.0, 10.0]$, $\beta_i \in [-10.0, 10.0]$, $\tau_i \in [0.0, 10.0]$, and $\lambda_i \in [0.0, 1.0]$. Note that we made the decay rate parameter λ_i flexible for analyzing this real gene expression profile.

Table 22.6 Predicted RNN model for SOS network

	uvrD	lexA	umuDC	recA	uvrA	uvrY	ruvA	polB
uvrD	6.424857	−7.213836	7.831464	7.814065	−9.167508	−2.723474	3.013275	1.299472
lexA	8.566925	−5.685499	−9.979448	1.986422	9.399351	9.869024	−9.644226	−6.941927
umuDC	−1.998766	−0.236688	−1.602020	8.958519	1.295936	1.663911	−1.294825	−3.710058
recA	9.891436	−2.182999	9.293353	−3.770621	9.769186	6.329382	−9.890675	9.283236
uvrA	9.534298	−4.073979	−0.893895	−0.808004	9.944793	9.976036	−9.575018	7.971381
uvrY	7.662362	−9.750357	9.312335	9.713831	−7.053784	9.263280	−7.632835	5.862812
ruvA	−9.222275	−6.405933	9.091333	−7.696163	−5.900112	−3.766886	−3.804311	−4.771628
polB	2.144721	−3.131107	−0.814872	4.217693	2.474628	5.033770	2.168423	−7.029893
β_i	−5.302547	−3.265813	−9.578692	−3.131999	−3.478737	−5.625459	−1.294883	−8.231654
τ_i	0.837486	0.147834	1.194170	0.156369	0.219462	0.217415	8.624435	4.009910
λ_i	0.699118	0.746625	0.981102	0.509149	0.997585	0.843420	0.939817	0.941208

Table 22.7 Predicted regulatory interactions in SOS network

	uvrD	lexA	umuDC	recA	uvrA	uvrY	ruvA	polB
uvrD	−							
lexA	−	−		+	+	+		
umuDC	−							
recA	+	−			+		−	+
uvrA	−	−			+		−	+
uvrY	−				−			
ruvA					−			
polB								

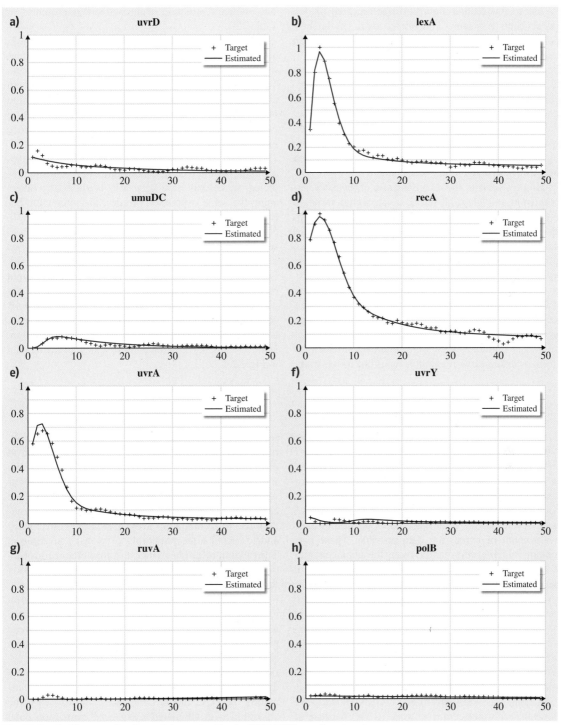

Fig. 22.4a–h Target and estimated dynamics for the SOS DNA repair network

The reconstruction algorithm was repeated for ten independent trial runs. The fitness score of the best models in different experimental runs ranged with an average 2.43×10^{-4} and standard deviation 1.56×10^{-4}; the average of the required reconstruction time was 89.5 ± 0.87s. The inferred RNN parameters in a typical algorithmic run are displayed in Table 22.6. The fitness score for this model was 2.11×10^{-4}. The actual dynamics of data set 1 and the dynamics generated by this network model are compared in Fig. 22.4. From Fig. 22.4 it is obvious that the predicted model captured the network dynamics very well.

However, due to the reconstruction of the SOS network from a single time series, a substantial variation in the predicted structure and parameter values was observed in different trials. Even the type of predicted regulation (synthesis or repression) varied from trial to trial. Moreover, all possible regulations (w_{ij}) were present in the predicted models as shown in Table 22.6. In order to form a unified estimation of the structure of the SOS network we filtered out the scattered interactions and retained only the robust regulations. In our predicted network structure we selected only those regulations as credible which were predicted at least nine times (with the same regulation type) out of ten trials.

The predicted structure of the network is presented in Table 22.7.

From the predicted regulations in Table 22.7, it is obvious that the reconstruction method was able to identify the essential regulations of the SOS repair network. The inhibition of uvrD, lexA, umuDC, recA, uvrA, and uvrY by lexA was identified accurately. The activation of lexA by recA was correctly inferred. The repression of lexA by umuDC is true according to the evidence in the scientific literature and online databases [22.77]. The other inferred regulations are either false positives or unknown interactions. However, some of these predicted regulations have been also predicted by others using the same expression profiles: the interaction between gene pairs lexA–uvrA was predicted in [22.50, 55], the interaction between recA and uvrA was presented in the reconstructed network of [22.55, 78], and the self-regulation of uvrA was identified in [22.55]. Such recurrent identification of these regulations might be coincidental or due to the noise bias present in the data. Furthermore, the repression of ruvA and polB could not be identified consistently in the reconstruction approach. The possible reason behind such failure is the low level expression of these genes, as can be seen in Fig. 22.4.

22.7 Conclusion

Gene regulatory networks are dynamic and distributed systems that govern the expression of genes in living cells and thereby regulate their behavior. Therefore, deciphering these complex molecular systems is essential to achieve the ultimate goal of systems and synthetic biology. High-throughput technologies such as DNA microarrays have made the collection of massive amount of expression data possible. These data contain valuable information about the characteristics of biological systems. Hence, reverse engineering of these systems from gene expression data has become a very popular research field. However, the field still faces many challenges, such as: precise modeling of the underlying process, insufficient amount of gene expression data, presence of significant error in the data, and a computationally efficient reverse engineering algorithm. In this chapter, we focused on these issues with some experimental study on GRN reconstruction.

Among the wide spectrum of GRN models, we chose the recurrent neural network (RNN) because of its adequate flexibility to represent different types of regulatory interactions and computational feasibility to

reconstruct large-scale GRNs. For reverse engineering of the RNN modeled genetic networks we presented an evolutionary algorithm. The reconstruction algorithm was developed using differential evolution (DE) – a versatile and reliable optimizer for real parameters. The canonical DE algorithm was enhanced using a random restart technique to avoid premature convergence and to increase the possibility of global convergence. The random restart heuristics also made the reconstruction algorithm robust against noisy gene expression data.

In order to investigate the capability of the algorithm to identify correct model parameters, we experimented using a well-known RNN model for artificial genetic networks. The algorithm exhibited adequate success in reverse engineering the network from a single time series with some error. When we supplied additional time dynamics to the algorithm, it was successful in estimating the exact model parameters. We also tried the algorithm with simulated noisy gene expression data. We found that if the noise level is not very high the algorithm can identify important regulations even in the presence of noise, and the effect of noise on the reconstruction algorithm can be minimized by supplying additional gene expression data. Even with 10% noise in the data the GRN was estimated with reasonable accuracy from multiple gene expression data set.

We also tried the algorithm developed using real gene expression data collected from the SOS DNA repair network in *E. coli*. Many of the known regulations in this response network were correctly predicted in this method. However, the algorithm failed to infer a few known regulations, perhaps due to the insufficient variation in the gene expression level. There were also a number of false positive regulations in the predicted SOS network. Nevertheless, this experiment validates the capability of the algorithm in identifying regulatory interactions through analysis of real gene expression data.

One of the very attractive features of the presented reconstruction algorithm is its efficiency. It can reconstruct the network within few seconds to minutes, depending of the amount of gene expression data sets used in inference. This feature will make this algorithm valuable in large-scale GRN estimation. However, for large-scale network estimation some enhancement of the algorithm will be necessary such as: identifying the sparse network architecture, taking advantage of the modular network-components, and incorporation of existing knowledge.

References

22.1   H. Kitano: Systems biology: A brief overview, Science **295**(5560), 1662–1664 (2002)

22.2   H. Kitano: Computational systems biology, Nature **420**(6912), 206–210 (2002)

22.3   A.P. Arkin, D.V. Schaffer: Network news: Innovations in 21st century systems biology, Cell **144**(6), 844–849 (2011)

22.4   J.R. Tejedor, J. Valcárcel: Gene regulation: Breaking the second genetic code, Nature **465**(7294), 45–46 (2010)

22.5   S. Kimura, K. Sonoda, S. Yamane, H. Maeda, K. Matsumura, M. Hatakeyama: Function approximation approach to the inference of reduced NGnet models of genetic networks, BMC Bioinformatics **9**, 23 (2008)

22.6   Z. Li, S.M. Shaw, M.J. Yedwabnick, C. Chan: Using a state-space model with hidden variables to infer transcription factor activities, Bioinformatics **22**(6), 747–754 (2006)

22.7   B.N. Kholodenko, A. Kiyatkin, F.J. Bruggeman, E. Sontag, H.V. Westerhoff, J.B. Hoek: Untangling the wires: A strategy to trace functional interactions in signaling and gene networks, Proc. Natl. Acad. Sci. USA **99**(20), 12841–12846 (2002)

22.8   T. Schlitt, A. Brazma: Current approaches to gene regulatory network modelling, BMC Bioinformatics **8**(6), S9 (2007)

22.9   T.S. Gardner, D. di Bernardo, D. Lorenz, J.J. Collins: Inferring genetic networks and identifying compound mode of action via expression profiling, Science **301**(5629), 102–105 (2003)

22.10   H. De Jong: Modeling and simulation of genetic regulatory systems: A literature review, J. Comput. Biol. **9**(1), 67–103 (2002)

22.11   M.A. Gibson, E. Mjolsness: Modeling the activity of single genes. In: *Computational Modeling of Genetic and Biochemical Networks*, ed. by J.M. Bower, H Bolouri (MIT, London 2001) pp. 3–48

22.12   M. Schena (Ed.): *DNA Microarrays: A Practical Approach* (Oxford Univ. Press, Oxford 1999)

22.13   P. Hegde, R. Qi, K. Abernathy, C. Gay, S. Dharap, R. Gaspard, J.E. Hughes, E. Snesrud, N. Lee, J. Quackenbush: A concise guide to cDNA microarray analysis, Biotechniques **29**(3), 548–562 (2000)

22.14   S. Liang, S. Fuhrman, R. Somogyi: REVEAL, a general reverse engineering algorithm for inference of genetic network architectures, Pac. Symp. Biocomput., Vol. 3 (1998) pp. 18–29

22.15   D. Jiang, C. Tang, A. Zhang: Cluster analysis for gene expression data: A survey, IEEE Trans. Knowl. Data Eng. **16**(11), 1370–1386 (2004)

22.16   M.B. Eisen, P.T. Spellman, P.O. Brown, D. Botstein: Cluster analysis and display of genome-wide expression patterns, Proc. Natl. Acad. Sci. USA **95**(25), 14863–14868 (1998)

22.17   P. Tamayo, D. Slonim, J. Mesirov, Q. Zhu, S. Kitareewan, E. Dmitrovsky, E.S. Lander, T.R. Golub: Interpreting patterns of gene expression with self-organizing maps: Methods and application to hematopoietic differentiation, Proc. Natl. Acad. Sci. USA **96**(6), 2907–2912 (1999)

22.18   G. Getz, E. Levine, E. Domany: Coupled two-way clustering analysis of gene microarray data, Proc. Natl. Acad. Sci. USA **97**(22), 12079–12084 (2000)

22.19   P. D'Haeseller, S. Liang, R. Somogyi: Genetic network inference: From co-expression clustering to reverse engineering, Bioinformatics **16**(8), 707–726 (2000)

22.20   J. Tegnér, M.K.S. Yeung, J. Hasty, J.J. Collins: Reverse engineering gene networks: Integrating genetic perturbations with dynamical modeling, Proc. Natl. Acad. Sci. USA **100**(10), 5944–5949 (2003)

22.21   M. Wahde, J. Hertz: Modeling genetic regulatory dynamics in neural development, J. Comput. Biol. **8**(4), 429–442 (2001)

22.22 P. D'haeseleer, X. Wen, S. Fuhrman, R. Somogyi: Linear modeling of mRNA expression levels during CNS development and injury, Pac. Symp. Biocomput., Vol. 4 (1999) pp. 41–52

22.23 N. Friedman, M. Linial, I. Nachman, D. Pe'er: Using Bayesian networks to analyze expression data, J. Comput. Biol. **7**(3–4), 601–620 (2000)

22.24 S.A. Kauffman: *The Origins of Order, Self-Organization and Selection in Evolution* (Oxford Univ. Press, Oxford 1993)

22.25 I. Shmulevich, E.R. Dougherty, W. Zhang: From boolean to probabilistic boolean networks as models of genetic regulatory networks, Proc. IEEE **90**(11), 1778–1792 (2002)

22.26 E.P. van Someren, L.F.A. Wessels, M.J.T. Reinders: Linear modeling of genetic networks from experimental data, Proc Intelligent Systems For Molecular Biology (ISMB 2000) (2000) pp. 355–366

22.27 E.P. van Someren, L.F.A. Wessels, M.J.T. Reinders: Genetic network models: A comparative study, Proc Proc. SPIE, Micro-Arrays: Opt. Technol. Inform., Vol. 4266 (2001) pp. 236–247

22.28 J. Vohradský: Neural network model of gene expression, FASEB J. **15**(3), 846–854 (2001)

22.29 M.A. Savageau: Biochemical systems analysis. I. Some mathematical properties of the rate law for the component enzymatic reactions, Theor. Biol. **25**(3), 365–369 (1969)

22.30 M.A. Savageau: Biochemical systems analysis. II. The steady-state solutions for an n-pool system using a power-law approximation, Theor. Biol. **25**(3), 370–379 (1969)

22.31 A. Arkin, J. Ross, H.H. McAdams: Stochastic kinetic analysis of developmental pathway bifurcation in phage lambda-infected *Escherichia coli* cells, Genetics **149**(4), 1633–1648 (1998)

22.32 H. Matsuno, A. Doi, M. Nagasaki, S. Miyano: Hybrid Petri net representation of gene regulatory network, Pac. Symp. Biocomput., Vol. 5 (2000) pp. 338–349

22.33 D.C. Weaver, C.T. Workman, G.D. Stormo: Modeling regulatory networks with weight matrices, Pac. Symp. Biocomput., Vol. 4 (1999) pp. 112–23

22.34 S. Kauffman, C. Peterson, B. Samuelsson, C. Troein: Random boolean network models and the yeast transcriptional network, Proc. Natl. Acad. Sci. USA **100**(25), 14796–14799 (2003)

22.35 K. Murphy, S. Mian: *Modelling Gene Expression Data Using Dynamic Bayesian Networks. Technical Report* (Univ. of California, Berkeley 1999)

22.36 I.M. Ong, J.D. Glasner, D. Page: Modelling regulatory pathways in *E. coli* from time series expression profiles, Bioinformatics **18**(1), S241–S248 (2002)

22.37 F.X. Wu, W.-J. Zhang, A.J. Kusalik: Modeling gene expression from microarray expression data with state-space equations, Pac. Symp. Biocomput., Vol. 9 (2004) pp. 581–592

22.38 J.M. Bower, H. Bolouri (Eds.): *Computational Modeling of Genetic and Biochemical Networks* (MIT, London 2004)

22.39 R. Callard, A.J.T. George, J. Stark: Cytokines, chaos, and complexity, Immunity **11**(5), 507–513 (1999)

22.40 J. Hasty, D. McMillen, F. Isaacs, J.J. Collins: Computational studies of gene regulatory networks: In numero molecular biology, Nat. Rev. Genet. **2**(4), 268–279 (2001)

22.41 M. Kabir, N. Noman, H. Iba: Reverse engineering gene regulatory network from microarray data using linear time-variant model, BMC Bioinformatics **11**(1), S56 (2010)

22.42 M.A. Savageau: *Biochemical Systems Analysis. A Study of Function and Design in Molecular Biology* (Addison-Wesley, Reading 1976)

22.43 N. Noman, H. Iba: Inferring gene regulatory networks using differential evolution with local search heuristics, IEEE/ACM Trans. Comput. Biol. Bioinform. **4**(4), 634–647 (2007)

22.44 S. Kimura, K. Ide, A. Kashihara, M. Kano, M. Hatakeyama, R. Masui, N. Nakagawa, S. Yokoyama, S. Kuramitsu, A. Konagaya: Inference of S-system models of genetic networks using cooperative coevolutionary algorithm, Bioinformatics **21**(7), 1154–1163 (2005)

22.45 V.N. Reddy, M.L. Mavrovouniotis, M.N. Liebmant: Petri net representations of metabolic pathways, 1st Int. Conf. Intell. Syst. Mol. Biol. (ISMB '93) (1993) pp. 328–336

22.46 D. Ruklisa, A. Brazma, J. Viksna: Reconstruction of gene regulatory networks under the finite state linear model, Genome Inform. **16**, 225–236 (2005)

22.47 M. Wahde, J. Hertz: Coarse-grained reverse engineering of genetic regulatory networks, Biosystems **55**(1–3), 129–136 (2000)

22.48 J. Vohradský: Neural model of the genetic network, J. Biol. Chem. **276**(39), 36168–36173 (2001)

22.49 E. Mjolsness, T. Mann, R. Castaño, B. Wold: From coexpression to coregulation: An approach to inferring transcriptional regulation among gene classes from large-scale expression data, Adv. Neural Inf. Process. Syst., Vol. 12 (1999) pp. 928–934

22.50 R. Xu, D.C. Wunsch II, R.L. Frank: Inference of genetic regulatory networks with recurrent neural network models using particle swarm optimization, IEEE/ACM Trans. Comput. Biol. Bioinform. **4**(4), 681–692 (2007)

22.51 P. Spirtes, C. Glymour, R. Scheines, S. Kauffman, V. Aimale, F. Wimberly: Constructing Bayesian network models of gene expression networks from microarray data, Proc. Atl. Symp. Comput. Biol. Genome Inf. Syst. Technol. (2000)

22.52 A. Ben-Dor, R. Shamir, Z. Yakhini: Clustering gene expression patterns, J. Comput. Biol. **6**(3/4), 281–297 (1999)

22.53   S. Kimura, S. Nakayama, M. Hatakeyama: Genetic network inference as a series of discrimination tasks, Bioinformatics **25**(7), 918–925 (2005)

22.54   A. Fujita, J. Ricardo Sato, H.M. Garay-Malpartida, P.A. Morettin, M.C. Sogayar, C.E. Ferreira: Time-varying modeling of gene expression regulatory networks using the wavelet dynamic vector autoregressive method, Bioinformatics **23**(13), 16230–1630 (2007)

22.55   B.-E. Perrin, L. Ralaivola, A. Mazurie, S. Bottani, J. Mallet, F. d'AlchéBuc: Gene networks inference using dynamic bayesian networks, Bioinformatics **19**(2), ii138–ii148 (2003)

22.56   N. Sugimoto, H. Iba: Inference of gene regulatory networks by means of dynamic differential bayesian networks and nonparametric regression, Genome Inform. **15**, 121–130 (2004)

22.57   H. Iba, E. Sakamoto: Inferring a system of differential equations for a gene regulatory network by using genetic programming, Congr. Evol. Comput. (CEC2001) (2001) pp. 720–726

22.58   S. Ando, H. Iba: Construction of genetic network using evolutionary algorithm and combined fitness function, Genome Inform. **14**, 94–103 (2003)

22.59   S. Kikuchi, D. Tominaga, M. Arita, K. Takahashi, M. Tomita: Dynamic modeling of genetic networks using genetic algorithm and S-sytem, Bioinformatics **19**(5), 643–650 (2003)

22.60   C. Spieth, F. Streichert, N. Speer, A. Zell: Optimizing topology and parameters of gene regulatory network models from time-series experiments, Proc. Genet. Evol. Comput. Conf. (2004) pp. 461–470

22.61   X. Yao: Evolving artificial neural networks, Proc. IEEE **87**(9), 1423–1447 (1999)

22.62   E. Keedwell, A. Narayanan: Discovering gene networks with a neural-genetic hybrid, IEEE/ACM Trans. Comput. Biol. Bioinform. **2**(3), 231–242 (2005)

22.63   H.W. Ressom, Y. Zhang, J. Xuan, Y.J.H. Wang, R. Clarke: Inference of gene regulatory networks from time course gene expression data using neural networks and swarm intelligence, IEEE Symp. Comput. Intell. Bioinform. Comput. Biol. (CIBCB) (2006) pp. 435–442

22.64   B.S. Mondal, A.K. Sarkar, M.M. Hasan, N. Noman: Reconstruction of gene regulatory networks using differential evolution, Proc. 13th Int. Conf. Comput. Inf. Technol. (ICCIT 2010) (2010) pp. 440–445

22.65   R. Storn: System design by constraint adaptation and differential evolution, IEEE Trans. Evol. Comput. **3**(1), 22–34 (1999)

22.66   R. Storn, K.V. Price: Differential evolution – a simple and efficient heuristic for global optimization over continuous spaces, J. Glob. Optim. **11**(4), 341–359 (1997)

22.67   K.V. Price, R.M. Storn, J.A. Lampinen: *Differential Evolution: A Practical Approach to Global Optimization* (Springer, Berlin, Heidelberg 2005)

22.68   N. Noman, H. Iba: Reverse engineering genetic networks using evolutionary computation, Genome Inform. **16**, 205–214 (2005)

22.69   F. Ghannadian, C.O. Alford, R. Shonkwiler: Application of random restart to genetic algorithms, J. Glob. Optim. **95**(1/2), 81–102 (1996)

22.70   G.N. Beligiannis, G.A. Tsirogiannis, P.E. Pintelas: Restartings: A technique to improve classic genetic algorithms' performance, J. Glob. Optim. **1**, 112–115 (2004)

22.71   C. Janion: Some aspects of the SOS response system – A critical survey, Acta Biochim. Pol. **48**(3), 599–610 (2001)

22.72   B. Michel: After 30 years of study, the bacterial sos response still surprises us, PLoS Biology **3**(7), e255 (2005)

22.73   J.W. Little, S.H. Edmiston, L.Z. Pacelli, D.W. Mount: Cleavage of the *Escherichia coli* lexA protein by the recA protease, Proc. Natl. Acad. Sci. USA **77**(6), 3225–3229 (1980)

22.74   G.C. Walker: Mutagenesis and inducible responses to deoxyribonucleic acid damage in *Escherichia coli*, Microbiol. Mol. Biol. Rev. **48**(1), 60–93 (1984)

22.75   M. Ronen, R. Rosenberg, B.I. Shraiman, U. Alon: Assigning numbers to the arrows: Parameterizing a gene regulation network by using accurate expression kinetics, Proc. Natl. Acad. Sci. USA **99**(16), 10555–10560 (2002)

22.76   U. Alon: Department of Molecular Cell Biology & Department of Physics of Complex Systems, Weizmann Institute of Science, Rehovot, Israel (2012) http://www.weizmann.ac.il/mcb/UriAlon/

22.77   T.S. Gardner, S. Shimer, J.J. Collins: Inferring microbial genetic networks, ASM News **70**(3), 121–126 (2004)

22.78   D.-Y. Cho, K.-H. Cho, B.-T. Zhang: Identification of biochemical networks by s-tree based genetic programming, Bioinformatics **22**(13), 1631–1640 (2006)

23. Structural Pattern Discovery in Protein–Protein Interaction Networks

Tamás Nepusz, Alberto Paccanaro

Most proteins in a cell do not act in isolation, but carry out their function through interactions with other proteins. Elucidating these interactions is therefore central for our understanding of cellular function and organization. Recently, experimental techniques have been developed, which have allowed us to measure protein interactions on a genomic scale for several model organisms. These datasets have a natural representation as weighted graphs, also known as protein–protein interaction (PPI) networks. This chapter will present some recent advances in computational methods for the analysis of these networks, which are aimed at revealing their structural patterns. In particular, we shall focus on methods for uncovering modules that correspond to protein complexes, and on random graph models, which can be used to de-noise large scale PPI networks. In Sect. 23.1, the state-of-the-art techniques and algorithms are described followed by the definition of measures to assess the quality of the predicted complexes and the presentation of a benchmark of the detection algorithms on four PPI networks. Section 23.2 moves beyond protein complexes and explores

other structural patterns of protein–protein interaction networks using random graph models.

Part D | 23

Biological processes in our cells are carried out by groups of biochemical compounds interacting with each other in elaborate ways. Proteins are probably the most important class of these compounds, consisting of chains of amino acids that are folded into a wide variety of complex fibrous or spherical shapes. Interactions between proteins are facilitated by specific domains on the amino acid chains that allow them to bind to each other. Some of these interactions are short-lived (i.e., a protein may be carrying another protein from one part of the cell to another), while others are stable and facilitate the formation of long-lived structures called protein complexes. The accurate knowledge of interactions and

complexes is crucial to deepen our understanding of biological processes and to assign a cellular function to proteins in the cell. It is thus no surprise that the web of interactions between proteins in living organisms has been the focus of scientific research for years in experimental [23.1–8] and computational (in silico) studies alike [23.9–15].

A natural representation of the interactions between proteins is an undirected graph, where vertices of the graph correspond to individual proteins and the edges between vertices denote pairwise interactions. Such networks are usually inferred via high-throughput experimental techniques such as yeast-2-hybrid (Y2H) experiments or co-affinity purification followed by mass spectrometry (AP-MS). Most frequently, the edges of this network are also labeled with confidence scores that represent how likely the existence of a given interaction is; these confidence scores are usually derived from the same experimental procedure that was used to produce the list of interacting protein pairs. Partial protein–protein interaction networks are now available for all major model organisms, ranging from the yeast *Saccharomyces cerevisiae* [23.1, 2, 7, 8, 12] to *Homo sapiens* [23.5, 6], opening up possibilities to analyze them using machine-learning algorithms in order to uncover their interesting structural patterns.

This chapter will first focus on the most intensively studied structural pattern, namely the modular architecture of PPI networks. We will see that PPI networks contain subsets of proteins that are connected to each other more densely than to the rest of the network and that these subsets show a good correspondence with protein complexes. In Sect. 23.1, we describe the state-of-the-art techniques and algorithms for detecting protein complexes in PPI networks, followed by the

definition of measures to assess the quality of the predicted complexes. To conclude the section, we present a benchmark of the described protein complex detection algorithms on four PPI networks of the yeast *Saccharomyces cerevisiae*, a well-known model organism in systems biology, genetics, and cell biology.

In Sect. 23.2, we move beyond protein complexes and explore other structural patterns of PPI networks using random graph models. First, we introduce the most promising random graph models for modeling PPI networks and describe how they can be fitted to a specific observed interaction dataset. After that, we will evaluate the predictive power of these graph models using standard machine-learning approaches (ROC and precision-recall curves) and present a case study that applies the degree-corrected stochastic block-model [23.16] to predict putative novel interactions between the proteins of *Homo sapiens*. Finally, we validate some of these predictions with the help of hand-curated and reviewed biological databases, such as the gene ontology (GO) [23.17] and the Kyoto Encyclopedia for Genes and Genomes (KEGG) [23.18].

23.1 Detecting Protein Complexes from Protein–Protein Interaction Networks

Protein complexes are groups of two or more associated polypeptide chains that bind together in a living organism and form a stable molecular structure that carries out a given biological function. They are the key components of many biological processes, and scientists generally think about a cell as a collection of modular protein complexes, each of which is responsible for an independent biological process. For instance, ribosome, one of the most well-known protein complexes, creates other proteins from amino acids and an RNA sequence that encodes the structure of the protein being built. This is achieved by a complex interplay of more than 70 proteins in eukaryotes [23.19]. The accurate description of protein complexes is thus one of the first steps towards understanding how different biological functions are carried out by the cells of a living organism.

Proteins in a complex are linked together by a dense network of PPI, and although there are quite a few interactions between proteins of different complexes, it is generally the case that the interaction density between members of the same complex is much higher than the interaction density between members of different complexes.

23.1.1 Complex Detection by Traditional Graph Clustering

The earliest algorithms for detecting protein complexes from pairwise interactions were mostly based on the application of a traditional graph clustering method to the graph representation of the set of interactions. The classical problem of graph clustering aims to decompose the vertex set V of a graph $G(V, E)$ into disjoint subsets $V_1, V_2, \ldots, V_k$ such that the intra-cluster edge densities (i. e., the number of edges within $V_1, V_2, \ldots, V_k$ divided by the number of theoretically possible edges within these sets) are maximal and the number of edges between clusters is minimal. However, since these methods were not designed specifically with PPI data in mind, they may fail to capture some peculiar properties of PPI networks. One such property is the fact that a protein may participate in more than one complex; in other words, protein complexes are not disjoint like clusters in the result of a graph clustering algorithm. Another property is that not all proteins participate in protein complexes, while graph clustering methods tend to classify each vertex into one of the clusters even when there is not enough evidence to support the membership of a vertex in *any* of the clusters. Despite this disadvantage, traditional graph clustering techniques are still being used to identify protein complexes, and no review would be complete without them. In this section, we will describe two of the most successful ones: restricted neighborhood search clustering (RNSC) [23.20] and Markov clustering (MCL) [23.21, 22].

Restricted Neighborhood Search Clustering

RNSC [23.20, 23] is a cost-based local search algorithm that starts from an initial random clustering and iteratively moves a randomly selected node from one cluster to another in order to improve the result. The quality of a clustering C is measured by two functions, the *naive* and the *scaled* cost function. Earlier stages of the algorithm use the naive cost function as it is faster to calculate, while later stages switch to the scaled cost function. The naive cost function for a graph $G(V, E)$ and its clustering C is defined as

$$C_n(G, C) = \frac{1}{2} \sum_{v \in V} \left[\#^1_{\text{out}}(G, C, v) + \#^0_{\text{in}}(G, C, v) \right],$$

$$(23.1)$$

where $\#^1_{\text{out}}(G, C, v)$ is the total number of edges incident on v that lead out of v's cluster in the clustering C, and $\#^0_{\text{in}}(G, C, v)$ is the number of vertices in the cluster of v that are *not* adjacent to v. Intuitively, a good clustering should contain only a small number of intercluster edges (which decreases the first term in the sum), and a large number of intra-cluster edges (which decreases the second term), therefore this cost function seems plausible. However, note that the cost function is biased towards small clusters. *King* [23.23] noted that for graphs with density less than 0.5, a single giant cluster including all the points will always have a higher cost than a clustering that puts every vertex into its own cluster. The scaled cost function resolves this problem by weighing the contribution of each cluster [23.23],

$$C_s(G, C) = \frac{n-1}{3} \sum_{v \in V} \frac{1}{N(v) \cup C_v}$$
$$\cdot \left[\#^1_{\text{out}}(G, C, v) + \#^0_{\text{in}}(G, C, v) \right], \quad (23.2)$$

where $N(v)$ is the neighborhood of v (i.e., the set of nodes adjacent to v) and C_v is the cluster in which v resides.

Local search algorithms typically suffer from the problem of getting stuck in suboptimal local minima. RNSC resolves this problem by occasionally inserting diversification moves into the optimization process.

A diversification move either places a subset of vertices into randomly selected clusters or destroys a cluster entirely and distributes its vertices among the remaining clusters. Both moves usually increase the cost of the clustering temporarily, but it may help the algorithm to climb out from local minima and allow it to decrease the cost by further optimization steps in later stages.

The RNSC algorithm also employs a tabu search heuristic [23.24] to prevent cycling around a local minimum. This is implemented by a fixed capacity first-in, first-out queue of vertices, called the *tabu queue*. When a vertex is moved, it enters the queue at one end and pushes all the vertices already in the queue towards the other end. Vertices within the tabu queue are not allowed to be moved between clusters until they leave the queue at the other end – which they will eventually do as the size of the tabu queue is fixed. The algorithm terminates after a fixed number of moves and it is restarted a fixed number of times from different random configurations. The final result will then be the clustering with the smallest scaled cost found among all the iterations in all trials.

The algorithm so far is a generic graph clustering algorithm. To incorporate domain-specific knowledge, *King* et al. [23.20] extended the algorithm by a postprocessing stage that excludes clusters that are smaller than a given size or sparser than a given density threshold; their recommendations keep predicted protein complexes containing at least four proteins and having a density larger than 0.7. In the final stage, the proteins in each cluster are evaluated for functional homogeneity using their annotations from the GO [23.17], and only those clusters are kept that contain at least one overrepresented gene ontology annotation. These steps allow some of the proteins to be left out entirely from protein complexes.

Although RNSC was shown to perform favorably on protein interaction data for the yeast *Saccharomyces cerevisiae* [23.20], it should be noted that the algorithm has a large number of parameters (including the search length, the number of repetitions, the frequency and length of diversification steps, the maximum number of clusters to consider, and the size of the tabu queue), all of which have to be tuned to the specific PPI network at hand. Even with careful tuning, the algorithm often converges to wildly different clusterings in different runs with the same parameter settings, which may require the evaluation of many different sets of predicted complexes to find the ones worthy of further small-scale follow-up experiments.

Part D | 23.1

Markov Clustering

MCL (MCL) [23.22] is based on the idea that a random walk on the vertices of the graph that is started from an arbitrary node is more likely to spend more time walking within a densely connected region (i. e., a cluster) of the graph than walking between different regions, assuming that the moves are only allowed along the edges and the random walker cannot jump arbitrarily between vertices. MCL deterministically approximates transition probabilities of such random walks by iteratively using two operators on row-stochastic matrices (also called Markov matrices, hence the name of the algorithm) until a well-defined set of clusters emerge.

A formal, step-by-step description of the algorithm on undirected, unweighted graphs without isolated vertices is as follows:

1. Let $M = (m_{ij})$ be the matrix of random walks on the graph $G(V, E)$; in other words, let m_{ij} be the probability that a random walker at node i moves to node j in one step. Since the random walker chooses each incident edge of node i with equal probability, $m_{ij} = A_{ij}/d_i$, where A_{ij} is 1 if vertices i and j are connected and zero otherwise, and d_i is the number of edges incident on vertex i. Note that each row of M sums up to 1, so M is row-stochastic.
2. Calculate the two-step random walk matrix by multiplying M with itself. This step is called the *expansion* step.
3. Inflate the probabilities of high-probability paths and deflate the probabilities of low-probability paths in M^2. This is achieved by raising each element of M^2 to a power $\gamma > 1$ and then re-normalizing the row sums of the result to 1 again to make the matrix row-stochastic. The re-normalized matrix is then stored in M'. This step is called the *inflation* step.
4. If the difference between M' and M is less than a given threshold, the algorithm has converged. Otherwise we assign M' to M and continue from step 2 with another expansion-inflation cycle.

The result of the above process is a doubly idempotent matrix M that remains the same after an expansion-inflation iteration. It can be proven that the above procedure converges quadratically around the equilibrium state and it is conjectured that the process always converges if the input graph is symmetric [23.22]. In practical applications, the convergence is noticeable after 310 iterations. It can also be shown that the graph constructed from the equilibrium state matrix M consists of different connected directed com-

ponents, each component having a star-like structure. The components are then interpreted as the clusters of the original graph.

The only parameter of the algorithm in the above description is the inflation exponent γ, whose values are typically between 1.2 and 5. The exponent controls the granularity of the obtained clustering; values close to 1 yield a few large clusters, while higher values generate a larger number of dense clusters. In practice, the algorithm also adds a pruning step after every expansion-inflation iteration, which removes entries close to zero from M; this ensures that the matrix remains sparse and matrix multiplications can be performed efficiently, enabling MCL to scale up to graphs containing millions of vertices and edges. The algorithm can also be extended naturally to directed and weighted graphs. In directed graphs, we consider only the outgoing edges when building the initial random walk matrix, and add loop edges to each vertex to ensure the convergence of the algorithm. Weights can be incorporated by making the random walk biased towards edges with larger weights; the probability of moving from vertex i to vertex j in a single step then becomes proportional to w_{ij} instead of A_{ij}, while keeping M row-stochastic. A post-processing step is commonly used to discard clusters with only one or two members or a density that is smaller than the expected density of real protein complexes.

MCL was first applied to identify sets of evolutionally related proteins [23.21] and it quickly became very popular in the bioinformatics community. It was subsequently applied for the identification of yeast protein complexes from high-throughput data by *Krogan* et al. [23.7]. However, note that MCL still does not solve the problem of overlapping protein complexes as it puts every protein into one and only one of the clusters.

23.1.2 Overlapping Clustering Methods for PPI Networks

Besides traditional graph clustering techniques, researchers have also proposed alternative algorithms that have been tailored to PPI networks. A common property of these methods is that they are able to put one protein into more than one predicted complex, or to leave out a protein entirely from all the predicted complexes if there is not enough evidence to support the assignment of that protein into any of the complexes. In this section, we will describe three such algorithms: molecular complex detection

(MCODE) [23.25], the k-clique percolation method (also called CFinder [23.26]), and clustering by overlapping neighborhood expansion (ClusterONE [23.27]).

Molecular Complex Detection

The MCODE [23.25] algorithm takes a different approach to classical graph clustering algorithms: instead of partitioning the entire vertex set of the graph, it tries to identify locally dense regions and disregards those parts of the graph that are not likely to belong to any of the clusters. The algorithm consists of three steps: vertex weighting, complex prediction, and post-processing. The vertex weighting step assigns a weight to all the vertices that quantifies how likely a given vertex is to be a core of a putative protein complex. The complex prediction stage selects highly weighted vertices from the previous step and grows protein complexes in its neighborhood. The post-processing step expands the protein complexes found in the previous step to allow overlaps between them.

The keystone of the algorithm is the vertex weighting scheme, which is based on two components: 1) the so-called core-clustering coefficient of the vertices and 2) the highest k-core of the immediate neighborhood of the vertices. To understand these measures, we need a few definitions first.

Definition 23.1 Clustering Coefficient
The clustering coefficient C_i of a vertex i is defined as the number of edges between neighbors of a vertex, divided by the number of theoretically possible edges between neighbors. Formally, given a vertex i with k_i neighbors and m edges between those neighbors, the clustering coefficient C_i is given by

$$C_i = \frac{2m}{k_i(k_i - 1)}.$$
(23.3)

One can think about the clustering coefficient as a local density measure; in fact, it is exactly the density of the subgraph spanned by the neighbors of a given vertex. Intuitively, vertices with high clustering coefficients are good candidates for proteins to be included in protein complexes as they have a densely connected neighborhood. However, many PPI networks contain hub proteins that interact with a diverse set of other proteins, potentially also including some that have only a single connection (the one towards the hub protein). These dangling links may rapidly decrease the clustering coefficient of a hub protein that could nevertheless have connections towards another, densely connected

region as well. *Bader* and *Hogue* [23.25] resolved this problem by introducing the *k-core indices* and the *core-clustering coefficient*.

Definition 23.2 k-Core
The k-core of a graph $G(V; E)$ is a subset of vertices $V_0 \subseteq V$ such that the degrees of the subgraph spanned by V_0 are all larger than or equal to k.

Definition 23.3 k-Core Index
The k-core index of a vertex v in a graph G is the largest k such that v is the member of the k-core of G but not the $(k+1)$-core of G. The k-core index of a graph G is the largest k such that the k-core of G is not empty.

It is easy to see that the k-core index of vertices with d links is at most d, and the k-core of G is smaller than or equal to the maximal vertex degree in the graph.

Definition 23.4 Core-Clustering Coefficient
The core-clustering coefficient of a vertex v in a graph G is the density of the highest k-core of the subgraph spanned by v and its neighbors.

Note that the core-clustering coefficient is similar to the clustering coefficient measure in the sense that it is also a local density measure, but it does not suffer from the negative effect of dangling links. Given a highly connected hub protein v, the highest k-core of its neighborhood will not include the dangling links, and the density of this vertex set will not be affected negatively. The MCODE algorithm uses the product of the core-clustering coefficient of v and the highest k-core index in the immediate neighborhood of v as the weight of vertex v.

Once the weights of the vertices have been determined, MCODE starts to grow protein complexes from highly weighted vertices as listed:

1. Select the vertex with the largest weight in the graph that has not yet been included in any of the predicted complexes. This vertex is called the *seed vertex* and it is the only initial member of the complex being grown. We denote the seed vertex with v and its weight with w_v.
2. Add those vertices in the neighborhood of v to the complex being grown that are not added to any other complex yet and have a weight larger than $(1 - \alpha)w_v$, where α is called the *vertex weight percentage*. Typical values of α are between 0 and 0.2.

3. Recursively keep on adding vertices adjacent to the complex being grown and not being in any other complex yet if their weight is larger than $(1 - \alpha)w_v$.
4. If there are no more vertices that have not been considered yet and have not been added to any of the predicted complexes, go back to step 1. Otherwise, start the post-processing stage.

The complex prediction stage gives us a list of nonoverlapping complex candidates. Every protein in the network is assigned to *at most* one of the complex candidates, but it may also have happened that a protein was not assigned to any of the complexes. The final post-processing stage first discards predicted complexes without a 2-core (i. e., a subgraph where all the degrees are larger than or equal to 2), then tries to introduce overlaps between the complexes by an operation called *fluffing* and trims unnecessary proteins from complexes by another step called *haircut*. Both the fluffing and the haircut steps are optional, but the former is required if we wish to see overlaps between predicted complexes.

The fluffing step considers every vertex in every predicted complex one by one. For each vertex v, the process adds the neighbors of v to the complex containing v if the density of the subgraph spanned by v and its neighbors is above the value of the fluff parameter of the algorithm. At this stage, it is possible that a vertex u becomes part of two or more complexes if it is part of two or more dense neighborhoods. The haircut step removes vertices from complexes that are connected to the rest of the complex with only a single edge, effectively reducing each complex to its 2-core.

One of the advantages of the MCODE algorithm is that it can be run in an exploratory analysis mode where the dense neighborhood of a selected seed vertex is extracted from the network. This is achieved by growing a single cluster from the seed vertex and performing the additional fluffing or haircut operations on it. This mode is very useful for researchers who are interested in the role of a particular protein within the cell and its interactions with other proteins. On the other hand, MCODE tends to be very strict in the sense that proteins are usually very tightly connected in MCODE clusters and it usually misses smaller molecular complexes, especially if the experimental data is noisy and there are a lot of missing interactions in the input data. It is also not able to take edge weights into account, therefore PPI networks with associated confidence information for each interaction have to be binarized and low-confidence edges have to be discarded before an MCODE analy-

sis, adding the optimal confidence threshold to the set of MCODE parameters to be tuned.

Complexes from k-Cliques: the CFinder Method

CFinder [23.26] was one of the first general-purpose overlapping clustering algorithms, which was later also applied successfully to biological networks [23.28]. The cluster definition of the algorithm is based on k-cliques, i. e., sets of vertices of size k where any pair of vertices is connected directly by an edge. The algorithm first constructs the k-clique reachability graph of the input graph G. The vertices of the k-clique reachability graph represent the k-cliques of the original graph, and two vertices in the reachability graph are connected by an edge if the corresponding k-cliques share at least $k - 1$ vertices. The clusters of the original graph are then derived from the connected components of the k-clique reachability graph such that each cluster becomes the union of k-cliques in one connected component of the reachability graph. This approach naturally includes overlaps between vertices of the original graph as a vertex may be a member of multiple k-cliques, and these k-cliques can end up in different connected components of the reachability graph. The only parameter of the algorithm is the value of k, which controls the granularity of the clustering; larger values of k yield densely connected clusters. $k = 2$ is equivalent to finding the connected components of the input graph, therefore k is always larger than 2 in practical applications of the algorithm.

One can easily recognize that it is enough to enumerate the maximal cliques of the original network that have at least k vertices instead of finding all k-cliques, as every subset of a maximal clique is also a clique, therefore a maximal clique of size n will be mapped to a connected subgraph consisting of $\binom{n}{k}$ vertices in the k-clique accessibility graph. Such subgraphs can be shrunk into a single vertex that will represent the whole maximal clique without affecting the connectivity properties of the k-clique accessibility graph.

The original CFinder publication [23.26] was not concerned with edge weights. Later on, a weighted extension of CFinder was proposed [23.29], which introduces a second parameter I, acting as an intensity threshold for the detected cliques. In the weighted CFinder algorithm, the product of the edge weights in a clique must exceed I in order to include that clique in the accessibility graph. This is computationally more prohibitive than the unweighted variant as we have to enumerate all k-subcliques of maximal cliques and check their intensities explicitly. In fact, the reference implementation of CFinder did not provide results of a PPI network. Nevertheless, the CFinder algorithm is still a very promising approach to detecting complexes from unweighted PPI networks and it is also applicable to weighted PPI data after the selection of an appropriate weight threshold.

Clustering by Overlapping Neighborhood Expansion

Clustering by overlapping neighborhood expansion (or ClusterONE in short) [23.27] is a recent overlapping clustering algorithm that grows clusters from seed proteins, similarly to the MCODE algorithm [23.25], and also allows researchers to explore the densely connected region around a given seed protein. The key difference is that ClusterONE explicitly strives to make use of the confidence scores (weights) that are usually associated to interactions in PPI datasets instead of discarding them after a thresholding step. The core of the algorithm is a quality measure called *cohesiveness*, which quantifies how likely it is for a set of proteins to be a good complex candidate based on structural considerations. The definition of cohesiveness for a set of vertices $V_0 \subseteq V$ in a graph $G(V, E)$ is

$$f(V_0) = \frac{w^{\text{in}}(V_0)}{w^{\text{in}}(V_0) + w^{\text{bound}}(V_0) + p|V_0|} , \quad (23.4)$$

where $w^{\text{in}}(V_0)$ is the total weight of internal edges in group V_0, $w^{\text{bound}}(V_0)$ is the total weight of edges that connect V_0 to the rest of the network, and $p|V_0|$ is a penalty term. The exact form of the cohesiveness function is based on the following considerations.

1. A good protein complex candidate should have many edges with large weights within the complex, hence $w^{\text{in}}(V_0)$ should be as large as possible.
2. Similarly, a good candidate should have only a few edges that connect the complex to the rest of the network, therefore $w^{\text{bound}}(V_0)$ should be small.
3. In PPI networks derived from experimental procedures, it is usually the case that not all the interactions are included in the network; some connections may be missing due to the limitations of the experiment. In the extreme case, it may happen that protein A in the PPI network is connected only to a putative complex with a single interaction, but in reality, protein A also interacts with other proteins not in the complex. Adding protein A to the complex is plausible if we know for sure that A interacts with a member of the complex and *with no other proteins*, but it is not justified if we know in

advance that many connections of A are likely to be uncharted yet. The penalty term $p|V_0|$ accounts for this by assuming that every protein has p missing connections in addition to the ones in the PPI network. We note that this definition could easily be extended to employ different p values for different proteins based on biological assumptions, thus a well-studied protein may have a lower p value since it is less likely to possess undiscovered interactions.

Note that $f(V_0)$ has a minimum value of zero (when there are no internal edges) and a maximum value of 1 (when there are no external edges, there is at least a single internal edge and $p = 0$), which leads to an intuitive probabilistic interpretation. Suppose we are selecting a random edge from the graph such that selection probabilities are proportional to edge weights. In this case, $f(V_0)$ is the conditional probability of selecting an edge with *two* internal endpoints, given that at least *one* endpoint is internal. An additional pleasant property of $f(V_0)$ is that when $f(V_0) > 1/3$, vertices of the subgraph have more internal weight than external weight on average, satisfying the conditions of being a community in the weak sense [23.30].

ClusterONE builds on the concept of cohesiveness and consists of two major steps: 1) finding a set of groups with high cohesiveness and 2) merging redundant cohesive groups and post-processing the results. However, we also note that the algorithm is not tied to the specific form of the cohesiveness function we introduced above: the same framework may be used in conjunction with any alternative local quality measure defined for sets of vertices in a graph [23.31–33].

A straightforward way to find a subgraph with a high cohesiveness is to adopt a greedy strategy: starting from a single seed vertex, one can extend the group one vertex at a time so that the newly added vertex always increases the cohesiveness of a group as much as possible. Removals are also allowed if removing a vertex from the group increases its cohesiveness. This approach yields locally optimal cohesive groups in the sense that the cohesiveness of the group cannot be improved any more by adding or removing a single vertex. The procedure to grow a locally optimal cohesive group from a seed vertex is as listed:

1. Select a seed vertex v_0 and let $V_0 = \{v_0\}$. Set the step number $t = 0$.
2. Calculate the cohesiveness of V_t and let $V_{t+1} = V_t$.

3. For every external vertex v incident on at least one boundary edge, calculate the cohesiveness of $V' = V_t \cup \{v\}$. If $f(V') > f(V_{t+1})$, let $V_{t+1} = V'$.

4. For every internal vertex v incident on at least one boundary edge, calculate the cohesiveness of $V'' = V_{t+1} \setminus \{v\}$. If $f(V'') > f(V_{t+1})$, let $V_{t+1} = V''$.

5. If $V_t \neq V_{t+1}$, increase t and return to step 2. Otherwise, declare V_t a locally optimal cohesive group.

This process is repeated from different seeds in order to form multiple, possibly overlapping groups as follows. We first select the vertex of G with the highest degree as the first seed and grow a locally optimal cohesive group around it. After a locally optimal cohesive group is found, the next seed vertex will always be the one that has the highest degree among those that have not yet been included in any other cohesive group. It may happen that the original seed vertex is not included in the detected cohesive group (because it was removed earlier in step 4); in this case, the vertex is declared an outlier. The process continues until all vertices are either included in at least one cohesive group or declared as outliers.

As an illustration of the above steps, let us return to the example graph in Fig. 23.1. Assuming $p = 0$ (i.e., no presumed undiscovered connections), the cohesive-

ness of the marked set is 10/15. In steps 3 and 4, the algorithm can either extend the current set by adding C, F, or G, or contract the set by removing A, B, D, or E. The best move for the algorithm is to add C to the set, since it converts three boundary edges to internal ones (in other words, it decreases w^{bound} by 3 and increases w^{in} by 3) and does not bring in any new boundary edges. After adding C, the cohesiveness of the group increases to 13/15 and the group becomes locally optimal, as adding F would result in a cohesiveness of 14/17 and adding G would yield 14/18.

The procedure outlined above will generate locally optimal cohesive groups that may overlap with each other to some extent. While the primary purpose of this procedure was, indeed, to generate overlapping clusters, some of the generated clusters will differ only in one or two proteins, which further complicates experimental validation of the results. It is, therefore, desirable to merge pairs of highly overlapping cohesive groups before declaring the clusters final. The algorithm uses the overlap score ω proposed in [23.25] to quantify the overlap between two protein sets A and B,

$$\omega(A, B) = \frac{|A \cap B|^2}{|A||B|} . \tag{23.5}$$

Pairs of groups having ω above a specified threshold are then merged in the predicted set of complexes. Finally, clusters smaller than a predefined size threshold or having a density less than a predefined density threshold are discarded.

23.1.3 Assessing the Quality of Complex Predictions

In this section, we compare the different techniques outlined in the previous section by benchmarking them on four PPI datasets for yeast. We first describe the datasets we are about to use in our tests and the set of reference complexes, then define some quality measures that will be employed to quantify how well a predicted set of complexes represents the reference complexes. Finally, we will test each of the algorithms and compare the quality scores obtained. We used the original implementations released by the authors of these algorithms and we followed the published protocols closely. Predicted complexes containing less than three proteins were discarded from the results of each algorithm unless the originally published protocol suggested otherwise: in the case of the RNSC algorithm, we therefore used a size threshold of 4 instead of 3.

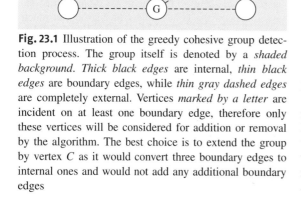

Fig. 23.1 Illustration of the greedy cohesive group detection process. The group itself is denoted by a *shaded background*. *Thick black edges* are internal, *thin black edges* are boundary edges, while *thin gray dashed edges* are completely external. Vertices *marked by a letter* are incident on at least one boundary edge, therefore only these vertices will be considered for addition or removal by the algorithm. The best choice is to extend the group by vertex C as it would convert three boundary edges to internal ones and would not add any additional boundary edges

Datasets and Reference Complexes

In our experiments, we have used the PPI networks for the yeast *Saccharomyces cerevisiae*, as published by *Krogan* et al. [23.7] and *Gavin* et al. [23.8], as well as the dataset of *Collins* et al. [23.12], which merges the Krogan and Gavin datasets using computational methods. The Krogan dataset was published in two variants in the original paper: the core dataset contained only highly reliable interactions (all having a confidence score larger than 0.273), while the extended dataset contained more interactions with less overall reliability (the smallest confidence score was 0.101). Both variants of the Krogan dataset were tested separately. Self-interactions and isolated proteins were discarded.

All the datasets that we used contained confidence values associated to each PPI. For the Gavin dataset, we re-scaled the original socio-affinity scores to the range [0, 1] for fairness to those algorithms that are not prepared for edge weights outside the range [0, 1]. Finally, in order to apply those algorithms that do not support edge weights (namely MCODE, RNSC, and CFinder), we first had to binarize the networks. This was done using the threshold values for the weights that were originally suggested by the authors of the datasets.

The set of reference complexes were extracted from the most recent version of the MIPS catalogue of protein complexes [23.34] (dated 18 May 2006). The MIPS dataset (Munich Information Center for Protein Sequences) is organized hierarchically: complexes in MIPS may consist of subcomplexes extending to at most five hierarchy levels deep. We considered MIPS categories containing at least 3 and at most 100 proteins as protein complexes, since these are plausible lower and upper bounds for protein complexes that are expected to appear in experimental PPI datasets. We also excluded MIPS category 550 and all its descendants, as these categories correspond to protein complexes that were predicted by computational methods and have not been confirmed experimentally.

In some cases, not all of the proteins in a given MIPS complex were found in a given benchmark dataset, and we cannot expect a protein complex detection algorithm to predict a complex that is only partially represented in the input data. To this end, MIPS complexes where less than half the proteins were represented in a given dataset were removed from the reference set for that dataset.

Clustering-Wise Sensitivity, Positive Predictive Value and Accuracy

In order to assess the performance of overlapping and nonoverlapping clustering algorithms within the same benchmark, we need to be able compare an arbitrary set of predicted complexes with a predefined gold standard complex set without relying on the assumption that a protein may belong to only one predicted and only one reference complex. The comparison is made difficult by the fact that often a match between a predicted complex and a gold standard one is only partial. Moreover, a gold standard complex can have a (partial) match with more than one predicted complex, and vice versa.

One of the first attempts to compare nonoverlapping and overlapping protein complex detection methods using a general quality score was published by *Brohée* and *van Helden* [23.35]. They introduced several measures, three of which will be used in our comparisons as well. These scores are the *clustering-wise sensitivity* (CWS), the *clustering-wise positive predictive value* (PPV) and the *geometric accuracy* (Acc).

CWS and PPV are based on the confusion matrix T between a predicted complex set and the set of reference complexes. Given n reference and m predicted complexes, let t_{ij} denote the number of proteins that are found both in reference complex i and predicted complex j, and let n_i denote the number of proteins in reference complex i. The CWS and PPV measures are then defined as

$$\text{CWS} = \frac{\sum_{i=1}^{n} \max_{j=1}^{m} t_{ij}}{\sum_{i=1}^{n} n_i}, \tag{23.6}$$

$$\text{PPV} = \frac{\sum_{j=1}^{m} \max_{i=1}^{n} t_{ij}}{\sum_{j=1}^{m} \sum_{i=1}^{n} t_{ij}}. \tag{23.7}$$

The Acc is the simply the geometric mean of CWS and PPV

$$\text{Acc} = \sqrt{\text{CWS} \times \text{PPV}}. \tag{23.8}$$

Note that the CWS and PPV measures are very similar but not entirely symmetric; the denominator of CWS includes the total size of reference complexes, while the denominator of PPV is simply the sum of elements in the confusion matrix T. When overlaps between complexes are allowed, n_i may be larger than, smaller than, or equal to the sum of row i in the confusion matrix (which we will denote with t_{i*} from now on), thus the denominator of CWS may also be larger than, smaller than, or equal to the denominator of PPV. This leads to a somewhat counter-intuitive behavior for these mea-

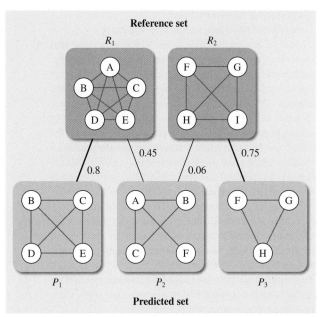

Fig. 23.2 Illustration of the maximum matching ratio between a reference and a predicted complex set. R_1 and R_2 are members of the reference set, while P_1, P_2 and P_3 are three predicted complexes. An edge connects a reference complex and a predicted complex if their overlap score is larger than zero. The maximum matching is shown by the *thick edges*. Note that P_2 was not matched to R_1 since P_1 provides a better match with R_1. The maximum matching ratio in this example is $(0.8 + 0.75)/2 = 0.775$

sures: in the case of a perfect agreement between the reference complexes and the set of predicted complexes, the CWS is equal to 1, but the positive predictive value may be lower when overlaps are present between the reference complexes. Since PPV is a component of the Acc score, this has a negative impact on the perceived performance of overlapping clustering algorithms when they are measured by the Acc score alone. It is even possible to construct a set of reference complexes such that a perfect clustering algorithm receives a lower score than a dummy algorithm that places every protein in a separate cluster; an example of this is given in the supplementary materials of [23.27].

The Maximum Matching Ratio

Recognizing the shortcomings of the Acc measure, *Nepusz* et al. introduced an alternative quality score called the *maximum matching ratio* (MMR) that is designed specifically for the comparison of overlapping clusterings [23.27]. The measure is based on the overlap score (23.5) and a maximum matching in a bipartite graph.

To calculate the MMR, we begin by building a bipartite graph where the two sets of nodes represent the reference and predicted complexes, respectively, and an edge connecting a reference complex with a predicted one is weighted by the overlap score between the two (Fig. 23.2). We then select the maximum weighted bipartite matching on this graph; that is, we choose a subset of edges such that each predicted and reference complex is incident on at most one edge in the chosen set and the sum of the weights of such edges is maximal. In this way, the chosen edges represent an optimal assignment between reference and predicted complexes in a way that no reference complex is assigned to more than one predicted complex and vice versa. The MMR between the reference and the predicted complex set is then given by the total weight of the selected edges divided by the number of reference complexes. This ratio measures how accurately the predicted complexes represent the reference complexes.

Note that the maximum matching ratio divides the weight of the maximum matching by the number of reference complexes instead of the number of predicted complexes. This is motivated by the fact that the gold standard sets are very often incomplete [23.36], therefore a predicted complex that does not match any of the reference complexes may belong to a valid but previously uncharacterized complex.

Figure 23.2 shows an example to illustrate this concept. The maximum matching ratio offers a natural, intuitive way to compare predicted complexes with a gold standard, and it explicitly penalizes cases when a reference complex is split into two or more parts in the predicted set, as only one of its parts is allowed to match the original reference complex.

Benchmarks

The standard procedure for evaluating the performance of a machine-learning algorithm on some dataset starts by dividing the dataset into a training and a testing set. The parameters of the algorithm are then tuned on the training set, and the optimal parameters are used to calculate the final performance score of the algorithm on the testing set. However, this procedure assumes that the input dataset can be naturally decomposed into problem *instances* such that 1) each instance is a complete input of the machine-learning algorithm on its own and 2) each instance is independent of others. Unfortunately, neither of these assumptions hold for graph clustering algorithms in biological contexts, where we usually have a single network to work with. The network itself is indivisible, since removing a predefined fraction of edges from a network changes its structural properties in a way that affects the outcome of a clustering algorithm significantly; in other words, removing some edges from a network is similar to adding noise to a feature vector in a standard machine-learning algorithm rather than to putting a set of problem instances aside in a testing set. Therefore, we cannot simply divide the input data into training and testing sets, and we cannot turn to the well-established methodology of k-fold cross-validation to evaluate the performance of a clustering method and to avoid the over-optimization of algorithm parameters to a given dataset or a given quality score [23.37]. To this end, we adopted the following counter-measures against overfitting in our benchmarks:

1. We tested each of the algorithms on four different datasets: three high-throughput experimental datasets [23.7, 8] and a computationally derived network that integrates the results of these studies [23.12].
2. We used more than one quality score to assess the performance of each algorithm: the fraction of matched complexes with a given overlap score threshold ($\omega \geq 0.25$), the Acc [23.35], and the maximum matching ratio that we proposed. These scores were combined into a composite quality score by taking the sum of the three individual scores.
3. For each algorithm *except* ClusterONE, the final results were obtained after having optimized the algorithm parameters to yield the best possible re-

sults as measured by the maximum matching ratio on the MIPS gold standard complex set [23.34], while the results for ClusterONE were obtained with one common parameter setting for all the datasets. Therefore, the scores of ClusterONE represent its performance when the method is adapted to a wider problem domain (i.e., detecting overlapping protein complexes from high-throughput experimental PPI networks in general), while the scores of other algorithms measure their performance when they are optimized to a *specific* dataset. It is thus rightly expected that these latter scores are optimistic estimates.

Figure 23.3 shows the results of our benchmarks, indicating that ClusterONE outperforms alternative approaches on the tested datasets, matching more complexes with a higher accuracy and providing a better one-to-one mapping with reference complexes in almost all the datasets. Only for the Gavin dataset did MCL achieve a better accuracy but a lower MMR, which can be explained by the bias of the accuracy score towards nonoverlapping clusterings, as discussed above. Also note that ClusterONE is more tolerant to the presence of low-confidence edges than other methods, as its maximum matching ratio on the Krogan extended dataset (which contains a high number of low-confidence interactions) is better than the maximum matching ratio of the runner-up MCL on the Krogan core dataset, which contains highly confident interactions only.

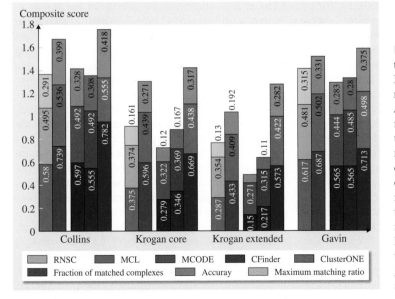

Fig. 23.3 Benchmark results of the tested algorithms on the MIPS gold standard. *Colors* correspond to the various algorithms, *shades of the same color* denote the individual components of the composite score of the algorithm (*dark* = fraction of matched complexes, *medium* = geometric accuracy, *light* = maximum matching ratio). The *numbers on the bars* show the exact scores for each component of the composite score. The total height of each column is the value of the composite score for a given algorithm on a given dataset. Larger scores are better

Part D | 23.1

So far in this chapter we have focused our analysis on the structure of PPI networks to the discovery of those densely connected submodules that correspond to protein complexes. These dense subgraphs are one of the peculiar structural properties of PPI networks; however, they are not the only ones.

In Sect. 23.2 we will try to capture the broader picture and provide models for the underlying organizational principles of PPI networks using random graphs. Moreover, we shall demonstrate how these models can be used for predicting putative novel interactions in PPI datasets.

23.2 Random Graph Models for Protein–Protein Interaction Networks

A random graph model is an algorithm that tells us how can one generate specific types of graphs using random numbers and a few predefined parameters. For instance, a simple random graph model could be specified as outlined here.

Take n vertices. Consider each pair of vertices once and for each such pair, add an edge between them with probability p.

The motivation behind the usage of random graphs for modeling PPI networks is as follows. Assume that we are able to find a well-fitting random graph model for PPI networks. This has two immediate advantages. First, we can use the generating mechanism of the random graph models to *reason* about how real PPI networks have evolved and what the general organizational principles behind them are, and second, we can use the random graph model to *make predictions* about the existence of yet uncharted connections between proteins as we can simply *ask* the model about the probability of the existence of every edge in the network. For instance, if it turns out that the above-mentioned simple model (which is one of the variants of the Erdős–Rényi random graph model [23.38]) is the best fitting model for PPI networks, we could simply infer that the connections between proteins are essentially distributed at random, and the probability of finding a new interaction between two proteins that are not known to interact yet is equal to p. Such a discovery (were it to be taken seriously) would probably put an end to all further research into the structure of protein interaction networks, but as we will see later, this is luckily not the case and there are still plenty of unanswered questions about how PPI networks evolved and why they look the way they do.

In the next subsections, we will introduce several random graph models that are thought to be good models for the structure of PPI networks. A comparative analysis will then follow; the main purpose of this analysis will be to assess the predictive power of these models with the underlying assumption that a well-

fitting random graph model should have the greatest predictive power among all of them. In particular, we will test the models on the PPI network of the yeast *Saccharomyces cerevisiae* and examine how accurately they are able to predict new connections in this network. Finally, we will present a case study where we apply the best model to identify putative connections in the interactome of *Homo sapiens*.

23.2.1 The Configuration Model

The configuration model was first proposed by *Molloy* and *Reed* [23.39], and was later applied to PPI networks by *Préulj* et al. [23.40] under the name *stickiness index based model*. The model assumes that the connections between proteins depend on the abundance of binding domains on the proteins themselves: two proteins are more likely to interact with each other if both of them have many and/or easily accessible binding domains. Formally, each protein has a stickiness index θ_i that determines the affinity of the protein to interact with others; the probability of an interaction between proteins i and j is then given by $\theta_i \theta_j$. Given a graph $G(V, E)$, one can easily calculate the likelihood of a given parameterization $\boldsymbol{\theta}$, i.e., the probability that the model generates G using these parameters

$$\mathcal{L}[\boldsymbol{\theta}|G(V, E)]$$
$$= \prod_{(i,j) \in V \times V} \left[\theta_i \theta_j A_{ij} + (1 - \theta_i \theta_j)(1 - A_{ij}) \right] ,$$
$$(23.9)$$

where A_{ij} is 1 if and only if vertices i and j are connected, zero otherwise, and the product iterates over *unordered* pairs of vertices.

When fitting the model to an observed network with n proteins, θ_i is usually set to $k_i / \sum_{j=1}^{n} k_j$, where k_i is the number of observed interactions that involve protein i. This ensures that the expected degree of each protein in the networks generated by the configuration model is equal to the observed degree of the protein in the original PPI network. Note that the model does not preserve the degree distribution on the level of individual networks; a protein with k_i observed interactions may have more or less interactions in any single generated random network, but it will have k_i interactions on average across the entire ensemble of random networks if we generate enough random instances. Also note that loop edges have a nonzero probability in this model, and explicitly disallowing loop edges would skew the expected degrees downwards, but the probability of loop edges is usually negligible for sparse networks.

23.2.2 Geometric Random Graphs

The configuration model was shown to be a good fit to PPI networks [23.40]. However, it has been noted in the literature that PPI networks may have a bias against connections between hub proteins (i. e., proteins with high degree) [23.41], and therefore the basic assumption of the configuration model may not be appropriate to describe connections between hubs. To this end, geometric random graphs [23.42] were suggested later as an alternative model for PPI networks [23.13].

The generation process of geometric random graphs first embeds the n vertices of the graph in an d-dimensional unit hypercube and connects vertex pairs with probabilities depending on the distances between them; more precisely, the probability of an edge between proteins i and j is equal to $f(\|x_i - x_j\|^2)$, where x_i is the position of the point corresponding to protein i and $\| \ldots \|^2$ denotes the Euclidean norm. The positions of the vertices and the function f can be considered as parameters of the model. In the simplest case, $f(x)$ is 1 if x is less than a predefined distance threshold δ and 0 otherwise – this function would connect all pairs of proteins that are closer to each other than δ. From a biological point of view, the model assumes that the proteins can be placed in an abstract d-dimensional *chemical space*, and proteins with similar chemical properties have a higher chance of interacting with each other. The likelihood of a given parameterization on a given graph $G(V, E)$ is given as

$$\mathcal{L}(X, f \mid G) = \prod_{(i,j)\in V\times V} \left\{ f(\|x_i - x_j\|^2) A_{ij} \right.$$
$$\left. + [1 - f(\|x_i - x_j\|^2)](1 - A_{ij}) \right\} \ . \tag{23.10}$$

The geometric random graph model is very flexible and it is naturally able to generate graphs with a clus-

tered structure. This property is very useful to us as we have already seen that PPI networks tend to contain tightly connected modules of proteins. However, placing the vertices in a d-dimensional space and choosing the function f in a way that produces a given PPI network with a high probability is a challenging task. The algorithm proposed in [23.13] and [23.15] consists of two major steps:

1. Estimate the position matrix X by applying multidimensional scaling in d dimensions to a matrix $M = (m_{ij})$ where m_{ij} is equal to the square root of the length of the shortest path from protein i to protein j in the network. The resulting configuration of the vertex positions minimizes the total squared elementwise difference between M and the distance matrix of the placed vertices in the d-dimensional space.
2. Assume that f is the mixture of three Gaussian curves,

$$f(x) = \sum_{i=1}^{3} m_i \left(\frac{1}{\sqrt{2\pi\sigma_i^2}} e^{-\frac{(x-\mu_i)^2}{2\sigma_i^2}} \right) \ ,$$

where μ_i and σ_i are the means and standard deviations of the corresponding Gaussian distributions and the m_i values are the mixing parameters. The optimal values of m_i, μ_i and σ_i for $i = 1, 2, 3$ are then found by expectation-maximization.

The application of the geometric random graph model to de-noise PPI datasets (i. e., to predict new protein interactions and to point out false positive interactions in experimental data) shows that the model has good predictive power for PPI prediction without using any additional data (e.g., protein sequences or the 3-D structure of the proteins) [23.15].

23.2.3 Stochastic Blockmodels

Stochastic blockmodels originate from the field of sociology [23.43, 44] where they have been applied to model social structures and to gain insights into the structure of observed relationships between individuals. However, the model has many applications outside the field of sociology; for instance, stochastic blockmodels have been used to predict yet uncharted connections in a network model of the visuotactile cortex of the macaque monkey [23.45, 46].

Stochastic blockmodels assume that each vertex of the graph corresponds to exactly one of d discrete types, and the probability of the existence of an edge depends on the types of the two endpoints of the edge. The two principal parameters of the model are a type vector $t = [t_1, t_2, \ldots, t_n]$ and a symmetric probability matrix $P = [p_{ij}]$ of size $d \times d$; in this setup, an edge is generated between vertices i and j with probability p_{t_i,t_j}. The likelihood of a given parameterization of a stochastic blockmodel for a graph $G(V, E)$ is given by

$$\mathcal{L}(d, t, P \mid G)$$
$$= \prod_{(i,j) \in V \times V} [p_{t_i,t_j} A_{ij} + (1 - p_{t_i,t_j})(1 - A_{ij})] .$$

(23.11)

Stochastic blockmodels can also exhibit a wide variety of structural patterns; for instance, high p_{ii} values generate clusters comprising of vertices of type i, while high p_{ij} $(i \neq j)$ values generate nearly complete bipartite subgraphs between vertices of type i and type j. These patterns are particularly important for protein interaction networks, as clusters may correspond to well-known protein complexes (to which we have dedicated the entire first half of this chapter), and nearly bipartite subgraphs may easily be generated by the lock-and-key mechanism of protein binding [23.47], where it is assumed that specific pairs of binding sites act like locks and keys, and a protein possessing a "key-type" binding site is, therefore, able to interact with almost any other protein that possesses the corresponding "lock-type" binding site. This is naturally represented in stochastic blockmodels via the type vector and the preference matrix – a high p_{ij} $(i \neq j)$ value in the preference matrix then identifies a putative lock–key pairing between proteins of type i and type j.

In order to fit a stochastic blockmodel to an observed network G, one has to find the optimal values for the number of types d, the type of each vertex (i. e., the vector t) and the connection probabilities between types (the matrix P). A possible way to do this is via maximum likelihood (ML) or maximum a posteriori (MAP) estimates. Assuming that d and the type vector t are known, one can estimate the probability matrix P by maximizing the likelihood $\mathcal{L}(P \mid G, k, t)$, and it is easy to show that the maximum likelihood estimate for P is

$$p_{rs} = \frac{\sum_{i=1}^{n} \sum_{j=1}^{n} A_{ij} \delta_{t_i r} \delta_{t_j s}}{\sum_{i=1}^{n} \delta_{t_i r} \sum_{j=1}^{n} \delta_{t_j s}} ,$$

(23.12)

where A_{ij} is 1 if and only if vertices i and j are connected (zero otherwise), and δ_{xy} is 1 if and only if $x = y$ (zero otherwise). Consequently, P can be treated as a quantity that depends on the type vector t and the graph G instead of a parameter on its own. The problem is then simplified to finding the optimal type vector

t and the number of required types, which is typically achieved using Markov chain Monte Carlo methods. The fitting algorithm that we propose to use will be described later.

One disadvantage of the stochastic blockmodel is the expected degree distribution of the network it generates; it can be shown that the degrees of the vertices are drawn from a mixture of Poisson distributions, and the mixing coefficients are determined by the relative frequencies of vertex types in the type vector t [23.48]. This distribution is able to mimic many other degree distributions from uniform to scale-free if the number of types is large enough, but increasing the number of types arbitrarily has the risk of overfitting the model to the observed network. In the most extreme case, the number of vertex types is equal to the number of vertices in the network (therefore each vertex has its own type), and the preference matrix is equal to the adjacency matrix of the original graph. Needless to say, such a configuration reproduces the original network perfectly, but does not tell us anything about the hidden structure of the network as it essentially treats each vertex as a unique entity. A possible solution to this problem is to control the number of vertex types using one of the well-known techniques for model selection (e.g., minimizing *Akaike*'s information criterion [23.49] or the Bayesian information criterion [23.50] instead of the log-likelihood), but it leaves us with a degree distribution that is inconsistent with the original network. To this end, the degree-corrected variant of stochastic blockmodels was introduced recently [23.16], which allows the generated networks to reproduce the degree distribution of the original network even with a small number of types, at the expense of some extra parameters.

The degree-corrected stochastic blockmodel combines the idea of the original blockmodel with the configuration model that we already described in Sect. 23.2.1. In this model, each vertex i has a type index t_i and an *activity* parameter θ_i, which is similar to the *stickiness* index in the configuration model. The probability matrix P is replaced by a rate matrix Ω. The *number* of edges between vertices i and j is then drawn from a Poisson distribution with rate parameter $\theta_i \theta_j \omega_{t_i,t_j}$; in other words, the number of edges between two vertices depends both on their individual activity levels and their affinity towards each other. Note that this model may generate multiple edges between two vertices, which lack a suitable interpretation in case of PPI networks, but one can simply collapse multiple edges into a single one in a post-processing step.

The likelihood function for degree-corrected stochastic blockmodels is [23.16]

$$
\begin{aligned}
&\mathcal{L}(d, t, \theta, \Omega) \\
&= \prod_{(i,j) \in V \times V} \frac{(\theta_i \theta_j \omega_{t_i, t_j})^{A_{ij}}}{A_{ij}!} \exp(-\theta_i \theta_j \omega_{t_i, t_j}) \\
&\times \prod_{i \in V} \frac{(\frac{1}{2} \theta_i^2 \omega_{t_i, t_i})^{A_{ii}/2}}{(A_{ii}/2)!} \exp\left(-\frac{1}{2} \theta_i \theta_i \omega_{t_i, t_i}\right).
\end{aligned}
$$

Note that parts of the vertex activity vector θ that belong to vertices within the same group are arbitrary to within a multiplicative constant, since multiplying all θ_i where i is in a given group g by a constant c can be counterbalanced by dividing $\omega_{t_i, j}$ and ω_{j, t_i} by c (where $j \neq t_i$) and ω_{t_i, t_i} by c^2. Without loss of generality, we can thus normalize θ_i such that the sum of θ_i values for all i within the same group is equal to 1. This allows one to re-write the likelihood function as

$$
\begin{aligned}
\mathcal{L}(d, t, \theta, \Omega) &= C \times \prod_{i \in V} \theta_i^{k_i} \times \prod_{r=1}^{d} \prod_{s=r}^{d} \omega_{rs}^{m_{rs}/2} \\
&\quad \cdot \exp\left(-\frac{1}{2} \omega_{rs}\right),
\end{aligned} \tag{23.13}
$$

where k_i is the degree of vertex i, C is a constant that depends solely on the adjacency matrix of the graph and not on the model parameters, and m_{rs} is the number of edge vertices of type r and s if $r \neq s$ and *twice* the number of edges between vertices of type r if $r = s$.

Similarly to the uncorrected stochastic blockmodel described above, one can factor out the rate matrix Ω and the vertex activity vector θ and simply treat them as dependent variables on t and the input graph G. The maximum likelihood estimates for Ω and θ are as [23.16]

$$
\hat{\theta}_i = \frac{k_i}{\sum_{j=1}^{d} m_{t_i j}}, \qquad \hat{\omega}_{ij} = m_{ij}, \tag{23.14}
$$

where m_{ij} is the number of edges between vertices of type i and j if $i \neq j$ and *twice* the number of edges between vertices of type i if $i = j$. The above choice yields an expected degree k_i for vertex i, thus the expected degree sequence of the generated graphs will match the degree sequence of the graph the model is being fitted to (neglecting loop and multiple edges).

Fitting the uncorrected and the degree-corrected stochastic blockmodels to a given network with a given type count d can then be achieved using Markov chain Monte Carlo methods. Note that in both cases, the only

parameter we really have to optimize is the type vector t; once it is known, all the remaining parameters (P in the uncorrected case or Ω and θ in the degree-corrected case) can be calculated quickly. To optimize t, we consider a Markov chain whose states consist of different type configurations, and propose a simple strategy to update the state of the Markov chain based on the Metropolis–Hastings algorithm as listed:

1. We choose a single vertex and select a new candidate type for this vertex.
2. We calculate the difference in the log-likelihood between the old and the new configuration. Since the changes introduced in the log-likelihood function by a single point mutation are local, not all the terms in the log-likelihood function are affected, and the difference can be calculated easily without re-evaluating the entire log-likelihood function.
3. When $\log \mathcal{L}_{\text{new}} > \log \mathcal{L}_{\text{old}}$, we accept the change unconditionally. Otherwise, we accept the change with probability $\exp(\log \mathcal{L}_{\text{new}} - \log \mathcal{L}_{\text{old}})$. If the change was rejected, the Markov chain stays in the same type configuration as in the previous step.
4. If the Markov chain did not converge to a stationary distribution yet, return to step 1.

Two points have to be clarified in the algorithm in order to make it fully specified. The first point is the problem of initialization, i.e., the starting configuration should we use for t to obtain a result with a high log-likelihood quickly. The second point is the problem of termination, i.e., the conditions we use to determine whether the Markov chain consisting of the type vectors in consecutive steps has converged or not. In the simplest case, the chain can simply be initialized by a random type vector. However, our experiments with the uncorrected stochastic blockmodel [23.46, 48] showed that the following, greedy initialization heuristic yields results in similar quality in a considerably shorter amount of time, as it brings the initial state of the Markov chain closer to the mode of the distribution:

1. First, we select a random type vector t.
2. Each vertex is then given the opportunity to change its type if it is able to increase its own contribution to the log-likelihood by doing so. Type changes for all the vertices are done in a synchronous manner; in other words, each vertex assumes that none of the other vertices change their types when they decide on their own new types they wish to belong to.
3. When the new configuration is equivalent to the old one, we consider the initialization to have converged

to a steady state and we start the Markov chain from the new configuration. Otherwise, the initialization algorithm returns to step 2.

The problem of termination can be solved by calculating the average log-likelihoods of two consecutive blocks of a large number of samples and stop the Markov chain when the difference between the two averages becomes smaller than a predefined threshold. This is motivated by the fact that the average log-likelihood of a block of samples is an estimator of the entropy of the distribution of likelihood values, and the entropy should be constant in a Markov chain that has converged to its stationary distribution. In our experiments, we use a block size of $2^{13} = 8192$ samples and an entropy difference threshold of 1.

Note that the fitting procedure described above can be used only if d (the number of types) is specified in advance. In most practical model fitting problems, d is unknown and must be chosen automatically. In our experiments, we have employed the following strategy to select the optimal d value:

1. Start from $d = 1$.
2. Run the optimization process outlined above with the given d until the Markov chain converges.
3. Let $\log \mathcal{L}_d$ denote the best log-likelihood encountered in step 2.
4. If $d < \sqrt{n}$, the square root of the number of proteins, increase d by one and go back to step 2.
5. For each d, calculate *Akaike*'s information criterion [23.49] for the model with the best log-likelihood as

$$\text{AIC}_d = 2r - 2 \log \mathcal{L}_d \,, \qquad (23.15)$$

where r is the number of parameters in the model. For the original blockmodel, $r = d(d+1)/2 + n + 1$, while for the degree-corrected blockmodel, $r = d(d+1)/2 + 2n + 1$.
6. Let $d = \text{argmin}_i \text{AIC}_i$.

Finally, note that using $d = 1$ (i.e., one vertex type only) for the degree-corrected stochastic blockmodel reproduces the configuration model as $\hat{\theta}_i = k_i/2m$ and $\hat{\omega}_{11} = 2m$ (where m is the number of edges), producing $p_{ij} = k_i k_j/2m$.

23.2.4 Hierarchical Random Graphs

The last random graph model that we are going to investigate in this chapter is the hierarchical random graph model [23.51]. This model has already been used to describe the structure of several real-world networks ranging from food webs to terrorist contact networks. One of the advantages of this model is that it avoids the problem of group count selection (unlike stochastic blockmodels). Instead of groups, the model uses a binary dendrogram, where the leaves of the dendrogram correspond to the vertices of the graph being generated. Since each intermediate node of the dendrogram joins two leaves or other intermediate nodes, it is easy to see that the dendrogram has n leaves and $n-1$ intermediate nodes. These intermediate nodes have associated probability values (denoted by a vector p_k, where k is an index in $[1; n-1]$ that identifies an intermediate node), and the probability of an edge between vertices i and j is equal to $p_{\text{LCA}(\mathcal{D}, i, j)}$, where $\text{LCA}(\mathcal{D}, i, j)$ is the lowest common ancestor node of leaves i and j in the dendrogram $\mathcal{D}$. The likelihood function of the model is then

$$\begin{aligned} \mathcal{L}[\mathcal{D}, \boldsymbol{p} \mid G(V, E)] \\ = \prod_{(i,j) \in V \times V} [p_{\text{LCA}(\mathcal{D}, i, j)} A_{ij} \\ + (1 - p_{\text{LCA}(\mathcal{D}, i, j)})(1 - A_{ij})] \,. \end{aligned} \qquad (23.16)$$

From a biological point of view, hierarchical random graph models assume that the proteins in a PPI network evolved from a remote common ancestor, and the affinity of two proteins towards each other depends on their ancestry; in particular, the node in the dendrogram where the two proteins diverged from each other. However, the model does not require that closely related proteins have a larger connection probability than remotely related proteins. In general, the idea of protein ancestry is well-known in biology, and several projects are dedicated to grouping proteins based on evolutionary relatedness. For instance, *protein families* in the SCOP (structural classification of proteins) database [23.52] *are clearly evolutionarily related*, and *superfamilies* contain *proteins that have low sequence identities, but whose structural and functional features suggest that a common evolutionary origin is probable* [23.53]. The basic assumption of hierarchical random graph models is, therefore, entirely plausible in the context of PPI networks.

The fitting procedure of hierarchical random graphs is very similar to the technique that we employed for stochastic blockmodels. It is easy to recognize that the probability vector $\boldsymbol{p} = (p_i)$ can be considered a dependent variable on the structure of the dendrogram as follows. Let L_i be the set of leaf nodes in the left subtree of intermediate node i, R_i be the set of leaf nodes in the right subtree and m_i is the number of edges going between L_i and R_i. The maximum likelihood estimate of p_i conditioned on the dendrogram is then given as $m_i/(|L_i||R_i|)$. Therefore, if we know the dendrogram, we can easily calculate the corresponding probabilities that maximize the log-likelihood. This makes it possible to employ Markov chain Monte Carlo optimization strategies again to find the optimal dendrogram structure and probability vector for a given network. The details of the process are given in [23.51].

23.2.5 Evaluating the Predictive Power of Random Graph Models

As mentioned earlier in Sect. 23.2, our primary motivation for using random graph models to analyze PPI networks is to facilitate the prediction of novel, yet uncharted interactions between proteins by fitting a model to the known part of a PPI network and then *asking* the fitted model for the probability of putative connections. In this section, we will, therefore, evaluate the predictive power of the random graph models presented in earlier sections.

The input data of our benchmarks will be the protein interaction dataset of *Collins* et al. [23.12], which combines two earlier experimental PPI datasets [23.7,8] into a self-consistent view of the interactome of the yeast *Saccharomyces cerevisiae*. Since it is based on two independent experimental datasets, the data of Collins et al. is thought to be one of the most accurate PPI networks for yeast. The dataset includes 5437 proteins and assigns a confidence score for each protein pair. Collins et al. also proposed a confidence score threshold that yields 9074 high-confidence interactions between 1622 proteins. These interactions can be turned into a graph containing 1622 vertices and 9074 edges. Since the geometric random graph model cannot handle disconnected networks, we extracted the largest connected component of this graph, leading to a network of 1004 proteins and 8323 interactions.

Since the fitting algorithms of random graph models require the whole network as input data, the standard machine learning approach (which divides the input data into training and testing sets) cannot be applied here either; for instance, it is not possible to calculate the multidimensional scaling required for fitting a geometric random graph model with only a fragment of the entire adjacency matrix. However, training and testing on the same dataset is likely to result in overfitting and, therefore, should be avoided. *Kuchaiev* et al. [23.15] circumvented the problem by training

the random graph models on the entire high confidence network and testing the model predictions on the set of mid and low-confidence edges, where they considered the mid-confidence edges as further positive examples. During the training process, mid-confidence edges are considered nonexistent, therefore a method that is prone to overfitting will learn the mid-confidence edges as negative examples and will have a lower performance in the test phase, where the mid-confidence edges are assumed to exist. The threshold for the confidence score of mid-confidence edges was set to 0.2, yielding 5352 mid-confidence edges and 489 831 negative testing examples.

Predictions for the geometric random graph model were obtained by calculating the probability of the existence of all test edges according to the models. For the stochastic blockmodels and the hierarchical random graph model, 10 000 samples were taken from the Markov chains after convergence, with sampling probability 0.1, and the final predicted probability for an edge were calculated by averaging the predicted probabilities for this edge across all the 10 000 samples. For instance, if an edge was predicted to exist with probability 0.7 by half of the samples and with probability 0.4 by the other half of the samples, the final predicted probability for that edge became 0.55. The geometric random graph model used five dimensions according to the recommendations of *Kuchaiev* et al. [23.15]. To account for the imbalance between the number of positive and negative examples in the test set, the results were evaluated both by ROC curves (Fig. 23.4) and by precision-recall curves (Fig. 23.5) [23.54].

Before drawing any conclusions on the difference between the AUC (area under the curve) values, we performed paired permutation tests on the curves [23.55] to assess the significance of these differences. All the differences between ROC curves were deemed significant (largest p-value $< 10^{-5}$), which allows us to conclude that the degree-corrected blockmodel shows superior performance to other alternative models in terms of predictive power, as measured by both the ROC and the precision-recall curves. Similarly, the configuration model showed the worst performance. The remaining three models were ranked differently by the two curves. The area under the ROC curve was larger for the geometric random graphs, while the precision-recall curve ranked the hierarchical random graph and the uncorrected stochastic blockmodel before geometric random graphs. The difference can be explained by looking at the precision-recall curves: the geometric random graph shows a consistently low

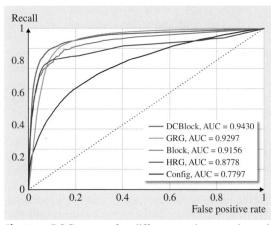

Fig. 23.4 ROC curves for different random graph models on the test set derived from the PPI network of *Collins* et al. [23.12]. Block = stochastic blockmodel, DCBlock = degree-corrected stochastic blockmodel, GRG = geometric random graph, HRG = hierarchical random graph, Config = configuration model

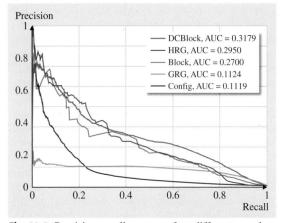

Fig. 23.5 Precision-recall curves for different random graph models on the test set derived from the PPI network of *Collins* et al. [23.12]. AUC = area under the curve, larger is better. Block = stochastic blockmodel, DCBlock = degree-corrected stochastic blockmodel, GRG = geometric random graph, HRG = hierarchical random graph, Config = configuration model

precision for almost any level of recall, while the performance of the other models generally improves as the recall rate is lowered. At a reasonable recall of 35%, the precision achieved by both stochastic blockmodels and the hierarchical random graph model is about 33%, which is more than twice as high as the

precision reported for geometric random graphs by *Kuchaiev* et al. [23.15] (precision = 15%, confirmed by our replicate).

These numbers also allow us to estimate the performance of these random graph models on human PPI data. Recent estimates of the size of the human PPI network indicate 154 000–369 000 interactions among 20 000–25 000 proteins [23.56]. At a recall of 35% and a precision of 33%, we can expect 53 900 novel interactions among only 163 000 predicted ones with stochastic blockmodels or hierarchical random graphs; by comparison, the same number of novel interactions would be expected among 359 000 interactions for the geometric random graph model.

23.2.6 Predicting Novel Interactions in the Human Interactome

The previous section has shown us that degree-corrected stochastic blockmodels outperform alternative random graph models in terms of predictive power. To show the potential of this approach in predicting new PPI, we will apply this graph model to a human protein interaction dataset derived from the BioGRID database [23.57]. The network contained 29 328 physical interactions among 8123 proteins and was constructed as listed:

1. We downloaded version 3.0.66 of the BioGRID database. Note that we intentionally used an older version of BioGRID because this allowed us to confirm some of the predictions by looking them up in a later version of BioGRID.
2. We selected all the physical interactions where both interactors correspond to *Homo sapiens* (species ID: 9606).
3. We kept those interactions annotated by an evidence code denoting a physical interaction between two proteins.
4. We extracted the largest connected component of the resulting network and used this for our experiments.

We fitted a degree-corrected stochastic blockmodel to the above network and used a predicted edge probability threshold of 0.6 to derive a set of predicted interactions, i. e., every connection that was not present in BioGRID but had a predicted probability larger than 0.6 was considered as a putative novel interaction. There were 217 such interactions in the entire network. We then used the GO project [23.17] and KEGG [23.18] to provide biological support for these interactions.

The GO project [23.17] aims to assemble a controlled vocabulary that can be used to annotate biological entities (genes, proteins, etc.) in three different aspects: biological process (BP), cellular component (CC), and molecular function (MF). Terms in the GO (GO) are then connected to each other via directed relations such as *is-a* or *part-of* to form three directed acyclic graphs (DAGs), one for each aspect of the tree. Figure 23.6 shows a subpart of the DAG corresponding to the cellular CC aspect, with all terms that the term *nucleoplasm* is directly or indirectly related to.

Each gene or protein may have multiple annotations in each of the three aspects. Since the GO is hierarchical and each term may have multiple subterms connected by *is-a* relations (e.g., nucleoplasm *is-a* nuclear part), GO annotations may be *direct* (when a GO term is directly assigned to an entity) or indirect (when the term is not assigned directly to an entity but such a relation can be inferred using the inference rules between the GO relations). For instance, the human gene *CDC25B* (and its corresponding protein Cdc25bp) is *directly* annotated by 9 BP, 6 CC and 4 MF terms in the GO at the time of writing, indicating its relatedness to cell division (term ID: GO:0051301 in the BP aspect), protein binding (term ID: GO:0005515 in the MF aspect), and the nucleoplasm (term ID: GO:0005654 in the CC aspect), among others. It is also annotated *indirectly* to superterms of these 19 terms, for instance:

- *CDC25B* is in a nuclear part (GO:0044428), because it is in the nucleoplasm (by direct annotation) and nucleoplasm *is-a* nuclear part.
- *CDC25B* is in an intracellular organelle part (as inferred above) and nuclear part *is-an* intracellular organelle part.
- *CDC25B* is in the nucleus (GO:0005634), because it is in a nuclear part (as inferred above), which is *part-of* the nucleus.

It is reasonable to assume that interacting proteins are situated in the same cellular component and/or participate in the same biological process, and the GO annotations and inference rules give us an opportunity to quantify what fraction of our predicted interacting protein pairs share annotations in the relevant aspects of the GO tree.

Similarly, KEGG [23.18] maintains (among others) an assignment between proteins and the biological pathways they participate in. We can assume that proteins in the same pathway correspond to the same biological process, therefore a shared KEGG pathway annotation for two proteins indicates their relatedness just like a shared GO annotation.

To this end, we calculated the fraction of predicted interaction pairs that 1) share at least one *direct* GO cellular component or biological process annotation or 2) participate in the same KEGG pathway in order to assess the biological plausibility of our predictions. We considered all the *Homo sapiens* pathways from KEGG and all the direct GO annotations ignoring cases when two proteins shared only the root GO terms (GO:0005575: cellular_component or GO:0008150: biological_process). Proteins not corresponding to any of the KEGG pathways were excluded from the KEGG score calculation, and proteins not having any annotations in the GO were excluded from the GO score calculation. In the following paragraphs, we report the fractions as percentages and also calculate the significance score of these fractions by comparing them with the same scores for randomly drawn protein pairs under the assumption of a hypergeometric model.

204 out of our 217 predicted interactions had GO annotations for both of the proteins involved. 171 out of these 204 (83.82%) of these pairs had at least one shared direct GO annotation in either the biological process (BP) or the cellular component (CC) aspect of the tree. The probability of observing such an extreme result by chance is less than 2.7×10^{-38} according to the hypergeometric test we have performed. However, since direct cellular component annotations may sometimes correspond to fairly broad categories (such as *intracellular*) due to lack of information about a given protein, and being in the same cellular component does not necessarily mean that the two proteins interact, we repeated the analysis with the BP aspect of the GO tree only. 187 of our predictions had associated BP annotations for both proteins of the interaction pair, and 82 (43.85%) of them shared at least one BP annotation. The probability of observing such a high number of shared BP annotations when drawing protein pairs at random is less than 4.2×10^{-46}. A similar analysis on the KEGG pathways showed that 111 out of the 217 predicted interactions had KEGG pathway information for both proteins involved, and 63 out of 111 (56.75%, p-value $< 4.6 \times 10^{-43}$) shared at least one KEGG pathway. These results together confirm the biological significance of our results.

Finally, we compared our predictions with a later version of BioGRID (version 3.1.72) that was published after we have finished our analyses on the BioGRID network and also with the most recent

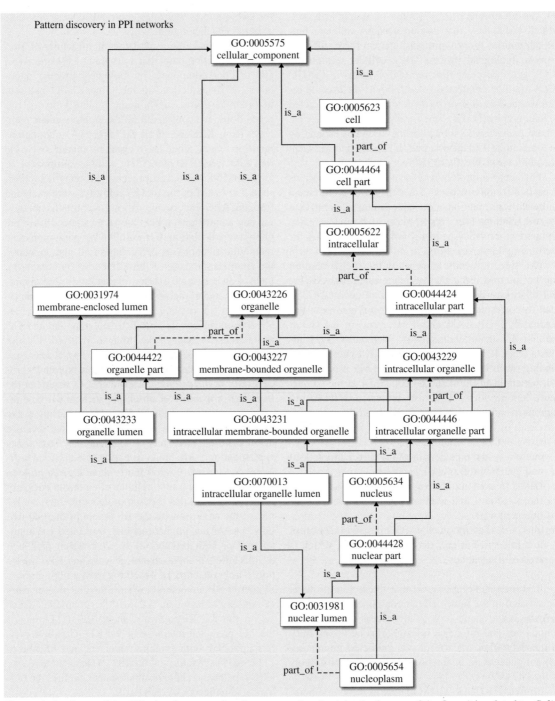

Fig. 23.6 A subpart of the GO, showing terms that the term *nucleoplasm* (at the *bottom* of the figure) is related to. *Solid edges* denote *is_a*, *dashed edges* denote *part_of* relations

version of the Human Protein Reference Database (HPRD [23.58]), an alternative repository of confirmed PPI in humans. Six out of the 217 predicted interactions (PIN1-UBC, ESR1-RELA, MYC-SMARCA2, MYC- SMARCC1, EP300-SMAD3 and SMAD2-UBC) were found in BioGRID 3.1.72 (p-value $< 4.8 \times 10^{-17}$), and the most recent version of HPRD provided confirmation for a further 28 interactions.

23.3 Conclusions

In this chapter, we have presented several computational methods to extract meaningful biological information from PPI networks. The structure we were able to unravel in these networks allowed us to detect the presence of protein complexes (sets of proteins that interact with each other more frequently than with the rest of the network) and even to pinpoint potentially novel interactions in PPI data that were probably overlooked by experimental methods.

We have presented several state-of-the-art techniques to extract protein complexes from PPI networks. Some of these methods (e.g., MCODE [23.25] or RNSC [23.20, 23]) work on the unweighted network only, while others can make use of the confidence scores associated to the interactions (e.g., MCL [23.21, 22] or ClusterONE [23.27]). More recent methods explicitly allow overlaps between the detected complexes [23.26, 27], which is an important feature of complexes found in real PPI networks. We have demonstrated that incorporating weights and the presence of overlaps into the clustering process improves the accuracy of the results of these methods.

In the second part of the chapter, we have given an overview of several random graph models that can be used to model the structure of PPI networks and also to make predictions about false negative or false positive interactions. The latter takes advantage of the fact that after having fitted a random graph model to an experimentally or computationally derived PPI network, we can simply *ask* the random graph model about the probability of the existence of every single connection in the network. Nonexistent connections with a high predicted probability are then candidates for false negative interactions, while existent connections with a very low predicted probability may be treated as false positives. We have shown that stochastic blockmodels [23.16, 43, 44, 46], especially its degree-corrected variant [23.16] have a larger predictive power than several alternative random graph models [23.13, 15, 40, 51] when evaluated on a PPI network derived from two experimental datasets [23.7, 8] and integrated using computational techniques [23.12].

However, despite all the recent advances both in the experimental techniques and the computational methods being used to analyze the produced data, there are still several open questions. For the problem of protein complex detection, it is important to note that some of the detected complexes are always bound to be the byproduct of random noise in the input data, and such noise is mostly unavoidable. Therefore, it would be important to devise computational techniques that are able to estimate the statistical significance of each detected complex in order to help biologists prioritize putative novel complexes for follow-up experiments. For the problem of modeling PPI networks, it is time to go back to the experimental data and see whether we can map the principles of the underlying assumptions of the best random graph models to the underlying biological processes. By being able to do so, we would be able to learn new facts about the biology of protein interactions. At the same time, we should check if all that we know about the biology of the problem is already included in our models and modify them, if necessary, in order to make them more realistic and thus more predictive.

References

23.1   P. Uetz, L. Giot, G. Cagney, T. Mansfield, R. Judson, J. Knight, D. Lockshon, V. Narayan, M. Srinivasan, P. Pochart, A. Qureshi-Emili, Y. Li, B. Godwin, D. Conover, T. Kalbfleisch, G. Vijayadamodar, M. Yang, M. Johnston, S. Fields, J. Rothberg: A comprehensive analysis of protein–protein interactions in Saccharomyces cerevisiae, Nature **403**(6770), 623–627 (2000)

23.2   T. Ito, K. Tashiro, S. Muta, R. Ozawa, T. Chiba, M. Nishizawa, K. Yamamoto, S. Kuhara, Y. Sakaki: Toward a protein–protein interaction map of the budding yeast: A comprehensive system to examine two-hybrid interactions in all possible combinations between the yeast proteins, Proc. Natl. Acad. Sci. USA **97**(3), 1143–1147 (2000)

23.3   L. Giot, J. Bader, C. Brouwer, A. Chaudhuri, B. Kuang, Y. Li, Y. Hao, C. Ooi, B. Godwin, E. Vitols, G. Vijayadamodar, P. Pochart, H. Machineni, M. Welsh, Y. Kong, B. Zerhusen, R. Malcolm, Z. Varrone, A. Collis, M. Minto, S. Burgess, L. McDaniel, E. Stimpson, F. Spriggs, J. Williams, K. Neurath, N. Ioime, M. Agee, E. Voss, K. Furtak, R. Renzulli, N. Aanensen, S. Carrolla, E. Bickelhaupt, Y. Lazovatsky, A. DaSilva, J. Zhong, C. Stanyon, R. Finley, K. White, M. Braverman, T. Jarvie, S. Gold, M. Leach, J. Knight, R. Shimkets, M. McKenna, J. Chant, J. Rothberg: A protein interaction map of Drosophila melanogaster, Science **302**(5651), 1727–1736 (2003)

23.4   S. Li, C. Armstrong, N. Bertin, H. Ge, S. Milstein, M. Boxem, P. Vidalain, J. Han, A. Chesneau, T. Hao, D. Goldberg, N. Li, M. Martinez, J. Rual, P. Lamesch, L. Xu, M. Tewari, S. Wong, L. Zhang, G. Berriz, L. Jacotot, P. Vaglio, J. Reboul, T. Hirozane-Kishikawa, Q. Li, H. Gabel, A. Elewa, B. Baumgartner, D. Rose, H. Yu, S. Bosak, R. Sequerra, A. Fraser, S. Mango, W. Saxton, S. Strome, S. Van Den Heuvel, F. Piano, J. Vandenhaute, C. Sardet, M. Gerstein, L. Doucette-Stamm, K. Gunsalus, J. Harper, M. Cusick, F. Roth, D. Hill, M. Vidal: A map of the interactome network of the metazoan C. elegans, Science **303**(5657), 540–543 (2004)

23.5   U. Stelzl, U. Worm, M. Lalowski, C. Haenig, F. Brembeck, H. Goehler, M. Stroedicke, M. Zenkner, A. Schoenherr, S. Koeppen, J. Timm, S. Mintzlaff, C. Abraham, N. Bock, S. Kietzmann, A. Goedde, E. Toksöz, A. Droege, S. Krobitsch, B. Korn, W. Birchmeier, H. Lehrach, E. Wanker: A human protein–protein interaction network: A resource for annotating the proteome, Cell **122**(6), 957–968 (2005)

23.6   J. Rual, K. Venkatesan, T. Hao, T. Hirozane-Kishikawa, A. Dricot, N. Li, G. Berriz, F. Gibbons, M. Dreze, N. Ayivi-Guedehoussou, N. Klitgord, C. Simon, M. Boxem, S. Milstein, J. Rosenberg, D. Goldberg, L. Zhang, S. Wong, G. Franklin, S. Li, J. Albala, J. Lim, C. Fraughton, E. Llamosas, S. Cevik, C. Bex, P. Lamesch, R. Sikorski, J. Vandenhaute, H. Zoghbi, A. Smolyar, S. Bosak, R. Sequerra, L. Doucette-Stamm, M. Cusick, D. Hill, F. Roth, M. Vidal: Towards a proteome-scale map of the human protein–protein interaction network, Nature **437**(7062), 1173–1178 (2005)

23.7   N. Krogan, G. Cagney, H. Yu, G. Zhong, X. Guo, A. Ignatchenko, J. Li, S. Pu, N. Datta, A. Tikuisis, T. Punna, J. Peregrin-Alvarez, M. Shales, X. Zhang, M. Davey, M. Robinson, A. Paccanaro, J. Bray, A. Sheung, B. Beattie, D. Richards, V. Canadien, A. Lalev, F. Mena, P. Wong, A. Starostine, M. Canete, J. Vlasblom, S. Wu, C. Orsi, S. Collins, S. Chandran, R. Haw, J. Rilstone, K. Gandi, N. Thompson, G. Musso, P. St. Onge, S. Ghanny, M. Lam, G. Butland, A. Altaf-Ui, S. Kanaya, A. Shilati-

fard, E. O'Shea, J. Weissman, C. Ingles, T. Hughes, J. Parkinson, M. Gerstein, S. Wodak, A. Emili, J. Greenblatt: Global landscape of protein complexes in the yeast Saccharomyces cerevisiae, Nature **440**(7084), 637–643 (2006)

23.8   A. Gavin, P. Aloy, P. Grandi, R. Krause, M. Boesche, M. Marzioch, C. Rau, L. Jensen, S. Bastuck, B. Dumpelfeld, A. Edelmann, M. Heurtier, V. Hoffman, C. Hoefert, K. Klein, M. Hudak, A. Michon, M. Schelder, M. Schirle, M. Remor, T. Rudi, S. Hooper, A. Bauer, T. Bouwmeester, G. Casari, G. Drewes, G. Neubauer, J. Rick, B. Kuster, P. Bork, R. Russell, G. Superti-Furga: Proteome survey reveals modularity of the yeast cell machinery, Nature **440**(7084), 631–636 (2006)

23.9   N. Pržulj, D. Corneil, I. Jurisica: Modeling interactome: Scale-free or geometric?, Bioinformatics **20**(18), 3508–3515 (2004)

23.10   L. Lu, Y. Xia, A. Paccanaro, H. Yu, M. Gerstein: Assessing the limits of genomic data integration for predicting protein networks, Genome Res. **15**(7), 945–953 (2005)

23.11   H. Yu, A. Paccanaro, V. Trifonov, M. Gerstein: Predicting interactions in protein networks by completing defective cliques, Bioinformatics **22**(7), 823–829 (2006)

23.12   S.R. Collins, P. Kemmeren, X.C. Zhao, J.F. Greenblatt, F. Spencer, F.C. Holstege, J.S. Weissman, N.J. Krogan: Toward a comprehensive atlas of the physical interactome of Saccharomyces cerevisiae, Mol. Cell Proteomics **6**, 439–450 (2007)

23.13   D. Higham, M. Rašajski, N. Pržulj: Fitting a geometric graph to a protein–protein interaction network, Bioinformatics **24**(8), 1093–1099 (2008)

23.14   H. Yu, P. Braun, M. Yildirim, I. Lemmens, K. Venkatesan, J. Sahalie, T. Hirozane-Kishikawa, F. Gebreab, N. Li, N. Simonis, T. Hao, J. Rual, A. Dricot, A. Vazquez, R. Murray, C. Simon, L. Tardivo, S. Tam, N. Svrzikapa, C. Fan, A. de Smet, A. Motyl, M. Hudson, J. Park, X. Xin, M. Cusick, T. Moore, C. Boone, M. Snyder, F. Roth, A. Barabási, J. Tavernier, D. Hill, M. Vidal: High-quality binary protein interaction map of the yeast interactome network, Science **322**(5898), 104–110 (2008)

23.15   O. Kuchaiev, M. Rašajski, D. Higham, N. Pržulj: Geometric de-noising of protein–protein interaction networks, PLoS Comp. Biol. **5**(8), e1000454 (2009)

23.16   B. Karrer, M.E.J. Newman: Stochastic blockmodels and community structure in networks, Phys. Rev. E **83**(1 Pt 2), 016107 (2011)

23.17   M. Ashburner, C. Ball, J. Blake, D. Botstein, H. Butler, J. Cherry, A. Davis, K. Dolinski, S. Dwight, J. Eppig, M. Harris, D. Hill, L. Issel-Tarver, A. Kasarskis, S. Lewis, J. Matese, J. Richardson, M. Ringwald, G. Rubin, G. Sherlock: Gene ontology: Tool for the unification of biology, Nat. Genet. **25**(1), 25–29 (2000)

23.18   M. Kanehisa, M. Araki, S. Goto, M. Hattori, M. Hirakawa, M. Itoh, T. Katayama, S. Kawashima, S. Okuda, T. Tokimatsu, Y. Yamanishi: KEGG for linking genomes to life and the environment, Nucl. Acids Res. **36**(Database issue), D480–4 (2008)

23.19   B. Alberts, A. Johnson, J. Lewis, M. Raff: *Molecular Biology of the Cell*, 4th edn. (Garland Science, New York 2002), Chap. 6, p. 342

23.20   A. King, N. Pržulj, I. Jurisica: Protein complex prediction via cost-based clustering, Bioinformatics **20**(17), 3013–3020 (2004)

23.21   A.J. Enright, S.V. Dongen, C.A. Ouzounis: An efficient algorithm for large-scale detection of protein families, Nucl. Acids Res. **30**(7), 1575–1584 (2002)

23.22   S. van Dongen: Graph clustering via a discrete uncoupling process, SIAM J. Matrix Anal. Appl. **30**, 121–141 (2008)

23.23   A. King: *Graph Clustering with Restricted Neighborhood Search*, Master's thesis (University of Toronto, Toronto 2004)

23.24   F. Glover, M. Laguna: *Tabu Search* (Kluwer Academic, Dordrecht 1997)

23.25   G.D. Bader, C.W. Hogue: An automated method for finding molecular complexes in large protein interaction networks, BMC Bioinformatics **4**, 2 (2003)

23.26   G. Palla, I. Derényi, I. Farkas, T. Vicsek: Uncovering the overlapping community structure of complex networks in nature and society, Nature **435**(7043), 814–818 (2005)

23.27   T. Nepusz, H. Yu, A. Paccanaro: Detecting overlapping protein complexes from protein–protein interaction networks, Nat. Methods **9**(5), 471–472 (2012)

23.28   B. Adamcsek, G. Palla, I. Farkas, I. Derényi, T. Vicsek: CFinder: Locating cliques and overlapping modules in biological networks, Bioinformatics **22**(8), 1021–1023 (2006)

23.29   I. Farkas, D. Ábel, G. Palla, T. Vicsek: Weighted network modules, New. J. Phys. **9**, 180 (2007)

23.30   F. Radicchi, C. Castellano, F. Cecconi, V. Loreto, D. Parisi: Defining and identifying communities in networks, Proc. Natl. Acad. Sci. USA **101**(9), 2658–2663 (2004)

23.31   A. Clauset: Finding local community structure in networks, Phys. Rev. E **72**, 026132 (2005)

23.32   J. Baumes, M. Goldberg, M. Magdon-Ismail: Efficient Identification of Overlapping Communities, LNCS **3495**, 27–36 (2005)

23.33   F. Luo, J.Z. Wang, E. Promislow: Exploring local community structures in large networks, Web Intell. Agent Syst. **6**(4), 387–400 (2008)

23.34   H.W. Mewes, C. Amid, R. Arnold, D. Frishman, U. Güldener, G. Mannhaupt, M. Münsterkötter, P. Pagel, N. Strack, V. Stümpflen, J. Warfsmann, A. Ruepp: MIPS: Analysis and annotation of proteins from whole genomes, Nucl. Acids Res. **32**(Database issue), D41–44 (2004)

23.35   S. Brohée, J. van Helden: Evaluation of clustering algorithms for protein–protein interaction networks, BMC Bioinformatics **7**, 488 (2006)

23.36   R. Jansen, M. Gerstein: Analyzing protein function on a genomic scale: The importance of gold-standard positives and negatives for network prediction, Curr. Opin. Microbiol. **7**(5), 535–545 (2004)

23.37   A.L. Boulesteix: Over-optimism in bioinformatics research, Bioinformatics **26**, 437–439 (2009)

23.38   P. Erdős;, A. Rényi: On random graphs, Publ. Math. **6**, 290–297 (1959)

23.39   M. Molloy, B. Reed: A critical point for random graphs with a given degree sequence, Random Struct. Algorithms **6**, 161–179 (1995)

23.40   N. Pržulj, D. Higham: Modelling protein–protein interaction networks via a stickiness index, J. R. Soc. Interface **3**(10), 711–716 (2006)

23.41   S. Maslov, K. Sneppen: Specificity and stability in topology of protein networks, Science **296**(5569), 910–913 (2002)

23.42   M.D. Penrose: *Random Geometric Graphs*, Oxford Studies in Probability, Vol. 5 (Oxford Univ. Press, Oxford 2003)

23.43   P. Holland, K.B. Laskey, S. Leinhardt: Stochastic blockmodels: Some first steps, Soc. Netw. **5**, 109–137 (1983)

23.44   T.A.B. Snijders, K. Nowicki: Estimation and prediction for stochastic blockmodels for graphs with latent block structure, J. Classif. **14**(1), 75–100 (1997)

23.45   L. Négyessy, T. Nepusz, L. Kocsis, F. Bazsó: Prediction of the main cortical areas and connections involved in the tactile function of the visual cortex by network analysis, Eur. J. Neurosci. **23**(7), 1919–1930 (2006)

23.46   T. Nepusz, L. Négyessy, G. Tusnády, F. Bazsó: Reconstructing cortical networks: Case of directed graphs with high level of reciprocity, Bolyai Soc. Math. Stud. **18**, 325–368 (2008)

23.47   J.L. Morrison, R. Breitling, D.J. Higham, D.R. Gilbert: A lock-and-key model for protein–protein interactions, Bioinformatics **22**(16), 2012–2019 (2006)

23.48   T. Nepusz: Data mining in complex networks: Fuzzy communities and missing link prediction. Ph.D. Thesis (Budapest University of Technology and Economics, Budapest 2008)

23.49   H. Akaike: Likelihood and the Bayes procedure. In: *Bayesian Statistics*, ed. by J.M. Bernardo, M.H. De Groot, D.V. Lindley, A.F.M. Smith (Valencia Univ. Press, Valencia 1980)

23.50   G.E. Schwarz: Estimating the dimension of a model, Ann. Stat. **6**(2), 461–464 (1978)

23.51   A. Clauset, C. Moore, M.E.J. Newman: Hierarchical structure and the prediction of missing links in networks, Nature **453**, 98–101 (2008)

23.52   A. Murzin, S. Brenner, T. Hubbard, C. Chothia: SCOP: A structural classification of proteins database for

Part D | 23

the investigation of sequences and structures, J. Mol. Biol. **247**(4), 536–540 (1995)

23.53   http://scop.mrc-lmb.cam.ac.uk/scop/intro.html (last accessed May 16, 2011)

23.54   J. Davis, M. Goadrich: The relationship between precision-recall and ROC curves, ICML '06: Proc. 23rd Int. Conf. Mach. Learn. (ACM, New York 2006) pp. 233–240

23.55   S. Swamidass, C. Azencott, K. Daily, P. Baldi: A CROC stronger than ROC: Measuring, visualizing and optimizing early retrieval, Bioinformatics **26**(10), 1348–1356 (2010)

23.56   G. Hart, A. Ramani, E. Marcotte: How complete are current yeast and human protein-interaction networks?, Genome Biol. **7**(11), 120 (2006)

23.57   C. Stark, B. Breitkreutz, T. Reguly, L. Boucher, A. Breitkreutz, M. Tyers: BioGRID: A general repository for interaction datasets, Nucl. Acids Res. **34**(Database issue), D535–9 (2006)

23.58   T. Keshava Prasad, R. Goel, K. Kandasamy, S. Keerthikumar, S. Kumar, S. Mathivanan, D. Telikicherla, R. Raju, B. Shafreen, A. Venugopal, L. Balakrishnan, A. Marimuthu, S. Banerjee, D. Somanathan, A. Sebastian, S. Rani, S. Ray, C. Harrys Kishore, S. Kanth, M. Ahmed, M. Kashyap, R. Mohmood, Y. Ramachandra, V. Krishna, B. Rahiman, S. Mohan, P. Ranganathan, S. Ramabadran, R. Chaerkady, A. Pandey: Human Protein Reference Database – 2009 update, Nucl. Acids Res. **37**(Database issue), D767–72 (2009)

24. Molecular Networks – Representation and Analysis

Miguel A. Hernández-Prieto, Ravi K.R. Kalathur, Matthias E. Futschik

Molecular networks, their representation and analysis have attracted increasing interest in recent years. Although the importance of molecular networks has been recognized for a long time, only the advent of new technologies during the last two decades has delivered the necessary data for a systematic study of molecular networks and their complex behavior. Especially the surge of genome-wide data as well as the increase in computational power have contributed to establishing network and systems biology as new paradigms. The conceptual framework is generally based on an integrated approach of computational and experimental methods. In this chapter, we introduce basic concepts and outline mathematical formalisms for representing and analyzing molecular networks. In particular, we review the study of transcriptional regulatory networks in prokaryotes and of protein interaction networks in humans as prime examples of network-orientated approaches to complex systems. The chapter is concluded with a discussion of current challenges and future directions of network biology.

24.1 Overview

Biological systems can range from a tiny bacterium in soil to large food webs across the oceans. Despite striking differences in size and appearance, they have one thing in common: they are complex networks made out of numerous single components interacting with each other. The cell can be seen as a leading example of a highly organized network with a plethora of single parts that are miraculously interwoven. While biological networks are truly fascinating, they are at the same time notoriously difficult to study due to their intrinsic complexity. For a long time, this was especially the case for molecular networks, as lack of technology limited the simultaneous measurement of the various components and interactions. This situation changed dramatically with technological breakthroughs in genomics, proteomics, and metabolomics, enabling the profiling of thousands of molecules.

In molecular biology, this motivated a shift of paradigms from reductionism, which places single components in the foreground, to systems biology, where we aim to study molecular processes as a whole [24.1, 2]. The key idea of systems biology is that the total system is more than the simple sum of its single components, i.e.. new properties of a system are emerging that are not exhibited by its single isolated components. The two main branches of systems biology are:

1. Generation of systems-wide in vitro and in vivo (and possibly quantitative) data
2. In silico representation and analysis of the biological systems.

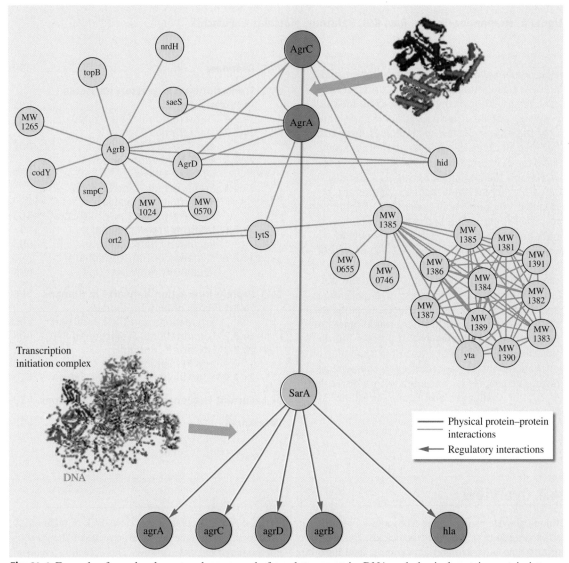

Fig. 24.1 Example of a molecular network composed of regulatory protein–DNA and physical protein–protein interactions. The network displays a model of a two-component signal-transduction cascade in *Staphylococcus aureus*. The sensor histidine kinase receptor AgrC, upon binding of a cyclic peptide, phosphorylates the response regulator AgrA, which in turn activates the transcription factor SarA. In conjunction with AgrA, SarA induces the expression of Agr virulence factors and the hla gene (RNAIII). Nodes represent genes or proteins in this network and are colored depending on their role: target genes in *green*, TF in *orange*, and the two-component system in *blue*. *Red edges* represent protein-DNA interactions, whereas *green* or *gray edges* represent PPIs. The *crystal structures* representing the different types of interactions are taken from PDB and are only for illustration purposes (i. e., showing equivalent structures only)

Besides providing a better understanding of complex molecular mechanisms, systems biology aims to establish a framework for integration and consolidation of different types of data. This is particularly important, as the current bottleneck in biomedical research is frequently not the generation of data, but their analysis and interpretation.

The application of systems biology can be manifold. Signaling pathways provide an excellent case how models of complex mechanisms can help biomedical research. The identification of one disease target in a pathway implies that other components of the same pathway may serve as alternative drug targets drastically increasing the possibilities for novel therapeutic interventions. For instance, the cholesterol synthesis pathway, whose end product is linked to cardiovascular diseases, can nowadays be targeted by multiple drugs. It should, however, be noted that although the term systems biology was coined only recently, many ideas of systems biology were introduced much earlier.

A major challenge of systems biology is the adequate representation of the system. Numerous components might have to be included for a consistent model. Especially for larger biological systems, the representation and analysis is computationally challenging. Here, concepts of network biology might help [24.3]. Its basis is the representation of biological systems as mathematical graphs. This leads to considerable simplification (see, e.g., Fig. 24.1). Molecular networks are conceived as connected sets of vertices, which can be analyzed using well-established tools of graph theory.

The importance of network-based approaches and their implementation will be illustrated on two specific types of networks in this chapter. First, we will review the study of transcriptional regulatory networks in prokaryotes. Transcriptional regulation in prokaryotes has been a focus of molecular biology since the seminal work by *Monod* and *Jacobs* elucidating the genetic regulation of the Lac operon [24.4]. In fact, much insight into molecular networks and their functioning has been gained by the study of transcriptional regulation in prokaryotes. We will thus introduce the underlying biology and give an overview of how to model regulatory networks.

As the second type of network, physical protein interaction networks in humans will be discussed. Proteins are of crucial importance for the correct functioning and orchestration of many cellular processes. Most proteins do not function alone, but within a cellular context network by interactions with other proteins. To obtain a better understanding, several large-scale interaction networks for human proteins have been constructed. Besides summarizing current approaches to obtain a human protein interactome, we present some network-based studies in cancer research. Especially for cancer, systems and network biology might provide promising tools to deal with the variability and heterogeneity of carcinogenesis.

It should be noted that other types of molecular networks might allow – or even demand – different kinds of representations and analysis tools. For instance, detailed knowledge of metabolic pathways has enabled the development of advanced mathematical frameworks, which can be used for a quantitative prediction instead of solely qualitative description [24.5]. For the types of networks that will be discussed here, the availability of high quality data is more restricted. Also, regulatory transcriptional networks and protein interaction networks exhibit a high degree of intrinsic complexity, making the formulation of exact models challenging.

24.2 Transcriptional Regulatory Networks in Prokaryotes

Transcriptional regulatory networks have drawn considerable attention for various reasons. Besides the experimental ease of studying prokaryotes in the laboratory and their general importance in biology and medicine, prokaryotes are prime examples of adaption and optimization of intrinsic cellular mechanisms to extrinsic conditions. Due to evolutionary pressure, the genomes of the first prokaryotic organisms have been extensively re-modeled to adapt to different environmental conditions. This adaptation is facilitated by horizontal gene transfer and a fast generation time [24.6]. As a consequence, prokaryotic organisms have successfully conquered most ecological niches on earth. Notably, the increasing number of sequenced genomes as well as of genome-wide expression studies has facilitated the study of prokaryotic adaption by means of bioinformatics and systems biology.

Before we discuss in more detail approaches to elucidating transcriptional regulatory networks, we briefly review the main components of prokaryotic transcriptional regulation in the next sections.

24.2.1 Transcription in Prokaryotes

Transcription describes the process by which RNA polymerases read and copy information from DNA to RNA. The prokaryotic RNA polymerase is a complex formed by several catalytic subunits and a single regulatory subunit known as the sigma (σ) factor. Transcription can generally be divided into four phases:

1. Preinitiation
2. Initiation
3. Elongation
4. Termination.

For regulation of gene expression, preinitiation and initiation are the most important phases.

During preinitiation, a sigma factor binds to an upstream sequence region (also called the promoter region) of the gene to be transcribed. Usually, several sigma factors exist in bacterial genomes. One is normally considered as the principal sigma factor responsible for expression of housekeeping genes, while the other sigma factors will regulate expression of genes required under specific conditions. After sigma factor binding, the RNA polymerase associates with promoter elements situated approximately at -10 and -35 bases upstream of the transcription start site.

Besides sequences recognized by sigma factors, the promoter can contain other specific sites that determine binding of different regulatory elements. In particular, short sequences can act as binding motifs for transcription factors (TF), which can be activators or repressors of gene expression. Thus, both sigma and TF regulators recognize specific DNA patterns in promoter regions; it is the combinations of these sequence elements that results in selectivity of gene expression. It is important to note that the initiation of transcription is not ensured by binding of sigma or TFs to the promoter, as cells have a limited pool of RNA polymerases and, therefore, subsequent assembly of RNA polymerases is rather a competitive process.

After binding of the RNA polymerase to the sigma factor and initiation of transcription, elongation continues until the RNA polymerase core releases the RNA transcript and dissociates from the DNA template. The dissociation can be provoked by binding of a so-called rho cofactor or formation of hairpin structures in rho-independent transcriptional terminators [24.7]. The synthesized RNA can then serve as template (in the form of messenger RNA) for protein synthesis or can have functionality by itself (e.g., in the form of ribosomal RNA).

24.2.2 Transcription Factors and Their Binding Sites

As mentioned earlier, TFs are the main determinants for transcription initiation besides sigma factors. Their binding to specific sequences in the promoter region is typically regulated in response to intracellular or environmental stimuli. The TF binding sites are normally $12-30$ nucleotides long. To understand the specificity of transcriptions, it is important to identify such binding sites. A common way is to search for a conserved short sequence (i. e., a consensus sequence) in the promoter region of genes under the control of the TF. Alternatively, one can determine the binding site experimentally. To this end, the promoter region is fused to a reporter gene and subsequently truncated or modified until the area responsible for the specific gene regulation is localized. Mutation analysis can then be used to confirm the TF factor binding site.

Frequently, our primary interest does not lie on regulation of a single gene but on the characterization of genome-wide effects of TFs. In this case, chromatin immunoprecipitation can be applied in case a specific antibody for the TF is available. The technique permits the coisolation of the TF and its bound DNA. Subsequently, the isolated double stranded DNA can then be labeled and hybridized onto a microarray (ChIP-chip) [24.8–10] or sequenced (ChIP-seq) [24.11, 12] in order to determine the promoters that were bound to the TF.

24.2.3 Gene Expression Control by Small RNAs

The importance of small noncoding RNAs (ncRNA) for regulation of gene expression has only recently come to be fully appreciated, despite the fact that evidence for regulatory ncRNA in bacteria was reported already over 30 years ago [24.13, 14]. The list of ncRNAs detected in prokaryotes has rapidly increased with the number of new genomes being sequenced. Approaches for their identification range from computational predictions [24.15] to experimental techniques such as high-throughput pyro-sequencing [24.16]. ncRNAs can influence gene expression in various ways. They can bind directly to specific mRNAs affecting their stability or translation rate or they can attach to proteins modifying their activity. Although the function of many ncRNAs is still unknown, their importance in gene regulation has become undisputed. In fact, cellular response to environmental changes seems to frequently depend on a tight coordination between ncRNAs and protein regulators. One intriguing example is the interplay of TF Fur (Ferric uptake regulator) and an ncRNA termed RhyB in order to ensure fast response to perturbations in extra-cellular iron concentrations [24.17].

24.2.4 Resources for Transcriptional Regulatory Interactions in Prokaryotes

Recent progress in sequencing and expression profiling has delivered us a wealth of data and has fueled the development of bioinformatics tools and online resources for the study of transcriptional regulation in prokaryotes. Unfortunately, a persistent challenge is that most studies and resources are biased towards a very small number of model organisms and thus may be of limited utility for the study of the great majority of prokaryotes. However, it can be expected that prokaryotes share at least a rudimentary basic regulatory network, as they appear to have derived from a common ancestor. From this common network, branches might have then evolved to adapt to specific external conditions [24.18]. Thus, genome level comparison of well-studied model organisms, such as *Escherichia coli* [24.19–21] and *Bacillus subtillis* [24.22, 23], might still be useful for interpretation of regulatory networks in less studied organisms.

Table 24.1 gives a short list of selected tools and resources for the study of transcriptional regulation in prokaryotes. It should be noted that this research field is highly dynamic, and new tools and databases are rapidly being developed. Thus, Table 24.1 should not be considered as a comprehensive overview, but rather a collection of initial pointers for the interested reader.

24.2.5 Inferring Transcriptional Regulatory Networks

Capturing the transcriptional regulatory network of a given organism should equip us with an explicit and comprehensive model of its transcriptional regulation. In the ideal case, this network would contain all regulatory components (such as TFs, ncRNAs, and their binding sites), as well as all the regulatory interactions between them and with their targets. Such a model could enable us to predict accurately transcript levels of target genes. Clearly, comprehensive regulatory networks would not only be valuable for microbiology, but could be of crucial importance for future bioengineering and synthetic biology, permitting accurate prediction of cellular responses to genetic manipulations. At present, however, we are still far from such models. Despite all the recent technological breakthroughs, deriving models for transcriptional regulatory networks remains

Table 24.1 Bioinformatic tools and on-line resources for the discovery, analysis, or visualization of transcription factor binding sites, regulatory motifs, ncRNAs, and transcriptional networks

Resource	URL	Description
TFBS/motif discovery tools		
AlignACE	http://atlas.med.harvard.edu/	Identification of DNA motifs
MEME	http://meme.sdsc.edu/	Identification of DNA motifs
Promoter prediction		
PPP	http://bioinformatics.biol.rug.nl/websoftware/ppp/	Prediction of promoter
PromEC	http://margalit.huji.ac.il/	Database of *E. coli* mRNA promoters
MOTIFATOR	http://www.motifator.nl	Mining and characterization of regulatory DNA motifs
ncRNAs prediction		
nocoRNAc	http://it.inf.uni-tuebingen.de/de/nocornac.php	Prediction and characterization of ncRNA
Databases		
MicrobesOnline	http://www.microbesonline.org	Compendium of genomes, expression data, operon predictions
RegPrecise	http://regprecise.lbl.gov/RegPrecise/index.jsp	Database of curated regulons in prokaryotic genomes
RegTransBase	http://regtransbase.lbl.gov	Curated database of regulatory interactions and database of TFBS
RegulonD	http://regulondb.ccg.unam.mx/	*E. coli* transcriptional regulatory network
Network analysis		
Bioconductor	http://www.bioconductor.org/	Network visualization and analysis
Cytoscape	http://www.cytoscape.org/	Network visualization and analysis
Pajek	http://pajek.imfm.si/	Network visualization and analysis

a formidable task even for simple processes in well-studied organism.

Commonly, transcriptional regulatory networks are represented as directed graphs comprised of nodes and edges. Nodes can represent TFs or ncRNAs and their target genes, while edges indicate regulatory interactions between the different components of the network [24.24]. To re-construct these networks, most approaches can be broadly classified into three types:

1. *Knowledge-based approaches*: Traditionally, models of transcriptional regulatory networks have been derived using the knowledge accumulated in scientific literature and databases. Here, an initial network model is constructed to simulate the system's behavior under different perturbations and is subsequently adjusted to consolidate it with experimental observations. Examples of this approach can be found later, where we discuss differential equations for modeling.

2. *Reverse engineering*: Recently, reverse engineering has become an promising alternative given the rapid increase of available expression data. Here, regulatory interactions are inferred directly from observed expression patterns of genes using correlation measures. The predicted transcriptional response is then compared with measured expression data and the network structure is iteratively improved. Reverse engineering methods might reduce the time needed to obtain accurate regulatory network models but also demand a large amount of data for inference [24.25]. Network inference can be direct or module-based and examples are given in the next section.

3. *Template-based methods*: Due to astonishing advances in sequencing technology, the number of available prokaryotic sequenced genomes has greatly increased over recent years. Notably, many of these newly sequenced genomes belong to organisms, of which only a limited knowledge of their biology exists. In such cases, transcriptional regulatory networks of a phylogenetically related organism might be used as a template for inference. Such methods are based on evolutionary conservation of regulation and can take advantage of abundant information on regulatory interaction from model organisms. A good example of this approach is the work by *Babu* and coworkers who inferred conserved regulatory networks of an organism by comparing its genes to known TFs and their gene targets in *E. coli* [24.26]. Homologs of

E. coli genes were identified by sequence comparison. If TF and its target are found to be conserved, their regulatory interaction is predicted to be also conserved. Clearly, template-based methods lose accuracy when applied to phylogenetically distant organisms. Here, additional analysis of potential binding sites for TFs might improve prediction of target genes [24.27].

Direct and Module-Based Network Inference

Direct and module-based methods use gene expression to detect and model the underlying regulatory interactions. Both methods assume that changes in gene expression are caused by changes in the expression of TFs. A direct inference approach was applied for *E. coli* by *Faith* and coworkers [24.28]. Initially, 179 measurements of microarrays for 69 different conditions from nine previous studies were collected and expanded by further in-house 266 microarray measurements for 121 selected conditions. The applied context likelihood of relatedness (CLR) algorithm is based on the previously developed concept of relevance networks [24.29, 30]. The CLR method assumes that the expression of a gene and of its potential regulator should vary in a coordinated manner over time and across environmental conditions. As a measure of correlation of expression, CLR calculates the mutual information between expression profiles. To assess significance of coexpression, the observed mutual information is compared to a background distribution of mutual information scores for all possible TF–gene pairs. Only pairs with mutual information scores that are significantly higher than the background distribution were subsequently included in the transcriptional regulatory network. The authors compared the detected interactions by the CLR algorithm with those annotated in the curated RegulonDB database and found that the filtering for significant interactions reduced the number of false positives considerably. At a 60% true positive rate, CLR predicted 1079 regulatory interactions, of which 741 were novel.

An alternative approach to capture relationships between regulators and their target genes is based on clustering of gene expression data. Clustering is commonly used to group genes into sets with similar expression patterns. Such sets of coexpressed genes, which are also called modules, may underlie the same transcriptional regulation and are the basic units of module networks. For these types of networks, we infer regulatory interactions between TF and whole modules instead of interactions between TF and their individual target genes. The usage of modules instead of single genes drastically reduces the number of variables in the model, and thus, the computational burden as well as the risk of over-fitting.

This approach was used by *Bonneau* and colleagues to study the transcriptional regulatory network of the poorly characterized *Halobacterium salinarum NRC-1* [24.31]. Genome-wide expression data for a variety of different environmental conditions and genetic mutations were clustered using the cMonkey algorithm [24.32]. This algorithm performs parallel *biclustering* of genes and conditions. The derived clusters (or modules) of genes can therefore be specific to conditions, in which the genes are coexpressed. This capacity of biclustering can allow subsequent identification of conditions in which a regulator is active. The final transcriptional regulatory network was derived by the *inferelator* method, which connects expression changes of individual or multiple TFs to expression changes in the detected modules using a regression algorithm [24.33].

24.2.6 Modeling of Transcriptional Regulatory Networks

Once the structure of a regulatory network has been inferred, its behavior can be modeled. A key step is the selection of the mathematical approach to be employed. This decision will mainly depend on the amount, type, and quality of available data and information. Several alternatives have been developed based on different mathematical frameworks such a differential equations [24.34], Boolean networks [24.35, 36], or Bayesian networks [24.37–39]. In the following section, we will briefly illustrate their use in modeling transcriptional networks.

Differential Equations. The usage of differential equations to describe regulatory relationships originated from the kinetic modeling of biochemical reactions [24.40]. A transcriptional regulatory network can be modeled by a system of linear differential equations with its rate parameters known or estimated. As exact parameters have typically to be estimated, it is important to limit the number of regulatory interactions. Therefore, a common assumption is that gene expression is determined by a small set of global transcriptional regulators. Also, the use of differential equations for modeling is computationally expensive. Thus, they are commonly used only for the modeling of small systems with a limited number of components and interactions.

Ropers and coworkers used piecewise-linear differential equations to simulate the adaption of *E. coli* during transition from a carbon-rich to a carbon-poor environment [24.41]. Their model consisted of six piecewise-linear differential equations which represented the expression of key global regulators over time during the response to carbon limitation. The use of piecewise-linear equations permitted qualitative analysis of the network dynamics even as quantitative information was scarce. For simulation, the open source software tool *Genetic Network Analyzer* was used [24.42]. The resulting network was extended in a later study to include also directionality of regulation and influence of metabolites on the network flux [24.43].

In a recent study, differential equations were used to model cell cycle regulation in *Caulobacter crescentus* [24.44]. The model described the dynamics of three global regulators (GcrA, DnaA, and CtrA) and other cell cycle proteins by 16 nonlinear ordinary differential equations. Although the model is of relative small scale, more than 40 parameters (rate constants, binding constants, and thresholds) had to be optimized, after initial values were estimated based on experimental observations from the literature. As an application of the trained model, the phenotypes of novel mutants could be predicted successfully. This shows that the modeling through differential equations is applicable for cases where sufficient information is available and the size of the system is small.

Boolean Networks

In the framework of Boolean networks, gene expression levels are binned to the values 1 or 0. Therefore, Boolean networks assume that a gene can be either expressed, i.e., in the *on* state described with the value 1, or not expressed, i.e., in the *off* state described with the value 0 [24.45]. During simulation, the level of a gene is derived from the levels of other genes via a Boolean function. In particular, the expression of a gene at a given time point depends on the expression of its regulators at the previous time point. The *on–off* assumption limits the capabilities of Boolean networks. Also, they fail to model TFs that regulate their own expression. Nevertheless, in cases where only qualitative knowledge is available and the size of the network is of moderate size, Boolean networks can explore the full state space.

The work by *Samal* and *Jain* provides a good example of how Boolean networks can assist in system level analysis and modeling [24.46]. Their study is based on a previously compiled data set for *E. coli* [24.47]. The constructed Boolean network comprised almost 600 genes and 100 metabolites. Studying the dynamical properties of the network, they found that the network states are highly robust to perturbation of single genes, while the system is still highly responsive to environmental changes.

Probabilistic Boolean Networks

Frequently, the lack of experimental information on certain regulatory entities makes it necessary to include probabilistic elements. Probabilistic elements were introduced to Boolean networks to overcome their deterministic nature [24.48]. Each entity is modeled by a family of Boolean functions, to which probabilities are assigned. The function is chosen with assigned probability to predict the particular expression value of the target gene at certain time point or condition. The stochastic model allows a greater range of possible solutions and can cope with uncertainty.

Chandrasekaran and *Price* integrated metabolic and transcriptional regulatory networks using an approach closely related to probabilistic Boolean networks [24.49]. Their method termed PROM (probabilistic regulation of metabolism) circumvents the tedious process involved in the determination of the Boolean rules between the regulator and its targets by automatically deriving interactions from high-throughput data. For *E. coli*, PROM was tested and compared with other state of the art methods that combine metabolic and gene regulatory data using Boolean rules. The main advantage of PROM resides in its automatism to calculate conditional probabilities between gene states from

expression data. The authors also demonstrated the capacity of their approach by genome-scale modeling the metabolic and regulatory network of *Mycobacterium tuberculosis* from available regulatory information and microarray data. Their model permitted the prediction of phenotypes for several TF knockout mutants, as well as the identification of genes candidates for drug targeting.

Bayesian Networks. Bayesian networks display conditional dependencies between variables represented as nodes. Each node is associated with a function determining the probability of the corresponding variable. A drawback of Bayesian networks is the failure to model regulatory feedback loops, since they can infer only directed acyclic graphs. This limitation is partially solved by the use of dynamical Bayesian networks (DBNs) [24.50].

Hodges et al. [24.51] used an implementation of the Bayesian network to analyze microarray gene expression data from *E. coli* and model the response to 27 genes linked to the reactive oxygen species (ROS) detoxification pathway defined in the EcoCyc database [24.52]. Expression data from over 300 measurements were utilized to score randomly initiated Bayesian networks. Subsequently, a consensus network was constructed from the networks with the highest posterior probability score. Even though the interactions of the consensus network did not fully match the known ROS pathway, it served as base for expansion. Some predictions of novel genes and interactions in the ROS pathway were successfully experimentally validated.

24.3 Protein Interaction Networks in Humans

Many – if not most – proteins do not function alone, but through interaction with other proteins in a defined physiological context. In fact, extensive experimental and computational analysis of protein–protein interactions (PPIs) in *S. cerevisae* showed that a large part of the proteome is organized in cluster structures (or so-called modules), in which proteins are highly connected [24.53]. The study of PPIs is, therefore, an essential prerequisite for understanding many cellular processes. Here, we present a brief review of the experimental detection and computational prediction of PPIs, as well as examples of relevant databases and network-orientated applications of PPI data. A graphical overview is presented in Fig. 24.2.

24.3.1 Types of Protein–Protein Interactions and Their Detection

PPIs can be broadly classified based on the complexes formed and the duration of interaction:

1. Interaction of identical polypeptide chains form homo-oligomers, whereas interactions of nonidentical chains form hetero-oligomers.
2. Permanent interactions persist until the complex is degraded, whereas transient interactions can form and break in vivo [24.54].

Various experimental techniques have been developed to detect different types of interactions. It is

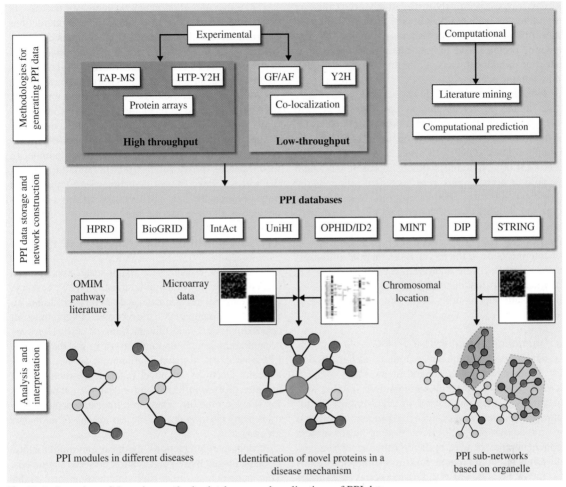

Fig. 24.2 Overview of detection methods, databases, and applications of PPI data

important to appreciate their characteristics as well as their strength and weakness, as different techniques can produce distinct sets of interactions even for the same set of proteins. In the following section, we introduce the yeast-two-hybrid (Y2H) and coimmunoprecipitation (Co-IP) approach, which have been used to screen for protein interactions in a high-throughput mode.

Yeast–Two–Hybrid (Y2H) Assay

Y2H is a well-established system, which was initially introduced in 1989 [24.55] to identify interactions between two selected proteins and has been now scaled up to cover the entire proteome of an organism. In the Y2H system, one protein (the so-called bait) is fused with the DNA binding domain of a yeast TF, while the second protein (prey) is fused with its activation domain.

Physical binding of bait and prey proteins reconstitutes a functional TF whose activity is monitored through a reporter gene. In the last decade, the technique has been successfully employed in large-scale mapping of PPIs for model organisms such as *S. cerevisiae* [24.56], *C. elegans* [24.57], and *D. melanogaster* [24.58], as well as for humans [24.59, 60]. In a high-throughput mode, large sets of yeast strains with bait and prey proteins are constructed. A common approach is to mate individual strains containing bait proteins with a pooled library of yeast strains containing prey proteins [24.61]. For positive read-outs, the interacting prey protein will be determined by sequencing. Besides its usability for large screens, an attractive feature of the Y2H assay is that weak transient interactions can be detected. However, false results may occur when additional proteins present in yeast positively or negatively affect binding of proteins assayed, or if bait or prey proteins itself have transcriptional activator or repressor activity. Also, interacting proteins have to be located to the nucleus for detection, which can cause difficulties in the screening of membrane proteins.

Notably, the Y2H system was applied in two large-scale studies to screen human proteins [24.59, 60]. *Stelzl* and coworkers used a combination of a fetal brain cDNA library and a set of full-length open reading frames (ORFs) to create over 11 000 Y2H clones [24.60]. Applying a pooling approach, more than 25 million protein pairs were tested resulting in the identification of over 3000 interactions. Independently, *Rual* and collaborators performed an Y2H screen with more than 8000 ORFs and detected ≈ 2800 interactions [24.59]. However, caution needs to be exercised in the interpretation of the reported interactions of human proteins, as they were measured in yeast (i. e., outside their native surroundings) and posttranscriptional modification altering binding might not be conserved in yeast.

Coimmunoprecipitation (Co-IP)

Coimmunoprecipitation is a commonly used method to test whether proteins are bound in the same complex in vivo. The target protein is immunoprecipitated with an antibody and fractionated by SDS-PAGE. Coimmunoprecipitated proteins are subsequently detected by autoradiography or Western blotting. Also, protein sequencing may be used to identify the interacting proteins. It is important to note that this technique assumes that interactions are preserved when a cell is lysed under nondenaturing conditions. To identify protein interaction in a high throughput mode, affinity purification is coupled with subsequent mass spectrometry [24.62,63]. On a system-wide level, this combination was first exercised for yeast [24.53]. A general observation was that a considerable part of the proteome can be organized in protein complexes. For instance, *Gavin* and colleagues were able to identify over 200 mostly novel complexes for an initial set of more than 1700 tagged yeast proteins [24.53]. Detailed analyses revealed that most protein complexes form a higher order network beyond the level of binary interactions. To inspect human complexes, *Ewing* chose an initial set of 338 bait proteins expressed in a human cell line (HEK293) [24.64]. Despite the rather small number of bait proteins, almost 25 000 interactions were detected. After filtering, ≈ 6500 interactions between over 2200 proteins remained, most of which were newly identified.

Note that identified interactions are not necessarily direct interactions, but might be indirect, interactions between proteins in the precipitated complex. This has consequences if we want to represent the complex by binary interactions in order to facilitate network analysis. As the internal structure of the complex is generally not known, two generic models are commonly employed: the *matrix* model postulates that all proteins in a complex interact with each other. Naturally, this postulation results in many interactions, especially for large complexes and thus potentially implies a large number of false positive binary interactions [24.65]. In contrast, the *spike* model assumes that direct interactions exist only between the bait and the coprecipitated proteins neglecting all other internal structures in the complex.

24.3.2 Computational Prediction of PPI

Despite rapid advances in large-scale experimental techniques applied in mapping the human PPI network, the coverage of experimentally determined PPI data remains scarce compared to the estimated total number of approximately 650 000 interactions [24.66]. However, computational methods might assist in the mapping of PPI networks. Computational approaches not only enable the discovery of novel putative interactions, but can also provide information for designing experiments for specific proteins. They are either based on predictions using existing data or on text mining of published literature.

For human interactions, the most important method is based on sequence conservation between organisms. This approach assumes that interactions are evolutionary conserved between orthologous proteins in different organisms. Initially, the concept of so-called *interologs* was introduced to examine the biological relevance of Y2H-derived interactions [24.67]. Although subsequent experimental validation only resulted in a limited accuracy (up to 30%) for predicted PPIs in *C. elegans*, several large sets of human PPIs have been computationally extrapolated from experimentally measured interactions in lower organisms, especially yeast, worms, flies, and mice [24.68–70]. For the identification of human orthologs for interacting proteins, the InParanoid algorithm has commonly been employed [24.69–71]. In a first application of the interolog concept to human PPIs, interactions from three model organisms (*S. cerevisiae*, *D. melanogaster*, and *C. elegans*) were utilized [24.69]. An interaction was predicted if both interacting proteins in a model organism have one or more human orthologs. Using this strategy, the authors generated a human interaction network comprising $\approx 71\,000$ interactions between ≈ 6000 human proteins. The generated map was then filtered using gene ontology annotation and coexpression in order to identify a core network of over 9500 interactions between ≈ 3500 unique proteins. A similar study was undertaken by *Perisco* et al. and led to the construction of the HomoMINT database [24.70]. Besides using interactions from lower organisms, they analyzed the domain composition of human proteins to refine the predictions of interaction. Instead of the InParanoid algorithm, *Brown* and *Jurisica* applied a BLAST and reciprocal best-hit approach to extrapolate interactions between organisms [24.72]. They created first an integrated interaction dataset from various model organisms and mapped it to human orthologs by aligning proteins from each model organism against human proteins stored in a SWISS-Prot database. As a next step, each top BLAST hit surpassing a predefined significance threshold was matched against the set of all protein sequences of the model organism. A protein was considered as potential ortholog, if it matched the original query protein in reverse direction. Following this method, the authors generated a human PPI map containing $\approx 25\,000$ interactions between ≈ 4000 proteins.

Ideally, we would like to accurately predict potential protein interactions directly from the primary sequences. Notably, several groups have undertaken this challenging task and have reported the successful prediction of PPIs based on sequence data only [24.73–77]. The different approaches commonly represent a protein sequence as a vector of features (such as the physicochemical properties of amino acids) and a protein pair as concatenation of the corresponding feature vectors. For subsequent classification and prediction, a support vector machine or other kernel-based method can be trained to distinguish between concatenated feature vectors of interacting and noninteracting protein pairs. The reported high accuracy, however, is somewhat surprising given that many PPI depend on the mutual recognition of detailed binding interfaces. Indeed, follow-up studies have shown that the reported performance depends strongly on the selection of training and test sets [24.77, 78]. Thus, the results need to be interpreted with caution. Nevertheless, the proposed prediction methods might help to derive an initial set of potential PPIs for experimental validation.

In addition to prediction based on existing data, potential PPIs can be computationally indicated by evaluating existing literature. A simple text-mining approach is the measurement of the frequency of protein names cooccurring in the same scientific text. If the frequency is higher than expected by chance, a functional association and potential physical interaction might be implicated. An early application of this approach generated a network of over 6600 interactions connecting ≈ 3700 human proteins based on text-mining of Medline abstracts [24.79]. However, one needs to keep in mind that the interactions deduced need not be physical. The advantages of such data mining techniques are that they are not biased to any particular study or experimental technique. A major drawback can be significant selection and detection bias towards well-studied proteins [24.80].

24.3.3 Literature Curation and PPI Databases

Besides experimental high-throughput approaches and computational prediction, the manual curation of literature is one main strategy to obtain large sets of human PPIs. Indeed, many interactions have been derived in dedicated small-scale experiments and have been reported in the scientific literature. To curate this information in a systematic way, several databases have been established including the Human Protein Reference Database (HPRD) [24.81], the Biological General Repository for Interaction Datasets (BioGRID) [24.82], IntAct [24.83], Database of Interaction Proteins (DIP) [24.84], the Biomolecular Interaction Network Database (BIND) [24.85], and the Mammalian Protein–Protein Interaction Database (MPPI) [24.86].

Notably, these databases are not synchronized with each other, and their data formats might also be incompatible. To merge publicly available protein interaction data, a time-consuming reformatting and mapping must be undertaken. To improve data exchange, *Hermjakob* et al. [24.87] proposed the molecular interaction (MI) extensible markup language (XML) format as a standard for the representation and exchange of protein interaction data. Notably, the PSI MI format is database-independent. This format can also help combine data from different sources.

To relieve users of cumbersome data preprocessing and merging, several databases have integrated both experimental as well as predicted PPIs sets. Examples are I2D [24.68], STRING [24.88] and UniHI [24.89]. For instance, the Unified Human Interactome database (UniHI) includes human interaction data from 14 different sources and comprises approximately $250\,000$ interactions between over $22\,000$ proteins [24.89]. The latest version of UniHI not only integrates the data from various sources, but also provides easy query options in order to extract a maximum of information. Furthermore, users can also map their own gene expression data onto the PPI network. Additionally, analyses can be performed to detect associations of a PPI network with biological processes, molecular functions, and cellular components. To illustrate the application of UniHI, we used expression data for a mutant p53 cell line [24.90]. The original study focused on the role of mutant p53 in promoting transformation and metastasis in breast cancer through aberrant protein interactions. UniHI enables one to derive a PPI network around p53 and to readily map the gene expression data onto the network. After filtering of interactions by differential expression, we

obtained a specific dysregulated p53-focused network. Using the tool for functional analysis in UniHI, we detected DNA damage response, cell cycle and regulation of apoptosis as processes that are significantly enriched in the network (Fig. 24.3). As this example illustrates, we can rapidly identify p53 interactions and functions that are potentially affected by dysregulation and that can serve as starting point for further study.

24.3.4 Network–Based Study of Cancer

For the study of pathogenic processes, the usage of PPI networks has become an attractive paradigm. Diseases frequently occur not due to malfunctioning of a single protein but of whole complexes or modules. An exam-

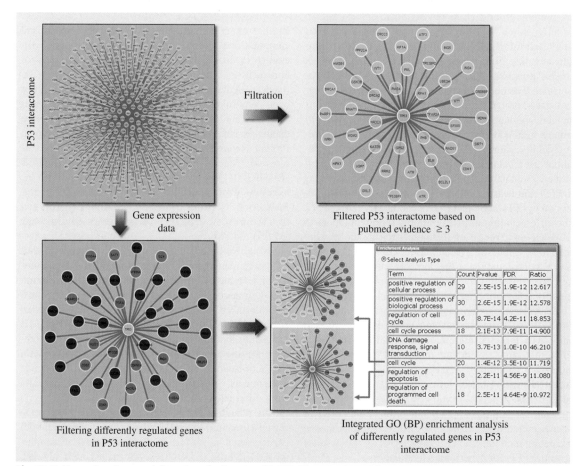

Fig. 24.3 Example of network-based analysis using UniHI. For proteins of interest such as p53, interaction partners can be queried and visualized. The derived networks can be subsequently filtered based on evidence (e.g., number of PubMed references reporting the interaction) or based on gene expression data. All networks can be readily inspected for enrichment in biological processes using an integrated tool

ple showing the importance of such complexes is Faconi anaemia, a genetic disease associated with an increased rate of leukaemia and bone marrow failure. Remarkably, the corresponding proteins of seven of the nine genes associated with Faconi anaemia form a physical complex involved in DNA repair. In fact, cancer per se can be seen as a classical example of how alterations in several genes and proteins need to occur for the disease to develop and progress. Genes involved in cancer can be generally assigned to three broad classes based on their functions:

1. *Oncogene* that can lead in their mutated form to aberrant growth signals
2. *Tumor-suppressor genes* that deter cells from unrestricted proliferation
3. *Stability genes* that keep genetic alterations to a minimum.

It is important to note that no single gene defect causes cancer. Rather, multiple mutations in oncogenes, tumor-suppressors, as well as stability genes have to occur to overcome the control mechanisms imbedded in cells [24.91, 92]. In support of this concept, recent sequencing of a large number of genes in breast and colorectal tumors revealed various mutations accumulated in a single tumor [24.93]. Eventually, the changes on the molecular level lead to a series of complex alterations in cell physiology, ranging from self-sufficiency in growth signals and evasion of apoptosis to tissue invasion and metastasis [24.91]. Again, each of these steps displays a high variability and is driven by multiple molecular mechanisms. To study this complexity and to find new targets in cancer treatment, network-based approaches promise to provide us with a powerful tool.

Identification of Novel Cancer-Associated Genes

PPI data have been utilized in various studies to identify novel genes associated with cancer. Such genes can aid in the early diagnosis and in the development of effective therapies. For instance, *Russo* et al. combined human PPI with other types of data to study prostate cancer [24.94]. Using gene expression and PPI data as well as pathway information, they identified the Met receptor tyrosine kinase as a metastatic biomarker and central regulator of prostate cancer progression. Another example of network-based approaches is the detection of GTP binding protein 4 (GTPBP4) as a marker for the breast cancer disease prognosis [24.95]. In this study, the authors first identified novel interactors of p53 in Drosophila and searched for corresponding

orthologs in humans. One of these novel interacting partners of p53 was GTPBP4. Knockdown of GTPBP4 in a cancer cell line led to activation of p53. Also, it was observed that expression levels of GTPBP4 were inversely correlated with the survival of the breast cancer patient. In the study of glioblastoma, a highly aggressive form of brain cancer, *Ladha* and coworkers used a set of up-regulated genes as seed proteins for network construction [24.96]. Analyses of the derived PPI network revealed that two proteins (casein kinase 2 catalytic subunit α and protein phosphatase 1 alpha) connected to network structures were potentially associated with cancer progression. Both proteins were shown to be up-regulated in a set of glioblastoma tumor samples. In a conceptually similar work, a serous ovarian cancer-related network was constructed [24.97]. Initially, 30 genes were obtained, which showed consistent up- or down-regulation of expression in different data sets. They were used as initial nodes for network construction. Subsequent integration of gene expression data with PPI data identified highly connected proteins (hubs) involved in early events of cell cycle progression, mismatch repair, and aneuploidy in ovarian cancer as well as novel cancer-related genes, whose precise involvement remains to be investigated.

As the studies show, the integration of PPI with complementary data is a common approach to detect cancer genes. A final example is a study of breast cancer, where the authors integrate PPI networks from different species with coexpression and genetic interactions, as well as functional gene annotation [24.98]. Starting from a set with four known breast cancer-associated genes (BRCA1, BRCA2, ATM, and CHEK2), the derived network was used for the prediction of novel breast cancer susceptibility genes. For one predicted gene, HMMR encoding a centrosome subunit, a functional association with BRCA1 was experimentally demonstrated.

Network-Based Analysis of Cancer-Related Mechanisms

In addition to prediction of novel cancer-associated genes, PPI networks have also been utilized to untangle cancer features. In a recent study [24.99], the authors analyzed features of prostate cancer using PPI networks and molecular profiles in adjacent normal cells. First, a specific set of genes was derived for normal cells, normal cells adjacent to tumor, and tumor cells. Second, PPI networks for these set of genes were analyzed. Notably, they identified three subnetworks consisting of pro-inflammatory cytokine, pro-metastatic chemokines, and growth factors. Furthermore, genes encoding cytokines and growth factors were found to be expressed in adjacent normal epithelial cells (i. e., in the tumor microenvironment) suggesting that the molecular state of adjacent normal cells might have a prognostic value.

An emerging application of molecular networks analysis is the use of PPI networks to improve disease classification. *Chuang* et al. employed a network-based classifier for the prognosis of breast cancer metastasis, which is the main cause of death among breast cancer patients [24.100]. Gene expression profiles of metastatic and nonmetastatic patients were mapped onto a human PPI network and subnetworks, whose expression levels correlate with metastasis, were identified. Here, the expression level of a subnetwork was defined as a function of expression levels of the genes included. Strikingly, it was found that classification of metastatic versus nonmetastatic tumor samples is more reproducible and more accurate if it is based on subnetworks than on individual gene expression signatures only.

Apart from disease classification, PPI networks have been employed in predicting cancer outcome using their modular structure [24.101]. A modular architecture can be imposed on the human protein interactome through inter-modular hubs (having low correlation of expression with interaction partners) and intra-modular hubs (having high correlation of coexpression with interaction partners). *Taylor* et al. studied the protein domain numbers, domain sizes, and linear motifs (i. e., post-translational modification and short binding motifs) of inter-modular and intra-modular hubs in the human protein interactome across 79 human tissues [24.101]. They identified that the number of domains is higher for inter-modular hubs, whereas domain sizes are larger in intra-modular hubs. Linear motifs are over-represented in inter-modular hubs. Further, they compared the modularity of the protein interactome in breast cancer patients with respect to disease outcome and found changes in modularity that were linked to the outcome of breast cancer. Such changes may provide a prognostic signature for breast cancer.

24.4 Current Challenges and Future Directions

Molecular interactions are crucial for the correct and efficient functioning of many cellular processes. Due to their biological relevance, the study of interaction networks has evoked increasing interest in their computational analysis. In parallel, systematic efforts and specialized databases have led to a dramatic growth in the number of chartered interaction data during recent years. This growth now offers an unprecedented wealth of interaction data to researchers for their investigations.

As a long-term perspective, we can aim to identify the full repertoire of molecular interactions for all components of a cell or even of a whole organism. Knowing this *interactome* will doubtlessly yield better insight into the molecular machinery of cells and disease mechanisms. Naturally, a combination of experimental and theoretical approaches is needed to accomplish such an ambitious goal and to consolidate the interactions obtained with gene expression, functional annotation, or other types of biological data.

At present, however, the available information about biological networks is still very limited and we have to work with crude approximations in network analyses. For instance, we would frequently prefer to utilize the activity of proteins instead of the transcript level for the modeling of molecular networks. For large net-

works, however, such analyses have remained beyond the current state of the art, as necessary information of e.g., protein abundances as well as of post-translational modifications (PTMs) and their effects is required, that is, information which is generally not yet available on a genome-wide level.

Also, molecular networks are inherently complex and difficult to model accurately because of their numerous components and multiple layers. For illustration: analyses of the human protein–protein interactome are typically confronted with the large number of predicted protein-coding genes ($\approx 25\,000$), unknown spatial and temporal localization of many proteins, as well as numerous PTMs that might alter PPIs. In particular, alterations in transient interactions can lead to various changes in network structures and thus contribute to a highly dynamic network. The stability of such interactions depends on cellular physiological conditions and the environment. This feature of molecular interactions should be always taken into account in the analysis and interpretation of interaction networks, especially as most presentations (such as Fig. 24.1) give a static – and potentially misleading – picture.

The dynamics of interactions is frequently regulated by post-translational modifications of proteins. In general, the effects of PTMs are manifold: they can influence protein size, hydrophobicity, and other physical-chemical properties; they can enhance, change, or block specific protein activities; and they can target proteins to specific subcellular localization. Currently there are over 80 000 experimentally characterized PTMs across different organisms reported in the SWISS-PROT database. Despite of their importance, however, PTMs are often not included in models of interaction networks due to missing data as well as their unknown effects. In general, the full structure of global molecular networks is largely unknown, as they consist of many interwoven networks of different type such as regulatory transcriptional, metabolic, or physical protein interaction networks.

Despite current limitations, recent studies have indicated that network-based approaches can contribute substantially to our knowledge of how biological systems function. In fact, molecular networks might be a natural extension of pathways models, a concept which has been highly successful in biology and medical research. For instance, pathway models helped to organize the large number of cancer-relevant genes in functional coherent groups [24.92]. In cancer research, pathway models were motivated by the observation that mutations within the same pathway frequently produce the same effects. For example, the control of a cell cycle can not only become defective by direct mutation of the retinoblastoma (Rb) gene but equally by other mutation in the associated Cdk/Rb/E2F pathway. The pathway concept allows here for consolidation of initially unrelated observations and can further reveal communalities of different types of cancers. Additionally, pathway-focused analyses have facilitated the interpretation of mutations in their functional context.

Despite the success of the pathway concept, it is important to note that pathways by no means derive naturally and that canonical pathway models frequently seem too simplistic. Notably, an inclusion of network-based concepts might contribute to our capability to capture molecular mechanisms. Indeed, it was pointed out by *Friedman* and *Perrimon* that the canonical view of signalling pathways as compartmentalized linear circuits is in strong contrast with the results from recent systematic screens [24.102]. A reason for the traditional simplistic picture of linear pathways might have been the dominance of developmental screens based on a qualitative readout. Such a setup may allow only the most important signalling proteins to be discovered. In contrast, large numbers of signalling modifiers have been identified when assays with sensitive quantitative readout were used. Thus, it can be argued that signalling transduction should be seen rather as quantitative information propagating through densely connected molecular interaction networks.

In future, we may therefore refer to specific networks (e.g., the *Wnt network*), similarly as we are now readily using the concept of pathways (e.g., the Wnt pathway). Clearly, much is still to be learned about the composition and regulation of molecular networks. However, we anticipate that such a step will constitute a crucial advancement towards a true understanding of the complexity lying within cells and organisms.

Part D | 24

References

24.1   H. Kitano: Computational systems biology, Nature **420**(6912), 206–210 (2002)

24.2   H. Kitano: Systems biology: A brief overview, Science **295**(5560), 1662–1664 (2002)

24.3   A.L. Barabasi, Z.N. Oltvai: Network biology: Understanding the cell's functional organization, Nat. Rev. Genet. **5**(2), 101–113 (2004)

24.4   F. Jacob, J. Monod: Genetic regulatory mechanisms in the synthesis of proteins, J. Mol. Biol. **3**, 318–356 (1961)

24.5   R. Steuer, B.H. Junker: Computational models of metabolism: Stability and regulation in metabolic networks, Adv. Chem. Phys. **42**, 105–251 (2008)

24.6   C. Médigue, T. Rouxel, P. Vigier, A. Hénaut, A. Danchin: Evidence for horizontal gene transfer in Escherichia coli speciation, J. Mol. Biol. **222**(4), 851–856 (1991)

24.7   M.D. Ermolaeva, H.G. Khalak, O. White, H.O. Smith, S.L. Salzberg: Prediction of transcription terminators in bacterial genomes, J. Mol. Biol. **301**(1), 27–33 (2000)

24.8   B. Ren, F. Robert, J.J. Wyrick, O. Aparicio, E.G. Jennings, I. Simon, J. Zeitlinger, J. Schreiber, N. Hannett, E. Kanin, T.L. Volkert, C.J. Wilson, S.P. Bell, R.A. Young: Genome-wide location and function of DNA binding proteins, Science **290**(5500), 2306–2309 (2000)

24.9   V.R. Iyer, C.E. Horak, C.S. Scafe, D. Botstein, M. Snyder, P.O. Brown: Genomic binding sites of the yeast cell-cycle transcription factors SBF and MBF, Nature **409**(6819), 533–538 (2001)

24.10   I. Simon, J. Barnett, N. Hannett, C.T. Harbison, N.J. Rinaldi, T.L. Volkert, J.J. Wyrick, J. Zeitlinger, D.K. Gifford, T.S. Jaakkola, R.A. Young: Serial regulation of transcriptional regulators in the yeast cell cycle, Cell **106**(6), 697–708 (2001)

24.11   D. Schmidt, M.D. Wilson, B. Ballester, P.C. Schwalie, G.D. Brown, A. Marshall, C. Kutter, S. Watt, C.P. Martinez-Jimenez, S. Mackay, I. Talianidis, P. Flicek, D.T. Odom: Five-vertebrate ChIP-seq reveals the evolutionary dynamics of transcription factor binding, Science **328**(5981), 1036–1040 (2010)

24.12   C. Kahramanoglou, A.S. Seshasayee, A.I. Prieto, D. Ibberson, S. Schmidt, J. Zimmermann, V. Benes, G.M. Fraser, N.M. Luscombe: Direct and indirect effects of H-NS and Fis on global gene expression control in Escherichia coli, Nucleic Acids Res. **39**(6), 2073–2091 (2011)

24.13   P. Stougaard, S. Molin, K. Nordstrom: RNAs involved in copy-number control and incompatibility of plasmid R1, Proc. Natl. Acad. Sci. USA **78**(10), 6008–6012 (1981)

24.14   G.G. Brownlee: Sequence of 6S RNA of E. coli, Nat. New Biol. **229**(5), 147–149 (1971)

24.15   A. Herbig, K. Nieselt: nocoRNAc: Characterization of noncoding RNAs in prokaryotes, BMC Bioinformatics **12**(1), 40 (2011)

24.16   R.C. Novais, Y.R. Thorstenson: The evolution of pyrosequencing for microbiology: From genes to genomes, J. Microbiol. Methods **86**(1), 1–7 (2011)

24.17   E. Masse, S. Gottesman: A small RNA regulates the expression of genes involved in iron metabolism in Escherichia coli, Proc. Natl. Acad. Sci. USA **99**(7), 4620–4625 (2002)

24.18   V. Daubin, M. Gouy, G. Perriere: A phylogenomic approach to bacterial phylogeny: Evidence of a core of genes sharing a common history, Genome Res. **12**(7), 1080–1090 (2002)

24.19   D. Thieffry, A.M. Huerta, E. Pérez-Rueda, J. Collado-Vides: From specific gene regulation to genomic networks: A global analysis of transcriptional regulation in Escherichia coli, Bioessays **20**(5), 433–440 (1998)

24.20   S.S. Shen-Orr, R. Milo, S. Mangan, U. Alon: Network motifs in the transcriptional regulation network of Escherichia coli, Nat. Genet. **31**(1), 64–68 (2002)

24.21   R. Dobrin, Q.K. Beg, A.L. Barabási, Z.N. Oltvai: Aggregation of topological motifs in the Escherichia coli transcriptional regulatory network, BMC Bioinformatics **5**, 10 (2004)

24.22   C.R. Harwood, I. Moszer: From gene regulation to gene function: Regulatory networks in bacillus subtilis, Comput. Funct. Genomics **3**(1), 37–41 (2002)

24.23   A.L. Sellerio, B. Bassetti, H. Isambert, M. Cosentino Lagomarsino: A comparative evolutionary study of transcription networks. The global role of feedback and hierachical structures, Mol. Biosyst. **5**(2), 170–179 (2009)

24.24   T.-M. Kim, P.J. Park: Advances in analysis of transcriptional regulatory networks, Wiley Interdiscip. Rev. **3**(1), 21–35 (2011)

24.25   T.S. Gardner, J.J. Faith: Reverse-engineering transcription control networks, Phys. Life Rev. **2**(1), 65–88 (2005)

24.26   M.M. Babu, N.M. Luscombe, L. Aravind, M. Gerstein, S.A. Teichmann: Structure and evolution of transcriptional regulatory networks, Curr. Opin. Struct. Biol. **14**(3), 283–291 (2004)

24.27   S. Tavazoie, J.D. Hughes, M.J. Campbell, R.J. Cho, G.M. Church: Systematic determination of genetic network architecture, Nat. Genet. **22**(3), 281–285 (1999)

24.28   J.J. Faith, B. Hayete, J.T. Thaden, I. Mogno, J. Wierzbowski, G. Cottarel, S. Kasif, J.J. Collins, T.S. Gardner: Large-scale mapping and validation of Escherichia coli transcriptional regulation from a compendium of expression profiles, PLoS Biology **5**(1), e8 (2007)

24.29   A. Butte, I. Kohane: Mutual information relevance networks: Functional genomic clustering using pairwise entropy measurements, Pac. Symp. Biocomput. (2000) p. 11

24.30   A.A. Margolin, I. Nemenman, K. Basso, C. Wiggins, G. Stolovitzky, R. Dalla Favera, A. Califano: ARACNE: An algorithm for the reconstruction of gene regulatory networks in a mammalian cellular context, BMC Bioinformatics **7**(Suppl 1), S7 (2006)

24.31   R. Bonneau, M.T. Facciotti, D.J. Reiss, A.K. Schmid, M. Pan, A. Kaur, V. Thorsson, P. Shannon, M.H. Johnson, J.C. Bare, W. Longabaugh, M. Vuthoori, K. Whitehead, A. Madar, L. Suzuki, T. Mori, D.E. Chang, J. Diruggiero, C.H. Johnson, L. Hood, N.S. Baliga: A predictive model for transcriptional control of physiology in a free living cell, Cell **131**(7), 1354–1365 (2007)

24.32   D.J. Reiss, N.S. Baliga, R. Bonneau: Integrated biclustering of heterogeneous genome-wide datasets for the inference of global regulatory networks, BMC Bioinformatics **7**, 280 (2006)

24.33   R. Bonneau, D.J. Reiss, P. Shannon, M. Facciotti, L. Hood, N.S. Baliga, V. Thorsson: The Inferelator: An algorithm for learning parsimonious regulatory networks from systems-biology data sets de novo, Genome Biol. **7**(5), R36 (2006)

24.34   T. Chen, H. He, G. Church: Modeling gene expression with differential equations, Pac. Symp. Biocomput. **4**, 29–40 (1999)

24.35   S.A. Kauffman: Metabolic stability and epigenesis in randomly constructed genetic nets, J. Theor. Biol. **22**(3), 437–467 (1969)

24.36   T. Akutsu, S. Miyano, S. Kuhara: Identification of genetic networks from a small number of gene expression patterns under the Boolean network model, Pac. Symp. Biocomput. **4**, 17–28 (1999)

24.37   S. Liang, S. Fuhrman, R. Somogyi: REVEAL: A general reverse engineering algorithm for inference of genetic network architectures, Pac. Symp. Biocomput. **3**, 18–29 (1998)

24.38 N. Friedman, M. Goldszmidt, A. Wyner: Data analysis with Bayesian networks: A bootstrap approach, Proc. 15th Conf. Uncertain. Artif. Intell. (UAI) (1999)

24.39 S. Imoto, T. Goto, S. Miyano: Estimation of genetic networks and functional structures between genes by using Bayesian networks and nonparametric regression, Pac. Symp. Biocomput. **7**, 175–186 (2002)

24.40 L. Glass, S.A. Kauffman: The logical analysis of continuous, nonlinear biochemical control networks, J. Theor. Biol. **39**(1), 103–129 (1973)

24.41 D. Ropers, H. de Jong, M. Page, D. Schneider, J. Geiselmann: Qualitative simulation of the carbon starvation response in Escherichia coli, Biosystems **84**(2), 124–152 (2006)

24.42 H. de Jong, J. Geiselmann, C. Hernandez, M. Page: Genetic Network Analyzer: Qualitative simulation of genetic regulatory networks, Bioinformatics **19**(3), 336–344 (2003)

24.43 V. Baldazzi, D. Ropers, Y. Markowicz, D. Kahn, J. Geiselmann, H. de Jong: The carbon assimilation network in Escherichia coli is densely connected and largely sign-determined by directions of metabolic fluxes, PLoS Comput. Biol. **6**(6), e1000812 (2010)

24.44 S. Li, P. Brazhnik, B. Sobral, J.J. Tyson: A quantitative study of the division cycle of Caulobacter crescentus stalked cells, PLoS Comput. Biol. **4**(1), e9 (2008)

24.45 S.A. Kauffman: The origins of order: Self-organization and selection in evolution, Biophys. J. **65**(6), 2 (1993)

24.46 A. Samal, S. Jain: The regulatory network of E. coli metabolism as a Boolean dynamical system exhibits both homeostasis and flexibility of response, BMC Syst. Biol. **2**(1), 21 (2008)

24.47 M.W. Covert, E.M. Knight, J.L. Reed, M.J. Herrgard, B.O. Palsson: Integrating high-throughput and computational data elucidates bacterial networks, Nature **429**(6987), 92–96 (2004)

24.48 I. Shmulevich, E.R. Dougherty, S. Kim, W. Zhang: Probabilistic Boolean networks: A rule-based uncertainty model for gene regulatory networks, Bioinformatics **18**(2), 261–274 (2002)

24.49 S. Chandrasekaran, N.D. Price: Probabilistic integrative modeling of genome-scale metabolic and regulatory networks in Escherichia coli and Mycobacterium tuberculosis, Proc. Natl. Acad. Sci. USA **107**(41), 17845–17850 (2010)

24.50 I.M. Ong, J.D. Glasner, D. Page: Modelling regulatory pathways in E. coli from time series expression profiles, Bioinformatics **18**(Suppl 1), S241–S248 (2002)

24.51 A.P. Hodges, P. Woolf, Y. He: Bayesian network expansion identifies new ROS and biofilm regulators, PLoS One **5**(3), e9513 (2010)

24.52 I.M. Keseler, J. Collado-Vides, A. Santos-Zavaleta, M. Peralta-Gil, S. Gama-Castro, L. Muñiz-Rascado, C. Bonavides-Martinez, S. Paley, M. Krummenacker, T. Altman, P. Kaipa, A. Spaulding, J. Pacheco, M. Latendresse, C. Fulcher, M. Sarker, A.G. Shearer, A. Mackie, I. Paulsen, R.P. Gunsalus, P.D. Karp: EcoCyc: A comprehensive database of Escherichia coli biology, Nucleic Acids Res. **39**(Suppl 1), D583–D590 (2011)

24.53 A.-C. Gavin, M. Bösche, R. Krause, P. Grandi, M. Marzioch, A. Bauer, J. Schultz, J.M. Rick, A.-M. Michon, C.-M. Cruciat, M. Remor, C. Höfert, M. Schelder, M. Brajenovic, H. Ruffner, A. Merino, K. Klein, M. Hudak, D. Dickson, T. Rudi, V. Gnau, A. Bauch, S. Bastuck, B. Huhse, C. Leutwein, M.-A. Heurtier, R.R. Copley, A. Edelmann, E. Querfurth, V. Rybin, G. Drewes, M. Raida, T. Bouwmeester, P. Bork, B. Seraphin, B. Kuster, G. Neubauer, G. Superti-Furga: Functional organization of the yeast proteome by systematic analysis of protein complexes, Nature **415**, 141–147 (2002)

24.54 I.M. Nooren, J.M. Thornton: Diversity of protein-protein interactions, EMBO Journal **22**(14), 3486–3492 (2003)

24.55 S. Fields, O. Song: A novel genetic system to detect protein–protein interactions, Nature **340**(6230), 245–246 (1989)

24.56 H. Yu, P. Braun, M.A. Yildirim, I. Lemmens, K. Venkatesan, J. Sahalie, T. Hirozane-Kishikawa, F. Gebreab, N. Li, N. Simonis, T. Hao, J.F. Rual, A. Dricot, A. Vazquez, R.R. Murray, C. Simon, L. Tardivo, S. Tam, N. Svrzikapa, C. Fan, A.S. de Smet, A. Motyl, M.E. Hudson, J. Park, X. Xin, M.E. Cusick, T. Moore, C. Boone, M. Snyder, F.P. Roth, A.L. Barabási, J. Tavernier, D.E. Hill, M. Vidal: High-quality binary protein interaction map of the yeast interactome network, Science **322**(5898), 104–110 (2008)

24.57 S. Li, C.M. Armstrong, N. Bertin, H. Ge, S. Milstein, M. Boxem, P.O. Vidalain, J.D. Han, A. Chesneau, T. Hao, D.S. Goldberg, N. Li, M. Martinez, J.F. Rual, P. Lamesch, L. Xu, M. Tewari, S.L. Wong, L.V. Zhang, G.F. Berriz, L. Jacotot, P. Vaglio, J. Reboul, T. Hirozane-Kishikawa, Q. Li, H.W. Gabel, A. Elewa, B. Baumgartner, D.J. Rose, H. Yu, S. Bosak, R. Sequerra, A. Fraser, S.E. Mango, W.M. Saxton, S. Strome, S. Van Den Heuvel, F. Piano, J. Vandenhaute, C. Sardet, M. Gerstein, L. Doucette-Stamm, K.C. Gunsalus, J.W. Harper, M.E. Cusick, F.P. Roth, D.E. Hill, M. Vidal: A map of the interactome network of the metazoan C. elegans, Science **303**(5657), 540–543 (2004)

24.58 L. Giot, J.S. Bader, C. Brouwer, A. Chaudhuri, B. Kuang, Y. Li, Y.L. Hao, C.E. Ooi, B. Godwin, E. Vitols, G. Vijayadamodar, P. Pochart, H. Machineni, M. Welsh, Y. Kong, B. Zerhusen, R. Malcolm, Z. Varrone, A. Collis, M. Minto, S. Burgess, L. McDaniel, E. Stimpson, F. Spriggs, J. Williams, K. Neurath, N. Ioime, M. Agee, E. Voss, K. Furtak, R. Renzulli, N. Aanensen, S. Carrolla, E. Bickelhaupt, Y. Lazovatsky, A. DaSilva, J. Zhong, C.A. Stanyon,

R.L. Finley Jr., K.P. White, M. Braverman, T. Jarvie, S. Gold, M. Leach, J. Knight, R.A. Shimkets, M.P. McKenna, J. Chant, J.M. Rothberg: A protein interaction map of Drosophila melanogaster, Science **302**(5651), 1727–1736 (2003)

24.59   J.F. Rual, K. Venkatesan, T. Hao, T. Hirozane-Kishikawa, A. Dricot, N. Li, G.F. Berriz, F.D. Gibbons, M. Dreze, N. Ayivi-Guedehoussou, N. Klitgord, C. Simon, M. Boxem, S. Milstein, J. Rosenberg, D.S. Goldberg, L.V. Zhang, S.L. Wong, G. Franklin, S. Li, J.S. Albala, J. Lim, C. Fraughton, E. Llamosas, S. Cevik, C. Bex, P. Lamesch, R.S. Sikorski, J. Vandenhaute, H.Y. Zoghbi, A. Smolyar, S. Bosak, R. Sequerra, L. Doucette-Stamm, M.E. Cusick, D.E. Hill, F.P. Roth, M. Vidal: Towards a proteome-scale map of the human protein–protein interaction network, Nature **437**(7062), 1173–1178 (2005)

24.60   U. Stelzl, U. Worm, M. Lalowski, C. Haenig, F.H. Brembeck, H. Goehler, M. Stroedicke, M. Zenkner, A. Schoenherr, S. Koeppen, J. Timm, S. Mintzlaff, C. Abraham, N. Bock, S. Kietzmann, A. Goedde, E. Toksöz, A. Droege, S. Krobitsch, B. Korn, W. Birchmeier, H. Lehrach, E.E. Wanker: A human protein–protein interaction network: A resource for annotating the proteome, Cell **122**(6), 957–968 (2005)

24.61   A.J. Walhout, M. Vidal: High-throughput yeast two-hybrid assays for large-scale protein interaction mapping, Methods **24**(3), 297–306 (2001)

24.62   D. Figeys, L.D. McBroom, M.F. Moran: Mass spectrometry for the study of protein–protein interactions, Methods **24**(3), 230–239 (2001)

24.63   T. Kocher, G. Superti-Furga: Mass spectrometry-based functional proteomics: From molecular machines to protein networks, Nat. Methods **4**(10), 807–815 (2007)

24.64   R.M. Ewing, P. Chu, F. Elisma, H. Li, P. Taylor, S. Climie, L. McBroom-Cerajewski, M.D. Robinson, L. O'Connor, M. Li, R. Taylor, M. Dharsee, Y. Ho, A. Heilbut, L. Moore, S. Zhang, O. Ornatsky, Y.V. Bukhman, M. Ethier, Y. Sheng, J. Vasilescu, M. Abu-Farha, J.P. Lambert, H.S. Duewel, I.I. Stewart, B. Kuehl, K. Hogue, K. Colwill, K. Gladwish, B. Muskat, R. Kinach, S.L. Adams, M.F. Moran, G.B. Morin, T. Topaloglou, D. Figeys: Large-scale mapping of human protein–protein interactions by mass spectrometry, Mol. Syst. Biol. **3**, 89 (2007)

24.65   G.D. Bader, C.W. Hogue: Analyzing yeast protein-protein interaction data obtained from different sources, Nat. Biotechnol. **20**(10), 991–997 (2002)

24.66   M.P. Stumpf, T. Thorne, E. de Silva, R. Stewart, H.J. An, M. Lappe, C. Wiuf: Estimating the size of the human interactome, Proc. Natl. Acad. Sci. USA **105**(19), 6959–6964 (2008)

24.67   A.J. Walhout, R. Sordella, X. Lu, J.L. Hartley, G.F. Temple, M.A. Brasch, N. Thierry-Mieg, M. Vidal: Protein interaction mapping in C. elegans using proteins involved in vulval development, Science **287**(5450), 116–122 (2000)

24.68   K.R. Brown, I. Jurisica: Unequal evolutionary conservation of human protein interactions in interologous networks, Genome Biol. **8**(5), R95 (2007)

24.69   B. Lehner, A.G. Fraser: A first-draft human protein-interaction map, Genome Biol. **5**(9), R63 (2004)

24.70   M. Persico, A. Ceol, C. Gavrila, R. Hoffmann, A. Florio, G. Cesareni: HomoMINT: An inferred human network based on orthology mapping of protein interactions discovered in model organisms, BMC Bioinformatics **6**(Suppl 4), S21 (2005)

24.71   S.E. O'Brien, D.G. Brown, J.E. Mills, C. Phillips, G. Morris: Computational tools for the analysis and visualization of multiple protein-ligand complexes, J. Mol. Graph Model **24**(3), 186–194 (2005)

24.72   K.R. Brown, I. Jurisica: Online predicted human interaction database, Bioinformatics **21**(9), 2076–2082 (2005)

24.73   S. Martin, D. Roe, J.L. Faulon: Predicting protein-protein interactions using signature products, Bioinformatics **21**(2), 218–226 (2005)

24.74   D.S. Han, H.S. Kim, W.H. Jang, S.D. Lee, J.K. Suh: PreSPI: A domain combination based prediction system for protein–protein interaction, Nucleic Acids Res. **32**(21), 6312–6320 (2004)

24.75   S.P. Kanaan, C. Huang, S. Wuchty, D.Z. Chen, J.A. Izaguirre: Inferring protein–protein interactions from multiple protein domain combinations, Methods Mol. Biol. **541**, 43–59 (2009)

24.76   Y. Guo, L. Yu, Z. Wen, M. Li: Using support vector machine combined with auto covariance to predict protein–protein interactions from protein sequences, Nucleic Acids Res. **36**(9), 3025–3030 (2008)

24.77   C.Y. Yu, L.C. Chou, D.T. Chang: Predicting protein-protein interactions in unbalanced data using the primary structure of proteins, BMC Bioinformatics **11**, 167 (2010)

24.78   Y. Park, E.M. Marcotte: Revisiting the negative example sampling problem for predicting protein–protein interactions, Bioinformatics **27**(21), 3024–3028 (2011)

24.79   A.K. Ramani, R.C. Bunescu, R.J. Mooney, E.M. Marcotte: Consolidating the set of known human protein-protein interactions in preparation for large-scale mapping of the human interactome, Genome Biol. **6**(5), R40 (2005)

24.80   M.E. Futschik, G. Chaurasia, H. Herzel: Comparison of human protein–protein interaction maps, Bioinformatics **23**(5), 605–6011 (2007)

24.81   T.S. Keshava Prasad, R. Goel, K. Kandasamy, S. Keerthikumar, S. Kumar, S. Mathivanan, D. Telikicherla, R. Raju, B. Shafreen, A. Venugopal, L. Balakrishnan, A. Marimuthu, S. Banerjee, D.S. Somanathan, A. Sebastian, S. Rani, S. Ray, C.J. Harrys Kishore, S. Kanth, M. Ahmed, M.K. Kashyap, R. Mohmood, Y.L. Ramachandra, V. Krishna,

B.A. Rahiman, S. Mohan, P. Ranganathan, S. Ramabadran, R. Chaerkady, A. Pandey: Human protein reference database – 2009 update, Nucleic Acids Res. **37**(Database issue), D767–D772 (2009)

24.82   C. Stark, B.J. Breitkreutz, A. Chatr-Aryamontri, L. Boucher, R. Oughtred, M.S. Livstone, J. Nixon, K. Van Auken, X. Wang, X. Shi, T. Reguly, J.M. Rust, A. Winter, K. Dolinski, M. Tyers: The BioGRID Interaction Database: 2011 update, Nucleic Acids Res. **39**(Database issue), D698–D704 (2011)

24.83   B. Aranda, P. Achuthan, Y. Alam-Faruque, I. Armean, A. Bridge, C. Derow, M. Feuermann, A.T. Ghanbarian, S. Kerrien, J. Khadake, J. Kerssemakers, C. Leroy, M. Menden, M. Michaut, L. Montecchi-Palazzi, S.N. Neuhauser, S. Orchard, V. Perreau, B. Roechert, K. van Eijk, H. Hermjakob: The IntAct molecular interaction database in 2010, Nucleic Acids Res. **38**(Database issue), D525–D531 (2010)

24.84   L. Salwinski, C.S. Miller, A.J. Smith, F.K. Pettit, J.U. Bowie, D. Eisenberg: The Database of Interacting Proteins: 2004 update, Nucleic Acids Res. **32**(Database issue), D449–D451 (2004)

24.85   R. Isserlin, R.A. El-Badrawi, G.D. Bader: The Biomolecular Interaction Network Database in PSI-MI 2.5, Database (Oxford) **2011**, baq037 (2011)

24.86   P. Pagel, S. Kovac, M. Oesterheld, B. Brauner, I. Dunger-Kaltenbach, G. Frishman, C. Montrone, P. Mark, V. Stümpflen, H.W. Mewes, A. Ruepp, D. Frishman: The MIPS mammalian protein–protein interaction database, Bioinformatics **21**(6), 832–834 (2005)

24.87   H. Hermjakob, L. Montecchi-Palazzi, G. Bader, J. Wojcik, L. Salwinski, A. Ceol, S. Moore, S. Orchard, U. Sarkans, C. von Mering, B. Roechert, S. Poux, E. Jung, H. Mersch, P. Kersey, M. Lappe, Y. Li, R. Zeng, D. Rana, M. Nikolski, H. Husi, C. Brun, K. Shanker, S.G. Grant, C. Sander, P. Bork, W. Zhu, A. Pandey, A. Brazma, B. Jacq, M. Vidal, D. Sherman, P. Legrain, G. Cesareni, I. Xenarios, D. Eisenberg, B. Steipe, C. Hogue, R. Apweiler: The HUPO PSI's molecular interaction format – a community standard for the representation of protein interaction data, Nat. Biotechnol. **22**(2), 177–183 (2004)

24.88   D. Szklarczyk, A. Franceschini, M. Kuhn, M. Simonovic, A. Roth, P. Minguez, T. Doerks, M. Stark, J. Muller, P. Bork, L.J. Jensen, C. von Mering: The STRING database in 2011: Functional interaction networks of proteins, globally integrated and scored, Nucleic Acids Res. **39**(Database issue), D561–D568 (2011)

24.89   G. Chaurasia, S. Malhotra, J. Russ, S. Schnoegl, C. Hänig, E.E. Wanker, M.E. Futschik: UniHI 4: New tools for query, analysis and visualization of the human protein–protein interactome, Nucleic Acids Res. **37**(Database issue), D657–D660 (2009)

24.90   J.E. Girardini, M. Napoli, S. Piazza, A. Rustighi, C. Marotta, E. Radaelli, V. Capaci, L. Jordan, P. Quinlan, A. Thompson, M. Mano, A. Rosato, T. Crook, E. Scanziani, A.R. Means, G. Lozano, C. Schneider, G. Del Sal: A Pin1/mutant p53 axis promotes aggressiveness in breast cancer, Cancer Cell **20**(1), 79–91 (2011)

24.91   D. Hanahan, R.A. Weinberg: The hallmarks of cancer, Cell **100**(1), 57–70 (2000)

24.92   B. Vogelstein, K.W. Kinzler: Cancer genes and the pathways they control, Nat. Med. **10**(8), 789–799 (2004)

24.93   L.D. Wood, D.W. Parsons, S. Jones, J. Lin, T. Sjöblom, R.J. Leary, D. Shen, S.M. Boca, T. Barber, J. Ptak, N. Silliman, S. Szabo, Z. Dezso, V. Ustyanksky, T. Nikolskaya, Y. Nikolsky, R. Karchin, P.A. Wilson, J.S. Kaminker, Z. Zhang, R. Croshaw, J. Willis, D. Dawson, M. Shipitsin, J.K. Willson, S. Sukumar, K. Polyak, B.H. Park, C.L. Pethiyagoda, P.V. Pant, D.G. Ballinger, A.B. Sparks, J. Hartigan, D.R. Smith, E. Suh, N. Papadopoulos, P. Buckhaults, S.D. Markowitz, G. Parmigiani, K.W. Kinzler, V.E. Velculescu, B. Vogelstein: The genomic landscapes of human breast and colorectal cancers, Science **318**(5853), 1108–1113 (2007)

24.94   A.L. Russo, K. Jedlicka, M. Wernick, D. McNally, M. Kirk, M. Sproull, S. Smith, U. Shankavaram, A. Kaushal, W.D. Figg, W. Dahut, D. Citrin, D.P. Bottaro, P.S. Albert, P.J. Tofilon, K. Camphausen: Urine analysis and protein networking identify met as a marker of metastatic prostate cancer, Clin. Cancer Res. **15**(13), 4292–4298 (2009)

24.95   A. Lunardi, G. Di Minin, P. Provero, M. Dal Ferro, M. Carotti, G. Del Sal, L. Collavin: A genome-scale protein interaction profile of Drosophila p53 uncovers additional nodes of the human p53 network, Proc. Natl. Acad. Sci. USA **107**(14), 6322–6327 (2010)

24.96   J. Ladha, S. Donakonda, S. Agrawal, B. Thota, M.R. Srividya, S. Sridevi, A. Arivazhagan, K. Thennarasu, A. Balasubramaniam, B.A. Chandramouli, A.S. Hegde, P. Kondaiah, K. Somasundaram, V. Santosh, S.M. Rao: Glioblastoma-specific protein interaction network identifies PP1A and CSK21 as connecting molecules between cell cycle-associated genes, Cancer Res. **70**(16), 6437–6447 (2010)

24.97   S.A. Bapat, A. Krishnan, A.D. Ghanate, A.P. Kusumbe, R.S. Kalra: Gene expression: Protein interaction systems network modeling identifies transformation-associated molecules and pathways in ovarian cancer, Cancer Res. **70**(12), 4809–4819 (2010)

24.98   M.A. Pujana, J.D. Han, L.M. Starita, K.N. Stevens, M. Tewari, J.S. Ahn, G. Rennert, V. Moreno, T. Kirchhoff, B. Gold, V. Assmann, W.M. Elshamy, J.F. Rual, D. Levine, L.S. Rozek, R.S. Gelman, K.C. Gunsalus,

R.A. Greenberg, B. Sobhian, N. Bertin, K. Venkate-san, N. Ayivi-Guedehoussou, X. Solé, P. Hernández, C. Lázaro, K.L. Nathanson, B.L. Weber, M.E. Cusick, D.E. Hill, K. Offit, D.M. Livingston, S.B. Gruber, J.D. Parvin, M. Vidal: Network modeling links breast cancer susceptibility and centrosome dysfunction, Nat. Genet. **39**(11), 1338–1349 (2007)

24.99   V. Trevino, M.G. Tadesse, M. Vannucci, F. Al-Shahrour, P. Antczak, S. Durant, A. Bikfalvi, J. Dopazo, M.J. Campbell, F. Falciani: Analysis of normal-tumor tissue interaction in tumors: Prediction of prostate cancer features from the mo-lecular profile of adjacent normal cells, PLoS One **6**(3), e16492 (2011)

24.100   H.Y. Chuang, E. Lee, Y.T. Liu, D. Lee, T. Ideker: Network-based classification of breast cancer metastasis, Mol. Syst. Biol. **3**, 140 (2007)

24.101   I.W. Taylor, R. Linding, D. Warde-Farley, Y. Liu, C. Pesquita, D. Faria, S. Bull, T. Pawson, Q. Morris, J.L. Wrana: Dynamic modularity in protein interaction networks predicts breast cancer outcome, Nat. Biotechnol. **27**(2), 199–204 (2009)

24.102   A. Friedman, N. Perrimon: Genetic screening for signal transduction in the era of network biology, Cell **128**(2), 225–231 (2007)

25. Whole-Exome Sequencing Data – Identifying Somatic Mutations

Roberta Spinelli, Rocco Piazza, Alessandra Pirola, Simona Valletta, Roberta Rostagno, Angela Mogavero, Manuela Marega, Hima Raman, Carlo Gambacorti-Passerini

The use of next-generation sequencing instruments to study hematological malignancies generates a tremendous amount of sequencing data. This leads to a challenging bioinformatics problem to store, manage, and analyze terabytes of sequencing data, often generated from extremely different data sources. Our project is mainly focused on sequence analysis of human cancer genomes, in order to identify the genetic lesions underlying the development of tumors. However, the automated detection procedure of somatic mutations and the statistical testing procedure to identify genetic lesions are still an open problem. Therefore, we propose a computational procedure to handle large-scale sequencing data in order to detect exonic somatic mutations in a tumor sample. The proposed pipeline includes several steps based on open-source software and the R language: alignment, detection of mutations, annotation, functional classification, and visualization of results. We analyzed Illumina whole-exome sequencing data from five leukemic patients and five paired controls plus one colon cancer sample and paired control. The results were validated by Sanger sequencing.

All cancers arise as a result of changes that occur in the DNA sequence of the genomes of cancer cells. These changes are collectively called somatic mutations to distinguish them from germline mutations that are inherited from parents and transmitted to offspring [25.1]. The genomic somatic mutations in a cancer cell include several events of DNA sequence changes: single-point mutations or substitutions of one base by another (SNVs); insertions or deletions of small (< 10 bp, indels) and large segments of DNA; genomic rearrangements such as inversions, translocations, and fusion genes in which a DNA segment is broken and rejoined from elsewhere in the genome; and copy number variations (CNVs), defined as copy number increases from the two copies present in the normal diploid genome, sometimes to several hundred copies (known as gene amplification), or copy number decreases that may result in complete absence of a DNA sequence from the cancer genome. The development of high-throughput sequencing technology provides a great opportunity to identify genetic alterations, structural changes, and molecular mechanisms involved in tumor genesis [25.2], allowing the identification of new therapeutic targets in diseases.

Our project is mainly focused on the sequence analysis of human cancer genomes in order to identify somatic point mutations underlying the development of tumors, by using the whole-exome sequencing approach. Somatic point mutations could cause deleterious DNA imbalance, significant alterations in gene expression, and profound effects on protein expression, leading to continuous acquisition of genetic variations in individual cells. Cells that acquire deleterious mutations or carrying alterations could gain the ability to proliferate and survive autonomously [25.1].

Whole-exome sequencing is a high-throughput technology to sequence all human exon regions. The exon is the coding part of a gene. The whole exome represents 2% of the human genome, and the protein-coding regions constitute about 1% of the human genome, totaling about 30 Mb, split across about 180 000 coding exons [25.3] with shortest exon of 2 bp and longest of 11 923 bp with median size of 120 bp [25.4]. Whole-exome sequencing had been widely applied to identify exonic lesions underlying tumor growth and to provide a molecular framework for the development of rational, biology-driven therapies in cancer [25.5], enabling systematic detection of rare variants occurring in a single individual or common variants shared across samples, allowing the identification of new therapeutic targets in cancer.

However, the automated detection procedure of somatic mutations is still an open problem. In order to identify somatic point mutations occurring in cancer genomes but not in paired normal genomes, we implemented a bioinformatic procedure using open-source software and the R language [25.6].

We applied our strategy to Illumina whole-exome sequencing data from chronic myeloid leukemia (CML) patients and atypical chronic myeloid leukemia (aCML) patients plus a colon cancer patient.

CML is a clonal bone marrow stem cell disorder in which proliferation of mature granulocytes (neutrophils, eosinophils, and basophils) and their precursors is the main finding. It is a myeloproliferative disease caused by a reciprocal translocation between the long arms of chromosome 9 and 22, t(9;22)(q34;q11.2), causing a Philadelphia (Ph) chromosome. The development of imatinib, a tyrosine kinase inhibitor targeting BCR-ABL1 kinase produced by t(9;22), has revolutionized treatment of CML. Although treatment with imatinib and other targeted therapies has dramatically improved survival, in some cases resistance to imatinib can potentially be developed by Ph+ cells persistent. Then, the disease can evolve into an aggressive and deadly acute leukemia called blast crisis (BC) [25.7]. The secondary genetic events associated with CML remain to be discovered.

aCML is a myeloproliferative disease clinically resembling chronic myeloid leukemia, but aCML cases lack the Philadelphia chromosome, which is the molecular hallmark of CML [25.8]. The identi-fication of specific genomic rearrangements, such as the Philadelphia chromosome, promyelocytic leukemia-retinoic acid receptor (PML-RAR) translocation, or fusions involving platelet-derived growth factor receptors (PDGFR), may have a profound impact on the prognosis and therapy of the underlying disorder. Unfortunately, specific cytogenetic changes have not been associated with aCML yet. This suggests that genetic lesions underlying the development of aCML remain to be discovered.

In this work we concentrated on the identification of damaging or deleterious point mutations characterizing each sample, in order to find recurrent mutations to be screened in larger CML and aCML datasets. Our pipeline allowed the identification of somatic mutations occurring in the cancer genomes but not in autologous normal lymphocytes according to our model. These mutations were confirmed by Sanger sequencing, indicating that our pipeline is able to effectively process high-throughput sequencing data and extract genes, or genomic regions, important for the discovery of new genes associated with diseases, suitable for further investigation. Sanger sequencing was established by Fredrick Sanger, who won his second Nobel Prize in 1980, and has became the standard because of its practicality. After Sanger sequencing of DNA from cancer cells and healthy nonneoplastic cells of patients, the results are two chromatograms, one for each population analyzed, from which it is possible to individuate the somatic mutation. This method essentially involves amplifying a single-stranded piece of DNA many times. The method employs special bases called dideoxynucleotides (ddNTPs). After many repeated cycles of amplification, this will result in all the possible lengths of DNA being represented and every piece of synthesized DNA containing a fluorescent label at its terminus. The reaction is set up so that a fluorescent ddNTP is present at every position in the DNA strand, so that every nucleotide in the strand can be determined [25.9]. A computer program can then compile the data into a colored graph showing the determined sequence.

25.1 Materials and Methods

Genomic DNA (gDNA) from leukemic cells (minimum content 80%) and normal lymphocytes was extracted from two CML and three aCML patients; furthermore, tumor and paired peripheral blood DNA was extracted from a patient affected by colon cancer. The samples were sequenced by Illumina Genome Analyzer IIX in paired-end mode by 76 bp reads from either end, using two lanes per sample. Genomic DNA was prepared for whole-exome sequencing according to the manufacturer's protocols [25.10, 11]. The genomic library was generated from 1 μg genomic DNA using the standard Illumina protocol with fragment size of 300 bp. After cluster amplification and sequencing, the images were acquired by Illumina Genome Analyzer Sequencing Control software (SCS 2.6) and analyzed by Real Time Analyzer software (RTA 1.6) to obtain base calls, quality metrics, and read calls, with default parameters and standard quality control. RTA 1.6 outputs the results in *Solexa/Illumina-FastQ* format [25.12].

To compare sequencing data from cancer cells with similar data from healthy nonneoplastic cells and to identify somatic mutations, we developed a computational strategy including several steps: alignment of sequences versus Human Genome Reference hg18 from the University of California, Santa Cruz (UCSC) genome browser; detection of single-nucleotide variations and small insertions or deletions in tumor and normal sample by the R language; identification of candidate somatic mutations, extracting the variations subset occurring only in cancer genomes; and annotation and functional prediction of somatic mutations (Fig. 25.1).

At each step, different bioinformatic tools were used: BioPython [25.13], SAMtools [25.14], R language, and tool to predict nonsynonimous/missense variants (SIFT) [25.15], and the genomic visualization of results was carried out using the Integrative Genomics Viewer (IGV) [25.16].

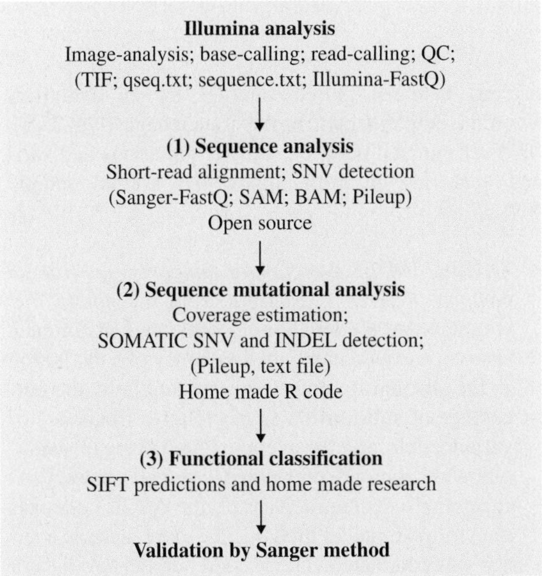

In our procedure we defined the somatic model by setting a number of parameters in order to ensure good quality control of the data generated. The parameters concern the minimum read coverage, the minimum frequency/percentage of substitution, and the minimum average Phred [25.17] quality score calculated at each genomic position [25.3].

25.1.1 Data Processing Pipeline

The data processing pipeline includes three main steps (Fig. 25.1).

Sequence Analysis
In the first step we describe the base-pair reads and the information at each chromosomal position of the sample to facilitate single-nucleotide polymorphism (SNP)/indel calling (Fig. 25.2):

1. *Preprocessing of the dataset of Illumina sequences.* We converted tumor and normal sequence data from *Solexa/Illumina-FastQ* format into *Sanger-FastQ* using the *SeqIO.convert* function in BioPython.
2. *Alignment of sequences.* We aligned 76 bp paired-end reads versus Human Genome Reference hg18 using the Burrows–Wheeler alignment (BWA) algorithm [25.18]. BWA is implemented in BWA tool and published by *Li* and *Durbin* [25.18]. BWA tool aligns relatively short sequences (queries) to a sequence database (target), such as the human genome reference; implements an algorithm based on the Burrows–Wheeler transform (BWT); and enables gapped global alignment queries. BWA is one of the fastest short-read alignment algorithms and supports paired-end reads. Using BWA we aligned the paired-end reads in the sequence alignment map (SAM) file and filtered the reads mapped in proper pairs in order to process further only reads that were correctly aligned. Moreover, we restricted the downstream analysis only to uniquely mapped reads. SAM is a text and standard format for storing large nucleotide sequence alignments.
3. *Conversion from alignment-oriented data to position-oriented data.* We converted the SAM files into binary alignment format (BAM, SAM binary) files using tools for sequence alignment map (SAMtools) [25.14]. SAMtools was implemented

Illumina analysis
Image-analysis; base-calling; read-calling; QC;
(TIF; qseq.txt; sequence.txt; Illumina-FastQ)

↓

(1) Sequence analysis
Short-read alignment; SNV detection
(Sanger-FastQ; SAM; BAM; Pileup)
Open source

↓

(2) Sequence mutational analysis
Coverage estimation;
SOMATIC SNV and INDEL detection;
(Pileup, text file)
Home made R code

↓

(3) Functional classification
SIFT predictions and home made research

↓

Validation by Sanger method

Fig. 25.1 The basic workflow of the computational procedure: from input of Illumina sequence data, preprocessing analysis, detection of somatic variations and functional annotation of coding variants, to candidate somatic mutations. Validation by Sanger method

Part D | 25.1

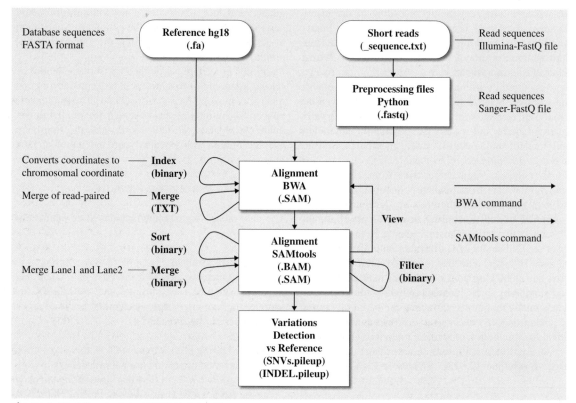

Fig. 25.2 Workflow to detect SNVs/indel calling

by *Li* et al. [25.14]. SAMtools is a set of utilities in C++ and Perl languages to generate and manipulate alignments. SAMtools imports SAM data and exports SAM or BAM format. Using SAMtools we sort, index, and merge the binary alignment-oriented data. Finally, we converted the binary alignment-oriented data into position-oriented *pileup* data (Fig. 25.2). The output pileup file describes the base-pair information at each chromosomal position and the variant calls in a text format. In pileup, each line consists of chromosome, 1-based coordinate, reference base, the number of reads covering the site (read coverage), read bases (match, mismatch, indel, strand, start, and end), and base qualities in ASCII code. The ASCII code transformed into a Phred quality score gives a measure of base-call accuracy and the probability of base-calling error.

Sequence Mutational Analysis

In the second step we identified the variations of sample versus human genome reference and the mis-

matches between paired samples by an analytical procedure implemented in the R language (Fig. 25.3). We discriminated between somatic mutations and variants occurring in both tumor and normal sample (SNPs).

1. *SNV and INDEL detection of sample versus Human Genome Reference*. Starting from the pileup file, we identified the single-nucleotide variants of each sample, and calculated the frequency of substitution as the absolute frequency of variant allele, the percentage of substitution as the relative frequency of variant allele, and the average Phred score of substitution as a measure of average base-call accuracy. At each single-nucleotide variant, the ASCII code was transformed into a Phred quality score, and the average was calculated. The R code outputs two lists of variations detected between the sample and the reference: SNVs and INDEL files. Each variation was described in a table by chromosome, 1-based coordinate, reference base, read coverage, read bases and base qualities in ASCII code, frequency of substitu-

tion, percentage of substitution, and average Phred score of substitutions.

2. *Somatic SNV and INDEL detection by paired analysis.* In this step we identified the hypothetical somatic mutations by paired mutational analysis. We discriminated between variants occurring in the cancer genome but not in paired healthy sample and variants occurring in both. In order to ensure high quality of the data and to obtain robustness of mutation detection, we filtered out variations in cancer sample using the following criteria: read coverage less than 20×, frequency of substitution less than 6, percentage of substitution less than 25%, and average Phred quality less than 30, corresponding to a probability of incorrect base call greater than 0.001 or accuracy equal to 99.9%. We accepted that each base was read by at least 20 individual sequences, each mutation was confirmed by at least 6 individual sequences, and the minimum frequency of substitution was set to 25% to find somatic mutations present not only as heterozygous or homozygous but also in subclones of each cancer sample. Finally, variations present in matched healthy sample with frequency lower than or equal to 10% are tolerated, otherwise being classified as SNPs. Two text files are generated as output; the first one list the somatic variations, and the second can be imported directly into SIFT for functional classification. Each somatic mutation is described in a table by chromosome, 1-physical position, reference base, read coverage, absolute frequency of mutation, percent of mutation, and average Phred score of mutation in cancer and in healthy sample.

Functional Classification of Coding Variant

In the third step we annotated and evaluated the functional effect of exonic hypothetical mutations, filtering out known polymorphisms (Fig. 25.4). We used SIFT genome [25.15] to predict whether an amino acid substitution affects protein function based on sequence homology. Using SIFT we extract coding variants from a large list of hypothetical somatic mutations providing the annotation and the functional prediction. Therefore, using SIFT annotation, we filtered out mutations predicted to be synonymous and polymorphic and confirmed the SIFT functional prediction by querying tools for functional annotation of genetic variants (ANNOVAR) (http://www.openbioinformatics.org/annovar/ [25.19]). ANNOVAR [25.20] is a free functional annotation tool

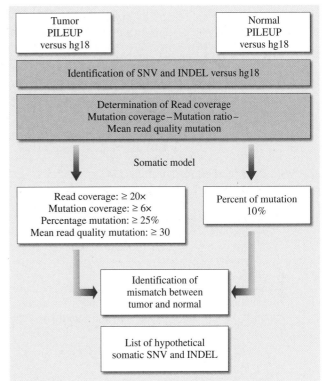

Fig. 25.3 Workflow to detect somatic SNVs/indel in tumor sample

for genetic variants for next-generation sequencing data that can identify whether SNPs cause protein coding changes and the amino acids that are affected. Moreover, using ANNOVAR we were able to flexibly use RefSeq genes to annotate the mutations at gene level, identify variants that are reported in dbSNP, or common SNPs (minor allele frequency (MAF) > 1%) in the 1000 Genome Project, and discover variants in conserved regions among 44 species or in segmental duplication regions. Moreover, we selected the annotated candidate genes for Sanger validation according to the following criteria: genes previously linked and associated with leukemia (data extracted by search in PUBMED), and genes know to participate in metabolic and biological process relevant to cancer, i.e., gene functional categories evidenced by ingenuity pathway analysis (IPA) [25.21]. Then all the critical data generated by our algorithm and SIFT were merged using dedicated R scripts. SIFT predictions (nucleotides position reference base and substitution, region belonging, dbSNP ID, SNP type, gene name, gene description, transcript ID, and protein family ID) were merged with the sequence data (coverage, absolute frequency of mu-

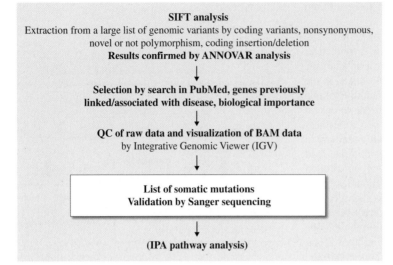

SIFT analysis
Extraction from a large list of genomic variants by coding variants, nonsynonymous, novel or not polymorphism, coding insertion/deletion
Results confirmed by ANNOVAR analysis

↓

Selection by search in PubMed, genes previously linked/associated with disease, biological importance

↓

QC of raw data and visualization of BAM data
by Integrative Genomic Viewer (IGV)

↓

List of somatic mutations
Validation by Sanger sequencing

↓

(IPA pathway analysis)

Fig. 25.4 Functional classification and selection of candidate genes; QC, quality control

tation, percent of mutation, and average Phred score of mutation in cancer, and the same fields for healthy sample). Finally, all selected somatic mutations were checked for quality control in pileup and BAM files and visualized using IGV, then sequenced by the Sanger method.

25.2 Results

The entire procedure was applied to whole-exome sequencing of five leukemia patients and five paired controls plus one solid tumor sample and paired control.

On average, 42 million paired-end reads and 5.4 gigabases (Gb) of sequences were generated per individual. The percentage of reads matching the reference human genome (hg18, NCBI36.1) was over 90%, with mean exon coverage of over 70-fold. For both the leukemia sample and the control, more than 90% of exons had mean coverage of over 20-fold. In the leukemia dataset we found several patterns of hypothetical somatic mutations.

In CML we identified 160 and 1389 somatic mutations for each patient, respectively. The number of mutations in coding regions was 32/160 and 181/1389, respectively. Of these, 22/32 and 40/181 were novel or not reported in the dbSNP database. In CML we detected a total of 33 novel mutations, being nonsynonymous and in exon regions, with minimum average Phred quality score of 30, corresponding to accuracy of 99.9%, minimum read depth of 20, confirmed by at least 6 individual sequences, and with minimum percent of mutation of 25% in 28 annotated genes. Additionally, candidate genes were selected for Sanger sequencing

analysis according to the criteria specified at step 3. This analysis allowed us to focus on four genes, three of which were validated by the Sanger sequencing method (data not shown).

In the aCML dataset we identified lists of hypothetical somatic mutations of 113, 146, and 517 single variants for each patient, respectively. The number of mutations in coding regions was 31/113, 42/146, and 55/517, respectively, and the numbers not reported in dbSNP were 27/31, 24/42, and 43/55. We detected in aCML a total of 69 novel nonsynonymous mutations located in exon regions, with accuracy of 99.9%, minimum read depth of 20, confirmed by at least 6 individual sequences, and with minimum percent of mutation of 25% in 56 annotated genes. The candidate genes were selected for Sanger sequencing according to the criteria of step 3. Then, we focused on 21 genes validated by the Sanger sequencing method (data not shown). Pathway analysis of the 21 genes performed by IPA confirmed that they correspond to cell cycle network function and cancer disease.

In the leukemia dataset, the remaining 59 genes may represent putative novel candidate genes involved in tumor etiology, and they are currently under analysis.

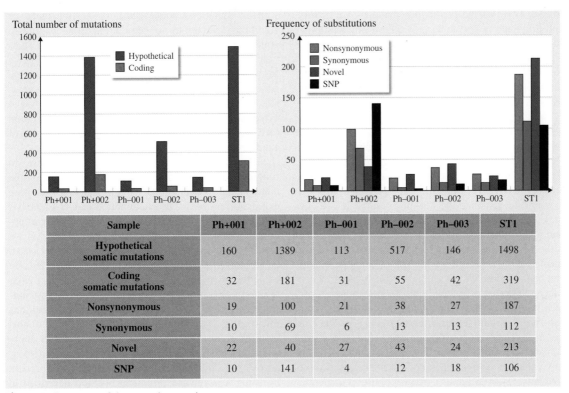

Fig. 25.5 Summary of the somatic mutations

Sample	Ph+001	Ph+002	Ph−001	Ph−002	Ph−003	ST1
Hypothetical somatic mutations	160	1389	113	517	146	1498
Coding somatic mutations	32	181	31	55	42	319
Nonsynonymous	19	100	21	38	27	187
Synonymous	10	69	6	13	13	112
Novel	22	40	27	43	24	213
SNP	10	141	4	12	18	106

However, due to the absence of any obvious link with oncogenesis, from a functional point of view approaching these genes is more challenging and will likely require functional biological models to gain insight into their role in cancer.

Unexpectedly, from the analysis on solid tumor exome, we identified a very long lists of hypothetical somatic variations: 1498 SNVs. The number of mismatches covering coding exon regions was 319, and 213 were not reported in the dbSNP. Globally, we found 144 novel coding SNVs, nonsynonymous, with accuracy of 99.9%, minimum read depth of 20, and minimum percent of substitution of 25% in 87 annotated genes. From this list, we selected 16 candidate genes according to the criteria specified in step 3. Their validation is in progress.

25.3 Summary and Future Prospects

We have developed and validated a computational method to manage large-scale sequencing data and to detect exonic somatic mutations in tumor samples.

For the analysis we used the hg18 human genome reference because at the time of the first round of analysis it was a complete and a stable version of available annotations. Today, most publications and bioinformatics tools still use hg18 for standard annotation purposes.

Our pipeline supports single-end and paired-end reads as input, and the output can be loaded directly into SIFT to predict the effect of somatic mutation. We also used ANNOVAR to confirm the functional annotation, gene annotation, exonic function, amino acid change, SNP annotation, and conservation analysis obtained by SIFT. Moreover, whole-exome precomputed PolyPhen v2, PhyloP, MutationTaster, and LRT (linear regression technique) [25.22] scores are available in ANNOVAR as a database, dbNSFP [25.23], giving more detailed annotation of nonsynonymous mutations in humans, in addition to SIFT. However, the

Part D | 25.3

overlap between the predictions made by LRT, SIFT, and PolyPhen is very low: only 5% of predictions are shared across all three methods [25.22]. dbNSFP is a lightweight database of human nonsynonymous SNPs for functional predictions to distinguish deleterious mutations from the massive number of nonfunctional variants [25.23].

We found genomic regions containing interesting genes potentially implicated in our diseases and suitable for further investigation. Recently, a new algorithm, Bambino [25.24], was published. Bambino is a tool to detect variations, somatic or germinal, and loss of heterozygosity (LOH) from single or pooled samples. It is useful to visualize the genomic results. It is a Java tool, being flexible and fast, but trivial to annotate and apply functional annotation. However, neither our procedure nor Bambino provides statistical testing to identify SNVs. Moreover, the list of somatic variants can be influenced by the settings used for variant detection and filtering of putative sites. In our somatic model we chose parameter values typically used in literature for whole-genome sequencing analysis [25.3]. The minimum frequency of substitution was set to 25% to find somatic mutations present not only as hetero- or homozygous but also in subclones of each cancer sample. Recently, GAMES [25.25], a new tool useful to identify and annotate mutations in next-generation sequencing, was published, but it is not implemented for somatic analysis.

In this work we concentrated on identification of damaging or deleterious mutations characterizing each sample, in order to find recurrent mutations to be screened in larger CML and aCML datasets. These somatic mutations could cause DNA imbalance or significant alterations in gene expression, leading to continuous acquisition of genetic variations in individual cells [25.1]. We analyzed a wide range of SNVs and indels in cancer samples and not the large genomic rearrangements or CNVs, because generally these abnormalities can be identified by classical cytogenetic, array-comparative genomic hybridization (aCGH), gene-expression profiling, and resequencing of candidate genes.

Further improvements in our pipeline are in progress to identify common aberrations in multiple datasets and to combine DNA data analysis with gene expression profiles. The combination of DNA analysis and gene expression profiling provides a powerful approach to understand functional effects. Moreover, we plan to improve the genomic view of results by circular visualization of tabular data (CIRCOS) [25.26]. By using this software we will be able to display a circular representation of the genome and explore the relationships between DNA events and the genomic positions involved to describe in synthesis the structural abnormalities of an individual. In addition, we will apply a new machine learning theory to discriminate molecular pathways involved in leukemic pathological process by using a novel and promising method described in *Barla* et al. [25.27]. The new pipeline can be applied to perform biomarker study in a leukemia and control dataset; moreover, we plan to apply a *modified* pipeline to a single group of leukemia patients to establish the correlation between the deregulated molecular pathways and the prognosis of patients from samples with known somatic mutations.

References

25.1   M.R. Stratton, P.J. Campbell, P.A. Futreal: The cancer genome, Nature **458**(7239), 719–724 (2009)

25.2   P.J. Campbell, P.J. Stephens, E.D. Pleasance, S. O'Meara, H. Li, T. Santarius, L.A. Stebbings, C. Leroy, S. Edkins, C. Hardy, J.W. Teague, A. Menzies, I. Goodhead, D.J. Turner, C.M. Clee, M.A. Quail, A. Cox, C. Brown, R. Durbin, M.E. Hurles, P.A. Edwards, G.R. Bignell, M.R. Stratton, P.A. Futreal: Identification of somatically acquired rearrangements in cancer using genome-wide massively parallel paired-end sequencing, Nat. Genet. **40**(6), 722–729 (2008)

25.3   S.B. Ng, E.H. Turner, P.D. Robertson, S.D. Flygare, A.W. Bigham, C. Lee, T. Shaffer, M. Wong, A. Bhattacharjee, E.E. Eichler, M. Bamshad, D.A. Nickerson, J. Shendure: Targeted capture and massively parallel sequencing of 12 human exomes, Nature **461**(7261), 272–276 (2009)

25.4   M.K. Sakharkar, V.T. Chow, P. Kangueane: Distributions of exons and introns in the human genome, In Silico Biol. **4**(4), 387–393 (2004)

25.5   Y. Jiao, C. Shi, B.H. Edil, R.F. de Wilde, D.S. Klimstra, A. Maitra, R.D. Schulick, L.H. Tang, C.L. Wolfgang, M.A. Choti, V.E. Velculescu, L.A. Diaz Jr., B. Vogelstein, K.W. Kinzler, R.H. Hruban, N. Papadopoulos: DAXX/ATRX, MEN1, and mTOR pathway genes are frequently altered in pancreatic neuroendocrine tumors, Science **331**(6021), 1199–1203 (2011)

25.6   R Core Team: *R: A Language and Enviroment for Statistical Computing* (R Foundation for Statistical Computing, Vienna 2012), available online at http://www.R-project.org/

25.7   Y. Chen, C. Peng, D. Li, S. Li: Molecular and cellular bases of chronic myeloid leukemia, Protein Cell **1**(2), 124–132 (2010)

25.8   S. Burgstaller, A. Reiter, N. Cross: BCR–ABL – negative chronic myeloid leukemia, Curr. Hematol. Malig. Rep. **2**(2), 75–82 (2007)

25.9   S.B. Primrose, R.M. Twyman: *Principles of Genome Analysis and Genomics* (Blackwell, Malden 2003)

25.10   Agilent Technologies: SureSelect Human All Exon Kit Illumina Paired-End Sequencing Library Prep Protocol Version 1.0.1 (2009)

25.11   Paired-End Sequencing Sample Preparation Guide, http://www.illumina.com

25.12   P.J. Cock, C.J. Fields, N. Goto, M.L. Heuer, P.M. Rice: The Sanger FASTQ file format for sequences with quality scores, and the Solexa/Illumina FASTQ variants, Nucleic Acids Res. **38**(6), 1767–1771 (2010)

25.13   P.J.A. Cock, T. Antao, J.T. Chang, B.A. Chapman, C.J. Cox, A. Dalke, I. Friedberg, T. Hamelryck, F. Kauff, B. Wilczynski, M.J.L. de Hoon: Biopython: Freely available Python tools for computational molecular biology and bioinformatics, Bioinformatics **25**(11), 1422–1423 (2009)

25.14   H. Li, B. Handsaker, A. Wysoker, T. Fennell, J. Ruan, N. Homer, G. Marth, G. Abecasis, R. Durbin, 1000 Genome Project Data Processing Subgroup: The sequence alignment/map format and SAMtools, Bioinformatics **25**(16), 2078–2079 (2009)

25.15   P. Kumar, S. Henikoff, P.C. Ng: Predicting the effects of coding nonsynonymous variants on protein function using the SIFT algorithm, Nat. Protoc. **4**(7), 1073–1081 (2009)

25.16   J.T. Robinson, H. Thorvaldsdottir, W. Winckler, M. Guttman, E.S. Lander, G. Getz, J.P. Mesirov: *Integrative genomics viewer*, Nat. Biotechnol. **29**, 24–26 (2011)

25.17   B. Ewing, P. Green: Base-calling of automated sequencer traces using *Phred*. II. Error probabilities, Genome Res. **8**(3), 186–194 (1998)

25.18   H. Li, R. Durbin: Fast and accurate long-read alignment with Burrows–Wheeler transform, Bioinformatics **26**(5), 589–595 (2010)

25.19   K. Wang, M. Li, H. Hakonarson: ANNOVAR: Functional annotation of genetic variants from next-generation sequencing data, Nucleic Acids Res. **38**, e164 (2010), available online at http://www.http://www.openbioinformatics.org/annovar/

25.20   K. Wang, M. Li, H. Hakonarson: ANNOVAR: Functional annotation of genetic variants from high-throughput sequencing data, Nucleic Acids Res. **38**(16), e164 (2010)

25.21   Ingenuity Systems: http://www.ingenuity.com/ (Ingenuity Systems, Inc., Redwood City)

25.22   S. Chun, J.C. Fay: Identification of deleterious mutations within three human genomes, Genome Res. **19**(9), 1553–1561 (2009)

25.23   X. Liu, X. Jian, E. Boerwinkle: dbNSFP: A lightweight database of human nonsynonymous SNPs and their functional predictions, Hum. Mutat. **32**(8), 894–899 (2011)

25.24   M.N. Edmonson, J. Zhang, C. Yan, R.P. Finney, D.M. Meerzaman, K.H. Buetow: Bambino: A variant detector and alignment viewer for next-generation sequencing data in the SAM/BAM format, Bioinformatics **27**(6), 865–866 (2011)

25.25   M.E. Sana, M. Iascone, D. Marchetti, J. Palatini, M. Galasso, S. Volinia: GAMES identifies and annotates mutations in next-generation sequencing projects, Bioinformatics **27**(1), 9–13 (2011)

25.26   M. Krzywinski, J.E. Schein, I. Birol, J. Connors, R. Gascoyne, D. Horsman, S.J. Jones, M.A. Marra: Circos: An information aesthetic for comparative genomics, Genome Res. **19**, 1639–1645 (2009), available online at http://mkweb.bcgsc.ca/circos/intro/genomic_data/

25.27   A. Barla, G. Jurman, R. Visintainer, M. Squillario, M. Filosi, S. Riccadonna, C. Furlanello: A machine learning pipeline for discriminant pathways identification, Proc. 8th Int. Meet. Comput. Intell. Methods Bioinf. Biostat., Gargnano (2011)

Part E

Bioinfor

Part E Bioinformatics Databases and Ontologies

Ed. by Francesco Masulli

26. Biological Databases

Mario Cannataro, Pietro H. Guzzi, Giuseppe Tradigo, Pierangelo Veltri

Biological databases constitute the data layer of molecular biology and bioinformatics and are becoming a central component of some emerging fields such as clinical bioinformatics, and translational and personalized medicine. The building of biological databases has been conducted either considering the different representations of molecular entities, such as sequences and structures, or more recently by taking into account high-throughput platforms used to investigate cells and organisms, such as microarray and mass spectrometry technologies. This chapter provides an overview of the main biological databases currently available and underlines open problems and future trends.

This chapter reports on examples of existing biological databases with information about their use and application for the life sciences. We cover examples in the areas of sequence, interactomics, and proteomics databases. In particular, Sect. 26.1 discusses sequence databases, Sect. 26.2 presents structure databases including protein contact maps, Sect. 26.3 introduces a novel class of databases representing the interactions among proteins, Sect. 26.4 describes proteomics databases, an area of biological databases that

is being continuously enriched by proteomics experiments, and finally Sect. 26.5 concludes the chapter by underlining future developments and the evolution of biological databases.

In recent years, the availability of high-performance computational platforms and communication networks has enabled the use of algorithms to support biological studies. Bioinformatics applications have allowed the study and evaluation of biological experiments in the areas of proteomics and genomics [26.1]. Indeed, thanks to such high-performance computational platforms, bioinformatics is supporting the pipelines of experiments: from data preparation, to data preprocessing and manipulation, and results extraction. Simulations of biological systems have also used computational platforms. Moreover, the increasing use of computational platforms is related to the necessity to design, create, and publish databases of biological data and information. Biological sequences as protein amino acids, microarray experimental results, spatial and geometrical molecule conformations, experimental configurations, spectral data resulting from mass spectrometry of tissues, and ontologies for classifying data and experiments are only limited examples of the data and information that biologists may found interesting to query, navigate, and use in their experiments.

Biological databases can be seen as large and available (web-based) biological information sources for experiments and data analysis tools. They can be classified depending on the information stored and their availability, modifiability, etc. We classify databases based on information type and based on their use for bioinformatics applications as well as by biologists. We consider sequence, structure, interactomics, proteomics, and genomics databases. Sequence databases contain information in the form of alphanumeric sequences representing proteins and genes, whereas structure databases contain information about complex molecules and their structures (for instance, protein structures) as well as their functionalities or potential interactions. Interactomics databases can be considered as recent arrivals, storing the results of complex molecular interactions (for instance, protein–protein interactions).

The available databases have been enriched during recent decades as new experimental results have become available and been approved by the international community. This is the case, for instance, of the Protein Data Bank (PDB) database that includes protein structures and information that are published as soon as they are validated. Today most of these databases are available online and publish their own application programming interfaces (APIs) as well as extensible markup language (XML)-based exchange formalism to maximize interoperability among different data sources and to maximize their use. Finally, databases are also created by single research groups, then quickly made available and published to the international research community, while the development and validation of large databases are carried out by the international community. Thus, biological databases can be seen as large libraries related to biology and life sciences data and information resulting from experiments conducted in laboratories worldwide as well as from simulation analysis and published papers. They contain information from research areas related to biology, chemistry, and life sciences, and most of them specialize in particular life science areas (e.g., proteomics or genomics) as well as in research analysis and simulation (microarray gene expression databases).

26.1 Sequence Databases

These databases store information about the primary sequence of proteins. Each sequence is generally annotated by several types of information, e.g., the name of the scientist who discovered the sequence, or about the posttranslational modification. The user can query these databases by using a protein identifier or a fragment of sequence in order to retrieve the most similar proteins.

26.1.1 The EMBL Nucleotide Sequence Database

The European Molecular Biology Laboratory (EMBL) nucleotide sequence database [26.2, 3], maintained at the European Bioinformatics Institute (EBI), collects nucleotide sequences and annotations from publicly available sources. The database is involved in an international collaboration, being synchronized with the DNA Data Bank of Japan (DDBJ) and GenBank (USA) (see next sections). The core data are the protein and nucleotide sequences. The annotations describe the following items:

1. Function(s) of the protein
2. Posttranslational modification(s)
3. Domains and sites
4. Disease(s) associated with deficiencies
5. Secondary structure.

Webin is the preferred tool for individual submissions of nucleotide sequences, including third-party annotations (TPAs) and alignments. Automated procedures are provided for submissions from large-scale sequencing projects and data from the European Patent Office. New and updated data records are distributed daily, and the whole EMBL nucleotide sequence database is released four times a year. Currently, it contains more than 180 000 000 000 nucleotides in more than 100 000 000 entries. The entries are structured as textual files, composed of lines. Different types of lines, each with their own format, are used to record the various data that make up the entry. Each line begins with a two-character line code, which indicates the type of data contained in the line; for instance, the code ID represent the identifiers of the protein, while the code AC represents the accession number. Data are accessible via file transfer protocol (ftp) and several web interfaces. Moreover, the web-based sequence retrieval system (SRS) links nucleotide data to other specialized databases maintained at the EBI. Many other tools for sequence comparison and sequence similarity search, such as FASTA or the basic local alignment search tool (BLAST), are available through web interfaces.

26.1.2 GenBank

The GenBank database [26.4, 5] stores information about nucleotide sequences, maintained by the National Center of Biotechnology Information (NCBI).

GenBank entries are structured as flat files (like the EMBL database) and share the same structure as EMBL and DDBJ. All entries are grouped following both taxonomic and biochemical criteria. In this way it is possible to distinguish entries belonging to taxonomic groups (e.g., bacteria, viruses, and primates) as well as expressed sequences tags (EST) or core nucleotide sequences. Compared with the EMBL flat file structure, the main difference in the GenBank flat files is in the abbreviations used in the structure and in their interpretation.

GenBank is accessible through a web interface. Through the ENTREZ system, the entries of GenBank are integrated with many data sources, enabling searching for information about proteins and their structures, as well as literature about the functions of genes.

26.1.3 Uniprot Database

After the introduction of the nucleotide sequence databases, the scientific community worked towards the development of databases storing information about protein sequences. This work was helped by the introduction of methods and technologies able to investigate proteins, leading to the accumulation of such data.

Historically, one of the first databases was Swiss-Prot [26.3]. The main characteristics of Swiss-Prot were:

1. The use of flat files organized in multiple subfields as the storage system
2. Avoidance of redundancies, achieved by employing a manual curation workflow
3. Extensive usage of annotation, providing useful information about each entry.

In 2002 the Swiss-Prot databases merged with two related projects: the Tr-EMBL database [26.3] (a supplement of Swiss-Prot that stores information about sequences that are candidates to be introduced into Swiss-Prot but are under verification) and the Protein Sequence Database that developed into the Protein Information Resource project (PSD–PIR [26.6]). The result of this process was the introduction of the UniProt [26.7] consortium, which is structured on three main knowledge bases:

1. UniProt (also referred to as the UniProt knowledge base), which is the main archive storing information about protein sequences and annotations extracted from Swiss-Prot, TrEMBL, and PSD-PIR
2. UniParc (Uniprot archive), which contains publicly available information about proteins extracted from the main archives
3. UniRef (Uniprot reference), a set of databases that organize entries of UniProt by their similarity sequence; e.g., UniRef90 groups entries of UniProt that present at least 90% sequence similarity into a single record.

The Uniprot database is publicly accessible at http://www.uniprot.org/, and it is freely downloadable.

26.2 Structure Databases

26.2.1 Protein Data Bank (PDB)

The Protein Data Bank (PDB) [26.8] is a worldwide archive of structure data of biological macromolecules. Such data are generated by crystallography and nuclear magnetic resonance (NMR) experiments. Each PDB entry is stored in a single flat file. There is an underlying ontology of about 1700 terms that define the macromolecular structure and crystallographic experiment. This ontology is called the macromolecular crystallographic information file (mmCIF) dictionary.

Although distributed as a flat file, PDB is based on a relational model. There are three distinct query interfaces:

1. Status Query
2. Search Lite
3. Search Field.

Search Lite has a single text field in which it is possible to write keywords. Search Field is a customizable query form allowing queries based on author citation, sequences (via FASTA algorithm), dates, and chemical formulas. Many interfaces present information related to the results. A query result browser interface allows detailed information to be browsed and a set of files storing the structures found to be downloaded. PDB files in XML format are currently being tested. Data are acquired from the research community by submission.

26.2.2 Databases of Structural Classifications

A structural domain of a protein is an element of its ternary structure that often folds independently of the rest of the protein chain and represents a biologically relevant module of the protein itself. Despite the large number of different proteins expressed in eukaryotic systems, there are many fewer different domains, structural motifs, and folds. Many domains are not unique to a protein that is produced by one gene or gene family but instead appear in a variety of proteins as a consequence of evolution, which has conserved spatial conformation better than primary sequence.

Consequently, several methods have been developed for structural classification of proteins, and a number of different databases have been introduced, such as the Structural Classification of Proteins (SCOP) [26.9] and CATH (class, architecture, topology, and homologous superfamily) databases [26.10].

SCOP

The Structural Classification of Proteins (SCOP, http://scop.mrc-lmb.cam.ac.uk/scop) database aims to order all the proteins whose structure has been published according to their structural domains. Protein domains in SCOP are hierarchically classified into *families, superfamilies, folds, and classes*. Firstly, proteins are grouped together into families on the basis of evolutionary or functional similarities, such as sequence alignment. Then, proteins whose sequences have low similarity but whose structure or functions are close are grouped into superfamilies. The secondary structure of the proteins in these two groups is analyzed, and when proteins in two groups, e.g., two families, have similar secondary structure, they have a common fold. Finally, folds are grouped into classes as follows:

1. All α, if the structure is essentially formed by α-helices
2. All β, if the structure is essentially formed by β-sheets

3. α/β, for those with α-helices and β-strands
4. α + β, for those in which α-helices and β-strands are largely segregated
5. Multidomain, for those with domains of different classes.

The SCOP database is available on the world wide web. The user has many options to browse its content. The main possibility is to start at the top of the hierarchy and then navigate through the levels from the root to the leaves, which are individual PDB entries. Alternatively, the user can search a protein starting from an amino acid sequence to retrieve the most similar proteins categorized in SCOP. The user can then download the found protein as a single PDB file.

CATH

CATH [26.10] stores a hierarchical classification of PDB structures obtained by NMR where crystal structures are solved at resolution higher than 4.0 Å (http://www.cathdb.info/latest/index.html). Protein structures are classified using a combination of automated and manual procedures on the protein domains. To divide multidomain protein structures into their constituent domains, a combination of automatic and manual techniques are used. The hierarchy is organize in four major levels: *class*, *architecture*, *topology* (fold family), and *homologous superfamily*. Class is determined according to the secondary structure composition; currently three major classes have been recognized: mainly α, mainly β, and α–β, which includes both alternating α/β structures and α + β structures. A fourth class contains protein domains which have low secondary structure content. The architecture level considers the overall shape of the domain structure as determined by the orientations of the secondary structures, being assigned manually. The topology level groups the structures depending on the overall shape and on the connectivity of the secondary structure by applying the SSAP (sequential structure alignment program for protein structure comparison) algorithm [26.11]. Finally, the homologous superfamily level groups protein domains which are thought to share a common ancestor and can therefore be described as homologous as recognized by SSAP. Currently, it contains 30 028 PDB structures. CATH contains PDB structures organized in a relational model. CATH can be searched by submitting a protein identifier, or by browsing the hierarchical structure. Moreover, the user can access data via ftp and download them.

26.2.3 Protein Contact Map

Protein contact maps are bidimensional data structures representing a view of the three-dimensional (3-D) structure of a protein. They are used to store the presence or absence of contacts among protein residue pairs. Two residues are said to be in contact if their mutual distance is lower than a certain distance threshold (i. e., 8 Å). Contact maps have a key role in most state-of-the-art protein structure prediction pipelines, i. e., the prediction of the three-dimensional space conformation of the amino acids composing a protein.

Indeed, contact information may be used to drive the computational folding process, to select structural templates, or to assess the quality of structural predictions. Thus, it is critical to develop accurate predictors of contact maps. Correct contact maps have been shown to lead to reasonably good 3-D structures [26.12, 13], and predicted contact maps have been used for driving protein folding in the ab initio case (that is, when a protein is folded without relying on homology to another protein of known structure), for selecting and ranking folded protein models, and for predicting folding times, protein domain boundaries, secondary structures, etc. Virtually no contact map databases exist (except for [26.14]), but many prediction tools are available [26.15–17]. In fact, in the case of an unknown protein for which no exact 3-D structure has been experimentally determined, a contact map can be directly predicted from its primary structure or derived from its predicted 3-D structure, while in the case of a known protein, the 3-D structure retrieved from a structure database (i. e., Protein Data Bank) can easily give the protein contact map.

26.3 Interactomics Databases

The accumulation of protein interaction data led to the introduction of several databases. Here we concentrate on *databases of experimentally determined interactions*, which include all databases storing interactions extracted from both literature and high-throughput experiments, and databases of *predicted interactions* that store data obtained by in silico predictions. Another important class that we report is constituted by *integrated databases* or metadatabases, i. e., databases that aim to integrate data stored in other publicly available datasets. Currently, there exist many databases that differ in various biological and information science aspects: the organism covered, the kind of interactions, the kind of interface, the query language, the file format, and the visualization of results.

Data produced in low- or high-throughput experiments are stored in *databases of experimentally determined interactions* after subsequent verification by a committee of database curators. Researchers can submit their own data directly to the databases, e.g., to Intact [26.18], or they can publish data in the literature and then the database curators will extract them, e.g., the MINT (Molecular INTeraction) database [26.19]. For a more complete description of interactomics databases see [26.20, 21].

For simpler organisms, such as yeasts, worms, or flies, the process of the whole coverage of the interaction network seems to be almost completed. This process led to the introduction of a huge amount of data that may be mined for many objectives. Conversely, the complexity of the interactomes of higher eukaryotes has prevented these experiments for humans. In this scenario, the need for the introduction of algorithms and tools able to use these experimental data to predict protein interactions arose. Thus, starting from existing databases of verified interactions, a number of algorithms have been developed to predict putative interactions that are accumulated into *databases of predicted interactions*. The common approach is based on the consideration that the interaction mechanisms are conserved through evolution; i. e., if two protein A and B interact in a simple organism, then the corresponding orthologous proteins A_1 and B_1 may interact in a complex organism. Thus, starting from the interacting proteins in a simple organism, predictions are made for other organisms.

Despite the existence of many databases, the resulting data present three main problems [26.22]: the low overlap among databases, the resulting lack of completeness with respect to the real interactome, and the absence of integration. Consequently, in order to perform an exhaustive data collection (e.g., for an experiment), researchers have to manually query different data sources. This problem is being addressed through the introduction of databases based on the integration of existing ones. Nevertheless, in the interactomics field, the integration of existing databases is not an easy task.

The integration of data from different laboratories and sources can be done through the adoption of an accepted system of interaction identifiers. It should be noted that, while in other biological database systems, such as sequence databases, there exists a common system of identifiers, and cross-referencing can be used to retrieve the same biological entity from different databases, PPI (protein-to-protein interaction) interactions are currently not identified by a unique identifier, but through the names of the corresponding partners.

Despite these problems, different approaches for data integration and building larger interaction maps have been pro posed. The rationale for these approaches is based on a three-step process:

1. Collection of data from different data sources
2. Transformation of data into a common model
3. Annotation and scoring of the resulting dataset.

All the existing databases go beyond storage of the interactions, also integrating them with functional annotations, sequence information, and references to corresponding genes. Finally, they generally provide some visualization that presents a subset of interactions in a comprehensive graph.

Nevertheless, currently there are some problems and characteristics that are common to almost all databases:

1. Errors in the databases
2. Lack of naming standards
3. Little overlap among interactions.

Any published dataset may contain errors, so any database may contain false interactions, often called false positives, i.e., proteins erroneously reported as interacting. This may be due, for instance, to technical (i.e., false positives that are due to the detection method) and biological problems (i.e., proteins that are reported to be interacting in vitro but that are never co-located).

In other biological database communities, such as those storing protein sequences or structures, there exist many projects providing common accepted identifiers for biological objects, or at least a system for cross-referencing the same object between almost all databases. In interactomics there is no such common identifier, and in general interactions are not identified

by a single code but rather by using the identifiers of the interacting proteins.

It has been noted [26.22] that existing databases present little overlap with respect to the dimension of the interactomes. Despite this, integration of databases remains an open problem due to the difficulties resulting from the absence of a naming standard.

Conversely, common aspects of existing datasets are:

1. Simple web-based interface for querying
2. Simple visualization of results in both tabular and graphical form
3. Data available for download in different formats.

It should be noted that almost all these databases offer the user the possibility of retrieving data and some annotations through a simple web-based interface. Despite this, querying of protein networks aims to go beyond the simple retrieval of a set of interactions stored in databases.

Databases can actually be queried through simple key-based searches, e.g., by inserting one or more protein identifiers. The output of such a query is in general a list of interacting protein pairs. These pairs share a protein, as specified in the query. Such an approach, despite its conceptual simplicity and easy practical use, presents some limitations. Let us consider, for instance, a researcher who wishes to compare patterns of interactions among species, or a researcher who wants to search for interactions related to a given biological compartment or biological process. The existing query interfaces, in general, do not enable such queries.

Thus, a more powerful querying system should provide a semantically more expressive language, e.g., enabling retrieval of all interaction patterns that share the same structure. Then, the query system should map the query, expressed in a high-level language (e.g., using a graph formalism), into suitable graph structures and search for them by applying appropriate algorithms. Unfortunately this problem is not easy from a computational point of view, and it requires:

1. Modeling of the PPI network in a suitable data structure
2. Appropriate algorithms for mapping, i.e., identification of the correspondence between the nodes in a subnetwork and those stored in the database [26.23].

26.4 Proteomics Databases

26.4.1 Global Proteome Machine Database

The Global Proteome Machine Database (http://www.thegpm.org/GPMDB/index.html) [26.24] was constructed to utilize information obtained from the different servers included in the Global Proteome Machine project (GPM), to validate peptide tandem mass spectrometry (MS/MS) spectra and protein coverage. GPM is a system for analyzing, storing, and validating proteomics information derived from tandem mass spectrometry. The system is based on a relational database on different servers for data analysis, and on a user-friendly interface to retrieve and analyze data. This database has been integrated into GPM server pages. The gpmDB data model is based on a modification of the Hupo-PSI minimum information about a proteomics experiment (MIAPE) [26.25] scheme. With respect to the proposed standard, the database is conceived to hold only the information needed in certain bioinformatics-related tasks, such as sequence assignment validation. Data are mainly held in a set of XML files: the database serves as an index to those files. This combination of a relational database with XML is called by the authors XML information about a proteomics experiment (XIAPE). The system is available both through a web interface and as a standalone application, allowing users to compare their experimental results with others previously observed by other scientists.

26.4.2 PeptideAtlas

PeptideAtlas [26.26] is a database that aims to annotate the human genome with protein-level information (http://www.peptideatlas.org/overview.php). It contains data coming from identified peptides analyzed by liquid chromatography tandem mass spectrometry (LC-MS/MS) and thus mapped onto the genome. PeptideAtlas is not a simple repository for mass spectrometry experiments, but uses spectra as a primary information source to annotate the genome, combining different information. Consequently, population of this database involves two main phases:

1. A proteomic phase in which samples are analyzed through LC-MS/MS, and resulting spectra are mined to identify the contained peptides
2. An in silico phase in which peptides are processed by applying a bioinformatic pipeline and each peptide is used to annotate a genome. Resulting derived data, both genomics and proteomics, are stored in the PeptideAtlas database.

Data submitted by researchers are organized in a relational model. PeptideAtlas is based on the Systems Biology Experiment Analysis Management System (SBEAMS) project, which is a framework for collecting, storing, and accessing data produced by a variety of different experiments. It combines a relational database management system back-end providing integrated access to remote data sources. User can query data through a web interface or can download a whole dataset, organized by the original publications.

26.4.3 NCI Repository

The National Cancer Institute Clinical Proteomics Databank (http://home.ccr.cancer.gov/ncifdaproteomics/ppatterns.asp), is not a database, but stores different datasets obtained by mass spectrometry. Currently it holds datasets obtained in different experimental conditions, coming from different mass spectrometry platforms, for both human and animals. It contains six different surface-enhanced laser desorption ionization time-of-flight (SELDI-TOF) mass spectrometry datasets. This technique is very similar to MALDI-TOF and generates spectra which have similar characteristics. Datasets are stored as flat files, each containing a whole SELDI-TOF spectrum. The datasets are freely available to download as flat files.

26.4.4 PRIDE

The proteomics identifications database (PRIDE) (http://www.ebi.ac.uk/pride) [26.27] is a database of protein and peptide identifications that may be annotated with supporting mass spectra. PRIDE stores information about a complete proteomic experiment, starting from the title and a brief description of the experiment itself. Each experiment is annotated by a description of the sample under analysis and of the instrumentation used to perform the analysis. The core element of each entry is the protein identifications, sorted by unique accession numbers and supported by a corresponding list of one or more peptide identifications. For each peptide identified, the database also stores the sequence and coordinates of the peptide within the protein for which it provides evidence. Optionally, an entry can contain a reference to any submitted mass spectra that form the evidence for the peptide identification, encoded in the versatile proteomics standard initiative (PSI) mzData format. Users can directly submit protein and peptide identification data to be published in peer-reviewed publications by using the PRIDE 2.1 XML schema. The PRIDE database currently contains 3178 experiments, 339 696 identified proteins, 2 145 505 identified peptides, 309 889 unique peptides, and 2 582 614 spectra. PRIDE is based on a relational database based on structured query language (SQL) and is currently available for ORACLE and MySQL. PRIDE provides the user with a web interface to retrieve data. Data can be exported in PRIDE XML schema, a format which embeds mzData as a subelement, or using the mzData XML schema.

26.4.5 2-D Gel Databases

Databases of data produced by using gel electrophoresis generally store both images and identification, the core data, and metadata relating to the experiment. Metadata are relative to the parameters of the experiment, while core data store the proteins contained in the associated image. In the following we present the SWISS-2DPAGE database.

SWISS–2DPAGE

The SWISS-2DPAGE (http://www.expasy.ch/ch2d/ch2d-top.html) [26.28] database was established in 1993 and is maintained collaboratively by the Swiss Institute of Bioinformatics (SIB) and the Biomedical Proteomics Research Group (BPRG) of the Geneva University Hospital. Current content includes identified spots and corresponding protein entries in 36 reference maps from human, mouse, *Arabidopsis thaliana*, *Dictyostelium discoideum*, *Escherichia coli*, *Saccharomyces cerevisiae*, and *Staphylococcus aureus*.

The protein entries in SWISS-2DPAGE are structured text files. Each entry is composed of defined lines, used to record various kinds of data. For standardization purposes, the format of SWISS-2DPAGE entries is similar to that used in the Swiss-Prot database, in addition to specific lines dedicated to the two-dimensional (2-D) polyacrylamide gel electrophoresis (PAGE) data:

1. The master line (MT) lists the reference maps where the entry has been identified
2. The images line (IM) lists the 2-D PAGE images available for the entry
3. The 2-D lines group different topics including the mapping procedure, spot coordinates, protein amino acid composition, protein expression levels, and modifications.

SWISS-2DPAGE is available through the ExPASy molecular biology server. The SWISS-2DPAGE top page provides text searches, and displays results with active links to other databases. It is also possible to get a local copy of SWISS-2DPAGE via ftp from the ExPASy ftp server. On the ExPASy webserver the data image associated with a protein entry displays the experimental location of the protein on the chosen map, in addition to a theoretical region computed from the protein sequence.

26.5 Conclusions

Biological databases have been developed as autonomous, specialized, but not integrated repositories of data; For instance, the first databases stored data regarding the different representations of DNA and proteins, such as sequences and structures. Since molecular medicine research needs information from different biological databases, the bioinformatics community started to develop integrated databases where integration is often obtained by cross-referencing common identifiers. Following the expansion of omics sciences, such as genomics, proteomics, and interactomics, and the diffusion of high-throughput experimental platforms, novel biological databases have been produced. Among them, mass spectrometry and gel electrophoresis data have improved proteomics databases, while data about protein interactions form the main interactomics protein–protein interaction databases. A future trend will be the further integration of this plethora of biological databases and especially the annotation of existing data with information contained in different knowledge bases and ontologies such as Gene Ontology (http://www.geneontology.org/).

References

26.1 R. Matthiesen: Methods, algorithms and tools in computational proteomics: A practical point of view, Proteomics **7**(16), 2815–2832 (2007)

26.2 EMBL Nucleotide Sequence (European Molecular Biology Laboratory, EMBL Heidelberg, Heidelberg) available online at http://www.ebi.ac.uk/embl

26.3 B. Boeckmann, A. Bairoch, R. Apweiler, M.-C.C. Blatter, A. Estreicher, E. Gasteiger, M.J. Martin, K. Michoud, C. O'Donovan, I. Phan, S. Pilbout, M. Schneider: The SWISS-PROT protein knowledgebase and its supplement TrEMBL in 2003, Nucleic Acids Res. **31**(1), 365–370 (2003)

26.4 GenBank database (National Center for Biotechnology Information, National Library of Medicine, Bethesda) USA available online at www.ncbi.nlm.nih.gov/genbank/

26.5 D.A. Benson, I. Karsch-Mizrachi, D.J. Lipman, J. Ostell, D.L. Wheeler: GenBank, Nucleic Acids Res. **36**, D25–D30 (2008)

26.6 W.C. Barker, J.S. Garavelli, P.B. Mcgarvey, C.R. Marzec, B.C. Orcutt, G.Y. Srinivasarao, L.S. Yeh, R.S. Ledley, H.W. Mewes, F. Pfeiffer, A. Tsugita, C. Wu: The PIR-international protein sequence database, Nucleic Acids Res. **27**(1), 39–43 (1999)

26.7 The UniProt Consortium: The Universal Protein Resource (UniProt) in 2010, Nucleic Acids Res. **38**(suppl 1), D142–D148 (2010)

26.8 H.M. Berman, J. Westbrook, Z. Feng, G. Gilliland, T.N. Bhat, H. Weissig, I.N. Shindyalov, P.E. Bourne: The protein data bank, Nucleic Acids Res. **28**(1), 235–242 (2000)

26.9 T.J.P. Hubbard, A.G. Murzin, S.E. Brenner, C. Chothia: SCOP: A structural classification of proteins database, Nucleic Acids Res. **25**(1), 236–239 (1997)

26.10 C.A. Orengo, A.D. Michie, S. Jones, D.T. Jones, M.B. Swindells, J.M. Thornton: CATH – a hierarchic classification of protein domain structures, Structure **5**(8), 1093–1108 (1997)

26.11 C. Orengo, W. Taylor: SSAP: Sequential structure alignment program for protein structure comparison. In: *Computer Methods for Macromolecular Sequence Analysis*, Methods in Enzymology, Vol. 266, ed. by S.P. Colowick, R.F. Doolittle, N.O. Kaplan (Academic, New York 1996) pp. 617–635

26.12 M. Vendruscolo, E. Kussell, E. Domany: Recovery of protein structure from contact maps, Fold. Des. **2**(5), 295–306 (1997)

26.13 I. Walsh, D. Baú, A.J.M. Martin, C. Mooney, A. Vullo, G. Pollastri: Ab initio and template-based prediction of multi-class distance maps by two-dimensional recursive neural networks, BMC Struct. Biol. **9**(1), 5 (2009)

26.14 P. Chen, C. Liu, L. Burge, M. Mohammad, B. Southerland, C. Gloster, B. Wang: IRCDB: A database of inter-residues contacts in protein chains, 1st Int. Conf. Adv. Databases (2009) pp. 1–6

26.15 D. Baú, A. Martin, C. Mooney, A. Vullo, I. Walsh, G. Pollastri: Distill: A suite of web servers for the prediction of one-, two- and three-dimensional structural features of proteins, BMC Bioinformatics **7**, 1–8 (2006)

26.16 R.M. MacCallum: Striped sheets and protein contact prediction, Bioinformatics **20**(suppl 1), i224–i231 (2004)

26.17 B. Rost, M. Punta: PROFcon: Novel prediction of long-range contacts, Bioinformatics **21**(9), 2960–2968 (2005)

26.18 H. Hermjakob, L. Montecchi-Palazzi, C. Lewington, S. Mudali, S. Kerrien, S. Orchard, M. Vingron, B. Roechert, P. Roepstorff, A. Valencia, H. Margalit, J. Armstrong, A. Bairoch, G. Cesareni, D. Sherman, R. Apweiler: IntAct: An open source molecular interaction database, Nucleic Acids Res. **1**(32), 452–455 (2004)

26.19 A. Zanzoni, L. Montecchi-Palazzi, M. Quondam, G. Ausiello, M. Helmer-Citterich, G. Cesareni: MINT: A Molecular INTeraction database, FEBS Lett. **513**(1), 135–140 (2002)

26.20 M. Cannataro, P.H. Guzzi, P. Veltri: Protein-to-protein interactions: Technologies, databases, and algorithms, ACM Comput. Surv. **43**, 1 (2010)

26.21 A. Batemen: NAR Database ISSUE, Nucleic Acids Res. **35**(Suppl. 1) (2007)

26.22 G. Chaurasia, Y. Iqbal, C. Hanig, H. Herzel, E.E. Wanker, M.E. Futschik: UniHI: An entry gate to the human protein interactome, Nucleic Acids Res. **35**(suppl1), D590–594 (2007)

26.23 S. Zhang, X.-S. Zhang, L. Chen: Biomolecular network querying: A promising approach in systems biology, BMC Syst. Biol. **2**(1), 5 (2008)

26.24 C. Robertson, J.P. Cortens, R.C. Beavis: Open source system for analyzing, validating, and storing protein identification data, J. Proteome Res. **3**(6), 1234–1242 (2004)

26.25 C.F. Taylor, H. Hermjakob, R.K. Julian, J.S. Garavelli, R. Aebersold, R. Apweiler: The work of the human proteome organisation's proteomics standards initiative (HUPO PSI), OMICS **10**(2), 145–151 (2006)

26.26 F. Desiere, E.W. Deutsch, N.L. King, A.I. Nesvizhskii, P. Mallick, J. Eng, S. Chen, J. Eddes, S.N. Loevenich, R. Aebersold: The PeptideAtlas project, Nucleic Acids Res. **34**(Suppl. 1), D655–D658 (2006)

26.27 P. Jones, R.G. Côté, L. Martens, A.F. Quinn, C.F. Taylor, W. Derache, H. Hermjakob, R. Apweiler: PRIDE: A public repository of protein and peptide identifications for the proteomics community, Nucleic Acids Res. **34**(Suppl. 1), D659–D663 (2006)

26.28   J.-C. Sanchez, D. Chiappe, V. Converset, C. Hoogland, P.-A. Binz, S. Paesano, R.D. Appel, S. Wang, M. Sennitt, A. Nolan, M.A. Cawthorne, D.F. Hochstrasser: The mouse SWISS-2D PAGE database: A tool for proteomics study of diabetes and obesity, Proteomics **1**(1), 136–163 (2001)

27. Ontologies for Bioinformatics

Andrea Splendiani, Michele Donato, Sorin Drăghici

This chapter provides an introduction to ontologies and their application in bioinformatics.

It presents and overview of the range of information artifacts that are denoted as ontologies in this field, from controlled vocabularies to rich axiomatizations. It then focuses on the conceptual nature of ontologies and introduces the role of upper ontologies in the conceptualization process.

Language and technologies that underpin the definition and usage of ontologies are then presented, with a particular focus on the ones derived from the semantic web framework. One objective of this chapter is to provide a concise and effective understanding of how technologies and concepts such as ontologies, RDF, OWL, SKOS, reasoning and Linked–Data relate to each other. The chapter is then complemented by a bioinformatics section (Sect. 27.4), both via an overview of the evolution of ontologies in this discipline, and via a more detailed presentation of a few notable examples such as gene ontologies (and the OBO family), BioPAX and pathway ontologies and UMLS. Finally, the chapter presents examples of a few areas where ontologies have found a significant usage in bioinformatics: data integration, information retrieval and data analysis (Sect. 27.5). This last section briefly lists some tools exploiting the information contained in biomedical ontologies when paired with the output of high–throughput experiments such as cDNA microarrays.

27.1 Defining Ontology

This chapter presents an introduction to ontologies and their application in bioinformatics. It can be argued that *ontologies* have long been an integral part of life sciences research. In biology, medicine, and in related sciences the use of classification systems (which can be seen as a particular type of ontology) dates back to the earlier days of these disciplines. The definition of anatomical parts and their relations can be dated back to Aristotle, while the classification of organisms was developed by Linnaeus in the early eighteenth century. Even the periodic table of elements, developed by Mendeleev later in the same century is, to a large extent, an *ontology* for chemistry.

It is, however, with the most recent developments of these areas of science that the use of ontologies has gained an even more critical dimension in research. Life sciences are characterized by a convergence of disciplines, which holds the promise for huge benefits to society, for instance via the convergence of basic research and medicine (translational medicine and personalized medicine).

Underpinning these promises is a growing amount of data (e.g., genome sequences) that has transformed biology into a data intensive science.

In this data intensive science, ontologies play several key roles:

1. In information retrieval from a large amount of data
2. In the annotation of information through consistent vocabularies (both in databases and in the scientific literature)
3. In the integration of heterogeneous datasets and information.

Ontologies also allow the development of analysis techniques that can be used to *explain* data in terms of the biological know-how on the observed systems.

The development of ontologies in life sciences (or, from an application point of view, in bioinformatics) sees the contribution of a variety of disciplines, both core disciplines such as medicine, biology, chemistry, and supporting disciplines such as computer science, mathematics (logics), and some areas of philosophy.

This is not surprising as both the variety of disciplines participating in the definition of ontologies and the variety of usages found by ontologies stems from a basic need that they address: the definition of a common language, which is shared across disciplines and whose semantics is *computable*.

27.1.1 What Is an Ontology?

In bioinformatics and computer sciences, the term *ontology* is commonly defined as *a formalization of a conceptualization*. A *conceptualization* is the definition of entities, classes and relations that hold in a given area of knowledge. As an example if we consider biological pathways, a conceptualization of biological pathways would list the entities (e.g., the p53 protein, a specific biochemical reaction), classes (e.g., protein, dimer, phosphorylation, biochemical reaction), and relations holding among them (e.g., a phosphorylation is a more specific kind of biochemical reaction, a protein

can have the role of catalyzer in a given biochemical reaction, etc.).

A conceptualization is *formalized* when such entities, classes, and relations are represented via a set of symbols to which a set of axioms can be associated, so that the truth of possible sentences in the representation reflects the truth of the entities that symbols represent.

For instance, if E is a symbol associated to the class of enzymes, C is a symbol associated to the class of catalysis, and *hasCatalyzer* a symbol associated to the relation which links a catalysis to its enzyme, we could formally express the notion that all catalysis processes have an enzyme (catalyzer) as

$$\forall c : c \in C, \exists e : e \in E \wedge \text{hasCatalyzer}(c, e) \, .$$

Without entering into the details of the theory of formal semantics, we can easily see how this formal assertion can be part of a formal theory of pathways.

Two important results stem from this representation. First, we can in general verify whether the theory is *consistent*. Second, we can deduce new relations among terms.

Some of the ontology languages that we will present later provide a set of *operators*, which balance the need for expressiveness with the computability of the resulting theories.

27.1.2 Defining a Conceptual Reference: Upper Ontologies

Defining a conceptualization of a domain is a nontrivial exercise, although its complexity may not be evident at first.

Let us consider our previous example in which we introduced *p53* as a part of our conceptualization of pathways. Which is the exact intended meaning of p53? Is it a representation of a consensus sequence, which is the canonical sequence for all genes encoding a p53 protein? Or is it the ensemble of all physical DNA that exists (and possibly existed or will exist) with the potentiality of encoding such protein? Is a nonviable version of the gene still a *p53* gene?

Often, such questions do not have a unique answer, as a conceptualization depends on a point of view, or a set of assumptions and references on which we base our understanding of the world. Some ontologies (known as *upper ontologies*) specify these assumptions and references, so that different conceptualizations developed on the basis of the same upper ontology are semantically interoperable.

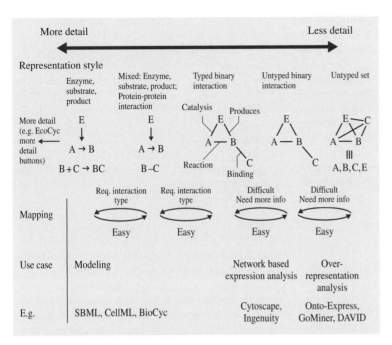

Fig. 27.1 Examples of different pathway modeling approaches, with different levels of detail (courtesy of the BioPAX Working Group, with modifications)

As an example, an upper ontology may require the partition of entities into entities that exist independently of time (e.g., physical objects) and entities that depend on time for their definition (e.g., events). A different upper ontology could postulate time as a dimension and consider all entities as belonging to the same four-dimensional space.

27.1.3 Many Artifacts Can Be Considered Ontologies

The definition of ontology that we have presented is very broad. According to this, different artifacts can be considered as ontologies, ranging from taxonomies to complete formal theories in a continuum of *ontological commitment*, that is, of how much of the domain knowledge is formally represented.

One aspect of the variability in biomedical ontologies stems from the level of detail in which reality can be represented. Building on the example of biological pathways, Fig. 27.1 presents different ways to represent pathway information, which vary according to the level of detail in which such pathways are described. Incidentally, different levels of details address different use cases, as reported in the figure.

Figure 27.1 displays different representations from the less detailed (far right) to the more detailed (left). At the extreme right, pathways are simply seen as untyped

sets, or equivalently as *labels* that characterize sets of genes. This is typical of much of the ontology first created for the annotation of biological processes (gene ontology, GO, which will be presented later). Beside annotation, such ontologies support a kind of analysis known as over-representation analysis (this will be discussed in more detail later).

Moving to the left, we see more detailed representations where pathways are seen as networks of interacting elements, then representations where these networks are typed, with types corresponding to the type of interactions and elements. These representations often correspond to a static view of pathways, mostly characterized in their topology and functional parts. They provide support for the description of pathways (e.g., pathway maps) and for network-based analysis approaches that can make use of the pathway topology.

To the far left, we find more detailed representations, which include a characterization of the dynamics of pathways and can be used as a base for dynamic modeling and simulations. As noted in Fig. 27.1, these different models are related to each other and often a mapping is possible from the most detailed to the less detailed representation.

While the level of detail is a feature characterizing biomedical ontologies, this is not to be confounded with the *ontological commitment* of these ontologies, which provides an orthogonal element of variability. While the notion that an entity participates in a pathway can be expressed with more or fewer details, the level of precision and formalization of this statement can also vary.

An ontology with a limited formalization could simply state *participatesIn(a, b)* to indicate that and entity *a* participates in a process *b*. A more precise characterization could, for instance, state that *a* participates in *b*, if and only if there is at least a time *t* such that *participatesIn(a, b)* is true.

It is clear that the more semantic content is formally represented in an ontology, the richer are the deductions and possibility to verify consistency that it supports.

27.1.4 On the Variety of Ontologies in Bioinformatics

The field of bioinformatics presents a wide range of artifacts representing conceptualizations of the life science domain, which can be broadly considered ontologies. One very common case is represented by ontologies that consist of large collections of classes, each class corresponding to a generic entity in the domain, and that are often organized according to a limited set of relations such as *is_a* or *part_of*, where often the ontological commitment of these relations is loosely defined (Fig. 27.2).

In these cases, ontologies are artifacts that are very similar (sometimes equivalent) to taxonomies or terminologies. Another common case is constituted by ontologies, like BioPAX, that have a limited number of classes and relations that represent in detail generic features of pathways (Fig. 27.6).

In this case, the ontology is very similar to a *schema* or a UML class model.

Often ontologies found in bioinformatics do not provide a deep formalization of the biological domain they

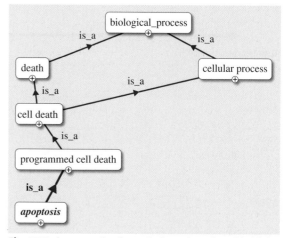

Fig. 27.2 An example of a typical ontology found in bioinformatics (fragment of GO biological process)

represent. Perhaps this is not surprising. As these ontologies are developed in a field which is becoming increasingly rich in information, an emphasis on *extensive* rather than *intensive* development of ontologies is natural: there is a natural tradeoff between the amount of knowledge that can be represented in an ontology (how *extensive* the ontology coverage is) and the precision and ontological commitment of the ontology (how *intensive* the definitions of entities are). In the biomedical domain, at this stage of the evolution of the science, the necessity for comprehensive, if somewhat sparse, characterization of the domain generally prevails over the need for detailed, but narrower, formalizations.

However, the development of ontologies that relate to bioinformatics is evolving, and more and more logic is being embedded in the definition of ontologies, as we will see in the case of ontologies like the Gene Ontology (GO) and the Open Biology Ontology (OBO).

27.2 Brief History of the Use of Ontologies in Bioinformatics

In this section, we provide a brief overview of the evolution of the ontologies commonly used in bioinformatics. Far from being extensive, this overview will link several topics presented in the remainder of the chapter, concerning what ontologies are available, their representations and their usage. Such has been the speed of development of ontologies in this area, that the evolution of ontologies is still an ongoing process, and current ontologies show the characteristics of their recent past.

As we have already mentioned, the use of some kind of ontology predates bioinformatics and is an integral part of several areas of the life sciences. It is, however, with modern data-intensive biology that the need for ontologies has become more crucial, and their definition and publishing have become more tightly linked to a computational environment.

The evolution of ontologies in bioinformatics has followed different, converging, paths. On the one side, existing terminologies (e.g., taxonomy, anatomy), already de facto ontologies, have been represented in a computational environment, and their design has, on some occasions, benefitted by the advantages of a proper formalization (e.g., consistency checking and automated classification) that can be made *actionable* by this environment (e.g., Galen).

When representing different domains of the life sciences (e.g., gene product functionalities or processes), the first approach has been to construct ontologies, which are essentially large terminological resources, often used in simple annotation of entities in databases. This is, for instance, the case of the famous GO, originally defined to annotate gene products in a cross-species database.

A large collection of ontologies has, therefore, emerged from terminology systems, gradually improving in their ontological commitment and coordination. The OBO foundry, which will present later, is now fostering the development of interoperable ontologies that share the same ontological framework. Services such as National Center for Biomedical Ontologies (NCBO) provide a well-supported access point to ontologies, embedding them in the web and Semantic Web frameworks, the latter representing a set of technologies that aim to combine web technologies and ontologies.

Ontologies have also been developed from the need to have a common representation of complex artifacts in specific areas of knowledge. Such a need was at first expressed as the need for exchange languages. While not explicitly ontologies, exchange languages allow us to represent a domain of knowledge in a (at least partially) computable way. However, the interpretation and hence the semantics of these languages rely on common shared understanding of the domain.

When such languages are used to exchange information across heterogeneous resources, which is inevitably followed by the need to integrate such information, the need to properly define the semantic of terms arises,

evolving in the evolution of ontologies. This is, for instance, the example of the BioPAX ontology, which will be introduced later.

The correlation and the convergence of these two *paths* in time can be perhaps exemplified by the development of ontologies to annotate experimental conditions. The need to properly qualify the experimental conditions has been first felt in the functional genomics community: with experiments measuring the expressions of thousands of genes at once, it was clear that there was a big scientific opportunity to have repositories in which the results of those experiments could be mined, to derive relations among such expression and conditions of interest (e.g., a specific stimulus to which a cell was reacting). A formal representation of experimental conditions was needed, which in the beginning lead to the microarray gene expression object model (MAGE-OM), serialized in MAGE-ML (an exchange language). To fill the values of properties of this model, a controlled vocabulary was introduced and named MAGE-ontology (in fact raising a few doubts as to its ontological status). This set of standards was complemented by MIAME (minimum information about a microarray experiment), which was available as a checklist.

The combination of object model with values derived from controlled vocabularies was later superseded by the use of a proper ontology to describe experimental conditions, as the objective was to model experimental conditions, rather than a software artifact representing experimental conditions. The ontology of biomedical investigation (OBI) was then born from the experience of the earlier object models and exchange languages, and was designed in accordance to basic formal ontology (BFO) principles, in common with OBO ontologies. A more extensive discussion on the evolution of ontologies in bioinformatics is available in the literature [27.1].

27.3 Languages and Systems to Represent Ontologies

By definition, ontologies are information artifacts that are represented through a formal language. The nature and expressiveness of this language, therefore, influence the extent of the ontological commitment that an ontology can have.

For ontologies in use in bioinformatics, two main languages are used to define ontologies: the OBO (open biomedical ontology) language, and the ontology web language (OWL). The first is confined to the biomedical

domain, the latter is of widespread and generic use and derives from the Semantic Web set of technologies.

OWL is getting increased acceptance in the biomedical community. New ontologies are been developed natively in OWL (e.g., the aforementioned OBI), and there is active research ongoing in mapping existing OBO ontologies to the more expressive OWL.

Because the Web is a primary medium for the publication and exchange of information, the publi-

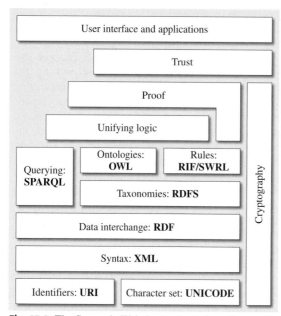

Fig. 27.3 The Semantic Web (technologies) stack

cation of ontologies that are represented in OWL is naturally accompanied by features derived from the Semantic Web. We will, therefore, briefly present OBO in the section dedicated to GO and OBO, and we will focus in this chapter on the relations between the Semantic Web, OWL, and other representations of ontologies.

27.3.1 The Semantic Web, RDF, and OWL

The *Semantic Web framework* is a set of technologies whose goal is to evolve the web into a distributed knowledge base. Whereas the web represents documents and links among documents, the Semantic Web is meant to represent entities (in documents), and the relations holding among them.

The set of technologies that form the Semantic Web is commonly illustrated through the Semantic Web stack (Fig. 27.3), where the layout of blocks representing technologies is a hint of the relations between them: technologies that *sit above* others are intended to be based on them.

It is out of the scope of this chapter to introduce all technologies presented in Fig. 27.3, and some of them (as well as their position in the stack) are still in evolution. However, we will introduce the basic building blocks that have a more direct implication in the definition of biomedical ontologies.

First, we should emphasize that the Semantic Web is a set of technologies, which do not necessarily exist in a monolithic space. At one extreme, some interpretation of the Semantic Web (e.g., linked-data) puts an emphasis on the *web side* of the Semantic Web; http-based resolvability of all identifiers used to represent entities is required, and a formal definition of their semantics is necessary. At the other extreme, some OWL ontologies may put less emphasis on the *web side* of the Semantic Web; they may use nonresolvable identifiers and even syntaxes that are not proper of this framework, but focus on the axiomatization of what they represent. Nevertheless, a basic set of technologies and principles are shared throughout *interpretations* of the Semantic Web.

A basic building block of the Semantic Web is the use of URIs, which are global unique identifiers. If we see the Semantic Web as a global distributed database, URIs are *the identifiers* of this global database. There is nothing in the Semantic Web architecture that dictates that such URIs should be resolvable (e.g., that an http request for such URIs would return a document). This is, however, a requirement for the set of best practices that go under the name of Linked Data and that de facto require the use of URLs in place of URIs.

A first immediate consequence of the representation of ontologies in Semantic Web language is that all identifiers are globally qualified,

GO:0051918 becomes

http://purl.obolibrary.org/obo/GO_0051918 .

Name clashes and to some extent authority of the definition (with its implication on versioning and validity of codes) are easily supported by the web architecture.

The second building block of the Semantic Web is the resource description framework (RDF). With some minor simplifications, this language specifies relations among entities through triples of the form

subject predicate object ,

where subject and predicate of the triple are URIs, while the object can be either a URI or a value. An example of a representation of an ontology in RDF is reported in Fig. 27.4.

The use of RDF has two immediate consequences in the definition of ontologies. First, RDF specifies a model and not a syntax (Fig. 27.4). RDF can be serialized in a variety of syntaxes (RDF-XML, in several versions, N3, Turtle, Ntriples, but even more *programming oriented* syntaxes such as JSON). An ontology

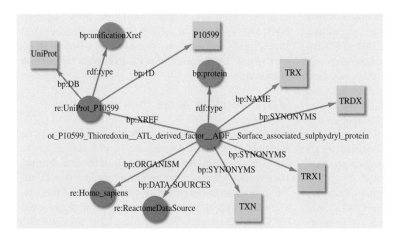

Fig. 27.4 An example of an RDF representation of detail of pathway information in RDF. Note how this RDF is represented as graph and not in a specific syntax. All statements of the form *subject predicate object* contribute to define a (*labeled*) graph. Objects that are data values are indicated in *squares* to distinguish them from objects identified by URIs (resources). Different *colors* of resources correspond to different meta-levels that will become clearer after the BioPAX ontology has been introduced

Part E | 27.3

represented in RDF can, therefore, be represented in different syntaxes. Second, RDF supports the modeling of knowledge as directed graphs, and the description of a domain of knowledge as concept maps, as well as the representation of networks, is easily expressible in RDF (although some of this intuitiveness is reduced by the encoding of OWL in RDF).

As RDF does not specify a syntax, it does not directly specify, to a large extent, a semantic. As subjects, predicates and nonvalue objects in RDF are identified by URIs, we can assign a semantic to each URI, which can be consistently interpreted. For instance, the property *rdfs:label* is used to associate an entity to the string that should be used to refer to it for in a human readable way. *rdfs:subPropertyOf* is used to express that a property $P1$ is a more specific property than $P2$ ($P1$ *rdfs:subPropertyOf* $P2$). Intuitively, we would expect that whenever we state x $P1$ y, we imply that also x $P2$ y is true.

When we referred to the properties *rdf:label* and *rdfs:subPropertyOf*, we used a syntactic shortcut for the full URI of the property: *http://www.w3.org/2000/01/rdf-schema#subPropertyOf*.

Several properties can be defined that start with the same string identified by RDFS, which is known as a namespace. Without entering into details, RDFS is, in fact, a collection of terms whose meaning is intended to express a conceptualization over entities in RDF. RDFS can be seen as weak ontology language, where the set of primitives provided (identified by URIs) are limited and associated with an imprecise semantics.

OWL is a more extensive set of constructs, with an associated semantic that is formally defined and largely derived from description logics. OWL provides a richer and better defined set of constructs of RDFS. OWL offers, for instance, relations as *disjunctions* or *property restrictions*, while classes, and subclasses in the OWL map to a set-theoretic interpretation that can be verified. To be precise, OWL exists in more flavors, which depend on which subset of OWL is used. Different flavors offer a different balance in the tradeoff between expressivity and computability.

Since OWL has been designed for the web, its semantics follows what is known as the *open world assumption*, which is often a source of confusion. The basic idea behind the open world assumption is that what is not known may be true (which seems to intuitively suit the nature of the web). In many applications and contexts, the *closed world assumption* per contra is more common, which implies that what is not known is false.

We present a simple example to highlight the differences between these two assumptions:

- Let us imagine than in a hypothetical simple ontology we state that *every catalyzer is a molecule*.
- We are presented with an instance of a pathway that has a photon as a catalyzer.

Our common understanding, which follows a closed world assumption, would be that this instance and the generic definition is inconsistent. In an open world assumption we would instead deduce that a photon is a molecule, as we did not explicitly state that photons and molecules are disjoint sets (this, and some other differences, often results in object-oriented models being erroneously translated into OWL ontologies, with very subtle consequences and major discrepancies between the intended and the resulting semantic of the ontology).

It should be clear at this point what it means that an ontology is expressed in OWL. It means that its semantics is represented using a set of constructs whose semantics is defined in the OWL language and that can be expressed in RDF in one of its serializations. The usage of OWL constructs could be minimal or extensive, and the use of OWL per se, therefore, does not have any implication on the amount of semantic content and ontological commitment of the ontology. If OWL may support a more committed representation than other languages, ontologies whose representation is ported to OWL from another language does not necessarily gain in its level of formalization, if not for minor practical details.

Finally, we note how there is a spectrum of ways in which terminologies can be expressed in OWL. They could be simply expressed in RDF/RDFS, for instance by representing relations as x is_a y as the corresponding RDF statement. They can make use of OWL constructs for their definition, and we would have

x is_a owl:class ,

y is_a owl:class ,

x owl:subClassOf y ,

where this representation would map to a semantics that could be computed to verify the consistency of the ontology, or in general to compute its hierarchy.

We could also use other dictionaries represented in RDF. An interesting example is offered by SKOS (simple knowledge organization system). SKOS provides a set of terms that model concepts and their relations (skos:concept, skos:narrower, skos:broader, . . .). In this case, we could assert that x *skos:broader* y. Interestingly, the semantics if SKOS is represented in OWL (*skos:concept rdf:type owl:Class*) is: an entity is an instance of a skos:concept, that is, at its time an instance of an owl:class.

27.3.2 Notes on Tools

It is worth concluding this section with a brief note on the tools used when dealing with ontologies in bioinformatics. Ontologies are generally designed through an ontology editor. Two common choices are OBO-Edit (which is suited for OBO ontologies), and Protege, a generic ontology editor, which is commonly used for OWL ontologies, but that can also import OBO ontologies.

Computation of inference on ontologies is performed through software called *reasoner*. Pellet is a reasoner (more precisely a DL reasoner) that is commonly used on OWL ontologies. It can be executed via a command line, or embedded into other applications (e.g., Protege). Another reasoner often used is FaCT++.

From a programming point of view, two of the most common APIs used to *handle* ontologies in OWL are the OWL-API and Jena, the latter being more oriented to the RDF representation of information than its semantics and in fact closer to the *Linked Data* approach. We refer to [27.2] for wider discussion on the Semantic Web, its role for ontology development and representation, and related tools and practices.

27.4 Overview of Ontologies in Bioinformatics

In this section, we provide a brief presentation of some ontologies that are of key usage in bioinformatics. It would be impossible to provide an extensive list of ontologies, or to sort them by relevance, as this ultimately depends on the specific area of research. In this selection, we present a few resources that can be considered archetypes for similar ontologies, or that, for historical reasons, or for the connectivity they provide, act as landmark in the bioinformatics ontology landscape.

27.4.1 Gene Ontology and the OBO Ontologies

Perhaps the most successful ontology in bioinformatics is the GO. Its success is evident in its wide adoption, in the range of use cases it supported, but also in its influence in shaping a long list of ontologies, which share some of its principles, design and tools and that are roughly grouped under the OBO umbrella.

GO was originally developed to support cross-species annotation of model organism databases and it is constituted of three independent ontologies: biological process, molecular function, and cellular component. At its core, and reflecting its past, these three ontologies are essentially terminologies, where each term identifies a class of, respectively, processes, functions, and components, and where these terms are related through relations as *is_a* (intended as *is a more specific term than*) and *part_of*.

Since it was first introduced, GO has been progressively enriched in terms of ontological commitment and new relations have been included. As shown later in this section, these improvements have not been isolated to GO, but they have been developed in a wider framework, whose goal it to foster of coordinated development of bio-ontologies.

GO is primarily developed in OBO language and made available in a variety of formats (including OWL). However, the expressivity of the OBO language limits the detail and the ontological commitment of the ontology. When representing the same ontology in more expressive languages, such as OWL, a few assumptions on the intended meaning of OBO constructs have to be formally introduced. The translation of GO to an expressive language as OWL is not only a syntactic operation, and the consistency and consequences of the implied formalization of the OBO semantics (and of GO itself) are still an active area of research.

To better understand the design of OBO, we use the following two examples of OBO definitions.

```
[Term]
id: GO:0048018
name: receptor agonist activity
namespace: molecular_function
def: Interacts with receptors such
    that the proportion of receptors
    in the active form is increased.
    [GOC:ceb, ISBN:0198506732]
synonym: receptor ligand activity
    BROAD [GOC:mtg_signaling]
is_a: GO:0005102 ! receptor binding
is_a: GO:0030546 ! receptor activator
    activity
relationship: part_of GO:0007165 !
    signal transduction
[Term]
id: GO:0048584
name: positive regulation of response
    to stimulus
namespace: biological_process
def: Any process that activates,
    maintains or increases the rate of
    a response to a stimulus. Response to
    stimulus is a change in state or
    activity of a cell or an organism
```

```
    (in termsof movement, secretion,
    enzyme production, gene expression,
    etc.) as a result of a stimulus.
    [GOC:jid]
comment: Note that this term is in the
    subset of terms that should not be
    used for direct gene product
    annotation. Annotations to this
    term will be removed during
    annotation QC.
subset: high_level_ annotation_qc
synonym: activation of response to
    stimulus NARROW []
synonym: stimulation of response to
    stimulus NARROW []
synonym: up regulation of response to
    stimulus EXACT []
synonym: up-regulation of response
    to stimulus EXACT []
synonym: upregulation of response to
    stimulus EXACT []
is_a: GO:0048518 ! positive regulation
    of biological process
is_a: GO:0048583 ! regulation of
    response to stimulus
intersection_of: GO:0065007 !
    biological regulation
intersection_of:
    positively_regulates GO:0050896 !
    response to stimulus
relationship: positively_regulates
    GO:0050896 ! response to stimulus
```

We note how in OBO a term is identified by a numeric code. These codes are semantically opaque and are not re-usable. Once defined, codes do not change meaning, but can be eventually made *obsolete* and eventually superseded by other terms. This is a very important feature of OBO and of the design of an ontology in general, which is maintained when GO (and other ontologies) are published in a Semantic Web framework. As can be seen in the example, the OBO ontology language presents constructs that highlight the origins of

GO as a terminology. Besides annotations, terms can have associated narrower and broader synonyms, which are typical of a terminology-oriented resource.

It is worth noting that GO only represents *terms* or classes. GO does not provide any information on the biological entities to which these terms may refer. The annotation of gene products with GO terms is part of independent projects, which release GO annotations (sometimes referred to as GOA). An annotation of a gene product with GO terms would be composed of the following fields.

DB, Object ID, DB Object Symbol, all required
(identifiers and descriptors of the gene product)

Qualifier (optional)

GO ID, Aspect, all required
(GO terms and ontology it belongs to)

DB:Reference (|DB:Reference), Evidence Code,
all required (evidence for the annotation)

With (or) From (optional)

DB Object Name, DB Object Synonym
(|Synonym), all optional, DB Object Type,
required, Taxon(|taxon), required
(additional information on the gene product)

Date, Assigned By, Assigned By, all required
(meta information on the annotation)

Annotation Extension (optional)

Gene Product Form ID (optional)

Some of the characteristics of a GO annotation require attention. Of particular interest is the *modifier* attributes, which specifies nature of the association between a gene product and its corresponding annotation: a possible value for this attribute is *not*, meaning that it was proved that the specific annotation does not apply to the specific gene. While this is a relatively infrequent and confined case, it is an emblematic example of how ignoring encoding details can result in a complete erroneous understanding of the intended annotation.

The OBO Foundry and the Coordinated Development of Bio-Ontologies

The OBO foundry is a collaborative effort to coordinate the development of biomedical ontologies as to reach a set of orthogonal, interoperable, reference ontologies for the biomedical domain. Ontologies that are part of the OBO foundry are meant to observe a set of principles that guarantee this interoperability, whose full list is

available on the OBO website. These principles address the following points.

- The ontology must be open (accessible) and available in standard formats (OBO, OWL).
- The ontology adhere to a defined policy for the provision of identifiers, source identification and versioning.
- The content of the ontology is well defined and orthogonal to other ontologies.
- The ontology is well-documented and provides definition for all terms.
- The development of ontology results from a community process.

In addition, a set of principles has been added in a later phase, principles that require a more precise ontological commitment from OBO foundry ontologies. For instance, the ontologies should provide text and formal definition for terms, ontologies should be developed within a defined upper ontology perspective and possibly using a defined set of relations from the OBO foundry. A few other principles are meant to uniform textual description and to provide more structure to the governance of the developed process.

Overall, the harmonization of the development of ontologies is built on three axes:

1. The involvement of the community, which is essential as the development of ontologies is intrinsically a social activity (it is after all the development of a common language).
2. The use of formal descriptions and in general of expressive formal languages, which support the validation of ontologies and the verification of their consistency. It should be noted that this is an ongoing task, due to the complexity of the field and the amount of legacy ontology.
3. The reference to a common ontology framework, which in essence implies the commitment to the use of a common ontology (BFO) and to use a basic set of relations (RO).

Basic Formal Ontology (BFO) and Upper Ontologies

BFO (basic formal ontology) is an upper level ontology based on a realistic perspective. It provides a coherent framework, or a conceptual point of view, for the conceptualization of a domain of knowledge.

In brief, and without entering its deep logical implications, BFO takes a realistic perspective. Intuitively, BFO assumes that knowledge is about a shared reality.

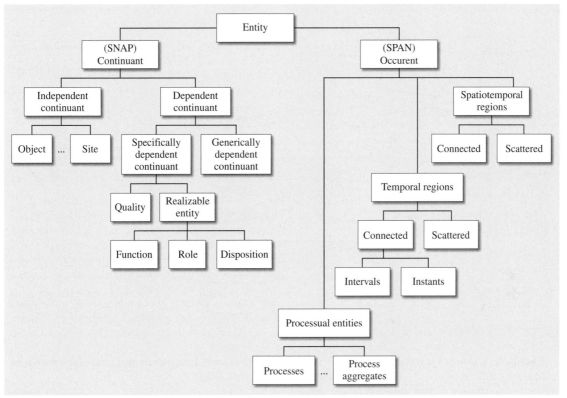

Fig. 27.5 BFO. A sketch of BFO class structure where the SNAP (*left*) and SPAN (*right*) branches are represented. A few classes have been omitted to simplify the representation

Hence, a BFO ontology primarily describes entities that exists in the real world. This is in contrast with other approaches where elements of the ontology can refer to cognitive, subjective, and even fictional entities.

As ontologies of interest for bioinformatics (and biomedical research in general) are primarily concerned with modeling experimental evidence, the *realistic* focus of BFO is an appealing characteristic.

BFO provides for a distinction of entities in two main groups: entities whose definition is independent from time (continuants or endurants), which are organized in an ontology called SNAP, and entities whose definition intrinsically relies on time (occurrents), which are organized in the SPAN ontology. A sketch of the SNAP and SPAN ontologies in provided in Fig. 27.5.

Continuants (SNAP entities in Fig. 27.5) are further divided in independent continuants, or things that can exist per se and dependent continuant, such as qualities or functions, which cannot exist without a dependent continuant *bearing* them. This kind of organization

of concepts, and the resulting clarity (and unambiguity) on primary entities, dependent entities, the relation between entities and properties, is the benefit of the adoption of an upper ontology.

The implications of this organization are not always intuitive. In Fig. 27.5 the term *generically dependent continuant*, for instance, is representing the class of continuants whose relation to their bearer (the entity in the real world they refer to) is not direct. An example of a generic dependent continuant is a map (e.g., a pathway map), which is something that depends on a substrate for it to be represented, but which is more general than the specific substrate itself representing it. A notable branch of BFO is the information artifact ontology (IAO), which specifically deals with information models.

While BFO provides a useful framework for the interoperability of ontologies, and ontologies designed according to BFO may be effective in supporting data integration across heterogeneous domains, its adoption currently presents some limitations.

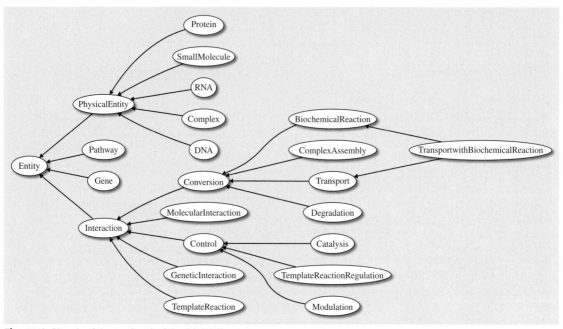

Fig. 27.6 Sketch of the top level of the BioPAX ontology

BFO lacks a proper formalization, hence it is not possible to verify in terms of consistency whether its intended semantics is present throughout the definition of compliant ontologies. Often an ontology can claim to be designed according to BFO just because some of its top-level entities to BFO terms. Lacking a proper formalization of BFO (and often of the ontology in question) it impossible to verify whether the resulting intended meaning of the whole ontology is consistent.

It is worth concluding this section by briefly mentioning other upper ontologies that the reader may encounter in bioinformatics. DOLCE (descriptive ontology for linguistics and cognitive engineering) is a top-level ontology that differs from BFO in that it is based on a *cognitive approach*. Terms in DOLCE are meant to refer to cognitive concepts, rather than entities in reality, and a few other differences distinguish the two ontologies: in DOLCE, for instance, qualities can exist independently from their bearer, as opposed to in BFO. DOLCE is also extensively defined through a first-order logic axiomatization.

Other upper ontologies worth mentioning are BioTOP (which provides bridges to SNOMED, BFO and the UMLS Semantic Network) and more general upper ontologies such as SUMO (suggested upper merged ontology). An interesting discussion of the role

of BFO in formalizing scientific knowledge exists in the literature [27.3].

The Relations Ontology (RO)

A relevant piece of ontological architecture within the OBO foundry framework is constituted by RO, which is intrinsically based on BFO in its definition. This ontology defines a limited number of relations, which are intended to hold within and across OBO ontologies. An example of a definition of RO relation (participates_in) is shown here:

```
<owl:ObjectProperty
  rdf:about="http://purl.obolibrary.org/
  obo/BFO_0000056">
    <rdfs:label xml:lang="en">
      participates in
    </rdfs:label>
    <obo:IAO_0000118 xml:lang="en">
      participates_in
    </obo:IAO_0000118>
    <rdfs:range
      rdf:resource="http://purl.
      obolibrary.org/obo/BFO_0000003"/>
```

```
<rdfs:domain
  rdf:resource="http://purl.
  obolibrary.org/obo/BFO_0000004"/>
</owl:ObjectProperty>
```

The relation is defined as holding among individuals of the classes BFO_0000003 (occurrent) that is the domain of the relation, and BFO_0000004 (independent continuant) which is its range. We note how this definition is meant to refer to a property holding among entities (not necessarily known) in the *real* world.

The adoption of RO provides immediate benefit in the integration of heterogeneous ontologies. However, since these definitions of relations may be retrofitted to previously roughly defined ontologies, a note of warning is necessary as some of the resulting semantics may be spurious or unintended.

27.4.2 BioPAX and Pathway Ontologies

BioPAX constitutes a different example of ontology in use in bioinformatics. Whereas OBO ontologies have evolved from terminologies and provide an extensive list of classes, BioPAX (biological pathway exchange language) originated from the need to share information on pathways across database providers and data consumers. As such it provides a concise set of classes that bridge the conceptualization of pathways found across different pathway databases.

BioPAX is an emblematic example of how the requirement for *exchange languages* has been at least in part an indirect requirement for a domain specific ontology.

A sketch of the top level of BioPAX is provided in Fig. 27.6.

As one can see from Fig. 27.6, classes and relations in the BioPAX ontology do not pertain to pathways themselves, but to the structure of pathways. They include terms such as *interaction*, *biochemical reaction*, *gene*, but not individual reactions or entities, which are treated in BioPAX as instances of the ontology. Compared to GO, BioPAX could be seen as a meta-pathway ontology, or an ontology on pathways.

A second observation is that the conceptualization of pathways that BioPAX proposes does not follow BFO, but is rather a conceptualization derived from biological network analysis practices. While BioPAX provides a partition between entities and processes that may seem to map to continuant and occurrent in BFO, some of its abstractions cross these two categories. An example is an *interaction* that encompasses both occurrents (e.g., a catalysis process) and continuants (e.g., the control role of an enzyme in a catalysis).

Also inherited from its exchange language commitment is a mixed reference between the domain formalized (pathways observed in molecular biology) and the information conveyed (annotations, references but even characteristics of pathway models). In BioPAX a set of *utility classes* (not shown in Fig. 27.6) are meant to represent entities that are needed for the representation of pathways, rather than being entities of pathways themselves.

BioPAX was originally meant to exchange pathway information, hence its emphasis on representation related classes. However, as pathways are complex entities that are very heterogeneous in nature, and since exchanging information cannot be distinguished from the integration of the exchanged information with other knowledge (or they would talk about duplication, rather than exchange of information), BioPAX quickly evolved as to include an ontological characterization of the pathway domain.

BioPAX is represented in OWL-DL (OWL description logic). Its semantics, however, make a limited use of OWL. First, given its fixed class structure, there is no need to compute the subsumption hierarchy of the ontology, which is a key feature of OWL. Also, in the typical use case, pathway information is explicitly exported as instances in BioPAX, which rule out the need to infer whether a specific pathway description is an instance of BioPAX.

Furthermore, in the pathway exchange scenario that BioPAX address, OWL does not provide direct support for the verification of a valid message. This is because it relies on an open world assumption, but also because a *message* is an information artifact, rather than something existing in the real world of pathways, which is what ontology is about. BioPAX relies on ad-hoc written libraries for these validation tasks. It could be argued that the semantics of BioPAX could have been largely expressed in other formalisms, such as UML.

However, BioPAX provides some important resources:

- A large collection of pathways expressed in a common syntax (RDF), which is the most common (if not the only shared) syntax across bio-ontologies. Some of these pathway resources have been published in a common framework at http://www.pathwaycommons.org/.

- An ontology that represents pathways according, primarily but not exclusively, to a network-analysis centric conceptualization.
- A shared conceptualization among database providers and consumers, which is the result of years of discussions on the definition of a common *data model*.

27.4.3 UMLS (Unified Medical Language System)

UMLS, the unified medical language system is primarily a terminological resource, which encompasses and unifies a range of terminologies (ontologies) found in bioinformatics and medical informatics as, for instance, GO or SNOMED.

UMLS is composed of three parts: a *Specialist* Lexicon, a meta-thesaurus, and the UMLS semantic network.

The Specialist Lexicon and related lexical tools provide information on lexical and morphological variants of terms, as well as tools such as normalizer, that have a typical usage in NLP (natural language processing) pipelines developed for the biomedical domain.

The meta-thesaurus is a multilingual terminological resource that encompasses terms from OBO SNOMED and other ontologies and code lists of biomedical interest. It unifies these terms on the basis of the concept they refer to, identified by a unique CUI (concept unique identifier). Concepts in the metathesaurus are linked to the Specialist Lexicon resources and to the UMLS Semantic Network.

The UMLS Semantic Network categorizes terms in semantic classes that are linked by relations (about 133 classes are linked by 53 classes). An extensive documentation is available on the NLM website (http://www.nlm.nih.gov/research/umls/) for a more detailed introduction to the UMLS resources.

27.5 Use of Ontologies in Bioinformatics

The spectrum of applications that ontologies support in bioinformatics is very varied. In particular, ontologies find significant usage in data integration, information retrieval, and data analysis. What follows is an overview of the panorama of such usages.

27.5.1 Data Integration

Different information systems, or databases, differ not only in the scope of the information they represent, but also in the way the same information is modeled. In many life sciences information resources, this inherent heterogeneity is even more acute, because of the different areas of study or disciplines that are involved. Perhaps specific to this area of science is the significant push toward the integration of data, which is at the basis of a new approach to medical research (translational medicine and personalized medicine).

The use of consistent terminologies across databases is a first step in making information from different databases interoperable. In large areas of bioinformatics, this is a step that has already yielded significant benefits. Much information in the life sciences is, in fact, extensive rather than intensive. In a genome annotation experiment, for instance, some tens of thousands of genes are annotated with the functions they carry, the biological processes they belong to, and the cellular component where their gene products are observed.

Such annotations are just simple terms and often represent all the know-how that is formalized on such a scale. Using common terminology to represent these functions, processes, and components allows a direct integration of annotations from different organisms.

A key step in the use of ontologies for data integration is the provision of a consistent set of terminologies, together with mechanisms that support their adoption and sharing. In the US, NCBO was founded to provide an access point to provide a coordinated set of ontologies, or terminologies, accessible to support biomedical investigation.

The NCBO BioPortal provides a repository of ontologies, which can be queried to find which terms or ontology represent some concepts. The system allows for the exploration of ontologies in a visual and interactive way, as well as the possibility to download ontologies. Additionally, the BioPortal also provides links among ontologies (cross products), as well as a space for discussion and annotation on ontologies, hence supporting the social aspect of ontology development.

The BioPortal also provides machine processable access points to ontologies, via a REST architecture, or via an experimental SPARQL endpoint (a query system for the Semantic Web), thus offering these ontologies for usage on the Semantic Web. Another example of the use of ontologies for data integration comes from the use of BioPAX in PathwayCommons.

Relating pathways representation from different databases cannot be re-conducted to matching terminologies, but it requires a more complex action of *schema* or *ontology* matching. In the area of bioinformatics and computational biology, we do not find reference to formalized data integration framework, such as local-as-view or global-as-view mapping. Often data integration is understood in terms of mapping to an exchange language, leaving the mapping implicit in the code of an exporter, rather than providing an explicit definition of relations among elements.

In the case of pathways, the previously introduced BioPAX is an initiative to map the content (or structure) of pathway databases to a common model, which is represented as an OWL ontology. BioPAX makes a peculiar use of OWL, for a set of reasons that are beyond the scope of this chapter.

PathwayCommons uses BioPAX to provide a coherent set of pathways, which are unified from the syntactic point of view and partially unified in their semantics. This comes in part from constructs that are proper to BioPAX, and in part from ad-hoc processing in PathwayCommons. It should be noted, however, that OWL is not used for unification of pathways in this context, and its scope of application is very limited for the way the ontology is structured, as has been discussed.

Ontologies can also provide the backbone for the integration of varied resources that contribute to the know-how of a specific area of science.

One of example of this is Neurocommons, a collaboration between Science Commons and the Teranode Corporation, that was originally designed as a prototype to demonstrate how Semantic Web technologies can be used for the aggregation and distribution of biomedical knowledge gathered from existing sources. Neurocommons allows for integration of heterogeneous sources through the conversion of the information that they provide to RDF/OWL, using a coherent identification scheme and re-using a number of shared ontologies. The primary purpose of Neurocommons was to focus on neurological diseases. However, the project effort is clearly towards a design that allows general use, therefore supporting other specializations. It includes the following sources of information:

- Entrez Gene
- The full set of OBO ontologies (including GO)
- The Gene Ontology Annotation, GALEN (the OWL version)
- Addgene plasmid catalog

- Links to the literature in the form of gene to article links from Entrez and GO
- The medical subject heading definitions and article associations from PubMed, as well as selected information associated with each article
- Information on homologs from Homologene.

Also, a selection of databases related to neuroscience has been included. Those sources include, but are not limited to, the Metadata associated with the Allen Brain Atlas images, NeuronDB, the Swanson 1998 rat portion of the Brain Architecture and Management System, and the PDSP Ki database.

Another example of a project-oriented resource is constituted by the DC-Thera directory (dc-research.eu), where ontologies are used to organize and link information on research assets of a translational medicine research project focusing on dendritic cells immunobiology. In these case ontologies, OBI are used to structure the relations among research assets (biomaterials, datasets), know-how (protocols, expertise), and the social dimension of research (participants and organizations).

27.5.2 Information Retrieval

Another use case for the use of ontologies in bioinformatics is information retrieval. The vast majority of the biomedical know-how is available in scientific literature. Indexing papers via subjects organized in a terminology (MESH) is a first step toward improving information retrieval. A deeper usage of ontologies is in the markup of papers with a formal representation of the entities and *facts* that they present in natural language. Such annotation of papers (and of biomedical information in general) improves information retrieval from such sources, as measured in terms of precision and recall.

Several systems provide entity recognition services that can be applied to the scientific literature, where they enrich it with standard identifiers (possibly from ontologies) for gene names, functions, and other entities. Such annotation can also be the basis for further automated processing meant to extract *facts* or the meaning of entire sentences from literature. We cite Whatizit and GoPubmed, respectively, as services that provide such markup and that leverage it to improve search.

Research is underway to embed entity markup in the process of writing (or publishing) a scientific paper, or to publish *facts* in a structured computable format, complementing natural language description of research findings.

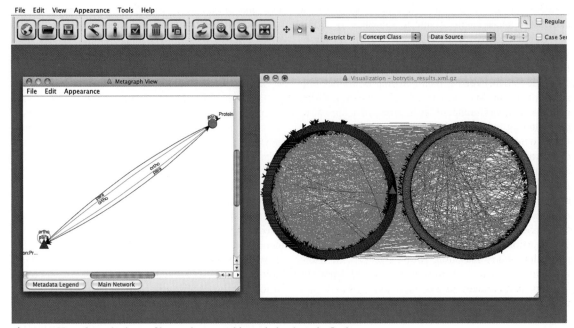

Fig. 27.7 Use of ontologies to filter an integrated knowledge base in Ondex

Ontologies such as CiTO (citation typing ontology) are used to represent the structure of citation among papers and allow for the computation of domain specific meta-information, such as the influence of papers in a research area.

A characteristic example of a system that makes an extensive use of ontologies for information retrieval is the already mentioned DC-Thera directory. In this system, ontologies provide multifaceted indexes for accessing specific information. Queries can refer to ontology terms and retrieve results that are relative to more specific terms, and search functionalities can recognize the lexical and morphological variants of such terms in free text. As different strategies have different precision and recall characteristics, such results are color coded for an intuitive representation of the quality of the retrieved information. The use of ontologies also allows the DC-Thera directory to dispatch queries to other systems (e.g., ArrayExpress Atlas) and to convey results in a unified framework. An extensive presentation of the DC-Thera directory, which touches the use of ontologies and Semantic Web technologies to manage research information, is available at [27.4].

Finally, ontologies can also be used to support interactive information retrieval. An example of this is provided by the Ondex project. In this system, hetero-geneous information on biological system is integrated in a concept network where concepts and relations are annotated through ontologies. In a visual interactive environment, these annotations allow selective visualization of the network. Figure 27.7 shows a screenshot from the OVTK (Ondex Visual Tool Kit). In this screenshot a graph representing classes and their relations (meta-graph, on the bottom left) can be used to filter the integrated network.

27.5.3 Knowledge Representation, Classification, and Reasoning

Although ontologies can support automated reasoning over a knowledge-base (a collection of facts), this usage in bioinformatics is still relatively scarce. In general, knowledge managements systems represent information on a given knowledge domain (facts and ontologies) and can act in a proactive way by providing deductions, detecting inconsistencies, and verifying the possibility of scenarios. In bioinformatics, and in the area of systems biology in particular, some of these tasks are often carried through specialized tools, rather than relying on the possibility offered by generic ontology languages.

As the bulk of ontologies used in bioinformatics are represented in OWL, the scope of inference supported by OWL is a first candidate for the evolution of knowledge management systems in this area.

As has been discussed, OWL allows us to represent a formalization over a domain in a fragment of first-order logic, which can be optimized as to fit a specific trade-off between expressivity and computability. A reasoner can than classify an OWL ontology. This means that it can compute the hierarchy of classes in the ontology, which is the basis to classify instances of the ontology, prove its consistency, or even querying a knowledge base: given a set of constraint which define a class, a reasoner can compute which individuals are compatible with this class definition, effectively answering a query.

It is easy to see how this inference is appealing for the Semantic Web, as it would allow for automatic classification of distributed information. However, this particular kind of reasoning has both potentialities and limitations.

The main limitations of OWL are related to the fact that it is based on logics and does not directly support inference over approximate information. Approximate information is often found in bioinformatics as many measures have a quantitative and qualitative nature (with the notable exception of nucleic sequences, which are by nature *digital* information). However, in large areas of the biomedical domain, approximate information is not an issue, as in the case of biomedical terminologies. OWL finds, in fact, a nice fit as a computational support for the definition of biomedical ontologies, where it promises both a computational support for checking consistency of terminologies (or across terminologies) and a system to compose terms from orthogonal terminologies.

As a note, other inference systems that are proposed in the Semantic Web framework are rule based systems, whose expressivity in general overlaps with that of OWL.

27.5.4 Data Analysis

One notable application of data analysis on ontologies in bioinformatics is the functional analysis of GO for the interpretation of the results of genomics experiments. Although specific, this kind of analysis touches some of the general issues in analyzing ontologies, such as dealing with relationships among entities and some issues bound to tests performed on all the elements of the ontology at once.

Traditionally, in the field of genomic studies, researchers analyzed a small number of genes at a time,

and data analysis techniques involved a search on the GO for quick and effective association of those genes to GO terms, allowing the profiling of the list of genes with respect to various biological processes, molecular functions, or cellular components. The usefulness of GO in this kind of investigation is clear: the researcher was able to exploit the knowledge built in the ontology with a simple search, validating his list of genes with a well-established and reliable source of expert knowledge. The process of validation, which previously consisted in a cumbersome and error prone literature search, could be resolved with a handful of queries to GO. Also, each specific process could be effortlessly related to all the parents in GO, allowing quick placement in the appropriate biological context.

However, the real advantage of an ontology in the area of genomic experiments strands out in the context of high-throughput experiments such as DNA microarrays or next-generation sequencing. When such high-throughput techniques started being widely available to the scientific community, researchers were presented for the first time in history with more genes than they could possibly investigate with manual querying techniques. When faced with the need to translate a list of thousands of genes in a (possibly smaller) list of important biological processes, molecular functions, or cellular components, the approach consisting in searching manually the existing literature was simply impossible to follow.

It is in this context that researchers in bioinformatics proposed automatic ontological analysis tools that helped the interpretation of those lists of genes and their relation with the underlying biological phenomena. In 2002, the first tool that addressed this problem was proposed, Onto-Express [27.5]. This tool, later included in a more comprehensive suite of tools for the analysis of genomic data, not only allowing for the automatic listing of GO terms associated with genes coming from a high-throughput experiment, but also allowing the performance of various statistical tests on each term, computing significance values for each functional category [27.6]. This approach became the de-facto standard for the analysis of genomic data in the context of the GO. Shortly after Onto-Express, many other tools were developed following the same basic approach: the identification of GO terms that are most significant in the biological phenomenon under analysis. At this point GO was not used as a mere *list* of terms and genes annotated with them, but as an input in a more refined type of analysis. This is achieved using the appropriate statistical models and techniques; the models used include the hypergeometric, binomial, chi-square, and Fisher's exact test. These models yield the probability that a certain term is over- or under-represented in the list of differentially expressed genes used as an input; various aspects of the input data, such as the size of the gene list and the interdependence among terms and genes, determine the most appropriate model that should be used for the specific experiment.

Here we will explain briefly the hypergeometric model for the assessment of the significance of a GO term. Let us suppose that a researcher is analyzing an organism with a specific microarray. The microarray is able to screen N genes simultaneously. Out of N, a subset T of genes are annotated with a specific term, while the subset NT (complementary to T with respect to N) are not annotated with this term. After analyzing the data from the microarray, the researcher determines, with the most appropriate analysis method, that K genes are relevant in the phenomenon in analysis. A number x out of K genes are annotated with T. The question that the researcher asks is *Given that I determined that x genes are annotated with this specific term, what is the probability of having x genes or more annotated with this term in a random list of x genes?* The idea here is that if a given term is either over- or under-represented in the list of differentially expressed genes, that term is probably meaningfully related to the condition under study. It is useful to mention that even though most people focus on the over-representation, the under-representation can also be meaningful [27.7].

The hypergeometric model answers this question by giving the probability that x or more genes annotated with a specific term will be determined as relevant just by chance; the lower this probability, the more the researcher is sure that the term is over-represented in the phenomenon in analysis. A similar reasoning can be made for the under-representation case.

The probability of having exactly x genes in the set T (with cardinality M) just by chance is

$$P(X = x | N, M, K) = \frac{\binom{M}{x}\binom{N-M}{K-x}}{\binom{N}{K}}.$$

From this, the probability of having x or more relevant genes associated with our term is

$$P(X \geq x | N, M, K) = 1 - \sum_{i=0}^{x-1} P(X = i | N, M, K).$$

Similar computations can be performed with other models, obtaining analogous results.

Another aspect that is treated differently among the various tools is the level of the ontology that is taken in account. While some of the tools limit themselves to a single term chosen by the user, as in the previous example, other tools perform a global analysis over the entire GO tree. This has an important consequence that raises a general issue when analyzing large ontologies: as all the terms are tested at once, a correction for multiple experiments is needed. This is due to the fact that when analyzing many terms at the same time, there is the chance of some terms resulting as significant while in reality they are not. This phenomenon is widely recognized in the literature, and several methods have been developed for dealing with it.

An extensive, albeit slightly outdated, comparison between many of the tools that perform this kind of analysis can be found in [27.8] with detailed information about statistical models used, the scope and abstraction level of the analysis, the details of the corrections for multiple comparisons used (if any), and performance estimates.

While some of these methods somewhat consider the dependencies between terms, for example using bootstrapping techniques, in 2006 two approaches were proposed that improve the scoring of terms taking in account the intrinsic dependency of such terms in the tree structure of GO; these approaches integrate the information about the parent–child relationships between various GO terms, refining the classical over-representation analysis.

The idea behind those two approaches, *Elim* and *Weight*, is that genes that are in common between a set of terms should not be considered in all the terms. In other words, if a node and one of his children share a subset of genes, this subset should not be counted both in the father node and in the child node.

The difference between the methods is in how this concept is applied. In *Elim*, the most straightforward methods, nodes are tested (with a statistical method of choice) bottom up, i.e., from the most specific node to the most general. If a term is reported as significant with the chosen statistical model, the genes belonging to this term are flagged for removal in the analysis of higher levels of the ontology; those genes, therefore, are not counted in the ancestors of the terms reported as being significant. The consequence of this approach is that more specific terms in a path from the root of the ontology to the leaves are reported.

The other approach, *Weight*, is slightly more complicated. As the name suggests, the method weights genes that are shared between terms depending on the

Table 27.1 Contingency matrix for Fisher's exact test on a term of GO

	Significant	Not significant	Total
In term	$|SigG \cap g(U)|$	$|\overline{SigG} \cap g(U)|$	$|g(U)|$
Not in term	$|SigG \cap \overline{g(U)}|$	$|\overline{SigG} \cap \overline{g(U)}|$	$|\overline{g(U)}|$
Total	$|SigG|$	$|\overline{SigG}|$	$|allGenes|$

Table 27.2 The first cell of the contingency matrix for Fisher's exact test on a term of the GO, after the application of the weight method

	Significant
In term	$\lceil \sum_{i \in \{SigG \cap g(U)\}} \text{weights}(i) \rceil$

significance of the terms. The basic principle of this approach is to assess the difference in significance between each node and his children. Again, the method starts from the leaves of the tree, and proceeds bottom up. Let us suppose that we are analyzing the term U. We check the significance of U and the significance of all its children. If there are no children more significant than U, genes in the children are down-weighted by a factor that is proportional to the ratio of the significance values of each pair (U, V), where V is a *less-significant* child of U. The ratio involves the use of an arbitrary increasing function $f(\bullet)$ responsible for the amount of weighting for the pairs (U, V), as in

$$\text{sigRatio}(\text{score}(U), \text{score}(V)) = \frac{f[\text{score}(U)]}{f[\text{score}(V)]} .$$

In the alternative case in which at least one child is more significant than the father, the genes in common between such children and *all the ancestors of U* are down-weighted, further decreasing the significance of U. Weights are updated at each step of the analysis: after an initialization step where all the weights are assigned the value 1, after every test the old weights are multiplied by the new weights appropriately.

The first issue that rises from this approach is that the counting of genes necessary for many statistical models used for over-representation analysis is no longer a proper counting when weights are introduced. Instead of counting the number of genes that are listed as interesting, after the application of *Weight* we use sums of weights for each term, rounded to the next integer. Let us assume, for example, that we choose to use Fisher's exact test for the assessment of the significance of a term. Let us define the set SigG as the set of genes reported to be significant, $g(U)$ as the set of genes annotated with node U. Similarly we define $\overline{SigG}$ as the set of genes that are not reported to be significant and

$\overline{g(U)}$ as the set of genes that do not belong to U. Without applying *Weight*, Fisher's exact test is performed on the following contingency matrix (Table 27.1).

In *Weight*, genes are associated with terms using weights, rather than a binary function that can only say that a gene is either associated with a term or not therefore the quantities in the contingency matrix have to be changed accordingly. In our example, the cell corresponding to genes that are significant *and* annotated with the term U would change as in Table 27.2. Changing all the other cells yields a contingency matrix that can be used for the computation of the significance of the term.

In addition to the single methods, it is possible to combine altogether the classical statistical p-value with the p-values yield by *Elim* and *Weight* with

$$\bar{p} = \exp\left(\frac{1}{3}\sum_{i=1}^{3}\text{pvalues}(i)\right) ,$$

where pvalues represents the vector containing the three p-values to combine.

Elim and *Weight* are available in the R package TopGO through the bioconductor repository. These issues are explained in more detail in the literature [27.7, Chap. 24].

With the evolution of high-throughput technologies, the analysis based on GO needs adjustments for the correction of issues linked to the characteristics of new data structures. One example is the use of next-gen sequencing (RNA-seq) data for the functional analysis of GO terms. RNA-seq allows for more precise measurements of RNA levels in a sample, in addition to many other benefits. However, the output coming from RNA-seq analyses has certain features that require the data analysis to be adjusted accordingly. In particular, the length of transcripts have been associated to the increase in statistical power for detecting differential expression; in other words, some genes have more probability to be selected as significant. Failure to take this effect in account yields biased results.

In [27.9] a method is presented for the correction of such effect. First (step 1), genes are selected as significant or not with classical methods. Second (step 2), the method quantifies the likelihood of a gene to be significant as a function of the transcript length, fitting a monotonic function on the DE/transcript length data.

In the third and final step, the function found in step 2 is incorporated in the statistical test of the GO term for which we are testing the significance.

27.6 Beyond Ontologies

Ontologies are a means to unify knowledge and make it interoperable, but we are still far from having a unified set of interoperable ontologies. This is not surprising, as our conceptualization of an area of knowledge depends on *use cases*, which are varied and require different assumptions and references. In the area of biomedical science, perhaps the OBO foundry proposes the most advanced effort in unifying ontologies, but this unification is still an ongoing task.

Perhaps it is emblematic to point out that, for biomedical information represented on the Semantic Web, there is not yet a de-facto agreement on the use of common URIs. In turn, there have been proposals such as LSID, LSRN, SharedNames, and now Identifers.org, which seem to have reached some consensus.

Currently, OWL represents the syntax most widely accepted to represent ontologies, however only subsets of this language are used in most ontologies, which makes their definition as *OWL ontologies* misleading. However, more and more logic is being introduced in the definition of ontologies, and the situation is evolving. A few experiments have been attempted in effectively using OWL features to classify protein or to verify the consistently of pathways.

It remains to be seen whether logics represent the best way to represent or extract semantics on the (bioinformatics) web. While in the biomedical world, and especially in the area of biomedical terminologies, this is relatively minor issue, in general web information is by nature inconsistent and approximate. Successful solutions that harness the *crowdsourcing* power of the web, for instance for information extraction from papers, may require different approaches to evaluate the truth of facts.

On a more technical side, in the original Semantic Web vision, OWL allows the definition of the semantics of facts expressed in RDF. OWL-2 has introduced a native XML serialization that shies away from its commitment to the web (though a canonical translation to RDF is mandatory).

RDF, on the other hand, has found a successful use case in linked data, which does not commit to the use of formal ontologies. Some tools like RelFinder, for instance, propose approaches to knowledge extraction that do not rely the definition of classes. Even more relevant, some aspects of RDF that are not meaningful in OWL, such as named graphs, are assuming an increasing importance in the way information is represented in RDF. Some alternative approaches to represent terminologies, like SKOS, are also gaining momentum. Outside the biomedical world, initiatives such as schema.org have been relying on other formats for annotations that, while related to RDF, clearly hint at strong use cases that are based on the consistent use of relations, rather than their axiomatization.

Still, new solutions are being proposed, which tie together the roles of RDF and OWL (SIO), and go even further embedding into this framework services and workflows (SADI).

A related question is which is the role of ontologies in bioinformatics. An intrinsic necessity in bioinformatics is data analysis. Biological entities, by definition, are difficult to define with clear boundaries. Sequence assemblies and microarray data, for example, have a semantic that intrinsically requires some degree of approximation and uncertainty. While current ontologies in bioinformatics are adequate to represent metadata and annotations, most of the information that is relevant for bioinformatics research will not be amenable to be formally represented in ontologies.

A computational intelligence is needed to make sense of the sheer amount of data that can connect events observed at the molecular scale to the (desired of not) features of organisms or entire ecosystems, and such computational intelligence needs ontologies that can express the common know-how of people. For this same reason, the development of ontologies is destined to be an ongoing process that will follow the evolution of science. Which approach best suits in defining such ontologies is, at this time, still the object of active research.

References

27.1 P. Romano, A. Splendiani: Applications of Semantic Web methodologies and techniques to biology and bioinformatics, LNCS **5224**, 200–239 (2008)

27.2 D. Allemang, J. Hendler: *Semantic Web for the Working Ontologist: Effective Modeling in RDFS and OWL* (Morgan Kaufmann, Burlington 2009)

27.3 B. Smith, W. Ceusters: Ontological realism as a methodology for coordinated evolution of scientific ontologies, Appl. Ontol. **5**, 139–188 (2010)

27.4 A. Splendiani, M. Gündel, J.M. Austyn, D. Cavalieri, C. Scognamiglio, M. Brandizi: Knowledge sharing and collaboration in translational research, and the DC-THERA Directory, Brief. Bioinforma. **12**(6), 562–575 (2011)

27.5 P. Khatri, S. Drăghici, G.C. Ostermeier, S.A. Krawetz: Profiling gene expression utilizing onto-express, Genomics **79**(2), 266–270 (2002)

27.6 S. Drăghici, P. Khatri, R.P. Martins, G.C. Ostermeier, S.A. Krawetz: Global functional profiling of gene expression, Genomics **81**(2), 98–104 (2003)

27.7 S. Drăghici: *Statistics and Data Analysis for Microarrays Using R and Bioconductor*, 2nd edn. (Chapman Hall/CRC, New York 2011)

27.8 P. Khatri, S. Drăghici: Ontological analysis of gene expression data: Current tools, limitations, and open problems, Bioinformatics **21**, 3587–3595 (2005)

27.9 M.D. Young, M.J. Wakefield, G.K. Smith, A. Oshlak: Gene ontology analysis for RNA-seq: Accounting for selection bias, Genome Biol. **11**(2), R14 (2010)

Part F

Part F Bioinformatics in Medicine, Health and Ecology

Ed. by Francesco Masulli

28. Statistical Signal Processing for Cancer Stem Cell Formation

Mónica F. Bugallo, Petar M. Djurić

Many mysteries related to the behavior of cancer stem cells (CSCs), their role in the formation of tumors, and the evolution of tumors with time remain unresolved. Biologists conduct experiments and collect data from them; these are then used for modeling, inference, and prediction of various unknowns that define the CSC system and are vital for its understanding. The aim of this chapter is to provide a summary of our progress in statistical signal processing models and methods for the advancement of the theory and understanding of cancer and the CSC paradigm. The chapter comprises three parts: model building, methods

for the forward problem, and methods for the inverse problem.

Human epithelial cancers remain incurable [28.1]. The ineffectiveness of standard anticancer drugs has recently been attributed to rare, highly drug-resistant tumor-driving cells, so-called cancer stem cells (CSCs) [28.2, 3]. The CSC concept of carcinogenesis suggests that effective anticancer therapy should target this biologically and functionally distinct cell population within the tumor. However, this new paradigm of cancer treatment requires novel criteria of drug effectiveness and suitable in vitro models for drug testing [28.4–6] and stem cell research [28.2, 7].

According to the most recent concept of carcinogenesis, cancer is thought to be due to uncontrolled stem cell division after genetic mutation or transformation of either an adult stem cell or dedifferentiated progenitor cell. Accumulated data suggest that not any cancer cell, but only a pathological pool of CSCs is responsible for tumor initiation, development, recurrence, invasion, and response to treatment in the majority of human cancer types (in contrast, the previous theory of carcinogenesis treated all cancer cells as equally malignant) [28.2, 3].

The studied biological system is described by a Markov chain with a large number of unknowns,

whereas the observations are characterized by sparsity and incompleteness. All these features introduce many challenges to the application of traditional models and methods, and motivate the research and progress in the theory of signal processing. The advances discussed in this chapter are not confined to the biological problem under consideration but can also be applied to many other problems in science and engineering.

From a signal processing point of view there are several important issues that need to be solved when dealing with the considered system. The first one is the building of the mathematical models for the studied phenomena on a scale that allows relatively easy interpretation of the results. This includes the construction of the chain of states of interest (and possibly some nuisance ones) that provides all the relevant interactions in the system. In general, one expects to build more than one model, in which case the problem of model selection needs to be resolved [28.8].

For a given model, one can consider two types of computational problems, usually referred to as the *forward* and *inverse* problems. The forward problem amounts to simulating the system from some starting values of the states and sequentially predicting their values in time. This presumes that all the transition probabilities and starting values are known or estimated by using inverse methods. For a Markov system with small number of states, this is relatively easy to accomplish. The problem becomes much more challenging when the state space is high dimensional [28.9]. If one wants to obtain the statistics of the random variables of the process using simulations, in general one runs many realizations or copies of the process [28.9–11]. This chapter discusses the implementation of the CSC system by an approach which allows for the inclusion of different effects and characteristics of open systems, and simulation by agent-based methods [28.12, 13].

The inverse problem consists of estimating the unknowns of the system based on experimental time-series measurements of some functions of the states. The unknowns include the unobservable states and the transition probabilities of the states. In most of the experiments related to CSC evolution, one deals with very high-dimensional systems, i.e., large number of states, and sparse time series of measurements, which are usu-ally short in duration and are characterized by many missing data. This entails that the inverse problem is ill-conditioned and requires inclusion of prior information. For this reason, a promising research line is based on the Bayesian methodology. It allows for incorporation of prior information about the unknowns and for its optimal combination with the information gained from experimental data.

From a biological point of view the reviewed models and simulation and estimation methods constitute a *microscope for biology* and an in silico laboratory. To that end, the developed models should mirror, as precisely as possible, the cellular communication that leads to different responses of the system [28.14,15] and should account for the sparse nature of the available data from in vivo experiments. The improved understanding obtained from models and methods should be used for design of better experiments, which in turn should provide more valuable information for pushing the advancement of knowledge about the studied biological phenomena.

28.1 The Biological Problem

Understanding the evolution and formation of cancer from CSCs is critical for the development of CSC-targeted anticancer therapies [28.16, 17]. Models for stem cell and CSC evolution have already been proposed [28.18–20]. However, existing models are either deterministic, which are known to be in-accurate in practical scenarios [28.21]; are applicable to well-stirred systems, which is clearly not the case in cell-mediated processes [28.22]; or assume two-dimensional schemes, which do not reflect in vivo systems [28.23]. At the present time, three-dimensional (3-D) cultures of floating cancer spheroids represent the most clinically relevant in vitro tumor model. It is known that tumor cells acquire multidrug resistance as they assemble into multicellular spheroids [28.7, 24]. Produced cancer spheroids differ in their size, density, growth rate, cell-to-cell adhesion, and other parameters, which can significantly affect drug penetration and response.

The processes that lead to formation and growth of tumors are mainly based on two types of cells, i.e., CSCs and daughter cells (i.e., other cells that are not CSCs). Figure 28.1a depicts the possible evolution of single cells in the system over time and the possible course of action depending on the types of cells in the system, whereas Fig. 28.1b illustrates the expected time/space evolution of the whole system. From Fig. 28.1a we see that CSCs can undergo either symmetrical divisions, in which case the outcomes are more CSCs, or divisions where the result is a CSC and a committed progenitor cell. The latter, after several additional symmetrical divisions, may become the building material of cancer spheroids. Figure 28.1b shows how the phenotype of stemness reduces with additional generations of differentiated cells. At the same time, the loss of stemness is replaced by the phenotype of differentiation. For simplicity and for better explanation of the intricacies of the problem, we consider that only CSCs exist at the beginning. The real system is more complicated and involves many more initial conditions, states, and transitions [28.7, 18]. This simple scheme can be clearly seen as a Markov chain. The CSCs have two unique properties:

1. They are long-lived cells.
2. They are capable of self-renewal, which means that they can divide symmetrically (from one stem cell, one gets two stem cells) or asymmetrically (from one stem cell, one gets one stem cell and one progenitor) (Fig. 28.1).

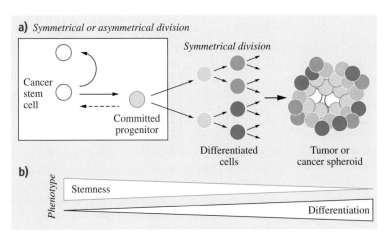

a) *Symmetrical or asymmetrical division*

Symmetrical division

Cancer stem cell

Committed progenitor

Differentiated cells

Tumor or cancer spheroid

b)

Phenotype

Stemness

Differentiation

Fig. 28.1a,b New paradigm of cancer formation from cancer stem cells. (**a**) Possible evolution of single cells. (**b**) Expected time/space evolution of the whole system

The daughter cells can only divide symmetrically and will eventually die. Cells make decisions based on many more different factors sensed by them, including density, quantity, or temperature.

28.2 The Model

We present a mathematical formulation of the system as follows. We assume that a CSC C_1 can participate in one of the following events

$$C_1 \xrightarrow{p_{01,t}} \emptyset , \tag{28.1}$$

$$C_1 \xrightarrow{p_{02,t}} 2C_1 , \tag{28.2}$$

$$C_1 \xrightarrow{p_{03,t}} C_1 + C_{2,1} , \tag{28.3}$$

where $\emptyset$ symbolizes death, $C_{2,1}$ is a daughter cell of first generation, and $p_{01,t}$, $p_{02,t}$, and $p_{03,t}$ are probabilities at a time instant t, where

$$p_{01,t} + p_{02,t} + p_{03,t} + p_{04,t} = 1 \tag{28.4}$$

with $p_{04,t}$ being the probability that C_1 is in quiescence state. The event described by (28.2), for example, represents symmetric division and can occur with probability $p_{02,t}$ in a time interval that separates $t-1$ and t. We note that this probability, in general, is a function of time.

A daughter cell of the j-th generation $C_{2,j}$, where $1 \le j < l$, participates in one of the following events

$$C_{2,j} \xrightarrow{p_{j1,t}} \emptyset , \tag{28.5}$$

$$C_{2,j} \xrightarrow{p_{j2,t}} 2C_{2,j+1} , \tag{28.6}$$

where $C_{2,j+1}$ is a cell from generation $j+1$, and the probabilities satisfy

$$p_{j1,t} + p_{j2,t} + p_{j3,t} = 1 , \tag{28.7}$$

where $p_{j3,t}$ is the probability that $C_{2,j}$ is in quiescence state.

Cells from the last generation $C_{2,\lambda}$ can either die within the interval from $t-1$ to t, i.e.,

$$C_{2,\lambda} \xrightarrow{p_{\lambda1,t}} \emptyset , \tag{28.8}$$

or stay alive with probability $p_{\lambda2,t}$, where

$$p_{\lambda1,t} + p_{\lambda2,t} = 1 . \tag{28.9}$$

The previous formulation defines events corresponding to fates of individual cells in the system. In order to simulate the system in a realistic way, one needs to account for communication among cells, attachment/detachment of cells for formation of tumors, as well as changes in the environment. It is clear that the whole system is per se open, in that no reasonable upper bound can be determined for the overall number of cells. It is thus evident that the structure of such a system requires resorting to simulation tools where random collections of objects with random states may be easily handled, a scenario quite reminiscent of the concept of random set theory (RST), which has recently received much attention in information fusion for distributed tracking [28.25] and communications [28.26].

28.3 The Forward Problem

We implement and simulate the behavior of CSCs and the processes that lead to formation and growth of tumors using concepts of RST and agent-based simulation tools. The system evolves from single cells to spheroids and aggregates (i. e., formations without CSCs). We consider that the formation of a spheroid is initiated by a CSC; i. e., one spheroid must contain at least one CSC. We also assume the hypothesis that from one CSC we may potentially get one spheroid irrespective of the number of symmetrical divisions that the original CSC will undergo. We use a compartmental approach; that is, we model the system as composed of two different compartments: the set of spheroids containing *useful* components (i. e., CSCs), and the set of *junk* aggregates with *useless* components (i. e., no CSCs),

$$Z_t = X_t \bigcup J_t , \tag{28.10}$$

where Z_t is the total number of sets in the system at time t, X_t is the set with the *useful* spheroids, i. e., those cell formations having a nonzero number of CSCs, while J_t is a set containing *useless* conglomerates, that is, cell aggregates that do not contain any CSCs. The X_t compartment can be viewed as a disjoint union of sets (spheroids)

$$X_t = \bigcup_{k=0}^{k_t} X_{k,t} , \tag{28.11}$$

where $k_t = |X_t|$ denotes the cardinality of X_t. We note that the cardinality varies with time; i. e., new spheroids can appear, or existing spheroids can disappear. The expression $k_t = |X_t| = 0$ means that the set of useful spheroids is empty (i. e., absence of stem cells). The set $X_{k,t}$ is a descriptor of the k-th useful spheroid, and it

can be expressed as

$$X_{k,t} = \left\{ \begin{pmatrix} x_{k,1,t} \\ x_{k,2,t} \\ \vdots \\ x_{k,\lambda,t} \\ l_{k,t} \end{pmatrix} \right\} , \tag{28.12}$$

where $x_{k,1,t}$ is the random variable modeling the number of CSCs in the k-th spheroid at time instant t, $x_{k,j,t}$, $j = 2, \cdots, n_k$ is the random variable modeling the number of daughter cells from the j-th generation in the k-th spheroid present at time instant t, λ is the last possible generation of cells, and $l_{k,t}$ represents the position of the group in the experiment domain. Note that, if $x_{k,1,t} = 0$, then $X_{k,t} = \emptyset$, which means that the k-th useful spheroid at t is junk since no new stem cells can be generated. At time $t + 1$, the set representing the k-th group has the form

$$X_{k,t+1} = \left\{ \begin{pmatrix} x_{k,1,t+1} \\ x_{k,2,t+1} \\ \vdots \\ x_{k,\lambda,t+1} \\ l_{k,t+1} \end{pmatrix} \right\} . \tag{28.13}$$

In order to derive the expressions that connect (28.12) and (28.13), we first note that we have two classes of cells with distinctive differentiating behaviors: the CSCs and the daughter cells. If at time instant t we have $x_{1,t}$ CSCs, those cells can participate in one of the following actions: die (with probability $p_{01,t}$), divide symmetrically (with probability $p_{02,t}$), divide asymmetrically (with probability $p_{03,t}$), or remain idle (with probability $p_{04,t} = 1 - p_{01,t} - p_{02,t} - p_{03,t}$). Note that we dropped the subindex k from $x_{1,t}$ for simplicity in the notation. By contrast, the daughter cells can only die, reproduce symmetrically towards the next generation (e.g., $C_{2,2} \xrightarrow{p_{22,t}} 2C_{2,3}$), or remain in a quiescence mode. Figure 28.2 displays a Markov chain representing the possible evolution of one CSC.

Denoting by $d_{1,t+1} \leq x_{1,t}$ the number of CSCs that will die in the interval $(t, t+1)$ from a particular spheroid, the number of possibly active CSCs (they may reproduce or do nothing) during $(t, t+1)$ is $x_{1,t} - d_{1,t+1}$. We represent by $s_{1,t+1} \leq x_{1,t} - d_{1,t+1}$ the number of CSCs that will reproduce symmetrically in the interval $(t, t+1)$, i. e., those CSCs that will produce new

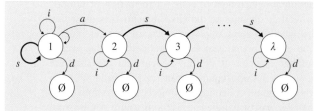

Fig. 28.2 Evolution of a CSC. The states represent the type of cell (1 is a CSC, 2 is a second generation of daughter cell, etc.). The *arrows* represent the possible actions at one time instant: *i*: idle; *d*: death; *s*: symmetrical division; *a*: asymmetrical division

CSCs. We can then write that the number of CSCs in the system evolves with time according to the expression

$$x_{1,t+1} = x_{1,t} + s_{1,t+1} - d_{1,t+1} . \qquad (28.14)$$

Equation (28.14) means that the new number of CSCs is the amount of CSCs in the previous time instant minus the ones that died plus the new ones that were generated through symmetrical division.

We next focus on the first generation of daughter cells. These cells can only die, reproduce symmetrically towards the next generation (i. e., $C_{2,2} \xrightarrow{p_{22,t}} 2C_{2,3}$), or remain in a quiescence mode. The number of cells in this group at time instant $t+1$ will be calculated as the number of cells of this same generation from the previous time instant that stay idle plus the cells that come from asymmetrical division of the CSCs minus the cells that either die or divide towards the next generation, i. e.,

$$x_{2,t+1} = x_{2,t} + a_{1,t+1} - s_{2,t+1} - d_{2,t+1} . \qquad (28.15)$$

For the next generations and up to the last one, we can write a common update equation since after the first generation of daughter cells only death, idleness, or symmetrical division towards future generation is possible. This can be formulated as

$$x_{j,t+1} = x_{j,t} + 2s_{j-1,t+1} - s_{j,t+1} - d_{j,t+1}$$
$$j = 3, \cdots, \lambda - 1 . \qquad (28.16)$$

Finally, the last generation is updated using the expression

$$x_{\lambda,t+1} = x_{\lambda,t} + 2s_{\lambda-1,t+1} - d_{\lambda,t+1} . \qquad (28.17)$$

In summary, the overall evolution of the k-th spheroid in the system will be given by

$$X_{k,t+1} = \left\{ \left(\begin{array}{l} x_{k,1,t+1} = x_{k,1,t} + s_{k,1,t+1} \\ \qquad - d_{k,1,t+1} \\ x_{k,2,t+1} = x_{k,2,t} + a_{k,1,t+1} \\ \qquad - s_{k,2,t+1} - d_{k,2,t+1} \\ x_{k,3,t+1} = x_{k,3,t} + 2s_{k,2,t+1} \\ \qquad - s_{k,3,t+1} - d_{k,3,t+1} \\ \qquad \vdots \\ x_{k,\lambda-1,t+1} = x_{k,\lambda-1,t} \\ \qquad + 2s_{k,\lambda-2,t+1} \\ \qquad - s_{k,\lambda-1,t+1} \\ \qquad - d_{k,\lambda-1,t+1} \\ x_{k,\lambda,t+1} = x_{k,\lambda,t} + 2s_{k,\lambda-1,t+1} \\ \qquad - d_{k,\lambda,t+1} \\ l_{k,t+1} = f(l_{k,t}, x_{k,t+1}) \end{array} \right) \right\} ,$$

where $f(\cdot)$ is a function that determines the evolution of the location of the spheroid depending on the previous location and the amount of cells in the spheroid $x_{k,t+1} = (x_{k,1,t+1}, x_{k,2,t+1}, \cdots, x_{k,\lambda,t+1})^{\top}$. We note that with this model we do not contemplate the case where a *useful* spheroid would break into two or more sets of cells, or the case where two spheroids (or spheroid and *junk* aggregates) would fuse.

The above equations show that the set sequence X_t can be easily simulated. As an example, the model leads to an algorithm wherein $X_{k,t+1}$ is generated from $X_{k,t}$ as follows:

CSC generation

1. Generate $(d_{k,1,t+1}, s_{k,1,t+1}, a_{k,1,t+1}, i_{k,1,t+1}) \sim MN(p_{01,t}, p_{02,t}, p_{03,t}, p_{04,t}; x_{k,1,t})$
   (the number of CSCs that die, divide symmetrically, divide asymmetrically, and remain idle, respectively. The notation $i_{k,1,t+1}$ is used to represent the number of CSCs that will stay in quiescence state from iteration time instant t to $t+1$. The symbol $MN(\cdot; \cdot)$ denotes a multinomial distribution.)
2. Update $x_{k,1,t+1} = x_{k,1,t} + s_{k,1,t+1} - d_{k,1,t+1}$
   (number of CSCs at $t+1$).

First generation of daughter cells

1. Generate $(d_{k,2,t+1}, s_{k,2,t+1}, i_{k,2,t+1}) \sim MN(p_{11,t}, p_{12,t}, p_{13,t}; x_{k,2,t})$
   (number of first generation that die, divide symmetrically, and remain idle, respectively)
2. Update $x_{k,2,t+1} = x_{k,2,t} + a_{k,1,t+1} + s_{k,2,t+1} - d_{k,2,t+1}$
   (number of first generation at $t+1$).

j-th generation of daughter cells, $j = 3, \cdots, \lambda - 1$

1. Generate $(d_{k,j,t+1}, s_{k,j,t+1}, i_{k,j,t+1}) \sim MN(p_{j1,t}, p_{j2,t}, p_{j3,t}; x_{k,j,t})$
   (number of j-th generation that die, divide symmetrically, and remain idle, respectively)
2. Update $x_{k,j,t+1} = x_{k,j,t} + 2s_{k,j-1,t+1} - s_{k,j,t+1} - d_{k,j,t+1}$
   (number of j-th generation at $t+1$).

λ-th generation of daughter cells

1. Generate $d_{k,\lambda,t+1} \sim Bi(p_{\lambda 1,t}; x_{k,\lambda,t})$
   (number of λ-th generation that die. The symbol Bi stands for binomial distribution.)
2. Update $x_{k,\lambda,t+1} = x_{k,\lambda,t} + 2s_{k,\lambda-1,t+1} - d_{k,\lambda,t+1}$
   (number of λ-th generation at $t+1$).

28.3.1 Forward Problem – Some Results

Using RST and a multinomial sampling approach for representation of cell growth, it is possible to derive the populations of cancer cells in each generation. However, we also need to deal with the locations of the spheroids. At a particular time instant, cell fates depend on a number of factors such as temperature, chemoattractant concentrations, as well as physical constraints due to the locations of neighboring cells. We consider only the latter (space constraints) as a location-determining factor.

We use the approach in [28.27] to simulate migration rules. We consider a square or cubical lattice with each side equal to n_{grid} points. Each cell can occupy at most one grid point. Daughter cells are assumed to have finite proliferation capacity (limited number of generations), whereas CSCs have infinite proliferation capacity. Time units are days. A cell can only proliferate if it is at least 1 day old, and if it is a daughter cell, it dies (with certainty) after it reaches a certain age (λ). Apart from this, there can be random events which also lead to death of daughter cells of any age, for which we assign some finite probability. At the beginning of each day, each cell can take a number of migration steps. At each possible step, a cell can choose with equal probability to move to any one of its free neighboring grid points or to stay idle at its present position. We specify some maximum possible migration steps per day for each cell. If no neighbor grid point is free, the cell stays idle.

After migration, for a CSC, the possible fates are remaining quiescent, death, symmetrical division, or asymmetrical division. For daughter cells, the possible fates are the same as above, except that there is no possibility for asymmetrical division. If no neighboring grid points are free for a CSC, its fate is being idle or death; otherwise it may have all the possible fates. Daughter cells also have similar fates, except if they have reached a dying age, when they have to die. For each possible fate, we assign probabilities that may be time-varying.

Figure 28.3 shows the results for two different migration rates for a $50 \times 50 \times 50$ 3-D lattice model. The simulation was initialized with three CSCs at three different locations. Figure 28.3a–c shows the results for a migration rate of 10 per day, while Fig. 28.3d–e shows results for no migration. The total cell population is pro-

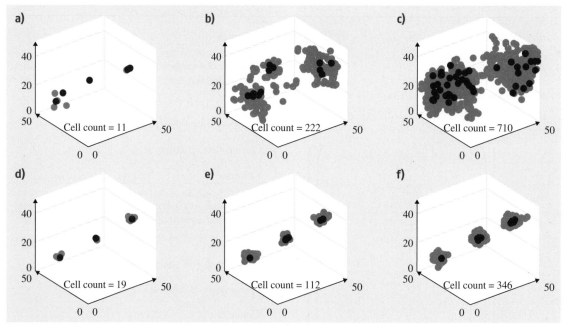

Fig. 28.3a–f Simulation snapshots for migration rates of 10 (**a–c**) and 0 (**d–f**) per day. Snapshots (**a**) and (**d**) were taken at $T = 5$ days, (**b**) and (**e**) at $T = 15$ days, and (**c**) and (**f**) at $T = 30$ days. CSCs are marked in *blue* and daughter cells in *red*

Table 28.1 Probabilities of the possible events

	Death	Idle	Migration	Asym. division	Sym. division
CSC	0	0.3	0.3	0.35	0.05
Daughters	0.1	0.3	0.3	0	0.3

vided in each subfigure. During each simulation, the probabilities for various cell fates remained unchanged over time (Table 28.1).

The simulation was carried out for a duration of 30 days, and the maximum number of daughter cell generations was $\lambda = 7$.

It is quite clear from the figure that, without migration, the increase in size of the tumor size (indicated by the cell count) is much slower. This suggests that in-

cluding a location function in our random set model is necessary so that it reflects reality better. Similar results have been observed in [28.27].

The above example demonstrates the potential of our approach to modeling tumor growth. In particular, the simulations show how a migration process can be simulated using simple random sampling rules and a lattice environment where proliferation and movement are limited by spatial constraints.

28.4 The Inverse Problem

In building the mathematical model of the stochastic evolution of a system involving cells, spheroids, and eventually inhibitors, we deal with experimental data including the amount of cells of a specific type, quantity of spheroids, density of the medium, and temperature. Some of these data are sparse. In particular, the experiment may start with K identical sets of cells that are allowed to evolve for various time periods. For each set, we get one observation when we stop its evolution. The first measurements y_1 will correspond to the outcome of the first set, y_2 to the second set, and so on. Thus, we can view the overall experiment as a set of subexperiments, where each of the subexperiments represents the evolution of a given set of cells. The goals of solving the inverse problem are to

1. Obtain an accurate model of the processes taking place in the experiment (there are several possible models [28.18, 28])
2. Estimate the parameters of the model.

Following the work in [28.28], and defining ρ as the ratio of new CSCs generated after each mitosis from the pool of stem cells that undergo cellular division ($\rho \geq 0.5$), α as the proliferation function, i.e., the ratio of active stem cells given the total number of stem cells, $x_{k,t}$ as the number of CSCs at time instant t $(t = 1, \cdots, T)$ in a specific subexperiment k $(k = 1, \cdots, t)$, and τ as the amount of time needed for a CSC to divide symmetrically, the rate of generation of

new stem cells is given by

$$\frac{(2\rho - 1) \cdot \alpha \cdot x_{k,t}}{\tau} = r \cdot x_{k,t} ,$$

where $r = (2\rho - 1) \cdot \alpha / \tau$ is the rate at which new stem cells are formed in the system. In real systems, the parameters ρ, α, and τ change with time, as do the conditions of the system. The evolution of the population of CSCs in a time interval Δt can be obtained according to

$$\frac{x_{k,t+\Delta t} - x_{k,t}}{\Delta t} = r \cdot x_{k,t} .$$

With this expression, we build a state-space model, where the state of the system is given by

$$x_{k,t} = ax_{k,t-1} + u_{k,t} , \tag{28.18}$$

where $a = 1 + (2 \cdot \rho - 1) \cdot \alpha / \tau$, $u_{k,t} \sim \mathcal{N}(0, \sigma_u^2)$ is a state noise that accounts for variability for not including factors such as density of cells in the system, quantity of different types of cells, temperature, foo, and loneliness, and Δt is normalized to one. We note that, in this section, $x_{k,t}$ denotes CSCs in the k-th subexperiment at time t.

As pointed out, the experiment is started with K identical sets of cells and conducted for various time periods. The observation equation for the k-th subexperiment is given by

$$y_{k,t} = \lfloor x_{k,t} \rfloor + v_{k,t} , \tag{28.19}$$

which represents the number of CSCs measured at time instant t and where the observation noise has a known discrete distribution centered at zero. The notation $\lfloor \cdot \rfloor$ represents the integer part of the argument. It is important to note that

1. From subexperiment 1 we get $y_{1,1}$, from subexperiment 2 we get $y_{2,2}$ (i.e., $y_{2,1}$ is a missing observation), and from subexperiment 3 we get $y_{3,3}$ (i.e., $y_{3,1}$ and $y_{3,2}$ are missing observations); that is, we are not able to get intermediate measurements for a specific subexperiment.
2. The overall model given by (28.18) and (28.19) is clearly nonlinear.

We know that $x_{1,0} = x_{2,0} = \cdots = x_{T,0}$, and the observations $y_{t,t}$, $t = 1, 2, \cdots, T$. The objective is to estimate the unknowns a, and $x_{k,t}$, $1 \le k \le t$, $t = 1, 2, \cdots, T$, which means that we want to estimate $k(k+1)/2$ unknowns with only k observations. In other words, our problem involves estimating a high-dimensional state with sparse observations. Once we count the CSCs, the subexperiment is completed. In other words, each subexperiment provides one number, which is the number of CSCs on the day when the counting takes place.

We find a solution for tracking the evolution of the unknowns within the Bayesian framework by using a particle filtering (PF) algorithm. We apply the PF methodology because of its effectiveness and flexibility when dealing with nonlinear and/or non-Gaussian models [28.29, 30], as is the case with our model [28.18, 28]. Rather than trying to adaptively provide point estimates of the unknown state of the system $x_{k,t}$, as almost all existing methods do, PF attempts to find approximations of the a posteriori distributions of the states. To that end, PF uses random grids which are composed of nodes, usually referred to as particles, with assigned weights, the latter being interpreted as probability masses. The particles and the weights form discrete random measures, which are basically probability mass functions. We denote them by $\chi_{k,t} = \{x_{k,t}^{(m)}, w_{k,t}^{(m)}\}_{m=1}^{M}$, where $x_{k,t}^{(m)}$ is the m-th particle and $w_{k,t}^{(m)}$ its associated weight. In general, standard PF (SPF) methods consist of three basic operations: particle generation (or in the language of biology, mutation), computation of weights (or fitness evaluation), and resampling (replication of the fit particles and removal of the ones that are not fit). Figure 28.4a shows the main three steps of PF methods, and Fig. 28.4b provides a flowchart of the generic PF algorithm.

Assuming that the prior of a is $p(a)$ defined on the interval $[1, 2)$, we implement the PF solution as follows (the bounds are calculated according to real parameters given in [28.18]. For the experiments we consider a uniform distribution for the prior).

[*Time series $k = 1$*]

1. Draw particles $a^{(m)}$ according to the prior $p(a)$,

$$a_1^{(m)} \sim p(a),$$

   where the subscript of a denotes that these are particles of a for the first time series.
2. For time instant $t = 1$, generate particles $x_{1,1}^{(m)}$, $m = 1, 2, \cdots, M$ according to

$$x_{1,1}^{(m)} \sim p(x_{1,1} | x_{1,0}, a_1^{(m)}).$$

3. Compute the weights of these particles by

$$w_{1,1}^{(m)} \propto p(y_{1,1} | x_{1,1}^{(m)}).$$

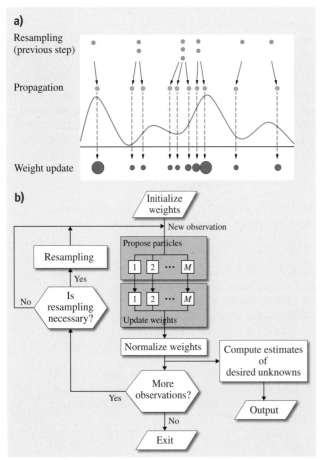

a)

Resampling (previous step)

Propagation

Weight update

b)

Fig. 28.4 (a) A pictorial description of particle filtering (PF). (b) Flowchart of PF

[*Time series k = 2*]

4. Draw particles $a_2^{(m)}$ from $p(a|y_{1,1}, x_{1,0})$ as follows:
   a) Choose the particle $x_{1,1}^{(m)}$ with probability $w_{1,1}^{(m)}$.
   b) Draw the particle of a for the second time series,

   $$a_2^{(m)} \sim p\left(a|x_{1,0}, x_{1,1}^{(m)}\right).$$

5. Given $a_2^{(m)}$, generate $x_{2,1}^{(m)}$ by

   $$x_{2,1}^{(m)} \sim p\left(x_{2,1}|a_2^{(m)}, x_{2,0}\right)$$

   and $x_{2,2}^{(m)}$ by

   $$x_{2,2}^{(m)} \sim p\left(x_{2,2}|a_2^{(m)}, x_{2,1}^{(m)}\right).$$

6. Compute the weights of the particles $(x_{2,1}^{(m)}, x_{2,2}^{(m)}, a_2^{(m)})$ by

   $$w_{2,2}^{(m)} \propto p\left(y_{2,2}|x_{2,2}^{(m)}\right).$$

[*Time series k = 3*]

1. Draw particles $a_3^{(m)}$ from $p(a|y_{1,1}, y_{2,2}, x_{1,0}, x_{2,0})$ as follows:
   a) Choose the particle $x_{2,2}^{(m)}$ with probability $w_{2,2}^{(m)}$.
   b) Draw particles of a for the third time series,

   $$a_3^{(m)} \sim p\left(a|x_{2,0}, x_{2,1}^{(m)}, x_{2,2}^{(m)}\right).$$

2. Given $a_3^{(m)}$, generate $x_{3,1}^{(m)}$ by

   $$x_{3,1}^{(m)} \sim p\left(x_{3,1}|a_3^{(m)}, x_{3,0}\right)$$

   followed by generation of $x_{3,2}^{(m)}$ by

   $$x_{3,2}^{(m)} \sim p\left(x_{3,2}|a_3^{(m)}, x_{3,1}^{(m)}\right)$$

   and generation of $x_{3,3}^{(m)}$ by

   $$x_{3,3}^{(m)} \sim p\left(x_{3,3}|a_3^{(m)}, x_{3,2}^{(m)}\right).$$

3. Compute the weights of the particles $(x_{3,1}^{(m)}, x_{3,2}^{(m)}, x_{3,3}^{(m)}, a_3^{(m)})$ by

   $$w_{3,3}^{(m)} \propto p\left(y_{3,3}|x_{3,3}^{(m)}\right).$$

   The processing continues in analogous fashion until $t = T$. We point out that one can develop alternative PF schemes to the one described before.

28.4.1 Inverse Problem – Simulation Results

We considered a set of 30 synthetically generated subexperiments, all of them starting with 100 CSCs. The noise of the state was modeled as zero-mean Gaussian with variance $\sigma_u^2 = 100$, and the parameter a was

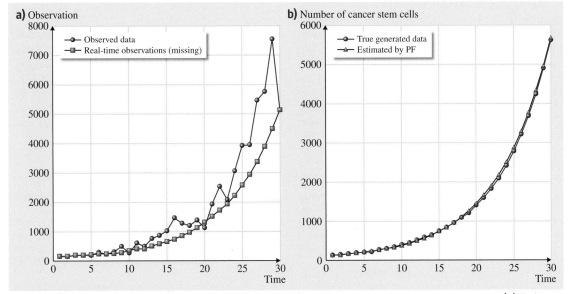

Fig. 28.5 (a) Observations for 30 experiments (last run and interpolation of the obtained measurements). (b) Estimates of the state

set to 1.15. The noise in the observation had a triangular probability mass function with lower limit -30, mode 0, and upper limit 30. In running the PF-based method, we used $M = 800$ particles. Note that in this case we have 30 observations available coming from 30 independent subexperiments, and we need to estimate 466 parameters. Also, Fig. 28.5a shows a comparison between the real (mostly nonobserved) sequence of observations from the last subexperiment (curve with square markers) and the interpolation of the true observed

subexperiments (curve with filled dot markers). It is evident that there is a great variability among both curves even for the considered low error of observations.

Figure 28.5b shows the results of estimating the number of CSCs for the 30-th subexperiment over time according to the proposed model and using the PF-based algorithm. It is apparent that the proposed method performs very accurately and the real and estimated curves are almost indistinguishable on the scale used in the figure.

28.5 Conclusions

The presented work is in the area of modeling, simulation, and in vitro experimentation for improved understanding of the behavior of CSCs (cellular level) and the processes that lead to formation and growth of tumors as well as metastasis (tissue level). The objective is to employ the developed models and methods as a *microscope for biology* and as an in silico laboratory. The improved understanding will be used for design of better experiments, which in turn may provide valuable information for expanding the knowledge about the studied biological phenomena. The motivation for this work is not only scientific. Cancer has become the second most fatal disease, only surpassed by cardiovascular diseases.

The chapter comprised three parts: model building, methods for the forward problem, and methods for the inverse problem. The addressed biological problem is characterized by a large number of unknowns and a set of sparse (in time) observations. The built model describes the behavior of CSCs and tumors, and we use this model to simulate the system by resorting to RST concepts and agent-based modeling. The flexibility of the proposed simulation model allows for inclusion of various effects (e.g., migration effects from differ-

ent spheroids or aggregates, indeterminacy of the next spheroid or aggregate location from the previous one, and so on) in a very easy and user-friendly way.

The solutions of the inverse problems help in building accurate models for simulations that address the forward problem. We apply a PF-based algorithm to estimate the population of CSCs over time in a system of high dimension and with sparse time-series measurements. The initial results on synthetically generated data show promising performance of the proposed method. Open issues that need to be addressed include:

1. Adding the rest of the cell compartments shown in Fig. 28.1b to the problem formulation, with particular emphasis on the 3-D nature of the model and the formation of the spheroids
2. A more accurate mathematical representation of the driving parameter a, which involves its time evolution in order to account for changes in cell density, medium composition, and interactions among cells
3. Improvement of the PF algorithm to improve the approximations of the posteriors of the unknowns, for instance, by applying PF in combination with population Monte Carlo sampling [28.31].

References

28.1   A. Kamb, S. Wee, C. Lengauer: Why is cancer drug discovery so difficult?, Nat. Rev. Drug Discov. **6**, 115–120 (2007)
28.2   P. Dalerba, R.W. Cho, M.F. Clarke: Cancer stem cells: Models and concepts, Annu. Rev. Med. **58**, 267–284 (2007)
28.3   M. Mimeault, R. Hauke, P.P. Mehra, S.K. Batra: Recent advances in cancer stem/progenitor cell research: Therapeutic implications for overcoming

resistance to the most aggressive cancers, J. Cell. Mol. Med. **11**, 981–1011 (2007)
28.4   R.E. Durand, P.L. Olive: Resistance of tumor cells to chemo- and radiotherapy modulated by the three-dimensional architecture of solid tumors and spheroids, Methods Cell Biol. **64**, 211–233 (2001)
28.5   G. Hamilton: Multicellular spheroids as an in vitro tumor model, Cancer Lett. **131**, 29–34 (1998)

28.6 R. Knuchel, F. Hofstadter, W. Jenkins, J.R. Masters: Sensitivities of monolayers and spheroids of the human bladder cancer cell line MGH-U1 to the drugs used for intravesical chemotherapy, Cancer Res. **49**, 1397–1401 (1989)

28.7 M. Al-Hajj, M.W. Becker, M. Wicha, I. Weissman, M.F. Clarke: Therapeutic implications of cancer stem cells, Curr. Opin. Genet. Dev. **14**, 43–47 (2004)

28.8 T. Toni, M.P.H. Stumpf: Simulation-based model selection for dynamical systems in systems and population biology, Bioinformatics **26**(1), 104–110 (2010)

28.9 D.J. Wilkinson: *Stochastic Modeling for System Biology* (Chapman Hall/CRC, New York 2006)

28.10 A. Arkin, J. Ross, H.H. McAdams: Stochastic kinetic analysis of developmental pathway bifurcation in phage lambda infected *Escherichia coli* cells, Genetics **149**, 633–648 (1998)

28.11 H.H. McAdams, A. Arkin: It's a noisy business: Genetic regulation at the nanomolar scale, Trends Genet. **15**, 65–69 (1999)

28.12 W.W. Wakeland, E.J. Gallaher, L.M. Macovsky, C.A. Aktipis: A comparison of system dynamics and agent-based simulation applied to the study of cellular receptor dynamics, Proc. 37th Hawaii Int. Conf. Syst. Sci. (IEEE, Bellingham 2004)

28.13 M. Woolridge: *An Introduction to Multi-Agent Systems* (Wiley, New York 2002)

28.14 M. Eyiyurekli, P. Manley, P.I. Lelkes, D.E. Breen: A computational model of chemotaxis-based cell aggregation, BioSystems **93**, 226–239 (2008)

28.15 E. Grinstein, P. Wernet: Cellular signaling in normal and cancerous stem cells, Cell. Signal. **19**, 2428–2433 (2007)

28.16 E.L. Bearer, J.S. Lowengrub, H.B. Frieboes, Y.-L. Chuang, F. Jin, S.M. Wise, M. Ferrari, D.B. Agus, V. Cristini: Multiparameter computational modeling of tumor invasion, Cander Res. **69**, 4493 (2009)

28.17 H.B. Frieboes, M.E. Edgerton, J.P. Fruehauf, F.R.A.J. Rose, L.K. Worrall, R.A. Gatenby, M. Ferrari, V. Cristini: Prediction of drug response in breast cancer using integrative experimental/ computational modeling, Cander Res. **69**, 4484 (2009)

28.18 R. Ganguly, I.K. Puri: Mathematical models for the cancer stem cell hypothesis, Cell Prolif. **39**, 3–14 (2006)

28.19 A.L. Garner, Y.Y. Lau, D.W. Jordan, M.D. Uhler, R.M. Gilgenbach: Implications of a simple mathematical model to cancer stem cell population dynamics, Cell Prolif. **39**, 15–28 (2006)

28.20 F. Michor: Mathematical models of cancer stem cells, J. Clin. Oncol. **26**, 2854–2861 (2008)

28.21 C.J. Morton-Firth, D. Bray: Predicting temporal fluctuations in an intracellular signaling pathway, J. Theor. Biol. **192**, 117–128 (1998)

28.22 T.G. Liou, E.J. Campbell: Nonisotropic enzyme inhibitor interactions: A novel nonoxidative mechanism for quantum proteolysis by human neutrophils, Biochemistry **34**, 16171–16177 (1995)

28.23 H. Song, S.K. Jain, R.M. Enmon, K.C. O'Connor: Restructuring dynamics of DU 145 and LNCaP prostate cancer spheroids, In Vitro Cell. Dev. Biol. **40**, 262–267 (2004)

28.24 N. Haraguchi, T. Utsunomiya, H. Inoue, F. Tanaka, K. Mimori, G.F. Barnard, M. Mori: Characterization of a side population of cancer cells from human gastrointestinal system, Stem Cells **24**, 506–513 (2006)

28.25 I.R. Goodman, R.P.S. Mahler, H.T. Nguyen: *Mathematics of Data Fusion* (Springer, Berlin, Heidelberg 1997)

28.26 E. Biglieri, M. Lops: Multiuser detection in a dynamic environment. Part I: User identification and data detection, IEEE Trans. Inf. Theory **53**, 3158–3170 (2007)

28.27 H. Enderling, L. Hlatky, P. Hahnfeldt: Migration rules: Tumours are conglomerates of self-metastases, Br. J. Cancer **100**, 1917–1925 (2009)

28.28 M. Loeffler, H.E. Wichmann: A comprehensive mathematical model of stem cell proliferation which reproduces most of the published experimental results, Cell Tissue Kinet. **13**, 543–561 (1980)

28.29 P.M. Djurić, J.H. Kotecha, J. Zhang, Y. Huang, T. Ghirmai, M.F. Bugallo, J. Míguez: Particle filtering, IEEE Signal Process. Mag. **20**(5), 19–38 (2003)

28.30 A. Doucet, S.J. Godsill, C. Andrieu: On sequential Monte Carlo sampling methods for Bayesian filtering, Stat. Comput. **10**, 197–208 (2000)

28.31 O. Cappé, A. Guillin, C.P. Robert: Population Monte Carlo, J. Comput. Graph. Stat. **13**, 927–929 (2004)

29. Epigenetics

Micaela Montanari, Marcella Macaluso, Antonio Giordano

In the past, the term *epigenetics* was used to describe all biological phenomena that do not follow normal genetic rules. Currently, it is generally accepted that epigenetics refers to the heritable modifications of the genome that do not involve changes in the primary DNA sequence. Important epigenetic events include DNA methylation, covalent post-transcriptional histone modifications, RNA-mediated silencing, and nucleosome remodeling. These epigenetic inheritances take place in the chromatin-mediated control of gene expression and are responsible for chromatin structure stability, genome integrity, modulation of the expression of tissue-specific genes, and embryonic development, which are essential mechanisms allowing the stable propagation of gene activity states from one generation of cells to the next. Importantly, during the past years, epigenetic events have emerged to be considered as key mechanisms in the regulation of critical biological processes and in the development of human diseases. From this point of view, the importance of epigenetic events in the control of both normal cellular processes and altered events associated with diseases has led to epigenetics being considered as a new frontier in cancer research.

Since the introduction of the term *epigenetics*, a number of biological events (such as imprinting, X-chromosome inactivation, and position-effect variegation in the fruit fly) that are not coded in the DNA sequence itself have been considered as epigenetic phenomena.

The importance of the epigenetic code is underlined by the fact that all cells in any given organism share an identical genome with other cell types exhibiting striking morphological and functional properties as well [29.1,2]. Therefore, it is obvious that the epigenetic events determine the identity and proliferation potential of different cells in the body.

Different types of epigenetic modifications are intimately linked and often act in a self-reinforcing manner in the regulation of different cellular processes [29.3]. DNA methylation and histone acetylation and methylation represent the most intensively studied epigenetic modifications, especially in the context of gene transcription and neoplastic process mechanisms. It has been reported that these epigenetic marks are dynamically linked in the epigenetic control of gene expression, suggesting that their deregulation plays an important role in tumorigenesis [29.4–6].

Physiologically, the presence of available transcription factors, co-activators, and repressors modulate the response of cells to the signaling pathways by determining which genes must be activated or repressed. Furthermore, DNA methylation, histone methylation, and deacetylation mediate gene transcription. Alterations of these mechanisms affect the expression of important genes involved in the regulation of cellular growth, differentiation and apoptosis triggering the neoplastic process [29.6,7]. Epigenetic modifications occur through different *pathways* by which a particular set of genes becomes available or less accessible to the transcription machinery. Moreover, the complexity of these processes is increased by the association of the DNA with nucleosome in the chromatin, which severely restricts the access of regulatory factors involved in the transcription [29.8].

Transcription in eukaryotic cells is strongly influenced by the manner in which DNA is packaged, and the chromatin structure itself is influenced by DNA methylation and DNA-histone interactions. These epigenetic mechanisms provide an additional layer of transcriptional control of gene expression beyond those associated with variation in the DNA sequence. It has been suggested that an epigenetic cross-talk, such as an interplay between DNA methylation and histone modifications, may be involved in the process of gene transcription and aberrant gene silencing in tumors [29.9–11]. All together, these modifications seem to integrate the DNA-based genome coding potential in guarantee the mitotic stability of transcription patterns as well as the maintenance of cell identity [29.12]. For these reasons, in recent years the scientists have begun to study a much more dynamic area of investigation, called epigenome, in which they correlate the existing different functional states, governed in part by epigenetic modifications, in time and space. These investigations have led to a distinction between *driver* and *passenger* epigenetic changes, giving then an additional layer of information for the understanding of neoplastic transformation [29.13]. Moreover, the permanent silencing of hypermethylated genes in tumors by so-called *epigenetic gatekeeper genes* has demonstrated not only that epigenetic rather than genetic changes are responsible for the transformation from a normal to a malignant cell, but also that they are an important driving force in this process, capable of making the affected cells unable to undergo the programmed cell death [29.5, 14].

The identification of epigenetic driver mutations and the full comprehension of *epigenomic* mutations will be helpful in developing more effective drugs capable of reversing those epigenetic changes that contribute to carcinogenesis and of restoring a normal gene expression.

29.1 DNA Methylation

DNA methylation represents one of the most intensely studied epigenetic mechanisms in mammals. The methylation of DNA is referred to as the covalent addition of a methyl group ($-CH_3$) to the 5-carbon (C_5) position of cytosine bases that are located 5o to a guanosine base. The covalent addition of a methyl group generally occurs in cytosine residues located within CpG dinucleotides. Usually, these dinucleotides are concentrated in large clusters, called CpG islands, between 0.2 and 2 kb long, containing a relatively high frequency of CpG dinucleotides, mostly in the promoter and/or first exon region [29.2, 15]. The whole genome is characterized by rather low overall CpG content, and both in normal tissues and germ line cells the majority of gene promoter-associated CpG islands remain unmethylated. Nevertheless, the density of CpG itself does not influence gene expression since this regulation only relies on the DNA methylation process. Normally, CpG islands are unmethylated in transcriptionally active genes, whereas silenced genes are characterized by methylation within the promoter region (e.g., tissue-specific or developmental genes) [29.2, 16].

Cancer exhibits aberrant CpG islands methylation when many hundreds of CpG islands in individual tumors acquire DNA methylation. Agents responsible for establishing and maintaining the methylation pattern are a class of enzymes called DNA methyltransferases (DNMTs), which catalyze the transfer of the methyl group from S-adenosyl-methionine onto cytosine. Five members of the DNMT family have been identified in mammals: DNMT1, DNMT2, DNMT3a, DNMT3b, and DNMT3L. However, only DNMT1, DNMT3a, and DNMT3b interplay produces the global cytosine methylation pattern [29.2]. These independently encoded proteins are classified as maintenance methylation enzyme (DNMT1) or as de novo methylation enzymes (DNMT3a and DNMT3b). DNMT1 is absolutely necessary for restoring full methylation arising as a result of the semiconservative replication from the hemimethylated DNA sites (maintenance activity) (Fig. 29.1). Maintenance methylation activity is necessary to preserve DNA methylation after every cellular DNA replication cycle. In fact, in absence of DNA methyltransferases, the replication machinery itself would produce unmethylated daughter strands, then leading to passive demethylation. On the other hand, DNMT3a and DNMT3b are highly methylated in embryonic cells and possess de novo methyltransferase activity rather than maintenance methyltransferase activity. This implies that these enzymes are responsible for initiating DNA methylation during early embryonic

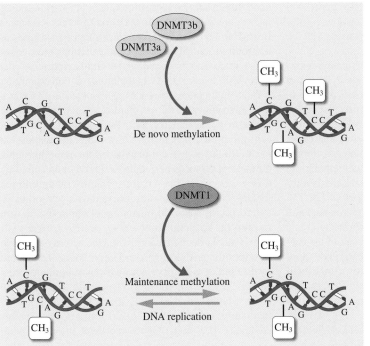

Fig. 29.1 De novo methylation and maintenance methylation

development by transferring a methyl group to previously unmethylated genomic regions.

DNMT1 exhibits a preference for hemimethylated substrates and appears to be involved in restoring the parental DNA methylation pattern in the newly synthesized DNA daughter strand, then ensuring the methylation status of CpG islands through multiple cell generations [29.17]. It has been reported that loss of DNMT1 function results in embryonic lethality in mice [29.18], then confirming the important role of this enzyme in proper cell functioning and in the cell development. Moreover, during embryogenesis and germ cell development de novo DNA methylation is carried out by DNMT3a and DNMT3b enzymes. The importance of the differential role played by these enzymes is demonstrated by the fact that DNMT3a knock-out mice die shortly after birth, while embryonic lethality is reported in the absence of DNMT3b in mice [29.19]. These observations suggest that DNMT3b may play a major role during early developmental stages, whereas DNMT3a may be the first player during the methylation process of DNA sequences, critical for the later development, or after birth [29.19].

During the mammalian development, the expression of various genes is controlled by DNA methylation. Moreover, reversible epigenetic alterations lead to selective use of genome information through transcription activation or inactivation of functional genes during gametogenesis, embryogenesis, development, and differentiation [29.20].

To understand cell development and differentiation, it is crucial to know how gene expression patterns are modulated and when and how they are established and maintained. The activities of replication and transcriptional machineries provide opportunities for epigenetic changes that are required for differentiation and development [29.21]. During differentiation changes in chromatin structure and nuclear organization establish heritable patterns of gene expression in response to signals even if similar or identical signaling pathways can produce different development outcome that cells interpret in very different ways depending on their developmental history [29.22].

With the exception of imprinted genes, widespread removal of epigenetic marks occurs following fertilization when maternal and paternal genomes undergo extensive demethylation to ensure the pluripotency of the developing zygote. Then, de novo methylation occurs just before the implantation [29.23]. However, changes in the methylation status, associated with cell differentiation, are established at different times during the embryonic development. Moreover, the methylation and demethylation reactions during germ cell differentiation and then after the fertilization are extensively involved in imprinting mechanisms [29.24, 25].

Imprinting refers to the selective inactivation mediated by epigenetic mechanisms (DNA methylation, in particular) of a parental specific allele, ensuring that in the offspring a specific locus is exclusively expressed from either maternal or paternal genomes [29.25, 26].

During early development, both maternal and paternal genomes undergo an impressive epigenetic reprogramming, most notably through DNA demethylation. After implantation, the methylation patterns are then reinitiated through de novo methylation activity [29.23]. Imprinted genes play critical roles in developmental and cellular processes; therefore, loss of imprinting (LOI) due to epigenetic alterations can result in loss of the normal expression pattern of specific parental allele, and, in cancer, it can lead to activation of grown-imprinted genes (e.g., insulin-like growth factor II, IGFII), as well as to silencing of tumor suppressor genes, and, then, to cancer development [29.26, 27].

In the past years different studies have demonstrated that DNA methylation causes changes in chromatin structure, modifying the interactions between the DNA and activating or repressing transcription factors (or complexes). Transcription factors often do not show a preference for unmethylated sequences. However, in gene transcriptional regulation CTCF (CCCTC-binding factor)-mediated, DNA methylation of target sequences is crucial. CTCF is a chromatin insulator that binds to imprinting control regions (ICRs) of imprinted genes (e.g., Igf2) and is essential for transcription of the INK/ARF locus. DNA methylation of corresponding CTCF-binding sites abolishes its association and therefore contributes to the permanent silencing of several genes at the respective loci. Interestingly, it was recently reported that in lung fibroblasts the expression of the Rb2/p130 gene, a member of the retinoblastoma tumor suppressor gene family is directly controlled by CTCF [29.28]. In these cells, CTCF by binding to the Rb2/p130 gene promoter is able to induce and/or maintain a specific local chromatin organization that in turn governs the transcriptional activity of Rb2/p130 gene itself. However, in lung cancer cells the activity of CTCF in controlling Rb2/p130 gene expression is impaired by BORIS, a CTCF-paralog, which by binding to the Rb2/p130 gene could trigger changes in the chromatin asset established by CTCF, thereby affecting CTCF regulatory activity on Rb2/p130 transcription. In this context, CTCF and BORIS, represent the main players whose *epigenetic activity* is crucial for explaining how alterations in the mechanism regulating Rb2/p130 gene expression may accelerate the progression of lung tumors, or favor the onset of recurrence after cancer treatment.

29.2 DNA Methylation and Cancer

DNA methylation may have major regulatory consequences in the regulation of gene expression during development and proper cell functioning. In fact, it has been reported that aberrant DNA methylation is tightly connected to a wide variety of human malignancies by playing a key role in the silencing of cancer-related genes and thereby affecting numerous cellular processes (including cell cycle checkpoint, apoptosis, signal transduction, and cell adhesion). Moreover, DNA methylation of promoter CpG islands is strongly associated with gene silencing and is known to be a frequent cause of expression loss not only in tumor suppressor genes but also in other genes involved in tumor formation [29.4, 5, 29]. However, the inactivation of certain tumor-suppressor genes occurs as a consequence of the gene promoter-associated (CpG island-specific) hypermethylation within the promoter regions, and numerous studies have shown that a broad range of genes is silenced by DNA methylation in different cancer types. On the other hand, the overall loss of 5-methyl-cytosine (global hypomethylation), which induces genomic instability, also contributes to cell transformation [29.5, 30, 31]. Both hyper and hypomethylation events can be observed in cancer. Generally, hypermethylation of CpG islands in promoters of genes that mainly act as tumor suppressor gene in normal cells has been shown as a critical hallmark during tumorigenesis [29.32], whereas a global decrease in methylated CpG content leading to genomic instability and, sometimes, to activation of silenced oncogenes has been observed in different tumors [29.33]. However, while it is clear that the hypermethylation of gene promoters is in turn associated with gene inactivation, the exact consequences of genome-wide hypomethylation are still under debate [29.34].

The global loss of methylcytosine is one of the earliest epigenetic modifications in cancer [29.35], even if DNA hypomethylation of individual genes in cancer cells is rather uncommon.

Although the mechanism responsible for genomic hypomethylation is unclear, it has been proposed that hypomethylation may be a result of deregulation of presumed demethylates enzymes or a consequence of the alteration in the DNMTs maintenance activity [29.36]. However, in cancer no active demethylating enzyme has been discovered and no clear evidence of any reduction in DNMTs activity has also been found. Despite the fact that the pathway leading to hypomethylation in cancer is still unknown, it is clear that hypomethylation is a hallmark of most cancer genomes and that leads to chromosomal instability [29.37].

In a normal cell, pericentromeric heterochromatin is highly methylated [29.38]. Satellite sequences, repetitive genomic sequences are silenced, thereby ensuring genomic integrity and stability. In a wide variety of tumors, however, this mechanism is disrupted and loss of DNA methylation of normally inactivated regions occurs [29.39]. As a consequence, there is a greater chance of undesired mitotic recombination. Transposable elements are then reactivated and can integrate at random sites in the genome, leading to mutagenesis and genomic instability [29.2].

Recently, it was suggested that the variants of DNMT3B, catalytically inactive, may play an important role in the hypomethylation process. In fact, these inactive DNMT3B variants may negatively regulate the DNA methyltransferase activity. DNMT3B variants may act by abolishing the DNA methylation mechanism either by titrating the DNMT3B binding partners, thus interfering with the DNA methylation machinery or by competing with the DNMT3B active form to bind to the target DNA sequences [29.40, 41].

DNA methylation abnormalities may occur in a very early, initial stage of neoplasia and may even affect the genes that are not directly related to tumorigenesis but which confer a *mutator phenotype* [29.42].

In cancer, despite global hypomethylation, certain genes undergo inactivation as a consequence of CpG islands' hypermethylation in regulatory regions, which are unmethylated in nonmalignant tissues [29.43]. Genes that acquire hypermethylation in regulatory regions are involved in a variety of important cellular pathways, which not only involves cell cycle related genes but also various genes associated with DNA repair processes and/or with genes whose pro-

tein products have pro-apoptotic functions [29.44, 45].

Nowadays, it is beyond any doubt that DNA methylation-associated silencing plays an important role in tumorigenesis and it is considered a hallmark of all types of human cancer. Moreover, it has been estimated that between 0.5–3% of all genes carrying CpG-rich promoter sequences may be silenced by DNA methylation in several types of cancer [29.46–48].

However, the mechanisms for aberrant CpG island methylation in cancer are unclear. It is generally observed that the hypermethylation of CpG islands is closely linked to modification of local chromatin structure and serve as a platform in triggering silencing of transcription. On the other hand, gene inactivity imposed by changes in chromatin structure or histone modification may predispose to DNA methylation. In addition, specific DNA sequences within CpG islands may be associated with the methylation process. However, so far, it is not clear whether these sequences are in vivo associated with DNA binding proteins that in some ways attract methylation.

In addition, recent genome-wide studies have revealed the presence of many genes whose methylation in GpG islands is considered to be a consequence of or in association with carcinogenesis (passenger methylation) leading to a new paradigm to discriminate them from the aberrant DNA methylation events, which are being causally involved in carcinogenesis (driver methylation) [29.49–51]. This fact clearly demonstrates the importance of carefully analyzing the role of any newly detected gene in terms of its role in carcinogenesis. Moreover, driver methylation is a methylation event that promotes tumorigenesis. Most likely, it is occurs during early stages of tumorigenesis and it is characterized by both tumor suppressor gene alteration and/or direct or indirect oncogenes activation [29.52].

However, all altered cellular phenotypes are not the immediate consequences of methylation-associated events, but they may be the result of different events that, in turn, may promote additional gene mutations.

Until now, genome-wide methylation-profiling studies have provided more detailed information regarding the methylation changes occurring in different diseases. However, the identification of critical driver methylation events, contributing to the transformed phenotype, versus passenger events still remains an open question [29.53].

29.3 Histone Modifications

In the eukaryotic cell, DNA is wrapped around an octamer of histone proteins forming a DNA–protein structure, termed chromatin. The basic chromatin structure unit, called nucleosome, consists of 146 bp of DNA wrapped around a histone core octamer composed by dimers of histone H2A, H2B, H3, and H4 [29.54]. Histone proteins are each composed of a structured globular central domain, which is in close contact to DNA, allowing the histones to interact together, and an unstructured amino-terminal domain whose 20–30 amino acids represent the histone *tail* [29.55, 56]. Histone tails extend from the central domain of the nucleosome and represent the sites where the post-translational modifications (such as methylation, acetylation, phosphorylation, ubiquitination) occur. It is accepted that histone tails bind the DNA (negatively charged) through charge interactions or contribute to chromatin compaction by mediating the interaction between nucleosomes [29.56, 57].

The packaging of DNA into nucleosome is critical to preserve both the genetic information and the integrity of the genome. In this context, the dynamic modulation of chromatin structure by a combination of different histone modifications in their globular and N-terminal domains represent a key component in the regulation of different cellular processes (such as gene transcription, DNA replication and repair, chromosome recombination, and segregation). Moreover, chromatin modifications (histone markings) can dynamically modulate the chromatin structure through mechanisms that are believed to facilitate the opening of chromatin throughout reversible modifications of core histones, then dictating the transcriptional status (active or repressed) of a gene [29.58, 59].

The chromatin structure surrounding unmethylated, transcriptionally active genes differs from that of methylated, silenced genes [29.60]. Methylation of specific residues on the tail of histone H3 is critical for maintaining the appropriate chromatin configuration. In fact, methylation of lysine 9 (K9) residues on histone 3 (H3) facilitates transcriptional repression, whereas methylation of lysine 4 on H3 is associated with transcriptionally active euchromatin. DNA methylation of promoter regions is associated with modification of repressive histone marks, including deacetylation of lysine residues and trimethylation of histones H3K9 and H3K27. Furthermore, the histone modifications seem to work in an ordered and consistent manner in regulating several key cellular processes, leading to the concept of the *histone code* [29.58, 61]. The different histone modifications present on histone tails are believed to generate a *code*, readable by different cellular machineries, which regulates chromatin function by affecting the nucleosome structural dynamics. Moreover, by defining the accessibility of the transcription machinery to genes and gating the accessibility of the genome to the other machineries, this histone code can, then, dictate the gene expression patterns, as well as the different cellular outcomes, adding a layer of complexity in the recruitment of the epigenetic modifiers and in the regulation of cellular processes [29.61]. It is clear that a complex interplay exists between the histone code and the susceptibility to DNA methylation, and importantly, defects disturbing this *interplay* may be crucial in leading to DNA hypermethylation in cancer cells.

In eukaryotes, in response to different stimuli, the regulation of gene expression requires chromatin modifiers and specific histone modifications, which assist and open the way for transcription factors to the DNA, leading to an *open* permissive chromatin, and to the activation of a cascade of events resulting in gene transcription. Conversely, histone modifications can also promote transcriptional repression by inducing a condensed chromatin state and closing DNA accessibility to transcription factors. Therefore, histone modifications can dictate the dynamic transitions between transcriptionally active or transcriptionally silent chromatin states.

Chromatin remodeling complexes and enzymes involved in post-translational modification of chromatin histone components play important roles in transcriptional regulation. Histone lysine acetylation has long been correlated with transcriptional activation [29.62, 63]. Indeed, through decreasing the positive charge of histone tails, acetylation is believed to weaken histone–DNA [29.64] or nucleosome–nucleosome interaction [29.65–68], thus relaxing the chromatin structure and inducing a chromatin conformational change that opens the way for the transcription machinery.

Acetylation is a reaction based on the introduction of an acetyl functional group into a chemical compound. Lysine acetylation is a reversible modification referred to the e-amino group of a lysine residue, controlled by two types of enzymes: histone acetyltransferases (HATs), which transfer an acetyl group from acetyl-CoA, that is specifically recognized and bound by the Arg/Gln-X-X-Gly-X-Gly/Ala segment of HATs, to the e-amino groups on the N-terminal tails of histones [29.69], and histone deacetylases (HDACs), which reverse this modification [29.70–72]. The dynamic equilibrium of lysine acetylation in vivo is governed by the opposing actions of histone acetyl transferases and histone deacetyl transferases. Deacetylation of histones by HDACs results in a reduction in the space between nucleosomes, leading to a closed (heterochromatin-like) chromatin conformation that diminishes accessibility for transcription factors and correlates with transcriptional repression [29.73, 74].

However, acetylated histones may serve as docking sites for recruitment of other transcription regulators influencing other histone modifications [29.61, 75]. The latter concept reinforces the *histone code hypothesis* that histones act as signaling platforms, integrating upstream signaling pathways to elicit appropriate nuclear response, depending on the modification status.

Histone methylation is a covalent modification that occurs on the side-chain nitrogen atoms of lysine and arginine of histones. Histone H3, followed by histone H4, represents one of the most heavily methylated histones. The methylation on lysine residues is mediated by the lysine-specific SET domain-containing histones methyltransferases (HMTs), which have a strong homology in a 140-amino acid catalytic domain known as the SET (su(var), enhancer of zeste, and trithorax) domain, and by the non-SET containing lysine HMTs [29.76]. Lysine methylation can occur in mono- (me), di- (2me), and tri- (3me) methylated forms.

Depending upon a particular lysine, methylation may serve as a marker of transcriptionally active euchromatin or transcriptionally repressed heterochromatin. For instance, histone H3 K9, H4 K20, and H3 K27 methylation are mainly involved in formation of heterochromatin, whereas histone H3 K4, H3 K36, and H3 K79 methylation correlate with euchromatin [29.77, 78]. In addition, arginine methylation of histones H3 and H4 induces a repressive chromatin state and recruits repressive complexes to transcription sites [29.77, 79]. On the other hand, lysine demethylation is associated with transcriptional activation [29.80]. Also, trimethylation of histone H3 Lys 9 (H3K9me3) is associated with the compaction and transcriptionally repressive features of pericentromeric heterochromatin, while dimethylation of histone H3 Lys 9 (H3K9me2) is a mark of euchromatic gene silencing [29.81].

Therefore, histone modifications are dynamic and tightly controlled processes that, in response to intrinsic or external stimuli, entail an adequate amount of organization and coordination in regulating cellular processes, such as gene transcription, DNA repair, and DNA repli-

cation. Consistent with the important roles that histone modifications play in key cellular processes, deregulation of histone marking has been intimately linked to a number of human malignancies [29.34, 82]. Moreover, aberrations in histone methyltransferases may alter the gene expression balance leading to alterations in critical cellular processes, resulting in cellular transformation and malignant outgrowth. In fact, alterations in these chromatin-based processes may lead to mutations in oncogenes, tumor suppressor genes, or DNA repair genes, resulting in genomic instability and oncogenic transformation. In addition, deregulation of histone modifications has been linked to abnormal cellular proliferation, invasiveness, metastatic progression, and therapy resistance, reinforcing the idea that deregulation of these processes contributes to cancer development and progression [29.82, 83]. Recently, the existence of a close and self-reinforcing crosstalk and interdependence between histone marks has been suggested [29.10, 84]. DNA methylation is associated with histone modifications (some of the lysine residues that are methylated in histones H3 and H4 are also found to be substrates for acetylation) and the interaction between these epigenetic modifications plays a critical role in regulating the genome function by changing chromatin architecture. Moreover, drastic alteration in one of these two epigenetic mechanisms inevitably affects the other. While some studies have suggested that DNA methylation patterns guide histone modifications during gene silencing; other studies argue that DNA methylation takes its cues primarily from histone modification states [29.10, 84]. Even if different studies have recently initiated disclosure of the mechanisms underlying these phenomena, it is still unclear which epigenetic event (DNA methylation or histone modification), and in which hierarchical order, upon a given exogenous or endogenous signal, initiates the gene silencing in tumor cells.

Chromatin inactivation mediated by histone deacetylation and DNA methylation has been reported as a critical component of estrogen receptor alpha (ER-α) gene silencing in human breast cancer cells. Several studies have revealed that chromatin remodeling enzymes can form different multimolecular complexes, which, by altering the local chromatin status, can modulate DNA accessibility and regulate gene expression on a specific promoter [29.85, 86]. In breast cancer cells, a different set of enzymatic activities is recruited in multimolecular complexes on the ER-α promoter, suggesting that the recruitment of SUV39H1, HDAC1, and histone acetyltransferase p300 by a retinoblastoma family member pRb2/p130 regulates the ER-α expression level in ER-α-positive breast cancer cells, and that further recruitment of DNMT1 (with the concomitant release of p300/CBP) is required for long-term ER-α gene silencing in ER-negative breast cancer cells [29.85, 86]. Interestingly, in these complexes the identity and the temporal specificity of recruited enzymatic activities can control the chromatin organization by inducing different acetylation and methylation levels, then causing important effects in the silencing or transcriptional regulation of the ER-alpha gene. In addition, it has been demonstrated that the sequence of epigenetic events for establishing and maintaining the silenced state of ER-α gene can be locus or pathway specific and that the remodeling of local chromatin structure of ER-α gene by pRb2/p130 multimolecular complexes may influence its susceptibility to specific DNA methylation [29.85, 86]. The specificity of the recruited enzymes may be a key element in determining a specific pattern of local chromatin remodeling that may dictate different *transcription modulation environments* closely correlated to ER-α expression in ER-α-positive and ER-α-negative breast cancer cells. In this context, the presence of a specific pRb2/p130-mediated multimolecular complex on the ER-α promoter strongly correlates with its methylation status, and the relationship between DNA methylation and chromatin modification may be viewed as a complex feedback loop.

29.4 Epigenetics and Cancer Therapy

The presence of epigenetic changes in many human malignancies has triggered an impressive quest for the development of *epigenetic drugs* and epigenetic therapies. Moreover, during the past years, a number of agents, including DNA methyltransferase inhibitors and HAT inhibitors, both of which aim to reactivate epigenetically silenced tumor suppressors, DNA repair genes, and other cancer-associated genes, have been investigated.

These *epigenetic approaches* were directed at modifying DNA methylation profiles and histone modification status in cancer cells, and are based on specific properties of different chemical agents affecting the activity of enzymes involved in the establishment and maintenance of epigenetic marking. Among these agents, DNA methyltransferase inhibitors and histone deacetylase inhibitors represent the most extensively studied epigenetic agents [29.87, 88].

5-azacytidine (5-aza-CR) and 5-aza-2-deoxycytidine (5-aza-CdR) are nucleoside analogs of cytosine, able to efficiently inhibit DNA methyltransferases and then lower DNA methylation levels, with subsequent expression reactivation of those genes (especially tumor suppressor genes) that were aberrantly silenced through epigenetic mechanism during progression of the disease. Moreover, during DNA replication, these molecules are incorporated in the cytosine position with consequent trapping and isolation of DNA methyltransferases [29.89].

Currently, these two DNA demethylating agents, 5-azacytidine and decitabine (5-aza-2′-deoxycytidine) have been approved by the US Food and Drug Administration (FDA) and are currently available for clinical use. However, these drugs were used in relatively high doses (even if more recent studies using significantly lower doses showed encouraging results) and their disadvantage consists in their myelotoxic effect, due to their incorporation into the DNA molecule rather than their DNA hypomethylation effect [29.90].

Different non-nucleoside agents, such as specific antisense oligonucleotides targeting DNMT enzymes, were tested in different models. However, even when less toxic and better tolerated, these agents failed to provide encouraging results since they showed no efficacy at all [29.91–93].

For these reasons, during the past years, investigations have focussed also on the creation and use of drugs able to inhibit the activity of histone deacetylase enzymes.

The histone deacetylase inhibitors (HDACis) represent a novel class of targeted drugs that alter the acetylation status of several cellular proteins, leading to the transcription modulation of specific genes whose expression causes inhibition of cancer cell growth [29.94]. Moreover, since treatment of cells with HDACis has been shown to induce differentiation, growth arrest, and/or apoptosis in a broad spectrum of transformed cells in culture and tumors in animals, new drugs have recently been translated and approved for clinical use in cancer patients [29.95, 96]. However, clinical responses have been shown to not always be reliable with the results observed in preclinical models. For these reasons, intense efforts have been made to develop small molecule inhibitors of HDACs, designed to specifically interfere with the catalytic domain, block substrate recognition, re-induce gene expression mainly in genes with tumor suppressor function, then reverting the aberrant epigenetic status of tumor cells, and inducing cell cycle arrest, differentiation, or apoptosis [29.97].

The most studied drug showing to selectively inhibit the class I and II mammalian HDACs (but not class III HDACs) is the trichostatin A (TSA). TSA was extensively used in vitro to alter gene expression by interfering with the removal of acetyl groups from histones, and therefore altering the ability of DNA transcription factors to access the DNA molecules inside chromatin. TSA promotes the expression of apoptosis-related genes, leading to cancerous cells surviving at lower rates, thus slowing the progression of cancer [29.98].

Vorinostat represents the first HDACi approved by the FDA for the treatment of patients with relapsed or refractory cutaneous T cell lymphoma [29.99]. At the cellular level, the antiproliferative effects of vorinostat are due to the increased acetylation of several proteins, including the core nucleosomal histones and other proteins (such as BCL6, p53, and HSP90), as well as in the increased expression level of p21, observed in different tumors and peripheral blood cells [29.100].

Unfortunately, although tremendous promise was shown in preclinical models, vorinostat has given a very limited clinical response in most cancer patients. For these reasons, new drugs, named as second generation drugs, rationally designed with improved properties and specificity, have been generated to overcome some limitations of the first generation HDACi and tested in preclinical trials for the cure of hematological malignancies.

Scientists are eager to know whether these new drugs may provide better clinical profiles for the cure of cancer patients. However, it still remains unclear whether selective inhibition of specific HDACs represents an advantage in cancer therapy, since the selectivity of specific HDACs in tumor maintenance and minimization of the side effects due to HDAC inhibition remain to be conclusively established. However, recently, the existence of an intimate crosstalk between different epigenetic mechanisms led scientists to investigate combinatorial therapies coupling DNA methylation inhibitors with HDAC inhibitors, in order to obtain more effective and less toxic drugs, to be useful tools in the cure not only of hematological malignancies but also in solid tumors.

29.5 Conclusions

Epigenetic therapies are based on the identification of drugs able to interfere with the activity of enzymes responsible for the epigenetic alterations occurring in tumor cells. However, the use of epigenetic drugs in a modality of treatment not based on a targeted approach, the use of inappropriate preclinical models, and the lack of specific biomarkers for directing therapeutic choices have represented big failures in cancer treatment.

In the recent years, the development of high-throughput and genome-wide methods has opened the opportunity to identify and then better characterize the epigenetic alterations affecting the epigenome of cancer cells; however, much work is still required to produce a treatment with satisfactory results also for solid tumors.

Further studies are necessary for understanding how tumors develop and progress, and to provide crucial information that will allow scientists to design novel and efficient anticancer strategies.

References

29.1   R. Taby, J.P. Issa: Cancer epigenetics, CA Cancer J. Clin. **60**(6), 376–392 (2010)

29.2   A. Portela, M. Esteller: Epigenetic modifications and human disease, Nat. Biotechnol. **28**(10), 1057–1068 (2010)

29.3   S. Winter, W. Fischle: Epigenetic markers and their cross-talk, Essay Biochem. **48**(1), 45–61 (2010)

29.4   P.A. Jones, S.B. Baylin: The fundamental role of epigenetic events in cancer, Nat. Rev. Genet. **3**(6), 415–428 (2002)

29.5 P.A. Jones, S.B. Baylin: The epigenomics of cancer, Cell **128**(4), 683–692 (2007)

29.6 N. Avvakumov, A. Nourani, J. Côté: Histone chaperones: Modulators of chromatin marks, Mol. Cell **41**(5), 502–514 (2011)

29.7 C. Cinti, M. Macaluso, A. Giordano: Tumor-specific exon 1 mutations could be the 'hit event' predisposing Rb2/p130 gene to epigenetic silencing in lung cancer, Oncogene **24**(38), 5821–5826 (2005)

29.8 L. Bai, A.V. Morozov: Gene regulation by nucleosome positioning, Trends Genet. **26**(11), 476–483 (2010)

29.9 R. Jaenisch, A. Bird: Epigenetic regulation of gene expression: How the genome integrates intrinsic and environmental signals, Nat. Genet. **33**, 245–254 (2003)

29.10 T. Vaissièe, C. Sawan, Z. Herceg: Epigenetic interplay between histone modifications and DNA methylation in gene silencing, Mutat. Res. **659**(1–2), 40–48 (2008)

29.11 J.P. Thomson, P.J. Skene, J. Selfridge, T. Clouaire, J. Guy, S. Webb, A.R. Kerr, A. Deaton, R. Andrews, K.D. James, D.J. Turner, R. Illingworth, A. Bird: CpG islands influence chromatin structure via the CpG-binding protein Cfp1, Nature **464**(7291), 1082–1086 (2010)

29.12 C. Lanzuolo, V. Orlando: The function of the epigenome in cell reprogramming, Cell Mol. Life Sci. **64**(9), 1043–1062 (2007)

29.13 E.T. Liu: Functional genomics of cancer, Curr. Opin. Genet. Dev. **18**(3), 251–256 (2008)

29.14 D. Van Heemst, P.M. den Reijer, R.G. Westendorp: Ageing or cancer: A review on the role of caretakers and gatekeepers, Eur. J. Cancer **43**(15), 2144–2152 (2007)

29.15 A.P. Bird: CpG-rich islands and the function of DNA methylation, Nature **321**(6067), 209–213 (1986)

29.16 R.A. Hinshelwood, J.R. Melki, L.I. Huschtscha, C. Paul, J.Z. Song, C. Stirzaker, R.R. Reddel, S.J. Clark: Aberrant de novo methylation of the p16INK4A CpG island is initiated post gene silencing in association with chromatin remodelling and mimics nucleosome positioning, Hum. Mol. Genet. **18**(16), 3098–3109 (2009)

29.17 A. Hermann, R. Goyal, A. Jeltsch: The Dnmt1 DNA-(cytosine-C5)-methyltransferase methylates DNA processively with high preference for hemimethylated target sites, J. Biol. Chem. **279**, 48350–48359 (2004)

29.18 E. Li, T.H. Bestor, R. Jaenisch: Targeted mutation of the DNA methyltransferase gene results in embryonic lethality, Cell **69**(6), 915–926 (1992)

29.19 M. Okano, D.W. Bell, D.A. Haber, E. Li: DNA methyltransferases Dnmt3a and Dnmt3b are essential for de novo methylation and mammalian development, Cell **99**(3), 247–257 (1999)

29.20 J. Turek-Plewa, P.P. Jagodziński: The role of mammalian DNA methyltransferases in the regulation of gene expression, Cell Mol. Biol. Lett. **10**, 631–647 (2005)

29.21 L. Chakalova, E. Debrand, J.A. Mitchell, C.S. Osborne, P. Fraser: Replication and transcription: Shaping the landscape of the genome, Nat. Rev. **6**(9), 669–677 (2005)

29.22 K.L. Arney, A.G. Fisher: Epigenetic aspects of differentiation, J. Cell Sci. **117**(19), 4355–4363 (2004)

29.23 W. Reik: Stability and flexibility of epigenetic gene regulation in mammalian development, Nature **447**(7143), 425–432 (2007)

29.24 J.M. Trasler: Epigenetics in spermatogenesis, Mol. Cell Endocrinol. **306**(1–2), 33–36 (2009)

29.25 M.V. Koerner, D.P. Barlow: Genomic imprinting-an epigenetic gene-regulatory model, Curr. Opin. Genet. Dev. **20**(2), 164–170 (2010)

29.26 S.K. Kota, R. Feil: Epigenetic transitions in germ cell development and meiosis, Dev. Cell **19**(5), 675–686 (2010)

29.27 M. Berdasco, M. Esteller: Aberrant epigenetic landscape in cancer: How cellular identity goes awry, Dev. Cell **19**(5), 698–711 (2010)

29.28 F.P. Fiorentino, M. Macaluso, F. Miranda, M. Montanari, A. Russo, L. Bagella, A. Giordano: CTCF and BORIS regulate Rb2/p130 gene transcription: A novel mechanism and a new paradigm for understanding the biology of lung cancer, Mol. Cancer Res. **9**(2), 225–233 (2011)

29.29 J.G. Herman, S.B. Baylin: Gene silencing in cancer in association with promoter hypermethylation, N. Engl. J. Med. **349**(21), 2042–2054 (2003)

29.30 M.F. Fraga, R. Agrelo, M. Esteller: Cross-talk between aging and cancer: The epigenetic language, Ann. N.Y. Acad. Sci. **1100**, 60–74 (2007)

29.31 F.I. Daniel, K. Cherubini, L.S. Yurgel, M.A. de Figueiredo, F.G. Salum: The role of epigenetic transcription repression and DNA methyltransferases in cancer, Cancer **117**(4), 677–687 (2011)

29.32 F. Chik, M. Szyf: Effects of specific DNMT gene depletion on cancer cell transformation and breast cancer cell invasion; toward selective DNMT inhibitors, Carcinogenesis **32**(2), 224–232 (2011)

29.33 I.P. Pogribny: Epigenetic events in tumorigenesis: Putting pieces together, Exp. Oncol. **32**(3), 132–136 (2010)

29.34 N. Sinčč, Z. Herceg: DNA methylation and cancer: Ghosts and angels above the genes, Curr. Opin. Oncol. **23**(1), 69–76 (2011)

29.35 A.P. Feinberg, B. Vogelstein: Hypomethylation of ras oncogenes in primary human cancers, Biochem. Biophys. Res. Commun. **111**(1), 47–54 (1983)

29.36 S.S. Palii, K.D. Robertson: Epigenetic control of tumor suppression, Crit. Rev. Eukaryot. Gene Expr. **17**(4), 295–316 (2007)

29.37 J. Veeck, M. Esteller: Breast cancer epigenetics: From DNA methylation to microRNAs, J. Mammary Gland Biol. Neoplasia **15**(1), 5–17 (2010)

29.38 M. Ehrlich: DNA hypomethylation, cancer, the immunodeficiency, centromeric region instability, facial anomalies syndrome and chromosomal rearrangements, Nutrition **132**(8 Suppl), 2424S–2429S (2002)

29.39 M. Ehrlich: DNA methylation and cancer-associated genetic instability, Adv. Exp. Med. Biol. **570**, 363–392 (2005)

29.40 K.R. Ostler, E.M. Davis, S.L. Payne, B.B. Gosalia, J. Expósito-Céspedes, M.M. Le Beau, L.A. Godley: Cancer cells express aberrant DNMT3B transcripts encoding truncated proteins, Oncogene **26**(38), 5553–5563 (2007)

29.41 D.J. Weisenberger, M. Velicescu, J.C. Cheng, F.A. Gonzales, G. Liang, P.A. Jones: Role of the DNA methyltransferase variant DNMT3b3 in DNA methylation, Mol. Cancer Res. **2**(1), 62–72 (2004)

29.42 K. Imai, H. Yamamoto: Carcinogenesis and microsatellite instability: The interrelationship between genetics and epigenetics, Carcinogenesis **29**(4), 673–680 (2008)

29.43 P.W. Laird, R. Jaenisch: DNA methylation and cancer, Hum. Mol. Genet. **3**, 1487–1495 (1994)

29.44 J. Felsberg, N. Thon, S. Eigenbrod, B. Hentschel, M.C. Sabel, M. Westphal, G. Schackert, F.W. Kreth, T. Pietsch, M. Loeffler, M. Weller, G. Reifenberger, J.C. Tonn: Promoter methylation and expression of MGMT and the DNA mismatch repair genes MLH1, MSH2, MSH6, and PMS2 in paired primary and recurrent glioblastomas, Int. J. Cancer **129**(3), 659–670 (2011)

29.45 K. Ramachandran, H. Miller, E. Gordian, C. Rocha-Lima, R. Singal: Methylation-mediated silencing of TMS1 in pancreatic cancer and its potential contribution to chemosensitivity, Anticancer Res. **30**(10), 3919–3925 (2010)

29.46 T.A. Rauch, X. Zhong, X. Wu, M. Wang, K.H. Kernstine, Z. Wang, A.D. Riggs, G.P. Pfeifer: High-resolution mapping of DNA hypermethylation and hypomethylation in lung cancer, Proc. Natl. Acad. Sci. USA **105**(1), 252–257 (2008)

29.47 S. Tommasi, D.L. Karm, X. Wu, Y. Yen, G.P. Pfeifer: Methylation of homeobox genes is a frequent and early epigenetic event in breast cancer, Breast Cancer Res. **11**(1), R14 (2009)

29.48 Y. Koga, M. Pelizzola, E. Cheng, M. Krauthammer, M. Sznol, S. Ariyan, D. Narayan, A.M. Molinaro, R. Halaban, S.M. Weissman: Genome-wide screen of promoter methylation identifies novel markers in melanoma, Genome Res. **19**(8), 1462–1470 (2009)

29.49 T. Ushijima: Detection and interpretation of altered methylation patterns in cancer cells, Nat. Rev. Cancer **5**, 223–231 (2005)

29.50 M. Weber, J.J. Davies, D. Wittig, E.J. Oakeley, M. Haase, W.L. Lam, D. Schübeler: Chromosome-wide and promoter-specific analyses identify sites of differential DNA methylation in normal and transformed human cells, Nat. Genet. **37**, 853–862 (2005)

29.51 T. Nakajima, S. Enomoto, T. Ushijima: DNA methylation: A marker for carcinogen exposure and cancer risk, Environ. Health Prev. Med. **13**, 8–15 (2008)

29.52 M.R. Estécio, J.P. Issa: Dissecting DNA hypermethylation in cancer, FEBS Letters **585**(13), 2078–2086 (2011)

29.53 P.A. Cowin, M. Anglesio, D. Etemadmoghadam, D.L. Bowtell: Profiling the cancer genome, Annu. Rev. Genomics Hum. Genet. **11**, 133–159 (2011)

29.54 R.D. Kornberg, Y. Lorch: Twenty-five years of the nucleosome, fundamental particle of the eukaryote chromosome, Cell **98**(3), 285–294 (1999)

29.55 T.H. Eickbush, E.N. Moudrianakis: The histone core complex: An octamer assembled by two sets of protein–protein interactions, Biochemistry **17**(23), 4955–4964 (1978)

29.56 A.P. Wolffe: Chromatin structure, Adv. Genome Biol. **5B**, 363–414 (1998)

29.57 P. Cheung, C.D. Allis, P. Sassone-Corsi: Signaling to chromatin through histone modifications, Cell **103**, 263–271 (2000)

29.58 B.D. Strahl, C.D. Allis: The language of covalent histone modifications, Nature **403**(6765), 41–45 (2000)

29.59 T. Kouzarides: Chromatin modifications and their function, Cell **128**(4), 693–705 (2007)

29.60 R. Margueron, D. Reinberg: Chromatin structure and the inheritance of epigenetic information, Nat. Rev. Genet. **11**(4), 285–296 (2010)

29.61 T. Jenuwein, C.D. Allis: Translating the histone code, Science **293**(5532), 1074–1080 (2001)

29.62 B.M. Turner: Histone acetylation and control of gene expression, J. Cell Sci. **99**, 13–20 (1991)

29.63 A.P. Wolffe, D. Pruss: Targeting chromatin disruption: Transcription regulators that acetylate histones, Cell **84**, 817–819 (1996)

29.64 P.A. Grant, D. Schieltz, M.G. Pray-Grant, D.J. Steger, J.C. Reese, J.R. Yates, J.L. Workman: A subset of TAF$_{II}$s are integral components of the SAGA complex required for nucleosome acetylation and transcriptional stimulation, Cell **94**(1), 45–53 (1998)

29.65 T.M. Fletcher, J.C. Hansen: The nucleosomal array: Structure/function relationships, Crit. Rev. Eukaryot. Gene Expr. **6**(2–3), 149–188 (1996)

29.66 L.C. Lutter, L. Judis, R.F. Paretti: Effects of histone acetylation on chromatin topology in vivo, Mol. Cell Biol. **12**(11), 5004–5014 (1992)

29.67 V.G. Norton, B.S. Imai, P. Yau, E.M. Bradbury: Histone acetylation reduces nucleosome core particle linking number change, Cell **57**(3), 449–457 (1998)

29.68   V.G. Norton, K.W. Marvin, P. Yau, E.M. Bradbury: Nucleosome linking number change controlled by acetylation of histones H3 and H4, J. Biol. Chem. **265**(32), 19848–19859 (1990)

29.69   P. Loidl: Histone acetylation: Facts and questions, Chromosoma **103**(7), 441–449 (1994)

29.70   J.E. Brownell, C.D. Allis: An activity gel assay detects a single, catalytically active histone acetyltransferase subunit in Tetrahymena macronuclei, Proc. Natl. Acad. Sci. USA **92**(14), 6364–6368 (1995)

29.71   S.E. Rundlett, A.A. Carmen, R. Kobayashi, S. Bavykin, B.M. Turner, M. Grunstein: HDA1 and RPD3 are members of distinct yeast histone deacetylase complexes that regulate silencing and transcription, Proc. Natl. Acad. Sci. USA **93**(25), 14503–14508 (1996)

29.72   A.J.M. De Ruijter, A.H. Van Gennip, H.N. Caron, S. Kemp, A.B.P. Van Kuilenburg: Histone deacetylases (HDACs): Characterization of the classical HDAC family, Biochem. J. **370**, 737–749 (2003)

29.73   H. Santos-Rosa, C. Caldas: Chromatin modifier enzymes, the histone code and cancer, Eur. J. Cancer **41**, 2381–2402 (2005)

29.74   M.A. Glozak, E. Seto: Histone deacetylases and cancer, Oncogene **26**, 5420–5432 (2007)

29.75   K.K. Lee, J.L. Workman: Histone acetyltransferase complexes: One size doesn't fit all, Mol. Cell Biol. **8**, 284–295 (2007)

29.76   M. Dalvai, K. Bystricky: The role of histone modifications and variants in regulating gene expression in breast cancer, J. Mammary Gland Biol. Neoplasia **15**, 19–33 (2010)

29.77   J.C. Black, J.R. Whetstine: Chromatin landscape, Epigenetics **6**(1), 9–15 (2011)

29.78   M. Lachner, R.J. O'Sullivan, T. Jenuwein: An epigenetic road map for histone lysine methylation, J. Cell Sci. **116**, 2117–2124 (2003)

29.79   M. Litt, Y. Qiu, S. Huang: Histone arginine methylations: Their roles in chromatin dynamics and transcriptional regulation, Biosci. Rep. **29**(2), 131–141 (2009)

29.80   K. Agger, J. Christensen, P.A.C. Cloos, K. Helin: The emerging functions of histone demethylases, Curr. Opin. Genet. Dev. **18**(2), 159–168 (2008)

29.81   A.H. Ting, K.M. McGarvey, S.B. Baylin: The cancer epigenome-components and functional correlates, Genes Dev. **20**(23), 3215–3231 (2006)

29.82   C. Sawan, T. Vaissière, R. Murr, Z. Herceg: Epigenetic drivers and genetic passengers on the road to cancer, Mutat. Res. **642**(1–2), 1–13 (2008)

29.83   P. Chi, C.D. Allis, G.G. Wang: Covalent histone modifications–miswritten, misinterpreted and mis-erased in human cancers, Nat. Rev. Cancer **10**(7), 457–469 (2010)

29.84   Y. Kondo: Epigenetic cross-talk between DNA methylation and histone modifications in human cancers, Yonsei Med. J. **50**(4), 455–463 (2009)

29.85   M. Macaluso, C. Cinti, G. Russo, A. Russo, A. Giordano: pRb2/p130-E2F4/5-HDAC1-SUV39H1-p300 and pRb2/p130-E2F4/5-HDAC1-SUV39H1-DNMT1 multimolecular complexes mediate the transcription of estrogen receptor-alpha in breast cancer, Oncogene **22**(23), 3511–3517 (2003)

29.86   M. Macaluso, M. Montanari, P.B. Noto, V. Gregorio, C. Bronner, A. Giordano: Epigenetic modulation of estrogen receptor-alpha by pRb family proteins: A novel mechanism in breast cancer, Cancer Res. **67**(16), 7731–7737 (2007)

29.87   J.M. Wagner, B. Hackanson, M. Lübbert, M. Jung: Histone deacetylase (HDAC) inhibitors in recent clinical trials for cancer therapy, Clin. Epigenet. **1**(3–4), 117–136 (2010)

29.88   C. Mund, F. Lyko: Epigenetic cancer therapy: Proof of concept and remaining challenges, BioEssays **32**(11), 949–957 (2010)

29.89   T.K. Kelly, D.D. De Carvalho, P.A. Jones: Epigenetic modifications as therapeutic targets, Nat. Biotechnol. **28**, 1069–1078 (2010)

29.90   J. Peedicayil: Epigenetic therapy – a new development in pharmacology, Indian J. Med. Res. **123**, 17–24 (2006)

29.91   A.J. Davis, K.A. Gelmon, L.L. Siu, M.J. Moore, C.D. Britten, N. Mistry: Phase I and pharmacologic study of the human DNA methyltransferase antisense oligodeoxynucleotide MG98 given as 21-day continuous infusion every 4 weeks, Investig. New Drugs **21**, 85–97 (2003)

29.92   D.J. Stewart, R.C. Donehowe, E.A. Eisenhaue, N. Wainman, A.K. Shah, C. Bonfils: A phase I pharmacokinetic and pharmacodynamic study of the DNA methyltransferase 1 inhibitor MG98 administered twice weekly, Ann. Oncol. **14**, 766–774 (2003)

29.93   R.B. Klisovic, W. Stock, S. Cataland, M.I. Klisovic, S. Liu, W. Blum: A phase I biological study of MG98, an oligodeoxynucleotide antisense to DNA methyltransferase 1, in patients with high-risk myelodysplasia and acute myeloid leukemia, Clin. Cancer Res. **12**, 2444–2449 (2008)

29.94   J.E. Bolden, M.J. Pearl, R.W. Johnstone: Anticancer activities of histone deacetylase inhibitors, Nat. Rev. Drug Discov. **5**(9), 769–784 (2006)

29.95   A. Mai, S. Massa, D. Rotili, I. Cerbara, S. Valente, R. Pezzi, S. Simeoni, R. Ragno: Histone deacetylation in epigenetics: An attractive target for anticancer therapy, Med. Res. Rev. **25**(3), 261–309 (2005)

29.96   L.S. Kristensen, H.M. Nielsen, L.L. Hansen: Epigenetics and cancer treatment, Eur. J. Pharmacol. **625**(1–3), 131–142 (2009)

29.97   F. Thaler, S. Minucci: Next generation histone deacetylase inhibitors: The answer to the search for optimized epigenetic therapies?, Exp. Opin. Drug Discov. **6**(4), 393–404 (2011)

29.98   S. Shankar, R.K. Srivastava: Histone deacetylase inhibitors: Mechanisms and clinical significance in cancer: HDAC inhibitor–induced apoptosis, Adv. Exp. Med. Biol. **615**, 261–298 (2008)

29.99 P.A. Marks: Discovery and development of SAHA as an anticancer agent, Oncogene **26**, 1351–1356 (2007)

29.100 V.M. Richon: Cancer biology: Mechanism of antitumour action of vorinostat (suberoylanilide hydroxamic acid), a novel histone deacetylase inhibitor, Br. J. Cancer **95**(S1), S2–S6 (2006)

30. Dynamics of Autoimmune Diseases

Hyeygjeon Chang, Alessandro Astolfi

Autoimmune diseases are due to the immune response of the human body, which reacts against substances or tissues of the body. Lupus is a systemic autoimmune disease, or autoimmune connective tissue disease, affecting any part of the human body. In this chapter we study the dynamics of autoimmune diseases using a control systems approach. We investigate how the drug scheduling framework previously developed by the authors can control the autoimmune phenomenon. The main purpose of this work is to demonstrate how available tools are capable of treating the autoimmune disease. We employ drug therapy as a control input and explore how it can contribute to treat autoimmune diseases. In particular, we study a model describing an autoimmune disease with a control input given by a lupus treatment drug, belimumab. We conduct additional modeling work since the models in the literature do not capture the explicit relation between autoimmunity and the drug therapy by belimumab. We also examine which part of the model can be controlled via manipulation of drug dosage and derive a control method with which to treat autoimmune inflammation related to autoreactive B cells.

Part F | 30

In Sect. 30.1 we give a brief introduction to lupus, because we study a model in which the control input is a newly developed and approved lupus treatment drug. In Sect. 30.2 we recall the model in [30.1] and conduct additional modeling work, since the current models do not capture the explicit relation between drug therapy and autoimmunity. We also examine which part of the model could be used as a control input and derive a control method. Section 30.3 describes the proposed control ideas and develops the control procedure for the model, which can treat autoimmune inflammation by means of controlling autoreactive B cells. Finally, we discuss future work and present further remarks in Sect. 30.4.

Autoimmune diseases are due to the immune response of the human body against substances or tissues of the body. In other words, the human body attacks some parts of itself. This is because the immune system regards some parts of the body as a pathogen and attacks them. The diseases may be restricted to certain organs, for example, in Chagas disease, while in other cases the diseases involve a particular tissue in different places, such as Goodpasture's disease, which affects the basement membrane in both the lung and the kidney. To treat autoimmune diseases we typically use immunosuppression medication which may decrease the overall immune response.

In this chapter we investigate the dynamics of autoimmune diseases using a control systems approach. In the last two decades, mathematical modeling for the dynamics of immune response and diseases has been researched; For example, mathematical models of the immune system with macrophages have been studied in [30.2, 3], and a similar problem with emphasis on diabetes has been dealt with in [30.4].

To deal with autoimmune disease we suggest the utilization of autoreactive B cells, which are essentially related to the autoimmune disease. Recently, progress on autoimmune disease modeling development has been explored in [30.1], based on biological modeling of the dynamics of autoreactive B cells in terms of a system of ordinary differential equations (ODEs). The model describes the dynamics of autoreactive B cells, autoantigen, and immune complexes. In [30.1] the model has been analytically studied, with identification of its possible steady states and investigation of their stability properties. In addition, a sensitivity analysis has been performed with regards to changes in parameter values and changes in the functional form of the equations, and the time-dependent behavior of the system has been analyzed during the disease development.

Meanwhile, taking a modeling approach different from the one used in [30.1], immune complex-related processes have been studied in [30.5]. The model developed in [30.5] incorporates several other components, such as T cells, plasma cells, macrophages, and complement proteins. The model is sophisticated, including 13 equations and 20 parameters. Note that the simplified model developed in [30.1] can provide an analytical explanation for the numerical behavior found in the biologically realistic model in [30.5].

In this chapter we employ the model in [30.1] to apply the control framework developed by the authors. The authors have studied nonlinear control theory with application to biological system; For example, a control method for human immunodeficiency virus (HIV)/acquired immunodeficiency syndrome (AIDS) dynamics based on bifurcation analysis has been reported in [30.6]. This control method has been applied to an HIV infection model and provides a way to drive the HIV patient state towards long-term nonprogressor status. A parameter adaptation in a reduced HIV dynamic model has been studied in [30.7]. Drug scheduling to boost the human immune response has been proposed in [30.8, 9]. This research could be applied to the dynamics of other infectious diseases such as malaria. Particularly, in [30.10], it has been discussed how understanding of mathematical models can be used to explain the experimental results reported in [30.11].

We investigate how the control framework developed by the authors can positively affect the autoimmune phenomenon. The main purpose of this work is to demonstrate how the available tools are capable of treating the autoimmune disease. We employ drug therapy as a control input to the human body and explore how it can contribute to the treatment of autoimmune diseases. We also describe validation of this treatment idea.

30.1 Overview of Systemic Lupus Erythematosus

Systemic lupus erythematosus (SLE or lupus) is a systemic autoimmune disease, or autoimmune connective tissue disease, affecting all parts of the human body. SLE is caused by a dysfunction of the immune system. As in other autoimmune diseases, in SLE the immune system attacks the cells and tissue of the body, resulting in inflammation and tissue damage. SLE can harm the heart, joints, skin, lungs, blood vessels, liver, kidneys, and nervous system. The progress of the disease is unpredictable, alternating between periods of illness and remission.

Typically, the B cells of the immune system produce antibodies that protect the body from invaders. In people with SLE, the B cells stay in the body longer than expected. These B cells (called autoreactive B cells) react against the body of patients with SLE. The autoreactive B cells produce a type of protein called an autoantibody. Antibodies react against foreign invaders, whereas au-

toantibodies attack the body of the subject with SLE and this can lead to inflamed body tissue.

Although the cause of the disease is not known, SLE appears to have a genetic component and can run in families. A number of factors may potentially trigger the disease, such as infections, stress, diet, toxins, and environmental conditions. Ultraviolet light, for example, sunlight, has been shown to trigger SLE in up to 70% of people affected by SLE. SLE affects everyone differently. Symptoms are wide-ranging and can change over time. This unpredictability can make SLE difficult to diagnose. An accurate diagnosis can take months or even years. The initial phase of SLE may be acute, appearing similar to an infection, or it may be a succession of vague, or seemingly disconnected, symptoms over time. Because the symptoms come and go and vary for each person, evaluation by a healthcare professional well acquainted with SLE is critical. This evaluation may include a physical examination, laboratory tests, and a complete medical history [30.12].

The rate of SLE varies considerably among countries, ethnicity, and gender. It changes over time [30.13]. In the USA the incidence of SLE is estimated to be about 53 per 100 000, translating to about 159 000 out of 300 million people in the USA being affected [30.13]. In Northern Europe the rate is about 40 per 100 000 people [30.14]. The survival rate of SLE for people in the USA, Canada, and Europe has risen from an approximate 4 year survival rate of 50% in the 1950s to a 15 year survival rate of 80% nowadays [30.14]. SLE occurs more commonly and with greater severity in those of non-European descent [30.14]; for example, the rate is as high as 159 per 100 000 among those of Afro-Caribbean descent [30.13]. SLE occurs in females nine times more frequently than in males, particularly in females of child-bearing age, i.e., between 15 and 35 years [30.13].

Currently SLE is treated by immunosuppression, mostly with cyclophosphamide and corticosteroids. However, there is no cure for SLE. SLE can be fatal, although such fatality is becoming rare.

SLE treatment involves preventing illness and decreasing the severity and duration of the illness. Treatment drugs include corticosteroids, antimalarial drugs, and cytotoxic drugs such as cyclophosphamide and mycophenolate. Nonsteroidal anti-inflammatory drugs might also be used. Medications such as prednisone, Cellcept, and Prograf have been used in the past [30.15].

Hydroxychloroquine (HCQ) was the last medication approved by the Food and Drug Administration (FDA), in 1955. HCQ is an FDA-approved antimalarial used for constitutional, cutaneous, and articular manifestations. HCQ has relatively few side-effects, and there is evidence that it improves survival among people who have SLE [30.16]. Some drugs approved for other diseases are used for SLE as well. A number of potential treatments are currently in clinical trials [30.17].

In November 2010, an FDA advisory panel recommended approving Benlysta (belimumab, previously known as LymphoStat-B) as a treatment for pain and flare-ups in lupus. Benlysta was approved by the US Food and Drug Administration (FDA) for treatment of SLE on March 9, 2011 [30.18]. It has subsequently been approved for use in Europe and Canada, and it is under evaluation for use in other autoimmune diseases. Belimumab successfully met the primary endpoints in its phase 3 clinical trials for SLE [30.19]. Benlysta is the first new drug approved to treat lupus in 56 years.

Benlysta is a prescription medication indicated for treatment of adult patients with active SLE who are receiving other lupus medicines. Benlysta is used along with other lupus medicines, and is given by intravenous (IV) infusion. It is not known if Benlysta is safe and effective in people with severe active lupus nephritis or severe active central nervous system lupus [30.12]. Benlysta is not a steroid. It contains belimumab, which belongs to a group of drugs called monoclonal antibodies. Belimumab is a fully human monoclonal antibody that inhibits B-lymphocyte stimulator (BLyS), also known as B cell activation factor of the tumor necrosis factor (TNF) family [30.20]. These are large, complex proteins produced in laboratories that copy the disease-fighting response of natural antibodies.

As already noted, in people with SLE, certain white blood cells called autoreactive B cells, which react against the body, stay in the body longer than they should. These autoreactive B cells produce the autoantibodies that attack the body. One of the important proteins for the growth of these autoreactive B cells is BLyS, which plays a key role in B lymphocyte differentiation, survival, and activation [30.21]. Benlysta binds to BLyS and prevents it from stimulating those B cells [30.22]. Adding Benlysta to other SLE treatments may help reduce the abnormal immune system activity that contributes to disease activity in SLE.

30.2 Disease Model with Control Input

There are several examples of mathematical models describing the development and progress of autoimmune diseases; see, for example [30.1, 23–25]. Throughout this work we study the effect of Benlysta and include its effect in the mathematical model. The main effect of Benlysta is on autoreactive B lymphocytes. Although the models of [30.23–25] describe the dynamic response of autoimmune diseases using a simple modeling approach, these models do not include autoreactive B lymphocytes explicitly.

The model in [30.24] has been constructed as a primary mathematical model for autoimmune diseases on the basis of the personal immune response and the target cell growth. While the simple model shows that these two functions are capable of capturing the essence of autoimmune diseases, explaining the characteristic symptom phases such as tolerance, repeated flare-ups, and dormancy, the model is preliminary and qualitative, and hence cannot be used for this chapter.

The spatiotemporal model in [30.25] has been constructed in order to illustrate a protective effect of autoimmunity, so the model is inconsistent with the context of the research of this chapter. The model supports the concept of autoimmunity as a defense against degenerative processes that operate by fighting against the threat of potential destructive activity originated or mediated within the organism. It is demonstrated that autoimmunity has a protective function after traumatic injuries to the central nervous system.

The model in [30.23] presents integration of human lupus nephritis biomarker data into a model of kidney inflammation. Lupus nephritis is a complex disease, but currently individual biomarkers are not sufficient to accurately describe disease activity. Biomarkers have been believed to be helpful to understand the disease activity with their integration into a pathogenic-based model of lupus nephritis. This model can fit clinical urine biomarker data from individual patients.

The model in [30.1] has been developed for the exploration of immune complex-mediated autoimmune inflammation and its clinical implications. It can characterize differences between normal individuals and those susceptible to such inflammation, the persistence of which is due to a positive feedback loop: the tissue damage caused by the inflammation can release autoantigen particles, which can stimulate autoreactive B cells, and this stimulation leads to the formation of further immune complexes and their subsequent deposition. The model includes autoreactive B cell-mediated dynamics, and it has been used to analyze the dynamic phenomena of SLE patients. In addition, the parameters of the model corresponding to SLE patients are estimated and available in [30.1] and its supplementary material.

Thus, we now consider the dynamic model in [30.1], namely

$$\dot{A} = \sigma_A^{max} \frac{I}{\theta_A + I} - \delta_A A - k_I A B \,, \tag{30.1}$$

$$\dot{I} = k_I A B - \delta_I I \,, \tag{30.2}$$

$$\dot{B} = \sigma_B + p_B f(h) B - \delta_B B \,, \tag{30.3}$$

where $f(h) = h/(\theta_B + h)$ and $h = \alpha_A A$.

The states describe the populations of specific cells in a unit volume of blood. In particular, A, I, and B describe the concentrations of autoantigen circulating in the human body, immune complexes due to the interaction between A and B, and autoreactive B cells responsible for the autoantigen, respectively.

The population of autoantigen (the state A) increases at a rate given by a saturating function of I proportional to σ_A^{max}, the maximal rate of autoantigen release due to the tissue damage resulting from the inflammation. It decreases at a constant rate proportional to δ_A, since autoantigens are cleared away by other mechanism than antibodies. Autoantigens interact with the corresponding autoantibodies and produce immune complexes (the state I) at a rate that is proportional to an effective rate k_I and the product of A and B. Immune complexes are cleared out at a constant rate δ_I.

The population of B cells increases at a rate σ_B since the cells are produced in bone marrow, and the cells die naturally at a rate proportional to δ_B. The B cells specific to the autoantigen proliferate at the effective rate $p_B f(h)$, where p_B is the maximal effective proliferation rate and h is the stimulation signal ($h = \alpha_A A$). The function $f(h)$ is assumed to be a saturation function of h.

The population of autoantibody is not separately modeled and is proportionally approximated by the level of B, since the dynamics of antibodies are comparatively fast. The model parameters σ_A^{max}, θ_A, δ_A, k_I, δ_I, σ_B, p_B, δ_B, θ_B, and α_A are positive and constant. The descriptions and values of the model parameters suggested in [30.1], and used in this chapter, are summarized in Table 30.1. For a detailed explanation of the model see [30.1] and its supplementary material.

Since the model parameters are based on clinical data (Table 30.2), the state variables in this chapter represent actual data. The effective rate of immune complexes formation (k_I) is the product of three terms: k_{asoc}, v, and β, corresponding to the association rate of the two molecules, their valence, and the proportionality factor between the autoantibodies and B cells, respectively.

With the parameters in Table 30.1, the system (30.1–30.3) is a bistable system [30.1] and the two stable equilibrium points are

Healthy state: $[0.0000, 0.0000, 40.0000]^\top$,

Disease state:

$[499.9906, 6.6562 \times 10^3, 3.1951 \times 10^6]^\top$.

To highlight this property, two trajectories of the systems are displayed in Figs. 30.1 and 30.2. The two initial points have been selected as $[A_0, I_0, B_0]^\top$, where $B_0 = \sigma_B/\delta_B$, $I_0 = A_0 B_0 k_I/\delta_I$, and $A_0 = 30$ or $A_0 = 50\,000$, respectively.

Table 30.1 Model parameters and their values [30.1]

Parameter	Physical description	Value
σ_A^{max}	Maximal autoantigen release rate by inflammation	2×10^6 (cells/day)
θ_A	Threshold for autoantigen release function	4 (pM)
δ_A	Autoantigen clearance rate	1 (day^{-1})
k_I	Effective rate of immune complex formation	5×10^{-4} (cell^{-1} day^{-1})
δ_I	Immune complex clearance rate	120 (day^{-1})
σ_B	Autoantigen-specific B cell production rate	20 (cells/day)
p_B	Maximal proliferation rate of B cells	1.5 (day^{-1})
δ_B	B cell death rate	0.5 (day^{-1})
θ_B	Threshold for B cell proliferation function	3 (pM)
α_A	Proportional rate of stimulation signal by autoantigen	1

Table 30.2 Parameters provided in the supplementary material of [30.1]

Parameter	Physical description	Range or value	Literature
δ_A	Autoantigen clearance rate	60–650 (day^{-1})	[30.26–28]
δ_I	Immune complex clearance rate	100–400 (day^{-1})	[30.27, 29, 30]
k_{asoc}	Autoantigen–antibody association rate	10^{-3} (pM^{-1} day^{-1})	[30.31, 32]
ν	Effective valence of autoantigen and antibodies	10	[30.30, 33]
β	Concentration of antibodies per B cell	10^{-3}–10^{-1} (pM/cell)	[30.34–36]
σ_B	Autoantigen-specific B cell production rate	1–100 (cells/day)	[30.37]
p_B	Maximal proliferation rate of B cells	1–3 (day^{-1})	[30.37]
δ_B	B cell death rate	0.33–1 (day^{-1})	[30.37]
θ_B	Threshold for B cell proliferation function	10^3–10^4 (pM)	[30.33, 38]

We integrate the model (30.1–30.3) from these initial points for 20 days. Figures 30.1 and 30.2 show that the trajectories converge to the stable equilibrium points, the *healthy state* and the *disease state*, respectively. While the initial state with $A_0 = 30$ is in the vicinity of the healthy state, the one with $A_0 = 50\,000$ represents an initial occurrence of autoantigen, such as an initial tissue damage by an invading external pathogen, which is a biologically realistic perturbation of the system (30.1–30.3).

To deal with drug dosage as a control input we modify the dynamic equation (30.3) of the model. A similar modification can be found in [30.39, 40]. In particular, we rewrite (30.3) as

$$\dot{B} = \eta[\sigma_B + p_B f(h)B] - \delta_B B . \tag{30.4}$$

The quantity η, which varies between 0 and 1, describes the effect of the drug. In the presence of a control input,

η can be rewritten as

$$\eta(t) = 1 - \eta^* u(t) ,$$

where η^* is the maximum effect of the drug and u is the control input, i.e., the drug dosage; for

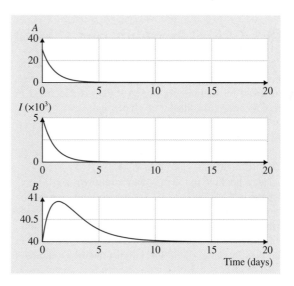

Fig. 30.1 State histories of the model (30.1–30.3) with initial state $[A_0, I_0, B_0]^\top$, where $B_0 = \sigma_B/\delta_B$, $I_0 = A_0 B_0 k_I/\delta_I$, and $A_0 = 30$. The state trajectory converges to the healthy state ▶

Part F | 30.2

Fig. 30.2 State histories of the model (30.1–30.3) with initial state $[A_0, I_0, B_0]^\top$, where $B_0 = \sigma_B/\delta_B$, $I_0 = A_0 B_0 k_I/\delta_I$, and $A_0 = 50\,000$. The state trajectory converges to the disease state ▶

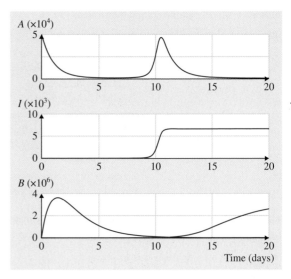

example, $\eta^* = 0.50$ indicates that the maximal effect of the drug is 50%. From a control perspective, the input u, which takes values between 0 and 1, represents the drug dose of belimumab. If $u = 1$, a patient receives the maximal drug therapy, while $u = 0$ means no medication. Currently we do not model the pharmacological dynamics of belimumab.

Given the system parameters in Table 30.1, we have bistability of system (30.1, 30.2, 30.4) with $u = 0$. Thus, we hope to drive the state into the region of attraction of the healthy state when a patient is treated. To this end we control the system state by manipulating the drug scheduling of belimumab.

During the drug treatment, the level of B cells should be maintained to some extent, since it is a key component of the human immune system. We consider this condition as a constraint in the control procedure. Belimumab suppresses the immune system, as immunosuppressives and corticosteroids do, thus there could be a risk of serious infections due to the drug treatment [30.41].

As mentioned before, and modeled in (30.4), the drug interferes with the activity of B cells, which implies that the control input has a direct effect on the concentration of B cells. In the positive feedback loop of autoimmune disease, the level of B cells is connected to

the formation of immune complexes. Thus, the dynamics of (30.2) is indirectly affected by the drug input, and the positive term $k_I AB$ on the right-hand side of (30.2) should be minimized.

Note that the level of I plays a key role in reducing the concentration of A, as a consequence of (30.1), leading to the suppression of the positive feedback loop of the disease. Accordingly, it is expected that the control input u can induce the healthy state in the model (30.1, 30.2, 30.4).

The basic control idea is to force the term AB to decrease as much as possible with a constraint on the concentration of the B cells.

30.3 Control Procedure and Computer Simulations

The suggested control idea can be implemented in the model (30.1, 30.2, 30.4) by applying the control steps proposed by the authors in [30.8, 42], as described in the following procedure.

30.3.1 Control Procedure

Initialization

Select a positive number T_s and a ratio $R_B \in [0, 1]$; T_s denotes the sampling time for the computation of the control input, and $R_B V_B$ denotes the lower bound on the concentration of B cells during the treatment procedure, where $V_B = \sigma_B/\delta_B$, i.e., the concentration of B cells in the healthy state. X_I is the initial condition of model (30.1–30.3).

Step 1: Integrate model (30.1, 30.2, 30.4) with initial condition X_I for T_s time instants with full medication and with no medication. Let $X_{F,fm} = [X_{F,fm}(1), X_{F,fm}(2), X_{F,fm}(3)]^\top$ and $X_{F,nm} = [X_{F,nm}(1), X_{F,nm}(2), X_{F,nm}(3)]^\top$ be the state vector of the model at the end of the integration period, with full medication and with no medication, respectively.

Step 2: If $X_{F,fm}(3) < R_B V_B$ and $X_{F,nm}(3) < R_B V_B$, then set $u = 0$.
If $X_{F,fm}(3) < R_B V_B$ and $X_{F,nm}(3) \geq R_B V_B$, then set $u = 0$.
If $X_{F,fm}(3) \geq R_B V_B$ and $X_{F,nm}(3) < R_B V_B$, then set $u = 1$.

If $X_{\mathrm{F,nm}}(3) \geq R_{\mathrm{B}} V_{\mathrm{B}}$, $X_{\mathrm{F}}, \mathrm{fm}(3) \geq R_{\mathrm{B}} V_{\mathrm{B}}$, and $X_{\mathrm{F,nm}}(1) X_{\mathrm{F,nm}}(3) \leq X_{\mathrm{F}}, \mathrm{fm}(1) X_{\mathrm{F}}, \mathrm{fm}(3)$, then set $u = 0$.

If $X_{\mathrm{F,nm}}(3) \geq R_{\mathrm{B}} V_{\mathrm{B}}$, $X_{\mathrm{F}}, \mathrm{fm}(3) \geq R_{\mathrm{B}} V_{\mathrm{B}}$, and $X_{\mathrm{F,nm}}(1) X_{\mathrm{F,nm}}(3) > X_{\mathrm{F}}, \mathrm{fm}(1) X_{\mathrm{F}}, \mathrm{fm}(3)$, then set $u = 1$.

Step 3: The input determined in step 2 is applied to the model (30.1, 30.2, 30.4) with initial point X_{I} for T_{s} time instants. Let X_{F} be the values of the state at the end of the integration period.

Step 4: Set $X_{\mathrm{I}} = X_{\mathrm{F}}$ and go to step 1.

In [30.12] the recommended dosage schedule for Benlysta is $10\,\mathrm{mg/kg}$ at 2 week intervals for the first three doses and at 4 week intervals thereafter. Thus, we assume T_{s} is 14 days. We also assume $R_{\mathrm{B}} = 0.75$, implying that the concentration of B cells is maintained at least over 75%, compared with the healthy state (although we might consult with a medical expert for the setting of the value of R_{B}). The control procedure is considered for 210 days, corresponding to a maximum of 15 doses, and the simulation is terminated at the 300th day.

Figure 30.3 shows the results of the application of the control strategy to the model (30.1, 30.2, 30.4). In this simulation it is assumed that the drug parameter η^* is 0.02, which implies a low efficiency of the drug. The initial state is $[A_0, I_0, B_0]^{\top}$, where $B_0 = \sigma_{\mathrm{B}}/\delta_{\mathrm{B}}$, $I_0 = A_0 B_0 k_{\mathrm{I}}/\delta_{\mathrm{I}}$, and $A_0 = 50\,000$. As already discussed, this initial condition is located in the region of attraction of the disease state. The simulation results in Fig. 30.3 indicate that the proposed control scheme can drive the initial state towards the healthy state even with low drug efficiency.

The control strategy with the same value of η does not work properly for the case of full-blown autoimmune response, described by the initial condition in Fig. 30.4. The figure presents simulation results with the initial state given by the disease state, i.e., $[499.9906, 6.6562 \times 10^3, 3.1951 \times 10^6]^{\top}$, describing a full-blown autoimmune disease. For this severe disease status, the drug with 2% effectivity can suppresses the B and I states only for the period of the control procedure (210 days), which is not sufficient to drive the patient state into the region of attraction of the healthy state. Higher levels of drug efficiency are needed for a subject with full-blown autoimmune disease, as indicated in Fig. 30.5.

Figure 30.5 shows the results of the control strategy with $\eta^* = 0.75$. The initial state is the same as in

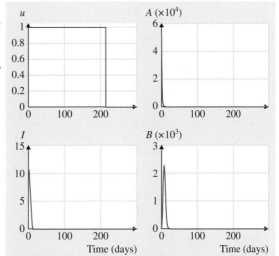

Fig. 30.3 Results of the application of the proposed control strategy to the model (30.1, 30.2, 30.4) with $\eta^* = 0.02$. The initial state is $[A_0, I_0, B_0]^{\top}$, where $B_0 = \sigma_{\mathrm{B}}/\delta_{\mathrm{B}}$, $I_0 = A_0 B_0 k_{\mathrm{I}}/\delta_{\mathrm{I}}$, and $A_0 = 50\,000$, describing a realistic perturbation of autoantigen

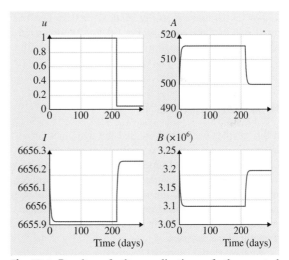

Fig. 30.4 Results of the application of the control strategy to the model (30.1, 30.2, 30.4) with $\eta^* = 0.02$. The initial state is the disease state, i.e., $[499.9906, 6.6562 \times 10^3, 3.1951 \times 10^6]^{\top}$, describing a full-blown autoimmune disease

Fig. 30.4. The control procedure leads the patient state towards the healthy state with only seven doses administered. Note that we cannot conclude that the control performance with this high effective drug is always bet-

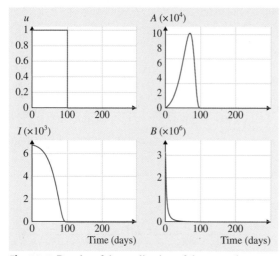

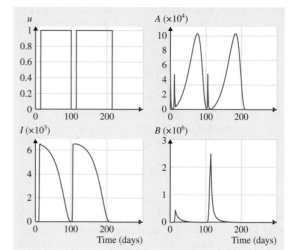

Fig. 30.5 Results of the application of the control strategy to the model (30.1, 30.2, 30.4), with $\eta^* = 0.75$. The initial state is given by the disease state

Fig. 30.6 Results of the application of the proposed control strategy to the model (30.1, 30.2, 30.4), with $\eta^* = 0.75$. The initial state is the same as in Fig. 30.3

ter than that with a low drug efficiency; for example, Fig. 30.6 shows some control results of the strategy contradicting this intuition.

For the simulation in Fig. 30.6, $\eta^* = 0.75$ and the initial state is the same as in Fig. 30.3. Although eventually driven into the region of attraction of the healthy

state, if compared with the simulation presented in Fig. 30.3, the states show high fluctuations, which is not a good sign, particularly from the biomedical point of view. This is mostly due to the control constraint on the B state, combined with the dosage interval $T_s = 14$ days.

30.4 Conclusions

In this chapter we have discussed the dynamics of autoimmune diseases using a control systems approach based on mathematical models. We have investigated how the control framework developed by the authors can counteract the autoimmune phenomenon. To this end, we have employed drug therapy as a control input and explored how it can contribute to the treatment of autoimmune diseases.

We have studied a model describing an autoimmune disease with a control input given by a lupus treatment drug, belimumab. We have conducted additional modeling work since the models in the literature do not capture the relation between autoimmunity and the drug therapy by belimumab. We have examined which part of the model could be controlled via the manipulation of drug dosage and derived a control method with which to treat an autoimmune inflammation controlling autoreactive B cells.

30.4.1 Discussion and Future Direction

In Sect. 30.3 we studied the role of η^*, with the assumption that u is either 0 (no medication) or 1 (full medication). η^* can be regarded as an individual-specific parameter, affected by the weight of the subject and their resistance (or sensitivity) to the drug. Note that Benlysta is for intravenous infusion only over a period of 1 h and must be reconstituted and diluted prior to administration [30.12]. Thus, we could consider a drug dosage input u varying between 0 and 1 using this dilution process.

By the definition of $\eta(t)$ [i.e., $1 - \eta^* u(t)$], η is implemented with the product of η^* and u, thus a weaker drug efficiency with $u = 1$ can be realized by the selection of $u \in [0, 1]$.

It is known that Benlysta decreases the disease activity of lupus more than any other medicines alone when given together with other lupus medicines, such as corticosteroids, antimalarials, nonsteroidal anti-inflammatory drugs, and immunosuppressants [30.12]. Considering this multi-input system and the recommended infusion scheduling of Benlysta, we plan to design a pharmacological dynamic model of Benlysta in the future.

Finally, the optimization of treatment of autoimmune disease can be an important topic of future research. Throughout the chapter we have discussed bistability properties. However, the autoimmune disease model (30.1–30.3) can have other properties, as suggested in [30.1]; for example, the system could have only one globally stable equilibrium representing a severe disease status. In such a case, we should design an optimal control to achieve a certain condition with minimal amount of drug.

Common adverse effects of belimumab include nausea, diarrhea, fever, and infusion-site reactions [30.43]. A greater number of deaths and serious infections were reported in patients treated with belimumab than in those treated with placebo [30.43]. Thus, it is desirable that drug treatment with belimumab should be optimized with regard to the total amount of drug administered and the duration of the drug treatment.

Part F | 30

References

30.1   A. Arazi, A.U. Neumann: Modeling immune complex-mediated autoimmune inflammation, J. Theor. Biol. **267**(3), 426–436 (2010)

30.2   P.S. Kim, P.P. Lee, D. Levy: Modeling regulation mechanisms in the immune system, J. Theor. Biol. **246**(1), 33–69 (2007)

30.3   S. Iwami, Y. Takeuchi, K. Iwamoto, Y. Naruo, M. Yasukawa: A mathematical design of vector vaccine against autoimmune disease, J. Theor. Biol. **256**(3), 382–392 (2009)

30.4   A.F.M. Marée, M. Komba, C. Dyck, M. Labecki, D.T. Finegood, L. Edelstein-Keshet: Quantifying macrophage defects in type 1 diabetes, J. Theor. Biol. **233**(4), 533–551 (2005)

30.5   M. Head, N. Meryhew, O. Runquist: Mechanism and computer simulation of immune complex formation, opsonization, and clearance, J. Lab. Clin. Med. **128**(1), 61–74 (1996)

30.6   H. Shim, N.H. Jo, H. Chang, J.H. Seo: A system theoretic study on a treatment of AIDS patient by achieving long-term nonprogressor, Automatica **45**, 611–622 (2009)

30.7   H. Chang, A. Astolfi: Control of HIV infection dynamics: Approximating high-order dynamics by adapting reduced-order model parameters, IEEE Control Syst. Mag. **28**, 28–39 (2008)

30.8   H. Chang, A. Astolfi: Activation of immune response in disease dynamics via controlled drug scheduling, IEEE Trans. Autom. Sci. Eng. **6**, 248–255 (2009)

30.9   H. Chang, A. Astolfi: Enhancement of the immune system in HIV dynamics by output feedback, Automatica **45**, 1765–1770 (2009)

30.10   H. Chang, A. Astolfi, H. Shim: A control theoretic approach to malaria immunotherapy with state jumps, Automatica **47**, 1271–1277 (2011)

30.11   M. Roestenberg, M. McCall, J. Hopman, J. Wiersma, A.J. Luty, G.J. van Gemert, M. van de Vegte-Bolmer, B. van Schaijk, K. Teelen, T. Arens, L. Spaarman, Q. de Mast, W. Roeffen, G. Snounou, L. Rénia, A. van der Ven, C.C. Hermsen, R. Sauerwein: Protection against a malaria challenge by sporozoite inoculation, N. Engl. J. Med. **361**(5), 468–477 (2009)

30.12   Human Genome Sciences: Benlysta Full Prescribing Information (2011) available at http://www.hgsi.com/images/Benlysta/pdf/benlysta_pi.pdf

30.13   N. Danchenko, J.A. Satia, M.S. Anthony: Epidemiology of systemic lupus erythematosus: A comparison of worldwide disease burden, Lupus **15**(5), 308–318 (2006)

30.14   A. Rahman, D.A. Isenberg: Systemic lupus erythematosus, N. Engl. J. Med. **358**(9), 929–939 (2008)

30.15   J. Wentworth, C. Davies: Systemic lupus erythematosus, Nat. Rev. Drug Discov. **8**(2), 103–104 (2009)

30.16   G.C. Tsokos: Systemic lupus erythematosus, N. Engl. J. Med. **365**(22), 2110–2121 (2011)

30.17   M. Ratner: Human genome sciences trial data wow lupus community, Nat. Biotechnol. **27**(9), 779–780 (2009)

30.18   U.S: Food and Drug Administration: FDA approves Benlysta to treat lupus (2011) press release

30.19   Human Genome Sciences: A study of Belimumab in subjects with systemic lupus erythematosus (SLE) (BLISS-52) (2011) available at http://clinicaltrials.gov/ct2/show/NCT00424476

30.20   C. Bossen, P. Schneider: BAFF, APRIL and their receptors: Structure, function and signaling, Semin. Immunol. **18**(5), 263–275 (2006)

30.21   J.E. Crowley, L.S. Treml, J.E. Stadanlick, E. Carpenter, M.P. Cancro: Homeostatic niche specification among naïve and activated B cells: A growing role for the BLyS family of receptors and ligands, Semin. Immunol. **17**(3), 193–199 (2005)

30.22   W.G. Halpern, P. Lappin, T. Zanardi, W. Cai, M. Corcoran, J. Zhong, K.P. Baker: Chronic administration of Belimumab, a BLyS antagonist, decreases tissue and peripheral blood B-lymphocyte populations in cynomolgus monkeys: Pharmacokinetic, pharmacodynamic, and toxicologic effects, Toxicol. Sci. **91**(2), 586–599 (2006)

30.23   P. Budu-Grajdeanu, R. Schugart, A. Friedman, D. Birmingham, B. Rovin: Mathematical framework for human SLE Nephritis: Disease dynamics and urine biomarkers, Theor. Biol. Med. Model. **7**(1), 14 (2010)

30.24   S. Iwami, Y. Takeuchi, Y. Miura, T. Sasaki, T. Kajiwara: Dynamical properties of autoimmune disease models: Tolerance, flare-up, dormancy, J. Theor. Biol. **246**(4), 646–659 (2007)

30.25   U. Nevo, I. Golding, A.U. Neumann, M. Schwartz, S. Akselrod: Autoimmunity as an immune defense against degenerative processes: A primary mathematical model illustrating the bright side of autoimmunity, J. Theor. Biol. **227**(4), 583–592 (2004)

30.26   P. Rumore, B. Muralidhar, M. Lin, C. Lai, C.R. Steinman: Haemodialysis as a model for studying endogenous plasma DNA: Oligonucleosome-like structure and clearance, Clin. Exp. Immunol. **90**(1), 56–62 (1992)

30.27   R.P. Kimberly, J.C. Edberg, L.T. Merriam, S.B. Clarkson, J.C. Unkeless, R.P. Taylor: In vivo handling of soluble complement fixing Ab/dsDNA immune complexes in chimpanzees, J. Clin. Invest. **84**(3), 962–970 (1989)

30.28   W. Emlen, M. Mannik: Clearance of circulating DNA–anti-DNA immune complexes in mice, J. Exp. Med. **155**(4), 1210–1215 (1982)

30.29   R.P. Kimberly, N.L. Meryhew, O.A. Runquist: Mononuclear phagocyte function in SLE. I. Bipartite Fc- and complement-dependent dysfunction, J. Immunol. **137**(1), 91–96 (1986)

30.30   W. Emlen, G. Burdick: Clearance and organ localization of small DNA anti-DNA immune complexes in mice, J. Immunol. **140**(6), 1816–1822 (1988)

30.31   S.M. Barbas, H.J. Ditzel, E.M. Salonen, W.P. Yang, G.J. Silverman, D.R. Burton: Human autoantibody recognition of DNA, Proc. Natl. Acad. Sci. USA **92**(7), 2529–2533 (1995)

30.32   E.R. Eivazova, J.M. McDonnell, B.J. Sutton, N.A. Staines: Specificity and binding kinetics of murine lupus anti-DNA monoclonal antibodies implicate different stimuli for their production, Immunology **101**(3), 371–377 (2000)

30.33   P.M. Rumore, C.R. Steinman: Endogenous circulating DNA in systemic lupus erythematosus.

Occurrence as multimeric complexes bound to histone, J. Clin. Invest. **86**(1), 69–74 (1990)

30.34   N.L. Bernasconi, E. Traggiai, A. Lanzavecchia: Maintenance of serological memory by polyclonal activation of human memory B cells, Science **298**(5601), 2199–2202 (2002)

30.35   E.A. Clutterbuck, P. Salt, S. Oh, A. Marchant, P. Beverley, A.J. Pollard: The kinetics and phenotype of the human B-cell response following immunization with a heptavalent pneumococcal–CRM conjugate vaccine, Immunology **119**(3), 328–337 (2006)

30.36   L.D. Erickson, B.G. Durell, L.A. Vogel, B.P. O'Connor, M. Cascalho, T. Yasui, H. Kikutani, R.J. Noelle: Short-circuiting long-lived humoral immunity by the heightened engagement of CD40, J. Clin. Invest. **109**(5), 613–620 (2002)

30.37   R.J. De Boer, A.S. Perelson, I.G. Kevrekidis: Immune network behavior I. From stationary states to limit cycle oscillations, Bull. Math. Biol. **55**(4), 745–780 (1993)

30.38   L. Raptis, H.A. Menard: Quantitation and characterization of plasma DNA in normals and patients with systemic lupus erythematosus, J. Clin. Invest. **66**(6), 1391–1399 (1980)

30.39   R. Zurakowski, A.R. Teel: A model predictive control based scheduling method for HIV therapy, J. Theor. Biol. **238**, 368–382 (2006)

30.40   B.M. Adams, H.T. Banks, M. Davidian, H. Kwon, H.T. Tran, S.N. Wynne, E.S. Rosenberg: HIV dynamics: Modeling, data analysis, and optimal treatment protocols, J. Comput. Appl. Math. **184**, 10–49 (2005)

30.41   National Institutes of Health U.S. Department of Health and Human Services: Intravenous Immunoglobulins (IVIGs) in Lupus Central Station, available at http://theodora.com/lupus_central_station/intravenous_immunoglobulins_ivigs.html

30.42   H. Chang, A. Astolfi: Immune response's enhancement via controlled drug scheduling, IEEE Proc. Conf. Decis. Control (2007) pp. 3919–3924

30.43   GlaxoSmithKline: GlaxoSmithKline and Human Genome Sciences announce FDA approval of Benlysta (Belimumab) for the treatment of systemic lupus erythematosus (2011) available at http://www.gsk.com/media/pressreleases/2011/2011_us_pressrelease_10017.htm

31. Nutrigenomics

Hylde Zirpoli, Mariella Caputo, Mario F. Tecce

The entire complex of molecular processes of the human organism results from endogenous physiological execution of the information encoded in the genome but is also influenced by exogenous factors, which include those originating from nutrition as major agents. The assimilation of nutrient molecules within the human body continuously allows homeostatic reconstitution of its qualitative and quantitative composition but also takes part in physiological changes of body growth and adaptation to particular situations. Nevertheless, in addition to replacing material and energetic losses, nutritional intake also provides bioactive molecules, which are selectively able to modulate specific metabolic pathways, noticeably affecting the risk of cardiovascular and neoplastic diseases, which are the major cause of mortality in developed countries. Numerous bioactive nutrients are being progressively identified and their chemopreventive effects are being described at clinical and molecular mechanism levels. All *omics* technologies (such as transcriptomics, proteomics, and metabolomics) allow systematic analyses to study the effect of dietary bioactive molecules on the totality of molecular processes.

Since each nutrient might also have specific effects on individually different genomes, nu-

trigenomic and nutrigenetic analysis data can be distinguished by two different observational views: 1) the effects of the whole diet and of specific nutrients on genes, proteins, metabolic pathways, and metabolites; and 2) the effects of specific individual genomes on the biological activity of nutritional intake and of specific nutrients. Nutrigenomic knowledge of physiologic status and disease risk will provide the development of better diagnostic procedures as well as new therapeutic strategies specifically targeted to nutritionally relevant processes.

31.1 Overview

The interaction of genomic and environmental factors is highly determinant in development or progression of most important human diseases. Nutritional intake is an environmental factor of major importance and its prominent role in disease etiology was firstly documented for monogenic diseases and then for multifactorial disorders [31.1].

Diet not only provides adequate nutrients to supply metabolic and energetic requirements for body com-

position homeostasis, but can also contribute to the improvement of human health through preventive effects and modulation of specific processes [31.2]. In this context it is important to highlight the role of bioactive nutrients, which are either essential (mandatorily to be taken from outside to avoid disease deficiencies, e.g., vitamins) and non-essential molecules (e.g., some polyphenols and polyunsaturated fatty acids). Bioactive nutrients are also known as chemopreventive molecules and their presence in nutritional intake affects human health, reducing disease risk with specific molecular mechanisms which are progressively revealed [31.3–5]. Both experimental and epidemiological studies confirmed that several eating patterns can decrease neoplastic and cardiovascular disease risk [31.4, 6]. Actually, nutritional research studies are growing more at molecular biology and genetics level than with population epidemiological studies [31.7]. The objective is to clarify how bioactive nutrients can act at the cellular and molecular level. Macronutrients and micronutrients are being evaluated as effective dietary signals, which influence both metabolic programming and cell homeostasis.

The purpose of nutritional genomics is to use diet to prevent or treat diseases and mainly delves into different aspects, nutrigenomics and nutrigenetics, first described by Brennan in 1975: *Nutrigenetics: New Concepts for Relieving Hypoglycemia*. In addition, understanding the molecular bases of these effects may also provide the way to new diagnostic and therapeutical procedures.

Nutrigenetics studies the interaction between nutritional intake and individual genetic variability and its effect on human health. Nutrigenetics can then provide recommendations about risks or benefits of specific diets or dietary components to each individual as *personalized nutrition*. Nutrigenomics, in parallel, focuses on the effect of individual bioactive nutrients on the human genome, proteome, and metabolome [31.8]. These are, respectively, the entire complex of genes, proteins, and metabolites of an organism, or of an organ, or a cell type. Each gene encodes a specific protein according to a number of mechanisms collectively defined as gene expression regulation processes. The metabolome is the complex of metabolites, molecules that are reagents and products of the very numerous biochemical reactions occurring in the organisms, each one regulated from an enzyme, a protein with specific catalytic activity. Moreover, nutrigenomics can be considered as being similar to pharmacogenomics, which systematically studies the effect of drugs on genomes. However, drugs are pure compounds, each administered in precise doses, usually at low concentrations, acting with a relatively high affinity and selectivity for a limited number of biological targets. By contrast, nutrigenomics must to handle the complexity and variability of diet, which includes a wide number of different nutrient molecules. Nutrients can reach high concentrations (μM to mM) without becoming toxic and can also bind to numerous targets with different affinities and specificities [31.9–11].

Micro and macronutrients can have important effects on gene and protein expression and on metabolism as fuel and co-factors. The molecular structure of a nutrient can influence specific signaling pathways. Small changes in structure can differentially activate metabolic steps and this molecular specificity explains why closely related nutrients can distinctly affect cellular homeostasis. One example is provided by the fatty acids class and their level of carbon chain unsaturation. The n–3 polyunsaturated fatty acids, in fact, have a positive preventive effect on cardiac arrhythmias [31.12], whereas saturated C16–18 fatty acids do not [31.13]. Furthermore, structurally similar *trans* fatty acids increase plasma levels of LDL-cholesterol [31.14]. Moreover, it is also important to highlight the strong relation between transcription factors and numerous nutrients as a direct mechanism to influence gene expression. In fact, dietary compounds affect numerous receptors, such as RXR, PPARs, LXR, and PXR, by changing the level of DNA transcription of specific genes [31.15].

Nutritional research is generally aimed at preventing chronic disorders, since dietary habits can play a decisive role in disease causation. A consistent part of data relating dietary intake to phenotypes and disease risk is provided by population studies using self-reported dietary questionnaires [31.16]. It is also possible to indirectly monitor the features of nutritional intake by measuring specific nutrient molecules in blood, urine, fat, or other tissues. This may also result in the identification of nutritional biomarkers, supplying a guidance on the relation between nutrition and health. Commonly used predictors of nutritional related diseases are altered serum lipid profiles (e.g., cholesterol, triglycerides), increased blood pressure, or reduced insulin sensitivity, which are used as biomarkers for pathophysiological changes and are also related to the individual genotype. A wider biomarkers profile should then help to characterize the health status better than an individual biomarker.

While a single biomarker concentration can be useful to indicate a normal or a pathological condition, a complete knowledge about conditions of all elements constituting the human body in different situations might obviously be much richer in potentialities. This is what new *omics* technologies now allow also in the field of nutrition. Differential comparisons technologies permit whole systematic analyses of all metabolites, or all proteins, or all RNA, or all genes identifying specific differences, for instance, between two examined conditions and possibly understanding all mechanisms that caused the change. In all research fields, identifying differences between two similar objects is the most common method of knowledge, allowing us to know many if not all the specific features of each one. This approach is currently being applied in the clinical field but it might be particularly effective in nutrition, since the effect of diet intake does not often produce big changes unless very long periods are observed. Nevertheless, even if the effects produced are small, they may be numerous and it could be necessary and more useful to monitor them all at the same time. For this reason, the most important analytical level of the *omics* technologies is the one where integration of all the levels is possible by the use of informatics.

These *system* profiles will completely facilitate the description of all aspects of the human body: from the macroscopical levels to each molecular component,

integrating the analyses of transcriptome, proteome, and metabolome, even at the nutritional level, and the crosstalk of different disciplines and expertise to build models that will integrate information about intake, gene polymorphisms, gene expression, phenotypes, diseases, effect biomarkers and susceptibility biomarkers. Therefore, while each *omics* has important perspectives since it analyzes simultaneously great numbers of elements, their combination through bioinformatic technologies in the *system biology* level should lead, by strong positive synergies, to even more important achievements, with an almost complete understanding and description of the totality of the mechanisms involved in nutritional effects. The complete development of this deep knowledge will certainly strongly improve the diagnosis and therapy of all nutritionally related diseases.

31.2 Effects of Bioactive Nutrients on Gene Expression

Through gene-expression profiling it is possible to characterize the basic molecular pathways of gene regulation affected by nutrients. In fact, eating patterns may be considered as endogenous cellular mediators that can specifically influence gene expression. Several technologies study, for instance, the comparison of the whole mRNA cellular expression patterns in different experimental conditions, comparing the effect of two distinct nutrient molecules. Among these techniques the most outstanding and effective is the employment of *microarrays*, also known as *DNA chips* [31.17].

A microarray is a set of thousands of microscopic spots of DNA oligonucleotides (10^{-12} mole), each representing a specific DNA sequence; the probes on this array are able to selectively hybridize to the corresponding sequence. In fact, a short section of a gene or another DNA element hybridizes to a complementary DNA (cDNA) or complementary RNA (cRNA) sample as a *target* in high-stringency conditions. This probe–target hybridization is then quantified by the detection of fluorophore-, silver-, or chemiluminescence-labeled targets. DNA microarrays can be used to measure changes in expression levels, to detect single nucleotide polymorphisms (SNPs), genotype, or sequence mutants. In the nutrition field, for instance, microarray analysis enables us to evaluate the effect of a specific diet or nutrient on the expression of the whole genome. Several studies reported gene-expression analyses according to caloric restriction, fasting or examination of the effects

of single nutrient deficiency [31.18–20]. This helped the investigation of mechanisms related to favorable or adverse actions of nutrients or diet. Furthermore, gene-expression profiling seeks to identify important genes that are altered in pathological states and that might act as biomarkers; in this way, these nutritional sensors might allow us, at an early and reversible stage of a disease, to gain useful information of prognostic importance.

Park et al. investigated the crucial role of microRNA (miRNA), which are non-coding RNA molecules, specifically expressed in liver tissue in diet-induced obese mice, using a miRNA microarray to determine miRNA differential expression in murine liver tissue. In conclusion of this study, more than 50 hepatic miRNAs were found modulated in diet-induced obese mice. Some of them regulated protein expression at translation level and others affected mRNA expression at transcriptional level, providing the evidence of a correlation between miRNAs and their targets in diet-induced obesity [31.21].

Moreover, to understand the mechanisms by which nutrients can be actively involved in cellular pathways, there are also other techniques, such as protein–DNA interaction profiling, ChIP-on-chip, or LA, which allow a systematic investigation of specific genomic protein-binding sites.

For instance, SREBP is a protein acting as a transcription factor that typically responds to diet changes and is mostly involved in cholesterol and fatty acid metabolism. Using a chromatin immunoprecipitation technique combined with genome tiling arrays (ChIP-on-chip), *Reed* et al. comprehensively investigated the effect of a certain diet on the promoters related to SREBP1 interaction and its binding partners, NFY and SP1, in a human hepatocarcinoma cell line [31.22]. They demonstrated a cooperation between SREBP1 with NFY and SP1 in many functional pathways related to diet derived effects. This example about the analysis of the binding of a transcription factor to the whole genome shows how identifying gene regulation in the specific condition is influenced by a certain nutritional intake [31.22].

Genetics contributes to determining an individual's risk of developing diseases. In particular, nutrigenetics, as a branch of genetics, examines genetic variations according to individual nutrient effects. For this reason, it is significant to relate SNPs, point of variations in DNA sequence, to diet responsiveness [31.23]. Population differences in SNPs can have, for example, a predictive role in risk assessment and life style recommendations. The goal of the Single Nucleotide Polymorphisms Consortium is to map all the important genome polymorphic sites and, in parallel, the challenge for molecular epidemiology is to link specific polymorphisms to disease predisposition [31.24]. As a result, nutrigenetics assesses an individual's genetic sensitivity to diet and incorporating studies of SNPs into metabolic and epidemiological studies might also help to characterize optimal diet formulations. Phenylketonuria (PKU) was the first evidence of a single-gene defect condition responding to dietary treatments, using a low-phenylalanine containing diet as nutrigenetic therapy [31.25]. As another example, in prostate cancer research, a study on the male Swedish population confirmed that frequent consumption of dietary omega 3 fatty acids was inversely associated with prostate cancer risk only in men carrying the variant allele +6364 A > G SNP in *COX-2* gene [31.26].

Other approaches to expand the knowledge about the effect of nutrition on genomes are aimed at understanding molecular mechanisms using in vivo transgenic or knockout models and in vitro experiments using *inducible expression systems, transdominant negative adenoviral constructs* and *RNA interference*. Recent studies investigated how polyunsaturated fatty acids (PUFAs) influence lipid metabolism using gene knockout models. PUFAs usually induce the expression of several genes related to fatty-acid oxidation. However, peroxisome proliferator-activated receptor-α (PPARα)-null mice lack this response suppressing lipogenesis [31.27]. This emphasizes the effects of dietary compounds on molecular pathways, directly affecting gene expression. Another approach to analyzing how nutrients can induce highly specific changes in the whole genome is *laser-capture microdissection* for single-cell gene-expression profiling. This technique enhances cell-specific information from in vivo nutrition experiments [31.28].

Other than in the nutrigenomics field, the use of primary cells and immortalized cell lines are additional tools for studying the effects of nutrients on individual gene expression; these applications permit the investigation of food components and essential nutrients as factors in the control of gene expression. For instance, dietary cholesterol exerts an intense inhibitory effect on the transcription of the gene for β-hydroxy-β-methyl-glutaryl (HMG)-CoA reductase or dietary PUFA suppress the hepatic messenger RNA (mRNA) production of fatty acid synthase for lipoproteinemia in adult and weanling rats [31.29, 30]. This ability to suppress the abundance of mRNA for lipogenic proteins is dependent on the degree of fatty acid unsaturation and the type [31.31]. The molecular effect of dietary n–3 PUFA are exerted by a synergic action that involves triacylglyceride (TG) lowering, anti-inflammatory activity, inflammation-resolving, and regulation of transcription factors, such as PPARs, NF-kB, and SREBPs [31.32].

Polyunsaturated fatty acids and more specifically n–3 docosahexaenoic and eicosapentaenoic acid (DHA and EPA, respectively) are well-known to be able to improve serum lipid composition and reduce cardiovascular risk. The research aimed to identify the molecular mechanisms accounting for these effects showed various interesting points. These polyunsaturated acids are certainly able to affect gene expressions in numerous cases. For example, Bouwens et al. performed microarray analysis on PBMC (peripheral blood mononuclear cells) RNA from 23 subjects who received 1.8 g EPA + DHA/d compared to 25 subjects who received high-oleic acid sunflower oil (HOSF) capsules. They found that high EPA + DHA intake changed the expression of 1040 genes, whereas HOSF intake changed the expression of only 298 genes. EPA + DHA intake resulted in a decreased expression of genes involved in inflammatory and atherogenic-related pathways, such as nuclear transcription factor kappaB signaling, eicosanoid synthesis, scavenger receptor activity, adipogenesis and hypoxia signaling [31.33].

Several bioactive nutrients are known to impact gene expression by activating or suppressing specific transcription factors. The most important group of transcription factors reported as mediating the effect of several nutrients and their metabolites on gene transcription is the superfamily of nuclear receptors, such as PPAR, SREBP, etc. PPARs are likely to have evolved as dietary lipid sensors. Accordingly, it can be hypothesized that PPARα is activated by changes in dietary fat load, for example by high fat feeding. Comparative microarray analysis of PPARα-dependent gene regulation induced by a synthetic PPARα agonist, by prolonged fasting or by high fat feeding indicated, although all treatments caused activation of PPARα, pronounced differences in the magnitude of PPARα-target gene induction [31.34].

Several studies aimed to identify a pre-disease or nutritionally specific state in blood cells. For instance, inter and intra-individual variations in blood cell gene expression were performed in healthy volunteers [31.35]. Expression patterns within a healthy person at different intervals over a 6-month period were highly concordant, but variation in gene expression between persons varied significantly. So it is important to consider this large inter-individual variation for distinguishing gene-expression signatures of healthy subjects from a pre-disease status. Once those pre-disease signatures are been defined and validated, nutrition interventions that focus on those pre-disease biomarkers can be developed, with the goal of pursuing healthful expression patterns and then better physical conditions.

To identify molecular markers of diet-related diseases, *Goyenechea* et al. evaluated, in peripheral blood mononuclear cells (PBMC), gene expression in relation to personalized obesity therapy. PBMC were obtained from obese men before and after an 8-week low-calorie diet (LCD) to lose weight [31.36]. Changes in gene expression before and after the LCD were firstly analyzed by a DNA-microarray platform and then validated by quantitative RT-PCR of selected species. Global gene expression data showed 385 differentially expressed transcripts after LCD. In addition, there was a decreased expression of genes related to oxidative stress and inflammation mechanisms. Interestingly, a lower IL8 gene expression was connected to higher fat mass decrease. These results cooperatively suggested that PBMC are a suitable RNA source and system model to perform nutrigenomics studies associated to obesity and IL8

gene expression may be a putative novel biomarker of changes in body fat quota.

As reported above, transcriptomics techniques can analyze expression levels of thousands of genes at the same time and within a single study, but these studies require substantial quantities of tissue for RNA isolation. The accessibility of human tissues is limited and has been a limit for this analysis. Nevertheless, biopsy samples from adipose tissue and muscle can be taken with low risk to investigate gene expression that can reflect a tissue-specific effect of a pathophysiological condition. For example, pro-inflammatory response observed in adipose tissue of obese people corresponds to an increased expression of pro-inflammatory genes in this tissue [31.37]. Another less-invasive possibility is to isolate RNA from circulating white blood cells. Disease-specific gene-expression patterns in blood cells have been identified for breast tumors and leukemia, and those patterns are now used as biomarkers and diagnostic tools for these diseases [31.38, 39].

A major point in transcriptomics analyses is data reproducibility and comparability with those resulting from similar studies. For this purpose, RNA quality and quantity must be strictly verified and all execution steps must be performed in highly controlled conditions, within the same microarray experiment and in conditions allowing the best normalization with other experiments. In order to enhance accuracy, pooling of samples is not desirable. On the contrary, it is better to increase the number of biological replicates to decrease false-positive rates, resulting in more reliable data. Another very important point is data processing. Since dietary studies generally result in small gene-expression changes, they require an even more reliable algorithm than with other differential studies. The most challenging part of these processes occurs after array data are obtained and have then to be interpreted in terms of biological meaning. Several commercial and non-commercial tools have been developed to obtain information about markedly changed genes, pathways, or related networks. The most common aim of transcriptome analysis is to find genes that are differentially expressed between the various experimental samples. Although early microarray papers used a simple *fold change* approach to generate lists of differentially expressed genes, most analyses now rely on more reliable statistical tests to identify differences in expression between groups [31.40].

31.3 Effect of Bioactive Nutrients on Metabolite Profiling

Metabolites are all the different organic and inorganic molecules produced by biochemical pathways of organisms, necessary for maintenance, growth, and physiological and pathological functions within cellular processes. Biochemical pathways are series of chemical reactions each catalyzed by specific protein enzymes. As all other proteins, even enzymes are strictly regulated in their expression. Therefore genome regulation, protein expression, and metabolite production are all highly related each other. Individual metabolite levels can then be considered as the last step of biological systems according to genetic or environmental factors and among these nutritional intake has a crucial role.

Metabolome, from the term *genome*, examines the complete set of metabolites in an organism, classifying and quantifying them individually in a biofluid, cell culture, or tissue sample through analytical methods. Their measures give information on how the enzymes and all other functional proteins act in cellular homeostatic mechanisms [31.41]. Nutrients can directly interact with our body at organ, cellular, and molecular levels. They usually come in complex mixtures, in which both the amount of a single compound and its interaction with multiple components are important, since this squarely influences their bioavailability and bioefficacy. In addition, metabolites have very different chemical and physical properties due to wide variations in atomic arrangements. In this context, several studies on metabolomes permitted the detection of a wide range of chemical species, from low and high molecular weight polar volatiles, non-polar lipids to inorganic species; since their concentrations can vary over nine orders of magnitude (pM–mM), this provides a hard challenge in metabolomics strategies [31.42].

Thus metabolomics allows the systematic investigation of small organic molecules and within nutrigenomics its main interest is to recognize how those molecules can reflect the effects of different diets, indicating an interactive and regulatory role of nutrition [31.43]. Metabolomics techniques can be essentially divided into two distinct parts. Firstly, a global analytical approach needs to examine biofluids, tissue extracts, or intact tissues by using ^{1}H nuclear magnetic resonance (NMR) spectroscopy, gas chromatography–mass spectrometry (GC–MS), and liquid chromatography–mass spectrometry (LC–MS); other methods include Fourier transform infrared (FT-IR) spectroscopy, high-performance liquid chromatography (HPLC), or capillary electrophoresis (CE)

in conjunction with diode or coulometric arrays and ion-cyclotron resonance Fourier transform (ICR-FT) MS. For example, using HPLC coulometric arrays, Kristal and co-workers investigated serotypes in mice undergoing dietary restriction compared to a normal diet condition [31.44].

High-resolution ^{1}H NMR spectroscopy, for example, measures metabolites in urine, blood plasma, or sera and in tissue extracts using a simple one-dimensional pulse sequence even if it can be used only for a limited number of analytes. Despite this disadvantage, NMR-based metabolomics, for instance, has been successful in distinguishing liver and kidney toxins in rats, mouse models of cardiac diseases, and mutants in yeast. Many metabolites take part in signaling, receptor binding, translocation and other reaction pathways and can directly be influenced by nutritional intake; thus by NMR analysis it is possible to study the metabolism interactions, since a perturbation at one point in the network, inducted, for example, by dietary signals, can affect another pathway, as a result of a highly connected hub [31.45].

Furthermore, alternative approaches are represented by GC–MS and LC–MS for measuring metabolic fingerprints of high-concentration metabolites, since these methods enable the detection of unique biomarkers. MS-based techniques are improving in both instrument design and data-processing software, in fact chromatography is making progress in the form of two-dimensional GC and more sensitive time-of-flight mass spectrometers. Rapid metabolic profiling of gender, age, and strains of mouse been enabled by the development of ultra-performance liquid chromatography (UPLC). Research is actually focused on examining the advantages of more recent mass spectrometry, such as ion cyclotron resonance (ICR)-FT-MS, matrix assisted laser desorption/ionization (MALDI)-MS, and capillary electrophoresis–MS. These types of metabolomics approaches suggest enormous chances for detection of the metabolic signatures profiling in according to different diets [31.45].

All dietary components, even without energetic value but with potential metabolic effects, must be included into metabolome analysis and may be critically important in human dietary studies that seek to use metabolomics techniques. The main idea of nutritional metabolomics is to detect and identify all endogenous human metabolites and exogenous components from food that coexist at least transiently in human biofluids. The multiple metabolites profile, corresponding to specific nutrition states, may help to characterize physiological or pathological conditions and certainly has more interest and a greater possibility of being used as biomarkers than a single molecule. Also the integration of this information with that from the other *omics* technologies will result in relevant information [31.46].

In this field, therefore, the biggest challenge is to match relate data from all these different approaches. *Griffin* used a combination of NMR, GC–MS, LC–MS, and ICR-FT-ICR MS with the aim of providing information about the highest number of metabolites in a tissue or biofluid [31.47]. A crucial step is to profile a metabolic phenotype, or *metabotype* related to a certain condition. Genetically determined metabotypes may provide the risk assessment for a certain disease phenotype, or the possible response to a given drug treatment, a nutritional intervention, or an environmental stimulus. *Gieger* et al. identified genetic variants linked to changes in the metabolic homeostasis of key lipids, carbohydrates, or amino acids [31.48]. They measured 363 serum metabolites in 284 male participants of the KORA study, finding correlation between SNPs and metabolite serum profiles. They identified four genetic variants in gene coding for several enzymes (FADS1, LIPC, SCAD, MCAD) where the corresponding metabolic phenotype (metabotype) clearly matches the biochemical pathways in which these enzymes are active.

As another aspect of metabolomics, lipidomics is a direct way to examine pathways and networks of cellular lipids. A recent study compared VLDL, IDL, LDL, HDL2, and HDL3 fraction profiles in insulin-resistant subjects *vs* healthy controls. The data showed that serum concentrations of specific triacylglycerides species, such as TG (16 : 0/16 : 0/18 : 1) or TG (16 : 0/18 : 1/18 : 0), may be more precise markers of insulin resistance than total serum TG concentration. These lipoprotein-specific changes may thus have an important role in regulating metabolism in peripheral tissues related to nutritional style [31.49].

A metabolic profile usually undergoes physiological variations. Therefore, it is a challenge to separate such variations from those due to nutrition related changes. Studies of acute dietary intake effect on urinary, plasma, and salivary metabolites show that consumption of a standard diet the day before sample collection can reduce biological variation in the urinary metabolic profiles of healthy individuals but not in plasma or salivary metabolic profiles [31.50]. In this context, it is also important to remember that all nutrients can also be subjected to metabolic conversions by gut microflora. Accordingly, large differences should be expected in urinary metabolomics profiling depending on the features of gut microflora.

The formation of microbial metabolites can be measured from complex food mixtures and isolated compounds with tailor-made approaches considering the conversion rate of the substrates and resistance of the substrate matrix. For example, in vitro microbial metabolism of flavanols, (+)-catechin and (−)-epicatechin was studied using a GC × GC-TOF/MS based metabolomic analysis to understand the effect of stereoisomerism on metabolic profiles [31.51].

Metabolomics plays a role in dietary assessment and identification of novel biomarkers that correspond to a certain dietary intake and can be associated to metabolites profile. For instance, ^{1}H NMR spectra of urine samples allowed identification of several metabolites matching the intake of specific dietary compositions [31.52].

In particular, within clinical applications related to dietary or lifestyle interventions, *Lankinen* et al. studied the effect of carbohydrates on serum metabolic profiles in subjects affected by metabolic syndrome (the typical condition in which serum dyslipidemia, obesity, hypertension, and altered glycaemia synergize in increasing the risk of cardiovascular diseases), as well as linking the metabolic profiles to specific gene expression pathways in adipose tissue [31.53]. Dietary carbohydrate modifications may alter pro-inflammatory serum lipid concentrations and may thus induce pro-inflammatory processes with adverse changes in insulin and glucose metabolism, which are part of the pathogenetic mechanisms involved in metabolic syndrome development. The same group, moreover, showed how lean fish and fatty fish vary serum lipidomic profiles in subjects with coronary heart disease [31.54]. Bioactive lipid species, such as ceramides, lysophosphatidylcholines, and diacylglycerols, were markedly reduced in the fatty fish exposed group. These changes are likely to be related to the protective effects of nutrients derived from fatty fish in atherosclerotic vascular diseases or insulin resistance.

Metabolomics was also applied to classify subjects according to fitness level during a recovery phase following strenuous physical exercise. Interestingly, nutritional modulation, obtained administrating a low carbohydrate-protein beverage, was shown to improve the metabolic profile of less fit subjects in the recovery phase [31.55].

Perspective developments in metabolomics and nutrition research will certainly include studies characterizing in more detail acute and chronic effects of diet on biofluid metabolomes, particularly regarding the effect of colonic microflora, low-residue diets, and the determination of the rates of change of human biofluid metabolomes, according to different dietary interventions. Similarly to all other nutrigenomic strategies, the major point will remain the combined bioinformatic integration of all metabolomic data from the different studies, with the different nutrients and with the different technologies.

31.4 Bioactive Effect of Nutrients on Protein Expression

All nutrient molecules, and particularly bioactive nutrients, affect cell metabolic functions in complex ways from gene expression to protein synthesis and degradation. *Nutriproteomics* is a recent branch of proteomics and systematically studies proteins structure and function, as well as protein–protein interactions, aiming to identify the molecular targets of dietary components [31.56].

This term was coined to make an analogy with genomics and to indicate the large-scale systematical analysis of proteins. Even in this case the goal is to identify differential elements in protein patterns under a certain condition, for example before and after a determinate dietary treatment. These analyses includes many different levels: *expression quantitative proteomics* (identification of proteins differently abundant in compared samples), *functional proteomics* (identification of specific protein–protein interactions), *post-translation modification proteomics* (identification of enzymatic or non-enzymatic protein modifications), *chemical proteomics* (identification of specific relations between proteins and specific chemical structures), and *structural proteomics* (identification of specific 3D protein structures). Technological advances in mass spectrometry enabled a great development in proteome research. The main challenges of proteomics concern the considerable dynamic range of protein concentrations and the variability of their physicochemical properties (solubility, size, hydrophobicity, etc.). Moreover proteomics in nutrition contributes to investigate all other *omics* analysis levels allowing their complete integration and synergy.

The transcription level of a gene generally gives a rough idea of its expression level into an encoded protein but this can be changed by many factors. In fact, a large amount of mRNA may be rapidly degraded or inefficiently translated, resulting in a low amount of protein. In addition, many proteins go through post-translational modifications that profoundly influence their activities; for instance some proteins are activated by phosphorylation. Methods such as phospoproteomics and glycoproteomics are, therefore, used to evaluate these post-translational modifications [31.56, 57]. Many transcripts result in more than one protein, through alternative splicing or alternative post-translational modifications. Finally, protein content is also adjusted by the degradation rate of these molecules.

Thus, proteomics analyzes at various levels the effect of dietary components, allowing the investigation of peptides as bioactive markers. Plasma proteomics is becoming important in nutrition and nutrigenomics research since it enables us to obtain information about a substantial number of proteins from easily accessible plasma samples [31.58]. Mass spectrometric techniques are able to identify proteins at a large scale and with high throughput, calculating their exact masses and their amino acid sequences [31.59]. The main ionization methods are electrospray and matrix-assisted laser desorption, which separate the ions by mass over charge. The major analyzers in proteomics are ion traps, triple-quadrupole (triple-Q), time-of-flight (ToF) tubes, orbitrap, and Fourier-transform ion-cyclotron resonance (FT-ICR) cells, with their specific advantages, such as high sensitivity and multiple-stage fragmentation for ion traps; high selectivity for triple-Q; high sensitivity and speed for ToF; very high mass accuracy and resolution for orbitrap and FT-ICR [31.60, 61]. Actually MS-based proteomic platforms also detect and quantify proteins present in low concentration both by depletion of abundant proteins or by selective enrichment. After depletion and/or enrichment, two-dimensional (2-D) gels, liquid chromatography (LC), or hybrid approaches (Gel-LC) are usually assayed as preseparation methods. Gel-based protein separation methods provide the advantage of physically preserving protein context and of generating real protein images. However, the dynamic range is limited, there is bias toward more soluble proteins, and there is a low degree of automation with low throughput [31.60, 61]. Differential imaging gel electrophoresis currently represents the main advanced technique for 2-D protein separation, based on multiplexed staining and co-processing of one control plus a maximum of two case samples. Selected differential protein spots are then detected, excised, digested with trypsin, and amended to LC-MS/MS for further analysis [31.60, 61]. Moreover, (multi)dimensional LC setups have been coupled online to MS analysis, with simple reversed-phase columns and combined strong cation exchange-reversed phase systems being the most frequently applied. These workflows are known as MudPIT (multidimensional protein identification technology) or shotgun proteomics [31.60, 61].

In the matter of protein quantification, the main strategies and steps are stable isotope rooted or label free, metabolic, or chemical labeling, providing relative or absolute quantitative information. Alternatively to differential imaging gel electrophoresis and compatible with online shotgun LC-MS/MS workflows, stable isotopes can be used to tag amino acid side chains, quantitatively comparing the results obtained to the MS method [31.60, 61].

Concerning clinical applications, proteomic techniques can investigate on nutritionally relevant biological pathways and on dietary interventions. For instance, to better clarify the molecular basis in functional gut disorders induced by environmental stress, a rat intestine proteomic catalog was obtained and stress effects on intestinal protein expression were evaluated [31.62]. Proteomics, also combined with gene expression analysis, was used in cancer prevention studies to identify innovative biomarkers. Thus, *Breikers* et al. analyzed several differentially expressed proteins in the colonic mucosa of healthy mice fed with a vegetable enriched diet [31.63]. Only six proteins showed altered expression levels in treatment mice *vs* control mice, which is likely to play a potential role in colorectal cancer protection associated with a vegetable enriched diet. Moreover, *Tan* et al. assessed sodium butyrate effects on growth inhibition of human HT-29 cancer cells by using a 2-D MS-based proteomic strategy [31.64]. Butyrate treatment induced altered protein expression, in particular related to the ubiquitin–proteasome pathway, suggesting that proteolysis may be a crucial mechanism by which butyrate regulates key proteins in control of the cell cycle, apoptosis, and differentiation. Combining gene and protein expression profiling in colonic cancer cells, *Herzog* et al. identified the flavonoid flavone, present in a variety of fruits and vegetables, as a potent apoptosis inducer through several heat-shock proteins, annexins, and cytoskeletal caspases [31.65]. Furthermore, *Daniel* et al. investigated the effect of a zinc-deficient diet, analyzing hepatic transcriptome,

proteome, and lipidome. By the combined *omics* analysis, they were able to identify factors in glucose and lipid metabolism, which affect liver lipid accumulation and hepatic inflammation [31.66]. Zhang et al. studied the protein profiling of fructose-induced fatty liver in hamsters. High fructose consumption is connected to development of fatty liver and dyslipidemia. Matrix-assisted laser desorption ionization–MS-based proteomic analysis of the liver tissue from those hamsters revealed a number of proteins whose expression levels were altered more than twofold after fructose administration. Identified proteins were grouped into categories as related to fatty acid metabolism, cholesterol, and triacylglycerol metabolism, molecular chaperones, enzymes in fructose catabolism, and proteins with housekeeping functions [31.67].

These nutritional intervention studies delve into specific gene/protein abundance changes. However, several food components may not only affect gene and protein expression but also target post-translational modifications. In this context, applied proteomic techniques can be useful to detect specific differences. For instance, the phosphorylation status of extracellular signal-regulated kinase (ERK) changes after exposure to diallyl disulfide, a dietary compound present in processed garlic, inducing cell cycle arrest [31.68]. Another example is given by Saleem and co-workers, who used a quantitative, integrated, phosphoproteomics approach to characterize the cellular responses of the yeast *Saccharomyces cerevisiae* to the fatty acid oleic acid, a molecule with broad human health implications and a potent inducer of peroxisomes. Using phosphoproteomic approaches, numerous phosphorylated peptides specific to oleate-induced and glucose-repressed conditions were identified and mapped to known signaling pathways. These include several transcription factors, two of which, Pip2p and Cst6p, must be phosphorylated for the normal transcriptional response of fatty acid-responsive loci encoding peroxisomal proteins [31.57].

All these proteomic approaches can provide a appropriate toolsets in nutritional studies to better recognize and compare the effects of whole diet or a single nutrient on protein expression and regulation in humans. A quantifiable change relating a normal or pathological condition with a modulation of an mRNA, a protein, or a metabolite concentration can be used as a molecular biomarker. Proteomics applied to nutrition research promises to provide specific and sensitive useful biomarkers with potentiality to monitor links between certain diet intakes and a clinical conditions. A protein concentration or its modification status can easily become a diagnostic tool, it being relatively easy to have precise and reproducible clinical diagnostic assays for these analytes [31.69]. Dietary dose levels of most nutrients are only weakly biologically active and may have several targets. In selecting a biomarker, the timing of its responses should be considered according to nutrient specific bioavailability and bioefficacy. Biomarkers may correlate with nutrient intake, but often their modulation is a combined result of intake, absorption, metabolism, and excretion. Also environmental factors and genetic predisposition may modulate the correlation between dietary intake and biomarkers. For all these reasons it can be expected that best results will not come from a single proteomic biomarker but from their multiple combination also with the information available from the other *omics* technologies.

31.5 Systems Biology and Bioinformatics Tools

The term *systems biology* was coined to indicate cross-disciplinary research in biology. Biochemical systems biology includes and put together several traditional disciplines, such as genomics, biochemistry and molecular biology, though mathematical and computational analysis, engineering practices and *omics* platform technologies, such as transcriptomics, proteomics and metabolomics [31.70].

The progress of high-throughput technologies has led to a substantial rise of functional genomics data. Different experimental strategies combine quantitative measurements of cellular components (mRNA, protein, and metabolite) with the development of mathematical and computational models as an equal partner [31.71, 72]. These very large amounts of information can be easily collected even at the nutritional level. Nevertheless all this information will be conveniently and efficiently used taking advantage of computational bioinformatics, which will provide a general view of data and *keys* to read the most relevant information for clinical perspectives [31.73].

A considerable challenge in systems biology is the development of data collections, data standards, and software tools for simulation, analysis and visualization of system components such as biochemical networks. The high-throughput molecular profiling methods need elaborated data processing and analysis and require signal processing and statistical analysis. Quantitative systems biology models have led to the development of the systems biology markup language (SBML) [31.74]. The SBML project contributed to improve computer-readable formats to describe biological processes with high fidelity and to outline biological connections by specific software tools. The graphical representation of biological processes was developed recently (systems biology graphical notation, SBGN) [31.75]. Actually, SBGN requires three complementary languages to characterize biological pathways and relationships among individual cellular components. The systems biology workbench (SWB) [31.76] is a framework that consents to several components to interact, exchange models via SBML, using programs for editing biochemical networks, tools for simulation, and to import and translate models. CellDesigner [31.77] is a helpful Java-based program for building and viewing biochemical networks, enabling to import models in SBML and display biochemical pathways, relied on in SBGN process diagram language. CellDesigner models can use either a built-in simulator or an external simulator, such as those provided by SWB. An example of an independent simulator is COPASI [31.78], which can simulate models based on ordinary differential equations (ODEs) as well as stochastic models by using Gillespie's algorithm. COPASI requires tools for the visual analysis of simulation data and carries out steady-state and metabolic control analyses.

All these recent tools join pre-existing traditional databases such as GenBank [31.79] or Protein Data Bank [31.80], which were started in the 1980s while systematical knowledge of genes and proteins was beginning. While these primary databases started collecting gene and protein sequence data, they now provide, in addition to experimentally derived information, the basis for the development of secondary databases and their integration. An example of such a secondary database is Pfam [31.81], which includes details about protein families and domains. Parallel to the progress of systems biology, a significant part of secondary biological databases is represented by the BioCyc project [31.82], which includes a database of human biological pathways [31.83], a database of interactions between small molecules and proteins [31.84], and databases of protein–protein interactions [31.85]. The gene ontology (GO) project collects three hierarchically structured vocabularies, clarifying interaction among gene products, biological processes, molecular functions and cellular components.

Cytoscape [31.86], an open-source software platform, visualizes molecular networks and combines these connections with gene expression profiles. The biological networks gene ontology (BiNGO), an open-source Java tool, identifies overrepresented classes in a pool of genes using GO. BiNGO provides either a list of genes or biological network graphs. It takes advantage of Cytoscape's program to generate an intuitive representation of the data [31.87].

These databases seek to organize and connect information about cellular components and all different pathways, editing higher-level knowledge and theories about the biological processes. While it is reasonable to expect much additional knowledge deriving from these research tools, also at the nutritional level, the results at the diagnostic level will also depend on the definition of analytical procedures with both reliability and applicability to clinical use.

31.6 Conclusions and Future Perspectives

Nutritional genomics is a growing research area and aims to improve general population dietary guidelines and individual nutrition recommendations. It mainly matches genome features with several environmental factors, including nutritional intake. The purpose of genetics is to investigate predisposition for all diseases with a genetic component and to provide the tools for its prevention. Nutrigenetics, in parallel with nutrigenomics, connects observational findings to molecular mechanisms and is aimed at describing nutritionally relevant genetic phenotypes, considering the biological variations. Nutrigenomic and nutrigenetics will also support pharmacogenomics and pharmacogenetics and will be integrated and applied in the therapy and prevention of many diseases, with individualized prescriptions and lifestyle modifications. Populations will be stratified into subgroups according to their genotypes and dietary histories to predict responsiveness to appropriate nutrient and drug interventions.

Another possible development in nutrition omics studies will likely regard the study of epigenetic changes. Epigenetics regard changes in gene expression caused by mechanisms other than changes in the underlying DNA sequence. These variations are studied and detected during gestation, neonatal development, puberty, and old age. Therefore a systematical analysis of epigenetic modifications in relation to quality and quantity of nutritional intake must also be collected and integrated with the other information, certainly with relevant perspectives.

The final goal of omics procedures is the complete definition in all molecular processes of an organism both from a static and from a dynamical view. There is wide evidence in the literature, which describes how both quality and quantity of nutrition can influence the risk of most important human disease, such as cardiovascular and neoplastic conditions, but details of molecular processes are only partially clarified. The use of these new technologies can then provide wide and complete information on these processes involved in the interplay between nutrition and health. Nutritional system biology development will certainly make it possible to identify new and more effective diagnostic biomarkers and target improved therapies.

It is conceivable that a detailed knowledge of the complex of molecular processes involved in the effects of nutrition will open important perspectives. This will happen when it will be possible to manage these information collecting reference profiles of data integrating all levels of omic analysis and will provide real applications of molecular and individually personalized medicine. This will include individually tailored nutrition plans to maximize disease prevention and highly sensitive and specific diagnostic biomarkers to assess nutrition related processes. While current biological technologies allow us to collect all the necessary data to reach these goals, the major future developments will be their integration with all areas of bioinformatics, including ontology systems and computational intelligence.

References

31.1   J.M. Ordovas, D. Corella: Nutritional genomics, Annu. Rev. Genomics Hum. Genet. **5**, 71–118 (2004)

31.2   B. Armstrong, R. Doll: Environmental factors and cancer incidence and mortality in different coun-

tries, with special reference to dietary practices, Int. J. Cancer **15**, 617–631 (1975)

31.3   M.L. Neuhouser, R.E. Patterson, I.B. King, N.K. Horner, J.W. Lampe: Selected nutritional

Part F | 31

biomarkers predict diet quality, Public Health Nutr. **6**, 703–709 (2003)

31.4   S. Eilat-Adar, U. Goldbourt: Nutritional recommendations for preventing coronary heart disease in women: Evidence concerning whole foods and supplements, Nutr. Metab. Cardiovasc. Dis. **20**(6), 459–466 (2010)

31.5   J.A. Milner: Molecular targets for bioactive food components, J. Nutr. **134**(9), 2492S–2498S (2004)

31.6   D. Eletto, A. Leone, M. Bifulco, M.F. Tecce: Effect of unsaturated fat intake from mediterranean diet on rat liver mRNA expression profile: Selective modulation of genes involved in lipid metabolism, Nutr. Metab. Cardiovasc. Dis. **15**, 13–23 (2005)

31.7   S.A. Lee: Gene-diet interaction on cancer risk in epidemiological studies, J. Prev. Med. Public Health **42**(6), 360–370 (2009)

31.8   M.A. Roberts, D.M. Mutch, J.B. German: Genomics: Food and nutrition, Curr. Opin. Biotechnol. **12**, 516–522 (2001)

31.9   W.E. Evans, J.A. Johnson: Pharmacogenomics: The inherited basis for interindividual differences in drug response, Annu. Rev. Genomics Hum. Genet. **2**, 9–39 (2001)

31.10   W.E. Evans, H.L. McLeod: Pharmacogenomics – drug disposition, drug targets, and side effects, N. Engl. J. Med. **348**, 538–549 (2003)

31.11   J. Kaput: Diet-disease gene interactions, Nutrition **20**, 26–31 (2004)

31.12   I.A. Brouwer, P.L. Zock, L.G. van Amelsvoort, M.B. Katan, E.G. Schouten: Association between n-3 fatty acid status in blood and electrocardiographic predictors of arrhythmia risk in healthy volunteers, Am. J. Cardiol. **89**, 629–631 (2002)

31.13   F.M. Sacks, M. Katan: Randomized clinical trials on the effects of dietary fat and carbohydrate on plasma lipoproteins and cardiovascular disease, Am. J. Med. **113**, S13–S24 (2002)

31.14   L. Ohlsson: Dairy products and plasma cholesterol levels, Food Nutr. Res. **54**, 5124 (2010)

31.15   G.A. Francis, E. Fayard, F. Picard, J. Auwerx: Nuclear receptors and the control of metabolism, Annu. Rev. Physiol. **65**, 261–311 (2002)

31.16   W.C. Willett: Nutritional epidemiology issues in chronic disease at the turn of the century, Epidemiol. Rev. **22**, 82–86 (2000)

31.17   G.A. Churchill: Fundamentals of experimental design for cDNA microarrays, Nat. Genet. **32**, 490–495 (2002)

31.18   C.K. Lee, D.B. Allison, J. Brand, R. Weindruch, T.A. Prolla: Transcriptional profiles associated with aging and middle age-onset caloric restriction in mouse hearts, Proc. Natl. Acad. Sci. USA **99**, 14988–14993 (2002)

31.19   S.X. Cao, J.M. Dhahbi, P.L. Mote, S.R. Spindler: Genomic profiling of short- and long-term caloric restriction effects in the liver of aging mice, Proc. Natl. Acad. Sci. USA **98**, 10630–10635 (2001)

31.20   J. Xiao, S. Gregersen, M. Kruhøffer, S.B. Pedersen, T.F. Ørntoft, K. Hermansen: The effect of chronic exposure to fatty acids on gene expression in clonal insulin-producing cells: Studies using high density oligonucleotide microarray, Endocrinology **142**, 4777–4784 (2001)

31.21   J.H. Park, J. Ahn, S. Kim, D.Y. Kwon, Y.T. Ha: Murine hepatic miRNAs expression and regulation of gene expression in diet-induced obese mice, Mol. Cells **31**(1), 33–38 (2010)

31.22   B.D. Reed, A.E. Charos, A.M. Szekely, S.M. Weissman, M. Snyder: Genome-wide occupancy of SREBP1 and its partners NFY and SP1 reveals novel functional roles and combinatorial regulation of distinct classes of genes, PLoS Genetics **4**(7), e1000133 (2008)

31.23   W.W. Grody: Molecular genetic risk screening, Annu. Rev. Med. **54**, 473–490 (2003)

31.24   R. Sachidanandam, D. Weissman, S.C. Schmidt, J.M. Kakol, L.D. Stein, G. Marth, S. Sherry, J.C. Mullikin, B.J. Mortimore, D.L. Willey, S.E. Hunt, C.G. Cole, P.C. Coggill, C.M. Rice, Z. Ning, J. Rogers, D.R. Bentley, P.Y. Kwok, E.R. Mardis, R.T. Yeh, B. Schultz, L. Cook, R. Davenport, M. Dante, L. Fulton, L. Hillier, R.H. Waterston, J.D. McPherson, B. Gilman, S. Schaffner, W.J. Van Etten, D. Reich, J. Higgins, M.J. Daly, B. Blumenstiel, J. Baldwin, N. Stange-Thomann, M.C. Zody, L. Linton, E.S. Lander, D. Altshuler: International SNP map working group: A map of human genome sequence variation containing 1.42 million single nucleotide polymorphisms, Nature **409**, 928–933 (2001)

31.25   N. Blau, F.J. van Spronsen, H.L. Levy: Phenylketonuria, Lancet **376**(9750), 1417–1427 (2010)

31.26   P. Terry, P. Lichtenstein, M. Feychting, A. Ahlbom, A. Wolk: Fatty fish consumption and risk of prostate cancer, Lancet **357**(9270), 1764–1766 (2001)

31.27   A. Morise, C. Thomas, J.F. Landrier, P. Besnard, D. Hermier: Hepatic lipid metabolism response to dietary fatty acids is differently modulated by PPARalpha in male and female mice, Eur. J. Nutr. **48**(8), 465–473 (2009)

31.28   T. Dreja, Z. Jovanovic, A. Rasche, R. Kluge, R. Herwig, Y.C. Tung, H.G. Joost, G.S. Yeo, H. Al-Hasani: Diet-induced gene expression of isolated pancreatic islets from a polygenic mouse model of the metabolic syndrome, Diabetologia **53**(2), 309–320 (2010)

31.29   C. Ness, C.M. Chambers: Feedback and hormonal regulation of hepatic 3-hydroxy-3-methylglutaryl coenzyme a reductase: The concept of cholesterol buffering capacity gene, Proc. Soc. Exp. Biol. Med. **224**(1), 8–19 (2000)

31.30   S.K. Cheema, M.T. Clandinin: Dietary fat-induced suppression of lipogenic enzymes in B/B rats during the development of diabetes, Lipids **35**(4), 421–425 (2000)

31.31 G. Schmitz, J. Ecker: The opposing effects of *n*-3 and *n*-6 fatty acids, Prog. Lipid Res. **47**(2), 147–155 (2008)

31.32 R.S. Chapkin, W. Kim, J.R. Lupton, D.N. McMurray: Dietary docosahexaenoic and eicosapentaenoic acid: Emerging mediators of inflammation, Prostaglandins Leukot. Essent. Fat. Acids **81**(2–3), 187–191 (2009)

31.33 M. Bouwens, O. van de Rest, N. Dellschaft, M.G. Bromhaar, L.C. de Groot, J.M. Geleijnse, M. Müller, L.A. Afman: Fish-oil supplementation induces antiinflammatory gene expression profiles in human blood mononuclear cells, Am. J. Clin. Nutr. **90**(2), 415–424 (2009)

31.34 M. Bünger, G.J. Hooiveld, S. Kersten, M. Müller: Exploration of PPAR functions by microarray technology – a paradigm for nutrigenomics, Biochim. Biophys. Acta **1771**(8), 1046–1064 (2007)

31.35 J.P. Cobb, M.N. Mindrinos, C. Miller-Graziano, S.E. Calvano, H.V. Baker, W. Xiao, K. Laudanski, B.H. Brownstein, C.M. Elson, D.L. Hayden, D.N. Herndon, S.F. Lowry, R.V. Maier, D.A. Schoenfeld, L.L. Moldawer, R.W. Davis, R.G. Tompkins, H.V. Baker, P. Bankey, T. Billiar, B.H. Brownstein, S.E. Calvano, D. Camp, I. Chaudry, J.P. Cobb, R.W. Davis, C.M. Elson, B. Freeman, R. Gamelli, N. Gibran, B. Harbrecht, D.L. Hayden, W. Heagy, D. Heimbach, D.N. Herndon, J. Horton, J. Hunt, K. Laudanski, J. Lederer, S.F. Lowry, R.V. Maier, J. Mannick, B. McKinley, C. Miller-Graziano, M.N. Mindrinos, J. Minei, L.L. Moldawer, E. Moore, F. Moore, R. Munford, A. Nathens, G. O'keefe, G. Purdue, L. Rahme, D. Remick, M. Sailors, D.A. Schoenfeld, M. Shapiro, G. Silver, R. Smith, G. Stephanopoulos, G. Stormo, R.G. Tompkins, M. Toner, S. Warren, M. West, S. Wolfe, W. Xiao, V. Young: Inflammation and Host Response to Injury Large-Scale Collaborative Research Program: Application of genome-wide expression analysis to human health and disease, Proc. Natl. Acad. Sci. USA **102**, 4801–4806 (2005)

31.36 E. Goyenechea, A.B. Crujeiras, I. Abete, J.A. Martínez: Expression of two inflammation-related genes (RIPK3 and RNF216) in mononuclear cells is associated with weight-loss regain in obese subjects, J. Nutrigenet. Nutrigenomics **2**(2), 78–84 (2009)

31.37 S.P. Weisberg, D. McCann, M. Desai, M. Rosenbaum, R.L. Leibel, A.W. Ferrante Jr: Obesity is associated with macrophage accumulation in adipose tissue, J. Clin. Invest. **112**, 1796–1808 (2003)

31.38 K.J. Martin, E. Graner, Y. Li, L.M. Price, B.M. Kritzman, M.V. Fournier, E. Rhei, A.B. Pardee: High-sensitivity array analysis of gene expression for the early detection of disseminated breast tumor cells in peripheral blood, Proc. Natl. Acad. Sci. USA **98**, 2646–2651 (2001)

31.39 A.R. Whitney, M. Diehn, S.J. Popper, A.A. Alizadeh, J.C. Boldrick, D.A. Relman, P.O. Brown: Individuality and variation in gene expression patterns in human blood, Proc. Natl. Acad. Sci. USA **100**, 1896–1901 (2003)

31.40 J. Quackenbush: *Computational Approaches to Analysis of DNA Microarray Data*, IMIA Yearbook Med. Inform. (Schattauer, Stuttgart 2006) pp. 91–103

31.41 P. Wai-Nang Lee, W. Vay Liang: Go. Nutrient-gene interaction: Tracer-based metabolomics, J. Nutr. **135**(12 Suppl), 3027S–3032S (2005)

31.42 M.J. Gibney, M. Walsh, L. Brennan, H.M. Roche, B. German, B. van Ommen: Metabolomics in human nutrition: Opportunities and challenges, Am. J. Clin. Nutr. **82**, 497–503 (2005)

31.43 K. Hollywood, D.R. Brison, R. Goodacre: Metabolomics: Current technologies and future trends, Proteomic **6**, 4716–4723 (2006)

31.44 U. Paolucci, K.E. Vigneau-Callahan, H. Shi, W.R. Matson, B.S. Kristal: Development of biomarkers based on diet-dependent metabolic serotypes: Characteristics of component-based models of metabolic serotypes, OMICS **8**(3), 221–238 (2004)

31.45 J.C. Lindon, E. Holmes, J.K. Nicholson: Metabonomics techniques and applications to pharmaceutical research and development, Pharm. Res. **23**(6), 1075–1088 (2006)

31.46 H. Tsutsui, T. Maeda, T. Toyo'oka, J.Z. Min, S. Inagaki, T. Higashi, Y. Kagawa: Practical analytical approach for the identification of biomarker candidates in prediabetic state based upon metabonomic study by ultraperformance liquid chromatography coupled to electrospray ionization time-of-flight mass spectrometry, J. Proteome Res. **9**(8), 3912–3922 (2010)

31.47 J.L. Griffin: Understanding mouse models of disease through metabolomics, Curr. Opin. Chem. Biol. **10**, 309–315 (2006)

31.48 C. Gieger, L. Geistlinger, E. Altmaier, M. Hrabé de Angelis, F. Kronenberg, T. Meitinger, H.W. Mewes, H.E. Wichmann, K.M. Weinberger, J. Adamski, T. Illig, K. Suhre: Genetics meets metabolomics: A genome-wide association study of metabolite profiles in human serum, PLoS Genetics **4**(11), e1000282 (2008)

31.49 A. Kotronen, V.R. Velagapudi, L. Yetukuri, J. Westerbacka, R. Bergholm, K. Ekroos, J. Makkonen, M.R. Taskinen, M. Oresic, H. Yki-Järvinen: Serum saturated fatty acids containing triacylglycerols are better markers of insulin resistance than total serum triacylglycerol concentrations, Diabetologia **52**(4), 684–690 (2009)

31.50 M.C. Walsh, L. Brennan, J.P.G. Malthouse, H.M. Roche, M.J. Gibney: Effect of acute dietary standardization on the urinary, plasma, and salivary metabolomic profiles of healthy humans, Am. J. Clin. Nutr. **84**(3), 531e9 (2006)

31.51 T.K. Mao, J. van de Water, C.L. Keen, H.H. Schmitz, M.E. Gershwi: Modulation of TNF-alpha secretion in peripheral blood mononuclear cells by cocoa flavanols and procyanidins, Dev. Immunol. **9**(3), 135–141 (2002)

31.52 E.M. Lenz, J. Bright, I.D. Wilson, A. Hughes, J. Morrisson, H. Lindberg, A. Lockton: Metabonomics, dietary influences and cultural differences: A 1H NMR-based study of urine samples obtained from healthy British and Swedish subjects, J. Pharm. Biomed. Anal. **36**(4), 841–849 (2004)

31.53 M. Lankinen, U. Schwab, P.V. Gopalacharyulu, T. Seppänen-Laakso, L. Yetukuri, M. Sysi-Aho, P. Kallio, T. Suortti, D.E. Laaksonen, H. Gylling, K. Poutanen, M. Kolehmainen, M. Oresic: Dietary carbohydrate modification alters serum metabolic profiles in individuals with the metabolic syndrome, Nutr. Metab. Cardiovasc. Dis. **20**(4), 249–257 (2010)

31.54 M. Lankinen, U. Schwab, A. Erkkilä, T. Seppänen-Laakso, M.L. Hannila, H. Mussalo, S. Lehto, M. Uusitupa, H. Gylling, M. Oresic: Fatty fish intake decreases lipids related to inflammation and insulin signaling – a lipidomics approach, PLoS One **4**(4), e5258 (2009)

31.55 A. Miccheli, F. Marini, G. Capuani, A.T. Miccheli, M. Delfini, M.E. Di Cocco, C. Puccetti, M. Paci, M. Rizzo, A. Spataro: The influence of a sports drink on the postexercise metabolism of elite athletes as investigated by NMR-based metabolomics, J. Am. Coll. Nutr. **28**(5), 553–564 (2009)

31.56 C.L. de Hoog, M. Mann: Proteomics, Annu. Rev. Genomics Hum. Genet. **5**, 267–293 (2004)

31.57 R.A. Saleem, R.S. Rogers, A.V. Ratushny, D.J. Dilworth, P.T. Shannon, D. Shteynberg, Y. Wan, R.L. Moritz, A.I. Nesvizhskii, R.A. Rachubinski, J.D. Aitchison: Integrated phosphoproteomics analysis of a signaling network governing nutrient response and peroxisome induction, Mol. Cell Proteomics **9**(9), 2076–2088 (2010)

31.58 S. Pan, R. Chen, D.A. Crispin, D. May, T. Stevens, M.W. McIntosh, M.P. Bronner, A. Ziogas, H. Anton-Culver, T.A. Brentnall: Protein alterations associated with pancreatic cancer and chronic pancreatitis found in human plasma using global quantitative proteomics profiling, J. Proteome Res. **10**(5), 2359–2376 (2011)

31.59 B. Cañas, D. López-Ferrer, A. Ramos-Fernández, E. Camafeita, E. Calvo: Mass spectrometry technologies for proteomics, Brief. Funct. Genomic Proteomic **4**(4), 295–320 (2006)

31.60 M. Kussmann, L. Krause, W. Siffert: Nutrigenomics: Where are we with genetic and epigenetic markers for disposition and susceptibility?, Nutr. Rev. **68**(Suppl. 1), S38–S47 (2010)

31.61 M. Kussmann, A. Panchaud, M. Affolter: Proteomics in nutrition: Status quo and outlook for biomark-

ers and bioactives, J. Proteome Res. **9**, 4876–4887 (2010)

31.62 L. Marvin-Guy, L.V. Lopes, M. Affolter, M.C. Courtet-Compondu, S. Wagnière, G.E. Bergonzelli, L.B. Fay, M. Kussmann: Proteomics of the rat gut: Analysis of the myenteric plexus-longitudinal muscle preparation, Proteomics **5**(10), 2561–2569 (2005)

31.63 G. Breikers, S.G. van Breda, F.G. Bouwman, M.H. van Herwijnen, J. Renes, E.C. Mariman, J.C. Kleinjans, J.H. van Delft: Potential protein markers for nutritional health effects on colorectal cancer in the mouse as revealed by proteomics analysis, Proteomics **6**(9), 2844–2852 (2006)

31.64 S. Tan, T.K. Seow, R.C. Liang, S. Koh, C.P. Lee, M.C. Chung, S.C. Hooi: Proteome analysis of butyrate-treated human colon cancer cells (HT-29), Int. J. Cancer **98**(4), 523–531 (2002)

31.65 A. Herzog, B. Kindermann, F. Döring, H. Daniel, U. Wenzel: Pleiotropic molecular effects of the pro-apoptotic dietary constituent flavone in human colon cancer cells identified by protein and mRNA expression profiling, Proteomics **4**(8), 2455–2464 (2004)

31.66 H. tom Dieck, F. Döring, D. Fuchs, H.P. Roth, H. Daniel: Transcriptome and proteome analysis identifies the pathways that increase hepatic lipid accumulation in zinc-deficient rats, J. Nutr. **135**(2), 199–205 (2005)

31.67 L. Zhang, G. Perdomo, D.H. Kim, S. Qu, S. Ringquist, M. Trucco, H.H. Dong: Proteomic analysis of fructose-induced fatty liver in hamsters, Metabolism **57**(8), 1115–1124 (2008)

31.68 L.M. Knowles, J.A. Milner: Diallyl disulfide induces ERK phosphorylation and alters gene expression profiles in human colon tumor cells, J. Nutr. **133**, 2901–2906 (2003)

31.69 C.D. Davis, J. Milner: Frontiers in nutrigenomics, proteomics, metabolomics and cancer prevention, Mutat. Res. **551**, 51–64 (2004)

31.70 G. Panagiotou, J. Nielsen: Nutritional systems biology: Definitions and approaches, Annu. Rev. Nutr. **29**, 329–339 (2009)

31.71 T. Ideker, T. Galitski, L. Hood: A new approach to decoding life: Systems biology, Annu. Rev. Genomics Hum. Genet. **2**, 343–372 (2001)

31.72 H. Kitano: Systems biology: A brief overview, Science **295**(5560), 1662–1664 (2002)

31.73 F. Desiere: Towards a systems biology understanding of human health: Interplay between genotype, environment and nutrition, Biotechnol. Annu. Rev. **10**, 51–84 (2004)

31.74 M. Hucka, A. Finney, B.J. Bornstein, S.M. Keating, B.E. Shapiro, J. Matthews, B.L. Kovitz, M.J. Schilstra, A. Funahashi, J.C. Doyle, H. Kitano: Evolving a lingua franca and associated software infrastructure for computational systems biology: The systems biology markup language (SBML) project, Syst. Biol. **1**(1), 41–53 (2004)

31.75   N. Le Novère, M. Hucka, H. Mi, S. Moodie, F. Schreiber, A. Sorokin, E. Demir, K. Wegner, M.I. Aladjem, S.M. Wimalaratne, F.T. Bergman, R. Gauges, P. Ghazal, H. Kawaji, L. Li, Y. Matsuoka, A. Villéger, S.E. Boyd, L. Calzone, M. Courtot, U. Dogrusoz, T.C. Freeman, A. Funahashi, S. Ghosh, A. Jouraku, S. Kim, F. Kolpakov, A. Luna, S. Sahle, E. Schmidt, S. Watterson, G. Wu, I. Goryanin, D.B. Kell, C. Sander, H. Sauro, J.L. Snoep, K. Kohn, H. Kitano: The systems biology graphical notation, Nat. Biotechnol. **27**(8), 735–741 (2009)

31.76   H.M. Sauro, M. Hucka, A. Finney, C. Wellock, H. Bolouri, J. Doyle, H. Kitano: Next generation simulation tools: The systems biology workbench and BioSPICE integration, OMICS **7**(4), 355–372 (2003)

31.77   A. Funahashi, M. Morohashi, H. Kitano, N. Tanimura: CellDesigner: A process diagram editor for gene-regulatory and biochemical networks, Biosilico **1**, 159–162 (2003)

31.78   P. Mendes, S. Hoops, S. Sahle, R. Gauges, J. Dada, U. Kummer: Computational modeling of biochemical networks using COPASI, Methods Mol. Biol. **500**, 17–59 (2009)

31.79   D.A. Benson, I. Karsch-Mizrachi, D.J. Lipman, J. Ostell, E.W. Sayers: GenBank, Nucleic Acids Res. **37**(1), D26–D31 (2010)

31.80   A. Kouranov, L. Xie, J. de la Cruz, L. Chen, J. Westbrook, P.E. Bourne, H.M. Berman: The RCSB PDB information portal for structural genomics, Nucleic Acids Res. **34**, D302–D305 (2006)

31.81   R.D. Finn, J. Mistry, J. Tate, P. Coggill, A. Heger, J.E. Pollington, O.L. Gavin, P. Gunasekaran, G. Ceric, K. Forslund, L. Holm, E.L. Sonnhammer,

S.R. Eddy, A. Bateman: The Pfam protein families database, Nucleic Acids Res. **38**(1), D211–D222 (2010)

31.82   R. Caspi, T. Altman, J.M. Dale, K. Dreher, C.A. Fulcher, F. Gilham, P. Kaipa, A.S. Karthikeyan, A. Kothari, M. Krummenacker, M. Latendresse, L.A. Mueller, S. Paley, L. Popescu, A. Pujar, A.G. Shearer, P. Zhang, P.D. Karp: The MetaCyc database of metabolic pathways and enzymes and the BioCyc collection of pathway/genome databases, Nucleic Acids Res. **38**(1), D473–D479 (2010)

31.83   L. Matthews, G. Gopinath, M. Gillespie, M. Caudy, D. Croft, B. de Bono, P. Garapati, J. Hemish, H. Hermjakob, B. Jassal, A. Kanapin, S. Lewis, S. Mahajan, B. May, E. Schmidt, I. Vastrik, G. Wu, E. Birney, L. Stein, P. D'Eustachio: Reactome knowledgebase of human biological pathways and processes, Nucleic Acids Res. **37**(1), D619–D622 (2009)

31.84   M. Kuhn, D. Szklarczyk, A. Franceschini, M. Campillos, C. von Mering, L.J. Jensen, A. Beyer, P. Bork: STITCH 2: An interaction network database for small molecules and proteins, Nucleic Acids Res. **38**(1), D552–D556 (2010)

31.85   B. Lehne, T. Schlitt: Protein-protein interaction databases: Keeping up with growing interactomes, Hum. Genomics **3**(3), 291–297 (2009)

31.86   P. Shannon, A. Markiel, O. Ozier, N.S. Baliga, J.T. Wang, D. Ramage, N. Amin, B. Schwikowski, T. Ideker: Cytoscape: A software environment for integratedmodels of biomolecular interaction networks, Genome Res. **13**(11), 2498–2504 (2003)

31.87   S. Maere, K. Heymans, M. Kuiper: BiNGO: A cytoscape plugin to assess overrepresentation of gene ontology categories in biological networks, Bioinformatics **21**(16), 3448–3449 (2005)

32. Bioinformatics and Nanotechnologies: Nanomedicine

Federico Ambrogi, Danila Coradini, Niccolò Bassani, Patrizia Boracchi, Elia M. Biganzoli

In this chapter we focus on the bioinformatics strategies for translating genome-wide expression analyses into clinically useful cancer markers with a specific focus on breast cancer with a perspective on new diagnostic device tools coming from the field of nanobiotechnology and the challenges related to high-throughput data integration, analysis, and assessment from multiple sources.

Great progress in the development of molecular biology techniques has been seen since the discovery of the structure of deoxyribonucleic acid (DNA) and the implementation of a polymerase chain reaction (PCR) method. This started a new era of research on the structure of nucleic acids molecules, the development of new analytical tools, and DNA-based analyses that allowed the sequencing of the human genome, the completion of which has led to intensified efforts toward comprehensive analysis of mammalian cell struc-

ture and metabolism in order to better understand the mechanisms that regulate normal cell behavior and identify the gene alterations responsible for a broad spectrum of human diseases, such as cancer, diabetes, cardiovascular diseases, neurodegenerative disorders, and others.

Part F | 32

32.1 Background

Technical advances such as the development of molecular cloning, Sanger sequencing, PCR, oligonucleotide microarrays and more recently the development of a variety of so-called next-generation sequencing (NGS) platforms has actually revolutionized translational research and in particular cancer research. Now, scientists can obtain a genome-wide perspective of cancer gene expression useful to discover novel cancer biomarkers for more accurate diagnosis and prognosis, and monitoring of treatment effectiveness. Thus, for instance, microRNA expression signatures have been shown to provide a more accurate method of classifying cancer subtypes than transcriptome profiling and allow classification of different stages in tumor progression, actually opening the field of personalized medicine (in which disease detection, diagnosis, and therapy are tailored to each individual's molecular profile) and predictive

medicine (in which genetic and molecular information is used to predict disease development, progression, and clinical outcome).

However, since these novel tools generate a tremendous amount of data and since the number of laboratories generating microarray data is rapidly growing, new bioinformatics strategies that promote the maximum utilization of such data, as well as methods for integrating gene ontology annotations with microarray data to improve candidate biomarker selection are necessary. In particular, the management and analysis of NGS data requires the development of informatics tools able to assemble, map, and interpret huge quantities of relatively or extremely short nucleotide sequence data.

As a paradigmatic example, a major pathology such as breast cancer can be considered. Breast cancer is the most common malignancy in women with a cumulative lifetime risk of developing the disease as high as one in every eight women [32.1]. Several factors are associated with this cancer such as genetics, life style, menstrual and reproductive history, and long-term treatment with hormones. Until now breast cancer has been hypothesized to develop, following a progression model similar to that described for colon cancer [32.2, 3], through a linear histological progression from adenosis, to ductal/lobular hyperplasia, to atypical ductal/lobular hyperplasia, to in situ carcinoma and finally to invasive cancer, corresponding to increasingly worse patient outcome. Molecularly, it has been suggested that this process is accompanied by increasing alterations of the genes that encode for tumor suppressor proteins, nuclear transcription factors, cell cycle regulatory proteins, growth factors, and corresponding receptors, which provide a selective advantage for the outgrowth of mammary epithelial cell clones containing such mutations [32.4].

Recent advances in genomic technology have improved our understanding of the genetic events that parallel breast cancer development. In particular, DNA microarray-based technology, with the simultaneous evaluation of thousands of genes, has provided researchers with an opportunity to perform comprehensive molecular and genetic profiling of breast cancer able to classify it into some clinically relevant subtypes and in the attempt to predict the prognosis or the response to treatment [32.5–8]. Unfortunately, the initial enthusiasm for the application of such an approach was tempered by the publication of several studies reporting contradictory results on the analysis of the same samples analyzed on different microarray platforms that arose the skepticism regarding the reliability and the reproducibility of this technique [32.9, 10]. In fact, despite the great theoretical potential for improving breast cancer management, the actual performance of predictors, built using genes' expression, is not as good as initially published, and the lists of genes obtained from different studies are highly unstable, resulting in disparate signatures with little overlap in their constituent genes. In addition, the biological role of individual genes in a signature, the equivalence of several signatures, and their relation to conventional prognostic factors are still unclear [32.11]. Even more incomplete and confusing is the information obtained when molecular genetics was applied to premalignant lesions; indeed, genome analysis revealed an unexpected morphological complexity of breast cancer, very far from the hypothesized multi-step linear process, but sug-

gesting a series of stochastic genetic events leading to distinct and divergent pathways towards invasive breast cancer [32.12], the complexity of which limits the application of really effective strategies for prevention and early intervention.

Therefore, despite the great body of information about breast cancer biology, improving our knowledge about the puzzling bio-molecular features of neoplastic progression is of paramount importance to better identify the series of events that, in addition to genetic changes, are involved in breast tumor initiation and progression and that enable premalignant cells to reach the six biological endpoints that characterize malignant growth (self-sufficiency in growth signals, insensitivity to growth-inhibitory signals, evasion of programmed cell death, limitless replicative potential, sustained angiogenesis, and tissue invasion and metastasis). To do that, instead of studying the single aspects of tumor biology, such as gene mutation or gene expression profiling, we must apply an investigational approach aimed to integrate the different aspects (molecular, cellular, and supracellular) of breast tumorigenesis.

At the molecular level, an increasing body of evidence suggests that gene expression alone is not sufficient to explain protein diversity and that epigenetic changes (i. e., heritable changes in gene expression that occur without changes in nucleotide sequences), such as alteration in DNA methylation, chromatin structure changes, and dysregulation of microRNA expression, may affect normal cells and predispose them to subsequent genetic changes with important repercussions in gene expression, protein synthesis, and ultimately cellular function [32.13–16]. At the cellular level, evidence indicates that to really understand cell behavior, we must consider also the microenvironment in which cells grow; an environment that recent findings indicate to have a relevant role in promoting and sustaining abnormal cell growth and tumorigenesis [32.17].

This picture is further complicated by the concept that among the heterogeneous cell population that makes up the tumor, there exists an approximately 1% of cells, also known as tumor initiating cells that are more likely derived from normal epithelial precursors (stem/precursor cells), and share with them a number of key properties including the capacity of self-renewal and the ability to proliferate and differentiate [32.18, 19]. When altered in their response to abnormal inputs from the local microenvironment, these stem/precursor cells can give rise to preneoplastic lesions [32.20]. In fact, similarly to bone marrow-derived stem cells, tissue-specific stem cells show remarkable plasticity within the microenvironment: they can enter a state of quiescence for decades (cell dormancy), but can become highly dynamic once activated by specific microenvironment stimuli from the surrounding stroma and are ultimately transformed in tumor initiating cells [32.21]. The stroma, in which the mammary gland is embedded, is composed of adipocytes, fibroblasts, blood vessels, and an extracellular matrix in which several cytokines and growth factors are present. While none of these cells are themselves malignant, they may acquire an abnormal phenotype and altered function due to their direct or indirect interaction with epithelial stem/precursor cells. Acting as an oncogenic agent, the stroma could provoke tumorigenicity in adjacent epithelial cells leading to the acquisition of genomic changes, at which epigenetic alterations also concur, that can accumulate over time and provoke silencing of more than 100 pivotal genes' encoding for proteins involved in tumor suppression, apoptosis, cell cycle regulation, DNA repair, and signal transduction [32.22]. Under these conditions, epithelial cells and the stroma co-evolve towards a transformed phenotype following a process that has not yet been worked out [32.23, 24].

Many of the soluble factors present in the stroma, essential for the normal mammary gland development, have been found to be associated with cancer initiation. This is the case of hormone steroids (estradiol and progesterone), which are physiological regulators of breast development and whose dysregulation may result in preneoplastic and neoplastic lesions [32.25–27]. In fact, through their respective receptors, in epithelial cells estrogens and progesterone may induce the synthesis of local factors that, on the one hand, trigger the activation of the stem/precursor cells and, on the other hand, exert a paracrine effect on endothelial cells, which in response to the vascular endothelial growth factor, trigger neoangiogenesis activation [32.21]. In addition, estrogens have been found implicated in the local modifications of tissue homeostasis associated to a chronic inflammation that may promote epithelial transformation due to the continued production of pro-inflammatory factors that favors generation of a pro-growth environment and fosters cancer development [32.28]; alternatively, transformed epithelial cells would enhance activation of fibroblasts through a vicious circle that supports the hypothesis according to which cancer should be considered as a never healing wound. Last but not least, very recent findings in animal models have clearly indicated that an early event occurring in the activation of estrogen-induced mammary carcinogenesis is represented by the altered expression of some oncogenic microRNAs (oncomir), suggesting a functional link between hormone exposure and epigenomic control [32.29].

Concerning the forecasted role of new nanobiotechnology applications, disclosing the bio-molecular events contributing to tumor initiation is, therefore, of paramount importance and to achieve this goal a convergence of advanced biocomputing tools for cancer biomarker discovery and multiplexed nanoprobes for cancer biomarker profiling is crucial. This is the one of the major tasks currently ongoing in medical research, namely the interaction of nanodevices with cells and tissues in vivo and their delivery to disease sites.

32.2 Biostatistics Supporting Bioinformatics for Biomarker Discovery

Biomarkers refer to genes, RNA, proteins, and miRNA expressions that can be correlated with a biological condition or may be important for prognostic or predictive aims as far as regards the clinical outcome. The discovery of biomarkers has a long history in translational research. In more recent years, microarrays have generated a great deal of work, promising the discovery of prognostic and predictive biomarkers able to change medicine as was known until then. Since the beginning, the importance of statistical methods in such a context was evident, starting from the seminal paper of Golub, which showed the ability of gene expression to classify tumors [32.30].

Although bioinformatics is the leading engine, referenced in biomolecular literature, providing informatics tool to handle massive omic data, the computational core is actually represented by biostatistics methodology aiming at extracting useful summary information.

Biostatistics cornerstones are represented by large sample and likelihood theories, hypothesis testing, experimental design, and exploratory multivariate techniques summarized in the genomic era according to class comparison, prediction, and discovery.

Actually, massive omic data and the idea of personalized medicine need to develop statistical theory according to new requirements. Even in the case of multivariate techniques, the problems usually faced using statistical techniques accounted for orders of magnitude of less data than those encountered with

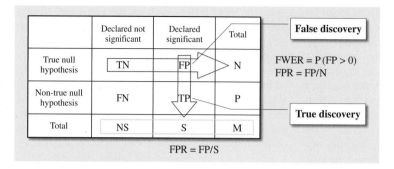

Fig. 32.1 Explanation of FDR (for further details, see [32.31])

Part F | 32.2

high-throughput technologies. A situation that NGS techniques will easily exacerbate.

In class comparison studies there is a predefined group which identifies samples and the interest is in evaluating if the groups express the transcripts of interest differently. Such studies are generally performed using a transcript by transcript analysis, performing thousands of statistical tests then correcting *p*-values to account for the desired percentages of false positives and negatives. In fact, the multiple comparison problem is the first concern as traditionally methods for family-wise control are generally too restrictive when accounting for thousands of tests. The false discovery rate (FDR) was a major breakthrough in such a context. The general concepts underlying FDR are outlined later (Fig. 32.1).

Another topic discussed regards the parametric assumptions underlying most of the statistical tests used. Permutation tests were much developed to face this issue and are now one of the standard tools available to researchers.

Jeffery and colleagues [32.32] performed a systematic comparison of 9 different methods for identifying genes differentially expressed across experimental groups, finding that different methods gave rise to very different lists of genes and that sample size and noise level strongly affected predictive performance of the methods chosen for evaluation. Also, evaluation of the accuracy of fold-change compared to ordinary and moderated *t* statistics was performed by *Witten* and *Tibshirani* [32.33], which discusses the issues of reproducibility and accuracy of gene lists returned by different methods, claiming that

a researcher's decision to use fold-change or a modified t-statistic should be based on biological, rather than statistical, considerations.

In this sense, the classical *limma-like* approach [32.34] has become a de facto standard in the analysis of high-throughput data: gene expression and miRNA

microarrays, proteomics, and serial analysis of gene expression (SAGE) generate an incredible amount of data which is routinely analyzed element-wise, without considering the multivariate nature of the problem. Akin to this, non-parametric multivariate analysis of variance (MANOVA) techniques have also been suggested to identify differentially expressed genes in the context of microarrays and qPCR-RT [32.35, 36], with the advantage of not making any distributional assumption on expression data and of being able to circumvent the dimensionality issue related to omic data ($n°$ of subjects $\ll n°$ of genes).

A well-known example of class comparison study was that of van't *Veer* and colleagues [32.37] in which a panel of genes, a signature, was claimed to be predictive of poor outcome at 5 years for breast cancer patients. In this case a group of patients relapsing at 5 years was compared in terms of gene expression to a group of patients not relapsing within 5 years.

In class discovery studies no predefined groups are available and the interest is in findings new groupings, usually called bioprofiles, using the available expression measures.

The standard statistical method to perform class discovery is cluster analysis that received great expansion due to gene expression studies. It is worth saying that cluster analysis is a powerful yet tricky method that should be applied taking care of outliers, stability of results, number of suspected profiles, and so on. These aspects are very hard to face with thousands of transcript to be analyzed. Even more subtle is the problem of the interpretation of the clusters obtained in terms of disease profiles and the definition of a rule to define the discovered profiles.

Alternatively, classical multivariate methods, such as principal components analysis (PCA), are gaining relevance for visualization of high-dimensional data (Fig. 32.2) through data reduction [32.38, 39]. Recent work on lung cancer highlighted different patterns of

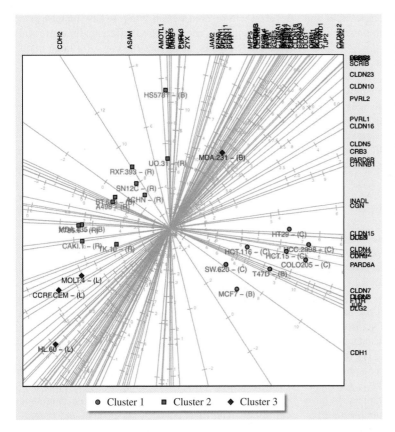

Fig. 32.2 PCA-based biplot for evaluating association between genes involved in cell-polarity pathway and a set of tumor cell-lines from the NCI60 panel. Expression values of 76 genes as profiled on 23 tumor cell lines (6 colon, 6 breast, 8 renal and 3 leukemia) were obtained from the well known NCI60 dataset [32.44,45], publicly available online on the ArrayExpress website with accession number E-GEOD-5720 (hybridization performed with human HG-U133A/133B Affymetrix GeneChip). To interpret results, let us use some examples. Cell lines labeled with *olive green points* (colon and ER+ breast cancer) are positively associated with the set of genes on the lower part of the *right vertical axis*, namely genes from CLDN15 and CDH1; on the other hand, cell lines labeled with *light blue squares* (renal lines and ER- breast cancer) are negatively associated with this set of genes, but show some mild positive association with genes on the *left side of the upper horizontal axis*, namely genes from CDH2 to ZYX

● Cluster 1   ■ Cluster 2   ◆ Cluster 3

expression between healthy subjects and mesotelioma-affected subjects using multivariate visualization techniques and evaluated relationships between gene-sets involved in different biological pathways via partial least squares regression [32.40,41]. It is relevant to note that all of these methods will have to face relevant challenges, in the perspective of being applied in the context of NGS data, for which a considerable amount of issues has already arisen [32.42]. The use of multivariate techniques is thus a principled way for making it through the steps of data dimensionality reduction, data normalization, and integration, which can lead to higher chances of driving the exploration of distinct bioprofiles [32.43].

The work of *Perou* and colleagues [32.5] is an important example of class discovery by cluster analysis in a major pathology such as breast cancer. In their work, the authors found genes distinguished between estrogen positive cancer with luminal characteristics and estrogen negative cancers. Among these two subgroups, one had a basal characterization and the other showed patterns of up-regulation for genes linked to oncogeneErb-B2. Repeated application of cluster anal-

ysis in different case series resulted in very similar groupings.

Notwithstanding the above-mentioned issues connected to cluster analysis, one of the major breakthroughs of genomic studies was actually believed to be the definition of genomic signatures/profiles by the repeated application of cluster analysis to different case series without the definition of a formal rule for class assignment of new subjects.

Profiles may then be correlated with clinical outcome as was done for breast cancer by *van't Veer* and colleagues [32.37]. Now, more than 10 years after this study, it is not yet clear which is the real contribution of microarray-based gene expression profiling to breast cancer prognosis. Of all the so-called *first-generation* signatures, only oncotype DX [32.46], a qRT-PCR based analysis of 21 genes, has reached level II of evidence to support tumor prognosis and has been included in the National Comprehensive Cancer Network guidelines, whereas the remaining signatures have only obtained level III of evidence so far [32.47]. Reasons for this are, among the others, a lack of stability in terms of genes that the lists are composed of and strong time-dependence, i. e., reduced prognostic value after 5 to 20 years of follow-up.

Another, and more important, issue for prognostic/prediction studies is connected to the design of the study itself. In fact, a prognostic study should be planned by defining a cohort that will be followed during time while a case control study may be only suggestive of signatures to be considered for clinical management. See letter [32.48] commenting reference [32.49] for concerns regarding such an issue.

32.2.1 Multiple Comparisons and the False Discovery Rate

Class comparison in genomic-wide studies is one of the most common and challenging applications since the advent of microarray technology. The first study on predictive signatures in breast cancer in 2002 [32.6] was mainly a class comparison study.

From the statistical viewpoint one of the first problems evidenced was the large number of statistical tests performed in such an analysis. In particular, the classical control for false positives, emphasizing the specificity of the screening appeared from the beginning to be too restrictive with the cost of false negatives too high.

To understand such an issue, let us suppose to have to compare the gene expression in a group of tumor tissues with that of a group of normal tissues. For each gene a statistical test controlling the probability of saying that a gene is different when in fact it is not (false positive, FP), is performed. Such an error is called type one error and its level is generally called α and fixed at a 5% level. The problem is that a test at α level is performed for each gene. Therefore, if the probability of making a mistake (FP) is 0.05, while the probability of not making a mistake is 0.95 (this is the probability of saying the gene is not differentially expressed when it is not, true negative), when performing, say, 1000 tests the probability of not making any mistake is 0.95^{1000}, which is practically 0. Accordingly, the probability of at least one FP is practically 1. How can the specificity of the experiment be controlled? A large number of procedures is available, the most simple and known is the Bonferroni correction. Let us see how it works.

In particular, if n tests are performed at α level, the probability of not having any false positive is $(1-\alpha)^n$, therefore the probability of making at least one false positive is $1-(1-\alpha)^n$, which can be approximated as $1-n\alpha$ (for small α). The Bonferroni correction originates from this. In fact, if the tests are performed at

level $\alpha_B = \alpha/n$, then we can expect to have no false positive among the genes declared differentially expressed at α level. This is, in fact, at the cost of a large number of false positives. In genomic experiments, when thousands of tests are performed, the Bonferroni significance level is so low that very few genes can easily pass the first screening probably paying too high costs in terms of genes declared not significantly differentially expressed when actually they have a differential expression. The balance between specificity and sensibility is a fairly old problem in screening problems, which is exacerbated with high-throughput data analysis.

One of the most common approaches applied in such a context is the proposal of *Benjamini* and *Hochberg* [32.50] called the false discovery rate (FDR) trying to control the number of false positives among the genes declared significant.

To better understand the novelty of FDR, let us suppose to have M genes to be considered in the high-throughput experiment, N of the M genes are truly differentially expressed while P are not. Performing the appropriate statistical test NS of the M genes are declared not different between groups under comparison while S are significantly different (Fig. 32.1). The type one error rate α (FPR) controls the number of FP with respect to N, while using the Bonferroni correction the probability that FP is greater than 0 is controlled. The FDR changes perspective and considers the columns of the table instead of the rows. FDR controls the number of FP with respect to S. If, for example, 10 genes are declared differentially expressed with an FDR of 20%, it is expected that 2 be false positives. This may allow greater flexibility in the managing of the screening phase of the analysis (see Fig. 32.3 for a graphical representation of results from a class comparison microarray study, with an application of FDR concepts). The problem first solved by *Benjamini* and *Hochberg* was basically how to estimate FDR and different proposals have appeared since then, for example the q-value of *Storey* [32.51].

In general, omic and multiplexed diagnostic technologies with the ability to produce vast amounts of biomolecular data, have vastly outstripped our ability to sensibly deal with this data deluge and extract useful and meaningful information for decision making. The producers of novel biomarkers assume that an integrated bioinformatics and biostatistics infrastructure exists to support the development and evaluation of multiple assays and their translation to the clinic. Actually, the best scientific practice for the use of high-throughput data is still to be developed. In this perspective, the existence of

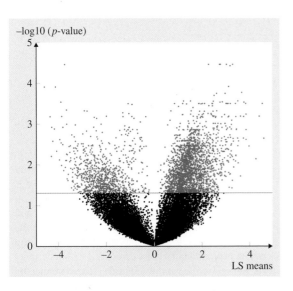

Fig. 32.3 Volcano plot of differential gene expression pattern between experimental groups. On the *x-axis* the least squares (LS) means (i. e., difference of mean expression on $\log_2$ scale between experimental groups) and on the *y-axis* $-\log_{10}$ transformed *p*-values corrected for multiplicity using the FDR method from *Benjamini* et al. [32.50] are reported. The *horizontal red line* corresponds to a cut-off for the significance level α at 0.05. *Points* above this threshold represent genes which are actually differentially expressed between experimental groups, and that are to be further investigated

advanced computational technologies for bioinformatics is irrelevant along the translational research process unless supporting biostatistical evaluation infrastructures exist to take advantage of developments in any technology.

In this sense, a key problem is the fragmentation of quantitative research efforts. The analysis of high dimensional data is mainly conducted by researchers with limited biostatistical experience using standard software without the knowledge of the underlying statistical principles of the methodology then exposing the results to a wide uncertainty not only due to sample size limitations. Moreover, so far, a large amount of biostatistical methods and software tools supporting bioinformatics analysis of genomic/proteomic data has been provided but reference standardized analysis procedures coping with suitable preprocessing and quality control approaches on raw data coming from omic and multiplex assays are still waiting for development. Formal initiatives for the integration of biostatistical research groups with functional genomics and proteomics labs are one of the major challenges in this context. In fact, besides the development of innovative biostatistics and bioinformatics tools, a major key of success lies in the ability to integrate different competencies. Such an integration cannot be simply demanded for the development of software, such as the ArrayTrack initiative, but needs to develop integrated skills assisted by a software platform able to outline the analysis plan. In such a context, different strategies can be adopted from open software, such as R and bioconductor, to commercial ones such as SAS/JMP genomics.

In a functional dynamic perspective, to the characterization of the bio-profiles of cancer affected patients, is added the complexity related to the prolonged follow-up of patients with the necessity of the registration of the event-history of possible adverse events (local recurrence and/or metastasis) before death, that may offer useful insight into disease dynamics to identify a subset of patients with worse prognosis and better response to the therapy. This makes it necessary to develop strategies for the integration of clinical and follow-up information with those deriving from genetic and molecular characterizations.

The evaluation and benchmarking of new analytical processes for the discovery, development, and clinical validation of new diagnostic/prognostic biomarkers is an extremely important problem especially in a fast growing area such as translational research based on functional genomics/proteomics. In fact, the presentation of overoptimistic results based on the unsuited application of biostatistical procedures can mask the true performance of new biomarker/bioprofiles and create false expectations about its effectiveness.

Guidelines for omic and cross-omic studies should be defined through the integration of different competencies coming from clinical-translational, bioinformatics, and biostatistics research competencies.

This integrated contribution from multidisciplinary research teams will have a major impact on the development of standard procedures that will standardize the results and make research more consistent and accurate according to relevant bioanalytical and clinical targets.

Part F | 32.2

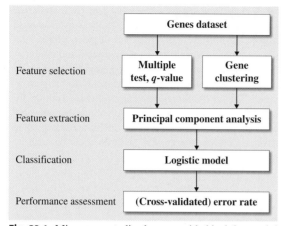

Fig. 32.4 Microarray studies have provided insight on global gene expression in cells and tissues with the expectation of prognostic assessments improvement. The identification of genes whose expression levels are associated with recurrence might also help better discriminating those subjects who are likely to respond to the various tailored systemic treatments. However, microarray experiments raised several questions to the statistical community about the design of the experiments, data acquisition and normalization, supervised and unsupervised analysis. All these issues are burdened by the fact that typically the number of genes being investigated far exceeds the number of patients. It is well-recognized that too large a number of predictor variables affects the performance of classification models: Bellman coined the term *curse of dimensionality* [32.52], referring to the fact that in the absence of simplifying assumptions, the sample size needed to estimate a function of several variables to a given degree of accuracy (i. e., to get a reasonably low-variance estimate) grows exponentially with the number of variables. To avoid this problem, feature selection and extraction issue play a crucial role in microarray analysis. This has led several researchers to find it judicious to filter out genes that do not change their expression level *significantly*, reducing the complexity of the data and improving the signal to noise ratio. However, the adopted *measure of significance* in filtering (the implicitly controlled error measure) is not often easy to interpret in terms of the simultaneous testing of thousands of genes. Moreover, gene expressions are usually filtered on a per-gene basis and seldom taking into account the correlation between different gene expressions. This filtering approach is commonly used in most current high-throughput experiments whose main objective is to detect differentially expressed genes (active genes) and, therefore, to generate hypotheses rather than to confirm them. All these methods, based on a *measure of significance*, select genes from a supervised perspective, i. e., accounting for the outcome of interest (the subject status). However, an unsupervised approach might be useful in order to reveal the pattern of associations among different genes making it possible to single out redundant information. The figure shows the data analysis pipeline developed in many papers dealing with expressions from high throughput experiments

Integration and standardization of approaches for assessment of diagnostic and prognostic performance is a key issue. Many of the clinical and translational research groups have chosen different approaches for biodata modeling, tailored to specific types of medical data. However, very few proper benchmarking studies of algorithm classes have been performed worldwide and fewer examples of best practice guidelines have been produced. Similarly, few studies have closely examined the criteria under which medical decisions are made. The integrating aspects of this theme relates to methods and approaches for inference, diagnosis, prognosis, and general decision making in the presence of heterogeneous and uncertain data.

A further priority is to ensure that research in biomarker analysis is designed and informed from the outset to integrate well with clinical practice (to facilitate widespread clinical acceptance) and that it exploits cross-over between methods and knowledge from different areas (to avoid duplication of efforts to facilitate rapid adoption of good practice in the development of this healthcare technology). Reference problems are related to the assessment of improved diagnostic and prognostic tools in the clinical setting, resorting to observational and experimental clinical studies from phase I to phase IV and the integration with studies on therapy efficacy which would involve bioprofile and biopattern analysis.

In this perspective, the integration of different omic data is a well-known issue that is receiving increasing attention in biomedical research [32.53, 54], and which questions the capability of researchers to make sense out of a huge amount of data with very different features. Since this integration can not only be seen as an IT problem, proper biostatistical approaches need to be taken into account that consider the multivariate nature of the problem in the light of exploiting the maximum prior information about the biological patterns underlying the clinical problem.

A critical review of microarray studies was performed earlier in a paper by *Dupuy* and *Simon* [32.10], in which a thorough analysis of the major limitations and pitfalls of 90 microarray studies published in 2004 concerning cancer outcome was done (see Fig. 32.4 for a general pipeline for high-throughput experiments). Integrated into this review was the attempt to write guidelines for statistical analysis and reporting of gene expression microarray studies. Starting from this work, it will be possible to extend the outlined criticisms to a wider range of omic studies, in order to produce updated guidelines useful for biomolecular researchers.

In the perspective of integrating omic data coming from different technologies, a comparison of microarray data with NGS platforms will be a relevant point [32.55–57]. Due to the lack of sufficiently standardized procedures for processing and analyzing NGS data, much attention will be given to the process of data generation and of quality control evaluation. Such an integration is crucial because, though capabilities of NGS platforms mostly outperform those of microarrays, protocols of management and data analysis are typically very time-consuming, thus making it impractical to be used for in-depth analysis of large samples.

Of note, one of the ultimate goals of biomedical research is to connect diseases to genes that specify their clinical features and to drugs capable of treating them. DNA microarrays have been used for investigating genome-wide expression of common diseases producing a multitude of gene signatures predicting survival, whose accuracy, reproducibility, and clinical relevance has, however, been debated [32.48, 49, 58, 59]. Moreover, the regulatory relationships between the signature genes have rarely been investigated, largely limiting their biological understanding. The genes, indeed, never act independently from each other. Rather, they form functional connections that coordinate their activity. Hence, it is fundamental that in each cell in every life stage, regulatory events take place in order to keep the healthy steady state. Any perturbation of a gene network, in fact, has a dramatic effect on our life, leading to disease and even death.

32.3 Nanotechnology in Human Healthcare

The prefix *nano* is from the Greek world meaning *dwarf*. Nanotechnology refers to the science of materials whose functional organization is on the nanometer scale, that is 10^{-9} m. Starting from ideas originating in physics in the 1960s and boosted by the need of miniaturization (i.e., speed) of the electronic industry the field has grown rapidly. Today, nanotechnology is gaining an important place in the medicine of the future. In particular, by using the patho-physiological conditions of diseased and inflamed tissues it is possible to target nanoparticles and with them drugs, genes, and diagnostic tools.

Moreover, the spatial and/or temporal contiguity of data from NGS and nanobiotech diagnostic approaches imposes the adoption of methods related to signal analysis which are still to be introduced in standard software, being related to statistical functional data analysis methods. Therefore, the extension of the multivariate statistical methodologies adopted so far is requested in a functional data context; a problem that has already been met in the analysis of mass spectrometry data from proteomic analyses.

Nanotechnology-based platforms for the high-throughput, multiplexed detection of proteins and nucleic acids actually promise to bring substantial advances in molecular diagnostics. Forecasted applications of nano-diagnostic devices are related to the assessment of the dynamics of cell process for a deeper knowledge of the ongoing etio-pathological process at the organ, tissue, and even single cell level.

NGS is a growing revolution in genomic nanobiotechnologies that parallelized the assay process, integrating reactions at the micro or nano scale on chip surfaces, producing thousands or millions of sequences at once. These technologies are intended to lower the costs of nucleic acid sequencing far beyond that possible with earlier methods.

Concerning cancer, a key issue is related to the improvement of early detection and prevention through the understanding of the cellular and molecular pathways of carcinogenesis. In such a way it would be possible to identify the conditions that are precursors of cancer before the start of the pathological process, unraveling its molecular origins. This should represent the next frontiers of bioprofiling to allow the strict monitoring and possible reversal of the neoplastic transformation through personalized preventive strategies. Advances in nanobiotechnology enables the visualization of changes in tissues and physiological process with a subcellular real-time spatial resolution. This is a revolution that can be compared to the *daguerreotype* pictures from current high-throughput multiplex approaches to the digital high resolution of next generation diagnostic devices.

Enormous challenges remain in managing and analyzing the large amounts of data produced. Such evolution is expected to have a strong impact in terms of personalized medical prevention and treatment with considerable effects on society. Therefore, the success will be strongly related to the capability of integrating data from multiple sources in a robust and sustainable research perspective, which could enhance the transfer of high-throughput molecular results to novel diagnostic and therapy application.

The new framework of nanobiotechnology approaches in biomedical decision support according to improved clinical investigation and diagnostic tools is emerging. There is a general need for guidelines for biostatistics and bioinformatics practice in the clinical translation and evaluation of new biomarkers from cross-omic studies based on hybridization, NGS, and high-throughput multiplexed nanobiotechnology assays. Specifically, the major topics concern: bioprofile discovery, outcome analysis in the presence of complex follow-up data, assessment of diagnostic, and prognostic values of new biomarkers/bioprofiles.

32.3.1 New Nanobiotechnological Perspectives in Diagnosis

Current molecular diagnostic technologies are not conceived to manage biological heterogeneity in tissue samples, in part because they require homogeneous preparation, leading to a loss of valuable spatial information regarding the cellular environment and tissue morphology. The development of nanotechnology has provided new opportunities for integrating morphological and molecular information and for the study of the association between observed molecular and cellular changes with clinical-epidemiological data.

Concerning specific approaches, bioconjugated quantum dots (QDs) [32.60–63] have been used to quantify multiple biomarkers in intact cancer cells and tissue specimens, allowing the integration of traditional histopathology versus molecular profiles for the same tissue [32.64–69]. Current interest is focused on the development of nanoparticles with one or multiple functionalities. For example, binary nanoparticles with two functionalities have been developed for molecular imaging and targeted therapy. Bioconjugated QDs, which have both targeting and imaging functions, can be used for targeted tumor imaging and for molecular profiling applications.

Nanoparticle material properties can be exploited to elicit clinical advantage for many applications, such as for medical imaging and diagnostic procedures. Iron oxide constructs and colloidal gold nanoparticles can provide enhanced contrast for magnetic resonance imaging (MRI) and computed tomography (CT) imaging, respectively [32.70, 71]. QDs provide a plausible solution to the problems of optical in vivo imaging due to the tunable emission spectra in the near-infrared region, where light can easily penetrate through the body without harm and their inherent ability to resist bleaching [32.72]. For ultrasound imaging, contrast relies on impedance mismatch presented by materials that are more rigid or flexible than the surrounding tissue, such as metals, ceramics, or microbubbles [32.73]. Continued advancements of these nano-based contrast agents will allow clinicians to image the tumor environment with enhanced resolution for a deeper understanding of disease progression and tumor location.

Additional nanotechnologically-based detection and therapeutic devices have been made possible using photolithography and nucleic acid chemistry [32.74, 75]. The same technology that enabled integrated circuitry, produced microelectromechanical systems (MEMS) for selective molecular sensing, sieving, and controlled drug release [32.76]. Microfluidic systems, also known as *lab-on-chip*, are fabricated by soft lithography of inexpensive polymers [32.77]. Micro- and nanoarrays, have experienced success for molecular diagnostic, genotyping, and biomarker-guided therapeutic targeting [32.76, 78, 79]. Moreover, advances in proteomics have been made possible due to the technical refinement of lithographic resolution [32.80]. Recent interest in nanowires [32.81, 82] and cantilever arrays [32.83–85] for biomarker detection has shown promise. The former are biologically gated transistors able to detect multiple, real-time, simultaneous molecular events. These innovative nanodevices equal or exceed the sensitivity of commercially available approaches [32.86] and are anticipated to be clinically available in the near future.

Highly sensitive biosensors that recognize genetic alterations or detect molecular biomarkers at extremely low concentration levels are crucial for the early detection of diseases and for early stage prognosis and therapy response. Nanowires have been used to detect several biomolecular targets such as DNA and proteins [32.82, 87]. The identification of DNA alterations is crucial to better understand the mechanism of a disease such as cancer and to detect potential genomic markers for diagnosis and prognosis. Other studies have reported the development of a three-dimensional gold nanowire platform for the detection of mRNA with enhanced sensitivity from cellular and clinical samples. Highly sensitive electrochemical sensing systems use peptide nucleic acid probes to directly detect specific mRNA molecules without PCR amplification steps [32.88–90].

The development of immuno and aptamer-based nanowire biosensors to detect cancer biomarkers such as VEGF [32.91] and CA125 [32.92], or SARS virus N-protein [32.93] has shown a great sensitivity, providing the potential use of these nanodevices for point-of-care diagnostic applications. To improve the diagnostic efficacy of the biosensors, a multiplexed approach is needed to accurately identify heterogeneous diseases such as cancer [32.94].

Cantilever nanosensors have also been used to detect minute amount of protein biomarkers. Label-free resonant microcantilever systems have been developed to detect the ng/mL level of alpha-fetoprotein, a potential marker of hepatocarcinoma, providing an opportunity for early disease diagnosis and prognosis [32.95]. Nanofabricated and functionalized devices such as nanowires and nanocantilevers are fast, multiplexed, and label-free methods that provide extraordinary potential for the future of personalized medicine.

The combination of data from multiple imaging techniques offers many advantages over data collected from a single modality. Potential advantages include: improved sensitivity and specificity of disease detection and monitoring, smarter therapy selection based on larger data sets, and faster assessment of treatment efficacy. The successful combination of imaging modalities, however, will be difficult to achieve with multiple contrast agents. Multimodal contrast agents stand to fill this niche by providing spatial, temporal, and/or functional information that corresponds with anatomic features of interest.

There is also great interest in the design of multifunctional nanoparticles, such as those that combine contrast and therapeutic agents. The integration of diagnostics and therapeutics, known as *theranostics*, is attractive because it allows the imaging of therapeutic delivery, as well as follow-up studies to assess treatment efficacy.

Finally, a key direction of research is the optimization of biomarker panels via principled biostatistics approaches for the quantitative analysis of molecular profiles for clinical outcome and treatment response prediction. The key issues that will need to be addressed are: (i) a panel of tumor markers will allow more accurate statistical modeling of the disease behavior than relying on single tumor markers; and (ii) the combination of tumor gene expression data and molecular information of the cancer microenvironment is necessary to define aggressive phenotypes of cancer, as well as for determining the response of early stage disease to treatment (chemotherapy, radiation, or surgery).

Currently, the major tasks in biomedical nanotechnology are (i) to understand how nanoparticles interact with blood, cells, and organs under in vivo physiological conditions and (ii) to overcome one of their inherent limitations, that is, their delivery to diseased sites or organs [32.96–98]. Another major challenge is to generate critical studies that can clearly link biomarkers with disease behaviors, such as the rate of tumor progression and different responses to surgery, radiation or drug therapy [32.99]. The current challenge is, therefore, related to the advancement of biostatistics and biocomputing techniques for the analysis of novel high-throughput biomarkers coming from nanotechnology applications. Current applications involve high-throughput analysis of gene expression data and for multiplexed molecular profiling of intact cells and tissue specimens. The advent of fast and low cost high-throughput diagnostic devices based on NGS approaches appears to be of critical relevance for improving the technology transfer to disease prevention and clinical strategies.

The development of nanomaterials and nanodevices offers new opportunities to improve molecular diagnosis, increasing our ability to discover and identify minute alterations in DNA, RNA, proteins, or other biomolecules. Higher sensitivity and selectivity of nanotechnology-based detection methods will permit the recognition of trace amounts of biomarkers which will open extraordinary opportunities for systems biology analysis and integration to elicit effective early detection of diseases and improved therapeutic outcomes; hence paving the way to achieving individualized medicine.

Effective personalized medicine depends on the integration of biotechnology, nanotechnology, and informatics. Bioinformatics and nanobioinformatics are cohesive forces that will bind these technologies together. Nanobioinformatics represents the application of information science and technology for the purpose of research, design, modeling, simulation, communication, collaboration, and development of nano-enabled products for the benefit of mankind. Within this framework a critical role is played by evaluation and benchmarking approaches according to a robust Health Technology Assessment approach; moreover the development of enhanced data analysis approaches for the integration of multimodal molecular and clinical data should be based on up to date and validated biostatisical approaches. Therefore, in the developing nanobiotechnology era, the role of biostatistical support to bioinformatics is definitely essential to prevent loss of money and suboptimal developments of biomarkers and diagnostic disease signature approaches of the past, which followed a limited assessment according to a strict business perspective rather than to social sustainability.

32.4 Discussion

Concerning the relevance and impact for national health systems, it is forecasted that current omic approaches based on nanobiotechnology will contribute to the identification of next generation diagnostic tests which could be focused on primary to advanced disease prevention by early diagnosis of genetic risk patterns, or the start or natural history of the pathological process of multifactor chronic disease by the multiplexed assessment of both direct and indirect, inner genetic, or environment causal factors.

A benefit of such a development would be finally related to the reduction of costs in the diagnostic process since nanobiotechological approaches seem best suited in the perspective of points-of-care POC diagnostic facilities which could be disseminated in large territories with a reduced number of excellence clinical facilities with reference diagnostic protocols. Nanomaterials are providing the small, disposable lab-on-chip tests that are leading this new approach to healthcare. A variety of factors are provoking calls for changes in how diagnosis is managed. The lack of infrastructure in the developing world can be added to the inefficiency and cost of many diagnostic procedures done in central labs, rather than by a local doctor. For the developed world, an increasingly elderly population is going to exacerbate demand on healthcare and any time-saving solutions will help deal with this new trend. POC devices are looking to reduce the dependence on lab tests and make diagnosis easier, cheaper, and more accessible for countries lacking healthcare infrastructure.

A key role in the overall framework will be played by data analysis under principled biostatistical approaches to develop suitable guidelines for data quality analysis, the following extraction of relevant information and communication of the results in an ethical and sustainable perspective for the individual and society.

The proper, safe and secure management of personalized data in a robust and shared bioethical reference framework is, indeed, expected to reduce the social costs related to unsuited medicalization through renewed preventive strategies. A strong biostatistical based Health Technology Assessment phase will be essential to avoid the forecasted drawbacks of the introduction of such a revolution in prevention and medicine.

To be relevant for national health services, research on biostatistics and bioinformatics applied to nano-biotechnology should exploit its transversal role across multiple applied translational research projects on biomarker discovery, development, and clinical validation until their release for routine application for diagnostic/prognostic aims. Objectives that would enable an accelerated framework for translational research since the involvement of quantitative support are listed here:

● Technological platforms for the developments in the fields of new diagnostic prevention and therapeutic tools. In the context of preventing and treating diseases, the objectives are to foster academic and industrial collaboration through technological platforms where multidisciplinary approaches using cutting edge technologies arising from genomic research may contribute to better healthcare and cost reduction through more precise diagnosis, individualized treatment, and more efficient development pathways for new drugs and therapies (such as the selection of new drug candidates), and other novel products of the new technologies.
● Patentable products: customized array and multiplex design with internal and external controls for optimized normalization. Validation by double checked expression results for genes or protein in the customized array and multiplex assays.

Patenting of validated tailor-made cDNA/proteomic arrays that encapsulate gene/protein signatures related to the response to the therapy with optimized cost/effectiveness properties.

A robust, multidisciplinary quantitative assessment framework in translational research is a global need, which should characterize any specific laboratory and clinical translation project. However, the quantitative assessment phase is rarely based on an efficient cooperation between biologists, biotechnologists, and clinicians with biostatisticians, with relevant skills in

this field. This represents a major limitation to the rapid transferability of basic research results to healthcare. Such a condition is solved in the context of pharmacology in the research and development of new drugs to their assessment in clinical trials, whereas, for diagnostic/prognostic biomarkers, this framework is still to be fully defined.

Such a gap is wasting resources and is malpractice in the use of biomarkers and related bioprofiles for clinical decision making in critical phases of chronic and acute major diseases like cancer and cardiovascular pathologies.

References

32.1   A. Jemal, R. Siegel, E. Ward, T. Murray, J. Xu, C. Smigal, M.J. Thun: Cancer statistics, CA – Cancer J. Clin. **56**, 106–130 (2006)

32.2   G. Arpino, R. Laucirica, R.M. Elledge: Premalignant and in situ breast disease: Biology and clinical implications, Ann. Intern. Med. **143**, 446–457 (2005)

32.3   B. Vogelstein, E.R. Fearon, S.R. Hamilton: Genetic alteration during colorectal tumor development, N. Engl. J. Med. **319**, 525–532 (1988)

32.4   D. Hanahan, R.A. Weinberg: The hallmarks of cancer, Cell **100**, 57–70 (2000)

32.5   C.M. Perou, T. Sørlie, M.B. Eisen, M. van de Rijn, S.S. Jeffrey, C.A. Rees, J.R. Pollack, D.T. Ross, H. Johnsen, L.A. Akslen, Ø. Fluge, A. Pergamenschikov, C. Williams, S.X. Zhu, P.E. Lønning, A. Børresen-Dale, P.O. Brown, D. Botstein: Molecular portraits of human breast tumours, Nature **406**, 747–752 (2000)

32.6   M.J. van de Vijver, Y.D. He, L.J. van't Veer, H. Dai, A.A.M. Hart, D.W. Voskuil, G.J. Schreiber, J.L. Peterse, C. Roberts, M.J. Marton, M. Parrish, D. Atsma, A. Witteveen, A. Glas, L. Delahaye, T. van der Velde, H. Bartelink, S. Rodenhuis, E.T. Rutgers, S.H. Friend, R. Bernards: A gene-expression signature as a predictor of survival in breast cancer, N. Engl. J. Med. **347**, 1999–2009 (2002)

32.7   Y. Wang, J.G.M. Klijn, Y. Zhang, A.M. Sieuwerts, M.P. Look, F. Yang, D. Talantov, M. Timmermans, M.E. Meijer-van Gelder, J. Yu, T. Jatkoe, E.M.J.J. Berns, D. Atkins, J.A. Foekens: Gene expression profiles to predict distant metastasis of lymph-node-negative primary breast cancer, The Lancet **365**, 671–679 (2005)

32.8   X. Ma, R. Salunga, J.T. Tuggle, J. Gaudet, E. Enright, P. McQuary, T. Payette, M. Pistone, K. Stecker, B.M. Zhang, Y. Zhou, H. Varnholt, B. Smith, M. Gadd, E. Chatfield, J. Kessler, T.M. Baer, M.G. Erlander, D.C. Sgroi: Gene expression profiles of human breast cancer progression, Proc. Natl. Acad. Sci. USA **100**, 5974–5979 (2003)

32.9   MAQC Consortium: The MicroArray Quality Control (MAQC) project shows inter- and intraplatform reproducibility of gene expression measurements, Nat. Biotechnol. **24**, 1151–1161 (2006)

32.10   A. Dupuy, R.M. Simon: Critical review of published microarray studies for cancer outcome and guidelines on statistical analysis and reporting, J. Natl. Cancer Inst. **99**, 147–157 (2007)

32.11   A.H. Sims, K.R. Ong, R.B. Clarke, A. Howell: High-throughput genomic technology in research and clinical management of breast cancer. Exploiting the potential of gene expression profiling: Is it ready for the clinic?, Breast Cancer Res. **8**, 214 (2006)

32.12   W. Boecker, H. Buerger, K. Schmitz, I.A. Ellis, P.J. van Diest, H.P. Sinn, J. Geradts, R. Diallo, C. Poremba, H. Herbst: Ductal epithelial proliferations of the breast: A biologic continuum? Comparative genomic hybridization and high-molecular-weight cytokeratin expression patterns, J. Pathol. **195**, 415–421 (2001)

32.13   J.G. Herman, S.B. Baylin: Gene silencing in cancer in association with promoter hypermethylation, N. Engl. J. Med. **349**, 2042–2054 (2003)

32.14   K.Y. Tai, S.G. Shiah, Y.S. Shieh, Y.R. Kao, C.Y. Chi, E. Huang, H.S. Lee, L.C. Chang, P.C. Yang, C.W. Wu: DNA methylation and histone modification regulate silencing of epithelial cell adhesion molecule for tumor invasion and progression, Oncogene **26**, 3989–3997 (2007)

32.15   S.L. Berger: Histone modifications in transcriptional regulation, Curr. Opin. Genet. Dev. **12**, 142–148 (2002)

32.16   M.V. Iorio, M. Ferracin, C.G. Liu, A. Veronese, R. Spizzo, S. Sabbioni, E. Magri, M. Pedriali, M. Fabbri, M. Campiglio, S. Ménard, J.P. Palazzo, A. Rosenberg, P. Musiani, S. Volinia, I. Nenci, G.A. Calin, P. Querzoli, M. Negrini, C.M. Croce: MicroRNA gene expression deregulation in human breast cancer, Cancer Res. **65**, 7065–7070 (2005)

32.17   M.J. Bissell, D. Radisky: Putting tumours in context, Nat. Rev. Cancer **1**, 46–54 (2001)

32.18   M. Al-Hajj, M.S. Wicha, A. Benito-Hernandez, S.J. Morrison, M.F. Clarke: Prospective identification of tumorigenic breast cancer cells, Proc. Natl. Acad. Sci. **100**, 3983–3988 (2003)

32.19   D. Ponti, A. Costa, N. Zaffaroni, G. Pratesi, G. Petrangolini, D. Coradini, S. Pilotti, M.A. Pierotti, M.G. Daidone: Isolation and in vitro propagation of tumorigenic breast cancer cells with stem/progenitor cell properties, Cancer Res. **65**, 5506–5511 (2005)

32.20   T. Reya, S.J. Morrison, M.F. Clarke, I.L. Weissman: Stem cells, cancer, and cancer stem cells, Nature **414**, 105–111 (2001)

32.21   J.B. Kim, R. Stein, M.J. O'Hare: Tumour-stromal interactions in breast cancer: The role of stroma in tumourigenesis, Tumour Biol. **26**, 173–185 (2005)

32.22   T.D. Tlsty, P.W. Hein: Know thy neighbor: Stromal cells can contribute oncogenic signals, Curr. Opin. Genet. Dev. **11**, 54–59 (2001)

32.23   V. Montel, E.S. Mose, D. Tarin: Tumor-stromal interactions reciprocally modulate gene expression patterns during carcinogenesis and metastasis, Int. J. Cancer **119**, 251–263 (2006)

32.24   M.M. Mueller, N.E. Fusenig: Friends or foes – bipolar effects of the tumour stroma in cancer, Nat. Rev. Cancer **4**, 839–849 (2004)

32.25   J.D. Yager, N.E. Davidson: Estrogen carcinogenesis in breast cancer, N. Engl. J. Med. **354**, 270–282 (2006)

32.26   C.L. Wilson, A.H. Sims, A. Howell, C.J. Miller, R.B. Clarke: Effects of oestrogen on gene expression in epithelium and stroma of normal human breast tissue, Endocr. Relat. Cancer **13**, 617–628 (2006)

32.27   J. Russo, I.H. Russo: The role of estrogen in the initiation of breast cancer, J. Steroid Biochem. Mol. Biol. **102**, 89–96 (2006)

32.28   L.M. Coussens, Z. Werb: Inflammation and cancer, Nature **420**, 860–867 (2002)

32.29   O. Kovalchuk, V.P. Tryndyak, B. Montgomery, A. Boyko, K. Kutanzi, F. Zemp, A.R. Warbritton, J.R. Latendresse, I. Kovalchuk, F.A. Beland, I.P. Pogribny: Estrogen-induced rat breast carcinogenesis is characterized by alterations in DNA methylation, histone modifications and aberrant microRNA expression, Cell Cycle **6**, 2010–2018 (2007)

32.30   T.R. Golub, D.K. Slonim, P. Tamayo, C. Huard, M. Gaasenbeek, J.P. Mesirov, H. Coller, M.L. Loh, J.R. Downing, M.A. Caligiuri, C.D. Bloomfield, E.S. Lander: Molecular classification of cancer: Class discovery and class prediction by gene expression monitoring, Science **286**, 531–537 (1999)

32.31   B. Walsh: *Multiple Comparisons: Bonferroni Corrections and False Discovery Rates*, Lecture Notes for EEB, Vol. 581 (Univ. Arizona, Tucson 2004)

32.32   I.B. Jeffery, D.G. Higgins, A.C. Culhane: Comparison and evaluation of methods for generating differentially expressed gene lists from microarray data, BMC Bioinformatics **7**, 359 (2006)

32.33   D.M. Witten, R. Tibshirani: A comparison of fold-change and the *t*-statistic for microarray data analysis, Stanford Univ. Tech. Report (2007)

32.34   G.K. Smyth: Linear models and empirical Bayes methods for assessing differential expression in microarray experiments, Stat. Appl. Genet. Biol. **3**, article 3 (2004)

32.35   J. Xu, X. Cui: Robustified MANOVA with applications in detecting differentially expressed genes from oligonucleotide arrays, Bioinformatics **24**, 1056–1062 (2008)

32.36   N. Bassani, F. Ambrogi, R. Bosotti, M. Bertolotti, A. Isacchi, E. Biganzoli: Non-parametric MANOVA methods for detecting differentially expressed genes in real-time RT-PCR experiments, Computational Intelligence Methods for Bioinformatics and Biostatistics 2009 – Revised Selected Papers (2010) pp. 56–69

32.37   L.J. van't Veer, H. Dai, M.J. van de Vijver, Y.D. He, A.A. Hart, M. Mao, H.L. Peterse, K. van der Kooy, M.J. Marton, A.T. Witteveen, G.J. Schreiber, R.M. Kerkhoven, C. Roberts, P.S. Linsley, R. Bernards, S.H. Friend: Gene expression profiling predicts clinical outcome of breast cancer, Nature **415**, 530–536 (2002)

32.38   S. Chapman, P. Schenk, K. Kazan, J. Manners: Using biplots to interpret gene expression in plants, Bioinformatics **18**, 202–204 (2001)

32.39   O. Alter, P.O. Brown, D. Botstein: Singular value decomposition for genome-wide expression data processing and modeling, Proc. Natl. Acad. Sci. USA **97**, 10101–10106 (2000)

32.40   C. Casarsa, N. Bassani, F. Ambrogi, G. Zabucchi, P. Boracchi, E. Biganzoli, D. Coradini: Epithelial-to-mesenchymal transition, cell polarity and stemness-associated features in malignant pleural mesothelioma, Cancer Lett. **302**, 136–143 (2011)

32.41   N. Bassani, F. Ambrogi, D. Coradini, E. Biganzoli: Use of biplots and partial least squares regression in microarray data analysis for assessing association between genes involved in different biological pathways, Computational Intelligence Methods for Bioinformatics and Biostatistcs 2010, Lecture Notes in Bioinformatics **6685**, 123–134 (2011)

32.42   D. Ghosh, Z.S. Qin: Statistical Issues in the Analysis of ChIP-Seq and RNA-Seq Data, Genes **1**, 317–334 (2010)

32.43   P.J.G. Lisboa, A. Vellido, R. Tagliaferri, F. Napolitano, M. Ceccarelli, J.D. Martin-Guerrero, E. Biganzoli: Data mining in cancer research, IEEE Comput. Intell. Mag. **5**, 14–18 (2010)

32.44   U. Scherf, D.T. Ross, M. Waltham, L.H. Smith, J.K. Lee, L. Tanabe, K.W. Kohn, W.C. Reinhold, T.G. Myers, D.T. Andrews, D.A. Scudiero, M.B. Eisen, E.A. Sausville, Y. Pommier, D. Botstein, P.O. Brown, J.N. Weinstein: A gene expression database for the molecular pharmacology of cancer, Nat. Genet. **24**, 236–244 (2000)

32.45 D.T. Ross, U. Scherf, M.B. Eisen, C.M. Perou, C. Rees, P. Spellman, V. Iyer, S.S. Jeffrey, M. Van de Rijn, M. Waltham, A. Pergamenschikov, J.C.F. Lee, D. Lashkari, D. Shalon, T.G. Myers, J.N. Weinstein, D. Botstein, P.O. Brown: Systematic variation in gene expression patterns in human cancer cell lines, Nat. Genet. **24**, 227–234 (2000)

32.46 S. Paik, S. Shak, G. Tang, C. Kim, J. Baker, M. Cronin, F.L. Baehner, M.G. Walker, D. Watson, T. Park, W. Hiller, E.R. Fisher, D.L. Wickerham, J. Bryant, N. Wolmark: A multigene assay to predict recurrence of tamoxifen-treaten, node-negative breast cancer, N. Engl. J. Med. **351**, 2817–2826 (2004)

32.47 P.E. Colombo, F. Milanezi, B. Weigelt, J.S. Reis-Filho: Microarrays in the 2010s: The contribution of microarray-based gene expression profiling to breast cancer classification, prognostication and prediction, Breast Cancer Res. **13**, 212 (2011)

32.48 E. Biganzoli, N. Lama, F. Ambrogi, L. Antolini, P. Boracchi: Prediction of cancer outcome with microarrays, Lancet **365**, 1683 (2005)

32.49 S. Michiels, S. Koscielny, C. Hill: Prediction of cancer outcome with microarrays: A multiple random validation strategy, Lancet **365**, 488–492 (2005)

32.50 Y. Benjamini, Y. Hochberg: Controlling the false discovery rate: A practical and powerful approach to multiple testing, J. R. Stat. Soc. B **57**, 289–300 (1995)

32.51 J.D. Storey: A direct approach to false discovery rates, J. R. Stat. Soc. B **64**, 479–498 (2002)

32.52 R. Bellman: *Adaptive Control Processes: A Guided Tour* (Princeton University Press, New Jersey 1961)

32.53 B. Palsson, K. Zengler: The challenges of integrating multi-omic data sets, Nat. Chem. Biol. **6**, 783 (2010)

32.54 D.B. Searls: Data integration: Challenges for drug discovery, Nat. Rev. Drug Discov. **4**, 45–58 (2005)

32.55 S.W. Roh, G.C.J. Abell, K. Kim, Y. Nam, J. Bae: Comparing microarrays and next-generation sequencing technologies for microbial ecology research, Trends Biotechnol. **28**, 291–299 (2010)

32.56 J.C. Marioni, C.E. Mason, S.M. Mane, M. Stephens, Y. Gilav: RNA-seq: An assessment of technical reproducibility and comparison with gene expression arrays, Genome Res. **18**, 1509–1517 (2008)

32.57 J. Juhila, T. Sipilä, K. Icay, D. Nicorici, P. Ellonen, A. Kallio, E. Korpelainen, D. Greco, I. Hovatta: MicroRNA expression profiling reveals MiRNA families regulating specific biological pathways in mouse frontal cortex and hippocampus, PLoS ONE **6**, e21495 (2011)

32.58 E.E. Ntzani, J.P. Ioannidis: Predictive ability of DNA microarrays for cancer outcomes and correlates: An empirical assessment, Lancet **362**, 1439–1444 (2003)

32.59 D. Dunkler, S. Michiels, M. Schemper: Gene expression profiling: Does it add predictive accuracy

to clinical characteristics in cancer prognosis?, Eur. J. Cancer **43**, 745–751 (2007)

32.60 W.C.W. Chan, S.M. Nie: Quantum dot bioconjugates for ultrasensitive nonisotopic detection, Science **281**, 2016–2018 (1998)

32.61 P. Alivisatos: The use of nanocrystals in biological detection, Nat. Biotechnol. **22**, 47–52 (2004)

32.62 X. Michalet, F.F. Pinaud, L.A. Bentolila, J.M. Tsay, S. Doose, J.J. Li, G. Sundaresan, A.M. Wu, S.S. Gambhir, S. Weiss: Quantum dots for live cells, in vivo imaging, and diagnostics, Science **307**, 538–544 (2005)

32.63 X. Gao, L. Yang, J.A. Petros, F.F. Marshall, J.W. Simons, S. Nie: In-vivo molecular and cellular imaging with quantum dots, Curr. Opin. Biotechnol. **16**, 63–72 (2005)

32.64 X. Gao, S. Nie: Molecular profiling of single cells and tissue specimens with quantum dots, Trends Biotechnol. **21**, 371–373 (2003)

32.65 Y. Xing, A.M. Smith, A. Agrawal, G. Ruan, S. Nie: Molecular profiling of single cancer cells and clinical tissue specimens with semiconductor quantum dots, Int. J. Nanomed. **1**, 473–481 (2006)

32.66 Y. Xing, Q. Chaudry, C. Shen, K.Y. Kong, H.E. Zhau, L.W. Chung, J.A. Petros, R.M. O'Regan, M.V. Yezhelyev, J.W. Simons, M.D. Wang, S. Nie: Bioconjugated quantum dots for multiplexed and quantitative immunohistochemistry, Nat. Protoc. **2**, 1152–1165 (2007)

32.67 M.V. Yezhelyev, A. Al-Hajj, C. Morris, A.I. Marcus, T. Liu, M. Lewis, C. Cohen, P. Zrazhevskiy, J.W. Simons, A. Rogatko, S. Nie, X. Gao, R.M. O'Regan: In situ molecular profiling of breast cancer biomarkers with multicolor quantum dots, Adv. Mater. **19**, 3146–3151 (2007)

32.68 A.A. Ghazani, J.A. Lee, J. Klostranec, Q. Ziang, R.S. Dacosta, B.C. Wilson, M.S. Tsao, W.C. Chan: High throughput quantification of protein expression of cancer antigens in tissue microarray using quantum dot nanocrystals, Nano Lett. **6**, 2881–2886 (2006)

32.69 M.V. Yezhelyev, X. Gao, Y. Xing, A. Al-Hajj, S. Nie, R.M. O'Regan: Emerging use of nanoparticles in diagnosis and treatment of breast cancer, Lancet Oncol. **7**, 657–667 (2006)

32.70 Y.X. Wang, S.M. Hussain, G.P. Krestin: Superparamagnetic iron oxide contrast agents: Physicochemical characteristics and applications in MR imaging, Eur. Radiol. **11**, 2319–2331 (2001)

32.71 Q.Y. Cai, S.H. Kim, K.S. Choi, S.Y. Kim, S.J. Byun, K.W. Kim, S.H. Park, S.K. Juhng, K.H. Yoon: Colloidal gold nanoparticles as a blood-pool contrast agent for X-ray computed tomography in mice, Invest. Radiol. **42**, 797–806 (2007)

32.72 X. Gao, W.C. Chan, S. Nie: Quantum-dot nanocrystals for ultrasensitive biological labeling and multicolor optical encoding, J. Biomed. Opt. **7**, 532–537 (2002)

Part F | 32

32.73   J.R. Lindner: Contrast ultrasound molecular imaging: Harnessing the power of bubbles, Cardiovasc. Res. **83**, 615–616 (2009)

32.74   S.E. Fodor: Light-directed spatially addressable parallel chemical synthesis, Sciences **251**, 767–773 (1991)

32.75   J.M. Nam, C.A. Mirkin: Bio-barcode-based DNA detection with PCR-like sensitivity, J. Am. Chem. Soc. **126**, 5932–5933 (2004)

32.76   J.T. Santini, A.C. Richards, R. Scheidt, M.J. Cima, R. Langer: Microchips as controlled drug delivery devices, Angew. Chem. Int. Edn. **39**, 2396–2407 (2000)

32.77   G.M. Whitesides, E. Ostuni, S. Takayama, X.Y. Jiang, D.E. Ingber: Soft lithography in biology and biochemistry, Annu. Rev. Biomed. Eng. **3**, 335–373 (2001)

32.78   D.A. LaVan, T. McGuire, R. Langer: Small-scale systems for in vivo drug delivery, Nat. Biotechnol. **21**, 1184–1191 (2003)

32.79   M.M. Orosco, C. Pacholski, M.J. Sailor: Real-time monitoring of enzyme activity in a mesoporous silicon double layer, Nature Nano. **4**, 255–258 (2009)

32.80   M.M. Cheng, G. Cuda, Y.L. Bunimovich, M. Gaspari, J.R. Heath, H.D. Hill, C.A. Mirkin, A.J. Nijdam, R. Terracciano, T. Thundat, M. Ferrari: Nanotechnologies for biomolecular detection and medical diagnostics, Curr. Opin. Chem. Biol. **10**, 11–19 (2006)

32.81   J.R. Heath, M.E. Phelps, L. Hood: NanoSystems biology, Mol. Imaging Biol. **5**, 312–325 (2003)

32.82   Y. Cui, Q. Wei, H. Park, C.M. Lieber: Nanowire nanosensors for highly sensitive and selective detetction of biological and chemical species, Science **293**, 1289–1292 (2001)

32.83   K.M. Hansen, H.-F. Ji, G. Wu, R. Datar, R. Cote, A. Majumdar, T. Thundat: Cantilever-based optical deflection assay for discrimination of DNA single-nucleotide mismatches, Anal. Chem. **73**, 1567–1571 (2001)

32.84   G. Wu, R.H. Datar, K.M. Hansen, T. Thundat, R.J. Cote, A. Majumdar: Bioassay of prostate-specific antigen (PSA) using microcantilevers, Nat. Biotechnol. **19**, 856–860 (2001)

32.85   K.S. Hwang, S.-M. Lee, S.K. Kim, J.H. Lee, T.S. Kim: Micro- and nanocantilever devices and systems for biomolecule detection, Annu. Rev. Anal. Chem. **2**, 77–98 (2009)

32.86   J.M. Perez, F.J. Simeone, Y. Saeki, L. Josephson, R. Weissleder: Viral-induced self-assembly of magnetic nanoparticles allows the detection of viral particles in biological media, J. Am. Chem. Soc. **125**, 10192–10193 (2003)

32.87   W.U. Wang, C. Chen, K.H. Lin, Y. Fang, C.M. Lieber: Label-free detection of small-molecule-protein interactions by using nanowaire nanosensors, Proc. Natl. Acad. Sci. USA **102**, 3208–3212 (2005)

32.88   Y.Y. Degenhardt, R. Wooster, R.W. McCombie, R. Lucito, S. Powers: High-content analysis of cancer genome DNA alterations, Curr. Opin. Genet. Dev. **18**, 68–72 (2008)

32.89   C.C. WU, F.H. Ko, Y.S. Yang, D.L. Hsia, B.S. Lee, T.S. Su: Label-free biosensing of a gene mutation using a silicon nanowire field-effect transistor, Biosens. Bioelectron. **25**, 820–825 (2009)

32.90   C.A. Pratilas, D.B. Solit: Therapeutic strategies for targeting BRAF in human cancer, Rev. Recent Clin. Trials **2**, 121–134 (2007)

32.91   H.S. Lee, K.S. Kim, C.J. Kim, S.K. Hahn, M.H. Jo: Electrical detection of VEGFs for cancer diagnoses using anti-vascular endothelial growth factor aptamer-modified Si nanowire FETs, Biosens. Bioelectron. **24**, 1801–1805 (2009)

32.92   M.A. Bangar, D.J. Shirale, W. Chen, N.V. Myung, A. Mulchandani: Single conducting polymer nanowire chemiresistive label-free immunosensor for cancer biomarker, Anal. Chem. **81**, 2168–2175 (2009)

32.93   F.N. Ishikawa, H.K. Chang, M. Curreli, H.I. Liao, C.A. Olson, P.C. Chen, R. Zhang, R.W. Roberts, R. Sun, R.J. Cote, M.E. Thompson, C. Zhou: Label-free, electrical detection of the SARS virus N-protein with nanowire biosensors utilizing antibody mimics as capture probes, ACS Nano **3**, 1219–1224 (2009)

32.94   G. Zheng, F. Patolsky, Y. Cui, W.U. Wang, C.M. Lieber: Multiplexed electrical detection of cancer markers with nanowire sensor arrays, Nat. Biotechnol. **23**, 1294–1301 (2005)

32.95   Y. Liu, X. Li, Z. Zhang, G. Zuo, Z. Cheng, H. Yu: Nanogram per milliliter-level immunologic detection of alpha-fetoprotein with integrated rotating-resonance microcantilevers for early-stage diagnosis of heptocellular carcinoma, Biomed. Microdevices **11**, 183–191 (2009)

32.96   R.K. Jain: Transport of molecules, particles, and cells in solid tumors, Annu. Rev. Biomed. Eng. **1**, 241–263 (1999)

32.97   R.K. Jain: Delivery of molecular and cellular medicine to solid tumors, Adv. Drug Del. Rev. **46**, 149–168 (2001)

32.98   R.K. Jain: The next frontier in molecular medicine: Delivery of therapeutics, Nat. Med. **4**, 655–657 (1998)

32.99   M.D. Wang, J.W. Simons, S. Nie: Biomedical nanotechnology with bioinformatics – the promise and current progress, Proc. IEEE **95**, 1386–1389 (2007)

33. Personalized Information Modeling for Personalized Medicine

Yingjie Hu, Nikola Kasabov, Wen Liang

Personalized modeling offers a new and effective approach for the study of pattern recognition and knowledge discovery, especially for biomedical applications. The created models are very useful and informative for analyzing and evaluating an individual data object for a given problem. Such models are also expected to achieve a higher degree of accuracy of prediction of outcome or classification than conventional systems and methodologies. Motivated by the concept of personalized medicine and utilizing transductive reasoning, personalized modeling was recently proposed as a new method for knowledge discovery in biomedical applications. Personalized modeling aims to create a unique computational diagnostic or prognostic model for an individual. Here we introduce an integrated method for personalized modeling that applies global optimization of variables (features) and an appropriate neighborhood size to create an accurate personalized model for an individual. This method creates an integrated computational system that combines different information processing techniques, applied at different stages of data analysis, e.g., feature selection, classification, discovering the interaction of genes, outcome prediction, personalized profiling and visualization, etc. It allows for adaptation, monitoring, and improvement of an individual's model and leads to improved accuracy and unique personalized

profiling that could be used for personalized treatment and personalized drug design.

Part F | 33

33.1 Medical Data Analysis

Contemporary medical and other data analysis and decision support systems predominantly use inductive global models for the prediction of a person's risk, or of the likely outcome of a disease for an individual [33.1,

2]. In such models, features are pre-processed to minimize learning function error (usually a classification error) in a global way to identify the patterns in large databases. Pre-processing is performed to constrain the features used for training global learning models. In general, global modeling is concerned with deriving a global formula (e.g., a linear regression function, a *black box neural network*, or a support vector machine (SVM)) from a large group of data samples. Once an optimal global model is trained, a set of features (variables) is selected and then applied to the whole problem space (i.e., all samples in the given dataset). Thus, the assumption is made that the global model is able to work properly on any new data sample. In clinical research, therapeutic treatment designed to target a disease is assumed to be suitable for any new patients anywhere at anytime. However, such global modeling based medical treatment systems are not always applicable to individual patients, as the molecular profiling information is not taken into account. Heterogeneity of diseases (e.g., cancer), means that there is different disease progress and different responses to the treatment, even when the patients have tumors in the same organ that are remarkably similar morphologically.

Statistic reports from the medical research community have shown that the treatment developed by such global modeling methods is only effective for approximately 70% of people, leaving the remainder of patients with no effective treatment [33.3]. In cases of aggressive diseases, e.g., cancer, any ineffective treatment of a patient (e.g., either a patient not being treated, or being incorrectly treated), can be the difference between life and death. Thus, more effective approaches are required that are capable of using a patient's unique information, such as protein, gene, or metabolite profile to design clinical treatment specific to the individual patient.

33.1.1 Why Personalized Modeling?

In order to develop an understanding of personalized modeling for medical data analysis and biomedical applications, we must answer the question: *why do we need personalized information modeling technologies*? For many common conditions a patient's health outcome is influenced by the complex interplay of genetic, clinical, and environmental factors [33.4]. With the advancement of microarray technologies collecting personalized genetic data on a genome-wide (or genomic) scale has become quicker and cheaper [33.5,6]. Such personalized genomic data may include: DNA sequence data (e.g., single nucleotide polymorphisms (SNPs)), gene and protein expression data. Many worldwide projects have already collected and published a vast amount of such personalized data. For example,

genome-wide association scan (GWAS) projects have so far been published for over 100 human traits and diseases and many have made data available for thousands of people (http://www.genome.gov/gwastudies).

The advance of molecular profiling technologies, including microarray messenger ribonucleic acid (mRNA) gene expression data, proteomic profiling, and metabolomic information make it possible to develop *personalized medicine* based on new molecular testing and traditional clinical information for treating individual patient. According to the United States Congress [33.7], the definition of *personalized medicine* is given as:

> the application of genomic and molecular data to better target the delivery of health care, facilitate the discovery and clinical testing of new products, and help determine a person's predisposition to a particular disease or condition.

Personalized medicine is expected to focus on the factors affecting each individual patient and to help fight chronic diseases. More importantly, it may allow the development of medical treatment tailored to an individual's needs.

Motivated by the concept of personalized medicine and utilizing transductive reasoning [33.8], personalized modeling was recently proposed as a new method for knowledge discovery in biomedical applications. For the purpose of developing medical decision support systems, it would be particularly useful to use the information from a data sample related to a particular patient (e.g., blood sample, tissue, clinical data, and/or DNA) and tailor a medical treatment specifically for her/him. This information can also be potentially useful for developing effective treatments for another part of the patient population.

In a broader sense, personalized modeling offers a new and effective approach for the study of pattern recognition and knowledge discovery. The models created are more useful and informative for analyzing and evaluating an individual data object for a given problem. Such models are also expected to achieve a higher degree of accuracy of outcome prediction or classification than conventional systems and methodologies [33.9]. In fact, being able to accurately predict an individual's disease risk or drug response and using such information for personalized treatment is a major goal of clinical medicine in the twenty-first century [33.10].

Personalized modeling has been reported as an efficient solution for clinical decision making systems [33.11], because its focus is not simply on the global problem space, but on the individual sample. For a new data vector, the whole (global) space usually contains much noise information that presents the learning algorithm working properly on this new data, though the same information might be valuable for other data samples. With personalized modeling, the noise (or redundant) information can be excluded within the local problem space that is only created for the observed data sample. This characteristic of personalized modeling makes it a more appropriate method for discovering more precise information specifically for the individual data sample than conventional models and systems.

33.1.2 Inductive Versus Transductive Reasoning

Inductive and transductive inference are two prevalent approaches used in the development of learning models and systems in artificial intelligence. The original theory of inductive inference was proposed by *Solomonoff* [33.12, 13] in the early 1960s and was used for predicting the new data based on observations of a series of given data. In the context of knowledge discovery, the inductive reasoning approach is concerned with the construction of a functional model based on the observations, e.g., predicting the next event (or data) based upon a series of historical events (or data) [33.2, 14]. Many statistical learning methods, such as, SVM, multilayer perception (MLP), and neural network models have been implemented and tested on inductive reasoning problems.

The inductive inference approach is widely used in the development of models and systems for data analysis and pattern discovery in computer science and engineering. This approach creates models based upon known historical data vectors that are applicable to the whole problem space. However, the inductive learning and inference approach is only efficient when the whole problem space (global space) is searched for the solution of a new data vector. Inductive models generally neglect any information related to the particular new data sample, which raises an issue about the suitability of a global model for analyzing new input data.

In contrast to inductive learning methods, transductive inference introduced by *Vapnik* [33.8] is a method

that creates a model to test a specific data vector (a testing data vector) based on the observation of a specific group of data vectors (training data). The models and methods created from transductive reasoning focus on a single point of the space (the new data vector), rather than on the whole problem space. Transductive inference systems emphasize the importance of the utilization of the additional information related to the new data point, which brings more relevant information to suit the analysis of the new data. Within the same given problem space, transductive inference methods may create different models, each of them specific for testing every new data vector.

Transductive inference systems have so far been applied to a variety of classification problems, such as heart disease diagnostics [33.15], promoter recognition in bioinformatics [33.16], and microarray gene expression data classification [33.17]. Other examples using transductive reasoning systems include: evaluating the predicting reliability in regression models [33.18], providing additional reliability measurement for medical diagnosis [33.19], transductive SVM for gene expression data analysis [33.20], and a transductive inference based radial basis function (TWRBF) method for medical decision support systems and time series prediction [33.21]. Most of these experimental results have shown that transductive inference systems outperform inductive inference systems, due to the former's ability to exploit the structural information of unknown data.

Some more sophisticated transductive inference approaches have been developed including: the transductive neural fuzzy inference system with weighted data normalization – TWNFI [33.11] and the transductive rbf neural network with weighted data normalization – TWRBF [33.21]. These methods create a learning model based on the neighborhood of new data vector and then use the trained model to calculate the output.

The transductive inference approach seems to be more appropriate to build learning models for clinical and medical applications, where the focus is not simply on the model, but on the individual patient's condition. Complex problems may require an individual or a local model that best fits a new data vector, e.g., a patient to be clinically treated; or a future time moment for a time-series data prediction, rather than a global model that does not take into account any specific information from the object data [33.11].

33.2 Global, Local, and Personalized Modeling Approaches

Global, local, and personalized modeling are currently the three main techniques for modeling and pattern discovery in the machine learning area. These three types of modeling techniques are derived from inductive and transductive inference and are the most commonly used learning techniques for building the models and systems for data analysis and pattern recognition [33.9, 22]. This section will investigate these three techniques for data analysis and model design.

- *Global modeling* creates a model from the data that covers the entire problem space. The model is represented by a single function, e.g., a regression function, a radial basis function (RBF), a MLP neural network, SVM, etc.
- *Local modeling* builds a set of local models from data, where each model represents a subspace (e.g., a cluster) of the whole problem space. These models can be a set of rules or a set of local regressions, etc.
- *Personalized modeling* uses transductive reasoning to create a specific model for each single data point (e.g., a data vector, a patient record) within a localized problem space.

To explain the concepts of global, local, and personalized modeling, we hereby present a comparative study in which each type of model will be applied to a benchmark gene expression dataset, namely colon cancer data [33.23] for cancer classification. This comparative study applies several popular algorithms for modeling development and investigates the performance using three modeling techniques on gene expression data. The data used in the comparative experiment originates from colon cancer data that consists of 62 samples of colon epithelial cells from colon cancer patients. 40 samples are collected from tumors and labeled as *diseased*, and 22 samples are collected from a healthy part of the colon of the same patient and are labeled as *normal*. Each sample is represented by 2000 genes selected out of a total of 6500 genes based on the confidence in measured expression levels. Since the goal of this experiment is to demonstrate the difference of classification performance generated by three modeling techniques, we simply select 15 out of 2000 genes by a signal-noise-to-ratio (SNR) method according to their statical scores for the purpose of reducing computational cost. SNR is a simple statistical algorithm and widely adopted to filter features. Let $\bar{x}_i$ and $\bar{y}_i$ denote the mean values of the i-th gene in the samples in class 1 and class 2, re-

spectively, σ_{xi} and σ_{yi} are the corresponding standard deviations. Then each feature's SNR score can be calculated as follows

$$\text{SNR}(i) = \frac{|\bar{x}_i - \bar{y}_i|}{\sigma_{xi} + \sigma_{yi}}, \quad i = 1, 2, \cdots, m, \quad (33.1)$$

where m is the number of features in the given dataset. The greater the SNR value, the more informative the feature. Therefore, the preprocessed subset used in the experiment presented here constitutes 62 samples. Each sample contains 15 top features (genes) selected based on their statistical SNR ranking scores. The subset is denoted as D_{colon15}.

As our interest in this experiment is mainly in the comparison of the classification performance obtained from three different modeling techniques, we applied a simple validation approach (*hold-out* method) to the classification on data D_{colon15}: the given data is split into training and testing data with a specified ratio, i.e., 70% of samples are used for training and the remaining 30% for testing.

33.2.1 Global Modeling

Linear and logistic regression modeling is one of the most popular global modeling techniques. It has been implemented in a variety of global methods for modeling gene expression data [33.24] and for modeling gene regulatory networks [33.25].

Multiple linear regression (MLR) is a global modeling technique that is among the simplest of all statistical learning algorithms. MLR analysis is a multivariate statistical technique that examines the linear correlations between a single dependent variable and two or more independent variables. For multiple linear regression analysis, the independent variable X is described by an m-dimensional vector $X = (x_1, x_2, \cdots, x_m)$, and an MLR model can be formulated as

$$y_i = \beta_0 + \beta_1 x_{i1} + \beta_2 x_{i2} + \cdots + \beta_m x_{im} \varepsilon_i,$$
$$i = (1, 2, \cdots, n), \quad (33.2)$$

where

- β is an m-dimensional parameter vector called effects or (regression coefficients)
- ε is the *residual* representing the deviations of the observed values y from their means $\bar{y}$, which are normally distributed with mean 0 and variance
- n is the number of observations.

For the purpose of investigating global modeling for classification problems, an MLR based approach is applied to the subset of colon cancer gene expression data ($D_{colon15}$). A global MLR based classifier is created from the training data (70%) analysis, which is given as

$$\mathcal{Y} = 0.1997 + 0.1354 \times X_1 + 0.70507 \times X_2$$
$$+ -0.42572 \times X_3 - 0.19511 \times X_4$$
$$+ 0.0943 \times X_5 - 0.6967 \times X_6 - 1.0139 \times X_7$$
$$+ 0.9246 \times X_8 + 0.1550 \times X_9 + 0.6190 \times X_{10}$$
$$+ 0.1793 \times X_{11} + 1.123 \times X_{12}$$
$$- 0.1615 \times X_{13} - 0.4789 \times X_{14}$$
$$- 0.4910 \times X_{15} , \tag{33.3}$$

where $\mathcal{Y}$ is an MLR model to predict the new input data vector (here to predict whether a patient sample is *diseased* or *normal*) and X_i, $i = 1, 2, \ldots, 15$ denotes each variable (feature).

Function 3 constitutes a global model to be used for evaluating the output for any new data vector in the 15-dimensional space regardless of where it is located. This global model extracts a *big* picture for the whole problem space, but lacks an individual profile [33.9]. It indicates to a certain degree the genes' importance: X_6, X_8, and X_{12} show a strong correlation to the corresponding output, while X_5, X_1, and X_9 are less important in terms of outcome prediction.

Figure 33.1 shows the prediction result from the global multilinear regression model over colon data with selected 15 genes. The results plotted in Fig. 33.1a,b demonstrate the inconsistent issue in microarray gene expression data analysis: the accuracy from testing data is significantly lower than that from training data, −95.3% versus 73.7%, when the threshold of disease distinction is set to 0.5.

33.2.2 Local Modeling

Unlike global models, local models are created to evaluate the output function especially within a subspace of the entire problem space (e.g., a cluster of data). Multiple local models can consist of the complete model across the entire problem space. Local models are usually based on clustering techniques. A cluster is a group of similar data samples, where similarity is measured predominantly as Euclidean distance in an orthogonal problem space. Clustering techniques can be found in the literature; classical *k*-means [33.26], self-organizing maps (SOM) [33.27, 28], fuzzy c-means clustering [33.29], hierarchical clustering for cancer

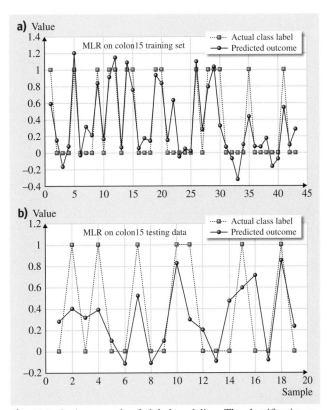

Fig. 33.1a,b An example of global modeling. The classification results from a multilinear regression model (MLR) over colon cancer gene data, where the *x*-axis is the sample index and the *y*-axis represents the value of the actual class label and predicted outcome for each sample. The *squares* represent the actual class labels of the samples, while the *filled circles* present the predicted outcome. **(a)** Classification result using a global MLR model on the $D_{colon15}$ training set (the training accuracy is 95.3%). **(b)** The classification result using a global MLR model on the $D_{colon15}$ testing set (the testing accuracy is 73.7%)

data analysis [33.23], and a simulated annealing procedure based clustering algorithm for finding globally optimal solution for gene expression data [33.30]. Fuzzy clustering is a popular algorithm used to implement local modeling for machine learning problems. The basic idea behind it is that one sample may belong to several clusters to a certain membership degree, and the sum of membership degree should be one.

Local learning models adapt to new data and discover local information and knowledge, which provide a better explanation for individual cases. However, these local modeling methods do not select specific subsets of features and precise neighborhood of samples for in-

a) Value

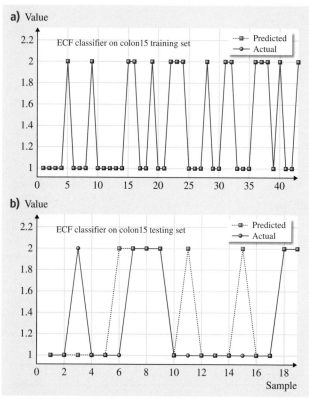

b) Value

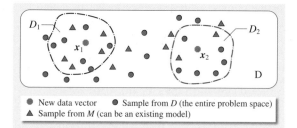

Fig. 33.3 An example of personalized space, where x_1 and x_2 represent two new input vectors, D is the entire (global) problem space, and D_1 and D_2 denote the two personalized spaces for x_1 and x_2, respectively

Fig. 33.2a,b An example of local modeling. The experimental results from a local modeling method (ECF) on the training and testing set from data ($D_{colon15}$), respectively. The *black solid line* represents the actual label of the sample, while the *broken line* is the predicted outcome. (**a**) A local modeling. The outcomes from ECF model on the training set of colon cancer data (70%), the training accuracy is 100%. (**b**) A local modeling. The outcomes from ECF model on the testing set of colon cancer data (30%), the testing accuracy is 79.0%

33.2.3 Personalized Modeling

The philosophy behind the proposed personalized modeling is the realization that every person is different, and preferably each individual should have their own personalized models and tailored treatment. In the context of medical research, it has become possible to utilize individual data for a person with the advance of technology, e.g., DNA, RNA, protein expression, clinical tests, inheritance, foods and drugs intake, and diseases. Such data is more readily obtainable nowadays and is easily measurable and storable in electronic data repositories with less cost.

In contrast to global and local modeling, personalized modeling creates a model for every new input data vector based on the samples that are closest to the new data vector in the given dataset. Figure 33.3 gives an example for personalized problem spaces. With a transductive approach, each individual data vector that represents a patient in any given medical area obtains a customized, local model that best fits the new data. This is contrary to using a global modeling approach where new data is matched to a model (function) averaged for the entire dataset. A global model may fail to take into account the specific information particular to individual data samples. Moreover, there are no efficient methods for identifying important features that assist complex disease classification, e.g., which genes, SNPs, proteins, and other clinical information contribute to the disease diagnosis. Hence, a transductive approach seems to be a step in the right direction when looking to devise personalized modeling useful for analyzing individual data samples, e.g., disease diagnosis, drug design, etc.

A personalized modeling framework (PMF) is initially designed for medical data analysis and knowledge discovery. However, PMF can be extended for solving

dividual samples that require a personalized modeling in the medical area. The evolving classification function (ECF) [33.31, 32] is a representative technique for local modeling. The classification result from ECF local model over dataset $D_{colon15}$ is shown in Fig. 33.2a,b. The classification accuracy from the ECF model on the training set (70% of the whole data) appeared excellent – 100% accurate, but the classification result from the testing set (30%) is only 78.95% (15 out of 19 samples are correctly classified). It seems that local modeling might not be an effective approach for analyzing this particular gene expression dataset. Moreover, it is difficult to optimism the parameters during the learning process.

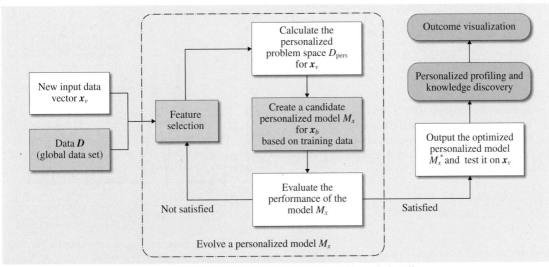

Fig. 33.4 A personalized modeling framework, PMF, for data analysis and knowledge discovery

various types of data analysis problems that require personalized modeling. PMF can be briefly described as follows:

1. Apply feature selection on the object data D (the global problem space) to identify which features are important to a new input vector x_v. The selected features are grouped into a candidate gene pool.
2. Select K_v nearest samples for x_v from D to form a local (personalized) problem space D_{pers}.
3. Create a personalized model candidate M_x specifically for x_v, which includes a learning function (usually a classifier or a clustering function) denoted by f.
4. Evaluate the candidate feature subset S by a learning function f based on their performance within the personalized problem space D_{pers}.
5. Optimize model M_x through an evolving approach until termination conditions are met. The output is the optimal or near-optimal solution to vector x_v. The solution includes an optimal personalized model M_x^* and a selected feature subset S^*.
6. Use the model M_x^* to test the new vector x_v and calculate the outcome y_v.
7. Create a personalized profile for the input vector x_v, visualize the outcome with the selected important features S^*, and provide an improvement scenario for data vector x_v for a given problem if it is possible.

An outline of PMF is depicted in Fig. 33.4.

The KNN method is probably the simplest technique to use for personalized modeling. In a KNN model, the K nearest samples for every new sample x_i are derived from the given dataset through a distance measurement (usually Euclidean distance), and the class label for the new sample x_i is assigned based on a voting scheme [33.33]. The classical KNN method calculates the output value y_i according to the determination made by the majority vote of its neighbors, i.e., the new data vector is assigned to the class most common amongst its k nearest neighbors.

The KNN algorithm is one of the most popular algorithms in machine learning, because it is simple to implement and works fast and effectively on many machine learning problems. However, the parameter selection is a critical factor impacting on the KNN classifier's performance, e.g., the choice of value for K. In general, more nearest neighbors (K) used in KNN method can reduce the effect of noise over the classification, but would make the boundaries between classes less distinct. If too few neighbors are selected, there can be insufficient information for decision making. Also, the performance of the KNN algorithm can be severely degraded by the presence of noisy features, which is a very common issue in biomedical data.

Weighted Nearest Neighbor Algorithms for Personalized Modeling: WKNN and WWKNN

In a weighted distance KNN algorithm (WKNN), the output y_i is calculated not only based on the output values (e.g., class label) y_j, but is also dependent on the weight w_j measured by the distance between the nearest neighbors and the new data sample x_i

$$y_i = \frac{\sum_{j=1}^{K_i} w_j \cdot y_j}{\sum_{j=1}^{K_i} w_j} \,, \qquad (33.4)$$

where y_i is the predicted output for the new vector x_i, y_j is the class label of each sample in the neighborhood of x_i, K_i is the number of K nearest samples to x_i, and w_j is the weight value calculated based on the distance from the new input vector x_j to its K nearest neighbors. The weight w_j can be calculated as

$$w_j = \frac{\max(d) - [d_j - \min(d)]}{\max(d)} \,, \qquad j = 1, \cdots, K \,, \qquad (33.5)$$

where the value of weights w_j ranges from $\frac{\min(d)}{\max(d)}$ to 1, $d = [d_1, d_2, \cdots, d_K]$ denotes the distance vector between the new input data d_i and the its K nearest neighboring samples, and $\max(d)$ and $\min(d)$ are the maximum and minimum values for vector d.

The distance vector d is computed as

$$d_j = \sqrt{\sum_{l=1}^{m} (x_{i,l} - x_{j,l})^2} \,, \qquad j = 1, \cdots, K \,, \qquad (33.6)$$

where m is the number of variables (features) representing the new input vector x_i within the problem space; $x_{i,l}$ and $x_{j,l}$ are the l-th variable values corresponding to the data vector x_i and x_j, respectively.

The output from a WKNN classifier for the new input vector x_i is a *personalized probability* that indicates the probability of vector x_i belonging to a given class. For a two-class classification problem, a WKNN classifier requires a threshold θ to determine the class label of x_i, i.e., if the output (*personalized probability*) is less than the threshold θ, then x_i is classified into the group with *small* class label, otherwise into the group with *big* class label. For example, in the case of a two-class problem, the output from the WKNN model for *sample#1* of data $D_{colon15}$ is 0.1444, so that this testing sample is classified into class **1** (*small* class label) when the threshold θ is set to 0.5.

The weighted distance and weighted variables K-nearest neighbors (WWKNN) is a personalized modeling algorithm that was introduced by *Kasabov* [33.9]. The main idea behind the WWKNN algorithm is: the K nearest neighbor vectors are weighted based on their distance to the new data vector x_i, and also the contribution of each variable is weighted according to their

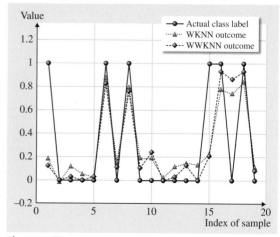

Fig. 33.5 The experimental results computed by two personalized models – WKNN and WWKNN on the colon cancer $D_{colon15}$ testing set (it contains 19 samples). The x axis is the sample index and the y axis shows value of the predicted outcome. $K = 15$ and the classification threshold is 0.5. Both of the models yielded 84.2% classification accuracy

importance within the local area where the new vector belongs [33.9]. In WWKNN, the assumption is made that the different variables have different importance to classifying samples into different classes when the variables are ranked in terms of their discriminative power of class samples over the whole m-*dimensional* space. Therefore, it will be more likely that the variables have different ranking scores if the discriminative power of the same variables is measured for a subspace (localized space) of the entire problem space. The calculation of Euclidean distance d_j between a new vector x_i and a neighbor x_j is mathematically formulated by

$$d_j = \sqrt{\sum_{l=1}^{K} c_{i,l} (x_{i,l} - x_{j,l})^2} \,, \qquad j = 1, \cdots, K \,, \qquad (33.7)$$

where $c_{i,l}$ is the coefficient weighing x_l in relation with its neighborhood of x_i, and K is the number of the nearest neighbors. The coefficient $c_{i,l}$ can be calculated by an SNR function that ranks variables across all vectors in the neighborhood set $D_{nbr}(x_i)$:

$$c_{i,l} = \{c_{i,1}, c_{i,2}, \cdots, c_{i,K}\} \,,$$

$$c_{i,l} = \frac{\left| \bar{x}_l^{\text{class1}} - \bar{x}_l^{\text{class2}} \right|}{\sigma_l^{\text{class1}} + \sigma_l^{\text{class2}}} \,, \qquad (33.8)$$

where: $\bar{x}_l^{\text{class}_i}$, $i = \{1, 2\}$ is the mean value of the l-th feature belonging to class i across the neighborhood $D_{\text{nbr}}(x_i)$ of x_j, and $\sigma_l^{\text{class}_i}$, $i = \{1, 2\}$ is the standard deviation of l-th feature belonging to class i across the neighborhood $D_{\text{nbr}}(x_i)$ of x_j.

Compared to a conventional KNN algorithm, the contribution of WWKNN lies in the new distance measurement: all variables are weighted according to their importance as discriminating factors in the neighborhood area (personalized subspace), which might provide more precise informa-tion for classification or prediction of the new data vector.

The experimental results from the classification of $D_{\text{colon}15}$ data using WKNN and WWKNN are illus-trated in Fig. 33.5; they show that WWKNN produced better predicting result for colon cancer data classifi-cation, as the predicted outcome from WWKNN both WKNN and WWKNN can create an outcome vector indicating the testing sample's probability of being dis-eased, which provides the important information for clinical decision making.

33.3 A Methodology to Build a Personalized Modeling System

We introduce a methodology for using the proposed PMF to build a personalized modeling system (PMS) to create the personalized model for each new in-put data sample based on its unique information. Given a dataset D pertaining to a bioinformatics prob-lem, $D = \{x_{ij}, y_i, i = 1, \cdots, n, j = 1, \cdots, m\}$, where x is a data sample, y is the responding outcome, n is the number of samples, and m denotes the number of features (variables). The proposed method aims to optimize a model M_x suitable for analyz-ing data, specific to every new input data vector x_v, e.g., to calculate y_v – the outcome of x_v. Data x_v contains a number of features that are related to the same scenario as the data samples in the global data D.

In order to obtain the optimal or near optimal personalized model M_x^* specifically for a new data sam-ple x_v, the proposed method aims to find the solutions to the following objectives:

1. Determine how many and which features (variables) S are most suitable for building the model M_x^* that is able to successfully predict the outcome for the new data vector x_v.
2. Determine the appropriate number K_v for the neigh-borhood of x_v to form a personalized problem space D_{pers}.
3. Identify K_v samples from the global data set D which have the pattern most similar to the data x_v, and use these K_v samples to form the neighborhood (a personalized problem space D_{pers}).
4. Calculate the importance of selected features S within the personalized problem space D_{pers}), based on their contribution to the outcome pre-diction of the data vectors in D_{pers}. Com-pute a weight vector w_v for all selected fea-tures S.
5. Create the optimal personalized model M_x^* with the optimized parameters obtained in steps 1–4.
6. Validate the obtained model M_x^* by calculating the outcome y_v for the new data x_v.
7. Profile the new input data x_v within its neigh-borhood D_{pers} using the most important features associated with a desired outcome.
8. If possible, provide the scenarios for improving the outcome for the new data vector x_v, which can be helpful for clinical use.

This is a method for determining a profile of a sub-ject (new input vector x_v) using an optimal personalized model M_x^* and for recommending the possible changes to the profile in relation to a scenario of interest in order to improve the outcome for x_v. The method comprises the following steps:

- Create a personalized profile for a new data vec-tor x_v.
- Compare each important feature of input data vec-tor x_v to the average value of important features of samples having the desired outcome.
- Determine which important features of input vec-tor x_v can be altered in order to improve the outcome.

Principally, the decision of which variables should be changed will be based on the observation of the weight vector W_x of features (i.e., the contribution of the fea-tures to the classification). The term *personalized profile* used here refers to an input vector x_v and to its predicted outcome and related information, such as the size of its neighborhood, its most important features specifically, etc.

Within the scope of PMS, the proposed method for building an optimal model M_x requires the following functional modules.

- A module for selecting most relevant V_v features (variables) S^* and ranking their weighter w_x by importance for x_v.

- The module for the selection of a number K_v of neighboring samples of x_v and for the selection of neighboring samples D_{pers}.
- A module for creating a prediction model M_x, defined by the a set of parameters P_v, such as K_v, V_v, D_{pers} which were derived in the previous modules.
- A module for calculating the final output y_v responding to the new data x_v.
- A module for the creation of personalized profile and the design of scenarios for potential improvement.

33.4 An Integrated Optimization Method for Implementing a PMS

There have been very few implementations of PMSs using computational intelligence to solve complex biomedical applications. In this section, we introduce an integrated method that has been recently developed for PMS implementations. The integrated method for personalized modeling (IMPM) [33.34] was developed based on the methodology described in Sect. 33.3. For every new individual sample (new data vector), all aspects of their personalized model (variables, neighboring samples, type of models and model parameters) are combined together to be optimized based on the accuracy of the outcome achieved within the local neighborhood of the sample. Next, a personalized model and personalized profile are derived that use the selected variables and the neighboring samples with known outcomes. The sample's profile is compared with average profiles of the other outcome classes in the neighborhood (e.g., positive outcome, or negative outcome of disease or treatment). The difference between the points and average profiles based on important variables that may need to be modified through treatment and can be utilized in personalized drug design.

33.4.1 A Detailed Description of IMPM

IMPM consists different functional modules and is summarized in Algorithm 33.1. Steps 2–5 are an iterative learning (training) process to optimize the local model M_x. The optimization continues until the termination criteria are reached, e.g., the maximum number of iterations is reached or a desired local accuracy of the model for a local data set D_x is achieved. The optimization of the parameters of the personalized model V_x, K_x, and D_x is global and is achieved through multiple runs of cEAP, which was described in the

previous section. The resulting competing personalized models for x form a population of such models that are evaluated over iterations (generations) using a fitness criterion – the best accuracy of outcome prognosis for the local neighborhood of new testing sample x. All variables and parameters of the personalized model form to an integrated single *chromosome* (Fig. 33.6) where variable values are optimized together as a global optimization.

Algorithm 33.1 The algorithm of IMPM

1: Data pre-processing stage:
   Include data collection, storage, update, etc.
2: Feature selection:
   Identify a subset of features (variables) V_x relevant to the new data sample x_v from all features V.
3: Local problem space creation:
   Select a number K_x of samples from the global dataset D to create a neighborhood D_x. D_x consists of a group of similar samples to x with the features from V_x.
4: Evaluate the V_x features within the local neighborhood D_x in terms of their contribution and obtaining a weight vector W_x.
5: Training model optimization:
   Optimize a local prognostic/classification model M_x that has a model parameter set P_x, a variable set V_x and local training data set D_x.
6: Testing the new data sample:
   Apply the optimized personalized model $M_x^*(P_x, V_x, D_x)$ on the new data sample x and output the prediction result.
7: Profiling:
   Generate a functional profile F_x for the new data sample x using the selected set V_x of variables,

along with the average profiles of the samples from D_x that belong to different outcome classes, e.g., F_i and F_j.

8: Perform a comparative analysis between F_x, F_i, and F_j to define what variables from V_x are the most important for the person x that make him/her very differential from the desired class. These variables may be used to define a personalised course of treatment, such as personalized medicine.

Initially, the assumption is made that all feature (variable) sets V have equal absolute and relative importance for a new sample x in relation to predicting its unknown output y

$$w_{v1} = w_{v2} = \cdots = w_{vq} = 1 \qquad (33.9)$$

and

$$w_{v1,\text{norm}} = w_{v2,\text{norm}} = \cdots = w_{vq,\text{norm}} = 1/q . \qquad (33.10)$$

The initial numbers for the variables V_x and K_x may be determined in a variety of different ways without departing from the scope of the method. For example, V_x and K_x may be initially determined by an assessment of the global dataset in terms of size and/or distribution of the data. The values of these parameters may be constrained according to the available data. For example, $V_{x\_\text{min}} = 3$ (the minimum three variables used in a personalized model) and $V_{x\_\text{max}} < K_x$ (the maximum variables used in a personalized model should be smaller than the number of samples in the neighborhood D_x of x), and usually $V_{x\_\text{max}} < 20$. The initial set of variables may include expert knowledge, i.e., variables which are referenced in the literature as highly correlated to the outcome of the problem (disease) in a general sense (over the whole population). Such variables, for example, are the BRCA genes in the study for breast cancer prediction [33.35]. For an individual patient the BRCA genes may interact with some other genes, which interaction will be specific for the person or a group of people and is likely to be discovered through local or/and personalized modeling only [33.9].

IMPM has a major advantage over global and local modeling methods, as its modeling process can start with all relevant variables available for a person, rather than with a pre-fixed set of variables in a global model. Such global models may be statistically representative for the whole population, but not necessarily representative for a single person in terms of optimal model and best profiling and prognosis for this person.

Fig. 33.6 A chromosome used in IMPM for the global optimization of the parameters (*genes*): the variables V_x to be selected, their corresponding weights W_x, number K of nearest neighbors to x_v, the neighborhood D_x with selected K samples $s_1 - s_K$, a local prognostic model M_x (e.g., classifier), and a parameter set P_m for the M_x

Selecting the initial number K_x of neighboring samples and the minimum and the maximum numbers K_{x_min} and K_{x_max} can also depend on the data available and on the problem in hand. A general requirement is that $K_{x\_\text{min}} > V_x$, and, $K_{x\_\text{max}} < cN$, where c is a ratio, e.g., 0.5, and N is the number of samples in the neighborhood D_x of x. Several formulas have been already suggested and experimented [33.8], e.g.:

- K_{x_min} equals the number of samples that belong to the class with a smaller number of samples when the data is imbalanced (one class has many more samples, e.g., 90%, than the another class) and the available data set D is of small or medium size (e.g., several tens or several hundreds of samples)
- $K_{x\_\text{min}} = \sqrt{N}$, where N is the total number of samples in the data set D.

At subsequent iterations of the method, the parameters V_x and K_x along with all other parameters are optimized via an optimization procedure, usually an evolutionary based algorithm, such as cEAP [33.36] that optimizes all or part of parameters from the *chromosome* in Fig. 33.6.

The closest K_x neighboring vectors to x from D are selected to form a new data set D_x. A local weighted variable distance measure is used to weigh the importance of each variable V_l ($l = 1, 2, \cdots, q$) to the accuracy of the model outcome calculation for all data samples in the neighborhood D_x. For example, the distance between x and z from D_x is measured as a local weighted variable distance

$$d_{x,z} = \frac{\sqrt{\sum_{l=1}^{q}(1 - w_{l,\text{norm}})(x_l - z_l)^2}}{q} , \qquad (33.11)$$

where w_l is the weight assigned to the variable V_l and its normalized value is calculated as

$$w_{l,\text{norm}} = \frac{w_l}{\sum_{i=1}^{q} w_i} . \qquad (33.12)$$

Part F | 33.4

Here the distance between a cluster center (in our case it is the vector x) and cluster members (data samples from D_x) is calculated not only based on the geometrical distance, as it is in the traditional nearest neighbor methods, but on the relative variable importance weight vector W_x in the neighborhood D_x as suggested in [33.9]. After a subset D_x of K_x data samples are selected based on the variables from V_x, the variables are ranked in descending order of their importance for prediction of the output y of the input vector x and a weighting vector W_x is obtained. Through an iterative optimization procedure the number of the variables V_x to be used for an optimized personalized model M_x will be reduced, and only the most appropriate variables that lead to the best local prediction accuracy for M_x will be selected. For weighting W_x (i.e., ranking) of the V_x variables, alternative algorithms can be used, such as the t-test, SNR, etc.

In the SNR algorithm, W_x are calculated as normalized coefficients and the variables are sorted in descending order: $V_1, V_2, \ldots, V_v$, where $w_1 \geq w_2 \geq \cdots \geq w_v$, using (33.8). This method is very fast, but evaluates the importance of the variables in the neighborhood D_x one by one and does not take into account a possible interaction between the variables, which might affect the model output.

A learning model, usually a classification or prediction model is applied to the neighborhood D_x of K_x data samples to derive a personalized model M_x using the already defined variables V_x, variable weights W_x, and a model parameter set P_x. A variety of classification or prediction models can be used here such as MLR, SVM, KNN, WKNN, WWKNN [33.9], TWNFI [33.11], etc. The outcome produced by the WKNN classifier for the new sample is calculated based on the weighted outcomes of the individuals in the neighborhood according to their distance to the new sample. In the WWKNN model [33.9] variables are ranked and weighted according to their importance for separating the samples of different classes in the neighborhood area in addition to the weighting according to the distance as in WKNN. In the TWNFI model [33.11], the number of variables in all personalized models is fixed, but the neighboring samples used to train the personalized neuro-fuzzy classification model are selected based on the variable weighted distance to the new sample the same as it is in the WWKNN.

The vector distance $d = (d_1, d_2, \cdots, d_K)$ is defined as the distances between the new input vector x and the nearest samples (x_j, y_j) for $j = 1$ to K_x; $\max(d)$ and $\min(d)$ are the maximum and minimum values in d, re-

spectively. The Euclidean distance d_j between vector x and a neighboring one x_j is calculated as

$$d_j = \sqrt{\sum_{l=1}^{V} (1 - w_l)(x_l - x_{jl})^2}\,, \tag{33.13}$$

where w_l is the coefficient weighing variable x_l in the neighborhood D_x of x (e.g., and w_l can be calculated by an SNR algorithm, refer to (33.1)).

When using the TWNFI classification or prediction model [33.11], the output y for the input vector x is calculated as

$$y = \frac{\sum_{l=1}^{m} \frac{n_l}{\delta_l^2} \prod_{j=1}^{P} \alpha_{lj} \exp\left(-\frac{w_j^2 (x_{ij} - m_{lj})^2}{2\sigma_{lj}^2}\right)}{\sum_{l=1}^{m} \frac{1}{\delta_l^2} \prod_{j=1}^{P} \alpha_{lj} \exp\left(-\frac{w_j^2 (x_{ij} - m_{lj})^2}{2\sigma_{lj}^2}\right)}\,, \tag{33.14}$$

where m is the number of the closest clusters to the new input vector x; each cluster l is defined as a Gaussian function G_l in a V_x dimensional space with a mean value m_l as a vector and a standard deviation δ_l as a vector too; $x = (x_1, x_2, \cdots, x_v)$; α_l (also a vector across all variables V) is membership degree to which the input vector x belongs to the cluster Gaussian function G_l; n_l is a parameter of each cluster [33.11].

A local accuracy (local error E_x) that estimates the personalized accuracy of the personalized prognosis (classification) for the data set D_x using model M_x is evaluated. This error is a local one, calculated in the neighborhood D_x, rather than a global accuracy, which is commonly calculated for the whole problem space D. Different methods can be used for calculating the error, such as: absolute error (AE), root-mean square error (RMSE), and area and the receiving operating characteristic curve (AUC).

We propose another method to calculate local error specific for model optimization

$$E_x = \frac{\sum_{j=1}^{K_x} (1 - d_{xj}) E_j}{K_x}\,, \tag{33.15}$$

where d_{xj} is the weighted Euclidean distance between sample x and sample S_j from D_x that takes into account the variable weights W_x (33.11); E_j is the error between what the model M_x calculates for the sample S_j from D_x and what its real output value is.

Based on a weighted distance measured by (33.15), the closer the data sample S_j to x is, the higher its contribution to the error E_x will be. The calculated personalized model M_x accuracy is then formulated as

$$A_x = 1 - E_x\,. \tag{33.16}$$

The best accuracy model obtained is stored for the purpose of future improvement and optimization. The optimization procedure iteratively returns to all previous procedures (steps 2–5) to select another set of parameter values for the parameter vector (Fig. 33.6) until the termination criteria are reached. The method also optimizes parameters P_x of the classification/prediction procedure. The output value y for the new input vector x is then calculated by the optimal model M_x^*. Next, a personalized profile F_x for the person can be assessed against possible desired outcomes for the scenario, and the possible ways to achieve an improved outcome can be designed, which is a major novelty of this method. The profile F_x for x is formed as a vector

$$F_x = (V_x, W_x, K_x, D_x, M_x, P_x, t), \tag{33.17}$$

where the variable t represents the time of the creation of the model M_x. At a future time $(t + \Delta t)$ the person's input data will change to x^* (due to changes in variables such as age, weight, protein expression values, etc.), or the data samples in the data set D may be updated and new data samples added. A new profile F_x^* derived at time $(t + \Delta t)$ may be different from the current one F_x.

The average profile F_i for every class C_i in the data D_x is a vector containing the average values of each variable of all samples in D_x from class C_i. The importance of each variable (feature) is indicated by its weighting in the weight vector W_x. The weighted distance from the person's profile F_x to the average class profile F_i (for each class i) is defined as

$$D(F_x, F_i) = \sum_{l=1}^{v} |V_{lx} - V_{li}| w_l , \tag{33.18}$$

where w_l is the weight of the variable V_l calculated for dataset D_x (33.12).

Assuming that F_d is the desired profile (e.g., normal outcome), the weighted distance $D(F_x, F_d)$ will be calculated as an aggregated indication of how much the person's profile should change to reach the average desired profile F_d:

$$D(F_x, F_d) = \sum_{l=1}^{v} |V_{lx} - V_{ld}| w_l . \tag{33.19}$$

A scenario for a person's improvement through changes made to variables (features) towards the desired average profile F_d can be produced as a vector of required variable changes, defined as

$$\Delta F_{x,d} = \Delta V_{lx,d} \mid l = 1, \cdots, v , \tag{33.20}$$

$$\Delta V_{lx,d} = |V_{lx} - V_{ld}|, \quad \text{with an importance of } w_l . \tag{33.21}$$

In order to find a smaller number of variables, as global markers that can be applied to the whole population X, procedures steps 2– 7 are repeated for every individual x. All variables from the derived sets V_x are then ranked based on their likelihood to be selected for all samples. The top m variables (most frequently selected for testing individual models) are considered as a set of global markers V_m. The procedures P2–P5 will be applied again with the use of V_m as initial variable set (instead of using the whole initial set V of variables). In this case, personalized models and profiles are obtained within a set of variable markers V_m that would make treatment and drug design more universal across the whole population X.

33.4.2 An Optimization Algorithm (cEAP) for PMS

The IMPM method employs a coevolutionary based algorithm for personalized modeling (cEAP) to optimize related parameters, selecting informative features and finding the appropriate neighborhood for personalized modeling [33.36]. Given a general objective optimization problem $f(x)$ to minimize (or maximize), $f(x)$ is subject to two constraints $g_i(x)$ and $h_j(x)$. A candidate solution is to minimize the objective function $f(x)$, where x represents a n-dimensional decision (or optimization) variable vector $X = \{x_i \mid i = 1, \cdots, n\}$ from the sample space Ω. The two constraints describe the dependence between decision variables and parameters involved in the problem, and must be satisfied in order to optimize $f(x)$. The constraints $g_i(x)$ and $h_j(x)$ are denoted as inequalities and equalities, respectively, and are mathematically formulated as:

$$g_i(x) \leq 0 \mid i = 1, \ldots, n , \tag{33.22}$$

$$h_j(x) = 0 \mid j = 1, \ldots, p . \tag{33.23}$$

The number of degrees of freedom is calculated by $n - p$. Note that the number of equality constraints must be smaller than the number of decision variables (i.e., $p < n$). The *overconstrained* issue occurs when $p \geq n$, because there are no degrees of freedom left for optimizing the objective function.

Algorithm 33.2 The optimization algorithm – cEAP

1: Initialize the subindividuals in the subcomponent for feature selection:
generate a probability vector p with l bits, $p_i = 0.5$, where $i \in 1, \cdots, l$.

2: Generate two subindividuals from the vector p, respectively:
$(G_a, G_b) = generate(p)$

3: Generate a pair of subindividuals K_a, K_b by a probability function f_p.

4: Generate a pair of subindividuals:
θ_a and θ_b using a probability function f_p'.

5: Recombine the above subindividuals from three subcomponents into two individuals:
$\alpha = G_a + K_a + \theta_a$, $\beta = G_b + K_b + \theta_b$.

6: Evaluate individuals α and β by a fitness function F_c, respectively.

7: Compete individual α and β:
winner, loser = *compete* (α, β).

8: Create new populations in three subcomponents:
(i) Use GA to create the new generation for feature selection subcomponent

if $G_a(i) \neq G_b(i)$ **then**
  if *winner*$(i) = 1$ **then** $p_i = p_i + \frac{1}{\mu}$
  else $p_i = p_i - \frac{1}{\mu}$.

(ii) Use ES (ES) to create the new generation for K and θ in the other subcomponents:
Keep the winner of K and θ to form the offspring K_a' and θ_a'; the other offspring K_b' and θ_b' are generated through a mutation performed by probability functions f_p and f_p'.

9: Check whether the termination criteria are reached:

if yes, then the winner individual represents the final solution ζ^*, including the selected features G^* and optimized parameters K^* and θ^*
else iterate the process from step 2.

The algorithm aims to find the optimal solution to an objective function. Given an objective function $f(x)$: for $x \in \Omega$, $\Omega \neq \emptyset$, a global minimum of the objective problem $f(x)$ can be mathematically defined as $f^* \triangleq f(x^*) > -\infty$, **only if**

$$\forall x \in \Omega : \quad f(x^*) \leq f(x), \tag{33.24}$$

where x^* denotes the minimum solution, Ω is the sample universe of x.

The optimization algorithm for selecting genes and optimizing the parameters of learning functions (e.g., a classifier threshold θ and the number of neighbors k_v) simultaneously. The basic idea underlying the cEAP algorithm is to coevolve the search in multiple search spaces (here for feature/variable selection and parameter optimization).

The objective of IMPM is to build personalized models for data analysis and knowledge discovery that are able to minimize the prediction accuracy of disease distinction and create a personalized profile for individual patient. Given a data $D = \{X, Y\} \mid X = x_{ij}$, $Y = y_i$, $i = 1 \ldots n$, $j = 1 \ldots m$, the objective is therefore defined to optimize a classifier that involves the selected features and related parameters

$$f(s^*) \leq f(s), \tag{33.25}$$

where f is a classification function, and s denotes an independent variables set. As s can be represented by the data vector X, Y with selected features and related parameters, (33.25) is rewritten as:

$$f(X, Y, \zeta_l^*) \leq f(X, Y, \zeta_l), \quad \mid \zeta \in \Omega, \ l = (1, 2, 3), \tag{33.26}$$

where ζ_l denotes the candidate solution from l different subcomponents. The final solution is obtained when (33.25) is fulfilled, i.e., ζ_l^* is taken as the desired solution to the problem of gene selection and parameter optimization when the classification error is less than or equal to the value at any other conditions.

For clarity, the pseudo code of the optimization algorithm cEAP is given in Algorithm 33.2.

33.5 Experiment

We present an experiment using personalized modeling with IMPM for diagnosis and profiling of cancer. A benchmark colon cancer gene expression dataset is used [33.23]. It consists of 62 samples, 40 collected from colon cancer patients and 22 from control subjects. Each sample is represented by 2000 gene expression variables. The objective is to create a diagnostic (classification) system that not only provides an accurate

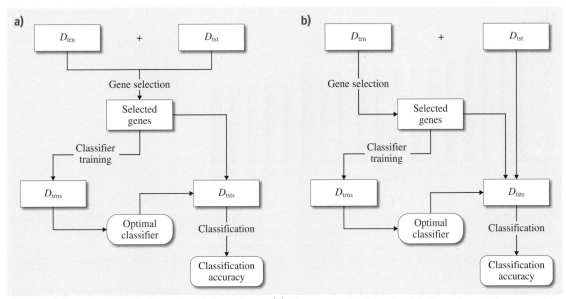

Fig. 33.7 (a) An example of a biased validation scheme. **(b)** The proposed unbiased validation scheme. The comparison between a biased and an unbiased verification scheme, where D_{trn} and D_{tst} are the training and testing set, and D_{trns} and D_{tsts} are the training and testing set with selected genes, respectively. In case **(a)** (biased verification scheme), the testing set is used twice in gene selection and classifier training procedure, which introduces a bias error from the gene selection stage into the final classification step. Whereas in case **(b)** (the unbiased scheme), the testing set is only used in the final classification (validation) stage, i.e., the testing set is independent all through gene selection and classifier training procedures

diagnosis, but also profiles the person to help define the best treatment.

An unbiased verification approach for personalized modeling data analysis should guarantee that generalization errors occur in either feature selection or classification procedures as little as possible. To this end, an efficient data sampling method should be used in the two procedures to maximally decrease the generalization error. In other words, the reliability and generalizability of the informative features should be evaluated on independent testing subsets, and then these features can be used for classification. The classification also needs to employ verification methods to estimate the bias error. Such a procedure is shown in Fig. 33.7b. For comparison, a simple example of biased validation schema is demonstrated in Fig. 33.7a.

Fig. 33.8 The evolution of feature (variable) selection for sample #32 from the colon cancer data (600 generations of GA optimization; the lighter the *color*, the higher the probability of the feature to be selected; each feature is represented as 1 bit on the *horizontal axis*; at the beginning all features are assigned equal probability to selected as 0.5) ▶

33.5.1 Personalized Modeling with IMPM for Colon Cancer Diagnosis and Profiling on Gene Expression Data

An example of a personalized model of colon cancer diagnosis and profiling of a randomly selected person is given in Fig. 33.8–Fig. 33.13 [33.34]. Figure 33.8 shows

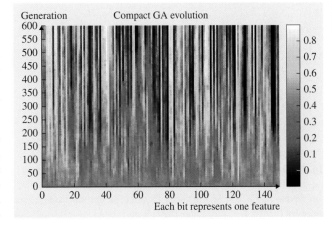

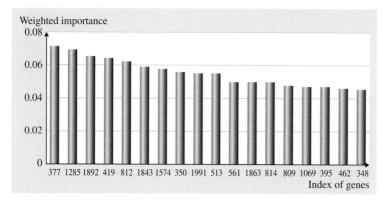

Fig. 33.9 The weighted importance of the selected features for sample #32 using weighted SNR based model

the evolution (GA) process of feature selection specifically for sample#32 from the colon cancer data through 600 generations. IMPM selects 18 genes (features) out of 2000 genes based the result from the GA optimization. Figure 33.9 illustrates the weighted importance of the selected 18 genes from Fig. 33.8. The weighted importance is calculated by a weighted SNR model ((33.12) and (33.1)). The larger the importance value, the more informative the gene.

Using the proposed IMPM, an optimized personalized model M_x for sample#32 from the colon cancer data is created. This personalized model M_x consists of the selected 18 informative genes, along with two parameters – classification threshold ($\theta = 0.40$) and the number of neighboring samples ($K = 18$), which are optimized specifically for sample#32. Figure 33.10 shows the data subset D_x with 18 samples (the neighborhood with an appropriate size) of sample#32 using top three selected genes (genes 377, 1285, and 1892). These neighboring samples are 61, 41, 12, 1, 38, 22, 26, 31, 34, 28, 19, 44, 6, 49, 57, 3, 8, 43.

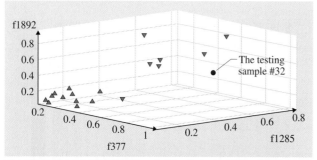

Fig. 33.10 Sample#32 (the *blue dot*) is plotted with its 18 neighboring samples selected by IMPM (*red triangles* – cancer samples and *green triangles* – control) in the 3-D space of the top three gene variables (genes 377, 1285, and 1892) from Fig. 33.9

The predicted outcome computed by the optimized personalized model M_x^* is 0.51, which successfully classifies sample#32 into the diseased class (class 2) (the classification threshold θ is optimized to 0.40 as a model parameter).

Using IMPM, a profile and a scenario of potential genome improvement for colon sample#32 was created as shown in Fig. 33.11. The desired average profile is the average gene expression level from a healthy samples group and the desired improvement value identifies the change of the gene expression level that this patient (sample#32) should follow in order to recover from the disease. For example, the expression level of gene 377 of sample#32 is 761.3790, while the average class profiles for class 1 (normal class) and class 2 (diseased class) are 233.8870 (for class 1) and 432.6468 (for class 2). The distance between the gene expression level of gene 377 for sample#32 and the desired average class 1 profile is 527.4920, i. e., a potential solution can be given to the colon cancer patient (sample#32) to decrease his/her gene 377 expression level from 761.3790 to 233.8870. The information in the generated profile can be used for designing personalized treatment for cancer patients.

To find a small number of variables (potential markers) for the whole population of colon cancer data, we have used the approach as follows: based on the experiment result for every sample, we selected 20 most frequently used genes as potential global markers. Table 33.1 lists these 20 global markers with their biological information. Here we use 20 selected genes as global markers. The number 20 is based on the suggestion in *Alon*'s work [33.23].

The next objective of our experiment is to investigate whether utilizing these 20 potential marker genes can lead to improved colon cancer classification accuracy and what classification algorithm will

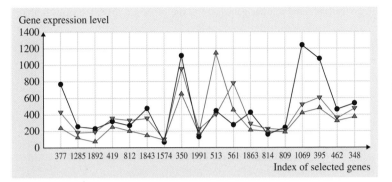

Fig. 33.11 The profile of sample#32 (*blue dots*) versus the average local profile of the control (*green*) and cancer (*red*) samples using the 18 selected genes from Fig. 33.9 as derived through the IMPM

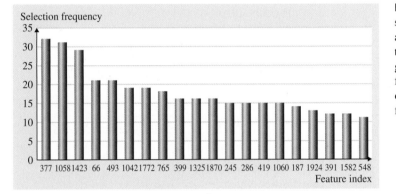

Fig. 33.12 The 20 most frequently selected genes using IMPM across all colon cancer data samples, where the *x*-axis represents the index of the gene in the data and the *y*-axis is the frequency of the gene as the marker of the optimized personalized models for which this gene has been selected

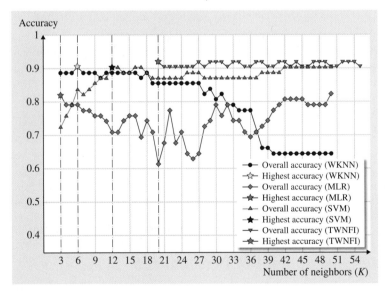

Fig. 33.13 A comparison of classification results obtained by four classification algorithms using 20 potential maker genes from Fig. 33.12, where the *x*-axis represents the size of the neighborhood and the *y*-axis is the average classification accuracy across all samples. The best accuracy is obtained with the use of the TWNFI classification algorithm (91.90%)

Part F | 33.5

perform best in the proposed IMPM. Four classification algorithms are tested as personalized models in this experiment, including WKNN, MLR, SVM, and TWNFI. All the classification results from four classifiers are validated based on leave-one-out cross

validation (LOOCV) across the whole dataset. Figure 33.13 shows the average accuracy obtained by these four algorithms with different size (K_x) of neighborhood. Table 33.2 summarizes the classification results from the four classification algorithms

Table 33.1 The 20 most frequently selected genes (potential marker genes) using the proposed IMPM across all colon cancer gene data samples (Fig. 33.12)

Index of Gene	GenBank Accession Number	Description of the Gene (from GenBank)
G377	Z50753	H.sapiens mRNA for GCAP-II/uroguanylin precursor
G1058	M80815	H.sapiens a-L-fucosidase gene, exon 7 and 8, and complete cds
G1423	J02854	Myosin regulatory light chain 2, smooth muscle ISOFORM (HUMAN)
G66	T71025	Human (HUMAN)
G493	R87126	Myosin heavy chain, nonmuscle (Gallus gallus)
G1042	R36977	P03001 Transcription factor IIIA
G1772	H08393	COLLAGEN ALPHA 2(XI) CHAIN (Homo sapiens)
G765	M76378	Human cysteine-rich protein (CRP) gene, exons 5 and 6
G399	U30825	Human splicing factor SRp30c mRNA, complete cds
G1325	T47377	S-100P PROTEIN (HUMAN)
G1870	H55916	PEPTIDYL-PROLYL CIS-TRANS ISOMERASE, MITOCHONDRIAL PRECURSOR (HUMAN)
G245	M76378	Human cysteine-rich protein (CRP) gene, exons 5 and 6
G286	H64489	Leukocyte antigen CD37 (Homo sapiens)
G419	R44418	Nuclear protein (Epstein-barr virus)
G1060	U09564	Human serine kinase mRNA, complete cds
G187	T51023	Heat shock protein HSP 90-BETA (HUMAN)
G1924	H64807	Placental folate transporter (Homo sapiens)
G391	D31885	Human mRNA (KIAA0069) for ORF (novel proetin), partial cds
G1582	X63629	H.sapiens mRNA for p cadherin
G548	T40645	Human Wiskott-Aldrich syndrome (WAS) mRNA, complete cds

Table 33.2 The best classification accuracy obtained by four classification algorithms on colon cancer data with 20 potential maker genes. Overall – overall accuracy; Class 1 – class 1 accuracy; Class 2 – class 2 accuracy

Classifier	Overall (%)	Class 1 (%)	Class 2 (%)	Neighborhood size
MLR (personalised (x))	82.3	90.0	68.2	3
SVM (personalised (x))	90.3	95.0	81.8	12
WKNN	90.3	95.0	81.8	6
TWNFI	**91.9**	95.0	85.4	20
Original publication [33.23]	87.1	–	–	–

using 20 selected potential marker genes. WKNN and a localized SVM yielded improved classification accuracy (90.3%) when compared to the global model [33.23]. However, the TWNFI classifier obtained the best classification performance (91.9%). Our results suggest that a small set of marker genes selected by IMPM could lead to improved cancer classification accuracy.

33.6 Conclusion and Future Development of Personalized Modeling Systems

When compared to global or local modeling, the proposed personalized modeling method (IMPM) has a major advantage. With personalized modeling methods, the modeling process starts with all relevant variables available for a person, rather than with a fixed number of features required by a global model. Such a global model may be statistically representative for a whole population (global problem space), but not necessarily representative for a single person in terms of best prognosis for this person. The proposed IMPM leads

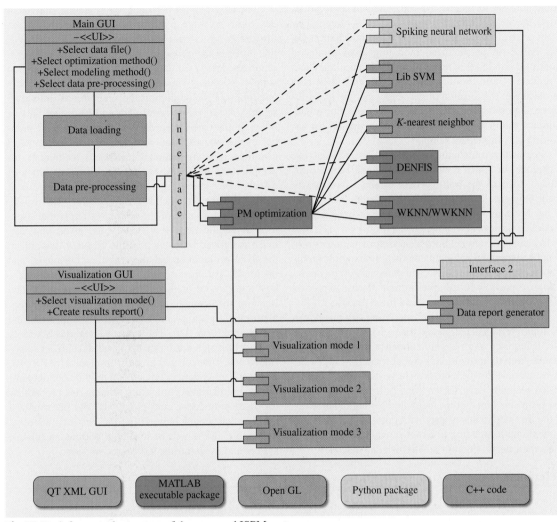

Fig. 33.14 A framework structure of the proposed ISPM system

to a better prognostic accuracy and a computed personalized profile. With global optimization, a small set of variables (potential markers) can be identified from the selected variable set across the whole population. This information can be utilized for the development of new, more efficient drugs. A scenario for outcome improvement is also created by IMPM, which can be utilized for the decision of efficient personalized treatment. We hope that this paper will motivate the biomedical applications of personalized modeling research.

Personalized modeling methods and systems will not substitute experts and current global or local modeling methods, but they are expected to derive information that is specifically relevant to a person and help individuals and clinicians make better decisions, thus saving lives, improving quality of life, and reducing cost of treatment. The IMPM method is capable of discovering more useful information, including selected informative genes and optimal disease classification parameters specifically for the observed patient sample, which are helpful for constructing the clinical decision support systems for cancer diagnosis and prognosis. For biological reference, some of our experimental findings have been reported in the literature, e.g., the selected genes of colon cancer data by our method are reported as biomarkers in other published papers.

In summary, personalized modeling offers a novel and integrated methodology that comprises different computational techniques for data analysis and knowledge discovery. Compared with the results obtained by other published methods, the new algorithms and methods based on personalized modeling have produced improved outcomes in terms of prediction accuracy and discovered more useful knowledge, because they take into account the location of new input sample in a subspace. The subspace (local space) excludes noise data samples and provides more precise information for analyzing new input data sample.

Personalized modeling is an adaptive and evolving technique, in which new data sample can be continuously added to the training dataset and subsequently contribute the learning process of personalized modeling. More importantly, the technique of personalized modeling offers a new tool to give a profile for each new individual data sample. Such a characteristic makes personalized modeling based methods promising for medical decision support systems and personalized medicine design, especially for complex human disease diagnosis and prognosis, such as cancer and brain disease.

However, as a personalized modeling system creates a unique (personalized) model for each testing data sample, it requires more computational power and performance time than traditional global modeling methods, especially to train the models on large data sets. The proposed methods have shown great potential for solving the problems that require individual testing. This study is the first step in this research direction and needs more in-depth understanding in bioinformatics for validating the experimental findings and knowledge discovery.

The next step for personalized modeling study is to develop a software platform, called an integrated optimization system for personalized modeling (ISPM) that utilizes the proposed novel personalized modeling methodology for data analysis and medical decision support systems. This platform offers a user-friendly environment for predicting the outcome of individual samples based on personal data and historical data of other similar cases, regardless of the type and the number of the available data and variables. It incorporates a variety of computational intelligent techniques for personalized modeling for solving different types of research problems. Figure 33.14 shows a framework structure of the proposed ISPM system.

The main feature of the ISPM system is that it optimizes all the factors related to the given objective in an integrated way, such as the features (variables), the local problem space, the classification model and its model parameters, with an objective function – best accuracy of predicted results for every individual input vector (sample, patient). ISPM includes a cross-platform class library, integrated development tools, and a cross-platform IDE. The system integrates a variety of modeling methods based on classical statistical algorithms and sophisticated models developed by KEDRI. The software system is expected to provide not only improved prediction accuracy, but reliable risk probability for disease diagnosis and prognosis, and personalized profiles that would help define the best actions (e.g., treatment). ISPM will be available both as an off-line system and as an on-line web based version.

References

33.1   J. Anderson, L.L. Hansen, F.C. Mooren, M. Post, H. Hug, A. Zuse, M. Los: Methods and biomarkers for the diagnosis and prognosis of cancer and other diseases: Towards personalized medicine, Drug Resist. Updates **9**(4/5), 198–210 (2006)

33.2   A.S. Levey, J.P. Bosch, J.B. Lewis, T. Greene, N. Rogers, D. Roth: A more accurate method to estimate glomerular filtration rate from serum creatinine: a new prediction equation. Modification of diet in renal disease study group, Ann. Intern. Med. **130**, 461–470 (1999)

33.3   A. Shabo: Health record banks: Integrating clinical and genomic data into patient-centric longitudinal and cross-institutional health records, Pers. Med. **4**(4), 453–455 (2007)

33.4   J.R. Nevins, E.S. Huang, H. Dressman, J. Pittman, A.T. Huang, M. West: Towards integrated clinico-genomic models for personalized medicine: Combining gene expression signatures and clinical factors in breast cancer outcomes prediction, Human Mol. Genet. **12**(2), R153–R157 (2003)

33.5   M.I. McCarthy, J.N. Hirschhorn: Genome-wide association studies: Potential next steps on a genetic journey, Human Mol. Genet. **17**(R2), R156–R165 (2008)

33.6   L.A. Hindorff, P. Sethupathy, H.A. Junkins, E.M. Ramos, J.P. Mehta, F.S. Collins, T.A. Manolio: Potential etiologic and functional implications of genome-wide association loci for human diseases and traits, Proc. Natl. Acad. Sci. USA **106**(23), 9362–9367 (2009)

33.7   The Congress of United States: S. 976 (110th): Genomics and Personalized Medicine Act of 2007 (2007)

33.8   V.N. Vapnik: *Statistical Learning Theory* (Wiley, New York 1998)

33.9   N. Kasabov: Global, local and personalized modelling and pattern discovery in bioinformatics: An integrated approach, Pattern Recognit. Lett. **28**(6), 673–685 (2007)

33.10   T.J. Jorgensen: From blockbuster medicine to personalized medicine, Pers. Med. **5**(1), 55–64 (2008)

33.11   Q. Song, N. Kasabov: TWNFI – a transductive neuro-fuzzy inference system with weighted data normalization for personalized modeling, Neural Netw. **19**(10), 1591–1596 (2006)

33.12   R. Solomonoff: A formal theory of inductive inference, part I, Inf. Control **7**(1), 1–22 (1964)

33.13   R. Solomonoff: A formal theory of inductive inference, part II, Inf. Control **7**(2), 224–254 (1964)

33.14   C. Bishop: *Neural Networks for Pattern Recognition* (Oxford Univ. Press, Cambridge 1995)

33.15   D. Wu, K.P. Bennett, N. Cristianini, J. Shawe-Taylor: Large margin trees for induction and transduction, Proc. Sixteenth Int. Conf. Mach. Learn. (ICML) (Morgan Kaufmann, San Francisco 1999) pp. 474–483

33.16   N. Kasabov, S. Pang: Transductive support vector machines and applications in bioinformatics for promoter recognition, Proc. Int. Conf. Neural Netw. Signal Process. (IEEE Press, Bellingham 2004)

33.17   M. West, C. Blanchette, H. Dressman, E. Huang, S. Ishida, R. Spang, H. Zuzan, J.A. Olson, J.R. Marks, J.R. Nevins: Predicting the clinical status of human breast cancer by using gene expression profiles, Proc. Natl. Acad. Sci. USA **98**(20), 11462–11467 (2001)

33.18   Z. Bosnic, I. Kononenko, M. Robnik-Sikonja, M. Kukar: Evaluation of prediction reliability in regression using the transduction principle, EUROCON 2003. The IEEE Region 8, Vol. 2 (2003) pp. 99–103

33.19   M. Kukar: Transductive reliability estimation for medical diagnosis, Artif. Intell. Med. **29**, 2003 (2002)

33.20   S. Pang, N. Kasabov: Inductive vs. transductive inference, global vs. local models: SVM, TSVM, and SVMT for gene expression classification problems, Neural Netw. 2004 IEEE Int. Joint Conf., Vol. 2 (2004) pp. 1197–1202

33.21   Q. Song, N. Kasabov: TWRBF: Transductive RBF neural network with weighted data normalization. In: *Neural Information Processing*, Lecture Notes in Computer Science, Vol. 3316, ed. by N. Pal, N. Kasabov, R. Mudi, S. Pal, S. Parui (Springer, Berlin, Heidelberg 2004) pp. 633–640

33.22   N. Kasabov: Soft computing methods for global, local and personalised modeling and applications in bioinformatics. In: *Soft Computing Based Modeling in Intelligent Systems*, ed. by V.E. Balas, J. Fodor, A. Varkonyi-Koczy (Springer, Berlin, Heidelberg 2009) pp. 1–17

33.23   U. Alon, N. Barkai, D.A. Notterman, K. Gish, S. Ybarra, D. Mack, A.J. Levine: Broad patterns of gene expression revealed by clustering analysis of tumor and normal colon tissues probed by oligonucleotide arrays, Proc. Natl. Acad. Sci. USA **96**, 6745–6750 (1999)

33.24   T. Furey, N. Cristianini, N. Duffy, D. Bednarski, M. Schummer, D. Haussler: Support vector machine classification and validation of cancer tissue samples using microarray expression data, Bioinformatics **16**(10), 906–914 (2000)

33.25   P. D'haeseleer, S. Liang, R. Somogyi: Genetic network inference: from co-expression clustering to reverse engineering, Bioinformatics **16**(8), 707–726 (2000)

33.26   S. Lloyd: Least squares quantization in PCM, IEEE Trans. Inf. Theory **28**(2), 129–137 (1982)

33.27   T. Kohonen: Self-organized formation of topologically correct feature maps, Biol. Cybern. **43**, 59–69 (1982)

33.28   T. Graepel, M. Burger, K. Obermayer: Self-organizing maps: Generalizations and new optimization techniques, Neurocomputing **21**, 173–190 (1998)

33.29   J.C. Bezdek: *Pattern Recognition with Fuzzy Objective Function Algorithms* (Kluwer Academic, Norwell 1982)

33.30   A.V. Lukashin, R. Fuchs: Analysis of temporal gene expression profiles: clustering by simulated annealing and determining the optimal number of clusters, Bioinformatics **17**(5), 405–414 (2001)

33.31   N. Kasabov: Evolving connectionist systems. In: *Methods and Applications in Bioinformatics, Brain Study and Intelligent Machines* (Springer, London 2002)

33.32   N. Kasabov, Q. Song: Denfis: Dynamic evolving neural-fuzzy inference system and its application for time-series prediction, IEEE Trans. Fuzzy Syst. **10**(2), 144–154 (2002)

33.33   T. Mitchell, R. Keller, S. Kedar-Cabelli: Explanation-based generalization: A unifying view, Mach. Learn. **1**(1), 47–80 (1986)

33.34   N. Kasabov, Y. Hu: Integrated optimisation method for personalised modelling and case studies for medical decision support, Int. J. Funct. Inform. Pers. Med. **3**(3), 236–256 (2010)

33.35   L.J. van't Veer, H. Dai, M.J. van de Vijver, Y.D. He, A.A.M. Hart, M. Mao, H.L. Peterse, K. van der Kooy, M.J. Marton, A.T. Witteveen, G.J. Schreiber, R.M. Kerkhoven, C. Roberts, P.S. Linsley, R. Bernards, S.H. Friend: Gene expression profiling predicts clinical outcome of breast cancer, Nature **415**(6871), 530–536 (2002)

33.36   Y. Hu, N. Kasabov: Coevolutionary method for gene selection and parameter optimization in microarray data analysis. In: *Neural Information Processing*, ed. by C.S. Leung, M. Lee, J.H. Chan (Springer, Berlin, Heidelberg 2009) pp. 483–492

34. Health Informatics

David Parry

Computers have been used in healthcare for many years for administrative, clinical, and research purposes. Health informatics is concerned with the use of data for the management of disease and the healthcare process. Increasingly health informatics is using data and approaches developed for bioinformatics and vice versa and there are many areas where computational intelligence has the potential to make a useful contribution to health informatics. Health informatics is both a practical profession and an area of research. This chapter deals with the organization of healthcare, areas of development of health informatics in recent times, and some active areas of research that may be relevant.

34.1 Healthcare

Healthcare is an important aspect of life for people around the world. However, the cost of healthcare in many countries is becoming unsustainable and in the extreme case around 18% of the US GDP is spent on healthcare. Healthcare is a very large employer in many countries, for example the British National Health Service is reputed to be the largest employer in Western Europe. Efficiency in healthcare requires increasing realization of the benefits in information technology, and this has become a central tenet of many reform schemes [34.1].

34.1.1 The Healthcare Environment

Healthcare can be divided into five different levels. Data moves between these levels but each one has its own requirements for collection and analysis of data. Understanding the needs of different care levels is important especially when computing professionals are new to the domain and may not realize that hospitals are not the sole environments for healthcare delivery.

Personal Care

Ultimately every person takes at least some responsibility for their own health and the health of their family. This can involve lifestyle decisions, self-monitoring of health status, and over the counter medications and devices. Information sources include health websites such as Medline Plus in the US, NHS Direct in the UK and commercial sites such as webMD. A large number of electronic personal health record sites have begun operating in the last few years, with Microsoft's Health Vault perhaps being the most prominent. These sites allow users to control and store medical information about themselves, which can also be made available to other users in the health system or other groups such as family or caregivers. These systems can take data directly from home care and lifestyle devices such as pedometers, blood pressure monitors, and blood sugar monitors.

Primary Care

Primary care includes family doctors – known as general practitioners in some parts of the world – dentists, physiotherapists, and nurses [34.2]:

> *The general practitioner is a specialist trained to work in the front line of a healthcare system and to take the initial steps to provide care for any health problem(s) that patients may have.*

Primary care informatics [34.3] is focused on making patient consultations more effective. Primary care workers may diagnose and treat the patient themselves, or refer them to other professionals or share care. Primary care physicians have a number of challenges in their work; he or she will often see the patient and make decisions alone, the consultation time is short, patients may present with a wide range of symptoms and no previous diagnosis or laboratory information, they may be dealing with chronic or acute illness and the seriousness of the complaint may not be obvious. In addition many primary healthcare systems are based around practitioners that are paid by a combination of fee for service and capitation (population-based) payment and targets for care – such as vaccination rates – that must be met. This means that primary care systems have to support both administration and clinical data, in an easy to use and efficient manner.

Other uses of data collected from primary care [34.4] include audit data, to assess the degree to which the practitioner is managing disease successfully, data related to prescriptions, and incidence of disease, especially infectious diseases and chronic disease. This data is especially rich and extensive, although care must be taken to preserve privacy and acknowledge that the primary purpose of the data collection was not large-scale modeling of populations, so that the data may need some normalization or cleaning.

Secondary Care

Secondary care is based in hospitals or specialized centers. In many health systems primary care practitioners act as *gatekeepers* to secondary care. Hospital informatics systems are often very well-developed and include administration systems, laboratory, and imaging systems (picture archiving and communication systems-PACS). One of the major issues for informatics in secondary care is the need to support communication between clinical staff, for example doctors, nurses, and allied staff who may have very different requirements. Secondary care units often deal with a very large number of patients, and informatics applications are regularly assessed in terms of their contribution to a reduction in the time taken to diagnose or discharge patients [34.5].

Tertiary Care

Tertiary care comprises specialized care that is not available at the secondary level. This can include specialized services such as liver transplantation, which may be provided on a regional or national basis. Tertiary care units tend to deal with small numbers of patients. Their informatics requirements include timely and accurate transfer of data from other units, including secondary care hospitals and assessment of the acuity – effectively a measurement of how sick the patient is. One popular means of assessing acuity is the acute physiology, age, chronic health evaluation (APACHE) III model [34.6]. These approaches attempt to calculate whether patients are likely to benefit from tertiary care based on a combination of physiological and pathological variables fed into a linear regression model. By having a widely accepted model, units can also compare outcomes and cost-effectiveness [34.7].

Transfer of information between these levels of care – both referral *to* and discharge *from* units – is a vital effort in health informatics. This includes information about diagnosis, symptoms, current therapy and appropriate care, as well as patient preferences. In many cases these levels of care are run by different organizations, and a funding model is applied, which may complicate patterns of referral. A large system implemented in the UK *Choose and Book* intended to support patient choice in referral has had limited success [34.8]. Ultimately, the different levels of care require a flexible and precise degree of semantic interoperability in order to fulfil this need [34.9].

Public Health

Public health deals with the population rather than the individual. Public health systems attempt to reduce the prevalence and impact of disease and ill-health and improve the well being of the whole population. Traditionally, public health has had a strong emphasis on infectious disease control, for example by identifying outbreaks of disease and responding to them by

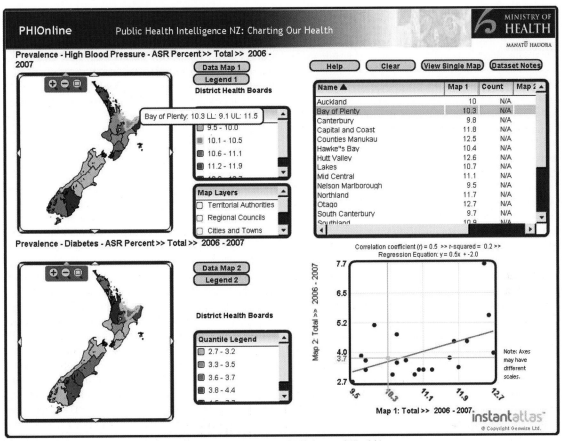

Fig. 34.1 Chronic disease mapping (courtesy of New Zealand Ministry of Health)

vaccination, public information, and even quarantine measures. Public health has more recently worked to control *lifestyle* diseases such as smoking related illness, accidents, and obesity. Public health informatics often uses visualization tools including geographical information systems (GIS) in order to track disease incidence and prevalence (Fig. 34.1).

Health informatics has traditionally been focused on clinical outcomes. Randomized controlled trials (RCT) are seen as the gold standard for assessing clinical interventions such as new drugs and therapies. This has been extended to systems – especially clinical decision support systems (CDSS) with researchers comparing clinical outcomes with and without the use of a particu-

lar system, in order to demonstrate benefit. However, this may not be appropriate for all informatics systems [34.10], as in practice clinical efficacy may not lead to actual use. More recently, the case for both more RCTs [34.11] and a wider view of system adoption, use, and usefulness [34.12] has been advocated. Computational intelligence research has traditionally focused more on demonstrating improvement in performance on well-known data sets than on ease of adoption or practical benefits, as befits a domain that is focussed on algorithm design and testing. In order to be credible to the user and funding community, health informatics research must always consider how improvement can be demonstrated in a clinical environment.

34.2 Clinical Uses of Information

According to the New Zealand Ministry of Health, clinical data collected should be available for; supporting clinical intervention, clinical governance, administration (in all parts of health), strategy, and policy development and research. The sixth use is patients using the clinical data for self-management. Health informatics as a discipline is related to a number of other domains (Fig. 34.2). A number of different aspects of health systems are examined in the following sections.

34.2.1 Health Information and Libraries

Health informatics has a relatively long history. Health information has been collected for many years. Initially data was collected on cause of death, and this became the international classification of disease (ICD), which was standardized in 1893. Development has continued to the present day and the current implementation, administered by the World health Organization is ICD-10. In 1879 Index Medicus was established by John Billings, which provided controlled indexing for medical research publications. This approach was computerized, and the bibliographic database MEDLARS was established in 1964. In 1997 PUBMED, which was set up by the National Library of Medicine in the USA, gave free public access via the web to this database [34.13]. Currently there are more than

20 million records available in PUBMED, which may include bibliographic data, an abstract, and a link to the full text of the article. Because of this history and vast resource, there has been a great deal of work in the area of concept hierarchies in particular the medical subject heading (MeSH), which is used for indexing. In order to improving mapping between databases and to make more sophisticated query handling approaches available the unified medical language system (UMLS) has also been developed. Tools for searching the database via the Entrez system have also been developed, such as ask Medline [34.14].

34.2.2 Coding and Vocabularies

Representing medical information in text format has occurred since doctors started making records, but more recently it was realized that standardization of such terms and concepts would be essential for decision support and information transfer. A number of schemes are in operation but the most comprehensive is SNOMED CT [34.15]. SNOMED was developed from a pathology vocabulary (SNOP) that came into being in 1965. SNOMED CT includes many thousands of concepts with a rich set of relations between them. SNOMED CT has been managed since 2007 by IHTSDO – the International Health Terminology Standards Development Organization.

34.2.3 Communication

Communication of health data is a key aspect of health informatics. Data needs to be coded, stored, transferred, and interpreted. In order to transfer clinical data effectively between practitioners a number of messaging systems have been developed of which the most widely used is Health Level 7 (HL7). HL7 was first developed in 1987 and as its name implies is designed to work at the upper (application) level of the OSI Model. HL7 has become increasingly sophisticated and now includes a clinical document architecture intended to support semantic interoperability [34.16]. Messaging systems are especially attractive in healthcare applications as data often needs to be transferred between providers. The CDA is essentially an XML-based wrapper for data that associates clinical observations and interventions with a particular patient and clinical event. HL7 systems can be used in primary and secondary or tertiary care. HL7 has been widely adopted in

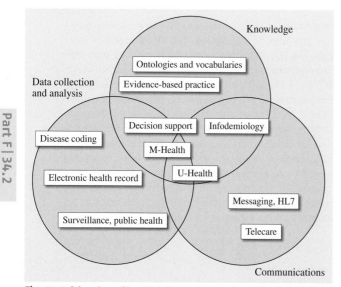

Fig. 34.2 Mapping of health informatics to other related disciplines

many countries and is particularly advanced in New Zealand [34.17].

Digital imaging and communications in medicine (DICOM) [34.18] is a messaging standard used for medical images and other multimedia data. It includes information related to the resolution of the image, suitable means of playback, and ensures that patient information is tightly bound to the image data.

34.2.4 Administration

Healthcare involves a great deal of administrative work, indeed the US spends around 31% of its health budget on administration [34.19], although this is around double that of Canada. Administrative tasks include ensuring that accurate records are kept of the activities of clinicians, the location and status of patients and payment status. Computer-based administrative systems have been present in secondary care for many years, and many of the original electronic data transfer systems were developed in order to make transfer of administrative details easier. Such standards – in the US mostly based around ASC X12N have been prescribed by legislation such as the health insurance portability and privacy act (HIPPA), but have not yet been fully adopted [34.20].

However, even what may be regarded as administration systems can provide valuable insights into patterns of disease and treatment. Perhaps the most famous example of this is the Dartmouth atlas [34.21], which by using data collected from administrative systems (Medicare) highlighted large variations in the degree of utilization of health services, indicating that resources may be being misallocated.

34.2.5 The Electronic Health Record (EHR)

Traditional hand-written medical notes are extremely information rich and support planning and communication, as well as acting as an information store. However paper-based records require storage and transport and are often not available at the point of care. They tend to be sequentially organized rather than problem-based. Their data is not usable by decision support systems or as the basis of an information sharing system.

The EHR contains information about an individual's health and treatment. Often seen as the key building-block of health informatics, electronic records attempt to capture all of the details that are recorded by clinical staff related to patient care. However paper records resist simple conversion to electronic format, as they are

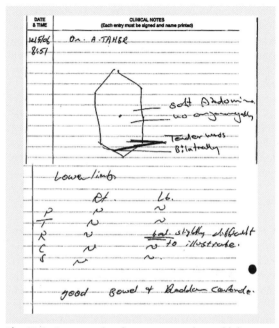

Fig. 34.3 An example of a paper record, combining text, graphics, and poor readability

often hard to read, combine images and text, and include input from many different authors (Fig. 34.3).

A large number of electronic health record systems are now available for both primary and secondary care. These systems may be characterized as part of health information systems, which can include all the IT sys-

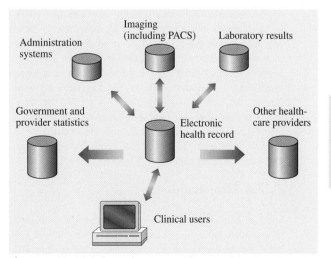

Fig. 34.4 Health information systems in secondary care, a system of systems

tems of a health organization (Fig. 34.4). Adoption of electronic health records has been patchy for various reasons, including cost [34.22], and rates of adoption around the world vary widely. For example, in primary care, some countries have over 95% adoption, but others are below 40% [34.23].

A large-scale effort to record and codify the semantics of the EHR continues based around the open EHR project. This systemizes what should be in a record and is based on an ontology of actions, and clinical state, etc. A good introduction to healthcare standards can be found in the review by *Marco* et al. [34.24].

34.2.6 Physiological Measurement

Data from physiological measurement systems are commonly recorded on computer systems. Automatic recording of data happens commonly in high risk areas such as intensive care units and operating theaters. This data can be collected and shared using standards such as DICOM. Relatively simple data such as *vital signs*, which include temperature, heart rate, blood pressure, and respiratory rate are collected routinely by nurses, and recent work has demonstrated that collecting this data via PDAs allows effective prediction of clinical outcome when large datasets are analyzed [34.25].

34.2.7 Decision Support

Decision support – using a computer system to assist with diagnosis or treatment decisions has had a long history, with one of the most famous expert-system approaches MYCIN [34.26] being developed in 1976. MYCIN was a system developed by Shortliffe to identify which antibiotic should be used for a particular infection based on clinical results. Many other decision-support systems have been developed since, and there have been systematic reviews [34.27] showing that practitioner performance has been improved by their use. Clinical decision support systems such as PRE-DICT [34.28] can use the data collected routinely in primary care and produce a risk score for the patient and assist with management (Fig. 34.5).

34.2.8 Error Avoidance

Clinical benefit has been shown by the use of health informatics tools in the area of avoiding clinical error. Clinical errors are an important cause of morbidity and mortality [34.29], and computerization has allowed improvements in record-keeping that can help to prevent such errors by making records more flexible and problem-focussed [34.30], rather than purely sequential. Computerized systems for drug prescription and dispensing have been shown to greatly reduce errors [34.31].

34.2.9 Infodemiology

Traditionally health informatics has concerned itself with recording and interpreting *traditional* data. However, the rise of the web, mobile devices, and ubiquitous connectivity means that new information sources are becoming available. *Eysenbach* has proposed that *infodemiology* [34.32] is the new science of using information sources, such as web 2.0 enabled technologies, and search patterns can be used to track the progress of epidemics and changing attitudes to health issues.

34.2.10 Telecare and Mobile Health (M–Health)

Traditional telecare has involved linking doctor and patient via video links or other approaches in order to allow consultations that would be difficult to organize because of geographical separation. A very simple approach has been taken by the Swinfen project [34.33], which uses email-based communication between resource-rich and resource-poor units. These approaches have widened to use mobile devices for novel approaches to supporting health, especially in the developing world. The mHealthAlliance involves a number of commercial and NGO partners working together to support work in this area. *Kaplan* [34.34] noted that the evidence for benefit is still relatively weak, and that mobile phones may need to be seen as community, rather than personal devices.

Ubiquitous healthcare [34.35] is perhaps the next frontier with smart clothing and smart houses monitoring health status and providing useful information for both patients and their caregivers. Privacy, reliability, and interpretation of such data remain key questions. RFID-based approaches have been suggested for monitoring activity in the elderly or infirm along with supporting activity especially in the case of memory loss or visual impairment [34.36].

34.2.11 Consumer E-Health

Internet-based health information services for health advice have become increasingly common. Many, such as

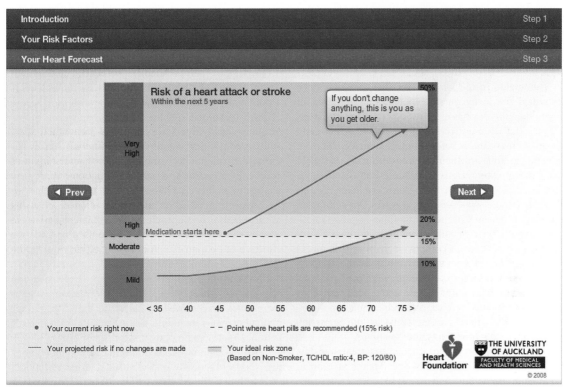

Fig. 34.5 Decision support – the PREDICT system for cardiovascular outcome in primary care

NHS direct (website) in the UK provide simple clinical algorithms to assist users to identify the severity of symptoms. Other health supporting initiatives include smoking cessation programs such as SMS-based systems [34.37], or online counseling for depression using a variety of multimedia and avatars.

Personal health records are one of the most exciting yet potentially troublesome areas of research in health record management [34.38]. There are large numbers of products available from Google Health and the Microsoft Health Vault, to provider-specific applications, which may make health data contained within their systems visible to the patient. Some countries such as Denmark have made the state-run health records available to individuals online, as well as other health related information [34.39]; the possibility of integrating personal health data with health advice.

Interoperability, privacy, and security of data remain key issues in terms of successful deployment of such systems.

34.2.12 Privacy and Security

Healthcare deals with personal and private information, and going back to the time of Hippocrates ($\approx$ 460–370 BC), clinicians have promised to keep secret information that is revealed to them by their patients. Health information systems have the potential to release private information by accident or by malicious attacks, if there is not sufficient security. In many countries there are specific laws covering the release and protection of medical data. It should be emphasized that this is not simply prudery – in some circumstances they release some medical information – especially that concerning reproductive health. However, there is also a requirement that data be available appropriately, even in emergencies, so particularly restrictive access policies may not be helpful. In many cases audit systems are used in large organizations to record any access to data and those who are unlikely to have a valid reason for access to particular records are liable to be contacted and even disciplined if found to be behaving unethically.

Part F | 34.2

34.3 Computational Intelligence in Health Informatics

A number of open and active research areas exist where computational intelligence interacts with health informatics.

Automated image analysis for diagnosis has been attempted for many years, especially in the area of mammography. Mammography (x-rays of the breast tissue) is a widely used screening method to detect early stages of breast cancer before clinical symptoms appear. Mammography is a difficult area to automate, partly because recognizing lesions is difficult, but also because very large numbers of mammograms are taken, as part of whole-population screening exercises. An automated approach needs an extremely low false-negative rate – otherwise there is no point screening, but also a very low false-positive rate, as otherwise very large numbers of healthy women will undergo unpleasant and possibly dangerous surgery. Recent developments in this area are using fuzzy segmentation and other approaches including Bayesian and support vector machine classifiers [34.40], but the complexity of the approach indicates that this problem is not yet solved.

Decision support for diagnosis and treatment has become more sophisticated [34.41]. This is starting to move from the area of population-based decision support to personalized decision making. This can include decision aids to help people decide what treatment they would prefer [34.42]. More radically, by using data from genomics and proteomics, we are in the era of personalized medicine, where an individual's genome is analyzed and potential outcomes calculated in order to provides personally optimal treatment.

Automated natural language processing and semantic annotation of documents and multimedia objects is a potentially fruitful area, especially with the success of such techniques on the web. The existence of sophisticated and large-scale ontologies and vocabularies, and the continuing use of free-text systems makes automated extraction of meaning both theoretically and practically important, and promising work is already happening in this area [34.43].

Supporting efficient and effective work practices in healthcare is particularly important. Large-scale observer-based studies such as WOMBAT [34.44] are effective but expensive. Automated systems including RFID-based approaches may be a way forward [34.45], but are currently computationally very expensive and rely on low-grade data. Health information systems themselves may provide clues on how to reduce clinical errors – including missed results from investigations [34.46] and incorrect selection of drugs [34.47].

Health outcome prediction remains very popular as it is such a practically important task. A general model of health and illness states still eludes description, although interesting and ambitious work is being done in this area [34.48].

The examination of social networks and E-health 2.0 to provide personal and family health support is becoming of interest but this is still in its early stages.

34.4 Conclusion

Health informatics is a discipline that supports and systemizes a great deal of practical activity. Large amounts of complex data are collected and used routinely. There have been great advances in the use of computational techniques in this area, but there is still much to do [34.49]. Many advances have occurred because of collaboration between clinicians, medical researchers, and computer scientists, and the application of computational intelligence in this area is only beginning.

References

34.1   B. Obama: Modern health care for all Americans, N. Engl. J. Med. **359**, 1537–1541 (2008)

34.2   F. Olesen, J. Dickinson, P. Hjortdahl: General practice–time for a new definition, BMJ **320**, 354–357 (2000)

34.3   S. de Lusignan: What is primary care informatics?, J. Am. Med. Inform. Assoc. **10**, 304–309 (2003)

34.4   S. Teasdale, D. Bates, K. Kmetik, J. Suzewits, M. Bainbridge: Secondary uses of clinical data in primary care, Inform. Prim. Care **15**, 157–166 (2007)

34.5 A.D. Mackinnon, R.A. Billington, E.J. Adam, D.D. Dundas, U. Patel: Picture archiving and communication systems lead to sustained improvements in reporting times and productivity: Results of a 5-year audit, Clin. Radiol. **63**, 796–804 (2008)

34.6 W.A. Knaus, D.P. Wagner, E.A. Draper, J.E. Zimmerman, M. Bergner, P.G. Bastos, C.A. Sirio, D.J. Murphy, T. Lotring, A. Damiano: The APACHE III prognostic system. Risk prediction of hospital mortality for critically ill hospitalized adults, Chest **100**, 1619–1636 (1991)

34.7 E. Golestanian, J.E. Scruggs, R.E. Gangnon, R.P. Mak, K.E. Wood: Effect of interhospital transfer on resource utilization and outcomes at a tertiary care referral center *, Crit. Care Med. **35**, 1470–1476 (2007)

34.8 R. Rabiei, P.A. Bath, A. Hutchinson, D. Burke: The national programme for IT in England: Clinician's views on the impact of the Choose and Book service, Health Inform. J. **15**, 167–178 (2009)

34.9 R. Lenz, M. Beyer, K.A. Kuhn: Semantic integration in healthcare networks, Int. J. Med. Inform. **76**, 201–207 (2007)

34.10 B. Kaplan: Evaluating informatics applications–clinical decision support systems literature review, Int. J. Med. Inform. **64**, 15–37 (2001)

34.11 J.L.Y. Liu, J.C. Wyatt: The case for randomized controlled trials to assess the impact of clinical information systems, J. Am. Med. Inform. Assoc. **18**, 173–180 (2011)

34.12 J.I. Westbrook, J. Braithwaite, A. Georgiou, A. Ampt, N. Creswick, E. Coiera, R. Iedema: Multimethod evaluation of information and communication technologies in health in the context of wicked problems and sociotechnical theory, J. Am. Med. Inform. Assoc. **14**, 746–755 (2007)

34.13 S.J. Pritchard, A.L. Weightman: Medline in the UK: Pioneering the past, present and future, Health Inform. Libr. J. **22**, 38–44 (2005)

34.14 P. Fontelo, F. Liu, M. Ackerman: askMEDLINE: A free-text, natural language query tool for MEDLINE/PubMed, BMC Med. Inform. Decis. Mak. **5**, 5 (2005)

34.15 K. Donnelly: SNOMED-CT: The advanced terminology and coding system for eHealth, Stud. Health Technol. Inform. **121**, 279–290 (2006)

34.16 R.H. Dolin, L. Alschuler, S. Boyer, C. Beebe, F.M. Behlen, P.V. Biron, A. Shabo Shvo: HL7 clinical document architecture, release 2, J. Am. Med. Inform. Assoc. **13**, 30–39 (2006)

34.17 D. Protti, T. Bowden, I. Johansen: Adoption of information technology in primary care physician offices in New Zealand and Denmark, part 4: Benefits comparisons, Inform. Prim. Care **16**, 291–296 (2008)

34.18 W.D. Bidgood Jr., S.C. Horii, F.W. Prior, D.E. Van Syckle: Understanding and using DICOM, the data interchange standard for biomedical imaging, J. Am. Med. Inform. Assoc. **4**, 199–212 (1997)

34.19 S. Woolhandler, T. Campbell, D.U. Himmelstein: Costs of health care administration in the United States and Canada, N. Engl. J. Med. **349**, 768–775 (2003)

34.20 W.E. Hammond: The making and adoption of health data standards, Health Aff. **24**, 1205–1213 (2005)

34.21 J.E. Wennberg: Unwarranted variations in healthcare delivery: Implications for academic medical centres, BMJ **325**, 961–964 (2002)

34.22 D. Gans, J. Kralewski, T. Hammons, B. Dowd: Medical groups' adoption of electronic health records and information systems, Health Aff. **24**, 1323–1333 (2005)

34.23 C. Schoen, R. Osborn, M.M. Doty, D. Squires, J. Peugh, S. Applebaum: A survey of primary care physicians in eleven countries, 2009: Perspectives on care, costs, and experiences, Health Affairs **28**, 1171–1183 (2009)

34.24 M. Eichelberg, T. Aden, J. Riesmeier, A. Dogac, B. Laleci: A survey and analysis of electronic healthcare record standards, ACM Comput. Surv. **37**(4), 277–315 (2005)

34.25 G.B. Smith, D.R. Prytherch, P. Schmidt, P.I. Featherstone, D. Knight, G. Clements, M.A. Mohammed: Hospital-wide physiological surveillance – A new approach to the early identification and management of the sick patient, Resuscitation **71**, 19–28 (2006)

34.26 E. H. Shortliffe: Mycin: A rule-based computer program for advising physicians regarding antimicrobial therapy selection (Stanford AI Lab Memo AIM-251 1974), p. 411

34.27 A.X. Garg, N.K.J. Adhikari, H. McDonald, M.P. Rosas-Arellano, P.J. Devereaux, J. Beyene, J. Sam, R.B. Haynes: Effects of computerized clinical decision support systems on practitioner performance and patient outcomes: A systematic review, J. Am. Med. Assoc. **293**, 1223–1238 (2005)

34.28 L. Bannink, S. Wells, J. Broad, T. Riddell, R. Jackson: Web-based assessment of cardiovascular disease risk in routine primary care practice in New Zealand: The first 18000 patients (PREDICT CVD-1), J. N. Z. Med. Assoc. **119**, U2313 (2006)

34.29 L. Kohn, J. Corrigan, M. Donaldson: *To err is human: Building a Safer Health System* (National Acadamies Press, Washington 2000)

34.30 L.L. Weed: Medical records that guide and teach, N. Engl. J. Med. **278**, 593–600 (1968)

34.31 E.G. Poon, C.A. Keohane, C.S. Yoon, M. Ditmore, A. Bane, O. Levtzion-Korach, T. Moniz, J.M. Rothschild, A.B. Kachalia, J. Hayes, W.W. Churchill, S. Lipsitz, A.D. Whittemore, D.W. Bates, T.K. Gandhi: Effect of bar-code technology on the safety of medication administration, N. Engl. J. Med. **362**, 1698–1707 (2010)

34.32   G. Eysenbach: Infodemiology: Tracking flu-related searches on the web for syndromic surveillance, AMIA Annu. Symp. Proc. (2006) pp. 244–248

34.33   P. Swinfen, R. Swinfen, K. Youngberry, R. Wootton: Low-cost telemedicine in Iraq: An analysis of referrals in the first 15 months, Telemed. Telecare **11**, 113 (2005)

34.34   W. Kaplan: Can the ubiquitous power of mobile phones be used to improve health outcomes in developing countries?, Glob. Health **2**, 9 (2006)

34.35   D.-O. Kang, K. Kang, H.-J. Lee, E.-J. Ko, J. Le: A context aware framework for u-healthcare in a wearable system, World Congr. Medical Phys. Biomed. Eng. 2006, Imaging the Future Medicine (COEX Seoul 2006)

34.36   J. Symonds, D. Parry, J. Briggs: An RFID-based system for assisted living: Challenges and solutions, Stud. Health Technol. Inform. **127**, 127–138 (2007)

34.37   A. Rodgers, T. Corbett, D. Bramley, T. Riddell, M. Wills, R.B. Lin, M. Jones: Do u smoke after txt? Results of a randomised trial of smoking cessation using mobile phone text messaging, Tob. Control **14**, 255–261 (2005)

34.38   D.C. Kaelber, A.K. Jha, D. Johnston, B. Middleton, D.W. Bates: A research agenda for personal health records (PHRs), J. Am. Med. Inform. Assoc. **15**, 729–736 (2008)

34.39   H. Andreassen, M. Bujnowska-Fedak, C. Chronaki, R. Dumitru, I. Pudule, S. Santana, H. Voss, R. Wynn: European citizens' use of E-health services: A study of seven countries, BMC Public Health **7**, 53 (2007)

34.40   A. RojasDomínguez, A.K. Nandi: Toward breast cancer diagnosis based on automated segmentation of masses in mammograms, Pattern Recognit. **42**, 1138–1148 (2009)

34.41   L. Ohno-Machado: Electronic health records and computer-based clinical decision support: Are we there yet?, J. Am. Med. Inform. Assoc. **18**, 109 (2011)

34.42   A. O'Connor, C. Bennett, D. Stacey, M. Barry, N. Col, K. Eden, V. Entwistle, V. Fiset, M. Holmes-Rovner, S. Khangura, H. Llewellyn-Thomas, D. Rovner: Decision aids for people facing health treatment or screening decisions, Cochrane Database Syst. Rev. **3**, CD001431 (2009)

34.43   J. Patrick, Y. Wang, P. Budd: An automated system for conversion of clinical notes into SNOMED clinical terminology, Proc. Fifth Australasian Symp. ACSW Frontiers, Vol. 68 (Australian Computer Society, Sidney 2007)

34.44   J.I. Westbrook, A. Ampt: Design, application and testing of the Work Observation Method by Activity Timing (WOMBAT) to measure clinicians' patterns of work and communication, Int. J. Med. Inform. **7**(8), S25–S33 (2009)

34.45   B. Houliston, D. Parry, A. Merry: Towards automated detection of anaesthetic activity, Methods Inf. Med. **50**(5), 464–471 (2011)

34.46   J. Callen, A. Georgiou, J. Li, J.I. Westbrook: The safety implications of missed test results for hospitalised patients: A systematic review, BMJ Qual. Saf. **20**, 194–199 (2011)

34.47   C.S. Webster, A.F. Merry, P.H. Gander, N.K. Mann: A prospective, randomised clinical evaluation of a new safety-orientated injectable drug administration system in comparison with conventional methods, Anaesthesia **59**, 80–87 (2004)

34.48   E.O. Voit: A systems-theoretical framework for health and disease: Inflammation and preconditioning from an abstract modeling point of view, Math. Biosci. **217**, 11–18 (2009)

34.49   A.D. Black, J. Car, C. Pagliari, C. Anandan, K. Cresswell, T. Bokun, B. McKinstry, R. Procter, A. Majeed, A. Sheikh: The impact of eHealth on the quality and safety of health care: A systematic overview, PLoS Medicine **8**, 1–16 (2011)

35. Ecological Informatics for the Prediction and Management of Invasive Species

Susan P. Worner, Muriel Gevrey, Takayoshi Ikeda, Gwenaël Leday, Joel Pitt, Stefan Schliebs, Snjezana Soltic

Ecologists face rapidly accumulating environmental data form spatial studies and from large-scale field experiments such that many now specialize in information technology. Those scientists carry out interdisciplinary research in what is known as ecological informatics. Ecological informatics is defined as a discipline that brings together ecology and computer science to solve problems using biologically-inspired computation, information processing, and other computer science disciplines such as data management and visualization. Scientists working in the discipline have research interests that include ecological knowledge discovery, clustering, and forecasting, and simulation of ecological dynamics by individual-based or agent-based models, as well as hybrid models and artificial life. In this chapter, ecological informatics techniques are applied to answer questions about alien invasive species, in particular, species that pose a biosecurity threat in a terrestrial ecological setting. Biosecurity is defined as the protection of a region's environment, flora and fauna, marine life, indigenous resources, and human and animal health. Because biological organisms can cause billions of dollars of impact in any country, good science, systems, and protocols that underpin a regulatory biosecurity system are required in order to facilitate international trade. The tools and techniques discussed in this chapter are designed to be used in a risk analysis procedure so that agencies in charge of biosecurity can prioritize scarce resources and effort and be better prepared to prevent unexpected incursions of dangerous invasive species. The methods are used to predict, (1) which species out of the many thousands might establish in a new area, (2) where those species might establish, and, (3) where they might spread over a realistic landscape so that their impact can be determined.

35.1 The Invasive Species Problem

As the flow of people and products steadily increases around the world so do many thousands of potential global invasive species. If allowed to cross national borders such species pose one of the greatest threats to a country's biodiversity, environment, economic activity, and health. A recent example of the potential impact that an invasive insect species, in particular, can have on a region is given by the exotic varroa mite, *Var-*

roa destructor. This small creature has invaded many countries this last decade, infesting beehives. It is easy to overlook such a species as its direct impact is on a bee-keeping industry that is judged small by national comparisons. However, often unappreciated is its indirect impact on pollination. Loss of pollinators can cause hundreds of millions of dollars of losses in primary production annually. While bees are commonly appreciated for their honey and other products, their major ecosystem service is as the main pollinator of many native plants, horticultural crops including vegetables and fruit, pasture clovers and vegetable seed [35.1]. With reduced pollination food production can fall dramatically causing escalating prices in developed nations and threats to food security in developing nations [35.2].

One of the most important activities that prevents invasive species impact involves proactive risk assessment that requires prediction and prioritization of species that pose a threat. In the following sections we discuss the use of ecological informatics

methods [35.3–5] to assist this process. The methods described here are

1. Self-organizing maps used to predict and prioritize biosecurity threats (in other words, identify *which* species are potential threats).
2. The use of neural networks and support vector machines to identify habitat suitability for particular species in novel areas (in other words *where* a species might establish).
3. An individual-based model IBM and agent-based models in geographic information systems (GISs) to project the spread of invasive species, or *where* it might find suitable habitat to determine possible impact.

The process of identifying where a species might find suitable resources for a self-sustaining population also assists the design of monitoring, detection, eradication, and containment programmes for alien invasive species.

35.2 Predicting and Prioritizing Species – Self-Organizing Maps (SOMS)

Despite the fact that invasive insect species have been studied for many decades, little progress has been made to identify and prioritize new threats before they arrive in a new country or region. When a large number of species, for example, more than 3000 global crop pests, have potential to cross the border of any country, it is difficult to know where to start to rank the risk posed by these potential invaders. Clearly, not all will get a chance to invade a new region and if they do, not all of them will find a suitable environment to establish a viable population. The sort of questions a risk assessor asks, are (1) which species out of a large list of potential invaders are more likely to establish if they arrive in the target region, and, (2) can those species that pose the greatest risk be identified?

The usual approach to risk assessment tends to be reactive. For example, agencies may be put on alert when an insect species or disease causes a problem somewhere else in the world. Agency personnel also take notice of species that are frequently intercepted at the border. They also carry out pest risk assessments on new commodities scheduled for importation to determine the likelihood of associated exotic species being introduced. The last thing any country needs is to be taken by surprise by a new incursion. Thus, recognizing the threat and preventing the species from breaching the border is the best way to avoid a new

dangerous incursion happening. The following analysis illustrates the application of a computational intelligence method combined with well-known ecological principles to rank many hundreds of well-known alien insect pest species with respect to their potential to establish in, for example, a new region, such as New Zealand (NZ).

The approach is based on the principle that the specific combination, assemblage, or profile of pest species already established in a particular region or country integrates information about the suitability of the environment for those species that can be used for predicting other species that might find that environment suitable [35.6]. In other words, certain species tend to co-occur because of factors such as suitable climate, ecological conditions, and presence of suitable host plants. If, in our analysis, we find, for example, that NZ shares a large number of its invasive pest species with a region in another part of the world, for example Spain, then one can infer that the two countries share some common characteristics. It is possible that some species that are in Spain but not yet established in NZ may pose a special threat. To find those regions that are most similar to each other with respect to their pest assemblage or the pest profiles, a large number of geographic regions and a large number of species need to be clustered. Many conventional clustering analy-

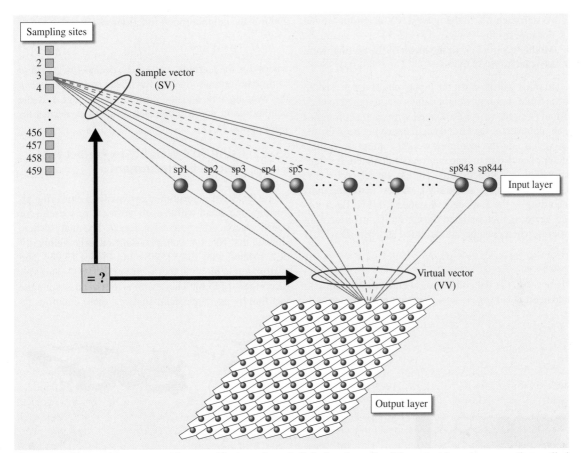

Fig. 35.1 Self-organizing map architecture. The input layer is linked to the cells of the output layer by connections called weights that define the virtual assemblages of the species. (After [35.6])

sis approaches can be used but cannot cope with high dimensional data.

A SOM was used to organize and visualize a large data set of the global distribution data of insects. First the SOM was used to indicate global geographic areas that share similar pest groupings [35.6, 7]. 844 pest species distributed over 459 geographic regions were used for this analysis. Data was extracted with permission from the CABI Crop Protection Compendium [35.8]. Figure 35.1 shows the vector or species profile for the third geographic area (out of 459) projected by the SOM onto a two-dimensional map according to the cluster to which they belong.

35.2.1 The SOM Algorithm

A detailed description of SOM can be found in [35.9, 10] and examples of its application to pest risk data

in [35.6] and [35.7]. The Kohonen SOM is an unsupervised learning algorithm that is a type of neural network. The SOM learns the patterns within the data in an unsupervised way. A SOM consists of two layers of artificial neurons, (1) the input layer that represents and the input data and the output layer or map, which is usually arranged in a two-dimensional structure. Every input neuron is connected to every output neuron, and each connection has a weight attached to it. The batch SOM algorithm can be summarized as follows:

1. Initialize the values of the virtual vectors (VV_i, $1 \leq i \leq c$) using random values
2. Repeat steps 1 to 6 until convergence
3. Read all the sample vectors (SV) one at a time
4. Compute the Euclidean distance between SV and VV

5. Assign each SV to the nearest VV according to the distance results
6. Modify each VV with the mean of the SV that were assigned to it [35.6].

In other words, when the input vectors are presented to the SOM, random weight values are assigned to each virtual (weight) vector associated with each neuron. For each input vector the Euclidean distance between the input vector and the incoming weight (output or virtual) vector of each output map neuron is calculated. Each input vector is assigned to the closest virtual vector (the winner, also known as the BMU or best matching unit) according to the Euclidean distance [35.11]. Each virtual vector is then updated during an iterative learning process, where weights are modified according to (35.1)

$$w_{i,j}(t+1) = w_{i,j}(t) + h(t)[x_i - w_{i,j}(t)] , \qquad (35.1)$$

where $w_{i,j}(t)$ is the connection weight from input i to map neuron j at time t, x_i is element i of input vector x,

and h is the neighborhood function, as defined in (35.2)

$$h(t) = \alpha \exp\{-d^2/[2\sigma^2(t)]\} , \qquad (35.2)$$

where α is the learning rate, which decays towards zero as time progresses, d is the Euclidean distance between the winning unit and the current unit j, and σ is the neighborhood width parameter, which also decays towards zero [35.11].

35.2.2 Using SOM Weights to Predict Pest Species Establishment

While the SOM algorithm is essentially a clustering algorithm, the detail within each cluster is very useful for questions concerning invasive insects. The analysis here showed that NZ, for example, shares a large number of pest species with Italy (59%) and France (58%), and a surprisingly high number with Turkey (46%) and Morocco (48%) [35.6]. These similarities exist despite the fact that by any superficial analysis, these countries do

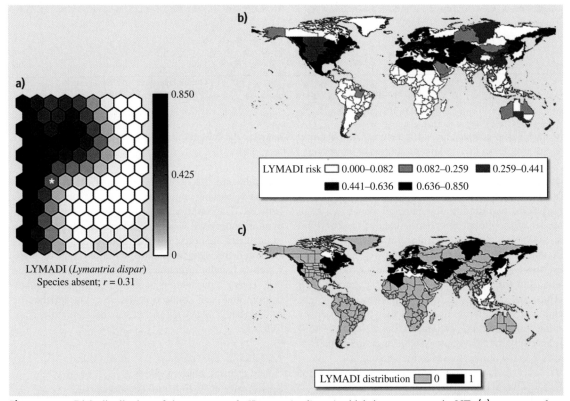

Fig. 35.2a–c Risk distribution of the gypsy moth (*Lymantria dispar*) which is not present in NZ, (**a**) represented on the SOM map (the *white asterisk* shows the cells where NZ is) and (**b**) represented on the world map. (**c**) The actual distribution (presence and absence) of the species on a world map. (After [35.6])

not appear to have analogous climates and conditions to NZ. Clearly, however, they require closer study to determine potential pathways by which serious invasive species could enter NZ. The most important result of the SOM analysis is that the SOM weights can be used to create a risk list where the weight assigned to each species (element in the vector of species) can be used as an index of the risk of those species most likely to establish in the target area (Fig. 35.2).

In this way, a subset of the 844 species can be targeted for more in-depth risk assessment. Clearly, no data set is complete and the impact of potentially inaccurate or incomplete data needs to be tested. Species profiles were bootstrapped 1000 times and the change

in each species rank, recorded [35.11]. For the top 50 most highly ranked species that are not established in NZ, their ranks changed on average only 14 places out of a possible 800, indicating considerable confidence in the method. The average rank over the 1000 bootstrap samples can now be used to organize the species to give the top 50 *least wanted* insect pest species [35.11]. Additionally, *Paini* et al. [35.12] evaluated the sensitivity of the SOM method by altering the original presence and absence data by introducing increasing amounts of error and comparing estimates of risk. They found that decisions based on the SOM analysis remained unaffected by alterations of up to 20% of data.

35.3 Species Distribution Models

While an analysis such as a SOM indicates which species might be a threat, risk assessors need to determine exactly where such species might establish themselves. If much biological information is available, more detailed mechanistic models can be created [35.13–15]. However, for those species about which we know little and where data are sparse, especially with respect to their environmental requirements, prediction of where they might establish or characterization of suitable habitat is more difficult. As climate change looms as a major driver of changes in species distributions it is not surprising that there is considerable international interest in characterizing suitable habitat for species for both invasive species research as well as conservation ecology.

The identification of areas that are most at risk from invasive species requires some assessment of where each species might establish a viable population if it has the chance to do so. Ecological niche models, or what are now generally known as species distribution models (SDMs), are commonly used for that purpose. For many potentially invasive species little is known about the species' environmental requirements such that detailed process or mechanistic models are not possible. Alternatively, discriminative or correlative species distribution models are based on the ecological principle that the geographic distribution and range of a species is largely structured by environmental predictors, particularly climatic factors, such as temperature and rainfall [35.16, 17]. Such models commonly use the current distribution of the species to infer the important environmental correlates that make the habitat suit-

able for sustained existence. Many species distribution models can utilize a range of additional environmental variables as well as interactions with other species. However, most often in invasive species research, where very little is known about an invading species, climate data are all that is available on a global scale (however, see [35.18, 19] for other data). There are currently a number of issues that need to be resolved to have greater confidence in such models, particularly to do with, (1) variable or feature selection, (2) absence data for discriminatory models, and, (3) model selection and performance criteria, and, (4) application of more innovative approaches such as adaptive models.

35.3.1 Variable or Feature Selection

The selection of relevant variables from a large number of predictors that explain the presence and/or absence of a species is often problematic when developing a prediction model. An optimal procedure would be to generate all possible combinations of input variables such that the subset of variables that best optimizes a criterion of acceptance is selected. When there are a large number of variables, an exhaustive search is difficult to implement. Consequently, many different techniques for variable selection have been developed [35.20].

Schliebs et al. [35.21] used a state of the art approach to solved this problem in an ecological informatics context. The approach comprised a quantum-inspired spiking neural network (QiSNN) framework for variable or feature selection. *Schliebs* et al. [35.22]

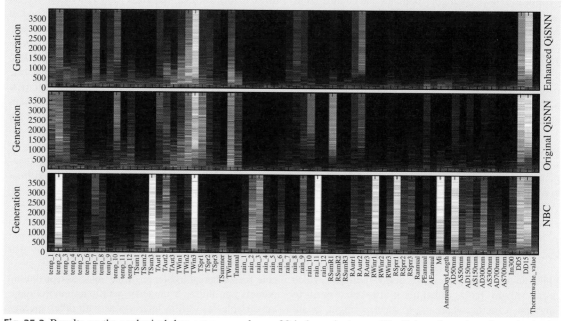

Fig. 35.3 Results on the ecological data set averaged over 30 independent runs. The lighter the *color* of a point in the diagram, the more often a specific feature was selected at the given generation. (After [35.22])

applied the QiSNN framework to a case study modeling Mediterranean fruit-fly (*Ceratitis capitata*), a serious global pest of fruit. The objectives of the study were to identify important climatic features relevant for predicting the presence/absence of this insect species. Results from the application of the QiSNN framework were compared to a classical naïve Bayesian classifier (NBC). Meteorological data, comprising monthly and seasonal temperature, rainfall, and soil moisture recordings for a range of geographical sites, were compiled from published results [35.7], and each site was cross referenced according to the known presence or absence of fruit fly populations around the world. Details of the approach can be found in *Schliebs* et al. [35.21] and in *Schliebs* et al. [35.22] where an enhanced QiSNN is used.

Figure 35.3 presents the results where the evolution of the average best feature subset is shown, and where the color indicates how often a specific variable was selected. Figure 35.4 shows that the enhanced QiSNN selected significantly fewer (9) variables than the original QiSNN (14) and 18 using NBC. Classification accuracy was similar for the three algorithms but the evolved feature subset for the enhanced QiSNN corresponds in general to what is known about *C. capitata* responses to climatic factors. For example, winter temperatures, autumn rainfall and degree-days, or the annual accumulated number of degrees of temperature above a threshold (5 and 15 °C in this study) were strong features. Figure 35.4 also shows that for all algorithms the number of variables selected decreases with increasing generations but faster for the enhanced QiSNN

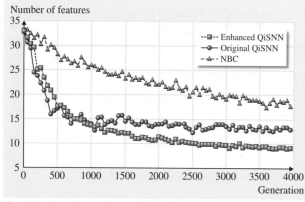

Fig. 35.4 In all algorithms the number of features decreases with increasing generations, the enhanced QiSNN being noticeably faster than the original QiSNN and NBC. All classifiers report a similar accuracy after the evolution of 4000 generations. (After [35.22])

compared with the original QiSNN and NBC. All classifiers reported a similar accuracy after the evolution of 4000 generations [35.22].

35.3.2 Absence Data for Discriminatory Models

One of the issues concerning species distribution models (SDMs) is that true absence data is generally not recorded. Field researchers are more interested in species presence but absence data is required for a full model correlative or discriminatory model. It is difficult to determine whether a species is truly absent in a region or whether it has simply not been detected or, maybe it has not managed to get to that locality. Many studies [35.23, 24] use *pseudo-absences*, which are simply random samples of the background (available) data. A study by *Chefaoui* and *Lobo* [35.25] showed that different methods of choosing pseudo-absences, whether they were chosen at random or based on a presence-only models altered the final results. Sometimes, oversampling the available background data can lead to an appearance of a high performing model when extremely unbalanced data is used. In other words a large number of absences are easy to fit and predict. This approach is also not appropriate when, (a) projecting habitat suitability globally, (b) when the world global environmental database comprises many thousands of points, and, (c) the target species is present in only a small proportion of locations. *Worner* et al. [35.26] developed a new procedure for choosing species absence data that reduces the problems often associated with this type of modeling. As an example we use the Asian tiger mosquito (*Aedes albopictus*). The Asian tiger mosquito can transmit a number of serious human and animal diseases [35.27], which would have serious consequences if it were to establish in many countries and particularly, NZ. The Asian tiger mosquito is recorded as present in only 2928 of potentially 580 000 locations for which global environmental data is available. A random selection of 580 000 locations to balance out the available presence locations would have varying effects on model output, depending on which locations are selected and, therefore, may lead to unreliable conclusions. To overcome this problem, we used one-class support vector machines [35.28] to select sites that are dissimilar (at various threshold levels) to the presence locations, out of the large number of possibilities. We chose one-class support vector machines (OCSVM) out of the several candidate methods because of their accuracy and ability to handle large datasets and creating a model in short

computational time. There have been many applications of one-class support vector machines in a number of disciplines, but very few in ecological fields [35.29–31].

While the analysis still resulted in many possible global absence locations, these were reduced to balance the number of presence points by grouping locations with similar environmental variables at each threshold into k clusters (Fig. 35.5), using k-means clustering [35.32]. The centroid of each cluster was used as a proxy for the environmental conditions that are dissimilar to species presence sites at the particular threshold chosen. Figure 35.6 shows absence locations derived from $k = 2928$ clusters required to balance the 2928 recorded absences for the mosquito. However, because the pseudo-absence points comprised the sample that represented environments most different from the presence locations (threshold = 0), we gradually changed the acceptance criteria of our OCSVM so that the sample of absence locations were brought closer to marginal presence points. That allowed us to model the data using more realistic absence locations with respect to the environmental conditions at those sites and also to characterize the environments we wished to model. This process avoided sampling bias that can be a consequence of randomly selected absences and improved the performance of all models in many cases.

35.3.3 The Multimodel Approach and Model Performance

While there are many different approaches and methodologies in this active area of research, few studies measure how different models perform given the same data. Those studies that compared model performance over a number of models [35.33, 34] used only one or two performance criteria and limited validation. However, a range of models can be tested in a multimodel approach to determine the best performing model out of several modeling approaches based on a number of performance measures and extensive validation [35.26] or, alternatively, a model ensemble may be used.

In the following example, the global habitat suitability of the Asian tiger mosquito in relation to environmental data was modeled using nine different modeling approaches. Results were compared and the *best* model was used to project the potential global distribution of this species (based on potential habitat suitability), including its potential distribution in NZ.

Among those models used for species distribution modeling that appear to perform particularly well are two types that belong to the area of computational in-

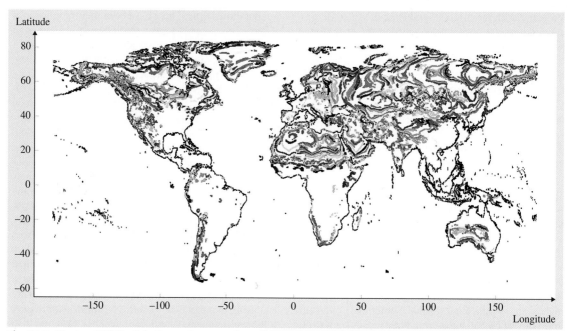

Fig. 35.5 Clusters (2928) of similar environmental conditions where the Asian tiger mosquito (*Aedes albopictus*) is likely to be absent are indicated by different colors. The *white areas* are more likely than the clusters to have environmental conditions suitable for establishment and are, therefore, likely to contain presence locations. (After [35.26])

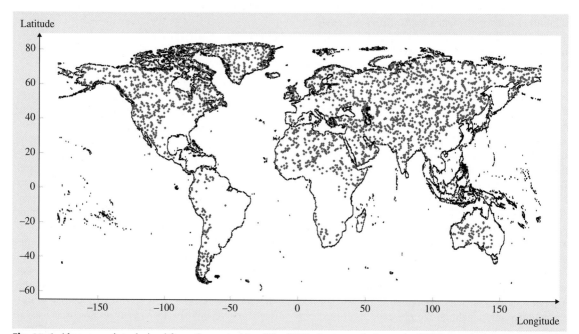

Fig. 35.6 Absence points derived from the centroids of the 2928 clusters to balance the 2928 recorded presences for the Asian tiger mosquito. (*Aedes albopictus*) (After [35.26])

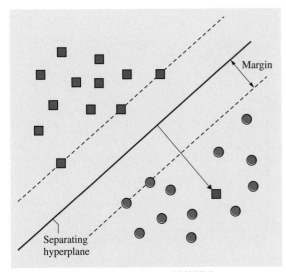

Fig. 35.7 Separating hyperplane with SVM

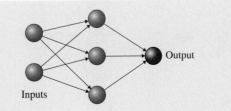

Fig. 35.8 A generic feed-forward neural network with one hidden layer

telligence, namely a back propagation artificial neural network (ANN) and a support vector machine (SVM).

Support Vector Machines

SVMs arose arisen from artificial neural network theory and have shown their capability in pattern recognition. SVMs were developed by *Vapnik* for binary classification [35.35]. The method attempts to find a hyperplane in the *gap* between two data groups, that is within a maximal margin (the distance from the hyperplane and the nearest points) (Fig. 35.7). Often, no separating hyperplane exists. This difficulty is tackled in two ways. First, by admitting some points on the wrong side of their margin and some on the wrong side of the hyperplane. A penalty error is defined as the distance to the hyperplane multiplied by the Lagrange multiplier $C > 0$. Second, the problem can be reconsidered in higher-dimensional space to make the separation easier. For nonlinear spaces, similarity functions or kernel functions are introduced.

A variety of kernel functions exist. In this study, only the *linear*, *radial basis*, and *polynomial* functions are implemented. The advantages of using SVMs for this type of modeling are that they work well when few observations are available and local minima tend not to occur. A disadvantage is that kernels tend to be black box functions.

Artificial Neural Networks (NNET)

Neural networks are designed to mimic networks of biological neurons [35.36]. The most common kind

of network is multilayer perceptron (MLP), whose information is passed forward (feed-forward networks), in other words, one direction from inputs (predictors) to outputs (response variable) without any feedback (Fig. 35.8). The other common type of neural network is where its learning takes place by backpropagation, in other words the error is backpropagated from outputs to inputs to update the neuron weights. The neural network is a mesh of neurons organized in several layers. These layers are joined or linked to each other, from left to right, taking outputs of a layer and using them as inputs of the next layer. Each layer filters information by amplifying or by reducing it. To do so, an activation function is used which in our study was the logistic function.

The advantages of a neural network (NNET) are the following:

1. Statistical learning can take place and no assumptions about the nature of the distribution of the data need to be made.
2. They have the ability to model complex nonlinear relationships and detect interactions between predictor variables.
3. The traditional protocol for training the network, if carried out appropriately, ensures that the resulting model has the capacity to generalize to new data (prevents overfitting).
4. Tolerance in relation to error in data.

The disadvantages are that the resulting model is computationally intense and complex and tends to be a *black box* unless variable contributions are analyzed.

For the example of Asian tiger mosquito, a random forests method [35.37] and stepwise regression analyses [35.26] was used for variable selection. Nine different models, a linear discriminant analysis, quadratic discriminant analysis, logistic regression, naïve Bayes, classification and regression tree, conditional tree, k nearest neighbors, support vector machine, and a neural network were compared. Model perfor-

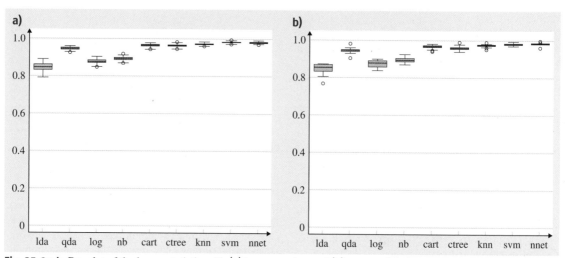

Fig. 35.9a,b Boxplot of the kappa statistic with (**a**) bootstrapping and (**b**) cross-validation resampling of the Asian tiger mosquito (*Aedes albopictus*) distribution data. Acronyms for the selected models are: *lda* for linear discriminant analysis, *qda* for quadratic discriminant analysis, *og* for logistic regression, *nb* for naive Bayes, *cart* for classification and regression tree, *ctree* for conditional tree, *knn* for *k* nearest neighbors, *svm* for support vector machine, and *nnet* for neural networks. Kappa values closer to 1 denote better model performance. (After [35.26])

mance was judged using nine performance criteria: Relative operating characteristic (ROC) curves [35.38], kappa index [35.39], measures derived from the confusion matrix, .632+ error [35.40], true skill statistic (TSS) (or the Hanssen–Kuiper skill score [35.41]) and model uncertainty (proportion of times a model predicts an index of suitability near 0.5 ± 1sd. SVM was selected as the best model with environmental variables comprising maximum temperature of the warmest month, minimum temperature of the coldest month, mean temperature of the wettest quarter, mean temperature of the driest quarter, mean temperature of warmest quarter, mean temperature of coldest quarter, annual precipitation, precipitation of wettest month, and coefficient of variation of precipitation seasonality, as significant variables. All models performed well over many of the performance criteria (Fig. 35.9) but the SVM model performed so well for this species during the validation step, the uncertainty of prediction for the model was insignificant. A projection of suitable habitat for the Asian tiger mosquito in NZ is shown in Fig. 35.10.

The field of species distribution modeling is multifaceted and continues to develop following a natural cycle of development, implementation, and evaluation [35.42] as new data and knowledge comes to hand. In this study, we have gone some way to improve methods for modeling presence-only data, model selection, and evaluation and assessing model uncertainty, three

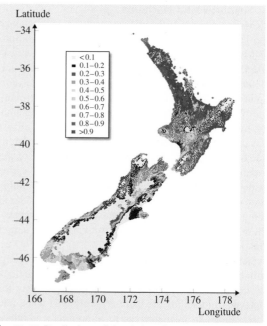

Fig. 35.10 Prediction of the Asian tiger mosquito (*Aedes albopictus*) distribution in NZ based on a support vector machine. The *legend* represents a scale of environmental suitability with *warmer colors* representing higher suitability for species establishment. (After [35.26])

areas called for by [35.42] in a review on species distribution models. However, there continues to be a need for more innovative research to solve the many issues in this field.

Adaptive Models

Most models discussed previously have drawbacks. Further advancements may lie in the area of adaptive models, which could provide incremental learning, and a facility for extracting knowledge (rules) as well as good generalization. *Soltic* et al. [35.43] described and compared a standard multiple regression model and an evolving dynamic neuro-fuzzy inference system (DENFIS) model as a type of SDM for estimating the potential establishment of pest insects. The two models were compared for their performance predicting the potential establishment of a pest in new locations using *Planococcus citri* (Risso), the citrus mealybug, as a case study [35.43]. The software environment Neu-Com (www.kedri.info) was used in this study for the analysis and prediction.

Given a domain data set $D = \{X_1, X_2, \ldots, X_k, Y\}$, where $X_i (i = 1, \ldots, k)$ are attributes of D, and Y is a discrete attribute to be estimated. Suppose attribute Y has m non-overlapping values $y_1, y_2, \ldots, y_m$ in D, and $d = x_1, x_2, \ldots, x_k, y$ is one transaction of D.

The target is to predict Y in terms of X by numerical function estimation $Y = f(X_1, \ldots, X_k)$.

The DENFIS model clusters data using an evolving clustering method (ECM) [35.44] and estimates f by DENFIS. The details of DENFIS can be found in [35.44, 45]. The model then fits a response surface as a function of predictors in environmental space $E = \{X_1, X_2, \ldots, X_k\}$, where $X_i (i = 1, \ldots, k)$ are data examples from D. It then uses the spatial pattern of predictor surfaces to predict the response in geographical space $G = \{g_1, g_2, \ldots, g_k\}$, where the examples are of type $g_i = (\text{latitude } i, \text{longitude } i)$ [35.43]. The DENFIS model is incrementally trainable on new data. *Soltic* et al. [35.43] implemented the following procedure to predict establishment potential as follows:

1. Apply a clustering algorithm to data from the problem space D.
2. For each cluster calculate the mean vector and establishment potential.
3. Use the establishment potential (P^C) and mean vector (X^C) to build the estimation function f.
4. Use f to make spatial predictions of the response (e.g., estimate the establishment potential for each location given in the original data set D).

The ECM was used for partitioning data D into 20 clusters. The DENFIS model was applied to the P^C and X^C from the ECM to obtain 15 rules, corresponding to the 15 rule nodes created during learning. Those rules cooperatively functioned as an estimate that can be used to predict the establishment potential of the citrus mealybug at each location. For example:

Rule 1:

if X_1 is $f(0.20 \quad 0.75)$ and
X_2 is $f(0.20 \quad 0.70)$ and
X_3 is $f(0.20 \quad 0.10)$ and
X_4 is $f(0.20 \quad 0.53)$ and
X_5 is $f(0.20 \quad 0.33)$ and
X_6 is $f(0.20 \quad 0.73)$ and
X_7 is $f(0.20 \quad 0.75)$ and
X_8 is $f(0.20 \quad 0.76)$ and
X_9 is $f(0.20 \quad 0.76)$ and
X_{10} is $f(0.20 \quad 0.72)$ and
X_{11} is $f(0.20 \quad 0.71)$ and
X_{12} is $f(0.20 \quad 0.69)$ and
X_{13} is $f(0.20 \quad 0.69)$ and
X_{14} is $f(0.20 \quad 0.71)$ and
X_{15} is $f(0.20 \quad 0.72)$ and
X_{16} is $f(0.20 \quad 0.71)$

then $Y = -2.45 - 27.88X_1 - 150.94X_2 - 1.27X_3$
$\quad - 4.04X_4 + 4.65X_5 - 59.00X_6$
$\quad + 85.32X_7 - 19.85X_8 - 29.54X_9$
$\quad + 72.00X_{10} + 45.41X_{11} - 129.34X_{12}$
$\quad + 203.15X_{13} + 11.39X_{14} + 12.75X_{15}$
$\quad - 6.59X_{16}$

The DENFIS model uses local generalization and, therefore, requires a larger data set than statistical models and when compared with predictions obtained by a multiple regression model both models gave similar predictions for the original data. However, the evolving method was able to accept new data for prediction and the fuzzy rules are useful to increase the understanding of the relationship between species establishment and environmental parameters. The DENFIS-based model can be used for online prediction applications as it is able to adapt its structure and produce output to accommodate the new input data. The model is also attractive in that it uses local rather than global clustering [35.43], and thus overcomes the problem of interpolation over large areas where the information about pest presence locations is better conserved in the estimation and may provide another solution to modeling species distributions without true absence data.

35.4 Individual-Based Models

Once a species arrives in NZ and establishes itself successfully, it would be useful to predict where it is likely to spread. Such a prediction would help the agencies in their search and destroy operations. Of course predicting where a species might spread is much like crystal ball gazing, but as described above it is possible to use information from species distribution modeling to identify a suitable habitat for a species to establish a viable population. A suitable climate is defined as that necessary for a species to complete its life cycle in that it provides appropriate food and a place to live and reproduce. While we can represent this environmental information spatially using GIS technology [35.14] (Fig. 35.10), the dispersal of the species needs to be modeled across this detailed landscape. *Pitt* et al. [35.13, 46] used several methods of spatial modeling to do this. First, the potential distribution of the gypsy moth, *Lymantria dispar*, a serious pest of plantation forestry, in NZ was modeled. *Pitt* et al. [35.13] used a life cycle model of individual development times to identify climates and regions in NZ where the gypsy moth can complete its life cycle (Fig. 35.11).

Those areas were combined with information on a topographic map and high resolution climate data to map the gypsy moth's potential distribution over the whole of NZ. Gypsy moth distribution under climate change was also modeled (Fig. 35.12).

However, the biological data that was used to construct the gypsy moth model described above is not often available. To explicitly model spread of an invasive species an individual-based stochastic dispersal model integrated with GIS was constructed [35.46]. Besides predicting spread, another important use for an individual-based stochastic dispersal model is that it can be used to run experiments that are impossible to carry out in real life. For example, one of the most difficult things to determine when a species invades a new area is to determine how long it has been there. Has it existed at such low abundance that it could not be detected? How much effort is required to detect a species that is in low abundance and sparsely spread around the area of interest? Such a dispersal model can be used to simulate the search to locate or detect a species in a complex environment. Such simulations will help determine appropriate sampling programs and thresholds

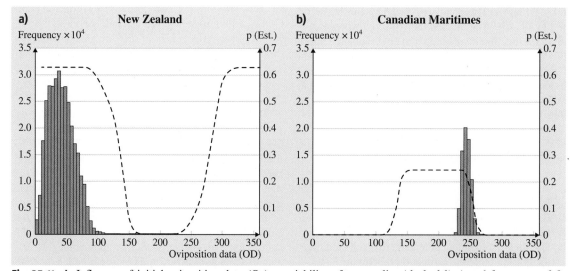

Fig. 35.11a,b Influence of initial oviposition date (O_0) on viability of seasonality (*dashed line*) and frequency of final oviposition achieved after 15 generations (*shaded bars*) using the same weather regime for (**a**) New Zealand and (**b**) Canadian Maritimes. Weather input was generated from 1971 to 2000 normals. (After [35.13])

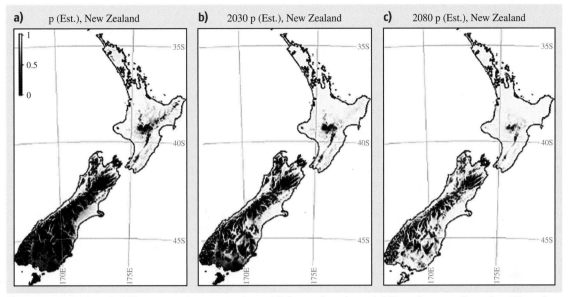

Fig. 35.12a–c NZ probability maps of gypsy moth establishment based on viability of seasonality using 50 stochastically different annual weather regimes for each of 500 simulation points. Interpolation between points was done by Kriging, with elevation as external drift. Using current temperature normals (**a**), 2030 normals (**b**), and 2080 normals (**c**). (After [35.13])

for detection in the event of a new incursion of a biosecurity threat.

Another question that such a model can help answer is: what is the best eradication strategy for a particular invasive species? Is it better to eradicate over large contiguous areas or to target a number of smaller foci throughout the species' current distribution? Such questions are impossible to answer by experimentation or even by experience in a real incursion. However, it is possible to simulate eradication strategies in the complex environment using GIS technology and environmental data that comprises vegetation cover, climate data, and land use information.

35.4.1 Modeling Dispersal Processes Within a Geographic Information System

While the ability to estimate the rate and direction of spread would greatly assist authorities in the design of sampling, monitoring and eradication programmes, modeling spread is difficult as data for model parameterization are seldom available for newly detected species. Many species can spread by multiple methods, including *random jumps* often meditated by humans. Traditional spread models are based on homogeneous

or uniform one-dimensional, abstract environments and are not terribly accurate, so *Pitt* et al. [35.46] modeled dispersal processes within a GIS using Argentine ant incursion into NZ as an example.

The *Pitt* et al. [35.46] model applied concepts taken from traditional theoretical population and spread models and individual-based modeling to the simulation of the dispersal of the Argentine ant *(Linepithema humile)* throughout NZ. This highly invasive species was already well-established before it was identified in NZ in 1990, so was largely left to disperse unhindered. As such, it provides a good example of the spread of an invasive species not confounded by eradication attempts. International data on Argentine ant distribution and spread were used to parameterize the model that was then used to simulate Argentine ant spread from its initial NZ invasion site.

35.4.2 Models of Dispersal

There are a number of methods that have been used to model dispersal. The most common is the use of analytic models, such as partial differential equations. Initially reaction diffusion equations were used by *Fisher* to model the diffusion of genes throughout a population [35.47], which inspired ecologists to use similar techniques for population dispersal. *Skellam* included logistic population growth to model oaks invading Europe [35.48]. These seminal works have since been greatly built upon [35.49], but analytic models can only include so much detail before they become intractable to solve. For this reason, analytic models tend to assume landscape homogeneity. IBMs also often do not include landscape heterogeneity [35.50] and unfortunately IBMs do not scale easily to simulate the dispersal of millions of individuals across geographically wide areas. Metapopulations split a population into spatially distinct areas and allows individuals to move between them [35.51]. These have the capacity to associate landscape heterogeneity with spatial regions, but tend to concentrate on localized regions or theoretical situations rather than on real landscapes at a broad geographic scale. Thus, *Pitt* et al. [35.46] created a framework for a metapopulation model that uses heterogeneous landscapes and can scale to encompass large geographic regions.

35.4.3 The MDIG Model

The advent of GIS and a subsequent abundance in geographic data from global navigation systems (GNS) and satellite imagery provides an ideal resource and environment for modeling spatial population dynamics.

Modeling in GIS has been neglected in the past [35.52] due to the computational resources needed, but it is now becoming popular. MDIG, an acronym for *modular dispersal in GIS*, is as its name suggests, a modular simulation framework for modeling dispersal within GIS. The model is spatially and temporally discrete, and for each time step it sequentially applies modules that update a species distribution map. Some of these modules are stochastic in nature, requiring simulations to be replicated to determine the range of possible dynamics. The program MDIG is implemented as an extension to the open source GIS, GRASS (geographic resources analysis support system) [35.14, 53], which in itself is a modular environment with each GIS tool implemented as a separate executable. MDIG's modular nature (Fig. 35.13) is based on observations that dispersal tends to be hierarchical, with different processes working at different levels [35.54, 55]. For example, most species spread contiguously as a result of localized population diffusion, but often they are also subject to long distance dispersal events that establish populations ahead of the invasion front [35.56]. MDIG has a module for local dispersal and for kernel-based dispersal, the latter being used to model long distance events. Sometimes a dispersal phenomenon is best described through multiple kernels being overlaid [35.57], and this can be modeled in MDIG by splitting a dispersal

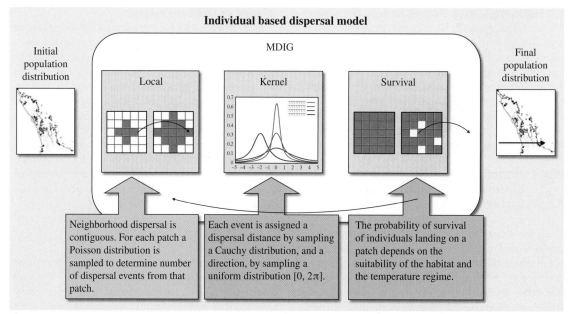

Fig. 35.13 MDIG model structure

event's probability between two kernels. In analyzing MDIG's dynamics, we only investigate the presence or absence of a population in a patch (represented as a cell in a raster map), but a module for local population growth exists and the modules described below are able to model the dispersal of individuals between patches. In this instance the population is stored in the cells of a distribution map instead of a presence/absence flag.

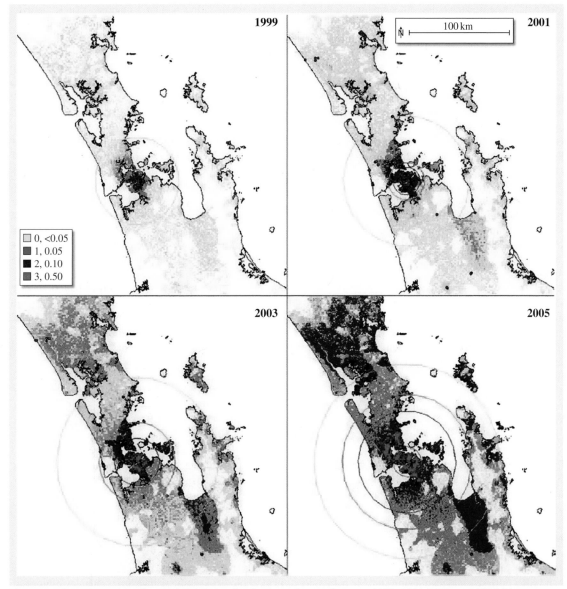

Fig. 35.14 Snapshots of the dispersal simulation for the Argentine ant for years 1999, 2001, 2003, and 2005. The map area is *colored* according to when the probability envelope predicted by MDIG simulations exceeds the thresholds (*red* > 0.5, *blue* > 0.1, *green* > 0.05, *yellow* > 0.05). The *colored circle* outlines indicate the boundary of a uniform spread model, the area of which is equivalent to the area encompassed by the thresholded probability envelope. *Red dots* with *black outlines* indicate observed Argentine ant occurrence sites. (After [35.46])

Neighborhood Dispersal

This module (Fig. 35.13) provides contiguous spread to neighboring patches using either *von Neumann* [35.58], *Moore* [35.59], or user defined dispersal neighborhoods. The neighborhood defines which nearby patches next to the current patch being processed will influence whether the current patch becomes occupied or not. So long as at least one neighborhood patch is present, the current one will become present in the next time-step.

Kernel Dispersal

Partial differential equations use a kernel function to describe the probability density of organisms landing a given distance from a site. This module (Fig. 35.13) uses a range of dispersal kernels to generate long-distance dispersal events, giving rise to sites that are not adjacent to the main population. This is often seen in recorded distribution data, resulting in stratified dispersal where individuals jump ahead of the advancement front and establish separate populations until the main population front catches up and envelopes them.

A prevalent and generic kernel form is

$$k(x) = \frac{1}{n} e^{-(\frac{x}{d})^s} , \qquad (35.3)$$

where x is the distance from the patch, d is the distance parameter, s is a dimensionless shape parameter, and n is a normalization constant to make the integral of the probability function sum 1. This kernel is equivalent to other probability distributions for certain values of s and is the one used by the simulations in this paper. A Poisson distribution is repeatedly sampled to generate the number of events that originate from each occupied patch.

Survival

To link a model to the underlying landscape there are multiple methods that can be used. One way is through module parameters that can be made dependent on the value of a map at the equivalent spatial location of the patch currently being processed (Fig. 35.13). This allows maps to be generated for parameters such as growth rates, population capacity, and dispersal event frequency, among others. Another way is to create a map representing habitat suitability that is used to evaluate the probability of a patch's population surviving to the next time-step. Those patches in suitable regions are given a much greater chance of survival than patches in less suitable areas. These habitat suitability maps can be generated in numerous ways – for example, through the utilization of classification tools and machine learning [35.26, 60], or expert opinion [35.61].

The ramifications of these new types of model are many. They can generate realistic predictions of spread using different spread models (Fig. 35.14) and be used in the design of appropriate, balanced sampling programmes. They allow designs with different sample sizes and spatial and temporal patterns to be tested over a realistic landscape, and experiments using different eradication treatments can be carried out. Such experiments are rarely possible in reality.

The GIS integration helps progress understanding of species spread in a complex environment as well as allowing predictions to be made about the direction and rate of spread. These predictions inform of invasion biology and will assist climate change studies by simulating the movement of species into new areas that become climatically suitable. Lastly, the simulations over the large spatial scales used in this study (Fig. 35.14), provide probabilistic estimates of establishments that may indicate hotspots where new populations are likely to develop. Additionally, however, simulating the spread of a species over such large spatial scales (Fig. 35.14), such as in this study, is important to help progress theoretical developments concerning the dispersal and spread of organisms over the heterogeneous environment.

35.5 Conclusions

Ranking a large number of species with respect to their biosecurity risk and identifying suitable habitat for species about which we have little information, as well as modeling their spread over the landscape is extremely difficult. The models featured in this chapter constitute useful advances to assess the risk of alien invasive species. The techniques used in this research fall within the emerging discipline of ecological informatics that concerns the application of computational intelligence in an ecological setting. However, as in many biological disciplines our knowledge is sometimes constrained and our data complex inaccurate and noisy [35.62], yet decisions must be made to protect national productive and natural ecosystems, human and animal health, and trade practices. Clearly, this is an active area of research and many issues remain. Some of the most important are that the very best methods of clustering high dimensional data and more comparative studies are required. In the area of habitat suitability modeling there is a dilemma in that while many models appear to perform exceptionally well they can often give quite different predictions despite extensive validation. Improved techniques to obtain the best model for a given dataset or a model ensemble are required. With respect to modeling spread across the realistic landscape population dynamics also need to be modeled to determine the effects and interactions

of resource distribution over a realistic spatial scale, dispersal distance, and reproductive strategies on the probability of establishment. In all of the applications illustrated here climate was either implicitly or explicitly a key feature of each model or analysis. Without a doubt, such models will continue to have application both in invasive species, and conservation research, particularly as both disciplines prepare for future climatic impacts that are likely to rapidly impact habitat suitability and the distribution of species of interest.

References

35.1   A. Ellis, K.S. Delaplane: Effects of nest invaders on honey bee (Apis mellifera) pollination efficacy, Agric. Ecosyst. Environ. **127**, 201–206 (2008)

35.2   T.D. Breeze, A.P. Bailey, K.G. Balcombe, S.G. Potts: Pollination services in the UK: How important are honeybees?, Agric. Ecosyst. Environ. **142**, 137–143 (2011)

35.3   M. Gevrey, S.P. Worner: Prediction of global distribution of insect pest species in relation to climate by using an ecological informatics method, J. Econ. Entomol. **99**, 979–986 (2006)

35.4   F. Recknagel: *Ecological Informatics: Understanding Ecology by Biologically Inspired Computation* (Springer, New York 2002)

35.5   F. Recknagel (Ed.): *Ecological Informatics. Scope, Techniques and Applications*, 2nd edn. (Springer, New York 2006) p.496

35.6   S.P. Worner, M. Gevrey: Modelling global insect species assemblages to determine risk of Invasion, J. Appl. Ecol. **43**, 858–867 (2006)

35.7   M. Gevrey, S.P. Worner, N. Kasabov, J. Pitt, J.-L. Giraudel: Estimating risk of events using SOM models: A case study on invasive species establishment, Ecol. Model. **197**, 361–372 (2006)

35.8   www.cabi.org/cpc

35.9   T. Kohonen: Self-organized formation of topologically correct feature maps, Biol. Cybern. **43**, 59–69 (1982)

35.10   T. Kohonen: *Self-Organizing Maps* (Springer, Berlin, Heidelberg 2001)

35.11   M.J. Watts, S.P. Worner: Estimating the risk of insect species invasion: Kohonen self-organising maps versus k-means clustering, Ecol. Model. **220**, 821–829 (2009)

35.12   D.R. Paini, S.P. Worner, M.B. Thomas, D.C. Cook: Using a self organising map to predict invasive species: Sensitivity to data errors and a comparison with expert opinion, J. Appl. Ecol. **47**, 290–298 (2010)

35.13   J. Pitt, J. Regniere, S.P. Worner: Risk assessment of gypsy moth, *Lymantria dispar* (L.) in New Zealand based on phenology modelling, Int. J. Biometeorol. **51**, 295–305 (2006)

35.14   W. Kresse, D. Danko (Eds.): *Springer Handbook of Geographic Information* (Springer, Berlin, Heidelberg 2012)

35.15   M. Kearney, B.L. Phillips, C.R. Tracy, K.A. Christian, G. Betts, W.P. Porter: Modelling species distributions without using species distributions: The cane toad in Australia under current and future climates, Ecography **31**, 423–434 (2008)

35.16   R.G. Pearson, T.P. Dawson: Predicting the impacts of climate change on the distribution of species: Are bioclimate envelope models useful?, Global Ecol. Biogeogr. **12**, 361–371 (2003)

35.17   A.T. Peterson: Predicting the geography of species' invasions via ecological niche modeling, Q. Rev. Biol. **78**, 419–433 (2003)

35.18   N. Roura-Pascual, C. Hui, T. Ikeda, G. Leday, D.M. Richardson, S. Carpintero, X. Espadaler, C. Gómez, B. Guénard, S. Hartley, P. Krushelnycky, P.J. Lester, M. McGeoch, S. Menke, J. Pedersen, J. Pitt, J. Reyes, N.J. Sanders, A.V. Suarez, Y. Touyama, D. Ward, P. Ward, J. Wetterer, S.P. Worner: Relative roles of climatic suitability and anthropogenic influence in determining the pattern of spread in a global invader, Proc. Natl. Acad. Sci. **108**, 220–225 (2011)

35.19   M.J. Watts, S.P. Worner: Comparing ensemble and cascaded neural networks that combine biotic and abiotic variables to predict insect species distribution, Ecol. Inform. **3**, 354–366 (2008)

35.20   I. Guyon, A. Elisseeff: An introduction to variable and feature selection, J. Mach. Learn. Res. **3**, 1157–1182 (2003)

35.21   S. Schliebs, M. Defoin-Platel, S. Worner, N. Kasabov: Quantum-Inspired Feature and Parameter Optimization of Evolving Spiking Neural Networks with a Case Study from Ecological Modelling, Proc. Int. Joint Conf. Neural Netw. (Atlanta, 2009) pp.2833–2840

35.22   S. Schliebs, M.D. Platel, S.P. Worner, N. Kasabov: Integrated feature and parameter optimisation for an evolving spiking neural network: Exploring heterogeneous probabilistic models, Neural Netw. **22**, 623–632 (2009)

35.23   A.H. Hirzel, J. Hausser, D. Chessel, N. Perrin: Ecological-niche factor analysis: How to compute habitat-suitability maps without absence data?, Ecology **83**, 2027–2036 (2002)

35.24   D. Stockwell, D. Peters: The GARP modelling system: Problems and solutions to automated spatial prediction, Int. J. Geogr. Inf. Sci. **13**, 143 (1999)

35.25   R.M. Chefaoui, J.M. Lobo: Assessing the effects of pseudo-absences on predictive distribution model performance, Ecol. Model. **210**, 478–486 (2008)

35.26   S.P. Worner, T. Ikeda, G. Leday, M. Joy: Surveillance tools for freshwater invertebrates, Ministry Agriculture Forestry NZ, Biosecurity Technical Paper 2010/21, 112 (2010)

35.27   N.G. Gratz: Critical review of the vector status of *Aedes albopictus*, Med. Vet. Entomol. **18**, 215–227 (2004)

35.28   B. Scholköpf, J.C. Platt, J. Shawe-Taylor, A.J. Smola, R.C. Williamson: Estimating the support of a high-dimensional distribution, Neural Comput. **13**, 1443–1471 (2001)

35.29   J.M. Drake, C. Randin, A. Guisan: Modelling ecological niches with support vector machines, J. Appl. Ecol. **43**, 424–432 (2006)

35.30   Q. Guo, M. Kelly, C.H. Graham: Support vector machines for predicting distribution of sudden oak death in California, Ecol. Model. **182**, 75–90 (2005)

35.31   W. Zuo, N. Lao, Y. Geng, K. Ma: GeoSVM: An efficient and effective tool to predict species' potential distributions, J. Plant Ecol. **1**, 143–145 (2008)

35.32   J.A. Hartigan, M.A. Wong: A k-means clustering algorithm, J. R. Stat. Soc. C **28**, 100–108 (1979)

35.33   P. Segurado, M.B. Araujo: An evaluation of methods for modelling species distributions, J. Biogeogr. **31**, 1555–1568 (2004)

35.34   J. Elith, C.H. Graham, R.P. Anderson, M. Dudík, S. Ferrier, A. Guisan, R.J. Hijmans, F. Huettmann, J.R. Leathwick, A. Lehmann, J. Li, L.G. LohmannL, B.A. Loiselle, G. Manion, C. Moritz, M. Nakamura, Y. Nakazawa, J.M. Overton, P.A. Townsend, S.J. Phillips, K. Richardson, R. Scachetti-Pereira, R.E. Schapire, J. Soberón, S. Williams, M.S. Wisz, N.E. Zimmermann: Novel methods improve prediction of species' distributions from occurrence data, Ecography **29**, 129–151 (2006)

35.35   V. Vapnik: *The Nature of Statistical Learning Theory*, 2nd edn. (Springer, Berlin, Heidelberg 1999)

35.36   S. Lek, J.F. Guégan: Artificial neural networks as a tool in ecological modelling, An Introduction. Ecol. Model. **120**, 65–73 (1999)

35.37   L. Breiman: Random forests, Mach. Learn. **45**, 5–32 (2001)

35.38   S.J. Mason, N.E. Graham: Areas beneath the relative operating characteristics (ROC) and relative operating levels (ROL) curves: Statistical significance and interpretation, Q. J. R. Meteorol. Soc. **128**, 2145–2166 (2002)

35.39   J.R. Landis, G.G. Koch: The measurement of observer agreement for categorical data, Biometrics **33**, 159–174 (1977)

35.40   C. Furlanello, S. Merler, C. Chemini, A. Rizzoli: *An Application of the Bootstrap 632 + Rule to Ecological Data*, Neural Nets, Vol. WIRN-97 (MIT Press, Cambridge 1998)

35.41   O. Allouche, A. Tsoar, R. Kadmon: Assessing the accuracy of species distribution models: Prevalence, kappa and the true skill statistic (TSS), J. Appl. Ecol. **43**, 1223–1232 (2006)

35.42   J. Elith, J.R. Leathwick: Species distribution models: Ecological explanation and prediction across space and time, Annu. Rev. Ecol. Evol. Syst. **40**, 677–697 (2009)

35.43   S. Soltic, S. Pang, L. Peacock, S.P. Worner: Evolving computation offers potential for estimation of pest establishment, Int. J. Comput. Syst. Signals **5**, 36–43 (2004)

35.44   N. Kasabov: *Evolving Connectionist Systems* (Springer, Berlin, Heidelberg 2002)

35.45   N. Kasabov, Q. Song: Dynamic evolving neural-fuzzy inference system and its application for time-series prediction, IEEE Trans. Fuzzy Syst. **10**, 144–154 (2002)

35.46   J.P.W. Pitt, S.P. Worner, A.V. Suarez: Predicting Argentine ant spread over the Heterogeneous landscape using a spatially-explicit stochastic model, Ecol. Appl. **19**, 1176–1186 (2009)

35.47   R.A. Fisher: The wave of advance of advantageous genes, Ann. Eugen. **7**, 355–369 (1937)

35.48   J.G. Skellam: Random dispersal in theoretical populations, Biometrika **38**, 196–218 (1951)

35.49   E.E. Holmes: Are diffusion models too simple? A comparison with telegraph models of invasion, Am. Nat. **142**, 403–419 (1993)

35.50   R.H. Gardner, E.J. Gustafson: Simulating dispersal of reintroduced species within heterogeneous landscapes, Ecol. Model. **171**, 339–358 (2004)

35.51   S. Hein, B. Pfenning, T. Hovestadt, H.-J. Poethke: Patch density, movement pattern, and realised dispersal distances in a patch-matrix landscape – a simulation study, Ecol. Model. **174**, 411–420 (2004)

35.52   D. Rhind: A GIS research agenda, Int. J. GIS **2**, 23–28 (1988)

35.53   M. Neteler, H. Mitasova: *Open Source GIS: A GRASS GIS Approach*, 2nd edn. (Kluwer Academic, Boston 2004)

35.54   E.C. Pielou: *Biogeography* (Wiley, New York 1979)

35.55   A.D. Cliff, P. Haggett, J.D. Ord, G.R. Versey: *Spatial Diffusion* (Cambridge Univ. Press, Cambridge 1981)

35.56   D. Mollison: Spatial contact models for ecological and epidemic spread, J. R. Stat. Soc. B **39**, 283–326 (1977)

35.57   J.S. Clark, M. Lewis, L. Horvath: Invasion by extremes: Population spread with variation in dispersal and reproduction, Am. Nat. **157**, 537–554 (2001)

35.58 J. von Neumann: The general and logical theory of automata. In: *Cerebral Mechanisms in Behavior – The Hixon Symposium*, ed. by L.A. Jeffress (Wiley, New York 1951)

35.59 A.J. Hoffman, R.R. Singleton: Moore graphs with diameter 2 and 3, IBM J. Res. Dev. **5**, 497–504 (1960)

35.60 M. Peng: Spatio-Temporal Modelling of Biological Invasions. Ph.D. Thesis (University of Auckland, NZ 2000)

35.61 S. Hartley, P.J. Lester: Temperature-dependent development of the Argentine ant, Linepithema humile (Mayr) (Hymenoptera: Formicidae): A degree-day model with implications for range limits in New Zealand, NZ Entomol. **26**, 91–100 (2003)

35.62 S. Worner: Use of models in applied entomology: The need for perspective, Environ. Entomol. **20**(3), 768–773 (1991)

Part G

Understa

Part G Understanding Information Processes in the Brain and the Nervous System

Ed. by Heike Sichtig

36. Information Processing in Synapses

Hiroshi Kojima

The synapse is a basic functional structure for information processing between neurons in the central nervous system, required for understanding of the functional properties of neural circuits and brain functions, and even the consciousness that emerges from them. There is now a wealth of experimental results concerning the detailed structure and functional properties of synapses on the basis of molecular biological, since the series of pioneering works by Bert Sakmann and Shosaku Numa in the 1980s that had been originally suggested and postulated by Bernard Katz and colleagues. With the introduction of more advanced research techniques, such as the patch-clamp method (in electrophysiology), two-photon confocal laser microscopy (in imaging), and molecular biological methods, into the research field of synaptic physiology, understanding of the functional significance of synapses has advanced enormously at the molecular level, with fruitful results. Furthermore, emerging new techniques with the invention of noninvasive whole-brain imaging methods (functional magnetic resonance imaging (fMRI), etc.) make it necessary for researchers to have deep understanding of the relationship between the microscopic physiological phenomena and the higher brain functions composed of and elicited from neural networks. Quantitative expressions of electrical and chemical signal transactions carried out within neural networks allow investigators in other fields such as engineering, computer science, and applied physics to treat these biological mechanisms mathematically for computational neuroscience. In this chapter, the physiology, biophysics, and pharmacology of the information processes of synaptic transmission in the central nervous system are presented to provide the necessary background knowledge to researchers in these fields. Especially, electrophysiological results regarding receptors and ion channels, playing important roles in synaptic transmission, are presented from the viewpoint of biophysics. Moreover, one of the most advanced techniques, namely fast multiple-point photolysis by two-photon laser beam activation of receptors in the membrane of a single neuron and neural tissues on the submicron level, is introduced together with results obtained by authors' experiments.

36.1 General View of Synaptic Transmission

Our present concepts concerning neurons have been developed over the last century from a convergence of several experimental fields such as physiology, pharmacology, anatomy, etc. Before the invention of the classical optical microscope, nervous tissue was thought to be an organ which transports fluid secreted by the central nervous system (brain and spinal cord) to the body periphery. This idea was postulated by the Greek physician *Galen* (*Claudius Galenus*, 129–200) [36.1]. The invention and development of the optical microscope revealed the true structure of neurons and other nervous tissues. However, the anatomical field concerning nervous tissue was not established until 1800, when *Camillo Golgi* (1843–1926) and *Santiago Ramón y Cajal* (1852–1934) described the detailed structure of these tissues in the central nervous system. *Golgi*, who invented the *Golgi staining method*, insisted that neural tissues were a continuous web-like endless structure (*reticular theory*) [36.2]. On the other hand, *Ramón y Cajal* proposed the hypothesis that neurons were separated from each other by membranes and were independent units of neural tissues (*neuron theory*) [36.3]. The invention of electron microscopy terminated the conflict between these two ideas by showing images of the synaptic cleft, the narrow space between two neurons. Moreover, it also enabled observation of the presynaptic vesicles in the presynaptic terminal. At that time, the term "synapse" was first introduced by *Sherrington* (1857–1952) in part III of the seventh edition of *Forster's Textbook of Physiology*, published in 1897 [36.4].

Later, ideas of two groups emerged: one is the group of physiologists led by the Australian physiologist *John C. Eccles* (1903–1997), and the other is the group of pharmacologists led by the British pharmacologist *Henry H. Dale* (1875–1968). *Eccles* argued that synaptic transmission is electrical [36.5]. However, *Dale* insisted that communication between two neurons is chemical in nature and the action potential that reached the presynaptic terminal of a neuron initiates the release of neurotransmitters which generate the potential in the postsynaptic neuron [36.6]. Through the development of electrical measurement techniques in the 1950s and 1960s, it was shown that most of the communication between neurons is conducted by chemical substances but that in some cases electrical communication also occurs. The latter is called the electrical synapse, in which two neurons are directly communicated through special channels (the gap junction) made up of proteins that serve as conduits between the cytoplasm of the two neurons.

In this chapter, the electrical synapse is not discussed in detail, even if it plays important roles in some of the physiological functions. We focus rather on the fundamental physiological mechanisms of the chemical synapses in order to make the discussion useful for scientists in the fields of engineering, computer science, and other related areas. Therefore, the basic concepts and significant principles of chemical synaptic transmission are presented to aid understanding in the context of information processing mechanisms in the nervous system. Readers who wish to know more details regarding current knowledge of the physiology, pharmacology, and molecular biology of the synaptic function can read other references [36.7].

36.2 Structure of the Synapse

36.2.1 General Features

We start our discussion by reviewing the morphological features of synapses. The basic feature of a synapse is a close apposition of specialized regions of the plasma membrane of two participating neurons to form a synaptic interface as shown in Fig. 36.1, which presents a typical synaptic connection, of which the postsynaptic element is the dendritic spine. The spine is thought to be a generally excitatory postsynaptic element of central neurons, whereas inhibitory synapse contacts are present on the proximal dendrite and soma. On the presynaptic side, a cluster of neurotransmitter-filled *synaptic vesicles* is associated with the presynaptic plasma membrane, and this area is called the *active zone*. On the postsynaptic membrane, which is sometimes observed as a high-electron-density scaffold called the *postsynaptic density* (PSD), an accumulation of neurotransmitter receptors and other machinery proteins are marked. The PSD (≈ 18 nm diameter) usually contains many granular particles such as calcium/calmodulin-dependent protein kinase II (CaMKII), PSD-95, glutamate receptor-interacting protein (GRIP), and other anchoring proteins. PSD-95 and GRIP have special domains of PDZ, SH3, etc. that make ion channels, receptors, and signaling proteins assemble together. The space between the pre- and postsynaptic membrane is called the *synaptic cleft*, having a space of around ≈ 20 nm and ≈ 500 nm for neuronal synapses and neuromuscular junctions, respectively. The basal lamina, which contains acetylcholinesterase (AChE), laminin, collagen, agrin, etc., is observed for the synaptic cleft of the neuromuscular junction. Some of these elements are discussed in later sections.

In typical cases, the presynaptic compartment is localized in the top (ending) of an axonal branch (called "*bouton*" in French or "button" in English), and the postsynaptic compartment is located at the surface of the cell body or of the dendrite. This organization reflects the general functional property of the neuron, according to which dendrites are the primary afferent elements (input) and axons are the primary efferent elements (output). However, there are exceptions to this general rule. Among these special cases are axonless olfactory bulb and retinal neurons that possess only short neurites that serve both effector (axonal) and receptive (dendritic) functions. Some olfactory bulb and thalamic neurons are provided with a standard axon, but their cell bodies and dendrites, which typically receive synaptic

inputs, are also capable of acting as presynaptic elements. Some of the morphological features of neurons and dendrites are related to the functional properties of neurons, such as integration of input signals by dendritic trees and soma. Figure 36.2 shows the variety of neurons that have unique dendritic trees and axons.

36.2.2 Synaptic Vesicles of the Presynaptic Site

Synaptic vesicles were thought to be linked to chemical transmission as soon as they were discovered independently by three groups in the early days of electron microscopy [36.8–10]. They have served ever since as one of the morphological hallmarks of chemical synapses. Synaptic vesicles are small, electron-lucent vesicles with a size range of $35 \sim 50$ nm that store nonpeptide neurotransmitters, such as acetylcholine (ACh), glutamate, gamma-aminobutyric acid (GABA), and glycine. At each synapse a few vesicles are in physical contact with the presynaptic plasma membrane at release sites (so-called docked vesicles), also called the

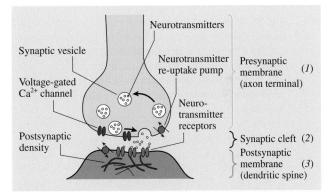

Fig. 36.1 Schematic representation of a typical chemical synapse. The synapse is divided into three parts according to functional and morphological criteria: (*1*) presynaptic membrane (axon terminal), (*2*) synaptic cleft, and (*3*) postsynaptic membrane (dendritic spine). The presynaptic terminal contains the synaptic vesicle, which includes neurotransmitters, the voltage-gated Ca^{2+} channel, the neurotransmitter re-uptake pump, etc. The synaptic cleft is the space between the presynaptic neuron and postsynaptic neuron. This was observed clearly by electron microscopy and strongly supported the *neuron hypotheses* of Ramón y Cajal. The postsynaptic membrane contains the neurotransmitter receptors, postsynaptic density, etc. The dendritic spine is generally the postsynaptic component of excitatory postsynaptic transmission (by Chloé Okuno)

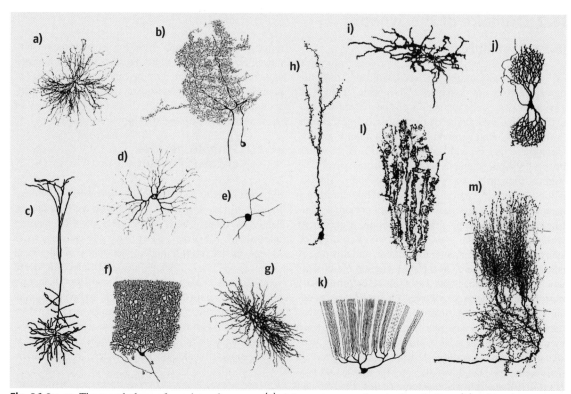

Fig. 36.2a–m The morphology of a variety of neurons: (**a**) alpha motoneuron in spinal cord of cat, (**b**) spiking interneuron in mesothoracic ganglion of locust, (**c**) layer 5 neocortical pyramidal cell in rat, (**d**) retinal ganglion cell in postnatal cat, (**e**) amacrine cell in retina of larval tiger salamander, (**f**) cerebellar Purkinje cell in human, (**g**) relay neuron in rat ventrobasal thalamus, (**h**) granule cell from olfactory bulb of mouse, (**i**) spiny projection neuron in rat striatum, (**j**) nerve cell in the nucleus of Burdach in human fetus, (**k**) Purkinje cell in mormyrid fish, (**l**) Golgi epithelial (glial) cell in cerebellum of normal–reeler mutant mouse chimera, and (**m**) axonal arborization of isthmotectal neurons in turtle (after *Mel* [36.12])

active zone as previously indicated. The synaptic vesicle was thought to correspond to a single unit, according to the quantal hypothesis proposed by Del Castillo and Katz based on the frog neuromuscular junction. Moreover, a decrease in the number of synaptic vesicles in the presynaptic terminal and an increase of the surface area of the presynaptic membrane supported their hypothesis of a transmitter release mechanism. At the central synapse, the active zone often forms a presynaptic grid where the docked synaptic vesicles are localized. It is organized into very dense masses which form ribbons or spherical dense bodies in specialized synapses of sensory organs. The main function of this dense matrix is suggested to be to provide a high concentration of synaptic vesicles in close proximity to release sites. Therefore, new vesicles can rapidly replace docked vesicles in order to undergo exocytosis. The molecular

composition of this matrix remains uncertain. However, some of the protein components, which share blocks of homology, have recently been identified. They are suggested to include Rab3-interacting modules (RIMs) and two very large proteins: Aczonin/Piccolo and Bassoon [36.11].

36.2.3 Proteins Associated with Exocytosis Machinery

Based on its many functions, the presynaptic membrane is thought to contain several specialized proteins in addition to the housekeeping proteins common to all plasma membranes. The detailed molecular mechanisms of transmitter release are discussed in Sect. 36.3.3, and more than 100 molecules are thought to be involved in this process. However, Fig. 36.3 indicates

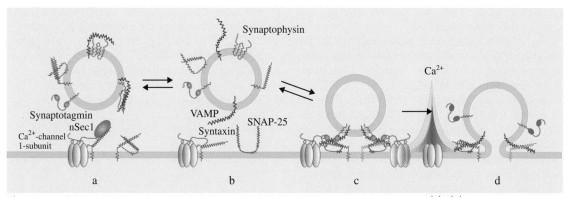

Fig. 36.3a–d Mechanisms of neurotransmitter release from the presynaptic membrane. (**a**), (**b**) Synaptic vesicle approaching the postsynaptic membrane. Several proteins are attached to the membrane of vesicles and the presynapse: synaptotagmin, VAMP, syntaxin, SNAP-25, and synaptophysin. nSec1 phosphorylated by PKC regulates the affinity to syntaxin. The Ca^{2+} ion channel is located in the postsynaptic membrane. (**c**) Docking and priming of vesicle; the Ca^{2+} channel is ready to be opened by the depolarization potential. (**d**) After opening of the Ca^{2+} channel, Ca^{2+} entry into the presynaptic terminal, and increase of the Ca^{2+} concentration, exocytosis is triggered by fusion between the membranes of the vesicle and presynaptic terminal. (SNAP: soluble NSF attachment protein) (Modified by Chloé Okuno after *Humeau* [36.13])

the simple mechanisms of exocytosis for the few of these specialized proteins that have been characterized. Two postsynaptic membrane proteins are the target proteins SNAREs (t-SNAREs; SNARE: SNAP REceptor), syntaxin and SNAP-25, which play a role in synaptic vesicle exocytosis. They act as the partners of the synaptic vesicle SNARE (v-SNARE) synaptobrevin/VAMP (vesicle associated membrane proteins) in the formation of the exocytotic fusion complex. However, neither syntaxin nor SNAP-25 is selectively concentrated in the presynaptic membrane, and it is suggested that other factors present either in the membrane or in close proximity to it are responsible for the selective exocytosis of synaptic vesicles at the synapse.

36.2.4 Voltage–Dependent Ca^{2+} Channels

Voltage-dependent Ca^{2+} channels, which are responsible for depolarization secretion coupling, play essential roles at sites of exocytosis, as described in Sect. 36.3.3. A direct connection between Ca^{2+} channels in the plasma membrane and docked vesicles is suggested by the interaction of these channels with a synaptic vesicle protein, synaptotagmin (Fig. 36.3). Such a link would allow ready-to-fuse vesicles to be exposed to transient, but high, peaks of cytosolic Ca^{2+} following depolarization-dependent channel opening when an action potential arrives at the presynaptic nerve terminal. Freeze–fracture views of the presynaptic membrane

characterize the presence of these arrays of large intramembranous particles. The shape of these arrays varies in different presynaptic membranes, but it generally reflects the distribution of docked synaptic vesicles. It has therefore been proposed that these particles represent Ca^{2+} channels [36.14].

36.2.5 Postsynaptic Compartment

A patch of plasma membrane containing a packed array of neurotransmitter receptors with other proteins and the *postsynaptic density* (PSD) constructs the postsynaptic compartment. The postsynaptic plasma membrane associated with the PSD has a characteristic appearance accompanied by the presence of a compact array of large intramembranous particles that would correspond to the postsynaptic receptors in freeze–fracture preparations. Within this membrane patch, two classes of receptors, namely *N*-methyl-D-aspartate (NMDA) and α-amino-3-hydroxy-5-methyl-4-isoxazolepropionic acid (AMPA) receptors, are arranged heterogeneously at glutamatergic synapses. NMDA nd AMPA receptors are localized in the central and peripheral regions in the postsynaptic membrane, respectively. Their detailed physiological functions are presented in Sect. 36.5.1. The postsynaptic membrane also includes other proteins that are required for information transmission and maintenance of the synaptic structures, such as the cell adhesion molecules of the presynaptic membrane, which include the neuroligins, syndecans, cadherins, and (at least in some synapses) densin-180 [36.15].

The PSD is a scaffold of molecular proteins that are different in molecular composition and morphology at excitatory and inhibitory synapses. Their major components include CaM kinase II, PSD-95, PSD-93, synapse-associated protein-102 (SAP-102), guanylate kinase-associated protein (GKAP), cysteine-rich interactor of PDZ3 (CRIPT), synaptic GTPase-activating protein (SynGAP), Homer, and gephyrin. These scaffold proteins regulate and modify synaptic functions by mediating receptor clustering, regulating receptor function, controlling receptor internalization and turnover, and linking the postsynaptic membrane to the cytoskeleton. Therefore, they have signaling roles with coordinating electrical responses generated by neurotransmitter-gated receptor ion channels with longer-lasting cellular responses.

36.2.6 Synaptic Cleft Between Pre- and Postsynaptic Membrane

The link between the pre- and the postsynaptic membrane is very tight, and generally the synaptic junction cannot be distributed. The intracellular space – the synaptic cleft – is occupied by a regular array of material of moderate electron density. The precise molecular nature of this material is not known at the central synapses, but many of the extracellular domains of the pre- and postsynaptic membranes should be included in the synaptic cleft. The only synapse for which some of the molecular components of the cleft have been identified and characterized is the neuromuscular junction, where the cleft is considerably wider (≈ 500 nm) than at central synapses (≈ 20 nm) and is occupied by a specialized basal lamina as described earlier. At this cholinergic synapse that is mediated by ACh receptors; it also includes the acetylcholine-dependent enzyme acetylcholinesterase (AChE) that quickly hydrates the ACh molecule into choline and acetate to terminate synaptic transmission. On the other hand, at the glutamatergic synapse, glutamate transporter located in the membrane and also glia cells take up glutamate molecules from the synaptic cleft and end the excitatory synaptic transmission smoothly.

The concentration of the neurotransmitters in the synaptic cleft is estimated to increase linearly to a maximum value of ≈ 1 mM in 20 μs and decay within a few milliseconds. The time course of the concentration profile is given in various forms and depends on several factors:

1. The number of synaptic vesicles released at a time,
2. The amounts of transmitter molecules in a vesicle,
3. The speed of the emission process of transmitter molecules from a vesicle,
4. The geometric structure of the synaptic cleft,
5. The diffusion speed of transmitter molecules,
6. The binding speed of transmitter molecule to receptor,
7. The binding speed of transmitter molecule to enzyme and transporter,
8. The kinetics of the reaction of enzyme or transporter.

The time course of the concentration of neurotransmitter molecules partly affects the waveform of the synaptic potential or current (under voltage-clamp conditions) and thus partly regulates the signaling mechanisms at the synapses. Most likely, the interaction of pre- and postsynaptic site through the synaptic cleft also mediates at least some aspects of retrograde signaling, through which the activation of postsynaptic receptors may modulate the efficacy of the presynaptic compartment. Finally these cognate receptors are selected, by these interactions, for axon that secretes a given neurotransmitter. In this chapter, such retrograde signaling mechanisms are not described.

36.2.7 Neuromuscular Junction as a Typical Model Synapse

The motor end-plate (in particular, the frog neuromuscular junction) has been the most widely used and studied synapse since *Bernard Katz* (1911–2003) and colleagues worked on it intensively [36.16–21]. Several features make the neuromuscular junction an especially favorable synapse for physiological studies and systematic correlative analyses of structure and function. Also, molecular biological investigation on the receptor protein (nicotinic ACh receptor) in the end-plate and its three-dimensional structure has been achieved by the combination of the patch-clamp technique invented by *Sakmann* et al. with molecular cloning by *Shosaku Numa* (1929–1992) [36.22, 23].

As shown in Fig. 36.4, the cholinergic axons that innervate a muscle fiber usually split into several branches, which form a rosette for mammals and run parallel to the long axis of the muscle fibers to contact with the muscle fibers in amphibians. Freeze–fracture views of the presynaptic plasma membrane underlying these sites reveal the presence of a double row of intramembranous particles. They are in precise register

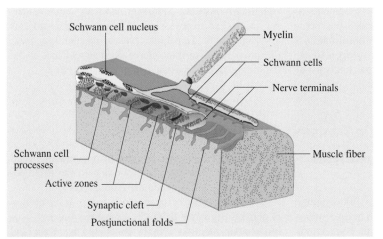

Fig. 36.4 Typical structure of the neuromuscular junction. The presynaptic terminal, i. e., the axon of a motoneuron whose cell body is located in the spinal cord, is covered by a myelin sheath and Schwann cells. The postsynaptic membrane is folded into muscle fiber called postjunctional folds, and at the top of the junctional folds ACh receptor channels are located. The opposite site of the junctional folds is the active zone of the presynaptic site. The postsynaptic membrane is called the end-plate. The action potentials propagate to the nerve terminal, and ACh molecules are released into the synaptic cleft. These ACh molecules bind to ACh receptors, which change their conformation to open ion channels. (Modified by Chloé Okuno after *Nicholls* [36.25])

with a double row of docked vesicles in the underlying cytoplasm and most likely represent Ca^{2+} channels, as described above. Simultaneously, these particles define the site where exocytosis occurs, as can be seen from freeze–fracture preparations [36.24].

The postsynaptic membrane of the motor end-plate is not flat; rather it is interrupted by deep infoldings that are in precise register with presynaptic active zones. The postsynaptic ionotropic receptors at this synapse [nicotinic acetylcholine receptors (nicotinic AChR)] are concentrated on the edges of the infoldings and are therefore in direct proximity to the release sites. One of the specialized features of the synaptic cleft of the end-plate is the *basal lamina*, which contains AChE, *agrin*, and a variety of other proteins that mediate the structural and functional connections between the pre- and postsynaptic compartments.

Nicotinic AChR has been studied molecular biologically, forming a pentamer composed of five subunits: two α-, one β-, one γ-, and one δ-subunit (for embryonic type), which is replaced by a ε-subunit for adult type [36.26]. The α-subunit contains a site that binds the ACh molecule with high affinity. *Arthur Karlin* and colleagues demonstrated that two extracellular binding sites for ACh are formed in a cleft between each α-subunit and its neighboring γ- and δ-subunits (or ε-subunit) [36.27, 28]. One molecule

of ACh binds to each of the two α-subunits for the channel to open. The inhibitory snake venom α-bungarotoxin (competitor) also binds to the α-subunit. The three-dimensional structural model of the AChR channel has been also investigated based on neutron scattering and electron diffraction images [36.29–31]. Investigation of this receptor molecular biology and its regulation, modulation, and synaptic plasticity by associated proteins is one of the major research topics in neuroscience today. The physiological features of the neuromuscular junction as revealed by recording electrical measurements from muscle fiber by the patch-clamp technique have been studied, with elucidation of the three-dimensional molecular structures of ACh receptor proteins. The results obtained from this investigation provide a fruitful example of the investigation of functional properties of other receptor channels such as glutamate and GABA. These results are discussed partly in Sect. 36.5.1.

36.2.8 Overview of the Structure of the Synapse

The description of the synaptic morphology summarized in this section makes it clear that the structural features of synapses are optimally designed to maximize the spatial precision and speed of signal transmission.

Spatial precision is achieved. Recent developments in molecular biological research have revealed the three-dimensional structure of receptors, the conformational change on agonist binding from closed to open to desensitized state, and the modulatory mechanisms of signaling molecules. Moreover, the Ca^{2+} binding sensor to trigger exocytosis and many of the proteins associated with transmitter release mechanisms from the presynaptic terminal have been elucidated in the membrane of both vesicle and presynaptic sites.

36.3 Physiology of Synaptic Transmission (Presynaptic Site)

36.3.1 Basic Properties of Different Types of Synapses

The average neuron receives more than 1000 synaptic inputs from other neurons, and the human brain contains at least 10^{11} neurons. Thus, 10^{14} synaptic connections are formed in the brain. The chemical synapse is the elemental functional unit that regulates a variety of functions in the central nervous system; for example, in the case of memory and learning, the efficacy of synaptic transmission becomes stronger or weaker for a long time after treatments that are thought to correspond to memory formation. Moreover, the transfer of information between neurons is regulated during memory formation, consolidation, and retrieval. Physiologically, this is termed *synaptic plasticity*, typical showing two forms: long-term potentiation (LTP) and long-term depression (LTD), for which molecular mechanisms have been studied since the discovery of LTP in the hippocampus [36.32]. Since the first measurements of long-term potentiation in the hippocampus, various types of synaptic efficacy change have been reported in various regions of the central nervous system in both in vivo and in vitro experiments [36.33, 34].

The physiological processes involved in chemical synaptic transmission are essentially categorized into three classes:

1. The processes of transmitter release from the presynaptic terminal,
2. The diffusion of neurotransmitters within the synaptic cleft toward the postsynaptic membrane,
3. The binding of the neurotransmitters to the receptor ion channels in the postsynaptic membrane and the generation of the graded electrical signals (induction of both fast and slow synaptic potential).

These three physiological phenomena proceed sequentially in a step-by-step manner during chemical synaptic transmission, and the time lag between the arrival of the action potential at the presynaptic nerve terminal and the onset of the synaptic potential is around $1-2$ ms. This time lag corresponding to the three sequential processes is called the *synaptic delay*.

Our basic idea of chemical synaptic transmission is derived primarily from classic studies on frog neuromuscular junction (NMJ) conducted by Katz and colleagues, as mentioned in the morphology section; later, they also used squid giant synapse for voltage clamping of the presynaptic site, which is large enough for insertion of two electrodes simultaneously. Their pioneering works with NMJ and squid giant synapse, and subsequent experiments concerning excitatory synaptic transmission in the central nervous system, revealed the fundamental mechanism of the chemical transmission, in which a series of processes (roughly grouped into the three elementary processes listed above) occur in a step-by-step (sequential) manner. Thus, we provide below the general mechanisms obtained from experiments with both NMJ and other preparations including excitatory synaptic transmission in the central nervous system. The mechanisms of the glutamatergic synapse, one of the excitatory synapses, could in principle also be understood in this way due to their similar mechanism. Moreover, the mechanism of inhibitory synaptic transmission is basically similar and where necessary is occasionally described as well:

1. A presynaptic neuron fires, and the action potential propagates along its axon towards the postsynaptic neuron to reach the presynaptic terminal or button.
2. The membrane potential of the presynaptic terminal is depolarized by these action potentials, and voltage-sensitive Ca^{2+} channels in the presynaptic membrane are activated.
3. The Ca^{2+} ions enter the presynaptic terminal through the Ca^{2+} channel and bind to *synaptotagmin*, a family of Ca^{2+} sensors that interact with SNAREs.
4. Exocytosis is triggered, and the neurotransmitters in the vesicles are delivered into the synaptic cleft in a quantal manner.

5. The neurotransmitters diffuse in the synaptic cleft towards the postsynaptic membrane.
6. These neurotransmitters bind to receptors such as AChR and glutamate-R in the postsynaptic membrane.
7. The receptor ion channels composed of several subunits of proteins elicit conformational (three-dimensional structural) changes such as open–close gating (including desensitization, which is another closed state with continuous presence of agonist in the vicinity of the receptors).
8. The cation-permeable (for the inhibitory synapse, anion-permeable) pore of this channel allows ions to pass across the membrane, and the membrane potential is depolarized.
9. An excitatory postsynaptic potential (EPSP) and/or end-plate potential (EPP) is produced in the post-synaptic cell, and when it reaches the threshold for action potential initiation, an impulse is fired at the initial segment of the neuron.
10. The final step in chemical synaptic transmission is neurotransmitter removal from the synaptic cleft. The mechanisms for transmitter removal include diffusion, degradation, and uptake into glial cells or nerve terminal; for example, at the neuromuscular junction, the action of acetylcholine molecules is terminated by the enzyme acetylcholinesterase (AChE), which hydrolyzes ACh to choline and acetate.
11. Finally, the vesicle membrane fuses with the postsynaptic membrane and is taken up into the presynaptic terminal by a mechanism called *endocytosis*.

These are the general detailed processes of the typical chemical transmission such as at NMJ and excitatory synapses in the central nervous system, even though each step and the structural components that participate vary from preparation to preparation. In the present chapter, the inhibitory synaptic transmission is not discussed, and readers who may be interested in these topics should read other references [36.35, 36].

36.3.2 Quantal Hypothesis and Quantal Analysis

In electrophysiological experiments, small electrical potentials are recorded by an electrode whose tip is inserted into a muscle fiber, and these small electrical signals, which are called miniature end-plate potentials (mEPPs), are found to occur spontaneously. *Fatt*

and *Katz* observed this phenomenon for the first time, and after careful inspection they considered that it was evoked by packets of thousands of neurotransmitter molecules in a vesicle, released spontaneously from the nerve terminal [36.17]. It was shown that the amplitude and frequency of the mEPP were less than 1 mV and around 1 Hz, respectively. On the other hand, we can observe larger synaptic potentials upon stimulation of a nerve fiber in the same preparation. It has been elucidated that end-plate potentials (EPPs) are evoked by a unit amount of ACh molecules (one vesicle) corresponding to a mEPP, whereas the larger-amplitude EPPs are caused by synchronous, multiquantum (multivesicle) transmitter release [36.37].

This observation led *Katz* and *Fatt* to propose the *quantum hypothesis*, which is one of the excellent examples of the application of a quantitative expression to a physiological process, i.e., the observed single quantum events and also as building blocks for the potentials evoked by stimulation [36.38].

The amplitude histograms of mEPP and evoked EPP recorded from mammalian neuromuscular junction by *Boyd* and *Martin* in 1956 are shown in Fig. 36.5 [36.39]. The mean amplitude and standard deviation of 78 spontaneous mEPPs are 0.4 and 0.086 mV, respectively, as shown in the inset of Fig. 36.5. The amplitude histogram of 198 evoked EPPs in low Ca^{2+} and high Mg^{2+} (12.5 mM) is shown in Fig. 36.5 with bin width of 0.1 mV. The histogram indicates the number of EPPs observed at each amplitude. The peaks of the histogram occur at 0 mV, at one (0.4 mV), two (0.8 mV), three (1.2 mV), four (1.6 mV), five (2.0 mV), six (2.4 mV), and seven (2.8 V) times the mean amplitude of the spontaneous mEPPs, indicating that the responses comprise 1, 2, 3, 4, 5, 6, and 7 quanta. The black bar indicates the number of failures with no response to stimulation. The number of quanta released per trial is defined as *m* and given by

$$m = \frac{\text{mean amplitude of evoked potentials}}{\text{mean amplitude of miniature potential}}.$$

(36.1)

In this experiment, $m = 0.933\,(\text{mV})/0.4\,(\text{mV})$, thus $m = 2.33$.

Statistical analysis is necessary to describe the quantal nature of the synaptic potential. *Del Castillo* and *Katz* [36.38] hypothesized that the motor nerve terminal contains thousands of quantal packets of acetylcholine molecules which correspond to *n* release sites, each of which has probability *p* of being released in response to a nerve impulse. Then, in a large number of trials,

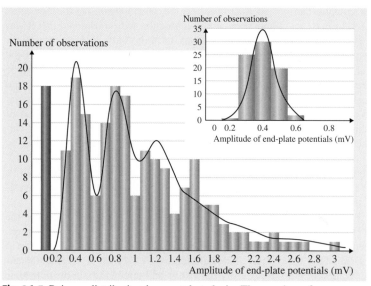

Fig. 36.5 Poisson distribution in quantal analysis. The number of responses on stimulating the presynaptic fiber was counted to plot in the histogram. The histogram has a number of peaks. The first with a black bar at 0 mV is the number of failures. The second peak, at 0.4 mV, represents the smallest evoked potential. The third, fourth, fifth, and sixth peaks correspond to 0.8, 1.2, 1.6, and 2.0 mV, respectively. The smallest evoked potential, which is called the unitary potential, has the same amplitude as the spontaneous miniature end-plate potential (mEPP) (*inset*). The *solid line* of the histogram shows the theoretical curve of the Poisson distribution. More precisely, the number of observations under each peak divided by the total number of observations in the experiment corresponds to the probability of each number of released quanta from the presynaptic terminal. The *inset* shows an amplitude histogram of the number of observations of spontaneous miniature end-plate potentials. Bar width is 0.1 mV. (Modified by Chloé Okuno after [36.39])

the mean number of quanta m released per trial would be given by $m = np$. The average response size of one quantum is defined as the quantal size q, so the mean amplitude of mEPPs is equal to

$$V = npq = mq .\qquad(36.2)$$

The number of times the response consisted of $0, 1, 2, 3, 4, \ldots$, or x quanta would be given by the binomial distribution

$$f(x) = nCxp^{x}(1 - p)^{n-x} .\qquad(36.3)$$

Under normal conditions, p may be assumed to be relatively large, as a large part of the synaptic population responds to an action potential. However, in Del Castillo and Katz's experimental conditions (low Ca^{2+} and high Mg^{2+} concentration), when p is small and n is quite large, the Poisson distribution is an appropriate approximation to the binomial distribution

$$f(x) = (m^{x}/x!)\exp(-m) .\qquad(36.4)$$

In the Poisson distribution, only m is a unique parameter to be determined, and if it is obtained, one

can calculate the probabilistic distribution experimentally.

It was postulated by *Del Castillo* and *Katz* that the amplitude histogram constructed from the experimental results shown in Fig. 36.5 could be fitted by a Poisson distribution under several physiological conditions [36.18, 38]. To test this hypothesis, it would be necessary to show that the experimental results (the amplitude distribution of evoked EPPs) can be fitted by a Poisson distribution. There are two ways to do this, i.e., fitting the amplitude distribution by a Poisson distribution directly, and direct comparison of the number of events against the prediction:

1. Comparison of m obtained by two different methods: one deduced from spontaneous miniature end-plate potentials, and another obtained from the probability of failure.
   From the hypothesis, $m = V/q$, and one can then calculate m as

$$m = \frac{(\text{mean amplitude of evoked potentials})}{(\text{mean amplitude of miniature potential})} .$$

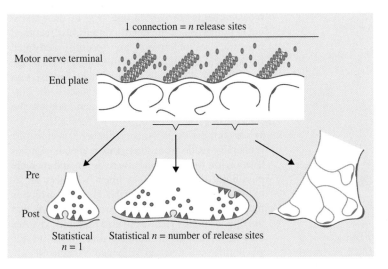

Fig. 36.6 The number of release sites is equal to the number of quanta. *Upper diagram* and *bottom diagram* show the neuromuscular junction and three types of synapses in the central nervous system, respectively. Statistical $n = 1$ (for *left part*) and $n = 4$ (*right part*) means the number of release sites likely to be active zones. (Modified by Chloé Okuno after [36.40])

The experimental results give

$$m = 0.933\,\mathrm{mV}/0.4\,\mathrm{mV} = 2.33 .\qquad(36.5)$$

On the other hand, the Poisson distribution gives the probability of failure by $f(0)$, where $x = 0$, as

$$f(0) = \frac{m^0}{0!}\exp(-m) = \exp(-m) .\qquad(36.6)$$

The experiment yielded

$$\frac{\text{number of failures (18)}}{\text{number of trials (198)}} ,$$

so

$$m = \ln(198/18) = 2.4 .$$

This is almost equal to the result of (36.5).

2. Direct comparison of the number of events (evoked EEPs) against those predicted by the Poisson distribution.

The value of m is obtained from (36.3), and this value is used for the calculation of the predicted values. The number of multiples of one quantum are $0, 1, 2, 3, 4, \ldots, 9$, and the corresponding values calculated from the Poisson distribution are $198[\exp(-2.33)] = 19, 44, 52, 40, 24, 11, 5, 2, 1, 0$. The values obtained by the physiological experiments for each multiple of one quantum are $18, 44, 55, 36, 25, 12, 5, 2, 1$, and 0, respectively, corresponding quite well. This indicates that the values predicted by the Poisson distribution and those obtained in the physiological experiment are almost equal, suggesting that EPPs are indeed composed of quanta and that each quantum is released independently.

36.3.3 Condition of Quantal Hypothesis and Quantal Analysis of the Synapse in the Central Nervous System

Del Castillo and *Katz* considered one vesicle in the nerve terminal as a possible candidate for the quantum of the quantal hypothesis [36.18, 38]. However, at present it is thought that one active zone of the presynaptic terminal corresponds to one quantum (Fig. 36.6) [36.41]. In general, quantal analysis of synapses at neuromuscular junction and sympathetic ganglion cells is reliable [36.42]. However, synapses in the central nervous system present many difficulties associated with the complex structures of the central synapses and dendritic trees. These difficulties are attributed to several factors when conducting experiments that have a variety of limitations due to the use of both in vivo and in vitro preparations and technical problems in carrying out the electrophysiology. The requirements for carrying out proper quantal analysis are as follows:

1. To avoid uncertainties in the variability of action potential generation or failure of spikes arriving at a particular synaptic terminal, exact stimulation of a single axon and measurement of the impulses at the presynaptic terminal are required.
2. Measurement of the synaptic events should be done at a sufficient close release site to avoid the possible filtering effect of dendritic cable properties [36.43].
3. The information regarding whether the release probability p is constant or not at all release sites is significant. This information determines the

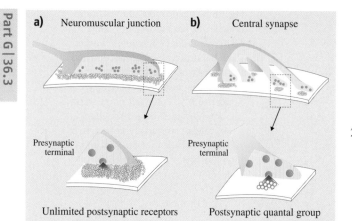

Fig. 36.7a,b Hypothesis of receptor saturation. (**a**) Hypothesis of receptor saturation of neuromuscular junctions. There are many ACh receptors at each release site. (**b**) Hypothesis that, at a synapse in the central nervous system, there are smaller numbers of ACh receptors in the postsynaptic site which is opposite to each release site. This causes saturation of the receptors during each quantal event. (Modified by Chloé Okuno after [36.44])

applicability of the binomial distribution to the experimental results (Fig. 36.5).

4. If the postsynaptic receptors are saturated by the released neurotransmitters, interpretation of the recorded data is quite difficult, and thus direct information on the variation in the quantal amplitude at a single release site is required (Fig. 36.7) [36.8, 44].

5. As there is a filtering effect of dendritic trees, the synaptic input at a variety of dendrite locations should at least be at the same electrotonic distance from the recording site, which is, generally, the soma of the cell [36.45].

6. Finally, the recording noise should be as small as possible to enable clear separation of the first peak in the evoked histogram and identification of failures [36.41, 44, 46, 47]. Quantal analysis is an interesting topic, and many works have already been conducted and reviews published [36.47, 48].

Detailed discussion of these problems is given later in order to understand and interpret properly the data of electrical measurements obtained from preparations in the central nervous system:

1. The synaptic potentials and/or currents generated at the synaptic sites of the dendritic trees which receive the inputs from presynaptic neurons and also far from the soma of the cells where the electrical measurements are carried out. Due to the passive cable properties of the dendritic trees, the recorded potentials and/or currents are generally deformed by these cable properties and filtering effects of the dendrites [36.43, 49]. To conduct quantal analysis on these synaptic signals, we need to discuss the correction of these deformed waves and estimate the filtering effects.

2. When stimulating presynaptic fibers by an electrode (generally made from glass and/or metals) for activation of the synapses, there is an unfavorable possibility that some other fibers will also be simultaneously stimulated by this electrode which is placed onto the presynaptic fibers. This means that not only a single synapse but also multiple synapses may be stimulated at the same time. To avoid this possibility, we need to stimulate a single presynaptic fiber and/or neuron by direct application of a patch pipette to the cell and/or by insertion of an intracellular electrode into the soma of these cells. However, these problems could be partially solved by recent advanced technology using photolysis, as discussed in a later section.

3. For the first time, *Jack* et al. and *Redman* suggested that postsynaptic receptors are saturated by released neurotransmitter at central synapses (Fig. 36.7) [36.46, 47]. Figure 36.7a shows a neuromuscular junction in which there are many postsynaptic receptors at each release site. Figure 36.7b presents the hypothesis of the central synapses in which there might be a limited number of receptors at each release site, showing the saturation of the receptors during each quantal event. Thus, it is hypothesized that the number of receptors at the subsynaptic membrane, rather than the amount of transmitter in each vesicle, is the rate-limiting factor for the amplitude of the excitatory postsynaptic currents (EPSCs) of the quantal event. In this case, little or no quantal variance at each release site is expected. A relatively fixed number of receptors at all active release sites on any particular neuron has been suggested [36.44, 50, 51]. This experiment would indicate that there is little or no quantal variance *across* release sites (Fig. 36.8).

4. Furthermore, it is *hypothesized* that the conductance associated with an individual synapse may have a variety of values depending on its electrotonic distance from the soma for central synapses. This hypothesis was deduced from experimental results showing that the charge measured at the soma between the proximal and distal synapses

has approximately the same value. This could be explained by assuming that larger conductance changes are produced at distal synapses than at proximal synapses. Therefore, the same amount of charge ultimately reaches the soma from the two inputs. Another possibility is that signals of the distal synapses are amplified by the activation of voltage-gated channels in the dendrites. Active properties of dendrites due to voltage-gated ion channels are discussed in Subsects. 36.5.2 and 36.5.3 in more detail [36.48].

There are some other requirements, in addition to addressing the described problems, in order to carry out proper quantal analysis, which are not discussed here. Studying the properties of central synapses suffers from many problems, because it is difficult to isolate and study individual synapses in central neurons. In spite of this, the quantal properties of synaptic transmission in central neurons with complex dendritic trees as shown in Fig. 36.2 will be the subject of intense investigation for many years in the future.

Including the aforementioned phenomenological results from electrophysiological experiments and quantitative analysis, we can summarize the substantial transmitter release processes, from the docking of vesicles against the postsynaptic membrane to exocytosis, finally ending up at endocytosis (Fig. 36.3), as follows:

1. Neurotransmitters are taken up into the vesicles and condensed within vesicles in the presynaptic terminal.
2. Vesicles move towards the active zones in the presynaptic membrane.
3. Proteins such as VAMP/synaptobrevin associated with the vesicle membrane and the proteins syntaxin and SNAP-25 in the presynaptic membrane combine with one another, resulting in the formation of a complex (SNARE-complex). This is the important process of vesicle docking to the nerve terminal membrane, and the ensembles of these docked vesicles are called the releasable pool. The four coils of these proteins (VAMP/synaptobrevin, syntaxin, and SNAP-25) remain parallel to one another to form a coil–coil complex, and the vesicles are tightly docked to the presynaptic membrane. This model with a coil–coil complex is called the zipper model.
4. The docked vesicles are ready to generate (execute) exocytosis by a process that consumes adenosine triphosphate (ATP). This pre-exocytosis process is

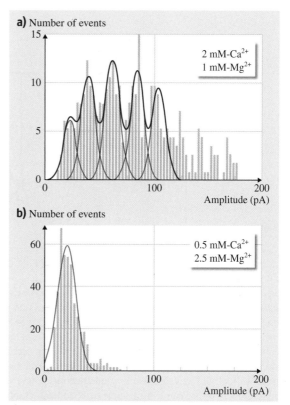

Fig. 36.8a,b Quantal analysis of IPSCs from dentate gyrus of rat hippocampus. (**a**) Histogram of the amplitude distribution of IPSCs from slice preparation in extracellular solution with high Ca^{2+} (2 mM) and low Mg^{2+} (1 mM). (**b**) Histogram of IPSCs from the same preparation with extracellular solution of different composition of low Ca^{2+} (0.5 mM) and high Mg^{2+} (2.5 mM). The Gaussian curve that fits histogram (**b**) fits well each of the five peaks in (**a**). The variance of each peak is small. This indicates that no multiple releases are summed up in the experiment and analysis. (Modified by Chloé Okuno after [36.51])

called priming, and the group of vesicles that are in the state of priming is called the readily releasable pool.

5. Depolarization of the membrane potential by the arrival of the action potential opens the voltage-gated Ca^{2+} channel and allows Ca^{2+} ions to enter the presynaptic terminal. The Ca^{2+} concentration around the region where the vesicles and active zones are located together increases to the order of 10^{-5}–10^{-6} M.
6. The Ca^{2+} ions that enter the terminal bind to the calcium sensor called synaptotagmin near the ac-

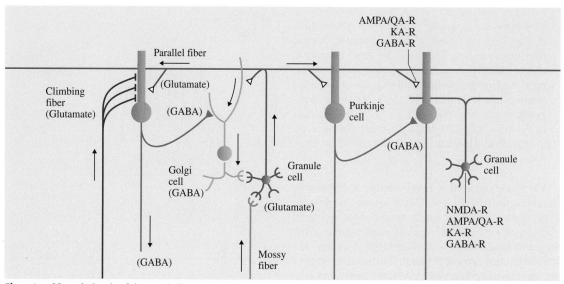

Fig. 36.9 Neural circuit of the cerebellar cortex. The structure of the cerebellar cortex is composed of three layers, from *bottom* to *top*: granular cell layer, Purkinje cell layer, and molecular layer where the dendrites of Purkinje cells extend toward the surface of the cortex. A Purkinje cell receives two excitatory inputs; one is the axon of a granule cell in the granular cell layer, and the other is a climbing fiber from the inferior olive. The sole output of the cerebellar cortex is the axon of a Purkinje cell that contacts to neuron of deep cerebellar nuclei. This synapse is mediated by the GABA receptor and is an inhibitory synapse by nature. Other significant neurons in the cerebellar cortex are Golgi cell, satellite cell, and basket cell; all of them are inhibitory interneurons. One Purkinje cell receives more than 10 000 synaptic contacts from parallel fibers whose soma are granule cells and one climbing fiber where multiple presynaptic buttons are attached to the soma and proximal dendrite of the Purkinje cell. Information flows continuously from the mossy fiber to the Purkinje cell via the granule cell toward cerebellar nuclei, and occasionally the information conveyed from the climbing fiber generates complex spikes at the Purkinje cell (KA-R: kainate receptor; QA-R: quisqualate receptor) (by Chloé Okuno)

tive zone and let vesicles stochastically fuse with the membrane in order to generate exocytosis. This type of calcium sensor, which is a calcium-binding protein, could be activated when the concentration of Ca^{2+} is increased to the order of 10^{-5}–10^{-6} M. The relationship between the Ca^{2+} concentration and the number of occurrences of exocytosis in not linear but rather can be fitted using a sigmoidal curve, indicating that 3–4 Ca^{2+} ions bind cooperatively to one calcium sensor. Such a calcium-sensor protein associated with the vesicle membrane is thought to be the tandem C2 domain of synaptotagmin-I and -II.

7. Dissociation of SNARE-complex and endocytosis. The time interval between the appearance of the synaptic response (onset of EPSP) and the entry of Ca^{2+} into the presynaptic terminal is around 0.1–0.2 ms. At present, more than 100 molecules are thought to participate in the process of exocytosis; in addition to synaptotagmin

and the SNARE-complex, Munc13, Munc18, complexin, and tomosyn play significant roles in the regulation of molecular mechanisms of transmitter release [36.52, 53].

Since Katz's formalization of the quantal theory of synaptic transmission, a central problem of the physiology of the synapse has been to identify the morphological and molecular correlates of the three key variables that characterize quantal neurotransmitter release, as already mentioned in the case of NMJ: the number of release sites N, the probability of a quantal release p, and the size of the quantal response q. At the frog NMJ, N and p are both large under natural conditions, which means that physiological concentrations of Ca^{2+} and Mg^{2+} are involved [36.38]. The squid giant synapse (SGS) has also been studied intensively, as the presynaptic terminal is exceptionally large and it is possible to insert several electrodes directly into this presynaptic terminal and simultaneously monitor the re-

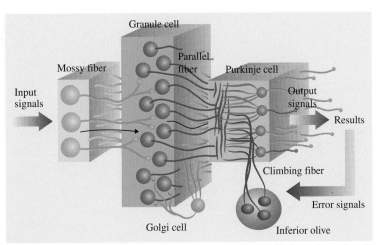

Fig. 36.10 Perceptron model of cerebellar cortex and vestibule-ocular reflex. The simple perceptron model is composed of three layers: mossy fiber layer, granule cell layer, and Purkinje cell layer. The synaptic efficacy between granule cell (parallel fiber) and the dendrite of Purkinje cell is modified according to the synaptic input from the neuron in the inferior olive whose axon is the climbing fiber. A pattern that should be learnt by the cortex is presented to the granule cell layer via the mossy fiber, and the output pattern from the perceptron is given by the signals of the Purkinje cell axon. If the output pattern is not correct, an error signal is conveyed by the climbing fiber in order to modify the synaptic strength between the parallel fiber and Purkinje cell dendrite. The model was first proposed by Marr, then modified by Albus, and finally confirmed experimentally by *Ito*. (Modified by Chloé Okuno after [36.54])

sponses from the postsynaptic neuron and the voltage of the presynaptic terminal. At this synapse (SGS), N and p are both very large to release sufficient neurotransmitter for quick movement of fins to escape from sudden danger. Furthermore, many types of synapses in the mammalian brain have also been investigated, such as the calyx of Held and climbing fiber–Purkinje cell synapses, which are similar to the frog NMJ in being high-N and high-p synapses.

On the other hand, many synapses receive a single synaptic contact from other cells, such as CA3 Schaffer collateral and CA1 pyramidal cells [36.55–57] in the hippocampus and granule cell axon (parallel fiber) and Purkinje cells in the cerebellum [36.58, 59].

To make it easy to understand the synaptic function with regard to its regulation and plasticity, we focus on the cerebellar Purkinje cell, with which other cerebellar neurons make synaptic contact. Figure 36.9 shows the neural circuit of the cerebellum cortex with the three typical neurons (Purkinje cell, granule cell, and Golgi cell) of all five neurons (the other two are stellite cell and basket cell). The Purkinje cell receives two major excitatory inputs from the climbing fiber and granule cell, whose axon is a parallel fiber. It is reported that the climbing fiber–Purkinje cell synapse is strong enough to generate action potentials (complex spikes) generated

by Ca^{2+} and Na^+ influx, whereas the parallel fiber–Purkinje cell synapse causes small but more frequent synaptic responses and also induces long-term depression (LTD), one of the significant aspects of synaptic plasticity. This LTD is suggested to be mediated by AMPA receptor channels but not by MNDA receptor channels, and to play a significant role in motor learning conducted by the cerebellum.

In the cerebellum, information is generally conveyed continuously by the action potentials propagating along the parallel fibers contacting to Purkinje cell. On the other hand, the occasional strong input through the climbing fiber to the Purkinje cell generates a strong prolonged Ca^{2+} depolarization (complex spike) in the dendrite. The Ca^{2+} entering the cell induces biochemical reactions that produce secondary messengers to change the state of the cell and, in some cases, generate slow synaptic potentials. This input from the climbing fiber to the Purkinje cell modifies the synaptic efficacy between parallel fiber and Purkinje cells, e.g., through LTD.

Figure 36.9a,b schematically represents the two functional hypotheses of the cerebellar cortex that have been proposed on the basis of available physiological and morphological results together with theoretical investigations. The English student David Marr (1945–

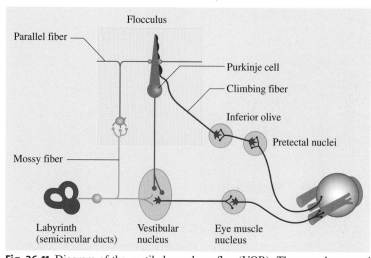

Fig. 36.11 Diagram of the vestibule-ocular reflex (VOR). The neural connection of the VOR in which eye movement is controlled by motoneurons innervating eye muscles in order to coincide with head movement (horizontal rotation) detected by the labyrinth in the ear. The signals detecting the head position from the labyrinth are sent to the vestibular nucleus, from where they are transmitted to a set of neurons controlling the eye muscles. At the same time, these signals are bypassed to a Purkinje cell through a parallel fiber. The synaptic strength between the parallel fiber and Purkinje cell is modified by occasional climbing fiber inputs that correspond to error signals or instructions from the retina. The instruction signals change the synaptic strength and induce plasticity (long-term potentiation (LTP)) that reduces the inhibition of Purkinje cell output to vestibular nucleus, resulting in a change in the signals from the labyrinth to the motoneurons. Then, eye movement is changed to coincide with the head movement by the modified signals to the muscle motoneurons. The neural circuit is thought to be located in the flocculus in the cerebellum. (Modified by Chloé Okuno after [36.54])

1980) at Cambridge University postulated a theoretical model focusing on the change in synaptic efficacy between the parallel fiber and Purkinje cell (long-term potentiation) and indicated that the function of the cerebellar cortex is equivalent to that of a simple perceptron that functions as a learning machine [36.60]. Figure 36.10 shows a schematic representation of a perceptron model composed of three cell layers (mossy fiber, granule cell layer, and Purkinje cell layer) of the cerebellar cortex. This hypothesis was later reevaluated and corrected (with the synaptic plasticity being not long-term potentiation but rather long-term depression) by *Albus* (1935–2011) [36.61]. Finally, a long-term change in synaptic efficacy was confirmed in rabbit cerebellum by *Ito*'s group based on in vivo electrophysiological experiments [36.62]. At present this is called the Marr–Albus–Ito theory of cerebellum.

LTD has been shown to be one of the most significant synaptic functions performed by the cerebellum,

and not only its physiological investigations but also its molecular biological mechanisms have been studied since its discovery in 1982. One of the most significant functional models postulated by Ito, who conducted elaborate experiments, is the vestibulo-ocular reflex (VOR) in the flocculus (Fig. 36.11) of the cerebellar cortex [36.62].

The cerebellar cortex is a unique anatomical structure composed of only five neurons with two excitatory inputs (mossy fiber and climbing fiber) and one inhibitory output (Purkinje cell axon to deep cerebellar nuclei). The neural circuit formed by these five neurons is conserved at any sagittal section of the whole cerebellar cortex, and thus various functional models have been proposed not only by physiologists but also theoretical neuroscientists, who have tried to investigate its function theoretically [36.63]. However, its detailed functional meaning remains elusive in spite of these intensive efforts.

36.4 Physiology of Synaptic Transmission (Postsynaptic Site)

The previous section discussed several presynaptic mechanisms involved in synaptic transmission, including the quantal nature of neurotransmitter release. In this section we deal with the postsynaptic processes of the synapse. First, the neuromuscular junction is introduced in order to understand the basic concept of postsynaptic processes, and then some aspects of the neuronal synapse in the central nervous system are discussed, focusing on the glutamatergic excitatory synapse.

36.4.1 Classical View of Synaptic Response

Synaptic transmission was first studied by electrical recording of synaptic potentials following the release of neurotransmitters due to action potential propagation to the terminal. The strength of the synaptic connection depends partly on the magnitude and time course of the synaptic potential, which is relevant to the conductance change of the postsynaptic neuron and the driving force for the synaptic current (36.8). Depending on the receptor ion, and the channel which is activated, these synaptic potentials are classified into two categories: the excitatory postsynaptic potential (EPSP), whose amplitude changes towards depolarization from the resting membrane potential (around -60 to -70 mV, RMP), and the inhibitory postsynaptic potential (IPSP), whose amplitude changes towards hyperpolarization. Although many significant features (the voltage dependence of the time course of the synaptic potentials, etc.) can be obtained from intracellular voltage recording of EPSPs and IPSPs, this recording technique is insufficient for observation of the conductance change during EPSP and IPSP, which has a direct relationship with synaptic inputs. Moreover, voltage-clamp recording gives the current change corresponding to the change in synaptic conductance.

One of the appropriate methods for analyzing synaptic input (current) and its information control mechanisms was and still is the voltage-clamp experiment using two microelectrodes. Furthermore, the patch-clamp technique, which is considered to be one of the voltage-clamp techniques and was developed in 1976 by *Sakmann* and *Neher*, is applicable to a variety of in vitro preparations such as acute slice preparations, cells in culture, and cloned ion channels (and receptors) expressed in oocyte membrane [36.64–68]. Before describing the results of experiments using the patch-clamp technique, experiments on neuromuscular junctions (NMJs) using the conventional two-electrode

voltage-clamp technique are presented as a typical example of the postsynaptic mechanism in order to describe the detailed mechanisms of excitatory synaptic function, on which experiments have been carried out by *Takeuchi* and *Takeuchi* [36.69, 70].

The end-plate potential is mediated by the ACh receptor in the membrane of the end-plate. Moreover, the end-plate potential that is generated at the neuromuscular junction was first and thoroughly examined and studied by *Fatt* and *Katz*, who concluded that the synaptic potential is generated by an inward ionic current confined to the end-plate region, which spreads passively into the muscle fiber away from the end-plate and also generates action potentials to initiate muscle contraction [36.16]. Here, the inward current corresponds to an influx of positive charge which depolarizes the inside of the membrane and generates the end-plate potential. This current that flow through the end-plate conductance composed of a number of the unitary currents of the ACh receptor channels has been investigated first by voltage-clamp experiments and then patch-clamp experiments since Takeuchi's first report. These experiment revealed that the end-plate current rises and decays more rapidly than the resultant end-plate potential.

The time course of the end-plate current that is an ensemble of many ACh receptor channels is directly determined by the properties of the ACh-gated receptor ion channels (36.10). Therefore, the end-plate potential exhibits a time lag behind the end-plate potential, as it takes time for an ionic current to charge or discharge the membrane capacitance around the end-plate. Figure 36.12a shows typical experimental results using the conventional voltage-clamp technique applied to the neuromuscular junction; here, the membrane at the end-plate is voltage-clamped using two electrodes. One electrode is for measurement of the voltage V_m, and the second electrode is for passing a current I_m into the end-plate. Both electrodes are connected to a voltage-clamp amplifier. The end-plate potential (upper trace of Fig. 36.12b) is measured when the voltage-clamp mode is not working, showing a relatively slow rise (≈ 1 ms) and slow decay (≈ 6 ms), whereas the end-plate current is measured under voltage-clamp conditions. The end-plate current rises and decays more rapidly than the end-plate potential, as shown in the lower trace of Fig. 36.12c. This rapid time course of the end-plate current is directly related to the change in the end-plate conductance due to the opening and closing of a number of ACh receptor channels, which is a typi-

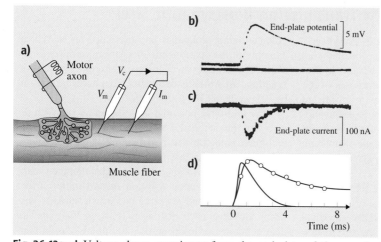

Fig. 36.12a–d Voltage-clamp experiment from the end-plate of the muscle fiber. (**a**) Experimental setup of a two-microelectrode voltage-clamp experiment. One electrode measures the membrane voltage V_m, and the second electrode passes the current I_m into the end-plate. The outputs of both electrodes are connected to an amplifier, which injects the current through the current electrode in order to hold the membrane V_m at the value V_c (holding potential). The motor axon is stimulated to evoke the synaptic current, which is measured at constant holding potential (after [36.75]) (**b**), (**c**) End-plate potential and end-plate current. The end-plate potential is measured when the voltage clamp is not functioning and it rises and decays slowly, whereas the synaptic current is recorded under the voltage-clamp conditions and its time course is faster in accordance with the conductance change. (**d**) Superimposed traces of the end-plate potential and the end-plate current. The end-plate potential is calculated from the end-plate current under the assumption that the membrane time constant is 25 ms. (Modified by Chloé Okuno after [36.70])

cal receptor channel whose molecular structure (amino acid sequence and subunit stoichiometry) has bee elucidated by both the patch-clamp technique and molecular biological methods during the last 20 years [36.71–74].

Takeuchi and *Takeuchi* also examined the relationship between the end-plate current amplitude and the holding potential of voltage-clamp recording, thus obtaining the simple (36.8). Equation (36.8) is similar to the expression for the ionic currents obtained by *Hodgkin* and *Huxley* [36.76]. However, the reversal potential of the end-plate current is equal to 0 mV, even though the reversal potential of Na^+ is about +55 mV. Thus, Takeuchi and Takeuchi in turn adjusted the concentration of the ionic compositions both outside and inside the membrane, simultaneously measuring the reversal potential for each concentration pair. Based on this experiment, they succeeded in elucidating the involvement of both Na^+ and K^+ ionic currents in the end-plate current. The detailed results are described in Sect. 36.4.2.

The conductance change is faster for the glutamate, gamma-aminobutyric acid (GABA), as well as the ACh receptor and is determined by many factors such as the diffusion speed of the transmitter, the binding rate of the

agonists to the receptor, and the removal of transmitters from the cleft. It is quite complicated to express the time course of the conductance change due to these factors. However, it is possible to describe the conductance change using the simple mathematical formula

$$g_{syn} = g_{syn}(\alpha^2 t)\exp(-\alpha t), \qquad (36.7)$$

where g_{syn} is the area under the curve. When $t = 0$, $g_{syn} = 0$; when $t = (1/\alpha)$, g_{syn} reaches its peak value.

The experimental approach for studying ACh receptor ion channels provides a typical example of investigations of ligand-gated ion channels. During the development of research on the ACh receptor, various techniques were also invented together.

36.4.2 A Ch–Gated Ion Channel Is Permeable to Both Sodium and Potassium

Here, we describe in more detail the voltage-clamp experiment that was introduced briefly in Sect. 36.4.1. The current flow (lower trace of Fig. 36.12c) through the membrane conductance under the voltage-clamp condition is obtained as the product of the membrane conductance and the electrochemical driving force on ions conducted through the channels. The end-plate current that underlies the excitatory postsynaptic potential is given as

$$I_{EPSC} = g_{EPSP}(V_m - E_{EPSP}),\qquad(36.8)$$

where I_{EPSC} is the end-plate current, g_{EPSP} is the conductance of the ACh-gated channels, V_m is the membrane potential, E_{EPSP} is the reversal potential and $(V_m - E_{EPSP})$ is chemical driving force produced from the conductance gradients of the ions passing through the ACh-gated channels. From (36.8), we can change the value of V_m in a step-by-step manner (for example, in 10 mV intervals) in a voltage-clamp experiment and observe the effect on the membrane current I_{EPSC}, which also changes its amplitude from negative to positive. If the membrane potential is held equal to the value of E_{EPSP} (the reversal potential), the net synaptic current could be zero because the electrochemical driving force balances the membrane potential.

The potential at which the net synaptic current is null is called the reversal potential of this type of ACh-gated ion channel. From Takeuchi's frog muscle experiments, the end-plate current actually reverses at 0 mV, because the ion channel is permeable to both sodium and potassium ions, which are able to move into and out of the cell simultaneously. The net synaptic current is the sum of the sodium and potassium ion fluxes through the ACh-gated channels. The diameter of the ionic pore of this channel is thought to be larger than that of voltage-gated channels such as sodium and potassium channels that are permeable to sodium and potassium ions, respectively, for the generation of the action potential. Figure 36.13 shows current traces for sodium ions (Na^+) only (Fig. 36.13a) and the total end-plate current (Fig. 36.13b) under voltage-clamp conditions. The upward and downward arrows for Na^+ and K^+ indicate the direction of each current that flows out of the end-plate and into the end-plate, respectively.

By introducing patch-clamp experiments for the measurement of the ACh receptor channel, we can observe the single-channel current that passes through the ACh-gated ion channel molecule. For the first time, *Neher* and *Sakmann* measured the current flow through this channel in a denervated muscle fiber and showed that the channels open and close in a step-like manner, generating a very small rectangular current whose duration and amplitude are approximately on the order of ms and pA, respectively [36.77–79]. This was the first observation of a current flow (functional factor) having a direct relationship with the conformational change

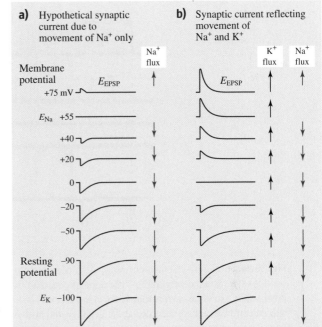

Fig. 36.13a,b End-plate current is carried by flow of both Na^+ and K^+ through the ACh receptor channel. (**a**) The synaptic current is measured at different holding potentials of the voltage-clamp experiment. When the Na^+ current alone is responsible for the membrane current, the reversal potential is equal to the equilibrium potential (+55 mV) for Na^+ calculated from the Nernst equation. The *arrows at the right* of each current indicate the direction of Na^+ flux at each holding potential. (**b**) When both Na^+ and K^+ flow through the ACh receptor channel, the total end-plate current is measured at each holding potential and the reversal potential is equal to 0 mV. The direction of each Na^+ and K^+ current flux is indicated by *arrows in the right column*. The net current is zero at the reversal potential. (Modified by Chloé Okuno after [36.75])

of a protein (structural change) at the molecular scale. The unitary current amplitude varies depending on the membrane potentials due to the electrochemical driving force generated by the membrane potential V_m and the reversal potential E_{EPSP}, as in the case of synaptic current in (36.8). A similar equation is given for the single-channel current of the ACh-gated channel:

$$i_{EPSC} = \gamma_{EPSP} \times (V_m - E_{EPSP}),\qquad(36.9)$$

where i_{EPSC} is the amplitude of the current flow through one channel and γ_{EPSP} is the conductance of a single channel. In (36.9), the single-channel conductance is a constant value which could be determined from the channel properties resulting from the protein struc-

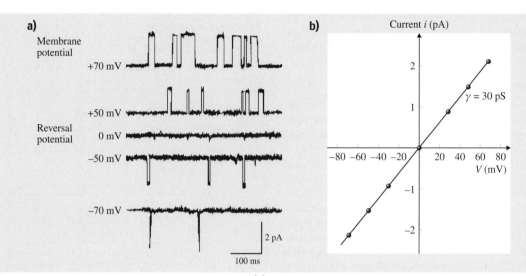

Fig. 36.14a,b Single-channel current of ACh receptor. (**a**) Single-channel currents of the ACh receptor recorded from outside of the patch configuration. The agonist concentration is $2\,\mu M$ which is added to the bath solution. At the holding potential of $0\,mV$ the current that flows in the channel is zero and the reversal potential is equal to $0\,mV$. Inward current and outward current are recorded at the negative and positive holding potential, respectively. (**b**) The membrane potential versus single-channel current is shown. The linear relationship indicates that the channel is a simple resistor with single-channel conductance γ of $30\,pS$. (Modified by Chloé Okuno after [36.75])

ture and subunit stoichiometry [36.74]. The duration of channel opening varies stochastically, and the amplitude of the single-channel current depends on the membrane potential; the linear relation between i_{EPSC} and V_m behaves as a simple resistor with constant conductance called the *single-channel conductance*. Figure 36.14a shows traces of the single-channel current of the ACh receptor channel under various holding potentials. The voltage across a patch of membranes was varied systematically during exposure to $2\,\mu M$ ACh. The current recorded in this experiment is inward at voltages negative to $0\,mV$ and outward at voltages positive to $0\,mV$, which gives the reversal potential. Figure 36.14b shows the relationship between the amplitude of the single-channel current and the holding potential. This linear relation shows that the channel behaves as a simple resistor with conductance of $30\,pS$. The opening and closing occurs randomly and abruptly, and nondeterministic and stochastic analysis should be used to describe such experimental data. From the single-channel recording, three parameters (the single-channel current amplitude, open time, and shut time) are obtained so that the kinetics of open–close gating of the ion channel can be obtained, providing a precise quantitative description of electrical signals in the brain and forming a bridge between brain function and a mo-

lecular biological phenomenon. This is one of the most prominent examples of quantitative expression of biological and physiological phenomena, which are generally thought to be qualitative sciences, and a typical achievement from the viewpoint of reductionists. Readers interested in the fields of single-channel kinetics and its mathematical treatment should consult [36.80]. In Sect. 36.4.3, we show how the macroscopic electrical signal (in the present case, the synaptic current) is built up based on the microscopic parameters obtained from the single-channel recording technique.

Moreover, experiments with other ligand-gated receptor channels (glutamate, GABA, etc.) have been conducted with the use of the patch-clamp technique applied to various (mainly in vitro) preparations, together with molecular biological experiments. The results obtained from these experiments are basically the same as those obtained with the ACh receptor mentioned above in this section and as presented in the next section.

36.4.3 Reconstruction of the Synaptic Current from the Single Channels

As described above, stimulation of the presynaptic fiber releases a large number of ACh molecules into the synaptic cleft, and these molecules diffuse within the

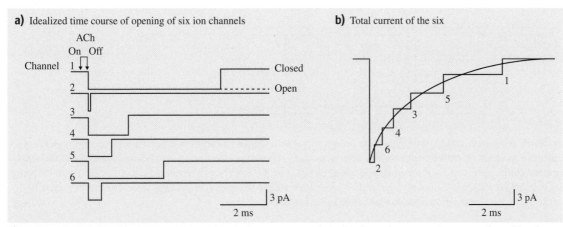

a) Idealized time course of opening of six ion channels

b) Total current of the six

Fig. 36.15a,b Relationship between the total end-plate current and single-channel currents that are activated by the release of ACh molecules. **(a)** Simulated behavior of six individual single channels that were opened at the same time when the acetylcholine concentration had decreased to zero. Opening is plotted downward. The channels stay open for a random length of time. **(b)** Sum of the six records. The total number of open channels decays exponentially with a time constant that is equal to the mean open time of the channels. (Modified by Chloé Okuno after [36.77])

cleft and bind to the ACh receptors in the postsynaptic membrane, causing more than 200 000 ACh receptor channels to open almost simultaneously. Such a rapid and large rise of the ACh concentration upon stimulation of the nerve terminal causes a large increase in the total conductance of the end-plate membrane g_{EPSP}, generating the rapid rise in the end-plate current as expressed in (36.8).

The concentration of ACh molecules in the synaptic cleft decreases to zero in less than 1 ms due to the action of an enzyme (ACh esterase, AChE) that cleaves the ACh molecule, and after this decrease of the ACh concentration, the channels begin to close in a stochastic manner as shown in Fig. 36.14. In Fig. 36.15a, the sum of the total number of channels (in this case, six) results in a decay curve that can be fitted by a single exponential function (Fig. 36.15b) contaminated with small, step-like changes (fluctuations). This corresponds to the decay phase of the end-plate current in which the open–close gating kinetics of the thousands of ACh receptor channels participates. The time constant of this single exponential function is equal to the mean open time of a single ACh receptor channel. The open–close kinetics of the individual channels is not visible, because the total end-plate current is so large that each of the single-channel current amplitudes is masked by the background noise.

When multiple populations of AChRs participate in the synaptic current or when the AChR has multiple conductance states, the decay phase of the synaptic current could be fitted by an exponential function with multiple components; For example, the decay phase of the synaptic current recorded from embryonic muscle fiber is slower than that of adult muscle fiber due to the replacement of the δ-subunit by the ε-subunit during development. Therefore, in the middle of development, the decay phase of the end-plate current can be fitted by a double exponential function with both fast and slow time constants.

Under these assumptions, the summed conductance of all open channels in a large population of ACh channels in the postsynaptic membrane is the total synaptic conductance, $g_{EPSP} = n\gamma$, where n is the average number of channels opened by ACh and γ is the conductance of a single ACh receptor channel. If a large number of channels (N) are located in the postsynaptic membrane and the probability that the channel is open (p_o) depends on the concentration of ACh molecules, the total end-plate current is given by

$$I_{EPSC} = Np_0\gamma(V_m - E_{EPSP}) \qquad (36.10)$$

or

$$I_{EPSC} = n\gamma(V_m - E_{EPSP}). \qquad (36.11)$$

Equations (36.10) and (36.11) indicate that the total end-plate current depends on four important factors:

1. The total number of ACh receptor channels N
2. The probability that a channel is open (p_o)
3. The conductance of each open channel (γ)
4. The driving force that acts on ions ($V_m - E_{EPSP}$).

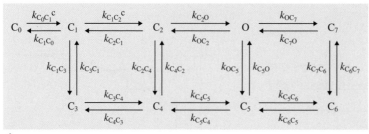

Fig. 36.16 Nine-state AMPA-type glutamate-receptor channel kinetics model of cerebellar Purkinje cells. AMPA-type glutamate receptor channel kinetics model with nine states, where C_0 is the unliganded closed state, C_1 is the singly liganded closed state, and C_2 is the doubly liganded closed state. C_3 to C_7 are desensitized (closed) states, with C_3 being singly liganded and C_4 to C_7 doubly liganded; O is the doubly liganded open (conducting) state. The rate constants $k_{C_0C_1}$ and $k_{C_1C_2}$ depend on the glutamate concentration (after [36.83])

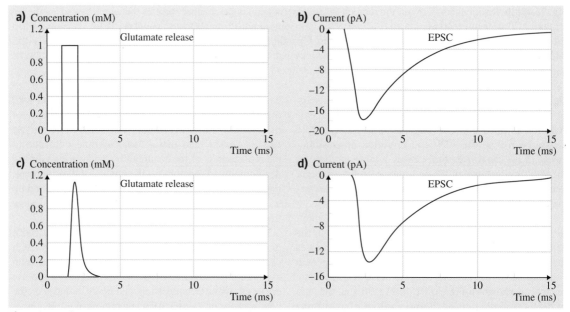

Fig. 36.17a–d Excitatory postsynaptic current (EPSC) waveforms calculated from the AMPA-receptor channel kinetics model in which the profiles of the concentration of glutamate molecules in the synaptic cleft are a step function and a complicated function. (**a**) The profile of the glutamate concentration in the synaptic cleft given by a step function with 1 ms duration and 1 mM amplitude. (**b**) Calculated waveform of the EPSC based on the AMPA/kainite kinetics model of Fig. 36.14 with the glutamate concentration in the synaptic cleft given in (**a**). (**c**) Profile of glutamate concentration in the synaptic cleft given by a model with detailed parameters in which the concentration of glutamate in a vesicle is 0.1 mM, the maximum concentration of binding factor is 0.001 mM, 9000 glutamates are involved in a vesicle, etc. (after *Destexhe* [36.84]). (**d**) Calculated waveform of the EPSC based on the AMPA/kainite kinetics model of Fig. 36.14 with the glutamate concentration in the synaptic cleft given in (**c**). The difference between the EPSC waveforms of (**b**) and (**d**) is not significant. (Modified by Chloé Okuno after [36.85])

When the experiment is conducted under voltage-clamp conditions, the holding potential V_m, γ, and the reversal potential E_{EPSP} are all time-independent constants, whereas n depends on time. If n is given as a function of time t, we can calculate I_{EPSC} as a function of time. n at time t can be estimated by calculation of the number in the open state at time t in a possible kinetics model that represents the reaction between the agonist

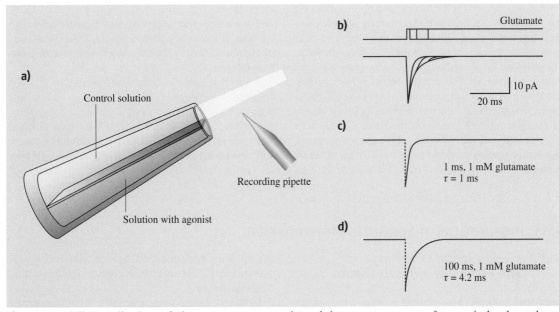

Fig. 36.18a–d Fast applications of glutamate receptor agonist and the current responses from an isolated membrane patch. (**a**) Schematic drawing of the double-barreled application pipette made from theta glass tubing used for the fast agonist application system. Two glass pipettes are continuously perfused with control solution and test solution with agonist (glutamate). The interface between the two solutions (one with and another without agonist) can be shifted across the isolated patch by a piezoelectric element to which the application pipette is attached (after [36.86]). (**b**) Superimposed current responses evoked by 1, 5, 10, and 100 ms of application of 1 mM glutamate (after [36.87]). (**c,d**) Deactivation (1 ms pulse) and desensitization (100 ms pulse) fitted by a single exponential function with time constant τ of 1.0 and 4.2 ms, respectively. (Modified by Chloé Okuno after [36.87])

(in this case, ACh) and the receptor as obtained from electrophysiological experiments. At present, a number of such kinetic models have been presented by many groups [36.81, 82].

Reconstruction of the excitatory postsynaptic current (EPSC) based on (36.10) or (36.11) and the kinetics models of the glutamate receptor channel, one of the ligand-gated receptor channels in the central nervous system, has been conducted by computer simulation. The many kinetics models that have already been proposed [36.45, 83, 87, 88] include, for example, a nine-state model for the glutamate receptor channel mediating the synaptic potential between parallel fiber and Purkinje cell in the cerebellar cortex of rat with one open state, three closed states, and five desensitized states, between which the forward and backward transition rate constants have been calculated experimentally [36.45] (Fig. 36.16). Here, the change in the agonist concentration in the synaptic cleft over time, represented as an α-function, which has a linear rise and an exponential decay with time constant τ, has been ex-

perimentally obtained [36.84, 89]. However, this time profile of the concentration of agonist in the synaptic cleft is substituted by the step (rectangular) function with duration of 1 ms and amplitude of 1 mM, because no difference in the calculated EPSC waveforms between using the α-function profile and the step (rectangular) function was observed, and the step function profile was adopted for simplicity of the calculation (Fig. 36.17).

These reconstructed EPSC waveforms are, for example, applicable to experiments of synaptic plasticity or the short-term change in synaptic strength efficacy attributed to a postsynaptic origin to estimate the possible change in the parameters (rate constants of the kinetics model) that could affect and/or be responsible for the changes in the EPSC waveform during these changes in synaptic efficacy [36.85], and also for the simulation of fast drug application experiments, etc.

Figure 36.18a indicates the experimental design for fast application of agonist to isolated patches, in which a double-barreled application pipette made from theta glass tubing is used for the fast agonist application system. The dual-channel theta glass tubing provides two glass pipettes that are continuously perfused with control solution and test solution with agonist (glutamate). The interface between the two solutions (one with and another without agonist) can be shifted across the isolated patch by using a piezoelectric element to which the application pipette is attached. Figure 36.18b–d shows an example of AMPA-receptor-mediated current activated by exposure of the outside of the membrane patch to glutamate pulses of 1 mM concentration for different durations (unit ms). The 20–80% rise time of the current is about $300\,\mu s$. The peak currents evoked by 1, 10, and 100 ms pulses are superimposed, but the decay time constants are clearly different (deactivation time constant about 2.5 ms, desensitization time constant about 10 ms) [36.86, 88].

The experimental results shown in Fig. 36.18 were simulated by computer calculations that can reconstruct the same current responses under the assumption that the kinetics model of the AMPA receptor channel shown in Fig. 36.16 is involved in the membrane patch. The calculated results (data not shown) match well with the experimental results (Fig. 36.18).

36.5 Integration of Synaptic Transmission

36.5.1 Integration of Excitatory and Inhibitory Synaptic Inputs

Compared with the neuromuscular junction, synaptic transmission in the central nervous system is more complex for several reasons:

1. Central neurons receive many inputs from other neurons, whereas muscle fiber receives only one fiber from a motoneuron in the spinal cord.
2. Muscle fiber receives only excitatory synaptic input, while a central neuron receives both excitatory and inhibitory inputs.
3. Muscle fiber receives only excitatory synaptic inputs mediated by the ACh receptor, and no other neurotransmitters are used for synaptic transmission. However, in the central nervous system, a variety of neurotransmitters are used to transmit and regulate information between neurons and also to change the physiological state of a single neuron through various inputs. These inputs activate not only excitatory receptors such as glutamate and ACh receptor but also inhibitory synaptic receptors, as well as metabotropic receptors which produce secondary messengers inducing changes in the cell's internal environment through a variety of biochemical reactions and as a result, in some cases, partly generate slow electric responses.

Therefore, at the first step of the sequential information processes, a single neuron integrates and converts diverse inputs from many neurons into one coordinated response. These inputs impinging onto a single neuron can be categorized into three major groups according to their physiological effect:

1. Excitatory inputs mediated by receptors (which generate EPSPs) activated by excitatory agonists (ACh, glutamate, etc.),
2. Inhibitory inputs (which generate IPSPs) mediated by receptors activated by other types of (inhibitory) agonists (GABA, glycine, etc.),
3. Slow synaptic responses generated indirectly by secondary messengers activated by metabotropic receptors (mGluR, $GABA_B$, etc.).

In addition to the variety of synaptic inputs that contribute to the information transfer process between neurons, the membrane (passive and active) properties of neurons are also significant factors to be considered in the integration of inputs by a single neuron.

Firstly, the passive membrane property of the neuron is affected in principle by two factors:

1. The time constant, which is determined by the electrical properties of the membrane, including the channel kinetics, determines the time course of the synaptic potential and thereby controls the temporal summation, the process by which consecutive synaptic potentials at the same location on the neuron are summed in the postsynaptic cell. Moreover, the membrane time constant also affects the *filtering effect* of a neuron, especially dendritic trees, whose membrane properties and morphology serve as a *low-pass filter* during the transmission of high-frequency components of electrical signals that are applied to the distal part of a dendrite.
2. The length constant, which is generally calculated by the equivalent cable equation of dendritic trees, determines the degree to which a depolarization cur-

rent decrease as it spreads passively [36.90–92]. The electrical signals spread in an electrotonic manner to the trigger zone with minimal decrease when the length constant is larger, whereas when the length constant is small, the electrical signals decay rapidly. The membrane potential at the trigger zone of a neuron receiving several inputs could reach the threshold and result in action potential generation. In general, the density of Na^+ channels is higher in the membrane of the trigger zone than in other parts of the neuron. This process is called spatial summation of a variety of synaptic inputs by a neuron on which many inputs impinge.

One of the major excitatory neurotransmitters in the mammalian central nervous system responsible for excitatory input is glutamate, which causes opening of glutamate-gated channels permeable to both Na^+ and K^+ (Ca^{2+} for NMDA receptor) with nearly equal permeability, as observed for the ACh receptor; therefore, the reversal potential of the glutamate receptor is about $0\,mV$ (Fig. 36.12) [36.93–95]. Furthermore, the glutamate receptor is divided into two classes: the metabotropic glutamate receptor (mGluR) and the ionotropic glutamate receptor, which is also pharmacologically categorized into three major groups: AMPA, kainate, and NMDA receptors or simply into two groups: non-NMDA (AMPA and kainate) and MNDA receptors [36.96–98]. The non-NMDA and MNDA receptors are pharmacologically blocked by CNQX and APV, respectively. The molecular stoichiometry and characterization of the single-channel currents of these receptors have been also elucidated by many groups [36.99–101].

The AMPA receptor channel contributes to the fast synaptic transmission that is regulated and modulated by a variety of proteins that induce receptor trafficking, phosphorylation, recycling (exocytosis and endocytosis of the subunits of the receptors), etc. [36.102–105]. The changes in efficacy and strength of the synapses by these modifications are thought to be a substrate and underlie the formation of memory, learning, and other behaviors relevant to synaptic functions [36.106]. The properties (kinetics and single-channel conductance) of AMPA receptor single-channel currents have been investigated using advanced experimental techniques since the development of the patch-clamp method. Historically, nonstationary noise analysis of AChR from frog neuromuscular junction was first introduced by *Katz* [36.37]. Later, *Anderson* and *Stevens* extended and generalized this work to estimate the single-channel

conductance and mean open lifetime of the channel by fitting of power spectra [calculated by fast Fourier transformation (FFT)] using a Lorentzian function [36.107]. Furthermore, nonstationary noise analysis was also developed at first to estimate the single-channel conductance of voltage-gated Na^+ channels of the node of Ranvier [36.108, 109]. This interesting analytical method was applied to the decay phase of synaptic currents to obtain the single-channel conductance in the postsynaptic membrane, where placement of a recording pipette for direct recording of the single-channel current is impossible because it is covered by a presynaptic button. With this method, detailed analysis of postsynaptic mechanisms could be carried out, for example, the change in single-channel conductance during synaptic plasticity or development, etc. [36.97, 110–112].

The postsynaptic sites of most neurons have both NMDA and non-NMDA glutamate receptors as excitatory receptors. (No NMDA receptor is observed in adult cerebellar Purkinje cells.) However, the NMDA receptor channel does not contribute to the fast EPSP because it is blocked by Mg^{2+}, which remains at the inside of the pore of the channel at the resting membrane potential. Thus, the waveform of the EPSP at the resting level depends largely on the activation of the non-NMDA (mainly AMPA) receptor channel. When the membrane potential increases by depolarization, the Mg^{2+} is driven out of the pore, and the current that is carried by Na^+, K^+, and Ca^{2+} flows through the NMDA channel pore. The properties of NMDA channels show that its open–close gating is relatively slower than that of the non-NMDA channel in response to glutamate and that it contributes to the late phase of EPSP [36.113]. This late phase of EPSP is small when a single action potential is reached to the presynaptic terminal. However, when a presynaptic axon fires repeatedly, the EPSPs are summed together (*temporal summation*) so that the postsynaptic neuron is depolarized enough for a physiologically important amount of Ca^{2+} to enter the neuron. This entry of Ca^{2+} triggers biochemical reactions and signal transduction pathways that contribute to, in the present case, modifications of the synaptic function that are thought to play a significant role in memory and learning and other physiological mechanisms in which Ca^{2+} is involved. The NMDA receptor simultaneously detects the excitations of both the presynaptic and postsynaptic neuron, where a necessary condition for synaptic plasticity is Hebb's rule. Here, the NMDA receptor works as a molecular switch to induce long-term plasticity as observed in some re-

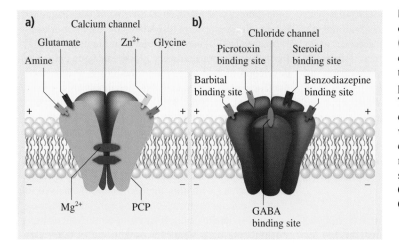

Fig. 36.19a,b Receptor ion channel of excitatory and inhibitory synapse. (**a**) MNDA-type glutamate receptor channel with binding sites for glutamate, glycine, Zn^{2+}, amine, and phenylcyclohexylpiperidine (PCP). The Mg^{2+} ion blocks the pore of the channel. Ca^{2+} is permeable together with Na^+ and K^+. (**b**) $GABA_A$ receptor channel with binding sites for muscimol, bicuculline, picrotoxin, steroid, benzodiazepine, and barbital. Cl^- is permeable. (Modified by Chloé Okuno after [36.114])

gions of the central nervous system [36.106]. A typical schematic representation of a NMDA receptor channel with several binding sites for other co-ligands (among which glycine is typical) is shown in Fig. 36.19a.

The other major inputs to the central neurons are mediated by neurotransmitter, GABA, and glycine [36.35]. These inputs are inhibitory synaptic inputs (IPSPs), which hyperpolarize the membrane potential of the neurons, and together with excitatory inputs determine the membrane potential of the neurons and the (temporally) summed potentials that passively spread to the trigger zone of the neurons.

GABA is a major inhibitory transmitter in the spinal cord and the central nervous system that operates two receptors: ionotropic $GABA_A$ that gates the Cl^- chan-

nel (Fig. 36.19b), and metabotropic $GABA_B$ receptors that couple to G-proteins and increase the K^+ permeability to inhibit voltage-gated Ca^{2+} channels in the neuron. The function of $GABA_A$ is controlled by three major classes of drugs: benzodiazepines, barbiturates, and alcohol, whose effects are to bind to portions of $GABA_A$ receptors in order to increase Cl^- flow through the channels in response to GABA. Picrotoxin (channel blocker) and bicuculline (competitor) are typical blockers of the $GABA_A$ receptor, and strychnine (competitor) is a blocker of the glycine receptor (Fig. 36.19b).

Figure 36.20 indicates the inhibitory action at the chemical synapse of the IPSP and inhibitory postsynaptic current (IPSC) when voltage-clamped at various holding potentials (shown to the left of each trace). The IPSC time course is similar to the change in the in-

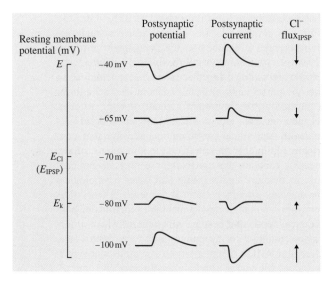

Fig. 36.20 Inhibitory postsynaptic potential (IPSP) and inhibitory postsynaptic current (IPSC) are reversed at $-70\,\mathrm{mV}$. IPSP is evoked as a small hyperpolarization potential at the resting membrane potential ($-65\,\mathrm{mV}$). The more depolarized the membrane potential, the more hyperpolarized the IPSP, whereas when the membrane potential is hyperpolarized, the IPSP is depolarized. The reversal potential ($-70\,\mathrm{mV}$) of the IPSP is equal to the equilibrium Cl^- potential (E_{Cl}) calculated from the Nernst equation. The IPSC is recorded by voltage-clamp measurements. The inward current corresponding to Cl^- outward flux (movement) is observed at the more negative holding potential, whereas the outward current corresponding to Cl^- inward flux (movement) is observed at the more positive holding potential. At the reversal potential (E_{Cl}), the postsynaptic current is zero. (Modified by Chloé Okuno after [36.75]) ◂

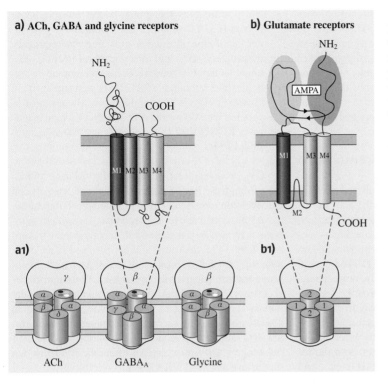

a) ACh, GABA and glycine receptors

b) Glutamate receptors

Fig. 36.21a,b Two typical families of ligand-gated ion channels. (**a**) Nicotinic ACh, GABA, and glycine receptor channels are pentamers composed of several types of subunits whose structure is formed of four transmembrane domains (M1–M4). The M2 domain of each subunit forms a channel, which are gathered together to form the pore of the channel. N- and C-terminal are located outside the cell. (**b**) Glutamate receptor channel is a tetramer composed of two different types of subunits. Each subunit has three transmembrane subunits (M1–M3), and the N-terminal is outside the cell whereas the C-terminal is inside the cell. (Modified by Chloé Okuno after [36.75])

hibitory synaptic conductance indicated in the EPSC case. At the resting membrane potential ($-65\,\text{mV}$), an action potential arriving at the presynaptic terminal produces a hyperpolarizing IPSP, which increases in amplitude when the membrane potential is depolarized to $-40\,\text{mV}$. On the other hand, when the membrane potential is hyperpolarized to $-70\,\text{mV}$, the amplitude of the IPSP becomes zero. This is the reversal potential of the IPSP that is equal to the equilibrium potential (calculated by the Nernst equation) of Cl^-, as the extracellular concentration of Cl^- is higher than its intracellular concentration. On further hyperpolarization of the holding potential at $-80\,\text{mV}$, the IPSP is inverted to a depolarizing postsynaptic potential. These voltage-clamp experiment results show that an inward current flows for membrane potentials more negative than the reversal potential, whereas outward current flows for membrane potentials that are positive with respect to the reversal potential, corresponding to an influx of Cl^-.

As in the case of the macroscopic current (IPSC), the single-channel currents through $GABA_A$ and glycine receptor channels have been measured using the patch-clamp technique. The reversal potential of the single-channel current is -60 to $-70\,\text{mV}$, a value similar to that obtained from whole-cell current recording

and the single-channel conductance of a $GABA_A$-gated channel and a glycine-activated channel, although depending on the preparations and compositions of the internal and external solutions, are $30\,\text{pS}$ and $46\,\text{pS}$, respectively.

The genes coding for the ionotropic glutamate, $GABA_A$, and glycine receptors have been cloned, and the GABA and glycine receptors are structurally similar to the ACh receptor channel. These three receptor channels are composed of five subunits. The GABA receptor channel is a pentamer and has stoichiometry of two α-, two β-, and one γ-subunit; these subunits are closely related to one another, so that the agonist GABA can bind to any of these subunits. The glycine receptor channel is also a pentamer, being composed of two β- and three α-subunits that are occupied with glycine. Figure 36.21 indicates the structures of these receptor channels schematically; each subunit has four transmembrane domains (M1–M4), and the M2 domain lines the channel pore. Detailed electron- and light-microscopic studies have shown that NMDA receptors (NMDARs) and AMPA receptors (AMPARs) are located in the postsynaptic density at the head of the spine, whereas inhibitory synapses typically form on the soma and dendritic shafts.

In the central nervous system, certain fast excitatory actions are mediated by the 5-HT$_3$ class of ligand-gated channels that are permeable to monovalent cations (mainly Na$^+$) and display a reversal potential of around 0 mV. Another family of transmitter-gated ion channels is referred to as receptors of adenosine triphosphate (ATP), serving as a neurotransmitter in some areas such as smooth muscle cells that are innervated by sympathetic ganglion cells and other neurons.

36.5.2 Modulation of Synaptic Transmission

It has been reported that there are a variety of types of metabotropic receptors which mainly activate G-protein (guanine nucleotide-binding protein) composed of three subunits (α-, β-, and γ-subunits), acting as secondary messengers in the cell. G-protein switches on and off when guanosine triphosphate (GTP) and guanosine diphosphate (GDP) bind to G-protein, respectively. The secondary messenger proteins indirectly regulate the activation of ion channels located away from the receptors and generate slow responses such as the slow EPSP or slow IPSP. As described briefly in an earlier section, the possible agonists for the metabotropic receptors are ACh, glutamate, GABA, 5-HT, etc., and their detailed mechanisms are described in other references [36.115]. About 1000 G-proteins are reported in the genome, and there are 7 transmembrane proteins. These are classified into three types depending on their roles in the signaling mechanism:

1. G$_s$ activates the cyclic adenosine monophosphate (cAMP)-dependent pathway by increasing the concentration of adenylate cyclase (AC) (the membrane-associated enzyme), as a result stimulating production of cAMP from ATP. cAMP acts as a secondary messenger that goes on to interact with and activate protein kinase A (PKA). PKA can then phosphorylate myriad downstream targets.
2. G$_{i/o}$ inhibits production of cAMP from ATP by decreasing the concentration of adenylate cyclase (AC) and then inhibits the activation of PKA. The activity of G$_{i/o}$ is blocked by pertussis toxin. The $\beta\gamma$-subunit (G$_{\beta\gamma}$) activates the K$^+$ channel (K$_V$3) that generates a slow hyperpolarization potential (slow IPSP) and inactivates the P/Q-type Ca^{2+} channel (Ca$_V$2.1) and N-type Ca^{2+} channel (Ca$_V$2.2). The muscarinic M$_{2,4}$ receptor, GABA$_B$ receptor, mGliR2-4, 6-8, adrenaline receptor, dopamine receptor, and serotonin receptor (5-HT$_{1A,B,D}$) are coupled with G$_{i/o}$ [36.116].
3. G$_{q/11}$ stimulates membrane-bound phospholipase C-beta (C$_\beta$PLC$_\beta$), which then cleaves PIP$_2$ (a minor membrane phosphoinositol) into two secondary messengers: IP$_3$ and diacylglycerol (DAG). IP$_3$ binds to the IP$_3$ receptor in the membrane of endoplasmic reticulum (ER). DAG activates protein kinase C (PKC). The muscarinic M$_1$ receptor is coupled with G$_{q/11}$ and activates PKC, which inactivates KCNQ/K$_V$7 (K$^+$ channel) and then generates slow EPSP.

G$_{12/13}$ is involved in Rho family GTPase signaling (through the RhoGEF superfamily) and controls cell cytoskeleton remodeling, thus regulating cell migration.

As described above, the secondary messenger pathways activated by a metabotropic receptor share a common logic. The neurotransmitters released from other neurons arrive at receptors in the postsynaptic membrane and activate transducer proteins. These transducer proteins activate primary enzymes that produce secondary messengers such as cAMP, IP$_3$, DAG, and in some cases gaseous secondary messengers (nitric oxide (NO) and carbon monoxide (CO)). These secondary messengers activate secondary effectors or directly act on the targets (PKC, PKA, etc.).

36.5.3 Integration by Active Membrane Property

Various receptor channels in each synapse contribute to the integration processes of a single neuron and their characteristics. The integration processes partly depend on the properties of the membrane in terms of whether they are passive, in which case electrical signals spread electrotonically, or active due to voltage-gated channels in the dendritic and somatic membrane.

As described in Sect. 36.5.1, the temporal and spatial summation of many synaptic inputs depends on the passive electrical properties of the membrane of neurons. However, many kinds of ion channels (not only voltage-gated ion channels such as Na$^+$, K$^+$, and Ca^{2+} but also ligand-gated ion channels) with a variety of gating properties are present in the membrane of dendrites and soma. These channels that collaborate with each other serve as local trigger regions and also amplify the small synaptic and/or slow potentials in order to transmit these signals to the initiation zone of the action potential without decrement [36.117, 118]. Although these ion channels in the dendritic membrane produce afferent signals that are conveyed from the distal portion of the dendrite to the soma, they are also used for generation of efferent signals called *back-propagating action potentials*, which travel from the soma to the dendrite. In pyramidal cells of certain areas of the cerebral cortex, these back-propagating action potentials are thought to be a feedback signal from the soma (trigger zone) to the distal dendrite so that the site of the synaptic input can monitor the activity or firing frequency of the neuron and thereby prevent divergence of the firing frequency. When a neuron fires with high frequency, the number of synaptic inputs to this neuron is decreased, and vice versa. The back-propagation action potentials, in some cases, contribute to spike timing-dependent plasticity (STDP) and modulate various functions of neural networks. Back-propagating action potentials are generally produced by fast regenerative Na^+ channels in the dendritic membrane. Therefore, a single neuron can be considered as a fine information processor by itself.

The Ca^{2+} channels in a cerebellar Purkinje cell contribute to the prolonged calcium action potentials generated at the dendritic trees of the Purkinje cell that enable increase of the Ca^{2+} concentration inside the cell and thereby trigger a cascade of biochemical reactions that induce physiological and molecular biological functions. In addition, these prolonged calcium spikes, which are induced (generated) after the regenerative fast sodium spike by the strong climbing fiber input, are relevant to the regulation of synaptic inputs, especially the long-term modification (long-term depression, long-term potentiation, and short-term potentiation) of the parallel fiber–Purkinje cell synapse that is thought to be the elementary process underlying motor learning and other important physiological functions performed by the cerebellum. Excess synaptic inputs to the Purkinje cell through the parallel fibers (axons of granule cells) are negatively regulated by Golgi cells to which parallel fibers and simultaneously collateral of Purkinje cell axons have excitatory and inhibitory inputs, respectively. These negative feedback regulations by Golgi cells are likely to be the feedback signals for pack-propagation action potentials in pyramidal cells (Fig. 36.9).

Individual neurons have different distributions of dendritic and somatic voltage-gated channels. In pyramidal cells, Na^+ channels permit action potentials generated at the axon hillock to propagate backwards into the dendritic tree. However, cerebellar Purkinje cells in which the density of Na^+ channels decreases rapidly with distance from the soma have more prominent features than pyramidal cells. Together with the unique morphology of this cell type, the Na-dependent

action potentials that are generated in the soma-axonal region do not invade the dendritic tree. However, slowly inactivating voltage-gated Ca^{2+} channels (mainly P/Q- and T-type Ca^{2+} channels) increase with distance from the soma, and a prolonged inward Ca^{2+} current that generates Ca-dependent action potentials is terminated once the intracellular Ca^{2+} concentration becomes sufficiently elevated to activate Ca-dependent K^+ channels. This unique physiological property of cerebellar Purkinje cells is apparently different from that of pyramidal cells [36.119, 120].

In principle, neural integration involves the summation of synaptic potentials that passively spread to the trigger zone of the neuron, and the cell at any given moment chooses one of two options: firing or not firing an action potential. The net effect of the many inputs to a single neuron is receipt of both excitatory inputs and inhibitory inputs together, which cause or prevent action potential initiation, respectively. These inputs depend on several factors: the location, size, and shape of the synapse, and its proximity to the soma.

As described in Sect. 36.5.1, the time constant of the passive membrane electrical property of the neurons determines the time course of the synaptic potentials so that the temporal summation affects the all-or-none determination of action potential firing. The length constant of the cable property of the neural membrane also determines the extent to which the synaptic currents electronically spread in order to reach the trigger zone without minimal decrement of potential. Since one excitatory synaptic input is not sufficient to fire an action potential, many excitatory inputs from many synaptic contact points must be added together to generate an action potential at the trigger zone (spatial summation). These two different types of summation are relevant to both the location and timing of synaptic inputs, generally being called spatiotemporal summation. The examples of Purkinje cells and pyramidal cells suggest, in addition to these two processes, that we add active propagation of the action potentials due to the active property of the membrane in the dendritic trees and soma towards the trigger zone, because these areas contain voltage-gated Na^+, K^+, and Ca^{2+} channels in addition to the neurotransmitter-operated receptor channels.

Besides the functional role of these voltage-gated channels, transmitter-gated receptor channels contribute to the excitability or metabolism of the synaptic target, as discussed in Sect. 36.5.2. With these many factors of chemical synapses, the interaction of neurons allows complex behaviors of biological systems [36.121].

36.5.4 Dynamical Integration of Synaptic Inputs to a Single Neuron

As discussed in the previous section, a neuron integrates synaptic signals from many thousands of inputs, whose patterns change both spatially and temporally. Understanding the integration of such a huge number of inputs and the mechanism of the information processing at the dendritic trees and/or somata is one of the central themes in experimental and computational neuroscience at the single-neuron level. Moreover, physiological and anatomical investigation of local neural circuits in the central nervous system is also one of the most significant topics for understanding the information processing of neural networks and the functional mechanisms of animal behavior. To understand these questions experimentally, the most popular technique is in general to carry out electrophysiological measurements by recording during stimulation of electrodes applied to in vitro slices and cultured preparations. This method has benefited from technological inventions and modifications (manipulators, microscope stages, microscope design, etc.) for many years. However, these experimental techniques in principle use conventional electrodes for direct stimulation of neural fibers and/or neurons to activate functional synapses that have synaptic contacts with other neurons, and it is almost impossible to apply and access more than three or four electrodes simultaneously for in vitro preparations (acute slices and neurons in culture) due to the small space around a preparation, mainly limited by the working distance between the surface of the preparation and the microscope objective. Moreover, there are other difficulties when stimulating a single fiber and/or neuron using conventional electrodes, due to the possible spread of injected currents to neighboring regions resulting in activation of synapses in which we are not interested. Thus, we cannot estimate which fiber is actually activated by the stimulation of a nerve fascicle.

Thus, there are three problems to be overcome that naturally arise from the difficulties of the conventional method of stimulation using electrodes in electrophysiological experiments:

1. Firstly, is it possible to stimulate a small point or region such as a single spine and/or synaptic active zone, synaptic button, or single fiber (axon)?
2. Second, is it possible to mimic or reproduce experimentally and artificially postsynaptic currents which have the same time courses and ampli-

tudes as those observed in naturally functioning synapses in the central and peripheral nervous system?
3. Finally, is it possible to reproduce the complicated synaptic input patterns (changing spatiotemporally) present in in vivo preparations for application to neurons in in vitro preparations?

Several experimental techniques have now been developed to overcome these problems. As already described in Sect. 36.4.3, for example, to activate receptor ion channels in a small region and mimic the postsynaptic response, application of a fast agonist (or chemical compound) by a glass theta tube quickly driven by a piezoelement to an isolated larger membrane patch was developed by *Colquhoun* et al. [36.86]. This membrane patch of out-side out patch-clamp configuration excised from the membranes of various parts of neurons has involved a comparative number of receptor ion channels so that it is sufficient to obtain the current response equivalent to the amplitude of miniature synaptic current. Such excised membrane patches are obtained from arbitrary neurons in the central nervous system, allowing investigation and characterization of the properties of the receptor channels from these neurons whose physiological function plays important roles in synaptic transmission.

A more advanced technology is photolysis (uncaging) of caged compounds, such as caged glutamate and caged GABA, by ultraviolet (UV) or infrared (IR) laser beams focused to diameters of less than a few microns in the focal plane of tissue preparations to activate the active zone of a single synapse and/or spine [36.122, 123]. This photolysis technique in which a laser beam is used to uncage caged compounds (4-methoxy-7-nitroindolinyl(MNI)-caged-L-glutamate, etc.) dissolved in solution (artificial cerebrospinal fluid (ACSF), etc.) around preparations is a powerful method when combined with electrophysiological measurements. The laser beam, e.g., an ultraviolet (UV) or infrared (IR) laser source, is introduced into the microscope optics, and before illumination of the surface of the preparation, steered using a galvano-mirror (or acoustooptical deflector composed of two crystals) to deflect the laser beam in two dimensions in the horizontal plane of the surface of the preparation [36.124, 125]. The laser beam is focused down to 200 nm (diameter) on the focal plane, as determined by theoretical limitations, which gives uncaging resolution of less than 1 μm. In particular, photolysis by an IR laser beam is called two-photon photolysis [36.126].

Receptor activation by both photolysis and double-barreled tubing makes it possible to estimate the properties of a single receptor (single-channel conductance, etc.) by nonfluctuation analysis, as developed for the first time for evaluation of the voltage-activated single sodium conductance in the node of Ranvier [36.108]. The photolysis method offers two important benefits that overcome problems 1 and 2 mentioned above: firstly, a detailed analytical approach for elucidation of the receptor channels at any area of the membrane of a neuron in any region of the brain for which one can prepare in vitro preparations, and secondly, stimulation of the receptors in a small region on the membrane of a neuron, from which one can obtain responses that are equivalent (in both amplitude and time course) to the miniature synaptic current.

With focused laser beam technology, stimulations (activations) of fine multiple points of dendrites and soma such as active zones, synaptic buttons, and spines can be carried out to study the mechanisms of synaptic sensitivities in dendrites and receptor mapping (distribution) in the membrane of neurons in combination with the photolysis (uncaging) method. However, the third problem mentioned above could not be overcome by this simple laser beam uncaging method, due to the technical difficulties of steering the laser beam rapidly and two-dimensionally in order to focus at multiple points in the same focal plane of the tissue preparation.

Recently, the introduction of a TeO_2 acoustooptical crystal (AOC) device as a deflector component allowed an ultraviolet laser beam to be deflected rapidly to perform uncaging of caged compounds such as caged glutamate at multiple points within time intervals of less than $\approx 40\,\mu s$ [36.124] (Fig. 36.22). However, in this technique the UV laser beam should be transmitted through the TeO_2 acoustooptical crystal deflector (AOCD), and necessarily, a high-energy UV laser source is required because of the large energy loss of the incident laser beam (mainly to heat) when passing through the solid crystal; For example, a 1.5 W pulsed UV laser (DPSS Corp., Bowie; 50–60 ns, $\lambda = 355$ nm pulses at a 100 kHz repetition rate with average power > 400 mW) is used, which is also quite expensive and difficult to handle for the present experimental purpose.

To overcome this difficulty, a deflector, which steers and modulates the UV laser beam for photolysis, has been alternatively developed by constructing a beam deflector with a conventional galvanometer mirror that is controlled by specially designed software. The galvanometer deflector (GMD), which has previously been used to scan the laser beam to take fluorescent im-

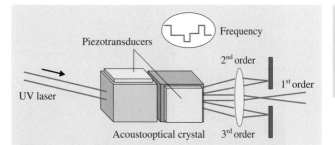

Fig. 36.22 Acoustooptical crystal deflector (AOCD) for ultraviolet laser beam steering in two-dimensions in the horizontal plane of the surface of the preparation plane. Introduction of the TeO_2 acoustooptical crystal (AOC) device into the deflector component allows ultraviolet laser beams to be deflected rapidly and performance of uncaging of caged compounds such as caged glutamate at multiple points within time intervals of less than $\approx 40\,\mu s$. Two crystal components are attached together, and two perpendicular directions of electric field are applied by piezotransducers controlled by computer-generated sequential pulses. (Modified by Chloé Okuno after [36.124])

ages in confocal microscopy, allows rapid access of a focused laser beam to many locations in a tissue preparation. The newly designed system could perform photolysis at over 1000 locations on the tissue preparation per second (corresponding to a 1 ms interval between two locations), considerably faster than the older uncaging system which was equipped with conventional galvanometer mirrors and mechanical shutters and was not originally designed to steer laser beams rapidly. The basic technical specification of the system, its available performance and limitations, and also a demonstration of the results obtained in applications to several topics in neuroscience are reported elsewhere [36.125, 127, 128] (Fig. 36.23a,b).

With these systems:

1. It is possible to perform photolysis in an arbitrary spatiotemporal dynamical pattern in order to emulate realistic neural input activity such as in in vivo preparations.
2. It allows investigators to study the properties of complicated synaptic transmission patterns, and synaptic plasticity such as spike timing-dependent plasticity (STDP) that is thought to be a substantial basis of memory and learning.
3. The mechanisms and significance of the dendritic compartmentalization of synaptic inputs in a single neuron, as mentioned in the previous section on dendritic integration, can be investigated.

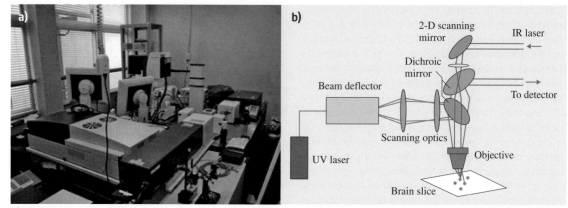

Fig. 36.23a,b Fast multiple-point photolysis system using a laser beam for uncaging of a caged compound. Photograph (**a**) and schematic diagram (**b**) showing the fast multiple-point uncaging system constructed from a conventional laser microscope (LSM510 meta; Zeiss) with an infrared laser (1.5 W, repetition rate 90 MHz, pulse width < 140 fs, 705–980 nm, Chameleon XR; Coherent), an ultraviolet laser (Enterprise II model 653; Coherent), and several visible-wavelength laser sources (Ar/ML 458/477/488/514 nm, G-HeNe 543 nm, 1 mW, R-HeNe 633 nm, 5 mW). The laser beam for uncaging is deflected by a galvanometer-based ultraviolet laser beam deflector (DuoScan). An electrophysiological setup for patch-clamp experiments can be seen in (**a**) (Modified by Chloé Okuno)

4. It is also able to investigate neural network-level topics, for example, the influence of the input rate (frequency) on the parameters which determine the membrane potential fluctuations, shaping of the local field potential by the synaptic input to a single cell, and long-range synaptic connectivity in the slices.

The generation of action potentials at the initiation site of a neuron results from the integration of many synaptic inputs to a single neuron and also from stochastic processes. When a neuron receives excitatory and inhibitory inputs with a complicated spatiotemporal dynamic pattern from other neurons, the integrated electrical signal reaches the initial segment, and if it exceeds the threshold, impulses fire in a stochastic manner and travel down the axon. Action potential initiation depends on the timing, location, and number of the synaptic inputs, which have different properties and transmission efficacies. However, the detailed mechanisms of synaptic integration at dendritic trees and somata are not yet perfectly understood due to the technical difficulties of such experiments.

36.A Summary of the Techniques Used for Investigation of Synaptic Functions

As described in this chapter, developments of experimental techniques and analytical methods play a key role in the investigation of synaptic functions. However, in spite of these technical developments, it has been shown that interpretation of the mechanisms of synaptic transmission in the central nervous system is difficult due to its complicated morphological features and the need to use living preparations [36.129]. We summarize here the significant techniques used for studying synaptic function and its regulation.

36.A.1 Noise Analysis and Single-Channel Current Recording

This technique, used for studying the properties of synaptic transmission and its receptor mechanisms, has developed with the progress of neurophysiology. Historically, the mechanisms of chemical transmission were studied through the use of extracellular and intracellular voltage recording from spinal motoneurons, since the development of this method in 1948. Stationary noise analysis was invented to estimate the properties of single channels by *Katz*, and then *Stevens* applied it for detailed description of single-channel conductance of the ACh receptor in the neuromuscular junction [36.107, 130]. Nonstationary noise analysis evolved from stationary noise analysis to estimate the properties of the receptor channel in the postsynaptic membrane [36.108, 110, 111].

In the late 1970s, the patch-clamp method was invented by *Hamill* and colleagues [36.64], and Numa introduced molecular biological techniques into the field of receptor mechanisms. These two techniques (patch-clamp and molecular biological techniques) promoted analysis of synaptic transmission enormously by enabling investigation of receptor mechanisms at the postsynaptic site. In particular, the single-channel recording techniques introduced by *Sakmann* and *Neher* and its statistical analysis were extended to postsynaptic processes such as neurotransmitter receptor mechanisms [36.74]. The quantitative description of single-channel behavior, which represents a typical achievement bridging between molecular (protein) structure and its function, is tractable by both mathematical treatment and computer calculations [36.80].

The invention and development of new methods of fluorescent optical recording by, for example, Shimomura (green fluorescent protein (GFP)) and Tsien (Ca^{2+} ion measurements), made it possible to visualize the processes of synaptic transmission.

36.A.2 Laser Photolysis and Rapid Multiple-Point Stimulation

As described in Sect. 36.5.3, when conducting conventional in vitro electrophysiological experiments, only a few recording and stimulating pipettes (electrodes) can be applied to the preparation (such as a slice, or cells in culture) under the objective of a microscope because of the narrow working space of the objective. Thus, it is impossible to stimulate many points and/or locations of dendritic trees almost simultaneously in a short time interval with the conventional technique. However, the recent development of the photolysis technique has overcome this experimental difficulty by enabling two significant developments:

1. By introducing the laser beam for photolysis, it becomes possible to stimulate a small region of nerve tissue such as a single spine and synaptic button.
2. By using specially designed laser beam deflectors it is also possible to stimulate many points of the

dendritic trees and soma of small size with constant intensity [36.124, 125, 131, 132].

Furthermore, this technique enables investigation of the integration and processing of the many synaptic inputs to a single neuron. The physiological and anatomical connections of neurons can also be elucidated by applying the laser photolysis method to local neural circuits, providing important advances in understanding of the functional mechanisms of animal behavior [36.133]. The fast multipoint uncaging technique also offers a bridge between experiments using in vitro preparations and in vivo experiments.

The application of this system to investigate the integration of dynamical patterned synaptic inputs by a single neuron allows one to model neurons having more realistic inputs as well as their synaptic transmission.

36.A.3 Fast Drug Application to Excised Membrane Patch and Fine Stimulation

For detailed analysis of synaptic transmission, there are several fundamental questions: (1) Is it possible to reproduce an elementary synaptic potential and/or current by an artificial method? and (2) Is it possible to activate a single synapse by presynaptic terminal stimulation? These questions are essential to reveal the functional characterization of receptor mechanisms and synaptic functions in the central nervous system. Fortunately, it has become possible to study these questions to some degree through the use of recent technological advances, applied to experiments using in vitro preparations.

Concerning question 1 above, in order to mimic and reproduce an elementary synaptic current and shed light on the mechanisms of the receptors and the pharmacological interactions with chemicals, the fast application of agonists (and/or chemicals) by theta tube to membrane patches isolated from neurons was introduced by *Sakmann* and *Jonas* in 1992 (Fig. 36.18) [36.88]. Using this method, the open–close gating kinetics of receptors has been elucidated and precise analyses of synaptic potentials and currents (in the case of voltage-clamp conditions) can be described in relation to elementary receptor mechanisms. Moreover, the single-channel conductance from neurons in various areas of the brain can be estimated with the use of fluctuation analysis as described in Sect. 36.A.1. Concerning question 2 above, the most direct way to investigate a single synapse is to stimulate only one synapse of a presynaptic cell and record the corresponding response from a target cell. The idea is to decrease the stimulus intensity so as to stimulate only a single axon of interest. Examples of such synapses are the Schaffer collateral synapses on CA1 pyramidal cells in the hippocampus and parallel fiber synapses onto Purkinje cells in the cerebellum. This method was developed to show that individual synapses in the hippocampus are very unreliable, which indicates that they release neurotransmitter with low probability and that hippocampal neurons receive synaptic inputs with diverse, individual properties. Moreover, as mentioned in Sect. 36.A.4, to activate a single synapse and spine, infrared two-photon laser uncaging methods are used to break the chemical bonds between the agonist and the chemicals, releasing free agonist in the vicinity of the postsynaptic density. The spatial resolution of activation by the UV and IR laser beam is less than $7\,\mu m$ and $3\,\mu m$, respectively. The chemical two-photon uncaging method is also useful for this purpose.

Another method for investigating single synapses is to study cell pairs that are connected to one another by a synaptic contact. A presynaptic cell is accessed and patch-clamped by an electrode for activation of a synapse, and measurement of the response of its target cell (postsynaptic neuron) is carried out simultaneously.

As long as each axon makes a synaptic contact with its postsynaptic cell, this method provides a reliable way to study single synapses. However, the difficulty with this approach is that the probability of finding two paired cells that have a synaptic contact with each other is very low, and simultaneous recording of paired cells is sometimes technically difficult and requires elaborate techniques and expensive electrophysiological setups with specially designed manipulators, etc.

36.A.4 Imaging Methods (Confocal Microscopy, Two-Photon Laser Microscopy)

One of the most important techniques used for visualization of synapses is laser microscopy. Confocal microscopy and two-photon laser scanning fluorescence microscopy are rapidly becoming common tools [36.123, 134]. The resolution required to observe synapses is close to that of light microscopy. Combined with computer image processing, confocal scanning microscopy enables reconstruction of images of synapses as three-dimensional structures at resolution below $1\,\mu m$. Against the many advantages of laser microscopy, its main disadvantage is the relatively long time needed for the scanner to form an image.

References

36.1   A.J. Brock (transl.): *Galen – On the Neural Faculties*, Loeb Classical Library (Harvard Univ. Press, Cambridge 1916)

36.2   C. Golgi: Sull sostanza grigia del cerevello, Gazz. Med. Lombarda **6**, 224–246 (1873)

36.3   S. Ramon y Cajal: The Croonian Lecture: La fine structure des centres nerveux, Proc. R. Soc. B **55**, 444–467 (1984)

36.4   C.G. Sherrington: The central nervous system. In: *A Textbook of Physiology*, 7th edn., ed. by M. Foster (Macmillan, London 2010)

36.5   J.C. Eccles: An electrical hypothesis of synaptic and neuromuscular transmission, Ann. N.Y. Acad. Sci. **47**, 429–455 (1946)

36.6   H.H. Dale, W. Feldberg: The chemical transmission of vagus effects to the stomach, J. Physiol. **81**, 320–380 (1934)

36.7   C. Ribrault, K. Sekimot, A. Triller: From the stochasticity of molecular processes to the variability of synaptic transmission, Nat. Rev. Neurosci. **20**, 375–387 (2011)

36.8   E.D. De Robertis, H.S. Bennett: A submicroscopic vesicular component of Schwann cells and nerve satellite cells, Exp. Cell Res. **6**(2), 543–545 (1954)

36.9   G.E. Palade: Electron microscopeobservation of innerneural and neuromuscular synapses, Anat. Record **118**, 335–336 (1954)

36.10   S.L. Palay: The morphology of synapses in the central nervous system, Exp. Cell Res. Suppl. **5**, 275–293 (1958)

36.11   T. Sudhof, R.H. Scheller: Mechanism and regulation of neurotransmitter release. In: *Synapses*, ed. by W.M. Cowan, T.C. Sudhof, C.F. Stevens (The Johns Hopkins Univ. Press, Baltimore 2003) pp. 177–216

36.12   B.W. Mel: Information processing in dendritic trees, Neural Comput. **6**, 103–1085 (1994)

36.13   Y. Humeau, F. Doussau, N.J. Grant, B. Poulain: How botulinum and tetanus neurotoxins block neurotransmitter release, Biochimie **82**(5), 427–446 (2000)

36.14   R. Robitaille, E.M. Adler, M.P. Charlton: Strategic location of calcium channel sat transmitter release sites of frog neuromuscular synapses, Neuron **5**, 773–779 (1990)

36.15   P. De Camilli, V. Haucke, K. Takei, E. Mugnaini: The structure of synapses. In: *Synapses*, ed. by W.M. Cowan, T.C. Sudhof, C.F. Stevens (The Johns Hopkins Univ. Press, Baltimore 2003) pp. 89–133

36.16 P. Fatt, B. Katz: An analysis of the end-plate potential recorded with an intra-cellular electrode, J. Physiol. (London) **115**, 320–370 (1951)

36.17 P. Fatt, B. Katz: Spontaneous subthreshold activity at motor nerve endings, J. Physiol. (London) **117**, 109–128 (1952)

36.18 J. Del Castillo, B. Katz: Statistical factors involved in neuromuscular facilitation and depression, J. Physiol. (London) **124**, 574–585 (1954)

36.19 B. Katz, R. Miledi: The study of synaptic transmission in the absence of nerve impulse, J. Physiol. (London) **192**, 407–436 (1967)

36.20 B. Katz, R. Miledi: The timing of calcium action during neuromuscular transmission, J. Physiol. (London) **192**, 535–544 (1967)

36.21 B. Katz, R. Miledi: Membrane noise produced by acetylcholine, Nature (London) **226**, 962–963 (1970)

36.22 B. Sakmann, C. Methfessel, M. Mishina, T. Takahashi, T. Takai, M. Kurasaki, K. Fukuda, S. Numa: Role of acetylcholine receptor subunits in gating of the channel, Nature **318**(6046), 538–543 (1985)

36.23 M. Mishina, T. Takai, K. Moto, M. Noda, T. Takahashi, S. Numa, C. Methfessel, B. Sakmann: Molecular distinction between fetal and adult forms of muscle acetylcholine receptor, Nature **321**(6068), 406–411 (1986)

36.24 J.E. Heuser, T.S. Reese, M.J. Dennis, Y. Jan, L. Yan, L. Evans: Synaptic vesicle exocytosis captured by quick freezing and correlated with quantal transmitter release, J. Cell Biol. **81**, 275–300 (1979)

36.25 J.G. Nicholls, A.R. Martin, B.G. Wallace, P.A. Fuchs: *From Neuron to Brain*, 4th edn. (Sinauer, Sunderland 2001)

36.26 T. Claudio, M. Ballivet, J. Patrick, S. Heinemann: Nucleotide and deduced amino acid sequences of Torpedo californica acetylcholine receptor γ-subunit, Proc. Natl. Acad. Sci. USA **80**, 1111–1115 (1983)

36.27 M.H. Akabas, C. Kaufmann, P. Archdeacon, A. Karlin: Identification of acetylcholine receptor-channel-lining residues in the entire M2 segment of the α-subunit, Neuron **13**, 919–927 (1994)

36.28 D.A. Karlin, M.H. Akabas: Toward a structural basis for the function of nicotinic acetylcholine receptors and their cousins, Neuron **15**, 1231–1244 (1995)

36.29 C. Toyoshima, N. Unwin: Ion channel of acetylcholine receptor reconstructed from images of postsynaptic membrane segment that determines the ion flow through the acetylcholine receptor-channel, Nature **336**, 247–250 (1988)

36.30 N. Unwin: Neurotransmitter action: Opening of ligandgated ion channels, Cell **72**(Suppl.), 31–41 (1993)

36.31 N. Unwin: Acetylcholine receptor-channel imaged in the open state, Nature **373**, 37–43 (1995)

36.32 T.V. Bliss, T. Lomo: Long-lasting potentiation of synaptic transmission in the dentate area of the anaesthetized rabbit following stimulation of the perforant path, J. Physiol. **232**(2), 331–356 (1973)

36.33 G.L. Collingridge, S.J. Kehl, H. McLennan: Excitatory amino acid in synaptic transmission in th Schaffer collateral-commissural pathway of the rat hippocampus, J. Physiol. **334**, 33–46 (1983)

36.34 J.T. Isaac, R.A. Nicoll, R.C. Malenka: Evidence for silent synapses: Implications for the expression of LTP, Neuron **15**(2), 427–434 (1995)

36.35 M. Farrant, R.A. Webster: GABA antagonists: Their use and mechanisms of action. In: *Neuromethods*, Vol. 12, ed. by A.A. Boulton, G.B. Baker, A.V. Juorio (Humana, Clifton 1989) pp. 161–219

36.36 O.P. Hamill, J. Bormann, B. Sakmann: Activation of multiple-conductance state chloride channels in spinal neurones by glycine and GABA, Nature **305**, 805–808 (1983)

36.37 B. Katz: *The Release of Neural Transmitter Substances* (Liverpool Univ. Press, Liverpool 1969)

36.38 J. Del Castillo, B. Katz: Quantal components of the end-plate potential, J. Physiol. (London) **124**, 560–573 (1954)

36.39 I.A. Boyd, A.R. Martin: The end-plate potential in mammalian muscle, J. Physiol. (London) **132**, 74–91 (1956)

36.40 H. Korn, D.S. Faber: Quantal analysis and synaptic efficacy in the CNS, Trends Neurosci. **14**, 439–445 (1991)

36.41 D.S. Faber, H. Korn: Unitary conductance change at teleost Mauthner cell glycinergic synapses: A voltage-clamp and phalmacologic analysis, J. Neurophysiol. **60**, 1982–1999 (1988)

36.42 U.J. McMahan, S.W. Kuffler: Visual identification of synaptic boutons on living ganglion cells and varicosities in postganglionic axons in the heart of the frog, Proc. R. Soc. B **177**, 485–508 (1971)

36.43 T.A. Benke, A. Lüthi, M.J. Palmer, M.A. Wikström, W.W. Anderson, J.T. Isaac, G.L. Collingridge: Mathematical modeling of non-stationary fluctuation analysis for studying channel properties of synaptic AMPA receptors, J. Physiol. **537**, 407–420 (2001)

36.44 F.A. Edwards: LTP is a long term problem, Nature (London) **350**, 271–272 (1999)

36.45 P. Jonas, G. Major, B. Sakmann: Quantal components of unitary EPSCs at the mossy fibre synapse on CA3 pyramidal cells of rat hippocampus, J. Physiol. **472**, 615–663 (1993)

36.46 J.J. Jack, S.J. Redman, K. Wong: The components of synaptic potentials evoked in cat spinal motoneurones by impulses in single group Ia afferents, J. Physiol. (London) **321**, 65–96 (1981)

36.47 S. Redman: Quantal analysis of synaptic potentials in neurons of the central nervous system, Physiol. Rev. **70**, 165–198 (1990)

36.48 D. Johnston, S.M.-S. Wu: *Foundation of Cellular Neurophysiology* (MIT Press, Cambridge 1995)

36.49 R.A. Silver, M. Farrant, S.G. Cull-Candy: Filtering of synaptic currents estimated from the time course of NMDA channel opening at rat cerebellar mossy fiber-granule cell synapse, J. Physiol. **494**, 85 (1996)

36.50   D.S. Faber, W.S. Young, P. Legendre, H. Korn: Intrinsic quantal variability due to stochastic properties of receptor-transmitter interactions, Science **258**, 1494–1498 (1992)

36.51   F.A. Edwards, A. Konnerth, B. Sakmann: Quantal analysis of inhibitory synaptic transmission in the dendate gyrus of rat hippocampal slices, J. Physiol. (London) **430**, 213–249 (1990)

36.52   S. Mochida, A.P. Few, T. Scheuer, W.A. Catterall: Regulation of presynaptic Cav2.1 channels by Ca^{2+} sensor proteins mediates short-term synaptic plasticity, Neuron **57**, 175–182 (2008)

36.53   S. Mochida: Activity-dependent regulation of synaptic vesicle exocytosis and presynaptic short-term lasticity, Neurosci. Res. **70**, 16–23 (2011)

36.54   M. Ito: *Physiology of Neuron* (Iwanami Press, Tokyo 1972), in Japanese

36.55   B.K. Andrasfalvy, J.C. Magee: Changes in AMPA receptor currents following LTP induction on rat CA1 pyramidal neurones, J. Physiol. **559**, 543–554 (2004)

36.56   B.K. Andrasfalvy, I. Mdody: Differences between the scaling of miniature IPSCs and EPSCs recorded in the dentrites of CA1 mouse pyramidal neurons, J. Physiol. **576**, 191–196 (2006)

36.57   A.A. Biro, N.B. Holderith, Z. Nusser: Quantal size is independent of the release probability at hippocampal excitatory synapses, J. Neurosci. **25**, 223–232 (2005)

36.58   B. Barbour, B.U. Keller, A. Marty: Prolonged presence of glutamate during excitatory synaptic transmission to cerebellar Purkinje cells, Neuron **12**, 1331–1343 (1994)

36.59   L. Cathala, S. Brickley, S. Cull-Candy, M. Farrant: Maturation of EPSCs and intrinsic membrane properties enhances precision at a cerebellar synapse, J. Neurosci. **23**, 6074–6085 (2003)

36.60   D. Marr: A theory of cerebellar cortex, J. Physiol. (London) **202**, 437–470 (1969)

36.61   J.S. Albus: A theory of cerebellar function, Math. Biosci. **10**, 25–61 (1971)

36.62   M. Ito: Cerebellar control of the vestibulo-ocular reflex around the flocculus hypothesis, Annu. Rev. Neurosci. **5**, 276–296 (1982)

36.63   E. De Schutter, J.M. Bower: Simulated responses of cerebellar Purkinje cell are independent of the dendritic location of granule cell synaptic inputs, Proc. Natl. Acad. Sci. USA **91**, 4736–4740 (1994)

36.64   O.P. Hamill, A. Marty, E. Neher, B. Sakmann, F.J. Sigworth: Improved patch-clamp techniques for hight-resolution current recording from cells and cell-free membrane patches, Pflügers Arch. **391**, 85–100 (1981)

36.65   F.A. Edwards, A. Konnerth, B. Sakmann, T. Takahashi: A thin slice preparation for patch clamp

recordings from neurons of the mammalian central nervous system, Pflügers Arch. **414**, 600–612 (1989)

36.66   B. Sakmann, E. Neher: *Single Channel Recording*, 2nd edn. (Plenum, New York 1995)

36.67   W. Waltz: *Patch Clamp Analysis*, 2nd edn. (Humana, New York 2007)

36.68   P. Molnar, J.J. Hickman: *Patch-Clamp Method and Protocols* (Humana, New York 2007)

36.69   A. Takeuchi, N. Takeuchi: Active phase of frog end-plate potential, J. Neurophysiol. **22**, 395–411 (1959)

36.70   A. Takeuchi, N. Takeuchi: On the permeability of end-plate membrane during the action of transmitter, J. Physiol. **154**, 52–67 (1960)

36.71   B. Hille: *Ionic Channels of Excitable Membranes*, 2nd edn. (Sinauer, Sunderland 1992) pp. 140–169

36.72   M. Noda, H. Takahashi, T. Tanabe, M. Toyosato, S. Kikyotani, Y. Furutani, T. Hirose, H. Takashima, S. Inayama, T. Miyata, S. Numa: Structural homology of Torpedo californica acetylcholine receptor subunits, Nature **302**, 528–532 (1983)

36.73   M. Noda, Y. Furutani, H. Takahashi, M. Toyosato, T. Tanabe, S. Shimizu, S. Kikyotani, T. Kayano, T. Hirose, S. Inayama, S. Numa: Cloning and sequence analysis of calf cDNA and human genomic DNA encoding α-subunit precursor of muscle acetylcholine receptor, Nature **305**, 818–823 (1983)

36.74   K. Imoto, C. Busch, B. Sakmann, M. Mishina, T. Konno, J. Nakai, H. Bujo, Y. Mori, K. Fukuda, S. Numa: Rings of negatively charged amino acids determine the acetylcholine receptor-channel conductance, Nature **335**, 645–648 (1988)

36.75   E.R. Kandel, J.H. Schwartz, T.M. Jessell: *Principles of Neural Science*, 4th edn. (McGraw-Hill, Hoboken 2000)

36.76   A.L. Hodgkin, A.F. Huxley: A quantitative description of membrane current and its application to conduction and excitation in nerve, J. Physiol. **117**, 500–544 (1952)

36.77   E. Neher, B. Sakmann: Single-channel currents recorded from membrane of denervated frog muscle fibers, Nature (London) **260**, 770–802 (1976)

36.78   D. Colquhoun, B. Sakmann: Fluctuations in the microsecond time range of the current through single acetylcholine receptor ion channels, Nature (London) **294**, 464–466 (1981)

36.79   D. Colquhoun, B. Sakmann: Fast events in single-channel currents activated by acetylcholine and its analogues at the frog muscle end-plate, J. Physiol. (London) **369**, 501–557 (1985)

36.80   D. Calquhoun, A.G. Hawkes: The principles of the stochastic interpretation of ion-channel mechanisms. In: *Single Channel Recording*, 2nd edn., ed. by B. Sakmann, E. Neher (Plenum, New York 1995) pp. 397–479

36.81   N. Spruston, P. Jonas, B. Sakmann: Dendritic glutamate receptor channels in rat hippocampal CA3 and CA1 pyramidal neurons, J. Physiol. **482**, 325–352 (1995)

36.82 R.A. Lester, J.D. Clements, G.L. Westbrook, C.E. Jahr: Channel kinetics determine the time course of NMDA receptor-mediated synaptic currents, Nature **346**, 565–567 (1990)

36.83 M. Hausser, A. Roth: Dendritic and somatic glutamate receptor channels in rat cerebellar Purkinje cells, J. Physiol. **501**, 77–95 (1997)

36.84 J.D. Clements: Transmitter timecourse in the synaptic cleft: Its role in central synaptic function, Trends Neurosci. **19**, 163–171 (1996)

36.85 H. Kojima, S. Katsumata: An analysis of synaptic transmission and its plasticity by glutamate receptor channel kinetics models and 2-photon laser photolysis, Lecture Notes in Computer Science **5506**, 88–94 (2009)

36.86 D. Colquhoun, P. Jonas, B. Sakmann: Action of brief pulses of glutamate on AMPA/kainite receptors in patches from different neurons of rat hippocampal slices, J. Physiol. (London) **458**, 261–287 (1992)

36.87 M. Hausser, A. Roth: Estimating the time course of excitatory postsynaptic conductance in neocortical pyramidal cells using a novel voltage jump method, J. Neurosci. **17**, 7606–7625 (1997)

36.88 P. Jonas, B. Sakmann: Glutamate receptor channels in isolated patches from CA1 and CA3 pyramidal cells of rat hippocampal slices, J. Physiol. **255**, 143–171 (1992)

36.89 J.D. Clements, R.A. Lester, G. Jahr, C.E. Tong, G.L. Westbrook: The time course of glutamate in the synaptic cleft, Science **258**, 1498–1501 (1992)

36.90 W. Rall, I. Segev: Space-clamp problems when voltage clamping branched neurons with intracellular microelectrodes. In: *Voltage and Patch Clamping with Intracellular Microelectrodes*, ed. by T.G. Smith Jr, H. Lecar, S.J. Redman, P. Gage (Am. Physiol. Soc., Bethesda 1985) pp. 191–215

36.91 M. London, M. Häusser: Dendritic computation, Annu. Rev. Neurosci. **28**, 503–532 (2005)

36.92 M. London, I. Segev: Synaptic scaling in vitro and in vivo, Nat. Neurosci. **4**, 853–855 (2001)

36.93 R. Dingledine, K. Borges, D. Bowie, S.F. Traynelis: The glutamate receptor ion channels, Pharmacol. Rev. **51**, 7–61 (1999)

36.94 S. Cull-Candy, L. Kelly, M. Farrant: Regulation of Ca^{2+}-permeable AMPA receptors: Synaptic plasticity and beyond, Curr. Opin. Neurobiol. **16**, 288–297 (2006)

36.95 M. Hollmann, S. Heinemann: Cloned glutamate receptors, Annu. Rev. Neurosci. **17**, 31–108 (1994)

36.96 B. Clark, M. Farrant, S.G. Cull-Candy: A direct comparison of the single-channel properties of synaptic and extrasynaptic NMDA receptors, J. Neurosci. **17**, 107–116 (1997)

36.97 T.A. Benke, A. Lüthi, J.T. Issac, G. Collingridge: Modulation of AMPA receptor unitary conductance by synaptic activity, Nature **393**, 793–797 (1998)

36.98 A. Harsch, H.P. Robinson: Postsynaptic variability of firing in rat cortical neurons: The roles of input synchronization and synaptic NMDA receptor conductance, J. Neurosci. **20**, 6181–6192 (2000)

36.99 K. Moriyoshi, M. Masu, T. Ishii, R. Shigemoto: Molecular cloning and characterization of the rat NMDA receptor, Nature **354**, 31–37 (1991)

36.100 M. Masu, Y. Tanabe, K. Tsuchida, R. Shigemoto, S. Nakanishi: Sequence and expression of a metabotropic glutamate receptor, Nature **349**, 760–765 (1991)

36.101 N. Armstrong, Y. Sun, G.Q. Chen, E. Gouaux: Structure of a glutamate-receptor ligand-binding core in complex with kainate, Nature **395**, 913–917 (1998)

36.102 M. Matsuzaki: Dendritic spine geometry is critical for AMPA receptor expression in hippocampal CA1 pyramidal neurons, Nat. Neurosci. **4**, 1086–1092 (2001)

36.103 J.J. Lawrence, L.O. Trussell: Long-term specification of AMPA receptor properties after synapse formation, J. Neurosci. **20**, 4864–4870 (2000)

36.104 A.A. Bagel, J.P. Kao, C.-M. Tng, S.M. Thompson: Long-term potentiation of exogeneous glutamate responses at single dendritic spines, Proc. Natl. Acad. Sci. USA **102**(40), 14434–14439 (2005)

36.105 D.S. Bredt, R.A. Nicoll: AMPA receptor trafficking at excitatory synapses, Neuron **40**, 361–379 (2003)

36.106 V. Derkach, A. Barria, T.R. Soderling: $Ca^{2+}/$ calmodulin-kinaseII enhances channel conductance of alpha-amino-3-hydroxy-5-methyl-4-isoxazolepropionate type glutamate receptors, Proc. Natl. Acad. Sci. USA **96**, 3269–3274 (1999)

36.107 C.R. Anderson, C.F. Stevens: Voltage clamp analysis of acetylcholine produced end-plate current fluctuations at frog neuromuscular junction, J. Physiol. (London) **235**, 655–691 (1973)

36.108 F.J. Sigworth: The variance of sodium current fluctuations at the node of Ranvier, J. Physiol. (London) **307**, 97–129 (1980)

36.109 F.J. Sigworth: The variance of sodium channels under conditions of reduced current at the node of Ranvier, J. Physiol. (London) **307**, 131–142 (1980)

36.110 H.P. Robinson, Y. Sahara, N. Kawai: Nonstationary fluctuation analysis and direct resolution of single channel currents at postsynaptic sites, Biophys. J. **59**, 295–304 (1991)

36.111 S.F. Traynelis, R.A. Silver, S.G. Cull-Candy: Estimated conductance of glutamate receptor channels activated during EPSCs at the rat cerebellar mossy fibret-granule cell synapse, Neuron **11**, 279–289 (1993)

36.112 H. Kojima, K. Ichikawa, L.V. Ileva, S. Traynelis: Properties of AMPA receptor channels during long-term depression in rat cerebellar Purkinje cells. In: *Slow Synaptic Responses and Modulation*, ed. by K. Kuba, H. Higashida, D. Brown, T. Yoshioka (Springer, Tokyo 2000)

36.113 D.J. Wyllie, P. Béhé, D. Colquhoun: Single-channel activations and concentration jumps: Comparison of recombinant NR1a/NR2A and NR1a/NR2D NMDA receptors, J. Physiol. **510**, 1–18 (1998)

36.114 N.R. Carlson: *Physiology of Behavior*, 10th edn. (Pearson Education, New York 2009)

36.115 C.M. Niswender, P.J. Conn: Metabotropic glutamate receptors: Physiology, Pharmacology, and disease, Annu. Rev. Pharmacol. Toxicol. **50**, 295–322 (2010)

36.116 J. Nabekura, S. Katsurabayashi, Y. Karazu, S. Shibata, A. Matsubara, S. Jinno, Y. Mizoguchi, A. Sasaki, H. Ishibashi: Developmental switch from GABA to glycine release in single central synaptic terminals, Nat. Neurosci. **7**, 17–23 (2004)

36.117 J.C. Magee, E.P. Cook: Somatic EPSP amplitude is independent of synapse location in hippocampal pyramidal neurons, Nat. Neurosci. **3**, 895–903 (2000)

36.118 J. Magee, D. Hoffmann, C. Colbert, D. Johnston: Electrical and calcium signaling in dendrites of hippocampal pyramidal neurons, Annu. Rev. Physiol. **60**, 327–346 (1998)

36.119 R. Llinas, M. Sugimori: Electrophysiological properties of in vitro Purkinje cell somata in mammalian cerebellar slices, J. Physiol. (London) **305**, 171–195 (1980)

36.120 R. Llinas, M. Sugimori: Electrophysiological properties of in vitro Purkinje cell dendrites in mammalian cerebellar slices, J. Physiol. (London) **305**, 197–203 (1980)

36.121 I. Llano, A. Marty, J.W. Johnson, P. Ascher, B.H. Gähwiler: Patch-clamp recording of amino acid-activated responses in 'organotypic' slice cultures, Proc. Natl. Acad. Sci. USA **85**, 3221–3225 (1988)

36.122 S.M. Thompson, J.P. Kao, R.H. Kramer: Flashy science: Controlling neural function with light, J. Neurosci. **9**, 10358–10365 (2005)

36.123 K. Svoboda, R. Yasuda: Principles of two-photon excitation microscopy and its applications to neuroscience, Neuron **50**, 823–839 (2006)

36.124 S. Shoham, D.H. O'Connor, D.V. Sarikov: Rapid neurotransmitter uncaging in spatially defined patterns, Nat. Methods **5**(11), 837–842 (2005)

36.125 H. Kojima, E. Simburger, C. Boucsein, T. Maruo, M. Tsukada, S. Okabe, A. Aertsen: Development of a system for patterned rapid photolysis and 2-photon confocal microscopy, IEEE Circuit Dev. Mag. **22**(6), 66–74 (2006)

36.126 M. Matsuzaki, G.C. Ellis-Davies, T. Nemoto, Y. Miyashita, M. Iino, H. Kasai: Dendric spine geometry is crtitical for AMPA receptor expression in hippocampal CA1 pyramidal neurons, Nat. Neurosci. **4**, 1086–1092 (2001)

36.127 M. Yoneyama, Y. Fukushima, H. Kojima, M. Tsukada: Analysis of the spatial-temporal characteristics of synaptic EPSP summation on the dendritic trees of hippocampal CA1 pyramidal neurons as revealed by laser uncaging stimulation, J. Jap. Neural Netw. Soc. **17**(1), 2–11 (2010), (in Japanese)

36.128 S. Toujoh, Y. Nakazato, T. Maruo, S. Katsumata, K. Sakai, H. Kojima: A system for rapid patterned photolysis by ultraviolet (UV) laser beam, Proc. Faculty Eng. (Tamagawa Univ.), Vol. 43 (2008) pp. 13–22

36.129 F.A. Edwards: Anatomy and electrophysiology of fast central synapses lead to a structural model fot long-term potentiation, Physiol. Rev. **75**, 759–787 (1995)

36.130 B. Katz, R. Miledi: The statistical nature of the acetylcholine potential and its molecular components, J. Physiol. (London) **224**, 665–699 (1972)

36.131 A. Losonczy, J.C. Magee: Integrative properties of radial oblique dendrites in hippocampal CA1 pyramidal neurons, Neuron **50**, 291–307 (2006)

36.132 C. Boucsein, M. Nawart, S. Rotter, A. Aertsen, D. Heck: Controlling Synaptic input pattern in vitro by dynamic photo stimulation, J. Neurophysiol. **94**, 2948–2958 (2005)

36.133 Y. Yoshimura, E.M. Callaway: Fine-scale specificity od cortical networks depends on inhibitory cell type and connectivity, Nat. Neurosci. **8**(11), 1552–1559 (2005)

36.134 S. Okabe: Nano-sacle fluorescent imaging analysis during synaptic dynamical changes, Biotechnol. J. **11–12**, 744–752 (2005), (in Japanese)

37. Computational Modeling with Spiking Neural Networks

Stefan Schliebs, Nikola Kasabov

This chapter reviews recent developments in the area of spiking neural networks (SNN) and summarizes the main contributions to this research field. We give background information about the functioning of biological neurons, discuss the most important mathematical neural models along with neural encoding techniques, learning algorithms, and applications of spiking neurons. As a specific application, the functioning of the evolving spiking neural network (eSNN) classification method is presented in detail and the principles of numerous eSNN based applications are highlighted and discussed.

37.1 Neurons and Brain

The brain is arguably the most complex organ of the human body. It contains approximately 10^{11} neurons, which are the elementary processing units of the brain. These neurons are interconnected and form a complex and very dense neural network. On average 1 cm^3 of brain matter contains 10^4 cell bodies and several kilometers of *wire*, i. e., connections between neurons in the form of branching cell extensions.

Like most cells in the human body, neurons maintain a certain ion concentration across their cell membrane. Therefore, the membrane contains ion pumps which actively transport sodium ions from the intracellular to the extra-cellular liquid. Potassium ions are pumped in the opposite direction from the outside to

the inside of the cell. In addition to the ion pumps, a number of specialized proteins, so-called ion channels, are embedded in the membrane. They allow a slow inward flow of sodium ions into the cell, while potassium ions leak outwards into the extra-cellular liquid. Thus, the ion streams at the channels have opposite directions to the ion pumps. Furthermore, since both ion streams differ in their strengths, an electrical potential exists across the cell membrane. The inside of the cell is negatively charged in relation the extra-cellular liquid.

A large variety of neural shapes and sizes exist in the brain. A typical neuron is illustrated in Fig. 37.1. The central part of the neuron is called the soma, in

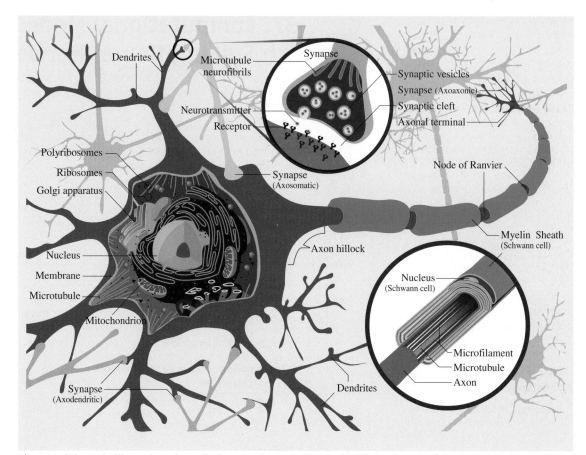

Fig. 37.1 Schematic illustration of a typical neuron in the human brain. The main part of the neuron is the soma containing the genetic information, the dendrites, and the axon, which are responsible for the reception and emission of electrical signals. Signal transmission occurs at the synapse between two neurons, see text for detailed explanations (after Wikipedia [37.1])

which the nucleus is located. It contains the genetic information of the cell, i. e., the DNA, from which genes are expressed and proteins constructed that are important for the functioning of the cell. The cell body has a number of cellular branch-like extensions known as dendrites. Dendrites are specialized for *receiving* electrical signals from other neurons that are connected to them. These signals are short pulses of electrical activity, also known as spikes or action potentials. If a neuron is stimulated by the spike activity of surrounding neurons and the excitation is strong enough, the cell triggers a spike. The spike is propagated via the axon, a long thin wire-like extension of the cell body, to the axonal terminals. These terminals in turn are connected to the dendrites of surrounding neurons and allow the transfer of information from one neuron to

the other. Thus an axon is responsible for *sending* information to other neurons connected to it. An axon may be covered by myelin sheaths that allow a faster propagation of electrical signals. These sheaths act as insulators and prevent the dissipation of the depolarization wave caused by an electrical spike triggered in the soma.

Information exchange between two neurons occurs at a synapse, which is a specialized structure that links two neurons together. A synapse is illustrated in the upper middle part of Fig. 37.1. The sending neuron is called pre-synaptic neuron, while the neuron receiving the signal is called post-synaptic. Sending information involves the generation of an action potential in the soma of the pre-synaptic cell. As described before, this potential is propagated through the axon of the neuron to the axonal terminals. These terminals contain the synapses in which neurotransmitter chemicals are stored. Whenever a spike is propagated through the axon, a portion of these neurotransmitters is released into a small gap between the two neurons also known as the synaptic cleft. The neurotransmitter diffuses into the cleft and interacts with specialized receptor proteins of the post-synaptic neuron. The activation of these receptors causes the sodium ion channels to open, which in turn results in the flow of sodium ions from the extracellular liquid into the post-synaptic cell. The ionic concentration across the membrane equalizes rapidly and the membrane depolarizes. Immediately after the depolarization the potassium channels open. As a consequence potassium ions stream outside the cell, which causes the re-polarization of the membrane. The process of de- and re-polarization, i. e., the action potential, lasts only around 2 ms, which explains the name spike or pulse.

A synaptic transmission can be either excitatory or inhibitory depending on the type of the transmitting synapse. Different neurotransmitters and receptors are involved in excitatory and inhibitory synaptic transmissions, respectively. Excitatory synapses release a transmitter called L-glutamate and increase the likelihood of the post-synaptic neuron triggering an action potential following stimulation. On the other hand, inhibitory synapses release a neurotransmitter called

GABA and decrease the likelihood of a post-synaptic potential.

The efficacy of a synapse, i. e., the strength of the post-synaptic response due to the neurotransmitter release in the synapse, is not fixed. The increase or decrease of the efficacy of a synapse is called *synaptic plasticity* and it enables the brain to learn and to memorize. There are several different possibilities to accomplish synaptic plasticity. One way is to change the time period of receptor activity in the post-synaptic neuron. Longer periods of receptor activity cause the ion channels to remain open for a longer time, which in turn results in a larger amount of ions flowing into the post-synaptic cell. Thus, the post-synaptic response increases. Short periods of receptor activity have the opposite effect.

Another way to change the synaptic efficacy is to increase or decrease the number of receptors, which would have a direct impact on the number of opened ion channels and as a consequence on the post-synaptic potential. The third possibility is a change of the amount of neurotransmitter chemicals released into the synaptic cleft. Here larger/smaller amounts would increase/decrease the synaptic efficacy.

Comprehensive information and details about the structure, functions, chemistry, and physiology of neurons can be found in a standard textbook on the matter by [37.2].

37.2 Models of Spiking Neurons

The remarkable information processing capabilities of the brain have inspired numerous mathematical abstractions of biological neurons. Spiking neurons represent the third generation of neural models, incorporating the concepts of time, neural, and synaptic state explicitly into the model [37.3]. Earlier artificial neural networks were described in terms of mean firing rates and used continuous signals for transmitting information between neurons. Real neurons, however, communicate by short pulses of electrical activity. In order to simulate

and describe biologically plausible neurons in a mathematical and formal way, several different models have been proposed in the recent past. Figure 37.2 illustrates

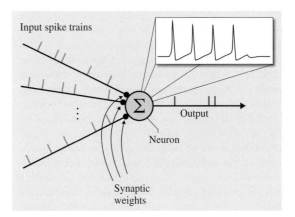

Fig. 37.2 Schematic illustration of a mathematical neuronal model. The model receives electrical stimulation in form of spikes through a number of connected pre-synaptic neurons. The efficacy of a synapse is modeled in the form of synaptic weights. Most models focus on the dynamics of the post-synaptic potential only. Output spikes are propagated via the axon to connected post-synaptic neurons ▶

schematically the mathematical abstraction of a biological neuron.

Neural modeling can be described on several levels of abstraction. On the microscopic level, the neuron model is described by the flow of ions through the channels of the membrane. This flow may, among other things, depend on the presence or absence of various chemical messenger molecules. Models at this level of abstraction include the Hodgkin–Huxley model [37.4] and the compartment models that describe separate segments of a neuron by a set of ionic equations.

On the other hand, the macroscopic level treats a neuron as a homogeneous unit, receiving and emitting spikes according to some defined internal dynamics. The underlying principles of how a spike is generated and carried through the synapse, dendrite, and cell body are not relevant. These models are typically known under the term integrate-and-fire models.

In the next sections the major neural models are discussed and their functions are explained. Since the macroscopic neuronal models are more relevant, the focus of the survey is put on these models. The only microscopic model presented here is the Hodgkin–Huxley model, due to its high significance for the research area of neuroscience.

37.2.1 Hodgkin–Huxley Model

This model dates back to the work of *Alan Lloyd Hodgkin* and *Andrew Huxley* in 1952 where they performed experiments on the giant axon of a squid [37.4]. Due to the significance of their contribution to neuroscience, both received the 1963 Nobel Prize in Physiology and Medicine. The model is a detailed description of the influences of the conductance of ion channels on the spike activity of the axon. The diameter of the squid's giant axon is approximately 0.5 mm and is visible to the naked eye. Since electrodes had to be inserted into the axon, its large size was a big advantage for biological analysis at that time.

Hodgkin and *Huxley* discovered three different ion currents in a neuron: a sodium, potassium, and a leak current. Voltage-dependent ion channels control the flow of ions through the cell membrane. Due to an active transport mechanism, the ion concentration within the cell differs from that in the extra-cellular liquid, resulting in an electrical potential across the cell membrane. In the mathematical model such a membrane is described as an electrical circuit consisting of a capacitor, resistors, and batteries that model the ion channels, (Fig. 37.3). The current I at a time t splits into the cur-

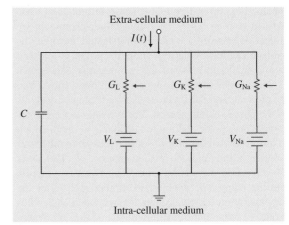

Fig. 37.3 Schematic illustration of the Hodgkin–Huxley model in the form of an electrical circuit (after [37.4]). The model represents the biophysical properties of the cell membrane of a neuron. The semipermeable cell membrane separates the interior of the cell from the extra-cellular liquid and thus acts as a capacitor. Ion movements through the cell membrane (in both directions) are modeled in the form of (constant and variable) resistors. In the diagram the conductance of the resistors $G_x = 1/R_x$ is shown. Three ionic currents exist: A sodium current (Na ions), potassium current (K ions), and a small leakage current (L) that is primarily carried by chloride ions

rent stored in the capacitor and the additional currents passing through each of the ion channels

$$I(t) = I_{cap}(t) + \sum_k I_k(t) \, , \qquad (37.1)$$

where the sum runs over all ion channels.

Substituting $I_{cap}(t) = C \, du/dt$ by applying the definition of the capacitance $C = Q/u$, where Q is the charge and u the voltage across the capacitor leads to

$$C \frac{du}{dt} = - \sum_k I_k(t) + I(t) \, . \qquad (37.2)$$

As mentioned earlier, in the Hodgkin–Huxley model three ion channels are modeled: A sodium current, potassium current, and a small leakage current that is primarily carried by chloride ions. Hence the sum in (37.1) consists of three different components that are formulated as

$$\sum_k I_k(t) = G_{Na} m^3 h(u - V_{Na})$$

$$+ G_K n^4 (u - V_K) + G_L (u - V_L) \, , \quad (37.3)$$

Table 37.1 Parameters of the Hodgkin–Huxley model. The membrane capacitance is $C = \mu F/cm^2$. The voltage scale is shifted in order to have a resting potential of zero

x	V_x (mV)	G_x (mS/cm^2)
Na	115	120
K	−12	36
L	10.6	0.3
x	$\alpha_x(u)$	$\beta_x(u)$
n	$\frac{0.1-0.01u}{\exp(1-0.1u)-1}$	$0.125\exp\left(-\frac{u}{80}\right)$
m	$\frac{2.5-0.1u}{\exp(2.5-0.1u)-1}$	$4\exp\left(-\frac{u}{18}\right)$
h	$0.07\exp\left(-\frac{u}{20}\right)$	$\frac{1}{\exp(3-0.1u)+1}$

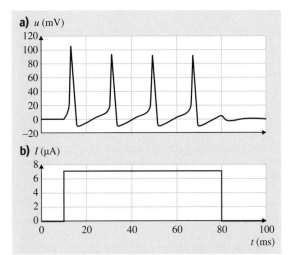

Fig. 37.4 (a) Evolution of the membrane potential u for a content input current I_0 using the Hodgkin–Huxley model. The current is switched on at time $t = 10$ ms for a duration of 70 ms. **(b)** The stimulus is strong enough to generate a spike train across the cell membrane (*upper diagram*). As soon as the input current vanishes ($I = 0$), the electrical potential returns to its resting potential ($u = 0$)

where V_{Na}, V_K, and V_L are constants called reverse potentials. Variables G_{Na} and G_K describe the maximum conductance of the sodium and potassium channel, respectively, while the leakage channel is voltage-independent with a conductance of G_L. The variables m, n, and h are gating variables whose dynamics are described by differential equations of the form

$$\frac{m}{dt} = \alpha_m(u)(1-m) - \beta_m(u)m , \tag{37.4}$$

$$\frac{n}{dt} = \alpha_n(u)(1-n) - \beta_n(u)n , \tag{37.5}$$

$$\frac{h}{dt} = \alpha_h(u)(1-h) - \beta_h(u)h , \tag{37.6}$$

where m and h control the sodium channel and variable n the potassium channel. Functions α_x and β_x, where $x \in \{m, n, h\}$, represent empirical functions of the voltage across the capacitor u, that need to be adjusted in order to simulate a specific neuron. Using a well-parameterized set of the above equations, *Hodgkin* and *Huxley* were able to describe a significant amount of data collected from experiments with the giant axon of a squid. The parameters discovered of the model are given in Table 37.1.

The dynamics of the Hodgkin–Huxley model are presented in Fig. 37.4. For the simulation, the parameter values from Table 37.1 are utilized. The membrane is stimulated by a constant input current $I_0 = 7\,\mu A$, switched on at time $t = 10$ ms for a duration of 70 ms. The current is switched off at time $t = 80$ ms. For $t < 10$ ms, no input stimulus occurs and the potential across the membrane stays at the resting potential. For 10 ms $\leq t \leq 80$ ms the current is strong enough to generate a sequence of spikes across the cell membrane. At time $t > 80$ ms and input cur-

rent $I = 0$, the electrical potential returns to its resting potential.

Additional reading on the Hodgkin–Huxley model can be found in the excellent review of [37.5], which also summarizes the historical developments of the model. A guideline for computer simulations of the model using the simulation platform GENESIS (general neural simulation system) can be found in [37.6].

37.2.2 Leaky Integrate–and–Fire Model (LIF)

The Hodgkin–Huxley model can reproduce electrophysiological measurements very accurately. Nevertheless, the model is computationally costly and simpler, more phenomenological models are required for the simulation of larger networks of spiking neurons. The leaky integrate-and-fire neuron (LIF) may be the best-known model for simulating spiking networks efficiently. The model has a long history and was first proposed by *Lapicque* in 1907, long before the actual mechanisms of action potential generation were known [37.7]. Discussions of this work can be found in [37.8, 9]. However, it was *Knight* who introduced the term *integrate-and-fire* in [37.10]. He called these models *forgetful*, but the term *leaky* quickly became more popular.

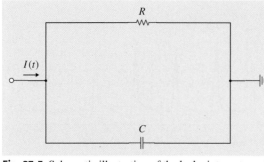

Fig. 37.5 Schematic illustration of the leaky integrate-and-fire model in the form of an electrical circuit. The model consists of a capacitor C in parallel with a resistor R, driven by a current $I = I(R) + I_{cap}$

Similar to the Hodgkin–Huxley model, the LIF model is based on the idea of an electrical circuit (Fig. 37.5). The circuit contains a capacitor with capacitance C and a resistor with a resistance R, where both C and R are assumed to be constant. The current $I(t)$ splits into two currents

$$I(t) = I_R + I_{cap} ,\tag{37.7}$$

where I_{cap} charges the capacitor and I_R passes through the resistor. Substituting $I_{cap} = C\,du/dt$ using the definition for capacity, and $I_R = u/R$ using Ohm's law, where u is the voltage across the resistor, one obtains

$$I(t) = \frac{u(t)}{R} + C\frac{du}{dt} .\tag{37.8}$$

Replacing $\tau_m = RC$ yields the standard form of the model

$$\tau_m \frac{du}{dt} = -u(t) + RI(t) .\tag{37.9}$$

The constant τ_m is called the membrane time constant of the neuron. Whenever the membrane potential u reaches a threshold ϑ, the neuron fires a spike and its potential is reset to a resting potential u_r. It is noteworthy that the shape of the spike itself is not explicitly described in the traditional LIF model. Only the firing times are considered to be relevant. Nevertheless, it is possible to include the shape of spikes as well [37.11].

A LIF neuron can be stimulated by either an external current I_{ext} or by the synaptic input current I_{syn} caused by pre-synaptic neurons. The external current $I(t) = I_{ext}(t)$ may be constant or represented by a function of time t. Figure 37.6 presents the dynamics of a LIF neuron stimulated by an input current $I_0 = 1.2$. The current is strong enough to increase the potential

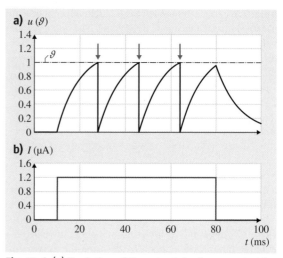

Fig. 37.6 (a) Evolution of the potential u for a constant input current I_0 using the leaky integrate-and-fire model. The membrane potential u is given in units of the threshold ϑ. The current is switched on at time $t = 10\,\mathrm{ms}$ for a duration of $70\,\mathrm{ms}$. **(b)** The stimulus is strong enough to generate a sequence of spike trains (*dark arrows*). As soon as the input current vanishes, the potential returns to its resting potential

u until the threshold ϑ is reached. As a consequence, a spike is triggered and the potential resets to $u_r = 0$. After the reset, the integration process starts again. At $t = 80\,\mathrm{ms}$, the current is switched off and the potential returns to its resting potential due to leakage.

If a LIF neuron is part of a network of neurons, it is usually stimulated by the activity of its pre-synaptic neurons. The resulting synaptic input current of a neuron i is the weighted sum over all spikes generated by pre-synaptic neurons j with firing times $t_j^{(f)}$

$$I(t) = I_{syn_i}(t) = \sum_j w_{ij} \sum_f \alpha\!\left(t - t_j^{(f)}\right) .\tag{37.10}$$

The weights w_{ij} reflect the efficacy of the synapse from neuron j to neuron i. Negative weights correspond to inhibitory synapses, while positive weights correspond to excitatory synapses. The time course of the post-synaptic current $\alpha(\cdot)$ can be defined in various ways. In the simplest form it is modeled by Dirac pulse $\delta(x)$, which has a non-zero function value for $x = 0$ and zero for all others. Thus the input current caused by a pre-synaptic neuron decreases/increases the potential u in a step-wise manner. More realistic models often employ different functions usually in the form $x \exp(-x)$, which is typically referred to as an α function.

In Fig. 37.7, a LIF neuron is stimulated by a spike train from a single pre-synaptic neuron. The post-synaptic current is modeled in the form of a Dirac pulse as described above. This results in a step-wise increase of the post-synaptic potential. If the potential reaches the threshold ϑ, a spike is triggered and the potential resets. Due to its simplicity, many LIF neurons can be connected to form large networks, while still allowing an efficient simulation.

Extensive additional information about the LIF model can be found in the excellent textbook [37.12] and in the two recent reviews by *Burkitt* [37.13, 14].

37.2.3 Izhikevich Model

Another neural model was proposed in [37.15]. It is based on the theory of dynamical systems. The model claims to be as biologically plausible as the Hodgkin–Huxley model while offering the computational complexity of LIF models. Depending on its parameter configuration, the model reproduces different spiking and bursting behavior of cortical neurons. Its dynamics are governed by two variables

$$\frac{dv}{dt} = 0.04v^2 + 5v + 140 - u + I , \tag{37.11}$$

$$\frac{du}{dt} = a(bv - u) , \tag{37.12}$$

where v represents the membrane potential of the neuron and u is a membrane recovery variable, which provides negative feedback for v. If the membrane potential reaches a threshold $\vartheta = 30\,\mathrm{mV}$, a spike is triggered and a reset of v and u occurs

$$\text{if } v \geq 30\,\mathrm{mV}, \quad \text{then } \begin{cases} v \leftarrow c \\ u \leftarrow u + d \end{cases} . \tag{37.13}$$

Variables a, b, c, d are parameters of the model. Depending on their setting, a large variety of neural characteristics can be modeled. Each parameter has an associated interpretation. Parameter a represents the decay rate of the membrane potential, b is the sensitivity of the membrane recovery, and c and d reset v and u, respectively.

In Fig. 37.8, the meaning of the parameters is graphically explained along with their effect on the dynamics of the model. For example, if we want to produce a regular spiking neuron, we would set $a = 0.02$, $b = 0.25$, $c = -65$, and $d = 8$.

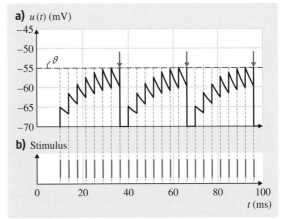

a) $u(t)$ (mV)

b) Stimulus

t (ms)

Fig. 37.7a,b Dynamics of the leaky integrate-and-fire model. The potential u increases due to the effect of pre-synaptic input spikes. If the membrane potential crosses a threshold ϑ, a spike is triggered (*straight dark arrows*) (**a**). The shape of this action potential is not explicitly described by the model, only the time of the event is of relevance (**b**). The synapse may have either an inhibitory or an excitatory effect on the post-synaptic potential that is determined by the sign of the synaptic weights

More information on this model can be found in the textbook on dynamical systems in neuroscience [37.16]. There are also a number of articles on the topic, e.g., the work on the suitability of mathematical models for simulation of cortical neurons [37.17] and the large-scale simulation of a mammalian thalamocortical system [37.18], which involves one million neurons and almost half a billion synapses.

37.2.4 Spike Response Model (SRM)

The spike response model (SRM) is a generalization of the LIF model and was introduced in [37.12]. In this model, the state of a neuron is characterized by a single variable u. A number of different kernel functions describe the impact of pre-synaptic spikes and external stimulation on u, but also the shape of the actual spike and its after-potential. Whenever the state u reaches a threshold ϑ from below, i.e., $u(t) = \vartheta$ and $du(t)/dt > 0$, a spike is triggered. In contrast to the LIF model, the threshold ϑ in SRM is not required to be fixed, but may depend on the last firing time $\hat{t}_i$ of neuron i. For example, the threshold might be increased after the neuron has spiked (also known as the refractory period) to avoid triggering another spike during that time.

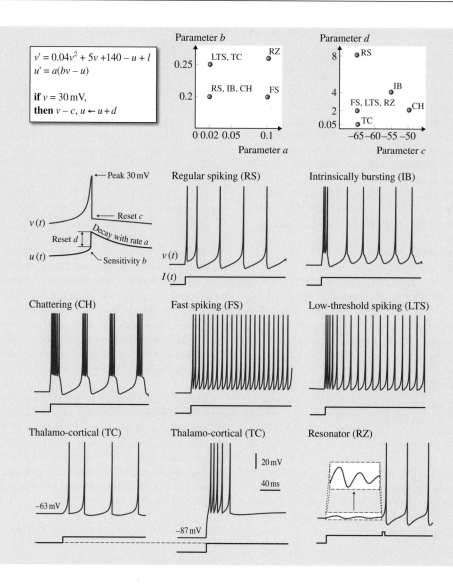

Let $u_i(t)$ be the state variable that describes neuron i at time t and $\hat{t}_i$ is the last time when the neuron emitted a spike, then the evolution of $u_i(t)$ can be formulated as

$$u_i(t) = \eta\left(t - \hat{t}_i\right) + \sum_j w_{ij} \sum_f \epsilon_{ij}\left(t - \hat{t}_i, t - t_j^{(f)}\right)$$

$$+ \int_0^\infty \kappa\left(t - \hat{t}_i, s\right) I_{\text{ext}}(t - s)\,\mathrm{d}s\,, \qquad (37.14)$$

where $t_j^{(f)}$ are the firing times of pre-synaptic neurons j, while w_{ij} represents the synaptic efficacy between neuron j and i.

Functions η, ϵ, and κ are response kernels. The first kernel, η, is the reset kernel. It describes the dynamics of an action potential and becomes non-zero each time a neuron fires. This kernel models the reset of the state u and its after-potential. A typical implementation is

$$\eta\left(t - \hat{t}_i\right) = \eta_0 \left\{ K_1 \exp\left(-\frac{t - \hat{t}_i}{\tau_m}\right) \right.$$

$$- K_2 \left[\exp\left(-\frac{t - \hat{t}_i}{\tau_m}\right) \right.$$

$$\left. \left. - \exp\left(-\frac{t - \hat{t}_i}{\tau_s}\right)\right]\right\} \theta(t - \hat{t}_i)\,, \qquad (37.15)$$

Part G | 37.2

Fig. 37.8 Dynamics of the Izhikevich model. Depending on the settings of the parameters a, b, c, and d, different neuron characteristics are modeled (generated by a freely available simulation tool provided by Izhikevich on http://www.izhikevich.org) ◄

where $\eta_0 = \vartheta$ equals the firing threshold of the neuron. The first term in (37.15) models the positive pulse with a decay rate τ_m and the second one is the negative spike after-potential with a decay rate τ_s, while K_1 and K_2 act as scaling factors. Function $\Theta(\cdot)$ is a step function known as the Heaviside function

$$\Theta(s) = \begin{cases} 0 & \text{if } s < 0 \\ 1 & \text{if } s \geq 0 \end{cases}, \qquad (37.16)$$

which ensures that the effect of the η kernel is zero if the neuron has not emitted a spike, i. e., $t < \hat{t}$. The shape of this kernel is presented in Fig. 37.9a, where $K_1 = 1$, $K_2 = 5$, $\tau_s = 0.005$, and $\tau_m = 0.01$ were used.

The second kernel determines the time course of a post-synaptic potential whenever the neuron receives an input spike. The kernel depends on the last firing time of the neuron $t - \hat{t}$ and on the firing times $t - t_j^{(f)}$ of the pre-synaptic neurons j. Due to the first dependence the post-synaptic neuron may respond differently to input spikes received immediately after a post-synaptic spike. A typical implementation of this kernel is e.g.,

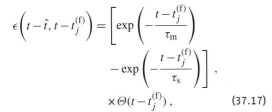

$$\epsilon\left(t - \hat{t}, t - t_j^{(f)}\right) = \left[\exp\left(-\frac{t - t_j^{(f)}}{\tau_m}\right)\right.$$
$$\left. - \exp\left(-\frac{t - t_j^{(f)}}{\tau_s}\right)\right]$$
$$\times \Theta(t - t_j^{(f)}), \qquad (37.17)$$

where $\Theta(\cdot)$ once more corresponds to the Heaviside function, the two exponential functions model a positive and a negative pulse with the corresponding decay rates, and $t_j^{(f)}$ is the spike time of a pre-synaptic neuron j. In (37.17), the first dependency of ϵ is neglected, which corresponds to a special case of the model, namely the simplified SRM. This simplified version of SRM is discussed in the next section. The time course of the ϵ kernel of (37.17), is presented in Fig. 37.9a. For the figure $\tau_s = 0.005$ and $\tau_m = 0.01$ were used. The implementations for the response kernels η and ϵ are adopted from the study on spike timing dependent plasticity in [37.19].

The third kernel function κ represents the linear response of the membrane to an input current I_{ext}. It depends on the last firing time of the neuron $t - \hat{t}$ and the time prior to t. It is used to model the time course of u due to external stimuli to the neuron.

A comprehensive discussion of the spike response model and its derivatives can be found in the excellent textbook [37.12] and also in [37.20].

Simplified Spike Response Model (SRM$_0$)
In a simplified version of SRM, the kernels ϵ and κ are replaced

$$\epsilon_0(s) = \epsilon_{ij}(\infty, s), \qquad (37.18)$$
$$\kappa_0(s) = \kappa_{ij}(\infty, s), \qquad (37.19)$$

which makes the kernels independent of the index j of pre-synaptic neurons and also of the last firing time $\hat{t}_i$ of the post-synaptic neuron. Using simple implementations of these kernel functions reduces the computational cost significantly. Hence, this model has been used to analyze the computational power of spiking neurons [37.21, 22], of network synchro-

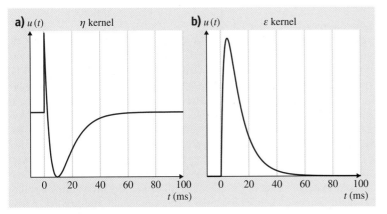

a) $u(t)$ η kernel   **b)** $u(t)$ ε kernel

t (ms)

Fig. 37.9a,b Shape of the response kernels η and ϵ. **(a)** A spike is triggered at time $t = t^{(f)} = 0$, which results in the activation of the η kernel. The shape of the spike and its after potential are modeled by this kernel function. **(b)** The neuron receives an input spike at time $t = 0$, which results in the activation of the ϵ kernel. If no further stimulus is received, the potential u returns to its resting potential

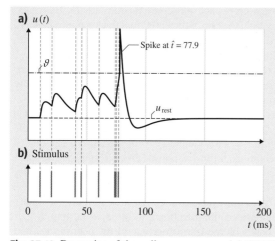

Fig. 37.10 Dynamics of the spike response model (SRM). In the post-synaptic neuron, spikes change the membrane potential described by the kernel function ϵ. If the membrane potential crosses a threshold ϑ, a spike is triggered. The shape of this action potential is modeled by the function η.

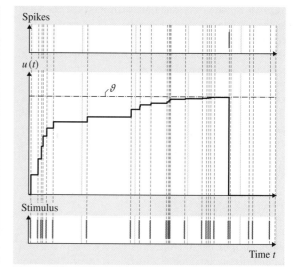

Fig. 37.11 Evolution of the post-synaptic potential (PSP) of the Thorpe neuronal model for a given input stimulus. If the potential reaches threshold ϑ, a spike is triggered and the PSP is set to 0 for the rest of the simulation, even if the neuron is still stimulated by incoming spike trains

nization [37.23] and collective phenomena of coupled networks [37.24].

The dynamics of the SRM_0 model are presented in Fig. 37.10. For the diagram, the ϵ and η kernels are defined by (37.15) and (37.17), respectively. The neuron receives a pre-synaptic stimulus in the form of several spikes which impact the potential u according to the response kernel ϵ. Due to the pre-synaptic activity, an action potential is triggered at time $t = 77.9$ ms, which results in the activation of the η kernel and the modeling of the spike shape and the after-potential. Figure 37.10 only presents excitatory synaptic activity.

37.2.5 Thorpe Model

A simplified LIF model was formally proposed in [37.25]. However, the general idea of the model can be traced back to publications from as early as 1990 [37.26]. This model lacks the post-synaptic potential leakage. The spike response of a neuron depends only on the arrival time of pre-synaptic spikes. The importance of early spikes is boosted and affects the post-synaptic potential more strongly than later spikes. This concept is very interesting due to the fact that the brain is able to compute even complex tasks quickly and reliably. For example, for the processing of visual data the human brain requires only approximately 150 ms [37.27], see also a similar study on rapid visual

categorization of natural and artificial objects [37.28]. Since it is known that this type of computation is partly sequential and several parts of the brain involving millions of neurons participate in the computation, it was argued in [37.29, 30] that each neuron has time and energy to emit only very few spikes that can actually contribute to the processing of the input. As a consequence, few spikes per neuron are biologically sufficient to solve a highly complex recognition task in real time.

Similar to other models, the dynamics of the Thorpe model are described by the evolution of the post-synaptic potential $u_i(t)$ of a neuron i as

$$u_i(t) = \begin{cases} 0 & \text{if fired} \\ \sum_{j \mid f(j) < t} w_{ji} m_i^{\text{order}(j)} & \text{else} \end{cases}, \quad (37.20)$$

where w_{ji} is the weight of a pre-synaptic neuron j, $f(j)$ is the firing time of j, and $0 < m_i < 1$ is a parameter of the model, namely the modulation factor. Function order(j) represents the rank of the spike emitted by neuron j. For example, a rank order(j) = 0 would be assigned if neuron j is the first among all pre-synaptic neurons of i that emits a spike. In a similar fashion, the spikes of all pre-synaptic neurons are ranked and then used in the computation of u_i. A neuron i fires a spike when its potential reaches a certain threshold ϑ. After emitting a spike, the potential resets to $u_i = 0$. Each neuron is allowed to emit only a single spike at most. The threshold $\vartheta = cu_{\max}$ is set to a fraction $0 < c < 1$ of the maximum potential $u_{\max}$ reachable for a neuron. Figure 37.11 presents the change of the post-synaptic potential for the Thorpe neural model if a series of input spikes stimulates the neuron through different synapses.

These simplifications allow a very fast real-time simulation of large networks. Due to its low com-putational costs this model was mainly used for studying image and speech recognition methods in-volving thousands of connected neurons [37.31, 32]. Many studies have investigated the Thorpe model, e.g., for face recognition [37.33, 34]. Additional studies uti-lizing this model are presented in Sect. 37.6, where principles and applications of the evolving spiking neural network architecture are discussed in the next sections.

37.3 Neural Encoding

This section addresses a fundamental question in neu-roscience: What is the code used by neurons to transmit information? Is it possible for an external observer to read and understand the message of neural activity? Traditionally, there are two main theories about neural encoding – pulse codes and rate codes. Both theories are discussed below.

37.3.1 Rate Codes

The first theory assumes that the mean firing rate of a neuron carries the most, maybe even all the informa-tion of a transmission. These codes are referred to as rate codes and have inspired the classical perceptron ap-proaches. The mean firing rate v is usually understood as the ratio of the average number of spikes n_{sp} observed over a specific time interval T, and T itself

$$v = \frac{n_{sp}}{T} \,. \tag{37.21}$$

This concept has been especially successful in the context of sensory or motor neural system, cf. e.g., the pioneering work by *Adrian* on the direct rela-tionship between the firing rate of stretch receptor neurons and the applied force in the muscles of frog legs [37.35]. Nevertheless, the idea of a mean firing rate has been repeatedly criticized (cf., e.g., [37.36]). The main argument is the comparably slow transmis-sion of information from one neuron to another, since each neuron has to integrate the spike activity of pre-synaptic neurons at least over a time T. Especially, the extremely short response times of the brain for certain stimuli, cannot be explained by the temporal averaging of spikes. For example, in [37.27] it was shown that the human brain can recognize a visual stimulus in approximately 150 ms. It is known that a moderate number of neural layers are involved in the processing of visual stimuli. If every layer had to wait a period T to receive the information from the previous layer, the recognition time would be much longer.

However, there is also another interpretation for the concept of the mean firing rate. It is defined as the av-erage spike activity over a population of neurons. The principle of this interpretation is explained in Fig. 37.12. A post-synaptic neuron receives stimulating inputs in the form of spikes emitted by a population of pre-synaptic neurons. This population produces a certain spike activity A, which is defined as the fraction of neurons being active within a short interval $[t, t + \Delta t]$

$$A = \frac{1}{\Delta t} \frac{n_{act}(t, t + \Delta t)}{N} \,, \tag{37.22}$$

where $n_{act}(t, t + \Delta t)$ denotes the number of active neu-rons in interval $[t, t + \Delta t]$, and N is the total number

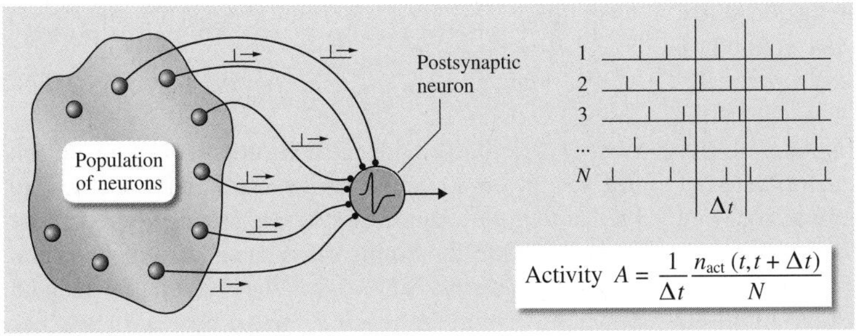

Activity $A = \dfrac{1}{\Delta t} \dfrac{n_{act}(t, t + \Delta t)}{N}$

Fig. 37.12 A neuron receives input spikes from a population of pre-synaptic neurons producing a certain activity A. The activity is defined as the fraction of neurons being active within a short interval $[t, t + \Delta t]$, di-vided by the population size N and the time period Δt. (After [37.12])

neuron in the population. The activity of a population may vary rapidly and thus allow fast responses of the neurons to changing stimuli [37.37, 38].

37.3.2 Pulse Codes

The second type of neural encoding is referred to as a spike or pulse code. These codes assume the precise spike time as the carrier of information between neurons. Experimental evidence for temporal correlations between spikes has been given through computer simulations [37.39], where integrate-and-fire models are investigated, but also through biological experiments, cf. the electrophysiological recordings and staining procedures in [37.40]; see also the in vivo measurements described in [37.41] in which spatio–temporal patterns of neuronal activity are analyzed in order to predict the behavior responses of rats.

A pulse code based on the timing of the first spike after a reference signal was discussed in [37.27]. This encoding is called time-to-first-spike and was inspired by the visual processing of the human eye. It was argued that each neuron has time to emit only a few spikes that can contribute to the overall processing of a stimulus. Indeed, it was also shown in [37.42] that a new stimulus is processed in the first 20–50 ms after its onset. Thus, earlier spikes carry most information about the stimulus. A specific neural model, namely the Thorpe model that boosts the importance of early spikes, was discussed already in Sect. 37.2.5.

Other pulse codes consider correlation and synchrony to be important. Neurons that represent a similar concept, object, or label are *labeled* by firing synchronously [37.43]. More generally, any precise spatio–temporal pulse pattern may be potentially meaningful and encode a particular information. Neurons that fire with a certain relative time delay may signify a certain stimulus.

As a practical example, the so-called rank order population encoding is presented in Sect. 37.6.1. Additional information about neural encoding in general can be found in the book by *Rieke* et al. [37.36].

37.4 Learning in SNN

This section presents some typical learning methods in the context of spiking neurons. A variety of problems impair the development of learning procedures for SNN. The explicit time dependence results in asynchronous information processing that commonly requires complex software and/or hardware implementations to simulate these neural networks. Additional difficulties are added by the fact that recurrent network topologies are commonly used in SNN and thus the formulation of a straightforward learning method, such as back-propagation for MLP, is not possible.

Similar to traditional neural networks, three different learning paradigms can be distinguished in SNN, which are referred to as unsupervised, reinforcement, and supervised learning. Reinforcement learning in SNN is probably the least common among the three. Some algorithms have been successfully applied in robotic applications [37.44], but were also theoretically analyzed in [37.45–47]. Unsupervised learning in the form of Hebbian learning is the most biologically realistic learning scenario. The so-called *spike-timing dependent plasticity* (STDP) belongs to this category and is discussed in the next section. Supervised techniques impose a certain input–output mapping on the network which is essential for practical applications of SNN.

Two methods are discussed in greater detail in the next sections. The learning algorithm employed in the eSNN architecture is discussed separately in Sect. 37.6.2. An excellent comparison of supervised learning methods developed for SNN can be found in [37.48].

37.4.1 STDP – Spike–Timing Dependent Plasticity

Spike-timing dependent plasticity is inspired by the experiments of Hebb published in his famous book *The Organization of Behavior* [37.49]. His essential postulate is often referred to as Hebb's law:

> *When an axon of cell A is near enough to excite cell B and repeatedly or persistently takes part in firing it, some growth process or metabolic change takes place in one or both cells such that A's efficiency, as one of the cells firing B, is increased.*

First experimental evidence that supports Hebb's postulate was given 20 years later in [37.50, 51]. Today, it is known that the change of synaptic efficacy in the brain is correlated to the timing of pre- and post-synaptic activity of a neuron [37.52–54]. Whenever the efficacy of a synapse is strengthened or weakened, we speak of

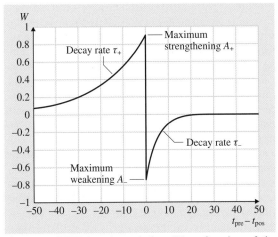

Fig. 37.13 STDP learning window W as function of the time difference $t_{pre} - t_{post}$ of pre- and post-synaptic spike times. The function presented is based on (37.23) using the following parameter setting: $A_+ = 0.9$, $A_- = -0.75$, $\tau_+ = 20$, and $\tau_- = 5$

long-term potentiation (LTP) or long-term depression (LTD), respectively. STDP is described by a function $W(t_{pre} - t_{post})$ that determines the fractional change of the synaptic weight in dependence of the difference between the arrival time t_{pre} of a pre-synaptic spike and the time t_{post} of an action potential emitted by the neuron. Function W is also known as the STDP window. Typical approximations of W are, e.g.,

$$W(t_{pre} - t_{post})$$
$$= \begin{cases} A_+ \exp\left(\dfrac{t_{pre} - t_{post}}{\tau_+}\right) & \text{if } t_{pre} < t_{post}\,, \\ A_- \exp\left(-\dfrac{t_{pre} - t_{post}}{\tau_+}\right) & \text{if } t_{pre} > t_{post}\,, \end{cases}$$
$$(37.23)$$

where parameters τ_+ and τ_- determine the temporal range of the pre- and post-synaptic time interval, while A_+ and A_- denote the maximum fractions of synaptic modification, if $t_{pre} - t_{post}$ is close to zero. Figure 37.13 presents the STDP window W according to (37.23).

The parameters for A_+, A_-, τ_+, and τ_- are adjusted according to the particular neuron to be modeled. The window W is usually temporally asymmetric, i.e., $A_+ \neq A_-$ and $\tau_+ \neq \tau_-$. However, there are also some exceptions, e.g., synapses of layer 4 spiny stellate neurons in the rat barrel cortex appear to have a symmetric window [37.55].

A study investigated the dynamics of synaptic pruning as a consequence of the STDP learning rule [37.56].

Synaptic pruning is a general feature of mammalian brain maturation and refines the embryonic nervous system by removing inappropriate synaptic connections between neurons, while preserving appropriate ones. Later studies extended this work by including apoptosis (genetically programmed cell death) into the analysis [37.57], and the identification of spatio–temporal patterns in the pruned network indicating the emergence of cell assemblies [37.58].

More information on STDP can be found in the excellent review on the matter in [37.59–62].

37.4.2 Spike-Prop

Traditional neural networks, like the multi-layer perceptron, usually employ some form of gradient based descent, i.e., error back-propagation, to modify synaptic weights in order to impose a certain input-output mapping on the network. However, the topological recurrence of SNN and their explicit time dependence do not allow a straightforward evaluation of the gradient in the network. Special assumptions need to be made to develop a version of back-propagation appropriate for spiking neurons.

In [37.63, 64] a back-propagation algorithm called spike-prop is proposed, which is suitable for training SNN. It is derived from the spike-response model discussed in Sect. 37.2.4. The aim of the method is to learn a set of desired firing times t_j^d of all output neurons j for a given input pattern presented to the network. Spike-prop minimizes the error E defined as the squared difference between all network output times t_j^{out} and desired output times t_j^d

$$E = \frac{1}{2} \sum_j \left(t_j^{out} - t_j^d\right)^2\,. \qquad (37.24)$$

The error is minimized with respect to the weights w_{ij}^k of each synaptic input

$$\Delta w_{ij}^k = -\eta \frac{dE}{dw_{ij}^k}\,, \qquad (37.25)$$

with η defining the learning rate of the update step.

A limitation of the algorithm is given by the requirement that each neuron is allowed to fire only once, which is similar to the limitations of the Thorpe neural model presented in Sect. 37.2.5. This simplification allows the error function defined in (37.24) to depend entirely on the difference between actual and desired spike time. Thus, only time-to-first-spike encoding is suitable in combination with spike-prop.

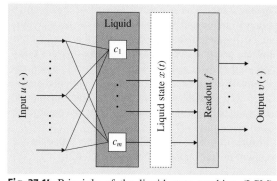

Fig. 37.14 Principle of the liquid state machine (LSM). The liquid transforms inputs u into a liquid state x, which in turn is mapped by a (linear) readout function f into the output v of the network. (After [37.68])

The algorithm was modified in a number of studies. In [37.65] a momentum term was included in the update of the weights, while [37.66] extended the method to learn additional neural parameters, such as synaptic delays, time constants, and neuron thresholds. An extension towards recurrent network topologies was presented in [37.67].

37.4.3 Liquid State Machine (LSM)

A very different approach to neural learning was proposed with the liquid state machine (LSM) introduced in [37.70]. The method is a specific form of reservoir

computing [37.71], that constructs a recurrent network of spiking neurons, for which all parameters of the network, i.e., synaptic weights, connectivity, delays, neural parameters, are randomly chosen and fixed during simulation. Such a network is also referred to as a *liquid*. If excited by an input stimulus, the liquid exhibits very complex non-linear dynamics that are expected to reflect the inherent information of the presented stimulus. The response of the network can be interpreted by a learning algorithm.

Figure 37.14 illustrates the principle of the LSM approach. As a first step in the general implementation of LSM a suitable liquid is chosen. This step determines, for example, the employed neural model along with its parameter configuration, as well as the connectivity strategy of the neurons, network size, and other network-related parameters. After creating the liquid, so-called liquid states $x(t)$ can be recorded at various time points in response to numerous different (training) inputs $u(t)$. Finally, a supervised learning algorithm is applied to a set of training examples of the form $(x(t), v(t))$ to train a readout function f, such that the actual outputs $f(x(t))$ are close to $v(t)$.

It was argued in [37.68] that LSM has universal computational power. A very appealing feature of the applied training method, i.e., the readout function, is its simplicity, since only a single layer of weights is actually modified, for which a linear training method is sufficient.

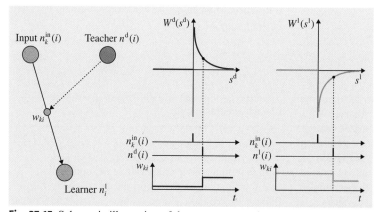

Fig. 37.15 Schematic illustration of the remote supervised method (ReSuMe). The synaptic change depends on the correlation of spike activities between input, teaching, and learning neurons. Spikes emitted by neuron input $n_k^{\text{in}}(i)$ followed by a spike of the teacher neuron $n^{\text{d}}(i)$ leads to an increase of synaptic weight w_{ki}. The value of w_{ki} is decreased, if $n_k^{\text{in}}(i)$ spikes before the learning neuron n_i^{l} is activated. The amplitude of the synaptic change is determined by two functions $W^{\text{d}}(s^{\text{d}})$ and $W^{\text{l}}(s^{\text{l}})$, where s^{d} is the temporal difference between the spike times of teacher neuron and input neuron, while s^{l} describes the difference between the spike times of learning neuron and input neuron. (After [37.69])

A specific implementation of the readout, the so-called remote supervised method (ReSuMe) introduced in [37.69], is presented here. The goal of ReSuMe is to impose a desired input–output spike pattern on an SNN, i.e., to produce target spike trains in response to a certain input stimulus. The method is based on the already presented STDP learning window as described in Sect. 37.4.1 for details, in which two opposite update rules for the synaptic weights are balanced. Additional teacher neurons are defined for each synapse, which remotely supervise the evolution of its synaptic weight. The teacher neuron is not explicitly connected to the network, but generates a reference spike signal which is used to update the connection weight in an STDP-like fashion. The post-synaptic neuron, whose activity is influenced by the weight update, is called the learning neuron.

Figure 37.15 illustrates the principle of ReSuMe. Let n_i^l denote the learning neuron which receives spike sequences from pre-synaptic neuron n_k^{in}, (i), the corresponding synaptic weight being w_{ki}, and neuron $n^d(i)$ being the teacher for weight w_{ki}. If input neuron n_k^{in}, (i) emits a spike that is followed by a spike of the teacher neuron $n^d(i)$, the synaptic weight w_{ki} is increased. On the other hand, if n_k^{in}, (i) spikes before the learning neuron n_i^l is activated, the synaptic weight is decreased. The amplitude of the synaptic change is determined by two functions $W^d(s^d)$ and $W^l(s^l)$, where s^d is the temporal difference between the spike times of teacher neuron and input neuron, while s^l describes the difference between the spike times of learning neuron and input neuron. Thus, the precise time difference of spiking activity defines the strength of the synaptic change.

A few studies on LSM can be found in the overview paper in [37.72] and the specific case study for isolated word recognition in [37.73]. More information on ReSuMe is available in [37.74–76].

37.5 Applications of SNN

Traditionally, SNN have been applied in the area of neuroscience to better understand brain functions and principles, the work by *Hodgkin* and *Huxley* [37.4] being among the pioneering studies in the field. A number of main directions for understanding the functioning of the nervous system are given in [37.77]. Here it is argued that a comprehensive knowledge about the anatomy of individual neurons and classes of cells, pathways, nuclei, and higher levels of organization is very important, along with detailed information about the pharmacology of ion channels, transmitters, modulators, and receptors. Furthermore, it is crucial to understand the biochemistry and molecular biology of enzymes, growth factors, and genes that participate in brain development and maintenance, perception, and behavior, learning, and diseases. A range of software systems for analyzing biologically plausible neural models exist, NEURON [37.78] and GENESIS [37.79], being the most prominent ones. Modeling and simulation are fundamental for the understanding of neural processes.

A number of large-scale studies have been undertaken recently to understand the complex behavior of ensembles of spiking [37.18, 80]. The review presented in [37.81] discusses challenges for implementations of spiking neural networks on FPGAs in the context of large-scale experiments.

SNN are also applied in many real-world applications. Notable progress has been made in areas such as speech recognition [37.73], learning rules [37.64], associative memory [37.82], and function approximation [37.83]. Other applications include biologically more realistic controllers for autonomous robots; see [37.84–86] for some interesting examples in this research area.

In Sect. 37.6 we focus on a few applications of the evolving spiking neural network architecture.

37.6 Evolving Spiking Neural Network Architecture

Based on [37.87], an evolving spiking neural network architecture (eSNN) was proposed in [37.88], which was initially designed as a visual pattern recognition system. Other studies have utilized eSNN as a general classification method, e.g., in the context of classifying water and wine samples [37.89]. The method is based on the already discussed Thorpe neural model, in which the importance of early spikes (after the onset of a certain stimulus) is boosted (Sect. 37.2.5). Synaptic plasticity is employed by a fast supervised one-pass learning algorithm that is explained as part of this section.

In order to classify real-valued data sets, each data sample, i. e., a vector of real-valued elements, is mapped into a sequence of spikes using a certain neural encoding technique. In the context of eSNN, the so-called rank order population encoding is employed, but other encoding may be suitable as well. The topology of eSNN is strictly feed-forward and organized in several layers. Weight modification only occurs on the connections between the neurons of the output layer and the neurons of either the hidden layer or the input layer. The weight modification occurs between the neurons of the output layer and the neurons of either hidden or input layer, only.

In Sect. 37.6.1 the encoding principle used in eSNN is presented, followed by the description of the one-pass learning method and the overall functioning of the eSNN method. Finally, a variety of applications based on the eSNN architecture are reviewed and summarized.

37.6.1 Rank Order Population Encoding

Rank order population encoding is an extension of the rank order encoding introduced in [37.25]. It allows the mapping of vectors of real-valued elements into a sequence of spikes. An implementation based on arrays of receptive fields was first described in [37.64]. Receptive fields allow the encoding of continuous values by using a collection of neurons with overlapping sensitivity profiles. Each input variable is encoded independently by a group of M one-dimensional receptive fields. For a variable n an interval $[I_{\min}^n, I_{\max}^n]$ is defined. The Gaussian receptive field of neuron i is given by its center μ_i

$$\mu_i = I_{\min}^n + \frac{2i - 3}{2} \frac{I_{\max}^n - I_{\min}^n}{M - 2} \qquad (37.26)$$

and width σ is

$$\sigma = \frac{1}{\beta} \frac{I_{\max}^n - I_{\min}^n}{M - 2} , \qquad (37.27)$$

with $1 \le \beta \le 2$. Parameter β directly controls the width of each Gaussian receptive field. Figure 37.16 depicts an example encoding of a single variable. For the diagram, $\beta = 2$ was used, the input interval $[I_{\min}^n, I_{\max}^n]$ was set to $[-1.5, 1.5]$, and $M = 5$ receptive fields were used.

More information on rank order coding strategies can be found in [37.90] and the accompanying article [37.34]. Very interesting is also the review on rapid spike-based processing strategies in the context of image recognition presented in [37.91], where most work on the Thorpe neural model and rank order coding is summarized. Rank order coding was also explored for speech recognition problems [37.92] and is a core part of the eSNN architecture.

37.6.2 One-Pass Learning

The aim of the learning method is to create output neurons, each of them labeled with a certain class label $l \in L$. The number and value of class labels depends on the classification problem to solve, i. e., L corresponds to the set of class labels of the given data set. After presenting a certain input sample to the network, the corresponding spike train is propagated through the SNN, which may result in the firing of certain output neurons. It is also possible that no output neuron is activated and the network remains silent. In this case, the classification result is undetermined. If one or more output neurons have emitted a spike, the neuron with the shortest response time among all activated output neurons is determined, i. e., the output neuron with the earliest spike time. The label of this neuron represents the classification result for the input sample presented.

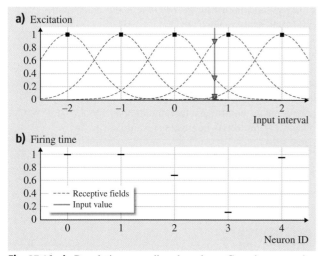

Fig. 37.16a,b Population encoding based on Gaussian receptive fields. (**a**) For an input value $v = 0.75$ (*thick straight*) the intersection points with each Gaussian is computed (*triangles*). (**b**) These points are then translated into spike time delays

Algorithm 37.1 Training an evolving spiking neural network

Require: m_l, s_l, c_l for a class label $l \in L$

1: initialize neuron repository $R_l = \{\}$
2: **for all** samples $X^{(i)}$ belonging to class l **do**
3:      $w_j^{(i)} \leftarrow (m_l)^{\text{order}(j)}, \forall j \mid j$ pre-synaptic neuron of i
4:      $u_{\max}^{(i)} \leftarrow \sum_j w_j^{(i)}(m_l)^{\text{order}(j)}$
5:      $\vartheta^{(i)} \leftarrow c_l u_{\max}^{(i)}$
6:      **if** $\min(d(w^{(i)}, w^{(k)})) < s_l, \quad w^{(k)} \in R_l$ **then**
7:          $w^{(k)} \leftarrow$ merge $w^{(i)}$ and $w^{(k)}$ according to (37.31)
8:          $\vartheta^{(k)} \leftarrow$ merge $\vartheta^{(i)}$ and $\vartheta^{(k)}$ according to (37.32)
9:      **else**
10:        $R_l \leftarrow R_l \cup \{w^{(i)}\}$
11:      **end if**
12: **end for**

The learning algorithm successively creates a repository of trained output neurons during the presentation of training samples. For each class label $l \in L$ an individual repository is evolved. The procedure is described in detail in Algorithm 37.1. For each training sample i with class label $l \in L$ a new output neuron is created and fully connected to the previous layer of neurons resulting in a real-valued weight vector $w^{(i)}$, with $w_j^{(i)} \in \mathbb{R}$ denoting the connection between the pre-synaptic neuron j and the created neuron i. In the next step, the input spikes are propagated through the network and the value of weight $w_j^{(i)}$ is computed according to the *order* of spike transmission through a synapse j, see line 6 in Algorithm 37.1

$$w_j^{(i)} = (m_l)^{\text{order}(j)} \forall j \mid j \text{ pre-synaptic neuron of } i . \tag{37.28}$$

Parameter m_l is the modulation factor of the Thorpe neural model. Differently labeled output neurons may have different modulation factors m_l. Function order(j) represents the rank of the spike emitted by neuron j. For example, a rank order(j) = 0 would be assigned, if neuron j is the first among all pre-synaptic neurons of i that emits a spike. In a similar fashion the spikes of all pre-synaptic neurons are ranked and then used in the computation of the weights.

The firing threshold $\vartheta^{(i)}$ of the created neuron i is defined as the fraction $c_l \in \mathbb{R}$, $0 < c_l < 1$, of the maximal possible potential $u_{\max}^{(i)}$, see lines 4 and 5 in

Algorithm 37.1,

$$\vartheta^{(i)} = c_l u_{\max}^{(i)} , \tag{37.29}$$

$$u_{\max}^{(i)} = \sum_j w_j^{(i)}(m_l)^{\text{order}(j)} . \tag{37.30}$$

The fraction c_l is a parameter of the model and for each class label $l \in L$ a different fraction can be specified.

The weight vector of the trained neuron is then compared the ones of neurons that are already stored neurons in the repository, cf. line 6 in Algorithm 37.1. If the minimal Euclidean distance between the weight vectors of the neuron i and an existing neuron k is smaller than a specified similarity threshold s_l, the two neurons are considered too *similar* and both the firing thresholds and the weight vectors are merged according to

$$w_j^{(k)} \leftarrow \frac{w_j^{(i)} + N w_j^{(k)}}{1 + N} ,$$
$$\forall j \mid j \text{ pre-synaptic neuron of } i \tag{37.31}$$

$$\vartheta^{(k)} \leftarrow \frac{\vartheta^{(i)} + N \vartheta^{(k)}}{1 + N} . \tag{37.32}$$

Integer N denotes the number of samples previously used to update neuron k. The merging is implemented as the (running) average of the connection weights, and the (running) average of the two firing thresholds. After the merging, the trained neuron i is discarded and the next sample processed. If no other neuron in the repository is similar to the trained neuron i, the neuron i is added to the repository as a new output neuron.

Figure 37.17 depicts the eSNN architecture. Due to the incremental evolution of output neurons, it is possible to accumulate knowledge as it becomes available. Hence, a trained network is able to learn new data without the need for re-training on already learnt samples. Real-world applications of the eSNN architecture are discussed in the next section.

37.6.3 Applications

The eSNN architecture is used in a variety of applications that are described and summarized here.

Visual Pattern Recognition

Among the earliest applications of eSNN is the visual pattern recognition system presented in [37.88], which extends the work of [37.93, 94] by including the online learning technique described before. In [37.88, 95] the method was studied on an image data set consisting of 400 faces of 40 different persons. The task here was to predict the class labels of presented images cor-

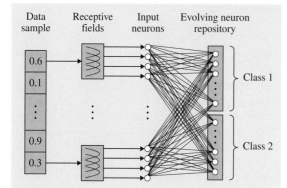

Fig. 37.17 Schematic illustration of the evolving spiking neural network architecture (eSNN). Real-valued vector elements are mapped into the time domain using rank order population encoding based on Gaussian receptive fields. As a consequence of this transformation input neurons emit spikes at pre-defined firing times, invoking the one-pass learning algorithm of the eSNN. The learning iteratively creates repositories of output neurons, one repository for each class. Here a two-class problem is presented. Due to the evolving nature of the network, it is possible to accumulate knowledge as it becomes available, without the requirement of re-training with already learnt samples

rectly. The system was trained on a subset of the data and then tested on the remaining samples of the data. Classification results were similar to [37.93, 94] with the additional advantages of the novel on-line learning method.

In a later study another processing layer was added to the system, which allows efficient multi-view visual pattern recognition [37.96]. The additional layer accumulates information over several different views of an image in order to reach a final decision about the associated class label of the frames. Thus, it is possible to perform an efficient on-line person authentication through the presentation of a short video clip to the system, although the audio information was ignored in this study.

The main principle of this image recognition method is briefly outlined here. The neural network is composed of four layers of Thorpe neurons, each of them grouping a set of neurons into several two-dimensional maps, so-called neural maps. Information in this network is propagated in a feed-forward manner, i.e., no recurrent connections exist. An input frame in the form of a gray-scale image is fed into the first neural layer (L_1), each pixel of the image corresponding to one neuron in a neural map of L_1. Several neural maps may exist in this layer. The map consists of *on* and *off*

neurons that are responsible for the enhancement of the high contrast parts of the image. Each map is configured differently and thus is sensitive to different gray scales in the image. The output of this layer is transformed into the spike domain using rank order encoding as described in [37.25]. As a consequence of this encoding, pixels with higher contrast are prioritized in the neural processing.

The second layer, denoted L_2, consists of orientation maps. Each map is selective for different directions, e.g., $0, 45, \ldots, 315°$, and is implemented by appropriately parameterized Gabor functions. It is noted that the first two layers are passive filters that are not subject to any learning process. In the third layer, L_3, the learning occurs using the one-pass learning method described in Sect. 37.6.2. Here neural maps are created and merged according to the rules of the learning algorithm. Finally, the fourth layer L_4, consists of a single neuron for each output class, which accumulates opinions about the class label of a certain sequence of input frames. The weights between L_3 and L_4 are fixed to a constant value, usually 1, and are not subject to learning. The first L_4 neuron that is activated by the stimuli presented determines the classification result for the input. After the activation of an L_4 neuron the system stops.

Experimental evidence about the suitability of this pattern recognition system is provided in [37.96] along with a comparison to other typical classification methods.

Auditory Pattern Recognition

A similar network, but in an entirely different context, was investigated in [37.97], where a text-independent speaker authentication system is presented. The classification task in this work consisted of the correct labeling of audio streams presented to the system.

Speech signals are split into temporal frames, each containing a signal segment over a short time period. The frames are first pre-processed using the mel-frequency cepstral coefficients (MFCCs) [37.98] and then used to invoke the eSNN. The MFCC frame is transformed into the spike domain using rank order encoding [37.25] and the resulting stimulus is propagated to the first layer of neurons. This layer, denoted L_1, contains two neural ensembles representing the speaker and the background model, respectively. While the former model is trained on the voice of a certain speaker, the latter is trained on the background noise of the audio stream. This system also collects opinions about the class label of the presented sequence of input frames, which is implemented by the second layer of the network. Layer L_2 consists of only two neurons, each of which accumulates information about whether a given frame corresponds to a certain speaker or to the background noise. Whenever an L_2 neuron is activated, the simulation of the network stops and the classification output is presented.

Audio-Visual Pattern Recognition

The two recognition systems presented above were successfully combined, forming an audio-visual pattern recognition method. Both systems are trained individually, but their output is propagated to an additional supra-modal layer. The supra-modal layer integrates incoming sensory information from individual modalities and cross-modal connections enable the influence of one modality upon the other. A detailed discussion of this system along with experimental evidence is given in [37.99, 100].

Case Study on Ecological Modeling

In [37.101, 102], the eSNN was applied on a real world data set in the context of an ecological modeling problem. For many invertebrate species little is known about their response to environmental variables over large spatial scales. That knowledge is important since it can help to identify critical locations in which a species that has the potential to cause great environmental harm might establish a new damaging population. The usual approach to determine the importance of a range of environmental variables that explain the global distribution of a species is to train or fit a model to its known distribution using environmental parameters measured in areas where the species is present and where it is absent.

Meteorological data that comprised 68 monthly and seasonal temperature, rainfall, and soil moisture variables for 206 global geographic sites were compiled from published records. These variables were correlated to global locations where the Mediterranean fruit-fly (*Ceratitis capitata*), a serious invasive species and fruit pest, was recorded at the time of the study, as either present or absent. Motivated by inadequate results [37.103–105] used a different method, namely the multi-layer perceptron (MLP); this study aimed to identify important features relevant for predicting the presence/absence of this insect species. The results obtained may also be of importance to evaluate the risk of invasion of certain species into specific geographical regions.

Taste Recognition

The last application of eSNN being discussed here investigates the use of SNN for taste recognition in a gustatory model. The classification performance of eSNN was experimentally explored based on water and wine samples collected from [37.106, 107]. The topology of the model consists of two layers. The first layer receives an input stimulus obtained from the mapping of a real-valued data sample into spike trains using a rank order population encoding (Sect. 37.6.1) The weights from the first neural layer are subject to training according to the already discussed one-pass learning method. Finally, the output of the second neural layer determines the class label of the presented input stimulus.

The method was investigated in a number of scenarios, where the size of the data sets and the number of class labels was varied. Generally, eSNN reported promising results on both large and small data sets, which has motivated an FPGA hardware implementation of the system [37.108].

References

37.1 Wikipedia: Neuron, available online at http://en.wikipedia.org/wiki/Neuron

37.2 E.R. Kandel: *Principles of Neural Science* (McGraw-Hill, Columbus 2000)

37.3 W. Maass: Networks of spiking neurons: The third generation of neural network models, Neural Netw. **10**(9), 1659–1671 (1997)

37.4 A.L. Hodgkin, A.F. Huxley: A quantitative description of membrane current and its application to conduction and excitation in nerve, J. Physiol. **117**(4), 500–544 (1952)

37.5 M. Nelson, J. Rinzel: The Hodgkin–Huxley model. In: *The Book of Genesis*, ed. by J.M. Bower, D. Beeman (Springer, Berlin, Heidelberg 1995) pp. 27–51

37.6 J.M. Bower, D. Beeman: *The Book of Genesis* (Springer, Berlin, Heidelberg 1995)

37.7 L. Lapicque: Recherches quantitatives sur l'excitation électrique des nerfs traitée comme une polarisation, J. Physiol. Pathol. Gen. **9**, 620–635 (1907)

37.8 L.F. Abbott: Lapicque's introduction of the integrate-and-fire model neuron (1907), Brain Res. Bull. **50**(5/6), 303–304 (1999)

37.9 N. Brunel, M.C.W. van Rossum: Lapicque's 1907 paper: From frogs to integrate-and-fire, Biol. Cybern. **97**(5), 337–339 (2007)

37.10 B.W. Knight: Dynamics of encoding in a population of neurons, J. Gen. Physiol. **59**, 734–766 (1972)

37.11 H. Meffin, A.N. Burkitt, D.B. Grayden: An analytical model for the "large, fluctuating synaptic conductance state" typical of neocortical neurons in vivo, J. Comput. Neurosci. **16**, 159–175 (2004)

37.12 W. Gerstner, W.M. Kistler: *Spiking Neuron Models: Single Neurons, Populations, Plasticity* (Cambridge Univ. Press, Cambridge 2002)

37.13 N. Burkitt: A review of the integrate-and-fire neuron model: I. Homogeneous synaptic input, Biol. Cybern. **95**(1), 1–19 (2006)

37.14 N. Burkitt: A review of the integrate-and-fire neuron model: II. Inhomogeneous synaptic input and network properties, Biol. Cybern. **95**(2), 97–112 (2006)

37.15 E.M. Izhikevich: Simple model of spiking neurons, IEEE Trans. Neural Netw. **14**(6), 1569–1572 (2003g)

37.16 E.M. Izhikevich: *Dynamical Systems in Neuroscience: The Geometry of Excitability and Bursting* (MIT Press, Cambridge 2006)

37.17 E.M. Izhikevich: Which model to use for cortical spiking neurons?, IEEE Trans. Neural Netw. **15**(5), 1063–1070 (2004)

37.18 E.M. Izhikevich, G.M. Edelman: Large-scale model of mammalian thalamocortical systems, Proc. Natl. Acad. Sci. USA **105**(9), 3593–3598 (2008)

37.19 T. Masquelier, R. Guyonneau, S.J. Thorpe: Spike timing dependent plasticity finds the start of repeating patterns in continuous spike trains, PLoS ONE **3**, e1377 (2008)

37.20 W. Maass, C.M. Bishop (Eds.): *Pulsed Neural Networks* (MIT Press, Cambridge 1999)

37.21 W. Maass: Lower bounds for the computational power of networks of spiking neurons, Neural Comput. **8**(1), 1–40 (1996)

37.22 W. Maass: Computing with spiking neurons. In: *Pulsed Neural Networks* (MIT Press, Cambridge 1999) pp. 55–85

37.23 W. Gerstner, J.L. van Hemmen, J.D. Cowan: What matters in neuronal locking?, Neural Comput. **8**(8), 1653–1676 (1996)

37.24 W.M. Kistler, R. Seitz, J.L. van Hemmen: Modeling collective excitations in cortical tissue, J. Phys. D **114**(3/4), 273–295 (1998)

37.25 S.J. Thorpe, J. Gautrais: *Rank Order Coding* (Plenum, New York 1998) pp. 113–118

37.26 S.J. Thorpe: Spike arrival times: A highly efficient coding scheme for neural networks. In: *Paralle Processing in Neural Systems and Computers*, ed. by R. Eckmiller, G. Hartmann, G. Hauske (Elsevier, Amsterdam 1990) pp. 91–94

37.27 S.J. Thorpe, D. Fize, C. Marlot: Speed of processing in the human visual system, Nature **381**, 520–522 (1996)

37.28 R. Van Rullen, S.J. Thorpe: Rate coding versus temporal order coding: What the retinal ganglion cells tell the visual cortex, Neural Comput. **13**(6), 1255–1283 (2001)

37.29 S.J. Thorpe, J. Gautrais: Rapid visual processing using spike asynchrony, Proc. Adv. Neural Inf. Process. Syst., Vol. 9 (NIPS), Denver 1996 (MIT Press, Cambridge 1996) pp. 901–907

37.30 S.J. Thorpe: How can the human visual system process a natural scene in under 150 ms? On the role of asynchronous spike propagation, ESANN 1997, 5th Eur. Symp. Artif. Neural Netw., D-Facto (1997)

37.31 A. Delorme, S.J. Thorpe: SpikeNET: An event-driven simulation package for modelling large networks of spiking neurons, Network **14**, 613–627 (2003)

37.32 S.J. Thorpe, R. Guyonneau, N. Guilbaud, J.-M. Allegraud, R. Van Rullen: SpikeNet: Real-time visual processing with one spike per neuron, Neurocomputing **58–60**, 857–864 (2004)

37.33 R. Van Rullen, J. Gautrais, A. Delorme, S. Thorpe: Face processing using one spike per neurone, Biosystems **48**(1–3), 229–239 (1998)

37.34 A. Delorme, L. Perrinet, S.J. Thorpe: Networks of integrate-and-fire neurons using rank order coding B: Spike timing dependent plasticity and emergence of orientation selectivity, Neurocomputing **38–40**, 539–545 (2001)

37.35 E.D. Adrian: The impulses produced by sensory nerve endings, J. Physiol. (London) **61**, 49–72 (1926)

37.36 F. Rieke, D. Warland, R.R. van Steveninck, W. Bialek: *Spikes: Exploring the Neural Code* (MIT Press, Cambridge 1999)

37.37 W. Gerstner: Population dynamics of spiking neurons: Fast transients, asynchronous states, and locking, Neural Comput. **12**(1), 43–89 (2000)

37.38 N. Brunel, F.S. Chance, N. Fourcaud, L.F. Abbott: Effects of synaptic noise and filtering on the frequency response of spiking neurons, Phys. Rev. Lett. **86**, 2186–2189 (2001)

37.39 R. Lestienne: Determination of the precision of spike timing in the visual cortex of anaesthetised cats, Biol. Cybern. **74**(1), 55–61 (1995)

37.40 M.P. Nawrot, P. Schnepel, A. Aertsen, C. Boucsein: Precisely timed signal transmission in neocortical networks with reliable intermediate-range projections, Front. Neural Circuits **3**(2), 1–11 (2009)

37.41 A.E.P. Villa, I.V. Tetko, B. Hyland, A. Najem: Spatiotemporal activity patterns of rat cortical neurons predict responses in a conditioned task, Proc. Natl. Acad. Sci. USA **96**(3), 1106–1111 (1999)

37.42 M.J. Tovee, E.T. Rolls, A. Treves, R.P. Bellis: Information encoding and the responses of single neurons in the primate temporal visual cortex, J. Neurophysiol. **70**(2), 640–654 (1993)

37.43 C. von der Malsburg: *The Correlation Theory of Brain Function*, Internal Report, Vol. 81-2 (Max-Planck-Institute for Biophysical Chemistry, Göttingen 1981)

37.44 R.V. Florian: A reinforcement learning algorithm for spiking neural networks, Proc. 7th Int. Symp.

Symb. Numer. Algorithms Sci. Comput. (SYNASC 2005) (IEEE, Los Alamitos 2005) pp. 299–306

37.45   R.V. Florian: Reinforcement learning through modulation of spike-timing-dependent synaptic plasticity, Neural Comput. **19**(6), 1468–1502 (2007)

37.46   H.S. Seung: Learning in spiking neural networks by reinforcement of stochastic synaptic transmission, Neuron **40**(6), 1063–1073 (2003)

37.47   X. Xie, H.S. Seung: Learning in neural networks by reinforcement of irregular spiking, Phys. Rev. E **69**(4), 041909 (2004)

37.48   A.J. Kasinski, F. Ponulak: Comparison of supervised learning methods for spike time coding in spiking neural networks, Int. J. Appl. Math. Comput. Sci. **16**, 101–113 (2006)

37.49   D.O. Hebb (Ed.): *The Organization of Behavior* (Wiley, New York 1949)

37.50   T.V.P. Bliss, T. Lomo: Long-lasting potentiation of synaptic transmission in the dentate area of the anaesthetized rabbit following stimulation of the perforant path, J. Physiol. **232**(2), 331–356 (1973)

37.51   T.V.P. Bliss, A.R. Gardner-Medwin: Long-lasting potentiation of synaptic transmission in the dentate area of the unanaesthetized rabbit following stimulation of the perforant path, J. Physiol. **232**(2), 357–374 (1973)

37.52   C.C. Bell, V.Z. Han, Y. Sugawara, K. Grant: Synaptic plasticity in a cerebellum-like structure depends on temporal order, Nature **387**, 278–281 (1997)

37.53   H. Markram, J. Lubke, M. Frotscher, B. Sakmann: Regulation of synaptic efficacy by coincidence of postsynaptic APs and EPSPs, Science **275**(5297), 213–215 (1997)

37.54   G.-Q. Bi, M.M. Poo: Synaptic modifications in cultured hippocampal neurons: Dependence on spike timing, synaptic strength, and postsynaptic cell type, J. Neurosci. **18**(24), 10464–10472 (1998)

37.55   V. Egger, D. Feldmeyer, B. Sakmann: Coincidence detection and changes of synaptic efficacy in spiny stellate neurons in rat barrel cortex, Nat. Neurosci. **2**, 1098–1105 (1999)

37.56   J. Iglesias, J. Eriksson, F. Grize, M. Tomassini, A.E.P. Villa: Dynamics of pruning in simulated large-scale spiking neural networks, Biosystems **79**(1–3), 11–20 (2005)

37.57   J. Iglesias, A.E.P. Villa: Neuronal cell death and synaptic pruning driven by spike-timing dependent plasticity, LNCS **4132**, 953–962 (2006)

37.58   J. Iglesias, A.E.P. Villa: Effect of stimulus-driven pruning on the detection of spatiotemporal patterns of activity in large neural networks, Biosystems **89**(1–3), 287–293 (2007)

37.59   G.-Q. Bi, M.-M. Poo: Synaptic modification by correlated activity: Hebb's postulate revisited, Annu. Rev. Neurosci. **24**(1), 139–166 (2001)

37.60   R. Kempter, W. Gerstner, J.L. van Hemmen: Hebbian learning and spiking neurons, Phys. Rev. E **59**(4), 4498–4514 (1999)

37.61   W. Gerstner, W.K. Kistler: Mathematical formulations of Hebbian learning, Biol. Cybern. **87**(5/6), 404–415 (2002)

37.62   W.M. Kistler: Spike-timing dependent synaptic plasticity: A phenomenological framework, Biol. Cybern. **87**(5/6), 416–427 (2002)

37.63   S.M. Bohte, J.N. Kok, J.A. La Poutré: SpikeProp: Backpropagation for networks of spiking neurons, ESANN Proc. (2000) pp. 419–424

37.64   S.M. Bohte, J.N. Kok, J.A. La Poutré: Error-backpropagation in temporally encoded networks of spiking neurons, Neurocomputing **48**(1–4), 17–37 (2002)

37.65   J. Xin, M.J. Embrechts: *Supervised Learning with Spiking Neural Networks* (IEEE, Bellingham 2001) pp. 1772–1777

37.66   B. Schrauwen, J. van Campenhout: Improving SpikeProp: Enhancements to an error-backpropagation rule for spiking neural networks, Proc. 15th ProRISC Workshop (2004) pp. 104–174

37.67   P. Tiño, A.J.S. Mills: Learning beyond finite memory in recurrent networks of spiking neurons, Neural Comput. **18**(3), 591–613 (2006)

37.68   T. Natschläger, W. Maass, H. Markram: The "liquid computer": A novel strategy for real-time computing on time series, Telematik **8**(1), 39–43 (2002)

37.69   F. Ponulak: *ReSuMe – New supervised learning method for Spiking Neural Networks*, Tech. Rep. (Institute of Control and Information Engineering, Poznan University of Technology 2005)

37.70   W. Maass, T. Natschläger, H. Markram: Real-time computing without stable states: A new framework for neural computation based on perturbations, Neural Comput. **14**(11), 2531–2560 (2002)

37.71   D. Verstraeten, B. Schrauwen, M. D'Haene, D. Stroobandt: An experimental unification of reservoir computing methods, Neural Netw. **20**(3), 391–403 (2007)

37.72   T. Natschläger, H. Markram, W. Maass: Computer models and analysis tools for neural microcircuits. In: *Neuroscience Databases: A Practical Guide*, ed. by R. Kötter (Kluwer Academic, Dordrecht 2003) pp. 123–138

37.73   D. Verstraeten, B. Schrauwen, D. Stroobandt: Isolated word recognition using a Liquid State Machine, Proc. ESANN (2005) pp. 435–440

37.74   A.J. Kasinski, F. Ponulak: Experimental demonstration of learning properties of a new supervised learning method for the spiking neural networks, LNCS **3696**, 145–152 (2005)

37.75   F. Ponulak, A.J. Kasinski: Generalization properties of spiking neurons trained with ReSuMe method, Proc. ESANN (2006) pp. 629–634

37.76   F. Ponulak: Analysis of the ReSuMe learning process for spiking neural networks, Appl. Math. Comput. Sci. **18**(2), 117–127 (2008)

37.77   N.T. Carnevale, M.L. Hines: *The NEURON Book* (Cambridge Univ. Press, University 2006)

37.78   M.L. Hines, N.T. Carnevale: Neuron: A tool for neuroscientists, Neuroscientist **7**(2), 123–135 (2001)

37.79   J.M. Bower, D. Beeman: GENESIS (simulation environment), Scholarpedia **2**(3), 1383 (2007)

37.80   B.P. Glackin, T.M. McGinnity, L.P. Maguire, Q. Wu, A. Belatreche: A novel approach for the implementation of large scale spiking neural networks on FPGA Hardware, LNCS **3512**, 552–563 (2005)

37.81   L.P. Maguire, T.M. McGinnity, B. Glackin, A. Ghani, A. Belatreche, J. Harkin: Challenges for large-scale implementations of spiking neural networks on FPGAs, Neurocomputing **71**(1–3), 13–29 (2007)

37.82   A. Knoblauch: Neural associative memory for brain modeling and information retrieval, Inf. Process. Lett. **95**(6), 537–544 (2005)

37.83   N. Iannella, L. Kindermann: Finding iterative roots with a spiking neural network, Inf. Process. Lett. **95**(6), 545–551 (2005)

37.84   D. Floreano, C. Mattiussi: Evolution of spiking neural controllers for autonomous vision-based robots, LNCS **2217**, 38–61 (2001)

37.85   D. Floreano, Y. Epars, J.-C. Zufferey, C. Mattiussi: Evolution of spiking neural circuits in autonomous mobile robots: Research articles, Int. J. Intell. Syst. **21**(9), 1005–1024 (2006)

37.86   X. Wang, Z.-G. Hou, A. Zou, M. Tan, L. Cheng: A behavior controller based on spiking neural networks for mobile robots, Neurocomputing **71**(4–6), 655–666 (2008)

37.87   N. Kasabov: *Evolving Connectionist Systems: The Knowledge Engineering Approach* (Springer, New York 2006)

37.88   S.G. Wysoski, L. Benuskova, N.K. Kasabov: Adaptive learning procedure for a network of spiking neurons and visual pattern recognition, LNCS **4179**, 1133–1142 (2006)

37.89   S. Soltic, S. Wysoski, N. Kasabov: Evolving spiking neural networks for taste recognition, Proc. IJCNN 2008 (IEEE, Bellingham 2008) pp. 2091–2097

37.90   L. Perrinet, A. Delorme, M. Samuelides, S.J. Thorpe: Networks of integrate-and-fire neuron using rank order coding A: How to implement spike time dependent Hebbian plasticity, Neurocomputing **38–40**, 817–822 (2001)

37.91   S.J. Thorpe, A. Delorme, R. van Rullen: Spike-based strategies for rapid processing, Neural Netw. **14**(6/7), 715–725 (2001)

37.92   S. Loiselle, J. Rouat, D. Pressnitzer, S. Thorpe: Exploration of rank order coding with spiking neural networks for speech recognition, Proc. IJCNN 2005 (IEEE, Bellingham 2005) pp. 2076–2080

37.93   A. Delorme, J. Gautrais, R. Van Rullen, S. Thorpe: SpikeNET: A simulator for modeling large networks of integrate and fire neurons, Neurocomputing **26–27**, 989–996 (1999)

37.94   A. Delorme, S.J. Thorpe: Face identification using one spike per neuron: Resistance to image degradations, Neural Netw. **14**(6/7), 795–803 (2001)

37.95   S.G. Wysoski, L. Benuskova, N. Kasabov: On-line learning with structural adaptation in a network of spiking neurons for visual pattern recognition, LNCS **4131**, 61–70 (2006)

37.96   S.G. Wysoski, L. Benuskova, N. Kasabov: Fast and adaptive network of spiking neurons for multi-view visual pattern recognition, Neurocomputing **71**(13–15), 2563–2575 (2008)

37.97   S.G. Wysoski, L. Benuskova, N. Kasabov: Text-independent speaker authentication with spiking neural networks, LNCS **4669**, 758–767 (2007)

37.98   L. Rabiner, B.-H. Juang: *Fundamentals of Speech Recognition* (Prentice-Hall, Upper Saddle River 1993)

37.99   S.G. Wysoski, L. Benuskova, N. Kasabov: Adaptive spiking neural networks for audiovisual pattern recognition, LNCS **4985**, 406–415 (2008)

37.100   S.G. Wysoski: Evolving Spiking Neural Networks for Adaptive Audiovisual Pattern Recognition. Ph.D. Thesis (Auckland University, Auckland 2008)

37.101   S. Schliebs, M. Defoin-Platel, S. Worner, N. Kasabov: Quantum-inspired feature and parameter optimisation of evolving spiking neural networks with a case study from ecological modeling, Proc. IJCNN (IEEE Computer Society, Washington 2009) pp. 2833–2840

37.102   S. Schliebs, M. Defoin-Platel, S. Worner, N. Kasabov: Integrated feature and parameter optimization for an evolving spiking neural network: Exploring heterogeneous probabilistic models, Neural Netw. **22**(5–6), 623–632 (2009)

37.103   S. Worner, G. Lankin, S. Samarasinghe, D. Teulon: Improving prediction of aphid flights by temporal analysis of input data for an artificial neural network, N. Z. Plant Prot. **55**, 312–316 (2002)

37.104   N. Cocu, R. Harrington, M.D. Rounsevell, S.P. Worner, M. Hulle: Geographical location, climate and land use influences on the phenology and numbers of the aphid, *Myzus persicae*, in Europe, J. Biogeogr. **32**(4), 615–632 (2005)

37.105   M.J. Watts, S.P. Worner: Using MLP to determine abiotic factors inuencing the establishment of insect pest species, Proc. 2006 Int. Joint Conf. Neural Netw. (IJCNN 2006) (IEEE, New York 2006) pp. 3506–3511

37.106   H.C. de Sousa, A. Riul Jr.: Using MLP networks to classify red wines and water readings of an electronic tongue, VII Braz. Symp. Neural Netw. (SBRN'02) (2002)

37.107   A. Riul, H.C. de Sousa, R.R. Malmegrim, D.S. dos Santos, A.C.P.L.F. Carvalho, F.J. Fonseca, O.N. Oliveira, L.H.C. Mattoso: Wine classification by taste sensors made from ultra-thin films and using neural networks, Sens. Actuators B **98**(1), 77–82 (2004)

37.108   A. Zuppicich, S. Soltic: FPGA Implementation of an evolving spiking neural network, LNCS **5506**, 1129–1136 (2009)

Part G | 37

38. Statistical Methods for fMRI Activation and Effective Connectivity Studies

Xingfeng Li, Damien Coyle, Liam Maguire, T. Martin McGinnity

Functional magnetic resonance imaging (fMRI) is a technique to indirectly measure activity in the brain through the flow of blood. fMRI has been a powerful tool in helping us gain a better understanding of the human brain since it appeared over 20 years ago. However, fMRI poses many challenges for engineers. In particular, to detect and interpret the blood oxygen level-dependent (BOLD) signals on which fMRI is based is a challenge; For example, fMRI activation may be caused by a local neural population (activation detection) or by a distant brain region (effective connectivity). Although many advanced statistical methods have been developed for fMRI data analysis, many problems are still being addressed to maximize the accuracy in activation detection and effective connectivity analysis. This chapter presents general statistical methods for activation detection and effective connectivity analysis in fMRI scans of the human brain. A linear regression model for activation detection is introduced (Sect. 38.2.1), and a detailed statistical inference method for activation detection (Sect. 38.2.2) when applying an autoregression model to correct residual terms of the linear model is presented in Sect. 38.2.3. We adopt a two-stage mixed model to combine different subjects (Sect. 38.3) for second-level data analysis. To estimate the variance for the mixed model, a modified expectation-maximization algorithm is employed. Finally, due to the false positives in the activation map, a Bonferroni-related threshold correction method is developed to control the false positives (Sect. 38.3.3). An fMRI dataset from retinotopic mapping experiments was employed to test the feasibility of the methods for both first- and second-level analysis.

In Sect. 38.4, we present a nonlinear system identification method (NSIM) for effective connectivity study. The mathematical theory of the method is presented in Sect. 38.4.1. An F statistical test is proposed to quantify the magnitude of the relationship based on two-connection and three-connection visual networks (Sect. 38.4.3). To circumvent the limitation of the model overfitting in NSIM, a model selection algorithm is suggested subsequently. In the model selection, we propose a nongreedy search method, e.g., least angle regression for effective connectivity analysis (Sects. 38.4.7 and 38.4.8). Three datasets obtained from standard block and random block designs are used to verify the method, and we outline some research directions and conclude our study in Sect. 38.5.

38.1 Functional Magnetic Resonance Imaging

38.1.1 Overview

The human brain is often considered as the most complex system known to humans and is no doubt the most complex biological system, hence it poses many difficulties in studies aimed at investigating and understanding its structural, functional, and dynamic complexity. This difficulty is mainly due to the fact that we lack effective tools to discover brain function in vivo. With the advent of functional magnetic resonance imaging (fMRI) in the early 1990s [38.1, 2], it is now possible not only to localize normal brain activation in response to a specific cognitive task but also to investigate effective connectivity from the response of cortex. Because fMRI offers high spatial image resolution, it has become an important tool to unveil human brain function for both neurology and physiology purposes. It greatly expands our ability to understand functional brains using different types of stimuli. However, in fMRI studies, we face one vital problem, namely how to extract interpretable information from a huge number of possible neuronal population responses. The brain is highly nonstationary and rarely in the same state. Although modern statisticians and signal processing engineers have made significant progress in detecting the information, e.g., activation and connectivity, many issues for activation detection and effective connectivity analysis remain open; For example, how to study brain function integration and segregation using a single model, instead of considering one brain system with two separate models, is a key issue. The objective of this chapter is to introduce these issues and present novel statistical solutions for activation and effective connectivity studies based on linear and nonlinear systems theory. Linear regression methods and mixed-effect models will be applied to address fMRI activation detection. In the study of fMRI effective connectivity, a nonlinear system identification approach is proposed, and a model selection algorithm to overcome the overfitting problem in the conventional nonlinear system identification method is also presented. Finally, we test these methods in a wide range of experiments and foresee applying these statistical methods for brain activation detection and connectivity studies in the future.

38.1.2 Background of BOLD-fMRI and fMRI Data Processing Methods

fMRI using blood oxygenation level-dependent (BOLD) contrast is currently the mainstay of neuroimaging in cognitive neuroscience [38.3]. BOLD contrast results from a complex interplay between cerebral blood flow (CBF), cerebral blood volume (CBV), blood oxygen extraction, and local metabolism which takes place as a consequence of neuronal activation. The end result is a decrease in the local concentration of venous deoxyhemoglobin [38.4]. The precise understanding of this physiological phenomenon is still under investigation [38.3, 5]. Despite its highly nonlinear nature [38.6, 7], most methods use linear systems theory [38.8, 9] to process the time course of fMRI. This is based on the assumption that the temporal hemodynamic response possesses linear characteristics and that the response is independent of prior response. Under this linear assumption, various methods have been proposed to analyze fMRI time series; For example, (cross-)correlation analysis was adopted [38.10] at the beginning of fMRI data processing technique development. In this method, a seed region of the fMRI time series is selected and correlation coefficients are calculated between the averaged seed-region time series and the time series from other voxels. Statistical tests/inferences for brain activation can be carried out based on these correlation coefficients. Later, linear regression methods such as the general linear model (GLM) were proposed [38.11, 12]. In the GLM method, the fMRI response model is predefined in the design matrix, and the nonlinear drifts are also included in the design matrix to exclude their effects. It should be mentioned that the exact cause of low-frequency drift has not been extensively explored [38.13], although several detrending methods are available to remove its effects for activation detection [38.14]. In the next section, we detail the GLM method for activation detection for both first- and second-level fMRI analysis.

38.2 GLM for Activation Detection

We begin with a linear model, since it has been extensively used for fMRI activation detection. Linear models are used in the situation in which the mean and the variances and covariances can be described as a linear function of unknown parameters [38.15]. The purpose of fMRI activation analysis is to find a significant relationship between system input (experimental design or external stimuli) and fMRI response (system output). As a result, applying a linear model is one of the methods commonly used to address this relationship. The details of the linear regression method for fMRI activation detection are given below.

38.2.1 Linear Model for First-Level Data Analysis

In the linear regression model for fMRI activation detection, we assume that the fMRI response Y within each run can be represented as [38.11, 16–18]

$$Y = X\beta + e \,, \tag{38.1}$$

where $X = [X_1, X_2]$ is the design matrix; X_1 can be obtained by convolving a box-car function with a Gaussian function or two-gamma function (Fig. 38.1c, d), while X_2 contains the drift terms, which can be modeled by polynomials (Fig. 38.2); β is a regression parameter, and $e \approx N(0, \sigma^2)$. The Euclidean distance between response and estimation (fitting) is defined in the following quadratic form:

$$D = (Y - X\beta)^{\mathrm{T}}(Y - X\beta) \,. \tag{38.2}$$

We want to obtain the parameter β value that minimizes the distance D, i.e., to minimize the mean square error (MSE) for the estimation. In this case, the least-squares solution is equivalent to solving the following well-known normal equation:

$$\frac{\partial D}{\partial \beta} = 0 = -2X^{\mathrm{T}}(Y - X\beta) \,. \tag{38.3}$$

From (38.3), it is easy to get

$$\hat{\beta} = (X^{\mathrm{T}}X)^{-1}X^{\mathrm{T}}Y \,. \tag{38.4}$$

It should be mentioned that at least one input function should be included in the design matrix X for the purpose of activation detection. This is because, in activation detection, we need to address the relationship between input stimulus or experimental design and output/response. In the linear model, the input function is employed to model the hemodynamic response function (HRF), which can be approximated by a two-gamma function [38.19], sine wave function [38.20], and Gaussian convolution with a box-car function [38.21]; For example, if the stimuli are changing periodically for a phase-encoded design, the sinusoidal function can be regarded as a brain system input; i.e., the periodical square block wave can be approximated by the fundamental frequency of the wave. Then, the fast Fourier transformation (FFT) method can be applied, because it is a powerful way to study experiments with periodically changing stimuli. The sine wave function (Fig. 38.1a) can be written as [38.17, 20, 22, 23]

$$f_{t,1} = a \cos(\omega t + \theta) \,, \tag{38.5}$$

where θ is the delay/onset or phase of the response, and can be estimated using the FFT method; ω is the angular frequency, $\omega = 2\pi f$, where f is the frequency of the stimulus (system input); a is the magnitude.

Another widely used function to model the BOLD HRF is the two-gamma function (Fig. 38.1b, c). To estimate the two-gamma function adaptively, we need to estimate the delay/onset of the function by FFT analysis (38.5) or cross-correlation analysis. Then, a two-gamma function is built according to the following equa-

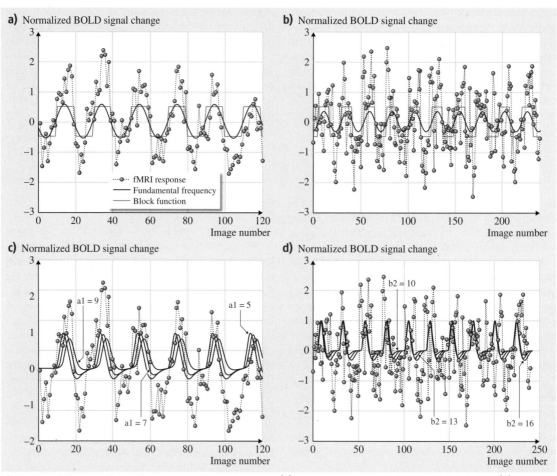

Fig. 38.1a–d Some typical fMRI responses (*dotted curves*): (**a**) box-car function for phase-encoded design, (**b**) averaged fMRI response for V1 brain area with standard block design (see Sect. 38.4.4), and two-gamma function with different parameters for (**c**) phase-encoded (**d**) and block design

tion [38.16, 24]:

$$f_{t,2} = \left(\frac{t}{d_1}\right)^{a_1} \exp\left(-\frac{t-d_1}{b_1}\right)$$
$$- c\left(\frac{t}{d_2}\right)^{a_2} \exp\left(-\frac{t-d_2}{b_2}\right), \qquad (38.6)$$

where $a_1 = 6, b_1 = 0.9, d_1 = a_1 \times b_1, c = 0.35, b_2 = 0.9$, $a_2 = 12$, and $d_2 = a_2 \times b_2$ are typical parameter values. As shown in Fig. 38.1c, d, variation of these parameters will lead to poorer or better estimation of the HRF. Figure 38.1c shows the result for parameter values of $a_1 = 5 : 2 : 10$, whereas in Fig. 38.1d, $a_2 = 10 : 3 : 17$ while all other parameters in (38.6) remain unchanged.

Furthermore, a block function (Fig. 38.1a,b) convolved with a Gaussian function [38.21] can also be em-

ployed to model the HRF. The function is expressed as

$$f_{t,3} = (\text{block}) \otimes \left\{\exp\left[-\left(\frac{t}{\sqrt{2}c}\right)^2\right]\right\}, \qquad (38.7)$$

where "block" is the block function (see, for example, Figs. 38.1a and 38.2a) and "$\otimes$" is the convolution operation. The full-width at half-maximum (FWHM) is determined according to: $\text{FWHM} = 2\sqrt{2\ln(2)}c$.

Apart from the system input or HRF, low-frequency fMRI drift terms can be modeled by polynomials [38.16, 17, 20], splines, and cosines. We use polynomials and include a maximum of 29th-order polynomials to model the drift in our retinotopic mapping experiment (Sect. 38.2.4). The shapes of the polynomials are given in Fig. 38.2 (used for fMRI responses in Fig. 38.1a,c). Figure 38.2a displays the curve of each polynomial against the time points, while the matrix format is given in Fig. 38.2b.

38.2.2 Statistical Inference for Activation Detection

After β is estimated from (38.4), the general linear hypothesis on β is written as [38.15]

$$H_0 : C\beta = h , \quad \text{versus} \quad H_0 : C\beta \neq h , \quad (38.8)$$

where $C = [c_1, c_2, \ldots, c_m]$ is the contrast matrix. In particular, we want to know whether there is an effect $C\beta \neq h$ or not, i. e., stimulus effect; That is

$$c_1\beta_1 + c_2\beta_2 + \cdots + c_m\beta_m = h . \quad (38.9)$$

In fMRI data analysis, we are mainly interested in the following two special cases. The first case is when a single component, say β_i, is zero. The contrast matrix is then the row vector $C = [0, \ldots, 0, 1, 0, \ldots, 0]$, where the "1" is in the ith position corresponding to the HRF (system input) in the design matrix; For example, if β_1 is the coefficient associated with the HRF at the first column of the design matrix in the GLM, then $C = [1, 0, \ldots, 0]$. In this case, we want to test whether the activation introduced by the experiment design is significant compared with the low-frequency drift; we thus consider the hypothesis (set $h = 0$)

$$H_0 : \beta_1 = 0 \quad \text{versus} \quad H_0 : \beta_1 \neq 0 \quad (38.10)$$

for single-input (task) activation detection. The second case is the hypothesis that two of the parameters, say β_1 and β_2 are equal. The second hypothesis (alternative hypothesis) is that β_1 and β_2 differ. In this case, the contrast matrix can be defined as $C = [1, -1, \ldots, 0]$; For example, we may be interested in the response difference between a red and a blue stimulus, where 1 corresponds to the model of the red stimulus and -1 corresponds to the model of the blue stimulus.

38.2.3 AR(1) Model for Error Term

The assumption of the independent error term in (38.1) often does not hold [38.25]. Therefore, temporal correction is needed to achieve more accurate estimation. One method is to employ an autoregression (AR) model. An AR(1) model to prewhiten the data for each fMRI time series, i. e., the prewhitening (38.1) in its matrix form, is

$$\tilde{Y}_t = \tilde{X}_t'\beta + \varepsilon_t , \quad (38.11)$$

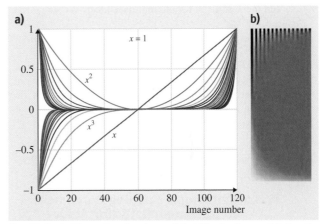

Fig. 38.2a,b Example of polynomial drifts in the model for the retinotopic mapping design with 120 fMRI image frames: (**a**) curve for each polynomial term, and (**b**) the corresponding matrix format

where $\tilde{Y}_1(1) = Y_1(1)$, $\tilde{Y}_t = (Y_t - \rho Y_{t-1})/\sqrt{1 - \rho^2}$; $\tilde{x}_1(1) = x_1(1)$, and $\tilde{x}_t = (x_t - \rho x_{t-1})/\sqrt{1 - \rho^2}$, $t = 2, \ldots, n$, where n is the number of fMRI time frames. In the same way as (38.4), the estimation of β is

$$\hat{\beta} = \tilde{X}^+\tilde{Y} , \quad (38.12)$$

where $\tilde{X}^+$ is the pseudoinverse of $\tilde{X} = (\tilde{x}_1, \ldots, \tilde{x}_n)'$, which is the transformed design matrix; $\tilde{Y} = (\tilde{Y}_1, \tilde{Y}_2, \ldots, \tilde{Y}_n)'$ is the transformed observation vector. The vector of residuals $R = (r_1, \ldots, r_n)'$ is

$$R = \tilde{Y} - \tilde{X}\hat{\beta} , \quad (38.13)$$

$$\hat{\sigma}^2 = \frac{R'R}{v} , \quad (38.14)$$

where $v = n - \text{rank}(\tilde{X})$ is the degrees of freedom (df). The autocorrelation coefficient is calculated using

$$\hat{\rho} = \frac{\sum\limits_{t=2}^{n} r_t r_{t-1}}{\sum\limits_{t=1}^{n} r_t^2} . \quad (38.15)$$

Then, an effect can be defined as

$$E = c\hat{\beta} , \quad (38.16)$$

where c is the contrast matrix $c = [1, 0, \ldots, 0]$ for one-input fMRI activation detection experiments, and 1 corresponds to the input/HRF. The estimated standard deviation is

$$S = \|c\tilde{X}^+\|\hat{\sigma} , \quad (38.17)$$

and the T statistic for the null hypothesis is

$$T = \frac{E}{S}, \qquad (38.18)$$

where T is often used to quantify the size of activation detected in the fMRI data. To detect more than one effect at the same time, that is, when the rank of the contrast matrix C is k, the T statistic is replaced by an F statistic defined as

$$F = \frac{E'(C\tilde{X}^+(C\tilde{X}^+)^{\mathrm{T}})^{-1}E}{(k\hat{\sigma}^2)}. \qquad (38.19)$$

38.2.4 Retinotopic Mapping Experiment for Activation Detection Study

This dataset was acquired using a retinotopic mapping stimuli experiment with a phase-encoded design from healthy subjects. This study was performed with the informed consent of the subjects and approved by the Montreal Neurological Institute Research Ethics Committee of McGill University (Montréal, Canada). A Siemens 1.5 T Magnetom scanner was used to collect both anatomical and functional images in the first experiment. All studies conformed to the Code of Ethics of the World Medical Association (Declaration of Helsinki), printed in the *British Medical Journal* (18 July 1964). Eleven normal subjects (age 32.55–4.5 years) were used in the first experiment [38.20, 26]. Briefly, anatomical images were acquired using a rectangular (14.5″ × 6.5″) head coil (circularly polarized transmit and receive) and a T_1-weighted sequence (repetition time (TR) = 22 ms, echo time (TE) = 10 ms, flip angle = 30°) giving 176 sagittal slices of 256×256 mm^2 image voxels. Functional scans for each subject were collected using a surface coil (circularly polarized, receive only) positioned beneath the subject's occiput. Each functional imaging session was preceded by a surface-coil anatomical scan (identical to the head-coil anatomical sequence, except that $80 \times 256 \times 256$ sagittal images of slice thickness 2 mm were acquired) in order to later co-register the data with the more homogeneous head-coil images. Functional scans were multislice T_2^*-weighted, gradient-echo, planar images (GE-EPI, TR = 3.0 s, TE = 51 ms, flip angle = 90°). The image volume consisted of 30 slices orthogonal to the calcarine sulcus. The field of view was 256×256 mm^2, and the matrix size was 64×64 with thickness of 4 mm, yielding voxel sizes of $4 \times 4 \times 4$ mm^3.

For functional data collection, phase-encoded designs [38.27–29] were used and each visual retinotopic experiment (phase-encoded design, traveling square wave) consisted of four acquisition runs for each eye (two eccentricity runs, two polar angle runs, two clockwise order runs, and two counterclockwise runs) and 128 image volumes acquired at 3 s intervals for the left and right eye of normal participants. Runs were alternated between the eyes in each case while the subject was performing a task to keep awake in the scanner. The eye not being stimulated was occluded with a black patch that excluded all light from the eye. Subjects monocularly viewed a stimulus back-projected into the bore of the scanner and viewed through an angled mirror. In addition, the middle temporal (MT) cortex or V5 cortex localizer experiment was conducted for seven normal subjects. The experiment consisted of two to five acquisition runs for both eyes using a checkerboard contrast stimulus [38.30]. During the MT localizer scanning sessions, subjects binocularly viewed a stimulus back-projected into the bore of the scanner and viewed through an angled mirror. In the data preprocessing, dynamic motion correction for functional image time series for each run and for different runs were realigned at the same time by using the fmr_preprocess function (provided in the MINC software package [38.31]) with default parameters for three-dimensional Gaussian low-pass filtering. The first eight scans of each functional run were discarded due to startup magnetization transients in the data, so that only 120 image volumes were used in each run. The MT area of each subject was defined based on T statistics of the BOLD response after different MT localizer runs were combined by using a random-effect model [38.16]. We localized MT in the regions which have a relatively larger T value ($T > 1.98$, $P < 0.05$), and we defined the common boundaries of different visual areas (from V1 to V4) by calculating the retinotopic visual field sign map [38.28, 29] information of each subject.

38.2.5 First-Level Activation Detection Results

Figure 38.3a (solid curve) shows one example fMRI response time series from the visual cortex from the retinotopic mapping experiment. A sine wave (38.5) as input and fourth-order polynomials were employed to construct the design matrix in the GLM (38.1). The T value was calculated as given in Sect. 38.2.3. Figure 38.3b shows the results with the same input but a different size of design matrix (the design matrix size is 12), i.e., different order of polynomials in the design matrix. It can be seen that the T value increases

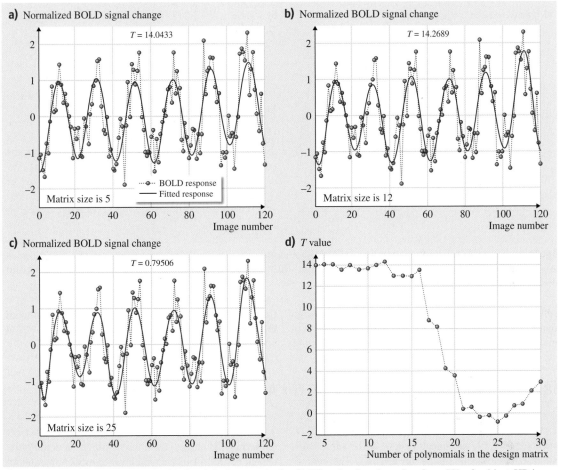

Fig. 38.3a–d The GLM method with fixed design matrix for fMRI activation detection from V1 of subject HP in response to the eccentric normal-order stimulus viewed with right eye (**a–c**). AR(1) model for modeling the error term e. The T value is decreased as the size of the design matrix is increased (**d**)

with the increase of the total number of polynomials in the design matrix. If we increase the polynomials to 25, however, the T value decreases to -0.79506, as shown in Fig. 38.3c. This is because the df is also increased with the increase of the design matrix. Before the total number of polynomials is 16 (Fig. 38.3d), R and v in (38.14) decrease at approximately the same speed; therefore, the T values from (38.18) do not change much. When the total number of polynomials becomes larger than 16, v decreases faster than R decreases; as a result, S from (38.17) increases, and this further leads to a T value decrease (via 38.18). Figure 38.3d displays the change in T value as the polynomials change in the design matrix. It is evident that the total number of polynomials is an important factor for fMRI activation

detection in the GLM, since the T values vary greatly as the size of the design matrix increases.

To overcome the limitation of GLM mentioned above (Fig. 38.3d), we proposed a least angle regression (LARS) [38.32–34] method for activation detection [38.17]. In the method, we provided several HRF models which worked as a dictionary for model selection to model the fMRI response caused by the experimental design. Instead of using fixed polynomial terms to model the low-frequency drift, we selected polynomials to fit the low-frequency drift adaptively. In this way, the limitation of the GLM, i.e., its being based on a fixed design matrix, can be overcome. Figure 38.4 shows an example of results using the LARS method with the same fMRI response as in Fig. 38.3a. LARS se-

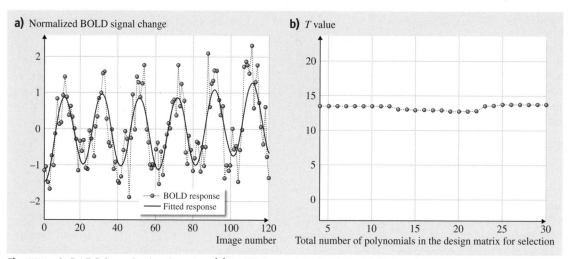

Fig. 38.4a,b LARS for activation detection. (**a**) LARS for model selection with 29th-order polynomials; LARS selected polynomial order of 23, 4, 3, and 5 in the design matrix. (**b**) T value change with the total number of polynomials for model selection (significant at $P < 0.05$)

lected polynomial orders of 23, 4, 3, and 5 in the design matrix. Figure 38.4b shows the total number of polynomials for LARS to model the slow drift. It is obvious that LARS produces small variation compared with GLM for fMRI activation detection (compare Fig. 38.3d with Fig. 38.4b).

38.3 General Linear Mixed Model for Activation Detection (Second-Level Data Analysis)

If we wish to study the individual effect (random effect) and fixed effect (experiment design) at the same time, mixed-effect models should be applied. The mixed-effect model has been extensively employed in the dairy industry and for longitudinal data analysis [38.35, 36] since the 1980s [38.37]. Recently, it has attracted great attention in the investigation of random effects in second- and high-level fMRI data analysis [38.16, 38–41]. In this section, we introduce the mixed-effect model and its application to fMRI analysis. Particularly, we point out the computational difficulties in applying this model to fMRI directly. To avoid this difficult, we employ the so-called two-stage summary statistics model [38.39] for the second/higher-level data analysis. Step-by-step details are provided to explain the method in the following section.

38.3.1 General Linear Mixed Model (GLMM)

From statistical theory, we know that the GLMM [38.42] can be expressed as

$$Y = X\beta + Zu + e,\qquad(38.20)$$

where X represents the design matrix for the fixed effect (experimental design) from the first level. Z denotes the design matrix for the second level to model the random effect (individual effect). Y is the fMRI response, where $u \approx N(0, G)$, $e \approx N(0, R)$, and $y \approx N(X\beta, ZGZ^T + R)$; G and R are variance matrices. The maximum-likelihood solution of this model is as follows [38.37, 43]:

$$\begin{pmatrix} X^T R^{-1} X & X^T R^{-1} Z \\ Z^T R^{-1} X & Z^T R^{-1} Z + G^{-1} \end{pmatrix} \begin{pmatrix} \hat{\beta} \\ \hat{u} \end{pmatrix}$$
$$= \begin{pmatrix} X^T R^{-1} y \\ Z^T R^{-1} y \end{pmatrix}.\qquad(38.21)$$

It is not difficult to see that it is computationally demanding to solve (38.21) because of the following two reasons: First, the first-level parameters (β and R) need to be updated prior to estimating the second-level parameters which are obtained from both the first- and second-level parameters. Second, when a new subject is scanned, all the first-level parameters/results need to be estimated again, and this is very computationally demanding for the whole brain, i. e., voxel by voxel. For these reasons, two-stage or mutlilevel models have been adopted in most fMRI studies [38.16, 18, 39]. The most striking advantage of the two-stage method is that, in the second-level analysis, the information from the first level does not need to updated. Due to its simplicity in computation, we present the two-stage model in the next section.

38.3.2 Two-Stage Model and Variance Estimation

In the two-stage models, one important task is to estimate the model variance for the mixed-effect model. For the second-level analysis, the linear mixed-effect model is expressed as [38.42, 44, 45]

$$E = Z \cdot \gamma + \eta, \qquad (38.22)$$

where $E = (E_1, \ldots, E_n)'$ and $S_0 = (S_1, \ldots, S_n)'$ are obtained from (38.17) and (38.18), and η is normally distributed with zero mean and variance $S_0^2 + \sigma_{\mathrm{random}}^2$ independently for $j = 1, \ldots, n$. In the second-level analysis, we often want to compare two groups, e.g., comparing normal controls with patients; we therefore define the design matrix Z in (38.22) as [38.36, 46]

$$Z = \begin{bmatrix} I_{n_1} & O_1 \\ O_2 & I_{n_2} \end{bmatrix},$$

where $I_{n_1} = [1, \ldots, 1]'_{1 \times n_1}$, $I_{n_2} = [1, \ldots, 1]'_{1 \times n_2}$, $O_2 = [0, \ldots, 0]'_{1 \times n_2}$, and $O_1 = [0, \ldots, 0]'_{1 \times n_1}$; n_1 and n_2 are the number of normal subjects and patients, respectively. In contrast, if we want to average different runs/subject, we can define the design matrix as

$$Z = \begin{bmatrix} 1 & 1 & \cdots & 1 \end{bmatrix}'_{1 \times n},$$

where n is the total number of subjects/runs to combine. After the design matrix has been determined, we need to estimate the covariance components of this model. An expectation-maximization (EM) algorithm [38.47, 48] is often applied to estimate the covariance for the two-stage model. Those interested in the EM algorithm to estimate variance can refer to [38.49–51] for details. A modified version of the EM algorithm has been proposed [38.16], and we introduce this method in the following section. In addition, we use the restricted maximum-likelihood [38.47, 49] (REML) method to estimate $\hat{\sigma}_{\mathrm{random}}^2$. In the algorithm, let $S = \mathrm{diag}(S_1, \ldots, S_{n^*})$ and I be the n^* identity matrix, where n^* is the total number of runs/subjects to

combine, then the variance matrix of the effects vector is (from (38.22))

$$V = S^2 + I \sigma_{\mathrm{random}}^2. \qquad (38.23)$$

In the numerical implementation of the algorithm, we subtract $\min(S_j)$ from S, and add it back after we have estimated the values. In this way, the error in the numerical analysis can be reduced. To estimate the random effect, we define the weighted residual matrix

$$R_V = V^{-1} - V^{-1} Z (Z'V^{-1}Z)^+ Z'V^{-1}. \qquad (38.24)$$

We start with an initial value of $\sigma_{\mathrm{random}}^2 = E'R_I E/v^*$, assuming that the fixed-effects variances are zero. The updated estimation is

$$\hat{\sigma}_{\mathrm{random}}^2 = \{ \sigma_{\mathrm{random}}^2 [p^* + \mathrm{tr}(S^2 R_V)] + \sigma_{\mathrm{random}}^4 E' R_V^2 E \}/n^*, \qquad (38.25)$$

where $p^* = \mathrm{rank}(Z)$. One then replaces $\sigma_{\mathrm{random}}^2$ with $\hat{\sigma}_{\mathrm{random}}^2$ in (38.23) and iterates (38.23–38.25) to convergence. Normally, 10 iterations appear to be sufficient for convergence [38.16]. Moreover, because of numerical noise, we smooth the random effect as follows:

$$\sigma_{\mathrm{fixed}}^2 = \sum_j \frac{v_j S_j^2}{v_{\mathrm{f}}}, \qquad (38.26)$$

where v_j is the degrees of freedom of S_j and $v_{\mathrm{f}} = \sum_j v_j$.

$$\tilde{\sigma}_{\mathrm{random}}^2 = \left[\mathrm{smooth} \left(\frac{\hat{\sigma}_{\mathrm{random}}^2}{\sigma_{\mathrm{fixed}}^2} \right) \right] \times \sigma_{\mathrm{fixed}}^2. \qquad (38.27)$$

Then, replacing $\sigma_{\mathrm{random}}^2$ by $\tilde{\sigma}_{\mathrm{random}}^2$ in (38.23), the estimation of γ is

$$\hat{\gamma} = (Z'V^{-1}Z)^+ Z'V^{-1} E. \qquad (38.28)$$

Its estimated variance matrix from the information matrix is [38.46, 52]

$$\hat{\mathrm{Var}}(\hat{\gamma}) = (Z'V^{-1}Z)^+. \qquad (38.29)$$

In cases where the variances of E are not homogeneous across the second-level analysis (e.g., where the same physician did not collect the data), the above equation should be replaced by [38.36, 45, 53]

$$\hat{\mathrm{Var}}(\hat{\gamma}) = (Z'V^{-1}Z)^+ Z'V^{-1}(E - Z\hat{\gamma}) \\ (E - Z\hat{\gamma})' V^{-1} Z (Z'V^{-1}Z)^+. \qquad (38.30)$$

Finally, the effect defined by a contrast matrix b in γ can be estimated by $E^* = b\hat{\gamma}$ with standard deviation

$$S^* = \sqrt{b \hat{\mathrm{Var}}(\hat{\gamma}) b'} \qquad (38.31)$$

Part G | 38.3

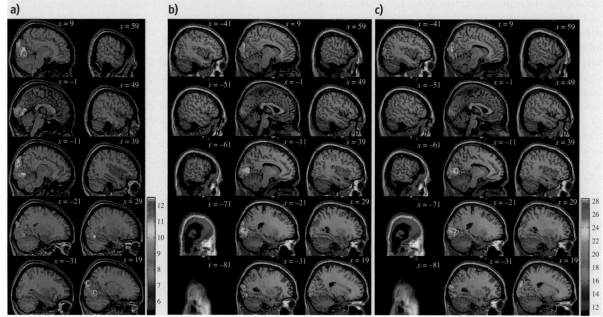

Fig. 38.5 (**a**) Single-run activation from one subject (from retinotopic mapping experiment). The activation is projected onto the structural MRI of this subject in Talairach space [38.54]; (**b**) EM algorithm to average activations from all 11 subjects; (**c**) modified EM (without smoothing) to average these subjects. FWE correction was applied to multiple comparisons, with FWE-corrected threshold of 5.1875 (**a**), 10.7242 (**b**), and 10.7242 (**c**). The x in the figure represents the x coordinate of Talairach space

and the T statistic is

$$T^* = \frac{E^*}{S^*} ,\tag{38.32}$$

with a nominal v^* df ($v^* = n^* - \mathrm{rank}(\mathbf{Z})$) used to detect the effect.

38.3.3 Threshold Correction Method (FDR and FWE)

Because there are false positives in comparing multiple images, this must be controlled over all tests. Bonferroni-related tests, e.g., the false discover rate (FDR) and familywise error rate (FWE), offer a solution to this problem [38.55, 56]. In this section, we introduce FDR and FWE methods for correction of fMRI datasets. Briefly, consider testing $H_1, H_2, \ldots, H_m$ based on the corresponding p values $P_1, P_2, \ldots, P_m$. Let $P_{(1)}, P_{(2)}, \ldots, P_{(m)}$ be the ordered p values, and denote by $H_{(i)}$ the null hypothesis corresponding to $P_{(i)}$, where m is the total number of (null) hypotheses tested simultaneously. Define the following Bonferroni-type multiple-testing procedure:

Let k be the largest i for which

$$P_{(i)} \le \frac{i}{m} q^* ;\tag{38.33}$$

then reject all $H_{(i)} i = 1, 2, \ldots, k$, where q^* is chosen to equal the significance level α. In this study, we set $\alpha = 0.05$; m is the total number of comparisons within the brain region mask (from 38.33).

The Bonferroni correction is

$$P_{(i)} \le \frac{i}{m} q^* .\tag{38.34}$$

The FWE correction is

$$P_{(i)} \le \frac{i}{m+1-i} q^* .\tag{38.35}$$

In this study, we applied both FDR and FWE methods to the activation maps. Due to space limitation, we show FWE correction figures (Fig. 38.5) only.

Figure 38.5a shows the activation mapping from a phase-encoded design (see Sect. 38.4.1 for details regarding the experiment) at the first level with the right eye open in one normal subject when the stimulus was polar angle reverse stimulus with counter-clockwise presentation [38.20, 26]. The FFT fundamental frequency was used to model the system input with two orders of polynomials to exclude slow drift effects. The error term was modeled based on an AR(1) model as presented in Sect. 38.2.3. The contrast matrix was defined as $c = [1, 0, 0]$, where 1 corresponds to the fundamental frequency to model the HRF. A T statistical test (38.18) was employed to quantify the magnitude of activation. The colored regions (Fig. 38.5a) show where the brain regions are significantly activated by the stimuli ($P < 0.05$, FWE-corrected threshold), and the color bar represents the corresponding magnitude.

In the second-level analysis, we adopted a standard EM algorithm to estimate model variance. The design matrix in (38.22) is defined as $Z = [1, 1, \ldots, 1]'_{1 \times 11}$ to combine 11 normal subjects. The results are given in Fig. 38.5b. In the results, the FWE method was applied to correct the threshold for multiple comparisons. Figure 38.5c displays the combination of these 11 subjects with the modified EM algorithm. It can be seen that

the modified EM algorithm produces stronger activation than the conventional EM algorithm for the mixed-effects model (comparing Fig. 38.5b with Fig. 38.5c). Figure 38.5c, b is in standard Talairach space [38.54], because it is a cross-subject analysis; the Montreal Neurological Institute (MNI) template in Talairach space was used to show the location of these significant activation regions.

So far, we have introduced the linear regression method for activation detection. In the first-level analysis, we presented the GLM with an AR(1) model for error term modeling. For the second-level data analysis, we employed the mixed model, which takes into account the individual (random) effect in the data analysis. As a result, it can model both fixed and random effects for fMRI activation detection. However, these models do not consider interaction between different, distant brain regions. In the next section, we present the effective connectivity method to address the regional influence (causality) between regions based on fMRI time series.

38.4 Effective Connectivity

Effective connectivity analysis assesses how one brain region influences another, distant brain region [38.57]. There are three approaches for modeling nonlinear brain activity in the study of fMRI effective connectivity. These methods can be categorized as white-box models, grey-box models, and black-box models [38.58]. White-box modeling is an off-building approach where the model is built before quantifying the connectivities. White-box modeling assumes that the structure of the neuron population is known, to address the problem of finding the parameters of the assumed structure. It is difficult to apply this identification method because the assumption of a known, stationary neuron population does not hold. Alternatively, grey-box models are often adopted when limited structural knowledge is available or assumed, and a model structure can be built on physical grounds, with a certain number of parameters to be estimated from the data. This could be a state-space model of a given order and structure; For example, using dynamic causal modeling (DCM) [38.59–62], an offline state-space model can be built at the neuron level, and the effective connectivity can be studied by identifying the known structure of models corresponding to an infinitely precise prior on the model. A major limitation of this method is that it uses one offline model for all the brain regions with-

out considering the variability of the fMRI response from different regions. For the black-box modeling method, physical insight is not available or assumed, but the chosen model structure belongs to families that are known to have good flexibility and therefore can be adapted more easily to the data. Current black-box model approaches to study effective connectivity include the Granger causality model (GCM) [38.63–67], the (sparse) multivariate autoregression (MAR) model [38.57, 68, 69], and autoregression and moving average (ARMA) models [38.70]. There are at least two limitations of applying these methods for effective connectivity studies; one is that they cannot be used in the case when the cerebral hemodynamic response does not reach a significant level [38.66]. The other is that these methods do not include the experimental design as input to the model for fMRI effective connectivity analysis. This is not optimal when knowledge of the experimental design is available [38.59]. To circumvent this limitation, we proposed a new nonlinear system identification method (black-box model method) to study effective connectivity [38.71].

In this section, we present the nonlinear system identification method and apply the method to two visual networks as shown in Fig. 38.6. These networks are based on previous studies [38.72, 73]. Figure 38.6a

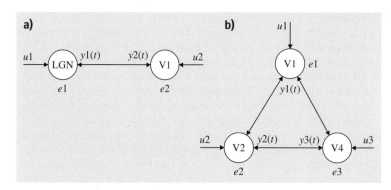

Fig. 38.6a,b Two examples of the visual network system: (**a**) LGN-V1 network; (**b**) three-connection ventral visual network

displays the lateral geniculate nucleus (LGN) and V1 network, while Fig. 38.6b shows the three-visual-region ventral network, i.e., the V1, V2, and V4 network. Firstly, we present the mathematical theoretical background of the method. Second, we provide an F statistical test for studying effective connectivity. Finally, we propose a model selection algorithm for nonlinear system identification-based effective connectivity analysis.

38.4.1 Nonlinear System Identification Method for Effective Connectivity

The physiological processes underlying the BOLD fMRI response can be modeled as a multiple-input, multiple-output (MIMO) system [38.6, 71]

$$\begin{cases} \dot{x}(t) = f(x(t), u(t), \theta) \\ y(t) = g(x(t), \theta) \end{cases}$$

and its discrete form is

$$\begin{cases} x(t+1) = f(x(t), u(t), \theta) \\ y(t) = g(x(t), \theta) \end{cases}, \qquad (38.36)$$

where f and g are nonlinear functions, and θ represents the set of model parameters. $y(t)$ is the BOLD response or nonlinear brain system output, $x(t)$ is the state variable of the system, and $u(t)$ is the system input. Under some mild assumptions, the discrete-time multivariate system (38.36) with p outputs and q inputs can be described by an autoregressive moving-average model with exogenous input (NARMAX) as follows [38.74, 75]:

$$y(t) = f_g[y(t-1), \ldots, y(t-n_y),$$
$$u(t-1), \ldots, u(t-n_u),$$
$$e(t-1), \ldots, e(t-n_e)] + e(t), \qquad (38.37)$$

where

$$y(t) = \begin{bmatrix} y_1(t) \\ \vdots \\ y_p(t) \end{bmatrix}, \quad u(t) = \begin{bmatrix} u_1(t) \\ \vdots \\ u_q(t) \end{bmatrix}, \quad e(t) = \begin{bmatrix} e_1(t) \\ \vdots \\ e_p(t) \end{bmatrix},$$

are the system output, input, and noise, respectively; n_y, n_u, and n_e are the maximum lags in the output, input, and noise respectively; $e(t)$ is a zero-mean independent sequence; f_g is a new nonlinear function which can be obtained from the nonlinear functions f and g. A special case of the general NARMAX model (38.37) is the nonlinear autoregressive with exogenous inputs (NARX) model

$$y(t) = f_g[y(t-1), \ldots, y(t-n_y),$$
$$u(t-1), \ldots, u(t-n_u)] + e(t). \qquad (38.38)$$

By applying the regression equation, the NARMAX model (38.37) and NARX model (38.38) can be approximated as [38.34, 76–78]

$$y(t) = \sum_{m=0}^{M} a_m P_m(t) + e(t), \quad t = 0, 1, \ldots, N,$$
$$(38.39)$$

where $P_0(t) = 1$; for $M \geq 1$, $P_m(t) = y_1 \cdots y_i u_1 u_2 \cdots u_j$, $i \geq 1$, $j \geq 0$; m is the number of nonlinear terms; M is the system order; N is the total number of time points (fMRI frames) in the time series; i is the number of connected regions; j is the number of inputs. Equation (38.4) denotes a general case where both input and output terms may be present, but it should be understood that some of the P_m may contain only input or output terms and cross-products; For example, for two stationary series of N values, the inputs u_{y_1} and y_2, the output y_1 of a closed-loop time-invariant nonlinear brain system can be described as [38.65]

$$y_1(t) = c_0 + \sum_{i=1}^{S_1} a_1(i) y_1(t-i) + \sum_{j=0}^{T_1} b_1(j) y_2(t-j)$$

$$+ \sum_{i=1}^{S_2} \sum_{j=1}^{S_2} a_2(i,j) y_1(t-i) y_1(t-j)$$

$$+ \sum_{i=0}^{T_2} \sum_{j=1}^{T_2} b_2(i,j) y_2(t-i) y_2(t-j)$$

$$+ \sum_{i=1}^{S_2} \sum_{j=0}^{T_2} c_2(i,j) y_1(t-i) y_2(t-j)$$

$$+ c_1 u_{y_1}(t) + e_{y_1} , \tag{38.40}$$

where the coefficients c_0, $\{a_1(i); b_1(j); c_1\}$, and $\{a_2(i,j); b_2(i,j); c_2(i,j)\}$ denote constant (zeroth-order), linear (first-order), and nonlinear (second-order) contributions to $y_1(t)$, respectively. u_{y_1} represents the experimental input, and e_{y_1} is the prediction error of $y_1(t)$. The model orders S_1 and S_2 are the maximum lags of the linear and nonlinear AR influences, respectively, while the maximum lags for linear and nonlinear exogenous effects are determined by the model orders T_1 and T_2. The model can be represented in the matrix form

$$Y = c_0 H_1 + H_{y_1} A_1 + H_{y_2} B_1 + H_{y_1 y_1} A_2$$
$$+ H_{y_2 y_2} B_2 + H_{y_1 y_2} C_2 + c_1 u_{y_1} + e_{y_1} , \tag{38.41}$$

where the vector $Y = [y_1(1), y_1(2), \ldots, y_1(N)]^T$ contains values of the output series, and $e_{y_1} = [e_{y_1}(1), e_{y_1}(2), \ldots, e_{y_1}(N)]$ is the prediction error series. $u_{y_1} = [u_{y_1}(1), u_{y_1}(2), \ldots, u_{y_1}(N)]$ is the experimental input time series; A_1, B_1, and C_1 are the first-order vector coefficients; A_2, B_2, and C_2 are the second-order vector coefficients. The matrices H_{y_1} and H_{y_2} contain the S_1 linear AR terms and the $(T_1 + 1)$ linear exogenous terms, respectively

$$H_{y_1} = \begin{bmatrix} y_1(0) & y_1(-1) & \cdots & y_1(1-S_1) \\ y_1(1) & y_1(0) & \cdots & y_1(2-S_1) \\ \vdots & \vdots & & \vdots \\ y_1(t-1) & y_1(t-2) & & y_1(t-S_1) \\ \vdots & \vdots & \cdots & \vdots \\ y_1(N-1) & y_1(N-2) & \cdots & y_1(N-S_1) \end{bmatrix},$$

$$H_{y_2} = \begin{bmatrix} y_2(1) & y_2(0) & \cdots & y_2(1-T_1) \\ y_2(2) & y_2(1) & \cdots & y_2(2-T_1) \\ \vdots & \vdots & & \vdots \\ y_2(t) & y_2(t-1) & & y_2(t-T_1) \\ \vdots & \vdots & \cdots & \vdots \\ y_2(N) & y_2(N-1) & \cdots & y_2(N-T_1) \end{bmatrix}.$$

The matrix $H_{y_1 y_1}$ contains the $S_2(S_2+1)/2$ quadratic AR terms given by the product of the terms of the matrix H_{y_1}. In the same way, the matrix $H_{y_2 y_2}$ contains the $(T_2+1)(T_2+2)/2$ quadratic exogenous terms, and the matrix contains the $S_2(T_2+1)$ cross-terms. Equation (38.41) can be written as

$$Y = W\beta + e_y , \tag{38.42}$$

where $W = [H_1, H_{y_1}, H_{y_2}, H_{y_1 y_1}, H_{y_2 y_2}, H_{y_1 y_2}, u_{y_1}]$, $\beta = [c_0, A_1^T, B_1^T, A_2^T, B_2^T, C_2^T, c_1]^T$. The coefficient matrix β can be estimated using least-squares estimation as $\hat{\beta} = \text{pinv}(W)Y$, where "pinv" is the Moore–Penrose pseudoinverse of the matrix.

Neglecting the nonlinear terms, i.e., $H_{y_1 y_1} A_2 + H_{y_2 y_2} B_2 + H_{y_1 y_2} C_2$, experimental input u_{y_1}, and considering only the first order of AR, i.e., AR(1), this leads to

$$Y = c_0 H_1 + H_{y_1} A_1 + H_{y_2} B_1 + e_{y_1}$$

or

$$y_1(t) = c_{01} + a_{11} y_1(t-1) + a_{12} y_2(t-1) + e_1(t) , \tag{38.43}$$

$$y_2(t) = c_{02} + a_{21} y_1(t-1) + a_{22} y_2(t-1) + e_2(t) . \tag{38.44}$$

This is the well-known two-connection linear Granger causality model (GCM) or fMRI effective connectivity estimation.

38.4.2 Granger Causality (GC) Tests

Instead of using conditional GC [38.79], we proposed a statistical test method to calculate causality as described in Sect. 38.2.2. This is because, once the coefficients of the model are determined, Granger causality tests [38.63,80,81] are derived based on F statistics. For simplicity and illustrative purposes, we take the nonlinear models (38.40) for example; the same principle can be applied for the linear system (38.43 and 38.44). The test for determining Granger cause (GC) is [38.80]:

1. y_2 is GC of y_1 if $b_1 = b_2 = c_2 = 0$ in (38.40) is *not true*. Given the data, we reach this conclusion if the condition $b_1 = b_2 = c_2 = 0$ is rejected.

2. Similarly, y_1 Granger causes of y_2 can be investigated by reversing the input–output roles of the two series.

Because the nonlinear term after c_2 is the combination of $y_1(t)$ and $y_2(t)$ in (38.40), it becomes a new predictor in the model; we call the causality introduced by this term an indirect GC, because this is a nonlinear indirect influence between two regions, as shown in (38.40). On the other hand, if at least one linear or nonlinear covariate (without interaction terms) exists between two regions, we say that there is a direct influence from one region to another; For example, if we have a nonlinear MAR model as in (38.40), direct influence from $y_2(t)$ to $y_1(t)$ means the direct influence from $y_2(t-1)$, without including the nonlinear interaction term $y_1(t-1)y_2(t-1)$, while indirect influence means influences from both $y_2(t-1)$ and $y_1(t-1)y_2(t-1)$ on $y_1(t)$.

Although t statistics can be applied to test these hypothesis (see Sect. 38.2.3 or [38.15, 46]), we introduce F statistics to detect significant relations (see below).

38.4.3 F Test for Nonlinear Connectivity Analysis

From (38.39) and (38.42), and considering the auxiliary system [38.82–84],

$$Y = W\beta + V, \tag{38.45}$$

where Y is $T \times n$, W is a $T \times K$ linear or nonlinear basis, β is $K \times n$; $V = e$ in (38.4) or $V = e_{y_1}$ in (38.40) or (38.42), and $E[V] = 0$; $E[VV'] = \sigma^2$. The coefficients $\hat{\beta}$ can be obtained from least squares as in (38.42) or (38.1), where the residuals are defined as $\hat{V} = Y - W\hat{\beta}$. Testing Granger causality is equivalent to testing whether the elements of β are zero. This can be done by Wald, likelihood ratio (LR), and the likelihood multiplier (LM) principle [38.85]. Partitioning the coefficients as $\beta = (\beta_1 : \beta_2)$ and $W = (W_1 : W_2)$ accordingly, we can write this test as

$$H_0 : \beta_2 = 0 \quad \text{versus} \quad H_1 : \beta_2 \neq 0,$$

with the maintained hypothesis given in (38.45). Defining R^2-type measures of goodness of fit

$$R_r^2 = 1 - |\hat{V}\hat{V}'| \, |\hat{V}_0 \hat{V}_0'|^{-1}, \tag{38.46}$$

where $\hat{V}_0$ is the residual from regression of Y on W_1 (that is, under H_0; this is the original system), while $\hat{V}$

results from the auxiliary system, the corresponding F-approximation to the likelihood ratio is (38.42) [38.82, 83].

$$\text{LMF} = \frac{1 - \left(1 - R_r^2\right)^{1/r}}{\left(1 - R_r^2\right)^{1/r}} \cdot \frac{Nr - q}{np}, \tag{38.47}$$

where $r = ((n^2 p^2 - 4)/(n^2 + p^2 - 5))^{1/2}$, $q = 1/2np - 1$, $N = T - k - p - 1/2(n - p + 1)$, k is the number of regressors in the original system (k is the column of W_1), n is the dimension of the system, T is the number of observations, and $p = ns$ (s is the column of W_2). LMF has an approximate $F(np, Ns - q)$ distribution (the F-approximation is exact for fixed regressors when $p \leq 2$ or $n \leq 2$). When $n = 1$, $R^2/(1 - R^2)(T - k - s)/s \approx F(s, T - k - s)$, where $R^2 = (\text{RSS}_0 - \text{RSS})/\text{RSS}_0$. RSS_0 and RSS are the residual sum of squares (RSS) of the original and auxiliary system, respectively, and $\text{RSS} = \sum_{i=1}^{T} V^2(i)$.

38.4.4 Datasets for Effective Connectivity Study

We chose three datasets to represent a wide variety of different experimental designs and different stimuli to test the robustness of our methods. The first experiment is based on a standard block design with six amblyopic subjects, and the third and fourth experiment are conducted using a random block design or counterbalanced block design [38.86] with seven amblyopic subjects, and the fourth experiment included eight normal subjects with random block design. The second, third, and fourth experiments were conducted within the constraints of the ethical clearance from the Medical Research Ethics Committee of the University of Queensland for MRI experiments on humans at the Centre for Magnetic Resonance (Brisbane, Australia). Informed written consent was gained from all participants prior to the commencement of the studies. All magnetic resonance images in the second, third, and fourth experiments were acquired using a 4 T Bruker MedSpec system. A transverse electromagnetic head coil was used for radiofrequency transmission and reception. Head movement was limited by foam padding within the head coil in all MRI scans. All studies conformed to the Code of Ethics of the World Medical Association (Declaration of Helsinki), printed in the *British Medical Journal* (18 July 1964). For details of the four fMRI experiments, data collections, and data preprocessing see references [38.20, 26, 86, 87]) or below.

The First Dataset: Standard Block Design for the LGN–V1 Network

Six amblyopic subjects were recruited in this experiment [38.87]. A standard block design composed of alternate stimulus and blank intervals (18 s stimulation, 18 s fixation, 10 blocks per run) was used. Each stimulus was presented in a two alternate forced-choice paradigm within a 3 s cycle; each stimulus presentation was for 800 ms with interstimulus interval of 200 ms and 1.2 s for response. To control for attentional modulation known to affect cortical and subcortical structures, subjects performed a two alternate forced-choice contrast discrimination task that involved discriminating subtle changes in the contrast of pairs of alternately presented stimuli within a stimulus cycle and responding with a button press. During the fixation epochs, dummy button presses were made. The contrast difference between alternately presented stimuli was detectable with all subjects performing the task with an average performance of $98.5\pm2\%$ with the amblyopic eye and $97.8\pm2\%$ with the fellow fixing eye, demonstrating that the targets were visible to each eye and properly imaged on their retinas. During the experimental paradigm, participants viewed the stimuli monocularly and a tight-fitting eye patch was used to occlude one eye.

For the fMRI experimental study, 252 T_2^*-weighted gradient-echo echoplanar images depicting BOLD contrast were acquired in each of 24 planes with TE = 30 ms, TR = 1500 ms, in-plane resolution 3.1×3.1 mm^2, and slice thickness 3 mm (0 mm gap). Two or three fMRI scans were performed in each session. In the same fMRI session, a high-resolution 3D T1 image was acquired using an MP-RAGE sequence with TI = 1500 ms, TR = 2500 ms, TE = 3.83 ms, and resolution of 0.9 mm^3. In the fMRI data preprocessing, the first 12 image volumes were cut because of magnetic instability; therefore we use only 240 image volumes in each run.

The Second Dataset: Random Block Design for the LGN–V1 Network

Seven amblyopic subjects (six subjects also participated in experiment 2) participated in this experiment [38.86]. Stimuli were radial sine wave gratings or Cartesian sine wave checkerboards (both 0.5 cycles/degree) whose contrast phase-reversed at 2 or 8 Hz. All stimuli were presented in a temporal Gaussian contrast envelope ($\sigma = 125$ ms). Four different stimulus conditions were used: achromatic (Ach), red–green (RG), blue–yellow (BY) stimuli, and a luminance (blank) condition in which only the fixation stimulus appeared. In the fix-

ation condition, a white ring surrounded the small black fixation spot. Stimuli were presented time-locked to the acquisition of fMRI time frames, i.e., every 3 s. Each stimulus was presented within a 500 ms time window in a temporal Gaussian contrast envelope ($\sigma = 125$ ms), with interstimulus interval of 500 ms. In the remaining 1.5 s the subjects' responses were recorded using an MR-compatible computer mouse. During the mean luminance (blank) condition an identical contrast discrimination task was performed for the fixation stimulus. The four stimulus types were presented in a random/counterbalanced block design (six presentations per block, duration = 18 s). Each block was repeated 10 times, giving a total of 240 presentations per scan, i.e., 12 min/scan. All results are based on data from two scans per experiment (480 presentations, 24 min).

For the fMRI studies, 241 T_2^*-weighted gradient-echo echoplanar images depicting BOLD contrast were acquired in each of 36 planes with TE = 30 ms, TR = 3000 ms, in-plane resolution of 3.6 mm^2, and slice thickness of 3 mm (0.6 mm gap). Two fMRI scans were performed in each session. In the same fMRI session, a high-resolution 3D T1 image was acquired using an MP-RAGE sequence with TI = 1500 ms, TR = 2500 ms, TE = 3.83 ms, and resolution of 0.9 mm^3. In the fMRI data preprocessing, the first two image volumes were cut because of magnetic instability; therefore we used only 239 image volumes in each run. The slices were taken parallel to the calcarine sulcus, and covered the entire occipital and parietal lobes and large dorsal-posterior parts of the temporal and frontal lobes in experiments 2 and 3. For more details regarding fMRI data preprocessing, see [38.87]. Identification of the early visual cortical areas of subjects in dataset 1 and 2, including V1, was performed in separate sessions with identical parameters except for the number of time frames (128), number of fMRI scans (1–4), and slice orientation (orthogonal to the calcarine for the retinotopic mapping experiments).

The Third Dataset: Random Block Design for the V1–V2–V4 Ventral Visual Network

Eight healthy observers were used as subjects (four female, mean age 41 years, age range 31–54 years), five of whom were naive to the purpose of the study [38.86]. The subjects were instructed to maintain fixation on the provided fixation point and trained prior to the scanning sessions to familiarize them with the task. All observers had normal or corrected-to-normal visual acuity. No participant had history of psychiatric or neurological disorder, head trauma, or substance abuse. The stimuli

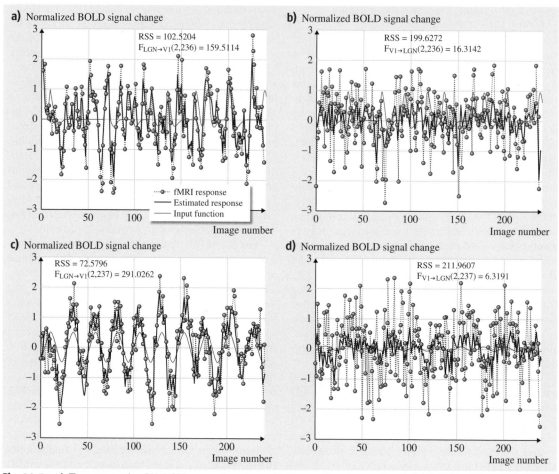

Fig. 38.7a–d Two-connection V1-LGN network system with blue–yellow input stimuli: (**a**) LGN feedforward influence on V1; V1 is the system output; (**b**) V1 feedback influence on LGN; the system output is LGN. (**c,d**) Results from standard block design for the feedforward (**c**) and feedback (**d**) influences

were the same as experiment 3; the fMRI and high-resolution 3D T1 image were acquired using the same procedure and parameters as in dataset 3. Two or three fMRI scans were performed in each session. The subject viewed the stimuli with both eyes.

38.4.5 Results from NARX for Effective Connectivity Study

In the effective connectivity analysis, the averaged fMRI time series from each region is often adopted to represent the response for the region. The conventional averaging method is not optimal because of its lack of robustness to response outliers. In this study, we used singular value decomposition (SVD) to obtain the aver-

age response for each region of interest (ROI). In short, suppose for each region that there are n voxels with m fMRI time points, and for all the response in this region, we can use X, which is an $m \times n$ matrix and can be expressed using SVD as

$$[S, V, D] = \text{svd}(X) , \tag{38.48}$$

where "svd" represents the SVD calculation from MATLAB; V is an $m \times n$ diagonal matrix with singular value on the diagonal elements in decreasing order. Keep the largest value in V and set all other values to zero, i.e., $V_{\text{new}}(1, 1) = V(1, 1)$, $V_{\text{new}}(i, i) = 0|_{i \neq 1}$, and perform an inverse calculation without the change matrix S and D, e.g.,

$$\text{avg} = S \times V_{\text{new}} \times D , \tag{38.49}$$

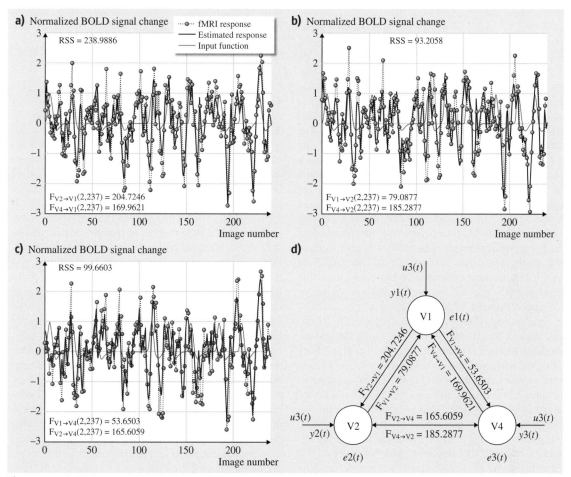

Fig. 38.8a–d Without model selection results using the NARX model. The system input is blue–yellow stimuli. The fMRI response is from the right hemisphere. Brain system output is V1 (**a**), V2 (**b**), and V4 (**c**). The thin solid curves in (**a–c**) denote the two-gamma function to model the system input. RSS indicates the residual sum of squares of the model. (**d**) Connectivity structure and strength of the network. RSS indicates the residual sum of squares of the model

where "avg" is the averaged response for this region. Figures 38.7 and 38.8 show the averaged response based on SVD decomposition for LGN (Fig. 38.7b,d), V1 (Figs. 38.1b,d, 38.7a,c, and 38.8a), V2 (Fig. 38.8b), and V4 (Fig. 38.8c).

We give the effective connectivity results based on two- and three-connection networks as shown in Fig. 38.6. The robust regional average method (38.48 and 38.49) is used to estimate the response for each region. Block and random block design data were employed to test the method for the LGN-V1 network. Datasets 1 and 2 for effective connectivity were used to test the algorithm. Based on the statistical method development in Sects. 38.4.1–38.4.3, we were able

to obtain the GC results as shown in Fig. 38.7 for the LGN-V1 network (Fig. 38.6a) using dataset 2. In this system, the two-gamma function with typical values (38.6) was used as the system input, and the system input stimuli were blue–yellow. The dotted curve in Fig. 38.7a represents the averaged V1 response from the right hemisphere of one amblyopic subject viewing the stimuli with the amblyopic eye. The solid curve denotes the predicted response, and the thin curve represents the system input. In Fig. 38.7, we use $F_{V1 \rightarrow LGN}$ to denote the influence from V1 to LGN, and $F_{LGN \rightarrow V1}$ to represent the influence from LGN to V1. In the NARX model for connectivity study, we employed AR(2) with two orders of nonlinearity to model the system. Using the statistical method developed in Sects. 38.4.1–38.4.3 and a two-gamma function to model the blue–yellow input (thin curves in Fig. 38.7), we obtained the LGN influence on V1 as $F(2, 236) = 159.5114$, significant at $P < 0.05$. Figure 38.7b displays the V1 influence on LGN, i. e., the feedback influence. As shown in Fig. 38.7a, the dotted curve denotes the SVD-averaged response from the LGN of the right hemisphere in the same subject. The feedback magnitude is $F(2, 236) = 16.3142$, significant at $P < 0.05$, suggesting that there is inhibitory feedback from V1 to LGN; however, it should be mentioned that the feedforward connection is nearly 10 times larger than the feedback influences for the particular response of this subject. In addition, the effective connectivity when viewing using an amblyopic eye for the standard block experiment design (dataset 1) was also studied. The results are given in Fig. 38.7c, d. As shown in Fig. 38.7a,b, there is a stronger feedforward connection between LGN and V1, but weak feedback inhibitory connection between V1 and LGN.

In addition, a three-connection network (Fig. 38.6b) was studied based on the random block design experiment (dataset 3 was adopted for the algorithm). Using the same statistical method as for LGN-V1 network connectivity analysis (Fig. 38.7), an example of the three-connection visual network (Fig. 38.6b) is presented in Fig. 38.8. The fMRI responses are from one normal subject with binocular stimuli (dataset 3). The dotted curves in Fig. 38.8a–c represent the right hemisphere averaged fMRI responses from V1, V2, and V4, respectively. We also adopted the two-gamma function (thin curves in Fig. 38.8a–c) as system input to model blue–yellow stimuli. As in Fig. 38.7, we use $F_{V1 \rightarrow v2}$ to denote the influence from V1 to V2 in Fig. 38.8, etc. In the NARX model for connectivity study, we employed AR(2) with two orders of nonlinearity to model the system. We found an influence of V2 and V4 on V1 of $F(2, 237) = 204.7246$, and $F(2, 237) = 169.9621$, respectively (significant at $P < 0.05$), while the influence from V1 and V4 to V2 was $F(2, 237) = 79.0877$ and $F(2, 237) = 185.2877$. We also found that the feedforward influence from V1 and V2 to V4 is $F(2, 237) = 53.6503$ and $F(2, 237) = 165.6059$. In addition, we investigated the influence of different input functions on the causality. We compare the box-car input function with the two-gamma function. We did not find significant differences between the box-car function and the two-gamma function as an input ($P < 0.05$) for the V1 $\rightarrow$ V4 feedforward influence, suggesting that there is no significant difference between the two-

gamma function and the box-car function as system input for effective connectivity study.

38.4.6 AR Model Order Selection for Effective Connectivity

The NARX method described in Sect. 38.3.1 is based on the assumption that the structure of the connectivity networks (the linear and nonlinear covariates in the AR model) is the same across the entire cortex. This assumption is not ideal, because the connectivity across different regions is very different and the variation of fMRI responses is consequently different. Additionally, an important question is not answered by this fMRI effective connectivity study method [38.71], namely how to deal with the large number of parameters that may be required to handle an arbitrary nonlinear dynamical system. Too many parameters can lead to overfitting and pose a high-dimensional nonlinear optimization problem to be solved [38.88], because the total number of covariates for the nonlinear model is increased as the model time lags increase along with the nonlinearity and brain regions included in the connectivity study; For example, suppose that the linear multiple fMRI time series are k-dimensional (number of regions) and the largest AR model order considered in the fitting is p, then the total number of subset vector AR models (without intercept) would be $2^{k^2 p}$, which results in a large number, even for low values of k and p [38.89]. This makes the nonlinear identification method very prone to overfitting the fMRI response, because in such cases, the nonlinear model contains unnecessary parameters, which leads to less effective parameter estimates, in particular for periodic fMRI data. Although this problem can be overcome using the subset selection method developed using control theory [38.90], this method depends on an orthogonal search algorithm, which may limit fMRI effective connectivity studies because the orthogonal search algorithm is computationally more demanding. To avoid the orthogonal procedure, we employ a least angle regression (LARS) method [38.32] for subset selection to study fMRI effective connectivity.

To apply LARS for model selection, the first step is to determine the system AR order. Although the AR order along with the nonlinearity can also be selected using the LARS algorithm, the search time for LARS is significantly reduced if the AR order of the system is determined prior to applying the LARS algorithm for model selection. In theory, the AR order for (38.39) can be estimated by Akaike information criterion (AIC) [38.91] or Schwartz's criterion (SC) [38.92]. If some range of p is known, then $p_{min} \leq p \leq p_{max}$ can be chosen to minimize the AIC or SC criteria; For instance, in phase-encoded or block design with 10 time points (fMRI time frames) on and 10 off, p_{max} can be set to be $p_{max} = 10$. These settings may be suitable, because the blood flow introduced by the first visual stimulus task within a block has finished cycling within each condition/block [38.93, 94]. We used stepwise least-squares estimation [38.95] and a software package [38.96] for the AR model to obtain the optimal order, where $p_{min} = 1$ and $p_{max} = 10$ in the software for the fMRI response. However, because there are 32 covariates for the second order of nonlinearity model for the three connections (Fig. 38.6b) used in this study, the fMRI time series is too short to determine the system AR order using the software; i.e., the stepwise [38.95] method and the software package failed to give any useful results for the phase-encoded design with 120 fMRI frames (the retinotopic mapping experiment). We therefore adopted the linear parts within this case and used the stepwise least-squares method to determine the system order for this experiment. For the other experiments which have a fMRI time frame of 239, so that the fMRI time series for these experiments are long enough to determine the system order, we included both linear and nonlinear bases to estimate the AR order. It was found that the best AR order is 2 for most of the systems. Therefore, we set AR = 2 for all the nonlinear systems. This also makes it easier to compare the results of the different datasets.

38.4.7 Model Selection for Effective Connectivity Study

The objective of model selection is to overcome the limitation of overfitting by the NARX method (38.39) for effective connectivity study. The NARX method can represent a standard Volterra series expansion to approximate a nonlinear biological system [38.97], but the nonlinear terms increase rapidly as the AR order and connected regions increase; For example, for a three-connection network (three-connection regions y_1, y_2, and y_3, as shown in Fig. 38.6b) with two orders of nonlinearity and lags (AR orders = 2) on the input of 1, the output y_1 can be expressed as

$$
\begin{aligned}
y_1(t) = {} & a_{10}y_{10} + f_1(y_1(t)) + f_2(y_2(t)) \\
& + f_3(y_3(t)) + f_4(y_1(t), y_2(t)) \\
& + f_5(y_1(t), y_3(t)) + f_6(y_2(t), y_3(t)) \\
& + f_7(y_1(t), y_2(t), y_3(t), u_1(t)) + e_1(t) ,
\end{aligned}
$$
$$(38.50)$$

where a_{10} is the coefficient for the constant drift y_{10} (intercept term) and

$$
\begin{aligned}
f_1(y_1(t)) = {} & a_1 y_1(t-1) + a_2 y_1(t-2) + a_3 y_1(t-1)^2 \\
& + a_4 y_1(t-2)^2 + a_5 y_1(t-1)y_1(t-2)
\end{aligned}
$$
$$(38.51)$$

is the nonlinear function for the mapping between output $y_1(t)$ and AR input $y_1(t-1)$ and $y_1(t-2)$. Similarly,

$$
\begin{aligned}
f_2(y_2(t)) = {} & b_1 y_2(t-1) + b_2 y_2(t-2) \\
& + b_3 y_2(t-1)^2 + b_4 y_2(t-2)^2 \\
& + b_5 y_2(t-1)y_2(t-2) ,
\end{aligned}
$$
$$(38.52)$$

$$
\begin{aligned}
f_3(y_3(t)) = {} & c_1 y_3(t-1) + c_2 y_3(t-2) \\
& + c_3 y_3(t-1)^2 + c_4 y_3(t-2)^2 \\
& + c_5 y_3(t-1)y_3(t-2) ,
\end{aligned}
$$
$$(38.53)$$

$$
\begin{aligned}
f_4(y_1(t), y_2(t)) = {} & d_1 y_1(t-1)y_2(t-1) \\
& + d_2 y_1(t-1)y_2(t-2) \\
& + d_3 y_1(t-2)y_2(t-1) \\
& + d_4 y_1(t-2)y_2(t-2) ,
\end{aligned}
$$
$$(38.54)$$

$$
\begin{aligned}
f_5(y_1(t), y_3(t)) = {} & g_1 y_1(t-1)y_3(t-1) \\
& + g_2 y_1(t-1)y_3(t-2) \\
& + g_3 y_1(t-2)y_3(t-1) \\
& + g_4 y_1(t-2)y_3(t-2) ,
\end{aligned}
$$
$$(38.55)$$

$$
\begin{aligned}
f_6(y_2(t), y_3(t)) = {} & h_1 y_2(t-1)y_3(t-1) \\
& + h_2 y_2(t-1)y_3(t-2) \\
& + h_3 y_2(t-2)y_3(t-1) \\
& + h_4 y_2(t-2)y_3(t-2)
\end{aligned}
$$
$$(38.56)$$

and

$$
\begin{aligned}
f_7 & (y_1(t), y_2(t), y_3(t), u(t)) \\
& = k_0 u_1(t) + k_1 y_1(t-1)u_1(t) \\
& + k_2 y_2(t-1)u_1(t) + k_3 y_3(t-1)u_1(t)
\end{aligned}
$$
$$(38.57)$$

in this study for model selection. Apart from the linear covariates as shown in the first two terms of (38.51–38.53), this model also includes nonlinear covariates. For second-order nonlinearity with AR order of 2, (38.50) includes 32 covariates, i.e., 6 covariates of the linear AR, 21 covariates of the nonlinearity resulting from the combination of these 6 linear AR covariates, 1 input, 3 first-order AR combinations with input (the last 3 interaction terms in (38.57)), and 1 constant drift. In addition, there is one input $u_1(t)$ and one constant baseline y_{10}. Because the total time series (fMRI image frames) is 239 image frames for the experiments (Sect. 38.4.4), the solution of (38.50) requires a lot of optimization and may be prone to overfitting. Therefore, it is necessary to develop model selection methods to determine the linear/nonlinear function f_g in (38.38) efficiently and accurately.

To handle the model overfitting problem, consider the term selection problem for the linear-in-parameters model (38.39). Let $y = [y(1), \ldots, y(n)]^T$ be a vector of measured fMRI response (system output) at n time instants, and $\phi_m = [P_m(1), \ldots, P_m(n)]^T$ be a vector formed by the mth candidate model term, where $m = 1, 2, \ldots, M$. Let $D = \{\phi_1, \ldots, \phi_M\}$ be a dictionary of the M candidate bases from (38.39). From the viewpoint of practical modeling and identification, the finite-dimensional set D is often redundant, as shown in (38.4). The model term selection problem is equivalent to finding a full-dimensional subset $D_p = \{x_0, \ldots, x_p\} = \{\phi_i, \ldots, \phi_{i_p}\}$ of $p(p \leq M)$ bases from the library D, where $x_k = \phi_{i_k}$, $i_k \in \{0, 1, \ldots, M\}$ and $k = 0, 1, \ldots, p$, so that the system output y can be satisfactorily approximated using a linear combination of $x_0, x_1, \ldots, x_p$ as below [38.90]

$$y(t) = \beta_0 x_0 + \beta_1 x_1 + \cdots + \beta_p x_p + e(t) \qquad (38.58)$$

or in a compact matrix form

$$y(t) = X\beta + e(t), \qquad (38.59)$$

where the matrix $X = [x_0, x_1, \ldots, x_p]$ is assumed to be of full column rank, $\beta = [\beta_0, \beta_1, \ldots, \beta_p]^T$ is a parameter vector, and $e(t)$ is the approximation error vector, where p is the total number of variables/covariates in (38.58), i.e., $p = 32$ for a three-connection model with second order of AR [for $y_1(t)$, $y_2(t)$, and $y_3(t)$] and nonlinearity (38.50). For a three-connection network (38.4), we set $\beta_0 = a_{1,0}$ and $x_0 = y_{10}$, and x_i equals the corresponding AR covariates in (38.51–38.57), i.e., $x_1 = y_1(t-1), \ldots, x_p = y_3(t-p_3)y_3(t-p_3)$; $p_1 = p_2 = p_3 = 2$; $Y = y_1(t) = y(t)$, which is an $n \times 1$ matrix, where n is the number of fMRI time frames. Unlike the orthogonal search method [38.90], the LARS method was used to determine matrix X and estimate parameters β in (38.59). The solution of (38.58) is

$$\hat{\beta} = \arg\min_{\beta} \left\| Y - \sum_{i=0}^{p} x_i \beta_i \right\|, \qquad (38.60)$$

where $\| \ \|$ is the L2 norm or Euclidean norm. To select the model for the connectivity study according to the fMRI responses, consider the optimization problem

$$\hat{\beta} = \arg\min_{\beta} \left\| Y - \sum_{i=0}^{p} x_i \beta_i \right\| \text{ subject to } \sum_{i=0}^{p} |\beta_i| \leq s, \qquad (38.61)$$

where $s \geq 0$ is the tuning parameter. If s is large enough, (38.61) will be the ordinary least-squares estimation algorithm. In (38.61), smaller values of s produce shrunken estimates $\hat{\beta}$, often with many components equal to zero; choosing s can be thought of as choosing the number of covariates/predictors to include in the regression model (38.58). There are several algorithms to do this. We describe the LARS method which employs a nongreedy search algorithm for model selection in six steps (for details see [38.32, 33]).

38.4.8 LARS Model Selection for Effective Connectivity

The six steps of the LARS method to select the model for studying fMRI effective connectivity are [38.98]:

1. Standardize the predictors x_i, $0 < i \leq p$ to have mean zero and variance 1, except the intercept term (constant predictor) x_0. Find the predictor x_i, where $0 \leq i \leq p$, most correlated with Y as the first predictor, x_1.

2. Calculate $\hat{Y} = x_1\beta_1 + r$, where $r = Y - \hat{Y}$ is the residual, and $\hat{Y}$ is the least-squares estimation of Y.

3. Find the predictor x_j most correlated with r, except for the first predictor x_1, which has been included in the second step.

4. Move β_j from 0 towards its least-squares coefficient [using the Moore–Penrose pseudoinverse (PINV) algorithm to estimate the coefficient] $\langle x_j, r \rangle$, until some other competitor x_k has as much correlation with the current residual as x_j.

5. Move (β_j, β_k) in the direction defined by the joint least-squares coefficient (using the PINV algorithm to estimate the coefficient) of the current residual on (x_j, x_k), until some other competitor x_l has as much correlation with the current residual.

6. Continue in this way until some criteria such as the AIC [38.91, 99], C_p-type [38.100] or cross-validation stopping rules are met. We compared Akaike's information criterion corrected (AICc) [38.99] and leave-one-out (LOO) cross-validation [38.101, 102] methods as the stopping criteria to determine the length of the nonlinear function f_g in (38.38). For the AICc method, we have $\text{AIC} = n(\log \hat{\sigma}^2 + 1) + 2(p+1)$ and $\text{AIC}_c = \text{AIC} + 2(p+1)(p+2)/n - p - 2$, where n is the total number of image frames used in the regression model; $\hat{\sigma}^2$ is the estimated variance, and p is the number of covariates. We used LOO for cross-validation and employed the predicted sum of

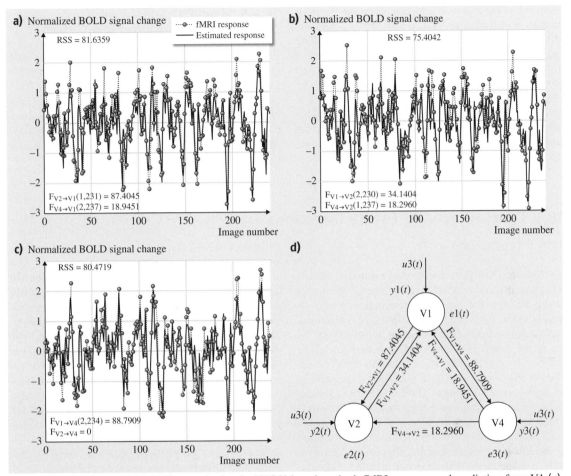

Fig. 38.9a–d LARS model selection results for the NARX-based method. fMRI response and prediction from V1 (**a**), V2 (**b**), and V4 (**c**). (**d**) Structure and strength of the model selection results. RSS indicates the residual sum of squares of the model

squares (PRESS) [38.103] as the stopping rule. For the case of $n \gg p$, which is an often encountered scenario and which will be considered in the present study, the calculation of PRESS can be significantly reduced to [38.104]

$$\text{PRESS}(p) \approx \left(\frac{1}{1 - p/n} \right)^2 \text{MSE}(p), \quad (38.62)$$

where $\text{MSE}(p) = (1/n) \sum_{i=1}^{n} (y(i) - \hat{y}(i))^2$, indicating the residual sum of squares (RSS).

Once the coefficients have been determined, an F-statistic test [38.71, 82, 105] can be used to calculate the regional influences (Granger causality [38.63]).

Based on the LARS for model section, we study the ventral visual network responses as shown in Fig. 38.8. The model bases were calculated from AR(2) with second order of nonlinearity, and the system input is blue–yellow stimulus. The fMRI responses are from the right hemisphere of this subject. Figure 38.9 shows the model selection results for the three-connection network (Fig. 38.6b). The dotted curves in Fig. 38.9a–c are the fMRI response from V1, V2, and V4. These responses are the same as in Fig. 38.8a–c. The thin solid curves in Fig. 38.9 represent the predicted response based on the LARS model selection algorithm. After the covariates have been selected using LARS, the statistical test outlined in Sect. 38.3.3 was applied to test the regional influence, as in the previous section in this chapter. We found the V2 and V4 to V1 feedback influences to be $F(1, 231) = 87.4045$ and $F(2, 237) = 18.9451$, respectively. The influences from V1 and V4 to V2 are $F(2, 230) = 34.1404$ and $F(1, 237) = 18.2960$. The feedforward influences from V1 and V2 to V4 are $F(2, 234) = 88.7909$ and $F = 0$. As in Figs. 38.7 and 38.8, we use $F_{V1 \to V2}$ to denote the influence from V1 to V2, and $F_{V2 \to V1}$ to represent the influence from V2 to V1, etc. From these influences, we can draw the network structure shown in Fig. 38.9d.

38.4.9 Second-Level Effective Connectivity Analysis and Threshold Correction

In theory, the second-level fMRI data analysis method for activation and threshold correction can be used for second-level effective connectivity directly. In practice, however, due to model selection in the second-level data analysis, more robust methods are needed to overcome the wide variability of the model and its associated coefficients [38.106]. This issue remains open and awaits further investigation.

38.5 Conclusions and Future Directions

38.5.1 Conclusions

We have presented general statistical methods for fMRI activation detection and effective connectivity studies in this chapter. For activation detection, we paid attention to the HRF variability at first level in general, and random effects from different runs/subjects for the mixed model in particular. One mixed model for both levels of fMRI activation detection could be applied to improve the statistical inference at the expense of more computation time. In the effective connectivity study, we proposed a black-box NARX method which overcomes the limitation of the fixed model method. We significantly improved the NARX method by introducing a model selection algorithm to study effective connectivity. The proposed method simplifies the nonlinear predictors and consequently reduces the complexity of the nonlinear model for effective connectivity study. Furthermore, it enables objective evaluation of the model fitting error and allows us to select the connected regions and models adaptively.

38.5.2 Future Directions

The current state of fMRI data analysis based on linear regression and nonlinear system identification methods is not completely satisfactory. The problems of unbalanced fMRI experimental design, HRF nonlinearity, too many covariates in the regression model, the appropriateness of the df in the statistical inferences, and expanding EM algorithm for variance estimation remain open to investigation. Apart from circumventing these limitations, we list some directions for future research here.

For activation detection, it is necessary to develop a method to overcome the variability of the HRF from different regions at the first level. One idea is to employ a model selection method to overcome this variability. The other is to apply nonlinear regression methods to fit the model. The implementation of nonlinear regression methods involves numerical analysis and optimization methods to achieve this goal. In second-level analysis for activation detection, one topic is to apply a recently developed variance estimation algorithm, i. e., the expanding EM [38.107] method, to estimate the variance for the analysis.

For effective connectivity study, we have not developed a model selection method for second-level analysis. Furthermore, due to model selection in the first-level analysis, we need to develop a robust method to overcome the high variability of models at the second/higher level.

In addition, the development of a single model for both activation detection and effective connectivity study is another research direction worth pursuing. Such an approach would have the advantage of including both segregation and integration within one model, therefore simplifying the computation. In this framework, a restricted input may be needed for activation detection.

Finally, it would be interesting to apply and validate these activation and connectivity study methods in clinical applications, e.g., in longitudinal Alzheimer's disease studies [38.36, 108].

References

38.1   S. Ogawa, T.M. Lee, A.S. Nayak, P. Glynn: Oxygenation-sensitive contrast in magnetic resonance image of rodent brain at high magnetic fields, Magn. Reson. Med. **14**(1), 68–78 (1990)

38.2   S. Ogawa, D.W. Tank, R. Menon, J.M. Ellermann, S.G. Kim, H. Merkle, K. Ugurbil: Intrinsic signal changes accompanying sensory stimulation: Functional brain mapping with magnetic resonance imaging, Proc. Natl. Acad. Sci. USA **89**(13), 5951–5955 (1992)

38.3   N.K. Logothetis: What we can do and what we cannot do with fMRI, Nature **453**(2), 869–878 (2008)

38.4   A.L. Vazquez, D.C. Noll: Nonlinear aspects of the BOLD response in functional MRI, NeuroImage **7**(2), 108–118 (1998)

38.5   N.K. Logothetis, J. Pauls, M. Augath, T. Trinath, A. Oeltermann: Neurophysiological investigation of the basis of the fMRI signal, Nature **412**(6843), 150–157 (2001)

38.6   K. Friston, A. Mechelli, R. Turner, C.J. Price: Nonlinear responses in fMRI: The balloon model, Volterra Kernels, and other hemodynamics, NeuroImage **12**, 466–477 (2000)

38.7   F.M. Miezin, L. Maccotta, J.M. Ollinger, S.E. Petersen, R.L. Buckner: Characterizing the hemodynamic response: Effects of presentation rate, sampling procedure, and the possibility of ordering brain activity based on relative timing, NeuroImage **11**(2), 735–759 (2000)

38.8   M.S. Cohen: Parametric analysis of fMRI data using linear systems methods, NeuroImage **6**(2), 93–103 (1997)

38.9   G. Boynton, S.A. Engel, G.H. Glover, D.J. Heeger: Linear systems analysis of functional magnetic resonance imaging in human V1, J. Neurosci. **16**(13), 4207–4221 (1999)

38.10 P.A. Bandettini, A. Jesmanowicz, E.C. Wong, J.S. Hyde: Processing strategies for time-course data sets in functional MRI of the human brain, Magn. Reson. Med. **30**(2), 161–173 (1993)

38.11 K. Friston, A.P. Holmes, K.J. Worsley, J.-B. Poline, C.D. Firth, R.S.J. Frackowiak: Statistical parametric maps in functional imaging: A general linear approach, Hum. Brain Mapp. **2**, 189–210 (1995)

38.12 K.J. Worsley, K.J. Friston: Analysis of fMRI time-series revisited–again, NeuroImage **2**(2), 173–181 (1995)

38.13 A.M. Smith, B.K. Lewis, U.E. Ruttimann, F.Q. Ye, T.M. Sinnwell, Y. Yang, J.H. Duyn, J.A. Frank: Investigation of low frequency drift in fMRI signal, NeuroImage **9**(5), 526–533 (1999)

38.14 J. Tanabe, D. Miller, J. Tregellas, R. Freedman, F.G. Meyer: Comparison of detrending methods for optimal fMRI preprocessing, NeuroImage **15**(4), 902–907 (2002)

38.15 R.R. Hocking: *Methods and Applications of Linear Models*, Wiley Series in Probability and Statistics (Wiley, New York 1996)

38.16 K. Worsley, C.H. Liao, J. Aston, V. Petre, G.H. Duncan, F. Morales, A.C. Evans: A general statistical analysis for fMRI data, NeuroImage **15**, 1–15 (2002)

38.17 X. Li, D. Coyle, L. Maguire, T.M. McGinnity, D.R. Watson, H. Benali: A least angle regression method for fMRI activation detection for phase-encoded experimental designs, NeuroImage **52**(2), 1390–1400 (2010)

38.18 M.W. Woolrich, T.E.J. Behrens, C.F. Beckmann, M. Jenkinson, S.M. Smith: Multilevel linear modelling for FMRI group analysis using Bayesian inference, NeuroImage **21**, 1732–1747 (2004)

38.19 G. Glover: Deconvolution of impulse response in event-related BOLD fMRI, NeuroImage **9**(4), 416–426 (1999)

38.20 X. Li, S.O. Dumoulin, B. Mansouri, R.F. Hess: Cortical deficits in human amblyopia: Their regional distribution and their relationship to the contrast detection deficit, Investig. Ophthalmol. Vis. Sci. **48**, 1575–1591 (2007)

38.21 A. Smith, K.D. Singh, A.L. Williams, M.W. Greenlee: Estimating receptive field size from fMRI data in human striate and extra-striate cortex, Cereb. Cortex, **11**, 1182–1190 (2001)

38.22 N. Lange, S.L. Zeger: Non-linear Fourier analysis of magnetic resonance functional neuroimage time series, Appl. Stat. **46**(1), 1–29 (1997)

38.23 X. Li, J. Tian, X. Wang, J. Dia, L. Ai: Fast orthogonal search method for modelling nonlinear hemodynamic response in fMRI. In: *SPIE, Medical Imaging 2004: Physiology, Function, and Structure from Medical* (San Diego, CA, USA 2004)

38.24 G. Golver: Deconvolution of impulse response in event-related BOLD fMRI, NeuroImage **9**(4), 416–426 (1999)

38.25 E. Bullmore, M.J. Brammer, S.C.R. Williams, S. Rabe-Hesketh, N. Janot, A.S. David, J.D.C. Mellers, R. Howard, P. Sham: Statistical methods of estimation and inference for functional MR images analysis, Magn. Reson. Med. **35**, 261–277 (1996)

38.26 X. Li, S.O. Dumoulin, B. Mansouri, R.F. Hess: The fidelity of the cortical retinotopic map in human amblyopia, Eu. J. Neurosci. **25**(5), 1265–1277 (2007)

38.27 S. Engel, G.H. Glover, B.A. Wandell: Retinotopic organization in human visual cortex and the spatial precision of functional MRI, Cereb. Cortex **7**, 181–192 (1997)

38.28 M. Sereno, A.M. Dale, J.B. Reppas, K.K. Kwong, J.W. Belliveau, T.J. Brady, B.R. Rosen, R.B. Tootell: Borders of multiple visual areas in humans revealed by functional magnetic resonance imaging, Science **268**, 889–893 (1995)

38.29   J. Warnking, M. Dojat, A. Guérie-Dugué, C. Delon-Martin, S. Olympieff, N. Richard, A. Chehikian, C. Segebarth: fMRI retinotopic mapping-step by step, NeuroImage **17**(4), 1665–1683 (2002)

38.30   S. Dumoulin, R.G. Bitter, N.J. Kabani, C.L. Baker, G.L. Goualher, G.B. Pike, A.C. Evans: A new anatomical landmark for reliable identification of human area V5/MT: A quantivative analysis of sulcal patterning, Cereb. Cortex **10**(5), 454–463 (2000)

38.31   All Brain image centre software source releases are stored on packages.bic.mni.mcgill.ca/

38.32   B. Efron, T. Hastie, I. Johnstone, R. Tibshirani: Least angle regression, Ann. Stat. **32**(2), 407–499 (2004)

38.33   T. Hastie, J. Taylor, R. Tibshirani, G. Walther: Forward stagewise regression and the monotone lasso, Electron. J. Stat. **1**, 1–29 (2007)

38.34   T. Hastie, R. Tibshirani, J. Friedman: *The Elements of Statistical Learning: Data Mining, Inference, and Prediction*, Springer Series in Statistics, 2nd edn. (Springer, New York 2009)

38.35   P.J. Diggle, P. Heagerty, K.Y. Liang, S. Zeger: *Analysis of Longitudinal Data*, Oxford Statistical Science Series, Vol. 25, 2nd edn., ed. by A.C. Atkinson (Oxford Univ. Press, Oxford 2003)

38.36   X. Li, D. Coyle, L. Maguire, D.R. Watson, T.M. McGinnity: Grey matter concentration and effective connectivity changes in Alzheimer's disease: A longitudinal structural MRI study, Neuroradiology **53**(10), 733–748 (2011)

38.37   C.R. Henderson: *Applications of Linear Models in Animal Breeding* (Canadian Cataloguing in Publication Data, Guelph 1984)

38.38   C.F. Beckmann, M. Jenkinson, S.M. Smith: General multilevel linear modeling for group analysis in FMRI, NeuroImage **20**, 1052–1063 (2003)

38.39   J.A. Mumford, T.E. Nichols: Power calculation for group fMRI studies accounting for arbitrary design and temporal autocorrelation, NeuroImage **39**(1), 261–268 (2008)

38.40   A. Roche, M. Mebastien, M. Keller, B. Thirion: Mixed-effect statistics for group analysis in fMRI: A nonparametric maximum likelihood approach, NeuroImage **38**, 501–510 (2007)

38.41   K.J. Friston, K.E. Stephan, T.E. Lund, A. Morcom, S. Kiebel: Mixed-effects and fMRI studies, NeuroImage **24**, 244–252 (2005)

38.42   N.E. Breslow, D.G. Clayton: Approximate inference in generalized linear mixed models, J. Am. Stat. Assoc. **88**(421), 9–25 (1993)

38.43   M. Lynch, B. Walsh: *Genetics and Analysis of Quantitative Traits* (Sinauer Associates, Sunderland 1998)

38.44   A.S. Bryk, S.W. Raudenbush: *Hierarchical linear models: Applications and data analysis methods*, Advanced Quantitative Techniques Techniques in the Social Sciences, ed. by C.D. Laughton, S. Robinson (SAGE, New Delhi 1992)

38.45   L.M. Sullivan, K.A. Dukes, E. Losina: Tutorial in biostatistics: An introduction to hierarbchical linear modelling, Stat. Med. **18**, 855–888 (1999)

38.46   G. Seber, A.J. Lee: *Linear Regression Analysis*, 2nd edn. (Wiley-Interscience, New York 2003) pp. 97–118

38.47   N. Laird, N. Lange, D. Stram: Maximum likelihood computations with repeated measures: Application of the EM algorithm, J. Am. Stat. Assoc. **82**(397), 97–105 (1987)

38.48   N.M. Laird, J.H. Ware: Random-effects models for longitudinal data, Biometrics **38**, 963–974 (1982)

38.49   A.P. Dempster, N.M. Laird, D.B. Rubin: Maximum likelihood from incomplete data via the EM algorithm, J. R. Stat. Soc. Ser. B **39**, 1–38 (1977)

38.50   J.A. Harville: Maximum likelihood approaches to variances component estimation and to related problems, J. Am. Stat. Assoc. **72**(358), 320–340 (1977)

38.51   C.E. McGulloch: Maximum likelihood algorithms for generalized linear mixed models, J. Am. Stat. Assoc. **92**(437), 162–170 (1997)

38.52   S. Searle, G. Casella, C. McCulloch: *Variance Components* (Wiley, New York 1992)

38.53   K.Y. Liang, S.L. Zeger: Longitudinal data analysis using generalized linear models, Biometrika **73**(1), 13–22 (1986)

38.54   J. Talairach, P. Tournoux: *Coplanar Stereotaxic Atlas of the Human Brain* (Thieme, Stuttgart 1998)

38.55   Y. Benjamini, Y. Hochberg: Controlling the false discovery rate: A practical and powerful approach to multiple testing, J. R. Stat. Soc. B **57**, 289–300 (1995)

38.56   Y. Benjamini, D. Yekutieli: The control of the false dicovery rate in multiple testing under dependency, Ann. Stat. **29**(4), 1165–1188 (2001)

38.57   L. Harrison, W.D. Penny, K.J. Friston: Multivariate autoregressive modeling of fMRI time series, NeuroImage **19**, 1477–1491 (2003)

38.58   J. Sjoberg, Q. Zhang, L. Ljung, A. Benveniste, B. Delyon, P.Y. Glorennec, H. Hjalmarsson, A. Juditsky: Nonlinear black-box modeling in system identification: A unified overview, Automatica **31**(12), 1691–1724 (1995)

38.59   K.J. Friston, L. Harrison, W. Penny: Dynamic causal modelling, NeuroImage **19**, 1273–1302 (2003)

38.60   W.D. Penny, K.E. Stephan, A. Mechelli, K.J. Friston: Comparing dynamic causal models, NeuroImage **22**, 1157–1172 (2004)

38.61   K. Stephan, L. Kasper, L.M. Harrison, J. Daunizeau, H.E. den Ouden, M. Breakspear, K.J. Friston: Nonlinear dynamic causal models for fMRI, NeuroImage **42**(2), 649–662 (2008)

38.62   T. Deneux, O. Faugeras: Using nonlinear models in fMRI data analysis: Model selection and activation detection, NeuroImage **32**, 1669–1689 (2006)

38.63   C. Granger: Investigating causal relations by econometric models and cross-spectral methods, Econometrica **37**, 424–438 (1969)

38.64 E. Pereda, R.Q. Quiroga, J. Bhattacharya: Nonlinear multivariate analysis of neurophysiological signals, Prog. Neurobiol. **77**, 1–37 (2005)

38.65 L. Faes, G. Nollo, K.H. Chon: Assessment of Granger causality by nonlinear model identification: Application to short-term cardiovascular variability, Ann. Biomed. Eng. **36**(3), 381–395 (2008)

38.66 A. Roebroeck, E. Formisano, R. Goebel: Mapping directed influence over the brain using Ganger causality and fMRI, NeuroImage **25**, 230–242 (2005)

38.67 G. Deshpande, K. Sathian, X. Hu: Effect of hemodynamic variability on Granger analysis of fMRI, NeuroImage **52**, 884–896 (2010)

38.68 P.J. Lahaye, J.B. Poline, G. Flandin, S. Dodel, L. Garneo: Functional connectivity: Study nonlinear delayed interactions between BOLD signals, NeuroImage **20**, 962–974 (2003)

38.69 P. Valdes-Sosa, J.M. Sanchez-Bornot, A. Lage-Castellanos, M. Vega-Hernandez, J. Bosch-Bayard, L. Melie-Carcia, E. Canales-Rodriguez: Estimating brain functional connectivity with spare multivariate autoregression, Philos. Trans. R. Soc. B **360**, 969–981 (2005)

38.70 E. Moller, B. Schack, N. Vath, H. Witte: Fitting of one ARMA model to multiple trials increases the time resolution of instantaneous coherence, Biol. Cybern. **89**, 303–312 (2003)

38.71 X. Li, G. Marrelec, R.F. Hess, H. Benali: A nonlinear identification method to study effective connectivity in functional MRI, Med. Image Anal. **14**(1), 30–38 (2010)

38.72 X. Li, K.T. Mullen, B. Thompson, R.F. Hess: Effective connectivity anomalies in human amblyopia, NeuroImage **54**(1), 505–516 (2011)

38.73 D. Felleman, D.C. Van Essen: Distributed hierarchical processing in the primate cerebral cortex, Cereb. Cortex **1**(1991), 1–47 (1991)

38.74 I. Leontaritis, S.A. Billings: Input-output parametric models for non-linear systems. Part 1 deterministic non-linear systems, Int. J. Control **41**, 303–328 (1985)

38.75 I. Leontaritis, S.A. Billings: Input-output parametric models for non-linear systems. Part 2: Stochastic non-linear systems, Int. J. Control **41**, 329–344 (1985)

38.76 S. Chen, S.A. Billings, W. Luo: Orthogonal least squares methods and their application to non-linear system identification, Int. J. Control **50**(5), 1873–1896 (1989)

38.77 K. Chon, M.J. Korenberg, N.H. Holstein-Rathlou: Application of fast orthogonal search to linear and nonlinear stochastic systems, Ann. Biomed. Eng. **25**, 793–801 (1997)

38.78 Q. Zhu, S.A. Billings: Fast orthogonal identification of non-linear stochastic models and radial basis function neural networks, Int. J. Control **64**(5), 871–886 (1996)

38.79 J. Geweke: Measures of conditional linear dependence and feedback between time series, J. Am. Stat. Assoc. **79**, 907–915 (1984)

38.80 C. Wernerheim: Cointegration and causality in the exports-GDP nexus: The post-war evidence for Canada, Empir. Econ. **25**, 111–125 (2000)

38.81 L. Oxley, D. Greasley: Vector autoregression, cointegration and causality: Testing for causes of the British industrial revolution, Appl. Econ. **30**, 1387–1397 (1998)

38.82 J.A. Doornik: Testing vector error autocorrelation and heteroscedasticity, Econometric Society 7th Congress (Tokio 1996)

38.83 D. Edgerton, G. Shukur: Testing autocorrelation in a system perspective tesing autocorrelation, Econ. Rev. **18**(4), 343–386 (1999)

38.84 J.F. Kiviet: On the rigour of some misspecification tests for modelling dynamic relationships, Rev. Econ. Stud. **53**(2), 241–261 (1986)

38.85 R.F. Engle: Wald, likelihood ratio, and Lagrange multiplier tests in econometrics. In: *Handbook of Econometrics*, Vol. 2, ed. by Z. Griliches, M.D. Intriligator (Elsevier, Amsterdam 1984) pp. 775–826

38.86 K.T. Mullen, S.O. Dumoulin, R.F. Hess: Color responses of the human lateral geniculate nucleus: Selective amplification of S-cone signals between the lateral geniculate nucleno and primary visual cortex measured with high-field fMRI, Euro. J. Neurosci. **28**, 1911–1923 (2008)

38.87 R.F. Hess, B. Thompson, G. Gole, K.T. Mullen: Deficient responsed from the lateral geniculate nucleus in humans with amblyopia, Eur. J. Neurosci. **29**, 1064–1070 (2009)

38.88 G. Kerschen, K. Worden, A.F. Vakakis, J.C. Golinval: Past, present and future of nonlinear system identification in stuctural dynamics, Mech. Syst. Signal Process. **20**, 505–592 (2006)

38.89 N. Hsu, H.L. Hung, Y.M. Chang: Subset selection for vector autoregressive processes using Lasso, Comput. Stat. Data Anal. **52**, 3645–3657 (2008)

38.90 S.A. Billings, H.L. Wei: An adaptive orthogonal search algorithm for model subset selection and non-linear system identification, Int. J. Control. **81**(5), 714–724 (2007)

38.91 H. Akaike: A new look at the statistical model identification, IEEE Trans. Autom. Control **19**(6), 716–723 (1974)

38.92 W.H. Greene: *Econometric Analysis*, 5th edn. (Prentice Hall, Upper Saddle River 2002)

38.93 H.K. Thompson, C.F. Starmer, R.E. Whalen, H.D. McIntosh: Indicator transit time considered as a gamma variate, Circ. Res. **14**(6), 502–515 (1964)

38.94 X. Li, J. Tian, R.K. Millard: Erroneous and inappropriate use of gamma fits to tracer-dilution curves in magnetic resonance imaging and nuclear medicine, Magn. Reson. Imaging **21**(9), 1095–1096 (2003)

38.95   A. Neumaier, T. Schneider: Estimation of parameters and eigenmodes of multivariate autoregressive models, ACM Trans. Math. Software **27**(1), 27–57 (2001)

38.96   T. Schneider, A. Neumaier: Algorithm 808: ARfit – A Matlab package for the estimation of parameters and eigenmodes of multivariate autoregressive models, ACM Trans. Math. Softw. **27**, 58–65 (2001), ARfit package available online at http://www.clidyn.ethz.ch/arfit/index.html

38.97   M.J. Korenberg, I.W. Hunter: The identification of nonlinear biological systems: Volterra kernel approaches, Ann. Biomed. Eng. **24**, 250–268 (1996)

38.98   X. Li, D. Coyle, L. Maguire, T.M. McGinnity: A model selection method for nonlinear system identification based fMRI effective connectivity analysis, IEEE Trans. Med. Imaging **30**(7), 1365–1380 (2011)

38.99   C. Hurvich, C.L. Tsai: Regression and time series model selection in small samples, Biometrika **76**(2), 297–307 (1989)

38.100   C. Mallows: Some comments on C_p, Technometrics **15**, 661–675 (1973)

38.101   M. Stone: Cross-validity choice and assessment of statistical predictor, J. R. Stat. Soc. **36**, 111–147 (1974)

38.102   P. Stoica, P. Eykhoff, P. Janssen, T. Soderstrom: Model-structure selection by cross-validation, Int. J. Control. **43**, 1841–1878 (1986)

38.103   D.M. Allen: The relationship between variable selection and data augmentation and a method for prediction, Technometrics **16**, 125–127 (1974)

38.104   A.J. Miller: *Subset Selection in Regression* (Chapman and Hall, London, 1990)

38.105   J. Durbin: Testing for serial correlation in least squares regression when some of the regressors are lagged dependent variables, Econometrica **38**, 410–421 (1970)

38.106   X. Li, D. Coyle, L. Maguire, T. McGinnity: A least trimmed square regression method for second level fMRI effective connectivity analysis, Neuroinformatics **11**, 105–118 (2013)

38.107   C. Liu, D.B. Rubin, Y.N. Wu: Parameter expansion to accelerate EM: The PX-EM algorithm, Biometrika **85**(4), 755–770 (1998)

38.108   X. Li, A. Messé, G. Marrelec, M. Pélégrini-Issac, H. Benali: An enhanced voxel-based morphometry method to investigate structural changes: Application to Alzheimer's disease, Neuroradiology **52**, 203–213 (2010)

39. Neural Circuit Models and Neuropathological Oscillations

Damien Coyle, Basabdatta S. Bhattacharya, Xin Zou, KongFatt Wong-Lin, Kamal Abuhassan, Liam Maguire

Degeneration of cognitive functioning due to dementia is among the most important health problems in the ageing population and society today. Alzheimer's disease (AD) is the most common cause of dementia, affecting more than 5 Mio. people in Europe with the global prevalence of AD predicted to quadruple to 106 Mio. by 2050. This chapter is focused on demonstrating models of neural circuitry and brain structures affected during neurodegeneration as a result of AD and how these model can be employed to better understand how changes in the physical basis in the electrochemical interactions at neuron/synapse level are revealed at the neural population level. The models are verified using known and observed neuropathalogical oscillations in AD. The thalamus plays a major role in generating many rhythmic brain oscillations yet, little is known about the role of the thalamus in neurodegeneration and whether or not thalamus atrophy is a primary or secondary phenomenon to hippocampal or neo cortical loss in AD. Neural mass models of thalamocortical networks are presented to investigate the role these networks have in the alterations of brain oscillation observed in AD. Whilst neural mass models offer many insights into thalamocortcial circuitry and rhythm generation in the brain, they are not suitable for elucidating changes synaptic processes and individual synaptic loss at the microscopic scale. There is significant evidence that AD is a synaptic disease. A model consisting of multiple Izhikevich type neurons elucidates now large scale networks of simple neurons can shed light on the relationship between synaptic/neuron degradation/loss and neural network oscillations. Focusing on thalamocortical circuitry may help explain oscillatory changes however the progression of AD is also usually associated with memory deficits, this implicates other brain structure such as the hippocampus. A hippocampal computational model that allows investigation of how the hippocampo-septal theta rhythms can be altered by beta-amyloid peptide (Aβ, a main marker of AD) is also described. In summary the chapter presents three different computational models of neural circuitry at different scales/brain regions and demonstrates how these models can be used to elucidate some of the vacuities in our knowledge of brain oscillations and how the symptoms associated with AD are manifested from the electrochemical interactions in neurobiology and neural populations.

Part G | 39

39.1 Modeling Neural Circuits at Multiple Scales

Although neuroscience is still predominantly based on experiments, computational modeling and theoretical analyses now form integral parts of neuroscience research [39.1] as they allow us to investigate and provide insights into and predictions about brain structure, function, and dynamics. There are many different spatial and temporal levels at which the neural system can be modeled, as well as justifications for these approaches [39.2]. Information signals in the brain can be encoded in various forms. For example, they can come in the shape of action potentials produced by a neuron, relative fidelity and strengths of synapses connecting neurons, timings of action potentials, or the frequency of the firing of action potentials. Information can be encoded in a single neuron (e.g., with a *high-level complex* feature) or distributed throughout a population of neurons. Here, we focus on a particular interesting type brain phenomenon: neural oscillations.

Although identified many years ago [39.3], only relatively recently has this phenomenon been modeled and its computational mechanisms and implications understood [39.4–7]. The study of brain oscillations have been surrounded by controversy. It was not known whether brain oscillations were an important computational element in the brain, or just epiphenomena. For example, in computational modeling, it is not difficult to produce oscillations in various neural circuits as long as there is some appropriate negative feedback. However, there is increasing evidence showing the importance of brain oscillations in the so-called *binding* phenomenon, where different brain regions that oscillate together can communicate better, thus allowing larger-scale computation over a wide region of the brain, which may also result in the formation of a more coherent brain state. With regard to memory processing in the brain, brain oscillations have been shown to be important. Computational modeling in this area will enable better understanding of the mechanisms of memory consolidation – a critical process in the formation of new memories. Limited models have been developed to capture the rapid associative learning properties of the hippocampus; however, computational models to date have generally addressed only systems-level consolidation (at the level of anatomical structures in the brain) [39.8].

This chapter presents examples of how computational models can be used to study the underlying information processing in the brain, in particular focusing on neuropathological correlates of neural oscillation abnormalities seen in several neurological disorders; we specifically study the case of Alzheimer's disease (AD). The goal of our research is to identify possible biomarkers in brain oscillatory activity (e.g., in electroencephalogram (EEG)) of people afflicted (early diagnosis) or likely to be afflicted (prediction) with AD. However, the chapter is written to be of interest to anyone working on computational model development and to provide a good source of information for anyone wishing to exploit computational models in understanding brain function and dysfunction.

AD is a complex brain disease. Different research studies have suggested that different sources, interactions, and feedback cycles are involved, each suggesting a role in AD pathology. One of the leading research directions is investigation of correlates of abnormal deposition of amyloid beta (Aβ) or plaques in AD. Recent research from Harvard University [39.9] suggests that a reduction in cortical thickness may be an indicator of abnormal Aβ deposition associated with AD. The recent report on AD diagnostic criteria [39.10] suggests that abnormal Aβ deposition may commence one to two decades prior to onset of clinical symptoms. Since EEG is a study of cortical neural populational behavior, it may be speculated that anomalous EEG in AD bears a causal relationship with such cortical thinning and the abnormal neuropathology. We further speculate that a computational modeling approach will help in better understanding the exact role of several Aβ peptide fragments in different types of neuronal oscillatory behavior and thus may provide a viable *combined biomarker* investigative direction in AD (as suggested in [39.10] and mentioned before). In Sect. 39.1 we present a model of hippocampus and use this low-level model to explore abnormal brain oscillations associated with disease and Aβ anomalies.

Modeling at microscopic levels to study brain network activity is complex and requires a significant level of neurophysiological knowledge and computational power, which is now feasible. Models at the single-neuron level and or single/multiple-compartment level for the soma, dendrites, and axon with various realistic conductances and ionic currents can help elucidate the importance of different physiological details to the overall network behavior. This is one of the most promising multiscale approaches in computational neuroscience. One of the most highly modeled brain regions is the hippocampus, which we discuss in detail later in the chapter.

Neurochemical and morphological studies have shown that many neurotransmitters may be released from both synaptic and nonsynaptic sites for diffusion to target cells more distant than those observed in regular synaptic transmission [39.11]; i.e., there are functional interactions among neurons even without synaptic contacts, hence a spiking neuron model would be insufficient to model many of the important interactions among neurons. Although nonsynaptic interaction has the potential to be an important contributor to the properties of neuronal networks and developing an understanding of the breakdown of the cytoskeleton in AD, this requires further investigation. Neuron–astrocyte interaction models have been investigated to study the role of astrocytes in regulating synapses. Astrocytes are emerging as potential contributors to early neuronal deficits in AD, network synchronization, and the production of oscillatory behavior. Further investigation is needed in this area [39.12–14].

Biophysical models such as the aforementioned conductance-based models and Hodgkin–Huxley-type models have many dynamical equations and parameters, and thus are computationally intensive and complex to develop and manipulate even if simplified to single compartments. Furthermore, single-compartmental models may prevent the incorporation of information related to the morphology of the cells and the spatial distribution of ionic conductance. In addition, some physiological information needed to construct these models may be incomplete, and related experiments are yet to be performed.

At the other extreme, lumped circuit models or neural mass models can be used to model the interactions of large neuronal populations. In neural mass models, spatial averages over neuronal populations are used rather than simulating a single cell in a distributed network, which results in a simplified network of interconnected neuronal populations that can approximate the essential properties of the original system [39.15–17]. In particular, *Lopes da Silva* [39.18–21] has proposed a lumped circuit model approach to understand the role of the neuronal and network properties of the thalamocortical system. The model is set at the mesoscopic scale (EEG thalmocortical rhythms are believed to be generated by neuronal populations (with between 10^4 and 10^7 neurons) in the thalamic and cortical tissues), which, according to Suffczínsky, has a number of advantages that [39.21]:

- Allow the establishment of the relationship between model parameters and both the cellular and synaptic properties of the modeled system
- Enable investigation of the system's dynamics
- Provide local models of the neurophysiological signals at the macroscopic level, which allows the model output to be compared with local field potentials (LFPs) or EEG
- Permit assumptions and application of system analysis methods to quantify the system behavior
- Produce computational efficiency.

Neural mass models of thalamocortical circuitry that produce alpha rhythms [39.21] can be used to develop a better understanding of the pathophysiology associated with Alzheimer's disease. Because Lopes da Silva's models focused on EEG alpha rhythms and event-related desynchronization and synchronization, they can be used to study many of the neurodynamical responses which are known to be associated with AD. Neural mass models can be verified by observing typical brain responses and ensuring that the models conform to produce similar characteristics to these processes. Both long-term and short-term characteristic brain responses can be modeled, and the associated and known oscillatory EEG correlates can be used to verify and validate the models. Here, we present a number of studies using a neural mass model.

Whilst neural mass models offer insights into thalamocortical circuitry and rhythm generation in the brain and provide a means which is computationally tractable, information linking single neuron activity or synaptic activity at the microscale to network activity at the macroscale (provided by neural mass models model) is not available. In contrast, detailed conductance-based or Hodgkin–Huxley-type models offer greater insight at the lower scales. However, these approaches are computationally intensive, involve large sets of model parameters, and are thus inconvenient for large-scale modeling of thalamocortical circuitry. *Izhikevich* and *Edelman* [39.22] have taken a compromise approach, at an intermediate level between the low-level and lumped models. Their network model consists of phenomenological simple spiking neuronal models that mimic conductance-based neuronal models with various currents [39.23]. Various neuronal types at different cortical layers were emulated in in vitro and in vivo recordings in animals' brains. Long-range connections of this large-scale model of the thalamocortical systems in the mammalian brain are informed by neuroanatomical studies such as retrograde/anterograde tracing or diffusion tensor imaging (DTI). The shape and connectivity (macroscopic levels) of the model are determined by the DTI and magnetic resonance imaging (MRI). Neuronal dynamics of individual neuron and dendritic tree formations are based on the simple phenomenological model proposed by *Izhikevich* [39.23, 24]. Short-term synaptic plasticity is incorporated based on a simple model, with different synaptic types determined by a small number of parameters. Equations proposed by *Izhikevich* in [39.24] are used to model the long-term conductance (weight) of each synapse in the model, which is simulated according to a spike-time-dependent plasticity (STDP), where the synapse is potentiated or depressed depending on the order of firing of the presynaptic neuron and the corresponding (dendritic) compartment of the postsynaptic neuron. For computational reasons, the density of neurons and synapses per mm^2 of cortical surface was necessarily reduced – fewer neurons and dendritic trees. Other neuromodulation systems including the dopaminergic reward systems and cholinergic systems (important for AD research) can also be incorporated.

Izhikevich and *Edelman* [39.22] have demonstrated that the model exhibited spatiotemporal dynamics and features similar to those observed from normal brain, even though these were not specifically built into the model, i. e., occurred spontaneously, and that adding or removing a spike of one neuron changed the state of the entire cortex after 0.5 s. In addition, regions of the brain exhibited collective waves of oscillations of local field potentials in the delta, alpha, and beta ranges similar to those recorded in humans (again very important for AD research).

This level of biological realism in the model may offer a range of benefits in developing better understanding of diseases such as AD. One way to deepen our understanding of how the synaptic and neuronal processes interact to produce the collective behavior of the brain is to develop large-scale anatomically detailed models of the mammalian brain; For example, by using DTI of patients with Alzheimer's disease or other neurological and psychiatric disorders, the influence of connectivity alone on brain dynamics may be investigated. A recent study by *de Jong* et al. [39.25] indicates that atrophy is considered a better marker for impaired functioning of brain regions than deposits of soluble proteins and plaques. Descriptions of cerebral atrophy patterns associated with onset and progression of AD are considered to be important, as they may elucidate the pathogenesis of AD and the contribution of various brain structures to cognitive decline. *de Jong*

et al. [39.25] observed reduced volumes of putamen and thalamus in Alzheimer's disease via a MRI study of 139 memory complainers (MC) and probable AD patients. The study showed that the left and right hippocampus, putamen, and thalamus of probable AD subjects are significantly smaller than those of MC. Also, the total brain volume and neocortical grey matter were shown to be significantly reduced in probable AD subjects, and the volumes of the left hippocampus, left putamen, and left thalamus in probable AD subjects correlate significantly with a range of cognitive test scores. The robust findings of smaller volumes of hippocampus, putamen, and thalamus in patients diagnosed with probable Alzheimer's disease strongly suggest that degenerative pathology affects these structures more or earlier in the process of Alzheimer's disease than other deep grey matter structures. Therefore, using this information to inform the macroscopic anatomy of the model via DTI and MRI, may simulate the effect of structural perturbations and atrophy in the thalamus and cortex due to AD on the global dynamics, and compare with results from human EEG data, for which a vast amount of information is available. To investigate these possibilities we present some preliminary data from a number of studies using simple phenomenological models based on an Izhichevich as well as a more detailed Hodgkin–Huxley-type model in this chapter.

The chapter is organized as follows. In Sect. 39.2, we present a brief review of AD and current research trends. In Sect. 39.3, we present a biophysical ion-channel-based model of the septohippocampal network which is implicated as the key area in generation of hippocampal theta oscillatory activity. We show that a decrease in the fast inactivating K^+ ionic channel current in the model can significantly change the theta band power. Furthermore, it seems to reduce the spiking threshold of hippocampal pyramidal cells, thus making them more excitable. In Sect. 39.4, we present a conductance-based spiking neuronal network of cortical neurons to study beta oscillations in AD and their correlation with cortical synaptic losses such as in AD-affected brains. We observe that a significant decrease in beta band power in the model occurs when the number of excitatory cells in the network is reduced by more than 13%, simulating neuronal loss associated with AD. Furthermore, a comparison with other frequency bands in a scaled-up version of the model also shows an early increase in the beta band frequency with a reduction of excitatory cell count in the network. This supports experimental studies showing the beta band to be affected in earlier stages of AD compared with the alpha and delta bands. In the final part of our work, presented in Sect. 39.5, we study a neural mass computational model (neural population level) of the thalamocorticothalamic circuitry to investigate alpha rhythm slowing as observed in the EEG of AD. Our results show a significant role of the inhibitory cell populations in the thalamus in slowing the alpha rhythm. Furthermore, a significant decrease in power within the alpha band is also observed

with decreased levels of synaptic connectivity in the model, simulating conditions of impaired neuronal connectivity in AD. The results of each of the three models are presented and discussed in the respective sections. We conclude the chapter in Sect. 39.6 with an overview of the presented models and implications of the results obtained in the context of AD, followed by possible directions for future work.

39.2 Alzheimer's Disease and Abnormal Brain Oscillations

In 1907, Alois Alzheimer reported an aggressive form of dementia that affected a 51-year-old female patient, Frau Auguste D., causing memory, language, and behavioral deficits. A brain autopsy revealed neuritic plaques, neurofibrillary tangles, and brain atrophy (shrinking of the cortical surface) [39.26, 27].

In 2011, more than a century later, definitive diagnosis of what is now known as *Alzheimer's disease* is still only possible by autopsy [39.28, 29]. This is an indication that, even with the remarkable advancement of scientific techniques and equipment, we still understand little in terms of the underlying neuronal mechanisms associated with onset and progression of AD to this day. According to current statistics, a total of 821 884 cases of dementia are reported in the UK, costing the economy 23 billion GBP per annum in health, social, and informal care costs as well as productivity losses [39.30]. Of these, AD accounts for 60–80% of cases, some examples of the other types of dementia being vascular dementia, mild cognitive impairment (MCI, speculated to be a precursor to AD), dementia with Lewy bodies, frontotemporal dementia, etc. [39.31]. In the USA, AD is the seventh leading cause of death, affecting 5.3 million people and costing the economy 172 billion USD per annum. Currently, treatment of AD involves medication that provides symptomatic relief and is effective in only half of the population that take the drugs [39.31–33]. The primary risk factor for AD is old age, and with increasing lifespans not only in developed but also in developing countries, incidence of AD is likely to increase further. The rising rate of the disease as well as the cost involved demand therapeutic drugs which can prevent the irreparable damage by targeting the underlying causes of the cascading neurodegeneration; the thrust now is on finding biomarkers associated with neuropathological changes present in the brain of AD patients several years or even decades prior to onset of cognitive deteri-

oration, thus aiding in preclinical diagnosis [39.34–37]. Furthermore, in the recently published modified diagnostic criteria for AD [39.10], two points are given special emphasis: an investigative approach that combines several biomarkers, and thorough validation of any proposed biomarker with post mortem studies.

The neuropathological hallmark of AD is the formation of neurofibrillary tangles (NFT) and neuritic plaques (NP); NFT are pathological tangles of hyperphosphorylated tau protein, while the main constituent of NP is β-amyloid peptide (Aβ). The direct consequence of such deposition is severe loss of neuronal synaptic connectivity and eventual neuronal death. Subsequently, severe atrophy (shrinking in area) of the cortex and hippocampus occurs, which is the post mortem structural hallmark of an AD-affected brain. The pathogenesis of degeneration in neuronal connectivity may be primarily attributed to abnormal deposition of Aβ in the AD-affected brain [39.38]. However, NFT and NP are also a part of the natural ageing process. Thus, one of the major challenges in the treatment of Alzheimer's disease is that early clinical symptoms of the disease cannot be distinguished from normal ageing or from other forms of dementia related to advancing age [39.39].

Recent studies have reported atrophy of the thalamus in AD [39.25, 40]. Indeed, the thalamocorticothalamic circuitry has long been implicated as the generator of brain oscillatory activity [39.41, 42]. This oscillatory behavior can be studied using electroencephalography (EEG), which is a low-cost, noninvasive measure of electrical activity along the (human/animal) scalp. Over the past few decades, EEG has been a popular tool in the quest for biomarkers in various pathological as well as neuropathological disorders including AD [39.43]. It is now well known that the EEG of AD patients shows a definite *slowing* (decrease in peak power) within the alpha frequency band (8–12 Hz) [39.44, 45], which is considered to be the hallmark of an AD-affected brain. Subsequently, other research has shown important correlates of cognitive as well as sensorimotor deficits in EEG of AD patients [39.46–53]. Such work reveals an overall decrease in the peak EEG power in AD, indicated by an increase in power within the theta (4–7.5 Hz) band during the early stages of the disease followed by a decrease in power within the beta (14–30 Hz) band as the disease progresses; a decrease within the alpha (8–10 Hz) band power and an increase within the delta (0.5–3 Hz) band power are observed in later stages of the disease [39.54]. Furthermore, longitu-

dinal studies show that the degree of EEG abnormality in AD patients is directly proportional to their cognitive deterioration [39.29]. Thus, it is speculated that one of the promising areas for identifying viable early markers for AD is through electroencephalography (EEG) studies and the underlying neuronal pathology reflected in these oscillatory changes in AD; this in turn will aid new diagnostic tools and methods for its early detection [39.55]. The remainder of the chapter focuses on understanding the underlying neural dynamics and physiology that result in altered EEG and brain oscillatory activities in Alzheimer's disease.

39.3 A Biophysical Model to Study Hippocampal Theta Rhythms in AD

The most prominent clinical symptom of AD is loss of memory. This is caused by severe neurodegeneration in the CA1 and CA2 regions of the hippocampus ("CA" is an abbreviated form of the Latin name *cornu ammonis* for hippocampus; it is subdivided into four regions: CA1–CA4). The hippocampal formation is among the earliest affected brain areas in AD. Furthermore, it has extensive connections with the entorhinal cortex [39.56], a region affected during the earliest stages of AD [39.57]. It is thus not surprising that there are several studies involving both biophysical as well as connectionist (large-scale network) models of the hippocampus to understand memory dysfunction in AD [39.58] (see [39.59] for a review). A characteristic oscillatory behavior displayed by the hippocampal formation arising out of its connections with the *medial septum* is within the theta band [39.60, 61] and is thought to be a correlate of memory [39.62] as well as other behavioral aspects [39.63]. EEG studies show that early stages of AD are often characterized by an increase in the hippocampal theta power [39.64–66]. To date, the cause of theta rhythm changes in AD remains unclear. This work aims to address this limitation using a computational model of the hippocampus and the associated medial septum, the goal being to investigate the abnormal deposition of Aβ in AD as an underlying neuropathological correlate of theta rhythm alteration in the early stages of the disease.

39.3.1 A Conductance-Based Model of the Septohippocampal Network

This work is based on a septohippocampal model presented in [39.67] consisting essentially of three cell pop-

ulations of the hippocampus CA1 region which participate in the generation of oscillatory activity: pyramidal cells (PY), basket interneurons (BI), and *stratum oriens* interneurons (which project to the *lacunosum moleculare*, hence abbreviated as OLM). A fourth cell population (MSGABA) represents the ionotropic alpha-amino-3-hydroxy-5-methyl-4-isoxazapropionic acid (AMPA) GABAergic cells (GABA$_A$) of the medial-septal region. The model is shown in Fig. 39.1. The cell populations are modeled as a network of 10 PY, 100 BI, 30 OLM, and 50 MSGABA neurons [39.67], while each neuron in the network is modeled using the Hodgkin–Huxley-type dynamical equations defined in (39.1–39.5) [39.68, 69]

$$\frac{dV_{sPY}}{dt} = -I_L - I_{Na} - I_K - I_{Ca} - I_A - I_{CT}$$
$$- I_h - \frac{g_c}{p}(V_s - V_d) - I_{syn,s} + I ,$$
$$\tag{39.1}$$

$$\frac{dV_{dPY}}{dt} = -I_L - I_{Ca} - I_{AHP} - I_A - I_{CT} - I_h$$
$$- \frac{g_c}{1-p}(V_d - V_s) - I_{syn,d} , \tag{39.2}$$

$$\frac{dV_{BI}}{dt} = -I_L - I_{Na} - I_K - I_{syn} + I , \tag{39.3}$$

$$\frac{dV_{OLM}}{dt} = -I_L - I_{Na} - I_K - I_{Ca} - I_h$$
$$- I_{AHP} - I_{syn} + I , \tag{39.4}$$

$$\frac{dV_{MSGABA}}{dt} = -I_L - I_{Na} - I_K - I_{KS} - I_{syn} + I . $$
$$\tag{39.5}$$

The pyramidal neurons are described by a two-compartmental model (39.1, 39.2), one for the soma

The PY cells innervate BI and OLM cells via neurotransmitter AMPA and AMPA+NMDA (*N*-methyl-D-aspartate), respectively. All other synaptic connections are mediated by GABA$_A$ neurotransmitter. The network is constructed using a sparse connectivity according to [39.67] (Sect. 39.A); i.e., the neurons are randomly coupled with a fixed average number of presynaptic inputs/postsynaptic outputs per neuron. To emulate the heterogeneity in real brain tissues, the injected DC current I for each neuron is not chosen to be identical. This is done by allowing I to follow a Gaussian distribution with mean I_μ and standard derivation I_σ. I_μ for the PY, BI, OLM, and MSGABA neural populations is chosen to be 5, 1.4, 0, and 2.2 μA/cm^2, respectively; $I_\sigma = 0.1$ μA/cm^2 for all populations. Values of the network connectivity parameters as well as all other parameters defined in (39.1–39.5) are consistent with those used in [39.67] (Sect. 39.A) and are defined in the Appendix.

39.3.2 Experimental Methods

We compute the LFP signal as a sum of the values of the synaptic currents of the pyramidal neurons [39.70]. This is under the assumption that pyramidal neurons contribute more to the overall signal due to their approximate open field arrangement. The network oscillations are investigated by analyzing a sum of all the membrane potentials (this can also be achieved using a summation of synaptic currents). The power spectrum is obtained by a fast Fourier transform with a 2 s-long Hamming window. The relative theta band power (% of total power) is calculated. A membrane potential noise that follows a Gaussian distribution with zero mean and 1.5 mV standard derivation is also introduced in some of the simulations. The heterogeneity and membrane noise are randomly generated in each trial that lasts for 10 s. Each presented result is obtained from simulations averaged over 15 trials of the model, representing different theta band recordings from individuals. We found that a higher number of trials does not alter the obtained average theta band power. The results from trials of the model run with normal parameter settings are considered as a *healthy* control group, whereas trials with alterations to parameters of various ionic channels simulate deficiencies and are considered as different *patient* groups. Various ionic channels are potentiated or suppressed to simulate the effects of Aβ. All of the results were obtained by adjusting the ionic currents in the pyramidal neurons only. The statistical significance of the differences between groups was eval-

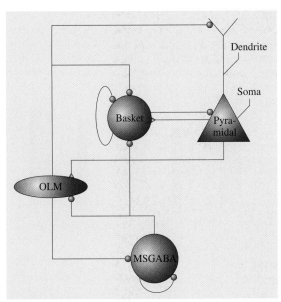

Fig. 39.1 Septohippocampal model architecture as used in [39.67]. The network consists of four types of neuronal populations: pyramidal, basket, OLM, and MSGABA. Inhibitory GABA$_A$-mediated synaptic connections are indicated by ⊙, and excitatory AMPA- and NMDA-mediated synaptic connections are indicated by △

(subscript "s" denotes soma) and the other for the dendrite (subscript "d" denotes dendrite) [39.68]. The soma compartment has spike-generating currents I_{Na} and I_K, and the dendrite contains a calcium-dependent potassium current I_{AHP}. Both the soma and dendrite contain leakage currents I_L, high-threshold L-type calcium currents I_{Ca}, and hyperpolarization-activated currents I_h. In addition, we model certain other currents which have been shown to be affected by Aβ. Thus, our model also contains an A-type potassium current I_A and a large-conductance calcium-dependent potassium current I_{CT} in both soma and dendrite. I is the injected direct current (DC). The somatic and dendritic membrane potentials are denoted by V_s and V_d, respectively, while the coupling conductance between the soma and dendrite is denoted by g_c, and p is defined as the ratio of the somatic area to the total cell area. The neurons of the BI, OLM, and MSGABA populations are single-compartment models, defined in (39.3–39.5), respectively. I_h in (39.4) denotes a hyperpolarization-activated current in an OLM cell, while I_{KS} in (39.5) denotes a slowly inactivating potassium current in a MSGABA cell. All other currents denote the same as for PY cells in (39.1) and (39.2).

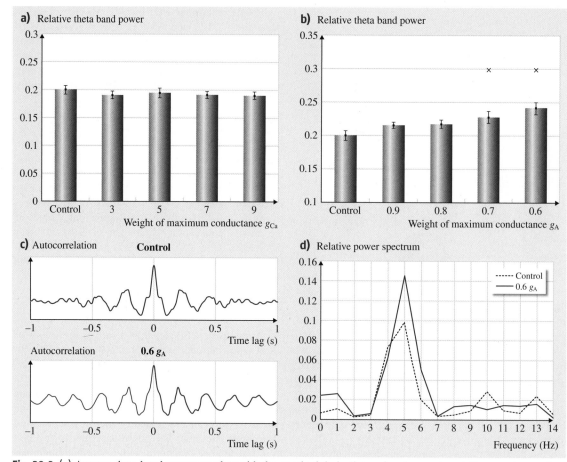

Fig. 39.2 (**a**) Average theta band power together with the standard derivations of each experiment. Increase in g_{Ca} does not induce changes in theta band power. (**b**) Theta band power increases with a decrease in g_A. "×" indicates that power is significantly larger than that obtained in control condition ($p < 0.05$). Error bar is standard error. (**c**) The autocorrelations of a summation of membrane potentials obtained in control and $0.6 \, g_A$ conditions. Theta rhythm is strengthened by decreased g_A. (**d**) More significant power spectrum peak in theta band in $0.6 \, g_A$ condition than in control condition. Both of the results are obtained in the same noisy and heterogeneity condition

uated using an ANOVA (one-way analysis of variance) test.

39.3.3 Results and Discussion

To better understand the spiking phases of different neuronal populations, noise in the membrane potential is removed. Our model shows that theta oscillation is generated by the spiking of different neuronal populations clustered at certain phases, assuming that a network theta oscillation begins with spikes from the pyramidal neurons. Then the OLM neurons are evoked via the excitatory synaptic connections from the pyramidal neurons. The basket neurons then gradually depolarize and produce a series of spikes. The spikes of basket neurons are inhibited by the spiking of MSGABA neurons. The slowly inactivating potassium current I_{KS} in the MSGABA neurons plays a very important role in the theta generation, being referred to as a *pacemaker* for theta rhythm [39.71].

It has been pointed out that the main cause of the loss of intracellular calcium homeostasis in AD patients is that Aβ can potentiate the L-type Ca^{2+} channels (I_{Ca}) [39.72], which causes a large influx of Ca^{2+} into the cells. The mechanism of Aβ increasing the influx of Ca^{2+} is still unclear. Aβ may form new cation channels

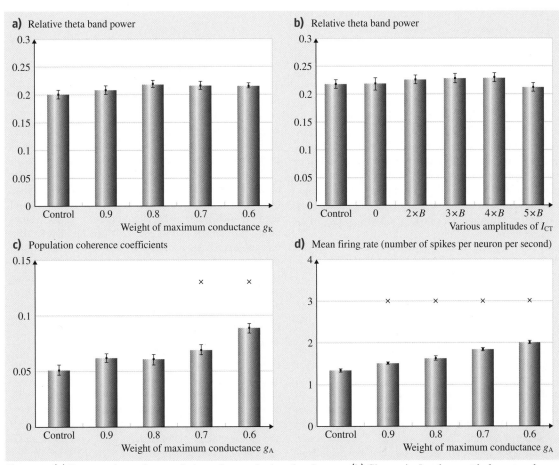

Fig. 39.3 (**a**) Decrease in g_K does not induce changes in theta band power. (**b**) Change in I_{CT} does not induce any change in theta band power. Both the completely blocked I_{CT} (0) and the potentiated I_{CT} (2B, 3B, 4B, 5B) are evaluated. (**c**) The pyramidal neuronal population coherence coefficients increase when g_A is decreased. "×" indicates that the coherence coefficient is significantly larger than that obtained in control condition ($p < 0.001$). (**d**) Pyramidal neuronal population firing rates increase with decrease in g_A. "×" indicates that the firing rate is significantly larger than that obtained in control condition ($p < 0.001$)

and/or alter the existing L-type Ca^{2+} channels. In our simulations, we emulate the effect of Aβ by increasing the maximum conductance of the L-type Ca^{2+} channels. The obtained theta band power with enhanced I_{Ca} is presented in Fig. 39.2a. It can be seen that changes in L-type Ca^{2+} channels do not cause a change in theta rhythm.

Aβ also blocks some K^+ ionic channels in pyramidal neurons, e.g., I_A and I_K [39.73, 74]. The experimental results showed that Aβ is more likely to block the channel from outside the neurons. Therefore, we emulate the effect of Aβ by decreasing the maximum conductance of I_A and I_K. Furthermore, it has been

shown that I_A has larger dendrite density compared with soma [39.75] and Aβ has much greater effect on the dendrite than I_A [39.76, 77]. Based on these findings, only I_A in the dendrites is reduced. The simulation results obtained in control conditions and with decreased I_A in the dendrite only are illustrated in Fig. 39.2b. Our simulation shows that the theta band power is significantly increased ($p < 0.05$) as I_A decreases. An example of the autocorrelation of the summation of all membrane potentials and the corresponding band power in control and $0.6\,g_A$ conditions is illustrated in Figs. 39.2c and 39.2d. It can be seen that theta oscillation and its power is significantly increased with low

g_A. Similar changes in theta band power due to I_K (via g_K) are not observed, as illustrated in Fig. 39.3a.

As AD disturbs the homeostasis of Ca^{2+}, the Ca^{2+}-activated BK channel (I_{CT}) is vulnerable to AD pathology. The BK channel can adjust the spike broadening during repetitive firing [39.78] and spiking frequency [39.79]. Previous research reveals that the activity of the BK channel is probably promoted by Aβ [39.80]. However, other research reports that the BK channel is suppressed in some cases [39.81–83]. Therefore, we simulated both increased and blocked I_{CT} in our simulations. I_{CT} is potentiated by increasing the fraction of Ca^{2+} influx, B (see Appendix). The simulation results are illustrated in Fig. 39.3b. It can be seen that neither blockage nor potentiation in I_{CT} can affect theta rhythm.

39.3.4 Mechanism Underlying Change in Theta Rhythm

The simulation results have shown that a decrease in I_A can significantly increase the theta band power. To evaluate whether this is due to an enhanced synchrony of neuronal populations, we calculate the population coherence coefficient [39.84]. The long time interval T ($T = 2$ s in our experiment) is first divided into small bins of $\tau = 1$ ms, and spike trains of the ith and jth neurons in the population are denoted as $X_i(l)$, $X_j(l) = 1$ or 0, $l = 1, \ldots, K$ ($K = T/\tau$), where "1" denotes spiking and "0" resting. The coherence coefficient κ_{ij} between the trains can be calculated as

$$\kappa_{ij} = \frac{\sum_{l=1}^{K} X_i(l) X_j(l)}{\sqrt{\sum_{l=1}^{K} X_i(l) \sum_{l=1}^{K} X_j(l)}} \,. \tag{39.6}$$

The whole population κ is obtained by averaging all of the combinations of i and j. κ is calculated for the control group and the group with decreased g_A. The obtained κ is illustrated in Fig. 39.3c. Consistent with our hypothesis, population synchrony is significantly increased as g_A decreases ($p < 0.05$).

The increased synchrony is probably caused by the enhanced excitability and firing rate of the pyramidal neurons. To support this hypothesis, the firing rates of the pyramidal neuronal population with various values of g_A are shown in Fig. 39.3d. It can be clearly seen that the decreased g_A has enhanced the excitability of the pyramidal neurons and their firing rates. Therefore, we suggest that, when I_A is decreased, the pyramidal neurons become more excitable. During the peak of each pyramidal population theta cycle, more pyramidal neurons spike simultaneously, which enhances the synchrony of the population.

In summary, our simulations have shown that a decrease in I_A in the pyramidal neurons induces an increase in theta band power by recruiting more pyramidal neurons to fire. In the next section, we explore the effects of the underlying neuropathology affecting cortical beta rhythms in AD. Further details of the model and an extended nonlinear dynamical systems analysis of the model can be found in [39.85, 86].

39.4 Neural Population Models to Study Cortical Rhythms in AD

Previous biological experiments on different types of cultured excitatory neurons show that cortical neuronal death is mediated by dysfunctional ionic behavior that might specifically contribute to the pathogenesis of β-amyloid peptide (Aβ)-induced neuronal death [39.87]. Furthermore, a number of studies reported an alteration of neuronal K^+ channel function after exposure to Aβ peptide fragments which led to remarkable perturbations in neuronal behavior. These channels provide a negative feedback to the membrane potential of neurons. Thus, K^+ channels regulate neuronal dynamics including the timing of interspike intervals (time between spikes), setting the resting potential, and keeping action potentials short [39.88]. Deregulation of K^+ channels has been investigated by a number of studies such as in [39.89–91]. In this section, we re-

view our earlier studies on investigating the effects of neuronal/synaptic loss and deregulation of negative feedback to the membrane potential of cortical excitatory neurons (which mainly results from Aβ-induced dysfunctional K^+ channels) on the oscillatory activity of cortical networks using two heterogeneous neuronal network models [39.38, 92], the goal being to provide better understanding of the underlying neural correlates of abnormal cortical oscillations within the beta frequency band in AD.

39.4.1 A Local Neural Network Model with Biophysical (Detailed) Neurons

We simulated a conductance-based neuronal network of 200 cells containing 160 excitatory (e-cells) and 40 inhibitory (i-cells) neurons in the baseline case. The cells are connected all-to-all, assuming that e–e synapses are weak within local networks [39.93]. Parameters and functional forms of the equations are adopted from [39.94]. Neurons were modeled by Hodgkin–Huxley dynamics as shown in (39.7) and (39.8). Both types of cells have a leak (L), transient sodium (Na), and delayed rectifier potassium (K) current. The e-cells have an additional after-depolarizing potential (AHP) resulting in a slow outward potassium current.

$$C\frac{\mathrm{d}V_i}{\mathrm{d}t} = -g_L(V_i - V_L) - g_K n^4(V_i - V_K)$$
$$- g_{Na}m^3h(V_i - V_{Na}) - I_{syn,i} + I_0 , \quad (39.7)$$

$$C\frac{\mathrm{d}V_e}{\mathrm{d}t} = -g_L(V_e - V_L) - g_K n^4(V_e - V_K)$$
$$- g_{Na}m^3h(V_e - V_{Na}) - g_{AHP}w(V_e - V_K)$$
$$- I_{syn,e} + I_0 . \quad (39.8)$$

The maximal conductances were $g_{Na} = 100\,\mathrm{mS/cm^2}$, $g_K = 80\,\mathrm{mS/cm^2}$, $g_L = 0.1\,\mathrm{mS/cm^2}$, and $g_{AHP} = 0.3\,\mathrm{mS/cm^2}$. Reversal potentials were $V_L = -67\,\mathrm{mV}$, $V_K = -100\,\mathrm{mV}$, and $V_{Na} = 50\,\mathrm{mV}$. The capacitances for e- and i-cells were $1\,\mathrm{AF/cm^2}$. The gating variables m, h, and n satisfy equations of the form

$$\frac{\mathrm{d}x}{\mathrm{d}t} = a_x(V)(1-x) - b_x(V)(x) \quad (39.9)$$

for $x = m, h, n$, where

$$a_m(V) = 0.32\frac{54+V}{1 - \exp\frac{-(V+54)}{4}} , \quad (39.10)$$

$$b_m(V) = 0.28\frac{V+27}{\exp\frac{(V+27)}{5} - 1} , \quad (39.11)$$

$$a_h(V) = 0.128\exp\frac{-(V+50)}{18} , \quad (39.12)$$

$$b_h(V) = \frac{4}{\left(1 + \exp\frac{-(V+27)}{5}\right)} , \quad (39.13)$$

$$a_n(V) = \frac{0.032(V+52)}{\left(1 - \exp\frac{-(V+52)}{5}\right)} , \quad (39.14)$$

$$b_n(V) = 0.5\exp\frac{-(57+V)}{40} . \quad (39.15)$$

The gating variable w is represented by

$$\frac{\mathrm{d}w}{\mathrm{d}t} = \frac{w_\infty(V) - w}{\tilde{\tau}_w(V)} , \quad (39.16)$$

$$w_\infty(V) = \frac{1}{\left(1 + \exp\left[\frac{-(V+35)}{10}\right]\right)} , \quad (39.17)$$

$$\tau_w(V) = \frac{400}{\left(3.3\exp\left[\frac{(V+35)}{20}\right] + \exp\left[\frac{-(V+35)}{20}\right]\right)} . \quad (39.18)$$

Synaptic currents were modeled by

$$I_{syn}, \alpha = g_{i\alpha}s_{i,tot}(V_\alpha - V_{in}) + g_{e\alpha}s_{e,tot}(V_\alpha - V_{ex}) \quad (39.19)$$

for $\alpha = $ e, i. Reversal potentials for AMPA and GABA$_A$ were $V_{ex} = 0\,\mathrm{mV}$ and $V_{in} = -80\,\mathrm{mV}$, respectively. The synaptic gates satisfy

$$s_{\alpha,tot} = \frac{1}{N_\alpha}\sum_{\alpha\text{-cells}} s_\alpha , \quad (39.20)$$

$$\frac{\mathrm{d}s_\alpha}{\mathrm{d}t} = a_\alpha\left(1 + \tanh\left(\frac{V_\alpha}{4}\right)\right)(1 - s_\alpha) - \frac{s_\alpha}{\tau_\alpha} , \quad (39.21)$$

where $a_e = 20\,/\mathrm{ms}$, $a_i = 1\,/\mathrm{ms}$, $\tau_e = 2.4\,\mathrm{ms}$, and $\tau_i = 12\,\mathrm{ms}$. The inhibitory GABA$_A$ conductances, g_{ie} and g_{ii}, are 5 and $10\,\mathrm{mS/cm^2}$, respectively. The excitatory conductances were $g_{ee} = 0.01\,\mathrm{mS/cm^2}$ and $g_{ei} = 0.05\,\mathrm{mS/cm^2}$ [39.94].

Gaussian noise generated by a Wiener process was added to the voltages at each integration step. The magnitude of the noise was $0.5\,\mathrm{mV^2/ms}$ for the e-cells. The equations were integrated using Euler's method with a time step of $0.025\,\mathrm{ms}$. The modeling time was $2000\,\mathrm{ms}$, and spike trains were analyzed after $1000\,\mathrm{ms}$, allowing a settlement period of $1000\,\mathrm{ms}$.

To achieve heterogeneity, the input currents (I_0) were varied in the range from 0.6 to $2.0\,\mu\mathrm{A/cm^2}$ for the e-cells and from 1 to $1.1\,\mu\mathrm{A/cm^2}$ for the i-cells. The release of neuromodulators such as acetylcholine (ACh) in the rest state is lower than that in the active states. Low ACh increases the AHP currents (I_{AHP}). This motivates the choice of relatively strong AHP currents (I_{AHP}) [39.94].

Hypothesis Testing

Based on the local network model, we have investigated the effect of excitatory circuit disruption on beta band power (13–30 Hz). To achieve our aim, we first ran the model with physiological (normal) values of all parameters for 25 trials. Then, we repeated the same procedure but with the loss rate of e-cells varied in the interval 6–19%, i.e., the number of e-cells (N_e) varied in the interval 130–150. We found that the mean value of beta band power becomes stable after 25 trials.

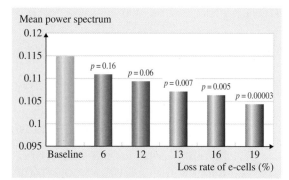

Fig. 39.4 Mean power spectra in beta frequency band (13–30 Hz) for a physiological case (baseline) and cases with reduced e-cells. *Brown bars* correspond to significant differences. Number of trials is 25 for each setup

The fast Fourier transform (FFT) technique was applied on the spiking train produced by each network setup to calculate the beta band power. One-way repeated-measures ANOVA was used to analyze the significance of the difference in beta band power between the normal (baseline) case and cases with abnormal ratio of e-cells to i-cells (corresponding to abnormal groups). *p*-Values less than 0.05 indicate significant difference.

Results: Effect of Excitatory Neuronal/Synaptic Loss on Beta Band Power

A significant decrease in beta band power is observed after the excitatory circuit loses more than 20 e-cells (the number of e-cells becomes ≤ 139, loss rate $\geq 13\%$, $p < 0.05$), as illustrated in Fig. 39.4. We refer to this point as a breakdown point, since the power spectrum of the beta rhythm exhibits a significant decrease after this point. In the network model, the death of each e-cell is associated with a synaptic loss of 319 (1%) excitatory synapses, since the network is fully connected (each e-cell receives 159 e-synapses, innervates 159 e-cells, and has one recurrent synapse).

In [39.95], it was emphasized that decreased activity in alpha, beta, and gamma waves is related to changes in excitatory circuit activity. The study [39.95] involved the analysis of a large EEG dataset using global field synchronization (GFS), a novel measure to quantify global EEG synchronization. A high GFS index for a certain frequency band reflects increased functional connectivity between brain processes. The patient results showed increased GFS values in the delta band, and decreased GFS values in alpha, beta, and gamma frequency bands, supporting the disconnection syndrome hypothesis [39.95, 96].

39.4.2 Local Neural Network Model with Simple Neurons

Spiking dynamics of neurons were simulated based on *Izhikevich*'s model of spiking neurons [39.23], which can reproduce the firing patterns of all known types of hippocampal, cortical, and thalamic neurons. The spiking neuron can be expressed in the form of ordinary differential equations defined in (39.22–39.24).

$$\frac{dV}{dt} = 0.04V^2 + 5V + 140 - u + I \,, \tag{39.22}$$

$$\frac{du}{dt} = a(bV - u) \,, \tag{39.23}$$

with the auxiliary after-spike resetting

$$\text{If } V \geq 30\,\text{mV then } V \leftarrow c, \quad u \leftarrow u + d \,, \tag{39.24}$$

where the dimensionless variables V and u represent the membrane potential and the recovery variable of the neuron, respectively. The recovery variable u provides negative feedback to V, and it corresponds to the inactivation of Na^+ ionic currents and activation of K^+ ionic currents [39.23].

Dimensionless parameters a–d (illustrated in Fig. 39.5) can be tuned to simulate the dynamics of inhibitory and excitatory neurons. Parameter b describes the sensitivity of the recovery variable u to the subthreshold fluctuations of the membrane potential V. Greater values of b couple V and u more strongly, resulting in possible subthreshold oscillations and low-threshold spiking dynamics [39.23].

Spiking networks of 1000 neurons of different types, fully and randomly connected to each other with no plasticity, were simulated to investigate two hypotheses about the underlying causes of abnormal cortical oscil-

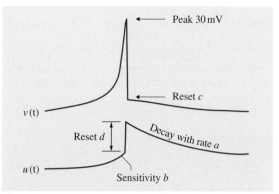

Fig. 39.5 Graphical representation of the influence of parameters a, b, c, and d on the spiking dynamics [39.23]

Table 39.1 Model setup for the investigated case studies. A number of parameters have been scaled to track the effect of changing such parameters on the EEG dynamics of the neuronal network

Case study	Number of e-cells	Number of i-cells	Loss rate (%) of e-cells	Parameter b for e-cells
1	764–794[a]	200	0.0075–0.045	0.2
2	800	200	0	0.195–0.1995[b]

[a] The loss rate of e-cells in case study 1 varies from 0.75% for group 1 to 4.5% for the last group. The length of step between each group in case study 1 is 0.25, i.e., the second group has 1% loss of e-cells. The total number of groups in case study 1 is 17 ([(794−764)/2] +1+ control group)

[b] Values of parameter b for e-cells were decreased from 0.1995 (group 1) to 0.195 (group 10). The length of step between each group in case study 2 is 0.0005. The normal value of b for e-cells is 0.2

lations in AD patients. The networks were stimulated by a random thalamic current at each time step. MATLAB software was used to simulate the networks in real time (resolution 1 ms). Setting the parameters of the model to physiological (normal) settings represents the physiological case. The ratio of excitatory to inhibitory neurons is 4 : 1, inspired by the anatomy of the mammalian cortex [39.91]. The modeling time was 30 000 ms, and spike trains were analyzed after 29 000 ms, allowing a settlement period (stability) of 29 000 ms.

Hypothesis Testing

The model parameters were varied to test two hypotheses. The different group comparisons and associated parameter alterations have been categorized into the two case studies described in Table 39.1. The first hypothesis is related to variations in the ratio of excitatory to inhibitory neurons in AD, while the second hypothesis is derived from AD observations of functional deficits in K^+ ionic channels in cortical neurons. Each hypothesis was investigated with a computational model as presented in Table 39.1; each model has a different setup, so we refer to these models as case studies; i. e., hypothesis 1 was investigated in case study 1, and hypothesis 2 was investigated in case study 2.

Hypothesis 1 relies on the loss of excitatory neurons that affects the ratio of excitatory to inhibitory neurons in the cortical network and decreases the excitatory current in the network as a possible cause of abnormal network oscillations. Hypothesis 2 links unbalanced cortical activity to an unbalanced negative feedback to the membrane potential. Specifically, it might result from dysfunctional K^+ channels in excitatory neurons during AD. Hypothesis 2 suggests that the enhancement of negative feedback in excitatory neurons may be involved in abnormal oscillatory activity. This results in high-threshold spiking dynamics and less spiking activity for excitatory neurons.

To determine the overall rhythmic activity of the model in its control (normal) context, 10 trials of the model were run, each with random input, therefore representing 10 individuals forming the healthy control group. The model parameters were then modified to test each hypothesis. Again the model was run 10 times for each test, with each test representing a group with a certain neuropathology. The FFT was applied on the spiking train produced by each model to calculate the power spectra within the following frequency bands: delta (1–3 Hz), theta (4–7 Hz), alpha (8–12 Hz), beta1 (13–18 Hz), beta2 (19–21 Hz), beta3 (22–30 Hz), gamma (31–50 Hz), and full (1–70 Hz). The categorization of the frequency bands is based on [39.97]. One-way repeated-measures ANOVA was used to analyze the statistical significance of the difference in frequency band power across the different groups outlined previously; p values smaller than 0.05 were considered statistically significant.

Results

The power spectrum averages of the alpha, beta3, gamma, and full frequency bands were significantly decreased when decreasing N_e from 794 to 764 (loss of e-cells increased from 0.75% to 4.5%). It is expected that reducing N_e will increase the inhibition in the network and slow down the spiking activity. From Table 39.2, we can see that the power spectrum is shifted to lower frequencies with a parallel decrease in the coherence of fast rhythms. The statisti-

Table 39.2 Mean power spectra after neuronal/synaptic loss

Frequency band	Min. value	Control value	Decrease
Delta	3.498	3.737	6.4
Theta	2.71	3.104	12.7
Alpha	3.178	4.236	25
Beta3	6.125	7.326	16.4
Gamma	13.15	15.034	12.5
Full	44.25	50.306	12

Table 39.3 Mean power spectra after varying parameter b for e-cells. Parameter b (39.23) for e-cells was varied from 0.1995 to 0.195. Control value is 0.2. Length of each step is 0.0005

Frequency band	Min. value	Control value	Decrease
Delta	3.208	3.737	14.2
Theta	2.544	3.104	18
Alpha	2.937	4.236	30.7
Beta1	2.671	3.579	25.4
Beta2	1.105	1.787	38.7
Beta3	5.359	7.326	26.9
Gamma	11.557	15.034	23.1
Full	39.326	50.306	21.8

cal analyses show that the significant decrease started in the beta3 band power ($N_e = 782$, 2.25% loss of e-cells). The delta, theta, beta1, and beta2 band powers were not significantly decreased.

As presented in Table 39.3, there are greater shifts for the alpha, beta2, beta3, and gamma bands than for slower frequency bands. One can see that the beta2 frequency band is the most affected. Again, decreasing parameter b in (39.23) for e-cells upregulates the negative feedback u to the membrane potential V (39.22), which results in high-threshold spiking dynamics and decreases the spiking activity of e-cells. This accounts for the activation of K^+ ionic currents and inactivation of Na^+ ionic currents [39.23].

39.4.3 Discussion: Cortical Rhythm Changes

The results are discussed in terms of structural and functional relevance in AD:

- *Structural Changes in AD*: It is commonly thought that an increase in theta band activity appears in the early stages of AD with a parallel decrease in beta activity, whereas delta activity increases later during the course of the disease [39.54]. However, the underlying neural causes of abnormal EEG dynamics in AD are still poorly understood. The research presented in this section is concerned with developing a better understanding of the underlying neural correlates of abnormal cortical beta oscillations in AD based on a computational modeling approach. It is emphasized in [39.95] that the decreased activity in alpha, beta, and gamma waves is related to changes in excitatory circuit activity. The computational model-based findings

presented here are in agreement with these experimental findings. Despite the heterogeneity of the network models in our work, we find that the power spectrum of beta rhythm is significantly decreased by excitatory neural and synaptic loss, but this decrease is not observed when the inhibitory circuitry is perturbed. Furthermore, the significant decrease begins in the beta3 band (upper beta) when N_e is 782 (2.25% loss of e-cells). Although the mean power spectrum of the full band is significantly decreased, slower bands (delta and theta) are the least affected by the decrease and most of the signals transitioned from fast bands toward slow bands.

- *Functional Changes in AD*: It was demonstrated that cortical excitatory neuronal death is mediated by the enhancement of outward K^+ current and that such enhancement might specifically contribute to the pathogenesis of Aβ-induced excitatory neuronal death [39.87]. The results in [39.87] showed that Aβ exposure in the delayed rectifier K^+ current I_K, an increase in maximal conductance, and a shift in its activation voltage relationship toward upregulation of this type of K^+ current provide more negative feedback to the membrane potential of e-cells, followed by neuronal death.

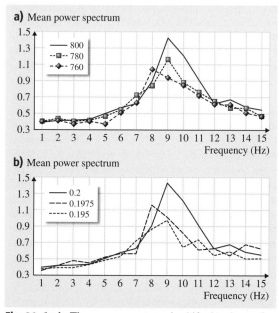

Fig. 39.6a,b The power spectrum is shifted to lower frequencies as a results of (**a**) excitatory neuron/synapse loss and (**b**) when decreasing parameter b for e-cells

We speculate that the neural mechanism that underlies the sequence of abnormal changes in EEG in AD can be described by inspecting the results in Table 39.3, which show a higher shift from upper frequency band powers, in particular, the power spectrum of beta2 and alpha rhythms (Fig. 39.6a,b). The significant decrease of the beta band power appears also (before other rhythms) when decreasing N_e, followed by a breakdown in other frequency bands on further decrement in N_e, simulating an increased rate of synaptic loss and subsequent neuronal death in AD. Further study to investigate compensation mechanisms can be found in [39.98, 99].

39.5 A Neural Mass Model to Study Thalamocortical Alpha Rhythms in AD

For an extended period, the only function of the thalamic structure was thought to be that of relaying sensory information to the cortex. Following seminal work during the 1930s and 1940s (see [39.100] for a review) and subsequent research in this area [39.19, 100–103], it is now known that the feedforward and feedback connections between the thalamic complex and the cortex play a central role in the generation of brain oscillations within the alpha band [39.41, 42, 104–106]. Furthermore, the thalamic reticular nucleus (TRN), a part of the thalamus consisting of a thin sheet of inhibitory cells surrounding the anterior and lateral parts of the thalamus, is thought to be vital for the oscillatory behavior of thalamic cells [39.107–111]. The TRN is controlled by inputs from various structures in the brainstem which are known to be affected in AD [39.57, 112]. It is speculated that the TRN afferents are corrupted in AD [39.113]. More recent research has shown thalamic atrophy in AD, caused by severe loss in synaptic connectivity in the thalamus [39.25, 40].

We have studied alpha rhythm slowing in AD using a classic computational model of the thalamocortical circuitry proposed by *Lopes da Silva* [39.19] to mimic thalamocortical alpha rhythms (alpha rhythm model) and subsequently used in epilepsy research [39.21, 114]. Our goal is to observe how depleted synaptic connectivity affects the model output oscillatory behavior within the alpha band. We speculate that such a direction will lead to better understanding of the abnormal alpha band activity – a hallmark in EEG of AD. Early results from our work with the alpha rhythm model (ARm) indicate significant slowing of alpha rhythm as a result of an increase in inhibitory connectivity [39.115, 116]. To bring the model up to date with recent research [39.117], we modified the structure of a single neuronal population in the ARm [39.118]. This modified ARm differed from the original in two aspects: first, a retinogeniculate connectivity parameter was introduced; second, the intramodel connectivity is now based on the most recent available experimental data [39.119–121]. Again, we see an increase in the inhibitory parameter as a direct correlate of alpha rhythm slowing in the model. Furthermore, the model is highly sensitive to the retinogeniculate connectivity; power within the alpha band is directly proportional to variations in this parameter [39.122] (results are discussed in context in Sect. 39.5.1, *Experimental Methods and Results*). More recently, we proposed a more biologically plausible thalamocorticothalamic circuitry [39.123] based on the physiological connectivity structure as mentioned in [39.124]. In the following sections, we first give a brief overview of the modified version of the alpha rhythm model as used in our work (Sect. 39.5.1), followed by the more recent thalamocorticothalamic model (Sect. 39.5.2) and a discussion of our results and possible implications (Sect. 39.5.3).

39.5.1 A Simple Thalamocortical Model

The concept of neural mass modeling was pioneered by *Freeman* [39.125], being in turn inspired by the pioneering works of *Wilson* and *Cowan* [39.126, 127] in mathematically describing the dynamics between the excitatory and inhibitory cell populations in the thalamus and the cortex. A cell population is considered as a single entity with an input and an output and is assumed to be representative of the combined behavior of around 10^4–10^7 neurons in the thalamic and cortical tissues, so densely packed that they may be considered as a continuum. Such a concept is appropriate in studying EEG dynamics as each EEG electrode records the behavior of a large number of cells at and near the point of recording. In his seminal work [39.19], *Lopes da Silva* tuned the model parameters to generate alpha rhythmic activity and validated the model with the EEG of a dog. Recently [39.118, 122], we modified the model to be consistent with more recent

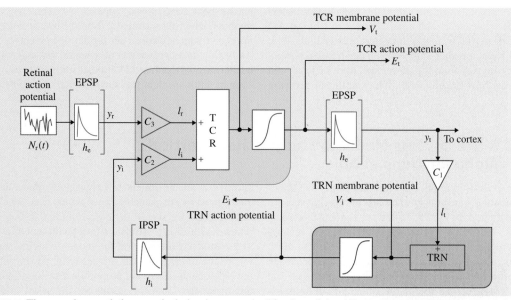

Fig. 39.7 The neural mass thalamocortical circuitry as a modification of the ARm and as simulated in our work with Simulink in MATLAB. The connectivity parameter values in the model are based on most recent experimental data (from the cat lateral geniculate nucleus, LGN) available from the literature [39.120, 121] and represent the proportion of excitatory or inhibitory synapses from respective afferents on a single dendritic terminal of a cell. The block N_r represents retinal input to the model in an eyes-closed resting condition, simulated by a stochastic signal with mean μ and standard deviation φ. All other blocks in the model are defined in (39.25) and (39.26). The parameter values for all blocks in the figure are provided in Table 39.4. IPSP: inhibitory post synaptic potential; EPSP: excitatory post synaptic potential; TCR: thalamocortical relay cells; TRN: thalamic reticular nucleus

Table 39.4 Values of the parameters defined in Eqs. (39.25–39.27) (after [39.118])

μ	φ	A_e	A_i	τ_e	τ_i	γ	v_0	e_0	C_1	C_2	C_3
pps	pps^2	mV	mV	ms	ms	mV^{-1}	mV	s^{-1}	%	%	%
550	120	3.25	22	10	20	0.56	6	25	35	28	10

research as well as to introduce experimentally derived parameter values into the model; the model is shown in Fig. 39.7. It consists of excitatory (forming excitatory synapses on the postsynaptic cell population) and inhibitory (forming inhibitory synapse with the postsynaptic cell population) cell populations simulating the TCRs and the TRN, respectively.

The extrinsic input N_r to the model is simulated by a Gaussian white noise having mean μ and standard deviation φ and represents the background firing rate (the unit of measure is spikes per second, sps) of the retinogeniculate neuronal populations corresponding to an awake relaxed state with eyes closed, the brain state associated with prominent alpha frequency in the EEG over the occipital cortex [39.19,21]. The functions $h_e(t)$ and $h_i(t)$ describe the excitatory and inhibitory synaptic

behavior of the TCR and TRN population, respectively, and are defined in (39.25) [39.128]. A sigmoid function S is used to transform the membrane potential V of a postsynaptic cell population into the firing rate E of the population and is defined in (39.26). The connectivity parameter C is the total number of synaptic contacts made by the presynaptic cell population as a percentage of the total number (including other afferent cell populations) of synaptic contacts made at the postsynaptic cell population dendrites. All parameter values are defined in Table 39.4. The output of the model V_t is the membrane potential of the TCR cell population. The model behavior is defined in (39.7).

$$h_e(t) = Aat\,e^{-at},$$
$$h_i(t) = Bbt\,e^{-bt}, \qquad (39.25)$$

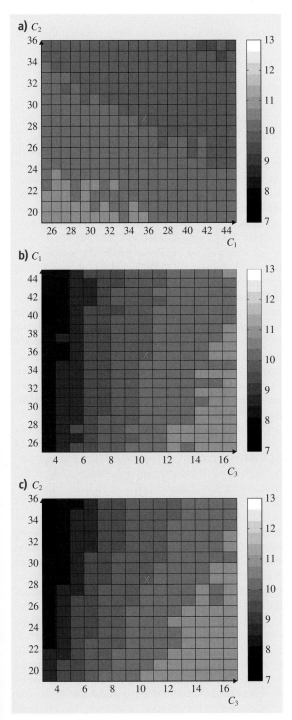

a) C_2

b) C_1

c) C_2

Fig. 39.8a–c Plots showing average dominant frequency behavior of the model output within the alpha (7.5–13 Hz) band. The basal value (a base value used to benchmark the model output oscillatory behavior representing that of a pathologically unaltered brain) of each parameter is as obtained from experimental studies in [39.119–121] and is presented in Table 39.4. The X in each plot identifies the output response when all three parameters are at their basal values. **(a)** Variation of the excitatory C_1 and inhibitory C_2 parameters in the thalamocortical pathway when the excitatory parameter C_3 in the retinal afferent pathway is at its basal value 10. **(b)** Variation of C_3 and C_2 when C_1 is at its basal value 35. **(c)** Variation of C_3 and C_1 when C_2 is at its basal value 28. ◄

where A and B are synaptic strengths and a and b are synaptic time constants.

$$S(v) = \frac{2e_0}{1 + e^{\gamma(v - v_0)}} \,, \tag{39.26}$$

where $2e_0$ is the maximum firing frequency, v_0 is the firing threshold of the population, and γ is the steepness parameter.

$$\ddot{y}_t = aAE_t - 2a\dot{y}_t - 2a^2 y_t \,,$$
$$E_t = S(V_t) \,,$$
$$V_t = I_r - I_i \,,$$
$$I_r = C_3 y_r \,,$$
$$I_i = C_2 y_i \,,$$
$$\ddot{y}_i = bBE_i - 2b\dot{y}_i - 2b^2 y_i \,,$$
$$E_i = S(V_i) \,,$$
$$V_i = I_t \,,$$
$$I_t = C_1 y_t \,,$$
$$\ddot{y}_r = AaC_3 N_r(t) - 2a\dot{y}_r - a^2 y_r \,. \tag{39.27}$$

Experimental Methods and Results

The values of the input mean μ and standard deviation φ were adjusted by trial simulations so as to obtain a dominant output frequency of 10.5–11 Hz within the alpha band while the connectivity parameters were set at their basal values. Model simulation was implemented using the fourth/fifth-order Runge–Kutta ordinary differential equation (ODE) solver within the Simulink environment in MATLAB. The total simulation time was 10 s with a sampling frequency of 250 Hz. The output from each simulation was clipped on the time axis so as to abstract the values from the start of the 4th second to the end of the 8th second as in [39.21, 129]. The output vector thus obtained was bandpass filtered between 6.5–14 Hz (to extract the alpha rhythm components) using a Butterworth filter of order 10. The power spectra of this filtered output was computed using a Welch periodogram with Hamming window of segment length 1/4th the size of the sampling frequency and overlap of 50% [39.115, 116, 130].

Results presented in Fig. 39.8a,b show overall slowing of the mean output frequency within the alpha band with increasing values of both the excitatory (C_1) and inhibitory (C_2) connectivity parameters in the thalamocortical loop, and a decrease of the excitatory parameter (C_3) in the retinogeniculate pathway. Thus, the output behavior is in agreement with various research reports suggesting the important role of (a) the TRN in modulating thalamic oscillatory behavior [39.107, 108, 110] and (b) the visual modality as a biomarker in detection of Alzheimer's disease [39.131]. In the next section we discuss the results corresponding to these parameters in a more biophysically plausible model of the thalamocortical circuitry.

39.5.2 The Thalamocorticothalamic Model

The thalamocorticothalamic model presented here consists of two modules, thalamic (Fig. 39.9a) and cortical (Fig. 39.9b), mutually connected as in [39.124] and as adapted in [39.123]. The basic structure of a single neuronal population in the modules remains the same as the modified ARm presented in Sect. 39.5.1. The cortical module is adapted from [39.117, 132], while the thalamocortical (thalamic TCR to cortical PY) connectivity parameter is as in [39.133]. The connectivity parameter values in the thalamic module are based on experimental data from the lateral geniculate nucleus of the cat [39.119] and rat [39.120, 121] (see [39.123] for a detailed discussion of assigning parameter values based on experimental findings). The values of the connectivity parameters in the thalamic module shown in Fig. 39.9a are presented in Table 39.5; all other parameter values are presented in Table 39.6. The parameters for the cortical module were sourced from [39.132] and are presented in Tables 39.7 and 39.8. In Tables 39.5 and 39.7, rows refer to the afferent cell populations while columns refer to the efferent population or extrinsic input. An "X" indicates that there is no connection between the two cell populations in the corresponding row and column. We study the model behavior corresponding to synaptic connectivity parameter variation in the thalamic module only (variation in the cortical

module is being considered as future work). Experimental methods in simulating the model are the same as mentioned in Sect. 39.5.1. Results are presented in the following section.

39.5.3 Results and Discussion

Figure 39.10 shows the behavior of the thalamic module with the variation of the total number of synapses T to each cell population and when it is disconnected from the cortical module. The basal (initial) value of $T = 100$ is assumed for all cell populations in the thalamus. We see that the power spectrum is fairly flat at the basal value of T, with similar power within the alpha subbands (α_1, 7–9 Hz; α_2, 9–11 Hz; α_3, 11–13 Hz). However, on varying T through ± 50 of its basal value, we see a slowing of the alpha rhythmic frequency with decreasing values of T, indicated by an increase in power within the α_1 and α_2 bands. Increasing the value of T brings about a decrease in the overall power within the alpha band, the relative power being higher within the upper alpha band.

Next, the modules were interconnected and the output of the thalamic module was analyzed by varying the synaptic connectivity parameters, one at a time, about their basal values while all other model parameters were held at their respective basal values as presented in Tables 39.4–39.7. Figure 39.10b,c shows the effects of varying the self-inhibitory loop synaptic connectivity parameters C_{isi} and C_{nsi} in the interneuron (IN) and TRN cell populations, respectively. Alpha rhythm slowing is observed with decreasing values of C_{isi} (IN) as the parameter is varied about $\pm 10\%$ of its basal value. On the other hand, alpha rhythm slowing is observed with increasing values of C_{nsi} (TRN) when the parameter is varied $\pm 10\%$ about its basal value. In Fig. 39.10d, we study the thalamic model behavior with variation of the retinogeniculate synaptic connectivity parameter in the TCR pathway, C_{tre}, about $\pm 1\%$ of its basal value. We observe alpha rhythm slowing with decreasing values of the parameter.

The results show remarkable agreement with experimental observations in AD and are discussed below:

- In Fig. 39.10a, the lowering of the synaptic count in the thalamic module may correspond to neurodegeneration and overall synaptic loss in the AD brain, which brings about a slowing (shift in peak alpha power from α_3 to α_1 and α_2 subbands) of the alpha rhythm activity recorded by EEG. On the other hand, increased synaptic connectivity may be

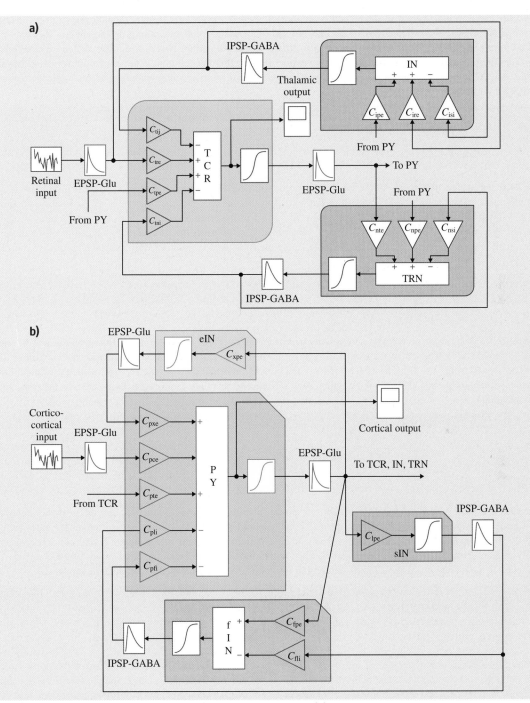

Fig. 39.9a,b Computational model of the (**a**) thalamic module and (**b**) cortical module of the thalamocorticothalamic model. The thalamic module has three cell populations: thalamocortical relay cells (TCR), interneurons (IN), and thalamic reticular nucleus (TRN). The cortical module has four cell populations: pyramidal cells (PY), excitatory interneurons (eIN), slow inhibitory interneurons (sIN), and fast inhibitory interneurons (fIN)

Table 39.5 Basal (initial) values of the synaptic connectivity parameters of the thalamic module shown in Fig. 39.9a. The values are expressed as a percentage of the total number of synapses T convergent on a cell population. The data are sourced from [39.119–121] and are based on in vitro studies of the cat and rat lateral geniculate nucleus. The efferent (*From*) cell population is mentioned in the top row, while the afferent (*To*) cell population is mentioned along the left column. An "X" indicates that there is no connectivity between the respective cell populations. In this work, we assume $T = 100$ for all thalamic cell populations

	Retinal	PY	TCR	TRN	IN
TCR	$C_{tre} = 7.1$	$C_{tpe} = 62$	X	$C_{tni} = 1/2$ of 30.9	$C_{tii} = 1/2$ of 30.9
TRN	X	$C_{npe} = 50$	$C_{nte} = 35$	$C_{nsi} = 15$	X
IN	$C_{ire} = 47.4$	$C_{ipe} = 29$	X	X	$C_{isi} = 23.6$

Table 39.6 Parameters used to model the synaptic response function, the sigmoid function, and the extrinsic noise input from the retina (with mean ξ_M and standard deviation ξ_D) in the thalamic module. The value of ξ_M is obtained from neurophysiological data corresponding to the resting firing rate in the brain. However, this is the case with eyes open, while our research is for the cases with eyes closed. Nevertheless, we consider this as a more biologically realistic option. Values of all other parameters are based on physiologically plausible data as used in [39.133], which in turn are sourced from [39.136]

H_e (mV)	H_i (mV)	τ_e (ms)	τ_i (ms)	ν (mV^{-1})	e_0 (s^1)	s_0 (mV)	ξ_M (pps)	ξ_D^2 (pps^2)
3.25	22	10	25	0.56	2.5	6	11	5

Table 39.7 Basal values of the synaptic connectivity parameters of the cortical module shown in Fig. 39.9b. The data are sourced from [39.132], which in turn are based on [39.136]. The thalamocortical connectivity parameter C_{pte} (from TCR to PY) is sourced from [39.133]. The efferent (*From*) cell population is mentioned in the top row, while the afferent (*To*) cell population is mentioned along the left column. An "X" indicates that there is no connectivity between the respective cell populations

	Cortico-cortical	TCR	PY	Ex-IN	sIn-IN	fIn-IN
PY	$C_{pce} = 1$	$C_{pte} = 80$	X	$C_{pxe} = 108$	$C_{pli} = 33.75$	$C_{pfi} = 108$
Ex-IN	X	X	$C_{xpe} = 135$	X	X	X
sIn-IN	X	X	$C_{lpe} = 33.75$	X	X	X
fIn-IN	X	X	$C_{fpe} = 40.5$	X	$C_{fli} = 13.5$	X

Table 39.8 Parameters used to model the synaptic response function, the sigmoid function, and the extrinsic noise input from neighboring cortical regions (with mean ξ_M and standard deviation ξ_D) in the cortical module. Values are as in [39.132] corresponding to the case for low-frequency oscillations. The mean and standard deviation are parameterized by trial and error so as to make the model oscillate within the alpha band with the basal values of the connectivity parameters as mentioned in Table 39.7 and operating in isolation (i. e., not connected to the thalamic module)

H_e (mV)	H_{is} (mV)	H_{if} (mV)	τ_e (ms)	τ_{is} (ms)	τ_{if} (ms)	ν (mV^{-1})	e_0 (s^1)	s_0 (mV)	ξ_M (pps)	ξ_D^2 (pps^2)
2.7	4.5	39	25	50	3.33	0.56	2.5	6	30	5

thought to be an attentive state of the brain which is believed to bring about a desynchronization (decrease in power) in the alpha band activity [39.42]. Furthermore, we speculate that the results might indicate a "threshold" of synaptic connectivity in the normal adult brain; when the connectivity slips below this threshold brought about by abnormal

neuropathology in AD, the brain slips into a "cognitively deficient" state associated with the disease. Further analysis (see [39.123]) has shown that the inhibitory cell populations, IN and TRN, play a major role in the overall behavior, as seen here; the TCR population does not show a significant effect. These observations conform to experimental stud-

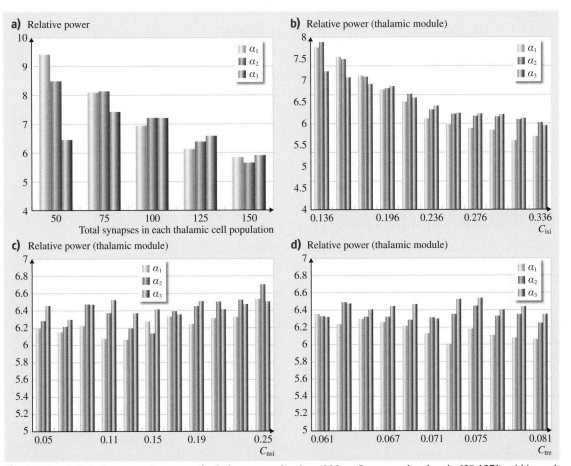

Fig. 39.10a–d Relative power (average of relative power density within a frequency band as in [39.137]) within each alpha subband when the (**a**) total number of synapses T of each cell population is varied from 50 to 150 at intervals of 50; (**b**) the parameter C_{isi} corresponding to the self-inhibitory loop in the IN cell population is varied across $\pm10\%$ of its basal value; (**c**) the parameter C_{nsi} corresponding to the self-inhibitory loop in the TRN cell population is varied across $\pm10\%$ of its basal value; (**d**) The parameter C_{tre} in the synaptic pathway from the retinogeniculate cells to the TCR cell population is varied across $\pm1\%$ of its basal value. All basal values are defined in Table 39.5

ies emphasizing the crucial role of the inhibitory cell population in brain rhythm activities in both functional and dysfunctional brain [39.41, 134, 135].

- Further emphasis of the crucial role of the inhibitory circuitry and pathway is observed in the thalamo-corticothalamic circuitry, as shown in Fig. 39.10b,c. A distinct slowing of alpha rhythm activity is seen with variations of the inhibitory synaptic connectivity. Moreover, the TRN circuitry behavior corresponding to the variation of C_{nsi} shown in Fig. 39.10b is in agreement with results obtained with our simple thalamocortical model presented in Sect. 39.5.1, albeit in a different pathway. Variation

of the connectivity parameter in the efferent pathways from the TCR to the TRN (C_{nti}) or IN (C_{iti}) cell population has little effect on the model oscillatory behavior within the alpha band (not shown here for brevity; refer to [39.123]). This is unlike the slowing of alpha rhythm observed with increase in values of the corresponding connectivity parameter C_2 in the simple thalamocortical model (Fig. 39.7) as shown in Fig. 39.8a. Thus, we may say that an advanced, biophysically plausible computational model such as that in Fig. 39.9 helps in better understanding of the specific neuronal pathways which significantly affect brain rhythm activities in AD.

Part G | 39.5

- The parameter C_{tre} (retinogeniculate connectivity) studied in Fig. 39.10d corresponds to C_3 in the simple thalamocortical model shown in Fig. 39.7. Comparing Figs. 39.10d and 39.8, we find an agreement in the model behavior – alpha rhythm slowing corresponds to a decrease in both C_{tre} and C_3. This result indicates an important role of aberrant synaptic connectivities in the sensory pathway in affecting brain rhythms within the alpha band. Indeed, the thalamus is believed to link perception to action [39.138], while visuomotor deficits have been reported in early-stage AD [39.139]. We speculate that further research in this direction with added biologically realistic features to the model will help to further validate the results with experimental studies. This may, in the near future, provide vital clues towards novel information enabling predictive diagnosis for AD and subsequent drug-based and therapeutic treatment opportunities. Additional studies using the thalamacortical neural mass model and extensions to the model can be found in [39.140–142].

39.6 Discussion and Conclusion

In this chapter, we have endeavored to present computational modeling of neuronal behavior at various levels of hierarchy, and in two vital brain parts, viz. the hippocampus (Sect. 39.3) and the thalamocortical circuitry (Sects. 39.5 and 39.6). In Sect. 39.3, we observed that the A-type potassium (K^+) current in the pyramidal neurons of the hippocampus induces an increase in theta band power by recruiting more pyramidal neurons to fire. Again, in Sect. 39.4, it is demonstrated that cortical excitatory neuronal death is mediated by enhancement of the outward K^+ current. These studies very much conform to experimental observations of cortical atrophy (neuronal death) and increased theta band power in Alzheimer's disease. Furthermore, results in Sect. 39.4 demonstrate that neuronal death and synaptic loss effect a significant decrease of beta and alpha band powers. In Sect. 39.5, we demonstrate a model at a higher level of hierarchy where a population of neurons is treated as a single entity, unlike the single-neuronal-level modeling in Sects. 39.3 and 39.4, thus giving an overview of the population dynamics of the thalamocortical neurons. Again, we observe that simulation of conditions of synaptic deficiency in the model brings about a decrease in the power within the alpha band. Specifically, the inhibitory cell population and related circuitry are demonstrated to play a significant role in slowing of the alpha rhythm, a hallmark of the Alzheimer's disease. Overall, the results presented in independent modeling studies in each of Sects. 39.3–39.5 show similar results and conform to experimental studies. Integrating these models in a combined multiscale modelling approach may help establish direct correlations between amyloid beta, potassium currents, synaptic loss, neuronal death, and brain dynamics. Such models may also improve our current understanding of brain functioning in normal conditions as well as in other neurological disorders or trauma.

39.A Appendix

In our work, all of the ionic currents are modeled by the Hodgkin–Huxley-type formalism, thus the dynamics of a gating variable x satisfies first-order kinetics, $dx/dt = \Phi_x[\alpha_x(1-x) - \beta_x x] = \Phi_x[x_\infty - x]/\tau_x$.

For the pyramidal neurons defined in (39.1) and (39.2): The sodium current

$$I_{Na} = g_{Na}m_\infty^3 h(V - E_{Na}),$$

where the fast activation gating variable is

$$m_\infty = \frac{\alpha_m}{\alpha_m + \beta_m},$$

$$\alpha_m = \frac{-0.1(V+33)}{\exp[-0.1(V+33)-1]},$$

$$\beta_m = 4\exp\frac{-(V+58)}{12},$$

$$\alpha_h = 0.07\exp\frac{-(V+50)}{10},$$

$$\beta_h = \frac{1}{\exp[-0.1(V+20)+1]}.$$

The potassium current

$$I_K = g_K n^4(V - E_K),$$

where

$$\alpha_n = \frac{-0.01(V+34)}{\exp[-0.1(V+34)-1]} \, ,$$

$$\beta_n = 0.125 \exp \frac{-(V+44)}{25} \, .$$

The calcium current

$$I_{\text{Ca}} = g_{\text{Ca}} m_\infty (V - E_{\text{Ca}}) \, ,$$

where

$$m_\infty = \frac{1}{\exp[\frac{-(V+20)}{9}]} \, .$$

The calcium-dependent potassium current

$$I_{\text{AHP}} = \frac{g_{\text{AHP}}[\text{Ca}^{2+}]}{([\text{Ca}^{2+}] + K_{\text{D}})(V - E_{\text{K}})} \, ,$$

where the intracellular calcium concentration $[\text{Ca}^{2+}]$ is governed by

$$\frac{\mathrm{d}[\text{Ca}^{2+}]}{\mathrm{d}t} = \frac{-[\text{Ca}^{2+}]}{\tau_{\text{Ca}}} - BI_{\text{Ca}} \, ,$$

$\tau_{\text{Ca}} = 1000$, $B = 0.002$, and $K_D = 30$.

The hyperpolarization-activated current

$$I_{\text{h}} = g_{\text{h}} H (V - E_{\text{h}}) \, .$$

The values of H_∞ and τ_H corresponding to different V can be found in [39.67].

The A-type transient potassium current

$$I_{\text{A}} = g_{\text{A}} a^3 b (V - E_{\text{K}}) \, ,$$

where

$$\alpha_a = \frac{-0.05(V+20)}{\exp \frac{-(V+20)}{15} - 1} \, ,$$

$$\beta_a = \frac{0.1(V+10)}{\exp \frac{(V+10)}{8} - 1} \, ,$$

$$\alpha_b = \frac{0.00015}{\exp \frac{(V+18)}{15}} \, ,$$

$$\beta_b = \frac{0.06}{\exp \frac{-(V+73)}{12} + 1} \, .$$

The large-conductance calcium-dependent potassium current

$$I_{\text{CT}} = g_{\text{CT}} c^2 d (V - E_{\text{K}}) \, ,$$

where

$$\alpha_c = \frac{-0.0077(V + V_{\text{shift}} + 103)}{\exp \frac{-(V + V_{\text{shift}} + 103)}{12} - 1} \, ,$$

$$\beta_c = 0.91 - \alpha_c \, ,$$

$$\alpha_d = \frac{1}{\exp \frac{(V+79)}{10}} \, ,$$

$$\beta_d = \frac{4}{\exp \frac{-(V-82)}{27} + 1} \, ,$$

$$V_{\text{shift}} = 40 \log \frac{[\text{Ca}^{2+}]}{13.805} \, ,$$

$\tau_{\text{Ca}} = 0.9$, $B = 0.06$. The values of the other parameters are $\Phi = 5$, $g_{\text{L}} = 0.1$, and $g_{\text{Ca}} = 0.5$ for soma and dendrite, $g_{\text{Na}} = 45$, $g_{\text{K}} = 18$, $g_{\text{A}} = 60$, and $g_{\text{CT}} = 140$ for the soma, and $g_{\text{AHP}} = 5$, $g_{\text{A}} = 30$, and $g_{\text{CT}} = 70$ for the dendrite; $E_{\text{L}} = -65$, $E_{\text{Na}} = 55$, $E_{\text{K}} = -80$, and $E_{\text{Ca}} = 120$. To emulate heterogeneity in the brain, the injected DC current I for each neuron is not chosen to be identical. This is achieved by allowing I to follow a Gaussian distribution with mean I_μ and standard deviation I_σ. For the pyramidal neuronal population, $I_\mu = 6$ and $I_\sigma = 0.1$. The units for all τ, E (or V), I, g, α (or β), K, and B are ms, mV, μA/cm^2, mS/cm^2, ms^{-1}, μM, and μMcm^2/(msμA). The rest are dimensionless constants.

The parameters for calculation of the ionic currents in the basket neuron defined in (39.3) are the same as those for the OLM neuron defined in (39.4). The sodium and potassium currents for both of these neuronal populations are defined in the same way as that of the pyramidal neuron with parameters

$$\alpha_m = \frac{-0.1(V+35)}{\exp[-0.1(V+35)-1]} \, ,$$

$$\beta_m = 4 \exp \frac{-(V+60)}{18} \, ,$$

$$\alpha_h = 0.07 \exp \frac{-(V+58)}{20} \, ,$$

$$\beta_h = \frac{1}{\exp[-0.1(V+28)+1]} \, ,$$

$$\alpha_n = \frac{-0.01(V+34)}{\exp[-0.1(V+34)-1]} \, ,$$

$$\beta_n = 0.125 \exp \frac{-(V+44)}{80} \, .$$

The calcium current

$$I_{\text{Ca}} = g_{\text{Ca}} m_\infty^2 (V - E_{\text{Ca}}) \, ,$$

where

$$m_\infty = \frac{1}{\exp \frac{-(V+20)}{9} + 1} \, .$$

The AHP current is defined the same as in pyramidal neurons with $\tau_{Ca} = 80$, $B = 0.002$, and $K_D = 30$. The hyperpolarization-activated current

$$I_h = g_h H(V - E_h),$$

where

$$H_\infty = \frac{1}{\exp\frac{(V+80)}{10} + 1},$$

$\tau_H = 200/\{\exp[(V + 70)/20] + \exp[-(V + 70)/20] + 5\}$. The other parameters are $\Phi = 5$, $g_L = 0.1$, $g_{Na} = 35$, $g_K = 9$, $g_{AHP} = 10$, $g_{Ca} = 1$, $g_h = 0.15$; $E_L = -65$, $E_{Na} = 55$, $E_K = -90$, $E_{Ca} = 120$, $E_h = -40$, $I_\mu = 0$, and $I_\sigma = 0.1$.

In the MSGABA neuronal model defined in (39.5), the sodium and potassium currents are defined in the same way as for the pyramidal neurons with parameters

$$\alpha_m = \frac{-0.1(V+33)}{\exp[-0.1(V+33)-1]},$$

$$\beta_m = 4 \exp\frac{-(V+58)}{18},$$

$$\alpha_h = 0.07 \exp\frac{-(V+51)}{10},$$

$$\beta_h = \frac{1}{\exp[-0.1(V+21)+1]},$$

$$\alpha_n = \frac{-0.01(V+38)}{\exp[-0.1(V+38)-1]},$$

$$\beta_n = 0.125 \exp\frac{-(V+48)}{80}.$$

The slowly inactivating potassium current

$$I_K = g_{KS} pq(V - E_K),$$

where

$$p_\infty = \frac{1}{\exp\frac{-(V+34)}{6.5} + 1},$$

$$q_\infty = \frac{1}{\exp\frac{(V+65)}{6.6} + 1},$$

$$\tau_q = 100 \left(\frac{1+1}{\exp\frac{-(V+50)}{6.8} + 1} \right),$$

and $\tau_p = 6$. The other parameters are $\Phi = 5$, $g_L = 0.1$, $g_{Na} = 50$, $g_K = 8$, $g_{KS} = 12$; $E_L = -50$, $E_{Na} = 55$, $E_K = -85$, $I_\mu = 2.2$, and $I_\sigma = 0.1$.

Network: The synaptic current is normalized by the number of presynaptic neurons. The AMPA and NMDA excitatory postsynaptic currents are described as

$$I_{AMPA} = g_{AMPA} s(V - E_{AMPA})$$

and

$$I_{NMDA} = g_{NMDA} B(V) s(V - E_{NMDA}),$$

respectively. s is updated as

$$ds/dt = \alpha[T](1-s) - \beta s,$$

$$[T] = \frac{T_{max}}{1 + \exp\frac{(-V_{pre}+V_p)}{K_p}},$$

$V_p = 2$, $V_K = 5$. $B(V)$ is calculated as

$$B(V) = \frac{1}{1 + \exp(-0.062V)\frac{[Mg^{2+}]}{3.5}},$$

where $[Mg^{2+}] = 1$ mM by default. The α and β for AMPA and NMDA are $\alpha = 1.1$, $\beta = 0.19$ and $\alpha = 0.072$, $\beta = 0.0066$, respectively. $E_{AMPA} = E_{NMDA} = 0$, $g_{AMPA} = 0.9$, and $g_{NMDA} = 0.45$. The GABA$_A$ inhibitory postsynaptic current is described as

$$I_{GABA_A} = g_{syn} s(V - E_{GABA_A}),$$

where the activation variable s is calculated by

$$ds/dt = \alpha F(V_{pre})(1-s) - \beta s.$$

V_{pre} is the presynaptic neuron membrane potential

$$F(V_{pre}) = \frac{1}{1 + \exp\left(\frac{-V_{pre}}{K}\right)}.$$

Parameters for GABA$_A$ synaptic connections between different neurons are as follows:

- Basket–pyramidal (b–p)
  $\alpha = 10$, $\beta = 0.07$, $K = 2$, $E_{GABA_A} = -80$, $g_{syn} = 1.38$
- OLM–basket (o–b)
  $\alpha = 20$, $\beta = 0.07$, $K = 2$, $E_{GABA_A} = -80$, $g_{syn} = 0.88$
- OLM–pyramidal (o–p)
  $\alpha = 20$, $\beta = 0.05$, $K = 2$, $E_{GABA_A} = -85$, $g_{syn} = 0.88$
- OLM–MSGABA (o–m)
  $\alpha = 20$, $\beta = 0.1$, $K = 0.5$, $E_{GABA_A} = -80$, $g_{syn} = 0.5$
- Basket–basket (b–b)
  $\alpha = 10$, $\beta = 0.1$, $K = 2$, $E_{GABA_A} = -75$, $g_{syn} = 0.125$
- MSGABA–OLM (m–o)
  $\alpha = 10$, $\beta = 0.1$, $K = 2$, $E_{GABA_A} = -75$, $g_{syn} = 0.5$

- MSGABA–MSGABA (m–m)
  $\alpha = 10,\ \beta = 0.1,\ K = 2,\ E_{GABA_A} = -75,\ g_{syn} = 0.25$

- MSGABA–basket (m–b)
  $\alpha = 10,\ \beta = 0.07,\ K = 2,\ E_{GABA_A} = -75,\ g_{syn} = 0.5.$

References

39.1 L.F. Abbott: Theoretical neuroscience rising, Neuron **60**, 489–4958 (2008)

39.2 A.V. Herz, T. Gollisch, C.K. Machens, D. Jaeger: Modeling single-neuron dynamics and computations: A balance of detail and abstraction, Science **314**, 80–85 (2006)

39.3 H. Berger, C. Gray: Über das Elektroenkephalogramm des Menschen, Arch. Psychiat. Nervenkr. **87**, 527–570 (1929)

39.4 G. Buzsáki: *Rhythms of the Brain* (Oxford Univ. Press, New York 2006)

39.5 G. Buzsáki, A. Draguhn: Neuronal oscillations in cortical networks, Science **304**, 1926–1929 (2004)

39.6 X.-J. Wang: Neurophysiological and computational principles of cortical rhythms in cognition, Physiol. Rev. **90**, 1195–1268 (2010)

39.7 M. Lengyel, Z. Huhn, P. Erdi: Computational theories on the function of theta oscillations, Biol. Cybern. **92**, 393–408 (2005)

39.8 R. Morris, G. Hitch, K. Graham, T. Bussey: Learning and Memory,. In: *Cognitive Systems. Information Processing Meets Brain Science*, ed. by R. Morris, L. Tarassenko, M. Kenward (Elsevier, London 2005)

39.9 A.J. Becker, T. Hedden, J. Carmasin, J. Maye, D.M. Rentz, D. Putcha, B. Fischl, D.N. Greve, G.A. Marshall, S. Salloway, D. Marks, R.L. Buckner, R.A. Sperling: Amyloid-β associated cortical thinning in clinically normal elderly, Ann. Neurol. **69**(6), 1032–1042 (2011)

39.10 C.R. Jack Jr., M.S. Albert, D.S. Knopman, G.M. McKhann, R.A. Sperling, M.C. Carrillo, B. Thies, C.H. Phelps: Introduction to the recommendations from the National Institute on Ageing and the Alzheimer's Association Workgroup on diagnostic guidelines for Alzheimer's disease, Alzheimers dement. **7**(3), 257–262 (2011)

39.11 E.S. Vizi: Role of high-affinity receptors and membrane transporters in nonsynaptic communication and drug action in the central nervous system, Pharmacol. Rev. **52**(1), 63–90 (2000)

39.12 A. Bacci, C. Verderio, E. Pravettoni, M. Matteoli: The role of glial cells in synaptic function, Philos. Trans. R. Soc. B **354**(1381), 403–409 (1999)

39.13 E.E. Benarroch: Neuron-astrocyte interactions: Partnership for normal function and disease in the central nervous system, Mayo Clin. Proc. **80**(10), 1326–1338 (2005)

39.14 A.J. Vincent, R. Gasperini, L. Foa, D.H. Small: Astrocytes in Alzheimer's disease: Emerging roles in calcium dysregulation and synaptic plasticity, J. Alzheimer's Dis. **22**(3), 699–714 (2010)

39.15 W.J. Freeman: Models of the dynamics of neural populations, Electroencephalogr. Clin. Neurophysiol. **34**, 9–18 (1978)

39.16 F.H. Loper da Silva, A. Hoeks, H. Smits, L.H. Zetterberg: Model of brain rhythmic activity, Kybernetic **15**, 27–37 (1974)

39.17 P. Robinson, C. Rennie, J. Wright, H. Bahramali, E. Gordon, D. Rowe: Prediction of electroencephalographic spectra from neurophysiology, Phys. Rev. E **63**, 021903 (2001)

39.18 F.H. Loper da Silva: Neural mechanisms underlying brain waves: From neural membranes to networks, Electroencephalogr. clin. Neurophysiol. **79**, 81–93 (1991)

39.19 F.H. Loper da Silva, T.H.M.T. van Lierop, C.F. Schrijer, W.S. van Leeuwen: Essential differences between alpha rhythms and barbiturate spindles: Spectra and thalamo-cortical coherences, Electroencephalogr. Clin. Neurophysiol. **35**, 641–645 (1973)

39.20 F.H. Loper da Silva, A.V. Rotterdam, P. Barts, E.V. Heusden, W. Burr: Models of neuronal populations: The basic mechanisms of rhythmicity. In: *Progress in Brain Research*, Vol. 45, ed. by M.A. Corner, D.F. Swaab (Elsevier, Amsterdam 1976) pp. 281–308

39.21 P. Suffczyński: Neural dynamics underlying brain thalamic oscillations investigated with computational models. Ph.D. Thesis (University of Warsaw, Warsaw 2000)

39.22 E.M. Izhikevich, G.M. Edelman: Large-scale model of mammalian thalamocortical systems, Proc. Natl. Acad. Sci. USA **105**(9), 3593–3598 (2008)

39.23 E.M. Izhikevich: Simple model of spiking neurons, IEEE Trans. Neural Netw. **14**, 1569–1572 (2003)

39.24 E.M. Izhikevich: *Dynamical Systems in Neuroscience: The Geometry of Excitability and Bursting*, 1st edn. (MIT Press, Cambridge 2007)

39.25 L.W. de Jong, K. van der Hiele, I.M. Veer, J.J. Houwing, R.G.J. Westendorp, E.L.E.M. Bollen, P.W. de Bruin, H.A.M. Middelkoop, M.A. van Buchem, J. van der Grond: Strongly reduced volumes of putamen and thalamus in Alzheimer's disease: An MRI study, Brain **131**, 3277–3285 (2008)

39.26 R. Kazman: Current research on Alzheimer's disease in a historical perspective. In: *Alzheimer's Disease: Causes(s), Diagnosis, Treatment and Care*, ed. by Z.S. Khachaturian, T.S. Radebaugh (CRC, New York 1996) pp. 15–29

39.27 S.S. Mirra, W.R. Markesbery: The neuropathology of Alzheimer's disease: Diagnostic features and standardization. In: *Alzheimer's Disease: Causes(s), Diagnosis, Treatment and Care*, ed. by Z.S. Khachaturian, T.S. Radebaugh (CRC, New York 1996) pp. 111–123

39.28   J.C. Troncoso, B.J. Crain, S.S. Sisodia, D.L. Price: Pathology, neurobiology, and animal models of Alzheimer's disease. In: *Alzheimer's Disease: Causes(s), Diagnosis, Treatment and Care*, ed. by Z.S. Khachaturian, T.S. Radebaugh (CRC, New York 1996) pp. 125–144

39.29   H. Adeli, S. Ghosh-Dastidar, N. Dadmehr: Alzheimer's Disease: Models of computation and analysis of EEGs, Clin. EEG Neurosci. **36**(3), 131–136 (2005)

39.30   R. Luengo-Fernandez, J. Leal, A. Gray: Dementia 2010, Tech. rep., Health Economics Research Centre, University of Oxford for the Alzheimer's Research Trust (2010), available online at http://www.alzheimers-research.org.uk

39.31   Alzheimer's Disease facts and figures. Tech. rep., Alzheimer's Association, http://www.alz.org (2010). Alzheimer's and Dementia, vol. 6

39.32   P.S. Aisen: Alzheimer's disease therapeutic research: The path forward, Biomed. Cent. **1**, 1–6 (2009)

39.33   P.S. Aisen: 100 years of Alzheimer research, 1st Int. Congr. Alzheimer's Dis. Adv. Neurotechnol. (2010), available online at http://meetingoftheminds2010.com/abstracts/

39.34   S. Andrieu: Primary prevention of dementia-review of current trials, methodological issues and perspectives, 1st Int. Congr. Alzheimer's Dis. Adv. Neurotechnol. (2010), available online at http://meetingoftheminds2010.com/abstracts

39.35   D. Schubert: Drugs for Alzheimer's Disease: Is the single-target approach going to work?, 1st Int. Congr. Alzheimer's Dis. Adv. Neurotechnol. (2010), available online at http://meetingoftheminds2010.com/abstracts/

39.36   B. Vellas, N. Coley, S. Andrieu: Disease modifying trials in Alzheimer's disease: Perspectives for the future, J. Alzheimer's Dis. **15**(2), 289–301 (2008)

39.37   W.H. Waugh: A call to reduce the incidence of Alzheimer's disease, J. Appl. Res. **10**(2), 53–57 (2010)

39.38   K. Abuhassan, D. Coyle, L. Maguire: Employing neuronal networks to investigate the pathophysiological basis of abnormal cortical oscillations in Alzheimer's disease, Proc. 33rd Annu. Conf. IEEE Eng. Med. Biol. Soc. (EMBC 2011) (2011) pp. 2065–2068

39.39   C.A. Raji, O.L. Lopez, L.H. Kuller, O.T. Carmichael, J.T. Becker: Age, Alzheimer disease, and brain structure, Neurology **73**, 1899–1905 (2009)

39.40   J.H. Xuereb, R.H. Perry, J.M. Candy, E.K. Perry, E. Marshall, J.R. Bonham: Nerve cell loss in the thalamus in Alzheimer's disease and Parkinson's disease, Brain **114**, 1363–1379 (1991)

39.41   M. Steriade, M. Deschenes: The thalamus as a neuronal oscillator, Brain Res. Rev. **8**, 1–63 (1984)

39.42   M. Steriade, P. Gloor, R.R. Llinas, F.H. Lopes da Silva, M.M. Mesulam: Basic mechanisms of cerebral rhythmic activities, Electroencephalogr. Clin. Neurophysiol. **76**, 481–508 (1990)

39.43   J. Dauwels, F. Vialatte, A. Cichocki: Diagnosis of Alzheimer's disease from EEG signals: Where are we standing?, NeuroImage **49**, 668–693 (2010)

39.44   A. Stoller: Slowing of the alpha-rhythm of the electroencephalogram and its association with mental deterioration and epilepsy, J. Ment. Sci. **95**, 972–984 (1949)

39.45   H. Soininen, K. Reinikainen, J. Partanen, E.L. Helkala, L. Paljarvi, P. Riekkinen: Slowing of elctroencephalogram and choline acetyltransferase activity in post mortem frontal cortex in definite Alzheimer's disease, Neuroscience **49**(3), 529–535 (1992)

39.46   P.M. Rossini, C. Percio, P. Pasqualetti, E. Cassetta, G. Binetti, G.D. Forno, F. Ferreri, G. Frisoni, P. Chiovenda, C. Miniussi, L. Parisi, M. Tombini, F. Vecchio, C. Babiloni: Conversion from mild cognitive impairment to Alzheimer's disease is predicted by sources and coherence of brain electroencephalography rhythms, Neuroscience **143**, 793–803 (2006)

39.47   C. Babiloni, F. Babiloni, F. Carducci, F. Cincotti, C.D. Percio, G.D. Pino, S. Maestrini, A. Priori, P. Tisei, O. Zanetti, P.M. Rossini: Movement-related electroencephalographic reactivity in Alzheimer disease, NeuroImage **12**, 139–146 (2000)

39.48   M. Karrasch, M. Laine, J.O. Rinne, P. Rapinoja, E. Sinerva, C.M. Krause: Brain oscillatory responses to an auditory-verbal working memory task in mild cognitive impairment and Alzheimer's disease, Int. J. Psychophysiol. **59**, 168–178 (2006)

39.49   B. Jelles, P. Scheltens, W.M. van der Flier, E.J. Jonkman, F.H. Lopes da Silva, C.J. Stam: Global dynamical analysis of the EEG in Alzheimer's disease: Frequency-specific changes of functional interactions, Clin. Neurophysiol. **119**, 837–841 (2008)

39.50   Y. Wada, Y. Nanbu, Z.Y. Jiang, Y. Koshino, T. Yamaguchi, T. Hashimoto: Electroencephalographic abnormalities in patients with presenile dementia of the Alzheimer type: Quantitative analysis at rest and during photic stimulation, Biol. Psychiatr. **41**, 217–225 (1997)

39.51   H. Soininen, J. Partanen, V. Jousmaki, E.L. Helkala, M. Vanhanen, S. Majuri, M. Kaski, P. Hartikainen, P. Riekkinen: Age-related cognitive decline and electroencephalogram slowing in down's syndrome as a model of Alzheimer's disease, Neuroscience **53**(1), 57–63 (1993)

39.52   P.N. Prinz, M.V. Vitiello: Dominant occipital (alpha) rhythm frequency in early stage Alzheimer's disease and depression, Electroencephalogr. Clin. Neurophysiol. **73**, 427–432 (1989)

39.53 C.M. Krause, L. Sillanmaki, M. Koivisto, C. Saarela, A. Haggqvist, M. Laine, H. Hamalainen: The effects of memory load on event-related EEG desynchronisation and synchronisation, Clin. Neurophysiol. **111**, 2071–2078 (2000)

39.54 J. Jeong: EEG dynamics in patients with Alzheimer's disease, Clin. Neurophysiol. **115**, 1490–1505 (2004)

39.55 W. de Haan, C.J. Stam, B.F. Jones, I.M. Zuiderwijk, B.W. van Dijk, P. Scheltens: Resting-state oscillatory brain dynamics in Alzheimer disease, J. Clin. Neurophysiol. **25**(4), 187–193 (2008)

39.56 G. Buzsáki: Hippocampus, Scholarpedia **6**(1), 1468 (2011)

39.57 H. Braak, E. Braak: Neuropathological stageing of Alzheimer-related changes, Acta Neuropathol. **82**, 239–259 (1991)

39.58 M.E. Hasselmo, J.L. McClelland: Neural models of memory, Curr. Opin. Neurobiol. **9**, 184–188 (1999)

39.59 H. Adeli, S. Ghosh-Dastidar, N. Dadmehr: Alzheimer's disease and models of computation: Imaging, classification and neural models, J. Alzheimer's Dis. **7**(3), 187–199 (2005)

39.60 G. Buzsaki: Theta oscillations in the hippocampus, Neuron **33**(3), 325–340 (2002)

39.61 D. Robbe, G. Buzsáki: Alteration of theta timescale dynamics of hippocampal place cells by a cannabinoid is associated with memory impairment, J. Neurosc. **29**, 12597–12605 (2009)

39.62 Y. Yamaguchi, Y. Aota, N. Sato, H. Wagatsuma, Z. Wu: Synchronization of neural oscillations as a possible mechanism underlying episodic memory: A study of theta rhythm in the hippocampus, J. Integr. Neurosci. **3**(2), 143–157 (2004)

39.63 M.E. Hasselmo: What is the function of hippocampal theta rhythm?–Linking behavioral data to phasic properties of field potential and unit recording data, Hippocampus **15**(7), 936–949 (2005)

39.64 N.V. Ponomareva, G.I. Korovaitseva, E.I. Rogaev: EEG alterations in non-demented individuals related to apolipoprotein E genotype and to risk of Alzheimer disease, Neurobiol. Ageing **29**, 819–827 (2008)

39.65 R. Ihl, T. Dierks, E.M. Martin, L. Frölich, K. Maurer: Topography of the maximum of the amplitude of EEG frequency bands in dementia of the Alzheimer type, Biol. Psychiatry **39**, 319–325 (1996)

39.66 R. Chiaramonti, G.C. Muscas, M. Paganini, T.J. Müller, A.J. Fallgatter, A. Versari, W.K. Strik: Correlations of topographical EEG features with clinical severity in mild and moderate dementia of Alzheimer type, Neuropsychobiology **36**, 153–158 (1997)

39.67 M. Hajos, W.E. Hoffmann, G. Orban, T. Kiss, P. Erdi: Modulation of septo-hippocampal θ activity by GABA$_A$ receptors: An experimental and computational approach, Neuroscience **126**(3), 599–610 (2004), http://geza.kzoo.edu/theta/supplmat.pdf

39.68 X. Wang: Calcium coding and adaptive temporal computation in cortical pyramidal neurons, J. Neurophysiol. **79**, 1549–1566 (1998)

39.69 E.N. Warman, D.M. Durand, G.L. Yuen: Reconstruction of hippocampal CA1 pyramidal cell electrophysiology by computer simulation, J. Neurophysiol. **71**, 2033–2045 (1994)

39.70 A. Mazzoni, K. Whittingstall, N. Brunel, N.K. Logothetis, S. Panzeri: Understanding the relationships between spike rate and delta/gamma frequency bands of LFPs and EEGs using a local cortical network model, NeuroImage **52**(3), 956–972 (2010)

39.71 X. Wang: Pacemaker neurons for the theta rhythm and their synchronization in the septohippocampal reciprocal loop, Neurophysiology **87**, 889–900 (2002)

39.72 N.J. Webster, M. Ramsden, J.P. Boyle, H.A. Pearson, C. Peers: Amyloid peptides mediate hypoxic increase of l-type Ca^{2+} channels in central neurones, Neurobiol. Ageing **27**, 439–445 (2006)

39.73 T.A. Good, D.O. Smith, R.M. Murphy: Beta-amyloid peptide blocks the fast-inactivating K^+ current in rat hippocampal neurons, J. Biophys. **70**, 296–304 (1996)

39.74 J.S. Qi, L. Ye, J.T. Qiao: Amyloid β-protein fragment 31–35 suppresses delayed rectifying potassium channels in membrane patches excised from hippocampal neurons in rats, Synapse **51**, 165–172 (2004)

39.75 D.A. Hoffman, J.C. Magee, C.M. Colbert, D. Johnston: K^+ channel regulation of signal propagation in dendrites of hippocampal pyramidal neurons, Nature **387**, 869–875 (1997)

39.76 C. Chen: Beta-amyloid increases dendritic Ca^{2+} influx by inhibiting the A-type K^+ current in hippocampal CA1 pyramidal neurons, Biochem. Biophys. Res. Commun. **338**, 1913–1919 (2005)

39.77 T.M. Morse, N.T. Carnevale, P.G. Mutalik, M. Migliore, G.M. Shepherd: Abnormal excitability of oblique dendrites implicated in early Alzheimer's: A computational study, Front. Neural Circuits **4**(16) (2010)

39.78 L.R. Shao, R. Halvorsrud, L. Borg-Graham, J.F. Storm: The role of BK-type Ca^{2+}-dependent K^+ channels in spike broadening during repetitive firing in rat hippocampal pyramidal cells, J. Phys. **521**(1), 135–146 (1999)

39.79 B. Lancaster, R.A. Nicoll: Properties of two calcium-activated hyperpolarizations in rat hippocampal neurones, J. Phys. **389**, 187–203 (1987)

39.80 H. Ye, S. Jalini, S. Mylvaganam, P. Carlen: Activation of large-conductance $Ca^{(2+)}$-activated $K^{(+)}$ channels depresses basal synaptic transmission in the hippocampal CA1 area in APP (swe/ind) TgCRND8 mice, Neurobiol. Ageing **31**, 591–604 (2010)

Part G | 39

39.81   J.S. Qi, J.T. Qiao: Suppression of large conductance Ca^{2+}-activated K^+ channels by amyloid beta-protein fragment 31–35 in membrane patches excised from hippocampal neurons, Sheng Li Xue Bao **53**, 198–204 (2001)

39.82   J.H. Jhamandas, C. Cho, B. Jassar, K. Harris, K.D. MacTavish, J. Easaw: Cellular mechanisms for amyloid β-protein activation of rat cholinergic basal forebrain neurons, J. Physiol. **86**, 1312–1320 (2001)

39.83   S. Chi, Z. Qi: Regulatory effect of sulphatides on BKCa channels, Brain J. Pharmacol. **149**, 1031–1038 (2006)

39.84   G.L. Gerstein, N.Y. Kiang: An approach to the quantitative analysis of electrophysiological data from single neurons, J. Biophys. **1**, 15–28 (1960)

39.85   X. Zou, D. Coyle, K. Wong-Lin, L. Maguire: Computational study of hippocampal-septal theta rhythm changes due to β-amyloid-altered ionic channels, PLoS One **6**(6), e21579 (2011)

39.86   X. Zou, D. Coyle, K. Wong-Lin, L. Maguire: Beta-amyloid induced changes in A-type K^+ current can alter hippocampo-septal network dynamics, J. Comput. Neurosci. **32**(3), 465–477 (2012)

39.87   S.P. Yu, Z.S. Farhangrazi, H.S. Ying, C.H. Yeh, D.W. Choi: Enhancement of outward potassium current may participate in β-amyloid peptide-induced cortical neuronal death, Neurobiol. Dis. **5**, 81–88 (1998)

39.88   S. Birnbaum, A. Varga, L.L. Yuan, A. Anderson, J. Sweatt, L. Schrader: Structure and function of Kv4-family transient potassium channels, Physiol. Rev. **84**, 803–833 (2004)

39.89   A. Pannaccione, F. Boscia, A. Scorzielloand, A. Adornetto, P. Castaldo, R. Sirabella, M. Taglialatela, G.D. Renzo, L. Annunziato: Up-regulation and increased activity of Kv3.4 channels and their accessory subunit mink-related peptide 2 induced by amyloid peptide are involved in apoptotic neuronal death, Mol. Pharmacol. **72**, 665–673 (2007)

39.90   Y. Pan, X. Xu, X. Tong, X. Wang: Messenger RNA and protein expression analysis of voltage-gated potassium channels in the brain of $A\beta25$-35-treated rats, J. Neurosci. Res. **77**, 94–99 (2004)

39.91   L. Plant, N. Webster, J. Boyle, M. Ramsden, D. Freir, C. Peers, H. Pearson: Amyloid β peptide as a physiological modulator of neuronal 'A'-type K^+ current, Neurobiol. Ageing **27**, 1673–1683 (2006)

39.92   K. Abuhassan, D. Coyle, L. Maguire: Simple spiking networks to investigate pathophysiological basis of abnormal cortical oscillations in Alzheimer's disease, Irish J. Med. Sci. **180**(Suppl. 2), S62 (2011)

39.93   S. Jones, D. Pinto, T. Kaper, N. Kopell: Alpha-frequency rhythms desynchronize over long cortical distances: A modeling study, J. Comput. Neurosci. **9**, 271–291 (2000)

39.94   O. Jensen, P. Goel, N. Kopell, M. Pohja, R. Hari, B. Ermentrout: On the human sensorimotor-cortex beta rhythm: Sources and modeling, NeuroImage **26**, 347–355 (2005)

39.95   T. Koenig, L. Prichep, T. Dierks, D. Hubl, L.O. Wahlund, E.R. Johan, V. Jelic: Decreased EEG synchronisation in Alzheimer's disease and mild cognitive impairment, Neurobiol. Ageing **26**, 165–171 (2005)

39.96   X. Li, D. Coyle, L. Maguire, D. Watson, T. McGinnity: Grey matter concentration and effective connectivity changes in Alzheimer's disease: A longitudinal structural mri study, Neuroradiology **53**(10), 733–748 (2011)

39.97   Y.M. Park, H.J. Che, C.H. Im, H.T. Jung, S.M. Bae, S.H. Lee: Decreased EEG synchronization and its correlation with symptom severity in Alzheimer's disease, Neurosci. Res. **62**, 112–117 (2008)

39.98   K. Abuhassan, D. Coyle, L.P. Maguire: Investigating the neural correlates of pathological cortical networks in alzheimer's disease using heterogeneous neuronal models, IEEE Trans. Biomed. Eng. **59**(3), 890–896 (2012)

39.99   K. Abuhassan, D. Coyle, L. P. Maguire: Compensating for synaptic loss in Alzheimer's disease: A computational study, submitted (2013)

39.100 P. Andersen, S.A. Andersson, T. Lomo: Some factors involved in the thalamic control of spontaneous barbiturate spindles, J. Phys. **192**, 257–281 (1967)

39.101 P. Andersen, S.A. Andersson, T. Lomo: Nature of thalamocortical relations during spontaneous barbiturate spindle activity, J. Phys. **192**, 257–281 (1967)

39.102 H. Jahnsen, R. Llinas: Electrophysiological properties of guinea-pig thalamic neurones: An in vitro study, J. Phys. **349**, 205–226 (1984)

39.103 H. Jahnsen, R. Llinas: Ionic basis for the electroresponsiveness and oscillatory properties of guinea-pig thalamic neurones in vitro, J. Phys. **349**, 227–247 (1984)

39.104 D.A. McCormick, T. Bal: Sleep and arousal: Thalamocortical mechanisms, Annu. Rev. Neurosci. **20**, 185–215 (1997)

39.105 M. Steriade, R.R. Llinás: The functional states of the thalamus and the associated neuronal interplay, Physiol. Rev. **68**, 649–745 (1988)

39.106 S.M. Sherman, R.W. Guillery: Thalamic relay function and their role in cortico-cortical communication: Generalizations from the visual system, Neuron **33**, 163–175 (2002)

39.107 F. Crick: Function of the thalamic reticular complex: The searchlight hypothesis, Proc. Natl. Acad. Sci. USA **81**, 4586–4590 (1984)

39.108 B. Amsallem, B. Pollin: Possible role of the nucleus reticularis thalami (nRT) in the control of specific, non-specific thalamic nuclei and cortex activity, Pain **18**, S-283 (1984)

39.109 M.E. Scheibel, A.B. Scheibel: The organisation of the nucleus reticularis thalami: A golgi study, Brain Res. **1**, 43–62 (1966)

39.110 U. Kim, M.V. Sanchez-Vives, D.A. McCormick: Functional dynamics of GABAergic inhibition in the thalamus, Science **278**, 130–134 (1997)

39.111 S.M. Sherman: A wake-up call from the thalamus, Nat. Neurosci. **4**(4), 344–346 (2001)

39.112 A.B. Scheibel: Thalamus. In: *Encyclopedia of the Neurological Sciences* (Elsevier, Amsterdam 2003) pp. 501–508

39.113 W.G. Tourtellotte, G.W.V. Hoesen, B.T. Hyman, R.K. Tikoo, A.R. Damasio: Afferents of the thalamic reticular nucleus are pathologically altered in Alzheimer's disease, J. Neuropathol. Exp. Neurol. **48**(3), 336 (1989)

39.114 F. Marten, S. Rodrigues, P. Suffczynski, M.P. Richardson, J.R. Terry: Derivation and analysis of an ordinary differential equation mean-field model for studying clinically recorded epilepsy dynamics, Phys. Rev. E **79**, 021911 (2009)

39.115 B.S. Bhattacharya, D. Coyle, L.P. Maguire: A computational modelling approach to investigate alpha rhythm slowing associated with Alzheimer's Disease, Proc. Brain Inspired Cogn. Sys. (BICS) (Madrid, Spain 2010) pp. 382–392

39.116 B.S. Bhattacharya, D. Coyle, L.P. Maguire: Thalamocortical circuitry and alpha rhythm slowing: An empirical study based on a classic computational model, Proc. Int. J. Neural Netw. (IJCNN) (Barcelona, Spain 2010) pp. 3912–3918

39.117 M. Ursino, F. Cona, M. Zavaglia: The generation of rhythms within a cortical region: Analysis of a neural mass model, NeuroImage **52**(3), 1080–1094 (2010)

39.118 B.S. Bhattacharya, D. Coyle, L.P. Maguire: Alpha and theta rhythm abnormality in Alzheimer's Disease: A study using a computational model. In: *From Brains to Systems: Brain-Inspired Cognitive Systems 2010*, Advances Experimental Medicine and Biology, Vol. 718, ed. by C. Hernandez, J. Gomez, R. Sanz, I. Alexander, L. Smith, A. Hussain, A. Chella (Springer, New York 2011) pp. 57–73

39.119 S.C.V. Horn, A. Erisir, S.M. Sherman: Relative distribution of synapses in the A-laminae of the lateral geniculate nucleus of the cat, J. Comp. Neurol. **416**, 509–520 (2000)

39.120 S.M. Sherman, R.W. Guillery: *Exploring the thalamus and its role in cortical functioning*, 2nd edn. (Academic, New York 2006)

39.121 E.G. Jones: *The Thalamus*, Vol. I and II, 1st edn. (Cambridge Univ. Press, Cambridge, UK 2007)

39.122 B.S. Bhattacharya, D. Coyle, L.P. Maguire: Assessing retino-geniculo-cortical connectivities in alzheimer's disease with a neural mass model, Proc. IEEE Symp. Ser. Comput. Intell. (SSCI) (Paris, France 2011) pp. 159–163

39.123 B.S. Bhattacharya, D. Coyle, L.P. Maguire: A thalamo-cortico-thalamic neural mass model to study alpha rhythms in Alzheimer's disease, Neural Netw. **24**(6), 631–645 (2011)

39.124 S.M. Sherman: Thalamus, Scholarpedia **1**(9), 1583 (2006)

39.125 W.J. Freeman: *Mass Action in the Nervous System* (Academic, New York 1975)

39.126 H.R. Wilson, J.D. Cowan: Excitatory and inhibitory interaction in localized populations of model neurons, J. Biophys. **12**, 1–23 (1972)

39.127 H.R. Wilson, J.D. Cowan: A mathematical theory of the functional dynamics of cortical and thalamic nervous tissue, Kybernetik **13**, 55–80 (1973)

39.128 B.H. Jansen, G. Zouridakis, M.E. Brandt: A neurophysiologically-based mathematical model of flash visual evoked potentials, Biol. Cybern. **68**, 275–283 (1993)

39.129 C.J. Stam, J. Pijn, P. Suffczyński, F.H. Lopes da Silva: Dynamics of the human alpha rhythm: Evidence for non-linearity?, Clin. Neurophysiol. **110**, 1801–1813 (1999)

39.130 J.L. Cantero, M. Atienza, A. Cruz-Vadell, A. Suarez-Gonzalez, E. Gil-Neciga: Increased synchronization and decreased neural complexity underlie thalamocortical oscillatory dynamics in mild cognitive impairment, NeuroImage **46**, 938–948 (2009)

39.131 P.J. Uhlhaas, J. Pantel, H. Lanfermann, D. Prvulovic, C. Haenschel, K. Maurer, D.E.J. Linden: Visual perceptual organization deficits in Alzheimer's dementia, Dement. Geriatr. Cogn. Disord. **25**(5), 465–475 (2008)

39.132 M. Zavaglia, L. Astolfi, F. Babiloni, M. Ursino: A neural mass model for the simulation of cortical activity estimated from high resolution EEG during cognitive or motor tasks, J. Neurosci. Methods **157**, 317–329 (2006)

39.133 R.C. Sotero, T.-N.J. Barreto, I.-Y. Medina: Realistically coupled neural mass models can generate EEG rhythms, Neural Comput. **19**, 479–512 (2007)

39.134 E. Başar: *Brain-Body-Mind in the Nebulous Cartesian System: A Holistic Approach by Oscillations*, 1st edn. (Springer, New York 2011)

39.135 E. Başar, M. Schurmann, C. Basar-Eroglu, S. Karakas: Alpha oscillations in brain functioning: An integrative theory, Int. J. Psychophys. **26**, 5–29 (1997)

39.136 B.H. Jansen, V.G. Rit: Electroencephalogram and visual evoked potential generation in a mathematical model of coupled cortical columns, Biol. Cybern. **73**, 357–366 (1995)

39.137 D.V. Moretti, C. Babiloni, G. Binetti, E. Cassetta, G.D. Forno, F. Ferreric, R. Ferri, B. Lanuzza, C. Miniussi, F. Nobili, G. Rodriguez, S. Salinari, P.M. Rossini: Individual analysis of EEG frequency and band power in mild Alzheimer's disease, Clin. Neurophysiol. **115**, 299–308 (2004)

39.138 R.W. Guillery: Branching thalamic afferents link action and perception, J. Neurophysiol. **90**(2), 539–548 (2003)

Part G | 39

39.139 W.J. Tippett, L.E. Sergio: Visuomotor integration is impaired in early stage Alzheimer's disease, Brain Res. **1102**, 92–102 (2006)

39.140 B.S. Bhattacharya, D. Coyle, L.P. Maguire, J. Stewart: Kinetic modelling of synaptic functions in the alpha rhythm neural mass model, Int. Conf. Artif. Neural Netw. (2012), 645–652

39.141 B.S. Bhattacharya, D. Coyle, L.P. Maguire: Assessing alpha band event-related synchronisation/ desynchronisation using a bio-inspired computational model, J. Univers. Comput. Sci. **18**(13), 1888–1904 (2012)

39.142 B. Sen Bhattacharya, Y. Cakir, N. Serap-Sengor, L. Maguire, D. Coyle: Model-based bifurcation and power spectral analyses of thalamocortical alpha rhythm slowing in Alzheimera's disease, Neurocomputing (2012) in press, available online

40. Understanding the Brain via fMRI Classification

Lavneet Singh, Girija Chetty

In this chapter, we present investigations on magnetic resonance imaging (MRI) of various states of brain by extracting the most significant features in order to classify brain images into normal and abnormal. We describe a novel method based on the wavelet transform to initially decompose the images, followed by the use of various feature selection algorithms to extract the most significant brain features from the MRI images. This chapter demonstrates the use of different classifiers to detect abnormal brain images from a publicly available neuroimaging dataset. A wavelet-based feature extraction followed by selection of the most significant features using principal component analysis (PCA)/quadratic discriminant analysis (QDA) with classification using learning-based classifiers results in a significant improvement in accuracy as compared with previously reported studies and to better understanding of brain abnormalities.

Part G | 40

40.1 Magnetic Resonance Imaging (MRI)

Magnetic resonance imaging (MRI) is an advanced technique used for medical imaging and clinical medicine, being an effective tool to study the various states of human brain. MRI images provide rich information of various states of brain which can be used to study, diagnose, and perform unparallel clinical analysis of brain to determine whether the brain is normal or abnormal and in which respect. The raw datasets extracted from the images are very large, and it is hard to analyze the data to reach any conclusions. In such cases, various image analysis tools can be used to analyze the MRI images, and to extract conclusive diagnostic information to classify the abnormalities of the brain. The level of detail in MRI images is increasing rapidly, and the technology is capable of providing two-dimensional (2-D) and three-dimensional (3-D) images of various organs inside the body.

MRI is often the medical imaging method of choice when soft tissue delineation is necessary. This is especially true for any attempt to classify brain tissues [40.1]. The most important advantage of MR imaging is that it is a noninvasive technique [40.2]. Use of computer technology in medical decision support is now widespread and pervasive across a wide range of medical areas such as cancer research, gastroenterology, heart diseases, brain tumors, etc. [40.3, 4]. Fully automatic classification of normal and diseased human brain based on MRI is of great importance for research and clinical studies. Recent work [40.2, 5] has shown that classification of abnormal conditions from human brain magnetic resonance (MR) images is possible via supervised techniques such as artificial neural networks and support vector machines (SVMs) [40.2], as well as unsupervised classification techniques such as self-organization map (SOM) [40.2] and fuzzy *c*-means clustering combined with feature extraction techniques [40.5]. Other supervised classification techniques, such as *k*-nearest neighbors (*k*-NN), also group pixels based on their similarities in each feature image [40.1, 6–8], and can be used to classify normal/pathological T2-weighted MRI images. We used supervised machine-learning algorithms to classify images into two categories: normal or abnormal.

There are various ways of using MRI images for diagnosis to produce high-resolution images for image segmentation and identification of brain tumors or abnormalities from MRI images. MRI images are also used to produce detailed, accurate pictures of different organs in different medical conditions. Use of high- and low-field MRI images can also allow the physician to see even very small tears and injuries to ligaments and muscles. MRI is based on absorption and emission of energy in the radio free range of the electromagnetic spectrum. The availability of accurate anatomical three-dimensional (3-D) models from 2-D medical image data provides precise information about spatial relationships between critical anatomical structures (e.g., eloquent cortical areas, vascular structures) and pathology, which are often indistinguishable by the naked eye [40.9]. The authors in [40.10] pro-

posed a Bayesian formulation for incorporating soft model assignments into the calculation of affinities, and the resulting model-aware affinities were integrated into the multilevel segmentation using a weighted aggregation algorithm, followed by application in the task of detecting and segmenting brain abnormalities and edema in multichannel magnetic resonance (MR) volumes. A key problem in medical imaging is automatically segmenting an image into its constituent heterogeneous process. Automatic segmentation has the potential to positively impact clinical medicine by freeing physicians from the burden of manual labeling and by providing robust, quantitative measurements as presented in [40.11].

In this chapter, we present the investigations being pursued in our research laboratory on magnetic resonance imaging (MRI) of various states of brain, by extracting the most significant features for classification into normal and abnormal brain images. We propose a novel method based on the wavelet transform to initially decompose the images, and then use various feature selection algorithms to extract the most significant features of brain from the MRI images. By using different classifiers to detect the abnormality of brain images from a publicly available neuroimaging dataset, we found that a principled approach involving wavelet-based feature extraction, followed by selection of the most significant features using the PCA/QDA technique, and classification using learning-based classifiers results in a significant improvement in accuracy as compared with previously reported studies. The rest of the chapter is organized as follows. The next section gives a brief background and review of previous work in this area, followed by a description of the materials and methods used in Sect. 40.3. The feature extraction, feature selection, and classifier techniques used are described in Sects. 40.4 and 40.5, and Sect. 40.6 presents some of the experimental work carried out. The paper concludes with outcomes of the experimental work using the proposed approach and outlines plans for future work.

40.2 Background

Two types of multiplicative noise often arise in several imaging modalities: speckle and Poisson noise. Both types are called multiplicative in the sense that their variance is not constant but depends on the parameters to be estimated [40.12]. The ultimate goal

of postscanning noise removal methods in MRI is to obtain piecewise-constant or slowly varying signals in homogeneous tissue regions while preserving tissue boundaries. In the literature, both statistical approaches and diffusion filter methods have been used to remove noise from digital images. An early study of *image improvement* in MRI using a statistical approach was described in [40.13].

Previous general reviews of wavelets in biomedical image processing, including some early work on functional MRI (fMRI) that measures brain activity by detecting associated changes in blood flow, are provided in [40.14–16]. Statistical issues in wavelet analysis of time series are addressed comprehensively in [40.17]. Reference [40.18] makes a detailed case for the general optimality of wavelets for analysis of fractal signals. The authors in [40.19] describe the implementation of wavelet methods in S-PLUS. Several

research groups have pioneered applications of wavelets to various issues in fMRI data analysis. The most popular application to date has been image compression or denoising [40.19, 20]. The authors in [40.21] explored two- and three-dimensional wavelet transforms as spatial filters of radioligand binding potential maps measured using positron emission tomography (PET) which uses a nuclear medical imaging technique that produces a three dimensional image of functional processes in the body. The work in [40.21] reported a technique for brain tissue classification or segmentation of structural MRI based on fuzzy clustering of wavelet coefficients.

40.3 Materials and Methods

40.3.1 Datasets

The input dataset consists of axial, T2-weighted, 256×256 pixel MR brain images (Fig. 40.1). These images are publicly available datasets downloaded from the Harvard Medical School website [40.22]. Only those sections of the brain in which lateral ventricles are clearly seen are considered in our study. The number of MR brain images in the input dataset is 60, of which 6 are of normal brain and 54 are of abnormal brain. The abnormal brain image set consists of images of brain affected by Alzheimer's and other diseases. The remarkable feature of a normal human brain is the symmetry that it exhibits in the axial and coronal images. Asymmetry in an axial MR brain image strongly indicates abnormality. Hence, symmetry in axial MRI images is

an important feature that needs to be considered in deciding whether the MR image at hand is of a normal or an abnormal brain. A normal and an abnormal T2-weighted MRI brain image are shown in Figs. 40.2 and 40.3, respectively. The lack of symmetry in the abnormal brain MR image is clearly seen in Fig. 40.3. Asymmetry beyond a certain degree is a sure indication of a diseased brain, and this has been exploited in our work for an initial classification at a gross level.

40.3.2 Decomposition of Images Using Wavelets

Wavelets are mathematical functions that decompose data into different frequency components and then

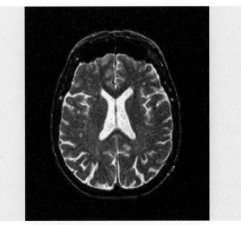

Fig. 40.2 T2-weighted axial MR image of normal brain after wavelet decomposition

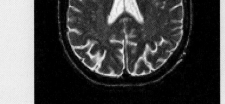

Fig. 40.1 T2-weighted axial MR brain image

Part G | 40.3

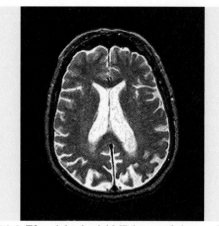

Fig. 40.3 T2-weighted axial MR image of abnormal brain

study each component at a resolution matched to its scale. Wavelets have emerged as powerful new mathematical tools for analysis of complex datasets. The Fourier transform provides a representation of an image based only on its frequency content. Hence, this representation is not spatially localized, while wavelet functions are localized in space. The Fourier transform decomposes a signal into a spectrum of frequencies, whereas the wavelet analysis decomposes a signal into a hierarchy of scales, starting from the coarsest. Hence, the wavelet transform, providing a representation of an image at various resolutions, is a better tool for feature extraction from images.

40.3.3 Discrete Wavelets Transform (DWT)

The DWT is an implementation of the wavelet transform using a discrete set of wavelet scales and translation, obeying some defined rules. For practical computations, it is necessary to discretize the wavelet transform. The scale parameters are discretized on a logarithmic grid. The translation parameter (τ) is then discretized with respect to the scale parameter; i.e., sampling is done on a dyadic (as the base of the logarithm is usually chosen as 2) sampling grid. The discretized scale and translation parameters are $s = 2^{-m}$ and $t = n2^{-m}$, where $m, n \in Z$, the set of all integers. Thus, the family of wavelet functions is represented in (40.1) and (40.2)

$$\psi_{m,n}(t) = 2^{m/2}\psi(2^m t - n) , \tag{40.1}$$

$$W\psi(a, b) = \int_{-\infty}^{\infty} f(x)\psi_{a,b}(t)\,dx . \tag{40.2}$$

In case of images, the DWT is applied to each dimension separately. This results in the image Y being decomposed into a first-level approximation component Y_a^1 and detailed components Y_h^1, Y_v^1, and Y_d^1 corresponding to horizontal, vertical, and diagonal details. Figure 40.3 depicts the process of an image being decomposed into approximate and detailed components.

The approximation component (Y_a) contains low-frequency components of the image, while the detailed components (Y_h, Y_v, and Y_d) contain high-frequency components. Thus,

$$Y = Y_a^1 + \{Y_h^1 + Y_v^1 + Y_d^1\} . \tag{40.3}$$

At each decomposition level, the length of the decomposed signals is half the length of the signal in the previous stage. Hence, the size of the approximation component obtained from the first-level decomposition of an $N \times N$ image is $N/2 \times N/2$, that of the second level is $N/4 \times N/4$, and so on. As the level of decomposition is increased, more compact but coarser approximations of the image are obtained. Thus, wavelets provide a simple hierarchical framework for interpreting image information.

40.4 Feature Selection

40.4.1 Quadratic Discriminant Analysis

Quadratic discriminant analysis (QDA) [40.23] describe the likelihood of a class as a Gaussian distribution and then uses posterior distribution estimates to estimate the class for a given test vector. This approach leads to the function

$$d_k(x) = (x - \mu_k)^T \sum_k^{-1} (x - \mu_k) + \log \sum k - 2 \log p(k) , \tag{40.4}$$

where $\sum_k$ is the covariance matrix, x is the test vector, μ_k is the mean vector, and $p(k)$ is the prior probability of the class k. The Gaussian parameters for each class can be estimated from the training dataset, so the values of $\sum_k$ and μ_k are replaced in the above formula by its estimates $\hat{\sum}_k$ and $\hat{\mu}_k$. However, when the number of training samples is small, compared with the number of dimensions of the training vector, the covariance estimation can be ill-posed. The approach to resolve the ill-posed estimation is to regularize the covariance matrix $\sum_k$.

40.4.2 Principal Component Analysis

Excessive features increase computation time and storage memory, which sometimes causes the classification process to become more complicated. This consequence is called the curse of dimensionality. A strategy is necessary to reduce the number of features used in classification. PCA is an efficient tool to reduce the dimension of a dataset consisting of a large number of interrelated variables while retaining the most sig-

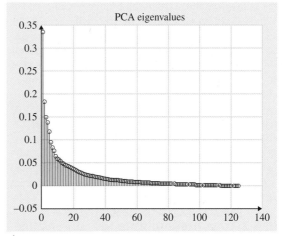

Fig. 40.4 PCA features of MRI images

nificant variations. It is achieved by transforming the dataset to a new set of ordered variables according to their degree of variance or importance, as shown in Fig. 40.4.

40.5 Classification Using Various Classifiers

40.5.1 Support Vector Machine (SVM) Classifiers

The support vector machine (SVM) is a well-known large-margin classifier proposed by *Vapnik* [40.24]. The basic concept of the SVM classifier is to find an optimal separating hyperplane which separates two classes. The decision function of the binary SVM is

$$f(x) = \sum_{i=1}^{N} (\alpha_i y_i K(x_i, x) + b), \qquad (40.5)$$

where b is a constant, $y_i \in \{-1, 1\}$, $0 \le \alpha_i \le C$, $I = 1, 2, \dots, N$ are nonnegative Lagrange multipliers, C is a cost parameter that controls the trade-off between allowing training errors and forcing rigid margins, x_i are the support vectors, and $K(x_i, x)$ is the kernel function.

We apply SVM to the multiclass problem using the one-against-one method. It was first introduced in [40.25], and the first use of this strategy for SVM was reported in [40.26, 27]. This method constructs $k(k - 1)/2$ classifiers, where each one trains data from two classes. For training data from the ith and jth classes, we solve the following binary classification problem.

In this study, we use the max win voting strategy suggested in [40.28]. If $\text{sign}((w^{ij})^T \theta(x) + b^{ij})$ says x is in the ith class, then the vote for the ith class is increased by one. Otherwise, the jth class is increased by one. Then, the largest vote determines the specific class of the variable x.

We used the LIBSVM software library for the experiments. LIBSVM is a general library for support vector classification and regression, being available at [40.29]. As mentioned above, there are different functions to map data to higher-dimensional spaces, and practically we need to select the kernel function $K(x^i; x^j) = \theta(xi)^T \theta(x^j)$. There are several types of kernels in use for all kinds of problems. Each kernel has different parameters for different problems; For example, some well-known problems with large numbers of features, such as text classification [40.30] and protein folding and image processing problems [40.31–33], are reported to be classified more correctly with a linear kernel. In our study, we uses the radial basis function (RBF) kernel which is real valued function whose value depends only on the distance from the origin. A learner with the RBF kernel usually performs no worse than others in terms of generalization ability. In this study, we carried out some simple comparisons and observed that, when using the RBF kernel, the performance is slightly better compared with the linear kernel $K(x^i; x^j) = \theta(xi)^T\theta(x^j)$ for all the problems we studied. Therefore, for the three datasets, instead of staying in the original space, a nonlinear mapping to a higher-dimensional space seems useful. Another important issue is the selection of parameters. For SVM training, a few parameters such as the penalty parameter C and the kernel parameter of the RBF function must be determined in advance. Choosing optimal parameters for support vector machines is an important step in SVM design. We use cross-validation on different parameters for model aselection.

40.5.2 Classification Using Self-Organizing Maps

Artificial neural networks (ANNs) [40.34, 35] are biologically inspired. They are composed of many nonlinear computational elements operating in parallel and are arranged in patterns similar to biological neural nets.

ANNs modify their behavior in response to their environment, learn from experience, and generalize from previous examples to new ones. ANNs have become the preferred technique for a large class of pattern recognition tasks. *Lippmann* [40.36] provides an excellent review on ANNs. A self-organizing map (SOM) is an unsupervised algorithm which has advantages over other networks; it can form similarity diagrams automatically, and can produce abstractions.

The self-organizing map is generalizable to nonvectorial, say symbolic, data, whereas other networks are not [40.37]. The support vector machine (SVM) is a machine-learning technique which originated from statistical theory and is used for classification of images. Its primary advantages are that it is able to model highly nonlinear systems and that the special properties of the decision surface ensure very good generalization. The SVM has been widely used in pattern recognition applications due to its computational efficiency and good generalization performance. Thus, ANN and SVM are very attractive for recognition and classification tasks. The results of both approaches are compared.

40.6 Experimental Results

40.6.1 Level of Wavelet Decomposition

We obtained wavelet coefficients of 60 brain MR images, each of size 256×256. Level 1 HAR wavelet decomposition of a brain MRI image produces 16 384 wavelet approximation coefficients, which is a sequence of rescaled *square-shaped* functions which together form a wavelet family or basis, while level 2 and level 3 produce 4096 and 1024 coefficients, respectively. The third level of wavelet decomposition greatly reduces the input vector size but results in lower classification percentage. With the first-level decomposition, the vector size (16 384) is too large to be given as an input to a classifier. Based on a systematic empirical analysis of the wavelet coefficients through simulation in MATLAB 7.10, we came to the conclusion that level 2 features are the most suitable for neural network, self-organizing map, and support vector machine classifiers, whereas level 1 and level 3 features result in lower classification accuracy. The second level of wavelet decomposition not only gives almost perfect results in the testing phase, but also has a reasonably manageable number of features (4096) that can be handled without much difficulty by the classifier. We also applied Daubechies-4 (DAUB4) mother wavelets which

are a family of orthogonal wavelets defining a discrete wavelet transform and characterized by a maximal number of vanishing moments for some given support to obtain the decomposition coefficients of MRI images at level 2 to provide a comparison of decomposition by two wavelet types in terms of classification accuracy.

40.6.2 Classification from Self-Organizing Maps

Feature extraction using wavelet decomposition, followed by dimensionality reduction with PCA, and classification through a self-organizing map-based neural network was employed. Wavelet coefficients of

Table 40.1 Comparison among different methods

Method	Recognition ratio (%)
SVM	91.32
k-NN	84.38
Bayesian naives	85.63
SVM-QDA (proposed method)	93.43
SVM-PCA (proposed method)	96.24
SOM-QDA (proposed method)	90.06
SOM-PCA (propsed method)	94.24

Table 40.2 Recognition ratio obtained using various feature selection and classification algorithms on HAR and Daubechies wavelets

Selection method (number of features)	Recognition ratio			
	SVM (%)	SVM–PCA (%)	SOM (%)	SOM–PCA (%)
HAR level 2 wavelets	91.32	96.24	88.56	94.24
DAUB4 level 2 wavelets	86.78	92.95	81.18	91.41

MR images were obtained using the wavelet toolbox of MATLAB 7.10. A program for the self-organizing neural network was written using MATLAB 7.10. Images marked as abnormal in the first stage were not considered in the second stage, to avoid expensive calculations for wavelet decomposition of these images. The second-level DAUB4 and HAR wavelet approximation coefficients were obtained and given as input to the self-organizing neural network classifier. The classification results are presented in Table 40.1 and Table 40.2. The number of MR brain images in the input dataset is 60, of which 6 are of normal brain and 54 are of abnormal brain. Experimentation was done with varying levels of wavelet decomposition. The final categories obtained after classification by the self-organizing map depend on the order in which the input vectors are presented to the network. Hence, we randomized the order of presentation of the input images. Our experiment was repeated, each time with a different order of input presentation, and we obtained the same classification percentage and normal–abnormal categories in all the experiments.

40.6.3 Classification Using Support Vector Machines

We implemented SVM in Weka 3.6.6, with the inputs being the wavelet-coded images, using the LIBSVM library for our classification. This is a 2-D classification

Table 40.3 Classification by SVM using various kernels

Kernel	SVM (%)		SVM–PCA (%)	
	HAR	DAUB4	HAR	DAUB4
Polynominal	91.32	86.78	96.24	92.95
RBF	76.33	71.45	86.41	91.68
Linear	71.56	69.44	73.63	70.81

technique. In this chapter, we treat the classification of MR brain images as a two-class pattern classification problem. In every wavelet-coded MR image, we apply a classifier to determine whether it is normal or abnormal. As mentioned previously, use of SVM involves training and testing the SVM, with a particular kernel function, which in turn has specific kernel parameters. The RBF and polynomial functions were used as SVM kernels for this classification. The linear kernel was also used for SVM training and testing, but it showed a lower classification rate than the polynomial and RBF kernels. The classification results with linear, polynomial, and RBF kernel are presented in Table 40.3. The classification accuracy is higher for the polynomial kernel in comparison with the RBF and linear kernels. We use the k-fold cross-validation technique to avoid overfitting, with $k = 7$ over the entire dataset. The parameters C from Eq. (40.5) and g have certain values. Both values were chosen experimentally using a cross-validation procedure on the training dataset. The best recognition ratio was achieved using parameter values of $g = 0.4$ and $C = 40$.

40.7 Conclusions and Further Work

In this chapter we present a principled approach for investigating brain abnormalities using wavelet-based feature extraction, PCA-based feature selection, and SVM/SOM-based classification. Experiments on a publicly available brain image dataset show that the proposed principled approach performs significantly better than other competing methods reported in the literature. The classification accuracy of more than 94% in case of self-organizing maps and 96% in case of the support vector machine demonstrates the utility of

the proposed method. In this chapter, we applied this method only to axial T2-weighted images at a particular depth inside the brain. The same method could be employed for T1-weighted, proton density, and other types of MR images. With the help of the described approaches, one could develop software for a diagnostic system for detection of brain disorders such as Alzheimer's, Huntington's, Parkinson's diseases, etc. Furthermore, the proposed approach uses reduced data by incorporating feature selection algorithms in the processing loop yet still provides improved recognition and accuracy. The combined SVM-QDA and SVM-PCA classifiers achieve much better results (93.43% and 96.24%) as compared with standalone SVM (91.32%) and other traditional classifiers reported in the literature. Further work will be pursued to classify different type of abnormalities, and to extract new features from MRI brain images depending on various parameters such as age, emotional state, and their feedback.

References

40.1   L.M. Fletcher-Heath, L.O. Hall, D.B. Goldgof, F.R. Murtagh: Automatic segmentation of non-enhancing brain tumors in magnetic resonance images, Artif. Intell. Med. **21**, 43–63 (2001)

40.2   L.M. Sandeep Chaplot, N.R. Patnaik: Jagannathan Classification of magnetic resonance brain images using wavelets as input to support vector machine and neuralnetwork, Biomed. Signal Process. Control **1**, 86–92 (2006)

40.3   F. Gorunescu: Data mining techniques in computer-aided diagnosis: Non-invasive cancer detection, Proc. World Acad. Sci. Eng. Technol. **25**, 427–430 (2007)

40.4   S. Kara, F. Dirgenali: A system to diagnose atherosclerosis via wavelet transforms, principal component analysis and artificial neural networks, Expert Syst. Appl. **32**, 632–640 (2007)

40.5   M. Maitra, A. Chatterjee: Hybrid multiresolution Slantlet transform and fuzzy c-means clustering approach for normal-pathological brain MR image segregation, Med. Eng. Phys. **30**(5), 615–623 (2007), doi:10.1016/j.medengphy.2007.06.009.

40.6   P. Abdolmaleki, F. Mihara, K. Masuda, L.D. Buadu: Neural networks analysis of astrocyticgliomas from MRI appearances, Cancer Lett. **118**, 69–78 (1997)

40.7   T. Rosenbaum, V. Engelbrecht, W. Krolls, F.A. van Dorsten, M. Hoehn-Berlage, H.-G. Lenard: MRI abnormalities in neuro-bromatosistype 1 (NF1): A study of men and mice, Brain Dev. **21**, 268–273 (1999)

40.8   C. Cocosco, A.P. Zijdenbos, A.C. Evans: A fully automatic and robust brain MRI tissue classification method, Med. Image Anal. **7**, 513–527 (2003)

40.9   X. Hu, K.K. Tan, D.N. Levin: Three-dimensional magnetic resonance images of the brain: Application to neurosurgical planning, J. Neurosurg. **72**, 433–440 (1990)

40.10 J.J. Corso, E. Sharon, S. Dube, S. El-Saden, U. Sinha, A. Yuille: Efficient multilevel brain tumor segmentation with integrated Bayesian model classification, IEEE Trans. Med. Imaging **27**(5), 623–640 (2008)

40.11 M.R. Patel, V. Tse: Diagnosis and staging of brain tumors, Semin. Roentgenol. **39**(3), 347–360 (2004)

40.12 A. Aldroubi, M. Unser: *Wavelets in Biology and Medicine* (RC Press, Boca Raton 1996)

40.13 A. Laine: Wavelets in temporal and spatial processing of biomedical images, Annu. Rev. Biomed. Eng. **2**, 511–550 (2000)

40.14 E.T. Bullmore, J. Fadili, M. Breakspear, R. Salvador, J. Suckling, M.J. Brammer: Wavelets and statistical analysis of functional magnetic resonance images of the human brain, Stat. Methods Med. Res. **12**, 375–399 (2003)

40.15 D.B. Percival, A.T. Walden: *Wavelet Methods for Time Series Analysis* (Cambridge Univ. Press, Cambridge 2000)

40.16 G.W. Wornell: *Signal Processing with Fractals: A Wavelet-Based Approach* (Prentice Hall, Upper Saddle River, NJ 1996)

40.17 A. Bruce, H.Y. Gao: *Applied Wavelet Analysis with S-PLUS* (Springer, New York 1996)

40.18 M.E. Alexander, R. Baumgartner, C. Windischberger, E. Moser, R.L. Somorjai: Wavelet domain denoising of time-courses in MR image sequences, Magn. Reson. Imaging **18**, 1129–1134 (2000)

40.19 S. Zaroubi, G. Goelman: Complex denoising of MR data via wavelet analysis: Application for functional MRI, Magn. Reson. Imaging **18**, 59–68 (2000)

40.20 Z. Cselenyi, H. Olsson, L. Farde, B. Gulyas: Wavelet-aided parametric mapping of cerebral dopamine D_2 receptors using the high affinity PET radioligand [^{11}C]FLB 457, NeuroImage **17**, 47–60 (2002)

40.21 V. Barra, J.Y. Boire: Tissue segmentation on MR images of the brainby possibilistic clustering on a 3D wavelet representation, J. Magn. Reson. Imaging **11**, 267–278 (2000)

40.22 K.A. Johnson, J.A. Becker: The whole brain atlas, available online at http://med.harvard.edu/AANLIB/

40.23 K. Fukunaga: *Introduction to Statistical Pattern Recognition*, 2nd edn. (Academic Press, New York 1990)

40.24 V. Vapnik: *The Nature of Statistical Learning Theory* (Springer, New York 1995)

40.25 S. Knerr, L. Personnaz, G. Dreyfus: Single-layer learning revisited: A step-wise procedure for building and training a neural network. In: *Neuro-computing: Algorithms, Architectures and Applications*, ed. by J. Fogelman (Springer, Berlin, Heidelberg 1990)

40.26 J. Friedman: Another approach to polychotomous classification, Tech. Rep. (Department of Statistics, Stanford University 1996)

40.27 U. Krebel: Pair-wise classification and support vector machines. In: *Advances in Kernel Methods—Support Vector Learning*, ed. by B. Scholkopf, C.J.C. Burges, A.J. Smola (MIT Press, Cambridge 1999) pp. 255–268

40.28 C.-J. Lin: Formulations of support vector machines: A note from an optimization point of view, Neural Comput. **13**(2), 307–317 (2001)

40.29 C.-C. Chang, C.-J. Lin: LIBSVM – A library for support vector machines, available online at http://www.csie.ntu.edu.tw/ cjlin/libsvm/

40.30 T. Joachims: The Maximum-Margin Approach to Learning Text Classifiers: Methods, Theory, and Algorithms. Ph.D. Thesis (Universität Dortmund, Dortmund 2000)

40.31 A. Mishra, L. Singh, G. Chetty: A novel image water marking scheme using extreme learning machine, Proc Proc. IEEE World Congr. Computational Intelligence (WCCI 2012) (Brisbane, Australia 2012)

40.32 L. Singh, G. Chetty: Hybrid approach in protein folding recognition using support vector machines, Proc. Int. Conf. Machine Learning Data Mining (MLDM 2012), 2012)

40.33 L. Singh, G. Chetty: Review of classification of brain abnormalities in magnetic resonance images using pattern recognition and machine learning, Proc. Int. Conf. Neuro Comput. Evolving Intelligence, NCEI 2012 (Springer, Auckland 2012)

40.34 P.D. Wasserman: *Neural Computing* (Van Nostrand Reinhold, New York 1989)

40.35 S. Haykin: *Neural Networks: A Comprehensive Foundation*, 2nd edn. (Pearson, Prentice Hall 1994)

40.36 R.P. Lippmann: An introduction to computing with neural nets, IEEE Acoust. Speech Signal Process. Mag. **4**(2), 4–22 (1987)

40.37 T. Kohonen: The self-organizing map, IEEE Proc. **78**, 1464–1477 (1990)

Part H
Advanced

Part H Advanced Signal Processing Methods for Brain Signal Analysis and Modeling

Ed. by Danilo Mandic

41. Nonlinear Adaptive Filtering in Kernel Spaces

Badong Chen, Lin Li, Weifeng Liu, José C. Príncipe

Recently, a family of online kernel-learning algorithms, known as the kernel adaptive filtering (KAF) algorithms, has become an emerging area of research. The KAF algorithms are developed in reproducing kernel Hilbert spaces (RKHS), by using the linear structure of this space to implement well-established linear adaptive algorithms and to obtain nonlinear filters in the original input space. These algorithms include the kernel least mean squares (KLMS), kernel affine projection algorithms (KAPA), kernel recursive least squares (KRLS), and extended kernel recursive least squares (EX-KRLS), etc. When the kernels are radial (such as the Gaussian kernel), they naturally build a growing RBF network, where the weights are directly related to the errors in each sample. The aim of this chapter is to give a brief introduction to kernel adaptive filters. In particular, our focus is on KLMS, the simplest KAF algorithm, which is easy to implement, yet efficient. Several key aspects of the algorithm are discussed, such as self-regularization, sparsification, quantization, and the mean-square

convergence. Application examples are also presented, including in particular the adaptive neural decoder for spike trains.

41.1 Overview

Adaptive filters find applications in a wide range of diverse fields such as communication, control, biomedicine, neuroinformatics, radar, sonar, acoustic, and speech processing, etc. [41.1–3]. In general, an adaptive filter is defined as a self-designing system whose free parameters (or structure) are self-adjusted by a certain recursive (or iterative) algorithm such that the system performs satisfactorily in an environment where knowledge of the relevant statistics is not available. Adaptive filters are usually classified into linear and nonlinear adaptive filters. Linear adaptive filters compute their output by using a linear combiner, other-

wise the adaptive filter is said to be nonlinear. Adaptive filters can also be classified into *supervised adaptive filters* and *unsupervised adaptive filters*. The supervised adaptive filters require a collection of desired (target) responses, and the training data are composed of the input-desired pairs. *Error-correction learning* and *memory-based learning* are two basic learning rules for a supervised adaptive filter. Unsupervised adaptive filters perform adjustment in a blind or self-organized manner, without the need for a desired signal. In this chapter our focus is on supervised adaptive filters.

41.1.1 Linear Adaptive Filters

Figure 41.1 shows the basic structure of an adaptive filter, where $u(i) \in \mathbb{R}^m$ is the input vector at iteration i, $w(i)$ is the free parameter vector or the weight vector, $d(i)$ is the desired signal, $y(i)$ is the filter output, and $e(i) = d(i) - y(i)$ is the error signal. At each iteration cycle, the weight vector $w(i)$ is adjusted by an adaptive algorithm so as to minimize a certain cost such as the mean square error (MSE) $J = E[e^2(i)]$. The commonly used linear adaptive filter is the transversal filter, whose output is simply a linear combination of the input

$$y(i) = w(i-1)^T u(i) \, . \tag{41.1}$$

The simplest and most popular adaptive algorithm is the least mean squares (LMS) algorithm, which is a stochastic gradient-based algorithm under the instantaneous cost function $J(i) = e^2(i)/2$. The weight update equation for LMS can be derived as

$$\begin{aligned} w(i) &= w(i-1) - \eta \frac{\partial}{\partial w(i-1)} \frac{1}{2} e^2(i) \\ &= w(i-1) - \eta e(i) \frac{\partial}{\partial w(i-1)} \\ &\quad (d(i) - w(i-1)^T u(i)) \\ &= w(i-1) + \eta e(i) u(i) \, , \end{aligned} \tag{41.2}$$

where $\eta > 0$ is the step size. The LMS algorithm is the workhorse of adaptive signal processing due to its simplicity of implementation, computational efficiency, and robust performance. The robustness of the LMS algorithm was explained in [41.4], where it was shown that a single realization of the LMS algorithm is optimal in the H^∞ sense. The step size η is critical to the performance of the LMS algorithm. In general, the choice of step size is a trade-off between the convergence rate and the asymptotic excess mean square error (EMSE).

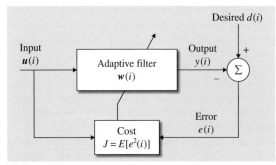

Fig. 41.1 Basic structure of an adaptive filter

The LMS algorithm suffers from two major drawbacks:

1. Slow rate of convergence
2. Sensitivity to the eigenvalue spread of the correlation matrix of the input vector.

One approach to overcome these limitations is to use the transform-domain adaptive algorithm, where the adaptation is performed in the frequency domain rather than the original time domain [41.5, 6]. Another approach is to use the *recursive least-squares* (RLS) algorithm, which recursively updates the estimated autocorrelation matrix of the input signal vector and the cross-correlation vector between the input vector and the desired response. The convergence rate of the RLS is usually an order of magnitude faster than the LMS algorithm. This improvement is, however, achieved at the expense of an increase in the computational complexity of the RLS filter.

The fast convergence rate does not mean a fast tracking capability in a nonstationary environment. In general, the RLS algorithm exhibits a worse tracking behavior than the LMS algorithm. In order to improve the tracking performance, an extended RLS (EX-RLS) algorithm was proposed in [41.7], where a *state-space* model is incorporated with the RLS. A more general linear filter based on a state-space model is the popular *Kalman filter* [41.8, 9].

41.1.2 Nonlinear Adaptive Filters

The approximation capability of a linear filter is very limited due to its inherent linearity. Nonlinear adaptive filters may achieve better results than conventional linear adaptive filters, especially when the underlying mapping between the input and output is nonlinear and complex. There have been many approaches to achieving a nonlinear adaptive filter. One simple way to implement a nonlinear adaptive filter is by using the cascade models such as the Wiener model (a linear filter followed by static nonlinearity) and the Hammerstein model (static nonlinearity followed by a linear filter) [41.10]. The modeling capability of this approach is, however, still limited, and the choice of the nonlinearity is highly problem dependent. The Volterra series and the Wiener series (an improvement over the Volterra series) can also be used to build a nonlinear adaptive filter [41.11, 12], but slow convergence and high complexity hinder their wide application. In recent years, nonlinear adaptive filters were also achieved in time-lagged neural networks such as the multilayer perceptrons (MLPs), radial basis function (RBF) networks, recurrent neural networks, and so on [41.13,14]. The use of the neural networks in adaptive filtering is, usually, hampered by some drawbacks such as topology specification and local minima.

41.1.3 Kernel Adaptive Filters

Kernel methods have been successfully used in nonlinear signal processing and machine learning due to their inherent advantages of convex optimization and universality in the space of L_2 functions. Typical examples of kernel methods include support vector machines (SVM) [41.15,16], kernel regularization networks [41.17], kernel principal component analysis (KPCA) [41.18], and kernel Fisher discriminant analysis (KFDA) [41.19], etc. These nonlinear algorithms show significant performance improvement over their linear counterparts. In kernel machine learning, one often uses a reproducing kernel Hilbert space (RKHS) $\mathcal{H}_k$ associated with a Mercer kernel (i. e., reproducing kernel) $\kappa(u, u')$ as the hypothesis space. The Mercer kernel is a continuous, symmetric, and positive-definite function $\kappa : U \times U \to R$, where $U \subseteq R^m$ denotes the m-dimensional input space. According to Mercer's theorem, any Mercer kernel $\kappa(u, u')$ induces a mapping φ from the input space U to a high (possibly infinite) dimensional feature space F (Fig. 41.2). In the feature space, the inner products can be calculated easily using the kernel evaluation (the so-called *kernel trick*)

$$\varphi(u)^{\mathrm{T}} \varphi(u') = \kappa(u, u') . \tag{41.3}$$

The key principle behind the kernel trick is that, as long as a linear algorithm in high dimensional feature space can be formulated in terms of inner products, a nonlinear algorithm can be developed by simply replacing the inner product with a Mercer kernel. It is worth noting that if identifying $\varphi(u) = \kappa(u, \cdot)$, the feature space F is essentially the same as the RKHS induced by the kernel κ.

The kernel adaptive filters (KAF) are a family of nonlinear adaptive filters recently developed by the kernel method [41.20]. In general, they are based on the following linear model in feature space (i. e., RKHS)

$$y(i) = \Omega(i - 1)^{\mathrm{T}} \varphi(i) , \tag{41.4}$$

where $\varphi(i) = \varphi(u(i))$ is the mapped feature vector from the input space U to the feature space F, $\Omega(i - 1)$ is the high-dimensional weight vector in the feature space. In form, the model (41.4) is very similar to the model of the transversal filter (41.1). However, one should keep in mind that the model (41.4) is essentially a nonlinear model in the original input space, since the kernel induced mapping φ is usually a nonlinear mapping. Now regarding the model (41.4) as a high-dimensional transversal filter, one can formulate the classical linear adaptive filtering algorithms (such as LMS, RLS, EX-RLS, etc.) in RKHS. If we can formulate these algorithms in terms of inner products, we will obtain nonlinear adaptive filters in input space, namely the kernel adaptive filters.

Selecting a proper Mercer kernel is important for a kernel adaptive filter. Commonly used kernels are listed in Table 41.1. Among these kernels, the Gaussian kernel is very popular and is usually a default choice due to its universal approximating capability, desirable smoothness, and numerical stability. In the spike train signal processing domain, the kernel function in the spike train space has also been defined [41.21,22].

Kernel adaptive filters have several desirable features:

1. If choosing a universal kernel (e.g., Gaussian kernel), they are universal approximators.
2. The performance surface is still quadratic so gradient descent learning does *not* suffer from local minima.
3. When pruned, the redundant features have moderate complexity in terms of computation and memory.

Table 41.1 Commonly used kernels in machine learning, with parameters $c > 0$, $\sigma > 0$ and $p \in N$

Kernel	Expression
Polynomial	$\kappa(u, u') = (c + u^{\mathrm{T}} u')^p$
Exponential	$\kappa(u, u') = \exp(u^{\mathrm{T}} u' / 2\sigma^2)$
Sigmoid	$\kappa(u, u') = \tanh(u^{\mathrm{T}} u' / \sigma + c)$
Gaussian	$\kappa(u, u') = \exp(-\|u - u'\|^2 / 2\sigma^2)$
Laplacian	$\kappa(u, u') = \exp(-\|u - u'\| / 2\sigma^2)$
Multiquadratic	$\kappa(u, u') = \sqrt{\|u - u'\|^2 + c}$
Inverse multiquadratic	$\kappa(u, u') = 1 / \sqrt{\|u - u'\|^2 + c}$

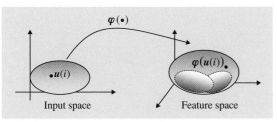

Fig. 41.2 Map $\phi(\cdot)$ from the input space to the feature space

Up to now, most linear adaptive filters were *kernelized*. Typical kernel adaptive filtering algorithms include the kernel least mean squares (KLMS) [41.23], kernel affine projection algorithms (KAPA) [41.24], kernel recursive least squares (KRLS) [41.25], and extended kernel recursive least squares (EX-KRLS) [41.26]. When

the kernel is radial (such as the Gaussian kernel), they naturally build a growing RBF network, where the weights are directly related to the errors at each sample. Recently, a book on kernel adaptive filtering was also published [41.20]. In the subsequent sections of this chapter, we mainly discuss the KLMS algorithm.

41.2 Kernel Least Mean Squares

41.2.1 Basic Algorithm

Suppose now the goal is to learn a continuous arbitrary input–output mapping $f : \mathbf{U} \to \mathbf{R}$ based on a sequence of input–output examples $\{u(i), d(i)\}$, $i = 1, 2, \ldots$, where $\mathbf{U} \subseteq \mathbf{R}^m$ is the input domain. Online learning with kernels sequentially finds an estimate of f such that f_i (the estimate at iteration i) is updated based on the last estimate f_{i-1} and current example $\{u(i), d(i)\}$.

Mapping the input data $\{u(i)\}$ into the feature space, one obtains the training data $\{\varphi(i), d(i)\}$, $i = 1, 2, \ldots$, where $\varphi(i) = \varphi(u(i))$. Then, using model (41.4) and performing the LMS algorithm on the new example sequence yields the KLMS algorithm [41.23]

$$\begin{cases} \boldsymbol{\Omega}(0) = \mathbf{0}, \\ e(i) = d(i) - \boldsymbol{\Omega}(i-1)^T \varphi(i), \\ \boldsymbol{\Omega}(i) = \boldsymbol{\Omega}(i-1) + \eta e(i)\varphi(i), \end{cases} \quad (41.5)$$

where $e(i)$ is the prediction error, and η is the step size. The KLMS is very similar to the LMS algorithm, except for the dimensionality (or richness) of the projection space. The learned mapping f_i is the composition of $\boldsymbol{\Omega}(i)$ and φ, that is $f_i = \boldsymbol{\Omega}(i)^T \varphi(\cdot)$. If identifying $\varphi(u) = \kappa(u, \cdot)$, one obtains the sequential learning rule in the original input space

$$\begin{cases} f_0 = 0, \\ e(i) = d(i) - f_{i-1}(u(i)), \\ f_i = f_{i-1} + \eta e(i)\kappa(u(i), \cdot). \end{cases} \quad (41.6)$$

At iteration i, given an input u, the output of the filter is

$$f_i(u) = \eta \sum_{j=1}^{i} e(j)\kappa(u(j), u). \quad (41.7)$$

From (41.5) one can see that, if choosing a radial kernel, KLMS produces a growing RBF network by allocating a new kernel unit for every new example with input

$u(i)$ as the center and $\eta e(i)$ as the coefficient. The algorithm of KLMS is summarized in Algorithm 41.1, and the network topology is illustrated in Fig. 41.3.

Algorithm 41.1 KLMS

Initialization:
  Choose kernel κ and step size η
  $a_1 = \eta d(1), \quad C(1) = \{u(1)\}, \quad f_1 = a_1(1)\kappa(u(1), \cdot)$
Computation:
  while $\{u(i), d(i)\}(i > 1)$ available **do**
    (1) Compute the filter output:
$$f_{i-1}(u(i)) = \sum_{j=1}^{i-1} a_j \kappa(u(i), u(j))$$
    (2) Compute the error: $e(i) = d(i) - f_{i-1}(u(i))$
    (3) Store the new center: $C(i) = \{C(i-1), u(i)\}$
    (4) Compute and store the coefficient: $a_i = \eta e(i)$
  end while

In KLMS, the kernel is usually chosen to be a Gaussian kernel. The kernel size (also known as the kernel bandwidth or smoothing parameter) σ in the Gaussian kernel is a crucial parameter that controls the degree

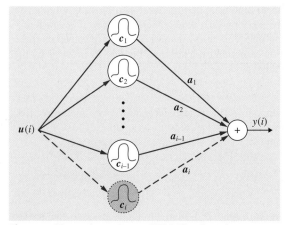

Fig. 41.3 Network topology of KLMS at iteration i

of smoothing and consequently has significant influence on the learning performance. At present, the kernel size is usually set manually or estimated in advance by *Silverman's* rule [41.27] based on the input sample distribution. The role of the step size η in KLMS in principle remains the same as the step size in traditional adaptive filters. Specifically, it controls the compromise between convergence speed and misadjustment. In practical use, choosing these two parameters (i.e., kernel size σ and step size η) usually requires experience and experiments.

41.2.2 Self-Regularization

The KLMS is a stochastic gradient algorithm that solves a least-squares problem in high-dimensional feature space. In most kernel machine learning algorithms, a certain regularization technique has been applied to obtain stable solutions that generalize appropriately. It can be shown that KLMS has a self-regularization property and does not need explicit regularization. Actually, the step size in KLMS plays a similar role to the regularization parameter in explicitly regularized cost functions. Some results are summarized below.

The concept of the regularization is strongly related to the solution norm constraint. One can show that under certain conditions, the weight vector $\boldsymbol{\Omega}(i)$ in KLMS will be upper-bounded at final iteration N. Let us assume the desired output data $\{d(i)\}$ are generated by a multiple linear regression model in feature space

$$d(i) = \boldsymbol{\Omega}^{*\mathrm{T}}\boldsymbol{\varphi}(i) + v(i) , \tag{41.8}$$

where $\boldsymbol{\Omega}^*$ denotes the unknown weight vector that needs to be estimated, and $v(i)$ stands for the modeling uncertainty or disturbance noise. Then the following theorem holds.

Theorem 41.1
Under the H^∞ stability condition, $\boldsymbol{\Omega}(N)$ is upper-bounded [41.23]

$$\|\boldsymbol{\Omega}(N)\| < \sqrt{N\varsigma_1\eta(\|\boldsymbol{\Omega}^*\|^2 + 2\eta\|v\|^2)} , \tag{41.9}$$

where ς_1 is the largest eigenvalue of the autocorrelation matrix $\boldsymbol{R}_\varphi = 1/N\boldsymbol{\Phi}\boldsymbol{\Phi}^{\mathrm{T}}$, $\boldsymbol{\Phi} = [\boldsymbol{\varphi}(1), \boldsymbol{\varphi}(2), \dots, \boldsymbol{\varphi}(N)]$, and $\boldsymbol{v} = [v(1), \dots, v(N)]^{\mathrm{T}}$.

The self-regularization property of KLMS can also be clearly shown by singular value analysis. Let the singular value decomposition (SVD) of $\boldsymbol{\Phi}$ be

$$\boldsymbol{\Phi} = \boldsymbol{P} \begin{bmatrix} \boldsymbol{S} & \boldsymbol{0} \\ \boldsymbol{0} & \boldsymbol{0} \end{bmatrix} \boldsymbol{Q}^{\mathrm{T}} , \tag{41.10}$$

where $\boldsymbol{P}$ and $\boldsymbol{Q}$ are orthogonal matrices, and $\boldsymbol{S} = \mathrm{diag}(s_1, \dots, s_r)$ with s_i the singular values, and r the rank of $\boldsymbol{\Phi}$. It is assumed that $s_1 \geq \dots \geq s_r > 0$ without loss of generality. Denote $\boldsymbol{d} = [d(1), \dots, d(N)]^{\mathrm{T}}$. Then the pseudo-inverse solution to the least squares optimization problem $\min_{\boldsymbol{\Omega}} \|\boldsymbol{d} - \boldsymbol{\Phi}^T\boldsymbol{\Omega}\|^2$ is

$$\boldsymbol{\Omega}_{\mathrm{PI}} = \boldsymbol{P}\mathrm{diag}(s_1^{-1}, \dots, s_r^{-1}, 0, \dots, 0)\boldsymbol{Q}^{\mathrm{T}}\boldsymbol{d} \tag{41.11}$$

If s_r is very small, the pseudo-inverse solution (41.11) becomes problematic as the solution approaches infinity. Tikhonov regularization [41.28] can be used to address this issue, where a regularization term $\lambda\|\boldsymbol{\Omega}\|^2$ is introduced in the least-squares cost. Solving the optimization $\min_{\boldsymbol{\Omega}}\{\|\boldsymbol{d} - \boldsymbol{\Phi}^{\mathrm{T}}\boldsymbol{\Omega}\|^2 + \lambda\|\boldsymbol{\Omega}\|^2\}$ yields the Tikhonov regularization solution

$$\boldsymbol{\Omega}_{\mathrm{TR}} = \boldsymbol{P}\mathrm{diag}\left(\frac{s_1}{s_1^2 + \lambda}, \dots, \frac{s_r}{s_r^2 + \lambda}, 0, \dots, 0\right)\boldsymbol{Q}^{\mathrm{T}}\boldsymbol{d}. \tag{41.12}$$

Comparing (41.11) and (41.12), one can observe that the Tikhonov regularization modifies the diagonal terms through the following regularization function (reg-function)

$$H_{\mathrm{TR}}(x) = \frac{x^2}{x^2 + \lambda} . \tag{41.13}$$

By Tikhonov regularization, we have that $H_{\mathrm{TR}}(s_r)s_r^{-1} \to 0$ if s_r is very small, and $H_{\mathrm{TR}}(s_r)s_r^{-1} \to s_r^{-1}$ if s_r is large (relative to λ). Attenuating the minor components usually yields a smaller norm solution, or in other words, a more stable solution. Another approach to attenuate the minor components is the truncated pseudo-inverse regularization [41.29], which is nothing but using the following hard cut-off reg-function

$$H_{\mathrm{PCA}}(x) = \begin{cases} 1 & \text{if } x > t \\ 0 & \text{if } x \leq t \end{cases} , \tag{41.14}$$

where t is the cut-off threshold. The subscript PCA in H_{PCA} indicates that this method is equivalent to applying the principal components analysis (PCA) technique.

Now we show why KLMS possesses a self-regularization property. Let $\boldsymbol{P}_j$ be the jth column of $\boldsymbol{P}$, M and ς_j be, respectively, the dimensionality and jth eigenvalue of $\boldsymbol{R}_\varphi$, and assume $\boldsymbol{\Omega}^* = \sum_{j=1}^{M} \boldsymbol{\Omega}_j^* \boldsymbol{P}_j$; we can prove the following result [41.23]

$$E[\boldsymbol{\Omega}(i)] = \sum_{j=1}^{M} [1 - (1 - \eta\varsigma_j)^i]\boldsymbol{\Omega}_j^*\boldsymbol{P}_j . \tag{41.15}$$

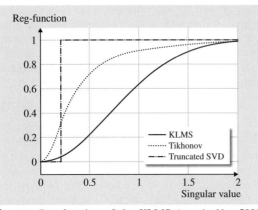

Fig. 41.4 Reg-function of the KLMS ($\eta = 1$, $N = 500$), Tikhonov ($\lambda = 0.1$), and truncated SVD ($t = 0.2$)

Replacing the optimal weight with the pseudo-inverse solution yields

$$E[\boldsymbol{\Omega}(i)] \approx \boldsymbol{P} \mathrm{diag}([1 - (1 - \eta\varsigma_1)^i]s_1^{-1}, \ldots,$$
$$[1 - (1 - \eta\varsigma_r)^i]s_r^{-1}, 0, \ldots, 0)\boldsymbol{Q}^{\mathrm{T}}\boldsymbol{d}$$
$$\overset{(a)}{=} \boldsymbol{P}\mathrm{diag}\left(\left[1 - \left(1 - \eta\frac{s_1^2}{N}\right)^i\right]s_1^{-1}, \ldots,\right.$$
$$\left.\left[1 - \left(1 - \eta\frac{s_r^2}{N}\right)^i\right]s_r^{-1}, 0, \ldots, 0\right)\boldsymbol{Q}^{\mathrm{T}}\boldsymbol{d},$$
$$\tag{41.16}$$

where (a) follows from $\varsigma_j = s_j^2/N$. Then the reg-function for KLMS can be obtained as

$$H_{\mathrm{KLMS}}(x) = \left[1 - \left(1 - \eta\frac{x^2}{N}\right)^N\right]. \tag{41.17}$$

One can easily verify that $H_{\mathrm{KLMS}}(s_r)s_r^{-1} \to 0$ if s_r is very small. Thus KLMS has the ability to automatically attenuate the small singular value components, i.e., has the self-regularization property. Figure 41.4 illustrates the reg-functions for three regularization approaches: KLMS, Tikhonov, and truncated SVD.

41.2.3 Sparsification

The main challenge of KAF algorithms is that their structure grows linearly with each sample, which results in high computational costs and memory requirements, especially for continuous adaptation scenarios. In order to curb the growth of the networks and to obtain a compact representation, a variety of sparsification techniques have been proposed, where the redundant data are removed and only the important input data are accepted as the new centers. As KAF algorithms are online learning algorithms, we focus the discussion here on online (or sequential) sparsification.

In general, sparsification uses a certain criterion function with thresholds to decide which data can be accepted as a new center. Typical sparsification criteria include the novelty criterion [41.30], prediction variance criterion [41.31], coherence criterion [41.32], approximate linear dependency (ALD) criterion [41.25], and the surprise criterion [41.33]. The prediction variance criterion is similar to ALD. The coherence criterion can be viewed as an approximation to ALD. In the following, we present a brief summary of three sparsification criteria: the novelty criterion, the ALD criterion and the surprise criterion.

Suppose the current learning system is

$$f_i(\cdot) = \sum_{j=1}^{m_i} a_j(i)\kappa(c_j, u), \tag{41.18}$$

where c_j is the j-th center, and $a_j(i)$ the j-th coefficient. Denote by $C(i) = \{c_j\}_{j=1}^{m_i}$ the center set or dictionary. When a new example $\{u(i+1), d(i+1)\}$ is presented, the learning system needs to judge whether $u(i+1)$ can be accepted as a new center.

Novelty Criterion

The novelty criterion (NC) first computes the distance of $u(i+1)$ to the present dictionary

$$\mathrm{dis}_1 = \min_{c_j \in C(i)} \|u(i+1) - c_j\|. \tag{41.19}$$

If dis_1 is smaller than some preset threshold $\delta_1(\delta_1 > 0)$, $u(i+1)$ will not be added to the dictionary. Otherwise, the algorithm computes the prediction error $e(i+1) = d(i+1) - f_i(u(i+1))$. Only if the magnitude of the prediction error is larger than another preset threshold $\delta_2(\delta_2 > 0)$, will $u(i+1)$ be accepted as a new center.

It is easy to prove that if the input domain $\mathbf{U}$ is a compact set, with NC sparsification the cardinality of the dictionary will be always finite and upper-bounded.

ALD Criterion

ALD uses the distance of the new input to the linear span of the present dictionary in feature space, which is

$$\mathrm{dis}_2 = \min_{\forall b} \left\| \varphi(u(i+1)) - \sum_{c_j \in C(i)} b_j\varphi(c_j) \right\|. \tag{41.20}$$

By straightforward calculation, one obtains

$$\text{dis}_2^2 = \kappa(u(i+1), u(i+1))$$
$$- h(i+1)^T G^{-1}(i) h(i+1) \,, \tag{41.21}$$

where

$$h(i+1) = [\kappa(u(i+1), c_1), \ldots,$$
$$\kappa(u(i+1), c_{m_i})]^T \,, \tag{41.22}$$

$$G(i) = \begin{bmatrix} \kappa(c_1, c_1) & \cdots & \kappa(c_{m_i}, c_1) \\ \vdots & \ddots & \vdots \\ \kappa(c_1, c_{m_i}) & \cdots & \kappa(c_{m_i}, c_{m_i}) \end{bmatrix} \,. \tag{41.23}$$

If dis_2 is smaller than some preset threshold δ_3, $u(i+1)$ will be discarded.

If regularization is added, (41.21) becomes

$$r(i+1) = \lambda + \kappa(u(i+1), u(i+1))$$
$$- h(i+1)^T (G(i) + \lambda I)^{-1} h(i+1) \,. \tag{41.24}$$

This quantity has already been defined in KRLS [41.25]. It is also the prediction variance in GPR [41.31].

ALD is computationally expensive, especially when the dictionary size m_i is very large. In order to simplify the computation, one can use the *nearest* center in the dictionary to approximate the overall distance, that is

$$\text{dis}_3 = \min_{\forall b, \forall c_j \in C(i)} \| \varphi(u(i+1)) - b\varphi(c_j) \| \,. \tag{41.25}$$

In this case we have

$$\text{dis}_3^2 = \min_{\forall c_j \in C(i)} \left[\kappa(u(i+1), u(i+1)) \right.$$
$$\left. - \frac{\kappa^2(u(i+1), c_j)}{\kappa(c_j, c_j)} \right] \,. \tag{41.26}$$

If $\kappa(c_j, c_j) = 1$ for all c_j, this distance measure is equivalent to the coherence measure. When the kernel κ is Gaussian, this distance measure is equivalent to dis_1 in NC.

Surprise Criterion

Surprise [41.33] is a subjective information measure of an exemplar $\{u, d\}$ with respect to a learning system $\mathcal{L}$. It is defined as the negative log-likelihood of the exemplar given the learning system's hypothesis on the data distribution

$$S_L(u, d) = -\ln p(u, d \mid L) \,, \tag{41.27}$$

where $p(u, d \mid L)$ is the subjective probability of (u, d) hypothesized by L. The surprise $S_L(u, d)$ measures how *surprising* the exemplar is to the learning system. For online application, the surprise of the new example $\{u(i+1), d(i+1)\}$ can be expressed as

$$S_{L(i)}(u(i+1), d(i+1))$$
$$= -\ln p(u(i+1), d(i+1) \mid L(i)) \,, \tag{41.28}$$

where $L(i)$ denotes the present learning system. To simplify notation one can write $S_{L(i)}(u(i+1), d(i+1))$ as $S(i+1)$.

According to the definition, if surprise $S(i+1)$ is large, the new datum $\{u(i+1), d(i+1)\}$ contains something new for the system to learn or it is suspicious. Otherwise, if surprise $S(i+1)$ is very small, the new datum is well expected by the learning system $L(i)$ and thus contains little information to be learned. Usually we can classify the new example into three categories

$$\begin{cases} \text{abnormal:} & S(i+1) > T_1 \,, \\ \text{learnable:} & T_1 \geq S(i+1) \geq T_2 \,, \\ \text{redundant:} & S(i+1) < T_2 \,, \end{cases} \tag{41.29}$$

where T_1 and T_2 are threshold parameters. The choice of the thresholds and learning strategies defines the characteristics of the learning system.

To evaluate the surprise, we have to estimate the posterior distribution $p(u(i+1), d(i+1) \mid L(i))$. The Gaussian process (GP) theory can be used to derive an analytical solution. In GP regression, the prior distribution of system outputs is assumed to be jointly Gaussian, i.e.,

$$[y(u(1)), \ldots, y(u(i))]^T \approx N(0, \sigma_v^2 I + G(i)) \,, \tag{41.30}$$

where

$$G(i) = \begin{bmatrix} \kappa(u(1), u(1)) & \cdots & \kappa(u(i), u(1)) \\ \vdots & \ddots & \vdots \\ \kappa(u(1), u(i)) & \cdots & \kappa(u(i), u(i)), \end{bmatrix} \tag{41.31}$$

and σ_v^2 is the noise variance. Under this prior assumption, the posterior distribution of the output given the input $u(i+1)$ and all past observation $\mathcal{D}(i)$ can be derived as

$$p(y(u(i+1)) \mid u(i+1), D(i))$$
$$\approx N(\bar{d}(i+1), \sigma^2(i+1)) \,, \tag{41.32}$$

which is again joint Gaussian, with

$$\bar{d}(i+1) = h(i+1)^{\mathrm{T}}[\sigma_v^2 \mathbf{I} + G(i)]^{-1} d(i) \quad (41.33)$$

$$\sigma^2(i+1) = \sigma_v^2 + \kappa(u(i+1), u(i+1))$$
$$\qquad - h(i+1)^{\mathrm{T}}[\sigma_v^2 \mathbf{I} + G(i)]^{-1} h(i+1), \quad (41.34)$$

where $h(i+1) = [\kappa(u(i+1), u(1)), \dots, \kappa(u(i+1), u(i))]^{\mathrm{T}}$, and $d(i) = [d(1), \dots, d(i)]^{\mathrm{T}}$.

Now assume $L(i) = D(i)$, i.e., the learning system memorizes all the past input–output pairs. By (41.32), the posterior joint probability density $p(u(i+1), d(i+1) \mid L(i))$ becomes

$$p(u(i+1), d(i+1) \mid L(i))$$
$$= p(d(i+1) \mid u(i+1), L(i)) \times p(u(i+1) \mid L(i))$$
$$= \frac{1}{\sqrt{2\pi}\sigma(i+1)} \exp\left(-\frac{(d(i+1) - \bar{d}(i+1))^2}{2\sigma^2(i+1)}\right)$$
$$\times p(u(i+1) \mid L(i)). \quad (41.35)$$

Hence, the surprise measure is

$$S(i+1) = -\ln p(u(i+1), d(i+1) \mid L(i))$$
$$= \ln \sigma(i+1) + \frac{(d(i+1) - \bar{d}(i+1))^2}{2\sigma^2(i+1)}$$
$$\qquad - \ln p(u(i+1) \mid L(i)) + \ln \sqrt{2\pi}. \quad (41.36)$$

The distribution $p(u(i+1) \mid L(i))$ is problem dependent. Usually we can assume $p(u(i+1) \mid L(i)) = p(u(i+1))$, i.e., the distribution of $u(i+1)$ is independent of the previous observations. To further simplify the computation, one can assume that the distribution $p(u(i+1))$ is uniform. By discarding the constant terms, the surprise measure can be simplified as

$$S(i+1) = \ln \sigma(i+1) + \frac{(d(i+1) - \bar{d}(i+1))^2}{2\sigma^2(i+1)}. \quad (41.37)$$

The complexity of computing the exact surprise measure for KLMS is $O(m_i^2)$ at iteration i, which may offset the advantage of KLMS over other KAF algorithms in terms of simplicity. The main challenge is how to simplify the computation of the prediction variance. To address this problem, we can use the equivalent relationship between prediction variance and the distance measure dis_2. By using dis_3 to approximate dis_2 and adding a regularization term λ, one obtains an approxi-

mation of the prediction variance

$$\sigma^2(i+1) \approx r(i+1) = \lambda + \kappa(u(i+1), u(i+1))$$
$$\qquad - \max_{\forall c_j \in C(i)} \frac{\kappa^2(u(i+1), c_j)}{\kappa(c_j, c_j)}. \quad (41.38)$$

The surprise measure can then be simply calculated as

$$S(i+1) = \frac{1}{2}\ln r(i+1) + \frac{e^2(i+1)}{2r(i+1)}. \quad (41.39)$$

In general, a new center will be added only if the example is learnable, i.e., $T_1 \geq S(i+1) \geq T_2$.

41.2.4 Quantized Kernel Least Mean Squares

Sparsification methods can effectively reduce the network size of KLMS. There is, however, a common drawback to these methods: the *redundant data* are purely discarded. Actually, the redundant data are still useful and can be, for example, utilized to update the coefficients (or weights) of the present network, although they are not accepted as new centers (update the topology). By doing so, one can expect to achieve better accuracy and a more compact network (with update of coefficients, fewer centers are needed to approximate the desired nonlinearity). This idea was already used by Platt in his resource-allocating networks [41.30]. There are two drawbacks to Platt's method:

1. For each redundant input (pattern), all the coefficients of the present network must be updated, which is computationally expensive.
2. As each pattern lies in a local region of the input space, learning performance may be negatively affected by the global update of the coefficients.

Thus a simple yet efficient way is to carry out a *local update* for each redundant input, i.e., only update the coefficients within the *responsive domain*. The *quantization* approach can be used to implement such a learning strategy: the input space is quantized, if the current quantized input has already been assigned a center, no new center will be added (from the viewpoint of sparsification, this input is redundant), but the coefficient of that center will be updated through merging a new coefficient (note only one coefficient is updated). Incorporating quantization into KLMS yields the quantized KLMS (QKLMS) algorithm [41.34].

The QKLMS algorithm can be obtained by just quantizing the feature vector $\varphi(i)$ in the weight update equation $\boldsymbol{\Omega}(i) = \boldsymbol{\Omega}(i-1) + \eta e(i)\boldsymbol{\varphi}(i)$ in (41.5), i.e.,

$$
\begin{cases}
\boldsymbol{\Omega}(0) = \mathbf{0}, \\
e(i) = d(i) - \boldsymbol{\Omega}(i-1)^{\mathrm{T}}\boldsymbol{\varphi}(i), \\
\boldsymbol{\Omega}(i) = \boldsymbol{\Omega}(i-1) + \eta e(i) Q[\boldsymbol{\varphi}(i)],
\end{cases}
\tag{41.40}
$$

where $Q[\cdot]$ denotes a quantization operator in $\mathbf{F}$. Since the dimensionality of the feature space is very high (possibly infinite), the quantization is usually performed in the original input space U. Hence, the learning rule for QKLMS can be

$$
\begin{cases}
f_0 = 0, \\
e(i) = d(i) - f_{i-1}(\boldsymbol{u}(i)), \\
f_i = f_{i-1} + \eta e(i)\kappa(Q[\boldsymbol{u}(i)], \cdot),
\end{cases}
\tag{41.41}
$$

where $Q[\cdot]$ is a quantization operator in U. For simplicity of notation, one may denote $\boldsymbol{\varphi}_q(i) = Q[\boldsymbol{\varphi}(i)]$ and $\boldsymbol{u}_q(i) = Q[\boldsymbol{u}(i)]$.

The network size of QKLMS can never be larger than the size of the quantization codebook (dictionary). A key problem in QKLMS is the design of the vector quantizer, including:

1. How to assign the code-vectors to the data.
2. How to find the *closest* code-vector representation.

In the literature, there is a variety of vector quantization (VQ) methods [41.35–38]. Most existing VQ methods, however, are not suitable for online implementation because the codebook must be supplied in advance (which is usually trained on an offline data set), and the computational burden is also heavy. In [41.34], a simple online VQ method is proposed, in which the codebook is trained directly from online samples and grows adaptively (see Algorithm 41.2).

Algorithm 41.2 Online VQ in U

Initialization:

Choose quantization size ε_U and initialize codebook $C(1) = \{\boldsymbol{u}(1)\}$.

Computation:

while $\{\boldsymbol{u}(i)\}(i > 1)$ available **do**
  (1) Compute the distance between $\boldsymbol{u}(i)$
  and $C(i-1)$:
    $\mathrm{dis}(\boldsymbol{u}(i), C(i-1))$
    $= \min\limits_{1 \le j \le \mathrm{size}(C(i-1))} \|\boldsymbol{u}(i) - C_j(i-1)\|$
  and determine the index of the closest center:
    $j^* = \arg\min\limits_{1 \le j \le \mathrm{size}(C(i-1))} \|\boldsymbol{u}(i) - C_j(i-1)\|$

(2) If $\mathrm{dis}(\boldsymbol{u}(i), C(i-1)) \le \varepsilon_U$, keep the codebook unchanged: $C(i) = C(i-1)$ and quantize $\boldsymbol{u}(i)$ to the closest code-vector: $\boldsymbol{u}_q(i) = C_{j^*}(i-1)$.
(3) Otherwise, update the codebook:
$C(i) = \{C(i-1), \boldsymbol{u}(i)\}$ and quantize $\boldsymbol{u}(i)$ as itself:
$\boldsymbol{u}_q(i) = \boldsymbol{u}(i)$
end while
(where $C_j(i-1)$ denotes the j-th element of the codebook $C(i-1)$)

If the kernel is Gaussian, we have

$$
\begin{aligned}
\|\boldsymbol{\varphi}(i) - \boldsymbol{\varphi}(j)\| &= \sqrt{(\boldsymbol{\varphi}(i) - \boldsymbol{\varphi}(j))^{\mathrm{T}}(\boldsymbol{\varphi}(i) - \boldsymbol{\varphi}(j))} \\
&= \sqrt{2 - 2\kappa(\boldsymbol{u}(i), \boldsymbol{u}(j))} \\
&= \sqrt{2 - 2\exp(-\|\boldsymbol{u}(i) - \boldsymbol{u}(j)\|^2/2\sigma^2)},
\end{aligned}
\tag{41.42}
$$

which implies that the distance in feature space F will be monotonically increasing with the distance in original input space U. In this case, VQ in U is actually equivalent to VQ in F, and one can identify $\kappa(\boldsymbol{u}_q(i), \cdot)$ with $\boldsymbol{\varphi}_q(i)$, where $\boldsymbol{\varphi}_q(i)$ is obtained by performing similar online VQ as in F but with quantization size

$$
\varepsilon_F = \sqrt{2 - 2\exp\left(\frac{-\varepsilon_U^2}{2\sigma^2}\right)}.
\tag{41.43}
$$

A summary of the QKLMS is given in Algorithm 41.3.

Algorithm 41.3 QKLMS

Initialization:

Choose step size η, kernel size σ, quantization size ε_U, and initialize the codebook and coefficient vector: $C(1) = \{\boldsymbol{u}(1)\}$, $\boldsymbol{a}(1) = [\eta d(1)]$.

Computation:

while $\{\boldsymbol{u}(i), d(i)\}(i > 1)$ available **do**
  (1) Compute the prediction error:
    $e(i)$
    $= d(i) - \sum\limits_{j=1}^{\mathrm{size}(C(i-1))} a_j(i-1)\kappa(C_j(i-1), \boldsymbol{u}(i))$
  (2) Compute the distance between $\boldsymbol{u}(i)$ and
  $C(i-1)$:
    $\mathrm{dis}(\boldsymbol{u}(i), C(i-1))$
    $= \min\limits_{1 \le j \le \mathrm{size}(C(i-1))} \|\boldsymbol{u}(i) - C_j(i-1)\|$
  and determine the index of the closest center:
    $j^* = \arg\min\limits_{1 \le j \le \mathrm{size}(C(i-1))} \|\boldsymbol{u}(i) - C_j(i-1)\|$
  (3)
  if $\mathrm{dis}(\boldsymbol{u}(i), C(i-1)) \le \varepsilon_U$, **then**
    $C(i) = C(i-1)$, $a_{j^*}(i) = a_{j^*}(i-1) + \eta e(i)$
  else $C(i) = \{C(i-1), \boldsymbol{u}(i)\}$, $\boldsymbol{a}(i) = [\boldsymbol{a}(i-1), \eta e(i)]$

endif
end while

QKLMS is somewhat similar to sparsified KLMS with a novelty criterion. The key difference between the two algorithms is that the QKLMS utilizes the *redundant* data to *locally update* the coefficient of the closest center. Intuitively, the coefficient update can enhance the *utilization efficiency* of that center and hence may yield better accuracy and a more compact network. When $\varepsilon_U = 0$, QKLMS reduces to standard KLMS.

In the QKLMS algorithm, the *redundant* data can also be used to *globally update* all the coefficients of the present network, using the LMS algorithm, just like Platt's approach in the resource-allocating network [41.30]. This variant of the algorithm, referred to as *QKLMS with global update* (QKLMS-GU), however, is more computationally intensive and also, may perform worse than *local* QKLMS. This has been confirmed by simulation examples [41.34]. One possible reason for this is that the local update (only the update the closest center) of QKLMS avoids the negative influence caused by far-away centers which are still under learning.

41.2.5 Mean Square Convergence

It is very important to investigate the convergence behavior of an adaptive filter. For classical linear adaptive filters, much research has been done in this area and significant results have been achieved. For most nonlinear adaptive filters, the convergence analysis is, in general, rather complicated and little studied. For kernel adaptive filters, the convergence analysis is, however, relatively tractable. The basic reason for this is that these filters are essentially linear in feature space, and hence their convergence analysis is much similar to that of classical linear adaptive filters.

In linear adaptive filtering theory, the *energy conservation relation* (ECR) is shown to be a powerful tool for mean square convergence analysis [41.39–42]. This fundamental relation can be easily extended into the feature space. In the following, we present some theoretical results on the mean square convergence of the QKLMS algorithm (which includes KLMS as a special case). The analysis is based on ECR in feature space. To simplify the derivation, the kernel κ is assumed to be Gaussian.

Energy Conservation Relation in Feature Space

Suppose the desired output data $\{d(i)\}$ are generated by the linear regression model (41.8) in feature space F.

Then the prediction error $e(i)$ will be

$$e(i) = d(i) - \boldsymbol{\Omega}(i-1)^{\mathrm{T}}\boldsymbol{\varphi}(i)$$
$$= (\boldsymbol{\Omega}^{*\mathrm{T}}\boldsymbol{\varphi}(i) + v(i)) - \boldsymbol{\Omega}(i-1)^{\mathrm{T}}\boldsymbol{\varphi}(i)$$
$$= \tilde{\boldsymbol{\Omega}}(i-1)^{\mathrm{T}}\boldsymbol{\varphi}(i) + v(i)$$
$$= e_a(i) + v(i), \tag{41.44}$$

where $\tilde{\boldsymbol{\Omega}}(i-1) = \boldsymbol{\Omega}^* - \boldsymbol{\Omega}(i-1)$ is the weight error vector in $\mathbb{F}$, $e_a(i) = \tilde{\boldsymbol{\Omega}}(i-1)^{\mathrm{T}}\boldsymbol{\varphi}(i)$ is called the *a priori* error at iteration i. Subtracting $\boldsymbol{\Omega}^*$ from both sides of the QKLMS weight update equation yields

$$\tilde{\boldsymbol{\Omega}}(i) = \tilde{\boldsymbol{\Omega}}(i-1) - \eta e(i)\boldsymbol{\varphi}_q(i). \tag{41.45}$$

Defining the *a posteriori* error $e_p(i) = \tilde{\boldsymbol{\Omega}}(i)^{\mathrm{T}}\boldsymbol{\varphi}(i)$, we have

$$e_p(i) = e_a(i) + (\tilde{\boldsymbol{\Omega}}(i)^{\mathrm{T}} - \tilde{\boldsymbol{\Omega}}(i-1)^{\mathrm{T}})\boldsymbol{\varphi}(i). \tag{41.46}$$

By incorporating (41.46) with (41.45)

$$e_p(i) = e_a(i) - \eta e(i)\boldsymbol{\varphi}_q(i)^{\mathrm{T}}\boldsymbol{\varphi}(i)$$
$$= e_a(i) - \eta e(i)\kappa(\boldsymbol{u}_q(i), \boldsymbol{u}(i)), \tag{41.47}$$

where the latter equality follows from $\boldsymbol{\varphi}_q(i) = \kappa(\boldsymbol{u}_q(i), \cdot)$ and the *kernel trick* (41.3). Combining (41.45) and (41.47) so as to eliminate the prediction error $e(i)$ yields

$$\tilde{\boldsymbol{\Omega}}(i) = \tilde{\boldsymbol{\Omega}}(i-1) + (e_p(i) - e_a(i))\frac{\boldsymbol{\varphi}_q(i)}{\kappa(\boldsymbol{u}_q(i), \boldsymbol{u}(i))}. \tag{41.48}$$

Squaring both sides of (41.48) and after some simple manipulations one obtains

$$\|\tilde{\boldsymbol{\Omega}}(i)\|^2 + \frac{e_a^2(i)}{\kappa(\boldsymbol{u}_q(i), \boldsymbol{u}(i))^2}$$
$$= \|\tilde{\boldsymbol{\Omega}}(i-1)\|^2 + \frac{e_p^2(i)}{\kappa(\boldsymbol{u}_q(i), \boldsymbol{u}(i))^2} + \beta_q, \tag{41.49}$$

where $\|\tilde{\boldsymbol{\Omega}}(i)\|^2 = \tilde{\boldsymbol{\Omega}}(i)^{\mathrm{T}}\tilde{\boldsymbol{\Omega}}(i)$, and β_q is

$$\beta_q = \frac{2(e_p(i) - e_a(i))}{\kappa(\boldsymbol{u}_q(i), \boldsymbol{u}(i))^2}$$
$$\times \left\{ \tilde{\boldsymbol{\Omega}}(i-1)^{\mathrm{T}}\boldsymbol{\varphi}_q(i)\kappa(\boldsymbol{u}_q(i), \boldsymbol{u}(i)) - e_a(i) \right\}. \tag{41.50}$$

Equation (41.49), referred to as the *energy conservation relation* (ECR) for QKLMS, shows how the error powers evolve during learning with QKLMS. If omitting the term β_q, (41.49) is very similar to the classical energy conservation relation for linear adaptive filters [41.39–42]. When the quantization size $\varepsilon_{\mathbb{U}}$ approaches zero, we have $\kappa(\boldsymbol{u}_q(i), \boldsymbol{u}(i)) \to 1$ and $\beta_q \to 0$, and in this case (41.49) becomes

$$\|\tilde{\boldsymbol{\Omega}}(i)\|^2 + e_a^2(i) = \|\tilde{\boldsymbol{\Omega}}(i-1)\|^2 + e_p^2(i), \quad (41.51)$$

which is the ECR for standard KLMS.

Substituting $e_p(i) = e_a(i) - \eta e(i) \kappa(\boldsymbol{u}_q(i), \boldsymbol{u}(i))$ into ECR (41.49), one can derive

$$\begin{aligned}
\|\tilde{\boldsymbol{\Omega}}(i)\|^2 &= \|\tilde{\boldsymbol{\Omega}}(i-1)\|^2 + \eta^2 e^2(i) \\
&\quad - 2\eta e(i) \tilde{\boldsymbol{\Omega}}(i-1)^{\mathrm{T}} \boldsymbol{\varphi}_q(i).
\end{aligned} \quad (41.52)$$

To study the mean square behavior of QKLMS, we take expectations of both sides of (41.52) and write

$$\begin{aligned}
E[\|\tilde{\boldsymbol{\Omega}}(i)\|^2] &= E[\|\tilde{\boldsymbol{\Omega}}(i-1)\|^2] + \eta^2 E[e^2(i)] \\
&\quad - 2\eta E[e(i) \tilde{\boldsymbol{\Omega}}(i-1)^{\mathrm{T}} \boldsymbol{\varphi}_q(i)], \quad (41.53)
\end{aligned}$$

where $E[\|\tilde{\boldsymbol{\Omega}}(i)\|^2]$ is called the *weight error power* (WEP) in F.

One can use (41.53) to analyze the mean square convergence of the QKLMS. Here, we give an assumption.

Assumption 41.1

The noise $v(i)$ is zero-mean, independent, identically distributed (i. i. d.), and independent of the *a priori* estimation error $e_a(i)$.

The above assumption is commonly used in convergence analysis for most adaptive filtering algorithms [41.2, 3]. Incorporating assumption A1 into (41.53) yields

$$\begin{aligned}
E[\|\tilde{\boldsymbol{\Omega}}(i)\|^2] &= E[\|\tilde{\boldsymbol{\Omega}}(i-1)\|^2] + \eta^2 \left(E[e_a^2(i)] + \sigma_v^2 \right) \\
&\quad - 2\eta E[e_a(i) \tilde{\boldsymbol{\Omega}}(i-1)^{\mathrm{T}} \boldsymbol{\varphi}_q(i)], \\
& \quad\quad\quad\quad\quad\quad\quad\quad\quad\quad\quad (41.54)
\end{aligned}$$

where σ_v^2 denotes the noise variance.

Sufficient Condition for Mean Square Convergence

From (41.54), it is easy to derive

$$\begin{aligned}
&E[\|\tilde{\boldsymbol{\Omega}}(i)\|^2] \le E[\|\tilde{\boldsymbol{\Omega}}(i-1)\|^2] \\
&\Leftrightarrow \eta^2 \left(E[e_a^2(i)] + \sigma_v^2 \right) \\
&\quad - 2\eta E\left[e_a(i) \tilde{\boldsymbol{\Omega}}(i-1)^{\mathrm{T}} \boldsymbol{\varphi}_q(i) \right] \le 0 \\
&\Leftrightarrow \eta \le \frac{2E\left[e_a(i) \tilde{\boldsymbol{\Omega}}(i-1)^{\mathrm{T}} \boldsymbol{\varphi}_q(i) \right]}{E[e_a^2(i)] + \sigma_v^2}. \\
&\quad\quad\quad\quad\quad\quad\quad\quad\quad\quad\quad\quad (41.55)
\end{aligned}$$

Thus, to ensure the monotonic decrease of the WEP in F, one should choose the step size η such that $\forall i$,

$$0 < \eta \le \frac{2E\left[e_a(i) \tilde{\boldsymbol{\Omega}}(i-1)^{\mathrm{T}} \boldsymbol{\varphi}_q(i) \right]}{E[e_a^2(i)] + \sigma_v^2}. \quad (41.56)$$

The existence of such a step size requires $\forall i$,

$$E\left[e_a(i) \tilde{\boldsymbol{\Omega}}(i-1)^{\mathrm{T}} \boldsymbol{\varphi}_q(i) \right] > 0. \quad (41.57)$$

Then a sufficient condition for mean square convergence (in the sense of the monotonic decrease of WEP) will be

$$\forall i, \begin{cases} E\left[e_a(i) \tilde{\boldsymbol{\Omega}}(i-1)^{\mathrm{T}} \boldsymbol{\varphi}_q(i) \right] > 0, & (C1) \\[2mm] 0 < \eta \le \dfrac{2E\left[e_a(i) \tilde{\boldsymbol{\Omega}}(i-1)^{\mathrm{T}} \boldsymbol{\varphi}_q(i) \right]}{E\left[e_a^2(i) \right] + \sigma_v^2}. & (C2) \end{cases}$$
$$(41.58)$$

Further, if $\boldsymbol{\varphi}(i)$ and $\tilde{\boldsymbol{\Omega}}(i-1)$ are statistically independent, we have

$$\begin{aligned}
&E\left[e_a(i) \tilde{\boldsymbol{\Omega}}(i-1)^{\mathrm{T}} \boldsymbol{\varphi}_q(i) \right] \\
&= E\left[\tilde{\boldsymbol{\Omega}}(i-1)^{\mathrm{T}} \boldsymbol{\varphi}(i) \boldsymbol{\varphi}_q(i)^{\mathrm{T}} \tilde{\boldsymbol{\Omega}}(i-1) \right] \\
&= E\left[\tilde{\boldsymbol{\Omega}}(i-1)^{\mathrm{T}} E(\boldsymbol{\varphi}(i) \boldsymbol{\varphi}_q(i)^{\mathrm{T}}) \tilde{\boldsymbol{\Omega}}(i-1) \right].
\end{aligned}$$
$$(41.59)$$

In this case, the condition $(C1)$ in (41.58) can be replaced by a more stringent condition

$$E\left(\boldsymbol{\varphi}(i) \boldsymbol{\varphi}_q(i)^{\mathrm{T}} \right) > 0, \quad (41.60)$$

which, in form, is similar to the *persistence of excitation* (PE) condition for the quantized regressor (QReg) algorithm given in [41.43].

Steady–State Mean Square Performance

Take the limit of (41.54) as $i \to \infty$ and write

$$\begin{aligned}
&\lim_{i \to \infty} E[\|\tilde{\boldsymbol{\Omega}}(i)\|^2] \\
&= \lim_{i \to \infty} E[\|\tilde{\boldsymbol{\Omega}}(i-1)\|^2] \\
&\quad + \eta^2 \left(\lim_{i \to \infty} E\left[e_a^2(i) \right] + \sigma_v^2 \right) \\
&\quad - 2\eta \lim_{i \to \infty} E\left[e_a(i) \tilde{\boldsymbol{\Omega}}(i-1)^{\mathrm{T}} \boldsymbol{\varphi}_q(i) \right]. \quad (41.61)
\end{aligned}$$

Suppose the WEP reaches the steady state, that is

$$\lim_{i \to \infty} E\left[\|\tilde{\boldsymbol{\Omega}}(i)\|^2 \right] = \lim_{i \to \infty} E\left[\|\tilde{\boldsymbol{\Omega}}(i-1)\|^2 \right].$$
$$(41.62)$$

Then one has

$$\eta^2 \left(\lim_{i \to \infty} E\left[e_a^2(i) \right] + \sigma_v^2 \right)$$
$$- 2\eta \lim_{i \to \infty} E\left[e_a(i)\tilde{\boldsymbol{\Omega}}(i-1)^{\mathrm{T}} \boldsymbol{\varphi}_q(i) \right] = 0 . \quad (41.63)$$

It follows that

$$\eta \left(\lim_{i \to \infty} E\left[e_a^2(i) \right] + \sigma_v^2 \right)$$
$$= 2 \lim_{i \to \infty} E\left[e_a(i)\tilde{\boldsymbol{\Omega}}(i-1)^{\mathrm{T}} \boldsymbol{\varphi}_q(i) \right]$$
$$= 2 \lim_{i \to \infty} E\left[e_a(i)\tilde{\boldsymbol{\Omega}}(i-1)^{\mathrm{T}} (\boldsymbol{\varphi}(i) + (\boldsymbol{\varphi}_q(i) - \boldsymbol{\varphi}(i))) \right]$$
$$= 2 \lim_{i \to \infty} E\left[e_a^2(i) \right]$$
$$+ 2 \lim_{i \to \infty} E\left[\tilde{\boldsymbol{\Omega}}(i-1)^{\mathrm{T}} \boldsymbol{\varphi}(i)\tilde{\boldsymbol{\Omega}}(i-1)^{\mathrm{T}} \right.$$
$$\left. (\boldsymbol{\varphi}_q(i) - \boldsymbol{\varphi}(i)) \right] . \quad (41.64)$$

Hence, the steady-state *excess mean square error* (EMSE) is

$$\lim_{i \to \infty} E\left[e_a^2(i) \right]$$
$$= \frac{\eta\sigma_v^2}{2-\eta}$$
$$- \frac{2 \lim_{i \to \infty} E\left[\tilde{\boldsymbol{\Omega}}(i-1)^{\mathrm{T}} \boldsymbol{\varphi}(i)\tilde{\boldsymbol{\Omega}}(i-1)^{\mathrm{T}} (\boldsymbol{\varphi}_q(i) - \boldsymbol{\varphi}(i)) \right]}{2-\eta} . \quad (41.65)$$

Note here that in the literature of adaptive filtering, the *a priori* error power is also referred to as the excess mean-square error

$$\left| E\left[\tilde{\boldsymbol{\Omega}}(i-1)^{\mathrm{T}} \boldsymbol{\varphi}(i)\tilde{\boldsymbol{\Omega}}(i-1)^{\mathrm{T}} (\boldsymbol{\varphi}_q(i) - \boldsymbol{\varphi}(i)) \right] \right|$$
$$\leq E\left[\|\tilde{\boldsymbol{\Omega}}(i-1)\|^2 \|\boldsymbol{\varphi}_q(i) - \boldsymbol{\varphi}(i)\| \right]$$
$$= E\left[\|\tilde{\boldsymbol{\Omega}}(i-1)\|^2 \sqrt{2 - 2\exp\left(-\frac{\|\boldsymbol{u}_q(i) - \boldsymbol{u}(i)\|^2}{2\sigma^2} \right)} \right]$$
$$\overset{(a)}{\leq} \sqrt{2 - 2\exp\left(-\frac{\varepsilon_{\mathbf{U}}^2}{2\sigma^2} \right)} E[\|\tilde{\boldsymbol{\Omega}}^*\|^2]$$
$$= \sqrt{2 - 2\exp\left(-\frac{1}{2}\gamma^2 \right)} E[\|\tilde{\boldsymbol{\Omega}}^*\|^2] , \quad (41.66)$$

where (a) follows from $\|\boldsymbol{u}_q(i) - \boldsymbol{u}(i)\| \leq \varepsilon_U$ and the monotonic decrease of WEP, γ is the *quantization factor* defined as $\gamma = \varepsilon_U/\sigma$, then we derive

$$\max\left\{ \frac{\eta\sigma_v^2 - 2\xi_\gamma}{2-\eta}, 0 \right\} \leq \lim_{i \to \infty} E\left[e_a^2(i) \right]$$
$$\leq \frac{\eta\sigma_v^2 + 2\xi_\gamma}{2-\eta} , \quad (41.67)$$

where $\xi_\gamma = \sqrt{2 - 2\exp(-1/2\gamma^2)} E[\|\boldsymbol{\Omega}^*\|_{\mathbb{F}}^2]$.

In (41.67), the lower and the upper bounds on the steady-state EMSE for QKLMS are provided. When the quantization factor $\gamma = 0$ (and hence $\xi_\gamma = 0$), one obtains the steady-state EMSE for KLMS

$$\lim_{i \to \infty} E\left[e_a^2(i) \right] = \frac{\eta\sigma_v^2}{2-\eta} , \quad (41.68)$$

which, interestingly, has no relation to the kernel size. The essential reason for this is perhaps that, with any kernel size, the feature space has the universal approximation property.

41.3 Application Examples

Kernel adaptive filters can be applied in a wide range of applications. In the following, we only give several examples illustrating their use.

41.3.1 Nonlinear Channel Equalization

The first example is nonlinear channel equalization. Suppose a binary signal $\{s_1, s_2, \ldots, s_L\}$ is fed into a nonlinear channel. The signal is further corrupted by additive i.i.d. Gaussian noise at the receiver end of the channel, and the observed signal is $\{r_1, r_2, \ldots, r_L\}$. The aim of channel equalization is to construct an *in-*

verse filter that recovers the original signal with as low an error rate as possible. This can be formulated as a regression problem with input–output training data $\{(r_{t+D}, r_{t+D-1}, \ldots, r_{t+D-l+1}), s_t\}$, where l is the time embedding length, and D is the equalization time lag.

In the experiment, the nonlinear channel is assumed to be

$$\begin{cases} z_t = s_t + 0.5s_{t-1} , \\ r_t = z_t - 0.9z_t^2 + v_t , \end{cases} \quad (41.69)$$

Table 41.2 Performance comparison of LMS, KLMS, and RN in nonlinear channel equalization

	LMS ($\eta = 0.005$)	KLMS ($\eta = 0.1$)	RN ($\lambda = 1.0$)
BER ($\sigma_v^2 = 0.01$)	0.162 ± 0.014	0.020 ± 0.012	0.008 ± 0.001
BER ($\sigma_v^2 = 0.16$)	0.177 ± 0.012	0.058 ± 0.008	0.046 ± 0.003
BER ($\sigma_v^2 = 0.64$)	0.218 ± 0.012	0.130 ± 0.010	0.118 ± 0.004

where $\{v_t\}$ is white Gaussian noise with variance σ_v^2. The average learning curves of the conventional LMS and the KLMS are plotted in Fig. 41.5. As expected, KLMS converges to a much smaller value of MSE due to its nonlinear nature. Testing was performed on a 5000-sample random test sequence. We compare the performance of LMS, KLMS, and the regularization network (RN), which serves as a batch-mode baseline. RN is a classic nonlinear modeling tool using a radial basis function (RBF) network topology specified by the kernel used [41.44]. The Gaussian kernel with kernel size $\sigma = \sqrt{5}$ is used in both KLMS and RN, and the time embedding length $l = 5$, equalization time lag $D = 2$. The results are summarized in Table 41.2, where each entry consists of the average and the standard deviation for 100 Monte Carlo independent tests. The results in Table 41.2 show that RN outperforms KLMS in terms of the bit error rate (BER) but not by much, which is surprising since one is a batch method and the other is online. They both outperform LMS substantially, as can be expected because the channel is nonlinear. The regularization parameter for RN and the learning rate of KLMS were set for best results.

41.3.2 Chaotic Time Series Prediction

Consider the Lorenz attractor, a nonlinear, three-dimensional dynamic system, whose states are governed by the following differential equations

$$\begin{cases} \dfrac{\mathrm{d}x}{\mathrm{d}t} = -\beta x + yz\,, \\[2mm] \dfrac{\mathrm{d}y}{\mathrm{d}t} = \delta(z - y)\,, \\[2mm] \dfrac{\mathrm{d}z}{\mathrm{d}t} = -xy + \rho y - z\,. \end{cases} \qquad (41.70)$$

Assume the parameters are set as $\beta = 8/3$, $\delta = 10$ and $\rho = 28$. The sample data are obtained using first order approximation with step size 0.01. The first state x is picked for the short term prediction task. The signal is preprocessed to be zero-mean and have unit-variance before the modeling. A segment of the processed Lorenz time series is shown in Fig. 41.6.

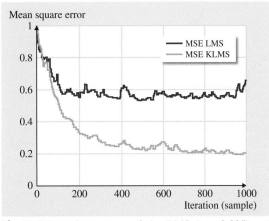

Fig. 41.5 Learning curves of the LMS ($\eta = 0.005$) and KLMS ($\eta = 0.1$) in the nonlinear channel equalization ($\sigma_v^2 = 0.16$)

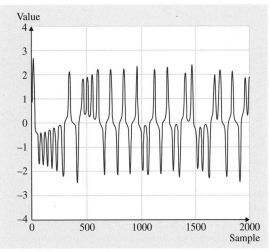

Fig. 41.6 A segment of the processed Lorenz time series

Now the goal is to predict the value of the current sample $x(i)$ using the previous five consecutive samples $\{x(i-5), x(i-4), \ldots, x(i-1)\}$. Define $u(i) = [x(i-5), x(i-4), \ldots, x(i-1)]^{\mathrm{T}}$. Then the problem is to learn the underlying mapping between the input vector $u(i)$ and the desired output $x(i)$. In this ex-

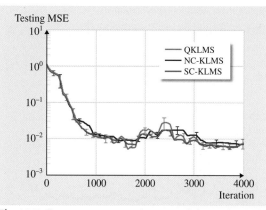

Fig. 41.7 Convergence curves in terms of the testing MSE for QKLMS, NC-KLMS, and SC-KLMS in Lorenz time series prediction. The parameters of the algorithms are chosen such that they produce almost the same testing MSE

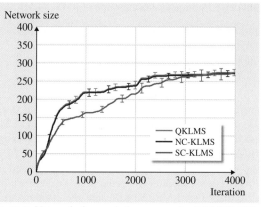

Fig. 41.9 Network size evolution curves for QKLMS, NC-KLMS, and SC-KLMS in Lorenz time series prediction. The parameters of the algorithms are chosen such that they produce almost the same final network size

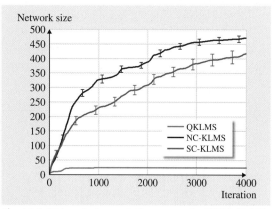

Fig. 41.8 Network size evolution curves for QKLMS, NC-KLMS, and SC-KLMS in Lorenz time series prediction. The parameters of the algorithms are chosen such that they produce almost the same testing MSE (Fig. 41.7)

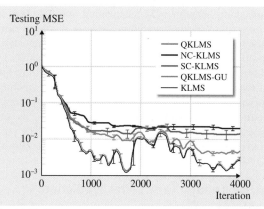

Fig. 41.10 Convergence curves in terms of the testing MSE for QKLMS, NC-KLMS, SC-KLMS, QKLMS-GU, and the standard KLMS in Lorenz time series prediction. The parameters of the algorithms (except KLMS) are chosen such that they produce almost the same final network size (Fig. 41.9)

ample, the performances of QKLMS, NC-KLMS, and SC-KLMS are compared. Here NC-KLMS and SC-KLMS denote, respectively, the novelty and surprise criterion-based sparsified KLMS. In the simulations, the kernel size is set as $\sigma = 0.707$, and unless mentioned otherwise, the step sizes involved are set as $\eta = 0.1$. The performance comparisons are presented in Figs. 41.7–41.10, where all the simulation results are averaged over 100 simulation runs with different segments of the signal. In Figs. 41.7 and 41.8, the parameters of the three algorithms are chosen such that they produce almost the same testing MSE (see Fig. 41.7), while in Figs. 41.9 and 41.10, the parameters are selected such

that the algorithms yield almost the same final network size (see Fig. 41.9). The testing MSE is calculated based on 200 test data (the filter is fixed in the testing phase). Table 41.3 gives the parameter settings. Simulation results clearly indicate that QKLMS exhibits much better performance, i.e., achieves either much smaller network size or much smaller testing MSE, than NC-KLMS and SC-KLMS. For further comparison purposes, in Fig. 41.10 we also plot the convergence curves of QKLMS-GU and standard KLMS. It is evident that QKLMS also performs better than QKLMS-GU and

Table 41.3 Parameter settings for different algorithms

	To produce the same testing MSE	To produce the same final network size
QKLMS	$\gamma = 1.0$	$\gamma = 0.2$
NC-KLMS	$\delta_1 = 0.1, \delta_2 = 0.001$	$\delta_1 = 0.142, \delta_2 = 0.001$
SC-KLMS	$\lambda = 0.005, T_1 = 300, T_2 = -0.5$	$\lambda = 0.01, T_1 = 300, T_2 = 0.0$

achieves almost the same testing MSE as KLMS. In the simulation, QKLMS and QKLMS-GU are set to the same quantization factor γ, and hence their network sizes are always identical. The step size for global update (using LMS) in QKLMS-GU is experimentally set to 0.002.

41.3.3 Regression and Filtering in Spike RKHS

In order to analyze spike trains – temporal patterns of action potentials – either an appropriate model for their stochasticity is required or tools that can quantify arbitrary stochasticity. Assumptions of the first kind have been widely deployed in neuroscience in the form of the Poisson process (PP) or renewal process models [41.45]. These assumptions enable summarizing the complicated spike train observations to a few intuitive statistical quantities and enable building computational tools, often with elegant analytic solutions. If one assumes a Poisson statistic, all the information is available in the rate function, since the rate function completely describes the process. Hence, neurons only need to decode the rate function; this fully supports the notion of rate code. In a similar way, a renewal process with a fixed interval distribution shape is also fully described by its mean rate function. However, recent evidence supports the notion of a temporal code – a neural coding scheme that assumes that there is more information than what is contained in the mean rate function [41.46]. Although it remains to be shown that these non-Poisson statistics are indeed used by the brain for decoding, these evidences suggest that the Poisson/renewal assumption is not always appropriate for analysis.

We submit that the *kernel* method [41.16] is especially useful for data types that do not exist in Euclidean space; therefore the reproducing kernel Hilbert space (RKHS) approach can be applied to PP with advantages and will yield a way to design linear optimal estimation models that will correspond to nonlinear models in the spike train domain [41.21]. Likewise, divergences can also be defined that capture the underlying probability law and measure robustly the statistical similarity of spike trains in statistical inference tests [41.47]. How-

ever, we cannot use directly the Gaussian kernel so common in kernel methods directly in spike trains. The key step is to define *characteristic spike kernels* that are Hermitian strictly positive definite functions in the spike train space yielding unique functions in RKHS for different PP. Since the Hilbert space is equipped with a metric, the distance between two point processes in the Hilbert space is also a point process divergence. As a matter of fact, the unique representation corresponds to the expectation of the spike train realizations in the Hilbert space of the point process and can be easily estimated empirically. Moreover, we can apply linear signal processing algorithms to the projected spike train, do our operations there (for simplicity, these operations have to be inner products), and obtain results back in the spike train space or in R^n. The details of all this methodology are beyond the scope of this chapter (see [41.21] and [41.47]), but we provide a simple example that illustrates the method of obtaining RKHS from spike trains.

Cross Intensity Kernels

If we define the spike train space as the ensemble of all possible spike trains, then a probability measure over the spike train space defines the point process. We will illustrate how one can design a positive definite function (a kernel) to capture the statistical properties of spike trains, which by itself defines a RKHS where signal processing operations can be performed because an inner product is implicitly defined once the kernel is selected. Given two point processes p_i and p_j, define the inner product between their conditional intensity functions as

$$I(p_i, p_j) = \langle \lambda_{p_i}(t|H_t^i), \lambda_{p_j}(t|H_t^j) \rangle_{L_2(T)}$$
$$= E[\int_T \lambda_{p_i}(t|H_t^i) \lambda_{p_j}(t|H_t^j) dt], \quad (41.71)$$

where the expectation is over the history and $\lambda(t \mid H)$ is the conditional intensity function defined as $\lambda(t|H_t) = \lim_{\Delta t \to 0} P\{\text{event in}[t, t+\Delta t]|H_t\}/\Delta t$ with H_t the history of the firings at time t. This inner product defines a family of cross intensity (CI) kernels depending upon the model imposed on the point process history H_t.

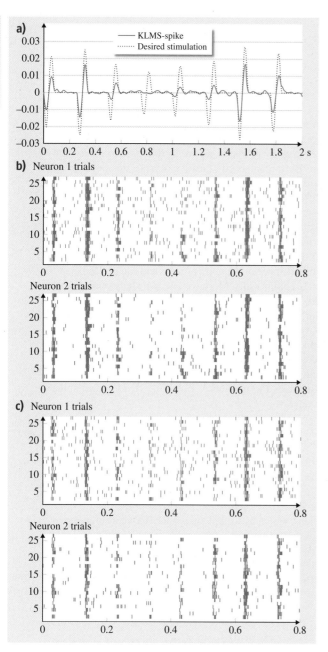

Fig. 41.11a–c The testing results of the adaptive KLMS decoder in RKHS of the spike time vectors. (**a**) The reconstructed and desired stimulations. (**b**) Raster plots of the neural response recorded from two output neurons (produced by the desired stimulation). (**c**) Raster plots of the neural response recorded from two output neurons (produced by the reconstructed stimulation) ◄

which is the simplest of the CI kernels called the memoryless kernel (mCI). One (among many) nonlinear cross intensity kernel (nCI) can be defined as

$$I_\sigma^*(p_i, p_j) = \int_T \kappa_\sigma(\lambda_{p_i}(t), \lambda_{p_j}(t))\,dt\,, \qquad (41.73)$$

where κ is a symmetric positive definite kernel, which is sensitive to nonlinear couplings in the time structure of the intensity functions (such as in renewal processes). On top of providing a bottom-up way of defining the RKHS, the advantage of the CI family is the simplicity of estimating the kernel from data. Let us place a smoothing function (e.g., exponential) $h(t) = e^{-\theta t}u(t)$ at each spike event location. Then the kernel becomes a Laplacian in the form $\hat{\lambda}_{p_i}(t) = \sum_{m=1}^{N_i} h(t - t_m^i)$, which directly yields

$$I(p_i, p_j) = \int_T \lambda_{p_i}(t)\lambda_{p_j}(t)\,dt$$

$$= \sum_{m=1}^{N_i}\sum_{n=1}^{N_j}\int_T h\left(t - t_m^i\right)h\left(t - t_n^j\right)\,dt$$

$$= \sum_{m=1}^{N_i}\sum_{n=1}^{N_j} \kappa\left(t_m^i - t_n^j\right)\,. \qquad (41.74)$$

This kernel has a free parameter θ that controls the inner product. The nCI kernel can also be implemented directly from the data if we use a Gaussian function in (41.74) to yield (σ is the free parameter)

$$I(p_i, p_j) = \int \exp\left(\frac{-\|\lambda_{s_i} - \lambda_{s_j}\|^2}{\sigma^2}\right)\,dt$$

$$= \sum_{n=1}^{N_i}\sum_{m=1}^{N_j} \exp\left(-\frac{\left|t_n^{N_i} - t_m^{N_j}\right|}{\sigma^2}\right)\,. \qquad (41.75)$$

We will illustrate the procedure using a simplifying assumption. If the PP belongs to the Poisson family, the conditional becomes the instantaneous intensity function, and the inner product simplifies to

$$I(p_i, p_j) = \int_T \lambda_{p_i}(t)\lambda_{p_j}(t)\,dt\,, \qquad (41.72)$$

In a sense, the free parameter is a continuous variable that links spike timing and rate methods, i.e., if the transformed events do not overlap a lot in time, one obtains spike time estimates, while if one uses a broader kernel there will be overlap between many transformed spikes, and the results are more in tune with the rate methods. However, here we have a clear mathematical view of the two types of processing, showing that, indeed, they only differ in the definition of the similarity metric in RKHS (i.e., the inner product). It is interesting that RKHS defined by mCI induces an RKHS that can isometrically embed the van Rossum distance.

Adaptive Neural Decoder for Spike Trains

The kernel adaptive filters can be used in neural systems modeling and tracking the neural plasticity. Based on the KLMS algorithm applied directly on the space of spike trains, a nonlinear adaptive neural decoder for somatosensory micro-stimulation was proposed in [41.22]. In order to effectively apply machine learning algorithms to neural decoding, one has to define an appropriate input space for the decoder. A variety of machine learning techniques have been applied on the discrete representations of spike trains (binned data) [41.48–50]. However, binned spike train representations (typically 50–100 ms in duration) usually impair the decoder time resolution of decoding models. For example, in the case of a somatosensory prosthesis, the time resolution of micro-stimulation signals used to create tactile sensation is small (0.2 ms), which requires a small bin size (less than the width of the stimulus pulse). With such a small bin size, the input space becomes sparse and dimensionality of the input space also becomes a problem. Instead of using a binned representation of spike trains, one can transform the vector of spike times into a RKHS where the inner product between two spike time vectors is defined by a nonlinear cross intensity kernel. Once the spike trains are transformed into RKHS functions, we can operate with them using our tool set of signal processing algorithms, because they now exist on a Hilbert space, which is a linear space with a well defined inner product.

In the following, we present some experimental results on KLMS-based neural decoders for somatosensory micro-stimulation. The kernel function is defined as in [41.22]. First, we consider a synthetic experiment, where the plant is simulated by *a neural circuit simulator* (CSIM). The network consists of 10 fully connected neurons. All neurons are modeled as leaky integrate-and-fire (LIF) units [41.51] with parameters: membrane time constant 30 ms, absolute refractory period 2 ms (excitatory neurons), threshold 15 mV (for a resting membrane potential assumed to be 0), reset voltage 14.3 mV, constant nonspecific background current 13.5 nA, input resistance 1 MΩ, and input noise

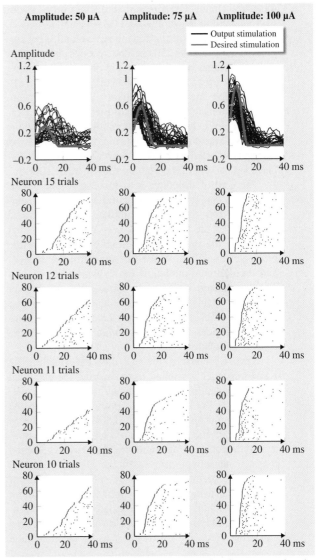

Fig. 41.12 The upper plots show the test results of reconstructing the micro-stimulation in a rat thalamus. The other plots represent input spike train rasters recorded from four channels in somatosensory cortical area (S1), which are sorted by the first spike time in order to show the spike train variability

9 nA. The postsynaptic current is modeled as an exponential decay $\exp(-t/\tau_s)$ with $\tau_s = 3$ ms. All neurons fire at a reasonable firing rate range (1–10 Hz) in the rest state. This neural circuit has one input and two output neurons, whose spike trains are recorded as the observed neural responses. The target spike trains are generated by feeding a periodic current stimulation into

the neural circuit. The stimulation repeats every 2 s with varying amplitude (see Fig. 41.11a). In each 2 s period, there are eight biphasic current pulses with different amplitudes separated by 250 ms. The decoder input is built from spike time vectors, which are obtained by sliding the window every 2 ms (window size: [−20 ms, 80 ms]). Each window includes the previous 20 ms spike train, which contains the information of the previous stimulation and helps to cancel the lasting influence of previous stimulus on the current neural firing pattern. The kernel size σ is selected based on the average pair wise distance $\|\lambda_{s_i} - \lambda_{s_j}\|$ of the training sample in the RKHS.

The goal of decoding is to reconstruct the stimulation that can produce the neural response indistinguishable from the desired stimulation. As illustrated in Fig. 41.11, the KLMS decoder is able to reconstruct the timing, the shape, and the amplitude pattern of the biphasic stimulation. A one-trial output of the decoder and the desired stimulation are shown in Fig. 41.11a. The comparison between the input spike train raster plots (Fig. 41.11b) and the elicited spike train raster plots (Fig. 41.11c) also illustrates that the decoded stimulation produces a spike train similar to that from the desired stimulation.

Next, we consider the real data experiment. The KLMS decoder is applied to reconstruct the electrical micro-stimulation imprinted in the ventral posterolateral nucleus (VPL) of the thalamus by decoding the elicited neural response recorded from somatosensory regions (S1). The neural data used were collected from a single chronically implanted rat with two 16-channel tungsten micro-wire arrays (Tucker Davis). Neuronal activity was recorded from arrays in S1 and in VPL of the thalamus using the Plexon multichannel acquisition processor. Action potentials were detected using a constant threshold and were sorted by a semi-automated clustering procedure (SortClient) using the first three

principal components of the detected waveforms. Prior to the recording session, anesthesia was induced by isofluorane followed by a Nembutal injection and maintained with isofluorane. Bipolar microstimulation (AM Systems Model 2200 Isolator) was applied to two adjacent electrodes in the thalamic array. The stimulations are always at 500 ms apart. The pulse duration is 0.25 ms and the pulse current amplitude is varied among three different levels (50, 75, and 100 μA), with 140 neural firing responses from each level randomly permuted throughout the recording. In practice the spike trains usually have a large variability. Thus it is essential to utilize multichannel spike trains to enhance the decoder robustness. However, if all channels are used as input, the computational cost will increase dramatically and the neuron channels that are not associated with stimuli will introduce noise into the decoder. Based on the dependence analysis results [41.52], only four channels (10, 11, 12, and 15) that are most highly dependent with the desired stimulation are selected as the input channels. The square wave stimulations are also preprocessed by passing through a *sinc* low pass filter with a Hamming window and a cut-off frequency of 500 Hz. Then the regression is performed from the four spiking channels to smoothed stimulation waveform. The input is again the vector of spike times, which is obtained by sliding the window every 0.1 ms (window size: 25 ms). As shown in Fig. 41.12, the multi-input single-output (MISO) KLMS decoder is able to reconstruct the shape and amplitude pattern of the desired stimulation with some deviations. The output deviation is caused by the large variability of the input spike trains with respect to the same desired stimulation. In general, the output deviation will decrease if the stimulation current amplitude is high, since the higher amplitude leads to the higher signal-noise-ratio (SNR) in input space.

41.4 Conclusion

Kernel adaptive filters are a new class of nonlinear adaptive filters which use Mercer kernels to map the input space into a high dimensional feature space and use the linear structure in feature space to implement nonlinear mappings in the original input space. Among various KAF algorithms, KLMS is the most computationally efficient, which is simply a stochastic gradient algorithm to solve least-squares problems in RKHS. The

good approximation ability of KLMS comes from the fact that the dimensionality of the transformed data is very high (possibly infinite), and the space spanned by the feature vectors is so large that the projection error of the desired signal could be rather small. KLMS has a self-regularization property, and its regularization and convergence behavior are controlled mainly by the step size parameter. In order to curb the linearly growing network of KLMS (as well as other KAF algorithms) and to obtain a compact model, some sparsification or quantization techniques can be applied.

Kernel adaptive filters are applicable in many situations where online nonlinear filters are a necessity, such as nonlinear channel equalization, nonlinear time series prediction, nonlinear system identification, and nonlinear adaptive control, etc. Recent studies show that KLMS is in particular useful in neural system modeling (e.g., neural encoding or decoding) and control.

References

41.1   B. Widrow, S.D. Stearns: *Adaptive Signal Processing* (Englewood Cliffs, NJ: Prentice-Hall 1985)

41.2   S. Haykin: *Adaptive Filtering Theory*, 3rd edn. (Prentice Hall, New York 1996)

41.3   A.H. Sayed: *Fundamentals of Adaptive Filtering* (Wiley, Hoboken 2003)

41.4   B. Hassibi, A.H. Sayed, T. Kailath: The H^∞ optimality of the LMS algorithm, IEEE Trans. Signal Process. **44**, 267–280 (1996)

41.5   S.S. Narayan, A.M. Peterson, M.J. Narashima: Transform domain LMS algorithm, IEEE Trans. Acoust. Speech Signal Process. **ASSP-31**, 609–615 (1983)

41.6   F. Beaufays: Transform-domain adaptive filters: An analytical approach, IEEE Trans. Signal Process. **43**, 422–431 (1995)

41.7   S. Haykin, A.H. Sayed, J.R. Zeidler, P. Yee, P.C. Wei: Adaptive tracking of linear time variant systems by extended RLS algorithm, IEEE Trans. Signal Process. **45**, 1118–1128 (1997)

41.8   A.H. Sayed, T. Kailath: A state-space approach to adaptive RLS filtering, IEEE Signal Process. Mag. **11**, 18–60 (1994)

41.9   B. Anderson, J. Moor: *Optimal Filtering* (Prentice-Hall, New York 1979)

41.10   S. Billings, S. Fakhouri: Identification of systems containing linear dynamics and static nonlinear elements, Automatica **18**, 15–26 (1982)

41.11   D. Gabor: Holographic model of temporal recall, Nature **217**, 584–585 (1968)

41.12   J.F. Barrett: The use of functionals in the analysis of non-linear physical systems, Int. J. Electron. **15**, 567–615 (1963)

41.13   S. Haykin: *Neural Networks: A Comprehensive Foundation*, 2nd edn. (Prentice-Hall, Upper Saddle River 1998)

41.14   J.C. Príncipe, B. de Vries, J.M. Kuo, P.G. de Oliveira: Modeling applications with the focused gamma net, Adv. Neural Inform. Process. Syst. **4**, 143–150 (1992)

41.15   V. Vapnik: *The Nature of Statistical Learning Theory* (Springer, New York 1995)

41.16   B. Scholkopf, A.J. Smola: *Learning with Kernels, Support Vector Machines, Regularization, Optimization and Beyond* (MIT Press, Cambridge 2002)

41.17   F. Girosi, M. Jones, T. Poggio: Regularization theory and neural networks architectures, Neural Comput. **7**, 219–269 (1995)

41.18   B. Scholkopf, A.J. Smola, K. Muller: Nonlinear component analysis as a kernel eigenvalue problem, Neural Comput. **10**, 1299–1319 (1998)

41.19   M.H. Yang: Kernel eigenfaces vs kernel fisherfaces: Face recognition using kernel methods, Proc. 5th IEEE ICAFGR (Washington, 2002) pp. 215–220

41.20   W. Liu, J.C. Príncipe, S. Haykin: *Kernel Adaptive Filtering: A Comprehensive Introduction* (Wiley, Hoboken 2010)

41.21   A.R.C. Paiva, I. Park, J.C. Príncipe: A reproducing kernel Hilbert space framework for spike train signal processing, Neural Comput. **21**, 424–449 (2009)

41.22   L. Li, I. Park, S. Seth, J.S. Choi, J.T. Francis, J.C. Sanchez, J.C. Príncipe: An adaptive decoder from spike trains to micro-stimulation using kernel least-mean-squares (KLMS), Mach. Learn. Signal Process. (MLSP), IEEE Int. Workshop (Beijing 2011) pp. 1–6

41.23   W. Liu, P. Pokharel, J. Príncipe: The kernel least mean square algorithm, IEEE Trans. Signal Process. **56**, 543–554 (2008)

41.24   W. Liu, J. Príncipe: Kernel affine projection algorithm, EURASIP J. Adv. Signal Process. **12**, 784292 (2008)

41.25   Y. Engel, S. Mannor, R. Meir: The kernel recursive least-squares algorithm, IEEE Trans. Signal Process. **52**, 2275–2285 (2004)

41.26   W. Liu, I. Park, Y. Wang, J.C. Príncipe: Extended kernel recursive least squares algorithm, IEEE Trans. Signal Process. **57**, 3801–3814 (2009)

41.27   B.W. Silverman: *Density Estimation for Statistics and Data Analysis* (Chapman Hall, New York 1986)

41.28   A. Tikhonov, V. Arsenin: *Solution of ill-posed Problems* (Winston, Washington 1977)

41.29   G. Golub, C. Loan: *Matrix Computations* (John Hopkins University Press, Washington, DC 1996)

41.30   J. Platt: A resource-allocating network for function interpolation, Neural Comput. **3**, 213–225 (1991)

41.31   L. Csato, M. Opper: Sparse online Gaussian process, Neural Comput. **14**, 641–668 (2002)

41.32   C. Richard, J.C.M. Bermudez, P. Honeine: Online prediction of time series data with kernels, IEEE Trans. Signal Process. **57**, 1058–1066 (2009)

41.33   W. Liu, I. Park, J.C. Príncipe: An information theoretic approach of designing sparse kernel adaptive filters, IEEE Trans. Neural Netw. **20**, 1950–1961 (2009)

41.34 B. Chen, S. Zhao, P. Zhu, J.C. Príncipe: Quantized kernel least mean square algorithm, IEEE Trans. Neural Netw. Learn. Syst. **23**(1), 22–32 (2012)

41.35 Y.Y. Linde, A. Buzo, R.M. Gray: An algorithm for vector quantizer design, IEEE Trans. Commun. **28**, 84–95 (1980)

41.36 P.A. Chou, T. Lookabaugh, R.M. Gray: Entropy-constrained vector quantization, IEEE Trans. Acoust. Speech Signal Process. **37**, 31–42 (1989)

41.37 T. Lehn-Schiøler, A. Hegde, D. Erdogmus, J.C. Príncipe: Vector quantization using information theoretic concepts, Nat. Comput. **4**, 39–51 (2005)

41.38 S. Craciun, D. Cheney, K. Gugel, J.C. Sanchez, J.C. Príncipe: Wireless transmission of neural signals using entropy and mutual information compression, IEEE Trans. Neural Syst. Rehabil. Eng. **19**, 35–44 (2011)

41.39 N.R. Yousef, A.H. Sayed: A unified approach to the steady-state and tracking analysis of adaptive filters, IEEE Trans. Signal Process. **49**, 314–324 (2001)

41.40 T.Y. Al-Naffouri, A.H. Sayed: Adaptive filters with error nonlinearities: Mean-square analysis and optimum design, EURASIP J. Appl. Signal Process. **4**, 192–205 (2001)

41.41 T.Y. Al-Naffouri, A.H. Sayed: Transient analysis of data-normalized adaptive filters, IEEE Trans. Signal Process. **51**, 639–652 (2003)

41.42 T.Y. Al-Naffouri, A.H. Sayed: Transient analysis of adaptive filters with error nonlinearities, IEEE Trans. Signal Process. **51**, 653–663 (2003)

41.43 W. Sethares, C.R. Johnson: A comparison of two quantized state adaptive algorithms, IEEE Trans. Acoust. Speech Signal Process. **37**, 138–143 (1989)

41.44 T. Poggio, F. Girosi: Networks for approximation and learning, Proc. IEEE **78**(9), 1481–1497 (1990)

41.45 D.R. Brillinger: Maximum likelihood analysis of spike trains of interacting nerve cells, Biol. Cybern. **59**, 189–200 (1988)

41.46 Z. Mainen, T. Sejnowski: Reliably of spike timing in neocortical neurons, Science **268**, 1503–1506 (1995)

41.47 I. Park: Capturing Spike Train Similarity Structure: A Point Process Divergence Approach. Ph.D. Thesis (Univ. of Florida, Gainesville 2010)

41.48 L. Paninski, J. Pillow, J. Lewi: Statistical models for neural encoding, decoding, and optimal stimulus design, Prog. Brain Res. **165**, 493–507 (2007)

41.49 E.N. Brown, L.M. Frank, D. Tang, M.C. Quirk, M.A. Wilson: A statistical paradigm for neural spike train decoding applied to position prediction from ensemble firing patterns of rat hippocampal place cells, J. Neurosci. **18**, 7411–7425 (1998)

41.50 J. Eichhorn, A. Tolias, E. Zien, M. Kuss, C.E. Rasmussen, J. Weston, N. Logothetis, B. Scholkopf: Prediction on spike data using kernel algorithms, Adv. Neural Inform. Process. Syst. **16**, 1367–1374 (2004)

41.51 W. Maass, T. Natschlager, H. Markram: Real-time computing without stable states: A new framework for neural computation based on perturbations, Neural Comput. **14**, 2531–2560 (2002)

41.52 S. Seth, A.J. Brockmeier, J.S. Choi, M. Semework, J.T. Francis, J.C. Príncipe: Evaluating dependence in spike train metric spaces, Int. Jt. Conf. Neural Netw. (2011)

42. Recurrence Plots and the Analysis of Multiple Spike Trains

Yoshito Hirata, Eric J. Lang, Kazuyuki Aihara

Spike trains are difficult to analyze and compare because they are point processes, for which relatively few methods of time series analysis exist. Recently, several distance measures between pairs of spike train windows (segments) have been proposed. Such distance measures allow one to draw recurrence plots, two-dimensional graphs for visualizing dynamical changes of time series data, which in turn allows investigation of many spike train properties, such as serial dependence, chaos, and synchronization. Here, we review some definitions of distances between windows of spike trains, explain methods developed on recurrence plots, and illustrate how these plots reveal spike train properties by analysis of simulated and experimental data.

42.1 Overview

Many experimental methods have been developed to measure the activity of multiple neurons simultaneously, making the analysis of their spike trains (i. e., datasets consisting of the times of action potentials) an increasingly important topic in neuroscience [42.1]. However, relatively few methods for analyzing multiple spike trains currently exist, possibly because spike trains are point processes, which are difficult to analyze mathematically. This paucity may partly reflect the historical development of time series analysis where the focus has been on data whose sampling intervals are constant, unlike spike train data. In this chapter, we review a new approach for analyzing multiple spike train data by using distances between spike trains, and recurrence plots.

In the past 15 years, several methods for measuring the distance between spike trains have been proposed [42.2–4]. Distances, in general, measure the dissimilarity between two spike trains. For example,

Victor and *Purpura* [42.2] measured distance in terms of the edits needed to transform one train into the other, where each edit is assigned a cost. (We particularly focus on the distance of Victor and Purpura here since the distance is a metric and its extension [42.4] provides a good reconstruction of information encoded by a neuron.) The distance between the two trains is then defined as the total minimal cost for converting one train into the other. The allowed operations include insertion, deletion, and shifting of spikes. For insertion or deletion of a spike, we assign a cost of 1. For shifting a spike, we assign a cost that is proportional to the time shifted. In *Hirata* and *Aihara* [42.4], this edit distance of *Victor* and *Purpura* [42.2] has been extended to allow analysis of spike train data from the viewpoint of dynamical systems theory. This extension also satisfies the three properties of a metric, namely, non-negativity (where the distance is zero if and only if two spike trains are identical), symmetry, and triangle inequality.

Recurrence plots [42.5, 6], the second element in the approach to be described in this chapter, have been developed over the last 25 years. Recurrence plots are two-dimensional graphs whose axes are times. For every pair of times, we evaluate whether or not the two corresponding states are similar. If they are similar, then a point is plotted at the corresponding place on the graph, otherwise nothing is plotted at that coordinate. Recurrence plots can provide information about the characteristics of the underlying dynamics of the data, including serial dependence, deterministic chaos, and synchronization.

The combination of using distances between spike trains with recurrence plots provides a useful method for analyzing multiple spike train data from the viewpoint of dynamical systems theory. In particular, synchronization defined by this approach is a broader notion than the conventional definition of synchronization for spike trains used in neuroscience, where spikes of the trains must occur within a fixed delay. Therefore, an approach using distances and recurrence plots can provide important dynamical information that other methods cannot provide.

42.2 Recurrence Plots

Let $x(t) \in X$ be the t-th point of time series ($t = 1, 2, \ldots, T$). Here, the space X might be a set of real-valued vectors or a more exotic space such as parts of spike trains. Let $d_X : X \times X \rightarrow \{0\} \cup R^+$ be a distance function for the space X. Let ε_X be a threshold for the distance. Then a recurrence plot of $x(t)$ is defined as follows [42.5]

$$R_X(i, j, \varepsilon_X) = \begin{cases} 1, & \text{if } d_X(x(i), x(j)) < \varepsilon_X , \\ 0, & \text{otherwise} . \end{cases} \quad (42.1)$$

Recurrence plots provide information about the nature and characteristics of the system. Examples of recurrence plots from three distinct dynamical systems are shown in Fig. 42.1. For uniform noise (Fig. 42.1a,d), points in the recurrence plot are distributed relatively uniformly and randomly. In the recurrence plots of periodic signals, the points form periodic patterns. This is shown for a sine wave (Fig. 42.1b,e) where the points fall on regularly spaced diagonal lines with the interline distance corresponding to the period of the signal. For signals displaying deterministic

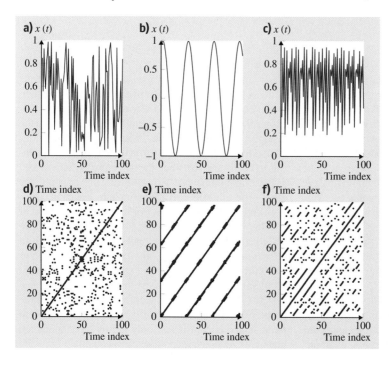

Fig. 42.1a–f Examples of recurrence plots. (**a–c**) Time series of the examples. (**d–f**) Their recurrence plots. The examples are, *from the left*, uniform noise, a periodic signal (sine wave), and a deterministic chaos (logistic map). In these examples, we choose the thresholds ε_X such that the recurrence rates, which are the probabilities that points are plotted, become 0.05

chaos, such as the logistic map (Fig. 42.1c,f), recurrence plots contain many diagonal line segments, since two nearby orbits will be separated by a certain distance some time in the future. Thus, the pattern of diagonal line segments reflects determinism in a recurrence plot.

Recurrence quantification analysis refers to the statistical analysis of patterns found in recurrence plots [42.6]. A typically used statistic is DET, which is the probability that plotted points are aligned diagonally.

Mathematically, DET can be defined as follows: First define the histogram $P(l, \varepsilon_X)$ of diagonal lines with length l by

$$Q(l, \varepsilon_X) = \sum_{i,j=1}^{T} [1 - R_X(i-1, j-1, \varepsilon_X)]$$
$$\times [1 - R_X(i+l, j+l, \varepsilon_X)]$$
$$\times \prod_{k=0}^{l-1} R_X(i+k, j+k, \varepsilon_X).$$

Then, DET can be defined as

$$\text{DET} = \frac{\sum_{l=l_{\min}}^{T} l Q(l, \varepsilon_X)}{\sum_{l=1}^{T} l Q(l, \varepsilon_X)}.$$

Usually, we set $l_{\min} = 2$.

For the examples shown in Fig. 42.1, DETs for the uniform noise, the periodic signal, and the deterministic chaos are 0.24, 0.99, and 0.73, respectively. Therefore, time series data that lack serial dependence, such as uniform random noise, tend to show smaller DET values compared with deterministic time series, such as periodic signals and those displaying deterministic chaos.

Recurrence plots can contain much information related to time series. When the space is limited to a set of real-valued vectors, then recurrence plots can contain almost all information of topology and distances related to a time series except for spatial scales, and thus it is possible to use them to reproduce the rough shape of the original time series [42.7, 8]. Additionally they can be used to estimate correlation dimension and correlation entropy [42.9, 10].

42.3 Univariate Analysis Using Recurrence Plots

Recently, we developed a series of methods for characterizing properties of time series as single systems. The properties we can characterize include serial dependence and deterministic chaos.

42.3.1 Serial Dependence

Serial dependence is the property that the future state of a system depends on the current or the past states. In a recurrence plot, serial dependence can be seen as points aligned diagonally. Therefore, it is useful to have a test for whether points tend to form diagonal lines or not. In *Hirata* and *Aihara* [42.11], we constructed such a test.

In this test we let p be the recurrence rate, or the probability that a point is plotted on a place of recurrence plot excluding the central diagonal line. Then, the probability that two points (i, j) and $(i+1, j+1)$ are aligned diagonally by chance is p^2. There are $(T-1)(T-2)/2$ independent pairs of two diagonally aligned places in a recurrence plot. Therefore, approximating the true distribution by the binomial distribution with probability p^2 and size $(T-1)(T-2)/2$ gives the mean number of diagonally aligned pairs of

points as $p^2(T-1)(T-2)/2$ with a standard deviation of $\sqrt{(1-p^2)p^2(T-1)(T-2)/2}$. Next, let n_d be the number of pairs where two points are actually aligned diagonally on the recurrence plot. If T is sufficiently large, then

$$z_d = \frac{n_d - p^2(T-1)(T-2)/2}{\sqrt{(1-p^2)p^2(T-1)(T-2)/2}} \tag{42.2}$$

follows the normal distribution of the mean 0 and standard deviation 1. Using this z_d we can evaluate the serial dependence of the underlying dynamics from the observed data.

If formula (42.2) is applied to the recurrence plots of Fig. 42.1 significance levels of 0.45, < 0.001, and < 0.001 are found for the random noise, periodic and deterministic chaos signals, respectively. Therefore, the proposed index works appropriately here.

42.3.2 Deterministic Chaos

Recurrence plots can be used to judge whether or not a particular time series displays deterministic chaos [42.12]. We will use the definition of determin-

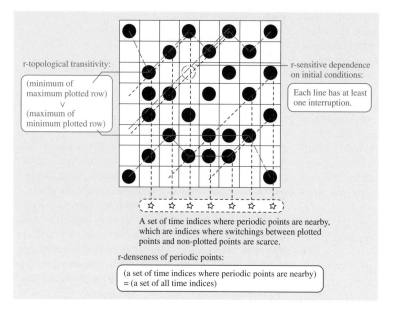

r-topological transitivity:

(minimum of
maximum plotted row)

∨

(maximum of
minimum plotted row)

r-sensitive dependence
on initial conditions:

Each line has at least
one interruption.

A set of time indices where periodic points are nearby,
which are indices where switchings between plotted
points and non-plotted points are scarce.

r-denseness of periodic points:

(a set of time indices where periodic points are nearby)
= (a set of all time indices)

Fig. 42.2 A schematic illustration of characteristics of deterministic chaos in a recurrence plot

istic proposed by Devaney, which consists of three conditions that need to be met: topological transitivity, denseness of periodic orbits, and sensitive dependence on initial conditions.

Topological transitivity is the condition that a trajectory exists from any open neighborhood to any open neighborhood. Denseness of periodic orbits means that in any open neighborhood, there is at least one periodic point. Sensitive dependence of initial conditions means that two nearby points can be separated by a certain distance, which is decided by the dynamics some time in the future. Since each of these conditions contains the notion of an arbitrary open neighborhood, we cannot apply them directly to a time series of finite length.

Therefore, we relaxed each of these conditions such that they can be checked with a time series with a finite length by using a recurrence plot by replacing arbitrary open neighborhoods in the original definitions with any neighborhoods defined by the threshold when we obtain the recurrence plot [42.12]. (See Fig. 42.2 for illustrative explanations.) Then, the relaxed condition of topological transitivity, or the r-topological transitivity, can be evaluated by testing whether the maximum of the minimal plotted row over all the columns is smaller than the minimum of the maximal plotted row over all the columns. (We eventually consider whether a point of any neighborhood used for defining a recurrence plot can be mapped to any neighborhood used for defining the recurrence plot.)

In the relaxed version of denseness of periodic orbits, or the r-denseness of periodic points, we focus on diagonal lines where the number of switchings between plotted places and non-plotted places is less than a specified chance level. We consider that these diagonal lines, which have distances of τ vertically or horizontally from the central diagonal line, correspond to periodic orbits of period τ. We include time indexes $\{i : R(i, i+\tau, \varepsilon_X) = 1\}$ and $\{j : R(j, j-\tau, \varepsilon_X) = 1\}$ of plotted points of the diagonal line segments in a set denoted by P which records points where a periodic point is nearby. After the calculations, if the set P becomes equal to the set of integers from 1 to T, which is a set of all time indexes for the recurrence plot, then it can be shown that we can find a periodic point from a neighborhood of any point in the time series, implying that periodic points are dense. Therefore, to test the r-denseness of periodic points, we test whether P is equal to a set of integers from 1 to T.

The relaxed version of sensitive dependence on initial conditions, or r-sensitive dependence on initial conditions, is much easier to verify. We check that every line such as $\{(i, i+\tau) \,|\, 0 < \tau < T/2\}$ has at least one interruption of the plotted points along it. The interruptions of plotted points mean that the two corresponding orbits are separated at least by ε_X.

Failure to meet all three relaxed criteria implies that the underlying dynamics is not consistent with deterministic chaos or that a given time series is too short. Therefore, we can use these relaxed criteria for an easy test of deterministic chaos. To illustrate, we apply these relaxed criteria to the examples of Fig. 42.1 in which the time series all had a length of 500. Only the recurrence plot of the logistic map satisfied all of the relaxed

criteria, implying that only the logistic map was consistent with Devaney's definition of deterministic chaos, i.e., consistent with the known underlying dynamics of these examples.

42.4 Multivariate Analysis Using Recurrence Plots

So far we have only discussed the use of recurrence plots in analyzing an individual signal; however, they may also provide information about the relationship between two signals as we will now describe. Recurrence plots have been extended to multivariate analysis in two ways. The first extension was proposed by *Zbilut* et al. [42.13] and called cross recurrence plots. Let $x(i) \in X$ be the i-th point of a time series $(i = 1, 2, \ldots, I)$. Let $y(j) \in X$ be the j-th point of a second time series $(j = 1, 2, \ldots, J)$. Therefore, $x(i)$ and $y(j)$ must be in the same space X. Let $d_X : X \times X \to \{0\} \cup R^+$ be a distance function for the space X. Let ε be a threshold for the distance. Then a cross recurrence plot of $x(t)$ and $y(t)$ is defined as follows [42.13, 14]

$$C(i, j) = \begin{cases} 1, & \text{if } d_X(x(i), y(j)) < \varepsilon , \\ 0, & \text{otherwise .} \end{cases} \quad (42.3)$$

The cross recurrence plot can be used for investigating whether the underlying dynamics of two time series are similar or not. If the two underlying dynamics are similar, then diagonal line segments will be observed. We may apply a method similar to that in Sect. 42.2 for obtaining the significance level of appearance of diagonal line segments here.

The other extension of recurrence plots to multivariate analysis is joint recurrence plots [42.15]. A joint recurrence plot is the product of two recurrence plots. Mathematically, a joint recurrence plot can be defined as follows: Let $R_X(i, j, \varepsilon_X)$ and $R_Y(i, j, \varepsilon_Y)$ be recurrence plots of $x(t)$ and $y(t)$, respectively. Then their joint recurrence plot is

$$J_{XY}(i, j, \varepsilon_X, \varepsilon_Y) = R_X(i, j, \varepsilon_X)R_Y(i, j, \varepsilon_Y) . \quad (42.4)$$

The advantage of joint recurrence plots is that we can compare two time series $x(t)$ and $y(t)$ whose spaces are different. Therefore, for example, by using a joint recurrence plot, we can compare spike trains with local field potentials directly.

In joint recurrence plots, the number of points remaining after converting recurrence plots to joint recurrence plots is an important index. Let p_X and p_Y be the probabilities that points are plotted on recurrence plots $R_X(i, j, \varepsilon_X)$ and $R_Y(i, j, \varepsilon_Y)$, respectively. These probabilities are called recurrence rates for $R_X(i, j, \varepsilon_X)$ and $R_Y(i, j, \varepsilon_Y)$. The probability that points are plotted on their joint recurrence plots is called the joint recurrence rate. If two time series are independent of each other, then the joint recurrence rate is expected to be $p_X p_Y$. If the two time series are related, or synchronized, then the joint recurrence rate is expected to be higher than $p_X p_Y$. Let T be the number of time points for time series $x(t)$ and $y(t)$. Then, in the recurrence plots and joint recurrence plots, the number of places where a point can be plotted in the upper triangle of the joint recurrence plot is $T(T-1)/2$. Then, if two time series satisfy the null-hypothesis that they are independent of each other, the number of points plotted in the upper triangle of their joint recurrence plots follows the binomial distribution of probability $p_X p_Y$ and size $T(T-1)/2$ [42.16]. Since $T(T-1)/2$ is, in general, sufficiently large, the number of points in the upper triangle can be approximated by the normal distribution with mean $p_X p_Y T(T-1)/2$ and standard deviation $\sqrt{p_X p_Y (1 - p_X p_Y) T(T-1)/2}$. Therefore, whether or not two time series are related can be easily quantified by the joint recurrence rate.

By further extending the analysis of joint recurrence plots, we even can infer network topology from multivariate time series; namely we can identify whether or not one of the systems drives the other, and whether or not there is a hidden common driver.

Inferring the asymmetry of coupling based on recurrence plots was proposed by *Romano* et al. [42.17]. To infer the asymmetry of coupling between two systems, we define the mean conditional probabilities of recurrence as follows

$$M_{\mathrm{CR}}(X \mid Y) = \frac{1}{T} \sum_{i=1}^{T} p(y(i) \mid x(i))$$

$$= \frac{1}{T} \sum_{i=1}^{T} \frac{\sum_{j=1}^{T} J_{XY}(i, j, \varepsilon_X, \varepsilon_Y)}{R_X(i, j, \varepsilon_X)} ,$$

$$(42.5)$$

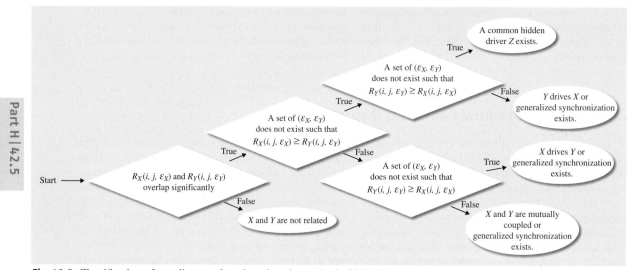

Fig. 42.3 Classification of coupling topology based on the method of [42.16]

$$M_{CR}(X \mid Y) = \frac{1}{T} \sum_{i=1}^{T} p(x(i) \mid y(i))$$

$$= \frac{1}{T} \sum_{i=1}^{T} \frac{\sum_{j=1}^{T} J_{XY}(i, j, \varepsilon_X, \varepsilon_Y)}{R_Y(i, j, \varepsilon_Y)} .$$

$$(42.6)$$

If X drives Y, then we have

$$M_{CR}(Y \mid X) < M_{CR}(X \mid Y) .$$

If Y drives X, then we have

$$M_{CR}(X \mid Y) < M_{CR}(Y \mid X) .$$

If the coupling is symmetric, then

$$M_{CR}(Y \mid X) = M_{CR}(X \mid Y) .$$

Inferring the existence of a hidden common third element was proposed by *Hirata* and *Aihara* [42.16]. To do this the first step is to use delay coordinates to reconstruct joint states of possible driver and driven systems. (In the analysis of spike trains, delay coordinates correspond to a window of spike trains.) Second, we check whether or not two time series are related by using the joint recurrence rate. If two time series are related, we check whether or not directional couplings from one system to the other system can be denied. If X drives Y, then we can choose the thresholds ε_X and ε_Y of their recurrence plots $R_X(i, j, \varepsilon_X)$ and $R_Y(i, j, \varepsilon_Y)$ so that $R_X(i, j, \varepsilon_X) \geq R_Y(i, j, \varepsilon_Y)$. If we take the contraposition of this statement, we can say that if we cannot choose the thresholds ε_X and ε_Y such that $R_X(i, j, \varepsilon_X) \geq R_Y(i, j, \varepsilon_Y)$, then X does not drive Y. Thus, we can deny the directional coupling from X to Y. In a similar way, we can also try to deny the directional coupling from Y to X. When some directional couplings are not denied, then directional couplings might be implied (see Fig. 42.3 for the classification).

42.5 Distances Between Spike Trains

So far the methods using recurrence plots have been constructed based on the space of real-valued vectors. However, if we can define a distance between spike trains, then we can apply the methods discussed above to multivariate spike trains. In this section, we discuss two methods for defining distances between spike trains.

A basic method for defining a distance between spike trains is the edit distance of *Victor* and *Purpura* described earlier [42.2]. To use the edit distance for investigating the underlying dynamics of spike trains, we set a time window of a fixed length. Then, the time window is moved by a fixed amount of time creating a set of spike train segments until the end of the spike train is reached. Then, for each pair of positions of the time window, the edit distance of the resulting train segments is calculated to see how the edit distance varies along the time axis.

However, the direct application of this edit distance of [42.2] to spike trains has a problem from the viewpoint of nonlinear dynamics because if a spike appears or disappears when the time window is moved, the distance jumps by 1 and does not change continuously.

To overcome this problem, we introduce two techniques [42.4]. The first technique is that we consider cases where we take into account a spike immediately before or after the time window. Therefore, there are four possibilities for each position of the time window. Since we calculate a distance between a pair of positions of the time window, there are 16 combinations of the possibilities as the total. For each combination, we calculate the edit distance by *Victor* and *Purpura* [42.2], find the minimum distance over 16 combinations, and declare the minimum distance as the local distance. By considering this local distance, we can define a distance between two positions of the time window whose

numbers of spikes are different by at most 2 only by shift.

Next, we connect these local distances to define a global distance between every pair of positions of the window. For this purpose, we use the method of Isomap, or isometric feature mapping [42.18], which was originally proposed for dimension reduction. First, a network is constructed by connecting neighboring nodes with edges and assigning each edge a local distance. Then, a global distance can be defined as the shortest path between every pair of nodes in the network. This shortest path corresponds to a geodesic distance. In our current setting, each node corresponds to a position of the time window, we consider all to all connections and assign each edge the local distance between two positions of the time window defined in the previous paragraph. By using the shortest paths on the graph, the global distances end up being defined using only the edges whose local distances are the result of shift of spike edits only, because edges whose local distances are defined by insertion and/or deletion of spikes have large values greater than 1.

42.6 Examples

First, we demonstrate how the global distance defined above is reasonable for analyzing spike trains from the viewpoint of nonlinear dynamics. We applied the global distance to an example of a cricket hair cell [42.4, 19]. A cricket hair cell was placed in a wind tunnel. Audio speakers were placed at each end of the tunnel and an oscillatory signal was fed to the speakers in anti-phase in order to generate wind in the tunnel. The oscillatory signal was generated by using the dynamics of *Rossler* chaos [42.20]. In sum, the wind represents the input to the hair cell and the latter's spikes are the response.

The results are shown in Fig. 42.4. We found that the reconstructed signal from the spike train of the hair cell correlated well with the change of the wind speed. In addition, we found that only the wind speed in one direction was reconstructed, meaning that the hair cell represents wind change for a particular direction. This finding agrees with previously published work [42.21].

Second, we applied the global distance to two integrate-and-fire neurons driven by *Lorenz* chaos [42.22]. The input of the Lorenz chaos and the spike trains generated by the two integrate-and-fire neurons

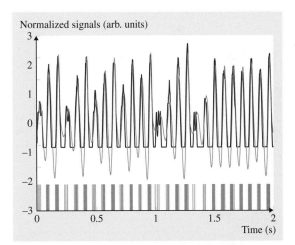

Fig. 42.4 Rossler chaos as an input for hair cell (*top, gray line*). The hair cell responded as a raster plot shown *at the bottom*. Then the proposed global distances were applied to the response of hair hell, which is represented as a spike train. The reconstructed signal (*top, black line*) correlates well with the change of the wind speed (*top, gray line*)

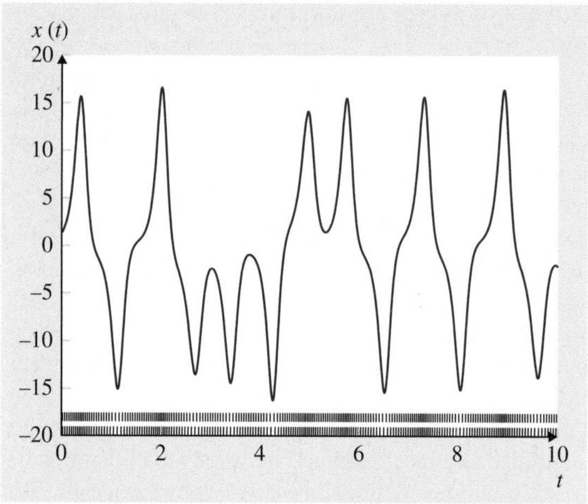

$x(t)$

Fig. 42.5 Lorenz chaos (*top*) as an input for two integrate-and-fire neurons. *At the bottom* we also show the spike trains generated from the two integrate-and-fire neurons with different parameters

to the first test of Fig. 42.3. By applying the subsequent tests listed in Fig. 42.3 we arrive at the conclusion that there is a hidden common driver of the two neurons. The existence of a hidden common driver reflects the fact that the Lorenz chaos has driven the two integrate-and-fire neurons! Therefore, the combination of the global distance for spike trains with recurrence plots works well.

Lastly, we calculated the global distances for spike trains of Purkinje cell complex spikes, which are triggered in a one-to-one manner by the activity of inferior olive neurons. To investigate the patterns of complex spikes simultaneous recordings were made from multiple Purkinje cells [42.23]. Typically, complex spikes of restricted subsets of Purkinje cells show synchronization based on traditional correlation analysis [42.24, 25]. In addition, models of inferior olivary neurons predict that these neurons should display deterministic chaotic behavior [42.26]. These reasons provided strong motivation for applying recurrence plot based analyses to complex spike trains.

First, we analyzed complex spike activity from individual Purkinje cells (see Fig. 42.7 for an example recurrence plot). The presence of serial dependence in the spike train is demonstrated by the tendency of points to form diagonal line segments (p-value < 0.001). In addition, this recurrence plot satisfies the three re-

are shown in Fig. 42.5. Since each integrate-and-fire neuron has a different parameter, their spikes do not necessarily occur within a fixed time delay. However, we can see diagonal line segments in their joint recurrence plot (Fig. 42.6), indicating that these neurons are synchronized in a sense of co-recurrences (p-value < 0.001), and thus that their activity is related according

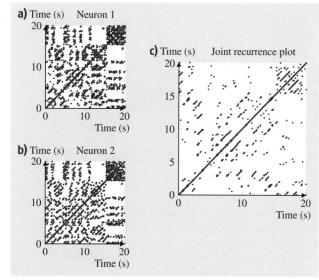

Fig. 42.6a–c Recurrence plots of the integrate-and-fire neurons (**a,b**) and their joint recurrence plot (**c**)

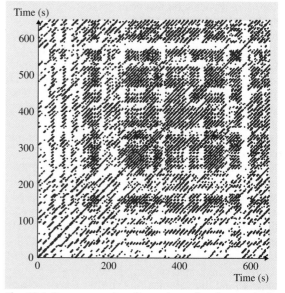

Fig. 42.7 Recurrence plot of complex spikes from a single Purkinje cell

laxed criteria of Devaney's definition of deterministic chaos, and thus may be the first evidence that a neuron in vivo shows firing patterns consistent with deterministic chaos.

Next, we analyzed the relationship between complex spike trains recorded simultaneously from two Purkinje cells using a joint recurrence plot (Fig. 42.8). We found that the activity of the two neurons is related (p-value < 0.001). When the method of [42.16] was applied to this dataset it denied both types of directional coupling (i.e., a true value at the second and third steps of the decision tree in Fig. 42.3), suggesting that there is a hidden common driver between the two neurons. In most instances this would be taken to be a synaptic input from an upstream region. Indeed, inferior olive neurons receive synaptic input from many areas, but they are also electrically coupled by gap junctions raising another possibility, which is that their activity is due to this coupling. Consistent with this last possibility, the block of gap junctions reduces complex spike synchrony [42.23].

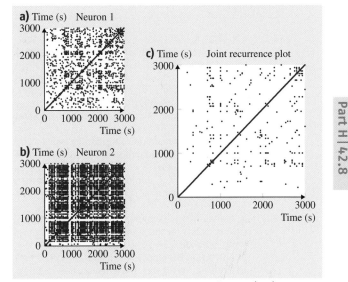

Fig. 42.8a–c Recurrence plots for neurons 1 and 2 (**a,b**), and their joint recurrence plot (**c**)

Part H | 42.8

42.7 Discussion

In this chapter, we analyzed spike trains using recurrence plots. The biggest advantage of this methodology is that recurrence plots can be applied to spike trains directly following the definition of a good distance measure between spike trains. In this chapter, the number of spike trains simultaneously applied was limited to two. However, by extending joint recurrence plots or modifying the method of *Nawrath* et al. [42.27], it would be possible to treat more than two neurons simultaneously. The methods using recurrence plots are still in their infancy. Therefore, it would be beneficial if methods using recurrence plots were further advanced so that they can treat large numbers of neurons simultaneously. This direction of research is in progress.

42.8 Conclusions

In this chapter, we have outlined a new approach for analyzing spike train data using distance measures and recurrence plots. Recently, numerous methods for using recurrence plots have been proposed. To take advantage of such recurrence plot based methods for analyzing spike trains it is necessary to define distance measures for the spike trains. After this, univariate analysis can test for serial dependence and deterministic chaos, and multivariate analysis with joint recurrence plots can test for synchronization between spike trains. Note that this is a more generalized type of synchronization being tested for than what is usually meant by standard definitions, since this generalized synchronization can occur even when the spikes of two spike trains are not locked within a fixed time delay. Moreover, directional couplings and hidden common causes may be inferred by using recurrence plots and a distance between spike trains. Therefore, recurrence plots will reveal not only nonlinear dynamics of neurons but also their unknown interactions.

References

42.1   E.N. Brown, R.E. Kass, P.P. Mitra: Multiple neural spike train data analysis: State-of-the-art and future challenges, Nat. Neurosci. **7**, 456–461 (2004)

42.2   J.D. Victor, K.P. Purpura: Metric-space analysis of spike trains: Theory, algorithms and application, Network **8**, 127–164 (1997)

42.3   M.C.W. van Rossum: A novel spike distance, Neural Comput. **13**, 751–763 (2001)

42.4   Y. Hirata, K. Aihara: Representing spike trains using constant sampling intervals, J. Neurosci. Methods **183**, 277–286 (2009)

42.5   J.-P. Eckmann, S.O. Kamphorst, D. Ruelle: Recurrence plots of dynamical systems, Europhys. Lett. **5**, 973–977 (1987)

42.6   N. Marwan, M.C. Romano, M. Thiel, J. Kurths: Recurrence plots for the analysis of complex systems, Phys. Rep. **438**, 237–329 (2007)

42.7   M. Thiel, M.C. Romano, J. Kurths, M. Rolfs, R. Kliegl: Generating surrogates from recurrences, Philos. Trans. R. Soc. A **366**, 545–557 (2008)

42.8   Y. Hirata, S. Horai, K. Aihara: Reproduction of distance matrices from recurrence plots and its applications, Eur. Phys. J. Spec. Top. **164**, 13–22 (2008)

42.9   P. Faure, H. Korn: A new method to estimate the Kolmogorow entropy from recurrence plots: Its application to neuronal signals, Physica D **122**, 265–279 (1998)

42.10   M. Thiel, M.C. Romano, P.L. Read, J. Kurths: Estimation of dynamical invariants without embedding by recurrence plots, Chaos **14**, 234–243 (2004)

42.11   Y. Hirata, K. Aihara: Statistical tests for serial dependence and laminarity on recurrence pltos, Int. J. Bifur. Chaos **21**, 1077–1084 (2011)

42.12   Y. Hirata, K. Aihara: Devaney's chaos on recurrence plots, Phys. Rev. **82**, 036209 (2010)

42.13   J.P. Zbilut, A. Giuliani, C.L. Webber Jr.: Detecting deterministic signals in exceptionally noisy environments using cross-recurrence quantification, Phys. Lett. A **246**, 122–128 (1998)

42.14   N. Marwan, J. Kurths: Nonlinear analysis of bivariate data with cross recurrence plots, Phys. Lett. A **302**, 299–307 (2002)

42.15   M.C. Romano, M. Thiel, J. Kurths, W. von Bloh: Multivariate recurrence plots, Phys. Lett. A **330**, 214–223 (2004)

42.16   Y. Hirata, K. Aihara: Identifying hidden common causes from bivariate time series: A method using recurrence plots, Phys. Rev. E **81**, 016203 (2010)

42.17   M.C. Romano, M. Thiel, J. Kurths, C. Grebogi: Estimation of the direction of the coupling by conditional probabilities of recurrence, Phys. Rev. E **76**, 036211 (2007)

42.18   J.B. Tenembaum, V. de Silva, J.C. Langford: A global geometric framework for nonlinear dimensionality reduction, Science **290**, 2319–2323 (2000)

42.19   H. Suzuki, K. Aihara, J. Murakami, T. Shimozawa: Analysis of neural spike trains with interspike interval reconstruction, Biol. Cybern. **82**, 305–311 (2000)

42.20   O.E. Rossler: An equation for hyperchaos, Phys. Lett. A **71**, 155–157 (1976)

42.21   E. Salinas, L.F. Abbott: Vector reconstruction from firing rates, J. Comput. Neurosci. **1**, 89–107 (1994)

42.22   E.N. Lorenz: Deterministic nonperiodic flow, J. Atmos. Sci. **20**, 130–141 (1963)

42.23   T.A. Blenkinsop, E.J. Lang: Block of inferior olive gap junctional coupling decreases Purkinje cell complex spike synchrony and rhythmicity, J. Neurosci. **26**, 1739–1748 (2006)

42.24   K. Sasaki, J.M. Bower, R. Llinas: Multiple Purkinje cell recording in rodent cerebellar cortex, Eur. J. Neurosci. **1**, 572–586 (1989)

42.25   E.J. Lang, I. Sugihara, J.P. Welsh, R. Llinás: Patterns of spontaneous Purkinje cell complex spike activity in the awake rat, J. Neurosci. **19**, 2728–2739 (1999)

42.26   N. Schweihofer, K. Doya, H. Fukai, J.V. Chiron, T. Furukawa, M. Kawato: Chaos may enhance information transmission in the inferior olive, Proc. Natl. Acad. Sci. USA **101**, 4655–4660 (2004)

42.27   J. Nawrath, M.C. Romano, M. Thiel, I.Z. Kiss, M. Wickramasinghe, J. Timmer, J. Kurths, B. Schelter: Distinguishing direct from indirect interactions in oscillatory networks with multiple time scales, Phys. Rev. Lett. **104**, 038701 (2010)

43. Adaptive Multiscale Time–Frequency Analysis

Naveed ur Rehman, David Looney, Cheolsoo Park, Danilo P. Mandic

Time–frequency analysis techniques are now adopted as standard in many applied fields, such as bio-informatics and bioengineering, to reveal frequency-specific and time-locked event-related information of input data. Most standard time–frequency techniques, however, adopt *fixed* basis functions to represent the input data and are thus suboptimal. To this cause, an empirical mode decomposition (EMD) algorithm has shown considerable prowess in the analysis of nonstationary data as it offers a fully data-driven approach to signal processing. Recent multivariate extensions of the EMD algorithm, aimed at extending the framework for signals containing multiple channels, are even more pertinent in many real world scenarios where multichannel signals are commonly obtained, e.g., electroencephalogram (EEG) recordings. In this chapter, the multivariate extensions of EMD are reviewed and it is shown how these extensions can be used to alleviate the long-standing problems associated with the standard (univariate) EMD algorithm. The ability of the multivariate extensions of EMD as a powerful real world data analysis tool is demonstrated via simulations on biomedical signals.

43.1 Data Driven Time Frequency Analysis

Empirical mode decomposition (EMD) is a data-driven method for the decomposition and time-frequency analysis of real world nonstationary signals [43.1]. The main advantages offered by EMD are its locality, adaptivity (data-driven decomposition), multiresolution, accurate time-frequency representation, and ability to capture oscillation of any type, e.g., nonharmonic signals. These properties have made EMD standard for processing signals exhibiting nonstationarity, a characteristic commonly associated with a wide range of real world signals–from biomedical to meteorological [43.2, 3]. Other established tools for time-frequency analysis, such as those based on the Fourier and wavelet transform, assume linearity and/or stationarity of input data as they adopt fixed basis functions to decompose the signal; as a result, these techniques are found to be unsuitable for many real world applications.

In scenarios involving real world nonstationary signals, there are instances when the convenience of choosing an a priori fixed set of basis functions, and their orthogonality (in strict sense), are less significant than the *sparsity and the local* nature of the expansion. To this end, it is important to consider a class of methods which may compromise the orthogonality and the choice of a fixed dictionary of basis functions, but provide a more physically meaningful representation of the signal in hand. This can be achieved by adopting a set of data-driven basis functions, where the expansion terms are chosen based on the input signal. This way, each expanded term is expected to carry significant information about the input signal in hand; otherwise, it is likely that a signal component will be spread over an unnecessary large number of terms (harmonics), causing unwanted distortion in the time-frequency spectrum, as experienced in the case of the short-time Fourier transform (STFT) and to some extent, the wavelet transform.

The EMD belongs to the above-mentioned class of fully data-driven methods which extract the basis functions from the data itself. Unlike Fourier or wavelet-based methods, EMD makes no prior assumptions on the data and instead uses a set of basis functions, also known as intrinsic mode functions (IMFs), which carry intrinsic time-scale information. This data-driven nature of EMD makes it more suitable for the processing of nonstationary signals.

Despite its initial success, EMD could not fulfill its potential in areas where underlying physical phenomena are represented by multichannel signals such as in bio-informatics and data fusion. Standard EMD applied channel-wise on multivariate signals yields suboptimal results due to the empirical nature of EMD and the fact that the interdependence between channels is ignored in such an analysis. To that cause, multivariate extensions of EMD have been developed with the expectation of gaining more insight into the dynamics and interdependence between the multiple channels of multivariate components.

This chapter gives a review of the recent multivariate extensions of the EMD algorithm and their advantages with regard to multivariate nonstationary data analysis [43.4–6]. Some important properties of such extensions, including their ability to exhibit wavelet-like dyadic filter bank structures for white Gaussian noise (WGN) and their capacity to align similar oscillatory modes from multiple channels are demonstrated. Using these properties, a noise-assisted EMD-based algorithm is presented in order to solve the mode-mixing problem in the original EMD algorithm [43.7]. Furthermore, simulations on multivariate EEG data for the classification of several motor imagery tasks, a key paradigm in brain computer interface (BCI), and phase synchrony will be presented. We show that MEMD is highly suited to monitor changing brain activity corresponding to motor imagery tasks due to its localized and adaptive nature, and the exploitation of common oscillatory modes within the multivariate data.

43.2 Empirical Mode Decomposition

EMD is a data-driven technique to decompose a signal, by means of an iterative process called the sifting algorithm, into a finite set of oscillatory components called intrinsic mode functions (IMFs), which represent the temporal modes (scales) present in the data [43.1]. Given an arbitrary time series $x(k)$, EMD decomposes it into a sum of IMFs $\{c_m(k)\}, m = 1, \ldots, M$ and the residual $r(k)$, that is,

$$x(k) = \sum_{m=1}^{M} c_m(k) + r(k) . \tag{43.1}$$

The residual $r(k)$, unlike $\{c_m(k)\}_{m=1}^{M}$, does not contain any oscillations and its physical meaning is a trend within the signal. By design, an IMF is a function that is characterized by the following two properties: the upper and lower envelopes are symmetric, and the number of zero-crossings and the number of extrema are exactly equal or they differ at most by one. The sifting algorithm is described for a signal $x'(k) = x(k)$ in Algorithm 43.1.

Algorithm 43.1 The sifting algorithm for EMD

1: Find the locations of all the extrema of $x'(k)$
2: Interpolate (using spline interpolation) between all the minima (or maxima) to obtain the signal envelope passing through the minima, $e_{\min}(k)$ (or $e_{\max}(k)$)
3: Compute the local mean $m(k) = (e_{\min}(k) + e_{\max}(k))/2$
4: Subtract the local mean from the signal to obtain the *oscillating* signal $d(k) = x'(k) - m(k)$
5: If the resulting signal $d(k)$ obeys the stopping criterion, it becomes the first IMF, otherwise set

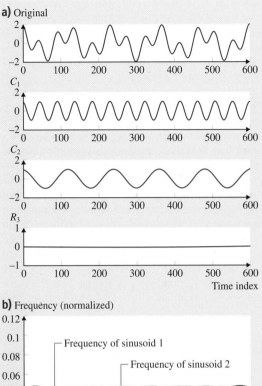

a) Original

b) Frequency (normalized)

Frequency of sinusoid 1

Frequency of sinusoid 2

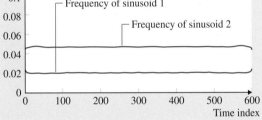

Fig. 43.1a,b EMD of two added sinusoids. (**a**) Intrinsic mode functions; (**b**) Hilbert spectrum

$x'(k) = d(k)$ and repeat the process from Step 1 until the first IMF is obtained

6: After finding an IMF, it is subtracted from the input signal. The residue of subtraction is fed back to the Step 1 of the algorithm as $x'(k)$

7: The algorithm is finished when the residual in the previous step becomes a monotonic function. The last residual is considered to be the trend $r(k)$

A stopping criterion in the sifting process of EMD is used to check whether the *local trend* $d(k)$, obtained

Part H | 43.2

in Step 5 of Algorithm 43.1, fulfills the IMF criteria. The choice of a suitable stopping criterion is important because if the data is over-sifted, this will result in over-decomposition of IMFs with uniform amplitude modulations, whereas, under-sifted IMFs will not satisfy mono-component criteria and the estimate of instantaneous frequency will be erratic. A robust criterion, based on the original definition of an IMF, stops the sifting process only after the condition of an IMF is met S consecutive times [43.8]; the condition checked for an IMF is that the difference between the number of extrema and zero crossings should not exceed one. It was also shown in [43.8] that the empirical range of S should be chosen between 4 and 8.

After obtaining the IMFs $\{c_m(k)\}_{m=1}^M$ from EMD, the Hilbert transform can be applied to each IMF separately and the instantaneous frequency can be computed. Given a signal $x(k)$ and its corresponding IMFs $\{c_m(k)\}_{m=1}^M$, application of the Hilbert transform to the decomposition given in (43.1) yields

$$x(k) = \sum_{m=1}^M a_m(k)e^{\iota\theta_m(k)} = \sum_{m=1}^M a_m(k)e^{\iota\int \omega_m(k)\,dt} \ .$$
(43.2)

The residue r is purposely omitted, as it is either a monotonic function or a constant. Note that (43.2) yields variable amplitude $a_m(k)$, and instantaneous frequency $w_m(k) = d\theta_m/dt$; where θ_m denotes instantaneous angle, making the corresponding expansion suitable for processing nonstationary data. The amplitude $a_m(k)$ and instantaneous frequency $w_m(k)$ can be plotted versus time index k to yield a time-frequency-amplitude representation of the entire signal known as the Hilbert–Huang spectrum, $H(k, w)$ [43.9].

To illustrate the operation of EMD, consider a signal consisting of two added sinusoids with different frequencies, shown in Fig. 43.1a (the original signal is in the first row, followed by the corresponding IMFs C1–C2, for convenience the last IMF is denoted by R3), and its Hilbert spectrum (Fig. 43.1b).

Observe the component separation performed by EMD, whereby the original signal is split into a set of IMFs (C1–C2) and residual (R3). The frequencies of the sinusoids composing the original signal are clearly visible in the time-frequency spectrum (Fig. 43.1b).

43.3 Multivariate Extensions of EMD

Given the popularity of EMD in real world nonstationary signals, efforts have been made recently to extend EMD to be able to process signals containing multiple data channels, that is, multivariate signals. These efforts have culminated in several multivariate frameworks for EMD, including complex extensions of EMD, trivariate EMD (TEMD), and a general multivariate EMD (MEMD). Of these algorithms, the most recent is MEMD, which is capable of processing multivariate signals containing any number of channels and which has already found numerous applications in phase synchrony, data fusion, and bio-informatics.

The main challenge in developing multivariate extensions of EMD is to find a suitable method to estimate the local mean of a multivariate signal. The problem arises due to the fact that, unlike the field of real numbers $\mathbb{R}$, the complex $\mathbb{C}$ and other higher dimensional fields are not ordered and, hence, it is not possible to define the local extrema directly [43.10]. Each multivariate extension of EMD, thus, has a unique way for defining the local mean and is classified based on that.

43.3.1 Complex Extensions of EMD

The most common complex extensions of EMD include complex EMD (CEMD) [43.11], rotation-invariant EMD (RI-EMD) [43.12], and bivariate EMD (BEMD) [43.4]. Complex EMD makes use of the analyticity of the signal to apply standard EMD component-wise on real and imaginary channels, whereas rotation invariant EMD and bivariate EMD estimate the local mean based on the envelopes obtained by taking projections of the input signal in multiple di-

rections. The following subsections will elaborate on each method in some detail.

43.3.2 Complex EMD

The complex EMD (CEMD) algorithm [43.11] operates by converting a general nonanalytic signal into two analytic signals, each corresponding to either the positive or the negative frequency components of the original signal. The standard EMD is then applied to the real part of the resulting analytic signals to obtain two sets of IMFs. These sets are then combined to form complex-valued IMFs. More precisely, let $x(k)$ be a complex-valued sequence and $X(e^{t\omega})$ its discrete Fourier transform (DFT). By processing signal $x(k)$ with the filter having a transfer function

$$H(e^{t\omega}) = \begin{cases} 1, & 0 < \omega \leq \pi \\ 0, & -\pi < \omega \leq 0 \end{cases} \tag{43.3}$$

the DFT of two analytic signals, denoted by $X_+(e^{t\omega})$ and $X_-(e^{t\omega})$, are generated, which correspond, respectively, to the positive and the negative frequency parts of $X(e^{t\omega})$. The subsequent application of the inverse Fourier transform, denoted by $\mathcal{F}^{-1}(\cdot)$, to $X_+(e^{t\omega})$ and $X_-(e^{t\omega})$ yields time series $x_+(k)$ and $x_-(k)$, defined as

$$x_+(k) = \mathcal{R}\{\mathcal{F}^{-1}[X_+(e^{t\omega})]\}, \tag{43.4}$$

$$x_-(k) = \mathcal{R}\{\mathcal{F}^{-1}[X_-(e^{t\omega})]\}, \tag{43.5}$$

where the operator $\mathcal{R}(\cdot)$ extracts the real component of a complex signal. Standard univariate EMD can then be

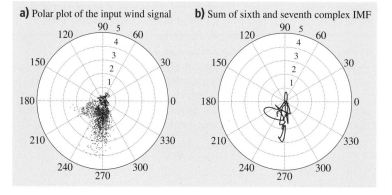

a) Polar plot of the input wind signal

b) Sum of sixth and seventh complex IMF

Fig. 43.2 A complex wind signal analyzed by the rotation-invariant EMD algorithm

applied to $x_+(k)$ and $x_-(k)$, to give

$$x_+(k) = \sum_{m=1}^{M^+} c_m(k) + r_+(k) , \qquad (43.6)$$

$$x_-(k) = \sum_{m=-1}^{-M^-} c_m(k) + r_-(k) , \qquad (43.7)$$

where symbols M^+ and M^- denote, respectively, the number of IMFs for the positive and the negative frequency parts, $c_m(k)$ are the IMFs, and $r_+(k)$ and $r_-(k)$ are residual signals for $x_+(k)$ and $x_-(k)$. The original complex signal $x(k)$ can then be reconstructed in terms of $x_+(k)$ and $x_-(k)$ as

$$x(k) = \{x_+(k) + \iota \mathcal{H}[x_+(k)]\}$$
$$+ \{x_-(k) + \iota \mathcal{H}[x_-(k)]\}^* , \qquad (43.8)$$

where $\mathcal{H}(\cdot)$ is the Hilbert transform operator and the symbol $(\cdot)^*$ denotes the complex conjugation operator.

For the m-th complex IMF $y_m(k)$, defined as

$$y_m(k) = \begin{cases} c_m(k) + \iota \mathcal{H}\big[c_m(k)\big], \\ \qquad\qquad m = 1, \ldots, M^+ , \\ \big\{c_m(k) + \iota \mathcal{H}\big[c_m(k)\big]\big\}^* , \\ \qquad\qquad m = -M^-, \ldots, -1 \end{cases} \qquad (43.9)$$

the original complex valued signal $x(k)$ can also be written as

$$x(k) = \sum_{m=-M^-, m \neq 0}^{m=M^+} y_m(k) + r(k) , \qquad (43.10)$$

where $r(k)$ represents the trend in the data and is represented in terms of the residuals of x_n^+ and x_n^- as

$$r(k) = \big[r_+(k) + r_-(k)\big] + \iota \mathcal{H}\big[r_+(k) - r_-(k)\big] . \qquad (43.11)$$

The so defined CEMD retains the generic structure of standard EMD; however, as the number of IMFs for $x_+(k)$ and $x_-(k)$ can in general be different, it is difficult to interpret the physical meaning of the extracted IMFs. Moreover, this approach is not suitable for extension to higher dimensions.

Rotation–Invariant EMD

The algorithm operates directly in $\mathbb{C}$ and defines the extrema of a complex signal as points where the angle of the first derivative of a signal becomes zero [43.12]. For a complex signal, $z(t) = x(t) + \iota y(t)$, it can be shown that this criterion is equivalent to $y'(t) = 0$, that is, the extrema of the imaginary part. Mathematically, for a complex signal $z(t) = x(t) + \iota y(t)$, the extrema locations are at

$$\angle z'(t) = 0 \Rightarrow \angle \{x'(t) + \iota y'(t)\} ,$$
$$\tan^{-1} \frac{y'(t)}{x'(t)} = 0 \Rightarrow y'(t) = 0 . \qquad (43.12)$$

As it is assumed that a local maximum is followed by a minimum, these sets can be interchanged. The spline interpolation is then performed on both components separately to obtain complex-valued envelopes, which are averaged to obtain the local mean. This method yields a single set of complex-valued IMFs, and the ambiguity at the zero frequency within the CEMD is avoided due to the direct operation in $\mathbb{C}$. Figure 43.2a shows the polar plot of a real wind data set, whereas Fig. 43.2b shows the contribution of the sixth and seventh IMF. It is clear that the complex IMFs have physical meaning as they can reveal the dynamics of teh original signal at different scales.

Bivariate EMD

The bivariate EMD (BEMD) algorithm [43.4] calculates the local mean envelopes based on the extrema of both (real and imaginary) components of a complex signal, yielding more accurate estimates than RI-EMD. It proceeds by projecting an input bivariate signal in V different directions, with each direction vector defined based on equidistant points along the unit circle. Next, the corresponding envelopes for each direction are obtained by interpolating the extrema of projected signals via component-wise spline interpolation; these envelopes are then averaged to obtain the local mean. Assuming four directions, the center of envelopes at a point in space is given by either of the following methods:

1. The barycenter of the four points
2. The intersection of two straight lines that pass though the middle of the horizontal and vertical tangents.

It can be noticed that while RI-EMD is a generic extension of standard EMD, it only uses projections in two directions to find the extrema, whereas BEMD can take projections in any number of directions, in turn, yielding a more accurate estimate of the local mean as compared to RI-EMD.

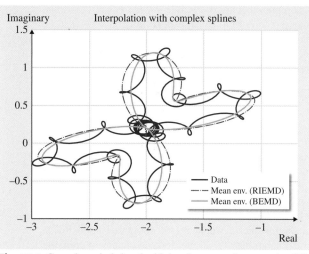

Fig. 43.3 Complex wind signal with local mean estimates using RI-EMD and BEMD

This issue is illustrated further in Fig. 43.3, which graphically shows the process of finding the local mean of a complex signal using RI-EMD and BEMD. In both cases, envelopes were calculated in multiple directions and then averaged to obtain the local mean. However, as expected, the estimates of BEMD were found to be more accurate than RI-EMD as it used the larger number of projections ($V = 4$) as compared to RI-EMD ($V = 2$).

43.3.3 Trivariate EMD

Trivariate EMD (TEMD) [43.5] extends the EMD algorithm to process trivariate signals directly. Estimation of the local mean of a trivariate signal is performed by taking signal projections along multiple directions in three-dimensional (3-D) spaces, using the rotation property of quaternions. If multiple direction vectors in 3-D space can be represented by points on the surface of a unit sphere (Fig. 43.4a), then the direction vectors chosen in TEMD are shown in Fig. 43.4b, which correspond to equidistant points taken along multiple longitudinal lines on the sphere, obtaining so-called *equi-longitudinal lines*. The projection of the input signal along these points (direction vectors) on an equi-longitudinal line can then be obtained by rotating the input signal along a rotation axis in the xy-plane and mapping it along the z-axis.

Since rotation axes are 3-D vectors, they can also be represented by a set of unit quaternions q in the xy plane, under an angle ϕ to the x-axis. Rotation axes, represented by a vector of quaternions q, can, therefore, be expressed as

$$q = 0 + \cos(\phi)\iota + \sin(\phi)j + 0\kappa .\tag{43.13}$$

Since a trivariate signal can also be represented as a pure quaternion $x(t)$, the projections of the input signal along multiple direction vectors on the sphere can be calculated by rotating the input signal about a set of

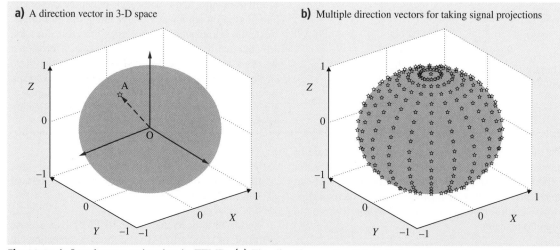

Fig. 43.4a,b Local mean estimation in TEMD: (**a**) The direction vector OA in 3-D space, which has unit norm, can also be represented by a point on the surface of a unit sphere. (**b**) Points on multiple longitudinal lines on a sphere, representing directions along which projections of the input signal can be taken by rotating the input signal along rotation axes represented by a set of unit quaternions q

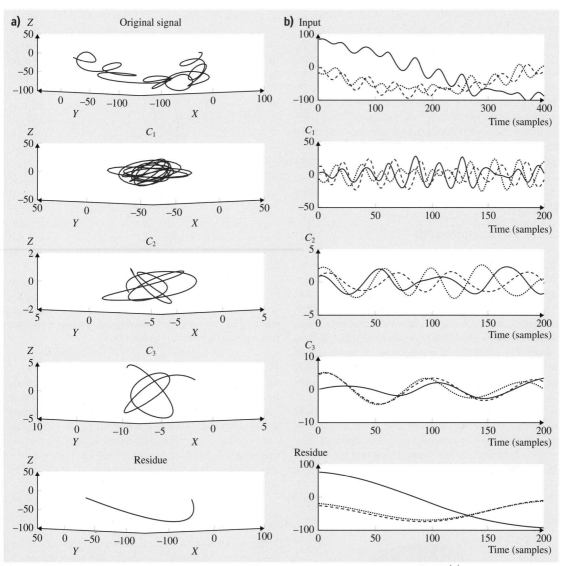

Fig. 43.5a,b A trivariate signal and its decomposition obtained using the proposed method. (**a**) 3-D plots of the input trivariate signal followed by the IMFs. (**b**) Time plots of the three components of the trivariate signal and their decomposition (*X*: *dotted line*, *Y*: *solid line*, *Z*: *dashed line*)

vectors q and taking its projection along the z-axis (κ), using

$$p_\theta^\phi = e^{q\theta} x(t)(e^{q\theta})^* \cdot \kappa, \qquad (43.14)$$

where symbol $\cdot$ denotes the dot product. To calculate the envelopes in multiple directions, angles ϕ and θ can be selected to have, respectively, N and K values between 0 and π. The range of π is necessary since both q and

$-q$ give projections in the same direction and also because the application of a unit quaternion q represents rotation by an angle 2θ.

The decomposition of a real world trivariate signal is next performed using TEMD; the corresponding results are shown in Fig. 43.5. The signal represents the 3-D orientation data generated by hand movements in a Tai Chi sequence, with a synthetically added mode for illustration purposes. The data was captured using an inertial 3-D sensor from Xsens Motion Technologies B.V. As shown in Fig. 43.5, the different 3-D rotating modes of the input trivariate signal are extracted, whereby the lower index IMFs contain higher frequency 3-D rotations and the higher index IMFs represent lower frequency rotating modes, as shown in Fig. 43.5a. The residue signal does not contain any 3-D rotating components. Time plots of the individual components of the input trivariate signal and their respective decomposition (IMFs) are also shown in Fig. 43.5b. The decomposition of individual components exhibits a mode alignment property, whereby common frequency modes in different input components are aligned in single IMFs.

43.3.4 Multivariate EMD

The multivariate empirical mode decomposition (MEMD) algorithm [43.6] extends EMD to process signals containing any number of data channels. (The free Matlab code for MEMD is available at: http://www.commsp.ee.ic.ac.uk/~mandic/research/emd.htm.) The rationale behind the MEMD algorithm is that it aims to extract local rotational components, if present, from the input signal; while BEMD and TEMD look for 2-D and 3-D rotational components, MEMD separates the rotational components in $\mathbb{R}^n$.

The MEMD operates by taking multiple real-valued projections of the input signal; the local extrema of the projected signals are interpolated component-wise to yield multidimensional envelopes of the signal. The envelopes obtained this way are then averaged

to give an estimate of the local mean. More specifically, if $e_{\{\theta_1,\theta_2,\dots,\theta_{n-1}\}}$ denotes the envelope in the direction represented by a vector $\boldsymbol{\theta} = \{\theta_1, \theta_2, \dots, \theta_{n-1}\}$ in $\mathbb{R}^n$, then the local mean can be estimated by using

$$m(t) = \frac{1}{2\pi^{n-1}} \int_{\theta_1=0}^{\pi} \int_{\theta_2=0}^{\pi} \cdots$$
$$\int_{\theta_{n-1}=0}^{2\pi} e_{\{\theta_1,\theta_2,\dots,\theta_{n-1}\}} \, \mathrm{d}\theta_1 \, \mathrm{d}\theta_2 \cdots \mathrm{d}\theta_{n-1} \,,$$

$$(43.15)$$

$$\approx \frac{1}{V_1 V_2 \cdots V_{n-1}}$$
$$\times \sum_{v_1=1}^{V_1} \sum_{v_2=1}^{V_2} \cdots \sum_{v_{n-1}=1}^{V_{n-1}} e_{\{\theta_{v_1},\theta_{v_2},\dots,\theta_{v_{n-1}}\}} \,,$$

$$(43.16)$$

where V_{n-1} denotes the number of direction vectors taken along the direction vector θ_{n-1}.

Since calculation of the local mean for multivariate signals can be considered as an approximation of the integral of all envelopes along multiple directions in $\mathbb{R}^n$, as shown in (43.16), the accuracy of the approximation is dependent on the uniformity of the chosen set of direction vectors. This is especially relevant if one only has a finite set of direction vectors. Therefore, the issue of choosing a suitable set of direction vectors for taking the signal projections in $\mathbb{R}^n$ needs special attention.

In the MEMD algorithm, the direction vectors are chosen as points generated by low-discrepancy quasi-Monte Carlo-based Hammersley sequences [43.13]. Figures 43.6 and 43.7 show, respectively, point sets on the surface of a sphere (2-sphere) and hypersphere (3-sphere), generated by the low-discrepancy Hammersley sequence. Observe that, as desired, the points generated by the low-discrepancy method are more uniformly distributed [43.14] as compared to the point set generated by the uniform angular sampling method in TEMD (Fig. 43.4b).

Consider a sequence of n-dimensional vectors $s(t) = \{s_1(t), s_2(t), \dots, s_n(t)\}$, representing a multivariate signal with n components, and $\boldsymbol{x}_{\theta_v} = \{x_1^v, x_2^v, \dots, x_n^v\}$ denoting a set of $v = 1, 2, \dots, V$ direction vectors along the directions given by the angles $\boldsymbol{\theta}_v = \{\theta_{v_1}, \theta_{v_2}, \dots \theta_{v_{n-1}}\}$ in $\mathbb{R}^n$. The proposed multivariate extension of EMD suitable for operating on general

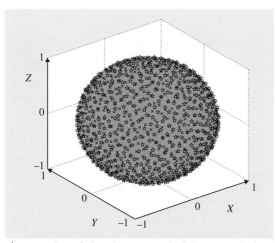

Fig. 43.6 Set of direction vectors in 3-D space obtained from a low-discrepancy Hammersley sequence

Fig. 43.7a−c Direction vectors for taking projections of a quaternion signal on a unit four-dimensional sphere (3-sphere) generated by using the Hammersley sequence. For visualization purposes, the point set is plotted on three unit spheres (2-spheres), defined respectively by the *WXY*, *XYZ*, and *WYZ*-axes. (**a**) Sampling a hypershere (3-sphere) using a low discrepancy sequence (*WXY*-axes shown), (**b**) sampling a hypershere (3-sphere) using a low discrepancy sequence (*XYZ*-axes shown), (**c**) sampling a hypershere (3-sphere) using a low discrepancy sequence (*WYZ*-axes shown) ▶

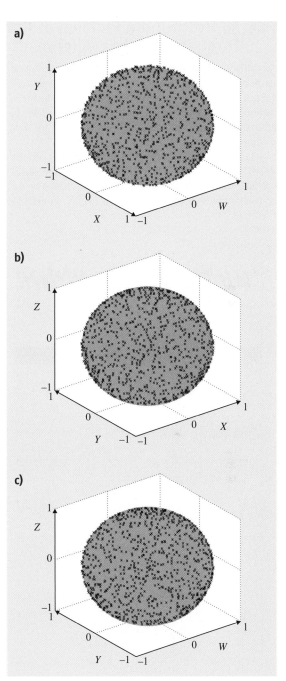

nonlinear and nonstationary multivariate time series is summarized in Algorithm 43.2.

Algorithm 43.2 Multivariate EMD

1: Find a point set for sampling an $(n - 1)$-sphere using the Hammersley sequence
2: Calculate a projection, denoted by $p_{\theta_v}(t)$, of the input signal $s(t)$ along the direction vector x_{θ_v}, for all v (the whole set of direction vectors), giving $\{p_{\theta_v}(t)\}_{v=1}^{V}$ as the set of projections
3: Find the time instants $\{t_{\theta_v}^i\}_{v=1}^{V}$ corresponding to the maxima of the set of projected signals $\{p_{\theta_v}(t)\}_{v=1}^{V}$
4: Interpolate $[t_{\theta_v}^i, s(t_{\theta_v}^i)]$ to obtain multivariate envelope curves $\{e_{\theta_v}(t)\}_{v=1}^{V}$
5: For a set of V direction vectors, the mean $m(t)$ of the envelope curves is calculated as

$$m(t) = \frac{1}{V} \sum_{v=1}^{V} e_{\theta_v}(t) \tag{43.17}$$

6: Extract *detail* $d(t)$ using $d(t) = s(t) - m(t)$. If $d(t)$ fulfills the stoppage criterion for a multivariate IMF, apply the above procedure to $s(t) - d(t)$, otherwise apply it to $d(t)$

The sifting process for a multivariate IMF can be stopped when all the projected signals fulfill any of the stoppage criteria adopted in standard EMD. One popular stopping criterion used in EMD stops the sifting when the number of extrema and the zero crossings differ at most by one for S consecutive iterations of the sifting algorithm [43.8].

43.4 Addressing the Problem of Nonuniqueness Through the Mode-Alignment Property of MEMD

The MEMD algorithm has the ability to align common scales in multivariate data: similar oscillatory scales are aligned in same-indexed IMFs from multiple channels. Such mode alignment helps to identify similar scales in different data sources, and hence, can be used for data fusion purposes [43.15, 16].

To illustrate the mode-alignment property of the MEMD algorithm, a synthetic hexavariate time series was analyzed; each component (variate), shown in the top row of Fig. 43.8 (denoted by U, V, W, X, Y, and Z),

was constructed from a set of four sinusoids. One sinusoid was made common to all components, whereas the remaining three sinusoidal components were combined so that the resulting signal had a common frequency mode in each UVX, $UVWY$, and $UWXZ$ component. A white Gaussian noise (WGN) signal was then added to first three components only. The MEMD algorithm was applied to the resulting hexavariate signal yielding multiple IMFs, as shown in Fig. 43.8. Observe that the sinusoid common to all components of the input

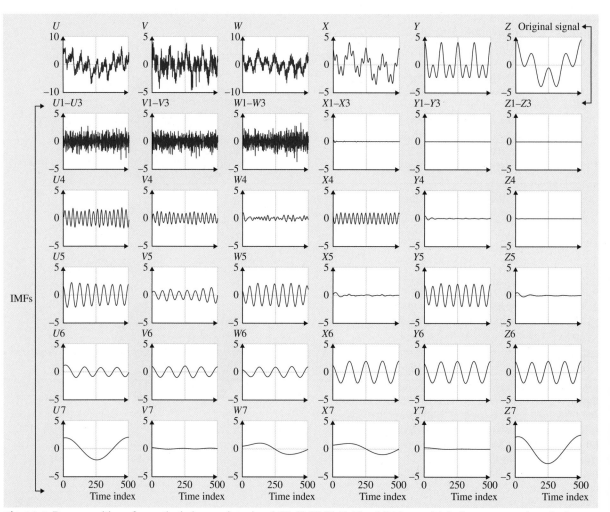

Fig. 43.8 Decomposition of a synthetic hexavariate signal (U, V, W, X, Y, Z) exhibiting multiple frequency modes, via the proposed multivariate EMD. Each IMF now carries a single frequency mode, illustrating the alignment of common scales within different components of a hexavariate signal

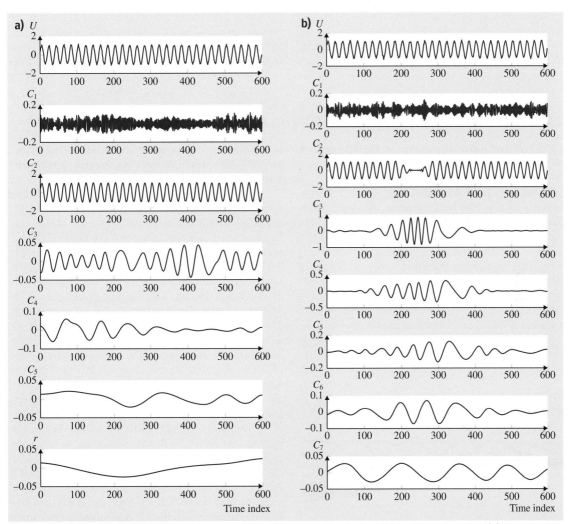

Fig. 43.9a,b EMD applied to a sinusoid corrupted by different realizations of white Gaussian noise (**a**) $M = 5$ IMFs obtained by applying EMD to a sinusoid corrupted by WGN (**b**) $M = 9$ (of which seven are shown) IMFs obtained from a signal corrupted by a different WGN realisation. Due to mode mixing, the original signal of interest (sinusoid) can be located in different modes for different decompositions, C_2 in (**a**) and C_2, C_3, and C_4 in (**b**)

signal is the sixth IMF, whereas the remaining three frequency modes were also accurately extracted in the respective IMFs, illustrating mode alignment between multiple data channels.

The mode alignment cannot be achieved by applying EMD to each channel separately due to the empirical nature [43.3]. To illustrate the empirical nature and the subsequent lack of mode alignment in its standard EMD algorithm, simulations were performed on a pair of signals consisting of a sinusoid of a certain frequency but corrupted with the different realizations

of WGN. In the case of the signal corrupted by the first realisation of WGN, $M = 5$ extracted IMFs from the EMD algorithm are shown in Fig. 43.9a. It should be noted that the original noise-free sinusoid corresponds to the second IMF, whereas the noise is mostly concentrated in the first IMF.

Figure 43.9b shows the IMFs obtained by applying EMD to the same sinusoid, as in the previous case but now corrupted by a different noise realization. In this case, a different number of IMF components ($M = 9$, only 7 of which are shown) were obtained and the original sinusoid is also distributed across several IMFs. This phenomenon, whereby similar oscillatory modes are found across different IMFs, is known as mode mixing. The mode-mixing problem hinders a multiscale com-

parison of IMFs from multiple channels and is a major obstacle in using EMD in fusion applications. Due to the mode-alignment property of multivariate extensions of EMD, however, this problem can be addressed [43.3].

43.5 Filter Bank Property of MEMD and Noise-Assisted MEMD

43.5.1 Filter Bank Property of MEMD-Based Decomposition

It is well-established that standard univariate EMD observes a dyadic filterbank structure for input white Gaussian noise (WGN) data [43.17]. The idea of a filter bank structure for multivariate inputs, however, is ambiguous since the concept of frequency is not clearly defined for multivariate signals. Even if the frequency response of individual channels of a multivariate signal are considered, the filter bank structure imposes an additional constraint on the frequency output of each multivariate IMF – *overlapping of the filter bands associated with the corresponding (same-index) IMFs from multiple channels*. It is vital for the IMFs obtained from MEMD to be physically meaningful, as any mismatch

in the frequency contents of the corresponding multi-channel IMFs would render their correlation estimates erroneous [43.7].

To perform the spectral analysis of MEMD, extensive simulations were carried out on multiple realizations of an 8-channel WGN process, which can be considered as a special case of fractional Gaussian noise (FGN), for $H = 0.5$. In simulations, $N = 500$ WGN realizations were used, each of length $K = 1000$, which were then ensemble averaged to yield an averaged power spectra. The stopping criterion used for MEMD is given in [43.8], with the value of $S = 5$. While the number of IMFs varied for different input realizations, it was never less than nine and, thus, the first nine IMFs are considered in the following analysis.

In the first case, the frequency response and the corresponding quasi-dyadic filter bank property of MEMD are illustrated by applying MEMD on $N = 500$ realizations of an 8-channel WGN; the power spectra of its resulting first nine IMFs are plotted in the top part of Fig. 43.10. Next, the same eight noise channels were separately processed via the standard EMD algorithm; the estimated power spectra of its IMFs are shown in the lower half of Fig. 43.10. It can be seen that the outputs from both the EMD and the MEMD algorithms follow a quasi-dyadic filter bank structure. However, overlapping of frequency bands of same-index IMFs, associated with different channels, is more prominent in the case of MEMD as compared to standard EMD. For a given number of noise realizations N, standard EMD failed to properly align the bandpass filters associated with the corresponding IMFs from different noise channels. Although this alignment is expected to become better with an increase in the number of noise realizations, MEMD-based spectra achieved much better results with the same number of ensembles.

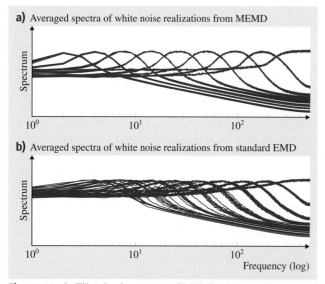

a) Averaged spectra of white noise realizations from MEMD

b) Averaged spectra of white noise realizations from standard EMD

Frequency (log)

Fig. 43.10a,b Filter bank property of MEMD: Averaged spectra of IMFs obtained for $N = 500$ realizations of 8-channel white Gaussian noise via MEMD (**a**) and the standard EMD (**b**). Overlapping of the frequency bands corresponding to the same-index IMFs is improved in both cases after averaging, but MEMD bands show much better alignment

43.5.2 Noise-Assisted MEMD Algorithm

Based on the quasi-dyadic filter bank structure of MEMD for multivariate WGN inputs, a noise assisted

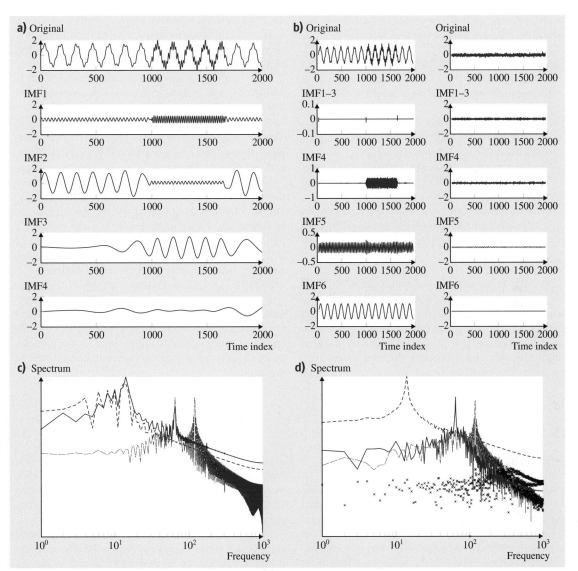

Fig. 43.11a–d NA-MEMD for reducing mode mixing: (**a**) IMFs of a synthetic signal obtained by applying standard EMD. (**b**) IMFs of synthetic signal obtained by applying NA-MEMD (*left column*); IMFs of one of the two noise channels (*right column*). (**c**) Spectrum of IMFs obtained from standard EMD. Mode mixing is evident due to overlapping of spectra from IMF1 and IMF2; and IMF2 and IMF3. (**d**) Spectrum of IMFs obtained from NA-MEMD; due to the added noise channels, the spectrum of the IMFs from the original signal do not overlap, in turn, reducing mode mixing. (See text for more detail)

MEMD algorithm (NA-MEMD) has been presented recently [43.7] which aims to provide a better decomposition of an input signal as compared to the standard EMD algorithm by reducing its inherent mode-mixing problem. It operates by adding extra channels containing multivariate independent WGN to the input signal, and then processes such a composite signal via MEMD. Incorporating extra noise channels imposes the desired dyadic filter bank structure on the data while also keeping the original input separate from the noise. Next, the IMF channels corresponding to the white noise are discarded, yielding the set of IMFs associated with only the original input signal. The details of the NA-MEMD method are outlined in Algorithm 43.3.

Algorithm 43.3 Noise–Assisted MEMD Algorithm

1: Create an uncorrelated Gaussian white noise time series (m-channel) of the same length as that of the input
2: Add the noise channels (m-channel) created in step 1 to the input multivariate (n-channel) signal, obtaining an ($n + m$)-channel signal
3: Process the resulting ($n + m$)-channel multivariate signal using the MEMD algorithm to obtain multivariate IMFs
4: From the resulting ($n + m$)-variate IMFs, discard the m channels corresponding to the noise, giving a set of n-channel IMFs corresponding to the original signal

---

In order to demonstrate the ability of NA-MEMD to reduce mode-mixing, a simulation was performed on a synthetic signal consisting of a combination of three predefined tones; two low frequency tones were added together along with a high frequency sinusoid which was only added between the time index 5000 and 8500. The resulting signal and its decomposition obtained from the standard EMD algorithm are shown in Fig. 43.11a. Mode mixing is evident since IMF1 contains multiple modes. Moreover, mode mixing can also be seen in IMF2 and IMF3.

The same signal was next processed using the NA-MEMD method with two extra noise channels ($m = 2$). The IMFs from the resulting trivariate signal are shown in Fig. 43.11b. Observe that the IMFs corresponding to the first channel are now free of mode mixing, as all the tones are decomposed as separate IMFs (IMF4, IMF5, and IMF6, respectively).

To analyze the effects of adding noise channels to the original synthetic signal, the spectra of IMFs obtained from the standard EMD and the NA-MEMD are plotted in Fig. 43.11c,d, respectively. It can be noticed that in the case of the standard EMD, the spectra of IMF1-IMF2 and IMF2-IMF3 overlap with each other, resulting in mode mixing in the time domain, as is evident in Fig. 43.11a. On the other hand, the power spectra of IMFs obtained from the NA-MEMD are well-separated in the frequency domain due to the quasi-filter bank structure imposed by incorporating extra WGN channels; the resulting decomposition in the time domain, therefore, does not suffer from the mode-mixing problem.

43.6 Applications

43.6.1 Phase Synchrony

Brain function requires the integration of numerous widely distributed functional areas in constant interaction with each other [43.18, 19]; the degree of integration between different brain locations can be characterized by sychronization features that reflect the similarity between the temporal structure of signal-pairs. Unlike standard measures such as coherence [43.20] or cross-correlation [43.21], which are limited by assumptions of linearity and stationarity, features based on phase synchrony – the temporal locking of phase information – can model higher-order synchronization statistics and have been widely used in the study of sleep [43.22] and epilepsy [43.23]. Typically, phase synchrony studies have employed the wavelet transform [43.24] and the Hilbert transform [43.25]. The wavelet transform is, however, based on a projection onto a fixed set of basis functions, which limits its time-frequency resolution and its analysis of nonstationary data. On the other hand, the Hilbert transform is only suitable for phase estimation if the data is first bandpass filtered so that it satisfies narrowband criteria, making the approach critically sensitive to the a priori selection of bandpass filter cutoffs so that synchrony events are likely to be missed [43.26].

The adaptive fashion in which EMD extracts signal components with well-defined phase information makes it highly suitable for the task at hand, and in [43.26] it was shown how the single channel algorithm can outperform standard techniques. As illustrated in this chapter, multivariate extensions of EMD exhibit desirable properties at the IMF level – mode alignment, greater robustness to mode mixing – that enable a more rigorous comparison between multichannel components. It was shown in [43.3] how bivariate EMD, and more recently multivariate EMD [43.27], exploit these properties to give a more localized estimate of phase synchrony and with greater accuracy in the presence of noise. The degree of phase synchrony between two channels of an MEMD decomposition is calculated as follows. The phase difference at the i-th IMF level is

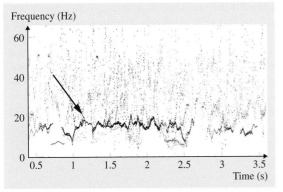

Fig. 43.12 Average phase synchrony spectrogram for EEG data. Note the increase in the phase coherence value at 1–3 s caused by a visual stimulus

calculated as ϕ_m, and its deviation from the δ distribution (perfect synchrony) can be quantified by estimating the phase coherence value (PCV) [43.25] as

$$\rho_n(k) = \frac{H_{\max} - H}{H_{\max}}, \tag{43.18}$$

where $H = -\sum_{n=1}^{N} p_n \ln p_n$; the Shannon entropy of the distribution of $\phi_m(k - W/2 : k + W/2)$ defined by a window of length W, N is the number of bins and p_n is the probability of $\phi_m(k - W/2 : k + W/2)$ within the

n-th bin [43.25]. The maximum entropy $H_{\max}$ is a function of W (see [43.25] for more details). The value of ρ is between 0 and 1, 1 indicating perfect synchrony and 0 a nonsynchronous state. The process is repeated for every IMF level, between every channel pair.

To illustrate how phase synchrony features can be used to detect changes in brain state, the MEMD-based approach was applied to multivariate EEG signals. A 4 s recording was made for a single subject, wherein the subject was presented with a visual stimulus for the time period 1–3 s. The stimulus was flashing at 20 Hz and produced the steady state evoked potential (SSVEP) response. The data was recorded according to the 10–20 system from electrodes Fp1, Fp2, F3, and F4 at 256 Hz. The average phase synchrony for several of the electrode pairs is plotted in Fig. 43.12 across time and frequency. Note that there is a clear increase in the phase coherence value at 20 Hz for the time period 1–3 s, indicating that the channels have become synchronized during the stimulus interval at the same frequency as the stimulus.

43.6.2 Classification of Motor Imagery Data

Recently brain–computer interface (BCI) based on the motor imagery response, that is, the imagination of a motor action without any real motor output, has been

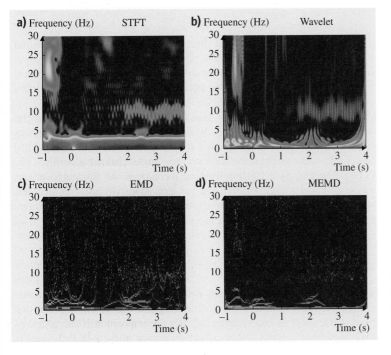

Fig. 43.13 Time-frequency representations of STFT, wavelet (Morlet), EMD, and MEMD for the *right-hand* motor imagery response. The results are the average of 'C4' and 'C6' electrode responses. Note the time-frequency localized MEMD spectrum compared to other methods

the subject of many studies [43.28]. The basis for motor imagery BCI is that the so-called *mu* rhythms (8–12 Hz) or sensorimotor rhythms (SMR) and *beta* rhythms (18–25 Hz) have been observed in the central region of the brain using both EEG and magnetoencephalography (MEG), when subjects plan and execute their hand or finger movements [43.29]. The MEMD algorithm is a natural choice for processing motor imagery data which is inherently nonstationary and multivariate. In this section, time-frequency analysis and classification performance of the MEMD algorithm is evaluated on the *BCI Competition IV Dataset I* [43.30] containing human-generated motor imagery data. During the recordings, subject a conducted *left hand* and *foot*, subject b *left hand* and *right hand*, subject f *left hand* and *foot*, and subject g *left hand* and *right hand* motor imagery tasks.

Figure 43.13 shows the spectra obtained by using the short-time Fourier transform (Fig. 43.13a), Wavelet transform (Fig. 43.13b), EMD (Fig. 43.13c), and MEMD (Fig. 43.13d). The motor imagery task lasted from 0–4 s and as expected, from 2 s after the cue sign was displayed, the spectrograms obtained using all the above methods illustrate ipsilateral increase, and event-related synchronization (ERS) in alpha and beta bands on the sensorimotor cortex during motor imagery. Note that among these four techniques, the MEMD approach was the most accurate in extracting the *mu* rhythm (approximately 10 Hz), which demonstrates its ability to perform a highly localised time-frequency estimation by exploiting multichannel information.

The classification performances of the above algorithms were also compared for all 200 motor imagery trials. For this purpose, input EEG data was first filtered into frequency bands ranging from 8 to 30 Hz, which contained *mu* and *beta* rhythms. The data collected for each electrode was then decomposed simultaneously using MEMD, whereas the standard EMD was applied independently channel-wise. IMFs decomposed using the standard EMD and MEMD were retained or omitted in an ad-hoc fashion to obtain bandpass

Table 43.1 Average classification rates of 20 different scenarios based on the different filter for four subjects. The Morlet wavelet was used for wavelet decomposition and the number of ensembles for EEMD was 100. Note that MEMD-CSP using $c_2(t)$ and $c_3(t)$ produced the best results for all subjects. Subject a conducted *left hand* and *foot*, subject b *left hand* and *right hand*, subject f *left hand* and *foot*, and subject g *left hand* and *right hand* motor imagery tasks

Subject	Algorithm	IMFs	Classification rate (%)
a	DFT		62.0 ± 11.2
	wavelet		66.2 ± 10.6
	EMD	$c_1(t) - c_3(t)$	57.0 ± 6.6
	MEMD	$c_2(t) - c_3(t)$	70.5 ± 11.2
b	DFT		57.6 ± 7.5
	wavelet		71.4 ± 6.5
	EMD	$c_1(t) - c_4(t)$	52.1 ± 5.7
	MEMD	$c_2(t) - c_3(t)$	75.6 ± 5.2
f	DFT		52.6 ± 6.9
	wavelet		52.9 ± 5.7
	EMD	$c_1(t) - c_4(t)$	52.2 ± 5.9
	MEMD	$c_2(t) - c_3(t)$	57.5 ± 13.4
g	DFT		86.9 ± 7.4
	wavelet		78.8 ± 9.4
	EMD	$c_1(t) - c_3(t)$	65.5 ± 10.8
	MEMD	$c_2(t) - c_3(t)$	91.9 ± 3.0

filtered signals. The common spatial patterns (CSP) algorithm [43.31] was used to extract features relevant to motor imagery responses. Table 43.1 shows the classification performances for four subjects using the discrete Fourier transform (DFT), Morlet wavelet, EMD, and the MEMD algorithm. It is obvious that the MEMD-CSP features obtained by applying CSP to the sum of selected IMFs always outperformed the other methods, where IMFs $c_2(t)$ and $c_3(t)$ were used for all subjects. On average, the MEMD-CSP feature yielded the highest classification performance of 73.9%, a 9.1% improvement over DFT, a 6.6% improvement over wavelet, and a 17.2% improvement over standard EMD.

43.7 Discussion and Conclusions

Recent extensions of the EMD algorithm to signals containing multiple channels have been reviewed and compared. More specifically, complex, trivariate and multivariate extensions of the EMD algorithm were presented with special emphasis on the multivariate EMD (MEMD) algorithm and its applications in bio-engineering. The benefits of MEMD with regards to its mode-alignment property for general inputs and filter bank property for multivariate WGN signals have been detailed. These properties of MEMD led to the development of the noise-assisted MEMD (NA-MEMD) algorithm which alleviates problems associated with the original EMD such as mode-mixing. Multivariate extensions of EMD and its variants, such as the NA-MEMD algorithm, are expected to become a standard in real world multivariate signal processing due to their data-driven nature, a characteristic necessary to process nonstationarity. Applications based on real world electroencephalogram recordings, such as estimating stimulus-induced neural synchronisation and the motor imagery response, support this claim.

References

43.1   N.E. Huang, Z. Shen, S. Long, M. Wu, H. Shih, Q. Zheng, N. Yen, C. Tung, H. Liu: The empirical mode decomposition and Hilbert spectrum for non-linear and non-stationary time series analysis, Proc. R. Soc. A **454**, 903–995 (1998)

43.2   N.E. Huang, Z. Wu: A review on Hilbert–Huang transform: Method and its applications to geophysical studies, Rev. Geophys. **46**, RG2006 (2008)

43.3   D. Looney, C. Park, P. Kidmose, M. Ungstrup, D.P. Mandic: Measuring phase synchrony using complex extensions of EMD, Proc. IEEE Stat. Signal Process. Symp. (2009) pp. 49–52

43.4   G. Rilling, P. Flandrin, P. Goncalves, J.M. Lilly: Bivariate empirical mode decomposition, IEEE Signal Process. Lett. **14**, 936–939 (2007)

43.5   N. Rehman, D.P. Mandic: Empirical mode decomposition for trivariate signals, IEEE Trans. Signal Process. **58**, 1059–1068 (2010)

43.6   N. Rehman, D.P. Mandic: Multivariate empirical mode decomposition, Proc. R. Soc. A **466**, 1291–1302 (2010)

43.7   N. Rehman, D.P. Mandic: Filterbank property of multivariate empirical mode decomposition, IEEE Trans. Signal Process. **59**, 2421–2426 (2011)

43.8   N.E. Huang, M. Wu, S. Long, S. Shen, W. Qu, P. Gloersen, K. Fan: A confidence limit for the empirical mode decomposition and Hilbert spectral analysis, Proc. R. Soc. A **459**, 2317–2345 (2003)

43.9   N.E. Huang, Z. Wu, S.R. Long, K.C. Arnold, K. Blank, T.W. Liu: On instantaneous frequency, Adv. Adapt. Data Anal. **1**, 177–229 (2009)

43.10   D.P. Mandic, V.S.L. Goh: *Complex Valued Nonlinear Adaptive Filters: Noncircularity, Widely Linear Neural Models* (Wiley, New York 2009)

43.11   T. Tanaka, D.P. Mandic: Complex empirical mode decomposition, IEEE Signal Process. Lett. **14**(2), 101–104 (2006)

43.12   M.U. Altaf, T. Gautama, T. Tanaka, D.P. Mandic: Rotation invariant complex empirical mode decomposition, Proc. IEEE Int. Conf. Acoust. (2007), Speech, Signal Process.

43.13   H. Niederreiter: *Random number generation and quasi-Monte Carlo methods*, CBMS-NSF Regional Conference Series in Applied Mathematics, Vol. 63 (Society for Industrial and Applied Mathematics, Philadelphia 1992)

43.14   J. Cui, W. Freeden: Equidistribution on the sphere, Siam J. Sci. Comput. **18**(2), 595–609 (1997)

43.15   N. Rehman, Y. Xia, D.P. Mandic: Application of multivariate empirical mode decomposition for seizure detection in EEG signals, Proc. 32rd Annu. Int. Conf. IEEE Eng. Med. Biol. Soc. (EMBC) (2010) pp. 1650–1653

43.16   D. Looney, D.P. Mandic: Multi-scale image fusion using complex extensions of EMD, IEEE Trans. Signal Process. **57**(4), 1626–1630 (2009)

43.17   P. Flandrin, G. Rilling, P. Goncalves: Empirical mode decomposition as a filter bank, IEEE Signal Process. Lett. **11**, 112–114 (2004)

43.18   K.J. Friston, K.M. Stephan, R.S.J. Frackowiak: Transient phase locking and dynamic correlations: Are they the same thing?, Human Brain Mapp. **5**, 48–57 (1997)

43.19   G. Tononi, M. Edelman: Consciousness and complexity, Science **282**, 1846–1851 (1998)

43.20   R. Srinivasan, D. Russell, G. Edelman, G. Tononi: Increased synchronization of neuromagnetic responses during conscious perception, J. Neurosci. **19**(13), 5435–5448 (1999)

43.21   V. Menon, W.J. Freeman, B.A. Cutillo, J.E. Desmond, M.F. Ward, S.L. Bressler, K.D. Laxer, N. Barbaro, A.S. Gevins: Spatio-temporal correlations in human gamma band electrocorticograms, Electroencephalogr. Clin. Neurophysiol. **98**(2), 89–102 (1996)

43.22   R. Ferri, F. Rundo, O. Bruni, M. Terzano, C. Stam: Dynamics of the EEG slow-wave synchronization during sleep, Clin. Neurophysiol. **116**(12), 2783–2795 (2005)

43.23   E.R. Kandel, J.H. Schwartz: *Principles of Neural Science*, 3rd edn. (Appleton and Lange, Norwalk 1985)

43.24   J. Lachaux, A. Lutz, D. Rudrauf, D. Cosmelli, M. Quyen, J. Martinerie, F. Varela: Estimating the time-course of coherence between single-trial brain signals: An introduction to wavelet coherence, Clin. Neurophysiol. **32**(3), 157–174 (2002)

43.25   P. Tass, M.G. Rosenblum, J. Weule, J. Kurths, A. Pikovsky, J. Volkmann, A. Schnitzler, H.-J. Freund: Detection of $n:m$ phase locking from noisy data: Application to magnetoencephalography, Phys. Rev. Lett. **81**(15), 3291–3294 (1998)

43.26   C.M. Sweeny-Reed, S.J. Nasuto: A novel approach to the detection of synchronisation in EEG based on empirical mode decomposition, J. Comput. Neurosci. **23**(1), 79–111 (2007)

43.27   A.Y. Mutlu, S. Aviyente: Multivariate empirical mode decomposition for quantifying multivariate phase synchronization, EURASIP J. Adv. Signal Process. **2011**, 1–13 (2011)

43.28   G. Pfurtscheller, C. Neuper, D. Flotzinger, M. Pregenzer: EEG-based discrimination between imagination of right and left hand movement, Electro-

encephalogr. Clin. Neurophysiol. **103**(6), 642–651 (1997)

43.29   G. Pfurtscheller, F.H. Lopes da Silva: Event-related EEG/MEG synchronization and desynchronization: Basic principles, Clin. Neurophysiol. **110**(11), 1842–1857 (1999)

43.30   B. Blankertz: BCI Competition IV (Berlin Institute of Technology), accessible online at http://www.bbci.de/competition/iv/

43.31   H. Ramoser, J. Muller-Gerking, G. Pfurtscheller: Optimal spatial filtering of single trial EEG during imagined hand movement, IEEE Trans. Rehabil. Eng. **8**(4), 441–446 (2000)

Part I

Information

Part I Information Modeling of Perception, Sensation and Cognition

Ed. by Lubica Benuskova

44. Modeling Vision with the Neocognitron

Kunihiko Fukushima

The *neocognitron*, which was proposed by *Fukushima* [44.1], is a neural network model capable of robust visual pattern recognition. It acquires the ability to recognize patterns through learning.

The neocognitron is a hierarchical network consisting of many layers of neuron-like cells. Its architecture was originally suggested from neurophysiological findings on visual systems of mammals. There are bottom-up connections between cells in adjoining layers. Some of these connections are variable and can be modified by learning. The neocognitron can acquire the ability to recognize patterns by learning. Since it has a large power of generalization, presentation of only a few typical examples of deformed patterns (or features) is enough for learning. It is not necessary to present all of the deformed versions of the patterns that might appear in the future. After learning, the neocognitron can recognize input patterns robustly, with little effect from deformation, changes in size, or shifts in location. It is even able to correctly recognize a pattern that has not been presented before, provided that it resembles one of the training patterns.

The principle of the neocognitron can be used in various kinds of pattern recognition systems, such as recognizing handwritten characters. Further

extensions and modifications of the neocognitron have been proposed to endow it with a function of selective attention, an ability to recognize partly occluded patterns, and so on.

44.1 Overview

In the visual systems of mammals, visual scenes are analyzed in parallel by separate channels. Loosely speaking, information concerning an object's shape is mainly analyzed through the temporal pathway in the cerebrum, while information concerning visual motion and location is mainly analyzed through the occipito-parietal pathway. The neocognitron is an artificial neural network, whose architecture was initially suggested from neurophysiological findings on the tem-

poral pathway: retina → LGN → area V1 (primary visual cortex) → area V2 → area V4 → IT (inferotemporal cortex).

In area V1, cells respond selectively to local features of a visual pattern, such as lines or edges in particular orientations [44.2, 3]. In areas V2 and V4, cells exist that respond selectively to complex visual features (e.g., [44.4–6]). In the inferotemporal cortex, cells exist that respond selectively to more complex features,

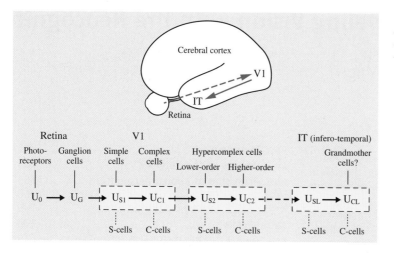

Fig. 44.1 Relation between the architecture of the neocognitron and the classical hypothesis of Hubel and Wiesel (after [44.1])

or even to human faces (e.g., [44.7–9]). Thus, the visual system seems to have a hierarchical architecture, in which simple features are first extracted from a stimulus pattern and then integrated into more complicated ones. In this hierarchy, a cell in a higher stage generally has a larger receptive field and is more insensitive to the location of the stimulus. This kind of physiological evidence suggested the network architecture for the neocognitron.

In the 1960s, Hubel and Wiesel classified cells in the visual cortex into simple, complex, lower-order hypercomplex, and higher-order hypercomplex cells. They hypothesized that visual information is processed hierarchically through simple cells → complex cells → lower-order hypercomplex cells → higher-order hypercomplex cells [44.2, 3]. They suggested that, in this hierarchy, the relation between lower-order hypercomplex cells to higher-order hypercomplex cells resembles that between simple cells to complex cells. Although classifying hypercomplex cells into lower-order and higher-order has not been popular among neurophysiologists recently, it is this hypothesis that suggested the original architecture of the neocognitron model when it was first proposed by *Fukushima* in 1980 [44.1].

In the neocognitron, there are two major types of cells, namely *S-cells* and *C-cells*. S-cells, which are named after simple cells, correspond to simple cells or lower-order hypercomplex cells. Similarly, C-cells, which are named after complex cells, correspond to complex cells or higher-order hypercomplex cells. As is shown in Fig. 44.1, the neocognitron consists of the cascaded connection of a number of modules, each of which consists of a layer of S-cells followed by a layer of C-cells.

44.2 Outline of the Network

The neocognitron is a multi-layered network, which consists of layers of S-cells and C-cells. These layers of S-cells and C-cells are arranged alternately in a hierarchical manner.

S-cells work as feature-extracting cells. They resemble the simple cells of the primary visual cortex in their response. Their input connections are variable and are modified through learning. After learning, each S-cell comes to respond selectively to a particular visual feature presented in its receptive field. The features extracted by S-cells are determined during the learning process. Generally speaking, *local* features, such as edges or lines in particular orientations, are extracted

in lower stages. More *global* features, such as parts of learning patterns, are extracted in higher stages.

C-cells, which resemble complex cells in the visual cortex, are inserted in the network to allow for positional errors in the features of the stimulus. The input connections of C-cells, which come from S-cells of the preceding layer, are fixed and invariable. Each C-cell receives excitatory input connections from a group of S-cells that extract the same feature, but from slightly different locations. The C-cell responds if at least one of these S-cells yields an output. Even if the stimulus feature shifts and another S-cell comes to respond instead of the first one, the same C-cell

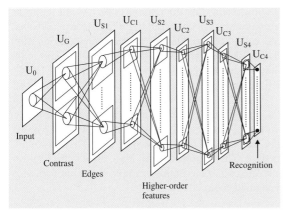

Fig. 44.2 A typical architecture of the neocognitron (after [44.10])

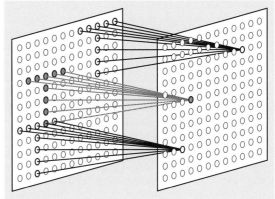

Fig. 44.3 An illustration of shared connections between two cell-planes. All cells in a cell-plane share the same set of input connections (after [44.1])

keeps responding. Thus, the C-cell's response is less sensitive to a shift in location of the input pattern. We can also express that C-cells have a blurring operation, because the response of a layer of S-cells is spatially blurred in the response of the succeeding layer of C-cells.

There are several versions of the neocognitron, which have slightly different architectures. Figure 44.2 shows a typical architecture of the network. Here we use notation like U_{Sl}, for example, to indicate the layer of S-cells of the l-th stage.

There are retinotopically ordered connections between cells of adjoining layers. Each cell receives input connections that lead from cells situated in a limited area on the preceding layer. Since cells in higher stages come to have larger receptive fields, the density of cells in each layer is designed to decrease with the order of the stage.

Each layer of the network is divided into a number of sub-layers, called *cell-planes*, depending on the feature to which cells respond preferentially. In Fig. 44.2, each rectangle drawn with thick lines represents a cell-plane. Incidentally, a cell-plane is a group of cells that are arranged retinotopically and share the same set of input connections [44.1]. Namely, all cells in a cell-plane share the same set of input connections, as illustrated in Fig. 44.3. In other words, the connections to a cell-plane have a translational symmetry. As a result, all cells in a cell-plane have identical receptive fields but at different locations. During learning, the modifica-

Fig. 44.4 An example of the response of a neocognitron that has been trained to recognize handwritten digits. The input pattern is recognized correctly as '5' ▶

tion of variable connections also progresses under the constraint of shared connections.

The lowest stage of the hierarchical network is the input layer U_0 consisting of two-dimensional array of cells, which correspond to photoreceptors of the retina. Stimulus patterns are presented to the input layer, U_0.

In the network shown in Fig. 44.2, a layer of contrast-extracting cells (U_G) follows the input layer. The cells of U_G correspond to retinal ganglion cells or lateral geniculate nucleus cells. Layer U_G consists of two cell-planes: one cell-plane consisting of cells with concentric on-center receptive fields and one cell-plane consisting of cells with off-center receptive fields. The former cells extract positive contrast in brightness, whereas the latter extract negative contrast from the images presented to the input layer.

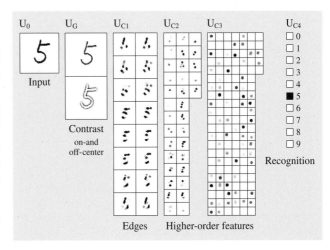

The output of U_G is then sent to U_{S1}. Each S-cell of U_{S1} resembles a simple cell in the primary visual cortex and responds selectively to an edge at a particular orientation. As a result, contours in the input image are decomposed into edges for every orientation in U_{S1}.

The hierarchical network has a number of stages, each of which consists of S-cell and C-cell layers. At each stage of the network, the output of layer U_{Sl} is fed into layer U_{Cl}. The response of layer U_{Cl} is then fed to U_{Sl+1}, the layer of S-cells of the next stage, where more global features are extracted.

S-cells at the highest stage (U_{SL}; $L = 4$ in the network of Fig. 44.2) are trained by supervised competitive learning using labeled training data. As the network learns varieties of deformed training patterns, more than one cell-plane per class is usually generated in U_{SL}. During the recognition phase, the label of the input stimulus is inferred from the response of U_{SL}. The C-cells at the highest stage (U_{CL}) show the inferred label.

Figure 44.4 shows an example of the response of a neocognitron that has learned to recognize handwritten digits. The responses of layers U_0, U_G, and layers of C-cells of all stages are displayed in series from left to right. The rightmost layer, U_{C4}, shows the final result of recognition. In this example, the input pattern is recognized correctly as '5'.

44.3 Principles of Robust Recognition

44.3.1 Toleration of Shift by C-Cells

In the whole network, with its alternate layers of S-cells and C-cells, the process of feature extraction by S-cells and toleration of shift by C-cells is repeated. During this process, local features extracted in lower stages are gradually integrated into more *global* features, as illustrated in Fig. 44.5.

Since small amounts of positional errors of local features are absorbed by the blurring operation by C-cells, each S-cell in a higher stage comes to respond robustly to a specific feature even if the feature is slightly deformed or shifted.

Let an S-cell in an intermediate stage of the network have already been trained to extract a global feature consisting of three local features of a training pattern A, as illustrated in Fig. 44.6. By the function of its presynaptic C-cells, the S-cell tolerates positional error of each local feature if the deviation falls within the dotted circle. Hence, the S-cell responds to any of the deformed patterns shown in Fig. 44.7b in a similar way as

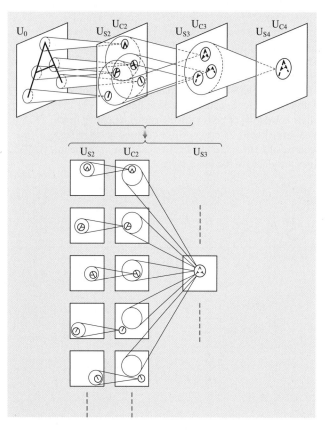

Fig. 44.5 The process of pattern recognition in the neocognitron. The lower half of the figure is an enlarged illustration of a part of the network (after [44.1]) ◄

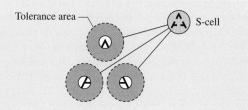

Fig. 44.6 Connections converging to an S-cell that has learned a global feature consisting of three local features of a training pattern A

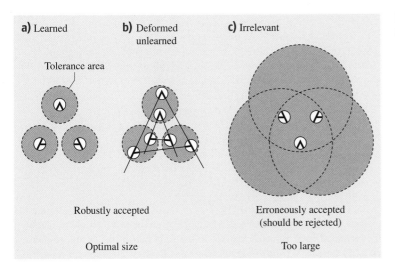

Fig. 44.7 Optimal size of the tolerance areas (after [44.11])

shown in Fig. 44.7a. The toleration of positional errors, however, should not be too large at this stage.

If large errors are tolerated at any one step, the network may come to respond erroneously, such as by recognizing a stimulus like Fig. 44.7c as an *A* pattern. Thus, tolerating positional error a little at a time at each stage, rather than all in one step, plays an important role in endowing the network with the ability to recognize even distorted patterns.

The C-cells in the highest stage work as recognition cells, which indicate the result of the pattern recognition. Each C-cell of the recognition layer at the highest stage integrates all the information of the input pattern, and responds only to one specific pattern. Since errors in the relative location of local features are tolerated in the process of extracting and integrating features, the same C-cell responds in the recognition layer at the highest stage, even if the input pattern is deformed, changed in size, or shifted in location. In other words, after having finished learning, the neocognitron can recognize input patterns robustly.

44.3.2 Blur by C-cells

The role of C-cells can also be understood from a different point of view. As illustrated in Fig. 44.8, the operation made by connections from S to C-cells can also be interpreted as a blurring operation, as well as an operation of tolerating shift.

Each S-cell measures the similarity between the stimulus feature and the feature that the S-cell has learned during the learning phase. As will be discussed later in Sect. 44.4.1, the similarity, which is defined by

the inner product of two feature vectors, is determined by the degree of overlap between the two vectors. The two patterns in the left and the center of Fig. 44.9a are perceived to be quite similar to each other when observed visually by human beings. S-cells, however, judge them completely differently, because their similarity defined by the inner product is zero. This is quite different from our natural feelings. If the patterns are blurred like in Fig. 44.9b, they will overlap greatly, and S-cells will also judge that they are similar to each other. This coincides with our natural feelings.

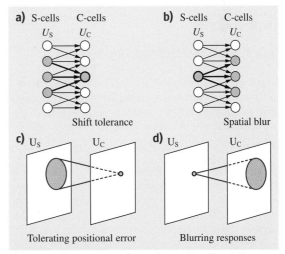

Fig. 44.8 Connections from S to C-cells: 1-D cross section (**a**), (**b**) and 3-D view (**c**), (**d**). Two different interpretations of the function of C-cells: tolerating shift (**a**), (**c**) and spatial blur (**b**), (**d**) (after [44.12])

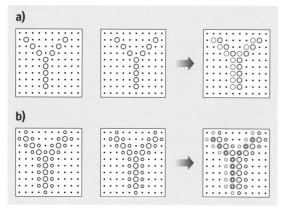

Fig. 44.9a,b Similarity between patterns, which is measured with the degree of overlap, is largely increased by blurring operation (after [44.12]). (**a**) There is no similarity (no overlap) between two patterns when they are not blurred. (**b**) There is a large similarity (overlap) between two patterns after blurring operation

If input patterns are blurred directly, however, fine details of the patterns are lost. Hence in the neocognitron, the blurring operation by C-cells is performed after extracting features by S-cells. Namely, responses of individual cell-plans of S-cells are blurred in the succeeding cell-planes of C-cells, as shown in Fig. 44.10.

We can summarize the function of C-cells that, by averaging their input signals, C-cells exhibit some level

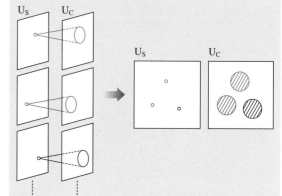

Fig. 44.10 Blurring operation for each cell-plane, namely, for each feature. In the right half of the figure, responses from different cell-planes are shown in different colors (after [44.12])

of translational invariance. As a result of averaging across location, C-cells encode a blurred version of their input. The blurring operation is essential for endowing the neocognitron with an ability to recognize patterns robustly, with little effect from deformation, change in size, or shift in the location of input patterns. The averaging operation is important, not only for endowing neural networks with an ability to recognize deformed patterns robustly, but also for smoothing additive random noise contained in the responses of S-cells.

44.4 S-cells

44.4.1 Feature Extraction by S-Cells

S-cells work as feature-extracting cells. Their input connections are variable and are modified through learning. After learning, each S-cell comes to respond selectively to a particular visual feature presented in its receptive field.

Input Signals to an S-Cell

To show the essence of the process of feature extraction, we watch the circuit converging to a single S-cell and analyze its behavior. Figure 44.11 shows the circuit. The S-cell of layer U_{Sl} receives excitatory signals directly from a group of C-cells, which are cells of the preceding layer U_{Cl-1}. It also receives an inhibitory signal through a V-cell, which accompanies the S-cell.

The V-cell receives fixed excitatory connections from the same group of C-cells as the S-cell does and

always responds with the average intensity of the output of the C-cells. The average is taken, not by an arithmetic mean, but by a root-mean-square.

Although several types of inhibitory mechanisms are used for S-cells depending on the versions of the neocognitron, here we discuss the case where the inhibitory signal from the V-cell works in a subtractive manner [44.13]. (See *Other Types of Inhibition* in Sect. 44.4.1).

Let a_n be the strength of the excitatory variable connection to the S-cell from the n-th C-cell, whose output is x_n. The output u of the S-cell is given by

$$u = \frac{1}{1-\theta} \cdot \varphi \left[\sum_n a_n x_n - \theta v \right], \qquad (44.1)$$

where $\varphi[\,]$ is a function defined by $\varphi[x] = \max(x, 0)$. Namely, $\varphi[\,]$ is a nonlinear function like a half-wave

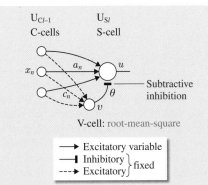

Fig. 44.11 Input connections converging to an S-cell. Subtractive inhibition (after [44.13])

rectifier. The strength of the inhibitory connection is θ, which determines the threshold of the S-cell $(0 < \theta < 1)$. The response of the V-cell is given by

$$v = \sqrt{\sum_n c_n x_n^2} \, , \tag{44.2}$$

where c_n is the strength of the fixed excitatory connection from the n-th C-cell.

We now use vector notation $x = (x_1, x_2, \ldots, x_n, \ldots)$ to represent the response of all pre-synaptic C-cells, from which the S-cell receives excitatory signals. We define the *weighted* inner product of arbitrary two vectors x and y by

$$(x, y) = \sum_n c_n x_n y_n \, , \tag{44.3}$$

where the strength of the input connections to the V-cell, c_n, is used as the weight for the inner product. We also define the norm of a vector, using the *weighted* inner product, by $\|x\| = \sqrt{(x, x)}$.

Renewing Connections to S-Cells
Generally speaking, S-cells of intermediate stages of the hierarchical network (U_{S2} and U_{S3} in the network of Fig. 44.2) are trained by unsupervised competitive learning. During the learning, S-cells compete with each other, and winners of the competition have their input connections renewed. Although concrete methods of choosing winners will be discussed later, here we discuss how the connections to the winners are renewed.

For the sake of simplicity, we discuss a case where training of the network is performed from lower stages to higher stages: after the training of a lower stage has

been completely finished, the training of the succeeding stage begins. For the training of S-cells of layer U_{Sl}, the response of C-cells of the preceding layer U_{Cl-1} works as a training stimulus.

Each time when a training pattern is presented to the input layer U_0, S-cells of U_{Sl} compete with each other, and several S-cells are selected as winners. The competition is made based on the strength of their responses.

Each S-cell usually becomes a winner several times during the training phase. Suppose an S-cell has become a winner at the t-th time. We use vector $X^{(t)}$ to represent the output of the C-cells pre-synaptic to this S-cell. Namely, $X^{(t)}$ is the training vector for this S-cell at this moment. Excitatory connection a_n is renewed through an auxiliary variable a_n', which increases in proportion to $X_n^{(t)}$. Namely, the amount of increase of a_n' is

$$\Delta a_n' = c_n X_n^{(t)} \, , \tag{44.4}$$

where c_n is the value of the fixed input connection to the inhibitory V-cell.

Let X be the sum of the training vectors that have made the S-cell a winner. Namely,

$$X = \sum_t X^{(t)} \, . \tag{44.5}$$

After having become winners for these training vectors, the strength of the auxiliary variable a_n' of this S-cell becomes

$$a_n' = \sum_t c_n X_n^{(t)} = c_n X_n \, . \tag{44.6}$$

The actual strength of the excitatory connection a_n is determined in proportion to a_n', but the total strength of all connections to a single S-cell, $\sqrt{\sum_n a_n^2 / c_n}$, is always kept constant.

To be more specific, the excitatory connection a_n is calculated from a_n' by

$$a_n = \frac{a_n'}{\sqrt{\sum_v a_v'^2 / c_v}} \, . \tag{44.7}$$

Since

$$\sqrt{\sum_v a_v'^2 / c_v} = \|X\|$$

holds, we have

$$a_n = \frac{c_n X_n}{\|X\|} \, . \tag{44.8}$$

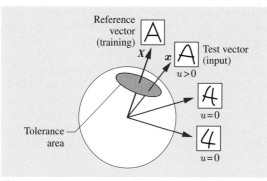

Fig. 44.12 Response of an S-cell in a multi-dimensional feature space (after [44.14])

Response of an S-cell

Using the weighted inner product defined by (44.3), we have

$$\sum_n a_n x_n = \frac{(X, x)}{\|X\|} \tag{44.9}$$

from (44.8). We also have

$$v = \|x\| \tag{44.10}$$

from (44.2).

Hence the response of the S-cell, which is given by (44.1), can also be expressed by

$$u = \|x\| \cdot \frac{\varphi[s - \theta]}{1 - \theta}, \tag{44.11}$$

where

$$s = \frac{(X, x)}{\|X\| \cdot \|x\|}. \tag{44.12}$$

In the multi-dimensional feature space, s shows a kind of similarity between x and X (Fig. 44.12). We call X, which is the sum of the training vectors, the reference vector of the S-cell. Using a neurophysiological term, we can also express that X is the preferred feature of the S-cell.

The second term $\varphi[s - \theta]/(1 - \theta)$ in (44.11) takes a maximum value 1 if the stimulus vector x is identical to the reference vector X, and becomes 0 when the similarity s is less than the threshold θ of the cell. In the multi-dimensional feature space, the area that satisfies $s < \theta$ becomes the tolerance area in feature extraction by the S-cell, and the threshold θ determines the size of the tolerance area. In other words, a non-zero response is elicited from the S-cell, if and only if the stimulus vector x is within a tolerance area around the reference vector X.

Figure 44.13 shows the equivalent circuit of an S-cell that has finished learning. The notes written in the margin of the figure show the values of connections and responses at several points in the equivalent circuit. These notes, however, do not show strict values: to help intuitive understanding, they show a simpler case of $c_n = 1$. (In Fig. 44.13, connections $(a_1, a_2, \ldots)$ are expressed by $X/\|X\|$, meaning $a_n = X_n/\|X\|$. To be more strict, they actually are $a_n = c_n X_n/\|X\|$.)

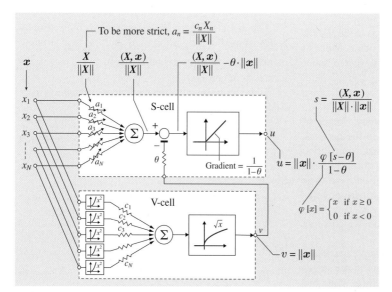

Fig. 44.13 Equivalent circuit of an S-cell that receives subtractive inhibition from the accompanying V-cell. The notes written in the *margin* show the values of connections and responses at several points in the equivalent circuit that have finished learning. These notes, however, do not show strict values: to help intuitive understanding, they show the case when $c_n = 1$ (after [44.13])

The selectivity of the response of an S-cell to a feature that is slightly different from its preferred feature (or from its reference vector) can thus be controlled by the threshold θ. A higher value of θ produces a smaller tolerance area. If the threshold is low, the radius of the tolerance area becomes large, and the S-cell responds even to features somewhat deformed from its reference vector.

Other Types of Inhibition

In the original neocognitron [44.1, 10], the inhibitory signal from the V-cell works in a shunting manner. If the strength of the inhibition is large enough, the shunting inhibition actually comes to work in a divisional manner. In most of the neocognitrons of previous versions, which used shunting inhibition, parameters were set in such a way that inhibition to S-cells works in the range of divisional inhibition.

When the inhibition from V-cells works in a divisional manner, response of an S-cell is given, not by (44.11), but by

$$u = \frac{\varphi[s - \theta]}{1 - \theta} . \tag{44.13}$$

If there is no background noise in the stimulus pattern, the characteristics of (44.13) are desirable for feature-extracting S-cells. The response of an S-cell is determined only by the similarity s between the input stimulus x and the training feature X. It is not affected by the strength of the input stimulus x. Hence S-cells can extract features robustly without being affected, say, by a gradual non-uniformity in thickness, darkness or contrast in an input pattern. If a stimulus pattern is contaminated by noise, however, interference from the background noise becomes serious.

Under noisy background, the neocognitron consisting of S-cells with subtractive inhibition can be much more insensitive to interference by noise [44.13]. Hence neocognitrons of recent versions use S-cells with subtractive inhibition.

44.4.2 Training S–Cells of Intermediate Stages

The neocognitron can be trained to recognize patterns through learning, where S-cells in the network have their input connections modified. Various training methods have been proposed to date.

Generally speaking, S-cells of intermediate stages of the hierarchical network (U_{S2} and U_{S3} in the network of Fig. 44.2) are trained by unsupervised competitive

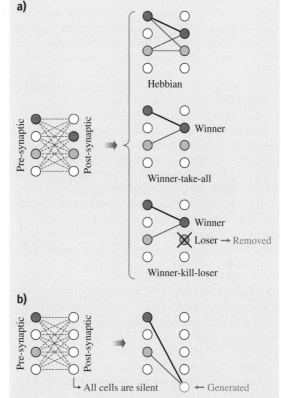

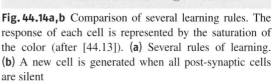

Fig. 44.14a,b Comparison of several learning rules. The response of each cell is represented by the saturation of the color (after [44.13]). **(a)** Several rules of learning. **(b)** A new cell is generated when all post-synaptic cells are silent

learning. The training methods for the highest stage are discussed later in Sect. 44.6.

As typical examples of unsupervised competitive learning, here we discuss two learning rules: *winner-take-all* and *winner-kill-loser*. Here, however, it should be noted that some other learning rules, such as those that accept incremental learning [44.15], have also been proposed.

Winner–Take–All

Figure 44.14 illustrates and compares several rules for unsupervised learning. The Hebbian rule, shown at the top of Fig. 44.14a, is one of the most commonly used learning rules for artificial neural networks [44.16]. During the learning phase, each synaptic connection is strengthened by an amount proportional to the product of the responses of the pre and post-synaptic cells.

In the *winner-take-all* rule, shown in the middle of Fig. 44.14a, post-synaptic cells compete with each other, and the cell from which the largest response is elicited becomes the winner. Only the winner can have its input connections renewed. The magnitude of the weight change is proportional to the response of the pre-synaptic cell. Incidentally, most conventional neocognitrons [44.1, 10] use this learning rule.

Once an S-cell has become a winner and has learned to respond to a feature, the S-cell usually loses its responsiveness to other features. When a different feature is presented, a different cell usually yields the maximum output and learns the second feature. Thus, a division-of-labor among the cells occurs automatically.

As is depicted in Fig. 44.14b, a new S-cell is generated if all cells are silent for a given training stimulus. The initial value of the input connections of the newly generated S-cell is proportional to the response of the pre-synaptic cells.

A Constraint Specific to the Neocognitron

When applying this learning rule to the neocognitron, a slight modification is required. As was mentioned before, each layer of the neocognitron is divided into cell-planes. All cells in a cell-plane share the same set of input connections. This condition of shared connections must be kept even during the learning phase, when input connections to S-cells are renewed.

When a winner is chosen in a given cell-plane, its input connections are renewed based on the responses of the C-cells pre-synaptic to it. Since all cells in the cell-plane share the same set of connections, all other cells in the cell-plane come to have the same connections as the winner. The winner thus works like a seed in crystal growth. Hence we call it a *seed-cell*.

In intermediate stages of the network, S-cells have small receptive fields, each of which covers only a part of stimulus patterns presented to the input layer. When a stimulus pattern is presented to the input layer in the recognition phase, a number of visual features have to be extracted in parallel by S-cells of a number of cell-planes.

To let a number of S-cells learn these features in parallel in the learning phase, the size of the competition areas of S-cells is set to be relatively small. When the response of an S-cell is the largest among S-cells situated in a certain small competition area, the S-cell becomes the winner and takes the place of a seed-cell.

The competition area of each S-cell has the shape of a *hypercolumn* [44.1]. A hypercolumn is defined here as a group of S-cells from all cell-planes in a layer whose

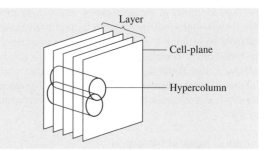

Fig. 44.15 Relation between cell-planes and hypercolumns within a layer (after [44.1])

receptive fields are situated approximately at the same location in the input layer. In other words, each hypercolumn contains all kinds of S-cells in it, and these S-cells extract features from approximately the same location in the input layer. If we rearrange the cell-planes of a layer in Fig. 44.2 and stack them as shown in Fig. 44.15, each columnar structure constitutes a hypercolumn.

Every time a training pattern is presented, all S-cells whose responses are the largest in their individual hypercolumns are chosen as candidates of seed-cells. If there is only one candidate from a cell-plane, the candidate is automatically chosen as the seed-cell for that cell-plane. If two or more candidates are chosen from a single cell-plane, the S-cell that yields the largest response among them is chosen as the seed-cell of the cell-plane. If no candidate appears in a cell-plane, no seed cell is selected from that cell-plane at this time.

Winner–Kill–Loser

The *winner-kill-loser* rule can train neocognitrons more efficiently than the winner-take-all rule, and an ability to robustly recognize patterns can be obtained with a smaller scale of the network.

The winner-kill-loser rule is illustrated at the bottom of Fig. 44.14a. It resembles the winner-take-all rule in the sense that only the winner learns the training stimulus. In the winner-kill-loser rule, however, not only does the winner learn the training stimulus, but the losers are also simultaneously removed from the network. Losers are defined as cells whose responses to the training stimulus are smaller than that of the winner, but whose activations are nevertheless greater than zero.

If a training stimulus elicits non-zero responses from two or more S-cells, it means that the preferred features of these cells resemble each other, and that they work redundantly in the network. To reduce this redundancy, only the winner has its input connections renewed to fit more to the training vector, while the other active cells, namely the losers, are removed from the network.

Since silent S-cells (namely, the S-cells whose responses to the training stimulus are zero) do not join the competition, they are not removed. These cells are expected to work toward extracting other features.

Similarly to the winner-take-all rule, a new S-cell is generated if all cells are silent for a given training stimulus, as depicted in Fig. 44.14b. The initial value of the input connections of the newly generated S-cell is proportional to the response of the pre-synaptic cells.

It should be noted here that the constraint discussed in *Constraint Specific to the Neocognitron* must be satisfied also under the winner-kill-loser rule. Namely, the constraint is also applied to losers. If there are non-silent cells at the location of the seed-cell, they are losers. The cell-planes, to which losers belong, are removed from the layer.

In the learning phase, a number of training stimuli are presented sequentially to the network. During this process, generation of new cells (cell-planes) and removal of redundant cells (cell-planes) occur repeatedly in the network. In particular, new cells are generated to cover areas of the multi-dimensional feature space that were not previously covered by existing cells. In the areas where similar cells exist in duplicate, redundant cells are removed. By repeating this process for a long enough time, the preferred features (reference vectors) of S-cells gradually become distributed uniformly over the multi-dimensional feature space.

The winner-kill-loser rule can thus produce a uniform distribution of feature vectors in the multi-dimensional feature space with a smaller number of feature vectors than the winner-take-all rule. This means that a large ability to recognize patterns can be obtained with a smaller scale of the network (namely, with a smaller number of cell-planes) by the winner-kill-loser than by the winner-take-all learning rule.

Dual Threshold for S-Cells

The neural networks' ability to recognize patterns robustly is influenced by the selectivity of feature-extracting cells, which is controlled by the threshold of the cells.

As shown in (44.11), a feature-extracting S-cell is active when $s < \theta$ and silent when $s < \theta$. This charac-

teristic is important for a successful self-organization by an unsupervised competitive learning. As mentioned above, winners are chosen from non-silent S-cells. If all S-cells are silent to the training vector, a new S-cell is generated. Generation (and also removal in the case of winner-kill-loser) of S-cells is repeated in the network, and the reference vectors of S-cells gradually come to distribute uniformly in the multi-dimensional feature space. Only a small number of S-cells come to respond to an individual feature, and the distance between adjacent vectors approaches approximately θ. This means that S-cells come to behave like grandmother cells. In order to produce a sufficient number of feature-extracting S-cells in a layer, the threshold of S-cells needs to be high enough during the learning.

In the recognition phase, however, a behavior like that of grandmother cells is not desirable for S-cells of a layer. If the threshold is high, a particular stimulus feature might elicit a response from only one cell. When the stimulus feature is slightly deformed, the S-cell stops responding, and another S-cell comes to respond instead of the first one. This decreases the ability of the network to recognize deformed patterns robustly.

If the threshold is low, however, S-cells respond even to features somewhat deformed from their reference vectors. This makes a situation like a population coding of features rather than the grandmother cell theory: many S-cells respond to a single feature if the response of an entire layer is observed. Even if the feature is slightly deformed, many of the S-cells still keep responding. Only a small number of S-cells change their responses. This situation of low threshold in the recognition phase usually endows the network with an ability of generalization and produces a better recognition rate of the neocognitron.

Hence it is desirable to use a dual threshold of S-cells for the learning and the recognition phases [44.14, 17]. In the recognition phase after having finished the learning, the threshold of S-cells is set to a lower value θ^R than the threshold θ^L for the learning.

Incidentally, if, by adding one more threshold for the learning, we use a triple threshold together with the winner-kill-loser rule, the learning can progress in a more stable manner [44.18]. The final number of cell-planes that has been created after learning is usually smaller with the triple threshold than with the dual threshold.

44.5 C-Cells

44.5.1 Averaging by C-Cells

As was discussed in Sect. 44.3, a C-cell has fixed excitatory connections from a group of S-cells of the corresponding cell-planes of S-cells. Through these connections, each C-cell averages the responses of S-cells whose receptive field locations are slightly deviated. In other words, S-cells' response is spatially blurred in the succeeding cell-planes of C-cells. The averaging operation is important, not only for endowing neural networks with an ability to recognize deformed patterns robustly, but also for smoothing additive random noise contained in the responses of S-cells.

In some versions of the neocognitron or related networks, the averaging is performed, not by arithmetic mean, but by root-mean-square [44.13] or a MAX operation [44.19].

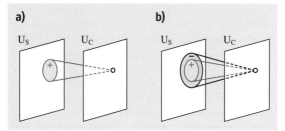

Fig. 44.16a,b Inhibitory surround in the connections to C-cells. (**a**) No inhibitory surround; (**b**) with inhibitory surround

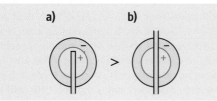

Fig. 44.17a,b Inhibitory surround in the connection to a C-cell produces a response like an end-stopped cell. Stilulus (**a**) produces a larger response than (**b**) (after [44.10])

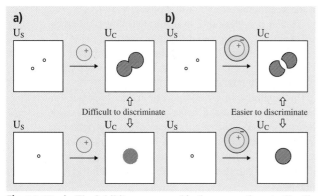

Fig. 44.18a,b The inhibitory surround in the connections to C-cells increases separation of the blurred responses produced by two independent features (after [44.10]). (**a**) No inhibitory surround; (**b**) with inhibitory surround

44.5.2 Inhibitory Surround in the Connections to C-Cells

The response of an S-cell layer U_{Sl} is spatially blurred in the succeeding C-cell layer U_{Sl}. In the original neocognitron, the input connections to a C-cell consist of only excitatory components of a circular spatial distribution, as shown in Fig. 44.16a.

The introduction of an inhibitory surround around the excitatory connections as shown in Fig. 44.16b increases the recognition rate of the neocognitron. The concentric inhibitory surround endows the C-cells with characteristics like those of end-stopped cells, and C-cells behave like hypercomplex cells in the visual cortex (Fig. 44.17). In other words, an end of a line elicits a larger response from a C-cell than a middle point of the line. Bend points and end points of lines are important features for pattern recognition. C-cells, whose input connections have inhibitory surrounds, thus participate in extraction of bend points and end points of lines while they are making a blurring operation. Incidentally, this kind of response is observed neurophysiologically in neurons of the primary visual cortex (e.g., [44.20]).

The inhibitory surrounds in the connections also have another benefit. The blurring operation by C-cells, which is usually effective for improving robustness against deformation of input patterns, sometimes makes it difficult to detect whether a lump of blurred response is generated by a single feature or by two independent features of the same kind (Fig. 44.18a). For example, a single line and a pair of parallel lines of a very narrow separation generate a similar response when they are blurred. The inhibitory surround in the connections to C-cells creates a non-responding zone between the two lumps of blurred responses (Fig. 44.18b). This silent zone makes the S-cells of the next stage easily detect the number of original features even after blurring.

The inhibitory surround in the input connections is usually introduced for C-cells of lower stages, but not for higher stages. In higher stages, an inhibitory surround seems to have little chance to display its real ability for two reasons:

1. Since there is little probability that a single input pattern has two identical global features at different locations, discrimination between one and two features is scarcely required.

2. Since spatial spread of the connections to a single cell becomes large enough to cover a great part of the preceding S-cell layer; most of the surrounding inputs come from outside of the boundary of a cell-plane and are treated as zero in the calculation.

44.6 The Highest Stage

At the highest stage of the network (U_{SL} and U_{CL}; $L = 4$ in the network of Fig. 44.2), the S-cells judges the final result of patterns recognition, and the C-cells yield the inferred label of the stimulus.

During the learning, S-cells of the highest stage are trained by supervised competitive learning using labeled training data. Although several methods have been proposed for learning, here we discuss two: *supervised winner-take-all* and *interpolating vectors*.

44.6.1 Supervised Winner–Take–All

This section discusses the *supervised winner-take-all* that is used for the learning of S-cells of the highest stage [44.10].

The learning rule resembles the competitive learning used to train S-cells of intermediate stages, but the class names of the training patterns are also used for learning. When each cell-plane first learns a training pattern, the class name of the training pattern is assigned to the cell-plane.

Every time a training pattern is presented during learning, competition occurs among all S-cells in the layer. If the winner of the competition has the same label as the training pattern, the winner becomes the seed-cell and learns the training pattern. However, if the winner has a wrong label (or if all S-cells are silent), a new cell-plane is generated. The new cell-plane hence learns the current training pattern simply by being assigned its corresponding label. Each cell-plane thus has a label indicating the category of the pattern. In the case of digit recognition, for example, the label indicates one of the ten digits.

During the recognition phase, the label of the maximally activated S-cell in the layer determines the final result of recognition. The C-cells at this stage yield the inferred label of the input stimulus.

In most neocognitrons of recent versions, the threshold value of S-cells of the highest stage is set to zero for both recognition and learning phases. Hence the process of finding the largest-output S-cell is equivalent to the process of finding the nearest reference vector in the multi-dimensional feature space. Each reference vector has its own territory determined by the Voronoi partition of the feature space (Fig. 44.19). The recognition process in the highest stage resembles the vector quantization [44.22, 23] in this sense.

Training vectors that are distant from class borders can be represented by a small number of reference vectors. Training vectors that are misclassified in the learning phase usually come from near class borders. If a particular training vector is misclassified in the learning, the reference vector of the winner, which caused a wrong recognition for this training vector, is not renewed this time. A new cell-plane is generated instead, and the misclassified training vector is adopted as the reference vector of the new cell-plane. Generation of a new reference vector causes a shift of decision borders in the feature space, and some of the training vectors,

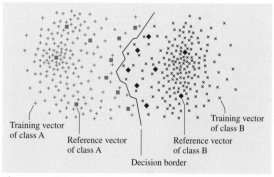

Fig. 44.19 Distribution of training and reference vectors in a multi-dimensional feature space (after [44.21])

which have been recognized correctly before, are now misclassified and additional reference vectors have to be generated again to readjust the borders. Thus, the decision borders are gradually adjusted to fit the real borders between classes. During this learning process, reference vectors come to be distributed more densely near the class borders, but their density remains low in the locations distant from class borders.

44.6.2 Interpolating Vectors

The method of *interpolating vectors* is another rule for analyzing the response of S-cells of the highest stage (U_{SL}) to classify input patterns [44.21]. Although it is slightly more complicated than the supervised winner-take-all rule, it usually produces a higher recognition rate. This section discusses the use of interpolating vectors in the learning and recognition phases.

Classification by Interpolating Vectors
In both training and recognition phases, responses of C-cells of the preceding stage (U_{CL-1}) become input signals to S-cells of the highest stage (U_{SL}). Here we use vector notation to represent these input signals.

Before explaining the method of creating reference vectors, we first discuss the recognition phase, where labeled reference vectors have already been produced. Namely, we assume a situation where a small number of reference vectors for each class have already been produced from a large number of training vectors. Each reference vector has a label of the class name.

We now discuss the process of classifying a test vector. The basic idea of the method of interpolating vectors is as follows. We assume a situation where vir-

tual vectors, which are named *interpolating vectors*, are densely placed along the line segments connecting every pair of reference vectors of the same label. From these interpolating vectors, we choose the one that has the largest similarity to the test vector. The label (or the class name) of the chosen vector is taken as the result of pattern recognition.

Actually, we do not need to generate an infinite number of interpolating vectors. We just assume line segments connecting all pairs of reference vectors of the same label. The line segments are assigned the same labels as the reference vectors on both sides. We then measure distances (based on similarity) to these line segments from the test vector and choose the nearest line segment (Fig. 44.20a). The label of the line segment shows the result of pattern recognition.

Mathematically, this process can be expressed as follows. Let X_i and X_j are two reference vectors of the same label. An interpolating vector $\boldsymbol{\xi}$ for this pair of reference vectors is given by their linear combination

$$\boldsymbol{\xi} = p\frac{X_i}{\|X_i\|} + q\frac{X_j}{\|X_j\|}, \quad (p+q=1). \tag{44.14}$$

Similarity s between the interpolating vector $\boldsymbol{\xi}$ and the test vector $\boldsymbol{x}$ takes a maximum value

$$s_{\max} = \sqrt{\frac{s_i^2 - 2s_i s_j s_{ij} + s_j^2}{1 - s_{ij}^2}} \tag{44.15}$$

at

$$p = \frac{s_i - s_j s_{ij}}{(s_i + s_j)(1 - s_{ij})},$$

$$q = \frac{s_j - s_i s_{ij}}{(s_i + s_j)(1 - s_{ij})}, \tag{44.16}$$

where

$$s_i = \frac{(X_i, \boldsymbol{x})}{\|X_i\|\|\boldsymbol{x}\|}, \quad s_j = \frac{(X_j, \boldsymbol{x})}{\|X_j\|\|\boldsymbol{x}\|},$$

$$s_{ij} = \frac{(X_i, X_j)}{\|X_i\|\|X_j\|}. \tag{44.17}$$

Incidentally, s_i and s_j are proportional to the responses of S-cells with X_i and X_j, respectively, because the threshold θ of S-cells of layer U_{SL} is always zero. To be more exact, from (44.11) and (44.12), the response of the i-th and j-th S-cells is given by

$$u_i = \frac{(X_i, \boldsymbol{x})}{\|X_i\|} = \|\boldsymbol{x}\|\frac{(X_i, \boldsymbol{x})}{\|X_i\|\|\boldsymbol{x}\|} = \|\boldsymbol{x}\|\, s_i,$$

$$u_j = \frac{(X_j, \boldsymbol{x})}{\|X_j\|} = \|\boldsymbol{x}\|\frac{(X_j, \boldsymbol{x})}{\|X_j\|\|\boldsymbol{x}\|} = \|\boldsymbol{x}\|\, s_j. \tag{44.18}$$

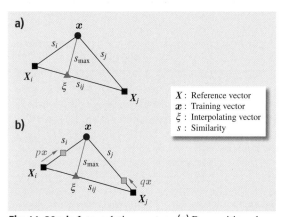

X : Reference vector
$\boldsymbol{x}$: Training vector
ξ : Interpolating vector
s : Similarity

Fig. 44.20a,b Interpolating vectors. (**a**) Recognition phase (after [44.21]). (**b**) Training reference vectors

Hence, if we have calculated s_{ij} in advance, we can easily obtain $\|x\| \cdot s_{max}$ from the responses of S-cells, u_i and u_j. Since the value of $\|x\|$ is the same for all S-cells that have receptive fields at the same location, we can easily find the pair of S-cells (reference vectors) that produces the maximum similarity s_{max}.

We can interpret that s_{max} represents similarity between test vector x and the line segment that connects a pair of reference vectors X_i and X_j (Fig. 44.20a). Among all line segments that connect every pair of reference vectors of the same label, we choose the one that has the largest similarity to the test vector. The label (or the class name) of the chosen line segment is taken as the result of pattern recognition. (If there exists only one reference vector of a class, similarity between the reference vector and the test vector is taken as s_{max}.)

Creating Reference Vectors

Every time a training vector is presented during the learning phase, we first try to classify it using interpolating vectors.

If the result of the classification is correct, we choose the line segment that shows the largest similarity to the training vector x. The two reference vectors, X_i and X_j on both sides of the line segment, learn the training vector x (Fig. 44.20b). The amounts of increase of X_i and X_j are px and qx, respectively, where p and q are given by (44.16). (If $p < 0$, namely, $q > 1$, we set $p = 0$ and $q = 1$. Similarly, if $p > 1$, namely, $q < 0$, we set $p = 1$ and $q = 0$.)

If the result of the classification is wrong, or if all S-cells are silent to the training vector, the training vector is adopted as a new reference vector and is assigned a label of the class name.

44.7 Networks Extended from the Neocognitron

Various extensions and modifications of the neocognitron have been proposed to endow it with further abilities or to make it more biologically plausible.

44.7.1 Selective Attention Model

Although the neocognitron has considerable ability to recognize deformed patterns, it does not always recognize patterns correctly when two or more patterns are presented simultaneously. The *selective attention model* has been proposed to eliminate these defects [44.24]. In the selective attention model, top-down (i.e., backward) connections were added to the neocognitron-type network, which had only bottom-up (i.e., forward) connections.

When a composite stimulus, consisting of two patterns or more, is presented, the model focuses its attention selectively on one of the patterns, segments it from the rest, and recognizes it. After the identification of the first segment, the model switches its attention to recognize another pattern. The model also has the function of associative recall. Even if noise or defects affect the stimulus pattern, the model can recognize it and recall the complete pattern from which the noise has been eliminated and defects corrected. These functions can be successfully performed even for deformed versions of training patterns, which have not been presented during learning.

This model has some similarity with the ART model [44.25], but the most important difference between the two is the fact that the selective attention model has the ability to accept patterns deformed in shape and shifted in location. With the selective attention model, not only the recognition of patterns, but also the filling-in process for defective parts of imperfect input patterns works on the deformed and shifted patterns themselves. The selective attention model can repair the deformed pattern without changing the basic shape and location of the deformed input pattern. The deformed patterns themselves can be repaired at their original locations, thus preserving their deformation.

The principles of this selective attention model can be extended to be used for several applications, such as the recognition and segmentation of connected characters in cursive handwriting of English words [44.26].

We can also design an artificial neural network that recognizes and segments a face and its components (e.g., eyes and mouth) from a complex background [44.27]. It consists of two channels of selective attention models with different resolutions. The high-resolution channel can analyze input patterns in detail, but usually lacks the ability to obtain global information because of the small receptive fields of the cells in it. On the other hand, the low-resolution channel, whose cells have large receptive fields, can capture global information, but only roughly. The network analyzes the object by the interaction of both channels. Even after having learned only a small number of facial front views, the network can recognize and segment faces, eyes, and mouths correctly from images containing a variety of faces against a complex background.

44.7.2 Recognition and Segmentation of Occluded Patterns

Human beings are often able to read or recognize a letter or word contaminated by ink stains that partly occlude the letter. If the stains are completely erased and the occluded areas of the letter are changed to white, however, we usually have difficulty in reading the letter, which now has some missing parts. For example, the patterns in Fig. 44.21a, in which the occluding objects are not visible, are almost illegible, but the patterns in Fig. 44.21b, in which the occluding objects are visible, are much easier to read.

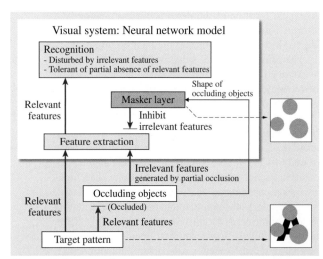

Fig. 44.21 (**a**) Patterns partly occluded by invisible masking objects are difficult to recognize. (**b**) It becomes much easier to recognize when the occluding objects are visible (after [44.28])

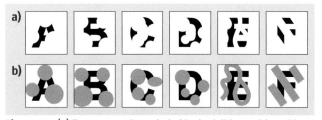

Fig. 44.22 Process of recognizing an occluded pattern (modified from [44.28])

Visual patterns have various local features, such as edges and corners. The visual system of animals extracts these features in its lower stages and tries to recognize a pattern using information of extracted local features. When a pattern is partly occluded, a number of new features, which do not exist in the original pattern, are generated.

If the occluding objects are visible, the visual system can easily distinguish relevant from irrelevant features and can ignore irrelevant features. Since the visual system has a large tolerance to a partial absence of relevant features, it can recognize the partly occluded patterns correctly, even though some relevant features are missing (Fig. 44.22). The same is true for the neocognitron model.

If the occluding objects are not visible, the visual system has difficulty in distinguishing which features are relevant to the original pattern and which are not. These irrelevant features largely disturb the correct recognition by the visual system.

To eliminate irrelevant features, which are usually generated near the contours of the occluding objects, a new layer U_M, named the *masker layer*, is added to a neocognitron as shown in Fig. 44.23 [44.28]. The masker layer detects and responds only to occluding objects. The shape of the occluding objects appears in U_M, in the same shape and at the same location as in input layer U_0. There are topographically ordered and slightly diverging inhibitory connections from layer U_M to all cell-planes of layer U_{S1}. The inhibitory signals from U_M thus suppress the responses to features irrelevant to the occluded pattern. Hence only local features relevant to

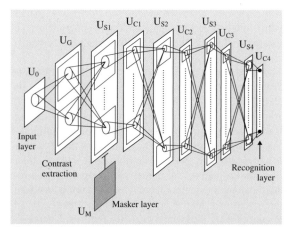

Fig. 44.23 If a layer U_M, which is called *masker layer*, is added to a neocognitron, the neocognitron can recognize partly occluded patterns correctly (after [44.28])

the occluded pattern are transmitted to higher stages of the network.

Figure 44.24 shows another example of stimuli, in which the perception is largely affected by the placement of occluding objects. The black parts of the patterns are actually identical in shape in the left and right figures. We feel as though different black patterns are occluded by gray objects. Namely, we perceive as though pattern 'R' were occluded in the left figure, while pattern 'B' is occluded in the right. The neocognitron with a masker layer can recognize these patterns correctly as 'R' and 'B', like human beings.

Fig. 44.24 Identical patterns are perceived differently by the placement of different gray objects (after 44.29)

By further adding top-down (i. e., backward) connections to the network, for example, the network comes to have the ability, not only to recognize occluded patterns correctly, but also to restore the occluded parts of the patterns [44.29, 30].

References

44.1 K. Fukushima: Neocognitron: A self-organizing neural network model for a mechanism of pattern recognition unaffected by shift in position, Biol. Cybern. **36**, 193–202 (1980)

44.2 D.H. Hubel, T.N. Wiesel: Receptive fields, binocular interaction and functional architecture in the cat's visual cortex, J. Physiol. (Lond.) **160**, 106–154 (1962)

44.3 D.H. Hubel, T.N. Wiesel: Receptive fields and functional architecture in nonstriate areas (18 and 19) of the cat, J. Neurophysiol. **28**, 229–289 (1965)

44.4 M. Ito, H. Komatsu: Representation of angles embedded within contour stimuli in area V2 of macaque monkeys, J. Neurosci. **24**, 3313–3324 (2004)

44.5 R. von der Hydt, E. Peterhans, G. Baumgartner: Illusory contours and cortical neuron responses, Science **224**, 1260–1262 (1984)

44.6 R. Desimone, S.J. Schein: Visual properties of neurons in area V4 of the macaque: Sensitivity to stimulus form, J. Neurophysiol. **57**, 835–868 (1987)

44.7 I. Fujita, K. Tanaka, M. Ito, K. Cheng: Columns for visual features of objects in monkey inferotemporal cortex, Nature **360**, 343–346 (1992)

44.8 C.J. Bruce, R. Desimone, C.G. Gross: Visual properties of neurons in a polysensory area in superior temporal sulcus of the macaque, J. Neurophysiol. **46**, 369–384 (1981)

44.9 S. Yamane, S. Kaji, K. Kawano: What facial features activate face neurons in the inferotemporal cortex of the monkey?, Exp. Brain Res. **73**, 209–214 (1988)

44.10 K. Fukushima: Neocognitron for handwritten digit recognition, Neurocomputing **51**, 161–180 (2003)

44.11 K. Fukushima: A neural network for visual pattern recognition, IEEE Comput. **21**, 65–75 (1988)

44.12 K. Fukushima: Analysis of the process of visual pattern recognition by the neocognitron, Neural Netw. **2**, 413–420 (1989)

44.13 K. Fukushima: Increasing robustness against background noise: Visual pattern recognition by a neocognitron, Neural Netw. **24**, 767–778 (2011)

44.14 K. Fukushima, M. Tanigawa: Use of different thresholds in learning and recognition, Neurocomputing **11**, 1–17 (1996)

44.15 K. Fukushima: Neocognitron capable of incremental learning, Neural Netw. **17**, 37–46 (2004)

44.16 D.O. Hebb: *Organization of Behavior* (Wiley, New York 1949)

44.17 K. Fukushima: Neocognitron trained with winner-kill-loser rule, Neural Netw. **23**, 926–938 (2010)

44.18 K. Fukushima, I. Hayashi, J. Léveillé: Neocognitron trained by winner-kill-loser with triple threshold, ICONIP 2011, Part II, Vol. 7063, ed. by B.-L. Lu, L. Zhang, J. Kwok (Springer, Berlin, Heidelberg 2011) pp. 628–637

44.19 M. Riesenhuber, T. Poggio: Hierarchical models of object recognition in cortex, Nat. Neurosci. **2**, 1019–1025 (1999)

44.20 G.A. Walker, I. Ohzawa, R.D. Freeman: Suppression outside the classical cortical receptive field, Vis. Neurosci. **17**, 369–379 (2000)

44.21 K. Fukushima: Interpolating vectors for robust pattern recognition, Neural Netw. **20**, 904–916 (2007)

44.22 R.M. Gray: Vector quantization, IEEE ASSP Magazine **1**, 4–29 (1984)

44.23 T. Kohonen: The self-organizing map, Proc. IEEE **78**, 1464–1480 (1990)

44.24 K. Fukushima: Neural network model for selective attention in visual pattern recognition and associative recall, Appl. Opt. **26**, 4985–4992 (1987)

44.25 G.A. Carpenter, S. Grossberg: ART 2: Self-organization of stable category recognition codes for analog input patterns, Appl. Opt. **26**, 4919–4930 (1987)

44.26 K. Fukushima, T. Imagawa: Recognition and segmentation of connected characters with selective attention, Neural Netw. **6**, 33–41 (1993)

44.27 K. Fukushima, H. Hashimoto: Recognition and segmentation of components of a face by a multiresolution neural network, Artificial Neural Networks – ICANN 97, Vol. 1327, ed. by W. Gerstner, A. Germond, M. Hasler, J.-D. Nicoud (Springer, Berlin, Heidelberg 1997) pp. 931–936

44.28 K. Fukushima: Recognition of partly occluded patterns: A neural network model, Biol. Cybern. **84**, 251–259 (2001)

44.29 K. Fukushima: Restoring partly occluded patterns: A neural network model, Neural Netw. **18**, 33–43 (2005)

44.30 K. Fukushima: Neural network model for completing occluded contours, Neural Netw. **23**, 465–582 (2010)

45. Information Processing in the Gustatory System

Alexandra E. D'Agostino, Patricia M. Di Lorenzo

Gustation is a sensory modality that is essential for survival. Information provided by the gustatory system enables identification and rejection of poisons and encourages ingestion of nutritious foods. In addition, fine taste discrimination, guided by postingestional feedback, is critical for the maintenance of energy homeostasis. The study of how gustatory information is processed by both the peripheral and the central nervous systems has been an active and fruitful area of investigation and has yielded some dramatic results in recent years. In this chapter, we will first discuss the general concept of neural coding of sensory information to provide a backdrop for the discussion of gustatory neural coding in particular. Next, the anatomy and organization of each level of the gustatory system from the tongue to the cortex will be presented in the context of various theories of sensory coding as they have been applied to gustation. Finally, we will review

the unifying ideas that characterize information processing in the taste system.

45.1 Neural Coding of Sensory Information – General Concepts

The idea of a neural code is as old as the study of neurophysiology itself. Investigators of the nervous system in the modern age discovered that neurons, the main cell type in the brain and periphery, emit electrical signals that are received and processed by other neurons, an event that constitutes communication of information from one cell to another. This unique electrical signal, known as the *action potential*, was first described in the late nineteenth century (see [45.1] for a review). It can be described as an abrupt and brief ($\approx 1\,\mathrm{ms}$) change in the voltage across the neuronal cell membrane that evokes a similar electrical event in neurons that are in physical proximity to it. It was later discovered that action potentials are produced in axons, a fine process emanating from the cell body of a neuron, and are actively propagated from the cell body to the terminal endings that closely abut other neurons in areas known as synapses. There, the action potential spurs a series of biochemical events that result in the release of a neurotransmitter substance. This chemical is what activates the *postsynaptic* neuron to eventually generate its own action potential through integration of incoming signals from many other presynaptic neurons. Thus, any given neuron integrates information from many sources and makes a singular decision (to generate its own action potential or not).

Fast forward to the mid-twentieth century. As the infant field of neuroscience was developing, theorists in computer science and engineering soon seized upon the deceptive simplicity of neuronal communication to devise models that might emulate the enormous computational power of the brain. The properties of the action potential made this venture particularly attractive to the philosophers who would later become the first computer scientists. Perhaps most importantly, the action potential was an all or none event. That is, in a given neuron, the signal was always the same amplitude and always had the same time course; but a neuron would either *fire*, i.e., generate and action potential, or not. In effect, the action potential was a binary event: it was a one or a zero. This analogy was brilliantly exploited by *McCulloch* and *Pitts* [45.2] in their treatise on how the properties of the nervous system could inform a burgeoning field of computer science (see [45.3] for a description of historical context).

Modeling computers after the nervous system has been fruitful for the field of computer science (e.g., the concept of parallel process was gleaned form neuroscience) but has also enabled a deeper understanding of how the nervous system works. For example, just as a *bit* of information can be represented in the pattern of ones and zeros within a computer, so too does the nervous system represent information in the distributed activity of action potentials across many neurons. However, unlike computers where the components that signal information are essentially all the same, neurons are diverse in their morphology (shape) and connectivity, and there is much processing that is graded, rather than all-or-none. This affords the nervous system a richness and complexity that can only be poorly approximated by any man-made device.

The neural representation of the energy that surrounds an organism in its environment is one of the most basic functions of the brain. This *energy*, defined as a sensory stimulus once it is represented and interpreted by the brain, comes in many forms, not all of which are *sensed* by any one species. That is, the sensory organs and brains afforded to members of each species are exquisitely well-suited to process the signals in its environment that enable it to survive. In the case of the gustatory system, one of the evolutionarily oldest sensory systems, the detection and evaluation of chemical candidates for ingestion is, as mentioned above, a critical function. However, how does the nervous system accomplish this feat? Most would argue that the first

step in the process of sensation is to transform environmental energy into biological energy. This process, known as *transduction*, precedes the conversion of the graded potentials that can be recorded from the sensory receptor cells to the stream of action potentials that constitute what we call the *neural code*.

The idea that the nervous system encodes information about a sensory stimulus carries with it several assumptions. As *Halpern* [45.4] has pointed out, the term *encode* poses the problem of who or what *decodes* or *reads* the code. Likewise, he argues that once decoded, sensory stimuli result in a *representation* or *reconstruction* of the world. This leads to the conundrum of who or what interprets the representation and how. As an alternative to this conceptualization, Halpern favors the idea of a *transformation* that eschews the idea of neural coding. *Tort* et al. [45.5] have also posited that sensation may literally be a series of transformations of inputs rather than a faithful encoding and representation of the sensory world. Such notions are appealing on several grounds. First, no decoding is required, i.e., information gleaned from neural signals generated by sensory signals can act directly on neural structures that modulate or generate behavior. Second, the concept of a transformation is in tune with the dynamic nature of neural activity, whereas the idea of a representation implies a static, or at least stable array of neurons that capture information about an encoded stimulus. Finally, transformation of sensory information is consistent with observations that sensory stimuli are processed along parallel and distributed neural pathways.

In spite of the caveats concerning the concept of neural coding of sensory stimuli, the term has widespread use both historically and in contemporary literature. In this review we will use the term *neural code* in deference to its common use, keeping in mind that there are perhaps better ways to conceptualize how the nervous system processes information.

In the gustatory system, transduction involves the transformation of the presence of a water-soluble chemical into neural activity in taste buds on the tongue. This deceptively simple process is then followed by activation of sensory *taste* nerves that carry signals to the brain. In the brain, taste-evoked activity is then transmitted across several structures, ending (but not really) in the gustatory cortex.

45.2 The Peripheral Gustatory System

45.2.1 Anatomy and Signal Transduction in the Periphery

Taste transduction occurs in taste buds located in the oropharyngeal area. There are about 10 000 taste buds in the human tongue, soft palate, and epiglottis [45.6]. Taste buds are found within protrusions called papillae. Fungiform papillae, characterized by their mushroom shape, are located on the rostral portion of tongue, towards the tip and along the sides. Each contains a number of taste buds located in a trench that surrounds the *mushroom top*. Foliate papillae consist of ridges located on the side of the tongue towards the back; taste buds are buried inside these ridges. Last, circumvallate papillae are dome-shaped and are located at the very back of the tongue, arranged in an inverted '*U*' shape; taste buds are embedded in the trenches surrounding the dome. Taste buds contain 50–150 taste receptor cells (TRCs) that are arranged like the sections of an orange. TRCs are replaced about every 12 days. At the base of the taste bud are 3–14 sensory nerve endings that receive the transduced signals from the TRCs and send action potentials to the sensory ganglia that then communicate with the brain (reviewed in [45.7]). Two-thirds of the tongue toward the tip, as well as the soft palate, are innervated by two branches of the facial nerve (cranial nerve VII): the chorda tympani (CT) nerve and the greater superior petrosal nerve, respectively. One-third of the tongue toward the back of the mouth is innervated by the lingual tonsillar branch of the glossopharyngeal nerve (cranial nerve X). The superior laryngeal branch of the vagus nerve (cranial nerve X) innervates the larynx, pharynx, and epiglottis, where taste buds are also present.

The chemicals that stimulate the taste system are commonly grouped into five primary *taste qualities*. Humans describe these as salty, sweet, bitter, umami (savory), or sour. Each primary taste is transduced by different receptor mechanisms (reviewed in [45.7, 8]. *Taste receptors* are molecules that are specialized to sense a taste stimulus and are embedded in the membranes of taste receptor cells. These can be specialized ion channels as in the case of salty [45.7] or other types of proteins. For example, receptor cells expressing the G protein coupled receptor (GPCR) heterodimer T1R2 + T1R3 respond to sweet tastes [45.9]. T2Rs, another class of GPCRs, are responsible for transducing bitter taste [45.10] and there are many variants that transduce the wide variety of bitter tastants (e.g., quinine, denatonium benzoate, etc.). Finally, the pro-

totypical stimulus for umami taste is monosodium glutamate (MSG). Putative umami GPCRs are the T1R1 + T1R3 heterodimer and mGluR-taste receptors (reviewed in [45.11]). Sour taste may be mediated by the intracellular proton concentration [45.12]. It is thought that high concentrations of extracellular protons are able to cross the membrane through ion channels and ion exchangers [45.8]. Some potential sour receptors include amiloride-sensitive epithelial sodium channels (amiloride is a chemical that blocks some types of ion channels [45.13]), members of the transient receptor potential family including PKD2L1 and PKD1L3 [45.14], and plasma membrane channels modulated by acidification of the cytoplasm [45.15].

45.2.2 Models of Peripheral Gustatory Coding

Historically, there have been two primary hypotheses related to gustatory coding in the periphery that have dominated the literature: the labeled line (LL) and the across-fiber pattern (fibers are axons; AFP) theories. The fact that taste receptor molecules that transduce each taste quality are found in their own dedicated groups of taste receptor cells has led many in the field to suggest that this division of labor carries over to the nervous system. That is, it has been argued that there are groups or types of nerve fibers (axons) that carry information exclusively about a single taste quality. Although peripheral nerve fibers are known to respond to more than one taste quality (that is, they are *broadly tuned*), the quality that evokes the largest response, the so-called *best stimulus*, predicts the relative response magnitude evoked by other taste qualities [45.16]. Based on these observations, many investigators contend that separate information channels connect peripheral receptors to the central nervous system (CNS). In contrast, and based on the same observations, the AFP theory asserts that taste qualities are represented by the activation across the population of fibers rather than in dedicated subsets of fibers as in the LL theory. In effect, all members of the array of broadly tuned nerve fibers can convey information about any taste stimulus.

One of the important predictions of the LL theory is that taste quality-specific receptors are innervated by taste quality-specific peripheral nerve fibers. In fact, there is evidence that stimulus-specific fiber types have unique behavioral functions. A good example of this can be seen in NaCl-best fibers. For example, sodium deprivation has been shown to affect responses to NaCl only in NaCl-best fibers [45.17,18]. A more recent study in marmosets showed that compounds that stimulated sucrose-best fibers were preferred, whereas compounds that stimulated quinine-best fibers were avoided [45.19].

Perhaps the most striking and convincing evidence for the LL theory comes from recent work using genetic deletion and substitution of the receptor molecules associated with each of the basic taste qualities. These studies have shown that if receptors for sweet, umami [45.20], bitter [45.10], sour [45.14], or salt [45.21] are genetically deleted, knockout mice show no response in the CT nerve to the respective taste qualities but respond normally to the other taste qualities. Moreover, if mice are genetically engineered to express bitter receptors in taste cells that normally would express sweet receptors, they will prefer bitter compounds [45.10]. In other experiments, *Zhao* and colleagues [45.20] expressed receptors for a normally tasteless (to a mouse) opioid in sweet-responsive TRCs and found that these mice could now taste and even prefer the opioid agonist. Conversely, when these same opioid receptors were expressed in bitter TRCs the animals rejected the opiate as though it were bitter. Collectively, these data strongly imply that the taste cells that house receptors for the various taste qualities are hard-wired to neural networks that orchestrate preference and rejection, as appropriate. Thus, at least in the periphery, the nervous system seems to be organized around quality-specific labeled lines.

There is also compelling evidence for the AFP theory. In spite of the frankly compelling studies using genetic engineering, decades of electrophysiological studies have consistently found that TRCs and peripheral nerves are broadly tuned across taste qualities (see, for example, [45.22, 23]). If a cell is broadly tuned, any response in that cell to a tastant would necessarily make its message ambiguous, that is, an objective observer would not be able to identify the stimulus that evoked the signal. The AFP theory solves this conundrum by proposing that the code for any given taste is the pattern of neural activity distributed across the population of taste-responsive cells. With that conceptualization, the patterns of activity generated by stimuli that taste alike should be more similar to each other than to patterns generated by stimuli that taste very different. In fact, dozens of studies in the literature show this to be true (see [45.24] for an example of early work).

Although at first blush it may seem that the preponderance of evidence at the present time favors the LL theory in the periphery, there are important caveats to keep in mind. It should be noted, for example, that taste quality-specific alterations in the responses of the CT nerve that have been reported in knockout mice have been derived from whole nerve recordings. It is entirely possible that single fiber recordings may reveal a different story. In addition, the CT is not the only nerve that subserves taste perception; information from other nerves may be conveyed with different mechanisms. Moreover, the LL and AF theories are not mutually exclusive, as has been noted in the literature [45.25]. Finally and most importantly, they are spatial codes and so ignore response dynamics. More about that later.

45.3 Taste Processing in the Brainstem

45.3.1 Anatomy

Gustatory coding becomes more complex and multifaceted at the level of the brainstem (see Fig. 45.1 for a diagram of the anatomy of the central gustatory pathways). The facial, glossopharyngeal, and vagus nerves that innervate the taste buds in the mouth all send their fibers to the rostral nucleus of the solitary tract (rNTS) in an orderly, topographical organization such that their fibers terminate from rostral (towards the nose) to caudal (towards the tail), respectively, with some overlap. In the rodent, the rNTS sends fibers to the parabrachial nucleus of the pons (PbN) as well as to the intermediate and lateral parvocellular divisions of the medullary reticular formation [45.26]. Ascending and descending fibers, however, originate from non-overlapping cell populations. The PbN provides ascending input to the insular cortex via the parvocellular subdivision of the ventral posteromedial nucleus of the thalamus (VPMpc) while a second pathway from the PbN sends information to the lateral hypothalamus (LH), central nucleus of the amygdala (CeA), bed nucleus of the stria terminalis (BNST) and the substantia innominata (SI) (reviewed in [45.27]). The two pathways from the PbN, i. e., the dorsal pathway ending in the cortex and the ventral pathway to the limbic system, are thought to mediate different taste functions. The dorsal path coordinates taste discrimination while the ventral path supports hedonic evaluation of taste (pleasure or disgust). In humans and Old World monkeys, the central gustatory pathway does not have

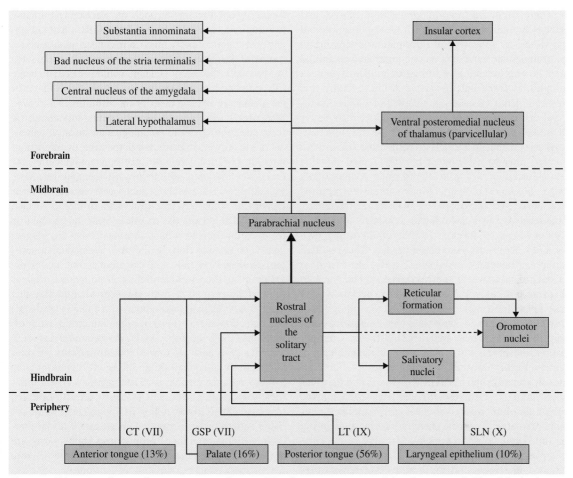

Fig. 45.1 Diagram of ascending gustatory pathways. *Blue boxes* represent peripheral receptor fields while brainstem nuclei are shown in *green boxes*. *Red* and *yellow boxes* indicate thalamocortical and ventral forebrain structures, respectively. *Thicker lines* indicate denser neural connections. Percentages in parentheses indicate the percentage of taste buds found in that receptor field. CT (VII) = chorda tympani branch of the facial nerve; GSP (VII) = greater superficial branch of the facial nerve; LT (IX) = lingual-tonsillar branch of the glossopharyngeal nerve; SLN (X) = superior laryngeal branch of the vagus nerve (after [45.29])

a synapse in the PbN; instead the rNTS projects directly to the thalamus [45.28].

45.3.2 Gustatory Information Processing in the Brainstem

In the study of information processing in the periphery as well as the brainstem, it is important to note that the basic data point upon which both the LL and AFP (or AFN in the central nervous system) are based involves the number of spikes counted over some response interval, as defined by the experimenter. Typically, this interval begins when the taste stimulus is flowed over the tongue and ends at a relatively arbitrary time thereafter. Sometimes, but not always, this period encompasses the entire stimulus presentation, but more often it ends prior to the stimulus offset. In nearly every study, the *baseline* period, defined as some measure of neuronal firing rate when either no stimulus or water is flowing over the tongue, is subtracted from the firing rate that occurs during taste stimulation. The response interval is on the order of seconds, most often 2–5 s, sometimes longer but rarely shorter. (These points and others have been discussed more fully in [45.30].)

There are at least two important implications, and perhaps limitations, of the choice of response interval and its use as a measure of information processing in the taste system. First, it is much longer than an animal needs to acquire enough information to identify a taste stimulus. *Halpern* and *Tapper* [45.31] have shown that a rat can identify a taste stimulus within 200 ms, an interval that is an order of magnitude shorter than nearly all studies of taste. *Lemon* and *Katz* [45.32], however, have argued that in most taste-responsive cells, there are very few spikes emitted by 200 ms and that the full appreciation of a taste stimulus may require much longer. In support of their argument are results reported by *Dinardo* and *Travers* [45.33] who show that behavioral reactivity to a taste stimulus may require a full 1.0 s or longer to develop. Further, *Halpern* and *Tapper*'s [45.31] results were based on data from rats that had been trained to avoid NaCl and were tested in an environment where the rat could expect NaCl to be presented, so both learning and expectation may have sped up recognition time. On the other hand, *Stapleton* et al. [45.34] have shown that gustatory cortical cells respond to taste stimuli within ≈ 90 ms, a result that is consistent with rapid stimulus identification. Clearly, further research is needed to resolve these issues. A second point, mentioned above, is that the characterization of a taste response using the average firing rate in an arbitrary interval necessarily ignores the response dynamics, i. e., the changes in firing rate during the timed course of the response that may be informative. *Katz* et al. [45.35] have shown that when response dynamics are considered, the proportion of cells that respond differentially to taste stimuli in the GC is nearly quadruple the proportion identified solely by changes in firing rate. Many investigators have noted that taste responses at all levels of the nervous system often show an initial phasic burst in firing rate followed by a later, more sustained, tonic increase in the firing rate above baseline. These changes, however, are only loosely predictive of taste quality and may be more closely associated with the response characteristics of individual cells (see [45.36], for a review, and the discussion below).

The use of taste-evoked firing rates in the formulation of taste coding theories of information processing (and more complex brain structures) has produced LL and AFP (now the across neuron pattern, ANP) theories of taste processing applied to the brainstem in the same way that they have been proposed in the periphery (see [45.29] for a review). However, these theories may not be as well-suited to the central nervous system, mainly because of the increase in the breadth of tuning as one ascends through the gustatory neuraxis. For example, as a result of convergent input from many peripheral nerve fibers, NTS cells generally respond to more taste qualities than peripheral nerve fibers [45.24]. Additionally, the receptive fields with respect to the area of the tongue that evokes a response in the NTS cells are quite large and increase during development [45.37]. Moreover, the problem of disambiguating taste intensity (concentration) from taste quality can be problematical if one only examines average the taste-evoked firing rate. Most NTS and PbN, and in the taste system in general, show monotonically increasing intensity-response functions. So, in a broadly tuned cell, there will necessarily be some concentration of more than one taste stimulus that evokes the identical firing rates, making the identification of which stimulus is on the tongue nearly impossible (but see below). This poses a serious challenge for the LL theory, but is not as vexing for the ANP theory, which relies on population coding rather than responses from stimulus-identified single cells. Interestingly, *Lemon* and *Smith* [45.38] used signal detection theory to show that the firing rates of individual NTS neurons could not accurately predict stimulus type. Instead, they demonstrated that the parallel activity of a small group of the NTS neurons was a more reliable predictor of taste quality. Their findings were at odds with the LL theory because neurons that contributed to the coding of a given taste stimulus had a variety of best stimuli. This so-called *distributed code*, as Lemon and Smith proposed, is a departure from the typical understanding of ANP theory in which activity across a large population of neurons encodes for different taste qualities.

There are some data that support the application of LL theory to taste processing in the brainstem. That is, there are reports of well-defined groups of narrowly tuned cells in both the NTS and PbN. For example, small populations of the NTS [45.39, 40] and PbN [45.41] cells respond exclusively to bitter tastes. Cells in this bitter-best group were generally located posteriorly in the PbN [45.41], suggesting the possibility of a topographic segregation of cells responsive to the various taste qualities. A similar topographic organization has also been found in the NTS, where *Harrer* and *Travers* [45.42] described a rostrocaudal separation of bitter and sweet tastes in the NTS. In fact, this kind of best-cell segregation may even be continued in the GC [45.43]. Further evidence for the LL theory in the brainstem comes from data on the intensity-response functions of taste cells in the NTS recorded in awake rats [45.44]. Specifically, although the problem of am-

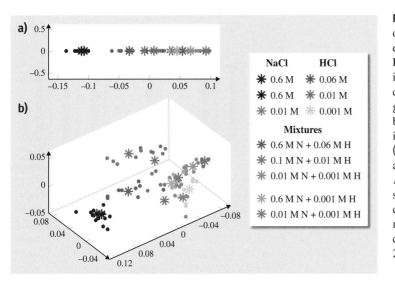

Fig. 45.2a,b Temporal coding analyses of one cell tested with three concentrations each of NaCl and HCl and five mixtures. For all plots, *dots* indicate the location of individual responses; *asterisks* indicate the centroid of the clusters of responses to a given taste stimulus. *Axes* are labeled in arbitrary units. Color coding of the stimuli is indicated in the *upper right* of the figure. **(a)** The one-dimensional response space created by MDS of the spike count distances, D^{count}. **(b)** The three-dimensional response space created by the MDS of the spike time distances $D^{spike}[q_{max}]$. For 11 stimuli, the maximum amount of information that can be conveyed is 3.46 bits; H_{max} for this cell was 2.88 (after [45.45])

biguity as described above remains, if one groups cells by their best stimulus, the intensity-response curve is steeper for a cell's best-stimulus than for its non-best stimuli (Fig. 45.2). In effect, these data suggest that neurons can distinguish between low concentrations of their best stimulus and high concentrations of non-best stimuli.

Another approach to the dilemma of resolving taste quality vs. intensity was proposed by *Chen* et al. [45.43]. Specifically, they used a measure of temporal coding, called metric space analysis [45.46,47], to evaluate the contribution of spike count vs. the temporal characteristics of a response to information processing of taste. In metric space analyses, the similarity of two spike trains are gauged by the *cost* of adding or deleting spikes as necessary, and moving spikes in time, such that the two spike trains are identical. This cost is called D^{spike} and is calculated at a variety of levels of temporal precision, called q. In sum, *Chen* et al. [45.43] showed that the precise timing of spikes within a taste-evoked spike train was a better predictor of taste quality than spike count alone, even when the concentration of the stimulus was varied. Furthermore, the temporal pattern of the taste-evoked spike train for a given taste quality was consistent across cells even when the stimulus was presented in varying concentrations in a mixture with a stimulus of a different taste quality (Fig. 45.3).

The idea of temporal coding of taste is based on the observation that different taste stimuli appear to produce different temporal patterns of response, as noted above (see [45.32,36] for reviews). Early investigations in this area quantified the time-dependent aspects of the

response [45.48–51]. In general, these approaches combined standardization of response magnitude with a variety of methods that compared sequences of increases and decreases in firing rate across the response interval; i.e. the *rate envelope* across the response interval. Although suggestive, these studies did not conclusively demonstrate that the rate envelope of the taste response can unambiguously identify a taste stimulus [45.49,52]. However, some evidence implied that the rate envelope of a taste response may signal the hedonic properties (pleasure or aversion) of taste stimuli [45.52].

More recent studies of temporal coding in the brainstem have focused on spike timing as a marker for taste quality. Using metric space analyses, several studies have shown that spike timing in the NTS cells can convey a significant amount of information about taste quality [45.53,54]) and intensity [45.43] above and beyond that of spike count alone. Furthermore, *Di Lorenzo* et al. [45.55] showed that temporal coding can be used by individual cells to convey information about the components of binary mixtures of different taste qualities, especially in the most broadly tuned cells.

Importantly, there is evidence that animals can actually use information conveyed by spike timing to make hedonic judgments as well as identify taste quality. A study by *Di Lorenzo* et al. [45.56] delivered lick-contingent 1.0 s electrical pulse trains into the NTS of rats while they were drinking plain water. The temporal patterns of these pulse trains mimicked the temporal pattern of a spike train evoked by either sucrose or quinine. Rats licked avidly for a pulse train based on the response to sucrose by avoided the pulse train that mim-

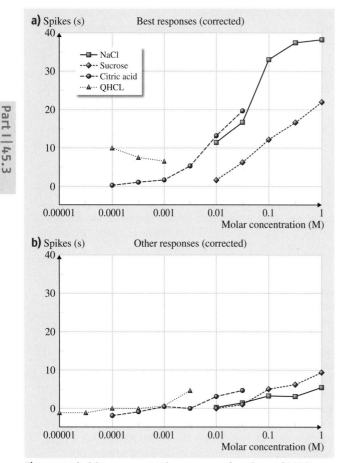

a) Spikes (s)   Best responses (corrected)

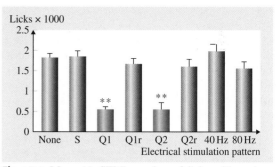

Licks × 1000

Fig. 45.4 Mean (± SEM) number of licks with various temporal patterns of electrical stimulation of the nucleus of the solitary tract. S = sucrose simulation pattern (1 s electrical pulse train based on the response to sucrose in an NTS cell); Q1 and Q2 = quinine simulation patterns based on electrophysiological responses of two different NTS cells to quinine; Q1r and Q2r = the quinine simulation patterns in which interpulse intervals are randomly shuffled. ** with a probability of $p > 0.01$ (after [45.56])

Fig. 45.3a,b Mean concentration-response functions of NTS neurons in chronic, behaving rats. (**a**) Neurons that responded to their best-stimuli presented in a concentration series. (**b**) Responses of other neurons to their non-best stimuli presented in the same concentration series. The responses of neurons to their best stimulus exhibited steeper response-concentration functions. $N = 52$ (after [45.44])

icked quinine (Fig. 45.4). When the temporal pattern of the pulse train associated with quinine was randomized, however, rats licked without hesitation. Follow-up studies from the same lab [45.55,57] showed that a conditioned aversion to the sucrose or quinine simulated patterns of electrical pulses generalized specifically to real sucrose or quinine, respectively, suggesting that the temporal pattern of pulses conveyed taste quality information.

In non-primates, the second relay for taste information in CNS is the parabrachial nucleus (PbN) of the pons. There are many complex connections between the

NTS and PbN that have been studied with the use of simultaneous recordings in each area. It was found that taste responses in relay NTS cells were larger than taste responses in non-relay NTS cells [45.58]. Interestingly, functionally connected cells in the NTS and PbN do not always share the same response profiles. For instance, it was found that among 13 functionally connected NTS-PbN pairs, seven of these pairs shared a best stimulus while the other pairs had differing best stimuli [45.59]. Furthermore, functionally connected pairs that shared a best stimulus were more effective at transmitting information about that stimulus. In contrast, functionally connected pairs not sharing a best stimulus did not show increased responses in the PbN to any particular stimulus. These results suggest that PbN cells are better at transmitting signals from the NTS cells with the same best-stimulus [45.59].

In addition to receiving information from relay cells in the NTS, PbN may also make use of temporal codes to relay taste information to the forebrain. A data set containing functionally connected pairs of NTS-PbN cells was reanalyzed to assess the importance of temporal codes [45.55]. The results indicated that the NTS input to PbN was important in the first 3 s of the PbN response during NaCl, HCl, and quinine trials. Metric space analysis was then run on isolated taste-responsive PbN cells to determine how much information spike count and spike timing contribute to the response. The analysis confirmed that spike count along with spike timing contributed more information than spike count alone. Importantly, the contribution of spike count and spike timing was found to be equivalent in the NTS and PbN, suggesting that temporal coding is maintained across these two areas [45.60]. Furthermore, a recent study from the same lab has provided evidence of temporal coding in PbN [45.61]. In agreement with research in the NTS, a positive correlation was found between breadth of tuning and information conveyed by temporal coding in isolated cells in PbN [45.61].

In rodents, the rostral NTS sends information about taste to the PbN and taste coding there is generally similar to that in the NTS. In particular, PbN may also make use of temporal codes to relay taste information to the forebrain [45.60, 61] and does so in the same proportion of cells and with the same amount of information. As in the NTS, in the PbN, more broadly tuned PbN cells are more likely to use temporal coding to convey taste information than more narrowly tuned cells [45.61]. Collectively, this research suggests that temporal coding is most evident when the spike count conveys an ambiguous message about taste identity.

Several studies have focused on how information is transferred and transformed from the NTS to PbN. Since not every taste-responsive cell in the NTS projects to PbN [45.58, 59], it is important to examine the specific characteristics of those NTS cells that provide direct input to PbN. In general, taste responses in the NTS-PbN relay cells are larger than taste responses in non-relay NTS cells [45.58, 59]. Studies of simultaneous recordings of taste-responsive cells in the NTS and PbN have shown that functionally connected cells in the NTS and the PbN do not always share the same response profiles. For instance, among 13 functionally connected NTS-PbN pairs, six responded best to different taste qualities [45.59]. However, in those functionally connected pairs of the NTS-PbN cells that did share a best stimulus, the NTS cell was more effective at driving spikes evoked by the best stimulus than non-best stimuli in the PbN cell [45.59]. Examination of how well PbN cells *followed* the time course of the functionally connected NTS cells suggested that the PbN target faithfully emulated the NTS input in the first 3 s of the taste response. After that, the firing patterns in the NTS and PbN became independent of each other [45.60].

45.3.3 Neural Circuitry in Relation to Information Processing About Taste in the Brainstem

Experiments showing that spike timing in the NTS is an important mechanism for communicating about taste begs the question of how the neural circuitry produces and interprets this type of information. How might neurons generate taste-evoked spike trains where the spikes consistently occur at specific times with respect to each other? One approach to answer this question is to examine the functional circuitry of the NTS. While it is known that the inputs to the NTS from the periphery are all glutaminergic, and thus exclusively excitatory [45.62, 63], once the taste signal reaches the NTS, there are extensive intranuclear connections that serve as an abundant source of GABAergic inhibition [45.64]. Additionally, feedback connections from the forebrain also provide inhibitory input to the NTS [45.65]. Indeed, about two thirds of the taste-responsive cells in the NTS receive some form of inhibitory influence [45.63, 66, 67] and inhibition is known to affect the breadth of tuning [45.63, 68, 69]. Since the breadth of tuning appears to predict the importance of spike timing in information processing about taste, a further examination of the role of inhibition in the NTS may be a fruitful strategy to uncover the mechanisms that generate reliable taste-evoked spike timing.

Studies have shown that the excitatory input to the NTS is altered in systematic ways by inhibition and that this may have relevance to taste coding. First, data from intracellular recordings in the NTS showed that tetanic stimulation of CT input generates inhibition that is potentiated by repeated stimulation of the CT nerve [45.70–72]. Extracellular recordings from single cells in the NTS also showed that electrical stimulation of the CT nerve generates a recurrent inhibitory influence [45.61, 68, 73]. In fact, *Lemon* and *Di Lorenzo* [45.68] showed that tetanic electrical stimulation of the CT nerve just prior to taste stimulation attenuated taste responses in some cells. To demonstrate the relevance of this input to taste coding, we presented brief (100 ms) pulses of taste stimuli, called *prepulses*, followed 1 s later by the same or a different taste stimulus [45.56]. The logic behind this experiment was that if the signal form the second taste stimulus arrived during the period of the recurrent inhibitory influence generated by the first taste stimulus, the response to the second stimulus should be suppressed compared to when it was presented without a prepulse. Over half of the cells showed significant changes as described: responses to the second stimulus were altered by taste prepulses, most often showing attenuation. Furthermore, when the prepulse-taste interval was lengthened to 5 s, the prepulse became ineffectual, suggesting that the recurrent inhibition had dissipated. Importantly, those cells that were most susceptible to

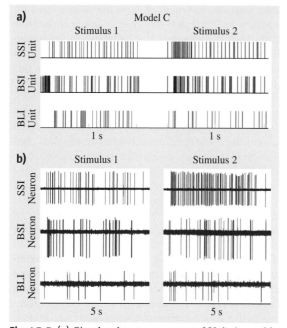

Fig. 45.5 (a) Simulated taste responses of Unit Assembly 2 to presentation of Stimulus 1 followed by Stimulus 2 (both at 25 Hz). Stimulus 2 evoked a phasic-tonic pattern of response in the selective short-inhibition unit that was not observed with presentation of stimulus one. Both broadly tuned units responded nearly equally well to both Stimulus 1 and 2. **(b)** Electrophysiological responses to a 5 s presentation of two different taste stimuli recorded from NTS neurons of each cell type. Similar to the network model, a phasic-tonic response patterns is apparent in the SSI neuron's response to its *best* stimulus (Stimulus 2) and was absent with presentation of Stimulus 1. Both broadly tuned units responded nearly equally well to both Stimulus 1 and 2. Abbreviations are: SSI, selective short-inhibition; BSI, broad short-inhibition; BLI, broad long-inhibition (after [45.56, 74, 75])

the effects of prepulses were those that were most broadly tuned, consistent with the idea that inhibition in the NTS may have a role in temporal coding.

To explore the relationship between inhibitory circuits in the NTS, the breadth of tuning of the NTS cells, and their role in temporal coding, we initiated a series of experiments using electrical stimulation of the CT nerve while recording from taste-responsive cells in the NTS[45.61, 76]. Initially, NTS cells with evoked responses to electrical stimulation of the CT were tested with exemplars of each of four primary taste qualities (NaCl, HCl, quinine, and sucrose) in

separate trials [45.77]. Next, the CT was electrically stimulated using a paired-pulse (10–2000 ms interpulse interval; blocks of 100 trials) paradigm. The majority (75.6%) of cells that showed CT-evoked responses also showed paired-pulse attenuation, defined as fewer spikes in response to the second pulse of the pair compared to the number of spikes evoked by the first pulse of the pair. A bimodal distribution of the peak of paired-pulse attenuation was found with modes at 10 and 50 ms in separate groups of cells. Early peak paired pulse attenuation had a short time course, dissipating by ≈ 50 ms; late peak paired pulse attenuation had a longer time course and was still evident at 500 ms. Cells with early peak paired pulse attenuation showed short CT-evoked response latencies and were narrowly tuned. Conversely, cells with late peak paired pulse attenuation showed long CT-evoked response latencies and were broadly tuned [45.76]. Further recordings revealed a third group of cells that showed early peak paired pulse attenuation but were also broadly tuned. To obtain a deeper understanding of how the circuitry of the NTS might be arranged so as to produce these results, we designed a computational model of the NTS circuitry. The results suggested that NTS cells may be organized into stimulus-specific cell assemblies that are dedicated to processing a single taste quality (Fig. 45.5). The architecture and final parameters of this artificial neural network were guided by a genetic algorithm [45.78]. Although the model was based on latency and paired pulse data alone, it accurately simulated both the breadth of tuning and the temporal patterns of firing that were recorded from actual NTS cells. Collectively, these data support the idea that inhibitory circuits in the NTS may constrain firing patters to generate the idiosyncratic temporal patterns that may identify various taste qualities.

45.3.4 Gustatory Coding in the Forebrain

The main projection of the taste-responsive portion of the PbN is to the ventromedial nucleus of the thalamus. Here, taste-responsive cells retain their broad sensitivity across taste qualities [45.79, 80]. In the rat, taste-evoked response magnitude is attenuated with respect to the PbN [45.79–81]. This de-amplification is accompanied by broader tuning in thalamic cells compared with cells in the PbN [45.80, 81]. In the monkey, electrophysiological data suggest that taste stimuli are processed similarly in the thalamus and brainstem [45.82]. Approximately 56% of the taste-responsive neurons in the thalamus project to the GN [45.83]. Thalamo-cortical relay neurons do not differ from non-relay neurons in their response properties, but do show larger overall response magnitudes [45.83], as with NTS-PbN relay cells [45.84].

Responses of GN neurons in rats are essentially multimodal and incorporate hedonics [45.35], attention [45.85], and memory [45.86, 87], as well as taste quality (reviewed in [45.88]). The multimodal nature of the GC makes its identification problematical. For example, most taste-responsive cells in the GC also respond to somatosensory input, making it difficult to tease apart the influence of the two modalities. Some recent work has suggested a topographical representation of taste quality in the GC, in support of a labeled line conceptualization of coding [45.43, 89]. However, changes in hedonic evaluation of a taste stimulus can alter the spatial representation of a taste quality in the GC [45.90].

The breadth of tuning in the GC has been, and continues to be, the subject of debate in the literature. Early studies in both anesthetized and chronic preparations found a sharpening of tuning in GC neurons [45.91, 92]. More recent studies in awake rats, however, have found that GC neurons are very broadly tuned, with the majority of neurons responding to three or more of the five primary taste qualities [45.34, 35]. The temporal dynamics of taste responses have also been shown to contribute information about taste quality in the cortex. For example, *Katz* et al. [45.35] found that when simply measuring the average firing rate over 2.5 s, only 14% of GC cells were taste responsive, but when they also considered the orderly changes in firing rate across the time course of the response, the percentage of taste-responsive GC cells increased to 41%. In addition, using the analyses of temporal dynamics, *Katz* et al. [45.35] were also able to differentiate somatosensory and chemosensory components of the response in GC neurons (Fig. 45.5). Their evidence suggested that the earliest portion of the response (< 200 ms) reflected purely somatosensory

input and was followed by a chemosensory response (0.2–1.0 s). Finally, the response reflected the hedonic value of the stimulus in the last 1 s of the response interval [45.35]. Katz et al.'s results seem to conflict with Stapleton et al.'s data suggesting that GC taste responses show chemosensory-related information in the initial ≈ 90 ms of the response. These contradictory findings may be related to differences in the methods of stimulus presentation. In the study by Katz et al. [45.35] rats received fluid passively through an intraoral cannula, while in Stapleton's study [45.34] rats that actively licked at a sipper tube. In this regard it is relevant that active acquisition of sensory stimuli is known to affect taste responses in the GC [45.93].

45.3.5 Conclusions

Here, we have presented an overview of the gustatory system in terms of anatomy, signal transduction and neural codes. In the peripheral gustatory system, evidence for the labeled line model of taste coding seems most compelling. That is, evidence points to the idea that each taste quality excites a subset of peripheral nerve fibers that then send information to the brain. Once in the brain, however, most cells are broadly tuned, although some remain stimulus-specific. As one ascends through each synapse in the gustatory system, reports are consistent in noting that the breadth of tuning increases. Studies of temporal coding suggest that those cells that are the most broadly tuned across taste qualities are those that are most likely to use the timing of spikes as a mechanism for communication. Also consistent are observations that within each gustatory-related structure, the cells that send information to the next structure respond to taste more vigorously than other cells. This begs the question as to what function the non-relay cells have and how they interact with inputs and outputs of each structure. These questions are at the forefront of research in the field of information processing in gustation.

References

45.1   S.M. Schuetze: The discovery of the action potential, Trends Neurosci. **6**, 164–168 (1983)

45.2   W.S. McCulloch, W. Pitts: A logical calculus of the ideas immanent in nervous activity, Bull. Math. Biophys. **5**, 115–133 (1943)

45.3   L. Kay: From logical neurons to poetic embodiments of mind: Warren S. McCulloch's Project in neuroscience, Sci. Context **14**(4), 591–615 (2001)

45.4   B.P. Halpern: Sensory coding, decoding, and representations. Unnecessary and troublesome constructs?, Physiol. Behav. **69**(1/2), 115–118 (2000)

45.5   A.B. Tort, A. Fontanini, M.A. Kramer, L.M. Jones-Lush, N.J. Kopell, D.B. Katz: Cortical networks produce three distinct 7–12 Hz rhythms during single sensory responses in the awake rat, J. Neurosci. **30**(12), 4315–4324 (2010)

45.6   I.J. Miller: Anatomy of the peripheral taste system. In: *In Handbook of Olfaction and Gustation*, ed. by R.L. Doty (Marcel Dekker, New York 1995) pp. 521–547

45.7   N. Chaudhari, S.D. Roper: The cell biology of taste, J. Cell Biol. **190**(3), 285–296 (2010)

45.8   S.D. Roper: Signal Transduction and information processing in mammalian taste buds, Eur. J. Physiol. **454**, 759–776 (2007)

45.9   G. Nelson, M.A. Hoon, J. Chandrashekar, Y. Zhang, N.J. Ryba, C.S. Zuker: Mammalian sweet taste receptors, Cell **106**, 381–390 (2001)

45.10   K.L. Mueller, M.A. Hoon, I. Erlenbach, J. Chandrashekar, C.S. Zuker, N.J. Ryba: The receptors and coding logic for bitter taste, Nature **434**(7030), 225–229 (2005)

45.11   J. Chandrashekar, M.A. Hoon, N.J. Ryba, C.S. Zuker: The receptors and cells for mammalian taste, Nature **444**, 288–294 (2006)

45.12   V. Lyall, R.I. Alam, D.Q. Phan, G.L. Ereso, T.H. Phan, S.A. Malik, M.H. Montrose, S. Chu, G.L. Heck, G.M. Feldman, J.A. DeSimone: Decrease in rat taste receptor cell intracellular pH is the proximate stimulus in sour taste transduction, Am. J. Physiol. Cell Physiol. **281**, C1005–C1013 (2001)

45.13   T.A. Gilbertson, S.D. Roper, S.C. Kinnamon: Proton currents through amiloride-sensitive Na^+ channels in isolated hamster taste cells: Enhancement by vasopressin and cAMP, Neuron **10**, 931–942 (1993)

45.14   A.L. Huang, X. Chen, M.A. Hoon, J. Chandrashekar, W. Guo, D. Tränkner, N.J. Ryba, C.S. Zuker: The cells and logic for mammalian sour taste detection, Nature **442**, 934–938 (2006)

45.15   W. Lin, C.A. Burks, D.R. Hansen, S.C. Kinnamon, T.A. Gilbertson: Taste receptor cells express pH-sensitive leak K^+ channels, J. Neurophysiol. **92**, 2909–2919 (2004)

45.16   M. Frank: An analysis of hamster afferent taste nerve response functions, J. Gen. Physiol. **61**(5), 588–618 (1973)

45.17   R.J. Contreras: Changes in gustatory nerve discharges with sodium deficiency: A single unit analysis, Brain Res. **121**(2), 373–378 (1977)

45.18   R.J. Contreras, M. Frank: Sodium deprivation alters neural responses to gustatory stimuli, J. Gen. Physiol. **73**(5), 569–594 (1979)

45.19   V. Danilova, G. Hellekant: Sense of taste in a New World monkey, the common marmoset. II. Link between behavior and nerve activity, J. Neurophysiol. **92**(2), 1067–1076 (2004)

45.20   G.Q. Zhao, Y. Zhang, M.A. Hoon, J. Chandrashekar, I. Erlenbach, N.J. Ryba, C.S. Zuker: The receptors for mammalian sweet and umami taste, Cell **115**, 255–266 (2003)

45.21   J. Chandrashekar, C. Kuhn, Y. Oka, D.A. Yarmolinsky, E. Hummler, N.J. Ryba, C.S. Zuker: The cells

45.22   and peripheral representation of sodium taste in mice, Nature **464**(7286), 297–301 (2010)

45.22   T.A. Gilberston, J.D. Boughter, H. Zhang, D.V. Smith: Distribution of gustatory sensitivities in rat taste cells: Whole-cell responses to apical chemical stimulation, J. Neurosci. **21**(13), 4931–4941 (2001)

45.23   A. Caicedo, K. Kim, S.D. Roper: Individual mouse taste cells respond to multiple chemical stimuli, J. Physiol. **544**, 501–509 (2002)

45.24   G.S. Doetsch, R.P. Erickson: Synaptic processing of taste-quality information in the nucleus tractus solitarius of the rate, J. Neurophysiol. **33**(4), 490–507 (1970)

45.25   D.V. Smith, R.L. Van Buskirk, J.B. Travers, S.L. Bieber: Coding of taste stimuli by hamster brain stem neurons, J. Neurophysiol. **50**(2), 541–558 (1983)

45.26   C.B. Halsell, S.P. Travers, J.B. Travers: Ascending and descending projections from the rostral nucleus of the solitary tract originate from separate neuronal populations, Neuroscience **72**(1), 185–197 (1996)

45.27   R.F. Lundy, R. Norgren: Activity in the hypothalamus, amygdala, and cortex generates bilateral and convergent modulation of pontine gustatory neurons, J. Neurophysiol. **91**, 1143–1157 (2004)

45.28   R.M. Beckstead, J.R. Morse, R. Norgren: The nucleus of the solitary tract in the monkey: Projections to the thalamus and brain stem nuclei, J. Comp. Neurol. **190**(2), 259–282 (1980)

45.29   A.C. Spector, S.P. Travers: The representation of taste quality in the mammalian nervous system, Behav. Cogn. Neurosci. Rev. **4**(3), 143–191 (2005)

45.30   P.M. Di Lorenzo, C.H. Lemon: Methodological considerations for electrophysiological recording and analysis of taste-responsive cells in the brain stem of the rat. In: *Methods in Chemosensory Research*, ed. by S.A. Simon, M.A.L. Nicolelis (CRC, New York 2001) pp. 293–324

45.31   B.P. Halpern, D.N. Tapper: Taste stimuli: Quality coding time, Science **171**(977), 1256–1258 (1971)

45.32   C.H. Lemon, D.B. Katz: The neural processing of taste, BMC Neuroscience **8**(Suppl. 3), S5 (2007)

45.33   L.A. Dinardo, J.B. Travers: Hypoglossal neural activity during ingestion and rejection in the awake rat, J. Neurophysiol. **72**(3), 1181–1191 (1994)

45.34   J.R. Stapleton, M.L. Lavine, R.L. Wolpert, M.A.L. Nicolelis, S.A. Simon: Rapid taste responses in the gustatory cortex during licking, J. Neurosci. **26**(15), 4126–4138 (2006)

45.35   D.B. Katz, S.A. Simon, M.A.L. Nicolelis: Dynamic and multimodal responses of gustatory cortical neurons in awake rats, J. Neurosci. **21**(12), 4478–4489 (2001)

45.36   R.M. Hallock, P.M. Di Lorenzo: Temporal coding in the gustatory system, Neurosci. Biobehav. Rev. **30**, 1145–1160 (2006)

45.37   M.B. Vogt, C.M. Mistretta: Convergence in mammalian nucleus of solitary tract during development and functional differentiation of salt taste circuits, J. Neurosci. **10**, 3148–3157 (1990)

45.38   C.H. Lemon, D.V. Smith: Influence of response variability on the coding performance of central gustatory neurons, J. Neurosci. **26**, 7433–7443 (2006)

45.39   U.K. Kim, D. Drayna: Genetics of individual differences in bitter taste perception: Lessons from the PTC gene, Clin. Genet. **67**, 275–280 (2005)

45.40   L.C. Geran, S.P. Travers: Single neurons in the nucleus of the solitary tract respond selectively to bitter taste stimuli, J. Neurophysiol. **96**(5), 2513–2527 (2006)

45.41   L.C. Geran, S.P. Travers: Bitter-responsive gustatory neurons in the rat parabrachial nucleus, J. Neurophysiol. **101**(3), 1598–1612 (2009)

45.42   M.I. Harrer, S.P. Travers: Topographic organization of Fos-like immunoreactivity in the rostral nucleus of the solitary tract evoked by gustatory stimulation with sucrose and quinine, Brain Res. **711**(1/2), 125–137 (1996)

45.43   X. Chen, M. Gabitto, Y. Peng, N.J. Ryba, C.S. Zuker: A gustotopic map of taste qualities in the mammalian brain, Science **333**(6047), 1262–1266 (2011)

45.44   K. Nakamura, R. Norgren: Gustatory responses of neurons in the nucleus of the solitary tract of behaving rats, J. Neurophysiol. **66**, 1232–1248 (1991)

45.45   J.-Y. Chen, J.D. Victor, P.M. Di Lorenzo: Temporal coding of intensity of NaCl and HCl in the nucleus of the solitary tract of the rat, J. Neurophysiol. **105**(2), 697–711 (2011)

45.46   J.D. Victor, K.P. Purpura: Nature and precision of temporal coding in visual cortex: A metric-space analysis, J. Neurophysiol. **76**, 1310–1326 (1996)

45.47   J.D. Victor, K.P. Purpura: Metric-space analysis of spike trains: Theory, algorithms and application, Network **8**, 127–164 (1997)

45.48   R.M. Bradley, H.M. Stedman, C.M. Mistretta: Superior laryngeal nerve response patterns to chemical stimulation of sheep epiglottis, Brain Res. **276**(1), 81–93 (1983)

45.49   T. Nagai, K. Ueda: Stochastic properties of gustatory impulse discharges in rat chorda tympani fibers, J. Neurophysiol. **45**(3), 574–592 (1981)

45.50   H. Ogawa, M. Sato, S. Yamashita: Variability in impulse discharges in rat chorda tympani fibers in response to repeated gustatory stimulations, Physiol. Behav. **11**(4), 469–479 (1973)

45.51   H. Ogawa, S. Yamashita, M. Sato: Variation in gustatory nerve fiber discharge pattern with change in stimulus concentration and quality, J. Neurophysiol. **37**(3), 443–457 (1974)

45.52   P.M. Di Lorenzo, J.S. Schwartzbaum: Coding of gustatory information in the pontine parabrachial

45.53   nuclei of the rabbit: Magnitude of neural response, Brain Res. **251**, 229–244 (1982)

45.53   P.M. Di Lorenzo, J.D. Victor: Taste response variability and temporal coding in the nucleus of the solitary tract of the rat, J. Neurophysiol. **90**, 1418–1431 (2003)

45.54   A.T. Roussin, J.D. Victor, J.-Y. Chen, P.M. Di Lorenzo: Variability in responses and temporal coding of tastants of similar quality in the nucleus of the solitary tract of the rat, J. Neurophysiol. **99**(2), 644–655 (2008)

45.55   P.M. Di Lorenzo, S. Leshchinskiy, D.N. Moroney, J.M. Ozdoba: Making time count: Functional evidence for temporal coding of taste sensation, Behav. Neurosci. **123**(1), 14–25 (2009)

45.56   P.M. Di Lorenzo, R.M. Hallock, D.P. Kennedy: Temporal coding of sensation: Mimicking taste quality with electrical stimulation of the brain, Behav. Neurosci. **117**(6), 1423–1433 (2003)

45.57   P.M. Di Lorenzo, J.D. Victor: Neural coding mechanisms for flow rate in taste-responsive cells in the nucleus of the solitary tract of the rat, J. Neurophysiol. **97**(2), 1857–1861 (2007)

45.58   Y.K. Cho, C.S. Li, D.V. Smith: Gustatory projections from the nucleus of the solitary tract to the parabrachial nuclei in the hamster, Chem. Senses **27**, 81–90 (2002)

45.59   P.M. Di Lorenzo, S. Monroe: Transfer of information about taste from the nucleus of the solitary tract to the parabrachial nucleus of the pons, Brain Res. **763**, 167–181 (1997)

45.60   P.M. Di Lorenzo, D. Platt, J.D. Victor: Information processing in the parabrachial nucleus of the pons: Temporal relationships of input and output, Ann. N.Y. Acad. Sci. **1170**, 365–371 (2009)

45.61   A.M. Rosen, J.D. Victor, P.M. Di Lorenzo: Temporal coding of taste in the parabrachial nucleus of the pons of the rat, J. Neurophysiol. **105**(4), 1889–1896 (2011)

45.62   C.S. Li, D.V. Smith: Glutamate receptor antagonists block gustatory afferent input to the nucleus of the solitary tract, J. Neurophysiol. **77**(3), 1514–1525 (1997)

45.63   D.V. Smith, C.S. Li: Tonic GABAergic inhibition of taste-responsive neurons in the nucleus of the solitary tract, Chem. Senses **23**(2), 159–169 (1998)

45.64   P.S. Lasiter, D.L. Kachele: Organization of GABA and GABA-transaminase containing neurons in the gustatory zone of the nucleus of the solitary tract, Brain Res. Bull. **21**, 623–636 (1988)

45.65   D.V. Smith, C.S. Li, B.J. Davis: Excitatory and inhibitory modulation of taste responses in the hamster brainstem, Ann. N.Y. Acad. Sci. **855**, 450–456 (1998)

45.66   M.S. King: Distribution of immunoreactive GABA and glutamate receptors in the gustatory portion of the nucleus of the solitary tract in rat, Brain Res. Bull. **60**(3), 241–254 (2003)

45.67 L. Wang, R.M. Bradley: Influence of GABA on neurons of the gustatory zone of the rat nucleus of the solitary tract, Brain Res. **616**(1–2), 144–153 (1993)

45.68 C.H. Lemon, P.M. Di Lorenzo: Effects of electrical stimulation of the chorda tympani nerve on taste responses in the nucleus of the solitary tract of the rat, J. Neurophysiol. **88**, 2477–2489 (2002)

45.69 A.M. Rosen, P.M. Di Lorenzo: Two types of inhibitory influences target different groups of taste-responsive cells in the nucleus of the solitary tract of the rat, Brain Res. **1275**, 24–32 (2009)

45.70 G. Grabauskas, R.M. Bradley: Ionic mechanism of GABAA biphasic synaptic potentials in gustatory nucleus of the solitary tract, Ann. N.Y. Acad. Sci. **855**, 486–487 (1998)

45.71 G. Grabauskas, R.M. Bradley: Potentiation of GABAergic synaptic transmission in the rostral nucleus of the solitary tract, Neuroscience **94**(4), 1173–1182 (1999)

45.72 G. Grabauskas, R.M. Bradley: Frequency-dependent properties of inhibitory synapses in the rostral nucleus of the solitary tract, J. Neurophysiol. **89**(1), 199–211 (2003)

45.73 C.H. Lemon, P.M. Di Lorenzo: Effects of electrical stimulation of the chorda tympani nerve on taste responses in the nucleus of the solitary tract of the rat, J. Neurophysiol. **88**, 2477–2489 (2002)

45.74 P.M. Di Lorenzo, C.H. Lemon, C.G. Reich: Dynamic coding of taste stimuli in the brain stem: Effects of brief pulses of taste stimuli on subsequent taste responses, J. Neurosci. **23**, 8893–8902 (2003)

45.75 A.M. Rosen, A.R. Roussin, P.M. Di Lorenzo: Water as an independent taste modality, Front. Neurosci. **4**, 175–185 (2010)

45.76 A.M. Rosen, P.M. Di Lorenzo: Two types of inhibitory influences target different groups of taste-responsive cells in the nucleus of the solitary tract of the rat, Brain Res. **1275**, 24–32 (2009)

45.77 A.M. Rosen, H. Sichtig, J.D. Schaffer, P.M. Di Lorenzo: Taste-specific cell assemblies in a biologically informed model of the nucleus of the solitary tract, J. Neurophysiol. **104**(1), 4–17 (2010)

45.78 L.J. Eshelman: The CHC adaptive search algorithm: How to have safe search when engaging in nontraditional genetic recombination. In: *Foundations of Genetic Algorithms*, ed. by G.J. Rawlins (Morgan Kaufman, San Mateo 1991) pp. 265–283

45.79 T. Nomura, H. Ogawa: The taste and mechanical response properties of neurons in the parvicellular part of the thalamic posteromedial ventral nucleus of the rat, Neurosci. Res. **3**(2), 91–105 (1985)

45.80 J.V. Verhagen, B.K. Giza, T.R. Scott: Responses to taste stimulation in the ventroposteromedial nucleus of the thalamus in rats, J. Neurophysiol. **89**(1), 265–275 (2003)

45.81 T.R. Scott, R.P. Erickson: Synaptic processing of taste-quality information in thalamus of the rat, J. Neurophysiol. **34**(5), 868–883 (1971)

45.82 T.C. Pritchard, R.B. Hamilton, R. Norgren: Neural coding of gustatory information in the thalamus of *Macaca mulatta*, J. Neurophysiol. **61**(1), 1–14 (1989)

45.83 H. Ogawa, T. Nomura: Receptive field properties of thalamo-cortical taste relay neurons in the parvicellular part of the posteromedial ventral nucleus in rats, Exp. Brain Res. **73**, 364–370 (1988)

45.84 S. Monroe, P.M. Di Lorenzo: Taste responses in neurons in the nucleus of the solitary tract that do and do not project to the parabrachial pons, J. Neurophysiol. **74**(1), 249–257 (1995)

45.85 A. Fontanini, D.B. Katz: State-dependent modulation of time-varying gustatory responses, J. Neurophysiol. **96**(6), 3183–3193 (2006)

45.86 S.E. Grossman, A. Fontanini, J.S. Wieskopf, D.B. Katz: Learning-related plasticity of temporal coding in simultaneously recorded amygdala-cortical ensembles, J. Neurosci. **28**(11), 2864–2873 (2008)

45.87 R. Gutierrez, S.A. Simon, M.A. Nicolelis: Licking-induced synchrony in the taste-reward circuit improves cue discrimination during learning, J. Neurosci. **30**(1), 287–303 (2010)

45.88 A. Carleton, R. Accolla, S.A. Simon: Coding in the mammalian gustatory system, Trends Neurosci. **33**(7), 326–334 (2010)

45.89 M. Sugita, Y. Shiba: Genetic tracing shows segregation of taste neuronal circuitries for bitter and sweet, Science **309**(5735), 781–785 (2005)

45.90 R. Accolla, B. Bathellier, C.C. Petersen, A. Carleton: Differential spatial representation of taste modalities in the rat gustatory cortex, J. Neurosci. **27**(6), 1396–1404 (2007)

45.91 T. Yamamoto, N. Yuyama, T. Kato, Y. Kawamura: Gustatory responses of cortical neurons in rats: I. Response characteristics, J. Neurophysiol. **51**, 616–635 (1984)

45.92 T. Yamamoto, R. Matsuo, Y. Kiyomitsu, R. Kitamura: Sensory inputs from the oral region to the cerebral cortex in behaving rats: An analysis of unit responses in cortical somatosensory and taste areas during ingestive behavior, J. Neurophysiol. **60**, 1303–1321 (1988)

45.93 A. Fontanini, D.B. Katz: Behavioral modulation of gustatory cortical activity, Ann. N.Y. Acad. Sci. **1170**, 403–406 (2009)

46. EEG Signal Processing for Brain–Computer Interfaces

Petia Georgieva, Filipe Silva, Mariofanna Milanova, Nikola Kasabov

This chapter is focused on recent advances in electroencephalogram (EEG) signal processing for brain computer interface (BCI) design. A general overview of BCI technologies is first presented, and then the protocol for motor imagery noninvasive BCI for mobile robot control is discussed. Our ongoing research on noninvasive BCI design based not on recorded EEG but on the brain sources that originated the EEG signal is also introduced. We propose a solution to EEG-based brain source recovering by combining two techniques, a sequential Monte Carlo method for source localization and spatial filtering by beamforming for the respective source signal estimation. The EEG inverse problem is previously studded assuming that the source localization is known. In this work for the first time the problem of inverse modeling is solved simultaneously with the problem of the respective source space localization.

46.1 Overview

Current computer input devices require some physical contact between the user and the input device, for example, a keyboard, mouse, or trackball. Other interfaces might include speech recognition to allow computer tasks and dictation to be performed by the user without any physical contact. Despite the wide range of computer input devices, it is clear that there may be circumstances where such devices are not suitable. For example, if the user has a motor-control disorder both speech and physical input may be impossible. This is how the idea of a brain computer interface (BCI) emerged, as an alternative communication channel between the human being and the external world. The objective is to record the brain activity variations associated with a thought or mental state and *translate* them into some kind of actuation or command over a target

output. Examples of possible uses for such an interface are control of a robot arm, wheelchair movement or writing on a monitor (BCI as a mental keyboard). Among various potential applications, prosthetic limb control is currently a major targeted application, allowing a paralyzed patient to gain motor ability in a BCI controlled limb. Potential users include those with severe head trauma or spinal cord injuries, as well as people with amyotrophic lateral sclerosis (ALS, or Lou Gehrig's disease), cerebral palsy, or other motor-control disorders. BCIs could be useful even for people with lesser, or no, motor impairment, as, for example, computer gamers.

This chapter will present recent advances towards the development of BCI systems that analyze the brain activity of a subject measured through the electroencephalogram (EEG). The work begins with a general overview of BCI technologies, which is followed by a more comprehensive analysis of the EEG-based BCI paradigms. Next, we focus on the main signal pro-

cessing challenges – EEG signal enhancement, feature extraction, and classification. Finally, we introduce our ongoing research on EEG source estimation by particle filters.

46.2 BCI Technologies

There are three main BCI technologies, which are invasive, partially invasive, and noninvasive. The *invasive BCIs* are based on detection of single neuron activity by intra-cortical electrodes implanted into the gray matter of the brain. Much of the related research has been done with animals (mainly rats and monkeys), although some tests with humans have also been done. For example, successful attempts were reported with patients who already have the electrodes from epilepsy monitoring trials. *Partially invasive BCIs* are based on recording of electrocorticographic (ECoG) implants inside the skull but outside the gray matter. Invasive and partially invasive BCIs are more reliable and less noisy, however implant placement by neurosurgery is required, which brings hardware and ethical problems. For example, important issues concerning the implant suitability for long-term use or their periodical replacement have still to be solved.

Noninvasive BCIs rely on recording of the brain activity at a macro level (in contrast to the single neuron level by invasive BCIs) using various noninvasive techniques. The most typical noninvasive BCI designs use an EEG to measure the potential difference between electrodes placed on the head surface. The EEG machine is a standard medical imaging device and has many advantages: it is safe, cheap, and relatively easy to use. Moreover, the subject preparation for experiments and the recordings are fast and, there-

fore, it is possible to perform real-time analysis of EEG components.

Alternative noninvasive BCI techniques such as magnetoencephalography (MEG) [46.1], functional magnetic resonance imaging (fMRI) [46.2], and functional near-infrared systems (fNIR) [46.3] are also subject of intensive research. In Fig. 46.1 time versus spatial resolutions of different techniques for brain imaging and the associated hardware complexity and price are depicted. Note that while the EEG technique has a temporal resolution similar to the real neuron activity (in the range of 1 ms), its spatial resolution is the worst among all techniques. Single photon emission computerized tomography (SPECT) and positron emission tomography (PET) do not provide resolution advantages and, additionally, they are more expensive, and therefore they have not yet been reported as BCI technologies. Note that fMRI has the closest spatial resolution to the real neuron activity (around $1\,\mathrm{mm}^3$). However, current technologies for recording MEG and fMRI are expensive, making them not quite practical for wider applications in the near future. fNIR is potentially cheaper but both fMRI and fNIR are based on changes in cerebral blood flow, which has an inherently slow response. Therefore, EEG signal processing is currently the most practical technology for BCI. Nevertheless, building a hybrid BCI that combines the two modalities, EEG and fMRI, and thus benefits from their respective favorable resolutions is an appealing still unexplored alternative.

BCI Sensors

The evolution over time of the electrical sensors used in BCI systems is summarized in Fig. 46.2a. The first BCI system was developed in 1977 by *Jacques Vidal* [46.4] and was based on visual evoked potentials (VEP). The real-time analysis of EEG signals suggested the use of waveform activity in the timeframe of N100–P200 components, with N and P indicating negative and positive peaks, and the numbers indicating the approximate latency in msec. Vidal's achievement was a proof of concept demonstration. Due to the limited computer processing capacity at that time (an XDS

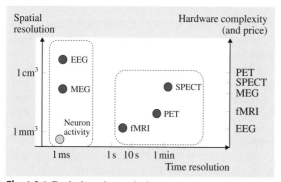

Fig. 46.1 Brain imaging techniques

Sigma 7 computer controlled the experiments and an IBM360 mainframe digitized the data) it was far from being practical. Online removal of ocular artifacts was also included in Vidal's system. A decade earlier, *Edmond Dewan*, 1967 [46.5], showed that subjects could learn to transmit Morse code messages using eye movements to modulate their brain waves.

Since 1992 experiments with invasive electrodes (ECoG and intra-cortical) have been performed in the BCI framework, mainly with animals (Fig. 46.2c). However, due to the high risk and hardware and ethical problems, invasive brain imaging is still less implemented. The noninvasive sensors (Fig. 46.2b) evolved from the classical EEG gel-based electrodes, through the near-IR electrodes, and very recently dry electrodes appeared on the market.

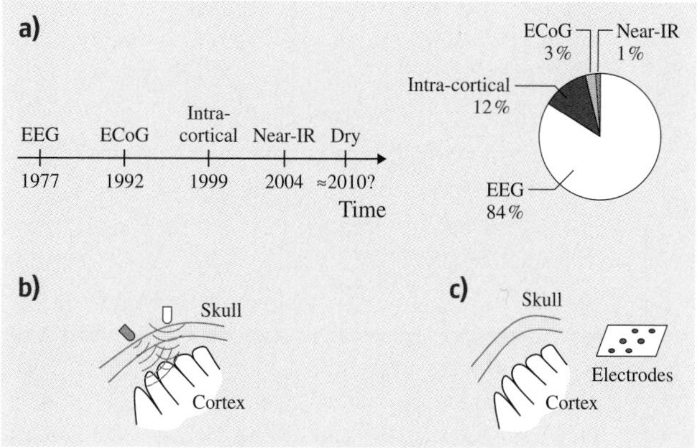

Fig. 46.2 (a) BCI electrodes technology evolution; (b) noninvasive BCI electrodes; (c) invasive BCI electrodes; BCI sensors

46.3 EEG-Based BCI

46.3.1 EEG Waves

The concept of interfacing a computer directly to the brain is a relatively new one, but analysis of brain waves has been reported since 1929 when Hans Berger first recorded EEG signals from a human source using electrodes and a galvanometer. This measured the potential difference between the electrodes placed on the scalp and a reference electrode, placed on the earlobe. The potential difference fluctuations gave an indication to the small current produced by the billions of neurons in the brain when they fire. The combination of the neurons firing in the brain can yield signals that are extremely complex but that have identifiable patterns based upon the activity in the brain. For example, sleeping will produce a substantially different EEG signal to that of a brain computing complex mathematical problems – this can be seen by performing a frequency analysis of the signals. The key concept for BCI designs is the frequency band segmentation. The EEG signal is composed of waves inside the $0–60\,\text{Hz}$ frequency band. Different brain activities and states can be identified based on the extracted frequency content of the recorded oscillations [46.6]. The main frequency bands of the EEG signal are listed in Fig. 46.3.

Delta waves (below $4\,\text{Hz}$) are associated with deep sleep. They are a high-amplitude, low-frequency wave and are generated by the lack of processing by neu-

rons. Delta waves can also be found when examining a comatose patient.

Theta waves ($4–8\,\text{Hz}$) are typical for dreamlike states and old memories, but can also be associated with anxiety, epilepsy, and traumatic brain injury.

The alpha band ($8–13\,\text{Hz}$) corresponds to a relaxed state recorded in the occipital brain zone. The amplitude of alpha waves ranges between 10 and $50\,\text{mV}$.

Sensorimotor (mu) rhythms ($8–12\,\text{Hz}$) are associated with the sensory-motor cortex and can be used to recognize intention or preparation of movement and also imaginary motor movement.

Beta ($13–30\,\text{Hz}$) waves are associated with alertness, arousal, concentration, and attention. Beta waves are fast but of low amplitude.

The gamma band ($30–50\,\text{Hz}$) is characteristic for mental activities such as perception, problem solving, and creativity.

Note that the alpha band and mu-rhythms cover the same frequency range, however, the respective waves are clearly identified at different brain zones.

For each particular brain activity there is one particular area that produces stronger electrical activity in one of the previously referred to frequency bands; similarly, internal artifacts are more relevant in some parts of the scalp than in others. Consequently, EEG signals are multi-channel signals, were each channel corresponds to a specific scalp location. The occipital area of the scalp is known to provide stronger electrical signals in

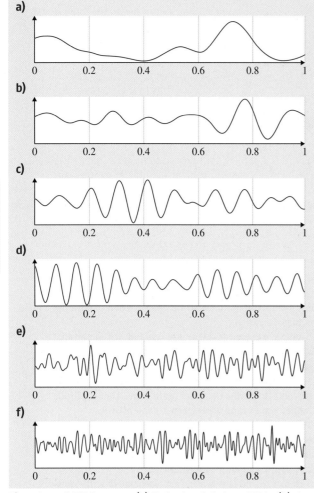

Fig. 46.3a–f EEG waves. (**a**) Delta band (below 4 Hz); (**b**) theta band (4–8 Hz); (**c**) alpha band (8–13 Hz); (**d**) mu-rhythm (8–12 Hz); (**e**) beta band (13–30 Hz); (**f**) gamma band (30–50 Hz)

the band in response to visual stimulation and perception of images.

46.3.2 EEG-Based BCI Paradigms

Among the various noninvasive BCI paradigms proposed, the *sensorimotor (mu) rhythms* (SMR) and the *visual evoked potentials* (VEP) are the most frequently reported.

SMR are spontaneous changes in the EEG oscillations in the 8–12 Hz range registered from the central region of the scalp (sensorimotor cortex) when

movement or imagination of movement is performed. At least a second before subjects initiate voluntary movement, the SMR over the hemisphere contralateral to the movement direction decrease in amplitude. SMR activity returns to baseline levels within a second after movement is initiated. These motor activity dependent changes are termed event related desynchronization (ERD) and event related synchronization (ERS). *Wolpaw* et al. [46.7] reported the first use of SMRs for cursor control.

In contrast to spontaneous SMR that do not require special stimuli to occur, VEP changes in the ongoing EEG are generated in response to visual stimulus. However, visual stimuli may comprise different components, such as color, texture, motion, objects, readability (text versus non-text), etc. Each of these components has an impact on the spatial dispersion of the VEP throughout the scalp, being observed differently in each EEG channel. Therefore, to focus the VEP production and analysis, the set of visual stimuli must be coherent, i.e., it should always stimulate the same brain areas. Typical VEP waves are denoted as N100, P100, P200, P300, and N300, with N and P indicating negative and positive peaks and the numbers indicating the approximate latency in milliseconds, for example, P300 stays for positive potentials over 300 ms. Early waves (N100, P100) correspond to perception of sensory stimulus, while latter waves (P300, N300) reflect higher cognitive processes like attention, learning tasks, decision-making, memory, etc. Various studies [46.8–10] showed that VEPs recorded from the human scalp contain a train of short latency wavelets, in the gamma band, precisely time locked to the stimulus and lasting approximately 100 ms. Furthermore, a more recent study [46.11] showed that the perception learning task of a picture set induced neural activity in the gamma band, highly synchronized between occipital electrodes. Finally, the analysis of EEG in response to coherent and noncoherent visual stimulus in a discrimination task evidenced a short-lasting occipital enhancement in the band around 300 ms after stimulus onset in response to coherent images only [46.12].

The most well-known BCI based on VIP is the so-called P300 BCI where a matrix of choices (letters of the alphabet, digits, and other symbols) is presented on screen and the EEG is recorded as these choices flash in succession. The positive potentials around 300 ms after an event significant to the subject is considered a *cognitive* potential since it is generated in tasks where

Table 46.1 BCI feature selection

Extraction method	Time	Frequency	Space	Examples
Time-frequency representation	yes	yes	yes	FT, wavelet
Cell firing rate	yes	yes	yes	—
Power spectral density (PSD)	no	yes	yes	Welch
Pattern matching	yes	no	yes	Correlation
Raw signal	yes	no	yes	Amplitude
Model parameters	yes	no	yes	AR, AAR, KF
Matrix transforms	yes	no	yes	ICA, PCA

the subject discriminates among stimuli. In the P300 spelling device, the subject focuses his/her attention on the desired symbol as the rows and columns of the matrix are repeatedly flashed to elicit VEPs.

An attempt for BCI design based on mental tasks such as solving a mathematical problem, mental counting, and imagining/rotating a 3D object was proposed recently [46.13].

BCI Challenges

Independently of the paradigms discussed above, the main challenges of EEG-based BCIs are related to:

1. The highly nonstationary nature of the EEG signal
2. The low signal to noise ratio (SNR) due to different sources of noise and physiological artifacts
3. Poor spatial resolution.

EEG signals are electric signals gathered at the scalp of an individual. These signals are a combination of signals from two different sources: 1) neural-cerebral activity, called features and 2) noncerebral origins, called artifacts. Internal (physiological) artifacts are artifacts caused by other body activities, such as eye motion, eye blinking, electrocardic activity, and electric activity resulting from muscle contraction. External (environmental) artifacts are artifacts created by external sources, such as power line induction or bad electrode contact. Currently there are many techniques for removing artifacts from EEG signals. Since the original signals received from the electrodes are very weak, in order to generate a useable signal they must be amplified substantially. Therefore, the EEG measurements are subject to pre-processing and filtering to remove the physiological and environmental artifacts and also the background brain activity (for example, jaw clenching) in order to isolate the event related features.

Apart from the EEG signal processing challenges, more technical problems such as effective electrode placement and the required impedance between scalp and electrodes have also been studied. In the past, the EEG was recorded on paper and an impedance of $5\,\mathrm{k\Omega}$ was required. However, with digital recording of the data it has been found that an impedance of $40\,\mathrm{k\Omega}$ will still produce usable results.

Further to this, the protocols and drivers of communication between the EEG machine, the computer, and eventually an output device often need special attention. Timing issues represent an additional challenge when real-time control of an external device is the goal of the BCI, as, for example, controlling a wheelchair. Last but not least, the multi-disciplinarity of BCI research requires knowledge and expertise in many different disciplines such as signal processing, computer science, computational neuroscience, and imbedded intelligent systems.

46.3.3 Feature Selection

After amplification and filtering of the row EEG data, the features that define the brain state or activity are selected and extracted. Due to the inherently large EEG signal variability that ranges from session to session for the same subject, through single versus multiple trials also for the same subject, to a clearly expressed

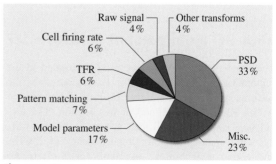

Fig. 46.4 Typical BCI features

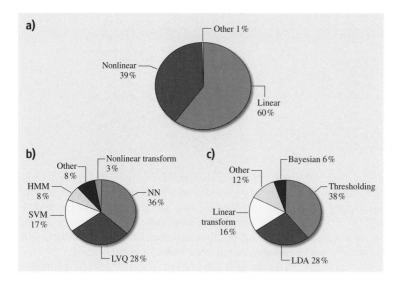

Fig. 46.5a–c Classification techniques for BCI. (**a**) Linear versus nonlinear approaches; (**b**) nonlinear techniques; (**c**) linear techniques

variability from subject to subject, features in time, space, and frequency are defined (see Fig. 46.4 and Table 46.1). The isolation and extraction of the most suitable features of interest is the key for BCI designs and it depends strongly on the BCI paradigm. Through training, the user can learn to control these detectable features.

For BCI-based on motor imagery tasks, SMRs are isolated by the power spectral density (PSD) method. PSD is the most widely applied technique for feature extraction, followed by classical linear methods such as auto regressive (AR) models, adaptive auto regressive (AAR) models, Kalman filtering (KF), and basic matrix transformation such as independent component analysis (ICA) and principal component analysis (PCA).

A combination of various features is an efficient approach to face the higher variability related with VEP BCI.

46.3.4 Classification Techniques

The next step of the BCI design is the feature classification and decoding of the brain state that reflects the subject's desire. The classification approaches are generally devised as nonlinear and linear, the latter being applied more often. The main linear and nonlinear techniques and their respective contribution in the BCI framework are summarized in Fig. 46.5. Among the linear techniques, simple thresholding and linear discriminant analysis (LDA) are generally the best classifiers. As with the nonlinear approach, neural networks (NN), learning vector quantization (LVQ), and nonlinear support vector machines (SVM) are implemented in more than 80% of cases. Although probabilistic techniques like Bayesian classifiers and hidden Markov models (HMM) are very powerful, they need longer execution time (particularly if Monte Carlo runs are required) and, therefore, are not suitable for real-time applications.

46.4 BCI Mobile Robot Control – IEETA Case Study

The protocol for motor imagery BCI designed in the Laboratory of Signal Processing (www.ieeta.pt) by the Institute of Electrical Engineering and Telematics of Aveiro (IEETA) is outlined in this section. The ambulatory EEG device, Trackit system LifeLines Ltd., has eight channels, a sampling frequency of 256 Hz, and a maximum voltage of 10 mV. The electrodes are located according to the standard 10/20 international system (Fig. 46.6) and their size is about

1.5 mm. EEG signals recorded from the central area (C3, C4, Cz), the frontal area (F3, F4), and the parietal area (P3, P4) are used for brain activity decoding.

The objective is to design a communication protocol between the brain and the Khepera mini-robot (5.7 cm diameter) during motor imagery tasks and to successfully control the robot movements over an improvised platform (Fig. 46.7).

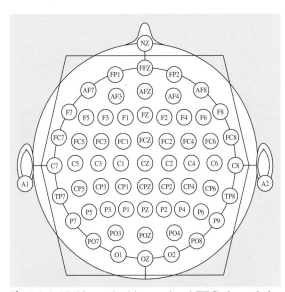

Fig. 46.6 10/20 standard international EEG electrode location system

Before real-time experiments, the user learns to modulate mu-rhythms with the help of visual feedback. The subject is given randomly generated instructions for performing one of the four motor imagery tasks – movements to the right, left, forward, and stopping the screen ball (Fig. 46.8). The BCI classifies the recorded EEG signals and the ball is moved according to the decoded intention.

The algorithm for signal acquisition, pre-processing, feature extraction, and classification of desired commands is summarized next in six steps.

Algorithm 46.1 EEG signal processing for BCI (Fig. 46.9)

Step 1: Initialization (fixation period)
Initial period of subject preparation and concentration (3–5 s) during which the personalized baseline EEG signal is recorded.

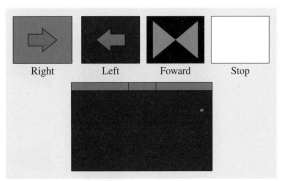

Fig. 46.8 Protocol of subject training

Step 2: Standard digital filter
The recorded EEG signals (C3, C4, Cz, P3, P4, F3, F4) are filtered by a standard digital infinite impulse response (IIR) Butterworth filter of the eighth order (1–40 Hz).

Step 3: Spatial surface Laplacian filter
For each hemisphere only one corresponding channel is considered obtained by subtracting the mean value of the three neighbor channels:

$$C\_LH = C3 - 1/3 * (F3 + P3 + Cz)$$
$$\text{(left hemisphere)}$$
$$C\_RH = C4 - 1/3 * (F4 + P4 + Cz)$$
$$\text{(right hemisphere)}$$

Step 4: Spectral power feature extraction
Sensorimotor (mu) rhythms (8–12 Hz) are extracted from the C_LH and C_RH equivalent signals and divided into segments of 128 sequential samples (0.5 s). The spectral power (P) for each segment per hemisphere is extracted.

Step 5: Event related de-synchronization (ERD)
ERD is computed for each channel C_LH and C_RH.

$$\text{ERD\%} = \frac{P - B}{B} 100 ,$$

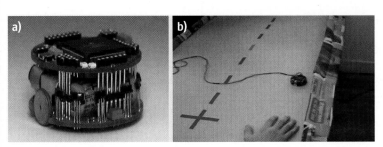

Fig. 46.7 (a) Khepera mini-robot; (b) BCI control of a mobile robot

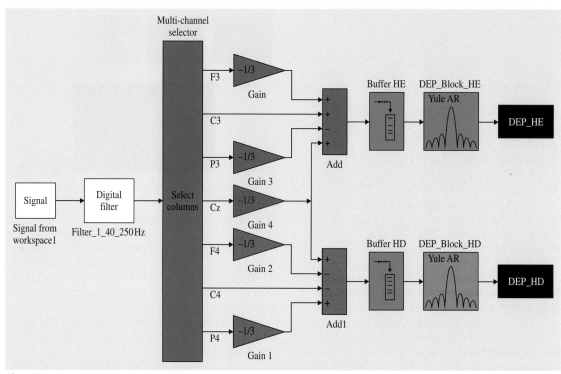

Fig. 46.9 Simulink EEG signal processing

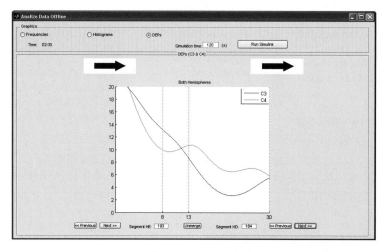

Fig. 46.10 Typical spectral power curves in left (C3) and right (C4) hemispheres related with imagery movement task *to the right*

where B is the mean power value of the baseline signal collected during the fixation period (step 1), P is the spectral power computed during motor imagery tasks (step 4).

Step 6: *Classification*

The classifier has two inputs, one for each ERD block.

Power attenuation is confirmed if ERD has a negative value and the following empirical rules drive the classifier:

- If only the right hemisphere signal verifies the ERD – the classifier output is "LEFT"
- If only the left hemisphere signal verifies the ERD – the classifier output is "RIGHT" (see Fig. 46.10).

Fig. 46.11 ERD/ERS due to movement imagery versus movement execution ▶

- If both signals verify the ERD – the output is "FORWARD"
- If neither of the signals verify the ERD – the output is "STOP".

In most of the cases studied, the motor imagery BCI can reach an accuracy of up to 70–75%. Other publications report similar results [46.14]. ERD/ERS due to imagination of a physical movement are normally less clearly identifiable compared with ERD/ERS provoked by real movement execution (Fig. 46.11). However, if the subject has some residual motor abilities there exist better human–computer communication techniques than motor imagery BCIs. On the other side, alternatives like VEP BCI are reported to achieve better accuracy; however their technical realization is more difficult and slow, since they require visual stimuli. Moreover, while motor imagery BCI can be automated and the subject

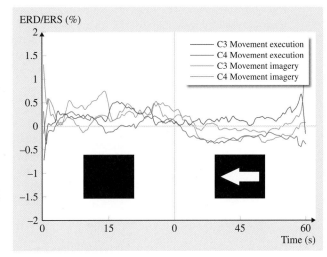

can do other cognitive tasks [46.15], P300 BCI requires the total attention of the subject during the stimulus presentation.

46.5 Source-Based BCI

Noninvasive BCI design based not on recorded EEG but on inner brain sources that originated the EEG is called source-based BCI [46.16–19]. The objective of source-based BCI is to find the localization and time patterns of the main brain activity sources by analyzing multichannel EEG recordings. The intuition behind this is that, hopefully, the recovered original sources will be less noisy and thus the essential information can be more reliably decoded. Additionally, we expect to reduce the dimensionality of the problem by extracting a fewer number of dominant sources than the number of surface electrodes.

In brain imaging, EEG source signal estimation is also known as the EEG inverse problem. The problem can be formulated as follows: using the measurements of electrical potential on the scalp recorded from multisensors, the goal is to build a reconstruction system able to estimate the location (the brain area) and the magnitude and directions of the dominative neural brain sources that most probably have originated the recorded EEG signal. Thus the problem can be divided into two stages: 1) *localization of the principal original sources inside the brain* and 2) *estimation of the source signal* (waveforms).

The problem of reconstructing the time pattern of the original source signals from a sensor array, can be

expressed as a number of related blind source separation (BSS) problems. *Choi* et al. [46.20] present a review of various BSS and independent component analysis (ICA) algorithms for static and dynamic models in their applications. Beamforming (BF) is also a popular analysis procedure for noninvasively recorded electrophysiological data sets. The goal is to use a set or recording sensors and combine the signals recorded at individual sites to increase the SNR, but focusing on a certain region in space (region-of-interest, ROI). In that sense, beamforming uses a different approach to image brain activities: the whole brain is scanned point by point. Thus, it is, in fact, a spatial filter designed to be fully sensitive to activity from the target location, while being as insensitive as possible to activity from other brain regions. This is achieved by constructing the spatial filter in an adaptive way, i. e., by taking into account the recorded data. More concretely, BF is carried out by weighting the EEG signals, thereby adjusting their amplitudes such as that when added together they form the desired source signal.

In this chapter we propose a solution to the EEG-based brain source recovering by combining two techniques, namely a sequential Monte Carlo (SMC) method for source localization and spatial filtering by BF for the respective source signal estimation based on

EEG measurements. The EEG inverse problem is intensively studded assuming that the source localization is known. In this work, for the first time the problem of inverse modeling is solved simultaneously with the problem of the respective source space localization.

46.5.1 Sequential Monte Carlo Problem Formulation

We consider the problem of the EEG source localization as an estimation problem and solve it within the SMC framework. To define the estimation problem, consider the evolution of the state sequence $\{x_k, k \in N\}$ of a target given by

$$x_k = f(x_{k-1}, w_{k-1}), \qquad (46.1)$$

where the state x at time k ($x_k \in \mathfrak{R}^{n_x}$) is possibly a nonlinear function f of the state x_{k-1} at the previous time $k-1$ and is also affected by the process noise sequence w_{k-1}; N is the set of natural numbers. The objective is to recursively estimate x_k from measurements

$$z_k = h(x_k, v_k), \qquad (46.2)$$

where h is a possibly nonlinear function and v_k is the measurement noise sequence. Expressions (46.1) and (46.2) are the state and the measurement equations of the general state-space transition model required by the SMC estimation [46.21]. In particular we seek filtered estimates of x_k based on the set of all available measurements up to time k. It is assumed that the observations are taken at discrete time points with a discretization time step T.

Within the Bayesian framework, the estimation problem is to recursively calculate some degree of belief in the state x_k at time k, given the data $z_{1:k}$ up to time k [46.22]. Thus it is required to construct the posterior probability density function (pdf) $p(x_k \mid z_{1:k})$. It is assumed that the initial pdf $p(x_0 \mid z_0) \equiv p(x_0)$ of the state vector, which is also known as the prior, is available (z_0 is the initial measurement). Then, in principle, the posterior conditional pdf $p(x_k \mid z_{1:k})$ may be obtained recursively, in two stages: *prediction* and *update*. Suppose that the required pdf $p(x_{k-1} \mid z_{1:k-1})$ at time $k-1$ is available. The prediction stage involves using the system model (46.1) to obtain the prior pdf of the state at time k via the Chapman–Kolmogorov equation

$$p(x_k \mid z_{1:k-1})$$
$$= \int p(x_k \mid x_{k-1}) p(x_{k-1} \mid z_{1:k-1}) dx_{k-1}, \qquad (46.3)$$

since (46.1) describes a Markov process of order one $p(x_k \mid x_{k-1}, z_{1:k-1}) = p(x_k \mid x_{k-1})$. The probabilistic model of the state evolution $p(x_k \mid x_{k-1})$ is defined by the system equation (46.1) and the known statistics of w_{k-1}. At time step k, a measurement z_k becomes available and this may be used to update the prior (update stage) via Bayes' rule

$$p(x_k \mid z_{1:k}) = \frac{p(z_k \mid x_k) p(x_k \mid z_{1:k-1})}{p(z_k \mid z_{1:k-1})}$$
$$= \frac{\text{likelihood} * \text{prior}}{\text{evidence}}, \qquad (46.4)$$

where $p(z_k \mid z_{1:k-1})$ is a normalizing constant defined by the measurement model (46.2) and the known statistics of v_k. Hence, the recursive update of $p(x_k \mid z_k)$ is proportional to

$$p(x_k \mid z_{1:k}) \propto p(z_k \mid x_k) p(x_k \mid z_{1:k-1}). \qquad (46.5)$$

In the update stage (46.4) the measurement is used to modify the prior density to obtain the required posterior density of the current state. The recurrence relations (46.3) and (46.4) form the basis for the optimal Bayesian framework.

In most real-life problems the recursive propagation of the posterior density cannot be performed analytically (the integral in (46.3) is intractable). Usually numerical methods are used and, therefore, a sample-based construction to represent the state pdf. The family of techniques that solve the estimation problem numerically are denoted as particle filtering methods (also known as nonparametric methods). Particle filters (PF) were first defined in the sequential Monte Carlo (SMC) framework and applied to object and video tracking. In the SMC framework, multiple particles (samples) of the state are generated, each one associated with a weight $W_k^{(l)}$ which characterizes the quality of a specific particle l, $l = 1, 2, \ldots, N$. Thus, a set of N weighted particles, drawn from the posterior conditional pdf, is used to map integrals to discrete sums. The posterior $p(x_k \mid z_{1:k-1})$ is approximated by the weighted sum of particles

$$\hat{p}(x_k \mid z_{1:k-1}) = \sum_{l=1}^{N} W_{k-1}^{(l)} \delta(x_k - x_k^l), \qquad (46.6)$$

and the update probability is

$$p(x_k \mid z_{1:k}) = \sum_{l=1}^{N} \hat{W}_k^{(l)} \delta(x_k - x_k^l), \qquad (46.7)$$

where

$$W_k^{(l)} = W_{k-1}^{(l)} p(z_k \mid x_k) = W_{\text{previous}} * , \text{ likelihood} \qquad (46.8)$$

and the normalized importance weights

$$\hat{W}_k^{(l)} = \frac{W_k^{(l)}}{\sum_{l=1}^{N} W_k^{(l)}} \,. \qquad (46.9)$$

New weights are calculated, putting more weight on particles that are important according to the *posterior* pdf (46.7). It is often impossible to sample directly from the posterior density function $p(x_k \mid z_{1:k})$. This difficulty is circumvented by making use of the importance sampling from a known *proposal distribution* $p(x_k \mid x_{k-1})$. During the prediction stage each particle is modified according to the state model (46.1). In the update stage, each particle's weight is re-evaluated based on the new data.

Particle Degeneracy Phenomenon

An inherent SMC problem is particle degeneracy, the case when a small set of particles (or even just one particle) have significant weights. An estimate of the measure of degeneracy [46.23] at time k is given as

$$N_{\text{eff}} = \frac{1}{\sum_{l=1}^{N} \left(W_k^{(l)} \right)^2} \,, \qquad (46.10)$$

where N_{eff} is the number of (effective) particles with significant weights. If the value of N_{eff} is very low, a resampling procedure can help to avoid degeneracy. A schematic representation of the resampling procedure is depicted in Fig. 46.1. Particles with small weights are eliminated, while particles with large weighs are replaced by a number of particles with smaller weights. There are two alternative mechanisms to schedule the resampling, either resampling at each iteration or resampling when the effective number of particles falls below a user-defined threshold N_{thres}. It is known that resampling reduces the variance of the particle population, making the estimation procedure susceptible to outliers with high importance weights and reducing the ability of the SMC to adapt to rapid changes in the states. Therefore, it is crucial to carefully choose N_{thres} in order to maintain the variety within the particle population and still counteract the tendency of focusing all the weight importance in one particle.

46.5.2 EEG Source Localization Model in State-Space

In order to apply the particle filter, outlined in the previous section, the state-space transition model of the source localization is first defined.

Source Model as a Current Dipole Model

Let us assume brain activity arises at a small zone of the cortex centered at location x_s and that the observation point x is some distance away from this zone. The primary current distribution can be approximated by an equivalent current dipole represented as a point source

$$J^p(x_s) = s\delta(x - x_s) \,, \qquad (46.11)$$

where $\delta(x)$ is the Dirac delta function, with moment

$$s \equiv \int J^p(x_s) \mathrm{d}x_s \,. \qquad (46.12)$$

The current dipole is an extension of the model of the paired-charges dipole in electrostatics. It is important to note that brain activity does not actually consist of discrete sets of physical current dipoles, but rather that the dipole is a convenient representation for coherent activation of a large number of pyramidal cells, possibly extending over a few square centimeters of gray matter. The current dipole model is the key of EEG processing since a primary current source of arbitrary extent can always be broken down into small regions, with each region represented by an equivalent current dipole.

Assuming that the electrical activity of the brain can be modeled by a number of dipoles, i.e., the measured multi-channel EEG signal signals $z_k \in \mathfrak{R}^{n_z}$ from n_z sensors at time k are produced by M dipoles, the forward EEG model is given by

$$z_k = \sum_{m=1}^{M} L_m(x_k(m)) s_k(m) + v_k \,, \qquad (46.13)$$

where $x_k(m)$ is a three-dimensional localization vector (space directions), $L_m(x_k(m)) \in \mathfrak{R}^{n_z \times 3}$ is the lead field matrix for dipole m, $s_k(m)$ is a three-dimensional moment vector of the m-th dipole (the source signal). By v_k the effect of noise in the measurements is simulated. $L_m(x_k(m))$ is a nonlinear function of the dipole localization, electrodes positions, and head geometry, [46.24]. Its three columns contain the activity that will be measured at the sensors due to a dipole source with unity moment in the x-, y-, and z-directions, respectively, and zero moment in the other directions. An analytical expression for the forward model exists if the dipole localization, electrode positions, and head geometry are known. The spherical head model is a simplification that preserves some important electrical characteristics of the head, while reducing the mathematical complexity of the problem. The different electric conductivities of the many layers between the brain and the measuring surface need to be known. The skull is typically assumed to be more resistive than the brain and scalp that, in turn, have similar conductivity properties [46.25].

In the framework of the dipole source localization problem, the states that need to be estimated are the geometrical positions of M dipoles

$$x_k = [x_k(1), \ldots, x_k(M)], \quad \text{where}$$
$$x_k(m) = [x(m), y(m), z(m)]^{\mathrm{T}} \quad \text{for } m = 1, \ldots, M.$$
$$(46.14)$$

Then the lead field matrix of M dipoles $L_m(x_k) \in \mathfrak{R}^{n_z \times 3M}$ is

$$L(x_k) = [L(x_k(1)), \ldots, L(x_k(M))]. \quad (46.15)$$

The vector of moments $s_k \in \mathfrak{R}^{3M \times 1}$ is $s_k = [s_k(1), \ldots, s_k(M)]^{\mathrm{T}}$, where each $s_k(m)$ consists of the brain source signals in each space direction, $s_k(m) = [s_x(m), s_y(m), s_z(m)]^{\mathrm{T}}$.

Equation (46.13) can be reformulated in matrix form as follows

$$z_k = L(x_k)s_k + v_k. \quad (46.16)$$

Expression (46.16) corresponds to the measurement equation (46.2) of the general transition model. As for the state equation (46.1), since it is unknown how the states (the geometrical positions of M dipoles) evolve over time, a random walk model (first-order Markov chain) is assumed in the source localization space,

$$x_k = x_{k-1} + w_k. \quad (46.17)$$

Equations (46.16) and (46.17) define the dipole source localization model in state space. The intuition behind the PF approach is to estimate the 3-D location (vector x) of the principle M dipoles (assuming M is known) that originated the underlying EEG recordings z_k. In the above model certain distributions for the process and the measurement noises are assumed and initial values for the states are chosen. The lead field matrix can then be computed, however the moments $s_k(m)$ are not known. In order estimate them the BF approach is used.

46.5.3 Beamforming as a Spatial Filter

BF deals with the estimation of the time patterns in three space directions of the m-th current dipole $s_k(m) = [s_x(m), s_y(m), s_z(m)]^{\mathrm{T}}$ located at $x_k(m) = [x(m), y(m), z(m)]^{\mathrm{T}}$ using the measurements of electrical potential on the scalp recorded from N sensors located at the surface of the head. The beamformer filter consists of weight coefficients (B) that when mul-

tiplied by the electrode measurements give an estimate of the dipole moment at time k

$$s_k = B^{\mathrm{T}} z_k, \quad (46.18)$$

where $B \in \mathfrak{R}^{n_z \times 3M}$ is the weighting matrix. The choice of the beamformer weights is based on the statistics of the signal vector z_k received at the electrodes. Basically, the objective is to optimize the beamformer response with respect to a prescribed criterion, so that the output s contains minimal contribution from noise and interference. There are a number of criteria for choosing the optimum weights. The method described below represents a linear transformation where the transformation matrix is designed according to the solution of a constrained optimization problem (early work on this is attributed to [46.26]).

The basic approach consists in the following: assuming that the desired signal and its direction are both unknown, accurate signal estimation can be provided by minimizing the output signal variance. To ensure that the desired signal is passed with a specific (unity) gain, a constraint may be used so that the response of the beamformer to the desired signal is

$$B^{\mathrm{T}} L(x_k) = I, \quad (46.19)$$

where I denotes the identity matrix. Minimization of contributions to the output due to interference is accomplished by choosing the weights to minimize the variance of the filter output

$$\mathrm{Var}\{y_k\} = \mathrm{tr}\{B^{\mathrm{T}} R_{zk} B\}, \quad (46.20)$$

where tr{} is the trace of the matrix in brackets nad R_{zk} is the covariance matrix of the EEG signals. In practice, R_{zk} will be estimated from the EEG signals during a given time window. Therefore, the filter is derived by minimizing the output variance subject to the constraint defined in (46.19). This constraint ensures that the desired signal is passed with unit gain. Finally, the optimal solution can be derived by constrained minimization using Lagrange multipliers [46.27] and it can be expressed as

$$B^{\mathrm{opt}} = R_{zk}^{-1} L^{\mathrm{T}}(x_k)\left(L^{\mathrm{T}}(x_k) R_{zk}^{-1} L(x_k)\right)^{-1}. \quad (46.21)$$

The response of the beamformer is often called the linearly constrained minimum variance (LCMV) beamformer. LCMV provides not only an estimate of source activity, but also its orientation, which is why it is classified as vector beamforming. The differences and similarities among beamformers based on this criterion for choosing the optimum weights are discussed in [46.28].

46.5.4 Experimental Results

In this section we test the algorithm with real EEG data. The solution of the EEG source localization problem requires a significant number of forward model evaluations. The proposed algorithm can require its evaluation at thousands of different source locations. In order to control the computational load, a discrete state-space is assumed, which consists of a finite number of states (dipoles). The dipoles are linearly distributed in a cube with limits $[-9; 9]$ cm. We use a linear grid $(8 \times 8 \times 8)$ to generate 512 dipoles. Based on the information for the localization of the grid of brain dipoles and the localization of the surface scalp electrodes the lead field matrix $L(x)$ is computed. The SMC algorithm for EEG source estimation is summarized below. The algorithm was inspired by related previous works [46.29–31].

Algorithm 46.2 Sequential Monte Carlo algorithm for EEG source localization

for run $= 1, 2, \ldots, $ MC

(repeat the same algorithm MC number of runs)

Initialization

I. $k = 0$, **for** $l = 1, 2, \ldots, N$

Generate N samples according to a chosen distribution $x_0^{(l)} \sim p(x_0)$ around the initial vector $x_0 = \min(D) + (\max(D) - \min(D)) * \text{rand}(1, N)$. Set initial weights $W_0^{(l)} = \frac{1}{N}$ (equal initial importance to all samples)

II. **for** $k = 1, 2, \ldots$

Prediction step

For $l = 1, 2, \ldots, N$ compute the state prediction according to the random walk state equation (46.17) $x_k = x_{k-1} + w_k$, where $w_k \sim N(0, Q)$ is the process (assumed Gaussian) noise, $E[w_k w'_{k+j}] = 0$ for $j \neq 0$. The covariance matrix Q of w_k is $Q = \sigma_w^2 I$, I denotes the unit matrix and σ_w is the standard deviation. σ_w is chosen as a percentage (0–50%) from the previously estimated state vector x_{k-1}.

Beamforming step

1. Compute the transfer function $L(x_k)$
2. apply the BF technique to define the spatial filter w using (46.21)
3. compute the amplitudes at time k of the source signal propagated in three directions, for all estimated sources y_k

Measurement Update

Evaluate the importance weights

for $l = 1, 2, \ldots, N$, on the receipt of a new measurement, compute the output according to the measurement equation (46.16) and compute the weights $W_k^{(l)} = W_{k-1}^{(l)} \text{Lic}(z_k \mid x_x^{(l)})$

The likelihood is calculated as

$\text{Lic}(z_k \mid x_x^{(l)}) \sim N(h(x_k^{(l)}), \sigma_v)$

$\text{Lic}(z_k \mid x_x^{(l)}) = \exp[-0.5 * (z_k - \text{EEG}_{\text{data}\_k})$

$R^{-1}(z_k - \text{EEG}_{\text{data}\_k})]$

$R = \text{cov}(\text{EEG}_{\text{data}}) = (\text{EEG}_{\text{data}})(\text{EEG}_{\text{data}})^{\text{T}} =$

1) **for** $l = 1, 2, \ldots, N$, normalize the weights

$\hat{W}_k^{(l)} = W_k^{(l)} / \sum_{l=1}^{N} W_k^{(l)}$

Output

2) Calculate the posterior mean $E[x_k \mid z_{1:k}]$ as

$\hat{x}_k = E[x_k \mid z_{1:k}] = \sum_{l=1}^{N} \hat{W}_k^{(l)} x_k^{(l)}$

Compute the effective sample size

$N_{\text{eff}} = \dfrac{1}{\sum_{l=1}^{N} (\hat{W}_k^{(l)})^2}$

Selection step (resampling) if $N_{\text{eff}} < N_{\text{tresh}}$

3) Multiply/suppress samples $x_k^{(l)}$ with high/low importance weights $\hat{W}_k^{(l)}$, in order to obtain N new random samples approximately distributed according to the posterior state distribution. The residual resembling algorithm, [46.23] is applied. This is a two-step process making use of sampling-importance-resampling scheme:

for $l = 1, 2, \ldots, N$ set $W_k^{(l)} = \hat{W}_k^{(l)} = \frac{1}{N}$.

It is assumed that the EEG signal has originated by activity in two principal dipoles, hence the dimension of the state vector at each time k is $x_k \in \mathfrak{R}^{6 \times 1}$ (three coordinates per dipole). The initial state vector $x_0 = [x_{10}, y_{10}, z_{10}, x_{20}, y_{20}, z_{20}]^{\text{T}}$ is chosen randomly

$\{d_{i0} : [x_{i0} \in [-9; 9] \text{cm}, \ y_{i0} \in [-9; 9] \text{cm}, $

$z_{i0} \in [9; 9]] \text{cm}\}, \quad i = 1, 2.$

The number of initially generated particles is $N = 500$ and the complete algorithm is executed 5 times (MC = 5 runs). Increasing N (the same stays also for MC) makes the estimation process rather time-consuming while decreasing N worsen the convergence properties of the algorithm.

The real EEG data correspond to visually evoked potential (VEP) signals extracted from 13 female subjects (20–28 years old). All participants had normal or corrected to normal vision and no history of neurological or psychiatric illness. Neutral, fearful, and disgusting faces of 16 different individuals (8 males and 8 females) were selected, giving a total of 48 different facial stimuli. Images of 16 different house fronts

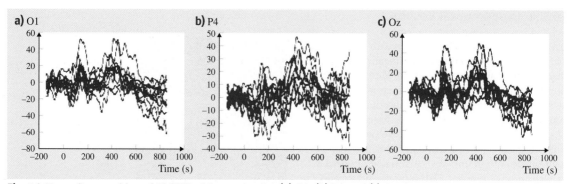

Fig. 46.12a–c Superposition of 10 VEP trials measured at (**a**) O1, (**b**) P4, and (**c**) Oz channels

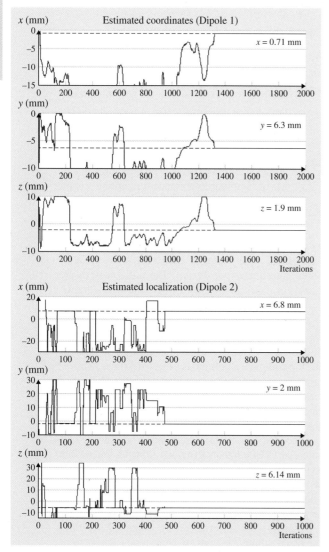

to be superimposed on each of the faces were selected from various internet sources. This resulted in a total of 384 gray scaled composite images (9.5 cm wide by 14 cm high) of transparently superimposed face and house with equivalent discriminability.

Participants were seated in a dimly lit room, where a computer screen was placed at a viewing distance of approximately 80 cm coupled to a PC equipped with software for the EEG recording. The images were divided into two experimental blocks. In the first, the participants were required to attend to the houses (ignoring the faces) and in the other they were required to attend to the faces (ignoring the houses). The participant's task was to determine, on each trial, if the current house or face (depending on the experimental block) was the same as the one presented on the previous trial. Stimuli were presented in sequence, for 300 ms each and were preceded by a fixation cross displayed for 500 ms. The inter-trial interval was 2000 ms.

EEG signals were recorded from 20 electrodes (Fp1, Fp2, F3, F4, C3, C4, P3, P4, O1, O2; F7, F8, T3, T6; P7, P8, Fz, Cz, Pz, Oz) according to the 10/20 international system. EOG signals were also recorded from electrodes placed just above the left supra orbital ridge (vertical EOG) and on the left outer canthus (horizontal EOG). VEP were calculated off-line, averaging segments of 400 points of digitized EEG (12 bit A/D converter, sampling rate 250 Hz).

These segments covered 1600 ms comprising a pre-stimulus interval of 148 ms (37 samples) and post-stimulus onset interval of 1452 ms. Before processing, EEG was visually inspected and those segments with excessive EOG artifacts were manually eliminated. Only trials with correct responses were included in

Fig. 46.13 Estimation of two source locations that produce P100 VEP peak in occipital channels ◀

the data set. The experimental setup was designed by [46.32] for their study on subject attention and perception using VEP signals.

Figure 46.12 represents ten enhanced (by principal component analysis) trials of three channels. In the reconstructed signals it is possible to identify a positive peak in the range of 100–160 ms (P100). P100 corresponds to the perception of the sensory stimulus, a brain activity that is known to happen in the primary visual cortex. The occipital channels (O1, Oz) that measure the brain activity around the visual cortex clearly represent the P100 peaks. The task of the SMC algorithm is to estimate the two strongest sources (d1 and d2) that may have produced the P100. The results of the estimation are summarized in Fig. 46.13. It is very interesting to observe that the final coordinates of d1 (0.71 mm, −6.3 mm, −1.9 mm) and d2 (6.8 mm, −2 mm, −6.14 mm) correspond to the zone of the primary visual cortex as illustrated in Fig. 46.14. Therefore, the proposed beamformer-based SMC suc-

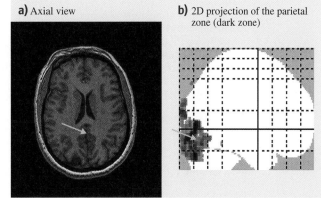

a) Axial view

b) 2D projection of the parietal zone (dark zone)

Fig. 46.14a,b Primary visual cortex zone. (**a**) Axial view; (**b**) 2-D projection of the parietal zone (*dark zone*)

cessfully estimated the space coordinates of the two strongest brain sources, producing the P100 peak, as located in the zone of the primary visual cortex.

46.6 Conclusions

This chapter has described recent efforts towards the development of EEG-based brain computer interface systems. In the first part, motor imagery noninvasive BCI for mobile robot control was discussed. In the second part an introduction to the underlying principles of source-based BCI was given. Different problems in developing BCI systems and in their applications arise when moving from the electrode-based domain to the source-based scale. The goal of the source-based approach is to obtain knowledge about brain activity and to answer fundamental questions about interacting regions. A combination of probabilistic (SMC) and deterministic (BF) techniques was proposed for the estimation of the principal activity zones in the brain, and its potential as a new direction in BCI design was

demonstrated. The method requires a priori knowledge about the number of active sources. The insights gained with this study can be relevant when optimizing the design and implementation of practical source-based BCI. However, there are a number of open issues that need to be studied. For example, the problem of the localization and number of measurement electrodes to be used in order to recover the active zones.

Creating a hybrid BCI paradigm where more than one noninvasive modality are combined such as, for example EEG, estimated brain sources, and fMRI seems the most logical continuation of the present research. In order to increase the information transfer rate and reach the accuracy of the invasive brain imaging techniques a multi-modal approach is required.

References

46.1 J. Mellinger, G. Schalk, C. Braun, H. Preissl, W. Rosenstiel, N. Birbaumer, A. Kübler: An MEG-based brain-computer interface, Neuroimage **36**(3), 581–593 (2007)

46.2 R. Sitaram, A. Caria, R. Veit, T. Gaber, G. Rota, A. Kuebler, N. Birbaumer: fMRI brain-computer interface: A tool for neuroscientific research and treatment, Comput. Intell. Neurosci. **2007**, 25487-1–25487-10 (2007), doi:10.1155/2007/25487

46.3 S.M. Coyle, T.E. Ward, C.M. Markham: Brain–computer interface using a simplified functional near-infrared spectroscopy system, J. Neural Eng. **4**, 219–226 (2007)

46.4 J.J. Vidal: Real-time detection of brain events in EEG, Proc. IEEE **65**(5), 633–641 (1977)

46.5 E.M. Dewan: Occipital alpha rhythm eye position and lens accommodation, Nature **214**, 975–977 (1967)

46.6   E. Niedermeyer, F. Lopes da Silva: *Electroen-cephalography* (Lippincott Williams and Wilkins, Philadelphia 1999)

46.7   J.R. Wolpaw, D.J. McFarland, G.W. Neat, C.A. Forneris: An EEG-based brain-computer interface for cursor control, Electroencephalogr. Clin. Neurophysiol. **78**(83), 252–259 (1991)

46.8   E. Basar: *EEG Brain Dynamics: Relation between EEG and brain evoked potentials* (Elsevier/North-Holland Biomedical, Amsterdam 1980)

46.9   E. Basar, B. Rosen, C. Basar-Eroglu, F. Greitschus: The associations between 40 Hz-EEG and the middle latency response of the auditory evoked potential, Int. J. Neurosci. **33**(1/2), 103–117 (1987)

46.10   R. Galambos: A comparison of certain gamma band (40-Hz) brain rhythms in cat and man. In: *Induced Rhythms in the Brain*, ed. by E. Başar, T.H. Bullock (Birkhäuser, Boston USA 1992) pp. 201–216

46.11   T. Gruber, M.M. Müller, A. Keil: Modulation of induced gamma band responses in a perceptual learning task in the human EEG, J. Cogn. Neurosci. **14**(5), 732–744 (2002)

46.12   C. Tallon-Baudry, O. Bertrand, C. Delpuech, J. Pernier: Stimulus specificity of phase-locked and non-phase-locked 40 Hz visual responses in human, J. Neurosci. **16**(13), 4240–4249 (1996)

46.13   J. Wang, N. Yan, H. Liu, M. Liu, C. Tai: Brain-Computer Interfaces Based on Attention and Complex Mental Tasks, Digital Human Modeling, LNCS **4561**, 467–473 (2007)

46.14   C. Neuper, G.R. Müller, A. Kübler, N. Birbaumer, G. Pfurtscheller: Clinical application of an EEG-based brain–computer interface: A case study in a patient with severe motor impairment, Clin. Neurophysiol. **114**(3), 399–409 (2003)

46.15   D. McFarland, J.R. Wolpaw: Brain computer inerfaces for communication and cotrol, Commun. ACM **54**(5), 60–66 (2011)

46.16   L. Qin, L. Ding, B. He: Motor imagery classification by means of source analysis for brain computer interface applications, J. Neural Eng. **1**, 133–141 (2004)

46.17   B. Kamousi, Z. Liu, B. He: An EEG inverse solution based brain-computer interface, Int. J. Bioelectromagn. **7**(2), 292–294 (2005)

46.18   Q. Noirhomme, R.I. Kitney, B. Macq: Signle-trial EEG source reconstruction for brain-computer interface, IEEE Trans. Biomed. Eng. **55**(5), 1592–1601 (2008)

46.19   M. Grosse-Wentrup, C. Liefhold, K. Gramann, M. Buss: Beamforming in non-invasive brain-computer interfaces, IEEE Trans. Biomed. Eng. **56**(4), 1209–1219 (2009)

46.20   S. Choi, A. Cichocki, H.-M. Park, S.-Y. Lee: Blind source separation and independent component analysis: A review, Neural Inf. Process. **6**(1), 1–57 (2005)

46.21   P. Brasnett, L. Mihaylova, D. Bull, N. Canagarajah: Sequential Monte Carlo tracking by fusing multiple cues in video sequences, Image Vis. Comput. Elsevier Sci. **25**(8), 1217–1227 (2007)

46.22   M. Arulampalam, S. Maskell, N. Gordon, T. Clapp: A tutorial on particle filters for online nonlinear/non-Gaussian Bayesian tracking, IEEE Trans. Signal Process. **50**(2), 174–188 (2002)

46.23   J. Liu, R. Chen: Sequential Monte Carlo methods for dynamic systems, J. Am. Statist. Assoc. **93**(443), 1032–1044 (1998)

46.24   Y. Salu, L.G. Cohen, D. Rose, S. Sato, C. Kufta, M. Hallet: An improved method for localizing electric brain dipoles, IEEE Trans. Biomed. Eng. **37**, 699–705 (1990)

46.25   Y. Lai, W. Van Drongelen, L. Ding, K.E. Hecox, V.L. Towle, D.M. Frim, B. He: Estimation of in vivo human brain-skull conductivity ratio from simultaneous extra- and intra-cranial electrical potential recordings, Clin. Neurophysiol. **116**, 456–465 (2005)

46.26   J. Capon: High-resolution frequency wavenumber spectrum analysis, Proc. IEEE **57**, 1408–1418 (1969)

46.27   B.D. Van Veen, W. van Drongelen, M. Yuchtman, A. Suzuki: Localization of brain electrical activity via linearly constrained minimum variance spatial filtering, IEEE Trans. Biomed. Eng. **44**(9), 867–880 (1997)

46.28   M.-X. Huang, J.J. Shih, R.R. Lee, D.L. Harrington, R.J. Thoma, M.P. Weisend, F. Hanion, K.M. Paulson, T. Li, K. Martin, G.A. Miller, J.M. Canive: Commonalities and differences among vectorized beamformers in electromagnetic source imaging, Brain Topogr. **16**, 139–158 (2004)

46.29   H.R. Mohseni, K. Nazarpour, E.L. Wilding, S. Saeid Sanei: The application of particle filters in single trial event-related potential estimation, Physiol. Meas. **30**, 1101–1116 (2009)

46.30   H.R. Mohseni, F. Ghaderi, E.L. Wilding, S. Saeid Sanei: A beamforming particle filter for EEG dipole source localization, Int. Conf. Acoust., Speech, Signal Process. (ICASSP) (2009) pp. 337–340

46.31   J.M. Anteli, J. Mingue: EEG source localization based on dynamic Bayesian estimation techniques, Int. J. Bioelectromagn. **11**(4), 179–184 (2009)

46.32   I.M. Santos, J. Iglesias, E.I. Olivares, A.W. Young: Differential effects of object-based attention on evoked potentials to fearful and disgusted faces, Neuropsychologia **46**(5), 1468–1479 (2008)

47. Brain-like Information Processing for Spatio-Temporal Pattern Recognition

Nikola Kasabov

Information processes in the brain, such as gene and protein expression, learning, memory, perception, cognition, consciousness are all spatio-and/or spectro temporal. Modelling such processes would require sophisticated information science methods and the best ones could be the brain-inspired ones, that use the same brain information processing principles. Spatio and spectro-temporal data (SSTD) are also the most common types of data collected in many domain areas, including engineering, bioinformatics, neuroinformatics, ecology, environment, medicine, economics, etc. However, there is lack of methods for the efficient analysis of such data and for spatio-temporal pattern recognition (STPR). The brain functions as a spatio-temporal information processing machine and deals extremely well with spatio-temporal data. Its organization and functions have been the inspiration for the development of new methods for SSTD analysis and STPR. Brain-inspired spiking neural networks (SNN) are considered the third generation of neural networks and are a promising paradigm for the creation of new intelligent ICT for SSTD. This new generation of computational models and systems is potentially capable of modeling complex information processes due to the ability to represent and integrate different information dimensions, such as time, space, frequency, and phase, and to deal with large volumes of data in an adaptive and self-organizing manner. This chapter reviews methods and systems of SNN for SSTD analysis and STPR, including single neuronal

models, evolving spiking neural networks (eSNN), and computational neurogenetic models (CNGM). Software and hardware implementations and some pilot applications for audio-visual pattern recognition, EEG data-analysis, cognitive robotic systems, BCI, neurodegenerative diseases, and others are discussed.

Part I | 47

47.1 Spatio and Spectro-Temporal Data Modeling and Pattern Recognition

Most problems in nature require spatio or/and spectro-temporal data (SSTD) that include measuring spatial or/and spectral variables over time. SSTD is described by a triplet (X, Y, F), where X is a set of independent variables measured over consecutive discrete time moments t, Y is the set of dependent output variables, and

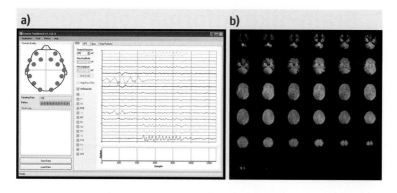

Fig. 47.1 (**a**) EEG SSTD recorded with the use of emotive EEG equipment (after [47.1]). (**b**) fMRI data (after [47.2])

F is the association function between whole segments (*chunks*) of the input data, each sampled in a time window d_t, and the output variables belonging to Y

$$F : X(d_t) \geq Y , \tag{47.1}$$

where

$$X(t) = (x_1(t), x_2(t), \ldots, x_n(t)), t = 1, 2, \ldots .$$

It is important for a computational model to capture and learn *whole* spatio and spectro-temporal patterns from data streams in order to predict most accurately future events for new input data. Examples of problems involving SSTD are: brain cognitive state evaluation based on spatially distributed EEG electrodes [47.3–8] (Fig. 47.1a); fMRI data [47.9] (Fig. 47.1b); moving object recognition from video data [47.10–12] (Fig. 47.15); spoken word recognition based on spectro-temporal audio data [47.13,14]; evaluating risk of disease, e.g., heart attack [47.15]; evaluating response of a disease to treatment based on clinical and environmental variables, e.g., stroke [47.16], prognosis of outcome of cancer [47.17], modeling the progression of a neurodegenerative disease, such as Alzheimer's disease [47.18, 19], and modeling and prognosis of the establishment of invasive species in ecology [47.20, 21]. The prediction of events in geology, astronomy, economics and many other areas also depends on accurate SSTD modeling.

The commonly used models for dealing with temporal information based on hidden Markov models (HMM) [47.22] and traditional artificial neural networks (ANN) [47.23] have limited capacity to achieve the integration of complex and long temporal spatial/spectral components because they usually either ignore the temporal dimension or over-simplify its representation. A new trend in machine learning is currently emerging and is known as *deep machine learning* [47.24, 24–27]. Most of the proposed models still learn SSTD by entering single time point frames rather than learning whole SSTD patterns. They are also limited in addressing adequately the interaction between temporal and spatial components in SSTD.

The human brain has the amazing capacity to learn and recall patterns from SSTD at different time scales, ranging from milliseconds to years and possibly to millions of years (e.g., genetic information, accumulated through evolution). Thus the brain is the ultimate inspiration for the development of new machine learning techniques for SSTD modeling. Indeed, brain-inspired spiking neural networks (SNN) [47.28–30] have the potential to learn SSTD by using trains of spikes (binary temporal events) transmitted among spatially located synapses and neurons. Both spatial and temporal information can be encoded in an SNN as locations of synapses and neurons and the time of their spiking activity, respectively. Spiking neurons send spikes via connections that have a complex dynamic behavior, collectively forming an SSTD memory. Some SNN employ specific learning rules such as spike-time-dependent-plasticity (STDP) [47.31] or spike driven synaptic plasticity (SDSP) [47.32]. According to the STDP a connection weight between two neurons increases when the pre-synaptic neuron spikes before the post-synaptic one. Otherwise, the weight decreases.

Models of single neurons as well as computational SNN models, along with their respective applications, have been already developed [47.29, 30, 33–36], including evolving connectionist systems and evolving spiking neural networks (eSNN), in particular where an SNN learns data incrementally by one-pass propagation of the data via creating and merging spiking neurons [47.37, 38]. In [47.38] an eSNN is designed to capture features and to aggregate them into audio and visual perceptions for the purpose of person authentication. It is based on four levels of feed-forward connected layers of spiking neuronal maps, similarly to the way the *cortex* works when learning and recognizing images or complex input stimuli [47.39]. It is an SNN realization of some computational models of vision, such as the five-level HMAX model inspired by the information processes in the cortex [47.39].

However, these models are designed for (static) object recognition (e.g., *a picture of a cat*), but not for moving object recognition (e.g., *a cat jumping to catch a mouse*). If these models are to be used for SSTD, they will still process SSTD as a sequence of static feature vectors extracted in single time frames. Although an eSNN accumulates incoming information carried in each consecutive frame from a pronounced word or a video, through the increase of the membrane potential of output spike neurons, they do not learn complex spatio/spectro-temporal associations from the data. Most of these models are deterministic and do not allow to model complex stochastic SSTD.

In [47.40, 41] a computational neurogenetic model (CNGM) of a single neuron and SNN are presented that utilize information about how some proteins and

genes affect the spiking activities of a neuron, such as fast excitation, fast inhibition, slow excitation, and slow inhibition. An important part of a CNGM is a dynamic gene regulatory network (GRN) model of genes/proteins and their interaction over time that affect the spiking activity of the neurons in the SNN. Depending on the task, the genes in a GRN can represent either biological genes and proteins (for biological applications) or some system parameters including probability parameters (for engineering applications). Recently some new techniques have been developed that allow the creation of new types of computational models, e.g., probabilistic spiking neuron models [47.42, 43], probabilistic optimization of features and parameters of eSNN [47.21, 44], reservoir computing [47.36, 45], and personalized modeling frameworks [47.46, 47]. This chapter reviews methods and systems for SSTD that utilize the above and some other contemporary SNN techniques along with their applications.

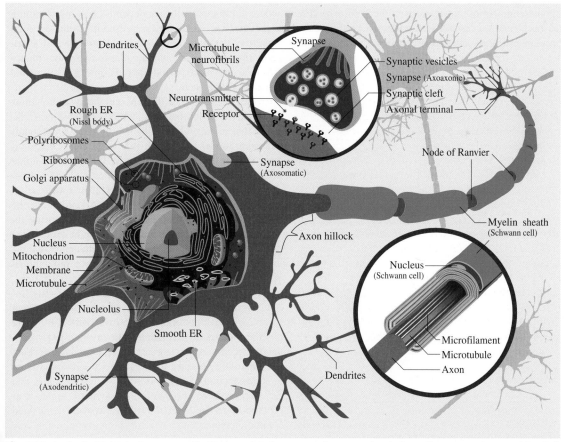

Fig. 47.2 A single biological neuron with the associated synapses is a complex information processing machine (after Wikipedia)

47.2 Single Spiking Neuron Models

47.2.1 A Biological Neuron

A single biological neuron and the associated synapses is a complex information processing machine, that involves short-term information processing, long-term information storage, and evolutionary information stored as genes in the nucleus of the neuron (Fig. 47.2).

47.2.2 Single Neuron Models

Some of the-state-of-the-art models of a spiking neuron include: early models by *Hodgkin* and *Huxley* [47.48]; more recent models by Maas, Gerstner, Kistler, Izhikevich and others, e.g., spike response models (SRM) [47.29, 30]; the integrate-and-fire model (IFM) [47.29, 30]; Izhikevich models [47.49–52], adaptive IFM, and others.

The most popular model for both biological modeling and engineering applications is the IFM. The IFM has been realized on software-hardware platforms for the exploration of patterns of activities in large scale

SNN under different conditions and for different applications. Several large scale architectures of SNN using IFM have been developed for modeling brain cognitive functions and engineering applications. Figure 47.3a,b illustrates the structure and the functionality of the leaky IFM (LIFM), respectively. The neuronal post-synaptic potential (PSP), also called membrane potential $u(t)$, increases with every input spike at a time t multiplied to the synaptic efficacy (strength) until it reaches a threshold. After that, an output spike is emitted and the membrane potential is reset to an initial state (e.g., 0). Between spikes, the membrane potential leaks, which is defined by a parameter.

An important part of a model of a neuron is the model of the synapses. Most neuronal models assume scalar synaptic efficacy parameters that are subject to learning, either on-line or off-line (batch mode). There are models of dynamics synapses (e.g., [47.43, 53, 54]), where the synaptic efficacy depends on synaptic parameters that change over time, representing both long-term memory (the final efficacy after learning) and short-term memory – the changes of the synaptic efficacy over a shorter time period not only during learning, but during recall as well.

One generalization of LIFM and the dynamic synaptic models is the probabilistic model of a neuron [47.42] as shown in Fig. 47.4a, which is also a biologically plausible model [47.30, 43, 55]. The state of a spiking neuron n_i is described by the sum $PSP_i(t)$ of the inputs received from all m synapses. When $PSP_i(t)$ reaches a firing threshold $\vartheta_i(t)$, neuron n_i fires, i.e., it emits a spike. Connection weights $(w_{j,i}, j = 1, 2, \ldots, m)$ associated with the synapses are determined during the learning phase using a learning rule. In addition to the connection weights $w_{j,i}(t)$, the probabilistic spiking neuron model has the following three probabilistic parameters:

1. A probability $p_{cj,i}(t)$ that a spike emitted by neuron n_j will reach neuron n_i at a time moment t through the connection between n_j and n_i. If $p_{cj,i}(t) = 0$, no connection and no spike propagation exist between neurons n_j and n_i. If $p_{cj,i}(t) = 1$ the probability for propagation of spikes is 100%.
2. A probability $p_{sj,i}(t)$ for the synapse $s_{j,i}$ to contribute to the $PSP_i(t)$ after it has received a spike from neuron n_j.
3. A probability $p_i(t)$ for the neuron n_i to emit an output spike at time t once the total $PSP_i(t)$ has

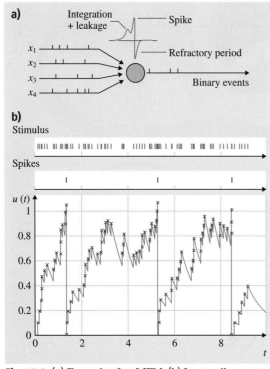

Fig. 47.3 (a) Example of an LIFM. **(b)** Input spikes, output spikes, and PSP dynamics of an LIFM

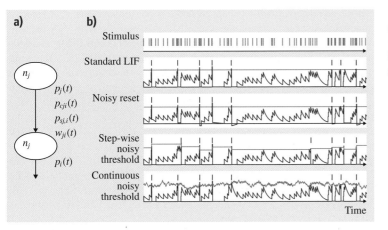

Fig. 47.4 (a) A simple probabilistic spiking neuron model (after [47.42]). **(b)** Different types of noisy thresholds have different effects on the output spikes (after [47.7, 56])

reached a value above the PSP threshold (a noisy threshold).

The total $PSP_i(t)$ of the probabilistic spiking neuron n_i is now calculated using the following formula [47.42]

$$PSP_i(t) = \sum_{p=t_0,\ldots,t} \left(\sum_{j=1,\ldots,m} e_j f_1(p_{cj,i}(t-p)) \right.$$
$$\left. f_2(p_{sj,i}(t-p))w_{j,i}(t) + \eta(t-t_0) \right),$$
(47.2)

where e_j is 1, if a spike has been emitted from neuron n_j, and 0 otherwise; $f_1(p_{cj,i}(t))$ is 1 with a probability $p_{cji}(t)$, and 0 otherwise; $f_2(p_{sj,i}(t))$ is 1 with a probability $p_{sj,i}(t)$, and 0 otherwise; t_0 is the time of the last spike emitted by n_i; and $\eta(t-t_0)$ is an additional term representing decay in PSP_i. As a special case, when all or some of the probability parameters are fixed to "1", the above probabilistic model will be simplified and will resemble the well-known IFM. A similar formula will be used when an LIFM is used as a fundamental model, where a time decay parameter is introduced.

It has been demonstrated that SNN that utilize the probabilistic neuronal model can learn better SSTD than traditional SNN with simple IFM, especially in a nosy environment [47.56, 57]. The effect of each of the above three probabilistic parameters on the ability of a SNN to process noisy and stochastic information was studied in [47.56]. Figure 47.4b presents the effect of different types of nosy thresholds on the neuronal spiking activity.

47.2.3 A Neurogenetic Model of a Neuron

A neurogenetic model of a neuron was proposed in [47.41] and studied in [47.40]. It utilizes information about how some proteins and genes affect the spiking activities of a neuron such as *fast excitation, fast inhibition, slow excitation, and slow inhibition*. Table 47.1 shows some of the proteins in a neuron and their relation to different spiking activities. For a real case application, apart from the GABAB receptor some other metabotropic and other receptors could be also included. This information is used to calculate the contribution of each of the different synapses, connected to a neuron n_i, to its post-synaptic potential $PSP_i(t)$

$$\varepsilon_{ij}^{\text{synapse}}(s) = A^{\text{synapse}}\left(\exp\left(-\frac{s}{\tau_{\text{decay}}^{\text{synapse}}}\right) \right.$$
$$\left. - \exp\left(-\frac{s}{\tau_{\text{rise}}^{\text{synapse}}}\right) \right),$$
(47.3)

where $\tau_{\text{decay/rise}}^{\text{synapse}}$ are time constants representing the rise and fall of an individual synaptic PSP, A is PSP's amplitude, $\varepsilon_{ij}^{\text{synapse}}$ represents the type of activity of the synapse between neuron j and neuron i that can be measured and modeled separately for a fast excitation, fast inhibition, slow excitation, and slow inhibition (it is affected by different genes/proteins). External inputs can also be added to model background noise, background oscillations or environmental information.

An important part of the model is a dynamic gene/protein regulatory network (GRN) model of the dynamic interactions between genes/proteins over time that affect the spiking activity of the neuron. Although biologically plausible, a GRN model is only a highly

Table 47.1 Neuronal action potential parameters and related proteins and ion channels in the computational neurogenetic model of a spiking neuron: AMPAR – (amino-methylisoxazole-propionic acid) AMPA receptor, NMDR – (N-methyl-D-aspartate acid) NMDA receptor, GABA$_A$R – (gamma-aminobutyric acid) GABA$_A$ receptor, GABA$_B$R – GABA$_B$ receptor, SCN – sodium voltage-gated channel, KCN – kalium (potassium) voltage-gated channel, CLC – chloride channel (after [47.40])

Different types of action potential of a spiking neuron used as parameters for its computational model	Related neurotransmitters and ion channels
Fast excitation PSP	AMPAR
Slow excitation PSP	NMDAR
Fast inhibition PSP	GABA$_A$R
Slow inhibition PSP	GABA$_B$R
Modulation of PSP	mGluR
Firing threshold	Ion channels SCN, KCN, CLC

simplified general model that does not necessarily take into account the exact chemical and molecular interactions. A GRN model is defined by:

1. A set of genes/proteins, $G = (g_1, g_2, \ldots, g_k)$
2. An initial state of the level of expression of the genes/proteins $G(t = 0)$
3. An initial state of a connection matrix $L = (L_{11}, \ldots, L_{kk})$, where each element L_{ij} defines the known level of interaction (if any) between genes/proteins g_j and g_i
4. Activation functions f_i for each gene/protein g_i from G. This function defines the gene/protein expression value at time $(t + 1)$ depending on the current values $G(t)$, $L(t)$ and some external information $E(t)$

$$g_i(t+1) = f_i(G(t).L(t), E(t)). \qquad (47.4)$$

47.3 Learning and Memory in a Spiking Neuron

47.3.1 General Classification

A learning process has an effect on the synaptic efficacy of the synapses connected to a spiking neuron and on the information that is memorized. Memory can be:

1. Short-term, represented as a changing PSP and temporarily changing synaptic efficacy
2. Long-term, represented as a stable establishment of the synaptic efficacy
3. Genetic (evolutionary), represented as a change in the genetic code and the gene/protein expression level as a result of the above short-term and long-term memory changes and evolutionary processes.

Learning in SNN can be:

1. Unsupervised – there is no desired output signal provided
2. Supervised – a desired output signal is provided
3. Semi-supervised.

Different tasks can be learned by a neuron, e.g.,

1. Classification
2. Input–output spike pattern association.

Several biologically plausible learning rules have been introduced so far, depending on the type of the information presentation:

1. Rate-order learning, which is based on the average spiking activity of a neuron over time [47.58–60]
2. Temporal learning, that is based on precise spike times [47.61–65]
3. Rank-order learning, that takes into account the order of spikes across all synapses connected to a neuron [47.65, 66].

Rate-order information representation is typical for cognitive information processing [47.58]. Temporal spike learning is observed in the auditory [47.13], the visual [47.67], and motor control information processing of the brain [47.61, 68]. Its use in neuroprosthetics is essential, along with applications for a fast, real-time recognition and control of sequence of related processes [47.69]. Temporal coding accounts for the precise time of spikes and has been utilized in several learning rules, the most popular being spike-time dependent plasticity (STDP) [47.31,70] and SDSP [47.32, 69]. Temporal coding of information in SNN makes use of the exact time of spikes (e.g., in milliseconds). Every spike matters and its time matters too.

47.3.2 The STDP Learning Rule

The STDP learning rule uses Hebbian plasticity [47.71] in the form of long-term potentiation (LTP) and depression (LTD) [47.31, 70]. Efficacy of synapses is strengthened or weakened based on the timing of post-synaptic action potential in relation to the pre-synaptic spike (an example is given in Fig. 47.5a). If the difference in the spike time between the pre-synaptic and post-synaptic neurons is negative (pre-synaptic neuron spikes first) then the connection weight between the two neurons increases, otherwise it decreases. Through STDP, connected neurons learn consecutive temporal associations from data. Pre-synaptic activity that precedes post-synaptic firing can induce long-term potentiation (LTP); reversing this temporal order causes long-term depression (LTD).

47.3.3 Spike Driven Synaptic Plasticity (SDSP)

SDSP is an unsupervised learning method [47.32, 69], a modification of the STDP that directs the change of the synaptic plasticity V_{w_0} of a synapse w_0 depending on the time of spiking of the pre-synaptic neuron and the post-synaptic neuron. V_{w_0} increases or decreases, depending on the relative timing of the pre-synaptic and post-synaptic spikes.

If a pre-synaptic spike arrives at the synaptic terminal before a postsynaptic spike within a critical time window, the synaptic efficacy is increased (potentiation). If the post-synaptic spike is emitted just before the pre-synaptic spike, synaptic efficacy is decreased (depression). This change in synaptic efficacy can be expressed as

$$\Delta V_{w_0} = \frac{I_{\text{pot}}(t_{\text{post}})}{C_p} \Delta t_{\text{spk}} \text{ if } t_{\text{pre}} < t_{\text{post}} , \qquad (47.5)$$

$$\Delta V_{w_0} = -\frac{I_{\text{dep}}(t_{\text{post}})}{C_d} \Delta t_{\text{spk}} \text{ if } t_{\text{post}} < t_{\text{pre}} , \qquad (47.6)$$

where Δt_{spk} is the pre-synaptic and post-synaptic spike time window.

The SDSP rule can be used to implement a supervised learning algorithm, when a teacher signal, which copies the desired output spiking sequence, is entered

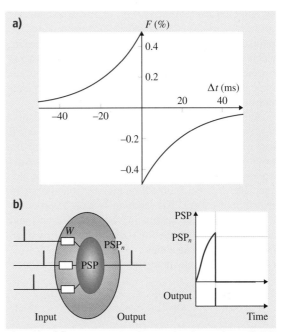

Fig. 47.5 (a) An example of synaptic change in an STDP learning neuron (after[47.31]). **(b)** Rank-order LIF neuron

along with the training spike pattern, but without any change of the weights of the teacher input.

The SDSP model is implemented as an VLSI analog chip [47.72]. The silicon synapses comprise bistability circuits for driving a synaptic weight to one of two possible analog values (either potentiated or depressed). These circuits drive the synaptic-weight voltage with a current that is superimposed on that generated by the STDP and which can be either positive or negative. If, on short time scales, the synaptic weight is increased above a set threshold by the network activity via the STDP learning mechanism, the bistability circuits generate a constant weak positive current. In the absence of activity (and hence learning) this current will drive the weight toward its potentiated state. If the STDP decreases the synaptic weight below the threshold, the bistability circuits will generate a negative current that, in the absence of spiking activity, will actively drive the weight toward the analog value, encoding its depressed state. The STDP and bistability circuits facilitate the implementation of both long-term and short-term memory.

47.3.4 Rank-Order Learning

The rank-order learning rule uses important information from the input spike trains – the rank of the first incoming spike on each synapse (Fig. 47.5b). It establishes a priority of inputs (synapses) based on the order of the spike arrival on these synapses for a particular pattern, which is a phenomenon observed in biological systems as well as an important information processing concept for some STPR problems, such as computer vision and control [47.65, 66].

This learning makes use of the extra information of spike (event) order. It has several advantages when used in SNN, mainly: fast learning (as it uses the extra information of the order of the incoming spikes) and asynchronous data entry (synaptic inputs are accumulated into the neuronal membrane potential in an asynchronous way). The learning is most appropriate for AER input data streams [47.10] as the events and their addresses are entered into the SNN *one by one*, in the order of their happening.

The post-synaptic potential of a neuron i at a time t is calculated as

$$PSP(i, t) = \sum \text{mod}^{\text{order}(j)} w_{j,i} , \qquad (47.7)$$

where mod is a modulation factor, j is the index for the incoming spike at synapse j, i and $w_{j,i}$ is the cor-

responding synaptic weight, and order(j) represents the order (the rank) of the spike at the synapse j, i among all spikes arriving from all m synapses to the neuron i. The order(j) has a value 0 for the first spike and increases according to the input spike order. An output spike is generated by neuron i if the PSP(i, t) becomes higher than a threshold PSPTh(i).

During the training process, for each training input pattern (sample, example) the connection weights are calculated based on the order of the incoming spikes [47.66]

$$\Delta w_{j,i}(t) = \text{mod}^{\text{order}(j,i(t))} . \qquad (47.8)$$

47.3.5 Combined Rank–Order and Temporal Learning

In [47.11] a method for a combined rank-order and temporal (e.g., SDSP) learning is proposed and tested on benchmark data. The initial value of a synaptic weight is set according to the rank-order learning based on the first incoming spike on this synapse. The weight is further modified to accommodate following spikes on this synapse with the use of a temporal learning rule – SDSP.

47.4 STPR in a Single Neuron

In contrast to the distributed representation theory and to the widely popular view that a single neuron cannot do much, some recent results showed that a single neuronal model can be used for complex STPR.

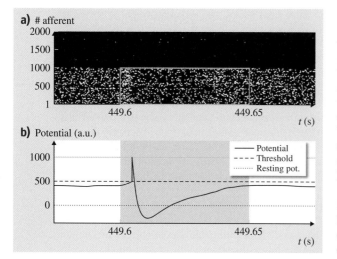

a) # afferent

b) Potential (a.u.)

— Potential
--- Threshold
······· Resting pot.

A single LIF neuron, for example, with simple synapses can be trained with the STDP unsupervised learning rule to discriminate a repeating pattern of synchronized spikes on certain synapses from noise (from [47.1]) – see Fig. 47.6.

Single neuron models have been introduced for STPR, for example, *Temportron* [47.74], *Chronotron* [47.75], *ReSuMe* [47.76], *SPAN* [47.77, 78]. Each of them can learn to emit a spike or a spike pattern (spike sequence) when a certain STP is recognized. Some of them can be used to recognize multiple STP per class and multiple classes [47.76–78]. Figure 47.7c,d shows the use of a single SPAN neuron for the classification of five STP belonging to five different classes [47.78]. The accuracy of classification is rightly lower for the class 1 (the neuron emits a spike at the very beginning of the input pattern) as there is no sufficient input data – Fig. 47.7d [47.78].

Fig. 47.6 A single LIF neuron with simple synapses can be trained with the STDP unsupervised learning rule to discriminate a repeating pattern of synchronized spikes on certain synapses from noise (after [47.73]) ◄

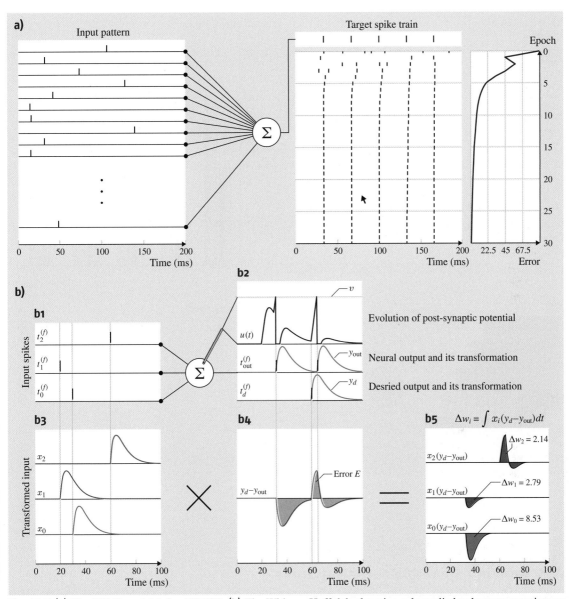

Fig. 47.7 (a) The SPAN model (after [47.78]). **(b)** The Widrow–Hoff delta learning rule applied to learn to associate an output spike sequence to an input STP (after [47.32, 78]). **(c,d)** The use of a single SPAN neuron for the classification of 5 STP belonging to five different classes (after [47.78]). The accuracy of classification is rightly lower for the class 1 – spike at the very beginning of the input pattern as there is no sufficient input data **(d)**

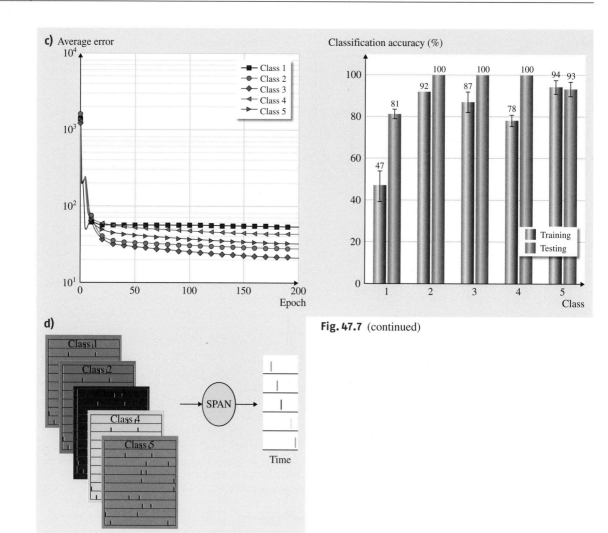

Fig. 47.7 (continued)

47.5 Evolving Spiking Neural Networks

Despite the ability of a single neuron to conduct STPR, a single neuron has a limited power and complex STPR tasks will require multiple spiking neurons.

One approach is proposed in the evolving spiking neural networks (eSNN) framework [47.37, 79]. eSNN evolve their structure and functionality in an on-line manner, from incoming information. For every new input pattern, a new neuron is dynamically allocated and connected to the input neurons (feature neurons). The neurons connections are established for the neuron to recognize this pattern (or a similar one) as a positive ex-

ample. The neurons represent centers of clusters in the space of the synaptic weights. In some implementations similar neurons are merged [47.37, 38]. This makes it possible to achieve very fast learning in an eSNN (only one pass may be necessary), both in a supervised and in an unsupervised mode.

In [47.77] multiple SPAN neurons are evolved to achieve a better accuracy of spike pattern generation than with a single SPAN – Fig. 47.8a.

In [47.69] the SDSP model from [47.32] was successfully used to train and test a SNN for 293 character

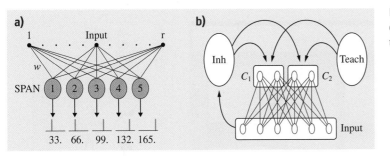

Fig. 47.8 (a) Multiple SPAN neurons (after [47.77]. (**b**)) Multiple SDSP trained neurons (after [47.69])

recognition (classes). Each character (a static image) is represented as a 2000 bit feature vector, and each bit is transferred into spike rates, with 50 Hz spike burst to represent 1 and 0 Hz to represent 0. For each class, 20 different training patterns are used and 20 neurons are allocated, one for each pattern (altogether 5860) (Fig. 47.8b) and trained for several hundreds of iterations.

A general framework of eSNN for STPR is shown in Fig. 47.9. It consists of the following blocks:

1. Input data encoding block
2. Machine learning block (consisting of several sub-blocks)
3. Output block.

In the input block continuous value input variables are transformed into spikes. Different approaches can be used:

1. Population rank coding [47.61] – Fig. 47.10a
2. Thresholding the input value, so that a spike is generated if the input value (e.g., pixel intensity) is above a threshold
3. Address event representation (AER) – thresholding the difference between two consecutive values of the same variable over time as it is in the artificial cochlea [47.14] and artificial retina devices [47.10] – Fig. 47.10b.

The input information is entered either on-line (for on-line, real-time applications) or as batch data. The *time* of the input data is in principle different from the internal SNN *time* of information processing.

Long and complex SSTD cannot be learned in simple one-layer neuronal structures as the examples in Fig. 47.8a,b. They require neuronal *buffers* as shown in Fig. 47.11a. In [47.80] a 3D buffer was used to store spatio-temporal *chunks* of input data before the data is classified. In this case, the size of the chunk (both in space and time) is fixed by the size of the reservoir. There are no connections between the layers in the buffer. Still, the system outperforms traditional classi-

fication techniques, as it is demonstrated on sign language recognition, where the eSNN classifier was applied [47.37, 38]. Reservoir computing [47.36, 45] has already become a popular approach for SSTD modeling and pattern recognition. In the classical view a *reservoir* is a homogeneous, passive 3D structure of probabilistically connected and fixed neurons that in principle has no learning and memory, nor has it an interpretable structure – Fig. 47.11b. A reservoir, such as a liquid state machine (LSM) [47.36, 81], usually uses *small world recurrent connections* that do not facilitate capturing explicit spatial and temporal components from the SSTD in their relationship, which is the main goal of learning SSTD. Despite difficulties with the LSM reservoirs, it was shown on several SSTD problems that they produce better results than using a simple classifier [47.7, 12, 36, 82]. Some publications demonstrated that probabilistic neurons are suitable for reservoir computing especially in a noisy environment [47.56,57]. In [47.83] an improved accuracy of the LSM reservoir structure on pattern classification of hypothetical tasks is achieved when STDP learning is introduced into the reservoir. The learning is based on comparing the liquid states for different classes and adjusting the connection weights so that same class inputs have closer connection weights. The method is illustrated on the phone recognition task of the TIMIT data base phonemes – spectro-temporal problem. 13 MSCC are turned into trains of spikes. The metric of separation between liquid states representing different classes is similar to Fisher's t-test [47.84].

After the presentation of an input data example (or a *chink* of data) the state of the SNN reservoir $S(t)$ is evaluated in an output module and used for classification purposes (both during training and recall phase). Different methods can be applied to capture this state:

1. Spike rate activity of *all* neurons at a certain time window: The state of the reservoir is represented as a vector of n elements (n is the number of neurons vector of n elements (n is the number of neurons in the reservoir), each element representing the spik-

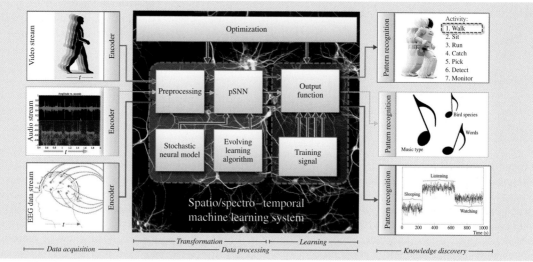

Fig. 47.9 The eSNN framework for STPR (after [47.85])

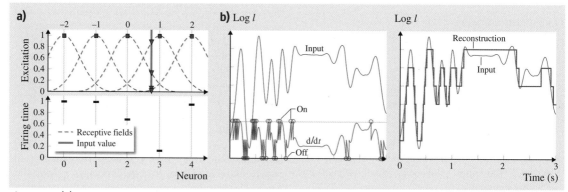

Fig. 47.10 (**a**) Population rank order coding of input information. (**b**) AER of the input information (after [47.10])

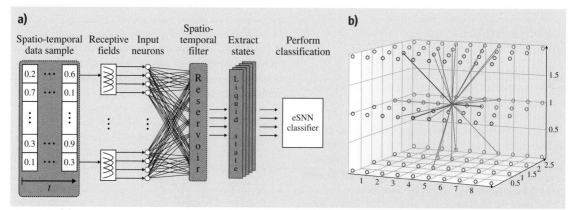

Fig. 47.11 (**a**) An eSNN architecture for STPR using a reservoir. (**b**) The structure and connectivity of a reservoir

ing probability of the neuron within a time window. Consecutive vectors are passed to train/recall an output classifier.

2. Spike rate activity of spatio-temporal clusters $C_1, C_2, \ldots C_k$ of close (both in space and time) neurons: The state $S_{C_i}(t)$ of each cluster C_i is represented by a single number, reflecting on the spiking activity of the neurons in the cluster in a defined time window (this is the internal SNN time, usually measured in ms). This is interpreted as the current spiking probability of the cluster. The states of all clusters define the current reservoir state $S(t)$. In the output function, the cluster states $S_{C_i}(t)$ are used differently for different tasks.

3. Continuous function representation of spike trains: In contrast to the above two methods that use spike rates to evaluate the spiking activity of a neuron or a neuronal cluster, here the train of spikes from each neuron within a time window, or a neuronal cluster, is transferred into a continuous value temporal function using a kernel (e.g., α-kernel). These functions can be compared and a continuous value error measured.

In [47.82] a comparative analysis of the three methods above is presented on a case study of Brazilian sign language gesture recognition (Fig. 47.18) using a LSM as a reservoir.

Different adaptive classifiers can be explored for the classification of the reservoir state into one of the output classes, including: statistical techniques, e.g., regression techniques; MLP; eSNN; nearest-neighbor techniques; and incremental LDA [47.86]. State vector transformation, before classification can be done with the use of adaptive incremental transformation functions, such as incremental PCA [47.87].

47.6 Computational Neurogenetic Models (CNGM)

Here, the neurogenetic model of a neuron [47.40, 41] is utilized. A CNGM framework is shown in Fig. 47.12 [47.19].

The CNGM framework comprises a set of methods and algorithms that support the development of computational models, each of them characterized by:

1. eSNN at the higher level and a gene regulatory network (GRN) at the lower level, each functioning at a different time-scale and continuously interacting between each other.
2. Optional use of probabilistic spiking neurons, thus forming an epSNN.
3. Parameters in the epSNN model are defined by genes/proteins from the GRN.
4. Ability to capture in its internal representation both spatial and temporal characteristics from SSTD streams.
5. The structure and the functionality of the model evolve in time from incoming data.
6. Both unsupervised and supervised learning algorithms can be applied in an on-line or in a batch mode.
7. A concrete model would have a specific structure and a set of algorithms depending on the problem and the application conditions, e.g., classification of SSTD; modeling of brain data.

The framework from Fig. 47.12 supports the creation of a multi-modular integrated system, where different modules, consisting of different neuronal types and genetic parameters, represent different functions (e.g., vision, sensory information processing, sound recognition, and motor-control) and the whole system works in an integrated mode.

The neurogenetic model from Fig. 47.12 uses as a main principle the analogy with biological facts about the relationship between spiking activity and gene/protein dynamics in order to control the learning and spiking parameters in an SNN when SSTD is learned. Biological support of this can be found in numerous publications (e.g., [47.40, 88–90]).

The Allen Human Brain Atlas [47.91] of the Allen Institute for Brain Science [47.92] has shown that at least 82% of human genes are expressed in the brain. For 1000 anatomical sites of the brains of two individuals 100 mln data points are collected that indicate gene expressions of each of the genes and underlies the biochemistry of the sites.

In [47.58] it is suggested that both the firing rate (rate coding) and spike timing as spatio-temporal patterns (rank order and spatial pattern coding) play a role in fast and slow, dynamic and adaptive sensorimotor responses, controlled by the cerebellar nuclei. Spatio-temporal patterns of a population of Purkinji cells are shaped by activities in the molecular layer of interneurons. In [47.88] it is demonstrated that the temporal spiking dynamics depend on the spatial structure of the neural system (e.g., different for the hippocampus

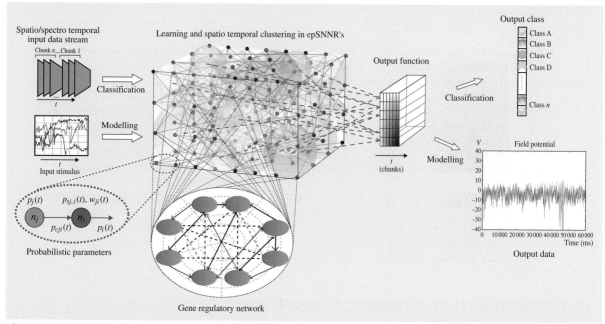

Fig. 47.12 A schematic diagram of a CNGM framework, consisting of: input encoding module; output function for SNN state evaluation; output classifier; GRN (optional module). The framework can be used to create concrete models for STPR or data modeling (after [47.19])

and the cerebellum). In the hippocampus the connections are scale free, e.g., there are hub neurons, while in the cerebellum the connections are regular. The spatial structure depends on genetic pre-determination and on the gene dynamics. Functional connectivity develops in parallel with structural connectivity during brain maturation. A growth-elimination process (synapses are created and eliminated) depend on gene expression [47.88], e.g., glutamatergic neurons issued from the same progenitors tend to wire together and form ensembles, also for the cortical GABAergic interneuron population. Connections between early developed neurons (mature networks) are more stable and reliable when transferring spikes than the connections between newly created neurons (thus the probability of spike transfer). Postsynaptic AMPA-type glutamate receptors (AMPARs) mediate most fast excitatory synaptic transmissions and are crucial for many aspects of brain function, including learning, memory, and cognition [47.40, 93].

Kasabov et al. [47.94] show the dramatic effect of a change of single gene, that regulates the τ parameter of the neurons, on the spiking activity of the whole SNN of 1000 neurons – see Fig. 47.13.

The spiking activity of a neuron may affect the expressions of genes as feedback [47.95]. As pointed out

in [47.90] on longer time scales of minutes and hours the function of neurons may cause changes of the expression of hundreds of genes transcribed into mRNAs and also in microRNAs, which makes the short-term, long-term, and the genetic memories of a neuron linked together in a global memory of the neuron and further – of the whole neural system.

A major problem with the CNGM from Fig. 47.12 is how to optimize the numerous parameters of the model. One solution could be to use evolutionary computation, such as PSO [47.57, 96] and the recently proposed quantum inspired evolutionary computation techniques [47.21, 44, 97]. The latter can deal with a very large dimensional space as each quantum-bit chromosome represents the whole space, each point to certain probability. Such algorithms are faster and lead to a close solution to the global optimum in a very short time. In one approach it may be reasonable to use same parameter values (same GRN) for all neurons in the SNN or for each of different types of neurons (cells) that will results in a significant reduction of the parameters to be optimized. This can be interpreted as the *average* parameter value for neurons of the same type. This approach corresponds to the biological notion to use one value (average)

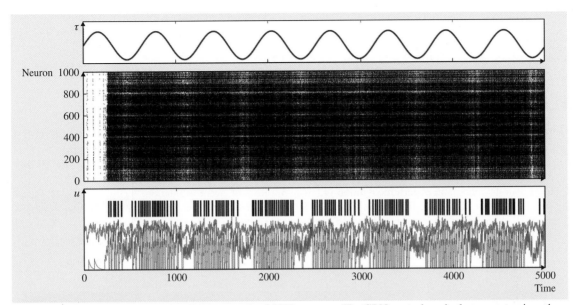

Fig. 47.13 A GRN interacting with a SNN reservoir of 1000 neurons. The GRN controls a single parameter, i.e., the τ parameter of all 1000 LIF neurons, over a period of 5 s. The *top diagram* shows the evolution of τ. The response of the SNN is shown as a raster plot of spike activity. A *black point in this diagram* indicates a spike of a specific neuron at a specific time in the simulation. The *bottom diagram* presents the evolution of the membrane potential of a single neuron from the network (*green curve*) along with its firing threshold ϑ (*red curve*). Output spikes of the neuron are indicated as *black vertical lines* in the same diagram (after [47.94])

of a gene/protein expression for millions of cells in bioinformatics.

Another approach to define the parameters of the probabilistic spiking neurons, especially when used in biological studies, is to use prior knowledge about the association of spiking parameters with relevant genes/proteins (neurotransmitter, neuroreceptor, ion channel, neuromodulator) as described in [47.19]. Combination of the two approaches above is also possible.

47.7 SNN Software and Hardware Implementations to Support STPR

Software and hardware realizations of SNN are already available to support various applications of SNN for STPR. Among the most popular software/hardware systems are [47.99–101]:

1. jAER [47.10, 102]
2. Software simulators, such as Brian [47.100], Nestor, NeMo [47.103], etc.
3. Silicon retina camera [47.10]
4. Silicon cochlea [47.14]
5. SNN hardware realization of LIFM and SDSP [47.72, 104–106]

Fig. 47.14 A hypothetical neuromorphic SNN application system (after [47.98]) ▶

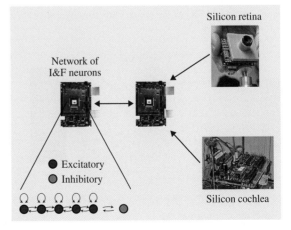

6. The SpiNNaker hardware/software environment [47.107, 108]
7. FPGA implementations of SNN [47.109]
8. The recently announced IBM LIF SNN chip.

Figure 47.14 shows a hypothetical engineering system using some of the above tools (from [47.11, 104]).

47.8 Current and Future Applications of eSNN and CNGM for STPR

The applications of eSNN for STPR are numerous. Here only few are listed:

1. Moving object recognition (Fig. 47.15) [47.10, 12].
2. EEG data modeling and pattern recognition [47.3–7, 111–113] directed to practical applications, such as: BCI [47.5], classification of epilepsy [47.112–114] – (Fig. 47.16).

3. Robot control through EEG signals [47.115] (Fig. 47.17) and robot navigation [47.116].
4. Sign language gesture recognition (e.g., the Brazilian sign language – Fig. 47.18) [47.82].
5. Risk of event evaluation, e.g., prognosis of establishment of invasive species [47.21] – Fig. 47.19; stroke occurrence [47.16], etc.
6. Cognitive and emotional robotics [47.19, 34].

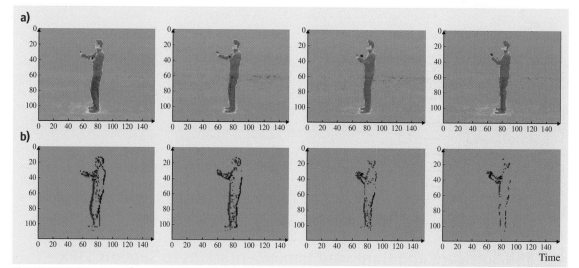

Fig. 47.15a,b Moving object recognition with the use of AER. (**a**) Disparity map of a video sample; (**b**) address event representation (AER) of the above video sample (after [47.10])

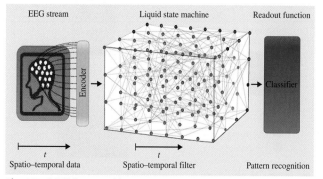

Fig. 47.16 EEG recognition system

Fig. 47.17 Robot control and navigation through EEG signals (from [47.110])

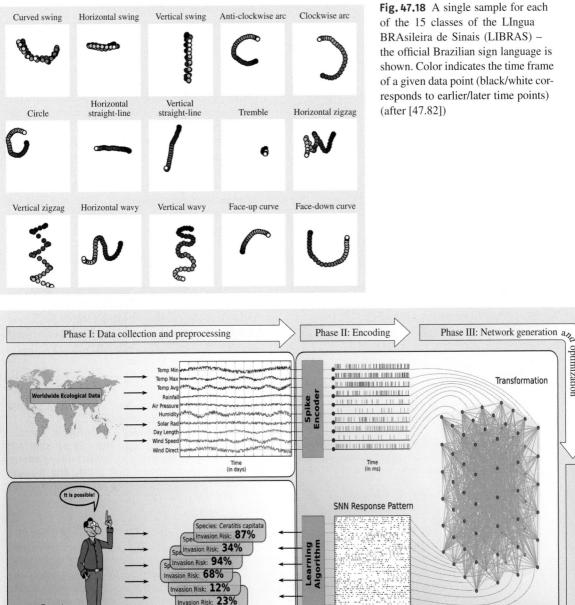

Fig. 47.18 A single sample for each of the 15 classes of the LIngua BRAsileira de Sinais (LIBRAS) – the official Brazilian sign language is shown. Color indicates the time frame of a given data point (black/white corresponds to earlier/later time points) (after [47.82])

Fig. 47.19 A prognostic system for ecological modeling (after [47.21])

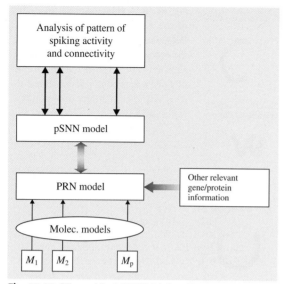

Fig. 47.20 Hierarchical CNGM (after 47.19)

7. Neurorehabilitation robots [47.117].
8. Modeling finite automata [47.118, 119].
9. Knowledge discovery from SSTD [47.120].

10. Neurogenetic robotics [47.121].
11. Modeling the progression or the response to treatment of neurodegenerative diseases, such as Alzheimer's disease [47.18, 19] – Fig. 47.20. The analysis of the GRN model obtained in this case could enable the discovery of unknown interactions between genes/proteins related to brain disease progression and how these interactions can be modified to achieve a desirable effect.
12. Modeling financial and economic problems (neuroeconomics) where at a *lower* level the GRN represents the dynamic interaction between time series variables (e.g., stock index values, exchange rates, unemployment, GDP, the price of oil), while the *higher* level epSNN states represents the state of the economy or the system under study. The states can be further classified into predefined classes (e.g., buy, hold, sell, invest, likely bankruptcy) [47.122].
13. Personalized modeling, which is concerned with the creation of a single model for an individual input data [47.17, 46, 47]. Here a whole SSTD pattern is taken as individual data rather than a single vector.

References

47.1 Emotiv: http://www.emotiv.com
47.2 The FMRIB Centre, University of Oxford, http://www. fmrib.ox.ac.uk
47.3 D.A. Craig, H.T. Nguyen: Adaptive EEG thought pattern classifier for advanced wheelchair control, Proc. Eng. Med. Biol. Soc. – EMBS'07 (2007) pp. 2544–2547
47.4 A. Ferreira, C. Almeida, P. Georgieva, A. Tomé, F. Silva: Advances in EEG-based biometry, LNCS **6112**, 287–295 (2010)
47.5 T. Isa, E.E. Fetz, K. Müller: Recent advances in brain-machine interfaces, Neural Netw. **22**(9), 1201–1202 (2009)
47.6 F. Lotte, M. Congedo, A. Lécuyer, F. Lamarche, B. Arnaldi: A review of classification algorithms for EEG-based brain–computer interfaces, J. Neural Eng. **4**(2), R1–R15 (2007)
47.7 S. Schliebs, N. Nuntalid, N. Kasabov: Towards spatio-temporal pattern recognition using evolving spiking neural networks, LNCS **6443**, 163–170 (2010)
47.8 B. Schrauwen, J. Van Campenhout: BSA, a fast and accurate spike train encoding scheme, Neural Netw. 2003, Proc. Int. Jt. Conf., Vol. 4 (IEEE 2003) pp. 2825–2830
47.9 D. Sona, H. Veeramachaneni, E. Olivetti, P. Avesani: Inferring cognition from fMRI brain images, LNCS **4669**, 869–878 (2007)

47.10 T. Delbruck: JAER open source project (2007) http://jaer.wiki.sourceforge.net
47.11 K. Dhoble, N. Nuntalid, G. Indivery, N. Kasabov: Online spatio-temporal pattern recognition with evolving spiking neural networks utilising address event representation, rank order, and temporal spike learning, Int. Joint Conf. Neural Netw. (IJCNN) (IEEE 2012)
47.12 N. Kasabov, K. Dhoble, N. Nuntalid, A. Mohemmed: Evolving probabilistic spiking neural networks for spatio-temporal pattern recognition: A preliminary study on moving object recognition, **7064**, 230–239 (2011)
47.13 A. Rokem, S. Watzl, T. Gollisch, M. Stemmler, A.V.M. Herz, I. Samengo: Spike-timing precision underlies the coding efficiency of auditory receptor neurons, J. Neurophys. **95**(4), 2541–2552 (2005)
47.14 A. van Schaik, L. Shih-Chii: AER EAR: A matched address event representation interface, Proc. IS-CAS – IEEE Int. Symp. Circuits Syst., Vol. 5 (2005) pp. 4213–4216
47.15 P.J. Cowburn, J.G.F. Cleland, A.J.S. Coats, M. Komajda: Risk stratification in chronic heart failure, Eur. Heart J. **19**, 696–710 (1996)
47.16 S. Barker-Collo, V.L. Feigin, V. Parag, C.M.M. Lawes, H. Senior: Auckland stroke outcomes study, Neurology **75**(18), 1608–1616 (2010)

47.17　N. Kasabov: Global, local and personalised modelling and profile discovery in Bioinformatics: An integrated approach, Pattern Recogn. Lett. **28**(6), 673–685 (2007)

47.18　R. Schliebs: Basal forebrain cholinergic dysfunction in Alzheimer's disease – interrelationship with β-amyloid, inflammation and neurotrophin signaling, Neurochem. Res. **30**, 895–908 (2005)

47.19　N. Kasabov, R. Schliebs, H. Kojima: Probabilistic computational neurogenetic framework: From modelling cognitive systems to Alzheimer's disease, IEEE Trans. Auton. Ment. Dev. **3**(4), 1–12 (2011)

47.20　C.R. Shortall, A. Moore, E. Smith, M.J. Hall, I.P. Woiwod, R. Harrington: Long-term changes in the abundance of flying insects, Insect Conserv. Divers. **2**(4), 251–260 (2009)

47.21　S. Schliebs, M. Defoin-Platel, S. Worner, N. Kasabov: Integrated feature and parameter optimization for evolving spiking neural network: Exploring heterogeneous probabilistic models, Neural Netw. **22**, 623–632 (2009)

47.22　L.R. Rabiner: A tutorial on hidden Markov models and selected applications in speech recognition, Proceedings of IEEE **77**(2), 257–285 (1989)

47.23　N. Kasabov: *Foundations of Neural Networks, Fuzzy Systems, and Knowledge Engineering* (MIT Press, Cambridge 1996) p. 550

47.24　I. Arel, D.C. Rose, T.P. Karnowski: Deep machine learning: A new frontier artificial intelligence research, Comput. Intell. Mag. **5**(4), 13–18 (2010)

47.25　I. Arel, D. Rose, B. Coop: DeSTIN: A deep learning architecture with application to high-dimensional robust pattern, Proc. 2008 AAAI Workshop Biologically Inspired Inspired Cognitive Architectures (BICA) (2008)

47.26　Y. Bengio: Learning deep architectures for AI, Found. Trends. Mach. Learn. **2**(1), 1–127 (2009)

47.27　I. Weston, F. Ratle, R. Collobert: Deep learning via semi-supervised embedding, Proc. 25th Int. Conf. Mach. Learn. (2008) pp. 1168–1175

47.28　W. Gerstner: Time structure of the activity of neural network models, Phys. Rev. **51**, 738–758 (1995)

47.29　W. Gerstner: What's different with spiking neurons?. In: *Plausible Neural Networks for Biological Modelling*, ed. by H. Mastebroek, H. Vos (Kluwer, Dordrecht 2001) pp. 23–48

47.30　G. Kistler, W. Gerstner: Spiking neuron models – single neurons. In: *Populations, Plasticity* (Cambridge Univ. Press, Cambridge 2002)

47.31　S. Song, K. Miller, L. Abbott: Competitive Hebbian learning through spike-timing-dependent synaptic plasticity, Nat. Neurosci. **3**, 919–926 (2000)

47.32　S. Fusi, M. Annunziato, D. Badoni, A. Salamon, D. Amit: Spike-driven synaptic plasticity: Theory, simulation, VLSI implementation, Neural Comput. **12**(10), 2227–2258 (2000)

47.33　A. Belatreche, L.P. Maguire, M. McGinnity: Advances in design and application of spiking neural networks, Soft Comput. **11**(3), 239–248 (2006)

47.34　F. Bellas, R.J. Duro, A. Faiña, D. Souto: Multilevel Darwinisb Brain (MDB): Artificial evolution in a cognitive architecture for real robots, IEEE Trans. Auton. Ment. Dev. **2**, 340–354 (2010)

47.35　S. Bohte, J. Kok, J. LaPoutre: Applications of spiking neural networks, Inf. Proc. Lett. **95**(6), 519–520 (2005)

47.36　W. Maass, T. Natschlaeger, H. Markram: Real-time computing without stable states: A new framework for neural computation based on perturbations, Neural Comput. **14**(11), 2531–2560 (2002)

47.37　N. Kasabov: *Evolving Connectionist Systems: The Knowledge Engineering Approach* (Springer, London 2007)

47.38　S. Wysoski, L. Benuskova, N. Kasabov: Evolving spiking neural networks for audiovisual information processing, Neural Netw. **23**(7), 819–835 (2010)

47.39　M. Riesenhuber, T. Poggio: Hierarchical model of object recognition in cortex, Nat. Neurosci. **2**, 1019–1025 (1999)

47.40　L. Benuskova, N. Kasabov: *Computational Neuro-Genetic Modelling* (Springer, New York 2007) p. 290

47.41　N. Kasabov, L. Benuskova, S. Wysoski: A computational neurogenetic model of a spiking neuron, IJCNN 2005 Conf. Proc., Vol. 1 (IEEE 2005) pp. 446–451

47.42　N. Kasabov: To spike or not to spike: A probabilistic spiking neuron model, Neural Netw. **23**(1), 16–19 (2010)

47.43　W. Maass, H. Markram: Synapses as dynamic memory buffers, Neural Netw. **15**(2), 155–161 (2002)

47.44　S. Schliebs, N. Kasabov, M. Defoin-Platel: On the probabilistic optimization of spiking neural networks, Int. J. Neural Syst. **20**(6), 481–500 (2010)

47.45　D. Verstraeten, B. Schrauwen, M. D'Haene, D. Stroobandt: An experimental unification of reservoir computing methods, Neural Netw. **20**(3), 391–403 (2007)

47.46　N. Kasabov, Y. Hu: Integrated optimisation method for personalised modelling and case study applications, Int. J. Funct. Inf. Personal. Med. **3**(3), 236–256 (2010)

47.47　N. Kasabov: *Data analysis and predictive systems and related methodologies – personalised trait modelling system*, NZ Patent PCT/NZ2009/000222 (2009)

47.48　A.L. Hodgkin, A.F. Huxley: A quantitative description of membrane current and its application to conduction and excitation in nerve, J. Physiol. **117**, 500–544 (1952)

47.49　E. Izhikevich: Simple model of spiking neurons, IEEE Trans. Neural Netw. **14**(6), 1569–1572 (2003)

Part I | 47

47.50 E.M. Izhikevich: Which model to use for cortical spiking neurons?, Neural Netw. **15**(5), 1063–1070 (2004)

47.51 E.M. Izhikevich, G.M. Edelman: large-scale model of mammalian thalamocortical systems, Proc. Natl. Acad. Sci. USA **105**, 3593–3598 (2008)

47.52 E. Izhikevich: Polychronization: Computation with spikes, Neural Comput. **18**, 245–282 (2006)

47.53 Z.P. Kilpatrick, P.C. Bresloff: Effect of synaptic depression and adaptation on spatio-temporal dynamics of an excitatory neural networks, Physica D **239**, 547–560 (2010)

47.54 W. Maass, A.M. Zador: Computing and learning with dynamic synapses. In: *Pulsed Neural Networks* (MIT Press, Cambridge 1999) pp.321–336

47.55 J.R. Huguenard: Reliability of axonal propagation: The spike doesn't stop here, Proc. Natl. Acad. Sci USA **97**(17), 9349–9350 (2000)

47.56 S. Schliebs, A. Mohemmed, N. Kasabov: Are probabilistic spiking neural networks suitable for reservoir computing?, Int. Jt. Conf. Neural Netw. (IJCNN) (IEEE 2011) pp.3156–3163

47.57 H. Nuzly, A. Hamed, N. Kasabov, S. Shamsuddin: Probabilistic evolving spiking neural network optimization using dynamic quantum inspired particle swarm optimization, Aust. J. Intell. Inf. Process. Syst. **11**(1), 1074 (2010), available online at http://cs.anu.edu.au/ojs/index.php/ajiips/article/viewArticle/1074

47.58 S.J. Thorpe: Spike-based image processing: Can we reproduce biological vision in hardware, LNCS **7583**, 516–521 (2012)

47.59 W. Gerstner, A.K. Kreiter, H. Markram, A.V.M. Herz: Neural codes: Firing rates and beyond, Proc. Natl. Acad. Sci. USA **94**(24), 12740–12741 (1997)

47.60 J.J. Hopfield: Neural networks and physical systems with emergent collective computational abilities, Proc. Natl. Acad. Sci. USA **79**, 2554–2558 (1982)

47.61 S.M. Bohte: The evidence for neural information processing with precise spike-times: A survey, Nat. Comput. **3**(2), 195–206 (2004)

47.62 J. Hopfield: Pattern recognition computation using action potential timing for stimulus representation, Nature **376**, 33–36 (1995)

47.63 H.G. Eyherabide, I. Samengo: Time and category information in pattern-based codes, Front. Comput. Neurosci. **4**, 145 (2010)

47.64 F. Theunissen, J.P. Miller: Temporal encoding in nervous rigorous definition, J. Comput. Neurosci. **2**(2), 149–162 (1995)

47.65 S. Thorpe, A. Delorme, R. VanRullen: Spike-based strategies for rapid processing, Neural Netw. **14**(6–7), 715–725 (2001)

47.66 S. Thorpe, J. Gautrais: Rank order coding, Comput. Neurosci. **13**, 113–119 (1998)

47.67 M.J. Berry, D.K. Warland, M. Meister: The structure and precision of retinal spiketrains, Proc. Natl. Acad. Sci. USA **94**(10), 5411–5416 (1997)

47.68 P. Reinagel, R.C. Reid: Precise firing events are conserved across neurons, J. Neurosci. **22**(16), 6837–6841 (2002)

47.69 J. Brader, W. Senn, S. Fusi: Learning real-world stimuli in a neural network with spike-driven synaptic dynamics, Neural Comput. **19**(11), 2881–2912 (2007)

47.70 R. Legenstein, C. Naeger, W. Maass: What can a neuron learn with spike-timing-dependent plasticity?, Neural Comput. **17**(11), 2337–2382 (2005)

47.71 D. Hebb: *The Organization of Behavior* (Wiley, New York 1949)

47.72 G. Indiveri, F. Stefanini, E. Chicca: Spike-based learning with a generalized integrate and fire silicon neuron, IEEE Int. Symp. Circuits Syst. (ISCAS 2010) (2010) pp.1951–1954

47.73 T. Masquelier, R. Guyonneau, S. Thorpe: Spike timing dependent plasticity finds the start of repeating patterns in continuous spike trains, PlosONE **3**(1), e1377 (2008)

47.74 R. Gutig, H. Sompolinsky: The tempotron: A neuron timing-based decisions, Nat. Neurosci. **9**(3), 420–428 (2006)

47.75 R.V. Florian: The chronotron: A neuron that learns to fire temporally-precise spike patterns, Nature Precedings (2010), available online at http://precedings.nature.com/documents/5190/version/1

47.76 F. Ponulak, A. Kasinski: Supervised learning in spiking neural networks with ReSuMe: Sequence learning, Neural Comput. **22**(2), 467–510 (2010)

47.77 A. Mohemmed, S. Schliebs, S. Matsuda, N. Kasabov: Evolving spike pattern association neurons and neural networks, Neurocomputing **107**, 3–10 (2013)

47.78 A. Mohemmed, S. Schliebs, S. Matsuda, N. Kasabov: SPAN: Spike pattern association neuron for learning spatio-temporal sequences, Int. J. Neural Syst. **22**(4), 1–16 (2012)

47.79 M. Watts: A decade of Kasabov's evolving connectionist systems: A Review, IEEE Trans. Syst. Man Cybern. C **39**(3), 253–269 (2009)

47.80 H. Nuzlu, N. Kasabov, S. Shamsuddin, H. Widiputra, K. Dhoble: An extended evolving spiking neural network model for spatio-temporal pattern classification, Proc. IJCNN (IEEE 2011) pp.2653–2656

47.81 E. Goodman, D. Ventura: Spatiotemporal pattern recognition via liquid state machines, Int. Jt. Conf. Neural Networks (IJCNN) '06 (2006) pp.3848–3853

47.82 S. Schliebs, H.N.A. Hamed, N. Kasabov: A reservoir-based evolving spiking neural network for on-line spatio-temporal pattern learning and recognition, 18th Int. Conf. Neural Inf. Proc. ICONIP 2011 (Springer, Shanghai 2011)

47.83 D. Norton, D. Ventura: Improving liquid state machines through iterative refinement of the reservoir, Neurocomputing **73**, 2893–2904 (2010)

47.84 R.A. Fisher: The use of multiple measurements in taxonomic problems, Ann. Eugen. **7**, 179–188 (1936)

47.85   EU FP7 Marie Curie project EvoSpike (2011–2012), http://ncs.ethz.ch/projects/evospike

47.86   S. Pang, S. Ozawa, N. Kasabov: Incremental linear discriminant analysis for classification of data streams, IEEE Trans. SMC-B **35**(5), 905–914 (2005)

47.87   S. Ozawa, S. Pang, N. Kasabov: Incremental learning of chunk data for on-line pattern classification systems, IEEE Trans. Neural Netw. **19**(6), 1061–1074 (2008)

47.88   J.M. Henley, E.A. Barker, O.O. Glebov: Routes, destinations and advances in AMPA receptor trafficking, Trends Neurosci. **34**(5), 258–268 (2011)

47.89   Y.C. Yu, R.S. Bultje, X. Wang, S.H. Shi: Specific synapses develop preferentially among sister excitatory neurons in the neocortex, Nature **458**, 501–504 (2009)

47.90   V.P. Zhdanov: Kinetic models of gene expression including non-coding RNAs, Phys. Rep. **500**, 1–42 (2011)

47.91   BrainMap Project: www.brain-map.org

47.92   Allen Institute for Brain Science: www.alleninstitute.org

47.93   Gene and Disease (2005) NCBI, http:/www.ncbi.nlm.nih.gov

47.94   N. Kasabov, S. Schliebs, A. Mohemmed: Modelling the effect of genes on the dynamics of probabilistic spiking neural networks for computational neurogenetic modelling, Proc. 6th Meet. Comp. Intell. Bioinfor. Biostat. (CIBB) 2011 (Springer 2011)

47.95   M. Barbado, K. Fablet, M. Ronjat, M. De Waard: Gene regulation by voltage-dependent calcium channels, Biochim. Biophys. Acta **1793**, 1096–1104 (2009)

47.96   A. Mohemmed, S. Matsuda, S. Schliebs, K. Dhoble, N. Kasabov: Optimization of spiking neural networks with dynamic synapses for spike sequence generation using PSO, Proc. Int. Joint Conf. Neural Netw. (IEEE, San Jose 2011) pp. 2969–2974

47.97   M. Defoin-Platel, S. Schliebs, N. Kasabov: Quantum-inspired evolutionary algorithm: A multimodel EDA, IEEE Trans. Evol. Comput. **13**(6), 1218–1232 (2009)

47.98   Neuromorphic Cognitive Systems Group, Institute for Neuroinformatics, ETH and University of Zurich, http://ncs.ethz.ch

47.99   R. Douglas, M. Mahowald: Silicon neurons. In: *The Handbook of Brain Theory and Neural Networks*, ed. by M. Arbib (MIT, Cambridge 1995) pp. 282–289

47.100 R. Brette, M. Rudolph, T. Carnevale, M. Hines, D. Beeman, J.M. Bower, M. Diesmann, A. Morrison, P.H. Goodman, F.C. Harris, M. Zirpe, T. Natschläger, D. Pecevski, B. Ermentrout, M. Djurfeldt, A. Lansner, O. Rochel, T. Vieville, E. Muller, A.P. Davison, S.E. Boustani, A. Destexhe: Simulation of networks of spiking neurons: A review

of tools and strategies, J. Comput. Neurosci. **23**, 349–398 (2007)

47.101 S. Furber, S. Temple: Neural systems engineering, Interface J. R. Soc. **4**, 193–206 (2007)

47.102 jAER Open Source Project: http://jaer.wiki.sourceforge.net

47.103 NeMo spiking neural network simulator, http://www.doc.ic.ac.uk/~akf/nemo/index.html

47.104 G. Indiveri, B. Linares-Barranco, T. Hamilton, A. Van Schaik, R. Etienne-Cummings, T. Delbruck, S. Liu, P. Dudek, P. Häfliger, S. Renaud: Neuromorphic silicon neuron circuits, Front. Neurosci. **5**, 1–23 (2011)

47.105 G. Indiveri, E. Chicca, R.J. Douglas: Artificial cognitive systems: From VLSI networks of spiking neurons to neuromorphic cognition, Cogn. Comput. **1**(2), 119–127 (2009)

47.106 G. Indiviery, T. Horiuchi: Frontiers in neuromorphic engineering, Front. Neurosci. **5**, 118 (2011)

47.107 A.D. Rast, X. Jin, F. Galluppi, L.A. Plana, C. Patterson, S. Furber: Scalable event-driven native parallel processing: The SpiNNaker neuromimetic system, Proc. ACM Int. Conf. Comput. Front. (ACM 2010) pp. 21–29

47.108 X. Jin, M. Lujan, L.A. Plana, S. Davies, S. Temple, S. Furber: Modelling spiking neural networks on SpiNNaker, Comput. Sci. Eng. **12**(5), 91–97 (2010)

47.109 S.P. Johnston, G. Prasad, L. Maguire, T.M. McGinnity: FPGA Hardware/software co-design methodology – towards evolvable spiking networks for robotics application, Int. J. Neural Syst. **20**(6), 447–461 (2010)

47.110 KEDRI: http://www.kedri.aut.ac.nz

47.111 R. Acharya, E.C.P. Chua, K.C. Chua, L.C. Min, T. Tamura: Analysis and automatic identification of sleep stages using higher order spectra, Int. J. Neural Syst. **20**(6), 509–521 (2010)

47.112 S. Ghosh-Dastidar, H. Adeli: A new supervised learning algorithm for multiple spiking neural networks with application in epilepsy and seizure detection, Neural Netw. **22**(10), 1419–1431 (2009)

47.113 S. Ghosh-Dastidar, H. Adeli: Improved spiking neural networks for EEG classification and epilepsy and seizure detection, Integr. Comput.-Aided Eng. **14**(3), 187–212 (2007)

47.114 A.E.P. Villa, Y. Asai, I. Tetko, B. Pardo, M.R. Celio, B. Schwaller: Cross-channel coupling of neuronal activity in parvalbumin-deficient mice susceptible to epileptic seizures, Epilepsia **46**(6), 359 (2005)

47.115 G. Pfurtscheller, R. Leeb, C. Keinrath, D. Friedman, C. Neuper, C. Guger, M. Slater: Walking from thought, Brain Res. **1071**(1), 145–152 (2006)

47.116 E. Nichols, L.J. McDaid, N.H. Siddique: Case study on self-organizing spiking neural networks for robot navigation, Int. J. Neural Syst. **20**(6), 501–508 (2010)

47.117 X. Wang, Z.G. Hou, A. Zou, M. Tan, L. Cheng: A behavior controller for mobile robot based on spiking

Part 1 | 47

neural networks, Neurocomputing **71**(4–6), 655–666 (2008)

47.118 D. Buonomano, W. Maass: State-dependent computations: Spatio-temporal processing in cortical networks, Nat. Rev. Neurosci. **10**, 113–125 (2009)

47.119 T. Natschläger, W. Maass: Spiking neurons and the induction of finite state machines, Theor. Comput. Sci. Nat. Comput. **287**(1), 251–265 (2002)

47.120 S. Soltic, N. Kasabov: Knowledge extraction from evolving spiking neural networks with rank order

population coding, Int. J. Neural Syst. **20**(6), 437–445 (2010)

47.121 Y. Meng, Y. Jin, J. Yin, M. Conforth: Human activity detection using spiking neural networks regulated by a gene regulatory network, Proc. Int. Jt. Conf. Neural Netw. (IJCNN) (IEEE, Barcelona 2010) pp. 2232–2237

47.122 R. Pears, H. Widiputra, N. Kasabov: Evolving integrated multi-model framework for on-line multiple time series prediction, Evol. Syst. **4**(2), 99–117 (2013)

48. Neurocomputational Models of Natural Language

Alistair Knott

In this chapter I review computational models of the neural circuitry which implements the human capacity for language. These models cover many different aspects of the language capacity, from representations of phonology and word forms to representations of sentence meanings and syntactic structures. The computational models discussed are neural networks: structures of simple units which are densely interconnected and can be active in parallel. I review the computational properties of the networks introduced and also empirical evidence from different sources (neural imaging, behavioral experiments, patterns of impairment following brain dysfunction) which supports the models described.

Part I | 48

Since the earliest days of neural network research, theorists have attempted to model the neural circuitry which supports the human language faculty. *Rosenblatt*'s [48.1] perceptron was designed to recognize individual letters; *Widrow* and *Hoff*'s [48.2] Adaline system was deployed in speech recognition among other tasks. Many of the pioneering PDP models of the 1980s had linguistic applications – for instance, *McClelland* and *Rumelhart*'s [48.3] interactive activation model of word reading, and *Rumelhart* and *McClelland*'s [48.4] model of past tense learning. These early models paved the way for a veritable industry of research into the neural basis of language processing,

combining empirical experimentation with computational simulations. In this chapter I will sketch some of the main results of this research program, summarizing what has been learned experimentally about language processing in the brain and describing some contemporary neural network models informed by these findings.

As is conventional in reviews of neural language circuitry, I will proceed from the outside inwards, working from the more peripheral sensory and motor representations of language in the brain to the more abstract, processed representations which are harder to investigate. In Sect. 48.1 I will consider the circuitry which supports auditory and articulatory processing. (I will restrict my attention to spoken language in the current chapter.) In Sect. 48.2 I will consider the circuitry which supports representation and processing of individual words; in Sect. 48.3 I will discuss some ideas about how the meanings of whole sentences are stored, and in Sect. 48.4 I will consider the circuitry involved in representing and processing the syntactic structure of sentences.

The neural network models I introduce in this chapter are models which learn to perform their allocated task through exposure to training data in one format or another. As a consequence, they are not just models of how the brain represents and processes language – they are also models of how children *acquire* language. In various places, therefore, I will also draw on empirical studies of child language acquisition to motivate the models I present.

48.1 Phonological Representations and Processing

48.1.1 Phonemes and Syllables

Articulatory movements are continuous in nature, so speech sounds vary along a continuum. However, there is good evidence that the neural circuitry which produces and interprets these sounds represents them discretely. The basic data bearing this out come from the linguistic discipline of *phonology*. Any given language has a fixed repertoire of discrete speech sounds called *phonemes*; speakers of a given language parse a speech stream into a sequence of phonemes. Different phonemes are available in different languages, for instance, the speech sounds which in English correspond to the phonemes *l* and *r* map onto a single phoneme in Japanese. Moreover, a given phoneme is produced in different ways in different contexts, due to a phenomenon called *coarticulation*: for instance, the phoneme *t* manifests itself differently in the phrases *hot porridge* and *light up*. During phonological development, infants become attuned to the phonological distinctions made in their mother tongue (see, e.g., [48.5]). What are the neural circuits which learn a repertoire of phonemes?

We know that phonemes must be represented both in the articulatory system and in the auditory system – and of course, the same sets of phonemes must be learned in each system. The neural areas we expect to be involved in articulation are in the inferior areas of motor cortex controlling the vocal apparatus, and also in associated inferior areas of premotor cortex, where complex articulatory movements are prepared (see, e.g., [48.6]). One influential idea is that phonemes correspond to planned articulatory gestures, i.e., planned movements of the lips and tongue to particular goal locations [48.7]. A discrete repertoire of phonemes can then be understood as a discrete set of articulatory plans. When an agent is speaking, there is a premium on speed, causing speakers

to overlap their articulatory gestures; for instance, when we produce the word *hot*, we do not wait until we finish producing the *h* sound before shaping our mouth for the *o* sound. Often gestures succeed each other so rapidly that some gestures are left incomplete, for instance, in the gesture producing the *t* phoneme in the phrase *hot porridge*, the tongue does not reach the same point it reaches in the word *light*. Nonetheless, the articulatory *intention* is the same in both cases. The hypothesis that phonemes correspond to planned or intended articulatory gestures is still subject to debate, but it can account for a good deal of the complexity of the mapping between phonemes and acoustic signals.

Of course, speakers can only overlap articulatory gestures to the extent that hearers can recover the intended sequence of gestures. A speaker must be able to *transmit* a phoneme sequence to a hearer in order for communication to take place. If phonemes are intended articulatory gestures, the hearer must be able to recover the speaker's articulatory intentions from the speech signal. Coarticulation means that the mapping cannot be expressed at the level of single phonemes: rather, hearers must learn to recognize the acoustic correlates of short *sequences* of phonemes. We must, therefore, envisage a neural circuit which learns to map acoustic signals onto planned sequences of articulatory gestures. This circuit obviously begins in the auditory cortex; there is evidence that it continues in the superior temporal sulcus (STS), by which time input speech signals are already represented in a more discrete phonological code (see, e.g., [48.8]).

How does the circuitry in between auditory cortex and STS learn to map from acoustic signals to phonemes? We know that this process happens during infancy (see again [48.5]). The standard proposal is that infants teach themselves the mapping by *babbling*. According to this proposal, infants produce articulatory gestures at random, generating reafferent acoustic signals which they perceive through audition, and learn to map these auditory signals back onto the motor plans which produced them. This idea requires that efferent copies of articulatory plans are available to the circuitry in STS, i.e., that there are strong connections between the articulatory system and the auditory system. Such connections have indeed been found. There are direct connections between STS and the inferior motor/premotor cortices, via a tract of white matter called the *arcuate fasciculus*, but also strong indirect connections between these two areas via a way-station in the inferior parietal cortex [48.9]. These tracts are better developed in humans than in chimpanzees or macaques [48.10], especially in the left hemisphere, which is specialized for language. In mature human speakers, there is good experimental evidence that auditory speech signals evoke articulatory representations in premotor cortex. For instance, *Wilson* et al. [48.11] found using fMRI that hearing a nonsense syllable activates the same premotor area activated when that syllable is pronounced. *Fadiga* et al. [48.12] found that in a condition when hearers received transcranial magnetic stimulation (TMS) over their motor cortex to amplify prepared motor commands, hearing words whose pronunciation required large tongue movements elicited more activity in tongue muscles than hearing words whose pronunciation required smaller tongue movements. These findings suggest that hearers do map speech sounds onto their associated articulatory movements. Additionally, *Ito* et al. [48.13] found that stretching the skin around a hearer's mouth to mimic the shape it assumes when pronouncing a particular vowel sound modulated the hearer's phonological interpretation of an acoustic stimulus presented in synchrony with the motor deformation. This finding is evidence that there is also a route from the articulatory system back to the auditory system. Corroborating evidence for this comes from an fMRI study by *Hickok* et al. [48.14], finding that speakers activate auditory representations in STS even when they cannot hear the sound of their own voice. In summary, we know something about the neural circuitry which allows infants to learn the discrete phonological representations which capture invariances between spoken and heard speech sounds. Interestingly, the circuitry implicated in learning to recognize the articulatory actions of others overlaps very significantly with the circuitry thought to be involved in the perception of other types of motor action, which also runs from primary sensory cortices through the STS, and then through the inferior parietal cortex to the premo-

tor cortex (see, e.g., [48.15]). The prevalent accounts of how this more general action-recognition circuitry is trained also assume a role for something like babbling [48.16–18].

Returning to the domain of speech, there are several connectionist models of the circuitry which learns associations between acoustic stimuli and articulatory plans (see, e.g., [48.19–21]). In these models, there are separate neural media representing articulatory plans and acoustic stimuli, which are linked by synaptic connections. During a simulated babbling phase, associations are learned between articulatory and acoustic representations using some variant of the Hebbian rule which associates units which are temporally correlated. These connections serve slightly different purposes at different points in development. To begin with, connections from articulatory to acoustic representations provide a training signal to allow a mapping to be learned, while in the mature system, these connections also serve to create auditory targets for articulatory gestures, which can be compared against the actual acoustic signals produced by articulations, to provide an error signal which can be corrected if necessary.

As already noted, the mapping between phonemes and acoustic signals must be expressed at the level of short *sequences* of phonemes rather than (or as well as) at the level of individual phonemes. (Indeed, the connectionist models just described operate at the level of phoneme sequences.) We, therefore, expect a distinct level of phonological representation in the brain, which groups phonemes into short sequences. Again, the linguistic discipline of phonology tells us that there is such a representation: phonemes are organized into *syllables*. The syllable can be defined in several ways, which roughly converge. In one definition it is a rhythmic unit of speech: a stream of speech consists of a stream of syllables, delivered in a roughly periodic manner. In another definition it is an acoustic unit, organized around a high-sonority sound (a vowel). A syllable consists of an *onset* followed by a *rhyme*, each of which is a short phoneme sequence. The onset is a (possibly empty) sequence of consonants; the rhyme is a vowel, followed by another (possibly empty) sequence of consonants. There is some evidence that syllables are the level at which articulatory and acoustic representations communicate with one another; see e.g., [48.22].

Figure 48.1 summarizes the proposals about the phonological representations introduced in this section. The labels *input syllable* and *output syllable* are somewhat misleading: as we have seen, superior temporal areas are active during output, and premotor areas are

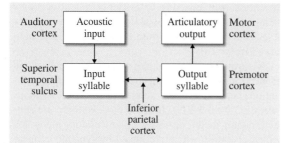

Auditory cortex	Acoustic input		Articulatory output	Motor cortex
Superior temporal sulcus	Input syllable	⟷	Output syllable	Premotor cortex
		Inferior parietal cortex		

Fig. 48.1 Phonological input and output representations

active during input. But the diagram at least roughs out the basic pathways involved in mapping between input and output representations.

48.1.2 Phonological Working Memory

When hearers apprehend a speech signal, the phonological representations they evoke can be maintained for a short time in working memory. There is good evidence that phonological representations are stored in a special-purpose working memory medium, which only interfaces with other cognitive mechanisms in certain well-delineated ways. The medium is called the *phonological buffer*, or the *phonological loop*; see classically [48.23]. The properties of this memory store are typically investigated in experiments where subjects are presented with a phonological stimulus taking the form of a sequence of syllables, which they must reproduce after a short interval. There are several key findings. Firstly, sequences containing syllables which phonologically resemble one another are harder to retain. This means that items in phonological working memory are likely to be active in parallel while they are being retained, in a way which causes similar items to interfere with one another. Secondly, sequences whose syllables form meaningful words are easier to retain. So there must be an interface between representations active in phonological working memory and semantic representations. Thirdly, there is evidence that subjects retain phonological stimuli in working memory by *rehearsing* these stimuli, either vocally or subvocally. If subjects have to perform a distractor task engaging the articulatory system (for instance counting to ten) while retaining a phonological stimulus, their ability to recall it is considerably impaired.

The dominant computational models of phonological working memory envisage a medium in which multiple phonological representations are active in parallel and compete for selection, with competition being biased by a separate mechanism representing a sequence of temporal contexts. For instance, in the model of *Burgess* and *Hitch* [48.24], input and output phonological representations are bidirectionally connected to one another through a *direct pathway* analogous to the arcuate fasciculus, but are also indirectly connected via a set of *item* units, which can be temporarily associated with both input and output phonemes, and also with a temporal context signal which can be played forward deterministically to produce a reference sequence of context representations. A phonological sequence is encoded in working memory by playing the context signal forward while the sequence is presented. At each time point, the current phoneme and the current context representation activate the item units. Item units compete so that only one remains active; this stores a temporary association between a temporal context and a phoneme. Winning item units also self-inhibit after a short time, ensuring that subsequent phonemes are stored by different item units. To recall a phonological sequence, the context signal is replayed, reactivating the item units, and biasing competition towards the production of a particular phoneme at each point in time. The direct connections from output phonemes back to input phonemes allow for a sequence of phonemes to be rehearsed: when a phoneme is produced (either vocally or subvocally), it reactivates its corresponding input phoneme, which can be reassociated with the current item and temporal context.

Burgess and Hitch's model uses single phonemes as its input and output units, but other models feature more structured phonological representations. For instance, in *Hartley* and *Houghton*'s [48.25] model, the phonological units which are sequenced by being associated with items represent whole syllables. Each syllable is stored as an onset and a rhyme unit, each of which in turn can represent multiple phonemes. Onset and rhyme units are directly associated with phonemes, but they are also associated with a template encoding general constraints on how onsets and rhymes can be formed, which help impose an order on the activated phonemes. Many of the errors found in phonological short-term recall reflect the syllabic structure of the stimulus; for instance, ordering errors tend to be of either onsets or rhymes, rather than of arbitrary phonemes (thus *dap, kon* is more likely to be misrecalled as *kap, don* than as *dan, kop*). Hartley and Houghton's model accounts well for these syllable-sized effects. It is also consistent with the suggestion argued for in Sect. 48.1.1, that the direct pathway from input to output phonological representations operates at the level of syllables.

Phonological working memory is thought to play a role in language development. It has been found that the capacity of infants' phonological working memory predicts their vocabulary size [48.26]. Computational models of this effect typically assume that phonological working memory increases the temporal window within which associations between phonological word forms and their meanings can be made. An infant rarely hears the word *dog* at exactly the same moment as she is looking at a dog. Models of phonological working memory, with their assumption that multiple phonological units can be active in parallel, allow a current meaning representation to be associated with several recent phonological forms, not just with the most recent one. Phonological working memory also correlates with syntactic development [48.27]; I will discuss models of this effect in Sect. 48.4.

While there is a direct mapping from input to output phonological representations at the level of syllables, there also appear to be working memory media in which input and output phonological representations are stored separately. Producing phonemes is ultimately a matter of articulation, i.e., of motor control. The motor system is hierarchically organized; like any high-level movements, complex articulatory movements need to be planned in advance, and there is likely to be a motor medium in which planned complex articulations are represented. Indeed, there is evidence for such a medium; for instance *Shallice* et al. [48.28] describe a patient whose working memory for digits is relatively normal, but who nonetheless has difficulty sequencing phonemes when speaking. The patient makes the same kinds of error whether repeating phonologically complex spoken words, reading complex words, or naming objects; since these tasks have different inputs, the problem appears to be in the output modality. At the same time, the patient does not have low-level articulatory difficulties; the problem is with planning complex phonological structures. Shallice et al. call the planning medium which appears to be damaged the *phonological output buffer*. To account for this pattern of dysfunction, we should probably assume that the phonemes (or syllables) which are produced as output by Burgess and Hitch's model are not directly pronounced, but provide input to a separate, specifically articulatory, phonological output buffer, which is also accessed by other tasks such as speaking or picture naming. Other patients display a converse pattern of dysfunction, showing serious impairments in phonological short-term memory, but intact abilities to generate speech and name objects (see, e.g., [48.29]). These

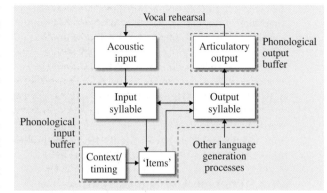

Fig. 48.2 A modified version of Burgess and Hitch's [48.24] model of phonological working memory

patients are diagnosed with damage to a *phonological input buffer*, which stores incoming phonological sequences and supports their repetition syllable by syllable. Burgess and Hitch's original network, or the modification of it proposed by Hartley and Houghton, are good models of the damaged circuitry in these patients.

A modified version of *Burgess* and *Hitch*'s [48.24] model of phonological working memory, updated to encode whole syllables rather than single phonemes, and extended with a separate phonological output buffer, is shown in Fig. 48.2. Note that the *direct* connection between input and output phonological representations hypothesized in Sect. 48.1.1 features in this diagram: it is the bidirectional link between input and output syllables. This is also the link through which subvocal repetition happens, which allows pre-articulatory outputs to reactivate phonological inputs. There is also an *indirect* link from phonological inputs to outputs through the *item* units, which could correspond to the indirect link between superior temporal and articulatory areas via the inferior parietal cortex. Whether the direct and indirect links between input and output phonology correspond to the arcuate fasciculus and the indirect pathway through inferior parietal cortex is not yet clear, but the inferior parietal cortex is certainly implicated in many studies of phonological short-term memory (see, e.g., [48.30, 31]).

A final point to note is that recall of a phonological sequence is sensitive to the way syllables are grouped when they are presented. For instance, if syllables are presented rhythmically in groups, the first and last syllables in each group are better recalled than the middle syllables (see [48.32] for a review). *Burgess and Hitch* [48.24] model this finding by assuming that the context representation is two-dimensional, with one dimension representing position of a syllable within a group and another representing serial position of the whole group. *Item* units now store the position of a syllables in a group, as well as the position of a group within a sequence of groups.

48.2 Lexical Representations and Processes

Words have phonological, semantic and syntactic aspects. In this section, I will discuss these aspects in turn.

48.2.1 Phonological Representations of Whole Words

The phonological structure of a word consists – for the main part – of a sequence of syllables. Word-sized phonological representations are basically representations of sequences of syllables which occur with particularly high frequency. In the input modality, word forms are regularities in the exposure language which infants become attuned to very early: 8-month-old infants can identify statistical regularities in syllable sequences after only a short period of *training* [48.33], long before they can reliably associate word forms with meanings. In the output modality, word forms are akin to motor schemata: sequences of articulations which are frequently performed, and through practice become compiled into encapsulated motor routines.

Whether we consider input or output modalities, word forms are evoked in some medium of working memory. Since we have already proposed separate input and output media for phonological working memory,

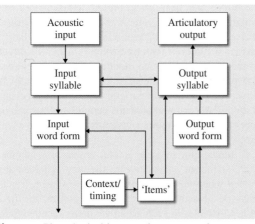

Fig. 48.3 Phonological input and output word representations

we expect to find separate input and output phonological representations of words. And indeed this seems to be the case. For instance, a patient described by *Auerbach* et al. [48.34] can produce words normally, and can understand written words, but cannot understand spoken words. Conversely, the patient described by *Shallice* et al. [48.28] can understand spoken words, but has difficulty producing words, especially when they are phonologically complex: this patient lacks word-sized phonological output representations. Figure 48.3 shows a model of phonological working memory extended with representations of whole word forms. In the remainder of this section, I will discuss some of the influential computational models of word forms and how they are accessed, in both the input and output modalities, and refer back to this figure.

Phonological Input Word Representations
One way of modeling phonological word representations in the input modality is to extend the model of the phonological input buffer by adding long-term synaptic connections encoding phonological patterns which occur particularly frequently. For instance, *Burgess* and *Hitch* [48.24] include long-term synaptic connections throughout their model to capture such common patterns. Note that their use of a two-dimensional context representation also allows the representation of multisyllable words. A multisyllable word has the same form as a rhythmically presented group of syllables in a working memory experiment. If we follow *Burgess* and *Hitch*'s [48.24] proposal that *item* units can store the position of syllables within a group, then we can think of phonological word representations as patterns of long-term synaptic connection which create a bias in the population of item units towards particular syllable sequences. This would explain why it is easier to retain real words in working memory than nonwords, all other things being equal. In this scheme, an active word representation is a particular pattern of activity in the item layer, denoting a particular syllable sequence. In the context of Fig. 48.3, we can think of these patterns as activating a separate layer of whole word representations, through the link connecting *items*

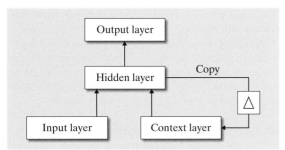

Fig. 48.4 A simple recurrent network

to *input word forms*. Other models of word representation in working memory have a similar flavor (see, e.g., [48.35]).

Another way of modeling phonological word representations in the input modality is to envisage a network which directly receives a stream of phonological inputs and learns to recognize sequential patterns in this stream. (This kind of model is provided for in Fig. 48.3 in the links directly connecting phonological inputs to input word forms.) The standard network for this purpose is a *simple recurrent network* (SRN; [48.36]), which is illustrated in Fig. 48.4. The classical SRN takes a sequence of inputs, and for each of these is trained to predict the next input in the sequence. The network takes a copy of its own hidden layer at the previous time point as an additional input, which allows it to condition its predictions on an exponentially decaying representation of past inputs, as well as on the current input. To configure a SRN for representing whole word forms, it is typically trained to generate a static representation as output for each word. Thus for each training word, we present a particular sequence of phonological input signals, and train the network to predict an unchanging output representation which declaratively represents that sequence. A well-known example of this kind of network is that of *Gaskell* and *Marslen-Wilson* [48.37]. Networks with this recurrent design have the interesting property of incrementality; at each point during the presentation of a phonological sequence, they predictively activate all the whole word form representations which start with the sequence provided so far. This mirrors experimental findings about the priming effects of partially-presented words (see, e.g., [48.38]). Gaskell and Marslen-Wilson's model is actually restricted to monosyllabic words; it takes as input fairly low-level phonological features, and generates as output the kind of static syllable representation proposed by Hartley and Houghton. So it could conceivably be a model of the circuitry which learns to

map from acoustic inputs to syllable structures (if syllables are thought of as copies of static articulatory plans passed back from the phonological output system. But SRNs have also been used to model the phonology of multisyllable words. For instance, *Sibley* et al.'s [48.39] recurrent network maps sequences of input phonemes straight onto static word forms, learning a large vocabulary of mainly polysyllabic words. Whatever their precise inputs and outputs, SRN-based models also capture the idea that a speech signal can activate multiple wordforms in parallel, in a medium in which they compete with one another. Broadly speaking, the recurrent circuitry in SRN models plays the same role as the additional temporal context dimension representing serial position within a chunk in Baddeley and Hitch's model.

A final interesting technique used in representing input phonological word forms is a *self-organizing map* (SOM; [48.40]). A SOM receives training inputs in some high-dimensional space, and learns to represent them in a lower dimensional (typically 2-D) space, in a way which positions similar inputs close to one another, and which devotes more space to inputs which appear frequently. If a SOM's inputs represent complex phonological sequences associated with whole words, it will learn a representation of word forms which places phonologically similar words close together, and which is biased towards recognition of frequently occurring words. A SOM is like an SRN in that it captures patterns in the input data, but it also has the effect of sharpening these patterns. *Li* et al. [48.41] use a SOM to represent word forms. The inputs to their network are declarative patterns representing sequences of up to three syllables, and encoding the onset, vowel and coda of each syllable separately [48.42]. (The vowel and coda together constitute the *rhyme* component of a syllable.) This encoding effectively bypasses much of the machinery discussed so far, and focusses on word-level encodings of phonology: a practical first step in many neural network applications.

Phonological Output Word Representations

Models of phonological word representations in the output modality are models of word production. The two dominant models in this area are those of *Dell* et al. [48.43] and of *Levelt* [48.44]. These models target slightly different levels: Dell et al.'s model only produces monosyllabic words, while Levelt's produces polysyllabic words as well.

Dell et al.'s model can be thought of as a model of the mapping from output syllables to articulations shown in Fig. 48.3. Monosyllabic word units are directly connected to phonemes, which are arranged in three banks: onsets, vowels and codas. A word like *cat* is connected to *c* in the onset units, *a* in the vowel units and *t* in the coda units. (Each consonant is represented twice, once as an onset and once as a coda.) Importantly, these connections are bidirectional, so that when word units activate, the phoneme units they activate also activate phonologically similar words. This allows the model to simulate phonological errors made by speakers, where they produce words other than the one intended: these errors tend to take the form of phonologically similar words. (Note that this model can probably also account for lexical effects in working memory tasks, although it was not designed for this purpose.)

Levelt's model of word forms maps polysyllabic words onto low-level phonological outputs. In relation to Fig. 48.3, this model encompasses all the stages from output word forms to articulatory outputs. Importantly, syllabification occurs at a fairly late point during this process. A word-form is associated directly with a sequence of phonemes, represented as a set of positions; for instance the word *escort* activates the phoneme *e* in position 1, the phoneme *s* in position 2, and so on. Particular subsequences of phonemes in turn activate particular syllables. The rationale for this structure is that syllables do not respect word boundaries; for instance, pronunciation of the words *escort us* creates the syllables *es-cor-tus*. There are many ways in which phonemes could be associated with particular serial positions in a sequence; for instance, we could use Burgess and Hitch's idea of a dynamically changing temporal context representation, together with item units associating contexts with phonemes, or we could use a SRN which maps a static word-form onto a sequence of phonemes. (For instance, *Dell* et al.'s earlier [48.45] model of word production takes the form of an SRN.) But the choice of representation has consequences on the kinds of psycholinguistic results the model can simulate.

Aside from its extension to polysyllabic words, there are several other differences between *Levelt*'s [48.44] model and that of Dell et al. [48.43]. One important difference is that Dell et al.'s model is geared towards simulating the kinds of errors which speakers make, while Levelt's model is geared towards accurately modeling the time course of speech production. Another difference – which relates to this difference in emphasis – is that Dell et al.'s model envisages feedback from phonemes to word forms, while *Levelt*'s model does not. *Dell* et al. [48.46] give a detailed discussion of these differences.

Note that while there is a direct pathway from input to output syllables in Fig. 48.3, there is no direct pathway from input to output word forms; the equivalences between phonological input and outputs are assumed to be expressed at a level lower than whole words.

48.2.2 Neural Representations of Word Meanings

A child learns to associate phonological word representations with meanings. What do these meaning representations look like? Linguistics can tell us something about them. For instance, in logical models of natural language semantics in the tradition of *Tarski* [48.47], word meanings are defined in relation to the role they play in the meanings of sentences. Thus, knowing the meaning of the verb *chase* involves being able to identify all the episodes in the world in which one thing chases another. In more recent empiricist models, the meaning of a word is defined statistically, by a vector of its co-occurrences with other words (see, e.g., [48.48]). These models can learn surprisingly intuitive and fine-grained semantic classifications of words from naturally-occurring text corpora; words with similar meanings are assigned similar representations. In fact, these statistically-defined meaning representations are learned naturally by the hidden layer of an Elman network, if it is given words from a text corpus as input, and trained to predict the next word (see [48.36] and much subsequent work).

Logical and empiricist conceptions of word meaning share the idea that meanings of words are tightly interconnected: the representation of one word's meaning makes reference to those of other words. However, word meanings also have to be *grounded*: at some point, we have to be able to tie word meanings to sensorimotor experience of the world. Researchers investigating neural representations of word meanings tend to focus on the meanings of concrete words. The main result from these studies is that concrete words activate sensorimotor areas of the brain – the same sensorimotor areas which are evoked by direct experience of the concepts they represent. For instance, concrete words with strong visual associations but weak motor associations (e.g., nouns denoting animals) tend to activate visual cortices, while words with the opposite associations (e.g., action verbs) tend to activate motor, premotor, and prefrontal cortices [48.49]. This is perhaps not surprising, but it is interesting that these sensorimotor responses are elicited very soon after presentation of a word, as early as 200 ms. Another study has shown that action verbs denoting actions of the mouth, arm, and leg activate the areas of motor and premotor cortex associated with these different effectors [48.50, 51], again within a very short interval of word presentation.

One common theme in all these studies is that each concrete word activates a widespread cortical region, encompassing sensorimotor areas but also many other areas. Word-sized semantic representations are not localized in any given area, but distributed through cortex – for concrete words, there is a bias towards the sensorimotor areas most associated with direct experience of the denoted concept. An interesting study bearing out this idea comes from Mitchell et al. [48.52]. This study makes use of the vector-based word meaning representations which can be derived from word co-occurrence statistics (discussed earlier in this section). Mitchell et al. found that these highly distributed word representations could be mapped to the distributed patterns of brain activity evoked in subjects reading these words. They recorded the fMRI signals associated with each word in a set of concrete nouns, and used these signals to train a function mapping the vector-based representation of a noun onto its associated fMRI signal. They found that this function could quite accurately predict the fMRI signal of an *unseen* word from its vector representation. These fMRI signals extended across the whole brain, but words with associations to particular sensory or motor modalities had predictable activations in the neural areas representing these modalities. Note that all words were of the same grammatical category (nouns), and that the vector-based input representations abstract completely away from the phonology of words, so the predicted fMRI activity really reflects the meanings of words, rather than their phonological form or grammatical class.

In summary: we know a little about the form which neural representations of word meanings take, at least in the case of concrete words. The meaning of a concrete word is represented by a highly distributed pattern of neural activity, which includes activity in the sensorimotor areas activated by direct experience of the denoted concept, but also activity in other cortical areas. The distributed nature of word meanings makes them hard to study; however, the technique of *Mitchell* et al. [48.52], using distributed meaning representations derived from co-occurrence statistics, may provide one way of making progress. In the diagrams in the remainder of this chapter, for clarity's sake, I will depict word

meanings as if they occupy a well-defined neural area of their own.

48.2.3 Mappings Between Word Forms and Word Meanings

I have described the neural representations of word forms, and of word meanings; I will now consider the circuitry which connects these two types of representation, and implements a mapping between word forms and word meanings. Figure 48.5 provides a diagram to refer to in this discussion.

As the figure shows, models of word interpretation and word production are not entirely symmetrical as regards this mapping. Models of interpretation often assume that word meanings are evoked directly from phonological inputs; for instance, *Gaskell* and *Marslen-Wilson*'s [48.37] incremental model of word interpretation discussed earlier learns not only to predict whole word forms from a phonological input sequence, but also whole word meanings (see the dotted lines in Fig. 48.5). On the other hand, during language production, there is often claimed to be an intermediate unit called a *lemma* intervening between word form and word meaning representations, as shown on the right-hand side of Fig. 48.5 (see especially [48.44]). A lemma is a linguistic representation of a word, which abstracts away from its form. *Levelt*'s model of word production has three stages: first a word-sized semantic concept is activated; this in turn activates a lemma; and the lemma activates the form of the word. The three

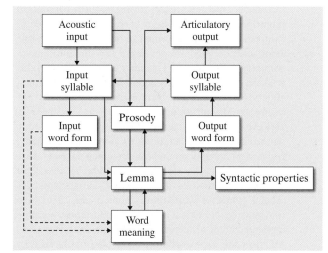

Fig. 48.5 The mappings between word forms/prosody and word meanings

stages are delineated by the dotted lines on the right of Fig. 48.5.

A lemma is an associative representation – something akin to a *convergence zone* in the parlance of *Damasio* and *Damasio* [48.53]. A lemma represents a particular word by connecting independently to a word meaning, and to a word form (a phonological sequence) and a *prosodic template* (a specification of which of the word's syllables are stressed). It also connects to a set of syntactic properties, indicating the word's part of speech, as well as properties such as gender for nouns, or tense for verbs. These syntactic associations are particularly close, being activated at the same stage of the word production pipeline as the lemma itself. Levelt gives several reasons for postulating this intermediary representation. One is the existence of *tip-of-the-tongue* states, in which a speaker has a particular word in mind, but cannot access its form (but can sometimes access its prosody). But it would be possible to obtain this effect even if word-sized semantic representations linked directly to output word forms, so it is not strong evidence. Better evidence for lemmas as an independent stratum comes from van *Turennout* et al. [48.54], who found in an ERP study that speakers can access syntactic properties of words (specifically gender) without activating word forms, but that the reverse is not possible.

Are lemmas specific to word production system, or do they also participate in the process of word interpretation? *Levelt* proposes that lemmas are also activated during word interpretation, in the same kind of incremental fashion proposed by Dell et al. for word forms and word meanings. This claim amounts to a claim that there are three routes from linguistic inputs to linguistic outputs: one using a sublexical phonological medium (connecting input to output syllables), one using a lexical medium (lemmas), and one through semantic representations. This idea is somewhat at odds with the neuropsychology and neuroimaging literature, which tends envisage a phonological route and a semantic route, but not a lexical one (see, e.g., [48.8]). However, there is some evidence for a route through lemma units. Levelt refers to a classic experiment by *Glaser* and *Düngelhoff* [48.55], in which subjects were asked to name a pictured object, on which a distractor word is superimposed. It was found that semantically related distractor words have more of an effect than semantically unrelated words; for instance, naming a pictured chair is delayed more by the distractor word *bed* than the distractor word *fish*. One might wonder why the effect is inhibitory rather than excitatory, given that the associative relationships between the concepts

chair and *bed* are presumably of equal strength. Levelt's proposal is that a distractor word can *directly* activate an output lemma (bed) to compete with the required output lemma (chair), while its facilitatory effect through associated semantic concepts is indirect. Indeed, if the experiment is set up so that the distractor word is not one of the possible words to be spoken, its net effect is facilitatory. These experiments provide further evidence for an independent layer of lemmas, and moreover, for Levelt's suggestion that the same lemmas participate in word interpretation and word production.

Considering how neural circuits are formed, it seems likely that lemmas have some existence as a representational medium in their own right. Connections between neural regions are profligate; if associations are learned between several distinct neural assemblies, it is likely that some additional population of neurons comes to participate in the resulting grouping, because it happens to be independently connected to each of these assemblies. In other words, postulating a layer of lemmas does not exclude the possibility of direct links between meanings and word forms.

Where in the brain is the circuitry which connects word forms and word meanings? Given that word meanings are widely distributed, we should expect this circuitry to be similarly widely distributed – however, there is evidence that the temporal cortex is particularly important in implementing these mappings, in both directions. Damage to the left superior temporal gyrus (somewhat informally termed *Wernicke's area*) often causes difficulty in understanding the meaning of words. It can also cause difficulty in producing meaningful words – in the classic (but rather rare) condition termed *Wernicke's aphasia*, patients produce sentences which are syntactically and phonologically well-formed, but whose words lack meaning (see, e.g., [48.56]). Other areas of temporal cortex are also implicated in the mapping between word forms and word meanings. For instance, *Damasio* et al. [48.57] found that damage to different areas of left temporal cortex correlated with deficits in naming different semantic types of object: damage to anterior, inferior, and posterior left temporal cortex correlated with impairments in naming people, animals, and tools, respectively. *Lu* et al. [48.58] found that damage to anterior temporal cortex correlated with impairments in naming actions.

Note that the circuitry which maps input word forms onto meanings must be somewhat distinct from that which maps meanings onto output word forms, according to our model of input and output word forms. We expect some overlap, to the extent that input and output syllable representations connect with one another. But there is still good evidence that input and output meaning-form mappings are somewhat distinct. For instance, *Kay* and *Ellis* [48.59] describe a patient who can understand concrete words, but has difficulty generating these same words in a picture naming task. This patient seems to have a relatively intact phonological output system, because he can repeat spoken words quite well – even those words he cannot spontaneously generate. This patient's deficit appears to be quite specific to the circuitry which maps semantic representations onto output word forms. *Auerbach* et al.'s [48.34] patient described earlier has the converse condition: he can produce words normally, but cannot understand spoken words. His deficit could be in the circuitry representing the form of incoming spoken words, but it could equally well be in the circuitry linking these form representations to meaning (or lemma) representations.

48.2.4 Morphology

Words often carry *inflections* signalling their syntactic properties. For instance, in the English sentence *He walks*, the verb *walks* is made up of the word stem *walk*, plus the inflection -*s*, which signals syntactic *agreement* with the subject *he*. *Walk* and -*s* are termed *morphemes*; they are phonological units in the word which carry their own meanings and/or syntactic properties. The branch of linguistics which investigates how words decompose into morphemes is called *morphology*.

Some inflections carry rich semantic information; for instance, tense inflections in English or subject agreement inflections in Romance languages (which carry enough information that pronominal subjects can often be omitted altogether). These inflections are very much like normal words; we can envisage them as having lemmas of their own. But inflections are normally also syntactically obligatory, even when they do not carry much information of their own; the English agreement inflection -*s* is a case in point. I have already briefly touched on the syntactic role of inflections in the discussion of lemmas in Sect. 48.2.3; in the current sec-

tion, I will just consider the nature of the neural circuitry which interprets and generates inflected words.

We must envisage circuitry in the input modality which decomposes an incoming word into its constituent morphemes, and circuitry in the output modality which assembles morphologically complex words. In either case, there is some evidence that frontal circuitry is involved. Patients with damage to left inferior frontal cortex (regions including, but not limited to *Broca's area* comprising left Brodmann areas BA44–45) often have difficulty generating inflected words, and often fail to notice inflection errors in incoming sentences, especially in cases where inflections do not carry much semantic information. There is interesting evidence that the production of inflected verbs involves a region of the left dorsolateral frontal cortex, part of BA9. *Shapiro* et al. [48.60] find that TMS over this area selectively impairs the production of inflected verbs, and *Shapiro* and *Caramazza* [48.61] describe a patient with damage to this same area who has a permanent selective deficit in producing inflected verbs.

The phonological form of inflected words is governed by a complex system of rules and exceptions. For instance, the morpheme -*s* is realized as *s* on the stem *walk* (to form *walks*), but as *es* on the stem *catch* (to form *catches*). Some of the most venerable connectionist models of language processing are models of circuitry which learns these rules from a training corpus of examples: for instance, *Rumelhart* and *McClelland* [48.62] proposed an influential model of the circuitry which learns the past tense form of verbs. Their network learns to map a phonological representation of a verb root onto a phonological representation of the appropriate past tense form, using a supervised training scheme. The network appeared to show a similar developmental profile in the task as real children: early in training it learned past tense forms on a word-by-word basis; later it learned general rules (and over-applied these rules, for instance producing *knowed* as the past tense of *know*); and finally it learned which words to treat as exceptions. While there have been many more sophisticated models of this process, the basic idea that a connectionist network can successfully learn a mixture of rules and exceptions is now very widely accepted.

48.3 Sentence–Sized Semantic Representations

Before we consider how syntactic processing is implemented in the brain, we must first consider how the meanings of sentences are represented. Syntac-

tic processing maps between sentences (represented as sequences of words) and their meanings. We have discussed how the brain represents word sequences in working memory. Before we discuss syntax, we must say something about how the brain represents sentence meanings in working memory, so we can define the computation which syntax performs. (We need to know what syntax *is* before we can look for it in the brain!) In this section, I will consider what we know about *semantic representations* in the brain. My approach will be to use formal theories of natural language semantics as a guide. In Sect. 48.3.1 I will introduce some concepts which have proved useful in formal models of sentence semantics. In Sects. 48.3.2–48.3.3 I will consider the cognitive representations which best correspond to these concepts, drawing mainly on neural theories of working memory and long-term memory. My proposals about cognitive representations of meaning will be broadly in line with those proposed in *cognitive linguistics* (see, e.g., [48.63, 64]), especially recent models which hold that entertaining the meaning of a sentence amounts to simulating the process of experiencing the episode it describes [48.65, 66]. However, the arguments I adduce for this conclusion, drawing on theories of memory, are largely my own. In Sect. 48.3.4 I will discuss various schemes for representing the relationships between concepts which are expressed in sentence meanings.

48.3.1 Formal Models of Sentence and Discourse Semantics

Formal semanticists sometimes refer to a sentence-sized meaning representation as an *episode*. The term *episode* is intended to encompass two ontologically distinct types of semantic entity: on the one hand a *state*, and on the other, an *event*. A state is a proposition which can be classed as true or false at any given single point in time; for instance, MY_CAT_IS_GRAY. An event describes an operation which brings about a change in the world; for instance, MY_CAT_ATE_MY_SANDWICH. In the standard model of semantics devised by *Reichenbach* [48.67] the meaning of any episode must be represented in relation to a *reference time*, which denotes *the time the speaker is currently talking about*. When a hearer assimilates an event, she must *update* the reference time, to accommodate the fact that events change the world: after an event takes place, certain things stop being true, and others start being true. On the other hand, when assimilating a state, the hearer should not update the reference time: states report facts which obtain at the reference time, and thus update our knowledge about this point in time, but not the temporal reference itself.

There are elegant formal theories about the nature of temporal update operations (e.g., [48.68]), and of the kinds of statements about time which natural language sentences can express (see, e.g., [48.69]). There are also elegant theories about how interpreting a sentence updates a hearer's knowledge of the world (see, e.g., [48.70, 71]). A key suggestion in both these bodies of theory is that the meaning of a sentence is best thought of not as a direct representation of some small part of the world, but as a *function*, which takes a representation of the world as input, and delivers a new representation as output. If the sentence denotes a state, the new representation provides new information about the current reference time; if it denotes an event, the new representation updates the reference time and provides information about the situation which obtains at the new time. The representations which are updated are often termed *discourse contexts*. These representations are rich, containing far more information than single sentences. Formally, a discourse context contains a set of *discourse referents* which are available to talk about, and a structure of *conditions*, which represent the sentence-sized things which have been said about these referents in the discourse so far. There may be many referents, and many conditions.

What are the neural equivalents of states, events, reference times, discourse contexts, referents and conditions? There are some very natural neural correlates of these concepts, which draw on models of human memory, as I will describe in the next two sections.

48.3.2 Episodic Long-Term Memory Representations

A good starting point is *Tulving*'s [48.72] well-known model of *episodic memory*. An agent's episodic memory is her memory for specific episodes she has experienced in her own life, either directly through perception, or indirectly through stories, films, and so on. Episodic memories are tied to particular places, and tend to have well-defined temporal structure: they take the form of narratives, in which one event leads to a subsequent event, and they are naturally recalled by recreating this structure. Computational models of episodic memory assume that this form of memory makes use of neural *context* representations which denote particular places and/or moments in time. We know something about these context representations; for instance, there is good evidence that spatial context representations are stored in the parahippocampal cortex, both in rodents [48.73, 74] and in humans [48.75,

76], and that temporal contexts are also stored in this parahippocampal area [48.77, 78]. Hippocampal representations of temporal contexts are a natural analog of Reichenbach's notion of temporal indeces, and the episodic memories which are linked to these contexts are natural analogs of the sentence-sized states which obtain at particular temporal indeces, or the events which transition between temporal indeces.

When a hearer interprets a narrative delivered in natural language, she ultimately encodes this narrative in episodic memory. Episodic memory is a form of *long-term memory*; if the hearer is asked about the events in the narrative some time later, she can query her episodic memory to retrieve the answers. In the standard model of episodic memory storage (see, e.g., [48.79, 80]), an episode is initially stored in the hippocampus and neighboring cortical areas, where it is associated with a spatiotemporal context representation to individuate it. Gradually, this hippocampal episode representation is transferred to (even) longer-term storage in a more widespread cortical network, in a process called *consolidation*.

48.3.3 Working Memory Episode and Context Representations

Of course, it is also important to consider how the meanings of the sentences in a narrative find their way *into* hippocampal storage in the first place. The standard proposal here is that in order to store an episode in the hippocampus, it must first be represented in *working memory*. There is good evidence that if an episode is actively elaborated in working memory, it is recalled better in the long term (see [48.81] and many subsequent studies). We have already discussed working memory for phonological sequences (see Sect. 48.1.2). However, the working memory medium in which episode representations are maintained is quite distinct from this phonological storage medium. It is semantic in nature and associated primarily with activity in the prefrontal cortex (PFC). Evidence for prefrontal involvement in the process of encoding episodic memories comes from several sources. For instance, both ERP and fMRI studies have shown that the level of PFC activity at the time a stimulus is encoded predicts how well it will later be retrieved [48.82, 83]. There is also evidence that material retrieved from long-term memory is re-established in prefrontal working memory (see [48.84] for a review).

An interesting recent proposal by *Baddeley* [48.85] is that the working memory medium which com-

municates with hippocampal long-term memory is specialized for holding semantic episode representations. Baddeley calls this working memory medium the *episodic buffer* and gives it a role in sentence processing. I will argue that the episodic buffer is the cognitive medium in which sentence meanings (*events* and *states*, in semanticists' terms) are maintained in working memory. This idea makes intuitive sense, but it also fits well with a broader model of updates in working memory, which I will outline below. Recall that semanticists think of events and states as update operations, which operate on rich, multifaceted *discourse context* representations. If we want to think of the episodic buffer as the medium which holds events and states, we must also propose some analog of discourse context representations and suggest how material in the episodic buffer can function to update these representations.

I will start by considering the analog of discourse context representations. To begin with, note that the episodic buffer is not the only semantic representation in working memory. There is also a more holistic working memory representation known as *cognitive set*, which Baddeley is at pains to distinguish from the episodic buffer. An agent's cognitive set specifies the rules which govern her actions at the current moment; formally, it defines a mapping from the stimuli she perceives to the actions she performs. The consensus is that cognitive set is also implemented by activity in the prefrontal cortex – see *Miller* [48.86] for a good articulation of this idea. Cognitive set can be thought of as an action-based representation of the agent's current situation – in the terminology of *Gibson* [48.87], we can think of it as an encoding of the actions which the situation *affords*. If we define actions broadly enough to include actions of attention to objects, cognitive set can also encompass the agent's working memory for the salient objects in the current situation. This kind of memory is also known to be prefrontal: there is evidence that working memory representations of task-relevant objects are maintained in PFC, both in monkeys (e.g., [48.88, 89]) and in humans (e.g., [48.90, 91]). I suggest that an agent's cognitive set is a good cognitive analog of semanticists' notion of *the current discourse context*. A cognitive set includes representations of recently encountered objects, and of the relevant properties of these objects. It does not represent these properties as explicitly as the discourse contexts of semanticists; they are not enumerated as conditions, but rather compiled into a rich set of action affordances. But their purpose is ultimately the same: semanticists envisage reasoning procedures which draw the kinds of conclusion about the current situation and the best actions to perform which are compiled into the cognitive set.

Just as semanticists talk about updates to the current discourse context, behavioral psychologists talk about updates to an agent's cognitive set. One of the interesting properties of cognitive set is that it can be changed suddenly (at least in humans and higher mammals). Updates come from new information about the environment. Often this information makes only minor changes to cognitive set. If we think of an agent's *set* as representing her current task or focus, many episodes she experiences have little or no impact on her cognitive set: one of the points of a task representation is to help the agent ignore incoming episodes which are not currently relevant. But sometimes, incoming episodes are important enough to update or change the agent's current goals. The interesting thing about these updates is that they are relatively fast and discrete: humans are very good at switching from one cognitive set to another (see [48.86] for a good model of this switching process). The fact that cognitive set supports update operations emphasizes its suitability as the cognitive analog of semanticists' notion of discourse context.

Now I will consider the relationship between cognitive set and the episodic buffer. If the episodic buffer holds events or states, which semanticists model as operations updating the current discourse context, and if cognitive set represents *the current discourse context*, we expect the contents of the episodic buffer to be implicated in bringing about updates to the cognitive set. More precisely, we expect the episodic buffer to implement a *function* mapping the agent's current cognitive set onto an updated set. A function implements a particular mapping from inputs to outputs, but it also has a domain of definition: it creates a new cognitive set *given the current set*. As regards the first of these roles, there are good accounts of the mechanisms through which current experiences update the PFC-based cognitive set; see in particular *Braver* and *Cohen* [48.92]. The mechanisms involve a special kind of reinforcement learning: an agent learns to update PFC in ways which have beneficial consequences some time later. As regards the second role, we must explain how an agent's current cognitive set influences or determines the material coming into the episodic buffer. If we think of an agent's cognitive set in Gibsonian terms, as the set of actions which are afforded in the current situation, then one natural way to think about the episodic buffer is as the medium in which these actions compete to be selected. This makes sense of the idea that the episodic buffer holds *just one episode*, but also shows how this one episode relates to the agent's wider representation of the current situation. Note that once an action is selected in the episodic buffer, it is also *executed*; it stops being a potential action, and becomes an actual one. In an agent interacting with the world, activating a representation in the episodic buffer actually changes the current situation, or, in the case of a perceptual action, the agent's knowledge about this situation. In either case, the change triggers a change in the agent's cognitive set.

In summary, many of the key concepts in a formal model of sentence semantics have natural correlates in a theory of working memory. Entertaining the meaning of a sentence is analogous to evoking an episode representation in the episodic buffer. Activating a particular representation of the current discourse context is analogous to adopting a particular cognitive set. If we accept these analogies, the proposal from formal semantics that sentence meanings are context update functions fits well with an account of how working memory representations bring about behavior in the world.

Note that the notion of a context update can also be connected to the account of long-term memory described in Sect. 48.3.2. We have to envisage particular points during experience of the world where information in the episodic buffer is transferred to longer-term hippocampal storage. Since there are representations of context in both working memory and episodic long-term memory, these transfer points should coincide with points at which working memory context is updated. If we consider a particular situation, where a particular episode is evoked in an agent's episodic buffer, we should envisage two operations: one which updates the agent's current cognitive set and another which encodes the update operation in episodic memory (perhaps updating the hippocampal context representation in the process). This second operation should obviously happen before the episodic buffer is cleared to set the stage for a decision about the next action.

48.3.4 Binding Mechanisms in Episode Representations

Having given a high-level picture of the semantic memory structures involved in representing a sentence, I will now consider the details of individual episode representations. For concreteness, I will focus on the format of episode representations in the episodic buffer.

An episode representation must combine several different conceptual elements: for instance a state like

MY_CAT_IS_GRAY combines at least the concepts ME, POSSESSION, CAT, PRESENT, GRAY (and possibly others too); an event like JOHN_JUMPED combines at least JOHN, JUMP, and PAST. Of course, an episode representation is more than just a bag of concepts: its component concepts are combined together in well-specified ways (which are reflected in the syntax of sentences which express it). A well-established idea in linguistics is that any episode representation takes the form of a set of role-filler pairs, identifying the type of the episode, and the roles played by the objects or entities which participate in it (see, e.g., [48.93] for a good presentation of this idea). So, for instance, we might represent JOHN_CHASED_MARY by associating the role ACTION_TYPE with the filler CHASE; actions of this type require AGENT and PATIENT participant roles, which in our case are filled by JOHN and MARY, respectively. There is a good consensus that we need a scheme of this kind to represent sentence meanings. A key question for computational neurolinguistics is how such a scheme can be implemented in neural circuitry. In computational terms, what is needed is a mechanism for expressing *binding* relationships between roles and fillers. This mechanism must be able to flexibly bind fillers to roles, and to express a fairly large number of binding relationships. Moreover, there must be some degree of hierarchy in the representation of fillers, which allows fillers to be units which are themselves specified by role-filler pairs. (For instance, in the episode MY_CAT_JUMPED, the AGENT role is occupied by an object whose TYPE is CAT, and whose POSSESSOR is ME.) The desiderata for a neural model of binding are well set out by *Jackendoff* [48.93]. There are many proposals about how a binding scheme can be implemented in neural hardware, and as yet no consensus, but I will briefly outline some of the key proposals.

The *binding-by-space* model represents a proposition in a set of units divided into groups, each of which is explicitly associated with a particular role; for instance, the first *n* units might be used to represent the filler of the AGENT role, the next *n* units to represent the filler of the TYPE role and so on. This model is something of a straw man. It requires each concept to be represented multiple times; for instance, we would need separate representations of CAT to fill AGENT and PATIENT roles. This is not just representationally expensive (there are lots of roles to fill, especially in nested propositions), it also fails to capture the commonalities between concepts associated with different roles. (A cat participating in an episode as an agent has much in common with a cat participating as a patient.)

The *binding-by-synchrony* model (see, e.g., [48.94]) makes reference to the periodic patterns of activity found in neural populations (revealed, for instance, in EEG signals). Different periodic patterns are found in different brain regions at different times; binding-by-synchrony typically refers to the theta cycle, which has a frequency of 4–8 Hz. The model is motivated from the observation that populations of neurons often fire at a particular phase in the brain's theta cycle. It proposes that roles and fillers are bound together by their activation at a particular phase of the theta cycle. This allows a single CAT concept to participate both as an AGENT and as a PATIENT. While there is now good consensus that precise timing of neural spikes is important (see, e.g., [48.95]), the idea of binding by synchrony has many problems; some of these are summarized by *Shadlen* and *Movshon* [48.96]. In any case, there are not enough phases in a theta cycle to hold the number of bindings required in representing the meaning of even moderately complex sentences.

The *binding-by-connection* model envisages that there are synaptic links between each role and each possible filler for that role, and that associating a filler with a particular role can be achieved by temporarily potentiating the appropriate connection, so that activating the filler activates the role and vice versa. This model also allows a single concept to be associated with different roles. The model has problems expressing the semantics of nested sentences featuring multiple copies of some roles – for instance, in *The boy chased the girl who ran*, the boy and the girl are both represented as agents (of different actions). But some sentence processing models put the topic of nested sentences to one side and employ binding-by-connection models quite effectively (see, e.g., [48.97]). Other models (e.g., [48.98]) resort to a binding-by-synchrony account to explain how multiple propositions can be stored.

The *neural blackboard* model [48.99] is an extension of the binding-by-connection idea, which envisages a pool of dedicated binding units, each of which is linked to every possible role and every possible filler, and can temporarily store an association between a particular role and filler. Dedicated binding units considerably reduce the number of connections which must be envisaged, and also allow for hierarchical objects to feature as fillers. However, it is unclear whether there are neural assemblies in the brain which have the right properties to function as binding units. There must be a relatively small pool of such units, and each of these must be fully connected to a wide range of other neural media. It is important that each binding unit can hold a connection between every possible role-filler pair – otherwise there will be certain bindings which simply cannot be expressed. What is more, there are some indications that the amount of neural circuitry required to implement binding units does not scale up to realistically sized sets of atomic concepts (see, e.g., [48.100]).

The *binding-by-serial-position* model *Knott* [48.101] makes reference to the proposal that experiencing an episode involves a canonical sequence of sensorimotor operations. For instance, experiencing a man grabbing a cup canonically involves attention to the agent first, then attention to the cup, and then a process of classifying or monitoring the grab action (see also [48.102, 103]). If this is the case, then we can store an episode in working memory as a prepared sensorimotor sequence, in which particular thematic roles are identified by particular serial positions. We know a great deal about how prepared attentional and motor actions are stored in the prefrontal cortex; see, e.g., [48.104, 105]: Knott's model assumes that episodes are held in prefrontal working memory in a similar format. There are limitations of this model: in particular, it does not yet address how nested propositions are stored in working memory. But the idea that entertaining a proposition involves rehearsing a sequence of representations does provide some interesting scope for an account of nested propositions, because it allows that representations of different propositions could be dominant in PFC at different points during this rehearsal process.

Another class of binding models use distributed representations to hold bindings between roles and their fillers. In these models, the representation of a binding relationship is held in a pattern of activation distributed over a whole ensemble of units, in which representations of the individual items being bound are hard to identify. An early example of the distributed approach is that of *McClelland* et al. [48.106], which can be termed *binding by query training*. In this model, a sentence is fed into an SRN-like network one word at a time, to create a *sentence gestalt* on the output layer. This gestalt then forms the input to a second network, which learns to answer questions about the binding of roles and fillers in the encoded sentence. The second network takes two inputs – the sentence gestalt itself and a single role or filler which functions as a query – and is trained to generate the item which is bound to this role or filler as output. Training updates the weights in the recurrent network as well as in the query network, so the recurrent network learns to generate sentence gestalts which support correct answers about the binding relationships between roles and fillers. A more complex model us-

ing a similar technique of query training is presented by *Rohde* [48.107], which supports the encoding of nested propositions. In this model, queries carry richer information: rather than just naming a role or a filler, queries are complete propositions with one missing element.

The *vector symbolic* model of [48.108, 109] capitalizes on the idea that the atomic components of episode representations are expressed in some distributed, high-dimensional feature space, as suggested in Sect. 48.2.2. I will describe Plate's model here. Plate's proposal is that binding two representations together involves performing an operation which takes their associated *n*-dimensional vectors as input and returns a new vector in the same *n*-dimensional space as output. The important property of the operation is that it returns a vector which does not closely resemble either of the representations being bound together; the *n*-dimensional space is assumed to be big enough to be able to hold a large number of atomic concepts and to represent the bindings of arbitrary role-filler pairs as distinguishable vectors. One attractive property of this scheme is that it naturally supports hierarchical binding operations; since the vector representing the result of a binding operation is in the same *n*-dimensional space as the concepts being bound, we can apply the same sort of operation to bind this derived representation with another concept. This binding model scales more naturally to large repertoires of concepts [48.100]. Assessed on its computational properties, it is an attractive model of neural binding. It is very hard to seek evidence for it, since representations are large distributed patterns of activity. One strong prediction is that the distributed representation of the binding of concepts A and B bears no resemblance to the representation of either A or B by themselves. Many studies of PFC-based working memory find that working memory representations of bound concepts tend to include identifiable representations of the individual concepts being bound (see, e.g., [48.88]). But these studies also tend to find that there are cells which encode specific combinations of bound concepts, which are what is predicted by the model. Perhaps better tests could be devised using the methodology of *Mitchell* et al. [48.52], which maps distributed word representations to whole fMRI brain scans.

A final attractive model of neural binding is IN-SOMnet (see, e.g., [48.110]). This scheme combines elements of distributed and blackboard binding models. It draws on a sophisticated symbolic scheme for representing sentence meanings, called *minimal recursion semantics* (*MRS*; [48.111]). In MRS, a sentence meaning is expressed as an unstructured set of *frames*. Each frame has a label or *handle* (h_n) and introduces one semantic element of the sentence, encoding the relation of this element to other semantic elements by binding relevant roles to the handles of other frames. (Thus, for instance, a frame with handle h_0 introducing the action concept HIT might bind the AGENT and PATIENT roles to the handles h_1 and h_2, which identify frames holding information about the agent and patient, respectively.) INSOMnet has a fixed-size set of assemblies encoding prototypical frame structures; for instance, frames introducing transitive and intransitive actions and frames introducing referential elements of different kinds might all be represented by different frame assemblies. Assemblies represent frame structures in a distributed manner; they are subsequently translated into localist representations by a *decoder* network. (The decoder network is constrained to use the same weights to decode each assembly, so similar assemblies are forced to use similar distributed schemes for representing frame structures.) Importantly, while each assembly represents a frame with a particular structure, it can represent many different token assemblies with this structure. (For instance, the same assembly might be able to represent a hit or a grasp action concept, because these concepts have a similar structure.) The front-end

of INSOMnet is a fairly standard SRN, which takes a sequence of words and produces a pattern of activation over the set of frame assemblies, which are then translated into a set of localist frame representations by the decoder network (aided by a mechanism for selecting particular frames which I will not discuss here). The whole network is trained in a supervised manner; for each sentence, a sequence of words is presented as input and the network is trained to produce the associated set of localist frame representations as output. (An interesting feature of INSOMnet is that it is trained on a very large corpus of sentence-meaning pairs, created automatically by a wide-coverage symbolic sentence interpretation system and postprocessed by hand.) As well as supervised training, the network also uses a self-organizing map to structure its set of assemblies into groups which represent similar frame structures. The trained network is very good at identifying the structure of the set of MRS frames representing unseen sentences, though somewhat less good at identifying the token concepts which fill these structures. Its strength is in its ability to learn a set of frame representations which collectively suffice to model the structure of a large range of real sentences in a semantically sophisticated way.

48.4 Syntactic Representations and Processes

In this section, I consider how syntactic processing is implemented in neural circuitry. I begin in Sect. 48.4.1 by defining the processing which will be under discussion and in Sect. 48.4.2 I give some broad ideas about the neural areas where this processing is implemented. In Sect. 48.4.3 I will discuss the dominant idea in connectionist models of syntax: that syntactic processing is carried out by circuitry in which a simple recurrent network plays an important role. In Sect. 48.4.4 I will discuss some limitations of standard SRNs as models of syntax and describe some ideas about how to extend or modify SRNs to overcome these limitations.

48.4.1 Conceptions of Syntax

The syntax of a natural language can be defined in two ways. On the one hand, we can think of the syntax of a given language as a specification of what counts as a *well-formed* sentence in that language. On the other hand, we can think of the syntax of a language as the principles which determine how the meanings of the words in a well-formed sentence are composed

together to yield the meaning of the whole sentence. These two conceptions of syntax are linked, in that they both require well-formed sentences to be represented as hierarchical structures, rather than as flat lists of words. The number of well-formed sentences in any language is infinite, for all practical purposes. This means that we must model well-formedness by defining a system of rules or principles which sentences must conform to, rather than by enumerating individual sentences. In traditional models of grammar, this system defines a sentence as being composed of *phrases* or groups of words. Phrases are defined recursively as composed of subphrases of their own, a definition terminating in elementary phrases consisting of single words. Phrases are also implicated in the conception of syntax which supports semantic interpretation of sentences; the phrase structure of a sentence is the guide which indicates how the meanings of its words combine together. In traditional models of compositional semantics, each phrase has a semantic interpretation, which is formed by combining the meanings of the phrases from which it is formed. The key question in theoretical syntax is, therefore, how to state the rules or principles which specify how sentences decompose into phrases.

One of the interesting debates concerns to what extent these rules abstract away from the lexical items which actually make up sentences. In one school of thought, some core set of syntactic rules is provided innately: infants arrive in the world *knowing* certain (quite abstract) facts about the syntax of all human languages and only have to learn about the syntactic idiosyncracies of the particular language they are exposed to. This idea is associated with the Chomskyan syntactic tradition (see, e.g., [48.112–114]). In this tradition, the core syntactic rules make no reference to words whatsoever. In another school of thought, infants learn all the syntactic principles of their native language by finding patterns in the sentences they are exposed to; in other words, they learn a more or less complete model of syntax from their exposure language. This idea is associated with constructivist linguists (see, e.g., [48.115, 116]). In this tradition, rules are created by a mechanism which takes concrete sentences as input, and there is nothing to prevent them making reference to specific words. Constructivist theories of grammar are good at modeling *surface* patterns in language, such as those found in idiomatic or semi-idiomatic constructions (e.g., *John let the cat out of the bag*; *Letting the cat out of the bag might be a good idea*), as well as graded notions of grammaticality (e.g., the fact that some sentence structures are *unusual, but not impossible*). They are also good at modeling the trajectory of syntactic development in infants; there is evidence that infants' earliest multiword constructions are structured around specific words, and that it takes some time for properly abstract syntactic constructions to develop (see again [48.116]).

48.4.2 Neural Areas Involved in Syntactic Processing

There is some evidence that the neural circuitry which represents well-formedness of sentences is distinct from that which maps word forms onto word meanings. Patients with damage to Broca's area and surrounding left inferior frontal cortex have difficulty generating word inflections, as noted in Sect. 48.2.4, but also have a wider difficulty generating syntactically well-formed sentences, and also in interpreting sentences, at least if they have complex syntactic structure (see, e.g., [48.117]). At the same time, if damage is limited to these areas, patients' knowledge of word meanings is often quite well preserved. The opposite pattern of damage is associated with damage to left posterior su-

perior temporal cortex, as already noted in Sect. 48.2.3. Some patients with damage in this area produce sentences which are fluent and syntactically well-formed, but which fail to convey any meaning at all; they are also unable to understand single words or to meaningfully produce single words. Damage in these cases is assumed to be to the circuitry which maps between word forms and word meanings, while the circuitry which encodes well-formed sentences remains intact. This classic double dissociation sees syntactic processing implemented in left inferior frontal cortex.

However, more recent work also implicates the left anterior superior temporal cortex in syntactic processing, particularly syntactic interpretation; in fact, syntactic interpretation deficits are more strongly associated with damage to this area than with damage to left frontal areas [48.118]. This area is activated more by syntactically well-formed utterances than by lists of words (see, e.g., [48.119]), and more activated by sentences reporting events than by nonlinguistic auditory stimuli depicting events [48.120]. In summary, there is syntactic circuitry in the brain in the left anterior temporal cortex as well as in the left inferior frontal cortex.

48.4.3 Basic SRN-Based Models of Syntax

In neural models of language, the key question concerns how syntactic knowledge is represented in neural circuitry. Again, we can think of syntax in two ways: either as a specification of well-formed sentences or as the principles which allow word meanings to be composed into full episode representations. I will give examples of models emphasizing each of these conceptions of syntax.

While traditional linguistics represents phrases symbolically, most connectionist models of syntax represent them subsymbolically or implicitly. Nearly all connectionist models have an empiricist flavor: they are set up to learn syntactic rules from training data, rather than hardwired to encode particular rules. The dominant idea in recent connectionist models is that an implicit representation of phrase structure is learned by something akin to a simple recurrent network (SRN). The network receives (or generates) sentences as flat sequences of words; through exposure to many sentences during training, it learns something about phrase structure in the weights which map its input and context units to its hidden layer.

To explain this idea, I will illustrate by describing *Elman*'s [48.36] original model of syntactic processing. This model emphasizes the conception of syntax as a specification of well-formed sentences. It takes the form of an SRN which receives the words in a sentence one by one, and in each case is trained to predict the next word. Of course there are many possible next words; what the network really learns to predict is the probability distribution for the next word, given the words in the sentence so far. This probability distribution embodies a graded notion of grammaticality, rather than a binary distinction between words which can legally occur next and words which cannot. The network represents which words are most likely to occur next, combining syntactic information with semantic information about what situations commonly occur (or are commonly reported). It also incorporates a model of commonly occurring surface patterns in language such as idioms.

How does a trained word-predicting SRN come to incorporate a model of phrase structure? To explain how this happens, it is useful to think of the SRN as a function which steps through a sequence of context representations when given a sequence of words as input. This sequence can be interpreted graphically as a trajectory defined in the n-dimensional space of the network's possible context unit activations. Information about phrase structure is present in the trajectories which the network learns. To illustrate, imagine a very simple language, in which a sentence is made up of two phrases P_1 and P_2 (in that order), and where there are no dependencies between these two phrases, so any instance of P_1 can be followed by any instance of P_2. An SRN trained on sentences from this language will learn to establish the same context representation after being presented with any sequence of words instantiating P_1, so that its subsequent predictions (about P_2) are properly independent of the particular instance of P_1 that it has just seen. The trained SRN represents the phrase boundary between P_1 and P_2 by having all its trajectories in context-unit space pass through the same point when this boundary is reached. (In more realistic situations where there is some degree of syntactic dependency between two phrases, so that particular varieties of P_1 require particular varieties of P_2, the SRN will learn different invariant points for the various different varieties, but the idea of a phrase boundary as a point where trajectories in context-unit space converge still carries over to these situations.) Importantly, the representations of phrase structure learned by an SRN are only approximations of the notion of phrase structure defined in symbolic models of grammar. I will return to this point in Sect. 48.4.4.

Other SRN-based models of syntax emphasize the role of syntax in representing the compositional seman-

tics of sentences, i.e., in defining a mapping between individual word meanings and the meanings of whole sentences. These models either receive or produce a sequence of words, but they are trained by associating word sequences with episode representations. In a model of sentence interpretation, the SRN receives a sequence of words as input and learns to produce an episode representation as output. We have already seen examples of this kind of network: the network models of *McClelland* et al. [48.106] and of *Mayberry* and *Miikkulainen* [48.110] both have this structure. In a model of sentence generation, the SRN takes an episode representation as input and produces a sequence of words as output: an example of this kind of model is *Chang* [48.97]. The purpose of the context layer in these networks is no longer to explicitly represent well-formed sentences, but to encode a mapping between word sequences and sentence meanings. However, these networks still learn something about the phrase structure of the training sentences they are exposed to, in the way already described.

One attractive feature of an SRN as a model of syntactic processing is that the circuitry it requires is amply attested in the frontal and temporal neural areas where syntactic processing is thought to happen. An SRN is a simple recurrent circuit, and there is evidence of recurrent loops throughout cortex (see, e.g., [48.121]): these loops pass through subcortical areas and return back to the same area of cortex from which they originated. There is also evidence that inferior frontal areas are involved not only in syntactic computations, but also in sequence processing more generally. Not all sequence processing depends on the inferior frontal cortex; for instance, Broca's aphasics can learn simple sequences of stimuli [48.122]. However, damage to inferior frontal cortex impairs the learning of sequences with hierarchical structure and of sequences of abstract symbols [48.123–125]. As I will discuss in the next section, a basic SRN has to be augmented with additional circuitry in order to allow it to learn more complex sequences of these kinds. In the remainder of this discussion, I will focus on computational arguments about the additional circuitry needed; ideas about the neural plausibility of this circuitry are still quite sketchy.

48.4.4 Limitations of Standard SRNs and Some Alternative Models

As noted above, the way a standard SRN represents phrase structure is just an approximation of the way phrase structure is represented in a symbolic model of syntax. In this section, I will discuss two of the computational shortcomings of SRNs.

Generalization to Unseen Patterns

For one thing, SRNs generalize quite poorly to patterns they have never encountered during training; the patterns they learn are strongly tied to the token words found in these patterns. The most problematic cases involve generalizations which allow a word encountered in one syntactic position to be used or understood in a new syntactic position. Imagine a language learner encounters a new word describing an animal, in sentences where this word always happens to occur in subject position. The learner should be able to understand (or generate) sentences where this word occurs in object position, even though she has never encountered it in that position. Syntactic formalisms which can generalize over syntactic positions in this way are said to possess *strong systematicity* [48.126]. A standard SRN exhibits little systematicity of this kind.

In some cases, this lack of strong systematicity is a good thing. The constructivist conception of syntax mentioned in Sect. 48.4.1 positively requires that syntactic constructions can be defined with reference to specific words and word combinations. But there are clearly also cases where properly abstract rules are required. So there are two questions: firstly, how can we create networks which exhibit strong systematicity, and secondly, how can we tell which constructions should retain reference to specific words and which should abstract away from them?

Most work has focussed on the first of these questions. There are now several network models which exhibit a good measure of strong systematicity. One of these is the network of *Chang* ([48.97], see also [48.127]). This network has an SRN at its core; the main innovation is that this SRN learns to sequence abstract semantic roles (AGENT, PATIENT, etc.) rather than actual words. This network uses the *binding-by-connection* scheme introduced in Sect. 48.3.4: a message is encoded by temporarily associating word meanings with particular roles. The words in an incoming sentence activate a sequence of role representations, and the SRN learns about the structure of these sequences. Chang et al.'s model has quite wide syntactic coverage: it can learn the right order for semantic roles for a wide variety of episode types. And because the rules it learns are framed in terms of semantic roles, it shows good strong systematicity. On the other hand, by generalizing away from individual words, the network loses an ability to encode idiomatic patterns which re-

fer to surface structures in language, which is one of the things a normal SRN does quite well.

Another model which achieves a form of strong systematicity is that of *Takac* et al. ([48.128], see also [48.101]). This model uses binding-by-serial-position (see again Sect. 48.3.4) to associate participants in an episode with semantic roles. The novel element in this network is that semantic roles can feature more than once in the canonical sequence representing an episode; for instance, the sequence of roles representing a transitive sentence could be summarized as AGENT, PATIENT, ACTION, AGENT, PATIENT. The doubling of AGENT and PATIENT roles reflects the fact that these participants are apprehended in different sensorimotor modalities when a transitive episode is observed. They must first be attended to as objects, to establish the context in which a motor action can be executed or monitored; during the course of this execution or monitoring they are reactivated, but this time in a motor modality. In Takac et al.'s model, producing a sentence which expresses an episode involves internally replaying the sensorimotor sequence characterizing the episode in a special mode where sensorimotor signals can have linguistic side-effects. When replaying the episode, there are multiple opportunities to pronounce the agent and the patient; what the syntactic network has to learn is which opportunity to take. Like Chang's network, Takac et al.'s network learns abstract word-ordering conventions and therefore exhibits strong systematicity. However, in Takac et al.'s model, a specific sequence of roles is provided in advance, and the network only needs to learn binary choices about when to pronounce the agent and the patient, while in Chang's network, a sequence of roles needs to be learned from scratch. Takac et al.'s network is, in fact, explicitly framed as a nativist model in the Chomskyan tradition: knowledge of the sensorimotor sequences associated with episodes is construed as innate linguistic knowledge and acquiring abstract word-ordering conventions just involves setting the values of a small number of parameters (see [48.38] for more about this interpretation of Chomskyan syntax). At the same time, Takac et al.'s network also provides a mechanism for learning surface structures in language. The network includes a simple SRN, which receives a semantic message as input and is trained to predict the next word in the sentence as output. Since messages have sequential structure, the SRN receives the message one element at a time rather than all at once. As long as the SRN can confidently predict the next word from the current semantic input, it can work autonomously, generating an *idiomatic* sequence of several words; when it can no longer do so, the next element in the message sequence is provided. This scheme allows the network to learn a range of idiomatic and semi-idiomatic structures, as well as abstract word-ordering conventions.

A final interesting approach to systematicity comes from *Frank* et al. [48.129]. The key suggestion in this model is that *episodes* are not a well-defined unit of representation in their own right, but should rather be thought of as components in larger representations of *situations*, which combine information about many episodes and their patterns of co-occurrence. In Frank et al.'s model, semantic representations are stored in a structure similar to a self-organizing map, which is trained on symbolic representations of whole situations, which somewhat resemble the *discourse context* representations discussed in Sect. 48.3.3. Symbolic situations are defined in relation to a microworld, in which a finite set of *n* episodes can occur. Each symbolic situation is represented in an *n*-dimensional vector whose components specify the truth value (1 or 0) of every one of these episodes. These vectors are presented as training data to the self-organizing map (which has a larger number of nodes). The network learns patterns of co-occurrence between the input episodes. Within the trained network, each episode is represented by a distinct vector, and Boolean combinations of episodes can be neatly derived from these basic vectors. Frank et al.'s sentence interpretation network is the simplest possible kind of SRN: it takes word sequences as input and is trained to generate appropriate situation representations as output. Nonetheless, it shows some degree of systematicity, being able to make good guesses about the semantics of unseen sentences, even when these feature unseen combinations of concepts. Interestingly, this network makes no attempt to represent the internal compositional structure of episodes. The previously described networks, by contrast, achieve systematicity precisely by representing this structure. One interesting possibility is that speakers make use of both kinds of episode representation, with each producing its own type of linguistic systematicity.

Representing Nested Phrase Structure

Another problem with using an SRN in a model of syntax is that SRNs are not good at representing long-distance syntactic dependencies, of the kind which are seen in deeply nested relative clauses (e.g., the agreement between *man* and *likes* in *The man [who my friend at Cadbury's plays snooker with] likes grappa*). These kind of dependencies are what motivate linguists

to think of syntactic representations as hierarchical. SRNs can learn something about phrases, as discussed in Sect. 48.4.3, but they are not explicitly designed to represent hierarchical phrase structures.

Encoding long-distance dependencies between words in an SRN is not impossible, but it requires reference to subtle features of context representations, reflecting words which occurred many iterations ago. Since information about the past fades exponentially in the context layer, these features are hard for learning algorithms to identify and exploit. Humans and SRNs, in fact, have difficulty with the same types of long-distance dependency; the hardest dependencies in each case are nested *center-embedded* clauses (*The man [who the woman [who danced] chased] sang*). So in some ways this deficit speaks in favor of SRNs as a cognitive model of syntax [48.130]. But simple SRNs also have difficulty with long-distance dependencies which pose no trouble to human speakers. Most modern SRN-based models include additional circuitry; for instance, adding additional recurrent loops [48.97] or using explicit representations of the recent words in a sentence [48.110]. But some networks go further, building in explicit representations of hierarchical, nested structures.

A key component in many of these latter networks is a *recursive auto-associative memory* (RAAM) network [48.131]. This is a network which takes a number of distinct vectors of size *n* as input and learns to reproduce these inputs on a similarly structured output layer, through a hidden layer whose size is the same as the input vectors (Fig. 48.6). After training, the hidden layer serves to compress the combined input vectors into a single vector which can be presented again to the network and combined with an additional vector. This property makes the RAAM a recursive data structure. Of course, compressing input vectors is a lossy operation, so there is a limit to the amount of recursive structure a RAAM can hold.

One interesting application of a RAAM is in a neural network implementation of a *stack*. A stack is a data

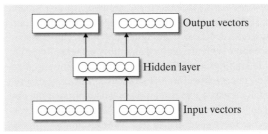

Fig. 48.6 A recursive auto-associative network

structure which holds a number of items; an item is added with a *push* operation and retrieved with a *pop* operation. The *pop* operation returns the item most recently pushed onto the stack. To implement a stack, a RAAM network needs two inputs and outputs, as shown in Fig. 48.6. The current state of the stack is represented by the RAAM's hidden layer; to represent an empty stack, this layer is initialized with an arbitrary conventional vector. Items to store in the stack are represented as input vectors, with the same dimensionality as the hidden layer. When a new item is pushed onto the stack, the RAAM is given this item and its own current state as inputs (and training outputs), and learns a new hidden layer which represents the new state of the stack; this process can then be repeated to push additional items onto the stack. When an item is popped from the stack, we can simply evoke the stack's current state in the hidden layer and read the popped item and the new state of the stack from the two vectors of the output layer. Of course, the limited capacity of a RAAM means that it can only store a certain number of items before it starts to make mistakes when popping items. But its performance degrades gradually; if it does not return the correct item, it is likely to return a similar one.

Early models of syntactic recursion used a RAAM to encode explicit representations of phrase structure. For instance, *Reilly* [48.132] trained a RAAM to encode a number of different hierarchical parse trees in its hidden layer. Trees were strictly binary-branching, so the RAAM just needed two vectors on its input and output; it recursively learned compressed representations of each binary node in each tree, starting from the simplest nodes and progressing to the more complex ones. In a separate learning phase, a standard SRN was trained to map sequences of words onto the parse tree representations learned by the RAAM. Subsequent models (e.g., [48.133, 134]) combined the RAAM and the SRN into a single network, but preserved the idea that parse trees are explicitly computed by a sentence processor.

An interesting alternative strategy is proposed in *Miikkulainen*'s [48.135] SPEC network. In this model, an SRN receives the words of a sentence one by one, but works in clause-sized units, mapping each clause in a complex sentence onto an assembly representing role-filler bindings for that particular clause. For instance, when interpreting *The girl [who liked the dog] saw the boy*, it creates a representation of *The girl liked the dog* (agent = girl, act = like, patient = dog) when processing the the nested clause and a representation of *The*

girl saw the boy (agent = girl, act = saw, patient = boy) when processing the matrix clause. Thus the output of the system for a complex sentence is a sequence of clause-sized semantic representations, rather than a single monolithic representation. The SRN itself never has to deal with the nested structure of clauses: this job is performed by two additional networks, which together act to store and load the SRN's context representations at clause boundaries. One, called the *segmenter network*, learns to recognize contexts where nested clauses begin and end. It takes the SRN's context and *current word* representations as input and learns to generate a control signal specifying what should happen at the next iteration of the SRN, as well as a signal which can be used as the SRN's context representation at this next iteration. At the start of a nested clause, the control signal indicates that the SRN's current context representation should be pushed onto a stack and replaced with a new context representation appropriate for processing the nested clause. At the end of a nested clause, the control signal indicates that the SRN's context representation should be replaced with the representation obtained by popping the stack. Where there is no clause boundary, it indicates that the context representation at the next iteration should be what the SRN would normally use (i. e., its current hidden layer). The stack is implemented by a RAAM network, as described above, configured to store vectors with the same dimensionality as the SRN's hidden layer. When a new context representation is pushed onto the stack, the RAAM learns to autoassociate this context representation with its own current hidden layer, creating a new hidden layer representing the new state of the stack. When the stack is popped to restore a previous context, this context (and the new state of the stack) is read from the two vectors of the RAAM's output layer. Dividing the task of sentence interpretation up between these three networks allows the model to learn good generalizations; the SRN learns generalizations which extend over all clauses, whether they are main clauses or nested ones, and the segmenter network learns generalizations about where clauses begin and end, regardless of their content or how deeply nested they are. The SRN and segmenter networks can learn what they need to know from a relatively small set of training sentences, at which point the SPEC network as a whole can successfully interpret a large space of complex sentences far greater than the size of its training set. At the same time, the RAAM's limited capacity results in psychologically realistic difficulties interpreting multiply center-embedded clauses.

48.5 Summary

Neural network models provide many insights into the neural circuitry implementing the human capacity for language. We have interesting models of phonological representations, word forms, and word meanings, and of the circuits which learn connections between word forms and word meanings. We also have interesting models of the representation of sentence-sized episodes, both in long-term memory and in working memory, and of the representation of the discourse contexts in which sentences are interpreted. Finally, we have interesting models of the syntactic machinery which implements the mapping between sequences of words and episode representations. There is a trend away from toy models towards models with wide syntactic coverage, echoing a similar trend in applied computational linguistics, where there is now a similar emphasis on machine learning techniques and naturalistic data sets. There is still real debate, however, in almost all the areas which have been discussed, especially in the areas of syntax and sentence/discourse semantics. I have described several radically different models of episode representations, and several radically different models of syntactic processing; all of them have their limitations, so the correct model may not bear much resemblance to any of the ones I have reviewed here. *Caveat lector!*

References

48.1 F. Rosenblatt: The perceptron: A probabilistic model for information storage and organization in the brain, Psychol. Rev. **65**(6), 386–408 (1958)

48.2 B. Widrow, M.E. Hoff: Adaptive switching circuits, IRE WESCON Conv. Record **4**, 96–104 (1960)

48.3 J. McClelland, D. Rumelhart: An interactive activation model of context effects in letter perception: Part 1. An account of basic findings, Psychol. Rev. **88**, 375–407 (1981)

48.4 D. Rumelhart, J. McClelland, PDP Research Group: *Parallel Distributed Processing: Explorations in the Microstructure of Cognition*, Vol. 1 (MIT Press, Cambridge 1986)

48.5 J. Werker, R. Tees: Influences on infant speech processing: Toward a new synthesis, Annu. Rev. Psychol. **50**, 509–535 (1999)

48.6 J. Bohland, F. Guenther: An fMRI investigation of syllable sequence production, NeuroImage **32**, 821–841 (2006)

48.7 C. Browman, L. Goldstein: Dynamics and articulatory phonology. In: *Mind as Motion: Explorations in the Dynamics of Cognition*, ed. by R. Port, T. van Gelder (MIT Press, Cambridge 1995) pp. 175–193

48.8 G. Hickok, D. Poeppel: The cortical organization of speech processing, Nat. Rev. Neurosci. **8**(5), 393–402 (2007)

48.9 M. Catani, D.K. Jones, D. Hffytche: Perisylvian language networks of the human brain, Ann. Neurol. **57**, 8–16 (2004)

48.10 J. Rilling, M. Glasser, T. Preuss, X. Ma, T. Zhao, X. Hu, T. Behrens: The evolution of the arcuate fasciculus revealed with comparative DTI, Nat. Neurosci. **11**, 426–428 (2008)

48.11 S. Wilson, A. Saygin, M. Sereno, M. Iacoboni: Listening to speech activates motor areas involved in speech production, Nat. Neurosci. **7**(7), 701–702 (2004)

48.12 L. Fadiga, L. Craighero, G. Buccino, G. Rizzolatti: Speech listening specifically modulates the excitability of tongue muscles: A TMS study, Eur. J. Neurosci. **15**, 399–402 (2002)

48.13 T. Ito, M. Tiede, D. Ostry: Somatosensory function in speech perception, Proc. Natl. Acad. Sci. USA **106**(4), 1245–1248 (2009)

48.14 G. Hickok, P. Erhard, J. Kassubek, A. Helms-Tillery, S. Naeve-Velguth, J. Struppf, P. Strick, K. Ugurbil: A functional magnetic resonance imaging study of the role of left posterior superior temporal gyrus in speech production: Implications for the explanation of conduction aphasia, Neurosci. Lett. **287**, 156–160 (2000)

48.15 A. Battaglia-Mayer, R. Caminiti, F. Lacquaniti, M. Zago: Multiple levels of representation of reaching in the parieto-frontal network, Cereb. Cortex **13**, 1009–1022 (2003)

48.16 M. Iacoboni, L. Koski, M. Brass, H. Bekkering, R. Woods, M. Dubeau, J. Mazziotta, G. Rizzolatti: Reafferent copies of imitated actions in the right superior temporal cortex, Proc. Natl. Acad. Sci. USA **98**(24), 13995–13999 (2001)

48.17 C. Keysers, D. Perrett: Demystifying social cognition: A Hebbian perspective, Trends Cogn. Sci. **8**(11), 501–507 (2004)

48.18 E. Oztop, M. Arbib: Schema design and implementation of the grasp-related mirror neuron system, Biol. Cybern. **87**, 116–140 (2002)

48.19 G. Westermann, E. Miranda: A new model of sensorimotor coupling in the development of speech, Brain Lang. **89**, 393–400 (2004)

48.20 F. Guenther, S. Ghosh, J. Tourville: Neural modeling and imaging of the cortical interactions underlying syllable production, Brain Lang. **96**, 280–301 (2006)

48.21 M. Garagnani, T. Wennekers, F. Pulvermüller: A neuroanatomically grounded Hebbian learning model of attention-language interactions in the human brain, Eur. J. Neurosci., **27**, 492–513 (2008)

48.22   L. Wheeldon, W. Levelt: Monitoring the time course of phonological encoding, J. Memory Lang. **34**, 311–334 (1995)

48.23   A. Baddeley, G. Hitch: Working memory. In: *The psychology of Learning and Motivation*, ed. by G. Bower (Academic Press, New York 1974) pp. 48–79

48.24   N. Burgess, G. Hitch: Memory for serial order: A network model of the phonological loop and its timing, Psychol. Rev. **106**, 551–581 (1999)

48.25   T. Hartley, G. Houghton: A linguistically constrained model of short-term memory for nonwords, J. Memory Lang. **35**, 1–31 (1996)

48.26   S. Gathercole, A. Baddeley: The role of phonological memory in vocabulary acquisition: A study of young children learning new names, Br. J. Psychol. **81**, 439–454 (1990)

48.27   A. Adams, S. Gathercole: Phonological working memory and speech production in preschool children, J. Speech Hear. Res. **38**, 403–414 (1995)

48.28   T. Shallice, R. Rumiati, A. Zadini: The selective impairment of the phonological output buffer, Cogn. Neuropsychol. **17**(6), 517–546 (2000)

48.29   T. Shallice, B. Butterworth: Short-term memory impairment and spontaneous speech, Neuropsychologia **15**, 729–735 (1977)

48.30   E. Paulesu, C. Frith, R. Frackowiak: The neural correlates of the verbal component of working memory, Nature **362**, 342–345 (1993)

48.31   R. Henson, N. Burgess, C. Frith: Recoding, storage, rehearsal and grouping in verbal short-term memory: An fMRI study, Neuropsychologia **38**(4), 426–440 (2000)

48.32   N. Burgess, G. Hitch: Computational models of working memory: Putting long-term memory into context, Trends Cogn. Sci. **9**(11), 535–541 (2005)

48.33   J. Saffran, R. Aslin, E. Newport: Statistical learning by 8-month-old infants, Science **274**, 1926–1928 (1996)

48.34   S. Auerbach, T. Allard, M. Naeser, M. Alexander, M. Albert: Pure word deafness: Analysis of a case with bilateral lesions and a defect at the pre-phonemic level, Brain **105**, 271–300 (1982)

48.35   S. Grossberg, C. Myers: The resonant dynamics of speech perception: Interword integration and duration-dependent backward effects, Psychol. Rev. **107**, 735–767 (2000)

48.36   J. Elman: Finding structure in time, Cogn. Sci. **14**, 179–211 (1990)

48.37   M. Gaskell, W. Marslen-Wilson: Integrating form and meaning: A distributed model of speech perception, Lang. Cogn. Process. **12**, 613–656 (1997)

48.38   P. Zwitserlood, H. Schriefers: Effects of sensory information and processing time in spoken-word recognition, Lang. Cogn. Process. **10**(2), 121–136 (1995)

48.39   D. Sibley, C. Kello, D. Plaut, J. Elman: Large-scale modeling of wordform learning and representation, Cogn. Sci. **32**, 741–754 (2008)

48.40   T. Kohonen: Self-organized formation of topologically correct feature maps, Biol. Cybern. **43**, 59–69 (1982)

48.41   P. Li, I. Farkas, B. MacWhinney: Early lexical development in a selforganizing neural network, Neural Networks **17**, 1345–1362 (2004)

48.42   P. Li, B. MacWhinney: PatPho: A phonological pattern generator for neural networks, Behav. Res. Methods Instrum. Comput. **34**, 408–415 (2002)

48.43   G. Dell, M. Schwartz, N. Martin, E. Saffran, D. Gagnon: Lexical access in aphasic and nonaphasic speakers, Psychol. Rev. **104**, 801–838 (1997)

48.44   W. Levelt: Perspective taking and ellipsis in spatial descriptions. In: *Language and Space*, ed. by P. Bloom, M. Peterson, L. Nadel, M. Garrett (MIT Press, Cambridge 1999) pp. 77–108

48.45   G. Dell, C. Juliano, A. Govindjee: Structure and content in language production: A theory of frame constraints in phonological speech errors, Cogn. Sci. **17**(2), 149–195 (1993)

48.46   G. Dell, N. Nozari, G. Oppenheim: Word production: Behavioral and computational considerations. In: *The Oxford Handbook of Language Production*, ed. by V. Ferreira, M. Goldrick, M. Miozzo (Oxford Univ. Press, Oxford 2011)

48.47   A. Tarski: The concept of truth in formalized languages. In: *Logic, Semantics and Mathematics*, ed. by J. Corcoran (Hackett, Indianapolis 1983)

48.48   P. Pantel, D. Lin: Discovering word senses from text, Proceedings of ACM SIGKDD Conference on Knowledge Discovery and Data Mining (2002) pp. 613–619

48.49   F. Pulvermüller, W. Lutzenberger, H. Preissl: Nouns and verbs in the intact brain: Evidence from event-related potentials and high-frequency cortical responses, Cereb. Cortex **9**(5), 497–506 (1999)

48.50   O. Hauk, I. Johnsrude, F. Pulvermüller: Somatotopic representation of action words in human motor and premotor cortex, Neuron **41**, 301–307 (2004)

48.51   F. Pulvermüller, O. Hauk, V. Nikulin, R. Ilmoniemi: Functional links between motor and language systems, Eur. J. Neurosci. **21**(3), 793–797 (2005)

48.52   T. Mitchell, S. Shinkareva, A. Carlson, K. Chang, V.R.M. Malave, M. Just: Predicting human brain activity associated with the meanings of nouns, Science **320**, 1191–1195 (2008)

48.53   A. Damasio, H. Damasio: Cortical systems for retrieval of concrete knowledge: The convergence zone framework. In: *Large-scale Neuronal Theories of the Brain*, ed. by C. Koch, J. Davis (MIT Press, Cambridge 1994)

48.54   M. van Turennout, P. Hagoort, C. Brown: Electrophysiological evidence on the time course of semantic and phonological processes in speech production, J. Exp. Psychol. Learn. **23**, 787–806 (1997)

48.55 W. Glaser, F. Düngelhoff: The time course of picture-word interference, J. Exp. Psychol. Hum. Percept. **10**, 640–654 (1984)

48.56 R. Brookshire: *Introduction to Neurogenic Communication Disorders* (Mosby-Year Book, St. Louis 1997)

48.57 H. Damasio, T. Grabowski, D. Tranel, R. Hichwa, A. Damasio: A neural basis for lexical retrieval, Nature **380**, 499–505 (1996)

48.58 L. Lu, B. Crosson, S. Nadeau, K. Heilman, L. Gonzalez-Rothi, A. Raymer, R. Gilmore, R. Bauer, S. Roper: Category-specific naming deficits for objects and actions: Semantic attribute and grammatical role hypotheses, Neuropsychologia **40**, 1608–1621 (2002)

48.59 J. Kay, A. Ellis: A cognitive neuopsychological case study of anomia: Implications for psychological models of word retrieval, Brain **110**, 613–629 (1987)

48.60 K. Shapiro, J. Shelton, A. Caramazza: Grammatical class in lexical production and morphological processing: Evidence from a case of fluent aphasia, Cogn. Neuropsychol. **17**(8), 665–682 (2000)

48.61 K. Shapiro, A. Caramazza: The representation of grammatical categories in the brain, Trends Cogn. Sci. **7**(5), 201–206 (2001)

48.62 D. Rumelhart, J. McClelland: On learning the past tense of English verbs. In: *Parallel Distributed Processing: Explorations in the Microstructure of Cognition*, Vol.1: Foundations (MIT Press, Cambridge 1986) pp. 216–271

48.63 G. Lakoff, M. Johnson: *Metaphors We Live By* (University of Chicago Press, Chicago and London 1980)

48.64 R. Langacker: *Cognitive Grammar: A Basic Introduction* (Oxford University Press, New York 2008)

48.65 L. Barsalou: Grounded cognition, Annu. Rev. Psychol. **59**, 617–645 (2008)

48.66 V. Gallese, G. Lakoff: The brain's concepts: The role of the sensory-motor system in conceptual knowledge, Cogn. Neuropsychol. **22**(3/4), 455–479 (2005)

48.67 H. Reichenbach: *Elements of Symbolic Logic* (University of California Press, Berkeley, 1947)

48.68 R. Reiter: *Knowledge in Action: Logical Foundations for Specifying and Implementing Dynamical Systems* (MIT Press, Cambridge 2001)

48.69 M. Steedman: Temporality. In: *Handbook of Logic and Language*, ed. by A. ter Meulen, J. van Benthem (North Holland, Amsterdam 1997) pp. 895–938

48.70 H. Kamp, U. Reyle: *From Discourse to Logic* (Kluwer Academic Publishers, Dordrecht 1993)

48.71 I. Heim: The semantics of definite and indefinite noun phrases. Ph.D. Thesis (University of Massachusetts, Amherst 1982)

48.72 E. Tulving: Episodic and semantic memory. In: *Organization of Memory*, ed. by E. Tulving,

W. Donaldson (Academic Press, New York 1972) pp. 381–403

48.73 H. Eichenbaum, A. Yonelinas, C. Ranganath: The medial temporal lobe and recognition memory, Ann. Rev. Neurosci. **30**, 123–152 (2007)

48.74 R. Diana, A. Yonelinas, C. Ranganath: Imaging recollection and familiarity in the medial temporal lobe: A three-component model, Trends Cogn. Sci. **11**(9), 379–386 (2007)

48.75 R. Epstein, N. Kanwisher: A cortical representation of the local visual environment, Nature **392**, 598–601 (1998)

48.76 R. Epstein, S. Higgins, K. Jablonski, A. Feiler: Visual scene processing in familiar and unfamiliar environments, J. Neurophysiol. **97**, 3670–3683 (2007)

48.77 M. Bar, E. Aminoff: Cortical analysis of visual context, Neuron **38**(2), 347–358 (2003)

48.78 M. Howard, M. Kahana: A distributed representation of temporal context, J. Math. Psychol. **46**, 269–299 (2002)

48.79 D. Marr: Simple memory: A theory for archicortex, Philos. Trans. R. Soc. B **262**(841), 23–81 (1971)

48.80 J. McClelland, B. McNaughton, R. O'Reilly: Why there are complementary learning systems in the hippocampus and neocortex: Insights from the successes and failures of connectionist models of learning and memory, Psychol. Rev. **102**, 419–457 (1995)

48.81 F. Craik, R. Lockhart: Levels of processing: A framework for memory research, J. Verbal Learn. Verbal Behav. **11**, 671–684 (1972)

48.82 M. Rugg: ERP studies of memory. In: *Electrophysiology of Mind: Event-Related Brain Potentials and Cognition*, ed. by M. Rugg, M. Coles (Oxford University Press, Oxford 1995)

48.83 M. Rotte, W. Koutstaal, D. Schacter, A. Wagner, B. Rosen, A. Dale, R. Buckner: Left prefrontal activation correlates with levels of processing during verbal encoding: An event-related fMRI study, NeuroImage **7**, S813 (2000)

48.84 R. Buckner: Functional-anatomic correlates of control processes in memory, J. Neurosci. **23**(10), 3999–4004 (2003)

48.85 A. Baddeley: The episodic buffer: A new component of working memory?, Trends Cogn. Sci. **4**(11), 417–423 (2000)

48.86 E. Miller: The prefrontal cortex and cognitive control, Nat. Rev. Neurosci. **1**, 59–65 (2000)

48.87 J. Gibson (Ed.): *The Perception of the Visual World* (Houghton Mifflin, Boston 1950)

48.88 G. Rainer, W. Asaad, E. Miller: Memory fields of neurons in the primate prefrontal cortex, Proc. Natl. Acad. Sci. USA **95**(25), 15008–15013 (1998)

48.89 G. Rainer, C. Rao, E. Miller: Prospective coding for objects in primate prefrontal cortex, J. Neurosci. **19**(13), 5493–5505 (1999)

48.90 S. Courtney, L. Petit, J. Maisog, L. Ungerleider, J. Haxby: An area specialized for spatial working memory in human frontal cortex, Science **279**(5355), 134–1351 (1998)

48.91   J. Haxby, L. Petit, L. Ungerleider, S. Courtney: Distinguishing the functional roles of multiple regions in distributed neural systems for visual working memory, NeuroImage **11**, 145–156 (2000)

48.92   T. Braver, J. Cohen: On the control of control: The role of dopamine in regulating prefrontal function and working memory. In: *Attention and Performance XVIII: Control of cognitive processes*, ed. by S. Monsell, J. Driver (MIT Press, Cambridge 2000) pp. 713–737

48.93   R. Jackendoff: *Foundations of Language: Brain, Meaning, Grammar, Evolution* (Oxford University Press, Oxford 2002)

48.94   L. Shastri, V. Ajjanagadde: From simple associations to systematic reasoning, Behav. Brain Sci. **16**(3), 417–494 (1993)

48.95   E. Izhikevich, J. Gally, G. Edelman: Spike-timing dynamics of neuronal groups, Cereb. Cortex **14**, 933–944 (2004)

48.96   M. Shadlen, J. Movshon: Synchrony unbound: A critical evaluation of the temporal binding hypothesis, Neuron **24**, 67–77 (1999)

48.97   F. Chang: Symbolically speaking: A connectionist model of sentence production, Cogn. Sci. **26**, 609–651 (2002)

48.98   J. Hummel, K. Holyoak: A symbolic-connectionist theory of relational inference and generalization, Psychol. Rev. **110**(2), 220–264 (2003)

48.99   F. van der Velde, M. de Kamps: Neural blackboard architectures of combinatorial structures in cognition, Behav. Brain Sci. **29**, 37–108 (2006)

48.100 T. Stewart, C. Eliasmith: Compositionality and biologically plausible models. In: *The Oxford Handbook of Compositionality*, ed. by M. Werning, W. Hinzen (Oxford University Press, New York 2012)

48.101 A. Knott: *Sensorimotor Cognition and Natural Language Syntax* (MIT Press, Cambridge 2012)

48.102 D. Ballard, M. Hayhoe, P. Pook, R. Rao: Deictic codes for the embodiment of cognition, Behav. Brain Sci. **20**(4), 723–767 (1997)

48.103 S. Goldin-Meadow, W. So, A. Özürek, C. Mylander: The natural order of events: How speakers of different languages represent events nonverbally, Proc. Natl. Acad. Sci. USA **105**(27), 9163–9168 (2008)

48.104 B. Averbeck, D. Lee: Prefrontal correlates of memory for sequences, J. Neurosci. **27**(9), 2204–2211 (2007)

48.105 B. Rhodes, D. Bullock, W. Verwey, B. Averbeck, M. Page: Learning and production of movement sequences: Behavioral, neurophysiological, and modeling perspectives, Hum. Mov. Sci. **23**, 699–746 (2004)

48.106 J. McClelland, M. St. John, R. Taraban: Sentence comprehension: A parallel distributed processing approach, Lang. Cogn. Process. **4**(3–4), 287–335 (1989)

48.107 D. Rohde: A Connectionist Model of Sentence Comprehension and Production. Ph.D. Thesis (Carnegie Mellon University, School of Computer Science, Pittsburgh 2002)

48.108 R. Gayler: Vector symbolic architectures answer Jackendoff's challenges for cognitive neuroscience, Proc. Jt. Int. Conf. Cogn. Sci, ed. by P. Slezak (2003) pp. 133–138

48.109 T. Plate: *Holographic Reduced Representations*, CSLI Lecture Notes Series, Vol. 150 (CSLI, Stanford 2003)

48.110 M. Mayberry, R. Miikkulainen: Incremental non-monotonic sentence interpretation through semantic self-organization. Technical Report AI08-12 (Department of Computer Sciences, The University of Texas at Austin, Austin 2008)

48.111 A. Copestake, D. Flickinger, I. Sag, C. Pollard: *Minimal Recursion Semantics: An Introduction. Manuscript* (CSLI, Stanford University 1999)

48.112 N. Chomsky: *Lectures on Government and Binding* (Foris, Dordrecht 1981)

48.113 N. Chomsky: Some notes on the economy of derivation and representation. In: *Principles and Parameters in Comparative Grammar*, Vol. 12, ed. by R. Freidin (MIT Press, Cambridge 1989)

48.114 N. Chomsky: *The Minimalist Program* (MIT Press, Cambridge 1995)

48.115 A. Goldberg: *Constructions. A Construction Grammar approach to Argument Structure* (Chicago Univ. Press, Chicago 1995)

48.116 M. Tomasello: *Constructing a Language: A Usage-Based Theory of Language Acquisition* (Harvard Univ. Press, Cambridge 2003)

48.117 E. Saffran: Aphasia and the relationship of language and brain, Semin. Neurol. **20**(4), 409–418 (2000)

48.118 N. Dronkers, D. Wilkins, R. Van Valin, B. Redfern, J. Jaeger: Lesion analysis of the brain areas involved in language comprehension, Cognition **92**(12), 145–177 (2004)

48.119 A. Friederici, A. Meyer, D. von Cramon: Auditory language comprehension: An event-related fMRI study on the processing of syntactic and lexical information, Brain Lang. **74**, 289–300 (2000)

48.120 C. Humphries, K. Willard, B. Buchsbaum, G. Hickok: Role of anterior temporal cortex in auditory sentence comprehension: An fMRI study, Neuroreport **12**(8), 1749–1752 (2001)

48.121 G. Alexander, M. DeLong, P. Strick: Parallel organization of functionally segregated circuits linking basal ganglia and cortex, Annu. Rev. Neurosci. **9**, 357–381 (1986)

48.122 T. Goschke, A. Friederici, S. Kotz, A. van Kampen: Procedural learning in Broca's aphasia: Dissociation between the implicit acquisition of spatio-motor and phoneme sequences, J. Cogn. Neurosci. **13**(3), 370–388 (2001)

48.123 C. Conway, M. Christiansen: Sequential learning in non-human primates, Trends Cogn. Sci. **5**(12), 539–546 (2001)

48.124 P. Dominey, M. Hoen, J.M. Blanc, T. Lelekov-Boissard: Neurological basis of language and sequential cognition: Evidence from simulation, aphasia and ERP studies, Brain Lang. **86**, 207–225 (2003)

48.125 M. Hoen, M. Pachot-Clouard, C. Segebarth, P. Dominey: When Broca experiences the Janus syndrome. An ER-fMRI study comparing sentence comprehension and cognitive sequence processing, Cortex **42**, 605–623 (2006)

48.126 R. Hadley: Systematicity in connectionist language learning, Mind Lang. **9**(3), 247–272 (1994)

48.127 F. Chang, G. Dell, K. Bock: Becoming syntactic, Psychol. Rev. **113**(2), 234–272 (2006)

48.128 M. Takac, L. Benuskova, A. Knott: Mapping sensorimotor sequences to word sequences: A connectionist model of language acquisition and sentence generation, Cognition **125**(2), 288–308 (2012)

48.129 S. Frank, W. Haselager, I. van Rooij: Connectionist semantic systematicity, Cognition **110**, 358–379 (2009)

48.130 M. Christiansen: The (non) necessity of recursion in natural language processing, Proceedings of the 14th Annual Cogn. Science Conference (Lawrence Erlbaum, Hillsdale 1992)

48.131 J. Pollack: Recursive distributed representations, Artif. Intell. **46**(1–2), 77–105 (1990)

48.132 R. Reilly: Connectionist technique for online parsing, Network **3**(1), 1–37 (1992)

48.133 G. Berg: A connectionist parser with recursive sentence structure and lexical disambiguation, Proc. 10th Natl. Conf. Artif. Intell., ed. by W. Swartout (MIT Press, Cambridge 1992) pp. 32–37

48.134 K. Ho, L. Chan: Confluent preorder parsing of deterministic grammars, Connect. Sci. **9**, 269–293 (1997)

48.135 R. Miikkulainen: Subsymbolic case-role analysis of sentences with embedded clauses, Cogn. Sci. **20**, 47–73 (1996)

Part J

Neuroinf

Part J Neuroinformatics Databases and Ontologies

Ed. by Shiro Usui (RIKEN) and Raphael Ritz (INCF)

49. Ontologies and Machine Learning Systems

Shoba Tegginmath, Russel Pears, Nikola Kasabov

In this chapter we review the uses of ontologies within bioinformatics and neuroinformatics and the various attempts to combine machine learning (ML) and ontologies, and the uses of data mining ontologies. This is a diverse field and there is enormous potential for wider use of ontologies in bioinformatics and neuroinformatics research and system development. A systems biology approach comprising of experimental and computational research using biological, medical, and clinical data is needed to understand complex biological processes and help scientists draw meaningful in-

ferences and to answer questions scientists have not even attempted so far.

49.1 Ontology in Computer Science

In modern computer science, ontology is a data model that represents knowledge within a domain (a part of the world), providing a common understanding about the type of objects and concepts that exist in the domain. The biological sciences are replete with descriptive terms, and ontologies are useful here to reach a common understanding of these terms. The explicit definition of concepts in an ontology supports the sharing and reuse of formally represented knowledge among systems [49.1] and helps people and machines to communicate semantically, and not just syntactically [49.2].

Data mining (DM) is a part of the larger overall process of knowledge discovery; it is the process of discovering meaningful patterns in data [49.3] and as such is wholly relevant to the field of biology with the massive quantities and types of data available. Ontolo-gies and DM work in parallel to identify and formalize knowledge – ontologies help in expressing the knowledge in a meaningful way while DM and ML help in extracting useful knowledge from data.

The current use of ontologies in biomedical sciences, however, is limited. The term ontology is used loosely in biomedicine and refers to a number of artifacts – such as controlled vocabularies, terminologies, and ontologies [49.4]. Ontologies have largely been used to facilitate interoperability among the various databases that contain datasets of biomedical experiments by indexing databases with standard terms to help locate and retrieve information. Ontologies have also been used for storing microarray experiment results and data; the microarray gene expression data ontology provides the common terminology and structure used for microarray experiments.

49.2 Ontologies in Bio- and Neuroinformatics

Gene ontology (GO) [49.4] is a preeminent example of the most common usage of ontologies in bioinformatics. GO provides a set of controlled, structured vocabularies to describe key domains of molecular and cellular biology and biological processes, including gene product attributes and biological sequences.

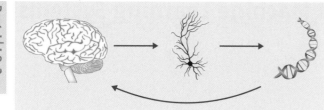

Fig. 49.1 BGO (brain-gene ontology) is concerned with the relationships between the brain and genes

Each ontology is structured in a classification that supports *is-a* and *part-of* relationships. The control in this controlled vocabulary arises from the commitment to use that ontology delivered vocabulary to describe attributes of classes of gene products in community-wide resources [49.5]. Collaborating databases provide data sets with links between database objects and GO terms – these are called annotations. A GO annotation is a link between a gene product type and a molecular function, biological process, or cellular component type. These annotations are what make the ontology useful and

able to support computational reasoning about the instances and GO terms. Observations from experiments and inferences drawn from such experiments are used to create annotations.

The brain-gene ontology (BGO) [49.6] is a biomedical ontology that integrates information from different disciplines such as neuroscience, bioinformatics, genetics, and computer and information sciences. BGO is focused on the gene-disease relationship and includes various concepts, facts, data, software simulators, graphs, videos, animations, and other information forms related to brain functions (as shown in Fig. 49.1), brain diseases, their genetic basis, and the relationship between all of them.

BGO [49.7] has been implemented in the Protégé ontology building environment [49.8]. BGO is based on GO [49.4] and the Unified Medical Language System [49.9]. In addition, knowledge acquired from biology domain experts, from other biological data sources such as Entrez Gene, Swissprot, and Interpro, and from literature databases such as PubMed has also been incorporated.

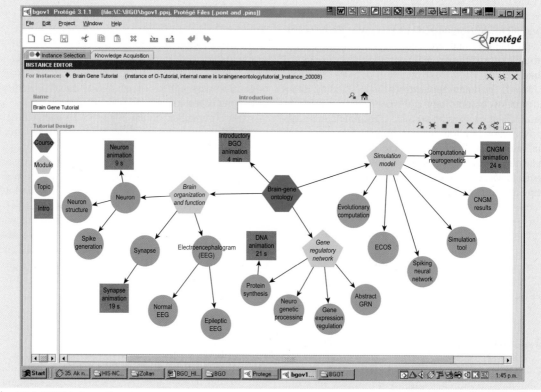

Fig. 49.2 BGO information structure

Table 49.1 General structure of CDO domains

Organism domain	Molecular domain	Medical domain	Nutritional domain	Biomedical informatics map
Human	Gene	Disease	Nutrients	Disease gene map
Group	Mutation	Clinical findings	Source	
Population group	Protein	Signs	Function	
Patient group		Symptoms		
		Laboratory tests		

The overall system as shown in Fig. 49.2 comprises three main parts:

1. Brain organization and functions, which contains information about neurons, their structure, the process of spike generation, and processes in synapses
2. Genes and gene regulatory networks (GRN) is divided into sections on neurogenetic processing, gene expression regulation, protein synthesis, and abstract GRNs
3. A simulation module that has sections on computational neurogenetic modeling, evolutionary computation, and evolving connectionist systems (ECOS).

ECOS [49.10] are modular connectionist-based systems that evolve their structure and functionality in a continuous, self-organized, adaptive way from incoming information, and are capable of processing both data and knowledge in a supervised or unsupervised manner.

One of the main applications of BGO is the integration between ontology and ML tools in relation to feature selection, classification, and prognostic modeling. Software machine learning environments such as NeuCom [49.6], WEKA [49.11], and Siftware [49.6] can be used to aid novel discoveries. By integrating results from ML with genetic information in BGO, a more complete understanding of the pathogenesis of brain

diseases is facilitated [49.12]. However, the discoveries from ML environments are currently entered back to the BGO manually, which is untenable.

Chronic disease ontology [49.6] (CDO) is a Protégé-based ontology [49.8], which contains information about genes involved in three diseases and their mutations, health and nutrition information, and life history data. The diseases are the top three common chronic diseases in developed countries–cardiovascular disease, type2 diabetes, and obesity. These diseases are thought to be mainly caused by interactions of common factors such as genes, nutrition, and life-style. Five domains – organism domain, molecular domain, medical domain, nutritional domain, and a biomedical informatics map exist in CDO. These five classes contain further subclasses and instances as shown in Table 49.1. Each subclass has a set of slots which provide information about each instance and have relationships among other slots, instances, and concepts [49.13]. There are about 76 genes in the ontology. Each gene instance has diverse information associated with the gene and has relationships with other domains. The population group of the organism domain contains information on 50 different population groups; the patient group contains individual patient data.

CDO makes it possible to input information on individual patients such as symptoms, gene maps, diet, and life history details, and generate a personalized

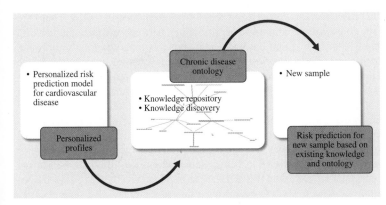

Fig. 49.3 Utilization of knowledge within the CDO

model of risks, profiles, and recommendations, based on the genes of interest and relevant diet components identified, as shown in Fig. 49.3. Large datasets can be imported into the ontology using the data-master plug-in while individual patient information, such as medical, genetic, clinical and nutritional, can be added manually. CDO has helped in identifying interrelationships in personalized risk evaluation for chronic diseases for individuals as well as for groups of individuals [49.13]. A personalized model is created for a single person, based on their personal data and the information in the ontology. A transductive neuro-fuzzy inference system with weighted data normalization [49.14] is used to evaluate personalized risk.

Ontologies in the biological domain have also been used supporting text mining and information retrieval. *Khelif, Dieng-Kuntz*, and *Barbry* [49.15] discuss the knowledge available in textual documents and the efficient detection and use of this knowledge, which is a challenging task. Biologists need tools to support them in the interpretation and validation of their experiments to facilitate future experiments. The authors propose an ontology-based approach for generation of semantic annotations and information retrieval and suggest that the proposed approach can probably be extended to other massive analysis of biological events. However, the authors do not present any experimental work. The proposed work is only suitable for microarray DNA experiment data. The proposed approach uses an annotation base and relations, and in certain cases these relations can be used interchangeably by the biologists

and there is a gap between system-based ontology and user-based ontology integration. To bridge the gap, biologists are required to strengthen the system-based ontologies by providing regular feedback of their ontology usage. The contextual information in the text documents is also very important in designing the ontologies in biological domain. The major problem, however, is that there are a number of text mining techniques consisting of different operators, but as yet there is no generic ontology framework for text mining in the biological domain.

Kuo, Lonie, Sonenberg, and *Paizis* [49.16] report on using a domain ontology-driven approach to DM using clinical data of patients undergoing treatment for chronic kidney disease. The authors acknowledge the challenges in mining such a multiple attribute dataset; the attributes in the dataset useful for DM are not apparent without domain knowledge. The reported work explores the use of medical domain ontology as a source of domain knowledge in both extracting and expressing knowledge in a useful format. Domain ontology has been used to categorize attributes in preparation for mining association rules in the data. The authors report that domain ontology driven DM can obtain more meaningful results than naive mining. However, determining the meaningfulness of the results is not an easy task – a strong association rule does not guarantee a useful rule. Also, although domain ontology has been used to provide domain expert knowledge to guide the mining process, a domain expert is still required to gauge the usefulness of the rules/results.

49.3 Data Mining Specific Ontologies

In this section, we present some recent work in ontologies for the DM domain.

Diamantini, Potena, and *Storti* [49.17] discuss the highly complex, iterative, and interactive, goal-driven and domain-dependent nature of the knowledge discovery in databases (KDD) and the DM process. The effective design of KDD process, the continuous development of new algorithms, their many characteristics, and the applicability to different kinds of data are all considered and the authors discuss issues in designing a successful knowledge discovery process. The main contribution of the work is the design of KDDONTO, an ontology for supporting both the discovery of suitable KDD algorithms and the composition of the KDD process. An ontology building methodology has been

proposed with the justification that previously proposed methodologies for ontology design were mainly borrowed from the software engineering field. There are a number of similarities between software engineering and ontology fields but the goals are totally different. For instance, the goal of software engineering is the development of implementable classes while the knowledge engineer looks for the formal representation of the domain. The proposed methodology satisfies the quality requirements of the formal ontology such as coherence, clarity, extensibility, minimization of encoding bias, and minimization of ontological commitment. For KDDONTO implementation, the authors have chosen the OWL-DL language. However, there are some issues yet to be resolved due to the expressive semantics restrictions in the properties of data and model classes. At present the KDDONTO implementation is formed of 95 classes, 31 relations, and more than 140 instances. Each step of the proposed methodology returns as output a valid ontology represented in a different language. The existing limitation of the work is that the KDD process composition using the proposed methodology is not automatic. However, authors plan to report the semi-automatic process composition in the future.

In order to help data miners consider the vast design space of possible processes thoroughly, *Bernstein*, *Provost*, and *Hill* [49.18] present an intelligent discovery assistant (IDA) that works with an explicit ontology of DM techniques to navigate space and present the user with a systematic enumeration of valid DM processes. IDA considers the characteristics of the data, the desired results, and works with the ontology to search for and enumerate valid plans to achieve the desired results. The processes may be ranked on criteria such as speed, accuracy, model comprehensibility, etc. The strength of the work is that the authors discuss how the proposed IDA tool can play an important role in knowledge sharing in the team of data miners. However, there are a few limitations in the work such as the fact that IDA provides a large number of valid plans to choose from but does not provide assistance in recommending a combination of processes to assist the user at this stage. Furthermore, there is no indication of which particular induction algorithm will be better to choose based on heuristic knowledge. IDA also lacks the incorporation of the characteristics of a dataset, which is a very important aspect before applying any DM technique. The ontology designed in this work is light-weight and does not contain the internal structure of DM operators [49.19]. The major gap is in the identification of interesting DM processes, which we think should be the primary objective of applying any DM technique.

Based on the need for a unifying framework for DM and an as yet unfulfilled proposal to describe the hierarchies of DM operations in terms of their signatures [49.20], *Panov*, *Dzeroski*, and *Soldatova* [49.19] put forward the idea of a DM ontology called OntoDM for the description of the DM domain. The authors' aim is to design OntoDM with sound theoretical foundations. They argue that ontologies proposed thus far are light-weight ontologies which can be easily developed and meet the specific purpose for which they were created but do not follow good practices of ontology development. They tend to be shallow, with no rigid relations between defined entities. The DM domain, unlike other domains, requires detailed inference

over data and, therefor, a rigid, heavy weight ontology is needed. In contrast to the work of *Bernstein* et al. [49.18], the proposed OntoDM covers the details of basic DM entities such as data types and datasets, DM tasks, and DM algorithms, components of DM algorithms, generalizations and constraints. OntoDM follows the philosophy of OBI (ontology for biomedical investigations) and EXPO ontology of scientific experiments. OBI's top-level ontology BFO (basic formal ontology) is used to define upper level classes. OBO RO (relational ontology) is also used to define the semantics of the relationships between the DM entities. The methodology is developed keeping in mind the complex entities and more popular research areas such as constraint based mining. The proposed ontology consists of three main components: classes, a hierarchical structure of classes, and relations between classes. The basic entities considered are based on a framework proposed by one of the authors in a prior work. The entities describe the different orthogonal dimensions of DM, and it is the authors' intention that different combinations of these entities can be used to describe the prevailing DM approaches. The entities are: dataset, data type, DM tasks, generalizations, DM algorithm, and components of DM algorithms such as distance functions, kernel functions, and features. Furthermore, it provides for defining more complex entities such as constraints and DM tasks. We find that the claim that OntoDM covers all the basic entities of DM and can be used for complex DM tasks needs proof of concept and further evaluation. The rapidly growing field of DM is incorporating intense nature inspired algorithms that are different from the traditional DM approaches and algorithms. We believe that such advancement should also be covered in the proposed ontology. Furthermore, the use of previous knowledge and the sharing of information have not been covered in the proposed ontology, which is an important aspect of KDD.

In *Panov*, *Soldatova*, and *Dzeroski* [49.21] the authors extend the work presented in [49.19] and present an updated version of OntoDM. The version described is updated in a number of ways. Alignment of the structure of the ontology with the top level structure of the OBI ontology introduced new entities in the ontology. For instance, the entity DM algorithm was split into three entities, each capturing a different aspect of the algorithm such as algorithm specification, algorithm implementation, and algorithm description. The set of relations used in the initial version is extended with relations defined in OBI ontology in order to express the relations between informational entities or entities that are realized in a process and processes. The authors have also extended the OBI classes with DM specific classes for defining complex entities such as DM scenarios and queries. However, the ontology presented is still in the early stages of development. There is a need to populate the proposed classes of DM entities with individuals and to refine the structure of OntoDM as needed in order to cover the various aspects of the DM domain.

An illustrative example of the use of ontology for DM has been provided in [49.22], which reports on an ontology-based DM system for discovering knowledge from incomplete data and the effectiveness of ontology in knowledge management. The results of applying ontologies for DM with incomplete data in a classroom environment are reported. The limitation is that a very simple example has been used for the demonstration of ontology development. As mentioned previously, there is a strong need for ontology development for complex DM tasks and procedures.

Hilario, Kalousis, Nguyen, and *Woznica* [49.23] present their vision of a DM ontology designed to support meta-learning for algorithm and model selection.

The authors argue that previous research has focused on aligning experiments and performance metrics and not much work has been done on explaining observations in terms of the internal logic and mechanism of learning algorithms. The authors extended the previously proposed Rice model, which related dataset descriptions to performance of algorithms, by adding algorithm features to dataset features as a parameter of algorithm selection. The key components of an algorithm include the structure and parameters of the models produced, the cost function used to quantify the appropriateness of a model, and the optimization strategy adopted to find the model parameter values that minimize the cost function. The authors discuss the next steps for the continuation of the proposed work. The first step is to gather interested data miners and ontology engineers to consolidate the core concepts of the DM ontology proposed, and the next step is to show how DM ontology can be used to improve algorithm selection through meta learning. However, there are still some ongoing issues in the proposed work. DM research has identified other components of bias for learning algorithms in addition to those described in this paper.

49.4 Discussion and Future Work

It is generally recognized that a standardized ontology framework makes data easily available for advanced methods of analysis and artificial intelligence algorithms. In the biomedical domain, GO and the subsequent OBO consortium have been instrumental in allowing the community of scientists to speak the same language, add to the knowledge by creating annotations, and use this in drawing inferences. The interested reader is referred to a detailed review [49.24] of trends in biomedical ontologies. The potential for wider use of ontologies in bioinformatics will be realized with greater use of ML methods.

Our review also found also that the use of ontologies with DM is either in a specific domain of interest, such as bioinformatics, or in the DM domain itself. There are ongoing research efforts both in the construction of light-weight DM ontologies and in the construction of top-level DM ontologies. Neither meet the needs of the KDD community entirely. Light-weight DM ontologies meet the specific purpose they were created for but do not follow good practices of ontology development. General purpose, top-level DM ontologies are not conceived for achieving specific support

requirements, like discovery of algorithms and process composition [49.17]. While such a general purpose top-level ontology is meant to be useful in supporting different activities, it ends up providing inefficient support in each activity. Ontologies, with their high level of abstraction, suffer as their construction and usage have been decoupled [49.25]. We believe that a coupling or integration of these two types of ontologies will open up new grounds for potentially rich areas of research into DM ontologies. Such integration is capable of providing excellent solutions to treat complex data and sophisticated DM methods and algorithms.

In the DM domain, ontologies are used to detect patterns in data, and further to retrieve facts or information. Knowledge in the ontology may be used to deduce features from the ontology, which help modify classical feature representations. We believe more work is required in enhancing an ML model with knowledge from ontology, resulting in a richer model.

We propose to develop a wholly-integrated ontology system capable of evaluating newly discovered knowledge and evolving in a recurring, automatic manner. The integrated ontology system will be composed of

ML methods, ontology, and associated knowledge base. To the best of our knowledge, there is no methodology that has dealt with these two types of ontologies in a single framework. This integration will ensure that usage will continuously add to the data and semantics of the ontology. For this to be realized a major challenge pertains to resolving the process of ontology augmentation – the knowledge, or additional facts, need to be confirmed as improving the accuracy of predictions on new data before they can be acknowledged as conceptual changes (new concepts or relationships), or as explication changes (changes to existing concepts or relationships) to the ontology [49.12]. We propose to use BGO in this exploration.

49.5 Summary

This chapter has reviewed current research in bio-neuro-informatics ontologies and found that, in general, biomedical ontologies are prolific and current ML efforts reuse existing ontologies in order to support text mining and/or other ML efforts. Research has concentrated on the solution of domain-specific problems or DM issues. We emphasize a need for a methodology that deals with the two types of ontologies in a single framework. Such an integrated solution can improve the analytical powers of modern DM systems. Our proposal for an integrated environment will also aid the coupling of construction and usage of ontologies and continued refinement, where usage will continuously add to the data and semantics of the ontology for significant advancement in computing research. Enhancement the decision-making process based on both ontology and ML is a rich and exciting area with substantial potential.

References

49.1 B. Chandrasekaran, J.R. Josephson, V.R. Benjamins: What are ontologies, and why do we need them?, Intell. Syst. Appl. **14**, 20–26 (1999)

49.2 A. Maedche, B. Motik, L. Stojanovic, R. Studer, R. Volz: Ontologies for enterprise knowledge management, Intell. Syst. IEEE **18**(2), 26–33 (2003)

49.3 I.H. Witten, E. Frank, M.A. Hall: *Data Mining: Practical Machine Learning Tools and Techniques* (Morgan Kaufmann, Burlington 2011)

49.4 M. Ashburner, C.A. Ball, J.A. Blake, D. Botstein, H. Butler, J.M. Cherry, A.P. Davis, K. Dolinski, S.S. Dwight, J.T. Eppig, M.A. Harris, D.P. Hill, L. Issel-Tarver, A. Kasarskis, S. Lewis, J.C. Matese, J.E. Richardson, M. Ringwald, G.M. Rubin, G. Sherlock: Gene ontology: Tool for the unification of biology, Nat. Genet. **25**(1), 25–29 (2000)

49.5 R. Stevens, P. Lord: Application of ontologies in bioinformatics. In: *Handbook on Ontologies*, ed. by S. Staab, R. Studer (Springer, Berlin, Heidelberg, 2009) pp. 735–756

49.6 http://www.kedri.aut.ac.nz (last accessed April 30, 2013)

49.7 N. Kasabov, V. Jain, P.C.M. Gottgtroy, L. Benuskova, F. Joseph: Brain-gene ontology, current version can be downloaded from http://www.kedri.aut.ac.nz/areas-of-expertise/neurocomputation-and-neuroinformatics/brain-gene-ontology

49.8 Protégé: http://protege.stanford.edu (last accessed April 30, 2013)

49.9 Unified Medical Language System http://www.nlm.nih.gov/research/umls/(lastaccessed 30 April 2013)

49.10 N. Kasabov: *Evolving Connectionist Systems The Knowledge Engineering Approach*, 2nd edn. (Springer, Berlin, Heidelberg 2007) p. 451

49.11 WEKA: http://www.cs.waikato.ac.nz/ml/weka (last accessed April 30, 2013)

49.12 N. Kasabov, V. Jain, P.C.M. Gottgtroy, L. Benuskova, S.G. Wysoski, F. Joseph: Evolving brain-gene ontology system (EBGOS): Towards integrating bioinformatics and neuroinformatics data to facilitate discoveries, Int. Joint Conf. Neural Netw. (IJCNN) 2007 (IEEE 2007) pp. 131–135

49.13 A. Verma, N. Kasabov, E. Rush, Q. Song: Ontology based personalized modeling for chronic disease risk analysis: An integrated approach, LNCS **5506**, 1204–1210 (2008)

49.14 Q. Song, N. Kasabov: TWNFI – a transductive neuro-fuzzy inference system with weighted data normalization for personalized modeling, Neural Netw. **19**, 1556–1591 (2006)

49.15 K. Khelif, R. Dieng-Kuntz, P. Barbry: An ontology-based approach to support text mining and information retrieval in the biological domain, J. Univers. Comput. Sci. **13**(12), 1881–1907 (2007)

49.16 Y. Kuo, A. Lonie, L. Sonenberg, K. Paizis: *Domain Ontology Driven Data Mining: A Medical Case Study*, ACM SIGKDD Workshop on Domain Driven DATA MINING (DDDM2007) (ACM, San Jose 2007)

49.17   C. Diamantini, D. Potena, E. Storti: KDDONTO: An ontology for discovery and composition of KDD algorithms, Third Generation Data Mining: Towards Service-Oriented Knowledge Discovery 19–24 (2009)

49.18   A. Bernstein, F. Provost, S. Hill: Toward intelligent assistance for a DATA MINING process: An ontology-based approach for cost-sensitive classification, IEEE Trans. Knowl. Data Eng. **17**(14), 503–518 (2005)

49.19   P. Panov, S. Dzeroski, L. Soldatova: OntoDM: An ontology of Data Mining, IEEE Int. Conf. DATA MINING Workshops (IEEE, Washington 2008) pp. 752–760

49.20   R. Ramakrishnan, R. Agrawal, J.-C. Freytag, T. Bollinger, C.W. Clifton, S. Dzeroski, J. Hipp, D. Keim, S. Kramer, H.-P. Kriegel, U. Leser, B. Liu, H. Mannila, R. Meo, S. Morishita, R. Ng, J. Pei, P. Raghavan, M. Spiliopoulou, J. Srivastava, V. Torra: Data mining: The next generation, Perspectives Workshop: Data Mining: The Next Generation, number 04292, Dagstuhl Seminar Proc., ed. by R. Agrawal, J.C. Freytag, R. Ramakrishnan (Internationales Begegnungs- and

Forschungszentrum für Informatik (IBFI), Schloss Dagstuhl 2005)

49.21   P. Panov, L.N. Soldatova, S. Džeroski: Towards an ontology of data mining investigations. In: *Discovery Science*, ed. by J. Gama (Springer Berlin, Heidelberg 2009) pp. 257–271

49.22   H. Wang, S. Wang: Ontology for data mining and its application to mining incomplete data, J. Database Manag. **19**(4), 81–90 (2008)

49.23   M. Hilario, A. Kalousis, P. Nguyen, W. Woznica: A DATA MINING ontology for algorithm selection and meta-mining, Third Generation Data Mining: Towards Service Oriented Towards Service-Oriented Knowledge Discovery (SoKD) (2009) p. 76

49.24   O. Bodenreider, R. Stevens: Bio-ontologies: Current trends and future directions, Brief. Bioinform. **7**(3), 256–274 (2006)

49.25   M. Hepp: Ontologies: State of the art, business potential, and grand challenges. In: *Data Management*, ed. by M. Hepp, P. De Leenheer, A. de Moor, Y. Sure (Springer, Berlin, Heidelberg 2007) pp. 3–24

50. Integration of Large-Scale Neuroinformatics – The INCF

Raphael Ritz, Shiro Usui

Understanding the human brain and its function in health and disease represents one of the greatest scientific challenges of our time. In the post-genomic era, an overwhelming accumulation of new data, at all levels of exploration from DNA to human brain imaging, has been acquired. This accumulation of facts has not given rise to a corresponding increase in the understanding of integrated functions in this vast area of research involving a large number of fields extending from genetics to psychology. Neuroinformatics is uniquely placed at the intersection between neuroscience and information technology, and emerges as an area of critical importance to facilitate the future conceptual development in neuroscience by creating databases which transcend different organizational levels and allow for the development of different computational models from the sub-cellular to the global brain level.

50.1 Neuroinformatics: A Megascience Issue Calling for Concerted Global Actions

The brain makes us perceive the world around us and generates the thoughts we have, the memories we store, and the emotions we express, that is, what makes us uniquely human. It controls our immediate and long-term response to the environment. The human brain can be considered as the most complex organ created during vertebrate evolution.

Many severe illnesses associated with brain function need to be alleviated or cured, such as the different forms of mental disorders, degeneration, stroke, and drug abuse. While patients and their families suffer, diseases of the nervous system represent more than one third of all costs for health care in society. The brain is also of interest in engineering, as it is, in many respects, the most effective information processor known to exist. The principles underlying cognitive brain functions are being applied to advance engineering and information technology in different areas, including developing advanced robot assistants for everyday life.

A large number of disciplines with different methodologies, approaches, and traditions address different aspects of brain function. Psychologists, linguists, and ethologists describe the ability of the nervous system to generate different aspects of behavior in man and animals. Molecular and cellular neuroscientists in turn describe the operation of the nervous system at the levels of cell, synapse, and microcircuit, and how the nervous system is put together during development. Other neuroscientists address the neural bases of many diseases of the brain. All of these 10–15 different subdisciplines of neuroscience contribute, each with its particular approach and methodology, to the understanding of the many different functions of the brain and its different disease mechanisms.

Using a great variety of sophisticated technologies, neuroscience has been dominated by the acquisition of experimental data and fact-finding that range from the genome to human brain imaging at all intervening analytical levels, with an ever increasing granularity. The resulting large quantities of data are heterogeneous and fractionated. While there has been an overwhelming accumulation of new facts, at all levels of exploration, there has not been a corresponding increase in insights into integrative brain function or disease mechanisms. One problem that can account for this short-coming is the difficulty for individual neuroscience researchers to study more than a fraction of the knowledge required to gain the necessary insights and to move conceptually from one organizational level to another; for example, from that of a gene expressed somewhere in the brain to its role for a given subroutine of the brain. This is where the field of neuroinformatics is of critical importance. It will facilitate the future conceptual development in neuroscience by creating databases which transcend different organizational levels, and allow for the development of different computational models from the subcellular to the global brain level. The International Neuroinformatics Coordinating Facility (INCF) has been formed to facilitate the development of the necessary infrastructure.

Neuroinformatics is uniquely placed at the intersection between neuroscience on the one hand, and information technology and computer science on the other. It brings together interdisciplinary contributions

to develop and apply the sophisticated tools and approaches needed for basic and clinical neuroscience. The resultant synergy achieved by combining these different disciplines will accelerate scientific and technological progress and lead to major medical, social, and economic benefits.

The three principal aims of INCF and of neuroinformatics are:

1. To facilitate the development of interoperable, structured neuroscience databases and to optimize the accumulation, storage, and sharing of vast amounts of primary data. The most immediate goal is to develop standards and to find mechanisms for making the large amount of varied data available to researchers in order to facilitate data sharing. In the near future, neuroinformatics databases will most likely have a similar role in neuroscience as bioinformatics databases have in the fields of genomics and proteomics.

2. To create infrastructure for the development of computational models of brain structure and function that can be validated by available data. As in all areas of science, an understanding of the systems and phenomena under investigation requires development of models that are not just descriptive, but also predictive and explanatory. In brain research, the systems and phenomena are among the most difficult to model: from molecule, cell, synapse, and microcircuit to perception, learning, memory, and reasoning. The only way to validate such models is through confrontation with the available detailed data sets of neuroscience.

3. To develop informatics tools for manipulating and managing different types of neuroscience data. In this field, the existence of very large bodies of data and of tools for navigating and manipulating these data will lead to breakthroughs in our understanding and to important commercial applications linked to human health.

To obtain all the evidence required to understand the operation of each part of the brain is beyond the capacity of any individual or country. This is due not only to the complexity of the brain and to the large number of neuroscientists active throughout the world, but also to the fact that evidence relevant to an understanding of how the brain works must be drawn jointly from the many different methodologies and disciplines described above. This is why neuroinformatics is considered a megascience issue that needs to be addressed

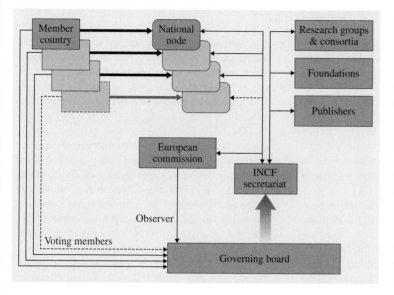

Fig. 50.1 The organization of INCF includes three main components. The Secretariat is the central facility of INCF, responsible for the execution of scientific and administrative activities and for the implementation of INCF work programs. It assists the national nodes (as defined below) in reaching their goals and plays a proactive role in furthering neuroinformatics. The Governing Board is the means by which the participating countries make collective decisions on all matters relating to INCF. The National Nodes are facilities, networks, or other mechanisms, funded by national sources. The nodes are established in order to coordinate and facilitate activities within a country and to provide an interface with the Secretariat. Furthermore, they participate in the formulation and implementation of INCF programs

in a coordinated manner, at both national and international levels.

Recognizing the enormous benefits of applying information technologies in the biomedical sciences, in January 1996, the OECD Megascience Forum established a Working Group focusing on neuroinformatics to strengthen international cooperation through the creation of new, shared internet-based capabilities for data and information management. In December 1998, the findings and recommendations were presented in the report *OECD Megascience Forum: Bioinformatics*. In this report, the then newly identified field of neuroinformatics was strongly endorsed, and additional steps were recommended in order to create a coordinated international neuroinformatics effort.

The subsequent OECD Global Science Forum continued the efforts with a further Neuroinformatics Working Group, established in January 2000, to examine issues of critical importance to furthering the development of neuroinformatics. It was recommended that, to fully realize the scientific, economic and social potential of neuroinformatics, the governments of the OECD countries should continue to individually support and develop national neuroinformatics pro-

grams and to jointly support a global neuroinformatics initiative to facilitate coordination of international research and resources in the field. These initiatives were considered critical to achieving global neuroinformatics capability across nations, scientific disciplines, and individual institutions. These recommendations were endorsed at a meeting with all ministers of research within the OECD in January 2004. As a result, the INCF was established in August 2005 with the proposed structure as shown in Fig. 50.1. In November 2005, after a competitive bid evaluated by a commission convened by the Secretary General of the Global Science Forum of the OECD, the decision was made to locate the Secretariat in Stockholm, on the premises of Karolinska Institutet with support from the Royal Institute of Technology. The founding Executive Director of INCF, Professor Jan Bjaalie, was appointed in April 2006. Since January 1st, 2011, Professor Sean Hill has taken over the role.

The Executive Director reports to the Governing Board, in which each member state has one vote, but may have two representatives present (e.g., one scientist and one governmental representative). The Chairman of the INCF Governing Board is Professor Sten Grillner.

50.2 Building Up of the INCF Organization

50.2.1 INCF Process for Planning Scientific Activities – Topical Workshops as the Basis for Actions

One of the most important stages for the development of the INCF organization has been the formation of a stepwise process for planning future actions in different target areas, as follows:

1. A topical workshop is held with leading experts in a target area.
2. The resulting workshop report provides recommendations for the development of a specific area and for the future role of INCF.
3. This is followed by an analysis by the Secretariat and discussions/confirmation at the next semi-annual INCF Governing Board meeting of the suitability and feasibility of the recommendations, in terms of service, standards, and training, to the neuroinformatics community at large.
4. Selected actions are launched.

As the basis for further actions and activities, INCF topical workshops are of critical importance for planning INCF scientific programs. They deal with selected problems, including technical issues of primary importance for neuroinformatics, or research topics coupled to databasing of neuroscience information, development of tools, and modeling of nervous system function. Ten workshops have been completed, with the contributions from more than 100 world leading experts from around the globe. The reports of these workshops are published on the INCF website as well as in Nature Precedings. A comprehensive survey of all INCF topical workshops is available at http://www.incf.org/programs/incf-topical-workshops.

50.2.2 INCF Mechanism for Carrying Out Activities – Oversight Committee, Task Force, and Reference Group

INCF has developed a mechanism for implementing prioritized recommendations resulting from the topical workshops. The mechanism actively involves opinion leaders and technical experts in three working groups with different functions: an Oversight Committee, a Task Force, and a Reference Group. Their roles are as follows:

- Upon the completion of the workshop report and evaluation of its recommendations, the Secretariat identifies two to four experts with appropriate expertise for intended actions and works with them to form an Oversight Committee of the INCF activities.
- The Oversight Committee consists of approximately ten world leading players in the relevant field of activity, reflecting different expertise and with wide geographic representation. The responsibility of this committee includes the determination of overall scope and process of the programmed actions. The Oversight Committee may set up one or more Task Force(s) to execute concrete action(s) and additionally establish a Reference Group.
- A Task Force is composed of approximately ten technical experts. Each Task Force is established for a specific task.
- The Reference Group is a more inclusive body of approximately 40–50 experts from INCF's internal network and the scientific community at large. Members of this group contribute to the preliminary products generated by the Task Force/Oversight Committee. The broad representation reflected in the Reference Group is essential for high impact of the INCF actions.

This sequential structure is used for each of the INCF Scientific Programs. It has been found to be particularly useful when establishing standards and guidelines where broad involvement of the community is necessary. The scheme is feasible for practical operations, as the respective deliverables of each program have to be specified (see the next section).

50.2.3 INCF Cooperative Partnerships – The Neuroscience Peer Review Consortium

Collaboration with strategic partners is important for INCF development as it allows INCF to be involved in coordinating infrastructure, standards, and guidelines for the field of neuroinformatics. By partnering with editors of scientific journals, INCF can facilitate scientific collaboration at the pre-publication stage, support development of non-traditional merit evaluation, and boost funding efficiency in scientific research enterprises. The Neuroscience Peer Review Consortium

(NPRC; http://nprc.incf.org) is an alliance of neuroscience journals that have agreed to accept manuscript reviews from other members of the Consortium. Its goals are to support efficient and thorough peer review of original research, speed up the publication of research reports, and reduce the burden on peer reviewers. The idea for such a consortium originated at the PubMed Plus Conference, a meeting of scientists, journal editors, and scholarly publishers, organized by the Society for Neuroscience in June 2007. As an independent, international neuroscience organization unaffiliated with a journal, INCF was deemed as a suitable, neutral venue to host the necessary infrastructure and provide technical support for theConsortium. NPRC has been operative since January 2008; there was a trial period of 1 year. With nearly 40 current member journals, including virtually all major ones in neuroscience, and the early indication of success, NPRC will continue and expand. The establishment of this Consortium has promoted awareness of the INCF in the neuroscience community, and has placed INCF in a better position for future collaboration with journals, such as establishing minimal requirements and standards for metadata.

50.3 INCF Scientific Programs

INCF programs represent long-term strategic undertakings to address issues of importance to the neuroscience community. Targeted to defined groups of stakeholders and clients, they span a relatively long period of time with the aim of solving identified problems through actions. INCF programs are multidisciplinary and organized under specific domain areas of neuroinformatics.

Within each program, diverse actions are carried out, including products and services provision, standards establishments and compliance, and forum and community development.

To date, four INCF programs have been launched: Digital Brain Atlasing, Ontologies of Neural Structures, Multi-Scale Modeling, and Standards for Data Sharing.

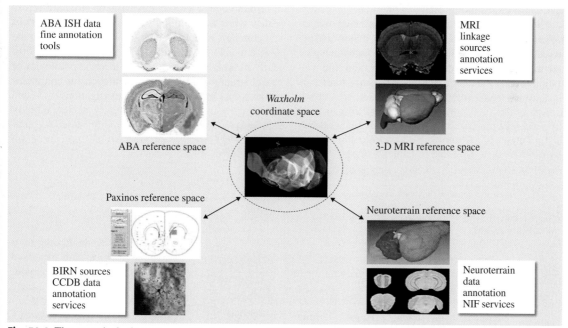

Fig. 50.2 The canonical atlas space, or Waxholm space (WHS), acts as the hub of a centralized infrastructure connecting several key reference spaces. Reference atlases mapped into this space are *normalized* and may share their associated data and services in a manner that is understandable to external sources/users

50.3.1 Program on Digital Atlasing

Digital brain atlases are an essential tool in neuroscience research. They function as references and as analytical tools, and provide a stable data integration framework to investigate normal and abnormal brain structure and function. Furthermore, web-accessible brain atlases and spatial indices promise to evolve into powerful tools for dynamic, multi-dimensional modes of scientific interaction.

The key aim of this INCF program is to coordinate and improve the impact of atlasing projects, with a focus on rodent brain. As a step towards fostering global interoperability, accessibility and sharing of databases, INCF started with a comprehensive analysis of the current and desired status of digital atlasing, which led to the vision and direction necessary to make the rapidly growing collection of multi-dimensional data such as images and gene expression widely accessible and usable to the international research community (First Task Force Report). Central to this endeavor is a new standardization in atlas mapping, proposed as Waxholm Space (named after the site of a key meeting in 2008). Waxholm Space is a coordinate-based reference space for the mapping and registration of neuroanatomical data, as illustrated in Fig. 50.2.

INCF has assembled a task force, consisting of leading experts who have been involved in major atlasing projects worldwide, to bring data into Waxholm Space and to establish procedures for registration and validation. From the onset, INCF has demonstrated its capacity and leadership for promoting international collaborations and partnerships among not only publicly funded projects but also private initiatives such as the Allen Brain Atlas (for which INCF also hosts the European mirror site). To date, Waxholm Space has been clearly defined and its construction has begun by setting a standardized acquisition procedure for passing data into it, using a high-resolution magnetic resonance imaging (MRI) dataset and companion Nissl-stained reconstructions. The next steps include the registration of key reference atlases into this space, as shown in Fig. 50.2, and development of a set of *best practices* for experimenters to ensure Waxholm Space compatibility.

The longer term goal of this program is to establish a broader INCF digital atlasing infrastructure to facilitate interoperability between atlases and data sharing. This infrastructure is envisioned as a collection of distributed services that support publication, discovery, and invocation of heterogeneous atlas resources, as detailed in Fig. 50.3.

50.3.2 Program on Ontologies of Neural Structures

The Program on Ontologies of Neural Structures aims to establish a platform for translation and clarification of terminologies, with an emphasis on structure across scales, while underlining the cross-mapping of indices and comparison of terminologies. Closely linked to the Program on Digital Atlasing, this initiative endeavors to reach all areas of neuroscience, since the ontologies of neural structures represent a foundation for communication across multiple levels of investigation.

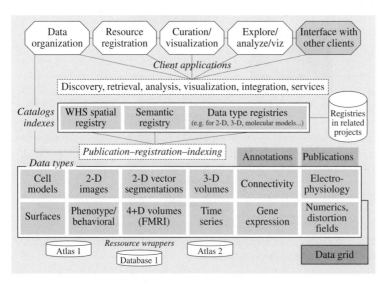

Fig. 50.3 INCF digital atlasing infrastructure resources will include multiple data types that are accessed via common mechanisms for each data type. Services and registries will be created to manage conversions between the sources and the INCF digital atlasing standards and to expose information to client applications

Three initial foci have been prioritized by the program's Oversight Committee: structural lexicon, neuronal registry, and technical infrastructure for the program. For the structural lexicon, a group of leading neuroanatomists worldwide have convened and pilot-tested a prototype wiki to serve as the initial venue for gathering consistent definitions of structural terms across scales. Reviewing the initially populated information has shown the feasibility of further populating the site with the goal to eventually open it to community input. The structural lexicon, in turn, will form the basis for more formal ontologies for cross species anatomy, as recommended by the infrastructure task force. In parallel, a neuron registry and a convention for naming neurons are being established, with particular consideration of the relation among morphological and functional characteristics. INCF has been bringing together various efforts that were started independently around the world, e.g., the Petilla standards for the description of cortical interneurons, but share common goals. As the technical support and closely coordinated with the above activities, INCF has been building the IT infrastructure technical platform to help the establishment of supporting services for making all knowledge usable to individuals.

Complementary to the consensus-building effort to harmonize semantics and ontologies, INCF has also facilitated coordinated technical developments in the creation of a lexicon of neuroanatomical structures through the leverage of existing resources and the INCF global network. Many of the essential features to achieve the objectives of this INCF program have already been implemented within the Brain-Info system (braininfo.rprc.washington.edu), a neuroanatomical resource established in the mid-1970s, and within the Neuroscience Information Framework (NIF; http://neuinfo.org). INCF played a critical role in brokering the sustained availability of BrainInfo to the community as a vital and dynamic resource via its US node. In addition, INCF has been supporting the continued population of this resource, using the SQL database and curatorial tools developed by BrainInfo and interacting with the INCF Japan node, to expand the utility of BrainInfo into an international neuroinformatics infrastructure. The NIF project is an initiative of the US National Institutes of Health Blueprint for Neuroscience Research. Its goal is also to provide a consistent framework for describing and locating neuroscience resources (tools, data, and knowledge). Through INCF, these efforts can be extended to the larger international community.

The long-term agenda of the INCF Program on Ontologies of Neural Structures is to provide a useful and practical means for data sharing and the re-use of data across research disciplines, with the consideration of state-of-the-art technological developments from molecular biology to cognitive science.

50.3.3 Program on Multi-Scale Modeling

Multi-scale modeling is an increasingly important tool in neuroscience research. As computational modeling techniques become integrated with experimental neuroscience, more knowledge can be extracted from existing experimental data. Quantitative models assist in generating experimentally testable hypotheses and in selecting informative experiments. Therefore, large-scale models are becoming an essential tool in bridging multiple levels of organization in the description and understanding of the nervous system. As an international organization advancing global collaborations in neuroscience research, this INCF program again concentrates on integrated actions to promote interoperable model construction and simulation sharing in neuroinformatics.

One major challenge in the field is that, because of the wide range of simulation tools being used in the community, it is unlikely that one laboratory can reproduce the results obtained by another group, even if they deposit the model in a database like ModelDB. This problem is due to the lack of appropriate simulator-independent language standards to communicate models constructed and published in the field of computational neuroscience. The systems biology community has led the way in proposing effective solutions to this problem like the SBML and CellML languages. The development of this standard language has fostered a strong growth of diverse software tools supporting systems biology and physiome modeling. Inspired by these examples, which are strongly embedded in the respective modeling communities, INCF has decided to foster the development of community-supported description languages for neuroscience models. The first INCF action focuses on a fast growing area in computational neuroscience, spiking networks of simple model neurons, as a particular level of modeling where the initiative is expected to exert a large impact. A community-based development effort has begun to establish a machine-readable declarative language standard that describes integrate-and-fire neural network models. Under INCF coordination, a proposal on language structure listing all model/connectivity/plasticity

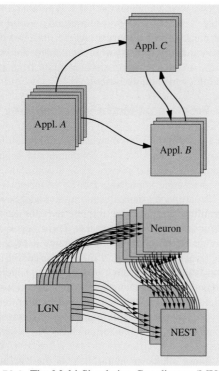

Fig. 50.4 The Multi-Simulation Coordinator (MUSIC) is a standard interface for run-time exchange of data among parallel applications in a cluster environment. It is designed for interconnecting large-scale neuronal network simulators with each other or with other tools. Data may consist of events, such as neuronal spikes, or graded continuous values, such as membrane voltages. A typical usage example is illustrated at the *top*, where three applications (A, B, and C) execute in parallel while exchanging data via MUSIC. The software interface promotes inter-operability by allowing models written for different simulators to be simulated together in a larger system. It enables re-usability of models or tools by providing a standard interface. As data are spread out over a number of processors, it is non-trivial to coordinate data transfer so that it reaches the correct destination at the correct time. The task for MUSIC is to relieve the applications of handling this complexity, as indicated at the *bottom*

types covered and including preliminary examples of implementation is being formed by technical experts worldwide. A first version of the standard language is expected to be fully defined by the end of 2010 and the first software implementations available by fall 2011. The formulation of a new standard to describe network models in a machine readable, declarative format, based on XML or similar technologies will have a profound impact on the modeling community as many researchers in the field still write their own simulation code for each modeling project. While this standard is defined, INCF has also liaised with the community to develop a minimal set of tools that facilitate the use of the standard, including updating network simulators to read/write the standard and a GUI-based tool to construct models.

While the action to establish standards will evidently assist the interoperability and sharing for future modeling work, many existing models would benefit from software tools for integration and interaction between different simulators in order to harness fully their usefulness. Accordingly, as a parallel project to the standardization initiative, INCF commissioned a simulator interoperability interface tool through the National Node of Sweden, following the recommendations of the INCF Workshop on Large-Scale Modeling. The Multi-Simulation Coordinator (MUSIC) is an API allowing large-scale neuron simulators using MPI internally to exchange data during runtime. As shown in Fig. 50.4, the software provides mechanisms to transfer massive amounts of event information and continuous values from one parallel application to another. In particular, it handles data transfer between applications that use different time steps and different data allocation strategies. Existing and future simulators can make use of MUSIC compliant general purpose tools and participate in multi-simulations. In addition, such a standard interface enables straightforward independent third-party development and community sharing of reusable and interoperable software tools for parallel processing.

MUSIC was released in March 2009 via the INCF Software Center (software.incf.org/software/music). Special effort was taken to facilitate easy adaptation with existing simulators; for example, MUSIC interconnects with both NEST and MOOSE (GENESIS), two widely used simulation environments. INCF is coordinating further efforts for the use of MUSIC with other major simulators such as NEURON.

50.3.4 Program on Standards for Data Sharing

Neuroscience data, particularly those in neuroinformatics related areas such as neuroimaging and electrophysiology, are associated with a rich set of descriptive information often called metadata. For data archive, storage, sharing, and re-use, metadata are of equal importance to primary data, as they define the methods and conditions of data acquisition (such as device characteristics, study/experiment protocol and parameters, behavioral paradigms, and subject/patient information), and statistical procedures. A further challenge for data sharing is the rapidly evolving nature of investigative methods and scientific applications.

The overall scope of this new program is to develop generic standards and tools to facilitate the recording, sharing, and reporting of metadata. It is expected that these efforts will greatly improve upon current practices for archiving and sharing neuroscience data. Two task forces have initially been created to focus on:

- Neuroimaging data and databases of neurological/psychiatric disorders and cognitive function
- Electrophysiological data and databases.

50.4 INCF Neuroinformatics Portal

The INCF Neuroinformatics Portal aims to function as the foremost channel for access to resources and interactions with the neuroscience community. Based on the recommendations from the INCF Workshop on Global Portal Services, a portal-of-portals service has been developed and is being improved to serve as an entry point to relevant neuroinformatics resources.

50.4.1 Overall Structure of the INCF Neuroinformatics Portal

The INCF Neuroinformatics Portal is designed as a *portal of portals* linking to and integrating with existing neuroinformatics portal services. Where necessary, it acts as a primary service provider as well. The overall structure of the portal is composed of five main categories. Firstly, about resources available to the community: research tools, training and funding opportunities, job offers, and supercomputer access. Secondly, a community section including a people directory, an event calendar, and information about other activities in the field such as competitions. Thirdly, the INCF News Room including all news around INCF, the INCF Blog, press material, and links to various content feeds. Fourthly, about INCF itself; the organization including the people involved, member countries, publications, and funding source. Last but not least there is a section covering all programs and program-related activities within INCF. The main portal is implemented using the web-based content management system Plone. The first release of the portal took place in November 2008 at the Society for Neuroscience Annual Meeting.

50.4.2 INCF Software Center

INCF has developed a Software Center as a primary resource to find and register neuroscience-related software tools. The INCF Software Center (http://software.incf.org) provides an interactive web interface allowing developers to provide information about their projects, thereby making their work discoverable to the neuroscience community at large. In addition, the Software Center can also be used to manage individual projects, including services for handling software releases, project wikis, documentation, issue trackers, and revision control systems. A major focus of INCF is on documentation and quality assessment. A reviewing system has been established to document a tool's usability and level of maturity, together with various means for the community to provide feedback on and contributions to individual projects. The Software Center was first released in July 2008 at the FENS Forum, the bi-annual conference of the Federation of European Neuroscience Societies.

50.4.3 Content Expansion and Coordination with INCF Programs

The principle of serving as a *portal of portals* linking to and integrating with existing portal services remains as the guidance for future development of the Portal. INCF is in discussion with various major resources to facilitate the coordination via technical implementations. A technical project in collaboration with the Neuroimaging Informatics Tools and Resources Clearinghouse (http://www.nitrc.org) has enabled content-sharing with the INCF Portal and Software Center, so that tools and other relevant information such as announcements of events or job posts registered at one site are also present in the other.

The basic structure of the INCF Neuroinformatics Portal has been completed. While further functions and improvements continue to be added, the focus of the next steps will shift to content development, in particular, the curation of relevant neuroinformatics resources and the expansion of neuroinformatics community section.

With the growth of the INCF programs, the INCF Neuroinformatics Portal will coordinate and interact with other activities within the INCF global network and establish closer links to distributed resources at the INCF National Nodes in member countries. Accordingly, the development of the INCF Neuroinformatics Portal will take into account the requirements of information technology derived from INCF programs, especially the infrastructural needs and tools for promoting adoption of INCF standards in the community.

50.5 INCF Global Network: National Neuroinformatics Nodes

INCF was conceived to operate as a central (Secretariat) as well as a distributed (National Nodes of INCF member countries) facility to coordinate and harmonize global neuroinformatics efforts. Implementation of INCF activities is carried out via the Secretariat and National Nodes, and National Nodes transmit international activities locally. For the past few years, INCF National Nodes have gradually been established and significant developments have been achieved based on the national needs, circumstances, and resources.

Among the current INCF member countries, the National Node of Japan (J-Node) will be taken as an example since it is the most advanced in terms of both structure and overall activity. It serves as a model for the development of a national node within the INCF global network. Fulfilling the mission to develop neuroinformatics in Asia, the Neuroinformatics Japan Center (NIJC) was organized at the RIKEN Brain Science Institute to act as the national interface for INCF in February 2006. To date, ten neuroscience databases have been established and are openly available through the J-Node Portal (www.neuroinf.jp) all based on the baseplatform XooNIps (xoonips.sourceforge.jp/index.php?ml_lang=en). The J-Node has also coordinated collaborative projects to set up the J-Node Platforms (Fig. 50.5).

To facilitate the development of these platforms, a committee of experts, drawn from the related fields throughout Japan, has been established. The Platform Committee members are responsible for accumulating

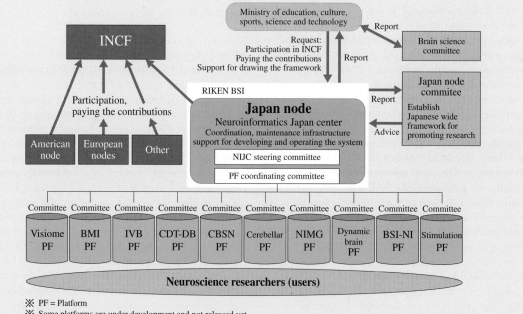

Fig. 50.5 Structure and function of the INCF National Node of Japan (J-Node) (courtesy of http://www.neuroinf.jp/)

data for each platform database, acting as moderators for maintaining the quality, and networking in the respective platform sub-domains, as follows:

- Visiome Platform
- Brain Machine Interface Platform
- Invertebrate Brain Platform
- Cerebellar Development Transcriptome Database
- Comprehensive Brain Science Network Platform
- Cerebellar Platform
- Neuroimaging Platform
- Dynamic Brain Platform
- BSI-Neuroinformatics Platform
- Simulation Server Platform.

Commencing with the symposium to mark the inauguration of J-Node, many outreach activities have been organized, including special sessions for INCF/J-Node at the Japan Neural Network Society, and a Joint UK-Japan Workshop organized by RIKEN and the Code Analysis Repository and Modeling for eNeuroscience (CARMEN) Project. A dissemination channel within Japan has been instituted through an email newsletter. Building on the strength, in 2010, J-Node has hosted the Third INCF Congress in cooperation with three national societies: the Japan Neuroscience Society, the Japanese Society for Neurochemistry, and the Japanese Neural Network Society.

To date, all current INCF countries have established respective national nodes. Although they are at different stages of maturity and of varied structures, better integration with Secretariat activities and coordination among individual national nodes have been reached via a number of measures:

- Descriptions and specifications of structure at more mature nodes have been gathered to help others to set up and coordinate their operations.
- A travel grant scheme has been established to facilitate communication between the Secretariat and National Nodes, and between the Nodes themselves.
- An INCF National Node workshop has been launched on an annual basis to comprehensively discuss issues related to local operations and to investigate concrete actions for improved integration of INCF worldwide components. The first workshop was held in October 2008.

As a result, INCF has begun to implement core infrastructure elements and other activities at the appropriate National Nodes.

50.6 Educational and Outreach Activities

Neuroinformatics requires the integration of knowledge from mathematics, physics, computer science, and engineering, together with detailed knowledge of the nervous system. It is essential that neuroinformaticians be able to communicate with researchers across the spectrum of all relevant disciplines. Therefore, training for the next generation of neuroinformaticians poses a specific challenge. Given the broad scope of neuroinformatics, only a few countries will be able to provide a wide range of training opportunities. International cooperation is thus required. INCF has spearheaded the efforts of coordination at the international level to examine issues on training in the field, particularly related to establishment of best practices for the different main areas of neuroinformatics for students with different undergraduate backgrounds, e.g., physics and engineering as opposed to biology and medicine. A series of workshops has investigated the current provision, needs for neuroinformatics training, good practice, and development of policies and standards for future curricula. A repository of existing courses has been established at the INCF Neuroinformatics Portal.

As an integrated action for promoting interaction and facilitating communication of the community, INCF has launched its annual congress with an interdisciplinary emphasis. The first INCF Congress in 2008 functioned successfully in terms of both the number of participants and the operation. Researchers from several distinct fields of neuroscience and technical development were attracted. Given the very positive response from attendees, it was decided to make the congress an annual event. With the 2009 congress held in Pilsen, the Czech Republic and the 2010 congress in Kobe, Japan, the 2011 congress in Boston, USA, and 2012 in Munich, Germany, this INCF outreach activity contributes positively to the profile and visibility of the organization.

50.7 Looking Ahead

In a short period, INCF has identified a number of areas of critical importance in neuroinformatics and initiated a diversity of actions. These are long-term undertakings for a sustainable neuroinformatics infrastructure with significant results over, not only the next few years, but also in a much longer perspective. The INCF programs are the initial steps of a strategy aiming at leading the neuroscience community down a path that facilitates integration between different levels of neuroscience from genes to behavior and the different disease mechanisms. Early successes have been obtained during the first years of operation, however, much is to follow. The overall goals of current INCF activities can be summarized as follows:

- To build up actions for fostering and coordinating interoperability, accessibility, and global sharing of databases.
- To develop and evolve a program for harmonizing semantics and ontologies as the fundamental framework connecting neuroscience research worldwide.
- To facilitate international standards and collaborations for time series data management and analysis.
- To establish integrated activities for promoting interoperable model construction and simulation sharing in neuroinformatics.
- To advocate and publicize the value of neuroinformatics in accelerating brain research via education and outreach to the neuroscience community.
- To promote opportunities for training the next generation of neuroinformaticians and to help member countries to increase the number of contributors to this interdisciplinary field.

- To develop further the management structure for effectively and efficiently serving the current needs and projected developments of the INCF scientific portfolio.

These goals represent a major enterprise and INCF must make a foremost effort in bringing these target areas from idea and plan to implementation and outcome in a way that make them useful for the neuroinformatics and neuroscience community. Although fundamentally a software problem, data integration is also an organizational challenge because there are different approaches being developed within the scientific community. INCF has arisen at a critical time and has positioned itself strategically to intermediate discussions vital to the development of standards. In addition to the four areas where INCF is active in establishing international agreements on standards, novel areas will be explored and actions are to be initiated in the future when the necessary resources become available. Such areas include molecular neuroscience and disorders of the nervous system.

More strategically, INCF intends to expand partnerships with journals and funding agencies to encourage collaborations at pre-publication stages and assist development of non-traditional merit evaluation in the field. INCF may instigate debate and initiative to boost funding efficiency by internationally coordinated funding schemes in neuroinformatics. Furthermore, INCF memberships with creative forms of participation will be promoted. As both opportunities and challenges arise beyond the immediate future, INCF will leverage its global network to reach a new horizon.

50.8 Concluding Remarks

The past few years witnessed the birth of an international organization. Within a short period of time, INCF has built up the organizational structure, developed a clear strategy and concrete objectives for the first 5-year period, and launched a series of scientific and technical activities that increasingly attract the attention of the neuroscience community at large.

This short history of INCF has already demonstrated that the development of dedicated databases and analytical and modeling tools, which extend from the genetic, anatomical and physiological levels to the behavioral

and cognitive levels, can be markedly facilitated if the responsibility for different aspects of the endeavor is shared among groups across the world, in an organized and coherent way with a consensus for long-term perspective. Obtaining a deeper understanding of the brain requires a large-scale, integrated global effort, and internationally collaborative research initiatives are better equipped to study the brain at multiple levels of complexity. INCF global coordination brings out added value and efficiency that are unachievable via individual efforts at the national level.

Further Reading

- J.G. Bjaalie, S. Grillner: Global neuroinformatics: The international neuroinformatics coordinating facility, J. Neurosci. **27**, 3613–3615 (2007)
- R. Williams: Mouse and rat brain digital atlasing systems, 1st INCF Workshop, ed. by J. Boline, M. Hawrylycz, R.W. Williams (Nature Precedings, Stockholm 2007), DOI: 10.1038/npre.2007.1046.1
- L. Swanson: Neuroanatomical nomenclature and taxonomy, 1st INCF Workshop, ed. by M. Bota, L. Swanson (Nature Precedings, Stockholm 2007), DOI: 10.1038/npre.2008.1780.1
- INCF: Business plan for the international neuroinformatics coordinating facility (2012), available online at http://www.incf.org/documents/incf-core-documents/INCF_BusinessPlan.pdf
- A. Lansner: Large-scale modeling of the nervous system, 1st INCF Workshop, ed. by M. Djurfeldt, A. Lansner (Nature Precedings, Stockholm 2006), DOI: 10.1038/npre.2007.262.1
- Ö. Ekeberg, M. Djurfeldt: MUSIC – Multisimulation Coordinator: Request for Comments, Nature Precedings (Stockholm 2008), DOI: 10.1038/npre.2008.1781.1
- P. Roland: NeuroImaging database integration, 1st INCF Workshop, ed. by L. Forsberg, P. Roland (Nature Precedings, Stockholm 2007), DOI: 10.1038/npre.2008.1781.1
- OECD: *Report of the OECD working group on neuroinformatics* (OECD, Paris 2002), available online at http://www.incf.org/med/Report_OECD_GSF_2002.pdf
- INCF: Program in international neuroinformatics (PIN) (2010) available online at http://www.incf.org/documents/incf-core-documents/INCF_PIN.pdf
- OECD: *Report of the OECD working group on biological informatics* (OECD, Paris 1999), available online at http://www.incf.org/documents/incf-core-documents/Report_OECD_MSF_1999.pdf
- INCF: Strategy overview (2008–2010), available online at http://www.incf.org/documents/incf-core-documents/INCFStrategyOverview
- INCF: Understanding for the international neuroinformatics coordinating facility, http://www.incf.org/documents/incf-core-documents/INCF_Understanding.pdf
- J. van Pelt: Sustainability of neuroscience databases, 1st INCF Workshop, ed. by J. Van Horn, J. van Pelt (Nature Precedings, Stockholm 2007), DOI: 10.1038/npre.2008.1983.1
- G. Shepherd: Global portal services for neuroscience, 1st INCF Workshop, ed. by J. van Pelt, G. Shepherd (Nature Precedings, Stockholm 2007), DOI: 10.1038/npre.2008.1779.1
- D. Willshaw: Needs for training in neuroinformatics, 1st INCF Workshop, ed. by D. Willshaw (Nature Precedings, Edinburgh 2008), DOI: 10.1038/npre.2008.2563.1

Part K Information Modeling for Understanding and Curing Brain Diseases

Ed. by Lubica Benuskova

Reinhard Schliebs

Alzheimer's disease (AD) is the most common neurodegenerative disorder in late life which is clinically characterized by dementia and progressive cognitive impairments with presently no effective treatment. This chapter summarizes recent progress achieved during the last decades in understanding the pathogenesis of AD. Basing on the pathomorphological hallmarks (senile amyloid plaque deposits, occurance of neurofibrillary tangles as hyperphosphorylated tau protein in cerebral cortex and hippocampus) and other consistent features of the disease (neurodegeneration, cholinergic dysfunction, vascular impairments), the mayor hypotheses of cause and development of the sporadic, not genetically inherited, AD are described. Finally, to reflect the disease in its entirety and internal connective relationships, the different pathogenetic hypotheses are tentatively combined to describe the interplay of the essential features of AD and their mutually influencing pathogenetic processes in a unified model. Such a unified approach may provide a basis to model pathogenesis and progression of AD by application of computational methods such as the recently introduced novel research framework for building probabilistic computational neurogenetic models (pCNGM) by *Kasabov* and coworkers [51.1].

51.1 Alzheimer's Disease

51.1.1 Epidemiology

In 1906/1907 the German psychiatrist Alois Alzheimer described a patient with progressive memory impairment, disordered cognitive function, altered personality and social behavior, including paranoia, delusions, and loss of social appropriateness. Following autopsy, in the patient's cerebral cortex and hippocampus he observed abundant extracellular amyloid-like protein deposits (neuritic plaques) and a number of pyramidal neurons containing intracellular inclusions of fibrillary structures (neurofibrillary tangles), which was accompanied by considerable neuronal, in particular cholinergic, cell loss [51.2]. During the following years similar cases were observed, and already in the German textbook of psychiatry edited by Emil Kraepelin in 1910, this novel brain disorder was named after *Alois Alzheimer*. Later it became evident that this kind of disorder also provides a common basis for impairments in cognition in elderly people in late life. The modern era of AD research, however, began with the identification of the β-amyloid sequence in the congophilic angiopathy of AD and Down's syndrome [51.3]. The finding that the amyloid plaques observed in both disorders contained largely the same peptide [51.4] and the cloning of the β-amyloid precursor protein gene [51.5–8] have led to major progress in understanding the pathology and biochemistry of AD.

AD represents the most common neurodegenerative disease with dementia and progressing cognitive deficits. Presently, in Germany there are about 1.2 million people suffering from AD, while worldwide there are estimates of more than 35 million AD patients [51.9]. Aging is the most important risk factor for AD (for a review, see [51.10]). At the age of 65 years about 1.2% of the population is afflicted, with an approximate doubling of incidence for every 5 years of age afterward. For those over the age of 90, the prevalence increases to nearly 40%. Due to ever-increasing life expectancy, the number of affected individuals is projected to rise to more than 2.3 million in Germany [51.9] and about 100 million worldwide by 2050 [51.11]. The increasing number of elderly people suffering from cognitive impairment and dementia will produce a considerable economic burden for the social communities in the near future, due to the growing costs of health-care and welfare systems. This challenging perspective indicates that detailed knowledge of the pathogenetic mechanisms of AD as well as pharmacotherapeutical strategies to combat AD are urgently required. It has been estimated that a delay in the onset of AD by 5 years may decrease the number of affected people by half.

51.1.2 Morphopathological Hallmarks

The pathology of AD is characterized by two major histopathological hallmarks such as senile plaques (syn. neuritic plaques, amyloid plaques) deposed extracellularly in cerebral cortical and hippocampal areas, and neurofibrillary tangles that occupy much of the cytoplasm of select cortical pyramidal neurons. These neuropathological hallmarks of AD are particularly prominent in areas such as the parietal and temporal cortices, the hippocampus, the entorhinal cortex, and the amygdala. In addition to senile plaques and neurofibrillary tangles, brains from AD patients also demonstrate astrocytic gliosis, reactive microglia, inflammation, as well as neuronal cell loss and synaptic dysfunction, which have been assumed to be consequences of the accumulation of the pathological protein components (for a review, see [51.12]).

Senile plaques are extracellular deposits that are mostly composed of β-amyloid (Aβ), a proteolytic fragment of the APP. The deposits are associated with neuronal terminals and degenerative swollen neurites and are surrounded by a web of astrocytic processes and microglia cells [51.3, 4, 13].

Neurofibrillary tangles, the second histopathological feature in AD, are pathological protein aggregates detectable in the cytoplasm of select cortical pyramidal neurons, which mostly consist of hyperphosphorylated tau, a microtubule-associated protein ([51.14–16], see also [51.17]). The tau protein is a highly soluble protein that functions as a modulator of the stability and flexibility of axonal microtubules, mediated by its degree of phosphorylation. Hyperphosphorylation of tau depresses its microtubule assembly activity and its binding to microtubules [51.18]. Abnormal hyperphosphorylated tau demonstrates a high tendency to aggregate and to form paired helical filaments (PHF) and straight filaments (SF), thus causing insoluble cytoplasmic inclusions, which disrupt the structure and function of the neurons [51.19]. However, recent studies have provided evidence that prefibrillar, oligomeric forms of tau, rather than PHF, cause the neurodegeneration observed in AD [51.20].

Many phosphokinases including glycogen synthase kinase 3β (GSK3β), cyclin-dependent kinase 5 (cdk5), and extracellular signal-related kinase 2 (ERK2), as well as the dual specificity tyrosine-phosphorylation-regulated kinase1A (DYRK1A) have been assumed to be involved in tau phosphorylation (for a review, see [51.21]).

The level of hyperphosphorylated tau in AD brains is about four–eight times higher as compared to age-matched normal brains. In the adult human brain, six tau isoforms are expressed through alternative splicing of a single tau gene on chromosome 17 (for a review, see [51.22]). Three isoforms contain three tandem microtubule binding repeats, while the other three forms contain four microtubule binding repeats. In AD all of the six tau isoforms contribute to the formation of PHF by hyperphosphorylation [51.23].

Tau pathology is also seen in a number of other human neurodegenerative disorders associated with neurodegeneration and dementia [51.22].

There is experimental evidence of a link between Aβ and tau pathology in AD with Aβ deposition preceding the tau pathology [51.21]. Several hypotheses have been suggested, including the *dual pathway* model of causality, whereby Aβ and tau are linked by separate mechanisms driven by a common upstream driver [51.24].

51.1.3 Genetic Factors Predisposing AD

A small subset of AD cases ($< 5\%$) results from an inherited autosomal dominant gene mutations and have an early-onset at ages between 40 and 60 years, while the majority of AD cases are sporadic with a complex aetiology due to interactions between environmental conditions and specific genetic features of the individual [51.23].

In patients with the early-onset familial form of the disease (FAD) that tends to be more aggressive, mutations in genes located on chromosome 21, 14, and 1, encoding the amyloid precursor protein (APP), presenilin1, and presenilin2, respectively, have been found (for reviews, see [51.12, 25]), while hitherto no tau gene mutations were observed in AD [51.23].

Moreover, a number of potential risk genes for AD have been described (for a review, see [51.21]), with ApoE being the most consistently associated risk gene. Carriers of the ApoE ε4-allele have a several times higher risk of developing AD as compared with those carrying the ApoE ε4-allele [51.26–28].

Inherited variants in the neuronal sortilin-related sorting receptor SORL1 have been found to be associated with late-onset AD [51.29]. Recently performed genome-wide association studies in late-onset AD patients revealed the identification of nine further novel genes that are linked to immune system function, cholesterol metabolism, and synaptic cell membrane processes, and which explain around 50% of late-onset AD genetics (for a review, see [51.30]).

51.2 Cholinergic Dysfunction in Alzheimer's Disease

Another consistent feature of AD is a progressive neuronal cell loss that is associated with region-specific brain atrophy. In particular, the cholinergic projection from the nucleus basalis of Meynert to areas of the cerebral cortex is the pathway that is very early and most severely affected in brains from AD patients. Changes in the cholinergic transmission in AD have been documented by assessing the major processes occurring at a cholinergic synapse. In the presynaptic compartment of the cholinergic neuron the neurotransmitter acetylcholine is synthesized from choline and acetyl-CoA by the catalytic action of the choline acetyltransferase (ChAT), and subsequently taken up into the vesicle by means of the vesicular acetylcholine trans-

porter (VAChT). Following the action potential-induced release into the synaptic cleft, acetylcholine binds to both post and presynaptically localized cholinergic muscarinic (mAChR) and nicotinic acetylcholine receptors (nAChR). While M1/M3-mAChR subtypes are mainly localized postsynaptically, parts of the M2/M4-mAChR subtypes act as autoreceptors on cholinergic presynaptic compartments to control acetylcholine release. Similarly, nAChRs are predominantly localized on both cholinergic and non-cholinergic nerve terminals to regulate transmitter release. In the synaptic cleft, acetylcholine undergoes a fast degradation by the catalytic action of the acetylcholinesterase (AChE, see also Fig. 51.1).

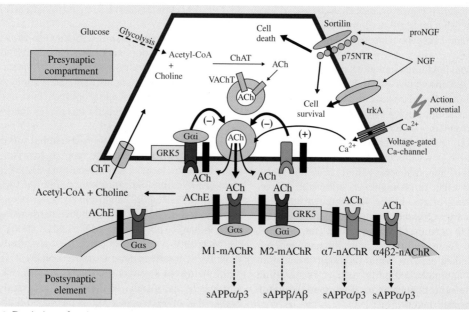

Fig. 51.1 *Depiction of main processes occurring at a cholinergic synapse in the brain.* In the presynaptic compartment of the cholinergic neuron the neurotransmitter acetylcholine is synthesized from choline and acetyl-CoA by the catalytic action of the choline acetyltransferase (ChAT) and subsequently taken up into the vesicle by means of the vesicular acetylcholine transporter (VAChT). Following the action potential-induced, Ca^{2+}-dependent release into the synaptic cleft, acetylcholine (ACh) binds to both post and presynaptically localized cholinergic muscarinic (mAChR) and nicotinic acetylcholine receptors (nAChR). While M1-mAChR subtypes are mainly localized postsynaptically, parts of the M2-mAChR subtypes act as autoreceptors on cholinergic presynaptic compartments to control ACh release. Similarly, nAChRs are predominantly localized on both cholinergic and non-cholinergic nerve terminals to regulate transmitter release. In the synaptic cleft, ACh undergoes a fast degradation by the catalytic action of the acetylcholinesterase (AChE). While activation of M1-mAChR and nAChR favor the non-amyloidogenic route of amyloid precursor protein (APP) processing, stimulation of M2-mAChR drives the APP processing toward the amyloidogenic path. Activated M2-mAChR are selectively desensitized by phosphorylation under the catalytic action of G protein-coupled receptor kinase 5 (GRK5). β-Amyloid ($A\beta$)-induced GRK5 deficiency leads to presynaptic M2-mAChR hyperactivity, which initiates a cascade of pathological events including enhanced $A\beta$ formation finally resulting in cholinergic dysfunction. For survival and maintenance, cholinergic neurons require neurotrophic support by the nerve growth factor (NGF) mediated through binding to both high- (trkA) and low-affinity (p75NTR) receptors which are nearly exclusively located on central cholinergic cells

For survival and maintenance, cholinergic neurons require neurotrophic support by the nerve growth factor (NGF) mediated through binding to both high- (trkA) and low-affinity (p75NTR) receptors, which are nearly exclusively located on central cholinergic cells (Sect. 51.4). Therefore, NGF and its receptors have additionally been used as markers to further characterize cholinergic dysfunction in AD.

Severe deficits of presynaptic cholinergic markers in the cerebral cortex of patients with early-onset AD were already observed in the late 1970s in a number of studies [51.31–34]. Biochemical and in situ hybridization studies reported a marked and region-dependent loss in ChAT activity (from 30 to 90%) and ChAT mRNA levels (about 50%) in the temporal lobe and frontal and parietal cortices of the AD brain being consistent with striking reductions in the number of basal forebrain cholinergic neurons. Accompanying the degenerations of cholinergic neurons are the presence of neuritic plaques in cholinoceptive cortical target regions, as well as within the basal forebrain nuclei (for references, see e.g., [51.35]). Neurochemical analysis have further revealed deficits in several other presynaptic cholinergic markers, such as acetylcholine synthesis and release, as well as a decreased number of nAChRs. While most studies did not detect changes in M1-

mAChRs, a decrease in the number of M2-mAChRs was reported [51.36].

The correlation of clinical dementia ratings with the reductions in a number of cortical cholinergic markers such as ChAT, M2-mAChR, and nAChR binding, as well as levels of acetylcholine [51.36–38], suggested an association of cholinergic hypofunction with cognitive deficits, which led to the formulation of the cholinergic hypothesis of memory dysfunction in senescence and in AD [51.39].

While there is no doubt that severe loss of cortical cholinergic innervation exists already in the initial stages of presenile (early-onset) as well as in the advanced stages of late-onset AD, the assumption of cholinergic denervation being an early and initial stage also in mild, late-onset AD has been debated (see e.g., [51.40]). Only mild losses of AChE activities have been observed in patients with mild cognitive impairment (MCI, a prodromal stage of AD), and early forms of AD, while in a number of brain regions studied in MCI patients no decrease in ChAT activity has been observed. Similarly, the number of ChAT-positive and VAChT-positive cells was unaltered in MCI as compared to non-demented controls. While the number of ChAT-positive neurons was unchanged, the trkA and p75NTR-containing neurons, which co-localize with ChAT, were significantly reduced in the nucleus basalis of subjects with MCI as compared to those with no cognitive deteriorations. These findings suggest

a downregulation of trkA and p75NTR receptors, thus a dysfunction of cholinergic neurons rather than cholinergic cell loss [51.41]. This is further emphasized by observations that other parameters of cholinergic function such as acetylcholine release, high-affinity choline uptake, and expression of mAChR and nAChR are also altered in MCI and early AD [51.35].

Gene expression analysis of single basal forebrain cholinergic neurons revealed that trkA, but not p75NTR, is reduced in MCI. The NGF precursor, proNGF, has been observed to be increased in the cortex of MCI and AD. As proNGF accumulates in the presence of reduced cortical trkA and sustained levels of p75NTR, a shift in the balance between cell survival and death molecules may occur in early AD. Similarly, BDNF and proBDNF, are reduced in the cortex of MCI, further depriving basal forebrain cholinergic cells of trophic support [51.41]. ProNGF is released from the cerebral cortex in an activity-dependent manner together with enzymes required to generate mature NGF. Thus, the upregulation of proNGF observed in AD may indicate a dysregulation in the maturation of NGF leading to enhanced vulnerability of the cholinergic system in AD [51.42]. These data further support the suggestion of a key role of the cholinergic system in the functional processes that lead to AD, while provoking the question whether the gradual progressive loss of cholinergic function is associated with $A\beta$ and tau pathology, as will be discussed in Sect. 51.3.2.

51.3 APP Metabolism and Its Regulation by Cholinergic Mechanisms

51.3.1 Processing of APP and Generation of $A\beta$

The major constituent of neuritic plaques is the 4 kDa $A\beta$ peptide that is derived from a much larger protein, the APP. APP belongs to a family of glycosylated type-I transmembrane proteins which are ubiquitously expressed but most abundantly in the brain. There are three major isoforms APP695, APP751, and APP770 (containing 696, 751, 770 amino acids, respectively) arising from alternative splicing. APP770 (and APP751) consists of a large extracellular domain, containing a 56 amino acid Kunitz protease inhibitor (KPI) region and glycosylation sites, a single membrane-spanning region, and a shorter intracellular carboxyl terminus. While APP751 and APP770 are expressed in most tissues, APP695 is predominantly expressed in neurons and lacks the KPI domain (for a review, see [51.43]).

The precursor protein contains the 39–42 amino acid sequence of $A\beta$ and is processed either by an α-secretory or β-secretory pathway yielding non-amyloidogenic cleavage products or potentially amyloidogenic $A\beta$ peptides, respectively. In order to generate $A\beta$ the APP must be cleaved by two proteases that have been termed β-and γ-secretases. The β-secretase cleavage results in a truncated APP that is denoted as secretory sAPP β, and in a remaining part consisting of 99 amino acids, the β-carboxy-terminal fragment, β CTF (12 kDa). The β CTF may undergo a further cleavage by the γ-secretase at the amino acid 711 or 713 to release either $A\beta$ (1–40), the predominant species, or to a lesser extent, the more amyloidogenic $A\beta$ (1–42) peptide, respectively, as well as the remaining intracellular domain of APP (AICD). Recently, further APP cleavage sites of the γ-secretase have been described: the ζ-site generating $A\beta$ (1–46) and the ε-site producing Aβ (1–49). An overview of the amyloidogenic route of APP processing is depicted in Fig. 51.2a.

The released AICD has been shown to regulate transcription of a number of genes including APP, glycogen synthase kinase (GSK) 3β, neprilysin, BACE1, p53, epidermal growth factor receptor (EGFR), and low density lipoprotein receptor-related protein 1 (LRP1), while AICD as the intracellular domain mediates interaction of APP with various cytosolic factors.

Recently, sAPPβ has been reported to undergo a further cleavage by a still unknown mechanism to release N-APP. N-APP was found to bind to the death receptor DR6 to cause axon pruning and neuronal cell death via caspase-6 and -3, respectively [51.44]. A novel function of sAPPβ as a regulator of transthyretin and Klotho gene expression has been detected recently [51.45].

The non-amyloidogenic secretory pathway includes the proteolytic cleavage of APP by the α-secretase between amino acid position 16 and 17 of the Aβ domain, which results in the secretion of a soluble ectodomain of APP, called sAPPα. The remaining membrane-anchored C-terminal fragment of APP, α CTF (83 amino acid residues) may further undergo a γ-secretase cleavage to produce a peptide called p3 and the remaining AICD (see also Fig. 51.2b).

sAPPα has been shown to be involved in mediating neuronal plasticity and survival, protection against excitoxicity, neural stem cell proliferation, and inhibition of stress-induced cdk5 activation; the physiological function of the rapidly degraded p3 peptide is still unclear [51.43].

The α, β and γ-secretases have recently been cloned and identified. The α-secretase is a zinc metalloproteinase and acts as a membrane-bound endoprotease to cleave APP within the plasma membrane. Three members of the ADAM (a disintegrin and metalloproteinase) family have been proposed to act as α-secretase: ADAM9, 10, and 17 (for reviews, see [51.43, 46].

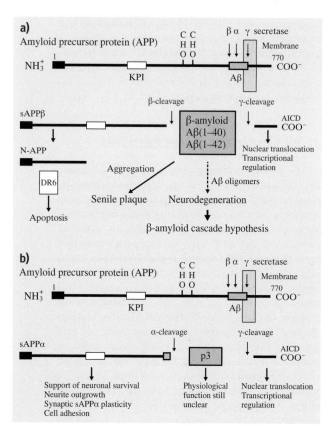

Fig. 51.2a,b *Processing of the amyloid precursor protein (APP).* (**a**) The APP contains the 40–42 amino acid sequence of Aβ, which is partly localized within the transmembrane domain of APP. To generate Aβ the APP is cleaved by two proteases termed as β-and γ-secretases. The β-secretase cleavage results in formation of secretory sAPPβ, and in a remaining part, the β-carboxy-terminal fragment, βCTF. The βCTF may undergo a further cleavage by the γ-secretase to release either Aβ (1–40), the predominant species, or to a lesser extent, the more amyloidogenic Aβ (1–42) peptide, respectively, as well as the remaining intracellular domain of APP (AICD). Aβ peptides may accumulate in the brain as synaptotoxic and cytotoxic oligomers and as fibrillar aggregates in the form of senile plaques. The released AICD has been shown to regulate transcription of a number of genes, while AICD as intracellular domain mediates interaction of APP with various cytosolic factors. Recently, sAPPβ was reported to undergo a further cleavage to release N-APP, which binds to the death receptor DR6 thereby causing axon pruning and neuronal cell death. (**b**) The non-amyloidogenic secretory pathway includes the proteolytic cleavage of APP by the α-secretase between amino acid position 16 and 17 of the Aβ domain which results in the secretion of the soluble ectodomain, sAPPα. The remaining membrane-anchored C-terminal fragment, αCTF may further undergo a γ-secretase cleavage to produce a peptide called p3 and the remaining AICD. sAPPα has been shown to be involved in mediating neuronal plasticity and survival, protection against excitoxicity, neural stem cell proliferation, and inhibition of stress-induced cdk5 activation. The physiological function of the rapidly degraded p3 peptide is still unclear [51.43] ◀

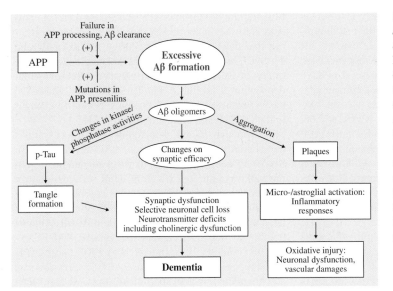

Fig. 51.3 *Amyloid cascade hypothesis of AD: modifying and detrimental effects of Aβ.* The amyloid cascade hypothesis states that the development of AD is initiated by abnormal cleavage of APP, resulting in an imbalance of production and clearance of Aβ. As a consequence, Aβ accumulates in the brain as synaptotoxic oligomers and as fibrillar aggregates in the form of senile plaques, thereby initiating a cascade of pathogenic events that ultimately result in the development of AD [51.47]. For more details, see text. (Graph adapted from [51.48] and [51.10])

The membrane-bound, transmembrane aspartyl protease β-site APP cleaving enzyme 1 (BACE1) has been identified as the β-secretase, while the γ-secretase activity is mediated by a high molecular weight complex consisting of presenilin 1 or 2, nicastrin, anterior pharynx-defective-1 (APH-1), and presenilin-enhancer-2 (PEN-2) (for reviews, see [51.21, 43, 46].

As the clinicopathological features of sporadic AD are indistinguishable from the inherited form of AD, significant efforts have been made to understand the physiological function of the genes tied to FAD. In particular, the development of transgenic mice carrying human FAD genes (for a review, see [51.49]) have essentially contributed to confirm the pathogenic effects of the mutated AD genes and to support the amyloid cascade hypothesis of AD, which has been proposed by [51.48] (Fig. 51.3). Changes in the metabolism of APP and formation and accumulation of Aβ peptides initiates a cascade of events that finally leads to neuronal dysfunction and cell death associated with neurotransmitter deficits and dementia (for reviews, see [51.48, 50]).

51.3.2 Cholinergic Control of APP Processing

The metabolism of APP underlies a strong physiological control with a balanced production and clearance of Aβ. Pathological changes in the processing of APP may favor the β-secretory pathway with the consequence of Aβ accumulation either as synaptotoxic oligomers or as fibrillar aggregates in the form of senile plaques. An

abundant number of investigations in the last decades provided compelling evidence that APP processing is controlled by neuronal signaling of several transmitter systems including cholinergic mechanisms. Particularly, G protein-coupled receptors have been disclosed to modify APP processing by affecting the proteolytic activities of the α, β, and γ-secretases (for a comprehensive review, see [51.50]).

This section focuses on the interaction of the cholinergic system and APP processing (for a review, see also [51.50–52]).

Control of APP by mAChR
The first evidence of a link between cholinergic dysfunction and APP processing was provided by observations of a co-localization of acetylcholinesterase (AChE) with Aβ deposits in AD brains [51.53, 54], which was further validated by a number of in vitro studies initiated by [51.55, 56]. Secretion of sAPPα was enhanced by electrical stimulation of tissue slices from rat brains, which could be blocked by the sodium-channel antagonist tetrodotoxin [51.57]. Selective activation of M1/M3-but not M2/M4-mAChR increased sAPPα secretion and decreased total Aβ formation both in vitro [51.58–61] and in vivo in AD patients [51.62, 63]. A similar effect was achieved by direct activation of protein kinase C (PKC) by phorbol ester, indicating that mAChR mediate their effects on APP processing through activation of the phosphatidyl inositol signaling pathway [51.64]. A recent cell culture study provided evidence that PKCα

and PKCε, but not PKCδ, are particularly involved in controlling sAPPα secretion [51.65]. Further studies revealed that other pathways downstream of the mAChR may also be involved in the α-secretase-mediated cleavage of APP, including protein kinase A (PKA), mitogen-activated protein kinase (MAPK), extracellular signal-regulated protein kinase (ERK), tyrosine kinase, and phosphatidylinositol 3-kinase (PI3K) [51.66, 67]. Activation of these signaling cascades shifts APP metabolism towards the α-secretase-mediated pathway with suppressing β-secretase-mediated Aβ formation (for a review, see [51.50]). However, a recent study has questioned the participation of the MEK/ERK pathway in PKC-stimulated and M1-mAChR-dependent formation of sAPPα [51.65].

Selective lesion of basal forebrain cholinergic cells in rat brains [51.68, 69] and administration of selective M1-mAChR agonists to mice [51.70] provided strong in vivo evidence that cortical APP processing is controlled by cholinergic activity originating from the basal forebrain. Scopolamine treatment of transgenic Tg2576 mice resulted in increased levels of fibrillar Aβ, and enhanced α-secretase activity, suggesting that chronic suppression of cortical muscarinic cholinergic transmission may alter the balance between α and β-secretory APP processing by favoring the amyloidogenic route [51.71].

While M2-/M4-mAChR have been observed to inhibit sAPPα release and enhance Aβ production [51.59, 72], studies in AD mouse models provided strong evidence that selective activation of M1-mAChRs drives the APP processing toward the non-amyloidogenic pathway. Thus, administration of the selective M1-mAChR agonist AF267B in the 3xTg-AD model resulted in reduced Aβ and tau pathologies in the hippocampus and cortex [51.73]. The decreased Aβ production indicates an M1-mAChR-mediated shift of APP processing toward the non-amyloidogenic pathway, mediated by an increase in PKC activation, ERK1 and ERK2 phosphorylation, and an increase in α-secretase ADAM17 expression, while the effect of AF267B on tau is obviously due to the reduced GSK3β activity [51.73]. These findings were further supported by genetic ablation studies. Deletion of the M1-mAChR in primary neurons has been shown to increase amyloidogenic APP processing in neurons as evidenced by decreased agonist-regulated shedding of the neuroprotective APP ectodomain sAPPα, and increased production of toxic Aβ peptides. In APP(Swe/Ind) transgenic mice, the loss of M1-mAChRs resulted in increased Aβ level and more amyloid plaques, thus favoring amyloidogenic APP processing [51.72]. This data is further supported by a recent study demonstrating that deletion of M1-mAChR in the 3xTgAD and Tg-SwDI mice increased plaque and tangle levels in the brains of 3xTgAD mice and elevated cerebrovascular deposition of fibrillar Aβ in Tg-SwDI mice, and led to tau hyperphosphorylation, presumably due to changes in the GSK3β and PKC activities [51.74].

Given that α and β-secretory APP processing is differentially regulated through mAChR subtypes, in SH-SY5Y neuroblastoma cells we examined the signaling pathways through which mAChR subtypes control β-secretase activity [51.75]. The expression of the β-secretase BACE1 was found to be differentially regulated in a subtype-specific manner by mAChR. Agonist binding to M1/M3-receptors upregulated BACE1 expression through activation of both PKC and MAPK signaling cascades. In contrast, BACE1 expression was downregulated by activation of M2-mAChR and PKA-mediated pathways [51.75]. These results may partly explain the observed deteriorations of AD patients after initial improvements by AChE-inhibitor or M1-mAChR agonist treatment.

Activated G protein-coupled receptors are desensitized by phosphorylation under the catalytic action of G protein-coupled receptor kinases [51.76]. GRK5, a member of the GRK protein family, is specifically involved in desensitizing activated Gi-coupled M2/M4-mAChR subtypes [51.77]. As M2/M4-mAChRs function primarily as presynaptic autoreceptors on cholinergic neurons, acting as feedback inhibitor of acetylcholine release [51.78], a functional loss of GRK5 should cause prolonged or persistent M2/M4-mAChR signaling leading to reduced acetylcholine release and cholinergic hypofunction. As GRK5 deficiency has been reported in AD [51.79], a role of GRK5 in promoting amyloidogenic APP processing via prolonged activation of M2/M4-muscarinic autoreceptors was suggested [51.80]. Indeed, in AD-like Tg2576 mice lacking one copy of Grk gene significantly increased Aβ accumulation including enhanced plaque load, and sAPPβ production was observed, which could be reversed by the administration of selective M2-mAChR antagonists [51.80].

As Aβ has been shown to be one of the main causes of the functional deficiency of GRK5 in AD [51.79], a positive feedback loop among Aβ, GRK5 deficiency, and cholinergic dysfunction has been postulated, by which each component can mutually promote each other, thus exacerbating both amyloid pathology and cholinergic dysfunction [51.80].

Control of APP by nAChR

Nicotine through action on nAChR has also been observed to modulate APP processing both in cell culture [51.81–84] and in vivo [51.85] by favoring the non-amyloidogenic pathway when treated with low doses of nicotine [51.86].

To reveal which subtype of neuronal nAChRs mediates the actions of nicotine on APP processing, subtype-selective suppression/transfection studies were performed in cell line cultures. Suppression of the α7-nAChR gene by siRNA indicated that α7-nAChR stimulation increases sAPPα formation by enhanced α-secretase activity, accompanied by activation of MAPK signaling, and improving antioxidant defenses [51.84].

By using cell lines expressing different subtypes of nAChRs, it has been detected that stimulation of both α4- and α7-nAChR by nicotinic compounds is involved in driving the APP processing toward the non-amyloidogenic pathway by increasing sAPPα secretion

and attenuating Aβ production [51.87]. While activation of α7 nAChR by nicotine and specific α7-nAChR agonists promotes non-amyloidogenic APP processing by decreasing the γ-secretase activity with no effects on α-secretase and β-secretase activity [51.88], the decrease in Aβ production by activation of α4β2 nAChR is mediated through regulation of BACE1 transcription, presumably via the ERK1/NFκB pathways [51.89].

These findings have also been validated by in vivo studies. Chronic treatment of transgenic Tg2576 APP mice with nicotine significantly reduced Aβ plaque deposition as compared to non-treated controls [51.90]. In a rat model of AD chronic nicotine was observed to restore normal Aβ levels and prevented short-term memory and LTP impairment [51.91]. Interestingly, tobacco-smoking elderly people demonstrated less Aβ deposition in the entorhinal cortex as compared to non-smokers [51.92].

51.4 Modulation of Cholinergic Function by Aβ

While impaired cholinergic neuronal signaling may induce pathologically enhanced Aβ production, there is also abundant evidence of the cytotoxic, modifying, and suppressing capacities of Aβ on cholinergic cells and function, suggesting the generation of vicious feedback loops. This section reviews the effect of Aβ on viability and signaling of cholinergic neurons including NGF signaling in cholinergic cells.

51.4.1 Cytotoxicity of Aβ on Cholinergic Cells

The most predominant Aβ peptides detected in Alzheimer plaques are Aβ (1–40) and Aβ (1–42). Aβ (1–42) is more fibrillogenic and displays higher neurotoxicity in vivo than Aβ (1–40) [51.93]. The insoluble, high-molecular weight fibrils are major components of the senile plaques in the Alzheimer brain and have been assumed for a long time to be the most toxic forms responsible for cholinergic neurodegeneration (see, e.g., [51.35, 94]). However, recent evidence indicates that soluble oligomers of Aβ (such as low-molecular weight monomers, oligomers and amyloid-derived diffusible ligands (ADDLs)), as well as protofibrils represent the main neurotoxic species leading to early neuronal dysfunction and memory deficits in AD (for reviews, see e.g., [51.95, 96]). In different cell and animal models prefibrillar assemblies of Aβ have

been shown to induce neurotoxicity (for reviews, see e.g., [51.51, 97], electrophysiological changes [51.98], modulation of synaptic plasticity [51.98, 99], and disruption of cognitive function [51.100, 101], which may explain why early onset of cholinergic dysfunction is in progress before there is considerable plaque formation in AD. Indeed, in AD brains, it has been observed that the severity of neurodegeneration correlates best with the pool of soluble Aβ rather than with the number of insoluble Aβ plaques [51.102]. This data fits well with observations in AD-like transgenic mouse models revealing decreases in cholinergic fibre density, in mAChR and nAChR binding levels in the cerebral cortex already before onset of plaque deposition [51.103–109]. Studies in BACE1 knock-out mice that do not produce any significant amount of Aβ [51.110] provided further evidence of Aβ-induced cholinotoxicity. BACE1 gene deletion rescued memory deficits and cholinergic dysfunction in AD mice which was associated with tremendously reduced Aβ levels [51.111].

In a recent study on brain autopsy in AD patients, the relationship between various Aβ oligomer assemblies found in AD brains with the levels of fibrillar Aβ and cholinergic synaptic function was examined, suggesting that only distinct Aβ oligomers induce impairment of cholinergic neurotransmission in AD [51.112].

51.4.2 Modulation of Cholinergic Transmission by Subtoxic Aβ

Aβ is also produced under normal conditions and secreted in the brain as a soluble peptide [51.113, 114], which raised the possibility that Aβ may also play a physiological role. This is supported by cell culture studies providing evidence of a modulatory role of soluble Aβ at subtoxic concentrations on cholinergic neurotransmission [51.35]. Soluble Aβ at pM to nM concentrations strongly inhibited the potassium-stimulated release of acetylcholine from hippocampal slices and reduced the high-affinity uptake of choline in synaptosomal preparations from cortex and hippocampus [51.115]. Choline deprivation has been suggested to render basal forebrain cholinergic neurons particularly vulnerable [51.116] and may initiate autocannibalism of cholinergic cells by cleavage of membrane phosphatidylcholine to replenish the choline availability [51.35].

Aβ peptides at subtoxic concentration also decreased the intracellular acetylcholine concentration in primary cultures [51.117] and resulted in reduced activity of ChAT but not AChE activity in cholinergic SN56 cells [51.118]. As exposure of primary neurons by Aβ reduced glucose uptake, the Aβ-mediated decline in intracellular acetylcholine might be due to the limited supply of acetyl-CoA, a prerequisite to acetylcholine synthesis.

AChE has been found in amyloid plaques. The association of AChE and Aβ alters the enzymatic properties of AChE and renders Aβ more toxic by accelerating assembly of Aβ into fibrils (for a review, see [51.119]).

Aβ at nanomolar concentration did not affect mAChR ligand binding, but impaired M1-mAChR associated signaling in primary septal and cortical cultures, which resulted in reduced inositol phosphate production and decreased calcium release from the intracellular pool [51.120].

Aβ exposure of primary cultures and brain slices have also been found to affect nAChR binding and signaling, but depending on the concentration of Aβ (see also Sect. 51.4.4).

In conclusion, in vitro data indicate that Aβ at physiological concentration (nM to pM) appears to play a potent negative modulator of acetylcholine synthesis and release, and interferes with normal cholinergic signaling mediated through mAChR and nAChR subtypes.

However, APP itself may also affect cholinergic transmission. APP was found to interact with the high-affinity choline uptake by mediating presynaptic localization and activity of the high-affinity choline transporter [51.121].

51.4.3 Effect of Aβ on mAChR

In AD brains the number of M1-mAChR has been reported to be mostly unaffected, while in some studies a decreased number of the M2-mAChR subtype was described as associated with the impaired coupling of the mAChR to heterotrimeric GTP-binding proteins (G proteins) [51.36].

While in vitro studies did not provide any evidence of a direct interaction of Aβ with mAChR binding sites, pathologically accumulating Aβ appears to affect mAChR signaling mainly by influencing downstream events, finally leading to loss of mAChRs and cholinergic dysfunction [51.103, 107, 109, 122]. Recently, a mechanism was proposed to explain how Aβ may interfere in mAChR signaling by interruption of the mAChR G-protein coupling [51.123].

Aβ accumulation triggers increased generation of reactive oxygen species inducing dimerization of the angiotensin type 2 receptor, which can further cross-link and lead to oligomerization of the angiotensin type 2 receptor dimers. The angiotensin type 2 oligomers sequester the G-protein $G\alpha_{,q/11}$, thus preventing the coupling of $G\alpha_{,q/11}$ to the M1-mAChR. Dysfunctional coupling of M1-mAChR and $G\alpha_{,q/11}$ is considered to mediate cholinergic deficits and cell loss, tau phosphorylation, and memory impairments [51.123].

Recently, in transgenic AD-like Tg2576 mice, Aβ-dependent inactivation of the JAK2/STAT3 axis was reported. This resulted in downregulation of ChAT activity and desensitization of the M1-mAChR, thus providing evidence of another mechanism of how enhanced level of Aβ can impair mAChR signaling and induce cholinergic dysfunction [51.124].

51.4.4 Effect of Aβ on nAChR

AD patients show a significant reduction in nAChRs, which has been documented by Western blotting [51.125, 126], immunohistochemical analysis [51.127, 128], or radioligand binding studies ([51.129]; for reviews, see [51.130, 131]). The most vulnerable neurons appear to be those expressing high levels of α7-nAChR [51.132]. As α7-type AChRs are highly expressed in brain regions relevant to memory functions, it has been suggested that α7 AChR has an important role in the development of AD (for a review, see [51.133]).

Exposure of PC12 cells to Aβ resulted in a significant decrease in nAChR, which led to the suggestion that Aβ can damage nAChR [51.134, 135]. Similarly, in rats chronically infused with Aβ peptides, significant decreases in the levels of nAChR subunits α7, α4, and β2, accompanied by an increased BACE expression, have been observed [51.136]. Both actions of Aβ could be reversed by chronic delivery of nicotine, suggesting inhibitory rather than damaging effects of Aβ on nAChRs [51.136].

On the other hand, there are reports that Aβ may affect nAChR by direct binding with high-affinity to nAChR, in particular to the α7 subtype [51.132, 137–140]. Other studies have been unable to confirm these findings [51.135, 141]. Aβ has been observed to act as an agonist of α7-nAChR, mediating the activation of the ERK2 MAP kinase signaling cascade [51.142]. Other groups have reported inhibitory actions of Aβ on α7 nAChR [51.143], which appears to depend on the concentration of Aβ; low concentration can activate, higher concentrations desensitize α7 nAChR [51.144]. This compares well with observations in triple transgenic mice overexpressing mutated human APP, presenilin-1 and tau (3xTg-AD), which demonstrate an age-dependent reduction in α7-nAChRs as compared to age-matched non-transgenic mice. The loss of α7-nAChRs is preceded by intracellular Aβ accumulation and is restricted to brain regions that develop Aβ pathology [51.139], a finding which mimics the situation detectable in brains of AD patients. Therefore, parts of the cholinergic deficits produced in Alzheimer's disease as well as in transgenic APP mice could be attributed to the suppression of cholinergic functions by Aβ peptides.

Recently, it was hypothesized that α7-nAChR may represent the missing link in understanding AD etiopathology [51.145]. This hypothesis stresses recent experimental findings that nicotinic agonists and Aβ may compete for the α7-nAChR binding site, whereas intracellular signaling cascades are activated that control either survival or cell death pathways, respectively [51.130]. Moreover, a recent study supports α7-nAChR as a mediator of Aβ-induced pathology in AD by demonstrating that both agonists and antagonists may modulate Aβ-induced tau phosphorylation through GSK-3β [51.146]. Deletion of the α7-nAChR gene in the PDAPP Alzheimer mouse model improved cognitive deficits and synaptic pathology, suggesting that blocking the α7-nAChR function may alleviate symptoms of AD and may represent a potential treatment strategy [51.147].

Recently, the functional effects of Aβ fibrils and oligomers on nAChRs were examined by measuring intracellular calcium levels in neuronal cells [51.148]. It was demonstrated that fibrillar Aβ exerts neurotoxic effects mediated partly through a blockade of α7-nAChRs, whereas oligomeric Aβ may act as a ligand activating α7-nAChRs, thereby stimulating downstream signaling pathways [51.148].

On the other hand, overexpression of APP (SWE) gene has been shown to influence the expression of nAChRs and resulted in neurotoxicity [51.149], while the disruption of cholesterol homeostasis as observed in AD brain may affect cholinergic synapses by decreasing the number of nAChR (see, e.g., [51.133]).

51.4.5 Aβ and Neurotrophin Signaling

For their maintenance and survival basal forebrain cholinergic cells require the neurotrophic support by the nerve growth factor (NGF) that is produced and released by cholinoceptive cells in the cortex. NGF exerts its action by binding to the high-affinity NGF-specific receptor tyrosine kinase, trkA, and the low-affinity, pan-neurotrophin receptor, p75NTR, which are expressed by basal forebrain cholinergic neurons, while neurons in other brain regions express little or no p75NTR. Receptor-bound NGF is internalized and retrogradely transported to the nucleus of basal forebrain cholinergic cells (for a review, see [51.150]). Transgenic mice that have been modified to produce anti-NGF antibodies exhibit degeneration of basal forebrain cholinergic cells and partly mimic AD-like pathology, suggesting that NGF has an important role for cholinergic cell viability [51.151].

While NGF expression is not altered in AD, dysfunction of cytoskeletal transport including reduced axonal transport of NGF has been assumed to play an important role in the development of AD pathology [51.150, 151], as both APP and tau are involved in axonal transport. Increased doses of APP markedly decreased retrograde transport of NGF and resulted in degeneration of forebrain cholinergic neurons [51.152].

Trophic deprivation has been observed to affect the cleavage of surface APP by BACE1, releasing sAPPβ, which is further cleaved to N-APP, which may bind to the death receptor 6 (DR6) to trigger degeneration [51.44]. On the other hand, APP may directly interact with p75NTR, which mediates death of basal forebrain cholinergic neurons [51.153].

Recently, it was discovered that the precursor of NGF, the proNGF, also displays biological activities, but that these activities are distinct from that of mature NGF ([51.17]; for a review, see [51.154]). proNGF may interact with sortilin as co-receptor for p75NTR and induce p75NTR-dependent apoptosis or may, to a lesser extent, bind to trkA and mediate trA-dependent neuronal survival [51.155]. In AD brains a reduced conversion of proNGF into mature NGF, and an increased NGF degradation has been observed [51.156], suggesting an imbalance of NGF/proNGF signaling in AD that may contribute to the cholinergic cell loss.

Because of the select expression of p75NTR by basal forebrain cholinergic neurons, a role of p75NTR in cholinergic dysfunctions in AD has been suggested (for a review, see [51.157]). There is abundant evidence from in vitro studies that p75NTR increases the susceptibility of cells to $A\beta$ toxicity, suggesting that p75NTR mediates $A\beta$ cytotoxicity by direct interaction with $A\beta$, or acting on signal transduction pathways mediated by p75NTR [51.158, 159]. These observations were further confirmed recently by a number of in vivo studies. $A\beta$ injected into the hippocampus of p75NTR/knock-out mice caused less neuronal death as compared to that observed in wild type (p75NTR+/+) mice [51.160]. Transgenic Thy1-hAPPLondon/Swe Alzheimer-like mice lacking wild type p75NTR did not display $A\beta$-associated basal forebrain cholinergic neuritic dystrophy and reduced cholinergic cortical fibre density as has been observed in transgenic Thy1-hAPPLondon/Swe mice with intact p75NTR [51.161]. The data provide evidence that p75NTR may be a major player in $A\beta$-associated cholinergic deficits, mediated through a p75NTR-activated c-Jun N-terminal kinase pathway [51.162].

Using another genetic approach (crossing the p75NTR knock-out mouse with the transgenic APPswe/PS1dE9 AD mouse), p75NTR has been shown to regulate $A\beta$ deposition by increasing neuronal $A\beta$ production [51.163]. The extracellular domain of p75NTR after shedding from the membrane (by an α-secretase activity) may bind and sequester $A\beta$ and thus inhibit $A\beta$ aggregation and reduce $A\beta$ deposition. Moreover, it may also block the interaction of p75NTR and $A\beta$ or proNGF by competitive binding, thus attenuating the p75NTR-dependent neurotoxicity and cell death [51.163]. On the other hand, p75NTR has been shown to be upregulated in two strains of AD transgenic mice compared to corresponding wild-type mice, which correlated with the age-dependent accumulation of $A\beta$ (1–42) level [51.164]. As activation of p75NTR signaling was observed to stimulate $A\beta$ production [51.163], the $A\beta$-induced p75NTR upregulation may initiate a vicious cycle that accelerates AD development [51.164].

51.5 Tau and Cholinergic Transmission

While no tau mutations have been described in the brains of AD patients, pathogenic mutations in the tau genes cause frontotemporal dementia [51.165], suggesting that posttranscriptional alterations in tau gene expression may contribute to the cognitive deficits in AD presumably also by interacting with the cholinergic transmission. Several studies have demonstrated that activation of nAChR results in a significant increase in tau phosphorylation, whereas mAChR activation may prevent tau phosphorylation (for reviews, see [51.130, 145]. Nicotine was found to induce tau phosphorylation at those sites that were also hyperphosphorylated in AD, presumably mediated through activation of the α7 subtype of nAChR [51.143]. This was further emphasized by observations in triple transgenic 3xTg-AD mice that develop age and regionally-dependent accumulation of both plaques and tangles as well as progressive deficits in cognition [51.166–168]. Chronic nicotine administration to one-month-old 3xTg-AD mice for five months did not change soluble $A\beta$ levels but resulted in a striking increase in phosphorylation and aggregation of tau, which appeared to be mediated by p38-MAP kinase [51.139].

Cholinergic basal forebrain neurons have been shown to demonstrate tau pathology both in patients with mild cognitive impairment and in AD patients. Single cell gene expression profiling in individual human cholinergic basal forebrain neurons revealed a shift in the ratio of three-tandem repeat tau to four-tandem repeat tau during the progression of AD but not during normal aging [51.169].

Basal forebrain pretangles and tangles have been observed prior to the pathology in the entorhinal/perirhinal cortex, indicating that abnormalities in cortical cholinergic axons and tau pathology within the basal forebrain cholinergic system occur very early in the course of life and increase in frequency in old age and AD [51.170, 171]. In a tau transgenic mouse model, tau pathology has been observed in both hippocampus and basal forebrain, which was associated with a significant reduction in medial septal cholinergic neurons and a decreased uptake and retrograde transport of NGF by cholinergic nerve terminals [51.172], suggesting that tau pathology may participate in cholinergic degeneration [51.172].

In conclusion, as Aβ may trigger tau phosphorylation, and Aβ pathology precedes tau pathology, tau-mediated changes in cholinergic cells may represent an indirect effect of pathogenic Aβ accumulation.

51.6 Neurovascular Function in AD

51.6.1 Cerebrovascular Abnormalities and Dysfunction in AD

In the brain, the cerebral blood flow is tightly regulated to assure adequate and timely blood supply to brain regions that have momentarily high energy demand because of enhanced neural activity, a phenomenon called functional hyperemia [51.174]. Already at very early stages of AD, changes in the cerebral blood flow, such as reduced blood supply at rest and altered per-

fusion to activated areas have been observed, which provided evidence to suggest a causal relationship between vascular mechanisms and the development of sporadic AD. This vascular hypothesis of AD was first formulated by [51.175] (for reviews, see [51.173, 176, 177], Fig. 51.4). Insufficient cerebral blood flow may induce hypoxia-sensitive pathways leading to inflammation with upregulation of pro-inflammatory cytokines and to oxidative stress with generation of reactive oxygen species, which may be detrimental to

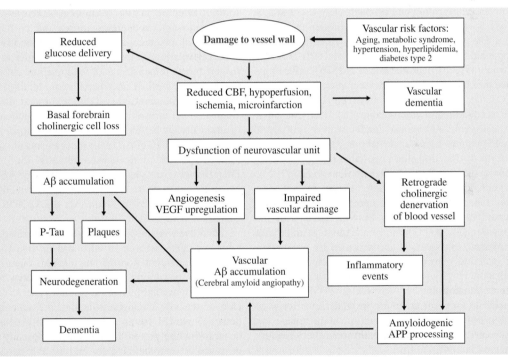

Fig. 51.4 *The vascular hypothesis of AD.* The vascular hypothesis for AD is based on findings that vascular risk factors damage the brain microvasculature, which results in chronic hypoperfusion, reduced cerebral blood flow, and reduced glucose supply [51.173]. Vascular synaptic damage may induce retrograde cell death of cholinergic neurons, which may trigger inflammation, favor amyloidogenic APP processing and cerebral amyloid angiopathy (CAA), and initiate tau pathology. Damage to vascular endothelial cells may induce angiogenesis and VEGF upregulation, which further promotes β-amyloidogenesis. As basal forebrain cholinergic neurons particularly respond sensitively to reduced glucose supply by synaptic dysfunction and degeneration, the cascade of events leading to the development of AD is further exacerbated. For details, see text

vascular integrity and function. Indeed, cerebrovascular abnormalities such as thickening of the microvascular basement membranes, decreased luminal diameter, and microvascular degeneration, in particular in the temporal-parietal cortex, have frequently been observed in AD patients [51.177].

In advanced AD cases $A\beta$ deposition occurs also in cerebral vessels (cerebral amyloid angiopathy CAA), which may result in smooth muscle degeneration and weakening of the vascular wall, impairing vasomotor function, and increasing the risk of cerebral hemorrhage [51.178]. Indeed, it has been suggested that the recently detected brain microbleeds (small dot-like lesions) in AD brains are the missing link between the amyloid cascade hypothesis and the vascular hypothesis [51.179]. The CAA observed during normal aging and in the majority of AD cases is likely to be caused by the failure of $A\beta$ elimination from the brain parenchyme. It has been suggested that receptor for advanced glycation end products (RAGE), and low density lipoprotein receptor related protein 1 (LRP1) play a role in controlling the $A\beta$ transport through the blood brain barrier (for a review, see [51.180]). A recent study in transgenic AD-like mice (Tg2576) further supported the *clearance hypothesis* demonstrating impaired perivascular solute drainage from the brain in aged mice as compared to younger ones [51.181]. Currently, in a transgenic AD mouse model with strong CAA pathology, it has been suggested that early perivascular astrocytic dysfunction play a particular role in impairing cerebrovascular and metabolic pathology [51.182].

Studies in transgenic mouse models of AD suggested that the compromised cerebral hemodynamics observed in AD appears to be associated with inflammation [51.183]. The local activation of microglia and reactive astrocytes is accompanied by production and secretion of proinflammatory cytokines such as interleukin-1β (IL-1β), tumor necrosis factor (TNF)-α, and transforming growth factor (TGF)-β1. Endothelial cells are known to respond sensitively to inflammatory stimuli by production of reactive oxygen species that may further exacerbate the vascular damages [51.184].

A number of risk factors are assumed to mediate the cerebrovascular dysfunctions and to trigger AD pathology, such as hypertension, hyperlipidemia, enhanced homocysteine levels, diabetes type 2, metabolic syndrome (obesity, hypertension, cardiovascular disease, atherosclerosis), as well as genetic factors (ε4 allele of ApoE), age, and life style [51.173, 185, 186]. However, it is still a matter of debate whether neurovascular dys-

function and vascular lesions play a causative role for the neurodegenerative processes as suggested by a number of reports (for a review, see [51.180]). However, regardless of that, cerebrovascular diseases appear to play an important role in determining the presence and severity of the clinical symptoms of AD [51.187].

51.6.2 Effect of $A\beta$ on Brain Vascular System

The microvascular degenerations observed in AD may also be the consequence of the vasoactive detrimental effects of $A\beta$. $A\beta$ is a potent vasoconstrictor in the brain, as has been shown in vivo and in vitro by application of exogenous $A\beta$ to normal blood vessels and to mouse cortices. On the other hand, $A\beta$ may cause degeneration of both the larger perforating arterial vessels as well as cerebral capillaries, presumably mediated through the induction of reactive oxygen species by activation of NADPH oxidase, which may subsequently severely affect regulation of cerebral blood vessels and brain perfusion, as well as impair the blood brain barrier (for reviews, see [51.174, 178, 188, 189]). Indeed, observations in transgenic AD-like mice revealed $A\beta$-mediated impairments of endothelium-dependent regulation of cortical microcirculation, abnormal vascular autoregulation, reduced cerebral blood flow, and attenuated cerebrovascular reactivity to functional hyperemia already before onset of any plaque load, further supporting the link of $A\beta$ to the mechanisms of vascular dysfunction (for comprehensive reviews, see [51.174, 177]. However, the view that elevated soluble $A\beta$ levels are sufficient to cause cerebrovascular dysfunction has also been challenged by studies on the Tg2576 transgenic mouse model [51.190].

Moreover, $A\beta$ peptides have been described to inhibit angiogenesis both in vitro and in vivo [51.191, 192], and deregulation of angiogenic factors may contribute to various neurological disorders including neurodegeneration (for a review, see [51.193]). One of the key angiogenic factors, the vascular endothelial growth factor (VEGF), a highly conserved heparin-binding protein [51.194], was originally found in vascular endothelial cells and is able to induce vascular endothelial cell proliferation, migration, and vasopermeability in many types of tissue [51.195].

Increased intrathecal levels of VEGF have also been observed in brains of AD patients as compared to age-matched healthy individuals [51.196–198], which has been correlated with the clinical severity of the disease [51.199]. However, the functional significance of VEGF upregulation in the pathogenesis and progression of AD is still a matter of debate. While VEGF and other angiogenic factors were found to be enhanced in AD [51.200–203], neovascularization has been observed only in the hippocampus of AD patients [51.203].

51.6.3 Effect of Ischemia and Hypoperfusion on APP Processing

There are reports that ischemia and hypoperfusion may trigger accumulation and cleavage of APP into Aβ, and its deposition in the brain, as well as hyperphosphorylation of tau and PHF formation (for a review, see [51.180]). The upregulation of VEGF in response to hypoxic, ischemic, or hypoglycemic stress [51.204–206], suggests its involvement also in the processing of APP. In turn, APP is also highly expressed in the endothelium of neoforming vessels [51.207], and inhibitors of β- and γ-secretases have been reported to inhibit angiogenesis and tumor growth [51.207], suggesting a role of APP metabolism also during angiogenesis. Recently, VEGF has been shown to also be involved in the induction of microglial-mediated inflammation by Aβ deposits via the microglial VEGF receptor subtype Flt-1 serving as a chemotactic receptor to mobilize microglial cells [51.199].

As vascular endothelial cells are also capable of expressing and to secreting APP [51.208], it has been hypothesized that VEGF may also be involved in the formation and deposition of Aβ. This hypothesis has been addressed by examining the effect of VEGF on APP processing in brain slice cultures, as well as in primary neuronal, astroglial, and vascular endothelial cells derived from AD-like transgenic Tg2576 mice [51.209]. The exposure of brain slices by VEGF resulted in an inhibition of the formation of soluble Aβ peptides, which was accompanied by a transient decrease in β-secretase activity, as compared to controls [51.209]. Similar studies in primary neurons, astrocytes, and endothelial cells expressing the Swedish mutation of human APP, have further provided evidence that VEGF affects APP processing but differentially acts in cells that form the neuron-glia-vascular unit [51.210].

51.6.4 Effect of Aβ on Cholinergic Function in the Brain Vascular System

There is a large body of evidence that cerebral blood flow and local glucose delivery is controlled by neuronal activity, known as neurometabolic and neurovascular coupling [51.177, 184, 211, 212]. Dysfunctions of the regulation of the cerebral blood circulation may af-

fect vital control mechanisms that ensure delivery of adequate amounts of substrate and maintain the homeostasis of the microenvironment of the neurovascular unit. The neurovascular unit defines the cellular interaction between brain capillary endothelial cells, the end feet of perivascular astrocytes, and neuronal axons [51.184]. Pial arteries at the surface of the brain are densely innervated by perivascular nerves that originate from autonomic and sensory ganglia, whereas intracerebral arterioles and capillaries receive afferents that originate from subcortical neuronal centers as well as from local cortical interneurons (for a comprehensive review, see [51.213]).

Based on findings in immunocytochemistry and electron microscopy, it is known that the cholinergic axons originating from the basal forebrain project not only to the cortical neuropile but also to arterioles, capillaries, and to perivascular astrocytes within the cerebral cortex [51.214]. Furthermore, there is physiological evidence that the central cholinergic pathways are involved in the regulation of cerebral cortical blood flow. Electrical or chemical stimulation of cholinergic basal forebrain neurons results in increased cerebral blood flow [51.215]. The involvement of ACh as a neurotransmitter in the control of regional cerebral blood flow has been further demonstrated by administration of cholinergic drugs [51.216]. While application mAChR antagonist decreased cerebral blood flow, the inhibition of AChE led to increased cerebral blood flow [51.217]. This response was found to be dependent on nitric oxide (NO) production and presumably to be mediated through the M5-mAChR subtype [51.216]. The basal forebrain cholinergic fibers can either directly affect the cerebrocortical microvasculature or innervate subpopulations of GABAergic interneurons releasing the vasodilators NO and VIP [51.218]. The interneurons appear to serve as a functional relay to adapt perfusion to locally increased neuronal activity [51.212].

On the other hand, damage to the neurovascular unit either by oxidative stress, inflammation, or Aβ accumulation may induce degeneration of vascular cholinergic nerve terminals and subsequent retrograde cell death of basal forebrain cholinergic neurons. The loss of cholinergic innervation of components of the neurovascular unit may affect APP processing with enhanced Aβ formation and deposition, microglia activation, and inflammation, thus suggesting a link between Aβ production, impairments in cerebrovascular function, and basal forebrain cholinergic deficits in AD (cholinergic-vascular hypothesis of AD; [51.219, 220]). Indeed, a semiquantitative immunohistochemical study on aged Tg2576 mice revealed an $A\beta$-mediated decrease in cholinergic innervation of cortical blood vessels [51.221], which has been assumed to contribute to the alterations of the cerebrovascular system observed in transgenic Tg2576 mice [51.209].

51.6.5 Effect of Glucose Deprivation on Cholinergic Neurons

AD patients feature decreased basal cerebral glucose utilization, presumably as a consequence of the cerebrovascular dysfunction and compromised cerebral hemodynamics [51.177]. Glucose deprivation has been suggested to render basal forebrain cholinergic neurons particularly vulnerable [51.222], as they require glucose not only for energy production but also to synthesize acetylcholine from choline and acetyl-CoA, which is generated by glucose degradation through glycolysis. As in cholinergic neurons energy production and acetylcholine synthesis compete for acetyl-CoA; reduced glucose supply with the consequence of a decline in acetyl-CoA synthesis may easily evoke energy deficits, leading to impairment of their function and structural integrity, as neurons have no energy reserves [51.222]. Interestingly, it has been proposed that these early abnormalities in brain glucose and energy metabolism in AD are caused by an *insulin-resistant brain state*, which appears to be responsible for the increased $A\beta$ accumulation, tau phosphorylation, and cognitive deficits (for a review, see [51.223]).

On the other hand, reduced glucose utilization in AD may also be caused by progressive $A\beta$ accumulation affecting expression and/or activity of key enzymes of the glycolysis in brain cells. In a mouse model of AD (Tg2576 mice) expression and activity of the phosphofructokinase (PFK), a key enzyme in regulation of glycolysis, was studied [51.224]. In a 24-month-old transgenic Tg2576 mouse cortex, but not in 7, 13, or 17-month-old mice, the copy number of PFK-C mRNA, the PFK protein level and PFK enzyme activity was significantly reduced as compared to non-transgenic litter mates, while the mRNA level of the other PFK isoforms did not differ between transgenic and non-transgenic tissue samples. In situ hybridization in brain sections from aged Tg2576 mice revealed reduced PFK-C mRNA expression in $A\beta$ plaque-associated neurons and upregulation in reactive astrocytes surrounding $A\beta$ deposits. The data demonstrate that long-lasting high $A\beta$ burden impairs cerebral cortical glucose metabolism by reducing PFK activity in $A\beta$ plaque-associated neurons and concomitant upregulation in reactive, plaque-surrounding astrocytes [51.224]. Therefore, $A\beta$-induced alterations in glucose metabolism may contribute to cholinergic dysfunction and cell loss.

In conclusion, the basal forebrain cholinergic system plays a significant role in neurovascular regulation and blood flow control, and vascular cholinergic deficits caused by damaged microvasculature may vice versa contribute to amyloidogenic APP metabolism, thus exacerbating AD pathology.

51.7 Interrelationship of Amyloid, Cholinergic Dysfunction, and Brain Vascular Impairments

For a variety of pathological features that have been attributed to play major roles in triggering AD, three major hypotheses of AD have been stressed.

51.7.1 Cholinergic Hypothesis of AD

The early occurrence of basal forebrain cholinergic cell loss in FAD and in advanced stages of sporadic AD, as well as the correlation of clinical dementia ratings with the impairments of cholinergic function have suggested a role of the cholinergic system in the pathogenesis of AD (Fig. 51.5). Indeed, both in vitro and in vivo studies provided evidence of a modulatory role of cholinergic signaling on APP metabolism, and impairments

in mAChR and nAChR signaling may drive the APP processing toward the amyloidogenic path and thus contribute to the amyloid pathology of AD.

Selective activation of M1/M3 but not M2/M4 muscarinic acetylcholine receptors (mAChR) increased $sAPP\alpha$ secretion and decreased total $A\beta$ formation, mediated through activation of PKC and/or MAPK signaling. Agonist action on nAChR has been shown to modulate APP processing by favoring the non-amyloidogenic pathway and to inhibit β-amyloid fibril formation. AChE was found to promote the aggregation of $A\beta$ and plaque formation by forming a complex with the growing fibrils. Vice versa, $A\beta$ may increase AChE around $A\beta$ plaques through the action of $A\beta$ on $\alpha7$-

nAChR. At nanomolar concentrations soluble $A\beta$ has been observed to inhibit markers of cholinergic function. Activation of α7-nAChR may induce degradation and clearance of $A\beta$ at least in transgenic APPswe mice.

NGF plays a maintaining role for cholinergic cells, but was also observed to modulate cholinergic control of APP processing and to influence APP metabolism by favoring the non-amyloidogenic pathway through TrkA receptors, but to increase neuronal $A\beta$ production via p75NTR.

Cholinergic dysfunctions may also be caused by impaired glucose utilization, damages of the microvasculature (retrograde cholinergic terminal degradation), and impaired NGF signaling (Fig. 51.5).

51.7.2 Amyloid Cascade Hypothesis

The discovery of $A\beta$ as the major component of senile plaques, one of the histopathological hallmarks of AD, provided the basis for the establishment of the amyloid cascade hypothesis of AD, stating that $A\beta$ accumulation is the essential event leading to neuronal dysfunction and cell death associated with neurotransmitter deficits and dementia (Fig. 51.3). This hypothesis was particularly supported by investigations of transgenic mice carrying human FAD genes. These mice mimic some of the key features of AD such as amyloid plaques, cognitive deficits, and impairments in neuronal signaling and microvasculature.

Both in vitro and in vivo studies demonstrated the detrimental and modifying effects of accumulating $A\beta$ on cholinergic synaptic events by blocking nAChR, impairing mAChR signaling, inhibiting AChE, and interfering with NGF signaling by binding to p75NTR.

$A\beta$ may damage vascular endothelial cells and demonstrates vasoactive properties by impairing perfusion, perivascular drainage, and cerebral blood flow, as well as inhibiting angiogenesis. It may trigger inflammation by activation of microglial and reactive astroglial cells, which secrete pro-inflammatory cytokines (IL-1β, TNFα, TGFβ), and mediates tau pathology, which finally leads to neurodegeneration and dementia. IL-1β and TNFα have been shown to selectively degenerate cholinergic basal forebrain cells, while IL-1β has been found to upregulate APP expression and to stimulate the amyloidogenic route of APP processing. Furthermore, IL-1β may also promote activity and expression of AChE.

51.7.3 Vascular Hypothesis of AD

The vascular hypothesis for AD is based on findings that vascular risk factors damage the neurovascular unit, which results in chronic hypoperfusion, reduced cerebral blood flow, and reduced glucose supply (Fig. 51.4). As brain capillary endothelial cells receive a cholinergic input from the basal forebrain, the vascular synaptic damages may induce retrograde cell death of cholinergic neurons, which may trigger inflammation and

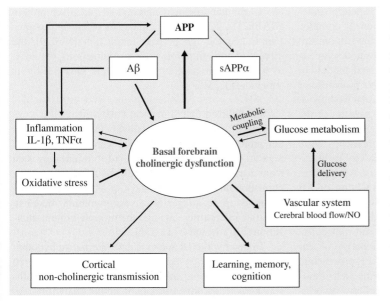

Fig. 51.5 *The cholinergic hypothesis of AD. The cholinergic hypothesis states that a dysfunction of basal forebrain cholinergic neurons causes the cognitive decline observed in AD patients, which is based on experimental findings that cholinergic neurotransmission has a fundamental role in learning and memory [51.39]. The hypothesis has been further validated by experimental observations that APP processing is controlled by cholinergic neuronal activity. Cholinergic dysfunction shifts the route of APP processing toward the amyloidogenic route, which again contributes to cholinergic deficits. Cholinergic dysfunctions may also be caused by impaired glucose utilization, damage to the microvasculature (retrograde cholinergic terminal degradation), and impaired NGF signaling (for details, see text)*

alterations in neurotrophin signaling, may favor amyloidogenic APP processing including CAA, and initiate tau pathology. Damage to vascular endothelial cells may induce angiogenesis and VEGF upregulation, which further promotes β-amyloidogenesis. As basal fore-brain cholinergic neurons particularly respond very sensitively to a reduced glucose supply by synaptic dysfunction and degeneration, the cascade of events leading to the development of AD is further exacerbated (Fig. 51.4).

51.8 Computational Modeling of the AD Brain

An understanding the pathogenesis of AD has been achieved by addressing a number of different approaches, ranging from investigations of the role and function of key proteins of AD pathology, exploring specific mechanisms and signaling cascades, neurophysiological and highly sophisticated neuroimaging techniques, up to the application of recently developed high-performance techniques of gene expression profiling and proteomic analyses, as well as genome-wide association studies. Consequently, during the last decades of research in neuroscience a vast number of experimental data and datasets have been accumulated and collected in various databases for genes, proteins, molecule structures, enzyme kinetics, etc., which require great computational effort and IT technology to be efficiently handled. On the other hand, computational approaches may provide useful tools not only to handle the abundant data available so far, but also to allow us to construct neuronal and brain networks to simulate the pathogenesis of the disease in its entirety and relevant connectivity, and to use the designed networks to computationally test for potential therapeutic strategies. Moreover, modeling the description of complex brain networks may also open new windows in our understanding of the basic mechanisms of the disease. In particular, solid and reliable computational models of the AD brain may help to elucidate the probabilities to which extent each of the hypotheses reviewed above contributes to the pathogenesis and progression of the disease.

In AD research computational approaches have been applied at various levels of complexity to investigate both single pathogenetic aspects of AD such as oligomerization, fibrillization, and aggregation of $A\beta$ and AD-related proteins, $A\beta$ binding motifs relevant to AD chemistry, structure – activity relationships and protein–protein interactions of AD relevant molecules, AD related cell signaling pathways, kinetics of enzyme substrate complexes relevant to APP processing, as well as to develop more complex systems such as molecular connectivity maps/protein interaction networks, neuronal and brain networks and their dynamics, gene–gene interactions and relevant proteomic changes, as well modeling of cognition/cognitive tasks.

In the following, a selection of recent advances made by applying computational approaches in AD-related research, will be presented; it is not intended to be complete or comprehensive.

Protein Aggregation

To understand the dynamics of the $A\beta$ peptide aggregation process involved in AD [51.225] and the differentiation between disease and non-disease protein aggregation [51.226], molecular-level simulation models have been applied. Computationally derived structural models of $A\beta$ and the interaction with possible aggregation inhibitors have been reported [51.227].

Protein–Protein Interactions

Examples of computational analysis and the prediction of protein–protein interaction include associations between the death cell receptor 6 (DR6) ectodomain and an N-terminal fragment of amyloid precursor protein (NAPP, [51.228]), in silico analysis of the apolipoprotein E and the $A\beta$ peptide interaction [51.229], the effects of mutations on protein interactions [51.230], as well as the regulation of tau protein kinase II by its subunits [51.231].

AD–Related Cell Signaling Pathways

To identify protein signal pathways between APP and tau proteins, recently a modified network-constrained regularization analysis was applied to microarray data from a transgenic mouse model and AD patients. The corresponding protein pathways were then constructed by applying an integer linear programming model to integrate microarray data and the protein–protein interaction (PPI) database [51.232].

To quickly model and visualize alternative hypotheses about uncertain pathway knowledge, a prototype tool for representing and displaying cell-signaling pathway knowledge, for carrying out simple qualitative reasoning over these pathways and for generating quantitative biosimulation codes, *Chalkboard*, was developed and tested for the network of APP processing [51.233].

Structure – Activity Relationship of AD–Relevant Molecules

Using computational chemistry a multitarget quantitative structure-activity relationship (QSAR) model has recently been developed being able to predict the results of 42 different experimental tests for GSK-3 inhibitors with heterogeneous structural patterns [51.234], while computational studies of Cu(II)/Met and Cu(I)/Met binding motifs revealed a low affinity binding site of Cu(II) in Aβ [51.235].

Kinetics of Enzyme Substrate Complexes Relevant to APP Processing

Combining bioinformatics, molecular dynamics, and density functional theory studies, enabled researchers to elucidate the mechanisms for the hydrolytic cleavage of Val-Ile and Ala-Thr peptide bonds of APP by the intramembrane aspartyl protease presenilin 1 [51.236].

Gene–Gene Interactions

Uncovering genes that are linked with AD can help to understand pathogenesis and progression of the disease. In genome-wide association studies millions of single nucleotide polymorphisms (SNP) have been analyzed in thousands of AD patients and compared with healthy individuals in order to correlate genomic changes with the disease and to reveal harmful mutations or susceptibility genes [51.237]. As any single genetic variant for AD may be dependent on other genetic variants (gene–gene interaction) and environmental factors (gene-environment interaction) the evaluation of genetic association studies represents a computational challenge. This has been addressed by the introduction of the multifactor dimensionality reduction (MDR) method proposed by [51.238], which was further improved by the log-linear model-based multifactor dimensionality reduction (LM MDR) method in sporadic AD [51.239], and grid computing to drive data intensive research [51.240]. Another computational strategy for simultaneously identifying multiple candidate genes for genetic human brain diseases from a brain-specific gene network-level perspective has been proposed by [51.241]. By integrating diverse genomic and proteomic datasets based on a Bayesian statistical model, a large-scale human brain-specific gene network has been established, which has been used to effectively identify multiple candidate genes for AD [51.241].

Gene–Protein Interactions, Posttranslational Modification of Gene Products

Microarray analyses are a valuable tool for large-scale screening of differential gene expressions that are presumably involved in AD pathogenesis, but cannot provide information about protein levels, posttranslational modifications of gene products, or activity changes, which are important for the understanding of the molecular processes of signal transduction pathways and intracellular interactions between proteins that influence cellular function. For example, toxicity studies of oligomeric Aβ (1−42) on cholinergic SN56.B5.G4 cells, revealed that gene products affected by Aβ (1−42) exposure were present in the endoplasmatic reticulum (ER), Golgi apparatus, and/or otherwise involved in protein modification and degradation (chaperones, ATF6) indicating a possible role of ER-mediated stress in Aβ-mediated toxicity. Moreover, a number of genes, which are known to be involved in AD (clusterin, acetylcholine transporter), were found to be affected following Aβ (1−42) exposure of cholinergic cells [51.242]. However, proteomic analyses of lysates of the cells exposed to Aβ (1−42) identified Aβ-mediated changes in protein levels and additionally in phosphorylation status, which provided complementary information to the microarray data. The levels of calreticulin and mitogen-activated protein kinase kinase 6c were upregulated in cholinergic cells exposed to Aβ (1−42), while γ-actin appeared downregulated. Decreased phosphorylation of phosphoproteins, such as the Rho GDP dissociation inhibitor, the ubiquitin carboxyl terminal hydrolase-1, and the tubulin α-chain isotype Mα6 have been observed [51.243]. The proteins identified have also been reported to be affected in brains of AD patients, suggesting a potential role of Aβ in influencing the integrity and functioning of the proteome in AD. The data suggest that Aβ (1−42) mediates its cytotoxic action by affecting proteins that also play key roles in a number of other physiological processes such as apoptosis, cytoskeletal changes, protein degradation of misfolded or damaged proteins, stress-related MAPK signaling, and GTPase function, indicating the Aβ-mediated disturbance of a complex mutually interacting network of signal transduction pathways. Pathological changes of one component of the network may have consequences on compartments of other cascades, finally resulting in either compensatory mechanisms, or when persistently overshooting, in pathogenic processes. In this respect, computational approaches may represent a very useful tool to model such complex networks because they are able to simulate the dynamics of such networks and the consequences of any disturbance. Assessment of particular signaling cascades reveals kinetic parameters that can be entered into a more complex network composed of a number of interacting pathways. This means that the complexity of the system can be steadily increased, finally addressing not only static but modeling also dynamic features of the system.

Molecular Connectivity Maps/Protein Interaction Networks

A computational framework was recently developed to build disease-specific drug-protein connectivity maps by integration of gene/protein and drug connectivity information based on protein interaction networks and literature mining, without requiring information on gene expression profiles [51.244]. Such molecular connectivity maps of various diseases appear to be adequate tools to computationally prove the therapeutic profile of candidate drugs before application in clinical trials [51.244].

Neuronal and Brain Networks

There are different organizational levels by which brain networks can be designed comprising either single neurons (microscale), a group of neurons, or anatomically distinct brain regions (macroscale; for a review, see e.g., [51.245]). A brain network consists of two basic elements, the nodes representing neurons or regions of interest and edges linking the nodes. The edges define either functional or structural associations among different neuronal elements of the brain. The small-world network introduced by [51.246] has been found to be an attractive model for the characterization of complex brain networks because it captures both segregated and integrated information processing of the brain [51.247]. Based on EEG, MEG and/or fMRI data, and using modern graph theory-based computational approaches the designed brain network can be used to map brain functional activities under both normal and pathological conditions. Analysis of AD patients revealed altered small-world characteristics, which reflect the aberrant neuronal connectivity in AD brain and may explain the cognitive deficits [51.245].

Neural networks have also been used for longitudinal studies in AD [51.248]. The proposed new type of neural network (mixed effects neural network) has been found to be a reliable tool for predicting non-linear disease trajectories and uncovering significant prognostic factors in longitudinal databases of AD [51.248].

A computational model of network dynamics describing the progression of the neuropathology from the hippocampus into neocortical structures in AD has been proposed by [51.249]. The model focuses on a final common breakdown in function, termed runaway synaptic modification, and proposes that the imbalance of variables regulating the influence of synaptic transmission on synaptic modification causes the pathogenic processes in the hippocampus and cortex. Memory deficits are described as being due to increased interference effects on recent memory caused by runaway synaptic modification [51.250].

Applying the framework of an associative memory model, the interplay between synaptic deletion and compensation, and memory deterioration was computationally studied in AD [51.251]. The simulation shows that a disease-induced synaptic deletion may be compensated by a local, dynamic mechanism, where each neuron maintains the profile of its incoming postsynaptic current, suggesting that the primary factor in the pathogenesis of cognitive deficiencies in AD is the failure of local neuronal regulatory mechanisms [51.251].

Recently, a novel research framework for building probabilistic computational neurogenetic models (pC-NGM) was proposed by [51.1]. The pCNGM represents a multilevel modeling framework inspired by the multilevel information processes in the brain. The framework comprises a set of several dynamic models including low (molecular) level models, a dynamic model of a protein regulatory network (PRN), and a probabilistic spiking neural network model (pSNN). It can be used for modeling both artificial cognitive systems and brain processes. Using data obtained from relevant genes/proteins a pCNGM may simulate their dynamic interaction which matches data related to brain development, higher-level brain function, or disorders such as AD [51.1].

Modeling of Cognition/Cognitive Tasks

There are a number of studies using computational modeling to simulate various cognitive and behavioral deficits observed in AD patients such as phonological dyslexia [51.252], decrease of performance in daily living activities [51.253], human path planning strategies [51.254], and memory impairments [51.255].

In conclusion, the recent advances in applying computational approaches to model the pathogenesis of AD by complex brain network methods provide some hope towards the understanding of the complexity of the disease as well as being able to computationally derive novel routes of treatment strategies and prove the therapeutic profile of candidate drugs.

51.9 General Conclusion: Mutual Interactions of Cholinergic, Aβ, and Vascular Pathologies

Taken together, the observations mentioned above indicate that a disbalance or disturbance of the homeostatic interrelationship between cholinergic function including NGF, APP metabolism, and control mechanisms of brain microvasculature (CBF, perfusion, glucose delivery) may result in deleterious consequences by favoring the amyloidogenic route of APP processing with enhanced plaque formation and induction of local neuroinflammatory events. Persisting β-amyloid-induced glial upregulation of pro-inflammatory cytokines then further contributes to damage cholinergic functions and

further promotes the amyloidogenic pathway. On the other hand, aging and life-style associated damage of the brain microvasculature may affect Aβ clearance and perivascular drainage, promote cerebrovascular Aβ deposition, inducing partial loss of cholinergic vascular innervation and changes in vascular function, angiogenesis and VEGF upregulation with consequences on APP processing and Aβ accumulation, which finally may develop into a detrimental vicious cycle.

The interplay of the essential features observed in AD (Aβ accumulation, cholinergic dysfunction, vascu-

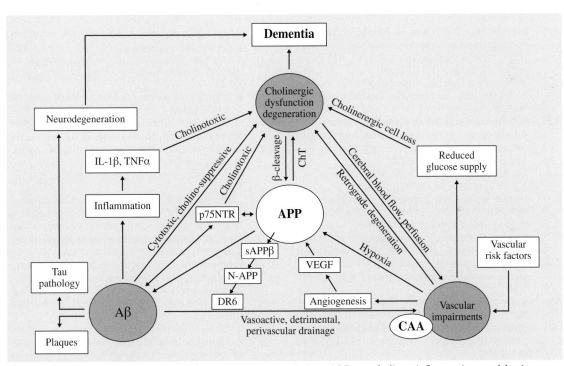

Fig. 51.6 *Interrelationship between cholinergic neurotransmission, APP metabolism, inflammation, and brain vascular system.* The interplay of the essential features observed in AD (Aβ accumulation, cholinergic dysfunction, vascular impairments, inflammation, neurodegeneration) and their mutually influencing processes is shown. A disbalance or disturbance of the homeostatic interrelationship between cholinergic function including NGF, APP metabolism, and control mechanisms of brain microvasculature (cerebral blood flow (CBF), perfusion, glucose delivery) may result in deleterious consequences by favoring the amyloidogenic route of APP processing with enhanced plaque formation and induction of local neuroinflammatory events. Persisting Aβ-induced glial upregulation of pro-inflammatory cytokines then further contributes to damage cholinergic functions as well as promotes the amyloidogenic pathway. On the other hand, aging, and life-style associated damage to the brain microvasculature may affect Aβ clearance and perivascular drainage, promote cerebrovascular Aβ deposition, inducing partial loss of cholinergic vascular innervation and changes in vascular function, angiogenesis, and VEGF upregulation with consequences on APP processing and Aβ accumulation, which finally may develop into a detrimental positive feedback loop, a vicious cycle

lar impairments, inflammation, neurodegeneration), and their mutually influencing processes are summarized in Fig. 51.6. We hypothesize that a disturbance of this homeostatic relationship may develop during late aging and may represent a major cause for the genesis of sporadic Alzheimer's disease. However, which kind of aging-related factors may play a role remains to be disclosed, but they would allow us to derive strategies for establishing prophylactic measures.

Computational approaches to model pathogenesis and progression of AD appear to be useful tools to reflect the disease in its entirety and internal connective relationships, and thus to open new views in our understanding of basic mechanisms of AD. The pCNGM research framework as recently introduced and described by [51.1], represents a promising and potential tool to model the AD brain in its complexity and dynamics at different operational levels, including functional neuron-glia-vascular units, gene–protein interactions, cell signaling cascades, regional association, and higher cognitive function. Thus computational approaches of modeling the AD brain may help to elucidate the probabilities to which extent each of the hypotheses reviewed in this chapter contribute to pathogenesis and progression of the disease.

References

51.1   N.K. Kasabov, R. Schliebs, H. Kojima: Probabilistic computational neurogenetic modeling: From cognitive systems to Alzheimer's disease, IEEE Trans. Auton. Mental Dev. **3**, 1–12 (2011)

51.2   A. Alzheimer: Über eine eigenartige Erkrankung der Hirnrinde, Allg. Z. Psychiatr. Psychisch-Gerichtl. Med. **64**, 146–148 (1907)

51.3   G.G. Glenner, C.W. Wong: Alzheimer's disease and Down's syndrome: Sharing of a unique cerebrovascular amyloid fibril protein, Biochem. Biophys. Res. Commun. **122**, 1131–1135 (1984)

51.4   C.L. Masters, G. Simms, N.A. Weinman, G. Multhaup, B.L. McDonald, K. Beyreuther: Amyloid plaque core protein in Alzheimer disease and down syndrome, Proc. Natl. Acad. Sci. USA **82**, 4245–4249 (1985)

51.5   J. Kang, H.G. Lemaire, A. Unterbeck, J.M. Salbaum, C.L. Masters, K.H. Grzeschik, G. Multhaup, K. Beyreuther, B. Müller-Hill: The precursor of Alzheimer's disease amyloid A4 protein resembles a cell-surface receptor, Nature **325**, 733–736 (1987)

51.6   D. Goldgaber, M.I. Lerman, O.W. McBride, U. Saffiotti, D.C. Gajdusek: (1987) Characterization and chromosomal localization of a cDNA encoding brain amyloid of Alzheimer's disease, Science **235**, 877–880 (2006)

51.7   N.K. Robakis, N. Ramakrishna, G. Wolfe, H.M. Wisniewski: Molecular cloning and characterization of a cDNA encoding the cerebrovascular and the neuritic plaque amyloid peptides, Proc. Natl. Acad. Sci. USA **84**, 4190–4194 (1987)

51.8   R.E. Tanzi, J.F. Gusella, P.C. Watkins, G.A. Bruns, P. St George-Hyslop, M.L. Van Keuren, D. Patterson, S. Pagan, D.M. Kurnit, R.L. Neve: Amyloid β protein gene: CDNA, mRNA distribution, and genetic linkage near the Alzheimer locus, Science **235**, 880–884 (1987)

51.9   H. Bickel: *Die Epidemiologie der Demenz*, Informationsblatt (Deutsche Alzheimer Gesellschaft e.V.,

Berlin 2010), available online at www.deutsche-alzheimer.de/

51.10   K. Herrup: Reimagining Alzheimer's disease—an age-based hypothesis, J. Neurosci. **30**, 16755–16762 (2010)

51.11   R.S. Turner: Alzheimer's disease, Semin. Neurol. **26**, 499–506 (2006)

51.12   R.J. Castellani, R.K. Rolston, M.A. Smith: Alzheimer disease, Dis. Mon. **56**, 484–546 (2010)

51.13   G.G. Glenner, C.W. Wong, V. Quaranta, E.D. Eanes: The amyloid deposits in Alzheimer's disease: Their nature and pathogenesis, Appl. Pathol. **2**, 357–369 (1984)

51.14   I. Grundke-Iqbal, K. Iqbal, Y.C. Tung, M. Quinlan, H.M. Wisniewski, L.I. Binder: Abnormal phosphorylation of the microtubule-associated protein tau (tau) in Alzheimer cytoskeletal pathology, Proc. Natl. Acad. Sci. USA **83**, 4913–4917 (1986)

51.15   I. Grundke-Iqbal, K. Iqbal, M. Quinlan, Y.C. Tung, M.S. Zaidi, H.M. Wisniewski: Microtubule-associated protein tau. A component of Alzheimer paired helical filaments, J. Biol. Chem. **261**, 6084–6089 (1986)

51.16   K. Iqbal, I. Grundke-Iqbal, A.J. Smith, L. George, Y.C. Tung, T. Zaidi: Identification and localization of a tau peptide to paired helical filaments of Alzheimer disease, Proc. Natl. Acad. Sci. USA **86**, 5646–5650 (1989)

51.17   R. Lee, P. Kermani, K.K. Teng, B.L. Hempstead: Regulation of cell survival by secreted proneurotrophins, Science **294**, 1945–1948 (2001)

51.18   K. Iqbal, C. Alonso Adel, S. Chen, M.O. Chohan, E. El-Akkad, C.X. Gong, S. Khatoon, B. Li, F. Liu, A. Rahman, H. Tanimukai, I. Grundke-Iqbal: Tau pathology in Alzheimer disease and other tauopathies, Biochim. Biophys. Acta **1739**, 198–210 (2005)

51.19   A. Alonso, T. Zaidi, M. Novak, I. Grundke-Iqbal, K. Iqbal: Hyperphosphorylation induces self-assembly of tau into tangles of paired heli-cal filaments/straight filaments, Proc. Natl. Acad. Sci. USA **98**, 6923–6928 (2001)

51.20 K.R. Patterson, C. Remmers, Y. Fu, S. Brooker, N.M. Kanaan, L. Vana, S. Ward, J.F. Reyes, K. Philibert, M.J. Glucksman, L.I. Binder: Characterization of prefibrillar tau oligomers in vitro and in Alzheimer disease, J. Biol. Chem. **286**, 23063–23076 (2011)

51.21 C. Ballard, S. Gauthier, A. Corbett, C. Brayne, D. Aarsland, E. Jones: Alzheimer's disease, Lancet **377**, 1019–1031 (2011)

51.22 M.S. Wolfe: Tau mutations in neurodegenerative diseases, J. Biol. Chem. **284**, 6021–6025 (2009)

51.23 M. Goedert, M.G. Spillantini: A century of Alzheimer's disease, Science **314**, 777–781 (2006)

51.24 S.A. Small, K. Duff: Linking Aβ and tau in late-onset Alzheimer's disease: A dual pathway hypothesis, Neuron **60**, 534–542 (2008)

51.25 L. Bertram, C.M. Lill, R.E. Tanzi: The genetics of Alzheimer disease: Back to the future, Neuron **68**, 270–281 (2010)

51.26 E.H. Corder, A.M. Saunders, W.J. Strittmatter, D.E. Schmechel, P.C. Gaskell, G.W. Small, A.D. Roses, J.L. Haines, M.A. Pericak-Vance: Gene dose of apolipoprotein E type 4 allele and the risk of Alzheimer's disease in late onset families, Science **261**, 921–923 (1993)

51.27 A.M. Saunders, W.J. Strittmatter, D. Schmechel, P.H. George-Hyslop, M.A. Pericak-Vance, S.H. Joo, B.L. Rosi, J.F. Gusella, D.R. Crapper-MacLachlan, M.J. Alberts, C. Hulette, B. Crain, D. Goldgaber, A.D. Roses: Association of apolipoprotein E allele epsilon 4 with late-onset familial and spo-radic Alzheimer's disease, Neurology **43**, 467–1472 (1993)

51.28 Y. Huang: Aβ-independent roles of apolipopro-tein E4 in the pathogenesis of Alzheimer's disease, Trends Mol. Med. **16**, 287–294 (2010)

51.29 E. Rogaeva, Y. Meng, J.H. Lee, Y. Gu, T. Kawarai, F. Zou, T. Katayama, C.T. Baldwin, R. Cheng, H. Hasegawa, F. Chen, N. Shibata, K.L. Lunetta, R. Pardossi-Piquard, C. Bohm, Y. Wakutani, L.A. Cupples, K.T. Cuenco, R.C. Green, L. Pinessi, I. Rainero, S. Sorbi, A. Bruni, R. Duara, R.P. Fried-land, R. Inzelberg, W. Hampe, H. Bujo, Y.Q. Song, O.M. Andersen, T.E. Willnow, N. Graff-Radford, R.C. Petersen, D. Dickson, S.D. Der, P.E. Fraser, G. Schmitt-Ulms, S. Younkin, R. Mayeux, L.A. Far-rer, P. St George-Hyslop: The neuronal sortilin-related receptor SORL1 is genetically associated with Alzheimer disease, Nat. Genet. **39**, 168–177 (2007)

51.30 K. Morgan: The three new pathways leading to AD, Neuropathol. Appl. Neurobiol. **37**, 353–357 (2011)

51.31 D.M. Bowen, C.B. Smith, P. White, A.N. Davison: Neurotransmitter-related enzymes and indices of hypoxia in senile dementia and other abiotro-phies, Brain **99**, 459–496 (1976)

51.32 P. Davies, A.J. Maloney: Selective loss of central cholinergic neurons in Alzheimer's disease, Lancet **2**, 1403 (1976)

51.33 E.K. Perry, P.H. Gibson, G. Blessed, R.H. Perry, B.E. Tomlinson: Neurotransmitter enzyme abnor-malities in senile dementia. Choline acetyltrans-ferase and glutamic acid decarboxylase activities in necropsy brain tissue, J. Neurol. Sci. **34**, 247–265 (1977)

51.34 E.K. Perry, R.H. Perry, G. Blessed, B.E. Tomlinson: Necropsy evidence of central cholinergic deficits in senile dementia, Lancet. **1**, 189 (1977)

51.35 D.S. Auld, T.J. Kornecook, S. Bastianetto, R. Quirion: Alzheimer's disease and the basal forebrain cholinergic system: Relations to β-amyloid pep-tides, cognition, and treatment strategies, Prog. Neurobiol. **68**, 209–245 (2002)

51.36 A. Nordberg: Neuroreceptor changes in Alzheimer disease, Cerebrovasc. Brain Metab. Rev. **4**, 303–328 (1992)

51.37 L.M. Bierer, V. Haroutunian, S. Gabriel, P.J. Knott, L.S. Carlin, D.P. Purohit, D.P. Perl, J. Schmeidler, P. Kanof, K.L. Davis: Neurochemical correlates of dementia severity in Alzheimer's disease: Relative importance of the cholinergic deficits, J. Neu-rochem. **64**, 749–760 (1995)

51.38 W. Gsell, G. Jungkunz, P. Riederer: Functional neurochemistry of Alzheimer's disease, Curr. Phar-macol. Des. **10**, 265–293 (2004)

51.39 R.T. Bartus: On neurodegenerative diseases, mod-els, and treatment strategies: Lessons learned and lessons forgotten a generation following the cholinergic hypothesis, Exp. Neurol. **163**, 495–529 (2000)

51.40 E.J. Mufson, S.E. Counts, S.E. Perez, S.D. Gins-berg: Cholinergic system during the progression of Alzheimer's disease: Therapeutic implications, Exp. Rev. Neurother. **8**, 1703–1718 (2008)

51.41 E.J. Mufson, S.E. Counts, M. Fahnestock, S.D. Gins-berg: Cholinotrophic molecular substrates of mild cognitive impairment in the elderly, Curr. Alzheimer Res. **4**, 340–350 (2007)

51.42 A.C. Cuello, M.A. Bruno, K.F. Bell: NGF-cholinergic dependency in brain aging, MCI and Alzheimer's disease, Curr. Alzheimer Res. **4**, 351–358 (2007)

51.43 Y.W. Zhang, R. Thompson, H. Zhang, H. Xu: APP processing in Alzheimer's disease, Mol. Brain **4**, 3 (2011), doi: 10.1186/1756-6606-4-3

51.44 A. Nikolaev, T. McLaughlin, D.D. O'Leary, M. Tessier-Lavigne: APP binds DR6 to trigger axon pruning and neuron death via distinct caspases, Nature **457**, 981–989 (2009)

51.45 H. Li, B. Wang, Z. Wang, Q. Guo, K. Tabuchi, R.E. Hammer, T.C. Südhof, H. Zheng: Soluble amy-loid precursor protein (APP) regulates transthyretin and Klotho gene expression without rescuing the essential function of APP, Proc. Natl. Acad. Sci. USA **107**, 17362–17367 (2010)

51.46   B. De Strooper, R. Vassar, T. Golde: The secretases: Enzymes with therapeutic potential in Alzheimer disease, Nat. Rev. Neurol. **6**, 99–107 (2010)

51.47   J. Hardy, D. Allsop: Amyloid deposition as the central event in the aetiology of Alzheimer's disease, Trends Pharmacol. Sci. **12**, 383–388 (1991)

51.48   J. Hardy: The amyloid hypothesis for Alzheimer's disease: A critical reappraisal, J. Neurochem. **110**, 1129–1134 (2009)

51.49   B.K. Harvey, C.T. Richie, B.J. Hoffer, M. Airavaara: Transgenic animal models of neurodegeneration based on human genetic studies, J. Neural Transm. **118**, 27–45 (2011)

51.50   A. Thathiah, B. De Strooper: The role of G protein-coupled receptors in the pathology of Alzheimer's disease, Nat. Rev. Neurosci. **12**, 73–87 (2011)

51.51   M. Pákáski, J. Kálmán: Interactions between the amyloid and cholinergic mechanisms in Alzheimer's disease, Neurochem. Int. **53**, 103–111 (2008)

51.52   R.H. Parri, T.K. Dineley: Nicotinic acetylcholine receptor interaction with beta-amyloid: Molecular, cellular, and physiological consequences, Curr. Alzheimer Res. **7**, 27–39 (2010)

51.53   M.M. Mesulam: Alzheimer plaques and cortical cholinergic innervation, Neuroscience **17**, 275–276 (1986)

51.54   M.A. Moran, E.J. Mufson, P. Gomez-Ramos: Colocalization of cholinesterases with β-amyloid protein in aged and Alzheimer's brain, Acta Neuropathol. **85**, 362–369 (1993)

51.55   R.M. Nitsch, B.E. Slack, R.J. Wurtman, J.H. Growdon: Release of Alzheimer amyloid precursor derivatives stimulated by activation of muscarinic acetylcholine receptors, Science **258**, 304–307 (1992)

51.56   J.D. Buxbaum, M. Oishi, H.I. Chen, R. Pinkas-Kramarski, E.A. Jaffe, S.E. Gandy, P. Greengard: Cholinergic agonists and interleukin 1 regulate processing and secretion of the Alzheimer β/A4 amyloid protein precursor, Proc. Natl. Acad. Sci. USA **89**, 10075–10078 (1992)

51.57   R.M. Nitsch, S.A. Farber, J.H. Growdon, R.J. Wurtman: Release of amyloid beta-protein precursor derivatives by electrical depolarization of rat hippocampal slices, Proc. Natl. Acad. Sci. USA **90**, 5191–5193 (1993)

51.58   H.A. Ensinger, H.N. Doods, A.R. Immel-Sehr, F.J. Kuhn, G. Lambrecht, K.D. Mendla, R.E. Muller, E. Mutschler, A. Sagrada, G. Walther: WAL 2014–a muscarinic agonist with preferential neuron-stimulating properties, Life Sci. **52**, 473–480 (1993)

51.59   S.A. Farber, R.M. Nitsch, J.G. Schulz, R.J. Wurtman: Regulated secretion of beta-amyloid precursor protein in rat brain, J. Neurosci. **15**, 7442–7451 (1995)

51.60   A. Walland, S. Burkard, R. Hammer, W. Troger: In vivo consequences of M1-receptor activation by talsaclidine, Life Sci. **60**, 977–984 (1997)

51.61   D.M. Müller, K. Mendla, S.A. Farber, R.M. Nitsch: Muscarinic M1 receptor agonists increase the secretion of the amyloid precursor protein ectodomain, Life Sci. **60**, 985–991 (1997)

51.62   C. Hock, A. Maddalena, A. Raschig, F. Müller-Spahn, G. Eschweiler, K. Hager, I. Heuser, H. Hampel, T. Müller-Thomsen, W. Oertel, M. Wienrich, A. Signorell, C. Gonzalez-Agosti, R.M. Nitsch: Treatment with the selective muscarinic m1 agonist talsaclidine decreases cerebrospinal fluid levels of A beta 42 in patients with Alzheimer's disease, Amyloid **10**, 1–6 (2003)

51.63   R.M. Nitsch: Treatment with the selective muscarinic m1 agonist talsaclidine decreases cerebrospinal fluid levels of Aβ42 in patients with Alzheimer's disease, Amyloid **10**, 1–6 (2003)

51.64   A.Y. Hung, C. Haass, R.M. Nitsch, W.Q. Qiu, M. Citron, R.J. Wurtman, J.H. Growdon, D.J. Selkoe: Activation of protein kinase C inhibits cellular production of the amyloid beta-protein, J. Biol. Chem. **268**, 22959–22962 (1993)

51.65   M. Cisse, U. Braun, M. Leitges, A. Fisher, G. Pages, F. Checler, B. Vincent: ERK1-independent α-secretase cut of β-amyloid precursor protein via M1 muscarinic receptors and PKCα/ε, Mol. Cell Neurosci. **47**, 223–232 (2011)

51.66   B.E. Slack, J. Breu, M.A. Petryniak, K. Srivastava, R.J. Wurtman: Tyrosine phosphorylation-dependent stimulation of amyloid precursor protein secretion by the m3 muscarinic acetylcholine receptor, J. Biol. Chem. **270**, 8337–8344 (1995)

51.67   R. Haring, A. Fisher, D. Marciano, Z. Pittel, Y. Kloog, A. Zuckerman, N. Eshhar, E. Heldman: Mitogen-activated protein kinase-dependent and protein kinase C-dependent pathways link the m1 muscarinic receptor to beta-amyloid precursor protein secretion, J. Neurochem. **71**, 2094–2103 (1998)

51.68   S. Roßner, U. Ueberham, R. Schliebs, J.R. Perez-Polo, V. Bigl: The regulation of amyloid precursor protein metabolism by cholinergic mechanisms and neurotrophin receptor signaling, Prog. Neurobiol. **56**, 541–569 (1998)

51.69   L. Lin, B. Georgievska, A. Mattsson, O. Isacson: Cognitive changes and modified processing of amyloid precursor protein in the cortical and hippocampal system after cholinergic synapse loss and muscarinic receptor activation, Proc. Natl. Acad. Sci. USA **96**, 12108–12113 (1999)

51.70   H. Seo, A.W. Ferree, O. Isacson: Cortico-hippocampal APP and NGF levels are dynamically altered by cholinergic muscarinic antagonist or M1 agonist treatment in normal mice, Eur J. Neurosci. **15**, 498–505 (2002)

51.71   W. Liskowsky, R. Schliebs: Muscarinic acetylcholine receptor inhibition in transgenic Alzheimer-like Tg2576 mice by scopolamine favours the amyloidogenic route of processing of amyloid precursor protein, Int. J. Devl. Neurosci. **24**, 149–156 (2006)

51.72 A.A. Davis, J.J. Fritz, J. Wess, J.J. Lah, A.I. Levey: Deletion of M1 muscarinic acetylcholine receptors increases amyloid pathology in vitro and in vivo, J. Neurosci. **30**, 4190–4196 (2010)

51.73 A. Caccamo, S. Oddo, L.M. Billings, K.N. Green, H. Martinez-Coria, A. Fisher, F.M. LaFerla: M1 receptors play a central role in modulating AD-like pathology in transgenic mice, Neuron **49**, 671–682 (2006)

51.74 R. Medeiros, M. Kitazawa, A. Caccamo, D. Baglietto-Vargas, T. Estrada-Hernandez, D.H. Cribbs, A. Fisher, F.M. Laferla: Loss of muscarinic M(1) receptor exacerbates Alzheimer's disease-like pathology and cognitive decline, Am. J. Pathol. **179**, 980–991 (2011)

51.75 T. Züchner, J.R. Perez-Polo, R. Schliebs: β-secretase BACE1 is differentially controlled through muscarinic acetylcholine receptor signaling, J. Neurosci. Res. **77**, 250–257 (2004)

51.76 T.A. Kohout, R.J. Lefkowitz: Regulation of G protein-coupled receptor kinases and arrestins during receptor desensitization, Mol. Pharmacol. **63**, 9–18 (2003)

51.77 J. Liu, I. Rasul, Y. Sun, G. Wu, L. Li, R.T. Premont, W.Z. Suo: GRK5 deficiency leads to reduced hippocampal acetylcholine level via impaired presynaptic M2/M4 autoreceptor desensitization, J. Biol. Chem. **284**, 19564–19571 (2009)

51.78 W. Zhang, A.S. Basile, J. Gomeza, L.A. Volpicelli, A.I. Levey, J. Wess: Characterization of central inhibitory muscarinic autoreceptors by the use of muscarinic acetylcholine receptor knock-out mice, J. Neurosci. **22**, 1709–1717 (2002)

51.79 Z. Suo, M. Wu, B.A. Citron, G.T. Wong, B.W. Festoff: Abnormality of G-protein-coupled receptor kinases at prodromal and early stages of Alzheimer's disease: An association with early beta-amyloid accumulation, J. Neurosci. **24**, 3444–3452 (2004)

51.80 S. Cheng, L. Li, S. He, J. Liu, Y. Sun, M. He, K. Grasing, R.T. Premont, W.Z. Suo: GRK5 deficiency accelerates β-amyloid accumulation in Tg2576 mice via impaired cholinergic activity, J. Biol. Chem. **285**, 41541–41548 (2010)

51.81 S. Efthimiopoulos, D. Vassilacopoulou, J.A. Ripellino, N. Tezapsidis, N.K. Robakis: Cholinergic agonists stimulate secretion of soluble full-length amyloid precursor protein in neuroendocrine cells, Proc. Natl. Acad. Sci. USA **93**, 8046–8050 (1996)

51.82 S.H. Kim, Y.K. Kim, S.J. Jeong, C. Haass, Y.H. Kim, Y.H. Suh: Enhanced release of secreted form of Alzheimer's amyloid precursor protein from PC12 cells by nicotine, Mol. Pharmacol. **52**, 430–436 (1997)

51.83 J. Seo, S. Kim, H. Kim, C.H. Park, S. Jeong, J. Lee, S.H. Choi, K. Chang, J. Rah, J. Koo, E. Kim, Y. Suh: Effects of nicotine on APP secretion and Aβ- or CT(105)-induced toxicity, Biol. Psychiatry **49**, 240–247 (2001)

51.84 X.L. Qi, A. Nordberg, J. Xiu, Z.Z. Guan: The consequences of reducing expression of the alpha7 nicotinic receptor by RNA interference and of stimulating its activity with an alpha7 agonist in SH-SY5Y cells indicate that this receptor plays a neuroprotective role in connection with the pathogenesis of Alzheimer's disease, Neurochem. Int. **51**, 377–383 (2007)

51.85 T. Utsuki, M. Shoaib, H.W. Holloway, D.K. Ingram, W.C. Wallace, V. Haroutunian, K. Sambamurti, D.K. Lahiri, N.H. Greig: Nicotine lowers the secretion of the Alzheimer's amyloid β-protein precursor that contains amyloid β-peptide in rat, J. Alzheimers Dis. **4**, 405–415 (2002)

51.86 D.K. Lahiri, T. Utsuki, D. Chen, M.R. Farlow, M. Shoaib, D.K. Ingram, N.H. Greig: Nicotine reduces the secretion of Alzheimer's β-amyloid precursor protein containing β-amyloid peptide in the rat without altering synaptic proteins, Ann. N. Y. Acad. Sci. **965**, 364–372 (2002)

51.87 M. Mousavi, E. Hellström-Lindahl: Nicotinic receptor agonists and antagonists increase sAPPalpha secretion and decrease Abeta levels in vitro, Neurochem. Int. **54**, 237–244 (2009)

51.88 H.Z. Nie, S. Shi, R.J. Lukas, W.J. Zhao, Y.N. Sun, M. Yin: Activation of α7 nicotinic receptor affects APP processing by regulating secretase activity in SH-EP1-α7 nAChR-hAPP695 cells, Brain Res. **1356**, 112–120 (2010)

51.89 H.Z. Nie, Z.Q. Li, Q.X. Yan, Z.J. Wang, W.J. Zhao, L.C. Guo, M. Yin: Nicotine decreases beta-amyloid through regulating BACE1 transcription in SH-EP1-α4β2 nAChR-APP695 cells, Neurochem. Res. **36**, 904–912 (2011)

51.90 E. Hellstrom-Lindahl, J. Court, J. Keverne, M. Svedberg, M. Lee, A. Marutle, A. Thomas, E. Perry, I. Bednar, A. Nordberg: Nicotine reduces Aβ in the brain and cerebral vessels of APPsw mice, Eur. J. Neurosci. **19**, 2703–2710 (2004)

51.91 M. Srivareerat, T.T. Tran, S. Salim, A.M. Aleisa, K.A. Alkadhi: Chronic nicotine restores normal Aβ levels and prevents short-term memory and E-LTP impairment in Aβ rat model of Alzheimer's disease, Neurobiol. Aging **32**, 834–844 (2011)

51.92 J.A. Court, M. Johnson, D. Religa, J. Keverne, R. Kalaria, E. Jaros, I.G. McKeith, R. Perry, J. Naslund, E.K. Perry: Attenuation of Aβ deposition in the entorhinal cortex of normal elderly individuals associated with tobacco smoking, Neuropathol. Appl. Neurobiol. **31**, 522–535 (2005)

51.93 A.M. Klein, N.W. Kowall, R.J. Ferrante: Neurotoxicity and oxidative damage of β-amyloid 1–42 versus β-amyloid 1–40 in the mouse cerebral cortex, Ann. N. Y. Acad. Sci. **893**, 314–320 (1999)

51.94 W. Härtig, A. Bauer, K. Brauer, J. Gosche, T. Hortobagyi, B. Penke, R. Schliebs, T. Harkany: Functional recovery of cholinergic basal forebrain neurons under disease conditions: Old problems, Rev. Neurosci. **13**, 95–165 (2002)

51.95   D.M. Walsh, D.J. Selkoe: Aβ oligomers – a decade of discovery, J. Neurochem. **101**, 1172–1184 (2007)

51.96   B.A. Yankner, T. Lu: Amyloid β-protein toxicity and the pathogenesis of Alzheimer disease, J. Biol. Chem. **284**, 4755–4759 (2009)

51.97   R. Schliebs, T. Arendt: The cholinergic system in aging and neuronal degeneration, Behav. Brain Res. **221**, 555–563 (2011)

51.98   T. Ondrejcak, I. Klyubin, N.W. Hu, A.E. Barry, W.K. Cullen, M.J. Rowan: Alzheimer's disease amyloid beta-protein and synaptic function, Neuromolecular Med. **12**, 13–26 (2010)

51.99   M.S. Parihar, G.J. Brewer: Amyloid beta as a modulator of synaptic plasticity, J. Alzheimers Dis. **22**, 741–763 (2010)

51.100 J.P. Cleary, D.M. Walsh, J.J. Hofmeister, G.M. Shankar, M.A. Kuskowski, D.J. Selkoe, K.H. Ashe: Natural oligomers of the amyloid-β protein specifically disrupt cognitive function, Nat. Neurosci. **8**, 79–84 (2005)

51.101 S. Lesné, M.T. Koh, L. Kotinilek, R. Kayed, C.G. Glabe, A. Yang, M. Gallagher, K.H. Ashe: A specific amyloid protein assembly in the brain impairs memory, Nature **440**, 352–357 (2006)

51.102 C.A. McLean, R.A. Cherny, F.W. Fraser, S.J. Fuller, M.J. Smith, K. Beyreuther, A.I. Bush, C.L. Masters: Soluble pool of Aβ amyloid as a determinant of severity of neurodegeneration in Alzheimer's disease, Ann. Neurol. **46**, 860–866 (1999)

51.103 J. Apelt, A. Kumar, R. Schliebs: Impairment of cholinergic neurotransmission in adult and aged transgenic Tg2576 mouse brain expressing the Swedish mutation of human beta-amyloid precursor protein, Brain Res. **953**, 17–30 (2002)

51.104 H.J. Lüth, J. Apelt, A. Ihunwo, R. Schliebs: Degeneration of β-amyloid-associated cholinergic structures in transgenic APPSW mice, Brain Res. **977**, 16–22 (2003)

51.105 M. Klingner, J. Apelt, A. Kumar, D. Sorger, O. Sabri, J. Steinbach, M. Scheunemann, R. Schliebs: Alterations in cholinergic and non-cholinergic neurotransmitter receptor densities in transgenic Tg2576 mouse brain with β-amyloid plaque pathology, Int. J. Devl. Neurosci. **21**, 357–369 (2003)

51.106 A. Bellucci, I. Luccarini, C. Scali, C. Prosperi, M.G. Giovannini, G. Pepeu, F. Casamenti: Cholinergic dysfunction, neuronal damage and axonal loss in TgCRND8 mice, Neurobiol. Dis. **23**, 260–272 (2006)

51.107 D. Van Dam, B. Marescau, S. Engelborghs, T. Cremers, J. Mulder, M. Staufenbiel, P.P. De Deyn: Analysis of cholinergic markers, biogenic amines, and amino acids in the CNS of two APP overexpression mouse models, Neurochem. Int. **46**, 409–422 (2005)

51.108 K.R. Bales, E.T. Tzavara, S. Wu, M.R. Wade, F.P. Bymaster, S.M. Paul, G.G. Nomikos: Cholinergic dysfunction in a mouse model of Alzheimer disease is reversed by an anti-Aβ antibody, J. Clin. Invest. **116**, 825–832 (2006)

51.109 E. Machová, J. Jakubík, P. Michal, M. Oksman, H. Iivonen, H. Tanila, V. Dolezal: Impairment of muscarinic transmission in transgenic APPswe/PS1dE9 mice, Neurobiol. Aging **29**, 368–378 (2008)

51.110 Y. Luo, B. Bolon, S. Kahn, B.D. Bennett, S. Babu-Khan, P. Denis, W. Fan, H. Kha, J. Zhang, Y. Gong, L. Martin, J.C. Louis, Q. Yan, W.G. Richards, M. Citron, R. Vassar: Mice deficient in BACE1, the Alzheimer's beta-secretase, have normal phenotype and abolished β-amyloid generation, Nat. Neurosci. **4**, 231–232 (2001)

51.111 M. Ohno, E.A. Sametsky, L.H. Younkin, H. Oakley, S.G. Younkin, M. Citron, R. Vassar, J.F. Disterhoft: BACE1 deficiency rescues memory deficits and cholinergic dysfunction in a mouse model of Alzheimer's disease, Neuron **41**, 27–33 (2004)

51.112 F. Bao, L. Wicklund, P.N. Lacor, W.L. Klein, A. Nordberg, A. Marutle: Different β-amyloid oligomer assemblies in Alzheimer brains correlate with age of disease onset and impaired cholinergic activity, Neurobiol. Aging **33**, 825.e1–825.e13 (2012)

51.113 C. Haass, D.J. Selkoe: Cellular processing of beta-amyloid precursor protein and the genesis of amyloid beta-peptide, Cell **75**, 1039–1042 (1993)

51.114 M. Shoji, T.E. Golde, J. Ghiso, T.T. Cheung, S. Estus, L.M. Shaffer, X.D. Cai, D.M. McKay, R. Tintner, B. Frangione: Production of the Alzheimer amyloid beta protein by normal proteolytic processing, Science **258**, 126–129 (1992)

51.115 S. Kar, S.P. Slowikowski, D. Westaway, H.T. Mount: Interactions between β-amyloid and central cholinergic neurons: Implications for Alzheimer's disease, J. Psychiatry Neurosci. **29**, 427–441 (2004)

51.116 R.J. Wurtman: Choline metabolism as a basis for the selective vulnerability of cholinergic neurons, Trends Neurosci. **15**, 117–122 (1992)

51.117 C. Hock, A. Maddalena, A. Raschig, F. Muller-Spahn, G. Eschweiler, K. Hager, I. Heuser, H. Hampel, T. Muller-Thomsen, W. Oertel, M. Wienrich, A. Signorell, C. Gonzalez-Agosti, M. Hoshi, A. Takashima, M. Murayama, K. Yasutake, N. Yoshida, K. Ishiguro, T. Hoshino, K. Imahori: Nontoxic amyloid β peptide 1–42 suppresses acetylcholine synthesis. Possible role in cholinergic dysfunction in Alzheimer's disease, J. Biol. Chem. **272**, 2038–2041 (1997)

51.118 W.A. Pedersen, M.A. Kloczewiak, J.K. Blusztajn: Amyloid β-protein reduces acetylcholine synthesis in a cell line derived from cholinergic neurons of the basal forebrain, Proc. Natl. Acad. Sci. USA **93**, 8068–8071 (1996)

51.119 D. Hicks, D. John, N.Z. Makova, Z. Henderson, N.N. Nalivaeva, A.J. Turner: Membrane targeting, J. Neurochem. **116**, 742–746 (2011)

51.120 J.F. Kelly, K. Furukawa, S.W. Barger, M.R. Rengen, R.J. Mark, E.M. Blanc, G.S. Roth, M.P. Mattson: Amyloid β-peptide disrupts carbachol-induced muscarinic cholinergic signal transduction in cortical neurons, Proc. Natl. Acad. Sci. USA **93**, 6753–6758 (1996)

51.121 B. Wang, L. Yang, Z. Wang, H. Zheng: Amyloid precursor protein mediates presynaptic localization and activity of the high-affinity choline transporter, Proc. Natl. Acad. Sci. USA **104**, 14140–14145 (2007)

51.122 J. Pavia, J. Alberch, I. Alvárez, A. Toledano, M.L. de Ceballos: Repeated intracerebroventricular administration of beta-amyloid (25–35) to rats decreases muscarinic receptors in cerebral cortex, Neurosci. Lett. **278**, 69–72 (2000)

51.123 A. Thathiah, B. De Strooper: G protein-coupled receptors, cholinergic dysfunction, and Aβ toxicity in Alzheimer's disease, Sci. Signal. **2**(93), re8 (2009)

51.124 T. Chiba, M. Yamada, J. Sasabe, K. Terashita, M. Shimoda, M. Matsuoka, S. Aiso: Amyloid-β causes memory impairment by disturbing the JAK2/STAT3 axis in hippocampal neurons, Mol. Psychiatry **14**, 206–222 (2009)

51.125 L. Burghaus, U. Schütz, U. Krempel, R.A. de Vos, E.N. Jansen Steur, A. Wevers, J. Lindstrom, H. Schröder: Quantitative assessment of nicotinic acetylcholine receptor proteins in the cerebral cortex of Alzheimer patients, Mol. Brain Res. **76**, 385–388 (2000)

51.126 M. Mousavi, E. Hellström-Lindahl, Z.Z. Guan, K.R. Shan, R. Ravid, A. Nordberg: Protein and mRNA levels of nicotinic receptors in brain of tobacco using controls and patients with Alzheimer's disease, Neuroscience **122**, 515–520 (2003)

51.127 C.M. Martin-Ruiz, J.A. Court, E. Molnar, M. Lee, C. Gotti, A. Mamalaki, T. Tsouloufis, S. Tzartos, C. Ballard, R.H. Perry, E.K. Perry: Alpha4 but not alpha3 and alpha7 nicotinic acetylcholine receptor subunits are lost from the temporal cortex in Alzheimer's disease, J. Neurochem. **73**, 1635–1640 (1999)

51.128 A. Wevers, H. Schröder: Nicotinic acetylcholine receptors in Alzheimer's disease, J. Alzheimers Dis. **1**, 207–219 (1999)

51.129 A. Nordberg: Nicotinic receptor abnormalities of Alzheimer's disease: Therapeutic implications, Biol. Psychiatry **49**, 200–210 (2001)

51.130 S.D. Buckingham, A.K. Jones, L.A. Brown, D.B. Sattelle: Nicotinic acetylcholine receptor signalling: Roles in Alzheimer's disease and amyloid neuroprotection, Pharmacol. Rev. **61**, 39–61 (2009)

51.131 S. Jürgensen, S. Ferreira: Nicotinic receptors, amaloid-β, and synaptic failure in Alzheimer's disease, J. Mol. Neurosci. **40**, 221–229 (2010)

51.132 R.G. Nagele, M.R. D'Andrea, W.J. Anderson, H.Y. Wang: Intracellular accumulation of beta-amyloid (1–42) in neurons is facilitated by the alpha 7 nicotinic acetylcholine receptor in Alzheimer's disease, Neuroscience **110**, 199–211 (2002)

51.133 F.J. Barrantes, V. Borroni, S. Vallés: Neuronal nicotinic acetylcholine receptor-cholesterol crosstalk in Alzheimer's disease, FEBS Lett. **584**, 1856–1863 (2010)

51.134 Z.Z. Guan, H. Miao, J.Y. Tian, C. Unger, A. Nordberg, X. Zhang: Suppressed expression of nicotinic acetylcholine receptors by nanomolar β-amyloid peptides in PC12 cells, J. Neural Transm. **108**, 1417–1433 (2001)

51.135 Q. Liu, H. Kawai, D.K. Berg: β-amyloid peptide blocks the response of α7-containing nicotinic receptors on hippocampal neurons, Proc. Natl. Acad. Sci. USA **98**, 4734–4739 (2001)

51.136 K.A. Alkadhi, K.H. Alzoubi, M. Srivareerat, T.T. Tran: Elevation of BACE in an Aβ rat model of Alzheimer's disease: Exacerbation by chronic stress and prevention by nicotine, Int. J. Neuropsychopharmacol. **28**, 1–11 (2011)

51.137 H.Y. Wang, D.H. Lee, M.R. D'Andrea, P.A. Peterson, R.P. Shank, A.B. Reitz: β-Amyloid(1–42) binds to α7 nicotinic acetylcholine receptor with high affinity, Implications for Alzheimer's disease pathology, J. Biol. Chem. **275**, 5626–5632 (2000)

51.138 H.Y. Wang, D.H. Lee, C.B. Davis, R.P. Shank: Amyloid peptide Aβ (1–42) binds selectively and with picomolar affinity to α7 nicotinic acetylcholine receptors, J. Neurochem. **75**, 1155–1161 (2000)

51.139 S. Oddo, A. Caccamo, K.N. Green, K. Liang, L. Tran, Y. Chen, F.M. Leslie, F.M. LaFerla: Chronic nicotine administration exacerbates tau pathology in a transgenic model of Alzheimer's disease, Proc. Natl. Acad. Sci. USA **102**, 3046–3051 (2005)

51.140 X.L. Qi, J. Xiu, K.R. Shan, Y. Xiao, R. Gu, R.Y. Liu, Z.Z. Guan: Oxidative stress induced by β-amyloid peptide (1–42) is involved in the altered composition of cellular membrane lipids and the decreased expression of nicotinic receptors in human SH-SY5Y neuroblastoma cells, Neurochem. Int. **46**, 613–621 (2005)

51.141 Z.Z. Guan, W.F. Yu, K.R. Shan, T. Nordman, J. Olsson, A. Nordberg: Loss of nicotinic receptors induced by β-amyloid peptides in PC12 cells: Possible mechanism involving lipid peroxidation, J. Neurosci. Res. **71**, 397–406 (2003)

51.142 K.T. Dineley, M. Westerman, D. Bui, K. Bell, K.H. Ashe, J.D. Sweatt: β-amyloid activates the mitogen-activated protein kinase cascade via hippocampal α7 nicotinic acetylcholine receptors: In vitro and in vivo mechanisms related to Alzheimer's disease, J. Neurosci. **21**, 4125–4133 (2001)

51.143 H.Y. Wang, W. Li, N.J. Benedetti, D.H. Lee: α7 nicotinic acetylcholine receptors mediate β-amyloid peptide-induced tau protein phosphorylation, J. Biol. Chem. **278**, 31547–31553 (2003)

51.144 K.T. Dineley, K.A. Bell, D. Bui, J.D. Sweatt: β-Amyloid peptide activates α7 nicotinic acetylcholine receptors expressed in *Xenopus oocytes*, J. Biol. Chem. **277**, 25056–25061 (2002)

51.145 M. Bencherif, P. Lippiello: α7 neuronal nicotinic receptors: The missing link to understanding Alzheimer's etiopathology?, Med. Hypotheses **74**, 281–285 (2010)

51.146 M. Hu, J.F. Waring, M. Gopalakrishnan, J. Li: Role of GSK-3β activation and α7 nAChRs in Aβ$_{1-42}$-induced tau phosphorylation in PC12 cells, J. Neurochem. **106**, 1371–1377 (2008)

51.147 G. Dziewczapolski, C.M. Glogowski, E. Masliah, S.F. Heinemann: Deletion of the α7 nicotinic acetylcholine receptor gene improves cognitive deficits and synaptic pathology in a mouse model of Alzheimer's disease, J. Neurosci. **29**, 8805–8815 (2009)

51.148 A.M. Lilja, O. Porras, E. Storelli, A. Nordberg, A. Marutle: Functional interactions of fibrillar and oligomeric amyloid-β with alpha7 nicotinic receptors in Alzheimer's disease, J. Alzheimers Dis. **23**, 335–347 (2011)

51.149 Y. An, X.L. Qi, J.J. Pei, Z. Tang, Y. Xiao, Z.Z. Guan: Amyloid precursor protein gene mutated at Swedish 670/671 sites in vitro induces changed expression of nicotinic acetylcholine receptors and neurotoxicity, Neurochem. Int. **57**, 647–654 (2010)

51.150 G. Niewiadomska, A. Mietelska-Porowska, M. Mazurkiewicz: The cholinergic system, nerve growth factor and the cytoskeleton, Behav. Brain Res. **221**, 515–526 (2011)

51.151 S. Capsoni, R. Brandi, I. Arisi, M. D'Onofrio, A. Cattaneo: A dual mechanism linking NGF/proNGF imbalance and early inflammation to Alzheimer's disease neurodegeneration in the AD11 Anti-NGF mouse model, CNS Neurol. Disord. Drug Targets, **10**, 635–647 (2011)

51.152 A. Salehi, J.D. Delcroix, P.V. Belichenko, K. Zhan, C. Wu, J.S. Valletta, R. Takimoto-Kimura, A.M. Kleschevnikov, K. Sambamurti, P.P. Chung, W. Xia, A. Villar, W.A. Campbell, L.S. Kulnane, R.A. Nixon, B.T. Lamb, C.J. Epstein, G.B. Stokin, L.S. Goldstein, C. Mobley: Increased APP expression in a mouse model of Down's syndrome disrupts NGF transport and causes cholinergic neuron degeneration, Neuron **51**, 29–42 (2006)

51.153 J. Fombonne, S. Rabizadeh, S. Banwait, P. Mehlen, D.E. Bredesen: Selective vulnerability in Alzheimer's disease: Amyloid precursor protein and p75(NTR) interaction, Ann. Neurol. **65**, 294–303 (2009)

51.154 K.K. Teng, S. Felice, T. Kim, B.L. Hempstead: Understanding proneurotrophin actions: Recent advances and challenges, Dev. Neurobiol. **70**, 350–359 (2010)

51.155 M. Fahnestock, G. Yu, M.D. Coughlin: ProNGF: A neurotrophic or an apoptotic molecule?, Prog. Brain Res. **146**, 101–110 (2004)

51.156 M.A. Bruno, W.C. Leon, G. Fragoso, W.E. Mushynski, G. Almazan, A.C. Cuello: Amyloid β-induced nerve growth factor dysmetabolism in Alzheimer disease, J. Neuropathol. Exp. Neurol. **68**, 857–869 (2009)

51.157 E.J. Coulson, L.M. May, A.M. Sykes, A.S. Hamlin: The role of the p75 neurotrophin receptor in choplinergic dysfunction in Alzheimer's disease, Neuroscientist **15**, 317–322 (2009)

51.158 C. Costantini, F. Rossi, E. Formaggio, R. Bernardoni, D. Cecconi, V. Della-Bianca: Characterization of the signaling pathway downstream p75 neurotrophin receptor involved in β-amyloid peptide-dependent cell death, J. Mol. Neurosci. **25**, 141–156 (2005)

51.159 E.J. Coulson: Does the p75 neurotrophin receptor mediate Aβ-induced toxicity in Alzheimer's disease?, J. Neurochem. **98**, 654–660 (2006)

51.160 A. Sotthibundhu, A.M. Sykes, B. Fox, C.K. Underwood, W. Thangnipon, E.J. Coulson: β-Amyloid$_{1-42}$ induces neuronal death through the p75 neurotrophin receptor, J. Neurosci. **28**, 3941–3946 (2008)

51.161 J.K. Knowles, J. Rajadas, T.V. Nguyen, T. Yang, M.C. LeMieux, L. Vander Griend, C. Ishikawa, S.M. Massa, T. Wyss-Coray, F.M. Longo: The p75 neurotrophin receptor promotes amyloid-beta (1–42)-induced neuritic dystrophy in vitro and in vivo, J. Neurosci. **29**, 10627–10637 (2009)

51.162 M. Yaar, S. Zhai, I. Panova, R.E. Fine, P.B. Eisenhauer, J.K. Blusztajn, I. Lopez-Coviella, B.A. Gilchrest: A cyclic peptide that binds p75(NTR) protects neurones from β amyloid (1–40)-induced cell death, Neuropathol. Appl. Neurobiol. **33**, 533–543 (2007)

51.163 Y.J. Wang, X. Wang, J.J. Lu, Q.X. Li, C.Y. Gao, X.H. Liu, Y. Sun, M. Yang, Y. Lim, G. Evin, J.H. Zhong, C. Masters, X.F. Zhou: p75NTR regulates Aβ deposition by increasing Aβ production but inhibiting Aβ aggregation with its extracellular domain, J. Neurosci. **31**, 2292–2304 (2011)

51.164 B. Chakravarthy, C. Gaudet, M. Ménard, T. Atkinson, L. Brown, F.M. Laferla, U. Armato, J. Whitfield: Amyloid-β peptides stimulate the expression of the p75(NTR) neurotrophin receptor in SHSY5Y human neuroblastoma cells and AD transgenic mice, J. Alzheimers Dis. **19**, 915–925 (2010)

51.165 M. Goedert, R. Jakes: Mutations causing neurodegenerative tauopathies, Biochim. Biophys. Acta **1739**, 240–250 (2005)

51.166 S. Oddo, A. Caccamo, M. Kitazawa, B.P. Tseng, F.M. LaFerla: Amyloid deposition precedes tangle formation in a triple transgenic model of Alzheimer's disease, Neurobiol. Aging **24**, 1063–1070 (2003)

51.167 L.M. Billings, S. Oddo, K.N. Green, J.L. McGaugh, F.M. LaFerla: Intraneuronal Aβ causes the onset of early Alzheimer's disease-related cognitive deficits in transgenic mice, Neuron **45**, 675–688 (2005)

51.168 M. Kitazawa, S. Oddo, T.R. Yamasaki, K.N. Green, F.M. LaFerla: Lipopolysaccharide-induced inflammation exacerbates tau pathology by a cyclin-dependent kinase 5-mediated pathway in a transgenic model of Alzheimer's disease, J. Neurosci. **25**, 8843–8853 (2005)

51.169 S.D. Ginsberg, S. Che, S.E. Counts, E.J. Mufson: Shift in the ratio of three-repeat tau and four-repeat tau mRNAs in individual cholinergic basal forebrain neurons in mild cognitive impairment and Alzheimer's disease, J. Neurochem. **96**, 1401–1408 (2006)

51.170 M. Mesulam: The cholinergic lesion of Alzheimer's disease: Pivotal factor or side show, Learn Mem. **11**, 43–49 (2004)

51.171 C. Geula, N. Nagykery, A. Nicholas, C.K. Wu: Cholinergic neuronal and axonal abnormalities are present early in aging and in Alzheimer disease, J. Neuropathol. Exp. Neurol. **67**, 309–318 (2008)

51.172 K. Belarbi, S. Burnouf, F.J. Fernandez-Gomez, J. Desmerciéres, L. Troquier, J. Brouillette, L. Tsambou, M.E. Grosjean, R. Caillierez, D. Demeyer, M. Hamdane, K. Schindowski, D. Blum, L. Buée: Loss of medial septum cholinergic neurons in THY-tau22 mouse model: What links with tau pathology?, Curr. Alzheimer Res. **8**, 633–638 (2011)

51.173 J.C. de la Torre: Vascular risk factor detection and control may prevent Alzheimer's disease, Ageing Res. Rev. **9**, 218–225 (2010)

51.174 C. Iadecola, L. Park, C. Capone: Threats to the mind: Aging, amyloid, and hypertension, Stroke **40**(3 Suppl), S40–44 (2009)

51.175 J.C. de la Torre, T. Mussivand: Can disturbed brain microcirculation cause Alzheimer's disease?, Neurol. Res. **15**, 146–153 (1993)

51.176 R.N. Kalaria: Vascular basis for brain degeneration: Faltering controls and risk factors for dementia, Nutr. Rev. **68**(Suppl.), S74–87 (2010)

51.177 N. Nicolakakis, E. Hamel: Neurovascular function in Alzheimer's disease patients and experimental models, Cereb. Blood Flow Metab. **31**, 1354–1370 (2011)

51.178 R.O. Weller, D. Boche, J.A.R. Nicoll: Microvasculature changes and cerebral amyloid angiopathy in Alzheimer's disease and their potential impact on therapy, Acta Neuropathol. **119**, 87–102 (2009)

51.179 C. Cordonnier, W.M. van der Flier: Brain microbleeds and Alzheimer's disease: Innocent observation or key player?, Brain **134**, 335–344 (2011)

51.180 B.V. Zlokovic: New therapeutic targets in the neurovascular pathway in Alzheimer's disease, Neurotherapeutics **5**, 409–414 (2008)

51.181 C.A. Hawkes, W. Härtig, J. Kacza, R. Schliebs, R.O. Weller, J.A. Nicoll, R.O. Carare: Perivascular drainage of solutes is impaired in the ageing mouse brain and in the presence of cerebral amyloid angiopathy, Acta Neuropathol. **121**, 431–443 (2011)

51.182 M. Merlini, E.P. Meyer, A. Ulmann-Schuler, R.M. Nitsch: Vascular β-amyloid and early astrocyte alterations impair cerebrovascular function and cerebral metabolism in transgenic arcAβ mice, Acta Neuropathol. **122**, 293–311 (2011)

51.183 D. Paris, J. Humphrey, A. Quadros, N. Patel, R. Crescentini, F. Crawford, M. Mullan: Vasoactive effects of Aβ in isolated human cerebrovessels and in a transgenic mouse model of Alzheimer's disease: Role of inflammation, Neurol. Res. **25**, 642–651 (2003)

51.184 C. Iadecola: Neurovascular regulation in the normal brain and in Alzheimer's disease, Nat Rev. Neurosci. **5**, 347–360 (2004)

51.185 A. Rocchi, D. Orsucci, G. Tognoni, R. Ceravolo, G. Siciliano: The role of vascular factors in late-onset sporadic Alzheimer's disease. Genetic and molecular aspects, Curr. Alzheimer Res. **6**, 224–237 (2009)

51.186 H.J. Milionis, M. Florentin, S. Giannopoulos: Metabolic syndrome and Alzheimer's disease: A link to a vascular hypothesis?, CNS Spectrums **13**, 606–613 (2008)

51.187 D.A. Snowdon, L.H. Greiner, J.A. Mortimer, K.P. Riley, P.A. Greiner, W.R. Markesbery: Brain infarction and the clinical expression of Alzheimer disease, The Nun Study, J. Am. Med. Assoc. **277**, 813–817 (1997)

51.188 S.L. Cole, R. Vassar: Linking vascular disorders and Alzheimer's disease: Potential involvement of BACE1, Neurbiol. Aging **30**, 1535–1544 (2009)

51.189 E.E. Smith, S.M. Greenberg: β-amyloid, blood vessels and brain function, Stroke **40**, 2601–2606 (2009)

51.190 H.K. Shin, P.B. Jones, M. Garcia-Alloza, L. Borrelli, S.M. Greenberg, B.J. Bacskai, M.P. Frosch, B.T. Hyman, M.A. Moskowitz, C. Ayata: Age-dependent cerebrovascular dysfunction in a transgenic mouse model of cerebral amyloid angiopathy, Brain. **130**(Pt 9), 2310–2319 (2007)

51.191 D. Paris, N. Patel, A. DelleDonne, A. Quadros, R. Smeed, M. Mullan: Impaired angiogenesis in a transgenic mouse model of cerebral amyloidosis, Neurosci. Lett. **366**, 80–85 (2004)

51.192 D. Paris, K. Townsend, A. Quadros, J. Humphrey, J. Sun, S. Brem, M. Wotoczek-Obadia, A. DelleDonne, N. Patel, D.F. Obregon, R. Crescentini, L. Abdullah, D. Coppola, A.M. Rojiani, F. Crawford, S.M. Sebti, M. Mullan: Inhibition of angiogenesis by Aβ peptides, Angiogenesis **7**, 75–85 (2004)

51.193 C. Ruiz de Almodovar, D. Lambrechts, M. Mazzone, P. Carmeliet: Role and therapeutic potential of VEGF in the nervous system, Physiol. Rev. **89**, 607–648 (2009)

51.194 F.Y. Sun, X. Guo: Molecular and cellular mechanisms of neuroprotection by vascular endothelial growth factor, J. Neurosci. Res. **79**, 180–184 (2005)

51.195 N. Ferrara, H.P. Gerber, J. LeCouter: The biology of VEGF and its receptors, Nat. Med. **9**, 669–676 (2003)

51.196 R.N. Kalaria, D.L. Cohen, D.R. Premkumar, S. Nag, J.C. LaManna, W.D. Lust: Vascular endothelial growth factor in Alzheimer's disease and experimental cerebral ischemia, Mol. Brain Res. **62**, 101–105 (1998)

51.197 E. Tarkowski, R. Issa, M. Sjogren, A. Wallin, K. Blennow, A. Tarkowski: Increased intrathecal levels of the angiogenic factors VEGF and TGF-beta in Alzheimer's disease and vascular dementia, Neurobiol. Aging **23**, 237–243 (2002)

51.198 S.P. Yang, D.G. Bae, H.J. Kang, B.J. Gwag, Y.S. Gho, C.B. Chae: Co-accumulation of vascular endothelial growth factor with β-amyloid in the brain of patients with Alzheimer's disease, Neurobiol. Aging **25**, 283–290 (2004)

51.199 J.K. Ryu, T. Cho, H.B. Choi, Y.T. Wang, J.G. McLarnon: Microglial VEGF receptor response is an integral chemotactic component in Alzheimer's disease pathology, J. Neurosci. **29**, 3–13 (2009)

51.200 A.I. Pogue, W.J. Lukiw: Angiogenic signaling in Alzheimer's disease, Neuroreport **15**, 1507–1510 (2004)

51.201 L. Thirumangalakudi, P.G. Samany, A. Owoso, B. Wiskar, P. Grammas: Angiogenic proteins are expressed by brain blood vessels in Alzheimer's disease, J. Alzheimer Dis. **10**, 111–118 (2006)

51.202 A.H. Vagnucci, W.W. Li: Alzheimer's disease and angiogenesis, Lancet **361**, 605–608 (2003)

51.203 B.S. Desai, J.A. Schneider, J.L. Li, P.M. Carvey, B. Hendey: Evidence of angiogenic vessels in Alzheimer's disease, J. Neural Transm. **116**, 587–597 (2009)

51.204 H.J. Marti, M. Bernaudin, A. Bellail, H. Schoch, M. Euler, W. Petit: Hypoxia-induced vascular endothelial growth factor expression precedes neovascularization after cerebral ischemia, Am. J. Pathol. **156**, 965–976 (2000)

51.205 I. Stein, M. Neeman, D. Shweiki, A. Itin, E. Keshet: Stabilization of vascular endothelial growth factor mRNA by hypoxia and hypoglycemia and coregulation with other ischemia-induced genes, Mol. Cell Biol. **15**, 5363–5368 (1995)

51.206 G.D. Yancopoulos, S. Davis, N.W. Gale, J.S. Rudge, S.J. Wiegand, J. Holash: Vascular-specific growth factors and blood vessel formation, Nature **407**, 242–248 (2000)

51.207 D. Paris, A. Quadros, N. Patel, A. DelleDonne, J. Humphrey, M. Mullan: Inhibition of angiogenesis and tumour growth by β- and γ-secretase inhibitors, Eur. J. Pharmacol. **514**, 1–15 (2005)

51.208 J.R. Ciallella, H. Figueiredo, V. Smith-Swintosky, J.P. McGillis: Thrombin induces surface and intracellular secretion of amyloid precursor protein from human endothelial cells, Thromb. Haemost. **81**, 630–637 (1999)

51.209 S. Bürger, M. Noack, L.P. Kirazov, E.P. Kirazov, C.L. Naydenov, E. Kouznetsova, Y. Yafai, R. Schliebs: Vascular endothelial growth factor (VEGF) affects processing of the amyloid precursor protein and

β-amyloidogenesis in brain slice cultures derived from transgenic Tg2576 mouse brain, Int. J. Dev. Neurosci. **27**, 517–523 (2009)

51.210 S. Bürger, Y. Yafai, M. Bigl, P. Wiedemann, R. Schliebs: Effect of VEGF and its receptor antagonist SU-5416, an inhibitor of angiogenesis, on processing of the β-amyloid precursor protein in primary neuronal cells derived from brain tissue of Tg2576 mice, Int. J. Dev. Neurosci. **28**, 597–604 (2010)

51.211 H. Girouard, C. Iadecola: Neurovascular coupling in the normal brain and in hypertension, stroke, and Alzheimer disease, J. Appl. Physiol. **100**, 328–335 (2006)

51.212 E. Hamel: Perivascular nerves and the regulation of cerebrovascular tone, J. Appl. Physiol. **100**, 1059–1064 (2006)

51.213 A.H. VanBeek, J.A. Claassen: The cerebrovascular role of the cholinergic neural system in Alzheimer's disease, Behav. Brain Res. **221**, 537–542 (2003)

51.214 E. Vaucher, E. Hamel: Cholinergic basal forebrain neurons project to cortical microvessels in the rat: Electron microscopic study with anterogradely transported *Phaseolus vulgaris* leucoagglutinin and choline acetyltransferase immunocytochemistry, J. Neurosci. **15**, 7427–7441 (1995)

51.215 E. Hamel: Cholinergic modulation of the cortical microvascular bed, Prog. Brain Res. **145**, 171–178 (2004)

51.216 A. Elhusseiny, E. Hamel: Muscarinic–but not nicotinic–acetylcholine receptors mediate a nitric oxide-dependent dilation in brain cortical arterioles: A possible role for the M5 receptor subtype, J. Cereb. Blood Flow Metab. **20**, 298–305 (2000)

51.217 E. Farkas, P.G. Luiten: Cerebral microvascular pathology in aging and Alzheimer's disease, Prog. Neurobiol. **64**, 575–611 (2001)

51.218 B. Cauli, X.K. Tong, A. Rancillac, N. Serluca, B. Lambolez, J. Rossier, E. Hamel: Cortical GABA interneurons in neurovascular coupling: Relays for subcortical vasoactive pathways, J. Neurosci. **24**, 8940–8949 (2004)

51.219 C. Humpel, J. Marksteiner: Cerebrovascular damage as a cause for Alzheimer's disease, Curr. Neurovasc. Res. **2**, 341–347 (2005)

51.220 J.A. Claassen, R.W. Jansen: Cholinergically mediated augmentation of cerebral perfusion in Alzheimer's disease and related cognitive disorders: The cholinergic-vascular hypothesis, J. Gerontol. A Biol. Sci. Med. Sci. **61**, 267–271 (2006)

51.221 E. Kouznetsova, R. Schliebs: Role of cholinergic system in β-amyloid related changes of perivascular innervation of cerebral microvessels in transgenic Tg2576 Alzheimer-like mice, J. Neurochem. **101**(Suppl. 1), 63 (2007)

51.222 A. Szutowicz, H. Bielarczyk, S. Gul, P. Zieliński, T. Pawełczyk, M. Tomaszewicz: Nerve growth factor and acetyl-l-carnitine evoked shifts in acetyl-CoA and cholinergic SN56 cell Vulnerability to neurotoxic inputs, J. Neurosci. Res. **79**, 185–192 (2005)

51.223 S.C. Correia, R.X. Santos, G. Perry, X. Zhu, P.I. Moreira, M.A. Smith: Insulin-resistant brain state: The culprit in sporadic Alzheimer's disease?, Ageing Res. Rev. **10**, 264–273 (2011)

51.224 M. Bigl, J. Apelt, K. Eschrich, R. Schliebs: Cortical glucose metabolism is altered in aged transgenic Tg2576 mice that demonstrate Alzheimer plaque pathology, J. Neural Transm. **110**, 77–94 (2003)

51.225 P. Ghosh, A. Kumar, B. Datta, V. Rangachari: Dynamics of protofibril elongation and association involved in Aβ42 peptide aggregation in Alzheimer's disease, BMC Bioinformatics **11**(Suppl 6), S24 (2010)

51.226 N.L. Fawzi, E.H. Yap, Y. Okabe, K.L. Kohlstedt, S.P. Brown, T. Head-Gordon: Contrasting disease and nondisease protein aggregation by molecular simulation, Acc. Chem. Res. **41**, 1037–1047 (2008)

51.227 A.R. George, D.R. Howlett: Computationally derived structural models of the beta-amyloid found in Alzheimer's disease plaques and the interaction with possible aggregation inhibitors, Biopolymers **50**, 733–741 (1999)

51.228 S.Y. Ponomarev, J. Audie: Computational prediction and analysis of the DR6-NAPP interaction, Proteins. **79**, 1376–1395 (2011)

51.229 J. Luo, J.D. Maréchal, S. Wärmländer, A. Gräslund, A. Perálvarez-Marín: In silico analysis of the apolipoprotein E and the amyloid beta peptide interaction: Misfolding induced by frustration of the salt bridge network, PLoS Comput. Biol. **6**(2), e1000663 (2010)

51.230 J.Y. Chen, E. Youn, S.D. Mooney: Connecting protein interaction data, mutations, Methods Mol. Biol. **541**, 449–461 (2009)

51.231 K.C. Chou, K.D. Watenpaugh, R.L. Heinrikson: A model of the complex between cyclin-dependent kinase 5 and the activation domain of neuronal Cdk5 activator, Biochem. Biophys. Res. Commun. **259**, 420–428 (1999)

51.232 Y. Huang, X. Sun, G. Hu: An integrated genetics approach for identifying protein signal pathways of Alzheimer's disease, Comput. Methods Biomech. Biomed. Engin. **14**, 371–378 (2011)

51.233 D.L. Cook, J.C. Wiley, J.H. Gennari: Chalkboard: Ontology-based pathway modeling and qualitative inference of disease mechanisms, Pac. Symp. Biocomput. **2007**, 16–27 (2007)

51.234 I. García, Y. Fall, G. Gómez, H. González-Díaz: First computational chemistry multi-target model for anti-Alzheimer, anti-parasitic, anti-fungi, and anti-bacterial activity of GSK-3 inhibitors in vitro, in vivo, and in different cellular lines, Mol. Divers. **15**, 561–567 (2011)

51.235 R. Gómez-Balderas, D.F. Raffa, G.A. Rickard, P. Brunelle, A. Rauk: Computational studies of Cu(II)/Met and Cu(I)/Met binding motifs relevant

for the chemistry of Alzheimer's disease, J. Phys. Chem. A **109**, 5498–5508 (2005)

51.236 R. Singh, A. Barman, R. Prabhakar: Computational insights into aspartyl protease activity of presenilin 1 (PS1) generating Alzheimer amyloid β-peptides (Aβ40 and Aβ42), J. Phys. Chem. B. **113**, 2990–2999 (2009)

51.237 M. Eisenstein: Genetics: Finding risk factors, Nature **475**, S20–S22 (2011)

51.238 M.D. Ritchie, L.W. Hahn, J.H. Moore: Power of multifactor dimensionality reduction for detecting gene–gene interactions in the presence of genotyping error, missing data, phenocopy, and genetic heterogeneity, Genet. Epidemiol. **24**, 150–157 (2003)

51.239 S.Y. Lee, Y. Chung, R.C. Elston, Y. Kim, T. Park: Log-linear model-based multifactor dimensionality reduction method to detect gene gene interactions, Bioinformatics. **23**, 2589–2595 (2007)

51.240 J. Andrade, M. Andersen, A. Sillén, C. Graff, J. Odeberg: The use of grid computing to drive data-intensive genetic research, Eur. J. Hum. Genet. **15**, 694–702 (2007)

51.241 B. Liu, T. Jiang, S. Ma, H. Zhao, J. Li, X. Jiang, J. Zhang: Exploring candidate genes for human brain diseases from a brain-specific gene network, Biochem. Biophys. Res. Commun. **349**, 1308–1314 (2006)

51.242 K. Heinitz, M. Beck, R. Schliebs, J.R. Perez-Polo: Toxicity mediated by soluble oligomers of β-amyloid (1–42) on cholinergic SN56.B5.G4 cells, J. Neurochem. **98**, 1930–1945 (2006)

51.243 S. Joerchel, M. Raap, M. Bigl, K. Eschrich, R. Schliebs: Oligomeric β-amyloid (1–42) induces the expression of Alzheimer disease-relevant proteins in cholinergic SN56.B5.G4 cells as revealed by proteomic analysis, Int. J. Dev. Neurosci. **26**, 301–308 (2006)

51.244 J. Li, X. Zhu, J.Y. Chen: Building disease-specific drug-protein connectivity maps from molecular interaction networks and PubMed abstracts, PLoS Comput. Biol. **5**(7), e1000450 (2009)

51.245 Y. He, Z. Chen, G. Gong, A. Evans: Neuronal networks in Alzheimer's disease, Neuroscientist **15**, 333–350 (2009)

51.246 D.J. Watts, S.H. Strogatz: Collective dynamics of 'small-world' networks, Nature **393**, 440–442 (1998)

51.247 C.J. Stam, B.F. Jones, G. Nolte, M. Breakspear, P. Scheltens: Small-world networks and functional connectivity in Alzheimer's disease, Cereb. Cortex **17**, 92–99 (2007)

51.248 R. Tandon, S. Adak, J.A. Kaye: Neural networks for longitudinal studies in Alzheimer's disease, Artif. Intell. Med. **36**, 245–255 (2006)

51.249 M.E. Hasselmo: A computational model of the progression of Alzheimer's disease, M.D. Computing **14**, 181–191 (1997)

51.250 M.E. Hasselmo: Neuromodulation and cortical function: Modeling the physiological basis of behavior, Behav. Brain Res. **67**, 1–27 (1995)

51.251 D. Horn, N. Levy, E. Ruppin: Neuronal-based synaptic compensation: A computational study in Alzheimer's disease, Neural Comput. **8**, 1227–1243 (1996)

51.252 L. Nickels, B. Biedermann, M. Coltheart, S. Saunders, J.J. Tree: Computational modelling of phonological dyslexia: How does the DRC model fare?, Cogn. Neuropsychol. **25**, 165–193 (2008)

51.253 A. Serna, V. Rialle, H. Pigot: Computational representation of Alzheimer's disease evolution applied to a cooking activity, Stud. Health Technol. Inform. **124**, 587–592 (2006)

51.254 M.A. Ahmadi-Pajouh, F. Towhidkhah, S. Gharibzadeh, M. Mashhadimalek: Path planning in the hippocampo-prefrontal cortex pathway: An adaptive model based receding horizon planner, Med. Hypotheses **68**, 1411–1415 (2007)

51.255 E. Ruppin, J.A. Reggia: A neural model of memory impairment in diffuse cerebral atrophy, Br. J. Psychiatry. **166**, 19–28 (1995)

52. Integrating Data for Modeling Biological Complexity

Sally Hunter, Carol Brayne

This chapter describes how information relating to the interactions between molecules in complex biological pathways can be identified from the scientific literature and integrated into maps of functional relationships. The molecular biology of the amyloid precursor protein (APP) in the synaptic processes involved in normal cognitive function and neurodegenerative disease is used as a case study. The maps produced are interpreted with reference to basic concepts of biological regulation and control. Direct and indirect feedback relationships between the amyloid precursor protein, its proteolytic fragments and various processes that contribute to processes involved in synaptic modifications are identified. The contributions of the amyloid precursor protein and its proteolytic fragments are investigated with reference to disease pathways in Alzheimer disease and new perspectives on disease progression are highlighted. Mapping functional relationships in complex biological pathways is useful to summarize the current knowledge base, identify further targets for research, and for empirical experimental design and interpretation of results.

52.1 Overview of the Brain

Neurobiology concerns the study of the complex processes in the brain that underlie cognitive function. Little is known about how the brain works to receive input from various sources in the environment, associate the various incoming data streams, and then integrate those streams into a conscious perception of the world. The brain is in a constant state of change, as new experiences build on past memories. How a life time of memories is recorded in the brain is not well understood.

Current theories suggest that memory is encoded by synaptic plasticity, i. e., modifications to synaptic connectivity and strength with follow-on changes in the behavior of neural networks [52.1, 2]. Long-term potentiation (LTP) is a state where synapses may be formed or strengthened in response to concomitant signal transmission and long-term depression (LTD) is a state where synapses are weakened or removed in response to non-concomitant signaling between neurons. Synaptic plasticity ultimately depends on the complex biochemical processes involved in signal transmission from one neuron to another and requires the coherent response of wide-ranging synaptic processes, from changes in cell adhesion to gene and protein expression [52.1].

Many techniques to study synaptic plasticity have been developed, from in vitro culture of neurons to whole animal studies. Experimental approaches aim to tease apart the mechanisms of synaptic processes by changing one step in a process, e.g., inhibiting a single neurotransmitter receptor type, and observing the consequent changes in animal or cell behavior. This reductionist approach, where the complex neuronal system is modeled in terms of isolated processes, has been essential in building the current knowledge base regarding memory formation. However, how the many synaptic processes relate to the emergence of normal cognitive function in the brain and ultimately the behavior of an organism as a whole is not known.

52.1.1 Overview of Alzheimer's Disease

While many people retain cognitive function into very old age, many suffer cognitive impairments to varying degrees. Many diseases involving the degeneration of neurons lead to dementia, with Alzheimer's disease (AD) being the most common form associated with aging. Despite intense research over the past century, there is no known AD disease mechanism or cure.

AD has two main forms depending on genetic status and age at disease onset. Familial AD (FAD), accounting for $< 1\%$ of cases is hereditary and has an early onset [52.3]. In contrast, sporadic AD (SAD), accounting for $> 99\%$ of cases, is associated primarily with aging and has a later onset [52.3]. Clinically, AD is characterized by progressive loss of memory, impaired cognitive function, and ultimately dementia [52.4]. These symptoms lead to severe impairments in activities of daily life with the result that individuals with AD require an increasing degree of support and care.

The neuropathological hallmarks of AD include synapse loss, intraneuronal tangles of paired helical filaments of hyperphosphorylated tau [52.5, 6], and microscopic extracellular senile and neuritic plaques containing aggregated amyloid-beta-protein ($A\beta$) [52.7]. $A\beta$ is a variable length proteolytic fragment of the amyloid precursor protein (APP) and various mutations within APP are associated with FAD. FAD is also associated with mutations in the presenilins (PS), which, as part of the γ-secretase enzyme complex, are involved in APP proteolysis. Additionally, Down's syndrome, with an extra copy of APP on chromosome 21, is associated with the development of AD-like dementia from middle age [52.3, 8]. The genetic and neuropathological data strongly suggest that APP and PS have central roles in disease progression but do not pinpoint the exact disease mechanism.

52.1.2 Models of Dementia

There are many models of disease progression in AD that attempt to explain the contributions of wide-ranging factors shown to be associated with the progression to dementia. These models include, amongst others, changes to cholesterol homeostasis [52.9–11], Ca^{2+} homeostasis [52.12], oxidative stress [52.13, 14], cell signaling cascades [52.15], and cell cycle [52.16]. While these models can be applied to the various selected factors associated with late onset SAD, they do not adequately explain the genetic contributions to disease seen in FAD. Theoretical models of disease progression in dementia underlie the design of hypotheses that are tested by experimentation and influence, either implicitly or explicitly, the interpretations of the resulting data.

The amyloid cascade hypothesis [52.17] has been the dominant model of disease progression in AD for over 25 years. It was derived from the evidence of aggregated $A\beta$ in senile and neuritic plaques and by interpretations of the genetic data from FAD and Down's syndrome where overexpression or altered processing of APP leads to increased expression of $A\beta$. The amyloid cascade hypothesis suggests that overproduction or a change in the ratio of the $A\beta$ sequence lengths, especially $A\beta(1–40)/A\beta(1–42)$, or an increased $A\beta$ aggregation state is neurotoxic and that the removal of $A\beta$ is an appropriate treatment strategy. As such, the amyloid cascade hypothesis is an infection-type model and has been modeled in an approximately linear way.

While the amyloid cascade hypothesis appears to adequately apply to familial forms of AD, data from the more common sporadic form (SAD) do not fit this model well. In particular, data from population studies showing that associations between amyloid load and dementia attenuate with increasing age [52.18, 19] support a more complex picture. An alternative approach is to view the brain as a dynamic, self-organized collection of individual processes which, through their multiple interactions and feedback loops, cause cognitive functions to emerge. This complex system can be broken down to more specific areas of study including, amongst others, genomics, epigenetics, proteomics, and neural connectivity. Systems biology is a holistic approach that aims to construct working models of complex biological systems by integrating basic quantitative data collected by experiment from all the individual processes involved. As such the brain may be best modeled as an iterative matrix [52.20].

Although researchers have integrated parts of the data available, for example, data relating to cholesterol metabolism has been integrated into various models describing the contributions of cholesterol to AD disease mechanisms [52.9–11] and data relevant to the amyloid beta protein (Aβ) has been integrated in the amyloid

cascade hypothesis [52.17], a fully integrative approach has not yet been attempted. Applying techniques from systems biology to investigate the interactions of APP and all its proteolytic fragments may allow alternative perspectives that are more consistent with the complex patterns of disease seen in AD. Integration of the volume of data regarding the roles of APP and PS in AD presents huge challenges. This chapter does not aim to meet this challenge but, rather, to examine how this might be done. Here we build a map of the interactions of APP with other molecules or cellular processes as identified in the literature and use this map to integrate the data to examine alternative perspectives on disease mechanisms in AD.

52.2 Identifying and Mapping the Interactions of APP

Several stages may be involved in building a map that describes the interactions of APP in normal cognition and AD disease pathways, including:

1. Identification of the interactions from the literature with each functionally different protein or peptide considered as a separate entity.
2. The interactions of each factor identified need to be characterized in terms of any process or modification that contributes to that interaction.
3. These interactions need to be described in terms of regulatory feedback loops so that potentially, the system can be computationally modeled dynamically over time.

Any model of cognition needs to replicate as far as possible normal and abnormal states and have understandable outputs that highlight targets for the development of treatments.

52.2.1 Identifying Interactions in the Literature

A vast amount of data relating to AD has been accumulated over the past century. Simple searches of Pubmed using MeSH terms retrieve thousands of papers that describe data relevant to the association of AD with APP and its proteolytic processing, as can be seen in Table 52.1. Functional interactions between APP and neuronal processes or other molecules can be identified from papers retrieved from literature searches.

Once interactions have been identified, each interaction needs to be characterized as far as possible

so that it can be integrated into models of disease progression.

Table 52.1 Accumulating evidence for APP involvement in AD

Pubmed Searches (07/04/2011)	Number of papers
Alzheimer Disease[Mesh]	53 274
Alzheimer disease	56 944
(*Alzheimer Disease*[Mesh]) AND *Amyloid beta-Protein Precursor*[Mesh]	10 308
Alzheimer amyloid precursor protein	12 205
(*Alzheimer Disease*[Mesh]) AND *Amyloid beta-Peptides*[Mesh]	7890
Alzheimer disease amyloid beta	13 147
(*Alzheimer Disease*[Mesh]) AND *Amyloid Precursor Protein Secretases*[Mesh]	1244
(((*ADAM9 protein, human* [Supplementary Concept]) OR *ADAM10 protein, human* [Supplementary Concept]) OR *tumor necrosis factor-alpha convertase* [Supplementary Concept]) AND *Alzheimer Disease*[Mesh]	42
Alzheimer alpha secretase	1595
((*BACE1 protein, human* [Supplementary Concept]) OR *BACE2 protein, human* [Supplementary Concept]) AND *Alzheimer Disease*[Mesh]	688
Alzheimer beta secretase	1825
(*Alzheimer Disease*[Mesh]) AND *Presenilins*[Mesh]	1733
Alzheimer gamma secretase	1697

52.2.2 Characterizing the Interactions

Within a cell as a whole, the probability that a biochemical reaction will occur between certain molecules and not others depends on many factors, including compartmentation, protein modifications, relative affinity, concentration, and half life.

Compartmentation

A cell is divided into different compartments at many levels, allowing a great degree of control over cellular processes. Organelles, such as the mitochondria, nucleus, endoplasmic reticulum, etc., isolate particular cellular processes from the rest of the cell within semipermeable membranes. Within these organelles, specific compartments can be defined by the close association of specific factors, such as more rigid cholesterol-rich lipid raft areas within a more fluid phospholipid membrane. In order to maintain cellular compartments, the cell must express the various components in the correct place and this involves the complex process of cellular trafficking.

Protein Modifications

After translation, proteins are often processed and/or modified before achieving an active form. Some proteins require proteolytic processing, where an immature protein is cleaved into a mature active form. There are also various reversible modifications that are fundamental to the regulation cellular processes such as i) glycosylation, the addition of sugar groups, ii) phosphorylation and dephosphorylation, the addition and removal of phosphate groups, and iii) acetylation and deacetylation, the addition or removal of acetyl groups. Phosphorylation and dephosphorylation, in particular, form a major mechanism by which cells can switch processes on or off or change the flow through a biochemical pathway.

Relative Affinity

The relative affinity of one protein for another contributes to the probability that they will react and the affinity depends on its conformation and charge distribution, which ultimately depend on the amino acid sequence. Protein conformation and charge distribution are altered by the protein modifications described above and by many other factors including pH, metal ion binding, and interactions with other cellular molecules.

Concentration

The concentration of the active form of a protein depends on many factors, including gene expression, protein synthesis, protein modification, trafficking and storage mechanisms, and protein degradation, amongst others. Concentration is usually tightly regulated and overexpression of active proteins can be disruptive to normal cellular processes. The outcomes of changing concentrations can be described by dose-response curves.

Half Life

The rate at which a protein is synthesized and degraded is its turnover and is characterized by its half life, i. e., the time it takes for half the amount of a particular protein to be degraded. The length of time a protein is active and available can contribute to the likelihood that it will be involved in a cellular reaction. The concentration of a protein with a short half life is more easily manipulated by the cell.

Once interactions involving APP have been identified and characterized; they need to be integrated in terms of feedback loops and regulatory mechanisms. This requires a basic understanding of the molecular biology of APP and its functional role(s) in synaptic plasticity.

52.3 The Amyloid Precursor Protein

52.3.1 Overview of APP

APP is a zinc-modulated, heparin-binding, type I transmembrane protein [52.21] that resembles a cell surface receptor [52.22]. It has a large extracellular N-terminal domain, a transmembrane domain, and an intracellular C-terminal domain. APP is ubiquitously expressed in several isoforms, with APP_{695} being the main isoform expressed by neurons. It has a high turnover, with a half life ranging from ≈ 1 h [52.23–25] to ≈ 4 h [52.26–

28]. APP has two homologs, (amyloid precursor like protein) APLP1 and APLP2 [52.21]. Evidence from studies with knock-out mice models [52.29–33] suggest that APP family members overlap functionally, making investigations into the exact function(s) of APP difficult. Regulated APP expression at the cell membrane involves many factors, which are summarized in Table 52.2.

Many of the factors in Table 52.2 also have signaling actions in other neural systems. Il-1β is involved in coor-

Table 52.2 Factors involved in the regulation of APP expression

Factor	Details of interaction	Reference
Glutamate NMDA signaling	Increased glutamate signaling via NMDA receptors increases expression of APP	[52.32]
Integrins	A3β1-integrin regulates APP trafficking and processing	[52.41]
Reelin integrins Dab1	Modulate APP trafficking and expression	[52.42]
PS	PS mediated trafficking of APP; mutant PS1 associated with APP retention in trans Golgi network	[52.40]
Il-1β	Increases expression of APP and β-cleavage in astrocytes and cell culture	[52.43, 44]
LRP1	Modulates APP trafficking and processing via interactions of their intracellular domains	[52.24, 45, 46]
Fibrillar peptides	Diverse fibril forming peptides, including Aβ and prion protein bind to APP and promote its expression	[52.47]
Lipids	Glycosphingolipids maybe involved in regulation of APP trafficking with effects on proteolytic processing	[52.48]
Cholesterol	Cholesterol depletion is associated with reduced trafficking of APP to lipid rafts and reduced processing via the β-pathway. Expression and processing of APP at the cell surface is increased by the cholesterol transporter protein ABCG1 in Down's syndrome	[52.39, 49]
OSBP1	Oxysterol binding Protein 1 (OSBP1) may link cholesterol metabolism with APP trafficking and affect APP processing	[52.50]
Insulin/IGF	Increased expression of APP at plasma membrane via a mechanism involving TK/MAPKK pathway; antagonized by Aβ; overactivation of IGF/insulin signaling is associated with inhibited trafficking of APP via negative feedback involving PI3K/Akt	[52.51–53]
P23/TMP21	Modulates APP trafficking independently from its effects on the γ-secretase complex	[52.54]
BDNF	BDNF enhanced expression of APP requires both Ras/MAPK and PI3K/Akt activation; down regulated by Aβ via MAPK	[52.55, 56]
NGF	APP expression is induced by NGF via Ras signaling	[52.44, 57]

Table 52.3 Interactions of full length N-terminal APP

Factor	Details of interaction	Reference
Heparins/heparin sulfates	APP has two zinc modulated, high-affinity binding sites for heparins. Heparins promote APP dimerization at the cell surface	[52.21, 58]
LRP	Transient interactions at cell surface	[52.59]
Reelin	Reelin interacts with APP and 3/β1-integrins and promotes neurite extension; APP endocytosis is reduced. Reelin signaling opposes the actions of Aβ	[52.41, 60]
Dab1	Increases cell surface expression of APP and promotes α-cleavage, this is modulated by reelin	[52.42]
CD74	CD74 is involved in trafficking and processing of MHC class II antigens and serves as a receptor for the cytokine MIF; interacts with APP and reduces expression of Aβ	[52.61, 62]

Table 52.4 Summary of the interactions of the APP cytoplasmic tail

Factor	Details of interaction	Reference
G-proteins	Full length and processed APP can potentially interact with G-proteins via the cytoplasmic tail and this can be altered by APP mutations around the G-protein binding site. This interaction has the potential to alter G-protein signaling with wide ranging effects including Ca^{2+} regulation and cell cycle pathways	[52.63–65]
Fe65	APP binds the transcription factor Fe65 at the YENPTY sequence in conjunction with TIP60 with effects on gene transcription, cytoskeleton and cell motility. Phosphorylation of T_{668} reduces the binding of Fe65 to YENPTY. Binding of Fe65 reduces $A\beta$	[52.66–70]
X11α/β	X11, part of the synaptic vesicle exocytotic machinery, binds to unphosphorylated APP and reduces $A\beta$ accumulation in transgenic mice	[52.71, 72]
Tyrosine kinases	Src, Abl, and Lyn bind to APP when phosphorylated at Y_{682} with affinity increased by phosphorylation of T_{668}; Overexpression of the NGF receptor TrkA associated with increased phosphorylation of Y_{682}	[52.71, 73]
Cholesterol	Interactions of cholesterol and APP may allow APP to react to cholesterol status of the cell	[52.74]
Numb	APP binds numb when Y_{682} is unphosphorylated and inhibits Notch signaling	[52.75]
SH domain adaptor proteins	Bind to APP when Y_{682} is phosphorylated; phosphorylation of T_{668} increases binding affinity	[52.71]
JNK/JIP1	JNK phosphorylation of APP at T_{668} modulated by JIP1	[52.76]

dinating responses of the immune system to injury, with both neurotoxic and neuroprotective effects [52.34], and may also have roles in synaptic plasticity [52.35]. The integrins are involved in many processes, including cell adhesion, signaling in the inflammatory pathway [52.36, 37], cell cycle activation, and gene transcription [52.38]. APP trafficking involves interactions with cholesterol homeostasis [52.39] and is associated with PS [52.40]. The involvement of glutamate signaling ties the expression of APP to synaptic plasticity.

While the data presented in Table 52.2 are in no way comprehensive, it can be seen that regulated APP expression and trafficking involves cross-talk between many systems involved in synaptic processes and that APP expression can be viewed as the outcome of all possible interactions at any one time.

52.3.2 Interactions of the Full Length Amyloid Precursor Protein

Extracellular Interactions of Full Length N-Terminal APP

At the cell surface, full length N-terminal APP interacts with many components of the extracellular matrix, (ECM), including heparins and heparin sulfates [52.21], laminin, collagen [52.77, 78], and β-1-integrin [52.22,

41, 42]. These interactions appear to underlie cellular processes such as cell adhesion, cell and neurite growth, synapse formation, and maintenance and plasticity [52.79].

Unprocessed APP is degraded or recycled via the endosomal or lysosomal pathways and may be recycled back to the membrane and processed within $\approx$30 min [52.25], with perhaps one third to one half being processed via the cleavage pathways as measured by secreted sAPPα/β [52.25]. Interactions of full length N-terminal APP are summarized in Table 52.3.

Interactions of the Full Length APP Cytoplasmic Tail

Interactions of the APP cytosolic domain with many binding proteins, presented in Table 52.4, are dependent on the phosphorylation state of the residues Y_{682} of the binding and signaling sequence GY$_{682}$ ENPTY and T_{668} of APP$_{695}$ [52.71, 80, 81]. Phosphorylation of Y_{682} appears increased in dementia [52.71] and phosphorylation of T_{668} has been linked to regulation of translocation of the amyloid precursor protein intracellular domain (AICD) to the nucleus with consequences for gene transcription and neurodegeneration [52.82, 83].

In general, phosphorylation of Y_{682} and T_{668} leads to differential binding of many proteins (reviewed

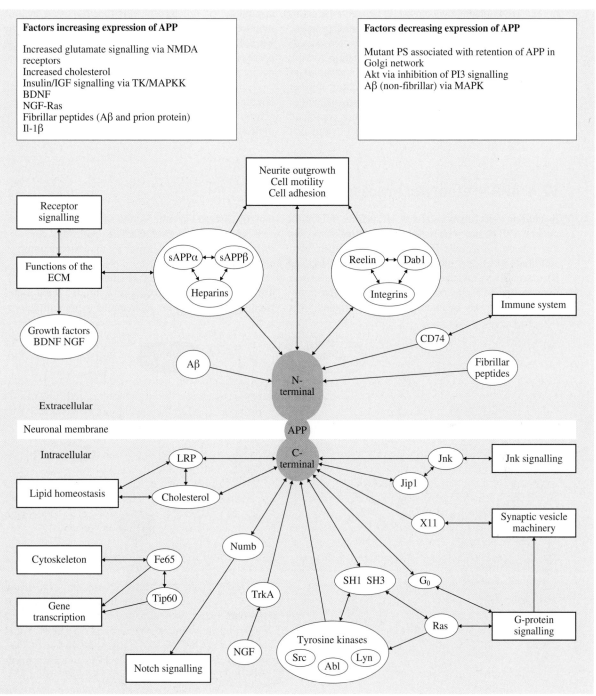

Fig. 52.1 Interactions of full length APP

in [52.71, 84]) and allows cross-talk between the cellular systems that share interactions with these proteins, including kinase cascades [52.15], the Notch signaling pathway [52.85], and G-protein signaling [52.63, 86]. This cross-talk involves differential activation of kinase/phosphatase cascades [52.15], adding further complexity to any model of APP in synaptic plasticity.

The data from Tables 52.3 and 52.4 are combined and mapped in Fig. 52.1. It is immediately obvious that any model of full length APP function must include interactions with multiple neuronal systems via the extracellular and intracellular domains. Each of these processes will have further regulatory mechanisms that do not involve APP and so are not considered here.

Evidence of two way interactions is shown with double arrow heads and evidence of single way interactions is shown with single arrow heads. The APP molecule is shown in gray.

52.4 APP Proteolytic Processing

In addition to its functions at the cell surface, full length mature APP is also processed via competing pathways that release proteolytic fragments, each with its own additional functional implications. APP processing has been extensively reviewed [52.22,87,88]. There are four main routes: APP may remain functionally active at the cell surface, it may be internalized and degraded, or it may be cleaved via either of the two competing α or β-cleavage pathways. Both cleavage pathways converge on an intramembrane cleavage by the PS-containing γ-secretase complex.

The possible cleavages are summarized in Fig. 52.2. The regulation and control of flow through these routes is complex and involves interactions with multiple neu-

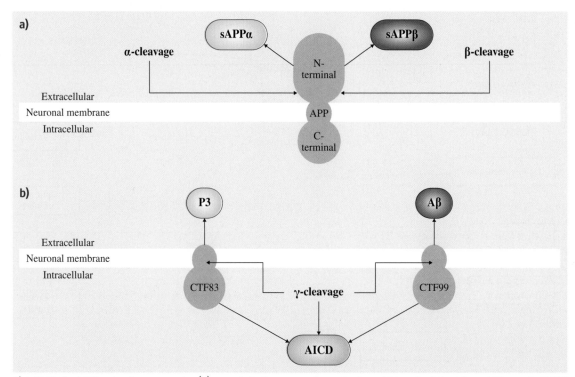

Fig. 52.2a,b APP main cleavage routes (**a**) α- and β-cleavages release the large N-terminal domains sAPPα and sAPPβ respectively to the extracellular space, leaving a membrane bound carboxy-terminal fragment (**b**) the intramembrane γ-cleavage releases the variable length P3 and Aβ peptides and the AICD. The Aβ and P3 peptides may be released extracellularly or retained intracellularly. The AICD translocates to the nucleus

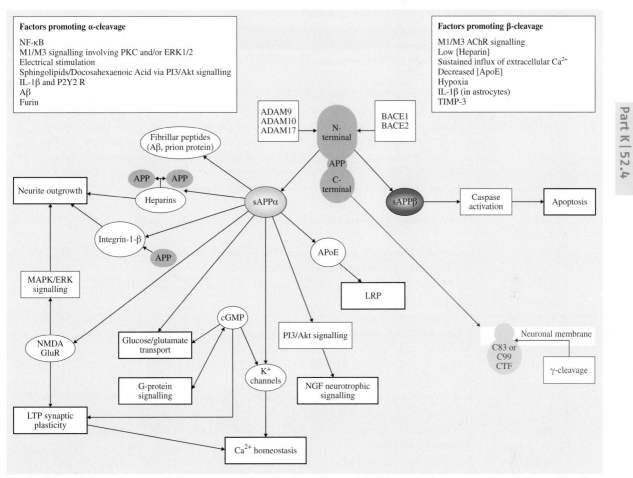

Factors promoting α-cleavage

NF-κB
M1/M3 signalling involving PKC and/or ERK1/2
Electrical stimulation
Sphingolipids/Docosahexaenoic Acid via PI3/Akt signalling
IL-1β and P2Y2 R
Aβ
Furin

Factors promoting β-cleavage

M1/M3 AChR signalling
Low [Heparin]
Sustained influx of extracellular Ca^{2+}
Decreased [ApoE]
Hypoxia
IL-1β (in astrocytes)
TIMP-3

Fig. 52.3 Regulation of α- and β-cleavages with functional outcomes. APP shown in *brown*, α-pathway shown in *yellow*, β-pathway shown in *blue*, CTF shown in *light brown*

ronal systems, of which only a few can be considered here. In Fig. 52.2, APP is shown in gray, the α-pathway is shown in yellow, the β-pathway is shown in blue, and the shared AICD is shown in orange.

52.4.1 α-Cleavage

α-cleavage occurs between residues Lys_{612} and Leu_{613} within the Aβ sequence of APP_{695}, releasing the large N-terminal sAPPα and leaving a membrane bound C83 C-terminal fragment (CTF) [52.22]. Several enzymes show α-secretase activity including a disintegrin and metalloproteinase (ADAM)9, ADAM10, ADAM17 [52.89–91], and possibly the matrix metalloproteinase (MMP) [52.92]. These proteases appear to be involved in many wide ranging processes in addition

to APP α-cleavage [52.93], introducing the possibility of competitive inhibition between these processes. Redundancy between the various ADAMs with respect to α-cleavage in APP makes investigations difficult [52.90, 91]. α-cleavage is both constitutive and regulated, with the various ADAMs responding in different ways depending on many factors [52.90]. Interactions involved in regulation of APP α-cleavage are summarized in Table 52.4 and mapped in Fig. 52.3.

It is clear from Table 52.5 that muscarinic acetylcholine (ACh) signaling, kinase cascades, and Ca^{2+} homeostasis are central to regulation of α-cleavage. In addition, α-cleavage is modulated by neuronal electrophysiology, with increases in electrical stimulation leading to up-regulation of α-cleavage, membrane composition, with lower cholesterol promoting α-cleavage,

Table 52.5 Factors involved in regulation of α-cleavage

Factor	Details of effect	Reference
Aβ	May up-regulate the expression of α-secretase via integrin receptors and MMP9 activity	[52.92]
NF-κB	NF-κB up-regulation by capacitive Ca^{2+} entry enhances sAPPα release via mAChR signaling; may reduce expression of BACE1	[52.94, 95]
mAChR	Increases in mAChR-M1 and M3 activation up regulate α-cleavage via PKC activation muscarinic up regulation of sAPPα secretion may involve the activation of a Src-TK leading to activation of PKCα and ERK1/2 Increased M2 activation decreases sAPPα secretion	[52.96–100]
Electrical stimulation	Non-specific electrical stimulation increases sAPP secretion	[52.100]
PKC	PKC activators enhance α-cleavage	[52.101, 102]
Lipids	Sphingolipids enhance α-cleavage via MAPK/ERK signaling; docosahexaenoic acid enhances α-cleavage via PI3/Akt signaling	[52.103, 104]
MAPK/ERK signaling	MAPK cascade may mediate the independent effects of PKC and tyrosine kinase in human astrocytes	[52.105]
Nucleotide signaling via P2Y2 Rs	G protein-coupled purine receptor, P2Y2 enhances the release of sAPPα in a time and dose dependent manner; probably mediated via ADAM10 and ADAM17	[52.106]
IL-1β	Enhanced α-cleavage by ADAM10/17 via up-regulation of P2Y2 receptors and may increase levels of ADAM10/17 by $\approx$3 fold	[52.107, 108]
IL-1	IL-1 up-regulates APP gene expression via PKC pathway and the upstream AP-1 binding site of the APP promoter	[52.109]
Furin	Furin enhances cleavage to active forms of ADAM10 and ADAM17 leading to enhanced α-cleavage	[52.110]
Nardilysin	N-arginine dibasic convertase may be involved in regulating the activity of ADAM17	[52.111]
Retinoic acid receptors	Retinoids increase gene expression of ADAM10 via retinoic acid receptors with retinoid X receptors and α cleavage: β cleavage is increased about 2–3 times	[52.112]
Disheveled	dvl-1 increases sAPPα production mediated via JNK and PKC/MAPK but not via p38 MAP kinase	[52.113]
Cholesterol transport	The cholesterol synthesis inhibitor, U18666a, increases APP at the cell surface and leads to increased processing by α-secretase	[52.114]
Purinergic receptor P2X7	Increased signaling promotes α-cleavage	[52.115]

and signaling by interleukins having multiple effects. The exact routes involved in these pathways between systems are not clear, making modeling difficult. In addition, the effects of Aβ on α-cleavage may be understood as direct feedback acting within the APP molecular system. The functions associated with the large N-terminal sAPPα, released by α-cleavage, are summarized in Table 52.6 and mapped in Fig. 52.3.

sAPPα is considered neuroprotective [52.22] and appears to be involved in promoting LTP, with effects on electrical behavior of neural networks [52.116] and ultimately cognition. Data in Table 52.6 suggest that it modulates glutamate and Ca^{2+} signaling with

Table 52.6 Functions of sAPPα

Factor	Functional association with sAPP α	Reference
LTP	Lowers stimulation frequency threshold for LTP and raises threshold for LTD via a mechanism involving cGMP, shifting the balance between LTP and LTD towards LTP	[52.117]
Ca^{2+}	Modulates Ca^{2+} signaling by activating high conductance K^{+} channels via a mechanism dependent on cGMP	[52.118]
Glutamate signaling	Suppresses NMDA currents rapidly and reversibly at concentrations of ≈ 0.011 nM, possibly involving cGMP and a protein phosphatase Reductions in sAPPα lead to reduced tetanically induced NMDA currents while increased sAPPα increased these currents and enhanced LTP	[52.119, 120]
Neurite outgrowth	Enhances neurite outgrowth via a mechanism involving membrane bound APP, NMDA receptors and MAPK/ERK signaling	[52.32]
Transthyetin	Neuroprotection in transgenic mice overexpressing mutant APP is associated with elevated levels of transthyretin	[52.121, 122]
Heparin/APP	Disrupts dimerization of APP by heparin	[52.123]
Integrins	Compete with APP for binding sites on integrin-β-1 and promote neurite outgrowth	[52.124]
Electrophysiology	Hippocampal and cortical electrophysiological processes are modulated by sAPPα	[52.116]
ApoE	Binds ApoE directly and is internalized via LRP	[52.22]
Glucose/glutamate transport	sAPPα enhances transport of glucose and glutamate in synapses and protects from oxidative stress via a mechanism involving cGMP	[52.125]
Neurotrophins (NGF)	sAPPα promotes NGF neurotrophic signaling via PI3/Akt signaling	[52.126]

consequences for synaptic plasticity and neurite outgrowth [52.32].

52.4.2 β-Cleavage

β-cleavage occurs between residues Met$_{596}$ and Asp$_{597}$ of APP$_{695}$ within the second heparin binding domain, releasing the N-terminal sAPPβ and a membrane bound C99 CTF [52.22, 127]. Two proteases are associated with β-cleavage, the β-site APP cleaving enzyme (BACE) 1 and to a lesser extent, BACE2 [52.22, 127]. BACE1 may additionally cleave 11 residues within the Aβ sequence [52.22, 127] and BACE2 may cleave an alternative θ-cleavage site between the phenylalanine residues, F$_{615}$ and F$_{616}$ of APP$_{695}$. BACE1 and BACE2 are differentially regulated and have different functions [52.128]. A third enzyme, the soluble carboxypeptidase B has also been linked to β-cleavage [52.129].

Many factors involved in the regulation of α-cleavage are also involved in regulating β-cleavage; they are summarized in Table 52.7 and mapped in Fig. 52.3. They allow a fine degree of coordination between the antagonistic pathways of APP processing.

The effects of cholesterol and Ca^{2+} appear quite straightforward, with increasing cholesterol and capacitative Ca^{2+} entry enhancing β-cleavage and decreases in both associated with enhanced α-cleavage. However the relationship between APP processing and ACh and Il-1β signaling is less clear. In neurons, increased Il-1β is associated with increased α-cleavage [52.107, 108], whereas in astrocytes, increased Il-1β may increase β-cleavage [52.43]. The effects of muscarinic ACh signaling are also not consistent between cell type/species. In the neuroblastoma cell line SK-SH-SY5Y, M1, and M3 activation increased BACE1 expression and promoted β-cleavage [52.130], whereas in the 3xTg-AD model M1 activation enhanced ADAM17 expression and promoted α-cleavage [52.96]. These results highlight the necessity of modeling molecular pathways in particular cell types and species, and generalization between cell types or species cannot be assumed.

Table 52.7 Factors involved in regulation of β-cleavage

Factor	Details of effect	Reference
Heparins	Proteolysis of immature BACE1 to its mature active form is promoted by low concentrations of heparin and inhibited at higher concentrations Certain heparin derivatives may act as inhibitors of BACE1 and have therapeutic potential	[52.131–133]
Muscarinic ACh signaling	Inhibition of muscarinic signaling promotes APP processing via β-pathway Increased M1/M3 signaling promoted β-cleavage via PKC; MEK/ERK and increased expression of BACE1; M2 activation suppressed BACE expression	[52.130, 134]
Ca^{2+}	Sustained increased influx of extracellular Ca^{2+} due to capacitive entry increases β-cleavage and inhibits α-cleavage	[52.135]
Cholesterol	Membrane cholesterol correlates with β-secretase activity and inhibition of β-secretase activity lead to increased membrane cholesterol levels. Moderate reductions in cholesterol enhance the co-expression of APP and BACE1 and promote the production of Aβ	[52.136, 137]
Lipid rafts	Association of APP and BACE1 with lipid rafts increases Aβ	[52.138]
ApoE	Decreased levels of ApoE lead to increased β-cleavage	[52.139]
Dab1	Dab1 is associated with decreased Aβ, possibly by reduced endocytosis and increased α-cleavage, also involves integrins and reelin	[52.42]
HIF-1α	Hypoxia increases BACE expression via increased HIF-1α expression and β-cleavage	[52.140]
Il-1β	Increases expression of APP and β-cleavage in astrocytes	[52.43]
TIMP-3	Promotes endocytosis and beta-secretase cleavage. Increases in TIMP-3 lead to decreased surface expression of ADAM10 and APP	[52.141]

From the data from Tables 52.2–52.7 and Figs. 52.1 and 52.2, it is clear that the decision to either enter into α or β-cleavage pathways or to remain at full length is a point of balance between many competing factors from wide ranging systems that are intimately involved in synaptic plasticity. This decision is dynamic and the outcome is dependent on the current state of the neuron and the cognitive demands being made upon it.

52.4.3 γ-Cleavage

γ-Cleavage Overview
Following α or β-cleavages, cleavage of the membrane bound carboxy-terminal fragments (CTF) (C83 or C99, respectively) by the γ-secretase complex occurs at various intramembrane sites to release the variable length 21–29 residue P3 fragment (α-pathway) and the variable length 38–46 residue Aβ peptide (β-pathway), with both pathways releasing the APP intracellular domain (AICD) [52.22, 87, 88, 142]. Variations in Aβ length are thought to underlie disease progression in AD, with the longer A$\beta_{(1-42)}$ being more prone to

aggregation than the more common A$\beta_{(1-40)}$ [52.88]; γ-cleavage is mapped in Fig. 52.4.

Regulation of γ-Cleavage
There are a number of alternative γ-secretase substrates, e.g., APLP1, APLP2, Notch, cadherins, the low-density lipoprotein receptor-related protein (LRP), and syndecan-1 [52.142, 143]. In addition to PS effects mediated by γ-secretase, some presenilin functions are independent of γ-secretase, so that in effect, γ-secretase may compete for presenilins with other γ-secretase independent PS functions, including cell adhesion, trafficking of various proteins [52.144], and Ca^{2+} homeostasis [52.142]. How the γ-secretase is regulated between the different substrates is not fully understood but may involve other binding proteins such as numb [52.75] and Rac1 [52.145], regulation of PS trafficking, including a possible reciprocal interaction with APP [52.136], and localization of PS within specific organelles and membrane compartments [52.146]. Various factors involved in the regulation of γ-cleavage are summarized in Table 52.8 and mapped in Fig. 52.4.

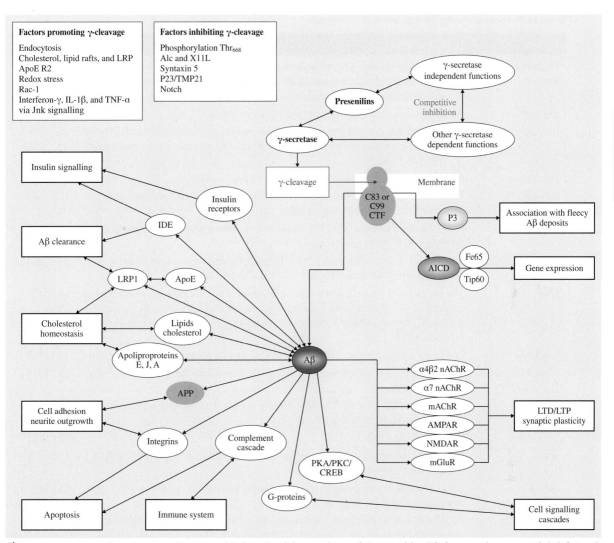

Fig. 52.4 Summary diagram of γ-cleavage with functional interactions of the peptides P3 from α-cleavage and Aβ from β-cleavage and the AICD. APP shown in *brown*, α-pathway shown in *yellow*, β-pathway shown in *blue*, shared AICD shown in *orange*

Contribution of the Presenilins

Mutations in PS1 and to a lesser extent PS2 [52.147] are associated with early onset FAD and, according to the amyloid cascade hypothesis, are thought to increase total Aβ and/or increase the level of $A\beta_{(1-42)}$ relative to $A\beta_{(1-40)}$ [52.17, 88, 148]. Mutations in APP also appear to affect the conformation of PS early in the secretory pathway resulting in increased total Aβ and increased ratio of $A\beta_{(1-42)}$ to $A\beta_{(1-40)}$ [52.149]. However, there is evidence of subtle differences in pathology between cases with different

PS mutations and SAD [52.150] and evidence that γ secretase assembly and Aβ production is differentially altered by various PS mutations [52.151]. These results suggest that the contributions of PS to disease progression are complex and may involve multiple pathways. Mutant PS is also associated with altered trafficking of various proteins, including APP [52.40], nAChRs and acetylcholinesterase [52.152], increased Ca^{2+} release from endoplasmic reticulum stores with a lowering of the neuronal threshold for LTP [52.153] and changes in Ca^{2+} homeostasis related to the changes

Table 52.8 Factors involved in regulation of γ-cleavage

Factor	Details of effect	Reference
Phosphorylation Thr$_{668}$	Conflicting evidence: Phosphorylation of T$_{668}$ decreases γ-cleavage and reduces levels of Aβ while c-Jun N-terminal kinase associated phosphorylation of T$_{668}$ promotes increased γ-cleavage	[52.155, 156]
Endocytosis	Aβ expression increases with increased endocytosis	[52.23]
Cholesterol	Aberrant cholesterol transport affects distribution of PS and enhances release of Aβ. Small reductions in cholesterol reduce Aβ by reducing the association of APP with lipid rafts	[52.39, 157, 158]
Lipid rafts	Association of γ-secretase with lipid rafts increases Aβ	[52.159]
ApoER2	Expression of ApoeR2 increases Aβ while decreasing endocytosis, may involve trafficking of APP to lipid rafts	[52.160]
IGF-1	IGF-1 promotes Aβ production through a secretase independent mechanism involving APP phosphorylation at Thr$_{668}$	[52.161]
Alc and X11L	Alc and X11L associate with C99 CTF and prevent γ-cleavage; Alc and APP co-localize in dystrophic neurites	[52.162]
LRP	The low density lipoprotein receptor related protein increases Aβ via increased trafficking of APP to lipid rafts and increased endocytosis. Also competes with APP for γ-secretase APP may reciprocally regulate ApoE and cholesterol metabolism via LRP	[52.163–165]
Oxidative stress	Oxidative stress increases expression of PS1 and its association with lipid rafts	[52.166]
Phospholipase D1 and APP	Both APP and Phospholipase D1 independently affect the trafficking of PS1 to the cell surface	[52.136]
JNK	Interferon-γ, IL-1β, and TNF-α promote γ-cleavage of APP and Notch via a JNK dependent MAPK pathway	[52.167]
Syntaxin 5	Reduces γ-cleavage	[52.168]
Rac-1	Modulates the affinity of γ-secretase for APP and Notch, inhibition of Rac-1 promotes the cleavage of Notch	[52.145]
Notch	Notch1 competes with APP for γ-cleavage and down regulates PS-1 gene expression	[52.169]
P23/TMP21	P23/TMP21 downregulates γ-secretase activity via interactions with the transmembrane helix	[52.170]

in muscarinic ACh and NMDA glutamate signaling that are independent of amyloid or tau accumulation [52.154].

FAD mutations that result in lost or altered function [52.147] coupled with the multiple interactions of PS both dependent on and independent from the γ-secretase complex are the basis of the presenilin hypothesis [52.143]. The interaction between APP and the γ-secretase complex can be understood as a second point of balance within the APP system.

Interactions of the Amyloid Beta Protein

While γ-cleavage releases both the P3 and Aβ peptides, there is little information regarding P3 and this will be considered later. In contrast, investigations into Aβ show interactions with multiple proteins in wide ranging cellular systems; these are summarized in Table 52.9 and mapped in Fig. 52.4.

In addition to the Aβ and P3 peptides, AICD is also released following γ-cleavage, but will not be considered further here.

Table 52.9 Interactions of $A\beta$

Factor	Functional association	Reference
HDL and apolipo-proteins	$A\beta$ binds apoA-I, apoA-II, apoE and apoJ; binding modulates $A\beta$ solubility. ApoE promotes polymerization of $A\beta$ into fibrils and enhances fibrillar $A\beta$ deposition in neuritic plaques in mice. The high affinity binding of $A\beta$ to ApoE reduces ability of ApoE to bind lipids	[52.171–174]
Lipid membranes, lipid rafts and cholesterol	Binding of $A\beta$ to acidic lipid molecules promotes $A\beta$ aggregation. $A\beta$ binds membrane gangliosides, sphingolipids and has high affinity for cholesterol. Cholesterol binding enhances $A\beta$ aggregation. $A\beta$ endocytosis may involve lipid rafts and clathryn coated pits. Aggregated $A\beta_{(1-40)}$ may affect lipid transport. $A\beta$ binds 24-hydroxycholesterol and affects membrane choline carriers	[52.175–180]
Integrins	$A\beta$ binds focal adhesion molecules and integrins; modulates integrin/FA signaling pathways involved in cell cycle activation and cell death. The αv integrin subunit is required for $A\beta$-associated suppression of LTP	[52.38, 181]
LTP	$A\beta$ suppresses LTP in hippocampal neurons via a mechanism involving $\alpha 4\beta 2$ nAChRs; $A\beta$ affects cascades downstream from NMDA glutamate receptor signaling	[52.182, 183]
Electrophysiology	$A\beta_{(1-40)}$ suppresses epileptiform activity in hippocampal neurons	[52.184]
AMPA receptors	AMPA glutamate receptor density is reduced by $A\beta$ oligomers via reduction of CamKII	[52.185, 186]
Insulin degrading enzyme IDE	$A\beta$ competes with insulin for the IDE and reduced IDE availability may contribute to dementia	[52.187]
Insulin signaling	$A\beta$ inhibits the autophosphorylation of insulin receptors and blocks the effect of insulin on the release of sAPPα	[52.52]
Complement cascade	$A\beta$ activates neuronal complement cascade to induce the membrane attack complex and reduces complement regulatory proteins, increasing complement-mediated cytotoxicity	[52.188]
$\alpha 7$-nAChR	$A\beta$ has high affinity for the $\alpha 7$-nAChR. Differential effects of $A\beta_{(1-40)}$ and $A\beta_{(1-42)}$ on $\alpha 7$-nAChR revealed by different effects on ACh release and Ca^{2+} influx. Disruption of signaling by $\alpha 7$-nAChR may be associated with $A\beta$-mediated increases in pre-synaptic Ca^{2+}	[52.189–191]
A4β2-nAChR	$A\beta_{(1-40)}$ and $A\beta_{(1-42)}$ reduced A4β2-nAChR and $\alpha 7$-nAChR currents. $A\beta_{(1-40)}$ but not $A\beta_{(1-42)}$ increased glutamatergic AMPA signaling via A4β2-nAChR is associated with reduced $A\beta$	[52.192, 193]
$\alpha 7$-nAChR and A4β2-nAChR	$A\beta_{(1-42)}$ has ≈ 5000 fold greater affinity for $\alpha 7$-nAChR than A4β2-nAChR	[52.194]
NMDA GluR	$A\beta$ promotes endocytosis of NMDA receptors in cortical neurons with the involvement of protein phosphatase 2B and the tyrosine phosphatase STEP	[52.195]
Neurite outgrowth	$A\beta_{(1-42)}$ reduces neurite outgrowth; reversed by the NMDA receptor antagonist memantine, $\alpha 7$-nAChR agonist PNU-282987 and the $\alpha 7$-nAChR antagonist methyllycaconitine	[52.196]
BDNF	$A\beta$ reduces BDNF-induced activation of MAPK/ERK and (PI3-K)/Akt pathways which may involve the docking proteins insulin receptor substrate-1 and Shc	[52.56]

Table 52.9 (continued)

Factor	Functional association	Reference
PKA/CREB/PKC	$A\beta$ inhibits PKA via increased persistence of its regulatory subunit PKAIIα, resulting in reduced CREB phosphorylation in response to glutamate $A\beta$ inhibits PKC	[52.197, 198]
Akt	$A\beta$ increases Akt phosphorylation in the short term via a mechanism involving α7-nAChR and NMDA receptors, with phosphorylation levels returning to baseline over the long term	[52.199]
G-protein signaling	$A\beta$ directly increases TNF-α at high levels and at low levels increases TNF-α release by altering G protein coupled receptor signaling at early stages of disease progression by effects on GPCR kinase 2/5	[52.200]

52.5 Integrating the Data So Far

52.5.1 Aβ in Synaptic Plasticity

A brief consideration of the effects of $A\beta$ on ACh and glutamate signaling will serve to illustrate the complexity involved in considering the contributions of APP and its proteolytic fragments to synaptic plasticity and ultimately normal cognition and disease pathways in AD.

The multiple interactions of $A\beta$ in both glutamate and ACh signaling have direct effects on synaptic plasticity. Defects in glutamatergic NMDA receptor function and signaling cascades in vivo in APP(V717I) transgenic mice point to decreased surface expression of NMDA-receptors as a mechanism involved in early synaptic defects [52.201, 202]. $A\beta$ also impairs AMPA receptor trafficking and function by reducing Ca^{2+}/calmodulin-dependent protein kinase II synaptic distribution [52.185].

$A\beta$ binds to α7-nAChRs with high affinity [52.194] and its removal may normalize ACh and NMDA receptor functions [52.203]. Reciprocally, α7-nAChRs may have a role in promoting the aggregation of $A\beta$ [52.204]. Transgenic mice show impaired cholinergic modulation of glutamate signaling before amyloid deposition and these effects are dependent on altered muscarinic ACh signaling [52.205].

Although both $A\beta_{(1-40)}$ and $A\beta_{(1-42)}$ reduced both α7-nAChR and A4β2-nAChR currents, only $A\beta_{(1-40)}$ increased the glutamatergic signaling via AMPA receptors [52.192]. In addition, the effects of $A\beta_{(1-40)}$ and $A\beta_{(1-42)}$ on α7-nAChR as measured by ACh release and Ca^{2+} influx are different [52.190]. These results suggest that the physiological effects of $A\beta_{(1-40)}$ and

$A\beta_{(1-42)}$ signaling are functionally different and may be involved in balancing glutamate and ACh signaling at different receptor types, with consequences for synaptic plasticity and cognition.

In addition to peptide length, there are differential effects arising from the aggregation state of $A\beta$. Protofibrils and mature fibrils show different effects on NMDA and non-NMDA glutamate receptor activation [52.206], and while both freshly soluble and aggregated $A\beta$ suppress LTP, the aggregates additionally affect neurotransmitter release [52.207]. Fibrillar forms of $A\beta$ interact with full length APP and promote APP expression, whereas non-fibrillar forms do not [52.47]. In terms of modeling the effects of $A\beta$ fragments, each fragment length and aggregation state should be considered as a separate entity.

Both $A\beta$ and PS have direct interactions with acetylcholinesterase (AChE) [52.152, 208], and AChE has been identified in association with heparin sulfates in senile plaques [52.209]. This evidence suggests that ACh signaling could be compromised by multiple mechanisms and suggests important roles for $A\beta$ in modulating both ACh [52.210] and glutamate signaling via direct and indirect mechanisms. This would allow reciprocal cross-talk between these neurotransmitters and contribute to coherence between them in maintaining a functional electrophysiological state at each synapse.

$A\beta$ appears to balance the effects of sAPPα, promoting LTD in the hippocampus while sAPPα promotes LTP. This fits well with the complex requirements of balance between LTP and LTD in synaptic plasticity. Considering the effects of ACh and glutamate signaling

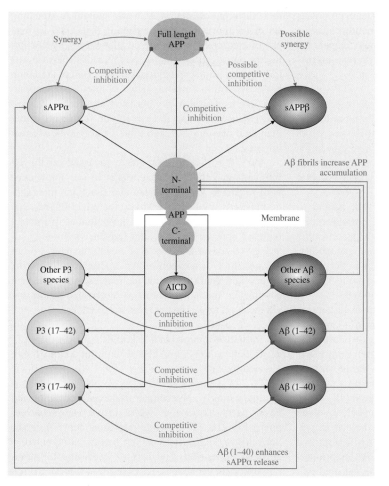

Fig. 52.5 Feedback loops between full length APP and its proteolytic fragments. *Red lines* signify inhibition, *green lines* indicate activation; full length (*brown solid* shading), alpha (*yellow shading*) or beta (*blue shading*) pathways; solid lines show interactions supported by evidence dotted lines indicate predicted interactions

on APP expression and the reciprocal effects of both sAPPα and Aβ on ACh and glutamate signaling, it is equally possible that either perturbations in APP processing drive changes in ACh and glutamate signaling or that changes in APP processing are driven by perturbations in ACh or glutamate signaling. The regulatory relationships between APP, its various proteolytic fragments, ACh, and glutamate may be central to disease progression in AD [52.210].

Neurotransmitter signaling is not the only system modulated by APP and its proteolytic fragments. These principles could be extended to investigations into the contributions of APP and its proteolytic fragments to other synaptic processes, including cholesterol homeostasis, the functions of the ECM in cell adhesion and neurite outgrowth, insulin signaling and metabolism, mitochondrial function, and the immune system.

52.5.2 Relationships Between APP and Its Proteolytic Fragments

The amyloid cascade hypothesis focuses on Aβ as a driving factor in disease initiation and progression. However, data from detailed investigations into the behavior of APP, PS, and its proteolytic fragments in synaptic plasticity reviewed above suggests a greater complexity than the cascade hypothesis can accommodate.

The various APP cleavage pathways appear to modulate each other by multiple mechanisms with feedback loops between Aβ, sAPPα, sAPPβ, and full length APP, as summarized in Fig. 52.5.

There is a constant presence but ever changing ratio of sAPPα, sAPPβ and full length APP in the extracellular matrix and, given that they share much sequence homology, there could be competition between them for binding sites. This will affect downstream processes as sAPPα may be around 100× more neuroprotective than sAPPβ [52.211]. To what degree they compete for exactly the same binding sites or have different interactions is not clear and requires investigation.

It is possible that APP, sAPPα, and sAPPβ work synergistically and that the ratios of full length APP and the various fragments are functionally more important than absolute levels in coordinating the responses of the various processes in the ECM. We need to know the affinity profiles for each fragment and interaction and how changes in these interactions affect synaptic processes over time.

sAPPα interactions, like those of Aβ, appear to have both neurotrophic and neurotoxic consequences. While sAPPα is clearly associated with neurotrophic responses, its association with neuronal excitability, like that of mutant PS [52.153], potentially contributes to excitotoxic neuronal death. Levels of both Aβ and sAPPα appear increased in Down's syndrome [52.49]. That sAPPα half life is almost doubled in transgenic mice [52.28] and levels of sAPPα show significant correlation with cognitive function in a Swedish FAD family [52.212] and in SAD [52.213] raises many questions concerning the exact roles of all the proteolytic fragments in progression to dementia. From a systems biology perspective, the balance between full length APP and its proteolytic fragments is intimately involved in the homeostatic regulation of neuronal electrophysiology where the inhibition of LTP by Aβ is a normal function and balances the role of sAPPα in promoting LTP. These relationships need to be clarified and quantified by experimentation.

Endocytosis enhances γ-cleavage and is dependent on the sequence, Y_{682}ENPTY of the APP intracellular domain [52.23]. Approximately 70% of Aβ released may be dependent on endocytosis and linked to synaptic activity [52.214]. Endocytosis allows a degree of dissociation between α and β cleavages and γ cleavage, and so sAPPα and sAPPβ levels may be independent from $Aβ_{(1-40/42)}$ and P3 levels, respectively. In functional terms this means that processing via the β-pathway has the potential to either:

1. Inhibit actions of sAPPα via competition with sAPPβ OR
2. Inhibit the actions of sAPPα AND initiate actions mediated by Aβ.

Increasingly, there is a distinction between the actions of $Aβ_{(1-40)}$ and $Aβ_{(1-42)}$, and this is supported by

their different effects on ACh and glutamate signaling reviewed above. It is also intriguing that $Aβ_{(1-40)}$ may protect neurons from the effects of $Aβ_{(1-42)}$ by disrupting the aggregation of $Aβ_{(1-42)}$ [52.215]. These results may be better interpreted in terms of different signaling pathways within a chaotic matrix of interdependent factors than as neurotoxicity per se. These results suggest that balance between the proteolytic fragments rather than their absolute concentration is likely to be of great importance.

Very few investigations have focussed on the P3 fragment and there are huge gaps in our knowledge with regard to its interactions and functions in synaptic plasticity. In a similar way to the effect of $Aβ_{(1-40)}$ on $Aβ_{(1-42)}$ [52.215], P3 has the potential to interact with Aβ species due to shared sequence homology, altering Aβ conformation and aggregation states. P3 has the potential to antagonize Aβ interactions, allowing a very fine control of cellular response via APP processing. In functional terms this means that processing via the α-pathway has the potential to either:

1. Initiate the actions of sAPPα OR
2. Initiate actions of sAPPα AND limit actions mediated by Aβ.

Interestingly, N-terminal truncated forms of Aβ have been associated with diffuse amyloid deposits [52.216, 217], raising questions concerning the behavior of P3 peptides in the processes involved in amyloid deposition. The lack of consistent terminology introduces problems, for example, while MeSH has a specific search term for Aβ, no term exists for the P3 fragment. P3 is variously described as N-truncated Aβ, Aβ (17–40/41), or P3, and extracting relevant information from the literature therefore requires a wider search and greater attention to the experimental methods section in each paper.

52.5.3 Relationships Between APP, Its Proteolytic Fragments, PS, and Wider Synaptic Systems

In the wider context, factors that promote maintenance of full length APP, α-cleavage or β-cleavage are often antagonistic. There is the potential for synergy between regulating factors so that the effects of changing one factor will depend greatly on the wider state of the neuron as a whole. Describing how functional outcomes, such as synaptic plasticity, are derived from the integration of multiple factors will not be straightforward.

It is impossible to detail the full extent of the known interaction of PS, APP, and its proteolytic fragments here, and there are likely to be other interactions that are as yet unknown. Understanding how the multiple processes reviewed above relate to the regulation of APP trafficking, processing, and interactions are central to understanding the way APP contributes to synaptic plasticity. APP can be visualized as part of a vast iterative matrix of synergistic and antagonistic interactions between interdependent systems, with each system contributing a partial pressure on all others over time. There may be direct and indirect relationships and focussing on one fragment or process involved in synaptic plasticity may be misleading.

APP as part of a wider matrix of interactions may have a central role in the temporal coherence of systems involved synaptic plasticity. Several factors make APP/PS ideally suited to this function. Firstly, a wide range of processes including functions of the ECM, the immune system, neurotransmitters, cholesterol homeostasis, and the cell cycle have the potential to interact with and regulate APP processing, allowing APP to monitor the current state of the cell. Secondly, the multiple processing pathways involving APP, ADAMs, BACEs, and PS assimilate signals from wide ranging cellular systems and communicate back to these systems via signals involving the APP fragments. Thirdly, the short half life allows neurons to be constantly aware of the current state of the cell and initiate future responses, with the potential to co-ordinate many cellular processes involved in synaptic plasticity over time. Fourthly, the potential for subtly different signals both from the α and β-pathways via alternate peptides and feedback relationships between the various peptides from the two major processing pathways allows the fine control required for integrating the many signaling pathways underlying mechanisms of synaptic plasticity.

52.6 Summary

While the work presented in this chapter is in no way comprehensive, mapping of the interactions and feedback loops involving APP highlights the complexity inherent in the processes of synaptic plasticity both in normal and diseased states. Indeed, the complexity identified here undermines the more reductionist models of disease causation and progression, such as the amyloid cascade hypothesis [52.17]. While models that focus on causal factors from other synaptic systems, such as cholesterol homeostasis [52.9–11], Ca^{2+} homeostasis [52.12], oxidative stress [52.13, 14], cell signaling cascades [52.15], and cell cycle [52.16] are perhaps more integrative in their approach, the contributions from APP and PS are to some extent neglected.

From a systems biology perspective, what unites all the possible disease pathways is imbalance in the regulation of and decoherence between the many processes involved in synaptic plasticity. There may be a number of factors that have the potential to initiate AD and which causal factor is specific to any one individual with AD may depend on the life history of that person. This suggests that there are many disease pathways that converge on a shared expression of synapse loss and neurodegeneration, and this may explain the heterogeneity seen in many aspects of AD. The challenge is to identify disease pathways in individuals with AD and develop appropriate treatments. This challenge is huge and is unlikely to be straightforward given the self-organizing nature of the brain. New computational models may have the potential to tease these pathways apart and contribute to our understanding of how synaptic plasticity works and how it fails in AD.

Currently, we do not have enough data to create a functional computational model from the maps of APP interactions. We need quantitative data that applies to each interaction and how the outcomes of those interactions behave over time. According to this perspective, experimental design must be modified to take account of the relative effects of various players in biochemical interactions in the neuron. Rather than focussing on observing the effects of changing just one part of the synaptic system, experimental design must try to account for changes in one aspect of the synaptic system while another is also changing. It could be that several synaptic systems need to be monitored across continua of relative changes to identify the outcomes of any synergistic and antagonistic relationships. Given the number of molecular players involved, new high through-put experimental techniques and new ways to analyze and integrate the data acquired from such experiments are needed.

In this chapter, we have demonstrated that mapping the network of interactions between molecules involved in complex biological processes is useful in many ways. Firstly, maps can summarize the current state of the knowledge base. This is important when trying to integrate data from wide ranging fields and to understand the relative contributions made by each factor in complex biological processes. Secondly, these maps can contribute to the identification of gaps in the knowledge base, allowing other targets for research to be identified. Thirdly, maps can contribute to alternative perspectives on processes in complex biological systems and generate new questions to be answered, thus feeding back into empirical experimental design and the interpretation of experimental results within the wider context of complex biological systems.

References

52.1   T. Miyashita, S. Kubik, G. Lewandowski, J.F. Guzowski: Networks of neurons, networks of genes: An integrated view of memory consolidation, Neurobiol. Learn. Mem. **89**(3), 269–284 (2008)

52.2   R.G. Morris, E.I. Moser, G. Riedel, S.J. Martin, J. Sandin, M. Day, C. O'Carroll: Elements of a neurobiological theory of the hippocampus: The role of activity-dependent synaptic plasticity in memory, Philos. Trans. R. Soc. Lond. B **358**(1432), 773–786 (2003)

52.3   D.J. Selkoe: Normal and abnormal biology of the β-amyloid precursor protein, Annu. Rev. Neurosci. **17**, 489–517 (1994)

52.4   J.C. Morris, A. Heyman, R.C. Mohs, J.P. Hughes, G. van Belle, G. Fillenbaum, E.D. Mellits, C. Clark: The Consortium to Establish a Registry for Alzheimer's Disease (CERAD). Part I. Clinical and neuropsychological assessment of Alzheimer's disease, Neurology **39**(9), 1159–1165 (1989)

52.5   H. Braak, E. Braak: Neuropathological stageing of Alzheimer-related changes, Acta Neuropathol. **82**(4), 239–259 (1991)

52.6   H. Braak, E. Braak: Diagnostic criteria for neuropathologic assessment of Alzheimer's disease, Neurobiol. Aging **18**(4 Suppl), S85–88 (1997)

52.7   S.S. Mirra, A. Heyman, D. McKeel, S.M. Sumi, B.J. Crain, L.M. Brownlee, F.S. Vogel, J.P. Hughes, G. van Belle, L. Berg: The Consortium to Establish a Registry for Alzheimer's Disease (CERAD). Part II. Standardization of the neuropathologic assessment of Alzheimer's disease, Neurology **41**(4), 479–486 (1991)

52.8   D.J. Selkoe: Physiological production of the beta-amyloid protein and the mechanism of Alzheimer's disease, Trends Neurosci. **16**(10), 403–409 (1993)

52.9   A.R. Koudinov, N.V. Koudinova: Cholesterol homeostasis failure as a unifying cause of synaptic degeneration, J. Neurol. Sci. **229–230**, 233–240 (2005)

52.10 J. Poirier: Apolipoprotein E and cholesterol metabolism in the pathogenesis and treatment of Alzheimer's disease, Trends Mol. Med. **9**(3), 94–101 (2003)

52.11 L. Puglielli, R.E. Tanzi, D.M. Kovacs: Alzheimer's disease: The cholesterol connection, Nat. Neurosci. **6**(4), 345–351 (2003)

52.12 M.P. Mattson: Calcium and neurodegeneration, Aging Cell **6**(3), 337–350 (2007)

52.13 G. Perry, M.A. Taddeo, A. Nunomura, X. Zhu, T. Zenteno-Savin, K.L. Drew, S. Shimohama, J. Avila, R.J. Castellani, M.A. Smith: Comparative biology and pathology of oxidative stress in Alzheimer and other neurodegenerative diseases: Beyond damage and response, Comp. Biochem. Physiol. C Toxicol Pharmacol. **133**(4), 507–513 (2002)

52.14 M.P. Mattson: Oxidative stress, perturbed calcium homeostasis, and immune dysfunction in Alzheimer's disease, J. NeuroVirol. **8**(6), 539–550 (2002)

52.15 T. Arendt: Synaptic plasticity and cell cycle activation in neurons are alternative effector pathways: The 'Dr. Jekyll and Mr. Hyde concept' of Alzheimer's disease or the yin and yang of neuroplasticity, Prog. Neurobiol. **71**(2–3), 83–248 (2003)

52.16 T. Arendt, M.K. Bruckner: Linking cell-cycle dysfunction in Alzheimer's disease to a failure of synaptic plasticity, Biochim. Biophys. Acta **1772**(4), 413–421 (2007)

52.17 J.A. Hardy, G.A. Higgins: Alzheimer's disease: The amyloid cascade hypothesis, Science **256**(5054), 184–185 (1992)

52.18 MRC CFAS: Pathological correlates of late-onset dementia in a multicentre, community-based population in England and Wales. Neuropathology Group of the Medical Research Council Cognitive Function and Ageing Study (MRC CFAS), Lancet **357**(9251), 169–175 (2001)

52.19 G.M. Savva, S.B. Wharton, P.G. Ince, G. Forster, F.E. Matthews, C. Brayne: Age, neuropathology, and dementia, N. Engl. J. Med. **360**(22), 2302–2309 (2009)

52.20 S. Hunter, R.P. Friedland, C. Brayne: Time for a change in the research paradigm for Alzheimer's disease: The value of a chaotic matrix modeling approach, CNS Neurosci. Ther. **16**(4), 254–262 (2010)

52.21 A.I. Bush, W.H. Pettingell Jr., M. de Paradis, R.E. Tanzi, W. Wasco: The amyloid beta-protein precursor and its mammalian homologues. Evidence for a zinc-modulated heparin-binding superfamily, J. Biol. Chem. **269**(43), 26618–26621 (1994)

52.22 P.R. Turner, K. O'Connor, W.P. Tate, W.C. Abraham: Roles of amyloid precursor protein and its fragments in regulating neural activity, plasticity and memory, Prog. Neurobiol. **70**(1), 1–32 (2003)

52.23 R.G. Perez, S. Soriano, J.D. Hayes, B. Ostaszewski, W. Xia, D.J. Selkoe, X. Chen, G.B. Stokin, E.H. Koo: Mutagenesis identifies new signals for beta-amyloid precursor protein endocytosis, turnover, and the generation of secreted fragments, including Abeta42, J. Biol. Chem. **274**(27), 18851–18856 (1999)

52.24 C.U. Pietrzik, T. Busse, D.E. Merriam, S. Weggen, E.H. Koo: The cytoplasmic domain of the LDL receptor-related protein regulates multiple steps in APP processing, EMBO J. **21**(21), 5691–5700 (2002)

52.25 E.H. Koo, S.L. Squazzo, D.J. Selkoe, C.H. Koo: Trafficking of cell-surface amyloid beta-protein precursor. I. Secretion, endocytosis and recycling as detected by labeled monoclonal antibody, J. Cell Sci. **109**(Pt 5), 991–998 (1996)

52.26 M.J. Savage, S.P. Trusko, D.S. Howland, L.R. Pinsker, S. Mistretta, A.G. Reaume, B.D. Greenberg, R. Siman, R.W. Scott: Turnover of amyloid beta-protein in mouse brain and acute reduction of its level by phorbol ester, J. Neurosci. **18**(5), 1743–1752 (1998)

52.27 A.W. Lyckman, A.M. Confaloni, G. Thinakaran, S.S. Sisodia, K.L. Moya: Post-translational processing and turnover kinetics of presynaptically targeted amyloid precursor superfamily proteins in the central nervous system, J. Biol. Chem. **273**(18), 11100–11106 (1998)

52.28 J. Morales-Corraliza, M.J. Mazzella, J.D. Berger, N.S. Diaz, J.H. Choi, E. Levy, Y. Matsuoka, E. Planel, P.M. Mathews: In vivo turnover of tau and APP metabolites in the brains of wild-type and Tg2576 mice: Greater stability of sAPP in the Beta-amyloid depositing mice, PLoS ONE **4**(9), e7134 (2009)

52.29 H. Zheng, M. Jiang, M.E. Trumbauer, D.J. Sirinathsinghji, R. Hopkins, D.W. Smith, R.P. Heavens, G.R. Dawson, S. Boyce, M.W. Conner, K.A. Stevens, H.H. Slunt, S.S. Sisoda, H.Y. Chen, L.H. Van der Ploeg: β-Amyloid precursor protein-deficient mice show reactive gliosis and decreased locomotor activity, Cell **81**(4), 525–531 (1995)

52.30 J.P. Steinbach, U. Muller, M. Leist, Z.W. Li, P. Nicotera, A. Aguzzi: Hypersensitivity to seizures in beta-amyloid precursor protein deficient mie, Cell Death Differ. **5**(10), 858–866 (1998)

52.31 C.S. von Koch, H. Zheng, H. Chen, M. Trumbauer, G. Thinakaran, L.H. van der Ploeg, D.L. Price, S.S. Sisodia: Generation of APLP2 KO mice and early postnatal lethality in APLP2/APP double KO mice, Neurobiol. Aging **18**(6), 661–669 (1997)

52.32 N. Gakhar-Koppole, P. Hundeshagen, C. Mandl, S.W. Weyer, B. Allinquant, U. Muller, F. Ciccolini: Activity requires soluble amyloid precursor protein alpha to promote neurite outgrowth in neural stem cell-derived neurons via activation of the MAPK pathway, Eur. J. Neurosci. **28**(5), 871–882 (2008)

52.33 S. Heber, J. Herms, V. Gajic, J. Hainfellner, A. Aguzzi, T. Rulicke, H. von Kretzschmar, C. von Koch, S.S. Sisodia, P. Tremml, H.P. Lipp, D.P. Wolfer, U. Muller: Mice with combined gene knock-outs reveal essential and partially redundant functions of amyloid precursor protein family members, J. Neurosci. **20**(21), 7951–7963 (2000)

52.34 G. Stoll, S. Jander, M. Schroeter: Detrimental and beneficial effects of injury-induced inflammation and cytokine expression in the nervous system, Adv. Exp. Med. Biol. **513**, 87–113 (2002)

52.35 M. Pickering, J.J. O'Connor: Pro-inflammatory cytokines and their effects in the dentate gyrus, Prog. Brain Res. **163**, 339–354 (2007)

52.36 C.W. Cotman, N.P. Hailer, K.K. Pfister, I. Soltesz, M. Schachner: Cell adhesion molecules in neural plasticity and pathology: Similar mechanisms, distinct organizations?, Prog. Neurobiol. **55**(6), 659–669 (1998)

52.37 G.J. Ho, R. Drego, E. Hakimian, E. Masliah: Mechanisms of cell signaling and inflammation in Alzheimer's disease, Curr. Drug Targets Inflamm. Allergy **4**(2), 247–256 (2005)

52.38 J. Caltagarone, Z. Jing, R. Bowser: Focal adhesions regulate Abeta signaling and cell death in Alzheimer's disease, Biochim. Biophys. Acta **1772**(4), 438–445 (2007)

52.39 C. Guardia-Laguarta, M. Coma, M. Pera, J. Clarimon, L. Sereno, J.M. Agullo, L. Molina-Porcel, E. Gallardo, A. Deng, O. Berezovska, B.T. Hyman, R. Blesa, T. Gomez-Isla, A. Lleo: Mild cholesterol depletion reduces amyloid-beta production by impairing APP trafficking to the cell surface, J. Neurochem. **110**(1), 220–230 (2009)

52.40 S. Gandy, Y.W. Zhang, A. Ikin, S.D. Schmidt, A. Bogush, E. Levy, R. Sheffield, R.A. Nixon, F.F. Liao, P.M. Mathews, H. Xu, M.E. Ehrlich: Alzheimer's presenilin 1 modulates sorting of APP and its carboxyl-terminal fragments in cerebral neurons in vivo, J. Neurochem. **102**(3), 619–626 (2007)

52.41 H.S. Hoe, K.J. Lee, R.S. Carney, J. Lee, A. Markova, J.Y. Lee, B.W. Howell, B.T. Hyman, D.T. Pak, G. Bu, G.W. Rebeck: Interaction of reelin with amyloid precursor protein promotes neurite outgrowth, J. Neurosci. **29**(23), 7459–7473 (2009)

52.42 H.S. Hoe, T.S. Tran, Y. Matsuoka, B.W. Howell, G.W. Rebeck: DAB1 and Reelin effects on amyloid precursor protein and ApoE receptor 2 trafficking and processing, J. Biol. Chem. **281**(46), 35176–35185 (2006)

52.43 J.G. Sheng, K. Ito, R.D. Skinner, R.E. Mrak, C.R. Rovnaghi, L.J. Van Eldik, W.S. Griffin: In vivo and in vitro evidence supporting a role for the inflammatory cytokine interleukin-1 as a driving force in Alzheimer pathogenesis, Neurobiol. Aging **17**(5), 761–766 (1996)

52.44   K.A. Chang, S.H. Kim, Y. Sakaki, H.S. Kim, C.W. Park, Y.H. Suh: Inhibition of the NGF and IL-1beta-induced expression of Alzheimer's amyloid precursor protein by antisense oligonucleotides, J. Mol. Neurosci. **12**(1), 69–74 (1999)

52.45   M.P. Marzolo, G. Bu: Lipoprotein receptors and cholesterol in APP trafficking and proteolytic processing, implications for Alzheimer's disease, Semin. Cell Dev. Biol. **20**(2), 191–200 (2009)

52.46   E. Waldron, C. Heilig, A. Schweitzer, N. Nadella, S. Jaeger, A.M. Martin, S. Weggen, K. Brix, C.U. Pietrzik: LRP1 modulates APP trafficking along early compartments of the secretory pathway, Neurobiol. Dis. **31**(2), 188–197 (2008)

52.47   A.R. White, F. Maher, M.W. Brazier, M.F. Jobling, J. Thyer, L.R. Stewart, A. Thompson, R. Gibson, C.L. Masters, G. Multhaup, K. Beyreuther, C.J. Barrow, S.J. Collins, R. Cappai: Diverse fibrillar peptides directly bind the Alzheimer's amyloid precursor protein and amyloid precursor-like protein 2 resulting in cellular accumulation, Brain Res. **966**(2), 231–244 (2003)

52.48   I.Y. Tamboli, K. Prager, E. Barth, M. Heneka, K. Sandhoff, J. Walter: Inhibition of glycosphingolipid biosynthesis reduces secretion of the beta-amyloid precursor protein and amyloid beta-peptide, J. Biol. Chem. **280**(30), 28110–28117 (2005)

52.49   G.H. Tansley, B.L. Burgess, M.T. Bryan, Y. Su, V. Hirsch-Reinshagen, J. Pearce, J.Y. Chan, A. Wilkinson, J. Evans, K.E. Naus, S. McIsaac, K. Bromley, W. Song, H.C. Yang, N. Wang, R.B. De-Mattos, C.L. Wellington: The cholesterol transporter ABCG1 modulates the subcellular distribution and proteolytic processing of beta-amyloid precursor protein, J. Lipid Res. **48**(5), 1022–1034 (2007)

52.50   C.V. Zerbinatti, J.M. Cordy, C.D. Chen, M. Guillily, S. Suon, W.J. Ray, G.R. Seabrook, C.R. Abraham, B. Wolozin: Oxysterol-binding protein-1 (OSBP1) modulates processing and trafficking of the amyloid precursor protein, Mol. Neurodegener. **3**, 5 (2008)

52.51   L. Gasparini, G.K. Gouras, R. Wang, R.S. Gross, M.F. Beal, P. Greengard, H. Xu: Stimulation of beta-amyloid precursor protein trafficking by insulin reduces intraneuronal beta-amyloid and requires mitogen-activated protein kinase signaling, J. Neurosci. **21**(8), 2561–2570 (2001)

52.52   X. Ling, R.N. Martins, M. Racchi, S. Craft, E. Helmerhorst: Amyloid beta antagonizes insulin promoted secretion of the amyloid beta protein precursor, J. Alzheimers Dis. **4**(5), 369–374 (2002)

52.53   D.W. Shineman, A.S. Dain, M.L. Kim, V.M. Lee: Constitutively active Akt inhibits trafficking of amyloid precursor protein and amyloid precursor protein metabolites through feedback inhibition of phosphoinositide 3-kinase, Biochemistry **48**(17), 3787–3794 (2009)

52.54   K.S. Vetrivel, P. Gong, J.W. Bowen, H. Cheng, Y. Chen, M. Carter, P.D. Nguyen, L. Placanica, F.T. Wieland, Y.M. Li, M.Z. Kounnas, G. Thinakaran: Dual roles of the transmembrane protein p23/TMP21 in the modulation of amyloid precursor protein metabolism, Mol. Neurodegener. **2**, 4 (2007)

52.55   Y. Ruiz-Leon, A. Pascual: Regulation of beta-amyloid precursor protein expression by brain-derived neurotrophic factor involves activation of both the Ras and phosphatidylinositide 3-kinase signalling pathways, J. Neurochem. **88**(4), 1010–1018 (2004)

52.56   L. Tong, R. Balazs, P.L. Thornton, C.W. Cotman: Beta-amyloid peptide at sublethal concentrations downregulates brain-derived neurotrophic factor functions in cultured cortical neurons, J. Neurosci. **24**(30), 6799–6809 (2004)

52.57   J.M. Cosgaya, M.J. Latasa, A. Pascual: Nerve growth factor and ras regulate beta-amyloid precursor protein gene expression in PC12 cells, J. Neurochem. **67**(1), 98–104 (1996)

52.58   S.S. Mok, G. Sberna, D. Heffernan, R. Cappai, D. Galatis, H.J. Clarris, W.H. Sawyer, K. Beyreuther, C.L. Masters, D.H. Small: Expression and analysis of heparin-binding regions of the amyloid precursor protein of Alzheimer's disease, FEBS Lett. **415**(3), 303–307 (1997)

52.59   G.W. Rebeck, R.D. Moir, S. Mui, D.K. Strickland, R.E. Tanzi, B.T. Hyman: Association of membrane-bound amyloid precursor protein APP with the apolipoprotein E receptor LRP, Brain Res. Mol. Brain Res. **87**(2), 238–245 (2001)

52.60   M.S. Durakoglugil, Y. Chen, C.L. White, E.T. Kavalali, J. Herz: Reelin signaling antagonizes beta-amyloid at the synapse, Proc. Natl. Acad. Sci. USA **106**(37), 15938–15943 (2009)

52.61   S. Matsuda, Y. Matsuda, L. D'Adamio: CD74 interacts with APP and suppresses the production of Abeta, Mol. Neurodegener. **4**, 41 (2009)

52.62   UniProt: P04233, available online at http://www.uniprot.org/uniprot/P04233

52.63   G.M. Shaked, S. Chauv, K. Ubhi, L.A. Hansen, E. Masliah: Interactions between the amyloid precursor protein C-terminal domain and G proteins mediate calcium dysregulation and amyloid beta toxicity in Alzheimer's disease, FEBS J. **276**(10), 2736–2751 (2009)

52.64   I. Nishimoto, T. Okamoto, Y. Matsuura, S. Takahashi, Y. Murayama, E. Ogata: Alzheimer amyloid protein precursor complexes with brain GTP-binding protein G(o), Nature **362**(6415), 75–79 (1993)

52.65   T. Okamoto, S. Takeda, Y. Murayama, E. Ogata, I. Nishimoto: Ligand-dependent G protein coupling function of amyloid transmembrane precursor, J. Biol. Chem. **270**(9), 4205–4208 (1995)

52.66   S.L. Sabo, A.F. Ikin, J.D. Buxbaum, P. Greengard: The Alzheimer amyloid precursor protein (APP) and

FE65, an APP-binding protein, regulate cell movement, J. Cell Biol. **153**(7), 1403–1414 (2001)

52.67 W.T. Kimberly, J.B. Zheng, S.Y. Guenette, D.J. Selkoe: The intracellular domain of the beta-amyloid precursor protein is stabilized by Fe65 and translocates to the nucleus in a notch-like manner, J. Biol. Chem. **276**(43), 40288–40292 (2001)

52.68 K. Ando, K.I. Iijima, J.I. Elliott, Y. Kirino, T. Suzuki: Phosphorylation-dependent regulation of the interaction of amyloid precursor protein with Fe65 affects the production of beta-amyloid, J. Biol. Chem. **276**(43), 40353–40361 (2001)

52.69 D. Santiard-Baron, D. Langui, M. Delehedde, B. Delatour, B. Schombert, N. Touchet, G. Tremp, M.F. Paul, V. Blanchard, N. Sergeant, A. Delacourte, C. Duyckaerts, L. Pradier, L. Mercken: Expression of human Fe65 in amyloid precursor protein transgenic mice is associated with a reduction in beta-amyloid load, J. Neurochem. **93**(2), 330–338 (2005)

52.70 D. Zhou, N. Zambrano, T. Russo, L. D'Adamio: Phosphorylation of a tyrosine in the amyloid-beta protein precursor intracellular domain inhibits Fe65 binding and signaling, J. Alzheimers Dis. **16**(2), 301–307 (2009)

52.71 R. Tamayev, D. Zhou, L. D'Adamio: The interactome of the amyloid beta precursor protein family members is shaped by phosphorylation of their intracellular domains, Mol. Neurodegener. **4**, 28 (2009)

52.72 J.H. Lee, K.F. Lau, M.S. Perkinton, C.L. Standen, S.J. Shemilt, L. Mercken, J.D. Cooper, D.M. McLoughlin, C.C. Miller: The neuronal adaptor protein X11alpha reduces Abeta levels in the brains of Alzheimer's APPswe Tg2576 transgenic mice, J. Biol. Chem. **278**(47), 47025–47029 (2003)

52.73 P.E. Tarr, C. Contursi, R. Roncarati, C. Noviello, E. Ghersi, M.H. Scheinfeld, N. Zambrano, T. Russo, L. D'Adamio: Evidence for a role of the nerve growth factor receptor TrkA in tyrosine phosphorylation and processing of beta-APP, Biochem. Biophys. Res. Commun. **295**(2), 324–329 (2002)

52.74 A.J. Beel, C.K. Mobley, H.J. Kim, F. Tian, A. Hadziselimovic, B. Jap, J.H. Prestegard, C.R. Sanders: Structural studies of the transmembrane C-terminal domain of the amyloid precursor protein (APP): Does APP function as a cholesterol sensor?, Biochemistry **47**(36), 9428–9446 (2008)

52.75 R. Roncarati, N. Sestan, M.H. Scheinfeld, B.E. Berechid, P.A. Lopez, O. Meucci, J.C. McGlade, P. Rakic, L. D'Adamio: The γ-secretase-generated intracellular domain of β-amyloid precursor protein binds Numb and inhibits Notch signaling, Proc. Natl. Acad. Sci. USA **99**(10), 7102–7107 (2002)

52.76 M.H. Scheinfeld, E. Ghersi, P. Davies, L. D'Adamio: Amyloid beta protein precursor is phosphorylated by JNK-1 independent of, yet facilitated by, JNK-

interacting protein (JIP)-1, J. Biol. Chem. **278**(43), 42058–42063 (2003)

52.77 D. Beher, L. Hesse, C.L. Masters, G. Multhaup: Regulation of amyloid protein precursor (APP) binding to collagen and mapping of the binding sites on APP and collagen type I, J. Biol. Chem. **271**(3), 1613–1620 (1996)

52.78 K.C. Breen: APP-collagen interaction is mediated by a heparin bridge mechanism, Mol. Chem. Neuropathol. **16**(1–2), 109–121 (1992)

52.79 C. Priller, T. Bauer, G. Mitteregger, B. Krebs, H.A. Kretzschmar, J. Herms: Synapse formation and function is modulated by the amyloid precursor protein, J. Neurosci. **26**(27), 7212–7221 (2006)

52.80 M.S. Lee, S.C. Kao, C.A. Lemere, W. Xia, H.C. Tseng, Y. Zhou, R. Neve, M.K. Ahlijanian, L.H. Tsai: APP processing is regulated by cytoplasmic phosphorylation, J. Cell Biol. **163**(1), 83–95 (2003)

52.81 T. Suzuki, T. Nakaya: Regulation of amyloid beta-protein precursor by phosphorylation and protein interactions, J. Biol. Chem. **283**(44), 29633–29637 (2008)

52.82 K.A. Chang, H.S. Kim, T.Y. Ha, J.W. Ha, K.Y. Shin, Y.H. Jeong, J.P. Lee, C.H. Park, S. Kim, T.K. Baik, Y.H. Suh: Phosphorylation of amyloid precursor protein (APP) at Thr668 regulates the nuclear translocation of the APP intracellular domain and induces neurodegeneration, Mol. Cell Biol. **26**(11), 4327–4338 (2006)

52.83 K.A. Chang, Y.H. Suh: Pathophysiological roles of amyloidogenic carboxy-terminal fragments of the beta-amyloid precursor protein in Alzheimer's disease, J. Pharmacol. Sci. **97**(4), 461–471 (2005)

52.84 H. Taru, T. Suzuki: Regulation of the physiological function and metabolism of AbetaPP by AbetaPP binding proteins, J. Alzheimers Dis. **18**(2), 253–265 (2009)

52.85 H.N. Woo, J.S. Park, A.R. Gwon, T.V. Arumugam, D.G. Jo: Alzheimer's disease and Notch signaling, Biochem. Biophys. Res. Commun. **390**(4), 1093–1097 (2009)

52.86 F. Sola Vigo, G. Kedikian, L. Heredia, F. Heredia, A.D. Anel, A.L. Rosa, A. Lorenzo: Amyloid-beta precursor protein mediates neuronal toxicity of amyloid beta through Go protein activation, Neurobiol. Aging **30**(9), 1379–1392 (2009)

52.87 D.J. Selkoe: Cell biology of the amyloid beta-protein precursor and the mechanism of Alzheimer's disease, Annu. Rev. Cell Biol. **10**, 373–403 (1994)

52.88 D.J. Selkoe: Alzheimer's disease: Genes, proteins, and therapy, Physiol. Rev. **81**(2), 741–766 (2001)

52.89 B.E. Slack, L.K. Ma, C.C. Seah: Constitutive shedding of the amyloid precursor protein ectodomain is up-regulated by tumour necrosis factor-alpha converting enzyme, Biochem J. **357**(3), 787–794 (2001)

52.90 T.M. Allinson, E.T. Parkin, A.J. Turner, N.M. Hooper: ADAMs family members as amyloid precursor protein alpha-secretases, J. Neurosci. Res. **74**(3), 342–352 (2003)

52.91 P. Yang, K.A. Baker, T. Hagg: The ADAMs family: Coordinators of nervous system development, plasticity and repair, Prog. Neurobiol. **79**(2), 73–94 (2006)

52.92 A.A. Talamagas, S. Efthimiopoulos, E.C. Tsilibary, M.E. Figueiredo-Pereira, A.K. Tzinia: Abeta(1–40)-induced secretion of matrix metalloproteinase-9 results in sAPPalpha release by association with cell surface APP, Neurobiol. Dis. **28**(3), 304–315 (2007)

52.93 M. Deuss, K. Reiss, D. Hartmann: Part-time alpha-secretases: The functional biology of ADAM 9, 10 and 17, Curr. Alzheimer Res. **5**(2), 187–201 (2008)

52.94 S. Choi, J.H. Kim, E.J. Roh, M.J. Ko, J.E. Jung, H.J. Kim: Nuclear factor-kappaB activated by capacitative Ca^{2+} entry enhances muscarinic receptor-mediated soluble amyloid precursor protein (sAPPalpha) release in SH-SY5Y cells, J. Biol. Chem. **281**(18), 12722–12728 (2006)

52.95 S. Rossner, M. Sastre, K. Bourne, S.F. Lichtenthaler: Transcriptional and translational regulation of BACE1 expression–implications for Alzheimer's disease, Prog. Neurobiol. **79**(2), 95–111 (2006)

52.96 A. Caccamo, S. Oddo, L.M. Billings, K.N. Green, H. Martinez-Coria, A. Fisher, F.M. LaFerla: M1 receptors play a central role in modulating AD-like pathology in transgenic mice, Neuron **49**(5), 671–682 (2006)

52.97 A. Fisher: M1 muscarinic agonists target major hallmarks of Alzheimer's disease – the pivotal role of brain M1 receptors, Neurodegener. Dis. **5**(3–4), 237–240 (2008)

52.98 R.M. Nitsch, B.E. Slack, S.A. Farber, J.G. Schulz, M. Deng, C. Kim, P.R. Borghesani, W. Korver, R.J. Wurtman, J.H. Growdon: Regulation of proteolytic processing of the amyloid beta-protein precursor of Alzheimer's disease in transfected cell lines and in brain slices, J. Neural. Transm. Suppl. **44**, 21–27 (1994)

52.99 R.M. Canet-Aviles, M. Anderton, N.M. Hooper, A.J. Turner, P.F. Vaughan: Muscarine enhances soluble amyloid precursor protein secretion in human neuroblastoma SH-SY5Y by a pathway dependent on protein kinase C_α, src-tyrosine kinase and extracellular signal-regulated kinase but not phospholipase C, Brain Res. Mol. Brain Res. **102**(1/2), 62–72 (2002)

52.100 S.A. Farber, R.M. Nitsch, J.G. Schulz, R.J. Wurtman: Regulated secretion of beta-amyloid precursor protein in rat brain, J. Neurosci. **15**(11), 7442–7451 (1995)

52.101 R. Etcheberrigaray, M. Tan, I. Dewachter, C. Kuiperi, I. Van der Auwera, S. Wera, L. Qiao, B. Bank, T.J. Nelson, A.P. Kozikowski, F. Van Leuven, D.L. Alkon: Therapeutic effects of PKC activators in Alzheimer's disease transgenic mice, Proc. Natl. Acad. Sci. USA **101**(30), 11141–11146 (2004)

52.102 S. Rossner, K. Mendla, R. Schliebs, V. Bigl: Protein kinase C_α and β1 isoforms are regulators of alpha-secretory proteolytic processing of amyloid precursor protein in vivo, Eur. J. Neurosci. **13**(8), 1644–1648 (2001)

52.103 N. Sawamura, M. Ko, W. Yu, K. Zou, K. Hanada, T. Suzuki, J.S. Gong, K. Yanagisawa, M. Michikawa: Modulation of amyloid precursor protein cleavage by cellular sphingolipids, J. Biol. Chem. **279**(12), 11984–11991 (2004)

52.104 G.P. Eckert, S. Chang, J. Eckmann, E. Copanaki, S. Hagl, U. Hener, W.E. Muller, D. Kogel: Liposome-incorporated DHA increases neuronal survival by enhancing non-amyloidogenic APP processing, Biochim. Biophys. Acta **1808**(1), 236–243 (2011)

52.105 C. Kim, C.H. Jang, J.H. Bang, M.W. Jung, I. Jool, S.U. Kim, I. Mook-Jung: Amyloid precursor protein processing is separately regulated by protein kinase C and tyrosine kinase in human astrocytes, Neurosci. Lett. **324**(3), 185–188 (2002)

52.106 J.M. Camden, A.M. Schrader, R.E. Camden, F.A. Gonzalez, L. Erb, C.I. Seye, G.A. Weisman: P2Y2 nucleotide receptors enhance alpha-secretase-dependent amyloid precursor protein processing, J. Biol. Chem. **280**(19), 18696–18702 (2005)

52.107 Q. Kong, T.S. Peterson, O. Baker, E. Stanley, J. Camden, C.I. Seye, L. Erb, A. Simonyi, W.G. Wood, G.Y. Sun, G.A. Weisman: Interleukin-1β enhances nucleotide-induced and α-secretase-dependent amyloid precursor protein processing in rat primary cortical neurons via up-regulation of the P2Y(2) receptor, J. Neurochem. **109**(5), 1300–1310 (2009)

52.108 Y. Tachida, K. Nakagawa, T. Saito, T.C. Saido, T. Honda, Y. Saito, S. Murayama, T. Endo, G. Sakaguchi, A. Kato, S. Kitazume, Y. Hashimoto: Interleukin-1 β up-regulates TACE to enhance α-cleavage of APP in neurons: Resulting decrease in Abeta production, J. Neurochem. **104**(5), 1387–1393 (2008)

52.109 D. Goldgaber, H.W. Harris, T. Hla, T. Maciag, R.J. Donnelly, J.S. Jacobsen, M.P. Vitek, D.C. Gajdusek: Interleukin 1 regulates synthesis of amyloid beta-protein precursor mRNA in human endothelial cells, Proc. Natl. Acad. Sci. USA **86**(19), 7606–7610 (1989)

52.110 E.M. Hwang, S.K. Kim, J.H. Sohn, J.Y. Lee, Y. Kim, Y.S. Kim, I. Mook-Jung: Furin is an endogenous regulator of alpha-secretase associated APP processing, Biochem. Biophys. Res. Commun. **349**(2), 654–659 (2006)

52.111 Y. Hiraoka, M. Ohno, K. Yoshida, K. Okawa, H. Tomimoto, T. Kita, E. Nishi: Enhancement of alpha-secretase cleavage of amyloid precursor protein by a metalloendopeptidase nardilysin, J. Neurochem. **102**(5), 1595–1605 (2007)

52.112 F. Tippmann, J. Hundt, A. Schneider, K. Endres, F. Fahrenholz: Up-regulation of the alpha-secretase ADAM10 by retinoic acid receptors and acitretin, FASEB J. **23**(6), 1643–1654 (2009)

52.113 A. Mudher, S. Chapman, J. Richardson, A. Asuni, G. Gibb, C. Pollard, R. Killick, T. Iqbal, L. Raymond, I. Varndell, P. Sheppard, A. Makoff, E. Gower, P.E. Soden, P. Lewis, M. Murphy, T.E. Golde, H.T. Rupniak, B.H. Anderton, S. Lovestone: Dishevelled regulates the metabolism of amyloid precursor protein via protein kinase C/mitogen-activated protein kinase and c-Jun terminal kinase, J. Neurosci. **21**(14), 4987–4995 (2001)

52.114 W. Davis Jr.: The cholesterol transport inhibitor U18666a regulates amyloid precursor protein metabolism and trafficking in N2aAPP "Swedish" cells, Curr. Alzheimer Res. **5**(5), 448–456 (2008)

52.115 C. Delarasse, R. Auger, P. Gonnord, B. Fontaine, J.M. Kanellopoulos: The purinergic receptor P2X7 triggers alpha-secretase-dependent processing of the amyloid precursor protein, J. Biol. Chem. **286**(4), 2596–2606 (2011)

52.116 M. Sanchez-Alavez, S.L. Chan, M.P. Mattson, J.R. Criado: Electrophysiological and cerebrovascular effects of the alpha-secretase-derived form of amyloid precursor protein in young and middle-aged rats, Brain Res. **1131**(1), 112–117 (2007)

52.117 A. Ishida, K. Furukawa, J.N. Keller, M.P. Mattson: Secreted form of beta-amyloid precursor protein shifts the frequency dependency for induction of LTD, and enhances LTP in hippocampal slices, Neuroreport. **8**(9/10), 2133–2137 (1997)

52.118 K. Furukawa, S.W. Barger, E.M. Blalock, M.P. Mattson: Activation of K$^+$ channels and suppression of neuronal activity by secreted beta-amyloid-precursor protein, Nature **379**(6560), 74–78 (1996)

52.119 K. Furukawa, M.P. Mattson: Secreted amyloid precursor protein alpha selectively suppresses N-methyl-d-aspartate currents in hippocampal neurons: Involvement of cyclic GMP, Neuroscience **83**(2), 429–438 (1998)

52.120 C.J. Taylor, D.R. Ireland, I. Ballagh, K. Bourne, N.M. Marechal, P.R. Turner, D.K. Bilkey, W.P. Tate, W.C. Abraham: Endogenous secreted amyloid precursor protein-alpha regulates hippocampal NMDA receptor function, long-term potentiation and spatial memory, Neurobiol. Dis. **31**(2), 250–260 (2008)

52.121 T.D. Stein, J.A. Johnson: Lack of neurodegeneration in transgenic mice overexpressing mutant amyloid precursor protein is associated with increased levels of transthyretin and the activation of cell survival pathways, J. Neurosci. **22**(17), 7380–7388 (2002)

52.122 R. Costa, F. Ferreira-da-Silva, M.J. Saraiva, I. Cardoso: Transthyretin protects against A-beta peptide toxicity by proteolytic cleavage of the peptide:

A mechanism sensitive to the Kunitz protease inhibitor, PLoS ONE **3**(8), e2899 (2008)

52.123 M. Gralle, M.G. Botelho, F.S. Wouters: Neuroprotective secreted amyloid precursor protein acts by disrupting amyloid precursor protein dimers, J. Biol. Chem. **284**(22), 15016–15025 (2009)

52.124 T.L. Young-Pearse, A.C. Chen, R. Chang, C. Marquez, D.J. Selkoe: Secreted APP regulates the function of full-length APP in neurite outgrowth through interaction with integrin beta1, Neural Dev. **3**, 15 (2008)

52.125 M.P. Mattson, Z.H. Guo, J.D. Geiger: Secreted form of amyloid precursor protein enhances basal glucose and glutamate transport and protects against oxidative impairment of glucose and glutamate transport in synaptosomes by a cyclic GMP-mediated mechanism, J. Neurochem. **73**(2), 532–537 (1999)

52.126 S. Jimenez, M. Torres, M. Vizuete, R. Sanchez-Varo, E. Sanchez-Mejias, L. Trujillo-Estrada, I. Carmona-Cuenca, C. Caballero, D. Ruano, A. Gutierrez, J. Vitorica: Age-dependent accumulation of soluble Abeta oligomers reverses the neuroprotective effect of sAPPalpha by modulating PI3K/Akt-GSK-3beta pathway in Alzheimer mice model, J. Biol. Chem. **286**(21), 18414–18425 (2011)

52.127 S.L. Cole, R. Vassar: The Alzheimer's disease beta-secretase enzyme, BACE1, Mol. Neurodegener. **2**, 22 (2007)

52.128 X. Sun, Y. Wang, H. Qing, M.A. Christensen, Y. Liu, W. Zhou, Y. Tong, C. Xiao, Y. Huang, S. Zhang, X. Liu, W. Song: Distinct transcriptional regulation and function of the human BACE2 and BACE1 genes, FASEB J. **19**(7), 739–749 (2005)

52.129 A. Matsumoto, K. Itoh, R. Matsumoto: A novel carboxypeptidase B that processes native beta-amyloid precursor protein is present in human hippocampus, Eur. J. Neurosci. **12**(1), 227–238 (2000)

52.130 T. Zuchner, J.R. Perez-Polo, R. Schliebs: Beta-secretase BACE1 is differentially controlled through muscarinic acetylcholine receptor signaling, J. Neurosci. Res. **77**(2), 250–257 (2004)

52.131 M. Beckman, R.M. Holsinger, D.H. Small: Heparin activates beta-secretase (BACE1) of Alzheimer's disease and increases autocatalysis of the enzyme, Biochemistry **45**(21), 6703–6714 (2006)

52.132 S.J. Patey, E.A. Edwards, E.A. Yates, J.E. Turnbull: Heparin derivatives as inhibitors of BACE-1, the Alzheimer's beta-secretase, with reduced activity against factor Xa and other proteases, J. Med. Chem. **49**(20), 6129–6132 (2006)

52.133 S.J. Patey, E.A. Edwards, E.A. Yates, J.E. Turnbull: Engineered heparins: Novel beta-secretase inhibitors as potential Alzheimer's disease therapeutics, Neurodegener. Dis. **5**(3/4), 197–199 (2008)

52.134 W. Liskowsky, R. Schliebs: Muscarinic acetylcholine receptor inhibition in transgenic Alzheimer-like Tg2576 mice by scopolamine favours the amyloidogenic route of processing of amyloid precursor protein, Int. J. Dev. Neurosci. **24**(2–3), 149–156 (2006)

52.135 N. Pierrot, P. Ghisdal, A.S. Caumont, J.N. Octave: Intraneuronal amyloid-beta$_{1-42}$ production triggered by sustained increase of cytosolic calcium concentration induces neuronal death, J. Neurochem. **88**(5), 1140–1150 (2004)

52.136 W.W. Liu, S. Todd, D.T. Coulson, G.B. Irvine, A.P. Passmore, B. McGuinness, M. McConville, D. Craig, J.A. Johnston: A novel reciprocal and biphasic relationship between membrane cholesterol and beta-secretase activity in SH-SY5Y cells and in human platelets, J. Neurochem. **108**(2), 341–349 (2009)

52.137 J. Abad-Rodriguez, M.D. Ledesma, K. Craessaerts, S. Perga, M. Medina, A. Delacourte, C. Dingwall, B. De Strooper, C.G. Dotti: Neuronal membrane cholesterol loss enhances amyloid peptide generation, J. Cell Biol. **167**(5), 953–960 (2004)

52.138 J.M. Cordy, I. Hussain, C. Dingwall, N.M. Hooper, A.J. Turner: Exclusively targeting beta-secretase to lipid rafts by GPI-anchor addition up-regulates beta-site processing of the amyloid precursor protein, Proc. Natl. Acad. Sci. USA **100**(20), 11735–11740 (2003)

52.139 J.C. Dodart, K.R. Bales, E.M. Johnstone, S.P. Little, S.M. Paul: Apolipoprotein E alters the processing of the beta-amyloid precursor protein in APP(V717F) transgenic mice, Brain Res. **955**(1–2), 191–199 (2002)

52.140 X. Zhang, K. Zhou, R. Wang, J. Cui, S.A. Lipton, F.F. Liao, H. Xu, Y.W. Zhang: Hypoxia-inducible factor 1alpha (HIF-1alpha)-mediated hypoxia increases BACE1 expression and beta-amyloid generation, J. Biol. Chem. **282**(15), 10873–10880 (2007)

52.141 H.S. Hoe, M.J. Cooper, M.P. Burns, P.A. Lewis, M. van der Brug, G. Chakraborty, C.M. Cartagena, D.T. Pak, M.R. Cookson, G.W. Rebeck: The metalloprotease inhibitor TIMP-3 regulates amyloid precursor protein and apolipoprotein E receptor proteolysis, J. Neurosci. **27**(40), 10895–10905 (2007)

52.142 K.S. Vetrivel, Y.W. Zhang, H. Xu, G. Thinakaran: Pathological and physiological functions of presenilins, Mol. Neurodegener. **1**, 4 (2006)

52.143 J. Shen, R.J. Kelleher III: The presenilin hypothesis of Alzheimer's disease: Evidence for a loss-of-function pathogenic mechanism, Proc. Natl. Acad. Sci. USA **104**(2), 403–409 (2007)

52.144 K. Uemura, A. Kuzuya, S. Shimohama: Protein trafficking and Alzheimer's disease, Curr. Alzheimer Res. **1**(1), 1–10 (2004)

52.145 J.H. Boo, J.H. Sohn, J.E. Kim, H. Song, I. Mook-Jung: Rac1 changes the substrate specificity of γ-secretase between amyloid precursor protein and Notch1, Biochem. Biophys. Res. Commun. **372**(4), 913–917 (2008)

52.146 K.S. Vetrivel, H. Cheng, S.H. Kim, Y. Chen, N.Y. Barnes, A.T. Parent, S.S. Sisodia, G. Thinakaran: Spatial segregation of gamma-secretase and substrates in distinct membrane domains, J. Biol. Chem. **280**(27), 25892–25900 (2005)

52.147 E.S. Walker, M. Martinez, A.L. Brunkan, A. Goate: Presenilin 2 familial Alzheimer's disease mutations result in partial loss of function and dramatic changes in Abeta 42/40 ratios, J. Neurochem. **92**(2), 294–301 (2005)

52.148 J. Hardy, D.J. Selkoe: The amyloid hypothesis of Alzheimer's disease: Progress and problems on the road to therapeutics, Science **297**(5580), 353–356 (2002)

52.149 L. Herl, A.V. Thomas, C.M. Lill, M. Banks, A. Deng, P.B. Jones, R. Spoelgen, B.T. Hyman, O. Berezovska: Mutations in amyloid precursor protein affect its interactions with presenilin/gamma-secretase, Mol. Cell Neurosci. **41**(2), 166–174 (2009)

52.150 C.L. Maarouf, I.D. Daugs, S. Spina, R. Vidal, T.A. Kokjohn, R.L. Patton, W.M. Kalback, D.C. Luehrs, D.G. Walker, E.M. Castano, T.G. Beach, B. Ghetti, A.E. Roher: Histopathological and molecular heterogeneity among individuals with dementia associated with Presenilin mutations, Mol. Neurodegener **3**, 20 (2008)

52.151 M. Bentahir, O. Nyabi, J. Verhamme, A. Tolia, K. Horre, J. Wiltfang, H. Esselmann, B. De Strooper: Presenilin clinical mutations can affect gamma-secretase activity by different mechanisms, J. Neurochem. **96**(3), 732–742 (2006)

52.152 M.X. Silveyra, G. Evin, M.F. Montenegro, C.J. Vidal, S. Martinez, J.G. Culvenor, J. Saez-Valero: Presenilin 1 interacts with acetylcholinesterase and alters its enzymatic activity and glycosylation, Mol. Cell Biol. **28**(9), 2908–2919 (2008)

52.153 I. Schneider, D. Reverse, I. Dewachter, L. Ris, N. Caluwaerts, C. Kuiperi, M. Gilis, H. Geerts, H. Kretzschmar, E. Godaux, D. Moechars, F. Van Leuven, J. Herms: Mutant presenilins disturb neuronal calcium homeostasis in the brain of transgenic mice, decreasing the threshold for excitotoxicity and facilitating long-term potentiation., J. Biol. Chem. **276**(15), 11539–11544 (2001)

52.154 Y. Wang, N.H. Greig, Q.S. Yu, M.P. Mattson: Presenilin-1 mutation impairs cholinergic modulation of synaptic plasticity and suppresses NMDA currents in hippocampus slices, Neurobiol. Aging **30**(7), 1061–1068 (2009)

52.155 C. Feyt, N. Pierrot, B. Tasiaux, J. Van Hees, P. Kienlen-Campard, P.J. Courtoy, J.N. Octave: Phosphorylation of APP695 at Thr668 decreases gamma-cleavage and extracellular Abeta, Biochem. Biophys. Res. Commun. **357**(4), 1004–1010 (2007)

52.156 V. Vingtdeux, M. Hamdane, M. Gompel, S. Begard, H. Drobecq, A. Ghestem, M.E. Grosjean, V. Kostanjevecki, P. Grognet, E. Vanmechelen, L. Buee, A. Delacourte, N. Sergeant: Phosphorylation of amyloid precursor carboxy-terminal fragments enhances their processing by a gamma-secretase-dependent mechanism, Neurobiol. Dis. **20**(2), 625–637 (2005)

52.157 M. Burns, K. Gaynor, V. Olm, M. Mercken, J. LaFrancois, L. Wang, P.M. Mathews, W. Noble, Y. Matsuoka, K. Duff: Presenilin redistribution associated with aberrant cholesterol transport enhances beta-amyloid production in vivo, J. Neurosci. **23**(13), 5645–5649 (2003)

52.158 M.P. Burns, U. Igbavboa, L. Wang, W.G. Wood, K. Duff: Cholesterol distribution, not total levels, correlate with altered amyloid precursor protein processing in statin-treated mice, Neuromol. Med. **8**(3), 319–328 (2006)

52.159 K.S. Vetrivel, H. Cheng, W. Lin, T. Sakurai, T. Li, N. Nukina, P.C. Wong, H. Xu, G. Thinakaran: Association of gamma-secretase with lipid rafts in post-Golgi and endosome membranes, J. Biol. Chem. **279**(43), 44945–44954 (2004)

52.160 R.A. Fuentealba, M.I. Barria, J. Lee, J. Cam, C. Araya, C.A. Escudero, N.C. Inestrosa, F.C. Bronfman, G. Bu, M.P. Marzolo: ApoER2 expression increases Abeta production while decreasing Amyloid Precursor Protein (APP) endocytosis: Possible role in the partitioning of APP into lipid rafts and in the regulation of gamma-secretase activity, Mol. Neurodegener. **2**, 14 (2007)

52.161 W. Araki, H. Kume, A. Oda, A. Tamaoka, F. Kametani: IGF-1 promotes beta-amyloid production by a secretase-independent mechanism, Biochem. Biophy. Res. Commun. **380**(1), 111–114 (2009)

52.162 Y. Araki, S. Tomita, H. Yamaguchi, N. Miyagi, A. Sumioka, Y. Kirino, T. Suzuki: Novel cadherin-related membrane proteins, Alcadeins, enhance the X11-like protein-mediated stabilization of amyloid beta-protein precursor metabolism, J. Biol. Chem. **278**(49), 49448–49458 (2003)

52.163 I.S. Yoon, E. Chen, T. Busse, E. Repetto, M.K. Lakshmana, E.H. Koo, D.E. Kang: Low-density lipoprotein receptor-related protein promotes amyloid precursor protein trafficking to lipid rafts in the endocytic pathway, FASEB Journal **21**(11), 2742–2752 (2007)

52.164 A. Lleo, E. Waldron, C.A. von Arnim, L. Herl, M.M. Tangredi, I.D. Peltan, D.K. Strickland, E.H. Koo, B.T. Hyman, C.U. Pietrzik, O. Berezovska: Low density lipoprotein receptor-related protein (LRP) interacts with presenilin 1 and is a competitive substrate of the amyloid precursor protein (APP) for gamma-secretase, J. Biol. Chem. **280**(29), 27303–27309 (2005)

52.165 Q. Liu, C.V. Zerbinatti, J. Zhang, H.S. Hoe, B. Wang, S.L. Cole, J. Herz, L. Muglia, G. Bu: Amyloid precursor protein regulates brain apolipoprotein E and cholesterol metabolism through lipoprotein receptor LRP1, Neuron **56**(1), 66–78 (2007)

52.166 A. Oda, A. Tamaoka, W. Araki: Oxidative stress upregulates presenilin 1 in lipid rafts in neuronal cells, J. Neurosci. Res. **88**(5), 1137–1145 (2009)

52.167 Y.F. Liao, B.J. Wang, H.T. Cheng, L.H. Kuo, M.S. Wolfe: Tumor necrosis factor-alpha, interleukin-1beta, and interferon-gamma stimulate gamma-secretase-mediated cleavage of amyloid precursor protein through a JNK-dependent MAPK pathway, J. Biol. Chem. **279**(47), 49523–49532 (2004)

52.168 K. Suga, A. Saito, T. Tomiyama, H. Mori, K. Akagawa: Syntaxin 5 interacts specifically with presenilin holoproteins and affects processing of betaAPP in neuronal cells, J. Neurochem. **94**(2), 425–439 (2005)

52.169 A. Lleo, O. Berezovska, P. Ramdya, H. Fukumoto, S. Raju, T. Shah, B.T. Hyman: Notch1 competes with the amyloid precursor protein for gamma-secretase and down-regulates presenilin-1 gene expression, J. Biol. Chem. **278**(48), 47370–47375 (2003)

52.170 R. Pardossi-Piquard, C. Bohm, F. Chen, S. Kanemoto, F. Checler, G. Schmitt-Ulms, P. St George-Hyslop, P.E. Fraser: TMP21 transmembrane domain regulates gamma-secretase cleavage, J. Biol. Chem. **284**(42), 28634–28641 (2009)

52.171 A.R. Koudinov, T.T. Berezov, A. Kumar, N.V. Koudinova: Alzheimer's amyloid beta interaction with normal human plasma high density lipoprotein: Association with apolipoprotein and lipids, Clin. Chim. Acta **270**(2), 75–84 (1998)

52.172 B.V. Zlokovic, S. Yamada, D. Holtzman, J. Ghiso, B. Frangione: Clearance of amyloid beta-peptide from brain: Transport or metabolism?, Nat. Med. **6**(7), 718–719 (2000)

52.173 D.M. Holtzman, A.M. Fagan, B. Mackey, T. Tenkova, L. Sartorius, S.M. Paul, K. Bales, K.H. Ashe, M.C. Irizarry, B.T. Hyman: Apolipoprotein E facilitates neuritic and cerebrovascular plaque formation in an Alzheimer's disease model, Ann. Neurol. **47**(6), 739–747 (2000)

52.174 S. Tamamizu-Kato, J.K. Cohen, C.B. Drake, M.G. Kosaraju, J. Drury, V. Narayanaswami: Interaction with amyloid beta peptide compromises the lipid binding function of apolipoprotein E, Biochemistry **47**(18), 5225–5234 (2008)

52.175 E. Terzi, G. Holzemann, J. Seelig: Interaction of Alzheimer beta-amyloid peptide (1–40) with lipid membranes, Biochemistry **36**(48), 14845–14852 (1997)

52.176 C. Hertel, E. Terzi, N. Hauser, R. Jakob-Rotne, J. Seelig, J.A. Kemp: Inhibition of the electrostatic interaction between beta-amyloid peptide and membranes prevents beta-amyloid-induced toxicity, Proc. Natl. Acad. Sci. USA **94**(17), 9412–9416 (1997)

52.177 Y. Verdier, M. Zarandi, B. Penke: Amyloid beta-peptide interactions with neuronal and glial cell plasma membrane: Binding sites and implications for Alzheimer's disease, J. Pept. Sci. **10**(5), 229–248 (2004)

52.178 L. Saavedra, A. Mohamed, V. Ma, S. Kar, E.P. de Chaves: Internalization of beta-amyloid peptide by primary neurons in the absence of apolipoprotein E, J. Biol. Chem. **282**(49), 35722–35732 (2007)

52.179 N.A. Avdulov, S.V. Chochina, U. Igbavboa, C.S. Warden, A.V. Vassiliev, W.G. Wood: Lipid binding to amyloid beta-peptide aggregates: Preferential binding of cholesterol as compared with phosphatidylcholine and fatty acids., J. Neurochem. **69**(4), 1746–1752 (1997)

52.180 Z. Kristofikova, V. Kopecky Jr., K. Hofbauerova, P. Hovorkova, D. Ripova: Complex of amyloid beta peptides with 24-hydroxycholesterol and its effect on hemicholinium-3 sensitive carriers, Neurochem. Res. **33**(3), 412–421 (2008)

52.181 Q. Wang, I. Klyubin, S. Wright, I. Griswold-Prenner, M.J. Rowan, R. Anwyl: αv integrins mediate beta-amyloid induced inhibition of long-term potentiation, Neurobiol. Aging **29**(10), 1485–1493 (2008)

52.182 M.N. Wu, Y.X. He, F. Guo, J.S. Qi: Alpha4beta2 nicotinic acetylcholine receptors are required for the amyloid beta protein-induced suppression of long-term potentiation in rat hippocampal CA1 region in vivo, Brain Res. Bull. **77**(2–3), 84–90 (2008)

52.183 G. Yamin: NMDA receptor-dependent signaling pathways that underlie amyloid beta-protein disruption of LTP in the hippocampus, J. Neurosci. Res. **87**(8), 1729–1736 (2009)

52.184 F.J. Sepulveda, C. Opazo, L.G. Aguayo: Alzheimer beta-amyloid blocks epileptiform activity in hippocampal neurons, Mol. Cell Neurosci. **41**(4), 420–428 (2009)

52.185 Z. Gu, W. Liu, Z. Yan: β-Amyloid impairs AMPA receptor trafficking and function by reducing Ca^{2+}/calmodulin-dependent protein kinase II synaptic distribution, J. Biol. Chem. **284**(16), 10639–10649 (2009)

52.186 D. Zhao, J.B. Watson, C.W. Xie: Amyloid beta prevents activation of calcium/calmodulin-dependent protein kinase II and AMPA receptor phosphorylation during hippocampal long-term potentiation, J. Neurophysiol. **92**(5), 2853–2858 (2004)

52.187 A. Perez, L. Morelli, J.C. Cresto, E.M. Castano: Degradation of soluble amyloid beta-peptides 1–40, 1–42, and the Dutch variant 1–40Q by insulin degrading enzyme from Alzheimer disease and control brains, Neurochem. Res. **25**(2), 247–255 (2000)

52.188 Y. Shen, T. Sullivan, C.M. Lee, S. Meri, K. Shiosaki, C.W. Lin: Induced expression of neuronal membrane attack complex and cell death by Alzheimer's beta-amyloid peptide, Brain Res. **796**(1–2), 187–197 (1998)

52.189 R.G. Nagele, M.R. D'Andrea, W.J. Anderson, H.Y. Wang: Intracellular accumulation of beta-amyloid (1–42) in neurons is facilitated by the alpha 7 nicotinic acetylcholine receptor in Alzheimer's disease, Neuroscience **110**(2), 199–211 (2002)

52.190 D.H. Lee, H.Y. Wang: Differential physiologic responses of alpha7 nicotinic acetylcholine receptors to beta-amyloid$_{1-40}$ and beta-amyloid$_{1-42}$, J. Neurobiol. **55**(1), 25–30 (2003)

52.191 J.J. Dougherty, J. Wu, R.A. Nichols: Beta-amyloid regulation of presynaptic nicotinic receptors in rat hippocampus and neocortex, J. Neurosci. **23**(17), 6740–6747 (2003)

52.192 H. Tozaki, A. Matsumoto, T. Kanno, K. Nagai, T. Nagata, S. Yamamoto, T. Nishizaki: The inhibitory and facilitatory actions of amyloid-beta peptides on nicotinic ACh receptors and AMPA receptors, Biochem. Biophys. Res. Commun. **294**(1), 42–45 (2002)

52.193 T. Kihara, S. Shimohama, M. Urushitani, H. Sawada, J. Kimura, T. Kume, T. Maeda, A. Akaike: Stimulation of alpha4beta2 nicotinic acetylcholine receptors inhibits beta-amyloid toxicity, Brain Res. **792**(2), 331–334 (1998)

52.194 H.Y. Wang, D.H. Lee, M.R. D'Andrea, P.A. Peterson, R.P. Shank, A.B. Reitz: β-Amyloid$_{(1-42)}$ binds to alpha7 nicotinic acetylcholine receptor with high affinity. Implications for Alzheimer's disease pathology, J. Biol. Chem. **275**(8), 5626–5632 (2000)

52.195 E.M. Snyder, Y. Nong, C.G. Almeida, S. Paul, T. Moran, E.Y. Choi, A.C. Nairn, M.W. Salter, P.J. Lombroso, G.K. Gouras, P. Greengard: Regulation of NMDA receptor trafficking by amyloid-beta, Nat. Neurosci. **8**(8), 1051–1058 (2005)

52.196 M. Hu, M.E. Schurdak, P.S. Puttfarcken, R. El Kouhen, M. Gopalakrishnan, J. Li: High content screen microscopy analysis of A beta 1–42-induced neurite outgrowth reduction in rat primary cortical neurons: Neuroprotective effects of alpha 7 neuronal nicotinic acetylcholine receptor ligands, Brain Res. **1151**, 227–235 (2007)

52.197 O.V. Vitolo, A. Sant'Angelo, V. Costanzo, F. Battaglia, O. Arancio, M. Shelanski: Amyloid beta -peptide inhibition of the PKA/CREB pathway and long-term potentiation: Reversibility by drugs that enhance cAMP signaling, Proc. Natl. Acad. Sci. USA **99**(20), 13217–13221 (2002)

52.198 A. Favit, M. Grimaldi, T.J. Nelson, D.L. Alkon: Alzheimer's-specific effects of soluble beta-amyloid on protein kinase C-alpha and -gamma degradation in human fibroblasts, Proc. Natl. Acad. Sci. USA **95**(10), 5562–5567 (1998)

52.199 J.J. Abbott, D.R. Howlett, P.T. Francis, R.J. Williams: Abeta$_{(1-42)}$ modulation of Akt phosphorylation via alpha7 nAChR and NMDA receptors, Neurobiol. Aging **29**(7), 992–1001 (2008)

52.200 Z. Suo, M. Wu, B.A. Citron, G.T. Wong, B.W. Festoff: Abnormality of G-protein-coupled receptor kinases at prodromal and early stages of Alzheimer's disease: An association with early beta-amyloid accumulation, J. Neurosci. **24**(13), 3444–3452 (2004)

52.201 I. Dewachter, R.K. Filipkowski, C. Priller, L. Ris, J. Neyton, S. Croes, D. Terwel, M. Gysemans, H. Devijver, P. Borghgraef, E. Godaux, L. Kaczmarek, J. Herms, F. Van Leuven: Deregulation of NMDA-receptor function and down-stream signaling in APP[V717I] transgenic mice, Neurobiol. Aging **30**(2), 241–256 (2009)

52.202 E.L. Schaeffer, W.F. Gattaz: Cholinergic and glutamatergic alterations beginning at the early stages of Alzheimer disease: Participation of the phospholipase A2 enzyme, Psychopharmacology **198**(1), 1–27 (2008)

52.203 H.Y. Wang, A. Stucky, J. Liu, C. Shen, C. Trocme-Thibierge, P. Morain: Dissociating beta-amyloid from alpha 7 nicotinic acetylcholine receptor by a novel therapeutic agent, S 24795, normalizes alpha 7 nicotinic acetylcholine and NMDA receptor function in Alzheimer's disease brain, J. Neurosci. **29**(35), 10961–10973 (2009)

52.204 L.R. Fodero, S.S. Mok, D. Losic, L.L. Martin, M.I. Aguilar, C.J. Barrow, B.G. Livett, D.H. Small: Alpha7-nicotinic acetylcholine receptors mediate an Abeta(1–42)-induced increase in the level of acetylcholinesterase in primary cortical neurones, J. Neurochem. **88**(5), 1186–1193 (2004)

52.205 Y. Goto, T. Niidome, H. Hongo, A. Akaike, T. Kihara, H. Sugimoto: Impaired muscarinic regulation of excitatory synaptic transmission in the APP-swe/PS1dE9 mouse model of Alzheimer's disease, Eur. J. Pharmacol. **583**(1), 84–91 (2008)

52.206 C. Ye, D.M. Walsh, D.J. Selkoe, D.M. Hartley: Amyloid beta-protein induced electrophysiological changes are dependent on aggregation state: N-methyl-d-aspartate (NMDA) versus non-NMDA receptor/channel activation, Neurosci. Lett. **366**(3), 320–325 (2004)

52.207 A.W. Schmid, D.B. Freir, C.E. Herron: Inhibition of LTP in vivo by beta-amyloid peptide in different conformational states, Brain Res. **1197**, 135–142 (2008)

52.208 N.C. Inestrosa, A. Alvarez, M.C. Dinamarca, T. Perez-Acle, M. Colombres: Acetylcholinesterase-amyloid-beta-peptide interaction: Effect of Congo Red and the role of the Wnt pathway, Curr. Alzheimer Res. **2**(3), 301–306 (2005)

52.209 R.N. Kalaria, S.N. Kroon, I. Grahovac, G. Perry: Acetylcholinesterase and its association with heparan sulphate proteoglycans in cortical amyloid deposits of Alzheimer's disease, Neuroscience **51**(1), 177–184 (1992)

52.210 R. Schliebs, T. Arendt: The significance of the cholinergic system in the brain during aging and in Alzheimer's disease, J. Neural Transm. **113**(11), 1625–1644 (2006)

52.211 K. Furukawa, B.L. Sopher, R.E. Rydel, J.G. Begley, D.G. Pham, G.M. Martin, M. Fox, M.P. Mattson: Increased activity-regulating and neuroprotective efficacy of alpha-secretase-derived secreted amyloid precursor protein conferred by a C-terminal heparin-binding domain, J. Neurochem. **67**(5), 1882–1896 (1996)

52.212 O. Almkvist, H. Basun, S.L. Wagner, B.A. Rowe, L.O. Wahlund, L. Lannfelt: Cerebrospinal fluid levels of alpha-secretase-cleaved soluble amyloid precursor protein mirror cognition in a Swedish family with Alzheimer disease and a gene mutation, Arch. Neurol. **54**(5), 641–644 (1997)

52.213 K. Sennvik, J. Fastbom, M. Blomberg, L.O. Wahlund, B. Winblad, E. Benedikz: Levels of alpha- and beta-secretase cleaved amyloid precursor protein in the cerebrospinal fluid of Alzheimer's disease patients, Neurosci. Lett. **278**(3), 169–172 (2000)

52.214 J.R. Cirrito, J.E. Kang, J. Lee, F.R. Stewart, D.K. Verges, L.M. Silverio, G. Bu, S. Mennerick, D.M. Holtzman: Endocytosis is required for synaptic activity-dependent release of amyloid-beta in vivo, Neuron **58**(1), 42–51 (2008)

52.215 K. Zou, D. Kim, A. Kakio, K. Byun, J.S. Gong, J. Kim, M. Kim, N. Sawamura, S. Nishimoto, K. Matsuzaki, B. Lee, K. Yanagisawa, M. Michikawa: Amyloid beta-protein (Abeta)$_{1-40}$ protects neurons from damage induced by Abeta$_{1-42}$ in culture and in rat brain, J. Neurochem. **87**(3), 609–619 (2003)

52.216 D.R. Thal, I. Sassin, C. Schultz, C. Haass, E. Braak, H. Braak: Fleecy amyloid deposits in the internal layers of the human entorhinal cortex are comprised of N-terminal truncated fragments of Abeta, J. Neuropathol. Exp. Neurol. **58**(2), 210–216 (1999)

52.217 L. Miravalle, M. Calero, M. Takao, A.E. Roher, B. Ghetti, R. Vidal: Amino-terminally truncated Abeta peptide species are the main component of cotton wool plaques, Biochemistry **44**(32), 10810–10821 (2005)

53. A Machine Learning Pipeline for Identification of Discriminant Pathways

Annalisa Barla, Giuseppe Jurman, Roberto Visintainer, Margherita Squillario, Michele Filosi, Samantha Riccadonna, Cesare Furlanello

Identifying the molecular pathways more prone to disruption during a pathological process is a key task in network medicine and, more generally, in systems biology. This chapter describes a pipeline that couples a machine learning solution for molecular profiling with a recent network comparison method. The pipeline can identify changes occurring between specific sub-modules of networks built in a case-control biomarker study, discriminating key groups of genes whose interactions are modified by an underlying condition. Different algorithms can be chosen to implement the workflow steps. Three applications on genome-wide data are presented regarding the susceptibility of children to air pollution, and early and late onset of Parkinson's and Alzheimer's diseases.

53.1 From Gene Profiling to Pathway Profiling

Nowadays, it is widely accepted as a consolidated fact that most known diseases are of systemic nature; their phenotypes can be attributed to the breakdown of a rather complex set of molecular interactions among a cell's components rather than imputed to the misfunctioning of a single entity such as a gene. A major aim of systems biology and, in particular, of its newly emerging discipline network medicine [53.1], is the understanding of the cellular wiring diagram at all possible levels of organization (from transcriptomics to signaling) of the functional design, the molecular pathways being a typical example. Molecular pathways functionally characterize genes by representing the interaction between a set of genes and network of molecules (proteins) in the cell. Thus, the systems biology approach is to shift the focus from a mere list of candidate genes selected as relevant for a given biological question, to important molecular pathways that functionally characterize such a question. The reconstruction of molecular pathways from high-throughput data is made feasible by

the recent advances in the theory of complex networks (e.g., [53.2–6]) and, in particular, in the reconstruction algorithms for inferring network topology and wiring starting from a collection of high-throughput measurements [53.7]. However, the tackled problem is hard (*a daunting task* [53.8]) and these methods are not flawless [53.9]: underdeterminacy is a major issue [53.10], and the ratio between the network dimension (number of nodes) and the number of available measurements to infer interactions plays a key role for the stability of the reconstructed structure. Although some effort has recently been put into facing this issue, the stability (and thus the reproducibility) of the process is still an open problem [53.8].

We propose a pipeline for machine learning driven determination of the disruption of important molecular pathways induced or inducing a condition starting from microarray measurements in a case/control experimental design. The problem of underdeterminacy in the inference procedure is avoided by focusing only on subnetworks, and the relevance of the studied pathways for the disease is judged in terms of discriminative relevance for the underlying classification problem. The profiling part of the pipeline, composed of a classifier and a feature selection method embedded within an adequate experimental procedure or data analysis protocol [53.11], is used to rank the genes with the highest discriminative power. These genes undergo an enrichment phase [53.12, 13] to identify the pathways that represent the functional dependencies among the genes. Such information would otherwise be lost by limiting the subnetwork analysis solely to the selected genes. The enrichment step is performed using the functional information stored in the gene ontology (GO) database [53.14]. GO is a vocabulary of *terms* that associate genes sharing similar molecular function, involved in the same biological process or acting in the same cellular compartment. In the present work, we refer to GO *terms* as pathways, assuming that genes annotated in each GO *term* interact among them and therefore constitute a molecular pathway. Finally, a network is inferred for both the case and the control samples on the selected pathways, and the two structures are compared to pinpoint the occurring differences and thus to detect the relevant pathway related variations.

A noteworthy point of this workflow is the independence from its ingredients: the classifier, the feature ranking algorithm, the enrichment procedure, the inference method, and the networks comparison function: they can all be swapped with alternative modules. In the workflow, a specific novelty is the use of a metric for the quantitative assessment of network difference. As discussed in [53.15], many classical distances (such as those of the edit family) have a relevant drawback in being local, that is focusing only on the portions of the network interested by the differences in the presence/absence of matching links. Other metrics can instead consider the global structure of the compared topologies; among such distances, the spectral ones – based on the list of eigenvalues of the Laplacian matrix of the underlying graph – are quite interesting, and, in particular, the Ipsen–Mikhailov [53.16] distance has been proven to be the most robust in a wide range of situations.

In what follows we will describe the newly introduced workflow in detail, providing three examples of application in problems of biological interest: the first tasks concerns the transcriptomics effects of exposure to environmental pollution on a cohort of children in Czech Republic, the second one investigates the molecular characteristics between Parkinson's disease (PD) at early and late stages, and the third regards the characterization of Alzheimer's disease (AD) at early and late stages. To strengthen the support of our proposal, the two problems will be dealt with by using different experimental conditions, i.e., varying the algorithms employed throughout the various steps of the workflow. In both cases, biologically meaningful considerations can be drawn, which are consistent with previous findings, showing the effectiveness of the proposed procedure in the assessment of the occurring subnetwork variations.

53.2 Methods

The proposed machine learning pipeline handles case/control transcription data through four main steps, from a profiling task output (a ranked list of genes) to the identification of discriminant pathways, see Fig. 53.1. Alternative algorithms can be used at each step of the pipeline; as an example in the profiling part different classifiers, regression, or feature selection methods can be adopted. In Sect. 53.3 we will describe the elementary steps used in the experiments.

Formally, we are given a collection of n subjects, each described by a p-dimensional vector x of measurements. Each sample is also associated with a phenotypical label $y = \{1, -1\}$, assigning it to a class (e.g., pollution versus no-pollution in Sect. 53.5.1). The dataset is therefore represented by an $n \times p$ gene expression data matrix $\mathbf{X}$, where $p \gg n$ and a corresponding labels vector $\mathbf{Y}$.

The matrix $\mathbf{X}$ is used to feed the profiling part of the pipeline. We choose a proper data analysis protocol [53.11] to ensure accurate and reproducible results and a prediction model. The model is built as a classifier (e.g., SRDA [53.17]) or a regression method (e.g., $\ell_1\ell_2$, [53.18]) coupled with a feature selection algorithm. Thus, we obtain a ranked list of genes from which we extract a gene signature $g_1, \ldots, g_k$ taking the top-k most discriminant genes. The choice is performed by finding a balance between the accuracy of the classifier and the stability of the signature [53.11].

Applying pathway enrichment techniques (e.g., GSEA or GSA [53.13, 19]), we retrieve for each gene

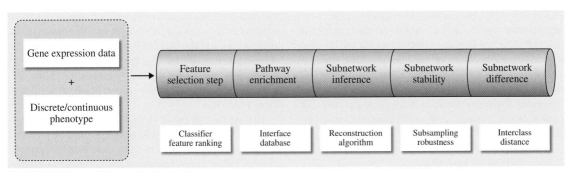

Fig. 53.1 Schema of the analysis pipeline

g_i the corresponding whole pathway $p_i = \{h_1, \ldots, h_t\}$, where the genes $h_j \neq g_i$ do not necessarily belong to the original signature $g_1, \ldots, g_k$. Extending the analysis to all the h_j genes of the pathway allows us to explore functional interactions that would otherwise be lost.

The subnetwork inference phase (e.g., WGCN or ARACNE [53.20][53.21]) requires a network to be reconstructed for each pathway p_i by the use of the steady state expression data of the samples of each class y. The network inference procedure is limited to the sole genes belonging to the pathway p_i in order to avoid the problem of intrinsic underdeterminacy of the task. As an additional caution against this problem, in the following experiments we limit the analysis to pathways having more than 4 nodes and less than 1000 nodes. For each p_i and for each y, we obtain a real-valued adjacency matrix, which is then binarized by choosing a thresh-

old on the correlation values. This choice requires the construction of a binary adjacency matrix $\mathbf{N}_{p_i, y, t_s}$ for each p_i, for each y and for a grid of threshold values $t_1, \ldots, t_T$. For each value t_s of the grid, we compute for each p_i both the distance D (e.g., the Ipsen–Mikhailov distance, for details see Sect. 53.3.4) between the case and control pathway graphs and the corresponding densities. We chose the t_s as providing the best balance between the average distance across the pathways p_i and the network density. For a fixed t_s and for each p_i, we obtain a score $D(\mathbf{N}_{p_i, 1, t_s}, \mathbf{N}_{p_i, -1, t_s})$ used to rank the pathways p_i. As an additional scoring indicator for $g_1, \ldots, g_k$, we also provide the difference between the weighted degree in the control ($y = -1$) and in the case ($y = 1$) network: $\Delta d(g_i) = d_{-1}(g_i) - d_1(g_i)$. A final step of biological relevance assessment of the ranked pathways concludes the pipeline.

53.3 Experimental Setup

The analysis pipeline presented is independent from the algorithms chosen for each step of the workflow. Here we give some details about the methods used for the experiments described in Sect. 53.5.

53.3.1 Feature Selection Step

In the feature selection step, we used spectral regression discriminant analysis (SRDA), as described in Sect. 53.5.1, and the $\ell_1\ell_2$ feature selection framework ($\ell_1\ell_2$FS) in the experiments detailed in Sects. 53.5.2 and 53.5.3, respectively. Both algorithms allow simultaneous binary classification and feature ranking, where features' weights are derived as (a function of) the coefficients of the separation rule.

SRDA belongs to the discriminant analysis algorithms family [53.17]. Its peculiarity is to exploit the

regression framework for improving the computational efficiency. Spectral graph analysis is used for solving only a set of regularized least squares problems avoiding the eigenvector computation. A score is assigned to each feature and can be interpreted as a feature weight, allowing feature ranking and selection directly. The regularization value α is the only parameter that needs to be tuned. The method is implemented in Python and it is available within the mlpy library, version 2.2.2 [53.22].

$\ell_1\ell_2$FS with double optimization is a feature selection method that can be tuned to give a minimal set of discriminative genes or larger sets including correlated genes [53.18,23]. The objective function is a linear model $f(x) = \beta x$, whose sign gives the classification rule that can be used to associate a new sample to one of the two classes. The sparse weight vector $\boldsymbol{\beta}$ is found by minimizing the $\ell_1\ell_2$ functional: $\| \mathbf{Y} - \boldsymbol{\beta} \mathbf{X} \|_2^2 +$

$\tau \parallel \boldsymbol{\beta} \parallel_1 + \mu \parallel \boldsymbol{\beta} \parallel_2^2$, where the least square error is penalized with the ℓ_1 and ℓ_2 norm of the coefficient vector $\boldsymbol{\beta}$. The training for selection and classification requires a careful choice of the regularization parameters for both $\ell_1 \ell_2$ and RLS. Indeed, model selection and statistical significance assessment is performed within two nested K-cross validation loops as in [53.24]. The framework is implemented in Python and uses the L1L2Py library [53.25].

53.3.2 Pathway Enrichment

The gene set enrichment analysis (GSEA) was performed by using WebGestalt, an online toolkit [53.12, 26]. This web-service takes as input a list of relevant genes/probe sets and performs a GSEA analysis [53.13] in several databases, as Kyoto Encyclopedia of Genes and Genomes (KEGG [53.27]) or GO, identifying the most relevant pathways and ontologies in the signatures. For our experiments we used the GO database and we selected the WebGestalt human genome as reference set, 0.05 as level of significance, 3 as the minimum number of genes, and the default hypergeometric test as statistical method.

53.3.3 Subnetwork Inference

As network reconstruction algorithms we used the weighted gene co-expression networks (WGCN), as described in Sect. 53.5.1, and the algorithm for the reconstruction of accurate cellular networks (ARACNE), in the experiments detailed in Sects. 53.5.2 and 53.5.3, respectively.

WGCN networks are based on the idea of using (a function of) the absolute correlation between the expression of a couple of genes across the samples to define a link between them. Soft thresholding techniques are then employed to obtain a binary adjacency matrix, where a suitable biologically motivated criterion (such as scale-free topology or some other prior knowledge) can be adopted [53.12, 20]. Due to the very small sample size, scale-freeness cannot be considered as a reliable criterion for threshold selection, so we adopted a different heuristics: for both networks in the two classes the selected threshold is the one maximizing the average Ipsen–Mikhailov distance on the selected pathways.

ARACNE is a recent method for inferring networks from the transcription level [53.28] to the metabolic level [53.29]. It was originally designed for handling the complexity of regulatory networks in mammalian

cells and it is able to address a wider range of network deconvolution problems. This information-theoretic algorithm removes the vast majority of indirect candidate interactions inferred by co-expression methods by using the data processing inequality property [53.30]. To be compliant, we adopt the same threshold criterion defined for WGCN.

Here we adopt the R bioconductor implementation both for WGCN (WGCNA package) and ARACNE (MiNET – mutual information networks package), keeping for the latter the default value for the data processing inequality tolerance parameter [53.21]. Moreover, the ARACNE implementation requires all the features to have non-zero variance on each class, and for consistency purposes we applied this filter in all experiments.

53.3.4 Subnetwork Stability and Difference

Although already fruitfully used even in a biological context [53.31], the problem of quantitatively comparing networks (e.g., using a metric instead of evaluating network properties) is a widely open issue affecting many scientific disciplines. The definition of the Ipsen–Mikhailov ϵ metric [53.16] follows the dynamical interpretation of an N-nodes network as N-atoms molecules connected by identical elastic strings, where the pattern of connections is defined by the adjacency matrix of the corresponding network [53.16]. The vibrational frequencies ω_i of the dynamical system are given by the eigenvalues of the Laplacian matrix of the network: $\lambda_i = \omega_i^2$, with $\lambda_0 = \omega_0 = 0$. The spectral density for a graph as the sum of Lorentz distributions is defined as

$$\rho(\omega) = K \sum_{i=1}^{N-1} \frac{\gamma}{(\omega - \omega_k)^2 + \gamma^2} ,$$

where γ is the common width specifying the half-width at half-maximum (HWHM, equal to half the interquartile range), and K is the normalization constant solution of

$$\int_0^\infty \rho(\omega) d\omega = 1 .$$

Then the spectral distance ϵ between two graphs G and H with densities $\rho_G(\omega)$ and $\rho_H(\omega)$ can be defined as

$$\sqrt{\int_0^\infty [\rho_G(\omega) - \rho_H(\omega)]^2 \, d\omega} .$$

To obtain a meaningful comparison of the value of ϵ on pairs of networks with a different number of nodes, we define the normalized version

$$\hat{\epsilon}(G, H) = \frac{\epsilon(G, H)}{\epsilon(F_n, E_n)},$$

where E_n, F_n indicate, respectively, the empty and the fully connected network on n nodes; they are the two most ϵ-distant networks for each n. The common width γ is set to 0.08 as in the original reference; being a multiplicative factor, it has no impact on comparing different values of the Ipsen–Mikhailov distance. The network analysis is implemented in R *igraph*.

Reconstruction Stability
A major issue affecting network inference is the high variability of network topology found for data perturbations, different parameter choices, and alternative methods. A quantitative measure of network stability will thus be used to evaluate the reliability of inferred topologies, derived from the metric introduced above. Given a dataset from which a (sub)network is inferred, a random subsampling (of a fraction p of the data) will be extracted from which the corresponding network will be reconstructed; repeating N times the subsampling/inferring procedure, a set of N nets will be generated for each initial dataset. Then all mutual

$$\binom{N}{2} = \frac{N(N-1)}{2}$$

Ipsen–Mikhailov distances are computed, and for each set of N graphs we build the corresponding distance histogram. Mean and variance of the constructed histograms will quantitatively assess the stability of the subnetwork inferred from the whole dataset; the lower the values, the higher the stability in terms of robustness to data perturbation (subsampling).

53.4 Data Description

Section 53.5 describes three different experiments. In the first experiment we used a genome-wide dataset created for investigating the effects of air pollution on children. In the second and third experiments we analyzed gene expression data on two neurodegenerative diseases: PD and AD. All the examples are based on the series matrix (preprocessed data) of publicly available datasets on the gene expression omnibus (GEO).

Children's Susceptibility to Air Pollution
The first dataset (GSE7543) collects data of children living in two regions of the Czech Republic that have different air pollution levels [53.32, 33]: 23 children recruited in the polluted area of Teplice and 24 children living in the cleaner area of Prachatice. The study was designed to have a representation of similar populations from both regions according to age, gender, and socioeconomic level. Blood samples were hybridized on Agilent Human 1 A 22 k oligonucleotide microarrays, corresponding to 17564 features.

Clinical Stages of Parkinson's Disease
For PD we consider two publicly available datasets from GEO: GSE6613 [53.34] and GSE20295 [53.35]. The former includes 22 controls and 50 whole blood samples from patients predominantly at early PD stages

while the latter is composed of 53 controls and 40 patients with late stage PD. Biological data were hybridized on the Affymetrix HG-U133A platform, estimating the expression of 22215 probe sets for each sample.

Clinical Stages of Alzheimer's Disease
For AD we analyzed two GEO datasets: GSE9770 and GSE5281 [53.36, 37]. The first includes 74 controls and 34 samples from non-demented patients with AD (since it is the earliest AD diagnosed, we will label it as early hereafter) and the second is composed of 74 controls and 80 samples from patients with late onset AD. The samples were extracted from six brain regions, differently susceptible to the disease: entorhinal cortex (EC), hippocampus (HIP), middle temporal gyrus (MTG), posterior cingulate cortex (PC), superior frontal gyrus (SFG), and primary visual cortex (VCX). The latter is known to be relatively spared by the disease, therefore we did not consider the samples within the VCX region. Overall, we analyzed 62 controls and 29 AD samples for GSE9770 and 62 controls and 68 AD samples for GSE5281. Biological data were hybridized on the Affymetrix HG-U133Plus2.0 platform, estimating the expression of 54713 probe sets for each sample.

53.5 Experimental Results

53.5.1 Air Pollution Experiment

The SRDA classification of the air pollution dataset was performed by applying a 100×5-fold cross validation (CV) schema, to obtain a gene signature, characteristic of the molecular differences between children in Teplice (polluted) and Prachatice (non-polluted). The signature consists of 50 probe sets, corresponding to 43 genes, achieving 76% of predictive accuracy. The enrichment analysis of the gene signature led to a functional characterization identifying 11 enriched ontologies in GO. We

then constructed the corresponding WGCN network for the selected pathways for both cases and controls.

Table 53.1 lists the enriched pathways and the total number of the genes belonging to each pathway. The list is ranked by the normalized Ipsen–Mikhailov distance $\hat{\epsilon}$, which provides a measure of the structural distance between the networks inferred for the two classes (see Sect. 53.3.4). The most disrupted pathway between the two conditions is GO:0043066 (*negative regulation of apoptotic process*) followed by GO:0001501 (*skeletal development*). Since the children under study

Table 53.1 Air pollution experiment: enriched pathways corresponding to the 43 most discriminant genes ranked by the normalized Ipsen–Mikhailov distance $\hat{\epsilon}$. The number of genes in the enriched pathway is also provided

Pathway ID	GO term	$\hat{\epsilon}$	# Genes
GO:0043066	*Negative regulation of apoptotic process*	0.257	21
GO:0001501	*Skeletal development*	0.149	89
GO:0009611	*Response to wounding*	0.123	16
GO:0007399	*Nervous system development*	0.093	252
GO:0016787	*Hydrolase activity*	0.078	718
GO:0005516	*Calmodulin binding*	0.076	116
GO:0007275	*Developmental process*	0.076	453
GO:0006954	*Inflammatory response*	0.048	180
GO:0005615	*Extracellular space*	0.038	417
GO:0007626	*Locomotory behavior*	0.000	5
GO:0006066	*Alcohol metabolic process*	0.000	8

Table 53.2 Air pollution experiment: list of Agilent probe sets in the signature with their corresponding Entrez gene symbol ID and GO pathway. The list is ranked according to the decreasing absolute value of the differential node degree Δd

Agilent ID	Gene symbol	Pathway ID	Δd
4701	NRGN	GO:0007399	−2.477
12235	DUSP15	GO:0016787	−1.586
8944	CLC	GO:0016787	−1.453
3697	ITGB5	GO:0007275	−1.390
4701	NRGN	GO:0005516	−1.357
12537	PROK2	GO:0006954	1.069
13835	OLIG1	GO:0007275	0.834
11673	HOXB8	GO:0007275	−0.750
16424	FKHL18	GO:0007275	−0.685
13094	DHX32	GO:0016787	−0.575
8944	CLC	GO:0007275	0.561
14787	MATN3	GO:0001501	0.495
15797	CXCL1	GO:0006954	0.467
15797	CXCL1	GO:0005615	0.338
11302	MYH1	GO:0005516	−0.194
15797	CXCL1	GO:0007399	0.131

are undergoing a rapid and constant growth causing changes to their skeleton and to the nervous system, the high differentiation between cases and controls of the GO:0001501 and GO:0007399 (*nervous system development*) together with the involvement of pathway GO:0007275 (*developmental process*) is biologically very sound. In particular, Fig. 53.2 shows the network of the GO:0007399 pathway, related to the nervous system development in the two cohorts. It is clear that several connections among the genes within this pathway are missing in the subjects living in the more polluted area (Teplice). Therefore, the development of the nervous system in these children is potentially at risk compared to those living in the less polluted city (Prachatice).

Another relevant pathway is GO:0006954, representing the response to infection or injury caused by

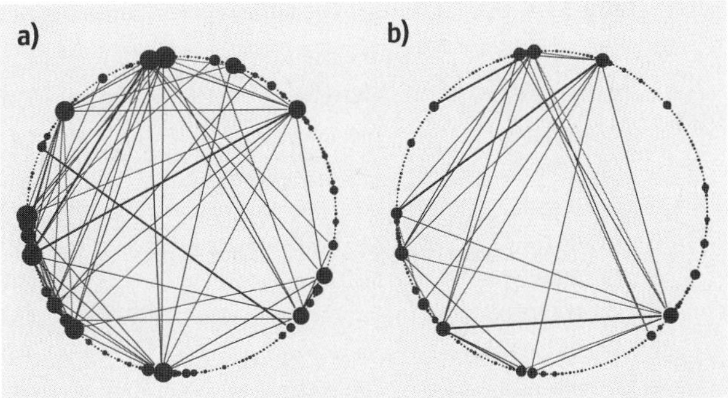

Fig. 53.2a,b Networks of the pathway GO:0007399 (*nervous system development*) (**a**) for Prachatice children (**b**) compared with Teplice children. Node diameter is proportional to the degree. Edge width is proportional to connection strength (estimated correlation)

Table 53.3 PD: most important pathways ranked by normalized Ipsen–Mikhailov distance $\hat{e}$. The Entrez gene symbol ID is also provided for the selected probe sets $g_1, \ldots, g_k$ in the corresponding pathway. The number of genes belonging to the pathway is also provided. Common elements between early and late stage PD are shown in *bold*

PD early			PD late		
Pathway ID	$\hat{e}$	**#Genes**	**Pathway ID**	$\hat{e}$	**#Genes**
GO:0005506	0.38	434	GO:0019226	0.31	20
GO:0006952	0.37	160	GO:0007611	0.16	34
GO:0045087	0.36	112	GO:0042493	0.15	109
GO:0042802	0.33	473	GO:0009725	0.11	27
GO:0006955	0.31	778	GO:0030424	0.10	93
GO:0006950	0.28	253	GO:0007267	0.09	264
GO:0020037	0.26	176	GO:0005516	0.09	215
GO:0005938	0.26	50	GO:0005096	0.09	252
GO:0005856	0.24	816	GO:0007610	0.08	40
GO:0003779	0.23	431[a]	**GO:0003779**	0.08	423[a]
GO:0030097	0.15	76	GO:0005624	0.08	616
GO:0009615	0.14	111	GO:0045202	0.08	278
GO:0051707	0.00	5	GO:0003924	0.07	294
			GO:0006928	0.07	166
			GO:0042995	0.07	231
			GO:0007268	0.06	201
			GO:0043234	0.06	233
			GO:0005525	0.05	450
			GO:0006412	0.05	466
			GO:0006836	0.05	42
			GO:0043005	0.05	51
			GO:0043025	0.04	82
			GO:0042221	0.00	16
			GO:0009266	0.00	6
			GO:0014070	0.00	13

The number of genes reported here for each pathway is computed after filtering features with zero variance on one class: it could be different between PD early and PD late since they are two separate datasets (e.g., GO:0003779)

Table 53.4 PD: GO terms. The table reports names for all the enriched GO pathway IDs reported in this section

PD early		PD late	
Pathway ID	**GO term**	**Pathway ID**	**GO term**
GO:0005506	*Iron ion binding*	GO:0019226	*Transmission of nerve impulse*
GO:0006952	*Defense response*	GO:0007611	*Learning or memory*
GO:0045087	*Innate immuno response*	GO:0042493	*Response to drug*
GO:0042802	*Identical protein binding*	GO:0009725	*Response to hormone stimulus*
GO:0006955	*Immune response*	GO:0030424	*Axon*
GO:0006950	*Response to stress*	GO:0007267	*cell–cell signaling*
GO:0020037	*Heme binding*	GO:0005516	*Calmodulin binding*
GO:0005938	*Cell cortex*	GO:0005096	*GTPase activator activity*
GO:0005856	*Cytoskeleton*	GO:0007610	*Behavior*
GO:0003779	*Actin binding*	GO:0003779	*Actin binding*
GO:0030097	*Hemopoiesis*	GO:0005624	*Membrane fraction*
GO:0009615	*Response to virus*	GO:0045202	*Synapse*
GO:0051707	*Response to other organism*	GO:0003924	*GTPase activity*
GO:0006915	*Apoptotic process*	GO:0006928	*Cellular component movement*
GO:0046983	*Protein dimerization activity*	GO:0042995	*Cell projection*
		GO:0007268	*Synaptic transmission*
		GO:0043234	*Protein complex*
		GO:0005525	*GTP binding*
		GO:0006412	*Translation*
		GO:0006836	*Neurotransmitter transport*
		GO:0043005	*Neuron projection*
		GO:0043025	*Neuronal cell body*
		GO:0042221	*Response to chemical stimulus*
		GO:0009266	*Response to temperature stimulus*
		GO:0014070	*Response to organic cyclic compound*
		GO:0007585	*Respiratory gaseous exchange*

chemical or physical agents. Several genes included in GO:0005516 bind (*calmodulin binding*) or interact with calmodulin, that is a calcium-binding protein involved in many essential processes, such as inflammation, apoptosis, nerve growth, and immune response. Other enriched terms are related to the capacity of an organism to defend itself (GO:0009611) and to the locomotion (GO:0007626).

Table 53.2 provides the subset of Agilent probe sets (together with their corresponding gene symbol and GO pathway) belonging to the signature and having a non-zero value of the differential node degree Δd (see Sect. 53.2). Since the Δd score is computed as the difference between the weighted degree in the two classes, the elements in Table 53.2 are those whose number of interactions varies most between the two conditions. Here we discuss a subset of the most biologically relevant genes. FKHL18, HOXB8, PROK2, DHX32, and MATN3 are directly involved in the de-

velopment. CLC is a key element in the inflammation and immune system. OLIG1 is a transcription factor that works in the oligodendrocytes within the brain. NRGN binds calcium and is a target for thyroid hormones in the brain. Finally, MYH1 encodes for myosin that is a major contractile protein that forms striated, smooth, and non-muscle cells. MYH1 isoforms show expression that is spatially and temporally regulated during development.

53.5.2 Parkinson's Disease Experiment

The $\ell_1\ell_2$ analysis of the two PD datasets was performed, respectively, within a ninefold nested CV loop for the early PD and an eightfold nested CV for late PD. The early PD signature consists of 77 probe sets, mapped on 70 genes, and associated to 62% accuracy. The late stage signature is composed of 94 probe sets corresponding to 90 genes and achieving 80% accuracy. Applying ARACNE, we constructed the relevance net-

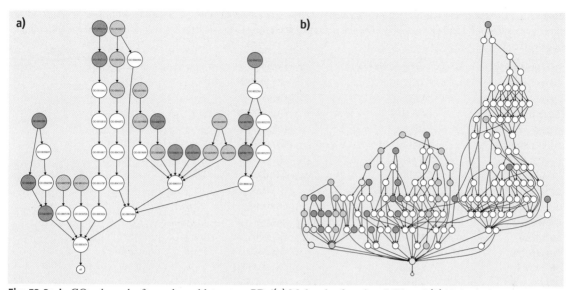

Fig. 53.3a,b GO subgraphs for early and late stage PD ((**a**) Molecular function (MF) and (**b**) biological processes (BP) domains). Selected nodes are represented in *light gray*, *gray*, and *dark gray* for late, early, and common nodes

work for both cases and controls for the 35 enriched pathways for late stage PD case and 42 pathways for early stage PD. Table 53.3 reports the most biologically relevant pathways, ranked for decreasing normalized Ipsen–Mikhailov distance $\hat{\epsilon}$. In Table 53.4 we report the names of the corresponding GO terms.

Having characterized the functional alteration of pathways for both early and late stage PD, we attempt a comparative analysis of the outcome, commenting on the most meaningful results from the biological viewpoint. We expected some common pathways between the two stages, especially within pathways that represent general processes and functions, but as commented in Sect. 53.2, the pipeline does not consider pathways having more than 1000 nodes, hence discarding the most general terms in the GO. Indeed, the only common pathway is GO:0003779, i.e., *actin binding*. Actin participates in many important cellular processes, including muscle contraction, cell motility, cell division and cytokinesis, vesicle and organelle movement, and cell signaling. Clearly, this term is strictly associated to the most evident movement-related symptoms in PD, including shaking, rigidity, slowness of movement, and difficulty with walking and gait. Nevertheless, in both early and late PD we note some alteration within the biological process class of *response to stimulus*: GO:0006950, GO:0009615, and GO:0051707 for early PD and GO:0042493, GO:0009725, GO:0042221, GO:0014070, and GO:0009266 for late PD. To better

visualize the results, we also plot the selected enriched pathways in Fig. 53.3. The two graphs represent the pathways in the molecular function and biological process domains, respectively. Despite the fact that only one pathway was found to be common in early and late AD, it is easy to note that the majority of selected pathways belong to common GO classes.

The pathways specific to early PD show a great involvement of the immune system, which is greatly stimulated by inflammation especially located in particular brain regions (mainly *substantia nigra*). Indeed,

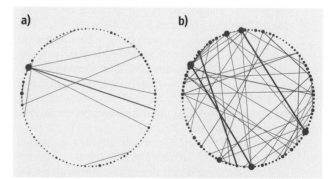

Fig. 53.4a,b Networks of the pathway GO:0045087 (*innate immune response*) (**a**) for PD early development patients (**b**) compared with healthy subjects. Node diameter is proportional to the degree. Edge width is proportional to connection strength (estimated correlation)

Table 53.5 PD experiment (early): list of Affymetrix probe sets in the early stage signature with their corresponding Entrez gene symbol ID and GO pathway. The list is ranked according to the decreasing absolute value of the differential node degree Δd

Affy probe set ID	Gene symbol	Pathway ID	Δd
200931_s_at	VCL	GO:0005856	−2.124
200931_s_at	VCL	GO:0003779	−2.107
213067_at	MYH10	GO:0003779	1.879
202887_s_at	DDIT4	GO:0006915	−1.872
201841_s_at	HSPB1	GO:0042802	−1.691
204439_at	IFI44L	GO:0006955	−1.585
201841_s_at	HSPB1	GO:0005856	−1.532
209480_at	HLA−DQB1	GO:0006955	−1.340
209116_x_at	HBB	GO:0005506	−1.008
203998_s_at	SYT1	GO:0042802	0.862
36711_at	MAFF	GO:0006950	−0.807
209116_x_at	HBB	GO:0020037	−0.599
36711_at	MAFF	GO:0046983	0.567
214414_x_at	HBA1/HBA2	GO:0020037	−0.376
205033_s_at	DEFA1/DEFA3	GO:0009615	−0.360
217232_x_at	HBB	GO:0005506	−0.239
205033_s_at	DEFA1/DEFA3	GO:0006955	−0.228
213067_at	MYH10	GO:0005938	0.183
214414_x_at	HBA1/HBA2	GO:0005506	−0.182
201841_s_at	HSPB1	GO:0006950	−0.154
217232_x_at	HBB	GO:0020037	−0.088
218197_s_at	OXR1	GO:0006950	−0.059
205033_s_at	DEFA1/DEFA3	GO:0006952	0.027

we identified GO:0006952, GO:0045087, GO:0006955, and GO:0030097. Figure 53.4 clearly shows that since in the early stages of PD the innate immune system is severely compromised, the body is highly subjected to the invasion and proliferation of microbes (like bacteria or viruses), resulting in a debilitated organism, less effective in fighting the consequent inflammation. In late stage PD, we detected several differentiated terms related to the central nervous system. Among others, we note GO:0019226, GO:0007611, GO:0007610, and GO:007268. These findings fit the late stage PD scenario, where cognitive and behavioral problems may arise with dementia.

Tables 53.5 and 53.6, respectively, list the subset of elements of the early and late PD signatures having non-zero differential node degree Δd. We recall that the elements in the two tables are those whose number of interactions varies most between the two case/control conditions.

Four common genes were identified between early and late stage PD: MYH10, SYT1, VCL, and HSPB1.

MYH10 is involved in several pathways: *actin binding* and *calmodulin binding*, *neural cell body*, *neuron projection*, and *cell cortex*. These pathways indicate that the damage mostly occurs in the neurons. In particular, actin binding and cell cortex affect the cytoskeleton and the muscular tissue. The calmodulin binding pathway indicates that other pre-processes, related to calmodulin and relevant for PD, might be damaged. These processes are related to the inflammation, metabolism, apoptosis, smooth muscle contraction, intracellular movement, short-term, long-term memory, nerve growth, and the immune response. Moreover, it is known that MYH10 is involved in the regulation of the actin cytoskeleton pathways and also in those related to the axon guidance. Mutations in this gene are known to be present in disease phenotypes affecting the heart and the brain [53.38]. The synaptotagmin SYT1, also involved in the calmodulin binding, is an integral membrane protein of synaptic vesicles, probably serving as a Ca(2+) sensor in the process of vesicular trafficking and exocytosis. Calcium binding

Table 53.6 PD experiment (late): list of Affymetrix probe sets in the late stage signature with their corresponding Entrez gene symbol and GO pathway. The list is ranked according to the decreasing absolute value of the differential node degree Δd

Affy probe set ID	Gene symbol	Pathway ID	Δd
213638_at	PHACTR1	GO:0045202	−3.255
213067_at	MYH10	GO:0003779	−3.252
213067_at	MYH10	GO:0005516	−2.597
204337_at	RGS4	GO:0005096	−2.194
213067_at	MYH10	GO:0043025	−2.107
213067_at	MYH10	GO:0043005	−1.696
214230_at	CDC42	GO:0003924	1.677
213638_at	PHACTR1	GO:0003779	−1.587
213067_at	MYH10	GO:0030424	−1.170
205857_at	SLC18A2	GO:0006836	−1.094
206552_s_at	TAC1	GO:0007268	−0.834
206552_s_at	TAC1	GO:0007267	0.809
205110_s_at	FGF13	GO:0007267	−0.804
203998_s_at	SYT1	GO:0005516	−0.787
208319_s_at	RBM3	GO:0006412	−0.759
201909_at	RPS4Y	GO:0006412	−0.688
200931_s_at	VCL	GO:0043234	−0.655
204337_at	RGS4	GO:0005516	−0.602
205105_at	MAN2A1	GO:0007585	−0.502
205857_at	SLC18A2	GO:0005624	−0.428
201841_s_at	HSPB1	GO:0006928	0.424
214230_at	CDC42	GO:0042995	−0.379
203998_s_at	SYT1	GO:0043005	−0.357
203998_s_at	SYT1	GO:0045202	−0.339
200931_s_at	VCL	GO:0003779	−0.311
200931_s_at	VCL	GO:0006928	−0.308
206836_at	SLC6A3	GO:0006836	−0.238
215342_s_at	RABGAP1L	GO:0005096	−0.211
211727_s_at	COX11	GO:0007585	0.188
203998_s_at	SYT1	GO:0007268	−0.159

to SYT1 participates in triggering neurotransmitter release at the synapse. This protein is, therefore, involved in the synaptic transmission and it predominantly works in the neuron projections and synapses. Vinculin (VCL) is a cytoskeletal protein associated with cell-cell and cell-matrix junctions, where it is thought to function as one of several interacting proteins involved in anchoring F-actin to the membrane. Defects in VCL are the cause of cardiomyopathy dilated type 1 W. This protein is involved in cell motility, proliferation, and differentiation and also in smooth muscle contraction, inflammation, and immune surveillance. VCL is located on a locus of chromosome 10 strongly associated with late onset AD [53.39]. HSPB1 is a heat shock protein induced by environmental stress and developmental changes. The encoded protein is involved in stress resistance and actin organization and translocates from the cytoplasm to the nucleus upon stress induction. This translocation occurs in order to modulate SP1-dependent transcriptional activity to promote neuronal protection [53.40]. Furthermore, defects in this gene cause two neuropathic diseases (i.e., Charcot–Marie–Tooth disease type 2F and distal hereditary motor neuropathy).

Besides the common genes, early stage PD (see Table 53.5) is characterized by several meaningful genes. HSPB1, IFI44L, MAFF, DEFA1/DEFA3, and OXR1 belong to pathways related to *response to stress*

Table 53.7 AD: most important pathways ranked by normalized Ipsen–Mikhailov distance $\hat{\varepsilon}$. Common pathways between early and late stage AD are shown in *bold*

AD early			AD late		
Pathway ID	$\hat{\varepsilon}$	# Genes	Pathway ID	$\hat{\varepsilon}$	# Genes
GO:0042598	0.21	16	GO:0040012	0.36	9
GO:0019787	0.16	116	**GO:0042598**	0.23	16
GO:0007417	0.10	199	**GO:0019226**	0.12	27
GO:0001508	0.14	31	GO:0030334	0.10	93
GO:0051246	0.15	121	GO:0045892	0.09	218
GO:0016874	0.12	735	GO:0042493	0.06	160
GO:0004842	0.11	368	GO:0042127	0.05	140
GO:0005768	0.08	490	GO:0008283	0.04	785
GO:0016567	0.07	206	GO:0005215	0.03	685
GO:0050877	0.06	31	GO:0008217	0.03	106
GO:0042552	0.05	36	GO:0007601	0.03	402
GO:0008015	0.04	103	GO:0007268	0.03	377
GO:0042391	0.04	67	GO:0007610	0.03	84
GO:0007399	0.04	806	GO:0008289	0.03	285
GO:0046982	0.03	364	**GO:0008015**	0.02	103
GO:0006633	0.02	109	GO:0016564	0.02	380
GO:0019226	0.00	27	GO:0008284	0.02	507
GO:0000267	0.00	4[a]	GO:0008285	0.02	578
			GO:0020037	0.02	265
			GO:0000267	0.00	5[a]
			GO:0050890	0.00	31

[a]Note: the number of genes reported here for each pathway is computed after filtering features with zero variance on one class: it could be different between AD early and AD late since they are two different datasets (e.g., GO:0000267)

and *to virus*. HLA-DQB1, HBB, HBA1/HBA2, and DEFA1/DEFA3 are related to *heme binding*, *iron ion binding*, and *immune response*. In particular, HBB encodes for hemoglobin beta that, together with another hemoglobin beta and two hemoglobin alpha, forms the adult hemoglobin. The work of [53.41] shows that the binding of amyloid-beta (Abeta) to the heme group (hemoglobins bond to iron) supports a unifying mechanism by which excessive Abeta induces heme deficiency, causes oxidative damage to macromolecules, and depletes specific neurotransmitters. Although Abeta is a known marker for AD, [53.42] also places it within a panel of PD biomarkers. DEFA1 and DEFA3 are both defensins, a family of microbicidal and cytotoxic peptides thought to be involved in host defense. They are abundant in the granules of neutrophils and also found in the epithelia of mucosal surfaces such as those of the intestine, respiratory tract, urinary tract, and vagina. Reference [53.43] presented some evidence for the recruitment of defensins in communication between the immune and nervous systems in the frog.

The late PD signature (Table 53.6) comprises some genes (RGS4, CDC42, RABGAPIL) that occur in pathways that are related to GTP, a purine nucleotide that can function either as source of energy for protein synthesis and in the signal transduction particularly with G-proteins. CDC42, a GTPase of the Rho subfamily, which regulates signaling pathways controlling diverse cellular functions including cell morphology, migration, endocytosis, and cell cycle progression. By interacting with other proteins, CDC42 is known to regulate the actin polymerization constituent both of the cytoskeleton and of the muscle cells. Specific to the brain is SLC6A3, a dopamine transporter which is a member of the sodium- and chloride-dependent neurotransmitter transporter family. This gene is associated with Parkinsonism-dystonia infantile [53.44]. Other significant genes in the list (MYH10, RGS4, PHACTR1, SYT1, VCL) are related to the actin and calmodulin binding, to the synaptic transmission, the neurotransmitter transport, the cell-cell signaling, the translation, and the cellular component movement.

53.5.3 Alzheimer's Disease Experiment

Classification and feature selection via $\ell_1\ell_2$, performed within a ninefold nested CV schema for AD early and eightfold for AD late, give, respectively, 90% accuracy and 95% with 50 probe sets for both cases. For AD, the same network analysis strategy as in the PD experiment was applied, inferring for both cases and controls 51 selected pathways for early stage AD and 34 for late stage AD. In Table 53.7 we summarize the main findings. In Table 53.8 we report the names of the GO terms corresponding to the selected pathways. Similarly to the PD analysis, we attempt a comparative analysis of the outcome for early and late stage AD having characterized the functional alteration of pathways for the two AD stages and comment the most meaningful results from the biological viewpoint.

Four common pathways were identified: GO:0019226, GO:0008015, GO:0000267, and GO:0042598.

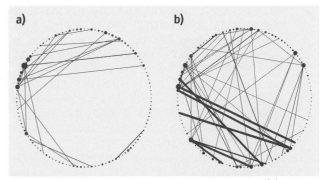

Fig. 53.5a,b Networks of the pathway GO:0019787 (**a**) for AD early development patients (**b**) compared with healthy subjects. Node diameter is proportional to the degree. Edge width is proportional to connection strength (estimated correlation)

The majority of pathways characterizing early stages of AD are related to the nervous system (GO:0007399, GO:0007417, GO:0042391, GO:0042552, GO:00508-

Table 53.8 AD: GO terms. The table reports names for all the 44 enriched GO pathway IDs reported in this Sect. 53.5.3

	AD early		AD late
Pathway ID	**GO term**	**Pathway ID**	**GO term**
GO:0042598	Vesicular fraction	GO:0040012	Regulation of locomotion
GO:0019787	Small conjugating	GO:0042598	Vesicular fraction
	Protein ligase activity	GO:0019226	Transmission of nerve impulse
GO:0007417	Central nervous system	GO:0030334	Regulation of cell migration
	Development	GO:0045892	Negative regulation of
GO:0001508	Regulation of action potential		Transcription, DNA-dependent
GO:0051246	Regulation of	GO:0042493	Response to drug
	Protein metabolic process	GO:0042127	Regulation of cell proliferation
GO:0016874	Ligase activity	GO:0008283	Cell proliferation
GO:0004842	Ubiquitin-protein ligase activity	GO:0005215	Transporter activity
GO:0005768	Endosome	GO:0008217	Regulation of blood pressure
GO:0016567	Protein ubiquitination	GO:0007601	Visual perception
GO:0050877	Neurological system process	GO:0007268	Synaptic transmission
GO:0042552	myelination	GO:0007610	Behavior
GO:0008015	Blood circulation	GO:0008289	Lipid binding
GO:0042391	Regulation of membrane potential	GO:0008015	Blood circulation
GO:0007399	Nervous system development	GO:0016564	transcription repressor activity
GO:0046982	Protein heterodimerization activity	GO:0008284	Positive regulation of
GO:0006633	Fatty acid biosynthetic process		Cell proliferation
GO:0019226	Transmission of nerve impulse	GO:0008285	Negative regulation of
GO:0000267	Cell fraction		Cell proliferation
GO:0050880	Regulation of blood vessel size	GO:0020037	Heme binding
GO:0045121	Membrane raft	GO:0000267	Cell fraction
GO:0008366	Axon ensheathment	GO:0050890	Cognition
GO:0019838	Growth factor binding	GO:0015630	Microtubule cytoskeleton

77, GO:0001508, and GO:0019226), and the blood (GO:0008015, GO:0050880). Most of the pathways characterizing late stage AD are related to the cell (GO:0008283, GO:0008284, GO:0008285, GO:0042127, GO:0030334) and to the nervous system (GO:0007268, GO:0007610, GO:0050890). Other relevant nodes are those related to the transcription regulation (GO:0016564, GO:0045892), the visual perception (GO:0007601), and the heme and lipid binding (i.e., GO:0020037 and GO:0008289).

Figure 53.5 displays the GO:0019787 node in early AD cases and controls. This pathway involves a class of enzymes that bind ubiquitin, a small regulatory protein that manages protein recycling and that was found in lesions associated with AD and PD. Figure 53.6 visualizes the enriched pathways in the molecular function and biological process domains. Despite the fact that only four pathways were found to be common in early and late AD, it is easy to note that the majority of selected pathways belong to common GO classes.

Tables 53.9 and 53.10 provide details of the network analysis results on early and late stage AD, respectively. The elements of the two signatures having non-zero Δd are listed for decreasing absolute value of the differential node degree score, thus giving top positions to genes that change the interaction network between the case/control condition most.

Table 53.9 reports the genes associated to the early stages of AD. UBE2D3 is an ubiquitin, targeting abnormal or short-lived proteins for degradation. It is a member of the E2 ubiquitin-conjugating enzyme family. This enzyme functions in the ubiquitination of the tumor-suppressor protein p53. It is also involved in several signaling pathways (BMP, TGF-β, TNF-α/NF-kB, and in the immune system), in the protein processing in the endoplasmic reticulum. PTGDS is an enzyme that catalyzes the conversion of prostaglandin H2 (PGH2) to prostaglandin D2 (PGD2). It functions as a neuromodulator as well as a trophic factor in the central nervous system; it is also involved in smooth muscle contraction/relaxation and is a potent inhibitor of platelet aggregation. This gene is preferentially expressed in the brain. Quantifying the protein complex of PGD2 and TTR in the cerebrospinal fluid may be useful in the diagnosis of AD, possibly in the early stages of the disease [53.45]. EGFR is a transmembrane glycoprotein that is a member of the protein kinase superfamily. This protein is a receptor for members of the epidermal growth factor family that binds to the epidermal growth factor. Binding of the protein to a ligand induces receptor dimerization and tyro-

sine autophosphorylation and leads to cell proliferation. This gene is involved in several pathways related to signaling, some types of cancer, to cell proliferation, migration and adhesion, and to axon guidance. It is expressed in pediatric brain tumors [53.46]. NTRK2 is member of the neurotrophic tyrosine receptor kinase (NTRK) family. This kinase is a membrane-bound receptor that upon neurotrophin binding phosphorylates itself and members of the MAPK pathway. Signaling through this kinase leads to cell differentiation. Mutations in this gene have been associated with obesity and mood disorders. SNPs in this gene are associated with AD [53.47].

The genes related to late stage AD are listed in Table 53.10. Even if SNCA is a known hallmark for PD, it also known to be expressed in late onset familial AD [53.48]. Other relevant genes are: SPEN, EIF2AK1, CAT, HBD, ATXN1, and XK. The first gene is a hormone inducible transcriptional repressor. Repression of transcription by this gene product can occur through interactions with other repressors by the recruitment of proteins involved in histone deacetylation or through sequestration of transcriptional activators. SPEN is involved in the Notch signaling pathway that is important for cell-cell communication since it involves gene regulation mechanisms that control multiple cell differentiation processes (*i.e., neuronal function and development, stabilization of arterial endothelial fate and angiogenesis, cardiac valve homeostasis*) during embryonic and adult life. EIF2AK1 acts at the level of translation initiation to downregulate protein synthesis in response to stress; therefore it seems to have a protective role diminishing the overproduction of proteins such as SNCA or beta amyloid. CAT encodes catalase, a key antioxidant enzyme in the body's defense against oxidative stress, therefore it acts against the oxidative stress present in the brain of AD patients. This gene, together with EIF2AK1, seems to fight against the disease. Like HBD (see Sect. 53.5.2), could display the same role [53.41]. ATXN1 is involved in the autosomal dominant cerebellar ataxias (ADCA), a heterogeneous group of neurodegenerative disorders characterized by progressive degeneration of the cerebellum brain stem and spinal cord. Therefore, because of the specific characteristics of these diseases (like the brain areas affected and the characteristics of the movement disorders), it might also play a role in AD. Finally, mutations of XK have been associated with McLeod's syndrome, an X-linked recessive disorder characterized by abnormalities in the neuromuscular and hematopoietic systems.

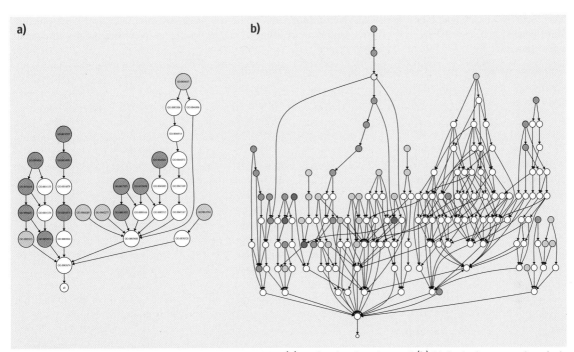

Fig. 53.6a,b GO subgraphs for early and late stage AD ((**a**) molecular function and (**b**) biological process domains). Selected nodes are represented in *light gray*, *gray*, and *dark gray* for late, early, and common nodes

Table 53.9 AD experiment (early): list of Affymetrix probe sets in the early stage signature with their corresponding Entrez gene symbol and GO pathway. The list is ranked according to the decreasing absolute value of the differential node degree Δd

Affy probe set ID	Gene symbol	Pathway ID	Δd
209116_x_at	HBB	GO:0050880	1.670
209116_x_at	HBB	GO:0008217	1.445
211748_x_at	PTGDS	GO:0006633	1.273
240383_at	UBE2D3	GO:0016874	−1.165
240383_at	UBE2D3	GO:0019787	−0.703
201061_s_at	STOM	GO:0045121	−0.662
240383_at	UBE2D3	GO:0051246	−0.613
201983_s_at	EGFR	GO:0046982	−0.476
221795_at	NTRK2	GO:0007399	−0.262
212226_s_at	PPAP2B	GO:0001568	0.259
201983_s_at	EGFR	GO:0005768	−0.256
211696_x_at	HBB	GO:0050880	−0.224
209072_at	MBP	GO:0008366	0.166
211696_x_at	HBB	GO:0008217	−0.149
212187_x_at	PTGDS	GO:0006633	−0.139
201185_at	HTRA1	GO:0019838	0.124
240383_at	UBE2D3	GO:0004842	0.120
209072_at	MBP	GO:0007417	0.113
240383_at	UBE2D3	GO:0016567	−0.047

Table 53.10 AD experiment (late): list of Affymetrix probe sets in the late stage signature with their corresponding Entrez gene symbol and GO pathway. The list is ranked according to the decreasing absolute value of the differential node degree Δd

Affy probe set ID	Gene symbol	Pathway ID	Δd
201996_s_at	SPEN	GO:0016564	1.590
211546_x_at	SNCA	GO:0040012	1.410
211546_x_at	SNCA	GO:0042493	1.310
201996_s_at	SPEN	GO:0045892	1.246
217736_s_at	EIF2AK1	GO:0020037	−1.066
201005_at	CD9	GO:0008285	0.725
210943_s_at	LYST	GO:0015630	0.706
204466_s_at	SNCA	GO:0042493	0.461
207827_x_at	SNCA	GO:0040012	0.434
206698_at	XK	GO:0005215	0.433
209184_s_at	IRS2	GO:0008283	0.208
212420_at	ELF1	GO:0016564	−0.203
207827_x_at	SNCA	GO:0042493	0.201
205592_at	SLCA4A1	GO:0005215	0.180
211922_s_at	CAT	GO:0008283	0.173
211922_s_at	CAT	GO:0020037	−0.094
203231_s_at	ATXN1	GO:0016564	−0.073
217736_s_at	EIF2AK1	GO:0008285	−0.072
204466_s_at	SNCA	GO:0040012	0.048
206834_at	HBD	GO:0008217	0.045
206834_at	HBD	GO:0020037	0.019

53.6 Conclusion

The theory of complex networks has recently proven to be a helpful tool to gain systematic and structural knowledge of cell mechanisms. To enhance the capabilities of the algorithms for the analysis of complex networks, we propose to couple them with a machine learning driven approach aimed at moving from global to local interaction scales. In other words, we focus on pathways that are most likely to change, for instance within particular pathological stages. Such a strategy is also better tailored to deal with situations where small sample size may affect the reliability of the network inference on a global scale. The method, demonstrated on three disease datasets of environmental pollution, PD and AD, was able to detect biologically meaningful differential pathways.

References

53.1   A.L. Barabasi, N. Gulbahce, J. Loscalzo: Network medicine: A network-based approach to human disease, Nat. Rev. Genet. **12**, 56–68 (2011)

53.2   S. Strogatz: Exploring complex networks, Nature **410**, 268–276 (2001)

53.3   M. Newman: The structure and function of complex networks, SIAM Review **45**, 167–256 (2003)

53.4   S. Boccaletti, V. Latora, Y. Moreno, M. Chavez, D.U. Hwang: Complex networks: Structure and dynamics, Phys. Rep. **424**(4–5), 175–308 (2006)

53.5   M. Newman: *Networks: An Introduction* (Oxford Univ. Press, Oxford 2010)

53.6   M. Buchanan, G. Caldarelli, P. De Los Rios, F. Rao, M. Vendruscolo (ed.): *Networks in Cell Biology* (Cambridge Univ. Press, Cambridge 2010)

53.7   F. He, R. Balling, A.P. Zeng: Reverse engineering and verification of gene networks: Principles, assumptions, and limitations of present methods and future perspectives, J. Biotechnol. **144**(3), 190–203 (2009)

53.8 A. Baralla, W. Mentzen, A. de la Fuente: Inferring gene networks: Dream or nightmare?, Ann. N.Y. Acad. Sci. **1158**, 246–256 (2009)

53.9 D. Marbach, R. Prill, T. Schaffter, C. Mattiussi, D. Floreano, G. Stolovitzky: Revealing strengths and weaknesses of methods for gene network inference, PNAS **107**(14), 6286–6291 (2010)

53.10 R. De Smet, K. Marchal: Advantages and limitations of current network inference methods, Nat. Rev. Microbiol. **8**, 717–729 (2010)

53.11 The MicroArray Quality Control Consortium, The MAQC–II Project: A comprehensive study of common practices for the development and validation of microarray-based predictive models, Nat. Biotechnol. **28**(8), 827–838 (2010)

53.12 B. Zhang, S. Horvath: A general framework for weighted gene co-expression network analysis, Stat. Appl. Genet. Mol. Biol. **4**(1), 17 (2005)

53.13 A. Subramanian, P. Tamayo, V.K. Mootha, S. Mukherjee, B.L. Ebert, M.A. Gillette, A. Paulovich, S.L. Pomeroy, T.R. Golub, E.S. Lander, J.P. Mesirov: Gene set enrichment analysis: A knowledge-based approach for interpreting genome-wide expression profiles, PNAS **102**(43), 15545–15550 (2005)

53.14 M. Ashburner, C.A. Ball, J.A. Blake, D. Botstein, H. Butler, J.M. Cherry, A.P. Davis, K. Dolinski, S.S. Dwight, J.T. Eppig, M.A. Harris, D.P. Hill, L. Issel-Tarver, A. Kasarskis, S. Lewis, J.C. Matese, J.E. Richardson, M. Ringwald, G.M. Rubin, G. Sherlock: Gene ontology: Tool for the unification of biology, The gene ontology consortium, Nat. Genet. **25**(1), 25–29 (2000)

53.15 G. Jurman, R. Visintainer, C. Furlanello: An introduction to spectral distances in networks, Proc. WIRN 2010 (2011) pp. 227–234

53.16 M. Ipsen, A. Mikhailov: Evolutionary reconstruction of networks, Phys. Rev. E **66**(4), 046109 (2002)

53.17 D. Cai, X. He, J. Han, SRDA: An efficient algorithm for large-scale discriminant analysis, IEEE Trans. Knowl. Data Eng. **20**, 1–12 (2008)

53.18 C. De Mol, S. Mosci, M. Traskine, A. Verri: A regularized method for selecting nested groups of relevant genes from microarray data, J. Comput. Biol. **16**, 1–15 (2009), . DOI 10.1089/cmb.2008.0171

53.19 B. Zhang, S. Kirov, J. Snoddy: WebGestalt: An integrated system for exploring gene sets in various biological contexts, Nucleic Acid Res. **33**, W741–W748 (2005)

53.20 W. Zhao, P. Langfelder, T. Fuller, J. Dong, A. Li, S. Horvath: Weighted gene coexpression network analysis: State of the art, J. Biopharm. Stat. **20**(2), 281–300 (2010)

53.21 P. Meyer, F. Lafitte, G. Bontempi: Minet: A R/Bioconductor package for inferring large transcriptional networks using mutual information, BMC Bioinformatics **9**(1), 461 (2008)

53.22 Mlpy website: http://mlpy.fbk.eu/

53.23 H. Zou, T. Hastie: Regularization and variable selection via the elastic net, J. R. Stat. Soc. B **67**(2), 301–320 (2005)

53.24 P. Fardin, A. Barla, S. Mosci, L. Rosasco, A. Verri, L. Varesio: The l1-l2 regularization framework unmasks the hypoxia signature hidden in the transcriptome of a set of heterogeneous neuroblastoma cell lines, BMC Genomics **10**, 474 (2009), DOI 10.1186/1471-2164-10-474

53.25 l1l2 website: http://slipguru.disi.unige.it/Software/L1L2Py

53.26 WebGestalt website: http://bioinfo.vanderbilt.edu/webgestalt/

53.27 M. Kanehisa, S. Goto, KEGG: kyoto encyclopedia of genes and genomes, Nucleic Acids Res. **28**(1), 27–30 (2000)

53.28 A. Margolin, I. Nemenman, K. Basso, C. Wiggins, G. Stolovitzky, R. Dalla-Favera, A. Califano: Aracne: An algorithm for the reconstuction of gene regulatory networks in a mammalian cellular context, BMC Bioinformatics **7**(7), S7 (2006)

53.29 I. Nemenman, G. Escola, W. Hlavacek, P. Unkefer, C. Unkefer, M. Wall: Reconstruction of metabolic networks from high-throughput metabolite profiling data, Ann. N.Y. Acad. Sci. **1115**, 102–115 (2007)

53.30 T. Cover, J. Thomas: *Elements of Information Theory* (Wiley, Hoboken 1991)

53.31 R. Sharan, T. Ideker: Modeling cellular machinery through biological network comparison, Nat. Biotechnol. **24**(4), 427–433 (2006)

53.32 D. van Leeuwen, M. van Herwijnen, M. Pedersen, L. Knudsen, M. Kirsch-Volders, R. Sram, Y. Staal, E. Bajak, J. van Delft, J. Kleinjans: Genome-wide differential gene expression in children exposed to air pollution in the Czech Republic, Mutat. Res. **600**(1–2), 12–22 (2006)

53.33 D. van Leeuwen, M. Pedersen, P. Hendriksen, A. Boorsma, M. van Herwijnen, R. Gottschalk, M. Kirsch-Volders, L. Knudsen, R. Sram, E. Bajak, J. van Delft, J. Kleinjans: Genomic analysis suggests higher susceptibility of children to air pollution, Carcinogenesis **29**(5), 977–983 (2008)

53.34 C.R. Scherzer, A.C. Eklund, L.J. Morse, Z. Liao, J.J. Locascio, D. Fefer, M.A. Schwarzschild, M.G. Schlossmacher, M.A. Hauser, J.M. Vance, L.R. Sudarsky, D.G. Standaert, J.H. Growdon, R.V. Jensen, S.R. Gullans: Molecular markers of early Parkinson's disease based on gene expression in blood, PNAS **104**(3), 955–960 (2007)

53.35 Y. Zhang, M. James, F. Middleton, R. Davis: Transcriptional analysis of multiple brain regions in Parkinson's disease supports the involvement of specific protein processing, energy metabolism and signaling pathways and suggests novel disease mechanisms, Am. J. Med. Genet. B **137B**, 5–16 (2005)

53.36 W. Liang, T. Dunckley, T. Beach, A. Grover, D. Mastroeni, K. Ramsey, R. Caselli, W. Kukull, D. Mckeel,

J. Morris, C. Hulette, D. Schmechel, E. Reiman, J. Rogers, D. Stephan: Neuronal gene expression in non-demented individuals with intermediate Alzheimer's disease neuropathology, Neurobiol. Aging **31**, 1–16 (2010)

53.37   W. Liang, E. Reiman, J. Valla, T. Dunckley, T. Beach, A. Grover, T. Niedzielko, L. Schneider, D. Mastroeni, R. Caselli, W. Kukull, J. Morris, C. Hulette, D. Schmechel, J. Rogers, D. Stephan: Alzheimer's disease is associated with reduced expression of energy metabolism genes in posterior cingulate neurons, PNAS **105**, 4441–4446 (2008)

53.38   K.Y. Kim, M. Kovács, S. Kawamoto, J.R. Sellers, R.S. Adelstein: Disease-associated mutations and alternative splicing alter the enzymatic and motile activity of nonmuscle myosins ii-b and ii-c, J. Biol. Chem. **280**(24), 22769–22775 (2005)

53.39   A. Grupe, Y. Li, C. Rowland, P. Nowotny, A.L. Hinrichs, S. Smemo, J.S.K. Kauwe, T.J. Maxwell, S. Cherny, L. Doil, K. Tacey, R. van Luchene, A. Myers, F.W.D. Vrièze, M. Kaleem, P. Hollingworth, L. Jehu, C. Foy, N. Archer, G. Hamilton, P. Holmans, C.M. Morris, J. Catanese, J. Sninsky, T.J. White, J. Powell, J. Hardy, M. O'Donovan, S. Lovestone, L. Jones, J.C. Morris, L. Thal, M. Owen, J. Williams, A. Goate: A scan of chromosome 10 identifies a novel locus showing strong association with late-onset Alzheimer disease, Am. J. Hum. Genet. **78**(1), 78–88 (2006), DOI 10.1086/498851

53.40   M.J. Friedman, S. Li, X.J. Li: Activation of gene transcription by heat shock protein 27 may contribute to its neuronal protection, J. Biol. Chem. **284**(41), 944–951 (2009)

53.41   H. Atamna, K. Boyle: Amyloid-beta peptide binds with heme to form a peroxidase: Relationship to the cytopathologies of Alzheimer's disease, PNAS **103**(9), 3381–3386 (2006)

53.42   M. Shi, J. Bradner, A.M. Hancock, K.A. Chung, J.F. Quinn, E.R. Peskind, D. Galasko, J. Jankovic, C.P. Zabetian, H.M. Kim, J.B. Leverenz, T.J. Montine, C. Ginghina, U.J. Kang, K.C. Cain, Y. Wang, J. Aasly, D. Goldstein, J. Zhang: Cerebrospinal fluid biomarkers for Parkinson disease diagnosis and progression, Ann. Neurol. **69**(3), 570–580 (2011)

53.43   G.N. Andrianov, A.D. Nozdrachev, I.V. Ryzhova: The role of defensins in the excitability of the peripheral vestibular system in the frog: Evidence for the presence of communication between the immune and nervous systems, Hear. Res. **230**(1–2), 1–8 (2007)

53.44   M.A. Kurian, J. Zhen, S.Y. Cheng, Y. Li, S.R. Mordekar, P. Jardine, N.V. Morgan, E. Meyer, L. Tee, S. Pasha, E. Wassmer, S.J.R. Heales, P. Gissen, M.E.A. Reith, E.R. Maher: Homozygous loss-of-function mutations in the gene encoding the dopamine transporter are associated with infantile Parkinsonism-dystonia, J. Clin. Invest. **119**(6), 1595–1603 (2009)

53.45   M.A. Lovell, B.C. Lynn, S. Xiong, J.F. Quinn, J. Kaye, W.R. Markesbery: An aberrant protein complex in csf as a biomarker of Alzheimer disease, Neurology **70**(23), 2212–2218 (2008)

53.46   A. Patereli, G.A. Alexiou, K. Stefanaki, M. Moschovi, I. Doussis-Anagnostopoulou, N. Prodromou, O. Karentzou: Expression of epidermal growth factor receptor and her-2 in pediatric embryonal brain tumors, Pediatr. Neurosurg. **46**(3), 188–192 (2010)

53.47   A. Cozza, E. Melissari, P. Iacopetti, V. Mariotti, A. Tedde, B. Nacmias, A. Conte, S. Sorbi, S. Pellegrini: SNPs in neurotrophin system genes and Alzheimer's disease in an Italian population, J. Alzheimers Dis. **15**(1), 61–70 (2008)

53.48   D.W. Tsuang, R.G. Rieske, K.M. Purganan, A.C. David, T.J. Montine, G.D. Schellenberg, E.J. Steinbart, E.C. Petrie, T.D. Bird, J.B. Leverenz: Lewy body pathology in late-onset familial Alzheimer's disease: A clinicopathological case series, J. Alzheimers Dis. **9**(3), 235–242 (2006)

54. Computational Neurogenetic Modeling: Gene–Dependent Dynamics of Cortex and Idiopathic Epilepsy

Lubica Benuskova, Nikola Kasabov

The chapter describes a novel computational approach to modeling the cortex dynamics that integrates gene–protein regulatory networks with a neural network model. Interaction of genes and proteins in neurons affects the dynamics of the whole neural network. We have adopted an exploratory approach of investigating many randomly generated gene regulatory matrices out of which we kept those that generated interesting dynamics. This naïve brute force approach served us to explore the potential application of computational neurogenetic models in relation to gene knock-out experiments. The knock out of a hypothetical gene for fast inhibition in our artificial genome has led to an interesting neural activity. In spite of the fact that the artificial gene/protein network has been altered due to one gene knock out, the dynamics of SNN in terms of spiking activity was most of the time very similar to the result obtained with the complete gene/protein network. However, from time to time the neurons spontaneously temporarily synchronized their spiking into coherent global oscillations. In our model, the fluctuations in the values of neuronal parameters leads to spontaneous development of seizure-like global synchronizations. These very same fluctuations also lead to termination of the seizure-like neural activity and maintenance of the inter-ictal normal periods of activity. Based on our model, we would like to suggest a hypothesis that parameter changes due to the gene–protein dynamics should also be included as a serious factor determining transitions in neural dynamics, especially when the cause of disease is known to be genetic.

Part K | 54

54.1 Overview

Properties of all cell types, including neurons, are determined by proteins they contain [54.1]. In turn, the types and levels of proteins are determined by differential transcription of genes in response to internal and external signals. Eventually, the properties of neurons determine the structure and dynamics of the whole neural network they are part of. In this chapter, we turn our attention to modeling the influence of genes and proteins upon the dynamics of mature cortex and the dynamic effects due to mutations of genes. The chapter is thus an extension and further development of a general framework for modeling brain functions and genetically caused dysfunctions by means of computational neurogenetic modeling [54.2].

54.1.1 Neurogenesis

Sophisticated mathematical and computer models of the general control of gene expression during early embryonic development of neural system in vertebrates have been developed [54.3]. Computational models based on the analogy of gene regulatory networks with artificial neural networks have been applied to model the steps in *Drosophila* early neurogenesis [54.4, 5]. The latter models attempted to elucidate how genes orchestrate the detailed pattern of early neural development. During early development, dynamics of changes in architecture and morphology of neural network parallels changes in gene expression in time.

54.1.2 Circadian Rhythms

One particular instance where the gene expression determines the neural dynamics is the circadian rhythm. A circadian rhythm is a roughly-24 h cycle in the physiological processes of plants and animals. The circadian rhythm partly depends on external cues such as sunlight and temperature, but otherwise it is determined by periodic expression patterns of the so-called clock genes [54.6, 7]. *Smolen* et al. [54.8] have developed a computational model to represent the regulation of core clock component genes in *Drosophila* (*per*, *vri*, *Pdp-1*, and *Clk*). To model the dynamics of gene expression, differential equations and first-order kinetics equations were employed for modeling the control of genes and their products. The model illustrates the ways in which negative and positive feedback loops within the gene regulatory network (GRN) cooperate to generate oscillations of gene expression. The relative amplitudes and phases of simulated oscillations of gene expressions resemble empirical data in most simulated situations. The model of *Smolen* et al. [54.8] shows that it is possible to develop detailed models of gene control of neural behavior provided that enough experimental data is available to adjust the model.

54.1.3 Neurodegenerative Diseases

Many diseases that affect the central nervous system and manifest cognitive symptoms have an underlying genetic cause – some are due to a mutation in a single gene, others are proving to have a more complex mode of inheritance [54.9]. A large group of disorders are neurodegenerative disorders (like Alzheimer's disease, Rett syndrome, Huntington disease, etc.), in which the underlying gene mutations lead to particular degenerative processes in the brain that progressively affect neural functions and eventually have fatal consequences. In the case of neurodegenerative diseases, the dynamics of neural networks degeneration is slow, usually lasting for years. The gene mutation is always there and particular degenerative changes in the brain tissue accumulate over time. It is a challenge for future computational neurogenetic-genetic models to attempt to model the onset and progression of these diseases using the integration of gene regulatory networks and artificial neural networks. Already for Alzheimer's disease, computational models of neural networks' dysfunction caused by experimentally identified biochemical factors that result from genetic abnormalities have been developed to gain insights into the neural symptoms of the disease [54.10–12]. All these neurodegenerative diseases (like Alzheimer's disease, Rett syndrome, Huntington disease, etc.) are characterized by the fact that once the gene mutation manifests itself, the disease progresses and its symptoms of cognitive impairment are always there.

54.1.4 Idiopathic Epilepsies

However, there are genetic diseases of the brain like, for instance, some genetically caused epilepsies, in which the main symptom – seizure – occurs only from time to time and between these episodes, in fact most of the time, the brain activity appears normal. In general, epilepsy is a disorder characterized by the occurrence of at least two unprovoked seizures [54.13]. Seizures are the manifestation of abnormal hypersynchronous discharges of neurons in the cerebral cortex. The clinical signs or symptoms of seizures depend on the location and extent of the propagation of the discharging cortical neurons. The prevalence of active epilepsy is about 1% [54.13], which, however, means about 70 million people worldwide are affected. Seizures are often a common manifestation of neurologic injury and disease, which should not be surprising because the main function of neurons is the transmission of electrical impulses. Thus, most epilepsies are not caused genetically. A particular case of nongenetic epilepsy is temporal lobe epilepsy (TLE). The most common pathological finding in TLE is hippocampal (temporal lobe) sclerosis that involves cell loss [54.14]. High-resolution MRI shows hippocampal atrophy in 87% of patients with TLE. Causes of the cell loss involve: past infections, trauma, and vascular and brain tissue malformations including tumors. Experimentally, the specific cell loss can be triggered by the injection of kainic acid [54.15]. Inhibitory neurons are more affected than excitatory neurons. Two broad classes of GABAergic inhibitory neurons can be distinguished: (1) dendritic-projecting interneurons mediating the slow inhibition through the $GABA_B$ and $GABA_A$ dendritic postsynaptic receptors [54.16, 17], and (2) interneurons that selectively innervate somas of other neurons and mediate the fast inhibition through $GABA_A$ type of postsynaptic receptors [54.15, 18]). In TLE, a selective reduction of the slow (dendritic) inhibition and an increase in the fast (somatic) inhibition have been identified [54.15]. There are many computational models of temporal lobe-like epileptic seizures based on the effect of abnormal values of inhibition parameters upon neural network dynamics [54.19–22].

There is quite a high percentage of pharmacoresistant epilepsies, i.e., 15–50% depending on age and definition [54.23]. In humans with intractable temporal lobe epilepsy (TLE), many of surviving inhibitory interneurons lose their PV content or PV immunoreactivity [54.24]. It has been proposed that efficient Ca^{2+} buffering by PV and its high concentration in PV-expressing inhibitory cells is a prerequisite for the proficient inhibition of cortical networks [54.25]. To investigate this hypothesis, Schwaller and co-workers used mice lacking PV ($PV^{-/-}$). These mice show no obvious abnormalities and do not have epilepsy [54.26]. However, the severity of generalized tonic-clonic seizures induced by pentylenetetrazol (PTZ) was significantly greater in $PV^{-/-}$ than in $PV^{+/+}$ animals. Extracellular single-unit activity recorded from over 1000 neurons *in vivo* in the temporal cortex revealed an increase of units firing regularly and a decrease of cells firing in bursts. In addition, control animals showed a lesser degree of synchronicity and mainly high frequency components above 65 Hz in the local field potential (LFP) spectrum compared to $PV^{-/-}$ mice. On the other hand, $PV^{-/-}$ mice were characterized by increased synchronicity and by abnormally high proportion of frequencies below 40 Hz [54.27]. In the hippocampus, PV deficiency facilitated the $GABA_A$ergic current reversal induced by

high-frequency stimulation, a mechanism implied in the generation of epileptic activity [54.28]. Through an increase in inhibition, the absence of PV facilitates hypersynchrony through the depolarizing action of GABA [54.26]. Thus there is a permanent change in the spectrum of the local field potential (LFP) of the PV gene KO mice. We developed a hierarchical spiking neural network model of LFP, in which each neuron has the values of parameters governed by an internal gene regulatory network [54.29]. For simplicity and because the measurements of LFP lasted only minutes, we assumed the gene expressions were constant. In spite of that, the removal of the gene for PV from the gene regulatory network affected all the parameters of neurons. We have evolved the gene interactions so that the resulting gene interaction matrix yielded such values of neuronal parameters that the resulting model LFP had similar changes of the frequency spectrum as in the experiment with $PV^{-/-}$ mice of *Villa* et al. [54.27].

Genetic contribution to etiology has been estimated to be present in about 40% of patients with epilepsy [54.30]. Pure Mendelian epilepsies, in which a single major locus can account for segregation of the disease trait are considered to be rare and probably account for no more than 1% of patients [54.30]. The common familial epilepsies tend to display complex inheritance, in which the pattern of familial clustering can be accounted for by the interaction of several loci together with environmental factors. Table 54.1 lists some types of genetically caused epilepsies, associated brain pathologies, symptoms, and putative mutated genes. This account is by far not complete.

54.1.5 Computational Models of Epilepsies

Let us consider in more detail childhood absence epilepsy (CAE). It is an idiopathic (i.e., arising from an unknown cause), generalized nonconvulsive epilepsy [54.31, 32]. The main feature is absence seizures. A typical absence is a nonconvulsive epileptic seizure, characterized by a briefly (4–20 s) lasting impairment of consciousness. This may happen up to about eight times each hour, and up to ≈ 200 times a day. Absence seizures occur spontaneously, i.e., they are not evoked by sensory or other stimuli. Absence is accompanied by a generalized, synchronous, bilateral, 2.5–4 Hz spike and slow-wave discharge (SWD) with large amplitudes in the electroencephalogram (EEG), see Fig. 54.1.

SWDs can start anywhere in the cortex and from there they quickly spread to the entire cortex and thalamus [54.33]. The origin can also be in the thala-

Table 54.1 Some idiopathic epilepsies, putative mutated genes, and affected functions of brain neurons in humans

Epilepsy	Mutated genes/chromosome location (if known)	Brain abnormality	Symptoms	References
Autosomal dominant nocturnal frontal lobe epilepsy (ADNFL)	α_4 subunit of the nicotinic AchR (CHRNA4)/20q β_2 subunit of the same receptor (CHRNB2)/1p	Reduced nAchR channel opening time and reduced conductance leading to hyperexcitability	Partial seizures during night that may generalize, arising from the frontal lobes, motor, tonic, postural type	[54.36–38]
Benign familial neonatal convulsions (BFNC1 and BFNC2)	EBN1 (K^+ channel gene KCQ2)/20q EBN2 (K^+ channel gene KCNQ3)/8q	Alteration of the gating properties of the K^+ channel leading to poor control of repetitive firing	Generalized epilepsy of newborns, seizures are frequent and brief, episodes resolve within a few days	[54.36–39]
Childhood absence epilepsy (CAE)	γ_2 subunit gene for the GABA$_A$ receptor gene GABRG2/5q gene CLCN2/3q	Fast and part of slow GABAergic inhibition is reduced, voltage-gated Cl^- channel function is impaired	Absence seizures (consciousness impaired) up to 200 times a day, bilateral 2–4 Hz spike and slow-wave EEG	[54.31, 32]
Generalized epilepsy and febrile seizures plus (GEFS+)	β_1 subunit of the Na^+ channel gene SCN1B/19q α_1 and α_2 subunits, gene SCN1A and gene SCN2A/2q GABRG2/5q	Normal inactivation kinetics of the Na^+ channel is reduced causing persistent Na^+ influx and hyperexcitability, reduced function of the GABA$_A$R	Childhood onset of febrile seizures, with febrile and afebrile generalized seizures continuing beyond 6 years of age	[54.36–38]
Intractable childhood epilepsy	α_1 subunit of the Na^+ channel, gene SCN1A/2q	Rapid recovery of the Na^+ channel from inactivation or very slow inactivation	Frequent intractable generalized tonic-clonic seizures	[54.39]
Juvenile absence epilepsy (JAE)	α_1/5q, α_5/15q, γ_2/5q subunit genes for the GABA$_A$ receptor gene (CLCN2)/3q	Fast and part of slow GABAergic inhibition is reduced, voltage-gated Cl channel function is impaired	Similar like CAE but the seizures start after year 10, seizures may be less frequent and last longer than few seconds	[54.39]
Juvenile myoclonic epilepsy (JME)	α_7 subunit of the nicotinic AChR (CHRNA7)/15q gene CLCN2/3q, β_4 subunit of Ca^{2+} channel, (CACNB4)/19p	Reduced function of the nicotinic AChR, voltage-gated Cl channel and voltage-gated Ca^{2+} channel have reduced conductance	Myoclonic jerks or seizures shortly after awakening, generalized tonic-clonic seizures, and sometimes absence seizures	[54.36]
Dravet syndrome, severe myoclonic epilepsy of infancy (SMEI)	α_1 subunit of the Na^+ channel, gene SCN1A/2q	Complete loss of activity of the Na^+ channel	Both generalized and localized seizures, clonic and myoclonic seizure types	[54.38, 39]
Lafora disease (progressive myoclonus epilepsy)	Laforin gene EPM2A/6q24 malin gene EPM2B/6p22.3	Presence of Lafora bodies (granules of accumulated carbohydrates)	Myoclonic jerking, ataxia, mental deterioration leading to dementia	[54.40]

mus [54.34]. Several gene mutations have been reported for CAE, and an extensive review can be found, for instance, in [54.31]. Some suspected gene mutations are linked to the receptor GABA$_A$ which mediates the so-called fast somatic inhibition and a smaller part of the slow dendritic inhibition in the cortex [54.16, 17]. Blockage of GABA$_A$ receptors by chemical agents leads to SWDs even in healthy brain tissue and in the corresponding model of the thalamo-cortical circuit [54.18, 34]. If excitatory and inhibitory cells generate high-frequency discharges in synchrony and if GABA$_B$ receptors are present (their time constants match with the period of SWDs), sufficient conditions are brought together to generate SWDs. In computational models, epileptic discharges can be evoked *chemically* or by a specifically manipulated input, and these seizures last as long as this altered input or chemical is present [54.20, 35]. However, in reality seizures occur *spontaneously* during the alert state of the brain (that is in CAE). The underlying cause is always there, i.e., the gene mutation, however, the seizures are not always present. When a spiking neural network model is used to model epilepsy, it can be shown that spiking neurons in a network exhibit spontaneous synchronizations when the overall excitation is permanently increased [54.22]. Depending on the level of total excitation of neurons the model network went from the nonbursting (asynchronous) activity through single synchronized bursts each lasting few milliseconds, and complex partly synchronized clusters of bursts, each lasting few seconds, to a continuous synchronization of firing. However in this model, the *ictal* and *inter-ictal* periods are unrealistically short, i.e., both periods lasting few seconds at most. While in CAE the seizures are indeed short-lasting, the inter-seizure periods are generally longer than just few seconds.

Although the above models cast invaluable insights into the mechanisms underlying epilepsies, according to us, these models do not satisfactorily explain spontaneous transitions to and from the seizures, neither the maintenance of inter-ictal and ictal states. Noise can be used to trigger transitions to and from abnormal activity, but cannot dramatically change the short duration of periodic and aperiodic phases of activity (being on the order of milliseconds – our own simulations of spiking networks). The reason is that these models work with constant values of parameters. We believe that if the parameter values were allowed to change dynamically, more realistic temporal dynamics could be achieved. *LeMasson, Marder* and *Abbott* [54.41, 42], realized that in neurons, like in other cells, there is a continuous turnover of ion channels and receptors that underlie neuronal signaling such that the properties of neurons are dynamically modified. They investigated the role of neuronal activity upon dynamic modification and maintenance of neuronal conductances to achieve a biologically plausible explanation of transitions between different neural activity modes in the stomatogastric ganglion. In their models the activity-dependent intracellular calcium concentration was used as a feedback element that leads to processes of insertion, removal, and modification of ion channels. These processes, happening probably at a local synaptic level, can be quite fast, taking just seconds or at most minutes, because they do not involve gene expression.

54.1.6 Outline of the Chapter

So far, due to the complexity of the whole issue, no attempt to bridge the slow dynamics of gene–protein-gene interactions with the fast dynamics of neural networks

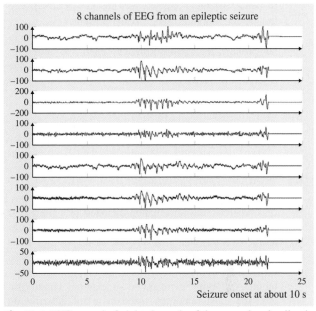

Fig. 54.1 EEG record of eight channels of the normal and epileptic slow-wave discharge (SWD) in childhood absence epilepsy. SWD have large amplitudes and frequency of 2.5–4 Hz

has been made. We think it can be done and we would like to elaborate a proposal of how to move forward in this direction. Thus, in this chapter we describe a computational neurogenetic model (CNGM) to investigate the influence of a slow gene–protein dynamics upon the fast neural dynamics. This approach is illustrated by means of a simple model of a spiking neural network, in which parameter values are linked to proteins that are the products of gene expression. Gene expressions change in time as a consequence of the dynamics of the internal gene–protein regulatory network. Thus, the values of neuronal parameters are not fixed but instead they vary in time due to the changes in expressions of genes coding for proteins that are behind the corresponding parameters of the neuronal functions (like amplitude of excitation or inhibition, resting firing threshold, etc.). In such a way, we can investigate different modes of gene interactions with normal or knock out genes, and their effects upon the neural dynamics. Based on these toy simulations we propose a hypothesis that spontaneous transitions to and from the ictal neural activity may be due to variations of neuronal parameters resulting from the underlying dynamics of gene–protein regulatory interactions. In other words, we would like to suggest a hypothesis why the seizures occur only from time to time and why between these episodes, in fact most of the time, the brain activity appears normal in spite of the genetic mutation always being present.

First we overview relevant facts from molecular biology and bioinformatics. Then we introduce the model that is a continuous extension of the discrete model of gene–protein dynamics introduced

in [54.43, 44]. Next we present some illustrative computer simulations with the new continuous model. In the discussion section, we outline directions for further development and applications of the neurogenetic approach to future modeling of the genetic brain disorders.

54.2 Gene Expression Regulation

54.2.1 Protein Synthesis

The term gene expression refers to the entire process whereby the information encoded in a particular (protein-coding) gene is decoded into a particular protein [54.1]. Regulation at any one of the various steps in this process can lead to differential gene expression in different cell types, different developmental stages of one cell type, in response to external conditions, etc. The most important mechanism for determining whether or not most genes are expressed is the control of transcription initiation [54.1].

After transcription has been initiated, RNA polymerase II, together with the necessary transcription elongation factors, travels along the DNA template and polymerizes ribonucleotides into a pre-messenger RNA (pre-mRNA) copy of the gene. The polymerase moves at a regular speed (approximately 30–50 nucleotides per second) and holds on to the DNA template efficiently, even if the gene is very long. At the end of the gene, the RNA polymerase falls off the DNA template and transcription terminates [54.45]. Each resulting pre-

mRNA consists of two types of segments – exons, that are segments translated into proteins, and introns – segments that are considered redundant and do not take part in the protein production. Removing the introns and ordering only the exon parts of the genes in the RNA sequence is called splicing and this process results in the production of a messenger RNA (mRNA) sequence. From one gene, many copies of mRNA are produced that are directly translated into proteins. Each protein consists of a sequence of amino acids, each of them coded by a base triplet of the transport RNA (tRNA), called a codon. Translation of mRNA sequence into the protein sequence (by means of tRNA) occurs on ribosomes, which are the protein synthesizing machines of cells. Elongation of a protein polypeptide proceeds at a rate of 3–5 amino acids added per second [54.45].

On average, a vertebrate gene is around 30 kb = 30 000 bases long, out of which the coding region is only about 1–2 kb = 1000–2000 bases long, that is 3–7%. However, huge deviations from the average can be observed. Thus from the initiation of transcription, it would take around 600–1000 s = 10–17 min to transcribe this average 30 kb gene. Translation of the coding part would take approximately 3–10 min. In total, the process of an average protein synthesis would last about 13–27 min. However, the process of protein preparation for its function is not over yet for the so-called proteins of the secretory pathways, like hormones, neurotransmitters, receptors, etc. [54.46]. After being synthesized on ribosomes, these proteins are transferred to endoplasmic reticulum (ER) to undergo the posttranslational modifications. After modification is completed in the ER, these proteins move via transport vesicles to the Golgi complex from where they are further sorted to several destinations, like axonal terminals, synapses, etc. The whole process is summarized in Fig. 54.2. Posttranslational modifications, sorting, and transport also take some time, probably on the order of minutes. Thus, from the initiation of transcription, it is fair to say, that it takes about 15 min

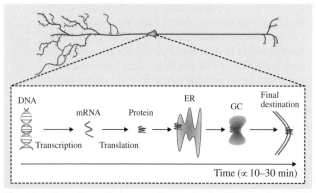

Fig. 54.2 Overview of the protein synthesis. ER = endoplasmic reticulum, GC = Golgi complex. Final destination for a protein from a secretory pathway can be, for instance, the plasmatic membrane, if it is a receptor

for an average protein of the secretory pathway to start functioning in its place, and it may take even more, depending on the size of the protein. It is also obvious that this time delay is different for different proteins. These temporal considerations will be important for the construction of the general gene–protein regulatory network.

54.2.2 Control of Gene Expression

Each gene has a promoter region. The binding of a nonspecific basal factor to promoter is a prerequisite for the basal level of the transcription of a gene. The differential level of transcription is controlled by transcription factors, activators and repressors, which bind to regulatory DNA sequences, located proximally to the promoter. Activators enhance gene expression while repressors suppress gene expression. Transcription can also be stimulated by control elements called enhancers that are more distant from the promoter region than the promoter-proximal elements. Enhancers include binding sites for several transcription factors. Thus, transcription from a single promoter may be regulated by binding of multiple transcription factors to alternative control elements, promoter-proximal elements and enhancers, permitting complex control of gene expression [54.1]. Moreover, there are molecules, which act as co-activators and co-repressors, that interact with activators and repressors to modify their activity. Whether or not a specific gene is expressed at a particular time is largely a consequence of the net effect of the activity of the number of transcription factors at that particular time.

To get from a particular protein in a cell to the control of expression of some gene takes many steps and thus some time. For instance, it is known that activation of the NMDA (*N*-methyl-D-aspartate) receptors in neurons causes de novo synthesis of BDNF (brain-derived neurotrophic factor) [54.47]. BDNF de novo synthesis was estimated by measuring the steady-state content of BDNF mRNA and protein at various times after NMDA treatment. NMDA elicited a time dependent increase in BDNF mRNA content, beginning at 3 h (2-fold) and lasting at least up to 8 h in vitro. However, a small, about 1.2-fold, increase was observed already after 1 h [54.47]. This means that it takes about an hour for the signal from the NMDA receptor to reach the genome, to initiate and carry out the transcription, and to synthesize enough proteins to be detected above the basal concentration. The NMDAR multiprotein complex contains 77 proteins out of which 19 participate in NMDAR signaling [54.48]. These signaling proteins mediate the effect by differential activation of different downstream effector pathways leading to the genome, for instance different mitogen-activated protein kinase (MAPK) pathways, depending on a particular signal [54.49]. Thus, in principle each protein in a cell has a way to influence transcription of genes through cascades of reactions leading to different transcriptional factors. This is the molecular basis for construction of different gene regulatory networks (GRNs) from gene expression data [54.50, 51].

54.3 Computational Neurogenetic Model

Here, we would like to mathematically formulate an extension of our computational neurogenetic model introduced in [54.43, 44], which was based on a discrete dynamics of gene regulatory network. Current extension takes into account different time scales and different time delays in the continuous dynamic system.

54.3.1 Gene–Protein Regulatory Network

Let us formulate a set of general equations for the gene–protein dynamic system (Fig. 54.3). As a first gross simplification, we will assume that every neuron has the same gene–protein regulatory network (GPRN) – that is, interactions between genes are the same in every neuron.

The following set of nonlinear delay differential equations (DDEs) was inspired by the mathematical model of *Chen and Aihara*, who also proved the general conditions of its stability and bifurcation for some simplifying assumptions [54.52]. Particular terms on the right-hand side of equations were inspired also by the *rough* network models from [54.53]. An equation similar to (54.2) for the protein (gene product) dynamics was used by [54.5] to model early neurogenesis in *Drosophila*. The linear difference form of (54.1) without decay was used by [54.54] to model mRNA levels during the rat CNS development and injury. In the following, we will consider the dynamics of genes and proteins to be continuous that is, we can describe their changes in a continuous time. We will speak about gene

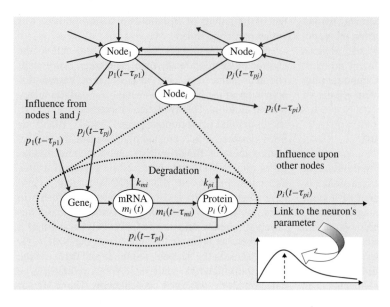

Fig. 54.3 Schematic illustration of the gene–protein regulatory network (GPRN). The edges represent regulatory interactions that lead either to up-regulation or down-regulation of expression of the target gene. These regulatory interactions are expressed as coefficients of the interaction matrix **W**. Protein properties and concentrations are linked to neuronal parameters, like, for instance, magnitude of excitation or inhibition. GPRN illustration was inspired by the network illustration in [54.52]

families rather than individual genes. By gene families we understand the set of genes that have a coordinated transcription control and code for different subunits of one protein. These subunits have to be synthesized in concert in order to produce a functional protein. We will represent the mRNA levels of all the relevant gene families with the vector $\boldsymbol{m} = (m_1, m_2, \ldots, m_N)$ and the corresponding protein levels with the vector $\boldsymbol{p} = (p_1, p_2, \ldots, p_N)$. The level of mRNA of the gene family i changes as

$$
\frac{\mathrm{d}m_i}{\mathrm{d}t} = A_{m_i}\sigma_{m_i}\left[\sum_{j=1}^{n} w_{ij}\, p_j(t - \tau_{p_j}) \right.
$$
$$
\left. + \sum_{k=1}^{K} v_{ik}x_k(t - \tau_{x_k}) + b_{m_i}\right] - \lambda_{m_i}m_i(t)\,,
$$

$$(54.1)$$

where $m_i(t)$ is the total level of mRNA for the i-th gene family at time t, σ_{m_i} is a nonlinear sigmoid regulation-expression (activation) function for the i-th gene family, A_{m_i} is the amplitude of this activation function, w_{ij} are the regulatory coefficients between the i-th and j-th gene families, while the regulatory interaction is mediated through proteins p_j, p_j is the level of the jth protein, τ_{p_j} is the delay, with which the j-th protein influences the transcription of the i-th gene family, v_{ik} is the influence of the k-th external factor upon the gene (hormone, drug, etc.), x_k is the concentration of the k-th external factor, τ_{x_k} is the delay with which the k-th external factor influences the transcription of the i-th gene

family, b_{m_i} is the bias, i. e., the basal expression level of the i-th gene family, and λ_{m_i} is the degradation rate of the mRNA of the i-th gene family.

Analogically, protein levels change as

$$
\frac{\mathrm{d}p_i}{\mathrm{d}t} = A_{p_i}\sigma_{p_i}\left[m_i(t - \tau_{m_i}) \right.
$$
$$
\left. + \sum_{k=1}^{K'} u_{ik}y_k(t - \tau_{y_k}) + b_{p_i}\right] - \lambda_{p_i}p_i(t)\,,
$$

$$(54.2)$$

where $p_i(t)$ is the level of a fully functional protein coded for by the i-th gene family, σ_{p_i} is a nonlinear sigmoid synthesis function for the i-th protein (note, we consider that one protein is coded for by only one gene family), A_{p_i} is the amplitude of this synthesis function, m_i is the total expression level of the i-th gene family, τ_{m_i} is the delay from initiation of transcription of the i-th gene family till the end of synthesis of the i-th protein (on the order of tens of minutes), u_{ik} is the influence of the k-th external factor upon the protein (hormone, drug, etc.), y_k is the concentration of the k-th external factor, τ_{y_k} is the delay with which the k-th external factor influences the i-th protein level, b_{p_i} is the bias, i. e., the basal level of the i-th protein, and λ_{p_i} is the degradation rate of the i-th protein.

If we, instead of gene families, worked with genes coding for n_s individual subunits, then the first term on the right-hand side of (54.2) would read $\sum_{j}^{n_s}$ (proportion of subunits) $\times m_j(t - \tau_{m_j})$ instead of $m_i(t - \tau_{m_i})$. Second terms in (54.1) and (54.2) enable us to investigate the effect of drugs and other factors like neurotransmitters or hormones, which are known to influence gene–protein interactions.

We solve (54.1) and (54.2) numerically and interpret one iteration as 1 s of the real time. Equations (54.1) and (54.2) are the so-called delay differential equations (DDEs). DDEs are similar to ordinary differential equations, but their evolution involves past values of the state variable. The solution of delay differential equations therefore requires knowledge of not only the current state, but also of the state a certain time previously [54.55]. For DDEs we must provide not just the value of the solution at the initial point, but also the history, that is the solution at times prior to the initial point [54.56]. As of MATLAB 6.5 (Release 13), the DDEs solver **dde23** is part of the official MATLAB release.

54.3.2 Proteins and Neural Parameters

Let P_j denotes the j-th parameter of a model neuron. Let $p_j \in (0, 1)$ be the normalized level of protein concentration obtained by the solution of (54.1) and (54.2) above. Then the value of parameter P_j is directly proportional to the concentration of the (functional) protein p_j, in such a way that

$$P_j(t) = p_j(t) \left(P_j^{\max} - P_j^{\min} \right) + P_j^{\min}, \qquad (54.3)$$

where $P_j^{\max}$ and $P_j^{\min}$ are maximal and minimal values of the j-th parameter, respectively. If $p_j \to 0$ then $P_j \to P_j^{\min}$, and if $p_j \to 1$ then $P_j \to P_j^{\max}$. Other, e.g., nonlinear relations between protein levels and parameter values are also possible. The linear relationship in (54.3) is justified by findings that protein complexes, which have clearly defined interactions between their subunits, have highly correlated levels with mRNA expression levels [54.57,58]. Subunits of the same protein complex show significant co-expression, both in terms of similarities of absolute mRNA levels and expression profiles, e.g., subunits of a complex have correlated patterns of expression over a time course [54.58]. This implies that there should be a correlation between mRNA and protein concentration, as these subunits have to be available in stoichiometric amounts for the complexes to function [54.57]. This is exactly the case of proteins in our model, which are receptors and ion channels, comprised of respective ratios of subunits. Equation (54.3) links the gene/protein dynamics to the dynamics of neural model. Values of neuronal parameters will not be constant anymore, but instead their values will depend on the levels of synthesized proteins [54.59, 60]. In such a way the system of (54.1) to (54.3) allows for investigation of how deleted or mutated genes can alter the activity of a neural network.

54.3.3 Thalamo–Cortical Model

We use the same spiking neural network as a thalamo-cortical model as we investigated in our previous studies of the cortical local field potential (LFP) and its dependence upon genes and proteins [54.43, 44]. The cortical local field potential (LFP) is calculated at each time instant as the total sum of current membrane potentials of all neurons in the network model of cortex, i.e., $\Phi(t) = \Sigma u_i(t)$.

Spiking Neuron

The spiking model of the cortical neuron is an integrate-and-fire neuron [54.61, 62]. The total somatic postsynaptic potential (PSP) of neuron i is denoted as $u_i(t)$. We update u_i every millisecond (as opposed to the gene–protein dynamics that is updated every second). When $u_i(t)$ reaches the firing threshold $\vartheta_i(t)$, the neuron i fires, i.e., emits a spike (Fig. 54.4). The moment of the threshold $\vartheta_i(t)$ crossing from below defines the firing time t_i of an output spike. The value of $u_i(t)$ is the weighted sum of all synaptic PSPs, $\varepsilon_{ij}\left(t - t_j - \Delta_{ij}^{ax}\right)$, such that

$$u_i(t) = \sum_{j \in \Gamma_i} \sum_{t_j \in F_j} J_{ij} \varepsilon_{ij} \left(t - t_j - \Delta_{ij}^{ax} \right). \qquad (54.4)$$

The weight of synaptic connection from neuron j to neuron i is denoted by J_{ij}. It takes positive (negative) values for excitatory (inhibitory) connections, respectively. Δ_{ij}^{ax} is an axonal delay between neurons i and j, which linearly increases with Euclidean distance between neurons. The positive kernel expressing an individual postsynaptic potential (PSP) evoked on neuron i when a presynaptic neuron j from the pool Γ_i fires at time t_j has a double exponential form, i.e.,

$$\varepsilon_{ij}^{\text{type}}(s) = A^{\text{type}} \left[\exp\left(-\frac{s}{\tau_{\text{decay}}^{\text{type}}} \right) - \exp\left(-\frac{s}{\tau_{\text{rise}}^{\text{type}}} \right) \right], \qquad (54.5)$$

where $\tau_{\text{decay/rise}}^{\text{type}}$ are time constants of the fall and rise of an individual PSP, respectively, A is the PSP's amplitude, and *type* denotes one of the following: *fast_excitation, fast_inhibition, slow_excitation, and*

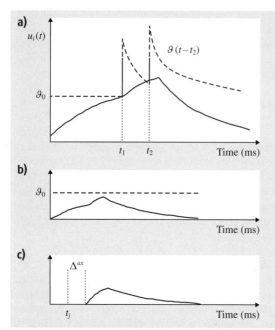

Fig. 54.4 (a) Suprathreshold summation of PSPs in the spiking neuron model. After each generation of postsynaptic spike there is a rise in the firing threshold that decays back to the resting value between the spikes. (b) Subthreshold summation of PSPs that does not lead to the generation of postsynaptic spike, but can still contribute to the generation of LFP. (c) PSP is generated after some delay of the presynaptic spike to travel from neuron j to neuron i

slow_inhibition. Fast excitation in excitatory synapses is mediated through the AMPA receptor-gated ion channels for sodium [54.18, 63]. Slow excitation in excitatory synapses is mediated through the NMDA receptor-gated ion channels for sodium and calcium [54.18,63]. Fast inhibition is mediated through the somatic GABA$_A$ receptor-gated ion channels for chloride, and slow inhibition is mostly mediated through the dendritic GABA$_B$ receptor-gated ion channels for potassium [54.16] as well as by the dendritic GABA$_A$ receptor-gated ion channels for chloride [54.17,21]. Immediately after firing the output spike at t_i, neuron's firing threshold $\vartheta_i(t)$ increases k-times and then returns to its resting value ϑ_0 in an exponential fashion

$$\vartheta_i(t - t_i) = k \times \vartheta_0 \exp\left(-\frac{t - t_i}{\tau_{\text{decay}}^{\vartheta}}\right), \tag{54.6}$$

where $\tau_{\text{decay}}^{\vartheta}$ is the time constant of the threshold decay. In such a way, absolute and relative refractory

Table 54.2 List of neuronal parameters and initial values used in computer simulations

Neuron's parameters	Value range
Fast excitation: amplitude rise/decay time constants	0.5–3.0 1–5 ms/5–10 ms
Slow excitation: amplitude rise/decay time constants	0.5–4.0 10–20 ms/30–50 ms
Fast inhibition: amplitude rise/decay time constants	4–8 5–10 ms/20–30 ms
Slow inhibition: amplitude rise/decay time constants	5–10 20–80 ms/50–150 ms
Resting firing threshold, decay time constant/rise k	17–25 5–50 ms/1–5 fold

periods are modeled. Table 54.2 contains the values of neuron's parameters used in our simulations. These values were inspired by experimental and computational studies [54.18, 21, 64, 65] and were further adjusted by experimentation.

We assume that all three parameters in (54.5) for the fast excitation, i. e., amplitude, rise, and decay time constants are proportional according to (54.3) to the concentration of the protein called amino-methylisoxazole-propionic acid receptor or AMPAR in short. All three parameters describing slow excitation are proportional to the concentration of the N-methyl-D-aspartate acid receptor or NMDAR. All three parameters describing fast inhibition are proportional to the concentration of the gamma-aminobutyric acid (GABA) receptor A or GABA$_A$R for short. The slow inhibition parameters are proportional to the levels of GABA receptor B (or GABA$_B$R). Concentration of the sodium voltage-gated channel (SCN) protein is inversely related to the firing threshold parameters, its resting value, and the decay constant. This inverse relationship is a trivial modification of the relation in (54.3). We could have made only the amplitudes dependent on the protein concentrations and we could have included more ion channel proteins to affect the firing threshold parameters. We do not claim particular assumptions above are the most appropriate. They do, however, serve our purpose to develop a model of SNN that has dynamically changing values of neuronal parameters that depend on levels of proteins which in turn depend on the dynamics of internal GPRN described by (54.1) and (54.2). Thus the values of parameters in Table 54.2 are initial values that change dynamically according to (54.3).

Spiking Neural Network

Figure 54.5 illustrates the architecture of our spiking neural network (SNN). Spiking neurons within the net-

Table 54.3 List of SNN parameters and values used in computer simulations

SNN Parameter	Value
Number of neurons	120
Proportion of inhibitory neurons	0.2
Probability of external input fiber firing	0.015
Peak/sigma of external input weight	5/1
Peak/sigma of lateral excitatory weights	10/4
Peak/sigma of lateral inhibitory weights	40/6
Probability of connection	0.5
Grid unit delay for excitatory/inhibitory spike propagation	1/2 ms

work that represents the cerebral cortex can be either excitatory or inhibitory. There can be as many as about 10–20% of inhibitory neurons positioned randomly on the rectangular grid of N neurons.

Lateral connections between neurons in the model cortex have weight values that decrease in strength with the distance from neuron i according to a Gaussian formula

$$J_{ij}(\text{dist}(i, j)) = \frac{J_0^{\text{exc/inh}}}{\sigma^{\text{exc/inh}}} \exp\left(-\frac{\text{dist}(i, j)^2}{\sigma^{\text{exc/inh}^2}}\right) \quad (54.7)$$

while the connections are established at random with the probability equal to 0.5. The same distribution of weights and connectivity probability is applied to the feedforward connections from the input layer that represents the thalamus. External inputs from the input layer are added to the right-hand side of (54.4) in each time step. Each external input has its own weights $J_i^{\text{ext\_input}}$ and the PSP evoked by the fast AMPAR-dependent excitation, i.e.,

$$u_i^{\text{ext\_input}}(t) = J_i^{\text{ext\_input}} \varepsilon_i^{\text{fast\_excitation}}(t). \quad (54.8)$$

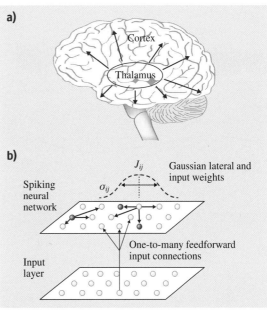

Fig. 54.5 (a) Simple neural network model of the thalamocortical (TC) system. **(b)** SNN represents the cortex and the input layer represents the thalamus. 20% of N neurons are inhibitory (*filled circles*). Units in both layers form a rectangular grid. The model does not have a feedback from the cortex to the thalamus

To stimulate the model cortex, we employed a uniformly random input with an average firing frequency of 15 Hz, since a tonic, low-frequency, nonperiodic and nonbursting firing of thalamocortical inputs with the frequency of 10–20 Hz is typical for the state of vigilance [54.66]. Table 54.3 contains the values of SNN parameters used in our simulations. Values of parameters like weight distributions and strengths and grid unit delays were adjusted by experimentation.

54.4 Dynamics of the Model

Figure 54.6 illustrates temporal evolution of variables at three levels, i.e., three dynamic systems that we combine into one integrated dynamic system – computational neurogenetic model.

The first dynamic system is the SNN, which similarly to biological neurons operates on the time scale of milliseconds (ms). Then we have changes in protein levels that lag behind the changes in gene expression/mRNA levels by the order of tens of minutes, even hours, due to the process of transcription, translation,

and post-translational modification. Protein levels are directly related to parameters of neuronal signaling, like excitation and inhibition. Gene expression levels (expressed as mRNA levels) change slowly – on the order of tens of minutes, even hours.

54.4.1 Estimation of Parameters

Ideally we would know all the parameters in (54.1)–(54.3), that is, ideally we would know all the delays, bi-

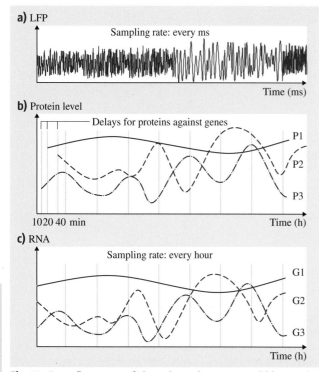

Fig. 54.6a–c Summary of three dynamic systems within one integrated dynamic system – computational neurogenetic model. (**a**) Local field potential (LFP) of SNN, (**b**) protein dynamics over time, (**c**) gene expression/mRNA dynamics over time

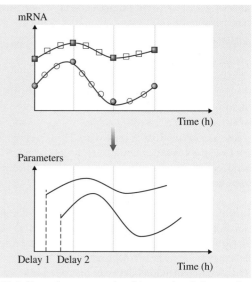

Fig. 54.7 From the measured and interpolated time evolution of mRNA levels, filled and empty symbols, respectively, we can derive the values of parameters for model neurons using (54.2) and (54.3)

ases, shapes of activation functions, amplitudes, degradation rates, and last but not least, the interaction matrix $W = \{w_{ij}\}$. We could infer the coefficients of W from the time course of mRNAs by reverse engineering. To do so, we would have to proceed in the following steps:

1. Obtain gene expression data (mRNA levels) from the relevant neural system for discrete sampling periods.
2. Then use the interpolation technique (for instance the extended Kalman filter) to interpolate missing values for mRNA levels for intermediate time intervals. In such a way, we can obtain the values of all m_i for, let us say, every minute.
3. Then we need to calculate the levels of proteins according to (54.2) to estimate the values of neuronal parameters according to (54.3), see Fig. 54.7.

Let us now discuss these steps in more detail. The first step, i. e., the interpolation, would enable us to work with the values of mRNAs and consequently parameter values in one-minute intervals. We can assume the

values of neuronal parameters being constant during these short intervals between interpolated values. Different methods can be used for data interpolation, for instance the Kalman filter [54.67, 68], evolutionary optimization [54.69], state space equations [54.70], etc. Then the biggest challenge is to estimate the delays from initiation of transcription of the gene families till the end of synthesis of relevant proteins in the gene–protein regulatory network, i. e., τ_{m_i}'s for (54.2). We need these delays for updating neuronal parameters to simulate SNN (or any other ANN model).

Proper updating of neuronal parameters is crucial for explaining changes in the SNN output – why they occur and when. We can make rough qualified estimates of these delays using the information of the length of the relevant proteins and the time, which is needed for their genes transcription and the subsequent protein translation [54.45] and posttranslational modifications [54.46]. After all this, the final computational challenge remains: the simulation of SNN activity in real time.

54.4.2 Simulation Plan

Another option is that instead of simulating the SNN for the whole time of evolving gene–protein dynamics, we can perform the simulations of SNN only at random or interesting time points of the gene–protein dynamics, as

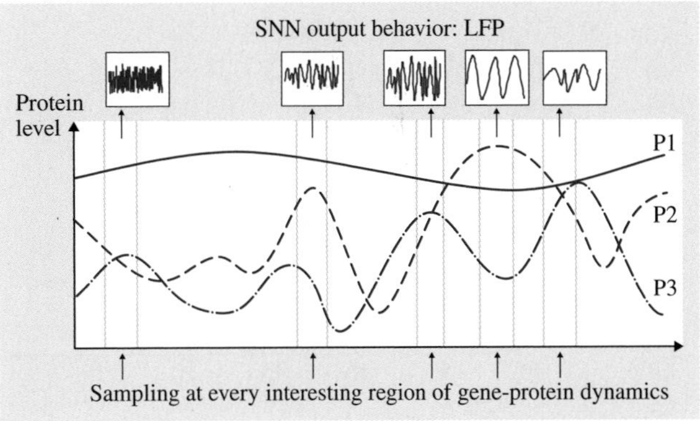

Fig. 54.8 Sampling the SNN output at interesting time intervals based on some heuristics. LFP means local field potential ▶

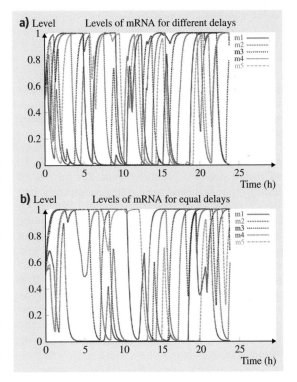

is illustrated in Fig. 54.8. For these random or interesting time points, the simulations of SNN will last only for some minutes of real time. Interesting intervals for sampling the SNN output can be based on some kind of heuristics – for instance based on the knowledge that at that particular time something has happened – for instance, there was a seizure. Otherwise this sampling can occur at intervals where the parameters have their extreme values, at intersections of values, etc.

54.5 Results of Computer Simulations

54.5.1 Dynamics of the Gene–Protein Regulatory Network

In order to illustrate the approach described above, we provide an example of an artificially created dynamic system described by (54.1)–(54.3). We have omitted the terms for external factors from both equations. Thus, we have numerically solved (54.1)–(54.3) (minus external factors terms) for several randomly generated interaction matrices Ws, with random coefficients $w_{ij} \in (-1, 1)$ and with randomly generated delays τ_{p_j}, $\tau_{m_i} \in (0\,\text{min}, 60\,\text{min})$. Other parameters had these values: $\sigma_{m_i}(x) = \sigma_{p_i}(x) = \tanh(x)$ for each i, $A_{m_i} = A_{p_i} = 0.01$ for each i, $b_{m_i} = b_{p_i} = 0.5$ for each i, and $\lambda_{m_i} = \lambda_{p_i} = 0.001$ for each i. These constants were chosen in such a way that the oscillations in gene and protein dynamics were on the orders of 1–2 or more hours [54.71]. The values of mRNAs and proteins were normalized to the interval $(0, 1)$. The total number of genes was $n = 10$, where five genes were directly related to the signal processing parameters and five genes were not. These five hypothetical genes directly affecting values of neural parameters were: AMPAR, NMDAR, GABAAR, GABABR, and SCN. For better clarity, we present only the curves for the five genes that are directly related to parameters. Initial conditions are always the same unless other-

Fig. 54.9a,b Effect of different delays upon the gene dynamics for the same GPRN interaction matrix W and the same initial conditions: (**a**) the delays are randomly generated from the interval 0–60 min, (**b**) all delays are equal to 30 min ▶

wise stated and that is $m_i(0) = 0.5$ and $p_i(0) = 0.5$ for all i.

In Fig. 54.9 we illustrate the effect of different delays upon the steady state gene dynamics for the same random W and the same initial conditions. The interval for numerical solving of (54.1) and (54.2) was equal to 1 s. We can see that delays can completely change

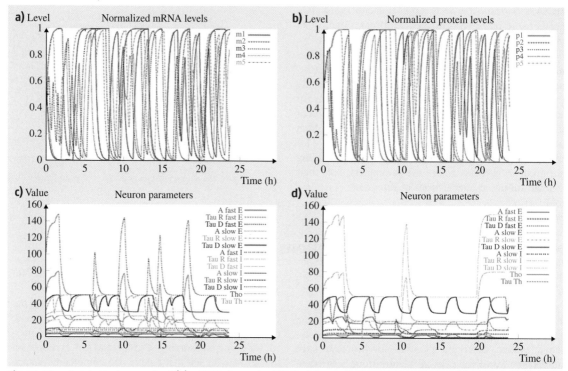

Fig. 54.10a–d A typical example of (**a**) mRNA dynamics (54.1), (**b**) protein dynamics (54.2), and (**c**) parameter dynamics (54.3) for one randomly generated GPRN interaction matrix **W** and one set of randomly generated delays. All delays are different and randomly generated from the interval (0 min, 60 min).(**d**) Temporal dynamics of the neuron's parameter values after the knock out of hypothetical genes for GABA$_A$R. Parameters related to fast inhibition are missing entirely and the other ones have changed their time course

the time course of mRNAs for the same values of other parameters. Therefore, in simulations of real experiments it would be very important to estimate the delays carefully.

In the next Fig. 54.10 we present a typical example of mRNA dynamics (54.1), protein dynamics (54.2), and related neuronal parameter dynamics for one of the randomly generated gene–protein regulatory matrices **W** and a set of randomly generated delays. All delays are different and randomly generated from the interval (0 min, 60 min). The time scale is in hours, and the whole simulation illustrates the gene and protein levels over 24 h. The update interval for the gene–protein dynamics was 1 s. We can see in Fig. 54.10a,b that the protein dynamics is essentially copying the gene dynamics, however with corresponding delays. The typical steady state of our model system resembles chaotic or quasi-periodic behavior. Constant or periodic steady states are rare.

54.5.2 Dynamics of GPRN After Gene Knock Out

In our artificial GPRN, we can *delete* or *mutate* genes one after another, and in various combinations, and observe the effect upon the parameter and consequently upon the SNN dynamics. In the model, we can manipulate also the genes that are not directly related to neuronal parameters, but that influence them through regulatory interactions **W**. In such a way, a computational model can aid an experimental search for various genetic conditions. For illustration, we will *knock out* one hypothetical gene, let us say that one responsible for GABA$_A$ receptors mediating fast inhibition in our model. GABA$_A$ receptor mRNA and protein levels as well as related parameters are all set to zero for the whole time of the simulation. The resulting parameter dynamics after GABA$_A$R gene deletion is illustrated in Fig. 54.10d. All the coefficients of the interaction matrix W are the same as for the dynamics illustrated in Fig. 54.10a–c, except of course all the links from knock out gene $GABA_AR$ are missing. We can see in Fig. 54.10d even with the naked eye that for the same initial conditions, the parameter dynamics governed by genes–protein network has changed compared to the complete GPRN. The curves corresponding to parameters related to fast inhibition are missing entirely, timing of others has changed. The prominent peaks in both graphs c and d belong to the decay constant of slow inhibition. After the knock out of $GABA_AR$ four peaks are missing in the dynamics of this slow inhibition parameter and a new peak appears after 20 h. The same holds for other parameters of slow inhibition. Although they have different values, they follow the same time courses because that is the time course of the concentration of NMDAR. We can conclude, not surprisingly, that the temporal evolution of all neural parameters has changed more or less as a consequence of the knock out of the hypothetical gene for $GABA_AR$.

54.5.3 Dynamics of the Cortex

The question arises now what the spiking activity of the SNN is like. Figure 54.11 illustrates the spiking of SNN over 1 min at some arbitrary time point of parameter dynamics of parameters illustrated in Fig. 54.10c. When the small artificial genome is complete, many random gene interaction matrices actually lead to an asynchronous spiking of SNN with very low frequencies and low spike count, like the one illustrated in Fig. 54.11.

54.5.4 Dynamics of the Cortex After Gene Knock Out

Now, we are interested in the effect of the gene knock out upon the SNN local field potential and spiking activity. We performed the SNN simulations and the LFP activity analysis for the same time intervals as in the case of a complete GPRN. Interestingly, most of the time, the spiking activity of SNN was low and asynchronous, like in the *normal* case illustrated in Fig. 54.11, in spite of the fact that the $GABA_AR$ gene and thus the fast inhibition were entirely missing. It seems that for this particular interaction matrix W there was a compensation for the missing fast inhibition most of the time. However, sometimes all the neurons in the SNN spontaneously synchronized. This behavior is illustrated in Fig. 54.12. Such behavior is reminiscent of a spontaneous epileptiform activity that spontaneously

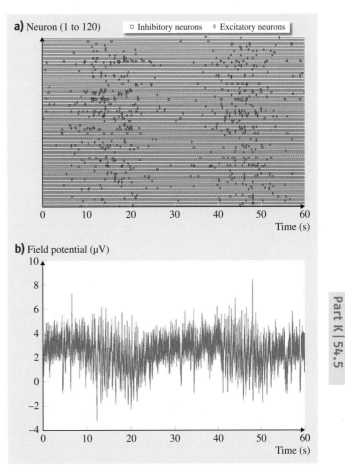

Fig. 54.11a,b Illustration of the output behavior of SNN for the complete artificial genome: (**a**) The spiking behavior. The x-axis is time and the y-axis is the index of a neuron. The spikes of excitatory neurons are marked by *red blobs* and the spikes of inhibitory neurons are marked by the *blue blobs*. Neurons spike on average with very low frequency of $0–0.5$ Hz. At any millisecond interval only $0–4$ neurons emit a spike. (**b**) The *graph* shows the corresponding local field potential (LFP), which is the sum of all membrane potentials of all neurons

arises and then ceases again. The SNN entered the synchronized state and left it spontaneously based on the underlying parameter dynamics. Most of the time however, the spiking was asynchronous like the one illustrated in Fig. 54.11, in spite of the fact that the gene for $GABA_AR$ was missing from the artificial genome.

To summarize, the observed behavior before and after the simulated knock out of the hypothetical gene for $GABA_AR$ from our artificial gene–protein regulatory network resulted from several randomly generated interaction matrices **W**. Most of the randomly generated regulatory matrices either produced no spiking in the SNN or the neurons were synchronized all the time. We did not perform any optimization of the interaction matrix **W** like we did, for instance, in our previous work by means of an evolutionary algorithm where the fitness function was particular spectrum of LFP or EEG [54.43, 44]. Rather, we adopted an exploratory approach of investigation of many randomly generated interaction matrices out of which we kept those that generated interesting dynamics. This naïve brute force approach served us to explore the potential application of computational neurogenetic model in relation to gene knock-out experiments.

54.6 Discussion and Future Directions

With the advancement of molecular research technologies more and more data and information will be made available about the genetic basis of neuronal functions and genetically based epilepsies and other brain genetic diseases. This information will then be utilized for the models of brain functions and diseases that include models of gene and protein interaction within neurons. Although the data are not yet available, we have started to conceive a new computational methodology that will be able to incorporate such knowledge into the existing models of neurons and neural networks. This new approach integrates knowledge from computer and information science, neuroscience, and molecular genetics. We call it computational neurogenetic modeling [54.43]. In this chapter we describe this novel computational approach to brain neural network modeling which integrates dynamic gene–protein networks with a neural network model. The bridging system is the protein network, in which individual proteins that are coded for by genes are related to neuronal parameters. Interaction of genes and proteins in model neurons thus affects the dynamics of the whole neural network through neuronal parameters, which are no longer constant, but change as a function of gene expression. For simulation of real experiments we would need real gene expression data to infer the parameters of GPRN. Then, through optimization of the gene interaction network, initial gene/protein expression values and neural parameters, particular target states of the neural network operation can be achieved. At present however, no such data are available. Therefore we have adopted an exploratory approach to demonstrate this new computational methodology by means of a simple neurogenetic model of a spiking neural network which generates LFP and we have shown some interesting behavior of this simple model.

Recently, *Thomas* et al. [54.72] adopted a kind of reverse approach. Using a computational model of dentate gyrus with mossy fiber sprouting they addressed these three questions: (1) What voltage-gated ion channels have the biggest influence on network excitability? (2) What changes in their electrophysiological properties have the biggest influence on network excitability? (3) What is the magnitude of these changes that leads to epileptiform activity? They call this approach a *genetic sensitivity* analysis to predict which genes are best positioned to increase risk as well as to predict *functionally* how variants in these genes might increase network excitability. Based on computer simulations they predicted that variants in sodium and delayed rectifier channels are likely to confer risk for development of epilepsy. This prediction is consistent with findings that mutations in genes coding these channels can cause generalized epilepsies (see, e.g., Table 54.1), although generalized epilepsy syndromes predominantly involve cortical and thalamic networks (like we simulated in this chapter albeit with only feedforward connectivity) and not dentate gyrus. Notable is the used model of the neural network. *Thomas* et al. [54.72] used a model of the dentate gyrus that contains morphologically realistic models of granule cells, and other excitatory and inhibitory neurons developed using the software NEURON [54.73, 74]. Neuron models had between 9 and 17 compartments describing the actual dendritic arborization and realistic conductances including the fast sodium and potassium channels that directly form the action potential, an A current, L-, N-, and T-type calcium channels, hyperpolarization-activated, cyclic nucleotide-gated (HCN) current, and slow-voltage and calcium-gated potassium channels. Similarly, a detailed neural network model of the relevant brain areas in connection with computational neurogenetic modeling will allow us to ask more detailed questions and simulate more detailed situations than we have shown in this chapter. At the time of writing of this chapter more than 600 of such detailed computational models were publicly available at the ModelDB website of the NEURON simulation environment [54.75].

54.6.1 Q and A for Neurogenetic Modeling

Computational neurogenetic modeling is a new area of research that has many open questions, for example:

1. *Which real neuronal parameters are to be included in an ANN model and how to link them to activities of genes/proteins?*

   In our model presented in this chapter, we have chosen those parameters that are relevant to neuronal excitation and inhibition as our goal was to model LFP and the consequence of gene knock out (or mutation) upon the changes in LFP that would resemble epileptiform activity. Thus, the general answer to this question would be to include those parameters that are relevant for the phenomenon modeled. We have linked the chosen parameters to the levels of proteins via (54.3). This linkage is based on the assumption that the magnitude of excitation/inhibition is proportional to the concentration of corresponding receptors for excitatory/inhibitory neurotransmitters, respectively. Or in other words, this assumption means that if the gene expression for particular receptor is increased/decreased, then there is probably an increased/decreased demand for it in the cell due to changes in synaptic transmission or due to effects of other genes. However, in addition to this relationship a more sophisticated relation is possible to model. Different variants of relevant genes are linked with variations in receptor or ion channels functions and that can be taken into account as well.

2. *Which genes/proteins are to be included in the model and how to represent the gene interaction over time within each neuron?*

   In our simple model we worked with the gene families that code for subunits of receptors or ion channels. We have also suggested how to extend the model to include individual genes. In addition to the genes that are directly related to neuronal parameters, there are also many other genes that influence them. These other genes can be found in the available literature and bioinformatics databases. For instance, the Ingenuity Pathway Analysis (IPA) system can be used to investigate interaction-based relationships between the genes and proteins based on their own Ingenuity Knowledge Base [54.76]. Then the choice of a particular computational model of gene–protein interaction should depend on the information that is available about the modeled system. There are many available theoretical models of gene–protein interaction networks, e.g., differential equations, stochastic models, weight matrices, Boolean models, etc.

3. *How can we integrate in time the activity of genes, proteins, and neurons in an ANN model, as it is known that neurons spike in millisecond intervals and the process of gene transcription and translation into proteins takes minutes or even hours?*

   This is really a computational issue that depends on the computational power available. Ideally one would simulate at the same time the internal gene–protein regulatory network and its effect upon neural parameters in the real time. An alternative solution is to run the GRN dynamics separately, store the results and feed the values of protein levels to the neural model only for some interesting intervals of gene–protein dynamics. We have adopted this approach in this chapter. As we did not know which periods of gene–protein dynamics were interesting, we just picked several intervals at random. This approach has the disadvantage that some interesting events in the dynamics of the model can be easily missed.

4. *How can we integrate internal and external variables in a CNGM (e.g., genes and neuronal parameters with external signals acting on the brain)?*

   The basis for this integration can be the second terms in (54.1) and (54.2). These terms express the effect of chemical agents be it drugs or upon the levels of mRNA and proteins. Our model can, in principle, be extended to include also the activity-dependent changes in neural parameters [54.42] and synaptic plasticity [54.20].

5. *How can we measure brain activity and the CNGM activity in order to validate the model?*

   Although our neural network is a model of LFP, since EEG is the sum of many LFPs [54.77], the EEG record like in Fig. 54.1 can serve as a validation of the model of epilepsy. For instance, we can see in Fig. 54.1 that there are about five epileptic slow-wave discharges starting at second 10, which then spontaneously cease. It seems another series starts between 20 and 25 s of the EEG record but the record suddenly ends. If we compare this behavior with the model spiking in Fig. 54.12, we can see it is, indeed, very similar in that each time there is an epileptiform behavior there are only few waves of global synchronization of neurons that spontaneously emerge and then cease and there is an inter-ictal interval of 20 s. Duration of periods between seizures is, however, highly variable in our

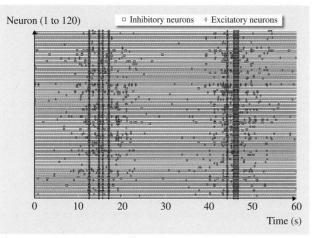

Fig. 54.12 After the knock out of the gene coding for GABA$_A$R, the SNN behavior was reminiscent of a spontaneous intra-ictal and inter-ictal activity. This is one particular illustration when the seizure-like synchronizations occurred in close succession within 1 min of simulated time. In general intervals between seizure-like activities varied widely

model and depends on the underlying GPRN dynamics that is relatively slow compared to neural dynamics.

6. *What is the effect of connectivity? How does it interact with the effects of GPRN?*

In each SNN simulation, the particular connectivity was different, albeit constructed according to the same statistics. It would be interesting to investigate whether connectivity can compensate for gene defects, or vice versa, or whether connectivity can lead to pathology even when the genome is complete with no mutations. We would need more information on variations of interneuronal connectivity between individuals and between the healthy subjects and subjects with epilepsy to be able to carry out this investigation.

7. *What useful information can be derived from CNGM?*

What happens if the nucleotide sequence of the gene encoding the protein is altered or deleted? Such changes in the DNA sequence, called mutations, can lead to the loss of the encoded protein or to a change in its structure, thereby altering its function. For instance, gene mutations may decrease or increase the activity of a given gene product or they may result in setting a new activity of the gene product [54.78]. Even when only one gene is mutated, the expression of other genes in the whole GRN

may be more or less affected. Sometimes a gene mutation matters and sometimes it does not. Everything depends on the function of its product and on its interactions with other genes in the GRN. Moreover, there can be mutations in the DNA sequence outside the gene region that correspond to the various gene expression regulation sites. While mutation of an activator-binding site leads to decreased expression of the linked gene, mutation of a repressor-binding site leads to increased expression of the gene. In our simple model we simulated the gene knock out in such a way, that the expression of the corresponding gene was set to zero and all the interactions to and from this gene were set to zero. In the case of the mutated gene, these interactions may remain intact and just the expression level can be modified, either increased or decreased. This will have an effect upon other genes in the network, and we can observe how the dynamics of the neural network changes. Thus, the computational neurogenetic approach has great potential in modeling gene dynamics and its changes due to DNA mutations and the consequences upon the neural dynamics.

These and many more questions remain to be addressed in the future. Although we are speaking about brain diseases and functions, having in mind mammals and higher vertebrates, the approach suggested in this chapter is applicable also to simpler organisms to aid the explanation of the genetic basis of their behaviors. Another system that can be used to validate our approach could be brain slices taken from normal and genetically modified brains.

54.6.2 Hypothesis

It is interesting to note that in our illustrative example that, in spite of the fact that the artificial gene–protein network has been altered due to one gene knock out, the dynamics of SNN in terms of spiking activity was most of the time very similar to the result obtained with the complete GPRN (Fig. 54.12). However, from time to time the neurons spontaneously temporarily synchronized their spiking as illustrated in Fig. 54.12. Thus, the knock out of a hypothetical gene for fast inhibition in our artificial genome has led to an epileptiform activity. In our model, the fluctuations in the values of neuronal parameters led to spontaneous development of seizure-like global synchronizations. These very same fluctuations also led to termination of the seizure-like neural activity and maintenance of the inter-ictal normal periods of activity. Our hypothesis does not rule out other mechanisms of seizure development, maintenance, and cessation, like for instance activity-dependent changes in neural parameters [54.42] and/or synaptic entraining [54.20], or other mechanism(s) like the processes in the thalamocortical recurrent loop, which we did not include in our model at all [54.34]. Based on our model, however, we would like to suggest a hypothesis that the parameter changes due to the gene–protein dynamics should also be included as a serious factor that determines transitions in neural dynamics, especially when the cause of disease is known to be genetic. This hypothesis can be tested, for instance, in slices or neuronal cultures from the wild type and gene knock out mice.

54.6.3 Robustness of Gene Regulatory Networks

Even in the presence of a mutated gene in the genome that is known to cause the brain disease, the neurons can still function normally provided a certain pattern of interaction between the genes is maintained [54.79]. It seems the expression regulation between genes is so robust that it can compensate, at least to some extent, for mutations in the genes. In fact, in our computer experiments, we observed that several different gene dynamics, i.e., several different regulatory matrices $\mathbf{W}$s can lead to very similar SNN LFPs and spiking activities. This observation relates to the hot topic of robustness versus evolvability of gene regulatory networks. It seems that our system manifests a certain robustness. We do not mean the robustness against noise or stochasticity in gene expression because our model system is fully deterministic. What we have discovered in our exploratory simulations is a certain robustness with respect to different regulatory interactions between the same set of genes leading to the same phenotype expressed as the spiking activity of the model SNN.

Biochemical parameters that determine the behavior of cellular systems – from proteins to genome-scale regulatory networks – change continually. These changes have two principal sources. One of them is genetic and consists of mutations. The other is not genetic and is related to noise internal to the organism and/or induced by environmental change. Much of the noise consists of stochastic variation in gene expression and expression regulation. Such noise makes all biochemical parameters affecting the cell's behavior appear to fluctuate randomly. Environmental change, such as a change in temperature, salinity, or nutrient availability, can simi-

larly affect many parameters at once in random manner. Such observations suggest that biological circuits are not fine-tuned to exercise their functions only for precise values of their biochemical parameters. Many biochemical parameters driving circuit behavior vary extensively and are thus not fine-tuned. Instead, biological circuits including gene regulatory networks must be able to function under a range of different parameters. In other words, they must be robust to parameter change. Current genetic and modeling research tries to elucidate the robustness of gene regulatory networks with respect to transcriptional mechanisms that cause robust versus stochastic gene expression and their relationship to phenotypic robustness and variability [54.80, 81].

In addition, it seems that organisms are robust to a great variety of genetic changes. *Aldana* et al. [54.82] addressed a problem of robustness and evolvability of the attractor landscape of genetic regulatory network models under the process of gene duplication followed by divergence. They showed that an intrinsic property of this kind of network is that, after the divergence of the parent and duplicate genes, with a high probability the previous phenotypes, encoded in the attractor landscape of the network, are preserved and new ones might appear. The above is true in a variety of network topologies and even for the case of extreme divergence in which the duplicate gene bears almost no relation with its parent. Their results indicate that networks operating close to the so-called *critical regime* exhibit the maximum robustness and evolvability simultaneously.

Ciliberti et al. (2007) argue that the topology of gene regulatory networks, that is who-interacts-with-whom, is the key to understanding their robustness to both mutations and noise [54.83]. The latter authors performed theoretical and computational investigation for a weight matrix model of transcriptional regulation networks, in which they explored millions of different network topologies. Topology is synonymous with the *structure* of the matrix $\mathbf{W}$, where each of $\mathbf{W}$'s nonzero entries corresponds to one regulatory interaction among the network's genes. Changes in topology correspond to the loss of a regulatory interaction, or to the appearance of a new regulatory interaction that was previously absent. The robust feature is the network's equilibrium gene expression pattern. Robustness to mutations corresponds to robustness of changes in regulatory interactions, either as a change in network topology, or as a change in the strength of regulatory interaction. Robustness to noise corresponds to robustness of equilibrium gene expression pattern to random changes in gene expression patterns. First, they showed that robustness to mutations and noise were correlated in these networks. They showed a skewed distribution, with a very small number of networks being vastly more robust than the rest. More importantly, they also showed that highly robust topologies can evolve from topologies with low robustness through gradual topological changes. Thus, they argue robustness is an evolvable property and that evolvability

of robust networks may be a general principle of evolutionary process. This result is general and thus applies to gene regulatory networks in the brain as well. It is congruent with the results of our exploratory simulations of spiking behavior of model SNN whose neuronal parameters of excitation and inhibition are dynamically varied due to the dynamics of internal gene–protein interaction network.

References

54.1   H. Lodish, A. Berk, S.L. Zipursky, P. Matsudaira, D. Baltimore, J. Darnell: *Molecular Cell Biology*, 5th edn. (Freeman, New York 2000), Chap. 10: Regulation of transcription initiation, pp. 341–403

54.2   L. Benuskova, N. Kasabov: *Computational Neurogenetic Modeling* (Springer, New York 2007)

54.3   H. Meinhardt: Different strategies for midline formation in bilaterians, Nat. Rev. Neurosci. **5**, 502–510 (2004)

54.4   M. Kerszberg, J.-P. Changeux: Molecular models of early neural development. In: *Modeling Neural Development*, ed. by A. van Ooyen (MIT Press, Cambridge 2003) pp. 1–26

54.5   G. Marnellos, E.D. Mjolsness: Gene network models and neural development. In: *Modeling Neural Development*, ed. by A. van Ooyen (MIT Press, Cambridge 2003) pp. 27–48

54.6   C. Lee, K. Bae, I. Edery: The Drosophila CLOCK protein undergoes daily rhythms in abundance, phosphorylation, and interactions with the PER-TIM complex, Neuron **21**, 857–867 (1998)

54.7   V. Suri, A. Lanjuin, M. Rosbash: TIMELESS-dependent positive and negative autoregulation in the Drosophila circadian clock, EMBO J. **18**, 675–686 (1999)

54.8   P. Smolen, P.E. Hardin, B.S. Lo, D.A. Baxter, J.H. Byrne: Simulation of Drosophila circadian oscillations, mutations, and light responses by a model with VRI, PDP-1, and CLK, Biophys. J. **86**, 2786–2802 (2004)

54.9   National Center for Biotechnology Information: The nervous system. In: *Genes and Disease* (National Center for Biotechnology Information, Bethesda 2011), available online at http://www.ncbi.nlm.nih.gov/books/NBK22197/ (last accessed 3 May 2013)

54.10 M.E. Hasselmo: Neuromodulation and the hippocampus: Memory function and dysfunction in a network simulation. In: *Disorder of Brain, Behavior and Cognition: The Neurocomputational Perspective*, ed. by J.A. Regia, E. Ruppin, D. Glanzman (Elsevier, Amsterdam 1999) pp. 3–18

54.11 D. Horn, N. Levy, E. Ruppin: Multimodular networks and semantic memory impairments. In: *Disorders of Brain, Behavior and Cognition: The Neurocomputational Perspective*, ed. by J.A. Reg-

gia, E. Ruppin, D. Glanzman (Elsevier, Amsterdam 1999) pp. 68–72

54.12 E.D. Menschik, L.H. Finkel: Cholinergic neuromodulation and Alzheimer's disease: From single cells to network simulation. In: *Disorders of Brain, Behavior and Cognition: The Neurocomputational Perspective*, ed. by J.A. Reggia, E. Ruppin, D. Glanzman (Elsevier, Amsterdam 1999) pp. 19–45

54.13 H.P. Goodkin: The founding of the American Epilepsy Society: 1936–1971, Epilepsia **48**(1), 15–22 (2007)

54.14 O. Devinsky: Diagnosis and treatment of temporal lobe epilepsy, Rev. Neurol. Dis. **1**(1), 2–9 (2004)

54.15 Y. Ben Ari, R. Cossart: Kainate, a double agent that generates seizures: Two decades of progress, Trends Neurosci. **23**, 580–587 (2000)

54.16 B.W. Connors, R.C. Malenka, L.R. Silva: Two inhibitory postsynaptic potentials, and $GABA_A$ and $GABA_B$ receptor-mediated responses in neocortex of rat and cat, J. Physiol. **406**, 443–468 (1988)

54.17 J.A. White, M.I. Banks, R.A. Pearce, N.J. Kopell: Networks of interneurons with fast and slow g-aminobutyric acid type A ($GABA_A$) kinetics provide substrate for mixed gamma-theta rhythm, Proc. Natl. Acad. Sci. USA **97**, 8128–8133 (2000)

54.18 A. Destexhe: Spike-and-wave oscillations based on the properties of $GABA_B$ receptors, J. Neurosci. **18**, 9099–9111 (1998)

54.19 R.D. Traub, R. Miles, R.K. Wong: Models of synchronized hippocampal bursts in the presence of inhibition. I. Single population events, J. Neurophysiol. **58**(4), 739–751 (1987)

54.20 B. Biswal, C. Dasgupta: Neural network model for apparent deterministic chaos in spontaneously bursting hippocampal slices, Phys. Rev. Lett. **88**, 88–102 (2002)

54.21 F. Wendling, F. Bartolomei, J.J. Bellanger, P. Chauvel: Epileptic fast activity can be explained by a model of impaired GABAergic dendritic inhibition, Eur. J. Neurosci. **15**, 1499–1508 (2002)

54.22 P. Kudela, P.J. Franaszczuk, G.K. Bergey: Changing excitation and inhibition in simulated neural networks: Effects on induced bursting behavior, Biol. Cybern. **88**, 276–285 (2003)

54.23 M.C. Picot, M. Baldy-Moulinier, J.P. Dau-rès, P. Dujols, A. Crespel: The prevalence of epilepsy and pharmacoresistant epilepsy in adults: A population-based study in a Western European country, Epilepsia **49**, 1230–1238 (2008)

54.24 L. Wittner, L. Eross, S. Czirjak, P. Halasz, T.F. Fre-und, Z. Magloczky: Surviving CA1 pyramidal cells receive intact perisomatic inhibitory input in the human epileptic hippocampus, Brain **128**, 138–152 (2005)

54.25 J. DeFelipe: Types of neurons, synaptic con-nections and chemical characteristics of cells immunoreactive for calbindin-D28K, parvalbumin and calretinin in the neocortex, J. Chem. Neu-roanat. **14**, 1–19 (1997)

54.26 B. Schwaller, I.V. Tetko, P. Tandon, D.C. Silveira, M. Vreugdenhil, T. Henzi, M.-C. Potier, M.R. Celio, A.E.P. Villa: Parvalbumin deficiency affects network properties resulting in increased susceptibility to epileptic seizures, Mol. Cell. Neurosci. **25**, 650–663 (2004)

54.27 A.E.P. Villa, Y. Asai, I.V. Tetko, B. Pardo, M.R. Celio, B. Schwaller: Cross-channel coupling of neuronal activity in parvalbumin-deficient mice susceptible to epileptic seizures, Epilepsia **46**(6), 359 (2005)

54.28 M. Vreugdenhil, J.G.R. Jefferys, M.R. Celio, B. Schwaller: Parvalbumin-deficiency facilitates repetitive IPSCs and related inhibition-based gamma oscillations in the hippocampus, J. Neu-rophysiol. **89**, 1414–1423 (2003)

54.29 L. Benuskova, N. Kasabov: Modeling brain dynam-ics using computational neurogenetic approach, Cogn. Neurodyn. **2**(4), 319–334 (2008)

54.30 R.M. Gardiner: Genetic basis of human epilepsies, Epilepsy Res. **36**, 91–95 (1999)

54.31 V. Crunelli, N. Leresche: Childhood absence epilepsy: Genes, channels, neurons and networks, Nat. Rev. Neurosci. **3**, 371–382 (2002)

54.32 C. Marini, L.A. Harkin, R.H. Wallace, J.C. Mul-ley, I.E. Scheffer, S.F. Berkovic: Childhood absence epilepsy and febrile seizures: A family with a GABA$_A$ receptor mutation, Brain **126**, 230–240 (2003)

54.33 H.K.M. Meeren, J.P.M. Pijn, E.L.J.M. van Lui-jtelaar, A.M.L. Coenen, F.H.L. da Silva: Cortical focus drives widespread corticothalamic networks during spontaneous absence seizures in rats, J. Neurosci. **22**, 1480–1495 (2002)

54.34 A. Destexhe, D.A. McCormick, T.J. Sejnowski: Tha-lamic and thalamocortical mechanisms underlying 3 Hz spike-and-wave discharges. In: *Disorders of Brain, Behavior and Cognition: The Neurocompu-tational Perspective*, ed. by J.A. Reggia, E. Ruppin, D. Glanzman (Elsevier Science, Amsterdam 1999) pp. 289–307

54.35 W.W. Lytton, D. Contreras, A. Destexhe, M. Steriade: Dynamic interactions determine partial thalamic quiescence in a computer network model of spike-

and-wave seizures, J. Neurophysiol. **77**, 1679–1696 (1997)

54.36 R.M. Gardiner: Genetic basis of human epilepsies, Epilepsy Res. **36**, 91–95 (1999)

54.37 M.H. Meisler, J. Kearney, R. Ottman, A. Escayg: Identification of epilepsy genes in human and mouse, Annu. Rev. Genet. **35**, 567–588 (2001)

54.38 O.K. Steinlein: Genetic mechanisms that underlie epilepsy, Nat. Rev. Neurosci. **5**, 400–408 (2004)

54.39 A.L. George: Inherited channelopathies associated with epilepsy, Epilepsy Curr. **4**, 65–70 (2004)

54.40 S. Ganesh, R. Puri, S. Singh, S. Mittal, D. Dubey: Recent advances in the molecular basis of Lafora's progressive myoclonus epilepsy, J. Hum. Gener. **51**, 1–8 (2006)

54.41 L.F. Abbott, K.A. Thoroughman, A.A. Prinz, V. Thiru-malai, E. Marder: Activity-dependent modifica-tions of intrinsic and synaptic conductances in neurons and rhytmic networks. In: *Modeling Neu-ral Development*, ed. by A. van Oojen (MIT Press, Cambridge 2003) pp. 151–166

54.42 G. LeMasson, E. Marder, L.F. Abbott: Activity-dependent regulations of conductances in model neurons, Science **259**, 1915–1917 (1993)

54.43 L. Benuskova, V. Jain, S.G. Wysoski, N. Kasabov: Computational neurogenetic modeling: A pathway to new discoveries in genetic neuroscience, Int. J. Neural. Syst. **16**, 215–227 (2006)

54.44 L. Benuskova, N. Kasabov: Modeling brain dynam-ics using computational neurogenetic approach, Cogn. Neurodyn. **2**, 319–334 (2008)

54.45 H. Lodish, A. Berk, S.L. Zipursky, P. Matsudaira, D. Baltimore, J. Darnell: *Molecular Cell Biology*, 5th edn. (Freeman, New York 2000), Chap. 4: Nu-cleic acids, the genetic code, and the synthesis of macromolecules, pp. 100–137

54.46 H. Lodish, A. Berk, S.L. Zipursky, P. Matsudaira, D. Baltimore, J. Darnell: *Molecular Cell Biology*, 5th edn. (Freeman, New York 2000), Chap. 17: Protein sorting: Organelle biogenesis and protein secre-tion. pp. 675–750

54.47 A.M. Marini, S.J. Rabin, R.H. Lipsky, I. Moc-cheti: Activity-dependent release of brain-derived neurotrophic factor underlies the neuroprotective effect of *N*-methyl-D-aspartate, J. Biol. Chem. **273**, 29394–29399 (1998)

54.48 H. Husi, M.A. Ward, J.S. Choudhary, W.P. Black-stock, S.G.N. Grant: Proteomic analysis of NMDA receptor-adhesion protein signaling complexes, Nat. Neurosci. **3**, 661–669 (2000)

54.49 T.P. Garrington, G.L. Johnson: Organization and regulation of mitogen-activated protein kinase signaling pathways, Curr. Op. Cell Biol. **11**, 211–218 (1999)

54.50 P. Baldi, S. Brunak: *Bioinformatics. A Machine Learning Approach*, 2nd edn. (MIT Press, Cambridge 2001)

54.51   J. Bower, H. Bolouri (Eds.): *Computational Modelling of Genetic and Biochemical Networks* (MIT Press, Cambridge 2001)

54.52   L. Chen, K. Aihara: Stability analysis of genetic regulatory networks with time delay, IEEE Trans. Circuits. Syst. – I: Fundam. Theory Appl. **49**, 602–608 (2002)

54.53   L.F.A. Wessels, E.P. van Someren, M.J.T. Reinders: A comparison of genetic network models, Proc. Pac. Symp. Biocomput., Vol. 6 (World Scientific, Singapore 2001) pp. 508–519

54.54   P. D'Haeseleer, X. Wen, S. Fuhrman, R. Somogyi: Linear modeling of mRNA expression levels during CNS development and injury, Proc. Pac. Symp. Biocomput., Vol. 4 (World Scientific Publ., Singapore 1999) pp. 41–52

54.55   J. Norbury, R.E. Wilson: Dynamics of constrained differential delay equations, J. Comput. Appl. Math. **125**, 201–215 (2000)

54.56   L.D. Drager, W. Layton: Initial value problems for nonlinear nonresonant delay differential equations with possibly infinite delay, Electron J. Differ. Equ. **24**, 1–20 (1997)

54.57   D. Greenbaum, C. Colangelo, K. Williams, M. Gerstein: Comparing protein abundance and mRNA expression levels on a genomic scale, Genome Biol. **4**, 117111–117118 (2003)

54.58   R. Jansen, D. Greenbaum, M. Gerstein: Relating whole-genome expression data with protein-protein interactions, Genome Res. **12**(1), 37–46 (2002)

54.59   M. Thoby-Brisson, J. Simmers: Transition to endogenous bursting after long-term decentralization requires de novo transcription in a critical time window, J. Neurophysiol. **84**, 596–599 (2000)

54.60   G. Turrigiano, G. LeMason, E. Marder: Selective regulation of current densities underlies spontaneous changes in the activity of cultured neurons, J. Neurosci. **15**, 3640–3652 (1995)

54.61   W. Gerstner, W.M. Kistler: *Spiking Neuron Models* (Cambridge Univ. Press, Cambridge 2002)

54.62   W. Maass, C.M. Bishop (Eds.): *Pulsed Neural Networks* (MIT Press, Cambridge 1999)

54.63   I.C. Kleppe, H.P.C. Robinson: Determining the activation time course of synaptic AMPA receptors from openings of colocalized NMDA receptors, Biophys. J. **77**, 1418–1427 (1999)

54.64   S. Charpier, H. Leresche, J.-M. Deniau, S. Mahon, S.W. Hughes, V. Crunelli: On the putative contribution of $GABA_B$ receptors to the electrical events occuring during spontaneous spike and wave discharges, Neuropharmacology **38**, 1699–1706 (1999)

54.65   R.A. Deisz: $GABA_B$ receptor-mediated effects in human and rat neocortical neurones in vitro, Neuropharmacology **38**, 1755–1766 (1999)

54.66   M. Beierlein, C.P. Fall, J. Rinzel, R. Yuste: Thalamocortical bursts trigger recurrent activity in neocortical networks: Layer 4 as a frequency-dependent gate, J. Neurosci. **22**, 9885–9894 (2002)

54.67   N. Kasabov, Z.S.H. Chan, V. Jain, I. Sidorov, D.S. Dimitrov: Gene regulatory network discovery from time-series gene expression data – a computational intelligence approach, LNCS **3316**, 1344–1353 (2004)

54.68   Z. Chan, N. Kasabov, L. Collins: A hybrid genetic algorithm and expectation maximization method for global gene trajectory clustering, J. Bioinform. Comput. Biol. **3**, 1227–1242 (2005)

54.69   D.J. Whitehead, A. Skusa, P.J. Kennedy: Evaluating an evolutionary approach for reconstructing gene regulatory networks, Proc. 9th Int. Conf. Simul. Synth. Living Syst. (ALIFE9) (MIT Press, Boston 2004)

54.70   F.X. Wu, W.J. Zhang, A.J. Kusalik: Modeling gene expression from microarray expression data with state-space equations, Proc. Pac. Symp. Biocomput., Vol. 9 (World Scientific, Singapore 2004) pp. 581–592

54.71   M.B. Elowitz, S. Leibler: A synthetic oscillatory network of transcriptional regulators, Nature **403**, 335–338 (2000)

54.72   E.A. Thomas, C.A. Reid, S.F. Berkovic, S. Petrou: Prediction by modeling that epilepsy may be caused by very small functional changes in ion channels, Arch. Neurol. **66**(10), 1225–1232 (2009)

54.73   V. Santhakumar, I. Aradi, I. Soltesz: Role of mossy fiber sprouting and mossy cell loss in hyperexcitability: A network model of the dentate gyrus incorporating cell types and axonal topography, J. Neurophysiol. **93**(1), 437–453 (2005)

54.74   R.J. Morgan, I. Soltesz: Nonrandom connectivity of the epileptic dentate gyrus predicts a major role for neuronal hubs in seizures, Proc. Natl. Acad. Sci. USA **105**(16), 6179–6184 (2008)

54.75   N. T. Carnevale, M.L. Hines: NEURON simulation environment (2011), available online at http://www.neuron.yale.edu/neuron/ (last accessed 3 May 2013)

54.76   Ingenuity Systems, Inc.: Ingenuity pathway analysis (IPA) (2011), available online at http://www.ingenuity.com/index.html (last accessed 3 May 2013)

54.77   H. Caspers: Mechanisms of EEG generation – historical and present aspects. In: *Basic Mechanisms of the EEG*, ed. by S. Zschocke, E.-J. Speckmann (Birkhauser, Boston 1993)

54.78   H. Lodish, A. Berk, S.L. Zipursky, P. Matsudaira, D. Baltimore, J. Darnell: *Molecular Cell Biology*, 5th edn. (Freeman, New York 2000), Chap. 8: Genetic analysis in cell biology, pp. 254–293

54.79   R. Morita, E. Miyazaki, C.G. Fong, X.-N. Chen, J.R. Korenberg, A.V. Delgado-Escueta, K. Yamakawa: JH8, A gene highly homologous to the mouse jerky gene, maps to the region for childhood absence epilepsy on 8q24, Biochem. Biophys. Res. Commun. **248**, 307–314 (1998)

54.80   L.T. MacNeil, J.M. Walhout: Gene regulatory networks and the role of robustness and stochasticity in the control of gene expression, Genome Res. **21**, 645–657 (2011)

54.81   A. Garg, K. Mohanran, A. Di Cara, G. De Micheli, I. Xenarios: Modeling stochasticity and robustness in gene regulatory networks, Bioinformatics **25**, i101–i109 (2009)

54.82   M. Aldana, E. Balleza, S. Kauffman, O. Resendiz: Robustness and evolvability in genetic regulatory networks, J. Theor. Biol. **245**, 433–448 (2007)

54.83   S. Ciliberti, O.C. Martin, A. Wagner: Robustness can evolve gradually in complex regulatory gene networks with varying topology, PLoS Comput. Biol. **3**(2), 0164–0173 (2007)

55. Information Methods for Predicting Risk and Outcome of Stroke

Wen Liang, Rita Krishnamurthi, Nikola Kasabov, Valery Feigin

Stroke is a major cause of disability and mortality in most economically developed countries. It is the second leading cause of death worldwide (after cancer and heart disease) [55.1, 2] and a major cause of disability in adults in developed countries [55.3]. Personalized modeling is an emerging effective computational approach, which has been applied to various disciplines, such as in personalized drug design, ecology, business, and crime prevention; it has recently become more prominent in biomedical applications. Biomedical data on stroke risk factors and prognostic data are available in a large volume, but the data are complex and often difficult to apply to a specific person. Individualizing stroke risk prediction and prognosis will allow patients to focus on risk factors specific to them, thereby reducing their stroke risk and managing stroke outcomes more effectively. This chapter reviews various methods–conventional statistical methods and computational intelligent modeling methods for predicting risk and outcome of stroke.

55.1 Background

The human brain is a control center, an intricate master computer that controls and integrates various organ systems of the body. The nerve cells of the brain are connected with other cells in every part of the body. They deliver messages from the brain to tell the organ systems and the brain itself how to function, communicate, form memories, or make decisions. They also send messages back to the brain, telling it what is happening throughout the body. The brain is a complex system which evolves its functions and structures during its lifetime [55.4]. In the brain, there are complex interactions between genes and neuronal functions.

However, abnormalities in some of these interactions might cause brain diseases, such as brain cancer, Parkinson's disease, and Alzheimer's disease, etc. Stroke is a prevalent brain disease and has become a major public health challenge and concern in New Zealand, as well as globally. As proposed by *Tobias* et al. [55.5], in New Zealand, over 7000 people each year will experience a stroke event, and at least three-quarters of this population will die or be dependent on others for health care 1 year after stroke.

Until now, many intelligent systems have been developed with the purpose of improving health care and providing better health care facilities at a reduced cost. However, a review of the literature on stroke occurrence and outcome shows that traditional predictive models using standard population statistics can only apply to a group of people and are unable to predict the degree of risk occurrence or disability level for either an individual person at risk of stroke or a stroke survivor. These conventional statistical methods of prediction employ only the most significant predictive variables so that less statistically significant personal information that may be *clinically* significant for a particular person but not for a group of people is certainly lost [55.6]. For that reason, the concept of personalized modeling may indeed be an ideal approach worth exploring and integrating into the medical system for diagnosis, prediction, and management. Personalized modeling is an emerging effective approach for knowledge discovery in biomedical applications. The principle of this computational intelligent approach is to create a personalized diagnostic or prediction model for an individual person based on his/her nearest neighbors of predictive variables that are pertinent to that person.

The objectives of this chapter are to provide a brief introduction to stroke and then review various information methods, including conventional statistical methods and computational intelligent modeling methods for predicting risk and outcome of stroke.

55.2 Literature Review

55.2.1 Stroke

What Is a Stroke?
The World Health Organization (WHO) defines stroke as [55.7]:

> *rapidly developing clinical signs of focal or global disturbance of cerebral function lasting more than 24 hours (unless interrupted by surgery or death) with no apparent cause other than of vascular origin.*

It is generally accepted that the lifetime risk of stroke occurrence is 1 in 6, at least as high as the risk for developing Alzheimer's disease [55.8].

Stroke has also a large physical, psychological and financial impact on patients/families, the health care system, and society [55.9, 10]. Lifetime costs per stroke patient range from US\$ 59.8K to US\$ 230K [55.10]. The majority (about 75%) of cases of stroke occur in people over the age of 65 years [55.11, 12], and about one third of patients die of stroke within a year of onset [55.13, 14]. Over half of the survivors remain dependent on others for everyday activities, often with significant adverse effects on caregivers [55.15].

Their family members are also affected due to the suffering of their loved ones as well as by the burden of caring for them, uncertain about future plans and anxious about increased financial burdens for the patient's treatment.

What Are the Risk Factors?
There are about 200 factors that can increase the risk of a stroke but the most important ones are:

- Elevated blood pressure
- Smoking
- Diabetes mellitus
- Increasing age
- Overweight (especially abdominal obesity)
- High blood cholesterol level
- Poor, unbalanced diet lacking fruit and vegetables
- Sedentary lifestyle
- Heart rhythm problems such as atrial fibrillation
- History of heart disease.

What Are the Symptoms?
The signs and symptoms of a stroke differ and depend on the area of the brain affected and the amount of brain tissue damaged. Small strokes may not cause any significant focal neurological symptoms (so-called silent strokes). However, when accumulated, they may lead to clinically significant consequences, such as vascular dementia. In general, stroke in the left side of the brain has clinical symptoms on the right side of the body, whereas the left side of the body is affected by a stroke in the right side of the brain.

According to the US National Institute of Neurological Disorders and Stroke (NINDS), the common symptoms of stroke generally come suddenly and may include:

- Sudden loss of consciousness: the victim may become stuporous or hard to arouse.
- Sudden loss of vision: difficulty with seeing in one or both eyes, such as blurred vision.
- Sudden headache: sudden onset severe headache that may be accompanied by vomiting or dizziness (loss of balance).
- Sudden trouble with muscle movements: difficulty with walking, moving the arm or leg on one side of body, carrying or picking up objects.
- Sudden trouble with speaking and understanding: may have problems with thinking or forming speech, such as when speaking, the words sound fine but do not make sense.

How Does Stroke Happen?
Stroke is a heterogeneous disorder that consists of two major pathological types (ischemic and hemorrhagic), and each type has different subtypes with different

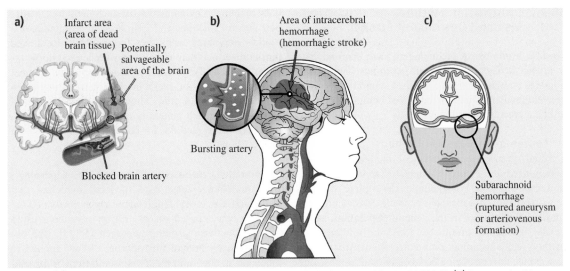

Fig. 55.1 (**a**) Ischemic stroke occurs when a blood vessel in the brain becomes blocked [55.16]. (**b**) Intracerebral hemorrhage occurs when blood vessels within the brain become damaged and burst within the brain [55.16]. (**c**) Subarachnoid hemorrhage occurs when a cerebral aneurysm ruptures; blood will fill the space surrounding the brain [55.16]

causes and outcomes. The two major types of stroke are described below (Fig. 55.1a–c).

Ischemic Stroke. Ischemic stroke is the most common type of stroke, accounting for almost 85% of all stroke cases. It occurs when there is a blood clot in the blood vessel of the brain that reduces or blocks the blood supply coming from the heart to the brain. As there is no nutrient/energy storage in the brain, it needs a constant supply of nutrients from the blood. The blood carries sugar and oxygen to the brain and takes away cellular waste and carbon dioxide. If an artery is blocked, the brain cells cannot receive the required level of oxygen and glucose needed, thus the affected cells begin to shut down. If there is no blood supply for as little as 7 s, the affected brain cells may die.

Based on different mechanisms of stroke, ischemic stroke consists of at least 4 subtypes: cardioembolic stroke, ischemic stroke due to large artery disease (such as atherosclerosis), ischemic stroke due to small artery disease (such as hypertension, intracranial arteritis), and ischemic stroke due to hematological disorders and other rare conditions.

Hemorrhagic Stroke. Hemorrhagic stroke accounts for up to 15% of all stroke cases. It often happens when an artery bursts and bleeds into the brain (intracerebral hemorrhage) or around the brain (subarachnoid hemorrhage).

Intracerebral Hemorrhage. Intracerebral hemorrhage is a type of stroke caused by the breaking of a diseased blood vessel in the brain, leading the blood to leak into the brain tissue. This may lead to direct (destroying brain cells) or indirect damage of the affected brain cells due to a sudden build-up in the intracranial pressure, each of which may lead to unconsciousness, lost neurological function, or even death. Intracerebral hemorrhage may be caused by different mechanisms (e.g., elevated blood pressure, amyloid angiopathy) in different parts of the brain (e.g., supratentorial, infratentorial hemorrhage), each of which carries a different prognosis and requires different management strategies.

Subarachnoid Hemorrhage. Subarachnoid hemorrhage is the result of a blood vessel bursting in the area between the brain and the thin tissues that surround the brain. This area is called the subarachnoid space, which is the area outside of the brain tissue. Typical symptoms of a patient with subarachnoid hemorrhage include loss of consciousness, vomiting, severe headache or neck pain, and neck stiffness. Subarachnoid hemorrhage most often results from a rupture of an intracranial aneurysm but it may also be caused by a rupture of the brain artery due to other causes (so-called nonaneurysmal subarachnoid hemorrhage). The management of aneurysmal subarachnoid hemorrhage is very much different from the management of nonaneurysmal subarachnoid hemorrhage.

55.2.2 Information Methods for Predicting Risk and Outcome of Stroke

To date, a number of technologies have been adopted to predict stroke occurrence and outcomes. These technologies can be divided into two major categories: conventional statistical methods and computational intelligent machine learning methods.

Conventional Statistical Methods

Currently conventional statistical methods are more widely used and commonly applied for stroke prediction data analysis. For example, descriptive statistics (e.g., frequency statistics) has been used to calculate the frequency of strokes in the general population, across gender, and ethnic groups etc. [55.17]; the correlation method (e.g., Spearman rank correlation) was utilized to compute the correlation between two different scales, such as the Barthel index and SF-36 [55.18]; logistic regression was applied to analyze the factors associated with the SF-36 subscales in order to discover which of these variables best discriminate between patients with low and high scores on the SF-36 subscales [55.19]; and one-way of variance and χ^2 (chi square) test have been adopted to assess the differences between different stroke outcomes [55.20].

However, conventional statistical methods have limitations in efficiency and prediction accuracy compared to machine learning methods. *Khosla* and his colleagues [55.21] present an integrated machine learning approach to compare the Cox proportional hazards model (one of the most commonly used conventional statistical methods in medical research) on the Cardiovascular Health Study (CHS) dataset for stroke risk prediction. Their research demonstrated that machine learning methods significantly outperform the Cox model in terms of stroke risk estimation.

Machine Learning Methods
In [55.22]:

> *Machine learning is the process of discovering and interpreting meaningful information, such as new correlations, patterns and trends by sifting through large amounts of data stored in repositories, using pattern recognition technologies as well as statistical and mathematical technique.*

In other worlds, machine learning is a process of using different analysis techniques to observe previously unknown, potentially meaningful information, and discover strong patterns and relationships from a large

dataset that can be applied most accurately to a particular person. *Kasabov* [55.23] classified computational models into three categories (e.g., global, local, and personalized), which have being widely used in the areas of data analysis and decision support in general, and in the areas of medicine and bioinformatics in particular. A review study shows that the personalized modeling approach generally outperforms the conventional statistical methods for prediction or classification of conditions.

Personalized Modeling. The concept of personalized modeling is one type of local modeling that is created for every single new input vector of the problem space based on its nearest neighbors using the transductive reasoning approach [55.23]. The basic philosophy behind this approach when applied to medicine is that every person is different from others, thus he/she needs and deserves a personalized model and treatment that best predicts possible outcomes for this person. This way of reasoning is much closer to a way of clinical decision making that is used by clinicians in their everyday practice as opposed to a group-based approach utilized by conventional statistical methods.

Personalized modeling is a novel and effective method that has been applied for evaluating and dealing with a variety of modeling problems. For instance, in the field of personalized health care, the knowledge uncovered by this approach has significantly contributed to prediction, diagnosis, and therapy for individual patients' diseases. In the articles by *Ginsburg* and *Mc-Carthy* [55.24] and *TEMU* [55.25], it is mentioned that providing a personalized therapy for an individual patient during the diagnosis time frame has proved to be very efficient and helpful.

Nowadays, the concept of personalized medicine is becoming a leading trend in medicine, health care, and life science. As presented by *Lesko* [55.26], from US Food and Drug Administration

> *Personalized medicine can be viewed ... as a comprehensive, prospective approach to preventing, diagnosing, treating, and monitoring disease in ways that achieve optimal individual healthcare decisions.*

As stated by the *Personalized Medicine Coalition* [55.27], traditional medicine is primarily based on the visible symptoms of the disease, but recently doctors have been able to integrate an individual patient's molecular profile to characterize various forms of cancer (e.g., breast cancer, brain cancer, and liver cancer, etc.) to make a decision about treatment. Furthermore, according to *Ginsburg* and *McCarthy* [55.24], the objective of personalized medicine is to determine a patient's disease at the molecular level, so the right therapies can be applied to the right people at the right time. The personalized medicine approach is being increasingly used to accommodate the individual patient's molecular/genetic profile.

Currently, there are several methods of personalized modeling, which are outlined below.

K Nearest Neighbor (KNN). The simplest method of personalized modeling was originally proposed by *Fix* and *Hodges* in 1951 [55.28]. KNN is a supervised learning algorithm that has been successfully used for classifying sets of samples based on nearest training samples in a multi-dimensional feature space.

The basic idea behind the KNN algorithm is as follows:

- Firstly, a set of pair features (e.g., $(x_1, y_1), \ldots, (x_n, y_n)$) are defined to specify each data point, and each of those data points are identified by the class labels $C = \{c_1, \ldots, c_n\}$.
- Secondly, a distance measure is chosen (e.g., Euclidean distance or Manhattan distance) to measure the similarity of those data points based on all their features.
- Finally, the k-nearest neighbors are found for a target data point by analyzing similarity and using the majority voting rule to determine which class the target data point belongs to.

Weighted Nearest Neighbor (WKNN). This is designed based on the transductive reasoning approach, which has been widely used to evaluate the output of a model focusing on solely an individual point of a problem space using information related to this point [55.29]. In the WKNN algorithm, each single vector requires a local model that is able to best fit each new input vector rather than a global model, thus each new input vector can be matched to an individual model without taking any specific information about existing vectors into account.

In contrast to the KNN algorithm, the basic idea behind the WKNN algorithm is that the output of a new input vector is calculated not only depending upon its k-nearest neighbor vectors, but also upon the distance between the existing vectors and the new input vector, which is represented as a weight vector w.

Weighted Distance and Weighted Variables K Nearest Neighbor (WWKNN). This is a novel personalized modeling algorithm, which was proposed by Kasabov in 2007. The basic idea behind this algorithm is as follows: the output of each new input vector is measured not only depending upon its k-nearest neighbors, but also upon the distance between the existing vectors and the new input vectors, and also the power of each vector which is weighted according to its importance within the subspace (local space) to which the new input vector belongs. We start with the assumption that all the variables from a data set are used and the distance of vectors is calculated in a V-dimensional space with all input variables having the same impact on the output variables. However, the different variables may vary in importance when classifying vectors if these variables are ranked by their discriminative power in classifying vectors over the entire V-dimensional Euclidean space. As a result, it can be seen that variables may have a different ranking when we measure the discriminative power of the same variables for a subspace of the problem space. The output of each new input vector can be calculated by using this type of ranking within the neighborhood of k-nearest neighbor vectors.

Evolving Spiking Neural Networks (eSNN). The brain is the center of the nervous system, it is an extremely complex organ. The cerebral cortex of the human brain contains roughly 15–33 billion neurons, depending on gender and age, linked with up to 10 000 synaptic connections each.

A typical neuron in the human brain is sketched in Fig. 55.2. A typical neuron can be divided into three functionally distinct parts, called dendrites, soma, and axon. Generally speaking, the dendrites play the role of the *input devices* that collect signals from other neurons and transmits them to the soma. Soma is the main part of the neuron containing the genetic information: if the total input exceeds a certain threshold, then an output signal is generated. The output signal is taken over by the *output devices*, the axons, which deliver the signal to other neurons or other parts of the body. The junction between two neurons is called a synapse. It is used to transfer signals between two neurons.

The remarkable information processing capabilities of the brain have inspired numerous mathematical abstractions of biological neurons. A neural network (NN) known as an artificial neural network (ANN) is defined as a hardware or software computational model that is inspired by the biological nervous systems, such

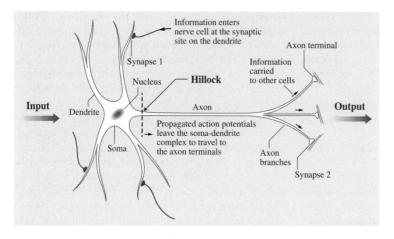

Fig. 55.2 Schematic drawing of a typical neuron [55.31]

as the brain, and processes information. Many ANN models have been successfully developed and applied across many disciplines, including classification, time series prediction, pattern recognition, and so on [55.30]. However, current ANN models do not provide good performance when applied to complex stochastic and dynamic processes such as modeling brain diseases. For this reason, new ANN models should be developed in order to become more accurate and efficient in knowledge discovery and information processing.

Maass [55.32] classifies past and current neural network models into three generations. Spiking neural networks (SNN) is the third generation of neural network models, which is a complex and biologically plausible connectionist model. In biological systems neurons are connected at synapses, and signals transfer information from one neuron to another. Quite a few models of SNN have been developed so far: Hodgkin–Huxley's model [55.33]; Spike response models [55.34–36]; integrate-and-fire models [55.35, 37]; Izhikevich models [55.38–40], etc.

SNN has been increasingly applied in the field of science and engineering as in other disciplines to solve complicated prediction and classification problems, such as speech recognition [55.41], audio and video analysis [55.42, 43], and financial forecasting [55.44]. In recent years, SNN has become a powerful computational tool that has been widely adopted for diagnosing and monitoring the prognosis of a disease, as evidenced by over 500 published papers each year featuring neural network applications in medicine [55.45]. In medical practice, SNN has been successfully used for diagnosing breast cancer [55.46], making prognosis for patients with congestive heart failure [55.47], and predicting the risk of death for lung cancer patients [55.48].

More recently, spiking neural networks have been successfully used for the prediction of functional outcome associated with clinical variables that are associated with rehabilitation in stroke [55.6]; early diagnosis of ischemic stroke [55.49] and classifying the gait patterns of post-stroke patients into homogenous groups [55.50].

Evolving SNN (eSNN), which is an SNN approach, was first proposed by *Wysoski* et al. in 2006 [55.51]. It consists a set of SNN that evolve their structure through incremental and fast one-pass learning from data.

Space and time can be viewed as the important aspects of all real-world phenomena. Spatio-temporal data (STD) contains information relating space and time. According to *Fayyad* and *Grinstein* [55.52], nowadays, approximately 80% of the available datasets have spatial components and are often related to some temporal aspects. Examples of such data are environmental, audio/visual, medical, and brain signals. So far, most available stroke data are STD, thus a great challenge is how to process these complex STD.

In general, classical statistical and computational techniques are insufficient when they are applied to spatio-temporal datasets for the following reasons:

- Spatio-temporal datasets are embedded into continuous space, whereas the classical datasets are often discrete.
- The patterns in these datasets are often local, but classical techniques normally focus on global patterns.
- They model either space and time separately, or mix both components in a simple way, missing to capture essential relations between variables in the STD.

eSNN is considered an emerging computational technique for the analysis of spatio-temporal datasets. As proposed by *Kasabov* [55.30], the development of a novel integrated evolving personalized modeling systems by utilizing novel technology such as evolving spiking neural networks (eSNN) might facilitate more precise decision-making to ensure patients receive optimal prognosis and treatment.

55.3 Conclusion

The human brain is a complex network of interconnected cells (neural networks) that develops its functionalities and structures during the lifetime of a person. There are complex interactions between genes, the environment, and neuronal function within the brain. However, some of these interactions might become abnormal and cause brain diseases.

Stroke is a prevalent brain disease and has become a major public health concern in New Zealand, as in many other countries worldwide. A large volume of biomedical data on stroke risk factors and prognosis is available, but the interpretation of this data is complex and challenging. An increasing number of studies have attempted to explore these complex data, but most of these studies applied conventional statistical methods. Literature reviews show that the traditional predictive models using conventional statistical methods can only apply to a group of people and are unable to predict the degree of risk occurrence or disability level for either an individual person at risk of stroke or a stroke survivor. Therefore, there is a growing need for utilizing computational models to study this data, especially applied

to estimating personalized risk to better understand the pathophysiology in individual and specific groups of stroke, and to achieve improved and reliable risk prediction for individuals.

Personalized modeling was recently proposed as a novel method for knowledge discovery in biomedical applications. The basic philosophy behind this computational intelligent approach when applied to medicine is that every person is different from others, thus he/she needs and deserves a personalized model and treatment that best predicts possible outcomes for this person. The principle of this approach is to create a personalized diagnostic or prediction model for an individual person based on his/her nearest neighbors of predictive variables that are pertinent for that person. For instance, employing eSNN might result in reducing the time, cost, medical error, and mortality rate. The concept of personalized modeling can be viewed as an ideal approach well worth exploring and integrating into the medical system for diagnosis, prediction, and prescription of various diseases, including stroke.

References

55.1   S.C. Johnston, S. Mendis, C.D. Mathers: Global variation in stroke burden and mortality: Estimates from monitoring, surveillance, and modelling, Lancet Neurol. **8**, 345–354 (2009)

55.2   P.M. Rothwell: The high cost of not funding stroke research: A comparison with heart disease and cancer, Lancet Neurol. **357**, 1612–1616 (2001)

55.3   M. Tobias, J. Cheung, H. McNaughton: *Modelling Stroke: A Multi-State Life Table Model*, Occasional Bulletin/Public Health Intelligence (Ministry of Health, Wellington 2002)

55.4   N. Kasabov: *Brain-, Gene-, and Quantum Inspired Computational Intelligence: Challenges and Opportunities* (Springer, Heidelberg 2007)

55.5   M. Tobias, J. Cheung, K. Carter, C. Anderson, V.L. Feigin: Stroke surveillance: Population-based estimates and projections for New Zealand, Aust. N. Z. J. Public Health **31**, 520–525 (2007)

55.6   W.J. Oczkowski, Susan Barreca: Neural network modeling accurately predicts the functional

outcome of stroke survivors with moderate disabilities, Arch. Phys. Med. Rehabil. **78**, 340–345 (1997)

55.7   K. Aho, P. Harmsen, S. Hatano, J. Marquardsen, V.E. Smirnov, T. Strasser: Cerebrovascular disease in the community: Results of a WHO collaborative study, Bull. World Health Organ. **58**, 113–130 (1980)

55.8   S. Seshadri, A. Beiser, M. Kelly-Hayes, C.S. Kase, R. Au, W.B. Kannel, P.A. Wolf: The lifetime risk of stroke: Estimates from the Framingham Study, Stroke **37**, 345–350 (2006)

55.9   K. Strong, C. Mathers, R. Bonita: Preventing stroke: Saving lives around the world, Lancet Neurol. **6**, 182–187 (2007)

55.10   J.J. Caro, K.F. Huybrechts, I. Duchesne: Management patterns and costs of acute ischemic stroke: An international study, Stroke **31**, 582–590 (2000)

55.11   R. Bonita, C.S. Anderson, J.B. Broad, K.D. Jamrozik, E.G. Stewart-Wynne, N.E. Anderson: Stroke incidence and case fatality in Australasia. A comparison of the Auckland and Perth population-based stroke registers, Stroke **25**, 552–557 (1994)

55.12   R. Bonita, J.B. Broad, R. Beaglehole: Changes in stroke incidence and case-fatality in Auckland, New Zealand, 1981–91, Lancet **342**, 1470–1473 (1993)

55.13   C.S. Anderson, K.D. Jamrozik, R.J. Broadhurst, E.G. Stewart-Wynne: Predicting survival for 1 year among different subtypes of stroke. Results from the Perth Community Stroke Study, Stroke **25**, 1935–1944 (1994)

55.14   R. Bonita, M.A. Ford, A.W. Stewart: Predicting survival after stroke: A three-year follow-up, Stroke **19**, 669–673 (1988)

55.15   C.S. Anderson, J. Linto, E.G. Stewart-Wynne: A population-based assessment of the impact and burden of caregiving for long-term stroke survivors, Stroke **26**, 843–849 (1995)

55.16   V. Feigin: *When Lightning Strokes: An Illustrated Guide to Stroke Prevention and Recovery* (Harper-Collins, Auckland 2004)

55.17   V. Feigin, K. Carter, M. Hackett, P.A. Barber, H. Mc-Naughton, L. Dyall, M.H. Chen, C. Anderson: Ethnic disparities in incidence of stroke subtypes: Auckland regional community stroke study, 2002–2003, Lancet Neurol. **5**, 130–139 (2006)

55.18   S.M. Lai, P.W. Duncan, J. Keighley: Prediction of functional outcome after stroke: Comparison of the orpington prognostic scale and the NIH stroke scale, Stroke **29**, 1838–1842 (1998)

55.19   M.L. Kauhanen: *Quality of life after stroke: Clinical, functional, psychosocial and cognitive correlates, Neurology* (University of Oulu, Oulu 1999)

55.20   V.L. Feigin, S. Barker-Collo, V. Parag, H. Senior, C.M. Lawes, Y. Ratnasabapathy, E. Glen: Auckland stroke outcomes study. Part 1: Gender, stroke types, ethnicity and functional outcomes 5 years poststroke., Neurology **75**, 1597–1607 (2010)

55.21   A. Khosla, Y. Cao, C.Y. Lin, H.K. Chiu, J. Hu, H. Lee: An integrated machine learning approach to stroke prediction, 16th ACM SIGKDD Conf. Knowl. Discov. Data Min. (2010)

55.22   D.T. Larose: *Discovering Knowledge in Data: An Introduction to Data Mining* (Wiley, Hoboken 2005)

55.23   N. Kasabov: *Evolving Connectionist Systems: The Knowledge Engineering Approach* (Springer, London 2007)

55.24   G.S. Ginsburg, J.J. McCarthy: Personalized medicine: Revolutionizing drug discovery and patient care, Trends in Biotechnol. **19**, 491–496 (2001)

55.25   TEMU: Personalised Medicine: Current Trends and Scientific Challenges, TEMU 2008 (TEMU, Crete 2008), available online at http://www.temu.gr/2008/sessions.html

55.26   L.J. Lesko: Personalized medicine: Elusive dream or imminent reality, Clin. Pharmacol. & Ther. **81**, 807–816 (2007)

55.27   Personalized Medicine Coalition: The case for personalized medicine (2006), available online at http://www.personalizedmedicinecoalition.org/communications/pmc_pub_11_06.php

55.28   E. Fix, J.L. Hodges: *Discriminatory analysis: Non-parametric discrimination: Consistency properties* (UASF School of Aviation Medicine, Randolph Field, 1951)

55.29   V. Vapnik: *Statistical Learning Theory* (Wiley-Interscience, New York 1998)

55.30   N. Kasabov: To spike or not to spike: A probabilistic spiking neuron model, Neural Netw. **23**, 16–19 (2010)

55.31   S. Barber. *AI: Neural Network for beginners (Part 1 of 3)* (2007), available online at http://www.codeproject.com/KB/recipes/NeuralNetwork_1.aspx

55.32   W. Maass: Networks of spiking neurons: The third generation of neural network models, Neural Netw. **10**, 1659–1671 (1997)

55.33   A.L. Hodgkin, A.F. Huxley: A quantitative description of membrane current and its application to conduction and excitation in nerve, J. Physiol. **117**, 500–544 (1952)

55.34   W. Gerstner: Time structure of the activity of neural network models, Phys. Rev. **51**, 738–758 (1995)

55.35   W. Gerstner, W.M. Kistler: *Neuron Models: Single Neurons, Populations, Plasticity* (Cambridge Univ. Press, Cambridge 2002)

55.36   W. Kistler, W. Gerstner: Reduction of Hodgkin-Huxley equations to a single variable threshold model, Neural Comput. **9**, 1015–1045 (1997)

55.37   W. Maass, C.M. Bishop: *Pulsed Neural Networks* (MIT Press, Cambridge 1999)

55.38   E.M. Izhikevich: Which model to use for cortical spiking neurons, IEEE Trans. Neural Netw. **15**, 1063–1070 (2004)

55.39   E.M. Izhikevich: *Dynamical Systems in Neuroscience* (MIT Press, Cambridge 2007)

55.40   E.M. Izhikevich, G.M. Edelman: Large-scale model of mammalian thalamocortical systems, Proc. Natl. Acad. Sci. USA **105**(9), 3593–3598 (2008)

55.41   W.C. Yau, D.K. Kumar, S.P. Arjunan: Visual recognition of speech consonants using facial movement features, Integr. Comput.-Aided Eng. **14**, 49–61 (2007)

55.42   C. Fyfe, W. Barbakh, W.C. Ooi, H. Ko: Topological mappings of video and audio data, Int. J. Neural Syst. **18**, 481–489 (2008)

55.43   N. Tsapatsoulis, K. Rapantzikos, C. Pattichis: An embedded saliency map estimator scheme: Application to video encoding, Int. J. Neural Syst. **17**, 289–304 (2007)

55.44   N.C. Schneider, D. Graupe: A modified lamstar neural network and its applications, Int. J. Neural Syst. **18**, 331–337 (2008)

55.45   V. Gant, R. Dybowski: Artificial Neural Networks: Practical Considerations for Clinical Applications. In: *Clinical Applications of Neural Networks*, ed.

by R. Dybowski, V. Gant (Cambridge Univ. Press, Cambridge 2001)

55.46 T. Kiyan, T. Yildirim: Breast cancer diagnosis using statistical neural networks, Turk. Symp. Artif. Intell. Neural Netw. (Istanbul 2003)

55.47 P.J. Cowburn, J.G.F. Cleland, A.J.S. Coats, M. Komajda: Risk stratification in chronic heart failure, Eur. Heart J. **19**, 696–710 (1996)

55.48 E. Bartfay, W.J. Mackillop, J.L. Pater: Comparing the predictive value of neural network models to logistic regression models on the risk of death for small-cell lung cancer patients, Eur. J. Cancer Care **15**, 115–124 (2006)

55.49 T. Anita, B. Surekha, S.N. Mishra: Early diagnosis of ischemia stroke using neural network, 2009

Proc. Int. Conf. Man-Machine Syst. (ICoMMS) (Batu Ferringhi, Penang 2009)

55.50 K. Kaczmarczyk, A. Wit, M. Krawczyk, J. Zaborski: Gait classification in post-stroke patients using artificial neural networks, Gait & Posture **30**, 207–210 (2009)

55.51 S.G. Wysoski, L. Benuskova, N. Kasabov: On-line learning with structural adaptation in a network of spiking neurons for visual pattern recognition, 2006 Int. Conf. Artif. Neural Netw., Lecture Notes in Computer Science (Heidelberg, Berlin 2006) pp. 61–70

55.52 U.M. Fayyad, G.G. Grinstein: *Information Visualization in Data Mining and Knowledge Discovery* (Morgan Kaufmann, Los altos 2001)

56. sEMG Analysis for Recognition of Rehabilitation Actions

Qingling Li, Zeng-Guang Hou, Yu Song

Surface electromyography (sEMG), a measurement of biomedical electronic signals from the muscle surface using electrodes, shows the motor status of the nerve–muscle system and motor instruction information. Active motion intention and the motor status of impaired stroke patients can be acquired by sEMG. Currently, sEMG is widely used in prosthetic arm control, rehabilitation robot control, exoskeletal power assist robot control, tele-operated robots, virtual reality, and so on. The application of sEMG to a rehabilitation robot is studied in this chapter. sEMG is used to build an information channel between the patient and the robot to obtain biological feedback control during rehabilitation training. It is of great significance for the improvement of patients' consciousness of active participation. It will also help to improve the evaluation and efficiency of rehabilitation. It establishes a general scheme for other applications related to the human–machine interface.

First, the generation mechanism and characteristic of the sEMG signal are presented in this chapter. Next, current feature analysis methods of sEMG are summarized. The advantages and disadvantages of each method are discussed. Then, an example of sEMG signal analysis for the recognition of rehabilitation actions is given. Finally, future work and discussions are given in the conclusion.

Part K | 56

56.1 Overview

An sEMG signal accompanied by generation of skeletal muscle motion is the direct manifestation of central nervous system regulation on the skin. It contains abundant motor function status information. According to the motor nerve conduction sequence, biomedical electronic signal occur ahead of muscle motion. This indicates *which muscle will act* and *how it will do*, which are all brain instructions. The sEMG signal is analyzed to predict the active motion intention of patients, which is very important for motor-impaired patients. This treatment mode fully inspires positive nerve conduction and encourages reverse nerve pathways and finally causes efficient therapy. Thus, effective sEMG feature information is particularly important. The analysis method of the sEMG signal for rehabilitation action recognition will be discussed next. Before that, *what sEMG is* and *what the characteristics of sEMG are* will be described.

56.2 What Is sEMG?

56.2.1 Generation Mechanism of sEMG

As the driving elements of joint movements, skeletal muscles are controlled by the brain. The corresponding muscles attached to the bone work together to result in limb coordination motions. Each skeletal muscle fiber is innervated by an α motor neuron in the spinal cord. The axon terminal of each α motor neuron is divided into a number of branches connecting muscle fibers. Thus, an exciting neuron causes innervated muscle contraction, which is called a motor unit (MU), i. e., an α motor neuron and all of the innervated skeletal muscle fibers. This is also the basic function unit of the neuromuscular system. Figure 56.1 shows the physiological structure.

An sEMG signal originates from an α motor neuron, which is excited by innervations or external stimulus. The nerve impulse passes to muscle fibers and generates a motor unit action potential (MUAP). During this course, muscle fibers of all MUs excite almost simultaneously, and the muscle fiber excitation of different MUs is mutually independent. So, sEMG acquired from the skin with electrode is actually an integration of all MUAP strings generated by the entire MU. There are two means for sEMG signal acquisition: the needle electrode and the surface electrode. The former is inserted into the muscle to detect MUAP, and the latter is placed on the skin for the integrated muscle motor potential. Compared with needle EMG, sEMG

is more susceptible to interference and has a lower signal-to-noise ratio due to indirect and non-specific measurement, a larger detection range, and lower spatial resolution. Its signal quality is easily influenced by many factors such as the electrode position, skin condition, hair condition, fat thickness, state of mind, circumstances of electromagnetic interference, and so on. However, sEMG measurement is non-invasive and does not need physician participation.

56.2.2 Characteristics of sEMG

sEMG signals of different patients have some differences and similarities.

1. sEMG is an AC voltage signal. Its amplitude is proportional to muscle strength and has approximately linear correlativity with relaxation and tension of the muscle.
2. The sEMG signal is so weak that MUAP amplitude fluctuates within a range of $100\,\mu V$ to $2\,mV$. The amplitude range of the superposition signal is from $2\,\mu V$ to $5\,mV$.
3. The sEMG signal is a low frequency signal. Its main frequency mainly resides in the range between $10-1000\,Hz$ and the largest frequency resides in the range between $50-150\,Hz$.
4. Another important characteristic of the sEMG signal is non-stationarity. That is to say its statistical characteristics vary with time because the sEMG signal is an averaging result of MUAP from muscle fibers with different intensity and propagation direction.
5. For the specified muscles of different individuals, regular sEMG signals are generated for the same action. On the other hand, there are differences between the sEMG signals of different muscles. These characteristics make it feasible to recognize limb actions and control rehabilitation robots using sEMG signals.

56.2.3 History of sEMG Analysis

The close relationship between muscle contraction and electricity was verified by Galvani in 1781. This new discovery launched a new era of electrophysiology and initiated discoveries and research regarding the bioelectricity signal and nerve impulse conduction. In 1894, Dubois-Reymond discovered the recording ability of

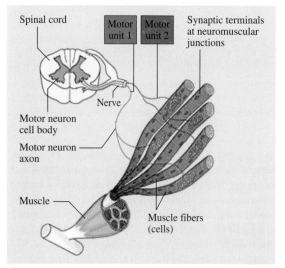

Fig. 56.1 Physiological structure of the motor unit [56.1]

electricity when a muscle contracts. He then proposed the concept of MUAP. In 1980, Marey recorded the first electrical activity of a muscle, called an electromyogram (EMG). This was followed by many inventions of detection instruments such as the electromyogram, the electroencephalogram (EEG), the electrocardiogram (ECG), the electrogastrogram (EGG), and so on. Applications on bioelectricity signal thus extend to many fields. Among these, sEMG is widely employed in rehabilitation medicine and sports science due to its many advantages, such as non-invasion, real-time, easy operation, multi-point measurement, and close relationship with muscle function and action status.

Presently, research on sEMG analysis mainly focuses on two parts: one is the physiology information and the other is the motion information. In the first part, the study object is the physiological and biochemical process during motion. It is used in disease diagnosis of the neuromuscular system, evaluation of motor function, ergonomics analysis, fatigue analysis, sports technical rationality, and so on. The second part studies motion recognition by analyzing the sEMG signal. It is widely used in human–machine interface, clinical rehabilitation, and so on. Analysis methods (sometimes called feature extraction methods) of sEMG include the frequency domain method and the time-frequency domain method. This will be elaborated in detail in the next section.

In this chapter, we focus on feature extraction methods and related applications on rehabilitation action recognition. This is the kernel technology of sEMG motion pattern recognition. Development history and related research on feature extraction methods will be reviewed. Among existing extraction methods, the time domain method is an initial and simple method. It takes the signal as a function of time and analyzes its statistic characteristics. This kind of method includes integration of sEMG, zero crossing, variance, Willison amplitude, waveform length, standard deviation, root mean square, histogram, and so on. Dorcas and Childress controlled an electronic prosthesis with a single degree of freedom using the amplitude and change rate of single channel sEMG in 1966 and 1969, respectively. Bernard Hudgins employed average absolute value, zero crossing, phase change number, wavelength, etc., to study multi-channel sEMG recognition of elbow joint flexion and extension. In 2005, Inhyuk Moon analyzed the amplitude of sEMG associated with the levator scapula muscle and then controlled an electric wheelchair to move forward, backward, or turn according to the double-threshold method. It can assist the

walking abilities of people affected by paralysis. In general, time domain-based statistical analysis methods have a lower motion recognition rate and, therefore, are not particularly suitable for sEMG because of the non-stationary property of its signals. Hence, parameter model methods, especially the autoregressive model (AR), are widely used because they can acquire the signal's short-term stationary property, are capable of describing the dynamic information of the signal and with higher resolution and linear complexity. In 1975, D. Graupe applied an autoregressive moving average model (ARMA) to sEMG analysis and then controlled a prosthesis accordingly. Experimental result showed that the sEMG signals were, as expected, stationary because of stationarity in short time intervals. This model realizes an 85% recognition rate of three upper limb motions. In 1995, Jer-JunnLuh applied a time-varying AR model to improve the estimation accuracy of the sEMG signal amplitude. After that, he developed a real-time sEMG classification system. The system judged the muscle contraction motion by the zero crossing method and extracted cepstrum coefficients of the AR model during the course of analyzing dual channel sEMG signals of five neck and shoulder motions. Other researchers transformed AR model parameters into reflection coefficients, cepstral coefficients, logarithmic area ratios, etc., to improve the degree of clustering separation. In 2002, Lamounier studied the on-line recognition algorithm of isometric and isotonic motion by the AR model method. A fourth-order model was used to analyze sEMG signals associated with the elbow flexion/extension and wrist pronation/supination motions. Final experimental results showed a 92% recognition accuracy. In 2005, Zhao Jingdong analyzed sEMG signals associated with the flexor extensor, the flexor hallucis longus muscle, during finger movements by using the AR model. It was combined with the integration of absolute values and inputted into a neural network to recognize four different motions and achieved high recognition accuracy.

In recent years, researchers have begun to explore time-frequency domain analysis methods. Typical methods include the short-time Fourier transform (STFT), the wavelet transform (WT), the wavelet packet transform (WPT), Wigner–Wille distribution (WVD), the complex cepstrum coefficient, the linear predictive coefficient (LPC), and so on. Englehart applied three time-frequency features (STFT, WT, and WPT) to sEMG action classification. The result showed that this kind of feature can increase the information amount of the pattern classification system. Researchers also extracted a WT-based feature set to improve recognition accuracy. Wang Rencheng analyzed sEMG signals associated with upper limbs by multi-scale decomposition and recognized hand gestures by variance of wavelet the coefficient. M. Khezri subtracted the average from WVD coefficients to obtain zero crossing. This feature vector had 91.3% average recognition accuracy for six hand gestures. Some comparative studies of this method and STFT, WT, etc., were also carried out to verify its recognition efficiency. Feature analysis of sEMG signals is an emerging technology that includes a variety of methods, which is explained in the following section.

56.3 sEMG Analysis Method

The sEMG signal has a typical non-stationary and non-linear characteristic. Many researchers are studying its analysis and recognition method, and some results have been obtained. It mainly includes five parts: the time domain analysis method, the frequency domain analysis method, the time-frequency domain analysis method, a high-order spectral method, and chaos and a fractal method.

56.3.1 Time Domain Analysis Method

Traditional analysis methods consider the sEMG signal as a random signal satisfying zero mean Gaussian distribution. This kind of method extracts features changing along time. It is called the time domain analysis method and includes following specific algorithms.

1. Integral of EMG (IEMG). This is an area of curves in unit time for an sEMG signal after rectifying and filtering. It reflects changes of signal intensity as time goes. In general, IEMG is used to analyze the contractile characteristics of muscles in unit time. It is calculated by

$$\text{IEMG} = \sum_{i=1}^{N} |x_i| . \tag{56.1}$$

2. Zero crossing (ZC). This is the number of zero crossings, which indicates the degree of fluctuation. Its calculation formula is

$$\text{ZC} = \sum_{i=1}^{N} \text{sgn}(-x_i x_{i+1}) ,$$

$$\text{sgn}(x) = \begin{cases} 1 & \text{if } x > 0 \\ 0 & \text{otherwise} . \end{cases} \tag{56.2}$$

3. Standard deviation (SD). This is the mean value of the difference between data and its average, also called mean variance and expressed by σ. It describes the discrete degree of sEMG signal data sets.

It is calculated as

$$\sigma = \sqrt{\frac{1}{N} \sum_{i=1}^{N} [x(i) - \bar{x}]^2} . \tag{56.3}$$

4. Root mean square (RMS). This is the square root of the average energy, so it is the common evaluation index in the time domain and represents an overall signal energy level

$$R = \sqrt{\frac{x_1^2 + x_2^2 + \ldots + x_N^2}{N}} . \tag{56.4}$$

5. Willison amplitude (WAMP). This is a method calculating the number of sEMG signal amplitude variations proposed by Willison in 1963. In 1988, some research results showed that the most suitable threshold interval was $50-100\,\mu\text{V}$. It is written as

$$\text{WAMP} = \sum_{i=1}^{N} f|x_i - x_{i+1}| ,$$

$$f(x) = \begin{cases} 1 & \text{if } x > \text{threshold} \\ 0 & \text{others} \end{cases} \tag{56.5}$$

6. v-order and log detector. In 1980, Hogan proposed a mathematical model to generate sEMG signals: $x(t) = s(t)n(t)$, where $n(t)$ is an ergodic Gaussian process and $s(t)$ is a function of muscle contraction, which is defined by $s(t) = f(m(t)) = \gamma m(t)^a$, where $m(t)$ is muscle contraction, γ and α are constants. In practice, α is in the range 1–1.75.

7. sEMG histogram. The sample number of various sEMG amplitudes is a valid feature because of different deviation degrees from baseline during muscle contraction. The feature extraction process divides the present threshold into equal parts and then counts the sample number of signals with different amplitudes, where the actual threshold and the number of segments are determined according to tests. It is also applied to the pathological study of muscles.

8. Peakness. This is the average peak value of the probability density curve and is also called the co-efficient of kurtosis. It is also used to determine whether or not it is a normal distribution. The peakness of normal distribution is 3. The peakness is larger than 3 when the curve peak is higher than the normal distribution, otherwise it is smaller than 3. Its formula is

$$bk = \frac{E(x-\mu)^4}{\sigma^4} \tag{56.6}$$

9. Skewness. This describes the average asymmetry degree of the probability density curve and is expressed by bs. $bs < 0$ shows us that the average data on the left-hand side is more discrete than that on the right-hand side. Otherwise, the right is more discrete. For normal distribution or any strictly symmetrical distribution, it is 0

$$bs = \frac{E(x-\mu)^3}{\sigma^3} . \tag{56.7}$$

10. Range analysis. Range is the difference between maximum and minimum values in a data set. It reflects the range of variation and the discrete amplitude of the data distribution. In the set, the difference between any two data values is smaller than the range. It is calculated as:

$$R = x_{max} - x_{min} . \tag{56.8}$$

Time domain analysis methods have some disadvantages as follows:

1. The sEMG signal is so weak and instable that it is easily influenced by external noise. This makes it difficult to extract time domain features.
2. Muscle contraction has no measurement standard and is hard to control. Excessive contraction generates the overlapping MUAP. It is the reason for acquisition inaccuracy and serious experimental error.
3. This kind of method cannot make full use of spectrum characteristics.

56.3.2 Frequency Domain Analysis Method

Traditional spectrum analysis methods transform sEMG signals with FFT. The frequency or power spectrum shows us changes at different frequency components. Compared with time domain methods, it has some advantages as follows.

1. During muscle fatigue, frequency domain features decrease linearly as time goes. Otherwise, there is a large difference between time domain features.
2. Frequency domain features are not influenced by subcutaneous fat thickness and limb circumference, but time domain features do not have this property.
3. The time sequence curve slope of frequency features has an apparent relationship to the load time, which is less dependent upon time features.

The disadvantages of FFT, which make it unsuitable for sEMG analysis, are as follows.

1. This method has a lack of time domain information for sEMG.
2. The precondition of FFT for analog signals requires all information in the time domain, even including future information. That is difficult.
3. A small signal change in the time domain will affect the entire frequency spectrum because of its non-linearity and non-stationarity, so this method is sensitive to noise.

Spectrum features of the sEMG signal are widely used in the diagnosis of muscle diseases and the detection of muscle fatigue. The common indicators are mean power frequency (MPF) and median frequency (MF), whose calculation formulas are

$$\text{MPF} = \frac{\int_0^\infty f\text{PSD}(f)\,df}{\int_0^\infty \text{PSD}(f)\,df} \tag{56.9}$$

$$\text{MF} = \frac{1}{2}\int_0^\infty \text{PSD}(f)\,df , \tag{56.10}$$

where $\text{PSD}(f)$ is power spectral density function. MPF is sensitive to spectral changes in the condition of low load. MF is the median value of muscle discharge frequency during the course of muscle contraction. It has a strong interference resistance ability and a wide applicable domain for the load level. There are other indicators, such as the slope of the MPF curve and that of MF curve. They show us the relative changes of MPF and MF during motions.

56.3.3 Time–Frequency Domain Analysis Method

The time-frequency domain analysis method combines the time domain and frequency domain which aims at dealing with non-stationarity of the signal. Nowadays, widely used methods include STFT, Wigner–Ville transform, Choi–Wiiliams distribution, wavelet transform, and so on.

1. STFT. This is the most basic time-frequency analysis method. Its fundamental principle is an intercepting signal with a sliding window function $g(t - \tau)$, where τ is used to move the window to cover the whole time domain. The signal within the window is assumed stationary and is analyzed by FFT. Frequency changes along the direction of time, which is the time-frequency distribution of the signal, will be obtained. It is defined as

$$ \text{STFT}(\tau, f) = \int_{-\infty}^{+\infty} x(t)g(t - \tau)e^{-j2\pi ft}\, dt \,, $$

(56.11)

where $g(t)$ is the window function and $x(t)$ is the signal. The product of $g(t - \tau)$ and $x(t)$ can realize the window interception and translation near τ. The window determined by $g(t)$ continuously moves along timer shaft to take $x(t)$ to be analyzed step by step.

This method is based on signal stationarity, so whether it is time domain windowing or frequency windowing, the window width must be very narrow. Another problem is that the narrow time window will mean higher time resolution and lower frequency resolution, and vice versa. This characteristic of time and frequency resolution restricts and contradicts each other. It is called the uncertainty principle.

2. Wigner–Ville transform. In order to overcome deficiencies of STFT, Ville developed the Wigner–Ville distribution. It can reflect power spectrum changes along time and has good anti-noise performance. The calculation formula of real discrete Wigner distribution is

$$ W_x(k, l) = 2 \sum_{m=-k}^{k} x(k + m)x^* $$
$$ \times (k - m)e^{-j(2\pi/M)lm} \,, $$

(56.12)

where k and l are time and frequency indices, respectively.

The definition of the Wigner–Ville transform is

$$ WV(t, w) = \int_{-\infty}^{\infty} s\left(t + \frac{\tau}{2}\right)s^*\left(t - \frac{\tau}{2}\right)e^{-jw\tau}\, d\tau \,, $$

(56.13)

where $s^*(t)$ is the complex conjugate of $s(t)$. t and τ are time and time delays, respectively.

The Wigner–Ville distribution is a distribution on both the time space and the frequency space. It has some advantages such as identity, invertible property, and so on. It is helpful to determine the time and frequency range of energy concentration and is sufficient for time-varying signal analysis. One main deficiency of the WVD is the cross-term interference caused by multi-component signal, which may be suppressed by smoothing window.

3. Wavelet transform. This is also called a mathematical microscope and was proposed by J. Morlet and A. Grossmann in 1984. The definition of the continuous wavelet transform is:

$$ \text{CWT}(a, b) = \frac{1}{\sqrt{a}} \int_{-\infty}^{\infty} x(t)\bar{\psi}\left(\frac{t - b}{a}\right) dt \,, $$

(56.14)

where $\bar{\psi}$ is the complex conjugate of ψ, which is the mother wavelet function satisfying admissible conditions. a and b are expansion and shift factors, respectively. The common mother wavelet function is the Morlet wavelet as

$$ \psi(t) = \pi^{-1/4} e^{j2\pi f_0 t} e^{-t^2/2} \,, $$

(56.15)

where f_0 is the center frequency of the Morlet wavelet.

The wavelet transform is the new development of the Fourier transform (FT). It inherits and develops the localization theory of FT and breaks the limitation of no resolution in the time domain. So, it has good localization characteristics in the time domain and in the frequency domain. The expansion and shift functions of WT can realize multi-scale signal analysis. That is a narrow window for high frequency and a wide window for low frequency. Variable resolution analysis is good for focusing on signal details. The application of WT is broad with an increasing trend, so the suitable wavelet base and scale parameter for sEMG analysis are two key problems. The global and local properties and insensitivity to noise are very suitable for non-stationary signals like sEMG.

56.3.4 High-Order Spectral Method

Traditional random signal processing technology is based on second-order statistics. It can completely reflect the probability structure of random signals with

Gaussian distribution. However, the sEMG signal during muscle force change is non-stationary. It is not a Gaussian signal and has rich phase spectrum information. The high-order spectral method can overcome the shortcoming of traditional methods and extract more useful information. For example, bispectral analysis can deal with non-Gaussian sEMG signal analysis, which is useful in the study of muscle recovery during muscle force change.

56.3.5 Non-Linear Kinetic Analysis Method

In recent years, a non-linear kinetic analysis method has been applied to sEMG signal analysis. The main idea is as follows:

1. Correlative dimension. As a kind of fractal dimension definition, this indicates the density in multi-dimensional space and the correlation degree between points of the system.
2. Lyapunov exponent. This is the average emission rate of an adjacent track as time goes. It is a statistical average. For the Lyapunov exponent, positive means that the phase volume of the system is divergent in a certain direction. On the contrary, negative means that it is convergent.
3. Entropy. This is a thermodynamic concept and is used to describe the confusion degree of a thermodynamic system. The sum of all positive Lyapunov exponents is defined as Koifmogrov entropy. This means the average exponential growth rate of an infinitesimal volume element in a certain direction.

4. Complexity. Its general idea originated in the 1960s, but it is used as a non-stationary feature after an algorithm proposed by Lempel and Ziv.

Introducing non-linear kinetic analysis methods into sEMG analysis is just a beginning. It needs more intensive study.

56.3.6 Chaos and Fractal Theory

Chaos is a definite resembling random process with aperiodicity, which is non-random, non-linear, and sensitive to initial conditions. It can be described by the Lyapunov exponent, power spectrum, phase plane, and so on. Because the sEMG signal is a non-linear coupling of a large number of motion units, this non-linear method may be useful for feature extraction. Some researchers have tried to introduce the chaos theory into this field.

Fractal theory gives a new approach to biological signal analysis, which is difficult to describe accurately with traditional methods. Some studies show that the fractal dimension of the sEMG signal monotonically increases with the rise of muscle contraction strength. This feature can be used in proportional myoelectric prostheses.

Currently, a non-linear kinetic method for sEMG signals is just at the primary exploration phase. These methods are limited in this field due to the large amount of calculations needed and because they are time-consuming and require complex calculations, and so on.

56.4 sEMG Analysis for Rehabilitation Action Recognition

In this section, sEMG analysis for rehabilitation action recognition is described in detail. The robot platform and control strategy of the method are introduced first. Next, the sEMG-based motion recognition algorithm is explained. Finally, the experimental results are given.

56.4.1 Rehabilitation Robot System

As shown in Fig. 56.2, a wearable exoskeleton system for upper limb rehabilitation is proposed. It is composed of an exoskeletal robotic arm, a mounting rack, and an electric control system. The robotic arm made of super duralumin is a two-link structure.

Its special structure makes it light but rigid enough to support a paralytic limb and it is easy to place sEMG electrodes on it. Five basic and important DOFs for upper limb motions of shoulder extension/flexion, shoulder abduction/adduction, elbow extension/flexion, wrist extension/flexion, and wrist pronation/supination are adopted.

sEMG directly reflects the motion and function status of relative muscles. In this research, sEMG is acquired using electrodes made by the German Otto Bock Company to understand patients' motion intention, which is designed with a built-in filter and adjustable gain, up to 10 000 times stronger than the myoelectric input signals.

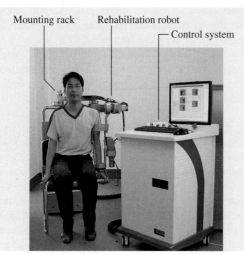

Fig. 56.2 Exoskeletal rehabilitation robot system

The robot system can provide multi-mode rehabilitation motions including passive and active motion modes. Therapists select the rehabilitation mode for hemiplegic patients at each recovery stage. In this chapter, the passive mode is discussed. Different from traditional passive rehabilitation methods, it is an autonomous passive rehabilitation motion mode, where *passive* means that a patient's affected limbs carry out exercises supported and detected by the robot system, and *autonomous* means motions executed according to The patient's intention. This method can inspire the patient's active motion sufficiently and make the patient's intention join the control and motion loop of the robot.

As we know, the clinical manifestation of hemiplegic patients is mostly unilaterally paralyzed. So, sEMG signals of the sound arm are used as control signals of the exoskeletal robot to retrain the malfunctioning one. The scheme flow of the autonomous passive rehabilitation strategy based on sEMG is shown in Fig. 56.3. It includes five steps. The first step is monitoring sEMG signals. When the motion is started,

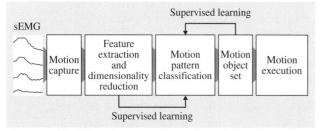

Fig. 56.3 sEMG-based autonomous passive training scheme

sEMG signals are acquired synchronously from multi-channels. The second step is analyzing acquired signals in order to extract features and reduce the data dimension. The third step is supervised learning and pattern classification, which includes classifier design, object set encoding, and classifier training. The fourth step is pattern recognition of upper limb motions according to multi-channel sEMG features. The last step is motion executing in terms of recognition result. The second and third steps are the core part of the autonomous passive rehabilitation strategy based on sEMG. In the next section, they are discussed in detail. We must notice that passive motions in this rehabilitation strategy include five upper limb movements that contain three crucial single joint movements: shoulder extension/flexion, shoulder abduction/adduction, elbow extension/flexion, and two basic composite movements for activities of daily life: eating and lifting trousers, which is very useful for motor function recovery of upper limbs.

56.4.2 sEMG Signal Analysis

Feature Extraction of sEMG Signals

Before analysis, preparation for sEMG acquirement is introduced briefly. The object's muscles for signal acquirement are selected based on upper limb rehabilitation movements. Four muscles of upper and fore arm, brachioradialis, biceps brachii, fore-deltoid, and mid-deltoid are selected. Electrodes are placed along the muscle fiber direction and over the muscle belly to obtain high quality sEMG signals. The signals' value is nearly zero and changes within the range of 0.02 v when the patient's arm is relaxed. The threshold method is used at the beginning discrimination of the valid motion to hinder the arm from vibrating confusingly. Too small a threshold would result in misjudgment; on the contrary, valid signals would be lost. In this research, the appropriate method is when two of four channel signals are greater than 0.05 v. The acquired signals are digitized by an A/D converter with 2000 Hz sample frequency.

The raw sEMG signal is processed by various methods in the time, frequency, and time-frequency domain. Among these, the AR model method is commonly used and confirmed to be effective. In the following paragraph, an improved AR model method for higher recognition rate and faster convergence speed is described.

The AR model is a very important analysis method for random signals in the time domain. It takes a random signal $x(n)$ as the response of white noise $w(n)$,

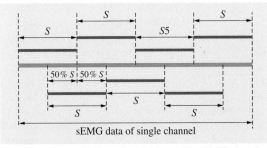

Fig. 56.4 Data splitting of each channel's sEMG signal

which applies to the linear system. If the parameters of white noise are confirmed, the character of the random signal can be described by AR model parameters. The sEMG signal is represented by an AR model on the condition of being short-time stationary. It is defined in the following equation

$$x(n) = w(n) - \sum_{k=1}^{p} a_k x(n-k) \,, \qquad (56.16)$$

where $x(n)$ is the n-th sample of the sEMG signal, $w(n)$ is the white noise signal, p is the order, and a_k ($k = 1, 2, \ldots, p$) is the k-th parameter of the AR model. p is a very important factor to exactly represent the sEMG signal. The final prediction error criterion (FPE) for determining model order is given by

$$\mathrm{FPE}(p) = \frac{N-1+p}{N-1-p} \sigma^2(p) \,, \qquad (56.17)$$

where N is the sample number and $\sigma^2(p)$ is RMS of the p order AR model. The minimum point of $\mathrm{FPE}(p)$ is the best order for the model. After testing the $\mathrm{FPE}(p)$ model from $p = 1$ to $p = 10$ for each channel of every motion, results indicate that the model with more than or equal to fourth order can describe the signal well. However, higher order can lead to more computational

cost and is does not improve the classification effect. The fourth-order AR model is the best choice for sEMG analysis.

As said before, extracting features of sEMG using the AR model is on the condition that the signal is short-term stationary. In this research, this assumption is obviously hard to be set up for the sample time of 1 s. A new improved AR method is proposed to consider both local short-time stationary and global characteristics for acquired signals. Different from the traditional method, it splits data into several segments, represented by $\{S_1, S_2, S_3 \cdots S_7\}$, and processes each of these with the fourth-order AR model. As shown in Fig. 56.4, each segment contains 512 data and 50% overlap each other. So, 28 features for every motion with a four-channel signal are as follows

$$R(i) = \left\{ a_{k(m)}^{(n)} | m = 1, 2 \ldots 7; \; n = 1, 2 \ldots 4 \right\} . \qquad (56.18)$$

As shown in Fig. 56.5, take shoulder abduction/adduction, for example, to compare the feature extracting result of the piecewise and global AR models. The former method results in a 112-dimension feature vector (4 channels × 7 segments × 4 features) and the latter is 16-dimension (4 channels × 4 features). It is obvious that the former result contains richer information than the latter. However, as the input of classifier, the 112-dimension vector would entail a lot of computation and training time. A dimension reduction of the feature vector using the principal component analysis method is described in the next paragraph.

Feature Reduction

As mentioned above, the valid dimension reduction method is helpful for gaining effective information to the maximum extent, saving calculation, and guarantee-

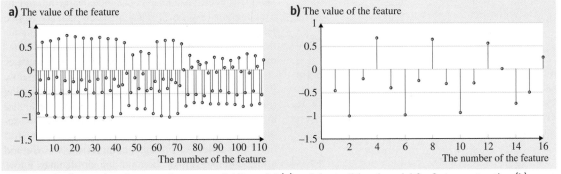

Fig. 56.5a,b Comparison between the proposed AR model (a) and the traditional model for feature extraction (b)

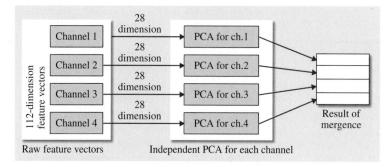

Fig. 56.6 Data comparison of sEMG feature vectors

ing convergence for net training. Principal component analysis (PCA), also called discrete Karhunen–Loeve (K–L) transformation, is widely used as a dimensionality reduction technique in data analysis. It has some important properties such as linear data compression, direct computation from the sample covariance matrix, and easy operations by multiplications. PCA performance is based on second-order statistics to transform correlated variables into uncorrelated ones and obtain principal components. Transform $\mathbf{x}$ linearly as follows

$$y = \mathbf{A}^{\top}\mathbf{x}; , \tag{56.19}$$

where $\mathbf{x} = (x_1, x_2, x_3, \cdots x_n)^{\top}$ is an n-dimension zero-mean random vector and $\mathbf{A} = [e_1, e_2, e_3, \ldots e_n]$, where $\{e_i\}, (i = 1 \ldots n)$ is the characteristic vector corresponding to characteristic value λ_i of correlation matrix $\mathbf{Rx}$, and they are all orthogonal to each other. So,

$$\begin{aligned} \mathbf{Ry} &= E[yy^{\top}] = E[\mathbf{A}^{\top}xx^{\top}\mathbf{A}] \\ &= \mathbf{A}^{\top}E[xx^{\top}]\mathbf{A} = \mathbf{A}^{\top}\mathbf{RxA} . \end{aligned} \tag{56.20}$$

It is clear that different component of y are uncorrelated, which is expressed as follows

$$E[y_i y_j] = \begin{cases} 0, & i \neq j \\ \lambda_i, & i = j \end{cases}. \tag{56.21}$$

Original signal x is reconstructed by the transformed result y as

$$x = (\mathbf{A}^{\top})^{-1}y = \sum_{i=1}^{n} y_i e_i . \tag{56.22}$$

In order to reduce dimension while keeping data information as much as possible, the characteristic vectors corresponding to larger characteristic values are reserved and the small ones are eliminated. Then, x is estimated as

$$\hat{x} = \sum_{i=1}^{m} y_i e_i . \tag{56.23}$$

The variance between $\hat{x}$ and x is

$$E[\|x - \hat{x}\|^2] = E\left[\left\|\sum_{i=m}^{n} y_i e_i\right\|^2\right] = \sum_{i=m}^{n} \lambda_i . \tag{56.24}$$

As shown in equation (56.24), the first m among total n principal components of y for signal reconstruction make truncation variance minimum, which is the integration of rejected $n - m$ characteristic values. The contribution rate of i-th principal component is defined as follows

$$K_i = \frac{\lambda_i}{\sum_{i=1}^{n} \lambda_i} = \frac{\lambda_i}{\mathbf{Tr}(\mathbf{Rx})} . \tag{56.25}$$

On the basis of the above theory, five upper limb motions are performed 50 times, respectively, and processed by the piecewise AR model method. 250 groups of characteristic vectors with 112 dimensions are used as the learning sample of PCA transformation. In this research, considering independency of signal source, independent PCA analysis is applied to each channel. This is explained in Fig. 56.6.

In the course of PCA analysis, the number of principal components is confirmed according to their contribution rate. The total of 28 principal components for each channel are calculated and figured in a histogram (Fig. 56.7). Principal components whose rate greater is than 90% are considered to contain the most information of the original signal. Accumulative contribution rates of the first three principal components for each channels are 95.54% (Fig. 56.7a), 95.28% (Fig. 56.7b), 95.24% (Fig. 56.7c) and 94.18% (Fig. 56.7d), respectively. As shown in Fig. 56.7, the higher the order of the principal components, the less the information content increases. On balance, the first three orders of principal components are adopted to be the projection direction of orthogonal linear transformation.

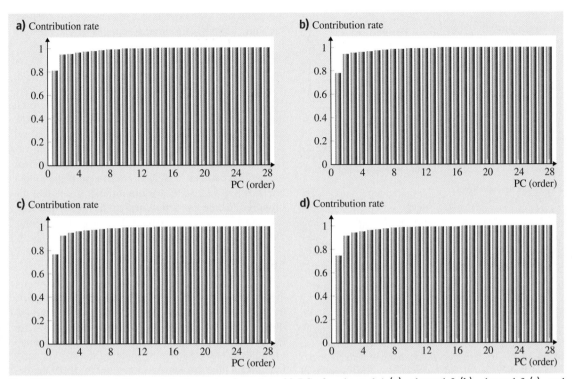

Fig. 56.7a–d Accumulative contribution rates from 1 to 28 PCs for channel 1 (**a**), channel 2 (**b**), channel 3 (**c**), and channel 4 (**d**)

Multi-Feature Fusion

The PCA method focuses on components with big change and rich local information. What is discarded may represent components with relative stability and small change, which always reflect the general trend of the signal. In this research, an area feature is fused into the features beforehand. In practical applications, a sample mean described as follows is used instead of area for the same sample length.

$$sv(i) = \frac{1}{N} \sum_{j=1}^{N} x^{(i)}(j) \,, \qquad (56.26)$$

where sv is the mean feature, i and j are indices of channel and sample, respectively, and N is the total sample count.

In a word, the final 16-dimension features of the sEMG signal is formed as shown in Fig. 56.8.

56.4.3 Feature Classification

In this research, a three-layer BPNN is applied to sEMG feature classification. The number of input and output nodes of the NN is determined according to feature dimension and recognized rehabilitation motions. The input layer contains 16 nodes for 16 dimensions' characteristic space, as stated in the previous paragraph. The output layer contains 5 nodes, as mentioned in the previous section. The number of nodes in the hidden layer is 10, determined with a trail-and-error method. The activation function of neural cells is the sigmoid function and the target error is 0.0001. Two different kinds of features extracted by AR and improved AR methods are applied to the neural network with 16 inputs,

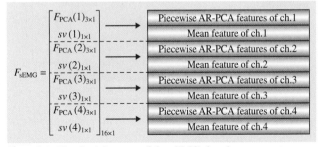

Fig. 56.8 The final features of the sEMG signal

Part K | 56.4

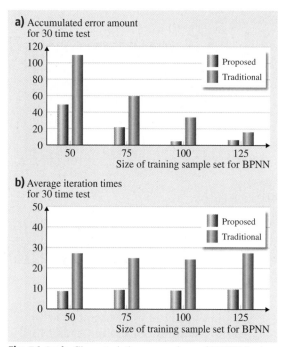

Fig. 56.9a,b Characteristics comparison of BPNN for two kinds of feature inputs. (**a**) Accumulated error amount of BPNN. (**b**) Average iteration times of BPNN, each for a 30 times test. The *grey bars* are for traditional AR model features and the *brown bars* for proposed features

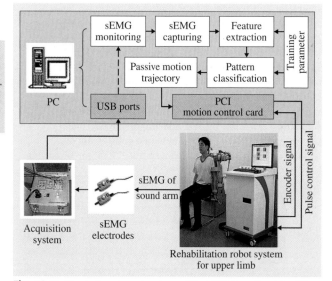

Fig. 56.10 Control system structure of the sEMG motion identification-based rehabilitation robot

5 outputs, and 10 hidden nodes to recognize upper limb motions. Meanwhile, various amounts of samples are used to train the network to observe its performance.

The training sample set is composed of 50 (10 for each motion, the rest may be deduced accordingly), 75, 100, and 125 groups of features, respectively. Among 50 groups of features, 10 groups are selected for each motion, and so on. To avoid instability of the BPNN, training sample set is sampled randomly from total set and used to train and test NN three times. Comparisons of the convergence speed and accumulative error recognition number of BPNN for different feature input are shown in Fig. 56.9; it is not difficult to see that the proposed feature extraction method has better recognition and generalization abilities than the traditional one, although they have the same dimension.

56.4.4 Experiments

As we know, rehabilitation for hemiplegic patients is a gradual process, and personalized motion therapy can improve treatment performance. The sEMG-based upper limb motion recognition discussed above is applied to an interactive passive rehabilitation strategy for early hemiplegia after stroke. As shown in Fig. 56.10 the detailed control structure is as follows: 4 electrodes are placed on the operator's sound arm. Weak current detected from the skin is filtered and amplified and then sent to computer via an A/D converter. The sEMG signal is analyzed for the purpose of feature extraction and pattern classification. Finally, the recognition result is used as input in an executable procedure to drive the exoskeletal robot with the programmed motion trajectory. Thus, the operator's affected arm is held by the exoskeleton to execute passive training according to intention.

A male and a female operator join the experiment in order to verify the effectiveness of proposed autonomous passive rehabilitation strategy based on motion recognition of sEMG. In this experiment (Fig. 56.11), an upper limb rehabilitation robot is worn on their left arm and the handle is held by hand. The sEMG electrodes worn on the right arm provide motion information for the robot to realize coordination training of bilateral limbs. The rehabilitation robot is set to help operators execute one training after motion recognition with the captured sEMG, and the human–machine system carries out shoulder extension/flexion, shoulder abduction/adduction, elbow extension/flexion, eating and lifting trousers 20 times, respectively. The

Table 56.1 Experimental results of sEMG HMI

Operators	Male		Female	
Motions	Correct (times)	Failed (times)	Correct (times)	Failed (times)
Shoulder abduction/ adduction	20	0	19	1
Shoulder extension/ flexion	19	1	20	0
Elbow extension/ flexion	20	0	19	1
Eating	19	1	18	2
Lifting trousers	18	2	19	1
Total	96	4	95	5
Accuracy rate	96%		95%	

Fig. 56.11 Experiment of upper limb complex

recognition results of the BP_Adaboost ensemble classifier are listed in Table 56.1.

The experimental results show us that the proposed AR method based on sEMG signal processing can make the rehabilitation robot *understand* the operator's active exercises intention accurately. sEMG is an effective human–machine interface for rehabilitation robot control. It can realize autonomous passive rehabilitation exercises for operators well. It is also good for helping patients to find correct motor sense and training coordination of bilateral limbs.

56.5 Conclusion

This chapter summarizes the analysis methods and application fields of sEMG signals. In the latter part, an sEMG analysis method based on the piecewise AR model for rehabilitation action recognition is proposed. Experimental results showed that the former algorithm is better than the traditional AR model method on two key indicators (recognition rate and convergence speed).

The sEMG signal is the most direct way to understand neuromuscular activities and a new direction of human–machine interaction. It has attracted many researchers, which has resulted in considerable achievements. However, there is no internationally recognized analysis method for the sEMG signal because of its non-linearity and non-stationarity. The most suitable algorithm for specific sEMG-based applications also needs future investigation.

Further Reading

- B. Bigland-Ritchie, E.F. Donovan, C.S. Roussos: Conduction velocity and EMG power spectrum changes in fatigue of sustained maximal efforts, J. Appl. Physiol. **51**(5), 1300–1305 (1981)
- M.A. Crary, G.D. Carnaby, M.E. Groher, E. Helseth: Functional benefits of dysphagia therapy using adjunctive sEMG biofeedback, Dysphagia. **19**(3), 160–164 (2004)
- D. Dorcas, R.N. Scott: A three state myoelectric control, Med. Biol. Eng. **4**, 367–372 (1966)
- D.A. Childress: A myoelectric three state controller using rate sensitivity, Proc. 8th IEEE Int. Conf. Med. Biol. Eng. (1969), S4–S576
- I. Moon, M. Lee, J. Chu: Wearable EMG-based HCI for electric-powered wheelchair users with motor disabilities, Proc. 2005 IEEE Int. Conf. Robot. Autom. (2005), pp. 2649–2654
- D. Graupe: Function separation of EMG signal via ARMA identification methods for prosthesis control purposes., IEEE Trans. Syst. Man Cybern. **5**(2), 252–259 (1975)
- D. Graupe: Multifunctional prosthesis and orthosis control via microcomputer identification of tem-

poral pattern differences in single site myoelectric signals, J. Biomed. Eng. **4**, 17–22 (1982)

- D. Graupe: Patient controlled electrical stimulation via EMG signature discrimination for providing certain paraplegics with primitive walking functions, J. Biomed. Eng. **5**, 220–226 (1983)

- J.J. Luh, G.C. Chang, C.K. Cheng, J.S. Lai, T.S. Kuo: Using time-varying autoregressive filter to improve EMG amplitudeestimator, Proc. 17th IEEE Int. Conf. Eng. Med. Biol. Soc. (1995) pp. 1343–1344

- G.C. Chang, W.J. Kang, J.J. Luh, C.K. Cheng, J.S. Lai, J.J.J. Chen, T.S. Kuo: Real-time implementation of electromyogram pattern recognition as a control command of man–machine interface, Med. Eng. Phys. **18**(7), 529–537 (1996)

- W.J. Kang, J.R. Shiu, C.K. Cheng, J.S. Lai, H.W. Tsao, T.S. Kuo: The application of cepstral coefficients and maximum likelihood method in EMG pattern recognition [movements classification], IEEE Trans. Biomed. Eng. **42**(8), 777–785 (1995)

- S.H. Park, S.P. Lee: EMG pattern recognition based on artificial intelligence techniques, IEEE Trans. Rehabil. Eng. **6**(4), 400–405 (1998)

- E. Lamounier, A. Soares, A. Andrade, R. Carrijo: A virtual prosthesis control based on neural networks for EMG pattern classification, Proc. Artif. Intell. Soft Comput. (2002)

- J.D. Zhao, Z.W. Xie, L. Jiang, H. Cai, H. Liu, G. Hirzinger: Levenberg-Marquardt based neural network control for a five-fingered prosthetic hand, Proc. 2005 IEEE Int. Conf. Robot. Autom. (2005), pp. 4482–4487

- K. Englehart, B. Hudgins, P.A. Parker, M. Stevenson: Classification of the myoelectric signal using time-frequency based representations, Med. Eng. Phys. **21**, 431–438 (1999)

- K. Englehart, B. Hudgin, P.A. Parker: A wavelet-based continuous classification scheme for multifunction myoelectric control, IEEE Trans. Biomed. Eng. **48**(3), 302–311 (2001)

- K. Englehart, B. Hudgin, A.D.C. Chan: Continuous multifunction myoelectric control using pattern recognition, Tech. Disabil. **15**(2), 95–103 (2003)

- K. Englehart, B. Hudgin: A robust, real-time control scheme for multifunction myoelectric control, IEEE Trans. Biomed. Eng. **50**(7), 848–854 (2003)

- J. Mingwen, W. Rencheng, W. Jingzhang: Motion identification of human hand based on wavelet transform of multi-channel EMG signal, Chin. J. Rehabil. Med. **21**(1), 22–24 (2006)

- M. Khezri, M. Jahed: An inventive quadratic time-frequency scheme based on Wigner–Ville distribution for classification of sEMG signals, Proc. 6th IEEE Int. Special Topic Conf. Inf. Technol. Appl. Biomed. (2007), pp. 261–264

References

56.1   D. DeWitt: Skeletal muscle (2013), available online at http://users.bergen.org/dondew/bio/AnP/AnP1/AnP1Tri2/FIGS/MUSCLE/muscle.html

Part L Nature Inspired Integrated Information Technologies

Ed. by Lubica Benuskova

57. Brain-Like Robotics

Richard J. Duro, Francisco Bellas, José A. Becerra Permuy

This chapter aims to provide an overview of what is happening in the field of brain like robotics, what the main issues are and how they are being addressed by different authors. It starts by introducing several concepts and theories on the evolution and operation of the brain and provides a basic biological and operational framework as background to contextualize the topic. Building on these foundations, the main body of the chapter is devoted to the different contributions within the robotics community that use brain-like models as a source of inspiration for controlling real robots. These contributions are addressed from two perspectives. On one hand the main cognitive architectures developed under a more or less strict brain-like point of view are presented, offering a brief description of each architecture as well as highlighting some of their main contributions. Then the point of view is changed and a more extensive review is provided of what is being done within three areas that we consider key for the future development of autonomous brain-like robotic creatures that can live and work in human environments interacting with other robots and human beings. These are: Memory, Attention and Emotions. This review is followed by a description of some of the current projects that are being car-

ried out or have recently finished within this field as well as of some robotic platforms that are currently being used. The chapter is heavily referenced in the hope that this extensive compilation of papers and books from the different areas that are relevant within the field are useful for the reader to really appreciate its breadth and beauty.

57.1 Natural and Artificial Brains

The information processing architectures of brains are some of the most impressive achievements of nature. However, these architectures cannot be contemplated in isolation from the bodies and environments within which they operate. In fact, these types of systems are the result of evolutionary processes that have taken place for millions of years in our world and which have involved generations and generations of brain–body–environment interactions leading to the current state of things, which is just an ephemeral point in the evolutionary timescale.

One of the merits of the solutions provided by nature is the very efficient use of resources. Brains employ, to make use of *Ritter* et al.'s words [57.1]:

components that are – as compared to transistor devices – slow, of low accuracy and of high manufacturing tolerances.

Nevertheless, even under these circumstances, they provide capabilities that allow their hosts to perform highly complex perception and control tasks in real time in a robust and adaptable manner. In fact, current technology is not capable of coming even close to these levels of performance. One would think that current massively parallel systems, with their huge number of very fast processors should reach a point where they would be able to perform at a level that is comparable to that of brains. However, this does not seem to be the case, as speed and massive processing capabilities do not appear to be enough to replicate what large numbers of slow and awkward neurons are capable of doing when paired with an appropriately linked body. Thus, it would seem that there is much more to *brain-like* capabilities than just speed and massive parallelism, and that other architectural and organizational issues should be contemplated and explored. To cite *Duch* [57.2]:

> It is the organization of the system, rather than the elementary unit, which is important. Biological properties and functions of complex organisms emerge from interactions of their cells, not from some special properties of elementary units.

There are numerous gaps in our knowledge of how brains operate and many processes are still not completely understood, especially in the information processing realm. *Sloman* [57.3] cited some of these: how brains see and understand many kinds of things in their environment, including 3-D spatial structure, affordances, and causal interactions, how brains decide what to do in diverse situations with conflicting pressures and tradeoffs, as well as how they plan and control actions, combining tasks where appropriate. On a different level, it is still unclear how brains remember things, learn about and take account of the mental states of others, enjoy or dislike things, sense the passage of time, and many others. Thus, the possibility of creating structures that attempt to mimic the way the brain works in order to obtain insights into some of these topics can be considered an interesting path towards the objective of understanding brain operation. This approach is based on what is called the machine hypothesis. This hypothesis has been very clearly stated by *Brooks* in his 2008 paper [57.4]. He wrote:

> Of all the hypotheses I've held during my 30-year career, this one in particular has been central to my research in robotics and artificial intelligence. I, you, our family, friends, and dogs – we all are machines. We are really sophisticated machines

> made up of billions and billions of biomolecules that interact according to well-defined, though not completely known, rules deriving from physics and chemistry. The bio molecular interactions taking place inside our heads give rise to our intellect, our feelings, our sense of self.

As a consequence, humans being a machine, there would be no reason why by trying to replicate the machine we could not gain insight into what is relevant from an architectural, organizational, or hardware point of view in order to achieve the information processing capabilities with the degree of robustness and adaptability that could be considered the trademark of real brains. In addition, a complementary and very interesting possibility would open up, that is, once the knowledge on these issues is gained, it should be possible to construct artificial systems that *would exhibit genuine human-level intelligence, emotions, and even consciousness* [57.4].

However, it has to be taken into account that the *human* or, in general, *animal* machine is made up not only of a brain, but also of a body, and it is the direct coupling of these two elements that provides the capabilities that animals display. Thus, it can be argued that the elucidation of the architecture of brains may benefit from insights gained from exploring the possibilities and limits of artificial control architectures for robot systems. Conversely, artificial control architectures for robots may benefit from this exploration of brain architecture in order to produce more robust, adaptable, and reliable systems. This is the realm of brain-like robotics (BLR).

This chapter is aimed at providing a general view of the problems and advances that are being made within the field of brain-like robotics. This is a very extensive field involving many different disciplines and considering multiple approaches. Here we will concentrate on some of the most important lines of research with the objective of allowing the reader to get a taste of the problem, what is involved, and of some of the solutions proposed. We do not claim to provide a complete review of this area, as it would require much more space than is available here, but by following the references that have been included, it should not be too hard for any researcher wishing to expand on what is presented here to do so.

Obviously, to address this subject requires asking and answering some very specific questions. On the one hand it is necessary to ascertain what is known about the structure of the brain and how it may work, at least in terms of models that can be handled within the robotics community. On the other hand, it would be of interest to decide what can be extracted from these models and how they can be implemented or, at least, what inspiration we can gain from them in order to implement cognitive structures within our robots that present the characteristics we are seeking. Finally, as this implementation and instantiation can become a very difficult problem and, in general, exponentially complex, it would be necessary to evaluate different approaches to the problem of how to produce these systems.

The chapter is divided into two main parts. The first one (Sect. 57.2) introduces concepts and theories on the evolution and operation of the brain and provides a basic biological and operational framework. The second part reviews different contributions within the robotics community that use brain-like models as a source of inspiration for controlling real robots. Obviously, this second part takes up much more space than the first one, as it is the main topic of the chapter. It is also divided into two parts. One of them (Sect. 57.3) is devoted to

reviewing the main cognitive architectures developed under a more or less strict brain-like point of view. A very brief description of each architecture is provided and some of their main contributions are highlighted. In the second part (Sect. 57.4), we concentrate on providing a more extensive review of what is being done within three areas that we consider essential for the future development of autonomous brain-like robotic creatures that can live and work in human environments interacting with other robots and human beings. These are: memory, attention, and emotions. Section 57.5 is an independent section that describes some of the current projects that are being carried out or have recently finished and robotic platforms that are currently being used within this field. Finally, in Sect. 57.6 we provide some conclusions of the work presented here. This chapter is heavily referenced, and, as mentioned above, we hope this extensive compilation of papers and books from the different areas that are relevant within the field will help the reader start their journey into what is becoming a very exciting topic.

57.2 Models of the Brain

The basic organization of the nervous system in terms of the main morphological divisions into spinal cord, hindbrain, midbrain, diencephalon, and telencephalon dates back to the early stages of vertebrate evolution over four hundred million years ago [57.5–7]. In fact, this organization is patent in the fossil record and already present in the earliest endocasts of jawless fish. Of course, as evolution progressed, variations within this basic plan of the nervous system appeared as vertebrates became more specialized to their niches. This led to different evolutionary paths such as increases in size and complexity of some brain regions over others [57.5, 8], as evidenced by the obvious differences between mammals, for which the cortex and cerebellum became huge structures, and reptiles for which this was not the case. Other examples can be mentioned such as the case of the telencephalon of birds, which is massively expanded compared to other non-mammalian tetrapods or that of some groups of sharks and rays with respect to other cartilaginous fish. Often, more than the creation of a new brain structure, changes occurred in terms of generating new functions from the evolution and adaptation of previously existing areas. An example of this is discussed by *Northcutt* and *Kaas* [57.9, 10] in terms of the likely evolution of the mammalian neocortex from areas in the dorsal pallium of jawed vertebrates that showed

similar connectivity. Thus, starting from a basic architecture or division of the brain into some basic modules, these seem to have evolved in different ways over different species as a consequence of the environments and activities or requirements they were exposed to within their niches. Sometimes this evolution led to an enlargement of certain areas and in others to the formation of new areas to cope with particular needs, whether by adapting and reusing previously existing hardware structures in new functions or generating the seeds for new structures. Consequently, as indicated by *Hodos* et al. [57.6, 7] the evolution of the brain cannot be taken as a linear process of increasing complexity from the earliest jawless fish to current mammals but rather as a tuning and adaptation process, where the complexity of the different brain areas is adapted to function and needs, as depicted in the diagram of Fig. 57.1.

From a functional point of view, it is very interesting to observe that the expansion or creation of new brain areas in different species, especially in vertebrates with complex brains, has not generally implied them taking over the functions of previously existing brain areas. Rather, these new structures usually complement the old ones by adding new or parallel functions and in general modulating existing structures as indicated by [57.5, 11, 12]. *Butler* and *Hodos* [57.7]

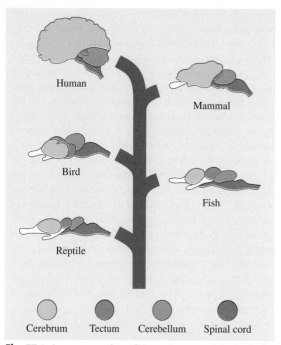

Fig. 57.1 A representation of the main parts of the brain for different evolutionary paths

provide an example of this in terms of the direct connections from the neocortex to brainstem and spinal cord motor neurons in mammals, which are not present in other types of vertebrates. These connections seem to be necessary for certain mammals to achieve more precise finger motions. This is, obviously, not necessary in other species such as birds or reptiles. Thus, these new cortical motor pathways evolved to improve precision when controlling manipulators as a complement to previously existing sub-cortical structures. It is important to note here that in experiments where these new motor pathways are blocked or severed, the older structures continue to perform their task without the modulation or complement of the newer structures and, consequently, the motions of the fingers are less precise and smooth.

Now, in terms of adaptation to new niches or functions it is relevant to emphasize that minor alterations in the connectional structures of brain areas can lead to very large functional changes. This is the case of spinal pattern generators. These systems originally evolved to produce undulatory movements in fish for the purpose of swimming. As limbed vertebrates started to move on dry ground, these same structures were adapted to produce the different gaits that were required. In fact, Cohen [57.13] suggests that this was achieved through

a number of very minor changes in the coupling and relative frequency of the oscillators. This argument can be extended to other systems, such as the peripheral sensory or motor apparatus, for which, according to [57.14] major functional changes may have been produced by limited reorganizations of central mechanisms.

There are two mechanisms that give rise to this flexibility for functional adaptation of preexisting structures. On the one hand it is necessary to consider developmental processes that allow certain areas to self-organize in order to adapt to changes in other structures from which they receive inputs or to which they send outputs as indicated by [57.15]. On the other hand, most neural circuits allow for reconfiguration into new operational modes through their modulation. Thus, new functionality may be achieved by adding modulatory inputs from other control structures [57.16, 17].

Whatever the mechanism, it seems that as brains evolved new structures appeared that mostly modulated the behavior of previous structures in order to achieve new functions and capabilities. This obviously gives credibility to the description of the brain as a layered control structure. In fact, as Prescott and his collaborators claim in [57.18]:

A substantial body of the neuroscience literature can be interpreted as demonstrating layered control systems in the vertebrate brain. In many ways the notion of layering is a common, often unspoken, assumption in contemporary neuroscience, however, the implications of the layered nature of the brain are not always acknowledged in a field dominated by the study of the mammalian cortex.

As shown in the schematic image of a human brain of Fig. 57.2 this view implies, to a certain extent, that different layers of the brain correspond to different evolutionary stages this particular type of brain went through, from a reptile-like simple brain to a more complex human brain with a neocortex.

This view of the nervous system as a layered structure can be traced back to neurologist *J.H. Jackson* in a 1884 lecture on the *evolution and dissolution of the nervous system* [57.19]. According to Jackson:

The doctrine of evolution implies the passage from the most organized to the least organized, or, in other terms, from the most general to the most special. Roughly, we say that there is a gradual adding on of the more and more special, a continual adding on of new organizations. But this adding on is at the same time a keeping down. The higher nervous

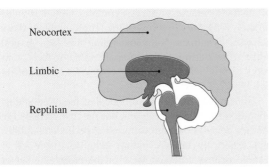

Fig. 57.2 A representation of different evolutionary layers of the human brain

arrangements evolved out of the lower keep down those lower, just as a government evolved out of a nation controls as well as directs that nation.

This view is obviously based on Darwinian evolution and was reached based on functional rather than morphological criteria. He postulated a division of the nervous system into lower, middle, and higher centers, the lower centers being more organized or fixed, automatic, and reflex than the higher centers, which are more modifiable or adaptable. In Jackson's view, the evolution of the nervous system is an incremental process where higher levels suppress or modify lower levels that are retained.

From a morphological perspective *von Baer* [57.20] had already suggested that the development of the brain took place from the center outwards and that as neurons moved to peripheral positions they became more specialized. In other words, more specialized systems are on the edges and more general systems are inside in an onion-like structure. This concentric organization has been recognized by authors such as *Magoun* [57.21] or *Berntson* et al. [57.22] and it is perfectly compatible with the Jacksonian view incorporating the real geometry of the system.

Other authors have expanded on the concept of layered architecture. For instance [57.23, 24] extended these ideas to development, that is, they have postulated that the brain matures by adding higher-level centers in a sequential manner. On the other hand, *Gallistel* [57.25] proposes viewing the nervous system as a *lattice hierarchy*. In this hierarchy lower-level centers are recruited as needed by higher-level centers. Each lower-level center can be recruited by more than one higher-level center. As a consequence of this structure, a lower-level center is activated by the combination of same level inputs, including sensory inputs and signals from higher-level centers that can potentiate or suppress the response of the center. This view is starting to look very similar to some of the approaches that have been followed in order to develop robot control systems, such as *Brooks'* subsumption architecture [57.26]. However, according to many authors (i. e., [57.27, 28]) this view is still incomplete as it does not take into account the feedback loops between lower and higher-level centers that are present in the nervous systems of vertebrates and which would lead to interpreting the system as heterarchical in its conception and make it difficult to determine what is lower or higher level in terms of control. Complicating matters even more, *Penfield* [57.29] proposed an organization whereby a group of central, sub-cortical brain structures, which were later identified by *Thompson* [57.30] as nuclei associated with the basal ganglia, coordinate and integrate the activity of both higher and lower-level centers in what has been termed a centrencephalic nervous system organization [57.22, 30].

Anyway, the fact remains that be it heterarchical or hierarchical, centrencephalic or a combination of these, it seems that the organization of the nervous system as a control structure can to a degree be described in terms of primitive neural centers (lower-level) and more modern neural centers (higher-level or cortex-related centers) that have evolved to make use, complement, and modulate these lower centers in order to achieve functionalities that were not contemplated in the original design. Obviously, all the levels of centers interact in complex ways and it is still an open question how this interaction can lead to the emergence of the complex behaviors that are observed in real animals. However, a picture is arising in the literature whereby, and to cite *Duch* [57.2] again:

It is the organization of the system, rather than the elementary unit, which is important. Biological properties and functions of complex organisms emerge from interactions of their cells, not from some special properties of elementary units.

In fact, recent results on brain organization show close interconnectivity of the majority of brain centers [57.31–33], providing clues on the pivotal role of structure at this level rather than the details of the modules. In addition, as is well known in other fields, especially in control, when a system exhibits large numbers of feedback connections, they can almost completely determine the behavior of the whole, making the details of the individual modules almost irrelevant. It is thus on the study of the architectures that can organize the computational contributions of the different modules, sensors, and actuators coherently and on how to harness these interactions where the emphasis should be placed.

In this line some authors (i. e., [57.34, 35]) view the brain as a dynamic system, more specifically as a multi-dimensional neurodynamical system. To cite [57.36]:

> *For decades, the cognitive and neural sciences have treated mental processes as though they involved passing discrete packets of information in a strictly feed-forward fashion from one cognitive module to the next or in a string of individuated binary symbols, like a digital computer. More recently, however, a growing number of studies, such as ours, support dynamical-systems approaches to the mind. In this model, perception and cognition are mathematically described as a continuous trajectory through a high-dimensional mental space; the neural activation patterns flow back and forth to produce non-linear, self-organized, emergent properties.*

This view implies that the brain is made up of many modules or centers that are active and influence each other constantly (they use the term resonate) leading to a global dynamic state of the brain. When some sensory or other type of signal or signals influence one or several of these modules, their dynamic state is changed and thus, their influence on other modules also changes, leading to a different global dynamic state for the brain. An example of this in the framework of the olfactory system performing its main task, which is to discriminate odorants, is described in [57.2]. The author considers a set of odorant receptive fields, which are basically feature detectors tuned to specific odorants, in the olfactory cortex. When an odor is perceived it will activate one or more of these modules in a resonant manner changing their dynamics. Their new contribution will be added to the global dynamic state of the brain and somehow this state will include the perception

of odor. Nevertheless, this is a complex field, still in its infancy, especially in terms of the reproduction of these complex dynamics. Notwithstanding this fact, several authors are starting to try to apply some of these concepts to producing brain-like behaviors and responses (i. e., [57.2, 37]).

To summarize, different models of the brain and its operation have been proposed, but the data obtained in the last decades through different experimental studies and, especially, brain scanning techniques have led to a more or less accepted view of the brain as a set of layered centers that are heavily interconnected. It seems that higher-level centers (peripheral modules or cortex) modulate and complement the behavior of lower-level or evolutionarily older brain centers. The term *modulation* has to be taken with care due to the heavy feedback occurring among all the modules or centers. The activity of the brain is not static, but rather it is a very complex dynamical system where somehow all of the modules participating in it influence the state of all the rest to a greater or lesser degree and it is through the learning processes that these influences are regulated in order to produce the appropriate states needed for the survival of the individual. How this is achieved is not yet known in detail, but we can postulate that through the instantiation of this brain view in artificial systems that include a body with sensors and actuators that can produce inputs for brain-like models we can gain insight into this operation. Also, by creating this type of brain-like robots we may be able to produce more robust and adaptable agents than more traditional approaches are achieving.

The following sections provide an overview of what is being done in the field of brain-like robotics, even though many of these approaches have not been included under this term and only claim to be *brain inspired*. We will show how many of the concepts and models stated in this section are implicitly or explicitly taken into account within what is an engineering realm.

57.3 Cognitive Robotics

We all envision future robots as highly versatile machines and very often very human-like machines. One can think of future robots as an advanced type of service robot able to accomplish different tasks in the real world, like helping at home or working in industry in a similar way to humans. Their cognitive capabilities will not limit their possible applications, which will depend only on their knowledge about a given task or on

their physical configuration (if we want our personal robot to paint a wall, it must be capable of grasping a brush). To achieve such versatility, the most important characteristic of those future robots will be their level of autonomy. They should be capable of learning from their embodied autonomous interaction with the real world, including other robots and people. Obviously, this learning should take place under changing conditions and, consequently, adaptability will be intrinsic to the learning process.

The required adaptive learning methods have been studied in the autonomous robotics field for decades, but the practical results have been very limited in real applications. The autonomy level achieved is still far from what would be expected from these future robots. Even though it is true that human and animal brains have been the main source of inspiration for years in this field, it is also true that the inspiration has been only partial and only considered specific aspects without using a complete brain model whose final computational implementation really reproduces the real processes occurring in brains. Most brain inspired implementations were based more on the perceived functional operation of the brain than on its structure, that is, more on psychological than on physiological aspects of the brain.

In the last decade, things have been changing in this field. Several authors have started to introduce in their robots the most advanced biological brain models taken from neurophysiology in an attempt to computationally reproduce the complex processes occurring within the brain with the aim of obtaining robots with a higher level of autonomy. That is, a group of autonomous robotics researchers are trying to computationally clone biological brains following what is known about the implementations of the biological sub-systems that make them up very closely. Their aim is to provide robots with the same functionalities that those brains show in humans or animals using the same operational principles and infrastructure (neurons and other underlying structures such as genes or proteins) as opposed to studying brain operation and trying to implement the resulting functionalities through an engineering process, which was the traditional approach. An example of addressing things this way that tries to jointly model a large set of the aspects that are known to influence the operation of these biological systems is that of computational neuro-genetic modeling, as described in [57.38] and [57.39]. In this model the authors use spiking neural networks (probabilistic spiking neural networks or pSNN) as the basic infrastructure and model the influence of some proteins and genes on the spiking behavior, thus creating a much richer set of behaviors and a more adaptable system. These types of protein regulatory models have been used for the study of brain developmental processes [57.40] as well as in robotics [57.41].

Thus, if this new approach works, the desired autonomy level and the adaptive learning capabilities required will arise naturally. These brain-like robotic approaches are typically included under the name of *cognitive robotics*, due to the fact that such robots are characterized by their cognition.

Cognition can be defined as [57.42]

the mental process of knowing, including aspects such as awareness, perception, reasoning, and judgment.

From a more psychological perspective, cognition is a term referring to

the mental processes involved in gaining knowledge and comprehension, including thinking, knowing, remembering, judging and problem-solving.

In computer science, cognition can be viewed as [57.38]

the process by which the system achieves robust adaptive, anticipatory, autonomous behavior, entailing embodied perception and action.

Therefore, a cognitive robot is characterized by its capacity of acquiring knowledge in an autonomous and adaptive way, that is, as commented above, a cognitive robot has adaptive learning capabilities and, as a consequence, its behavior is really autonomous.

A point that must be commented on here is that, although there is no explicit distinction between human or animal brain models and functionalities, in most brain-related cognitive robotics work, as we will see in the following sections, the different authors tend to use the human brain as their main reference. In fact, most of the research in this field has human-like robots as its final application platform. This is a consequence of the desired final objective: if we try to develop robots that live with humans and interact with them in a natural fashion, it seems useful to use the human brain and body as a reference. It is well known that human and animal brains are not so different in biological terms, but their high-level functionalities are. As remarked in [57.43] cognitive robots

are mainly inspired by human neurocognitive and psychological abilities, where typical human cognitive tasks such as perception, learning, memory, emotions, reasoning, decision-making, behavior, language, consciousness, etc. are in some way modeled and used as a source of inspiration in order to enhance the capabilities of autonomous mobile robots.

Thus, most authors use the human brain model and its capabilities as the target, but are very aware at the same time that reaching the autonomy level of most animals is, nowadays, a very ambitious quest.

Cognitive architectures are the computational implementation of a cognitive model [57.44], and as such, constitute the substrate for all cognitive functionalities, like perception, attention, action selection, learning, reasoning, etc. [57.45]. It seems obvious that developing a complete cognitive architecture for a brain-like autonomous robot is a very complex issue. Anyway, several approaches can be found in the literature that propose general architectures. However, when they are analyzed in detail, it is usually the case that some specific cognitive functionalities are studied in depth and others are barely touched or even missed in most cases. Nevertheless, after reviewing these proposals an overall view of what has been done in cognitive robotics and in the application of brain-like systems can be obtained, including the main strengths and the bottlenecks of the approach. This is the objective of this section and the next one. The main cognitive architectures developed under the cognitive robotics paradigm will be described in this section, focusing the discussion on their success and suitability in their application to real robots. Section 57.4, on the other hand, reviews a set of cognitive functionalities that have been chosen as key to the development of future robots that can achieve the level of autonomy and interactive capabilities that will be required from them in the complex human populated world where they will have to operate.

57.3.1 Context

As in almost every application field where artificial intelligence is used, in the field of robotics one can find cognitive architectures constructed using symbolic approaches, sub-symbolic (usually also bio-inspired) approaches, and hybrid approaches. Although each one has its own advantages and disadvantages, it is not easy for a purely symbolic system to mimic some structures and dynamics of brains, their adaptability, and their plasticity. On the other hand, it is not easy for a purely sub-symbolic system to implement some high-level cognitive features like reasoning either. Thus, when trying to build a complete cognitive architecture the most common solution adopted by the researchers in the field has been to implement some kind of hybrid system, at least initially. A system of this type can be obtained employing a bottom-up approach, that is, starting with a sub-symbolic system and adding typical symbolic features, or employing a top-down procedure, starting with a symbolic cognitive architecture and adding sub-symbolic components later.

Here we are considering approaches with a brain-like inspiration, which mimic brains from an operational and/or structural point of view. However, it is possible to design a brain-like cognitive architecture from that point of view without using any bio-inspired low-level support. It is also true that bio-inspiration is not necessarily the same as brain-like inspiration, as it is possible to use bio-inspired tools, like artificial neural networks (ANNs), and not model any part of a brain. Finally, brain-like and mind-like are not the same either, as it is possible to model some human mental capabilities without paying any attention to how these capabilities are produced in human brains. So, to be precise, brain-like cognitive architectures try to reproduce some brain capabilities, usually human brain capabilities, but not necessarily, while at the same time mimicking internal structures and dynamics of real brains, with or without bio-inspired support. The first option, that is, with bio-inspired support, is preferred for coherence, but it is not 100% feasible due to the current state of computational hardware and bioscience knowledge. In fact, the models of the brain that are used in most implementations of cognitive architectures are not too precise and are often high-level in terms of modules and interactions between modules. They are usually loosely based on the view of the brain as a set of layered centers where higher-level centers modulate and complement the behavior of lower-level, or evolutionarily older brain centers. However, the very high degree of interconnection of these centers found in brains, as well as the details of these interconnections in dynamic terms have seldom been considered in real implementations and are still an open field of research.

In what follows, we will briefly describe the most well-known brain-like cognitive architectures that have been developed, or for which there is work-in-progress, and that have been applied to real robots. We will first consider cognitive architectures that were not originally created for their implementation in real robots. Most of these were designed from a purely symbolic perspective, but were later modified towards a more bio-inspired point of view when implementing some components of the architecture (top-down design). In fact, some of them were even completely re-implemented. We will then describe a set of cognitive architectures specifically designed for robots. These are usually, but not always, characterized by a high level of bio-inspiration but at the same time usually lack some high-level cognitive features present in the first group of cognitive architectures.

Although there is no standardized cognitive model used by all architectures, they share some basic cognitive functionalities and characteristics. One of the most frequent is the use of several types of memory elements, specialized on different types of information, and the existence of mechanisms that control the flow of information between them.

Usually, and following theoretical concepts of neuroscience and cognitive psychology, memory is segmented into the following types, although not every architecture implements all of them:

- *Sensory memory* (SM): it contains the information perceived by the perceptual system through the sensors. A selective attention mechanism filters the information that is transferred to the short-term memory.
- *Short-term memory* (STM): it stores the conscious information transferred from the sensory memory for a short period of time, in which it undergoes a process of consolidation until it is permanently stored in the long-term memory.
- *Working memory*: it can be considered as a short-term memory cache that actively maintains information relevant to the current task for a short period of time.
- *Long-term memory* (LTMZ): it is a very complex memory sub-system that preserves a large amount of information for a very long time. It is typically divided into:
  - *Non-declarative*: also known as *procedural* or *implicit* memory. It is related with implicit knowledge, such as perceptions and actions (non-conscious processes like the tuning of motor skills).
  - *Declarative*: it is related to *explicit* knowledge and can be divided into:
    - *Semantic*: it stores general knowledge about the world, including facts and properties of objects.
    - *Episodic*: it stores facts, events, or episodes that occurred during a lifetime contextualized in time and space.

Other cognitive functionalities that are present in most of the cognitive architectures we will review in the next section are perception, reasoning/decision-making/action-selection, learning, attention, and emotions. Due to the fact that the last three elements are not always present, that there are many alternatives and different ways of implementing each one of these functionalities, that these ways are conditioned by how memory is implemented, and that all the elements interact through the contents of the memory, the descriptions of the architectures will mainly be focused on memory, with the particularities of the other components mentioned when relevant.

57.3.2 Non-Native Robotic Cognitive Architectures

Brain-like cognitive architectures that were not originally created to be applied within real robots, but that were modified with this objective at a later stage, will be described in this section. Most of them were initially focused on a functional approach, mimicking high-level brain capabilities and functional areas, but without using any bio-inspired support. Nevertheless, all of them are being modified towards a more bio-inspired implementation and towards a more complete model, including low-level sensing and acting, the later being necessary to implement the cognitive architecture in a real robot operating in real time.

SOAR [57.46] is a classical example of symbolic cognitive architecture with psychological roots [57.47], although current work [57.48] goes further in brain-likeness splitting long-term memory into procedural, semantic and episodic memories, incorporating specific learning algorithms (still in development) for each one, including reinforcement learning to update procedural memory, and adding the concept of emotions as a behavior modulator. Nevertheless, all cognitive procedures still rely on symbolic representations, excluding perception (visual, for instance) that uses a neural representation. There exists some preliminary work-in-progress in order to implement the architecture in real robots [57.49]. The suitability of this cognitive architecture for its use in robotics is unknown and it depends on the advances achieved in all these recently opened research lines.

ACT-R is another symbolic-born cognitive architecture inspired by the work of Allen Newell that has recently been modified to be more bio-inspired. It has two main types of memory: procedural (production rules) and declarative (chunks of information as the result of applying production rules). Recent versions have incorporated a modified *EPIC* (another cognitive architecture) in order to model perception and motor operation [57.50], capabilities that were not present in the original architecture. The procedural and declarative memories have been split into several modules that can be mapped to known brain sub-systems (basal ganglia in the first case and temporal lobes and hippocampus in the second one), and new structures connecting procedural and declarative areas have also been created to store temporarily active information to simulate cortical activity [57.51]. There is an, apparently discontinued, connectionist implementation named *ACT-RN* where chunks are made of associative memories and procedural knowledge is spread out over several memories with connections between them [57.52]. *ACT-R* has been employed to model human memory, to solve some complex problems, to model natural language, to mimic human behavior or to predict patterns of brain activity in cognitive neuroscience, but its use in real robots has been marginal [57.53].

Still following the same philosophy, *LIDA* was designed using a brain-like approach from a functional point of view. Its guiding principle is that every autonomous agent's life can be viewed as a continuous sequence of iterations of sensing, input processing (including modeling the world) and acting cognitive cycles [57.54]. According to this principle, it establishes a cognitive cycle integrating several types of memories and processes. In this cognitive cycle there are three phases: understanding, attention, and learning/action selection. In the understanding phase, the data flows from the sensors to low-level feature detectors in the sensory memory, its output goes to higher-level feature detectors in the perceptual associative memory, and then to the workspace, where it cues both transient episodic memory and declarative memory, producing local associations. These associations are combined with the percept to generate a situational model. The attention phase begins with the formation of coalitions of the most salient portions of the situational model, which compete for a place in the current conscious contents. These conscious contents are then broadcast to the perceptual associative memory, transient episodic memory, and procedural memory, initiating the different types of learning that create new entities and associations and reinforces the relevant old ones. In parallel with learning, and using the conscious contents, possible action schemes are instantiated from procedural memory and sent to action selection, where they compete to be the behavior selected for this cognitive cycle, which will trigger the sensory-motor memory to produce the right actions. Recently, its authors established a very detailed correlation between modules and functions in *LIDA* and different areas of the brain (there are several diagrams and animations showing this in their webpage at http://ccrg.cs.memphis.edu/tutorial). There are several papers theorizing on how *LIDA* could be applied to real robots [57.55,56]. Some of them represent work-in-

progress to really incorporate *LIDA* to a real robot, such as [57.57], where a simplified version is implemented as a secondary controller that works in the form of an advisory sub-system. To summarize, the original design philosophy of LIDA seems more appropriate for implementation in real robots than that of *SOAR* or *ACT-R*, but its problem seems to be the computational resources needed to operate in real time.

MicroPsi [57.58] is based on a psychologically inspired cognitive theory called *Psi* [57.59]. *MicroPsi* models demands (water, energy, affiliation, certainty, competence, physical integrity) as values that should be kept in a particular target range. Whenever the value of some of these demands deviate from their target range, an urge arises and generates a motivation to act. Those motives are modeled as (urge, goal) pairs, where a goal is a statement that the agent will try to make true in order to satisfy the urge. When an urge is being satisfied, that is, the demand's level approaches its target range, pleasure appears. The set active motives of an agent in a particular instant of time make up its intention. There are four modulators that conform an emotional system, acting on the perception, cognition/action selection and action execution processes: activation, resolution level, certainty, and selection threshold. Activation measures the degree of rapid activity versus reflective activity. The resolution level determines how accurately the agent tries to perceive the world. Certainty is one of the demands and indicates the confidence in the agent's knowledge. Finally, the selection threshold determines how willing the agent is to change its choice of which goals to focus on. Everything in *MicroPsi*, current control, and memory, is stored using a very simple model of neuron with a step-like activity function. These neurons are structured as quads, which are structures with a central neuron that has four peripheral neurons connected to it. These can be activated by the central neuron and have special meanings: sur (is part of something), sub (has something as part), por (points to something in the future), and ret (points to something in the past). Quads are connected hierarchically to build sensory/perception schemas and behaviors/motor schemas. These schemas are also connected as triplets made up of a sensory schema, a motor schema and another sensory schema that represents the predicted sensing after acting. MicroPsi has three ways of choosing an action: automatisms for well-known situations, application of a known set of strategies when there are no automatisms but there are valid rules to derive a course of action, and planning when current strategies are not valid. Experiments were performed using real robots in 2005 and 2006 (the robot was a *robotized* Mac Mini), but it seems that this line of application is stalled, at least there do not seem to be new publications and the original web page is not available anymore. Again, it may be that the mechanism is not designed to operate in real time in a physical robot with the necessary speed and robustness.

4D/RCS is another cognitive architecture [57.60] that combines perception, world modeling, planning and action although it lacks, for instance, learning. It contains three hierarchies, which can be interconnected, for perception, modeling and action, and where each memory/processing unit presents a particular spatiotemporal scope. Although the initial implementations are not brain-like at all, recent work [57.61] goes in this direction linking architectural components with brain functions and implementing part of the perceptual hierarchy using a particular type of units (cortical computational units) that emulate cortical columns of neurons in real brains. This architecture, although only in its original form, has been employed in many robots, as it was originally conceived to be used in military unmanned vehicles.

OpenCogPrime is based on the systems theory of intelligence outlined in [57.62] characterized by combining multiple AI paradigms (conceptual blending, probabilistic logic networks, economic attention networks, probabilistic evolutionary learning ...). Cognitive processes embodying these different paradigms interoperate using a common neural-symbolic knowledge store called Atomspace. This knowledge store includes different types of memory that can be used by the different cognitive processes: declarative, procedural, sensory, episodic, attentional (used just by the economic attention networks process), and intentional (for storing goal related information in a hierarchical way). It is possible to convert information stored in one type of memory to another one, although this process can be computationally costly. *OpenCogPrime* has been used in software agents in virtual environments and there is an effort, still in a very early stage, to implement it in real robots. For instance, *OpenCogPrime* is being integrated with *DeSTIN* to handle lower-level sensorimotor intelligence. *DeSTIN* is another cognitive architecture focused on perception, which like *4D/RCS* uses several hierarchical spatiotemporal networks (for carrying out inference on sensing processes, acting and obtaining feedback from the environment, which is necessary to perform reinforcement learning when acting) with the main objective of making the whole perception process scalable. The network nodes use probabilistic

pattern recognition and can receive inputs from nodes of different networks. In the context of robotics, there is also some work trying to fuse *OpenCogPrime* with neural networks and evolution [57.63]. The most important problem of *OpenCogPrime* is, even to a higher degree than in other architectures, real-time operation and the internal coherence of information, due to the simultaneous execution of several cognitive processes using different information representations. This problem is augmented because low-level sensing and acting was not originally considered and the integration with specific methods for this purpose increases the computational requirements and complexity.

57.3.3 Native Robotic Cognitive Architectures

Most brain-like cognitive architectures specifically designed for real robots have been implemented in the first decade of this twenty-first century and can be included (to a greater or lesser degree) within the cognitive developmental robotics (CDR) sub-field. Following [57.64] all the above architectures are non-developmental because *a developmental architecture requires not only a specification of processors and their interconnections, but also their online, incremental, automatic generation from real-time experience*. In this sense, CDR is a new paradigm in cognitive robotics characterized by the assertion that [57.65]:

> *Existing approaches often explicitly implement a control structure in the robot's brain that was derived from a designer's understanding of the robot's physics. According to CDR, the structure should reflect the robot's own process of understanding through interactions with the environment.*

The main feature of a CDR system is development, which implies *the progressive acquisition of predictive anticipatory capabilities by a system over its lifetime through experiential learning* [57.66].

Specifically, CDR architectures should display the following features [57.66–68]:

1. They should address the dynamics of the neural element in different regions of the brain, the structure of these regions, and especially the connectivity and interaction between these regions.
2. A developmental system should be a model generator, rather than a model fitter [57.69] organizing unlabeled sensory signals into categories without a priori knowledge or external instruction.

3. Physical embodiment: the agent's physical body specifies the constraints on the interaction between the agent and its environment that generate the rich contents of its process or consequences. It also provides a meaningful structure to the interaction with environment and is the physical infrastructure to form cognition and action [57.70].

4. They should have some minimal set of innate behaviors or reflexes in order to explore and survive in its initial environmental niche.

5. They must display adaptive capabilities in order to improve and increase the basic set of innate behaviors.

In [57.70], the authors review in depth the main developments in the field and their final robotic application. They concentrate on a set of salient research issues:

1. Basic structures for embodied motion, based on the study of body representation through the analysis of emergence of fetal sensorimotor development and other synthetic approaches.

2. Mechanisms for achieving dynamic motions of the whole body from basic skills to voluntary movements.

3. Body/motor representations and spatial perception to link individual and social development.

4. Social aspects of development: early communications, action execution and understanding, vocal imitation, joint attention, and empathy development.

CDR puts more emphasis on human/humanoid cognitive development, and most of the architectures belonging to this field have been applied in the context of humanoid robots. In this sense, several robots have been designed for CDR research. Some examples are SAIL [57.71], DEV [57.72], iCub [57.73], and Acroban [57.74], etc.

Focusing now on specific cognitive architectures, even those approaches that are not explicitly classified within the CDR sub-field by most authors in the literature share many of its principles. One of them is IMA (intelligent machine architecture) [57.75, 76], which runs in a robot called ISAC that was designed to be human-friendly in human–robot interactions (it has a stereo vision system, two pneumatically-powered arms with 6 DOF, a voice synthesizer, and can understand voice commands). This cognitive architecture is multi-agent based and it makes use of short-term memory for spatio-temporal data coming from sensors, long-term memory in the form of procedural, semantic and episodic memories, and working memory used

by the self-agent to store temporary information about the tasks being performed coming from short-term and long-term memory. The main agents present in the architecture are the perception agents, which are linked to sensors and populate the short-term memory, control agents, which take commands from the self-agent and send control signals to the actuators, the self-agent [57.77], which executes tasks using procedural and declarative knowledge and controls behavior using attention, emotion [57.78] and the working memory, and the human agent, which interprets human intention in human–robot interaction processes. Most of the implementation is not bio-inspired with the recent exception of some controllers in the procedural memory [57.79]. Figure 57.3 contains a block diagram representing the IMA architecture in the left image and a photo of the ISAC robot in the right image.

Another example of cognitive architecture that can be considered a precursor of CDR is *Shanahan's* global workspace cognitive architecture [57.80]:

> *It realizes cognitive function through topographically organized maps of neurons, which can be thought of as a form of analogical representation whose structure is close to that of the sensory input of the robot whose actions they mediate.*

It is based on the global workspace theory [57.81–83] which basically obtains a sequence of states in time that result from the interaction of multiple parallel processes to facilitate anticipation and planning, leading to a cognition based action selection strategy [57.84]. The architecture proposed by *Shanahan* contains [57.85]:

> *Analogues of a variety of brain structures and systems, including multiple motor-cortical populations (that compete for access to the global workspace), internal sensorimotor loops (capable of rehearsing trajectories through sensorimotor space), the basal ganglia (to carry out action selection), and the amygdala (to guide action selection through affect).*

From these background biological structures, high-level concepts like imagination, emotion, and consciousness emerge in a natural way, all of them supported by the global workspace theory. Its working cycle is divided into two sensorimotor loops, a reactive first-order loop devoted to the actuation motivated by the current time requirements of the world, and a higher-order loop that controls the action selection. To do this, the architecture includes analogies of basal ganglia and amygdala that compute the action salience by using models or abstractions of the robot's sensorimotor space where the

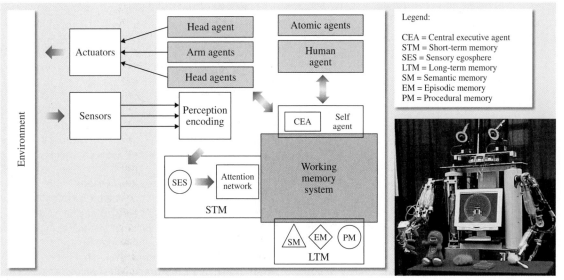

Fig. 57.3 IMA architecture and ISAC robot

system tries out trajectories in an off-line mode as described in [57.86]. The top-level components of this architecture are displayed in Fig. 57.4. The real robot application of the architecture has been very limited. Basically, the authors have used a Khepera robot in simulation with the main objective of studying its basic operation and expected features.

Brain-based devices (BBD) [57.87, 88] is the name for a set of theoretical brain models that have been created to investigate nervous system function with the requirement of being run in real robots. As usual in CDR, the fundamental premise is that an organism's brain is closely coupled to its body, which actively interacts with its environment, and cognition cannot be fully understood and modeled without taking into account that fact. The first models were run in software simulations but real robots have been used as platforms for BBD since 1992. In particular, a NOMAD has been used since 2000 and a Segway since 2004. Both are wheeled and camera equipped robots, the latter with outdoor capabilities. The different versions of BBD are called Darwin followed by a roman numeral, the latest being Darwin XI, and were used to experiment with different concepts. So, the architecture is different in each case but it is always based on artificial nervous systems with tens of thousands of neurons and hundreds of thousands of synapses. This artificial nervous system usually runs in a Beowulf cluster with tens of CPUs, although some of the experiments with the Segway robot were run in the real robot. Also, some of the

experiments [57.89] employed other techniques in order to carry out complex tasks that could not be solved at that time using just plain BBD. The neural model employed since Darwin III is called synthetic neural modeling [57.90], which uses a very biologically inspired and massive artificial neural model that includes synaptic changes, organization into neuronal groups and layers emulating different brain regions. The last experiments (Darwin X and XI) are the most complex ones and a model of a medial temporal lobe is implemented (using one hundred thousand neurons and more than one million synapses) and its role in spatial, episodic and associative memory is studied. The neural model corresponding to Darwin XI is displayed in Fig. 57.5 while Fig. 57.6 contains a photo of the real robot used in the experiments.

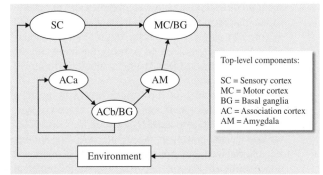

Fig. 57.4 Top-level components of Shanahan's architecture

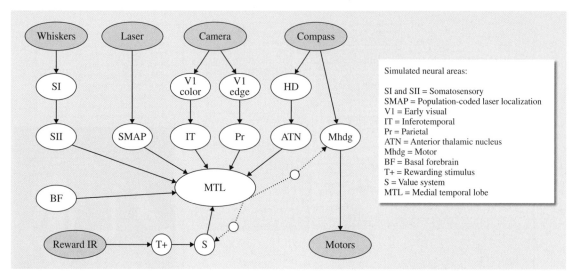

Simulated neural areas:

SI and SII = Somatosensory
SMAP = Population-coded laser localization
V1 = Early visual
IT = Inferotemporal
Pr = Parietal
ATN = Anterior thalamic nucleus
Mhdg = Motor
BF = Basal forebrain
T+ = Rewarding stimulus
S = Value system
MTL = Medial temporal lobe

Fig. 57.5 Darwin XI neural model. *Dark ellipses* are sensors and actuators. *Dotted lines* represent plastic synapses where learning occurs

The architectures presented above are not usually included within the CDR paradigm in the different reviews although they share many concepts and ideas with this approach and one could argue that they can really be taken as CDR architectures. Anyway, it is not our objective to discuss this here, and in what follows we will concentrate on and present more extensively a set of representative architectures that are usually considered CDR. As we will see, most of them are based on a brain model that is very similar to the one that was introduced in Sect. 57.2.

Self-Aware and Self-Effecting Architecture (SASE)

Weng et al. have been working on the *SASE* architecture since 1999 [57.91]. In fact, it was the first cognitive architecture to include autonomous development for its main modules, the attention selector, model generator, motor mapping and the motivational system [57.69]. A core idea of *SASE* is that the brain senses the internal environment and acts on it, being the body and the real world both considered as the external environment for the architecture.

The brain model and the learning processes included in *SASE* are biologically supported by neuroanatomy, and there is a correspondence between the cortical areas and the components of the architecture. The authors have formalized the architecture's principles and basic concepts by dividing it into six types of incremental capabilities [57.64]:

Fig. 57.6 Darwin XI robot

1. Type 1: it is defined as an observation-driven Markov decision process of order k. Basically, this architecture contains a regressor R that performs a mapping between the feature input space and the primed context output space. This regressor is developed from real-time experience. In addition, this architecture has a value system V (motivational system) in charge of selecting the most appropriate context from the set of primed ones. The authors apply Q-learning within this value system.

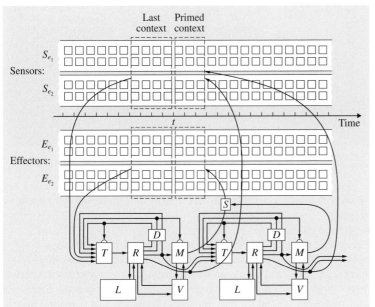

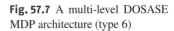

Fig. 57.7 A multi-level DOSASE MDP architecture (type 6)

2. Type 2: starting from the previous architecture, the authors include a spatial and temporally specific attentional system *T*, called *sensory mapping* that allows the brain to actively select responses from specific areas.

3. Type 3: it includes a motor mapping system M that can generate a representation for frequently practiced action sequences. With this system, the architecture can test the consequences of action sequences without actually carrying them out.

4. Type 4: an agent with this architecture is considered a self-aware and self-effecting (SASE) agent. Its main feature is that it is able to sense the external and internal voluntary decisions and use them in the attention selector *T*. Furthermore, it provides the agent with pre-motor areas where information related with the control of the effectors is stored and can be sensed, so the robot can *talk to itself*. This type of architecture allows the agent to perform

external, internal, and mixed reasoning corresponding to the attention in which the attention module T attends to external, internal or both, respectively,

which implies that the agent can accomplish non-associative learning, classical learning, and instrumental conditioning learning. Finally, a SASE agent shows autonomous planning capabilities.

5. Type 5: the authors call this architecture type developmental observation-driven SASE MDP (DOSASE MDP), which is characterized by being programmed without an explicit knowledge of the tasks the agent will learn and by being inaccessible once the robot starts execution. As a consequence the human teacher can interact with the architecture only through the environment.

6. Type 6: it is made up of several levels of type-5 architectures to provide the architecture with sensory and motor hierarchies that integrate different sensors and effectors feeding responses from low-level systems as sensory inputs of higher-level systems as in most layered architectures. A diagram of this architecture is displayed in Fig. 57.7 (extracted from [57.64]).

Directly associated with *SASE* is the autonomous learning method the authors propose, communicative learning, which requires grounded language acquisition and teaching using the acquired language.

The main elements of the *SASE* architecture have been implemented and tested in real humanoid robots (SAIL1, SAIL2 and DEV). Experiments have been carried out focusing on the attention selector *T*, the learning regressor *R* using the IHDR method [57.92], communicative learning [57.71], task transfer [57.64] and action chaining [57.92]. All these experiments study the developmental capabilities of the proposed architecture [57.69]. Figure 57.8 shows one of the experiments carried out with the SAIL robot in an

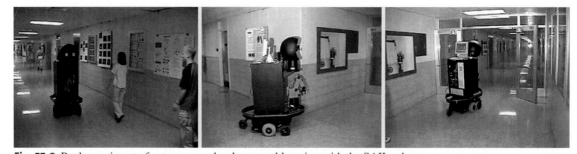

Fig. 57.8 Real experiment of autonomous developmental learning with the SAIL robot

autonomous navigation task using the vision-based sensorimotor skills acquired through online real-time developmental learning (extracted from [57.69]).

In the last few years, *Weng* et al. have focused their work on Where-What networks, which are SASE networks whose aim is to study visual attention, object recognition and tracking from a developmental point of view. These networks have been applied to a vision system that purports to be general purpose and whose aim is to extract objects from complex backgrounds [57.93].

iCub Cognitive Architecture

The iCub is a humanoid robot of small size (that of a 3–4 year-old child) developed within the RobotCub EU project [57.94]. It is being extensively used by the CDR community as the main humanoid platform for research, mainly in human-like early development. *Vernon*, *Von Hofsten*, and *Fadiga* have published a roadmap that summarizes and formalizes the design of a humanoid robot, in this case the one followed in the case of iCub addressing the main topics that any researcher should take into account [57.95]. A cognitive architecture has been developed for this robot within the same project and its main elements are summarized in [57.96]:

> *Gaze control, reaching, and locomotion constitute the initial simple goal-directed actions. Episodic and procedural memories are included to effect a simplified version of internal simulation in order to provide capabilities for prediction and reconstruction, as well as generative model construction bootstrapped by learned affordances. A very simple process of homeostatic self-regulation governed by the affective state provides elementary action selection. Finally, all the various components of the cognitive architecture operate concurrently so that a sequence of states representing cognitive behavior emerges from the interaction of many separate*

> *parallel processes rather than being dictated by some state-machine as in the case of most cognitive architectures.*

As we can see, it constitutes a very ambitious project that comprises the main neurological aspects of mental development. Specifically, it is made up of thirteen interconnected blocks [57.96]: Exogenous Salience, Endogenous Salience, Egosphere, and Attention Selection constitute the perception system, Gaze Control, Vergence, Reach & Grasp and Locomotion make up the actions system, the Episodic and Procedural Memory together constitute the main elements for anticipation and adaptation, the Affective State comprises the motivational system and regulates the autonomous capabilities of the robot together with the Action Selection system and, finally, it contains an Interface component.

The iCub cognitive architecture considers actions as the basic element of cognitive behaviors. In this context, actions are different from reactive movements because they are intentional, guided by a motivation and are executed monitored by a predictive system or model. The authors highlight that the main feature of the iCub cognitive architecture is self-development, which is implemented from a set of *innate perceptuo-motor and cognitive skills* [57.97], whose modulation and coordination is achieved through three basic elements: auto-associative memory, action selection and motivation. The low-level layers of the iCub architecture are displayed in Fig. 57.9.

As in most CDR architectures, a key aspect is the development of sensorimotor coordination and mapping through autonomous interaction. The authors have achieved very interesting advances in the development of a layered control system for iCub [57.94] that contains: an architecture for the generation of discrete and rhythmic movements where trajectories can be modulated by high-level commands and sensory feed-back, an attention system, which includes sensory input process-

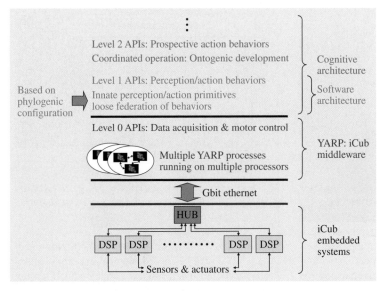

Fig. 57.9 A layered representation of the iCub architecture (courtesy of Prof. Giorgio Metta)

ing, eye–neck coordination, eye movements, a robust task-space reaching controller, and methods for learning internal models and a grasping module that allows the coordination of looking for a potential target, reaching for it (placing the hand close to the target) and attempting a grasping motion (or another basic action). In fact, manipulation is a very important task in iCub design, and very relevant work has been carried out on this topic.

In addition, the architecture includes a sub-system to simulate events including a prediction element for action selection. This system is based son *Shanahan's* global working memory [57.86] that, as explained by the authors,

uses an associative memory that receives efferent (motor) input produces afferent (sensory) output. This feeds into a motor hetero-associative memory that in turn produces (simulated) efferent (motor) output. This output is connected recurrently back to the sensory associative memory and also back to the modulation circuit.

Another basic element in the iCub architecture is the computational model of affordances, allowing for a developmental mixing of perception and action in a single representation that allows the

creation of a computation model of imitation and interaction between humans and robots by evaluating the automatic construction of models from experience, their correction via feedback, timing and synchronization.

Consequently, learning is a crucial aspect in this model. On the one hand, learning by observation is considered in the architecture, but as a higher-level capacity that requires basic capabilities like object and action recognition. On the other hand, autonomous learning constitutes the background of learning affordances, focused on problems like object location or executing simple actions over those objects. As remarked in [57.94]:

After the system has acquired the capability to coordinate movements with respect to sensory information, it can start interacting with objects and understanding its interface – how to grab the object, what are the effects of certain applied actions. Then, the system may start recognizing and interpreting other agents interacting with similar objects, learning other object affordances and interpreting activities.

The iCub cognitive architecture uses Bayesian networks for learning affordances that operate over the previously commented set of innate skills and over the knowledge acquired by the robot through object manipulation. The affordances model has been tested in a real world experiment with the iCub humanoid robot [57.94] showing the capability of the model for autonomously capturing the basic relations between the object's shape, hand velocity and selected actions. Figure 57.10 displays the iCub robot in one of these experiments devoted to developmental object recognition.

The authors have applied the architecture to more complex tasks related to imitation and communication

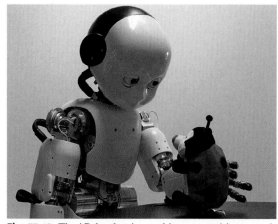

Fig. 57.10 The iCub robot in an object recognition experiment (courtesy of Lorenzo Natale)

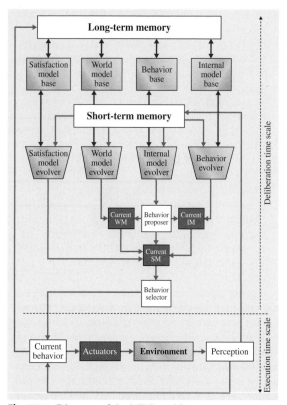

Fig. 57.11 Diagram of the MDB architecture

with the real robot. Specifically, they have carried out a very interesting human–robot interaction experiment where iCub had to recognize and interpret the gestures of people interacting with it [57.94].

Multi-Level Darwinist Brain (MDB)

The multi-level Darwinist brain (*MDB*) is a cognitive architecture that was first presented in [57.98]. It follows a developmental approach for the automatic acquisition of knowledge in a real robot through its interaction with the environment. The background idea of the *MDB* of applying artificial evolution for knowledge acquisition takes inspiration from classical bio-psychological theories that relate the brain and its operation through a Darwinist process. All of these theories lead to the same concept of cognitive structure based on the brain adapting its neural connections in real time through evolutionary or selectionist processes [57.99].

Typical CDR low-level topics like fetal sensorimotor development, voluntary movement acquisition, spatial perception, body/motor representation or understanding are beyond the scope of this architecture, which is more focused on higher-level features that a real robot must exhibit like adaptive motivation, open-ended lifelong learning or autonomous knowledge acquisition. In this sense, the authors pay attention to the computational implementation problems that arise when using a developmental approach in real-time operation [57.100].

The *MDB* architecture is based on a cognitive model that uses three basic models (W:world, I:internal and S:satisfaction) that must be obtained at execution time as the agent interacts with the world in a developmental

fashion. To be able to carry out this modeling process, information must be extracted from the real data the agent has after each interaction with the environment. If the sensorial information changes (dynamic environment, hardware failures), the world and internal models can be updated or replaced without any consequence in terms of the architecture. In addition, if the motivation that guides the robot changes, the satisfaction model would change while the action selection method remains unchanged. Thus, for every interaction of the agent with its environment, two processes must be solved: the modeling of functions W, I, and S and the optimization of the action using the models available at that time trying to maximize the predicted satisfaction provided by the satisfaction model.

MDB includes the following elements, displayed in the architecture diagram of Fig. 57.11 (extracted from [57.99]):

1. *Behavior structures*: they generalize the concept of single action used in the cognitive model. A behavior represents a decision element able to provide

actions or sequences of actions according to the particular sensorial inputs [57.101].

2. *Memory elements*: a short-term and a long-term memory are required in the learning processes, the first one involved in the storage of relevant input data from the sensors and the second one in the storage of relevant models. The interplay between them is discussed in [57.102].

3. *Time scales*: the *MDB* is structured into two different time scales, one devoted to the execution of the actions in the environment (reactive part) and the other dealing with learning the models and behaviors (deliberative part).

4. *Neuroevolution*: the authors decided to use ANNs as the representation for the models. Consequently, the acquisition of knowledge in *MDB* is a neuroevolutionary process, with an evolutionary algorithm devoted to learning the parameters of the ANN.

The architecture has been tested in real robots in two main experiments, one devoted to the learning of basic skills in a hexapod robot [57.103] and the other one focused on a more complex psychological experiment of induced behavior [57.104] and behavior reuse through the long-term memory [57.102] carried out in wheeled and legged robots (shown in Fig. 57.12). Both experiments followed the developmental principles of *MDB*. It is now being improved from a computational point of view [57.100] to make it more efficient and its elements are being made more brain-like in order to be able to consider more complex scenarios.

Epigenetic Robotics Architecture (ERA)

The epigenetic robotics architecture (*ERA*) developed by *Morse* et al. at the University of Plymouth is presented as a homogeneous and intuitive framework to achieve ongoing development, scalability, concept use and transparency [57.105].

ERA is based on the conceptual spaces theory of concepts [57.106], being the use of conceptual structures the main objective of the architecture designers [57.107]. As remarked by the authors in [57.105]:

Conceptual spaces are postulated as a way to represent knowledge on a level that resides in between the symbolic approaches on the one hand, and connectionist modeling on the other.

A conceptual space

is a collection of one or more domains (like color, shape, or tone), where a domain is postulated as

Fig. 57.12a,b Induced behavior of the MDB using (**a**) a wheeled and (**b**) a legged robot

a collection of inseparable sensory-based quality dimensions with a metric.

This type of conceptual knowledge performs a classification of the sensorial inputs into concepts using a distance metric.

With this conceptual background, the architecture is constructed associating multiple self-organizing maps, introduced by the authors as a more powerful and *rich* representation of concepts [57.105]:

The ERA architecture comprises multiple SOMs, each receiving input from a different sensory modality, and each with a single winning unit. Each of these winning units is then associated to the winning unit of a special hub SOM using a bidirectional connection weighted with positive Hebbian learning.

The left image in Fig. 57.13 displays a representation of a basic *ERA* unit.

This architecture is under development but it has been applied to humanoid robots in several examples focused on conditioned learning, various forms of priming, the relation between movement and orientation selectivity, sustained inattentional blindness, or showing that body posture is central to the linking of linguistic and visual information [57.108–111]. The right image of Fig. 57.13 shows an experiment for learning object names carried out with the iCub robot. Continuous development of the architecture takes place within the *iTalk* project [57.112].

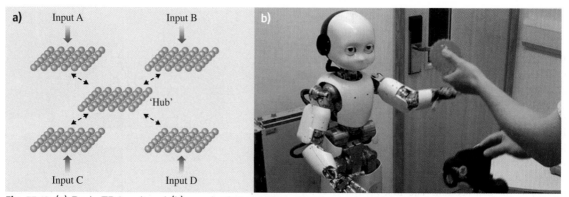

Fig. 57.13 (a) Basic ERA unit and **(b)** a real robot experiment having to do with learning the names of objects (both images courtesy of Anthony Morse)

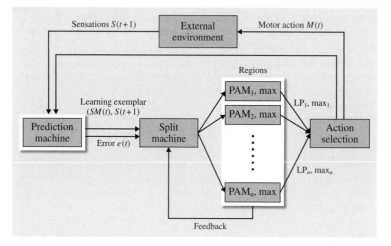

Fig. 57.14 R-IAC main elements and operational structure

Intelligent Adaptive Curiosity (IAC)

Although it is not exactly a cognitive architecture, *IAC* must be considered when reviewing relevant CDR systems. It was initially introduced as a developmental mechanism allowing a robot to self-organize developmental trajectories of increasing complexity without pre-programming the particular developmental stages [57.113]. It has been improved with the *R-IAC* version, that allows a robot *to learn actively, fast and correctly forward kinematic models in an unprepared sensorimotor space* and with the self-adaptive goal generation – robust intelligent adaptive curiosity (*SAGG-RIAC*) algorithm focused on inverse kinematic models [57.114]. Fig. 57.14 contains a block diagram of the *IAC* and *R-IAC* operational structure (extracted from [57.113]).

IAC has been tested in real robots [57.115] while *R-IAC* and *SAGG-RIAC* mainly in simulated ones. Fig-

ure 57.15 contains an image taken from a real robot experiment carried out with the AIBO robot where the authors show how IAC allows the robot to autonomously generate a developmental sequence.

We must highlight the ongoing work carried out in Flower's research group at the INRIA research institute focused on the

> *study of developmental constraints that allow for efficient open-ended learning of novel skills, in particular, the constraints that guide exploration in large sensorimotor spaces, like mechanisms of intrinsically motivated exploration and learning, mechanisms for social learning, morphological computation, and maturational constraints.*

Their work in language acquisition, as a developmental process grounded in action, is very remark-

Fig. 57.15 AIBO robot in a developmental experiment with the IAC architecture ▶

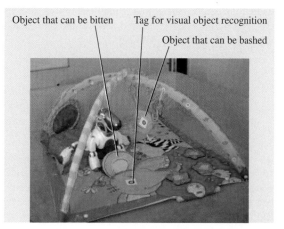

Object that can be bitten   Tag for visual object recognition

Object that can be bashed

able [57.116]. They are developing a humanoid robot called Acroban as the main real platform for testing their algorithms [57.74].

As a conclusion to this review of cognitive architectures and revisiting the dynamic and distributed brain model that was introduced in Sect. 57.2, we must say that things have been changing in the last decade and different authors are starting to propose new cognitive architectures in line of with the brain model introduced in that section. The first approaches were mainly symbolic with low biological inspiration but trying to replicate the cognitive functionalities displayed by humans or animals. These initial approaches to cognitive robotics served as a basis to the emergence of the CDR field, were most of the authors decided to break with the original engineering tendency and started developing non-symbolic, dynamic, and distributed architectures with the principles commented in Sect. 57.3.

Moreover, this new field has taken real robots as the only valid platform for testing cognitive architectures and, as a consequence, the number of robots that have been physically constructed to support this approach is remarkable.

57.4 Some Key Areas

From the previous revision of some the most relevant brain inspired cognitive architectures for robots developed in the last decade, one can extract the conclusion that the main cognitive functionalities such as perception, attention, memory, emotions, learning, reasoning, decision-making, or social development have been taken into account to different extents and using different approaches. The main reason behind this heterogeneity is the complexity of each one of these topics, and as a consequence, researchers usually concentrate their efforts on one of them before proceeding with others. Thus, there remains a lot of work to be done in the sense of unifying the research lines on these topics and on the joint consideration of several of them in order to be able to address the well-known interdependence between these cognitive functionalities in the brain.

Out of all of the basic functionalities that go into making a cognitive system present the properties and function of a brain based individual, in this section we will concentrate on three of them: memory, attention, and emotions. We have selected these three topics for discussion due to their remarkable brain-like inspiration background and due to the high relevance that, in our opinion, the advances in these topics will have in

the achievement of the next type of autonomous robot commented in the introduction of Sect. 57.3.

57.4.1 Memory

One of the most studied aspects of human brains in cognitive robotics is memory. Several authors assert that cognitive functions cannot arise without a memory system supporting them [57.117, 118]. Generally speaking, a robot that does not remember past actions or situations is bound to repeat previous mistakes [57.119], so learning is clearly compromised. As commented in [57.43]:

> *Human psychological research indicates that the knowledge of his/her personal history enhances one person's ability to accomplish several cognitive capabilities in the context of sensing, reasoning and learning.*

More specifically, memory is required for the main cognitive processes occurring in the brain. Memory plays a basic role in learning, and although it is possible to learn without explicitly using previous experience, the retention of learned patterns is clearly improved using memory. In planning and decision-making, memory is used as the substrate for the creation of models where predictions can be carried out. Finally, action selection is guided by previously selected actions and their consequences. In this sense [57.120]:

> *The most pleasurable action – meaning that the internal values have been influenced most positively – can be the action that is selected as the next action. The memory therefore can have an impact – positive or negative – on the stability of such agent internal values like the energy level, extending the lifespan of the agent.*

Advanced auxiliary processes such as attention, visual processing, and reasoning are naturally strongly coupled to the cognitive memory system [57.121]. Even emotions affect memory by modulating the behavior of the neural cells, either by inhibiting or facilitating the memory consolidation process [57.122, 123]. Consequently, all of these processes are interdependent, being the complete memory system characterized by complex relationships among them.

A classical discussion in the autonomous robotics field is that of the practical relevance of having a complex memory system in a real robot, as opposed to simpler and more reactive control methods [57.124]. It has been clearly shown throughout decades of research that purely reactive approaches can support simple tasks in a very efficient way, even achieving the tremendous success of introducing autonomous robots in homes [57.125]. These robots lack adaptive learning capabilities, so they are forced to repeat the same reactive behaviors during their lifetime and, consequently, the higher level of autonomy commented at the beginning of Sect. 57.3 is far from their possibilities. The practical consequences of using a brain-like memory system in real robots are precisely summarized in [57.126]:

- Focus attention on the most relevant features of the current task
- Support generalization of tasks without explicitly programming the robot
- Guide perceptual processes by limiting the perceptual search space
- Provide a focused memory to prevent the robot from being confused by occlusions
- Provide robust operation in the presence of distracting irrelevant events.

The traditional view of memory in neurosciences as a modular and hierarchical structure embraces most of the memory system proposals in cognitive robotics [57.49, 76, 127, 128]. Recently, new trends emphasize the distributed nature of cortical organization [57.129] based on dynamic, constructive, and self-referential remembering processes [57.130]. This new approach breaks with the traditional computer science view where memory is based on the database metaphor (using the storage-and-retrieval concept) [57.131]. These ideas are under development in neuroscience and their impact on cognitive robotics is still marginal, but they follow the same brain-like model commented on in Sect. 57.2, so their potential is very high.

Hierarchical Memory Systems

Starting with the classical view of brain-like memory, it must be pointed out that two main models with slight differences have been considered. The first one is based on the Atkinson–Shiffrin model [57.132], where memory can be divided into three basic types: sensory memory, short-term memory, and long-term memory (defined in Sect. 57.3.1).

In general, the whole memory system works on the basis of three different processes [57.133]: the information from the sensory system of the organism (external and internal sensors) is encoded, then stored, and finally retrieved (if not forgotten). Anyway, it is yet unclear and controversial among scientists, how exactly memory works [57.134]. In fact, this model that includes a sensory memory as an isolated element is controversial in the neuroscience community that tends to include it within the short-term memory. We can find this first model in cognitive robotics mainly in works focused on different aspects of the memory system, without paying direct attention to the sensory memory like in [57.133, 135] and [57.136]. Sensory memory has been studied and implemented on the basis of its main function acting as a filter for the short-term memory through the attentional mechanisms that will be analyzed in the next section.

The second model, widely used in cognitive robotics, is based on the existence of three memory elements: short-term and long-term memories as in the previous model, and a working memory [57.137]. Working memory can be considered a short-term memory cache that actively maintains information relevant to the current task for a short period of time [57.76]. As commented in [57.126],

> *working memory has been described as a system that stores a small number of "chunks" of information, protecting them from interference from other processing systems and positioning them so as to directly influence the generation of behavior.*

This memory element has generated much attention within the cognitive robotics field because it is used to store transient information during perception, reasoning, or planning for short periods of time, and it is fundamental for information selection, and consequently, determines the information that is stored and is used in the remaining cognitive functions. Some authors consider that working memory includes the short-term memory and other processing mechanisms required to use the short-term memory properly [57.138, 139].

To illustrate the relevance of working memory in cognitive robotics we must highlight some references, mainly the line developed at the Vanderbilt University School of Engineering. In 2005, *Phillips* and *Noelle* presented a biologically inspired working memory model for robots [57.126], which has been applied in the ISAC humanoid robot by professor *Kawamura* and their collaborators [57.76]. Their working memory system is based on data structures called *chunks*, which are able to store different items such as percepts, behaviors, or other task-related information [57.140]. The authors use a working memory neural network with temporal-difference learning *to learn the appropriate chunks to load given the current state of the system*. Their working memory system is implemented using the working memory toolkit [57.141]. Real experiments with the humanoid robot ISAC were carried out concentrating on the analysis of the ability to choose the appropriate behavior and percept chunk to accomplish a task (handshaking, reaching, waving, etc.) or the ability to use a higher-level decision rule to distinguish among similar performances.

In [57.48], the authors develop a working memory that acts as a global short-term memory. It uses relational symbolic structures that can describe complex situations among multiple entities, like current mission information, choices of tactics, etc. In this work, its authors use a mobile robot in navigation-based tasks such as path following, dynamic obstacle avoidance, replanning when a path is blocked, station keeping, and communicating with other robots about obstacles and unfriendly entities.

Regarding short-term and long-term memory, they have been considered in most of the memory systems developed in cognitive robotics. Their main differences lie in their capacity and duration limits. As established in [57.139]:

A duration difference means that items in short-term storage decay from this sort of storage as a function of time. A capacity difference means that there is a limit in how many items short-term storage can hold.

As commented above, short-term memory is typically devoted to the storage and basic processing of percepts obtained from the perceptual system of the robot [57.48, 140] and, after a generalization procedure, the most relevant information is transferred to the LTM. For example, in [57.135] the authors represent STM and LTM as finite state machines, having each memory a set of states (S_i). The LTM represents the recent stable configuration of features in the environment and the transfer of feature points from STM to LTM depends on their frequency. This policy means that spurious features should be quickly forgotten, while persistent features will be transferred to LTM. With this system, the movements of several objects are tracked in a dynamic environment using a wheeled robot equipped with a camera. From another point of view, in [57.99], the authors propose an interplay mechanism between STM and LTM where the LTM stores models and the STM percepts, the STM management strategy being the the one that decides whether a model is relevant and novel according to the percepts they generalize. This mechanism has been successfully applied to real dynamic robotics tasks using Pioneer II and *Aibo Robots* [57.99]

An interesting approach is that of considering the working memory as the element involved in this interplay between STM and LTM. In this sense, in [57.126], the authors develop a model where working memory is directly connected to the perceptual system and transfers relevant information to the STM and LTM, which are not directly connected. This scheme has been adopted in the ISAC robot by *Kawamura* et al. [57.76] and by [57.46] in the SOAR cognitive architecture, both of them providing promising results in real robotic tasks.

Long-term memory is the most complex memory element and it has received much attention in cognitive robotics research. Most of the authors agree on the division of long-term memory into declarative and non-declarative [57.142]:

- *Non-declarative*: also known as *procedural* or *implicit* memory. It is related with implicit knowledge, such as perceptions and actions (non-conscious processes like the tuning of motor skills).
- *Declarative*: it is related with *explicit* knowledge and can be divided into:
  - *semantic*: storing general knowledge about the world, including facts and properties of objects.

- *episodic*: storing facts, events, or episodes occurring during a lifetime contextualized in time and space.

Conscious, declarative, or explicit memories are mediated by the hippocampus and related cortical areas, whereas implicit emotional memory involves the amygdala and related areas [57.143]. Episodic memory is mainly related to the medial temporal lobe and hippocampal structures [57.144].

Most cognitive architectures (SOAR, ACT-R, OpenCog, ISAC, . . .), as remarked in the previous section, include a procedural memory element that contains the basic knowledge of how to select and perform basic actions or behaviors. Episodic memory provides the capacity to remember specific events or objects in the environment, while semantic memory stores accumulative knowledge of the world, that is, some generalized representation of the different episodes experienced [57.135]. Only a few architectures consider semantic memory systems appropriately due to the complexity of transforming events or facts into knowledge, that is, the classical problem of assigning meanings to the percepts. Particularizing to robotic applications, the number of relevant approaches is even smaller [57.135] (ISAC).

The case of episodic memory is different, the LTM module being the one that has been studied more in depth and implemented in real robots (SOAR, LIDA, SMRITI, . . .). It allows an agent to adapt its behavior to the environment according to previous experiences. As remarked in [57.119], where its authors present one of the most sophisticated models for episodic memory using the EIPROME framework:

> *Episodic memory is oriented towards the past in a way in which no other kind of memory system is. It is the only memory system that allows people to consciously re-experience their past.*

In this work, the authors revise the main cognitive architectures in robotics that use an episodic memory sub-system, reaching the conclusions that all of them have the following problems in common:

- Only applicable in highly limited domains
- Inappropriate for realizing higher psychological functionality of episodic memory
- Only consider actions, no perceptual and executive information
- Mostly handle short sequences
- Do not use one-shot learning

- Exhibit gap of terminology of episodic memory among different disciplines.

Most of the authors that have included this type of memory in their architectures have tried to implement three basic functionalities: encoding (where only salient events are processed), storage (consolidation and forgetting are very relevant issues), and retrieval [57.43, 120]. These functionalities have been basically implemented within a computational perspective, although often trying to mimic the neurological processes that occur in brains. For example, the salience of an event is strongly influenced by the emotional state of a person [57.145], and this connection between episodic memory and emotions has been studied in different works [57.76, 120, 146].

Distributed and Dynamical Memory Systems

Although its application to real robots is still marginal, different researchers differ from the classical hierarchical view of memory in neurobiology and propose distributed and dynamical memory systems. As remarked in [57.147]:

> *Memory is, in this context, considered to be fundamentally associative, following Hebbian principles, and in the low-level sense, rather than at the level of, for example, episodic memory. This principle emphasizes the distributed nature of cortical organization as the substrate for cognitive behaviors, rather than the competing modular view: in so doing it provides the basis of a common framework for both perception and action.*

As commented above, the researchers that have been working with brain inspired memory models in robotics have devoted much effort to the computational implementation of the neurological models so as to make them practical, that is, operational in real robots. Recent discussions in cognitive and neuropsychology diverge from the classical data base metaphor of memories and propose dynamic, constructive, and self-referential remembering processes. In this sense, *Rosenfield* [57.130] presented an approach to memory on the basis of clinical case studies that proposes the following basic concepts [57.131]:

- There is no memory but the process of remembering.
- Memories do not consist of static items which are stored and retrieved but they result from a construction process.

- The body is the point of reference for all remembering events.
- Body, time, and the concept of *self* are strongly interrelated.

In [57.147] the authors propose a memory-based cognitive framework where learning is considered as the adaptation of memory. They use the network memory theory [57.129] that proposes that distributed networks of neurons make up the basic substrate of memory, perception, and action. As remarked by the authors:

Each of these distributed networks encode associative relationships, which may be considered as basic neural elements, and are termed cognits [57.148]. These are proposed to be arranged in two informal and overlapping hierarchies, the first based on the primary sensory cortical regions, and the second on primary motor regions. Behavior emerges from this system through the activation-based interaction of cognits across, and at all levels of, these hierarchies – this active process may be described as cognition.

The authors apply this framework to a real-robot case in a preliminary study.

This approach to memory is still in its initial stages in terms of application within real robot architectures. In fact, to make it feasible, these architectures would have to become much more brain-like in their implementations and not just in their functionalities presenting a much more coherent neural based organization and, especially, considering much more precise models of interactions between and within brain areas.

57.4.2 Attention

Attention could be defined as a type of categorization of sensorial spaces by means of the relevance of the different regions of the space in terms of their importance for performing a given task [57.149]. Natural organisms have highly complex attentional systems coupled with their perceptual system and high-level cognitive systems, which select the relevant information for a given task, motivation or objective. Real robots will have to coexist with these natural organisms in their environment, so they are influenced by the same dynamic sensorial world, a world that can easily overwhelm any cognitive architecture if some order is not introduced. Usually, up to now, most of the work carried out in the development of artificial attentional systems has been applied in static and controlled environments, so it must

be clearly stated that current attentional systems are far from the required ones. Anyway, this is a very active research field [57.150, 151] where new systems and improvements are continuously being implemented and tested.

As we have seen in the review of the main cognitive architectures performed in Sect. 57.3, different kinds of attentional systems have been implemented within them, some included as a part of the memory system [57.54, 76] and some as an intrinsic part of the perceptual system [57.93, 94]. However, in general, most of the attentional systems one can find in the literature, including those described in this section have not been developed in the framework of any cognitive architecture. In fact, most work on attention has been carried out in the context of image processing and visual systems, without taking into account any brain-like interactions. That is, attention in natural systems implies the integration of multiple sensorial sources with a high degree of sensitivity. In robotics, attentional systems consider mainly visual information [57.152, 153] as the only sensorial input. Whenever multi-modal attentional systems are addressed in the literature, this is usually done in an ad hoc manner and generally using visual and auditive information streams [57.154–156]). The typical problems that can be found in multi-modal attentional systems are that they take into account sensorial data sources in a homogeneous way and that they lack the possibility of combining the different elements of attentional systems using novel schemes allowing the designer to decide the relationship among the different elements in any particular case. Examples of multi-modal attentional systems for robots are arising like [57.157] or [57.158–160].

Focusing on visual attention for robots, researchers have taken the human visual attention mechanisms and their related eye motion control (foveation) as inspiration [57.161]. In this sense, as remarked in [57.153]:

About 90% of the visual information leads to the Lateral Geniculate Nucleus (LGN). From the LGN, the information is transferred to the primary visual cortex (V1). ... From V1, the information is transmitted to the "higher" brain areas V2–V4, infero temporal cortex (IT), the middle temporal area (MT or V5) and the posterior parietal cortex (PP).

It is well-known that the visual information is processed in parallel in the brain, with functionally separated areas for processing color, shape, motion ... These areas have been modeled as processors, agents, or cells that cooperate and compete, leading to the concept of consciousness emergence from their dynamic relations [57.162, 163]. Consequently [57.153]:

> there is not a single brain area that controls attention but a network of areas. Several areas have been verified to be involved in attentional processes but the accurate task and behavior of each area as well as the interplay among them still remain open questions.

Thus, attentional systems developed for robots have adopted models taken from the computer vision field [57.164–166], which are based on the above commented knowledge of the operation of the human brain. These models use, basically, the same elements and operational scheme as described in detail in [57.153]:

1. Several features of the original image are processed in parallel creating a saliency map, which represents the relevance of the pixels, or as collections of nodes of an artificial neural network. Different features are represented in different maps that are combined, usually, through a weighing function to create the saliency map.
2. The saliency map contains a trajectory of image regions where selected image regions are local maxima in the saliency map. Here a winner-take-all (WTA) network is typically used.
3. The most salient region is determined to establish the focus of attention (FOA).
4. An inhibition of return (IOR) method is applied; a mechanism for avoiding re-exploring previously scanned positions. It prevents the FOA from permanently staying at the most salient region.

In the application of these elements and steps to autonomous robots the following main research topics have arisen:

1. *Establishing the saliency operator:* different authors have proposed different approaches to calculate the saliency map [57.166–168]. This is the most *classical* problem in attention, extensively studied in computer vision.
2. *Covert versus overt attention:* covert attention assumes that the robotic head or eye is static, simplifying the analysis of the image. In this sense, the models used in computer vision can be directly applied to robots, and remarkable results have been achieved in the identification of the most salient stimuli and in visual search in the image. Anyway, visual attention in nature is overt, and different authors have started to propose solutions in this trend

for cognitive robots. As remarked in [57.152], in overt attention:

> the content of the visual field (VF) changes after every head-eye movement. This makes the "saliency map" calculated prior to the movement partially obsolete. Moreover, each head movement causes a corresponding shift in the camera and the image coordinate systems. Shifts of the camera and the image coordinate systems between two successive attention cycles make the implementation of the inhibition of return (IOR) complicated. Finally, head movement often results in situations where only a partial view of different stimuli are perceivable. Such incomplete perception makes the identification of task-relevant information uncertain and might severely affect the visual search behavior.

Visual attention in overt approaches performs a parallel processing of the input information, where a robot-centric perspective is used following the concept of ego-centric saliency map [57.157, 169, 170].

3. *Bottom-up* versus *top-down analysis:* the former is related to seeking salient physical stimuli (for example: certain bright or colored regions, oriented edges, optical flows, etc.) and the latter to task-oriented attention, implying higher-level cognitive capabilities. Most of the initial attentional systems in robotics perform a bottom-up analysis, but as concluded in [57.152] cognitive robots require a top-down attentional system, because:

> In reality, the top-down influence plays a major role in the primates' visual attention mechanism. In case of the primates the top-down influence in attentional selection, however, is the result of a complex interaction among knowledge, emotion, personality, and reasoning.

Works like [57.171] and [57.172], implement a combination of top-down and bottom-up attention in a real robot.

4. *Unit of attention:* most computer vision models use *space* as the basic attentional unit [57.164, 165, 167]. Recently, attention models have become more focused on objects as the basic unit [57.173–175] or even on the mutual relationships between these two units [57.176, 177], which seems to be the process that occurs in human brains.

Typical robotic applications where attentional systems have been tested can be summarized into four main groups [57.153]:

- Active vision: it is basically the problem of deciding the position of the visual system of the robot according to the task or objective and, in its more general form, is highly coupled with overt attention. It has been applied to robots in classical works like [57.178] or [57.179] and in more recent examples of humanoid robotic vision [57.180, 181].

- Object manipulation: grasp and manipulation of objects is a very active field in cognitive robotics, mainly with humanoid robots, because it implies a wide range of cognitive skills from low to high level [57.150, 182].

- Robot navigation: controlling the motion of robots in real time implies an active control system that exploits an attentional system with quick response. In this field, researchers have been developing attentional systems for years (highly simplified versions, with low brain-like inspiration) [57.183]. Recent examples of attentional systems are scarce.

- Human–robot interaction: paying attention to humans has been the main research field of humanoid robots, because it implies from low to high-level cognitive capabilities, from detecting salient regions in a human (face, colors, expressions, etc.) to social interactions with behavior adaptation. Examples within this application area can be found in [57.184] or [57.185].

Recent trends based on modeling neuromodulation and action selection like [57.186–189] must be highlighted here. For example, in [57.187] the authors design and implement

an integrated neural architecture, modeled on human executive attention, which is used to control both automatic (reactive) and willed action selection in a simulated robot.

In [57.186] a strategy for controlling autonomous robots that is based on principles of neuromodulation in the mammalian brain is presented. As the authors mention in this work:

the neuromodulatory systems provide a foundation for cognitive function in higher organisms; Attention, emotion, goal-directed behavior, and decision-making all derive from the interaction between the neuromodulatory systems, and brain areas such as the amygdala, frontal cortex, and hippocampus. Therefore, understanding neuromodulatory function may provide control and action selection algorithms for autonomous robots that effectively interact with the environment.

In both works, robotic experiments are carried out analyzing the model's response to simulated lesions.

To conclude, in [57.152] its authors present two very complete reviews of visual attentional systems for robots. They conclude that, nowadays, there are five basic open issues that must be solved to achieve a real world operative attentional system for cognitive robots: overt shift of attention (which implies studying the change of reference frame, the development of dynamic IOR and considering the partial appearance of objects), integrated space and object-based analysis, optimal learning strategy for visual search, autonomy in the operation of the attentional system and on-line learning.

57.4.3 Emotions

The relationship between emotion and behaviors in human beings has been studied in several areas of knowledge, such as psychology, sociology, neurophysiology, and brain science [57.190–197]. In particular, the relationship between emotional expressions and social behavior has been discussed in sociology, where it is considered that emotion plays a very important role in social interactions. It is mainly this fact that led to the study of emotions in the field of robotics and artificial cognition.

There are three main reasons to incorporate the use of emotions in robotics. The first one is to obtain sociable robots, that is, robots closer to humans that are capable of recognizing human emotions and expressing their own emotions in order to improve human–robot interaction. The second one is to incorporate emotions to the robot behavior architecture, not to make behaviors more expressive, but to organize them and to modulate them in a more complete cognitive architecture. Finally, some researchers are also interested in obtaining useful feedback to understand human cognition processes by means of modeling those processes in robots. Often researchers are interested in more than one of these aspects and it is common to find research work that addresses two or even all of them.

Regarding sociable robots, several human-like robots (or robotic heads) able to show facial expressions have been developed [57.198–202]; probably Kismet [57.203] is the best known. Concrete facial expressions, or their number, are usually specified explicitly by the designer in advance, and consequently, the robots learn to mimic humans but are not able to adapt to non-specified situations from an emotional point of view. *Breazeal* et al. proposed a developmental model that enables a robot to derive the relationship between motor commands for its facial expressions and those of the caregiver's by imitating facial expressions during the robot's motor babbling [57.204]. *Kobayashi* et al. went a step beyond [57.205] and proposed a method for learning states rather than expressions, so that the robot categorizes a user's facial expressions under given emotional labels and is able to evoke the right emotional label later. *Watanabe* et al. [57.206] also proposed a communications model that enables a robot to associate facial expressions to internal states through intuitive parenting by users who mimic or exaggerate a robot's facial expression, like humans do with babies. The robot strengthens the connection between its internal state and the facial expression associated with a particular state. The internal state of a robot and its facial expressions change dynamically depending on the external stimuli. After learning, facial expressions and internal states are classified and made to mutually correspond by strengthened connections. More recently, *Lee* et al. [57.207] presented a system to learn combinations of unit behaviors, using simulated annealing, in order to allow a robot to express new complex emotions, which are linear combinations of a set of predefined basic emotions. They call unit behaviors the physical elements of a robot employed to express emotions (eyes, arms, cheeks, brows, etc.). *Han* et al. [57.208] address a similar problem, but using fuzzy Kohonen clustering networks, in an attempt to generate a continuous of emotional behaviors as a combination of four basic emotions.

Regarding the role of emotions in the organization of behaviors, many authors propose explicit functions for emotions in the context of a global cognitive architecture [57.209, 210], such as coordinating behaviors, shifting behavioral hierarchies, social communication, shortcut cognitive processing, and facilitating storage and recall of memories. *Ortony* et al. [57.211] establish three levels of information processing in a cognitive architecture: reactive, routine, and reflective, and they analyze the interplay between affect, motivation, and cognition at those three levels. *Arbib* [57.212] also points to the fact that emotions, although present in living beings, are not always beneficial, so it is very important to determine how the brain maximizes the benefits of emotions minimizing their occasional inappropriateness and how to take into account this fact when designing robots. *Arkin* et al. incorporate emotions to their previously elaborated schema theory in [57.213]. In this work they use schema theory as a tool to implement and study emotion-driven behaviors in robots performing in multi-agent environments, exemplifying the approach using a Hermes II robot that implements four different behaviors of a praying mantis: prey acquisition (motivated by hunger), predator avoidance (motivated by fear), mating (motivated by sex desire), and search for an habitat (they called it *chantlitaxia*).

The importance of emotions is also being studied in the field of cognitive developmental robotics [57.65] from both the point of view of neuroscience and that of developmental psychology. From a neuroscience point of view, it seems that there are, at least, two emotional processes that play a very important role in cognitive development [57.214], in particular in memorizing (the emotional stimulus of the amygdala affects the pathway from the cortex to the hippocampus, strengthening the memory of the events before and after the stimulus [57.215, 216]), and reward prediction (where the dopamine neuron in basal ganglia plays an important role [57.217]). From a developmental psychology point of view, several researchers try to mimic in robots the peekaboo game of children. Peekaboo is treated as one of the communication styles in which infants adjust to the emotions and affections of their caregivers, detect regular social behavior in communications, and predict certain behaviors [57.218]. *Mirza* et al. [57.219] propose a model in which this game emerges as a robot's behavior based on its own experience and the stimulus from the environment. This work is extended in [57.220] incorporating a deeper interaction history and using a humanoid robot (the iCub). The authors in [57.214] propose a communications model for a robot to acquire peekaboo abilities based on reward prediction. The system keeps the sensor data in short-term memory and transfers them to long-term memory when the value of the internal state corresponding to the emotion is increased, so future sensor data is compared to memory contents to predict the response of the caregiver.

To summarize, the study of emotions and their implementation in robotic architectures is a completely open field. As indicated throughout this section, several approaches and implementations have been studied in the last ten years and there has been a lot of speculation on the particular role emotions could play other than simplifying communications and interaction with humans through emotion detection or emotion simulation. In fact, much work in this line has gone into making robots *seem* emotionally competent for the benefit of humans, but a lot less into exploring the mechanisms and advantages of including emotions within robotic cognitive architectures as an intrinsic part of their processing or decision-making processes.

57.5 Cognitive Robotic Platforms and Research Projects

Several projects have arisen in the last ten years that are related, one way or another, with the study of cognition, embodiment and development, often over real agents/robots, with the hope of increasing the learning and adaptability capabilities of agents and, at the same time, our knowledge about how these things work in real brains and bodies in nature.

Meyer and other authors [57.221] have been involved in a biological inspired project called Psikharpax, the main objective of which is to build a rat-like robot with a neural control architecture that will permit some of the capacities of autonomy and adaptation exhibited by real rats. The alleged reason to use a rat as inspiration is the reticence expressed by some authors such as [57.222] and [57.223] about the complexity of trying to mimic human intelligence taking into account the current knowledge about the human brain, and the convenience of starting with a less ambitious task such as to understand and reproduce other (and simpler) animal capabilities.

Several robotic projects have been started with the aim of carrying out research in the field of human–robot interaction and how robots can learn from humans (and mimic them) in those kinds of processes. In this line, a social robotic platform (Roillo) for investigating human–robot communications is presented in [57.224]. The platform is centered in timing in non-verbal communications and its role in the expression of attention, intention, and emotion. The experiments are performed in the context of children's playrooms, and the robot is designed accordingly, simple but with the capability of expressing emotions and with the appearance of a toy. Another project, funded by the European Seventh Framework Programme (FP7) and using relatively simple robots, is SERA [57.225], which is also focused on social interactions between humans and artificial systems (robots, agents, etc.). The same idea is also behind the development of several humanoids by the Humanoid Robotics Group at MIT (such as Coco, Cog, and Kismet).

There are also several more ambitious and long-term global projects. For instance, in 2002 the UK Computing Research Committee initiated a discussion of possible long-term projects (they called them Grand Challenges in Computer Research [57.226]) to address deep scientific issues rather than purely practical goals. This was materialized in 2004 as a list of detailed proposals. One of those proposals, called Architecture of Brain and Mind [57.227], is about attempting to understand and model mechanisms of the brain and mind in the context of an ambitious project to design a robot with a significant sub-set of the capabilities of a young child. Although this is a long-term project with neither a specific deadline nor funds, several UK research proposals have been inspired by it.

The European Union, throughout the sixth and the seventh Framework Programmes, has funded two initiatives called *Cognitive Systems* and *Cognitive Systems, Interaction, Robotics* [57.228] that, although their objectives do not explicitly bind the creation of more advanced cognitive systems to biological concepts, include projects and scientific research in this direction. Perhaps the most relevant example is the RobotCub project [57.229], the main goal of which is to study cognition through the implementation of a humanoid robot, the iCub [57.230]. This robot presents the size of a 3.5 year old child and numerous sensors and actuators (53 motors that move the head, arms, hands, waist, and legs, microphones, video cameras, accelerometers, gyroscopes, etc.). The *age* of the robot is a consequence of the philosophy behind the RobotCub project, as it is assumed that manipulation plays a fundamental role in the development of cognitive capabilities and manipulation skills are developed during the first years of growth. Both the design and the software of the iCub are open and publicly available. This has contributed to its fast dissemination within the research community [57.73, 231, 232]. Not only are there several research lines in RobotCub focusing on different aspects of the iCub robot, but also other robotics projects have emerged, usually also within FP7, that use this same robot and that share some of the objectives of the RobotCub project, like development (MACSi [57.233]), embodiment and complex object manipulation (Darwin [57.234]), embodiment and social learning (ITALK [57.235]), embodiment and cumulative learning (IM-CLeVeR [57.236]), and human–robot interaction and communications (EFAA [57.237], CHRIS [57.238] or ROSSI [57.239]).

An earlier project with similar objectives within the FP6 program, was Feelix Growing [57.240], where several research groups in Europe were involved in several robotic projects with the common aim of adapting robotic behaviors to human mood/emotions and mimicking those human emotions to make more human-friendly robotic behaviors. One of the robots employed in this project was Nao [57.241], another humanoid robot of 58 cm height, 4.3 kg weight and many sensors and actuators (it has a total of 25 DOF), which was developed with the aim of being widely used by the research community around the world. Nao is nowadays the robot used in the RoboCup [57.242] soccer standard platform league.

The Socially-Synergistic Intelligence Group has developed two other baby robotic platforms within the JST ERATO program in Japan with the aim of being used in the cognitive developmental robotics field. One of them, CB^2 [57.243], is similar in conception to iCub, that is, large, with many DOF and expensive, whereas the other one [57.244] is more like Nao, smaller (3 kg versus 22 kg in the case of iCub and 33 kg in the case of CB^2) and simpler, in order to decrease the cost (for example, the iCub costs between 200 000 € (v1.1) and 250 000 € (v2.0) plus taxes, versus Nao that costs 12 000 € plus taxes with all the bells and whistles), preserving enough flexibility to carry out the desired experiments, mainly related with embodiment.

The LIRA-Lab in Italy has also developed a humanoid robot with 18 DOF called BabyBot [57.245]. This robot is not as complex as other humanoids mentioned before, but the objective was the same: research on cognition within the developmental paradigm. Finally, Acroban [57.246] at Inria in France, is a very recent humanoid project whose objectives include both development and human–robot interaction.

Most robotic projects nowadays consider embodiment and development as key issues and employ humanoid robots in their research, as there is more interest in human-like brains than in any other life-like brains, even if the complexity of a human brain is huge and the task of emulating its capabilities will not be achieved in the short term. In this direction, although there are a few research groups with the necessary resources to implement their own humanoid robots, most of the groups use Nao or, if they can afford it, iCub.

57.6 Concluding Remarks

This chapter has provided an overview of a research field that is starting to explode in terms of number of research papers and projects that are focusing on creating really autonomous robots that are able to operate in human populated environments and interact with humans. In this line, the creation of more or less brain-like approaches to the control of these robots is becoming ubiquitous. However, the field is still in its infancy and there are a lot of avenues of research that are completely open and often unexplored. It is important to take into account that the brain itself, which is the object that all of the researchers are attempting to clone, is not well known. Of course great progress has been made in its study in the last two or three decades, especially with new scanning techniques. Nonetheless, this progress is still not detailed enough to permit a unified vision of the operation of the brain or of the precise role of each one of its components or their interrelations. As a consequence, the models of the brain that are used in robotics are usually very basic (despite the complexity they may display) and even naive in many instances. They generally address the most salient architectural features in terms of a layered structure with blocks that modulate other blocks and often hybridize sub-symbolic approaches with symbolic processing structures in order to achieve some perceived functions of the brain. Nevertheless, in the last few years, there has been a trend towards making these elements and structures more biologically plausible and some authors have started to base some of these blocks on computational neural structures that mimic the image of the behavior of some brain areas that is currently available. Some of them have even gone into modeling the influence of genes and proteins on the behavior of neurons. This approach, however, is still in its very early stages and, although some results have been produced, they are still very far from the objective of a brain-like robot controller.

Many topics are still open both from a neurophysiological and from a computational point of view, including how memory functions, how attention comes about, or the role of emotion on the operation of the brain. In addition, there is a lot of work to be done on how learning in these systems takes place and can be controlled. In fact, without a precisely agreed model on the dynamics of the operation of natural brains it is hard to envision any of these topics coming to a conclusion soon. What is really happening now is a sort of a coevolutionary process whereby research into brain operation is being used to produce ever more biologically credible analogs of the brain for the control of robots, and these are being used to try to dilucidate some of the characteristics of the operation of brains. This is not, however, a balanced cooperation yet as roboticists are borrowing heavily from neuroscientists, whereas the feedback the latter are getting is still very limited due to the inherent problems of having to make things work in real systems, which often implies just concentrating on one aspect or simplifying some of the premises.

Nonetheless, progress is being made in this exciting field and, as more researchers become involved, this progress, will accelerate, and perhaps in not such a distant future we will be able to produce robots that operate and think like humans and that are able to live and cooperate with humans and, hopefully, this will be good.

References

57.1   H. Ritter, J.J. Steil, C. Noelker, F. Roethling, P.C. McGuire: Neural architectures for robot intelligence, Rev. Neurosci. **14**(1/2), 121–143 (2003)

57.2   W. Duch: Brain-inspired conscious computing architecture, J. Mind Behav. **26**(1/2), 1–21 (2005)

57.3   S. Sloman: *Grand challenge 5: The Architecture of Brain and Mind: Integrating Low-Level Neuronal Brain Processes with High-Level Cognitive Behaviours in a Functioning Robot, Technical Report, COSY-TR-0607* (School of Computer Science, University of Birmingham UK 2006)

57.4   R. Brooks: I, Rodney Brooks, am a robot, IEEE Spectr. **45**(6), 71–75 (2008)

57.5   H. Jerison: *Evolution of the Brain and Intelligence* (Academic, New York 1973)

57.6   W. Hodos: Some perspectives on the evolution of intelligence and the brain. In: *Animal Mind–Human Mind*, ed. by D.R. Griffin (Springer, Berlin, Heidelberg 1982) pp. 33–56

57.7   A.B. Butler, W. Hodos: *Comparative Vertebrate Neuroanatomy* (Wiley-Liss, New York 1996)

57.8   P.H. Harvey, J.R. Krebs: Comparing brains, Science **249**, 140–146 (1990)

57.9   R.G. Northcutt: Evolution of the telencephalon in nonmammals, Annu. Rev. Neurosci. **4**, 301–350 (1981)

57.10   R.G. Northcutt, J.H. Kaas: The emergence and evolution of mammalian neocortex, Trends Neurosci. **18**, 373–379 (1995)

57.11   S. Walker: *Animal Thought* (Routledge Kegan Paul, London 1983)

57.12   M.G. Belekhova, N.P. Veselkin: Telencephalization and transfer of function in the central nervous system of vertebrates in light of current data, J. Evol. Biochem. Physiol. **21**, 357–365 (1985)

57.13   A.H. Cohen: Evolution of the vertebrate central pattern generator for locomotion. In: *Neural Control of Rhythmic Movements in Vertebrates*, ed. by A.H. Cohen, S. Rossignol, S. Grillner (Wiley, New York 1988) pp. 129–166

57.14   W. Wilczynski: Central neural systems subserving a homoplasous periphery, Am. Zool. **24**, 755–763 (1984)

57.15   T.W. Deacon: Rethinking mammalian brain evolution, Am. Zool. **30**, 629–705 (1990)

57.16   P.A. Getting: Emerging principles governing the operation of neural networks, Annu. Rev. Neurosci. **12**, 185–204 (1989)

57.17   R.M. Harris-Warrick: Modulation of neural networks for behaviour, Annu. Rev. Neurosci. **14**, 39–57 (1991)

57.18   T.J. Prescott, P. Redgrave, K. Gurney: Layered control architectures in robots and vertebrates, Adapt. Behav. **7**, 99–127 (1999)

57.19   J.H. Jackson: Evolution and dissolution of the nervous system, Br. Med. J. **I**, 662 (1884)

57.20   K.E. von Baer: *Über Entwicklungsgeschichte der Thiere: Beobachtung und Reflexion*, Vol. 2 (Königsberg, Bornträger 1828) pp. 1828–1888

57.21   H.W. Magoun: *The Waking Brain* (Thomas, Springfield, Illinois 1958)

57.22   G.G. Berntson, S.T. Boysen, J.T. Cacioppo: Neurobehavioral organization and the cardinal principle of evaluative bivalence, Annu. New York Acad. Sci. **702**, 75–102 (1993)

57.23   P. Teitelbaum, T. Schallert, I.Q. Whishaw: Sources of spontaneity in motivated behaviour. In: *Handbook of Behavioural Neurobiology*, Vol. 6, ed. by P. Teitelbaum, E. Satinoff (Plenum, New York 1983) pp. 23–65

57.24   J.W. Rudy, S. Stadler-Morris, P. Albert: Ontogeny of spatial navigation behaviours in the rat: Dissociation of *proximal* – and *distal* – cue-based behaviours, Behav. Neurosci. **101**, 62–73 (1987)

57.25   C.R. Gallistel: *The Organization of Action: A New Synthesis* (Lawrence Erlbaum, Hillsdale, NJ 1980)

57.26   R. Brooks: A robust layered control system for a mobile robot, IEEE J. Robot. Autom. **2**(1), 14–23 (1986)

57.27   J. Kien: From command fibers to command systems to consensus are these labels really useful anymore, Behav. Brain Sci. **9**, 732–733 (1986)

57.28   A.H. Cohen: The role of heterarchical control in the evolution of central pattern generators, Brain, Behav. Evol. **40**, 112–124 (1992)

57.29   W. Penfield: *The Mystery of the Mind* (Princeton Univ., Princeton, NJ 1975)

57.30   R. Thompson: Centrencephalic theory, the general learning system, and subcortical dementia, Annu. New York Acad. Sci. **702**, 197–223 (1993)

57.31 M.P. Young: Special issue on brain-structure-function relationships, advances from neuroinformatics, Philos. Trans. R. Soc. Lond. B **355**, 3–6 (2000)

57.32 A. Meyering, H. Ritter: Learning to recognize 3d-hand postures from perspective pixel images, Artif. Neural Netw. **2**, 821–824 (1992)

57.33 M.P. Young, J.W. Scannel: Analysis and modelling of the organization of the mammalian cerebral cortex. In: *Experimental and Theoretical Advances in Biological Pattern Formation*, Vol. 259 of NATO ASI Series A, ed. by H.G. Othmer, P.K. Maini, J.D. Murray (Plenum, New York 1993) pp. 369–384

57.34 W. Duch: Platonic model of mind as an approximation to neurodynamics. In: *Brain-Like Computing and Intelligent Information Systems*, ed. by S. Amari, N. Kasabov (Springer, Singapore 1997) pp. 491–512

57.35 W.J. Freeman: *Neurodynamics, An Exploration in Mesoscopic Brain Dynamics* (Springer, London 2000)

57.36 M. Spivey, M. Grosjean, G. Knoblich: Continuous attraction toward phonological competitors, Proc. Natl. Acad. Sci. USA **102**(29), 10393–10398 (2005)

57.37 T.D. Frank, A. Daffertshofer, C.E. Peper, P.J. Beek, H. Haken: Towards a comprehensive theory of brain activity: Coupled oscillator systems under external forces, Physica D **144**, 62–86 (2000)

57.38 N. Kasabov, L. Benuskova, S. Wysoski: Biologically plausible computational neurogenetic models: Modeling the interaction between genes/proteins, neurons and neural networks, J. Comput. Theoret. Nanosci. **2**(4), 569–575 (2005)

57.39 N. Kasabov, R. Schliebs, H. Kojima: Probabilistic computational neurogenetic modelling: From cognitive systems to Alzheimer's disease, IEEE Trans. Auton. Ment. Dev. **3**(4), 300–311 (2011)

57.40 Y. Meng, Y. Jin, J. Yin, M. Conforth: Human activity detection using spiking neural networks regulated by a gene regulatory network, Proc. Int. Joint Conf. Neural Netw. (IJCNN) (2010) pp. 2232–2237

57.41 A.F. Morse, J. de Greeff, T. Belpeame, A. Cangelosi: Epigenetic robotic architecture (ERA), IEEE Trans. Auton. Ment. Develop. **2**(4), 325–339 (2010)

57.42 W. Morris (Ed.): *The American Heritage Dictionary of the English Language*, 4th edn. (Houghton Mifflin Harcourt, Boston 2000)

57.43 E.C. de Castro, R.R. Gudwin: An episodic memory for a simulated autonomous robot, Proc. Robot. 2010 (2010) pp. 1–7

57.44 M.D. Byrne: Cognitive architecture. In: *The Human-Computer Interaction Handbook*, ed. by J.A. Jacko, A. Sears (L. Erlbaum Associates Inc., Hillsdale, NJ, USA 2002) pp. 97–117

57.45 B. van Heuveln: *What is Cognitive Robotics?* (Cognitive Science Department, Rensselaer Polytechnic Institute, Troy 2010), available online at http://www.cogsci.rpi.edu/public_html/heuveb/Teaching/CognitiveRobotics/Cognitive%20Robotics.html

57.46 J.E. Laird, A. Newell, P.S. Rosenbloom: SOAR: An architecture for general intelligence, Artif. Intell. **33**(1), 1–64 (1987)

57.47 A. Newell: *Unified Theories of Cognition* (Harvard University, Cambridge 1994)

57.48 J.E. Laird: Toward cognitive robotics, SPIE Def. Sens. Conf. (2009)

57.49 J.E. Laird: Extending the soar cognitive architecture, Proc. 2008 Conf. Artif. Gen. Intell., ed. by P. Wang, B. Goertzel, S. Franklin (IOS, Amsterdam 2008) pp. 224–235

57.50 J.R. Anderson, C. Lebiere: *The Atomic Components of Thought* (Lawrence Erlbaum Associates, Mahwah, NJ 1998)

57.51 J.R. Anderson, D. Bothell, M.D. Byrne, S. Douglass, C. Lebiere, Y. Qin: An integrated theory of the mind, Psychol. Rev. **111**(4), 1036–1060 (2004)

57.52 C. Lebiere, J.R. Anderson: A connectionist implementation of the ACT-R production system, 15th Annu. Conf. Cogn. Sci. Soc. (1993) pp. 635–640

57.53 B.J. Best, C. Lebiere: Cognitive agents interacting in real and virtual worlds. In: *Cognition and Multi-Agent Interaction: From Cognitive Modeling to Social Simulation*, ed. by R. Sun (Cambridge Univ. Press, Cambridge 2006) pp. 186–218

57.54 D. Friedlander, S. Franklin: LIDA and a Theory of Mind, 1st Conf. Artif. Gen. Intell. (AGI) (Amsterdam 2008) pp. 137–148

57.55 S. Franklin: Cognitive robots: Perceptual associative memory and learning, IEEE Int. Workshop on Robot and Human Interact. Commun. (ROMAN) (2005) pp. 427–433

57.56 S. Franklin, M. Ferkin: Using broad cognitive models to apply computational intelligence to animal cognition. In: *Applications of Computational Intelligence in Biology*, Vol. 122, ed. by T. Smolinski, M. Milanova, A.-E. Hassanien (Springer Berlin, Heidelberg 2008) pp. 363–394

57.57 D.L. De Luise, G. Barrera, S. Franklin: Robot localization using consciousness, J. Pattern Recognit. Res. **6**, 1–23 (2011)

57.58 J. Bach: The micropsi agent architecture, 5th Int. Conf. Cognit. Model. (ICCM) (Bamberg, Germany 2003) pp. 15–20

57.59 J. Bach: *Principles of Synthetic Intelligence PSI: An Architecture of Motivated Cognition*, 1st edn. (Oxford Univ., Inc., New York 2009)

57.60 J.S. Albus, A.M. Meystel: *Engineering of Mind: An Introduction to the Science of Intelligent Systems* (Wiley, New York 2001)

57.61 J.S. Albus: Reverse engineering the brain. In: *AAAI Fall Symp. Biol. Inspired Cogn. Archit.* (Arlington, Virginia, USA 2008) pp. 5–14

57.62 B. Goertzel: *The Hidden Pattern: A Patternist Philosophy of Mind* (Brown Walker, FL 2006)

57.63 B. Goertzel, H. de Garis: XIA-MAN: An extensible, integrative architecture for intelligent humanoid robotics, AAAI Fall Symp. Biol. Inspired Cogn. Archit. (2008) pp. 65–74

57.64 J. Weng: On developmental mental architectures, Neurocomputing **70**(13–15), 2303–2323 (2007)

57.65 M. Asada, K.F. MacDorman, H. Ishiguro, Y. Kuniyoshi: Cognitive developmental robotics as a new paradigm for the design of humanoid robots, Robot. Auton. Syst. **37**, 185–193 (2001)

57.66 D. Vernon, G. Metta, G. Sandini: The iCub cognitive architecture: Interactive development in a humanoid robot, Proc. ICDL 2007 (2007) pp. 122–127

57.67 J.L. Krichmar, G.N. Reeke: The Darwin brain-based automata: Synthetic neural models and real-world devices. In: *Modelling in the Neurosciences: From Biological Systems to Neuromimetic Robotics*, ed. by G.N. Reeke, R.R. Poznanski, K.A. Lindsay, J.R. Rosenberg, O. Sporns (Taylor and Francis, Boca Raton 2005) pp. 613–638

57.68 J.L. Krichmar, G.M. Edelman: Principles underlying the construction of brain-based devices, Proc. AISB '06, Vol. 2 (2006) pp. 37–42

57.69 J. Weng: Developmental robotics: Theory and experiments, Int. J. Humanoid Robot. **1**(2), 199–236 (2004)

57.70 M. Asada, K. Hosoda, Y. Kuniyoshi, H. Ishiguro, T. Inui, Y. Yoshikawa, M. Ogino, C. Yoshida: Cognitive Developmental Robotics: A Survey, IEEE Trans. Auton. Ment. Dev. **1**(1), 12–34 (2009)

57.71 Y. Zhang, J. Weng: Grounded auditory development by a developmental robot, Proc. INNS-IEEE Int. Jt. Conf. Neural Netw. (2001) pp. 1059–1064

57.72 S. Zeng, J. Weng: Online-learning and attention-based approach to obstacle avoidance using a range finder, J. Intell. Robotic Syst. **50**(3), 219–239 (2007)

57.73 G. Metta, G. Sandini, D. Vernon, L. Natale, F. Nori: The iCub humanoid robot: An open platform for research in embodied cognition, PerMIS: Perform. Metr. Intell. Syst. Workshop (2008)

57.74 O. Ly, P.-Y. Oudeyer: Acroban the humanoid: Playful and compliant physical child-robot interaction, ACM SIGGRAPH'2010 Emerg. Technol. (2010)

57.75 K. Kawamura, W. Dodd, P. Ratanaswasd, R.A. Gutierrez: Development of a robot with a sense of self, IEEE Int. Symp. Comput. Intell. Robot. Autom. (CIRA) (2005) pp. 211–217

57.76 K. Kawamura, S.M. Gordon, P. Ratanaswasd, E. Erdemir, J.F. Hall: Implementation of cognitive control for a humanoid robot, Int. J. Humanoid Robot. **5**(4), 547–586 (2008)

57.77 P. Ratanaswasd, W. Dodd, K. Kawamura, D.C. Noelle: Modular behavior control for a cognitive robot, 12th Int. Conf. Adv. Robot. (ICAR) (2005) pp. 713–718

57.78 P. Ratanaswasd, C. Garber, A. Lauf: Situation-based stimuli response in a humanoid robot, 5th Int. Conf. Dev. Learn. (2006)

57.79 B. Ulutas, E. Erdemir, K. Kawamura: Application of a hybrid controller with non-contact impedance to a humanoid robot, Int. Workshop Var. Struct. Syst. (VSS) (2008) pp. 378–383

57.80 M.P. Shanahan: A cognitive architecture that combines internal simulation with a global workspace, Conscious. Cogn. **15**, 433–449 (2006)

57.81 B.J. Baars: *A Cognitive Theory of Consciousness* (Cambridge Univ. Press, Cambridge 1998)

57.82 B.J. Baars: *In the Theater of Consciousness* (Oxford Univ., New York 1997)

57.83 B.J. Baars: The conscious access hypothesis: Origins and recent evidence, Trends Cogn. Sci. **6**(1), 47–52 (2002)

57.84 B.J. Baars, S. Franklin: How conscious experience and working memory interact, Trends Cogn. Sci. **7**, 166–172 (2003)

57.85 M.P. Shanahan: Consciousness, emotion, and imagination: A brain-inspired architecture for cognitive robotics, Proc. AISB 2005 Symp. Next Gener. Approaches Mach. Conscious (2005) pp. 26–35

57.86 M.P. Shanahan, B.J. Baars: Global workspace theory emerges unscathed, Behav. Brain Sci. **30**(5/6), 524–525 (2007)

57.87 J.L. Krichmar, G.M. Edelman: Brain-based devices: Intelligent systems based on principles of the nervous system, IEEE/RSJ Int. Conf. Intell. Robotic Syst. (IROS), Vol. 1 (2003) pp. 940–945

57.88 G.M. Edelman: Learning in and from brain-based devices, Science **318**(5853), 1103–1105 (2007)

57.89 J.G. Fleischer, B. Szatmáry, D.B. Hutson, D.A. Moore, J.A. Snook, G.M. Edelman, J.L. Krichmar: A neurally controlled robot competes and cooperates with humans in Segway soccer, IEEE Int. Conf. Robot. Autom. (ICRA) (2006) pp. 3673–3678

57.90 G.N. Reeke Jr., O. Sporns, G.M. Edelman: Synthetic neural modeling: The 'Darwin' series of recognition automata, Proc. IEEE **78**(9), 1498–1530 (1990)

57.91 J. Weng, C. Evans, W.S. Hwang, Y.B. Lee: The developmental approach to artificial intelligence: Concepts, developmental algorithms and experimental results, Proc. NSF Des. Manuf. Grantees Conf. (1999)

57.92 J. Weng, W. Hwang: Online image classification using IHDR, Int. J. Doc. Anal. Recognit. **5**(2/3), 118–125 (2002)

57.93 J. Weng, M. Luciw: Online learning for attention, recognition, and tracking by a single developmental framework, Proc. 23rd IEEE Conf. Comput. Vis. Pattern Recognit. 4th IEEE Online Learn. Comput. Vis. (2010) pp. 1–8

57.94 G. Metta, L. Natale, F. Nori, G. Sandini, D. Vernon, L. Fadiga, C. von Hofsten, K. Rosander, M. Lopes, J. Santos-Victor, A. Bernardino, L. Montesano:

The iCub humanoid robot: An open-systems platform for research in cognitive development, Neural Netw. **23**(8–9), 1125–1134 (2010)

57.95 D. Vernon, C. von Hofsten, L. Fadiga: A Roadmap for Cognitive Development in Humanoid Robots. In: *Cognitive Systems Monographs*, Vol. 11, ed. by R. Dillmann, Y. Nakamura, S. Schaal, D. Vernon (Springer, Berlin, Heidelberg 2010)

57.96 D. Vernon: Enaction as a conceptual framework for developmental cognitive robotics, Paladyn J. Behav. Robot. **1**(2), 89–98 (2010)

57.97 D. Vernon, G. Metta, G. Sandini: A survey of artificial cognitive systems: Implications for the autonomous development of mental capabilities in computational agents, IEEE Trans. Evol. Comput. **11**(2), 151–180 (2007)

57.98 R.J. Duro, J. Santos, F. Bellas, A. Lamas: On line Darwinist cognitive mechanism for an artificial organism, Proc. Suppl. Book SAB2000 (2000) pp. 215–224

57.99 F. Bellas, R.J. Duro, A. Faiña, D. Souto: Multilevel Darwinist brain (MDB): Artificial evolution in a cognitive architecture for real robots, IEEE Trans. Auton. Ment. Dev. **2**(4), 340–354 (2010)

57.100 B. Santos-Diez, F. Bellas, A. Faiña, R.J. Duro: Lifelong learning by evolution in robotics: Bridging the gap from theory to reality, Proc. Int. Symp. Evol. Intell. Syst. (2010) pp. 48–53

57.101 R.J. Duro, F. Bellas, J.A. Becerra: Evolutionary architecture for lifelong learning and real-time operation in autonomous robots. In: *Evolving Intelligent Systems: Methodology and Applications*, ed. by P. Angelov, D.P. Filev, N. Kasabov (Wiley, New Jersey 2010)

57.102 F. Bellas, J.A. Becerra, R.J. Duro: Internal and external memory in neuroevolution for learning in non-stationary problems, Lect. Notes Artif. Intell. **5040**, 62–72 (2008)

57.103 F. Bellas, R.J. Duro: Multilevel Darwinist brain in robots, initial implementation, Proc. 1st Int. Conf. Inform. Control, Autom. Robot. (2004) pp. 25–32

57.104 F. Bellas, J.A. Becerra, R.J. Duro: Induced behaviour in a real agent using the multilevel Darwinist brain, Lect. Notes Comput. Sci. **3562**, 425–434 (2005)

57.105 A.F. Morse, J. De Greeff, T. Belpaeme, A. Cangelosi: Epigenetic robotics architecture (ERA), IEEE Trans. Auton. Ment. Dev. **2**(4), 325–339 (2010)

57.106 P. Gardenfors: Conceptual spaces as a framework for knowledge representation, Mind and Matter **2**(2), 9–27 (2004)

57.107 J. Parthemore, A.F. Morse: Reclaiming symbols: An enactive account of the inter-dependence of concepts and experience, Pragmat. Cogn. **18**(2), 273–312 (2010)

57.108 A.F. Morse, T. Belpaeme, A. Cangelosi, L.B. Smith: Thinking with your body: Modelling spatial biases in categorization using a real humanoid robot, 32nd Annu. Conf. Cogn. Sci. Soc. (2010)

57.109 A.F. Morse, R. Lowe, T. Ziemke: Manipulating space: Modelling the role of transient dynamics in inattentional blindness, Connect. Sci. **21**(4), 275–295 (2009)

57.110 A.F. Morse, M. Aktius: Dynamic liquid association: Complex learning without implausible guidance, Neural Netw. **22**(1), 875–889 (2009)

57.111 A.F. Morse, T. Ziemke: Action, detection, and perception: A computational model of the relation between movement and orientation selectivity in the cerebral cortex, Proc. CogSci. 2009 – 31st Annu. Conf. Cogn. Sci. Soc. (2009)

57.112 A. Cangelosi, T. Belpaeme, G. Sandini, G. Metta, L. Fadiga, G. Sagerer, K. Rohlfing, B. Wrede, S. Nolfi, D. Parisi, C. Nehaniv, K. Dautenhahn, J. Saunders, K. Fischer, J. Tani, D. Roy: The ITALK project: Integration and transfer of action and language knowledge, Proc. 3rd ACM/IEEE Int. Conf. Hum. Robot Interact. (HRI 2008) (2008)

57.113 A. Baranes, P.-Y. Oudeyer: R-IAC: Robust intrinsically motivated exploration and active learning, IEEE Trans. Auton. Mental Dev. **1**(3), 155–169 (2009)

57.114 A. Baranes, P.-Y. Oudeyer: Intrinsically motivated goal exploration for active motor learning in robots: A Case Study, Proc. IEEE/RSJ Int. Conf. Intell. Robotic Syst. (2010)

57.115 P.-Y. Oudeyer, F. Kaplan, V. Hafner: Intrinsic motivation systems for autonomous mental development, IEEE Trans. Evol. Comput. **11**(2), 265–286 (2007)

57.116 F. Kaplan, P.-Y. Oudeyer, B. Bergen: Computational models in the debate over language learnability, Infant Child Dev. **17**(1), 55–80 (2008)

57.117 J.M. Fuster: *The Prefrontal Cortex: Anatomy, Physiology, and Neuropsychology of the Frontal Lobe*, 2nd edn. (Lippincott, Williams Wilkins 1997)

57.118 B.R. Postle: Working memory as an emergent property of the mind and brain, Neuroscience **139**(1), 23–38 (2006)

57.119 S. Jockel, M. Weser, D. Westhoff, J. Zhang: Towards an episodic memory for cognitive robots, Proc. 6th Int. Cognit. Rob. Workshop 18th Eur. Conf. Artif. Intell. (ECAI) (IOS, July 21–22 2008) pp. 68–74

57.120 T. Deutsch, A. Gruber, R. Lang, R. Velik: Episodic memory for autonomous agents, Conf. Hum. Syst. Interact., ed. by Łukasz Piatek (Rzeszow University of Information, Technology and Management, Kraków 2008) pp. 621–626

57.121 P. Kronberg: *Evolving short-term memory for behavior-based robots*, Master Thesis (Chalmers Univ. Technol., Göteborg 2003)

57.122 S.A. Christianson (Ed.): *Handbook of Emotion and Memory: Current Research and Theory* (Lawrence Erlbaum Associates, New Jersey 1992)

57.123 E.R. Kandel, J.H. Schwartz, T.M. Jessell: *Principles of Neural Science*, 4th edn. (McGraw-Hill, New York 2000)

57.124 R.A. Brooks: *Elephants Don't Play Chess, Designing Autonomous Agents: Theory and Practice from Biology to Engineering and Back* (MIT, Cambridge USA 1990)

57.125 Roomba vacuum cleaner website: http://www.irobot.com (2011)

57.126 J. Phillips, D. Noelle: A biologically inspired working memory framework for robots, Proc. IEEE Intl. Workshop Robot. Hum. Interact. Commun. (RO-MAN) (2005)

57.127 J.R. Anderson, M. Matessa: The rational analysis of categorization and the ACT-R architecture. In: *Rational Models of Cognition*, ed. by M. Oaksford, N. Chater (Oxford Univ. Press, Oxford 1998) pp. 197–217

57.128 B. Goertzel, H. de Garis: XIA-MAN: An extensible, integrative architecture for intelligent humanoid robotics, Proc. BICA-08 (2008) pp. 86–90

57.129 J.M. Fuster: Network memory, Trends Neurosci. **20**(10), 451–459 (1997)

57.130 I. Rosenfield: *The Strange, Familiar, and Forgotten. An Anatomy of Consciousness* (Vintage Books, New York 1993)

57.131 K. Dautenhahn: Embodiment and interaction in socially intelligent life-like agents. In: *Computation for Metaphors, Analogy, and Agents*, Lecture Notes In Computer Science, Vol. 1562, ed. by C.L. Nehaniv (Springer, Berlin, Heidelberg 1999) pp. 102–141

57.132 R. Atkinson, R. Shiffrin: Human memory: A proposed system and its control processes. In: *The Psychology of Learning and Motivation*, Vol. 2, ed. by K.W. Spence, J.T. Spence (Academic Press, New York 1968) pp. 89–195

57.133 P.A. Vargas, W.C. Ho, M. Lim, S. Enz, R. Aylett: To forget or not to forget: Towards a roboethical memory control, Proc. AISB-2009 (2009) pp. 18–23

57.134 J.M. Levenson: Epigenetic mechanisms: A common theme in vertebrate and invertebrate memory formation, Cell. Mol. Life Sci. **63**, 1009–1016 (2006)

57.135 F. Dayoub, T. Duckett, G. Cielniak: Toward an object-based semantic memory for long-term operation of mobile service robots. In: *Workshop on Semantic Mapping and Autonomous Knowledge Acquisition* (IROS, Taipei, Taiwan 2010)

57.136 C. Gurrin, H. Lee, J. Hayes: iForgot: A model of forgetting in robotic memories, Proc. HRI 2010 – 5th ACM/IEEE Int. Conf. Hum.-Robot. Interact. (2010) pp. 93–94

57.137 A.D. Baddeley, G. Hitch: Working memory. In: *The Psychology of Learning and Motivation: Advances in Research and Theory*, Vol. 8, ed. by G.H. Bower (Academic Press, New York 1974) pp. 47–89

57.138 M. Solms, O. Turnbull: *The Brain and the Inner World* (Karnac/Other, Cathy Miller Foreign Rights Agency, London, England 2002)

57.139 N. Cowan: What are the differences between long-term, short-term, and working memory? Prog. Brain Res. **169**, 323–338 (2008)

57.140 S. Gordon, J. Hall: System integration with working memory management for robotic behavior learning, Proc. 5th Int. Conf. Dev. Learn. (ICDL'06) (2006)

57.141 D.M. Wilkes, M. Tugcu, J.E. Hunter, D. Noelle: Working memory and perception, Proc. IEEE Int. Workshop Robot. Hum. Interact. Commun. (2005) pp. 686–691

57.142 E. Tulving: *Episodic and Semantic Memory, in Organization of Memory* (Academic, NewYork 1972) pp. 89–101

57.143 S. Halpern: *Can't Remember what I Forgot: The Good News From the Front Lines of Memory Research* (Harmony Books, New York 2008)

57.144 L.R. Squire, B. Knowlton, G. Musen: The structure and organization of memory, Annu. Rev. Psychol. **44**, 453–495 (1993)

57.145 A. Baddeley: *Human Memory: Theory and Practice* (Psychology, UK 1997)

57.146 W. Dodd, R. Gutierrez: The role of episodic memory and emotion in a cognitive robot, Proc. 14th Annu. IEEE Int. Workshop Robot. Hum. Interact. Commun. (RO-MAN) (2005) pp. 692–697

57.147 P. Baxter, W. Browne: Memory as the substrate of cognition: A developmental cognitive robotics perspective, Proc. 10th Int. Conf. Epigenet. Robot. (2010) pp. 19–26

57.148 J.M. Fuster: The Cognit: A network model of cortical representation, Int. J. Psychophysiol. **60**, 125–132 (2006)

57.149 M.M. Mesulam: Attention, confusional states, and neglect. In: *Principles of Behavioral and Cognitive Neurology*, ed. by M.M. Mesulam (Oxford Univ. Press, New York 2000) pp. 174–256

57.150 J.L. Tsotsos: *A Computational Perspective on Visual Attention* (MIT Press, Cambridge 2011)

57.151 L. Paletta, J.L. Tsotsos: Attention in cognitive systems, 5th Int. Workshop Atten. Cognit. Syst., WAPCV 2008 (Springer, Berlin, Heidelberg 2009)

57.152 M. Begum, F. Karray: Visual attention for robotic cognition: A survey, IEEE Trans. Auton. Mental Dev. **3**(1), 1–14 (2010)

57.153 S. Frintrop, E. Rome, H.I. Christensen: Computational visual attention systems and their cognitive foundations: A survey, ACM Trans. Appl. Percept. **7**(1), 6 (2010)

57.154 S.N. Wrigley, G.J. Brown: A computational model of auditory selective attention, IEEE Trans. Neural Netw. **15**(5), 1151–1163 (2004)

57.155 S. Lang, M. Kleinehagenbrock, S. Hohenner, J. Fritsch, G.A. Fink, G. Sagerer: Providing the basis for human–robot-interaction: A multi-modal attention system for a mobile robot, Proc. Int. Conf. Multimodal Interfaces (2003)

57.156 S.W. Ban, M. Lee: Biologically motivated visual selective attention for face localization, Proc. 2nd

Int. Workshop Atten. Perform. Comput. Vis. (WAPCV 2004), Vol. 3368, ed. by N.L. Paletta, J.K. Tsotsos, E. Rome, G.W. Humphreys (Springer, Berlin, Heidelberg 2004) pp. 196–205

57.157 J.L. Crespo, A. Faiña, R.J. Duro: An adaptive detection/attention mechanism for real time robot operation, Neurocomputing **72**, 850–860 (2009)

57.158 R. Arrabales, A. Ledezma, A. Sanchis: A cognitive approach to multimodal attention, J. Phys. Agents **3**(1), 53–64 (2009)

57.159 A. Haasch, N. Hofemann, J. Fritsch, G. Sagerer: A multi-modal object attention system for a mobile robot, IROS 2005 Proc. (2005) pp. 2712–2717

57.160 L.M. Garcia, A.A.F. Oliveira, R.A. Grupen, D.S. Wheeler, A.H. Fagg: Tracing patterns and attention: Humanoid robot cognition, IEEE Intell. Syst. Appl. **15**(4), 70–77 (2000)

57.161 P.J. Burt: Attention mechanisms for vision in a dynamic world, 9th Int. Conf. Pattern Recognit., Vol. 2 (1988) pp. 977–987

57.162 B.J. Baars: *A Cognitive Theory of Consciousness* (Cambridge Univ. Press, New York 1993)

57.163 D.C. Dennett: *Consciousness Explained* (Little, Brown and Co, Boston 1991)

57.164 C. Koch, S. Ullman: Shifts in selective visual attention: Toward the underlying neural circuitry, Hum. Neurobiol. **4**, 219–227 (1985)

57.165 L. Itti, C. Koch, E. Niebur: A model of saliency-based visual attention for rapid scene analysis, IEEE Trans. Pattern Anal. Mach. Intell. **20**(11), 1254–1259 (1998)

57.166 S. Frintrop: *VOCUS: A visual attention system for object detection and goal-directed search*, Lecture Notes in Artificial Intelligence (LNAI), Vol. 3899 (Springer, Berlin, Heidelberg 2006)

57.167 V. Navalpakkam, L. Itti: Top-down attention selection is fine- grained, J. Vis. **6**, 1180–1193 (2006)

57.168 M.R. Heinen, P.M. Engel: Visual selective attention model for robot vision, Latin Am. Robot. Symp. Intell. Robot. Meet. (2008) pp. 29–34

57.169 J.M. Canas, M.M. Casa, T. Gonzalez: An overt visual attention mechanism based on saliency dynamics, Int. J. Intell. Comput. Med. Sci. Image Process **2**, 93–100 (2008)

57.170 J. Ruesch, M. Lopes, A. Bernardino, J. Hornstein, J.S. Victor, R. Pfeifer: Multi modal saliency-based bottom-up attention: A framework for the humanoid robot icub, Proc. IEEE Int. Conf. Robot. Autom. (2008) pp. 962–967

57.171 L. Zhang: Vision attention learning model and its application in robot, Proc. 7th Asian Control Conf. (2009) pp. 970–975

57.172 T. Xu, K. Kühnlenz, M. Buss: Autonomous behavior-based switched top-down and bottom-up visual attention for mobile robots, IEEE Trans. Robot. **26**(5), 947–954 (2010)

57.173 Y. Sun, R. Fisher: Object-based visual attention for computer vision, Artif. Intell. **146**, 77–123 (2003)

57.174 T. Wu, J. Gao, Q. Zhao: A computational model of object-based selective visual attention mechanism in visual information acquisition, Proc. IEEE Conf. Inform. Acquis. (2004) pp. 405–409

57.175 Y. Yu, G.K. Mann, R.G. Gosine: A novel robotic visual perception method using object-based attention, Proc. 2009 IEEE Int. Conf. Robot. Biomim. (2009)

57.176 M. Begum, G. Mann, R. Gosine, F. Karray: Object and space-based visual attention: An integrated framework for autonomous robots, Proc. IEEE/RSJ Int. Conf. Intell. Robotic Syst., Nice (2008) pp. 301–306

57.177 L.J. Lanyon, S.L. Denham: A model of active visual search with object-based attention guiding scan paths, Neural Netw. **17**, 873–897 (2004)

57.178 B. Mertsching, M. Bollmann, R. Hoischen, S. Schmalz: The neural active vision system NAVIS. In: *Handbook of Computer Vision and Applications*, Vol. 3, ed. by B. Jähne, H. Haussecke, P. Geissler (Academic, San Diego 1999) pp. 543–568

57.179 M. Bollmann, R. Hoischen, M. Jesikiewicz, C. Justkowski, B. Mertsching: Playing domino: A case study for an active vision system. In: *Computer Vision Systems*, ed. by H. Christensen (Springer, New York 1999) pp. 392–411

57.180 S. Vijayakumar, J. Conradt, T. Shibata, S. Schaal: Overt visual attention for a humanoid robot, Proc. Int. Conf. Intell. Robot. Auton. Syst. (IROS 2001) (2001) pp. 2332–2337

57.181 A. Dankers, N. Barnes, A. Zelinsky: A reactive vision system: Active-dynamic saliency, Proc. 5th Int. Conf. Comput. Vis. Syst. (ICVS 2007) (2007)

57.182 S. Jeong, M. Lee, H. Arie, J. Tani: Developmental learning of integrating visual attention shifts and bimanual object grasping and manipulation tasks, Proc. IEEE 9th Int. Conf. Dev. Learn. (ICDL2010) (2010) pp. 165–170

57.183 J.J. Clark, N.J. Ferrier: Modal control of an attentive vision system, Proc. 2nd Int. Conf. Comput. Vis. (Tampa, Florida, US 1988)

57.184 C. Breazeal: A context-dependent attention system for a social robot, Proc. Int. Jt. Conf. Artif. Intell. (IJCAI 99) (Stockholm 1999) pp. 1146–1151

57.185 Y. Nagai: From bottom-up visual attention to robot action learning, IEEE 8th Int. Conf. Dev. Learn. (2009)

57.186 B.R. Cox, J.L. Krichmar: Neuromodulation as a robot controller: A brain inspired design strategy for controlling autonomous robots, IEEE Rob. Autom. Mag. **16**(3), 72–80 (2009)

57.187 J. Garforth, S.L. McHale, A. Meehan: Executive attention, task selection and attention-based learning in a neurally controlled simulated robot, Neurocomputing **69**(16–18), 1923–1945 (2006)

57.188 W.H. Alexander, O. Sporns: An embodied model of learning, plasticity, and reward, Adapt. Behav. **10**, 143–159 (2002)

57.189 T.J. Prescott, F.M. Montes Gonzalez, K. Gurney, M.D. Humphries, P. Redgrave: A robot model of the basal ganglia: Behavior and intrinsic processing, Neural Netw. **19**, 31–61 (2006)

57.190 D. Sperber, D. Wilson: *Relevance: Communication and Cognition* (Blakwell, Oxford 1986)

57.191 R.R. Cornelius: *The Science of Emotion: Research and Tradition in the Psychology of Emotions* (Prentice-Hall, Inc, Englewood Cliffs, NJ, US 1996)

57.192 R. Buck: *The Communication of Emotion* (Guilford, New York 1984)

57.193 D. Keltner, J. Haidt: Social functions of emotions. In: *Emotions: Current Issues and Future Directions*, ed. by T.J. Mayne, G.A. Bonanno (Guilford, New York 2001) pp. 192–213

57.194 D. Keltner, A.M. Kring: Emotion, social function, and psychopathology, Rev. Gen. Psychol. **2**(3), 320–342 (1998)

57.195 M. Ridley: *The Origins of Virtue* (Penguin, London 1997)

57.196 M.W. Eysenck: *Psychology: An Integrated Approach* (Longman, Harlow, Essex 1998)

57.197 L. Cosmides, J. Tooby: Evolutionary psychology and the emotions. In: *Handbook of Emotions*, Vol. 7, 2nd edn., ed. by M. Lewis, J.M. Haviland-Jones (Wiley, Chichester 2000) pp. 91–115

57.198 D. Matsui, T. Minato, K.F. MacDorman, H. Ishiguro: Generating natural motion in an android by mapping human motion, IEEE/RSJ Int. Conf. Intell. Robot. Syst. (IROS) (2005) pp. 3301–3308

57.199 H. Miwa, T. Okuchi, K. Itoh, H. Takanobu, A. Takanishi: A new mental model for humanoid robots for human friendly communication introduction of learning system, mood vector and second order equations of emotion, IEEE Int. Conf. Robot. Autom. (ICRA), Vol. 3 (2003) pp. 3588–3593

57.200 H. Miwa, K. Itoh, M. Matsumoto, M. Zecca, H. Takanobu, S. Rocella, M.C. Carrozza, P. Dario, A. Takanishi: Effective emotional expressions with expression humanoid robot WE-4RII: Integration of humanoid robot hand RCH-1, IEEE/RSJ Int. Conf. Intell. Robots Syst. (IROS), Vol. 3 (2004) pp. 2203–2208

57.201 T. Hashimoto, S. Hitramatsu, T. Tsuji, H. Kobayashi: Development of the face robot SAYA for rich facial expressions, Int. Jt. Conf. SICE-ICASE (2006) pp. 5423–5428

57.202 L. Canamero, J. Fredslund: I show you how I like you – can you read it in my face?, IEEE Trans. Syst. Man Cyber. A **31**(5), 454–459 (2001)

57.203 C. Breazeal: Toward sociable robots, Robot. Auton. Syst. **42**(3), 167–175 (2003)

57.204 C. Breazeal, D. Buchsbaum, J. Gray, D. Gatenby, B. Blumberg: Learning from and about others: Towards using imitation to bootstrap the social understanding of others by robots, Artif. Life **11**(1), 31–62 (2005)

57.205 H. Kobayashi, Y. Ichikawa, M. Senda, T. Shiiba: Realization of realistic and rich facial expressions by face robot, IEEE/RSJ Int. Conf. Intell. Rob. Syst. (IROS), Vol. 2 (2003) pp. 1123–1128

57.206 A. Watanabe, M. Ogino, M. Asada: Mapping facial expression to internal states based on intuitive parenting, J. Robot. Mechatron **19**(3), 315–323 (2007)

57.207 D. Lee, H.S. Ahn, J.Y. Choi: A general behavior generation module for emotional robots using unit behavior combination method, The 18th IEEE Int. Symp. Robot Hum. Interact. Commun. (RO-MAN) (2009) pp. 375–380

57.208 M.-J. Han, C.-H. Lin, K.-T. Song: Autonomous emotional expression generation of a robotic face, IEEE Int. Conf. Syst., Man and Cybern. (SMC) (2009) pp. 2427–2432

57.209 E.T. Rolls: On The brain and emotion, Behav. Brain Sci. **23**(2), 219–228 (2000)

57.210 R.W. Levenson: Blood, sweat, and fears: The autonomic architecture of emotion, Annu. New York Acad. Sci. **1000**, 348–366 (2003)

57.211 A. Ortony, D.A. Norman, W. Revelle: Affect and proto-affect in effective functioning. In: *Who Needs Emotions: The Brain Meets the Machine*, ed. by J.M. Fellous, M.A. Arbib (Oxford Univ., New York 2005)

57.212 M.A. Arbib: Beware the passionate robot. In: *Who Needs Emotions? The Brain Meets the Machine*, ed. by J.M. Fellous, M.A. Arbib (Oxford Univ., New York 2005)

57.213 R.C. Arkin, K. Ali, A. Weitzenfeld, F. Cervantes-Pèrez: Behavioral models of the praying mantis as a basis for robotic behavior, Robot. Auton. Syst. **32**(1), 39–60 (2000)

57.214 M. Ogino, T. Ooide, A. Watanabe, M. Asada: Acquiring peekaboo communication: Early communication model based on reward prediction, IEEE 6th Int. Conf. Dev. Learn. (ICDL) (2007) pp. 116–121

57.215 R. Paz, J.G. Pelletier, E.P. Bauer, D. Paré: Emotional enhancement of memory via amygdala-driven facilitation of rhinal interactions, Nat. Neurosci. **9**(10), 1321–1329 (2006)

57.216 J.L. McGaugh: *Memory and Emotion* (Orion, London 2003)

57.217 W. Schultz, P. Dayan, P.R. Montague: A neural substrate of prediction and reward, Science **275**(5306), 1593–1599 (1997)

57.218 P. Rochat, J.G. Querido, T. Striano: Emerging sensitivity to the timing and structure of proto-conversation in early infancy, Dev. Psychol. **35**(4), 950–957 (1999)

57.219 N.A. Mirza, C.L. Nehaniv, K. Dautenhahn, R. Te Boekhorst: Grounded sensorimotor interaction histories. In an information theoretic metric space for robot ontogeny, Adapt. Behav. **15**, 167–187 (2007)

57.220 N.A. Mirza, C.L. Nehaniv, K. Dautenhahn, R. te Boekhorst: Developing social action capabilities in a humanoid robot using an interaction history architecture, 8th IEEE-RAS Int. Conf. Humanoid Robots (2008) pp. 609–616

57.221 J.-A. Meyer, A. Guillot, B. Girard, M. Khamassi, P. Pirim, A. Berthoz: The Psikharpax project: Towards building an artificial rat, Robot. Auton. Syst. **50**(4), 211–223 (2005)

57.222 H.L. Dreyfus: *What Computers Still Can't Do: A Critique of Artificial Reason* (MIT, Cambridge, USA 1992)

57.223 R.A. Brooks: *Cambrian Intelligence: The Early History of the New AI* (MIT, Cambridge, USA 1999)

57.224 M.P. Michalowski, S. Sabanovic, P. Michel: Roillo: Creating a social robot for playrooms, 15th IEEE Int. Symp. Robot Hum. Interact. Commun. (ROMAN) (2006) pp. 587–592

57.225 Project SERA (Social Engagement with Robots and Agents): http://project-sera.eu/ (last accessed 5 May 2013)

57.226 UK Computing Research Committee (Grand Challenges in Computer Research): http://www.ukcrc.org.uk/grand-challenge/ (last accessed 5 May 2013)

57.227 UK Computing Research Committee (Grand Challenge 5 Architecture of Brain and Mind), http://www.cs.stir.ac.uk/gc5/ (last accessed 5 May 2013)

57.228 Seventh Framework Programme (FP7), Information and Communication Technologies Challenge 2: Cognitive Systems and Robotics, http://cordis.europa.eu/fp7/ict/programme/challenge2_en.html (last accessed 5 May 2013)

57.229 RobotCub – An Open Framework for Research in Embodied Cognition, http://www.robotcub.org (last accessed 5 May 2013)

57.230 iCub – An Open Source Cognitive Humanoid Robotic Platform, http://www.icub.org (last accessed 5 May 2013)

57.231 N.G. Tsagarakis: iCub: The design and realization of an open humanoid platform for cognitive and neuroscience research, Adv. Robot. **21**(10), 1151–1175 (2007)

57.232 G. Sandini, G. Metta, D. Vernon: The iCub cognitive humanoid robot: An open-system research platform for enactive cognition. In: *50 Years of Artificial Intelligence*, Vol. 4850, ed. by M. Lungarella, F. Iida, J. Bongard, R. Pfeifer (Springer, Berlin, Heidelberg 2007) pp. 358–369

57.233 MACSi – Motor Affective Cognitive Scaffolding for the iCub, http://macsi.isir.upmc.fr (last accessed 5 May 2013)

57.234 Darwin Project – Dexterous Assembler Robot Working with embodied Intelligence, http://darwin-project.eu (last accessed 5 May 2013)

57.235 ITALK – Integration and Transfer of Action and Language Knowledge in Robots, www.italkproject.org (last accessed 5 May 2013)

57.236 IM-CLeVeR – Intrinsically Motivated Cumulative Learning Versatile Robots, http://www.im-clever.eu (last accessed 5 May 2013)

57.237 EFAA – Experimental Functional Android Assistant, http://efaa.upf.edu (last accessed 5 May 2013)

57.238 CHRIS – Cooperative Human Robot Interaction Systems, http://www.chrisfp7.eu (last accessed 5 May 2013)

57.239 ROSSI – Emergence of communication in RObots through Sensorimotor and Social Interaction, http://www.rossiproject.eu (last accessed 5 May 2013)

57.240 Feelix Growing Project, http://www.feelix-growing.org (last access date 07/28/2011)

57.241 RoboCup, http://www.robocup.org (last accessed 5 May 2013)

57.242 BabyBot Humanoid, LIRA Lab, http://www.liralab.it/babybotmain.htm (last accessed 5 May 2013)

57.243 T. Minato, Y. Yoshikawa, T. Noda, S. Ikemoto, H. Ishiguro, M. Asada: CB2: A child robot with biomimetic body for cognitive developmental robotics, 7th IEEE-RAS Int. Conf. Humanoid Robots (2007) pp. 557–562

57.244 T. Minato, F. DallaLibera, S. Yokokawa, Y. Nakamura, H. Ishiguro, E. Menegatti: A baby robot platform for cognitive developmental robotics, Workshop Synerg. Intell. 2009 IEEE/RSJ Int. Conf. Intell. Robots Syst. (IROS) (2009)

57.245 The Acroban Humanoid Project, INRIA Ensta-ParisTech research team on developmental and social robotics, http://flowers.inria.fr/acroban.php (last accessed 5 May 2013)

57.246 NAO Humanoid Robot, Aldebaran Robotics, http://www.aldebaran-robotics.com (last accessed 5 May 2013)

58. Developmental Learning for User Activities

Xiao Huang , Juyang Weng , Zhengyou Zhang

This chapter presents a brain-inspired developmental learning system. A personal computer *lives* with the human user as long as the power is on. It can develop and report some activities of the user like a *shadow* machine, a virtual machine that runs in the background while the human user is doing its regular activities, on the computer or off the computer. The goal of the teacher of this *shadow machine* is to enable it to observe human users' status, recognize users' activities, and provide the taught actions as desired reports. Both visual and acoustic contexts are used by this *shadow machine* to infer the user's activities (e.g., in an office). A major challenge is that the system must be applicable to open domains – without a handcrafted environmental model. That is, there is no handcrafted constraint on office lighting, size, setting, nor requirements of the use of a head-mounted close-talk microphone. A room microphone sits somewhere near the computer. The distance between the sound sources and the microphone varies significantly. This system is designed to respond to its sensory inputs. A more challenging issue is to make the system adapt to different users and different environments. Instead of building all the world knowledge in advance (which is intractable), the

system's adaptive capability enables it to learn sensorimotor association (which is tractable). The real-time prototype system has been tested in different office environments.

58.1 Overview

Demonstrated by human cognitive and behavioral development from infancy to adulthood, autonomous development is nature's approach to human intelligence [58.1–3].

58.1.1 Developmental Learning

Inspired by human mental development from conception, a developmental robot is one that autonomously develops its mental skills and task performance capabilities from interactions with the environment. Figure 58.1 illustrates the paradigm of autonomous mental development [58.3].

As far as we know, Cresceptron (1993) [58.4, 5] was the first developmental model for visual learning from complex natural backgrounds. By developmental, we mean that the internal representation is fully emergent from interactions with the environment, without

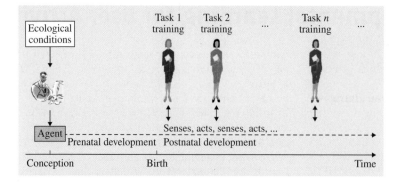

Fig. 58.1 The paradigm of autonomous mental development by machines, inspired by human mental development. No task is given during the programming (i. e., conception) time. The tasks that the agent learns during lifetime are determined after birth by other users

allowing a human to manually instantiate a task-specific representation. However, the although the features used by Cresceptron are acquired from incremental experience, they are not based on statics limited by a finite amount of resources.

In this chapter, we present a developmental learning system for user activities, using the visual and auditory context of the user of interest. This system has the following characteristics of developmental learning:

1. Open environment: we do not handcraft a model about the environment so the environment is open, although the capabilities of the system can be limited.
2. The use of incremental hierarchical discriminant regression (IHDR) to automatically learn features in an automatically generated hierarchy of feature spaces, organized as a tree structure. A potential of developmental learning is to incrementally learn to perform multiple tasks. Toward this goal, this system interactively allows the user to define different events, such as *rest*, *moving near door*, *moving in the office*, and *out*, as the actions taught at the system's action output end.

58.1.2 Context-Aware Systems

Context-aware systems [58.6] have drawn increasing attention from researchers and engineers. Context is defined as *the physical and social situation in which computational devices are embedded*. Context-awareness is the key component of next generation human–computer interaction technology, which tends to measure the information about *where*, *what*, *when*, and *who*. A few prototype context-aware systems have been implemented in the last decade. *Shafer*, *Brumitt* and *Cadiz* [58.7] create an *EasyLiving* home environment which is similar to the intelligent office

environment [58.8]. In [58.9, 10], a driver's behaviors are modeled so that the intelligent assistance systems can improve the safety of driving. Wearable computing [58.11] is another hot area, where intelligent devices are embedded in clothes, watches, or glasses to monitor and help users.

The goal of context-aware computing is to provide users with services that are appropriate to their particular situational information. The system detects whether users are in certain environments (room, office, car, etc.) and what their status is (in, out, online, offline, awake, sleeping, etc.). However, some context-aware systems rely on keyboard and mouse activities, which do not reflect actual user activities. For example, one current context-aware system may set the presence status to offline after 15 min of keyboard inactivity. If the user has left the office immediately after working on the computer, his presence status is still shown as *active*. Our work aims at computing the user presence and activity reliably in real time. In order to build such a system, we integrate audio-visual contextual information and other aspects of a user's information such as past states.

A significant portion of previous work related to context-aware computing focusing on recognizing human activities based on a single modality input in a specific environment. *Aggarwal* and *Cai* [58.12] review studies in human motion analysis. There are two popular probabilistic approaches in visual activity recognition: the hidden Markov model (HMM) and the Bayesian belief network (BBN). An earlier attempt to apply HMMs to activity recognition is found in [58.13]. Since then, a lot of extensions to HMMs have tried to model different human activities. Variable-length HMM [58.14] is applied to exercise behavior recognition. *Brand* and *Oliver* use coupled-HMM [58.15] to detect interactions between multiple people. Entropic-HMM [58.16] is introduced to detect and recognize activities in video. *Bobick* and *Ivanov* [58.17] apply stochastic context-free parsing to recognize activity sequences generated by low-level HMMs. BBN is another popular approach. *Buxton* and *Gong* [58.18] adopt BBN for visual surveillance. *Madbhushi* and *Aggarwal* [58.19] use BBN to recognize actions including sitting down, standing up, hugging, etc.

A limitation of the above studies is that there have been few studies on human activity recognition using multimodal context information. Recently, a lot of results have been reported to combine visual and audio cues to improve speech recognition or human identification. *Bengio* [58.20] propose an asynchronous HMM for audiovisual speech recognition and user authentication. This work takes advantage of the inherently close *temporal coupling* of speech and lip movements as modalities, and it requires a large amount of high-quality video and acoustic training data. For example, after training conducted with 185 recordings from 37 subjects, performance with asynchronous HMM exceeded that of an HMM (i. e., yielding 88.6% correct for nine digits at 10 dB signal-to-noise ratio). However, in the context-awareness system discussed here, these modes are not as closely aligned temporally. As such, this is a more challenging problem than processing closely-coupled multimodal data. Furthermore, most of these audio-visual systems require specific lighting conditions or ask users to wear close-talk microphones. A user does not need a close-talk microphone nor to adjust to the settings (i. e., lighting/size). In other words, our system requires activity recognition through natural interactions, which is a very difficult problem.

There were some early attempts to develop audio-visual context awareness through natural interactions. *Clarkson* and *Petland* [58.21] combine audio and visual contexts to classify simple activities such as crossing roads and passing through doors. *Oliver* [58.22] proposes a layered HMM architecture (LHMM), which integrates information from multimodal inputs to recognize six activities in an office, used to infer six distinct human activities in an office environment from users' audio-visual activities and mouse input after 1 hour of training. High accuracy was also reported for activity classification (over 99%), although generalization across variations in office conditions (e.g., changes in lighting) is known to be a limitation with this type of approach.

A more challenging issue is the adaptive capability of context-aware systems. These systems should be able to work in different environments and adapt to different users. Most current user activity recognition systems are limited to a constrained setting. These constraints are not acceptable in practical applications. Our goal is to build a context-aware system for open domains – working at any office room without restricted lighting/size/setting requirements. A reliable system for a user in one environment is not useful for other users in different settings. For example, if we can detect the phone ring signal in an office, it would be helpful to infer whether a user is going to have a phone conversation, and then to decide whether the system should provide assistance to the user. However, different rooms have different telephones, and different users have different cell phones. It is impossible to build a system to detect all kinds of phone ring signals (world knowledge) using pre-trained classifiers. A useful system must adapt to its user and learn user-centered signals on-the-fly in real time. Here we define world knowledge as all the patterns that need to be classified while user-centered knowledge is patterns related to a specific user. As shown in the above example, recognizing world knowledge is an intractable task, while recognizing user-centered knowledge is possible.

In order to tackle the above challenges, we propose a real-time audio-visual user activity recognition system to monitor the behavior of a single human in an office. The system has the following features:

1. Recognition of human activities through natural interactions. Instead of using a close-talk microphone, a room microphone is used to capture acoustic signals, so that the user can act naturally in the office. A new type of feature for acoustic signals is developed to improve acoustic signal classification. The feature is robust, since the distance between the sound source and the microphone varies from 10 mm to 2.5 m.
2. There is no impractical constraint. It is applicable to open domains (any regular office). The system is robust to environment changes (lighting/size, etc.).
3. An observation-driven Markov decision process (ODMDP) is implemented to automatically generate models for acoustic signals. As the engine of ODMDP, the IHDR algorithm is implemented to learn new acoustic signals online. The system can adapt to new settings, which is the major difference between this work and [58.22]. This novel component is crucial for adaptation to unknown environments as a consumer product.
4. A layered architecture is implemented to integrate both visual and audio cues to infer human activities with different granularities.

5. A real-time system is successfully tested in different environments.

It is worth noting that our goal is to build an adaptive context-aware system to recognize user activities. In this chapter, only the audio component is adaptive, while the visual component and high-level reasoning are still predefined.

In what follows, we first describe the system and challenges we have to face in Sect. 58.2. System architecture is discussed in Sect. 58.3. Then we present the experimental results and summarize our work.

58.2 System Overview and Challenges

A typical setting of a context-aware user activity recognition system is shown in Fig. 58.2. A table and a chair are put in an office. On the table, we can find a telephone, a personal computer, a video camera, and a microphone. The sensors we use to collect context information are:

1. USB camera: a Logitech QuickCam camera sampled at 15 FPS (FPS: frames per second) is used to detect human motions. The resolution is 640×480.
2. Microphone: a built in microphone associated with the camera.

However, we do not handcraft such an environmental model. The context-aware system response to its sensory inputs with the taught actions.

The goal of the human teacher is to teach this sensorimotor system to detect human activities in an office. The system overview is shown in Fig. 58.3. Two kinds of sensory inputs are used: auditory and visual. An audition pattern classifier discriminates four kinds of auditory patterns: *phone ring*, *conversation*, *uncertain*

noise, and *silence*. Our ultimate goal is to detect human activity and status and then provide proper assistance. The major activities in an office include face-to-face conversation and phone-conversation, which is why we define the above four auditory patterns. Based on the same goal, a motion detector captures motion information by computing the difference between two consecutive images. Sequences of submotion activities are classified into four motion patterns: *rest*, *moving near door*, *moving in the office*, and *out*. The outputs from the above two components are integrated to infer human activities in the office. There are four kinds of activities in total: *conversation*, *other activity*, *rest*, and *nobody around*. It is worth noting that the number of different activities in an office is not necessary limited to the above definition. Our system can easily learn to recognize more activities, which is explained in later sections.

In order to work in an open domain, we choose motion as the major visual feature, since motion is insensitive to office lighting/size/setting. In contrast, many vision systems rely on color or intensity, which is vulnerable to setting changes. No close-talk microphone is required. We use a room microphone, which has a capturing distance up to 2.5 m. Most speech recognition systems require a close-talk microphone to guarantee reliable performance. Here the difficulty lies in that the distance between sound sources and the microphone varies. Few studies discuss this important issue.

Another technical difficulty is to fuse different context inputs. In this system, the sampling rate of acoustic signals is 50 Hz while the sampling rate of visual signals is 15 Hz. A *moving in the office* pattern can be detected in 0.1 s, while to detect *nobody around* the system has to wait for 2 s to confirm the status. Psychological studies show that human behaviors are hierarchical [58.23]. In order to model human activities with different granularities, we implement LHMM. Low-level HMMs are used for motion pattern classifica-

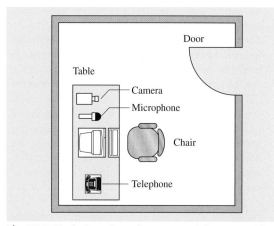

Fig. 58.2 Typical setting of a user activity recognition system

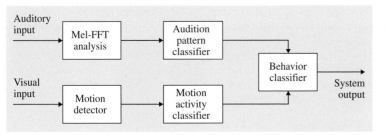

Fig. 58.3 Overview of the user activity recognition system

tion while high-level HMMs are used for human activity classification.

Adaptation to users and environments is a more challenging issue. As we mentioned in Sect. 58.1, if the system can detect a user's sound or detect his phone ring signal (office telephone or cell phone), it would be easy to infer whether he is in the office and what his status is (face-to-face conversation, phone conversation, silent, etc.). Since each user has a unique voice and there are thousands of different phone ring tones in this world, it is impossible to have a system to recognize all these acoustic signals. If the system could adapt to each user and the related environment, we would have a tractable solution. We propose ODMDP based on the IHDR algorithm for acoustic signal classification and learning. HMMs have been used for speech recognition because

of their dynamic modeling capability for sequential signals. However, one limitation of HMM is that it is only a computational model in the sense that it is not designed to be generated automatically from observations. Engineers have to manually design the system for given settings in advance, which affects its adaptive capability in unknown environments. For example, if a new type of sensory input is added to the system, the designer has to create new HMMs and specify parameters for each HMM, which is inconvenient, especially for unprofessional users. In contrast, ODMDP organizes the learned patterns in tree structure. New patterns are added as new leaf nodes. One tree is enough to model all acoustic signals related to a user. Furthermore, its coarse-to-fine tree structure enables efficient learning in real time.

58.3 System Architecture

The detailed architecture of the user activity recognition system is shown in Fig. 58.4. Mel analysis [58.24] is applied to raw auditory signals to extract Mel-FFT

(fast Fourier transform) features, which are fed into an IHDR tree to classify four kinds of auditory patterns. Sequences of motion activities are classified by

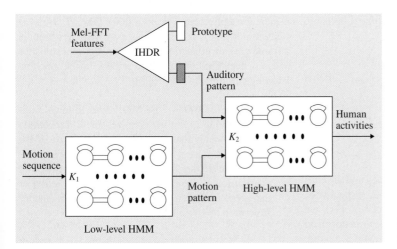

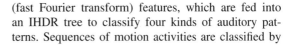

Fig. 58.4 Detailed architecture of the user activity recognition system

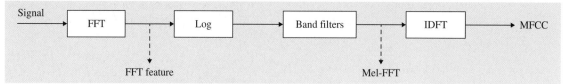

Fig. 58.5 The procedure to extract FFT, Mel-FFT, and MFCC features

low-level HMMs into four motion patterns. High-level HMMs integrate outputs from the above two components to infer human activities.

58.3.1 Acoustic Signal Classification

Feature Extraction

With the audition component, we try to discriminate the following four signals: conversation, phone ring, silence (background noise), and uncertain noise (radio, music, flapping, clapping, etc.). In most speech processing systems, Mel-frequency cepstral coefficients (MFCCs) [58.24] are the dominating feature because MFCC captures the properties of human speech signals. The procedure to calculate the MFCC feature is shown in Fig. 58.5.

First, we apply FFT to obtain features in the frequency space. Usually, we calculate the logarithmic energy of the FFT coefficients. Critical band filters are applied to derive the appropriate frequency components for computing MFCC. This implementation is described in (58.1),

$$Y(i) = \sum_{k=F_{is}}^{F_{ie}} \log |Z(k)| H_i \left(k \frac{2\pi}{N'} \right) , \tag{58.1}$$

where $\log |Z(k)|$ is the k-th log energy of the Fourier coefficient and N' is the length of the i-th triangular band. H_i is a triangular filter

$$H_i(f) = \begin{cases} 1 - |f|, & f \leq 1 \\ 0, & f > 1 . \end{cases} \tag{58.2}$$

The variable f is defined in (58.3)

$$f = \frac{k - F_{is}}{F_{is} - F_{ie}} , \tag{58.3}$$

where F_{is} and F_{ie} are the starting frequency and the ending frequency of the ith triangular band, respectively. $F_{is} \leq k \leq F_{ie}$. After we obtain a sequence of $Y(i)$, we can apply IDFT (inverse discrete Fourier transform) to this sequence, and then we can obtain the MFCC feature

using (58.4)

$$c(w) = \frac{1}{N'} \sum_{k=0}^{N'-1} Y(k) e^{jk(2\pi/N')w} , \tag{58.4}$$

where w is the index of the MFCC feature.

In 58.1, $Y(i)$ denotes the sum of the weighted $\log |S(k)|$ within the i-th critical band filter. $Y(i)$ is sometimes called the weighted log energy in the i-th critical band. In this paper, we call $Y(i), i = 1, 2, \ldots N'$ the Mel-FFT feature. In Sect. 58.4, we compare the FFT feature, the Mel-FFT feature, and the MFCC feature. It turns out that Mel-FFT is the best choice for this application. We all know that MFCC is designed to capture the property of human speech signals, while in this study our goal is to classify different acoustic signals. Mel-FFT represents properties in the frequency space, while MFCC has another IDFT procedure after calculating Mel-FFT. In other words, MFCC represents properties in a pseudo time space. Our argument in this study is that these acoustic signals are better separated in the frequency space.

58.3.2 Observation–Driven Markov Decision Process

Our architecture follows the ODMDP, introduced as a mental architecture by *Weng* 2007 [58.25]. We start with the Markov decision process (MDP), which is a framework originated from mathematics and statistics and has been studied and used widely in the machine learning community [58.26, 27].

● *Definition 1 (MDP)*. The MDP is as follows. Suppose $S = \{1, 2, \ldots, n\}$ is a set of n predefined symbolic states that is used to model a part of the world. The state s_t at time t is a random variable taking one of the values in S. Its prior probability distribution is $P(s_0)$. The action a_t is the action of the agent at time t. Let H_t be the random history from time $t = 0$ up to time $t - 1$:

$$H_t = \{s_{t-1}, s_{t-2}, \ldots, s_0, a_{t-1}, a_{t-2}, \ldots, a_0\} .$$

If its conditional state transitional probability $P(s_t|H_t)$ satisfies

$$P(s_t|H_t) = P(s_t|l_t),$$

where l_t is the short last k frames of the history

$$l_t = \{s_{t-1}, s_{t-2}, \ldots, s_{t-k}, a_{t-1}, a_{t-2}, \ldots, a_{t-k}\},$$

we call it the k-th order MDP [58.26, 27]. A limitation of MDP is that it assumes the state is directly observable. In many applications, the state of the world is not directly observable by the agent, or is partially observable (with noise).

- *Definition 2 (Partially observable MDP).* If the state s_t of the world is not totally observable to the agent. Instead, there is an observation x_t at time t that depends on the state s_t by an observation probability $P(x_t|s_t)$, the process is called partially observable MDP or POMPD [58.27] (or HMM [58.28]). Now let us give the definition of ODMDP.
- *Definition 3 (ODMDP).* Let $x_t \in X$ and $p_t \in P$ be the observations and outcome covariates (i.e., random vectors) at time t, respectively. Let H_t be the random vector of the entire history

$$H_t = \{x_t, x_{t-1}, \ldots, x_0, p_{t-1}, \ldots, p_0\}.$$

If its state transitional probability $P(s_t|H_t)$ satisfies

$$P(p_t|H_t) = P(p_t|l_t),$$

where l_k is the last k observations

$$l_t = \{x_t, x_{t-1}, \ldots, x_{t-k}, p_{t-1}, \ldots, p_{t-k}\},$$

we call the process the k-th order ODMDP [58.29].

The following are the major differences between POMDP (or HMM) and ODMDP:

1. POMDP is world-centered, where each state corresponds to a modeled object or event of the world (e.g., a corner). ODMDP is mind-centered (from sensors), where each state corresponds to an observation from the environment (e.g., a view of a corner with other background objects).
2. The states s_t of POMDP are hand-specified but states of ODMDP can be automatically generated (developed) on-the-fly. With POMDP, the meaning of each state must be specified so that the initial estimates of the three probability distributions ($P(s_0)$, $P(s_t|s_{t-1})$, and $P(x_t|s_t)$) can be provided.
3. In POMDP, there are two layers of probability: the state transition probability $P(s_t|x_t, s_{t-1})$ and the state observation probability $P(x_t|s_t)$, while ODMDP has only one layer of probability: $P(p_t|l_t)$, making a more efficient learning algorithm possible.

58.3.3 Incremental Hierarchical Discriminant Regression

The engine of the ODMDP model is IHDR. After extracting the features of acoustic signals, how do we classify them? Classification and regression problems can be formulated as a complicated function which maps high-dimensional input and the current state to low-dimensional output signals. The decision tree [58.30, 31] has been a popular method for function approximation. We use the IHDR [58.32, 33] tree algorithm for acoustic signal classification. In order to build such a tree, two types of clusters are incrementally updated at each node of the tree: y-clusters and x-clusters. The y-clusters are clusters in the output space Y and x-clusters are those in the input space X. The y-clustering result affects the distribution of x-clustering because IHDR in principal is a supervised learning approach. This doubly-clustering mechanism is shown clearly in Fig. 58.6. The samples come one by one. For each new sample (x, y), y finds the nearest y-cluster in Euclidean distance and updates the center of the y-cluster. This y-cluster indicates which corresponding x-cluster the input (x, y) belongs to. Then, the x part of (x, y) is used to update the statistics of the x-cluster (the mean vector and the covariance matrix). It is worth noting that ODMDP is a general model. In the (x, y) pair described here, x corresponds to $x(t)$ and y corresponds to $p(t)$ in Definition 3. In order to more effectively explain how to apply IHDR to acoustic signal classification, we adopt (x, y) as the notation in this section. There is a maximum of q clusters at each node. The idea is that each node models a region of the space using q Gaussians. The centers of these q x-clusters are denoted by

$$M = \{m_1, m_2, \ldots, m_q | c_i \in X, i \\ = 1, 2, \ldots, q\}. \tag{58.5}$$

The q centers span a $q - 1$-dimensional hyperplane, which is the discriminating feature subspace. In this sense, the original d-dimensional input space is mapped to a $q - 1$-dimensional discriminant subspace. We only conduct linear discriminant analysis (LDA) [58.34] in the very-low dimensional subspace, which saves tremendous computational cost. This projection from the original input space to the low-dimensional subspace is shown in Fig. 58.7. There are three Gaussians

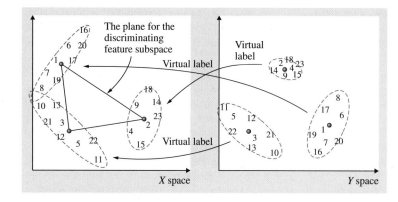

Fig. 58.6 Y-clusters in space Y and the corresponding X-clusters in space X. Each sample is indicated by a number which denotes the order of arrival. The first and the second order statistics are updated for each cluster. The first statistic gives the position of the cluster, while the second statistic gives the size, shape, and orientation of each cluster

($q = 3$). Three center vectors are m_1, m_2, and m_3, respectively. Noting that the dimension of m_i could be very high, we define scatter vectors s_i

$$s_i = m_i - \bar{m} \,, \tag{58.6}$$

where $\bar{m}$ is the mean of all center vectors. Let S be the set that contains these scatter vectors: $S = \{s_i | i = 1, 2, 3\}$. The discriminant subspace spanned by S (denoted by $\mathrm{span}(S)$) consists of all the possible linear combinations from vectors in S. We can see a two-dimensional subspace passes the head tips of the center vectors. By applying Gram–Schmidt orthogonalization (GSO), we can obtain two orthogonalized

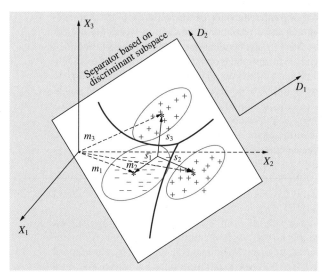

Fig. 58.7 We use three ($q = 3$) Gaussians to model the input space. The linear variety (hyperplane) that passes through the head points of the center of each Gaussian. It can be represented by $\bar{m} + \mathrm{span}(S)$. We can use two orthogonal vectors D_1 and D_2 to determine this hyperplane. The dimension of the subspace is $q - 1 = 2$

discriminant vectors D_1 and D_2. Now the separators (the thicker arcs) based on D_1 and D_2 can partition these three classes in the subspace.

Each Gaussian is represented by its first two-order statistics: mean and covariance matrix. The mean is updated incrementally as follows

$$\bar{x}^{(n+1)} = \frac{n - \mu}{n+1}\bar{x}^n + \frac{1 + \mu}{n+1}x_{n+1} \,, \tag{58.7}$$

where x_{n+1} is the $(n + 1)$-th sample, $\bar{x}^{(n+1)}$ is the mean after this sample is trained, and μ is a parameter. If $\mu > 0$, the new input has more weight than old inputs. We called this implementation the amnesic average. The covariance matrix can also be updated incrementally by using the amnesic average (58.8):

$$\Gamma_x^{(n+1)} = \frac{n - 1 - \mu}{n}\Gamma_x^{(n)}$$
$$+ \frac{1 + \mu}{n}(x_{n+1} - \bar{x}^{(n+1)})(x_{n+1} - \bar{x}^{(n+1)})^T \,, \tag{58.8}$$

where $\Gamma_x^{(n+1)}$ is covariance matrix in X space after $(n + 1)$-th updating.

Each leaf node generates quite a few samples (numbers in Fig. 58.4), which represent different patterns (model generator). In the testing phase, if a sample is reached, the label associated with it would be the output (computational model). A more detailed description of the algorithm can be found in [58.35]. The advantages of IHDR include:

1. IHDR maps high dimensional data into low-dimensional subspace. LDA is conducted in the discriminating subspace, which saves a lot of computational power.
2. IHDR organizes the learning knowledge in a coarse to fine way, which speeds up the retrieval proce-

dure. The time complexity is $N \log N$, where N is the number of samples in the tree.

3. IHDR intrinsically has an incremental online learning capability to adapt to new signals since the sufficient statistics are updated incrementally. This is why we use IHDR instead of traditional HMMs to classify auditory patterns. For practical applications, the system has to work in different offices, which have different users and different telephones. In order to make HMMs adapt to new settings, a bank of new HMMs has to be created manually. In contrast, one IHDR is enough to handle all types of auditory signals related to a user. It is also important that the adaptation can be incrementally conducted in real time. Another advantage of IHDR is that it does not require decorrelated features, which are necessary for HMMs employing diagonal covariance matrices.

58.3.4 HMMs for Motion Pattern Classification

In order to recognize human motion activities, we have to find the moving subject and his/her location. We implement a motion detector to find the bounding box of a moving body with two consecutive images. First, we calculate the difference between these two images. Then a low-pass filter is used to remove noise in the difference image. The bounding box of the moving body is determined by the histogram of the image [58.36].

Now we know the size and the location of the bounding box (denoted by a rectangle R_b). The user can specify the door region of an office (denoted by a convex polygon P_d). Let us first define submotion patterns. If the bounding box is within the door region ($R_b \subseteq P_d$), we can treat it as subpattern 1 (S_{p1}: the user is in the door region). If there is overlap between R_b and P_d ($R_b \neq P_d$ and $R_b \cap p_d > \Phi$), we treat it as subpattern 2 (S_{p2}: the user moves near the door region). Note that Φ means an empty set. If there is no overlap between R_b and P_d ($R_b \cap P_d = \Phi$), we treat it as subpattern 3 (S_{p3}: moving in the office). If there is no motion, we treat it as subpattern 4 (S_{p4}: stop moving). Each submotion pattern covers about 0.1 s. It is obvious that these subpatterns (motion features) are robust. They can tolerate lighting changes, and the size and the setting of the office do not affect these features too much, which makes them applicable to open domains. After obtaining a sequence of motion subpatterns, we use discrete HMM to classify human motion behaviors. An HMM is denoted by $\lambda = (A, B, \pi)$, where A is state transition probability matrix, B is the observation symbol probability matrix,

and π is the initial state distribution. Specification of an HMM involves the choice of the number of states N and the number of observations M. With training data, we can calculate λ by using the Baum–Welch algorithm [58.28]. Given a model λ and a sequence of observation $O = \{O_1, O_2, \ldots, O_T\}$, we can calculate the likelihood of the sequence $P(O|\lambda)$ using the forward algorithm.

Letting K be the number of motion/activity patterns, we usually need to generate a bank of K HMMs ($\lambda_k (1 \leq k \leq K)$). The likelihood of the observation sequence in each model is $L_k = P(O|\lambda_k)$. Suppose the maximal likelihood is $L_{\max}$ and the minimal likelihood is $L_{\min}$, the normalized likelihood is

$$L'_k = \frac{L_k - L_{\min}}{L_{\max} - L_{\min}} . \tag{58.9}$$

HMMs choose the pattern $\hat{k}$ with the largest normalized likelihood as output

$$\hat{k} = \arg \max_k \{L'_k\} . \tag{58.10}$$

58.3.5 Integration Component

The system is able to recognize four types of human activities: *conversation*, *other activity*, *rest*, and *nobody around*. Integrating the above two low-level components to infer human activities is difficult because different modalities have different updating frequencies and they can be either related or unrelated. High-level HMMs are necessary for reasoning human activities based on information of low-level components. Usually the reasoning is conducted with a larger time granularity (for example, 2 s), while in motion pattern classification and auditory pattern classification, the granularity is less than 1 s. The outputs from low-level components are labels, which can be fed into the integration HMMs. Since the vision component outputs four types of patterns (as does the audition component), the combination of these two components gives 16 types of observations. Let us first define the observation and the output for high-level HMMs. Suppose the output from the motion component is $L_v = \{1, 2, 3, 4\}$ and the output from the audition component is $L_a = \{1, 2, 3, 4\}$. Then the observation for high-level HMMs is $o^h = L_v \cdot 4 + L_a$. For example, if the motion component output is *moving in office* ($L_v = 2$) and the audition is *conversation* ($L_a = 2$), then the observation for the integration component is $o^h = L_v \cdot 4 + L_a = 2 \cdot 4 + 2 = 10$. With sequences of o^h, we can implement the Baum–Welch algorithm [58.28] to train high-level HMMs.

58.4 Experimental Results

We conducted experiments for each of these three components. In the data recording phase, we asked each subject to step in/out the office, move around, rest, and talk. We recorded 40 min activities and manually annotated the data to obtain the ground truth. About a half of the data was used for training and the other half was used for testing.

58.4.1 Experimental Results of the Auditory Component

We trained the system with nine kinds of male conversation signals, three kinds of female conversation signals, 16 kinds of phone ring signals, background noise, and uncertain noise (radio, music, flapping, clapping) using IHDR. Some signals are shown in Fig. 58.8.

1. *Comparison of different feature vectors.* First, we compared three kinds of features: FFT, Mel-FFT, and MFCC. The auditory data were digitized at 11 025 Hz by a normal sound blaster card. The number of Mel-FFT feature vectors was about 12 000. Each feature vector covered 256 ms. The overlap between two consecutive feature vectors was 56 ms. The number of training samples was about 6000 for each feature, the number of testing samples was 5174. We can see in Table 58.1 that Mel-FFT is better than FFT and MFCC. The recognition rate is about 90%. The dimensions of FFT, Mel-FFT, and MFCC feature vectors are 512, 21, and 21, respectively. The corresponding time to train each feature vector using IHDR is 115 ms, 8.6 ms, and 8.5 ms, respectively. In order to build a real-time online learning system, FFT is not acceptable since the system has to incrementally learn about 50 FPS. Considering training time and recognition rate, we chose Mel-FFT as the feature of acoustic signals. Note that all of these signals are captured by a room microphone. The distance between the sound source and the microphone changes from 10 mm to 2.5 m, which is a very difficult problem because most context-aware systems use a close-talk microphone.

2. *Online learning.* As we mentioned earlier, we used IHDR rather than HMMs for acoustic signals classification. The reason is that one IHDR tree can adapt to new signals, new users, and new environments. In contrast, using HMM the designer has to create new models for new signals. In order to verify the adaptive capability of IHDR, we tested the online learning component. We divided the dataset into three sets. Set A consists of conversation signals and eight types of phone ring signals. Set B consists of conversation signals and seven types of phone rings signals. Set C consists of one type of phone ring signal, which is in A but not in B. We first trained set B offline then tested it on set A. The confusion matrix is shown in Table 58.2. A confusion ma-

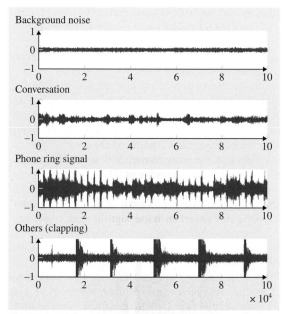

Fig. 58.8 Different acoustic signals. From *top* to *bottom*: background noise, conversation, phone ring signal, and other noise (clapping). The *x*-axis denotes the number of raw signal points and the *y*-axis denotes the normalized energy

Table 58.1 Comparison of different features for acoustic signals

Feature types	Dimension	Training time	Correct	Total	Recog. Rate
FFT	512	115 ms	4314	5174	83.38%
MelFFT	21	8.6 ms	4641	5174	89.70%
MFCC	21	8.5 ms	4467	5174	86.33%

Table 58.2 Confusion matrix of offline training (C = conversation, P = phone)

Data	C	P	Total	Rate
C	4064	254	4318	94.12%
P	275	1212	1487	81.50%

Table 58.3 Confusion matrix after online training (C = conversation, P = phone)

Data	C	P	Total	Rate
C	4071	247	4318	94.28%
P	154	1333	1487	89.64%

Table 58.4 Recognition rate of each auditory set. (C = conversation; UN = uncertain noise; P = phone; S = silence)

Data	C	UN	P	S	Total	Rate
C	2275	188	6	4	2470	92.11%
UN	216	2104	0	0	2320	90.69%
P	8	14	1768	5	1795	98.50%
S	0	0	0	1141	1141	100%

trix contains information about actual and predicted classifications done by a classification system. The performance of such systems is commonly evaluated using the data in the matrix. Using Table 58.2 as an example, there are totally 1487 phone ring testing samples. 275 samples are misclassified as *conversation* while other 1212 samples are correctly classified as *phone*. The recognition rate of phone ring signal is only 81.5%.

In the next experiment, using the offline generated IHDR tree, we trained the system with set C and then tested it on set A again. The result of online learning is shown in Table 58.3. The recognition rate of phone ring signals is improved from 81.50% to 89.64%, which proves the adaptive capability of IHDR.

Then we tested all four kinds of acoustic signals. The confusion matrix is shown in Table 58.4. Conversation and uncertain noise signals were confusing sometimes (188 conversation feature vectors were recognized as uncertain noise) because uncertain noise covers a very large feature space. The overall recognition rate was 94.33%. Compared to [58.37], the performance gained about 0.5%. We need to note that the test is source-dependent. In other words, the testing set and the training set come from the same source. If we test the system with signals of a telephone that we never train, the performance would drop. That is why IHDR is important for this application since it can incrementally learn new auditory patterns. HMMs are not necessarily the best choice for acoustic signal classification since we do not need to recognize the meaning of each signal. Moreover, IHDR is very efficient for online learning because one IHDR tree is enough to model all user-centered contexts.

58.4.2 Experimental Results of the Motion Component

Figure 58.9 shows a motion sequence, which is classified as *moving near the door*. The first row is the video sequence while the second row shows the bounding box

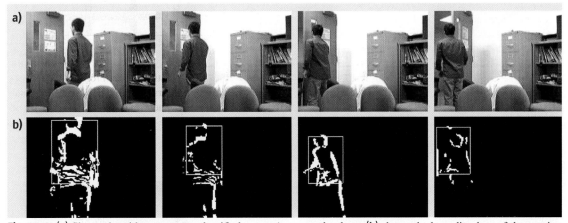

Fig. 58.9 (a) Shows the video sequence classified as *moving near the door*; **(b)** shows the bounding box of the moving object

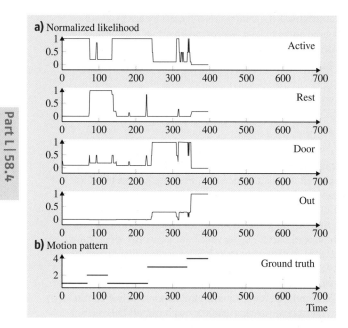

of the moving object. We fed motion subpattern sequences into low-level HMMs. In this experiment, the parameters of the low-level HMMs are: $N_s = 6$, $M = 4$ and $K = 4$. The normalized likelihood of each motion pattern is shown in the first four plots of Figure 58.10. The x-axis is the time line. In this figure, the video covers about 400 s. The granularity of each HMM output is

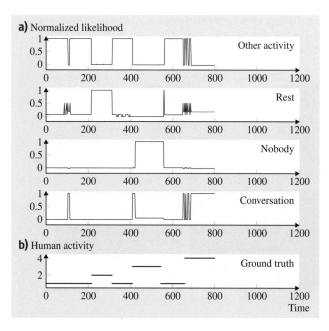

Fig. 58.10a,b Likelihood of motion patterns over time. (**a**) The normalized likelihood of each motion pattern is shown in the first four plots. (**b**) The fifth plot shows the ground truth, where pattern 1 is *active* (*moving in the room*); pattern 2 is *rest*; pattern 3 is *door* (*moving in the door region*), and pattern 4 is *out* ◄

Table 58.5 Recognition rate of human activities (N = nobody; OA = other activity; R = rest; C = conversation)

Data	N	O	R	C
N	100%	0	0	0
O	0	94.31%	0	5.69%
R	0	0	100%	0
C	0	4.52%	0	95.48%

1 s. If $L'_k = 1$, then pattern k is reported. The ground truth of activity sequences is shown in the fifth plot, which goes as follows: the user firstly moved in the room (pattern 1), rested for a while (2), moved around the door (3), and then went out (4). The motion behaviors are correctly recognized. A mistake occurs at step 320, the system classified pattern (3) as pattern (1) when the user moved near the boundary of the door and the rest of the room.

58.4.3 Experimental Results of the High-Level Reasoning

The specifications of the high level HMMs are: $N_s = 4$, $M = 16$, and $K = 4$. The normalized likelihood of each human activity is shown in the first four plots of Figure 58.11. The ground truth of activity sequences is in the fifth plot, which goes like this: the user moved around in the office (activity 1), rested (2), moved around again (1), talked for a while (3), moved to the door (1), and went out (4). The x-axis is the time line and in this figure it covers about 800 s. *Nobody* and *rest* are perfectly classified, while for a short period of time, *conversation* and *other activity* are misclassified because sometimes the user moved and talked at the same time. The recognition rate of human activities is shown in Table 58.5. About 4% of *conversation* is incorrectly recognized as *other activity*, which is consistent

Fig. 58.11a,b Likelihood of human activities over time. (**a**) The normalized likelihood of each human activity is shown in the first four plots. (**b**) The fifth plot shows the ground truth, where pattern 1 is *other activity*; pattern 2 is *rest*; pattern 3 is *nobody* and pattern 4 is *conversation* ◄

with the results in Figure 58.11. The initial result is very promising since we tested the system in uncontrolled environments (different offices, different users, different lighting conditions, etc.). The distance between sound sources and the microphone varied because the user moved around. It would be easy to extend this system to other context-awareness systems (intelligent car/room, etc.).

58.5 Summary and Discussion

In this chapter, a multimodal context-aware system is used as an example of developmental learning systems. It is trained to detect user activities in an office, although other types of actions can also be taught, depending on the user's preference. In principle, the system is able to classify its sensory inputs into actions required by the user desire. We adopted the ODMDP model to automatically generate models for acoustic signals related to a user. New signals can be learned on-the-fly. For example, IHDR can learn its owner's sound and the phone ring signal in his/her office in real time. Both visual and acoustic contexts are integrated using LHMM to generate the user-taught actions.

Compared with the brain, this example of a developmental learning system has several major limitations. First, although the acoustic features are inspired by known physiological studies (e.g., Mel sensitivity profile), the model of the acoustic features is handcrafted, different from the features in IHDR which are automatically learned. Second, the detection of motion information uses human handcrafted features. Future work will be to study how to use a developmental network [58.38] that does not have any dedicated temporal component for dealing with time. Third, we would like to study how visual attention mechanisms [58.39] can be used in application-oriented systems.

References

58.1   J. Piaget: *The Construction of Reality in the Child* (Basic Books, New York 1954)

58.2   J. Elman, E.A. Bates, M.H. Johnson, A. Karmiloff-Smith, D. Parisi, K. Plunkett: *Rethinking Innateness: A Connectionist Perspective On Development* (MIT, Cambridge 1997)

58.3   J. Weng, J. McClelland, A. Pentland, O. Sporns, I. Stockman, M. Sur, E. Thelen: Autonomous mental development by robots and animals, Science **291**(5504), 599–600 (2001)

58.4   J. Weng, N. Ahuja, T.S. Huang: Learning recognition and segmentation of 3-D objects from 2-D images, Proc. IEEE 4th Int. Conf. Comput. Vis. (Michigan State Univ., East Lansing 1993) pp. 121–128

58.5   J. Weng, N. Ahuja, T.S. Huang: Learning recognition using the Cresceptron, Int. J. Comput. Vis. **25**(2), 109–143 (1997)

58.6   T. Moran, P. Dourish: Introduction to this special issue on context-aware computing, Hum. Comput. Interact. **16**, 87–95 (2001)

58.7   S. Shafer, B. Brumitt, J. Cadiz: Interaction issues in context-aware interactive environments, Hum. Comput. Interact. **16**, 363–378 (2001)

58.8   N. Oliver, E. Horvitz: Selective perception policies for guiding sensing and computation in multimodal systems: A comparative analysis, Proc. Int. Conf. Multimodal Interfaces (Vancouver 2003) pp. 3–8

58.9   N. Oliver, A. Pentland: Driver behavior recognition and prediction in a smartcar, Proc. SPIE Aerosense2000 'Enhanc. Synth. Vis.' (Orlando, Florida 2000)

58.10   K. Torkkola, N. Massey, B. Leivian, C. Wood, J. Summers, S. Kundalkar: Classification of critical driving events, Proc. Int. Conf. Mach. Learn. Appl. (ICMLA) (Los Angeles, CA, USA 2003) pp. 81–85

58.11   F. Sparacino, A. Pentland, G. Davenport: Wearable performance, 1st Int. Symp. Wearable Comput. (Cambridge 1997)

58.12   J.K. Aggarwal, Q. Cai: Human motion analysis: A review, Comput. Vis. Image Underst. **73**(3), 428–440 (1999)

58.13   J. Yamato, J. Ohya, K. Ishii: Recognizing human action in time-seqential images using hidden Markov model, Proc. Int. Conf. Comput. Vis. Pattern Recognit. (NTT Hum. Interface Labs, Yokosuka 1992) pp. 379–385

58.14   A. Galata, N. Johnson, D. Hogg: Learning variable length Markov models of behaviour, Comput. Vis. Image Underst. **81**(3), 398–413 (2001)

58.15   M. Brand, N. Oliver, A. Pentland: Coupled hidden Markov models for modeling interacting processes, Proc. Int. Conf. Comput. Vis. Pattern Recognit. (1996) pp. 994–999

58.16   M. Brand, V. Kettnaker: Discovery and segmentaion of activities in video, IEEE Trans. Pattern Anal. Mach. Intell. **22**(8), 844–851 (2000)

58.17   Y. Ivanov, A. Bobick: Recognition of visual activities and interactions by stochastic parsing, IEEE Trans. Pattern Anal. Mach. Intell. **22**(8), 852–872 (2000)

58.18   H. Buxton, S. Gong: Advanced Visual Surveillance using Bayesian Networks, Proc. Int. Conf. Comput. Vis. (Cambridge 1995) pp. 111–123, June

58.19   A. Madabhushi, J. Aggarwal: A Bayesian approach to human activity recognition, Proc. 2nd Int. Workshops Vis. Surveill. (Washington D.C. 1999) pp. 25–30

58.20   S. Bengio: An asynchronous hidden Markov model for audio-visual speech recognition, Proc. Adv. Neural Inf. Process. Syst. (2003) pp. 1213–1220

58.21   B. Clarkson, A. Pentland: Unsupervised clustering of ambulatory audio and video, Int. Jt. Conf. Acoust., Speech Signal Proces., ICASSP'99 (1999) pp. 3037–3040

58.22   N. Oliver, E. Horvitz, A. Garg: Layered representation for human activity recognition, Proc. Int. Conf. Multimodal Interfaces (2002) pp. 3–8

58.23   J. Zacks, B. Tersky: Event structure in perception and cognition, Psychol. Bull. **127**(1), 3–21 (2001)

58.24   J. Deller, J. Proakis, J. Hansen: *Discrete-Time Processing of Speech Signals* (Inst. Electr. Electron. Eng., New York 2000)

58.25   J. Weng: On developmental mental architectures, Neurocomputing **70**(13–15), 2303–2323 (2007)

58.26   M.L. Puterman: *Markov Decision Processes* (Wiley, New York 1994)

58.27   L. Kaelbling, M. Littman, A. Moore: Reinforcement learning: A survey, J. Artif. Intell. Res. **4**, 237–285 (1996)

58.28   L.R. Rabiner: A tutorial on hidden Markov models and selected applications in speech recognition, Proc. IEEE **77**(2), 257–286 (1989)

58.29   D.R. Cox: Statistical analysis of time series: Some recent developments, Scand. J. Stat. **8**(2), 93–115 (1981)

58.30   J. Quinlan: *C4.5: Programs for Machine Learning* (Morgan Kaufmann, San Mateo, CA 1993)

58.31   L. Breiman, J. Friedman, R. Olshen, C. Stone: *Classification and Regression Trees* (Chapman Hall, New York 1993)

58.32   W. Hwang, J. Weng: Hierarchical discriminant regression, IEEE Trans. Pattern Anal. Mach. Intell. **22**(11), 1277–1293 (2000)

58.33   X. Huang, J. Weng: Locally balanced incremental hierarchical discriminant regression, 4th Int. Conf. Intell. Data Eng. Autom. Learn. (Hong Kong 2003)

58.34   R. Duta, P. Hart, D. Stork: *Pattern Classification*, 2nd edn. (Wiley, New York 2001)

58.35   W. Hwang, J. Weng: An online training and online testing algorithm for OCR and image orientation classification using hierarchical discriminant regression, Proc. 4th IAPR Int. Workshop Document Anal. Syst. (Rio De Janeiro, Brazil 2000)

58.36   W. Pratt: *Digital Image Processing* (John Wiley, New York 1991)

58.37   X. Huang, J. Weng, Z. Zhang: Office presence detection using multimodal context information, Proc. Int. Conf. Acoust., Speech Signal Proces. (ICASSP 2004) (Montreal, Quebec, Canada, USA 2004)

58.38   J. Weng: Why have we passed "neural networks do not abstract well"?, Nat. Intell.: INNS Mag. **1**(1), 13–22 (2011)

58.39   M. Luciw, J. Weng: Where what network 3: Developmental top-down attention with multiple meaningful foregrounds, Proc. IEEE Int. Jt. Conf. Neural Netw. (Barcelona, Spain 2010) pp. 4233–4240

59. Quantum and Biocomputing – Common Notions and Targets

Mika Hirvensalo

Biocomputing and quantum computing are both relatively novel areas of information processing sciences under the umbrella *natural computing* established in the late twentieth century. From the practical point of view one can say that in both bio and quantum paradigms, the purpose is to replace the traditional media of computing by an alternative. Biocomputing is based on an appropriate treatment of biomolecules, and quantum computing is based on the physical realization of computation on systems so small that they must be described by using quantum mechanics. The efficiency of the proposed biomolecular computing is based on massive parallelism, which is implementable by already existing technology for small instances. In a sense, also quantum computing involves parallelism. From time to time, there are proposals or attempts to create a uniform approach to both biocomputational and quantum parallelism. The main purpose of this article is the explain why this a very challenging task. For this aim, we present the usual mathematical formalism needed to speak about quantum computing and

compare quantum parallelism to its biomolecular counterpart.

59.1 Overview

In 1994 *Adleman* aroused a lot of attention by describing a biomolecular solution to the traveling salesman problem [59.1] (see also [59.2]). It is by no means exaggerated to say that Adleman actually set the establishment of a new kind of science, even though the ideas of bio-inspired computational models are far older. For example, the theoretical properties of artificial neural networks had been studied decades before, and a notable exposition was published by *Minsky* and *Papert* in 1969 [59.3]. However, the previous studies on bio-inspired computing seemed to focus on analogies of biological processes, and that is exactly where *Adleman* took one step further: he proposed that instead

of simulation, it could be useful to utilize directly the biochemical processes to perform computation.

By a coincidence, another branch of new science gained a lot of attention in 1994, too. In that year Shor published his famous polynomial-time algorithm for integer factorization [59.4]. Shor's study was remarkable for several reasons. First, since antiquity, there has been an unsuccessful quest for an efficient procedure for integer factorization. Another reason making Shor's discovery important is a very practical one. The security of the broadly used RSA encryption system is based on the assumption that no efficient method for factoring integers exists. The third important feature of Shor's results was that the factoring method was designed for *quantum computers* and could not be directly described in terms of traditional *Turing machines* [59.5].

It is also true that *quantum computing* did not begin with Shor's article. The first ideas of quantum information were coined by *von Neumann* in 1927 [59.6, 7], but the first ideas of the quantum computer were introduced as late as early 1980s by *Benioff* [59.8, 9], and very notably by *Feynman* [59.10], who actually suggested that quantum computers may be more efficient that the traditional ones. The theory was further developed by *Deutsch*, who introduced quantum Turing machines [59.11] and quantum networks [59.12]. Also the first examples of the superior efficiency of quantum computers were given by *Deutsch* and *Jozsa* [59.13] in 1992. In 1994, *Simon* presented a more usable example [59.14], and Shor actually built his factoring procedure on Simon's algorithm. In 1997 *Bernstein* and *Vazirani* established quantum complexity theory using an improved version of Deutsch's quantum Turing machine [59.15].

Since the early days, the theory of quantum computing has been under strong development, but many basic questions still remain unresolved. For instance, we know that factoring integers would be feasible by quantum computers, but we cannot *prove* it unfeasible for classical computers. All we know is that no-one has discovered any feasible factoring for classical computers. In fact we do not know for sure of *any* problems for which quantum computers could provably be more powerful. This lack of knowledge becomes understandable when noticing that many basic questions on classical computing remain unresolved, too: as of 2013, we do not know whether polynomial-time nondeterministic computing any more powerful than its deterministic counterpart. This so-called P versus NP-problem is generally acknowledged as one of

the most difficult problems in contemporary mathematics [59.16], and there is no reason to believe that analogous problems on the relations between quantum and classical computing were any simpler to resolve.

Quantum computing has an important common feature with molecular computing: in both paradigms, the idea is to perform computation on non-traditional hardware. To run quantum algorithms requires a quantum computer, i. e., a computer capable of storing and handling quantum information. For that purpose, the information should be presented by using physical systems so small that the quantum effects occur. From time to time, the analogies between quantum and biomolecular computing encourage authors to submit an idea of joint computing model. However, the success of such models has been very limited so far, and the main purpose of this article is to explain why to a reader only weakly familiar with quantum computing concepts. This will be done by introducing the basic notions of quantum computing in a superficial way and pointing out the essential differences between the two computing paradigms. One single and perhaps the most essential difference between the computational paradigms can be introduced without presenting any deeper structures. The objects of molecular computing (DNA molecules) are microscopic to humans, but yet macroscopic from the quantum computing perspective. Hence the information in molecular computing is treated as classical information, whereas the starting point of quantum computing is the *quantumness* of information.

As a secondary purpose, we describe briefly the four types of existing quantum algorithms and some restrictions of quantum computing to present an idea of what can be achieved by using quantum computing. The types presented cover almost all known quantum algorithms.

59.2 Biomolecular and Quantum Parallelism

Adleman designed and expressed his algorithm for the traveling salesman problem (actually Adleman's formulation was a problem which should be called the Hamiltonian path problem) by using biomolecular operations. Anyone interested can learn details about those operations in [59.1], but roughly speaking, *Adleman*'s procedure can be described as follows. The problem itself is to decide whether in a given directed graph (*city map*) there is a path beginning and ending at fixed vertices and visiting every vertex exactly once. Adle-

man's solution was to encode each vertex and edge into a single-stranded DNA-sequence in such a way, that if there is an edge e from vertex c_1 to c_2, then *half* of the strand encoding c_1 is complementary to a half of the strand encoding e, and the latter half of e is complementary to a half encoding c_2. As there may be multiple incoming and outgoing edges, there may also be multiple encodings of cities. In a test tube containing multiple copies of DNA strands encoding both vertices and edges, the single (let us call them *lower*) strands $c_1, c_2, \ldots$ tend to form longer strands $c_{i_1} c_{i_2} c_{i_3} \ldots$ bounded by the *upper* (single) strands $e_{i_1} e_{i_2} \ldots$ (e_{i_1} extends over c_{i_1} and c_{i_2}, e_{i_2} over c_{i_2} and c_{i_3}, etc.) The encoding ensures that c_{i_j} and $c_{i_{j+1}}$ will be adjacent only if there is an edge from c_{i_j} to $c_{i_{j+1}}$. This means that the chemical tendency of DNA sticking to its complementary counterpart will generate various double-stranded DNA sequences encoding paths in the graph that was originally encoded.

The problem is then to detect whether a strand encoding a desired path exists. Using electrophoresis it is possible to filter out DNA strands of wrong length, and additional existing techniques suffice to filter out exactly the desired paths, if there are any. The crucial point is that it is known how to the duplicate the existing DNA. Once the encodings are available, they can be duplicated to the extent guaranteeing that the above procedure will detect a desired solution with high probability, if any exists.

The above description emphasizes that Adleman's DNA-based solution actually utilizes heavy parallelism. A test tube containing multiple copies of vertex and edge encodings acts like a nondeterministic device generating potential paths, and the problem is to filter out the desired path – a solution, if any exists. An existing biotechnique is sufficient to do the filtering.

In quantum computing, there also occurs parallelism – quantum parallelism, which will be described in the rest of this chapter. The presentation here will be merely informal, but the notions will be defined in the next chapter. It is not necessary to focus on any specific NP-complete problem, and we choose to study a general version.

59.2.1 A General NP–Complete Problem

- Input: $N \in \mathbb{N}$ and $f : \{0, 1\}^N \to \{0, 1\}$ a polynomial-time (in N) computable function.
- Output:

$$\begin{cases} 1, & \text{if there is } x \in \{0, 1\}^N \text{ so that } f(x) = 1 , \\ 0, & \text{otherwise} . \end{cases}$$

Here and hereafter, $\{0, 1\}^N$ stands for the bit strings of length N, so a general NP-complete problem is typically a search problem: one has to decide whether there is an N-bit string x so that $f(x) = 1$. A solution to this problem can evidently be obtained via exhaustive search, but that is computationally expensive: try all 2^N candidates $x \in \{0, 1\}^N$ and check if any of them satisfies $f(x) = 1$. By assumption, any value $f(x)$ can

be computed in polynomial time (in N), but there are exponentially many (2^N) possibilities to be checked.

In quantum computing, it is possible to form a state

$$\sum_{x \in \{0,1\}^N} \frac{1}{\sqrt{2^N}} |x\rangle , \tag{59.1}$$

so-called *superposition* of all bit strings $x \in \{0, 1\}^N$, and then to compute function f on all possible inputs simultaneously by a cost of single computation to obtain

$$\sum_{x \in \{0,1\}^N} \frac{1}{\sqrt{2^N}} |x\rangle \, |f(x)\rangle . \tag{59.2}$$

This is what quantum parallelism means: all values $x \in \{0, 1\}^N$ occur, in a sense, parallel in (59.1), and all values $f(x)$ are, again in a sense, computed simultaneously.

However, this so-called quantum parallelism is very different from the biomolecular parallelism described earlier. In (59.1) there are no 2^N physical systems each consisting of N bits, but only a single physical system of N bits in a *state*, which allows, in a sense, an interpretation as any $x \in \{0, 1\}^N$. Quantum parallelism is not comparable to DNA computing parallelism, and the incomparability is underlined by the physical interpretation of (59.2). Observation of (59.2) will give any pair $(x, f(x))$, each with probability $\left| 1/\sqrt{2^N} \right|^2 = \frac{1}{2^N}$, but on a measurement, (59.2) is destroyed irreversibly.

It is worth noticing that the quantum parallelism, as described above, is not far apart from *probabilistic parallelism*: Toss N times a fair coin to obtain a random bit string $x \in \{0, 1\}^N$, then compute $f(x)$. A string x with the property $f(x) = 1$ (if any exists) will be found exactly with the same probability as observing (59.2) would give. Hence the problem with straightforward quantum parallelism (59.2) is the same as with the *probabilistic parallelism*. If there are only a few, or even only one string x (let us call it *solution*) such that $f(x) = 1$, then the cases *solution exists* and *no solution* cannot be distinguished from each other with any better probability than $\frac{1}{2^N}$.

However, as *Deutsch* and *Josza*'s [59.13], *Simon*'s [59.14], and *Shor*'s [59.4] discoveries demonstrated, quantum computing offers possibilities to solve *some* problems more efficiently than any known classical procedure allows. But it may be useful to underline right now that it is strongly believed (although not proved) that quantum computers cannot solve NP-complete problems in polynomial time.

59.3 Quantum Computing Preliminaries

In this chapter, the basics of quantum information will be presented only very superficially, and a reader desiring more detailed exposition is advised to consult [59.17,18] or [59.19]. To understand the formalism, it is necessary to accept the fact that quantum mechanics is a stochastic theory, meaning that in general, the complete description of the system, *the state*, cannot in general result into any deterministic description, only a probability distribution over potential outcomes. The probabilistic structure of quantum mechanics is very well studied, and lots of results are already available. We could ask, for instance, whether there could be a deterministic theory lying under quantum mechanics, and the probability distribution is only due to unknown boundary values (hidden variables). For instance, when tossing a classical coin, one could imagine that if the initial circumstances are known precisely enough, one could be always predict the outcome.

A deep investigation has shown that the quantum randomness cannot emerge from any deterministic procedure as described above, but randomness in inherently inseparable feature of quantum mechanics (see [59.20], for instance).

59.3.1 Hilbert Space Basic Structure

The *Hilbert space* formalism of quantum computing is usually based on *pure states* and requires some basic notions. *An n-level quantum system* means a (quantum) physical system with n different states which are mutually distinguishable with certainty. *An n-dimensional Hilbert space* H_n is a complex vector space $\mathbb{C}^n$ equipped with a Hermitian inner product $\langle x|y \rangle = x_1^* y_1 + \ldots + x_n^* y_n$. The inner product induces a norm $\|x\| = \sqrt{\langle x|x \rangle}$. For any element $(x_1, \ldots, x_n) \in \mathbb{C}^n$, a *A ket-vector* is defined as column vector ($n \times 1$-matrix)

$$|x\rangle = \begin{pmatrix} x_1 \\ x_2 \\ \vdots \\ x_n \end{pmatrix},$$

and a bra-vector as $1 \times n$-matrix (row vector)

$$\langle x| = \left(x_1^*, x_2^*, \ldots, x_n^* \right).$$

Usually $\mathbb{C}^n$ is identified with the space of ket-vectors ($n \times 1$-matrices), and we say that H_n is the *state space* of

the quantum system. The mathematical description of an n-level quantum system is based on n-dimensional Hilbert space in the following way: an orthonormal basis $\{|x_1\rangle, \ldots, |x_n\rangle\}$ is fixed as a *computational basis*, and a general state of the system is presented as a *superposition* of computational states

$$\alpha_1 |x_1\rangle + \cdots + \alpha_n |x_n\rangle, \tag{59.3}$$

so that $|\alpha_1|^2 + |\alpha_2|^2 + \cdots + |\alpha_n|^2 = 1$. In other words, a general state of an n-level system can be represented as unit-length vectors in H_n. Basically any basis for H_n could be chosen for representation (59.3), but some bases may preferred because of the physical implementation. Hence the term *computational basis* should not be understood as any mathematical definition, but as a chosen reference basis. States of a computational basis are also called *basis states* and complex coefficients α_i *amplitudes*.

An *observable* of quantum system H_n is a collection of mutually orthogonal subspaces $\{V_1, \ldots, V_k\}$ so that $H_k = V_1 \oplus \cdots \oplus V_k$. The intuitive meaning of the notion is that each subspace V_i refers to a physical property the system can have. For example, the computational basis itself induces an observable $\{L(x_1), \ldots, L(x_n)\}$, where $L(x_i)$ stands for the subspace generated by x_i.

The *minimal interpretation* of quantum physics is an axiom connecting the mathematical structure to the real world. For this representation, it is sufficient to introduce the minimal interpretation in the following way: let $\{V_1, \ldots, V_k\}$ be an observable and $x = \alpha_1 x_1 + \cdots + \alpha_k x_k$ a presentation of state x so that $x_i \in V_i$ and $\|x_i\| = 1$ for each i. Then the probability that quantum system in state x is seen to have property V_i is

$$\mathbb{P}(i) = |\alpha_i|^2. \tag{59.4}$$

It may be worth mentioning here that usually an extra element is associated to an observable: a real number λ_i to each subspace V_i. Number λ_i is the observable value, and equation (59.4) should read as

$$\mathbb{P}(\lambda_i) = |\alpha_i|^2,$$

meaning that the probability that the measured value of the observable is λ_i equals $|\alpha_i|^2$, However, when just studying quantum computation, it is not usually necessary to address explicit values $\lambda_i \in \mathbb{R}$, but it is enough to identify λ_i with its index i.

According to the *projection postulate*, the quantum system *collapses* to the observed state, and the superposition is irreversibly lost. That is, if property i was observed, then the state immediately after the observation is x_i. The projection postulate is among the most problematic features in quantum mechanics, but this article cannot be extended to treat that specifically.

Example 59.1: A two-level quantum system is referred as to a *quantum bit*, or *qubit* for short. We fix an orthonormal computational basis $|0\rangle = (1, 0)^T$, $|1\rangle = (0, 1)^T$ for H_2 (T stands for transposition), and a general state of a quantum bit is a vector

$$\alpha|0\rangle + \beta|1\rangle , \tag{59.5}$$

where $|\alpha|^2 + |\beta|^2 = 1$ (meaning that the length of (59.5) is 1). Let $C = \{L(|0\rangle), L(|1\rangle)\}$ be an observable consisting of two subspaces generated by $|0\rangle$ and $|1\rangle$, respectively, and $C' = \{L(|0'\rangle), L(|1'\rangle)\}$ another observable, where $|0'\rangle = \frac{1}{\sqrt{2}}(|0\rangle + |1\rangle)$ and $|1'\rangle = \frac{1}{\sqrt{2}}(|0\rangle - |1\rangle)$. If a quantum bit is in state (59.5), and observable C is measured, then 0 is seen with probability $|\alpha|^2$ and 1 with probability $|\beta|^2$.

Hence a qubit state (59.5) may look like a generalized probability distribution, but that is not the case. It is perfectly possible to measure also observable C' in state (59.5), and the outcome may be totally different. In fact,

$$\frac{1}{\sqrt{2}}|0\rangle + \frac{1}{\sqrt{2}}|1\rangle = 1 \cdot |0'\rangle + 0 \cdot |1'\rangle , \tag{59.6}$$

so measuring observable C results in 0 with probability $\frac{1}{2}$ and 1 with probability $\frac{1}{2}$. On the other hand, measuring observable C' results in $0'$ with probability 1 and $1'$ with probability 0. This is to emphasize that the state of a quantum system cannot be treated as a probability distribution. In fact, for every state of a finite-level (excluding the trivial case $n = 1$) quantum system there is a nontrivial (i. e., with more than 1 potential values) observable so that a single value will be observed with probability 1.

Based on the above definitions, we can now clarify the role of the *computational basis* a little bit. In fact, observing state (59.3) would generally require specification of the observable to be measured, but it is traditional to use the terminology *observing a state*, if the observable is induced by the computational basis.

59.3.2 Compound Systems

The states of a quantum system consisting of two distinguishable subsystems can be presented by using a *tensor product construction*. For the purposes of this article, it is not necessary to define the notion of tensor product exactly, it is enough just to know that the tensor product is (essentially) associative and distributive, but a non-commutative product of vectors obeying the obvious scalar rules. For more details, see [59.17] or [59.19]. It is also worth emphasizing that the counterpart of the tensor product in *concrete* objects such as matrices is the *Kronecker product*.

Now if H_m and H_n are the state spaces of m- and n-level quantum systems with computational bases $\{|x_1\rangle, \ldots, |x_m\rangle\}$ and $\{|y_1\rangle, \ldots, |y_n\rangle\}$, then the state space of the compound system is mn-dimensional if tensor product $H_m \otimes H_n$, whose computational basis can be chosen as

$$\{|x_i\rangle \otimes |y_j\rangle \mid (i, j) \in \{1, \ldots, m\} \times \{1, \ldots, n\}\} .$$

It is common to use shorthand notations $|x_i\rangle \otimes |y_j\rangle = |x_i\rangle|y_j\rangle = |x_i, y\rangle_j$, (even $|x_i y\rangle_j$ is used if there is no danger of confusion) so the state of the compound system can be represented as

$$\sum_{i=1}^{m} \sum_{j=1}^{n} \alpha_{ij} |x_i, y\rangle_j ,$$

where

$$\sum_{i=1}^{m} \sum_{j=1}^{n} |\alpha_{ij}|^2 = 1 . \tag{59.7}$$

It is clear that the observables of subsystems give raise to observables of the compound system.

State (59.7) is called *decomposable* if it can be presented as a product state

$$\left(\sum_{i=1}^{m} \alpha_i |x_i\rangle\right) \left(\sum_{j=1}^{n} \beta_j |y_j\rangle\right) ,$$

otherwise, the state is called *entangled*.

Example 59.2: A two-qubit state

$$\frac{1}{2}|00\rangle + \frac{1}{2}|01\rangle + \frac{1}{2}|10\rangle + \frac{1}{2}|11\rangle$$

is decomposable, as

$$\frac{1}{2}|00\rangle + \frac{1}{2}|01\rangle + \frac{1}{2}|10\rangle + \frac{1}{2}|11\rangle$$
$$= \frac{1}{\sqrt{2}}(|0\rangle + |1\rangle)\frac{1}{\sqrt{2}}(|0\rangle + |1\rangle) .$$

On the other hand, a two-qubit state

$$\frac{1}{\sqrt{2}}|00\rangle + \frac{1}{\sqrt{2}}|11\rangle \qquad (59.8)$$

is entangled, since assumption

$$\frac{1}{\sqrt{2}}|00\rangle + \frac{1}{\sqrt{2}}|11\rangle$$
$$= (\alpha_0|0\rangle + \alpha_1|1\rangle)(\beta_0|0\rangle + \beta_1|1\rangle)$$
$$= \alpha_0\beta_0|00\rangle + \alpha_0\beta_1|01\rangle + \alpha_1\beta_0|10\rangle + \alpha_1\beta_1|11\rangle$$

leads into equations $\alpha_0\beta_0 = \alpha_1\beta_1 = \frac{1}{\sqrt{2}}$ and $\alpha_0\beta_1 = \alpha_1\beta_0 = 0$, which are clearly impossible.

Entangled state (59.8) is historically and philosophically of great interest. Indeed, *Bohm* used [59.21] a state analogous to it to reformulate an apparent paradox of quantum mechanics introduced by *Einstein, Podolsky,* and *Rosen* [59.22]. That formulation eventually led *Bell* to present a resolution of the paradox [59.23] (for an exposition, see [59.20]). State (59.8) is called *an EPR state*, and a pair of qubits in state (59.8) an *EPR pair* for the aforementioned reasons.

The minimal interpretation implies directly that if the state (59.8) is observed (i.e., observable generated by the computational basis is measured), then we will see "00" with probability of $\frac{1}{2}$, and "11" with probability of $\frac{1}{2}$, too. Hence the quantum bits in an EPR state (59.8) are perfectly correlated; when observed, they always have the same value, which, however, can be 0 or 1, either with probability $\frac{1}{2}$.

It has been experimentally demonstrated that the correlation of the EPR pairs as described above is detectable even if the two physical systems (quantum bits) are spatially separated by 144 km [59.24]. However, the correlation over distance should not be surprising or anything specific to quantum physics; it is left to the reader to describe a non-quantum bipartite system with distant correlations analogous to the EPR state.

Whereas the correlation itself is not specific to quantum mechanics, the *violation of Bell inequalities* is [59.20], for instance. Violation of Bell inequalities has been experimentally detected over a physical distance of 144 km [59.24].

The mathematical description of compound systems with more than 2 subsystems is again based on tensor product construction. In this article, we will not focus on details, but will merely present an example.

Example 59.3: A system of N quantum bits has its description in a state space $H_2 \otimes \cdots \otimes H_2$, a Hilbert space

isomorphic to H_{2^N}. A general state of H_{2^N} can be described as

$$\sum_{x \in \{0,1\}^N} \alpha_x |x\rangle ,$$

where

$$\sum_{x \in \{0,1\}^N} |\alpha_x|^2 = 1 .$$

State

$$\frac{1}{\sqrt{2}}(|0\rangle + |1\rangle) \cdots \frac{1}{\sqrt{2}}(|0\rangle + |1\rangle)$$
$$= \frac{1}{\sqrt{2^N}} \sum_{x \in \{0,1\}^N} |x\rangle \qquad (59.9)$$

presents a uniformly distributed superposition over all basis states $|x\rangle$. It is worth noticing that presentation (59.9) shows that the state is clearly decomposable.

59.3.3 Quantum Operations

It was explained in the previous sections, in a very simplified way, how to present quantum information in pure states. It is, however, clear that the stagnant pictures of quantum states are not sufficient for using the theory. Instead, it is necessary to describe how quantum systems change in time. For most quantum computing models, and also for this article, it is sufficient to describe *closed* quantum system transformations, which will be mathematically formalized as follows: a *(closed) quantum system state transformation* is a *unitary* mapping $H_n \to H_n$. A linear mapping U is *unitary*, if $U^*U = UU^* = I$ (identity mapping), where U^* is the complex conjugate of the transpose of U. A closed quantum system state transformation is also called a *quantum gate*, see [59.25] for a study on quantum gates.

Example 59.4:

$$H = \frac{1}{\sqrt{2}} \begin{pmatrix} 1 & 1 \\ 1 & -1 \end{pmatrix} .$$

If is straightforward to verify that $H^* = H$, and that $H^*H = HH^* = HH = I$, meaning that H is unitary. H is hence a *unary* quantum gate, i.e., a gate on one qubit. The action of H on computational basis $\{|0\rangle = (1, 0)^T, |1\rangle = (0, 1)^T\}$ is given by

$$H|0\rangle = \frac{1}{\sqrt{2}}(|0\rangle + |1\rangle) \text{ and } H|1\rangle = \frac{1}{\sqrt{2}}(|0\rangle - |1\rangle).$$

Gate H is called a *Hadamard transform* or a *Walsh transform*.

Example 59.5: Mapping

$$C = \begin{pmatrix} 1 & 0 & 0 & 0 \\ 0 & 1 & 0 & 0 \\ 0 & 0 & 0 & 1 \\ 0 & 0 & 1 & 0 \end{pmatrix}$$

can be easily verified to be unitary. C is a binary gate called *controlled not*, and its name is justified by computing its action. First, matrix presentations of $|00\rangle$, $|01\rangle$, $|10\rangle$, $|11\rangle$ are obtained by using the Kronecker product

$$|00\rangle = |0\rangle \otimes |0\rangle = \begin{pmatrix} 1 \\ 0 \end{pmatrix} \otimes \begin{pmatrix} 1 \\ 0 \end{pmatrix} = \begin{pmatrix} 1 \\ 0 \\ 0 \\ 0 \end{pmatrix},$$

$$|01\rangle = \begin{pmatrix} 1 \\ 0 \end{pmatrix} \otimes \begin{pmatrix} 0 \\ 1 \end{pmatrix} = \begin{pmatrix} 0 \\ 1 \\ 0 \\ 0 \end{pmatrix},$$

$$|10\rangle = |1\rangle \otimes |0\rangle = \begin{pmatrix} 0 \\ 1 \end{pmatrix} \otimes \begin{pmatrix} 1 \\ 0 \end{pmatrix} = \begin{pmatrix} 0 \\ 0 \\ 1 \\ 0 \end{pmatrix},$$

and

$$|11\rangle = \begin{pmatrix} 0 \\ 1 \end{pmatrix} \otimes \begin{pmatrix} 0 \\ 1 \end{pmatrix} = \begin{pmatrix} 0 \\ 0 \\ 0 \\ 1 \end{pmatrix}.$$

The action of C is then easy to compute: $C|00\rangle = |00\rangle$, $C|01\rangle = |01\rangle$, $C|10\rangle = |11\rangle$, and $C|11\rangle = |10\rangle$, meaning that the second qubit is flipped exactly when the first, the control bit, equals 1.

It is possible to establish quantum computing on quantum gates only, and that would lead into *quantum circuit formalism* [59.17, 18]. On the other hand, there are also other possible ways to establish the formalism of quantum computing. In fact, for any classical model for computing, there is a canonical way of transforming it into a quantum version. For example, for the definitions of quantum finite automata, see [59.26, 27] and for quantum Turing machines, see [59.17] or [59.19]. It must, however, be emphasized that unitary mappings are invertible by definition, and therefore all quantum computing models based on them are reversible. For quantum Turing machines the reversibility does not

bring any disadvantage, as it is well known that all computation can be made reversible by introducing extra space [59.28]. On the other hand, many unitary models of finite automata are strictly weaker than the traditional one [59.26, 27], just because the transformation into a reversible machine would require extra space. The unitarity can be relaxed by using *open* system transforms, but representing them would require too much space in this article. See [59.29] for an automaton model with open time evolution.

Let us now revisit quantum parallelism and add more details. If $f : \{0, 1\}^N \to \{0, 1\}$ is computable in polynomial time, there is also a polynomial-size quantum circuit computing f. In fact, an algorithm computing f can be efficiently turned into a quantum circuit that computes f [59.17]. Usually it is necessary to add some auxiliary bits to bypass the reversibility requirement, but those extra bits are not usually written down explicitly. This implies that it is possible to construct a unitary mapping U_f by using a polynomial number of simple quantum gates (selected from a finite set) with the following action

$$U_f|x\rangle|0\rangle = |x\rangle|f(x)\rangle,$$

where x is a sequence (register) of N quantum bits. Applying the Hadamard transform to N first quantum bits in state

$$|\mathbf{0}\rangle|0\rangle$$

will result into state

$$\frac{1}{\sqrt{2^N}} \sum_{x \in \{0,1\}^N} |x\rangle|0\rangle,$$

(59.9), and a further application of U_f will lead into state

$$\frac{1}{\sqrt{2^N}} \sum_{x \in \{0,1\}^N} |x\rangle|f(x)\rangle, \qquad (59.10)$$

an equally balanced superposition over all potential pairs $(x, f(x))$. This is the state (59.2) of a previous example. From the computational complexity point of view, it is important to realize that the state (59.10) can be generated by N Hadamard gate actions plus the number of quantum gates required to implement U_f (polynomial in N). This is exactly what quantum parallelism means: by a polynomial number of actions it is possible to generate state (59.10) extending over exponentially many basis states.

Unfortunately (59.10) is only a mathematical description of a state of a physical system. In particular, the exponentially many values do not exist physically observable to us, but observing (59.10) will give only a single pair $(x, f(x))$, each with probability $\frac{1}{2^N}$, and observation will make (59.10) to collapse into state $|x\rangle|f(x)\rangle$.

59.4 Quantum Algorithms

The last example of the previous section clearly justifies the following question: Why should we regard quantum parallelism any better than a simple probability distribution? In fact, we could obtain equally good results just by selecting N random bits to form a bit string x, then to compute $f(x)$. If necessary, we could even invent a notation for probability distribution. Let us agree that notation

$$\sum_{x \in \{0,1\}^N} \frac{1}{2^N}[x, 0]$$

stands for a probability distribution over $N+1$ bit strings x, 0, where each $x \in \{0, 1\}^N$ occurs with a probability $\frac{1}{2^N}$. Then, computing f results into

$$\sum_{x \in \{0,1\}^N} \frac{1}{2^N}[x, f(x)],$$

and any pair $(x, f(x))$ is seen with a probability of $\frac{1}{2^N}$, as in the case of (59.10).

The answer to that question is that the straightforward use of (59.10) is obviously not the only possible strategy. It should be noted that the amplitudes can be negative as well, and consequently it may be possible to design quantum computing in such a way that the desirable basis states would gain more *visibility* because their amplitudes would sum up, and nondesirable ones could cancel each other. Should that happen, we would call the former *constructive interference* and the latter *destructive interference*.

Example 59.6: State $|0\rangle$ turns into $\frac{1}{\sqrt{2}}(|0\rangle + |1\rangle)$, if affected by the Hadamard transform. If the affected state were observed, one would see 0 and 1, both with probability $\frac{1}{2}$. On the other hand, if the state is not observed, but another Hadamard transform is applied, we get the following

$$H \frac{1}{\sqrt{2}}(|0\rangle + |1\rangle)$$

$$= \frac{1}{\sqrt{2}}(H|0\rangle + H|1\rangle)$$

$$= \frac{1}{\sqrt{2}}\left(\frac{1}{\sqrt{2}}(|0\rangle + |1\rangle)\right)$$

$$+ \frac{1}{\sqrt{2}}\left(\frac{1}{\sqrt{2}}(|0\rangle - |1\rangle)\right)$$

$$= \left(\frac{1}{2} + \frac{1}{2}\right)|0\rangle + \left(\frac{1}{2} - \frac{1}{2}\right)|1\rangle = |0\rangle,$$

which demonstrates in detail how the amplitudes $\frac{1}{2}$ and $\frac{1}{2}$ sum up to 1, and $\frac{1}{2}$ and $-\frac{1}{2}$ cancel each other.

A parallel to classical information processing could be as follows: the Hadamard transform may be interpreted as a fair coin toss. Beginning either from $|0\rangle$ or $|1\rangle$, one reaches state $1/\sqrt{2}(|0\rangle \pm |1\rangle)$, where 0 and 1 are both seen with probability $\frac{1}{2}$ (single coin toss). If the coin is tossed twice in the classical settings, then again 0 and 1 are seen both with 50% probability. But in this example, the second *coin toss* returns the state into $|0\rangle$, and hence 0 is seen with 100% probability. This is a feature that is clearly impossible with classical information.

Powerful quantum algorithms, such as Shor's factoring algorithm are indeed all algorithms utilizing quantum interference in a clever manner. Unfortunately, the interference behavior of a quantum algorithm is quite difficult to control in practice, and consequently only a few families of quantum algorithms are known to date.

The known quantum algorithm families enclosing almost all known quantum algorithms are:

1. Quantum algorithms based on Fourier transforms
2. Amplitude amplification methods
3. Quantum random walks
4. Adiabatic quantum algorithms.

59.4.1 Quantum Algorithms Based on Fourier Transforms

This class of quantum algorithms usually provide an apparent exponential speedup over their classical counterparts. The algorithms in this class attempt to construct a superposition

$$\sum_x \alpha_x |x\rangle$$

Hence it is not possible to use quantum parallelism, at least not in this straightforward way, to resolve efficiently NP-complete problems.

whose *amplitudes* α_x form a periodic or almost periodic sequence. In many cases, it is then possible to perform the discrete Fourier transform *on amplitudes* (exponentially many) with only a polynomial number of operations on quantum bits (see, e.g., [59.17]). This structure takes care of interfering quantum computational paths by using centuries old knowledge on discrete Fourier transforms.

The approach has been very successful. The algorithm by *Deutsch–Jozsa* [59.13], *Simon*'s algorithm [59.14], and *Shor*'s algorithm [59.4] are all quantum algorithms that control their amplitudes by a structure given by discrete Fourier transforms. The most prominent examples of exponential speed-ups are provided by Fourier transform-based quantum algorithms. For details, see [59.17] or [59.19].

59.4.2 Amplitude Amplification Methods

The amplitude amplification method was presented by *Grover* in 1996 [59.30], and it has been extended thereafter. The basic form of Grover's method involves a function $f : \{0, 1\}^N \to \{0, 1\}$ assumed to be computable in polynomial time, and is applied to a superposition

$$\sum_{x \in \{0,1\}^N} \alpha_x |x\rangle .$$

The purpose is to use quantum interference to increase the joint squared absolute values of those amplitudes α_x, for which $f(x) = 1$, to make such values x more likely to be observed. Grover presented an iterative procedure for that purpose. An iteration step typically involves one evaluation of f.

Among all quantum algorithms, the amplitude amplification methods deserve the first right to be called *quantum-most* methods, as all the other ones have an analog or a counterpart in classical computing. Grover's method does not have any; it is a method purely originating from quantum computing purposes.

The most remarkable consequence of Grover's method is that a general NP-complete problem can be solved (with a high probability) by quantum computers using only $O(\sqrt{N})$ evaluations of function f. When comparing this to $O(N)$ evaluations in the classical case, this is an essential improvement, but not yet not an exponential one: $\sqrt{2^N} = (\sqrt{2})^N$ is again an exponential function, although with a smaller base. For details, see [59.17] or [59.19].

59.4.3 Quantum Random Walks

Quantum random walks is a straightforward analog of classical random walks. It is known that quantum random walks sometimes have exponentially faster hitting times than their classical counterparts [59.31], and the technique can be attempted for a great variety of computational problems. This article is too narrow to address quantum random walks in detail, but for an exposition, see [59.32].

59.4.4 Adiabatic Quantum Algorithms

Adiabatic quantum algorithms should not be called algorithms, but just a technique for designing quantum algorithms. It, or at least its generality can be loosely compared to classical evolutionary algorithms. Adiabatic quantum computing can be adapted to any computational problem.

Adiabatic quantum computing is based on the *adiabatic theorem*, which says that if a *Hamiltonian operator* H_0 is transformed into another Hamiltonian H_1 slowly enough, then the ground state $|x_0\rangle$ of H_0 is transformed into the ground state $|x_1\rangle$ of H_1, as well. The terminology is not explained here, but the reader is advised to consult [59.33]. Instead of detailed definitions and descriptions, we just mention that the adiabatic computation has been shown to be equally as powerful as quantum computing based on quantum gates [59.33]. The technique of adiabatic computing just provides an advantage that the algorithm design can be circumvented in some cases.

59.4.5 Restrictions of Quantum Computing

Theoretical computer science seems to suffer from powerful absolute limitations of computational models. It has been previously mentioned that question $P \neq NP$ is waiting for resolution, but there are many analogous unsolved problems. From a quantum computing point of view, the most important such problem is probably: Is polynomial-time quantum computing more powerful than its classical counterpart, at least for some instances? Very likely this problem is at least as difficult as the P versus NP problem, and in the sight of the present understanding, there is no apparent route how to even approach these problems.

However, there are easy ways to obtain *relativized* lower bounds for computational complexity. For relativization in computing, we refer to [59.34], but the basic results are easy to state as follows: a general NP-complete problem described in an earlier section is *relativized* in the sense that no structure of function f is available. I should apologize to complexity theory experts for the simplifications in this explanation (but on second thought, I will not do that), but the intuitive idea in a relativized lower bound is the following: if *nothing* is known of the structure of function f (but the values are chosen arbitrarily), then to decide about the existence of $x \in \{0, 1\}^N$ so that $f(x) = 1$ will inherently take 2^N evaluations of f, since for any process asking less values, there is a possibility to introduce f assigning a *wrong* value to the non-queried string. This certainly gives reasons to believe that P $\neq$ NP, but does not constitute any proof of that, since no polynomial-time computable function exists without a structure.

Relative lower bounds are known for quantum computing, too. It is, for example, possible to say, that for a general NP-complete problem, quantum computing offers no polynomial time solution. Instead, at least $\sqrt{2^N}$ computational steps are needed for a solution (this is to say that Grover's method is asymptotically optimal). There are various techniques for obtaining relativized lower bounds for quantum computing, and as the most notable ones we can mention the *polynomial technique* [59.35] and the *adversary technique* [59.36].

59.4.6 Physical Realization of Quantum Algorithms

As of 2013, quantum algorithms have been under development almost for three decades, and many things are known about them. In a previous section, we listed four general families of quantum algorithms and mentioned some techniques for proving lower bounds for quantum computing. Even though the development may seem modest in some sense, there are enough interesting quantum algorithms to justify the quantum computer development project. Unfortunately, to build a quantum computer has turned out to be a very challenging task.

In principle, any quantum physical two-state system could serve as a quantum bit. Unfortunately such a system is always very vulnerable to external disturbances, and consequently in many realizations, the lifetime of a qubit is only a tiny fraction of a second. The following physical realizations (among others) have been proposed. Cold trapped ions [59.37], nuclear spin [59.38], and photon polarization [59.39] are all potential implementations of quantum bits, but the most advanced quantum computer (with respect to the number of quantum bits) in modern technology allows us to hold only 12 quantum bits [59.40]. Quantum factoring algorithm for N-bit integers will require approximately $2N$ qubits [59.41]. This implies that the current quantum computers cannot handle enough bits to perform universal computation to truly challenge RSA or other public-key cryptosystems used currently.

Hence we have to conclude that with modern technology, we cannot yet realize very much quantum computing. The most important lesson the 30-year lasting research on quantum computing provides us is, therefore, new insights into the theory of computing and relations between the theory of computation and physical world.

59.5 Biological Applications of Quantum Computing

Even though large-scale quantum computers do not exist yet, we already know various potential applications. A fast integer factoring algorithm would be very influential, even though its influences may not be called entirely positive. Quantum algorithms providing an exponential speed-up over known classical ones are almost all designed by using quantum Fourier transform, and hence they are applicable only for problems having a suitable periodic structure. Such structures do not typically exist in biological problems, and therefore they are not very likely to have an exponential speed-up on biological problems.

On the other hand, a quantum computer would provide a quadratic speed-up on all search problems, and there are various potential applications. For instance, protein folding problems are typical search problems that could benefit from a quadratic speed-up, but as this would apply for any search problem, we are not going to list especially biological search problems.

Instead, we will conclude this article by pointing out a specific problem that quantum computers are good at and which may have consequences in biology, as well. This specific problem also encloses the circle: quantum computers are good at simulating quantum physics, just as *Feynman* explained in his article [59.10]. In general, simulating a quantum mechanical system of N particles with classical computers seems to lead into an exponential slowdown in the simulation efficiency, and Feynman proposed that a quantum computer could be used to avoid the slowdown.

As quantum mechanics governs microsystems, and large biomolecules are built of smaller particles, one could expect that quantum mechanics to have an important explanatory value on biomolecular processes [59.42]. Perhaps it is so, but there is not much existing research on this topic. One reason for this is that structures like DNA are so complex from the physical perspective that even the modeling becomes extremely hard, to say nothing of the explicit solutions. Another reason is that the help provided by computers will not lead very far – as long as we do not have quantum computers.

References

59.1   L.M. Adleman: Molecular computation of solutions to combinatorial problems, Science **266**(11), 1021–1024 (1994)

59.2   R.J. Lipton: DNA solution of hard computational problems, Science **268**(5210), 542–545 (1995)

59.3   M.L. Minsky, S.A. Papert: *Perceptrons* (MIT Press, Cambridge 1969)

59.4   P.W. Shor: Algorithms for quantum computation: Discrete log and factoring, Proc. 35th Annu. IEEE Symp. Found. Comput. Sci. (1994) pp. 20–22

59.5   A.M. Turing: On computable numbers, with an application to the entscheidungsproblem, Proc. Lond. Math. Soc. 2(42), 230–265 (1936)

59.6   J. von Neumann: Thermodynamik quantum-mechanischer Gesamheiten, Nachr. Ges. Wiss. Gött. **1**, 273–291 (1927)

59.7   J. von Neumann: *Mathematische Grundlagen der Quantenmechanik* (Springer, Berlin 1932)

59.8   P.A. Benioff: The computer as a physical system: A microscopic quantum mechanical Hamiltonian model of computers as represented by Turing machines, J. Stat. Phys. **22**(5), 563–591 (1980)

59.9   P.A. Benioff: Quantum mechanical Hamiltonian models of discrete processes that erase their own histories: Application to turing machines, Int. J. Theor. Phys. 21(3/4), 177–202 (1982)

59.10   R.P. Feynman: Simulating physics with computers, Int. J. Theor. Phys. **21**(6/7), 467–488 (1982)

59.11   D. Deutsch: Quantum theory, the Church–Turing principle and the universal quantum computer, Proc. R. Soc. A **400**, 97–117 (1985)

59.12   D. Deutsch: Quantum computational networks, Proc. R. Soc. A **425**, 73–90 (1989)

59.13   D. Deutsch, R. Jozsa: Rapid solutions of problems by quantum computation, Proc. R. Soc. A **439**, 553 (1992)

59.14   D.R. Simon: On the power of quantum computation, Proc. 35th Annu. IEEE Symp. Found. Comput. Sci. (1994) pp. 116–123

59.15   E. Bernstein, U. Vazirani: Quantum complexity theory, SIAM J. Comput. **26**(5), 1411–1473 (1997)

59.16   Clay Mathematics Institute: http://www.claymath.org/millennium/P_vs_NP/

59.17   M. Hirvensalo: *Quantum Computing*, 2nd edn. (Springer, Berlin, Heidelberg 2004)

59.18   M. Hirvensalo: Mathematics for quantum information processing. In: *Handbook of Natural Computing*, ed. by G. Rozenberg, T. Bäck, J. Kok (Springer, Berlin, Heidelberg 2011)

59.19   M.A. Nielsen, I.L. Chuang: *Quantum Computation and Quantum Information* (Cambridge Univ. Press, Cambridge 2000)

59.20   M. Hirvensalo: EPR Paradox and Bell Inequalities, Bulletin EATCS **92**, 115–139 (2007)

59.21   D. Bohm: *Quantum Theory* (Prentice-Hall, Englewood Cliffs 1951) pp. 614–619

59.22   A. Einstein, B. Podolsky, N. Rosen: Can quantum-mechanical description of physical reality be considered complete?, Phys. Rev. **47**, 777–780 (1935)

59.23   J.S. Bell: On the Einstein–Podolsky–Rosen paradox, Physics **1**, 195–200 (1964)

59.24   R. Ursin, F. Tiefenbacher, T. Schmitt-Manderbach, H. Weier, T. Scheidl, M. Lindenthal, B. Blauensteiner, T. Jennewein, J. Perdigues, P. Trojek, B. Ömer, M. Fürst, M. Meyenburg, J.G. Rarity, Z. Sodnik, C. Barbieri, H. Weinfurter, A. Zeilinger: Free-space distribution of entanglement and single photons over 144 km, Nat. Phys. **3**(7), 481–486 (2007)

59.25   A. Barenco, C.H. Bennett, R. Cleve, D.P DiVincenzo, N. Margolus, P. Shor, T. Sleator, J. Smolin, H. Weinfurter: Elementary gates for quantum computation, Phys. Rev. A **52**(5), 3457–3467 (1995)

59.26   A. Kondacs, J. Watrous: On the power of quantum finite state automata, Proc. 38th Annu. Symp. Found. Comput. Sci. (1997) pp. 66–75

59.27   C. Moore, J.P. Crutchfield: Quantum automata and quantum grammars, Theor. Comput. Sci. **237**(1/2), 275–306 (2000)

59.28   C.H. Bennett: Logical reversibility of computation, IBM J. Res. Dev. **17**, 525–532 (1973)

59.29   M. Hirvensalo: Quantum automata with open time evolution, Int. J. Nat. Comput. Res. **1**, 70–85 (2010)

59.30   L.K. Grover: A fast quantum-mechanical algorithm for database search, Proc. 28th Annu. ACM Symp. Theory Comput. (1996) pp. 212–219

59.31   J. Kempe: Discrete quantum walks hit exponentially faster, Probab. Theory Relat. Fields **133**(2), 215–235 (2005)

59.32   J. Kempe: Quantum random walks – an introductory overview, Contemp. Phys. **44**(4), 307–327 (2003)

59.33   D. Aharonov, W. van Dam, J. Kempe, Z. Landau, S. Lloyd, O. Regev: Adiabatic quantum computation is equivalent to standard quantum computation, SIAM J. Comput. **37**, 166–194 (2007)

59.34   C.H. Papadimitriou: *Computational Complexity* (Addison-Wesley, Reading 1994)

59.35   R. Beals, H. Buhrman, R. Cleve, M. Mosca, R. Wolf: Quantum lower bounds by polynomials, Journal ACM **48**(4), 778–797 (2001)

59.36   A. Ambainis: Quantum lower bounds by quantum arguments, J. Comput. System Sci. **64**(4), 750–767 (2002)

59.37   J.I. Cirac, P. Zoller: Quantum computations with cold trapped ions, Phys. Rev. Lett. **74**, 4091–4094 (1995)

59.38   I.L. Chuang, N. Gershenfeld, M. Kubinec: Experimental implementation of fast quantum searching, Phys. Rev. Lett. **80**, 3408–3411 (1998)

59.39   J.L. O'Brien: Optical quantum computing, Science **318**, 1567–1570 (2007)

59.40   C. Negrevergne, T.S. Mahesh, C.A. Ryan, M. Ditty, F.-Y. Cyr-Racine, W. Power, N. Boulant, T. Havel, D.G. Cory, R. Laflamme: Benchmarking quantum control methods on a 12-qubit system, Phys. Rev. Lett. **96**, 170501 (2006)

59.41   S. Beauregard: Circuit for Shor's algorithm using $2n+3$ qubits, Quantum Inform. Comput. **3**(2), 175–185 (2003)

59.42   V.V. Nelayev, K.N. Dovzhik, V.V. Lyskouski: Quantum effects in biomolecular structures, Rev. Adv. Mater. Sci. **20**, 42–47 (2009)

60. Brain, Gene, and Quantum Inspired Computational Intelligence

Nikola Kasabov

This chapter discusses opportunities and challenges for the creation of methods of computational intelligence (CI) and more specifically – artificial neural networks (ANN), inspired by principles at different levels of information processing in the brain: cognitive, neuronal, genetic, and quantum, and mainly, the issues related to the integration of these principles into more powerful and accurate CI methods. It is demonstrated how some of these methods can be applied to model biological processes and to improve our understanding in the subject area; generic CI methods being applicable to challenging generic AI problems. The chapter first offers a brief presentation of some principles of information processing at different levels of the brain and then presents brain inspired, gene inspired, and quantum inspired CI. The main contribution of the chapter, however, is the introduction of methods inspired by the integration of principles from several levels of information processing, namely:

1. A computational neurogenetic model that in one model combines gene information related to spiking neuronal activities.
2. A general framework of a quantum spiking neural network (SNN) model.
3. A general framework of a quantum computational neurogenetic model (CNGM).

Many open questions and challenges are discussed, along with directions for further research.

60.1 Levels of Information Processing in the Brain

The brain is a dynamic information processing system that evolves its structure and functionality in time through information processing at different levels – Fig. 60.1: quantum, molecular (genetic), single neuron, ensemble of neurons, cognitive, evolutionary.

Principles from each of these levels have been already used as inspiration for CI methods, and more specifically – for methods of ANN. The chapter focuses on the interaction between these levels and mainly on how this interaction can be modeled and how it can be used in principle to improve existing CI methods and for a better understanding of brain, gene, and quantum processes.

At the quantum level, particles (atoms, ions, electrons, etc.), which make every molecule in the material world, move continuously, being in several states at the same time, and are characterized by probability, phase, frequency, and energy.

At a molecular level, RNA and protein molecules evolve in a cell and interact in a continuous way, based on the information stored in the DNA and on external factors, and affect the functioning of a cell (neuron) under certain conditions.

At the level of a neuron, the internal information processes and the external stimuli cause the neuron to produce a signal that carries information to be transferred to other neurons.

At the level of neuronal ensembles, all neurons operate in a *concert*, defining the function of the ensemble, for instance the perception of a spoken word.

At the level of the whole brain, cognitive processes take place, such as language and reasoning, and global information processes are manifested, such as consciousness.

At the level of a population of individuals, species evolve through evolution, changing the genetic DNA code for a better adaptation.

The information processes at each level shown in Fig. 60.1 are very complex and difficult to understand, but much more difficult to understand is the interaction between the different levels. It may be that understanding the interaction through its modeling would be a key to understanding each level of information processing in the brain and perhaps the brain as a whole. Using principles from different levels in one ANN CI model and modeling their relationship can lead to a next generation of ANN as more powerful tools to understand the brain and to solve complex problems.

Some examples of CI models that combine principles from different levels shown in Fig. 60.1 are: computational neurogenetic models [60.1–3], quantum inspired CI and ANN [60.4, 5], and evolutionary models [60.6, 7]. Suggestions are made that modeling of higher cognitive functions and consciousness in particular can be achieved if principles from quantum information processing are considered [60.8, 9]. There are many issues and open questions to be addressed when creating CI methods that integrate principles from different levels; some of these are presented in this chapter.

In Sect. 60.2 models inspired by information processes in the brain, which include local learning evolving connectionist systems (ECOS) and SNN are discussed briefly. Section 60.3 presents CI methods inspired by genetic information processes, mainly models of gene regulatory networks (GRN). In Sect. 60.4, the issue of combining neuronal with genetic information processing is discussed and the principles of CNGM are presented. Section 60.5 presents some ideas behind quantum inspired CI. Section 60.6 presents a model of a quantum inspired SNN and offers a theoretical framework for the integration of principles from quantum, -genetic, and neuronal information processing. Section 60.7 concludes the chapter with more open questions and challenges for the future.

6. Evolutionary (population/generation) processes

5. Brain cognitive processes

4. System information processing (e.g., neural ensemble)

3. Information processing in a cell (neuron)

2. Molecular information processing (genes, proteins)

1. Quantum information processing

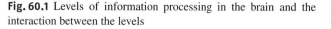

Fig. 60.1 Levels of information processing in the brain and the interaction between the levels

60.2 CI and ANN Models Inspired by Neuronal and Cognitive Processes in the Brain

Many CI methods, in particular ANN, are brain inspired (using some principles from the brain), or brain-like (more biologically plausible models, usually developed to model a brain function) [60.1, 10–15]. Examples are: models of single neurons and neural network ensembles [60.16–22], cognitive ANN models [60.14, 15, 23, 24], etc.

These models have been created with the goals of:

- Modeling and understanding brain functions.
- Creating powerful methods and systems of CI for solving complex problems in all areas of science and the humanity.

In this section we present only two groups of models, namely ECOS and SNN, as they will be used in other sections to create models that incorporate principles from other levels of information processing.

60.2.1 Local, Knowledge-Based Learning Evolving Connectionist Systems – Weakly Brain Inspired Models

ECOS are adaptive, incremental learning and knowledge representation systems that evolve their structure and functionality, where there is a connectionist architecture in the core of a system that consists of neurons (information processing units) and connections between them [60.25]. ECOS is a CI system based on neural networks, but using other techniques of CI, that operates continuously in time and adapts its structure and functionality through continuous interaction with the environment and with other systems. The adaptation is defined through:

1. A set of evolving rules.
2. A set of parameters (*genes*) that are subject to change during the system operation.
3. An incoming continuous flow of information, possibly with unknown distribution.
4. Goal (rationale) criteria (also subject to modification) that are applied to optimize the performance of the system over time.

ECOS learning algorithms are inspired by brain-like information processing principles, e.g.,

1. They evolve in an open space, where the dimensions of the space can change.

2. They learn via incremental learning, possibly in an on-line mode.
3. They learn continuously in a lifelong learning mode.
4. They learn both as individual systems and as an evolutionary population of such systems.
5. They use constructive learning and have evolving structures.
6. They learn and partition the problem space locally, thus allowing for a fast adaptation and tracing the evolving processes over time.
7. They evolve different types of knowledge representation from data, mostly a combination of memory-based and symbolic knowledge.

Many ECOS have been suggested so far, where the structure and the functionality of the models evolve through incremental, continuous learning from incoming data, sometimes in an on-line mode, and through interaction with other models and the environment. Examples are: growing SOMs [60.17], growing gas [60.26], RAN [60.27], growing RBF networks [60.28, 29], FuzzyARTMAP [60.14], EFuNN [60.25, 30, 31], DENFIS [60.32], and many more.

A block diagram of EFuNN is given in Fig. 60.2. It is used to model GRN in Sect. 60.5. At any time of the EFuNN continuous incremental learning, rules can be derived from the structure, which rules represent clusters of data and local functions associated with these clusters

IF < data is in cluster N_{cj},

   defined by a cluster center N_j,

   a cluster radius R_j

   and a number of examples

   N_{jexamp} in this cluster >

THEN < the output function is F_c >   (60.1)

In the case of DENFIS, first-order local fuzzy rule models are derived incrementally from data, for example,

IF < the value of $\times 1$ is in the area defined by

   a Gaussian membership function with a center

   at 0.1 and a standard deviation of 0.05 >,

AND < the value of $\times 2$ is in the area defined

   by a Gaussian function with parameters

   (0.25, 0.1) respectively >

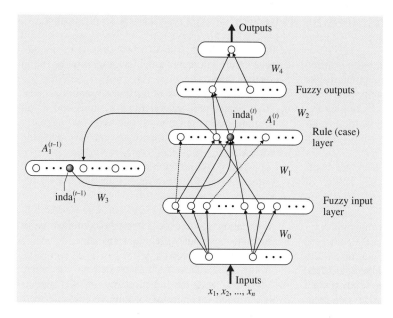

Fig. 60.2 An EFuNN (evolving fuzzy neural networks) architecture with a short term memory and feedback connections [60.33]. It is used in Sect. 60.5 to model GRN with inputs being the expression of genes at a time (t) and the outputs being the expression of genes/proteins at time $(t + \mathrm{d}t)$

THEN $<$ the output y is calculated by the formula

$$y = 0.01 + 0.7 \times 1 + 0.12 \times 2 > \qquad (60.2)$$

In the case of EFuNN, local simple fuzzy rule models are derived, for example,

IF $\times 1$ is (Medium 0.8) and $\times 2$ is (Low 0.6)
THEN y is (High 0.7), radius $R = 0.24$;

$$N_{\mathrm{examp}} = 6 , \qquad (60.3)$$

where: low, medium and high are fuzzy membership functions defined for the range of each of the variables $\times 1$, $\times 2$, and y; the number and the type of the membership functions can either be deduced from the data through learning algorithms, or can be predefined based on human knowledge [60.34, 35]; R is the radius of the cluster; and N_{examp} is the number of examples in the cluster.

A further development of the EFuNN and the DEN-FIS local ECOS models is the transductive weighted neuro-fuzzy inference engine (TWNFI) [60.30, 36]. In this approach, for every new vector (sample/example S) a *personalized* model is developed from existing nearest samples, where each of the variables is normalized in a different subrange of [0,1] so that they have a different influence on the Euclidean distance from (60.1), therefore they are ranked in terms of their importance to the output calculated for any new sample individually. Samples are also weighted in the model based on their distance to the new sample, where in the Euclidean

distance formula variables are also weighted. Each personalized model can be represented as a rule (or a set of rules) that represents the personalized profile for the new input vector. The TWNFI model is evolving as new data samples, added to a data set, can be used in any further personalized model development. This includes using different sets of variables and features [60.30, 36].

ECOS have been applied to both model brain functions and as general CI tools [60.30]. In one application, an ECOS was trained to classify EEG data measured from a single person's brain, into four classes representing four perceptual states – hearing, seeing, both, and nothing [60.30]. In another application, ECOS were used to model emerging acoustic clusters, when multiple spoken languages are learned [60.30].

ECOS have been applied to a wide range of CI applications, such as adaptive classification of gene expression data, adaptive robot control, adaptive financial data modeling, adaptive environmental, and social data modeling [60.30].

ECOS are used in Sect. 60.3 for building GRN models.

60.2.2 Spiking Neural Networks – Strongly Brain Inspired Models

Spiking models of a neuron and of neural networks – SNN, have been inspired and developed to mimic more biologically the spiking activity of neurons in the brain when processing information.

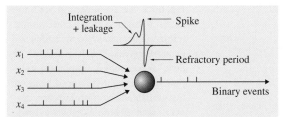

Fig. 60.3 A general representation of a spiking neuron model (after [60.13])

One model – the spike response model (SRM) of a neuron [60.31, 37] is described below and extended in Sect. 60.4 to a CNGM.

A neuron i receives input spikes from presynaptic neurons $j \in \Gamma_i$, where Γ_i is a pool of all neurons presynaptic to neuron i. The state of the neuron i is described by the state variable $u_i(t)$ that can be interpreted as a total postsynaptic potential (PSP) at the membrane of soma (Fig. 60.3). When $u_i(t)$ reaches a firing threshold $\vartheta_i(t)$, neuron i fires, i.e., emits a spike. The value of the state variable $u_i(t)$ is the sum of all postsynaptic potentials, i.e.,

$$u_i(t) = \sum_{j \in \Gamma_i} \sum_{t_j \in F_j} J_{ij} \left(t - t_j - \Delta_{ij}^{ax} \right). \tag{60.4}$$

The weight of the synaptic connection from neuron j to neuron i is denoted by J_{ij}. It takes positive (negative) values for excitatory (inhibitory) connections, respectively. Depending on the sign of J_{ij}, a presynaptic spike generated at time t_j increases (or decreases) $u_i(t)$ by an amount $\varepsilon_{ij}(t - t_j - \Delta_{ij}^{ax})$. Δ_{ij}^{ax} is an axonal delay between neurons i and j which increases with Euclidean distance between neurons.

The positive kernel $\varepsilon_{ij}(t - t_j - \Delta_{ij}^{ax}) = \varepsilon_{ij}(s)$ expresses an individual postsynaptic potential (PSP) evoked by a presynaptic neuron j on neuron i. A double exponential formula can be used

$$\varepsilon_{ij}^{\text{synapse}}(S) A^{\text{synapse}} \left(\exp \left(\frac{s}{\tau_{\text{decay}}^{\text{synapse}}} \right) - \exp \left(-\frac{s}{\tau_{\text{rise}}^{\text{synapse}}} \right) \right). \tag{60.5}$$

The following notations are used above: $\tau_{\text{decay/rise}}^{\text{synapse}}$ are time constants of the rise and fall of an individual PS, A is the PSP's amplitude, and *synapse* represents the type of the activity of the synapse from the neuron j to neuron i that can be measured and modeled separately for *fast_excitation, fast_inhibition,*

slow_excitation, and slow_inhibition, all integrated in formula [60.13]. These types of PSPs are based on neurobiology [60.38] and will be the basis for the development of the computational neurogenetic model in Sect. 60.4, where the different synaptic activities are represented as functions of different proteins (neurotransmitters and neuroreceptors).

External inputs from the input layer are added at each time step, thus incorporating the background noise and/or the background oscillations. Each external input has its own weight $J_{ik}^{\text{ext\_input}}$ and amount of signal $\varepsilon_k(t)$, such that

$$u_i^{\text{ext\_input}}(t) = J_{ik}^{\text{ext\_input}} \varepsilon_{ik}(t). \tag{60.6}$$

It is optional to add some degree of Gaussian noise to the right-hand side of the equation above to obtain a stochastic neuron model instead of a deterministic one.

SNN models can be built with the use of the above spiking neuron model. Spiking neurons within an SNN can be either excitatory or inhibitory. Lateral connections between neurons in an SNN may have weights that decrease in value with distance from neuron i for instance, according to a Gaussian formula, while the connections between neurons themselves can be established at random.

SNN can be used to build biologically plausible models of brain functions. Examples are given in [60.13, 31, 37, 38]. Figure 60.4 graphically shows an application of an SNN to model brain functions that connect signals from the thalamus to the temporal cortex (from [60.13]).

Other applications of SNN include image recognition. In [60.39] an adaptive SNN model is developed where new SNN submodules (maps) are created incrementally to accommodate new data samples over time. For example, a new submodule of several spiking neurons and connections evolves when a new class of objects (e.g., a new face in the case of a face recognition problem) is presented to the system for learning at any time of this process. When there are no active inputs presented to the system, the system merges close spiking neuronal maps depending on their similarity.

Developing new methods for learning in evolving SNN is a challenging direction for future research with a potential for applications in both computational neuroscience and pattern recognition, e.g., multimodal information processing – speech, image, odor, gestures, etc.

SNN are extended to CNGM in Sect. 60.4.

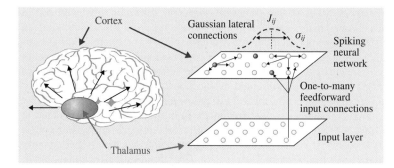

Fig. 60.4 An example of a SNN to model a function of the cortex with internal inputs from the thalamus and external input stimuli. About 20% of $N = 120$ neurons are inhibitory neurons that are randomly positioned on the grid (*filled circles*). External input is random with a defined average frequency (e.g., between 10–20 Hz) (after [60.13])

60.2.3 Open Questions

Further development of brain-like or brain inspired ANN requires some that some questions be addressed:

- How much should an ANN mimic the brain in order to become an efficient CI model?

- How is a balance between structure definition and learning achieved in ANN?
- How can ANN evolve and optimize their parameters and input features over time in an efficient way?
- How can incremental learning in ANN be applied without the presentation of an input signal (e.g., *sleep* learning)?

60.3 Gene Inspired Methods of Computational Intelligence

60.3.1 The Central Dogma in Molecular Biology and GRN

The central dogma of molecular biology states that DNA, which resides in the nucleus of a cell or a neuron, transcribes into RNA and then translates into proteins, which process is continuous, evolving, so that proteins, called transcription factors, cause genes to transcribe, etc. [60.40, 41] (Fig. 60.5).

The DNA is a long, double stranded sequence (a double helix) of millions or billions of 4 base molecules (nucleotides) denoted as A, C, T, and G, which are chemically and physically connected to each other through other molecules. In the double helix, they

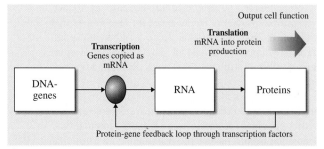

Fig. 60.5 The genes in the DNA transcribe into RNA and then translate into proteins that define the function of a cell. (The central dogma of molecular biology)

make pairs such that every A from one strand is connected to a corresponding T on the opposite strand and every C is connected to a G. A gene is a sequence of hundreds and thousands of bases as part of the DNA that is translated into protein. Only less than 5% of the DNA of the human genome constitutes genes, the other part is a noncoding region that contains useful information as well.

The DNA of each organism is unique and resides in the nucleus of each of its cells. But it is the proteins that are expressed from the genes and define the function of the cell that make a cell alive. The genes and proteins in each cell are connected in a dynamic *GRN* consisting of regulatory pathways.

Normally, only a few hundreds of genes are expressed as proteins in a particular cell. At the transcription phase, one gene is transcribed in many RNA copies and their number defines the expression level of this gene [60.40, 41]. Some genes may be over-expressed, resulting in too much protein in the cell, some genes may be under-expressed resulting in too little protein; in both cases the cell may be functioning in a wrong way, which may be causing a disease. Abnormal expression of a gene can be caused by a gene mutation – a random change in the code of the gene, where a base molecule is either inserted or deleted, or altered into another base molecule. Drugs can be used to stimulate or suppress the expression of certain genes and proteins, but how that will affect indirectly the other genes related to the targeted one must be evaluated and this is where computational modeling of GRN can help.

It is always difficult to establish the interaction between genes and proteins. The question *What will happen with a cell or the whole organism if one gene is under-expressed or missing?* is now being attempted by the use of a technology called *knock-out gene technology*. This technology is based on the removal of a gene sequence from the DNA and letting the cell/organism to develop, where parameters are measured and compared with the parameters when the gene was not missing.

60.3.2 GRN–ANN Models

Modeling GRN is the task of creating a dynamic interaction network between genes that defines the next time expression of genes based on their previous time expression. A detailed discussion of the methods for GRN modeling can be found in [60.41, 43, 44]. Models of GRN, derived from gene expression RNA data, have been developed with the use of different mathematical and computational methods, such as: statistical correlation techniques; evolutionary computation; ANN; differential equations, both ordinary and partial; Boolean models; kinetic models; state-based models; and others [60.41].

A model of GRN, trained on time-course data is presented in [60.42] where the human response to fibroblast serum data is used (Fig. 60.6) and a GRN is extracted from it (Fig. 60.7). The method uses a genetic algorithm to select the initial cluster centers of the time course clustered gene expression values and then applies a Kalman filter to derive the gene connecting equations.

In [60.44] a GRN-ECOS is proposed and applied on small-scale cell line gene expression data. An ECOS is evolved with inputs being the expression level of a certain number of selected genes (e.g., 4) at a time moment (t) and the outputs being the expression level of the same or other genes/proteins at the next time moment ($t + dt$). After an ECOS is trained on time course gene expression data, rules are extracted from the ECOS and linked between each other in terms of their creation in the model, thus representing the GRN. The rule nodes in an ECOS capture clusters of input genes that are related to the output genes/proteins at the next time moment. Figure 60.7 shows an example of EFuNN used for modeling GRN [60.33, 44].

The rules extracted from an EFuNN model, for example, represent the relationship between the gene expression of a group of genes $G(t)$ at a time moment t

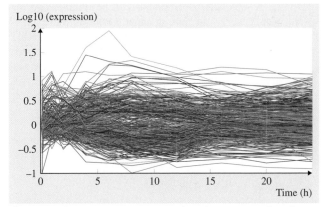

Fig. 60.6 Time-course gene expression data representing the response of thousands of genes of fibroblast to serum (after [60.42])

and the expression of the genes at the next time moment $G(t + dt)$, e.g.,

IF $g_{13}(t)$ is High (0.87) and $g_{23}(t)$ is Low (0.9)

THEN $g_{87}(t + dt)$ is High (0.6) and

$$g_{103}(t + dt) \text{ is Low.} \quad (60.7)$$

Through modifying a threshold for rule extraction one can extract stronger or weaker patterns of a dynamic relationship.

Adaptive training of an ECOS makes incremental learning of a GRN possible, as well as adding new genes to the GRN.

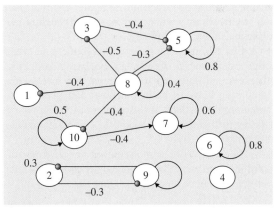

Fig. 60.7 A GRN obtained with the use of the method from [60.42] on the data from Fig. 60.5 after the time gene expression series are clustered into 10 clusters. The nodes represent gene clusters while the arcs represent the dynamic relation (interaction) between these gene groups over consecutive time moments

A set of DENFIS models can be trained, one for each gene g_i, so that an input vector is the expression vector $G(t)$ and the output is a single variable $g_i(t + \mathrm{d}t)$. DENFIS allows for a dynamic partitioning of the input space. Takagi–Sugeno fuzzy rules, which represent the relationship between gene g_i with the rest of the genes, are extracted from each DENFIS model, e.g.,

IF g_1 is $(0.63, 0.70, 0.76)$ and

g_2 is $(0.71, 0.77, 0.84)$ and

g_3 is $(0.71, 0.77, 0.84)$ and

g_4 is $(0.59, 0.66, 0.72)$

THEN $g_5 = 1.84 - 1.26g_1 - 1.22g_2$
$$+ 0.58g_3 - 0.03g_4. \tag{60.8}$$

The ECOS structure from Fig. 60.2 can be used in a multilevel, hierarchical way, where the transcription process is represented in one ECOS and translation in another ECOS, which inputs are connected to the outputs of the first one, using feedback connections to represent transcription factors.

Despite the variety of different methods used so far for modeling GRN and for systems biology in general, there is no single method that will suit all requirements to model a complex biological system, especially to meet the requirements for adaptation, robustness, and information integration.

In the next section GRN modeling is integrated with SNN to model the interaction between genes/proteins in relation to activity of a spiking neuron and an SNN as a whole.

60.4 Computational Neurogenetic Models

60.4.1 General Notions

With the advancement of molecular and brain research technologies more and more data and information are being made available about the genetic basis of some neuronal functions (see, for example, the brain-gene map of a mouse [60.45] and the brain-gene ontology BGO in [60.46]).

This information can be utilized to create biologically plausible ANN models of brain functions and diseases that include models of gene interaction. This area integrates knowledge from computer and information science, brain science, and molecular genetics and it is here called CNGM [60.2].

A CNGM integrates genetic, proteomic, and brain activity data and performs data analysis, modeling, prognosis, and knowledge extraction that reveals the relationship between brain functions and genetic information. Let us look at this process as a process of building mathematical function or a computational algorithm as follows.

A future state of a molecule M' or a group of molecules (e.g., genes and proteins) depends on its current state M and on an external signal Em

$$M' = Fm(M, Em). \tag{60.9}$$

A future state N' of a neuron or an ensemble of neurons will depend on its current state N and on the state of the molecules M (e.g., genes) and on external signals En

$$N' = Fn(N, M, En). \tag{60.10}$$

Finally, a future neuronal state C' of the brain will depend on its current state C and also on the neuronal N and the molecular M state, and on the external stimuli Ec

$$C' = Fc(C, N, M, Ec). \tag{60.11}$$

The above set of equations (or algorithms) is a general one and in different cases it can be implemented differently, e.g., one gene – one neuron/brain function; multiple genes – one neuron/brain function, no interaction between genes; multiple genes – multiple neuron/brain functions, where genes interact in a GRN and neurons also interact in a neural network architecture; multiple genes – complex brain/cognitive function/s, where genes interact within GRN and neurons interact in several hierarchical neural networks.

Several CNGM models have been developed so far, varying from modeling a single gene in a biologically realistic ANN model [60.3] to modeling a set of genes forming an interaction GRN [60.13,43]. In the next section we give an example of a CNGM that combines SNN and GRN into one model [60.13].

60.4.2 A Computational Neurogenetic Model that Integrates GRN Within an SNN Model

The main idea behind the model proposed in [60.2] is that interaction of genes in neurons affect the dynam-

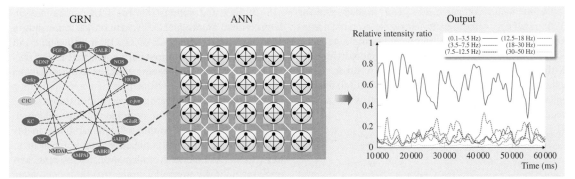

Fig. 60.8 A CNGM, where a GRN is used to represent the interaction of genes, and a SNN is employed to model a brain function. The model output is compared against real brain data for validation of the model and for verifying the derived gene interaction GRN after model optimization is applied [60.13]

ics of the whole ANN through neuronal parameters, which are no longer constant but change as a function of gene/protein expression. Through optimization of the GRN, the initial gene/protein expression values, and the ANN parameters, particular target states of the ANN can be achieved, so that the ANN can be tuned to model real brain data in particular.

This idea is illustrated in Fig. 60.8. The behavior of the SNN is evaluated by means of the local field potential (LFP), thus making it possible to attempt modeling the role of genes in different brain states, where EEG data is available to test the model. A standard FFT signal processing technique is used to evaluate the SNN output and to compare it with real human EEG data. A broader theoretical and biological background of CNGM construction is given in [60.13].

In general, we consider two sets of genes – a set G_{gen} that relates to general cell functions and a set G_{spec} that defines specific neuronal information-processing functions (receptors, ion channels, etc.). The two sets together form a set $G = \{G_1, G_2, \ldots, G_n\}$. We assume that the expression level of each gene is a nonlinear function of expression levels of all the genes in G

$$g_j(t + \Delta t') = \sigma \left(\sum_{k=1}^{n} w_{jk} g_k(t) \right). \quad (60.12)$$

In [60.13] it is assumed that:

1. One protein is coded by one gene.
2. The relationship between the protein level and the gene expression level is nonlinear.
3. Protein levels lie between the minimal and maximal values. Thus, the protein level is expressed by

$$p_j(t + \Delta t) = \left(p_j^{\text{max}} - p_j^{\text{min}} \right)$$
$$\times \sigma \left(\sum_{k=1}^{n} w_{jk} g_k(t) \right) + p_j^{\text{min}}. \quad (60.13)$$

The delay constant introduced in the formula corresponds to the delay caused by the gene transcription, mRNA translation into proteins and posttranslational protein modifications, and also the delay caused by gene transcription regulation by transcription factors.

Some proteins and genes are known to affect the spiking activity of a neuron represented in an SNN model by neuronal parameters, such as *fast_excitation, fast_inhibition, slow_excitation,* and *slow_inhibition* (Sect. 60.2). Some neuronal parameters and their correspondence to particular proteins are summarized in Table 60.1.

Besides genes directly affecting the spiking dynamics of a neuron, a GRN model can include other genes relevant to a problem in hand, e.g., modeling a brain function or a brain disease. In [60.13] these genes/proteins are c-jun, mGLuR3, Jerky, BDNF, FGF-2, IGF-I, GALR1, NOS, and S100beta [60.13].

The goal of the CNGM in Fig. 60.8 is to achieve a desired SNN output through optimization of the model parameters. The LFP of the SNN, defined as $\text{LFP} = (1/N)\Sigma u_i(t)$, by means of FFT is evaluated in order to compare the SNN output with the EEG signal analyzed in the same way. It has been shown that brain LFPs in principle have the same spectral characteristics as EEG [60.47].

In order to find an optimal GRN within the SNN model, so that the frequency characteristics of the LFP of the SNN model are similar to the brain EEG

Table 60.1 Neuronal parameters and related proteins (PSP; AMPAR: (amino-methylisoxazole-propionic acid) ampa receptor; NMDAR: (*N*-methyl-D-aspartate acid) NMDA receptor; GABRA: (gamma-aminobutyric acid) GABA$_A$ receptor; GABRB: GABA$_B$ receptor; SCN: sodium voltage-gated channel; KCN: kalium (potassium) voltage-gated channel; CLC: chloride channel; PV: parvalbumin)

Neuronal parameter amplitude and time constants of	Protein
Fast excitation PSP	AMPAR
Slow excitation PSP	NMDAR
Fast inhibition PSP	GABRA
Slow inhibition PSP	GABRB
Firing threshold	SCN, KCN, CLC
Late excitatory PSP through GABRA	PV

characteristics, the following evolutionary computation procedure is used:

1. Generate a population of CNGMs, each with randomly, but constrained, generated values of coefficients for the GRN matrix **W**, initial gene expression values $g(0)$, initial values of SNN parameters $P(0)$, and different connectivity.
2. Run each SNN model over a period of time T and record the LFP.
3. Calculate the spectral characteristics of the LFP using FFT.
4. Compare the spectral characteristics of SNN LFP to the characteristics of the target EEG signal. Evaluate the closeness of the LFP signal for each SNN to the target EEG signal characteristics. Proceed further according to the standard GA algorithm to find a SNN model that matches the EEG spectral characteristics better than previous solutions.
5. Repeat steps 1 to 4 until the desired GRN and SNN model behavior is obtained.
6. Analyze the GRN and the SNN parameters for significant gene patterns that cause the SNN model to manifest similar spectral characteristics as the real data.

The proposed CNGM modeling framework can be used to find patterns of gene regulation related to brain functions. In [60.13] some preliminary results of analysis performed on real human interictal EEG data are presented. The model performance and the real EEG data are compared for the following relevant to the problem subbands: delta (0.5–3.5 Hz), theta (3.5–7.5 Hz), alpha (7.5–12.5 Hz), beta 1 (12.5–18 Hz), beta 2

(18–30 Hz), and gamma (above 30 Hz). This particular SNN had an evolved GRN with only 5 genes out of 16 (s100beta, GABRB, GABRA, mGLuR3, c-jun), all other genes having constant expression values. A GRN is obtained that has a meaningful interpretation and can be used to model what will happen if a gene/protein is suppressed by administering a drug, for example.

In *evolving CNGM* new genes can be added to the GRN model at a certain time, in addition to the new spiking neurons and connections created incrementally, as is the case in *evolving SNN*. Developing new evolving CNGM to model brain functions and brain diseases such as epilepsy, Alzheimer's, Parkinson's disease, schizophrenia, mental retardation, and others is a challenging problem for future research [60.13, 43].

60.4.3 Open Questions

Some questions emerged from the first CNGM experiments:

- How many different GRNs would lead to similar LFPs and what do they have in common?
- What neuronal parameters should be included in an ANN model and how can they be linked to activities of genes/proteins?
- What genes/proteins should be included in the model and can the gene interaction be represented over time within each neuron?
- How can the output activity of the ANN and the genes be integrated in time, as it is known that neurons spike in millisecond intervals and the process of gene transcription and translation into proteins takes minutes?
- How can a CNGM be created and evaluated in a situation of insufficient data?
- How can brain activity and the CNGM activity be measured in order to validate the model?
- What useful information (knowledge) can be derived from a CNG model?
- How can a CNGM model be adapted incrementally in a situation of new incoming data about brain functions and genes related to them?

Integrating principles from gene and neuronal information processing in a single ANN model raises many other, more general, questions that need to be addressed in the future, for example:

- Is it possible to create a truly adequate CNGM of the whole brain? Would gene-brain maps help in this respect [60.3]?

- How can dynamic CNGM be used to trace over time and predict the progression of a brain diseases, such as epilepsy and Parkinson's?
- How can CNGM be used to model gene mutation effects?

- How can CNGM be used to predict drug effects?
- How can CNGM help us to understand brain functions better, such as memory and learning?
- What CI problems can be efficiently solved with the use of a brain-gene inspired ANN?

60.5 Quantum Inspired CI

60.5.1 Quantum Level of Information Processing

At the quantum level, particles (e.g., atoms, electrons, ions, photons, etc.) are in a complex evolving state all the time. The atoms are the material that everything is made of. They can change their characteristics due to the frequency of external signals. Quantum computation is based upon physical principles from the theory of quantum mechanics [60.48].

One of the basic principles is the *linear superposition* of states. At a macroscopic or classical level a system exists only in a single basis state as energy, momentum, position, spin, and so on. However, at a microscopic or quantum level a quantum particle (e.g., atom, electron, positron, ion) or a quantum system is in a superposition of all possible basis states. At the microscopic level any particle can assume different positions at the same time moment, can have different values of energy, can have different spins, and so on. This *superposition* principle is counterintuitive because in classical physics one particle has only one position, energy, spin, etc.

If a quantum system interacts in any way with its environment, the superposition is assumed to be destroyed and the system *collapses* into one single real state as in the classical physics (Heisenberg). This process is governed by a probability amplitude. The square of the intensity for the probability amplitude is the quantum probability to observe the state.

Another quantum mechanics principle is *entanglement* – two or more particles, regardless of their location, are in the same state with the same probability function. The two particles can be viewed as *correlated*, undistinguishable, *synchronized*, coherent. An example is a laser beam consisting of millions of photons having the same characteristics and states.

Quantum systems are described by a probability density ψ that exists in a Hilbert space. The Hilbert space has a set of states $|\varphi_i\rangle$ forming a basis. A system can exist in a certain quantum state $|\psi\rangle$, which is defined as

$$|\psi\rangle = \sum c_i |\varphi_i\rangle, \quad \sum |c_i|^2 = 1; , \qquad (60.14)$$

where the coefficients c_i may be complex. $|\psi\rangle$ is said to be in a superposition of the basis states $|\varphi_i\rangle$. For example, the quantum inspired analog of a single *bit* in classical computers can be represented as a *qu-bit* in a quantum computer

$$|x\rangle = a|0\rangle + b|1\rangle; , \qquad (60.15)$$

where $|0\rangle$ and $|1\rangle$ represent the states 0 and 1, and a and b their probability amplitudes, respectively. The *qu-bit* is not a single value entity, but is a function of parameters whose values are complex numbers. After the loss of coherence the *qu-bit* will *collapse* into one of the states $|0\rangle$ or $|1\rangle$ with the probability a^2 for the state $|0\rangle$ and probability b^2 for the state $|1\rangle$.

The state of a quantum particle (represented, for example, as a *qu-bit)* can be changed by an operator called a *quantum gate*. A quantum gate is a reversible gate and can be represented as a unitary operator U acting on the *qu-bit* basis states. The defining property of a unitary matrix is that its conjugate transpose is equal to its inverse. Several quantum gates have been introduced, such as the NOT gate, controlled NOT gate, rotation gate, Hadamard gate, etc. [60.49–52].

60.5.2 Why Quantum Inspired CI?

Quantum mechanical computers and quantum algorithms try to exploit the massive quantum parallelism which is expressed in the principle of *superposition*. The principle of superposition can be applied to many existing methods of CI, where instead of a single state (e.g., a parameter value, or a finite automaton state, or a connection weight, etc.) a superposition of states will be used, described by a wave probability function, so that all these states will be computed in parallel, resulting in an increased speed of computation by many orders of magnitude [60.5, 8, 9, 49–57].

Quantum mechanical computers were proposed in the early 1980s and a description was formalized in the late 1980s. These computers, when implemented, are expected to be superior to classical computers in various specialized problems. Much effort has been made to extend the principal ideas of quantum mechanics to other fields of interest. There are well-known quantum algorithms such as Shor's quantum factoring algorithm [60.58] and Grover's database search algorithm [60.50, 54].

The advantage of quantum computing is that while a system is *uncollapsed* it can carry out more computing than a collapsed system, because, in a sense, it is computing in *many universes* at once. The above quantum principles have inspired research in both computational methods and brain study.

New theories (some of them speculative at this stage) have already been formulated. For example, *Penrose* [60.8, 9] argues that solving the quantum measurement problem is prerequisite for understanding the mind and that consciousness emerges as a macroscopic quantum state due to a coherence of quantum-level events within neurons.

60.5.3 Quantum Inspired Evolutionary Computation and Connectionist Models

Quantum inspired methods of evolutionary computation (QIEC) and other techniques were proposed and discussed in [60.51, 55]. They include genetic programming [60.59], particle swarm optimizers [60.60], finite automata and Turing machines, etc.

In QIEC, a population of n *qu-bit* individuals at time t can be represented as

$$Q(t) = \left\{ q_1^t, q_2^t, \ldots q_n^t \right\} , \qquad (60.16)$$

where n is the size of the population.

Evolutionary computing with *qu-bit* representation has a better characteristic of population diversity than other representations, since it can represent linear superposition of states probabilistically. The *qu-bit* representation leads to a quantum parallelism of the system as it is possible to evaluate the fitness function on a superposition of possible inputs. The output obtained is also in the form of superposition, which needs to be *collapsed* to obtain the actual solution.

Recent research activities have focussed on using quantum principles for ANN [60.4, 5, 61–63]. Considering quantum ANN seems to be important for at least two reasons. There is evidence for the role that quantum processes play in the living brain. *Penrose* argued that a new physics binding quantum phenomena with general relativity can explain such mental abilities as *understanding*, *awareness*, and *consciousness* [60.9]. The second motivation is the possibility that the field of classical ANN could be generalized to the promising new field of quantum computation [60.53]. Both considerations suggest a new understanding of mind and brain functions, as well as new unprecedented abilities in information processing. *Ezhov* and *Ventura* consider quantum neural networks as the next natural step in the evolution of neurocomputing systems [60.4].

Several quantum inspired ANN models have been proposed and illustrated on small examples. In [60.63] QIEA is used to train a MLP ANN. *Narayanan* and *Meneer* simulated classical and quantum inspired ANN and compared their performances [60.5]. Their work suggests that there are, indeed, certain types of problems for which quantum neural networks will prove superior to classical ones.

Other relevant work includes quantum decision making, quantum learning models [60.64], quantum networks for signal recognition [60.62], and quantum associative memory [60.61, 65]. There are also recent approaches to quantum competitive learning where the quantum system's potential for excellent performance is demonstrated on real-world data sets [60.66, 67].

60.6 Towards the Integration of Brain, Gene, and Quantum Information Processing Principles: A Conceptual Framework for Future Research

60.6.1 Quantum Inspired SNN

In Sect. 60.4 we presented a CNGM that integrated principles from neuronal information processing and gene information processing in the form of integrating SNN with GRN. Following some ideas from QI-ANN, we

can expect that *QI-SNN* and *QI-CNGM* would open new possibilities for modeling gene–neuron interactions related to brain functions and to new efficient AI applications.

The CNGM from Sect. 60.4 linked principles of information processing in gene/protein molecules with neuronal spiking activity, and then – to the information processing of a neuronal ensemble, that is measured as local field potentials (LFP). How the quantum information processes in the atoms and particles (ions, electrons, etc.), that make the large gene/protein molecules, relate to the spiking activity of a neuron and to the activity of a neuronal ensemble, is not known yet and it is a challenging question for the future.

What is known at present, is that the spiking activity of a neuron relates to the transmission of ions and neurotransmitter molecules across the synaptic clefts and to the emission of spikes. Spikes, as carriers of information, are electrical signals made of particles that are emitted in one neuron and transmitted along the nerves to many other neurons. These particles are characterized by their quantum properties. So, quantum properties may influence, under certain conditions, the spiking activity of neurons and of the whole brain, as brains obey the laws of quantum mechanics (as everything else in the material world does).

Similarly to a chemical effect of a drug to the protein and gene expression levels in the brain, which may affect the spiking activity and the functioning of the whole brain (modeling of these effects is subject of the computational neurogenetic modeling), external factors like radiation, light, high frequency signals, etc., can influence the quantum properties of the particles in the brain through gate operators. According to *Penrose* [60.9] microtubules in the neurons are associated with quantum gates, even though what constitutes a quantum gate in the brain is still a highly speculative topic.

So, the question is: *Is it possible to create an SNN model and a CNGM that incorporate some quantum principles?*

A *QI-SNN* can be developed as an extension of the concept of evolving SNN [60.39] using the superposition principle, where instead of many SNN maps, each representing one object (e.g., a face), there will be a single SNN, where both connections and neurons are represented as particles, being in many states at the same time defined as probability wave function. When an input vector is presented to the QI-SNN, the network collapses in a single SNN defining the class of the recognized input vector.

60.6.2 A Conceptual Framework of a QI-CNGM

Here we extend the concept of CNGM (60.9)–(60.11) by introducing the level of quantum information pro-

cessing. This results in a conceptual and hypothetical QI-CNGM.

The following is a list of equations that include quantum particle states and functions (hypothetical at this stage) into (60.9)–(60.11) and (60.18)–(60.20), starting with a new (60.17) that is concerned only with the level of quantum particle states.

A future state Q' of a particle or a group of particles (e.g. ions, electrons, etc.) depends on the current state Q and on the frequency spectrum Eq of an external signal, according to the Max Planck constant

$$Q' = Fq\,(Q, Eq)\,. \tag{60.17}$$

A future state of a molecule M' or a group of molecules (e.g., genes, proteins) depends on its current state M, on the quantum state Q of the particles, and on an external signal Em:

$$M' = Fm\,(Q, M, Em)\,. \tag{60.18}$$

A future state N' of a spiking neuron or an ensemble of neurons will depend on its current state N, on the state of the molecules M, on the state of the particles Q, and on external signals En

$$N' = Fn\,(N, M, Q, En)\,. \tag{60.19}$$

Finally, a future neuronal state C' of the brain will depend on its current state C and also on the neuronal N, on the molecular M, and on the quantum Q states of the brain:

$$C' = Fc\,(C, N, M, Q, Ec)\,. \tag{60.20}$$

The above hypothetical model of integrated function representations is based on the following assumptions:

- A large number of atoms are characterized by the same quantum properties, possibly related to the same gene/protein expression profile of a large number of neurons characterized by spiking activity that can be represented as a function.
- A large neuronal ensemble can be represented by a single LFP function.
- A cognitive process can be represented, at an abstract level, as a function Fc that depends on all lower levels of neuronal, genetic, and quantum activities.

60.6.3 Open Questions

Several reasons can be given in support of the research on integrating principles from quantum, molecular, and brain information processing into future CI models:

- This may lead to a better understanding of neuronal, molecular, and quantum information processes.
- This may lead to new computer devices – a million times faster and more accurate than the current ones.

- At the nanolevel of microelectronic devices, quantum processes would have a significant impact and new methods of computation would be needed anyway.

60.7 Conclusions and Directions for Further Research

This chapter presents some CI models inspired by principles from different levels of information processing in the brain – including neuronal level, gene/protein level, and quantum level, and argues that CI models that integrate principles from different levels of information processing would be useful tools for a better understanding of brain functions and for the creation of more powerful methods and systems of computational intelligence.

Many open questions need to be answered in the future, some of these are:

- How do quantum processes affect the functioning of a living system in general?
- How do quantum processes affect cognitive and mental functions?
- Is it true that the brain is a quantum machine – working in a probabilistic space with many states (e.g., thoughts) being in a superposition all the time and it is only when we formulate our thought through speech or writing that the brain *collapses* in a single state?
- Is fast pattern recognition in the brain, involving far away segments, a result of both parallel spike transmissions and particle entanglement?
- Is communication between people and between living organisms in general a result of entanglement processes?

- How does the energy in the atoms relate to the energy of the proteins, the cells, and the whole brain?
- Would it be beneficial to develop different QI computational intelligence techniques, such as QI-SVM, QI-GA, QI-decision trees, QI-logistic regression, QI-cellular automata, and QI-ALife?
- How do we implement QI computational intelligence algorithms in order to benefit from their high speed and accuracy? Should we wait for the quantum computers to be realized many years from now, or we can implement them efficiently on specialized computing devices based on classical principles of physics?

Further directions in our research are:

- Building a brain–gene-quantum ontology system that integrates facts, information, knowledge, and CI models of different levels of information processing in the brain and their interaction.
- Building novel brain, gene, and quantum inspired CI models, studying their characteristics, and interpreting the results.
- Applying the new methods to solving complex CI problems in neuroinformatics and brain diseases, bioinformatics and cancer genetics, multimodal information processing, and biometrics.

References

60.1 C. Bishop: *Neural Networks for Pattern Recognition* (Oxford Univ. Press, Oxford, UK 1995)

60.2 N. Kasabov, L. Benuskova: Computational neurogenetics, Int. J. Theor. Comput. Nanosci. **1**(1), 47–61 (2004)

60.3 G. Marcus: *The Birth of the Mind: How a Tiny Number of Genes Creates the Complexity of the Human Mind* (Basic, New York 2004)

60.4 A. Ezhov, D. Ventura: Quantum neural networks. In: *Future Directions for Intelligent Systems and Information Sciences*, ed. by N. Kasabov (Springer, Berlin, Heidelberg 2000) pp. 213–234

60.5 A. Narayanan, T. Meneer: Quantum artificial neural network architectures and components, Inf. Sci. **128**, 199–215 (2000)

60.6 D.B. Fogel: *Evolutionary Computation – Toward a New Philosophy of Machine Intelligence* (IEEE, New York 1995)

60.7 X. Yao: Evolutionary artificial neural networks, Int. J. Neural Syst. **4**(3), 203–222 (1993)

60.8 R. Penrose: *Shadows of the Mind. A Search for the Missing Science of Conscious* (Oxford Univ. Press, Oxford 1994)

60.9 R. Penrose: *The Emperor's New Mind* (Oxford Univ. Press, Oxford 1989)

60.10 S. Amari, N. Kasabov: *Brain-like Computing and Intelligent Information Systems* (Springer, New York 1998)

60.11 M. Arbib: *Brains, Machines and Mathematics* (Springer, Berlin 1987)

60.12 M. Arbib (Ed.): *The Handbook of Brain Theory and Neural Networks* (MIT, Cambridge 2003)

60.13 L. Benuskova, N. Kasabov: *Towards Computational Neurogenetic Modelling* (Springer, New York 2007)

60.14 G. Carpenter, S. Grossberg, N. Markuzon, J.H. Reynolds, D.B. Rosen: Fuzzy ARTMAP: A neural network architecture for incremental supervised learning of analogue multi-dimensional maps, IEEE Trans. Neural Netw. **3**(5), 698–713 (1991)

60.15 G. Carpenter, S. Grossberg: *Pattern Recognition by Self-Organizing Neural Networks* (The MIT, Cambridge, USA 1991)

60.16 N. Kasabov: Foundations of neural networks. In: *Fuzzy Systems and Knowledge Engineering* (MIT Press, MA 1996)

60.17 T. Kohonen: *Self-Organizing Maps* (Springer, Cambridge 1997)

60.18 E. Rolls, A. Treves: *Neural Networks and Brain Function* (Oxford Univ. Press, Oxford 1998)

60.19 F. Rosenblatt: *Principles of Neurodynamics* (Spartan Books, New York 1962)

60.20 D.E. Rumelhart, G.E. Hinton, R.J. Williams (Eds.): Learning Internal Representations by Error Propagation, Parallel Distrib, Processing: Explorations in the Microstructure of Cognition (MIT/Bradford Books, Cambridge 1986)

60.21 G.A. Rummery, M. Niranjan: *On-line Q-learning Using Connectionist System* (Cambridge Univ. Press, Cambridge 1994), 166 pp., CUED/F-INENG/TR

60.22 S. Schaal, C. Atkeson: Constructive incremental learning from only local information, Neural Comput. **10**, 2047–2084 (1998)

60.23 S. Grossberg: *Studies of Mind and Brain* (Reidel, Boston 1982)

60.24 J.G. Taylor: *The Race for Consciousness* (MIT, Cambridge 1999)

60.25 N. Kasabov: Evolving fuzzy neural networks – algorithms, applications and biological motivation. In: *Methodologies for the Conception, Design and Application of Soft Computing*, ed. by T. Yamakawa, G. Matsumoto (World Scientific, Singapore 1998) pp. 271–274

60.26 B. Fritzke: A growing neural gas network learns topologies, Adv. Neural Inf. Process. Syst. **7**, 625–632 (1995)

60.27 J. Platt: A resource allocating network for function interpolation, Neural Comput. **3**, 213–225 (1991)

60.28 J. Freeman, D. Saad: On-line learning in radial basis function networks, Neural Comput. **9**(7), 1601–1622 (1997)

60.29 T. Poggio: Regularization theory, radial basis functions and networks. In: *From Statistics to Neural Networks: Theory and Pattern Recognition Applications*, NATO ASI Series, Vol. 136, ed. by V. Cherkassky, J.H. Friedman, H. Wechsler (NATO, Les Arcs 1994) pp. 83–104

60.30 N. Kasabov: *Evolving Connectionist Systems: The Knowledge Engineering Approach* (Springer, London 2007)

60.31 W. Maass, C.M. Bishop (Eds.): *Pulsed Neural Networks* (The MIT, Cambridge 1999)

60.32 N. Kasabov, Q. Song: DENFIS: Dynamic, evolving neural-fuzzy inference systems and its application for time-series prediction, IEEE Trans. Fuzzy Syst. **10**, 144–154 (2002)

60.33 N. Kasabov: Evolving fuzzy neural networks for online supervised/unsupervised, knowledge – based learning, SMC B: Cybern. **31**(6), 902–918 (2001)

60.34 T. Yamakawa, H. Kusanagi, E. Uchino, T. Miki: A new effective algorithm for neo fuzzy neuron model, Proc. Fifth IFSA World Congr. (IFSA, Seoul, Korea 1993) pp. 1017–1020

60.35 L.A. Zadeh: Fuzzy Sets, Inf. Control. **8**, 338–353 (1965)

60.36 Q. Song, N. Kasabov: TWNFI – a transductive neuro-fuzzy inference system with weighted data normalisation for personalised modelling, Neural Netw. **19**(10), 1591–1596 (2006)

60.37 W. Gerstner, W.M. Kistler: *Spiking Neuron Models* (Cambridge Univ. Press, Cambridge 2002)

60.38 A. Destexhe: Spike-and-wave oscillations based on the properties of $GABA_B$ receptors, J. Neurosci. **18**, 9099–9111 (1998)

60.39 S. Wysoski, L. Benuskova, N. Kasabov: On-line learning with structural adaptation in a network of spiking neurons for visual pattern recognition, Artificial Neural Networks – ICANN 2006 **4131**, 61–70 (2006)

60.40 C. Brown, M. Shreiber, B. Chapman, G. Jacobs: Information science and bioinformatics. In: *Future Directions of Intelligent Systems and Information Sciences*, ed. by N. Kasabov (Physica, Heidelberg 2000) pp. 251–287

60.41 D.S. Dimitrov, I. Sidorov, N. Kasabov: Computational biology. In: *Handbook of Theoretical and Computational Nanotechnology*, Vol. 1, ed. by M. Rieth, W. Schommers (American Scientific, Stevenson Ranch 2004), Chap. 21

60.42 Z. Chan, N. Kasabov, L. Collins: A two-stage methodology for gene regulatory network extraction from time-course gene expression data, Expert Syst. Appl. **30**(1), 59–63 (2006)

60.43 H. Chin, S. Moldin (Eds.): *Methods in Genomic Neuroscience* (CRC, Boca Raton 2001)

60.44 N. Kasabov, S.H. Chan, V. Jain, I. Sidirov, S.D. Dimitrov: Gene Regulatory Network Discovery from Time-Series Gene Expression Data – A Computational Intelligence Approach, LNCS **3316**, 1344–1353 (2004)

60.45   Allen Brain Institute: http://alleninstitute.org
60.46   The Knowledge Engineering and Discovery Research Institute (KEDRI), Auckland University of Technology: http://www.kedri.info
60.47   W. Freeman: *Neurodynamics* (Springer, London 2000)
60.48   R.P. Feynman, R.B. Leighton, M. Sands: *The Feynman Lectures on Physics* (Addison-Wesley Publishing Company, Massachusetts 1965)
60.49   T. Hey: Quantum computing: An introduction. In, Comput. Control Eng. J. **10**(3), 105–112 (1999)
60.50   T. Hogg, D. Portnov: Quantum optimization, Inf. Sci. **128**, 181–197 (2000)
60.51   J.-S. Jang, K.-H. Han, J.-H. Kim: Quantum-inspired evolutionary algorithm-based face verification, LNCS **2724**, 2147–2156 (2003)
60.52   S.C. Kak: *Quantum Neural Computation, Research Report* (Louisiana State Univ., Baton Rouge 1995)
60.53   M. Brooks: *Quantum Computing and Communications* (Springer, Berlin, Heidelberg 1999)
60.54   L.K. Grover: A fast quantum mechanical algorithm for database search, STOC '96: Proc. Twenty-Eighth Ann. ACM Symp. Theory Comput. (ACM, New York, USA 1996) pp. 212–219
60.55   K.-H. Han, J.-H. Kim: Quantum-inspired evolutionary algorithms with a new termination criterion, H gate, and two phase scheme, IEEE Trans. Evol. Comput. **8**(2), 156–169 (2004)
60.56   G.E. Hinton: Connectionist learning procedures, Artif. Intell. **40**, 185–234 (1989)
60.57   M.A. Perkowski: Multiple-valued quantum circuits and research challenges for logic design and computational intelligence communities, IEEE Comput. Intell. Soc. Mag. **2005**, 6–12 (2005)

60.58   P.W. Shor: Polynomial-time algorithms for prime factorization and discrete logarithms on a quantum computer, SIAM J. Comput. **26**, 1484–1509 (1997)
60.59   L. Spector: *Automatic Quantum Computer Programming: A Genetic Programming Approach* (Kluwer Academic, Boston 2004)
60.60   J. Liu, W. Xu, J. Sun: Quantum-Behaved Particle Swarm Optimization with Mutation Operator, 17th IEEE Int. Conf. Tools Artif. Intell. (ICTAI'05) (2005)
60.61   C.A. Trugenberger: Quantum pattern recognition, Quantum Inf. Process. **1**, 471–493 (2002)
60.62   X.-Y. Tsai, H.-C. Huang, S.-J. Chuang: Quantum NN vs. NN in signal recognition, in: ICITA'05, Proc. Third Int. Conf. Inf. Technol. Appl. (ICITA'05), Vol. 2 (IEEE Computer Society, Washington, DC, USA 2005) pp. 308–312
60.63   G.K. Venayagamoorthy, S. Gaurav: Quantum-inspired evolutionary algorithms and binary particle swarm optimization for training MLP and SRN neural networks, J. Theor. Comput. Nanosci. **2**, 561–568 (2005)
60.64   N. Kouda, N. Matsui, H. Nishimura, F. Peper: Qu-bit neural network and its learning efficiency, Neural Comput. Appl. **14**, 114–121 (2005)
60.65   D. Ventura, T. Martinez: Quantum associative memory, Inf. Sci. Inf. Comput. Sci. **124**, 273–296 (2000)
60.66   D. Ventura: Implementing competitive learning in a quantum system. In, Proc. Int. Jt. Conf. Neural Netw. (IEEE 1999)
60.67   G. Xie, Z. Zhuang: A quantum competitive learning algorithm, Liangzi Dianzi Xuebao/Chin. J. Quantum Electron. (China) **20**, 42–46 (2003)

61. The Brain and Creativity

Francesco C. Morabito, Giuseppe Morabito, Matteo Cacciola, Gianluigi Occhiuto

Modern abstract art is considered complex and the extraction of meaning from some works of art is largely controversial. However, some artists have explicitly tried to produce paintings in accordance with specific goals. This means that behind their artwork there is a project realized through creativity. Their paintings clearly reflect those efforts and are able to show the emergence of complex ideas reproducing a non-linear and uncertain world. This chapter investigates the link between brain-states of a subject's perception of art with the complexity of the art. More than 25 paintings of famous artists of modern art are studied and evaluated. The concept of artistic complexity, C_A has been introduced as a metric for assessing the complexity of paintings of different artists. The results achieved have been compared to the saliency maps earlier introduced in computer vision as computational models of bottom-up VA. The measure proposed is based on an interplay between top-down and bottom-up approaches,

manifesting the difficulty of the human brain in extracting invariants from some abstract representations. The intriguing relationships shown may offer a paradigm for testing novel computational models on brain-like machines. The methodologies described are likely to be of interest for multimedia quality assessment as metrics able to emulate the integral mechanisms of human visual systems as well as to correlate well with visual perception of quality.

Part L | 61

61.1 Creativity and Perception of Art

Creativity is one of the most relevant qualities of the human brain. The arts are remarkable activities requiring creativity, in addition to education and technical knowledge. Recently, a number of researchers have hypothesized interesting relationships between art and neuroscience, aiming to extract from this interplay cues about neural correlates of creativity [61.1, 2]. In this work, the study of artistic oeuvres, namely exceptional paintings, is driven by the idea of objectively measuring their complexity. The visual quality of some aspects of the images has been measured in different technological contexts, namely in multimedia, and complexity is one such aspects. The quantitative analysis carried out in this work has net yet been corroborated by any fMRI experiments; it merely exploits some image processing concepts. Accordingly, the paintings are treated as images, i.e., as two-dimensional vectors.

Investigating the relationships between the perception and information processing of art through visual attention (VA), thus, between the creative brain and the fine arts, can be useful to advance knowledge of the psycho-physiological mechanisms underlying the interaction between VA and quality perception. It has been suggested that only the unexpected at one stage of processing is transmitted to the next stage [61.3]. The sensory cortex may have evolved to adapt to and to relax in front of statistical regularities [61.4]. At higher levels of abstraction, non-repetitive and novel stimulus are also relevant to learning and memory formation [61.5]. At a behavior level, it is well known that the strongest attractor of attention are stimuli that pop-up from their neighbors in space (or time), like a transverse bar embedded within an array of horizontal bars [61.6], like in the artistic works of Piet Mondrian, or an abrupt onset of a bright spot in a regular display. Many computational techniques are still using purely bottom-up VA models even though they are known to not perform well in complex scenarios. *Korner* and *Matsumoto* [61.7] earlier proposed a model of neocortical computation based on bidirectional processing: the recognition process functionally consists of a recurrent two-way multi-hierarchical procedure that is initiated by a forward categorization (generated by a retinotopic input representation in a metric of the external world) that activates an initial hypothesis at each level of the processing hierarchy. This non-retinotopical representation provides top-down predictive feedback to verify the hypothesis. The *top-down bottom-up* process creates a consistent description of the input in terms of already memorized knowledge. For the categorization step, bottom-up processing can be sufficient; for recognition and in complex segmentation problems, bidirectional processing is required. In the hierarchical model of the cortex, the interpretation of a scene is done through a layered collection of cells that encode the building blocks and lead to an invariant representation of the objects in the scene. The interpretation of the scene is obtained from the sensory input data flowing up to the cortical hierarchy through successive sub-processing steps, but also from our learned invariant representations that await the data while simultaneously guiding the recognition process from above. Art and, particularly, modern art, through its explicit tendency toward abstraction, stands in between the top-hierarchic level of the invariant rep-

resentation and the essential raw sensory data of the incoming input. A color-field painting brings us into the viewing experience, and the lack of categories guided from the raw sensory input needs to activate some mid-hierarchical levels of representation in the Korner–Matsumoto scheme. It implies a strong activation of the essential features of the piece at hand, like the detection of edges (Mondrian) or the suggested movement of dynamical effect (Kandinsky), in our visual system. In our brain's attitude of problem-solving and semantic extraction, the mid-hierarchical sub-representations are connected to some invariant representations in the top-hierarchical levels: this way, the meaning is generated from our internal memory stores in an interplay with the top-down process. The artistic abstractions involve the viewer's own memory (as well as *familiarity*, according to *Forsythe* et al. [61.8]) and active imagination to a greater degree [61.9]. Abstraction requires more top-down processing than a bottom-up feed-forward pattern of activation. In terms of neuroinformatics and machine learning, this is achieved by sparse representation and unsupervised learning algorithms. Through the mechanisms of art generation and fruition it is possible to gain insight on how information is efficiently handled by the brain. Complexity is just one aspect of visual quality. Since recent years have witnessed an increased interest in easy-to-use and accurate image and video quality assessment tools in order to improve the perceptual quality of multimedia content in a variety of applications, the techniques presented here can be of interest with regards to quality of service (QoS) of delivered content [61.10]. In the next sections, the concept of artistic complexity is introduced and discussed and several results regarding paintings are proposed. The resulting metric is then compared to the saliency metric of computational bottom-up models.

61.2 Picture Complexity Metrics

Perceptual metrics are relevant in image and video processing systems, in terms of both complexity and quality. These metrics aim to emulate the mechanism of perception and assessment of the human visual system and of the brain. One relevant property of the perception system is VA. In turn, VA is intuitively correlated to the complexity of the scene under inspection also in terms of some invariants that can be extracted for the scene interpretation. The major aspects of VA derive from higher cognitive processing, deployed in order to reduce the complexity of scene analysis. Accordingly,

the extraction of visual information implies shifting the focus of attention across the scene visualized to the most significant objects and their evident correlations. Not all the objects in a scene draw the attention of the viewer to the same degree: common objects are easily recognized and categorized in a scene independently from the perspective and possible partial occlusions. A visual scene, in particular a picture, is gradually inspected by shifting the focus point through saccadic movements in a search of the most relevant information in the context. From an evolutionary perspective, the most significant stimuli in a scene are favored over the less relevant ones. VA is directed by bottom-up and top-down mechanisms. Bottom-up attention is signal-driven and task-independent. It is dominated by low-level features that can be experienced as visually salient and contrasted to the background. The low-level features, pertinent to pictures, are colors, edges, kinds of geometric objects, and orientations. Top-down attention is rather driven by higher-level cognitive factors and personal preferences and attitudes, like familiarity with the scene under analysis. Top-down attention modulates bottom-up attention and can redirect the focus to specific cues. Bottom-up attention is dominant in cases of sudden stimuli and its importance is gradually reduced in the observation of a picture after the slower impact of top-down attention tends to dominate. For example, in the observation of a figurative painting, the parts of interest are quickly defined as they correspond to expected and known objects and the role of bottom-up mechanisms are limited to predict visually salient locations. Complexity is one of various subjective image characteristics useful for categorizing or qualifying pictures. As an example, anthropogenic objects are simple to categorize while nonsense shapes are complex and difficult to associate to some invariants, thus requiring a more relevant involvement of top-down mechanisms. If the picture contains a variety of patterns at different levels of resolution (scales), it is reasonable to characterize it as complex. On the surface of a painting we perceive forms, which may be abstract or representational, well-defined or ambiguous. These forms are defined through colors and textures; they have well-defined relative positions to each other and within the whole picture. These elemental traits are the essential blocks of the whole image and are properly fused together by the brain in the visual perception act. The criteria commonly used to define visual complexity are:

1. The *level of details*, i.e., the varied constituents of an image, thus, the quantity of visual elements present and their size.
2. The *number of colors*, i.e., the variety of picture elements, since a greater variety of colors contributes to a greater complexity because it requires a more complicated description.
3. The *redundancy*, which has a negative correlation with complexity, which concerns the order of a picture (this feature is largely used for making an objective measurement of complexity, which is the basis of file-compression techniques).
4. The *amount of work or difficulty in perceiving* the picture.
5. The *depth*, since the perception of the three dimensions involves a building-up of visual elements and is more complex than the perception of two dimensions.

The adoption of these intuitively reasonable criteria can be a starting point for defining a *quantitative* measure of complexity. This metric could allow a comparison and a ranking of different pictures. The metric proposed here allows multi-scale analysis of a painting based on VA mechanisms and is related to the concept of neural complexity introduced earlier by *Tononi* et al. [61.11]. The metric can be generalized by introducing a parametric form of it in terms of *Tsallis* entropy [61.12]. Some preliminary versions of the present study have been published in [61.13, 14].

61.3 Artistic Complexity and the Brain

The challenge of defining a quantitative metric for complexity of paintings has been the subject of many quite controversial efforts. In 1928, *Birkhoff* introduced the concept of aesthetic measure (AM), defined as the ratio between order and complexity [61.15]. In 1965, *Bense* reinterpreted AM from an information theory viewpoint [61.16]. From a theoretical information viewpoint, the process of creating a painting can be seen as a transformation from an initial uncertainty, expressed by the Shannon entropy of the *repertoire* (palette), to the algorithmic information content of the final image, measured by the Kolmogorov complexity. In recent papers [61.17, 18], modern art has been proposed as a paradigm of complexity; in particular, the oeuvres of the Russian artist Wassily Kandinsky have been analyzed. In the present work, we enlarge the focus of the complexity assessment to other kinds of artistic paintings: three categories of paintings are considered, namely, modern, figurative, and geometric art. As a representative of figurative artists, Caravaggio is considered, while not only Kandinsky but also the well-known works of Jackson Pollock have been studied. The geometric paintings are basically due to Piet Mondrian. In particular, in the case of Kandinsky and Pollock, the results presented here agree well with the *intuitive* judgment of experts in terms of complexity evaluation.

Scientific objectivity has been proven to be an essential instrument for interpreting the content of abstract paintings. For example, the possibility of carrying out a fractal analysis of Pollock's drip paintings is well known in the related literature [61.19]. The concept of neural complexity (C_N) was proposed early for measuring the interplay between two fundamental aspects of brain organization, namely segregation of local areas and their global integration during perception [61.20]. The original C_N formulation is in terms of Shannon entropy, and, thus, for images, of first-order statistics.

61.4 Quantitative Measure of Complexity

Different artists roughly belong to so-called *currents* that are certainly related to the historical time of the artists' life. In figurative arts, the object of representation can be considered *simple*, in the sense that it corresponds to our ordinary perception of the surrounding world. In contrast, in abstract art, it is somehow not as easy to understand the underlying message of the artist. In some oeuvres of abstract painters there is an explicit attempt to favor the emergence of a meaning in the observer. The ability of integrating the information conveyed by the painting and to discover its meaning is a cognitive act. The painting is a static object that, in terms of computer analysis, can be represented as an image. The original color image corresponding to the painting is actually stored as the collection of the three-channel RGB sub-images. The measure of complexity carried out here and the results presented have been carried out on a luminance matrix of pixels. Thus, instead of the three-color RGB chrominance signals, the Y_{709} luminance signal has been considered [61.21]. However, it is easy to show that this apparent limitation is quite irrelevant: the behavior of the complexity curve shows that considering a luminance signal does not make any relevant difference with respect to the consideration of the curve averaged on the three channels. The *repertoire*, i.e., the palette of colors, is assumed to be given by the normalized histogram of the luminance values of the image, $X(i, j)$. The creative process consists in the selection of a specialized final product (the painting) among every possible realizations, given the distribution of probability of the repertoire. The Shannon entropy, $H_S(X)$, of the image is defined by

$$H_S(X) = - \sum_{i=1}^{256} p_i \cdot \log_2(p_i) \,, \tag{61.1}$$

where p_i is the i-th bin of the image histogram and X is the image corresponding to the painting. In a 8 bit representation, each pixel of the image may assume $2^8 = 256$ gray levels. By considering a bipartition of the image X into a j-th sub-image (window) and its complement $(X - X_j)$, the uncertainty about the state of a subset X_j of X, which is accounted for by the state of the rest of the system $(X - X_j)$, is called *mutual information*, and it is given by

$$\begin{aligned} \mathrm{MI}(X_j, X - X_j) = &\, H_S(X_j) \\ &+ H_S(X - X_j) - H_S(X) \,, \end{aligned} \tag{61.2}$$

where $X = X_j U(X - X_j) \cdot H_s(X_j)$ and $H_s(X - X_j)$ are the entropies of the two complementary subsets and $H_s(X)$ is the entropy of the system considered as a whole. *MI* is positive and symmetric and it zeroes if X_j and $(X - X_j)$ are statistically independent. *MI* is high if both the entropies of X_j and $X - X_j$ are high and they share a large fraction of it. The image treated here consists of a number of elementary components (subsets of different size), which interact with each other only spatially. In this case, the two subsets are a subimage of fixed size extracted from the whole image and the remaining counterpart. According to this perspective, a painting can be considered complex if it is both highly informative (in the Shannon sense) and highly integrated (coherent). As in [61.20], it is possible to estimate the average integration for subsets of the image of increasing size, namely, at multiple spatial scales. If a painting is composed of multiple *segregated* elements, the average integration for small subsets is low. This implies that such elements have independent meaning in the context and provide separated sources of information. If the painting, as a whole, shows an organized emerging effect, the average integration for large subsets is high. It becomes thus possible to quantitatively measure the coexistence of the two aspects by defining a suitable measure of *artistic complexity* (C_A). The technique looks at the different scales (sub-parts of the image) and then averages over all the scales, thus yielding novel insights on both the distinguishing traits of works of art and the way how the brain can process them. C_A can be computed by integration over all subset sizes k of the average mutual information at the different scales (all of the possible bipartitions of the image X).

The mathematical expression of the C_A is reminiscent of the expression of C_N [61.11, 20]

$$C_N(X) = \sum_{k=2}^{N/2} \left\langle MI\left(X_j^k, X - X_j^k\right)\right\rangle,\qquad (61.3)$$

where $N/2$ is the dimension of the largest squared subset extracted from the original image, $\langle \bullet \rangle$ denotes an average over all subsets of size k of the given image. The definition of C_A can be easily extended to the general Tsallis entropy (TE) which is a non-extensive parametric definition of entropic content useful to extract suitable features depending on the parameter. TE, introduced by *C. Tsallis* (1988), has the ability to catch long-range interactions between sub-parts of a system [61.12, 22, 23]. Given a random variable X, which attains N possible different values $(x_1, \ldots, x_N)$, TE is defined as

$$H_T(X) = \frac{1}{\ln(2)}\frac{1}{q-1} \times \left(1 - \sum_{i=1}^{256} p(x_i)^q\right);\quad q \in R,\qquad (61.4)$$

where $p(\bullet)$ is the probability density function and q is a real number that characterizes the degree of non-extensivity. The usual Shannon entropy, H_S, can be obtained in the limit $q \to 1$. TE is expressed in bits through the normalization by $\ln(2)$.

61.5 Artistic Complexity of Paintings

To examine the complexity of works of art, different categories of painting styles are considered. Figure 61.1 shows the three cases considered here, namely, one of figurative style, one of geometric abstract art, and one of abstract art. Table 61.1 reports the results achieved by computing the C_A on different paintings of differ-

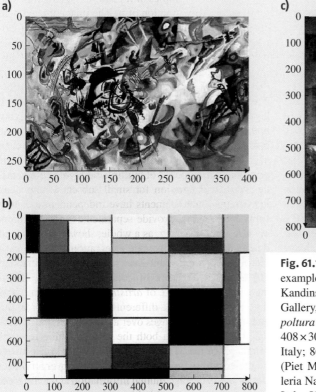

Fig. 61.1a–c The paintings of the three styles taken as examples of the analysis: (a) *Composition VII* (Wassily Kandinsky), 1913, oil on canvas, 200×300 cm, Tretyakov Gallery, Moscow; 400×267 pixels – JPEG format; (b) *Sepoltura di Santa Lucia* (Caravaggio), 1608, oil on canvas, 408×300 cm, Chiesa di Santa Lucia alla Badia, Siracusa, Italy; 867×1180 pixels–JPEG format; (c) *Composition A* (Piet Mondrian), 1923, oil on canvas, 91.5×92 cm, Galleria Nazionale d'Arte Moderna e Contemporanea, Rome, Italy; 889×866 pixels–JPEG format

Part L | 61.5

Table 61.1 Values of C_A computed as the area under the complexity curve normalized to the size of the painting (N) and to the maximum value of the entropy. Two values of q have been considered for several paintings of different styles; **bold** indicates the maximum values of complexity

Normalized area under the curve (normalized by maximum entropy)	Tsallis ($q = 0.5$)	Shannon ($q = 1$)
Dominant curve, Kandinsky	0.6748	0.8069
Suonatore di liuto, Caravaggio	0.5458	0.7251
On points, Kandinsky	0.6794	0.8179
Bacco, Caravaggio	0.5979	0.7591
Composition IX, Kandinsky	0.5681	0.7420
Composition VIII, Kandinsky	0.5262	0.5892
Composition VII, Kandinsky	**0.7812**	**0.8845**
Composition VI, Kandinsky	0.7696	0.8788
Composition X, Kandinsky	0.6191	0.7726
Gelb rot blau, Kandinsky	0.7057	0.8323
Improvisation 19, Kandinsky	0.6400	0.8030
Improvisation 28, Kandinsky	0.7697	0.8823
Improvisation 34, Kandinsky	0.6655	0.7759
Improvisation 6, Kandinsky	0.6818	0.8131
Improvisation dreamy, Kandinsky	0.7437	0.8681
Le petit rond rouge, Kandinsky	0.5054	0.6797
Composition A, Mondrian	0.5194	0.6686
On white II, Kandinsky	0.5102	0.6673
Las meninas, Picasso	0.7661	0.8728
Number 1 (Lavender mist), Pollock	**0.8523**	**0.9293**
Number 81, Pollock	**0.8321**	**0.9159**
Primo acquerello, Kandinsky	0.7721	0.8687
Sepoltura Santa Lucia, Caravaggio	0.5929	0.7789
Transverse line, Kandinsky	0.6673	0.7861
Venice, Claude Monet	0.5832	0.7814

ent artists. Figure 61.2 illustrates the behavior of the C_A for three paintings. The C_A curve was obtained by considering successive decomposition of different size of the original image X, in terms of fixed-size $k \times k$ non-overlapping blocks (windows); thus, the MI between each single block and its complement is calculated by using (61.2). Then, increasing k from 2 to $N/2$, C_A is obtained by the double-level average in (61.3). The curve for *Composition VII* shows a monotonic increase of MI, which can be understood by considering that the image has no well-defined geometric objects, thus no evident structure at well-defined k size. On the contrary, most of the sub-images include neighborhoods with structures exhibiting flatter distribution. The whole palette of colors is used and there are sub-blocks of various colors at different scales. The curve of MI for *Composition A* shows a different behavior, which can be related to the presence of rectangular structures of different sizes and colors. The complexity values are low

and so is C_A. Caravaggio's paintings also show some oscillations in MI trends and an intermediate level of C_A. The interpretation of this result is that only some portions (subsets) of the image have high entropy values while vast areas are uniform, thus showing peaked histograms. Indeed, the entropy of a *flat* region of uniform color is very low and corresponds to a minimum of MI. On the contrary, windows including different details, color variations, and small parts generate high values of MI and contribute to increase complexity. C_A is then a measure of complexity that is able to take into account both local complexity (by measuring the entropy on the local histogram of intensity) and the change of complexity at varying scales. Further insights on the metric can be gained through 2-D representations of the MI computed on different bipartitions of the painting for a fixed window size. Figure 61.3 shows the different representations of the painting *On Points* of Kandinsky at five different levels of resolu-

a) Bits

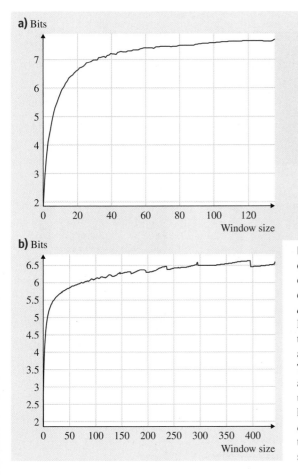

b) Bits

c) Bits

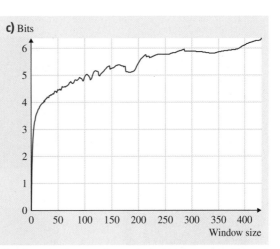

Fig. 61.2a–c Curves of complexity (representing the trend of averaged *MI* for different window sizes) in the case of the paintings representative of the three styles considered, namely, *Composition VII* by V. Kandinsky, *Sepoltura di Santa Lucia* by Caravaggio, and *Composition A* by P. Mondrian. (**a**) The first *curve* rises quickly for a relative low window size, thus indicating a complex structure at the low scale; the maximum value is well above 7. The *curve* is quite smooth; this behavior reflects the absence of evident geometric objects or well-defined figurative parts popping out from the background and a homogeneous complexity at all scales. (**b,c**) The other *curves* attain lower levels of C_A and there are oscillations due to the presence of objects and/or geometric structures

tion (scales). The map closely resembles the original painting at small scales. At large scales, however, a defocused representation of the object is still present. The original presence of geometric objects gradually disappears. The 2-D representations of *MI* for bipartitions of different window size basically allow recovery of the main segregated aspects of the painting. One of the apparent limitations of the above complexity measures is that it attributes high values of complexity to fully random images, like snow images. Recently, scientific approaches to complexity have attempted to retain the intuitive, common sense notion of complexity by emphasizing the idea that complex systems are neither completely regular nor completely random. Actually, C_A confirms that any system of elements (image or subimage) arranged in homogeneous way is not complex. In contrast, a fully random image implies both a high value of C_A and a fast growing *MI* curve. This limitation, however, can be easily overcome by looking at the statistics of the *MI* values; in particular, in Figure 61.4, for the painting *Composition VII*, it is shown that although a shuffled image has comparable complexity, the variance of the *MI* values is largely different for the original and the shuffled images. The joint use of C_A and the second-order statistics of *MI* distribution can allow discrimination of structural complexity from randomness. The 2-D representation of the *MI* distribution can be useful to extract relevant features about the painting. Furthermore, the presence of similar features at different scales is the hallmark of specially relevant structures in the painting. Peculiar objects in the paintings correspond to attracting bumps for human gaze and attention as relevant visual stimuli. In other words, they may correspond to salient parts of the scene.

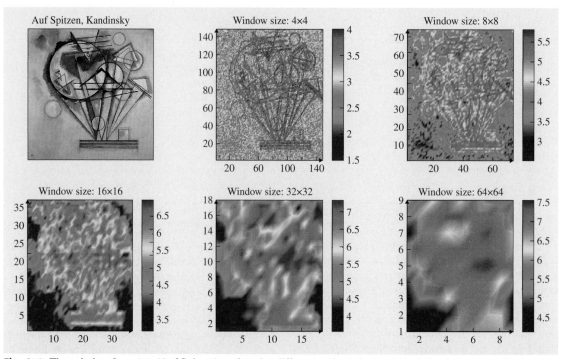

Fig. 61.3 The painting *On points* (Auf Spitzen) analyzed at different scales (windows); both the defocusing effect at large scales (64×64) and the ability to catch details at small scales (4×4) are evident. Each sub-image represents a 2-D map of the *MI*

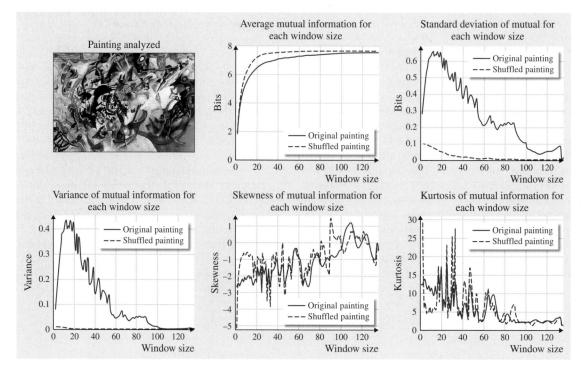

61.6 Saliency Maps of Paintings

Visual saliency is defined as [61.24]:

> *The distinct subjective perceptual quality which makes some items in the world stand out from their neighbors and immediately grab our attention.*

It represents a broad term that conveys the idea that some specific parts of a scene are pre-attentively distinctive and it is able to create visual arousal within the early stages of the visual system. The term *pop-out* was coined to describe the visual saliency process occurring at the pre-attentive stage [61.25, 26]. The fruition of a work of art, however, implies a modulation of this *bottom-up* approach through a *top-down* mechanism that is influenced by many aspects; for example the education and familiarity of the observer. It is thus a highly subjective task, with particular regard to the aesthetic appreciation of the object inspected by the visual sense. However, early stages of visual processing give rise to a sort of perceptual quality that makes some stimuli stand out from among other items or locations. Our brain has the evolutionarily determined ability to realize a real-time, automatic, apparently effortless processing over the entire visual field where salient visual locations attract VA.

High complexity has been proposed as a suitable descriptor that can be used as a measure of local saliency [61.25,26]; however, it fails in the case of noise and self-similar images. In any case, this kind of local descriptor is not appropriate for analyzing complexity of the whole and its relationship with local complexity. In our specific case, the paintings, it is interesting to compare the computation approach to measuring VA based on saliency and the C_A approach.

The principle behind computing salience is the iterative detection of locations whose visual attributes significantly differ from the surrounding image attributes. This means extracting a number of simple features represented in the early stages of cortical visual processing, namely, color, edges, orientation, and luminance changes [61.24].

Many efficient computational techniques to extract saliency maps have been proposed in the last decade [61.17, 18], most of which are the evolution of a basic approach due to *Itti* and *Koch*. The underlying

hypothesis is that a 2-D saliency map can provide an efficient control strategy for the deployment of attention based on bottom-up cues. The painting, i. e., the input image, is decomposed through several feature detection mechanisms, which operate in parallel over the image. This multi-scale low-level feature extraction stage generates maps that encode for spatial contrast in each of the feature channels. The extraction of saliency implies competition. Then, the resulting single conspicuity maps are combined into a unique saliency map, which is the representation of the topographical encoding for saliency, irrespective of the feature channel where the

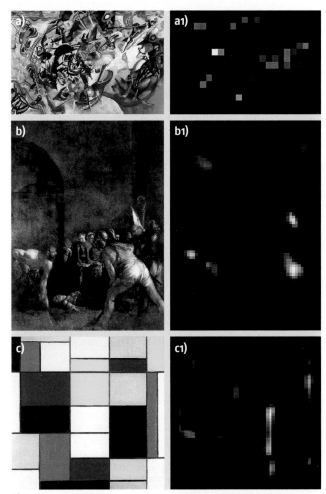

Fig. 61.4 The statistics of the MI distributions allow us to distinguish the behavior of the complexity of a painting from one of a shuffled image with the same histogram. Both the variance (standard deviation) and the skewness of the *MI* maps show relevant differences ◀

Fig. 61.5 The saliency maps extracted from the three paintings analyzed: **(a,c)** original paintings (see Fig. 61.1); **(a1)**–**(c1)** locations and intensity of the saliency regions. It is highlighted that more complex paintings include a higher number of salient areas

stimuli appeared salient. The saliency map is scanned by attention through a double mechanism; a winner-take-all network, which detects the areas of highest saliency, and inhibition of return, which suppresses the last attended location from the saliency map, in order to focus attention onto the next most salient location. The resulting procedure was applied to the analysis of the paintings, thus extracting the related saliency maps. The results are reported in Fig. 61.5, for paintings representative of the three styles mentioned above. However, many paintings have been analyzed through the technique. The general conclusion is that the saliency map of the paintings whose artistic complexity is high includes a higher number of extracted salient locations. This result can be intuitively understood if related to the fruition experience of paintings; an artwork that displays familiar human shapes is more easily recognized by conscious image-recognition areas than an impressionist work, whose unrealistic patchwork of brush strokes and mottled coloring distract conscious vision. In the most complex work of the abstract artist Kandinsky, namely *Composition VII*, the presence of numerous

motifs abstracted from natural objects or pure abstracts, which are pictorial constructions of unusual form and colors, are organized into confounding visual structures that can be experienced simultaneously. At the simple bottom-up level implemented through the saliency map, this effect is already appreciated by the more numerous areas selected through the multi-step procedure. In turn, this can be read as a cross-check of the objectivity of C_A as a measure of complexity. By extending the comparison between the saliency areas and the parts of the paintings corresponding to minima or maxima of the *MI* maps at selected scales, in the three cases presented here, it is possible to note a good match among the relevant pop-ups. By using the concept of TE, at different values of the parameter q various different features at more scales can be observed. This reflects the ability of TE to catch long-range interactions and fractal-type structures on the images (or their sub-parts). The non-extensivity of the system is justified by the presence of correlations between pixels of the same object in the image with respect to luminance values and space filling.

61.7 Concluding Remarks

Modern abstract art is considered complex and the extraction of meaning from some works of art is largely controversial. However, some artists have explicitly tried to produce paintings in accordance with specific goals. This means that behind their artwork there is a project realized through creativity. Their paintings clearly reflect those efforts and are able to show the emergence of complex ideas reproducing a non-linear and uncertain world.

The concept of C_A has been introduced as a metric for assessing complexity of paintings of different artists. The results achieved have been compared to the saliency

maps earlier introduced in computer vision as computational models of bottom-up VA. The proposed measure is based on an interplay between top-down and bottom-up approaches manifesting the difficulty of the human brain in extracting invariant from some abstract representations. The intriguing relationships shown may offer a paradigm for testing novel computational models on brain-like machines. The methodologies described can presumably be of interest for multimedia quality assessment as metrics able to emulate the integral mechanisms of human visual systems to correlate well with the visual perception of some aspects of quality.

References

61.1   M. Livingstone: *Vision and Art: The Biology of Seeing* (Harry N. Abrams, New York 2002)

61.2   S. Zeki: Artistic creativity and the brain, Science **293**, 51–52 (2001)

61.3   R. Rao, D. Ballard: Predicting coding in visual cortex: A functional interpretation of some extra-classical receptive-field effects, Nat. Neurosci. **2**(1), 79–87 (1999)

61.4   S. David, W. Vinje, J. Gallant: Natural stimulus statistics alter the receptive field structure of V1 neurons, J. Neurosci. **24**(31), 6991–7006 (2004)

61.5   W. Shultz, A. Dickinson: Neuronal coding of prediction errors, Annu. Rev. Neurosci. **23**, 473–500 (2000)

61.6   A. Treisman, G. Gelade: A feature-integration theory of attention, Cogn. Psychol. **12**(1), 97–136 (1980)

61.7   E. Korner, G. Matsumoto: Cortical architecture and self-referential control for brain-like computation, IEEE Eng. Med. Biol. Mag. **21**(5), 121–133 (2002)

61.8   A. Forsythe, G. Mulhern, M. Sawey: Confounds in pictorial sets: The role of complexity and familiarity in basic-level picture processing, Behav. Res. Methods **40**(1), 116–129 (2008)

61.9 L. Itti, P. Baldi: Bayesian surprise attracts human attention, Vis. Res. **49**, 1295–1306 (2009)

61.10 F. Porikli: Multimedia Quality Assessment, IEEE Signal Process. Mag. **28**(6), 164–177 (2011)

61.11 G. Tononi, O. Sporns, G. Edelman: A measure for brain complexity: Relating functional segregation and integration in the nervous system, Proc. Natl. Acad. Sci. USA **91**, 5033–5037 (1994)

61.12 C. Tsallis: Possible generalization of Boltzmann–Gibbs statistics, J. Stat. Phys. **52**, 479–487 (1988)

61.13 F.C. Morabito: Artistic Complexity and brain: Quantitative measurement of creativity, Tri-Soc. Newsl. **8**(2), 10–11 (2010), INNS/ENNS/JNNS

61.14 F.C. Morabito, M. Cacciola, G. Occhiuto: Creative brain and abstract art: A quantitative study on Kandinskij paintings, Proc. IJCNN (2011) pp. 2387–2394

61.15 G. Birkhoff: *Aesthetic Measure* (Harvard Univ., Cambridge, USA 1933)

61.16 M. Bense: *Einführung in die Informationstheoretische Ästhetik* (Rowohlt Taschenbuch, Hamburg 1969)

61.17 L. Itti, C. Koch, E. Niebur: A model of saliency based visual attention for rapid scene analysis, IEEE Trans. Pattern Anal. Mach. Intell. **20**(11), 1254–1259 (1998)

61.18 L. Zhang, M.H. Tong, T.K. Marks, H. Shan, G.W. Cottrell: SUN: A Bayesian framework for saliency using natural statistics, J. Vis. **8**(7), 1–20 (2008)

61.19 K. Jones-Smith, H. Mathur: Fractal analysis: Revisiting Pollock's drip paintings, Nature **444**, E9–E10 (2006)

61.20 G. Tononi, O. Sporns, G.M. Edelman: Complexity and coherency: Integrating information in the brain, Trends Cogn. Sci. **2**(12), 474–484 (1998)

61.21 International Telecommunication Union: Parameter values for the HDTV standards for production and international programme exchange (2002), available online at www.itu.int/dms_pubrec/itu-r/rec/bt/R-REC-BT.709-5-200204-I!!PDF-E.pdf

61.22 O.A. Rosso, M.T. Martin, A. Plastino: Brain electrical activity analysis using wavelet-based informational tools (II): Tsallis non-extensivity and complexity measures, Physica A **320**, 497–511 (2003)

61.23 M. Gell-Mann, C. Tsallis (Eds.): *Nonextensive Entropy-Interdisciplinary Applications* (Oxford Univ. Press, New York 2004)

61.24 T. Kadir, M. Brady: *Saliency, Scale and Image Description* (University of Oxford, UK 2000)

61.25 L. Itti, C. Koch: Computational modelling of visual attention, Nature **2**, 194–203 (2001)

61.26 B. Julesz: *Dialogues on Perception* (MIT Press, Cambridge 1995)

62. The Allen Brain Atlas

Michael Hawrylycz, Lydia Ng, David Feng, Susan Sunkin, Aaron Szafer, Chinh Dang

The Allen Brain Atlas is an online publicly available resource that integrates gene expression and connectivity data with neuroanatomical information for the mouse, human, and non-human primate. Launched in 2004 by the Allen Institute for Brain Science, the portal currently receives about 45 000 unique users each month. More than one petabyte of in situ hybridization imagery and over 240 million microarray data points from six adult human brains representing 3700 tissue samples have been generated to date. As one of the most comprehensive gene expression resources for the nervous system, scientists regularly use these resources to study the expression profile of genes in the various regions of the brain. Additional usage includes searching for biomarkers, correlating gene expression to neuroanatomy, and other large-scale correlative data analysis. This chapter reviews the resources available and describes how they were constructed to enable development of visualization and search tools to analyze the massive amount of data generated. Finally, examples are provided on how these tools can be leveraged for scientific discovery.

Part L | 62

62.1 The Portal

The Allen Brain Atlas portal [62.1] contains a growing collection of data and software that integrates gene expression and connectivity with neuroanatomy. There are currently seven major gene expression and connectivity atlases that span across species and developmental time points and three additional data sets focused on gene expression in sleep, diversity of gene expression across mouse strains, and gene expression profiles of glioblas-toma in human. While the portal (Fig. 62.1) serves as the entry point for these data resources, it also provides an integrated environment to communicate and provide information to the scientific community across all of the resources. The portal is updated three times each year. An announcement accompanying each release highlights newly released data, new application features and the next data-release date.

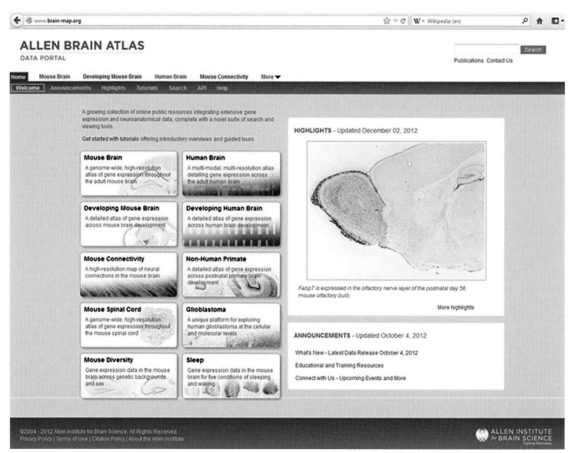

Fig. 62.1 The Allen Brain Atlas main portal. Users can readily access and search all data sets and tools from a convenient project oriented interface. In addition to the main data sets, the portal presents daily data analysis highlights and news announcements

62.2 Mouse Atlas Resources

There are four primary large-scale atlases associated with the mouse. They are the Allen Mouse Brain Atlas, the Allen Developing Mouse Brain Atlas, the Allen Mouse Brain Connectivity Atlas, and the Allen Spinal Cord Atlas. The inaugural project of the Allen Institute for Brain Science, the Allen Mouse Brain Atlas [62.2] is a genome-wide 3-D atlas of gene expression in the adult mouse brain. The atlas consists of over 600 terabytes of high resolution in situ hybridization images of the C57Bl/6J P56 mouse brain. Building upon the foundations of the adult mouse data pipeline [62.3,4], the Allen Developing Mouse Brain Atlas profiles the changes of gene expression during development. Effectively a 4-D

atlas, the data set comprises gene expression profiles of over 2000 genes key to neurodevelopment in seven different stages ranging from early embryonic to adult. The atlas is a framework to explore when and where genes are activated or deactivated as the mouse brain develops. The Allen Mouse Brain Connectivity Atlas examines the longer range mesoconnectivity of neural connections in the mouse brain. When completed, the atlas will contain axonal projections mapped from ≈ 300 regions of the adult mouse brain and diverse neuronal populations defined by ≈ 100 transgenic mice genetically engineered to target specific cell types [62.5–7]. Complementing atlases associated with the mouse brain, the

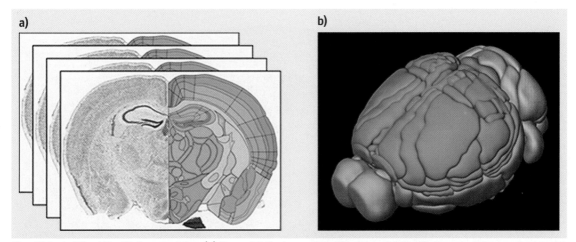

Fig. 62.2a,b The Allen Reference Atlas. (**a**) Several 2-D plates of the Nissl based Allen Reference Atlas are shown with updated annotation in the cerebral cortex. (**b**) 3-D reconstruction formed from the symmetric annotation

Allen Spinal Cord Atlas is a genome-wide gene expression map of the juvenile and adult mouse spinal cord. Similar to other mouse gene expression atlases, the Allen Spinal Cord Atlas uses in situ hybridization to assay gene expression across all anatomic segments of the spinal cord.

62.2.1 Common Reference Space in the Mouse Brain

The Allen Reference Atlas was originally created as an annotated Nissl histology reference space for the Allen Mouse Brain Atlas [62.8]. It was designed for three purposes: to serve as a reference resource for gene expression pattern comparison to neuroanatomical structures, to provide standard neuroanatomical ontology structural labels, and to act as a template for the development of 3-D computer graphic models of the adult mouse brain and the automated informatics annotation tools. Since the initial construction, it has been refined and utilized in other Allen Institute resources, mainly the Allen Mouse Brain Connectivity Atlas and future work on adult mouse systems. The initial 2-D annotation and creation of the 3-D reference volume [62.3, 8] was updated in 2011 with increased delineation in the cerebral cortex, hippocampus, and cerebellum. Currently, the Allen Reference Atlas contains over 738 annotated structures. In addition to incorporating the additional structures into the 3-D volume, another major improvement was also made. The original 3-D model was based on one

hemisphere (the annotated hemisphere). During the atlas update, the reconstructed volume was aligned with a sagittal sectioned specimen in addition to the original coronal sections. Reflecting one hemisphere to the other side of the volume resulted in an updated, symmetric, fully-annotated reference space with a more consistent and deeper level of annotation (Fig. 62.2).

Unlike the adult mouse, reference atlas for the Allen Developing Mouse Brain Atlas was created to provide a novel neuroanatomical framework based on geno-architectonic data. With the expertise of Luis Puelles, MD, PhD (University of Murcia, Spain), the developing mouse brain reference atlas consists of seven atlases of different stages of the developing mouse brain with a common developmental ontology. The developmental stages include four embryonic time points (E11.5, E13.5, E15.5, and E18.5) and three postnatal time points (P4, P14, and P56). With the exception of the annotation and ontology, the steps to create the 3-D models for each of the seven time points is similar to that of the adult mouse.

Although the ontologies and structural delineations differ between the adult mouse of the Allen Reference Atlas and the developing stages of the developing mouse brain reference atlas, the two atlases share a common 3-D volume for the P56 stage. This common 3-D volume enables ongoing development to spatially map and integrate the two atlases, thereby allowing cross ontology comparisons of anatomical structure delineations (Fig. 62.3).

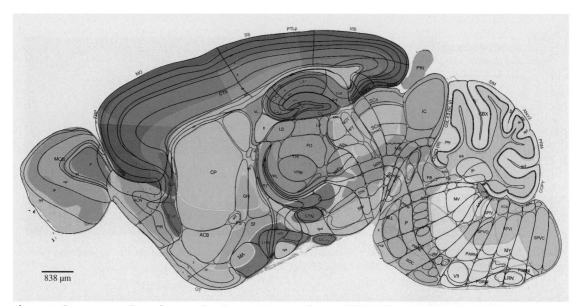

Fig. 62.3 Common coordinate framework and annotation used for the C57Bl6/6J mouse for both adult mouse and final stage developing mouse atlas. Relationship and intersection of the Allen Reference Atlas (*drawn in colors*) and the P56 adult stage of the Allen Developing Mouse Brain Atlas (*drawn in black polygons*)

62.2.2 Gene Expression and Connectivity Mapping to 3-D in the Mouse Brain

The reference atlases and 3-D models provide a framework for gene expression data and axonal projections from the Allen Brain Atlas resources based upon the adult P56 mouse to be automatically annotated and co-registered. In the Allen Mouse Brain Atlas, an automated informatics pipeline [62.3, 4] was developed to process the images containing the expression of ≈ 20 000 genes for the pipeline and enables online anatomic structural search, visualization, and data mining of the in situ hybridization data. Using the same architectural pipeline, new informatics modules were developed to detect and map axonal projections for the Allen Mouse Brain Connectivity Atlas. Unlike the bright field images of the Allen Mouse Brain Atlas and Allen Developing Mouse Brain Atlas, the data from the Allen Mouse Brain Connectivity Atlas contains fluorescent images of axonal projections targeting different anatomic regions or various cell types. This informatics pipeline (Fig. 62.4) consists of a preprocessing module, a 3-D reference module, an alignment module, a detection module, and a gridding module. Although the modules are different, the output of the pipeline is similar to that of the Allen Mouse Brain Atlas in that quantified signal values are obtained at the level of a ref-

erence atlas 3-D voxel and annotated according to the Allen Reference Atlas ontology.

All seven developmental stages of gene expression data in the Allen Developing Mouse Brain Atlas were mapped into the respective stages of the reference atlas. An exception is the gene expression data from the P28 stage, which was mapped into the P56 atlas. An informatics pipeline, similar to that of the Allen Mouse Brain Atlas, contains the modules to preprocess the images, segment, expressing signal, and annotate grid voxels. To enable temporal searches and visualization of data across time points, an additional step was introduced to create a master reference space and co-register the data sets into this master reference space.

62.2.3 Integrated Search and Visualization

Basic visualization of one or more experiments is standardized across all mouse resources. At the core of this is the *Experiment Image Viewer* (EIV), which allows for zooming and panning of high resolution images. An unlimited number of experiments and corresponding reference atlases can be viewed simultaneously. The user can drag-and-drop experiment windows, customize the number of columns of experiments displayed and resize experiment windows. When working with such large data sets, it is crucial to be able to simul-

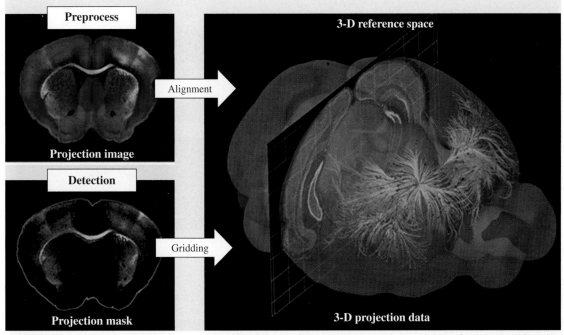

Fig. 62.4 The informatics pipeline of the Allen Brain Atlas resources. A preprocessing module performs general imaging histogram normalization and other preliminary operations. A detection module is used to segment image characteristics of interest, especially gene expression or anatomic projection. An alignment module registers images into a common coordinate framework, while a gridding module collapses gene expression or other data to a standardized grid. The result is that all experimental information can be analyzed in a common coordinate framework

taneously view similar neuroanatomical regions across many experiments. A cross-plane and cross-time point *synchronize* feature in this viewer zooms all experiment windows to the same approximate zoom level and brain position based on the linear alignment of the images to the corresponding reference atlas (Fig. 62.5). These features allow users to efficiently compare and analyze many experiments at the same time.

The Brain Explorer tool is a 3-D desktop visualization application that displays gene expression and other data to be viewed within a three dimensional environment. Similar to standardized features in 2-D experiment visualization, basic navigation and 3-D visualization in the Brain Explorer application are also standardized. In this desktop application, users can view spatially registered data from the adult mouse, developing mouse, and mouse connectivity atlases. There are three panes in the Brain Explorer application window. The main pane shows 3-D brain anatomy and reconstructed experimental data at the voxel level. For the Allen Developing Mouse Brain Atlas, multiple experiments at different time points can be viewed si-

multaneously. Within this main pane, basic zoom, pan, and rotate functionalities are available. Informatics values, location of the voxels, and correspondence point to the original 2-D images from the web application are given when selecting a voxel for a particular experiment. The upper right hand pane shows the structures of the ontology in hierarchical or alphabetical order of the corresponding reference atlas. Because these individual panes are interconnected, users can control what structure(s) and experimental gene expression voxel data are displayed in the main window. A third pane in the lower right shows which experiments are currently displayed in the main pane. In addition to being able to view the experimental data within their respective environment in the Brain Explorer application, the registration of the Allen Mouse Brain Atlas and the Allen Mouse Brain Connectivity Atlas into the Allen Reference Atlas enables visualization within the same environment. This feature is an important component when searching for potential correlations between gene expression and axonal projections. This concept is illustrated in Fig. 62.6.

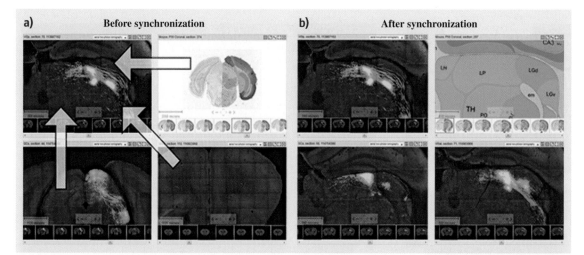

Fig. 62.5a,b Image synchronization feature. (**a**) A selection of experiments from the Allen Mouse Brain Connectivity Atlas is shown along with a plate from the Allen Reference Atlas in the *upper right*. After clicking the synchronization button in the window targeted by the arrows in (**a**), the frame of reference is aligned so that all images and the atlas plate are seen at a common frame of reference (**b**). This mechanism is enabled through image alignment and registration

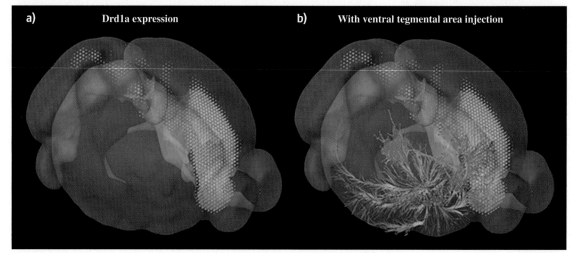

Fig. 62.6a,b Viewing 3-D data in the Brain Explorer desktop application. (**a**) Gene expression for *Drd1a* is shown as *bright yellow* in the striatum. Expression values are collapsed to 100 μm voxels and represented as spheres pseudocolored by expression intensity. (**b**) Projection connectivity data is overlaid with axonal projections from a primary injection in the ventral tegmental area. Basic correspondence of gene expression and projection connectivity can be viewed within the 3-D context

Another important component of this large data set is the ability to search and mine the data. Because of the inherent differences in data modalities between gene expression and axonal projection data, searches are implemented differently in these atlases. In the gene expression data sets of the Allen Mouse Brain Atlas and Allen Developing Mouse Brain Atlas, the core search is the Gene Search. Users input one or more genes into the search box and the results displayed are based on the experiments available for those genes. Users can also perform a Differential Search that identifies experiments with enriched gene expression in

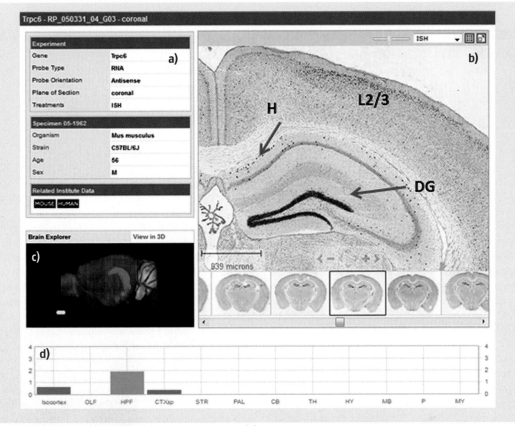

Fig. 62.7a–d *Tpc6* in the primary image viewer. Panel (**a**) provides basic experimental annotation and metadata. The primary in situ hybridization (ISH) image viewer is shown in (**b**) with pronounced expression seen in the dentate gyrus (DG) of the hippocampus indicated by the *arrow*. There is also expression in the hilar interneurons (H) and layer 2/3 of the cortex (L2/3). A rotatable 3-D perspective of the geometry of gene expression (**c**), and high level expression summary statistics by region (**d**) complete the attributes of the primary image viewer

particular brain structures in contrast to other structures entered. Differential Search is an on-the-fly service that computes differential gene expression fold change calculations over all experiments in the atlas. The result of the search displays experiments sorted by highest to lowest expression fold change between the target structure(s) and contrast structure(s). Another useful search functionality between these two atlases is the *genes like me* search or *NeuroBlast*. This search allows users to mine for experiments within the atlas with similar gene expression patterns to an experiment with a gene expression pattern of interest. While the expression pattern of interest can be found in the whole brain, the search can also be limited to certain brain regions, effectively limiting the expression pattern to a region of interest.

One way users can search in the online application for gene expression in the adult mouse brain is by accessing a curated list of genes whose expression is enriched in one of 26 regions. These lists have been validated by manual examination and are therefore called *marker* genes.

In the Allen Mouse Brain Connectivity Atlas, the most basic search is for experiments with injection sites in one or more brain regions. Users can also browse experiments based on the anatomic locations of the injection sites. The search can be further refined by specifying afferent structure(s), which limits the search results to experiments with detected signal in those structures. By default, the search returns experiments with projection signal from either hemisphere, but the results can be limited to a particular hemisphere. A cor-

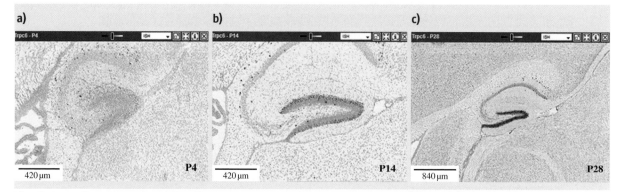

Fig. 62.8a–c Gene expression of *Trpc6* in the dentate gyrus of the hippocampus at P4 (**a**), P14 (**b**), and P28 (**c**). The cation channel is implicated in neuronal development. A scale bar for each image is shown in the lower left

relative search, similar to the NeuroBlast search, is also available in this data set. This search allows exploration of experiments with similar projection patterns to the selected experiment of interest.

62.2.4 Analyzing and Understanding Mouse Gene Expression Data

We next consider a use-case data analysis scenario for the Allen Mouse Brain atlas. By searching through the adult mouse list for enrichment in the dentate gyrus of the hippocampus one finds the transient receptor potential cation channel (*Trpc6*) illustrated in the main online image browser shown in Fig. 62.7. The hippocampus is one of the most widely studied areas in the brain because of its important functional role in memory processing and learning, as well as its remarkable neuronal cell plasticity.

Trpc6 is well known for its expression in the hippocampus [62.9] and in particular for its involvement with excitatory synapse formation and dendritic growth [62.10–12]. This gene is also available in the Allen Developing Mouse Brain Atlas and shows the interesting emergence of expression in the more mature neuronal layers of the dentate gyrus (P14 and P28) with less expression in the mature layers of the dentate gyrus in P4 (Fig. 62.8).

Starting with the image series for *Trpc6* users may employ the correlation search tool *NeuroBlast* to search for other genes in the adult mouse exhibiting a similar expression pattern. This tool and its utility for finding related voxel-wise expression patterns and has been de-

scribed elsewhere [62.13]. The most correlated genes (*Crlf1*, *Prdm5*, *Lct*, *Prox1*, *Tnip2*, *Slc26a10*, *Cacna1a*, *Tnfrsf25*, *Rras2*, *Vwa3b*, *Glis3*, *Ppfia2*, *Pcgf6*, *Dsp*, *Sesn1*, and *Tdo2*) have correlations with *Trpc6* in the range (0.662, 0.775). Although the input image series was in the coronal plane, cross-brain alignment is also able to return high correlates in the sagittal plane as shown in Fig. 62.9. Several of these genes, such as *Dsp*, *Tdo2*, and *Prox1*, exhibit hippocampus specific function related to spatial learning or neurogenesis [62.14–16].

As demonstrated by *Hawrylycz* et al. [62.13] the dual problem to finding genes of high spatial correlation is to investigate how gene expression on average varies from one spatial location to another. The Anatomic Gene Expression Atlas (AGEA) presents an online atlas of spatial correlation maps over an input set of genes, and in the present case the set of all coronal ISH image series in the Allen Mouse Brain Atlas. As shown in Fig. 62.10, users can spatially navigate by selecting a voxel (Fig. 62.10a) in the three orthogonal planes (coronal, sagittal, and horizontal, the latter not displayed). The result is a spatial correlation volume that can be navigated as well (Fig. 62.10b). Higher intensity color represents regions that on average will have high expression to the original seed point chosen. There is additionally a *Gene Finder* algorithm that can return specific high spatial correlates to the spatial correlation volume. In this case the algorithm returns *Tprc6* and many of the higher correlates described. Using both NeuroBlast and AGEA is an effective way of searching for spatially related gene expression patterns and returning relevant genes.

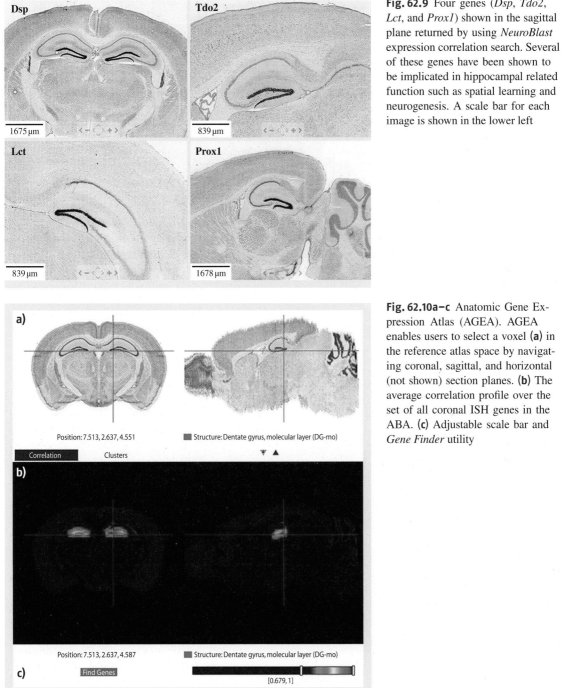

Fig. 62.9 Four genes (*Dsp*, *Tdo2*, *Lct*, and *Prox1*) shown in the sagittal plane returned by using *NeuroBlast* expression correlation search. Several of these genes have been shown to be implicated in hippocampal related function such as spatial learning and neurogenesis. A scale bar for each image is shown in the lower left

Fig. 62.10a–c Anatomic Gene Expression Atlas (AGEA). AGEA enables users to select a voxel (**a**) in the reference atlas space by navigating coronal, sagittal, and horizontal (not shown) section planes. (**b**) The average correlation profile over the set of all coronal ISH genes in the ABA. (**c**) Adjustable scale bar and *Gene Finder* utility

62.3 Human and Non-Human Primate Atlas Resources

In addition to the mouse data sets, there are three large human and non-human primate data sets within the Allen Brain Atlas portal. The Allen Human Brain Atlas is a multi-modal atlas that maps gene expression in the adult human brain [62.17]. The atlas contains genome-wide microarray gene expression sampled from ≈ 900 brain regions across both hemispheres and mapped into the 3-D reconstructed magnetic resonance imaging (MRI) volume of the donor brain as well as MNI (Montreal Neurological Institute) template space [62.18]. High resolution in situ hybridization data is available for selected sets of genes on specific brain regions. Complementing the adult human data set, the BrainSpan Atlas *of the Developing Human Brain* is a foundational resource for studying transcriptional mechanisms in human brain development. This atlas profiles gene expression in sixteen cortical and subcortical structures across the full course of human development using RNA sequencing and exon microarrays. More detailed

structural sampling (500+ different structures) is done in four mid-gestational prenatal male and female specimens using microarrays. Similar to the strategy for in situ hybridization in the Allen Human Brain Atlas, the high-resolution in situ hybridization data in this project is done on selected sets of genes in developing and adult human brains. Finally, the National Institutes of Health (NIH) Blueprint Non-Human Primate Atlas is a data set containing gene expression in the developing rhesus macaque brain. This data set and the BrainSpan atlas of the developing human brain are designed to complement one another in studying human brain development.

62.3.1 Informatics of Adult Human Microarray, Anatomic Mapping, and Search and Visualization

Unlike the fully automated data pipeline developed for the mouse atlases, the Allen Human Brain Atlas data

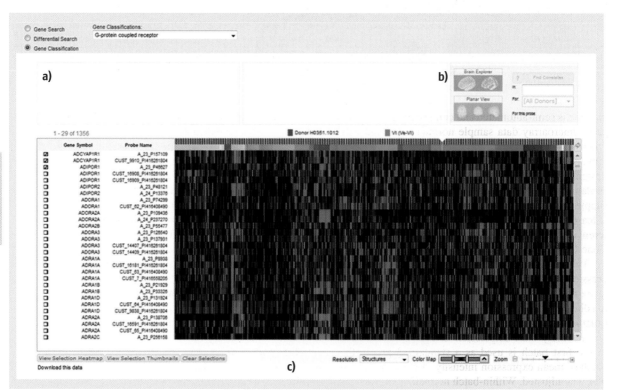

Fig. 62.11a–c Heat map rendering of gene expression in the Allen Human Brain Atlas. **(a)** Gene class selected from G-protein coupled receptors. **(b)** Differential and correlative search option bar. **(c)** Heat map rendering of a set of probes from the G-protein coupled receptor class with color bar range adjustment. Data from several brains is simultaneously presented

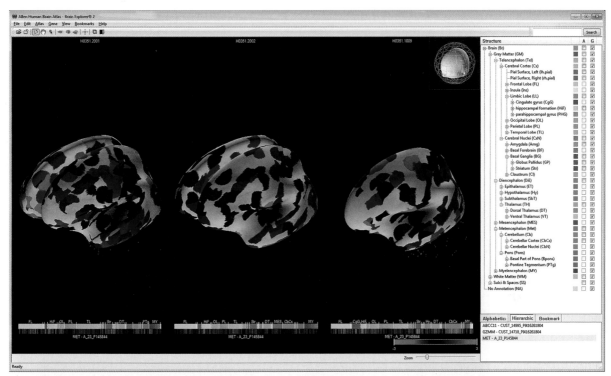

Fig. 62.12 Visualizations in the Allen Human Brain Atlas of MET (met proto-oncogene) on an inflated representation of the human cortex for three human brain specimens. MET has a strong dorsal-ventral expression gradient as can be seen from higher red expression in occipital cortex in each of the three brains. The human brain ontology is shown on the *right*

pipeline is semi-automated and composed of three main parts: microarray data sample normalization and MRI processing and registration of microarray data. A custom Agilent array [62.19] containing 60 000 probes was designed and used throughout this project. This array contains 44 000 Agilent Whole Human Genome probes supplemented with an additional 16 000 probes where at least two different probes, and where possible, on different exons are available for 93% of genes with Entrez Gene IDs.

There are three steps to normalizing the gene expression microarray data. First, normalization is done at the batch level. This includes standard preprocessing to adjust for probe related technical biases including GC (guanine-cytosine) content, array spatial probe position, and intensity bias. Expression data from the controls of the first batch is used as a standard. For all other batches, mean expression intensity of the control samples was adjusted. Within-batch normalization was then performed using a 75% centering algorithm where expression distribution of all samples in the batch were normalized to have the same 75th percentile expression

values. Next, all samples within the brain were normalized using a cross-batch normalization algorithm. This includes aligning brain specific controls between the brains as well as an extra step to compensate for the differences in sampling technique between macro and laser capture microdissection samples. The online resources provide a detailed description of this normalization process. The final step in the normalization of the microarray data is cross-brain normalization whereby data is aligned to the 75th percentile of expression values of the control samples of each brain to the first brain. This step allows for comparison of data across the different subjects within the data set. The normalized expression data forms the core of search and visualization of this data that is displayed as a heat map in the web application. After a search, using tools for differential expression, or correlative search, the heat map displays normalized microarray values using a color map (Fig. 62.11). The default heat map colors are in the green-red scale where green is relatively low expression and red is relatively high expression. Each row in the heat map represents a probe while the col-

Part L | 62.3

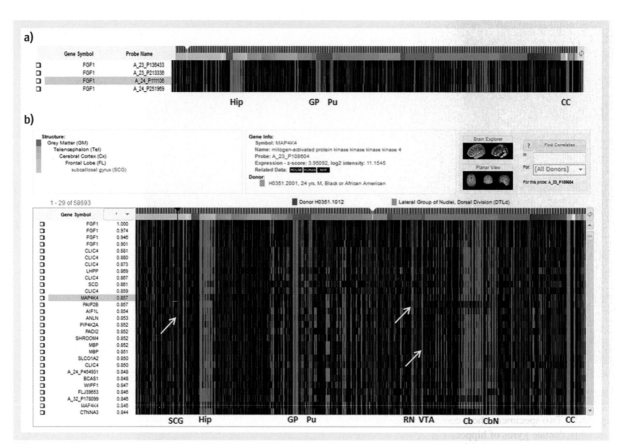

Fig. 62.13a,b Heat map representation of gene expression data in the Allen Human Brain Atlas. (**a**) Searching the MAP3K gene class for fibroblast growth factors shows a consistent pattern for FGF1. There is agreement amongst the probes for high expression in the striatum (GP, Pu) and white matter CC, with down regulation in most areas of the hippocampus (Hip). (**b**) Performing a correlation search yields the results shown. These high correlations (Pearson > 0.7) also show high expression in the cerebellar nuclei (CbN), red nucleus (RN), ventral tegmental area (VTA), and possibly subcallosal gyrus (SCG). These results can be confirmed by downloading the actual data from the website. *White arrows* show structures with common expression

umn represents a sample from a particular region of the brain.

MRI was obtained for each brain used in the data set. All T1-weighted volumes were anatomically labeled using FreeSurfer's *recon-all* pipeline [62.18] and the entire volume was mapped into MNI space. To enable a unified navigation and visualization of the histology sections, microarray data, and MRI, manual registration was done on each brain. Briefly, the process involves constructing virtual slab images from 2×3 Nissl sections where RNA samples were taken. These virtual slab images were then placed in the context of the T1-weighted volume. Because of significant deformation, registration landmarks of the gene expression sampling points were manually placed on the virtual slab and MR

volume using BioImage Suite [62.20]. When needed, other slab based images (blockface and 6×8 histology images) were registered to the virtual slab image. Results from the registration processes allowed the construction of the 2-D and 3-D MR-based navigation maps where all the histology data, gene expression data for a particular gene, and MRI are interconnected. These views are presented within both the web application and the Brain Explorer desktop application (Fig. 62.12).

62.3.2 Understanding Gene Expression in the Human Brain

The fibroblast growth factors (FGF) are a family of growth factors that form part of the MAPK signaling

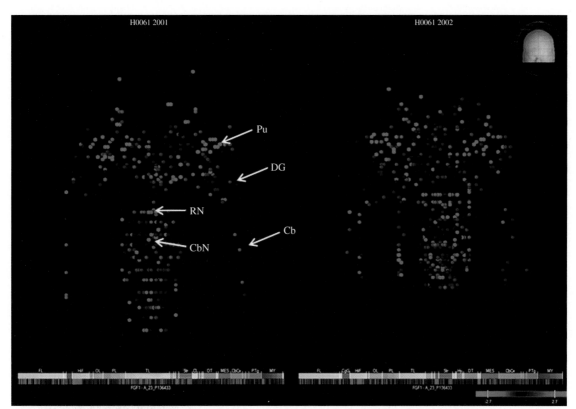

Fig. 62.14 Two specimens show consistent expression profiles in coronal view. Highlighted structures are labeled putamen (Pu), dentate gyrus of hippocampus (DG), cerebellum (CB), cerebellar nuclei (CbN), and red nucleus (RN) of the midbrain. This expression profile can be consistently seen in each of the specimens online

cascade [62.21]. These genes are involved in a variety of functions from development to wound healing. During CNS development FGFs play important roles in neurogenesis, axon growth, and differentiation. In addition, FGFs are major determinants of neuronal survival both during development and during adulthood. FGF1 and FGF2 seem to be involved in the regulation of synaptic plasticity and processes attributed to learning and memory [62.22], and are implicated in the regulation of synaptic plasticity and processes attributed to learning and memory, at least in the hippocampus [62.23].

A search for the MAPK cascade from the Allen Human Brain Atlas home page enables us to identify the heat map expression for FGF1 as shown in Fig. 62.13a. Amongst other features in Fig. 62.13a there is pronounced expression in the striatum in globus pallidus (GP) and putamen (Pu) and down regulation in hippocampus (Hip). This gene is also highly expressed in the corpus callosum (CC). Using the correlative search tool, we can find a much larger set of human expression

probes with the same profile shown in Fig. 62.13b. The results for the unique subset of 301 genes returned with correlation at least 0.7 clearly identify up regulation in the cerebellar nuclei (CbN), red nucleus (RN), and ventral tegmental area (VTA) and possibly the subcallosal gyrus (SCG), shown with arrows in Fig. 62.13b, as well as pronounced down regulation in the cerebellum (Cb). These patterns would be much less likely to be perceived when examining the set of probes for FGF1 alone in Fig. 62.13a despite their high consistency.

The resulting gene set is expectedly enriched in gene ontology (GO) [62.24] functional categories of cytoskeletal and membrane organization, but perhaps more curiously highly significant for neuron and axon ensheathment and myelination (BH corrected p-value $< 3.5 \times 10^{-5}$). One gene found among the set is MYB, myelin basic protein is an oligodendrocyte marker and believed to be important in myelination [62.25]. This in combination with high corpus collusum expression suggests that the returned gene set is richly expressing oligodendrocytes. Other gene families represented by this set include kinesins, transmembrane proteins, G-protein coupled receptors, and zinc fingers, and suggest analogous roles to FGF1 for these genes. Interestingly, the only other related family member which appears in this list is FGFR2, the fibroblast growth factor receptor 2 with correlation 0.738.

Finally, the Allen Human Brain Explorer gives a three dimensional perspective on gene expression of FGF1. In Fig. 62.14 coronal views of two specimens are shown with several structures of interest labeled in common with Fig. 62.13. The coronal view illustrates the symmetry of gene expression as well as its reproducibility between two specimens. Other specimens can be viewed in the online application.

62.4 Beyond the Atlases

While each resource within the Allen Brain Atlas portal is a standalone application, when combined, they provide unprecedented opportunities to mine and analyze data in the mammalian brain. The Allen Brain Atlas portal provides capabilities to access data across projects as well as programmatic access to the huge amount of data generated.

62.4.1 Navigating Across Atlases and Projects

The portal environment provides a cross-project search tool. Searching on genes, gene family, species, experimental data modalities, gender, age, and anatomical structures [62.26] is supported. The text-based search allows for data discovery in the Allen Brain Atlas database and the query can be further refined in the user interface by selecting or deselecting search facets. Another convenient data discovery mechanism is the *Related Data* present within each of the atlases. The underlying search mechanism is the same as the text-based search across all atlases. However, this functionality is available within each of the atlases and the result of the search is context-sensitive based on the view of the experimental data the user is currently accessing. The context sensitivity, initially implemented at just the gene level, has since been expanded to including age, sex, anatomic structures, and experimental data modalities.

62.4.2 Application Programming Interface

As with most large scale databases, the ability to programmatically access the data is critical for large scale data mining and analysis. The Allen Brain Atlas application programming interface (API) enables all of the data within the portal to be accessed and downloaded. Using the Representational State Transfer (REST) software architecture [62.27], data can be requested in

JavaScript object notation (JSON), extensible markup language (XML), or comma-separated values (CSV) formats using the standard HTTP GET method. Services available for the Allen Mouse Brain Atlas and Allen Developing Mouse Brain Atlas include:

- ISH experiments and Nissl series metadata
- Images at multiple resolutions
- Grid-level gene expression values registered to the reference atlas space
- Reference atlas ontologies and 3-D reference models
- Image-to-image synchronization service
- Gene expression search services including gene search, differential expression, and *NeuroBlast*
- Summarized expression level and density values for computed structures.

In the Allen Human Brain Atlas, BrainSpan Atlas of the Developing Human Brain, and NIH Blueprint Non-Human Primate Atlas, services in the API include:

- Experiment metadata and normalized gene expression values for all microarray and RNA sequencing data
- Raw microarray gene expression data
- Microarray gene expression values and samples mapped to MNI space in the Allen Human Brain Atlas
- Search services including gene search, differential expression and *NeuroBlast* search
- MRI and DTI files and metadata for donors
- Ontology for brain structures
- Search and download of all high resolution ISH and other histological images.

In addition to the API, example applications demonstrating some of the most common data retrieval methods are also provided along with source code. For certain examples, source code in multiple programming languages are also available. While the examples in this library will continue to grow, the current library includes the following examples:

- Retrieving the adult mouse ontology and summarized gene expression data for the gene *Adora2a* and visualization of the data using a sunburst chart
- Retrieving the adult mouse expression grid data for two data sets and computing the fold change of the structures between the data sets
- Retrieving all adult mouse expression data and structure information to compute correlations between gene expression and structures of the atlas, computed structure similarity network based on gene expression is then visualized using a Javascript library
- An application that uses the open source Visualization Toolkit and Qt application user interface to visualize expression data and anatomic structure in 3-D

- Mechanism to download the annotated images for the reference atlas and manually annotated gene expression data from the developing mouse and map the gene expression values to the reference atlas map
- Downloading two sets of microarray gene expression values from two different genes from the Allen Human Brain Atlas and plot the data in a scatterplot
- Downloading Allen Human Brain Atlas expression data and identifying the Brodmann areas of the samples using the Statistical Parametric Mapping (SPM) Matlab package
- Downloading the raw microarray data and reprocess and perform correlation analyzing using the Weighted Gene Co-expression Network Analysis (WGCNA) [62.28]
- A UI driven query builder that builds the HTTP URLs to retrieve the data from the Allen Brain Atlas without the need to learn the URL syntax created from the Allen Brain Atlas API.

62.5 Conclusions

Large scale data generation and integration is a cornerstone of modern science and information technology. These concepts have come to the forefront with emerging big data science and its need for effective data mining and discovery. Neuroinformatics and information sciences form an essential component in ongoing design considerations for the Allen Brain Atlas resources. To maximize capacity for user scientific utility it is essential to consider aspects of

both neuroinformatics tool development and appropriate user experience. Techniques employed from these disciplines can play a role not only in the development of the resources but also scientific data analysis and knowledge discovery. This chapter presented an overview of how the Allen Institute for Brain Science has employed these methods. Further information is available in the references and online documents.

References

62.1 Allen Brain Institute: Brain Atlas, available online at www.brain-map.org

62.2 E.S. Lein, M.J. Hawrylycz, N. Ao, M. Ayres, A. Bensinger, A. Bernard, A.F. Boe, M.S. Boguski, K.S. Brockway, E.J. Byrnes, et al.: Genome-wide atlas of gene expression in the adult mouse brain, Nature **445**, 168–176 (2007)

62.3 L.L. Ng, S.D. Pathak, C.L. Kuan, C. Lau, H. Dong, A.J. Sodt, C.N. Dang, B. Avants, P. Yushkevich, J.C. Gee, et al.: Neuroinformatics for genome-wide 3-D gene expression mapping in the mouse brain, IEEE Trans. Comput. Biol. Bioinforma. **4**, 382–393 (2007)

62.4 L.L. Ng, S.M. Sunkin, D. Feng, C. Lau, C. Dang, M.J. Hawrylycz: Large-scale neuroinformatics for in situ hybridization data in the mouse brain, Int. Rev. Neurobiol. **104**, 159–182 (2012)

62.5 L. Madisen, T.A. Zwingman, S.M. Sunkin, S.W. Oh, H.A. Zariwala, H. Gu, L.L. Ng, R.D. Palmiter, M.J. Hawrylycz, A.R. Jones, et al.: A robust and high-throughput cre reporting and characterization system for the whole mouse brain, Nat. Neurosci. **13**, 133–140 (2010)

62.6 L. Madisen, T. Mao, H. Koch, J.M. Zhuo, A. Berenyi, S. Fujisawa, Y.W. Hsu, A.J. Garcia III, X. Gu, S. Zanella, et al.: A toolbox of cre-dependent optogenetic transgenic mice for light-induced activation and silencing, Nat. Neurosci. **15**, 793–802 (2012)

62.7   H. Zeng, L. Madisen: Mouse transgenic approaches in optogenetics, Prog. Brain Res. **196**, 193–213 (2012)

62.8   H.W. Dong: *The Allen Reference Atlas: A Digital Color Brain Atlas of the C57BL/6J Male Mouse* (Wiley, Hoboken 2008)

62.9   P. Xu, J. Xu, Z. Li, Z. Yang: Expression of TRPC6 in renal cortex and hippocampus of mouse during postnatal development, PLoS ONE **7**, e38503 (2012)

62.10   J. Zhou, W. Du, K. Zhou, Y. Tai, H. Yao, Y. Jia, Y. Ding, Y. Wang: Critical role of TRPC6 channels in the formation of excitatory synapses, Nat. Neurosci. **11**, 741–743 (2008)

62.11   Y. Tai, S. Feng, R. Ge, W. Du, X. Zhang, Z. He, Y. Wang: TRPC6 channels promote dendritic growth via the CaMKIV-CREB pathway, J. Cell Sci. **121**, 2301–2307 (2008)

62.12   K. Leuner, W. Li, M.D. Amaral, S. Rudolph, G. Calfa, A.M. Schuwald, C. Harteneck, T. Inoue, L. Pozzo-Miller: Hyperforin modulates dendritic spine morphology in hippocampal pyramidal neurons by activating Ca^{2+}-permeable TRPC6 channels, Hippocampus **23**, 40–52 (2013)

62.13   M. Hawrylycz, L. Ng, D. Page, J. Morris, C. Lau, S. Faber, V. Faber, S. Sunkin, V. Menon, E. Lein, et al.: Multi-scale correlation structure of gene expression in the brain, Neural Netw. **24**, 933–942 (2011)

62.14   H. Hagihara, K. Toyama, N. Yamasaki, T. Miyakawa: Dissection of hippocampal dentate gyrus from adult mouse, J. Vis. Exp. **17**(33), 1543 (2009)

62.15   J. Hauser, T.A. Sontag, O. Tucha, K.W. Lange: The effects of the neurotoxin DSP4 on spatial learning and memory in Wistar rats, Atten. Defic. Hyperact. Disord. **4**, 93–99 (2012)

62.16   A. Lavado, O.V. Lagutin, L.M. Chow, S.J. Baker, G. Oliver: Prox1 is required for granule cell maturation and intermediate progenitor maintenance during brain neurogenesis, PLoS Biol. **8**, e1000460 (2010)

62.17   M.J. Hawrylycz, E.S. Lein, A.L. Guillozet-Bongaarts, E.H. Shen, L. Ng, J.A. Miller, L.N. van de Lagemaat, K.A. Smith, A. Ebbert, Z.L. Riley, et al.: An anatomically comprehensive atlas of the adult human brain transcriptome, Nature **489**, 391–399 (2012)

62.18   A.M. Dale, B. Fischl, M.I. Sereno: Cortical surface-based analysis. I. Segmentation and surface reconstruction, NeuroImage **9**, 179–194 (1999)

62.19   www.genomics.agilent.com

62.20   www.bioimagesuite.org

62.21   R. Seger, E.G. Krebs: The MAPK signaling cascade, FASEB J. **9**, 726–735 (1995)

62.22   B. Reuss, O. von Bohlen und Halbach: Fibroblast growth factors and their receptors in the central nervous system, Cell Tissue Res. **313**, 139–157 (2003)

62.23   S. Zechel, S. Werner, K. Unsicker, O. von Bohlen und Halbach: Expression and functions of fibroblast growth factor 2 (FGF-2) in hippocampal formation, Neuroscientist **16**, 357–373 (2010)

62.24   http://david.abcc.ncifcrf.gov/

62.25   C.M. Deber, S.J. Reynolds: Central nervous system myelin: Structure, function, and pathology, Clin. Biochem. **24**, 113–134 (1991)

62.26   S.M. Sunkin, L. Ng, C. Lau, T. Dolbeare, T.L. Gilbert, C.L. Thompson, M. Hawrylycz, C. Dang: Allen Brain Atlas: An integrated spatio-temporal portal for exploring the central nervous system, Nucleic Acids Res. **41**(41), D996–D1008 (2012)

62.27   http://en.wikipedia.org/wiki/Representational_state_transfer

62.28   P. Langfelder, S. Horvath: WGCNA: An R package for weighted correlation network analysis, BMC Bioinformat. **9**, 559 (2008)

Glossary

A

Adaptation

The process of structural and functional changes of a system in order to improve its performance in a changing environment.

Alzheimer's disease (AD)

A brain disorder that is clinically characterized by a global decline of cognitive function that progresses slowly and leaves end-stage patients in custodial care. All of the currently used drugs are of limited benefit, because they have only modest symptomatic effects. Other drugs are used to manage mood disorder, agitation, and psychosis in later stages of the disease, but no treatment with a strong disease-modifying effect is currently available.

Approximate reasoning

A process of achieving approximate, imprecise solutions and/or conclusions often based on inexact facts and uncertain rules.

ART (Adaptive resonance theory)

Refers to both a cognitive and computational theory of the brain.

Artificial life

A modeling paradigm that assumes that many individuals are governed by the same or similar rules to grow, die, and communicate with each other. Ensembles of such individuals exhibit repetitive patterns of behavior.

Artificial neural network (ANN)

Biologically inspired computational model which consists of processing elements (neurons) and connections between them with coefficients (weights) bound to the connections. Training and recall algorithms are also attached to the structure.

Atom

The smallest particle of a chemical element that retains its chemical properties. Most atoms are composed of three types of subatomic particles which govern their properties: electrons (with a negative charge), protons (with a positive charge), and neutrons (without charge).

Automatic speech recognition system (ASRS)

A computer system which aims at providing enhanced access to machines via voice commands. Some speech recognition systems need to be trained by the voice of the intended user and are known as *speaker dependent* systems. Other systems do not need to be trained and are known as *speaker independent*.

B

Backpropagation training algorithm

An algorithm for supervised learning in artificial neural networks. During a training phase, after input data is entered and propagated forwards to the output of the model, the difference between the current output value and the expected output value is propagated backwards through the network as an error to adjusts the connection weights so that the next time the same data is entered, the error will be smaller. A gradient descent rule is used for finding the optimal connection weights w_{ij} that minimize the global error E. A change of a weight Δw_{ij} at a cycle $(t+1)$ is in the direction of the negative gradient of the error E.

Bayesian probability

The following formula, which represents the conditional probability between two events C and A, is known as Bayes' formula (Tamas Bayes, eighteenth century): $p(A|C) = p(C|A) . p(A)/p(C)$. Using Bayes' formula involves difficulties, mainly concerning the evaluation of the prior probabilities $p(A)$, $p(C)$, $p(C|A)$. In practice (for example, in statistical pattern recognition), the latter is assumed to be of a Gaussian type. Bayes' theorem also assumes that if the condition C consists of condition elements $C1, C2, \ldots, Ck$ they are independent (which may not be the case in some applications).

Blue Brain Project

A research project hosted by EPFL in Lausanne, aiming at the development of biologically adequate brain models (http://bluebrainproject.epfl.org).

Brain Atlas

A repository of data and knowledge and software tools to explore different brain structures and functions and genes related to them produced by the Allen Brain Science Institute (http://www.brain-map.org).

C

Calcium ions

Ca^{2+} ions are stored in the synapses and enter the neuron via voltage-gated calcium channels and the NMDA receptor-channel complex. The intracellular calcium concentration Ca^{2+} is the principal trigger for the induction of LTD/LTP.

Catastrophic forgetting

A phenomenon which represents the inability of a learning system to retain previously learned information. An example is an ANN model that *forgets* what has been learned from previous data examples, when they are no longer presented to it, but other, new examples are presented instead.

Cellular automata

A set of regularly connected simple finite automata. The simple automata communicate and compute together when working on a single global task. Cellular automata may be able to grow, to shrink, and to reproduce, thus providing a flexible environment for computation with arbitrary complexity. They are also called *non-von Neumann* models because of their difference from the standard digital von Neumann computer organization.

Central dogma of molecular biology

It states that DNA, which resides in the nucleus of a cell, transcribes into RNA, and then RNA translates into proteins; a process which is continuous, evolving, so that some proteins cause genes to transcribe, etc.

Center-of-gravity defuzzification method (COG)

A method for defuzzification, e.g., transforming a membership function B of an output fuzzy variable in a fuzzy system into a crisp value y such that y is the geometrical center of the area occupied by B.

Cerebellum

A part of the brain, below the cortex, that regulates body movements, reflexes, balance, posture, etc.

Cerebral cortex

Three quarters of all 10^{10} neurons in the human brain form a 4–6 mm thick cerebral cortex that constitutes a heavily folded brain surface. The cerebral cortex is thought to be a seat of cognitive functions, like perception, imagery, memory, learning, thinking, etc. The cortex cooperates with evolutionary older subcortical nuclei that are located in the middle of the brain, in and around the so-called brain stem.

Chaos

A complex behavior of a *nonlinear dynamical system* that follows some underlying rules. The state of the system at time (t) depends on its states at previous time moments and there is no periodically repetitive pattern of exactly the same states of the system over time.

Chaotic attractor

An area or points from the phase space of a chaotic process where the process often goes through time, but without repeating the same trajectory.

Classification problem

A generic AI problem which arises when it is necessary to associate an object with some already existing groups, clusters, or classes of objects.

Clustering

The process of finding groups (called clusters) of similar data elements. A cluster is defined by its center and its members' data elements. Similarity between data elements can be measured in different ways, e.g.,

Euclidean distance, correlation, and Hamming distance. Usually there is no previous knowledge of the groups' definitions. The membership of elements to clusters can be *crisp*, where one element can belong to only one cluster, or fuzzy – each element may belong to several clusters with a different membership degree, all of them summing to 1. This is one of the most important unsupervised learning methods in the area of machine learning. Well-known techniques for clustering are k-means, fuzzy c-means, etc. Data clustering is widely used for data mining.

Computational intelligence (CI)

CI encompasses methods for information processing based on learning, reasoning, dealing with incomplete and uncertain data, and their numerous applications in almost all areas of science, engineering, and human activities. These methods include probabilistic methods, neural networks, rule-based and fuzzy systems, evolutionary computation, hybrid systems, and biologically inspired methods. Many CI methods of are inspired by nature and human intelligence.

Computational neurogenetic modeling (CNGM)

This is a computational modeling approach that integrates genetic, proteomic, and brain activity data into a complex neural network system to perform data analysis, modeling, prognosis, and knowledge extraction that reveals the relationship between network performance and genetic information. It can be used for both brain/gene data and engineering applications.

Computer tomography (CT)

This is based on the classical x-ray principle. X-rays reflect the relative density of the tissue through which they pass. If a narrow x-ray beam is passed through the same point at many different angles, it is possible to construct a cross-sectional visual image of the brain. A 3-D x-ray technique is called CAT (computerized axial tomography). CT is noninvasive and shows only the anatomical structure of the brain, not its function.

Conditional probabilities

The probability $p(A|C)$ defines the probability of the event A to occur, given that the event C has occurred. It is described by the formula:
$p(A|C) = p(A \wedge C)/p(C)$, where $p(A \wedge C)$ is the probability of the two events A and C to be observed together and $p(C)$ is the probability of the event C to happen.

Connectionist production system

A connectionist (neural network) system that implements production rules of the form IF C THEN A, where C is a set of conditions and A is a set of actions.

Control

The process of acquiring information for the current state of an object or a process and emitting control signals in order to keep the object/process in desirable states.

D

Data analysis

Data analysis aims at answering important questions about a process (or an object) under investigation. Some exemplar questions are: What are the statistical parameters of the data representing the process, e.g. mean, standard deviation, distribution? What is the nature of the process, random, chaotic, periodic, stable, etc.? How is the available data distributed in the problem space, e.g., clustered into groups, sparse, covering only patches of the problem space and, therefore, not enough to rely on fully when solving the problem, uniformly distributed? Is there missing data? Is there a critical obstacle which could make the process of solving the problem by using data impossible? What other methods can be used either in addition to, or instead of, methods based on data?

Data, information, and knowledge

Data are the numbers, the characters, the quantities operated on by a computer. *Information* is the ordered, structured, interpreted data. *Knowledge* is the theoretical or practical understanding of a subject, gained experience, true and justified belief, the way we do things. For example, the number 34 is data; *34 degrees of temperature in Auckland today* is information; the expression *IF temperature is too high THEN risk of stroke increases* is a piece of knowledge.

Decision support system

An intelligent system that supports the human decision making process. Such a system analyzes available data in a given problem space and suggests decisions. Examples are automated trading systems on the Internet, systems that grant loans through electronic submissions, and medical decision support systems for cardiovascular event prediction.

Defuzzification

The process of calculating a single output numerical value for a fuzzy variable in a fuzzy system when a fuzzy membership function for this variable is given.

Destructive learning

A learning technique, usually in artificial neural network models, that modifies an initial neural network architecture, e.g., removes connections, for the purpose of better future learning.

Discrete Fourier transform (DFT)

The transformation of a discrete input function (usually in the time domain) into another function in the frequency domain.

Distance between data points

The distance between two data points a and b in an n-dimensional geometrical space can be measured in several ways. The most widely used formulas are: Hamming distance, $D_{ab} = \Sigma |a_i - b_i|$ and Euclidean distance, $E_{ab} = \sqrt{\Sigma (a_i - b_i)^2 / n}$.

Distributed representation

A way of encoding information, usually in a neural network model, where a concept or a value for a variable is represented by the collective activation of a group of neurons.

DNA (deoxyribonucleic acid)

This is a chemical chain, present in the nucleus of each cell of an organism. It consists of pairs of small chemical molecules (bases) ordered in a double helix, which are: adenine (A), cytidine (C), guanidine (G), and thymidine (T), linked together by a deoxyribose sugar phosphate nucleic acid backbone. Almost all cells in an organism contain the same DNA information, but different parts of the DNA, different genes, express in different parts of the organism and produce different proteins.

E

Electroencephalography (EEG)

An EEG is a recording of electrical signals from the brain made by attaching surface electrodes to the subject's scalp. These electrodes record electric signals naturally produced by the brain, called brainwaves. EEGs allow researchers to follow electrical potentials across the surface of the brain and observe changes over split seconds of time.

Elitism (in genetic algorithms and other evolutionary optimization algorithms)

The fittest members of a population at generation (t) is copied unmodified into the population of the next generation ($t + 1$). The intention of this strategy is to reduce the chance of losing the best genotypes. If elitism were a principle of human evolution, we would still have Leonardo da Vinci among artists and scientists nowadays.

Epilepsy

This is a brain disorder characterized by the occurrence of at least two unprovoked seizures. Seizures are the manifestation of abnormal hypersynchronous discharges of neurons in the cerebral cortex. The clinical signs or symptoms of seizures depend on the location and extent of the propagation of the discharging cortical neurons.

Evolutionary computation

A computational paradigm that uses principles from natural evolution, such as genetic representation, mutation, survival of the fittest, population of individuals, and generations of populations. Evolutionary computation is mainly used as a population-generation based optimization technique where the best or close to it solution to a problem is achieved through evaluating many individual solutions in a population over generations.

Evolutionary programming

Evolutionary algorithms applied for the automatic creation or optimization of computer programs.

Evolutionary strategies

Strategies that use evolutionary algorithms to represent a solution to a problem as a single chromosome and evaluate different mutations of this chromosome over generations through a fitness function. This process is carried out until a satisfactory solution is found.

Evolving connectionist systems (ECOS)

Artificial neural networks proposed by Kasabov (1998) that develop (evolve) their structure and functionality from incoming data in an adaptive, incremental way.

Evolving intelligent systems (EIS)

Intelligent systems that are characterized by adaptation and incremental evolving of knowledge. The methods used in such systems are mainly based on neural networks, but may include many other techniques from the area of computational intelligence.

Expert system

Knowledge-based systems that provide expertise, similar to that of human experts in a restricted application area, for the solution of problems in that area. An expert system consists of the following main blocks: knowledge base, data base, inference engine, explanation module, user interface, and knowledge acquisition module.

Explanation in an intelligent system

This is a desirable property for many AI systems. It means tracing, in a contextually comprehensible way, the process of inferring the solution and reporting it. Explanation is easier for AI symbolic systems where sequential inference takes place.

F

Fast Fourier transformation (FFT)

This is a fast algorithm for discrete Fourier transformation (DFT). A nonlinear transformation applied on time series data to transform the signal taken within a small portion of time (the time scale domain) into a vector in the frequency scale domain.

Feed-forward neural network

A neural network in which there are no connections back from the output to the input neurons.

Finite automaton

A computational model represented by a set X of inputs, a set Y of outputs, a set Q of internal states, and two functions f_1 (state transfer function) and f_2 (output function): $f_1: X \times Q \to Q$, i.e., $[x, q(t)] \to q(t+1)$, $f_2: X \times Q- \to Y$, i.e., $[x, q(t)] \to y(t+1)$, where $x \in X$, $q \in Q$, $y \in Y$, t, and $(t+1)$ represent two consecutive time moments.

Fractal

An object which occupies a fraction (called *embedding space*) of a standard space (i. e., space with integer numbers for dimensions). The dimensionality of fractal can be, e.g., 2.4, rather than the standard 3-D.

Functional MRI (fMRI)

This combines visualization of brain anatomy with the dynamic image of brain activity into one comprehensive scan. This noninvasive technique measures the ratio of oxygenated to deoxygenated haemoglobin which have different magnetic properties. Active brain areas have higher levels of oxygenated haemoglobin than less active areas. An fMRI can produce images of brain activity at the time scale of a second with very precise spatial resolution of about 1–2 mm. Thus, fMRI provides both an anatomical and functional view of the brain.

Fuzzification

The process of finding the membership degree $\mu A(x')$ to which a value x' of a fuzzy variable x belongs to a fuzzy set A defined on the same universe as the variable x.

Fuzzy clustering

A procedure of clustering data into possibly overlapping clusters, such that each of the data elements may belong to each of the clusters to a certain degree. The procedure aims at finding the cluster centers V_i ($i = 1, 2, \ldots, c$) and the cluster membership functions μ_i which define to what degree each of the n data elements belongs to the i-th cluster. The number of clusters c is either defined a priori or chosen by the clustering procedure (evolving clustering). The result of a clustering procedure can be represented as a fuzzy relation $\mu_{i,k}$, such that: (i) $\Sigma \mu_{i,k} = 1$, for each $k = 1, 2, \ldots, n$ (the total membership of an instance to all the clusters equals 1); (ii) $\Sigma \mu_{i,k} > 0$, for each $i = 1, 2, \ldots, c$ (there are no empty clusters).

Fuzzy control

The application of fuzzy logic to control problems. A fuzzy control system is a fuzzy system applied to solve a control problem.

Fuzzy expert system

An expert system to which methods of fuzzy logic are applied. Fuzzy expert systems use fuzzy data, fuzzy rules, and fuzzy inference, in addition to the standard ones implemented in ordinary expert systems.

Fuzzy logic

A logic system that is based on fuzzy relations and fuzzy propositions, the latter being defined on the basis of fuzzy sets.

Fuzzy neural network

An artificial neural network model that can be interpreted as a fuzzy system. The model can have neurons that represent fuzzy concept (e.g., small, large, fuzzy cluster). The connections between these neurons can be represented and interpreted as fuzzy relations, e.g. IF temperature is extremely high THEN risk of stroke is high.

Fuzzy propositions

Propositions which contain fuzzy variables with their fuzzy values. The truth value of a fuzzy proposition X is A is given by a membership function μ_A.

Fuzzy relations

Fuzzy relations link two fuzzy sets or two fuzzy variables in a predefined manner. Fuzzy relations make it possible to represent ambiguous relationships, such as: *the grades of the 3rd and 2nd year classes are similar*, *team A performed slightly better than team B*, or *the more fat you eat, the higher your risk of heart attack*.

G

GenBank

A repository of genes and their functions across species and diseases, maintained by the NCBI (http://www.ncbi.nlm.nih.gov/Genbank/index.html).

Gene expression atlas

A repository of gene expression data across species and diseases (http://expression.gnf.org/cgi-bin/index.cgi).

Gene ontology

An ontology knowledge repository system designed to produce a controlled vocabulary that can be applied to all organisms even if knowledge of genes and proteins is changing.

Gene regulatory network (GRN)

A biological or computational network of genes connected between each other according to their interaction in time.

Gene-brain ontology

An ontology knowledge repository system that includes knowledge, data and known relationships between brain structures and functions and genes that are related to them.

Generalization

The process of matching new input data to a model, system, or in principle, to an existing set of problem knowledge, in order to obtain an output value (e.g., solution) that corresponds to this input data.

Genes

Parts of a DNA sequence that are transcribed into RNA and translated into proteins or alternatively produce microRNA (not translated into proteins). Genes are the carrier of information that is passed from one generation of species to another in an evolutionary process.

Genetic algorithms (GA)

These are algorithms for solving complex multivariant combinatorial and organizational problems by

employing methods of evolutionary computation that are analogous to evolution in nature. There are several general steps that a genetic algorithm cycles through: generate a population of individuals; evaluate the fitness (goodness) of each individual; select the best individuals; perform cross-over operation between these individuals; mutate individuals, if necessary. These steps are repeated all over again until an acceptable solution is found or the time for performing the algorithm has expired.

Glutamate neurotransmitters

Molecules released in the synapses during afferent activity that bind to AMPA, NMDA, and metabotropic glutamate (mGlu) receptors to produce postsynaptic response.

Goodness function (fitness function)

A function that can be used to measure the appropriateness of an individual element from a population of individuals at a certain generation over time. An individual element would represent a possible solution to a problem, e.g., the shortest path from one city to another, or a set of genes to diagnose cancer.

H

Hebbian learning law

A generic learning principle which states that a synapse connecting two neurons i and j increases its strength w_{ij} if the two neurons i and j are repeatedly and simultaneously activated by input stimuli.

Hopfield network

A fully connected feedback neural network which is an auto-associative memory. It can be trained to recognize input patterns and to recover them if they are presented only partially at a later stage. It is named after its inventor John Hopfield.

Human Brain Project

An EU funded long-term programme aiming at understanding the brain, simulating brain functions and structures and creating brain-like computer systems (see the Foreword by K. Meier).

I

Image filtering

A transformation of an original image through a set of operations that use the original pixel intensities of the image and apply a two-dimensional array of numbers, which is known as a kernel. This process is also called convolution.

Independent component analysis

The process of separating independent components of multidimensional time series data, such as signal and noise.

Inference in an AI system

The process of matching new data from a domain space to the knowledge existing in an AI system and obtaining output values.

Information entropy

A measure for the level of uncertainty (or unpredictability) associated with a random variable. The more unpredictable an event is, the higher the information entropy.

Information retrieval

The process of retrieving relevant information from a data base.

Information science

This is the area of science that develops methods and systems for information and knowledge processing regardless of the domain specificity of this information. Information science includes the following subject areas: data collection and data communication (sensors and networking); information storage and retrieval (data base systems); methods for information processing (information theory); creating computer programs and information systems (software engineering and system development); acquiring, representing, and processing knowledge (knowledge–based systems); and creating intelligent systems and machines (artificial intelligence, knowledge engineering).

Information

Collection of structured data. In its broad meaning it includes knowledge as well as simple meaningful data.

Initialization in ANN

The process of setting the connection weights in an ANN to some initial values before starting the training algorithm.

Intelligent system (IS)

An information system that manifests features of intelligence, such as learning, generalization, reasoning, adaptation, knowledge discovery, and applies these to complex tasks such as decision making, adaptive control, pattern recognition, speech, image and multimodal information processing, etc.

Interaction (human–computer)

Communication between a user and a computer system.

K

Knowledge engineering

The area of science and engineering that deals with data, information and knowledge representation in machines, information processing, and finally, knowledge elucidation and knowledge discovery.

Knowledge

Concise presentation of facts, skills, previous experience, principles, definitions, etc., that is

interpretable under different conditions. Knowledge resides in the human brain. As a term it is used to represent information in a computer system that can be interpreted by humans.

Knowledge-based neural networks (KBNN)

These are prestructured ANN that allow for data and machine knowledge manipulation, including learning from data, rule insertion, rule extraction, adaptation, and reasoning. KBNN have been developed either as a combination of symbolic AI systems and ANN, or as a combination of fuzzy logic systems and ANN, or as other hybrid systems. Rule insertion and rule extraction are typical operations for a KBNN to accommodate existing knowledge along with data, and to produce an explanation of what the system has learned.

Kohonen self-organizing map (SOM)

A self-organized ANN that uses unsupervised learning to map multidimensional input vectors into low-dimensional matrix known as map. The concept of SOM was first introduced and developed by the Finish scientist Prof. Teuvo Kohonen.

L

Laws of inference in fuzzy logic

The way fuzzy propositions are used to make inference over new facts. The following are the two most used laws illustrated on two fuzzy propositions A and B. (a) Generalized modus ponens: $A \rightarrow B$, and $A' \therefore B'$, where $B' = A'o(A \rightarrow B)$; (b) generalized modus tollens (law of the contrapositive): $A \rightarrow B$, and $B', \therefore A'$, where $A' = (A \rightarrow B)oB'$.

Learning vector quantization algorithm (LVQ)

A supervised learning algorithm, which is an extension of the unsupervised Kohonen self-organized network learning algorithm.

Learning

Process of obtaining new information (possibly interpretable as knowledge) from data and existing information.

Linear transformation

Transformation $f(x)$ of a variable x such that f is a linear function of x, for example, $f(x) = 2x + 1$.

Linguistic variable

A variable that takes fuzzy values that have linguistic meaning, e.g. the variable *temperature* can take a fuzzy value of *very high temperature*.

Local representation in a neural network

A way of encoding information in an ANN in which every neuron represents one concept or one variable.

Logic system

An abstract system that consists of four parts: an alphabet – a set of basic symbols from which more complex sentences (constructions) can be made; syntax – a set of rules or operators for constructing sentences (expressions) or other more complex structures from the alphabet elements; semantics – define the meaning of the constructions in the logic system; laws of inference – a set of rules or laws for constructing semantically equivalent but syntactically different sentences; this set of laws is also called a set of inference rules.

Long-term depression (LTD)

A process of a long-lasting decrease in the strength of synaptic transmission, produced by low-frequency stimulation of presynaptic afferents. The majority of synapses in many brain regions and in many species that express LTP also express LTD.

Long-term potentiation (LTP)

This is a process of a long-lasting increase in synaptic efficacy, produced by high-frequency stimulation of presynaptic afferents or by pairing presynaptic stimulation with postsynaptic depolarization.

M

Machine learning

An area of information and computer science concerned with the methods for accumulating, changing, and updating information and obtaining machine knowledge through algorithms.

Magnetic resonance imaging (MRI)

This uses the properties of magnetism. A large cylindrical magnet creates a magnetic field around the subject's head. Detectors measure local magnetic fields caused by alignment of atoms in the brain with the externally applied magnetic field. The degree of alignment depends upon the structural properties of the scanned tissue. MRI provides a precise anatomical image of both surface and deep brain structures.

Magnetoencephalography (MEG)

This measures millisecond-long changes in magnetic fields created by the brain's electrical currents. MEG machines use a noninvasive, whole-head, 248-channel, *super-conducting-quantum-interference-device* (SQUID) to measure small magnetic signals reflecting changes in the electrical signals in the human brain.

Mel-scale filter bank transformations

The process of filtering a signal through a set of frequency bands represented by triangular filter functions similar to the functions that describe the function of the human inner ear.

Membership function

Generalized characteristic function which defines the degree to which an object from a universe belongs to a given fuzzy concept.

Memory capacity of an ANN

The maximum number m of patterns that can be learned properly in a network.

Mental retardation

A developmental deficit, beginning in childhood, which results in significant limitation of intellect and cognition and poor adaptation to the demands of everyday life.

Methods for feature extraction

Methods used for reducing the dimensionality of raw data by transforming it from the original space into a space of selected features.

Microarray for gene expression

A device that evaluates the level of transcription (expression) of a predefined set of genes in a single biological cell or a piece of tissue. The five principal steps in the microarray technology are: tissue collection, RNA extraction, microarray gene expression evaluation, scanning and image processing, and data analysis.

Monitoring

The process of interpreting continuous input information and recommending intervention if appropriate.

Moving averages

A moving average of a time series is calculated by using $MA_t = (\Sigma S_{t-i})/n$, for $i = 1, 2, \ldots, n$, where n is the number of the data points, S_{t-i} is the value of the series at a time moment $(t-i)$, and MA_t is the moving average at a time moment t. Moving averages are often used in an information system as input features in addition to, or in substitution of, the real values of a time series.

Multilayer perceptron (MLP)

An ANN that consists of an input layer, at least one intermediate or *hidden* layer, and one output layer where the neurons from each layer are fully connected to the neurons from the next layer. In some particular applications they may be partially connected.

Mutation

A random change in the value of a gene; this relates to both biological genes and to gene parameters of an evolutionary algorithm.

N

Neural networks (NN)

See *artificial neural network*.

Neurotransmitters

Molecules that are produced in neurons in the brain and reside in synapses. When a synapse receives a spike, the synapse transfers neurotransmitters across the synaptic cleft so they can bind to receptors in the postsynaptic membrane that causes ion gates to open and to receive ions that change the membrane potential of the postsynaptic neuron. It is estimated that there are about 50 different neurotransmitters acting in the human brain. Neurotransmitters control the opening of the channels in the neuronal membrane and are vital for neuronal functions, including learning, memory, emotions, and decision making. The three major categories of substances that act as neurotransmitters are: (1) amino acids (primarily glutamate, GABA, aspartic acid and glycine); (2) peptides (vasopressin, somatostatin, neurotensin, etc.); and (3) monoamines (norepinephrine, dopamine and serotonin) plus acetylcholine. There are also other categories like opioids, tachykinins, and so on. The vast majority of neurotransmitters is produced in evolutionary older subcortical nuclei.

Noise
A random value without meaning that is added to the general function that describes the underlying behavior of a process or a signal.

Nonlinear dynamical system
A system whose next state on the time scale can be expressed by a nonlinear function of its previous time states.

Nonlinear transformation
Transformation f of a variable x, where f is a nonlinear function of x, for example, $f(x) = 1/(1 + e^{-xc})$, where c is a constant.

Normalization
Transforming data from its original range into another, predefined range, e.g., [0, 1].

Nyquist sampling frequency
A Nyquist sampling frequency for a particular signal is defined as twice the highest frequency contained within the signal (e.g., if $F_{signal} = 10.025$ Hz then $F_{NyqSampling} = 22.050$ Hz). When a signal is sampled at Nyquist frequency, the numeric sequence obtained completely determines the signal.

O

Ontology systems
This is both a data and a knowledge repository. Ontology is defined in the artificial intelligence literature as a specification of a conceptualization. Ontology specifies at a higher level the classes of concepts that are relevant to the domain and the relations that exist between these classes. Ontology captures the intrinsic conceptual structure of a domain along with the data that is available. For any given domain, the ontology forms the heart of the knowledge representation.

Optimization
The process of finding such values for the parameters of an object, system, or a process that would minimize an objective (cost) for this object/system/process.

Overfitting
A phenomenon that indicates that an ANN has approximated (or learned) a set of data examples too

closely, and as a result the network cannot generalize well on new examples.

P

Pattern matching
The process of matching a feature vector to already existing ones and finding the best match among them.

Phonemes
A basic distinctive unit of speech sound in a specified language.

Photon
In physics, a photon is a quantum of electromagnetic field, for instance, light. The term photon was coined by Gilbert Lewis in 1926. A photon can be perceived as a wave or a particle, depending on how it is measured. The photon is an elementary particle. Its interactions with electrons and atomic nuclei account for a great many of the features of matter, such as the existence and stability of atoms, molecules, and solids.

Planning
An important biological process and also AI-problem which is about generating a sequence of actions in order to achieve a given goal when a description of the current situation is available.

Positron emission tomography (PET)
This is used to study living brain activity. This noninvasive method involves an on-site use of a machine called a cyclotron to label specific drugs or analogs of natural body compounds (such as glucose or oxygen) with small amounts of radioactivity. The labeled compound (a radiotracer) is then injected into the bloodstream, which carries it into the brain. Radiotracers break down, giving off subatomic particles (positrons). By surrounding the subject's head with a detector array, it is possible to build up images of the brain showing different levels of radioactivity, and therefore, cortical activity.

Prediction
Generating information for possible future development of a process from data that represents its past and present development.

Principle component analysis (PCA)
A statistical procedure for finding a smaller number of m components $Y = (y_1, y_2, \ldots, y_m)$ (aggregated variables, eigenvectors) that can represent a function $F(x_1, x_2, \ldots, x_n)$ of n variables, where $n > m$, to a desired degree of accuracy Θ, i. e., $F = M \cdot Y + \Theta$, where M is a matrix that must be found through PCA.

Probability automata
Finite automata whose transitions are defined as probabilities. They are also known as stochastic automata.

Probability theory

The theory is based on the following three axioms: Axiom 1 defines the probability $p(E)$ of an event E as a real number in the closed interval [0, 1], i.e. $0 \leq p(E) \leq 1$. A probability $p(E) = 1$ indicates a certain event while $p(E) = 0$ indicates an impossible event. Axiom 2 is expressed as $\sum p(E_i) = 1$, $E_1 \cup E_2 \cup \ldots \cup E_k = U$, where U denotes the problem space (universe) as an union as subspaces. Axiom 3 indicates that if two independent events E_1 and E_2 cannot occur simultaneously, the probability of one or the other happening is the sum of their probabilities, i.e., $p(E_1 \vee E_2) = p(E_1) + p(E_2)$, where E_1 and E_2 are mutually exclusive events.

Production system

A computer system consisting of three main parts: (a) a list of facts, considered a working memory (the facts being called *working memory elements*); (b) a set of production inference rules, considered the production memory; and (c) an inference engine which is a reasoning procedure.

Productions

Transformation rules that are applied for obtaining one sequence of characters from another.

Propositional logic

A logic system that can be dated back to Aristotle (384–322 B.C.). There are three types of symbols in the propositional logic: propositional symbols (the alphabet), connective symbols, and symbols denoting the meaning of the sentences. There are rules in propositional logic to construct syntactically correct sentences (called well-formed formulas) and rules to evaluate the semantics of the sentences. A proposition represents a statement about the world, for example: *The temperature is over 40*. The semantic meaning of a proposition is expressed by two possible semantic symbols – true and false. Statements or propositions can be only *true* or *untrue* (false), nothing in between.

Proteins

Biological molecules that result from RNA translation. Proteins provide the majority of the structural and functional components of a cell. The area of molecular biology that deals with all aspects of proteins is called proteomics. A protein is a sequence of amino acids, each of them defined by a group of three nucleotides (codons). There are 20 amino acids all together, denoted by letters (A,C-H,I,K-N,P-T,V,W,Y). The length of a protein in number of amino acids is from tens to several thousands. Each protein is characterized by some characteristics, for example: structure, function, charge, acidity, hydrophilicity, and molecular weight. An initiation codon defines the start position of a gene in an mRNA where the translation of the mRNA into protein begins. A stop codon defines the end position.

Pruning in ANN

Technique where during the training procedure of the ANN weak connections (i.e., connections that have weights around 0) and the neurons connected by them are gradually removed.

R

Recall process

The process of using a trained ANN where new data is entered and results are calculated.

Recurrent fuzzy rule

A fuzzy rule that uses in its antecedent part one or more previous time-moment values of the output fuzzy variable.

Recurrent networks

ANN with feedback connections from neurons in one layer to neurons in a previous layer.

Reinforcement learning

A learning method that is based on presenting input data x to a learning system and observing the produced output. If this output is evaluated as *good*, then a *reward* is given to the learning system, e.g., connection weights of a neural network model increase in values, otherwise the system is *punished*, e.g., connection weights decrease.

RNA (ribonucleic acid)

A transcribed copy of part of an DNA that has a similar structure to the DNA, but here thymidine (T) is substituted by uridine (U) nucleotide. In the pre-RNA only segments that contain genes are extracted from the DNA. Each gene consists of two types of segments – exons, that are segments translated into proteins, and introns – segments that are considered redundant and do not take part in the protein production. Removing the introns and ordering only the exon parts of the genes in a sequence is called splicing and this process results in the production of messenger RNA (or mRNA) sequences. mRNAs are directly translated into proteins.

Roulette wheel selection (in genetic algorithms)

A selection strategy according to which each individual from a population of individuals at a certain generation is assigned a sector in an imaginary roulette wheel, with the size of the sector depending on the fitness of the individual. The size of the sector represents the probability of the individual to be selected when a random number is generated. Therefore, the fitter the individual, the higher the chance of it being selected for cross-over with other selected individuals to produce the population of individuals for the next generation.

S

Sampling

A process of selecting a subset of data from a larger data set. Sampling can be applied on continuous time-series data (e.g., speech data can be sampled at a frequency of 22 kHz), or on static data where only a smaller subset of the data is selected at a time for processing.

Schizophrenia

A brain disorder that has typical characteristic symptoms such as: delusions, hallucinations, and various thinking and perceptual disorders. Schizophrenic withdrawal from reality can manifest itself in many peculiar ways. Disorder is accompanied by serious deterioration of the previous level of functioning in such areas as work, social relations, and self-care.

Sensitivity to initial conditions

A characteristic of a chaotic process which means that a slight difference in the initial values of some parameters will result in different trends in a future development of the chaotic process. A set of algorithms which computationally embody ART. Developed by Gail A. Carpenter and Stephen Grossberg.

Spatio-temporal ANN

These networks can learn and represent spatio-temporal patterns from data.

Spatio-temporal data

data that is characterized by spatially distributed variables that are measured over time, e.g., electroencephalogram data.

Spike time dependent plasticity (STDP)

A method for learning in spiking neural networks that modifies the connection weight between two neurons, so that if the presynaptic neuron spikes first and then, within a certain time interval, spikes the postsynaptic neuron – the connection weight increases, otherwise – the connection weight decreases.

Spiking neural networks

Biological or artificial neural networks that consist of spiking neurons and connections. The information is represented as trains of spikes (binary events over time).

Spiking neuron

A biological or artificial neuron model that receives binary input signals (spikes) over time from many inputs (dendrites). It emits a spike (action potential) when the cumulative input (the membrane potential) of this neurons has reached a threshold. After that the neuronal membrane potential is set to a reset value and the process continues.

Stability/plasticity dilemma

The ability of a system to keep the balance between retaining previously learned information and patterns and learning new information/patterns.

Statistical analysis methods

Methods used for discovering repetitiveness in data based on probability estimation.

Subcortical structure of the brain

This consists of brain areas excluding the cortex, such as: basal ganglia, thalamus, hypothalamus, amygdala, and dozens of other groups of neurons with more or less specific functions in operations of the whole brain.

Supervised learning

A process of inferring a function from a set of training data with known outputs (labels). The training data set consists of data items each of which contains values for attributes (features) – independent variables, labeled by the desired value(s) for the dependant variables. Supervised learning can be viewed as approximating a mapping between a domain and a solution space of a problem: $X \rightarrow Y$, when samples (examples) of (input vector–output vector) pairs (x, y) are known, and $x \in X$, $y \in Y$, $x = (x_1, x_2, \ldots, x_n)$, $y = (y_1, y_2, \ldots, y_m)$.

Supervised training algorithm for an ANN

Training of an ANN when the training examples comprise input vectors x and the desired output vectors y; training is performed until the neural network *learns* to associate each input vector x to its corresponding and desired output vector y to a desired accuracy.

Synaptic efficacy

The level of concentration of ions in a synapse that can be transmitted to the postsynaptic neuronal membrane through ion channels that become open after certain neurotransmitters bind to them.

Synaptic plasticity

The process of changing synaptic efficacy through LTP/LTD learning.

System biology

An approach to treat and understand complex biological systems in their entirety, i. e., at a system level. It involves the integration of different data, knowledge, data analysis approaches, and tools. One of the major challenges of systems biology is the identification of the logic and dynamics of gene-regulatory and biochemical networks. The most feasible application of systems biology is to create a detailed model of a cell regulation to provide system-level insights into mechanism-based drug discovery.

T

Test error

An error that is calculated for a learning system that is trained with a set of training data. When a test (or validation) data set, for which the results are known, is applied in a recall procedure the test error is calculated.

Time alignment

A process where a sequence of input vectors recognized in a system over time are aligned to represent a meaningful output (e.g., a phoneme, word, or trend in stock).

Time-series prediction

Prediction of time series events.

Training error

The error of a learning system that is evaluated on the data used for training.

Training of a neural network

A procedure for presenting training examples to a neural network, which results in changing the network's connection weights according to a certain learning law.

Tree (refers to decision trees)

Directed graph in which one of the nodes, called *root*, has no incoming arcs, but from which each node in the tree can be reached by exactly one path and the end nodes (also called leaves) represent outputs of the system (e.g., decisions).

Turing test

Test of the ability of a digital computer to demonstrate intelligent behavior or, more precisely, whether it can imitate a human. The test was first described by the British mathematician Alan Turing in his 1950 paper *Computing Machinery and Intelligence*. The Turing test has been highly influential in the area of artificial intelligence and at the same time has been very controversial. The idea of the test is that an interrogator communicates with an entity in written form and, based on the reactions of the entity, decides whether it is another human or a computer. If a computer can trick the interrogator into believing that it is a human, then the machine has passed the test.

Type-2 fuzzy inference system

A fuzzy rule based system that uses type-2 fuzzy rules.

Type-2 fuzzy set

A fuzzy set to which elements belong with a membership degree that is represented not by a single number but by an interval of min–max membership degrees.

U

Universal function approximator (for ANN)

A theorem that was proved by Hornik (1989), Cybenko (1989), and Funahashy (1989). It states that an MLP with one hidden layer can approximate any continuous function to any desired accuracy, subject to sufficient number of hidden nodes. As a corollary, any Boolean function of n Boolean variables can be approximated by an MLP.

Unsupervised learning algorithm

A learning procedure where only input vectors x are supplied to a learning system (e.g., a neural network). The system learns some internal characteristics, e.g., clusters, for the whole set of input vectors presented to it. An example of such an algorithm is the self-organizing maps.

V

Validation

Process of testing how good the solutions produced by a system are. The solutions are usually compared to the results obtained either by experts or by other systems.

Vector quantization

A process of representing data from n-dimensional problem space into the m-dimensional one, where $m < n$, in a way that preserves the distance between data examples (points) from the original space.

Vigilance

Parameter in an ART network that controls the degree of mismatch between the new patterns and the learned patterns that the system can tolerate.

W

Wavelet transformation

A nonlinear transformation that can be used to represent slight changes of a time series within a chosen time interval.

Acknowledgements

A.2 Information Processing at the Cellular Level: Beyond the Dogma
by Alberto Riva

The author wishes to thank Prof. Silvano Riva for reviewing the contents of this chapter. All figures were obtained from Wikipedia, and are licensed under Creative Commons.

A.7 Understanding Evolving Bacterial Colonies
by Leonie Z. Pipe

I extend sincere thanks to the staff of the *Journal of the Physical Society of Japan*, and to Professor Harry Swinney, for their gracious permission to reproduce Fig. 7.1 from *Ohgiwari* et al. (*Morphological changes in growth phenomena of bacterial colony patterns*, J. Phys. Soc. Japan 61(3), 816–822 (1992) and Fig. 77.1 from *Zhang* et al. (*Collective motion and density fluctuations in bacterial colonies*, Proc. Natl. Acad. Sci. USA 107(31), 13626–13630 (2010)), respectively. Special thanks are due to Professor Nikola Kasabov, who read and commented on the manuscript.

B.10 Bioinformatic Methods to Discover *Cis*-regulatory Elements in mRNAs
by Stewart G. Stevens, Chris M. Brown

This work was partially supported by a Human Frontier Science Program grant to Ian Macara, Anne Spang, and C.M.B. (RGP0031_2009) and a University of Otago Research Grant to C.M.B.

C.16 Phylogenetic Cladograms: Tools for Analyzing Biomedical Data
by Mones S. Abu-Asab, Jim DeLeo

We would like to thank Heike Sichtig for her constant and consistent support, encouragement, and coaching throughout our writing of this chapter, and also for her help in designing the cartoon, which was drawn by Ann Aiken, who we also wish to acknowledge and thank.

C.18 Kernel Methods and Applications in Bioinformatics
by Yan Fu

This work was supported by the National Natural Science Foundation of China under grant no. 30900262.

D.19 Path Finding in Biological Networks
by Lore Cloots, Dries De Maeyer, Kathleen Marchal

This work is supported by the KU Leuven Research Council [GOA/08/011, CoE EF/05/007 – SymBioSys, NATAR C1895-PF/10/10]; the agency for Innovation by Science and Technology [SBO-BioFrame]; Interuniversity Attraction Poles [P6/25 – BioMaGNet]; and Research Foundation – Flanders [IOK-B9725-G.0329.09]. The authors would also like to thank Dr. Kristof Engelen for valuable comments on the manuscript.

D.21 Computational Methods for Analysis of Transcriptional Regulation
by Yue Fan, Mark Kon, Charles DeLisi

This work was partially supported by NIH grant 1R21CA13582-01 and NIH grant 1R01GM080625-01A1.

D.25 Whole-Exome Sequencing Data – Identifying Somatic Mutations
by Roberta Spinelli, Rocco Piazza, Alessandra Pirola, Simona Valletta, Roberta Rostagno, Angela Mogavero, Manuela Marega, Hima Raman, Carlo Gambacorti-Passerini

We are grateful to the patients. This work was supported by AIRC 2010 (IG-10092); PRIN 2008 program; Fondazione Cariplo (2009-2667); Lombardy Region (ID-16871; ID14546A).

E.27 Ontologies for Bioinformatics
by Andrea Splendiani, Michele Donato, Sorin Drăghici

A.S. has been supported by the BBSRC and in particular the Ondex project (BBSRC BB/F006039/1).

M.D. and S.D. have been supported by the following grants: NIH RO1 RDK089167-01, NIH 2R42GM087013-02 and NSF DBI-0965741.

Any opinions, findings, and conclusions or recommendations expressed in this material are those of the author(s) and do not necessarily reflect the views of the BBSRC, NSF, NIH, or any other of the funding agencies.

Acknowl.

F.29 Epigenetics
by Micaela Montanari, Marcella Macaluso,
Antonio Giordano

The authors apologize to scientists whose relevant publications were not cited because of space limitations.

G.36 Information Processing in Synapses
by Hiroshi Kojima

I would like to thank Chloé Okuno who illustrated many figures that apparear in this article.

G.38 Statistical Methods for fMRI Activation and Effective Connectivity Studies
by Xingfeng Li, Damien Coyle, Liam Maguire,
T. Martin McGinnity

This study is currently supported under the CNRT award by the Northern Ireland Department for Employment and Learning through its *Strengthening the All-Island Research Base* initiative. The data collection in this study was supported by CIHR grants to Robert F. Hess (#MOP53346) and Kathy T. Mullen (#MOP-10819).

G.39 Neural Circuit Models and Neuropathological Oscillations
by Damien Coyle, Basabdatta S. Bhattacharya,
Xin Zou, KongFatt Wong-Lin, Kamal Abuhassan,
Liam Maguire

The authors apologize to scientists whose relevant publications were not cited because of space limitations. This work was supported under the Computational Neuroscience Research Team project by the N. Ireland Department for Education and Learning "Strengthening the All-island Research Base" funding and The Centre of Excellence in Intelligent Systems supported by ILEX and Invest Northern Ireland.

H.42 Recurrence Plots and the Analysis of Multiple Spike Trains
by Yoshito Hirata, Eric J. Lang, Kazuyuki Aihara

The research of Y.H. and K.A. was partially supported by the Aihara Innovate Mathematical Modeling Project, the Japan Society for the Promotion of Science (JSPS) through its *Funding Program for World-Leading Innovative R&D on Science and Technology (FIRST Program)*, initiated by the Council for Science and Technology Policy (CSTP). Y.H. was also partially supported by Grants in Aid for Young Scientists (B), Grant No. 21700249 from the Japanese Ministry of Education, Culture, Sports, Science, and Technology, and Grant No. 23700261 from the Japan Society for the Promotion and Science (JSPS). The research of E.J.L. was supported by National Science Foundation (IOS-1051858), and Irma T. Hirschl/Monique Weill-Caulier Trust.

I.46 EEG Signal Processing for Brain–Computer Interfaces
by Petia Georgieva, Filipe Silva,
Mariofanna Milanova, Nikola Kasabov

This work was supported by the Portuguese Foundation for Science and Technology under the grant SFRH/BSAB/1092/2010 and the Institute of Electronic Engineering and Telematics of Aveiro (IEETA), Portugal.

I.47 Brain-like Information Processing for Spatio-Temporal Pattern Recognition
by Nikola Kasabov

The author acknowledges discussions with G. Indivery and also with: A.Mohemmed, T.Delbruck, S-C.Liu, N.Nuntalid, K.Dhoble, S.Schliebs, R.Hu, R.Schliebs, H.Kojima, F.Stefanini. The work in this chapter was sponsored by the Knowledge Engineering and Discovery Research Institute, KEDRI (www.kedri.info) and the EU FP7 Marie Curie International Incoming Fellowship project PIIF-GA-2010-272006 *EvoSpike*, hosted by the Institute for Neuroinformatics – the Neuromorphic Cognitive Systems Group, at the University of Zurich and ETH Zurich (http://ncs.ethz.ch/projects/evospike). Diana Kassabova helped with the proofreading. This chapter is based on the material published as: N. Kasabov: Evolving spiking neural networks and neurogenetic systems for spatio- and spectro-temporal data modelling and pattern recognition. In: *IEEE WCCI 2012*, LNCS Vol. 7311, ed. by J. Liu et al. (Springer, Berlin, Heidelberg 2012), pp. 234-260.

K.53 A Machine Learning Pipeline for Identification of Discriminant Pathways
by Annalisa Barla, Giuseppe Jurman,
Roberto Visintainer, Margherita Squillario,
Michele Filosi, Samantha Riccadonna,
Cesare Furlanello

This work at FBK was supported by the European Union [FP7 HiperDART] and the Autonomous Province of Trento [CancerAtlas Trentino, ENVIROCHANGE].

K.54 Computational Neurogenetic Modeling: Gene-Dependent Dynamics of Cortex and Idiopathic Epilepsy

by Lubica Benuskova, Nikola Kasabov

We thank our former research assistant Dr. Simei Gomes Wysoski for his excellent programming work.

K.56 sEMG Analysis for Recognition of Rehabilitation Actions

by Qingling Li, Zeng-Guang Hou, Yu Song

This research is supported in part by the National Natural Science Foundation of China (Grants 61005070, 61175076, 61225017), and the International S&T Cooperation Project of China (Grant 2011DFG13390).

L.60 Brain, Gene, and Quantum Inspired Computational Intelligence

by Nikola Kasabov

The work presented here is an extended version of the previously published by Springer chapter: N. Kasabov, Brain-, Gene-, and Quantum Inspired Computational Intelligence: Challenges and Opportunities, in: W. Duch and J. Manzduk (eds.) Challenges in Computational Intelligence, ISBN: 978-3-540-71983-0, 193–219, Springer 2007.

Acknowl.

About the Authors

Mones S. Abu-Asab

National Institutes of Health
National Eye Institute
Bethesda, MD, USA
mones@mail.nih.gov

Chapter C.16

Mones S. Abu-Asab holds a PhD degree in Phylogenetic Systematics and is currently working on ultra structural pathology at the National Eye Institute. Recently, he has been advocating the application of phylogenetic analysis to high-throughput heterogeneous omics data. His analyses have shown that parsimony phylogenetics is a multidimensional tool that can be utilized for disease modeling, profiling, and subtyping, as well as biomarker discovery.

Kamal Abuhassan

University of Ulster
Intelligent Systems Research Centre
Derry, Northern Ireland, UK
Abuhassan-K@email.ulster.ac.uk

Chapter G.39

Kamal Abuhassan is a PhD research student at the Intelligent Systems Research Centre (ISRC), University of Ulster, UK. His main research interests include the investigation abnormal brain oscillations with biomarkers in Alzheimer's Disease and computational modeling of neuronal networks. He has more than six years of experience in developing SOA and J2EE applications using JAVA and IBM tools.

Kazuyuki Aihara

The University of Tokyo
Institute of Industrial Science
Tokyo, Japan
aihara@sat.t.u-tokyo.ac.jp

Chapter H.42

Kazuyuki Aihara received the PhD degree in Electronic Engineering in 1982 from the University of Tokyo, Japan. Currently, he is Professor of the Institute of Industrial Science and Director of the Collaborative Research Center for Innovative Mathematical Modelling at the University of Tokyo. His research interests include mathematical modeling of complex systems and nonlinear time series analysis of complex data.

Federico Ambrogi

University of Milan
Department of Clinical Sciences and
Community Health, Section of Medical
Statistics, Biometry and Bioinformatics
"Giulio A. Maccacaro"
Milan, Italy
federico.ambrogi@unimi.it

Chapter F.32

Federico Ambrogi has a PhD in biomedical statistics. He worked from 1998 to 2001 as a Consultant Researcher for the support decision systems at NUNATAC, Milan, Italy. From 2001 to 2005 he was a Research Fellow at the Istituto Nazionale Tumori, Milan, Italy. Since 2006, he is a Researcher at the University of Milan. His main interests are survival regression models for clinical decision support and multivariate techniques for bioprofile identification of complex diseases.

Periklis Andritsos

University of Toronto
Faculty of Information (iSchool)
Toronto, ONT, Canada
periklis.andritsos@utoronto.ca

Chapter C.13

Periklis Andritsos is an Assistant Professor in the Faculty of Information (iSchool) at the University of Toronto, Canada. His research focuses on the analysis of Big Data repositories and, more specifically, the structure discovery in order to facilitate design and speed up querying. He has made contributions in the fields of clustering categorical data, structure characterization and discovery in unstructured data and information extraction.

Authors

Alessandro Astolfi

Imperial College London
Electrical and Electronic Engineering
London, UK
a.astolfi@ic.ac.uk

Chapter F.30

Alessandro Astolfi graduated in Electrical Engineering from the University of Rome (1991) and then joined ETH-Zurich where he obtained a MSc in Information Theory in 1995 and the PhD degree in 1995. In 1996 he was awarded a Ph.D. from the University of Rome "La Sapienza". Since 1996 he has been with the Electrical and Electronic Engineering Department of Imperial College, London (UK), where he is currently Professor in Non-linear Control Theory. Since 2005 he is also Professor at Dipartimento di Informatica, Sistemi e Produzione, University of Rome Tor Vergata. His research interests are focused on mathematical control theory and control applications. He has authored more than 100 journal papers, 20 book chapters.

Annalisa Barla

University of Genova
DIBRIS
Genova, Italy
annalisa.barla@unige.it

Chapter K.53

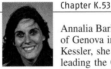

Annalia Barla received her PhD in Computer Science from the University of Genova in 2005. After a post-doc position at the Fondazione Bruno Kessler, she joined the Department of Informatics in Genova in 2008, leading the Computational Biology unit. Since 2012 she is Assistant Professor. Her research is focused on regularization methods for the understanding of molecular data and their functional characterization.

Niccolò Bassani

University of Milan
Department of Clinical Sciences and
Community Health
Milano, Italy
niccolo.bassani@unimi.it

Chapter F.32

Niccolò Bassani obtained his MD in Biostatistics in 2008 and entered the PhD in Biomedical Statistics by the beginning of 2010 at the University of Milan. He joined the BBCTR as a fellow researcher in 2006, and is finishing the PhD course with a study of the reliability of microRNA microarray platforms. His main research interests are multivariate visualization techniques and genomic high-throughput data analysis.

José A. Becerra Permuy

University of A Coruña
Department of Computer Science
Ferrol, Spain
jose.antonio.becerra.permuy@udc.es

Chapter L.57

J.A. Becerra received the BS and MS degrees in Computer Science from the University of A Coruña, Spain, in 1999, and a PhD in Computer Science from the same university in 2003. He is currently an Associate Professor and member of GII at the University of A Coruña. His research activities are mainly related to autonomous robotics, evolutionary algorithms and parallel computing.

Francisco Bellas

Universidade da Coruña
Escola Politecnica Superior, Department
of Computer Science
Ferrol, Spain
francisco.bellas@udc.es

Chapter L.57

Francisco Bellas is a Professor Titular at the University of A Coruña, Spain. He received the BS and MS degree in Physics from the University of Santiago de Compostela, Spain, in 2001, and a PhD in Computer Science from the University of A Coruña in 2003. He is a member of the Integrated Group for Engineering Research at the University of A Coruña. His main research activities are focused to evolutionary robotics, collective intelligence and neuroevolutionary algorithms.

Lubica Benuskova   Chapter K.54   For biographical profile, please see section "About the Part Editors" on page XV.

Basabdatta S. Bhattacharya

University of Lincoln
School of Engineering
Lincoln, UK
bbhattacharya@lincoln.ac.uk

Chapter G.39

Basabdatta Sen Bhattacharya is currently a Lecturer of Electrical Engineering at the University of Lincoln, UK. Dr. Sen Bhattacharya received her B.Eng. in Electrical Engineering from NIT Silchar (India) in 1992, M. Eng. in Electronics Engineering from Jadavpur University, Kolkata (India) in 2002 and PhD in Computer Science from University of Manchester (UK) in 2008. Prior to that, she has worked as post-doctoral researcher at the University of Ulster, UK and the University of Bordeaux, France. Her research interests include computational modeling of the thalamocortical circuitry and the visual pathway.

Elia M. Biganzoli

University of Milan
Department of Clinical Sciences and
Community Health, Section of Medical
Statistics, Biometry and Bioinformatics
"Giulio A. Maccacaro"
Milan, Italy
elia.biganzoli@unimi.it

Chapter F.32

Elia Biganzoli has a PhD in Medical Statistics. His career started 1989–1995 as Research Fellow at Lepetit Research Centre, Marion Merrel Down USA. From 1995 he was Senior Biostatistician at the Istituto Nazionale Tumori, Milano and since 2007 he is an Associate Professor of Biostatistics at the University of Milan. His main research fields cover statistical methods for survival analysis and high-throughput bioassays. He authored more than 100 scientific publications in books and international journals.

Veselka Boeva

Technical University of Sofia, Branch
Plovdiv
Department of Computer Systems and
Technologies
Plovdiv, Bulgaria
vboeva@tu-plovdiv.bg

Chapter C.14

Veselka Boeva holds a PhD degree in Computer Science from the Technical University of Sofia-branch Plovdiv, Bulgaria. She is currently an Associate-Professor in Computer Science at the Department of Computer Systems and Technologies of the same university. Her main research interests and activities are in the areas of artificial intelligence, decision making, bioinformatics and software systems. She is currently performing research dealing with computational analysis of gene expression data and development of expert location systems. She has more than 50 publications in refereed journals and conference proceedings.

Patrizia Boracchi

University of Milan
Department of Clinical Sciences and
Community Health, Section of Medical
Statistics, Biometry and Bioinformatics
"Giulio A. Maccacaro"
Milano, Italy

Fondazione IRCCS Istituto Nazionale
Tumori
patrizia.boracchi@unimi.it

Chapter F.32

Patrizia Boracchi received her PhD in Medical Statistics from the University of Milan in 1989. Since 1991 she is Technical Collaborator and Researcher (2001) at the Institute of Medical Statistics and Biometry, University of Milan. Her main research interests are statistical methods in survival analysis and flexible regression modeling and competing risks. She collaborates in planning clinical trials and observational studies. She has authored more than 120 scientific publications in international journals.

Carol Brayne

University of Cambridge
Department of Public Health and
Primary Care
Cambridge, UK
cb105@medschl.cam.ac.uk

Chapter K.52

Carol Brayne is Professor of Public Health Medicine in Department of Public Health and Primary Care in the University of Cambridge, UK. She graduated in medicine from the Royal Free Hospital School of Medicine, University of London. After gaining membership she moved on to training in epidemiology with a Training Fellowship with the Medical Research Council. Since the mid eighties her main research area has been longitudinal studies of older people. She is lead principal investigator in the group of MRC CFA Studies and has been responsible for training programs in epidemiology. She is Director of the Cambridge Institute of Public Health at the University of Cambridge.

Chris M. Brown   Chapter B.10   For biographical profile, please see section "About the Part Editors" on page XV.

Mónica F. Bugallo   Chapter F.28

Stony Brook University
Electrical and Computer Engineering
Stony Brook, NY, USA
monica@ece.sunysb.edu

Mónica F. Bugallo received the PhD degree in Computer Engineering from the University of A Coruña, Spain, in 2001. From 1998 to 2000 she was with the Departamento de Electrónica y Sistemas at the Universidade da Coruña, Spain and joined the Department of Electrical and Computer Engineering at Stony Brook University in 2001, where she is currently an Associate Professor for digital communications and signal processing. Her research interests lie in the area of statistical signal processing and its applications to different disciplines including sensor networks and biology.

Matteo Cacciola   Chapter L.61

Uuniversity Mediterranea of Reggio
Calabria
DICEAM
Reggio Calabria, Italy
matteo.cacciola@unirc.it

Matteo Cacciola is a post-doc researcher at the University Mediterranea of Reggio Calabria, Italy. He owns a MD in Electronic Engineering (2003) and PhD in Computer, Biomedic and Telecommunication Engineering (2008) from the "Mediterranea" University. His research activity is focused on soft computing applied to regularization of inverse electric and magnetic problems. He co-authored more than 100 papers and was awarded the Best ENNS Poster Award 2008.

Colin Campbell   Chapter C.12

University of Bristol
Department of Engineering Mathematics
Bristol, UK
c.campbell@bristol.ac.uk

Dr. Colin Campbell gained a First Class Honors degree in Physics from Imperial College, London and a Doctorate from the Department of Mathematics, King's College, London. He is currently a Reader and Deputy Director of the Intelligent Systems Laboratory in the Merchant Venturer's School of Engineering, University of Bristol, UK. His research interests are machine learning, including probabilistic graphical models and kernel-based methods and algorithm design. He is interested in the application of machine-learning techniques in bioinformatics, particularly medical bioinformatics. His past research has been funded by the EPSRC, the Medical Research Council, Cancer Research UK, the EU and PASCAL2.

Mario Cannataro   Chapter E.26

University Magna Graecia of Catanzaro
Department of Medical and Surgical
Sciences
Catanzaro, Italy
cannataro@unicz.it

Mario Cannataro is Associate Professor of Computer Engineering at the Magna Græcia University of Catanzaro, Italy, since 2002. He received his Laurea Degree cum Laude in Computer Engineering from the University of Calabria, Rende, Italy, in 1993. His current research interests are on bioinformatics, health informatics, grid and parallel computing. Dr. Cannataro has published three books and more than 150 papers in international journals and conference proceedings. He is a member of ACM, ACM SIGHIT, IEEE Computer Society, and Italian Society for Bioinformatics.

Mariella Caputo   Chapter F.31

Università di Salerno
Dipartimento di Farmacia
Fisciano, Italy
macaputo@unisa.it

Mariella Caputo received the Laurea degree in Pharmaceutical Chemistry from the University of Naples and the PhD in Biochemistry from the University of Salerno. Her research activities include the analyses of molecular effects of n-3 and n-6 polyunsaturated fatty acids in cell cultures to identify molecular mechanisms involved in the action of chemopreventive nutrients.

Hyeygjeon Chang   Chapter F.30

Kookmin University
School of Electrical Engineering
Seoul, Korea
hchang@kookmin.ac.kr

Hyeygjeon Chang received the BS and the MS degrees from the School of Electrical Engineering of Seoul National University, in 1998 and 2004, respectively, and the PhD degree from the Electrical and Electronic Engineering Department of Imperial College, London, in 2009. After postdoctoral research experiences at the Boston University and CNRS, he is currently working at CPSE, Imperial College London as research associate. His research interests include nonlinear control theory, nonlinear systems analysis, and theoretical biology.

Badong Chen   Chapter H.41

Xi'an Jiaotong University
Institute of Artificial Intelligence and Robotics
Xi'an, P. R. China
chenbd@mail.xjtu.edu.cn

Badong Chen received his PhD degree in Computer Science and Technology from Tsinghua University, China, in 2008. He was a Postdoctoral Associate at the University of Florida from Oct., 2010 to Sept., 2012. He is currently a Professor at Xi'an Jiaotong University, China. His research interests are in signal processing and machine learning, and their applications in cognition and neuroscience.

Sixue Chen   Chapter A.5

University of Florida
Department of Biology
Gainesville, FL, USA
schen@ufl.edu

Dr. Sixue Chen is an Associate Professor and Director of Proteomics Facility at the University of Florida. His current research activities include elucidating molecular networks controlling glucosinolate metabolism, proteomics and metabolomics of guard cell signalling networks, as well as investigating plant interaction with pathogens and other environmental factors.

Girija Chetty   Chapters C.17, G.40

University of Canberra
Faculty of Information Sciences and Engineering
Bruce, Canberra, ACT, Australia
girija.chetty@canberra.edu.au

Girija Chetty is an Assistant Professor in Software Engineering at the University of Canberra, Australia. She received her Bachelors and Masters (Research) degrees in Electrical Engineering and Computer Science from India and her Doctorate in Information Sciences and Engineering from Australia. She has published more than 100 articles and papers in Internationally peer reviewed Journals. Her research interests are in the area of image processing, information fusion, machine learning, pattern recognition as well as software engineering.

Yoonjoo Choi   Chapter B.11

Dartmouth College
Department of Computer Science
Hanover, NH, USA
yoonjoo@cs.dartmouth.edu

Yoonjoo Choi majored in philosophy, Chinese philosophy, physics and mathematics. He did his DPhil in Statistics at the University of Oxford working on protein loop structure prediction. He is currently a postdoctoral researcher at Dartmouth College, Hanover, NH, working on computational immunology. His research interests are protein structure prediction, protein loop modeling, antibody complementarity determining region prediction, antibody complex modeling and protein deimmunization.

Feng Chu   Chapter C.15

Siemens Ptd Ltd
The Siemens Center
Singapore
chufeng67@gmail.com

Feng Chu received the BEng degree from Zhejiang University, Hangzhou, China and the MEng degree from Huazhong University of Science and Technology, Wuhan, China in 1995 and 2002, respectively. He earned his PhD degree at Nanyang Technological University in Singapore in 2006. Since 2005, he has been also working for Siemens Pte Ltd in Singapore as a research and development engineer. His research interests include computational intelligence, data mining, and their applications, in bioinformatics, computational finance, etc.

Lore Cloots

KU Leuven
Centre of Microbial and Plant Genetics
Heverlee, Belgium
lore.cloots@biw.kuleuven.be

Chapter D.19

In 2008 I started my PhD in bioinformatics at the Centre of Microbial and Plant Genetics, KU Leuven and have been involved in developing and applying a query-based biclustering method for expression analysis. Furthermore, I have developed strategies to generate molecular interaction networks from high throughput data and implemented a novel network-based gene prioritization method in the context of eQTL studies.

Danila Coradini

University of Milan
Department of Clinical Sciences and
Community Health, Fondazione IRCCS
Istituto Nazionale Tumori
Milano, Italy
danila.coradini@libero.it

Chapter F.32

Danila Coradini, PhD, is a cellular and molecular biologist with a 30 year long experience in basic and translational studies on clinical tumors. Especially focused on breast cancer, she addressed to evaluate the biologic predictivity on prognosis and treatment responses of hormone-related markers. More recently, taking advantage from the information provides by new high throughput technologies, she focused on the study of the earliest events occurring in breast carcinogenesis.

Damien Coyle

University of Ulster
Intelligent System Research Centre
Derry, Northern Ireland, UK
dh.coyle@ulster.ac.uk

Chapters G.38, G.39

Damien Coyle is a lecturer/senior lecturer at the School of Computing and Intelligent Systems, University of Ulster, UK. His research interests include brain–computer interfaces, computational intelligence and neuroscience, neuroimaging and biosignal processing. He is a recipient of the IEEE Computational Intelligence Society's Outstanding Doctoral Dissertation Award and the International Neural Network Society's Young Investigator of the Year Award and is a University of Ulster Distinguished Research Fellow and a Royal Academy of Engineering/Leverhulme Trust Senior Research Fellow.

Alexandra E. D'Agostino

Stony Brook University
Department of Neurobiology
Stony Brook, NY, USA
alexandra.dagostino@stonybrook.edu

Chapter I.45

Alexandra D'Agostino received her MS in Behavioral Neuroscience from Binghamton University in 2011. Her master's work investigated the neural coding of taste mixtures in the nucleus of the solitary tract. She is currently a graduate student at Stony Brook University working toward her PhD in Neuroscience.

Joyce D'Mello

Auckland University of Technology
Knowledge Engineering and
Discovery Research Institute
Auckland, New Zealand
joyce.dmello@aut.ac.nz

Joyce D'Mello is the Administrative Manager at the Knowledge Engineering and Discovery Research Institute, KEDRI, at Auckland University of Technology and contributes to the functioning of a high-performing team.

Shaojun Dai

Northeast Forestry University
Alkali Soil Natural Environmental Science
Center
Harbin, China
daishaojun@hotmail.com

Chapter A.5

Dr. Shaojun Dai obtained his PhD degree from Northeast Forestry University (NEFU), Harbin, China, in 2002. He had postdoctoral training at the Institute of Botany, Chinese Academy of Sciences in 2006, and at the University of Florida in 2009. Now, he is a Professor in NEFU. Dr. Dai's laboratory mainly focuses on functional proteomics of plants in response to stress.

Chinh Dang   Chapter L.62

Allen Institute for Brain Science
Department of Technology
Seattle, WA, USA
chinhda@alleninstitute.org

Since joining the Allen Institute in early 2004, Chinh Dang leads the Technology Team responsible for the Allen Brain Atlas portal development as well as the development of software and hardware infrastructure to support high-throughput data generation. Prior to joining the Allen Institute, Chinh held several positions in the bioinformatics and biotech industries developing enterprise-level software for large-scale data management, analysis, and visualization. Chinh holds a BS in Biology from the University of Denver and a MS in Information Systems from the University of Colorado.

Dries De Maeyer   Chapter D.19

KU Leuven
Department of Microbial and Molecular Systems
Heverlee, Belgium
dries.demaeyer@biw.kuleuven.be

Dries De Maeyer received his MSc in Bio-Engineering from the KU Leuven where he obtained an additional degree in informatics. Currently he is a PhD student at the Centre of Microbial and Plant Genetics where he performs research into computational techniques on physical interaction networks of model organisms to uncover the mechanisms between genotypes and phenotypes.

Charlotte M. Deane   Chapter B.11

University of Oxford
Department of Statistics
Oxford, UK
deane@stats.ox.ac.uk

Charlotte Deane is a Professor in the Department of Statistics at Oxford University, Oxford, UK. Her research work is focussed on proteins, their interactions, their structures and their evolution. She works closely with several industrial partners include UCB, Roche, GSK and Pfizer and several of her software packages are commercially licensed.

Jim DeLeo   Chapter C.16

National Institutes of Health Clinical Center
Laboratory for Informatics Development
Bethesda, MD, USA
jdeleo@nih.gov

Jim DeLeo is a computer scientist who develops innovative computational methods for biomedical, space, and defense applications. Most of his career he has been at his present position with the National Institutes of Health (NIH) in Bethesda, MD, where he is engaged in several collaborative biomedical projects with various scientists at the NIH, and with other governmental agencies, universities and industry.

Charles DeLisi   Chapter D.21

Boston University
Program in Bioinformatics
Boston, MA, USA
delisi@bu.edu

Charles DeLisi is the Arthur G.B. Metcalf Professor of Science and Engineering, as well as the founding Chair of the all-University PhD Program in Bioinformatics at Boston University. He is the recipient of numerous awards including the Presidential Citizens Medal from President Clinton for initiating the Human Genome Project.

Patricia M. Di Lorenzo   Chapter I.45

Binghamton University
Department of Psychology
Binghamton, NY, USA
diloren@binghamton.edu

Patricia M. Di Lorenzo earned a PhD in Biopsychology from the University of Rochester, NY (1981). Her research probes issues of neural coding in the brainstem gustatory system using behavioral, electrophysiological and computational techniques. She is currently a Professor of Psychology and Director of the Integrative Neuroscience Program at Binghamton University, Binghamton, NY.

Authors

Petar M. Djurić

Stony Brook University
Department of Electrical and
Computer Engineering
Stony Brook, NY, USA
djuric@ece.sunysb.edu

Chapter F.28

Petar M. Djurić received his BS and MS degrees from the University of Belgrade, in 1981 and 1986, respectively, and his PhD degree in Electrical Engineering from the University of Rhode Island, in 1990. From 1981 to 1986 he was a Research Associate with the Institute of Nuclear Sciences, Belgrade. Since 1990 he has been with Stony Brook University, where he is a Professor in the Department of Electrical and Computer Engineering. He works in the area of statistical signal processing, and his primary interests are in the theory of signal modeling, detection, and estimation. He is a Member-at-Large of the Board of Governors of the Signal Processing Society and a Fellow of IEEE.

Michele Donato

Wayne State University
Department of Computer Science
Detroit, MI, USA
dw2237@wayne.edu

Chapter E.27

Michele Donato earned a Master's Degree in Computer Engineering from the University of Pisa, Italy, and is currently a PhD student in Bioinformatics in the Computer Science Department of the Wayne State University in Detroit, MI. His research interests are systems biology, signaling pathways, and biomedical ontologies.

Stijn van Dongen

EMBL – European Bioinformatics Institute
Wellcome Trust Genome Campus
Hinxton, Cambridge, UK
stijn@ebi.ac.uk

Chapter B.9

Stijn van Dongen received his PhD in Mathematics in 2000 from the University of Utrecht in the Netherlands. He works on algorithms, software and data analysis in the field of computational biology. His areas of interest include network analysis and network clustering, microRNA signal detection in microarray data, and methods to analyse next generation and high-throughput sequencing data.

Sorin Drǎghici

Wayne State University
Department of Computer Science
Detroit, MI, USA
sorin@wayne.edu

Chapter E.27

Sorin Drǎghici holds the Robert J. Sokol MD Endowed Chair in Systems Biology in the Department of Obstetrics and Gynecology, and is a Professor in the Department of Clinical and Translational Science and the Department of Computer Science, as well as the Head of the Intelligent Systems and Bioinformatics Laboratory at Wayne State University, Detroit, MI. He is also the chief of the Bioinformatics and Data Analysis Section in the Perinatology Research Branch of the National Institute for Child Health and Development. He earned a PhD in Computer Science from the University of St. Andrews, UK.

Richard J. Duro

Universidade da Coruña
Department of Computer Science
Ferrol, Spain
richard@udc.es

Chapter L.57

Richard J. Duro received a MS degree in Physics from the University of Santiago de Compostela, Spain, in 1989, and a PhD in Physics from the same University in 1992. He is currently a Full Professor in the Department of Computer Science and head of the Integrated Group for Engineering Research at the University of A Coruña. His research interests include higher order neural network structures, signal processing and autonomous and evolutionary robotics.

Anton J. Enright   Chapter B.9

EMBL – European Bioinformatics Institute
Wellcome Trust Genome Campus
Hinxton, Cambridge, UK
aje@ebi.ac.uk

Dr. Anton Enright originally trained in Genetics at Trinity College in Dublin before embarking on a PhD in Computational Biology at the EMBL – European Bioinformatics Institute and the University of Cambridge, UK. After completing his PhD in 2001 he worked at Memorial Sloan–Kettering Cancer Center in New York before returning to Cambridge to start his laboratory at the Wellcome Trust Sanger Institute in 2004. He returned to the EMBL-EBI in 2008 and his laboratory focuses on the analysis of non-coding RNAs in animal genomes.

Yue Fan   Chapter D.21

Boston University
Department of Mathematics and Statistics
Boston, MA, USA
yue@bu.edu

Yue Fan obtained his PhD in Mathematics from Boston University in 2012. His current primary research interest is in machine learning and its applications to computational biology and bioinformatics.

Valery Feigin   Chapter K.55

AUT University
National Institute for Stroke and
Applied Neurosciences
Northcote, Auckland, New Zealand
valery.feigin@aut.ac.nz

Valery Feigin is a Professor of Neurology and Epidemiology and Director of the National Institute for Stroke and Applied Neurosciences at AUT University, Auckland. His research interests include stroke prevention, epidemiology and management as well as epidemiology of traumatic brain injury. We currently run several observational and experimental epidemiological studies in stroke and traumatic brain injury in New Zealand, and are also involved in international collaborative research projects in stroke and traumatic brain injury.

David Feng   Chapter L.62

Allen Institute for Brain Science
Department of Technology
Seattle, WA, USA
davidf@alleninstitute.org

David Feng joined the informatics team at the Allen Institute in 2010. He develops visualization tools and informatics-driven web applications in addition to supporting the Institute's data analysis pipeline. His background includes computer graphics, scientific data visualization, and image analysis. David received a BS and MS from Northwestern University and a PhD in Computer Science from The University of North Carolina at Chapel Hill.

Michele Filosi   Chapter K.53

Fondazione Bruno Kessler
Predictive Models for Biomedicine and
Environment
Povo, Italy
filosi@fbk.eu

Michele Filosi is currently enrolled in the Biomolecular Science PhD School at the University of Trento (Italy). His project is focused on application of Computer Science to Biology with a particular interest in analysis and pathway detection in biological networks.

Alexandru G. Floares   Chapter D.20

SAIA, OncoPredict
Cancer Institute Cluj-Napoca
Cluj-Napoca, Romania
alexandru.floares@oncopredict.com;
alexandru.floares@saia-institute.org

Alexandru G. Floares is a medical doctor (neurologist) and a computer scientist. He received his PhD degree in Biophysics in 2000, from Iasi University of Medicine and Pharmacy, Romania. His research interests include developing artificial intelligence based OMICS tests in cancer, and methodologies for automatic modeling complex systems, including gene regulatory networks. He is a member of Institute of Electrical and Electronics Engineers (IEEE) and New York Academy of Science.

Yan Fu

Chinese Academy of Sciences
Academy of Mathematics and Systems
Science, Haidian District
Beijing, China
yfu@amss.ac.cn

Chapter C.18

Yan Fu obtained his PhD degree in Computer Science from the Institute of Computing Technology, Chinese Academy of Sciences (CAS), and is now an Associate Professor at the Academy of Mathematics and Systems Science. His research interests are mainly in application of statistical and machine-learning methods to biological problems, with a current focus on computational proteomics and mass spectrometry.

Kunihiko Fukushima

Tokyo, Japan
fukushima@m.ieice.org

Chapter I.44

Kunihiko Fukushima received a BEng and a PhD degree in Electrical Engineering from Kyoto University, Japan. He was a Professor at Osaka University, and later at several universities. Prior to his Professorship, he was a Senior Research Scientist at the NHK Science Research Laboratories. He was the founding President of Japanese Neural Network Society (JNNS) and a founding member of International Neural Network Society (INNS).

Cesare Furlanello

Fondazione Bruno Kessler
Povo, Italy
furlan@fbk.eu

Chapter K.53

Cesare Furlanello is Senior Researcher at FBK, directing the MPBA unit. He graduated in Mathematics at the University of Padua in 1986. He is a member of the Steering Committee of the FDA SEQC initiative for NGS data, and of the Board of Directors of the Functional Genomics Data Society. He is an investigator in the FANTOM5 RIKEN project and a Wistar Adjunct Professor.

Matthias E. Futschik

University of Algarve
IBB/Centre for Biomedical and
Structural Biomedicine
Faro, Portugal
mfutschik@ualg.pt

Chapter D.24

Matthias E. Futschik obtained a PhD in Information Science from the University of Otago, New Zealand. Currently, he is Principal Investigator at the Centre for Molecular and Structural Biomedicine of the University of Algarve, Portugal. His research interests comprise various topics in the field of bioinformatics and systems biology with applications in stem cell biology, synthetic biology, and human diseases.

Carlo Gambacorti-Passerini

University of Milano Bicocca
Department of Health Science
Monza, Italy
carlo.gambacorti@unimib.it

Chapter D.25

Professor Carlo Gambacorti-Passerini is an oncologist and hematologist known for his contributions to cancer research. He is presently Professor of Internal Medicine at the University of Milan Bicocca in Italy and Director of the Clinical Research Unit at S. Gerardo Hospital, Monza, Italy. His main scientific contribution relates to the preclinical and clinical development of imatinib.

Petia Georgieva

University of Aveiro
Department of Electronics
Telecommunications and Informatics
(DETI)
Aveiro, Portugal
petia@ua.pt

Chapter I.46

Petia Georgieva received her PhD in Electrical Engineering from the Technical University of Sofia, Bulgaria (1997). She is a Lecturer at the University of Aveiro and the Head of Signal Processing Lab, Institute of Electronics Engineering and Telematics of Aveiro (IEETA). Her research interests are in data mining and machine learning with a strong focus on brain study applications like the brain–machine interface and brain neural activity recovering. Dr. Georgieva is a Senior member of International Neural Network Society (INNS).

Muriel Gevrey   Chapter F.35

University of Maryland
Chesapeake Biological Laboratory
Solomons, MD, USA
mgevrey@gmail.com

Muriel Gevrey has accomplished her PhD in Biomathematics and was then involved in an invasive species subject during her New-Zealand post-doctoral fellowship. Since that, she has participated in several European projects dealing with the Water Framework Directive. She has recently spent one year as a scientific visitor at the Chesapeake Biological Laboratory, USA, working on nitrogen removal by storm water management structures.

Antonio Giordano   Chapter F.29

Temple University
Biology – Center for Biotechnology
Philadelphia, PA, USA
antonio.giordano@temple.edu

Antonio Giordano is a "Chiara fama" Professor at the Department of Pathology and Oncology at the University of Siena. He is also the Director of the Sbarro Institute for Cancer Research and Molecular Medicine and Center for Biotechnology at Temple University, College of Science & Technology, Philadelphia. Dr. Giordano made significant contributions to the field of cancer research by discovering pRb2/p130, a member of retinoblastoma family proteins, and studying its role during cancer initiation and progression. He has published over 400 articles on his work in the field of cell cycle, gene therapy and genetics of cancer as well as several book chapters in oncology and pathology field.

Janice I. Glasgow   Chapter C.13

Queen's University
School of Computing
Kingston, ONT., Canada
janice@cs.queensu.ca

Janice Glasgow is a Professor and Research Chair in the School of Computing at Queen's University, Kingston, Canada. She is the Director of the Computational Imagery Laboratory and a co-director of the Molecular Scene Analysis Laboratory. She is a co-founder of the company Molecular Mining. Janice Glasgow's primary research interests include the development of knowledge representation and reasoning tools for computational imagery.

Pietro H. Guzzi   Chapter E.26

University Magna Graecia of Catanzaro
Surgical and Medical Sciences
Catanzaro, Italy
hguzzi@unicz.it

Pietro H. Guzzi is an Assistant Professor of Computer Engineering at the University "Magna Graecia" of Catanzaro, Italy, since 2008. He received his PhD in Biomedical Engineering in 2008, from Magna Graecia University of Catanzaro and his Laurea degree in Computer Engineering in 2004 from the University of Calabria, Rende, Italy. His research interests comprise bioinformatics, the analysis of proteomics data, and the analysis of protein interaction networks. Pietro is an ACM member and serves the scientific community as reviewer for many conferences.

Michael Hawrylycz   Chapter L.62

Allen Institute for Brain Science
Department of Modeling, Analysis, and Theory
Seattle, WA, USA
mikeh@alleninstitute.org

Mike Hawrylycz joined the Allen Institute in 2003. He is responsible for the direction of the data analysis and annotation effort. Hawrylycz has worked in a variety of applied mathematics and computer science areas, addressing challenges in consumer and investment finance, electrical engineering and image processing, and computational biology and genomics. Hawrylycz received his PhD in Applied Mathematics from the Massachusetts Institute of Technology, Cambridge. He subsequently was a post-doctoral researcher in the Computer Research and Applications Group at the Los Alamos National Laboratory.

Authors

Miguel A. Hernandez-Prieto   Chapter D.24

University of Algarve
IBB/Centre for Biomedical and
Structural Biomedicine
Faro, Portugal
mprieto@ualg.pt

Miguel A. Hernández-Prieto is a Postdoctoral Fellow at the Systems Biology group (SysBioLab), Universidade do Algarve, Portugal. In 2009, he received a PhD in Biochemistry from the University of Umeå, Sweden. He combines genetics and computational biology approaches to discover and understand mechanisms that allow cyanobacteria to adapt to different environmental conditions.

Yoshito Hirata   Chapter H.42

The University of Tokyo
Institute of Industrial Science
Tokyo, Japan
yoshito@sat.t.u-tokyo.ac.jp

Yoshito Hirata obtained BEng and MEng from the Department of Mathematical Engineering and Information Physics, The University of Tokyo, and his PhD from the School of Mathematics and Statistics, University of Western Australia. He is a project Associate Professor at the Institute of Industrial Science, The University of Tokyo. He is interested in developing nonlinear time series analysis and applying it to real data.

Mika Hirvensalo   Chapter L.59

University of Turku
Department of Mathematics
Turku, Finland
mikhirve@utu.fi

Mika Hirvensalo owns a PhD in Mathematics obtained 2003 from the University of Turku. Between 2003 and 2006 he was a post doc, lecturer of mathematics at the University of Turku Since 2008 he serves as a reader/adjunct professor. His current research interests are models of computing, decidability/undecidability, complexity of computation, and quantum automata.

Zeng-Guang Hou   Chapter K.56

Chinese Academy of Sciences
Institute of Automation, State Key
Laboratory of Management and Control
for Complex Systems
Beijing, China
zengguang.hou@ia.ac.cn

Zeng-Guang Hou is a Professor and Deputy Director of the State Key Laboratory of Management and Control for Complex Systems, Institute of Automation, Chinese Academy of Sciences. He currently serves as the Editor of Computational Intelligence Society E-Letter, and as an Editor Editorial Board Member of the International Journal of Intelligent Systems Technologies and Applications, and the Journal of Intelligent and Fuzzy Systems.

Yingjie Hu   Chapter F.33

Auckland University of Technology
Knowledge Engineering and
Discovery Research Institute
Auckland, New Zealand
rhu@aut.ac.nz

Dr. Yingjie Hu received his Master degree (Hons) and PhD in Computer and Information Sciences from Auckland University of Technology (AUT), New Zealand. He is currently a research fellow in KEDRI at AUT and is involved in several projects of data analysis. His research interests are in the areas of personalized modeling, data mining, neural networks and image recognition.

Xiao Huang   Chapter L.58

Bellevue, WA, USA

Xiao Huang received the BS degree in Electrical Engineering from Beijing Jiaotong University, Beijing, China, in 1997, the MEng degree in Automation from the Chinese Academy of Sciences, Beijing, China, in 2000, and the PhD degree in Computer Science from Michigan State University, Lansing, Michigan, USA, in 2005. He is currently a research software design engineer at the Microsoft Bing Data Mining team. His research interests include data mining and machine learning.

Sally Hunter

University of Cambridge
Public Health and Primary Care
Cambridge, UK
seh66@medschl.cam.ac.uk

Chapter K.52

Sally Hunter is a researcher with the Cambridge City over-75s Cohort Study (CC75C) (CC75C) with interests in neuropathology and the Amyloid Precursor Protein proteolytic system. She graduated from the Open University with a degree in the Life Sciences and following a short contract with the Cambridge Brain Bank she went on to complete a Master's degree, also with the Open University. Since 2008 she has worked on various neuropathological studies with the CC75C collaboration.

Hitoshi Iba

The University of Tokyo
Graduate School of Information, Science and Technology
Tokyo, Japan
iba@iba.t.u-tokyo.ac.jp

Chapter D.22

Hitoshi Iba received the PhD degree from the University of Tokyo, Japan, in 1990. From 1990 to 1998, he was with the ElectroTechnical Laboratory (ETL), Ibaraki, Japan. He has been with the University of Tokyo, since April 1998 where he is currently a Professor at the Graduate School of Information Science and Technology, University of Tokyo. His research interests include evolutionary computation, genetic programming, bioinformatics, foundation of artificial intelligence, machine learning, and robotics. Dr. Iba is an Associate Editor of the IEEE Transactions on Evolutionary Computation.

Takayoshi Ikeda

University of Otago
Dean's Department
Wellington, Wellington South, New Zealand
tak.ikeda@otago.ac.nz

Chapter F.35

Takayoshi Ikeda is a research fellow and biostatistician at the University of Otago, Wellington, New Zealand, providing biostatistical assistance to fellow staff and postgraduate students. His interests in medical research include tobacco, cancer and infectious disease modeling. He also has experience in species distribution modeling looking at the potential risk of freshwater invertebrate species invading New Zealand.

Hiroko Imasato

Fuzzy Logic Systems Institute (FLSI)
Fukuoka, Japan
imasato@flsi.or.jp

Chapter A.3

Hiroko Imasato received the MEng and the PhD degrees from Kyushu Institute of Technology in 2004 and 2008, respectively. Since 1982, she had been a laboratory medical technologist at the University Hospital of Occupational and Environmental Health (UOEH). Since 2008, she has been a Senior Researcher at Fuzzy Logic Systems Institute (FLSI). Her research interest lies on dielectrophoresis and clinical laboratory tests.

Grant H. Jacobs

BioinfoTools
Dunedin, New Zealand
gjacobs@bioinfotools.com

Chapter B.8

Since doctoral training exploring the structure and function of DNA-binding proteins (MRC Laboratory of Molecular Biology, Cambridge, UK), Dr. Jacobs has worked at the University of Otago, New Zealand as post-doc and research fellow: He then founded BioinfoTools, offering computational biology to research groups and companies.
Among his interests are computational studies of gene regulation including chromatin and genome structure.

Igor Jurisica   Chapter C.13

University of Toronto
Department of Computer Science
Toronto, Ontario, Canada
juris@ai.utoronto.ca

Igor Jurisica, Tier I Canadian Research Council in Integrative Cancer Informatics, is a Senior Scientist at Ontario Cancer Institute, Professor at the University of Toronto and Visiting Scientists at IBM's CAS. He is also an Adjunct Professor at the School of Computing, Queen's University and Computer Science at York University. His research focuses on integrative computational biology and the representation, analysis and visualization of high-dimensional data to identify prognostic/predictive signatures, drug mode of action and in silico repurposing drugs.

Giuseppe Jurman   Chapter K.53

Fondazione Bruno Kessler
Predictive Models for Biomedicine and
Environment
Povo, Italy
jurman@fbk.eu

Giuseppe Jurman earned his PhD in Mathematics (Algebra) at the University of Trento (Italy) in 1998. After 2 years at ANU Canberra, he moved to the Fondazione Bruno Kessler in Trento, in the Predictive Models for Biomedicine and Environment unit. His research field is computational biology, and, in particular, the application of complex networks and machine learning to -omics data.

Ravi K.R. Kalathur   Chapter D.24

University of Algarve
DIBB/Centre for Biomedical and Structural
Biomedicine
Faro, Portugal
rkkalathur@ualg.pt

Ravi Kiran Reddy Kalathur accomplished his PhD in Bioinformatics at the IGBMC, University of Strasbourg, France. At present, he is a Postdoctoral Fellow at the University of Algarve and studies the molecular networks underlying the development and progression of T-cell acute lymphoblastic leukemia.

Nikola Kasabov   Chapters 1, F.33, G.37, I.46, I.47, J.49, K.54, K.55, L.60

Auckland University of Technology
KEDRI – Knowledge Engineering and
Discovery Research Institute
Auckland, New Zealand
nkasabov@aut.ac.nz

Professor Nikola K. Kasabov is Fellow of IEEE, RSNZ and a Distinguished Visiting Fellow of the Royal Academy of Engineering UK. He obtained his Masters and PhD degrees from the Technical University of Sofia, Bulgaria. He is the Director and the Founder of the Knowledge Engineering and Discovery Research Institute (KEDRI) and Professor of Knowledge Engineering at the School of Computing and Mathematical Sciences at the Auckland University of Technology, New Zealand. He has published more than 450 papers, books and patents in the areas of informatics, computational intelligence, neural networks, bioinformatics, neuroinformatics. He is Past-President of the International Neural Network Society (INNS, 2009, 2010), Governor of INNS, Past President of the Asia Pacific Neural Network Assembly (APNNA), Distinguished IEEE CIS Lecturer, EU Marie Curie Fellow and Visiting Professor at the Institute for Neuroinformatics ETH/UZH Zurich, Guest Professor Shanghai Jiao Tong University. He received the INNS Gabor Award (2012), Bayer Innovation Award (2007), APPNA Excellence Award (2005), RSNZ Science and Technology Medal (2002), APPNA Outstanding Achievement Award (2012), and numerous IEEE best paper awards.

Diana A. Kassabova

Knowledge Engineering Consulting Ltd.
Auckland, New Zealand
diana.kassabova@gmail.com

Diana Kassabova is a consultant and editor on technical writings, including postgraduate disserations, scientific books, and technical manuals.
Contribution: Assistance to the Editor in the preparation of selected chapters.

Sebastian Kelm   Chapter B.11

University of Oxford
Department of Statistics
Oxford, UK
kelm@stats.ox.ac.uk

Sebastian Kelm obtained his BSc in Biochemistry and MSc in Bioinformatics and Systems Biology from Imperial College London. He completed his doctorate in the prediction of membrane protein structure at the University of Oxford's Department of Statistics, where he now works as a postdoctoral researcher. His research interests include the structure, function and dynamics of membrane proteins and antibodies.

Alvin T. Kho   Chapter A.4

Boston Children's Hospital
Boston, MA, USA
alvin_kho@hms.harvard.edu

Alvin T. Kho is a staff scientist at the Boston Children's Hospital Informatics Program. His research focuses on multi-scalar mathematical modeling of developing biological systems to identify molecular programs that under pathogenesis in such systems. He has worked on the systematic connections between central nervous system and pulmonary cancers in relation to their cognate developing systems. Most recently he is investigating the developmental bases (e.g., fetal programming) for paediatric and adult chronic lung diseases. He has a mathematics PhD degree from the University of Massachusetts at Amherst.

Alistair Knott   Chapter I.48

University of Otago
Department of Computer Science
Dunedin, New Zealand
alik@cs.otago.ac.nz

Alistair Knott is an Associate Professor at the Department of Computer Science, at the University of Otago. His interests are in computational linguistics and computational neuroscience. He has recently published a book about the relationship between language and the sensorimotor system.

Hiroshi Kojima   Chapter G.36

Tamagawa University
Intelligent Information Systems
Tokyo, Japan
hkojima@lab.tamagawa.ac.jp

Hiroshi Kojima holds a PhD degree in Medical Sciences from the Graduate School of Medicine, Kyoto University, Japan obtained in 1986. Since 2005, he has been working as a Full Professor at Tamagawa University, Japan. Present interests are mechanisms and regulation of synaptic transmission.

Mark Kon   Chapter D.21

Boston University
Department of Mathematics and Statistics
Boston, MA, USA
mkon@bu.edu

Mark Kon received a PhD in Mathematics from MIT, Cambridge, MA, and Bachelor's degrees in Mathematics, Physics, and Psychology from Cornell University, Ithaka, NY. He has had appointments at Columbia University as Assistant and Associate Professor of Computer Science and Mathematics, and at MIT as graduate instructor. He has been Departmental Director of graduate studies at Boston University, and is affiliated with the Bioinformatics Graduate Program.

Elena Kostadinova   Chapter C.14

Technical University of Sofia, Plovdiv Branch
Department of Computer Systems and Technologies
Plovdiv, Bulgaria
elli@tu-plovdiv.bg

Elena Kostadinova received her BSc (2006) and MSc (2008) degrees in Computer Engineering from the Technical University of Sofia, branch Plovdiv, Bulgaria. She is currently a PhD student in the Department of Computer Systems and Technologies of the same university. Her research interests are directed to the development of methods for gene expression data analysis, including missing values imputation, gene clustering and data integration.

Authors

Rita Krishnamurthi

AUT University
National Institute for Stroke and
Applied Neurosciences
Auckland, New Zealand
rita.krishnamurthi@aut.ac.nz

Chapter K.55

Dr. Krishnamurthi is a senior research fellow at AUT University working in stroke and applied neurosciences. Her current research activities include stroke epidemiology, interventions for primary and secondary stroke prevention, prognostic modeling of stroke risk factors and outcomes, and contributing to the Global Burden of Diseases Stroke panel.

Eric J. Lang

New York University
Department of Physiology & Neuroscience
School of Medicine
New York, NY, USA
eric.lang@nyumc.org

Chapter H.42

Eric J. Lang earned his MD and PhD from New York University, where he is now an Associate Professor of Physiology and Neuroscience. He did his post-doctoral work at Laval University, Quebec. His research focuses on understanding the mechanisms underlying the patterns of neuronal activity in the cerebellum, and how these patterns relate to the cerebellum's role in motor control.

Gwenaël Leday

Vrije Universiteit Amsterdam
Department of Mathematics
Amsterdam, The Netherlands
g.g.r.leday@vu.nl

Chapter F.35

Gwenaël G.R. Leday is currently a PhD student in Statistics at the VU University in Amsterdam, Netherlands. His research interests are in the applications of statistical methods to the field of ecology and biology. His PhD project aims to develop statistical methods for the (integrative) analysis of molecular data such as DNA and mRNA.

Lin Li

Philips Research North America
Briarcliff Manor, NY, USA
lin-li@philips.com

Chapter H.41

Lin Li received her PhD in Electrical and Computer Engineering from the University of Florida in 2012. She is currently a member of research staff with Philips Research North America, NY. Her research interests include machine learning, computational neuroscience, and big data analytics. She is a member of IEEE Engineering in Medicine & Biology Society since 2010.

Qingling Li

China University of Mining &
Technology, Beijing
Department of Mechanical & Electrical
Engineering
Beijing, China
doudouhit@163.com

Chapter K.56

Qingling Li received the BSc and the MSc degree in precision instrument and the PhD degrees in mechanical engineering from Harbin Institute of Technology, China, in 2002, 2004, and 2009, respectively. From 2009 to 2011, she did her post-doctoral research at the Institute of Automation, Chinese Academy of Sciences. Since 2011, he has been at the China University of Mining & Technology, Beijing, where she is currently a lecturer of mechanical engineering. Her current research interests include rehabilitation robots, sEMG signal analysis, and, and assistive technologies for the disabled.

Xingfeng Li

University of Ulster
Department of Computing and
Engineering
Derry, Northern Ireland, UK
x.li@ulster.ac.uk

Chapter G.38

Dr. Xingfeng Li has been a research fellow at the University of Ulster, UK, since 2009. He holds a PhD in Pattern Recognition and Machine Intelligence from the Institute of Automation, Chinese Academy of Sciences (2004). He worked at McGill University and Paris 6th University to analyze functional MRI data. His current research is focused on algorithm development for fMRI, DTI, and PET data processing.

Wen Liang

Auckland University of Technology
School of Computing and Mathematical Science
Auckland, New Zealand
lliang@aut.ac.nz

Chapters F.33, K.55

Wen Liang (Linda) received her Master's degree at AUT in 2009 with First Class Honors. She is now a PhD student at AUT under the supervision of Prof. Nik Kasabov and Prof. Valery Feigin. Her research interests are in the areas of personalized modeling of brain data and spiking neural networks. Linda was awarded the Vice-Chancellor's Doctoral Scholarship by the AUT University in 2011, and the Graduate Assistantship Award 2010 from the Faculty of Design and Creative Technologies.

Hongye Liu

Harvard Medical School/Boston Children's Hospital
Informatics Program
Boston, MA, USA
hongye.liu@gmail.com;
hongye.liu@childrens.harvard.edu

Chapter A.4

Hongye Liu is a Bioinformatician in the Children's Hospital Boston. She graduated with a PhD from MIT in Engineering. Dr. Liu is interested in understanding of cancers in both lung and the CNS. She has developed a development-based framework for lung cancer classification and prognostics and unraveled the tissue-specific microRNA regulation shared in CNS development and medulloblastom.

Weifeng Liu

Jump Trading
Chicago, IL, USA
weifeng@ieee.org

Chapter H.41

Weifeng Liu received the PhD degree in Electrical and Computer Engineering from the University of Florida in 2008. He was a senior scientist with Amazon.com, Seattle, WA, from 2008 to 2012. He is currently a senior researcher with Jump Trading, Chicago, IL. His research interests include machine learning, adaptive signal processing and their applications to e-commerce, business, and finance.

David Looney

Imperial College London
Communication and Signal Processing Research Group, Department of Electrical and
Electronic Engineering
London, UK
david.looney06@imperial.ac.uk

Chapter H.43

David Looney received the BEng degree in Electronic Engineering from University College Dublin, Ireland. In 2011. he received his PhD degree in Signal Processing from Imperial College, London, UK, where he is currently a Research Associate. His research interests are mainly in the areas of data fusion, exploratory data analysis and wearable solutions for health monitoring.

Irina Luludachi

SAIA Institute
Cluj-Napoca, Romania
irina.luludachi@saia.ro

Chapter D.20

Irina Luludachi is a computer scientist who received her Master Degree in Applied Mathematics from the Babes-Bolyai University of Cluj-Napoca, Romania, in 2010. Her research interests include cancer Computational Biology and Bioinformatics.

Marcella Macaluso

Temple University
Biology – Center for Biotechnology
Philadelphia, PA, USA
macaluso@temple.edu

Chapter F.29

Marcella Macaluso is an Associate Professor of Biology and Director of the Epigenetic and Genetic Program at Sbarro Health Research Organization (S.H.R.O.). Dr Macaluso's research focuses on understanding the molecular mechanisms underlying epigenetic and genetic alterations in human cells leading to cancer formation and progression. Her interest is especially focused on the pRb family proteins and their role in normal and cancer cells.

Authors

Liam Maguire

University of Ulster
Intelligent Systems Research Centre
Derry, Northern Ireland, UK
lp.maguire@ulster.ac.uk

Chapters G.38, G.39

Liam received MEng and PhD degrees in Electrical and Electronic Engineering from the Queen's University of Belfast and is currently Head of the School of Computing and Intelligent Systems at the University of Ulster where he leads the bio-inspired systems and neuro-engineering research team. His research interests are in developing new bio-inspired intelligent systems. He is the author of over 200 research papers including 70 journal papers. He has an established track record of securing research funding and has supervised 15 PhD and 3 MPhil students to completion.

Danilo P. Mandic

Chapter H.43

For biographical profile, please see section "About the Part Editors" on page XV.

Kathleen Marchal

Ghent University
Department of Plant Biotechnology and Bioinformatics
Gent, Belgium
kamar@psb.ugent.be

Chapter D.19

Kathleen Marchal is Associate Professor at the Department of Plant Biotechnology and Bioinformatics, Ghent University and at the Department of Microbial and Molecular Systems, KU Leuven. She is also part of the VIB department Plant Systems Biology. Her research focuses on the development of computational methods for systems biology and data integration, focusing on network inference and network-based omics integration.

Manuela Marega

Justus-Liebig-Universität
Institut für Neuropathologie
Gießen, Germany
*manuela.marega@patho.med.
uni-giessen.de*

Chapter D.25

Dr. Manuela Marega graduated in Medical Biotechnology from the University of Trieste in 2006 and received her PhD in Experimental Haematology from the University of Milano-Bicocca in 2010. She moved to the Institute of Neuropathology, Gießen, Germany, where she holds a position as post-doc. Her research activity focuses on the relationship between microenvironment and the tumour, comparing different tumour model.

T. Martin McGinnity

University of Ulster
School of Computing and Intelligent Systems, Computer Science Research Institute
Derry, Northern Ireland, UK
tm.mcginnity@ulster.ac.uk

Chapter G.38

T. Martin McGinnity holds a 1st Class (Hons.) degree in Physics (1975) and a PhD degree from the University of Durham, UK (1979). He is a Professor of Intelligent Systems Engineering, in the University of Ulster, Northern Ireland, and the Director of the Intelligent Systems Research Centre. He is the author or co-author of more than 275 research papers.

Mariofanna Milanova

University of Arkansas at Little Rock
Department of Computer Science
Little Rock, AR, USA
mgmilanova@ualr.edu

Chapter I.46

Dr. Mariofanna Milanova is a Professor of Computer Science in the Department of Computer Science at the University of Arkansas at Little Rock. She received her PhD degree in Computer Science in 1995 from the Technical University, Sofia, Bulgaria and did her post-doctoral research in visual perception at the University of Paderborn, Germany. Her main research interests are in the areas of artificial intelligence, biomedical signal processing, computer vision, machine learning, and security based on biometric research.

Angela Mogavero   Chapter D.25

Università di Milano Bicocca
Department of Health Science
Monza, Italy
angela.mogavero@gmail.com

Angela Mogavero received her PhD in Experimental Hematology from the University of Milano-Bicocca, Italy (2010). Her doctoral research was devoted to detect the epigenetic silencing of the proapoptotic gene BIM in glucocorticoid poor-responsive pediatric acute lymphoblastic leukemia and in anaplastic large cell lymphoma. Currently, her postdoctoral fellowship interests include molecular biology and treatment of colorectal cancer.

Micaela Montanari   Chapter F.29

Temple University
Biology-Center for Biotechnology
Philadelphia, PA, USA
montanar@temple.edu

Micaela Montanari is an Adjunct Assistant Professor of Biology and Director of the Hormone-Signalling and Cancer Program at SHRO. She received her PhD in Oncology at the Catholic University of Sacred Hearth, Rome, Italy. Her research focuses on understanding the molecular mechanisms underlying estrogen signals alterations in human cells leading to breast cancer formation and progression, and on disclosing the genetic and epigenetic mechanisms responsible for the regulatory disruption of cellular proliferation, survival and differentiation in the neoplastic process.

Francesco C. Morabito

University Mediterranea
DICEAM
Reggio Calabria, Italy
morabito@unirc.it

Chapter L.61

Francesco Carlo Morabito is a Full Professor of Electrical Engineering at the University Mediterranea of Reggio Calabria, Italy, where he was former Dean of the Faculty of Engineering and is now serving as Vice-Rector for Internationalization. He served for 12 years as Governor of International Neural Network Society and is now President of the Italian Neural Network Society, Foreign Member of the Royal Academy of Doctors, Spain. He is author/co-author of more than 300 papers in international journals/conference proceedings in various fields of engineering (radar data processing, nuclear fusion, biomedical signal processing, non-destructive testing and evaluation, computational intelligence). He is co-editor of 10 books and holds 3 international patents.

Giuseppe Morabito

University of Pavia
Pavia, Italy
*giuseppe.
morabito01@universitadipavia.it*

Chapter L.61

Giuseppe Morabito graduated in Electronic Engineering from the University Mediterranea of Reggio Calabria (summa cum laude) and is now going towards his MSc degree in Bio-Engineering at the University of Pavia (Italy). His main research interests are in the fields of biomedical signal processing, particularly EEG, measures of complexity and early detection of Alzheimer's Disease.

Tamás Nepusz   Chapter D.23

Eötvös Loránd University
Department of Biological Physics
Budepast, Hungary
nepusz@hal.elte.hu

Tamás Nepusz received his PhD in Computer Science from the Budapest University of Technology and Economics (2008) and is currently a Research Associate at the Eötvös Loránd University, Budapest, Hungary. His research interests include machine-learning algorithms and random-graph models. He also spent two years at Royal Holloway, University of London, supported by the Newton International Fellowship of the Royal Society.

Authors

Authors

Lydia Ng

Allen Institute for Brain Science
Department of Technology
Seattle, WA, USA
lydian@alleninstitute.org

Chapter L.62

Lydia Ng joined the Allen Institute in 2004 and currently leads the atlas development team responsible for the design and implementation of the Institute's web applications as well as visualization and data mining tools. Her research background includes image processing and analysis, image registration, data analysis and mining. Lydia received a BEng in Electrical Engineering and a BSc in Computer Science from the University of New South Wales (Sydney, Australia) and a PhD in Electronics from Macquarie University (Sydney, Australia).

Nasimul Noman

University of Tokyo
Graduate School of Engineering
IBA Laboratory
Department of Electrical Engineering and
Information Systems
Tokyo, Japan
noman@iba.t.u-tokyo.ac.jp,
noman@univdhaka.edu

Chapter D.22

Nasimul Noman received the BSc and MSc degrees in Computer Science from the University of Dhaka, Bangladesh and his PhD degree in Frontier Informatics from the Graduate School of Frontier Sciences, University of Tokyo, Japan. He is a faculty member in the Department of Computer Science and Engineering, University of Dhaka, since March 2002 and also a joint researcher in the Department of Information Science in Technology at Tokyo University. His research interests include evolutionary computation, computational biology and bioinformatics.

Gianluigi Occhiuto

University Mediterranea
DICEAM
Reggio Calabria, Italy
gianluigi.occhiuto@unirc.it

Chapter L.61

Gianluigi Occhiuto received the Master's Degree in Electronic Engineering from the "Mediterranea" University of Reggio Calabria in October 2008. He has been enrolled as a PhD candidate in Information Engineering of the Mediterranea University of Reggio Calabria since the end of 2009. His research interests are on the study of main complexity measures and the application signal-processing tools to biomedical signals.

Alberto Paccanaro

University of London
Department of Computer Science
Egham, Surrey, UK
alberto@cs.rhul.ac.uk

Chapter D.23

Alberto Paccanaro is Reader in Computational Biology in the Department of Computer Science at Royal Holloway, University of London. He received his PhD in Computer Science from the University of Toronto in 2002, specializing in machine learning. From 2002 to 2006 he was at Queen Mary University of London and Yale University. His research interests are in applying and developing machine learning and pattern recognition techniques for solving problems in molecular biology. His recent research focuses on methods for analysis and inference in large scale biological networks.

Leon Palafox

The University of Tokyo
School of Electrical Engineering
Tokyo, Japan
leon@iba.t.u-tokyo.ac.jp

Chapter D.22

Leon Palafox received the BSc degree in Electronic Engineering from the National Autonomous University of Mexico (UNAM) and the MSc degree in Electronic Engineering Graduate from the School of Engineering, University of Tokyo, Japan. He is currently pursuing his PhD in Electrical Engineering and Information Sciences at the University of Tokyo. His research interests include evolutionary computation, machine learning and bioinformatics.

Cheolsoo Park

University California, San Diego
Department of BioEngineering
La Jolla, CA, USA
charles586@gmail.com

Chapter H.43

Cheolsoo Park received his PhD degree in Adaptive Nonlinear Signal Processing and Brain–Computer Interface from Imperial College London, UK in 2012. Currently, he is working as a postdoctoral researcher at the University California, San Diego, USA. His research interests are in brain–computer interface, time–frequency analysis and statistical signal processing.

David Parry

Auckland University of Technology
School of Computing and Mathematical
Sciences
Auckland, New Zealand
dave.parry@aut.ac.nz

Chapter F.34

Dave Parry is a Senior Lecturer and Director of the AUT Radio Frequency Identification (RFID) Laboratory (AURA) in the AUT School of Computing and Mathematical Sciences. His research interests include health informatics, ontology-based information retrieval, and RFID applications for pervasive computing. Current research projects include the use of RFID in healthcare and healthcare IT for developing nations.

Michael G. Paulin

University of Otago
Department of Zoology
Dunedin, New Zealand
mike.paulin@otago.ac.nz

Chapter A.6

Mike Paulin is Associate Professor in Zoology at the University of Otago, New Zealand. His main research interest is the coevolution of brains and bodies for agile movement. He has previously worked at the University of Auckland (Mathematics); University of Southern California (Neurobiology and Engineering); California Institute of Technology (Computational Neuroscience and Engineering) and NASA Jet Propulsion Laboratory (Robotics).

Russel Pears

Auckland University of Technology
Department of Computing and
Mathematical Sciences
Auckland, New Zealand
rpears@aut.ac.nz

Chapter J.49

Russel is a Senior Lecturer in the School of Computing and Mathematical Sciences, Auckland University of Technology, New Zealand. He has published widely in the Machine Learning area and is currently involved in a number of research projects in the Machine Learning area including Data Stream Mining, Pattern Mining and the interface between Machine Learning systems and Ontologies.

Rocco Piazza

University of Milano–Bicocca
Department of Health Science
Monza, Italy
rocco.piazza@unimib.it

Chapter D.25

Rocco Piazza qualified in Medicine at the University of Pavia. He received his PhD degree in Biochemistry from Pavia University and completed his post-doctoral work at the National Cancer Institute of Milan. In 2005, he completed a 4 years fellowship in clinical haematology in Monza and is now researcher at the University of Milano-Bicocca in the field of high-throughput sequencing of oncohematological disorders.

Leonie Z. Pipe

Auckland University of Technology
Knowledge Engineering and Discovery
Research Institute (KEDRI)
Auckland, New Zealand
leonie.pipe@gmail.com

Chapter A.7

Leonie's PhD thesis, entitled Mathematical and Numerical Modelling of Colony Growth on High-Nutrient Surfaces, was completed in 2009. Currently, she is involved with the KEDRI institute as part of the Artificial Bacterial Intelligence group, which aims to better understand the cooperative group behaviour of microorganisms in colonies, with potential applications to optimization problems and robotics.

Alessandra Pirola

University of Milano–Bicocca
Department of Health Science
Monza, Italy
alessandra.pirola@unimib.it

Chapter D.25

Alessandra Pirola received the Master's degree in Biology from the University of Milan, Italy, in 2004. She has been working at the University of Milano-Bicocca in Professor Gambacorti-Passerini Molecular Oncology's lab since 2007 as a Research Technician. Her research activities include processing of biological samples derived from haematological patients for molecular analysis and now she works mainly on the high-throughput sequencing instrument Illumina Genome Analyzer IIx for libraries preparation.

Authors

Joel Pitt

Chapter F.35

Lincoln University
c/0 Bio-Protection Research Centre
Lincoln, New Zealand
joel@joelpitt.com

Joel is a multidisciplinary scientist who has contributed to the fields of bioinformatics, artificial intelligence, and ecology. His PhD from Lincoln University, New Zealand involved the development of a spatially explicit stochastic simulation model for investigating large-scale spread of invasive species across real landscapes.

José C. Príncipe

Chapter H.41

University of Florida
Department of Electrical and
Computer Engineering
Gainesville, FL, USA
principe@cnel.ufl.edu

José C. Príncipe is currently a Distinguished Professor of Electrical and Biomedical Engineering at the University of Florida, Gainesville, USA. He is Founder and Director of the University of Florida Computational Neuro-Engineering Laboratory (CNEL). Dr. Principe is an IEEE fellow and AIMBE fellow He is involved in biomedical signal processing, in particular, the electroencephalogram (EEG) and the modeling and applications of adaptive systems.

Hima Raman

Chapter D.25

University of Milano Bicocca
Department of Health Science
Monza, Italy
ram.hima@gmail.com

Hima Raman received the Master degree in Microbiology in 2005 from Mahatma Gandhi University, Kerala, India and the Bachelor degree in Zoology from University of Calicut, Kerala, India. In 2006 she received the Post Graduate Diploma in Bioinformatics from STG Global, Bangalore, India.

Naveed ur Rehman

Chapter H.43

COMSATS Institute of Information
Technology
Islamabad, Pakistan
naveed.rehman@comsats.edu.pk

Naveed ur Rehman received the BEng degree in Electrical Engineering from National University of Sciences and Technology, Pakistan. He completed his PhD degree in Signal Processing in December 2011 at the Imperial College, London, UK. He is currently working as an Assistant Professor in COMSATS Institute of Information Technology, Islamabad, Pakistan. His research interests are multivariate time-frequency algorithms, nonlinear methods and their applications in biomedical engineering, communications, and renewable energy.

Samantha Riccadonna

Chapter K.53

Fondazione Edmund Mach
Computational Biology Department
S. Michele all'Adige, Italy
samantha.riccadonna@fmach.it

Samantha Riccadonna received her PhD in Information and Communication Technologies from the University of Trento, Italy, in 2009. Then she joined the MPBA research unit of the Fondazione Bruno Kessler and in 2013 she moved to the Computational Biology Department of the Fondazione Edmund Mach, Italy. Her research interests are in computational biology, with a special focus in pathway analysis.

Raphael Ritz

Chapter J.50

For biographical profile, please see section "About the Part Editors" on page XV.

Alberto Riva

Chapter A.2

University of Florida
Molecular Genetics and Microbiology
Gainesville, FL, USA
ariva@ufl.edu

Alberto Riva is a computer scientist with a background in knowledge engineering, artificial intelligence and bioengineering. His research activity focuses on bioinformatics, and in particular on developing methods and tools for the representation of biomedical knowledge and its use in automated reasoning for computational biology applications.

Roberta Rostagno

University Milano Bicocca
Department of Health Science
Monza, Italy
roberta.rostagno@unimib.it

Chapter D.25

Dr. Roberta Rostagno received the MSc in Chemistry from the University of Milano. She is research fellow at the University of Milano-Bicocca. Her research activities include molecular modeling, studies on protein–ligand and protein–protein interaction and molecular dynamics calculation.

Reinhard Schliebs

University of Leipzig
Medical Faculty, Paul Flechsig Institute for
Brain Research
Leipzig, Germany
schre@medizin.uni-leipzig.de

Chapter K.51

Professor Reinhard Schliebs holds a Diploma in Chemistry (1972), a doctor's degree in Physical Chemistry (1975), and a DSc (Habilitation) in Biochemistry (1987) all from Karl Marx University of Leipzig, Germany. From 1975 to 1979 he was Research Assistant in the Department of Chemistry, and since 1979 Research Group leader and Professor at the Paul Flechsig Institute for Brain Research, University of Leipzig. Currently, his research interests include neurochemical and pathogenetic aspects of Alzheimer's disease.

Stefan Schliebs

Auckland University of Technology
School of Computing and Mathematical
Sciences
Auckland, New Zealand
stefan.schliebs@aut.ac.nz

Chapters F.35, G.37

Stefan Schliebs received the MSc degree in Computer Science from the University of Leipzig, Germany, and the PhD in Machine Learning from the Auckland University of Technology, New Zealand. His research interests include the study of neural networks, reservoir computing and evolutionary algorithms. Currently, he works on the recognition of temporal patterns in multi-dimensional time series for applications in human action detection systems.

Filipe Silva

University of Aveiro, Campus Universitário
de Santiago
Department of Electronics,
Telecommunications and Informatics
Aveiro, Portugal
fmsilva@ua.pt

Chapter I.46

Filipe Silva obtained his PhD degree in Electrical and Computer Engineering from the University of Porto in 2002. Currently, he is Assistant Professor at the Department of Electronics, Telecommunications and Informatics of the University of Aveiro, Portugal. His research interests include humanoid robotics, artificial biped locomotion systems and EEG-based brain-computer interfaces.

Lavneet Singh

University of Canberra
Bruce, Canberra, ACT, Australia
lavneet.singh@canberra.edu.au

Chapters C.17, G.40

Lavneet Singh is currently working as Research Associate and is working towards his PhD at the Faculty of ISE, University of Canberra, Australia. He did his Bachelor's and Master's degree at the Uttar Pradesh Technical University, Lucknow, India in Computer Sciences and Engineering. His research interests include image processing, machine learning, pattern recognition, artificial Intelligence, bioinformatics, and software engineering.

Snjezana Soltic

Manukau Institute of Technology
Engineering Centre of Excellence
Manukau, New Zealand
ssoltic@manukau.ac.nz

Chapter F.35

Snjezana Soltic received the BEng and MSc degrees in Electrical Engineering from University of Zagreb and the PhD degree in Computing and Mathematical Sciences from Auckland University of Technology, New Zealand. Her research interests are in neuro-computing, biomedicine and optimization. Currently, she is focusing on optimization of multi-band spectra for the design of the next generation of energy-saving lighting.

Authors

Authors

Yu Song

Beijing Jiaotong University
Department of Automation
Beijing, China
songyu@bjtu.edu.cn

Chapter K.56

Yu Song received his BEng degree in Mechanical Engineering from Harbin Institute of Technology, China, in 2001, his MEng and PhD degrees in Mechanical Engineering from the State Key Laboratory of Robotics and System, Harbin Institute of Technology, Harbin, China in 2004 and 2008, respectively. From 2008 to 2010 he was a postdoctoral research fellow at the State Key Laboratory of Intelligent Technology and System, Department of Computer Science, Tsinghua University, Beijing, China. Since 2011, he has been with the Department of Automation, Beijing Jiaotong University Beijing, China where he currently is a Lecturer His research interests cover intelligent robotics and computer vision. He is a member of the IEEE.

Roberta Spinelli

University of Milano–Bicocca
Department of Health Science
Monza, Italy
roberta.spinelli@unimib.it

Chapter D.25

Dr. Roberta Spinelli graduated in Mathematics from the University of Milan, Italy, in 2003. She received her Master in Statistics from the University of Milano-Bicocca in 2004. From 2004 to 2009 she worked as a researcher at ITB-CNR in microarray analysis to develop statistical procedures for genomic and transcriptomic profiles of several cancer pathologies and congenital diseases. Since 2010, she has been working on whole-exome and RNA sequencing of haematological patients at the University of Milano-Bicocca in the Molecular Oncology laboratory.

Andrea Splendiani

Rothamsted Research
Harpenden, Hertfordshire, UK
andrea.splendiani@intellileaf.com,
andrea.splendiani@rothamsted.ac.uk

Chapter E.27

Andrea Splendiani earned a Master's Degree in Information Technology from the Politecnico di Milano, Italy and a PhD in Computer Science from the University of Milano-Bicocca, Italy. His research interests are in use and development of web and semantic technologies to support data integration, user interaction and analytics. He has been active in systems biology, biomedical ontologies, bioinformatics and functional genomics. He is currently and independent consultant (IntelliLeaf Ltd) and a Bioinformatics Scientist at Rothamsted Research.

Margherita Squillario

University of Genova
DIBRIS
Genova, Italy
margherita.squillario@unige.it

Chapter K.53

Margherita Squillario is currently enrolled in the Informatics PhD School at the University of Genova, Italy. Her Bachelor's degree was on Biotechnology while her Master's degree on Bioinformatics. She is interested in the analysis of high-throughput data (e.g, microarray, NGS): gene signatures, their functional characterization and the prior knowledge integration. She also focuses on the analysis of data concerning neurodegenerative diseases.

Stewart G. Stevens

University of Otago
Department of Biochemistry
Dunedin, New Zealand
stewart.stevens@otago.ac.nz

Chapter B.10

Stewart Stevens started his career with Fujfilm researching and developing image processing algorithms. He is currently working on a PhD while jointly reading an Mb ChB at the University of Otago. His research is on RNA regulatory elements effecting post-transcriptional gene regulation.

Susan Sunkin

Allen Institute for Brain Science
Scientific Program Management
Seattle, WA, USA
susans@alleninstitute.org

Chapter L.62

Susan Sunkin joined the Allen Institute in 2004 She holds a PhD in Molecular Genetics, Biochemistry, and Microbiology from the University of Cincinnati, OH. She did her post-doctoral training at the University of Cincinnati and the Seattle Biomedical Research Institute. Prior to 2004, she managed aspects of the genomics production environments at Celltech R & D, Inc. and Genelex Corporation.

Aaron Szafer

Allen Institute for Brain Science
Department of Technology
Seattle, WA, USA
aarons@alleninstitute.org

Chapter L.62

Aaron Szafer joined the Allen Institute in 2010 and is the Technical Program Manager for the Human and Non-Human Primate projects. His background includes design, anaysis and optimization of software and communication networks, as well as research on the physics of MRI, mesoscopic systems and elementary particles. Aaron holds a PhD in Theoretical Physics from Yale University, New Haven, CT.

Mario F. Tecce

Università di Salerno
Dipartimento di Farmacia
Fisciano, Italy
tecce@unisa.it

Chapter F.31

Mario Felice Tecce, MD, PhD, is a Professor of Biochemistry at the University of Salerno and coordinates the Molecular Nutrition Laboratory within the Pharmacyl Department. His current research is aimed to the identification of relevant molecular processes related to bioactive nutrients and their role in neoplastic and cardiovascular prevention. Previously, he was a senior researcher in the pharmaceutical industry.

Shoba Tegginmath

Auckland University of Technology
Computing and Mathematical Sciences
Auckland, New Zealand
stegginm@aut.ac.nz

Chapter J.49

Shoba is the Postgraduate Programme Leader and a Senior Lecturer in the School of Computing and Mathematical Sciences, Auckland University of Technology, New Zealand. Shoba's research interests lie in the area of efficient storage and retrieval of data; in databases, data warehouses and ontologies for the semantic web. Her current research is on using ontologies to improve machine learning techniques.

Giuseppe Tradigo

University Magna Graecia of Catanzaro
Department of Medical and Surgical Sciences
Catanzaro, Italy
gtradigo@unicz.it

Chapter E.26

Giuseppe Tradigo is a PhD student at the Biomedical and Informatics Engineering PhD course at Magna Graecia University of Catanzaro, Italy. His research interests are protein structure prediction, clinical and biological data analysis and clinical engineering systems. He is currently investigating the role of contact maps in the protein 3-D structure prediction pipeline.

Elena Tsiporkova

Sirris, The Collective Center for the Belgian Technological Industry
Department of ICT & Software Engineering
Brussels, Belgium
elena.tsiporkova@sirris.be

Chapter C.14

Elena Tsiporkova is currently a member of the ICT & Software Engineering Group of Sirris. She holds an MSc in Mathematics/Informatics from the University of Plovdiv, Bulgaria (1985), and a PhD in Mathematics from the University of Ghent, Belgium (1995). She has a rather extensive R&D experience, both in an academic and an industrial environment in the field of data analysis, decision making support systems, knowledge engineering and multimedia. She has more than 50 publications in refereed journals and conference proceedings.

Shiro Usui

Chapter J.50

For biographical profile, please see section "About the Part Editors" on page XV.

Simona Valletta

University of Milano Bicocca
Department of Health Science
Monza, Italy
s.valletta@campus.unimib.it

Chapter D.25

Simona Valletta received her Master's degree in Biotechnology from the University of Florence in 2009. Actually she is a PhD student in Experimental Hematology and works in Professor Gambacorti-Passerini's Molecular Oncology laboratory at the University of Milan Bicocca. Her research concerns mutational analysis in Philadelphia negative Chronic Myeloid Leukemia (aCML) by next-generation sequencing.

Authors

Pierangelo Veltri   Chapter E.26

University Magna Graecia of Catanzaro
Department of Medical and Surgical
Sciences
Catanzaro, Italy
veltri@unicz.it

Pierangelo Veltri is an Associate Professor of Computer Engineering at the Magna Graecia University of Catanzaro, Italy. His research explores bioinformatics, health informatics, and databases. He obtained his PhD In 2002 from the University of Orsay, Paris 13. He works in bioinformatics since 2002 and is co-author of more than 70 papers published in international journals and conference proceedings. For 2012 he is editor of ACM SIGHIT, the newsletter of the special interest group in health informatics.

Roberto Visintainer   Chapter K.53

Fondazione Bruno Kessler
Predictive Models for Biomedicine and
Environment
Povo, Italy
visintainer@fbk.eu

Roberto Visintainer received his PhD in Computer Science from the University of Trento, Italy, in 2013. His research interests are mainly in computational biology with a specific focus on biological network theory. In particular, he is dealing with the problem of reproducibility and stability in network inference.

Lipo Wang   Chapter C.15

Nanyang Technological University
School of Electrical and Electronic
Engineering
Singapore
elpwang@ntu.edu.sg

Dr. Lipo Wang is an elected member of the AdCom (2010–2012) of the IEEE Computational Intelligence Society (CIS) and served as IEEE CIS Vice President for Technical Activities (2006–2007). He is an elected member of the Board of Governors of the International Neural Network Society (2011–2013) and was President of the Asia-Pacific Neural Network Assembly (APNNA) in 2002/2003.

Juyang Weng   Chapter L.58

Michigan State University
Department of Computer Science and
Engineering
East Lansing, MI, USA
weng@cse.msu.edu

Juyang Weng received the BS degree in Computer Science from Fudan University, Shanghai, China, in 1982, and MSc and PhD degrees in Computer Science from the University of Illinois at Urbana-Champaign, in 1985 and 1989, respectively. He is currently a Professor of Computer Science and Engineering at Michigan State University, East Lansing. He is also a faculty member of the Cognitive Science Program and the Neuroscience Program at Michigan State University. Since the work of Cresceptron (ICCV 1993), he expanded his research interests in biologically inspired systems, especially the autonomous development of a variety of mental capabilities by robots and animals, including perception, cognition, behaviors, motivation, and abstract reasoning skills.

KongFatt Wong-Lin   Chapter G.39

University of Ulster
Intelligent Systems Research Centre
Derry, Northern Ireland, UK
k.wong-lin@ulster.ac.uk

KongFatt Wong-Lin is a Lecturer in Computational Neuroscience at the University of Ulster. He received his PhD in Physics (Computational Neuroscience) fromt Brandeis University, and was a Research Associate at Princeton University affiliated with the Program in Applied and Computational Mathematics, and the Princeton Neuroscience Institute. His research interests are in computational modeling and mathematical analysis in systems and cognitive neurosciences.

Susan P. Worner   Chapter F.35

Lincoln University
Bio-Protection Research Centre
Lincoln, New Zealand
worner@lincoln.ac.nz

Susan P. Worner is Associate Professor and Project Leader in the Bio-Protection Research Centre, Lincoln University, New Zealand, a National Centre of Research Excellence. Her research interests are the ecoclimatic assessment and risk prediction of invasive insect species and the application of ecological informatics, including computational intelligence and spatial modeling, to assess potential establishment and the spread of invasive species.

Wei Xie

Institute for Infocomm Research
Singapore
wxie@i2r.a-star.edu.sg

Chapter C.15

Wei Xie received his BEng degree in Communication Engineering in 2002 with first class honor, and the MEng degree in Information Engineering in 2004 from Nanyang Technological University. He is now working in the Institute for Infocomm Research, Singapore. His research interests include biomedical signal processing, image processing and pattern recognition.

Takeshi Yamakawa

Fuzzy Logic Systems Institute (FLSI)
Kitakyushu, Japan
yamakawa@flsi.or.jp

Chapter A.3

Takeshi Yamakawa, Professor Emeritus of Kyushu Institute of Technology (KIT), is now the Founding Director of the Fuzzy Logic Systems Institute. He received the BEng degree from KIT in 1969, the MEng. and the PhD degrees from Tohoku University in 1971 and 1974, respectively. His research interests lie on soft computing, epilepsy treatment, dielectrophoresis, electrochemistry, and chip fabrication.

Zhengyou Zhang

Microsoft
Microsoft Research
Redmond, WA, USA
zhang@microsoft.com

Chapter L.58

Zhengyou Zhang is a Research Manager and Principal Researcher with Microsoft Research, Redmond, WA, USA. He received the BS degree from Zhejiang University, Hangzhou, China, the MS degree from the University of Nancy, France, and the PhD degree and the Doctorate of Science (Habilitation à diriger des recherches) both from the University of Paris XI, Paris, France. His research interests include computer vision, audio and speech signal processing, multimedia, human–computer interaction, and autonomous mental development. He has published more than 200 papers and 5 monographs. He is a Fellow of the IEEE.

Hylde Zirpoli

Università di Salerno
Dipartimento di Farmacia
Fisciano, Italy
hzirpoli@unisa.it

Chapter F.31

Hylde Zirpoli is a Post-Doctoral Research Fellow at Columbia Univeristy Medical Center, NY. She earned her PhD degree in Biochemistry and Pathology on the action mechanism of drugs from the University of Salerno. Her current research activity is focused on the effect of n-3 polyunsaturated fatty acids acute treatment in cardiovascular diseases, specifically adopting ex vivo and in vivo ischemia/reperfusion models.

Xin Zou

The Universities of Greenwich and Kent
at Medway
Medway School of Pharmacy
Chatham, Kent, UK
x.zou@kent.ac.uk

Chapter G.39

Dr. Xin Zou owns a PhD in Statistical Signal Processing from the University of Birmingham, UK. He is currently a bioinformatician working in the School of Pharmacy, Kent University. His research interests are bioinformatics, statistics, pattern recognition, computational neuroscience.

Authors

Detailed Contents

Part A Understanding Information Processes in Biological Systems

6 Pattern Formation and Animal Morphogenesis

Detailed Cont.

Part C Machine Learning Methods for the Analysis, Modeling and Knowledge Discovery from Bioinformatics Data

17 Protein Folding Recognition

18 Kernel Methods and Applications in Bioinformatics

Part D Modeling Regulatory Networks: The Systems Biology Approach

Detailed Cont.

Part E   Bioinformatics Databases and Ontologies

Part F Bioinformatics in Medicine, Health and Ecology

28 **Statistical Signal Processing for Cancer Stem Cell Formation**

Detailed Cont.

Part G Understanding Information Processes in the Brain and the Nervous System

36 Information Processing in Synapses

Part I Information Modeling of Perception, Sensation and Cognition

Part J Neuroinformatics Databases and Ontologies

Part K Information Modeling for Understanding and Curing Brain Diseases

51 Alzheimer's Disease

Detailed Cont.

Detailed Cont.

Subject Index

Subject Index

Subject Index

Recently Published Springer Handbooks

Springer Handbook of Bio-/Neuro-Informatics (2014)
ed. by Kasabov, 1230 p., 978-3-642-30573-3

Springer Handbook of Nanomaterials (2013)
ed. by Vajtai, 1222 p., 978-3-642-20594-1

Springer Handbook of Lasers and Optics (2nd) (2012)
ed. by Träger, 1694 p., 978-3-642-19408-5

Springer Handbook of Geographic Information (2012)
ed. by Kresse, Danko, 1120 p., 978-3-540-72678-4

Springer Handbook of Medical Technology (2011)
ed. by Kramme, Hoffmann, Pozos, 1500 p., 978-3-540-74657-7

Springer Handbook of Metrology and Testing (2011)
ed. by Czichos, Saito, Smith, 1229 p., 978-3-642-16640-2

Springer Handbook of Crystal Growth (2010)
ed. by Dhanaraj, Byrappa, Prasad, Dudley, 1816 p., 978-3-540-74182-4

Springer Handbook of Nanotechnology (3rd) (2010)
ed. by Bhushan, 1961 p., 978-3-642-02524-2

Springer Handbook of Automation (2009)
ed. by Nof, 1812 p., 978-3-540-78830-0

Springer Handbook of Mechanical Engineering (2009)
ed. by Grote, Antonsson, 1576 p., 978-3-540-49131-6

Springer Handbook of Robotics (2008)
ed. by Siciliano, Khatib, 1611 p., 978-3-540-23957-4

Springer Handbook of Experimental Solid Mechanics (2008)
ed. by Sharpe, 1096 p., 978-0-387-26883-5